S0-BRR-497

Third Edition

GROUNDWATER CHEMICALS

Desk Reference

John H. Montgomery

CRC LEWIS PUBLISHERS

Boca Raton   London   New York   Washington, D.C.

Library of Congress Cataloging-in-Publication Data

Montgomery, John H. (John Harold), 1955–
    Groundwater chemicals desk reference / John H. Montgomery.—3rd ed.
        p.   cm.
    Includes bibliographical references and index.
    ISBN 1-56670-498-7
    1. Groundwater—Pollution—Handbooks, manuals, etc. 2. Pollutants—Handbooks,
manuals, etc. I. Title.
    TD426 .M66 2000
    628.1′61—dc21   00-037077
                                                                                            CIP

This book contains information obtained from authentic and highly regarded sources. Reprinted material is quoted with permission, and sources are indicated. A wide variety of references are listed. Reasonable efforts have been made to publish reliable data and information, but the author and the publisher cannot assume responsibility for the validity of all materials or for the consequences of their use.

Neither this book nor any part may be reproduced or transmitted in any form or by any means, electronic or mechanical, including photocopying, microfilming, and recording, or by any information storage or retrieval system, without prior permission in writing from the publisher.

The consent of CRC Press LLC does not extend to copying for general distribution, for promotion, for creating new works, or for resale. Specific permission must be obtained in writing from CRC Press LLC for such copying.

Direct all inquiries to CRC Press LLC, 2000 N.W. Corporate Blvd., Boca Raton, Florida 33431.

Trademark Notice: Product or corporate names may be trademarks or registered trademarks, and are used only for identification and explanation, without intent to infringe.

© 2000 by CRC Press LLC
Lewis Publishers is an imprint of CRC Press LLC

No claim to original U.S. Government works
International Standard Book Number 1-56670-498-7
Library of Congress Card Number 00-037077
Printed in the United States of America 1 2 3 4 5 6 7 8 9 0
Printed on acid-free paper

Preface

The proliferation of research publications and continuing public awareness dealing with the fate, transport, and remediation of hazardous substances in the environment has stimulated this third edition of the *Groundwater Chemicals Desk Reference*.

Since the publication of the second edition, additional data have been critically reviewed. New data included in this edition are bioconcentration factors, aquatic mammalian toxicity values, degradation rates, corresponding half-lives in various environmental compartments, ionization potentials, aqueous solubility of miscellaneous compounds, and biological, chemical, and theoretical oxygen demand values for various organic compounds. The following data field fields have been added to each chemical profile: sources of contamination, entropy of fusion, and Merck reference citation. New tables include quantitative and qualitative nature of various petroleum fuels, Henry's law constants, soil adsorption constants, and toxicity data for a wide variety of chemicals. These data should prove useful to the environmental regulating community and environmental consultants, especially for conducting risk-based contamination assessments.

The Registry of Toxic Effects of Chemical Substances (RTECS) Number Index was deleted for three reasons. First, the Chemical Abstract Service (CAS) Registry Number Index is commonly used to search the documented literature for the compound of interest. Second, the CAS Number Index can be used to obtain the RTECS Number for a particular substance. Third, the RTECS Number Index was deleted to conserve space.

The presentation of data remains unchanged in this edition to make it easy for the reader to locate information. The Environmental Fate section has been expanded significantly. This section is subdivided into the following categories: Biological, Soil, Plant, Surface Water, Groundwater, Photolytic, and Chemical/Physical. When available, photolytic, hydrolysis, biodegradation, and volatilization half-lives of chemicals in various media are included. To conserve space, references are no longer given at the end of each chemical profile. All references now appear at the end of the main text, before the tables.

The book is based on more than 2,600 references. Most of the citations reviewed from the documented literature and included in this book pertain to the fate and transport of chemicals in the subsurface environment and toxicity data for mammals and aquatic species (approximately 700 new references). Every effort has been made to select accurate information and present it in a clear, consistent format. The publisher and author would appreciate hearing from readers regarding corrections and suggestions for material that might be included for use in future editions.

The author is grateful to the staff of CRC Press, in particular Arline Massey, Carol Whitehead, and Christine Andreasen, for their invaluable contributions and suggestions during the preparation of this book. The author also extends thanks to the many anonymous reviewers for their comments and suggestions on draft proofs.

John H. Montgomery

Introduction

The compounds profiled in this book include solvents, herbicides, insecticides, fumigants, and other hazardous substances most commonly found in the groundwater environment; the organic Priority Pollutants promulgated by the U.S. Environmental Protection Agency (U.S. EPA) under the Clean Water Act of 1977 (40 Code of Federal Regulations (CFR) 136, 1977); and compounds most commonly found in the workplace.

The compound headings are those commonly used by the U.S. EPA and many agricultural organizations. Positional and/or structural prefixes set in italic type are not integral parts of the chemical names and hence are disregarded in alphabetizing. These include *asym-*, *sym-*, *n-*, *sec-*, *cis-*, *trans-*, α-, β-, γ-, *o-*, *m-*, *p-*, *N-*, *S-*, etc.

Synonyms: These are listed alphabetically following the convention used for the compound headings. Compounds in boldface type are the Chemical Abstracts Service (CAS) names listed in the eighth or ninth Collective Index. If no synonym appears in boldface type, then the compound heading is the CAS assigned name. Synonyms include chemical names, common or generic names, trade names, registered trademarks, government codes and acronyms. All synonyms found in the literature are listed.

Synonyms were retrieved from several references, primarily from the Registry of Toxic Effects of Chemical Substances (RTECS, 1985).

Beneath the synonyms is the structural formula, a graphic representation of atoms or group(s) of atoms relative to one other. This is given for every compound regardless of its complexity. The limitation of structural formulas is that they depict these relationships in two dimensions.

Chemical Abstracts Service (CAS) Registry Number: This unique identifier is assigned by the American Chemical Society to chemicals recorded in the CAS Registry System. This number is used to access various chemical databases including the Hazardous Substances Data Bank (HSDB), CAS Online, Chemical Substances Information Network, and many others. This entry is also useful to conclusively identify a substance regardless of the assigned name.

Department of Transportation (DOT) designation: This is a four-digit number assigned by the U.S. Department of Transportation (DOT) to hazardous materials and is identical to the United Nations identification number (which is preceded by the letters UN). This number is required on shipping papers, on placards or orange panels on tanks and on a label or package containing the material. These numbers are widely used by personnel responding to emergency situations, e.g., overturned tractor trailers, to identify transported material quickly and easily. The DOT designations and appropriate responses to each chemical or compounds are cross-referenced in the Emergency Response Guidebook. Additional information is provided in this book, which may be obtained from the U.S. Department of Transportation, Research and Special Programs Administration, Materials Transportation Bureau, Washington, D.C. 20590.

DOT label: This label is the hazard classification assigned by the U.S. Department of Transportation. Federal regulations require it on all containers shipped.

Molecular formula: This is arranged by carbon, hydrogen, and remaining elements in alphabetical order in accordance with the system developed by Hill (1900). Molecular formulas are useful for identifying isomers (i.e., compounds with identical molecular formulas) and are required to calculate the formula weight of a substance.

Formula weight: This is calculated to the nearest hundredth using the empirical formula and the 1981 Table of Standard Atomic Weights as reported in Weast (1986). Formula weights are required for many calculations, such as converting weight/volume units, e.g., mg/L or g/L, to molar units (mol/L); with density for calculating molar volumes; and estimating Henry's law constants.

Registry of Toxic Effects of Chemical Substances (RTECS) number: Many compounds are assigned a unique accession number consisting of two letters followed by seven numerals. This number is needed to quickly and easily locate additional toxicity and health-based data that are cross-referenced in the RTECS (1985). For additional information, contact NIOSH, U.S. Department of Health and Human Services, Mail Stop C-13, 4676 Columbia Parkway, Cincinnati, OH 45226-1998 (toll free: 1-800-35-NIOSH; fax: 513-533-8573).

Merck Reference: The two sets of numbers refer to the edition and monograph number as found in the 10th edition of "The Merck Index" (Windholz et al., 1983). For example, the entry 10, 1063 refers to an entry in the 10th edition as monograph number 1063.

Physical state, color, and odor: The appearance, including the color and physical state (solid, liquid, or gas) of a chemical at room temperature (20-25 °C) is provided. If the compound can be detected by the olfactory sense, the odor is noted. Unless noted otherwise, the information provided in this category is for the pure substance and was obtained from many sources (CHRIS, 1984; HSDB, 1989; Hawley, 1981; Keith and Walters, 1992; Sax, 1984; Sax and Lewis, 1987; Toxic and Hazardous Industrial Chemicals Safety Manual for Handling and Disposal with Toxicity and Hazard Data, 1986; Sittig, 1985; Verschueren, 1983; Windholz et al., 1983). If available, odor thresholds in air and/or taste thresholds in water are given. The odor threshold is the lowest compound concentration in air that can be detected by the olfactory sense and it is usually reported as parts per billion (ppb) as micrograms per cubic meter ($\mu g/m^3$). Sometimes the odor threshold in water is given and is reported as $\mu g/kg$. Taste thresholds in water are usually given in mass per unit volume (e.g., $\mu g/L$).

Boiling point: This is defined as the temperature at which the vapor pressure of a liquid equals the atmospheric pressure. Unless otherwise noted, all boiling points are reported at 1.0 atmosphere pressure (760 mmHg). Although not used in environmental assessments, boiling points for aromatic compounds have been found to be linearly correlated with aqueous solubility (Almgren et al., 1979). Boiling points are also useful for assessing entry of toxic substances into the body. Body contact with high-boiling liquids is the most common means of entry of toxic substances into the body. Inhalation is the most common means of entry into the body for low-boiling liquids (Shafer, 1987).

Diffusivity in water: Molecular diffusion is defined as the transport of molecules (e.g., organic compounds) in either liquid or gaseous states. Typically, molecular diffusion is not a major factor under most environmental conditions. However, in saturated aquifers with low pore water velocities (i.e., <0.002 cm/sec), diffusion can be a contributing factor in the transport of organic compounds.

Very few experimentally determined diffusivities of organic substances in water are available. If experimentally determined diffusivity values are not available, Hayduk and Laudie (1974) recommend the following equation for estimating this parameter:

$$D = (1.326 \times 10^{-4})/(\eta^{1.14} V^{0.589}) \qquad [1]$$

where D is the diffusivity of the substance in water (cm^2/sec), η is the viscosity of water (centipoise, cps), and V is the molar volume of the solute (cm^3/mol). The molar volume is easily determined if the liquid density of the solute is available. Molar volume may also be determined using the LeBas incremental method as described in Lyman et al. (1982). The latter method is required to determine molar volumes of substances that are solids at ordinary temperatures. More recent methods for estimating molar volume using molecular group contributions are described by Baum (1998).

Diffusivity values are reported in a modified exponential form. For example, the experimentally determined diffusivity of benzene in water is 1.09×10^{-5} cm^2/sec, but this value is reported as 1.09 (x 10^{-5} cm^2/sec).

Dissociation constant: In an aqueous solution, an acid (HA) will dissociate into the carboxylate anion (A$^-$) and hydrogen ion (H$^+$) and may be represented by the general equation:

$$HA_{(aq)} \rightleftharpoons H^+ + A^-$$ [2]

At equilibrium, the ratio of the products (ions) to the reactant (non-ionized electrolyte) is related by the equation:

$$K_a = ([H^+] [A^-]/[HA])$$ [3]

where K_a is the dissociation constant. This expression shows that K_a increases if there is increased ionization and vice versa. A strong acid (weak base) such as hydrochloric acid ionizes readily and has a large K_a, whereas a weak acid (or stronger base) such as benzoic acid ionizes to a lesser extent and has a lower K_a. The dissociation constants for weak acids are sometimes expressed as K_b, the dissociation constant for the base, and both are related to the dissociation constant for water by the expression:

$$K_w = K_a + K_b$$ [4]

where K_w is the dissociation constant for water (10^{-14} at 25 °C), K_a is the acid dissociation constant, and K_b is the base dissociation constant.

The dissociation constant is usually expressed as $pK_a = -\log_{10}K_a$. Equation [4] becomes:

$$pK_w = pK_a + pK_b$$ [5]

When the pH of the solution and the pK_a are equal, 50% of the acid will have dissociated into ions. The percent dissociation of an acid or base can be calculated if the pH of the solution and the pK_a of the compound are known (Guswa et al., 1984):

For organic acids: $\alpha_a = [100/(1 + 10^{(pH-pKa)})]$ [6]

For organic bases: $\alpha_b = [100/(1 + 10^{(pKw-pKb-pH)})]$ [7]

where α_a is the percent of the organic acid that is nondissociated, α_b is the percent of the organic base that is nondissociated, pK_a is the $-\log_{10}$ dissociation constant for an acid, pK_w is the $-\log_{10}$ dissociation constant for water (14.00 at 25 °C), pK_b is the $-\log_{10}$ dissociation constant for base ($pK_b = pK_w - pK_a$), and pH is the $-\log_{10}$ hydrogen ion activity (concentration) of the solution.

Because ions tend to remain in solution, the degree of dissociation will affect processes such as volatilization, photolysis, adsorption, and bioconcentration (Howard, 1989).

Heat of fusion: This value, normally reported in kcal/mol, is also referred to as the heat of melting. For solids, the heat of fusion is required to estimate the solubility of the solute to account for crystal lattice interactions. The theoretical basis for introducing this value into the estimation of aqueous solubility of organic solids is explained by Irmann (1965) and

Yalkowsky and Valvani (1979). Heat of fusion data is available in many texts including Dean (1987), Weast (1986), and CHRIS (1984).

Henry's law constant: Sometimes referred to as the air-water partition coefficient, the Henry's law constant is defined as the ratio of the partial pressure of a compound in air to the concentration of the compound in water at a given temperature under equilibrium conditions. If the vapor pressure and solubility of a compound are known, this parameter can be calculated at 1.0 atm (760 mmHg) as follows:

$$K_H = Pfw/760S \qquad [8]$$

where K_H is Henry's law constant (atm·m^3/mol), P is the vapor pressure (mmHg), S is the solubility in water (mg/L), and fw is the formula weight (g/mol).

Henry's law constant can also be expressed in dimensionless form and may be calculated using one of the following equations:

$$K_{H'} = K_H/RK$$

or

$$K_{H'} = S_a/S \qquad [9]$$

where $K_{H'}$ is Henry's law constant (dimensionless), R is the ideal gas constant (8.20575 x 10^{-5} atm·m^3/mol·K), K is the temperature of water (degrees Kelvin), S_a is the solute concentration in air (mol/L), and S is the aqueous solute concentration (mol/L). Many papers present Henry's law constants as H (M/atm) representing the solubility of the chemical. The relationship between K_H and H is:

$$K_H = 1/(H \times 1000) \qquad [10]$$

In addition, Henry's law constants may be expressed as a mole fraction ratio between the air and aqueous phases (Leighton and Calo, 1981):

$$K'_{AW} = y/x \qquad [11]$$

where y is the mole fraction of the chemical in air and x is the mole fraction of the chemical in water. The partition coefficients $K_{H'}$ and K'_{AW} are related by the expression:

$$K_{H'} = K'_{AW} V_w/V_a \qquad [12]$$

where V_w is the molar volume of water (m^3/mol) and V_a is the molar volume of air (m^3/mol). At standard temperature (25 °C) and pressure (1 atm), V_w and V_a are 1.81 x 10^{-5} and 2.45 x 10^{-2} m^3/mol, respectively.

It should be noted that estimating Henry's law constant assumes the gas obeys the ideal gas law and the aqueous solution behaves as an ideally dilute solution. The solubility and vapor pressure data inputted into the equations are valid only for the pure compound and must be in the same standard state at the same temperature.

The major drawback in estimating Henry's law constant is that both the solubility and the vapor pressure of the compound are needed in equation [8]. If one or both parameters are unknown, an empirical equation based on quantitative structure-activity relationships (QSAR) may be used to estimate Henry's law constants (Nirmalakhandan and Speece, 1988). In this QSAR model, only the structure of the compound is needed. From this, connectivity indexes (based on molecular topology), polarizability (based on atomic contributions) and the propensity of the compound to form hydrogen bonds can easily be determined. These parameters, when regressed against known Henry's law constants for 180 organic compounds, yielded an empirical equation that explained more than 98% of the variance in the data set having an average standard error of only 0.262 logarithm units.

Henry's law constant may also be estimated using the bond or group contribution method developed by Hine and Mookerjee (1975). The constants for the bond and group contributions were determined using experimentally determined Henry's law constants for 292 compounds. The authors found that those estimated values significantly deviating from observed values (particularly for compounds containing halogen, nitrogen, oxygen and sulfur substituents) could be explained by "distant polar interactions," i.e., interactions between polar bonds or structural groups.

A more recent study for estimating Henry's law constants using the bond contribution method was provided by Meylan and Howard (1991). In this study, the authors updated and revised the method developed by Hine and Mookerjee (1975) based on new experimental data that have become available since 1975. Bond contribution values were determined for 59 chemical bonds based on known Henry's law constants for 345 organic compounds. A good statistical fit [correlation coefficient (r^2) = 0.94] was obtained when the bond contribution values were regressed against known Henry's law constants for all compounds. For selected chemicals classes, r^2 increased slightly to 0.97.

Russell et al. (1992) conducted a similar study using the same data set from Hine and Mookerjee (1975). They developed a computer-assisted model based on five molecular descriptors which was related to the compound's bulk, lipophilicity, and polarity. They found that 63 molecular structures were highly correlative with the log of Henry's law constants $(r^2 = 0.96)$.

Henry's law constants provided an indication of the relative volatility of a substance. According to Lyman et al. (1982), if $K_H < 10^{-7}$ atm·m^3/mol, the substance has a low volatility. If K_H is $> 10^{-7}$ but $< 10^{-5}$ atm·m^3/mol, the substance will volatilize slowly. Volatilization becomes an important transfer mechanism in the range $10^{-5} < H < 10^{-3}$ atm·m^3/mol. Values of $K_H > 10^{-3}$ atm·m^3/mol indicate volatilization will proceed rapidly.

The rate of volatilization will also increase with an increase in temperature. ten Hulscher et al. (1992) studied the temperature dependence of Henry's law constants for three chlorobenzenes, three chlorinated biphenyls, and six polynuclear aromatic hydrocarbons. They observed that over the temperature range of 10 to 55 °C, Henry's law constant was doubled for every 10 °C increase in temperature. This temperature relationship should be considered when assessing the role of chemical volatilization from large surface water bodies whose temperatures are generally higher than those typically observed in groundwater.

Henry's law constants are reported in a modified exponential form. For example, the experimentally determined Henry's law constant of benzene is 5.48 x 10^{-3} atm·m^3/mol; however, this value is reported as 5.48 (x 10^3 atm·m^3/mol).

Interfacial tension with water: With few exceptions, most organic compounds entering the aqueous environment are non-miscible liquids. The interfacial tension between the compound and water (i.e., groundwater, surface water bodies, etc.) is numerically equivalent to the free surface energy that is formed at the interface. Compounds with high interfacial tension values relative to water are easy to separate after mixing and are not likely to form emulsions (Lyman et al., 1982). The interfacial tension of organic compounds can be used to calculate the spreading coefficient to determine whether it forms a lens or macromolecular film with water (Demond and Lindner, 1993):

$$s_{OW} = \gamma_{W(O)} - \gamma_{O(W)} - \gamma_{OW} \qquad [13]$$

where s_{OW} is the spreading coefficient of the organic liquid at the air-water interface, $\gamma_{W(O)}$ is the surface tension of water saturated with the organic liquid, $\gamma_{O(W)}$ is the surface tension of the organic liquid saturated with water and γ_{OW} is the interfacial tension between organic liquid and water. Organic liquids with spreading coefficients greater than zero form a thin layer on the water surface. Conversely, the organic liquid forms a lens if the spreading coefficient is negative.

The units of interfacial tension are identical for surface tension, i.e., dyn/cm. Interfacial tension values of organic compounds range from zero for completely miscible liquids (e.g., acetone, methanol, ethanol) up to the surface tension of water at 25 °C which is 72 dyn/cm (Lyman et al., 1982). Interfacial tension values may be affected by pH, surface-active agents, and dissolved gases (Schowalter, 1979). Most of the interfacial tension values reported in this book were obtained from Dean (1987), Demond and Lindner (1993), CHRIS (1984), and references cited therein.

Ionization potential: The ionization potential of a compound is defined as the energy required to remove a given electron from the molecule's atomic orbit (outermost shell) and is expressed in electron volts (eV). One electron volt is equivalent to 23,053 cal/mol.

Knowing the ionization potential of a contaminant is required to determine the appropriate photoionization lamp for detecting that contaminant or family of contaminants in air. Photo-ionization instruments are equipped with a radiation source (UV lamp), pump, ionization chamber, an amplifier (detector), and a recorder (either digital or meter). Generally, compounds with ionization potentials less than the radiation source (UV lamp rating) being used will readily ionize and will be detected by the instrument. Conversely, compounds with ionization potentials more than the lamp rating will not ionize and will not be detected by the instrument.

Bioconcentration factor, log BCF: The bioconcentration factor is defined as the steady-state concentration of the chemical in an organism (kg of chemical/kg of organism) to the concentration of the chemical in the organism's environment (kg of chemical/kg of medium or kg of chemical/m^3 of medium). Bioconcentration factors have also been determined for volatile organic compounds in vegetation. The bioconcentration factors were determined by measuring the concentration of the compound in leaves (mass/unit volume) to the concentration of the compound in air (mass/unit volume) (Hiatt, 1998). Generally, high bioconcentration factors tend to be associated with very lipophilic (low aqueous solubility) compounds. Conversely, low bioconcentration factors are associated with compounds having high aqueous solubilities.

Logarithmic bioconcentration factors have been shown to be correlated with the logarithmic octanol/water partition coefficient in aquatic organisms (Davies and Dobbs, 1984; de Wolf et al., 1992; Isnard and Lambert, 1988) and fish (Davies and Dobbs, 1984; Kenaga, 1980; Isnard and Lambert, 1988; Neely et al., 1974; Ogata et al., 1984; Oliver and Niimi, 1985). In addition, bioconcentration factors are well correlated by a linear solvation energy relationship (commonly known as LSER) that includes the intrinsic solute molecular volume and solvatochromic parameters that measure hydrogen bond acceptor basicity and donor acidity of the compound (Park and Cho, 1993).

Soil sorption coefficient, log K_{oc}: The soil/sediment partition or sorption coefficient is defined as the ratio of adsorbed chemical per unit weight of organic carbon to the aqueous solute concentration. This value provides an indication of the tendency of a chemical to partition between particles containing organic carbon and water. Compounds that bind strongly to organic carbon have characteristically low solubilities, whereas compounds with low tendencies to adsorb onto organic particles have high solubilities.

Nonionizable chemicals (e.g., hydrocarbons, ethers, alcohold) that sorb onto organic materials in an aquifer (i.e., organic carbon) are retarded in their movement in groundwater. The sorbing solute travels at linear velocity that is lower than the groundwater flow velocity by a factor of R_d, the retardation factor. If the K_{oc} of a compound is known, the retardation factor may be calculated using the following equation from Freeze and Cherry (1974) for unconsolidated sediments:

$$R_d = V_w/V_c = [1 + (BK_d/n_e)] \qquad [14]$$

where R_d is the retardation factor (unitless), V_w is the average linear velocity of groundwater (e.g., ft/day), V_c is the average linear velocity of contaminant (e.g., ft/day), B is the average soil bulk density (g/cm^3), n_e is the effective porosity (unitless), K_d is the distribution (sorption) coefficient (cm^3/g).

By definition, K_d is defined as the ratio of the concentration of the solute on the solid to the concentration of the solute in solution. This can be represented by the Freundlich equation:

$$K_F \text{ (or } K_d) = VM_S/MM_L = C_S/C_L^n \qquad [15]$$

where V is the volume of the solution (cm^3), M_S is the mass of the sorbed solute (g), M is the mass of the porous medium (g), M_L is mass of the solute in solution (g), C_S is the concentration of the sorbed solute (g/cm^3), C_L is the concentration of the solute in the solution (g/cm^3), and n is a constant.

Values of n are normally between 0.7 and 1.1 although values of 1.6 have been reported (Lyman et al., 1982). If n is unknown, it is assumed to be unity and a plot of C_S versus C_L will be linear. The distribution coefficient is related to K_{oc} by the equation:

$$K_{oc} = K_d/f_{oc} \qquad [16]$$

where f_{oc} is the fraction of naturally occurring organic carbon in soil.

Sometimes K_d is expressed on a naturally occurring organic-matter basis and is defined as:

$$K_{om} = K_d/f_{om} \qquad [17]$$

where f_{om} is the fraction of naturally occurring organic matter in soil. The relationship between K_{oc} and K_{om} is defined as:

$$K_{om} = 0.58K_{oc} \qquad [18]$$

where the constant 0.58 is assumed to represent the fraction of carbon present in the soil or sediment organic matter (Allison, 1965).

For fractured rock aquifers in which the porosity of the solid mass between fractures is insignificant, Freeze and Cherry (1974) report the retardation equation as:

$$R_d = V_w/V_c = [1 + (2K_A/b)] \qquad [19]$$

where K_A is the distribution coefficient (cm) and b is the aperture of fracture (cm).

To calculate the retardation factors for ionizable compounds such as acids and bases, the fraction of un-ionized acid (α_a) or base (α_b) needs to be determined (see **Dissociation constant**). According to Guswa et al. (1984), if it is assumed only the un-ionized portion of the acid is adsorbed onto the soil, the retardation factor for the acid becomes:

$$R_a = [1 + (\alpha_a BK_d/n_e)] \qquad [20]$$

However, for a base they assume that the ionized portion is exchanged with a monovalent ion and the un-ionized portion of the base is adsorbed hydrophobically. Therefore, the retardation factor for the base is:

$$R_b = \{1 + [(\alpha_b BK_d)/n_e] + [CECB(1-\alpha_b)]/(100\sum z^+n_e)\} \qquad [21]$$

where CEC is the cation exchange capacity of the soil (cm^3/g) and $\sum z^+$ is the sum of all positively charged particles in the soil (milliequivalents/cm^3). Guswa et al. (1984) report that the term $\sum z^+$ is approximately 0.001 for most agricultural soils.

Correlations between K_{oc} and bioconcentration factors in fish and beef have shown a log-log linear relationship (Kenaga, 1980) as well as solubility of organic compounds in water (Abdul et al., 1987; Means et al., 1980). Moreover, the log K_{oc} has been shown to be related to molecular connectivity indices (Govers et al., 1984; Gerstl and Helling, 1987; Koch, 1983; Meylan et al., 1992; Sabljić and Protić, 1982; Sabljić, 1984, 1987), solvatochromic parameters (Cho and Park, 1995), and high performance liquid chromatography (HPLC) capacity factors (Haky and Young, 1984; Hodson and Williams, 1988; Szabo et al., 1990, 1990a).

In instances where experimentally determined K_{oc} values are not available, they can be estimated using recommended regression equations as cited in Lyman et al. (1982) or Meylan et al. (1992). All the K_{oc} estimations are based on regression equations in which the aqueous solubility or the K_{ow} of the substance is known.

Various classes of compounds are not represented in recommended regression equations for estimating K_{oc} values because experimentally determined K_{oc} values are not available. These classes of compounds include, but are not limited to, alcohols, aldehydes, esters, ether, saturated hydrocarbons, and ketones. Generally, these types of compounds are very soluble in water (i.e., mg/L or greater) and low K_{ow} values suggesting its adsorption to soil will be nominal. Nevertheless, these compounds will adsorb to some degree on activated carbon. Removal of these compounds and other hazardous substances from water by adsorption on powdered or granular activated carbon is commonly used as a primary or secondary remedial alternative.

Adsorption isotherms of compounds onto activated carbon or other materials are usually approximated by the Freundlich equation:

$$\log x/m = \log K_F + (1/n) \log C \qquad [22]$$

where x/m is the amount of solute adsorbed (e.g., mg/g of carbon), C is the equilibrium concentration (e.g., mg/L), and K_F and $1/n$ are empirical constants characteristic of the compound and adsorption material. When n is unity, the adsorption isotherm is linear. Freundlich constants (K_F) for a variety of organic compounds are summarized in Table 12.

Octanol/water partition coefficient, log K_{ow}: The K_{ow} of a substance is the n-octanol/water partition coefficient and is defined as the ratio of the solute concentration in the water-saturated n-octanol phase to the solute concentration in the n-octanol-saturated water phase. Values of K_{ow} are therefore unitless.

The partition coefficient has been recognized as a key parameter in predicting the environmental fate of organic compounds. The log K_{ow} has been shown to be linearly correlated with log BCF in aquatic organisms (Davies and Dobbs, 1984; de Wolf et al., 1992; Isnard and Lambert, 1988), in fish (Davies and Dobbs, 1984; Kenaga, 1980; Isnard and Lambert, 1988; Neely et al., 1974; Ogata et al., 1984; Oliver and Niimi, 1985), log soil/sediment partition coefficients (K_{oc}) (Chiou et al., 1979; Kenaga and Goring, 1980), log of the solubility of organic compounds in water (Banerjee et al., 1980; Chiou et al., 1977, 1982; Hansch et al., 1968; Isnard and Lambert, 1988; Miller et al., 1984, 1985; Tewari et ai., 1982; Yalkowsky and Valvani, 1979, 1980), molecular surface area (Camilleri et al., 1988; Funasaki et al., 1985; Miller et al.,

1984; Woodburn et al., 1992; Yalkowsky and Valvani, 1979, 1980), molar refraction (Yoshida et al., 1983), molecular connectivity indices (Govers et al., 1984; Patil, 1991; Woodburn et al., 1992), reversed-phase liquid chromatography (RPLC) retention factors (Khaledi and Breyer, 1989; Woodburn et al., 1992), RPLC capacity factors (Braumann, 1986; Minick et al., 1989), isocratic RPLC capacity factors (Hafkenscheid and Tomlinson, 1983); HPLC capacity factors (Brooke et al., 1986; Carlson et al., 1975; DeKock and Lord, 1987; Eadsforth, 1986; Hammers et al., 1982; Harnisch et al., 1983; Kraak et al., 1986; Miyake et al., 1982, 1987, 1988; Szabo et al., 1990, 1990a, 1900b), HPLC retention times (Burkhard and Kuehl, 1986; Mirrlees et al., 1976; Sarna et al., 1984; Veith and Morris, 1978; Webster et al., 1985), reversed-phase thin layer chromatography retention parameters (Bruggeman et al., 1982), gas chromatography retention indices (Valkó et al., 1984), distribution coefficients (Campbell et al., 1983), solvatochromic parameters (Sadek et al., 1985), biological responses (Schultz et al., 1989), log of the n-hexane/water and L-a-phosphatidycholine dimyristol partitioning coefficients (Gobas et al., 1988), molecular descriptors and physicochemical properties (Bodor et al., 1989; Warne et al., 1990), molecular structure (Suzuki, 1991), LSER's (Kamlet et al., 1988), and substituent constants which are based on empirically derived atomic or group constants and structural factors (Hansch and Anderson, 1967; Hansch et al., 1972), including some intramolecular interactions (Wang et al., 1997). Variables needed for employing the LSER method have recently been presented by Hickey and Passino-Reader (1991).

For ionizable compounds (e.g., acids, amines and phenols), K_{ow} values are a function of pH. Unfortunately, many investigators have neglected to report the pH of the solution at which the K_{ow} was determined. If a K_{ow} value is used for an ionizable compound for which the pH is known, both values should be noted.

Melting point: The melting point of a substance is defined as the temperature at which a solid substance undergoes a phase change to a liquid. The reverse process, the temperature at which a liquid freezes to a solid, is called the freezing point. For a given substance, the melting point is identical to the freezing point.

Unless noted otherwise, all melting points are reported at the standard pressure of 1.0 atm (760 mmHg). Although the melting point of a substance is not directly used in predicting its behavior in the environment, it is useful for determining the phase in which the substance would be found under typical conditions.

Solubility in organics: The presence of small quantities of solvents can enhance a compound's solubility in water (Nyssen et al., 1987). Consequently, its fate and transport in soils, sediments and groundwater will be changed due to the presence of these cosolvents, e.g., soils contaminated with compounds having low water solubilities tend to remain bound to the soil by adsorbing onto organic carbon and/or by interfacial tension with water. A solvent introduced to an unsaturated soil environment (e.g., a surface spill, leaking aboveground tank, etc.) may come in contact with existing soil contaminants. As the solvent interacts with the existing contamination, it may mobilize it, thereby facilitating its migration. Consequently, the organic solvent can facilitate the leaching of contaminants from the soil to the water table. Therefore, the presence of cosolutes must be considered when predicting the fate and transport of contaminants in the unsaturated zone, the water table and surface water bodies.

Solubility in water: The water solubility of a compound is defined as the saturated concentration of the compound in water at a given temperature and pressure. This parameter is perhaps the most important factor in estimating a chemical's fate and transport in the aquatic environment. Compounds with high water solubilities tend to desorb from soils and sediments (i.e., they have low K_{oc} values), are less likely to volatilize from water, and are susceptible to biodegradation. Conversely, compounds with low solubilities tend to adsorb onto soils and sediments (have high K_{oc}), volatilize more readily from water and bioconcentrate in aquatic organisms. The more soluble compounds commonly enter the water table more readily than their less soluble counterparts.

The water solubility of a compound varies with temperature, pH (particularly, ionizable compounds such as acids and bases) and other dissolved constituents, e.g., inorganic salts (electrolytes) and organic chemicals including naturally occurring organic carbon, such as humic and fulvic acids. At a given temperature, the variability or discrepancy of water-solubility measurements documented by investigators may be attributed to one or more of the following: (1) purity of the compound; (2) analytical method employed; (3) particle size (for solid solubility determinations only); (4) adsorption onto the container and/or suspended solids; (5) time allowed for equilibrium conditions to be reached, (6) losses due to volatilization; and (7) chemical transformations (e.g., hydrolysis).

The water solubility of chemical substances has been related to log BCF, log K_{oc} (Abdul et al., 1987; Means et al., 1980), log K_{ow} (Chiou et al., 1977, 1982; Hansch et al., 1968; Miller et al., 1984, 1985; Yalkowsky and Valvani, 1979, 1980), HPLC capacity factors (Hafkenscheid and Tomlinson, 1981; Whitehouse and Cooke, 1982), molecular descriptors and physico-chemical properties (Warne et al., 1990), log K_{om} (Chiou et al., 1983), total molecular surface area (Amidon et al., 1975; Hermann, 1972; Lande and Banerjee, 1981; Lande et al., 1985; Valvani et al., 1976), the compound's molecular structure or quantitative structure-property relationship (Nirmalakhandan and Speece, 1988a, 1989; Patil, 1991; Sutter and Jurs, 1996; Suzuki, 1989), boiling points (Almgren et al., 1979; Yaws et al., 1997) and for homologous series of hydrocarbons or classes of organic compounds-carbon number (Bell, 1973; Krzyzanowska and Szeliga, 1978; Mitra et al., 1977; Robb, 1966) and molar volumes (Lande and Banerjee, 1981). With the exception of the molecular structure-solubility relationship, regression equations generated from the other relationships have demonstrated a log-log linear relationship for these properties. The reported regression equations are useful in estimating the solubility of a compound in water if experimental values are not available. In addition, the solubility of a compound may be estimated from experimentally determined Henry's law constants (Kamlet et al., 1987) or from measured infinite dilution activity coefficients (Wright et al., 1992).

Unless otherwise noted, all reported solubilities were determined using distilled and/or deionized water. For some compounds, solubilities were determined using groundwater, natural seawater or artificial seawater.

Solubility concentrations can be expressed many ways, including molarity (mol/L), molality (mol/kg), mole fraction, weight percent, mass per unit volume (e.g., g/L), etc. The conversion formulas for solutions having different concentration units are presented in Table 1.

Methods for estimating the aqueous solubilities of organic solutes can be found in Lyman et al. (1982), Yalkowsky and Banerjee (1992), and Baum (1998).

Specific density: The specific density, also known as relative density, is defined as:

$$\rho = d_s/d_w \qquad [23]$$

where d_s is the density of a substance (g/mL or g/cm^3) and d_w is the density of distilled water (g/mL or g/cm^3). Values of specific density are unitless and are reported in the form ρ at T_s/T_w where ρ is the specific density of the substance, T_s is the temperature of substance at the time of measurement (°C) and T_w is the water temperature (°C).

For example, the value 1.1750 at 20/4 °C indicates a specific density of 1.1750 for the substance at 20 °C with respect to water at 4 °C. At 4 °C, the density of water is exactly 1.0000 g/mL (g/cm^3). Therefore, the specific density of a substance is equivalent to the density of the substance relative to the density of water at 4 °C.

The density of a hydrophobic substance enables it to sink or float in water. Density values are especially important for liquids migrating through the unsaturated zone and encountering the water table as "free product." Generally, liquids that are less dense than water "float" on the water table. Conversely, organic liquids that are more dense than water commonly "sink" through the water table, e.g., dense nonaqueous phase liquids such as chloroform, dichloroethane, and tetrachloroethylene.

Hydrophilic substances, on the other hand, behave differently. Acetone, which is less dense than water, does not float on water because it is freely miscible with water in all proportions. Therefore, the solubility of a substance must be considered in assessing its behavior in the subsurface.

Environmental fate: Chemicals released in the environment are susceptible to several degradation pathways, including chemical (i.e., hydrolysis, oxidation, reduction, dealkylation, dealkoxylation, decarboxylation, methylation, isomerization, and conjugation), photolysis or photooxidation and biodegradation. Compounds transformed by one or more of these processes may result in the formation of more toxic or less toxic substances. In addition, the transformed product(s) will behave differently from the parent compound due to changes in their physicochemical properties. Many researchers focus their attention on transformation rates rather than the transformation products. Consequently, only limited data exist on the transitional and resultant end products. Where available, compounds that are transformed into identified products as well as environmental fate rate constants and/or half-lives are listed.

In addition to chemical transformations occurring under normal environmental conditions, abiotic degradation products are also included. Types of abiotic transformation processes or treatment technologies fall into two categories: physical and chemical. Types of physical processes used in removing or eliminating hazardous wastes include sedimentation, centrifugation, flocculation, oil/water separation, dissolved air flotation, heavy media separation, evaporation, air stripping, steam stripping, distillation, soil flushing, chelation, liquid-liquid extraction, supercritical extraction, filtration, carbon adsorption, absorption, reverse osmosis, ion exchange, and electrodialysis. This information can be useful in evaluating abiotic degradation as a possible remedial measure. Chemical processes include neutralization, precipitation, hydrolysis (acid or base catalyzed), photolysis or irradiation, oxidation-reduction, oxidation by hydrogen peroxide, alkaline chlorination, electrolytic oxidation, catalytic dehydrochlorination, and alkali metal dechlorination.

If available, experimentally determined hydrolysis and photolytic half-lives of chemicals are provided. The half-life of a chemical is the time required for the parent chemical to reach one-half or 50% of its original concentration.

Chemicals will undergo photolysis if they can absorb sunlight. Photolysis can occur in air, soil, water and plants. The rate of photolysis is dependent upon the pH, temperature, presence of sensitizers, sorption to soil, depth of the compound in soil and water. Lyman et al. (1982) present an excellent overview of the photolysis process.

The rate of chemical hydrolysis is highly dependent upon the compound's solubility, temperature and pH. Since other environmental factors such as photolysis, adsorption, volatility (i.e., Henry's law constants) and adsorption can affect the rate of hydrolysis, these factors are virtually eliminated by performing hydrolysis experiments under carefully controlled laboratory conditions. The hydrolysis half-lives reported in the literature were calculated using experimentally determined hydrolysis rate constants.

Most of the abiotic chemical transformation products reported in this book are limited to only three processes: hydrolysis, photooxidation, and chemical oxidation-reduction. These processes are the most widely studied and reported in the literature. Detailed information describing the above technologies, their availability/limitation and company sources is available (U.S. EPA, 1987).

Vapor density: The vapor density of a substance is defined as the ratio of the mass of vapor per unit volume. An equation for estimating vapor density is readily derived from a varied form of the ideal gas law:

$$PV = MRK/fw \qquad [24]$$

where P is the vapor pressure (atm), V is the volume (L), M is the mass (g), R is the ideal gas constant (8.20575×10^{-2} atm·L/mol·K) and K is the temperature (degrees Kelvin). Recognizing that the density of a substance is defined as:

$$d_s = M/V \qquad [25]$$

and substituting this equation into the equation [24], rearranging, and simplifying yields the vapor density (g/L) of a substance, d_v:

$$d_s = Pfw/RK \qquad [26]$$

At standard temperature (293.15 K) and pressure (1 atm), equation [26] simplifies to:

$$d_v = fw/24.47 \qquad [27]$$

The specific vapor density of a substance relative to air is determined using:

$$p_v = fw/24.47p_{air} \qquad [28]$$

where p_v is the specific vapor density of a substance (unitless) and p_{air} is the vapor density of air (g/L).

The specific vapor density, p_v, is simply the ratio of the vapor density of the substance to that of air under the same pressure and temperature. According to Weast (1986), the vapor density of dry air at 20 °C and 760 mmHg is 1.204 g/L. At 25 °C, the vapor density of air decreases slightly to 1.184 g/L. Calculated specific vapor densities are reported relative to air (set equal to 1) only for compounds that are liquids at room temperature (i.e., 20-25 °C).

Vapor pressure: The vapor pressure of a substance is defined as the pressure exerted by the vapor (gas) of a substance when it is under equilibrium conditions. It provides a semi-quantitative rate at which it will volatilize from soil and/or water. The vapor pressure of a substance is a required input parameter for calculating the air-water partition coefficient (see **Henry's law constant**), which in turn is used to estimate the volatilization rate of compounds from groundwater to the unsaturated zone and from surface water bodies to the atmosphere.

FIRE HAZARDS

Flash point: The flash point is defined as the minimum temperature at which a substance releases ignitable flammable vapors in the presence of an ignition source, e.g., spark or flame. Flash points may be determined by two methods: Tag closed cup via American Society for Testing and Materials (ASTM) method D56 or Cleveland open cup via ASTM method D93. Unless otherwise noted, all flash point values represent closed cup method determinations. Flash point values determined by the open cup method are slightly higher (about 10-15 °C) than those determined by the closed cup method; however, the open cup method is more representative of actual environmental conditions.

A material with a flash point of ≤100 °F is considered dangerous, whereas a material having a flash point >200 °F is considered to have a low flammability (Sax, 1984). Substances with flash points within this temperature range are considered to have moderate flammabilities.

Lower explosive limit: The minimum concentration (vol % in air) of a flammable gas or vapor required for ignition or explosion to occur in the presence of an ignition source (see also **Flash point**).

Upper explosive limit: The maximum concentration (vol % in air) of a flammable gas or vapor required for ignition or explosion to occur in the presence of an ignition source (see also **Flash point**).

HEALTH HAZARD DATA

Immediately Dangerous to Life or Health (IDLH): According to NIOSH (1997), the IDLH level ". . . for the purpose of respirator selection represents a maximum concentration from which, in the event of respirator failure, one could escape within 30 minutes without experiencing any escape-impairing or irreversible health effects." Concentrations are typically reported in parts per million (ppm) or milligrams per cubic meter (mg/m^3).

Exposure limits: The permissible exposure limits (PELs) in air, set by the Occupational Health and Safety Administration (OSHA), can be found in the Code of Federal Regulations (General Industry Standards for Toxic and Hazardous Substances, 1977). Unless noted otherwise, the PELs are 8-h time-weighted average (TWA) concentrations.

Also included are recommended exposure limits (RELs) published by NIOSH (1997) and/or threshold limit values (TLVs) published by the American Conference of Governmental Industrial Hygienists (ACGIH). Unless noted otherwise, the NIOSH RELs are TWA concentrations for up to a 10-h workday during a 40-h workweek. The short-term exposure limit (STEL) is a 15-min TWA that should be exceeded at any time during the workday. Recommended ceiling values are concentrations that should never be exceeded at any time during the day.

The ACGIH's TLVs, are subdivided into three exposure classes. The TLVs, which are updated annually, are defined as follows:

Threshold Limit Value-Time Weighted Average (TLV-TWA) - the TWA concentration for a normal 8-h workday and a 40-h workweek, to which nearly all workers may be repeatedly exposed, day after day, without adverse effect.

Threshold Limit Value-Short Term Exposure Limit (TLV-STEL) - the concentration to which workers can be exposed continuously for a short period of time without suffering from (1) irritation; (2) chronic or irreversible tissue damage; or (3) narcosis of sufficient degree to increase the likelihood of accidental injury, impair self-rescue, or materially reduce work efficiency, provided that the daily TLV-TWA is not exceeded. It is not a separate independent exposure limit; rather, it supplements the TWA limit where there are recognized acute toxic effects from a substance whose toxic effects are primarily of a chronic nature. STELs are recommended only where toxic effects have been reported from high short-term exposures in either humans or animals.

A STEL is defined as a 15-minute time-weighted average exposure which should not be exceeded at any time during a workday even if the 8-h TWA is within the TLV. Exposures at the STEL should not be longer than 15 minutes and should not be repeated more than four times per day. There should be at least 60 minutes between successive exposures at the STEL. An averaging period other than 15 minutes may be recommended when warranted by observed biological effects.

Threshold Limit Value-Ceiling - the concentration that should not be exceeded during any part of the working exposure.

For additional information from OSHA, write to Technical Data Center, U.S. Department of Labor, Washington, DC 20210. The NIOSH Pocket Guide to Chemical Hazards is available in electronic formats (e.g., CD-ROM and diskettes) from Industrial Hygiene Services, Inc., 941 Gardenview Office Parkway, St. Louis, MO 63141 (toll free: 1-800-732-3015; fax: 314-993-3193), Micromedex, Inc., 6200 South Syracuse Way, Suite 300, Englewood, CO (toll free: 1-800-525-9083; fax: 303-486-6464), or Praxis Environmental Systems, Inc., 251 Nortontown Road, Guilford, CT (telephone: 203-458-7111; fax: 203-458-7121). The ACGIH's address is 1300 Kemper Meadow Drive, Cincinnati, OH 45240-1634 (telephone: 513-742-2020; fax: 513-742-3355; http://www.acgih.org).

Symptoms of exposure: Effects of exposure caused by inhalation of gases, ingestion of liquids or solids, contact with eyes or skin are provided. This information should only be used as a guide to identify potential effects of exposure. If exposure to a chemical is suspected or known, seek immediate medical attention. Additional information on the symptoms and effects of chemical exposure can be obtained from Patnaik (1992), Sax and Lewis (1987) and CHRIS (1984).

Toxicology: Information on toxicity to aquatic life was obtained primarily from the peer-reviewed articles, Hartley and Kidd (1987), and the Chemical Hazard Response Information System (CHRIS, 1984). Information on toxicity to rats, mice, and other mammals were also obtained from the documented literature and from Hartley and Kidd (1987) and RTECS (1985). The absence of toxicity data does not imply that toxic effects do not exist.

Drinking water standard: Drinking water standards established by the U.S. EPA are given in maximum contaminant level goals (MCLGs) or maximum contaminant levels (MCLs). The MCLG is a non-enforceable concentration of a drinking water contaminant that is protective of adverse human health effects and allows an adequate margin of safety. The MCL is the maximum permissible concentration of a drinking water contaminant that is delivered to any user of a public water supply system. The reader should be aware that many states may have adopted more stringent drinking water standards than those promulgated or regulated by the U.S. EPA. For additional information on drinking water regulations, contact the U.S. EPA, Office of Water, Washington, D.C. (telephone: (202) 260-7571). The Safe Drinking Water Hotline is 1-800-426-4791 and is open Monday through Friday, 8:30 a.m. to 5:00 p.m. Eastern Standard Time.

Uses: Descriptions of specific uses are based on one or more of the following sources - HSDB (1989), CHRIS Manual (1984), Sittig (1985), and Verschueren (1983). This information is useful in attempting to identify potential sources of the industrial and environmental contamination.

REFERENCES

Abdul, S.A., Gibson, T.L., and Rai, D.N. Statistical correlations for predicting the partition coefficient for nonpolar organic contaminants between aquifer organic carbon and water, *Haz. Waste Haz. Mater.*, 4(3):211-222, 1987.

Allison, L.E. Organic carbon, in *Methods of Soil Analysis, Part 2.*, Black, C., Evans, D., White, J., Ensminger, L., and Clark, F., Eds. (Madison, WI: American Society of Agronomy, 1965), pp. 1367-1378.

Almgren, M., Grieser, F., Powell, J.R., and Thomas, J.K. A correlation between the solubility of aromatic hydrocarbons in water and micellar solutions, with their normal boiling points, *J. Chem. Eng. Data*, 24(4):285-287, 1979.

Amidon, G.L., Yalkowsky, S.H., Anik, S.T., and Valvani, S.C. Solubility of nonelectrolytes in polar solvents. V. Estimation of the solubility of aliphatic monofunctional compounds in water using a molecular surface area approach, *J. Phys. Chem.*, 79(21):2239-2246, 1975.

Ashton, F.M. and Monaco, T.J. *Weed Science* (New York: John Wiley & Sons, 1991), 466 p.

Banerjee, S., Yalkowsky, S.H., and Valvani, S.C. Water solubility and octanol/water partition coefficients of organics. Limitations of the solubility-partition coefficient correlation, *Environ. Sci. Technol.*, 14(10):1227-1229, 1980.

Baum, E.J. *Chemical Property Estimation: Theory and Application* (Boca Raton, FL: CRC Press, 1998), 386 p.

Bell, G.H. Solubilities of normal aliphatic acids, alcohols and alkanes in water, *Chem. Phys. Lipids*, 10:1-10, 1973.

Bodor, N., Gabanyi, Z., and Wong, C.-K. A new method for the estimation of partition coefficient, *J. Am. Chem. Soc.*, 111(11):3783-3786, 1989.

Braumann, T. Determination of hydrophobic parameters by reversed-phase liquid chromatography: theory, experimental techniques, and application in studies on quantitative structure-activity relationships, *J. Chromatogr.*, 373:191-225, 1986.

Brooke, D.N., Dobbs, A.J., and Williams, N. Octanol:water partition coefficients (P): measurement, estimation, and interpretation, particularly for chemicals with P >10^5, *Ecotoxicol. Environ. Saf.*, 11(3):251-260, 1986.

Bruggeman, W.A., Van Der Steen, J., and Hutzinger, O. Reversed-phase thin-layer chromatography of polynuclear aromatic hydrocarbons and chlorinated biphenyls. Relationship with hydrophobicity as measured by aqueous solubility and octanol-water partition coefficient, *J. Chromatogr.*, 238:335-346, 1982.

Burkhard, L.P. and Kuehl, D.W. *n*-Octanol/water partition coefficients by reverse phase liquid chromatography/mass spectrometry for eight tetrachlorinated planar molecules, *Chemosphere*, 15(2):163-167, 1986.

Camilleri, P., Watts, S.A., and Boraston, J.A. A surface area approach to determine partition coefficients, *J. Chem. Soc., Perkin Trans. 2* (September 1988), pp. 1699-1707.

Campbell, J.R., Luthy, R.G., and Carrondo, M.J.T. Measurement and prediction of distribution coefficients for wastewater aromatic solutes, *Environ. Sci. Technol.*, 17(10):582-590, 1983.

Carlson, R.M., Carlson, R.E., and Kopperman, H.L. Determination of partition coefficients by liquid chromatography, *J. Chromatogr.*, 107:219-223, 1975.

Chiou, C.T., Freed, V.H., Schmedding, D.W., and Kohnert, R.L. Partition coefficients and bioaccumulation of selected organic chemicals, *Environ. Sci. Technol.*, 11(5):475-478, 1977.

Chiou, C.T., Peters, L.J., and Freed, V.H. A physical concept of soil-water equilibria for nonionic organic compounds, *Science (Washington, D.C.)*, 206(4420):831-832, 1979.

Chiou, C.T., Porter, P.E., and Schmedding, D.W. Partition equilibria of nonionic organic compounds between organic matter and water, *Environ. Sci. Technol.*, 17(4):227-231, 1983.

Chiou, C.T., Schmedding, D.W., and Manes, M. Partitioning of organic compounds in octanol-water systems, *Environ. Sci. Technol.*, 16(1):4-10, 1982.

Cho, E.H. and Park, J.H. Prediction of soil sorption coefficients for organic nonelectrolytes from solvatochromic parameters, *Bull. Korean Chem. Soc.*, 16(3):290-293, 1995.

CHRIS Hazardous Chemical Data, Vol. 2, U.S. Department of Transportation, U.S. Coast Guard, U.S. Government Printing Office, November, 1984.

Davies, R.P. and Dobbs, A.J. The prediction of bioconcentration in fish, *Water Res.*, 18(10):1253-1262, 1984.

DeKock, A.C. and Lord, D.A. A simple procedure for determining octanol-water partition coefficients using reverse phase high performance liquid chromatography (RPHPLC), *Chemosphere*, 16(1):133-142, 1987.

Demond, A.H. and Lindner, A.S. Estimation of interfacial tension between organic liquids and water, *Environ. Sci. Technol.*, 27(12):2318-2331, 1993.

de Wolf, W., de Bruijn, J.H.M., Sienen, W., and Hermens, J.L.M. Influence of biotransformation on the relationship between bioconcentration factors and octanol-water partition coefficients, *Environ. Sci. Technol.*, 26(6):1197-1201, 1992.

Eadsforth, C.V. Application of reverse-phase H.P.L.C. for the determination of partition coefficients, *Pestic. Sci.*, 17(3):311-325, 1986.

Freeze, R.A. and Cherry, J.A. *Groundwater* (Englewood Cliffs, NJ: Prentice-Hall, 1974), 604 p.

Funasaki, N., Hada, S., and Neya, S. Partition coefficients of aliphatic ethers - molecular surface area approach, *J. Phys. Chem.*, 89(14):3046-3049, 1985.

Gerstl, Z. and Helling, C.S. Evaluation of molecular connectivity as a predictive method for the adsorption of pesticides in soils, *J. Environ. Sci. Health*, B22(1):55-69, 1987.

Gobas, F.A.P.C., Lahittete, J.M., Garofalo, G., Shiu, W.-Y., and Mackay, D. A novel method for measuring membrane-water partition coefficients of hydrophobic organic chemicals: comparison with 1-octanol-water partitioning, *J. Pharm. Sci.*, 77(3):265-272, 1988.

Govers, H., Ruepert, C., and Aiking, H. Quantitative structure-activity relationships for polycyclic aromatic hydrocarbons: correlation between molecular connectivity, physico-chemical properties, bioconcentration and toxicity in *Daphnia Pulex*, *Chemosphere*, 13(2):227-236, 1984.

Guidelines Establishing Test Procedures for the Analysis of Pollutants, U.S. Code of Federal Regulations, 40 CFR 136, 44(233):69464-69575.

Guswa, J.H., Lyman, W.J., Donigan, A.S., Jr., Lo, T.Y.R., and Shanahan, E.W. Groundwater contamination and emergency response guide (Park Ridge, NJ: Noyes Publications, 1984), 490 p.

Hafkenscheid, T.L. and Tomlinson, E. Estimation of aqueous solubilities of organic non-electrolytes using liquid chromatographic retention data, *J. Chromatogr.*, 218:409-425, 1981.

Hafkenscheid, T.L. and Tomlinson, E. Correlations between alkane/water and octan-1-ol/water distribution coefficients and isocratic reversed-phase liquid chromatographic capacity factors of acids, bases and neutrals, *Int. J. Pharm.*, 16:225-239, 1983.

Haky, J.E. and Young, A.M. Evaluation of a simple HPLC correlation method for the estimation of the octanol-water partition coefficients of organic compounds, *J. Liq. Chromatogr.*, 7(4):675-689, 1984.

Hammers, W.E., Meurs, G.J., and De Ligny, C.L. Correlations between chromatographic capacity ratio data on Lichrosorb RP-18 and partition coefficients in the octanol-water system, *J. Chromatogr.*, 247:1-13, 1982.

Hansch, C. and Anderson, S.M. The effect of intramolecular hydrophobic bonding on partition coefficients, *J. Org. Chem.*, 32:2853-2586, 1967.

Hansch, C., Leo, A., and Nikaitani, D. On the additive-constitutive character of partition coefficients, *J. Org. Chem.*, 37(20):3090-3092, 1972.

Hansch, C., Quinlan, J.E., and Lawrence, G.L. The linear free-energy relationship between partition coefficients and aqueous solubility of organic liquids, *J. Org. Chem.*, 33(1):347-350, 1968.

Harnisch, M., Mockel, H.J., and Schulze, G. Relationship between log P_{ow} shake-flask values and capacity factors derived from reversed-phase high performance liquid chromatography for *n*-alkylbenzene and some OECD reference substances, *J. Chromatogr.*, 282:315-332, 1983.

Hartley, D. and Kidd, H., Eds. *The Agrochemicals Handbook*, 2nd ed. (England: Royal Society of Chemistry, 1987.

Hawley, G.G. *The Condensed Chemical Dictionary* (New York: Van Nostrand Reinhold, 1981), 1135 p.

Hayduk, W. and Laudie, H. Prediction of diffusion coefficients for nonelectrolytes in dilute aqueous solution, *Am. Inst. Chem. Eng.*, 20(3):611-615, 1974.

Hazardous Substances Data Bank. National Library of Medicine, Toxicology Information Program, 1989.

Hermann, R.B. Theory of hydrophobic bonding. II. The correlation of hydrocarbon solubility in water with solvent cavity surface area, *J. Phys. Chem.*, 76(19):2754-2759, 1972.

Hiatt, M.H. Bioconcentration factors for volatile organic compounds in vegetation, *Anal. Chem.*, 70(5):851-856, 1998.

Hickey, J.P. and Passino-Reader, D.R. Linear solvation energy relationships: "rules of thumb" for estimation of variable values, *Environ. Sci. Technol.*, 25(10):1753-1760, 1991.

Hill, E.A. On a system of indexing chemical literature: adopted by the Classification Division of the U.S. Patent Office, *J. Am. Chem. Soc.*, 22(8):478-494, 1900.

Hine, J. and Mookerjee, P.K. The intrinsic hydrophobic character of organic compounds. Correlations in terms of structural contributions, *J. Org. Chem.*, 40(3):292-298, 1975.

Hodson, J. and Williams, N.A. The estimation of the adsorption coefficient (K_{oc}) for soils by high performance liquid chromatography, *Chemosphere*, 19(1):67-77, 1988.

Howard, P.H. *Handbook of Environmental Fate and Exposure Data for Organic Chemicals - Volume I. Large Production and Priority Pollutants* (Chelsea, MI: Lewis Publishers, 1989), 574 p.

Irmann, F. Eine einfache Korrelation zwischen Wasserlöslichkeit und Struktur von Kohlenwasserstoffen und Halogenkohlenwasserstoffen, *Chemie. Ing. Techn.*, 37:789-798, 1965.

Isnard, S. and Lambert, S. Estimating bioconcentration factors from octanol-water partition coefficient and aqueous solubility, *Chemosphere*, 17(1):21-34, 1988.

Kamlet, M.J., Doherty, R.M., Abraham, M.H., Carr, P.W., Doherty, R.F., and Taft, R.W. Linear solvation energy relationships. 41. Important differences between aqueous solubility relationships for aliphatic and aromatic solutes, *J. Phys. Chem.*, 91(7):1996-2004, 1987.

Kamlet, M.J., Doherty, R.M., Carr, P.W., Mackay, D., Abraham, M.H., and Taft, R.W. Linear solvation energy relationships. 44. Parameter estimation rules that allow accurate prediction of octanol/water partition coefficients and other solubility and toxicity properties of polychlorinated biphenyls and polycyclic aromatic hydrocarbons, *Environ. Sci. Technol.*, 22(5):503-509, 1988.

Kenaga, E.E. Correlation of bioconcentration factors of chemicals in aquatic and terrestrial organisms with their physical and chemical properties, *Environ. Sci. Technol.*, 14(5):553-556, 1980.

Kenaga, E.E. and Goring, C.A.I. Relationship between water solubility, soil sorption, octanol-water partitioning and concentration of chemicals in biota, in *Aquatic Toxicology, ASTM STP 707*, Eaton, J.G., Parrish, P.R., and Hendricks, A.C., Eds. (Philadelphia, PA: American Society for Testing and Materials, 1980), pp. 78-115.

Khaledi, M.G. and Breyer, E.D. Quantitation of hydrophobicity with micellar liquid chromatography, *Anal. Chem.*, 61(9):1040-1047, 1989.

Koch, R. Molecular connectivity index for assessing ecotoxicological behaviour of organic compounds, *Toxicol. Environ. Chem.*, 6(2):87-96, 1983.

Kraak, J.C., Van Rooij, H.H., and Thus, J.L.G. Reversed-phase ion-pair systems for the prediction of *n*-octanol-water partition coefficients of basic compounds by high-performance liquid chromatography, *J. Chromatogr.*, 352:455-463, 1986.

Krzyzanowska, T. and Szeliga, J. A method for determining the solubility of individual hydrocarbons, *Nafta*, 28:414-417, 1978.

Lande, S.S. and Banerjee, S. Predicting aqueous solubility of organic nonelectrolytes from molar volume, *Chemosphere*, 10(7):751-759, 1981.

Lande, S.S., Hagen, D.F., and Seaver, A.E. Computation of total molecular surface area from gas phase ion mobility data and its correlation with aqueous solubilities of hydrocarbons, *Environ. Toxicol. Chem.*, 4(3):325-334, 1985.

Leighton, D.T., Jr. and Calo, J.M. Distribution coefficients of chlorinated hydrocarbons in dilute air-water systems for groundwater contamination applications, *J. Chem. Eng. Data*, 26(4):382-385, 1981.

Lyman, W.J., Reehl, W.F., and Rosenblatt, D.H. *Handbook of Chemical Property Estimation Methods: Environmental Behavior of Organic Compounds* (New York: McGraw-Hill, 1982).

Means, J.C., Wood, S.G., Hassett, J.J., and Banwart, W.L. Sorption of polynuclear aromatic hydrocarbons by sediments and soils, *Environ. Sci. Technol.*, 14(2):1524-1528, 1980.

Meylan, W. and Howard, P.H. Bond contribution method for estimating Henry's law constants, *Environ. Toxicol. Chem.*, 10(10):1283-1293, 1991.

Meylan, W., Howard, P.H., and Boethling, R.S. Molecular topology/fragment contribution method for predicting soil sorption coefficients, *Environ. Sci. Technol.*, 26(8):1560-1567, 1992.

Miller, M.M., Ghodbane, S., Wasik, S.P., Tewari, Y.B., and Martire, D.E. Aqueous solubilities, octanol/water partition coefficients, and entropies of melting of chlorinated benzenes and biphenyls, *J. Chem. Eng. Data*, 29(2):184-190, 1984.

Miller, M.M., Wasik, S.P., Huang, G.-L., Shiu, W.-Y., and Mackay, D. Relationships between octanol-water partition coefficient and aqueous solubility, *Environ. Sci. Technol.*, 19(6):522-529, 1985.

Minick, D.J., Brent, D.A., and Frenz, J. Modeling octanol-water partition coefficients by reversed-phase liquid chromatography, *J. Chromatogr.*, 461:177-191, 1989.

Mirrlees, M.S., Moulton, S.J., Murphy, C.T., and Taylor, P.J. Direct measurement of octanol-water partition coefficient by high-pressure liquid chromatography, *J. Med. Chem.*, 19(5):615-619, 1976.

Mitra, A., Saksena, R.K., and Mitra, C.R. A prediction plot for unknown water solubilities of some hydrocarbons and their mixtures, *Chem. Petro-Chem. J.*, 8:16-17, 1977.

Miyake, K., Kitaura, F., Mizuno, N., and Terada, H. Phosphatidylcholine-coated silica as a useful stationary phase for high-performance liquid chromatographic determination of partition coefficients between octanol and water, *J. Chromatogr.*, 389(1):47-56, 1987.

Miyake, K., Mizuno, N., and Terada, H. Effect of hydrogen bonding on the high-performance liquid chromatographic behaviour of organic compounds. Relationship between capacity factors and partition coefficients, *J. Chromatogr.*, 439:227-235, 1988.

Miyake, K. and Terada, H. Determination of partition coefficients of very hydrophobic compounds by high-performance liquid chromatography on glyceryl-coated controlled-pore glass, *J. Chromatogr.*, 240(1):9-20, 1982.

Neely, W.B., Branson, D.R., and Blau, G.E. Partition coefficient to measure bioconcentration potential of organic chemicals in fish, *Environ. Sci. Technol.*, 8(13):1113-1115, 1974.

NIOSH Pocket Guide to Chemical Hazards, U.S. Department of Health and Human Services, U.S. Government Printing Office, 1997, 440 p.

Nirmalakhandan, N.N. and Speece, R.E. QSAR model for predicting Henry's constant, *Environ. Sci. Technol.*, 22(11):1349-1357, 1988.

Nirmalakhandan, N.N. and Speece, R.E. Prediction of aqueous solubility of organic compounds based on molecular structure, *Environ. Sci. Technol.*, 22(3):328-338, 1988a.

Nirmalakhandan, N.N. and Speece, R.E. Prediction of aqueous solubility of organic compounds based on molecular structure. 2. Application to PNAs, PCBs, PCDDs, etc., *Environ. Sci. Technol.*, 23(6):708-713, 1989.

Nyssen, G.A., Miller, E.T., Glass, T.F., Quinn, C.R., II, Underwood, J., and Wilson, D.J. Solubilities of hydrophobic compounds in aqueous-organic solvent mixtures, *Environ. Monit. Assess.*, 9(1):1-11, 1987.

Ogata, M., Fujisawa, K., Ogino, Y., and Mano, E. Partition coefficients as a measure of bioconcentration potential of crude oil in fish and sunfish, *Bull. Environ. Contam. Toxicol.*, 33(5):561-567, 1984.

Oliver, B.G. and Niimi, A.J. Bioconcentration factors of some halogenated organics for rainbow trout: limitations in their use for prediction of environmental residues, *Environ. Sci. Technol.*, 19(9):842-849, 1985.

Park, J.H., and Cho, E.H. Estimation of bioconcentration factors in fish for organic nonelectrolytes using the linear solvation energy relationship, *Bull. Korean Chem. Soc.*, 14(4):457-461, 1993.

Patil, G.S. Correlation of aqueous solubility and octanol-water partition coefficient based on molecular structure, *Chemosphere*, 22(8):723-738, 1991.

Patnaik, P. *A Comprehensive Guide to the Hazardous Properties of Chemical Substances* (New York: Van Nostrand Reinhold, 1992), 763 p.

RTECS (Registry of Toxic Effects of Chemical Substances), U.S. Department of Health and Human Services, National Institute for Occupational Safety and Health, 1985, 2050 p.

Robb, I.D. Determination of the aqueous solubility of fatty acids and alcohols, *Aust. J. Chem.*, 18:2281-2285, 1966.

Russell, C.J., Dixon, S.L., and Jurs, P.C. Computer-assisted study of the relationship between molecular structure and Henry's law constant, *Anal. Chem.*, 64(13):1350-1355, 1992.

Sabljić, A. Predictions of the nature and strength of soil sorption of organic pollutants by molecular topology, *J. Agric. Food Chem.*, 32(2):243-246, 1984.

Sabljić, A. On the prediction of soil sorption coefficients of organic pollutants from molecular structure: application of molecular topology model, *Environ. Sci. Technol.*, 21(4):358-366, 1987.

Sabljić, A. and Protić, M. Relationship between molecular connectivity indices and soil sorption coefficients of polycyclic aromatic hydrocarbons, *Bull. Environ. Contam. Toxicol.*, 28(2):162-165, 1982.

Sadek, P.C., Carr, P.W., Doherty, R.M., Kamlet, M.J., Taft, R.W., and Abraham, M.H. Study of retention processes in reversed-phase high-performance liquid chromatography by the use of the solvatochromic comparison method, *Anal. Chem.*, 57(14):2971-2978, 1985.

Sarna, L.P., Hodge, P.E., and Webster, G.R.B. Octanol-water partition coefficients of chlorinated dioxins and dibenzofurans by reversed-phase HPLC using several C_{18} columns, *Chemosphere*, 13(9):975-983, 1984.

Sax, N.I. *Dangerous Properties of Industrial Materials* (New York: Van Nostrand Reinhold, 1984), 3124 p.

Sax, N.I. and Lewis, R.J., Sr. *Hazardous Chemicals Desk Reference* (New York: Van Nostrand Reinhold, 1987), 1084 p.

Schowalter, T.T. Mechanics of secondary hydrocarbon migration and entrapment, *Assoc. Pet. Geol. Bull.*, 63(5):723-760, 1979.

Schultz, T.W., Wesley, S.K., and Baker, L.L. Structure-activity relationships for di and tri alkyl and/or halogen substituted phenols, *Bull. Environ. Contam. Toxicol.*, 43(2):192-198, 1989.

Shafer, D. *Hazardous Materials Training Handbook* (Madison, CT: Bureau of Law and Business, 1987), 206 p.

Sittig, M. *Handbook of Toxic and Hazardous Chemicals and Carcinogens* (Park Ridge, NJ: Noyes Publications, 1985), 950 p.

Sutter, J.M. and Jurs, P.C. Prediction of aqueous solubility for a diverse set of heteroatom-containing organic compounds using a quantitative structure-property relationship, *J. Chem. Inf. Comput. Sci.*, 36(1):100-107, 1996.

Suzuki, T. Development of an automatic estimation system for both the partition coefficient and aqueous solubility, *J. Comput.-Aided Mol. Des.*, 5:149-166, 1991.

Szabo, G., Prosser, S.L., and Bulman, R.A. Adsorption coefficient (K_{oc}) and HPLC retention factors of aromatic hydrocarbons, *Chemosphere*, 21(4/5):495-505, 1990.

Szabo, G., Prosser, S.L., and Bulman, R.A. Determination of the adsorption coefficient (K_{oc}) of some aromatics for soil by RP-HPLC on two immobilized humic acid phases, *Chemosphere*, 21(6):777-788, 1990a.

Szabo, G., Prosser, S.L., and Bulman, R.A. Prediction of the adsorption coefficient (K_{oc}) for soil by a chemically immobilized humic acid column using RP-HPLC, *Chemosphere*, 21(6):729-739, 1990b.

ten Hulscher, Th.E.M., van der Velde, L.E., and Bruggeman, W.A. Temperature dependence of Henry's law constants for selected chlorobenzenes, polychlorinated biphenyls and polycyclic aromatic hydrocarbons, *Environ. Toxicol. Chem.*, 11(11):1595-1603, 1992.

Tewari, Y.B., Miller, M.M., Wasik, S.P., and Martire, D.E. Aqueous solubility of octanol/water partition coefficient of organic compounds at 25.0 °C, *J. Chem. Eng. Data*, 27(4):451-454, 1982.

Threshold Limit Values and Biological Exposure Indices for 1987-1988 (Cincinnati, OH: American Conference of Governmental Industrial Hygienists, 1987), 114 p.

Toxic and Hazardous Industrial Chemicals Safety Manual for Handling and Disposal with Toxicity and Hazard Data (Tokyo, Japan: International Technical Information Institute, 1986), 700 p.

U.S. EPA. A compendium of technologies used in the treatment of hazardous wastes, Office of Research and Development, U.S. EPA Report-625/8-87-014, 1987, 49 p.

Valkó, K., Papp, O., and Darvas, F. Selection of gas chromatographic stationary phase pairs for characterization of the 1-octanol-water partition coefficient, *J. Chromatogr.*, 301:355-364, 1984.

Valvani, S.C., Yalkowsky, S.H., and Amidon, G.L. Solubility of nonelectrolytes in polar solvents. VI. Refinements in molecular surface area computations, *J. Phys. Chem.*, 80(8):829-835, 1976.

Veith, G.D. and Morris, R.T. A rapid method for estimating log P for organic chemicals, U.S. EPA Report-600/3-78-049, 1978, 15 p.

Verschueren, K. *Handbook of Environmental Data on Organic Chemicals* (New York: Van Nostrand Reinhold, 1983), 1310 p.

Wang, R., Fu, Y., and Lai, L. A new atom-additive method for calculating partition coefficients, *J. Chem. Inf. Comput. Sci.*, 37(3):615-621, 1997.

Warne, M. St.J., Connell, D.W., Hawker, D.W., and Schüürmann, G. Prediction of aqueous solubility and the octanol-water partition coefficient for lipophilic organic compounds using molecular descriptors and physicochemical properties, *Chemosphere*, 21(7):877-888, 1990.

Weast, R.C., Ed. *CRC Handbook of Chemistry and Physics*, 67th ed. (Boca Raton, FL: CRC Press, 1986), 2406 p.

Webster, G.R.B., Friesen, K.J., Sarna, L.P., and Muir, D.C.G. Environmental fate modeling of chlorodioxins: determination of physical constants, *Chemosphere*, 14(6/7):609-622, 1985.

Whitehouse, B.G. and Cooke, R.C. Estimating the aqueous solubility of aromatic hydrocarbons by high performance liquid chromatography, *Chemosphere*, 11(8):689-699, 1982.

Windholz, M., Budavari, S., Blumetti, R.F., and Otterbein, E.S., Eds. *The Merck Index*, 10th ed. (Rahway, NJ: Merck, 1983), 1463 p.

Woodburn, K.B., Delfino, J.J., and Rao, P.S.C. Retention of hydrophobic solutes on reversed-phase liquid chromatography supports: Correlation with solute topology and hydrophobicity indices, *Chemosphere*, 24(8):1037-1046, 1992.

Worthing, C.R. and Walker, S.B., Eds. *The Pesticide Manual - A World Compendium*, 9th ed. (Great Britain: British Crop Protection Council, 1991), 1141 p.

Wright, D.A., Sandler, S.I., and DeVoll, D. Infinite dilution activity coefficients and solubilities of halogenated hydrocarbons in water at ambient temperature, *Environ. Sci. Technol.*, 26(9):1828-1831, 1992.

Yalkowsky, S.H. and Banerjee, S. *Aqueous Solubility-Methods of Estimation for Organic Compounds* (New York: Marcel Dekker, 1992), 264 p.

Yalkowsky, S.H. and Valvani, S.C. Solubilities and partitioning 2. Relationships between aqueous solubilities, partition coefficients, and molecular surface areas of rigid aromatic hydrocarbons, *J. Chem. Eng. Data*, 24(2):127-129, 1979.

Yalkowsky, S.H. and Valvani, S.C. Solubility and partitioning I: solubility of nonelectrolytes in water, *J. Pharm. Sci.*, 69(8):912-922, 1980.

Yaws C.L., Hopper, J.R., Sheth, S.D., Han, M., and Pike, R.W. Solubility and Henry's law constant for alcohols in water, *Waste Manage.*, 17(8):541-547, 1997.

Yoshida, K., Tadayoshi, S., and Yamauchi, F. Relationship between molar refraction and octanol/water partition coefficient, *Ecotoxicol. Environ. Saf.*, 7(6):558-565, 1983.

Abbreviations and Symbols

A	amp
Å	angstrom
α	alpha
α_a	percent of acid that is nondissociated
α_b	percent of base that is nondissociated
$\approx$	approximately equal to
ACGIH	American Conference of Governmental Industrial Hygienists
ASTM	American Society for Testing and Materials
ATSDR	Agency for Toxic Substances and Disease Registry
asym-	asymmetric
atm	atmosphere
ATSDR	Agency for Toxic Substances and Disease Registry
b	aperture of fracture
B	average soil bulk density (g/cm^3)
β	beta
BAF	bioaccumulation factor
BCF	bioconcentration factor
BOD	biological oxygen demand $(g\ O_2/g\ chemical)$
C	equilibrium concentration (e.g., mg/L)
$°C$	degrees Centigrade (Celsius)
Ca	calcium (as in Ca-montmorillonite)
cal	calorie
CAS	Chemical Abstracts Service
CEC	cation exchange capacity (meq/L unless noted otherwise)
CERCLA	Comprehensive Environmental Response, Compensation and Liability Act
CFR	U.S. Code of Federal Regulations
CHRIS	Chemical Hazard Response Information System
cis-	stereochemical opposite of *trans-*
cm	centimeter
C_L	concentration of solute in solution
CO_2	carbon dioxide
COD	chemical oxygen demand $(g\ O_2/g\ chemical)$
cps	centipoise
C_S	concentration of sorbed solute
Cu	copper (as in Cu-montmorillonite)
d	day or days
D	diffusivity
DOT	Department of Transportation (U.S.)
d_s	density of a substance
d_v	vapor density
d_w	density of water
dyn	dyne
δ	delta
EC_{50}	concentration necessary for 50% of aquatic or plant species tested to show abnormal behavior or visible injury, respectively
EPA	(U.S.) Environmental Protection Agency
et al.	and others
eV	electron volts
$°F$	degrees Fahrenheit
FA	fulvic acid
f_{oc}	fraction of organic carbon
fw	formula weight

γ	gamma
g	gram
GAC	granular activated carbon
gal	gallon
GC/MS	gas chromatography/mass spectrometry
>	greater than
≥	greater than or equal to
h	hour(s)
HPLC	high performance liquid chromatography
HSDB	Hazardous Substances Data Bank
Hz	hertz
IARC	International Agency for Research on Cancer
IC_{50}	concentration necessary to reduce growth or immobilization 50%
IDLH	immediately dangerous to life or health
IP	ionization potential
k	rate constant
K	Kelvin (°C + 273.15)
K_a	acid dissociation constant
K_A	distribution coefficient (cm)
K_b	base dissociation constant
kcal	kilocalories
K_d	distribution coefficient (cm^3/g)
kg	kilogram
K_F	Freundlich adsorption coefficient (various units)
K_H	Henry's law constant ($atm-m^3/mol-K$)
$K_{H'}$	Henry's law constant (dimensionless)
km	kilometer
K_{oc}	soil/sediment partition coefficient (organic carbon basis)
K_{om}	soil/sediment partition coefficient (organic matter basis)
K_{ow}	n-octanol/water partition coefficient
kPa	kilopascal
K_w	dissociation constant for water (10^{-14} at 25 °C)
<	less than
≤	less than or equal to
L	liter
lb	pound
LC_{50}	lethal concentration necessary to kill 50% of the aquatic species tested
LC_{100}	lethal concentration necessary to kill 100% of the aquatic species tested
LD_{50}	lethal dose necessary to kill 50% of the mammals tested
LOEL	lowest-observed effect level
m	meter
m	molality
m-	meta (as in m-dichlorobenzene)
M	molarity (moles/liter)
M	mass
MCL	maximum contaminant level
MCLG	maximum contaminant level goal
meq	milliequivalents
mg	milligram
min	minute(s)
mL	milliliter

M_L	mass of sorbed solute
mM	millimolar per liter (mmol/L)
mmHg	millimeters of mercury
mmol	millimole
mol	mole
M_S	mass of solute in solution
mV	millivolt
Na	sodium (as in Na-montmorillonite)
N	normality (equivalents/liter)
n-, N-	normal (as in n-propyl, N-nitroso)
η	viscosity
ND	non-detect
n_e	effective porosity
ng	nanogram
NIOSH	National Institute for Occupational Safety and Health
nm	nanometer
NOEC	no-observed effect concentration
NOEL	no-observed effect level
o-	ortho (as in o-dichlorobenzene)
OSHA	Occupational Safety and Health Administration
ρ	specific density (unitless)
p	pico
p-	para (as in p-dichlorobenzene)
P	pressure
Pa	pascal
P_{air}	vapor density of air
PEL	permissible exposure limit
pg	picogram
pH	$-\log_{10}$ hydrogen ion activity (concentration)
pK_a	$-\log_{10}$ dissociation constant of an acid
pK_b	$-\log_{10}$ dissociation constant of a base
pK_w	$-\log_{10}$ dissociation constant of water
ppb	parts per billion (μg/L)
pph	parts per hundred
ppm	parts per million (mg/L)
p_v	specific vapor density
QSAR	quantitative structure-activity relationships
R	ideal gas constant (8.20575×10^{-5} atm·m^3/mol)
R_a	retardation factor for an acid
R_b	retardation factor for a base
RCRA	Resource Conservation and Recovery Act
R_d	retardation factor
REL	recommended exposure limit
RP-HPLC	reverse-phase high performance liquid chromatography
rpm	revolutions per minute
RTECS	Registry of Toxic Effects of Chemical Substances
S	solubility
S_a	solute concentration in air (mol/L)
SARA	Superfund Amendments and Reauthorization Act
sec-	secondary (as in sec-butyl)
sec	second

s_{OW}	spreading coefficient of organic liquid at air-water table interface (dyn/cm)
sp.	species
spp.	species (plural)
STEL	short-term exposure limit
sym	symmetric
t-	tertiary (as in *t*-butyl; but *tert*-butyl)
ThOD	theoretical oxygen demand
TLV	threshold limit value
TOC	total organic carbon (mg/L)
trans-	stereochemical opposite of *cis-*
T_s	temperature of a substance
T_w	temperature of water
TWA	time-weighted average
μ	micro (10^{-6})
μg	microgram
unsym	unsymmetric
U.S. EPA	U.S. Environmental Protection Agency
UV	ultraviolet
V, vol	volume
V	molar volume
V_a	molar volume of air (e.g., m^3/mol)
V_c	average linear velocity of contaminant (e.g., ft/day)
V_w	average linear velocity of groundwater (e.g., ft/day)
V_w	molar volume of water (e.g., m^3/mol)
w	mass fraction
W	watt
wk	week(s)
wt	weight
x	mole fraction
x/m	amount of solute adsorbed (e.g. mg/g of absorbent)
X_a	effective mole fraction
yr	year(s)
λ	wavelength
γ_{OW}	interfacial tension between organic liquid and water (dyn/cm)
$\gamma_{O(W)}$	surface tension of organic liquid saturated with water (dyn/cm)
$\gamma_{W(O)}$	surface tension of water saturated with organic liquid (dyn/cm)
z^+	positively charged species (milliequivalents/cm^3)

Contents

Tables

ACENAPHTHENE

Synonyms: Ethylenenaphthalene; 1,8-Ethylenenaphthalene; 1,8-Hydroacenaphthylene; Peri-ethylenenaphthalene.

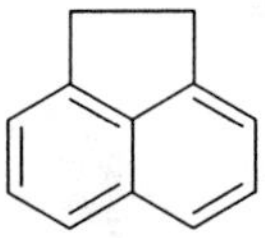

CAS Registry Number: 83-32-9
Molecular formula: $C_{12}H_{10}$
Formula weight: 154.21
RTECS: AB1000000
Merck reference: 10, 22

Physical state, color, and odor:
White crystalline solid or orthorhombic bipyramidal needles from alcohol. Odor threshold concentration ranges from 0.02 to 0.22 ppm (Keith and Walters, 1992). In Wisconsin, the taste and odor threshold concentration in water that is non-toxic to humans is 20 μg/L (ATSDR, 1995).

Melting point (°C):
96.2 (Weast, 1986)
93.0 (Pearlman et al., 1984)
89.9 (Casellato et al., 1973)

Boiling point (°C):
279 (Weast, 1986)

Density (g/cm³):
1.0242 at 90/4 °C (Weast, 1986)

Diffusivity in water (x 10^{-5} cm²/sec):
0.64 at 20 °C using method of Hayduk and Laudie (1974)

Dissociation constant, pK_a:
>15 (Christensen et al., 1975)

Entropy of fusion (cal/mol·K):
13.99 (Finke et al., 1977)

Heat of fusion (kcal/mol):
4.95 (Tsonopoulos and Prausnitz, 1971)
5.13 (Osborn and Douslin, 1975)
5.23 (Wauchope and Getzen, 1972)

Henry's law constant (x 10^{-5} atm·m³/mol):
14.6 (Mackay et al., 1979)
15.5 (Mackay and Shiu, 1981)
24.1 (Warner et al., 1987)
19 (Petrasek et al., 1983)

6.36 at 25 °C (Fendinger and Glotfelty, 1990)
16.0 at 25 °C (Shiu and Mackay, 1997)
3.47, 6.21, 10.8, 18.3, and 28.2 at 4.1, 11.0, 18.0, 25.0, and 31.0 °C, respectively (Bamford et al., 1998)

Ionization potential (eV):
7.75 ± 0.05 (Lias, 1998)

Bioconcentration factor, log BCF:
2.58 (bluegill sunfish, Veith et al., 1980)

Soil sorption coefficient, log K_{oc}:
1.25 (Mihelcic and Luthy, 1988a)
3.59, 3.79 (RP-HPLC immobilized humic acids, Szabo et al., 1990)
5.38 (average, Kayal and Connell, 1990)
4.58, 5.03 (San Francisco, CA mudflat sediments, Maruya et al., 1996)

Octanol/water partition coefficient, log K_{ow}:
3.92 (Banerjee et al., 1980; Veith et al., 1980)

Solubility in organics (g/L):
Methanol (17.9), ethanol (32.3), propanol (40.0), chloroform (400.0), benzene, or toluene (200.0), glacial acetic acid (32) (Windholz et al., 1983)

Solubility in water:
3.47 mg/kg at 25 °C (Eganhouse and Calder, 1976)
3.93 mg/L at 25 °C (Mackay and Shiu, 1977)
4.16 mg/L at 25 °C (Walters and Luthy, 1984)
2.42 mg/kg at 25 °C. In seawater (salinity 35 g/kg): 0.214, 0.55, and 1.84 mg/kg at 15, 20, and 25 °C, respectively (Rossi and Thomas, 1981)
47.8 μmol/L at 25 °C (Banerjee et al., 1980)
In mg/kg (°C): 3.57 (22.0), 4.60, 4.72 and 4.76 (30.0), 5.68, 5.73 and 6.00 (34.5), 6.8, 7.0 and 7.1 (39.3), 9.3, 9.4 and 9.4 (44.7), 12.4, 12.4 and 12.5 (50.1), 15.8, 15.9 and 16.3 (55.6), 25.9 and 27.8 (64.5), 22.8, 23.4 and 23.7 (65.2), 30.1, 33.6 and 34.3 (69.8), 35.2 (71.9), 39.1 and 40.1 (73.4), 39.3 and 40.8 (74.7) (Wauchope and Getzen, 1972)
23.3 μmol/kg at 25.0 °C (Vesala, 1974)

Vapor pressure (x 10^{-3} mmHg):
2.30 at 25 °C (Banerjee et al., 1990)
149 at 65 °C, 231 at 70 °C, 351 at 75 °C, 529 at 80 °C, 787 at 85 °C, 1,151 at 90 °C, 1,388 at 92.5 °C, 1,463 at 93.195 °C (Osborn and Douslin, 1975)
2.15 at 25 °C (Sonnefeld et al., 1983)

Environmental fate:
Biological. When acenaphthene was statically incubated in the dark at 25 °C with yeast extract and settled domestic wastewater inoculum, significant biodegradation with rapid adaptation was observed. At concentrations of 5 and 10 mg/L, 95 and 100% biodegradation, respectively, were observed after 7 d (Tabak et al., 1981). A *Beijerinckia* sp. and a mutant strain (*Beijerinckia* sp. strain B8/36) cooxidized acenaphthene to the following metabolites: 1,2-ace-naphthenediol, acenaphthenequinone, and a compound tentatively identified as 1,2-dihydroxy-acenaphthylene (Schocken and Gibson, 1984). The fungus *Cunninghamella elegans* ATCC

36112 degraded approximately 64% acenaphthene added within 72 h of incubation. Metabolites identified and their respective yields were 6-hydroxyacenaphthenone (24.8%), 1,2-ace-naphthenedione (19.9%), *trans*-1,2-dihydroxyacenaphthene (10.3%), 1,5-dihydroxyace-naphthene (2.7%), 1-acenaphthenol (2.4%), 1-acenaphthenone (2.1%), and *cis*-1,2-dihydroxy-acenaphthene (1.8%) (Pothuluri et al., 1992). In a soil-water system, acenaphthene did not biodegrade under anaerobic conditions. Under denitrification conditions, acenaphthene (water concentration 400 μg/L) degraded to non-detectable levels in 40 d. In both studies, the acclimation period was approximately 2 d (Mihelcic and Luthy, 1988).

Contaminated soil from a manufactured coal gas plant that had been exposed to crude oil was spiked with acenaphthene (400 mg/kg soil) to which Fenton's reagent (5 mL 2.8 M hydrogen peroxide; 5 mL 0.1 M ferrous sulfate) was added. The treated and nontreated soil samples were incubated at 20 °C for 56 d. Fenton's reagent did not promote the mineralization of acenaphthene by indigenous microorganisms to any appreciable extent. The mineralization of acenaphthene was enhanced only 1.2-fold when compared with the nontreated control sample. The amounts of acenaphthene recovered as carbon dioxide after treatment with and without Fenton's reagent were 20 and 17%, respectively (Martens and Frankenberger, 1995).

Based on aerobic soil column test data, the reported half-life of acenaphthene in soil ranged from 295 h to 102 d (Kincannon and Lin, 1985).

Photolytic. Based upon a rate constant of 0.23/h, the photolytic half-life of acenaphthene in water is 3 h (Fukuda et al., 1988). Behymer and Hites (1985) determined the effect of different substrates on the rate of photooxidation of acenaphthene using a rotary photoreactor equipped with a 450 W medium pressure mercury lamp (λ = 300-410 nm). The photolytic half-lives of acenaphthene absorbed onto silica gel, alumina, and fly ash were 2.0, 2.2, and 44 h, respectively. The estimated photooxidation half-life of acenaphthene in the atmosphere via OH radicals is 0.879 to 8.79 h (Atkinson, 1987).

Chemical/Physical. Ozonation in water at 60 °C produced 7-formyl-1-indanone, 1-indanone, 7-hydroxy-1-indanone, 1-indanone-7-carboxylic acid, indane-1,7-dicarboxylic acid and indane-1-formyl-7-carboxylic acid (Chen et al., 1979). Wet oxidation of acenaphthene at 320 °C yielded formic and acetic acids (Randall and Knopp, 1980).

Acenaphthene will not hydrolyze because it has no hydrolyzable functional group (Kollig, 1993).

Toxicity:

LC_{50} (contact) for earthworm (*Eisenia fetida*) 49 μg/cm^2 (Neuhauser et al., 1985).

LC_{50} (10-d) for *Rhepoxynius abronius* 2.31 mg/g organic carbon (Swartz et al., 1997).

LC_{50} (96-h) for channel catfish 1,720 μg/L, rainbow trout 670 μg/L, brown trout 580 μg/L, fathead minnows 1,600 μg/L (Holcombe et al., 1983), bluegill sunfish 1,700 μg/L (Spehar et al., 1982).

LC_{50} (96-h) for *Cyprinodon variegatus* in natural seawater 2.2 ppm (Heitmuller et al., 1981).

LC_{50} (72-h) for rainbow trout 800 μg/L, brown trout 600 μg/L, fathead minnows 1,700 μg/L (LeBlanc, 1980).

LC_{50} (48-h) for rainbow trout 1,130 μg/L, brown trout 650 μg/L (Holcombe et al., 1983), *Daphnia magna* 41 mg/L (LeBlanc, 1980), *Cyprinodon variegatus* 2.37 ppm (Heitmuller et al., 1981).

LC_{50} (24-h) for *Daphnia magna* >280 mg/L (LeBlanc, 1980), rainbow trout 1,570 μg/L, brown trout 840 μg/L (Holcombe et al., 1983), *Cyprinodon variegatus* 3.7 ppm (Heitmuller et al., 1981).

Carcinogenicity in animals has not been determined. Mutagenicity results were inconclusive (Patnaik, 1992).

Final acute values for freshwater and saltwater organisms were determined to be 80.01 and 140.8 μg/L, respectively. Acute toxicity values determined from individual toxicity tests for freshwater and saltwater organisms ranged from 120.0 to 2,045 and 160 to 16,440 μg/L, respectively (U.S. EPA, 1993).

Knobloch et al. (1969) reported acute oral LD_{50} values of 2.1 and 10 g/kg for mice and rats, respectively.

LD_{50} (intraperitoneal) for rats 600 mg/kg (Reshetyuk et al., 1970).

Heitmuller et al. (1981) reported a NOEC of 1.0 ppm.

Drinking water standard: No MCLGs or MCLs have been proposed (U.S. EPA, 1996).

Source: Detected in groundwater beneath a former coal gasification plant in Seattle, WA at a concentration of 180 μg/L (ASTR, 1995). Acenaphthene is present in tobacco smoke, asphalt, combustion of aromatic fuels containing pyridine (Verschueren, 1983). Present in diesel fuel and corresponding aqueous phase (distilled water) at concentrations of 100-600 mg/L and 4-14 μg/L, respectively (Lee et al., 1992).

Thomas and Delfino (1991) equilibrated contaminant-free groundwater collected from Gainesville, FL with individual fractions of three individual petroleum products at 24-25 °C for 24 h. The aqueous phase was analyzed for organic compounds via U.S. EPA approved test method 625. Average acenaphthene concentrations reported in water-soluble fractions of unleaded gasoline, kerosene, and diesel fuel were 1, 2, and 6 μg/L, respectively.

Acenaphthene occurs naturally in coal tar. Based on analysis of 7 coal tar samples, acenaphthene concentrations ranged from 350 to 12,000 ppm (EPRI, 1990). Detected in 1-yr aged coal tar film and bulk coal tar at concentrations of 5,800 and 5,900 mg/kg, respectively (Nelson et al., 1996).

Acenaphthene was detected in a diesel-powered medium duty truck exhaust at an emission rate of 19.3 μg/km (Schauer et al., 1999) and is a component in cigarette smoke.

Acenaphthene was detected in soot generated from underventilated combustion of natural gas doped with toluene (3 mole %) (Tolocka and Miller, 1995).

Uses: Manufacture of dye intermediates, pharmaceuticals, insecticides, fungicides, and plastics; chemical research. Derived from coal tar and petroleum refining.

ACENAPHTHYLENE

Synonym: Cyclopenta[*d,e*]naphthalene.

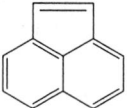

CAS Registry Number: 208-96-8
Molecular formula: $C_{12}H_8$
Formula weight: 152.20
RTECS: AB1254000

Physical state:
Solid

Melting point (°C):
92-93 (Weast, 1986)
96.2 (Lande et al., 1985)

Boiling point (°C):
280 (Aldrich, 1988)

Density (g/cm³):
0.8988 at 16/4 °C (Keith and Walters, 1992)

Diffusivity in water (x 10^{-5} cm²/sec):
0.66 at 20 °C using method of Hayduk and Laudie (1974)

Dissociation constant, pK_a:
>15 (Christensen et al., 1975)

Henry's law constant (x 10^{-4} atm·m³/mol):
1.14 at 25 °C (Warner et al., 1987)
1.125 at 25 °C (Fendinger and Glotfelty, 1990)
0.23, 0.42, 0.74, 1.25, and 1.93 at 4.1, 11.0, 18.0, 25.0, and 31.0 °C, respectively (Bamford et al., 1998)

Ionization potential (eV):
8.12 ± 0.01 (Lias, 1998)
8.29 (Rav-Acha and Choshen, 1987)

Bioconcentration factor, log BCF:
2.58 (Isnard and Lambert, 1988)

Soil sorption coefficient, log K_{oc}:
3.75, 3.83 (RP-HPLC immobilized humic acids, Szabo et al., 1990)

Octanol/water partition coefficient, log K_{ow}:
4.07 (Yoshida et al., 1983)

Solubility in organics:
Soluble in ethanol, ether, and benzene (U.S. EPA, 1985)

Solubility in water (mg/L at 25 °C):
3.93 (quoted, Verschueren, 1983)
16.1 (Walters and Luthy, 1984)

Vapor pressure (x 10^{-3} mmHg):
29.0 at 20 °C (Sims et al., 1988)
6.68 at 25 °C (Sonnefeld et al., 1983)

Environmental fate:
   Biological. When acenaphthylene was statically incubated in the dark at 25 °C with yeast extract and settled domestic wastewater inoculum, significant biodegradation with rapid adaptation was observed. At concentrations of 5 and 10 mg/L, 100 and 94% biodegradation, respectively, were observed after 7 d (Tabak et al., 1981). A *Beijerinckia* sp. and a mutant strain was able to cooxidize acenaphthylene to the following metabolites: acenaphthenequinone and a compound tentatively identified as 1,2-dihydroxyacenaphthylene. When acenaphthylene was incubated with a mutant strain (*Beijerinckia* sp. strain B8/36) one metabolite formed which was tentatively identified as *cis*-1,2-acenaphthenediol (Schocken and Gibson, 1984). Acenaphthylene in an unacclimated agricultural sandy loam soil (30.4 mg/kg) was incubated at 10 and 20 °C. After 60 d, acenaphthylene was not detected. The estimated biodegradation half-lives for acenaphthylene in aerobic soil ranged from 12 to 121 d (Coover and Sims, 1987).
   Kincannon and Lin (1985) studied the biodegradation of acenaphthylene in a various soil columns containing several mixture of sludges and all containing nitrogen and phosphorus amendments. In a Derby soil column containing a sludge mixture, acenaphthylene decreased in concentration from 130 to 42 mg/kg after 97 d. In a Derby soil column containing a wood-preserving sludge, acenaphthylene decreased in concentration from 958 to 35 mg/kg after 203 d. In a Derby soil column containing an oil sludge, acenaphthylene decreased in concentration from 772 to 244 mg/kg after 76 d. In a Masham soil column containing an oil sludge mixture, acenaphthylene decreased in concentration from 661 to 9.2 mg/kg after 76 d. Based on this study, the investigators reported biodegradation half-lives ranging from 12.3 to 121.3 d (Kincannon and Lin, 1985).
   Soil. Bossert and Bartha (1986) reported that acenaphthylene in a Nixon sandy loam soil (1 g/kg) completely disappeared in <4 months. They concluded volatilization was more important than biodegradation in the disappearance of acenaphthylene from soil.
   Photolytic. Based on data for structurally similar compounds, acenaphthylene may undergo photolysis to yield quinones (U.S. EPA, 1985). In a toluene solution, irradiation of ace-naphthylene at various temperatures and concentrations all resulted in the formation of dimers. In water, ozonation products included 1,8-naphthalene dialdehyde, 1,8-naphthalene anhydride, 1,2-epoxyacenaphthylene, 1-naphthoic acid, and 1,8-naphthaldehyde. In methanol, ozonation products included 1,8-naphthalene dialdehyde, 1,8-naphthalene anhydride, methyl 8-formyl-1-naphthoate, and dimethoxyacetal 1,8-naphthalene dialdehyde (Chen et al., 1979). Acenaphth-ylene reacts with photochemically produced OH radicals and ozone in the atmosphere. The rate constants and corresponding half-life for the vapor-phase reaction of acenaphthylene with OH radicals (500,000/cm^3) at 25 °C are 8.44 x 10^{-11} cm^3/molecule·sec and 5 h, respectively. The rate constants and corresponding half-life for the vapor-phase reaction of acenaphthylene with ozone (7 x 10^{11}/cm^3) at 25 °C are 2.52 x 10^{-16} cm^3/molecule·sec and 1 h, respectively. The overall atmospheric half-life was estimated to range from 0.191 to 1.27 h (Atkinson, 1987).
   Behymer and Hites (1985) determined the effect of different substrates on the rate of photooxidation of acenaphthylene using a rotary photoreactor equipped with a 450 W medium

pressure mercury lamp (λ = 300-410 nm). The photolytic half-lives of acenaphthylene absorbed onto silica gel, alumina, fly ash, and carbon black were 0.7, 2.2, 44, and 170 h, respectively.

Chemical/Physical. Ozonation in water at 60 °C produced 1,8-naphthalene dialdehyde, 1,8-naphthalene anhydride, 1,2-epoxyacenaphthylene, 1-naphthoic acid, and 1,8-naphthaldehydic acid (Calvert and Pitts, 1966).

Acenaphthylene will not hydrolyze because it has no hydrolyzable functional group.

Exposure limits: No individual standards have been set. As a constituent in coal tar pitch volatiles, the following exposure limits have been established (mg/m^3): NIOSH REL: TWA 0.1 (cyclohexane-extractable fraction), IDLH 80; OSHA PEL: TWA 0.2 (benzene-soluble fraction); ACGIH TLV: TWA 0.2 (benzene solubles).

Toxicity:
Carcinogenicity in animals and humans is not known (Patnaik, 1992).

Source: Detected in groundwater beneath a former coal gasification plant in Seattle, WA at concentrations ranging from nondetect (method detection limit 5 μg/L) to 250 μg/L (ASTR, 1995). Based on analysis of 7 coal tar samples, acenaphthylene concentrations ranged from 260 to 18,000 ppm (EPRI, 1990).

Detected in a distilled water-soluble fraction of used motor oil at concentrations ranging from 4.5 to 4.6 μg/L (Chen et al., 1994).

Acenaphthylene was detected in a diesel-powered medium duty truck exhaust at an emission rate of 70.1 μg/km (Schauer et al., 1999). Acenaphthylene was also detected in soot generated from underventilated combustion of natural gas doped with toluene (3 mole %) (Tolocka and Miller, 1995).

Uses: Research chemical. Derived from industrial and experimental coal gasification operations where the maximum concentrations detected in gas, liquid, and coal tar streams were 28, 4.1, and 18 mg/m^3, respectively (Cleland, 1981).

ACETALDEHYDE

Synonyms: Acetic aldehyde; Aldehyde; Ethanal; Ethyl aldehyde; NCI-C56326; RCRA waste number U001; UN 1089.

CAS Registry Number: 75-07-0
DOT: 1089
DOT label: Flammable liquid
Molecular formula: C_2H_4O
Formula weight: 44.05
RTECS: AB1925000
Merck reference: 10, 31

Physical state, color, and odor:
Colorless, mobile, fuming liquid or gas with a penetrating, pungent odor; fruity odor when diluted. Odor threshold concentration is 0.05 ppm (Amoore and Hautala, 1983).

Melting point (°C):
-121 (Weast, 1986)
-123.5 (Windholz et al., 1983)

Boiling point (°C):
20.8 (Weast, 1986)

Density (g/cm³):
0.7834 at 18/4 °C (Weast, 1986)
0.788 at 16/4 °C (Windholz et al., 1983)

Diffusivity in water (x 10⁻⁵ cm²/sec):
1.23 at 20 °C using method of Hayduk and Laudie (1974)

Dissociation constant, pK_a:
14.15 at 0 °C (HSDB, 1989)

Flash point (°C):
-37.8 (NIOSH, 1997)
-40 (open cup, Hawley, 1981)

Lower explosive limit (%):
4.0 (NIOSH, 1994)

Upper explosive limit (%):
57 (Sax and Lewis, 1987)
60 (NIOSH, 1994)

Heat of fusion (kcal/mol):
0.775 (Riddick et al., 1986)

Henry's law constant (x 10^{-5} atm·m³/mol):
6.61 at 25 °C (Buttery et al., 1969)
6.71 at 25 °C (Zhou and Mopper, 1990)
1.81 at 5 °C, 2.54 at 15 °C, 8.77 at 25 °C, 15.24 at 35 °C (Betterton and Hoffmann, 1988)
7.69 at 25 °C (Snider and Dawson, 1985; Benkelberg et al., 1995)

Ionization potential (eV):
10.21 (Franklin et al., 1969)
10.229 ± 0.007 (Lias, 1998)

Soil sorption coefficient, log K_{oc}:
Unavailable because experimental methods for estimation of this parameter for aldehydes are lacking in the documented literature. However, its miscibility in water suggests its adsorption to soil will be nominal (Lyman et al., 1982).

Octanol/water partition coefficient, log K_{ow}:
0.52 (Sangster, 1989)

Solubility in organics:
Miscible with acetone, alcohol, benzene, ether, gasoline, solvent naphtha, toluene, turpentine, and xylene (Hawley, 1981)

Solubility in water:
Miscible (Palit, 1947)

Vapor density:
1.80 g/L at 25 °C, 1.52 (air = 1)

Vapor pressure (mmHg):
740 at 20 °C (NIOSH, 1994)
900 at 25 °C (Lide, 1990)

Environmental fate:
Biological. Heukelekian and Rand (1955) reported a 5-d BOD value of 1.27 g/g which is 69.8 of the ThOD value of 1.82 g/g.

Photolytic. Photooxidation of acetaldehyde in nitrogen oxide-free air using radiation between 2900-3500 Å yielded hydrogen peroxide, alkyl hydroperoxides, carbon monoxide, and lower molecular weight aldehydes. In the presence of nitrogen oxides, photooxidation products include ozone, hydrogen peroxide, and peroxyacyl nitrates (Kopczynski et al., 1974). Anticipated products from the reaction of acetaldehyde with ozone or OH radicals in the atmosphere are formaldehyde, and carbon dioxide (Cupitt, 1980). Reacts with nitrogen dioxide forming peroxyacyl nitrates, formaldehyde, and methyl nitrate (Altshuller, 1983). Irradiation in the presence of chlorine yielded peroxyacetic acid, carbon monoxide, and carbon dioxide (Hanst and Gay, 1983). Synthetic air containing gaseous nitrous acid and exposed to artificial sunlight (λ = 300-450 nm) photooxidized acetaldehyde into formic acid, methyl nitrate, and peroxyacetal nitrate (Cox et al., 1980).

The room-temperature photooxidation of acetaldehyde in the presence of oxygen with continuous irradiation (λ >2200 Å) resulted in the following by-products: methanol, carbon monoxide, carbon dioxide, water, formaldehyde, formic acid, acetic acid, CH_3OOCH_3, and probably $CH_3C(O)OOH$ (Johnston and Heicklen, 1964).

Rate constants reported for the reaction of acetaldehyde and OH radicals in the atmosphere:

9.6 x 10^{12} cm³/mol·sec at 300 K (Hendry and Kenley, 1979), 1.5 x 10^{-11} cm³/molecule·sec (Morris et al., 1971), 1.622 x 10^{-11} cm³/molecule·sec (Sabljić and Güsten, 1990), 1.6 x 10^{-11} cm³/molecule·sec (Niki et al., 1978; Baulch et al., 1984); with NO_3: 3.02 x 10^{-15} cm³/molecule·sec at 298 K (Atkinson, 1985), 1.40 x 10^{-15} cm³/molecule·sec (Atkinson and Lloyd, 1984), 2.5 x 10^{-15} cm³/molecule·sec (Atkinson, 1985), 2.59 x 10^{-15}, 3.15 x 10^{-15}, and 2.54 x 10^{-15} cm³/molecule·sec at 298, 299, and 300 K, respectively (Atkinson, 1991); with ozone: 3.4 x 10^{-20} cm³/molecule·sec (Stedman and Niki, 1973).

Chemical/Physical. Oxidation in air yields acetic acid (Windholz et al., 1983). In the presence of sulfuric, hydrochloric, or phosphoric acids, polymerizes explosively forming trimeric paraldehyde (Huntress and Mulliken, 1941; Patnaik, 1992). In an aqueous solution at 25 °C, acetaldehyde is partially hydrated, i.e., 0.60 expressed as a mole fraction, forming a gem-diol (Bell and McDougall, 1960). Acetaldehyde decomposes above 400 °C, forming carbon monoxide and methane (Patnaik, 1992).

Prolonged exposure to air may result in the formation of explosive peroxides; easily undergoes polymerization (NIOSH, 1997).

At an influent concentration of 1,000 mg/L, treatment with granular activated carbon resulted in an effluent concentration of 723 mg/L. The adsorbability of the carbon used was 22 mg/g carbon (Guisti et al., 1974).

Exposure limits: Potential occupational carcinogen. NIOSH REL: IDLH 2,000 ppm; OSHA PEL: TWA 200 ppm (360 mg/m³); ACGIH TLV: TWA 100 ppm (180 mg/m³), STEL 150 ppm (270 mg/m³).

Symptoms of exposure: Conjunctivitis, central nervous system, eye and skin burns, and dermatitis are symptoms of ingestion. Inhalation may cause irritation of the eyes, nose, throat, and mucous membranes. At high concentrations headache, sore throat, and paralysis of respiratory muscles may occur (Windholz et al., 1983; Patnaik, 1992). An irritation concentration of 90.00 mg/m³ in air was reported by Ruth (1986).

Toxicity:

LC_{50} (48-h) for red killifish 1,820 mg/L (Yoshioka et al., 1986), hamsters 17,000 ppm/4-h, mice 1,400 ppm/4-h, rats 37 gm/m³/30-min.

Acute oral LD_{50} for rats 1,930 mg/kg (quoted, RTECS, 1985).

Source: Manufactured by oxidizing ethanol with sodium dichromate and sulfuric acid or from acetylene, dilute sulfuric acid, and mercuric oxide catalyst.

Acetaldehyde was detected in diesel fuel at a concentration of 41,800 µg/g (Schauer et al., 1999).

Acetaldehyde occurs naturally in many plant species including Merrill flowers (*Telosma cordata*), in which it was detected at a concentration of 1,026 ppm (Furukawa et al., 1993). In addition, acetaldehyde was detected in witch hazel leaves (160 ppm), orange juice (3-15 ppm), tangerines (0-2 ppm), pineapples (0.61-1.4 ppm), celery leaves, coffee seeds, cantaloupes, soybeans, carrot roots, tomatoes, tobacco leaves, apples, peaches, black currant, fishwort, peppermint, rice plants, and caraway (Duke, 1992).

Uses: Manufacture of acetic acid, acetic anhydride, aldol, aniline dyes, 1-butanol, 1,3-butylene glycol, cellulose acetate, chloral, 2-ethylhexanol, paraldehyde, pentaerythritol, peracetic acid, pyridine derivatives, terephthalic acid, trimethylolpropane, flavors, perfumes, plastics, synthetic rubbers, disinfectants, drugs, explosives, antioxidants, yeast; silvering mirrors; hardening gelatin fibers.

ACETIC ACID

Synonyms: Acetic acid (aqueous solution); Ethanoic acid; Ethylic acid; Glacial acetic acid; Methanecarboxylic acid; Pyroligneous acid; UN 2789; UN 2790; Vinegar acid.

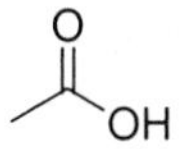

CAS Registry Number: 64-19-7
DOT: 2789 (glacial, >80% acid), 2790 (10-80% acid)
DOT label: Corrosive
Molecular formula: $C_2H_4O_2$
Formula weight: 60.05
RTECS: AF1225000
Merck reference: 10, 47

Physical state, color, and odor:
Clear, colorless, corrosive liquid with a strong vinegar-like odor. Lower and upper odor thresholds in air are 5 and 80 ppm, respectively (Keith and Walters, 1992). Very sour taste.

Melting point (°C):
16.6 (Weast, 1986)

Boiling point (°C):
117.9 (Weast, 1986)

Density (g/cm³):
1.0492 at 20/4 °C (Weast, 1986)

Diffusivity in water (x 10^{-5} cm²/sec):
1.19 at 25 °C (quoted, Hayduk and Laudie, 1974)

Dissociation constant, pK_a:
4.74 (Windholz et al., 1983)
4.733 (Korman and La Mer, 1936)
4.75 (Weast, 1986)

Flash point (°C):
39.5 (NIOSH, 1994)

Lower explosive limit (%):
4.0 (NFPA, 1984)

Upper explosive limit (%):
19.9 at 93-94 °C (NFPA, 1984)

Heat of fusion (kcal/mol):
2.801 (Riddick et al., 1986)

Henry's law constant (x 10^{-7} atm·m³/mol):
1.82 at 25 °C (Khan et al., 1995)

2.44 at 25 °C (Johnson et al., 1996)

1 at pH 4 (quoted, Gaffney et al., 1987)

1.81 at the pH range 1.6 to 1.9 (Khan and Brimblecombe, 1992)

1.08 at 25 °C (value concentration dependent, Servant et al., 1991)

133, 122, 6.88, and 1.27 at pH values of 2.13, 3.52, 5.68, and 7.14, respectively (25 °C, Hakuta et al., 1977)

Ionization potential (eV):
10.69 ± 0.03 (Franklin et al., 1969)

10.37 (Gibson et al., 1977)

10.65 ± 0.02 (Lias, 1998)

Soil sorption coefficient, log K_{oc}:
0.00 (Meylan et al., 1992)

Octanol/water partition coefficient, log K_{ow}:
-0.29, -0.30 (Sangster, 1989)

-0.17 (Leo et al., 1971)

-0.53 (Onitsuka et al., 1989)

-0.31 (Collander, 1951)

Solubility in organics:
Miscible with alcohol, carbon tetrachloride, glycerol (Windholz et al., 1983)

Solubility in water:
Miscible (NIOSH, 1994)

Vapor density:
2.45 g/L at 25 °C, 2.07 (air = 1)

Vapor pressure (mmHg):
11.4 at 20 °C, 20 at 30 °C (Verschueren, 1983)

Environmental fate:
   Biological. Near Wilmington, NC, organic wastes containing acetic acid (representing 52.6% of total dissolved organic carbon) were injected into an aquifer containing saline water to a depth of about 1,000 feet. The generation of gaseous components (hydrogen, nitrogen, hydrogen sulfide, carbon dioxide, and methane) suggests acetic acid and possibly other waste constituents, were anaerobically degraded by microorganisms (Leenheer et al., 1976).

   Plant. Based on data collected during a 2-h fumigation period, EC_{50} values for alfalfa, soybean, wheat, tobacco, and corn were 7.8, 20.1, 23.3, 41.2, and 50.1 mg/m^3, respectively (Thompson et al., 1979).

   Photolytic. A photooxidation half-life of 26.7 d was based on an experimentally determined rate constant of 6×10^{-13} cm^3/molecule·sec at 25 °C for the vapor-phase reaction of acetic acid with OH radicals in air (Atkinson, 1985). In an aqueous solution, the rate constant for the reaction of acetic acid with OH radicals was determined to be 2.70×10^{-17} cm^3/molecule·sec (Dagaut et al., 1988).

   Chemical/Physical. Ozonolysis of acetic acid in distilled water at 25 °C yielded glyoxylic acid which was oxidized readily to oxalic acid. Oxalic acid was further oxidized to carbon dioxide. Ozonolysis accompanied by UV irradiation enhanced the removal of acetic acid (Kuo et al., 1977).

At an influent concentration of 1,000 mg/L, treatment with granular activated carbon resulted in an effluent concentration of 760 mg/L. The adsorbability of the carbon used was 48 mg/g carbon (Guisti et al., 1974).

Reacts slowly with alcohols forming acetate esters (Morrison and Boyd, 1971).

Exposure limits: NIOSH REL: TWA 10 ppm (25 mg/m^3), STEL 15 ppm (37 mg/m^3), IDLH 50 ppm; OSHA PEL: TWA 10 ppm; ACGIH TLV: TWA 10 ppm, STEL 15 ppm.

Symptoms of exposure: Produces skin burns. Causes eye irritation on contact. Inhalation may cause irritation of the respiratory tract. Acute toxic effects following ingestion may include corrosion of mouth and gastrointestinal tract, vomiting, diarrhea, ulceration, bleeding from intestines and circulatory collapse (Patnaik, 1992; Windholz et al., 1983). An irritation concentration of 25.00 mg/m^3 in air was reported by Ruth (1986).

Toxicity:

LC$_{50}$ (48-h static or semi-static conditions) for *Hyale plumulosa* 310 mg/L in seawater, red killifish (*Oryzias latipes*) 350 and 94 mg/L in seawater and freshwater, respectively (Onitsuka et al., 1989).

LC$_{50}$ static bioassay values for fathead minnows in Lake Superior water maintained at 18-22 °C after 1, 24, 48, and 96 h are >315, 122, 92, and 88 mg/L, respectively (Mattson et al., 1976).

LC$_0$ (24-h) and LC$_{100}$ (24-h) for creek chub in Detroit river area 100 and 200 mg/L, respectively (Gillette et al., 1952).

Acute oral LD$_{50}$ for rats 3,530 mg/kg; LC$_{50}$ (inhalation) for mice 5,620 ppm/1 h (quoted, RTECS, 1985).

Source: Present in domestic sewage effluent at concentrations ranging from 2.5 to 36 mg/L (Verschueren, 1983).

Acetic acid occurs naturally in many plant species including Merrill flowers (*Telosma cordata*), in which it was detected at a concentration of 2,610 ppm (Furukawa et al., 1993). In addition, acetic acid was detected in cacao seeds (1,520-7,100 ppm), celery, blackwood, blueberry juice (0.7 ppm), pineapples, licorice roots (2 ppm), grapes (1,500-2,000 ppm), onion bulbs, oats, horse chestnuts, coriander, ginseng, hot peppers, linseed (3,105-3,853 ppm), ambrette, and chocolate vines (Duke, 1992).

Uses: Manufacture of acetate rayon, acetic anhydride, acetone, acetyl compounds, cellulose acetates, chloroacetic acid, ethyl alcohol, ketene, methyl ethyl ketone, vinyl acetate, plastics and rubbers in tanning; laundry sour; acidulate and preservative in foods; printing calico and dyeing silk; solvent for gums, resins, volatile oils and other substances; manufacture of nylon and fiber, vitamins, antibiotics and hormones; production of insecticides, dyes, photographic chemicals, stain removers; latex coagulant; textile printing.

ACETIC ANHYDRIDE

Synonyms: Acetic acid anhydride; Acetic oxide; Acetyl anhydride; Acetyl ether; Acetyl oxide; Ethanoic anhydrate; Ethanoic anhydride; UN 1715.

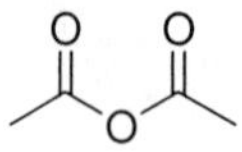

CAS Registry Number: 108-24-7
DOT: 1715
DOT label: Corrosive
Molecular formula: $C_4H_6O_3$
Formula weight: 102.09
RTECS: AK1925000
Merck reference: 10, 48

Physical state, color, and odor:
Colorless, very mobile liquid with a very strong, acetic acid-like odor. Odor threshold concentration is 0.13 ppm (Amoore and Hautala, 1983).

Melting point (°C):
-68 to -73 (Verschueren, 1983)

Boiling point (°C):
139.55 (Weast, 1986)

Density (g/cm³):
1.0820 at 20/4 °C (Weast, 1986)
1.080 at 15/4 °C (Windholz et al., 1983)

Flash point (°C):
48.9 (NIOSH, 1994)
54.4 (Windholz et al., 1983)

Lower explosive limit (%):
2.7 (NIOSH, 1994)

Upper explosive limit (%):
10.3 (NIOSH, 1994)

Heat of fusion (kcal/mol):
2.51 (Dean, 1987)

Ionization potential (eV):
≈10.00 (Lias et al., 1998)

Soil sorption coefficient, log K_{oc}:
Unavailable because experimental methods for estimation of this parameter for anhydrides and its acetic acid (hydrolysis product) are lacking in the documented literature. However, its high solubility in water and low K_{ow} for its hydrolysis product (acetic acid) suggest its adsorption to soil will be nominal (Lyman et al., 1982).

Octanol/water partition coefficient, log K_{ow}:
-0.31, -0.17 (acetic acid, Leo et al., 1971)

Solubility in organics:
Miscible with acetic acid, alcohol, and ether (Hawley, 1981)

Solubility in water:
12 wt % at 20 °C (NIOSH, 1994)

Vapor density:
4.17 g/L at 25 °C, 3.52 (air = 1)

Vapor pressure (mmHg):
3.5 at 20 °C, 5 at 25 °C, 7 at 30 °C (Verschueren, 1983)

Environmental fate:
   Chemical/Physical. Slowly dissolves in water forming acetic acid. In ethanol, ethyl acetate is formed (Windholz et al., 1983).

Exposure limits: NIOSH REL: ceiling 5 ppm (20 mg/m^3), IDLH 200 ppm; OSHA PEL: 5 ppm; ACGIH TLV: ceiling 5 ppm.

Symptoms of exposure: Severe eye and skin irritant (NIOSH, 1997). An irritation concentration of 20.00 mg/m^3 in air was reported by Ruth (1986).

Toxicity:
   Acute oral LD$_{50}$ for rats 1,780 mg/kg; LC$_{50}$ (inhalation) 1,000 ppm/4-h (quoted, RTECS, 1985).

Uses: Preparation of acetyl compounds and cellulose acetates; detection of rosin; dehydrating and acetylating agent in the production of pharmaceuticals, dyes, explosives, perfumes, and pesticides; organic synthesis.

ACETONE

Synonyms: Chevron acetone; Dimethylformaldehyde; Dimethylketal; Dimethyl ketone; DMK; Ketone propane; β-Ketopropane; Methyl ketone; Propanone; **2-Propanone**; Pyroacetic acid; Pyroacetic ether; RCRA waste number U002; UN 1090.

CAS Registry Number: 67-64-1
DOT: 1090
DOT label: Flammable liquid
Molecular formula: C_3H_6O
Formula weight: 58.08
RTECS: AL3150000
Merck reference: 10, 58

Physical state, color, and odor:
Clear, colorless, volatile, liquid with a sweet, fragrant odor. Odor threshold concentration is 13 ppm (Amoore and Hautala, 1983). Sweetish taste.

Melting point (°C):
-95.35 (Weast, 1986)

Boiling point (°C):
56.2 (Weast, 1986)

Density (g/cm³ at 20/4 °C):
0.7899 (Weast, 1986)

Diffusivity in water (x 10^{-5} cm²/sec, 25 °C):
1.28 (quoted, Hayduk and Laudie, 1974)
1.17 (mole fraction = 10^{-5}) (Gabler et al., 1996)

Flash point (°C):
-18 (NIOSH, 1997)

Lower explosive limit (%):
2.5 (NIOSH, 1997)

Upper explosive limit (%):
12.8 (NIOSH, 1997)

Dissociation constant, pK_a:
≈20 (Gordon and Ford, 1972)

Heat of fusion (kcal/mol):
1.366 (Dean, 1987)
1.360 (Riddick et al., 1986)

Henry's law constant (x 10^{-5} atm·m³/mol):
3.85 at 25 °C (Snider and Dawson, 1985)

3.30 at 25 °C (Butler and Ramchandani, 1935)
3.67 at 25 °C (Buttery et al., 1969)
3.70 at 25 °C (Hoff et al., 1993)
3.46 at 25 °C (Zhou and Mopper, 1990)
3.57 at 25 °C (Burnett, 1963)
4.00 at 25 °C (Vitenberg et al., 1975)
3.70 at 25 °C (Benkelberg et al., 1995)
2.27 at 14.9 °C, 3.03 at 25 °C, 7.69 at 35.1 °C, 11.76 at 44.9 °C (Betterton, 1991)

Ionization potential (eV):
9.68 (Gibson, 1977)
9.703 ± 0.006 (Lias, 1998)

Soil sorption coefficient, log K_{oc}:
Although experimental methods for estimation of this parameter for ketones are lacking in the documented literature, an estimated value of -0.588 was reported by Ellington et al. (1993). Its miscibility in water and low K_{oc} and K_{ow} values suggest that acetone adsorption to soil will be nominal (Lyman et al., 1982).

Octanol/water partition coefficient, log K_{ow}:
-0.24, -0.48 (Sangster, 1989)
-0.23 (Collander, 1951)

Solubility in organics:
Soluble in ethanol, benzene, and chloroform (U.S. EPA, 1985). Miscible with dimethyl-formaldehyde, chloroform, ether, and most oils (Windholz et al., 1983).

Solubility in water:
Miscible (Palit, 1947). A saturated solution in equilibrium with its own vapor had a concentration of 440.6 g/L at 25 °C (Kamlet et al., 1987).

Vapor density:
2.37 g/L at 25 °C, 2.01 (air = 1)

Vapor pressure (mmHg):
180 at 20 °C (ACGIH, 1986; NIOSH, 1997)
235 at 25 °C (Howard, 1990)

Environmental fate:
 Biological. Following a lag time of 20-25 h, acetone degraded in activated sludge (30 mg/L) at a rate constant ranging from 0.016 to 0.020/h (Urano and Kato, 1986). Bridié et al. (1979) reported BOD and COD values 1.85 and 1.92 g/g using filtered effluent from a biological sanitary waste treatment plant. These values were determined using a standard dilution method at 20 °C for a period of 5 d. Similarly, Heukelekian and Rand (1955) reported a 5-d BOD value of 0.85 g/g which is 38.5% of the ThOD value of 2.52 g/g.
 Photolytic. Photolysis of acetone in air yields carbon monoxide and free radicals, but in isopropanol, pinacol is formed (Calvert and Pitts, 1966). Photolysis of acetone vapor with nitrogen dioxide via a mercury lamp gave peroxyacetyl nitrate as the major product with smaller quantities of methyl nitrate (Warneck and Zerbach, 1992).
 Reported rate constants for the reaction of acetone with OH radicals in the atmosphere and in water are 2.16 x 10^{-13} and 1.80 x 10^{-13} cm³/molecule·sec, respectively (Wallington and Kurylo,

1987; Wallington et al., 1988a).

Chemical/Physical. Hypochlorite ions, formed by the chlorination of water for disinfection purposes, may react with acetone to form chloroform. This reaction is expected to be significant within the pH range of 6-7 (Stevens et al., 1976).

Acetone will not hydrolyze because it has no hydrolyzable functional group (Kollig, 1993).

At an influent concentration of 1,000 mg/L, treatment with granular activated carbon resulted in an effluent concentration of 782 mg/L. The adsorbability of the carbon used was 43 mg/g carbon (Guisti et al., 1974).

Complete combustion in air yields carbon dioxide and water vapor.

Exposure limits: NIOSH REL: TWA 250 ppm (590 mg/m^3), IDLH 2,500 ppm; OSHA PEL: TWA 1,000 ppm (2,400 mg/m^3); ACGIH TLV: TWA 750 ppm (1,780 mg/m^3), STEL 1,000 ppm.

Symptoms of exposure: Inhalation of acetone at high concentrations may produce headache, mouth dryness, fatigue, nausea, dizziness, muscle weakness, speech impairment, and dermatitis. Ingestion causes headache, dizziness, and drowsiness (Patnaik, 1992). Prolonged contact with skin may produce erythema and dryness (Windholz et al., 1983). An irritation concentration of 474.67 mg/m^3 in air was reported by Ruth (1986).

Toxicity:

EC$_{50}$ (96-h) for rainbow trout 5.54 mg/L (Mayer and Ellersieck, 1986).

EC$_{50}$ (48-h) and EC$_{50}$ (24-h) values for *Spirostomum ambiguum* were 9.7 and 9.35 g/L, respectively (Nałecz-Jawecki and Sawicki, 1999).

EC$_{50}$ (24-h) for rainbow trout 6.01 mg/L (Mayer and Ellersieck, 1986).

LC$_{50}$ (14-d) for *Poecilia reticulata* 6,368 mg/L (Könemann, 1981), 6,100 mg/L (Majewski et al., 1978).

LC$_{50}$ (12-d) for grass shrimp embryos (*Palaemonetes pugio*) 6.94 g/L (Rayburn and Fisher, 1997).

LC$_{50}$ (4-d) for grass shrimp embryos (*Palaemonetes pugio*) 6.78 g/L (Rayburn and Fisher, 1997).

LC$_{50}$ (96-h) for fathead minnows 8,120 mg/L (Veith et al., 1983), and rainbow trout 5.54 mL/L at pH 7.4 and 12 °C (Mayer and Ellersieck, 1986).

LC$_{50}$ (48-h) for Mexican axolotl and clawed toad 20.0 and 24.0 mg/L, respectively (Sloof and Baerselman, 1980).

LC$_{50}$ (48-h) and LC$_{50}$ (24-h) values for *Spirostomum ambiguum* were 22.1 and 26.2 g/L, respectively (Nałecz-Jawecki and Sawicki, 1999).

LC$_{50}$ (24-h) for rainbow trout 6.01 mL/L at pH 7.4 and 12 °C (Mayer and Ellersieck, 1986).

LC$_{50}$ (8-h inhalation) for rats 50,100 mg/m^3 (Possani et al., 1959).

LC$_{100}$ (4-h) for protozoan (*Paramecium caudatum*) 2.9 volume % (Rajini et al., 1989).

LC$_{100}$ (10-min) for protozoan (*Paramecium caudatum*) 2.9 volume % (Rajini et al., 1989).

Acute oral LD$_{50}$ for rats 5,800 mg/kg, mice 3,000 mg/kg (quoted, RTECS, 1985). Kimura et al. (1971) reported acute oral LD$_{50}$ values of 1.8, 4.5, 7.3, and 6.8 g/kg for newborn, immature, young adult, and old adult rats, respectively.

Source: Naturally occurs in blood and urine at low concentrations. Reported in cigarette smoke (1,100 ppm) and gasoline exhaust (2.3 to 14.0 ppm) (Verschueren, 1983).

Acetone occurs naturally in a many plant species including cuneate Turkish savory (*Satureja cuneifolia*), catmint (*Nepeta racemosa*), Guveyoto (*Origanum sipyleum*), and Topukcayi shoots (*Micromeria myrtifolia*) at concentrations of 20, 2, 2, and 0.1, respectively (Baser et al., 1992, 1993; Ozek et al., 1992; Tumen, 1991). Acetone was also detected in Turkish calamint

(*Calamintha nepeta subsp. glandulosa*) (Kirimer et al., 1992), pineapples, cauliflower leaves, tea leaves, West Indian lemongrass, jimsonweed, soybeans, carrots, bay leaves, hop flowers, apples, tomatoes, water mint leaves, alfalfa, pears, rice plants, white mulberries, clover-pepper, and roses (Duke, 1992).

Acetone was detected in diesel fuel at a concentration of 22,000 μg/g (Schauer et al., 1999).

Uses: Intermediate in the manufacture of many chemicals including acetic acid, chloroform, methyl isobutyl ketone, methyl isobutyl carbinol, methyl methacrylate, bisphenol-A; paint, varnish, and lacquer solvent; spinning solvent for cellulose acetate; to clean and dry parts for precision equipment; solvent for potassium iodide, potassium permanganate, cellulose acetate, nitrocellulose, acetylene; delustrant for cellulose acetate fibers; specification testing for vulcanized rubber products; extraction of principals from animal and plant substances; ingredient in nail polish remover; manufacture of rayon, photographic films, explosives; sealants and adhesives; pharmaceutical manufacturing; production of lubricating oils; organic synthesis.

ACETONITRILE

Synonyms: Cyanomethane; Ethanenitrile; Ethyl nitrile; Methanecarbonitrile; Methyl cyanide; NA 1648; NCI-C60822; RCRA waste number U003; UN 1648; USAF EK-488.

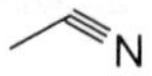

CAS Registry Number: 75-05-8
DOT: 1648
DOT label: Flammable liquid and poison
Molecular formula: C_2H_3N
Formula weight: 41.05
RTECS: AL7700000
Merck reference: 10, 62

Physical state, color, and odor:
Colorless liquid with an ether-like or pungent odor of vinegar. Odor threshold concentration is 40 ppm (Keith and Walters, 1992).

Melting point (°C):
-41.0 (Stull, 1947)
-45.7 (Weast, 1986)

Boiling point (°C):
81.6 (Weast, 1986)

Density (g/cm³):
0.7820 at 20.00/4 °C, 0.7712 at 30.00/4 °C, 0.7603 at 40.00/4 °C, 0.7492 at 50.00/4 °C (Ku and Tu, 1998)

Diffusivity in water (x 10⁻⁵ cm²/sec):
1.23 at 20 °C using method of Hayduk and Laudie (1974)

Dissociation constant, pK$_a$:
29.1 (Riddick et al., 1986)

Flash point (°C):
5.6 (open cup, NIOSH, 1997)
12.8 (Windholz et al., 1983)
6 (open cup, NFPA, 1984)

Lower explosive limit (%):
3.0 (NFPA, 1984)

Upper explosive limit (%):
16 (NIOSH, 1994)

Heat of fusion (kcal/mol):
1.952 (Riddick et al., 1986)

Henry's law constant (x 10⁻⁵ atm·m³/mol at 25 °C):
2.04 (Snider and Dawson, 1985)

1.85 (Hamm et al., 1984)
1.89 (Benkelberg et al., 1995)

Ionization potential (eV):
12.20 ± 0.01 (Lias, 1998)

Soil sorption coefficient, log K_{oc}:
Although experimental methods for estimation of this parameter for nitriles are lacking in the documented literature, an estimated value of -0.714 was reported by Ellington et al. (1993). Its miscibility in water and low K_{oc}, and K_{ow} values suggest that acetonitrile adsorption to soil will be nominal (Lyman et al., 1982).

Octanol/water partition coefficient, log K_{ow}:
-0.34 (Hansch and Anderson, 1967)
-0.54 (Tanii and Hashimoto, 1984)

Solubility in organics:
Miscible with acetamide solutions, acetone, carbon tetrachloride, chloroform, 1,2-dichloroethane, ether, ethyl acetate, methanol, methyl acetate, and many unsaturated hydrocarbons (Windholz et al., 1983). Immiscible with many saturated hydrocarbons (Keith and Walters, 1992).

Solubility in water:
Miscible (Riddick et al., 1986). A saturated solution in equilibrium with its own vapor had a concentration of 139.1 g/L at 25 °C (Kamlet et al., 1987).

Vapor density:
1.68 g/L at 25 °C, 1.42 (air = 1)

Vapor pressure (mmHg):
73 at 20 °C (NIOSH, 1994)
88.8 at 25 °C (Banerjee et al., 1990)
87.02 at 25.00 °C (Hussam and Carr, 1985)
86.6 at 25 °C (Hoy, 1970)
100 at 27.0 (Stull, 1947)
115 at 30 °C (Verschueren, 1983)

Environmental fate:
   Photolytic. A rate constant of 4.94 x 10^{-14} cm^3/molecule·sec at 24 °C was reported for the vapor-phase reaction of acetonitrile and OH radicals in air (Harris et al., 1981). Reported rate constants for the reaction of acetonitrile and OH radicals in the atmosphere and in water are 1.90 x 10^{-14} and 3.70 x 10^{-14} cm^3/molecule·sec, respectively (Kurylo and Knable, 1984). The estimated lifetime of acetonitrile in the atmosphere is estimated to range from 6 to 17 months (Arijs and Brasseur, 1986).
   Chemical/Physical. The estimated hydrolysis half-life of acetonitrile at 25 °C and pH 7 is >150,000 yr (Ellington et al., 1988). No measurable hydrolysis was observed at 85 °C at pH values 3.26 and 6.99. At 66.0 °C (pH 10.42) and 85.5 °C (pH 10.13), the hydrolysis half-lives based on first-order rate constants were 32.2 and 5.5 d, respectively (Ellington et al., 1987). The presence of hydroxide or hydronium ions facilitates hydrolysis transforming acetonitrile to the intermediate acetamide which undergoes hydrolysis forming acetic acid and ammonia (Kollig, 1993).

At an influent concentration of 1,000 mg/L, treatment with granular activated carbon resulted in an effluent concentration of 28 mg/L. The adsorbability of the carbon used was 194 mg/g carbon (Guisti et al., 1974).

Burns with a luminous flame (Windholz et al., 1983).

Exposure limits: NIOSH REL: TWA 20 ppm (34 mg/m^3), IDLH 500 ppm; OSHA PEL: TWA 40 ppm (70 mg/m^3); ACGIH TLV: TWA 40 ppm, STEL 60 ppm.

Symptoms of exposure: Inhalation may cause nausea, vomiting, asphyxia, and tightness of the chest. Symptoms of ingestion may include gastrointestinal pain, vomiting, nausea, stupor, convulsions, and weakness (Patnaik, 1992). An irritation concentration of 875.00 mg/m^3 in air was reported by Ruth (1986).

Toxicity:

EC_{50} (24-h), EC_{50} (48-h), LC_{50} (24-h), and LC_{50} (48-h) values for *Spirostomum ambiguum* were 7.35, 6.32, 17.2, and 15.2 g/L, respectively (Nałecz-Jawecki and Sawicki, 1999).

LC_{50} (inhalation) for cats 18 gm/cm^3, guinea pigs 5,655 ppm/4-h, mice 2,693 ppm/1 h, rabbits 2,828 ppm/4-h (quoted, RTECS, 1985).

Acute oral LD_{50} in guinea pigs 177 mg/kg, rats 2,730 mg/kg, mice 269 mg/kg (quoted, RTECS, 1985).

Uses: Preparation of acetamidine, acetophenone, α-naphthaleneacetic acid, thiamine; dyeing and coating textiles; extracting fish liver oils, fatty acids, and other animal and vegetable oils; recrystallizing steroids; solvent for polymers, spinning fibers, casting and molding plastics; manufacture of pharmaceuticals; chemical intermediate for pesticide manufacture; catalyst.

2-ACETYLAMINOFLUORENE

Synonyms: AAF; 2-AAF; 2-Acetamidofluorene; 2-Acetaminofluorene; Acetoaminofluorene; 2-(Acetylamino)fluorene; *N*-Acetyl-2-aminofluorene; FAA; 2-FAA; ***N*-9*H*-Fluoren-2-ylacetamide**; 2-Fluorenylacetamide; *N*-2-Fluorenylacetamide; *N*-Fluoren-2-acetylacetamide; *N*-Fluorenyl-2-acetamide; RCRA waste number U005.

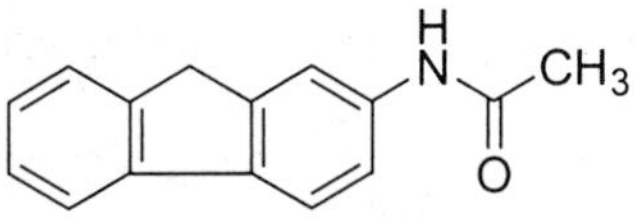

CAS Registry Number: 53-96-3
Molecular formula: $C_{15}H_{13}NO$
Formula weight: 223.27
RTECS: AB9450000
Merck reference: 10, 4058

Physical state and color:
Tan crystalline solid or needles

Melting point (°C):
194 (Weast, 1986)

Diffusivity in water (x 10^{-5} cm²/sec):
0.52 at 20 °C using method of Hayduk and Laudie (1974)

Soil sorption coefficient, log K_{oc}:
3.20 (calculated, Mercer et al., 1990)

Octanol/water partition coefficient, log K_{ow}:
3.28 (Mercer et al., 1990)

Solubility in organics:
Soluble in acetone, acetic acid, alcohol (Weast, 1986), glycols, and fat solvents (Windholz et al., 1983)

Solubility in water:
10.13 mg/L at 26.3 °C (Ellington et al., 1987)

Environmental fate:
   Biological. In the presence of suspended natural populations from unpolluted aquatic systems, the second-order microbial transformation rate constant determined in the laboratory was reported to be 4.8±2.8 x 10^{-12} L/organism·h (Steen, 1991).
   Chemical/Physical. Based on first-order rate constants determined at 85.5 °C, hydrolysis half-lives at pH values of 2.49, 2.97, 7.34, 9.80, 10.25, and 10.39 were 4.2, 12, 41, 13, 7.2, and 1.9 d, respectively (Ellington et al., 1987). Releases toxic nitrogen oxides when heated to decomposition (Sax and Lewis, 1987).

Exposure limits: Potential occupational carcinogen. Because no standards have been established, NIOSH (1997) recommends the most reliable and protective respirators be used, i.e., a self-contained breathing apparatus that has a full facepiece and is operated under positive-pressure or a supplied-air respirator that has a full facepiece and is operated under pressure-demand or under positive-pressure in combination with a self-contained breathing apparatus operated under pressure-demand or positive-pressure.

OSHA recommends that worker exposure to this chemical is to be controlled by use of engineering control, proper work practices, and proper selection of personal protective equipment. Specific details of these requirements can be found in CFR 1910.1003-1910.1016.

Toxicity:
Acute oral LD_{50} for mice 1,020 mg/kg (quoted, RTECS, 1985).

Uses: Biochemical research.

ACROLEIN

Synonyms: Acraldehyde; Acrylaldehyde; Acrylic aldehyde; Allyl aldehyde; Aqualin; Aqualine; Biocide; Crolean; Ethylene aldehyde; Magnacide H; NSC 8819; Propenal; **2-Propenal**; Prop-2-en-1-al; 2-Propen-1-one; RCRA waste number P003; Slimicide; UN 1092.

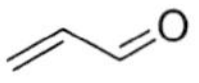

Note: Normally inhibited to prevent polymerization (Keith and Walters, 1992).

CAS Registry Number: 107-02-8
DOT: 1092 (inhibited), 2607 (stabilized dimer)
DOT label: Flammable liquid and poison
Molecular formula: C_3H_4O
Formula weight: 56.06
RTECS: AS1050000
Merck reference: 10, 123

Physical state, color, and odor:
Colorless to yellow, watery liquid with a very sharp, pungent, irritating odor. Odor threshold concentrations ranged from 0.16 ppm (Amoore and Hautala, 1983) to 0.11 mg/kg (Guadagni et al., 1963).

Melting point (°C):
-86.9 (Weast, 1986)

Boiling point (°C):
52.7 (Standen, 1963)

Density (g/cm³):
0.8410 at 20/4 °C (Weast, 1986)
0.8389 at 25/4 °C (Riddick et al., 1986)

Diffusivity in water (x 10^{-5} cm²/sec):
1.12 at 20 °C using method of Hayduk and Laudie (1974)

Flash point (°C):
-25 (Weiss, 1986)
-18 (open cup, Aldrich, 1988)

Lower explosive limit (%):
2.8 (NIOSH, 1994)

Upper explosive limit (%):
31 (NIOSH, 1994)

Henry's law constant (x 10^{-6} atm·m³/mol at 25 °C):
4.4 (quoted, Howard, 1989)
135 (Snider and Dawson, 1985)

Interfacial tension with water (dyn/cm):
35 at 20 °C (estimated, CHRIS, 1984)

Ionization potential (eV):
10.11 ± 0.01 (Lias, 1998)

Bioconcentration factor, log BCF:
2.54 (bluegill sunfish, Veith et al., 1980)

Soil sorption coefficient, log K_{oc}:
Although experimental methods for estimation of this parameter for unsaturated aldehydes are lacking in the documented literature, an estimated value of -0.588 was reported by Ellington et al. (1993). Its high solubility in water and low K_{oc}, and K_{ow} values suggest that acrolein adsorption to soil will be low (Lyman et al., 1982).

Octanol/water partition coefficient, log K_{ow}:
-0.01 (Sangster, 1989)
0.90 (Veith et al., 1980)

Solubility in organics:
Soluble in ethanol, ether, and acetone (U.S. EPA, 1985)

Solubility in water (wt %):
19.7 at 0 °C, 20.9 at 10.0 °C, 22.9 at 20.0 °C, 23.0 at 30 °C, 24.2 at 40.0 °C, 24.5 °C at 53.0 °C (Stephenson, 1993b)

Vapor density:
2.29 g/L at 25 °C, 1.94 (air = 1)

Vapor pressure (mmHg):
210 at 20 °C (NIOSH, 1997)
265 at 25 °C (Howard, 1989)
273 at 25 °C (Banerjee et al., 1990)

Environmental fate:
   Biological. Microbes in site water converted acrolein to β-hydroxypropionaldehyde (Kobayashi and Rittman, 1982). This product also forms when acrolein is hydrated in distilled water (Burczyk et al., 1968) which can revert back to acrolein. This suggests water, not site microbes, is primarily responsible for the formation of the aldehyde. When 5 and 10 mg/L of acrolein were statically incubated in the dark at 25 °C with yeast extract and settled domestic wastewater inoculum, complete degradation was observed after 7 d (Tabak et al., 1981). Activated sludge was capable of degrading acrolein at concentrations of 2,300 ppm but no other information was provided (Wierzbicki and Wojcik, 1965).
   The half-life of acrolein in unsterilized supply water samples from an irrigation area was 29 h versus 43 h in thymol-treated water. A half-life of 43 h was also reported for acrolein in buffered distilled water at identical pH. These data suggest that biotransformation occurred in these aquatic systems. At higher aqueous concentrations (6.0 to 50.5 mg/L), the marked decrease in pH suggests that carboxylic acids were formed as end products (Bowmer and Higgins, 1976).
   Bridié et al. (1979) reported BOD and COD values 0.00 and 1.72 g/g using filtered effluent from a biological sanitary waste treatment plant. These values were determined using a standard

dilution method at 20 °C for a period of 5 d. The ThOD for acrolein is 2.00 g/g.

Surface Water. In canal water, the initial acrolein concentration of 100 μg/L was reduced to 90, 50 and 30 μg/L at 5, 10, and 15 miles downstream. No explanation was given for the decrease in concentration, e.g., volatilization, chemical hydrolysis, dilution, etc. (Bartley and Gangstad, 1974).

Photolytic. Photolysis products include carbon monoxide, ethylene, free radicals, and a polymer (Calvert and Pitts, 1966). Anticipated products from the reaction of acrolein with ozone or OH radicals in the atmosphere are glyoxal, formaldehyde, formic acid, and carbon dioxide (Cupitt, 1980). The major product reported from the photooxidation of acrolein with nitrogen oxides is formaldehyde with a trace of glyoxal (Altshuller, 1983). Osborne et al. (1962) reported that acrolein was stable at 30 °C and UV light (λ = 313 nm) in the presence and absence of oxygen.

Rate constants of 1.90-2.53 x 10^{-13} cm³/molecule·sec (Atkinson, 1985), 1.99 x 10^{-11} cm³/molecule·sec (Atkinson, 1990), and 2.29 x 10^{-11} cm³/molecule·sec (Sabljić and Güsten, 1990) were reported for the gas-phase reaction of acrolein and OH radicals. Acrolein reacts with ozone and NO_3 radicals in gas-phase at rates of 6.4 x 10^{-19} cm³/molecule·sec (Atkinson and Carter, 1984) and 1.15 x 10^{-15} cm³/molecule·sec, respectively (Sabljić and Güsten, 1990).

Chemical/Physical. Wet oxidation of acrolein at 320 °C yielded formic and acetic acids (Randall and Knopp, 1980). May polymerize in the presence of light and explosively in the presence of concentrated acids (Worthing and Hance, 1991) forming disacryl, a white plastic solid (Humburg et al., 1989; Windholz et al., 1983).

In distilled water, acrolein is hydrolyzed to β-hydroxypropionaldehyde (Burczyk et al., 1968; Reinert and Rodgers, 1987; Kollig, 1993). The estimated hydrolysis half-life in water is 22 d (Burczyk et al., 1968). Bowmer and Higgins (1976) reported a disappearance half-life of 69 and 34 d in buffered water at pH values of 5 and 8.5, respectively.

In a large tank containing non-turbulent water, acrolein disappeared at a first-order rate constant of 0.83/d which corresponds to a half-life of 0.83 d (Bowmer et al., 1974).

At an influent concentration of 1,000 mg/L, treatment with granular activated carbon resulted in an effluent concentration of 694 mg/L. The adsorbability of the carbon used was 61 mg/g carbon (Guisti et al., 1974).

Exposure limits: NIOSH REL: TWA 0.1 ppm (0.25 mg/m³), STEL 0.3 ppm (0.8 mg/m³), IDLH 2 ppm; OSHA PEL: TWA 0.1 ppm; ACGIH TLV: TWA 0.1 ppm, STEL 0.3 ppm.

Symptoms of exposure: Strong lachrymator and nasal irritant. Eye contact may damage cornea. Skin contact may cause delayed pulmonary edema (Patnaik, 1992). An irritation concentration of 1.25 mg/m³ in air was reported by Ruth (1986). Inhalation of acrolein at a concentration of 153 ppm for 10 min resulted in death (quoted, Verschueren, 1983).

Toxicity:
LC_{50} (96-h) for bluegill sunfish 90 μg/L (Spehar et al., 1982).

LC_{50} (48-h) for oysters 560 μg/L, shrimp 100 μg/L (Worthing and Hance, 1991), *Daphnia magna* 83 μg/L (LeBlanc, 1980).

LC_{50} (24-h) for bluegill sunfish 79 μg/L, brown trout 46 μg/L (Burdick et al., 1964), *Daphnia magna* 23 μg/L (LeBlanc, 1980), rainbow trout 150 μg/L, mosquito fish 390 μg/L, shiners 40 μg/L (Worthing and Hance, 1991).

LC_{50} (4-h inhalation) for Sprague-Dawley rats (combined sexes) 8.3 ppm (Ballantyne et al., 1989a).

LC_{50} (1-h inhalation) for Sprague-Dawley rats (combined sexes) 26 ppm (Ballantyne et al., 1989a).

Acute oral LD_{50} for rats 46 mg/kg (Ashton and Monaco, 1991), 25.1 mg/kg, mice 40 mg/kg, rabbits 7 mg/kg (quoted, RTECS, 1985), bobwhite quail 19 mg/kg, mallard ducks 9.1 mg/kg (Worthing and Hance, 1991).

LD_{50} (inhalation) for mice 66 ppm/6-h, rats 300 mg/m^3/30-min (quoted, RTECS, 1985).

Acute percutaneous LD_{50} for rats 231 mg/kg (Worthing and Hance, 1991).

In 90-d feeding trials, the NOEL in rats is 5 mg/kg daily (Worthing and Hance, 1991).

Source: Reported in cigarette smoke (150 ppm) and gasoline exhaust (0.2 to 5.3 ppm) (quoted, Verschueren, 1983). May be present as an impurity in 2-methoxy-3,4-dihydro-2*H*-pyran (Ballantyne et al., 1989a).

Acrolein was detected in diesel fuel at a concentration of 3,400 μg/g (Schauer et al., 1999).

Uses: Intermediate in the manufacture of many chemicals (e.g., glycerine, 1,3,6-hexanediol, β-chloropropionaldehyde, 1,2,3,6-tetrahydrobenzaldehyde, β-picoline, nicotinic acid), pharmaceuticals, polyurethane, polyester resins, liquid fuel, slimicide; herbicide; anti-microbial agent; control of aquatic weeds in irrigation canals and ditches; warning agent in gases.

ACRYLAMIDE

Synonyms: Acrylamide monomer; Acylic amide; Ethylene carboxamide; Propenamide; **2-Propenamide**; RCRA waste number U007; UN 2074.

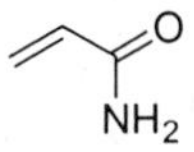

CAS Registry Number: 79-06-1
DOT: 2074
DOT label: Poison
Molecular formula: C_3H_5NO
Formula weight: 71.08
RTECS: AS3325000
Merck reference: 10, 125

Physical state, color, and odor:
Colorless, odorless solid or flake-like crystals

Melting point (°C):
84-85 (Weast, 1986)

Boiling point (°C):
125 at 25 mmHg (Windholz et al., 1983)

Density (g/cm³):
1.05 at 25/4 °C (CHRIS, 1984)
1.122 at 30/4 °C (Windholz et al., 1983)

Diffusivity in water (x 10^{-5} cm²/sec):
1.49 at 30 °C using method of Hayduk and Laudie (1974)

Flash point (°C):
137 (NIOSH, 1994)

Henry's law constant (x 10^{-9} atm·m³/mol):
3.03 at 20 °C (approximate - calculated from water solubility and vapor pressure)

Ionization potential (eV):
9.5 (Lias et al., 1998)

Soil sorption coefficient, log K_{oc}:
Although experimental methods for estimation of this parameter for amides are lacking in the documented literature, an estimated value of -0.989 was reported by Ellington et al. (1993). Its high solubility miscibility in water and low K_{oc}, and K_{ow} values suggest that acrylamide adsorption to soil will be nominal (Lyman et al., 1982). This is in agreement with the findings of Brown et al. (1980). These investigators concluded that owing to the neutral hydrophilic nature of acrylamide, no significant adsorption was observed when various adsorbents were used including natural sediments, clays, peat, sewage sludge, cationic and anionic resins.

Octanol/water partition coefficient, log K_{ow}:
-0.78 (Sangster, 1989)
-0.90 (Fujisawa and Masuhara, 1981)

Solubility in organics:
At 30 °C (g/L): acetone (631), benzene (3.46), chloroform (26.6), ethanol (862), ethyl acetate (126), heptane (0.068), methanol (1,550) (Windholz et al., 1983)

Solubility in water (g/L):
2,155 at 30 °C (quoted, Windholz et al., 1983)
2,050 (quoted, Verschueren, 1983)

Vapor pressure (x 10^{-3} mmHg):
7 at 20 °C (NIOSH, 1994)

Environmental fate:
   Biological. Bridié et al. (1979) reported BOD and COD values 0.05 and 1.33 g/g using filtered effluent from a biological sanitary waste treatment plant. These values were determined using a standard dilution method at 20 °C for a period of 5 d. When a sewage seed was used in a separate screening test, a BOD value of 0.92 g/g was obtained. In a treatment plant, a BOD value of 0.40 g/g was reported after 10 d (Mills et al., 1953). The ThOD for acrylamide is 1.35 g/g.
   Brown et al. (1980) observed that when acrylamide (0.5 mg/L) was added to estuarine water and river water with and without sediment, no acrylamide was detected after 7 d. The percentage of acrylamide remaining in seawater and sediment, sewage work effluent and sewage work effluent, and sludge were 25, 62, and 35%, respectively.
   Soil. Under aerobic conditions, ammonium ions is oxidized to nitrite ions and nitrate ions. The ammonium ions produced in soil may volatilize as ammonia or accumulate as nitrite ions in sandy or calcareous soils (Abdelmagid and Tabatabai, 1982).
   Chemical/Physical. Readily polymerizes at the melting point or under UV light. In the presence of alkali, polymerization is a violent reaction. On standing, may turn to yellowish color (Windholz et al., 1983).
   Acrylic acid and ammonium ions (Abdelmagid and Tabatabai, 1982; Brown and Rhead, 1979; Kollig, 1993) were reported as hydrolysis products. The hydrolysis rate constant at pH 7 and 25 °C was determined to be <2.1 x 10^{-6}/h, resulting in a half-life of <37.7 yr (Ellington et al., 1988). The hydrolysis half-lives are reduced significantly at varying pHs and temperature. At 88.0 °C and pH values of 2.99 and 7.04, the half-lives were 2.3 and 6.0 d, respectively (Ellington et al., 1986).
   Decomposes between 175 and 300 °C (NIOSH, 1994).

Exposure limits (mg/m³): Potential occupational carcinogen. NIOSH REL: TWA 0.03, IDLH: 60; OSHA PEL: TWA 0.3; ACGIH TLV: TWA 0.03.

Toxicity:
   LC_{50} (24-h static bioassay) for brown trout yearlings 400 mg/L (Woodiwiss and Fretwell, 1974).
   Acute oral LD_{50} for mice 107 mg/kg, quail 186 mg/kg, rats 124 mg/kg (quoted, RTECS, 1985).

Drinking water standard (final): MCLG: zero; MCL: lowest concentration obtained using conventional treatment techniques (U.S. EPA, 1996).

Uses: Synthesis of dyes; flocculents; polymers or copolymers as plastics, adhesives, soil conditioning agents; sewage and waste treatment; ore processing; permanent press fabrics.

ACRYLONITRILE

Synonyms: Acritet; Acrylon; Acrylonitrile monomer; An; Carbacryl; Cyanoethylene; ENT 54; Fumigrain; Miller's fumigrain; Nitrile; Propenenitrile; **2-Propenenitrile**; RCRA waste number U009; TL 314; UN 1093; VCN; Ventox; Vinyl cyanide.

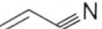

Note: Inhibited with hydroquinone monomethyl ether to prevent polymerization (Aldrich, 1990).

CAS Registry Number: 107-13-1
DOT: 1093 (inhibited)
DOT label: Flammable liquid and poison
Molecular formula: C_3H_3N
Formula weight: 53.06
RTECS: AT5250000
Merck reference: 10, 127

Physical state, color, and odor:
Clear, colorless to pale yellow, watery, volatile liquid with a sweet irritating odor resembling peach pits. Odor threshold concentration is 17 ppm (Amoore and Hautala, 1983).

Melting point (°C):
-83 (Verschueren, 1983)

Boiling point (°C):
77.5-79 (Weast, 1986)

Density (g/cm³):
0.8060 at 20/4 °C (Weast, 1986)
0.8004 at 25.00/4 °C (Aralaguppi et al., 1999a)

Diffusivity in water (x 10^{-5} cm²/sec):
1.12 at 20 °C using method of Hayduk and Laudie (1974)

Flash point (°C):
-1 (NIOSH, 1994)

Lower explosive limit (%):
3.05 ± 0.5 (Standen, 1963)

Upper explosive limit (%):
17.0 ± 0.5 (Standen, 1963)

Heat of fusion (kcal/mol):
1.489 (Riddick et al., 1986)

Henry's law constant (x 10^{-4} atm·m³/mol):
1.10 at 25 °C (Howard, 1989)

Ionization potential (eV):
10.91 ± 0.01 (Lias, 1998)

Bioconcentration factor, log BCF:
1.68 (bluegill sunfish, Veith et al., 1980)

Soil sorption coefficient, log K_{oc}:
1.10 (Captina silt loam), 1.01 (McLaurin sandy loam) (Walton et al., 1992)

Octanol/water partition coefficient, log K_{ow}:
0.00 (Sangster, 1989)
0.25 (Hansch and Leo, 1985)
0.09 (Tanii and Hashimoto, 1984)
1.20 (Veith et al., 1980)

Solubility in organics:
Soluble in ether, acetone, benzene (U.S. EPA, 1985), carbon tetrachloride, and toluene (Yoshida et al., 1983a). Miscible with alcohol and chloroform (Meites, 1963).

Solubility in water:
7.2% at 0 °C, 7.35% at 20 °C, 7.9% at 40 °C (WHO, 1983)
15.6 wt % at 20 °C (Riddick et al., 1986)
79,000 mg/L at 25 °C (Klein et al., 1957)
110 g/kg at 60.3 °C (Fordyce and Chapin, 1947)

Vapor density:
2.17 g/L at 25 °C, 1.83 (air = 1)

Vapor pressure (mmHg):
82.5 at 20 °C (Riddick et al., 1986)
137 at 30 °C (Verschueren, 1983)
107.8 at 25 °C (Howard, 1989)

Environmental fate:
Biological. Degradation by the microorganism *Nocardia rhodochrous* yielded ammonium ion and propionic acid, the latter being oxidized to carbon dioxide and water (DiGeronimo and Antoine, 1976). When 5 and 10 mg/L of acrylonitrile were statically incubated in the dark at 25 °C with yeast extract and settled domestic wastewater inoculum, complete degradation was observed after 7 d (Tabak et al., 1981). Heukelekian and Rand (1955) reported a 5-d BOD value of 1.09 g/g which is 60.0% of the ThOD value of 1.81 g/g.

Soil. Donberg et al. (1992) reported acrylonitrile (≤100 mg/kg) underwent aerobic biodegradation to non-detectable levels in a London soil after 2 d.

Photolytic. In an aqueous solution at 50 °C, UV light photooxidized acrylonitrile to carbon dioxide. After 24 h, the concentration of acrylonitrile was reduced 24.2% (Knoevenagel and Himmelreich, 1976). A rate constant of 4.06 x 10^{-12} cm^3/molecule·sec at 26 °C was reported for the vapor-phase reaction of acrylonitrile and OH radicals in air (Harris et al., 1981).

Chemical/Physical. Ozonolysis of acrylonitrile in the liquid phase yielded formaldehyde and the tentatively identified compounds glyoxal, an epoxide of acrylonitrile and acetamide (Munshi et al., 1989). In the gas phase, cyanoethylene oxide was reported as an ozonolysis product (Munshi et al., 1989a). Anticipated products from the reaction of acrylonitrile with ozone or OH radicals in the atmosphere are formaldehyde, formic acid, HC(O)CN and cyanide ions (Cupitt, 1980).

Incineration or heating to decomposition releases toxic nitrogen oxides (Sittig, 1985) and cyanides (Lewis, 1990). Wet oxidation of acrylonitrile at 320 °C yielded formic and acetic acids (Randall and Knopp, 1980). Polymerizes readily in the absence of oxygen or on exposure to visible light (Windholz et al., 1983). If acrylonitrile is not inhibited with methyl-hydroquinone, it may polymerize spontaneously or when heated in the presence of alkali (NIOSH, 1997).

The hydrolysis rate constant for acrylonitrile at pH 2.87 and 68 °C was determined to be 6.4 x 10^{-3}/h, resulting in a half-life of 4.5 d. At 68.0 °C and pH 7.19, no hydrolysis or disappearance was observed after 2 d. However, when the pH was raised to 10.76, the hydrolysis half-life was calculated to be 1.7 h (Ellington et al., 1986). Acrylonitrile hydrolyzes to acrylamide which undergoes further hydrolysis forming acrylic acid and ammonia (Kollig, 1993).

Exposure limits (ppm): Potential occupational carcinogen. NIOSH REL: TWA 1, 15-min ceiling 1, IDLH 85; OSHA PEL: TWA 2, 15-min ceiling 10; ACGIH TLV: TWA 2 (proposed).

Symptoms of exposure: Eye and skin irritant. Inhalation may cause asphyxia and headache. Ingestion and skin absorption may cause headache, lightheadedness, sneezing, weakness, nausea, and vomiting (Patnaik, 1992).

Toxicity:
   LC_{50} (96-h) for bluegill sunfish 10 mg/L (Spehar et al., 1982), brown shrimp 6 mg/L (Verschueren, 1983).
   LC_{50} (48-h) for red killifish 600 mg/L (Yoshioka et al., 1986), brown shrimp 20 mg/L, zebra fish 15 mg/L (Slooff, 1979), *Daphnia magna* 7.6 mg/L (LeBlanc, 1980).
   LC_{50} (24-h) for *Daphnia magna* 13 mg/L (LeBlanc, 1980).
   LC_{50} (1-h) for brown shrimp 3,600 mg/L (Verschueren, 1983); CD-1 mice 4,524 mg/m^3 (Willhite, 1981).
   LC_{100} (24-h) for all freshwater fish 100 mg/L (quoted, Sloof, 1979).
   Acute oral LD_{50} for mice 27 mg/kg, guinea pigs 50 mg/kg, rabbits 93 mg/kg (quoted, RTECS, 1985), rats 81 mg/kg (quoted, Slooff, 1979).

Drinking water standard (tentative): MCLG: zero (U.S. EPA, 1996).

Uses: Copolymerized with methyl acrylate, methyl methacrylate, vinyl acetate, vinyl chloride, or 1,1-dichloroethylene to produce acrylic and modacrylic fibers and high-strength fibers; ABS (acrylonitrile-butadiene-styrene) and acrylonitrile-styrene copolymers; nitrile rubber; cyano-ethylation of cotton; synthetic soil block (acrylonitrile polymerized in wood pulp); manufacture of adhesives; organic synthesis; grain fumigant; pesticide; monomer for a semi-conductive polymer that can be used similar to inorganic oxide catalysts in dehydrogenation of *t*-butyl alcohol to isobutylene and water; pharmaceuticals; antioxidants; dyes and surfactants.

ALDRIN

Synonyms: Aldrec; Aldrex; Aldrex 30; Aldrite; Aldrosol; Altox; Compound 118; Drinox; ENT 15949; Hexachlorohexahydro-*endo,exo*-dimethanonaphthalene; **1,2,3,4,10,10-Hexachloro-1,4,-4a,5,8,8a-hexahydro-1,4:5,8-dimethanonaphthalene**; 1,2,3,4,10,10-Hexachloro-1,4,4a,5,8,-8a-hexahydro-1,4-*endo,exo*-5,8-dimethanonaphthalene; 1,2,3,4,10,10-Hexachloro-1,4,4a,5,8,-8a-hexahydro-*exo*-1,4-*endo*-5,8-dimethanonaphthalene; 1,4,4a,5,8,8a-Hexahydro-1,4-*endo,exo*-5,8-dimethanonaphthalene; HHDN; NA 2761; NA 2762; NCI-C00044; Octalene; OMS 194; RCRA waste number P004; Seedrin; Seedrin liquid.

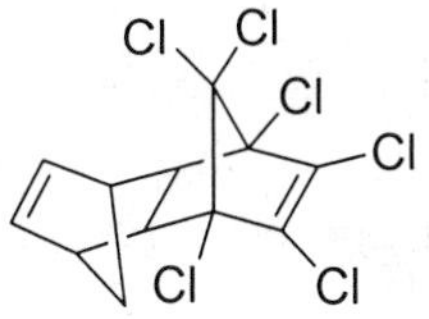

CAS Registry Number: 309-00-2
DOT: 2761 (additional numbers may exist and may be provided by the supplier)
DOT label: Poison
Molecular formula: $C_{12}H_8Cl_6$
Formula weight: 364.92
RTECS: IO2100000
Merck reference: 10, 220

Physical state, color, and odor:
White, odorless crystals when pure; technical grades are tan to dark brown with a mild chemical odor. The odor threshold in water is 17μg/kg (Sigworth, 1964).

Melting point (°C):
104 (Weast, 1986)
107 (Sims et al., 1988)

Boiling point (°C):
145 at 2 mmHg (Hayes, 1982)

Density (g/cm³):
1.70 at 20/4 °C (Hayes, 1982)

Diffusivity in water (x 10⁻⁵ cm²/sec):
0.45 at 20 °C using method of Hayduk and Laudie (1974)

Flash point (°C):
Not applicable (NIOSH, 1994)

Lower explosive limit (%):
Not applicable (NIOSH, 1994)

Upper explosive limit (%):
Not applicable (NIOSH, 1994)

Henry's law constant (x 10^{-6} atm·m^3/mol):
1.4 (Eisenreich et al., 1981)
496 at 25 °C (Warner et al., 1987)

Bioconcentration factor, log BCF:
4.10 (*Chlorella fusca*, Freitag et al., 1982; Geyer et al., 1984)
4.26 (activated sludge), 3.44 (golden ide) (Freitag et al., 1985)
3.50 (freshwater fish), 3.50 (fish, microcosm) (Garten and Trabalka, 1983)
4.13 freshwater clam (*Corbicula manilensis*) (Hartley and Johnston, 1983)

Soil sorption coefficient, log K_{oc}:
2.61 (Kenaga, 1980a)
4.68 (Geescroft/Rothamsted Farm soil, Lord et al., 1980)
4.69 (Batcombe silt loam, Briggs, 1981)
5.38 (Taichung soil: pH 6.8, 25% sand, 40% silt, 35% clay) (Ding and Wu, 1995)

Octanol/water partition coefficient, log K_{ow}:
5.52 (Travis and Arms, 1988)
6.496 (de Bruijn et al., 1989)
7.4 (Briggs, 1981)

Solubility in organics:
50 g/L in alcohol at 25 °C (quoted, Meites, 1963)

Solubility in water (μg/L):
27 at 25-29 °C (Park and Bruce, 1968)
17 at 25 °C (Weil et al., 1974)
105 at 15 °C, 180 at 25 °C, 350 at 35 °C, 600 at 45 °C (particle size $\leq$5 μ, Biggar and Riggs, 1974)

Vapor pressure (x 10^{-5} mmHg):
7.5 at 20 °C (quoted, Windholz et al., 1983)
2.31 at 20 °C (quoted, Martin, 1972)
15.2, 17.3 at 25 °C (Hinckley et al., 1990)
24.9 at 25 °C (Bidleman, 1984)

Environmental fate:
Biological. Dieldrin is the major metabolite from the microbial degradation of aldrin by oxidation or epoxidation (Lichtenstein and Schulz, 1959; Korte et al., 1962; Kearney and Kaufman, 1976). Microorganisms responsible for this reaction were identified as *Aspergillus niger*, *Aspergillus flavus*, *Penicillium chrysogenum*, and *Penicillium notatum* (Korte et al., 1962). Dieldrin may further degrade to photodieldrin (Kearney and Kaufman, 1976). A pure culture of the marine alga, namely *Dunaliella* sp., degraded aldrin to dieldrin and the diol 23.2 and 5.2%, respectively (Patil et al., 1972). In four successive 7-d incubation periods, aldrin (5 and 10 mg/L) was recalcitrant to degradation in a settled domestic wastewater inoculum (Tabak et al., 1981). Incubation with a mixed anaerobic population resulted in a degradation yield of 87% in 4 d. Two dechlorinated products were reported (Maule et al., 1987).

Soil. Patil and Matsumura (1970) reported 13 of 20 soil microorganisms were able to degrade aldrin to dieldrin under laboratory conditions. Harris and Lichtenstein (1961) studied the volatilization of aldrin (4 ppm) in Plainfield sand and quartz sands. Air was passed over the soil at a wind speed of 1 L/min for 6 h at 22 °C and 38% relative humidity. No volatilization was observed when both soils were dry. When the soils were wet, however, the amount of aldrin that volatilized from wet Plainfield and quartz sands were 4 and 2.19%, respectively. When the relative humidity was increased to 100%, 4.52 and 7.33% of the aldrin volatilized from Plainfield and quartz sands.

Aldrin was found to be very persistent in an agricultural soil. Fifteen years after application at a rate of 20 lb/acre, 5.8% of the applied amount was recovered as dieldrin and 0.2% was recovered as photodieldrin (Lichtenstein et al., 1971). Patil and Matsumura (1970) reported 13 of 20 soil microorganisms were able to degrade aldrin to dieldrin under laboratory conditions.

Plant. Photoaldrin and photodieldrin formed when aldrin was codeposited on bean leaves and exposed to sunlight (Ivie and Casida, 1971a). Dieldrin and 1,2,3,4,7,8-hexachloro-1,4,4a,6,7,7a-hexahydro-1,4-*endo*-methyleneindene-5,7-dicarboxylic acid were identified in aldrin-treated soil on which potatoes were grown (Klein et al., 1973).

Surface Water. Under oceanic conditions, aldrin may undergo dihydroxylation at the chlorine free double bond to produce aldrin diol (Verschueren, 1983). In a river die-away test using raw water from the Little Miami River in Ohio, 26, 60, and 80% of aldrin present degraded after 2, 4, and 6 wk, respectively (Eichelberger and Lichtenberg, 1971).

Mackay and Wolkoff (1973) estimated an evaporation half-life of 10.1 d from a surface water body that is 25 °C and 1 m deep. Singmaster (1975) studied the rate of volatilization of aldrin (1 ng/L) in a flask filled with 0.9 L water obtained from California. The flask was gently stirred and an air stream was passed over the air-water interface. He reported volatilization half-lives of 0.38, 0.59, and 0.60 h from San Francisco Bay, American River, and Sacramento River, respectively.

Photolytic. Aldrin exhibits low absorption at wavelengths greater than 290 nm (Gore et al., 1971). Photolysis of 0.33 ppb aldrin in San Francisco Bay water by sunlight produced photodieldrin with a reported photolysis half-life of 1.1 d (Singmaster, 1975). Aldrin on silica gel plates in the presence of photosensitizers and exposed to sunlight produced photoaldrin (Ivie and Casida, 1971). Photoaldrin also formed in a 77% yield when a benzene solution containing aldrin and benzophenone as a sensitizer was exposed to UV light (λ = 268-356 nm) (Rosen and Carey, 1968). Photodegradation of aldrin by sunlight yielded the following products after 1 month: dieldrin, photodieldrin, photoaldrin, and a polymeric substance at yields of 4.1%, 24.1%, 9.6%, and 59.7%, respectively (Rosen and Sutherland, 1967). Photolysis of solid aldrin using a high pressure mercury lamp with a Pyrex filter (λ >300 nm) yielded a polymeric substance with small amounts of photoaldrin, dieldrin, hydrochloric acid, and carbon dioxide (Gäb et al., 1974). When aldrin (166.8 mg) adsorbed on silica gel surface was irradiated by a mercury high pressure lamp (λ >230 nm), however, high yields of epoxides were formed. Photoproducts identified and their respective conversion yields were dieldrin (63.9%), photoaldrin (5.1%), photodieldrin (2.7%), photoketoaldrin (4.7%), polymer and polar products (23.6%) (Gäb et al., 1975).

Sunlight and UV light can convert aldrin to photoaldrin (Georgacakis and Khan, 1971). Oxygen atoms can also convert aldrin to dieldrin (Saravanja-Bozanic et al., 1977). When aldrin vapor (5 mg) in a reaction vessel was irradiated by a sunlamp for 45 h, 14-34% degraded to dieldrin (50-60 μg) and photodieldrin (20-30 μg). However, when the aldrin vapor concentration was reduced to 1 μg and irradiation time extended to 14 d, 60% degraded to dieldrin (0.63 μg), photodieldrin (0.02 μg) and photoaldrin (0.02 μg) (Crosby and Moilanen, 1974). When an aqueous solution containing aldrin was photooxidized by UV light at 90-95 °C, 25, 50, and 75% degraded to carbon dioxide after 14.1, 28.2, and 109.7 h, respectively (Knoevenagel and Himmelreich, 1976). Aldrin in a hydrogen peroxide solution (5 μM) was irradiated by UV light (λ = 290 nm). After 12 h, the aldrin concentration was reduced 79.5%. Dieldrin, photoaldrin, and an unidentified compound were reported as the end products (Draper and Crosby, 1984). After a short-term exposure to sunlight (<1 h), aldrin on silica gel chromatoplates was converted to photoaldrin. Photodecomposition was accelerated by several photosensitizing agents (Ivie and Casida, 1971).

When an aqueous solution of aldrin (0.07 μM) in natural water samples collected from California and Hawaii were irradiated (λ <220 nm) for 36 h, 25% was photooxidized to dieldrin. By comparison, no loss was reported when aldrin in deionized water was subject to UV light for 10 h. However, when aldrin in sterile paddy water containing 0.1% acetone or acetaldehyde was irradiated by UV light for 60 min, dieldrin formed at yields of 24 and 26%, respectively (Ross and Crosby, 1975). Dieldrin also formed as the major product when a film of aldrin was irradiated at 254 nm (Roburn, 1963).

Chemical/Physical. In an aqueous solution containing peracetic acid, aldrin was transformed to dieldrin in the dark (Ross and Crosby, 1975). Oxidation of aldrin by oxygen atoms yielded dieldrin (Saravanja-Bozanic et al., 1977). The hydrolysis rate constant for aldrin at pH 7 and 25 °C was determined to be 3.8 x 10^{-5}/h, resulting in a half-life of 760 d (Ellington et al., 1988). At higher temperatures, the hydrolysis half-lives decreased significantly. At 68 °C and pH values of 3.03, 6.99, and 10.70, the calculated hydrolysis half-lives were 1.9, 2.9, and 2.5 d, respectively (Ellington et al., 1986). No disappearance of aldrin was observed in water after 2 wk at pH 11 and 85 °C (Kollig, 1993).

In hexane and water containing 10% acetone, ozone readily oxidized aldrin to dieldrin (Hoffman and Eichelsdoerfer, 1971). When heated to decomposition, toxic chlorides are released (Lewis, 1990).

Exposure limits (mg/m³): Potential occupational carcinogen. NIOSH REL: 0.25, IDLH 25; OSHA PEL: TWA 0.25; ACGIH TLV: TWA 0.25.

Symptoms of exposure: Inhalation may cause nausea, vomiting, asphyxia, and tightness of the chest. Symptoms of ingestion may include gastrointestinal pain, vomiting, nausea, stupor, convulsions, and weakness (Patnaik, 1992).

Toxicity:

EC_{50} (48-h) for mature *Cypridopsis vidua* 18 μg/L (pH 7.4, 21 °C), first instar *Daphnia pulex* 28 μg/L (pH 7.1, 15 °C), first instar *Simocephalus serrulatus* 28 μg/L (pH 7.1, 15 °C), (Mayer and Ellersieck, 1986).

LC_{50} (96-h) for American eel 5 ppb, mummichog 4-8 ppb, striped killifish 17 ppb, Atlantic silverside 13 ppb, striped mullet 100 ppb, bluehead 12 ppb, northern puffer 36 ppb, striped bass 10 μg/L, pumpkinseed 20 μg/L, white perch 42 μg/L (Verschueren, 1983), fathead minnows 28 μg/L, bluegill sunfish 13 μg/L (Henderson et al., 1959), rainbow trout 17.7 μg/L, coho salmon 45.9 μg/L, chinook 7.5 μg/L (Katz, 1961), rainbow trout 10 μg/L (Macek and McAllister, 1970), fathead minnow 8.2 μg/L (pH 7.1, 18 °C), black bullhead 19 μg/L (pH 7.1, 24 °C), channel catfish 53 μg/L (pH 7.4, 18 °C), bluegill sunfish 5.6 μg/L (pH 7.4, 24 °C), largemouth bass 5 μg/L (pH 8, 18 °C), tadpole 68 μg/L, chinook salmon 14.3 μg/L (flow-through test, pH 7.2, 15 °C), immature *Palaemonetes kadiakensis* 50 μg/L (pH 7.1, 21 °C), second-year class *Pteronarcys californica* 1.3 μg/L (pH 7.1, 15 °C), rainbow trout 2.6 μg/L (pH 7.1, 18 °C), Fowlers tadpole 68 μg/L (pH 7.1, 16 °C) (Mayer and Ellersieck, 1986).

LC_{50} (48-h) for mosquito fish 36 ppb (Verschueren, 1985), red killifish 220 μg/L (Yoshioka et al., 1986).

LC_{50} (24-h) for bluegill sunfish 260 ppb (Verschueren, 1983), rainbow trout 4.5 μg/L (pH 7.1, 18 °C), fathead minnow 15 μg/L (pH 7.1, 18 °C), black bullhead 22 μg/L (pH 7.1, 24 °C), channel catfish 53 μg/L (pH 7.4, 18 °C), bluegill sunfish 130 μg/L (pH 7.4, 24 °C), largemouth bass 19 μg/L (pH 8, 18 °C), immature *Palaemonetes kadiakensis* 120 μg/L (pH 7.1, 21 °C), second-year class *Pteronarcys californica* 30 μg/L (pH 7.1, 15 °C), Fowlers tadpole >180 μg/L (pH 7.1, 16 °C) (Mayer and Ellersieck, 1986).

Acute oral LD_{50} for male and female rats: 39, 60 mg/kg (Windholz et al., 1983), wild birds

7.2 mg/kg, cats 10 mg/kg, chickens 10 mg/kg, ducks 520 mg/kg, dogs 65 mg/kg, guinea pigs 33 mg/kg, hamsters 100 mg/kg, mice 44 mg/kg, pigeons 56.2 mg/kg, quail 42.1 mg/kg, rabbits 50 mg/kg (quoted, RTECS, 1985).

Drinking water standard: No MCLGs or MCLs have been proposed (U.S. EPA, 1996).

Uses: Insecticide and fumigant.

ALLYL ALCOHOL

Synonyms: AA; Allyl al; Allylic alcohol; 3-Hydroxypropene; Orvinylcarbinol; Propenol; Propenol-3; Propen-1-ol-3; 1-Propenol-3; 1-Propen-3-ol; 2-Propenol; **2-Propen-1-ol**; Propenyl alcohol; 2-Propenyl alcohol; RCRA waste number P005; UN 1098; Vinyl carbinol.

CAS Registry Number: 107-18-6
DOT: 1098
DOT label: Flammable liquid; poison
Molecular formula: C_3H_6O
Formula weight: 58.08
RTECS: BA5075000
Merck reference: 10, 279

Physical state, color, and odor:
Colorless liquid with a pungent, mustard-like odor. Odor threshold concentration is 1.1 ppm (Amoore and Hautala, 1983).

Melting point (°C):
-129 (Weast, 1986)

Boiling point (°C):
97.1 (Weast, 1986)

Density (g/cm³):
0.8540 at 20/4 °C (Weast, 1986)
0.85511 at 15/4 °C (Riddick et al., 1986)

Diffusivity in water (x 10^{-5} cm²/sec):
1.10 at 20 °C using method of Hayduk and Laudie (1974)

Flash point (°C):
23.9, 21.1 (open cup, Windholz et al., 1983)

Lower explosive limit (%):
2.5 (NIOSH, 1994)

Upper explosive limit (%):
18.0 (NIOSH, 1994)

Henry's law constant (x 10^{-6} atm·m³/mol):
5.00 at 25 °C (Hine and Mookerjee, 1975)

Ionization potential (eV):
9.67 ± 0.03 (Lias, 1998)

Soil sorption coefficient, log K_{oc}:
0.51 (calculated, Mercer et al., 1990)

Octanol/water partition coefficient, log K_{ow}:
0.17 (Leo et al., 1971)

Solubility in organics:
Miscible with alcohol, chloroform, ether, and petroleum ether (Windholz et al., 1983)

Solubility in water:
Miscible (Gunther et al., 1968)

Vapor density:
2.37 g/L at 25 °C, 2.01 (air = 1)

Vapor pressure (mmHg):
20 at 20 °C, 32 at 30 °C (Verschueren, 1983)
28.1 at 25 °C (Banerjee et al., 1990)

Environmental fate:
Biological. Bridié et al. (1979) reported BOD and COD values 1.79 and 2.12 g/g using filtered effluent from a biological sanitary waste treatment plant. These values were determined using a standard dilution method at 20 °C for a period of 5 d. The ThOD for allyl alcohol is 2.21 g/g.

Photolytic. Atkinson (1985) reported a rate constant of 2.59 x 10^{11} cm³/molecule·sec at 298 K. Based on an atmospheric OH concentration of 1.0 x 10^6 molecule/cm³, the reported half-life of allyl alcohol is 0.35 d. The reaction of allyl alcohol results in the OH addition to the C=C bond (Grosjean, 1997).

Chemical/Physical. Will slowly polymerize over time into a viscous liquid (Windholz et al., 1983). Polymerization may be caused by elevated temperatures, oxidizers, or peroxides (NIOSH, 1997).

Irradiation of an aqueous solution at 50 °C for 24 h resulted in a 13.9% yield of carbon dioxide (Knoevenagel and Himmelreich, 1976).

At an influent concentration of 1,010 mg/L, treatment with granular activated carbon resulted in an effluent concentration of 789 mg/L. The adsorbability of the carbon used was 24 mg/g carbon (Guisti et al., 1974).

Exposure limits: NIOSH REL: TWA 2 ppm (5 mg/m³), STEL 4 ppm (10 mg/m³), IDLH 20 ppm; OSHA PEL: TWA 2 ppm; ACGIH TLV: TWA 2 ppm, STEL 4 ppm.

Symptoms of exposure: Inhalation may cause severe irritation of mucous membranes. Ingestion may cause irritation of intestinal tract (Patnaik, 1992). At a concentration of 6.25 ppm, slight eye irritation was reported (Verschueren, 1983). An irritation concentration of 12.50 mg/m³ in air was reported by Ruth (1986).

Toxicity:
LC_{50} (inhalation) for mice 500 mg/m³/2-h, rats 76 ppm/8-h; LD_{50} for rats 64 mg/kg, mice 96 mg/kg, rabbits 71 mg/kg (quoted, RTECS, 1985).

Uses: Manufacture of acrolein, allyl compounds, glycerol, plasticizers, resins, military poison gas; contact pesticide for weed seeds and certain fungi; intermediate for pharmaceuticals and other organic compounds; herbicide.

ALLYL CHLORIDE

Synonyms: Chlorallylene; 3-Chloroprene; 1-Chloropropene-2; 1-Chloro-2-propene; 3-Chloro-propene; 3-Chloroprene-1; **3-Chloro-1-propene**; 1-Chloro-2-propylene; 1-Chloropropylene-2; 3-Chloropropylene; α-Chloropropylene; 3-Chloro-1-propylene; NCI-C04615; UN 1100.

CAS Registry Number: 107-05-1
DOT: 1100
DOT label: Flammable liquid
Molecular formula: C_3H_5Cl
Formula weight: 76.53
RTECS: UC7350000
Merck reference: 10, 283

Physical state, color, and odor:
Colorless, yellow, purple, or brown liquid with a pungent, unpleasant, garlic-like odor. Odor threshold concentration is 1.2 ppm (Amoore and Hautala, 1983).

Melting point (°C):
-134.5 (Weast, 1986)

Boiling point (°C):
45 (Weast, 1986)

Density (g/cm³):
0.9376 at 20/4 °C (Weast, 1986)

Diffusivity in water (x 10^{-5} cm²/sec):
0.99 at 20 °C using method of Hayduk and Laudie (1974)

Flash point (°C):
-31.7 (NIOSH, 1994)
-29 (open cup, Hawley, 1981)

Lower explosive limit (%):
2.9 (NFPA, 1984)

Upper explosive limit (%):
11.1 (NFPA, 1984)

Henry's law constant (x 10^{-2} atm·m³/mol):
1.08 at 25 °C (calculated, Dilling, 1977)

Ionization potential (eV):
10.05 ± 0.01 (Lias, 1998)

Soil sorption coefficient, log K_{oc}:
1.68 using method of Kenaga and Goring (1980)

Octanol/water partition coefficient, log K_{ow}:
1.79 using method of Hansch et al. (1968)

Solubility in organics:
Miscible with alcohol, chloroform, ether, and petroleum ether (Windholz et al., 1983)

Solubility in water:
3,600 mg/L at 20 °C (Krijgsheld and van der Gen, 1986)

Vapor density:
3.13 g/L at 25 °C, 2.64 (air = 1)

Vapor pressure (mmHg):
295 at 20 °C (NIOSH, 1994)
340 at 20 °C, 440 at 30 °C (Verschueren, 1983)
360 at 25 °C (quoted, Nathan, 1978)

Environmental fate:
   Biological. Bridié et al. (1979) reported BOD and COD values 0.23 and 0.86 g/g using filtered effluent from a biological sanitary waste treatment plant. These values were determined using a standard dilution method at 20 °C and stirred for a period of 5 d. When a sewage seed was used in a separate screening test, a BOD value of 0.42 g/g was obtained. The ThOD for allyl chloride is 1.67 g/g.
   Photolytic. Anticipated products from the reaction of allyl chloride with ozone or OH radicals in the atmosphere are formaldehyde, formic acid, chloroacetaldehyde, chloroacetic acid, and chlorinated hydroxy carbonyls (Cupitt, 1980).
   Chemical/Physical. Hydrolysis under alkaline conditions will yield allyl alcohol (Hawley, 1981). The estimated hydrolysis half-life in water at 25 °C and pH 7 is 69 d (Mabey and Mill, 1978).
   The evaporation half-life of allyl chloride (1 mg/L) from water at 25 °C using a shallow-pitch propeller stirrer at 200 rpm at an average depth of 6.5 cm is 26.6 min (Dilling, 1977).
   When heated to decomposition, hydrogen chloride gas is produced (CHRIS, 1984).

Exposure limits: NIOSH REL: TWA 1 ppm (3 mg/m^3), STEL 2 ppm (6 mg/m^3), IDLH 250 ppm; OSHA PEL: TWA 1 ppm; ACGIH TLV: STEL 2 ppm.

Symptoms of exposure: Irritation of eyes and respiratory passages (Windholz et al., 1983). An irritation concentration of 75.00 mg/m^3 in air was reported by Ruth (1986).

Toxicity:
   LC_{50} (14-d) for *Poecilia reticulata* 1,213 µg/L (Hermens et al., 1985).
   Acute oral LD_{50} for rats 64 mg/kg; LD_{50} (skin) for rabbits 2,066 mg/kg (quoted, RTECS, 1985).

Uses: Preparation of epichlorohydrin, glycerol, allyl compounds, pharmaceuticals; thermosetting resins for adhesives, plastics, varnishes; glycerol and insecticides.

ALLYL GLYCIDYL ETHER

Synonyms: AGE; Allyl-2,3-epoxypropyl ether; 1-Allyloxy-2,3-epoxypropane; 1,2-Epoxy-3-allyloxypropane; Glycidyl allyl ether; NCI-C56666; **[(2-Propenyloxy)methyl]oxirane**; UN 2219.

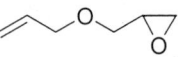

CAS Registry Number: 106-92-3
DOT: 2219
Molecular formula: $C_6H_{10}O_2$
Formula weight: 114.14
RTECS: RR0875000

Physical state, color, and odor:
Colorless liquid with a strong, pleasant odor. An odor threshold value of 47 mg/m^3 was reported (Verschueren, 1983).

Melting point (°C):
-100 (Verschueren, 1983)

Boiling point (°C):
154 (NIOSH, 1994)

Density (g/cm^3):
0.9698 at 20/4 °C (Sax and Lewis, 1987)

Diffusivity in water (x 10^{-5} cm^2/sec):
0.80 at 20 °C using method of Hayduk and Laudie (1974)

Flash point (°C):
57.2 (NIOSH, 1994)

Henry's law constant (x 10^{-6} atm·m^3/mol):
3.83 at 20 °C (approximate - calculated from water solubility and vapor pressure)

Ionization potential (eV):
10.04 (Mallard and Linstrom, 1998)

Soil sorption coefficient, log K_{oc}:
Unavailable because experimental methods for estimation of this parameter for epoxy ethers are lacking in the documented literature. However, its very high solubility in water suggests its adsorption to soil will be nominal (Lyman et al., 1982).

Octanol/water partition coefficient, log K_{ow}:
0.63 using method of Hansch et al. (1968)

Solubility in organics:
Miscible with toluene (Keith and Walters, 1992)

Solubility in water:
141 g/L (quoted, Verschueren, 1983)

Vapor density:
4.67 g/L at 25 °C, 3.94 (air = 1)

Vapor pressure (mmHg):
2 at 20 °C (NIOSH, 1994)
3.6 at 20 °C, 5.8 at 30 °C (Verschueren, 1983)

Environmental fate:
   Biological. Bridié et al. (1979) reported BOD and COD values 0.06 and 1.99 g/g using filtered effluent from a biological sanitary waste treatment plant. These values were determined using a standard dilution method at 20 °C for a period of 5 d. The ThOD for allyl glycidyl ether is 2.11 g/g.

Exposure limits: NIOSH REL: TWA 5 ppm (22 mg/m^3), STEL 10 ppm (44 mg/m^3), IDLH 50 ppm; OSHA PEL: ceiling 10 ppm; ACGIH TLV: TWA 5 ppm, STEL 10 ppm.

Symptoms of exposure: Irritation of eyes, nose, skin, and respiratory system (NIOSH, 1994). An irritation concentration of 1,144.00 mg/m^3 in air was reported by Ruth (1986).

Toxicity:
   LC$_{50}$ (96-h) for goldfish 30 mg/L (quoted, Verschueren, 1983).
   LC$_{50}$ (24-h) for goldfish 78 mg/L (quoted, Verschueren, 1983).
   Acute oral LD$_{50}$ for mice 390 mg/kg, rats 922 mg/kg (Verschueren, 1983).

Uses: Ingredient in epoxy resins.

4-AMINOBIPHENYL

Synonyms: *p*-Aminobiphenyl; 4-Aminodiphenyl; *p*-Aminodiphenyl; Anilinobenzene; Biphenylamine; 4-Biphenylamine; *p*-Biphenylamine; **[1,1′-Biphenyl]-4-amine**; Paraminodiphenyl; 4-Phenylaniline; *p*-Phenylaniline; Xenylamine.

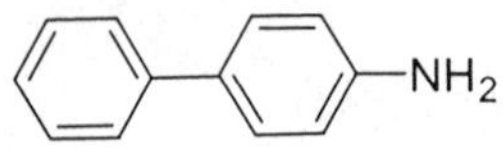

CAS Registry Number: 92-67-1
Molecular formula: $C_{12}H_{11}N$
Formula weight: 169.23
RTECS: DU8925000
Merck reference: 10, 1233

Physical state, color, and odor:
Colorless to yellow-brown crystals with a floral-like odor. Becomes purple on exposure to air.

Melting point (°C):
53-54 (Weast, 1986)
50-52 (Sittig, 1985)

Boiling point (°C):
302 (Weast, 1986)

Density (g/cm³):
1.160 at 20/20 °C (Sax and Lewis, 1987)

Diffusivity in water (x 10^{-5} cm²/sec):
0.59 at 20 °C using method of Hayduk and Laudie (1974)

Dissociation constant, pK_a:
4.27 at 25 °C (Dean, 1987)

Henry's law constant (x 10^{10} atm·m³/mol):
3.89 at 25 °C (calculated, Mercer et al., 1990)

Ionization potential (eV):
7.49 (Farrell and Newton, 1966)

Soil sorption coefficient, log K_{oc}:
2.03 (calculated, Mercer et al., 1990)

Octanol/water partition coefficient, log K_{ow}:
2.86 (Sangster, 1989)

Solubility in organics:
Soluble in alcohol, chloroform, and ether (Weast, 1986)

Solubility in water:
842 mg/L at 20-30 °C (Mercer et al., 1990)

Vapor pressure (mmHg):
6×10^{-5} at 20-30 °C (Mercer et al., 1990)

Exposure limits: Known human carcinogen. Because no standards have been established, NIOSH (1997) recommends the most reliable and protective respirators be used, i.e., a self-contained breathing apparatus that has a full facepiece and is operated under positive-pressure or a supplied-air respirator that has a full facepiece and is operated under pressure-demand or under positive-pressure in combination with a self-contained breathing apparatus operated under pressure-demand or positive-pressure.

OSHA recommends that worker exposure to this chemical is to be controlled by use of engineering control, proper work practices, and proper selection of personal protective equipment. Specific details of these requirements can be found in CFR 1910.1003-1910.1016.

Symptoms of exposure: Headache, dizziness, lethargy, dyspnea, ataxia, weakness, urinary burning (NIOSH, 1994)

Toxicity:
Acute oral LD_{50} for rats 200 mg/kg, mice 50 mg/kg (quoted, Verschueren, 1983).

Uses: Detecting sulfates; formerly used as a rubber antioxidant; cancer research.

2-AMINOPYRIDINE

Synonyms: Amino-2-pyridine; α-Aminopyridine; *o*-Aminopyridine; **2-Pyridinamine**; α-Pyridinamine; α-Pyridylamine; 2-Pyridylamine; UN 2671.

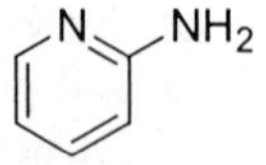

CAS Registry Number: 504-29-0
DOT: 2671
DOT label: Poison
Molecular formula: $C_5H_6N_2$
Formula weight: 94.12
RTECS: US1575000
Merck reference: 10, 478

Physical state, color, and odor:
Colorless crystals, leaflets, or powder with a characteristic odor

Melting point (°C):
57-58 (Weast, 1986)

Boiling point (°C):
210.6 (Windholz et al., 1983)

Density (g/cm³):
1.073 at 20/4 °C (calculated, Lyman et al., 1982)

Diffusivity in water (x 10^{-5} cm²/sec):
0.84 at 20 °C using method of Hayduk and Laudie (1974)

Dissociation constant, pK_a:
6.86 at 25 °C (Dean, 1973)

Flash point (°C):
68 (NIOSH, 1994)
92 (Dean, 1987)

Ionization potential (eV):
8.85 (Lias et al., 1998)

Soil sorption coefficient, log K_{oc}:
Unavailable because experimental methods for estimation of this parameter for pyridines are lacking in the documented literature. However, its high solubility in water and low K_{ow} suggest its adsorption to soil will be nominal (Lyman et al., 1982).

Octanol/water partition coefficient, log K_{ow}:
-0.22 (quoted, Verschueren, 1983)

Solubility in organics:
Soluble in acetone, alcohol, benzene, and ether (Weast, 1986)

Solubility in water:
Miscible (NIOSH, 1994)

Vapor pressure (mmHg):
25 at 25 °C (NIOSH, 1994)

Environmental fate:
  Soil. When radio-labeled 4-aminopyridine was incubated in moist soils (50%) under aerobic conditions at 30°C, the amount of $^{14}CO_2$ released from an acidic loam (pH 4.1) and an alkaline, loamy sand (pH 7.8) was 0.4 and 50%, respectively (Starr and Cunningham, 1975).
  Chemical/Physical. Releases toxic nitrogen oxides when heated to decomposition (Sax and Lewis, 1987).

Exposure limits: NIOSH REL: TWA 0.5 ppm (2 mg/m³), IDLH 5 ppm; OSHA PEL: 0.5 ppm; ACGIH TLV: TWA 0.5 ppm.

Toxicity:
  LC_{50} (48-h) for red killifish 63 mg/L (Yoshioka et al., 1986).

Uses: Manufacture of pharmaceuticals, especially antihistamines.

AMMONIA

Synonyms: Amfol; Ammonia anhydrous; Ammonia gas; Anhydrous ammonia; Nitrosil; R 717; Spirit of Hartshorn; UN 1005.

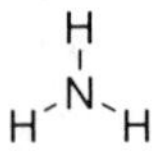

$$H-\underset{\underset{H}{|}}{N}-H$$

CAS Registry Number: 7664-41-7
DOT: 1005 (anhydrous or >50%), 2672 (12-44% solution), 2073 (>44% solution)
DOT label: Liquefied compressed gas
Molecular formula: H_3N
Formula weight: 17.04
RTECS: BO0875000
Merck reference: 10, 500

Physical state, color, and odor:
Colorless gas with a penetrating, pungent, suffocating odor. Odor threshold concentration is 5.2 ppm (Amoore and Hautala, 1983).

Melting point (°C):
-77.8 (NIOSH, 1994)

Boiling point (°C):
-33.3 (NIOSH, 1994)

Density (g/cm³):
0.77 at 0/4 °C (Hawley, 1981)

Diffusivity in water (x 10^{-5} cm²/sec):
1.10 at 0 °C using method of Hayduk and Laudie (1974)

Dissociation constant, pK_a:
9.247 at 25 °C (as ammonium hydroxide, Gordon and Ford, 1972)

Flash point (°C):
Not applicable (NIOSH, 1997)

Lower explosive limit (%):
15 (NFPA, 1984)

Upper explosive limit (%):
28 (NFPA, 1984)

Henry's law constant (x 10^{-5} atm·m³/mol at 25 °C):
1.28 (Holzwarth et al., 1984)
1.32 (Hales and Drewes, 1979)
1.64 (Clegg and Brimblecombe, 1989)
0.67, 0.89, 1.02, and 1.24 at 5.7, 9.7, 13.7, and 17.4 °C, respectively (Dasgupta and Dong, 1986)

Ionization potential (eV):
10.15 (Gibson et al., 1977)
10.2 (Franklin et al., 1969)

Soil sorption coefficient, log K_{oc}:
0.49 (calculated, Mercer et al., 1990)

Octanol/water partition coefficient, log K_{ow}:
0.00 (Mercer et al., 1990)

Solubility in organics:
Soluble in chloroform, ether, methanol (16 wt % at 25 °C), and ethanol (10 and 20 wt % at 0 and 25 °C, respectively) (Windholz et al., 1983)

Solubility in water:
895 g/L at 0 °C, 531 g/L at 20 °C, 440 g/L at 28 °C (quoted, Verschueren, 1983)
In wt %: 38 at 15 °C, 34 at 20 °C, 31 at 25 °C, 28 at 30 °C, 18 at 50 °C (quoted, Windholz et al., 1983)

Vapor density:
0.7714 g/L, 0.5967 (air = 1) (Windholz et al., 1983)

Vapor pressure (mmHg):
7,600 at 25.7 °C (Sax and Lewis, 1987)
6,460 at 20 °C (NIOSH, 1994)

Environmental fate:
   Chemical/Physical. Reacts violently with acetaldehyde, ethylene oxide, ethylene dichloride (Patnaik, 1992). Reacts with acids forming water-soluble salts.

Exposure limits: NIOSH REL: TWA 25 ppm (18 mg/m³), STEL 35 ppm (27 mg/m³), IDLH 300 ppm; OSHA PEL: STEL 50 ppm; ACGIH TLV: TWA 25 ppm, STEL 35 ppm.

Symptoms of exposure: Very irritating to eyes, nose and respiratory tract. An irritation concentration of 72.00 mg/m³ in air was reported by Ruth (1986). Exposure to 3,000 ppm for several min may result in serious blistering of skin, lung edema, and asphyxia leading to death (Patnaik, 1992). Ingestion may cause bronchospasm, difficulty in breathing, chest pain, and pulmonary edema. Contact with liquid ammonia or aqueous solutions may cause vesiculation or frostbite (NIOSH, 1994).

Toxicity:
   LC_{50} (60-d) for freshwater clam (*Sphaerium novaezelandiae*) 3.8 mg/L [as (N)/L] (Hickey and Martin, 1999).
   LC_{50} (96-h) for guppy fry 1.26-74 mg/L, coho salmon (flow-through bioassay) 0.45 mg/L, cut throat trout (flow-through bioassay) 0.5-0.8 mg/L (Verschueren, 1983), euryhaline amphipod (*Corophium* sp.) 5.5 mg/L (Hyne and Everett, 1998), Atlantic salmon (*Salmo salar*) as non-ionized ammonia (as μg/L NH_3-N) ranged from 31 (2.1 °C) to 111 (17.1 °C) at pH 6.0 and from 30 (1.8 °C) to 146 (12.5 °C) at pH 6.4 (Knoph, 1992).
   LC_{50} (96-h), LC_{50} (72-h), and LC_{50} (48-h) values for juvenile shrimp (*Penaeus vannamei*), as ammonia-N, were 70.9, 85.3, and 110.6 mg/L, respectively (Frías-Espericueta et al., 1999).
   Calculated inhalation LC_{50} values for rats at exposure periods of 10, 20, 40, and 60 min were

40,300, 28,595, 20,300, and 20,300 ppm, respectively (Appelman et al., 1982).

Acute oral LD_{50} for rats 250 mg/kg; LC_{50} (inhalation) for mice 4,230 ppm/h, rats 2,000 ppm/4-h, rabbits 7 gm/m^3/4-h (quoted, RTECS, 1985).

Source: Ammonia is released as a combustion product of coal, fuel oil, natural gas, wood, butane, and propane (Verschueren, 1983).

Ammonia naturally occurs in soybean (8,600 ppm), evening-primrose seeds (2,300-2,455 ppm), lambsquarter, and tobacco leaves (Duke, 1992).

Uses: Manufacture of acrylonitrile, hydrazine hydrate, hydrogen cyanide, nitric acid, sodium carbonate, urethane, explosives, synthetic fibers, fertilizers; refrigerant; condensation catalyst; dyeing; neutralizing agent; synthetic fibers; latex preservative; fuel cells, rocket fuel; nitrocellulose; nitroparaffins; ethylenediamine, melamine; sulfite cooking liquors; developing diazo films; yeast nutrient.

n-AMYL ACETATE

Synonyms: Acetic acid amyl ester; **Acetic acid pentyl ester**; Amyl acetate; Amyl acetic ester; Amyl acetic ether; *n*-Amyl ethanoate; Banana oil; Birnenoel; Pear oil; Pentacetate; Pentacetate 28; 1-Pentanol acetate; 1-Pentyl acetate; *n*-Pentyl acetate; *n*-Pentyl ethanoate; Primary amyl acetate; UN 1104.

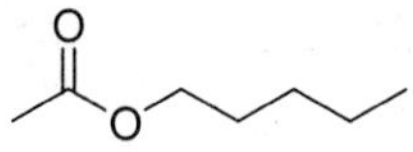

CAS Registry Number: 628-63-7
DOT: 1104
DOT label: Flammable liquid
Molecular formula: $C_7H_{14}O_2$
Formula weight: 130.19
RTECS: AJ1925000

Physical state, color, and odor:
Colorless liquid with a sweet, banana-like odor. Odor threshold concentration is 54 ppb (Amoore and Hautala, 1983).

Melting point (°C):
-70.8 (Weast, 1986)

Boiling point (°C):
149.25 (Weast, 1986)

Density (g/cm³):
0.8756 at 20/4 °C (Weast, 1986)

Diffusivity in water (x 10⁻⁵ cm²/sec):
0.70 at 20 °C using method of Hayduk and Laudie (1974)

Flash point (°C):
25 (NIOSH, 1994)
16-21 (NFPA, 1984)

Lower explosive limit (%):
1.1 (NIOSH, 1994)

Upper explosive limit (%):
7.5 (NIOSH, 1994)

Henry's law constant (x 10⁻⁴ atm·m³/mol):
3.57 at 25 °C (Kieckbusch and King, 1979)

Interfacial tension with water (dyn/cm):
50 at 20 °C (estimated, CHRIS, 1984)

Soil sorption coefficient, log K$_{oc}$:
Unavailable because experimental methods for estimation of this parameter for aliphatic esters are lacking in the documented literature

Octanol/water partition coefficient, log K_{ow}:
2.23 (Leo et al., 1969)

Solubility in organics:
Miscible with alcohol and ether (Hawley, 1981)

Solubility in water:
1.8 g/L at 20 °C (quoted, Verschueren, 1983)
In wt %: 0.29 at 0 °C, 0.22 at 19.7 °C, 0.16 at 30.6 °C, 0.16 at 39.5 °C, 0.10 at 50.0 °C, 0.10 at
    60.3 °C, 0.17 at 70.2 °C, 0.17 at 80.1 °C (Stephenson and Stuart, 1986)
1,730 mg/L at 25 °C (McBain and Richards, 1946)

Vapor density:
5.32 g/L at 25 °C, 4.49 (air = 1)

Vapor pressure (mmHg):
4.1 at 25 °C (Abraham, 1984)

Environmental fate:
    Chemical/Physical. Hydrolyzes in water forming acetic acid and 1-pentanol.
    At an influent concentration of 985 mg/L, treatment with granular activated carbon resulted
in an effluent concentration of 119 mg/L. The adsorbability of the carbon used was 175 mg/g
carbon (Guisti et al., 1974).

Exposure limits: NIOSH REL: TWA 100 ppm (525 mg/m³), IDLH 1,000 ppm; OSHA PEL:
TWA 100 ppm; ACGIH TLV: TWA 100 ppm with intended TWA and STEL values of 50 and
100 ppm, respectively.

Symptoms of exposure: Irritating to eyes and respiratory tract. At concentrations of 1,000
ppm, inhalation may cause headache, somnolence, and narcotic effects (Patnaik, 1992). An
irritation concentration of 53.00 mg/m³ in air was reported by Ruth (1986).

Toxicity:
    LC_0 (24-h) and LC_{100} (24-h) for creek chub in Detroit river water were 50 and 120 mg/L,
respectively (Gillette et al., 1952).
    Acute oral LD_{50} for rats 6,500 mg/kg (quoted, RTECS, 1985).

Uses: Solvent for lacquers and paints; leather polishes; flavoring agent; photographic film;
extraction of penicillin; nail polish; printing and finishing fabrics; odorant.

sec-AMYL ACETATE

Synonyms: 2-Acetoxypentane; 1-Methylbutyl acetate; 1-Methylbutyl ethanoate; **2-Pentanol acetate**; 2-Pentyl acetate; *sec*-Pentyl acetate.

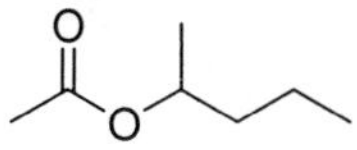

CAS Registry Number: 626-38-0
DOT: 1104
DOT label: Flammable liquid
Molecular formula: $C_7H_{14}O_2$
Formula weight: 130.19
RTECS: AJ2100000

Physical state, color, and odor:
Clear, colorless liquid with a fruity odor. Odor threshold concentration is 2.0 ppb (Amoore and Hautala, 1983).

Melting point (°C):
-78.4 (NIOSH, 1994)

Boiling point (°C):
134 (Weast, 1986)

Density (g/cm^3):
0.862-0.866 at 20/20 °C (Hawley, 1981)

Diffusivity in water (x 10^{-5} cm^2/sec):
0.69 at 20 °C using method of Hayduk and Laudie (1974)

Flash point (°C):
31.7 (NIOSH, 1994)

Lower explosive limit (%):
1 (NIOSH, 1994)

Upper explosive limit (%):
7.5 (NIOSH, 1994)

Henry's law constant (x 10^{-4} atm·m^3/mol):
7.7 at 25 °C (approximate - calculated from water solubility and vapor pressure)

Interfacial tension with water (dyn/cm):
44.1 at 20 °C (estimated, CHRIS, 1984)

Soil sorption coefficient, log K$_{oc}$:
Unavailable because experimental methods for estimation of this parameter for aliphatic esters are lacking in the documented literature

Octanol/water partition coefficient, log K_{ow}:
5.26 using method of Hansch et al. (1968)

Solubility in organics:
Soluble in alcohol and ether (Weast, 1986)

Solubility in water:
2.2 g/L at 25 °C (Montgomery, 1989)

Vapor density:
5.32 g/L at 25 °C, 4.49 (air = 1)

Vapor pressure (mmHg):
10 at 35.2 °C (estimated, Weast, 1986)

Environmental fate:
   Chemical/Physical. Slowly hydrolyzes in water forming acetic acid and 2-pentanol.

Exposure limits: NIOSH REL: TWA 125 ppm (650 mg/m^3), IDLH 1,000 ppm; OSHA PEL: TWA 125 ppm; ACGIH TLV: TWA 125 ppm with intended TWA and STEL values of 50 and 100 ppm, respectively.

Symptoms of exposure: Irritating to eyes, nose, and respiratory tract (NIOSH, 1994)

Uses: Solvent for nitrocellulose and ethyl cellulose; coated paper, lacquers; cements; nail enamels, leather finishes; textile sizing and printing compounds; plastic wood.

ANILINE

Synonyms: Aminobenzene; Aminophen; Aniline oil; Anyvim; **Benzenamine**; Benzidam; Blue oil; C.I. 76000; C.I. oxidation base 1; Cyanol; Krystallin; Kyanol; NCI-CO3736; Phenylamine; RCRA waste number U012; UN 1547.

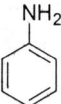

CAS Registry Number: 62-53-3
DOT: 1547
DOT label: Poison
Molecular formula: C_6H_7N
Formula weight: 93.13
RTECS: BW6650000
Merck reference: 10, 681

Physical state, color, and odor:
Colorless, oily liquid with a faint ammonia-like odor and burning taste. Darkens on exposure to air or light. The lower and upper odor thresholds are 2 and 128 ppm, respectively (Keith and Walters, 1992).

Melting point (°C):
-6.3 (Weast, 1986)
-5.98 (Dean, 1987)

Boiling point (°C):
184 (Weast, 1986)

Density (g/cm³):
1.02166 at 20.00/4 °C (Tsierkezos et al., 2000)

Diffusivity in water (x 10⁻⁵ cm²/sec):
1.05 at 25 °C (quoted, Hayduk and Laudie, 1974)

Dissociation constant, pK$_a$:
4.630 at 25 °C (Gordon and Ford, 1972)

Flash point (°C):
70 (NFPA, 1984; Dean, 1987)
76 (Windholz et al., 1983)

Lower explosive limit (%):
1.3 (NFPA, 1984)

Upper explosive limit (%):
11 (NFPA, 1984)

Heat of fusion (kcal/mol):
2.519 (Dean, 1987)

Henry's law constant (x 10^{-6} atm·m³/mol at 25 °C):
22.1, 101, 95.9, and 103 at pH values of 2.93, 7.30, 8.88, and 9.07, respectively (Hakuta et al., 1977)
1.99 (Jayasinghe et al., 1992)

Interfacial tension with water (dyn/cm at 20 °C):
5.77 (Harkins et al., 1920)

Ionization potential (eV):
7.72 ± 0.002 (Lias, 1998)

Bioconcentration factor, log BCF:
0.60 (*Chlorella fusca*, Freitag et al., 1982; Geyer et al., 1984)
0.78 (mosquito fish, Lu and Metcalf, 1975)
0.41 (*Brachydanio rerio*, Kalsch et al., 1991; Devillers et al., 1996)
2.70 (activated sludge, Freitag et al., 1985)
1.87 (uptake-phase data), 2.77 (elimination-phase data) (*Daphnia magna*, Dauble et al., 1986)
1.96 (*Scenedesmus quadricauda*, Hardy et al., 1985)

Soil sorption coefficient, log K_{oc}:
1.96 (river sediment), 3.4 (coal wastewater sediment) (Kopinke et al., 1995)
1.65 (Batcombe silt loam, Briggs, 1981)
2.11 (Hagerstown clay loam), 2.61 (Palouse silt loam) (Pillai et al., 1982)
K_d values of 2.11 and 3.11 cm³/g for H-montmorillonite and Na-montmorillonite, respectively (pH 9.0, Bailey et al., 1968)

Octanol/water partition coefficient, log K_{ow}:
0.90 (Fujita et al., 1964; Mirrlees et al., 1976; Campbell and Luthy, 1985)
1.09 (Geyer et al., 1984)
0.781 (Klein et al., 1988)
0.940 (de Bruijn et al., 1989; Brooke et al., 1990)
0.91 at 25 °C (Andersson and Schräder, 1999)
0.93 (Könemann et al., 1979)
0.89 at pH 7.4 (El Tayar et al., 1984)
0.79, 0.96 (Garst and Wilson, 1984)

Solubility in organics:
Miscible with alcohol, benzene, chloroform (Windholz et al., 1983)

Solubility in water:
34.1 g/L (Fu et al., 1986)
27.2, 35.4, and 47.8 g/L at 4, 25, and 40 °C, respectively (Moreale and Van Bladel, 1979)
36.65 g/L at 25 °C (Hill and Macy, 1924)
38.0 and 49.0 mL/L water at 25 and 60 °C, respectively (Booth and Everson, 1949)

Vapor density:
3.81 g/L at 25 °C, 3.22 (air = 1)

Vapor pressure (mmHg at 25 °C):
0.6 (Sonnefeld et al., 1983)
0.49 (Banerjee et al., 1990)

Environmental fate:

Biological. Under anaerobic conditions using a sewage inoculum, 10% of the aniline present degraded to acetanilide and 2-methylquinoline (Hallas and Alexander, 1983). In a 56-d experiment, [^{14}C]aniline applied to soil-water suspensions under aerobic and anaerobic conditions gave $^{14}CO_2$ yields of 26.5 and 11.9%, respectively (Scheunert et al., 1987). A bacterial culture isolated from the Oconee River in North Georgia degraded aniline to the intermediate catechol (Paris and Wolfe, 1987). Aniline was mineralized by a soil inoculum in 4 d (Alexander and Lustigman, 1966).

Silage samples (chopped corn plants) containing aniline were incubated in an anaerobic chamber for 2 wk at 28 °C. After 3 d, aniline was biologically metabolized to formanilide, propioanilide, 3,4-dichloroaniline, 3- and 4-chloroaniline (Lyons et al., 1985). Various micro-organisms isolated from soil degraded aniline to acetanilide, 2-hydroxyacetanilide, 4-hydroxy-aniline, and two unidentified phenols (Smith and Rosazza, 1974). In activated sludge, 20.5% mineralized to carbon dioxide after 5 d (Freitag et al., 1985). In the presence of suspended natural populations from unpolluted aquatic systems, the second-order microbial transformation rate constant determined in the laboratory was reported to be 1.1 ± 0.8 x 10^{-11} L/organism·h (Steen, 1991).

Heukelekian and Rand (1955) reported a 5-d BOD value of 1.55 g/g which is 64.3% of the ThOD value of 2.41 g/g. In activated sludge inoculum, following a 20-d adaptation period, 94.5% COD removal was achieved. The average rate of biodegradation was 19.0 mg COD/g·h (Pitter, 1976).

Soil. A reversible equilibrium is quickly established when aniline covalently bonds with humates in soils forming imine linkages. These quinodal structures may oxidize to give nitrogen-substituted quinoid rings. The average second-order rate constant for this reaction in a pH 7 buffer at 30 °C is 9.47 x 10^{-5} L/g·h (Parris, 1980). In sterile soil, aniline partially degraded to azobenzene, phenazine, formanilide, and acetanilide and the tentatively identified compounds nitrobenzene and *p*-benzoquinone (Pillai et al., 1982).

Surface Water. Aniline degraded in pond water containing sewage sludge to catechol, which further degraded to carbon dioxide. Intermediate compounds identified in minor degradative pathways include acetanilide, phenylhydroxylamine, *cis,cis*-muconic acid, β-ketoadipic acid, levulinic acid, and succinic acid (Lyons et al., 1984).

Photolytic. A carbon dioxide yield of 46.5% was achieved when aniline adsorbed on silica gel was irradiated with light (λ >290 nm) for 17 h (Freitag et al., 1985). Products identified from the gas-phase reaction of ozone with aniline in synthetic air at 23 °C were nitrobenzene, formic acid, hydrogen peroxide, and a nitrated salt having the formula: $[C_6H_5NH_3]^+NO_3^-$ (Atkinson et al., 1987).

Irradiation of an aqueous solution at 50 °C for 24 h resulted in a 28.5% yield of carbon dioxide (Knoevenagel and Himmelreich, 1976).

A second-order rate constant of 6.0 x 10^{-11} cm^3/molecule·sec at 26 °C was reported for the vapor-phase reaction of aniline and OH radicals in air at room temperature (Atkinson, 1985).

Chemical/Physical. Alkali or alkaline earth metals dissolve with hydrogen evolution and the formation of anilides (Windholz et al., 1983). Laha and Luthy (1990) investigated the redox reaction between aniline and a synthetic manganese dioxide in aqueous suspensions at the pH range 3.7-6.5. They postulated that aniline undergoes oxidation by loss of one electron forming cation radicals. These radicals may undergo head-to-tail, tail-to-tail, and head-to-head couplings forming 4-aminophenylamine, benzidine, and hydrazobenzene, respectively. These compounds were further oxidized, in particular, hydrazobenzene to azobenzene at pH 4 (Laha and Luthy, 1990).

Kanno et al. (1982) studied the aqueous reaction of aniline and other substituted aromatic hydrocarbons (toluidine, 1-naphthylamine, phenol, cresol, pyrocatechol, resorcinol, hydro-quinone and 1-naphthol) with hypochlorous acid in the presence of ammonium ion. They reported that the aromatic ring was not chlorinated as expected but was cleaved by chloramine forming cyanogen chloride (Kanno et al., 1982). The amount of cyanogen chloride formed increased at lower pHs. At pH 6, the greatest amount of cyanogen chloride was formed when the reaction mixture contained ammonium ion and hypochlorous acid at a ratio of 2:3 (Kanno et al., 1982). When aniline in an aqueous solution containing nitrite ion was ozonated, nitrosobenzene, nitrobenzene, 4-aminodiphenylamine, azobenzene, azoxybenzene, benzidine, phenazine (Chan and Larson, 1991), 2-, 3-, and 4-nitroaniline formed as products (Chan and Larson, 1991a). The yields of nitroanilines were higher at a low pH (6.25) than at high pH (10.65) and the presence of carbonates inhibited their formation (Chan and Larson, 1991a).

Aniline will not hydrolyze because it has no hydrolyzable functional group (Kollig, 1993).

At an influent concentration of 1,000 mg/L, treatment with granular activated carbon resulted in an effluent concentration of 251 mg/L. The adsorbability of the carbon used was 150 mg/g carbon (Guisti et al., 1974).

Exposure limits: Potential occupational carcinogen. NIOSH REL: IDLH 100 ppm; OSHA PEL: TWA 5 ppm (19 mg/m^3); ACGIH TLV: TWA 2 ppm (10 mg/m^3).

Symptoms of exposure: Absorption through skin may cause headache, weakness, dizziness, ataxia, and cyanosis (Patnaik, 1992).

Toxicity:

EC_{50} (96-h) for *Zenopus laevis* 370 mg/L (Davis et al., 1981).

LC_{50} (28-d) for *Brachydanio rerio* 39 mg/L during embryo larval life stages (van Leeuwen et al., 1990).

LC_{50} (21-d) for *Daphnia magna* 47 μg/L (Gersich and Milazzo, 1988).

LC_{50} (7-d) for *Oncorhynchus mykiss* 8.2 mg/L (Abram and Sims, 1982).

LC_{50} (96-h) for *Carassius auratus* 187 mg/L (Holcombe et al., 1987), rainbow trout (*Salmo gairdneri*) 20 mg/L (Spehar et al., 1982), Japanese medaka (*Oryzias latipes*) 108 mg/L (Holcombe et al., 1995).

LC_{50} (48-h) for *Daphnia pulex* 0.1 mg/L, *Lymnaea stagnalis* 800 mg/L (Sloof et al., 1983), red killifish 1,820 mg/L (Yoshioka et al., 1986).

LC_{50} (inhalation) for mice 175 ppm/7-h (quoted, RTECS, 1985).

Acute oral LD_{50} for mice is 464 mg/kg, wild birds 562 mg/kg, dogs 195 mg/kg, quail 750 mg/kg, rats 250 mg/kg (quoted, RTECS, 1985).

Source: Detected in distilled water-soluble fractions of regular gasoline (87 octane) and Gasohol at concentrations of 0.55 and 0.20 mg/L, respectively (Potter, 1996).

Based on analysis of 7 coal tar samples, aniline concentrations ranged from ND to 13 ppm (EPRI, 1990).

Aniline in the environment may originate from the anaerobic biodegradation of nitrobenzene (Razo-Flores et al., 1999).

Uses: Manufacture of dyes, resins, varnishes, medicinals, perfumes, photographic chemicals, shoe blacks, chemical intermediates; solvent; vulcanizing rubber; isocyanates for urethane foams; explosives; petroleum refining; diphenylamine; phenolics; fungicides; herbicides.

o-ANISIDINE

Synonyms: 2-Aminoanisole; *o*-Aminoanisole; 1-Amino-2-methoxybenzene; 2-Anisylamine; 2-Methoxy-1-aminobenzene; 2-Methoxyaniline; *o*-Methoxyaniline; **2-Methoxybenzenamine**; *o*-Methoxyphenylamine; UN 2431.

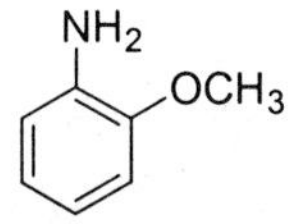

CAS Registry Number: 90-04-0
DOT: 2431
DOT label: Poison
Molecular formula: C_7H_9NO
Formula weight: 123.15
RTECS: BZ5410000
Merck reference: 10, 689

Physical state, color, and odor:
Colorless, yellow to reddish liquid with an amine-like odor. Becomes brown on exposure to air.

Melting point (°C):
6.2 (Weast, 1986)
5 (Windholz et al., 1983)

Boiling point (°C):
224 (Weast, 1986)

Density (g/cm³):
1.0923 at 20/4 °C (Weast, 1986)

Diffusivity in water (x 10⁻⁵ cm²/sec):
0.82 at 20 °C using method of Hayduk and Laudie (1974)

Dissociation constant, pK_a:
4.09 at 25 °C (Dean, 1973)

Flash point (°C):
118 (open cup, NFPA, 1984)

Henry's law constant (x 10⁻⁶ atm·m³/mol):
1.25 at 25 °C (approximate - calculated from water solubility and vapor pressure)

Ionization potential (eV):
7.46 (Farrell and Newton, 1966)

Soil sorption coefficient, log K_{oc}:
Unavailable because experimental methods for estimation of this parameter for anilines are lacking in the documented literature

Octanol/water partition coefficient, log K_{ow}:
0.95 (Leo et al., 1971)
1.18 (Camilleri et al., 1988)
1.23 (HPLC, Unger et al., 1978)

Solubility in organics:
Soluble in acetone and benzene (Weast, 1986). Miscible with alcohol and ether (Windholz et al., 1983).

Solubility in water:
1 wt % at 20 °C (NIOSH, 1994)

Vapor density:
5.03 g/L at 25 °C, 4.25 (air = 1)

Vapor pressure (mmHg):
<0.1 at 25 °C (NIOSH, 1994)

Exposure limits: Potential occupational carcinogen. NIOSH REL: TWA 0.5 mg/m³, IDLH 50 mg/m³; OSHA PEL: TWA 0.5 mg/m³; ACGIH TLV: TWA 0.1 ppm (adopted).

Symptoms of exposure: Absorption and inhalation may cause headache and dizziness (Patnaik, 1992).

Toxicity:
Acute oral LD_{50} for rats 2,000 mg/kg, wild birds 422 mg/kg, mice 1,400 mg/kg, rabbits 870 mg/kg (quoted, RTECS, 1985).

Uses: Manufacture of azo dyes.

p-ANISIDINE

Synonyms: 4-Aminoanisole; *p*-Aminoanisole; 1-Amino-4-methoxybenzene; *p*-Aminometh-oxybenzene; *p*-Aminomethylphenyl ether; 4-Anisidine; *p*-Anisylamine; 4-Methoxy-1-amino-benzene; 4-Methoxyaniline; *p*-Methoxyaniline; **4-Methoxybenzenamine**; *p*-Methoxybenzen-amine; 4-Methoxyphenylamine; *p*-Methoxyphenylamine.

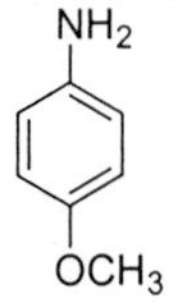

CAS Registry Number: 104-94-9
DOT: 2431
DOT label: Poison
Molecular formula: C_7H_9NO
Formula weight: 123.15
RTECS: BZ5450000
Merck reference: 10, 689

Physical state, color, and odor:
Yellow to light brown solid or crystals with a characteristic amine-like odor

Melting point (°C):
57.2 (Weast, 1986)
57-60 (Aldrich, 1988)

Boiling point (°C):
243 (Weast, 1986)
246 (Windholz et al., 1983)
240-243 (Aldrich, 1988)

Density (g/cm³):
1.096 at 20/4 °C (Aldrich, 1988)

Diffusivity in water (x 10^{-5} cm²/sec):
0.73 at 20 °C using method of Hayduk and Laudie (1974)

Dissociation constant, pK_a:
4.49 at 25 °C (Dean, 1973)

Ionization potential (eV):
7.44 (NIOSH, 1994)
7.82 (Franklin et al., 1969)

Soil sorption coefficient, log K_{oc}:
Unavailable because experimental methods for estimation of this parameter for anilines are lacking in the documented literature

Octanol/water partition coefficient, log K_{ow}:
0.95 (Leo et al., 1971)
0.83 (Garst and Wilson, 1984)

Solubility in organics:
Soluble in acetone, alcohol, benzene, and ether (Weast, 1986)

Solubility in water:
3.3 mg/L at 20-25 °C using method of Kenaga and Goring (1980)

Vapor pressure (mmHg):
6×10^{-3} at 25 °C (NIOSH, 1994)

Environmental fate:
   Chemical/Physical. Releases toxic nitrogen oxides when heated to decomposition (Sax and Lewis, 1987).

Exposure limits: NIOSH REL: TWA 0.5 mg/m^3, IDLH 50 mg/m^3; OSHA PEL: TWA 0.5 mg/m^3; ACGIH TLV: TWA 0.1 ppm (adopted).

Symptoms of exposure: Anemia and cyanosis (Patnaik, 1992)

Toxicity:
   Acute oral LD$_{50}$ for rats 1,400 mg/kg, mice 810 mg/kg, rabbits 2,900 mg/kg (quoted, RTECS, 1985).

Uses: Azo dyestuffs; chemical intermediate.

ANTHRACENE

Synonyms: Anthracin; Green oil; Paranaphthalene; Tetra olive N2G.

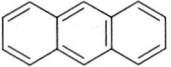

CAS Registry Number: 120-12-7
Molecular formula: $C_{14}H_{10}$
Formula weight: 178.24
RTECS: CA9350000
Merck reference: 10, 704

Physical state, color, and odor:
White to yellow crystalline flakes or crystals with a bluish or violet fluorescence and a weak aromatic odor. Impurities (naphthacene, tetracene) impart a yellowish color with green fluorescence.

Melting point (°C):
216.2-216.4 (Aldrich, 1988)
219.5 (Casellato et al., 1973)

Boiling point (°C):
339.9 (Dean, 1973)

Density (g/cm³):
1.283 at 25/4 °C (Weast, 1986)
1.24 at 20/4 °C (Weiss, 1986)

Diffusivity in water (x 10⁻⁵ cm²/sec):
0.59 at 20 °C using method of Hayduk and Laudie (1974)

Flash point (°C):
121.1 (Weiss, 1986)

Lower explosive limit (%):
0.6 (Weiss, 1986)

Dissociation constant, pK_a:
>15 (Christensen et al., 1975)

Entropy of fusion (cal/mol·K):
14.1 (Tsonopoulos and Prausnitz, 1971; Wauchope and Getzen, 1972)

Heat of fusion (kcal/mol):
6.93 (Wauchope and Getzen, 1972)

Henry's law constant (x 10⁻⁵ atm·m³/mol):
140 (Petrasek et al., 1983)

6.51 at 25 °C (Southworth, 1979)

1.93 at 25 °C (Fendinger and Glotfelty, 1990)

7.56 at 25 °C (Shiu and Mackay, 1997)

4.88 at 25 °C (Alaee et al., 1996)

1.23, 2.09, 3.45, 5.57, and 8.25 at 4.1, 11.0, 18.0, 25.0, and 31.0 °C, respectively (Bamford et al., 1998)

Ionization potential (eV):

7.55 (Krishna and Gupta, 1970)

7.58 (Yoshida et al., 1983a)

7.43 (Cavalieri and Rogan, 1985)

Bioconcentration factor, log BCF:

3.83 (activated sludge), 2.96 (golden ide) (Freitag et al., 1985)

2.58 (*Daphnia pulex*, Southworth et al., 1978)

2.95 (bluegill sunfish, Spacie et al., 1983)

2.21 (goldfish, Ogata et al., 1984)

3.89 (*Chlorella fusca*, Freitag et al., 1982; Geyer et al., 1984)

Apparent values of 3.8 (wet wt) and 5.5 (lipid wt) for freshwater isopods including *Asellus aquaticus* (L.) (van Hattum et al., 1998)

Soil sorption coefficient, log K_{oc}:

4.27 (aquifer sands, Abdul et al., 1987)

4.41 (average silt fraction from Doe Run and Hickory Hill sediments, Karickhoff et al., 1979)

4.205 (Nkedi-Kizza et., 1985)

4.50 (humic acid, Landrum et al., 1984)

4.42, 4.53 (RP-HPLC immobilized humic acids, Szabo et al., 1990)

4.44 (flint aquifer material, Abdul and Gibson, 1986)

4.93 (Gauthier et al., 1986)

5.76 (average, Kayal and Connell, 1990)

4.11 (fine sand, Enfield et al., 1989)

4.73, 5.18, 5.86 (San Francisco, CA mudflat sediments, Maruya et al., 1996)

6.7 (average value using 8 river bed sediments from the Netherlands, van Hattum et al., 1998)

Average K_d values for sorption of anthracene to corundum (α-Al_2O_3) and hematite (α-Fe_2O_3) were 0.0666 and 0.226 mL/g, respectively (Mader et al., 1997)

Octanol/water partition coefficient, log K_{ow}:

4.45 (Hansch and Fujita, 1964; DeKock and Lord, 1987)

4.54 (Miller et al., 1985)

4.34 (Mackay, 1982)

4.63 (Bruggeman et al., 1982)

4.68 at 25 °C (de Maagd et al., 1998)

Solubility in organics:

Ethanol (14.9 g/L), methanol (14.3 g/L), benzene (16.1 g/L), carbon disulfide (32.3 g/L), carbon tetrachloride (11.6 g/L), chloroform (11.8 g/L), and toluene (8.0 g/L) (Windholz et al., 1983); 6.6, 21, and 16 mM at 25 °C in isooctane, butyl ether, and pentyl ether, respectively (Anderson et al., 1980).

N,N-dimethylformamide, g/kg (°C): 13.3414 (29.8), 16.9352 (34.8), 19.9337 (39.6), 22.5539 (44.2), 27.1358 (49.6). In 1,4-dioxane, g/kg (°C): 2.0787 (29.8), 3.7332 (34.8), 5.4112 (39.6), 8.3659 (44.2), 13.6541 (49.6). In ethylene glycol, g/kg (°C): 0.4384 (64.8), 0.7955

(78.8), 1.0680 (86.8), 1.5346 (97.8), 2.3934 (110.8), 3.3500 (123.8), 6.5100 (146.2), 8.5000 (159.8) (Cepeda et al., 1989).

Solubility in water:
At 20 °C: 180, 118, 107 and 126 nmol/L in distilled water, Pacific seawater, artificial seawater and 35% NaCl, respectively (Hashimoto et al., 1984)
44.6 μg/kg at 25 °C, 57.0 μg/kg at 29 °C. In seawater (salinity = 35.0 g/kg): 31.1 μg/kg at 25 °C (May et al., 1978a)
75 μg/L at 27 °C (Davis et al., 1942; Klevens, 1950)
73 μg/L at 25 °C (Mackay and Shiu, 1977)
70 μg/L at 23 °C (Pinal et al., 1991)
30 μg/L at 25 °C (Schwarz and Wasik, 1976)
112.5 μg/L at 25 °C (Sahyun, 1966)
In mg/kg: 0.119-0.125 at 35.4 °C, 0.148-0.152 at 39.3 °C, 0.206-0.210 at 44.7 °C, 0.279 at 47.5 °C, 0.297-0.302 at 50.1 °C, 0.389-0.402 at 54.7 °C, 0.480-0.525 at 59.2 °C, 0.62-0.72 at 64.5 °C, 0.64-0.67 at 65.1 °C, 0.92 at 69.8 °C, 0.90-0.97 at 70.7 °C, 0.91 at 71.9 °C, 1.13-1.26 at 74.7 °C (Wauchope and Getzen, 1972)
In μg/kg: 12.7 at 5.2 °C, 17.5 at 10.0 °C, 22.2 at 14.1 °C, 29.1 at 18.3 °C, 37.2 at 22.4 °C, 43.4 at 24.6 °C, 55.7 at 28.7 °C (May et al., 1978)
In nmol/L: 131 at 8.6 °C, 137 at 11.1 °C, 144 at 12.2 °C, 154 at 14 °C, 166 at 15.5 °C, 181 at 18.2 °C, 222 at 20.3 °C, 234 at 23.0 °C, 230 at 25.0 °C, 267 at 26.2 °C, 325 at 28.5 °C, 390 at 31.3 °C. In 0.5 M NaCl: 93 at 8.6 °C, 101 at 8.6 °C, 122 at 11.7 °C, 147 at 19.2 °C, 168 at 21.5 °C, 204 at 25.0 °C, 192 at 25.3 °C, 202 at 27.1 °C, 246 at 30.2 °C (Schwarz, 1977)
41 μg/L at 20 °C (Kishi and Hashimoto, 1989)
250 nmol/L at 25 °C (Akiyoshi et al., 1987; Wasik et al., 1983)
70 μg/L at 23 °C (Pinal et al., 1990)
93 μg/L at 25 °C (de Maagd et al., 1998)
48.8 μg/L at 25 °C (Etzweiler et al., 1995)
At pH 9 containing humic acids (wt %) derived from Sagami Bay: 470 μg/L (0.02), 172 μg/L (0.04), 157 μg/L (0.06), 154 μg/L (0.09), 343 μg/L (0.12) (Shinozuka et al., 1987)
69.8 μg/L at 25 °C (Walters and Luthy, 1984)
53.9, 72.4, 99.3, 133, 181, and 248 nmol/L at 4.6, 8.8, 12.9, 17.0, 21.1, and 25.3 °C, respectively. In seawater (salinity = 36.5 g/kg): 37.9, 50.7, 68.4, 97.6, 131, and 182 nmol/L at 4.6, 8.8, 12.9, 17.0, 21.1, and 25.3 °C, respectively (Whitehouse, 1984)
58 μg/L at 25 °C (Vadas et al., 1991)
44.3 and 34 μg/L at 25 °C (Billington et al., 1988)
In mole fraction (x 10^{-9}): 1.284 at 5.20 °C, 1.637 at 9.70 °C, 1.769 at 10.00 °C, 2.245 at 14.10 °C, 2.537 at 16.60 °C, 2.941 at 18.30 °C, 3.760 at 22.40 °C, 3.821 at 23.20 °C, 4.387 at 24.60 °C, 5.630 at 28.70 °C, 5.781 at 29.30 °C (May et al., 1983)

Vapor pressure (x 10^{-5} mmHg):
0.60 at 25 °C (Wasik et al., 1983)
19.5 at 25 °C (Radding et al., 1976)
51.7, 75 at 25 °C (Hinckley et al., 1990)
669 at 85.25 °C, 1,020 at 90.15 °C (Macknick and Prausnitz, 1979)
8.6 at 65.7 °C, 10.5 at 67.10 °C, 11.8 at 68.75 °C (Bradley and Cleasby, 1953)
0.43 at 25 °C (McVeety and Hites, 1982)
33,000 at 127 °C (Eiceman and Vandiver, 1983)
0.56 at 25 °C (de Kruif, 1980)
48 at 25 °C (Bidleman, 1984)
1.25, 1.91, and 2.91 at 95, 100, and 105 °C, respectively (Kelley and Rice, 1964)

0.86, 4.31, 12.15, 26.63, 46.50, 153.00, and 193.50 at 27.70, 39.70, 47.60, 54.60, 60.10, 72.70, and 74.10 °C, respectively (Oja and Suuberg, 1998)

Environmental fate:

Biological. Catechol is the central metabolite in the bacterial degradation of anthracene. Intermediate by-products included 3-hydroxy-2-naphthoic acid and salicylic acid (Chapman, 1972). Anthracene was statically incubated in the dark at 25 °C with yeast extract and settled domestic wastewater inoculum. Significant biodegradation with gradual adaptation was observed. At concentrations of 5 and 10 mg/L, biodegradation yields at the end of 4 wk of incubation were 92 and 51%, respectively (Tabak et al., 1981). A mixed bacterial community isolated from seawater foam degraded anthraquinone, a photodegradation product of anthracene, to traces of benzoic and phthalic acids (Rontani et al., 1975). In activated sludge, only 0.3% mineralized to carbon dioxide after 5 d (Freitag et al., 1985).

Contaminated soil from a manufactured coal gas plant that had been exposed to crude oil was spiked with anthracene (400 mg/kg soil) to which Fenton's reagent (5 mL 2.8 M hydrogen peroxide; 5 mL 0.1 M ferrous sulfate) was added. The treated and nontreated soil samples were incubated at 20 °C for 56 d. Fenton's reagent greatly enhanced the mineralization of anthracene by indigenous microorganisms. The amounts of anthracene recovered as carbon dioxide after treatment with and without Fenton's reagent were 24 and <1%, respectively (Martens and Frankenberger, 1995).

Soil. In a 14-d experiment, [^{14}C]anthracene applied to soil-water suspensions under aerobic and anaerobic conditions gave $^{14}CO_2$ yields of 1.3 and 1.8%, respectively (Scheunert et al., 1987). The reported half-lives for anthracene in a Kidman sandy loam and McLaurin sandy loam are 134 and 50 d, respectively (Park et al., 1990).

Surface Water. The removal half-lives for anthracene in a water column at 25 °C in midsummer sunlight were 10.5 h for deep, slow, slightly turbid water; 21.6 h for deep, slow, muddy water; 8.5 h deep, slow, clear water; 3.5 h for shallow, fast, clear water, and 1.4 h for very shallow, fast, clear water (Southworth, 1977).

Photolytic. Oxidation of anthracene adsorbed on silica gel or alumina by oxygen in the presence of UV-light yielded anthraquinone. This compound further oxidized to 1,4-dihydroxy-9,10-anthraquinone. Anthraquinone was also formed by the oxidation of anthracene in diluted nitric acid or nitrogen oxides (Nikolaou et al., 1984) and in the dark when adsorbed on fly ash (Korfmacher et al., 1980). Irradiation of anthracene (2.6 mM) in cyclohexanone solutions gave 9,10-anthraquinone as the principal product (Korfmacher et al., 1980). Photocatalysis of anthracene and sulfur dioxide at -25 °C in various solvents yielded anthracene-9-sulfonic acid (Nielsen et al., 1983).

A carbon dioxide yield of 16.0% was achieved when anthracene adsorbed on silica gel was irradiated with light (λ >290 nm) for 17 h (Freitag et al., 1985). The photolytic half-life of anthracene in water ranged from 0.58 to 1.7 h (Southworth, 1979).

Behymer and Hites (1985) determined the effect of different substrates on the rate of photooxidation of anthracene (25 μg/g substrate) using a rotary photoreactor. The photolytic half-lives of anthracene using silica gel, alumina, and fly ash were 1.9, 0.5, and 48 h, respectively. Anthracene (5 mg/L) in a methanol-water solution (1:1 v/v) was subjected to a high pressure mercury lamp or sunlight. Based on a rate constant of 2.3 x 10^{-2}/min, the corresponding half-life is 30 min (Wang et al., 1991).

In a 5-m deep surface water body, the calculated half-lives for direct photochemical transformation at 40 °N latitude, in the midsummer during midday were 5.2 and 4.5 with and without sediment-water partitioning, respectively (Zepp and Schlotzhauer, 1979).

Chemical/Physical. In urban air from St. Louis, MO, anthracene reacted with NO_x to form 9-nitroanthracene (Ramdahl et al., 1982).

Anthracene will not hydrolyze in water (Kollig, 1995).

Exposure limits: Anthracene is a potential human carcinogen. No individual standards have been set; however, as a constituent in coal tar pitch volatiles, the following exposure limits have been established (mg/m^3): NIOSH REL: TWA 0.1 (cyclohexane-extractable fraction), IDLH 80; OSHA PEL: TWA 0.2 (benzene-soluble fraction); ACGIH TLV: TWA 0.2 (benzene solubles).

Toxicity:
Intraperitoneal LD$_{50}$ for mice is >430 mg/kg (Salamone, 1981).

Drinking water standard: No MCLGs or MCLs have been proposed (U.S. EPA, 1996).

Source: Concentrations in 8 diesel fuels ranged from 0.026 to 40 mg/L with a mean value of 6.275 mg/L (Westerholm and Li, 1994). Lee et al. (1992) reported concentration ranges of 100-300 mg/L and 0.04-2 μg/L in diesel fuel and corresponding aqueous phase (distilled water), respectively. Schauer et al. (1999) reported anthracene in diesel fuel at a concentration of 5 μg/g and in a diesel-powered medium-duty truck exhaust at an emission rate of 12.5 μg/km. Anthracene was detected in a distilled water-soluble fraction of used motor oil at concentrations ranging from 1.1 to 1.3 μg/L (Chen et al., 1994).

Thomas and Delfino (1991) equilibrated contaminant-free groundwater collected from Gainesville, FL with individual fractions of three individual petroleum products at 24-25 °C for 24 h. The aqueous phase was analyzed for organic compounds via U.S. EPA approved test method 625. Average anthracene concentrations reported in water-soluble fractions of kerosene and diesel fuel were 12 and 25 μg/L, respectively. Anthracene was ND in the water-soluble fraction of unleaded gasoline.

Based on analysis of 7 coal tar samples, anthracene concentrations ranged from 400 to 8,600 ppm (EPRI, 1990).

Uses: Dyes; starting material for the preparation of alizarin, phenanthrene, carbazole, 9,10-anthraquinone, 9,10-dihydroanthracene, and insecticides; in calico printing; as component of smoke screens; scintillation counter crystals; organic semiconductor research; wood preservative.

ANTU

Synonyms: Anturat; Bantu; Chemical 109; Krysid; **1-Naphthalenylthiourea**; 1-(1-Naphthyl)-2-thiourea; α-Naphthylthiourea; *N*-1-Naphthylthiourea; α-Naphthylthiocarbamide; Rattrack.

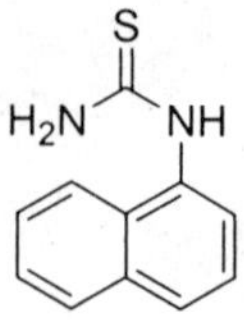

CAS Registry Number: 86-88-4
DOT: 1651
Molecular formula: $C_{11}H_{10}N_2S$
Formula weight: 202.27
RTECS: YT9275000
Merck reference: 10, 747

Physical state, color, and odor:
Colorless to gray, odorless solid

Melting point (°C):
198 (Weast, 1986)

Boiling point (°C):
Decomposes (NIOSH, 1994)

Density (g/cm³):
1.895 using method of Lyman et al. (1982)

Diffusivity in water (x 10^{-5} cm²/sec):
0.56 at 20 °C using method of Hayduk and Laudie (1974)

Flash point (°C):
Not applicable because ANTU is noncombustible (NIOSH, 1997)

Lower explosive limit (%):
Not applicable (NIOSH, 1997)

Upper explosive limit (%):
Not applicable (NIOSH, 1997)

Solubility in organics:
4.3 and 86 g/L in acetone and triethylene glycol, respectively (Windholz et al., 1983)

Solubility in water:
600 mg/L at 20 °C (quoted, Windholz et al., 1983)

Environmental fate:
Chemical/Physical. The hydrolysis rate constant for ANTU at pH 7 and 25 °C was

determined to be 8 x 10^{-5}/h, resulting in a half-life of 361 d (Ellington et al., 1988). At 85 °C, hydrolysis half-lives of 3.1, 1.2, and 0.6 d were observed at pH values of 3.26, 7.17, and 9.80, respectively (Ellington et al., 1987).

Emits toxic fumes of nitrogen and sulfur oxides when heated to decomposition (Lewis, 1990).

Exposure limits (mg/m³): NIOSH REL: TWA 0.3, IDLH 100; OSHA PEL: TWA 0.3; ACGIH TLV: TWA 0.3 mg/m³ (adopted).

Symptoms of exposure: Vomiting, dyspnea, cyanosis, course pulmonary rales after ingestion of large doses (NIOSH, 1997)

Toxicity:

Acute oral LD_{50} for Norwegian rats 6-8 mg/kg (Cremlyn, 1991; Hartley and Kidd, 1987).

Use: Rat poison. Banned in Britain due to carcinogenic impurities such as β-naphthylamine (Cremlyn, 1991).

BENZENE

Synonyms: Annulene; Benxole; Benzol; Benzole; Benzolene; Bicarburet of hydrogen; Carbon oil; Coal naphtha; Coal tar naphtha; Cyclohexatriene; Mineral naphthalene; Motor benzol; NCI-C55276; Nitration benzene; Phene; Phenyl hydride; Pyrobenzol; Pyrobenzole; RCRA waste number U019; UN 1114.

CAS Registry Number: 71-43-2
DOT: 1114
DOT label: Flammable liquid
Molecular formula: C_6H_6
Formula weight: 78.11
RTECS: CY1400000
Merck reference: 10, 1063

Physical state, color, and odor:
Clear, colorless to light yellow watery liquid with an aromatic or gasoline-like odor. The reported odor threshold concentrations in air ranged from 0.19 to 0.84 ppm (Keith and Walters, 1992; Young et al., 1996).

Melting point (°C):
5.533 (Standen, 1964)

Boiling point (°C):
80.100 (Standen, 1964)

Density (g/cm³):
0.87891 at 20.00/4 °C (Tsierkezos et al., 2000)
0.8784 at 20/4 °C, 0.8680 at 30/4 °C, 0.8572 at 40/4 °C (Sumer et al., 1968)
0.8728 at 25.00/4 °C (Aminabhavi and Banerjee, 1999)
0.87378 at 25/4 °C (Kirchnerová and Cave, 1976)
0.8630 at 35.00/4 °C, 0.8520 at 45.00/4 °C (Sastry et al., 1999)

Diffusivity in water (x 10^{-5} cm²/sec):
1.02 at 20 °C (Witherspoon and Bonoli, 1969)
1.09 at 25 °C (quoted, Hayduk and Laudie, 1974)
At 25 °C: 1.13 and 1.06 at mole fractions of 10^{-5} and 4 x 10^{-5}, respectively (Gabler et al., 1996)

Flash point (°C):
-11 (NIOSH, 1997)

Lower explosive limit (%):
1.2 (NIOSH, 1997)

Upper explosive limit (%):
7.8 (NIOSH, 1997)

Dissociation constant, pK_a:
≈ 37 (Gordon and Ford, 1972)

Entropy of fusion (cal/mol·K):
8.5 (Tsonopoulos and Prausnitz, 1971)

Heat of fusion (kcal/mol):
2.370 (Tsonopoulos and Prausnitz, 1971)

Henry's law constant (x 10^{-3} atm·m^3/mol):
5.56 at 23 °C (Anderson, 1992)
5.56 at 25 °C (Vitenberg et al., 1975; Mackay et al., 1979)
3.30, 3.88, 4.52, 5.28 and 7.20 at 10, 15, 20, 25 and 30 °C, respectively (Ashworth et al., 1988)
2.86, 3.75, 4.54, 5.96 and 7.31 at 10, 15, 20, 25 and 30 °C, respectively (Perlinger et al., 1993)
Distilled water: 1.73, 2.20, 2.38, 3.70 and 4.75 at 2.0, 6.0, 10.0, 18.2 and 25.0 °C, respectively;
    natural seawater: 2.63 and 6.04 at 6.0 and 25.0 °C, respectively (Dewulf et al., 1995)
4.76 at 25 °C (Nielsen et al., 1994)
3.77 at 25 °C (Allen et al., 1997)
5.88 at 25 °C (Hoff et al., 1993)
5.29, 6.77, 8.79, 12.0, and 14.3 at 25, 30, 40, 45, and 50 °C, respectively (Robbins et al., 1993)
4.79 at 22.0 °C (mole fraction ratio, Leighton and Calo, 1981)
5.62 at 25.0 °C (Ettre et al., 1993)
6.25 at 25 °C (Wasik and Tsang, 1970)
At 25 °C (NaCl concentration, mol/L): 4.41 (0), 4.55 (0.1), 5.09 (0.3), 5.56 (0.5), 6.12 (0.7),
    6.95 (1.0); Salinity, NaCl = 37.34 g/L (°C): 4.35 (15), 5.73 (20), 7.03 (25), 8.19 (30), 9.79
    (35), 11.93 (40), 13.66 (45) (Peng and Wan, 1998).
5.43 at 25 °C (Alaee et al., 1996)
1.90 and 4.74 at 5.7 and 24.9 °C, respectively (Dewulf et al., 1999a)

Interfacial tension with water (dyn/cm):
28.7 at 25 °C (Murphy et al., 1957)
34.1 at 25 °C (Donahue and Bartell, 1952)
33.63 at 25 °C (Jańczuk et al., 1993)
35.03 at 20 °C (Harkins et al., 1920)

Ionization potential (eV):
9.25 (Lo et al., 1986)
9.56 (Yoshida et al., 1983a)
9.38 (Krishna and Gupta, 1970)

Bioconcentration factor, log BCF:
0.54 (eels, Ogata and Miyake, 1978)
3.32 green alga, *Selenastrum capricornutum* (Casserly et al., 1983)
1.10 (fathead minnow, Veith et al., 1980)
0.63 (goldfish, Ogata et al., 1984)
3.23 (activated sludge, Freitag et al., 1985)
1.48 (algae, Geyer et al., 1984)
2.35 (*Daphnia pulex*) (Trucco et al., 1983)

Soil sorption coefficient, log K_{oc}:
1.69 (aquifer sands, Abdul et al., 1987)
1.92 (Schwarzenbach and Westall, 1981)
1.96, 2.00 (Overton silty clay loam, Rogers et al., 1980)
1.58, 1.64, 1.73 (various Norwegian soils, Seip et al., 1986)

1.50 (Woodburn silt loam, Chiou et al., 1983)

2.16, 2.53, 2.73 (Cohansey sand), 2.09, 2.31, 3.01 (Potomac-Raritan-Magothy sandy loam) (Uchrin and Mangels, 1987)

1.42 (estuarine sediment, Vowles and Mantoura, 1987)

2.10, 2.40 (Allerod), 2.30 (Borris, Brande), 2.55 (Brande), 2.15 (Finderup), 2.65, 2.68 (Gunderup), 2.48 (Herborg), 2.92 (Rabis), 2.40, 2.50 (Tirstrup), 2.05 (Tylstrup), 2.70 (Vasby), 2.38, 2.78 (Vejen), 2.85, 2.95, 2.28 (Vorbasse) (Larsen et al., 1992)

1.74 (Captina silt loam), 1.81 (McLaurin sandy loam) (Walton et al., 1992)

From crude oil: 0.68 (Grimsby silt loam), 1.00 (Vaudreil sandy loam), 1.54 (Wendover silty clay), 0.53 (Rideau silty clay) (Nathwani and Phillips, 1977)

1.73 (river sediment), 1.70 (coal wastewater sediment) (Kopinke et al., 1995)

1.80 (Piwoni and Banerjee, 1989)

1.82, 1.87 (RP-HPLC immobilized humic acids, Szabo et al., 1990)

2.82 (glaciofluvial, sandy aquifer, Nielsen et al., 1996)

1.48, 1.45, and 1.52 for Oakville sand (A horizon), Oakville sand (B horizon), and Pipestone sand, respectively (Maraqa et al., 1998)

1.39 (Mt. Lemmon soil, Hu et al., 1995)

1.84, 1.86, 1.87, 1.88, 1.90, 1.87 and 1.90 at 2.3, 3.8, 6.2, 8.0, 13.5, 18.6, at 25.0 °C, respectively, for a Leie River (Belgium) clay (Dewulf et al., 1999a)

1.74 (muck), 1.31 (Eustis sand) (Brusseau et al., 1990)

Octanol/water partition coefficient, log K_{ow}:
2.13 (Hansch and Fujita, 1964; DeKock and Lord, 1987)

2.11 (Mackay, 1982)

1.56 (Rogers and Cammarata, 1969)

2.15 (Campbell and Luthy, 1985; Leo et al., 1971)

2.12 (Veith et al., 1980)

1.95 (Eadsforth, 1986)

2.186 (de Bruijn et al., 1989)

2.20 (Hammers et al., 1982)

2.16, 2.28 (Suntio et al., 1988)

2.08 (Maraqa et al., 1998)

1.97, 1.96, 2.05, 2.01, 2.04, and 1.97 at 2.2, 6.0, 10.0, 14.1, 18.7, and 24.8 °C, respectively (Dewulf et al., 1999a)

Solubility in organics:
Miscible with ethanol, ether, acetic acid, acetone, chloroform, carbon tetrachloride (U.S. EPA, 1985), carbon disulfide, oils (Windholz et al., 1983), and hexane (Corby and Elworthy, 1971)

Solubility in water:
0.181 wt % at 25 °C (Lo et al., 1986)

0.153 wt % at 0 °C (Hill, 1922)

1,820 mg/L at 22 °C (Chiou et al., 1977)

1,740 mg/L at 25 °C (Andrews and Keefer, 1949)

1,750 mg/L at 25 °C (Banerjee et al., 1980)

1,760 mg/L at 25 °C (Etzweiler et al., 1995)

1,850 mg/kg at 30 °C (Gross and Saylor, 1931)

1,755 mg/L at 25 °C (McDevit and Long, 1952)

1,780 mg/kg at 25 °C (McAuliffe, 1963, 1966)

1,790 mg/L at 25 °C (Bohon and Claussen, 1951; Wasik et al., 1983)

1,791 μg/kg at 25 °C (May et al., 1978a)

1,800 mg/L at 25 °C (Howard and Durkin, 1974; Klevens, 1950)

In wt %: 0.175 at 20 °C, 0.180 at 25 °C, 0.190 at 30 °C (quoted, Stephen and Stephen, 1963)

1,755 mg/kg at 25 °C (Polak and Lu, 1973)

1,740 mg/kg at 25 °C, 1,391 mg/kg in artificial seawater (34.472 mg NaCl/kg) at 25 °C (Price, 1976)

1,710 mg/L at 20 °C (Freed et al., 1977)

1,796 mg/L at 20 °C (Hayashi and Sasaki, 1956)

21.7 mM at 25.00 °C (Keeley et al., 1988)

1,860 mg/kg at 25 °C (Stearns et al., 1947)

1,000 mg/L in fresh water at 25 °C, 1,030 mg/L in salt water at 25 °C (Krasnoshchekova and Gubergrits, 1975)

0.18775 wt % at 23.5 °C (Schwarz, 1980)

23.3 mM at 25 °C (Ben-Naim and Wilf, 1980)

17.76 mM in 0.5 M NaCl at 25 °C (Wasik et al., 1984)

24.2 mM at 35 °C (Hine et al., 1963)

1.74 g/kg at 25 °C (Chey and Calder, 1972)

20.7, 20.2, 20.7, 21.8, and 22.8 mM at 5, 15, 25, 35, and 45 °C, respectively (Sanemasa et al., 1982)

32 mM at 25 °C (Hogfeldt and Bolander, 1963)

22 mM at 25 °C (Taha et al., 1966)

23.6 and 24.3 mmol/kg at 30 and 35 °C, respectively (Saylor et al., 1938)

1,510 mg/L at 25 °C (McBain and Lissant, 1951)

1,779.5 mg/L at 25 °C (Mackay and Shiu, 1973)

1,800 and 2,200 mg/kg at 25 and 50 °C, respectively (Griswold et al., 1950)

2,170 mg/L at 25 °C (Worley, 1967)

1,650 mg/L (Coutant and Keigley, 1988)

22.0 mmol/kg at 25 °C (Morrison and Billett, 1952)

1.84, 1.85, 1.81, 1.81, 1.77, 1.77, 1.79, 1.79, and 1.76 g/kg at 4.5, 6.3, 7.1, 9.0, 11.8, 12.1, 15.1, 17.9, and 20.1 °C, respectively. In artificial seawater: 1.323, 1.376, 1.347, 1.318, and 1.296 g/kg 0.19, 5.32, 10.05, 14.96, and 20.04 °C, respectively (Brown and Wasik, 1974)

1.76 g/L at 25 °C (Brady and Huff, 1958)

In g/kg: 1.79, 1.77, 1.80, 1.83, 1.92, 2.03, 2.14, 2.34, and 2.57 at 9.4, 16.8, 24.0, 31.0, 38.0, 44.7, 51.5, 58.8, and 65.4 °C, respectively (Alexander, 1959)

10^{-4} mole fraction (°C): 3.95 (17.0), 3.97 (22.0), 3.99 (26.0), 4.02 (29.0), 4.12 (32.0), 4.20 (35.0), 4.39 (40.5), 4.40 (42.0), 4.45 (44.0), 4.57 (46.0), 4.78 (51.0), 5.03 (56.0), 5.31 (61.0), 5.42 (63.0) (Franks et al., 1963)

21.8 mM at 20 °C (Corby and Elworthy, 1971)

In wt % (°C): 1.283 (153), 1.913 (178), 2.902 (204), 3.790 (225), 4.471 (241), 5.073 (154) (Guseva and Parnov, 1963)

1,765 mg/L at 25 °C (Leinonen and Mackay, 1973)

2.0403 mL/L at 25 °C (Sada et al., 1975)

24.4 mmol/kg at 25.0 °C (Vesala, 1974)

15.4 mM at 25.00 °C (Sanemasa et al., 1985)

4.03×10^{-4} at 25 °C (mole fraction, Li et al., 1993)

1.1 and 1.9 mL/L at 25 and 60 °C, respectively (Booth and Everson, 1949)

In mole fraction ($\times 10^{-3}$): 4.232 at 0.20 °C, 4.159 at 6.20 °C, 4.147 at 11.20 °C, 4.080 at 14.20 °C, 4.062 at 16.90 °C, 4.073 at 18.60 °C, 4.129 at 25.00 °C, 4.193 at 25.80 °C (May et al., 1983)

Vapor density:
3.19 g/L at 25 °C, 2.70 (air = 1)

Vapor pressure (mmHg):
60 at 15 °C, 76 at 20 °C, 118 at 30 °C (quoted, Verschueren, 1983)
95.2 at 25 °C (Mackay and Leinonen, 1975)
397 at 60.3 °C, 556 at 70.3 °C, 764 at 80.3 °C, 1,031 at 90.3 °C, 1,397 at 100.3 °C (Eon et al., 1971)
146.8 at 35 °C (Hine et al., 1963)
95 at 25 °C (Milligan, 1924)
93.56 at 25.00 °C (Hussam and Carr, 1985)

Environmental fate:
  Biological. A mutant of *Pseudomonas putida* dihydroxylyzed benzene into *cis*-benzene glycol, accompanied by partial dehydrogenation, yielding catechol (Dagley, 1972). Bacterial dioxygenases can cleave catechol at the *ortho* and *meta* positions to yield *cis,cis*-muconic acid and α-hydroxymuconic semialdehyde, respectively (Chapman, 1972). Pure microbial cultures hydroxylated benzene to phenol and two unidentified phenols (Smith and Rosazza, 1974). Muconic acid was reported to be the biooxidation product of benzene by *Nocardia corallina* V-49 using hexadecane as the substrate (Keck et al., 1989).
  In activated sludge, 29.2% of the applied benzene mineralized to carbon dioxide after 5 d (Freitag et al., 1985). In anoxic groundwater near Bemidji, MI, benzene was anaerobically biodegraded to phenol (Cozzarelli et al., 1990). When benzene was statically incubated in the dark at 25 °C with yeast extract and settled domestic wastewater inoculum, significant biodegradation with rapid adaptation was observed. At concentrations of 5 and 10 mg/L, 49 and 37% biodegradation, respectively, were observed after 7 d. After 14 d of incubation, benzene demonstrated complete dissimilation (Tabak et al., 1981). Based on a first-order degradation rate constant of 0.2/yr, the half-life of benzene is 100 d (Zoeteman et al., 1981).
  Estimated half-lives of benzene (1.4 μg/L) from an experimental marine mesocosm during the spring (8-16 °C), summer (20-22 °C), and winter (3-7 °C) were 23, 3.1, and 13 d, respectively (Wakeham et al., 1983).
  Bridié et al. (1979) reported BOD and COD values 2.18 and 2.15 g/g using filtered effluent from a biological sanitary waste treatment plant. These values were determined using a standard dilution method at 20 °C and stirred for a period of 5 d. The ThOD for benzene is 3.08 g/g.
  Photolytic. A photooxidation rate constant of 6 x 10^{-11} cm^3/molecule·sec at room temperature was reported for the vapor-phase reaction of benzene with OH radicals in air (Atkinson, 1985). The reported rate constant and half-life for the reaction of benzene and OH radicals in the atmosphere are 8.2 x 10^{-10} M/sec and 6.8 d, respectively (Mill, 1982). Major photooxidation products in air include nitrobenzene, nitrophenol, phenol, glyoxal, butanedial, formaldehyde, carbon dioxide, and carbon monoxide (Nojima et al., 1975; Finlayson-Pitts and Pitts, 1986).
  Groundwater. Nielsen et al. (1996) studied the degradation of benzene in a shallow, glacio-fluvial, unconfined sandy aquifer in Jutland, Denmark. As part of the *in situ* microcosm study, a cylinder that was open at the bottom and screened at the top was installed through a cased borehole approximately 5 m below grade. Five liters of water was aerated with atmospheric air to ensure aerobic conditions were maintained. Groundwater was analyzed weekly for approximately 3 months to determine benzene concentrations over time. The experimentally determined first-order biodegradation rate constant and corresponding half-life following a 6-d lag phase were 0.5/d and 1.39 d, respectively.
  Surface Water. Mackay and Wolkoff (1973) estimated an evaporation half-life of 37.3 min from a surface water body that is 25 °C and 1 m deep.
  Chemical/Physical. Titanium dioxide suspended in an aqueous solution and irradiated with UV light (λ = 365 nm) converted benzene to carbon dioxide at a significant rate (Matthews, 1986). Irradiation of benzene in an aqueous solution yields mucondialdehyde. Photolysis of benzene vapor at 1849-2000 Å yields ethylene, hydrogen, methane, ethane, toluene, and a

polymer resembling cuprene. Other photolysis products reported under different conditions include fulvene, acetylene, substituted trienes (Howard, 1990), phenol, 2-nitrophenol, 4-nitrophenol, 2,4-dinitrophenol, 2,6-dinitrophenol, nitrobenzene, formic acid, and peroxyacetyl nitrate (Calvert and Pitts, 1966). Under atmospheric conditions, the gas-phase reaction with OH radicals and nitrogen oxides resulted in the formation of phenol and nitrobenzene (Atkinson, 1990). A carbon dioxide yield of 40.8% was achieved when benzene adsorbed on silica gel was irradiated with light (λ >290 nm) for 17 h.

Kanno et al. (1982) studied the aqueous reaction of benzene and other aromatic hydrocarbons (toluene, xylene, and naphthalene) with hypochlorous acid in the presence of ammonium ion. They reported that the aromatic ring was not chlorinated as expected (forming chlorobenzene) but was cleaved by chloramine forming cyanogen chloride (Kanno et al., 1982). The amount of cyanogen chloride formed was inversely proportional to the pH of the solution. At pH 6, the greatest amount of cyanogen chloride was formed when the reaction mixture contained ammonium ion and hypochlorous acid at a ratio of 2:3 (Kanno et al., 1982). Benzene vapor reacted with nitrate radicals in purified air forming nitrobenzene (Chiodini et al., 1993).

At an influent concentration of 416 mg/L, treatment with granular activated carbon resulted in an effluent concentration of 21 mg/L. The adsorbability of the carbon used was 80 mg/g carbon (Guisti et al., 1974). Similarly, at influent concentrations of 10.0, 1.0, 0.1, and 0.01 mg/L, the adsorption capacities of the granular activated carbon used were 40, 1.0, 0.03, and 0.0007 mg/g, respectively (Dobbs and Cohen, 1980).

Benzene will not hydrolyze because it does not contain a hydrolyzable functional group (Kollig, 1995).

Exposure limits (ppm): Known human carcinogen. NIOSH REL: TWA 0.1, STEL 1, IDLH 500; OSHA PEL: TWA 1, STEL 5; ACGIH TLV: TWA 0.5, STEL 2.5 (adopted).

Symptoms of exposure: Hallucination, distorted perception, euphoria, somnolence, nausea, vomiting, and headache. Narcotic at air concentrations of 200 ppm. At higher concentrations, convulsions may occur. Eye, nose, and respiratory irritant (Patnaik, 1992). An irritation concentration of 9,000.00 mg/m³ in air was reported by Ruth (1986).

Toxicity:
EC_{50} (72-h) for *Selenastrum capricornutum* 29 mg/L (Galassi et al., 1988).
EC_{10} and EC_{50} concentrations inhibiting the growth of alga *Scenedesmus subspicatus* in 96 h were both >1,360 mg/L (Geyer et al., 1985).
LC_{50} (contact) for earthworm (*Eisenia fetida*) 98 μg/cm² (Neuhauser et al., 1985).
LC_{50} (14-d) for *Poecilia reticulata* 63.5 mg/L (Könemann, 1981).
LC_{50} (96-h) for bass (*Marone saxatilis*) 5.8-10.9 mg/L (quoted, Verschueren, 1983), coho salmon 9 mg/L (Moles et al., 1979), juvenile rainbow trout 5.3 mg/L (deGraeve et l., 982), *Salmo gairdneri* 5.9 mg/L, *Poecilia reticulata* 28.6 mg/L (Galassi et al., 1988).
LC_{50} (48-h) for Mexican axolotl 370 mg/L, clawed toad 190 mg/L (Sloof et al., 1983), *Ischnura elegans* nymphs 10 mg/L (Sloof, 1983a), *Daphnia magna* 31.2 mg/L (Bobra et al., 1983) and 200 mg/L (LeBlanc, 1980).
LC_{50} (24-h) for *Daphnia magna* 250 mg/L (LeBlanc, 1980), rainbow trout 9.2 μg/L, channel catfish 425 μg/L, bluegill sunfish 102 (pH 8) to 740 (pH 7.4) (Mayer and Ellersieck, 1986).
LC_{50} (3-h) for *Chlorella vulgaris* 312 mg/L (Hutchinson et al., 1980).
LC_{50} (60-min) for brown trout yearlings (static bioassay) 12 mg/L (Woodiwiss and Fretwell, 1974).
LC_{50} (inhalation) for mice 9,980 ppm, rats 3,306 ppm/7-h (quoted, RTECS, 1985).
LC_{50} (4-h inhalation) 44,500 mg/m³ for rats (Drew and Fouts, 1974).
Acute oral LD_{50} for mice 4,700 ppm, rats 3,306 mg/kg (quoted, RTECS, 1985).

Drinking water standard (final): MCLG: zero; MCL: 5 μg/L (U.S. EPA, 1996).

Source: Detected in distilled water-soluble fractions of 87 octane gasoline (24.0 mg/L), 94 octane gasoline (80.7 mg/L), Gasohol (32.3 mg/L), No. 2 fuel oil (0.50 mg/L), jet fuel A (0.23 mg/L), diesel fuel (0.28 mg/L), military jet fuel JP-4 (17.6 mg/L) (Potter, 1996), new motor oil (0.37-0.40 μg/L), and used motor oil (195-198 μg/L) (Chen et al., 1994). The average volume percent and estimated mole fraction in American Petroleum Institute PS-6 gasoline are 2.082 and 0.2969, respectively (Poulsen et al., 1992). Schauer et al. (1999) reported benzene in a diesel-powered medium-duty truck exhaust at an emission rate of 2,740 μg/km.

Thomas and Delfino (1991) equilibrated contaminant-free groundwater collected from Gainesville, FL with individual fractions of three individual petroleum products at 24-25 °C for 24 h. The aqueous phase was analyzed for organic compounds via U.S. EPA approved test method 602. Average benzene concentrations reported in water-soluble fractions of unleaded gasoline, kerosene, and diesel fuel were 8.652, 0.349, and 0.200 mg/L, respectively. When the authors analyzed the aqueous-phase via U.S. EPA approved test method 610, average benzene concentrations in water-soluble fractions of unleaded gasoline, kerosene, and diesel fuel were lower, i.e., 1.107, 0.073, and 0.066 mg/L, respectively.

Benzene is produced from petroleum refining, coal tar distillation, coal processing, and coal coking (Verschueren, 1983).

Uses: Manufacture of ethylbenzene (preparation of styrene monomer), dodecylbenzene (for detergents), cyclohexane (for nylon), nitrobenzene, aniline, maleic anhydride, biphenyl, benzene hexachloride, benzene sulfonic acid, phenol, dichlorobenzenes, insecticides, pesticides, fumigants, explosives, aviation fuel, flavors, perfume, medicine, dyes, and many other organic chemicals; paints, coatings, plastics and resins; food processing; photographic chemicals; nylon intermediates; paint removers; rubber cement; antiknock gasoline; solvent for fats, waxes, resins, inks, oils, paints, plastics, and rubber.

BENZIDINE

Synonyms: Azoic diazo component 112; Benzidine base; *p*-Benzidine; 4,4′-Bianiline; *p,p*′-Bianiline; **(1,1′-Biphenyl)-4,4′-diamine**; 4,4′-Biphenyldiamine; *p,p*′-Biphenyldiamine; 4,4′-Biphenylenediamine; *p,p*′-Biphenylenediamine; C.I. 37225; C.I. azoic diazo component 112; 4,4′-Diaminobiphenyl; *p,p*′-Diaminobiphenyl; 4,4′-Diamino-1,1′-biphenyl; 4,4′-Diaminodiphenyl; *p*-Diaminodiphenyl; *p,p*′-Diaminodiphenyl; 4,4′-Dianiline; *p,p*′-Dianiline; 4,4′-Diphenylenediamine; *p,p*′-Diphenylenediamine; Fast corinth base B; NCI-C03361; RCRA waste number U021; UN 1885.

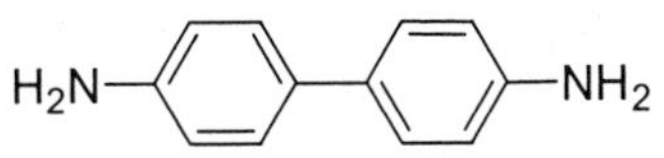

CAS Registry Number: 92-87-5
DOT: 1885
DOT label: Poison
Molecular formula: $C_{12}H_{12}N_2$
Formula weight: 184.24
RTECS: DC9625000
Merck reference: 10, 1077

Physical state, color, and odor:
Grayish-yellow powder or white to pale reddish, odorless crystals. Darkens on exposure to air or light.

Melting point (°C):
128 (Weast, 1986)
117.2 (ACGIH, 1986)

Boiling point (°C):
401.7 (Sax, 1984)
400 (Patnaik, 1992)

Density (g/cm³):
1.250 at 20/4 °C (Shriner et al., 1978)

Diffusivity in water (x 10^{-5} cm²/sec):
0.57 at 20 °C using method of Hayduk and Laudie (1974)

Flash point (°C):
Combustible solid, but difficult to burn (NIOSH, 1997)

Dissociation constant, pK$_a$ (at 25 °C):
$pK_1 = 4.70$, $pK_2 = 3.63$ (Dean, 1973)

Henry's law constant (x 10^{-11} atm·m³/mol):
3.88 at 25 °C (estimated, Howard, 1989)

Ionization potential (eV):
6.88 (Mallard and Linstrom, 1998)

Bioconcentration factor, log BCF:
1.74 (mosquito fish), 2.66 (mosquito), 2.81 (snail), 3.42 (algae) (Lu et al., 1977)
1.90 (golden ide), 2.93 (algae), 3.08 (activated sludge) (Freitag et al., 1985)

Soil sorption coefficient, log K_{oc}:
5.72 (average using 4 soils, Graveel et al., 1986)
3.46 (Meylan et al., 1992)

Octanol/water partition coefficient, log K_{ow}:
1.34 (Mabey et al., 1982; Hansch and Leo, 1985)
1.63 (Hassett et al., 1980)

Solubility in organics:
Soluble in ethanol (U.S. EPA, 1985) and ether (1 g/50 mL) (Windholz et al., 1983)

Solubility in water (mg/L):
400 at 12 °C, 9,400 at 100 °C (quoted, Verschueren, 1983)
500 at 25 °C (Bowman et al., 1976)
360 at 24 °C (Hassett et al., 1980)
520 at 25 °C (Shriner et al., 1978)

Vapor density:
7.50 g/L presumably at 20 °C (Sims et al., 1988)

Vapor pressure (mmHg):
Based on the specific vapor density value of 6.36 (Sims et al., 1988), the vapor pressure was calculated to be 0.83 at 20 °C.

Environmental fate:
   Biological. In activated sludge, <0.1% mineralized to carbon dioxide after 5 d (Freitag et al., 1985). Kincannon and Lin (1985) reported a half-life of 76 d when benzidine in sludge was applied to a sandy loam soil.
   Soil. Benzidine was added to different soils and incubated in the dark at 23 °C under a carbon dioxide-free atmosphere. After 1 yr, 8.3 to 11.6% of the added benzidine degraded to carbon dioxide primarily by microbial metabolism and partially by hydrolysis (Graveel et al., 1986). Tentatively identified biooxidation compounds using GC/MS include hydroxy-benzidine, 3-hydroxybenzidine, 4-amino-4'-nitrobiphenyl, N,N'-dihydroxybenzidine, 3,3'-di-hydroxybenzidine and 4,4'-dinitrobiphenyl (Baird et al., 1977). Under aerobic conditions, the half-life was estimated to be 2 to 8 d (Lu et al., 1977).
   In the presence of hydrogen peroxide and acetylcholine at pH 11 and 20 °C, benzidine oxidized to 4-amino-4'-nitrobiphenyl (Aksnes and Sandberg, 1957).
   Photolytic. A carbon dioxide yield of 40.8% was achieved when benzene adsorbed on silica gel was irradiated with light (λ >290 nm) for 17 h (Freitag et al., 1985).
   Chemical/Physical. Benzidine is not subject to hydrolysis (Kollig, 1993).

Exposure limits: Known human carcinogen. Because no standards have been established, NIOSH (1997) recommends the most reliable and protective respirators be used, i.e., a self-contained breathing apparatus that has a full facepiece and is operated under positive-pressure or a supplied-air respirator that has a full facepiece and is operated under pressure-demand or under positive-pressure in combination with a self-contained breathing apparatus operated under pressure-demand or positive-pressure.

OSHA recommends that worker exposure to this chemical is to be controlled by use of engineering control, proper work practices, and proper selection of personal protective equipment. Specific details of these requirements can be found in CFR 1910.1003-1910.1016.

Toxicity:
  LC_{50} (48-h) for red killifish 57.5 mg/L (Yoshioka et al., 1986).
  Acute oral LD_{50} for mice 214 mg/kg, rats 309 mg/kg (quoted, RTECS, 1985).

Source: Benzidine can enter the environment by transport, use, and disposal, or by dyes and pigments containing the compound. A photodegradation product of 3,3'-dichlorobenzidine.
  Based on analysis of 7 coal tar samples, benzidine was ND (EPRI, 1990).

Uses: Organic synthesis; manufacture of azo dyes, especially Congo Red; detection of blood stains; stain in microscopy; laboratory reagent in determining cyanide, sulfate, nicotine, and some sugars; stiffening agent in rubber compounding.

BENZO[*a*]ANTHRACENE

Synonyms: BA; B(*a*)A; Benzanthracene; **Benz[*a*]anthracene**; 1,2-Benzanthracene; 1,2-Benz[*a*]anthracene; 2,3-Benzanthracene; Benzanthrene; 1,2-Benzanthrene; Benzoanthracene; 1,2-Benzoanthracene; Benzo[*a*]phenanthrene; Benzo[*b*]phenanthrene; 2,3-Benzophenanthrene; Naphthanthracene; RCRA waste number U018; Tetraphene.

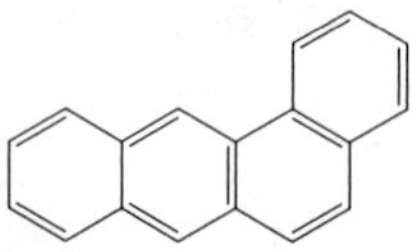

CAS Registry Number: 56-55-3
Molecular formula: $C_{18}H_{12}$
Formula weight: 228.30
RTECS: CV9275000
Merck reference: 10, 1059

Physical state and color:
Colorless leaflets or plates with a greenish-yellow fluorescence

Melting point (°C):
162 (Weast, 1986)
161.1 (Casellato et al., 1973)
156.9 (Murray et al., 1974)

Boiling point (°C):
435 with sublimation (Weast, 1986)
437.6 (Aldrich, 1988)
400 (Sims et al., 1988)

Density (g/cm³ at 20 °C):
1.274 (HSDB, 1989)
1.2544 (Mailhot and Peters, 1988)

Diffusivity in water (x 10^{-5} cm²/sec):
0.52 at 20 °C using method of Hayduk and Laudie (1974)

Dissociation constant, pK_a:
>15 (Christensen et al., 1975)

Henry's law constant (x 10^{-6} atm·m³/mol):
8.0 (Southworth, 1979)
1.48, 3.06, 6.22, 12.0, and 20.8 at 4.1, 11.0, 18.0, 25.0, and 31.0 °C, respectively (Bamford et al., 1998)

Ionization potential (eV):
8.01 (Franklin et al., 1969)
7.45 (Yoshida et al., 1983a)
7.54 (Cavalieri and Rogan, 1985)

Bioconcentration factor, log BCF:
4.00 (*Daphnia pulex*, Southworth et al., 1978)
4.00 (fathead minnow, Veith et al., 1979)
4.01 (*Daphnia magna*, Newsted and Giesy, 1987)
4.39 (activated sludge), 3.50 (algae), 2.54 (golden ide) (Freitag et al., 1985)
Apparent values of 4.1 (wet wt) and 5.9 (lipid wt) for freshwater isopods including *Asellus aquaticus* (L.) (van Hattum et al., 1998)

Soil sorption coefficient, log K_{oc}:
5.0 (Meylan et al., 1992)
5.81 (average coarse silt fraction from Doe Run and Hickory Hill sediments, Karickhoff et al., 1979)
5.30 (humic acid, Landrum et al., 1984)
6.30 (average, Kayal and Connell, 1990)
7.2 (average value using 8 river bed sediments from the Netherlands, van Hattum et al., 1998)

Octanol/water partition coefficient, log K_{ow}:
5.61 (Radding et al., 1976)
5.91 at 25 °C (Yoshida et al., 1983; de Maagd et al., 1998)

Solubility in organics:
Soluble in ethanol, ether, acetone, and benzene (U.S. EPA, 1985)

Solubility in water:
10 μg/L at 25 °C (Klevens, 1950)
14 μg/L at 25 °C (Mackay and Shiu, 1977)
9.4 and 12.2 μg/kg at 25 and 29 °C, respectively. In seawater (salinity = 35.0 g/kg): 5.6 μg/kg at 25 °C (May et al., 1978a)
11 μg/L at 27 °C (Davis et al., 1942)
44 μg/L at 24 °C (practical grade, Hollifield, 1979)
5.7 μg/L at 20 °C (Smith et al., 1978)
16.8 μg/L at 25 °C (Walters and Luthy, 1984)
13.0 μg/L at 25 °C (de Maagd et al., 1998)
94.4 μg/L at 25 °C (Billington et al., 1988)
21.3, 18.9, 18.8, 20.6, 27.6, and 37.4 nmol/L at 3.7, 8.0, 12.4, 16.7, 20.9, and 25.0 °C, respectively. In seawater (salinity = 32.1 g/kg): 118, 65.4, 57.2, 38.5, 36.7, and 40.9 nmol/L at 3.7, 8.0, 12.4, 16.7, 20.9, and 25.0 °C, respectively (Whitehouse, 1984).
In mole fraction (x 10^{-9}): 2.359 at 6.90 °C, 2.983 at 10.70 °C, 2.849 at 11.10 °C, 3.780 at 14.30 °C, 4.403 at 18.10 °C, 4.995 at 19.30 °C, 6.313 at 23.60 °C, 6.605 at 23.10 °C, 6.794 at 25.00 °C, 9.785 at 29.50 °C, 10.02 at 29.70 °C (May et al., 1983).

Vapor pressure (x 10^{-9} mmHg):
5 at 20 °C (Pupp et al., 1974)
210 at 25 °C (Sonnefeld et al., 1983)
2,250 at 25 °C (Bidleman, 1984)
30.5 at 25 °C (Banerjee et al., 1990)
110.3 at 25 °C (Murray et al., 1974)
54.8 at 25 °C (de Kruif, 1980)

Environmental fate:
Biological. In an enclosed marine ecosystem containing planktonic primary production and

heterotrophic benthos, the major metabolites were water soluble and could not be extracted with organic solvents. The only degradation product identified was benzo[*a*]anthracene-7,12-dione (Hinga and Pilson, 1987). Under aerobic conditions, *Cunninghanella elegans* degraded benzo[*a*]anthracene to 3,4-, 8,9-, and 10,11-dihydrols (Kobayashi and Rittman, 1982; Riser-Roberts, 1992).

A strain of *Beijerinckia* oxidized benzo[*a*]anthracene producing 1-hydroxy-2-anthranoic acid as the major product. Three other metabolites identified were 2-hydroxy-3-phenanthroic acid, 3-hydroxy-2-phenanthroic acid, and *cis*-1,1-dihydroxy-1,2-dihydrobenzo[*a*]anthracene (Gibson et al., 1975; Mahaffey et al., 1988).

In a marine microcosm containing Narragansett Bay sediments, the polychaete *Mediomastis ambesita* and the bivalve *Nucula anulata*, benzo[*a*]anthracene degraded to carbon dioxide, phenols, and quinones (Hinga et al., 1980).

In activated sludge, <0.1% mineralized to carbon dioxide after 5 d (Freitag et al., 1985). When benzo[*a*]anthracene (5 and 10 mg/L) was statically incubated in the dark at 25 °C with yeast extract and settled domestic wastewater inoculum, no significant biodegradation was observed (Tabak et al., 1981).

Ye et al. (1996) investigated the ability of *Sphingomonas paucimobilis* strain U.S. EPA 505 (a soil bacterium capable of using fluoranthene as a sole source of carbon and energy) to degrade 4, 5, and 6-ringed aromatic hydrocarbons (10 ppm). After 16 h of incubation using a resting cell suspension, 72.9% of benzo[*a*]anthracene had degraded via ring cleavage.

Soil. The reported half-lives for benzo[*a*]anthracene in a Kidman sandy loam and McLaurin sandy loam are 261 and 162 d, respectively (Park et al., 1990).

Surface Water. In a 5-m deep surface water body, the calculated half-lives for direct photochemical transformation at 40 °N latitude, in the midsummer during midday were 4.8 and 22.8 h with and without sediment-water partitioning, respectively (Zepp and Schlotzhauer, 1979).

Photolytic. Benzo[*a*]anthracene-7,12-dione formed from the photolysis of benzo[*a*]anthracene ($\lambda = 366$ nm) in an air-saturated, acetonitrile-water solvent (Smith et al., 1978).

A carbon dioxide yield of 25.3% was achieved when benzo[*a*]anthracene adsorbed on silica gel was irradiated with light ($\lambda > 290$ nm) for 17 h (Freitag et al., 1985). Behymer and Hites (1985) determined the effect of different substrates on the rate of photooxidation of chrysene using a rotary photoreactor. The photolytic half-lives of benzo*a*]anthracene using silica gel, alumina, and fly ash were 4.0, 2.0, and 38 h, respectively.

Benzo[*a*]anthracene (12.5 mg/L) in a methanol-water solution (2:3 v/v) was subjected to a high pressure mercury lamp or sunlight. Based on a rate constant of 2.51×10^{-2}/min, the corresponding half-life is 0.46 h (Wang et al., 1991).

Chemical/Physical. Benzo[*a*]anthracene-7,12-dione and a monochlorinated product were formed during the chlorination of benzo[*a*]anthracene. At pH 4, the reported half-lives at chlorine concentrations of 0.6 and 10 mg/L were 2.3 and <0.2 h, respectively (Mori et al., 1991). When an aqueous solution containing benzo[*a*]anthracene (16.11 μg/L) was chlorinated for 6 h using chlorine (6 mg/L), the concentration was reduced 53% (Sforzolini et al., 1970).

Benzo[*a*]anthracene will not hydrolyze because it does not contain a hydrolyzable functional group (Kollig, 1993).

Exposure limits: Potential occupational carcinogen. No individual standards have been set; however, as a constituent in coal tar pitch volatiles, the following exposure limits have been established (mg/m^3): NIOSH REL: TWA 0.1 (cyclohexane-extractable fraction), IDLH 80; OSHA PEL: TWA 0.2 (benzene-soluble fraction); ACGIH TLV: TWA 0.2 (benzene solubles).

Toxicity:
LD$_{50}$ for mice by intravenous injection 10 mg/kg (Patnaik, 1992).

Drinking water standard: No MCLGs or MCLs have been proposed (U.S. EPA, 1996).

Source: Concentrations in 8 diesel fuels ranged from 0.018 to 5.9 mg/L with a mean value of 0.93 mg/L (Westerholm and Li, 1994). Identified in Kuwait and South Louisiana crude oils at concentrations of 2.3 and 1.7 ppm, respectively (Pancirov and Brown, 1975).

Based on analysis of 7 coal tar samples, benzo[*a*]anthracene concentrations ranged from 600 to 5,100 ppm (EPRI, 1990). Detected in 1-yr aged coal tar film and bulk coal tar at concentrations of <1,500 and 850 mg/kg, respectively (Nelson et al., 1996).

Uses: Organic synthesis. Not manufactured commercially but is derived from industrial and experimental coal gasification operations where the maximum concentrations detected in gas, liquid, and coal tar streams were 28, 4.1, and 18 mg/m^3, respectively (Cleland, 1981).

BENZO[*b*]FLUORANTHENE

Synonyms: **Benz[*e*]acephenanthrylene**; 3,4-Benz[*e*]acephenanthrylene; 2,3-Benzfluoranthene; 3,4-Benzfluoranthene; Benzo[*b*]fluoranthene; Benzo[*e*]fluoranthene; 2,3-Benzofluoranthene; 3,4-Benzofluoranthene; 3,4-Benzo[*b*]fluoranthene; B(*b*)F.

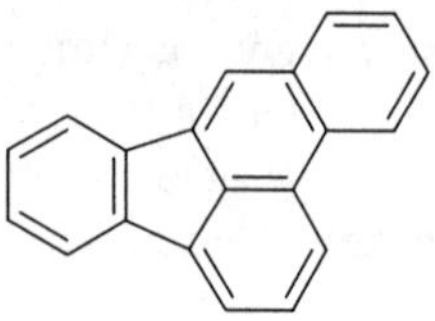

CAS Registry Number: 205-99-2
Molecular formula: $C_{20}H_{12}$
Formula weight: 252.32
RTECS: CU1400000

Physical state:
Solid

Melting point (°C):
168 (Weast, 1986)
163-165 (Aldrich, 1988)
161.6 (Casellato et al., 1973)

Diffusivity in water (x 10^{-5} cm^2/sec):
0.49 at 20 °C using method of Hayduk and Laudie (1974)

Dissociation constant, pK$_a$:
>15 (Christensen et al., 1975)

Henry's law constant (x 10^{-7} atm·m^3/mol):
5.03 at 25 °C (de Maagd et al., 1998)
2.47, 5.03, 11.74, 14.90, 20.53, and 36.52 at 10.0, 20.0, 35.0, 40.1, 45.0, and 55.0 °C, respectively (ten Hulscher et al., 1992)

Bioconcentration factor, log BCF:
4.00 (*Daphnia magna*, Newsted and Giesy, 1987)
0.96 (*Polychaete* sp.), 0.23 (*Capitella capitata*) (Bayona et al., 1991)
Apparent values of 4.3 (wet wt) and 6.0 (lipid wt) for freshwater isopods including *Asellus aquaticus* (L.) (van Hattum et al., 1998)

Soil sorption coefficient, log K$_{oc}$:
6.26, 6.70, 6.82 (San Francisco, CA mudflat sediments, Maruya et al., 1996)
7.3 (average value using 8 river bed sediments from the Netherlands, van Hattum et al., 1998)

Octanol/water partition coefficient, log K$_{ow}$:
6.06 (Mabey et al., 1982)
6.40 (Bayona et al., 1991)

Solubility in organics:
Soluble in most solvents (U.S. EPA, 1985)

Solubility in water:
1.2 μg/L at 25 °C (U.S. EPA, 1980a)

Vapor pressure (mmHg):
5 x 10^{-7} at 20 °C (U.S. EPA, 1982)

Environmental fate:
   Biological. Ye et al. (1996) investigated the ability of *Sphingomonas paucimobilis* strain U.S. EPA 505 (a soil bacterium capable of using fluoranthene as a sole source of carbon and energy) to degrade 4, 5, and 6-ringed aromatic hydrocarbons (10 ppm). After 16 h of incubation using a resting cell suspension, only 12.5% of benzo[*b*]fluoranthene had degraded. It was suggested that degradation occurred via ring cleavage resulting in the formation of polar metabolites and carbon dioxide.
   Soil. The reported half-lives for benzo*b*]fluoranthene in a Kidman sandy loam and McLaurin sandy loam are 294 and 211 d, respectively (Park et al., 1990).
   Photolytic. The atmospheric half-life was estimated to range from 1.43 to 14.3 h (Atkinson, 1987).
   Chemical/Physical. Benzo[*b*]fluoranthene will not hydrolyze because it has no hydrolyzable functional group (Kollig, 1993).

Exposure limits: Potential occupational carcinogen. No individual standards have been set;however, as a constituent in coal tar pitch volatiles, the following exposure limits have been established (mg/m^3): NIOSH REL: TWA 0.1 (cyclohexane-extractable fraction), IDLH 80; OSHA PEL: TWA 0.2 (benzene-soluble fraction); ACGIH TLV: TWA 0.2 (benzene solubles).

Drinking water standard: No MCLGs or MCLs have been proposed (U.S. EPA, 1996).

Source: Benzo[*b*]fluoranthene and benzo[*k*]fluoranthene were detected in 8 diesel fuels at concentrations ranging from 0.0027 to 3.1 mg/L with a mean value of 0.266 mg/L (Westerholm and Li, 1994). Also present in low octane gasoline (0.16-0.49 mg/kg), high octane gasoline (0.26-1.34 mg/kg), used motor oil (2.8-141.0 mg/kg), and bitumen (40-1,600 ppb), cigarette smoke (3 μg/1,000 cigarettes), and gasoline exhaust (19-48 μg/L) (quoted, Verschueren, 1983).
   Based on analysis of 7 coal tar samples, benzo[*b*]fluoranthene concentrations ranged from 490 to 3,100 ppm (EPRI, 1990).

Use: Produced primarily for research purposes. Derived from industrial and experimental coal gasification operations where the maximum concentrations detected in gas, liquid, and coal tar streams were 0.38, 0.033, and 3.2 mg/m^3, respectively (Cleland, 1981).

BENZO[*k*]FLUORANTHENE

Synonyms: 8,9-Benzfluoranthene; 8,9-Benzofluoranthene; 11,12-Benzofluoranthene; 11,12-Benzo[*k*]fluoranthene; B(*k*)F; 2,3,1′,8′-Binaphthylene; Dibenzo[*b,jk*]fluorene.

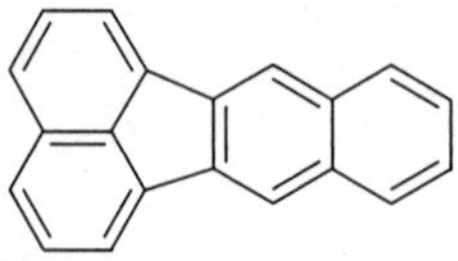

CAS Registry Number: 207-08-9
Molecular formula: $C_{20}H_{12}$
Formula weight: 252.32
RTECS: DF6350000

Physical state and color:
Pale yellow needles

Melting point (°C):
198-217 (Murray et al., 1974)
217 (Weast, 1986)

Boiling point (°C):
480 (Pearlman et al., 1984)
481 (Bjørseth, 1983)

Diffusivity in water (x 10^{-5} cm^2/sec):
0.49 at 20 °C using method of Hayduk and Laudie (1974)

Dissociation constant, pK_a:
>15 (Christensen et al., 1975)

Henry's law constant (x 10^{-7} atm·m^3/mol):
4.24 at 25 °C (de Maagd et al., 1998)
2.17, 4.24, 10.56, 13.62, 19.54, and 39.77 at 10.0, 20.0, 35.0, 40.1, 45.0, and 55.0 °C, respectively (ten Hulscher et al., 1992)

Bioconcentration factor, log BCF:
4.00 (*Daphnia magna*, Newsted and Giesy, 1987)
1.15 (*Polychaete* sp.), 0.26 (*Capitella capitata*) (Bayona et al., 1991)
Apparent values of 4.4 (wet wt) and 6.1 (lipid wt) for freshwater isopods including *Asellus aquaticus* (L.) (van Hattum et al., 1998)

Soil sorption coefficient, log K_{oc}:
5.99 (average, Kayal and Connell, 1990)
5.97, 6.01, 6.70, 6.94 (San Francisco, CA mudflat sediments, Maruya et al., 1996)
7.4 (average value using 8 river bed sediments from the Netherlands, van Hattum et al., 1998)

Octanol/water partition coefficient, log K_{ow}:
6.85 (quoted, Mills et al., 1985)

6.40 (Bayona et al., 1991)
6.11 at 25 °C (de Maagd et al., 1998)

Solubility in organics:
Soluble in most solvents (U.S. EPA, 1985)

Solubility in water:
1.09 μg/L at 25 °C (de Maagd et al., 1998)

Vapor pressure (mmHg):
9.59 x 10^{-11} at 25 °C (Radding et al., 1976)

Environmental fate:
   Soil. Based on aerobic soil die-away test data, the half-life in soil ranged from 910 d to 5.86 yr (Bossert et al., 1984).
   Photolytic. The atmospheric half-life was estimated to range from 1.1 to 11 h (Atkinson, 1987).
   Chemical/Physical. Benzo[*k*]fluoranthene will not hydrolyze because it has no hydrolyzable functional group (Kollig, 1995).

Exposure limits: Potential occupational carcinogen. No individual standards have been set; however, as a constituent in coal tar pitch volatiles, the following exposure limits have been established (mg/m^3): NIOSH REL: TWA 0.1 (cyclohexane-extractable fraction), IDLH 80; OSHA PEL: TWA 0.2 (benzene-soluble fraction); ACGIH TLV: TWA 0.2 (benzene solubles).

Drinking water standard: No MCLGs or MCLs have been proposed (U.S. EPA, 1996).

Source: Benzo[*b*]fluoranthene and benzo[*k*]fluoranthene were detected in 8 diesel fuels at concentrations ranging from 0.0027 to 3.1 mg/L with a mean value of 0.266 mg/L (Westerholm and Li, 1994). Also present in gasoline (9 μg/L), bitumen (34-1,140 μg/L), crude oil (<1 ppm) (quoted, Verschueren, 1983), and coal (32.5 g/kg) (Lao et al., 1975).
   Based on analysis of 7 coal tar samples, benzo[*k*]fluoranthene concentrations ranged from 350 to 3,000 ppm (EPRI, 1990).

Use: Produced primarily for research purposes. Derived from industrial and experimental coal gasification operations where the maximum concentrations detected in liquid and coal tar streams were 0.017 and 1.6 mg/m^3, respectively (Cleland, 1981).

BENZOIC ACID

Synonyms: Benzenecarboxylic acid; Benzeneformic acid; Benzenemethanoic acid; Benzoate; Carboxybenzene; Dracylic acid; NA 9094; Phenylcarboxylic acid; Phenylformic acid; Retarder BA; Retardex; Salvo liquid; Salvo powder; Tenn-plas.

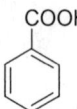

CAS Registry Number: 65-85-0
DOT: 9094
Molecular formula: $C_7H_6O_2$
Formula weight: 122.12
RTECS: DG0875000
Merck reference: 10, 1093

Physical state, color, and odor:
Colorless to white needles, scales or powder with a faint benzoin or benzaldehyde-like odor

Melting point (°C):
121.7 (Stull, 1947)
122.13 (Weast, 1986)

Boiling point (°C):
249.2 (Weast, 1986)

Density (g/cm³):
1.2659 at 15/4 °C (Weast, 1986)
1.316 at 28/4 °C (Standen, 1964)

Diffusivity in water (x 10^{-5} cm²/sec):
0.79 at 20 °C using method of Hayduk and Laudie (1974)

Flash point (°C):
121 (Aldrich, 1988)

Diffusivity in water (x 10^{-5} cm²/sec):
0.90 at 25 °C (undissociated molecule, Noulty and Leaist, 1987)

Dissociation constant, pK_a:
4.21 at 25 °C (Dean, 1973)
4.20 (Clarke and Cahoon, 1987; Muccini et al., 1999)

Entropy of fusion (cal/mol·K):
9.82 (Pacor, 1967)
10.5 (Andrews et al., 1926)
10.88 (Furukawa et al., 1951; Ginnings and Furukawa, 1953)
11 (David, 1964)

Heat of fusion (kcal/mol):
4.32 (Dean, 1987)

Henry's law constant (x 10^{-8} atm·m^3/mol):
7.02 (calculated, U.S. EPA, 1980a)

Ionization potential (eV):
9.73 ± 0.09 (Franklin et al., 1969)

Bioconcentration factor, log BCF:
0.48 (algae, Geyer et al., 1984)
3.11 (activated sludge, Freitag et al., 1985)
2.00 (mosquito, Lu and Metcalf, 1975)

Soil sorption coefficient, log K_{oc}:
1.48-2.70 (average 2.26 for 10 Danish soils, Løkke, 1984)

Octanol/water partition coefficient, log K_{ow}:
1.68, 1.94 (Sangster, 1989)
1.87 (Leo et al., 1971)
2.03 (Lu and Metcalf, 1975)
1.81-1.88 (Fujita et al., 1964)
2.18 (Garst and Wilson, 1984)

Solubility in organics:
Soluble in chloroform (222 g/L), ether (333 g/L), acetone (333 g/L), carbon tetrachloride (33 g/L), benzene (100 g/L), carbon disulfide (33 g/L), volatile and fixed oils (Windholz et al., 1983).

Solubility in water (g/L):
3.00, 3.20, and 3.40 at 18, 23.5, and 25 °C, respectively (quoted, Stephen and Stephen, 1963)
1.70 at 0 °C, 2.10 at 10 °C, 2.90 at 20 °C, 3.40 at 25 °C, 4.20 at 30 °C, 6.00 at 40 °C, 8.50 at 50 °C, 12.00 at 60 °C, 17.70 at 70 °C, 27.50 at 80 °C, 45.50 at 90 °C, 68.00 at 95 °C (quoted, Standen, 1964)
1.80 at 4 °C, 2.70 at 18 °C, 22.00 at 75 °C (Hodgman et al., 1961)

Vapor pressure (x 10^{-4} mmHg):
45 at 25 °C (Howard, 1989)
8.18 at 25 °C (Sonnefeld et al., 1983)
4.5 at 20 °C (Klöpffer et al., 1988)

Environmental fate:
Biological. Benzoic acid may degrade to catechol if it is the central metabolite whereas, if protocatechuic acid (3,4-dihydroxybenzoic acid) is the central metabolite, the precursor is 3-hydroxybenzoic acid (Chapman, 1972). Other compounds identified following degradation of benzoic acid to catechol include *cis,cis*-muconic acid, (+)-muconolactone, 3-oxoadipate enol lactone, and 3-oxoadipate (Verschueren, 1983). Pure microbial cultures hydroxylated benzoic acid to 3,4-dihydroxybenzoic acid, 2- and 4-hydroxybenzoic acid (Smith and Rosazza, 1974). In a methanogenic enrichment culture, 91% of the added benzoic acid anaerobically biodegraded to carbon dioxide and methane (Healy and Young, 1979). In activated sludge, 65.5% mineralized to carbon dioxide after 5 d (Freitag et al., 1985).

Heukelekian and Rand (1955) reported a 5-d BOD value of 1.38 g/g which is 70.0% of the ThOD value of 1.97 g/g. In activated sludge inoculum, following a 20-d adaptation period, 99.0% COD removal was achieved in 5 d. The average rate of biodegradation was 88.5 mg

COD/g·h (Pitter, 1976).

Photolytic. Titanium dioxide suspended in an aqueous solution and irradiated with UV light (λ = 365 nm) converted benzoic acid to carbon dioxide at a significant rate (Matthews, 1986). An aqueous solution containing chlorine and irradiated with UV light (λ = 350 nm) converted benzoic acid to salicylaldehyde and other unidentified chlorinated compounds (Oliver and Carey, 1977). A carbon dioxide yield of 10.2% was achieved when benzoic acid adsorbed on silica gel was irradiated with light (λ >290 nm) for 17 h (Freitag et al., 1985).

Chemical/Physical. At an influent concentration of 1,000 mg/L, treatment with granular activated carbon resulted in an effluent concentration of 89 mg/L. The adsorbability of the carbon used was 183 mg/g carbon (Guisti et al., 1974). Ward and Getzen (1970) investigated the adsorption of aromatic acids on activated carbon under acidic, neutral, and alkaline conditions. The amount of benzoic acid (10^{-4} M) adsorbed by carbon at pH values of 3.0, 7.0, and 11.0 were 49.7, 11.2, and 2.5%, respectively.

In water, benzoic acid reacted with hydroxy radicals at a rate constant of 3.2 x 10^9/M·sec (Mabury and Crosby, 1996).

The evaporation rate of benzoic acid at 20 °C is 6.9 x 10^{-11} mol/cm^2·h (Gückel et al., 1982).

Toxicity:

LC_{50} for *Tetrahymena pyriformis* 12.6 mg/L (Muccini et al., 1999).

LC_{50} (48-h) for red killifish 910 mg/L (Yoshioka et al., 1986).

Acute oral LD_{50} for mice 1,940 mg/kg, cats 2,000 mg/kg, dogs 2,000 mg/kg, rats 2,530 mg/kg (quoted, RTECS, 1985).

Source: Naturally occurs in cranberries, ligonberries (1,360 ppm), peppermint leaves (20-200 ppb), tea leaves, cassia bark, carob, blessed thistle, purple foxglove, jasmine, hyacinth, apples, tobacco leaves, daffodils, autumn crocus, prunes, anise seeds, ripe cloves, and wild black cherry tree bark (Duke, 1992; quoted, Verschueren, 1983).

Schauer et al. (1999) reported benzoic in diesel fuel at a concentration of 1,260 μg/g.

Uses: Preparation of sodium and butyl benzoates, benzoyl chloride, phenol, caprolactum, and esters for perfume and flavor industry; plasticizers; manufacture of alkyl resins; preservative for food, fats, and fatty oils; seasoning; tobacco; dentifrices; standard in analytical chemistry; anti-fungal agent; synthetic resins and coatings; pharmaceutical and cosmetic preparations; plasticizer manufacturing (to modify resins such as polyvinyl chloride, polyvinyl acetate, phenol-formaldehyde).

BENZO[*ghi*]PERYLENE

Synonyms: 1,12-Benzoperylene; 1,12-Benzperylene; B(*ghi*)P.

CAS Registry Number: 191-24-2
Molecular formula: $C_{22}H_{12}$
Formula weight: 276.34
RTECS: DI6200500

Physical state:
Solid

Melting point (°C):
222 (Cleland and Kingsbury, 1977)
278-280 (Fluka, 1988)
275-277 (Murray et al., 1974)

Boiling point (°C):
>500 (Aldrich, 1988)
525 (Pearlman et al., 1984)

Diffusivity in water (x 10^{-5} cm²/sec):
0.49 at 20 °C using method of Hayduk and Laudie (1974)

Dissociation constant, pK_a:
>15 (Christensen et al., 1975)

Henry's law constant (x 10^{-7} atm·m³/mol):
2.66 at 25 °C (de Maagd et al., 1998)
1.88, 2.66, 5.13, 5.33, 6.51, and 8.59 at 10.0, 20.0, 35.0, 40.1, 45.0, and 55.0 °C, respectively
 (ten Hulscher et al., 1992)

Ionization potential (eV):
7.17 ± 0.02 (Lias, 1998)
7.24 (Franklin et al., 1969)

Bioconcentration factor, log BCF:
4.45 (*Daphnia magna*, Newsted and Giesy, 1987)
Apparent values of 3.7 (wet wt) and 5.4 (lipid wt) for freshwater isopods including *Asellus aquaticus* (L.) (van Hattum et al., 1998)

Soil sorption coefficient, log K_{oc}:
6.56, 6.78 (San Francisco, CA mudflat sediments, Maruya et al., 1996)
6.6 (average value using 8 river bed sediments from the Netherlands, van Hattum et al., 1998)

Octanol/water partition coefficient, log K_{ow}:
7.10 (Bruggeman et al., 1982; Mackay et al., 1980)
6.22 at 25 °C (de Maagd et al., 1998)

Solubility in organics:
Soluble in most solvents (U.S. EPA, 1985)

Solubility in water (μg/L at 25 °C):
0.137 (de Maagd et al., 1998)
0.26 (Mackay and Shiu, 1977)
0.83 (Wise et al., 1981)

Vapor pressure (x 10^{-10} mmHg at 25 °C):
1.01 (Radding et al., 1976)
1.04 (Murray et al., 1974)

Environmental fate:
  Biological. Based on aerobic soil die away test data at 10 to 30 °C, the estimated half-lives ranged from 590 to 650 d (Coover and Sims, 1987).
  Groundwater. Based on aerobic soil die away test data at 10 to 30 °C, the estimated half-lives ranged from 3.23 to 3.56 yr (Coover and Sims, 1987).
  Photolytic. The atmospheric half-life was estimated to range from 0.321 to 3.21 h (Atkinson, 1987). Behymer and Hites (1985) determined the effect of different substrates on the rate of photooxidation of benzo[*ghi*]perylene using a rotary photoreactor. The photolytic half-lives of benzo*ghi*]perylene using silica gel, alumina, and fly ash were 7.0, 22, and 29 h, respectively.
  Chemical/Physical. Benzo[*ghi*]perylene will not hydrolyze because it has no hydrolyzable functional group (Kollig, 1995).

Exposure limits: Potential occupational carcinogen. No individual standards have been set; however, as a constituent in coal tar pitch volatiles, the following exposure limits have been established (mg/m^3): NIOSH REL: TWA 0.1 (cyclohexane-extractable fraction), IDLH 80; OSHA PEL: TWA 0.2 (benzene-soluble fraction); ACGIH TLV: TWA 0.2 (benzene solubles).

Drinking water standard: No MCLGs or MCLs have been proposed (U.S. EPA, 1996).

Source: Detected in 7 of 8 diesel fuels at concentrations ranging from 0.008 to 0.35 mg/L with a mean value of 0.113 mg/L (Westerholm and Li, 1994). Identified in Kuwait and South Louisiana crude oils at concentrations of <1 and 1.6 ppm, respectively (Pancirov and Brown, 1975). Also present in fresh motor oil (120 μg/kg) and used motor oil (108.8-289.4 mg/kg) (quoted, Verschueren, 1983).
  Based on analysis of 7 coal tar samples, benzo[*ghi*]perylene concentrations ranged from ND to 1,900 ppm (EPRI, 1990).

Use: Research chemical. Derived from industrial and experimental coal gasification operations where the maximum concentration detected in coal tar streams was 2.7 mg/m^3 (Cleland, 1981).

BENZO[a]PYRENE

Synonyms: Benzo[*d,e,f*]chrysene; 1,2-Benzopyrene; 3,4-Benzopyrene; 6,7-Benzopyrene; Benz[*a*]pyrene; Benzo[α]pyrene; 1,2-Benzpyrene; 3,4-Benzpyrene; 3,4-Benz[*a*]pyrene; 3,4-Benzypyrene; BP; B(*a*)P; 3,4-BP; RCRA waste number U022.

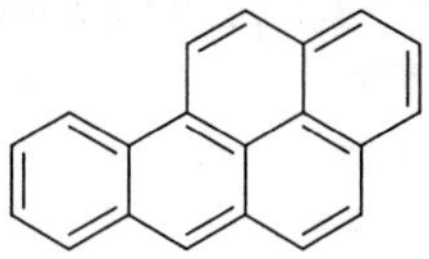

CAS Registry Number: 50-32-8
Molecular formula: $C_{20}H_{12}$
Formula weight: 252.32
RTECS: DJ3675000
Merck reference: 10, 1106

Physical state, color, and odor:
Odorless, yellow, orthorhombic or monoclinic crystals. Solution in concentrated sulfuric acid is orange-red with a green fluorescence (Keith and Walters, 1992).

Melting point (°C):
179-179.3 (Weast, 1986)
181.3 (Casellato et al., 1973)
176.4 (Murray et al., 1974)

Boiling point (°C):
495 (Aldrich, 1988)

Density (g/cm³):
1.351 (Kronberger and Weiss, 1944)

Diffusivity in water (x 10^{-5} cm²/sec):
0.50 at 20 °C using method of Hayduk and Laudie (1974)

Dissociation constant, pK_a:
>15 (Christensen et al., 1975)

Entropy of fusion (cal/mol·K):
9.27 (Hinckley et al., 1990)

Henry's law constant (x 10^{-7} atm·m³/mol):
3.36 at 25 °C (de Maagd et al., 1998)
2.17, 3.36, 7.30, 9.08, 10.86, and 23.89 at 10.0, 20.0, 35.0, 40.1, 45.0, and 55.0 °C, respectively (ten Hulscher et al., 1992)

Ionization potential (eV):
7.12 ± 0.01 (Lias, 1998)
7.23 (Cavalieri and Rogan, 1985)

Bioconcentration factor, log BCF:
3.69 (bluegill sunfish, Spacie et al., 1983)

3.90 (*Daphnia magna*, McCarthy, 1983)
4.00 (activated sludge), 3.53 (algae), 2.68 (golden ide) (Freitag et al., 1985)
3.51 (bluegill sunfish, Devillers et al., 1996)
Apparent values of 4.4 (wet wt) and 6.1 (lipid wt) for freshwater isopods including *Asellus aquaticus* (L.) (van Hattum et al., 1998)

Soil sorption coefficient, log K_{oc}:
5.95 (Meylan et al., 1992)
5.95 (humic acids, Landrum et al., 1984)
6.33 (humic acids, McCarthy and Jimenez, 1985)
6.26 (average, Kayal and Connell, 1990)
7.01 (Rotterdam Harbor sediment, Hageman et al., 1995)
7.4 (average value using 8 river bed sediments from the Netherlands, van Hattum et al., 1998)

Octanol/water partition coefficient, log K_{ow}:
6.00 (Sangster, 1989)
5.99 (Mallon and Harrison, 1984)
6.04 (Radding et al., 1976)
6.13 at 25 °C (de Maagd et al., 1998)
6.50 (Bruggeman et al., 1982; Landrum et al., 1984)
5.81 (Zepp and Scholtzhauer, 1979)

Solubility in organics:
Soluble in benzene, toluene and xylene; sparingly soluble in ethanol and methanol (Windholz et al., 1983)

Solubility in water:
3.8 μg/L at 25 °C (Mackay and Shiu, 1977)
4 μg/L at 25 °C (Schwarz and Wasik, 1976)
1.82 μg/L at 25 °C (de Maagd et al., 1998)
1.2 ng/L at 22 °C (Smith et al., 1978)
0.505 μg/L in Lake Michigan water at $\approx$ 25 °C (Eadie et al., 1990)
2.64, 3.00, 3.74, 4.61, and 6.09 nmol/L at 8.0, 12.4, 16.7, 20.9, and 25.0 °C, respectively. In seawater (salinity = 36.7 g/kg): 1.40, 2.07, 2.50, 3.43, and 4.45 nmol/L at 8.0, 12.4, 16.7, 20.9, and 25.0 °C, respectively (Whitehouse, 1984)
4.8 nmol/L (Mill et al., 1981)
3.0, 3.5, 4.0, 4.0 and 4.5 μg/L at 27 °C (Davis et al., 1942)
4.0 μg/L at 27 °C (Davis and Parke, 1942)
19 nmol/L at 25 °C (Barone et al., 1967)
1.6 μg/L at 25 °C (Billington et al., 1988)
In mole fraction (x 10^{-10}): 0.3998, 0.5712, 0.8139, 1.157, and 1.635 at 10.00, 15.00, 20.00, 25.00, and 30.00 °C, respectively (May et al., 1983)

Vapor pressure (x 10^{-9} mmHg):
5.49 at 25 °C (Radding et al., 1976)
5.6 at 25 °C (Murray et al., 1984)
5 at 25 °C (Smith et al., 1978)
509 at 20 °C (Sims et al., 1988)
2.4 at 25 °C (McVeety and Hites, 1988)
113 at 25 °C (Bidleman, 1984)
543, 840 at 25 °C (Hinckley et al., 1990)

3.9 at 25 °C (extrapolated from vapor pressures determined at higher temperatures, Tesconi and Yalkowsky, 1998)

Environmental fate:

Biological. Benzo[a]pyrene was biooxidized by *Beijerinckia* B836 to *cis*-9,10-di-hydroxy-9,10-dihydrobenzo[a]pyrene. Under nonenzymatic conditions, this metabolite mono-dehydroxylated to form 9-hydroxybenzo[a]pyrene (Verschueren, 1983). Under aerobic conditions, *Cunninghanella elegans* degraded benzo[a]pyrene to *trans*-7,8-dihydroxy-7,8-dihydrobenzo[a]pyrene (Kobayashi and Rittman, 1982), 3-hydroxybenzo[a]pyrene, 9-hydroxybenzo[a]pyrene and vicinal dihydrols including *trans*-9,10-dihydroxy-9,10-dihydrobenzo[a]pyrene (Cerniglia and Gibson, 1980; Gibson et al., 1975). The microorganisms *Candida lipolytica* and *Saccharomyces cerevisiae* oxidized benzo[a]pyrene to *trans*-7,8-dihydroxy-7,8-dihydrobenzo[a]pyrene, 3- and 9-hydroxybenzo[a]pyrene (Cerniglia and Crow, 1980; Wiseman et al., 1978) whereas 3-hydroxybenzo[a]pyrene was the main degradation product by the microbe *Neurospora crassa* (Lin and Kapoor, 1979).

After a 30-d incubation period, the white rot fungus *Phanerochaete chrysosporium* converted benzo[a]pyrene to carbon dioxide. Mineralization began between the third and sixth day of incubation. The production of carbon dioxide was highest between 3-18 d of incubation, after which the rate of carbon dioxide produced decreased until the 30th day. It was suggested that the metabolism of benzo[a]pyrene and other compounds, including *p,p'*-DDT, TCDD, and lindane, was dependent on the extracellular lignin-degrading enzyme system of this fungus (Bumpus et al., 1985). In activated sludge, <0.1% mineralized to carbon dioxide after 5 d (Freitag et al., 1985).

Contaminated soil from a manufactured coal gas plant that had been exposed to crude oil was spiked with benzo[a]anthracene (400 mg/kg soil) to which Fenton's reagent (5 mL 2.8 M hydrogen peroxide; 5 mL 0.1 M ferrous sulfate) was added. The treated and nontreated soil samples were incubated at 20 °C for 56 d. Fenton's reagent greatly enhanced the mineralization of benzo[a]anthracene by indigenous microorganisms. The amounts of anthracene recovered as carbon dioxide after treatment with and without Fenton's reagent were 17 and 2%, respectively (Martens and Frankenberger, 1995).

Ye et al. (1996) investigated the ability of *Sphingomonas paucimobilis* strain U.S. EPA 505 (a soil bacterium capable of using fluoranthene as a sole source of carbon and energy) to degrade 4, 5 and 6-ringed aromatic hydrocarbons (10 ppm). After 16 h of incubation using a resting cell suspension, 33.3% of benzo[a]pyrene had degraded. In the presence of 5 ppm benzo[b]fluoranthene, biodegradation was reduced about 30%. It was suggested that biodegradation occurred via ring cleavage resulting in the formation of polar metabolites and carbon dioxide.

Soil. Lu et al. (1977) studied the degradation of benzo[a]pyrene in a model ecosystem containing Drummer silty clay loam. Samples were incubated at 27.6 °C for 1, 2, and 4 wk before extraction with acetone for TLC analysis. After 4 wk, only 8.05% of benzo[a]pyrene degraded forming one polar compound and two unidentified compounds. The reported half-lives for benzo[a]pyrene in a Kidman sandy loam and McLaurin sandy loam are 309 and 229 d, respectively (Park et al., 1990).

Surface Water. In a 5-m deep surface water body, the calculated half-lives for direct photochemical transformation at 40 °N latitude, in the midsummer during midday were 3.2 and 13 d with and without sediment-water partitioning, respectively (Zepp and Schlotzhauer, 1979). The volatilization half-life of benzo[a]pyrene from surface water (1 m deep, water velocity 0.5 m/sec, wind velocity 1 m/sec) using experimentally determined Henry's law constants is estimated to be 1,500 h (Southworth, 1979).

Photolytic. Coated glass fibers exposed to air containing 100-200 ppb ozone yielded benzo[a]pyrene-4,5-oxide. At 200 ppb ozone, conversion yields of 50 and 80% were observed after 1 and 4 h, respectively (Pitts et al., 1990). Free radical oxidation and photolysis of benzo[a]pyrene at a wavelength of 366 nm yielded the following tentatively identified products: benzo[a]pyrene-1,6-quinone, benzo[a]pyrene-3,6-quinone, and benzo[a]pyrene-6,12-quinone (Smith et al., 1978).

In a solution containing oxygen, photolysis yields a mixture of 6,12-, 1,6-, and 3,6-diones. Nitration by nitrogen dioxide forms 6-nitro-, 1-nitro-, and 3-nitrobenzo[a]pyrene. When benzo[a]pyrene in methanol (1 g/L) was irradiated at 254 nm in a quartz flask for 1 h, the solution turned pale yellow. After 2 h, the solution turned yellow and back to clear after 4 h of irradiation. After 4 h, 99.67% of benzo[a]pyrene was converted to polar compounds. One of these compounds was identified as a methoxylated benzo[a]pyrene (Lu et al., 1977). A carbon dioxide yield of 26.5% was achieved when benzo[a]pyrene adsorbed on silica gel was irradiated with light (λ >290 nm) for 17 h (Freitag et al., 1985).

In a 5-m deep surface water body, the calculated half-lives for direct photochemical transformation at 40 °N latitude, in the midsummer during midday were 3.2 and 13 d with and without sediment-water partitioning, respectively (Zepp and Schlotzhauer, 1979).

Behymer and Hites (1985) determined the effect of different substrates on the rate of photo-oxidation of benzo[a]pyrene using a rotary photoreactor. The photolytic half-lives of benzo[a]pyrene using silica gel, alumina, and fly ash were 4.7, 1.4, and 31 h, respectively.

Benzo[a]pyrene (2.5 mg/L) in a methanol-water solution (3:7 v/v) was subjected to a high pressure mercury lamp or sunlight. Based on a rate constant of 3.22 x 10^{-2}/min, the corresponding half-life is 0.35 h (Wang et al., 1991).

Chemical/Physical. Ozonolysis to benzo[a]pyrene-1,6-quinone or benzo[a]py-rene-3,6-quinone followed by further oxidation to benzanthrone dicarboxylic anhydride was reported (IARC, 1983).

In a simulated atmosphere, direct epoxidation by ozone led to the formation of benzo[a]pyrene-4,5-oxide. Benzo[a]pyrene reacted with benzoyl peroxide to form the 6-benzoyloxy derivative (Nikolaou et al., 1984). It was reported that benzo[a]pyrene adsorbed on fly ash and alumina reacted with sulfur dioxide (10%) in air to form benzo[a]pyrene sulfonic acid (Nielsen et al., 1983). Benzo[a]pyrene coated on a quartz surface was subjected to ozone and natural sunlight for 4 and 2 h, respectively. The compounds 1,6-quinone, 3,6-quinone, and the 6,12-quinone of benzo[a]pyrene were formed in both instances (Rajagopalan et al., 1983).

When benzo[a]pyrene adsorbed from the vapor phase onto coal fly ash, silica and alumina was exposed to nitrogen dioxide, no reaction occurred. However, in the presence of nitric acid, nitrated compounds were produced (Yokley et al., 1985). Chlorination of benzo[a]pyrene in polluted humus poor lake water gave 11,12-dichlorobenzo[a]pyrene and 1,11,12-, 3,11,12- or 3,6,11-trichlorobenzo[a]pyrene, representing 99% of the chlorinated products formed (Johnsen et al., 1989). When an aqueous solution containing benzo[a]pyrene (53.14 μg/L) was chlorinated for 6 h using chlorine (6 mg/L), the concentration was reduced 98% (Sforzolini et al., 1970).

Benzo[a]pyrene will not hydrolyze in water because it does not contain a hydrolyzable functional group (Kollig, 1993).

Exposure limits: Potential occupational carcinogen. No individual standards have been set; however, as a constituent in coal tar pitch volatiles, the following exposure limits have been established (mg/m^3): NIOSH REL: TWA 0.1 (cyclohexane-extractable fraction), IDLH 80; OSHA PEL: TWA 0.2 (benzene-soluble fraction); ACGIH TLV: TWA 0.2 (benzene solubles).

Toxicity:

LC$_{50}$ (96-h) for *Daphnia pulex* 5 μg/L (Trucco et al., 1983).

LD$_{50}$ for mice (subcutaneous) 50 mg/kg (quoted, RTECS, 1985).

LD$_{50}$ for mice (intraperitoneal) 232 mg/kg (Salamone, 1981).

Drinking water standard (final): MCLG: zero; MCL: 0.2 μg/L (U.S. EPA, 1996).

Source: Identified in Kuwait and South Louisiana crude oils at concentrations of 2.8 and 0.75 ppm, respectively (Pancirov and Brown, 1975). Emitted to the environment from coke production, coal refuse and forest fires, motor vehicle exhaust, and heat and power (utility) generation (Suess, 1976). Benzo[*a*]pyrene is produced from combustion of tobacco and fuels. It is also a component of gasoline (133-143 μg/L), fresh motor oil (20-100 g/kg), used motor oil (83.2-242.4 mg/kg), asphalt ($\leq$0.0027 wt %), coal tar pitch ($\leq$1.25 wt %), cigarette smoke (25 μg/1,000 cigarettes), and gasoline exhaust (quoted, Verschueren, 1983). Benzo[*a*]pyrene was also detected in liquid paraffin at an average concentration of 25 μg/kg (Nakagawa et al., 1978).

Benzo[*a*]pyrene was reported in a variety of foodstuffs including raw and cooked meat (ND to 12 ppb), fish (0.3-6.9 ppb), vegetables oils (ND-4), fruits (ND to 6.2 ppb) (quoted, Verschueren, 1983).

Based on analysis of 7 coal tar samples, benzo[*a*]pyrene concentrations ranged from 500 to 6,400 ppm (EPRI, 1990).

Use: Produced primarily for research purposes. Derived from industrial and experimental coal gasification operations where the maximum concentrations detected in gas, liquid, and coal tar streams were 5.0, 0.036, and 3.5 mg/m^3, respectively (Cleland, 1981).

BENZO[*e*]PYRENE

Synonyms: 1,2-Benzopyrene; 1,2-Benzpyrene; 4,5-Benzopyrene; 4,5-Benzo[*e*]pyrene; B(*e*)P.

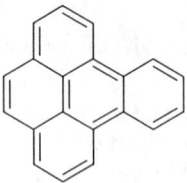

CAS Registry Number: 192-97-2
Molecular formula: $C_{20}H_{12}$
Formula weight: 252.32
RTECS: DJ4200000
Merck reference: 10, 1107

Physical state:
Crystalline, solid, prisms, or plates

Melting point (°C):
178-179 (Windholz et al., 1983)
178.8 (Murray et al., 1974)
179 (Bjørseth, 1983)

Boiling point (°C):
493 (Bjørseth, 1983)

Density (g/cm³):
0.8769 at 20/4 °C (Windholz et al., 1983)

Diffusivity in water (x 10^{-5} cm²/sec):
0.50 at 20 °C using method of Hayduk and Laudie (1974)

Dissociation constant, pK$_a$:
>14 (Schwarzenbach et al., 1993)

Entropy of fusion (cal/mol·K):
10.11 (Hinckley et al., 1990)

Henry's law constant (x 10^{-7} atm·m³/mol):
4.84 at 25 °C (approximate - calculated from water solubility and vapor pressure)

Ionization potential (eV):
7.41 (Lias et al., 1998)
7.62 (Cavalieri and Rogan, 1985)

Bioconcentration factor, log BCF:
4.56 (freshwater isopod *A. aquaticus*, van Hattum and Montanes, 1999)

Soil sorption coefficient, log K$_{oc}$:
7.20 (Broman et al., 1991)
6.19 (San Francisco, CA mudflat sediments, Maruya et al., 1996)

Octanol/water partition coefficient, log K_{ow}:
6.75 using method of Yalkowsky and Valvani (1979)

Solubility in organics:
Slightly soluble in methanol (Patnaik, 1992)

Solubility in water:
In ng/kg: 325 at 8.6 °C, 358 at 14.0 °C, 444 at 17.0 °C, 394 at 17.5 °C, 459 at 20.0 °C, 479 at 20.2 °C, 535 at 23.2 °C, 507 at 23.0 °C, 643 at 29.2 °C, 681 at 31.7 °C. In seawater: 3.32 ng/kg at 25 °C. In NaCl solution (salinity = 30 g/kg), nmol/L: 0.082 at 8.9 °C, 0.088 at 10.8 °C, 0.101 at 15.6 °C, 0.104 at 19.2 °C, 0.113 at 21.7 °C, 0.135 at 25.3 °C, 0.142 at 27.1 °C, 0.166 at 30.2 °C (Schwarz, 1977)
29 nmol/L at 25 °C (Barone et al., 1967)
3.5 μg/L at 27 °C (Davis et al., 1942)

Vapor pressure (x 10^{-9} mmHg):
2.4 (McVeety and Hites, 1988)
5.54 at 25 °C (Pupp et al., 1974)
5.55 at 25 °C (Murray et al., 1974)
644 at 25 °C (Hinckley et al., 1990)

Environmental fate:
    Chemical/Physical. Benzo[e]pyrene will not hydrolyze because it has no hydrolyzable functional group.

Source: Detected in 8 diesel fuels at concentrations ranging from 0.047 to 2.1 mg/L with a mean value of 0.113 mg/L (Westerholm and Li, 1994). Identified in Kuwait and South Louisiana crude oils at concentrations of 0.5 and 2.5 ppm, respectively (Pancirov and Brown, 1975).
    Benzo[e]pyrene is produced from the combustion of tobacco and petroleum fuels. It also occurs in low octane gasoline (0.18-0.87 mg/kg), high octane gasoline (0.45-1.82 mg/kg), used motor oil (92.2-278.4 mg/kg), asphalt ($\leq$0.0052 wt %), coal tar pitch ($\leq$0.70 wt %), cigarette smoke (3 μg/1,000 cigarettes), and gasoline exhaust (quoted, Verschueren, 1983).

Uses: Chemical research.

BENZYL ALCOHOL

Synonyms: Benzal alcohol; Benzene carbinol; **Benzene methanol**; Benzoyl alcohol; α-Hydroxytoluene; NCI-C06111; Phenol carbinol; Phenyl carbinol; Phenyl methanol; Phenyl methyl alcohol; α-Toluenol.

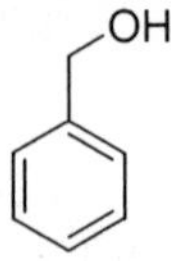

CAS Registry Number: 100-51-6
Molecular formula: C_7H_8O
Formula weight: 108.14
RTECS: DN3150000
Merck reference: 10, 1130

Physical state, color, and odor:
Colorless, hygroscopic liquid with a faint, pleasant, aromatic odor. Odor threshold concentration is 5.5 ppm (Keith and Walters, 1992).

Melting point (°C):
-15.3 (Stull, 1947)
-11 to -9 (Fluka, 1988)

Boiling point (°C):
204.7 (Stull, 1947)
205.3 (Weast, 1986)

Density (g/cm³):
1.044 at 20/4 °C (Fluka, 1988)
1.0424 at 25.00/4 °C, 1.0383 at 30.00/4 °C, 1.0313 at 40.00/4 °C, 1.0232 at 50.00/4 °C, 1.0153 at 60.00/4 °C, 1.0075 at 70.00/4 °C (Abraham et al., 1971)

Diffusivity in water (x 10⁻⁵ cm²/sec):
0.93 at 25 °C (quoted, Hayduk and Laudie, 1974)

Flash point (°C):
93 (NFPA, 1984)

Heat of fusion (kcal/mol):
4.074 (Riddick et al., 1986)

Interfacial tension with water (dyn/cm at 22.5 °C):
4.75 (Demond and Lindner, 1993)

Ionization potential (eV):
9.14 ± 0.05 (Franklin et al., 1969)

Soil sorption coefficient, log K_{oc}:
1.193 (Australian soil, Broggs, 1981)
<0.70 (Apison, Fullerton and Dormant soils, Southworth and Keller, 1986)

Octanol/water partition coefficient, log K$_{ow}$:
1.00 (Sangster, 1989)
1.10 (Fujita et al., 1964)
1.16 (Garst and Wilson, 1984)

Solubility in organics:
Miscible with ether and absolute alcohol (Windholz et al., 1983)

Solubility in water:
42,900 mg/L at 25 °C (Banerjee, 1984)
3.66 wt % at 20 °C (quoted, Stephen and Stephen, 1963)
In wt %: 4.8 at 0 °C, 4.7 at 9.8 °C, 4.3 at 20.1 °C, 4.3 at 29.6 °C, 4.6 at 40.2 °C, 5.2 at 50.0 °C
   (Stephenson and Stuart, 1986)
34 mM at 25 °C (Southworth and Keller, 1986)

Vapor density:
4.42 g/L at 25 °C, 3.73 (air = 1)

Vapor pressure (mmHg):
0.11 at 25 °C (Riddick et al., 1986)

Environmental fate:
   Biological. Heukelekian and Rand (1955) reported a 5-d BOD value of 1.55 g/g which is 61.5% of the ThOD value of 2.52 g/g.
   Chemical/Physical. Slowly oxidizes in air to benzaldehyde (Huntress and Mulliken, 1941). Benzyl alcohol will not hydrolyze because it has no hydrolyzable functional group (Kollig, 1993).

Toxicity:
   LC$_{50}$ static bioassay values for fathead minnows in Lake Superior water maintained at 18-22 °C after 1, 24, 48, 72, and 96 h are 770, 770, 770, 480, and 460 mg/L, respectively (Mattson et al., 1976).
   LC$_{50}$ (96-h, static bioassay) for bluegill sunfish 15 mg/L (quoted, Verschueren, 1983).
   Acute oral LD$_{50}$ for wild bird 100 mg/kg, mouse 1,580 mg/kg, rat 1,230 mg/kg, rabbit 1,040 mg/kg (quoted, RTECS, 1985), adult and neonatal mice 1,000 mg/kg (McCloskey et al., 1986).

Source: Benzyl alcohol naturally occurs in tea (900 ppm), daffodils (165-330 ppm), hyacinths (64-920 ppm), jasmine (120-228 ppm) rosemary (7-32 ppm), hyssop (0.1-30 ppm), tangerines (1-2 ppm), blueberries (0.01-0.08 ppm in fruit juice), ylang-ylang, colocynth, licorice, roselle, tomatoes, spearmint, sweet basil, apricots, tuberose (Duke, 1992), and small-flowered oregano shoots (2 ppm) (Baser et al., 1991).

Uses: Manufacture of esters for use in perfumes, flavors, soaps, lotions, ointments; photographic developer for color movie films; dying nylon filament, textiles, and sheet plastics; solvent for dyestuffs, cellulose, esters, casein, waxes, etc.; heat-sealing polyethylene films; bacteriostat, insect repellant; emulsions; ballpoint pen inks and stencil inks; surfactant.

BENZYL BUTYL PHTHALATE

Synonyms: BBP; **1,2-Benzenedicarboxylic acid butyl phenylmethyl ester**; Benzyl *n*-butyl phthalate; Butyl benzyl phthalate; *n*-Butyl benzyl phthalate; Butyl phenylmethyl 1,2-benzene-dicarboxylate; NCI-C54375; Palatinol BB; Santicizer 160; Sicol 160; Unimoll BB.

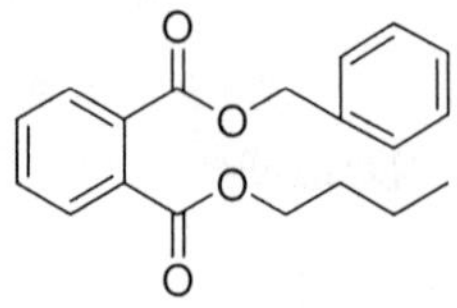

CAS Registry Number: 85-68-7
Molecular formula: $C_{19}H_{20}O_4$
Formula weight: 312.37
RTECS: TH9990000

Physical state and odor:
Clear, oily liquid with a faint odor

Melting point (°C):
-35 (Fishbein and Albro, 1972)

Boiling point (°C):
370 (Verschueren, 1983)
377 (Fishbein and Albro, 1972)

Density (g/cm³):
1.12 at 20/4 °C (Weiss, 1986)
1.111 at 25/4 °C (Standen, 1968)

Diffusivity in water (x 10^{-5} cm²/sec):
0.48 at 20 °C using method of Hayduk and Laudie (1974)

Flash point (°C):
197 (Weiss, 1986)
110 (Aldrich, 1988)

Henry's law constant (x 10^{-6} atm·m³/mol):
1.3 at 25 °C (calculated, Howard, 1989)

Bioconcentration factor, log BCF:
2.89 (bluegill sunfish, Veith et al., 1980)
2.80 (Barrows et al., 1980)
2.82 (bluegill sunfish, Gledhill et al., 1980)

Soil sorption coefficient, log K_{oc}:
4.23 (Broome County, NY composite soil, Russell and McDuffie, 1986)
4.05 (Gledhill, 1980)

Octanol/water partition coefficient, log K_{ow}:
4.05 (Veith et al., 1980)

4.73 (Ellington and Floyd, 1996)
4.80 (Hirzy et al., 1978)
4.91 (Leyder and Boulanger, 1983)
4.77 (Gledhill et al., 1980)

Solubility in water (mg/L):
2.82 at 20 °C (Leyder and Boulanger, 1983)
2.69 at 25 °C (Howard et al., 1985)
0.710 at 24 °C (practical grade, Hollifield, 1979)
2.0 at 25 °C (Russell and McDuffie, 1986)

Vapor density:
12.76 g/L at 25 °C, 10.78 (air = 1)

Vapor pressure (x 10^{-6} mmHg):
8.6 at 20 °C (Verschueren, 1983)
8.25 at 25 °C (Banerjee et al., 1990)

Environmental fate:
Biological. In anaerobic sludge diluted to 10%, benzyl butyl phthalate biodegraded to monobutyl phthalate, which subsequently degraded to phthalic acid. After 40 d, >90% of applied amount degraded (Shelton et al., 1984). When benzyl butyl phthalate (5 and 10 mg/L) was statically incubated in the dark at 25 °C with yeast extract and settled domestic wastewater inoculum, complete biodegradation with rapid adaptation was observed after 7 d (Tabak et al., 1981). In activated sludge, the half-life was 2 h (Saeger and Tucker, 1976). Gledhill et al. (1980) reported half-lives of 2 and <4 d for benzyl butyl phthalate in river water and a lake water microcosm.
Surface Water. The biological half-life of benzyl butyl phthalate in river water was determined to be 2 d (Saeger and Tucker, 1976).
Photolytic. Gledhill et al. (1980) reported the photolytic half-life is >100 d.
Chemical/Physical. Benzyl butyl phthalate initially hydrolyzes to butyl hydrogen phthalate. This compound undergoes further hydrolysis yielding *o*-phthalic acid, *n*-butanol, and benzyl alcohol (Kollig, 1993). Gledhill et al. (1980) reported the hydrolysis half-life is >100 d.

Symptoms of exposure: Toxic symptoms include nausea, somnolence, hallucination, and dizziness (Patnaik, 1992).

Toxicity:
EC_{50} (48-h) for *Daphnia magna* 3.7 mg/L (Gledhill et al., 1980).
LC_{50} (14-d) for fathead minnows 2.25 mg/L (flow through test) (Gledhill et al., 1980).
LC_{50} (4-d) for fathead minnows (flow through test) 2.32 mg/L (Gledhill et al., 1980).
LC_{50} (96-h) for bluegill sunfish 43 mg/L (Spehar et al., 1982), *Cyprinodon variegatus* 440 ppm using natural seawater (Heitmuller et al., 1981).
LC_{50} (72-h) for *Cyprinodon variegatus* 430 ppm (Heitmuller et al., 1981).
LC_{50} (48-h) for *Daphnia magna* 92 mg/L (LeBlanc, 1980), *Cyprinodon variegatus* 420 ppm (Heitmuller et al., 1981).
LC_{50} (24-h) for *Daphnia magna* >160 mg/L (LeBlanc, 1980), *Cyprinodon variegatus* 380 ppm (Heitmuller et al., 1981).
Acute oral LD_{50} for guinea pigs 13,750 mg/kg, mice 4,170 mg/kg, rats 2,330 mg/kg (quoted, RTECS, 1985).
Heitmuller et al. (1981) reported a NOEC of 360 ppm.

Drinking water standard: No MCLGs or MCLs have been proposed (U.S. EPA, 1996).

Source: Detected in distilled water-soluble fractions of new and used motor oil at concentrations of 8.6-13 and 14-17 μg/L, respectively (Chen et al., 1994)

Uses: Plasticizer used in polyvinyl chloride formulations; additive in polyvinyl acetate emulsions, ethylene glycol, and ethyl cellulose; organic synthesis.

BENZYL CHLORIDE

Synonyms: Benzyl chloride anhydrous; **(Chloromethyl)benzene**; Chlorophenylmethane; α-Chlorotoluene; Ω-Chlorotoluene; NCI-C06360; RCRA waste number P028; Tolyl chloride; UN 1738.

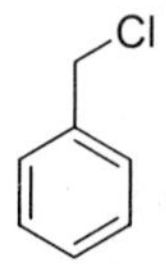

CAS Registry Number: 100-44-7
DOT: 1738
DOT label: Corrosive material
Molecular formula: C_7H_7Cl
Formula weight: 126.59
RTECS: XS8925000
Merck reference: 10, 1136

Physical state, color, and odor:
Colorless to pale yellowish-brown liquid with a pungent, aromatic, irritating odor. Odor threshold concentration is 47 ppb (Keith and Walters, 1992).

Melting point (°C):
-39 (Weast, 1986)
-43 to -48 (Windholz et al., 1983)

Boiling point (°C):
179.3 (Weast, 1986)

Density (g/cm³):
1.1002 at 20/4 °C (Weast, 1986)

Diffusivity in water (x 10^{-5} cm²/sec):
0.81 at 20 °C using method of Hayduk and Laudie (1974)

Flash point (°C):
67.3 (NIOSH, 1997)

Lower explosive limit (%):
1.1 (NIOSH, 1997)

Henry's law constant (x 10^{-4} atm·m³/mol):
3.57 at 20 °C (Hovorka and Dohnal, 1997)

Interfacial tension with water (dyn/cm):
30 at 20 °C (estimated, CHRIS, 1984)

Ionization potential (eV):
9.10-9.30 Lias et al., 1998)

Soil sorption coefficient, log K_{oc}:
2.28 using method of Chiou et al. (1979)

Octanol/water partition coefficient, log K_{ow}:
2.30 (Leo et al., 1971)

Solubility in organics:
Miscible with alcohol, chloroform, and ether (Windholz et al., 1983)

Solubility in water:
493 mg/L at 20 °C (Howard, 1989)

Vapor density:
5.17 g/L at 25 °C, 4.37 (air = 1)

Vapor pressure (mmHg):
1.3 at 25 °C (Banerjee et al., 1990)

Environmental fate:
   Biological. When incubated with raw sewage and raw sewage acclimated with hydrocarbons, benzyl chloride degraded forming non-chlorinated products (Jacobson and Alexander, 1981).
   Chemical/Physical. Anticipated products from the reaction of benzyl chloride with ozone or OH radicals in the atmosphere are chloromethyl phenols, benzaldehyde and chlorine radicals (Cupitt, 1980).
   Slowly hydrolyzes in water forming hydrochloric acid and benzyl alcohol. The estimated hydrolysis half-life in water at 25 °C and pH 7 is 15 h (Mabey and Mill, 1978). The hydrolysis rate constant for benzyl chloride at pH 7 and 59.2 °C was determined to be 0.0204/min, resulting in a half-life of 34 min (Ellington et al., 1986).
   May polymerize in contact with metals except nickel and lead (NIOSH, 1997). When heated to decomposition, hydrogen chloride gas may be released (CHRIS, 1984).

Exposure limits: NIOSH REL: 15-min ceiling 1 ppm (5 mg/m³), IDLH 10 ppm; OSHA PEL: TWA 1 ppm; ACGIH TLV: TWA 1 ppm (adopted).

Symptoms of exposure: Eye contact may cause corneal injury. Exposure to fumes may cause irritation to eyes, nose, skin, and throat (NIOSH, 1997; Patnaik, 1992). An irritation concentration of 41.00 mg/m³ in air was reported by Ruth (1986).

Toxicity:
   LC_{50} (14-d semi-static) for *Poecilia reticulata* 391 μg/L (Hermens et al., 1985).
   LC_{50} (96-h) for fathead minnows 6 mg/L, white shrimp 3.9 mg/L (Curtis et al., 1979), fathead minnows 5.0 mg/L (Curtis et al., 1978).
   LC_{50} (48-h) for fathead minnows 7.3 mg/L, white shrimp 4.4 mg/L (Curtis et al., 1979).
   LC_{50} (24-h) for fathead minnows 11.6 mg/L, white shrimp 7.1 mg/L (Curtis et al., 1979), fathead minnows 12.5 mg/L (Curtis et al., 1978).
   LC_{50} (inhalation) for mice 80 ppm/2-h, rats 150 ppm/2-h (quoted, RTECS, 1985).
   Acute oral LD_{50} for mice 1,624 mg/kg, rats 1,231 mg/kg; LD_{50} (subcutaneous) for rats 1 gm/kg (quoted, RTECS, 1985).

Drinking water standard: As of October 1996, benzyl chloride was listed for regulation but no MCLGs or MCLs have been proposed (U.S. EPA, 1996).

Uses: Manufacture of perfumes, benzyl compounds (e.g., benzyl alcohol, benzyl acetate), pharmaceutical products, artificial resins, dyes, photographic developers, synthetic tannins; gasoline gum inhibitors; quaternary ammonium compounds; penicillin precursors; chemical intermediate.

α-BHC

Synonyms: Benzene hexachloride-α-isomer; α-Benzene hexachloride; ENT 9232; α-HCH; α-Hexachloran; α-Hexachlorane; α-Hexachlorcyclohexane; α-Hexachlorocyclohexane; 1,2,3,4,-5,6-Hexachloro-α-cyclohexane; **1α,2α,3β,4α,5β,6β-Hexachlorocyclohexane**; α-1,2,3,4,5,6-Hexachlorocyclohexane; α-Lindane; TBH.

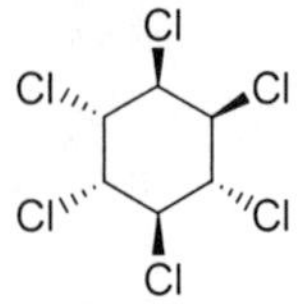

CAS Registry Number: 319-84-6
DOT: 2761
DOT label: Poison
Molecular formula: $C_6H_6Cl_6$
Formula weight: 290.83
RTECS: GV3500000

Physical state, color, and odor:
Brownish to white crystalline solid or powder with a phosgene-like odor (technical grade). The odor threshold concentration is 88 μg/kg (Sigworth, 1964).

Melting point (°C):
156-161 (Aldrich, 1988)
159.8 (Horvath, 1982)

Boiling point (°C):
288 (Weast, 1986)

Density (g/cm³):
1.870 at 20/4 °C (Horvath, 1982)

Diffusivity in water (x 10^{-5} cm²/sec):
0.50 at 20 °C using method of Hayduk and Laudie (1974)

Henry's law constant (x 10^{-6} atm·m³/mol):
5.3 at 20 °C (approximate - calculated from water solubility and vapor pressure)

Bioconcentration factor, log BCF:
3.20-3.38 (fish tank), 2.85 (Lake Ontario) (rainbow trout, Oliver and Niimi, 1985)
2.49 (*Chlamydomonas*, Canton et al., 1977)
2.79 (freshwater fish), 0.00 (fish, microcosm) (Garten and Trabalka, 1983)
3.04 (*Brachydanio rerio*, Devillers et al., 1996)
3.90-4.04 (*Artemia salina*), 4.23 (*Lebistes reticulatus*) (lipid basis, Canton et al., 1978)

Soil sorption coefficient, log K_{oc}:
3.279 (Karickhoff, 1981)
3.30 (Meylan et al., 1992)
3.01, 3.10, 3.30, 3.31, 3.34 (lateritic soil), 3.12, 3.33 (Kari soil), 3.14 (sandy soil), 3.19, 3.46 (alluvial soil), 3.32 (Pokkali soil) (Wahid and Sethunathan, 1979)

Octanol/water partition coefficient, log K_{ow}:
3.81 (Kurihara et al., 1973)
3.46 (Geyer et al., 1987)
3.72 (Schwarzenbach et al., 1983)
3.776 (de Bruijn et al., 1989)

Solubility in organics:
Soluble in ethanol, benzene, chloroform (Weast, 1986), cod liver oil, and octanol (Montgomery, 1993)

Solubility in water (mg/L):
1.4 (salt water, Canton et al., 1978)
At 28 °C: 1.48, 1.77 (0.05 μ particle size), 1.21, 2.03 (0.1 μ particle size) (Kurihara et al., 1973)
2.00 at 25 °C (Weil et al., 1974)
1.63 at 20 °C (Brooks, 1974)

Vapor pressure (x 10^{-5} mmHg):
2.5 at 20 °C (Balson, 1947)
2.15 at 20 °C (Sims et al., 1988)
173 at 25 °C (Hinckley et al., 1990)

Environmental fate:
Biological. Clostridium sphenoides biodegraded α-BHC to δ-3,4,5,6-tetrachloro-1-cyclo-hexane (Heritage and MacRae, 1977). In four successive 7-d incubation periods, α-BHC (5 and 10 mg/L) was recalcitrant to degradation in a settled domestic wastewater inoculum (Tabak et al., 1981).

Soil. Under aerobic conditions, indigenous microbes in contaminated soil produced pentachlorocyclohexane. However, under methanogenic conditions, α-BHC was converted to chlorobenzene, 3,5-dichlorophenol, and the tentatively identified compound 2,4,5-trichloro-phenol (Bachmann et al., 1988).

Surface Water. Hargrave et al. (2000) calculated BAFs as the ratio of the compound tissue concentration [wet and lipid weight basis (ng/g)] to the concentration of the compound dissolved in seawater (ng/mL). Average log BAF values for α-BHC in ice algae and phyto-plankton collected from the Barrow Strait in the Canadian Archipelago were 6.18 and 6.08, respectively.

Photolytic. When an aqueous solution containing α-BHC was photooxidized by UV light at 90-95 °C, 25, 50, and 75% degraded to carbon dioxide after 4.2, 24.2, and 40.0 h, respectively (Knoevenagel and Himmelreich, 1976). In basic, aqueous solutions, α-BHC de-hydrochlorinates forming pentachlorocyclohexene, which is further transformed to trichloro-benzenes. In a buffered aqueous solution at pH 8 and 5 °C, the calculated hydrolysis half-life is 26 yr (Ngabe et al., 1993).

Chemical/Physical. Emits very toxic chloride fumes when heated to decomposition (Lewis, 1990). α-BHC will hydrolyze via *trans*-dehydrochlorination of the axial chlorines resulting in the formation of hydrochloric acid and the intermediate 1,3,4,5,6-pentachlorocyclohexene. The intermediate will undergo further hydrolysis resulting in the formation of 1,2,3-trichloro-benzene, 1,2,4-trichlorobenzene, and hydrochloric acid (Kollig, 1993).

Toxicity:
EC_{50} (96-h) for guppies (*Lebistes reticulatus*) 1.31 mg/L (Canton et al., 1978).
LC_{10} (35-d) for *Lebistes reticulatus* 0.5 mg/L (Canton et al., 1978).
LC_{50} (96-h) for *Poecilia reticulata* 1.49, *Brachydanio rerio* 1.11, *Paracheirodon axelrodi*

1.52 (Oliverira-Filho and Paumgarten, 1997), *Lebistes reticulatus* >1.4 mg/L, *Artemia salina* 0.5 mg/L (Canton et al., 1978).

LC_{50} (72-h) values of 1.58, 1.11, and 1.52 mg/L for *Poecilia reticulata*, *Brachydanio rerio*, and *Paracheirodon axelrodi*, respectively (Oliverira-Filho and Paumgarten, 1997).

LC_{50} (48-h) values of 1.95, 1.11, and 1.52 mg/L for *Poecilia reticulata*, *Brachydanio rerio*, and *Paracheirodon axelrodi*, respectively (Oliverira-Filho and Paumgarten, 1997).

LC_{50} (24-h) values of 2.62, 1.41, and 1.64 mg/L for *Poecilia reticulata*, *Brachydanio rerio*, and *Paracheirodon axelrodi*, respectively (Oliverira-Filho and Paumgarten, 1997).

Acute oral LD_{50} for rats 177 mg/kg (quoted, RTECS, 1985).

Use: Not produced commercially in the U.S. and its sale is prohibited by the U.S. EPA.

β-BHC

Synonyms: *trans*-α-Benzene hexachloride; β-Benzene hexachloride; Benzene-*cis*-hexachloride; ENT 9233; β-HCH; β-Hexachlorobenzene; **1α,2β,3α,4β,5α,6β-Hexachlorocyclohexane**; β-Hexachlorocyclohexane; 1,2,3,4,5,6-Hexachloro-β-cyclohexane; 1,2,3,4,5,6-Hexachloro-*trans*-cyclohexane; β-1,2,3,4,5,6-Hexachlorocyclohexane; β-Isomer; β-Lindane; TBH.

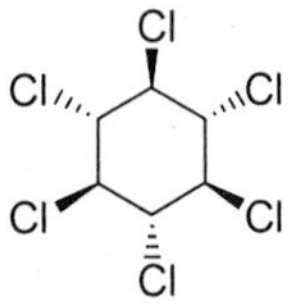

CAS Registry Number: 319-85-7
DOT: 2761
DOT label: Poison
Molecular formula: $C_6H_6Cl_6$
Formula weight: 290.83
RTECS: GV4375000

Physical state:
Although β-BHC is a solid at room temperature, the odor threshold concentration is 0.32 $\mu g/kg$ (Sigworth, 1964).

Melting point (°C):
314.5 (Horvath, 1982)
311.7 (Standen, 1964)

Boiling point (°C):
60 at 0.58 mmHg (Horvath, 1982)
Sublimes at 760 mmHg (U.S. EPA, 1980a)

Density (g/cm³):
1.89 at 19/4 °C (Weast, 1986)

Diffusivity in water (x 10⁻⁵ cm²/sec):
0.50 at 20 °C using method of Hayduk and Laudie (1974)

Henry's law constant (x 10⁻⁷ atm·m³/mol):
2.3 at 20 °C (calculated)

Bioconcentration factor, log BCF:
2.82 (brown trout, Sugiura et al., 1980)
3.08 (activated sludge), 2.26 (algae), 2.65 (golden ide) (Freitag et al., 1985)
2.86 (freshwater fish), 2.97 (fish, microcosm) (Garten and Trabalka, 1983)
3.16, 3.18 (*Brachydanio rerio*, Devillers et al., 1996)

Soil sorption coefficient, log K_{oc}:
3.462 (silt loam, Chiou et al., 1979)
3.322 (Karickhoff, 1981)
3.553 (Reinbold et al., 1979)
3.50 (Meylan et al., 1992)
Ca-Staten peaty muck: 3.55 and 3.54 at 20 and 30 °C, respectively; Ca-Venado clay: 3.12 and

3.07 at 19.8 and 30 °C, respectively (Mills and Biggar, 1969)
3.06, 3.17, 3.27, 3.41, 3.43, 3.50 (lateritic soil), 3.27, 3.38 (Kari soil), 3.26 (sandy soil), 3.29, 3.32 (alluvial soil), 3.50 (Pokkali soil) (Wahid and Sethunathan, 1979)

Octanol/water partition coefficient, log K_{ow}:
3.80 (Kurihara et al., 1973)
4.50 (Geyer et al., 1987)
3.842 (de Bruijn et al., 1989)

Solubility in organics:
Soluble in ethanol, benzene, and chloroform (Weast, 1986)

Solubility in water:
130, 200 ppb at 28 °C (0.1 μm particle size, Kurihara et al., 1973)
240 μg/L at 25 °C (Weil et al., 1974)
700 ppb at 25 °C (Brooks, 1974)
5 mg/L at 20 °C (Chiou et al., 1979)
2.7 mg/L at 20 °C (Mills and Biggar, 1969)

Vapor pressure (x 10^{-7} mmHg):
2.8 at 20 °C (Balson, 1947)
4.66 at 25 °C (Banerjee et al., 1990)

Environmental fate:
Biological. No biodegradation of β-BHC was observed under denitrifying and sulfate-reducing conditions in a contaminated soil collected from the Netherlands (Bachmann et al., 1988). In four successive 7-d incubation periods, β-BHC (5 and 10 mg/L) was recalcitrant to degradation in a settled domestic wastewater inoculum (Tabak et al., 1981).
Chemical/Physical. Emits very toxic fumes of chloride, hydrochloric acid and phosgene when heated to decomposition (Lewis, 1990). β-BHC will not hydrolyze to any reasonable extent (Kollig, 1993).

Toxicity:
LC_{50} (96-h) values of 1.66, 1.52, and 1.10 mg/L for *Poecilia reticulata, Brachydanio rerio*, and *Paracheirodon axelrodi*, respectively (Oliverira-Filho and Paumgarten, 1997).
LC_{50} (72-h) values of 2.18, 1.63, and 1.10 mg/L for *Poecilia reticulata, Brachydanio rerio*, and *Paracheirodon axelrodi*, respectively (Oliverira-Filho and Paumgarten, 1997).
LC_{50} (48-h) values of 2.68, 1.63, and 1.10 mg/L for *Poecilia reticulata, Brachydanio rerio*, and *Paracheirodon axelrodi*, respectively (Oliverira-Filho and Paumgarten, 1997).
LC_{50} (24-h) values of 3.14, 1.78, and 1.70 mg/L for *Poecilia reticulata, Brachydanio rerio*, and *Paracheirodon axelrodi*, respectively (Oliverira-Filho and Paumgarten, 1997).
Acute oral LD_{50} for rats 6,000 mg/kg (quoted, RTECS, 1985).

Use: Insecticide.

δ-BHC

Synonyms: δ-Benzene hexachloride; ENT 9234; δ-HCH; δ-Hexachlorocyclohexane; δ-1,2,-3,4,5,6-Hexachlorocyclohexane; δ-(aeeeee)-1,2,3,4,5,6-Hexachlorocyclohexane; **1α,2α,3α,-4β,5β,6β-Hexachlorocyclohexane**; 1,2,3,4,5,6-Hexachloro-δ-cyclohexane; δ-Lindane; TBH.

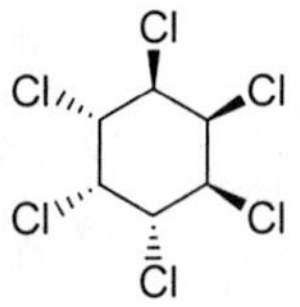

CAS Registry Number: 319-86-8
DOT: 2761
DOT label: Poison
Molecular formula: $C_6H_6Cl_6$
Formula weight: 290.83
RTECS: GV4550000

Physical state and odor:
Solid with a faint musty-like odor

Melting point (°C):
141.8 (Horvath, 1982)

Boiling point (°C):
60 at 0.34 mmHg (Horvath, 1982)

Density (g/cm³):
≈ 1.87 (Hawley, 1981)

Diffusivity in water (x 10^{-5} cm²/sec):
0.50 at 20 °C using method of Hayduk and Laudie (1974)

Henry's law constant (x 10^{-7} atm·m³/mol):
2.5 at 20 °C (approximate - calculated from water solubility and vapor pressure)

Bioconcentration factor, log BCF:
2.45 (*Oncorhynchas mykiss*, Devillers et al., 1996)

Soil sorption coefficient, log K_{oc}:
3.279 (Karickhoff, 1981)

Octanol/water partition coefficient, log K_{ow}:
4.14 (Kurihara et al., 1973)
2.80 (Geyer et al., 1987)

Solubility in organics:
Soluble in ethanol, benzene, and chloroform (Weast, 1986)

Solubility in water (ppm):
31.4 at 25 °C (Weil et al., 1974)

At 28 °C: 8.64, 11.6 (0.05 μ particle size), 10.7, 15.7 (0.1 μ particle size) (Kurihara et al., 1973)
21.3 at 25 °C (Brooks, 1974)
7 at 20 °C (Worthing and Hance, 1991)

Vapor pressure (x 10^{-5} mmHg):
1.7 at 20 °C (Balson, 1947)
3.52 at 25 °C (Banerjee et al., 1990)

Environmental fate:
   Biological. Dehydrochlorination of δ-BHC by a *Pseudomonas* sp. under aerobic conditions was reported by Sahu et al. (1992). They also reported that when deionized water containing δ-BHC was inoculated with this species, the concentration of δ-BHC decreased to undetectable levels after 8 d with concomitant formation of chloride ions and δ-pentachlorocyclohexane. In four successive 7-d incubation periods, δ-BHC (5 and 10 mg/L) was recalcitrant to degradation in a settled domestic wastewater inoculum (Tabak et al., 1981).
   Chemical/Physical. δ-BHC dehydrochlorinates in the presence of alkalies. Although no products were reported, the hydrolysis half-lives at pH values of 7 and 9 are 191 d and 11 h, respectively (Worthing and Hance, 1991).

Toxicity:
   LC$_{50}$ (96-h) values of 2.83, 1.58, and 0.84 mg/L for *Poecilia reticulata*, *Brachydanio rerio*, and *Paracheirodon axelrodi*, respectively (Oliverira-Filho and Paumgarten, 1997).
   LC$_{50}$ (72-h) values of 3.27, 1.58, and 1.42 mg/L for *Poecilia reticulata*, *Brachydanio rerio*, and *Paracheirodon axelrodi*, respectively (Oliverira-Filho and Paumgarten, 1997).
   LC$_{50}$ (48-h) values of 3.79, 1.87, and 2.19 mg/L for *Poecilia reticulata*, *Brachydanio rerio*, and *Paracheirodon axelrodi*, respectively (Oliverira-Filho and Paumgarten, 1997).
   LC$_{50}$ (24-h) values of 4.05, 2.64, and 2.59 mg/L for *Poecilia reticulata*, *Brachydanio rerio*, and *Paracheirodon axelrodi*, respectively (Oliverira-Filho and Paumgarten, 1997).
   Acute oral LD$_{50}$ for rats 1,000 mg/kg (quoted, RTECS, 1985).

Use: Insecticide.

BIPHENYL

Synonyms: Bibenzene; **1,1′-Biphenyl**; Diphenyl; Lemonene; Phenylbenzene.

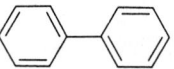

CAS Registry Number: 92-52-4
Molecular formula: $C_{12}H_{10}$
Formula weight: 154.21
RTECS: NU8050000
Merck reference: 10, 3326

Physical state, color, and odor:
White scales with a pleasant but peculiar odor. Odor threshold concentration is 0.83 ppb (Amoore and Hautala, 1983).

Melting point (°C):
71 (Weast, 1986)
68.6 (Parks and Huffman, 1931)

Boiling point (°C):
255.9 (Weast, 1986)
254-255 (Windholz et al., 1983)

Density (g/cm³):
0.8660 at 20/4 °C (Weast, 1986)
1.18 at 0/4 °C (Verschueren, 1983)

Diffusivity in water (x 10^{-5} cm²/sec):
0.62 at 20 °C using method of Hayduk and Laudie (1974)

Dissociation constant, pK_a:
>14 (Schwarzenbach et al., 1993)

Flash point (°C):
113 (NIOSH, 1997)

Lower explosive limit (%):
0.6 at 111 °C (NIOSH, 1997)

Upper explosive limit (%):
5.8 at 155 °C (NIOSH, 1997)

Entropy of fusion (cal/mol·K):
12.20 (Miller et al., 1984)
13.0 (Smith, 1979; Spaght et al., 1932; Ueberreiter and Orthmann, 1950)

Heat of fusion (kcal/mol):
4.18 (Miller et al., 1984)
4.52 (Wauchope and Getzen, 1972)

Henry's law constant (x 10^{-4} atm·m^3/mol):
1.93 at 25 °C (Fendinger and Glotfelty, 1990)
4.05 at 25 °C (Mackay et al., 1979)
3.08 at 25 °C (Mackay and Shiu, 1981; Shiu and Mackay, 1997)
4.74, 3.44, 6.28, 10.76, and 12.97 at 2.0, 6.0, 10.0, 18.0, and 25.0 °C, respectively (Dewulf et al., 1999)

Ionization potential (eV):
8.16 (Lias, 1998)
8.27 ± 0.01 (Franklin et al., 1969)
8.30 (Krishna and Gupta, 1970)

Bioconcentration factor, log BCF:
2.53 (fish, Kenaga and Goring, 1980)
3.12 (rainbow trout, Veith et al., 1979)
2.73 (*Chlorella fusca*, Freitag et al., 1982; Halfon and Reggiani, 1986)
3.41 (activated sludge), 2.45 (golden ide) (Freitag et al., 1985)
2.45 (fish), 3.41 (activated sludge) (Halfon and Reggiani, 1986)

Soil sorption coefficient, log K_{oc}:
3.03 (Kishi et al., 1990)
3.52 (Apison soil), 2.95 (Fullerton soil), 2.94 (Dormont soil) (Southworth and Keller, 1986)
3.76, 3.85, 4.10 (glaciofluvial, sandy aquifer, Nielsen et al., 1996)
Average K_d values for sorption of biphenyl to corundum (α-Al$_2$O$_3$) and hematite (α-Fe$_2$O$_3$) were 0.231 and 0.983 mL/g, respectively (Mader et al., 1997)

Octanol/water partition coefficient, log K_{ow}:
3.16 (Rogers and Cammarata, 1969)
4.09 (Rogers and Cammarata, 1969; Johnsen et al., 1989)
4.04 (Banerjee et al., 1980; Rogers and Cammarata, 1969)
3.76 (Geyer et al., 1984; Miller et al., 1984)
3.95 (Ruepert et al., 1985)
4.00 (DeKock and Lord, 1987)
4.008 (de Bruijn et al., 1989)
3.89 (Camilleri et al., 1988; Eadsforth, 1986; Woodburn et al., 1984)
4.11 (Garst and Wilson, 1984)
3.91 (Tipker et al., 1988)
3.79 (Rapaport and Eisenreich, 1984)
4.10 (Bruggeman et al., 1982)

Solubility in organics:
Soluble in alcohol, benzene, and ether (Weast, 1986)

Solubility in water:
39.1 μmol/L at 25 °C (Banerjee et al., 1980)
43.5 μmol/L at 25 °C (Miller et al., 1984)
7.45 mg/kg at 25 °C in distilled water, 4.76 mg/kg in artificial seawater (salinity 35 g/kg) at 25 °C (Eganhouse and Calder, 1976)
In mg/L: 2.83 at 0.4 °C, 2.97 at 2.4 °C, 3.38 at 5.2 °C, 3.64 at 7.6 °C, 4.06 at 10.0 °C, 4.58 at 12.6 °C, 5.11 at 14.9 °C, 5.27 at 15.9 °C, 7.48 at 25.0 °C, 7.78 at 25.6 °C, 9.64 at 30.2 °C, 9.58 at 30.4 °C, 11.0 at 33.3 °C, 11.9 at 34.9 °C, 12.5 at 36.0 °C, 17.2 at 42.8 °C (Bohon

and Claussen, 1951)

5.94 mg/L at 25 °C (Andrews and Keefer, 1949)

7.0 mg/L at 25 °C (Mackay and Shiu, 1977)

2.64, 7.08, 8.88, 13.8, 22.1, and 37.2 mg/kg at 0.0, 25.0, 30.3, 40.1, 50.1, and 60.5 °C, respectively (Wauchope and Getzen, 1972)

42 μmol/L at 29 °C (Stucki and Alexander, 1987)

6.92 mg/L (Morehead et al., 1986)

At 20 °C: 38, 23.9, 23.9, and 26.3 μmol/L in doubly distilled water, Pacific seawater, artificial seawater, and 35% NaCl, respectively (Hashimoto et al., 1984)

In NaCl (g/kg) at 25 °C, μg/kg: 6.08 (13.25), 5.46 (26.24), 4.62 (39.05), 4.16 (46.28), 4.13 (51.62), 3.54 (63.97), 3.45 (63.97) (Paul, 1952)

7.2 mg/L at 25 °C (Billington et al., 1988)

45.7 μmol/L at 25 °C (Akiyoshi et al., 1987)

45.70 μmol/L at 30 °C (Yalkowsky et al., 1983)

49.1 μmol/kg at 25.0 °C (Vesala, 1974)

6.99 mg/L at 22 °C (Coyle et al., 1997)

Vapor pressure (x 10^{-4} mmHg):

145,500 at 127 °C (Eiceman and Vandiver, 1983)

100 at 25 °C (Mackay et al., 1982)

5.84 at 20.70 °C, 8.87 at 24.00 °C, 15.4 at 29.15 °C (Bradley and Cleasby, 1953)

530 at 25 °C (Bidleman, 1984)

1,200 at 53.0 °C, 2,600 at 61.0 °C, 6,900 at 71.0 °C (Sharma and Palmer, 1974)

497 at 25 °C (Foreman and Bidleman, 1985)

97.5 at 25 °C (Bright, 1951)

251 at 25 °C (Hinckley et al., 1990)

88.1 at 25 °C (extrapolated from vapor pressures determined at higher temperatures, Tesconi and Yalkowsky, 1998)

Environmental fate:

Biological. Reported biodegradation products include 2,3-dihydro-2,3-dihydroxybiphenyl, 2,3-dihydroxybiphenyl, 2-hydroxy-6-oxo-6-phenylhexa-2,4-dienoate, 2-hydroxy-3-phenyl-6-oxohexa-2,4-dienoate, 2-oxopenta-4-enoate, phenylpyruvic acid (Verschueren, 1983), 2-hydroxybiphenyl, 4-hydroxybiphenyl, and 4,4'-hydroxybiphenyl (Smith and Rosazza, 1974). The microbe *Candida lipolytica* degraded biphenyl into the following products: 2-, 3-, and 4-hydroxybiphenyl, 4,4'-dihydroxybiphenyl, and 3-methoxy-4-hydroxybiphenyl (Cerniglia and Crow, 1981). With the exception of 3-methoxy-4-hydroxybiphenyl, these products were also identified as metabolites by *Cunninghanella elegans* (Dodge et al., 1979). In activated sludge, 15.2% mineralized to carbon dioxide after 5 d (Freitag et al., 1985).

Under aerobic conditions, *Beijerinckia* sp. degraded biphenyl to *cis*-2,3-dihydro-2,3-dihydroxybiphenyl. In addition, *Oscillatoria* sp. and *Pseudomonas putida* degraded biphenyl to 4-hydroxybiphenyl and benzoic acid, respectively (Kobayashi and Rittman, 1982).

Surface Water. The evaporation half-life of biphenyl in surface water (1 m depth) at 25 °C is estimated to be 7.52 h (Mackay and Leinonen, 1975).

Groundwater. Nielsen et al. (1996) studied the degradation of biphenyl in a shallow, glaciofluvial, unconfined sandy aquifer in Jutland, Denmark. As part of the *in situ* microcosm study, a cylinder that was open at the bottom and screened at the top was installed through a cased borehole approximately 5 m below grade. Five liters of water was aerated with atmospheric air to ensure aerobic conditions were maintained. Groundwater was analyzed weekly for approximately 3 months to determine biphenyl concentrations over time. The experimentally determined first-order biodegradation rate constant and corresponding half-life were 0.2/d and 3.47 d, respectively.

Photolytic. A carbon dioxide yield of 9.5% was achieved when biphenyl adsorbed on silica gel was irradiated with light (λ >290 nm) for 17 h (Freitag et al., 1985). Irradiation of biphenyl (λ >300 nm) in the presence of nitrogen monoxide resulted in the formation of 2- and 4-nitrobiphenyl (Fukui et al., 1980). Biphenyl (16.2 mg/L) in a methanol-water solution (3:7 v/v) was subjected to a high pressure mercury lamp or sunlight. Based on a rate constant of 5.1 x 10^{-4}/min, the corresponding half-life is 22.61 h (Wang et al., 1991).

Chemical/Physical. The aqueous chlorination of biphenyl at 40 °C over a pH range of 6.2 to 9.0 yielded 2-chlorobiphenyl and 3-chlorobiphenyl (Snider and Alley, 1979). In an acidic aqueous solution (pH 4.5) containing bromide ions and a chlorinating agent (sodium hypochlorite), 4-bromobiphenyl formed as the major product. Minor products identified include 2-bromobiphenyl, 2,4- and 4,4'-dibromobiphenyl (Lin et al., 1984).

Biphenyl will not hydrolyze because it has no hydrolyzable functional group.

Exposure limits: NIOSH REL: TWA 1 mg/m^3 (0.2 ppm), IDLH 100 mg/m^3; OSHA PEL: TWA 1 mg/m^3; ACGIH TLV: TWA 0.2 ppm (adopted).

Symptoms of exposure: Irritation of throat, eyes; headache, nausea, fatigue, numbness in limbs (Hamburg et al., 1989). An irritation concentration of 7.50 mg/m^3 in air was reported by Ruth (1986).

An acceptable daily intake reported for humans is 125 μg/kg body weight (Worthing and Hance, 1991).

Toxicity:

EC_{50} (48-h) for *Daphnia pulex* 1.6 mg/L (Passino-Reader et al., 1997).

LC_{50} (48-h) for *Daphnia magna* 4.7 mg/L (LeBlanc, 1980).

LC_{50} (24-h) for *Daphnia magna* 27 mg/L (LeBlanc, 1980).

Acute oral LD_{50} for rats 3,280 mg/kg (quoted, RTECS, 1985).

Source: Lao et al. (1975) reported biphenyl was present in coal tar at a concentration of 2.72 mg/g.

Uses: Heat transfer liquid; fungistat for oranges; plant disease control; manufacture of benzidine; organic synthesis.

BIS(2-CHLOROETHOXY)METHANE

Synonyms: BCEXM; Bis(2-chloroethyl)formal; Bis(β-chloroethyl)formal; Dichlorodiethyl formal; Dichlorodiethyl methylal; Dichloroethyl formal; Di-2-chloroethyl formal; **1,1′-[Methylenebis(oxy)]bis(2-chloroethane)**; 1,1′-[Methylenebis(oxy)]bis(2-chloroformaldehyde); Bis-(β-chloroethyl)acetal ethane; Formaldehyde bis(β-chloroethylacetal); RCRA waste number U024.

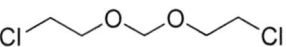

CAS Registry Number: 111-91-1
DOT: 1916
DOT label: Poison
Molecular formula: $C_5H_{10}Cl_2O_2$
Formula weight: 173.04
RTECS: PA3675000

Physical state and color:
Colorless liquid

Melting point (°C):
-32.8 (Hawley, 1981)

Boiling point (°C):
218.1 (Webb et al., 1962)

Density (g/cm³):
1.2339 at 20/20 °C (Hawley, 1981)

Diffusivity in water (x 10^{-5} cm²/sec):
0.72 at 20 °C using method of Hayduk and Laudie (1974)

Flash point (°C):
110 (open cup, Hawley, 1981)

Henry's law constant (x 10^{-7} atm·m³/mol):
3.78 (calculated, U.S. EPA, 1980a)

Soil sorption coefficient, log K_{oc}:
2.06 using method of Kenaga and Goring (1980)

Octanol/water partition coefficient, log K_{ow}:
1.26 (calculated, Leo et al., 1971)

Solubility in water:
81,000 mg/L at 25 °C using method of Moriguchi (1975)

Vapor density:
7.07 g/L at 25 °C, 5.97 (air = 1)

Vapor pressure (mmHg):
1 at 53 °C (Weast, 1986)

Environmental fate:
   Biological. Using settled domestic wastewater inoculum, bis(2-chloroethoxy)methane (5 and 10 mg/L) did not degrade after 28 d of incubation at 25 °C (Tabak et al., 1981).

Toxicity:
   Acute oral LD_{50} for rats 65 mg/kg (quoted, RTECS, 1985).

Uses: Manufacture of insecticides, polymers; degreasing solvent; intermediate for polysulfide rubber.

BIS(2-CHLOROETHYL) ETHER

Synonyms: BCEE; Bis(β-chloroethyl) ether; Chlorex; 1-Chloro-2-(β-chloroethoxy)ethane; Chloroethyl ether; 2-Chloroethyl ether; (β-Chloroethyl) ether; DCEE; Dichlorodiethyl ether; 2,2′-Dichlorodiethyl ether; β,β′-Dichlorodiethyl ether; Dichloroether; Dichloroethyl ether; α,α′-Dichloroethyl ether; Di(β-chloroethyl) ether; Di(2-chloroethyl) ether; *sym*-Dichloroethyl ether; 2,2′-Dichloroethyl ether; Dichloroethyl oxide; ENT 4504; **1,1′-Oxybis(2-chloroethane)**; RCRA waste number U025; UN 1916.

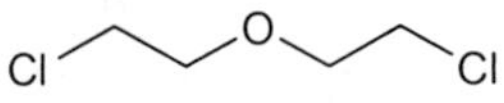

CAS Registry Number: 111-44-4
DOT: 1916
DOT label: Poison
Molecular formula: $C_4H_8Cl_2O$
Formula weight: 143.01
RTECS: KN0875000
Merck reference: 10, 3050

Physical state, color, and odor:
Clear, colorless liquid with a strong, fruity, or chlorinated-like odor. The low odor and high odor threshold concentrations were 90.0 and 2,160 mg/m^3, respectively (Ruth, 1986).

Melting point (°C):
-47 (Aldrich, 1988)
-50 (NIOSH, 1997)

Boiling point (°C):
178.5 (Dean, 1973)

Density (g/cm³):
1.2199 at 20/4 °C (Weast, 1986)

Diffusivity in water (x 10^{-5} cm²/sec):
0.80 at 20 °C using method of Hayduk and Laudie (1974)

Flash point (°C):
55 (NFPA, 1984)

Heat of fusion (kcal/mol):
2.070 (Riddick et al., 1986)

Henry's law constant (x 10^{-5} atm·m³/mol):
1.3 (Schwille, 1988)

Ionization potential (eV):
9.85 (Franklin et al., 1969)

Bioconcentration factor, log BCF:
1.04 (bluegill sunfish, Veith et al., 1980)

Soil sorption coefficient, log K_{oc}:
1.88 (Wilson et al., 1981)

Octanol/water partition coefficient, log K_{ow}:
1.12 (Veith et al., 1980)
1.29 (Hansch and Leo, 1985)

Solubility in organics:
Soluble in acetone, ethanol, benzene, and ether (Weast, 1986)

Solubility in water:
17,400 mg/L (Hake and Rowe, 1963).
10,200 mg/L at 20 °C (Du Pont, 1966)
In wt %: 1.14 at 0 °C, 1.09 at 9.6 °C, 1.04 at 20.0 °C, 1.03 at 31.0 °C,1.05 at 40.0 °C, 1.11 at
    50.1 °C, 1.28 at 70.6 °C, 1.36 at 80.9 °C, 1.51 at 91.7 °C (Stephenson, 1992)
17,195 mg/L at 25 °C (Veith et al., 1975)

Vapor density:
5.84 g/L at 25 °C, 4.94 (air = 1)

Vapor pressure (mmHg):
0.71 at 20 °C, 1.4 at 25 °C (Verschueren, 1983)
1.55 at 25 °C (Howard, 1989)

Environmental fate:
    Biological. When 5 and 10 mg/L of bis(2-chloroethyl) ether were statically incubated in the
dark at 25 °C with yeast extract and settled domestic wastewater inoculum, complete
degradation was observed after 7 d (Tabak et al., 1981).
    Soil. Based on data obtained from a 97-d soil column study, the estimated half-life of bis(2-
chloroethyl) ether is soil is approximately 16.7 d (Kincannon and Lin, 1985).
    Surface Water. Using the method of Mackay and Wolkoff (1973), the calculated
volatilization half-life from a surface water body 1 m deep is 5.8 d at 25 °C.
    Chemical/Physical. The hydrolysis rate constant for bis(2-chloroethyl) ether at pH 7 and 25
°C was determined to be 2.6 x 10^{-5}/h, resulting in a half-life of 3.0 yr. Products of hydrolysis
include 2-(2-chloroethoxy)ethanol, bis(2-hydroxyethyl) ether, hydrochloric acid, and/or 1,4-
dioxane (Ellington et al., 1988; Enfield and Yates, 1990; Kollig, 1993).
    Emits chlorinated acids when incinerated (Sittig, 1985).

Exposure limits: Potential occupational carcinogen. NIOSH REL: TWA 5 ppm (30 mg/m³),
STEL 10 ppm (60 mg/m³), IDLH 100 ppm; OSHA PEL: ceiling 15 ppm (90 mg/m³); ACGIH
TLV: TWA 30 mg/m³, STEL 60 mg/m³.

Symptoms of exposure: Eye contact may cause conjunctival irritation and corneal injury.
Extreme irritation to the eyes and nasal passages was observed by male volunteers when
exposed to concentrations ranging from 550 to 1,000 ppm (Canadian Environmental Protection
Act, 1993). Ingestion of low concentrations may cause nausea and vomiting (Patnaik, 1992).
Symptoms of inhalation include irritation of nose and throat (NIOSH, 1997). An irritation
concentration of 600.00 mg/m³ in air was reported by Ruth (1986).

Toxicity:
    LC_{50} (contact) for earthworm (*Eisenia fetida*) 19 μg/cm² (Neuhauser et al., 1985).

LC_{50} (96-h) for bluegill sunfish 600 mg/L (Buccafusco et al., 1981).

LC_{50} (48-h) for *Daphnia magna* 240 mg/L (LeBlanc, 1980).

LC_{50} (24-h) for *Daphnia magna* 340 mg/L (LeBlanc, 1980).

LC_{50} (inhalation, 0.75-h) for rats 117 gm/m³; LC_{50} (inhalation, 4-h) for rats 5,850 gm/m³ (Smyth and Carpenter, 1948; Carpenter et al., 1949).

Acute oral LD_{50} for mice 112 mg/kg, rats 75 mg/kg (quoted, RTECS, 1985).

Source: Bis(2-chloroethyl) ether does not occur naturally in the environment. In Canada, this compound enters the environment as a by-product from chlorination of waste streams containing ethylene, propylene (Environment Canada, 1993) or ethyl ether (Verschueren, 1983).

Uses: Scouring and cleaning textiles; fumigants; processing fats, waxes, greases, cellulose esters; dewaxing agent for lubricating oils; preparation of insecticides, butadiene, pharmaceuticals; solvent in paints, varnishes and lacquers; selective solvent for production of high-grade lubricating oils; fulling, wetting and penetrating compounds; finish removers; spotting and dry cleaning; soil fumigant; acaricide; organic synthesis.

BIS(2-CHLOROISOPROPYL) ETHER

Synonyms: BCIE; BCMEE; Bis(β-chloroisopropyl) ether; Bis(2-chloro-1-methylethyl) ether; Bis(2-chloro-3-methylethyl) ether; 1-Chloro-2-(β-chloroisopropoxy)propane; 2-Chloroisopropyl ether; β-Chloroisopropylether; (2-Chloro-1-methylethyl) ether; DCIP; Dichlorodiisopropyl ether; Dichloroisopropyl ether; 2,2′-Dichloroisopropyl ether; NCI-C50044; **2,2′-Oxybis(1-chloropropane)**; RCRA waste number U027; UN 2490.

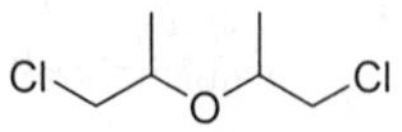

CAS Registry Number: 108-60-1
DOT label: Poison and corrosive material
Molecular formula: $C_6H_{12}Cl_2O$
Formula weight: 171.07
RTECS: KN1750000

Physical state and color:
Colorless to brown oily liquid. The reported odor threshold concentration is 320 ppb (Verschueren, 1983).

Melting point (°C):
-97 (Verschueren, 1983)

Boiling point (°C):
78-79 (quoted, Kawamoto and Urano, 1989)

Density (g/cm³):
1.103 at 20/4 °C (Weast, 1986)
1.1127 at 25/4 °C (Standen, 1964)

Diffusivity in water (x 10⁻⁵ cm²/sec):
0.68 at 20 °C using method of Hayduk and Laudie (1974)

Flash point (°C):
85 (Hawley, 1981)

Henry's law constant (x 10⁻⁴ atm·m³/mol):
1.1 (Pankow and Rosen, 1988)
0.24 at 25 °C (Kawamoto and Urano, 1989)

Soil sorption coefficient, log K_{oc}:
Although experimental methods for estimation of this parameter for ethers are lacking in the documented literature, an estimated value of 2.39 was reported by Ellington et al. (1993).

Octanol/water partition coefficient, log K_{ow}:
2.48 (Kawamoto and Urano, 1989)

Solubility in organics:
Soluble in acetone, ethanol, benzene, and ether (Weast, 1986)

Solubility in water:
In wt %: 0.409 at 9.5 °C, 0.245 at 19.2 °C, 0.237 at 31.0 °C, 0.218 at 40.3 °C, 0.182 at 51.1 °C, 0.209 at °C, 0.265 at 80.7 °C, 0.241 at 91.4 °C (Stephenson, 1992)
1,700 mg/L (quoted, Kawamoto and Urano, 1989)

Vapor density:
6.99 g/L at 25 °C, 5.91 (air = 1)

Vapor pressure (mmHg at 20 °C):
0.85 (Verschueren, 1983)
0.56 (quoted, Kawamoto and Urano, 1989)

Environmental fate:
Biological. When bis(2-chloroisopropyl) ether (5 and 10 mg/L) was statically incubated in the dark at 25 °C with yeast extract and settled domestic wastewater inoculum, complete biodegradation was achieved after 14 d (Tabak et al., 1981).

Chemical/Physical. Kollig (1993) reported that bis(2-chloroisopropyl) ether is subject to hydrolysis forming hydrochloric acid and the intermediate (2-hydroxyisopropyl-2-chloro-isopropyl) ether. The intermediate compound undergoes further hydrolysis yielding bis(2-hydroxyisopropyl) ether. Van Duuren et al. (1972) reported a hydrolysis half-life of 21 h at 25 °C and pH 7.

At an influent concentration of 1,008 mg/L, treatment with granular activated carbon resulted in non-detectable concentrations in the effluent. The adsorbability of the carbon used was 20 mg/g carbon (Guisti et al., 1974).

Symptoms of exposure: Exposure to vapors may cause eye and respiratory tract irritation (Patnaik, 1992).

Toxicity:
LC_{50} (4-h inhalation) for rats 12.8 mg/L (Worthing and Hance, 1991).
LC_{50} (48-h) for carp >40 mg/L (Hartley and Kidd, 1987).
Acute oral LD_{50} for male rats 240 mg/kg, female mice 536 mg/kg (Worthing and Hance, 1991).
Acute percutaneous LD_{50} for rats >2,000 mg/kg (Worthing and Hance, 1991).

Drinking water standard: No MCLGs or MCLs have been proposed (U.S. EPA, 1996).

Source: A waste by-product in the manufacture of propylene glycol (Verschueren, 1983)

Uses: Chemical intermediate in the manufacturing of dyes, resins, and pharmaceuticals; solvent and extractant for fats, waxes, and greases; textile manufacturing; agent in paint and varnish removers, spotting and cleaning agents; a combatant in liver fluke infections; preparation of glycol esters in fungicidal preparations and as an insecticidal wood preservative; apparently used as a nematocide in Japan but is not registered in the U.S. for use as a pesticide.

BIS(2-ETHYLHEXYL) PHTHALATE

Synonyms: BEHP; **1,2-Benzenedicarboxylic acid bis(2-ethylhexyl) ester**; Bioflex 81; Bioflex DOP; Bis(2-ethylhexyl)-1,2-benzenedicarboxylate; Compound 889; DAF 68; DEHP; Di(2-ethylhexyl)ortho phthalate; Di(2-ethylhexyl) phthalate; Dioctyl phthalate; Di-*sec*-octyl phthalate; DOP; Ergoplast FDO; Ethylhexyl phthalate; 2-Ethylhexyl phthalate; Eviplast 80; Eviplast 81; Fleximel; Flexol DOP; Flexol plasticizer DOP; Goodrite GP 264; Hatcol DOP; Hercoflex 260; Kodaflex DOP; Mollan 0; NCI-C52733; Nuoplaz DOP; Octoil; Octyl phthalate; Palatinol AH; Phthalic acid bis(2-ethylhexyl) ester; Phthalic acid dioctyl ester; Pittsburgh PX-138; Platinol AH; Platinol DOP; RC plasticizer DOP; RCRA waste number U028; Reomol D 79P; Reomol DOP; Sicol 150; Staflex DOP; Truflex DOP; Vestinol 80; Witicizer 312.

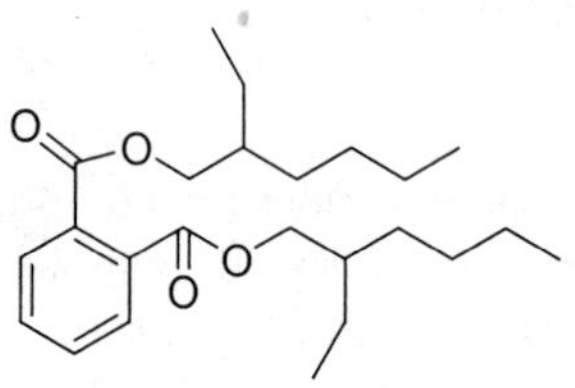

Note: In the documented literature, bis(2-ethylhexyl) phthalate is sometimes referred to as dioctyl phthalate. Not to be confused with di-n-octyl phthalate.

CAS Registry Number: 117-81-7
Molecular formula: $C_{24}H_{38}O_4$
Formula weight: 390.57
RTECS: TI0350000
Merck reference: 10, 1248

Physical state, color, and odor:
Colorless, oily liquid with a very faint odor

Melting point (°C):
-55 (Verschueren, 1983)
-46 (Standen, 1968)

Boiling point (°C at 5 mmHg):
386.9 (Fishbein and Albro, 1972)
230 (Howard, 1989)

Density (g/cm³):
0.987 at 14.9/4 °C, 0.983 at 20.1/4 °C, 0.980 at 25.0/4 °C, 0.976 at 30.0/4 °C (De Lorenzi et al., 1997)

Diffusivity in water (x 10^{-5} cm²/sec):
0.39 at 20 °C using method of Hayduk and Laudie (1974)

Flash point (°C):
207 (Aldrich, 1988)
196 (open cup, Broadhurst, 1972)
215 (open cup, NFPA, 1984)

Lower explosive limit (%):
0.3 at 245 °C (NFPA, 1984)

Henry's law constant (x 10^{-5} atm·m^3/mol):
1.1 at 25 °C (calculated, Howard, 1989)

Bioconcentration factor, log BCF:
3.73 (*Chlorella fusca*, Freitag et al., 1982; Geyer et al., 1984)
5.02 (mosquito larvae, Metcalf et al., 1973)
2.32 (*Daphnia magna*, Brown and Thompson, 1982)
3.37, 3.40, 3.42 (*Mytilus edulis*, Brown and Thompson, 1982a)
3.48 (activated sludge), 1.60 (golden ide) (Freitag et al., 1985)
1.04 (*Crassostrea virginica*, Wofford et al., 1981)
2.06 (bluegill sunfish, Barrows et al., 1980)
2.49 (freshwater fish), 2.11 (fish, microcosm) (Garten and Trabalka, 1983)
4.13 (scud), 3.49 (midge larvae), 3.72 (water flea), 3.36 (mayfly), 2.36 (sowbug) (Sanders et al., 1973)
2.19, 2.95 (*Pimephales promelas*, Devillers et al., 1996)
0.95 (*Oncorhynchus mykiss*, Tarr et al., 1990)
3.14 (*Pimephales promelas*, Mayer and Sanders, 1973)
2.90 (fathead minnow, 28-d exposure), 4.32 (snail) (quoted, Callahan et al., 1979)

Soil sorption coefficient, log K_{oc}:
5.0 (Neely and Blau, 1985)
4.94 (Broome County, NY composite soil, Russell and McDuffie, 1986)
5.15 (Carter and Suffett, 1983)
4.48 (Missouri River sediments), 4.95 (Mississippi River sediments), 5.41 (river sediments east of Lorenzo, IL) (Williams et al., 1995)

Octanol/water partition coefficient, log K_{ow}:
4.20 (Mackay, 1982)
3.98 (Kenaga and Goring, 1980)
5.11 (Geyer et al., 1984)
5.0 (Klein et al., 1988)
5.03 (Harnisch et al., 1983)
7.453 (de Bruijn et al., 1989)
7.137 (Brooke et al., 1990)
4.88 (Wams, 1987)
7.27 (Ellington and Floyd, 1996)

Solubility in organics:
Miscible with mineral oil and hexane (U.S. EPA, 1985)

Solubility in water (μg/L):
400 at 25 °C (Wolfe et al., 1980)
41 at 20 °C (Leyder and Boulanger, 1983)
At 25 °C: 340, 300 (well water), 160 (natural seawater) (Howard et al., 1985)
285 at 24 °C (technical grade, Hollifield, 1979)
47 at 25 °C (Klöpffer et al., 1982)
360 (DeFoe et al., 1990)
300 at 25 °C (Wams, 1987)

Vapor density:
15.96 g/L at 25 °C, 13.48 (air = 1)

Vapor pressure (x 10^{-9} mmHg):
200 at 20 °C (Hirzy et al., 1978)
50 at 68 °C, 5,000 at 120 °C (Gross and Colony, 1973)
6,450 at 25 °C (Howard et al., 1985)
100 at 20 °C (Broadhurst, 1972)
62 at 25 °C (Giam et al., 1980)
8.25 at 20 °C (Riederer, 1990)
340 at 25 °C (Wams, 1987)
143 at 25 °C (Hinckley et al., 1990)
1,125 at 40 °C (Goodman, 1997)

Environmental fate:
Biological. Bis(2-ethylhexyl) phthalate degraded in both amended and unamended calcareous soils from New Mexico. After 146 d, 76 to 93% degraded (mineralized) to carbon dioxide. No other metabolites were detected (Fairbanks et al., 1985). In a 56-d experiment, [^{14}C]bis(2-ethylhexyl) phthalate applied to soil-water suspensions under aerobic and anaerobic conditions gave $^{14}CO_2$ yields of 11.6 and 8.1%, respectively (Scheunert et al., 1987).

When bis(2-ethylhexyl) phthalate was statically incubated in the dark at 25 °C with yeast extract and settled domestic wastewater inoculum, no degradation was observed after 7 d. Over a 21-d period, however, gradual adaptation did occur resulting in 95 and 93% losses at concentrations of 5 and 10 mg/L, respectively (Tabak et al., 1981).

Half-lives of 19 h and 4-5 wk were reported for bis(2-ethylhexyl) phthalate in activated sludge and river water, respectively (Saeger and Tucker, 1976).

In the presence of suspended natural populations from unpolluted aquatic systems, the second-order microbial transformation rate constant determined in the laboratory was reported to be 4.2 ± 0.7 x 10^{-15} L/organism·h (Steen, 1991).

Photolytic. The estimated photolytic half-life of bis(2-ethylhexyl) phthalate in water is 143 d (Wolfe et al., 1980).

Chemical/Physical. Hydrolyzes in water to *o*-phthalic acid (via the intermediate 2-ethyl-hexyl hydrogen phthalate) and 2-ethylhexyl alcohol (Kollig, 1993; Wolfe et al., 1980). Although no pH value was given, the reported hydrolysis rate constant under alkaline conditions is 1,400/M·yr (Ellington et al., 1993; Kollig, 1993).

Klöpffer et al. (1982) estimated an evaporation half-life of approximately 140 d from a 21-cm deep vessel.

Pyrolysis of bis(2-ethylhexyl) phthalate in the presence of polyvinyl chloride at 600 °C for 10 produced the following compounds: methylindene, naphthalene, 1-methylnaphthalene, 2-methylnaphthalene, biphenyl, dimethylnaphthalene, acenaphthene, fluorene, methylacenaphth-ene, methylfluorene, phenanthrene, anthracene, methylphenanthrene, methylanthracene, meth-ylpyrene or fluoranthene, and 17 unidentified compounds (Bove and Dalven, 1984).

At influent concentrations of 1.0, 0.1, 0.01, and 0.001 mg/L, the granular activated carbon adsorption capacities were 11,300, 340, 10, and 0.32 mg/g, respectively (Dobbs and Cohen, 1980).

Exposure limits (mg/m^3): Potential occupational carcinogen. NIOSH REL: TWA 5, STEL 10, IDLH 5,000; OSHA PEL: TWA 5; ACGIH TLV: TWA 5 (adopted).

Symptoms of exposure: Ingestion of 10 mL may cause gastrointestinal pain, hypermobility, and diarrhea (Patnaik, 1992).

Toxicity:

EC_{50} (48-h) for *Chironomus plumosus* >72 mg/L (Mayer and Ellersieck, 1986).

LC_{50} (contact) for earthworm (*Eisenia fetida*) >25,000 μg/cm^2 (Neuhauser et al., 1985).

LC_{50} (96-h) for bluegill sunfish >770 mg/L (Spehar et al., 1982), *Cyprinodon variegatus* >550 ppm (Heitmuller et al., 1981).

LC_{50} (72-h) for *Cyprinodon variegatus* >550 ppm (Heitmuller et al., 1981).

LC_{50} (48-h) for *Daphnia pulex* 113 μg/L (Passino and Smith, 1987), *Daphnia magna* 11 mg/L (LeBlanc, 1970), *Cyprinodon variegatus* >550 ppm (Heitmuller et al., 1981).

LC_{50} (24-h) for *Daphnia magna* >68 mg/L (LeBlanc, 1980), *Cyprinodon variegatus* >550 ppm (Heitmuller et al., 1981).

LD_{50} (dermal) for guinea pigs 10 g/kg (quoted, Autian, 1973).

Acute oral LD_{50} for guinea pigs 26 gm/kg, mice 30 gm/kg, rats 30,600 mg/kg, rabbits 34 gm/kg (quoted, RTECS, 1985).

Heitmuller et al. (1981) reported a NOEC of 550 ppm.

Drinking water standard (final): MCLG: zero; MCL: 6 μg/L (U.S. EPA, 1996).

Source: Detected in distilled water-soluble fractions of new and used motor oil at concentrations of 17-21 and ND-1.2 μg/L, respectively (Chen et al., 1994). Also may leach from plastic products used in analytical laboratories (e.g., tubing, containers).

Uses: Plasticizer in PVC flooring, films, and sheets; in vacuum pumps.

BROMOBENZENE

Synonyms: Monobromobenzene; NCI-C55492; Phenyl bromide; UN 2514.

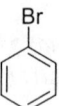

CAS Registry Number: 108-86-1
DOT: 2514
DOT label: Flammable liquid
Molecular formula: C_6H_5Br
Formula weight: 157.01
RTECS: CY9000000
Merck reference: 10, 1376

Physical state, color, and odor:
Clear, colorless liquid with an aromatic odor. The reported odor threshold is 4.6 ppm (Mateson, 1955).

Melting point (°C):
-30.8 (Weast, 1986)

Boiling point (°C):
156 (Weast, 1986)

Density (g/cm³):
1.5017 at 15/4 °C, 1.4952 at 20/4 °C, 1.4815 at 30/4 °C (Windholz et al., 1983)
1.49466 at 20.00/4 °C (Tsierkezos et al., 2000)

Diffusivity in water (x 10^{-5} cm²/sec, 25 °C):
1.03, 0.99, and 0.96 at mole fractions of 10^{-6}, 2 x 10^{-6}, and 5 x 10^{-6}, respectively (Gabler et al., 1996)

Flash point (°C):
51 (Windholz et al., 1983)

Entropy of fusion (cal/mol·K):
10.48 (Stull, 1937)
10.55 (Masi and Scott, 1975)

Heat of fusion (kcal/mol):
2.54 (Dean, 1987)

Henry's law constant (x 10^{-3} atm·m³/mol):
2.40 at 25 °C (Valsaraj, 1988)
2.527, 3.277, and 5.685 at 30.0, 35.0, and 44.8 °C, respectively (Hansen et al., 1993, 1995)
1.66 at 20 °C (Hovorka and Dohnal, 1997)
2.47 at 25 °C (Shiu and Mackay, 1997)

Interfacial tension with water (dyn/cm):
38.1 at 25 °C (Donahue and Bartell, 1952)

Ionization potential (eV):
8.98 ± 0.02 (Franklin et al., 1969)
9.41 (Yoshida et al., 1983a)

Bioconcentration factor, log BCF:
3.18 (activated sludge), 2.28 (algae), 1.70 (golden ide) (Freitag et al., 1985)

Soil sorption coefficient, log K_{oc}:
2.33 using method of Chiou et al. (1979)

Octanol/water partition coefficient, log K_{ow}:
3.01 (Garst and Wilson, 1984; Watarai et al., 1982)
2.99 (Hansch and Anderson, 1967)
2.96 (Kenaga and Goring, 1980)
2.98 (Wasik et al., 1981, 1983)

Solubility in organics:
At 25 °C (wt %): alcohol (10.4) and ether (71.3). Miscible with benzene, chloroform, petroleum ethers (Windholz et al., 1983), and many other organic solvents (Patnaik, 1992).

Solubility in water:
500 mg/L at 20 °C (quoted, Verschueren, 1983)
446 mg/L at 30 °C (Chiou et al., 1977)
409 mg/L at 25 °C (Valsaraj, 1988)
446 mg/kg at 30 °C (Gross and Saylor, 1931)
3.62 mM at 25 °C (Andrews and Keefer, 1950)
2.92 mM at 35 °C (Hine et al., 1963)
2.3 mM at 25 °C (Yalkowsky et al., 1979)
2.62 mM at 25 °C (Wasik et al., 1981, 1983)
2.84 mmol/kg at 25.0 °C (Vesala, 1974)

Vapor density:
6.42 g/L at 25 °C, 5.41 (air = 1)

Vapor pressure (mmHg):
3.3 at 20 °C (Verschueren, 1983)
3.8 at 25 °C (Valsaraj, 1988)
4.14 at 25 °C (Mackay et al., 1982)
7.48 at 35 °C (Hine et al., 1963)

Environmental fate:
Biological. In activated sludge, 34.8% of the applied bromobenzene mineralized to carbon dioxide after 5 d (Freitag et al., 1985).

Photolytic. A carbon dioxide yield of 19.7% was achieved when bromobenzene adsorbed on silica gel was irradiated with light (λ >290 nm) for 17 h (Freitag et al., 1985). Irradiation of bromobenzene in air containing nitrogen oxides gave phenol, 4-nitrophenol, 2,4-dinitrophenol, 4-bromophenol, 3-bromonitrobenzene, 3-bromo-2-nitrophenol, 3-bromo-4-nitrophenol, 3-bromo-6-nitrophenol, 2-bromo-4-nitrophenol, and 2,6-dibromo-4-nitrophenol (Nojima et al., 1980).

Chemical/Physical. Bromobenzene will not hydrolyze to any reasonable extent. In the laboratory, no change in concentration was observed after 29 d at 85 °C in 0.1M sodium hydroxide and 0.1M hydrochloric acid (Kollig, 1995).

Toxicity:

Acute oral LD_{50} for rats 2,699 mg/kg, mice 2,700 mg/kg (quoted, RTECS, 1985).

Drinking water standard: As of October 1996, bromobenzene was listed for regulation but no MCLGs or MCLs have been proposed (U.S. EPA, 1996).

Source: Storm water runoff, waste motor oils, improper disposal of laboratory solvent containing bromobenzene (Verschueren, 1983)

Uses: Preparation of phenyl magnesium bromide used in organic synthesis; solvent for fats, waxes, and oils; motor oil additive; crystallizing solvent; chemical intermediate.

BROMOCHLOROMETHANE

Synonyms: CB; CBM; Chlorobromomethane; Halon 1011; Methylene chloromethane; Mil-B-4394-B; UN 1887.

$$\begin{array}{c} Br \\ | \\ H{-}C{-}Cl \\ | \\ H \end{array}$$

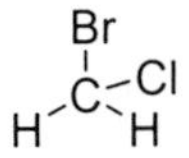

CAS Registry Number: 74-97-5
DOT: 1887
Molecular formula: CH_2BrCl
Formula weight: 129.39
RTECS: PA5250000

Physical state, color, and odor:
Clear, colorless liquid with a sweet, chloroform-like odor

Melting point (°C):
-87.5 (Riddick et al., 1986)

Boiling point (°C):
68.1 (Weast, 1986)

Density (g/cm³):
1.9344 at 20/4 °C (Weast, 1986)
1.991 at 20 °C (Aldrich, 1988)
1.9229 at 25 °C (Dreisbach, 1959)

Diffusivity in water (x 10^{-5} cm²/sec):
1.13 at 20 °C using method of Hayduk and Laudie (1974)

Flash point (°C):
Non applicable (NIOSH, 1997)

Lower explosive limit (%):
Not applicable (NIOSH, 1997)

Upper explosive limit (%):
Not applicable (NIOSH, 1997)

Henry's law constant (x 10^{-3} atm·m³/mol):
1.44 at 25 °C (approximate - calculated from water solubility and vapor pressure)

Ionization potential (eV):
10.77 ± 0.01 (Franklin et al., 1969)

Soil sorption coefficient, log K_{oc}:
1.43 using method of Chiou et al. (1979)

Octanol/water partition coefficient, log K_{ow}:
1.41 (Tewari et al., 1982; Wasik et al., 1981)

Solubility in organics:
Soluble in acetone, alcohol, benzene, and ether (Weast, 1986)

Solubility in water (25 °C):
129 mM (Tewari et al., 1982; Wasik et al., 1981)
15 g/kg H_2O (O'Connell, 1963)

Vapor density:
5.29 g/L at 25 °C, 4.47 (air = 1)

Vapor pressure (mmHg):
115 at 20 °C (NIOSH, 1997)
141.07 at 24.05 °C (Kudchadker et al., 1979)

Environmental fate:
   Biological. When bromochloromethane (5 and 10 mg/L) was statically incubated in the dark at 25 °C with yeast extract and settled domestic wastewater inoculum for 7 d, 100% biodegradation with rapid adaptation was observed (Tabak et al., 1981).
   Chemical/Physical. Although no products were identified, the estimated hydrolysis half-life in water at 25 °C and pH 7 is 44 yr (Mabey and Mill, 1978).

Exposure limits: NIOSH REL: TWA 200 ppm (1,050 mg/m³), IDLH 2,000 ppm; OSHA PEL: TWA 200 ppm; ACGIH TLV: TWA 200 ppm (adopted)

Toxicity:
   LC_{50} (inhalation) for mice 15,850 mg/m³/8-h, rats 28,800 ppm/15-min (quoted, RTECS, 1985).
   Acute oral LD_{50} for rats 5,000 mg/kg, mice 4,300 mg/kg (quoted, RTECS, 1985).

Drinking water standard: No MCLGs or MCLs have been proposed (U.S. EPA, 1996).

Uses: Fire extinguishing agent; organic synthesis.

BROMODICHLOROMETHANE

Synonyms: BDCM; Dichlorobromomethane; NCI-C55243.

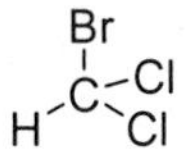

$$
\begin{array}{c}
\text{Br} \\
| \\
\text{H} - \text{C} - \text{Cl} \\
| \\
\text{Cl}
\end{array}
$$

CAS Registry Number: 75-27-4
Molecular formula: $CHBrCl_2$
Formula weight: 163.83
RTECS: PA5310000

Physical state and color:
Clear, colorless liquid

Melting point (°C):
-57.1 (Weast, 1986)

Boiling point (°C):
90.1 (Dean, 1973)

Density (g/cm³ at 20/4 °C):
1.980 (Weast, 1986)

Diffusivity in water (x 10⁻⁵ cm²/sec):
0.98 at 20 °C using method of Hayduk and Laudie (1974)

Flash point (°C):
None (Dean, 1987)

Henry's law constant (x 10⁻⁴ atm·m³/mol):
2.12 at 25 °C (Warner et al., 1987)
16 at 20 °C (Nicholson et al., 1984)
16, 26, and 40 at 20, 30, and 40 °C, respectively (Tse et al., 1992)
In seawater (salinity 30.4‰): 5.52, 10.51, and 18.97 at 0, 10, and 20 °C, respectively (Moore et al., 1995)

Ionization potential (eV):
10.88 ± 0.05 (Franklin et al., 1969)

Soil sorption coefficient, log K_{oc}:
1.77 (estimated, Ellington et al., 1993)
1.79 (estimated, Schwille, 1988)

Octanol/water partition coefficient, log K_{ow}:
1.88 (quoted, Mills et al., 1985)
2.10 (Hansch and Leo, 1979; Mabey et al., 1982)

Solubility in organics:
Soluble in acetone, ethanol, benzene, chloroform, and ether (Weast, 1986)

Solubility in water (mg/L):
4,500 at 0 °C (quoted, Schwille, 1988)
4,700 at 22 °C (Mabey et al., 1982)
2,968 at 30 °C (McNally and Grob, 1984)
3,031.9 at 30 °C (McNally and Grob, 1983)

Vapor density:
6.70 g/L at 25 °C, 5.66 (air = 1).

Vapor pressure (mmHg):
50 at 20 °C (Dreisbach, 1952)

Environmental fate:
    Biological. Bromodichloromethane showed significant degradation with gradual adaptation in a static-culture flask-screening test (settled domestic wastewater inoculum) conducted at 25 °C. At concentrations of 5 and 10 mg/L, percent losses after 4 wk of incubation were 59 and 51, respectively. At a substrate concentration of 5 mg/L, 8% was lost due to volatilization after 10 d (Tabak et al., 1981).
    Chemical/Physical. The estimated hydrolysis half-life in water at 25 °C and pH 7 is 137 yr (Mabey and Mill, 1978). Reported products of hydrolysis include carbon monoxide, hydrochloric and hydrobromic acids (Ellington et al., 1993; Kollig, 1993).
    At influent concentrations of 1.0, 0.1, 0.01, and 0.001 mg/L, the granular activated carbon adsorption capacities at pH 5.3 were 7.9, 1.9, 0.47, and 0.12 mg/g, respectively (Dobbs and Cohen, 1980).

Toxicity:
    Acute oral LD_{50} for rats 916 mg/kg, mice 450 mg/kg (quoted, RTECS, 1985).

Drinking water standard (proposed): MCLG: zero; MCL: 0.1 mg/L. Total for all trihalomethanes cannot exceed a concentration of 0.08 mg/L (U.S. EPA, 1996).

Source: By-product in chlorination of drinking water and use of fire extinguishers (Verschueren, 1983)

Uses: Component of fire extinguisher fluids; solvent for waxes, fats, and resins; degreaser; flame retardant; heavy liquid for mineral and salt separations; chemical intermediate; laboratory use.

BROMOFORM

Synonyms: Methenyl tribromide; Methyl tribromide; NCI-C55130; RCRA waste number U225; **Tribromomethane**; UN 2515.

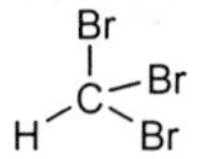

$$\underset{H}{\overset{Br}{\underset{}{}}}\overset{Br}{\underset{Br}{C}}-Br$$

CAS Registry Number: 75-25-2
DOT: 2515
DOT label: Poison
Molecular formula: $CHBr_3$
Formula weight: 252.73
RTECS: PB5600000
Merck reference: 10, 1389

Physical state, color, and odor:
Clear, colorless to yellow liquid with a chloroform-like odor. Odor threshold concentration from water is 0.3 mg/kg (Verschueren, 1982).

Melting point (°C):
8.05 (Riddick et al., 1986)
8.3 (Weast, 1986)

Boiling point (°C):
149.5 (Weast, 1986)

Density (g/cm³ at 20/4 °C):
2.9031 (Dean, 1987)
2.8899 (Weast, 1986)
2.89165 (Kudchadker et al., 1979)

Diffusivity in water (x 10⁻⁵ cm²/sec):
0.95 at 20 °C using method of Hayduk and Laudie (1974)

Flash point (°C):
Noncombustible liquid (NIOSH, 1997)

Lower explosive limit (%):
Not applicable (NIOSH, 1997)

Upper explosive limit (%):
Not applicable (NIOSH, 1997)

Henry's law constant (x 10⁻⁴ atm·m³/mol):
5.6 (Pankow and Rosen, 1988)
4.35 at 25 °C (Wright et al., 1992)
4.3 at 20 °C (Nicholson et al., 1984)
4, 7, and 12 at 20, 30, and 40 °C, respectively (Tse et al., 1992)
In seawater (salinity 30.4‰): 1.41, 2.88, and 5.22 at 0, 10, and 20 °C, respectively (Moore et al., 1995)

5.32 at 25 °C (Warner et al., 1987)
4.31 at 20 °C (Hovorka and Dohnal, 1997)
2.33, 4.09, and 6.93 at 10, 20, and 30 °C, respectively (Munz and Roberts, 1987)

Interfacial tension with water (dyn/cm at 20 °C):
40.85 (Demond and Lindner, 1993)

Ionization potential (eV):
10.51 ± 0.02 (Franklin et al., 1969)

Soil sorption coefficient, log K_{oc}:
2.45 (Abdul et al., 1987)
2.10 (Hutzler et al., 1986)

Octanol/water partition coefficient, log K_{ow}:
2.30 (quoted, Mills et al., 1985)
2.38 (Valsaraj, 1988)

Solubility in organics:
Soluble in ligroin (Weast, 1986). Miscible with benzene, chloroform, ether, petroleum ether, acetone, and oils (Windholz et al., 1983).

Solubility in water:
3,010 mg/kg at 15 °C, 3,190 mg/kg at 30 °C (Gross and Saylor, 1931)
3,130 mg/L at 25 °C (Valsaraj, 1988)
3,931 mg/L at 30 °C (McNally and Grob, 1984)
3,180 mg/L at 30 °C (Horvath, 1982)

Vapor density:
10.33 g/L at 25 °C, 8.72 (air = 1)

Vapor pressure (mmHg):
4 at 20 °C (quoted, Munz and Roberts, 1987)
5.4 at 25 °C (Mackay et al., 1982)

Environmental fate:
  Biological. Bromoform showed significant degradation with gradual adaptation in a static-culture flask-screening test (settled domestic wastewater inoculum) conducted at 25 °C. At concentrations of 5 and 10 mg/L, percent losses after 4 wk of incubation were 48 and 35, respectively (Tabak et al., 1981).
  Surface Water. Kaczmar et al. (1984) estimated the volatilization half-life of bromoform from rivers and streams to be 65.6 d.
  Chemical/Physical. The estimated hydrolysis half-life in water at 25 °C and pH 7 is 686 yr (Mabey and Mill, 1978). Products of hydrolysis include carbon monoxide and hydrobromic acid (Kollig, 1993). When an aqueous solution containing bromoform was purged with hydrogen for 24 h, only 5% of the bromoform reacted to form methane and minor traces of ethane. In the presence of colloidal platinum catalyst, the reaction proceeded at a much faster rate forming the same end products (Wang et al., 1988). In an earlier study, water containing 2,000 ng/μL of bromoform and colloidal platinum catalyst was irradiated with UV light. After 20 h, about 50% of the bromoform had reacted. A duplicate experiment was performed but the concentration of bromoform was increased to 3,000 ng/μL and 0.1 g zinc was added. After 14

h, only 0.1 ng/μL bromoform remained. Anticipated transformation products include methane and bromide ions (Wang and Tan, 1988).

Photolysis of an aqueous solution containing bromoform (989 μmol) and a catalyst [Pt(colloid)/Ru(bpy)$^{2+}$/MV/EDTA] yielded the following products after 25 h (μmol detected): bromide ions (250), methylene bromide (475), and unreacted bromoform (421) (Tan and Wang, 1987).

Bromoform (0.11 mM) reacted with OH radicals in water (pH 8.5) at a rate of 1.3 x 10^8/M·sec (Haag and Yao, 1992).

At influent concentrations of 1.0, 0.1, 0.01, and 0.001 mg/L, the adsorption capacities of the granular activated carbon used were 19.6, 5.9, 1.8, and 0.52 mg/g, respectively (Dobbs and Cohen, 1980).

Exposure limits: NIOSH REL: TWA 0.5 ppm (5 mg/m^3), IDLH 850 ppm; OSHA PEL: 0.5 ppm; ACGIH TLV: TWA 0.5 ppm (adopted).

Symptoms of exposure: Inhalation may cause respiratory irritation (Patnaik, 1992).

Toxicity:

LC$_{50}$ (96-h) for bluegill sunfish 29 mg/L (Spehar et al., 1982), *Cyprinodon variegatus* 18 ppm using natural seawater (Heitmuller et al., 1981).

LC$_{50}$ (72-h) for *Cyprinodon variegatus* 18 ppm (Heitmuller et al., 1981).

LC$_{50}$ (48-h) for *Daphnia magna* 46 mg/L (LeBlanc, 1980), *Cyprinodon variegatus* 19 ppm (Heitmuller et al., 1981).

LC$_{50}$ (24-h) for *Daphnia magna* 56 mg/L (LeBlanc, 1980), *Cyprinodon variegatus* 19 ppm (Heitmuller et al., 1981).

Acute oral LD$_{50}$ for rats 1,147 mg/kg, mice 1,400 mg/kg (quoted, RTECS, 1985); for male rats 1,399, female rats 1,147 mg/kg (Chu et al., 1980); male mice 1,400 mg/kg, female mice 1,550 mg/kg (Bowman et al., 1978).

Heitmuller et al. (1981) reported a NOEC of 2.9 ppm.

Drinking water standard (proposed): MCLG: zero; MCL: 0.1 mg/L. Total for all trihalomethanes cannot exceed a concentration of 0.08 mg/L (U.S. EPA, 1996).

Uses: Solvent for waxes, greases, and oils; separating solids with lower densities; component of fire-resistant chemicals; geological assaying; medicine (sedative); gauge fluid; intermediate in organic synthesis.

4-BROMOPHENYL PHENYL ETHER

Synonyms: 4-Bromodiphenyl ether; *p*-Bromodiphenyl ether; **1-Bromo-4-phenoxybenzene**; 1-Bromo-*p*-phenoxybenzene; 4-Bromophenyl ether; *p*-Bromophenyl ether; *p*-Bromophenyl phenyl ether; Phenyl-4-bromophenyl ether; Phenyl-*p*-bromophenyl ether.

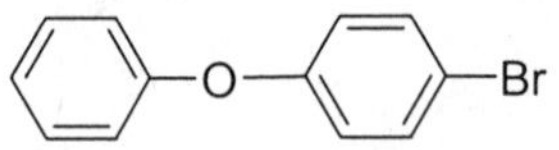

CAS Registry Number: 101-55-3
Molecular formula: $C_{12}H_9BrO$
Formula weight: 249.20

Physical state:
Liquid

Melting point (°C):
18.7 (Weast, 1986)

Boiling point (°C):
310.1 (Weast, 1986)
305 (Aldrich, 1988)

Density (g/cm³):
1.4208 at 20/4 °C (Weast, 1986)

Diffusivity in water (x 10^{-5} cm²/sec):
0.63 at 20 °C using method of Hayduk and Laudie (1974)

Flash point (°C):
>110 (Aldrich, 1988)

Henry's law constant (x 10^{-4} atm·m³/mol):
1 (Pankow and Rosen, 1988)

Soil sorption coefficient, log K_{oc}:
4.94 using method of Karickhoff et al. (1979)

Octanol/water partition coefficient, log K_{ow}:
5.15 (Walton, 1985)

Solubility in organics:
Soluble in ether (Weast, 1986)

Vapor density:
10.19 g/L at 25 °C, 8.60 (air = 1)

Vapor pressure (mmHg):
1.5 x 10^{-3} at 20 °C (calculated, Dreisbach, 1952)

Environmental fate:

Biological. Using settled domestic wastewater inoculum, 4-bromophenyl phenyl ether (5 and 10 mg/L) did not degrade after 28 d of incubation at 25 °C (Tabak et al., 1981).

Toxicity:

LC_{50} (96-h) for bluegill sunfish 5.9 mg/L (Spehar et al., 1982).

LC_{50} (48-h) for *Daphnia magna* 0.36 mg/L (LeBlanc, 1980).

LC_{50} (24-h) for *Daphnia magna* 0.46 mg/L (LeBlanc, 1980).

Use: Research chemical.

BROMOTRIFLUOROMETHANE

Synonyms: Bromofluoroform; F 13B1; Freon 13-B1; Halocarbon 13B1; Halon 1301; Monobromotrifluoromethane; Refrigerant 13B1; Trifluorobromomethane; Trifluoromonobromomethane; UN 1009.

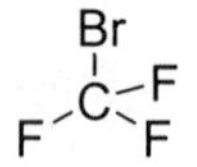

$$\overset{\displaystyle Br}{\underset{\displaystyle F}{\underset{|}{C}}} \!\!\! \begin{matrix} F \\ F \end{matrix}$$

CAS Registry Number: 75-63-8
DOT: 1009
Molecular formula: $CBrF_3$
Formula weight: 148.91
RTECS: PA5425000

Physical state, color, and odor:
Colorless gas with an ether-like odor

Melting point (°C):
-168 (Horvath, 1982)

Boiling point (°C):
-58 to -57 (Aldrich, 1988)

Density (g/cm³):
1.538 at 20/4 °C (Horvath, 1982)

Diffusivity in water (x 10^{-5} cm²/sec):
0.90 at 20 °C using method of Hayduk and Laudie (1974)

Flash point (°C):
Nonflammable gas (NIOSH, 1997)

Lower explosive limit (%):
Not applicable (NIOSH, 1997)

Upper explosive limit (%):
Not applicable (NIOSH, 1997)

Henry's law constant (atm·m³/mol):
0.500 at 25 °C (Hine and Mookerjee, 1975)

Ionization potential (eV):
11.40 (Lias, 1998)

Soil sorption coefficient, log K_{oc}:
2.44 using method of Chiou et al. (1979)

Octanol/water partition coefficient, log K_{ow}:
1.54 (Hansch and Leo, 1979)

Solubility in organics:
Soluble in chloroform (Weast, 1986)

Solubility in water:
0.03 wt % at 20 °C (NIOSH, 1997)

Vapor density:
6.09 g/L at 25 °C, 5.14 (air = 1)

Vapor pressure (mmHg):
>760 at 20 °C (NIOSH, 1997)

Exposure limits: NIOSH REL: TWA 1,000 ppm (6,100 mg/m^3), IDLH 40,000 ppm; OSHA PEL: TWA 1,000 ppm.

Symptoms of exposure: Exposure to 10% in air for 3 min caused lightheadedness and paresthesia (Patnaik, 1992).

Toxicity:
LC$_{50}$ (inhalation) for mice 381 gm/m^3, rats 416 gm/m^3 (quoted, RTECS, 1985).

Uses: Chemical intermediate; metal hardening; refrigerant; fire extinguishers.

1,3-BUTADIENE

Synonyms: Biethylene; Bivinyl; Butadiene; Buta-1,3-diene; α,γ-Butadiene; Divinyl erythrene; NCI-C50602; Pyrrolylene; Vinylethylene.

CAS Registry Number: 106-99-0
DOT: 1010
DOT label: Flammable liquid
Molecular formula: C_4H_6
Formula weight: 54.09
RTECS: EI9275000
Merck reference: 10, 1476

Physical state, color, and odor:
Colorless gas with a mild, aromatic odor. Odor threshold concentration is 160 ppb (4 mg/m³) (Keith and Walters, 1992).

Melting point (°C):
-108.9 (Weast, 1986)

Boiling point (°C):
-4.4 (Weast, 1986)
-4.50 (Howard, 1989)

Density (g/cm³):
0.6211 at 20/4 °C and 0.6149 at 25/4 °C at saturation pressure (Dreisbach, 1959)
0.6789 at 25/4 °C (Hayduk and Minhas, 1987)

Diffusivity in water (x 10⁻⁵ cm²/sec):
0.95 at 20 °C using method of Hayduk and Laudie (1974)

Dissociation constant, pKₐ:
>14 (Schwarzenbach et al., 1993)

Flash point (°C):
-76 (NIOSH, 1997)

Lower explosive limit (%):
2.0 (NIOSH, 1997)

Upper explosive limit (%):
12.0 (NIOSH, 1997)

Heat of fusion (kcal/mol):
1.908 (Dean, 1987)

Henry's law constant (x 10⁻² atm·m³/mol):
6.3 at 25 °C (Hine and Mookerjee, 1975)

Interfacial tension with water (dyn/cm at 22 °C):
67 (estimated, CHRIS, 1984)

Ionization potential (eV):
9.18 (Krishna and Gupta, 1970)

Soil sorption coefficient, log K_{oc}:
2.08 (calculated, Mercer et al., 1990)

Octanol/water partition coefficient, log K_{ow}:
1.99 (Hansch and Leo, 1979)

Solubility in organics:
In *n*-heptane (mole fraction): 0.668, 0.360, and 0.210 at 4.00, 25.00, and 50.00 °C, respectively (Hayduk and Minhas, 1987)

Solubility in water:
735 ppm at 25 °C (McAuliffe, 1966)
At 37.8 °C, the reported mole fraction solubilities at 517 and 1,034 mmHg are 8×10^{-5} and 1.6×10^{-4}, respectively (Reed and McKetta, 1959)
11 mM at 25 °C and 520 mmHg (Fischer and Ehrenberg, 1948)
6.8×10^{-4} at 4.00 °C (mole fraction, Hayduk and Minhas, 1987)

Vapor density:
2.29 g/L at 25 °C, 1.87 (air = 1)

Vapor pressure (mmHg):
1,840 at 21 °C (Sax and Lewis, 1987)
2,105 at 25 °C (Wilhoit and Zwolinski, 1971)

Environmental fate:
Surface Water. The estimated volatilization half-life of 1,3-butadiene in a model river 1 m deep, flowing 1 m/sec and a wind speed of 3 m/sec is 3.8 h (Lyman et al., 1982).

Photolytic. The following rate constants were reported for the reaction of 1,3-butadiene and OH radicals in the atmosphere: 6.9×10^{-11} cm^3/molecule·sec (Atkinson et al., 1979) and 6.7×10^{-11} cm^3/molecule·sec (Sabljić and Güsten, 1990). Atkinson and Carter (1984) reported a rate constant of 6.7-8.4 x cm^3/molecule·sec for the reaction of 1,3-butadiene and ozone in the atmosphere. Photooxidation reaction rate constants of 2.13×10^{-13} and 7.50×10^{-18} cm^3/molecule·sec were reported for the reaction of 1,3-butadiene and NO$_3$ (Benter and Schindler, 1988; Sabljić and Güsten, 1990). The half-life in air for the reaction of 1,3-butadiene and NO$_3$ radicals is 15 h (Atkinson et al., 1984a).

Chemical/Physical. Will polymerize in the presence of oxygen if no inhibitor is present (Hawley, 1981).

Exposure limits: Potential occupational carcinogen. NIOSH REL: IDLH 2,000 ppm; OSHA PEL: TWA 1,000 ppm (2,200 mg/m^3); ACGIH TLV: TWA 2 ppm (adopted).

Symptoms of exposure: An asphyxiant. Inhalation may cause hallucinations, distorted perception, and eye, nose, and throat irritation. At high concentrations drowsiness, light-headedness, and narcosis may occur (Patnaik, 1992). Contact of liquid with skin may result in frostbite (NIOSH, 1997).

Toxicity:

LC$_{50}$ (inhalation) for mice 270 gm/m^3/2-h, rats 285 gm/m^3/4-h (quoted, RTECS, 1985).

Uses: Synthetic rubbers and elastomers (styrene-butadiene, polybutadiene, neoprene); organic synthesis (Diels-Alder reactions); latex paints; resins; chemical intermediate.

BUTANE

Synonyms: *n*-Butane; Diethyl; Methylethylmethane; UN 1011.

CAS Registry Number: 106-97-8
DOT: 1011
DOT label: Flammable gas
Molecular formula: C_4H_{10}
Formula weight: 58.12
RTECS: EJ4200000
Merck reference: 10, 1483

Physical state, color, and odor:
Colorless gas with a natural gas-like odor. Odor threshold concentration is 2,700 ppm (Amoore and Hautala, 1983).

Melting point (°C):
-138.4 (Weast, 1986)
-135.0 (Stull, 1947)

Boiling point (°C):
-0.5 (Weast, 1986)

Density (g/cm³):
0.6012 at 0/4 °C, 0.5788 at 20/4 °C (Weast, 1986)
0.57287 at 25/4 °C (Riddick et al., 1986)

Diffusivity in water (x 10⁻⁵ cm²/sec):
0.89 at 20 °C (Witherspoon and Bonoli, 1969)
0.97 at 25 °C (quoted, Hayduk and Laudie, 1974)

Dissociation constant, pK_a:
>14 (Schwarzenbach et al., 1993)

Flash point (°C):
-72 (Kuchta et al., 1968)

Lower explosive limit (%):
1.6 (NFPA, 1984)

Upper explosive limit (%):
8.4 (NFPA, 1984)

Heat of fusion (kcal/mol):
1.050 (Parks and Huffman, 1931)
1.114 (Riddick et al., 1986)

Henry's law constant (atm·m³/mol):
0.930 at 25 °C (Hine and Mookerjee, 1975)

Interfacial tension with water (dyn/cm at 22 °C):
65 (estimated, CHRIS, 1984)

Ionization potential (eV):
10.63 ± 0.03 (Franklin et al., 1969)

Soil sorption coefficient, log K_{oc}:
Unavailable because experimental methods for estimation of this parameter for aliphatic hydrocarbons are lacking in the documented literature

Octanol/water partition coefficient, log K_{ow}:
2.89 (Hansch and Leo, 1979)

Solubility in organics:
   At 17 °C (mL/L): chloroform (25,000), ether (30,000) (Windholz et al., 1983). At 10 °C (mole fraction): acetone (0.2276), aniline (0.04886), benzene (0.5904), 2-butanone (0.3885), cyclohexane (0.6712), ethanol (0.1647), methanol (0.04457), 1-propanol (0.2346), 1-butanol (0.2817). At 25 °C (mole fraction): acetone (0.1108), aniline (0.03241), benzene (0.2851), 2-butanone (0.1824), cyclohexane (0.3962), ethanol (0.07825), methanol (0.03763), 1-propanol (0.1138), 1-butanol (0.1401) (Miyano and Hayduk, 1986).
   Mole fraction solubility in 1-butanol: 0.140, 0.0692, and 0.0397 at 25, 30, and 70 °C, respectively; in chlorobenzene: 0.274, 0.129, and 0.0800 at 25, 30, and 70 °C, respectively and in octane: 0.423, 0.233, and 0.152 at 25, 30, and 70 °C, respectively (Hayduk et al., 1988).
   Mole fraction solubility in 1-butanol: 0.139 and 0.0725 at 25 and 70 °C, respectively; in chlorobenzene: 0.269 and 0.131 at 25 and 70 °C, respectively; and in carbon tetrachloride: 0.167 at 70 °C (Blais and Hayduk, 1983).

Solubility in water:
61.4 mg/kg at 25 °C (McAuliffe, 1963, 1966)
At 0 °C, 0.0327 and 0.0233 volume of gas dissolved in a unit volume of water at 19.8 and 29.8
   °C, respectively (Claussen and Polglase, 1952)
7 mM at 17 °C and 772 mmHg (Fischer and Ehrenberg, 1948)
1.09 mM at 25 °C (Barone et al., 1966)
3.21, 1.26, and 0.66 mM at 4, 25, and 50 °C, respectively (Kresheck et al., 1965)

Vapor density:
2.38 g/L at 25 °C, 2.046 (air = 1)

Vapor pressure (mmHg):
1,820 at 25 °C (Wilhoit and Zwolinski, 1971)

Environmental fate:
   Biological. In the presence of methane, *Pseudomonas methanica* degraded butane to 1-butanol, methyl ethyl ether, butyric acid, and 2-butanone (Leadbetter and Foster, 1959). 2-Butanone was also reported as a degradation product of butane by the microorganism *Mycobacterium smegmatis* (Riser-Roberts, 1992). Butane may biodegrade in two ways. The first is the formation of butyl hydroperoxide which decomposes to 1-butanol followed by oxidation to butyric acid. The other pathway involves dehydrogenation yielding 1-butene, which may react with water forming 1-butanol (Dugan, 1972). Microorganisms can oxidize alkanes under aerobic conditions (Singer and Finnerty, 1984). The most common degradative pathway involves the oxidation of the terminal methyl group forming the corresponding alcohol

(1-butanol). The alcohol may undergo a series of dehydrogenation steps forming butanal followed by oxidation forming butyric acid. The fatty acid may then be metabolized by β-oxidation to form the mineralization products, carbon dioxide, and water (Singer and Finnerty, 1984).

Photolytic. Major products reported from the photooxidation of butane with nitrogen oxides under atmospheric conditions are: acetaldehyde, formaldehyde, and 2-butanone. Minor products included peroxyacyl nitrates and methyl, ethyl and propyl nitrates, carbon monoxide, and carbon dioxide. Biacetyl, *tert*-butyl nitrate, ethanol, and acetone were reported as trace products (Altshuller, 1983; Bufalini et al., 1971). The amount of *sec*-butyl nitrate formed was about twice that of *n*-butyl nitrate. 2-Butanone was the major photooxidation product with a yield of 37% (Evmorfopoulos and Glavas, 1998). Irradiation of butane in the presence of chlorine yielded carbon monoxide, carbon dioxide, hydroperoxides, peroxyacid, and other carbonyl compounds (Hanst and Gay, 1983). Nitrous acid vapor and butane in a "smog chamber" were irradiated with UV light. Major oxidation products identified included 2-butanone, acetaldehyde, and butyraldehyde. Minor products included peroxyacetyl nitrate, methyl nitrate, and other unidentified compounds (Cox et al., 1981).

The rate constant for the reaction of butane and OH radicals in the atmosphere at 300 K is 1.6×10^{12} cm^3/mol·sec (Hendry and Kenley, 1979). Based upon a photooxidation rate constant of 2.54×10^{-12} cm^3/molecule·sec with OH radicals in summer daylight, the atmospheric lifetime is 54 h (Altshuller, 1991). At atmospheric pressure and 298 K, Darnall et al. (1978) reported a rate constant of 2.35-4.22×10^{-12} cm^3/molecule·sec for the same reaction. A rate constant of 1.28×10^{11} L/mol·sec was reported for the reaction of butane with OH radicals in air at 298 K, respectively (Greiner, 1970). At 296 K, a rate constant of 6.5×10^{-17} cm^3/molecule·sec was reported for the reaction of butane with NO_3 (Atkinson, 1990).

Chemical/Physical. Complete combustion in air gives carbon dioxide and water. Butane will not hydrolyze because it has no hydrolyzable functional group.

Exposure limits: NIOSH REL: TWA 800 ppm (1,900 mg/m^3); ACGIH TLV: TWA 800 ppm (adopted).

Symptoms of exposure: High concentrations may cause narcosis (Patnaik, 1992).

Toxicity:
    LC$_{50}$ (inhalation) for mice 680 gm/m^3/2-h, rats 658 gm/m^3/4-h (quoted, RTECS, 1985).

Source: Present in gasoline ranging from 4.31 to 5.02 vol % (Verschueren, 1983).
    Schauer et al. (1999) reported butane in a diesel-powered medium-duty truck exhaust at an emission rate of 3,830 μg/km.

Uses: Manufacture of synthetic rubbers, ethylene; raw material for high octane motor fuels; solvent; refrigerant; propellant in aerosols; calibrating instruments; organic synthesis.

1-BUTANOL

Synonyms: Butan-1-ol; *n*-Butanol; Butyl alcohol; *n*-Butyl alcohol; Butyl hydroxide; Butyric alcohol; CCS 203; 1-Hydroxybutane; Methylolpropane; NA 1120; NBA; Propylcarbinol; Propylmethanol; RCRA waste number U031; UN 1120.

CAS Registry Number: 71-36-3
DOT: 1120
DOT label: Flammable liquid
Molecular formula: $C_4H_{10}O$
Formula weight: 74.12
RTECS: EO1400000
Merck reference: 10, 1513

Physical state, color, and odor:
Colorless liquid with a characteristic odor similar to fusel oil. Odor threshold concentration is 0.83 ppm (Amoore and Hautala, 1983). The low odor and high odor threshold concentrations were 0.36 and 150.0 mg/m^3, respectively (Ruth, 1986).

Melting point (°C):
-89.5 (Weast, 1986)

Boiling point (°C):
117.2 (Weast, 1986)

Density (g/cm^3):
0.8098 at 20/4 °C (Weast, 1986)
0.8095 at 20.00/4 °C, 0.8057 at 25.00/4 °C (Shan and Asfour, 1999)
0.80240 at 30.00/4 °C (Nikam et al., 2000)

Diffusivity in water (x 10^{-6} cm^2/sec, 25 °C):
9.7 (quoted, Hayduk and Laudie, 1974)
9.6 (Hao and Leaist, 1996)

Dissociation constant, pK$_a$:
20.89 (Riddick et al., 1986)

Flash point (°C):
28.9 (NIOSH, 1997)
36-38 (Windholz et al., 1983)

Lower explosive limit (%):
1.4 (NIOSH, 1997)

Upper explosive limit (%):
11.2 (NIOSH, 1997)

Heat of fusion (kcal/mol):
2.240 (Riddick et al., 1986)

Henry's law constant (x 10^{-6} atm·m^3/mol at 25 °C):
7.90 (Snider and Dawson, 1985)
8.81 (Buttery et al., 1969; Amoore and Buttery, 1978)
8.33 (Butler et al., 1935)
7.14 (Burnett, 1963)
7.11 (Chaintreau et al., 1995)

Interfacial tension with water (dyn/cm at 25 °C):
1.8 (Donahue and Bartell, 1952)
1.9 (Murphy et al., 1957)

Ionization potential (eV):
10.04 (Franklin et al., 1969)

Soil sorption coefficient, log K_{oc}:
0.50 (Gerstl and Helling, 1987)

Octanol/water partition coefficient, log K_{ow}:
0.88 (Leo et al., 1971)
0.83 (Collander, 1951)
0.785 (Tewari et al., 1982; Wasik et al., 1981)

Solubility in organics:
Miscible with alcohols (e.g., methanol, ethanol), ether, and many other organic solvents (Windholz et al., 1983)

Solubility in water:
77,085 mg/L at 20 °C (Mackay and Yeun, 1983)
74,700 mg/L at 25 °C in Lake Superior water having a hardness and alkalinity of 45.5 and 42.2 mg/L as $CaCO_3$, respectively (Veith et al., 1983)
74.5 g/L at 25 °C (Stockhardt and Hull, 1931)
63.3 g/L at 25.0 °C (Tewari et al., 1982; Wasik et al., 1981)
6.4 wt % at 20 °C (Palit, 1947)
74 g/kg at 25 °C (De Santis et al., 1976)
In wt %: 7.497 at 22.60 °C, 7.407 at 23.70 °C, 7.318 at 24.85 °C, 7.31 at 25.00 °C, 7.202 at 26.4 °C, 7.090 at 29.18 °C, 7.016 at 29.18 °C (Butler et al., 1933)
77 g/L (Price et al., 1974)
6,700 mg/kg at 37.7 °C (McCants et al., 1953)
73 g/kg at 25 °C (Donahue and Bartell, 1952)
73 g/kg at 26.7 °C (Skrzec and Murphy, 1954)
70 g/kg at 25 °C (Petriris and Geankopolis, 1959)
91 mL/L at 25 °C (Booth and Everson, 1948)
0.9919 M at 18 °C (Fühner, 1924)
In wt %: 10.33 at 0 °C, 8.98 at 9.6 °C, 8.03 at 20.0 °C, 7.07 at 30.8 °C, 6.77 at 40.1 °C, 6.54 at 50.0 °C, 6.35 at 60.1 °C, 6.73 at 70.2 °C, 7.04 at 80.1 °C, 7.26 at 90.6 °C (Stephenson and Stuart, 1986)
7.2 wt % at 37.8 °C (Jones and McCants, 1954)
74.1 g/kg at 25.0 °C (Hansen et al., 1949)
In g/kg: 95.5 at 5.0 °C, 89.1 at 10.0 °C, 82.1 at 15.0 °C, 78.1 at 20.0 °C, 73.5 at 25.0 °C, 70.8 at 30.0 °C, 68.3 at 35.0 °C, 66.0 at 40.0 °C, 64.6 at 45.0 °C, 65.2 at 50.0 °C, 67.3 at 55.0 °C, 68.9 at 80.0 °C, 87.4 at 97.9 °C, 127.3 at 114.5 °C, 134.6 at 116.9 °C, 197.3 at 123.3

°C, 272.6 at 124.83 °C, 328.2 at 125.1 °C, 304.4 at 125.15 °C (Hill and Malisoff, 1926)
In g/kg: 127.2 at -18.01 °C, 97.9 at -3.11 °C, 60.3 at 40.0 °C, 60.3 at 65.0 °C, 64.7 at 81.0 °C,
    97.9 at 107.72 °C, 127.2 at 117.4 °C, 151.5 at 120.30 °C, 175.1 at 122.45 °C (Jones, 1929)
65 and 61 g/kg at 26 and 50 °C, respectively (Othmer et al., 1942)
74 g/L at 25 °C (Amoore and Buttery, 1978)

Vapor density:
3.03 g/L at 25 °C, 2.56 (air = 1)

Vapor pressure (mmHg):
6 at 20 °C (NIOSH, 1997)
4.4 at 20 °C, 6.5 at 25 °C, 10 at 30 °C (Verschueren, 1983)

Environmental fate:
Biological. 1-Butanol degraded rapidly, presumably by microbes, in New Mexico soils releasing carbon dioxide (Fairbanks et al., 1985). Bridié et al. (1979) reported BOD and COD values 1.71 and 2.46 g/g using filtered effluent from a biological sanitary waste treatment plant. These values were determined using a standard dilution method at 20 °C for a period of 5 d. Heukelekian and Rand (1955) reported a similar 5-d BOD value of 1.66 g/g which is 64.0% of the ThOD value of 2.59 g/g. In activated sludge inoculum, following a 20-d adaptation period, 98.8% COD removal was achieved. The average rate of biodegradation was 84.0 mg COD/g·h (Pitter, 1976).

Photolytic. An aqueous solution containing chlorine and irradiated with UV light (λ = 350 nm) converted 1-butanol into numerous chlorinated compounds which were not identified (Oliver and Carey, 1977).

Reported rate constants for the reaction of 1-butanol and OH radicals in the atmosphere: 6.8 x 10^{-10} cm$_3$/molecule·sec at 292 K (Campbell et al., 1976), 8.31 x 10^{-12} cm^3/molecule·sec (Wallington and Kurylo, 1987), Reported rate constants for the reaction of methanol and OH radicals in the atmosphere: 8.3 x 10^{-12} cm^3/molecule·sec at 298 K (Atkinson, 1990); with OH radicals in aqueous solution: 2.2 x 10^9 L/mol·sec (OH concentration 10^{-17} M) (Anbar and Neta, 1967). Based on an atmospheric OH concentration of 1.0 x 10^6 molecule/cm^3, the reported half-life of 1-butanol is 0.96 d (Grosjean, 1997).

Chemical/Physical. Complete combustion in air yields carbon dioxide and water vapor. Burns with a strongly luminous flame (Windholz et al., 1983).

1-Butanol will not hydrolyze because it has no hydrolyzable functional group (Kollig, 1993).

At an influent concentration of 1,000 mg/L, treatment with granular activated carbon resulted in an effluent concentration of 466 mg/L. The adsorbability of the carbon used was 107 mg/g carbon (Guisti et al., 1974).

Exposure limits: NIOSH REL: ceiling 50 ppm (150 mg/m^3), IDLH 1,400 ppm; OSHA PEL: TWA 100 ppm (300 mg/m^3); ACGIH TLV: ceiling 50 ppm (adopted), ceiling 25 ppm (intended change for 1999).

Symptoms of exposure: Inhalation may cause irritation to eyes, nose, and throat (coughing). Chronic exposure to high concentrations may cause photophobia, blurred vision, and lacrimation (Patnaik, 1992; Windholz et al., 1983). An irritation concentration of 75.00 mg/m^3 in air was reported by Ruth (1986).

Toxicity:
EC$_{50}$ (48-h) and EC$_{50}$ (24-h) values for *Spirostomum ambiguum* were 875 and 823 mg/L, respectively (Nałecz-Jawecki and Sawicki, 1999).

LC$_{50}$ (96-h) for fathead minnows 1,730 mg/L (Veith et al., 1983).

LC$_{50}$ (48-h) and LC$_{50}$ (24-h) values for *Spirostomum ambiguum* were 1,097 and 3,365 mg/L, respectively (Nałecz-Jawecki and Sawicki, 1999).

LC$_0$ (24-h) and LC$_{100}$ (24-h) for creek chub in Detroit river water were 1,000 and 1,400 mg/L, respectively (Gillette et al., 1952).

LC$_{50}$ static bioassay values for fathead minnows in Lake Superior water maintained at 18-22 °C after 1, 24, 48, and 72 h were identical at a concentration of 1,950 mg/L and after 96 h, the LC$_{50}$ was 1,910 mg/L (Mattson et al., 1976).

Acute oral LD$_{50}$ for wild birds 2,500 mg/kg, mice 5,200 mg/kg, rats 790 mg/kg, rabbits 384 mg/kg (quoted, RTECS, 1985).

Source: Butanol naturally occurs in white mulberries and papaya fruit (Duke, 1992).

Uses: Preparation of butyl esters, glycol ethers, and di-*n*-butyl phthalate; solvent for resins and coatings; hydraulic fluid; ingredient in perfumes and flavors; gasoline additive.

2-BUTANONE

Synonyms: Butanone; Ethyl methyl ketone; Meetco; MEK; Methyl acetone; Methyl ethyl ketone; RCRA waste number U159; UN 1193; UN 1232.

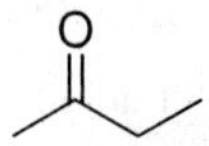

CAS Registry Number: 78-93-3
DOT: 1193
DOT label: Flammable liquid
Molecular formula: C_4H_8O
Formula weight: 72.11
RTECS: EL6475000
Merck reference: 10, 5945

Physical state, color, and odor:
Clear, colorless liquid with a sweet, mint-like odor. Odor threshold concentration is 2.0 ppm (Keith and Walters, 1992).

Melting point (°C):
-86.9 (Dean, 1973)

Boiling point (°C):
79.6 (Weast, 1986)

Density (g/cm³):
0.8054 at 20/4 °C (Weast, 1986)
0.7997 at 25 °C (Riddick et al., 1986)

Diffusivity in water (x 10⁻⁵ cm²/sec):
0.94 at 20 °C using method of Hayduk and Laudie (1974)

Dissociation constant, pKₐ:
14.7 (Riddick et al., 1986)

Flash point (°C):
-9 (NFPA, 1984)

Lower explosive limit (%):
1.4 at 93 °C (NIOSH, 1997)

Upper explosive limit (%):
11.4 at 93 °C (NIOSH, 1997)

Heat of fusion (kcal/mol):
2.017 (Riddick et al., 1986)

Henry's law constant (x 10⁻⁵ atm·m³/mol):
4.65 at 25 °C (Buttery et al., 1969)
5.56 at 25 °C (Vitenberg et al., 1975)
1.05 at 25 °C (Snider and Dawson, 1985)

10.0 at 30 °C (Friant and Suffet, 1979)
5.15 at 25 °C (Chaintreau et al., 1995)
5.26 at 25 °C (Rohrschneider, 1973)
7.0 at 25 °C (Hawthorne et al., 1985)
1.98 at 25 °C (Zhou and Mopper, 1990)
28, 39, 19, 13, and 11 at 10, 15, 20, 25, and 30 °C, respectively (Ashworth et al., 1988)

Interfacial tension with water (dyn/cm at 25 °C):
1.0 (Murphy et al., 1957; Demond and Lindner, 1993)
3.0 (Lyman et al., 1982)

Ionization potential (eV):
9.54 (NIOSH, 1997)
9.53 (Gibson, 1977)

Soil sorption coefficient, log K_{oc}:
1.47 (Captina silt loam), 1.53 (McLaurin sandy loam) (Walton et al., 1992)

Octanol/water partition coefficient, log K_{ow}:
0.26, 0.50 (Sangster, 1989)
0.29 (Hansch and Anderson, 1967; Leo et al., 1971)
0.69 (Wasik et al., 1981)
0.25 (Collander, 1951)

Solubility in organics:
Miscible with acetone, ethanol, benzene, and ether (U.S. EPA, 1985)

Solubility in water:
353 g/L at 10 °C, 190 g/L at 90 °C (quoted, Verschueren, 1983)
24.00 wt % at 20 °C (Palit, 1947; Riddick et al., 1986)
In wt %: 27.33 at 20 °C, 25.57 at 25 °C, 24.07 at 30 °C (Ginnings et al., 1940)
In wt %: 35.7 at 0 °C, 31.0 at 9.6 °C, 27.6 at 19.3 °C, 24.5 at 29.7 °C, 22.0 at 39.6 °C, 20.6 at
    49.7 °C, 18.0 at 60.6 °C, 18.2 at 70.2 °C (Stephenson, 1992)
268 g/L (Price et al., 1974)
1.89 mol/L at 25.0 °C (Wasik et al., 1981)

Vapor density:
2.94 g/L at 25 °C, 2.49 (air = 1)

Vapor pressure (mmHg):
77.5 at 20 °C (Verschueren, 1983)
71.2 at 20 °C (Standen, 1967)
90.6 at 25 °C (Ambrose et al., 1975)
92.64 at 25.00 °C (Hussam and Carr, 1985)

Environmental fate:
Biological. Following a lag time of approximately 5 h, 2-butanone degraded in activated sludge (30 mg/L) at a rate constant ranging from 0.021 to 0.025/h (Urano and Kato, 1986).
Bridié et al. (1979) reported BOD and COD values 2.03 and 2.31 g/g using filtered effluent from a biological sanitary waste treatment plant. These values were determined using a standard dilution method at 20 °C for a period of 5 d. The ThOD for 2-butanone is 2.44 g/g.

Photolytic. Synthetic air containing gaseous nitrous acid and exposed to artificial sunlight (λ = 300-450 nm) photooxidized 2-butanone into peroxyacetyl nitrate and methyl nitrate (Cox et al., 1980). The OH-initiated photooxidation of 2-butanone in a smog chamber produced peroxyacetyl nitrate and acetaldehyde (Cox et al., 1981). Reported rate constants for the reaction of acetone with OH radicals in the atmosphere and in water are 1.15×10^{-13} and 1.50×10^{-13} cm^3/molecule·sec, respectively (Wallington and Kurylo, 1987; Wallington et al., 1988a). The rate constant for the reaction of 2-butanone and OH radicals in the atmosphere at 300 K is 2.0×10^{12} cm^3/mol·sec (Hendry and Kenley, 1979). Cox et al. (1981) reported a photooxidation half-life of 2.3 d for the reaction of 2-butanone and OH radicals in the atmosphere.

Chemical/Physical. 2-Butanone will not hydrolyze because it has no hydrolyzable functional group (Kollig, 1993).

At an influent concentration of 1,000 mg/L, treatment with granular activated carbon resulted in an effluent concentration of 532 mg/L. The adsorbability of the carbon used was 94 mg/g carbon (Guisti et al., 1974).

Exposure limits: NIOSH REL: TWA 200 ppm (590 mg/m^3), STEL 300 ppm (885 mg/m^3), IDLH 3,000 ppm; OSHA PEL: TWA 200 ppm; ACGIH TLV: TWA 200 ppm, STEL 300 ppm (adopted).

Symptoms of exposure: Inhalation may cause irritation of eyes and nose and headache. Narcotic at high concentrations (Patnaik, 1992). An irritation concentration of 590.00 mg/m^3 in air was reported by Ruth (1986).

Toxicity:

LC$_{50}$ (96-h) for fathead minnows 3,200 mg/L (Veith et al., 1983), *Cyprinodon variegatus* >400 ppm using natural seawater (Heitmuller et al., 1981).

LC$_{50}$ (72-h) for *Cyprinodon variegatus* >400 ppm (Heitmuller et al., 1981).

LC$_{50}$ (48 and 24-h) for *Daphnia magna* >520 mg/L (LeBlanc, 1980), *Cyprinodon variegatus* >400 ppm (Heitmuller et al., 1981).

Acute oral LD$_{50}$ for rats 2,737 mg/kg, mouse 4,050 mg/kg; LD$_{50}$ (skin) for rabbit 13 gm/kg (quoted, RTECS, 1985).

Heitmuller et al. (1981) reported a NOEC of 400 ppm.

Drinking water standard: No MCLGs or MCLs have been proposed (U.S. EPA, 1996).

Source: Improper disposal of cleaning fluids, adhesives, paints, and lacquers, and laboratory solvent. Leaches from PVC cement used to join tubing (Wang and Bricker, 1979). Also present in cigarette smoke (500 ppm) and exhaust from gasoline-powered engines (<0.1-2.6 ppm) (Verschueren, 1983).

Uses: Solvent in nitrocellulose coatings, vinyl films and "Glyptal" resins; paint removers; cements and adhesives; organic synthesis; manufacture of smokeless powders and colorless synthetic resins; preparation of 2-butanol, butane, and amines; cleaning fluids; printing; catalyst carrier; acrylic coatings.

1-BUTENE

Synonyms: α-Butylene; Ethylethylene.

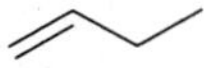

CAS Registry Number: 106-98-9
DOT: 1012
Molecular formula: C_4H_8
Formula weight: 56.11
Merck reference: 10, 1488

Physical state, color, and odor:
Colorless gas with a weak, aromatic odor. Faint odor recognized at a concentration range of 21.5-25.4 ppb (Verschueren, 1983). The low odor and high odor threshold concentration were identical at a concentration of 54.96 mg/m³ (Ruth, 1986).

Melting point (°C):
-185.3 (Weast, 1986)

Boiling point (°C):
-6.3 (Weast, 1986)

Density (g/cm³):
0.5951 at 20/4 °C (Weast, 1986)

Diffusivity in water (x 10^{-5} cm²/sec):
0.91 at 20 °C using method of Hayduk and Laudie (1974)

Dissociation constant, pK_a:
>14 (Schwarzenbach et al., 1993)

Flash point (°C):
-79 (Hawley, 1981)

Lower explosive limit (%):
1.6 (NFPA, 1984)

Upper explosive limit (%):
10.0 (NFPA, 1984)

Heat of fusion (kcal/mol):
0.92 (Dean, 1987)

Henry's law constant (atm·m³/mol):
0.25 at 25 °C (Hine and Mookerjee, 1975)

Ionization potential (eV):
9.6 (Franklin et al., 1969)
9.26 (Collin and Lossing, 1959)

Soil sorption coefficient, log K_{oc}:
Unavailable because experimental methods for estimation of this parameter for aliphatic hydrocarbons are lacking in the documented literature

Octanol/water partition coefficient, log K_{ow}:
2.40 (Hansch and Leo, 1979)

Solubility in organics:
Soluble in alcohol, benzene, and ether (Weast, 1986)

Solubility in water:
222 ppm at 25 °C (McAuliffe, 1966)

Vapor density:
2.29 g/L at 25 °C, 1.94 (air = 1)

Vapor pressure (mmHg at 25 °C):
2,708 at 25 °C (estimated from Antoine equation, Stull, 1947)
2,230 at 25 °C (Wilhoit and Zwolinski, 1971)

Environmental fate:
   Biological. Biooxidation of 1-butene may occur yielding 3-buten-1-ol, which may further oxidize to give 3-butenoic acid (Dugan, 1972). Washed cell suspensions of bacteria belonging to the genera *Mycobacterium*, *Nocardia*, *Xanthobacter*, and *Pseudomonas* and growing on selected alkenes metabolized 1-butene to 1,2-epoxybutane (Van Ginkel et al., 1987).
   Photolytic. Products identified from the photoirradiation of 1-butene with nitrogen dioxide in air are epoxybutane, 2-butanone, propanal, ethanol, ethyl nitrate, carbon monoxide, carbon dioxide, methanol, and nitric acid (Takeuchi et al., 1983).
   The following rate constants were reported for the reaction of 1-butene and OH radicals in the atmosphere: 1.0 x 10^{-17} cm^3/molecule·sec (Bufalini and Altshuller, 1965); 2.70 x 10^{-11} cm^3/molecule·sec (Atkinson et al., 1979); 3.14 x 10^{-11} cm^3/molecule·sec (Atkinson, 1990; Sabljić and Güsten, 1990). Reported photooxidation reaction rate constants for the reaction of 1-butene and ozone are 1.23 x 10^{-17}, 1.0 x 10^{-17}, 1.03 x 10^{-17} cm^3/molecule·sec (Adeniji et al., 1981). Based on the reaction of 1-butene and OH radicals gas phase, the atmospheric lifetime was estimated to be 5.5 h in summer sunlight.
   Chemical/Physical. Complete combustion in air yields carbon dioxide and water.

Symptoms of exposure: Narcotic at high concentrations (Patnaik, 1992)

Source: In exhaust of gasoline-powered engines (1.8 vol % of total exhaust hydrocarbons) and component of gasoline (Verschueren, 1983)

Uses: Polybutylenes; polymer and alkylate gasoline; intermediate for butyl and pentyl aldehydes, alcohols, maleic acid, and other organic compounds.

2-BUTOXYETHANOL

Synonyms: Butyl cellosolve; Butyl glycol; Butyl glycol ether; Butyl oxitol; Dowanol EB; Ektasolve; Ethylene glycol monobutyl ether; Ethylene glycol mono-*n*-butyl ether; Glycol monobutyl ether; Jeffersol EB.

CAS Registry Number: 111-76-2
DOT: 2369
DOT label: Poison and combustible liquid
Molecular formula: $C_6H_{14}O_2$
Formula weight: 118.18
RTECS: KJ8575000
Merck reference: 10, 1532

Physical state, color, and odor:
Clear, colorless, oily liquid with a mild, ether-like odor. Odor threshold concentration is 0.10 ppm (Amoore and Hautala, 1983).

Melting point (°C):
-77 (NIOSH, 1997)

Boiling point (°C):
171 (Weast, 1986)
171 at 743 mmHg (Aldrich, 1988)

Density (g/cm³):
0.9015 at 20/4 °C (Weast, 1986)

Diffusivity in water (x 10⁻⁵ cm²/sec):
0.75 at 20 °C using method of Hayduk and Laudie (1974)

Flash point (°C):
62 (NIOSH, 1997)

Lower explosive limit (%):
1.1 at 93 °C (NIOSH, 1997)

Upper explosive limit (%):
12.7 at 93 °C (NIOSH, 1997)

Henry's law constant (x 10⁻⁶ atm·m³/mol):
2.36 (approximate - calculated from water solubility and vapor pressure)

Ionization potential (eV):
10.00 (NIOSH, 1997)

Soil sorption coefficient, log K_{oc}:
Unavailable because experimental methods for estimation of this parameter for cellosolves are lacking in the documented literature

Octanol/water partition coefficient, log K_{ow}:
0.45 using method of Hansch et al. (1968)

Solubility in organics:
Soluble in alcohol, ether (Weast, 1986), and mineral oil (Windholz et al., 1983)

Solubility in water:
Miscible (Price et al., 1974)

Vapor density:
4.83 g/L at 25 °C, 4.08 (air = 1)

Vapor pressure (mmHg):
0.8 at 20 °C (NIOSH, 1997)

Environmental fate:
   Biological. Bridié et al. (1979) reported BOD and COD values 0.71 and 2.20 g/g using filtered effluent from a biological sanitary waste treatment plant. These values were determined using a standard dilution method at 20 °C for a period of 5 d. When a sewage seed was used in a separate screening test, a BOD value of 0.20 g/g was obtained. The ThOD for 2-butoxy-ethanol is 2.31 g/g.
   Chemical/Physical.   At an influent concentration of 1,000 mg/L, treatment with granular activated carbon resulted in an effluent concentration of 441 mg/L. The adsorbability of the carbon used was 112 mg/g carbon (Guisti et al., 1974).

Exposure limits: NIOSH REL: TWA 5 ppm (24 g/m³), IDLH 700 ppm; OSHA PEL: TWA 50 ppm (240 mg/m³); ACGIH TLV: TWA 20 ppm (adopted).

Symptoms of exposure: An 8-h exposure to 200 ppm may cause nausea, vomiting, and headache (Patnaik, 1992).

Toxicity:
   LC_{50} (96-h) for brown shrimp 775 mg/L (Blackmann, 1974).
   LC_{50} (48-h) for brown shrimp 800 mg/L (Blackmann, 1974).
   LC_{50} (24-h static bioassay) for bluegill sunfish at 23 °C is 983 mg/L (quoted, Verschueren, 1983).
   LC_{50} (4-h inhalation) values for Fischer-344 rats were 486 and 450 ppm for males and females, respectively (Dodd et al., 1983).
   LC_{50} (1-h inhalation) for guinea pigs >633 mg/L (Gingell et al., 1998).
   Acute oral LD_{50} for rats is 450 mg/kg (Patnaik, 1992), guinea pigs 1,400 mg/kg (Gingell et al., 1998).

Uses: Dry cleaning; solvent for nitrocellulose, cellulose acetate, resins, oil, grease, albumin; perfume fixative; coating compositions for paper, cloth, leather; lacquers.

n-BUTYL ACETATE

Synonyms: Acetic acid butyl ester; 1-Butanol acetate; Butyl acetate; 1-Butyl acetate; Butyl ethanoate; *n*-Butyl ethanoate; UN 1123.

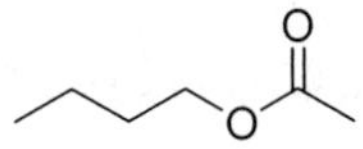

CAS Registry Number: 123-86-4
DOT: 1123
DOT label: Flammable liquid
Molecular formula: $C_6H_{12}O_2$
Formula weight: 116.16
RTECS: AF7350000
Merck reference: 10, 1508

Physical state, color, and odor:
Colorless liquid with a fruity odor. Odor threshold concentrations ranged from 7 to 20 ppm (Keith and Walters, 1992).

Melting point (°C):
-77.9 (Weast, 1986)

Boiling point (°C):
126.5 (Weast, 1986)
124-126 (Aldrich, 1988)

Density (g/cm³):
0.88145 at 20.00/4 °C (Lee and Tu, 1999)

Diffusivity in water (x 10^{-5} cm²/sec):
0.75 at 20 °C using method of Hayduk and Laudie (1974)

Flash point (°C):
22.2 (NIOSH, 1997)
36.6 (open cup, Hawley, 1981)

Lower explosive limit (%):
1.7 (NIOSH, 1997)

Upper explosive limit (%):
7.6 (NIOSH, 1997)

Henry's law constant (x 10^{-4} atm·m³/mol at 25 °C):
2.81 (Kieckbusch and King, 1979)
24.8, 25.2, 18.3, 8.56, and 2.68 at pH values of 2.93, 5.69, 6.22, 6.58, and 8.80, respectively (Hakuta et al., 1977)

Interfacial tension with water (dyn/cm at 25 °C):
14.5 (Donahue and Bartell, 1952)

Ionization potential (eV):
9.56 ± 0.03 (Franklin et al., 1969)

Soil sorption coefficient, log K_{oc}:
Unavailable because experimental methods for estimation of this parameter for aliphatic esters are lacking in the documented literature

Octanol/water partition coefficient, log K_{ow}:
1.82 (Tewari et al., 1982; Wasik et al., 1981)

Solubility in organics:
Miscible with alcohol and ether (Windholz et al., 1983)

Solubility in water:
57.7 mM at 25.0 °C (Tewari et al., 1982; Wasik et al., 1981)
57.4, 58.5, 61.2, and 61.9 mM at 30, 40, 50, and 60 °C, respectively (Gandhi et al., 1998)
6.8 g/L (Price et al., 1974)
0.84 wt % at 25 °C (Lo et al., 1986)
In wt %: 0.96 at 0 °C, 0.76 at 9.1 °C, 0.64 at 19.7 °C, 0.52 at 30.3 °C, 0.50 at 39.6 °C, 0.50 at
    50.0 °C, 0.50 at 60.2 °C, 0.47 at 70.2 °C, 0.48 at 80.1 °C, 0.48 at 90.5 °C (Stephenson and
    Stuart, 1986)

Vapor density:
4.75 g/L at 25 °C, 4.01 (air = 1)

Vapor pressure (mmHg):
10 at 20 °C (NIOSH, 1997)
11.0 at 25 °C (Abraham, 1984)

Environmental fate:
Biological. Heukelekian and Rand (1955) reported a 5-d BOD value of 0.52 g/g which is 23.5% of the ThOD value of 2.21 g/g.

Photolytic. *n*-Butyl acetate reacts with OH radicals in the atmosphere at a rate constant of 4.15×10^{-12} cm^3/molecule·sec at 296 K (Wallington et al., 1988b).

Chemical/Physical. Hydrolyzes in water forming 1-butanol and acetic acid. Estimated hydrolysis half-lives at 20 °C: 11.4 d at pH 9.0, 114 d at pH 8.0, and 3.1 yr at pH 7.0 (Mabey and Mill, 1978).

At an influent concentration of 1,000 mg/L, treatment with granular activated carbon resulted in an effluent concentration of 154 mg/L. The adsorbability of the carbon was 169 mg/g carbon (Guisti et al., 1974).

Exposure limits: NIOSH REL: TWA 150 (710 mg/m^3), STEL 200 ppm (950 mg/m^3), IDLH 1,700 ppm; OSHA PEL: TWA 150 ppm; ACGIH TLV: TWA 150 ppm, STEL 200 ppm (adopted).

Symptoms of exposure: Exposure to 200-400 ppm may cause moderate eye and throat irritation and headache. May be narcotic at higher concentrations (Verschueren, 1983; Patnaik, 1992). An irritation concentration of 473.33 mg/m^3 in air was reported by Ruth (1986).

Toxicity:
LC$_{50}$ (96-h static bioassay) for bluegill sunfish 100 ppm, *Menidia beryllina* 185 ppm (quoted,

Verschueren, 1983).

LC_{50} (inhalation) for mice 6 gm/m^3/2-h, rats 2,000 ppm/4-h (quoted, RTECS, 1985).

Acute oral LD_{50} for rats 14 mg/kg, mice 7,060 mg/kg, rabbits 7,400 mg/kg (quoted, RTECS, 1985).

Uses: Manufacture of artificial leathers, plastics, safety glass, photographic films, lacquers; as a solvent in the production of perfumes, natural gums, and synthetic resins; solvent for nitrocellulose lacquers; dehydrating agent.

sec-BUTYL ACETATE

Synonyms: Acetic acid 2-butoxy ester; Acetic acid *sec*-butyl ester; **Acetic acid 1-methyl-propyl ester**; 2-Butanol acetate; 2-Butyl acetate; *sec*-Butyl alcohol acetate; 2-Butyl acetate; 1-Methylpropyl acetate; UN 1123.

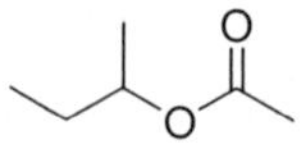

CAS Registry Number: 105-46-4
DOT: 1123
DOT label: Flammable liquid
Molecular formula: $C_6H_{12}O_2$
Formula weight: 116.16
RTECS: AF7380000
Merck reference: 10, 1509

Physical state, color, and odor:
Colorless liquid with a pleasant odor

Melting point (°C):
-37.8 (NIOSH, 1997)

Boiling point (°C):
112 (Weast, 1986)

Density (g/cm³):
0.8758 at 20/4 °C (Weast, 1986)
0.865 at 25/4 °C (Windholz et al., 1983)

Diffusivity in water (x 10^{-5} cm²/sec):
0.74 at 20 °C using method of Hayduk and Laudie (1974)

Flash point (°C):
16.7 (NIOSH, 1997)
31 (open cup, Windholz et al., 1983)

Lower explosive limit (%):
1.7 (NIOSH, 1997)

Upper explosive limit (%):
9.8 (NIOSH, 1997)

Henry's law constant (x 10^{-4} atm·m³/mol):
1.91 at 20 °C (approximate - calculated from water solubility and vapor pressure)

Interfacial tension with water (dyn/cm at 17 °C):
58 (estimated, CHRIS, 1984)

Ionization potential (eV):
9.91 ± 0.03 (Franklin et al., 1969)

Soil sorption coefficient, log K_{oc}:
Unavailable because experimental methods for estimation of this parameter for aliphatic esters are lacking in the documented literature

Octanol/water partition coefficient, log K_{ow}:
1.66 using method of Hansch et al. (1968)

Solubility in organics:
Soluble in acetone, alcohol, and ether (Weast, 1986)

Solubility in water (wt %):
1.330 at 0 °C, 0.879 at 9.6 °C, 0.869 at 19.5 °C, 0.753 at 29.7 °C, 0.663 at 39.9 °C, 0.629 at 50.0 °C, 0.613 at 60.1 °C, 0.605 at 70.5 °C, 0.622 at 80.2 °C, 0.604 at 90.5 °C (Stephenson and Stuart, 1986)

Vapor density:
4.75 g/L at 25 °C, 4.01 (air = 1)

Vapor pressure (mmHg):
10 at 20 °C (NIOSH, 1997)

Environmental fate:
   Photolytic. The rate constant for the reaction of *sec*-butyl acetate and OH radicals in the atmosphere at 300 K is 3.4 x 10^{12} cm^3/mol·sec (Hendry and Kenley, 1979).
   Chemical/Physical. Slowly hydrolyzes in water forming *sec*-butyl alcohol and acetic acid.

Exposure limits: NIOSH REL: TWA 200 ppm (950 mg/m^3), IDLH 1,700 ppm; OSHA PEL: TWA 200 ppm; ACGIH TLV: TWA 200 ppm (adopted).

Symptoms of exposure: Exposure to vapors may cause irritation to eyes and respiratory passages (Patnaik, 1992).

Uses: Solvent for nitrocellulose lacquers; nail enamels, thinners; leather finishers.

tert-BUTYL ACETATE

Synonyms: Acetic acid *tert*-butyl ester; **Acetic acid 1,1-dimethylethyl ester**; *t*-Butyl acetate; *t*-Butyl ethanoate; Texaco lead appreciator; TLA; UN 1123.

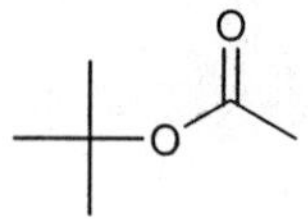

CAS Registry Number: 540-88-5
DOT: 1123
DOT label: Flammable liquid
Molecular formula: $C_6H_{12}O_2$
Formula weight: 116.16
RTECS: AF7400000
Merck reference: 10, 1510

Physical state, color, and odor:
Colorless liquid with a fruity odor. Odor threshold concentration in water is 4 ppm (CHRIS, 1994).

Boiling point (°C):
97-98 (Weast, 1986)

Density (g/cm³):
0.8665 at 20/4 °C (Weast, 1986)
0.8593 at 25/4 °C (Windholz et al., 1983)

Diffusivity in water (x 10⁻⁵ cm²/sec):
0.74 at 20 °C using method of Hayduk and Laudie (1974)

Flash point (°C):
15 (Dean, 1987)
22.2 (NIOSH, 1997)

Lower explosive limit (%):
1.5 (NIOSH, 1997)

Soil sorption coefficient, log K_{oc}:
Unavailable because experimental methods for estimation of this parameter for aliphatic esters are lacking in the documented literature

Solubility in organics:
Miscible with alcohol and ether (Windholz et al., 1983)

Solubility in water (wt %):
1.170 at 0 °C, 1.000 at 9.2 °C, 0.803 at 19.2 °C, 0.703 at 29.6 °C, 0.620 at 40.0 °C, 0.573 at 50.0 °C, 0.526 at 60.5 °C, 0.538 at 70.5 °C, 0.499 at 80.5 °C (Stephenson and Stuart, 1986)

Vapor density:
4.75 g/L at 25 °C, 4.01 (air = 1)

Environmental fate:

Chemical/Physical. Hydrolyzes in water to *tert*-butyl alcohol and acetic acid. The estimated hydrolysis half-life at 25 °C and pH 7 is 140 yr (Mabey and Mill, 1978).

Exposure limits: NIOSH REL: TWA 200 ppm (950 mg/m^3), IDLH 1,500 ppm; OSHA PEL: TWA 200 ppm; ACGIH TLV: TWA 200 ppm (adopted).

Symptoms of exposure: Exposure to vapors may cause irritation to eyes and respiratory passages; narcotic at high concentrations (Patnaik, 1992).

Uses: Gasoline additive; solvent.

sec-BUTYL ALCOHOL

Synonyms: Butanol-2; Butan-2-ol; **2-Butanol**; *sec*-Butanol; 2-Butyl alcohol; Butylene hydrate; CCS 301; Ethylmethyl carbinol; 2-Hydroxybutane; Methylethylcarbinol; SBA.

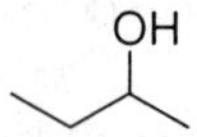

$$OH$$

CAS Registry Number: 78-92-2
DOT: 1120
DOT label: Flammable liquid
Molecular formula: $C_4H_{10}O$
Formula weight: 74.12
RTECS: EO1750000
Merck reference: 10, 1514

Physical state, color, and odor:
Colorless liquid with a pleasant odor. Odor threshold concentration is 2.6 ppm (Amoore and Hautala, 1983).

Melting point (°C):
-114.7 (Windholz et al., 1983)

Boiling point (°C):
99.5 (Weast, 1986)

Density (g/cm³):
0.8063 at 20/4 °C (Weast, 1986)
0.8024 at 25.00/4 °C (Aralaguppi et al., 1999a)

Diffusivity in water (x 10⁻⁶ cm²/sec):
9.4 at 25 °C (Hao and Leaist, 1996)

Dissociation constant, pK_a:
>14 (Schwarzenbach et al., 1993)

Flash point (°C):
23.9 (NIOSH, 1997)
31 (open cup, Windholz et al., 1983)

Lower explosive limit (%):
1.7 at 100 °C (NFPA, 1984)

Upper explosive limit (%):
9.8 at 100 °C (NFPA, 1984)

Henry's law constant (x 10⁻⁵ atm·m³/mol at 25 °C):
1.03 (Butler, 1935)
0.79 (Snider and Dawson, 1985)

Interfacial tension with water (dyn/cm at 25 °C):
2.1 (quoted, Freitas et al., 1997)

Ionization potential (eV):
10.10 (NIOSH, 1997)

Soil sorption coefficient, log K_{oc}:
Unavailable because experimental methods for estimation of this parameter for aliphatic alcohols are lacking in the documented literature

Octanol/water partition coefficient, log K_{ow}:
0.61 (Hansch and Anderson, 1967)
0.81 (Yonezawa and Urushigawa, 1979)

Solubility in organics:
Miscible with acetone, alcohol, benzene, and ether (Windholz et al., 1983)

Solubility in water (wt %):
20 at 20 °C (Palit, 1947)
22.5 at 25 °C (De Santis et al., 1976)
26.0 at 0 °C, 23.5 at 10.0 °C, 19.6 at 20.0 °C, 17.0 at 29.9 °C, 15.1 at 40.0 °C, 14.0 at 50.0 °C, 13.4 at 60.3 °C, 13.3 at 70.1 °C, 13.6 at 80.1 °C, 14.5 at 90.2 °C (Stephenson and Stuart, 1986)

Vapor density:
3.03 g/L at 25 °C, 2.56 (air = 1)

Vapor pressure (mmHg):
12 at 20 °C, 24 at 30 °C (Verschueren, 1983)

Environmental fate:
Biological. Bridié et al. (1979) reported BOD and COD values 2.15 and 2.49 g/g using filtered effluent from a biological sanitary waste treatment plant. These values were determined using a standard dilution method at 20 °C for a period of 5 d. The ThOD for *sec*-butyl alcohol is 2.59 g/g. In activated sludge inoculum, following a 20-d adaptation period, 98.5% COD removal was achieved. The average rate of biodegradation was 55.0 mg COD/g·h (Pitter, 1976).
Photolytic. The estimated half-life of *sec*-butyl alcohol for the reaction of OH radicals in air ranges from 129 d to 23 yr (Anbar and Neta, 1967).
Chemical/Physical. sec-Butyl alcohol will not hydrolyze in water (Kollig, 1993).

Exposure limits: NIOSH REL: TWA 100 ppm (305 mg/m³), STEL 150 ppm (455 mg/m³), IDLH 2,000 ppm; OSHA PEL: TWA 150 ppm; ACGIH TLV: TWA 100 ppm (adopted).

Symptoms of exposure: Inhalation may cause irritation to eyes. Narcotic at high concentrations. May irritate skin on contact (Patnaik, 1992).

Toxicity:
LC_{50} (24-h) for goldfish 4,300 mg/L (quoted, Verschueren, 1983).
Acute oral LD_{50} for rats 6,480 mg/kg, rabbits 4,893 mg/kg (quoted, Verschueren, 1983).

Uses: Manufacture of flotation agents, esters (perfumes and flavors), dyestuffs, wetting agents; ingredient in industrial cleaners and paint removers; preparation of methyl ethyl ketone; solvent in lacquers; in hydraulic brake fluids; organic synthesis.

tert-BUTYL ALCOHOL

Synonyms: *t*-Butanol; *tert*-Butanol; *t*-Butyl alcohol; *t*-Butyl hydroxide; 1,1-Dimethylethanol; **2-Methyl-2-propanol**; NCI-C55367; TBA; Trimethylcarbinol; Trimethyl methanol; UN 1120.

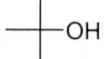

CAS Registry Number: 75-65-0
DOT: 1120
DOT label: Flammable liquid
Molecular formula: $C_4H_{10}O$
Formula weight: 74.12
RTECS: EO1925000
Merck reference: 10, 1515

Physical state, color, and odor:
Colorless liquid or crystals with a camphor-like odor. Odor threshold concentration is 47 ppm (Amoore and Hautala, 1983).

Melting point (°C):
25.5 (Weast, 1986)

Boiling point (°C):
82.3 (Weast, 1986)

Density (g/cm³):
0.78581 at 20/4 °C, 0.78086 at 25/4 °C (Windholz et al., 1983)
0.77502 at 30.00/4 °C (Nikam et al., 2000)

Diffusivity in water (x 10^{-6} cm²/sec):
8.7 at 25 °C (Hao and Leaist, 1996)

Dissociation constant, pK_a:
≈ 19 (Gordon and Ford, 1972)

Flash point (°C):
11.1 (NIOSH, 1997)
4 (Aldrich, 1988)

Lower explosive limit (%):
2.4 (NIOSH, 1997)

Upper explosive limit (%):
8.0 (NIOSH, 1997)

Entropy of fusion (cal/mol·K):
5.44 (Yalkowsky and Valvani, 1980)

Heat of fusion (kcal/mol):
1.587 (Riddick et al., 1986)

Henry's law constant (x 10^{-5} atm·m^3/mol at 25 °C):
1.19 (Butler et al., 1935)
1.44 (Snider and Dawson, 1985)

Ionization potential (eV):
9.70 (NIOSH, 1997)

Soil sorption coefficient, log K_{oc}:
Unavailable because experimental methods for estimation of this parameter for alcohols are lacking in the documented literature. However, its miscibility in water and low K_{ow} suggest its adsorption to soil will be nominal (Lyman et al., 1982).

Octanol/water partition coefficient, log K_{ow}:
0.37 (Hansch and Anderson, 1967)
0.59 (Yonezawa and Urushigawa, 1979)

Solubility in organics:
Miscible with alcohol and ether (Windholz et al., 1983)

Solubility in water:
Miscible (NIOSH, 1997; Palit, 1947). A saturated solution in equilibrium with its own vapor had a concentration of 316.2 g/L at 25 °C (Kamlet et al., 1987).

Vapor pressure (mmHg):
42 at 25 °C, 56 at 30 °C (Verschueren, 1983)

Environmental fate:
  Biological. Bridié et al. (1979) reported BOD and COD values 0.02 and 2.49 g/g using filtered effluent from a biological sanitary waste treatment plant. These values were determined using a standard dilution method at 20 °C for a period of 5 d. The ThOD for *tert*-butyl alcohol is 2.59 g/g. In activated sludge inoculum, 98.5% COD removal was achieved. The average rate of biodegradation was 30.0 mg COD/g·h (Pitter, 1976).
  Photolytic. Wallington (1988c) reported a rate constant of 1.07 x 10^{-12} cm^3/molecule·sec at 298 K. Based on an atmospheric OH concentration of 1.0 x 10^6 molecule/cm^3, the reported half-life of *tert*-butyl alcohol is 8.6 d (Grosjean, 1997).
  Chemical/Physical. May react with strong mineral acids (e.g., hydrochloric) or oxidizers releasing isobutylene (NIOSH, 1997).
  tert-Butyl alcohol will not hydrolyze because it has no hydrolyzable functional group (Kollig, 1993).
  At an influent concentration of 1,000 mg/L, treatment with granular activated carbon resulted in effluent concentration of 705 mg/L. The adsorbability of the carbon used was 59 mg/g carbon (Guisti et al., 1974).

Exposure limits (ppm): NIOSH REL: TWA 100, STEL 150, IDLH 1,600; OSHA PEL: TWA 100; ACGIH TLV: TWA 100 (adopted).

Symptoms of exposure: Ingestion may cause headache, dizziness, dry skin, and narcosis. Inhalation may cause drowsiness and mild irritation to eyes and nose (Patnaik, 1992).

Toxicity:
  LC_{50} (7-d) for *Poecilia reticulata* 3,550 mg/L (Könemann, 1981).

LC_{50} (24-h) for goldfish 4,300 mg/L (Verschueren, 1983).

LC_0 (24-h) and LC_{100} (24-h) for creek chub in Detroit river water were 3,000 and 6,000 mg/L, respectively (Gillette et al., 1952).

Acute oral LD_{50} for rats 3,500 mg/kg, rabbits 3,559 mg/kg (quoted, RTECS, 1985).

Source: Detected in a distilled water-soluble fraction of 94 octane unleaded gasoline at a concentration of 3.72 mg/L (Potter, 1996)

Uses: Denaturant for ethyl alcohol; manufacturing flavors, perfumes (artificial musk), flotation agents; solvent; paint removers; octane booster for unleaded gasoline; dehydrating agent; chemical intermediate.

n-BUTYLAMINE

Synonyms: 1-Aminobutane; Butylamine; **1-Butanamine**; Monobutylamine; Mono-*n*-butyl-amine; Norvalamine; UN 1125.

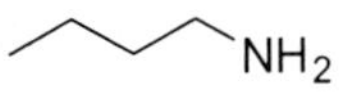

$$\text{CH}_3\text{CH}_2\text{CH}_2\text{CH}_2\text{—NH}_2$$

CAS Registry Number: 109-73-9
DOT: 1125
DOT label: Flammable liquid
Molecular formula: $C_4H_{11}N$
Formula weight: 73.14
RTECS: EO2975000
Merck reference: 10, 1516

Physical state, color, and odor:
Clear, colorless liquid with an ammonia-like odor. Odor threshold concentration is 1.8 ppm (Amoore and Hautala, 1983).

Melting point (°C):
-49.1 (Weast, 1986)

Boiling point (°C):
77.8 (Weast, 1986)

Density (g/cm³):
0.7414 at 20/4 °C (Weast, 1986)
0.7327 at 25/4 °C (Windholz et al., 1983)

Diffusivity in water (x 10^{-5} cm²/sec):
0.89 at 20 °C using method of Hayduk and Laudie (1974)

Dissociation constant, pK$_a$:
9.33 at 20 °C (Gordon and Ford, 1972)
10.685 at 25 °C (Dean, 1987)

Flash point (°C):
-12.2 (NIOSH, 1997)
<-10 (Mitchell et al., 1999)
-1 (open cup, Windholz et al., 1983)

Lower explosive limit (%):
1.7 (NIOSH, 1997)

Upper explosive limit (%):
9.8 (NIOSH, 1997)

Henry's law constant (x 10^{-5} atm·m³/mol at 25 °C):
1.52 (Butler and Ramchandani, 1935)
1.72 (Christie and Crisp, 1967)
2.15 (Amoore and Buttery, 1978)

Ionization potential (eV):
8.71 ± 0.03 (Franklin et al., 1969)

Soil sorption coefficient, log K_{oc}:
1.88 (Meylan et al., 1992)

Octanol/water partition coefficient, log K_{ow}:
0.81 (Leo et al., 1971)
0.68 (Collander, 1951)
0.88 (Sandell, 1962)

Solubility in organics:
Soluble in alcohol and ether (Weast, 1986)

Solubility in water:
Miscible (NIOSH, 1997). A saturated solution in equilibrium with its own vapor had a concentration of 667.0 g/L at 25 °C (Kamlet et al., 1987).

Vapor density:
2.99 g/L at 25 °C, 2.52 (air = 1)

Vapor pressure (mmHg at 20 °C):
82 (NIOSH, 1997)
72 (Verschueren, 1983)

Environmental fate:
  Chemical/Physical. Reacts with mineral acids forming water-soluble salts.
  At an influent concentration of 1,000 mg/L, treatment with granular activated carbon resulted in effluent concentration of 480 mg/L. The adsorbability of the carbon used was 103 mg/g carbon (Guisti et al., 1974).

Exposure limits: NIOSH REL: ceiling 5 ppm (15 mg/m³), IDLH 300 ppm; ACGIH TLV: ceiling 2.5 ppm (adopted).

Symptoms of exposure: Irritates eyes and respiratory tract at concentrations of 5 to 10 ppm. Contact with skin may cause severe burns (Verschueren, 1983; Patnaik, 1992). An irritation concentration of 30.00 mg/m³ in air was reported by Ruth (1986).

Toxicity:
  LC_{50} (96-h static bioassay) for bluegill sunfish 32 ppm, *Menidia beryllina* 24 ppm (quoted, Verschueren, 1983).
  LC_{50} (inhalation) for mice 800 gm/m³/2-h, rats 4,000 ppm/4-h (quoted, RTECS, 1985).
  Acute oral LD_{50} for guinea pigs 430 mg/kg, mice 430 mg/kg, rats 366 mg/kg (quoted, RTECS, 1985).

Uses: Intermediate for dyestuffs, emulsifying agents, pharmaceuticals, rubber chemicals, insecticides, and synthetic tanning agents; preparation of isocyanates for coatings.

n-BUTYLBENZENE

Synonyms: Butylbenzene; 1-Phenylbutane.

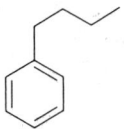

CAS Registry Number: 104-51-8
DOT: 2709
Molecular formula: $C_{10}H_{14}$
Formula weight: 134.22
RTECS: CY9070000
Merck reference: 10, 1522

Physical state and color:
Colorless liquid

Melting point (°C):
-88 (Weast, 1986)

Boiling point (°C):
183.31 (Wilhoit and Zwolinski, 1971)

Density (g/cm³):
0.8601 at 20/4 °C (Weast, 1986)

Diffusivity in water (x 10⁻⁵ cm²/sec):
0.68 at 20 °C using method of Hayduk and Laudie (1974)

Dissociation constant, pK_a:
>14 (Schwarzenbach et al., 1993)

Flash point (°C):
71 (open cup, Windholz et al., 1983)
59 (Aldrich, 1988)

Lower explosive limit (%):
0.8 (Sax and Lewis, 1987)

Upper explosive limit (%):
5.8 (Sax and Lewis, 1987)

Henry's law constant (x 10⁻³ atm·m³/mol):
5.35, 8.17, 11.0, 16.7, and 21.4 at 10, 15, 20, 25, and 30 °C, respectively (Perlinger et al., 1993)

Interfacial tension with water (dyn/cm at 20 °C):
39.6 (Demond and Lindner, 1993)

Ionization potential (eV):
8.69 ± 0.01 (Franklin et al., 1969)

Soil sorption coefficient, log K_{oc}:
3.39 (Schwarzenbach and Westall, 1981)
3.40 (estuarine sediment, Vowles and Mantoura, 1987)

Octanol/water partition coefficient, log K_{ow}:
4.643 (Klein et al., 1988)
4.29 (Brooke et al., 1990; Schantz and Martire, 1987)
4.377 (Brooke et al., 1990; de Bruijn et al., 1989)
4.26 (Camilleri et al., 1988; Hammers et al., 1982; Hansch and Leo, 1979)
4.28 (Tewari et al., 1982; Wasik et al., 1981, 1983)
4.18 (Bruggeman et al., 1982)
4.44 (Hammers et al., 1982)

Solubility in organics:
Miscible with alcohol, benzene, ether (Windholz et al., 1983), and many other organic solvents

Solubility in water:
In mg/kg at 25 °C: 11.8 (distilled water), 7.09 (artificial seawater) (Sutton and Calder, 1975)
73.0 μmol/L at 25.0 °C (Andrews and Keefer, 1950a)
In μmol/L (°C): 96.6 (15.0), 101.8 (20.0), 102.5 (25.0) (Owens et al., 1986)
434 μmol/L at 25.0 °C (Tewari et al., 1982)
15.4 and 17.7 mg/L at 25 °C (Mackay and Shiu, 1977)
64.2 μmol/L in 0.5 M NaCl at 25 °C (Wasik et al., 1984)
103 μmol/L at 25 °C (Wasik et al., 1981, 1983)
50.0 mg/L at 25 °C (Klevens, 1950)
1.77 x 10^{-6} at 25 °C (mole fraction, Li et al., 1993)

Vapor density:
5.49 g/L at 25 °C, 4.63 (air = 1)

Vapor pressure (mmHg):
1.03 at 25 °C (Mackay et al., 1982)

Environmental fate:
    Biological. n-Butylbenzene is subject to cometabolism. The reported oxidation product is phenylacetic acid (Pitter and Chudoba, 1990).
    Chemical/Physical. n-Butylbenzene will not hydrolyze because it has no hydrolyzable functional group (Kollig, 1995).

Drinking water standard: No MCLGs or MCLs have been proposed (U.S. EPA, 1996).

Source: Evaporation and/or dissolution of gasoline, naphtha, coal tar, and asphalt

Uses: Pesticide manufacturing; plasticizer; solvent for coatings; surface active agents; polymer linking agent; ingredient in naphtha; organic synthesis.

sec-BUTYLBENZENE

Synonyms: (1-Methylpropyl)benzene; 2-Phenylbutane; UN 2709.

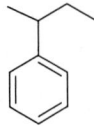

CAS Registry Number: 135-98-8
DOT: 2709
Molecular formula: $C_{10}H_{14}$
Formula weight: 134.22
RTECS: CY9100000
Merck reference: 10, 1523

Physical state and color:
Colorless liquid

Melting point (°C):
-75.5 (Weast, 1986)
-82.7 (Windholz et al., 1983)

Boiling point (°C):
173 (Weast, 1986)
173.34 (Wilhoit and Zwolinski, 1971)

Density (g/cm³ at 20/4 °C):
0.8621 (Weast, 1986)
0.8608 (Windholz et al., 1983)

Diffusivity in water (x 10^{-5} cm²/sec):
0.68 at 20 °C using method of Hayduk and Laudie (1974)

Dissociation constant, pK$_a$:
>14 (Schwarzenbach et al., 1993)

Flash point (°C):
52 (Windholz et al., 1983)
62.7 (open cup, Hawley, 1981)
45 (Aldrich, 1988)

Lower explosive limit (%):
0.8 (Sax and Lewis, 1987)

Upper explosive limit (%):
6.9 (Sax and Lewis, 1987)

Henry's law constant (x 10^{-2} atm·m³/mol):
1.14 at 25 °C (Hine and Mookerjee, 1975)

Ionization potential (eV):
8.68 ± 0.01 (Franklin et al., 1969)

Soil sorption coefficient, log K_{oc}:
2.95 using method of Kenaga and Goring (1980)

Octanol/water partition coefficient, log K_{ow}:
4.24 using method of Hansch et al. (1968)

Solubility in organics:
Miscible with alcohol, benzene, and ether (Windholz et al., 1983)

Solubility in water (25 °C):
17.6 mg/kg (distilled water), 11.9 mg/kg (artificial seawater) (Sutton and Calder, 1975)
171 mg/L at (Andrews and Keefer, 1950a)
87.6 μmol/L in 0.5 M NaCl (Wasik et al., 1984)

Vapor density:
5.49 g/L at 25 °C, 4.63 (air = 1)

Vapor pressure (mmHg):
1.1 at 20 °C (Verschueren, 1983)
1.81 at 25 °C (Mackay et al., 1982)

Environmental fate:
   Chemical/Physical. sec-Butylbenzene will not hydrolyze because it has no hydrolyzable functional group (Kollig, 1995).

Toxicity:
   Acute oral LD_{50} for rats 2,240 mg/kg (quoted, RTECS, 1985).

Drinking water standard: No MCLGs or MCLs have been proposed (U.S. EPA, 1996).

Uses: Solvent for coating compositions; plasticizer; surface-active agents; organic synthesis.

tert-BUTYLBENZENE

Synonyms: *t*-Butylbenzene; **(1,1-Dimethylethyl)benzene**; 2-Methyl-2-phenylpropane; Pseudobutylbenzene; Trimethylphenylmethane; UN 2709.

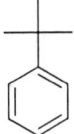

CAS Registry Number: 98-06-6
DOT: 2709
Molecular formula: $C_{10}H_{14}$
Formula weight: 134.22
RTECS: CY9120000
Merck reference: 10, 1524

Physical state and color:
Colorless liquid

Melting point (°C):
-57.8 (Weast, 1986)

Boiling point (°C):
169 (Weast, 1986)
169.15 (Wilhoit and Zwolinski, 1971)

Density (g/cm³):
0.8665 at 20/4 °C (Weast, 1986)

Diffusivity in water (x 10^{-5} cm²/sec):
0.68 at 20 °C using method of Hayduk and Laudie (1974)

Dissociation constant, pK$_a$:
>14 (Schwarzenbach et al., 1993)

Flash point (°C):
60 (open cup, Windholz et al., 1983)
34 (Aldrich, 1988)

Lower explosive limit (%):
0.7 at 100 °C (Sax and Lewis, 1987)

Upper explosive limit (%):
5.7 at 100 °C (Sax and Lewis, 1987)

Henry's law constant (x 10^{-2} atm·m³/mol):
1.17 at 25 °C (Hine and Mookerjee, 1975)

Ionization potential (eV):
8.68 ± 0.01 (Franklin et al., 1969)

Soil sorption coefficient, log K_{oc}:
2.83 using method of Kenaga and Goring (1980)

Octanol/water partition coefficient, log K_{ow}:
4.11 (Leo et al., 1971)
4.07 (Nahum and Horvath, 1980)

Solubility in organics:
Soluble in acetone (Weast, 1986) but miscible with alcohol, benzene, and ether (Windholz et al., 1983)

Solubility in water (25 °C):
29.5 mg/kg in distilled water, 21.2 mg/kg in seawater (Sutton and Calder, 1975)
196 μmol/L (Andrews and Keefer, 1950a)
134 μmol/L in 0.5 M NaCl (Wasik et al., 1984)

Vapor density:
5.49 g/L at 25 °C, 4.63 (air = 1)

Vapor pressure (mmHg):
2.14 at 25 °C (Mackay et al., 1982)
1.5 at 20 °C (Verschueren, 1983)

Environmental fate:
   Photolytic. At 25 °C, a rate constant of 4.58 x 10^{-12} cm^3/molecule·sec was reported for the gas-phase reaction of *tert*-butylbenzene with OH radicals (Ohta and Ohyama, 1985).
   Chemical/Physical. tert-Butylbenzene will not hydrolyze because it has no hydrolyzable functional group (Kollig, 1995).

Drinking water standard: No MCLGs or MCLs have been proposed (U.S. EPA, 1996).

Uses: Polymerization solvent; polymer linking agent; organic synthesis.

n-BUTYL MERCAPTAN

Synonyms: Butanethiol; **1-Butanethiol**; *n*-Butanethiol; Butyl mercaptan; *n*-Butyl thioalcohol; 1-Mercaptobutane; NCI-C60866; Thiobutyl alcohol; UN 2347.

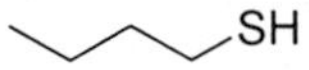

$$\diagdown\diagup\diagdown\diagup\text{SH}$$

CAS Registry Number: 109-79-5
DOT: 2347
Molecular formula: $C_4H_{10}S$
Formula weight: 90.18
RTECS: EK6300000
Merck reference: 10, 1549

Physical state, color, and odor:
Colorless liquid with a strong garlic, cabbage, or heavy skunk-like odor. Odor threshold concentrations ranged from 1 ppb (CHRIS, 1984) to 0.97 ppm (Amoore and Hautala, 1983).

Melting point (°C):
-115.7 (Weast, 1986)

Boiling point (°C):
98.4 (Weast, 1986)

Density (g/cm³):
0.8337 at 20/4 °C (Weast, 1986)
0.83679 at 25/4 °C (Windholz et al., 1983)
0.8412 at 20/4 °C (Hawley, 1981)

Diffusivity in water (x 10^{-5} cm²/sec):
0.84 at 20 °C using method of Hayduk and Laudie (1974)

Flash point (°C):
1.7 (NIOSH, 1997)
12 (Aldrich, 1988)

Heat of fusion (kcal/mol):
2.500 (Dean, 1987)

Henry's law constant (x 10^{-3} atm·m³/mol):
7.04 at 20 °C (approximate - calculated from water solubility and vapor pressure)

Interfacial tension with water (dyn/cm at 20 °C):
30 (CHRIS, 1984)

Ionization potential (eV):
9.14 ± 0.02 (Franklin et al., 1969)

Soil sorption coefficient, log K_{oc}:
Unavailable because experimental methods for estimation of this parameter for aliphatic mercaptans are lacking in the documented literature

Octanol/water partition coefficient, log K_{ow}:
2.28 (Sangster, 1989)

Solubility in organics:
Soluble in alcohol and ether (Weast, 1986).

Solubility in water:
590 mg/L at 22 °C (quoted, Verschueren, 1983)

Vapor density:
3.69 g/L at 25 °C, 3.11 (air = 1)

Vapor pressure (mmHg):
35 at 20 °C (NIOSH, 1997)
55.5 at 25 °C (Wilhoit and Zwolinski, 1971)

Environmental fate:
  Chemical/Physical. Releases toxic sulfur oxide fumes when heated to decomposition (Sax and Lewis, 1987).

Exposure limits: NIOSH REL: 15-min ceiling 0.5 ppm (1.8 mg/m^3), IDLH 500 ppm; OSHA PEL: TWA 10 ppm (35 mg/m^3); ACGIH TLV: TWA 0.5 ppm (adopted).

Toxicity:
  Acute oral LD_{50} for rats 1,500 mg/kg (quoted, RTECS, 1985).

Uses: Chemical intermediate; solvent.

CAMPHOR

Synonyms: 2-Bornanone; 2-Camphanone; Camphor-natural; Camphor-synthetic; Formosa camphor; Gum camphor; Japan camphor; 2-Keto-1,7,7-trimethylnorcamphane; Laurel camphor; Matricaria camphor; Norcamphor; 2-Oxobornane; Synthetic camphor; **1,7,7-Trimethylbicyclo[2.2.1]heptan-2-one**; UN 2717.

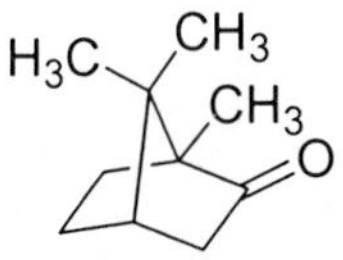

CAS Registry Number: 76-22-2
DOT: 2717
DOT label: Flammable solid
Molecular formula: $C_{10}H_{16}O$
Formula weight: 152.24
RTECS: EX1225000
Merck reference: 10, 1710

Physical state, color, and odor:
Colorless to white crystalline semi-solid with a penetrating, fragrant, or aromatic odor. Odor threshold concentration is 0.27 ppm (Amoore and Hautala, 1983).

Melting point (°C):
174-179 (Verschueren, 1983)

Boiling point (°C):
Sublimes at 204 (Weast, 1986)

Density (g/cm³):
0.990 at 25/4 °C (Weast, 1986)

Diffusivity in water (x 10⁻⁵ cm²/sec):
0.78 at 25 °C using method of Hayduk and Laudie (1974)

Flash point (°C):
65.6 (NIOSH, 1997)

Lower explosive limit (%):
0.6 (NIOSH, 1997)

Upper explosive limit (%):
3.5 (NIOSH, 1997)

Entropy of fusion (cal/mol·K):
3.609 (Frandsen, 1931)

Heat of fusion (kcal/mol):
1.635 (Dean, 1987)

Henry's law constant (x 10^{-5} atm·m^3/mol):
3.00 at 20 °C (approximate - calculated from water solubility and vapor pressure)

Ionization potential (eV):
8.76 (NIOSH, 1997)

Soil sorption coefficient, log K_{oc}:
Unavailable because experimental methods for estimation of this parameter for cyclic ketones are lacking in the documented literature

Octanol/water partition coefficient, log K_{ow}:
2.42 using method of Hansch et al. (1968)

Solubility in organics:
At 25 °C (g/L): acetone (2,500), alcohol (1,000), benzene (2,500), chloroform (2,000), ether (1,000), glacial acetic acid (2,500), oil of turpentine (667). Also soluble in aniline, carbon disulfide, decalin, methylhexalin, nitrobenzene, petroleum ether, tetralin, higher alcohols, in fixed and volatile oils (Windholz et al., 1983).

Solubility in water:
≈ 1.25 g/L at 25 °C (quoted, Windholz et al., 1983)
0.170 wt % at 20-25 °C (Fordyce and Meyer, 1940)

Vapor pressure (mmHg):
0.2 at 20 °C (NIOSH, 1997)

Exposure limits (mg/m^3): NIOSH REL: TWA 2, IDLH 200; OSHA PEL: TWA 2, ceiling 4 ppm (adopted).

Symptoms of exposure: Vapors may irritate eyes, mucous membranes and throat. Ingestion may cause headache, nausea, vomiting, and diarrhea (NIOSH, 1997; Patnaik, 1992). An irritation concentration of 10.62 mg/m^3 in air was reported by Ruth (1986) for synthetic camphor.

Toxicity:
    LC$_{50}$ static bioassay values for fathead minnows after 1, 24, and 96 h are 145, 112, and 110 mg/L, respectively (Mattson et al., 1976).
    LD$_{50}$ (intraperitoneal) for mice 3,000 mg/kg (quoted, RTECS, 1985).

Source: Major component in pine oil (Verschueren, 1983) Also present in a variety of rosemary shoots (330-3,290 ppm) (Soriano-Cano et al., 1993), anise-scented basil leaves (1,785 ppm) (Brophy et al., 1993), Iberian savory leaves (2,660 ppm) (Arrebola et al., 1994), African blue basil shoots (7,000 ppm), Greek sage (160-5,040 ppm), Montane Mountain mint (3,395-3,880 ppm), yarrow leaves (45-1,780 ppm), and coriander (100-1,300 ppm) (Duke, 1992).

Uses: Plasticizer for cellulose esters and ethers; manufacture of plastics, cymene, incense, celluloid; in lacquers, explosives, and embalming fluids; pyrotechnics; moth repellent; preservative in pharmaceuticals and cosmetics; odorant/flavorant in household, pharmaceutical, and industrial products; tooth powders.

CARBARYL

Synonyms: Arylam; Carbamine; Carbatox; Carbatox 60; Carbatox 75; Carpolin; Carylderm; Cekubaryl; Crag sevin; Denapon; Devicarb; Dicarbam; ENT 23969; Experimental insecticide 7744; Gamonil; Germain's; Hexavin; Karbaspray; Karbatox; Karbosep; Methyl carbamate-1-naphthalenol; Methyl carbamate-1-naphthol; Methyl carbamic acid 1-naphthyl ester; *N*-Methyl-1-naphthylcarbamate; *N*-Methyl-α-naphthyl carbamate; *N*-Methyl-α-naphthylurethan; NA 2757; NAC; **1-Naphthalenol methyl carbamate**; 1-Naphthol-*N*-methyl carbamate; 1-Naphthyl methyl carbamate; 1-Naphthyl-*N*-methyl carbamate; α-Naphthyl-*N*-methyl carbamate; OMS 29; OMS 629; Panam; Ravyon; Rylam; Seffein; Septene; Sevimol; Sevin; Sok; Tercyl; Toxan; Tricarnam; UC 7744; Union Carbide 7744.

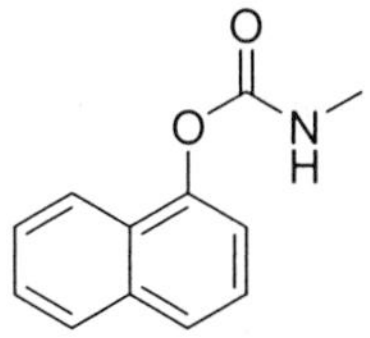

CAS Registry Number: 63-25-2
DOT: 2757
DOT label: Poison
Molecular formula: $C_{12}H_{11}NO_2$
Formula weight: 201.22
RTECS: FC5950000
Merck reference: 10, 1766

Physical state and color:
Colorless solid or white to light tan crystals. The reported odor threshold concentration in air and taste threshold concentration in water are 37 and 44 ppb, respectively (Young et al., 1996).

Melting point (°C):
142 (Worthing and Hance, 1991)
145.1 (NIOSH, 1997)

Density (g/cm³):
1.232 at 20/20 °C (Windholz et al., 1983)

Diffusivity in water (x 10^{-5} cm²/sec):
0.56 at 20 °C using method of Hayduk and Laudie (1974)

Flash point (°C):
195 (ACGIH, 1986)

Henry's law constant (x 10^{-5} atm·m³/mol):
1.27 at 20 °C (approximate - calculated from water solubility and vapor pressure)

Bioconcentration factor, log BCF:
1.86 (*Chlorella fusca*, Freitag et al., 1982; Geyer et al., 1984)
1.95 (activated sludge), 1.48 (golden ide) (Freitag et al., 1985)
0.00 (fish, microcosm) (Garten and Trabalka, 1983)
0.95 (*Pseudorasbora parva*, Devillers et al., 1996)

Soil sorption coefficient, log K_{oc}:
2.42, 2.55, and 2.59 in Commerce, Tracy, and Catlin soils, respectively (McCall et al., 1981)
2.36 (Kenaga and Goring, 1980)
2.02 (Batcombe silt loam, Briggs, 1981)
2.12, 2.51, 2.40, and 2.48 in Bolpur, Nalhati, Barokodali, and Dudherkuthi soils, respectively
    (Jana and Das, 1997)
2.62 and 2.55 at 18 and 30 °C, respectively (Borg El-Arab soil, Aly et al., 1980)

Octanol/water partition coefficient, log K_{ow}:
2.36, 2.56 (Leo et al., 1971)
2.32 (Briggs, 1981; Geyer et al., 1984)
2.31 (Bowman and Sans, 1983a)
2.34 (Swann et al., 1983)

Solubility in organics:
Moderately soluble in acetone, cyclohexanone, *N,N*-dimethylformamide (400-450 g/kg), and
isophorone (Windholz et al., 1983; Worthing and Hance, 1991)

Solubility in water:
0.4105 mol/m^3 at 25 °C (Swann et al., 1983)
120 mg/L at 30 °C (quoted, Windholz et al., 1983)
40 mg/L at 30 °C (Gunther et al., 1968)
72.4, 104, and 130 mg/L at 10, 20, and 30 °C, respectively (Bowman and Sans, 1985)
350 μmol/L at 25 °C (LaFleur, 1979)
67 ppm at 12 °C, 114 ppm at 24 °C (Wauchope and Haque, 1973)
50 mg/L at 20 °C (Fühner and Geiger, 1977)

Vapor pressure (x 10^{-3} mmHg):
5 at 25 °C (Wright et al., 1981)
2 at 40 °C (Meister, 1988)

Environmental fate:
   Biological. Carbaryl degraded completely after 4 wk of incubation in the dark in Holland
March canal water (Sharom et al., 1980). Fourteen soil fungi metabolized [methyl-^{14}C]carbaryl
via hydroxylation forming 1-naphthyl-*N*-hydroxymethyl carbamate, 4-hydroxy-1-naphthyl-
methyl carbamate and 5-hydroxy-1-naphthylmethyl carbamate (Bollag and Liu, 1972).
Carbaryl was degraded by a culture of *Aspergillus terreus* to 1-naphthyl carbamate. The
half-life was determined to be 8 d (Liu and Bollag, 1971). When ^{14}C-carbonyl-labeled carbaryl
(200 ppm) was added to five different soils and incubated at 25 °C for 32 d, evolution of ^{14}CO$_2$
varied from 2.2-37.4% (Kazano et al., 1972).
   Soil. The rate of hydrolysis of carbaryl in flooded soil increased when the soil was pretreated
with the hydrolysis product, 1-naphthol (Rajagopal et al., 1986). Carbaryl is hydrolyzed in both
flooded and nonflooded soils but the rate is slightly higher under flooded conditions (Rajagopal
et al., 1983). When ^{14}C-carbonyl-labeled carbaryl (200 ppm) was added to five different soils
and incubated at 25 °C for 32 d, evolution of ^{14}CO$_2$ varied from 2.2-37.4% (Kazano et al.,
1972). Metabolites identified in soil included 1-naphthol (hydrolysis product) (Ramanand et al.,
1988a; Sud et al., 1972), hydroquinone, catechol, pyruvate (Sud et al., 1972), coumarin, carbon
dioxide (Kazano et al., 1972), 1-naphthyl carbamate, 1-naphthyl *N*-hydroxymethyl carbamate,
5-hydroxy-1-naphthylmethyl carbamate, 4-hydroxy-1-naphthylmethyl carbamate, and 1-naphth-
ylhydroxymethyl carbamate (Liu and Bollag, 1971, 1971a). 1-Naphthol was readily degraded
by soil microorganisms (Sanborn et al., 1977).

Sud et al. (1972) discovered that a strain of *Achromobacter* sp. utilized carbaryl as the sole source of carbon in a salt medium. The organism grew on the degradation products 1-naphthol, hydroquinone, and catechol. 1-Naphthol, a metabolite of carbaryl in soil, was recalcitrant to further degradation by a bacterium tentatively identified as an *Arthrobacter* sp. under anaerobic conditions (Ramanand et al., 1988a). Carbaryl or its metabolite 1-naphthol at normal and ten times the field application rate had no effect on the growth of *Rhizobium* sp. or *Azotobacter chroococcum* (Kale et al., 1989). The half-lives of carbaryl under flooded and non-flooded conditions were 13-14 and 23-28 d, respectively (Venkateswarlu et al., 1980).

When carbaryl was applied to soil at a rate of 1,000 L/ha, more than 50% remained in the upper 5 cm (Meyers et al., 1970). The half-lives of carbaryl in a sandy loam, clay loam and an organic amended soil under non-sterile conditions were 96-1,462, 211-2,139, and 51-4,846 d, respectively, while under sterile conditions the half-lives were 67-5,923, 84-9,704, and 126-4,836 d, respectively (Schoen and Winterlin, 1987).

Liu and Bollag (1971) reported that the fungus *Gliocladium roseum* degraded carbaryl to 1-naphthyl *N*-hydroxymethyl carbamate, 4-hydroxy-1-naphthylmethyl carbamate, and 1-naphthylhydroxymethyl carbamate.

Rajagopal et al. (1984) proposed degradation pathways of carbaryl in soil and in microbial cultures included the following compounds: 5,6-dihydrodihydroxy carbaryl, 2-hydroxy carbaryl, 4-hydroxy carbaryl, 5-hydroxy carbaryl, 1-naphthol, *N*-hydroxymethyl carbaryl, 1-naphthyl carbamate, 1,2-dihydroxynaphthalene, 1,4-dihydroxynaphthalene, *o*-coumaric acid, *o*-hydroxybenzalpyruvate, 1,4-naphthoquinone, 2-hydroxy-1,4-naphthoquinone, coumarin, γ-hydroxy-γ-*o*-hydroxyphenyl-α-oxobutyrate, 4-hydroxy-1-tetralone, 3,4-dihydroxy-1-tetralone, pyruvic acid, salicylaldehyde, salicylic acid, phenol, hydroquinone, catechol, carbon dioxide, and water. When carbaryl was incubated at room temperature in a mineral salts medium by soil-enrichment cultures for 30 d, 26.8 and 31.5% of the applied insecticide remained in flooded and nonflooded soils, respectively (Rajagopal et al., 1984a). A *Bacillus* sp. and the enrichment cultures both degraded carbaryl to 1-naphthol. Mineralization to carbon dioxide was negligible (Rajagopal et al., 1984a).

Groundwater. According to the U.S. EPA (1986), carbaryl has a high potential to leach to groundwater.

Plant. In plants, the *N*-methyl group may be subject to oxidation or hydroxylation (Kuhr, 1968). The presence of pinolene (β-pinene polymer) in carbaryl formulations increases the amount of time carbaryl residues remain on tomato leaves and decreases the rate of decomposition. The half-life in plants range from 1.3 to 29.5 d (Blazquez et al., 1970).

Surface Water. In a laboratory aquaria containing estuarine water, 43% of dissolved carbaryl was converted to 1-naphthol in 17 d at 20 °C (pH 7.5-8.1). The half-life of carbaryl in estuarine water without mud at 8 °C was 38 d. When mud was present, both carbaryl and 1-naphthol decreased to less than 10% in the estuarine water after 10 d. Based on a total recovery of only 40%, it was postulated that the remainder was evolved as methane (Karinen et al., 1967). The rate of hydrolysis of carbaryl increased with an increase in temperature (Karinen et al., 1967) and in increases of pH values above 7.0 (Rajagopal et al., 1984). The presence of a micelle [hexadecyltrimethylammonium bromide (HDATB), 3×10^{-3} M] in natural waters greatly enhanced the hydrolysis rate. The hydrolysis half-lives in natural water samples with and without HDATB were 0.12-0.67 and 9.7-138.6 h, respectively (González et al., 1992). In a sterilized buffer solution, a hydrolysis half-life of 87 h was observed (Ferreira and Seiber, 1981). In the dark, carbaryl was incubated in 21 °C water obtained from the Holland Marsh drainage canal. Degradation was complete after 4 wk (Sharom et al., 1980).

In pond water, carbaryl degraded very rapidly to 1-naphthol. The latter was further degraded, presumably by *Flavobacterium* sp., into hydroxycinnamic acid, salicylic acid, and an unidentified compound (HSDB, 1989). Four d after carbaryl (30 mg/L and 300 μg/L) was added to Fall Creek water, >60% was mineralized to carbon dioxide. At pH 3, however, <10%

was converted to carbon dioxide (Boethling and Alexander, 1979). Under these conditions, hydrolysis of carbaryl to 1-naphthol was rapid. The authors could not determine how much carbon dioxide was attributed to biodegradation of carbaryl and how much was due to the biodegradation of 1-naphthol (Boethling and Alexander, 1979). Hydrolysis half-lives of carbaryl in filtered and sterilized Hickory Hills (pH 6.7) and U.S. Department of Agriculture Number 1 pond water (pH 7.2) were 30 and 12 d, respectively (Wolfe et al., 1978).

Six brooks and three rivers in Maine contaminated by carbaryl originated from forest spraying. The disappearance half-lives of carbaryl in Clayton, Carry, Squirrel, Burntland, Mud and Wyman Brooks were 15, 19, 77, 111, 42, and 33-83 h, respectively. Disappearance half-lives of 58-200, 21-71, and 24 h were reported for Presque Isle, Machia, and Penobscot Rivers, respectively (Stanley and Trial (1980).

Photolytic. Based on data for phenol, a structurally related compound, an aqueous solution containing 1-naphthoxide ion (3 x 10^{-4} M) in room light would be expected to photooxidize to give 2-hydroxy-1,4-naphthoquinone (Tomkiewicz et al., 1971). 1-Naphthol, methyl isocyanate and other unidentified cholinesterase inhibitors were reported as products formed from the direct photolysis of carbaryl by sunlight (Wolfe et al., 1976). In an aqueous solution at 25 °C, the photolysis half-life of carbaryl by natural sunlight or UV light ($\lambda = 313$ nm) is 6.6 d (Wolfe et al., 1978a).

A photolysis half-life of 1.88 d was observed when carbaryl in an acidic (pH 5.5), buffered aqueous solution was irradiated with UV light ($\lambda > 290$ nm) (Wolfe et al., 1976).

Peris-Cardells et al. (1993) studied the photocatalytic degradation of carbaryl (200 μg/L) in aqueous solutions containing a suspended catalyst (titanium dioxide) in a continuous flow system equipped with a UV lamp. The investigators concluded that the degradation yield depends on the ratio of the insecticide and catalyst. The rate of degradation was independent of pH and temperature. Under optimal experimental conditions, >99% degradation was achieved in less than 60 sec of irradiation in solutions containing a concentration of 0.2 mg/L carbaryl.

Chemical/Physical. Ozonation of carbaryl in water yielded 1-naphthol, naphthoquinone, phthalic anhydride, *N*-formyl carbamate of 1-naphthol (Martin et al., 1983), naphthoquinones and acidic compounds (Shevchenko et al., 1982).

Hydrolysis and photolysis of carbaryl forms 1-naphthol (Wauchope and Haque, 1973; Rajagopal et al., 1984, 1986; Miles et al., 1988; MacRae, 1989; Ramanand et al., 1988a; Lewis, 1989; Somasundaram et al., 1991) and 2-hydroxy-1,4-naphthoquinone (Wauchope and Haque, 1973), respectively. In aqueous solutions, carbaryl hydrolyzes to 1-naphthol (Boethling and Alexander, 1979; Vontor et al., 1972), methylamine, and carbon dioxide (Vontor et al., 1972), especially under alkaline conditions (Wolfe et al., 1978). At pH values of 5, 7, and 9, the hydrolysis half-lives at 27 °C were 1,500, 15, and 0.15 d, respectively (Wolfe et al., 1978).

Miles et al. (1988) studied the rate of hydrolysis of carbaryl in phosphate-buffered water (0.01 M) at 26 °C with and without a chlorinating agent (10 mg/L hypochlorite solution). The hydrolysis half-lives at pH 7 and 8 with and without chlorine were 3.5 and 10.3 d and 0.05 and 1.2 d, respectively (Miles et al., 1988). The reported hydrolysis half-lives of carbaryl in water at pH values of 7, 8, 9, and 10 were 10.5 d, 1.3 d, 2.5 h, and 15.0 min, respectively (Aly and El-Dib, 1971). The hydrolysis half-lives of carbaryl in a sterile 1% ethanol/water solution at 25 °C and pH values of 4.5, 6.0, 7.0, and 8.0, were 300, 58, 2.0, and 0.27 wk, respectively (Chapman and Cole, 1982).

Products reported from the combustion of carbaryl at 900 °C include carbon monoxide, carbon dioxide, ammonia, and oxygen (Kennedy et al., 1972).

In water, carbaryl reacted with hydroxy radicals at a rate constant of 3.4 x 10^9/M·sec (Mabury and Crosby, 1996). Carbaryl degradation followed first-order kinetics and it was more reactive than another closely related carbamate, carbofuran.

Carbaryl is stable to light, heat, and in slightly acidic and alkaline solutions (Hartley and Kidd, 1987).

Exposure limits (mg/m³): NIOSH REL: TWA 5, IDLH 100; OSHA PEL: TWA 5; ACGIH TLV: TWA 5 (adopted).

Symptoms of exposure: Symptoms may include nausea, vomiting, diarrhea, abdominal cramps, miosis, lachrymation, excessive salivation, nasal discharge, sweating, muscle twitching, convulsions, and coma (Patnaik, 1992). An acceptable daily intake reported for humans is 0.01 mg/kg body weight (Worthing and Hance, 1991).

Toxicity:

EC_{50} (48-h) for *Daphnia magna* 5.6 μg/L, *Simocephalus serrulatus* 11 μg/L (Mayer and Ellersieck, 1986).

EC_{50} (24-h) for *Cancer magister* 600 ppb, *Hemigrapsis oregonensis* 270 ppb, *Mytilus edulis* larvae 2.30 ppm, *Crassostrea gigas* larvae 2.20 ppm (Stewart et al., 1967).

EC_{50} (5-min) for *Photobacterium phosphoreum* 5.0 mg/L (Somasundaram et al., 1990).

LC_{50} (contact) for earthworm (*Eisenia fetida*) 14 μg/cm² (Neuhauser et al., 1985).

LC_{50} (8-d) for bluegill sunfish 10 mg/L, rainbow trout 1.3 mg/L, sheepshead minnow 2.2 mg/L (Worthing and Hance, 1991).

LC_{50} (96-h) for *Channa punctatus* 2.0 mg/L (Saxena and Garg, 1978), *Gammarus italicus* 28 μg/L, *Echinogammarus tibaldii* 6.5 μg/L (Pantani et al., 1997), catfish 15.8 mg/L, goldfish, 13.2 mg/L, fathead minnows 14.6 mg/L, carp 5.3 mg/L, bluegill sunfish 6.8 mg/L, rainbow trout 4.3 mg/L, coho salmon 0.76 mg/L, perch 0.75 mg/L (Macek and McAllister, 1970), *Pteronarcella badia* 13.7 μg/L, *Pteronarcys californica* 4.8 μg/L, *Gammarus fasciatus* 26 μg/L, *Gammarus lacustris* 22 μg/L, *Gammarus pseudolimnaeus* 7.1 μg/L, *Palaemonetes kadiakensis* 5.6 μg/L, *Claassenia sabulosa* 5.6 μg/L, coho salmon 1,150-4,340 μg/L, cutthroat trout 970-7,100 μg/L, Atlantic salmon 250-4,500 μg/L, brown trout 6,300 μg/L, brook trout 680-4,560 μg/L (Mayer and Ellersieck, 1986).

LC_{50} (72-h) for catfish (*Mystus vittatus*) 17.5 ppm (Arunachalam et al., 1980).

LC_{50} (48-h) for red killifish 32 mg/L (Yoshioka et al., 1986), carp 11-13 mg/L, goldfish >10 mg/L, medaka 2.8 mg/L, pond loach 13 mg/L, rainbow trout 8.5 mg/L, guppy 3.4 mg/L (quoted, RTECS, 1985), *Daphnia pulex* 6.4 μg/L (Mayer and Ellersieck, 1986), freshwater fish (*Channa punctatus*) 8.71 ppm (Rao et al., 1985).

LC_{50} (24-h) for *Pteronarcella badia* 18.5 μg/L, *Pteronarcys californica* 30 μg/L, rainbow trout 1,340 μg/L (flow-through), 3,500 μg/L (static bioassay), *Asellus brevicaudus* 340 μg/L *Gammarus fasciatus* 50 μg/L, *Gammarus lacustris* 40 μg/L, *Gammarus pseudolimnaeus* 10.5-22 μg/L, *Palaemonetes kadiakensis* 120 μg/L, *Claassenia sabulosa* 12 μg/L, coho salmon 1,350-6,520 μg/L, cutthroat trout 2,750-7,400 μg/L, Atlantic salmon 500-4,900 μg/L, brown trout 6,840 μg/L, brook trout 770-10,000 μg/L (Mayer and Ellersieck, 1986), *Paramecium multimicronucleatum* 28 ppm (Edmiston et al., 1985), Protozoan (*Spirostomum teres*) 3.34 mg/L (Twagilimana et al., 1998).

LC_{50} (17-h) for *Paramecium multimicronucleatum* 34 ppm (Edmiston et al., 1985).

LC_{50} (13-h) for *Paramecium multimicronucleatum* 46 ppm (Edmiston et al., 1985).

LC_{50} (9-h) for *Paramecium multimicronucleatum* 65 ppm (Edmiston et al., 1985).

LC_{50} (7-h) for *Paramecium multimicronucleatum* 105 ppm (Edmiston et al., 1985).

LC_{50} (30-min) for Protozoan (*Spirostomum teres*) 0.49 mg/L (Twagilimana et al., 1998).

Acute oral LD_{50} for wild birds 56 mg/kg, cats 150 mg/kg, guinea pigs 250 mg/kg, gerbils 491 mg/kg, hamsters, 250 mg/kg, mice, 212 mg/kg, rabbits 710 mg/kg (quoted, RTECS, 1985); male rats 250 mg/kg, female rats 500 mg/kg (Worthing and Hance, 1991).

Acute percutaneous LD_{50} for rats 4,000 mg/kg, rabbits >2,000 mg/kg (Worthing and Hance, 1991).

A NOEL of 200 mg/kg diet was reported for rats in 2-yr feeding trials (Worthing and Hance, 1991).

Drinking water standard: No MCLGs or MCLs have been proposed (U.S. EPA, 1996).

Uses: Contact and stomach insecticide with slight systemic properties used against Alticinae, Cicadellidae, Coleoptera, Dermateptera, Jassidae, Lepidoptera, Miridae, *Tipula* spp. cotton, vegetables, and fruit crops (Worthing and Hance, 1991).

CARBOFURAN

Synonyms: Bay 70143; **2,3-Dihydro-2,2-dimethyl-7-benzofuranol methyl carbamate**; 2,2-Dimethyl-7-coumaranyl *N*-methyl carbamate; 2,2-Dimethyl-2,3-dihydro-7-benzofuranyl-*N*-methyl carbamate; ENT 27164; Furadan; Methyl carbamic acid 2,3-dihydro-2,2-dimethyl-7-benzofuranyl ester; NIA 10242; OMS 864.

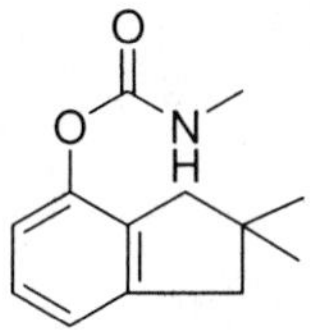

CAS Registry Number: 1563-66-2
DOT: 2757
DOT label: Poison
Molecular formula: $C_{12}H_{15}NO$
Formula weight: 221.26
RTECS: FB9450000
Merck reference: 10, 1786

Physical state, color, and odor:
Odorless, white to grayish crystalline solid

Melting point (°C):
150-153 (Windholz et al., 1983)
151 (Bowman and Sans, 1983a)
153-154 (Worthing and Hance, 1991)

Boiling point (°C):
152 (NIOSH, 1997)

Density (g/cm³):
1.18 at 20/20 °C (Verschueren, 1983)

Diffusivity in water (x 10⁻⁵ cm²/sec):
0.54 at 20 °C using method of Hayduk and Laudie (1974)

Henry's law constant (x 10⁻⁸ atm·m³/mol):
3.88 at 30 °C (approximate - calculated from water solubility and vapor pressure)

Soil sorption coefficient, log K_{oc}:
2.20 (Commerce soil), 1.98 (Tracy soil), 2.02 (Catlin soil) (McCall et al., 1981)

Octanol/water partition coefficient, log K_{ow}:
1.60 (Kenaga and Goring, 1980)
1.63 (Bowman and Sans, 1983a)
1.23-1.41 (Worthing and Hance, 1991)

Solubility in organics (g/L):
Methylene chloride (>200), 2-propanol (20-50) (Worthing and Hance, 1991)

194   **Carbofuran**

Solubility in water:
415 ppm (Kenaga and Goring, 1980)
700 ppm at 25 °C (quoted, Windholz et al., 1983)
291, 320, and 375 mg/L at 10, 20, and 30 °C, respectively (Bowman and Sans, 1985)
320 mg/L at 19.0 °C (Bowman and Sans, 1979)

Vapor pressure (mmHg):
2×10^{-5} at 33 °C (Verschueren, 1983)

Environmental fate:
  Biological. Carbofuran or their metabolites (3-hydroxycarbofuran and 3-ketocarbofuran) at normal and 10 times the field application rate had no effect on *Rhizobium* sp. However, in a nitrogen-free culture medium, *Azotobacter chroococcum* growth was inhibited by carbofuran, 3-hydroxycarbofuran, and 3-ketocarbofuran (Kale et al., 1989). Under *in vitro* conditions, 15 of 20 soil fungi degraded carbofuran to one or more of the following compounds: 3-hydroxy-carbofuran, 3-ketocarbofuran, carbofuran phenol, and 3-hydroxyphenol (Arunachalam and Lakshmanan, 1988). *Pseudomonas* sp. and *Achromobacter* sp. were also capable of degrading carbofuran (Felsot et al., 1981). Derbyshire et al. (1987) reported that an enzyme, isolated from the microorganism *Achromobacter* sp., hydrolyzed carbofuran via the carbamate linkage forming 2,3-dihydro-2,2-dimethyl-7-benzofuranol. The optimum pH and temperature for the degradation of carbofuran and two other pesticides (aldicarb and carbaryl) were 9.0-10.5 and 45 and 53 °C, respectively. Degradation of carbofuran in poorly drained clay-muck, well drained clay-muck, and poorly drained clay field test plots from British Columbia was very low. After 2 successive years of treatment, carbofuran concentrations began to increase (Williams et al., 1976).
  In a small watershed, carbaryl was applied to corn seed farrows at a rate of 5.03 kg/ha active ingredient. Carbaryl was stable up to 166 d, but after 135 d, 95% had disappeared. The long lag time suggests that carbaryl degradation was primarily due to microbial degradation (Caro et al., 1974).
  Soil. Carbofuran is relatively persistent in soil, especially in dry, acidic, and low temperature soils (Ahmad et al., 1979; Caro et al., 1973; Fuhremann and Lichtenstein, 1980; Gorder et al., 1982; Greenhalgh and Belanger, 1981; Ou et al., 1982). In alkaline soil with high moisture content, microbial degradation of carbofuran was more important than leaching and chemical degradation (Gorder et al., 1982; Greenhalgh and Belanger, 1981).
  Carbofuran phenol is formed from the hydrolysis of carbofuran at pH 7.0. Carbofuran phenol was also found to be the major biodegradation product by *Azospirillum lipoferum* and *Streptomyces* sp. isolated from a flooded alluvial soil (Venkateswarlu and Sethunathan, 1984). The hydrolysis of carbofuran to carbofuran phenol was catalyzed by the addition of rice straw in an anaerobic flooded soil where it accumulated (Venkateswarlu and Sethunathan, 1979). The rate of transformation of carbofuran in soil increased with repeated applications (Harris et al., 1984). In an alluvial soil, carbaryl and its analog carbosulfan 2,3-dihydro-2,2-dimethyl-7-benzofuranyl [(di-*n*-butyl)aminosulfenyl methyl carbamate] both degraded faster at 35 °C than at 25 °C with carbosulfan degrading to carbofuran (Sahoo et al., 1990). An enrichment culture isolated from a flooded alluvial soil (Ramanand et al., 1988) and a bacterium tentatively identified as an *Arthrobacter* sp. (Ramanand et al., 1988a) readily mineralized carbofuran to carbon dioxide at 35 °C. Mineralization was slower at lower temperatures (20-28 °C). Under anaerobic conditions, carbofuran did not degrade (Ramanand et al., 1988a). The reported half-lives in soil are 1-2 months (Hartley and Kidd, 1987); 11-13 d at a pH value of 6.5 and 60-75 d for a granular formulation (Ahmad et al., 1979).
  Rajagopal et al. (1984) proposed degradation pathways of carbofuran in both soils and microbial cultures included the following compounds: 3-hydroxycarbofuran, 3-ketocarbofuran,

carbofuran phenol, 3-hydroxycarbofuran phenol, 3-ketocarbofuran phenol, 6,7-dihydroxycarbo-furan phenol, 3,6,7-trihydroxycarbofuran phenol, 3-keto-6,7-dihydroxycarbofuran phenol, and carbon dioxide. In soils, microorganisms degraded carbofuran to carbofuran phenol (Ou et al., 1982), then to 3-hydroxycarbofuran and 3-ketocarbofuran (Kale et al., 1989; Ou et al., 1982).

Hydrolyzes in soil and water to carbofuran phenol, carbon dioxide, and methylamine (Rajagopal et al., 1986; Seiber et al., 1978; Somasundaram et al., 1989, 1991). Hydrolysis of carbofuran occurs in both flooded and nonflooded soils, but the rate is slightly higher under flooded conditions (Venkateswarlu et al., 1977), especially when the soil is pretreated with the hydrolysis product, carbofuran phenol (Rajagopal et al., 1986). In addition, the hydrolysis of carbofuran was found to be pH dependent in both deionized water and rice paddy water. At pH values of 7.0, 8.7, and 10.0, the hydrolysis half-lives in deionized water were 864, 19.4, and 1.2 h, respectively. In paddy water, the hydrolysis half-lives at pH values of 7.0, 8.7, and 10.0 were and 240, 13.9, and 1.3, respectively (Seiber et al., 1978).

The half-lives for carbofuran in soil incubated in the laboratory under aerobic conditions ranged from 6 to 93 d with an average half-life of 37 d (Getzin, 1973; Venkateswarlu et al., 1977). In flooded soils, Venkateswarlu (1977) reported a half-life of 44 d. In field soils, half-lives ranging from 31 to 117 d with an average half-life of 68 d were reported (Caro et al., 1973; Mathur et al., 1976; Talekar et al., 1977). The average mineralization half-life for carbofuran in soil was 535 d (Getzin, 1973). Ou et al. (1982) reported the following disappearance half-lives of carbofuran in soil: Webster clay loam: 14-107 d, Cecil loamy sand: 26-261 d, Sharpsburg silty clay loam: 53-227 d, Grenada silt loam: 30-73 d, Yolo silt loam: 29-155 d, and Houston clay loam: 5-109 d. Mineralization half-lives of carbofuran incubated in soil at 27 °C at soil-water tensions of 0.1 and 0.33 bars were 279 and 350 for Cecil loamy sand, 216 and 236 d for Webster clay loam, 273 and 442 d for Grenada silt loam, 839 and 944 d for Sharpsburg silty clay loam, 271 and 295 d for Yolo silt loam, and 317 and 302 d for Houston clay loam, respectively.

Gorder et al. (1982) studied the persistence of carbofuran in an alkaline cornfield soil having a high moisture content. They concluded that microbial degradation was more important than leaching in the dissipation from soil. Similarly, Greenhalgh and Belanger (1981) reported that microbial degradation was more important than chemical attack in a humic soil and a sandy loam soil.

Groundwater. According to the U.S. EPA (1986), carbofuran has a high potential to leach to groundwater.

Plant. Carbofuran is rapidly metabolized in plants to nontoxic products (Cremlyn, 1991). Metcalf et al. (1968) reported that carbofuran undergoes hydroxylation and hydrolysis in plants, insects, and mice. Hydroxylation of the benzylic carbon gives 3-hydroxycarbofuran, which is subsequently oxidized to 3-ketocarbofuran. In carrots, carbofuran initially degraded to 3-hydroxycarbofuran. This compound reacted with naturally occurring angelic acid in carrots forming a conjugated metabolite identified as 2,3-dihydro-2,2-dimethyl-7-(((methylamino)car-bonyl)oxy)-3-benzofuranyl (Z)-2-methyl-2-butenoic acid (Sonobe et al., 1981). Metabolites identified in three types of strawberries (Day-Neutral, Tioga and Tufts) were 2,3-dihydro-2,2-dimethyl-3-hydroxy-7-benzofuranyl-N-methyl carbamate, 2,3-dihydro-2,2-dimethyl-3-oxo-7-benzofuranyl-N-methyl carbamate, 2,3-dihydro-2,2-dimethyl-3-benzofuranol, 2,3-dihydro-2,2-dimethyl-3,7-benzofuranol and 2,3-dihydro-2,2-dimethyl-3-oxo-7-benzofuranol (Archer et al., 1977). Oat plants were grown in two soils treated with [^{14}C]carbofuran. Most of the residues recovered in oat leaves were in the form of carbofuran and 3-hydroxycarbofuran. Other metabo-lites identified were 3-ketocarbofuran, a 3-keto-7-phenol, and 3-hydroxy-7-phenol (Fuhremann and Lichtenstein, 1980).

Photolytic. 2,3-Dihydro-2,2-dimethylbenzofuran-4,7-diol and 2,3-dihydro-3-keto-2,2-di-methylbenzofuran-7-yl carbamate were formed when carbofuran dissolved in water was irradiated by sunlight for 5 d (Raha and Das, 1990).

In cornfield soils, carbofuran losses were high but in autoclaved soils, virtually no losses were observed. This insecticide appears to be poorly absorbed onto soil matrix and therefore, is readily leachable (Felsot et al., 1981).

Surface Water: Sharom et al. (1980) reported that the half-lives for carbofuran in sterilized and non-sterilized water collected from the Holly Marsh in Ontario, Canada were 2.5 to 3 wk at pH values of 7.8-8.0 and 8.0, respectively. The half-lives observed in distilled water were 2 and 3.8 wk at pH values of 7.0-7.2 and 6.8, respectively. They reported that chemical degradation of dissolved carbofuran was more significant than microbial degradation.

Chemical/Physical. Releases toxic nitrogen oxides when heated to decomposition (Sax and Lewis, 1987; Lewis, 1990).

In water, carbofuran reacted with hydroxy radicals at a first-order rate constant of 2.2 x 10^9/M·sec (Mabury and Crosby, 1996).

The hydrolysis half-lives of carbofuran in a sterile 1% ethanol/water solution at 25 °C and pH values of 4.5, 5.0, 6.0, 7.0, and 8.0 were 170, 690, 690, 8.2, and 1.0 wk, respectively (Chapman and Cole, 1982).

Exposure limits (mg/m^3): NIOSH REL: TWA 0.1; ACGIH TLV: TWA 0.1 (adopted).

Toxicity:

EC_{50} (5-min) for *Photobacterium phosphoreum* 20.5 mg/L (Somasundaram et al., 1990).

LC_{50} (96-h) for *Gammarus italicus* 12 μg/L, *Echinogammarus tibaldii* 4.6 μg/L (Pantani et al., 1997), rainbow trout 490 μg/L, brown trout 560 μg/L, lake trout 164 μg/L, fathead minnow 872-1,990 μg/L, channel catfish 248 μg/L, bluegill sunfish 88 μg/L, yellow perch 253 μg/L (Mayer and Ellersieck, 1986).

LC_{50} (96-h semi-static bioassay) for *Tigriopus brevicornis* females 59.9 μg/L (Forget et al., 1998).

LC_{50} (24-h) for rainbow trout 850 μg/L, brown trout 842 μg/L, lake trout 164 μg/L, fathead minnow 883-2,240 μg/L, channel catfish 372 μg/L, bluegill sunfish 101 μg/L, yellow perch 407.50 μg/L (Mayer and Ellersieck, 1986).

Acute oral LD_{50} for rats 11 mg/kg (quoted, Verschueren, 1983).

Drinking water standard (final): MCLG: 40 μg/L; MCL: 40 μg/L (U.S. EPA, 1996).

Uses: Systematic insecticide, nematocide, and acaricide.

CARBON DISULFIDE

Synonyms: Carbon bisulfide; Carbon bisulphide; Carbon disulphide; Carbon sulfide; Carbon sulphide; Dithiocarbonic anhydride; NCI-C04591; RCRA waste number P022; Sulphocarbonic anhydride; UN 1131; Weeviltox.

CAS Registry Number: 75-15-0
DOT: 1131
DOT label: Flammable liquid
Molecular formula: CS_2
Formula weight: 76.13
RTECS: FF6650000
Merck reference: 10, 1795

Physical state, color, and odor:
Clear, water-white to pale yellow liquid; ethereal odor when pure; technical and reagent grades have strong, foul odors. Odor threshold concentration is 0.11 ppm (Amoore and Hautala, 1983).

Melting point (°C):
-111.5 (Weast, 1986)

Boiling point (°C):
46.2 (Weast, 1986)

Density (g/cm³):
1.2632 at 20/4 °C (Weast, 1986)
1.2632 at 15/4 °C, 1.2559 at 25/4 °C, 1.2500 at 30/4 °C (Standen, 1964)

Diffusivity in water (x 10^{-5} cm²/sec):
1.18 at 20 °C using method of Hayduk and Laudie (1974)

Flash point (°C):
-30 (NIOSH, 1997)

Lower explosive limit (%):
1.3 (NIOSH, 1997)

Upper explosive limit (%):
50.0 (NIOSH, 1997)

Heat of fusion (kcal/mol):
1.049 (Dean, 1987)

Henry's law constant (atm·m³/mol):
24.25 at 24 °C (Elliott, 1989)

Interfacial tension with water (dyn/cm at 25 °C):
48.36 (Harkins et al., 1920)
48.1 (Donahue and Bartell, 1952)

Ionization potential (eV):
10.08 (Franklin et al., 1969)

Soil sorption coefficient, log K_{oc}:
2.38-2.55 using method of Kenaga and Goring (1980)
1.84 (estimated, Ellington et al., 1993)

Octanol/water partition coefficient, log K_{ow}:
1.84, 2.16 (calculated, Leo et al., 1971)

Solubility in organics:
Soluble in ethanol, chloroform, and ether (Weast, 1986)

Solubility in water:
2,300 mg/L at 22 °C (quoted, Verschueren, 1983)
2,200 mg/L at 22 °C (ACGIH, 1986)
2,000 mg/L at 20 °C (quoted, WHO, 1979)
0.210 wt % at 20 °C (quoted, Riddick et al., 1986)
0.294% at 20 °C (quoted, Windholz et al., 1983)
0.286 wt % at 25 °C (Lo et al., 1986)

Vapor density:
3.11 g/L at 25 °C, 2.63 (air = 1)

Vapor pressure (mmHg):
297 at 20 °C (NIOSH, 1997)
360 at 25 °C (Lide, 1990)
430 at 30 °C (Verschueren, 1983)

Environmental fate:
   Chemical/Physical. In alkaline solutions, carbon disulfide hydrolyzes to carbon dioxide and hydrogen sulfide (Peyton et al., 1976) via the intermediate carbonyl sulfide (Ellington et al., 1993; Kollig, 1993). In an aqueous alkaline solution containing hydrogen peroxide, dithioper-carbonate, sulfide, elemental sulfur, and polysulfides may be expected by-products (Elliott, 1990). In an aqueous, alkaline solution (pH ≥8), carbon disulfide reacted with hydrogen per-oxide forming sulfate and carbonate ions. However, when the pH is lowered to 7 and 7.4, colloidal sulfur is formed (Adewuyi and Carmichael, 1987). Forms a hemihydrate which decomposes at -3 °C (Keith and Walters, 1992).
   An aqueous solution containing carbon disulfide reacts with sodium hypochlorite forming carbon dioxide, sulfuric acid and sodium chloride (Patnaik, 1992).
   Burns with a bright blue flame releasing carbon dioxide and sulfur dioxide (Windholz et al., 1983).
   Oxidizes in the troposphere forming carbonyl sulfide. The atmospheric half-lives of carbon disulfide and carbonyl sulfide were estimated to be approximately 2 yr and 13 d, respectively (Khalil and Rasmussen, 1984).

Exposure limits: NIOSH REL: TWA 1 ppm (3 mg/m³), STEL 10 ppm, IDLH 500 ppm; OSHA PEL: TWA 20 ppm, ceiling 30 ppm, 100 ppm peak for 30 min; ACGIH TLV: TWA 10 ppm (adopted).

Symptoms of exposure: Dizziness, headache, poor sleep, fatigue, nervousness; anorexia, weight loss; polyneuropathy, burns, dermatitis. Contact with skin causes burning pain, erythema, and exfoliation (NIOSH, 1997).

Toxicity:

Acute oral LD_{50} for rats 3,188 mg/kg, guinea pigs 2,125 mg/kg, mice 2,780 mg/kg, rabbits 2,550 mg/kg (quoted, RTECS, 1985).

Uses: Manufacture of viscose rayon, cellophane, flotation agents, ammonium salts, carbon tetrachloride, carbanilide, paints, enamels, paint removers, varnishes, tallow, textiles, rocket fuel, soil disinfectants, electronic vacuum tubes, herbicides; grain fumigants; solvent for fats, resins, phosphorus, sulfur, bromine, iodine, and rubber; petroleum and coal tar refining; solvent and eluant for organics adsorbed on charcoal for air analysis.

CARBON TETRACHLORIDE

Synonyms: Benzinoform; Carbona; Carbon chloride; Carbon tet; ENT 4705; Fasciolin; Flukoids; Freon 10; Halon 104; Methane tetrachloride; Necatorina; Necatorine; Perchloromethane; R 10; RCRA waste number U211; Tetrachloormetaan; Tetrachlorocarbon; **Tetrachloromethane**; Tetrafinol; Tetraform; Tetrasol; UN 1846; Univerm; Vermoestricid.

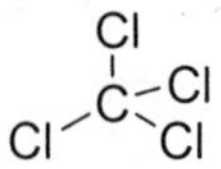

CAS Registry Number: 56-23-5
DOT: 1846
DOT label: Poison
Molecular formula: CCl_4
Formula weight: 153.82
RTECS: FG4900000
Merck reference: 10, 1799

Physical state, color, and odor:
Clear, colorless, heavy, watery liquid with a strong, sweetish, distinctive odor resembling ether. Odor threshold concentration is 21.4 ppm (Keith and Walters, 1992).

Melting point (°C):
-22.99 (Horvath, 1982)

Boiling point (°C):
76.54 (Horvath, 1982)

Density (g/cm³):
1.59472 at 20/4 °C (Standen, 1964)
1.5844 at 25/4 °C (Kirchnerová and Cave, 1976)

Diffusivity in water (x 10⁻⁵ cm²/sec):
0.90 at 20 °C using method of Hayduk and Laudie (1974)

Flash point (°C):
Noncombustible (Rogers and McFarlane, 1981)

Heat of fusion (kcal/mol):
0.581 (Riddick et al., 1986)
0.601 (Dean, 1987)

Henry's law constant (x 10⁻² atm·m³/mol):
3.02 at 25 °C (Warner et al., 1987)
2.4 at 20 °C (Roberts and Dändliker, 1983)
10.2 at 37 °C (Sato and Nàkajima, 1979)
3.04 at 24.8 °C (Gossett, 1987)
0.2226 at 20 °C (Roberts et al., 1985)
2.63 at 25 °C (Tancréde and Yanagisawa, 1990)
3.78 at 30 °C (Jeffers et al., 1989)
2.78 at 25 °C (Hoff et al., 1993)

3.33 at 25 °C (Wright et al., 1992)

2.22 at 25 °C (Pearson and McConnell, 1975)

1.48, 1.91, 2.32, 2.95, and 3.78 at 10, 15, 20, 25, and 30 °C, respectively (Ashworth et al., 1988)

2.04, 3.37, 3.82, and 4.52 at 20, 30, 35, and 40 °C, respectively (Tse et al., 1992)

2.40 and 3.68 in distilled water and seawater at 25 °C, respectively (Hunter-Smith et al., 1983)

Distilled water: 0.894, 0.963, 1.098, 1.946, and 2.568 at 2.0, 6.0, 10.0, 18.2, and 25.0 °C, respectively; natural seawater: 1.32 and 3.32 at 6.0 and 25.0 °C, respectively (Dewulf et al., 1995)

3.27, 4.49, and 6.26 at 27.6, 35.0, and 45.0 °C, respectively (Hansen et al., 1993, 1995)

2.38 at 22.0 °C (mole fraction ratio, Leighton and Calo, 1981)

1.42, 2.36, and 3.82 at 10, 20, and 30 °C, respectively (Munz and Roberts, 1987)

0.938 and 2.57 at 5.7 and 24.9 °C, respectively (Dewulf et al., 1999a)

Interfacial tension with water (dyn/cm at 25 °C):
43.7 (Donahue and Bartell, 1952)
43.26 (Harkins et al., 1920)
45.0 (20 °C, Freitas et al., 1982)

Ionization potential (eV):
11.47 ± 0.01 (Franklin et al., 1969)

Bioconcentration factor, log BCF:
1.24 (rainbow trout), 1.48 (bluegill sunfish) (Veith et al., 1980)
2.48 (algae, Geyer et al., 1984)
2.68 (activated sludge, Freitag et al., 1985)

Soil sorption coefficient, log K_{oc}:
2.35 (Abdul et al., 1987)
2.62 (Chin et al., 1988)
2.16 (Captina silt loam), 1.69 (McLaurin sandy loam) (Walton et al., 1992)
2.10 (extracted peat, Rutherford et al., 1992)
1.78 (normal soils), 2.01 (suspended bed sediments) (Kile et al., 1995)
2.20, 2.24, 2.43, 2.25, 2.28, 2.27, and 2.62 at 2.3, 3.8, 6.2, 8.0, 13.5, 18.6, and 25.0 °C, respectively, for a Leie River (Belgium) clay (Dewulf et al., 1999a)

Octanol/water partition coefficient, log K_{ow}:
2.83 (Chou and Jurs, 1979)
2.73 (Banerjee et al., 1980)
2.73, 2.74, and 2.73 at 25, 30, and 40 °C, respectively (Bhatia and Sandler, 1995)
2.39, 2.38, 2.50, 2.44, 2.47, and 2.38 at 2.2, 6.0, 10.0, 14.1, 18.7, and 24.8 °C, respectively (Dewulf et al., 1999a)

Solubility in organics:
Miscible with ethanol, benzene, chloroform, ether, carbon disulfide (U.S. EPA, 1985), petroleum ether, solvent naphtha, and volatile oils (Yoshida et al., 1983a)

Solubility in water:
785 mg/L at 20 °C (Pearson and McConnell, 1975)
970 mg/L at 0 °C (quoted, Dean, 1973)
800 mg/L at 25 °C (Valsaraj, 1988)

757 mg/L at 25 °C (Banerjee et al., 1980)
770 mg/kg at 15 °C, 810 mg/kg at 30 °C (Gross and Saylor, 1931)
In m/kg: 970 at 0 °C, 830 at 10 °C, 800 at 20 °C, 850 at 30 °C (Rex, 1906)
770 mg/kg at 25 °C (Gross, 1929)
805 mg/L at 20 °C (Howard, 1990)
0.8 g/kg at 25 °C (McGovern, 1943)
0.080 wt % at 25 °C (Lo et al., 1986)
780 mg/L at 23-24 °C (Broholm and Feenstra, 1995)
1.0 mL/L at 25 °C (Booth and Everson, 1949)
In wt %: 0.089 at 0 °C, 0.063 at 10.0 °C, 0.060 at 20.5 °C, 0.072 at 31.0 °C, 0.068 at 41.3 °C,
    0.078 at 52.5 °C, 0.096 at 64.0 °C, 0.115 at 75.0 °C (Stephenson, 1992)

Vapor density:
6.29 g/L at 25 °C, 5.31 (air = 1)

Vapor pressure (mmHg):
56 at 10 °C, 90 at 20 °C, 113 at 25 °C, 137 at 30 °C (Verschueren, 1983)
115 at 25 °C (Rogers and McFarlane, 1981)

Environmental fate:
Biological. An anaerobic species of *Clostridium* biodegraded carbon tetrachloride by reductive dechlorination yielding trichloromethane (chloroform), dichloromethane and unidentified products (Gälli and McCarty, 1989). Chloroform also formed by microbial degradation of carbon tetrachloride using denitrifying bacteria (Smith and Dragun, 1984).

Carbon tetrachloride (5 and 10 mg/L) showed significant degradation with rapid adaptation in a static-culture flask-screening test (settled domestic wastewater inoculum) conducted at 25 °C. Complete degradation was observed after 14 d of incubation (Tabak et al., 1981).

Photolytic. Anticipated products from the reaction of carbon tetrachloride with ozone or OH radicals in the atmosphere are phosgene and chlorine radicals (Cupitt, 1980). Phosgene is hydrolyzed readily to hydrogen chloride and carbon dioxide (Morrison and Boyd, 1971).

Chemical/Physical. Under laboratory conditions, carbon tetrachloride will partially hydrolyze in aqueous solutions forming chloroform and carbon dioxide (Smith and Dragun, 1984). Complete hydrolysis yields carbon dioxide and hydrochloric acid (Ellington et al., 1993; Kollig, 1993). The estimated hydrolysis half-life in water at 25 °C and pH 7 is 7,000 yr (Mabey and Mill, 1978) and 40.5 yr (Jeffers et al., 1989; Ellington et al., 1993).

Carbon tetrachloride slowly reacts with hydrogen sulfide in aqueous solution yielding carbon dioxide via the intermediate carbon disulfide. However, in the presence of two micaceous minerals (biotite and vermiculite) and amorphous silica, the rate transformation increases. At 25 °C and a hydrogen sulfide concentration of 0.001 M, the half-lives of carbon tetrachloride were calculated to be 2,600, 160, and 50 d for the silica, vermiculite, and biotite studies, respectively. In all three studies, the major transformation pathway is the formation of carbon disulfide. This compound is then hydrolyzed to carbon dioxide (81-86% yield) and HS⁻. Minor intermediates detected include chloroform (5-15% yield), carbon monoxide (102% yield), and a non-volatile compound tentatively identified as formic acid (3-6% yield) (Kriegman-King and Reinhard, 1992).

Matheson and Tratnyek (1994) studied the reaction of fine-grained iron metal in an anaerobic aqueous solution (15 °C) containing carbon tetrachloride (151 μM). Initially, carbon tetrachloride underwent rapid dehydrochlorination forming chloroform, which further degraded to methylene chloride and chloride ions. The rate of reaction decreased with each dehydro-chlorination step. However, after 1 h of mixing, the concentration of carbon tetrachloride decreased from 151 to approximately 15 μM. No additional products were identified although the authors concluded that environmental circumstances may exist where degradation of methylene chloride may occur. They also reported that reductive dehalogenation of carbon tetrachloride and other chlorinated hydrocarbons used in this study appears to take place in conjunction with the oxidative dissolution or corrosion of the iron metal through a diffusion-limited surface reaction.

The evaporation half-life of carbon tetrachloride (1 mg/L) from water at 25 °C using a shallow-pitch propeller stirrer at 200 rpm at an average depth of 6.5 cm is 29 min (Dilling et al., 1977).

Carbon tetrachloride reacts with OH radicals in the atmosphere at a calculated rate of $<2 \times 10^6$/M·sec (Haag and Yao, 1992).

At influent concentrations of 1.0, 0.1, 0.01, and 0.001 mg/L, the adsorption capacities of the granular activated carbon used were 11, 1.6, 0.24, and 0.04 mg/g, respectively (Dobbs and Cohen, 1980).

Emits toxic chloride and phosgene fumes when heated to decomposition (Lewis, 1990).

Exposure limits: Potential occupational carcinogen. NIOSH REL: STEL (1 h) 2 ppm (12.6 mg/m^3), IDLH 200 ppm; OSHA PEL: TWA 10 ppm, ceiling 25 ppm, 5-min/4-h peak 200 ppm; ACGIH TLV: TWA 5 ppm, ceiling 10 ppm (adopted).

Symptoms of exposure: Central nervous system depression, nausea, vomiting, skin irritation (NIOSH, 1997). Repeated exposure may result in liver damage (Worthing and Hance, 1991).

Toxicity:

EC_{50} (24-h) for *Daphnia magna* 74.5 mg/L (Lilius et al., 1995).

LC_{50} (contact) for earthworm (*Eisenia fetida*) 160 μg/cm^2 (Neuhauser et al., 1985).

LC_{50} (14-d) for *Poecilia reticulata* 67 mg/L (Könemann, 1981).

LC_{50} (96-h) for bluegill sunfish 27 mg/L (Spehar et al., 1982).

LC_{50} (48-h) for red killifish 617 mg/L (Yoshioka et al., 1986), *Daphnia magna* 35 mg/L (LeBlanc, 1980).

LC_{50} (24-h) for *Daphnia magna* 35 mg/L (LeBlanc, 1980).

Acute oral LD_{50} for rats 2,800 mg/kg, guinea pigs 5,760 mg/kg, mice 8,263 mg/kg, rabbits 5,760 mg/kg (quoted, RTECS, 1985).

Drinking water standard (final): MCLG: zero; MCL: 5 μg/L (U.S. EPA, 1996).

Source: Carbon tetrachloride is used in fumigant mixtures such as 1,2-dichloroethane (Granosan) because it reduces the fire hazard (Worthing and Hance, 1991).

Uses: Preparation of dichlorodifluoromethane, refrigerants, aerosols, and propellants; metal degreasing; agricultural fumigant; chlorinating unsaturated organic compounds; production of semiconductors; solvent for fats, oils, rubber, etc; dry cleaning operations; industrial extractant; spot remover; in fire extinguishers; veterinary medicine (anthelminitic); organic synthesis.

CHLORDANE

Synonyms: A 1068; Aspon-chlordane; Belt; CD 68; Chlordan; γ-Chlordan; γ-Chlordane; Chloridan; Chlorindan; Chlorkil; Chlorodane; Chlortox; Corodane; Cortilan-neu; Dichloro-chlordene; Dowklor; ENT 9932; ENT 25552-X; HCS 3,260; Kypchlor; M 140; M 410; NA 2762; NCI-C00099; Niran; Octachlor; 1,2,4,5,6,7,8,8-Octachlor-2,3,3a,4,7,7a-hexahydro-4,7-methanoindane; Octachlorodihydrodicyclopentadiene; 1,2,4,5,6,7,8,8-Octachloro-2,3,3a,4,7,-7a-hexahydro-4,7-methanoindene; **1,2,4,5,6,7,8,8-Octachloro-2,3,3a,4,7,7a-hexahydro-4,7-methano-1H-indene**; 1,2,4,5,6,7,8,8-Octachloro-3a,4,7,7a-hexahydro-4,7-methyleneindane; Octachloro-4,7-methanohydroindane; Octachloro-4,7-methanotetrahydroindane; 1,2,4,5,6,7,8,-8-Octachloro-4,7-methano-3a,4,7,7a-tetrahydroindane; 1,2,4,5,6,7,8,8-Octachloro-3a,4,7,7a-tetrahydro-4,7-methanoindan; 1,2,4,5,6,7,8,8-Octachloro-3a,4,7,7a-tetrahydro-4,7-methanoin-dane; Octaklor; Octaterr; OMS 1437; Orthoklor; RCRA waste number U036; SD 5532; Shell SD-5532; Synklor; Tatchlor 4; Topichlor 20; Topiclor; Topiclor 20; Toxichlor; Velsicol 1068.

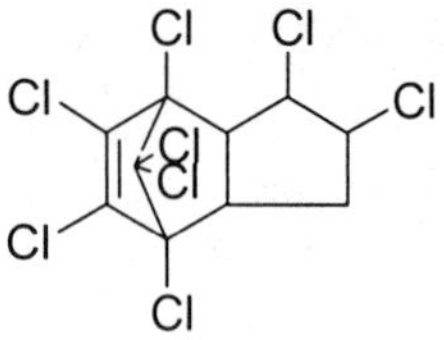

Note: Chlordane is a mixture of cis- and trans-chlordane and other complex chlorinated hydro-carbons including heptachlor (≤7%) and nonachlor. According to Brooks (1974), technical chlordane has the approximate composition: trans-chlordane (24%), four chlordene isomers ($C_{10}H_6Cl_8$) (21.5%), cis-chlordane (19%), heptachlor (10%), nonachlor (7%), pentachlorocyclo-pentadiene (2%), hexachlorocyclopentadiene (>1%), octachlorocyclopentene (1%), $C_{10}H_{7-8}Cl_{6-7}$ (8.5%) and unidentified compounds (6%).

CAS Registry Number: 57-74-9
DOT: 2762
DOT label: Poison
Molecular formula: $C_{10}H_6Cl_8$
Formula weight: 409.78
RTECS: PB9800000
Merck reference: 10, 2046

Physical state, color, and odor:
Colorless to amber to yellowish-brown, viscous liquid with an aromatic, slight pungent odor similar to chlorine

Melting point (°C):
103-109 (NIOSH, 1997)

Boiling point (°C):
175 at 2 mmHg (Roark, 1951)

Density (g/cm³):
1.59-1.63 at 20/4 °C (Melnikov, 1971)

Diffusivity in water (x 10^{-5} cm²/sec):
0.43 at 20 °C using method of Hayduk and Laudie (1974)

Flash point (°C):
Noncombustible but may be utilized in flammable solutions (NIOSH, 1997)

Lower explosive limit (%):
0.7 (kerosene solution, Weiss, 1986)

Upper explosive limit (%):
5 (kerosene solution, Weiss, 1986)

Henry's law constant (x 10^{-5} atm·m³/mol):
4.8 at 25 °C (Warner et al., 1987)

Bioconcentration factor, log BCF:
4.58 (freshwater fish), 3.92 (fish, microcosm) (Garten and Trabalka, 1983)
3.74 (white suckers, Roberts et al., 1977)
3.38-4.18 for green alga (*Scenedesmus quadricauda*) (Glooschenko et al., 1979)
4.20 (*Oncorhynchus mykiss*, Devillers et al., 1996)
2.03 (frogs), 2.51 (bluegill sunfish), 3.00 (goldfish) (Khan et al., 1979)

Soil sorption coefficient, log K_{oc}:
5.15, 5.57 (Chin et al., 1988)
4.85, 4.88 (sand), 4.72, 4.78 (silt) (Johnson-Logan et al., 1992)

Octanol/water partition coefficient, log K_{ow}:
6.00 (Travis and Arms, 1988)
6.21 (Ellington and Stancil, 1988)

Solubility in organics:
Miscible with acetone, cyclohexanone, deodorized kerosene, ethanol, 2-propanol, trichloroethylene (Worthing and Hance, 1991)

Solubility in water:
56 ppb at 25 °C (Sanborn et al., 1976)
1.85 ppm at 25 °C (Weil et al., 1974)
9 ppb at 25 °C (NAS, 1977)
34 μg/L at 24 °C (Johnson-Logan et al., 1992)

Vapor density:
16.75 g/L at 25 °C, 14.15 (air = 1)

Vapor pressure (x 10^{-6} mmHg):
10 at 25 °C (Sunshine, 1969)

Environmental fate:
   Biological. In four successive 7-d incubation periods, chlordane (5 and 10 mg/L) was recalcitrant to degradation in a settled domestic wastewater inoculum (Tabak et al., 1981).
   Soil. The actinomycete *Nocardiopsis* sp. isolated from soil extensively degraded pure *cis*- and *trans*-chlordane to dichlorochlordene, oxychlordane, heptachlor, heptachlor *endo*-epoxide,

chlordene chlorohydrin, and 3-hydroxy-*trans*-chlordane. Oxychlordane slowly degraded to 1-hydroxy-2-chlorochlordene (Beeman and Matsumura, 1981). In Hudson River, NY sediments, the presence of adsorbed chlordane suggests it is very persistent in this environment (Bopp et al., 1982). The reported half-life in soil is approximately 1 yr (Hartley and Kidd, 1987).

The percentage of chlordane remaining in a Congaree sandy loam soil after 14 yr was 40% (Nash and Woolson, 1967).

Chlordane did not degrade in settled domestic wastewater after 28 d (Tabak et al., 1981). Chlordane was metabolized by the fungus *Aspergillus niger* but it was not the main carbon source (Iyengar and Rao, 1973).

Plant. Alfalfa plants were sprayed with technical chlordane at a rate of 1 lb/acre. After 21 d, 95% of the residues had volatilized (Dorough et al., 1972)..

Surface Water. Hargrave et al. (2000) calculated BAFs as the ratio of the compound tissue concentration [wet and lipid weight basis (ng/g)] to the concentration of the compound dissolved in seawater (ng/mL). Average log BAF values for chlordane in ice algae and phytoplankton collected from the Barrow Strait in the Canadian Archipelago were <5.07 and <5.22, respectively.

Photolytic. Chlordane should not undergo direct photolysis because it does not absorb UV light at wavelengths greater than 280 nm (Gore et al., 1971).

Chemical/Physical. In an alkaline medium or solvent, carrier, diluent or emulsifier having an alkaline reaction, chlorine will be released (Windholz et al., 1983). Technical grade chlordane that was passed over a 5% platinum catalyst at 200 °C resulted in the formation of tetrahydro-dicyclopentadiene (Musoke et al., 1982).

Chlordane is subject to hydrolysis via the nucleophilic substitution of chlorine by hydroxyl ions to yield 2,4,5,6,7,8,8-heptachloro-3a,4,7,7a-tetrahydro-4,7-methano-1*H*-indene which is resistant to further hydrolysis (Kollig, 1993). The hydrolysis half-life at pH 7 and 25 °C was estimated to be >197,000 yr (Ellington et al., 1988).

Chlordane (1 mM) in methyl alcohol (30 mL) underwent dechlorination in the presence nickel boride (generated by the reaction of nickel chloride and sodium borohydride). The catalytic dechlorination of chlordane by this method yielded a pentachloro derivative as the major product having the empirical formula $C_{10}H_9Cl_5$ (Dennis and Cooper, 1976).

In aquatic sediments, chlordane was nearly complete after 6 d (Oloffs et al., 1972, 1973).

Emits very toxic chloride fumes when heated to decomposition (Lewis, 1990).

Exposure limits (mg/m³): Potential occupational carcinogen. NIOSH REL: TWA 0.5, IDLH 100; OSHA PEL: TWA 0.5; ACGIH TLV: TWA 0.5 (adopted).

The acceptable daily intake for humans is 0.5 μg/kg (Worthing and Hance, 1991).

Symptoms of exposure: Blurred vision, confusion, ataxia, delirium, coughing, abdominal pain, nausea, vomiting, diarrhea, irritability, tremor, convulsions, anuria (NIOSH, 1997)

Toxicity:

EC_{50} (96-h) for goldfish 0.5 mg/L (Hartley and Kidd, 1987).

EC_{50} (48-h) for *Daphnia pulex* 29 μg/L, *Simocephalus serrulatus* 22 μg/L (Mayer and Ellersieck, 1986).

LC_{50} (96-h) for rainbow trout 8.2-59 μg/L (Mayer and Ellersieck, 1986), bluegill sunfish 41 μg/L (technical), 62 μg/L (72% emulsifiable concentrate), *Daphnia magna* 97 μg/L (technical), 156 μg/L (72% emulsifiable concentrate) (Randall et al., 1979), fish (*Cyprinodon variegatus*) 24.5 μg/L, fish (*Lagodon rhomboides*) 6.4 μg/L, decapod (*Penaeus duorarum*) 0.4 μg/L, decapod (*Palaemonetes pugio*) 4.8 μg/L (quoted, Reish and Kauwling, 1978), *Pteronarcys californica* 15 μg/L (Sanders and Cope, 1968).

LC_{50} (48-h) for red killifish 245 μg/L (Yoshioka et al., 1986), *Daphnia magna* 590 μg/L

(Worthing and Hance, 1991).

LC_{50} (24-h) for rainbow trout 20-200 μg/L (Mayer and Ellersieck, 1986).

LC_{50} (inhalation) for cats 100 mg/m^3/4-h (quoted, RTECS, 1985).

Acute oral LD_{50} for rats 283 mg/kg, chickens 220 mg/kg, hamsters 1,720 mg/kg, mice 145 mg/kg, rabbits 100 mg/kg (quoted, RTECS, 1985), bobwhite quail 83 mg/kg diet, mallard ducks 795 mg/kg diet (Worthing and Hance, 1991).

Acute percutaneous LD_{50} for rabbits >200 but <2,000 mg/kg (Worthing and Hance, 1991).

In 2-yr feeding trials, the NOEL for dogs was 3 mg/kg diet (Worthing and Hance, 1991).

Drinking water standard (final): MCLG: zero; MCL: 2 μg/L (U.S. EPA, 1996).

Uses: Non-systemic contact and ingested insecticide and fumigant. Treating termites and other insect pests. Also used as a wood preservative, treating underground cables, and reducing earthworm populations in lawns (Worthing and Hance, 1991).

cis-CHLORDANE

Synonyms: α-Chlordane; α-1,2,4,5,6,7,8,8-Octachloro-3a,4,7,7a-tetrahydro-4,7-methanoindan.

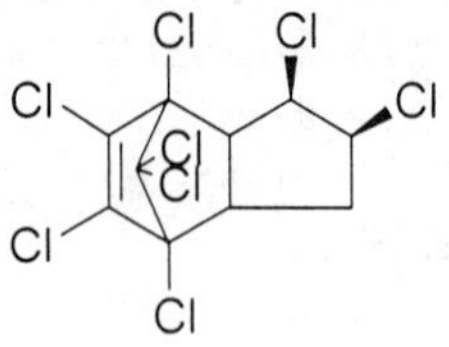

CAS Registry Number: 5103-74-2
DOT: 2762
DOT label: Flammable liquid
Molecular formula: $C_{10}H_6Cl_8$
Formula weight: 409.78
RTECS: PC0175000

Physical state:
Solid

Melting point (°C):
107.0-108.8 (Callahan et al., 1979)

Boiling point (°C):
175 (technical grade containing both *cis* and *trans* isomers, Sims et al., 1988)

Diffusivity in water (x 10^{-5} cm²/sec):
0.43 at 20 °C using method of Hayduk and Laudie (1974)

Henry's law constant (x 10^{-4} atm·m³/mol):
8.75 and 41.3 at 23 °C in distilled water and seawater, respectively (Atlas et al., 1982)

Bioconcentration factor, log BCF:
3.62-4.34 (fish tank), 6.15 (Lake Ontario) (rainbow trout, Oliver and Niimi, 1985)
3.30 (*B. subtilis*, Grimes and Morrison, 1975)
3.68 freshwater clam (*Corbicula manilensis*) (Hartley and Johnston, 1983)
4.45 (*Oncorhynchus mykiss*, Devillers et al., 1996)

Soil sorption coefficient, log K_{oc}:
5.40 (river sediments, Oliver and Charlton, 1984)
5.57 (Chin et al., 1988)
4.77 (Meylan et al., 1992)
4.94 (muck), 4.83 (loam), 4.55 (loess), 4.73 (clay), 4.48 (sand) (Erstfeld et al., 1996)

Octanol/water partition coefficient, log K_{ow}:
5.93 using method of Kenaga and Goring (1980)

Solubility in organics:
Miscible with aliphatic and aromatic solvents (Windholz et al., 1983)

Solubility in water:
51 μg/L at 20-25 °C (Geyer et al., 1984)

Vapor pressure (mmHg):
3.6 x 10^{-5} at 25 °C (Hinckley et al., 1990)

Environmental fate:
    Groundwater. According to the U.S. EPA (1986), *cis*-chlordane has a high potential to leach to groundwater.
    Photolytic. Irradiation of *cis*-chlordane by a 450-W high-pressure mercury lamp gave photo-*cis*-chlordane (Ivie et al., 1972).
    Chemical/Physical. In an alkaline medium or solvent, carrier, diluent or emulsifier having an alkaline reaction, chlorine will be released (U.S. EPA, 1985). The hydrolysis half-lives of *cis*-chlordane at pH values of 10.18 (84.0 °C) and 10.85 (65.0 °C) were 1.92 and 16.8 h, respectively (Ellington et al., 1987).
    Chlordane (0.009 mM) reacted with OH radicals in water (pH 3.3) at a rate of 6-170 x 10^8/M·sec (Haag and Yao, 1992).
    Emits toxic chloride fumes when heated to decomposition (Lewis, 1990).

Symptoms of exposure: Blurred vision, confusion, ataxia, delirium, coughing, abdominal pain, nausea, vomiting, diarrhea, irritability, tremor, convulsions, anuria (NIOSH, 1997)

Toxicity:
    EC_{50} (96-h) for goldfish 0.5 mg/L (Hartley and Kidd, 1987).
    LC_{50} (96-h) for rainbow trout 90 μg/L (Hartley and Kidd, 1987), bluegill sunfish 7.09 μg/L (Mayer and Ellersieck, 1986).
    LC_{50} (24-h) for bluegill sunfish 28.3 μg/L (Mayer and Ellersieck, 1986).
    Acute oral LD_{50} for rats 365-590 mg/kg (Hartley and Kidd, 1987).

Drinking water standard: See chlordane.

Use: Insecticide.

trans-CHLORDANE

Synonyms: β-Chlordane; 1,2,4,5,6,7,8,8-Octachloro-3a,4,7,7a-tetrahydro-4,7-methanoindan.

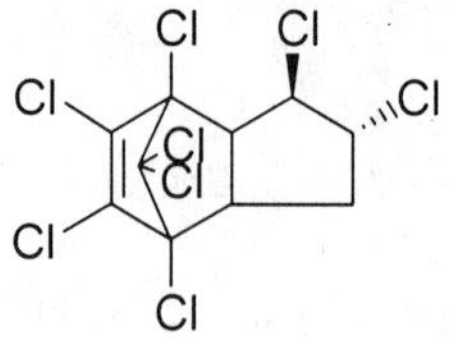

CAS Registry Number: 5103-71-9
DOT: 2762
DOT label: Flammable liquid
Molecular formula: $C_{10}H_6Cl_8$
Formula weight: 409.78
RTECS: PB9705000

Physical state:
Solid

Melting point (°C):
103.0-105.0 (Callahan et al., 1979)

Boiling point (°C):
175 (technical grade containing both *cis* and *trans* isomers, Sims et al., 1988)

Diffusivity in water (x 10^{-5} cm²/sec):
0.43 at 20 °C using method of Hayduk and Laudie (1974)

Henry's law constant (x 10^{-3} atm·m³/mol at 23 °C):
1.34 and 5.59 in distilled water and seawater, respectively (Atlas et al., 1982)

Bioconcentration factor, log BCF:
4.18-4.30 (fish tank), 4.88 (Lake Ontario) (rainbow trout, Oliver and Niimi, 1985)
3.36 (*B. subtilis*, Grimes and Morrison, 1975)
3.64 freshwater clam (*Corbicula manilensis*) (Hartley and Johnston, 1983)

Soil sorption coefficient, log K_{oc}:
5.48 (Niagara River sediments, Oliver and Charlton, 1984)
An average value of 6.3 was experimentally determined using 10 suspended sediment samples
collected from the St. Clair and Detroit Rivers (Lau et al., 1989).
5.03 (muck), 5.04 (loam), 4.67 (loess), 4.92 (clay), 4.67 (sand) (Erstfeld et al., 1996)

Octanol/water partition coefficient, log K_{ow}:
8.69, 9.65 using method of Kenaga and Goring (1980)

Solubility in organics:
Miscible with aliphatic and aromatic solvents (U.S. EPA, 1985)

Vapor pressure (mmHg):
5.03 x 10^{-5} at 25 °C (Hinckley et al., 1990)

Environmental fate:

Photolytic. Irradiation of *trans*-chlordane by a 450-W high-pressure mercury lamp gave photo-*trans*-chlordane (Ivie et al., 1972).

Chemical/Physical. In an alkaline medium or solvent, carrier, diluent, or emulsifier having an alkaline reaction, chlorine will be released (Windholz et al., 1983).

Chlordane (0.009 mM) reacted with OH radicals in water (pH 3.3) at a rate of 6-170 x 10^8/M·sec (Haag and Yao, 1992).

Emits toxic chloride fumes when heated to decomposition (Lewis, 1990).

Toxicity:

LC_{50} (96-h) and LC_{50} (24-h) values for bluegill sunfish were 50.5 and 210 μg/L, respectively (Mayer and Ellersieck, 1986).

Drinking water standard: See chlordane.

Use: Insecticide.

CHLOROACETALDEHYDE

Synonyms: 2-Chloroacetaldehyde; Chloroacetaldehyde monomer; 2-Chloroethanal; 2-Chloro-1-ethanal; Monochloroacetaldehyde; RCRA waste number P023; UN 2232.

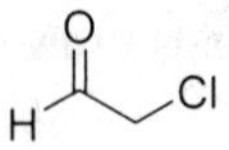

CAS Registry Number: 107-20-0
DOT: 2232
DOT label: Poison
Molecular formula: C_2H_3ClO
Formula weight: 78.50
RTECS: AB2450000
Merck reference: 10, 2075

Physical state, color, and odor:
Clear, colorless liquid with an irritating, acrid odor

Melting point (°C):
-19.5 (40% aqueous solution, NIOSH, 1997)

Boiling point (°C):
85-85.5 at 748 mmHg (Weast, 1986)
85 (40% solution, Hawley, 1981)

Density (g/cm³ at 20/4 °C):
1.19 (40% solution, NIOSH, 1997)
1.236 (Aldrich, 1988)

Diffusivity in water (x 10^{-5} cm²/sec):
1.15 at 20 °C using method of Hayduk and Laudie (1974)

Flash point (°C):
87.8 (40% aqueous solution, NIOSH, 1997)
53 (Aldrich, 1988)

Ionization potential (eV):
10.61 (NIOSH, 1997)

Soil sorption coefficient, log K_{oc}:
Unavailable because experimental methods for estimation of this parameter for halogenated aldehydes are lacking in the documented literature. However, its miscibility in water suggests its adsorption to soil will be nominal (Lyman et al., 1982).

Octanol/water partition coefficient, log K_{ow}:
Unavailable because experimental methods for estimation of this parameter for halogenated aldehydes are lacking in the documented literature

Solubility in organics:
Soluble in ether (Weast, 1986), acetone, and methanol (Hawley, 1981)

Solubility in water:
>50 wt %, forms a hemihydrate (Hawley, 1981)

Vapor density:
3.21 g/L at 25 °C, 2.71 (air = 1)

Vapor pressure (mmHg):
100 at 20 °C (NIOSH, 1997)

Environmental fate:
Chemical/Physical. Polymerizes on standing (Windholz et al., 1983).

Exposure limits: NIOSH REL: ceiling 1 ppm (3 mg/m^3), IDLH 45 ppm; OSHA PEL: ceiling 1 ppm; ACGIH TLV: ceiling 1 ppm (adopted).

Symptoms of exposure: Severe irritation and blurred vision on exposure to vapors. Skin contact with 40% aqueous solution can cause skin burn and tissue damage (Patnaik, 1992). An irritation concentration of 9.00 mg/m^3 in air was reported by Ruth (1986).

Toxicity:
Acute oral LD$_{50}$ for rats 75 mg/kg, mice 69 mg/kg (quoted, RTECS, 1985).

Uses: Removing barks from trees; manufacture of 2-aminothiazole; chemical intermediate; organic synthesis.

α-CHLOROACETOPHENONE

Synonyms: CAF; CAP; Chemical mace; 2-Chloroacetophenone; Ω-Chloroacetophenone; Chloromethyl phenyl ketone; **2-Chloro-1-phenylethanone**; CN; Mace; NCI-C55107; Phenacyl chloride; Phenyl chloromethyl ketone; Tear gas; UN 1697.

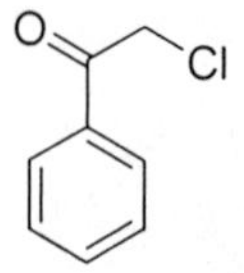

CAS Registry Number: 532-27-4
DOT: 1697
DOT label: Poison
Molecular formula: C_8H_7ClO
Formula weight: 154.60
RTECS: AM6300000
Merck reference: 10, 2082

Physical state, color, and odor:
Colorless to gray crystalline solid with a sharp, penetrating, irritating odor. Odor threshold concentration is 0.035 ppm (Amoore and Hautala, 1983). The low odor and high odor threshold concentrations were 102 and 150 $\mu g/m^3$, respectively (Ruth, 1986).

Melting point (°C):
56.5 (Weast, 1986)
58-59 (Windholz et al., 1983)
54-56 (Aldrich, 1988)

Boiling point (°C):
247 (Weast, 1986)
244-245 (Windholz et al., 1983)

Density (g/cm³):
1.324 at 15/4 °C (Weast, 1986)

Diffusivity in water (x 10⁻⁵ cm²/sec):
0.69 at 15 °C using method of Hayduk and Laudie (1974)

Flash point (°C):
117.9 (NIOSH, 1997)

Ionization potential (eV):
9.5 (Franklin et al., 1969)
9.44 (NIOSH, 1997)

Soil sorption coefficient, log K_{oc}:
Unavailable because experimental methods for estimation of this parameter for halogenated aromatic ketones are lacking in the documented literature. However, its miscibility in water suggests its adsorption to soil will be nominal (Lyman et al., 1982).

Octanol/water partition coefficient, log K_{ow}:
Unavailable because experimental methods for estimation of this parameter for halogenated aromatic ketones are lacking in the documented literature

Solubility in organics:
Soluble in acetone (Weast, 1986). Freely soluble in alcohol, benzene, and ether (Windholz et al., 1983)

Vapor pressure (x 10^{-3} mmHg):
5.4 at 20 °C (Windholz et al., 1983)
4 at 20 °C, 14 at 30 °C (Verschueren, 1983)

Environmental fate:
   Chemical/Physical. Releases toxic chloride fumes when heated to decomposition (Sax and Lewis, 1987).

Exposure limits: NIOSH REL: TWA 0.3 mg/m^3 (0.05 ppm), IDLH 15 mg/m^3; OSHA PEL: TWA 0.05 ppm (0.3 mg/m^3); ACGIH TLV: TWA 0.05 ppm (adopted).

Symptoms of exposure: An irritation concentration of 0.05 mg/m^3 in air was reported by Ruth (1986).

Toxicity:
   Acute oral LD$_{50}$ for rats 50 mg/kg, rabbits 118 mg/kg, mice 139 mg/kg, guinea pigs 158 mg/kg (quoted, RTECS, 1985).

Uses: Riot control agent.

4-CHLOROANILINE

Synonyms: 1-Amino-4-chlorobenzene; 1-Amino-*p*-chlorobenzene; 4-Aminochlorobenzene; *p*-Aminochlorobenzene; 4-Chloraniline; *p*-Chloraniline; *p*-Chloroaniline; **4-Chlorobenzamine**; *p*-Chlorobenzamine; 4-Chlorophenylamine; *p*-Chlorophenylamine; NCI-C02039; RCRA waste number P024; UN 2018; UN 2019.

CAS Registry Number: 106-47-8
DOT: 2018 (solid); 2019 (liquid)
DOT label: Poison
Molecular formula: C_6H_6ClN
Formula weight: 127.57
RTECS: BX0700000
Merck reference: 10, 2087

Physical state, color, and odor:
Yellowish-white solid with a mild, sweetish odor. Odor threshold concentration is 287 ppm (Keith and Walters, 1992).

Melting point (°C):
72.50 (Martin et al., 1979)
70 (Morrison and Boyd, 1971)

Boiling point (°C):
232 (Weast, 1986)

Density (g/cm³):
1.429 at 19/4 °C (Weast, 1986)

Diffusivity in water (x 10⁻⁵ cm²/sec):
0.75 at 20 °C using method of Hayduk and Laudie (1974)

Dissociation constant, pK$_a$:
4.15 (Weast, 1986)
3.98 (Howard, 1989)

Flash point (°C):
>1,205 (Weiss, 1986)

Lower explosive limit (%):
Not pertinent (Weiss, 1986)

Upper explosive limit (%):
Not pertinent (Weiss, 1986)

Entropy of fusion (cal/mol·K):
13.8 (Tsonopoulos and Prausnitz, 1971)

Heat of fusion (kcal/mol):
3.75 (Tsonopoulos and Prausnitz, 1971)

Henry's law constant (x 10^{-5} atm·m^3/mol):
1.07 at 25 °C (calculated, Howard, 1989)

Ionization potential (eV):
7.77 (Farrell and Newton, 1966)
8.00 (Kinoshita, 1962)

Bioconcentration factor, log BCF:
2.41 (*Chlorella fusca*, Freitag et al., 1982; Geyer et al., 1984)
-0.10, 0.23 (*Cyprinus carpio*, Devillers et al., 1996)
1.11 (golden orfe), 2.42 (algae), 2.45 (activated sludge) (Freitag et al., 1982)
0.91 (zebrafish, Kalsch et al., 1991; Zok et al., 1991)

Soil sorption coefficient, log K_{oc}:
2.42 (quoted, Hodson and Williams, 1988)
1.98 (Spodosol, pH 3.9), 2.05 (Speyer soil 2/2, pH 5.8), 3.10 (Alfisol, pH 7.5), 3.13 (Speyer soil
    2.1, pH 7.0), 3.18 (Entisol, pH 7.9) (Rippen et al., 1982)
2.75 (Rao and Davidson, 1980)
1.96 (Meylan et al., 1992)
2.60, 2.70, 2.82, 2.91 (Van Bladel and Moreale, 1977)

Octanol/water partition coefficient, log K_{ow}:
1.83 (Leo et al., 1971)
2.78 (Hammers et al., 1982)
1.88 (Hammers et al., 1982; Garst and Wilson, 1984)
2.02 (Geyer et al., 1984)

Solubility in organics:
Soluble in ethanol and ether (Weast, 1986). Freely soluble in acetone and carbon disulfide
(Windholz et al., 1983).

Solubility in water (g/L):
3.9 at 20-25 °C (Kilzer et al., 1979)
3.1 at 4 °C, 4.7 at 25 °C, 7.8 at 40 °C (Moreale and Van Bladel, 1979)
2.8 at 20 °C (Hafkenscheid and Tomlinson, 1983)

Vapor pressure (x 10^{-2} mmHg):
1.5 at 20 °C, 5 at 30 °C (Verschueren, 1983)
2.5 at 25 °C (Piacente et al., 1985)

Environmental fate:
   Biological. In an anaerobic medium, the bacteria of the *Paracoccus* sp. converted 4-chloroaniline to 1,3-bis(*p*-chlorophenyl)triazene and 4-chloroacetanilide with product yields of 80 and 5%, respectively (Minard et al., 1977). In a field experiment, [^{14}C]4-chloroaniline was applied to a soil at a depth of 10 cm. After 20 wk, 32.4% of the applied amount was recovered

in soil. Metabolites identified include 4-chloroformanilide, 4-chloroacetanilide, 4-chloronitrobenzene, 4-chloronitrosobenzene, 4,4'-dichloroazoxybenzene, and 4,4'-dichloroazobenzene (Freitag et al., 1984).

In a 56-d experiment, [^{14}C]4-chloroaniline applied to soil-water suspensions under aerobic and anaerobic conditions gave $^{14}CO_2$ yields of 3.0 and 2.3%, respectively (Scheunert et al., 1987). Silage samples (chopped corn plants) containing 4-chloroaniline were incubated in an anaerobic chamber for 2 wk at 28 °C. After 3 d, 4-chloroaniline was biologically metabolized to 4-chloroacetanilide and another compound, tentatively identified as 4-chloroformanilide (Lyons et al., 1985). In activated sludge, 22.7% mineralized to carbon dioxide after 5 d (Freitag et al., 1985).

In activated sludge inoculum, 98.0% COD removal was achieved. The average rate of biodegradation was 16.7 mg COD/g·h (Pitter, 1976).

Soil. 4-Chloroaniline covalently bonds with humates in soils to form quinoidal structures followed by oxidation to yield a nitrogen-substituted quinoid ring. A reaction half-life of 13 min was determined with one humic compound (Parris, 1980). Catechol, a humic acid monomer, reacted with 4-chloroaniline yielding 4,5-bis(4-chlorophenylamino)-3,5-cyclohexadiene-1,2-dione (Adrian et al., 1989).

Photolytic. Under artificial sunlight, river water containing 2-5 ppm 4-chloroaniline photodegraded to 4-aminophenol and unidentified polymers (Mansour et al., 1989). Photooxidation of 4-chloroaniline (100 μM) in air-saturated water using UV light (λ >290 nm) produced 4-chloronitrobenzene and 4-chloronitrosobenzene. About 6 h later, 4-chloroaniline completely reacted leaving dark purple condensation products (Miller and Crosby, 1983). A carbon dioxide yield of 27.7% was achieved when 4-chloroaniline adsorbed on silica gel was irradiated with light (λ >290 nm) for 17 h (Freitag et al., 1985).

A rate constant of 8.3 x 10^{-11} cm^3/molecule·sec was reported for the gas-phase reaction of 4-chloroaniline and OH radicals in air (Wahner and Zetzsch, 1983).

Chemical/Physical. 4-Chloroaniline will not hydrolyze to any reasonable extent (Kollig, 1993).

Toxicity:

EC$_{10}$ and EC$_{50}$ concentrations inhibiting the growth of alga *Scenedesmus subspicatus* in 96 h were 0.4 and 2.4 mg/L, respectively (Geyer et al., 1985).

LC$_{50}$ (96-h static bioassay) for rainbow trout 14 mg/L, fathead minnows 12 mg/L, channel catfish 23 mg/L, bluegill sunfish 2 mg/L (quoted, Verschueren, 1983), Japanese medaka (*Oryzias latipes*) 37.7 mg/L (Holcombe et al., 1995).

LC$_{50}$ (48-h) for red killifish 219 mg/L (Yoshioka et al., 1986).

Acute oral LD$_{50}$ for wild birds 100 mg/kg, guinea pigs 350 mg/kg, mice 100 mg/kg, quail 237 mg/kg, rats 310 mg/kg; LD$_{50}$ (inhalation) for mice 250 mg/m^3/6-h.

Uses: Dye intermediate; pharmaceuticals; agricultural chemicals.

CHLOROBENZENE

Synonyms: Benzene chloride; Chlorbenzene; Chlorbenzol; Chlorobenzol; MCB; Mono-chlorbenzene; Monochlorobenzene; NCI-C54886; Phenyl chloride; RCRA waste number U037; UN 1134.

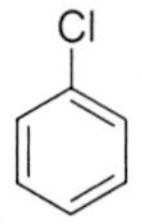

CAS Registry Number: 108-90-7
DOT: 1134
DOT label: Flammable liquid
Molecular formula: C_6H_5Cl
Formula weight: 112.56
RTECS: CZ0175000
Merck reference: 10, 2090

Physical state, color, and odor:
Colorless, watery liquid with a sweet almond or mothball-like odor. The reported odor threshold concentrations ranged from 0.19 to 0.21 ppm (Keith and Walters, 1992; Young et al., 1996).

Melting point (°C):
-45.6 (Weast, 1986)

Boiling point (°C):
132 (Weast, 1986)

Density (g/cm^3):
1.10646 at 20.00/4 °C (Tsierkezos et al., 2000)
1.0904 at 35.00/4 °C, 1.0790 at 45.00/4 °C (Sastry et al., 1999)
1.1009 at 25/4 °C (Kirchnerová and Cave, 1976)

Diffusivity in water (x 10^{-5} cm^2/sec, 25 °C):
1.04, 1.00, and 0.98 at mole fractions of 2 x 10^{-6}, 4 x 10^{-6}, and 8 x 10^{-6}, respectively (Gabler et al., 1996)

Flash point (°C):
27.8 (NIOSH, 1997)

Lower explosive limit (%):
1.3 (NIOSH, 1997)

Upper explosive limit (%):
9.6 (NIOSH, 1997)

Entropy of fusion (cal/mol·K):
10.02 (Stull, 1937)

Heat of fusion (kcal/mol):
2.28 (Dean, 1987)

Henry's law constant (x 10^{-3} atm·m^3/mol):
3.93 at 25 °C (Warner et al., 1987)
3.6 (Pankow and Rosen, 1988)
3.70 at 25 °C (Mackay et al., 1979; Valsaraj, 1988)
6.21 at 37 °C (Sato and Nakajima, 1979)
2.44, 2.81, 3.41, 3.60, and 4.73 at 10, 15, 20, 25, and 30 °C, respectively (Ashworth et al., 1988)
2.84 at 20 °C (Hovorka and Dohnal, 1997)
3.11 at 25 °C (Shiu and Mackay, 1997)
3.85 at 25 °C (Hoff et al., 1993)
3.18 at 23.0 °C (mole fraction ratio, Leighton and Calo, 1981)
3.38 at 25.0 °C (Ramachandran et al., 1996)
1.11, 1.54, 1.81, 2.80, and 3.79 at 2.0, 6.0, 10.0, 18.0, and 25.0 °C, respectively (Dewulf et al., 1999)

Interfacial tension with water (dyn/cm at 20 °C):
37.41 (Harkins et al., 1920)

Ionization potential (eV):
9.07 (Lias, 1998)
9.14 (Yoshida et al., 1983a)

Bioconcentration factor, log BCF:
3.23 (activated sludge), 1.70 (algae), 1.88 (golden ide) (Freitag et al., 1985)
2.65 (fathead minnow, Veith et al., 1979)
3.69 green alga, *Selenastrum capricornutum* (Casserly et al., 1983)

Soil sorption coefficient, log K_{oc}:
1.92 (Woodburn silt loam, Chiou et al., 1983)
2.07 (Lincoln sand, Wilson et al., 1981)
2.59 (Roberts et al., 1980)
1.91 (Mt. Lemmon soil, Hu et al., 1995)
2.50 (Captina silt loam), 2.17 (McLaurin sandy loam) (Walton et al., 1992)
2.22 (field sample), 2.41 (average of 6 measurements), 2.70 (field sample) (Schwarzenbach and Westall, 1981)
2.37 (sandy soil, Van Gestel and Ma, 1993)
2.10 (muck), 1.89 (Eustis sand) (Brusseau et al., 1990)

Octanol/water partition coefficient, log K_{ow}:
2.84 (Fujita et al., 1964; Mirrlees et al., 1976; Garst and Wilson, 1984; Pereira et al., 1988)
2.81 (Mirrlees et al., 1976)
2.98 (Tewari et al., 1982; Wasik et al., 1981, 1983; Paschke et al., 1998)
2.71 (Schwarzenbach and Westall, 1981)
2.898 (Brooke et al., 1990; de Bruijn et al., 1989)
2.96 at 25 °C (Andersson and Schräder, 1999)
2.784 (Brooke et al., 1990)
2.83 (Hammers et al., 1982; Yoshida et al., 1983)
2.65 (Campbell and Luthy, 1985)

Solubility in organics:
Soluble in ethanol, benzene, carbon tetrachloride, chloroform, ether, and methylene chloride

Solubility in water:
502 mg/L at 25 °C (Banerjee, 1984)
295 mg/L at 25 °C (Tewari et al., 1982; Wasik et al., 1981)
4.43 mM at 25 °C (Wasik et al., 1983)
471.7 mg/L at 25 °C (Aquan-Yuen et al., 1979)
500 mg/L at 25 °C, 488 mg/kg at 30 °C (Andrews and Keefer, 1950)
446 mg/kg at 30 °C (Gross and Saylor, 1931)
448 mg/L at 30 °C (Freed et al., 1977)
503 mg/L at 25 °C (Yalkowsky et al., 1979)
534 mg/kg at 25 °C (Chey and Calder, 1972)
474.0 mg/L at 30 °C (McNally and Grob, 1983, 1984)
0.035 wt % at 25 °C (Lo et al., 1986)
4.11 mmol/kg at 25.0 °C (Vesala, 1974)
3.78 mM at 25.0 °C (Sanemasa et al., 1987)
496 mg/L at 25 °C (Paschke et al., 1998)
0.046, 0.045, and 0.050 wt % at 10.0, 20.0, and 30.0 °C, respectively (Schwarz and Miller, 1980)

Vapor density:
4.60 g/L at 25 °C, 3.88 (air = 1)

Vapor pressure (mmHg):
9 at 20 °C (NIOSH, 1997)
11.86 at 25 °C (Mackay et al., 1982)

Environmental fate:
Biological. In activated sludge, 31.5% of the applied chlorobenzene mineralized to carbon dioxide after 5 d (Freitag et al., 1985). A mixed culture of soil bacteria or a *Pseudomonas* sp. transformed chlorobenzene to chlorophenol (Ballschiter and Scholz, 1980). Pure microbial cultures isolated from soil hydroxylated chlorobenzene to 2- and 4-hydroxychlorobenzene (Smith and Rosazza, 1974). Chlorobenzene was statically incubated in the dark at 25 °C with yeast extract and settled domestic wastewater inoculum. At a concentration of 5 mg/L, biodegradation yields at the end of 1 and 2 wk were 89 and 100%, respectively. At a concentration of 10 mg/L, significant degradation with gradual adaptation was observed. Complete degradation was not observed until after the 3rd week of incubation (Tabak et al., 1981).

Heukelekian and Rand (1955) reported a 5-d BOD value of 0.03 g/g which is 1.5% of the ThOD value of 2.00 g/g.

Surface Water. Estimated half-lives of chlorobenzene (1.0 μg/L) from an experimental marine mesocosm during the spring (8-16 °C), summer (20-22 °C), and winter (3-7 °C) were 21, 4.6, and 13 d, respectively (Wakeham et al., 1983).

Photolytic. Under artificial sunlight, river water containing 2-5 ppm chlorobenzene photo-degraded to phenol and chlorophenol. The lifetimes of chlorobenzene in distilled water and river water were 17.5 and 3.8 h, respectively (Mansour et al., 1989). In distilled water containing 1% acetonitrile exposed to artificial sunlight, 28% of chlorobenzene in solution photolyzed to phenol, chloride ion, and acetanilide with reported product yields of 55, 112, and 2%, respectively (Dulin et al., 1986).

Titanium dioxide suspended in an aqueous solution and irradiated with UV light (λ = 365 nm) converted chlorobenzene to carbon dioxide at a significant rate (Matthews, 1986). Products identified as intermediates in this reaction include three monochlorophenols, chlor-ohydroquinone and hydroxyhydroquinone as intermediates (Kawaguchi and Furuya, 1990).

Photooxidation of chlorobenzene in air containing nitric oxide in a Pyrex glass vessel and a quartz vessel gave 3-chloronitrobenzene, 2-chloro-6-nitrophenol, 2-chloro-4-nitrophenol, 4-chloro-2-nitrophenol, 4-nitrophenol, 3-chloro-4-nitrophenol, 3-chloro-6-nitrophenol, and 3-chloro-2-nitrophenol (Kanno and Nojima, 1979). A carbon dioxide yield of 18.5% was achieved when chlorobenzene adsorbed on silica gel was irradiated with light (λ >290 nm) for 17 h. The sunlight irradiation of chlorobenzene (20 g) in a 100-mL borosilicate glass-stoppered Erlenmeyer flask for 28 d yielded 1,060 ppm monochlorobiphenyl (Uyeta et al., 1976).

When an aqueous solution containing chlorobenzene (190 μM) and a nonionic surfactant micelle (Brij 58, a polyoxyethylene cetyl ether) was illuminated by a photoreactor equipped with 253.7-nm monochromatic UV lamps, phenol, hydrogen and chloride ions formed as the major products. Although not measured, it was reported that aromatic aldehydes, organic acids, and carbon dioxide would form from the photoreaction of benzene in water under similar conditions. A duplicate experiment was conducted using an ionic micelle (triethylamine, 5 mM), which serves as a hydrogen source. Products identified were phenol and benzene (Chu and Jafvert, 1994).

Chemical/Physical. Anticipated products from the reaction of chlorobenzene with ozone or OH radicals in the atmosphere are chlorophenols and ring cleavage compounds (Atkinson et al., 1985; Cupitt, 1980).

In the absence of oxygen, chlorobenzene reacted with Fenton's reagent forming chloro-henols, dichlorobiphenyls, and phenolic polymers as major intermediates. With oxygen, chloro-enzoquinone, chlorinated and nonchlorinated diols formed (Sedlak and Andren, 1991).

Based on an assumed base mediated 1% disappearance after 16 d at 85 °C and pH 9.70 (pH 11.26 at 25 °C), the hydrolysis half-life was estimated to be >900 yr (Ellington et al., 1988).

Toxic fumes of phosgene and hydrogen chloride may form when exposed to an open flame (CHRIS, 1984). Chlorobenzene is stable up to 700 °C but in combination with other chlorinated compounds, it is stable up to 900 °C (Graham et al., 1986).

At influent concentrations of 10.0, 1.0, 0.1, and 0.01 mg/L, the adsorption capacity of the granular activated carbon used at pH 7.4 were 890, 91, 9.3, and 0.95 mg/g, respectively (Dobbs and Cohen, 1980).

Exposure limits: NIOSH REL: Awaiting OSHA ruling to determine if the recommended TWA exposure limit of 75 ppm is protective of human health, IDLH 1,000 ppm; OSHA PEL: TWA 75 ppm (350 mg/m³); ACGIH TLV: TWA 10 ppm (adopted).

Symptoms of exposure: Inhalation of vapors may cause drowsiness, incoordination, and liver damage. May irritate eyes and skin (Patnaik, 1992). An irritation concentration of 933.33 mg/m³ in air was reported by Ruth (1986).

Toxicity:
EC_{50} (96-h) and EC_{50} (3-h) concentrations that inhibit the growth of 50% of *Selenastrum capricornutum* population are 12.5 and 33.0 mg/L, respectively (Calamari et al., 1983).

EC_{50} (3-h) for *Selenastrum capricornutum* 33.0 mg/L (Calamari et al., 1983).

LC_{50} (contact) for earthworm (*Eisenia fetida*) 29 μg/cm² (Neuhauser et al., 1985).

LC_{50} for goldfish 1.8 mL/kg (Verschueren, 1983), 89.7 and 164 mg/L (soil porewater concentration) for earthworm (*Eisenia andrei*) and 252 and 482 mg/L (soil porewater concentration) for earthworm (*Lumbricus rubellus*) (Van Gestel and Ma, 1993).

LC_{50} for male rats 13,490 mg/m³ (Bonnet et al., 1982), female mice 8,581 mg/m³ (Bonnet et al., 1979).

LC_{50} (14-d) for *Poecilia reticulata* 19.1 mg/L (Könemann, 1981).

LC_{50} (7-d) for *Micropterus salmoides* 50 μg/L (Birge et al., 1979).

LC_{50} (96-h) for bluegill sunfish 16 mg/L (Spehar et al., 1982).

LC$_{50}$ (96-h static flow-through system) for *Oncorhynchus mykiss* 7.36 mg/L (Hodson et al., 1984), 4.7 mg/L (Dalich et al., 1982), *Cyprinodon variegatus* 10 ppm using natural seawater (Heitmuller et al., 1981).

LC$_{50}$ (48-h) for *Daphnia magna* 86 mg/L (LeBlanc, 1980), *Cyprinodon variegatus* 8.9 ppm (Heitmuller et al., 1981).

LC$_{50}$ (24-h) for *Daphnia magna* 140 mg/L (LeBlanc, 1980), *Cyprinodon variegatus* >20 ppm (Heitmuller et al., 1981).

Acute oral LD$_{50}$ for rats 2,910 mg/kg, guinea pigs 5,060 mg/kg, rabbits 2,250 mg/kg (quoted, RTECS, 1985).

Drinking water standard (final): MCLG: 0.1 mg/L; MCL: 0.1 mg/L (U.S. EPA, 1996).

Uses: Preparation of phenol, 4-chlorophenol, chloronitrobenzene, aniline, 2-, 3-, and 4-nitro-chlorobenzenes; carrier solvent for methylene diisocyanate and pesticides; solvent for paints; insecticide, pesticide, and dyestuffs intermediate; heat transfer agent.

o-CHLOROBENZYLIDENEMALONONITRILE

Synonyms: 2-Chlorobenzalmalononitrile; *o*-Chlorobenzalmalononitrile; 2-Chlorobenzylidene-malononitrile; 2-Chlorobnm; **[(2-Chlorophenyl)methylene]propanedinitrile**; CS; β,β-Dicyano-*o*-chlorostyrene; NCI-C55118; OCBM; USAF KF-11.

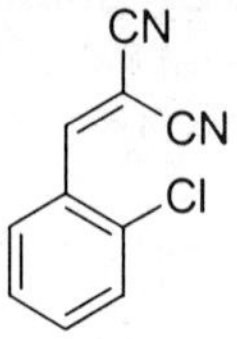

CAS Registry Number: 2698-41-1
Molecular formula: $C_{10}H_5ClN_2$
Formula weight: 188.61
RTECS: OO3675000
Merck reference: 10, 2096

Physical state, color, and odor:
White crystalline solid with a pepper-like odor

Melting point (°C):
95-96 (Windholz et al., 1983)

Boiling point (°C):
310-315 (Windholz et al., 1983)

Density (g/cm³):
1.472 using method of Lyman et al. (1982)

Diffusivity in water:
Not applicable - reacts with water

Lower explosive limit:
For a dust, 25 mg/m³ is the minimum explosive concentration in air (NIOSH, 1997).

Henry's law constant (atm·m³/mol):
Not applicable - reacts with water

Soil sorption coefficient, log K_{oc}:
Not applicable - reacts with water

Octanol/water partition coefficient, log K_{ow}:
Not applicable - reacts with water

Solubility in organics:
Soluble in acetone, benzene, 1,4-dioxane, ethyl acetate, and methylene chloride (Windholz et al., 1983)

Solubility in water:
Not applicable - reacts with water

Vapor pressure (mmHg):
3 x 10^{-5} at 20 °C (NIOSH, 1997)

Environmental fate:
  Chemical/Physical. Hydrolyzes in water forming 2-chlorobenzaldehyde and malononitrile (Verschueren, 1983).

Exposure limits: NIOSH REL: ceiling 0.05 ppm (0.4 mg/m^3), IDLH 2 mg/m^3; OSHA PEL: TWA 0.05 ppm; ACGIH TLV: ceiling 0.05 ppm (adopted).

Symptoms of exposure: An irritation concentration of 1.52 mg/m^3 in air was reported by Ruth (1986).

Toxicity:
  After 12, 24, 48, and 96 h, observed LC_{50} concentrations were 1.28, 0.45, 0.42, and 0.22 mg/L, respectively (Abram and Wilson, 1979).
  Acute oral LD_{50} for rats 178 mg/kg, guinea pigs 212 mg/kg, rabbits 143 mg/kg, mice 282 mg/kg (quoted, RTECS, 1985).

Uses: Riot control agent.

p-CHLORO-*m*-CRESOL

Synonyms: Aptal; Baktol; Baktolan; Candaseptic; *p*-Chlor-*m*-cresol; Chlorocresol; 4-Chloro-cresol; *p*-Chlorocresol; 4-Chloro-*m*-cresol; 6-Chloro-*m*-cresol; 4-Chloro-1-hydroxy-3-methyl-benzene; 2-Chlorohydroxytoluene; 2-Chloro-5-hydroxytoluene; 4-Chloro-3-hydroxytoluene; 6-Chloro-3-hydroxytoluene; **4-Chloro-3-methylphenol**; *p*-Chloro-3-methylphenol; 3-Methyl-4-chlorophenol; Ottafact; Parmetol; Parol; PCMC; Peritonan; Preventol CMK; Raschit; Raschit K; Rasenanicon; RCRA waste number U039.

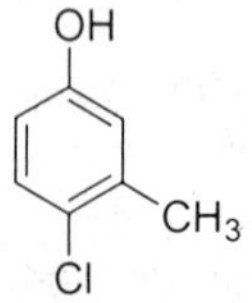

CAS Registry Number: 59-50-7
DOT: 2669
DOT label: Poison
Molecular formula: C_7H_7ClO
Formula weight: 142.59
RTECS: GO7100000
Merck reference: 10, 2102

Physical state, color, and odor:
Colorless, white, or pinkish crystals with a slight phenolic odor. On exposure to air it slowly becomes light brown.

Melting point (°C):
66-68 (Weast, 1986)
63-65 (Fluka, 1988)

Boiling point (°C):
235 (Weast, 1986)

Diffusivity in water (x 10^{-5} cm^2/sec):
0.71 at 20 °C using method of Hayduk and Laudie (1974)

Dissociation constant, pK_a:
9.549 at 25 °C (Dean, 1973)

Henry's law constant (x 10^{-6} atm·m^3/mol):
2.5 at 20 °C (calculated, Mabey et al., 1982)

Soil sorption coefficient, log K_{oc}:
2.89 using method of Karickhoff et al. (1979)

Octanol/water partition coefficient, log K_{ow}:
2.18 (Hansch and Leo, 1979)
3.10 (Leo et al., 1971)

Solubility in organics:
Soluble in ethanol, ether (Weast, 1986), benzene, chloroform, acetone, petroleum ether, fixed oils, terpenes, and aqueous alkaline solutions (Windholz et al., 1983)

Solubility in water (mg/L):
3,846 at 20 °C (quoted, Windholz et al., 1983)
3,990 at 25 °C and pH 5.1 (Blackman et al., 1955)

Vapor pressure (mmHg):
No data was found, however, a value of 5 x 10^{-2} at 20 °C was assigned by analogy (Mabey et al., 1982).

Environmental fate:
Biological. When *p*-chloro-*m*-cresol was statically incubated in the dark at 25 °C with yeast extract and settled domestic wastewater inoculum, significant biodegradation with rapid adaptation was observed. At concentrations of 5 and 10 mg/L, 78 and 76% biodegradation, respectively, were observed after 7 d (Tabak et al., 1981).

Toxicity:
LD_{50} for rats (subcutaneous) 400 mg/kg (quoted, RTECS, 1985).
Acute oral LD_{50} for rats 1,830 mg/kg (quoted, RTECS, 1985).

Uses: External germicide; preservative for gums, glues, paints, inks, textiles, and leather products; topical antiseptic (veterinarian).

CHLOROETHANE

Synonyms: Aethylis; Aethylis chloridum; Anodynon; Chelen; Chlorethyl; Chloridum; Chloryl; Chloryl anesthetic; Ether chloratus; Ether hydrochloric; Ether muriatic; Ethyl chloride; Hydrochloric ether; Kelene; Monochlorethane; Monochloroethane; Muriatic ether; Narcotile; NCI-C06224; UN 1037.

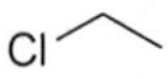

$$Cl \diagup\diagdown$$

CAS Registry Number: 75-00-3
DOT: 1037
DOT label: Flammable gas/liquid
Molecular formula: C_2H_5Cl
Formula weight: 64.52
RTECS: KH7525000
Merck reference: 10, 3729

Physical state, color, and odor:
Colorless gas or liquid with a pungent ether-like odor. Odor threshold concentration is 4.2 ppm (Amoore and Hautala, 1983).

Melting point (°C):
-136.4 (Weast, 1986)
-139.0 (Stull, 1947)

Boiling point (°C):
12.3 (Weast, 1986)

Density (g/cm³):
0.9028 at 15/4 °C, 0.8970 at 20/4 °C (Standen, 1964)
0.8706 at 25/4 °C (Dreisbach, 1959)

Diffusivity in water (x 10^{-5} cm²/sec):
1.07 at 20 °C using method of Hayduk and Laudie (1974)

Flash point (°C):
-50 (NIOSH, 1997)
-43 (open cup, Windholz et al., 1983)

Lower explosive limit (%):
3.8 (NIOSH, 1997)

Upper explosive limit (%):
15.4 (NIOSH, 1997)

Heat of fusion (kcal/mol):
1.064 (Riddick et al., 1986)

Henry's law constant (x 10^{-3} atm·m³/mol):
11.1 at 25 °C (Gossett, 1987)

9.3 (Pankow and Rosen, 1988)
7.59, 9.58, 11.0, 12.1, and 14.3 at 10, 15, 20, 25, and 30 °C, respectively (Ashworth et al., 1988)

Ionization potential (eV):
10.97, 11.01 (Horvath, 1982)

Soil sorption coefficient, log K_{oc}:
0.51 using method of Chiou et al. (1979).

Octanol/water partition coefficient, log K_{ow}:
1.43 (Hansch et al., 1975; Valvani et al., 1981)
1.54 (Leo et al., 1971)

Solubility in organics:
Soluble in ethanol, ether (U.S. EPA, 1985); miscible with chlorinated hydrocarbons such as chloroform, carbon tetrachloride, and tetrachloeoethane

Solubility in water:
3,330 mg/L at 0 °C (quoted, Verschueren, 1983)
0.57 wt % at 17.5 °C (quoted, Stephen and Stephen, 1963)
0.455 wt % at 0 °C (Konietzko, 1984)
4.5 g/kg at 0 °C (McGovern, 1943)
5,710 mg/L at 20 °C (Mackay and Shiu, 1981)
89 mM at 20 °C (Fischer and Ehrenberg, 1948)
89.0 mM (Fühner, 1924)

Vapor density:
2.76 kg/m^3 at 20 °C (Konietzko, 1984)

Vapor pressure (mmHg):
1,011 at 20 °C (Standen, 1964)
1,444 at 30 °C (Verschueren, 1983)
766 at 12.5 °C (Howard, 1990)
1,199 at 25 °C (quoted, Nathan, 1978)

Environmental fate:
Photolytic. The rate constant for the reaction of chloroethane and OH radicals in the atmosphere at 300 K is 2.3 x 10^{11} cm^3/mol·sec (Hendry and Kenley, 1979). At 296 K, a photooxidation rate constant of 3.9 x 10^{-13} cm^3/molecule·sec was reported (Howard and Evenson, 1976). The estimated tropospheric lifetime is 14.6 d (Nimitz and Skaggs, 1992).
Chemical/Physical. Under laboratory conditions, chloroethane hydrolyzed to ethanol (Smith and Dragun, 1984). An estimated hydrolysis half-life in water at 25 °C and pH 7 is 38 d, with ethanol and hydrochloric acid being the expected end-products (Mabey and Mill, 1978). In the atmosphere, formyl chloride is the initial photooxidation product (U.S. EPA, 1985). In the presence of water, formyl chloride hydrolyzes to hydrochloric acid and carbon monoxide (Morrison and Boyd, 1971).
Burns with a smoky, greenish flame releasing hydrogen chloride (Windholz et al., 1983).
In the laboratory, the evaporation half-life of chloroethane (1 mg/L) from water at 25 °C using a shallow-pitch propeller stirrer at 200 rpm at an average depth of 6.5 cm is 23.1 min (Dilling, 1977).

At influent concentrations of 1.0, 0.1, and 0.01 mg/L, the granular activated carbon adsorption capacities at pH 5.3 were 0.59, 0.07, and 0.007 mg/g, respectively (Dobbs and Cohen, 1980).

Exposure limits: Potential occupational carcinogen. NIOSH REL: IDLH 3,800 ppm; OSHA PEL: TWA 1,000 ppm (2,600 mg/m^3); ACGIH TLV: TWA 100 ppm (adopted).

Symptoms of exposure: May cause stupor, eye irritation, incoordination, abdominal cramps, anesthetic effects, cardiac arrest, and unconsciousness (Patnaik, 1992)

Toxicity:
LC$_{50}$ (inhalation) for mice 146 gm/m^3/2-h, rats 160 gm/m^3/2-h (quoted, RTECS, 1985).

Drinking water standard: As of October 1996, chloroethane has been listed for regulation but no MCLGs or MCLs have been proposed (U.S. EPA, 1996).

Uses: Intermediate for tetraethyl lead and ethyl cellulose; topical anesthetic; organic synthesis; alkylating agent; refrigeration; analytical reagent; solvent for phosphorus, sulfur, fats, oils, resins, and waxes; insecticides.

2-CHLOROETHYL VINYL ETHER

Synonyms: 2-Chlorethyl vinyl ether; **(2-Chloroethoxy)ethene**; RCRA waste number U042; Vinyl 2-chloroethyl ether; Vinyl β-chloroethyl ether.

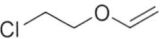

CAS Registry Number: 110-75-8
Molecular formula: C_4H_7ClO
Formula weight: 106.55
RTECS: KN6300000
Merck reference: 10, 2110

Physical state and color:
Colorless liquid

Melting point (°C):
-70.3 (Sax, 1984)
-69.7 (Dean, 1987)

Boiling point (°C):
108 (Weast, 1986)
109 (Standen, 1970)
110 (Dean, 1987)

Density (g/cm³ at 20/4 °C):
1.0475 (Weast, 1986)
1.0493 (Standen, 1970)

Diffusivity in water (x 10^{-5} cm²/sec):
0.87 at 20 °C using method of Hayduk and Laudie (1974)

Flash point (°C):
16 (Aldrich, 1988)
27 (open cup, NFPA, 1984)

Henry's law constant (x 10^{-4} atm·m³/mol):
2.5 (Pankow and Rosen, 1988)

Soil sorption coefficient, log K_{oc}:
0.82 (estimated, Schwille, 1988)

Octanol/water partition coefficient, log K_{ow}:
1.28 (calculated, Leo et al., 1971)

Solubility in organics:
Soluble in ethanol and ether (Weast, 1986)

Solubility in water:
6,000 mg/L at 20 °C (Standen, 1970)

Vapor density:
4.36 g/L at 25 °C, 3.68 (air = 1)

Vapor pressure (mmHg):
26.75 at 20 °C (U.S. EPA, 1980a)

Environmental fate:
  Biological. When 2-chloroethyl vinyl ether was statically incubated in the dark at 25 °C with yeast extract and settled domestic wastewater inoculum, significant biodegradation with rapid adaptation was observed. At concentrations of 5 and 10 mg/L, complete degradation was observed after 21 d (Tabak et al., 1981).
  Chemical/Physical. Chlorination of 2-chloroethyl vinyl ether to α-chloroethyl ethyl ether or β-chloroethyl ethyl ether may occur in water treatment facilities. The *alpha* compound is very unstable in water and decomposes almost as fast as it is formed (Summers, 1955). Although stable in sodium hydroxide solutions, in dilute acid solutions hydrolysis yields acetaldehyde and chlorohydrin (Windholz et al., 1983). At pH 7 and 25 °C, the hydrolysis half-life is 175 d (Jones and Wood, 1964).
  At influent concentrations of 10.0, 1.0, 0.1, and 0.01 mg/L, the granular activated carbon adsorption capacities at pH 5.4 were 25, 3.9, 0.6, and 0.1 mg/g, respectively (Dobbs and Cohen, 1980).

Symptoms of exposure: Exposure to vapors may cause irritation of eyes, nose, and lungs (Patnaik, 1992).

Toxicity:
  LC_{50} (contact) for earthworm (*Eisenia fetida*) 33 μg/cm^2 (Neuhauser et al., 1985).
  LC_{50} (96-h) for bluegill sunfish 350 mg/L (Spehar et al., 1982).
  Acute oral LD_{50} for rats 250 mg/kg (quoted, RTECS, 1985).

Uses: Anesthetics, sedatives, and cellulose ethers; copolymer of 95% ethyl acrylate with 5% 2-chloroethyl vinyl ether is used to produce an acrylic elastomer.

CHLOROFORM

Synonyms: Formyl trichloride; Freon 20; Methane trichloride; Methenyl chloride; Methenyl trichloride; Methyl trichloride; NCI-C02686; R 20; R 20 (refrigerant); RCRA waste number U044; TCM; Trichloroform; **Trichloromethane**; UN 1888.

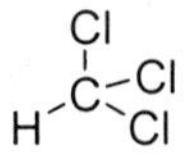

$$\overset{\displaystyle Cl}{\underset{\displaystyle H}{\overset{|}{C}}}\!\!-\!Cl$$

CAS Registry Number: 67-66-3
DOT: 1888
DOT label: Poison
Molecular formula: $CHCl_3$
Formula weight: 119.38
RTECS: FS9100000
Merck reference: 10, 2111

Physical state, color, and odor:
Clear, colorless, volatile liquid with a strong, sweet, pleasant ether-like odor. The reported odor threshold concentrations in air ranged from 7.5 to 1,000 ppm (Ruth, 1986; Keith and Walters, 1992; Young et al., 1996). The taste threshold concentration in water is 1.2 ppm (Young et al., 1996).

Melting point (°C):
-63.5 (McGovern, 1986)

Boiling point (°C):
61.7 (Weast, 1986)

Density (g/cm³):
1.4832 at 20/4 °C (Weast, 1986)
1.4890 at 20/4 °C, 1.48069 at 25/4 °C (Standen, 1964)

Diffusivity in water (x 10⁻⁵ cm²/sec):
1.00 at 20 °C using method of Hayduk and Laudie (1974)

Flash point (°C):
Noncombustible (NIOSH, 1997)

Heat of fusion (kcal/mol):
2.28 (Riddick et al., 1986)

Henry's law constant (x 10⁻³ atm·m³/mol):
3.39 at 25 °C (Warner et al., 1987)
3.23 (Valsaraj, 1988)
5.3 at 20 °C (Roberts and Dändliker, 1983)
3.18 at 25 °C (Dilling, 1977)
3.57 at 25°C (Gossett, 1987)
7.27 at 37 °C (Sato and Nakajima, 1979)
5.76 at 20 °C (Roberts et al., 1985)

3.00 at 20 °C (Nicholson et al., 1984)

5.54 at 30 °C (Jeffers et al., 1989)

4.00 at 25 °C (Hoff et al., 1993)

1.720, 2.33, 33.2, 42.1, and 55.4 at 10, 15, 20, 25, and 30 °C, respectively (Ashworth et al., 1988)

4.90 at 25 °C in seawater (Hunter-Smith et al., 1983)

In seawater (salinity 30.4‰): 1.25, 2.11, and 3.48 at 0, 10, and 20 °C, respectively (Moore et al., 1995)

Distilled water: 1.24, 1.43, 1.72, 2.79, and 3.75 at 2.0, 6.0, 10.0, 18.2, and 25.0 °C, respectively; natural seawater: 1.75 and 4.38 at 6.0 and 25.0 °C, respectively (Dewulf et al., 1995)

2.80 at 25 °C (McConnell et al., 1975)

8.33 at 25 °C (Tancréde and Yanagisawa, 1990)

3.85 at 25 °C (Wright et al., 1992)

3.08 at 20 °C (Hovorka and Dohnal, 1997)

3.20 at 24.9 °C (mole fraction ratio, Leighton and Calo, 1981)

1.78, 2.97, and 4.81 at 10, 20, and 30 °C, respectively (Munz and Roberts, 1987)

1.33 and 3.69 at 5.7 and 24.9 °C, respectively (Dewulf et al., 1999a)

Interfacial tension with water (dyn/cm at 25 °C):
31.6 (Donahue and Bartell, 1952)

32.63 (Harkins et al., 1920)

Ionization potential (eV):
11.42 ± 0.03 (Franklin et al., 1969)

11.50 (Horvath, 1982)

Bioconcentration factor, log BCF:
0.78 (bluegill sunfish, Veith et al., 1980)

2.84 (green algae, Davies and Dobbs, 1984)

Soil sorption coefficient, log K_{oc}:
1.80 (Potomac-Raritan-Magothy sand), 1.94 (Cohansey sand) (Uchrin and Michaels, 1986)

1.88 (Lincoln sand, Wilson et al., 1981)

1.44 (soil, sand, and loess), 1.98 (weathered shale), 2.79 (unweathered shale) (Grathwohl, 1990)

1.60 (Hutzler et al., 1983)

1.77 (Eerd soil), 1.91 (peat soil) (Loch et al., 1986)

1.57 (Captina silt loam), 1.46 (McLaurin sandy loam) (Walton et al., 1992)

1.63, 1.65, 1.63, 1.66, 1.69, 1.65, and 1.70 at 2.3, 3.8, 6.2, 8.0, 13.5, 18.6, and 25.0 °C, respectively, for a Leie River (Belgium) clay (Dewulf et al., 1999a)

Octanol/water partition coefficient, log K_{ow}:
1.97 (Hansch and Anderson, 1967; Hansch et al., 1975)

1.94 (Hansch and Leo, 1979)

1.90 (Veith et al., 1980)

1.95 (Mackay, 1982)

2.00, 2.01, and 2.00 at 25, 35, and 50 °C, respectively (Bhatia and Sandler, 1995)

1.83, 1.81, 1.90, 1.87, 1.92, and 1.86 at 2.2, 6.0, 10.0, 14.1, 18.7, and 24.8 °C, respectively (Dewulf et al., 1999a)

Solubility in organics:
Miscible with ethanol, ether, benzene, and ligroin (U.S. EPA, 1985)

Solubility in water:
In g/kg: 10.62 at 0 °C, 8.95 at 10 °C, 8.22 at 20 °C, 7.76 at 30 °C (Rex, 1906)
8,200 mg/L at 20 °C (Pearson and McConnell, 1975)
7,950 mg/L at 25 °C (Kenaga, 1975)
7,222 mg/L at 25 °C (Banerjee et al., 1980)
7,840 mg/L at 25 °C (Dilling, 1977)
8,520 mg/kg at 15 °C, 7,710 mg/kg at 30 °C (Gross and Saylor, 1931)
2,524 mg/L at 30 °C (McNally and Grob, 1984)
0.82 wt % at 25 °C (Lo et al., 1986)
7.9 g/kg at 25 °C (McGovern, 1943)
8,700 mg/L at 23-24 °C (Broholm and Feenstra, 1995)
1.11×10^{-3} at 25 °C (mole fraction, Li et al., 1993)
In wt %: 1.02 at 0 °C, 0.93 at 9.5°C, 0.82 at 19.6°C, 0.79 at 29.5°C, 0.74 at 39.3 °C, 0.77 at
    49.2 °C, 0.79 at 59.2 (Stephenson, 1992)

Vapor density:
4.88 g/L at 25 °C, 4.12 (air = 1)

Vapor pressure (mmHg):
160 at 20 °C, 245 at 30 °C (Verschueren, 1983)
150.5 at 20 °C (McConnell et al., 1975)
198 at 25 °C (Warner et al., 1987)
246 at 25 °C (Howard, 1990)

Environmental fate:
Biological. An anaerobic species of *Clostridium* biodegraded chloroform (a metabolite of carbon tetrachloride) by reductive dechlorination yielding methylene chloride and unidentified products (Gälli and McCarty, 1989). Chloroform showed significant degradation with gradual adaptation in a static-culture flask-screening test (settled domestic wastewater inoculum) conducted at 25 °C. At concentrations of 5 and 10 mg/L, complete degradation was observed at the end of the third subculture period (28 d). The amount lost due to volatilization after 10 d was 6-24% (Tabak et al., 1981).
    Heukelekian and Rand (1955) reported a 5-d BOD value of 0.0 g/g which is 0.0% of the ThOD value of 0.136 g/g.
    Photolytic. Complete mineralization was reported when distilled deionized water containing chloroform (118 ppm) and 0.1 wt % titanium dioxide as a catalyst was irradiated with UV light. Mineralization products included carbon dioxide and hydrochloric acid (Pruden and Ollis, 1983). Photocatalyzed mineralization of chloroform in the presence of titanium dioxide as catalyst occurred at a rate of 4.4 ppm/min/gm catalyst (Ollis, 1985). Titanium dioxide suspended in an aqueous solution and irradiated with UV light (λ = 365 nm) converted chloroform to carbon dioxide at a significant rate. Intermediate compounds were not identified (Matthews, 1986).
    An aqueous solution containing 300 ng/μL chloroform and colloidal platinum catalyst was irradiated with UV light. After 15 h, only 10 ng/μL chloroform remained. A duplicate experiment was performed but 0.1 g zinc was added to the system. At approximately 2 h, 10 ng/μL chloroform remained and 210 ng/μL methane was produced (Wang and Tan, 1988).
    Photolysis of an aqueous solution containing chloroform (314 μmol) and the catalyst [Pt(colloid)/Ru(bpy)$^{2+}$/MV/EDTA] yielded the following products after 15 h (μmol detected): chloride ions (852), methane (265), ethylene (0.05), ethane (0.52), and unreacted chloroform (10.5) (Tan and Wang, 1987). In the troposphere, photolysis of chloroform via OH radicals may yield formyl chloride, carbon monoxide, hydrogen chloride, and phosgene as the principal

products (Spence et al., 1976). Phosgene is hydrolyzed readily to hydrogen chloride and carbon dioxide (Morrison and Boyd, 1971).

Chloroform reacted with OH radicals in water (pH 8.5) at a rate of 5.4 x 10^7/M·sec. At pH 3, the reaction rate was 5.0 x 10^7/M·sec (Haag and Yao, 1992).

Chemical/Physical. Matheson and Tratnyek (1994) studied the reaction of fine-grained iron metal in an anaerobic aqueous solution (15 °C) containing chloroform (107 μM). Initially, chloroform underwent rapid dehydrochlorination forming methylene chloride and chloride ions. As the concentration of methylene chloride increased, the rate of reaction appeared to decrease. After 140 h, no additional products were identified. The authors reported that reductive dehalogenation of chloroform and other chlorinated hydrocarbons used in this study appears to take place in conjunction with the oxidative dissolution or corrosion of the iron metal through a diffusion-limited surface reaction.

The volatilization half-life of chloroform (1 mg/L) from water at 25 °C using a shallow-pitch propeller stirrer at 200 rpm at an average depth of 6.5 cm is 20.2 min (Dilling, 1977). The predicted volatilization half-life of chloroform from a 1 m deep steam flowing 1 m/sec with a wind velocity of 3 m/sec at 20 °C is 4 h (Smith et al., 1990).

The experimental half-life for hydrolysis of chloroform in water at 25 °C is approximately 15 months (Dilling et al., 1975). The estimated hydrolysis half-lives in water at 25 °C and pH 7 are 3,500 yr (Mabey and Mill, 1978) and 1,849.6 yr (Jeffers et al., 1989). End products of hydrolysis include carbon monoxide and hydrochloric acid (Kollig, 1993).

When chloroform is heated to decomposition, phosgene gas is formed (NIOSH, 1997).

At influent concentrations of 1.0, 0.1, 0.01, and 0.001 mg/L, the granular activated carbon adsorption capacities at pH 5.3 were 2.6, 0.48, 0.09, and 0.02 mg/g, respectively (Dobbs and Cohen, 1980).

Exposure limits: Potential occupational carcinogen. NIOSH REL: STEL (1 h) 2 ppm (9.78 mg/m^3), IDLH 500 ppm; OSHA PEL: ceiling 50 ppm (240 mg/m^3); ACGIH TLV: TWA 10 ppm (adopted).

Symptoms of exposure: Dizziness, lightheadedness, dullness, hallucination, nausea, headache, fatigue, and anesthesia (Patnaik, 1992). An irritation concentration of 20.48 g/m^3 in air was reported by Ruth (1986).

Toxicity:

EC$_{50}$ (24-h) for *Daphnia magna* 573 mg/L (Lilius et al., 1995).

LC$_{50}$ (contact) for earthworm (*Eisenia fetida*) 111 μg/cm^2 (Neuhauser et al., 1985).

LC$_{50}$ (14-d) for *Poecilia reticulata* 102 mg/L (Könemann, 1981).

LC$_{50}$ (48-h) for red killifish 1,260 mg/L (Yoshioka et al., 1986), zebra fish 100 mg/L (Slooff, 1979), *Daphnia magna* 29 mg/L (LeBlanc, 1980), *Crassostrea virginica* larvae 1 mg/L (quoted, Verschueren, 1983).

LC$_{50}$ (24-h) for *Daphnia magna* 29 mg/L (LeBlanc, 1980).

Acute oral LD$_{50}$ for rats 23 mg/kg (Patnaik, 1992), 908 mg/kg, mice 36 mg/kg, and guinea pigs 820 mg/kg (quoted, RTECS, 1985).

Drinking water standard (tentative): MCLG: zero; MCL: 0.1 mg/L. Total for all trihalomethanes cannot exceed a concentration of 0.08 mg/L (U.S. EPA, 1996).

Uses: Manufacture of fluorocarbon refrigerants, fluorocarbon plastics, and propellants; solvent for natural products; analytical chemistry; cleansing agent; soil fumigant; insecticides; preparation of chlorodifluoromethane, methyl fluoride, salicylaldehyde; cleaning electronic circuit boards; in fire extinguishers.

2-CHLORONAPHTHALENE

Synonyms: β-Chloronaphthalene; RCRA waste number U047.

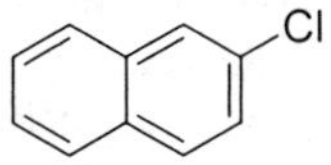

CAS Registry Number: 91-58-7
Molecular formula: $C_{10}H_7Cl$
Formula weight: 162.62
RTECS: QJ2275000
Merck reference: 10, 2120

Physical state and color:
Off-white monoclinic plates or leaflets

Melting point (°C):
61 (Weast, 1986)

Boiling point (°C):
256 (Weast, 1986)

Density (g/cm³):
1.1377 at 71/4 °C (Dean, 1987)

Diffusivity in water (x 10^{-5} cm²/sec):
0.65 at 20 °C using method of Hayduk and Laudie (1974)

Flash point (°C):
Nonflammable (Sittig, 1985)

Henry's law constant (x 10^{-4} atm·m³/mol):
3.31 at 25 °C (Shiu and Mackay, 1997)

Ionization potential (eV):
8.11 (Lias et al., 1998)

Soil sorption coefficient, log K_{oc}:
3.93 using method of Karickhoff et al. (1979)

Octanol/water partition coefficient, log K_{ow}:
3.98 (Sangster, 1989)

Solubility in organics:
Soluble in chloroform and carbon disulfide (Windholz et al., 1983)

Solubility in water:
<1 mg/L at 30 °C (McNally and Grob, 1984)

Vapor pressure (mmHg):
2.88 x 10^{-2} at 25.00 °C (Lei et al., 1999)

Environmental fate:

Biological. Reported biodegradation products include 8-chloro-1,2-dihydro-1,2-dihydroxy-naphthalene and 3-chlorosalicylic acid (Callahan et al., 1979). When 2-chloronaphthalene was statically incubated in the dark at 25 °C with yeast extract and settled domestic wastewater inoculum, complete biodegradation was observed after 7 d (Tabak et al., 1981).

Chemical/Physical. The hydrolysis rate constant for 2-chloronaphthalene at pH 7 and 25 °C was determined to be 9.5 x 10^{-6}/h, resulting in a half-life of 8.3 yr (Ellington et al., 1988). At 85.5 °C, hydrolysis half-lives of 255, 156, and 244 d were reported at pH values of 2.93, 7.10, and 9.58, respectively (Ellington et al., 1977).

Toxicity:

Acute oral LD_{50} for rats 2,078 mg/kg, mice 886 mg/kg (quoted, RTECS, 1985).

Uses: Chlorinated naphthalenes were formerly used in the production of electric condensers, insulating electric condensers, electric cables, and wires; additive for high pressure lubricants.

p-CHLORONITROBENZENE

Synonyms: **1-Chloro-4-nitrobenzene**; 4-Chloronitrobenzene; 4-Chloro-1-nitrobenzene; 4-Nitrochlorobenzene; *p*-Nitrochlorobenzene; PCNB; PNCB; UN 1578.

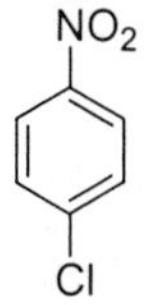

CAS Registry Number: 100-00-5
DOT: 1578
DOT label: Poison
Molecular formula: $C_6H_4ClNO_2$
Formula weight: 157.56
RTECS: CZ1050000
Merck reference: 10, 2121

Physical state, color, and odor:
Yellow, crystalline solid with a sweet odor

Melting point (°C):
83.6 (Weast, 1986)

Boiling point (°C):
242 (Weast, 1986)

Density (g/cm³):
1.520 at 18/4 °C (Verschueren, 1983)

Diffusivity in water (x 10^{-5} cm²/sec):
0.76 at 20 °C using method of Hayduk and Laudie (1974)

Flash point (°C):
127.3 (NIOSH, 1997)

Entropy of fusion (cal/mol·K):
7.98 (Marchidan and Ciopec, 1978)

Henry's law constant (x 10^{-5} atm·m³/mol):
9.2 at 20-30 °C (approximate - calculated from water solubility and vapor pressure)

Ionization potential (eV):
9.96 (NIOSH, 1997)

Bioconcentration factor, log BCF:
2.00 (*Oncorhynchus mykiss*, Devillers et al., 1996)

Soil sorption coefficient, log K_{oc}:
K_d = 44 mL/g on a Cs^+-kaolinite (Haderlein and Schwarzenbach, 1993)

Octanol/water partition coefficient, log K_{ow}:
2.39 (Fujita et al., 1964)

Solubility in organics:
Soluble in acetone and alcohol (Weast, 1986)

Solubility in water (20 °C):
2,877 μmol/L (Eckert, 1962)
3.40 g/L (Hafkenscheid and Tomlinson, 1983)

Vapor pressure (mmHg):
0.2 at 30 °C (NIOSH, 1997)

Environmental fate:
Biological. Under aerobic conditions, the yeast *Rhodosporidium* sp. metabolized 4-chloronitrobenzene to 4-chloroacetanilide and 4-chloro-2-hydroxyacetanilide as final major metabolites. Intermediate compounds identified include 4-chloronitrosobenzene, 4-chlorophenylhydroxylamine, and 4-chloroaniline (Corbett and Corbett, 1981).

Under continuous flow conditions involving feeding, aeration, settling and reflux, a mixture of *p*-chloronitrobenzene and 2,4-dinitrochlorobenzene was reduced 61-70% after 8-13 d by *Arthrobacter simplex*, a microorganism isolated from industrial waste. A similar experiment was conducted using two aeration columns. One column contained *A. simplex*, the other a mixture of *A. simplex* and microorganisms isolated from soil (*Streptomyces coelicolor*, *Fusarium* sp., probably *aquaeductum* and *Trichoderma viride*). After 10 d, 89.5-91% of the nitro compounds was reduced. *p*-Chloronitrobenzene was reduced to 4-chloroaniline and six unidentified compounds (Bielaszczyk et al., 1967).

Photolytic. An aqueous solution containing *p*-chloronitrobenzene and a titanium dioxide (catalyst) suspension was irradiated with UV light (λ >290 nm). 2-Chloro-5-nitrophenol was the only compound identified as a minor degradation product. Continued irradiation caused further degradation yielding carbon dioxide, water, hydrochloric and nitric acids (Hustert et al., 1987).

Irradiation of *p*-chloronitrobenzene in air and nitrogen produced 4-chloro-2-nitrophenol and *p*-chlorophenol, respectively (Kanno and Nojima, 1979).

Chemical. Although no products were identified, *p*-chloronitrobenzene (1.5 x 10^{-5} M) was reduced by iron metal (33.3 g/L acid washed 18-20 mesh) in a carbonate buffer (1.5 x 10^{-2} M) at pH 5.9 and 15 °C. Based on the pseudo-first-order disappearance rate of 0.0336/min, the half-life was 20.6 min (Agrawal and Tratnyek, 1996).

Exposure limits (mg/m³): Potential occupational carcinogen. NIOSH REL: IDLH 100; OSHA PEL: TWA 1.

Symptoms of exposure: Anoxia, unpleasant taste, anemia (NIOSH, 1997)

Toxicity:
EC_{50} (15-min) for *Photobacterium phosphoreum* 18.1 mg/L (Yuan and Lang, 1997).
IC_{50} (24-h) for river bacteria 27.4 mg/L (Yuan and Lang, 1997).
Acute oral LD_{50} for mice 650 mg/kg, rats 420 mg/kg (quoted, RTECS, 1985).
LD_{50} (skin) for rats 16 gm/kg (quoted, RTECS, 1985).

Uses: Intermediate for dyes; rubber and agricultural chemicals; manufacture of *p*-nitrophenol.

1-CHLORO-1-NITROPROPANE

Synonyms: Chloronitropropane; Korax; Lanstan.

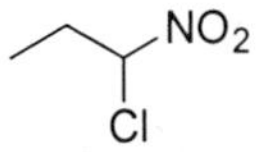

CAS Registry Number: 600-25-9
Molecular formula: $C_3H_6ClNO_2$
Formula weight: 123.54
RTECS: TX5075000

Physical state, color, and odor:
Colorless liquid with an unpleasant odor

Boiling point (°C):
170.6 at 745 mmHg (Hawley, 1981)

Density (g/cm³):
1.209 at 20/20 °C (Weast, 1986)

Diffusivity in water (x 10^{-5} cm²/sec):
0.87 at 20 °C using method of Hayduk and Laudie (1974)

Flash point (°C):
62.3 (open cup, NIOSH, 1997)

Henry's law constant (atm·m³/mol):
0.157 at 20 °C (approximate - calculated from water solubility and vapor pressure)

Ionization potential (eV):
9.90 (NIOSH, 1997)

Soil sorption coefficient, log K_{oc}:
3.34 using method of Chiou et al. (1979)

Octanol/water partition coefficient, log K_{ow}:
4.25 using method of Kenaga and Goring (1980)

Solubility in organics:
Soluble in alcohol and ether (Weast, 1986).

Solubility in water:
6 mg/L at 20 °C (quoted, Verschueren, 1983)

Vapor density:
5.05 g/L at 25 °C, 4.26 (air = 1)

Vapor pressure (mmHg):
6 at 25 °C (NIOSH, 1997)

242   **1-Chloro-1-nitropropane**

Exposure limits: NIOSH REL: TWA 2 ppm (10 mg/m^3), IDLH 100 ppm; OSHA PEL: TWA 20 ppm (100 mg/m^3); ACGIH TLV: TWA 2 ppm (adopted).

Toxicity:

Acute oral LD$_{50}$ for mice 510 mg/kg (quoted, RTECS, 1985).

Use: Fungicide.

2-CHLOROPHENOL

Synonyms: 1-Chloro-2-hydroxybenzene; 1-Chloro-*o*-hydroxybenzene; 2-Chlorohydroxy-benzene; *o*-**Chlorophenol**; 1-Hydroxy-2-chlorobenzene; 1-Hydroxy-*o*-chlorobenzene; 2-Hydroxychlorobenzene; RCRA waste number U048; UN 2020; UN 2021.

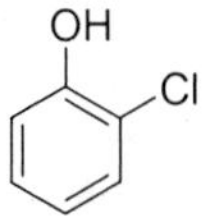

CAS Registry Number: 95-57-8
DOT: 2020 (liquid); 2021 (solid)
DOT label: Poison
Molecular formula: C_6H_5ClO
Formula weight: 128.56
RTECS: SK2625000
Merck reference: 10, 2125

Physical state, color, and odor:
Pale amber liquid with a slight phenolic odor. The reported odor threshold concentrations ranged from 0.088 to 20 ppb (Keith and Walters, 1992; Young et al., 1996). The taste threshold concentration in water is 0.14 ppb (Young et al., 1996).

Melting point (°C):
9.00 (Martin et al., 1979)
7.0 (Stull, 1947)

Boiling point (°C):
174.9 (Weast, 1986)
174.5 (Stull, 1947)

Density (g/cm³):
1.2634 at 20/4 °C (Weast, 1986)
1.257 at 25/4 °C (Krijgsheld and van der Gen, 1986a)

Diffusivity in water (x 10^{-5} cm²/sec):
0.87 at 20 °C using method of Hayduk and Laudie (1974)

Dissociation constant, pK_a:
8.48 at 25 °C (Dean, 1973)
8.30 (Hoigné and Bader, 1983)

Flash point (°C):
64 (NFPA, 1984)

Entropy of fusion (cal/mol·K):
10.58 (Poeti et al., 1982)

Heat of fusion (kcal/mol):
2.57 (Tsonopoulos and Prausnitz, 1971)

Henry's law constant (x 10^{-7} atm·m³/mol):
5.6 at 25 °C (estimated, Howard, 1989)

Ionization potential (eV):
9.28 (Franklin et al., 1969)

Bioconcentration factor, log BCF:
2.33 (bluegill sunfish, Barrows et al., 1980; Veith et al., 1980)

Soil sorption coefficient, log K_{oc}:
1.71 (Brookstone clay loam, Boyd, 1982)
3.69 (fine sediments), 3.60 (coarse sediments) (Isaacson and Frink, 1984)
2.60 (Meylan et al., 1992)
2.00 (coarse sand), 1.36 (loamy sand) (Kjeldsen et al., 1990)
K_d = <0.1 mL/g on a Cs^+-kaolinite (Haderlein and Schwarzenbach, 1993)

Octanol/water partition coefficient, log K_{ow}:
2.16 (Banerjee et al., 1980; Veith et al., 1980)
2.25 (Menges et al., 1990)
2.17 (Hansch and Leo, 1979)
2.19 (Leo et al., 1971)
2.15 (Fujita et al., 1964)
2.29 (Kishino and Kobayashi, 1994)

Solubility in organics:
Soluble in ethanol, benzene, ether (Weast, 1986), and caustic alkaline solutions (Windholz et al., 1983)

Solubility in water:
28,500 mg/L at 20 °C (quoted, Verschueren, 1983)
24,650 mg/L at 20 °C (Mulley and Metcalf, 1966)
22,000 mg/L at 25 °C (Roberts et al., 1977)
11,350 mg/L at 25 °C (Banerjee et al., 1980)
0.2 M at 25 °C (Caturla et al., 1988)
20.838 and 22.660 g/L at 15.4 and 24.6 °C, respectively (Achard et al., 1996)

Vapor density:
5.25 g/L at 25 °C, 4.44 (air = 1)

Vapor pressure (mmHg at 25 °C):
1.42 (Howard, 1989)
2.25 (quoted, Nathan, 1978)

Environmental fate:
   Biological. Chloroperoxidase, a fungal enzyme isolated from *Caldariomyces fumago*, reacted with 2-chlorophenol yielding traces of 2,4,6-trichlorophenol, 2,4- and 2,6-dichlorophenols (Wannstedt et al., 1990). When 2-chlorophenol was statically incubated in the dark at 25 °C with yeast extract and settled domestic wastewater inoculum, significant biodegradation with rapid adaptation was observed. At concentrations of 5 and 10 mg/L, 86 and 83% biodegradation, respectively, were observed after 7 d (Tabak et al., 1981). Biodegradation rates of 10 and 8 μmol/L·d were reported for chlorophenol in saline water and acclimated

sulfidogenic sediment cultures (Häggblom and Young, 1990).

In activated sludge inoculum, 95.6% COD removal was achieved in 6 h. The average rate of biodegradation was 25.0 mg COD/g·h (Pitter, 1976).

Soil. In laboratory microcosm experiments kept under aerobic conditions, half-lives of 7.2 and 1.7 d were reported for 2-chlorophenol in an acidic clay soil (<1% organic matter) and slightly basic sandy loam soil (3.25% organic matter) (Loehr and Matthews, 1992). In a non-sterile clay loam soil, a loss of 91% was reported when 2-chlorophenol was incubated in a non-sterile clay loam at 0 °C. Non-detectable levels of 2-chlorophenol was reported in sediments obtained from a stream at 20 °C after 10-15 d (Baker et al., 1980).

Surface Water. Hoigné and Bader (1983) reported a 2-chlorophenol reacts with ozone at a rate constant of 1,100/M·sec at the pH range of 1.8 to 4.0.

Photolytic. Monochlorophenols exposed to sunlight (UV radiation) produced catechol and other hydroxybenzenes (Hwang et al., 1986). Titanium dioxide suspended in an aqueous solution and irradiated with UV light ($\lambda = 365$ nm) converted 2-chlorophenol to carbon dioxide at a significant rate (Matthews, 1986). In a similar experiment, irradiation of an aqueous solution containing 2-chlorophenol and titanium dioxide with UV light ($\lambda > 340$ nm) resulted in the formation of chlorohydroquinone and trace amounts of catechol. Hydroxylation of both of these compounds forms the intermediate hydroxyhydroquinone, which degrades quickly to unidentified carboxylic acids and carbonyl compounds (D'Oliveira et al., 1990).

Irradiation of an aqueous solution at 296 nm and pH values from 8 to 13 yielded different products. Photolysis at a pH nearly equal to the dissociation constant (undissociated form) yielded pyrocatechol. At an elevated pH, 2-chlorophenol is almost completely ionized; photolysis yielded cyclopentadienic acid (Boule et al., 1982). Irradiation of an aqueous solution at 296 nm containing hydrogen peroxide converted 2-chlorophenol to catechol and 2-chlorohydroquinone (Moza et al., 1988). In the dark, nitric oxide (10^{-3} vol %) reacted with 2-chlorophenol forming 4-nitro-2-chlorophenol and 6-nitro-2-chlorophenol at yields of 36 and 30%, respectively (Kanno and Nojima, 1979).

Chemical/Physical. Wet oxidation of 2-chlorophenol at 320 °C yielded formic and acetic acids (Randall and Knopp, 1980). Wet oxidation of 2-chlorophenol at elevated pressure and temperature yielded the following products: acetone, acetaldehyde, formic, acetic, maleic, oxalic, muconic, and succinic acids (Keen and Baillod, 1985).

The volatilization half-lives of 2-chlorophenol in stirred and static water maintained at 23.8 °C were 1.35 and 1.60 h, respectively (Chiou et al., 1980).

In an aqueous phosphate buffer solution at 27 °C, a reaction rate 9.2×10^6/M·sec was reported for the reaction with singlet oxygen (Tratnyek and Hoigné, 1991).

2-Chlorophenol will not hydrolyze to any reasonable extent (Kollig, 1993).

Toxicity:

EC_{50} (48-h) for *Daphnia magna* 3.91 mg/L (Keen and Baillod, 1985).

LC_{50} (contact) for earthworm (*Eisenia fetida*) 2.2 μg/cm^2 (Neuhauser et al., 1985).

LC_{50} (96-h) for bluegill sunfish 6.6 mg/L, fathead minnows 11-13 mg/L (Spehar et al., 1982).

LC_{50} (48-h) for fathead minnows 9.7 mg/L (Spehar et al., 1982), *Daphnia magna* 2.6 mg/L (LeBlanc, 1980).

LC_{50} (24-h) for fathead minnows 16 ppm (quoted, Verschueren, 1983), *Daphnia magna* >22 mg/L (LeBlanc, 1980).

Acute oral LD_{50} for mice 345 mg/kg, rats 670 mg/kg (quoted, RTECS, 1985).

Drinking water standard: No MCLGs or MCLs have been proposed (U.S. EPA, 1996).

Uses: Component of disinfectant formulations; chemical intermediate for phenolic resins;

solvent for polyester fibers, antiseptic (veterinarian); preparation of 4-nitroso-2-methylphenol and other compounds.

4-CHLOROPHENYL PHENYL ETHER

Synonyms: 4-Chlorodiphenyl ether; *p*-Chlorodiphenyl ether; **1-Chloro-4-phenoxybenzene**; 1-Chloro-*p*-phenoxybenzene; 4-Chlorophenyl ether; *p*-Chlorophenyl ether; *p*-Chlorophenyl phenyl ether; Monochlorodiphenyl oxide.

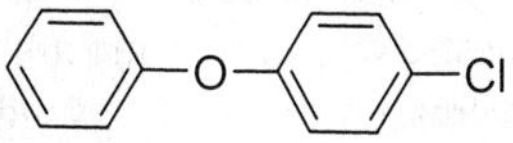

CAS Registry Number: 7005-72-3
Molecular formula: $C_{12}H_9ClO$
Formula weight: 204.66

Physical state:
Liquid

Melting point (°C):
-8 (U.S. EPA, 1980a)

Boiling point (°C):
284-285 (Weast, 1986)

Density (g/cm³):
1.2026 at 15/4 °C (Weast, 1986)
1.193 at 20/4 °C (Aldrich, 1988)

Diffusivity in water (x 10^{-5} cm²/sec):
0.64 at 20 °C using method of Hayduk and Laudie (1974)

Flash point (°C):
110 (Aldrich, 1988)

Henry's law constant ((x 10^{-4} atm·m³/mol):
2.2 (Pankow and Rosen, 1988)

Bioconcentration factor, log BCF:
2.867 (rainbow trout, Branson, 1977)

Soil sorption coefficient, log K_{oc}:
3.60 using method of Kenaga and Goring (1980)

Octanol/water partition coefficient, log K_{ow}:
4.08 (Branson, 1978)

Solubility in water:
3.3 mg/L at 25 °C (Branson, 1978)

Vapor density:
8.36 g/L at 25 °C, 7.06 (air = 1)

Vapor pressure (mmHg):
2.7 x 10^{-3} at 25 °C (calculated, Branson, 1978)

Environmental fate:

Biological. 4-Chlorophenyl phenyl ether (5 and 10 mg/L) did not significantly biodegrade following incubation in settled domestic wastewater inoculum at 25 °C. Percent losses reached a maximum after 2-3 wk but decreased thereafter suggesting a deadaptive process was occurring (Tabak et al., 1981). In activated sludge, a half-life of 4.0 h was measured (Branson, 1978).

Photolytic. In a methanolic solution irradiated with UV light (λ >290 nm), dechlorination of 4-chlorophenyl phenyl ether resulted in the formation of diphenyl ether (Choudhry et al., 1977). Photolysis of an aqueous solution containing 10% acetonitrile with UV light (λ = 230-400 nm) yielded 4-hydroxybiphenyl ether and chloride ion (Dulin et al., 1986).

Use: Research chemical.

CHLOROPICRIN

Synonyms: Acquinite; Chlor-o-pic; Chloropicrine; Dolochlor; G 25; Larvacide 100; Microlysin; NA 1583; NCI-C00533; Nitrochloroform; Nitrotrichloromethane; Picclor; Picfume; Picride; Profume A; PS; S 1; **Trichloronitromethane**; Triclor; UN 1580.

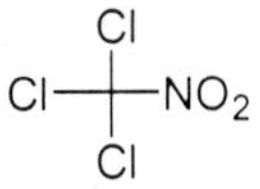

$$Cl—\underset{\underset{Cl}{|}}{\overset{\overset{Cl}{|}}{C}}—NO_2$$

CAS Registry Number: 76-06-2
DOT: 1580
DOT label: Poison
Molecular formula: CCl_3NO_2
Formula weight: 164.38
RTECS: PB6300000
Merck reference: 10, 2129

Physical state, color, and odor:
Colorless to pale yellow, oily liquid with a sharp, penetrating odor. Odor threshold concentration is 0.78 ppm (Amoore and Hautala, 1983).

Melting point (°C):
-64.5 (Weast, 1986)
-64 (Worthing and Hance, 1991)

Boiling point (°C):
111.8 (Weast, 1986)
112.4 (Worthing and Hance, 1991)

Density (g/cm³):
1.6558 at 20/4 °C, 1.6483 at 25/4 °C (Windholz et al., 1983)

Diffusivity in water (x 10⁻⁵ cm²/sec):
0.89 at 20 °C using method of Hayduk and Laudie (1974)

Henry's law constant (x 10⁻³ atm·m³/mol):
2.06 at 25 °C (Kawamoto and Urano, 1989)

Interfacial tension with water (dyn/cm at 20 °C):
32.3 (estimated, CHRIS, 1984)

Soil sorption coefficient, log K_{oc}:
0.82 using method of Chiou et al. (1979)

Octanol/water partition coefficient, log K_{ow}:
1.03 (Kawamoto and Urano, 1989)

Solubility in organics:
Miscible with acetone, benzene, carbon disulfide, carbon tetrachloride, ether, and methanol (Worthing and Hance, 1991)

Solubility in water:
2,270 mg/L at 0 °C (Gunther et al., 1968)
1.621 g/L at 25 °C (quoted, Windholz et al., 1983)
2,300 mg/L (quoted, Kawamoto and Urano, 1989)

Vapor density:
6.72 g/L at 25 °C, 5.67 (air = 1)

Vapor pressure (mmHg):
16.9 at 20 °C, 33 at 30 °C (Verschueren, 1983)
23.8 at 25 °C (quoted, Kawamoto and Urano, 1989)
18.3 at 20 °C (Meister, 1988)

Environmental fate:
Biological. Four *Pseudomonas* sp., including *Pseudomonas putida* (ATCC culture 29607) isolated from soil, degraded chloropicrin by sequential reductive dechlorination. The proposed degradative pathway is chloropicrin → nitrodichloromethane → nitrochloromethane → nitromethane + small amounts of carbon dioxide. In addition, a highly water soluble substance tentatively identified as a peptide, was produced by a nonenzymatic mechanism (Castro et al., 1983).

Photolytic. Photodegrades under simulated atmospheric conditions to phosgene and nitrosyl chloride. Photolysis of nitrosyl chloride yields chlorine and nitrous oxide (Moilanen et al., 1978; Woodrow et al., 1983). When aqueous solution of chloropicrin (10^{-3} M) is exposed to artificial UV light (λ <300 nm), protons, carbon dioxide, hydrochloric and nitric acids are formed (Castro and Belser, 1981).

Chemical/Physical. Releases very toxic fumes of chlorides and nitrogen oxides when heated to decomposition (Sax and Lewis, 1987). Reacts with alcoholic sodium sulfite solutions and ammonia to give methanetrisulfonic acid and guanidine, respectively (Sittig, 1985).

Exposure limits: NIOSH REL: TWA 0.1 ppm (0.7 mg/m^3), IDLH 2 ppm; OSHA PEL: 0.1 ppm; ACGIH TLV: TWA 0.1 ppm (adopted).

Symptoms of exposure: An irritation concentration of 2.10 mg/m^3 in air was reported by Ruth (1986).

Toxicity:
LC_{50} (inhalation) for mice 1,500 mg/m^3/10-min (quoted, RTECS, 1985).
Acute oral LD_{50} for rats 250 mg/kg (quoted, RTECS, 1985).

Drinking water standard: No MCLGs or MCLs have been proposed although chloropicrin has been listed for regulation (U.S. EPA, 1996).

Uses: Disinfecting cereals and grains; fumigant and soil insecticide; dyestuffs; odorant in methyl bromide fungicide; rat exterminator; organic synthesis; war gas.

CHLOROPRENE

Synonyms: Chlorobutadiene; 2-Chlorobutadiene-1,3; 2-Chlorobuta-1,3-diene; **2-Chloro-1,3-butadiene**; β-Chloroprene; Neoprene; UN 1991.

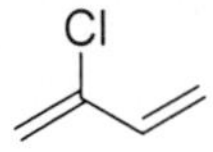

CAS Registry Number: 126-99-8
DOT: 1991
DOT label: Flammable liquid
Molecular formula: C_4H_5Cl
Formula weight: 88.54
RTECS: EI9625000

Physical state, color, and odor:
Colorless liquid with a pungent, ether-like odor. The odor threshold is 0.40 mg/m^3 (CHRIS, 1984).

Melting point (°C):
-103 (NIOSH, 1997)

Boiling point (°C):
59.4 (Weast, 1986)

Density (g/cm^3):
0.9583 at 20/4 °C (Weast, 1986)

Diffusivity in water (x 10^{-5} cm^2/sec):
0.93 at 20 °C using method of Hayduk and Laudie (1974)

Flash point (°C):
-20 (NIOSH, 1997)

Lower explosive limit (%):
4.0 (NIOSH, 1997)

Upper explosive limit (%):
20.0 (NIOSH, 1997)

Henry's law constant (x 10^{-2} atm·m^3/mol):
3.20 using method of Hine and Mookerjee (1975)

Ionization potential (eV):
8.79 (NIOSH, 1997)

Soil sorption coefficient, log K_{oc}:
1.74 (estimated, Ellington et al., 1993)

Octanol/water partition coefficient, log K_{ow}:
2.06 (estimated, Ellington et al., 1993)

Solubility in organics:
Soluble in acetone, benzene, and ether (Weast, 1986)

Vapor density:
3.62 g/L at 25 °C, 3.06 (air = 1)

Vapor pressure (mmHg):
188 at 20 °C (NIOSH, 1997)
174 at 25 °C (Boublik et al., 1984)
118 at 10 °C, 275 at 30 °C, 200 at 20 °C (Verschueren, 1983)

Environmental fate:
Chemical/Physical. Anticipated products from the reaction of chloroprene with ozone or OH radicals in the atmosphere are formaldehyde, 2-chloroacrolein, OHCCHO, ClCOCHO, $H_2CCHCCIO$, chlorohydroxy acids, and aldehydes (Cupitt, 1980).

Chloroprene will polymerize at room temperature unless inhibited with antioxidants (NIOSH, 1997). 2-Chlorobutadiene is resistant to hydrolysis under neutral and alkaline conditions (Carothers et al., 1931).

Chloroprene is subject to hydrolysis forming 3-hydroxypropene and hydrochloric acid. The reported hydrolysis half-life at 25 °C and pH 7 is 40 yr (Kollig, 1993).

Exposure limits: Potential occupational carcinogen. NIOSH REL: 15-min ceiling 1 ppm (3.6 mg/m³), IDLH 300 ppm; OSHA PEL: TWA 25 ppm (90 mg/m³); ACGIH TLV: TWA 10 ppm (adopted).

Symptoms of exposure: Irritation of eyes, skin, and respiratory system; dermatitis; nervousness (NIOSH, 1997).

Toxicity:
Acute oral LD_{50} for mice 260 mg/kg, rats 900 mg/kg (quoted, RTECS, 1985).

Use: Manufacture of neoprene.

CHLORPYRIFOS

Synonyms: Brodan; Chlorpyrifos-ethyl; Detmol U.A.; *O,O*-Diethyl-*O*-3,5,6-trichloro-2-pyridyl phosphorothioate; Dowco-179; Dursban; Dursban F; ENT 27311; Eradex; Lorsban; NA 2783; OMS 971; **Phosphorothionic acid *O,O*-diethyl *O*-(3,5,6-trichloro-2-pyridyl) ester**; Pyrinex.

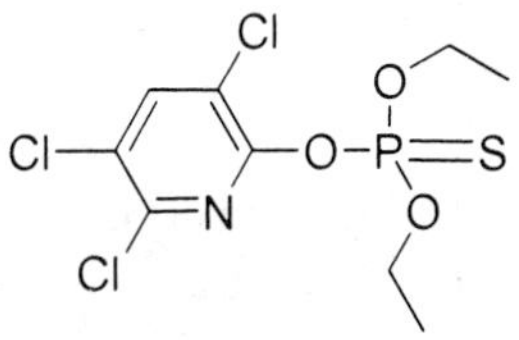

CAS Registry Number: 2921-88-2
DOT: 2783
DOT label: Poison
Molecular formula: $C_9H_{11}Cl_3NO_3PS$
Formula weight: 350.59
RTECS: TF6300000
Merck reference: 10, 2167

Physical state, color, and odor:
Colorless to white granular crystals or amber-colored oil with a mercaptan-like odor

Melting point (°C):
42.3 (NIOSH, 1997)

Boiling point (°C):
Decomposes at 160 (NIOSH, 1997)

Density (g/cm³):
1.398 at 43.5/4 °C (Verschueren, 1983)

Henry's law constant (x 10^{-6} atm·m³/mol):
4.16 at 25 °C (Fendinger and Glotfelty, 1990)
3.13 and 4.87 in distilled water and 33.3‰ NaCl at 20 °C, respectively (Rice et al., 1997a)

Bioconcentration factor, log BCF:
2.67 (freshwater fish), 2.50 (fish, microcosm) (Garten and Trabalka, 1983)
2.68 (*Mytilus edulis*, Serrano et al., 1997)
3.23 (*Poecilia reticulata*) (Welling and de Vries, 1992)

Soil sorption coefficient, log K_{oc}:
3.86 (Commerce soil), 3.77 (Tracy soil), 3.78 (Catlin soil) (McCall et al., 1981)
4.13 (Kenaga and Goring, 1980)
3.27 (average of 2 soil types, Kanazawa, 1989)
3.34-3.79 (average = 3.66 in 4 sterilized Iowa soils, Felsot and Dahm, 1979)

Octanol/water partition coefficient, log K_{ow}:
5.11 (Chiou et al., 1977)
4.80 (DeKock and Lord, 1987)
5.267 (de Bruijn et al., 1989)

Solubility in organics (25 °):
6.5, 7.9, 6.3, and 0.45 kg/kg in acetone, benzene, chloroform, and methanol, respectively
(Worthing and Hance, 1991)

Solubility in water (ppm):
1.12 at 24 °C (Felsot and Dahm, 1979)
0.4 at 23 °C (Chiou et al., 1977)
0.45 at 10 °C, 0.73 at 20 °C, 1.3 at 30 °C (Bowman and Sans, 1985)
0.7 at 19.0 °C (Bowman and Sans, 1979)

Vapor pressure (mmHg):
5.03×10^{-5} at 25 °C (Hinckley et al., 1990)

Environmental fate:
  Biological. From the first-order biotic and abiotic rate constants of chlorpyrifos in estuarine
water and sediment/water systems, the estimated biodegradation half-lives were 3.5-41 and
11.9-51.4 d, respectively (Walker et al., 1988).
  Soil: Hydrolyzes in soil to 3,5,6-trichloro-2-pyridinol (Somasundaram et al., 1991). The half-
lives in a silt loam and clay loam were 12 and 4 wk, respectively (Getzin, 1981). In another
study, Getzin (1981a) reported the hydrolysis half-lives in a Sultan silt loam at 5, 15, 25, 35,
and 45 °C were >20, >20, 8, 3, and 1 d, respectively. The only breakdown product identified
was the hydrolysis product 3,5,6-trichloro-2-pyridinol. Degrades in soil forming oxychlor-
pyrifos, 3,5,6-trichloro-2-pyridinol (hydrolysis product), carbon dioxide, soil-bound residues,
and water-soluble products (Racke et al., 1988). Leoni et al. (1981) reported that the major
degradation product of chlorpyrifos in soil is 3,5,6-trichloro-2-pyridinol. The major factors
affecting the rate of degradation include chemical hydrolysis in moist soils, clay-catalyzed
hydrolysis on dry soil surfaces, microbial degradation, and volatility (Davis and Kuhr, 1976;
Felsot and Dahm, 1979; Miles et al., 1979; Chapman and Harris, 1980; Getzin, 1981; Chapman
et al., 1984; Miles et al., 1983, 1984; Getzin, 1985; Chapman and Chapman, 1986). Getzin
(1981) reported that catalyzed hydrolysis and microbial degradation were the major factors of
chlorpyrifos disappearance in soil. The reported half-lives in sandy and muck soils were 2 and 8
wk, respectively (Chapman and Harris, 1980).
  Plant: Although no products were identified, the half-life of chlorpyrifos in Bermuda grasses
was 2.9 d (Leuck et al., 1975). The concentration and the formulation of application of
chlorpyrifos will determine the rate of evaporation from leaf surfaces. Reported foliar half-lives
on tomato, orange, and cotton leaves were 15-139, 1.4-96, and 5.5-57 h, respectively (Veierov
et al., 1988).
  Surface Water: In an estuary, the half-life of chlorpyrifos was 24 d (Schimmel et al., 1983).
  Photolytic. 3,5,6-Trichloro-2-pyridinol formed by the photolysis of chlorpyrifos in water.
Continued photolysis yielded chloride ions, carbon dioxide, ammonia, and possibly poly-
hydroxychloropyridines. The following photolytic half-lives in water at north 40° latitude were
reported: 31 d during midsummer at a depth of 10^{-3} cm; 345 d during midwinter at a depth of
10^{-3} cm; 43 d at a depth of 1 m; 2.7 yr during midsummer at a depth of 1 m in river water
(Dilling et al., 1984). The combined photolysis-hydrolysis products identified in buffered,
distilled water were *O*-ethyl *O*-(3,5,6-trichloro-2-pyridyl)phosphorothioate, 3,5,6-trichloro-2-
pyridinol, and five radioactive unknowns (Meikle et al., 1983).
  Chemical/Physical. Hydrolysis products include 3,5,6-trichloro-2-pyridinol, *O*-ethyl *O*-
hydrogen-*O*-(3,5,6-trichloro-2-pyridyl)phosphorthioate, and *O,O*-dihydrogen-*O*-(3,5,6-trichlor-
o-2-pyridyl)phosphorothioate. Reported half-lives in buffered distilled water at 25 °C at pH
values of 8.1, 6.9, and 4.7 are 22.8, 35.3, and 62.7 d, respectively (Meikle and Youngson,
1978).

The hydrolysis half-life in three different natural waters was approximately 48 d at 25 °C (Macalady and Wolfe, 1985). Freed et al. (1979) reported hydrolysis half-lives of 120 and 53 d at pH 6.1 and pH 7.4, respectively, at 25 °C. At 25 °C and a pH range of 1-7, the hydrolysis half-life was about 78 d (Macalady and Wolfe, 1983). However, the alkaline hydrolysis rate of chlorpyrifos in the sediment-sorbed phase were found to be considerably slower (Macalady and Wolfe, 1985). Over the pH range of 9-13, 3,5,6-trichloro-2-pyridinol and *O,O*-diethyl phosphorothioic acid formed as major hydrolysis products (Macalady and Wolfe, 1983). The hydrolysis half-lives of chlorpyrifos in a sterile 1% ethanol/water solution at 25 °C and pH values of 4.5, 5.0, 6.0, 7.0, and 8.0 were 11, 11, 7.0, 4.2, and 2.7 wk, respectively (Chapman and Cole, 1982).

Chlorpyrifos is stable to hydrolysis over the pH range of 5-6 (Mortland and Raman, 1967). However, in the presence of a Cu(II) salt (i.e., cupric chloride) or when present as the exchangeable Cu(II) cation in montmorillonite clays, chlorpyrifos is completely hydrolyzed via first-order kinetics in <24 h at 20 °C. It was suggested that chlorpyrifos decomposition in the presence of Cu(II) was a result of coordination of molecules to the copper atom with subsequent cleavage of the side chain containing the phosphorus atom forming 3,5,6-trichloro-2-pyridinol and *O,O*-ethyl-*O*-phosphorothioate (Mortland and Raman, 1967).

Emits toxic fumes of hydrogen chloride, ethyl sulfide, diethyl sulfide and nitrogen oxides if exposed to fire. At 130 °C, chlorpyrifos undergoes exothermic decomposition which could lead to higher temperature (Turf & Ornamental Chemicals Reference, 1991).

Chlorpyrifos is stable in neutral and weakly acidic solutions (Hartley and Kidd, 1987).

Exposure limits (mg/m³): NIOSH REL: TWA 0.2, STEL 0.6; ACGIH TLV: TWA 0.2 (adopted). The acceptable daily intake for humans is 10 μg/kg body weight (Worthing and Hance, 1991).

Symptoms of exposure: May cause pain and slight eye irritation. Excessive exposure may cause headache, dizziness, nausea, muscle twitching, incoordination, abdominal cramps, diarrhea, sweating, blurred vision, pinpoint pupils, salivation, tearing, tightness in chest, excessive urination, convulsions (Turf & Ornamental Chemicals Reference, 1991).

Toxicity:

EC_{50} (24-h) for brine shrimp (*Artemia* sp.) 2 mg/L, for estuarine rotifer (*Brachionus plicatilis*) 1.7 mg/L (Guzzella et al., 1997).

EC_{50} (5-min) for *Photobacterium phosphoreum* 46.3 mg/L (Somasundaram et al., 1990).

LC_{50} (96-h) for *Pteronarcys californica* 10 μg/L, *Pteronarcella badia* 0.38 μg/L, *Classenia sabulosa* 0.57 μg/L (Sanders and Cope, 1968), juvenile Gulf toadfish (*Ospanus beta*) 520 μg/L (Hansen et al., 1988), *Gammarus fasciatus* 0.32 μg/L (quoted, Verschueren, 1983), rainbow trout 15-550 μg/L, brook trout 200 μg/L, lake trout 73-227 μg/L, channel catfish 280 μg/L, bluegill sunfish 1.7-4.2 μg/L, *Gammarus lacustris* 0.11 μg/L, fathead minnow 678 μg/L (Mayer and Ellersieck, 1986), *Pimephales promelas* 122.2 μg/L (Jarvinen et al., 1988); estuarine mysid (*Mysidopsis bahia*) 35 ng/L, sheepshead minnow (*Cyprinodon variegatus*) 136 μg/L, longnose killifish (*Fundulus similis*) 4.1 μg/L, Atlantic silverside (*Menidia menidia*) 1.7 μg/L, striped mullet (*Mugil cephalus*) 5.4 μg/L (Schimmel et al., 1983).

LC_{50} (48-h) for red killifish 1,350 mg/L (Yoshioka et al., 1986), *Gammarus pulex* 0.08 μg/L, *Simocephalus* sp. 0.8 μg/L, *Asellus aquaticus* 4.3 μg/L (Van Wijgaarden et al., 1993).

LC_{50} (24-h) for *Brachionus calyciflorus* 11.85 μg/L, *Brachionus plicatilis* 10.67 μg/L (Ferrando and Andreu-Moliner, 1991), *Gammarus lacustris* 0.76 μg/L, rainbow trout 7.1-51 μg/L, brook trout 400 μg/L, fathead minnow 712 μg/L, lake trout 195-419 μg/L, channel catfish 410 μg/L, bluegill sunfish >10 μg/L (Mayer and Ellersieck, 1986).

LC_{50} for flies (*Drosophila melanogaster*) ranged from 52 to 99 ng/cm² (Ringo et al., 1995).

Acute oral LD_{50} for chickens 25.4 mg/kg, ducks 76 mg/kg, guinea pigs 504 mg/kg, quail 13.3 mg/kg, rabbits 1,000 mg/kg (quoted, RTECS, 1985), mice 152 mg/kg, rats 169 mg/kg (Berteau and Deen, 1978), female rats 382 mg/kg (Turf & Ornamental Chemicals Reference, 1991).

LD_{50} (inhalation) for mice 94 mg/kg, rats 78 mg/kg (Berteau and Deen, 1978); 4-h exposure for rats >1.5 mg/L (Turf & Ornamental Chemicals Reference, 1991).

LD_{50} (skin) for rats 202 mg/kg, rabbits 2,000 mg/kg (quoted, RTECS, 1985).

Acute percutaneous LD_{50} (solution) for rabbits ≈4,000 mg/kg (Worthing and Hance, 1991).

In 2-yr feeding trials, NOELs based on blood plasma cholinesterase activity for rats and dogs were 30 and 10 μg/kg daily, respectively (Worthing and Hance, 1991).

Drinking water standard: No MCLGs or MCLs have been proposed (U.S. EPA, 1996).

Use: Insecticide used for control of Coleoptera, Diptera, Homoptera, and Lepidoptera in soil or foliage of citrus, coffee, cotton, maize, and sugar beets (Worthing and Hance, 1991). Controls pests infesting lawns, golf courses, parks, perennial turf grasses, flowers, shrubs, evergreens, vines, shade and flowering trees (Turf & Ornamental Chemicals Reference, 1991).

CHRYSENE

Synonyms: Benz[*a*]phenanthrene; Benzo[*a*]phenanthrene; Benzo[α]phenanthrene; 1,2-Benzo-phenanthrene; 1,2-Benzphenanthrene; 1,2-Dibenzonaphthalene; 1,2,5,6-Dibenzonaphthalene; RCRA waste number U050.

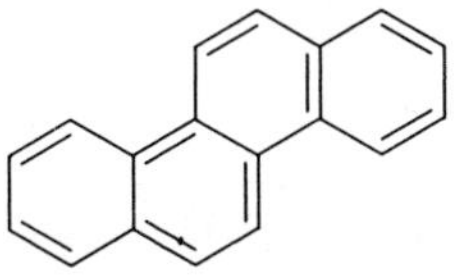

CAS Registry Number: 218-01-9
Molecular formula: $C_{18}H_{12}$
Formula weight: 228.30
RTECS: GC0700000
Merck reference: 10, 2235

Physical state:
Orthorhombic, bipyramidal plates from benzene exhibiting strong fluorescence under UV light

Melting point (°C):
255-256 (Weast, 1986)
258.2 (Casellato, 1973)

Boiling point (°C):
448 (Weast, 1986)

Density (g/cm³):
1.274 at 20/4 °C (Weast, 1986)

Diffusivity in water (x 10⁻⁵ cm²/sec):
0.63 at 20 °C using method of Hayduk and Laudie (1974)

Dissociation constant, pK_a:
>15 (Christensen et al., 1975)

Entropy of fusion (cal/mol·K):
14.9 (Yalkowsky, 1981)

Henry's law constant (x 10⁻⁷ atm·m³/mol):
1.97, 6.91, 18.8, 52.3, and 118 at 4.1, 11.0, 18.0, 25.0, and 31.0 °C, respectively (Bamford et al., 1998)

Ionization potential (eV):
7.85 ± 0.15 (Franklin et al., 1969)

Bioconcentration factor, log BCF:
3.79 (*Daphnia magna*, Newsted and Giesy, 1987)
1.17 (*Polychaete* sp.), 0.79 (*Capitella capitata*) (Bayona et al., 1991)
Apparent values of 4.3 (wet wt) and 6.0 (lipid wt) for freshwater isopods including *Asellus aquaticus* (L.) (van Hattum et al., 1998)

Soil sorption coefficient, log K_{oc}:
6.27 (average, Kayal and Connell, 1990)
5.98 (San Francisco, CA mudflat sediments, Maruya et al., 1996)
6.8 (average value using 8 river bed sediments from the Netherlands, van Hattum et al., 1998)

Octanol/water partition coefficient, log K_{ow}:
5.81 at 25 °C (de Maagd et al., 1998)
5.91 (Bruggeman et al., 1982; Yoshida et al., 1983)
5.78 at 25 °C (Andersson and Schräder, 1999)

Solubility in organics:
769 mg/L in absolute alcohol at 25 °C (Windholz et al., 1983)

Solubility in water:
2.0 μg/L at 25 °C (Mackay and Shiu, 1977)
1.8 μg/kg at 25 °C, 2.2 μg/kg at 29 °C (May et al., 1978a)
1.5 μg/L at 27 °C (Davis et al., 1942)
17 μg/L at 24 °C (practical grade, Hollifield, 1979)
6 μg/L at 25 °C (Klevens, 1950)
3.27 μg/L at 25 °C (Walters and Luthy, 1984)
1.6 μg/L at 25 °C (Vadas et al., 1991)
1.50 μg/L at 25 °C (de Maagd et al., 1998)
1.02 μg/L at 25 °C (Billington et al., 1988)
2.1 μg/L at 23 °C (Pinal et al., 1991)
In mole fraction (x 10^{-10}): 0.5603 at 6.65 °C, 0.6313 at 11.00 °C, 1.105 at 20.40 °C, 1.326 at
    24.00 °C, 1.491 at 25.30 °C, 1.744 at 28.85 °C (May et al., 1983)

Vapor pressure (x 10^{-9} mmHg):
6.3 at 25 °C (Mabey et al., 1982)
630 at 20 °C (Sims et al., 1988)
4.3 at 25 °C (de Kruif, 1980)

Environmental fate:
Biological. When chrysene was statically incubated in the dark at 25 °C with yeast extract and settled domestic wastewater inoculum, significant biodegradation with varied adaptation rates was observed. At concentrations of 5 and 10 mg/L, 59 and 38% biodegradation, respectively, were observed after 28 d (Tabak et al., 1981).

Contaminated soil from a manufactured coal gas plant that had been exposed to crude oil was spiked with chrysene (400 mg/kg soil) to which Fenton's reagent (5 mL 2.8 M hydrogen peroxide; 5 mL 0.1 M ferrous sulfate) was added. The treated and nontreated soil samples were incubated at 20 °C for 56 d. Fenton's reagent greatly enhanced the mineralization of chrysene by indigenous microorganisms. The amounts of chrysene recovered as carbon dioxide after treatment with and without Fenton's reagent were 25 and 21%, respectively. Pretreatment of the soil with a surfactant (10 mM sodium dodecylsulfate) before addition of Fenton's reagent, increased the mineralization rate 32% as compared to nontreated soil (Martens and Frankenberger, 1995).

Ye et al. (1996) investigated the ability of *Sphingomonas paucimobilis* strain U.S. EPA 505 (a soil bacterium capable of using fluoranthene as a sole source of carbon and energy) to degrade 4, 5, and 6-ringed aromatic hydrocarbons (10 ppm). After 16 h of incubation using a resting cell suspension, 31.5% of chrysene had degraded. It was suggested that degradation occurred via ring cleavage resulting in the formation of polar metabolites and carbon dioxide.

Soil. The reported half-lives for chrysene in a Kidman sandy loam and McLaurin sandy loam are 371 and 387 d, respectively (Park et al., 1990).

Surface Water. In a 5-m deep surface water body, the calculated half-lives for direct photochemical transformation at 40 °N latitude, in the midsummer during midday were 13 h and 68 d with and without sediment-water partitioning, respectively (Zepp and Schlotzhauer, 1979).

Photolytic. Based on structurally related compounds, chrysene may undergo photolysis to yield quinones (U.S. EPA, 1985) and/or hydroxy derivatives (Nielsen et al., 1983). The atmospheric half-life was estimated to range from 0.802 to 8.02 h (Atkinson, 1987). Behymer and Hites (1985) determined the effect of different substrates on the rate of photooxidation of chrysene using a rotary photoreactor. The photolytic half-lives of chrysene using silica gel, alumina, and fly ash were 100, 78, and 38 h, respectively.

Chrysene (5.0 mg/L) in a methanol-water solution (1:1 v/v) was subjected to a high pressure mercury lamp or sunlight. Based on a rate constant of 7.07 x 10^{-3}/min, the corresponding half-life is 1.63 h (Wang et al., 1991).

Chemical/Physical. A monochlorochrysene product was formed during the chlorination of chrysene in aqueous solutions at pH 4. The reported half-lives at chlorine concentrations of 0.6 and 10 mg/L were >24 and 0.45 h, respectively (Mori et al., 1991).

Chrysene will not hydrolyze because it has no hydrolyzable functional group (Kollig, 1993).

Exposure limits: Potential occupational carcinogen. No individual standards have been set; however, as a constituent in coal tar pitch volatiles, the following exposure limits have been established (mg/m^3): NIOSH REL: TWA 0.1 (cyclohexane-extractable fraction), IDLH 80; OSHA PEL: TWA 0.2 (benzene-soluble fraction); ACGIH TLV: TWA 0.2 (benzene solubles).

Drinking water standard: No MCLGs or MCLs have been proposed (U.S. EPA, 1996).

Source: Detected in groundwater beneath a former coal gasification plant in Seattle, WA at a concentration of 10 μg/L (ASTR, 1995). Identified in Kuwait and South Louisiana crude oils at concentrations of 6.9 and 17.5 ppm, respectively (Pancirov and Brown, 1975). Also present in high octane gasoline (6.7 mg/kg), fresh motor oil (0.56 mg/kg), used motor oil (86.2-236.6 mg/kg), bitumen (1.64-5.14 ppm), gasoline exhaust (27-318 μg/m^3), cigarette smoke 60 μg/1,000 cigarettes), and South Louisiana crude oil (17.5 ppm) (quoted, Verschueren, 1983).

Based on analysis of 7 coal tar samples, chrysene concentrations ranged from 620 to 5,100 ppm (EPRI, 1990).

Use: Organic synthesis. Derived from industrial and experimental coal gasification operations where the maximum concentrations detected in gas, liquid, and coal tar streams were 7.3, 0.16, and 8.6 mg/m^3, respectively (Cleland, 1981).

CROTONALDEHYDE

Synonyms: 2-Butenal; Crotenaldehyde; Crotonal; Crotonic aldehyde; **1,2-Ethanediol dipropanoate**; Ethylene glycol dipropionate; Ethylene dipropionate; Ethylene propionate; β-Methylacrolein; NCI-C56279; Propylene aldehyde; RCRA waste number U053; Topanel.

CAS Registry Number: 4170-30-3
DOT: 1143
DOT label: Flammable liquid
Molecular formula: C_4H_6O
Formula weight: 70.09
RTECS: GP9499000
Merck reference: 10, 2585

Physical state, color, and odor:
Clear, colorless to straw-colored liquid with a pungent, irritating, suffocating odor. Odor threshold concentration is 7 ppm (Keith and Walters, 1992).

Melting point (°C):
-74 (Weast, 1986)
-76.5 (Windholz et al., 1983)

Boiling point (°C):
104-105 (Weast, 1986)
99 (Verschueren, 1983)

Density (g/cm³):
0.853 at 20/20 °C (Weast, 1986)
0.8477 at 20.5/4 °C (Huntress and Mulliken, 1941)

Diffusivity in water (x 10^{-5} cm²/sec):
0.99 at 20 °C using method of Hayduk and Laudie (1974)

Flash point (°C):
7.2 (NIOSH, 1997)
13 (open cup, Windholz et al., 1983)

Lower explosive limit (%):
2.1 (NIOSH, 1997)

Upper explosive limit (%):
15.5 (NIOSH, 1997)

Henry's law constant (x 10^{-5} atm·m³/mol):
1.92 (Buttery et al., 1971)

Ionization potential (eV):
9.73 ± 0.01 (Franklin et al., 1969)

Soil sorption coefficient, log K_{oc}:
Unavailable because experimental methods for estimation of this parameter for aldehydes are lacking in the documented literature. However, its high solubility in water suggests its adsorption to soil will be nominal (Lyman et al., 1982).

Octanol/water partition coefficient, log K_{ow}:
Unavailable because experimental methods for estimation of this parameter for aliphatic aldehydes are lacking in the documented literature

Solubility in organics:
Miscible with alcohol, benzene, gasoline, kerosene, solvent naphtha, and toluene (Hawley, 1981)

Solubility in water (wt %):
19.2 at 5 °C, 18.1 at 20 °C (quoted, Windholz et al., 1983)

Vapor density:
2.86 g/L at 25 °C, 2.41 (air = 1)

Vapor pressure (mmHg):
19 at 20 °C (NIOSH, 1997)

Environmental fate:
  Biological. Heukelekian and Rand (1955) reported a 10-d BOD value of 1.30 g/g which is 56.8% of the ThOD value of 2.29 g/g.
  Chemical/Physical. Slowly oxidizes in air forming crotonic acid (Windholz et al., 1983). At elevated temperatures, crotonaldehyde may polymerize (NIOSH, 1997).
  Crotonaldehyde undergoes addition of water across the CH=CH bond yielding 3-hydroxy-1-butanal (Kollig, 1995).
  At an influent concentration of 1,000 mg/L, treatment with granular activated carbon resulted in effluent concentration of 544 mg/L. The adsorbability of the carbon used was 92 mg/g carbon (Guisti et al., 1974).

Exposure limits: NIOSH REL: TWA 2 ppm (6 mg/m^3), IDLH 50 ppm; OSHA PEL: TWA 2 ppm; ACGIH TLV: ceiling 0.3 ppm (adopted).

Symptoms of exposure: Irritation of eyes and respiratory system (NIOSH, 1997). An irritation concentration of 23.01 mg/m^3 in air was reported by Ruth (1986).

Toxicity:
  LC$_{50}$ (96-h) for bluegill sunfish 3.5 mg/L, *Menidia beryllina* 1.3 mg/L (quoted, Verschueren, 1983).
  Acute oral LD$_{50}$ for mice 104 mg/kg, rats 206 mg/kg (quoted, RTECS, 1985).

Source: Reported in exhaust of gasoline-powered automobile at concentrations ranging from 100 to 900 ppb (quoted, Verschueren, 1983)

Uses: Preparation of 1-butanol, butyraldehyde, 2-ethylhexanol, quinaldine; chemical warfare; insecticides; leather tanning; alcohol denaturant; solvent; warning agent in fuel gases; purification of lubricating oils; organic synthesis.

CYCLOHEPTANE

Synonyms: Heptamethylene; Suberane; UN 2241.

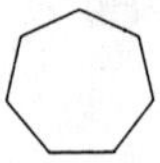

CAS Registry Number: 291-64-5
DOT: 2241
Molecular formula: C_7H_{14}
Formula weight: 98.19
RTECS: GU3140000

Physical state:
Oily liquid

Melting point (°C):
-12 (Weast, 1986)

Boiling point (°C):
118.5 (Weast, 1986)

Density (g/cm³ at 20/4 °C):
0.8098 (Weast, 1986)
0.8011 (Riddick et al., 1986)

Diffusivity in water (x 10^{-5} cm²/sec):
0.79 at 20 °C using method of Hayduk and Laudie (1974)

Dissociation constant, pK_a:
>14 (Schwarzenbach et al., 1993)

Flash point (°C):
15 (Sax and Lewis, 1987)
6 (Aldrich, 1988)

Lower explosive limit (%):
1.1 (NFPA, 1984)

Upper explosive limit (%):
6.7 (NFPA, 1984)

Heat of fusion (kcal/mol):
0.450 (Dean, 1987)

Henry's law constant (x 10^{-2} atm·m³/mol):
9.35 (approximate - calculated from water solubility and vapor pressure)

Ionization potential (eV):
9.9 (Lias, 1998)

Soil sorption coefficient, log K_{oc}:
Unavailable because experimental methods for estimation of this parameter for alicyclic hydrocarbons are lacking in the documented literature

Octanol/water partition coefficient, log K_{ow}:
2.64 using method of Hansch et al. (1968)

Solubility in organics:
Soluble in alcohol, benzene, chloroform, ether, and ligroin (Weast, 1986)

Solubility in water:
30 mg/kg at 25 °C (McAuliffe, 1966)
27.1 mg/L at 30 °C, 17.0 mg/L in artificial seawater (34.5 parts NaCl per 1,000 parts water) at 30 °C (Groves, 1988)

Vapor density:
4.01 g/L at 25 °C, 3.39 (air = 1)

Vapor pressure (mmHg):
21.7 (extrapolated, Boublik et al., 1986)

Environmental fate:
Biological. Cycloheptane may be oxidized by microbes to cycloheptanol, which may further oxidize to give cycloheptanone (Dugan, 1972).
Photolytic. The following rate constants were reported for the reaction of cycloheptane and OH radicals in the atmosphere: 1.31×10^{-12} cm³/molecule·sec at 298 K (Atkinson, 1985) and 1.25×10^{-11} cm³/molecule·sec (Atkinson, 1990).
Chemical/Physical. Cycloheptane will not hydrolyze because it has no hydrolyzable functional group.

Uses: Organic synthesis; gasoline component.

CYCLOHEXANE

Synonyms: Benzene hexahydride; Hexahydrobenzene; Hexamethylene; Hexanaphthene; RCRA waste number U056; UN 1145.

CAS Registry Number: 110-82-7
DOT: 1145
DOT label: Flammable liquid
Molecular formula: C_6H_{12}
Formula weight: 84.16
RTECS: GU6300000
Merck reference: 10, 2717

Physical state, color, and odor:
Colorless liquid with a sweet, chloroform-like odor. Odor threshold concentration is 410 ppb (Keith and Walters, 1992).

Melting point (°C):
6.5 (Weast, 1986)
6.6 (Stull, 1947)

Boiling point (°C):
80.7 (Weast, 1986)

Density (g/cm³):
0.7785 at 20/4 °C (Weast, 1986)

Diffusivity in water (x 10^{-5} cm²/sec):
0.84 at 20 °C (Witherspoon and Bonoli, 1969)
0.90 at 25 °C (Bonoli and Witherspoon, 1968)

Dissociation constant, pK_a:
≈ 45 (Gordon and Ford, 1972)

Flash point (°C):
-18 (NIOSH, 1997)

Lower explosive limit (%):
1.3 (NIOSH, 1997)

Upper explosive limit (%):
8.0 (NFPA, 1984)

Heat of fusion (kcal/mol):
0.640 (Riddick et al., 1986)

Henry's law constant (x 10^{-1} atm·m³/mol):
1.03, 1.26, 1.40, 1.77, and 2.23 at 10, 15, 20, 25, and 30 °C, respectively (Ashworth et al., 1988)

0.54, 0.69, 0.82, 1.43, and 1.79 at 2.0, 6.0, 10.0, 18.0, and 25.0 °C, respectively (Dewulf et al., 1999)

Interfacial tension with water (dyn/cm at 25 °C):
50.0 (Donahue and Bartell, 1952)

Ionization potential (eV):
9.88 (Franklin et al., 1969)
11.00 (Yoshida et al., 1983a)

Soil sorption coefficient, log K_{oc}:
Unavailable because experimental methods for estimation of this parameter for alicyclic hydrocarbons are lacking in the documented literature

Octanol/water partition coefficient, log K_{ow}:
3.44 (Leo et al., 1975; Yoshida et al., 1983)

Solubility in organics:
Miscible with acetone, benzene, chloroform, carbon tetrachloride, ethanol, and ethyl ether (Windholz et al., 1983).
In methanol, g/L: 344 at 15 °C, 384 at 20 °C, 435 at 25 °C, 503 at 30 °C, 600 at 35 °C, 740 at 40 °C (Kiser et al., 1961).

Solubility in water:
66.5 mg/kg at 25 °C (Price, 1976)
55.0 mg/kg at 25 °C (McAuliffe, 1963, 1966)
57.5 mg/L at 25 °C (Mackay and Shiu, 1975)
52 mg/kg at 23.5 °C (Schwarz, 1980)
At 25 °C: 58.4 mg/L, 40.1 mg/L in 3.3% NaCl solution (Groves, 1988)
100 mg/L at 20 °C (Korenman and Aref'eva, 1977)
80 mg/L at 25 °C (McBain and Lissant, 1951)
In mg/kg: 0.008 at 25 °C, 0.017 at 56 °C, 0.028 at 94 °C, 0.0517 at 127 °C, 0.146 at 162 °C, 1.785 at 220.5 °C (Guseva and Parnov, 1963a).
56.7 mg/L at 25 °C (Leinonen and Mackay, 1973)
627 μmmol/L at 25 °C (Sanemasa et al., 1987)

Vapor density:
3.44 g/L at 25 °C, 2.91 (air = 1)

Vapor pressure (mmHg):
78 at 20 °C (NIOSH, 1997)
97.61 at 25 °C (Wilhoit and Zwolinski, 1971)

Environmental fate:
  Biological. Microbial degradation products reported include cyclohexanol (Dugan, 1972; Verschueren, 1983), 1-oxa-2-oxocycloheptane, 6-hydroxyheptanoate, 6-oxohexanoate, adipic acid, acetyl-CoA, succinyl-CoA (Verschueren, 1983), and cyclohexanone (Dugan, 1972; Keck et al., 1989).
  Photolytic. The following rate constants were reported for the reaction of cyclohexane and OH radicals in the atmosphere: 5.38 x 10^{12} cm^3/mol·sec at 295 K (Greiner, 1970); 6.7 x 10^{-12} cm^3/molecule·sec at 300 K (Darnall et al., 1978); 6.69 x 10^{-12} cm^3/molecule·sec at 298 (DeMore

and Bayes, 1999); 7.0 x 10^{-12} cm^3/molecule·sec (Atkinson et al., 1979); 7.49 x 10^{-12} cm^3/molecule·sec (Atkinson, 1990). A photooxidation reaction rate constant of 1.35 x 10^{-16} cm^3/molecule·sec was reported for the reaction of cyclohexane with NO$_3$ in the atmosphere (Atkinson, 1991).

   Chemical/Physical. Cyclohexane will not hydrolyze because it has no hydrolyzable functional group.

Exposure limits: NIOSH REL: TWA 300 ppm (1,050 mg/m^3), IDLH 1,300 ppm; OSHA PEL: TWA 300 ppm; ACGIH TLV: TWA 300 ppm (adopted) with proposed TWA and STEL values of 200 and 400 ppm, respectively.

Symptoms of exposure: Irritation of the eyes and respiratory system (Patnaik, 1992) at a concentration of 300 ppm (quoted, Verschueren, 1983). An irritation concentration of 1,050.00 mg/m^3 in air was reported by Ruth (1986).

Toxicity:
   LC$_{50}$ (7-d) for *Poecilia reticulata* >84 mg/L (Könemann, 1981).
   LC$_{50}$ (96-h static bioassay) for fathead minnows 93 mg/L (Mattson et al., 1976).
   Acute oral LD$_{50}$ for mice 813 mg/kg, rats 12,705 mg/kg (quoted, RTECS, 1985).

Source: Schauer et al. (1999) reported cyclohexane in a diesel-powered medium-duty truck exhaust at an emission rate of 210 μg/km.

Uses: Manufacture of nylon; solvent for cellulose ethers, fats, oils, waxes, resins, bitumens, crude rubber; paint and varnish removers; extracting essential oils; glass substitutes; solid fuels; fungicides; gasoline and coal tar component; organic synthesis.

CYCLOHEXANOL

Synonyms: Adronal; Anol; Cyclohexyl alcohol; Hexahydrophenol; Hexalin; Hydralin; Hydro-phenol; Hydroxycyclohexane; Naxol.

CAS Registry Number: 108-93-0
Molecular formula: $C_6H_{12}O$
Formula weight: 100.16
RTECS: GV7875000
Merck reference: 10, 2719

Physical state, color, and odor:
Colorless to pale yellow, viscous, hygroscopic liquid with a camphor-like odor. Odor threshold concentration is 0.15 ppm (Amoore and Hautala, 1983).

Melting point (°C):
25.1 (Weast, 1986)
23 (Hawley, 1981)
20-22 (Aldrich, 1988)

Boiling point (°C):
161.1 (Weast, 1986)

Density (g/cm³):
0.9624 at 20/4 °C (Weast, 1986)
0.9449 at 25/4 °C (Sax and Lewis, 1987)
0.937 at 37/4 °C (Hawley, 1981)

Diffusivity in water (x 10^{-5} cm²/sec):
0.86 at 20 °C using method of Hayduk and Laudie (1974)

Dissociation constant, pK$_a$:
>14 (Schwarzenbach et al., 1993)

Flash point (°C):
67.8 (NIOSH, 1997)

Heat of fusion (kcal/mol):
0.406 (Dean, 1987)

Henry's law constant (x 10^{-6} atm·m³/mol):
5.74 at 25 °C (Hine and Mookerjee, 1975)

Interfacial tension with water (dyn/cm):
3.92 at 16.2 °C (Demond and Lindner, 1993)
3.1 at 25 °C (Murphy et al., 1957)

Ionization potential (eV):
10.00 (NIOSH, 1997)

Soil sorption coefficient, log K_{oc}:
Unavailable because experimental methods for estimation of this parameter for alicyclic alcohols are lacking in the documented literature

Octanol/water partition coefficient, log K_{ow}:
1.23 (Hansch and Anderson, 1967)

Solubility in organics:
Miscible with aromatic hydrocarbons, ethanol, ethyl acetate, linseed oil, and petroleum solvents (Windholz et al., 1983)

Solubility in water:
3.82 wt % at 20 °C (Palit, 1947)
33.70 g/L at 26.7 °C (Skrez and Murphy, 1954)
43 g/kg at 30 °C (Patnaik, 1992)
34 mL/L at 25.0 °C (Booth and Everson, 1948, 1949)
33.8 mL/L at 60 °C (Booth and Everson, 1949)
39.2 g/kg at 25 °C (Hansen et al., 1949)
In g/kg: 50.0 at 7.2 °C, 47.8 at 9.4 °C, 45.8 at 9.7 °C, 44.1 at 11.2 °C, 45.5 at 12.0 °C, 42.3 at 14.2 °C, 42.9 at 15.2 °C, 40.9 at 16.3 °C, 39.5 at 20.6 °C, 38.2 at 20.8 °C, 37.5 at 24.6 °C, 35.2 at 27.55 °C, 35.7 at 28.7 °C, 33.7 at 31.85 °C, 34.1 at 33.6 °C, 32.6 at 40.4 °C, 31.8 at 40.45 °C, 31.9 at 45.8 °C, 51.4 at 121.95 °C, 92.2 at 156.9 °C, 150.0 at174.3 °C, 192 at 179.4 °C, 324 at 184.72 °C (Sidgwick and Sutton, 1930)
37,500 mg/L at 25 °C (Etzweiler et al., 1995)

Vapor density:
4.09 g/L at 25 °C, 3.46 (air = 1)

Vapor pressure (mmHg):
1 at 20 °C (NIOSH, 1997)
3.5 at 34 °C (Verschueren, 1983)
14.9 at 68.1 °C, 30.0 at 80.1 °C (Steele et al., 1997)

Environmental fate:
Biological. Reported biodegradation products include cyclohexanone (Dugan, 1972; Verschueren, 1983), 2-hydroxyhexanone, 1-oxa-2-oxocycloheptane, 6-hydroxyheptanonate, 6-oxohexanoate, and adipate (Verschueren, 1983). In activated sludge inoculum, following a 20-d adaptation period, 96.0% COD removal was achieved. The average rate of biodegradation was 28.0 mg COD/g·h (Pitter, 1976).

Photolytic. A photooxidation reaction rate constant of 1.74×10^{-13} cm^3/molecule·sec was estimated for the reaction of cyclohexanol with OH radicals in the atmosphere (Atkinson, 1987).

Chemical/Physical. Cyclohexanol will not hydrolyze in water because it does not contain a hydrolyzable functional group (Kollig, 1993).

Complete combustion in air yields carbon dioxide and water vapor.

Exposure limits: NIOSH REL: TWA 50 ppm (200 mg/m^3), IDLH 400 ppm; OSHA PEL: TWA 50 ppm; ACGIH TLV: TWA 50 ppm (adopted).

Symptoms of exposure: May irritate eyes, nose, and throat. Ingestion can cause nausea, trembling, gastrointestinal disturbances (Patnaik, 1992), and narcosis (NIOSH, 1997). An irritation concentration of 200.00 mg/m^3 in air was reported by Ruth (1986).

Toxicity:

LC$_{50}$ (96-h) for fathead minnows 704 mg/L (Veith et al., 1983), bluegill sunfish 1,100 mg/L, *Menidia beryllina* 720 mg/L (quoted, Verschueren, 1983).

LC$_{50}$ static bioassay values for fathead minnows in Lake Superior water maintained at 18-22 °C after 1 and 24-96 h were 1,550 and 1.033 mg/L, respectively (Mattson et al., 1976).

Acute oral LD$_{50}$ for rats 2,060 mg/kg (quoted, RTECS, 1985).

LD$_{50}$ (subcutaneous) for mice 2,480 mg/kg (quoted, RTECS, 1985).

Uses: In paint and varnish removers; solvent for lacquers, shellacs and resins; manufacture of adipic acid, caprolactum, benzene, cyclohexene, chlorocyclohexane, cyclohexanone, nitro-cyclohexane, and solid fuel for camp stoves; fungicidal formulations; polishes; plasticizers; soap and detergent manufacturing (stabilizer); emulsified products; blending agent; recrystallizing steroids; germicides; plastics; organic synthesis.

CYCLOHEXANONE

Synonyms: Anone; Cyclohexyl ketone; Hexanon; Hytrol O; Ketohexamethylene; Nadone; NCI-C55005; Pimelic ketone; Pimelin ketone; RCRA waste number U057; Sextone; UN 1915.

CAS Registry Number: 108-94-1
DOT: 1915
DOT label: Flammable liquid
Molecular formula: $C_6H_{10}O$
Formula weight: 98.14
RTECS: GW1050000
Merck reference: 10, 2720

Physical state, color, and odor:
Clear, colorless to pale yellow, oily liquid with a peppermint-like odor. Odor threshold concentration is 120 ppb (Keith and Walters, 1992).

Melting point (°C):
-16.4 (Weast, 1986)
-32.1 (Windholz et al., 1983)
-26 (Verschueren, 1983)
-47 (Aldrich, 1988)

Boiling point (°C):
155.6 (Weast, 1986)

Density (g/cm³):
0.9478 at 20/4 °C (Weast, 1986)
0.9412 at 25.00/4 °C (Aralaguppi et al., 1999)

Diffusivity in water (x 10⁻⁵ cm²/sec):
0.87 at 20 °C using method of Hayduk and Laudie (1974)

Flash point (°C):
35.2 (NIOSH, 1997)
46 (Aldrich, 1988)

Lower explosive limit (%):
1.1 at 100 °C (NIOSH, 1997)

Upper explosive limit (%):
9.4 (NFPA, 1984)

Henry's law constant (x 10⁻⁵ atm·m³/mol):
1.2 at 25 °C (Hawthorne et al., 1985)

Interfacial tension with water (dyn/cm at 25 °C):
3.9 (quoted, Freitas et al., 1997)

Ionization potential (eV):
9.14 ± 0.01 (Franklin et al., 1969)

Soil sorption coefficient, log K_{oc}:
Unavailable because experimental methods for estimation of this parameter for alicyclic ketones are lacking in the documented literature. However, its high solubility in water suggests its adsorption to soil will be nominal (Lyman et al., 1982).

Octanol/water partition coefficient, log K_{ow}:
0.81 (Leo et al., 1971)

Solubility in organics:
Soluble in acetone, alcohol, benzene, chloroform, and ether (Weast, 1986)

Solubility in water:
15 g/L 10 °C, 50 g/L at 30 °C (quoted, Windholz et al., 1983)
23 g/L at 20 °C, 24 g/L at 31 °C (quoted, Verschueren, 1983)
In wt %: 13.7 at 0 °C, 11.5 at 9.8 °C, 9.7 at 19.5 °C, 8.2 at 29.8 °C, 7.5 at 40.1 °C, 7.0 at 50.2
  °C, 6.7 at 60.5 °C, 6.5 at 71.1 °C, 6.8 at 80.2 °C, 6.9 at 90.7 °C (Stephenson, 1992)

Vapor density:
4.01 g/L at 25 °C, 3.39 (air = 1)

Vapor pressure (mmHg):
4 at 20 °C, 6.2 at 30 °C (Verschueren, 1983)
5 at 20 °C (NIOSH, 1997)
4.5 at 25 °C (Banerjee et al., 1990)

Environmental fate:
  Biological. In activated sludge inoculum, 96.0% COD removal was achieved. The average rate of biodegradation was 30.0 mg COD/g·h (Pitter, 1976).
  Photolytic. Atkinson (1985) reported an estimated photooxidation rate constant of 1.56 x 10^{-11} cm^3/molecule·sec for the reaction of cyclohexanone and OH radicals in the atmosphere at 298 K.
  Chemical/Physical. Cyclohexanone will not hydrolyze because it has no hydrolyzable functional group.
  At an influent concentration of 1,000 mg/L, treatment with granular activated carbon resulted in effluent concentration of 332 mg/L. The adsorbability of the carbon used was 134 mg/g carbon (Guisti et al., 1974).

Exposure limits: NIOSH REL: TWA 25 ppm (100 mg/m^3), IDLH 700 ppm; OSHA PEL: TWA 50 ppm (200 mg/m^3); ACGIH TLV: TWA 25 ppm (adopted).

Symptoms of exposure: May irritate eyes and throat. Contact with eyes may cause cornea damage (Patnaik, 1992). An irritation concentration of 100.00 mg/m^3 in air was reported by Ruth (1986).

Toxicity:
  EC_{50} (48-h) and EC_{50} (24-h) values for *Spirostomum ambiguum* were 442 mg/L (Nałecz-Jawecki and Sawicki, 1999).
  LC_{50} (96-h) for fathead minnows 527 mg/L (Veith et al., 1983).

LC_{50} (48-h) and LC_{50}(24-h) values for *Spirostomum ambiguum* were 3.57 and 3.66 g/L, respectively (Nałecz-Jawecki and Sawicki, 1999).

Acute oral LD_{50} for mice 1,400 mg/kg, rats 1,535 mg/kg (quoted, RTECS, 1985).

Uses: Solvent for cellulose acetate, crude rubber, natural resins, nitrocellulose, vinyl resins, waxes, fats, oils, shellac, rubber, DDT and other pesticides; preparation of adipic acid and caprolactum; wood stains; paint and varnish removers; degreasing of metals; spot remover; lube oil additive; leveling agent in dyeing and delustering silk.

CYCLOHEXENE

Synonyms: Benzene tetrahydride; Tetrahydrobenzene; 1,2,3,4-Tetrahydrobenzene; UN 2256.

CAS Registry Number: 110-83-8
DOT: 2256
DOT label: Flammable liquid
Molecular formula: C_6H_{10}
Formula weight: 82.15
RTECS: GW2500000
Merck reference: 10, 2721

Physical state, color, and odor:
Colorless liquid with a sweet odor. Odor threshold concentration is 0.18 ppm (Amoore and Hautala, 1983).

Melting point (°C):
-103.5 (Weast, 1986)

Boiling point (°C):
83 (Weast, 1986)

Density (g/cm³):
0.8102 at 20/4 °C (Weast, 1986)
0.7823 at 50/4 °C (Windholz et al., 1983)
0.81096 at 20/4 °C, 0.80609 at 25/4 °C (Dreisbach, 1959)

Diffusivity in water (x 10⁻⁵ cm²/sec):
0.88 at 20 °C using method of Hayduk and Laudie (1974)

Dissociation constant, pK$_a$:
>14 (Schwarzenbach et al., 1993)

Flash point (°C):
-11.7 (NIOSH, 1997)

Heat of fusion (kcal/mol):
0.787 (Dean, 1987)

Henry's law constant (x 10⁻² atm·m³/mol):
3.85 at 25 °C (Nielsen et al., 1994)

Ionization potential (eV):
8.72 (Franklin et al., 1969)
9.18 (Collin and Lossing, 1959)
9.20 (Rav-Acha et al., 1987)

Soil sorption coefficient, log K_{oc}:
Unavailable because experimental methods for estimation of this parameter for alicyclic hydrocarbons are lacking in the documented literature

Octanol/water partition coefficient, log K_{ow}:
2.86 (Hansch and Leo, 1979)

Solubility in organics:
Soluble in acetone, alcohol, benzene, and ether (Weast, 1986)

Solubility in water:
130 mg/L at 25 °C (McBain and Lissant, 1951)
213 mg/kg at 25 °C (McAuliffe, 1966)
281, 286 mg/kg at 23.5 °C (Schwarz, 1980)
In 1 mM nitric acid: 4.95, 3.46, and 2.49 mM at 30, 35, and 40 °C, respectively (Natarajan and Venkatachalam, 1972).

Vapor density:
3.36 g/L at 25 °C, 2.84 (air = 1)

Vapor pressure (mmHg):
67 at 20 °C (NIOSH, 1997)
88.8 at 25 °C (estimated using Antoine equation, Dreisbach, 1959)

Environmental fate:
Biological. Cyclohexene biodegrades to cyclohexanone (Dugan, 1972; Verschueren, 1983).

Photolytic. The following rate constants were reported for the reaction of cyclohexene with OH radicals in the atmosphere: 6.80×10^{-11} cm^3/molecule·sec (Atkinson et al., 1979), 6.75×10^{-11} cm^3/molecule·sec at 298 K (Sabljić and Güsten, 1990), 5.40×10^{-11} cm^3/molecule·sec at 298 K (Rogers, 1989), 1.0×10^{-10} cm^3/molecule·sec at 298 K (Atkinson, 1990; with ozone in the gas-phase: 1.69×10^{-16} cm^3/molecule·sec at 298 K (Japar et al., 1974), 2.0×10^{-16} at 294 K (Adeniji et al., 1981), 1.04×10^{-16} cm^3/molecule·sec (Atkinson et al., 1983), 1.04×10^{-16} at 298 K (Atkinson and Carter, 1984); with NO_3 in the atmosphere: 5.26×10^{-13} cm^3/molecule·sec (Sabljić and Güsten, 1990); 5.3×10^{-13} cm^3/molecule·sec at 298 K (Atkinson, 1990), and 5.28×10^{-13} cm^3/molecule·sec at 295 K (Atkinson, 1991).

Chemical/Physical. Gaseous products formed from the reaction of cyclohexene with ozone were (% yield): formic acid (12), carbon monoxide (18), carbon dioxide (42), ethylene (1), and valeraldehyde (17) (Hatakeyama et al., 1987).

Cyclohexene reacts with chlorine dioxide in water forming 2-cyclohexen-1-one (Rav-Acha et al., 1987).

Exposure limits: NIOSH REL: TWA 300 ppm (1,015 mg/m^3), IDLH 2,000 ppm; OSHA PEL: TWA 300 ppm; ACGIH TLV: TWA 300 ppm (adopted).

Symptoms of exposure: Irritation of eyes, skin, and respiratory tract. Inhalation of high concentrations may cause drowsiness (NIOSH, 1997; Patnaik, 1992).

Uses: Manufacture of adipic acid, hexahydrobenzoic acid, maleic acid, 1,3-butadiene; catalyst solvent; oil extraction; component of coal tar; stabilizer for high octane gasoline; organic synthesis.

CYCLOPENTADIENE

Synonyms: 1,3-Cyclopentadiene; Pentole; R-pentine; Pyropentylene.

CAS Registry Number: 542-92-7
Molecular formula: C_5H_6
Formula weight: 66.10
RTECS: GY1000000
Merck reference: 10, 2732

Physical state, color, and odor:
Colorless liquid with a turpentine-like odor. Odor threshold concentration is 1.9 ppm (Amoore and Hautala, 1983).

Melting point (°C):
-97.2 (Weast, 1986)
-85 (Windholz et al., 1983)

Boiling point (°C):
40.0 (Weast, 1986)
41.5-42 (Windholz et al., 1983)

Density (g/cm³):
0.8021 at 20/4 °C (Weast, 1986)
0.7966 at 25/4 °C (Windholz et al., 1983)

Diffusivity in water (x 10^{-5} cm²/sec):
0.99 at 20 °C using method of Hayduk and Laudie (1974)

Dissociation constant, pK_a:
15 (Gordon and Ford, 1972)
16.0 (Streitwieser and Nebenzahl, 1976)

Flash point (°C):
25.0 (open cup, NIOSH, 1997)

Henry's law constant (x 10^{-2} atm·m³/mol):
5.1 at 25 °C (approximate - calculated from water solubility and vapor pressure)

Ionization potential (eV):
8.56 (NIOSH, 1997)
8.97 (Franklin et al., 1969)

Soil sorption coefficient, log K_{oc}:
Unavailable because experimental methods for estimation of this parameter for alicyclic hydrocarbons are lacking in the documented literature

Octanol/water partition coefficient, log K_{ow}:
2.34 using method of Hansch et al. (1968)

Solubility in organics:
Miscible with acetone, benzene, carbon tetrachloride, and ether. Soluble in acetic acid, aniline, and carbon disulfide (Windholz et al., 1983).

Solubility in water:
10.3 mM at 20-25 °C (Streitwieser and Nebenzahl, 1976)

Vapor pressure (mmHg):
400 at 20 °C (NIOSH, 1997)

Vapor density:
2.70 g/L at 25 °C, 2.28 (air = 1)

Environmental fate:
Biological. Cyclopentadiene may be oxidized by microbes to cyclopentanone (Dugan, 1972).
Chemical/Physical. Dimerizes to dicyclopentadiene on standing (Windholz et al., 1983).

Exposure limits: NIOSH REL: TWA 75 ppm (200 mg/m^3), IDLH 750 ppm; OSHA PEL: TWA 75 ppm; ACGIH TLV: TWA 75 ppm (adopted).

Symptoms of exposure: Irritation of eyes and nose (Patnaik, 1992)

Toxicity:
Acute oral LD$_{50}$ (dimeric form) for rats is 820 mg/kg (Patnaik, 1992).

Uses: Manufacture of resins; chlorinated insecticides; organic synthesis (Diels-Alder reaction).

CYCLOPENTANE

Synonyms: Pentamethylene; UN 1146.

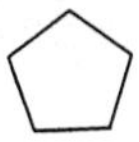

CAS Registry Number: 287-92-3
DOT: 1146
DOT label: Flammable liquid
Molecular formula: C_5H_{10}
Formula weight: 70.13
RTECS: GY2390000
Merck reference: 10, 2734

Physical state and color:
Colorless, mobile liquid

Melting point (°C):
-93.9 (Weast, 1986)
-95 (Huntress and Mulliken, 1941)

Boiling point (°C):
49.2 (Weast, 1986)

Density (g/cm³):
0.7457 at 20/4 °C (Weast, 1986)
0.74059 at 15/4 °C (Huntress and Mulliken, 1941)
0.74394 at 25/4 °C (Riddick et al., 1986)

Diffusivity in water (x 10^{-5} cm²/sec):
0.93 at 20 °C (Witherspoon and Bonoli, 1969)
1.04 at 25 °C (Bonoli and Witherspoon, 1968)

Dissociation constant, pK_a:
≈ 44 (Gordon and Ford, 1972)

Flash point (°C):
-7 (Sax and Lewis, 1987)
-37.2 (NIOSH, 1997)

Lower explosive limit (%):
1.1 (NIOSH, 1997)

Upper explosive limit (%):
8.7% (NIOSH, 1997)

Heat of fusion (kcal/mol):
0.1455 (Riddick et al., 1986)

Henry's law constant (atm·m³/mol):
0.164, 0.240, and 0.300 at 27.9, 35.8, and 45.0 °C, respectively (Hansen et al., 1995)

Interfacial tension with water (dyn/cm at 20 °C):
28 (CHRIS, 1984)

Ionization potential (eV):
10.53 ± 0.05 (Franklin et al., 1969)

Soil sorption coefficient, log K_{oc}:
Unavailable because experimental methods for estimation of this parameter for alicyclic hydrocarbons are lacking in the documented literature

Octanol/water partition coefficient, log K_{ow}:
3.00 (Leo et al., 1975)

Solubility in organics:
Miscible with ether and other hydrocarbon solvents (Windholz et al., 1983), such as pentane, hexane, and cyclohexane.
In methanol, g/L: 680 at 5 °C, 860 at 10 °C, 1,400 at 15 °C. Miscible at higher temperatures (Kiser et al., 1961).

Solubility in water:
In mg/kg: 160 at 25 °C, 163 at 40.1 °C, 180 at 55.7 °C, 296 at 99.1 °C, 372 at 118.0 °C, 611 at 137.3 °C, 792 at 153.1 °C (Price, 1976)
156 mg/kg at 25 °C (McAuliffe, 1963, 1966)
164 mg/L at 25 °C, 128 mg/L in artificial seawater (34.5 parts NaCl per 1,000 parts water) at 25 °C (Groves, 1988)

Vapor density:
2.87 g/L at 25 °C, 2.42 (air = 1)

Vapor pressure (mmHg):
400 at 31.0 °C (estimated, Weast, 1986)

Environmental fate:
Biological. Cyclopentane may be oxidized by microbes to cyclopentanol, which may further oxidize to cyclopentanone (Dugan, 1972).
Photolytic. The following rate constants were reported for the reaction of octane and OH radicals in the atmosphere: 3.7×10^{12} cm^3/mol·sec at 300 K (Hendry and Kenley, 1979); 5.40×10^{-12} cm^3/molecule·sec (Atkinson, 1979); 4.83×10^{-12} cm^3/molecule·sec at 298 K (DeMore and Bayes, 1999); 6.20×10^{-12}, 5.24×10^{-12}, and 4.43×10^{-12} cm^3/molecule·sec at 298, 299, and 300 K, respectively (Atkinson, 1985), 5.16×10^{-12} cm^3/molecule·sec at 298 K (Atkinson, 1990).
Chemical/Physical. Cyclopentane will not hydrolyze because it has no hydrolyzable functional group. Complete combustion in air yields carbon dioxide and water.
At elevated temperatures, rupture of the ring occurs forming ethylene and presumably allene and hydrogen (Rice and Murphy, 1942).

Exposure limits: NIOSH REL: 600 ppm (1,720 mg/m^3); ACGIH TLV: TWA 600 ppm (adopted).

Symptoms of exposure: Exposure to high concentrations may produce depression of central nervous system. Symptoms include excitement, loss of equilibrium, stupor, and coma (Patnaik, 1992).

Source: Component of high octane gasoline (quoted, Verschueren, 1983).

Schauer et al. (1999) reported cyclopentane in a diesel-powered medium-duty truck exhaust at an emission rate of 410 μg/km.

Uses: Solvent for cellulose ethers and paints; azeotropic distillation agent; motor fuel; extractions of fats and wax; shoe industry; organic synthesis.

CYCLOPENTENE

Synonym: UN 2246.

CAS Registry Number: 142-29-0
DOT: 2246
Molecular formula: C_5H_8
Formula weight: 68.12
RTECS: GY5950000

Physical state and color:
Colorless liquid

Melting point (°C):
-135 (Weast, 1986)
-135.08 (Dreisbach, 1959)

Boiling point (°C):
44.2 (Weast, 1986)

Density (g/cm³):
0.77199 at 20/4 °C, 0.76653 at 25/4 °C (Dreisbach, 1959)

Diffusivity in water (x 10^{-5} cm²/sec):
0.95 at 20 °C using method of Hayduk and Laudie (1974)

Dissociation constant, pK_a:
>14 (Schwarzenbach et al., 1993)

Flash point (°C):
-28.9 (Sax and Lewis, 1987)
-34 (Aldrich, 1988)

Heat of fusion (kcal/mol):
0.804 (Dean, 1987)

Henry's law constant (x 10^{-2} atm·m³/mol):
6.3 at 25 °C (Hine and Mookerjee, 1975)

Ionization potential (eV):
9.01 (Franklin et al., 1969)
9.27 (Collin and Lossing, 1959)

Soil sorption coefficient, log K_{oc}:
Unavailable because experimental methods for estimation of this parameter for alicyclic hydrocarbons are lacking in the documented literature

Octanol/water partition coefficient, log K_{ow}:
2.45 using method of Hansch et al. (1968)

Solubility in organics:
Soluble in alcohol, benzene, ether, and petroleum (Weast, 1986)

Solubility in water:
535 mg/kg at 25 °C (McAuliffe, 1966)
In 1 mM nitric acid: 9.21, 8.97, and 8.71 mM at 30, 35, and 40 °C, respectively (Natarajan and Venkatachalam, 1972)

Vapor density:
2.78 g/L at 25 °C, 2.35 (air = 1)

Vapor pressure (mmHg):
380 at 25 °C (estimated using Antoine equation, Dreisbach, 1959)

Environmental fate:
Biological. Cyclopentene may be oxidized by microbes to cyclopentanol, which may further oxidize to cycloheptanone (Dugan, 1972).

Photolytic. The following rate constants were reported for the reaction of cyclopentene with OH radicals in the atmosphere: 6.39×10^{-11} cm^3/molecule·sec (Atkinson et al., 1983), 4.99×10^{-11} cm^3/molecule·sec at 298 K (Rogers, 1989), 4.0×10^{-10} cm^3/molecule·sec (Atkinson, 1990) and 6.70×10^{-11} cm^3/molecule·sec (Sabljić and Güsten, 1990); with ozone in the atmosphere: 8.13×10^{-16} at 298 K (Japar et al., 1974) and 9.69×10^{-16} cm^3/molecule·sec at 294 K (Adeniji et al., 1981); with NO$_3$ in the atmosphere: 4.6×10^{-13} cm^3/molecule·sec at 298 K (Atkinson, 1990) and 5.81×10^{-13} cm^3/molecule·sec at 298 K (Sabljić and Güsten, 1990).

Chemical/Physical. Gaseous products formed from the reaction of cyclopentene with ozone were (% yield): formic acid (11), carbon monoxide (35), carbon dioxide (42), ethylene (12), formaldehyde (13), and butyraldehyde (11). Particulate products identified include succinic acid, glutaraldehyde, 5-oxopentanoic acid, and glutaric acid (Hatakeyama et al., 1987).

At elevated temperatures, rupture of the C-C bond occurs forming molecular hydrogen and cyclopentadiene (95% yield) as the principal products (Rice and Murphy, 1942).

Toxicity:
Acute oral LD$_{50}$ for rats is 1,656 mg/kg (quoted, RTECS, 1985).

Uses: Cross-linking agent; organic synthesis.

2,4-D

Synonyms: Agrotect; Amidox; Amoxone; Aqua-kleen; BH 2,4-D; Brush-rhap; B-Selektonon; Chipco turf herbicide D; Chloroxone; Crop rider; Crotilin; D 50; 2,4-D acid; Dacamine; Debroussaillant 600; Decamine; Ded-weed; Ded-weed LV-69; Desormone; Dichlorophenoxy-acetic acid; **(2,4-Dichlorophenoxy)acetic acid**; Dicopur; Dicotox; Dinoxol; DMA-4; Dormone; Emulsamine BK; Emulsamine E-3; ENT 8538; Envert 171; Envert DT; Esteron; Esteron 76 BE; Esteron 44 weed killer; Esteron 99; Esteron 99 concentrate; Esteron brush killer; Esterone 4; Estone; Farmco; Fernesta; Fernimine; Fernoxone; Ferxone; Foredex 75; Formula 40; Hedonal; Herbidal; Ipaner; Krotiline; Lawn-keep; Macrondray; Miracle; Monosan; Moxone; NA 2765; Netagrone; Netagrone 600; NSC 423; Pennamine; Pennamine D; Phenox; Pielik; Planotox; Plantgard; RCRA waste number U240; Rhodia; Salvo; Spritz-hormin/2,4-D; Spritz-hormit/2,4-D; Super D weedone; Transamine; Tributon; Trinoxol; U 46; U-5043; U 46DP; Vergemaster; Verton; Verton D; Verton 2D; Vertron 2D; Vidon 638; Visko-rhap; Visko-rhap drift herbicides; Visko-rhap low volatile 4L; Weedar; Weddar-64; Weddatul; Weed-b-gon; Weedez wonder bar; Weedone; Weedone LV4; Weed-rhap; Weed tox; Weedtrol.

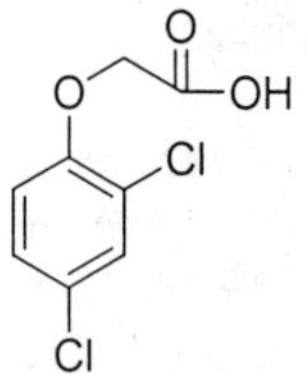

CAS Registry Number: 94-75-7
DOT: 2765
DOT label: Poison
Molecular formula: $C_8H_6Cl_2O_3$
Formula weight: 221.04
RTECS: AG6825000
Merck reference: 10, 2790

Physical state, color, and odor:
Odorless, white to pale yellow, powder or prismatic crystals

Melting point (°C):
140-141 (Weast, 1986)
140.5 (Worthing and Hance, 1991)
138.2-138.8 (Crosby and Tutass, 1966)
136 (Riederer, 1990)
138 (Windholz et al., 1983)

Boiling point (°C):
160 at 0.4 mmHg (Weast, 1986)

Density (g/cm³):
1.416 at 25/4 °C (Verschueren, 1983)
1.57 at 30/4 °C (Bailey and White, 1965)

Diffusivity in water (x 10⁻⁵ cm²/sec):
0.57 at 20 °C using method of Hayduk and Laudie (1974)

Dissociation constant, pK$_a$:
2.73 (Nelson and Faust, 1969)
2.64, 2.80 at room temperature, 3.22 at 60 °C (Bailey and White, 1965)
2.87 (Cessna and Grover, 1978)
2.90 (Jafvert et al., 1990)

Flash point:
Noncombustible solid (NIOSH, 1997)

Henry's law constant (x 10^{-5} atm·m^3/mol):
6.72 and 0.84 at pH values of 1 and 7, respectively (Rice et al., 1997a)

Bioconcentration factor, log BCF:
0.78 (*Chlorella fusca*, Freitag et al., 1982; Geyer et al., 1984)
1.30 (activated sludge, Freitag et al., 1985)
0.00 (fish, microcosm) (Garten and Trabalka, 1983)

Soil sorption coefficient, log K$_{oc}$:
1.68 (Commerce soil), 1.88 (Tracy soil), 1.76 (Catlin soil) (McCall et al., 1981)
1.30 (includes salts, Kenaga and Goring, 1980)
2.04-2.35 (quoted, Hodson and Williams, 1988)
1.70-2.73 (average = 2.18 for 10 Danish soils, Løkke, 1984)
2.05 (Spodosol, pH 3.9), 2.11 (Speyer soil, pH 5.8), 2.16 (Alfisol, pH 7.5) (Rippen et al., 1982)
1.77 (Lubbeek II sand loam), 2.33 (Lubbeek II sand), 1.83 (Lubbeek I silt loam), 1.82 (Lubbeek III silt loam), 2.22 (Stookrooie II loamy sand), 1.80 (Fleron silty clay loam), 1.76 (Bullingen silt loam), 2.79 (Spa silty clay loam), 2.74 (Bernard-Fagne silt loam), 2.48 (Stavelot silt loam), 2.29 (Meerdael silt loam), 2.52 (Soignes silt loam), 1.77 (Heverlee II sandy loam), 1.73 (Nodebais silt loam), 2.91 (Zolder sand - A$_1$ horizon), 2.71 (Zolder sand - A$_2$ horizon) (Moreale and Van Bladel, 1980)

Octanol/water partition coefficient, log K$_{ow}$:
2.81 (Leo et al., 1971)
1.57, 4.88 (Geyer et al., 1984)
2.65 (Hansch and Leo, 1985)
2.50 (Riederer, 1990)
2.59 (Freese et al., 1979)
2.14, 2.16 (Jafvert et al., 1990)
-1.56 and 1.26 in dissociated (0.1 M Na$_2$CO$_3$) and undissociated (0.5 M H$_2$SO$_4$) water phases, respectively (Wang et al., 1994)

Solubility in organics:
At 25 °C (g/L): carbon tetrachloride (1), ethyl ether (270), acetone (850), and ethyl alcohol (1,300) (Bailey and White, 1965)

Solubility in water:
890 ppm at 25 °C (Chiou et al., 1977)
400 mg/L at 20 °C (Riderer, 1990)
725 ppm at 25 °C (Bailey and White, 1965)
530 mg/L at 17 °C, 2.36 mM at 25 °C (Gunther et al., 1968)
2,940 μmol/L at 25 °C (LaFleur, 1979)
620 mg/L at 20 °C (Fühner and Geiger, 1977)

Vapor pressure (x 10^{-3} mmHg):
4.7 at 20 °C (Riderer, 1990)

Environmental fate:

Biological. 2,4-D degraded in anaerobic sewage sludge to 4-chlorophenol (Mikesell and Boyd, 1985). In moist nonsterile soils, degradation of 2,4-D occurs via cleavage of the carbon-oxygen bond at the 2-position on the aromatic ring (Foster and McKercher, 1973). In filtered sewage water, 2,4-D underwent complete mineralization but degradation was much slower in oligotrophic water, especially when 2,4-D was present in high concentrations (Rubin et al., 1982). In a primary digester sludge under methanogenic conditions, 2,4-D did not display any anaerobic biodegradation after 60 d (Battersby and Wilson, 1989).

Faulkner and Woodstock (1964) reported the soil microorganism *Aspergillus niger* degraded 2,4-D to 2,4-dichloro-5-hydroxyphenoxyacetic acid.

Soil. In moist soils, 2,4-D degraded to 2,4-dichlorophenol and 2,4-dichloroanisole as intermediates followed by complete mineralization to carbon dioxide (Wilson and Cheng, 1978; Smith, 1985; Stott, 1983). 2,4-Dichlorophenol was reported as a hydrolysis metabolite (Somasundaram et al., 1989, 1991; Somasundaram and Coats, 1991). In a soil pretreated with its hydrolysis metabolite, 80% of the applied [^{14}C]2,4-D mineralized to $^{14}CO_2$ within 4 d. In soils not treated with the hydrolysis product (2,4-dichlorophenol), only 6% of the applied [^{14}C]2,4-D degraded to $^{14}CO_2$ after 4 d (Somasundaram et al., 1989). Steenson and Walker (1957) reported that the soil microorganisms *Flavobacterium peregrinum* and *Achromobacter* both degraded 2,4-D yielding 2,4-dichlorophenol and 4-chlorocatechol as metabolites. The microorganisms *Gloeosporium olivarium*, *Gloeosporium kaki*, and *Schisophyllum communs* also degraded 2,4-D in soil forming 2-(2,4-dichlorophenoxy)ethanol as the major metabolite (Nakajima et al., 1973). Microbial degradation of 2,4-D was more rapid under aerobic conditions (half-life = 1.8-3.1 d) than under anaerobic conditions (half-life = 69-135 d) (Liu et al., 1981). In a 5-d experiment, [^{14}C]2,4-D applied to soil water suspensions under aerobic and anaerobic conditions gave $^{14}CO_2$ yields of 0.5 and 0.7%, respectively (Scheunert et al., 1987). Degradation was observed to be lowest at low redox potentials (Gambrell et al., 1984). The reported degradation half-lives for 2,4-D in soil ranged from 4 d in a laboratory experiment (McCall et al., 1981a) to 15 d (Jury et al., 1987). Degradation half-lives were determined in six soils: Catlin (1.5 d), Cecil (3.0 d), Commerce (5 d), Fargo (8.5 d), Keith (3.9 d), Walla Walla (2.5 d) (McCall et al., 1981a). Residual activity in soil is limited to approximately 6 wk (Hartley and Kidd, 1987).

After one application of 2,4-D to soil, the half-life was reported to be approximately 80 d but can be as low as 2 wk after repeated applications (Cullimore, 1971). The half-lives for 2,4-D in soil incubated in the laboratory under aerobic conditions ranged from 4 to 34 d with an average of 16 d (Altom and Stritzke, 1973; Foster and McKercher, 1973; Yoshida and Castro, 1975). In field soils, the disappearance half-lives were lower and ranged from approximately 1 to 15 d with an average of 5 d (Radosevich and Winterlin, 1977; Wilson and Cheng, 1976; Stewart and Gaul, 1977). Under aerobic conditions, the mineralization half-lives of 2,4-D in soil ranged from 11 to 25 d (Ou et al., 1978; Wilson and Cheng, 1978). The half-lives of 2,4-D in a sandy loam, clay loam, and an organic amended soil under non-sterile conditions were 722-2,936, 488-3,609, and 120-1,325 d, respectively (Schoen and Winterlin, 1987). Disappearance half-lives of 2,4-D were determined in two soils following a 28-d incubation period. In a Cecil loamy sand (Typic Hapludult), half-lives ranged from 3.9-9.4 and 6.8-115 d at 25 and 35 °C, respectively. In a Wenster sandy clay loam (Typic Haplaquoll), half-lives ranged from 7.0-254 and 6.7 to 176 d at 25 and 35 °C, respectively. The disappearance half-lives of 2,4-D generally increased with an increase of soil moisture content (Ou, 1984).

Groundwater. According to the U.S. EPA (1986), 2,4-D has a high potential to leach to groundwater.

Plant. Reported metabolic products in bean and soybean plants include 4-*O*-β-glucosides of 4-hydroxy-2,5-dichlorophenoxyacetic acid, 4-hydroxy-2,3-dichlorophenoxyacetic acid, *N*-(2,4-dichlorophenoxyacetyl)-L-aspartic acid and *N*-(2,4-dichlorophenocyacetyl)-L-glutamic acid. Metabolites identified in cereals and strawberries include 1-*O*-(2,4-dichlorophenoxyacetyl)-β-D-glucose and 2,4-dichlorophenol, respectively (Verschueren, 1983). In alfalfa, the side chain in the 2,4-D molecule was found to be lengthened by two and four methylene groups resulting in the formation of (2,4-dichlorophenoxy)butyric acid and (2,4-dichlorophenoxy)hexanoic acid, respectively. In several resistant grasses, however, the side chain increased by one methylene group forming (2,4-dichlorophenoxy)propionic acid (Hagin and Linscott, 1970).

2,4-D was metabolized by soybean cultures forming 2,4-dichlorophenoxyacetyl derivatives of alanine, leucine, phenylalanine, tryptophan, valine, aspartic and glutamic acids (Feung et al., 1971, 1972, 1973). On bean plants, 2,4-D degraded via β-oxidation and ring hydroxylation to form 2,4-dichloro-4-hydroxyphenoxyacetic acid, 2,3-dichloro-4-hydroxyphenoxyacetic acid (Hamilton et al., 1971), and 2-chloro-4-hydroxyphenoxyacetic acid. 2,5-Dichloro-4-hydroxy-phenoxyacetic acid was the predominant product identified in several weed species, as well as in smaller quantities of 2-chloro-4-hydroxyphenoxyacetic acid in wild buckwheat, yellow foxtail, and wild oats (Fleeker and Steen, 1971).

Esterification of 2,4-D with plant constituents via conjugation formed the β-D-glucose ester of 2,4-D (Thomas et al., 1964).

Photolytic. Photolysis of 2,4-D in distilled water using mercury arc lamps ($\lambda = 254$ nm) or by natural sunlight yielded 2,4-dichlorophenol, 4-chlorocatechol, 2-hydroxy-4-chlorophenoxy-acetic acid, 1,2,4-benzenetriol, and polymeric humic acids. The half-life for this reaction is 50 min (Crosby and Tutass, 1966). A half-life of 2-4 d was reported for 2,4-D in water irradiated at 356 nm (Baur and Bovey, 1974).

Bell (1956) reported that the composition of photodegradation products formed were dependent upon the initial 2,4-D concentration and pH of the solutions. 2,4-D undergoes reductive dechlorination when various polar solvents (methanol, butanol, isobutyl alcohol, *t*-butyl alcohol, octanol, ethylene glycol) are irradiated at wavelengths between 254-420 nm. Photoproducts formed included 2,4-dichlorophenol, 2,4-dichloroanisole, 4-chlorophenol, 2- and 4-chlorophenoxyacetic acid (Que Hee and Sutherland, 1981).

Irradiation of a 2,4-D sodium salt solution by a 660 W mercury discharge lamp produced 2,4-dichlorophenol within 20 min. Further irradiation resulted in further decomposition. The irradiation times required for 50% decomposition of the 2,4-D sodium salt at pH values of 4.0, 7.0, and 9.0 are 71, 50, and 23 min, respectively (Aly and Faust, 1964).

Surface Water. In filtered lake water at 29 °C, 90% of 2,4-D (1 mg/L) mineralized to carbon dioxide. The half-life was <5 d. At low concentrations (0.2 mg/L), no mineralization was observed (Wang et al., 1984). Subba-Rao et al. (1982) reported that 2,4-D in very low concentrations mineralized in one of three lakes tested. Mineralization did not occur when concentrations were at the picogram level. A degradation rate constant of 0.058/d at 29 °C was reported. At original concentrations of 100 mg/L and 100 μg/L in autoclaved water, the amount of 2,4-D remaining after 56 d of incubation were 81.6 and 79.6%, respectively (Wang et al., 1994).

In clear water and muddy water, hydrolysis half-lives of 18->50 d and 10-25 d, respectively, were reported (Nesbitt and Watson, 1980).

Chemical/Physical. In a helium pressurized reactor containing ammonium nitrate and polyphosphoric acid at temperatures of 121 and 232 °C, 2,4-D was oxidized to carbon dioxide, water and hydrochloric acid (Leavitt and Abraham, 1990). Carbon dioxide, chloride, aldehydes, oxalic and glycolic acids, were reported as ozonation products of 2,4-D in water at pH 8 (Struif et al., 1978). Reacts with alkalies, metals, and amines forming water soluble salts (Hartley and Kidd, 1987).

In water, 2,4-D reacted with hydroxy radicals at a first-order rate constant of 1.6 x 10^9/M·sec

(Mabury and Crosby, 1996).

When 2,4-D was heated at 900 °C, carbon monoxide, carbon dioxide, chlorine, hydrochloric acid, and oxygen were produced (Kennedy et al., 1972, 1972a). In liquid ammonia containing metallic sodium or lithium, 2,4-D degraded completely (Kennedy et al., 1972a). Total mineralization of 2,4-D was observed when a solution containing the herbicide and Fenton's reagent (ferrous ions and hydrogen peroxide) was subjected to UV light (λ = 300-400 nm). One intermediate compound identified was oxalic acid (Sun and Pignatello, 1993). Emits very toxic chloride fumes when heated to decomposition (Lewis, 1990).

The solubilities of the calcium and magnesium salts of 2,4-D acid at 25 °C are 9.05 and 25.1 mM, respectively (Aly and Faust, 1964).

2,4-D will not hydrolyze to any reasonable extent (Kollig, 1993).

Ward and Getzen (1970) investigated the adsorption of aromatic acids on activated carbon under acidic, neutral, and alkaline conditions. The amount of 2,4-D (10^{-4} M) adsorbed by carbon at pH values of 3.0, 7.0, and 11.0 were 60.1, 18.8, and 14.3%, respectively.

Exposure limits (mg/m^3): NIOSH REL: TWA 10, IDLH 100; OSHA PEL: TWA 10; ACGIH TLV: TWA 10 (adopted).

Toxicity:

EC$_{50}$ (24-h) for *Daphnia magna* 249 mg/L, *Daphnia pulex* 324 mg/L (Lilius et al., 1995).

EC$_{50}$ (5-min) for *Photobacterium phosphoreum* 100.7 mg/L (Somasundaram et al., 1990).

LC$_{50}$ (96-h) for carp 5.1-20 mg/L (Spehar et al., 1982), cutthroat trout 24.5-172 mg/L, rainbow trout 110 mg/L, lake trout 62-120 mg/L, fathead minnow 133 mg/L, bluegill sunfish 180 mg/L (Mayer and Ellersieck, 1986), Japanese medaka (*Oryzias latipes*) 2,780 mg/L (Holcombe et al., 1995).

LC$_{50}$ (48-h) for bluegill sunfish 900 ppb, for rainbow trout 1.1 ppm (Edwards, 1977).

LC$_{50}$ (24-h) for cutthroat trout 32-185 mg/L, lake trout 44.5-127.5 mg/L (Mayer and Ellersieck, 1986).

Acute oral LD$_{50}$ for chickens 541 mg/kg, dogs 100 mg/kg, guinea pigs 469 mg/kg, hamsters 500 mg/kg, mice 368 mg/kg, rats 370 mg/kg (quoted, RTECS, 1985), chicks 420 mg/kg (Morgulis et al., 1998).

Drinking water standard (final): MCLG: 70 μg/L; MCL: 70 μg/L (U.S. EPA, 1996).

Uses: Systemic herbicide; weed killer and defoliant.

Synonyms: 1,1-Bis(4-chlorophenyl)-2,2-dichloroethane; 1,1-Bis(*p*-chlorophenyl)-2,2-dichloroethane; 2,2-Bis(4-chlorophenyl)-1,1-dichloroethane; 2,2-Bis-(*p*-chlorophenyl)-1,1-dichloroethane; DDD; 4,4'-DDD; 1,1-Dichloro-2,2-bis(*p*-chlorophenyl)ethane; 1,1-Dichloro-2,2-di(4-chlorophenyl)ethane; 1,1-Dichloro-2,2-di(*p*-chlorophenyl)ethane; Dichlorodiphenyldichloroethane; 4,4'-Dichlorodiphenyldichloroethane; *p,p'*-Dichlorodiphenyldichloroethane; **1,1'-(2,2-Dichloroethylidene)bis[4-chlorobenzene]**; Dilene; ENT 4225; ME-1700; NA 2761; NCI-C00475; OMS 1078; RCRA waste number U060; Rhothane; Rhothane D-3; Rothane; TDE; 4,4'-TDE; *p,p'*-TDE; Tetrachlorodiphenylethane.

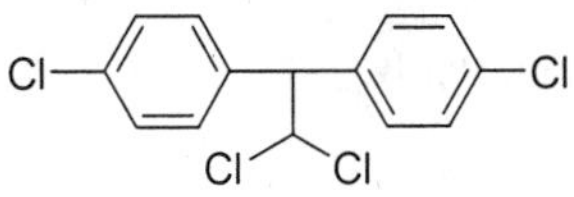

CAS Registry Number: 72-54-8
DOT: 2761
DOT label: Poison
Molecular formula: $C_{14}H_{10}Cl_4$
Formula weight: 320.05
RTECS: KI0700000
Merck reference: 10, 3044

Physical state and color:
Crystalline white solid

Melting point (°C):
112 (Verschueren, 1983)
107-109 (Aldrich, 1988)

Boiling point (°C):
193 (Sax, 1985)

Density (g/cm³):
1.476 at 20/4 °C (Weiss, 1986)

Diffusivity in water (x 10^{-5} cm²/sec):
0.45 at 20 °C using method of Hayduk and Laudie (1974)

Flash point (°C):
Not pertinent (Weiss, 1986)

Lower explosive limit (%):
Not pertinent (Weiss, 1986)

Upper explosive limit (%):
Not pertinent (Weiss, 1986)

Henry's law constant (x 10^{-5} atm·m³/mol):
2.16 (calculated, U.S. EPA, 1980a)

Bioconcentration factor, log BCF:
4.92 (fish, microcosm) (Garten and Trabalka, 1983)
2.79 (alga), 3.65 (snail), 3.43 (carp) (quoted, Verschueren, 1983)

Soil sorption coefficient, log K_{oc}:
5.12 (Taichung soil: pH 6.8, % sand: 25, % silt: 40, % clay: 35) (Ding and Wu, 1995)
5.38 (Jury et al., 1987)
6.6 (average using 9 suspended sediment samples from the St. Clair and Detroit Rivers, Lau et al., 1989)

Octanol/water partition coefficient, log K_{ow}:
5.99 (Callahan et al., 1979)
5.061 (Rao and Davidson, 1980)
5.80 (DeKock and Lord, 1987)
6.217 (de Bruijn et al., 1989)

Solubility in water (ppb):
160 at 24 °C (quoted, Verschueren, 1983)
20 at 25 °C (Weil et al., 1974)
50 at 15 °C, 90 at 25 °C, 150 at 35 °C, 240 at 45 °C (particle size $\leq$5 μ) (Biggar and Riggs, 1974)

Vapor density:
17.2 ng/L at 30 °C (Spencer and Cliath, 1972)

Vapor pressure (x 10^{-6} mmHg):
1.02 at 30 °C (Spencer and Cliath, 1972)
4.68 at 25 °C (Bidleman, 1984)
8.25, 12.2, at 25 °C (Hinckley et al., 1990)

Environmental fate:
Biological. It was reported that *p,p'*-DDD, a major biodegradation product of *p,p'*-DDT, was degraded by *Aerobacter aerogenes* under aerobic conditions yielding 1-chloro-2,2-bis(*p*-chlorophenyl)ethylene, 1-chloro-2,2-bis(*p*-chlorophenyl)ethane, and 1,1-bis(*p*-chlorophenyl)ethylene. Under anaerobic conditions, however, four additional compounds were identified: bis(*p*-chlorophenyl)acetic acid, *p,p'*-dichlorodiphenylmethane, *p,p'*-dichlorobenzhydrol and *p,p'*-dichlorobenzophenone (Fries, 1972). Under reducing conditions, indigenous microbes in Lake Michigan sediments degraded DDD to 2,2-bis(*p*-chlorophenyl)ethane and 2,2-bis(*p*-chlorophenyl)ethanol (Leland et al., 1973). Incubation of *p,p'*-DDD with hematin and ammonia gave 4,4'-dichlorobenzophenone, 1-chloro-2,2-bis(*p*-chlorophenyl)ethylene, and bis(*p*-chlorophenyl)acetic acid methyl ester (Quirke et al., 1979). Using settled domestic wastewater inoculum, *p,p'*-DDD (5 and 10 mg/L) did not degrade after 28 d of incubation at 25 °C (Tabak et al., 1981).
Soil: In Hudson River, NY sediments, the presence of adsorbed *p,p'*-DDD in core samples suggests it is very persistent in this environment. The estimated half-life ranged from 4.2 to 4.5 yr (Bopp et al., 1982).
Chemical/Physical. The hydrolysis rate constant for *p,p'*-DDD at pH 7 and 25 °C was determined to be 2.8 x 10^{-6}/h, resulting in a half-life of 28.2 yr (Ellington et al., 1987). At 85 °C, the hydrolysis half-lives were 5.3 d, 43 min, and 130 min at pH values 7.22, 9.67, and 10.26, respectively (Ellington et al., 1987). 2,2-Bis(4-chlorophenyl)-1-chloroethene and hydrochloric acid were reported as hydrolysis products (Kollig, 1993).

Toxicity:

EC$_{50}$ (48-h) for *Daphnia pulex* 3.2 μg/L, *Daphnia magna* 9.0 μg/L, *Simocephalus serrulatus* 4.9 μg/L, *Cypridopsis vidua* 45 μg/L (Mayer and Ellersieck, 1986).

LC$_{50}$ (96-h) for *Gammarus lacustris* 0.64-0.86 μg/L, *Palaemonetes kadiakensis* 0.68 μg/L, *Asellus breviacaudus* 10.0 μg/L (quoted, Verschueren, 1983).

LC$_{50}$ (48-h) for *Simocephalus serrulatus* 4.5 μg/L, *Daphnia pulex* 3.2 μg/L (Sanders and Cope, 1966).

Acute oral LD$_{50}$ for rats 113 mg/kg (quoted, RTECS, 1985), 60.8 μg/roach and 86.4 μg/roach for male and female, respectively (Gardner and Vincent, 1978).

Uses: Dusts, emulsions, and wettable powders for contact control of leaf rollers and other insects on vegetables and tobacco.

p,p'-DDE

Synonyms: 2,2-Bis(4-chlorophenyl)-1,1-dichloroethene; 2,2-Bis(*p*-chlorophenyl)-1,1-dichloroethene; 1,1-Bis(4-chlorophenyl)-2,2-dichloroethylene; 1,1-Bis(*p*-chlorophenyl)-2,2-dichloroethylene; DDE; 4,4'-DDE; DDT dehydrochloride; 1,1-Dichloro-2,2-bis(*p*-chlorophenyl)-ethylene; Dichlorodiphenyldichloroethylene; *p,p'*-Dichlorodiphenyldichloroethylene; **1,1'-(Dichloroethenylidene)bis(4-chlorobenzene)**; NCI-C00555.

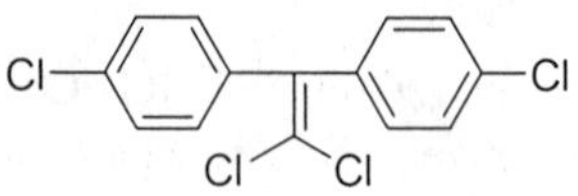

CAS Registry Number: 72-55-9
DOT: 2761
DOT label: Poison
Molecular formula: $C_{14}H_8Cl_4$
Formula weight: 319.03
RTECS: KV9450000

Physical state and color:
White crystalline powder

Melting point (°C):
88-90 (Leffingwell, 1975)
112 (Melnikov, 1971)

Diffusivity in water (x 10^{-5} cm²/sec):
0.46 at 20 °C using method of Hayduk and Laudie (1974)

Henry's law constant (x 10^{-3} atm·m³/mol at 23 °C):
1.22 and 3.65 in distilled water and seawater, respectively (Atlas et al., 1982)

Bioconcentration factor, log BCF:
4.00-4.15 (fish tank), 7.26 (Lake Ontario) (rainbow trout, Oliver and Niimi, 1985)
4.08 (fish, Metcalf et al., 1975)
4.71 (freshwater fish), 4.44 (fish, microcosm) (Garten and Trabalka, 1983)
4.91 (*Oncorhynchus mykiss*, Devillers et al., 1996)
4.72 (*Crassostrea virginica*, Schimmel and Garnas, 1981)
4.05 (*Oedogonium cardiacum*), 4.56 (snail), 4.77 (mosquito), 4.08 (*Gambusia affinis*) (Metcalf et al., 1975)

Soil sorption coefficient, log K_{oc}:
5.386 (Reinbold et al., 1979)
4.42 (Taichung soil: pH 6.8, % sand: 25, % silt: 40, % clay: 35) (Ding and Wu, 1995)
6.6 (average using 10 suspended sediment samples collected from the St. Clair and Detroit Rivers, Lau et al., 1989)
K_d = 2.5 mL/g (Barcelona coastal sediments, Bayona et al., 1991).

Octanol/water partition coefficient, log K_{ow}:
5.83 (Travis and Arms, 1988)
5.69 (Freed et al., 1977, 1979a)

5.766 (Kenaga and Goring, 1980)
5.89 (Burkhard et al., 1985a)
6.20 (DeKock and Lord, 1987)
6.956 (de Bruijn et al., 1989)

Solubility in organics:
Soluble in fats and most solvents (IARC, 1974)

Solubility in water (ppb):
40 at 20 °C, 1.3 μg/L at 25 °C (Metcalf et al., 1973a)
14 at 20 °C (Weil et al., 1974)
40 at 20 °C (Chiou et al., 1977)
55 at 15 °C, 120 at 25 °C, 235 at 35 °C, 450 at 45 °C (particle sizes ≤5 μ) (Biggar and Riggs, 1974)
65 at 24 °C (Hollifield, 1979)

Vapor density:
109 ng/L at 30 °C (Spencer and Cliath, 1972)

Vapor pressure (x 10^{-6} mmHg):
6.49 at 30 °C (Spencer and Cliath, 1972)
13 at 30 °C (Wescott et al., 1981)
15.7 at 25 °C (Bidleman, 1984)
7.4 at 25 °C (Wescott and Bidleman, 1981)
14.0, 20.3 at 25 °C (Hinckley et al., 1990)

Environmental fate:
  Biological. In four successive 7-d incubation periods, *p,p'*-DDE (5 and 10 mg/L) was recalcitrant to degradation in a settled domestic wastewater inoculum (Tabak et al., 1981).
  Photolytic. When an aqueous solution of *p,p'*-DDE (0.004 μM) in natural water samples from California and Hawaii were irradiated (maximum λ = 240 nm) for 120 h, 62% was photo-oxidized to *p,p'*-dichlorobenzophenone (Ross and Crosby, 1985). In an air-saturated distilled water medium irradiated with monochromic light (λ = 313 nm), *p,p'*-DDE degraded to *p,p'*-dichlorobenzophenone, 1,1-bis(*p*-chlorophenyl)-2-chloroethylene (DDMU), and 1-(4-chloro-phenyl)-1-(2,4-dichlorophenyl)-2-chloroethylene (*o*-chloro DDMU). Identical photoproducts were also observed using tap water containing Mississippi River sediments (Miller and Zepp, 1979). The photolysis half-life under sunlight irradiation was reported to be 1.5 d (Mansour et al., 1989).
  When *p,p'*-DDE in water was irradiated at 313 nm, a quantum yield of 0.3 was achieved. A photolysis half-life of 0.9 d in summer and 6.1 d in winter by direct sunlight at 40° latitude was observed. Photolysis products included DDMU (yield 20%), *o*-chloro DDMU (yield 15%), and a dichlorobenzophenone (Zepp et al., 1976, 1977). Quantum yields of 0.26 and 0.24 were reported for the photolysis of *p,p'*-DDE in hexane at wavelengths of 254 and 313 nm, respectively (Mosier et al., 1969; Zepp et al., 1977). Singmaster (1975) reported a photolytic half-life of 1.1 d when *p,p'*-DDE (0.84 μg/L) in San Francisco Bay, CA water was subjected to sunlight.
  When *p,p'*-DDE in a methanol solvent was photolyzed at 260 nm, a dichlorobenzophenone, a dichlorobiphenyl, DDMU, and 3,6-dichlorofluorenone (yield 10%) formed as the major products (Plimmer et al., 1970).
  Chemical/Physical. May degrade to bis(chlorophenyl)acetic acid in water (Verschueren, 1983), or oxidize to *p,p'*-dichlorobenzophenone using UV light as a catalyst (HSDB, 1989).

Chlordane is resistant to hydrolysis. At pH 5 and 27 °C, the estimated half-life in >120 yr (Wolfe et al., 1977). This is in agreement with results reported by Eichelberger and Lichtenberg (1971). They found no change in *p,p'*-DDE concentration in water over an 8-wk period.

Toxicity:

LC_{50} (96-h) for rainbow trout 32 μg/L, Atlantic salmon 96 μg/L, bluegill sunfish 240 μg/L (Mayer and Ellersieck, 1986).

Acute oral LD_{50} for rats 880 mg/kg (quoted, RTECS, 1985).

Source: Agricultural runoff degradation of *p,p'*-DDT (quoted, Verschueren, 1983)

Uses: Military product; chemical research.

Synonyms: Agritan; Anofex; Arkotine; Azotox; 2,2-Bis(4-chlorophenyl)-1,1,1-trichloro-ethane; 2,2-Bis(*p*-chlorophenyl)-1,1,1-trichloroethane; α,α-Bis(*p*-chlorophenyl)-β,β,β-tri-chloroethane; 1,1-Bis(*p*-chlorophenyl)-2,2,2-trichloroethane; Bosan Supra; Bovidermol; Chlor-ophenothan; Chlorophenothane; Chlorophenotoxum; Citox; Clofenotane; DDT; 4,4′-DDT; Dedelo; Deoval; Detox; Detoxan; Dibovan; Dichlorodiphenyltrichloroethane; *p,p*′-Dichlorodi-phenyltrichloroethane; 4,4′-Dichlorodiphenyltrichloroethane; Dicophane; Didigam; Didimac; Diphenyltrichloroethane; Dodat; Dykol; ENT 1506; Estonate; Genitox; Gesafid; Gesapon; Gesarex; Gesarol; Guesapon; Gyron; Havero-extra; Ivoran; Ixodex; Kopsol; Mutoxin; NCI-C00464; Neocid; Parachlorocidum; PEB1; Pentachlorin; Pentech; PPzeidan; RCRA waste number U061; Rukseam; Santobane; Trichlorobis(4-chlorophenyl)ethane; Trichlorobis(*p*-chlorophenyl)ethane; 1,1,1-Trichloro-2,2-bis(*p*-chlorophenyl)ethane; 1,1,1-Trichloro-2,2-di(4-chlorophenyl)ethane; 1,1,1-Trichloro-2,2-di(*p*-chlorophenyl)ethane; **1,1′-(2,2,2-Trichloroeth-ylidene)-bis[4-chlorobenzene]**; Zeidane; Zerdane.

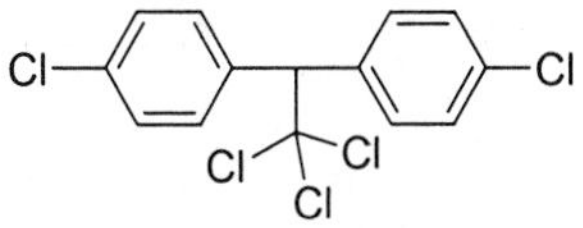

CAS Registry Number: 50-29-3
DOT: 2761
DOT label: Poison
Molecular formula: $C_{14}H_9Cl_5$
Formula weight: 354.49
RTECS: KJ3325000
Merck reference: 10, 2823

Physical state, color, and odor:
Colorless crystals, white powder, or waxy solid, odorless to slightly fragrant powder. Odor threshold concentration is 200 ppb (Keith and Walters, 1992) and in water, 350 μg/kg (Sigworth, 1964).

Melting point (°C):
108.5 (Bowman et al., 1960)
108-109 (Weast, 1986)

Boiling point (°C):
260 (Weast, 1986)
185 (U.S. EPA, 1980a)

Density (g/cm³):
1.56 at 15/4 °C (Weiss, 1986)

Diffusivity in water (x 10⁻⁵ cm²/sec):
0.37 at 20 °C using method of Hayduk and Laudie (1974)

Flash point (°C):
72.3-77.3 (NIOSH, 1997)

Henry's law constant (x 10^{-5} atm·m³/mol):
3.8 (Eisenreich et al., 1983)
48.9 (Jury et al., 1984)
10.3 (Jury et al., 1984a)
5.2 (Jury et al., 1983)
1.29 at 23 °C (Fendinger et al., 1989)

Bioconcentration factor, log BCF:
3.97 (algae, Geyer et al., 1984)
4.15 (activated sludge), 3.28 (golden ide) (Freitag et al., 1985)
4.53 (freshwater fish), 4.49 (fish, microcosm) (Garten and Trabalka, 1983)
4.81, 4.86, 4.95, 4.97, 4.99 (*Oncorhynchus mykiss*, Devillers et al., 1996)
3.67 (alga), 3.41 (*Daphnia pulex*), 3.56 (snail), 3.38 (carp) (quoted, Verschueren, 1983)

Soil sorption coefficient, log K_{oc}:
5.146 (silt soil loam, Chiou et al., 1979)
5.14 (Schwarzenbach and Westall, 1981)
5.20 (Commerce soil), 5.17 (Tracy soil), 5.18 (Catlin soil) (McCall et al., 1981)
6.26 (marine sediments, Pierce et al., 1974)
5.39 (Rao and Davidson, 1980)
5.68 (Taichung soil: pH 6.8, % sand: 25, % silt: 40, % clay: 35) (Ding and Wu, 1995)
6.7 (average using 7 suspended sediment samples from the St. Clair and Detroit Rivers, Lau et al., 1989)
5.77 (Mivtahim soil), 5.48 (Gilat soil), 5.22 (Neve Yaar soil), 4.98 (Malkiya soil), 5.40 (Kinneret sediment), 5.30 (Kinneret-G sediment) (Gerstl and Mingelgrin, 1984)

Octanol/water partition coefficient, log K_{ow}:
6.36 (Chiou et al., 1982)
6.19 (DeKock and Lord, 1987; Freed et al., 1977, 1979; Johnsen et al., 1989)
5.76 (Travis and Arms, 1988)
5.98 (Mackay and Paterson, 1981)
5.38 (Kenaga, 1980)
6.16, 6.17, 6.22, 6.44 (Brooke et al., 1986)
6.28 (Geyer et al., 1984)
4.89 (Wolfe et al., 1977)
5.44 (Gerstl and Mingelgrin, 1984; Burkhard et al., 1985a)
6.914 (Brooke et al., 1990; de Bruijn et al., 1989)
6.307 (Brooke et al., 1990)
6.38 (Hammers et al., 1982)
6.00 at 25 °C (Paschke et al., 1998)
5.94 (Ellgehausen et al., 1981)

Solubility in organics (g/L):
Acetone (580), benzene (780), benzyl benzoate (420), carbon tetrachloride (450), chlorobenzene (740), cyclohexanone (1,160), ethyl ether (280), gasoline (100), isopropanol (30), kerosene (80-100), morpholine (750), peanut oil (110), pine oil (100-160), tetralin (610), tributyl phosphate (500) (Windholz et al., 1983).

Solubility in water:
5.5 ppb at 25 °C (Weil et al., 1974)
1.2 μg/L at 25 °C (Bowman et al., 1960)

7 μg/L at 20 °C (Nisbet and Sarofim, 1972)
5.4 μg/L at 24 °C (Chiou et al., 1986)
4 μg/L at 24-25 °C (Hollifield, 1979; Chiou et al., 1979)
5.9 ppb at 2 °C, 37.4 ppb at 25 °C, 45 ppb at 37.5 °C (Babers, 1955)
10-100 ppb at 22 °C (Roeder and Weiant, 1946)
2 ppb (Kapoor et al., 1973)
In ppb: 17 at 15 °C, 25 at 25 °C, 37 at 35 °C, 45 at 45 °C (particle size ≤5 μ) (Biggar and
    Riggs, 1974)
260 ppb at 25 °C (NAS, 1977)
At 20-25 °C: 40 ppb (particle size ≤5 μ), 16 ppb (particle size ≤0.05 μ) (Robeck et al., 1965)
7.7 μg/L at 20 °C (Friesen et al., 1985)
3.0 μg/L at 20 °C (Ellgehausen et al., 1980)
40 μg/L at 20 °C (Ellgehausen et al., 1981)
3 nmol/L at 25 °C (LaFleur, 1979)
3.54 μg/L in Lake Michigan water at ≈ 25 °C (Eadie et al., 1990)
4.5 ppb at 25 °C (Gerstl and Mingelgrin, 1984)
5.1 μg/L at 25 °C (Paschke et al., 1998)

Vapor density:
13.6 ng/L at 30 °C (Spencer and Cliath, 1972)

Vapor pressure (x 10^{-7} mmHg):
1 at 25 °C (quoted, Mackay and Wolkoff, 1973)
1.5 at 20 °C (Balson, 1947)
7.26 at 30 °C (Spencer and Cliath, 1972)
1.29 at 20 °C, 4.71 at 50 °C, 6.76 at 100 °C (Webster et al., 1985)
1.40 at 30 °C (Wescott and Bidleman, 1981)
2.2, 4.3, 9.3, 40, 150, 480, 1,500, and 4,500 at 20, 25, 30, 40, 50, 60, 70, and 80 °C, respec-
    tively (Rothman, 1980).
150, 533, 5,850, and 14,900 at 50.1, 60.1, 80.4, and 90.2 °C, respectively (Dickenson, 1956)
62.3 at 25 °C (Hinckley et al., 1990)
0.5 at 25 °C (extrapolated from vapor pressures determined at higher temperatures, Tesconi and
    Yalkowsky, 1998)

Environmental fate:
Biological. In four successive 7-d incubation periods, *p,p'*-DDT (5 and 10 mg/L) was recalcitrant to degradation in a settled domestic wastewater inoculum (Tabak et al., 1981).

Castro (1964) reported that iron(II) porphyrins in dilute aqueous solution was rapidly oxidized by DDT to form the corresponding iron(III) chloride complex (hematin) and DDE, respectively. Incubation of *p,p'*-DDT with hematin and ammonia gave *p,p'*-DDD, *p,p'*-DDE, bis(*p*-chlorophenyl)acetonitrile, 1-chloro-2,2-bis(*p*-chlorophenyl)ethylene, 4,4'-dichlorobenzo-phenone, and the methyl ester of bis(*p*-chlorophenyl)acetic acid (Quirke et al., 1979).

In 1 d, *p,p'*-DDT reacted rapidly with reduced hematin forming *p,p'*-DDD and unidentified products (Baxter, 1990). The white rot fungus *Phanerochaete chrysosporium* degraded *p,p'*-DDT yielding the following metabolites: 1,1-dichloro-2,2-bis(4-chlorophenyl)ethane (*p,p'*-DDD), 2,2,2-trichloro-1,1-bis(4-chlorophenyl)ethanol (dicofol), 2,2-di-chloro-1,1-bis(4-chloro-phenyl)ethanol, and 4,4'-dichlorobenzophenone and carbon dioxide (Bumpus et al., 1985; Bumpus and Aust, 1987). Mineralization began between the third and sixth day of incubation. The production of carbon dioxide was highest between 3-18 d of incubation, after which the rate of carbon dioxide produced decreased until the 30th day. It was suggested that the metabolism of *p,p'*-DDT was dependent on the extracellular lignin-degrading enzyme system of this fungus (Bumpus et al., 1985). White rot fungi, namely *Phanerochaete chrysosporium*, *Pleurotus sajorcaju*, *Pleurotus florida*, and *Pleurotus eryngii*, biodegraded *p,p'*-DDT (50 μM) at degradation yields of 68.99-77.75, 78.23-91.70, 74.74-77.97, and 46.92-65.98%, respectively. The experiments were carried out in a culture incubated at 30 °C for 20 d (Arisoy, 1998).

Mineralization of *p,p'*-DDT by the white rot fungi *Pleurotus ostreatus*, *Phellinus weirri*, and *Polyporus versicolor* was also demonstrated (Bumpus and Aust, 1987). Fries (1972) reported that *Aerobacter aerogenes* degraded *p,p'*-DDT under aerobic conditions forming *p,p'*-DDD, *p,p'*-DDE, 1-chloro-2,2-bis(*p*-chlorophenyl)ethylene, 1-chloro-2,2-bis(*p*-chlorophenyl)ethane, and 1,1-bis(*p*-chlorophenyl)ethylene. Under anaerobic conditions the same organism produced four additional compounds. These were bis(*p*-chlorophenyl)acetic acid, *p,p'*-dichlorodiphenylmethane, *p,p'*-dichlorobenzhydrol, and *p,p'*-dichlorobenzophenone. Other degradation products of *p,p'*-DDT under aerobic and anaerobic conditions in soils using various cultures not previously mentioned include 1,1-bis(*p*-chlorophenyl)-2,2,2-trichloroethanol (Kelthane) and 4-chlorobenzoic acid (Fries, 1972).

Under aerobic conditions, the amoeba *Acanthamoeba castellanii* (Neff strain ATCC 30.010) degraded *p,p'*-DDT to *p,p'*-DDE, *p,p'*-DDD, and dibenzophenone (Pollero and dePollero, 1978).

Incubation of *p,p'*-DDT with hematin and ammonia gave *p,p'*-DDD, *p,p'*-DDE, bis(*p*-chlorophenyl)acetonitrile, 1-chloro-2,2-bis(*p*-chlorophenyl)ethylene, 4,4'-dichlorobenzophenone, and the methyl ester of bis(*p*-chlorophenyl)acetic acid (Quirke et al., 1979).

Chacko et al. (1966) reported DDT dechlorinated to DDD by six actinomycetes (*Norcardia* sp., *Streptomyces albus*, *Streptomyces antibioticus*, *Streptomyces auerofaciens*, *Streptomyces cinnamoneus*, *Streptomyces viridochromogenes*) but not by 8 fungi. The maximum degradation observed was 25% in 6 d.

Thirty-five microorganisms isolated from marine sediment and marine water samples taken from Hawaii and Houston, TX were capable of degrading *p,p'*-DDT. *p,p'*-DDD was identified as the major metabolite. Minor transformation products included 2,2-bis(*p*-chlorophenyl)-ethanol, 2,2-bis(*p*-chlorophenyl)ethane, and *p,p'*-DDE (Patil et al., 1972).

In a 42-d experiment, [^{14}C]*p,p'*-DDT applied to soil water suspensions under aerobic and anaerobic conditions gave $^{14}CO_2$ yields of 0.8 and 0.7%, respectively (Scheunert et al., 1987). Similarly, Matsumura et al. (1971) found that *p,p'*-DDT was degraded by numerous aquatic microorganisms isolated from water and silt samples collected from Lake Michigan and its tributaries in Wisconsin. The major metabolites identified were TDE, DDNS, and DDE. *p,p'*-DDT was metabolized by the following microorganisms under laboratory conditions to *p,p'*-DDD: *Escherichia coli* (Langlois, 1967), *Aerobacter aerogenes* (Plimmer et al., 1968; Wedemeyer, 1966), and *Proteus vulgaris* (Wedemeyer, 1966). In addition, *p,p'*-DDT was degraded to *p,p'*-DDD, *p,p'*-DDE, and dicofol by *Trichoderma viride* (Matsumura and Boush, 1968) and to *p,p'*-DDD and *p,p'*-DDE by *Ankistrodemus amalloides* (Neudorf and Khan, 1975).

Jensen et al. (1972) studied the anaerobic degradation of *p,p'*-DDT (100 mg) in 1 L of sewage sludge containing *p,p'*-DDD (4.0%) and *p,p'*-DDE (3.1%) as contaminants. The sludge was incubated at 20 °C for 8 d under a nitrogen atmosphere. The parent compound degraded rapidly (half-life = 7 h) forming *p,p'*-DDD, *p,p'*-dichlorodiphenylbenzophenone (DBP), 1,1-bis(*p*-chlorophenyl)-2-chloroethylene (DDMU), and bis(*p*-chlorophenyl)acetonitrile. After 48 h, the original amount of *p,p'*-DDD added to the sewage sludge had completely reacted. In a similar study, Pfaender and Alexander (1973) observed the cometabolic conversion of DDT (0.005%) in unamended sewage sludge to give DDD, DDE, and DBP. When the sewage sludge was amended with glucose (0.10%), the rate of DDD formation was enhanced. However, with the addition of diphenylmethane, the rate of formation of both DDD and DBP was reduced. The diphenylmethane-amended sewage sludge showed the greatest abundance of bacteria capable of cometabolizing DDT, whereas the unamended sewage showed the fewest number of bacteria. Zoro et al. (1974) also reported that *p,p′*-DDT in untreated sewage sludge was converted to *p,p′*-DDD, especially in the presence of sodium dithionate, a widely used reducing agent.

In an *in vitro* fermentation study, rumen microorganisms metabolized both isomers of [^{14}C]DDT (*o,p′*- and *p,p′*-) to the corresponding DDD isomers at a rate of 12%/h. With *p,p′*-DDT, 11% of the ^{14}C detected was an unidentified polar product associated with microbial and substrate residues (Fries et al., 1969). In another *in vitro* study, extracts of *Hydrogenomonas* sp. cultures degraded DDT to DDD, 1-chloro-2,2-bis(*p*-chlorophenyl)ethane (DDMS), DBP, and several other products under anaerobic conditions. Under aerobic conditions containing whole cells, one of the rings is cleaved and *p*-chlorophenylacetic acid is formed (Pfaender and Alexander, 1972).

Soil. *p,p′*-DDD and *p,p′*-DDE are the major metabolites of *p,p′*-DDT in the environment (Metcalf, 1973). In soils under anaerobic conditions, *p,p′*-DDT is rapidly converted to *p,p′*-DDD via reductive dechlorination (Johnsen, 1976) and very slowly to *p,p′*-DDE under aerobic conditions (Guenzi and Beard, 1967; Kearney and Kaufman, 1976). The aerobic degradation of *p,p′*-DDT under flooded conditions is very slow with *p,p′*-DDE forming as the major metabolite. Dicofol was also detected in minor amounts (Lichtenstein et al., 1971). In addition to *p,p′*-DDD and *p,p′*-DDE, 2,2-bis(*p*-chlorophenyl)acetic acid (DDA), bis(*p*-chlorophenyl)methane (DDM), *p,p′*-dichlorobenzhydrol (DBH), DBP, and *p*-chlorophenyl-acetic acid (PCPA) were also reported as metabolites of *p,p′*-DDT in soil under aerobic conditions (Subba-Rao and Alexander, 1980).

The anaerobic conversion of *p,p′*-DDT to *p,p′*-DDD in soil was catalyzed by the presence of ground alfalfa or glucose (Burge, 1971). Under flooded conditions, *p,p*-DDT was rapidly converted to TDE via reductive dehalogenation and other metabolites (Guenzi and Beard, 1967; Castro and Yoshida, 1971). Degradation was faster in flooded soil than an upland soil and was faster in soils containing high organic matter (Castro and Yoshida, 1971). Other reported degradation products under aerobic and anaerobic conditions by various soil microbes include 1,1′-bis(*p*-chlorophenyl)-2-chloroethane, 1,1′-bis(*p*-chlorophenyl)-2-hydroxyethane, and *p*-chlorophenyl acetic acid (Kobayashi and Rittman, 1982). It was also reported that *p,p′*-DDE formed by hydrolyzing *p,p′*-DDT (Wolfe et al., 1977). The clay-catalyzed reaction of DDT to form DDE was reported by Lopez-Gonzales and Valenzuela-Calahorro (1970). They observed that DDT adsorbed of sodium bentonite clay surfaces was transformed more rapidly than on the corresponding hydrogen-bentonite clay. In 1 d, *p,p′*-DDT reacted rapidly with reduced hematin forming *p,p′*-DDD and unidentified products (Baxter, 1990). In an Everglades muck, *p,p′*-DDT was slowly converted to *p,p′*-DDD and *p,p′*-DDE (Parr and Smith, 1974). The reported half-life in soil is 3,800 d (Jury et al., 1987).

Oat plants were grown in two soils treated with [^{14}C]*p,p′*-DDT. Most of the residues remained bound to the soil. Metabolites identified were *p,p′*-DDE, *o,p′*-DDT, TDE, DBP, dicofol, and DDA (Fuhremann and Lichtenstein, 1980).

The half-lives for *p,p′*-DDT in field soils ranged from 106 d to 15.5 yr with an average half-life of 4.5 yr (Lichtenstein and Schulz, 1959; Lichtenstein et al., 1960; Nash and Woolson, 1967; Lichtenstein et al., 1971; Stewart and Chisolm, 1971; Suzuki et al., 1977). The average half-life of *p,p′*-DDT in a variety of anaerobic soils was 692 d (Burge, 1971; Glass, 1972; Guenzi and Beard, 1976).

p,p′-DDT is very persistent in soil. The percentage of the initial dosage (1 ppm) remaining after 8 wk of incubation in an organic and mineral soil were 76 and 79%, respectively, while in sterilized controls 100 and 92% remained, respectively (Chapman et al., 1981).

Photolytic. Photolysis of *p,p′*-DDT in nitrogen-sparged methanol solvent by UV light (λ = 260 nm) produced DDD and DDMU. But photolysis of *p,p′*-DDT at 280 nm in an oxygenated methanol solution yielded a complex mixture containing the methyl ester of 2,2-bis(*p*-chlorophenyl)acetic acid (Plimmer et al., 1970). *p,p′*-DDT in an aqueous solution containing suspended titanium dioxide as a catalyst and irradiated with UV light (λ >340 nm) formed chloride ions. Based on the amount of chloride ions generated, carbon dioxide and hydrochloric acid were reported as the end products (Borello et al., 1989).

When an aqueous solution containing *p,p'*-DDT was photooxidized by UV light at 90-95 °C, 25, 50, and 75% degraded to carbon dioxide after 25.9, 66.5, and 120.0 h, respectively (Knoevenagel and Himmelreich, 1976). When *p,p'*-DDT in distilled water and San Francisco Bay, CA water was subjected to sunlight for 1 wk, percent losses were 0 and 50, respectively (Singmaster, 1975). In a similar study, Leffingwell (1975) studied the photolysis of *p,p'*-DDT (10 ppm) suspension in water using sunlight and a mercury lamp. The presence of triphenylamine, diphenylanthracene, or azobis(isobutryonitrile) accelerated the rate of reaction. Photolysis of *p,p'*-DDT in cyclohexane using UV light (λ = 310 nm) did not occur unless triphenylamine or *N,N*-diethylaniline was present. Photoproducts DDD, DDE, and DDMU formed at yields of 6, 15, and 16%, respectively (Miller and Narang, 1970).

When *p,p'*-DDT on quartz was subjected to UV radiation (2537 Å) for 2 d, 80% of *p,p'*-DDT degraded to 4,4'-dichlorobenzophenone, 1,1-dichloro-2,2-bis-(*p*-chlorophenyl)ethane, and 1,1-dichloro-2,2-bis(*p*-chlorophenyl)ethene. Irradiation of *p,p'*-DDT in a hexane solution yielded 1,1-dichloro-2,2-bis(*p*-chlorophenyl)ethane and hydrochloric (Mosier et al., 1969).

Chemical/Physical. In alkaline solutions and temperatures >108.5 °C, *p,p'*-DDT undergoes dehydrochlorination via hydrolysis releasing hydrochloric acid to give the noninsecticidal *p,p'*-DDE (Hartley and Kidd, 1987; Kollig, 1993; Worthing and Hance, 1991). This reaction is also catalyzed by ferric and aluminum chlorides and UV light (Worthing and Hance, 1991).

Castro (1964) reported that iron(II) porphyrins in dilute aqueous solution was rapidly oxidized by DDT to form the corresponding iron(III) chloride complex (hematin) and DDE, respectively. Incubation of *p,p'*-DDT with hematin and ammonia gave *p,p'*-DDD, *p,p'*-DDE, bis(*p*-chlorophenyl)acetonitrile, 1-chloro-2,2-bis(*p*-chlorophenyl)ethylene, 4,4'-dichlorobenzophenone, and the methyl ester of bis(*p*-chlorophenyl)acetic acid (Quirke et al., 1979).

When *p,p'*-DDT was heated at 900 °C, carbon monoxide, carbon dioxide, chlorine, hydrochloric acid, and other unidentified substances were produced (Kennedy et al., 1972, 1972a). Emits hydrochloric acid and chlorine when incinerated (Sittig, 1985).

At 33 °C, 35% relative humidity and a 2 mile/h wind speed, the volatility losses of *p,p'*-DDT as a thick film, droplets on glass, droplets on leaves, and formulation film on glass after 48 h were 97.1, 46.0, 40.6, and 5.5%, respectively (Que Hee et al., 1975).

Mackay and Wolkoff (1973) estimated an evaporation half-life of 3.7 d from a surface water body that is 25 °C and 1 m deep.

Exposure limits (mg/m³): Potential occupational carcinogen. NIOSH REL: TWA 0.5, IDLH 500; OSHA PEL: TWA 1; ACGIH TLV: TWA 1 (adopted).

Toxicity:

EC$_{50}$ (48-h) for *Daphnia magna* 4.7 μg/L, *Daphnia pulex* 0.36 μg/L, *Simocephalus serrulatus* 2.7 μg/L (Mayer and Ellersieck, 1986).

LC$_{50}$ (96-h) for *Salmo gairdneri* 7 μg/L, *Salmo trutta* 2 μg/L, *Perca flavescens* 9 μg/L, *Oncorhynchus kisutch* 4 μg/L, *Ictalurus melas* 5 μg/L, *Ictalurus punctatus* 16 μg/L, *Pimephales promelas* 19 μg/L, *Lepomis microlophus* 5 μg/L *Lepomis machrochirus* 8 μg/L, *Micropterus salmoides* 2 μg/L (Macek and McAllister, 1970), *Pteronarcys californica* 7.0 μg/L, *Pteronarcella badia* 1.9 μg/L, *Claassenia sabutosa* 3.5 μg/L (Sanders and Cope, 1968), *Gammarus italicus* 7.0 μg/L, *Echinogammarus tibaldii* 3.9 μg/L (Pantani et al., 1997), carp 250 μg/L, goldfish 70 μg/L, medaka 10 μg/L, pond loach 240 μg/L (Spehar et al., 1982), bluegill sunfish 3.4 μg/L (technical), 9.0 μg/L (25% emulsifiable concentrate), *Daphnia magna* 1.1 μg/L (technical), 1.7 μg/L (25% emulsifiable concentrate) (Randall et al., 1979), northern pike

2.7 μg/L, coho salmon 11.7 μg/L, cutthroat trout 6.7 μg/L, brown trout 1.8 μg/L, goldfish 23.1 μg/L, carp 9.7 μg/L river shiner 5.8 μg/L, fathead minnow 7.8 μg/L, black bullhead 5.0 μg/L, green sunfish 6.5-10.9 μg/L, walleye 3.8 μg/L, largemouth bass 1.5 μg/L, redear sunfish 15 μg/L, 1 wk old Western chorus tadpole 800 μg/L (Mayer and Ellersieck, 1986).

LC$_{50}$ (96-h) static lab bioassay for *Crangon septemspinosa* (sand shrimp) 0.6 μg/L, *Palaemonetes vulgaris* 2 μg/L, *Pagurus longicarpus* (hermit crab) 6 μg/L (Eisler, 1969).

LC$_{50}$ (48-h) for red killifish 100 μg/L (Yoshioka et al., 1986), *Simocephalus serrulatus* 2.5 μg/L, *Daphnia pulex* 0.36 μg/L (Sanders and Cope, 1966).

LC$_{50}$ (24-h) for fish (*Fundulus heteroclitus*) 250 μg/L (quoted, Reish and Kauwling, 1978), northern pike 5.5 μg/L, coho salmon 18.5 μg/L, cutthroat trout 9.9 μg/L, brown trout >6.5 μg/L, goldfish 15.1 μg/L, carp 14 μg/L, river shiner 6.7 μg/L, fathead minnow 14.0 μg/L, black bullhead 31.5 μg/L, green sunfish 16.9-19.8 μg/L, walleye 4.4 μg/L, largemouth bass 2.9 μg/L, redear sunfish 19 μg/L, 1 wk old Western chorus tadpole 1,400 μg/L (Mayer and Ellersieck, 1986).

Acute oral LD$_{50}$ for dogs 150 mg/kg, frogs 7,600 μg/mg, guinea pigs 150 mg/kg, monkeys 200 mg/kg, mice 135 mg/kg, rats 87 mg/kg, rabbits 250 mg/kg (quoted, RTECS, 1985), 6.0 μg/roach and 11.7 μg/roach for male and female, respectively (Gardner and Vincent, 1978).

Acute percutaneous LD$_{50}$ for female rats 2,510 mg/kg (Worthing and Hance, 1991).

In 160-d feeding trials, the NOEL in rats was 1 mg/kg (Worthing and Hance, 1991).

Uses: Use as an insecticide is now prohibited; chemical research; nonsystemic stomach and contact insecticide.

DECAHYDRONAPHTHALENE

Synonyms: Bicyclo[4.4.0]decane; Dec; Decalin; Decalin solvent; Dekalin; Naphthalane; Naphthane; Perhydronaphthalene; UN 1147.

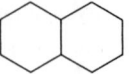

CAS Registry Number: 91-17-8
DOT: 1147
Molecular formula: $C_{10}H_{18}$
Formula weight: 138.25
RTECS: QJ3150000
Merck reference: 10, 2830

Physical state, color, and odor:
Water-white liquid with a methanol-like odor

Melting point (°C):
-43 (*cis*), -30.4 (*trans*) (Weast, 1986)

Boiling point (°C):
195.6 (*cis*), 187.2 (*trans*) (Weast, 1986)

Density (g/cm³):
0.8965 at 20/4 °C (*cis*), 0.8699 at 20/4 °C (*trans*) (Weast, 1986)

Diffusivity in water (x 10⁻⁵ cm²/sec at 20 °C):
0.68 and 0.67 for *cis* and *trans* isomers, respectively, using method of Hayduk and Laudie (1974)

Dissociation constant, pK$_a$:
>14 (Schwarzenbach et al., 1993)

Flash point (°C):
58 (commercial mixture, Windholz et al., 1983)
58 (*cis*), 52 (*trans*) (Dean, 1987)

Lower explosive limit (%):
0.7 at 100 °C (Sax and Lewis, 1987)

Upper explosive limit (%):
4.9 at 100 °C (Sax and Lewis, 1987)

Entropy of fusion (cal/mol·K):
14.19 (McCullough et al., 1957a)

Heat of fusion (kcal/mol):
2.268 and 3.445 for the *cis* and *trans* isomers, respectively (Riddick et al., 1986)

Henry's law constant (x 10^{-2} atm·m³/mol):
7.00, 8.37, 10.6, 11.7, and 19.9 at 10, 15, 20, 25, and 30 °C, respectively (Ashworth et al., 1988)

Interfacial tension with water (dyn/cm at 20 °C):
51.24 (*cis*), 50.7 (*trans*) (Demond and Lindner, 1993)
53.24 (*cis*), 51.29 (*trans*) (Fowkes, 1980)

Soil sorption coefficient, log K_{oc}:
Unavailable because experimental methods for estimation of this parameter for alicyclic hydrocarbons are lacking in the documented literature

Octanol/water partition coefficient, log K_{ow}:
4.00 using method of Hansch et al. (1968)

Solubility in organics:
Soluble in acetone, alcohol, benzene, ether, and chloroform (Weast, 1986). Miscible with propanol, isopropanol, and most ketones and ethers (Windholz et al., 1983).

Solubility in water (25 °C):
889 μg/kg (Price, 1976)
<0.2 mL/L (Booth and Everson, 1948)

Vapor density:
5.65 g/L at 25 °C, 4.77 (air = 1)

Vapor pressure (mmHg):
195.77 and 187.27 at 25 °C for *cis* and *trans* isomers, respectively (Wilhoit and Zwolinski, 1971)

Environmental fate:
   Photolytic. The following rate constants were reported for the reaction of decahydro-naphthalene and OH radicals in the atmosphere: 1.96 x 10^{-11} and 2.02 x 10^{-11} cm³/molecule·sec at 299 K for *cis* and *trans* isomers, respectively (Atkinson, 1985). A photooxidation reaction rate constant of 2.00 x 10^{-11} was reported for the reaction of decahydronaphthalene (mixed isomers) and OH radicals in the atmosphere at 298 K (Atkinson, 1990).
   Chemical/Physical. Decahydronaphthalene will not hydrolyze because it has no hydrolyzable functional group.

Toxicity:
   Acute oral LD_{50} for rats 4,170 mg/kg (quoted, RTECS, 1985).

Uses: Solvent for naphthalene, waxes, fats, oils, resins, rubbers; motor fuel and lubricants; cleaning machinery; substitute for turpentine; shoe-creams; stain remover.

DECANE

Synonyms: *n*-Decane; Decyl hydride; UN 2247.

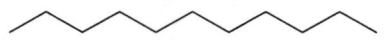

CAS Registry Number: 124-18-5
DOT: 2247
Molecular formula: $C_{10}H_{22}$
Formula weight: 142.28
RTECS: HD6550000

Physical state and color:
Clear, colorless liquid. The odor threshold is 11.3 mg/m^3 (Laffort and Dravnieks, 1973).

Melting point (°C):
-29.7 (Weast, 1986)
-30.0 (Stephenson and Malanowski, 1987)

Boiling point (°C):
174.1 (Dreisbach, 1959)

Density (g/cm^3):
0.73005 at 20/4 °C (Dreisbach, 1959)
0.7262 at 25.00/4 °C (Aralaguppi et al., 1999)

Diffusivity in water (x 10^{-5} cm^2/sec):
0.60 at 20 °C using method of Hayduk and Laudie (1974)

Dissociation constant, pK$_a$:
>14 (Schwarzenbach et al., 1993)

Flash point (°C):
46.1 (Sax and Lewis, 1987)

Lower explosive limit (%):
0.8 (Sax and Lewis, 1987)

Upper explosive limit (%):
5.4 (Sax and Lewis, 1987)

Heat of fusion (kcal/mol):
6.864 (Riddick et al., 1986)

Henry's law constant (atm·m^3/mol):
0.187 at 25 °C (approximate - calculated from water solubility and vapor pressure)

Interfacial tension with water (dyn/cm):
51.2 at 20 °C (Girifalco and Good, 1957)

51.30 at 25 °C (Jańczuk et al., 1993)
52.26 at 20 °C (Fowkes, 1980)

Ionization potential (eV):
9.65 (Lias, 1998)

Soil sorption coefficient, log K_{oc}:
Unavailable because experimental methods for estimation of this parameter for aliphatic hydrocarbons are lacking in the documented literature

Octanol/water partition coefficient, log K_{ow}:
6.69 (Burkhard et al., 1985a)

Solubility in organics:
Miscible with hexane (Corby and Elworthy, 1971) and many other hydrocarbons (e.g., hexane, cyclohexane).
In methanol, g/L: 62 at 5 °C, 68 at 10 °C, 74 at 15 °C, 81 at 20 °C, 89 at 25 °C, 98 at 30 °C, 109 at 35 °C, 120 at 40 °C (Kiser et al., 1961).

Solubility in water:
2.20 x 10^{-5} mL/L at 25 °C (Baker, 1959)
52 μg/kg at 25 °C (McAuliffe, 1969)
19.8 μg/kg at 25 °C (Franks, 1966)
9 μg/L at 20 °C (distilled water), 0.087 mg/L at 20 °C (seawater) (quoted, Verschueren, 1983)

Vapor density:
5.82 g/L at 25 °C, 4.91 (air = 1)

Vapor pressure (mmHg):
2.7 at 20 °C (Verschueren, 1983)
1.35 at 25 °C (Wilhoit and Zwolinski, 1971)

Environmental fate:
   Biological. Decane may biodegrade in two ways. The first is the formation of decyl hydroperoxide, which decomposes to 1-decanol, followed by oxidation to decanoic acid. The other pathway involves dehydrogenation to 1-decene, which may react with water giving 1-decanol (Dugan, 1972). Microorganisms can oxidize alkanes under aerobic conditions (Singer and Finnerty, 1984). The most common degradative pathway involves the oxidation of the terminal methyl group forming the corresponding alcohol (1-decanol). The alcohol may undergo a series of dehydrogenation steps, forming decanal, followed by oxidation forming decanoic acid. The fatty acid may then be metabolized by β-oxidation to form the mineralization products, carbon dioxide and water (Singer and Finnerty, 1984). Hou (1982) reported 1-decanol and 1,10-decanediol as degradation products by the microorganism *Corynebacterium.*
   Photolytic. A photooxidation reaction rate constant of 1.16 x 10^{-11} cm³/molecule·sec was reported for the reaction of decane with OH in the atmosphere (Atkinson, 1990).
   Chemical/Physical. Complete combustion in air yields carbon dioxide and water vapor. Decane will not hydrolyze because it has no hydrolyzable functional group.

Toxicity:
   LC_{50} (96-h) for *Cyprinodon variegatus* >500 ppm using natural seawater (Heitmuller et al.,

1981).

LC$_{50}$ (72-h) for *Cyprinodon variegatus* >500 ppm (Heitmuller et al., 1981).

LC$_{50}$ (48-h) for *Daphnia magna* 18 mg/L (LeBlanc, 1980), *Cyprinodon variegatus* >500 ppm (Heitmuller et al., 1981).

LC$_{50}$ (24-h) for *Daphnia magna* 23 mg/L (LeBlanc, 1980), *Cyprinodon variegatus* >500 ppm (Heitmuller et al., 1981).

LC$_{50}$ (inhalation) for mice 72,300 gm/kg/2-h (quoted, RTECS, 1985).

Heitmuller et al. (1981) reported a NOEC of 500 ppm.

Source: Major constituent in paraffin (quoted, Verschueren, 1983)

Uses: Solvent; standardized hydrocarbon; manufacturing paraffin products; jet fuel research; paper processing industry; rubber industry; organic synthesis.

DIACETONE ALCOHOL

Synonyms: DAA; Diacetone; Diacetonyl alcohol; Diketone alcohol; Dimethylacetonylcarbinol; 4-Hydroxy-2-keto-4-methylpentane; 4-Hydroxy-4-methylpentanone-2; 4-Hydroxy-4-methylpentan-2-one; **4-Hydroxy-4-methyl-2-pentanone**; 2-Methyl-2-pentanol-4-one; Pyranton; Pyranton A; UN 1148.

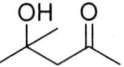

CAS Registry Number: 123-42-2
DOT: 1148
DOT label: Flammable liquid
Molecular formula: $C_6H_{12}O_2$
Formula weight: 116.16
RTECS: SA9100000
Merck reference: 10, 2928

Physical state, color, and odor:
Colorless liquid with a mild, pleasant odor

Melting point (°C):
-44 (Weast, 1986)

Boiling point (°C):
164 (Weast, 1986)
167.9 (Windholz et al., 1983)

Density (g/cm³):
0.9387 at 20/4 °C (Weast, 1986)
0.9306 at 25/4 °C (Windholz et al., 1983)

Diffusivity in water (x 10^{-5} cm²/sec):
0.78 at 20 °C using method of Hayduk and Laudie (1974)

Flash point (°C):
51.7 (NIOSH, 1997)
66 (reagent grade), 48 (commercial grade), 13 (commercial grade - open cup) (Windholz et al., 1983)

Lower explosive limit (%):
1.8 (NIOSH, 1997)

Upper explosive limit (%):
6.9 (NIOSH, 1997)

Soil sorption coefficient, log K_{oc}:
Unavailable because experimental methods for estimation of this parameter for ketones are lacking in the documented literature. However, its miscibility in water suggests its adsorption to soil will be nominal (Lyman et al., 1982).

Solubility in organics:
Soluble in alcohol and ether (Weast, 1986)

Solubility in water:
Miscible (NIOSH, 1997).

Vapor density:
4.75 g/L at 25 °C, 4.01 (air = 1)

Vapor pressure (mmHg):
1 at 20 °C (NIOSH, 1997)
1.7 at 30 °C (Verschueren, 1983)

Environmental fate:
Photolytic. Grosjean (1997) reported rate constant of 4.0 x 10^{-12} cm^3/molecule·sec at 298 K for the reaction of OH radicals in the atmosphere. Based on a OH concentration of 1.0 x 10^6 molecule/cm^3, the reported half-life of diacetone alcohol is 2.0 d (Grosjean, 1997).

Exposure limits: NIOSH REL: TWA 50 ppm (240 mg/m^3), IDLH 1,800 ppm; OSHA PEL: TWA 50 ppm; ACGIH TLV: TWA 50 ppm (adopted).

Symptoms of exposure: May cause irritation of eyes, nose, throat, and skin (Patnaik, 1992). An irritation concentration of 240.00 mg/m^3 in air was reported by Ruth (1986).

Toxicity: LC_{50} (24-h) for goldfish >5,000 mg/L (quoted, Verschueren, 1983); acute oral LD_{50} for rats 4,000 mg/kg, mice 3,950 mg/kg (quoted, RTECS, 1985).

Uses: Solvent for celluloid, cellulose acetate, fats, oils, waxes, nitrocellulose and resins; wood preservatives; rayon and artificial leather; imitation gold leaf; extraction of resins and waxes; in antifreeze mixtures and hydraulic fluids; laboratory reagent; preservative for animal tissue; dyeing mixtures; stripping agent for textiles.

DIBENZ[*a,h*]ANTHRACENE

Synonyms: 1,2:5,6-Benzanthracene; DBA; 1,2,5,6-DBA; DB[*a,h*]A; 1,2:5,6-Dibenzanthracene; 1,2:5,6-Dibenz[*a,h*]anthracene; 1,2:5,6-Dibenzoanthracene; Dibenzo[*a,h*]anthracene; RCRA waste number U063.

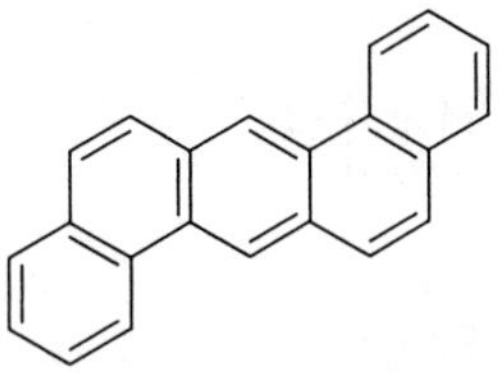

CAS Registry Number: 53-70-3
Molecular formula: $C_{22}H_{14}$
Formula weight: 278.36
RTECS: HN2625000
Merck reference: 10, 2978

Physical state and color:
White, monoclinic or orthorhombic crystals or leaflets

Melting point (°C):
271 (Casellato et al., 1973)
262-265 (Fluka, 1988)

Boiling point (°C):
524 (Verschueren, 1983)

Density (g/cm³):
1.282 (IARC, 1973)

Diffusivity in water (x 10^{-5} cm²/sec):
0.46 at 20 °C using method of Hayduk and Laudie (1974)

Dissociation constant, pK_a:
>15 (Christensen et al., 1975)

Henry's law constant (x 10^{-6} atm·m³/mol):
1.70 at 25 °C (approximate - calculated from water solubility and vapor pressure)

Ionization potential (eV):
7.28 ± 0.29 (Franklin et al., 1969)

Bioconcentration factor, log BCF:
4.00 (*Daphnia magna*, Newsted and Giesy, 1987)
4.63 (activated sludge), 3.39 (algae), 1.00 (fish) (Freitag et al., 1985)
Apparent values of 3.7 (wet wt) and 5.5 (lipid wt) for freshwater isopods including *Asellus aquaticus* (L.) (van Hattum et al., 1998)

Soil sorption coefficient, log K_{oc}:
6.22 (Abdul et al., 1987)

5.75 (Illinois soil), 5.91 (Illinois and North Dakota sediments), 6.07 (Iowa sediment), 6.23 (Georgia and Missouri sediments), 6.36 (Iowa loess), 6.38 (Illinois sediment), 6.42 (South Dakota sediment), 6.43 (Indiana sediment), 6.43, 6.47 (Illinois sediment), 6.48 (West Virginia soil), 6.49 (Kentucky sediment) (Hassett et al., 1980; Means et al., 1980)

Octanol/water partition coefficient, log K_{ow}:
6.50 (Abdul et al., 1987; Means et al., 1980)
6.36 (Chiou et al., 1982)
5.97 (Sims et al., 1988)
6.58 (Burkhard et al., 1985a)

Solubility in organics:
Soluble in petroleum ether, benzene, toluene, xylene, and oils. Slightly soluble in alcohol and ether (Windholz et al., 1983).

Solubility in water:
0.5 μg/L at 27 °C (Davis et al., 1942)
2.49 μg/L at 25 °C (Means et al., 1980)
2.15 nmol/L at 25 °C (Klevens, 1950)

Vapor pressure (mmHg):
2.78 x 10^{-12} at 25 °C (de Kruif, 1980)

Environmental fate:
   Biological. In activated sludge, <0.1% of the applied dibenz[*a,h*]anthracene mineralized to carbon dioxide after 5 d (Freitag et al., 1985). Based on aerobic soil die away test data, the estimated half-lives ranged from 361 to 940 d (Coover and Sims, 1987).
   Ye et al. (1996) investigated the ability of *Sphingomonas paucimobilis* strain U.S. EPA 505 (a soil bacterium capable of using fluoranthene as a sole source of carbon and energy) to degrade 4, 5, and 6-ringed aromatic hydrocarbons (10 ppm). After 16 h of incubation using a resting cell suspension, only 7.8% of dibenz[*a,h*]anthracene had degraded. It was suggested that degradation occurred via ring cleavage resulting in the formation of polar metabolites and carbon dioxide.
   Soil. The reported half-lives for dibenz[*a,h*]anthracene in a Kidman sandy loam and McLaurin sandy loam are 361 and 420 d, respectively (Park et al., 1990).
   Photolytic. A carbon dioxide yield of 45.3% was achieved when dibenz[*a,h*]anthracene adsorbed on silica gel was irradiated with light (λ >290 nm) for 17 h (Freitag et al., 1985). The photooxidation half-life in the atmosphere was estimated to range from 0.428 to 4.28 h (Atkinson, 1987).
   Chemical/Physical. Dibenz[*a,h*]anthracene will not hydrolyze because it does not contain a hydrolyzable functional group (Kollig, 1993).

Exposure limits: Potential occupational carcinogen. No individual standards have been set; however, as a constituent in coal tar pitch volatiles, the following exposure limits have been established (mg/m^3): NIOSH REL: TWA 0.1 (cyclohexane-extractable fraction), IDLH 80; OSHA PEL: TWA 0.2 (benzene-soluble fraction); ACGIH TLV: TWA 0.2 (benzene solubles).

Toxicity:
   LD_{50} (intravenous) for rats is 10 mg/kg (Patnaik, 1992).

Drinking water standard: No MCLGs or MCLs have been proposed (U.S. EPA, 1996).

Source: Constituent in coal tar, cigarette smoke (4 μg/1,000 cigarettes), and exhaust condensate of gasoline engine (96 μg/g) (quoted, Verschueren, 1983).

Based on analysis of 7 coal tar samples, dibenz[*a,h*]anthracene was not detected.

Use: Research chemical. Although not produced commercially in the U.S., dibenz[*a,h*]anthracene is derived from industrial and experimental coal gasification operations where the maximum concentrations detected in gas and coal tar streams were 0.0061 and 3.4 mg m^3, respectively (Cleland, 1981).

DIBENZOFURAN

Synonyms: Biphenylene oxide; Diphenylene oxide.

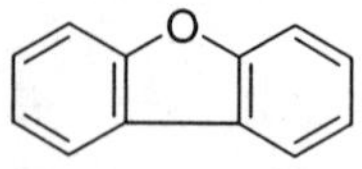

CAS Registry Number: 132-64-9
Molecular formula: $C_{12}H_8O$
Formula weight: 168.20

Physical state and color:
Colorless crystals

Melting point (°C):
86-87 (Weast, 1986)
82 (Banerjee et al., 1980)

Boiling point (°C):
287 (Weast, 1986)

Density (g/cm³):
1.0886 at 99/4 °C (Weast, 1986)

Diffusivity in water (x 10⁻⁵ cm²/sec):
0.63 at 20 °C using method of Hayduk and Laudie (1974)

Entropy of fusion (cal/mol·K):
12.90 (Rordorf, 1989)

Heat of fusion (kcal/mol):
4.6845 (Rordorf, 1989)

Henry's law constant (x 10⁻⁵ atm·m³/mol):
5.82 at 25 °C (approximate - calculated from water solubility and vapor pressure)

Ionization potential (eV):
8.59 (Franklin et al., 1969)

Bioconcentration factor, log BCF:
3.13 (fathead minnow, Veith et al., 1979)

Soil sorption coefficient, log K_{oc}:
4.54, 4.58 (clayey till, Broholm et al., 1999)

Octanol/water partition coefficient, log K_{ow}:
4.17 (Banerjee et al., 1980)
4.12 (Leo et al., 1971)
4.31 (Doucette and Andren, 1988)

Solubility in organics:
Soluble in acetic acid, acetone, ethanol, and ether (Weast, 1986)

Solubility in water (mg/L):
10.03 at 25 °C (Banerjee et al., 1980)
1.65, 4.22, and 6.96 at 4.0, 25.0, and 39.8 °C, respectively (Doucette and Andren, 1988a)

Vapor pressure (mmHg):
2.63×10^{-3} at 25 °C (Rordorf, 1989)

Environmental fate:
Soil. The estimated half-lives of dibenzofuran in soil under aerobic and anaerobic conditions were 7 to 28 and 28 to 112 d, respectively (Lee et al., 1984).
Groundwater. Based on aerobic acclimated and unacclimated groundwater die away test data, the estimated half-life of dibenzofuran in groundwater ranged from 8.54 to 34.9 d (Lee et al., 1984).
Photolytic. The estimated half-life for the reaction of dibenzofuran with OH radicals in the atmosphere ranged from 1.9 to 19 h (Atkinson, 1987).
Chemical/Physical. It was suggested that the chlorination of dibenzofuran in tap water accounted for the presence of chlorodibenzofuran (Shiraishi et al., 1985). Dibenzofuran will not hydrolyze because it has no hydrolyzable functional group (Kollig, 1995).

Toxicity:
LC_{50} (96-h) for *Cyprinodon variegatus* 1.8 ppm using natural seawater (Heitmuller et al., 1981).
LC_{50} (72-h) for *Cyprinodon variegatus* 3.1 ppm (Heitmuller et al., 1981).
LC_{50} (48-h) for *Daphnia magna* 1.7 mg/L (LeBlanc, 1980), *Cyprinodon variegatus* >3.2 ppm (Heitmuller et al., 1981).
LC_{50} (24-h) for *Daphnia magna* 7.5 mg/L (LeBlanc, 1980), *Cyprinodon variegatus* >3.2 ppm (Heitmuller et al., 1981).
Heitmuller et al. (1981) reported a NOEC of 1.0 ppm.

Source: Based on analysis of 7 coal tar samples, dibenzofuran concentrations ranged from 170 to 4,000 ppm (EPRI, 1990).
Schauer et al. (1999) reported dibenzofuran in diesel fuel at a concentration of 29 μg/g and in a diesel-powered medium-duty truck exhaust at an emission rate of 28.7 μg/km.

Use: Research chemical. Derived from industrial and experimental coal gasification operations where the maximum concentration detected in coal gas tar streams was 12 mg/m^3 (Cleland, 1981).

1,4-DIBROMOBENZENE

Synonym: *p*-Dibromobenzene.

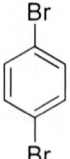

CAS Registry Number: 106-37-6
DOT: 2711
DOT label: Combustible liquid
Molecular formula: $C_6H_4Br_2$
Formula weight: 235.91
RTECS: CZ1791000
Merck reference: 10, 2992

Physical state, color, and odor:
Colorless liquid with a pleasant, aromatic odor

Melting point (°C):
87.3 (Weast, 1986)

Boiling point (°C):
218-219 (Weast, 1986)
219 (Dean, 1987)
220.4 (Windholz et al., 1983)
225 (Hawley, 1981)

Density (g/cm³):
1.9767 at 25/4 °C (Hawley, 1981)
1.841 at 87-89 °C (Aldrich, 1988)

Diffusivity in water (x 10^{-6} cm²/sec):
8.30 at 25 °C (mole fraction = 2 x 10^{-7}) (Gabler et al., 1996)

Flash point (°C):
None (Dean, 1987)

Entropy of fusion (cal/mol·K):
13.3 (Ueberreiter and Orthman, 1950)

Henry's law constant (x 10^{-4} atm·m³/mol):
5.0 at 25 °C (Hine and Mookerjee, 1975)

Ionization potential (eV):
8.82 (Lias, 1998)

Bioconcentration factor, log BCF:
2.36 (wet weight based), -0.11 (lipid based) (*Gambusia affinis*, Chaisuksant et al., 1997)

Soil sorption coefficient, log K_{oc}:
3.20 using method of Chiou et al. (1979)

Octanol/water partition coefficient, log K_{ow}:
3.79 (Watarai et al., 1982)
3.89 (Gobas et al., 1988)
4.41, 4.35, 4.27, 4.19, and 4.12 at 5, 15, 25, 35, and 45 °C, respectively (Shiu et al., 1997)

Solubility in organics:
Miscible with acetone, alcohol, benzene, carbon tetrachloride, ether, and heptane (Hawley, 1981)

Solubility in water:
16.5 mg/L at 25 °C (Andrews and Keefer, 1950)
0.112 mM at 35 °C (Hine et al., 1963)
20.0 mg/L at 25 °C (Mackay and Shiu, 1977)
1.914, 3.044, 4.75, 7.56, and 11.7 mg/L at 5, 15, 25, 35, and 45 °C, respectively (Shiu et al., 1997)

Vapor pressure (mmHg):
0.161 at 25 °C (Mackay et al., 1982)
0.134 at 35 °C (Hine et al., 1963)

Toxicity:
Acute oral LD_{50} for mice 120 mg/kg (quoted, RTECS, 1985).

Uses: Solvent for oils; ore flotation; motor fuels; organic synthesis.

DIBROMOCHLOROMETHANE

Synonyms: Chlorodibromomethane; CDBM; NCI-C55254.

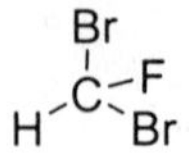

Br
|
C–F
H   Br

CAS Registry Number: 124-48-1
Molecular formula: CHBr₂Cl
Formula weight: 208.28
RTECS: PA6360000

Physical state and color:
Clear, colorless to pale yellow, heavy liquid

Melting point (°C):
-23 to -21 (Dean, 1973)

Boiling point (°C):
116 (Hawley, 1981)
122 (Horvath, 1982)

Density (g/cm³):
2.451 at 20/4 °C (Weast, 1986)

Diffusivity in water (x 10⁻⁵ cm²/sec):
0.97 at 20 °C using method of Hayduk and Laudie (1974)

Flash point (°C):
Noncombustible (Aldrich, 1988)

Henry's law constant (x 10⁻⁴ atm·m³/mol):
78.3 at 25 °C (Warner et al., 1987)
8.7 at 20 °C (Nicholson et al., 1984)
3.8, 4.5, 10.3, 11.8, and 15.2 at 10, 15, 20, 25, and 30 °C, respectively (Ashworth et al., 1988)
8, 14, and 22 at 20, 30, and 40 °C, respectively (Tse et al., 1992)
In seawater (salinity 30.4‰): 2.78, 5.56, and 10.13 at 0, 10, and 20 °C, respectively (Moore et al., 1995)

Ionization potential (eV):
10.59 (HNU, 1986)

Soil sorption coefficient, log K_{oc}:
1.92 (Schwille, 1988)

Octanol/water partition coefficient, log K_{ow}:
2.24 (Hansch and Leo, 1979)
2.08 (quoted, Mills et al., 1985)

Solubility in organics:
Miscible with oils, dichloropropane, and isopropanol (U.S. EPA, 1985)

Solubility in water (mg/L at 30 °C):
2,509 (McNally and Grob, 1984)
1,049.9 (McNally and Grob, 1983)

Vapor density:
8.51 g/L at 25 °C, 7.19 (air = 1)

Vapor pressure (mmHg):
76 at 20 °C (Schwille, 1988)

Environmental fate:
   Biological. Dibromochloromethane showed significant degradation with gradual adaptation in a static-culture flask-screening test (settled domestic wastewater inoculum) conducted at 25 °C. At concentrations of 5 and 10 mg/L, percent losses after 4 wk of incubation were 39 and 25, respectively. At a substrate concentration of 5 mg/L, 16% was lost due to volatilization after 10 d (Tabak et al., 1981).
   Surface Water. The estimated volatilization half-life of dibromochloromethane from rivers and streams is 45.9 h (Kaczmar et al., 1984).
   Photolytic. Water containing 2,000 ng/μL of dibromochloromethane and colloidal platinum catalyst was irradiated with UV light. After 20 h, dibromochloromethane degraded to 80 ng/μL bromochloromethane, 22 ng/μL methyl chloride, and 1,050 ng/μL methane. A duplicate experiment was performed but 1 g zinc was added. After about 1 h, total degradation was achieved. Presumed transformation products include methane, bromide, and chloride ions (Wang and Tan, 1988).
   Chemical/Physical. The estimated hydrolysis half-life in water at 25 °C and pH 7 is 274 yr (Mabey and Mill, 1978). Hydrogen gas was bubbled in an aqueous solution containing 18.8 μmol bromodichloromethane. After 24 h, only 18% of the bromodichloromethane reacted to form methane and minor traces of ethane. In the presence of colloidal platinum catalyst, the reaction proceeded at a much faster rate forming the same end products (Wang et al., 1988).
   At influent concentrations of 1.0, 0.1, 0.01, and 0.001 mg/L, the granular activated carbon adsorption capacities at pH 5.3 were 4.8, 2.2, 1.0, and 0.46 mg/g, respectively (Dobbs and Cohen, 1980).

Drinking water standard (proposed): MCL: 60μg/L. Total for all trihalomethanes cannot exceed a concentration of 0.08 mg/L (U.S. EPA, 1996).

Toxicity:
   Acute oral LD_{50} for rats 848 mg/kg, mice 800 mg/kg (quoted, RTECS, 1985).

Uses: Manufacture of fire extinguishing agents, propellants, refrigerants, and pesticides; organic synthesis.

1,2-DIBROMO-3-CHLOROPROPANE

Synonyms: BBC 12; 1-Chloro-2,3-dibromopropane; 3-Chloro-1,2-dibromopropane; DBCP; Dibromochloropropane; Fumagon; Fumazone; Fumazone 86; Fumazone 86E; NCI-C00500; Nemabrom; Nemafume; Nemagon; Nemagon 20; Nemagon 20G; Nemagon 90; Nemagon 206; Nemagon soil fumigant; Nemanax; Nemapaz; Nemaset; Nematocide; Nematox; Nemazon; OS 1987; Oxy DBCP; RCRA waste number U066; SD 1897; UN 2872.

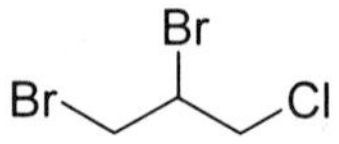

CAS Registry Number: 96-12-8
DOT: 2872
Molecular formula: $C_3H_5Br_2Cl$
Formula weight: 236.36
RTECS: TX8750000
Merck reference: 10, 2994

Physical state, color, and odor:
Yellow to brown liquid with a pungent odor at high concentrations

Melting point (°C):
5 (NIOSH, 1997)

Boiling point (°C):
196 (Windholz et al., 1983)

Density (g/cm³):
2.093 at 14/4 °C (Windholz et al., 1983)
2.05 at 20/4 °C (Hawley, 1981)

Diffusivity in water (x 10^{-5} cm²/sec):
0.81 at 20 °C using method of Hayduk and Laudie (1974)

Flash point (°C):
76.7 (open cup, NIOSH, 1997)

Henry's law constant (x 10^{-4} atm·m³/mol):
2.49 at 20 °C (approximate - calculated from water solubility and vapor pressure)

Soil sorption coefficient, log K_{oc}:
2.11 (Kenaga and Goring, 1980)
1.49-2.16 (Panoche clay loam, Biggar et al., 1984)
2.48, 2.55 (Fresno, CA aquifer solids, Deeley et al., 1991)

Octanol/water partition coefficient, log K_{ow}:
2.63 using method of Hansch et al. (1968).
2.49 and 2.51 were estimated using fragment contribution methods of Boto et al. (1984) and Viswanadhan et al. (1989), respectively.

Solubility in organics:
Miscible with oils, dichloropropane, and isopropanol (Windholz et al., 1983)

Solubility in water:
1,270 ppm (Kenaga and Goring, 1980)

Vapor density:
9.66 g/L at 25 °C, 8.16 (air = 1)

Vapor pressure (mmHg):
0.8 at 21 °C (Verschueren, 1983)

Environmental fate:
   Biological. Soil water cultures converted 1,2-dibromo-3-chloropropane to 1-propanol, bromide, and chloride ions. Precursors to the alcohol formation include allyl chloride and allyl alcohol (Castro and Belser, 1968).
   Soil. Biodegradation is not expected to be significant in removing 1,2-dibromo-3-chloropropane. In aerobic soil columns, no degradation was observed after 25 d (Wilson et al., 1981). The reported half-life in soil is 6 months (Jury et al., 1987).
   Groundwater. According to the U.S. EPA (1986), 1,2-dibromo-3-chloropropane has a high potential to leach to groundwater. Deeley et al. (1991) calculated a half-life of 6.1 yr for 1,2-dibromo-3-chloropropane in a Fresno, CA aquifer (pH 7.8 and 21.1 °C).
   Chemical/Physical. 1,2-Dibromo-3-chloropropane is subject to both neutral and base-mediated hydrolysis (Kollig, 1993). Under neutral conditions, the chlorine or bromine atoms may be displaced by hydroxyl ions. If nucleophilic attack occurs at the carbon-chlorine bond, 2,3-dibromopropanol is formed which reacts further to give 2,3-dihydroxybromopropane via the intermediate epibromohydrin. 2,3-Dihydroxybromopropane will undergo hydrolysis via the intermediate 1-hydroxy-2,3-propylene oxide which further reacts with water to give glycerol. If the nucleophilic attack occurs at the carbon-bromine bond, 2-bromo-3-chloropropanol is formed which further reacts forming the end product glycerol (Kollig, 1993). If hydrolysis of 1,2-dibromo-2-chloropropane occurs under basic conditions, the compound will undergo dehydrohalogenation forming 2-bromo-3-chloropropene and 2,3-dibromo-1-propene as intermediates. Both compounds are subject to further attack forming 2-bromo-3-hydroxy-propene as the end product (Burlinson et al., 1982; Kollig, 1993). The hydrolysis half-life at pH 7 and 25 °C was calculated to be 38 yr (Burlinson et al., 1982; Ellington et al., 1986).
   The rate constants for the reaction of 1,2-dibromo-3-chloropropane with ozone and OH radicals in the atmosphere at 296 K are <5.4 x 10^{-20} and 4.4 x 10^{-13} cm^3/molecule·sec (Tuazon et al., 1986). The smaller rate constant for the reaction with ozone indicates that the reaction with ozone is not an important atmospheric loss of 1,2-dibromo-3-chloropropane. The calculated photolytic half-life and tropospheric lifetime for the reaction with OH radicals in the atmosphere are 36 and 55 d, respectively. The compound 1-bromo-3-chloropropan-2-one was tentatively identified as a product of the reaction of 1,2-dibromo-3-chloropropane with OH radicals. In water, 1,2-dibromo-3-chloropropane (0.045 mM) reacted with OH radicals (pH 2.7). Reaction rates were 3.2 x 10^8/M·sec and 4.2 x 10^8/M·sec (Haag and Yao, 1992).
   Emits toxic chloride and bromide fumes when heated to decomposition (Lewis, 1990).

Exposure limits: Potential occupational carcinogen. Because no standards have been established, NIOSH (1997) recommends the most reliable and protective respirators be used, i.e., a self-contained breathing apparatus that has a full facepiece and is operated under positive-pressure or a supplied-air respirator that has a full facepiece and is operated under pressure-demand or under positive-pressure in combination with a self-contained breathing apparatus operated under pressure-demand or positive-pressure. OSHA, however, recommends a PEL TWA of 1 ppb.

Symptoms of exposure: An irritation concentration of 1.93 mg/m^3 in air was reported by Ruth (1986).

Toxicity:

   Acute oral LD$_{50}$ for chickens 60 mg/kg, guinea pigs 150 mg/kg, mice 257 mg/kg, rats 170 mg/kg, rabbits 180 mg/kg (quoted, RTECS, 1985).

Drinking water standard (final): MCLG: zero; MCL: 0.2 μg/L (U.S. EPA, 1996).

Uses: Used in the U.S. as a soil fumigant and nematocide until its ban in 1977 (Tuazon et al., 1986); organic synthesis.

DIBROMODIFLUOROMETHANE

Synonyms: Difluorodibromomethane; Freon 12-B2; Halon 1202; UN 1941.

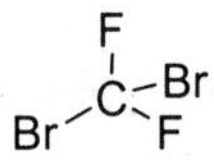

$$Br \overset{\displaystyle F}{\underset{\displaystyle F}{\overset{|}{\underset{|}{C}}}} Br$$

CAS Registry Number: 75-61-6
DOT: 1941
Molecular formula: CBr_2F_2
Formula weight: 209.82
RTECS: PA7525000

Physical state, color, and odor:
Colorless liquid or gas with a characteristic odor

Melting point (°C):
-146.2 (NIOSH, 1997)
-142 to -141 (Aldrich, 1988)

Boiling point (°C):
24.5 (Weast, 1986)
22-23 (Aldrich, 1988)
23-24 (Dean, 1987)

Density (g/cm³):
2.3063 at 15/4 °C (Horvath, 1982)
2.288 at 15/4 °C (Hawley, 1981)
2.297 at 20/4 °C (Aldrich, 1988)

Diffusivity in water (x 10⁻⁵ cm²/sec):
0.93 at 20 °C using method of Hayduk and Laudie (1974)

Flash point (°C):
Noncombustible (NIOSH, 1997)

Ionization potential (eV):
11.07 (NIOSH, 1997)

Solubility in organics:
Soluble in acetone, alcohol, benzene, and ether (Weast, 1986)

Vapor density:
8.58 g/L at 25 °C, 7.24 (air = 1)

Vapor pressure (mmHg):
688 at 20 °C (calculated using the reported constants for the Antoine equation) (Kudchadker et al., 1979)

Exposure limits: NIOSH REL: TWA 100 ppm (860 mg/m³), IDLH 2,000 ppm; OSHA PEL: TWA 100 ppm.

Symptoms of exposure: May cause headache, drowsiness, and excitement (Patnaik, 1992)

Toxicity:

A 15-min exposure to 6,400 and 8,000 ppm were fatal to rats and mice, respectively (Patnaik, 1992).

Uses: Synthesis of dyes; quaternary ammonium compounds; pharmaceuticals; fire-extinguishing agent.

DI-*n*-BUTYL PHTHALATE

Synonyms: 1,2-Benzenedicarboxylate; **1,2-Benzenedicarboxylic acid dibutyl ester**; *o*-Benzenedicarboxylic acid dibutyl ester; Benzene-*o*-dicarboxylic acid di-*n*-butyl ester; Butyl phthalate; *n*-Butyl phthalate; Celluflex DPB; DBP; Dibutyl-1,2-benzenedicarboxylate; Dibutyl phthalate; Elaol; Hexaplas M/B; Palatinol C; Phthalic acid dibutyl ester; Polycizer DP; PX 104; RCRA waste number U069; Staflex DBP; Witicizer 300.

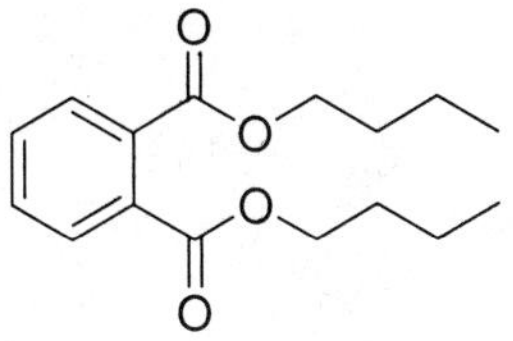

CAS Registry Number: 84-74-2
DOT: 9095
Molecular formula: $C_{16}H_{22}O_4$
Formula weight: 278.35
RTECS: TI0875000
Merck reference: 10, 1559

Physical state, color, and odor:
Colorless to pale yellow, oily, viscous liquid with a mild, aromatic odor

Melting point (°C):
-35 (Verschueren, 1983)
-40 (Standen, 1968)

Boiling point (°C):
340 (Weast, 1986)
335 (Weiss, 1986)

Density (g/cm³):
1.0465 at 21/4 °C (Fishbein and Albro, 1972)
1.042 at 25/4 °C (Standen, 1968)

Diffusivity in water (x 10^{-5} cm²/sec):
0.49 at 20 °C using method of Hayduk and Laudie (1974)

Flash point (°C):
157 (NFPA, 1984)
159 (open cup, Broadhurst, 1972)

Lower explosive limit (%):
0.5 at 235 °C (NFPA, 1984)

Upper explosive limit (%):
2.5 (calculated, Weiss, 1986)

Henry's law constant (x 10^{-5} atm·m³/mol):
6.3 (Petrasek et al., 1983)

Bioconcentration factor, log BCF:
3.61 (*Chlorella pyrenoidosa*, Yan et al., 1995)
3.15 (fish, Mayer and Sanders, 1973)
1.07 (sheepshead minnow), 1.50 (American oyster), 1.22 (brown shrimp) (Wofford et al., 1981)
4.36 green alga, *Selenastrum capricornutum* (Casserly et al., 1983)

Soil sorption coefficient, log K_{oc}:
3.14 (Russell and McDuffie, 1986)
3.00, 2.98, and 2.60 for Apison, Fullerton, and Dormont soils, respectively (Southworth and Keller, 1986)

Octanol/water partition coefficient, log K_{ow}:
4.31 (Doucette and Andren, 1988)
4.50 (Ellington and Floyd, 1996)
4.57 (Leyder and Boulanger, 1983)
4.79 (Howard et al., 1985)
4.72 (DeFoe et al., 1990)
5.20 (Wang et al., 1992)
4.01 at 25 °C (Andersson and Schräder, 1999)

Solubility in organics:
Soluble in benzene and ether (Weast, 1986); very soluble in acetone (Windholz et al., 1983)

Solubility in water:
9.40 mg/L (DeFoe et al., 1990)
13 ppm at 25 °C (Fukano and Obata, 1976)
10.1 mg/L at 20 °C (Leyder and Boulanger, 1983)
11.2 mg/L at 25 °C (Howard et al., 1985)
100 mg/L at 22 °C (Nyssen et al., 1987)
1.300 mg/L at 20-25 °C (Narragansett Bay water, 1.8 mg/L dissolved organic carbon, Boehm and Quinn, 1973)
1.83×10^{-3} wt % at 23.5 °C (Schwarz, 1980)
1.32×10^{-3}, 1.11×10^{-3}, and 1.15×10^{-3} wt % at 10.0, 20.0, and 30.0 °C, respectively (Schwarz and Miller, 1980)
9.2 mg/L at 25 °C (Russell and McDuffie, 1986)
0.013 wt % at 20-25 °C (Fordyce and Meyer, 1940)

Vapor density:
11.38 g/L at 25 °C, 9.61 (air = 1)

Vapor pressure (x 10^{-5} mmHg at 25 °C):
1.4 (Giam et al., 1980)
7.3 (Banerjee et al., 1990; Howard et al., 1985)
4.2 (Hinckley et al., 1990)
10.4 at 25 °C (extrapolated from vapor pressures determined at higher temperatures, Tesconi and Yalkowsky, 1998)

Environmental fate:
  Biological. In anaerobic sludge, di-*n*-butyl phthalate degraded as follows: monobutyl phthalate to *o*-phthalic acid to protocatechuic acid followed by ring cleavage and mineralization. More than 90% of di-*n*-butyl phthalate degraded in 40 d (Shelton et al., 1984). Di-*n*-butyl

phthalate followed the same biodegradation pathway by five strains of microorganisms isolated from a coke-plant wastewater treatment plant sludge. It was proposed, however, that proto-catechuic acid further degraded to a tricarboxylic acid before mineralizing to carbon dioxide and water (Jianlong et al., 1995). In activated sludge obtained from a coke-plant wastewater treatment plant, di-*n*-butyl phthalate degraded rapidly via biological oxidation. The rate of degradation followed first-order kinetics. The rate constant was independent of the initial concentration when the initial di-*n*-butyl phthalate concentration was <200 mg/L. The biodegradation half-life of di-*n*-butyl phthalate in activated was approximately 46 days (Wang et al., 1997). Engelhardt et al. (1975) reported that a variety of microorganisms were capable of degrading of di-*n*-butyl phthalate and suggested the following degradation scheme: di-*n*-butyl phthalate to mono-*n*-butyl phthalate to *o*-phthalic acid to 3,4-dihydroxybenzoic acid and other unidentified products. Di-*n*-butyl phthalate was degraded to benzoic acid by tomato cell suspension cultures (*Lycopericon lycopersicum*) (Pogány et al., 1990).

In a static-culture-flask screening test, di-*n*-butyl phthalate showed significant biodegradation with rapid adaptation. The ester (5 and 10 mg/L) was statically incubated in the dark at 25 °C with yeast extract and settled domestic wastewater inoculum. After 7 d, 100% biodegradation was achieved (Tabak et al., 1981). At 30 °C, microorganisms (*Pseudomonas* sp.) isolated from soil degraded di-*n*-butyl phthalate with a half-life of 3 d (Kurane et al., 1977).

In the presence of suspended natural populations from unpolluted aquatic systems, the laboratory determined second-order microbial transformation rate constant was $3.1 \pm 0.8 \times 10^{-11}$ L/organism·h (Steen, 1991).

Soil. Under aerobic conditions using a fresh-water hydrosol, mono-*n*-butyl phthalate and phthalic acid were produced. Under anaerobic conditions, however, phthalic acid was not formed (Verschueren, 1983).

Inman et al. (1984) reported degradation yields of approximately 90% within 80 d under both aerobic and anaerobic conditions. They also reported half-lives in soil ranging from 11 to 53 d.

Photolytic. An aqueous solution containing titanium dioxide and subjected to UV radiation (λ >290 nm) produced hydroxyphthalates and dihydroxyphthalates as intermediates (Hustert and Moza, 1988).

Chemical/Physical. Pyrolysis of di-*n*-butyl phthalate in the presence of polyvinyl chloride at 600 °C gave the following compounds: indene, methylindene, naphthalene, 1-methylnaphtha-lene, 2-methylnaphthalene, biphenyl, dimethylnaphthalene, acenaphthene, fluorene, methylace-naphthene, methylfluorene, and six unidentified compounds (Bove and Dalven, 1984).

Under alkaline conditions, di-*n*-butyl phthalate will initially hydrolyze to *n*-butyl hydrogen phthalate and *n*-butanol. The monoester will undergo further hydrolysis forming *o*-phthalic acid and *n*-butanol (Kollig, 1993). Wolfe et al. (1980) reported a hydrolysis half-life of 10 yr.

At influent concentrations of 10.0, 1.0, 0.1 and 0.01 mg/L, adsorption capacities using granular activated carbon were 610, 220, 77, and 28 mg/g, respectively (Dobbs and Cohen, 1980).

Exposure limits (mg/m³): NIOSH REL: TWA 5, IDLH 4,000; OSHA PEL: TWA 5; ACGIH TLV: TWA 5 (adopted).

Symptoms of exposure: Ingestion at a dose level of 150 mg/kg may cause nausea, vomiting, hallucination, dizziness, distorted vision, lacrimation, and conjunctivitis (Patnaik, 1992).

Toxicity:
EC_{50} (48-h) for *Chironomus plumosus* 4.7 mg/L (Mayer and Ellersieck, 1986).
LC_{50} (contact) for earthworm (*Eisenia fetida*) 1,360 $\mu g/cm^2$ (Neuhauser et al., 1985).
LC_{50} (96-h) for bluegill sunfish 1.2 mg/L (Spehar et al., 1982), *Gammarus pseudolimnaeus*

2.1 mg/L (Mayer and Ellersieck, 1986); LC_{50} (48-h) for red killifish 15 mg/L (Yoshioka et al., 1986).

LC_{50} (24-h) for grass shrimp larvae 10-50 mg/L (Laughlin et al., 1978), *Gammarus pseudolimnaeus* 7 mg/L (Mayer and Ellersieck, 1986).

Acute oral LD_{50} for guinea pigs 10 gm/kg, mice 5,289 mg/kg, rats 8,000 mg/kg (quoted, RTECS, 1985).

Drinking water standard: No MCLGs or MCLs have been proposed (U.S. EPA, 1996).

Source: Detected in distilled water-soluble fractions of new and used motor oil at concentrations of 38-43 and 15-23 μg/L, respectively (Chen et al., 1994). Leaching from flexible plastics in contact with water. Laboratory contaminant.

Uses: Manufacture of plasticizers, insect repellents, printing inks, paper coatings, explosives, adhesives, cosmetics, safety glass; organic synthesis.

1,2-DICHLOROBENZENE

Synonyms: Chloroben; Chloroden; Cloroben; DCB; 1,2-DCB; *o*-DCB; 1,2-Dichlorbenzene; *o*-Dichlorbenzene; 1,2-Dichlorbenzol; *o*-Dichlorbenzol; *o*-Dichlorobenzene; 1,2-Dichloro-benzol; *o*-Dichlorobenzol; Dilantin DB; Dilatin DB; Dizene; Dowtherm E; NCI-C54944; ODB; ODCB; Orthodichlorobenzene; Orthodichlorobenzol; RCRA waste number U070; Special termite fluid; Termitkil; UN 1591.

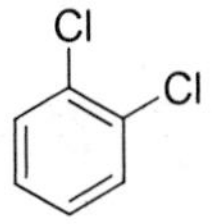

CAS Registry Number: 95-50-1
DOT: 1591
Molecular formula: $C_6H_4Cl_2$
Formula weight: 147.00
RTECS: CZ4500000
Merck reference: 10, 3040

Physical state, color, and odor:
Clear, colorless to pale yellow liquid with a pleasant, aromatic odor. The reported odor threshold concentrations in air ranged from 200 ppb to 4.0 ppm (Keith and Walters, 1992; Young et al., 1996). The taste threshold concentration in water is 200 ppb (Young et al., 1996).

Melting point (°C):
-16 to -14 (Fluka, 1988)
-17.00 (Martin et al., 1979)

Boiling point (°C):
180.5 (Weast, 1986)
179.5 (Standen, 1964)

Density (g/cm³):
1.3048 at 20/4 °C (Weast, 1986)
1.30024 at 25/4 °C (Kirchnerová and Cave, 1976)

Diffusivity in water (x 10⁻⁶ cm²/sec, 25 °C):
9.4, 8.9, and 8.7 at mole fractions of 2 x 10⁻⁷, 4 x 10⁻⁷, and 8 x 10⁻⁷, respectively (Gabler et al., 1996)

Flash point (°C):
66 (NIOSH, 1997)

Lower explosive limit (%):
2.2 (NIOSH, 1997)

Upper explosive limit (%):
9.2 (NIOSH, 1997)

Entropy of fusion (cal/mol·K):
12.1 (Yalkowsky and Valvani, 1980)

Heat of fusion (kcal/mol):
3.19 (Dean, 1987)

Henry's law constant (x 10^{-3} atm·m^3/mol):
1.9 (Pankow and Rosen, 1988)
1.2 at 20 °C (Oliver, 1985)
2.83 at 37 °C (Sato and Nakajima, 1979)
1.63, 1.43, 1.68, 1.57, and 2.37 at 10, 15, 20, 25, and 30 °C, respectively (Ashworth et al., 1988)
1.39 at 20 °C (Hovorka and Dohnal, 1997)
1.92 at 25 °C (Shiu and Mackay, 1997)

Ionization potential (eV):
9.06 (NIOSH, 1997)

Bioconcentration factor, log BCF:
1.95 (bluegill sunfish) (Barrows et al., 1980; Veith et al., 1980)
2.43, 2.75 (*Oncorhynchus mykiss*, Oliver and Niimi, 1983)
3.94 (Atlantic croakers), 4.46 (blue crabs), 3.79 (spotted sea trout), 3.82 (blue catfish) (Pereira et al., 1988)
4.17 green alga, *Selenastrum capricornutum* (Casserly et al., 1983)

Soil sorption coefficient, log K_{oc}:
2.255 (Willamette silt loam, Chiou et al., 1979)
2.43 (Marlette soil - B+ horizon), 2.45 (Marlette soil - A horizon) (Lee et al., 1989)
2.51 (Woodburn silt loam) (Chiou et al., 1983)
2.59 (Appalachee soil, Stauffer and MacIntyre, 1986)
2.69 (peaty soil, Friesel et al., 1984)
2.70 (Piwoni and Banerjee, 1989)
3.10 (Captina silt loam), 2.90 (McLaurin sandy loam) (Walton et al., 1992)
3.02 (Tinker), 2.83 (Carswell), 2.45 (Barksdale), 2.91 (Blytheville), 3.51 (Traverse City), 3.29 (Borden), 2.85 (Lula) (Stauffer et al., 1989)
2.46 (normal soils), 2.70 (suspended bed sediments) (Kile et al., 1995)
3.29, 3.36, 3.49 (glaciofluvial, sandy aquifer, Nielsen et al., 1996)
2.44 (muck), 2.38 (Eustis sand) (Brusseau et al., 1990)

Octanol/water partition coefficient, log K_{ow}:
3.38 (Leo et al., 1971; Wasik et al., 1981)
3.40 (Banerjee et al., 1980)
3.55 (Könemann et al., 1979)
3.433 (de Bruijn et al., 1989)
3.34 (Hammers et al., 1982)
3.49 (Pereira et al., 1988)
3.51, 3.41, 3.29, 3.20, and 3.09 at 5, 15, 25, 35, and 45 °C, respectively (Bahadur et al., 1997)
3.75 (Veith et al., 1980)
3.56 (Garst, 1984)
3.45 at 25 °C (Paschke et al., 1998)
3.70 at 25 °C (Andersson and Schräder, 1999)

Solubility in organics:
Miscible with alcohol, ether, benzene (Windholz et al., 1983), and many other organic solvents,

particularly chlorinated compounds (e.g., carbon tetrachloride, methylene chloride, chloroform, 1,1,1-trichloroethane).

Solubility in water:
137 mg/L at 25 °C (Banerjee, 1984)
156 mg/L at 25 °C (Banerjee et al., 1980)
148 mg/L at 20 °C (Chiou et al., 1979)
92.7 mg/L at 25 °C (Yalkowsky et al., 1979)
0.628 mM at 25 °C (Miller et al., 1984)
145 mg/L at 25 °C (Etzweiler et al., 1995)
142.3 mg/L at 30 °C (McNally and Grob, 1984)
149.4 mg/L at 30 °C (McNally and Grob, 1983)
0.017 wt % at 25 °C (Lo et al., 1986)
97 mg/L at 25 °C (Paschke et al., 1998)
In wt %: 0.005 at 0 °C, 0.0031 at 19.5 °C, 0.0017 at 40.0 °C, 0.0024 at 50.0 °C, 0.0054 at 60.5 °C, 0.0055 at 70.7 °C, 0.0091 at 80.0 °C, 0.0083 at 90.5 °C (Stephenson, 1992)
0.0162 and 0.0126 wt % at 10.0 and 20.0 °C, respectively (Schwarz and Miller, 1980)
134 mg/L at 20 °C, 145 mg/L at 25 °C, 171 mg/L at 30 °C, 183 mg/L at 35 °C, 194 mg/L at 40 °C, 203 mg/L at 45 °C, 223 mg/L at 55 °C, 232 mg/L at 60 °C (Klemenc and Löw, 1930)

Vapor density:
6.01 g/L at 25 °C, 5.07 (air ≈ 1)

Vapor pressure (mmHg):
1.03 at 20 °C (Stull, 1984)
1.9 at 30 °C (Verschueren, 1983)
62 at 100 °C (Bailey and White, 1965)
1.5 at 25 °C (Mackay et al., 1982)

Environmental fate:
  Biological. Pseudomonas sp. isolated from sewage samples produced 3,4-dichloro-*cis*-1,2-dihydroxycyclohexa-3,5-diene. Subsequent degradation of this metabolite yielded 3,4-dichlorocatechol, which underwent ring cleavage to form 2,3-dichloro-*cis,cis*-muconate, followed by hydrolysis to form 5-chloromaleylacetic acid (Haigler et al., 1988). When 1,2-dichlorobenzene was statically incubated in the dark at 25 °C with yeast extract and settled domestic wastewater inoculum, significant biodegradation with gradual acclimation was followed by a deadaptive process in subsequent subcultures. At a concentration of 5 mg/L, 45, 66, 48, and 29% losses were observed after 7, 14, 21, and 28-d incubation periods, respectively. At a concentration of 10 mg/L, only 20, 59, 32, and 18% losses were observed after 7, 14, 21, and 28-d incubation periods, respectively (Tabak et al., 1981).
  Groundwater. Nielsen et al. (1996) studied the degradation of 1,2-dichlorobenzene in a shallow, glaciofluvial, unconfined sandy aquifer in Jutland, Denmark. As part of the *in situ* microcosm study, a cylinder that was open at the bottom and screened at the top was installed through a cased borehole approximately 5 m below grade. Five liters of water was aerated with atmospheric air to ensure aerobic conditions were maintained. Groundwater was analyzed weekly for approximately 3 months to determine 1,2-dichlorobenzene concentrations over time. The experimentally determined first-order biodegradation rate constant and corresponding half-life following a 13-d lag phase were 0.06/d and 11.55 d, respectively.
  Photolytic. Titanium dioxide suspended in an aqueous solution and irradiated with UV light (λ = 365 nm) converted 1,2-dichlorobenzene to carbon dioxide at a significant rate (Matthews, 1986). The sunlight irradiation of 1,2-dichlorobenzene (20 g) in a 100-mL borosilicate glass-

stoppered Erlenmeyer flask for 56 d yielded 2,270 ppm 2,3′,4′-trichlorobiphenyl (Uyeta et al., 1976).

When an aqueous solution containing 1,2-dichlorobenzene (190 μM) and a nonionic surfactant micelle (Brij 58, a polyoxyethylene cetyl ether) was illuminated by a photoreactor equipped with 253.7-nm monochromatic UV lamps, photoisomerization took place yielding 1,3- and 1,4-dichlorobenzene as the principal products. The half-life for this reaction, based on the first-order photodecomposition rate of 1.35 x 10^{-3}/sec, is 8.6 min (Chu and Jafvert, 1994).

Chemical/Physical. Anticipated products from the reaction of 1,2-dichlorobenzene with ozone or OH radicals in the atmosphere are chlorinated phenols, ring cleavage products and nitro compounds (Cupitt, 1980). Based on an assumed base-mediated 1% disappearance after 16 d at 85 °C and pH 9.70 (pH 11.26 at 25 °C), the hydrolysis half-life was estimated to be >900 yr (Ellington et al., 1988).

When 1,2-dichlorobenzene in hydrogen-saturated deionized water was exposed to a slurry of palladium catalyst (1%) at room temperature, benzene formed via the intermediate chloro-benzene. The reaction rate decreased in the order of MCM-41 (mesoporous oxide having a silicon: aluminum ratio of 35) > alumina > Y (dealuminated zeolite having a silicon:aluminum ratio of 15). It appeared the reaction rate was directly proportional to the surface area of the support catalyst used (Schüth and Reinhard, 1997).

At influent concentrations of 1.0, 0.1, 0.01, and 0.001 mg/L, the granular activated carbon adsorption capacities at pH 5.5 were 129, 47, 17, and 64 mg/g, respectively (Dobbs and Cohen, 1980).

Exposure limits: NIOSH REL: ceiling 50 ppm (300 mg/m³), IDLH 200 ppm; OSHA PEL: ceiling 50 ppm; ACGIH TLV: TWA 25 ppm, ceiling 50 ppm (adopted).

Symptoms of exposure: Lacrimation, depression of central nervous system, anesthesia, and liver damage (Patnaik, 1992). An irritation concentration of 150.00 mg/m³ in air was reported by Ruth (1986).

Toxicity:

Concentrations that reduce the fertility of *Daphnia magna* in 2 wk for 50% (EC_{50}) and 16% (EC_{16}) of the population are 0.55 and 0.37 mg/L, respectively (Calamari et al., 1983).

EC_{50} (96-h) and EC_{50} (3-h) concentrations that inhibit the growth of 50% of *Selenastrum capricornutum* population are 2.2 and 10.0 mg/L, respectively (Calamari et al., 1983).

IC_{50} (24-h) for *Daphnia magna* 0.78 mg/L (Calamari et al., 1983).

LC_{50} (contact) for earthworm (*Eisenia fetida*) 21 μg/cm² (Neuhauser et al., 1985).

LC_{50} (14-d) for *Poecilia reticulata* 5.85 mg/L (Könemann, 1981).

LC_{50} (96-h) for bluegill sunfish 5.6 mg/L, (Spehar et al., 1982), fathead minnows 57 mg/L, grass shrimp 9.4 mg/L (Curtis et al., 1979), *Cyprinodon variegatus* 9.7 ppm using natural sea-water (Heitmuller et al., 1981).

LC_{50} (72-h) for *Cyprinodon variegatus* 9.7 ppm (Heitmuller et al., 1981).

LC_{50} (48-h) for red killifish 68 mg/L (Yoshioka et al., 1986), fathead minnows 76 mg/L, grass shrimp 10.3 mg/L (Curtis et al., 1979), *Daphnia magna* 2.4 mg/L (LeBlanc, 1980), *Salmo gairdneri* 2.3 mg/L, *Brachydanio rerio* 6.8 mg/L (Calamari et al., 1983), *Cyprinodon variegatus* 9.3 ppm (Heitmuller et al., 1981).

LC_{50} (24-h) for *Daphnia magna* 2.4 mg/L (LeBlanc, 1980), grass shrimp 14.3 mg/L (Curtis et al., 1979), fathead minnows 120.0 mg/L (Curtis et al., 1978), *Cyprinodon variegatus* 9.7-13 ppm (Heitmuller et al., 1981).

Acute oral LD_{50} for mice 4,386 mg/kg, rats 500 mg/kg, rabbits 500 mg/kg (quoted, RTECS, 1985).

Heitmuller et al. (1981) reported a NOEC of 9.7 ppm.

Drinking water standard (final): MCLG: 0.6 mg/L; MCL: 0.6 mg/L (U.S. EPA, 1996).

Uses: Preparation of 3,4-dichloroaniline; solvent for a wide variety of organic compounds and for oxides of nonferrous metals; solvent carrier in products of toluene diisocyanate; intermediate for dyes; fumigant; insecticide for termites; degreasing hides and wool; metal polishes; degreasing agent for metals, wood and leather; industrial air control; disinfectant; heat transfer medium.

1,3-DICHLOROBENZENE

Synonyms: 1,3-DCB; *m*-DCB; 1,3-Dichlorbenzene; *m*-Dichlorbenzene; 1,3-Dichlorbenzol; *m*-Dichlorbenzol; *m*-Dichlorobenzene; 1,3-Dichlorobenzol; *m*-Dichlorobenzol; RCRA waste number U071; UN 1591.

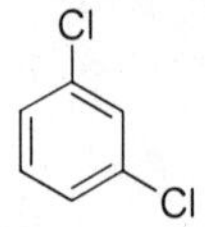

CAS Registry Number: 541-73-1
Molecular formula: $C_6H_4Cl_2$
Formula weight: 147.00
RTECS: CZ4499000
Merck reference: 10, 3039

Physical state and color:
Colorless liquid. The reported odor threshold concentrations ranged from 20 to 77 ppb (Keith and Walters, 1992; Young et al., 1996). The taste threshold concentration in water is 190 ppb (Young et al., 1996).

Melting point (°C):
-24.70 (Martin et al., 1979)

Boiling point (°C):
174 (Miller et al., 1984)
173 (Weast, 1986)
172 (Verschueren, 1983)

Density (g/cm³):
1.2881 at 20/4 °C, 1.2799 at 25/4 °C (Standen, 1964)

Diffusivity in water (x 10^{-5} cm²/sec):
0.82 at 20 °C using method of Hayduk and Laudie (1974)

Flash point (°C):
63 (Aldrich, 1988)

Lower explosive limit (%):
2.02 (estimated, Weiss, 1986)

Upper explosive limit (%):
9.2 (estimated, Weiss, 1986)

Entropy of fusion (cal/mol·K):
12.2 (Yalkowsky and Valvani, 1980)

Heat of fusion (kcal/mol):
3.021 (Weast, 1986)

Henry's law constant (x 10^{-3} atm·m^3/mol):
3.6 (Pankow and Rosen, 1988)
2.63 at 25 °C (Warner et al., 1987)
4.63 at 37 °C (Sato and Nakajima, 1979)
1.8 at 20 °C (Oliver, 1985)
2.94 at 25 °C (Hoff et al., 1993)
2.21, 2.31, 2.94, 2.85, and 4.22 at 10, 15, 20, 25, and 30 °C, respectively (Ashworth et al., 1988)
2.14 at 20 °C (Hovorka and Dohnal, 1997)

Ionization potential (eV):
9.12 (Franklin et al., 1969)

Bioconcentration factor, log BCF:
1.82 (bluegill sunfish) (Barrows et al., 1980; Veith et al., 1980)
2.62, 2.87 (rainbow trout, Oliver and Niimi, 1983)
1.99 (fathead minnow, Carlson and Kosian, 1987)
3.60 (Atlantic croakers), 3.86 (blue crabs), 3.25 (spotted sea trout), 3.40 (blue catfish) (Pereira et al., 1988)

Soil sorption coefficient, log K_{oc}:
2.23 (log K_{om} for a Woodburn silt loam soil, Chiou et al., 1983)
4.60 (Niagara River sediments, Oliver and Charlton, 1984)
K_d = 1.4 mL/g on a Cs$^+$-kaolinite (Haderlein and Schwarzenbach, 1993)

Octanol/water partition coefficient, log K_{ow}:
3.38 (Leo et al., 1971)
3.43 (Miller et al., 1985)
3.48 (Miller et al., 1984; Wasik et al., 1981)
3.44 (Banerjee, 1984)
3.53 (Watarai et al., 1982)
3.72 at 13 °C, 3.55 at 19 °C, 3.48 at 28 °C, 3.42 at 33 °C (Opperhuizen et al., 1988)
3.60 (Könemann et al., 1979)
3.525 (de Bruijn et al., 1989)
3.46 (Hammers et al., 1982)
3.50 (Pereira et al., 1988)

Solubility in organics:
Soluble in ethanol, acetone, ether, benzene, carbon tetrachloride, ligroin (U.S. EPA, 1985), and many organic solvents

Solubility in water:
143 mg/L at 25 °C (Banerjee, 1984)
133 mg/L at 25 °C (Banerjee et al., 1980)
0.847 mM at 25 °C (Miller et al., 1984)
0.01465 wt % at 23.5 °C (Schwarz, 1980)
In wt %: 0.0118 at 10.0 °C, 0.0101 at 20.0 °C, 0.0135 at 30.0 °C (Schwarz and Miller, 1980)
125.5 mg/L at 30 °C (McNally and Grob, 1983, 1984)
700 μmol/kg at 25.0 °C (Vesala, 1974)
In mg/L: 111 at 20 °C, 123 at 25 °C, 140 at 30 °C, 150 at 35 °C, 167 at 40 °C, 177 at 45 °C, 196 at 55 °C, 201 at 60 °C (Klemenc and Löw, 1930)

Vapor density:
6.01 g/L at 25 °C, 5.07 (air = 1)

Vapor pressure (mmHg at 25 °C):
1.9 (Warner et al., 1987)
2.3 (Mackay et al., 1982)
2.15 (Banerjee et al., 1990)

Environmental fate:
Biological. When 1,3-dichlorobenzene was statically incubated in the dark at 25 °C with yeast extract and settled domestic wastewater inoculum, significant biodegradation with gradual acclimation was followed by a deadaptive process in subsequent subcultures. At a concentration of 5 mg/L, 59, 69, 39, and 35% losses were observed after 7, 14, 21, and 28-d incubation periods, respectively. At a concentration of 10 mg/L, percent losses were virtually unchanged. After 7, 14, 21, and 28-d incubation periods, percent losses were 58, 67, 31, and 33, respectively (Tabak et al., 1981).

Photolytic. The sunlight irradiation of 1,3-dichlorobenzene (20 g) in a 100-mL borosilicate glass-stoppered Erlenmeyer flask for 56 d yielded 520 ppm trichlorobiphenyl (Uyeta et al., 1976).

When an aqueous solution containing 1,3-dichlorobenzene (190 μM) and a non-ionic surfactant micelle (Brij 58, a polyoxyethylene cetyl ether) was illuminated by a photoreactor equipped with 253.7-nm monochromatic UV lamps, photoisomerization took place yielding 1,2- and 1,4-dichlorobenzene as the principal products. The half-life for this reaction, based on the first-order photodecomposition rate of 1.40 x 10^{-3}/sec, is 8.3 min (Chu and Jafvert, 1994).

Chemical/Physical. Anticipated products from the reaction of 1,3-dichlorobenzene with atomspheric ozone or OH radicals are chlorinated phenols, ring cleavage products and nitro compounds (Cupitt, 1980). Based on an assumed base-mediated 1% disappearance after 16 d at 85 °C and pH 9.70 (pH 11.26 at 25 °C), the hydrolysis half-life was estimated to be >900 yr (Ellington et al., 1988). 1,3-Dichlorobenzene (0.17-0.23 mM) reacted with OH radicals in water (pH 8.7) at a rate of 5.0 x 10^9/M·sec (Haag and Yao, 1992).

At influent concentrations of 1.0, 0.1, 0.01, and 0.001 mg/L, the granular activated carbon adsorption capacities at pH 5.1 were 118, 42, 15, and 5.1 mg/g, respectively (Dobbs and Cohen, 1980).

Toxicity:
LC_{50} (14-d) for *Poecilia reticulata* 7.4 mg/L (Könemann, 1981).

LC_{50} (96-h) for fathead minnows 7.8 mg/L (Veith et al., 1983), bluegill sunfish 5.0 mg/L (Spehar et al., 1982), *Cyprinodon variegatus* 7.8 ppm using natural seawater (Heitmuller et al., 1981).

LC_{50} (72-h) for *Cyprinodon variegatus* 8.0 ppm (Heitmuller et al., 1981).

LC_{50} (48-h) for *Daphnia magna* 11 mg/L (LeBlanc, 1980), *Cyprinodon variegatus* 8.0 ppm (Heitmuller et al., 1981).

LC_{50} (24-h) for *Daphnia magna* 42 mg/L (LeBlanc, 1980), *Cyprinodon variegatus* 8.5 ppm (Heitmuller et al., 1981).

Heitmuller et al. (1981) reported a NOEC of 4.2 ppm.

Drinking water standard: No MCLGs or MCLs have been proposed (U.S. EPA, 1996).

Uses: Fumigant and insecticide; organic synthesis.

1,4-DICHLOROBENZENE

Synonyms: 4-Chlorophenyl chloride; *p*-Chlorophenyl chloride; 1,4-DCB; *p*-DCB; Dichloricide; 4-Dichlorobenzene; *p*-Dichlorobenzene; 4-Dichlorobenzol; *p*-Dichlorobenzol; Evola; NCI-C54955; Paracide; Para crystals; Paradi; Paradichlorobenzene; Paradichlorobenzol; Paradow; Paramoth; Paranuggetts; Parazene; Parodi; PDB; PDCB; Persia-Perazol; RCRA waste number U072; Santochlor; UN 1592.

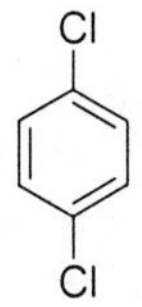

CAS Registry Number: 106-46-7
DOT: 1592
Molecular formula: $C_6H_4Cl_2$
Formula weight: 147.00
RTECS: CZ4550000
Merck reference: 10, 3041

Physical state, color, and odor:
Colorless to white crystals with a penetrating, mothball-like odor. Odor threshold concentrations ranged from 4.5 ppb to 30 ppm (Keith and Walters, 1992; Young et al., 1996). The taste threshold concentration in water is 11 ppb (Young et al., 1996).

Melting point (°C):
53.10 (Martin et al., 1979)

Boiling point (°C):
174.4 (Dean, 1973)

Density (g/cm³):
1.2475 at 20/4 °C (Weast, 1986)

Diffusivity in water (x 10⁻⁶ cm²/sec, 25 °C):
9.9 and 9.3 at mole fractions of 2 x 10^{-7} and 4 x 10^{-7}, respectively (Gabler et al., 1996)

Flash point (°C):
65.6 (NIOSH, 1997)

Lower explosive limit (%):
2.5 (NIOSH, 1997)

Entropy of fusion (cal/mol·K):
13.3 (Ueberreiter and Orthman, 1950)
13.4 (Yalkowsky and Valvani, 1980)

Heat of fusion (kcal/mol):
4.34 (Wauchope and Getzen, 1972)
4.54 (Miller et al., 1984)

Henry's law constant (x 10^{-3} atm·m³/mol):
3.1 (Pankow and Rosen, 1988)
2.72 at 25 °C (Warner et al., 1987)
1.5 at 20 °C (Oliver, 1985)
2.12, 2.17, 2.59, 3.17, and 3.89 at 10, 15, 20, 25, and 30 °C, respectively (Ashworth et al., 1988)
1.86 at 20 °C (Hovorka and Dohnal, 1997)
2.41 at 25 °C (Shiu and Mackay, 1997)

Ionization potential (eV):
8.95 (Horvath, 1982)
9.07 (Yoshida et al., 1983a)

Bioconcentration factor, log BCF:
2.71, 2.86, 2.95 (rainbow trout, Oliver and Niimi, 1985)
2.00, 1.70, and 2.75 for algae, fish, and activated sludge, respectively (Freitag et al., 1985)
1.78 (bluegill sunfish, Veith et al., 1980)
2.33 determined for (rainbow trout during a 4-d exposure under static test conditions (Neely et al., 1974)
3.91 (Atlantic croakers), 4.53 (blue crabs), 4.09 (spotted sea trout), 3.51 (blue catfish) (Pereira et al., 1988)
2.47 (*Jordanella floridae*, Devillers et al., 1996)
2.04 (fathead minnow, Carlson and Kosian, 1987)
3.26 (guppy, Könemann and van Leeuwen, 1980)
1.89 (wet weight based) and -0.60 (lipid based) for *Gambusia affinis* (Chaisuksant et al., 1997)
2.90, 3.36, 3.43, 3.60, 3.73, and 3.79 for olive, holly, grass, ivy, mock orange, and pine leaves, respectively (Hiatt, 1998)

Soil sorption coefficient, log K_{oc}:
2.44 (Woodburn silt loam, Chiou et al., 1983)
2.48 (Garyling soil), 2.85 (average using five soils) (Hutzler et al., 1983)
2.60 (Lincoln sand) (Wilson et al., 1981)
2.63 (Friesel et al., 1984)
2.78, 2.87, 3.14 (Schwarzenbach and Westall, 1981)
2.92 (humic acid polymers) (Chin and Weber, 1989)
2.96 (Eerd soil) (Loch et al., 1986)
2.82 (Apison soil), 2.93 (Fullerton soil), 2.45 (Dormont soil) (Southworth and Keller, 1986)
3.29, 3.38, 3.53 (glaciofluvial, sandy aquifer, Nielsen et al., 1996)
2.57 (sandy soil, Van Gestel and Ma, 1993)
Average K_d values for sorption of 1,4-dichlorobenzene to corundum (α-Al_2O_3) and hematite (α-Fe_2O_3) were 0.00451 and 0.0105 mL/g, respectively (Mader et al., 1997)

Octanol/water partition coefficient, log K_{ow}:
3.39 (Leo et al., 1971)
3.37 (Banerjee et al., 1980; Garst and Wilson, 1984; Wasik et al., 1981)
3.38 (Chiou et al., 1977)
3.53 (Mackay, 1982)
3.62 (Könemann et al., 1979)
3.444 (Brooke et al., 1990; de Bruijn et al., 1989)
3.355 (Brooke et al., 1990)
3.41 (Hammers et al., 1982)

3.40 (Campbell and Luthy, 1985)
3.47 (Pereira et al., 1988)
3.42, 3.35, 3.23, 3.12, and 3.03 at 5, 15, 25, 35, and 45 °C, respectively (Bahadur et al., 1997)
3.52 (quoted, Howard et al., 1989)

Solubility in organics:
Soluble in ethanol, acetone, ether, benzene, carbon tetrachloride, ligroin (U.S. EPA, 1985),
carbon disulfide, and chloroform (Windholz et al., 1983)

Solubility in water:
65.3 mg/L at 25 °C (Banerjee, 1984)
74 mg/L at 25 °C (Banerjee et al., 1980)
77 mg/kg at 30 °C (Gross and Saylor, 1931)
76 mg/L at 25 °C, 79.1 mg/kg at 25 °C (Andrews and Keefer, 1950)
87.15 mg/L at 25 °C (Aquan-Yuen et al., 1979)
80 mg/L at 25 °C (Gunther et al., 1968)
90.6 mg/L at 25 °C (Yalkowsky et al., 1979)
0.21 mM at 25 °C (Miller et al., 1984)
94.4 mg/L at 30 °C (McNally and Grob, 1984)
92.13 mg/L at 30 °C (McNally and Grob, 1983)
In mg/kg: 77.8 at 22.2 °C, 83.4 at 24.6 °C, 86.9 at 25.5 °C, 92.6 at 30.0 °C, 102 at 34.5 °C,
    121 at 38.4 °C, 159 at 47.5 °C, 173 at 50.1 °C, 210 at 59.2 °C, 218 at 60.7 °C, 230 at 65.1
    °C , 237 at 65.2 °C, 281 at 73.4 °C (Wauchope and Getzen, 1972)
581.6 μmol/kg at 25.0 °C (Vesala, 1974)
48.6, 63.0, 81.4, 104.5, and 130 mg/L at 5, 15, 25, 35, and 45 °C, respectively (Shiu et al.,
    1997)
156 and 163 mg/L at 55 and 65 °C, respectively (Klemenc and Löw, 1930)

Vapor pressure (mmHg at 25 °C):
0.4 (Standen, 1964)
0.7 (Mackay et al., 1982)

Environmental fate:
   Biological. In activated sludge, <0.1% degraded (mineralized) to carbon dioxide after 5 d
(Freitag et al., 1985). When 1,4-dichlorobenzene was statically incubated in the dark at 25 °C
with yeast extract and settled domestic wastewater inoculum, significant biodegradation with
gradual acclimation was followed by a deadaptive process in subsequent subcultures. At a
concentration of 5 mg/L, 55, 61, 34, and 16% losses were observed after 7, 14, 21, and 28-d
incubation periods, respectively. At a concentration of 10 mg/L, only 37, 54, 29, and 0% losses
were observed after 7, 14, 21, and 28-d incubation periods, respectively (Tabak et al., 1981).
   Surface Water. Estimated half-lives of 1,4-dichlorobenzene (1.5 μg/L) from an experimental
marine mesocosm during the spring (8-16 °C), summer (20-22 °C), and winter (3-7 °C) were
18, 10, and 13 d, respectively (Wakeham et al., 1983).
   Groundwater. Nielsen et al. (1996) studied the degradation of 1,4-dichlorobenzene in a
shallow, glaciofluvial, unconfined sandy aquifer in Jutland, Denmark. As part of the *in situ*
microcosm study, a cylinder that was open at the bottom and screened at the top was installed
through a cased borehole approximately 5 m below grade. Five liters of water was aerated with
atmospheric air to ensure aerobic conditions were maintained. Groundwater was analyzed
weekly for approximately 3 months to determine 1,4-dichlorobenzene concentrations over time.
The experimentally determined first-order biodegradation rate constant and corresponding half-
life following a 22-d lag phase were 0.05/d and 13.86 d, respectively.

Under anaerobic conditions, 1,4-dichlorobenzene in a plume of contaminated groundwater was found to be very persistent (Barber, 1988).

Photolytic. Under artificial sunlight, river water containing 2-5 ppm of 1,4-dichlorobenzene photodegraded to chlorophenol and phenol (Mansour et al., 1989). A carbon dioxide yield of 5.1% was achieved when 1,4-dichlorobenzene adsorbed on silica gel was irradiated with light (λ >290 nm) for 17 h (Freitag et al., 1985). Irradiation of 1,4-dichlorophenol in air containing nitrogen oxides gave 2,5-dichloro-6-phenol (major product), 2,5-dichloronitrobenzene, 2,5-dichlorophenol and 2,5-dichloro-4-nitrophenol (Nojima and Kanno, 1980). The sunlight irradiation of 1,4-dichlorobenzene (20 g) in a 100-mL borosilicate glass-stoppered Erlenmeyer flask for 56 d yielded 1,860 ppm 4,2′,5′-trichlorobiphenyl (Uyeta et al., 1976).

When an aqueous solution containing 1,2-dichlorobenzene (190 μM) and a nonionic surfactant micelle (Brij 58, a polyoxyethylene cetyl ether) was illuminated by a photoreactor equipped with 253.7-nm monochromatic UV lamps, photoisomerization took place, yielding 1,3- and 1,4-dichlorobenzene as the principal products. The half-life for this reaction, based on the first-order photodecomposition rate of 1.34 x 10^{-3}/sec, is 8.6 min (Chu and Jafvert, 1994). A room temperature rate constant of 3.2 x 10^{-13} cm^3/molecule·sec was reported for the vapor-phase reaction of 1,4-dichlorobenzene with OH radicals (Atkinson, 1985).

Chemical/Physical. Anticipated products from the reaction of 1,4-dichlorobenzene with ozone or OH radicals in the atmosphere are chlorinated phenols, ring cleavage products, and nitro compounds (Cupitt, 1980).

Based on an assumed base-mediated 1% disappearance after 16 d at 85 °C and pH 9.70 (pH 11.26 at 25 °C), the hydrolysis half-life was estimated to be more than 900 yr (Ellington et al., 1988).

At influent concentrations of 1.0, 0.1, 0.01, and 0.001 mg/L, granular activated carbon adsorption capacities at pH 5.1 were 121, 41, 14, and 4.6 mg/g, respectively (Dobbs and Cohen, 1980).

Exposure limits: Potential occupational carcinogen. NIOSH REL: IDLH 150 ppm; OSHA PEL: TWA 75 ppm (450 mg/m^3); ACGIH TLV: TWA 10 ppm (adopted).

Symptoms of exposure: Repeated inhalation of high concentrations of vapors may cause head-ache, weakness, dizziness, nausea, vomiting, diarrhea, loss of weight, and injury to kidney and liver (Patnaik, 1992). An irritation concentration of 240.00 mg/m^3 in air was reported by Ruth (1986).

Toxicity:

Concentrations that reduce the fertility of *Daphnia magna* in 2 wk for 50% (EC_{50}) and 16% (EC_{16}) of the population are 0.93 and 0.64 mg/L, respectively (Calamari et al., 1983).

EC_{50} (96-h) and EC_{50} (3-h) concentrations that inhibit the growth of 50% of *Selenastrum capricornutum* population are 1.6 and 5.2 mg/L, respectively (Calamari et al., 1983).

IC_{50} (24-h) for *Daphnia magna* 1.6 mg/L (Calamari et al., 1983).

LC_{50} (14-d) for *Poecilia reticulata* 4 mg/L (Könemann, 1981).

LC_{50} (96-h) for fathead minnows 4.0 mg/L (Veith et al., 1983), bluegill sunfish (*Lepomis machrochirus*) 4.3 mg/L, fathead minnows 30 mg/L (Spehar et al., 1982) and 34.5 mg/L (Curtis et al., 1978), *Cyprinodon variegatus* 7.4 ppm using natural seawater (Heitmuller et al., 1981).

LC_{50}: 17.8 and 51 mg/L (soil porewater concentration) for the earthworm *Eisenia andrei* and 26 and 229 mg/L (soil porewater concentration) for the earthworm *Lumbricus rubellus* (Van Gestel and Ma, 1993).

LC_{50} (72-h) for *Cyprinodon variegatus* 7.4 ppm (Heitmuller et al., 1981).

LC_{50} (48-h) for fathead minnows 35.4 mg/L, grass shrimp 129 mg/L (Curtis et al., 1979), *Daphnia magna* 11 mg/L (LeBlanc, 1980), 2.2 mg/L (Canton et al., 1985), *Salmo gairdneri*

1.18 mg/L, *Brachydanio rerio* 4.25 mg/L (Calamari et al., 1983), *Cyprinodon variegatus* 7.2 ppm (Heitmuller et al., 1981).

LC$_{50}$ (24-h) for *Daphnia magna* 42 mg/L (LeBlanc, 1980), fathead minnows 35.4 mg/L (Curtis et al., 1979), fathead minnows 34.0 mg/L (Curtis et al., 1978), *Cyprinodon variegatus* 7.5-10 ppm (Heitmuller et al., 1981).

Acute oral LD$_{50}$ for mice 2,950 mg/kg, rats 500 mg/kg, rabbits 2,830 mg/kg (quoted, RTECS, 1985).

Heitmuller et al. (1981) reported a NOEC of 5.6 ppm.

Drinking water standard (final): MCLG: 75 μg/L; MCL: 75 μg/L (U.S. EPA, 1996).

Uses: Moth and bird repellent; general insecticide, fumigant and germicide; space odorant; manufacture of 2,5-dichloroaniline and dyes; pharmacy; agriculture (fumigating soil); disinfectant, urinal deodorizer, air freshener, and chemical intermediate in the manufacture of 1,2,4-trichlorobenzene and polyphenylene sulfide.

3,3'-DICHLOROBENZIDINE

Synonyms: C.I. 23060; Curithane C126; DCB; 4,4'-Diamino-3,3'-dichlorobiphenyl; 4,4'-Di-amino-3,3'-dichlorodiphenyl; Dichlorobenzidine; Dichlorobenzidine base; *m,m'*-Dichlorobenzi-dine; **3,3'-Dichloro-1,1'-(biphenyl)-4,4'-diamine**; 3,3'-Dichlorobiphenyl-4,4'-diamine; 3,3'-Dichloro-4,4'-biphenyldiamine; 3,3'-Dichloro-4,4'-diaminobiphenyl; 3,3'-Dichloro-4,4'-di-amino-(1,1-biphenyl); RCRA waste number U073.

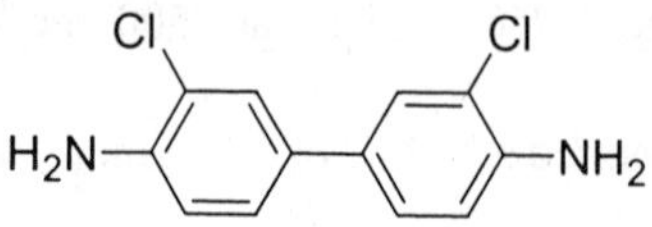

Note: Normally found as the dihydrochloride salt, 3,3'-dichlorobenzidine dihydrochloride ($C_{12}H_{10}Cl_2N_2 \cdot HCl$).

CAS Registry Number: 91-94-1
Molecular formula: $C_{12}H_{10}Cl_2N_2$
Formula weight: 253.13
RTECS: DD0525000
Merck reference: 10, 3043

Physical state, color, and odor:
Colorless to grayish-purple crystals with a mild odor

Melting point (°C):
132-133 (Windholz et al., 1983)

Boiling point (°C):
420 (NIOSH, 1997)

Diffusivity in water (x 10^{-5} cm^2/sec):
0.51 at 20 °C using method of Hayduk and Laudie (1974)

Dissociation constant, pK_a:
5.4, 3.3 (Korenman and Nikolaev, 1974)

Henry's law constant (x 10^{-8} atm·m^3/mol):
4.5 at 25 °C (estimated, Howard, 1989)

Bioconcentration factor, log BCF:
3.49 (activated sludge), 2.97 (algae), 2.79 (golden orfe) (Freitag et al., 1985)
2.70 (bluegill sunfish, Appleton and Sikka, 1980)

Soil sorption coefficient, log K_{oc}:
4.35 (Meylan et al., 1992)

Octanol/water partition coefficient, log K_{ow}:
3.51 (Banerjee et al., 1980)

Solubility in organics:
Soluble in ethanol, benzene, and glacial acetic acid (Windholz et al., 1983)

Solubility in water:
3.11 mg/L at 25 °C (Banerjee et al., 1980)
4.0 mg/L at 22 °C (dihydrochloride, U.S. EPA, 1980a)
3.99 ppm at pH 6.9 (Appleton and Sikka, 1980)

Vapor pressure (mmHg):
10^{-5} at 22 °C (assigned by analogy, Mabey et al., 1982)
4.2 x 10^{-7} at 25 °C (estimated, Howard, 1989)

Environmental fate:
  Biological. In activated sludge, 2.7% mineralized to carbon dioxide after 5 d (Freitag et al., 1985). Sikka et al. (1978) reported 3,3′-dichlorobenzidine is resistant to degradation by indigenous aquatic microbial communities over a 4-wk period. Under aerobic and anaerobic conditions, 3,3′-dichlorobenzidine is mineralized very slowly (Boyd et al., 1984; Chung and Boyd, 1987).
  Photolytic. An aqueous solution subjected to UV radiation caused a rapid degradation (half-life <10 min) to monochlorobenzidine, benzidine, and several unidentified, brightly-colored, water-insoluble chromophores (Banerjee et al., 1978). In a similar experiment, 3,3′-dichloro-benzidine in an aqueous solution was subjected to radiation at λ=310 nm for approximately 15 min. During the period of irradiation, concentrations of 3,3′-dichlorobenzidine decreased rapidly over the period of irradiation. 3-Chlorobenzidine formed as a transient intermediate which underwent dechlorination forming a benzidine, a stable photoproduct. Depending upon the wavelength used, the benzidine yields ranged from 8 to 12% of the total 3-chlorobenzidine transformed (Nyman et al., 1997). A carbon dioxide yield of 41.2% was achieved when 3,3′-dichlorobenzidine adsorbed on silica gel was irradiated with light (λ >290 nm) for 17 h (Freitag et al., 1985).
  Chemical/Physical. 3,3′-Dichlorobenzidine will not hydrolyze to any reasonable extent (Kollig, 1993).

Exposure limits: Potential occupational carcinogen. Because no standards have been established, NIOSH (1997) recommends the most reliable and protective respirators be used, i.e., a self-contained breathing apparatus that has a full facepiece and is operated under positive-pressure or a supplied-air respirator that has a full facepiece and is operated under pressure-demand or under positive-pressure in combination with a self-contained breathing apparatus operated under pressure-demand or positive-pressure.
  OSHA recommends that worker exposure to this chemical is to be controlled by use of engineering control, proper work practices, and proper selection of personal protective equipment. Specific details of these requirements can be found in CFR 1910.1003-1910.1016.

Symptoms of exposure: May cause irritation of eyes, nose, throat, and skin (Patnaik, 1992)

Toxicity:
  LC_{50} (48-h) for bluegill sunfish 2 mg/L (Sikka et al., 1978).
  Acute oral LD_{50} for rats as free base or dihydrochloride are 7,070 and 3,820, respectively (Gerade and Gerade, 1974).

Uses: Intermediate in the manufacture of azo dyes and pigments for printing inks, textiles, paints, plastics, and crayons; curing agent for isocyanate-terminated polymers and resins; rubber compounding ingredient; analytical determination of gold; formerly used as chemical intermediate for direct red 61 dye.

DICHLORODIFLUOROMETHANE

Synonyms: Algofrene type 2; Arcton 6; Difluorodichloromethane; Electro-CF 12; Eskimon 12; F 12; FC 12; Fluorocarbon 12; Freon 12; Freon F-12; Frigen 12; Genetron 12; Halon; Halon 122; Isceon 122; Isotron 2; Kaiser chemicals 12; Ledon 12; Propellant 12; R 12; RCRA waste number U075; Refrigerant 12; Ucon 12; Ucon 12/halocarbon 12; UN 1028.

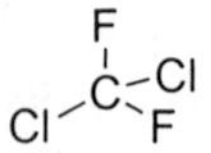

$$Cl-\underset{\underset{F}{|}}{\overset{\overset{F}{|}}{C}}-Cl$$

CAS Registry Number: 75-71-8
DOT: 1028
DOT label: Nonflammable gas
Molecular formula: CCl_2F_2
Formula weight: 120.91
RTECS: PA8200000
Merck reference: 10, 3048

Physical state, color, and odor:
Colorless gas with an ethereal odor

Melting point (°C):
-158 (Weast, 1986)

Boiling point (°C):
-29.8 (Weast, 1986)

Density (g/cm³):
1.329 at 20/4 °C (Verschueren, 1983)
1.311 at 25/4 °C (Horvath, 1982)

Diffusivity in water (x 10^{-5} cm²/sec):
0.93 at 20 °C using method of Hayduk and Laudie (1974)

Flash point (°C):
Nonflammable (Weiss, 1986)

Lower explosive limit (%):
Nonflammable (Weiss, 1986)

Upper explosive limit (%):
Nonflammable (Weiss, 1986)

Henry's law constant (x 10 atm·m³/mol):
4.0 at 20 °C (Pearson and McConnell, 1975)
1.72, 2.63, and 3.91 at 10, 20, and 30 °C, respectively (Munz and Roberts, 1987)

Ionization potential (eV):
11.75 (NIOSH, 1997)
12.31 ± 0.05 (Franklin et al., 1969)

Soil sorption coefficient, log K_{oc}:
2.56 using method of Kenaga and Goring (1980)

Octanol/water partition coefficient, log K_{ow}:
2.16 (Hansch et al., 1975)

Solubility in organics:
Soluble in acetic acid, acetone, chloroform, ether (Weast, 1986), and ethanol (ITII, 1986)

Solubility in water (mg/L at 25 °C):
280 (Pearson and McConnell, 1975)
301 (quoted, Munz and Roberts, 1987)

Vapor density:
4.94 g/L at 25 °C, 4.17 (air = 1)

Vapor pressure (mmHg):
4,250 at 20 °C, 5,776 at 30 °C (Verschueren, 1983)
4,870 at 25 °C (Jordan, 1954)
4,306 at 20 °C (McConnell et al., 1975)

Environmental fate:
Surface Water. Estimated half-lives of dichlorodifluoromethane from an experimental marine mesocosm during the spring (8-16 °C) and winter (3-7 °C) were 20 and 13 d, respectively (Wakeham et al., 1983).

Exposure limits: NIOSH REL: TWA 1,000 ppm (4,950 mg/m³), IDLH 15,000 ppm; OSHA PEL: TWA 1,000 ppm; ACGIH TLV: TWA 1,000 ppm (adopted).

Toxicity:
LC_{50} (inhalation) for guinea pigs 80 pph/30-min, mice 76 pph/30-min, rats, 80 pph/30-min, rabbits 80 pph/30-min (quoted, RTECS, 1985).
LD_{50} for rats >1 g/kg (quoted, Verschueren, 1983).

Drinking water standard: No MCLGs or MCLs have been proposed although dichlorodifluoromethane has been listed for regulation (U.S. EPA, 1996).

Uses: Refrigerant; aerosol propellant; plastics; blowing agent; low temperature solvent; chilling cocktail glasses; freezing foods by direct contact; leak-detecting agent.

1,3-DICHLORO-5,5-DIMETHYLHYDANTOIN

Synonyms: Dactin; DCA; DDH; **1,3-Dichloro-5,5-dimethyl-2,4-imidazolidinedione**; Dichlorodimethylhydantoin; Halane; NCI-C03054; Omchlor.

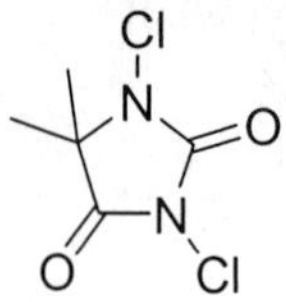

CAS Registry Number: 118-52-5
Molecular formula: $C_5H_6Cl_2N_2O_2$
Formula weight: 197.03
RTECS: MU0700000
Merck reference: 10, 3049

Physical state, color, and odor:
White powder or solid with a chlorine-like odor

Melting point (°C):
132 (NIOSH, 1997)

Boiling point:
Sublimes at 100 °C (Keith and Walters, 1992)

Density (g/cm³):
1.5 at 20/20 °C (Windholz et al., 1983)

Diffusivity in water:
Not applicable - reacts with water

Flash point (°C):
174.6 (NIOSH, 1997)

Henry's law constant (atm·m³/mol):
Not applicable - reacts with water

Soil sorption coefficient, log K_{oc}:
Not applicable - reacts with water

Octanol/water partition coefficient, log K_{ow}:
Not applicable - reacts with water

Solubility in organics (wt % at 25 °C):
Benzene (9.2), carbon tetrachloride (12.5), chloroform (14), 1,2-dichloroethane (32.0), methylene chloride (30.0), 1,1,2,2-tetrachloroethane (17.0) (Windholz et al., 1983).

Solubility in water (wt %):
0.21 at 25 °C, 0.60 at 60 °C (quoted, Windholz et al., 1983)

Environmental fate:
Chemical/Physical. Reacts with water (pH 7.0) releasing hypochlorous acid. At pH 9, nitrogen chloride is formed (Windholz et al., 1983).

Exposure limits (mg/m^3): NIOSH REL: TWA 0.2, STEL 0.4, IDLH 5; OSHA PEL: TWA 0.2, STEL/C 0.4 (adopted).

Uses: Chlorinating agent; industrial deodorant, disinfectant; intermediate for amino acids, drugs and insecticides; polymerization catalyst; stabilizer for vinyl chloride polymers; household laundry bleach; water treatment; organic synthesis.

1,1-DICHLOROETHANE

Synonyms: Chlorinated hydrochloric ether; 1,1-Dichlorethane; *asym*-Dichloroethane; Ethylidene chloride; Ethylidene dichloride; 1,1-Ethylidene dichloride; NCI-C04535; RCRA waste number U076; UN 2362.

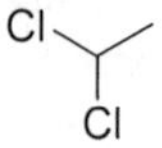

CAS Registry Number: 75-34-3
DOT: 2362
DOT label: Flammable liquid
Molecular formula: $C_2H_4Cl_2$
Formula weight: 98.96
RTECS: KI0175000
Merck reference: 10, 3756

Physical state, color, and odor:
Clear, colorless, oily liquid with a chloroform-like odor

Melting point (°C):
-97.4 (Dean, 1973)

Boiling point (°C):
57.3 (Weast, 1986)

Density (g/cm³):
1.1757 at 20/4 °C (Weast, 1986)
1.1830 at 15/4 °C, 1.60010 at 30/4 °C (Standen, 1964)

Diffusivity in water (x 10^{-5} cm²/sec):
0.98 at 20 °C using method of Hayduk and Laudie (1974)

Flash point (°C):
-16.7 (NIOSH, 1997)
13.3 (McGovern, 1943)

Lower explosive limit (%):
5.4 (NIOSH, 1997)

Upper explosive limit (%):
11.4 (NIOSH, 1997)

Heat of fusion (kcal/mol):
1.881 (Riddick et al., 1986)

Henry's law constant (x 10^{-3} atm·m³/mol):
4.3 (Pankow and Rosen, 1988)
5.45 at 25 °C (Warner et al., 1987)
5.56 at 25 °C (Gossett, 1987)
9.43 at 37 °C (Sato and Nakajima, 1979)

5.88 at 25 °C (Wright et al., 1992)

7.76 at 30 °C (Jeffers et al., 1989)

3.68, 4.54, 5.63, 6.25, and 7.76 at 10, 15, 20, 25, and 30 °C, respectively (Ashworth et al., 1988)

4.6, 7.0, and 10.2 at 20, 30, and 40 °C, respectively (Tse et al., 1992)

Distilled water:1.63, 2.06, 2.03, 3.75, and 5.05 at 2.0, 6.0, 10.0, 18.2, and 25.0 °C, respectively; natural seawater: 2.42 and 5.80 at 6.0 and 25.0 °C, respectively (Dewulf et al., 1995)

4.60 at 20 °C (Hovorka and Dohnal, 1997)

1.81 and 4.87 at 5.7 and 24.9 °C, respectively (Dewulf et al., 1999a)

Ionization potential (eV):
11.06 (NIOSH, 1997)

Soil sorption coefficient, log K_{oc}:
1.46 (Ellington et al., 1993)
1.66 (Jury et al., 1990)
1.79 (Roy et al., 1987)
1.43, 1.46, 1.43, 1.48, 1.50, 1.49 and 1.55 at 2.3, 3.8, 6.2, 8.0, 13.5, 18.6, at 25.0 °C, respectively, for a Leie River (Belgium) clay (Dewulf et al., 1999a)

Octanol/water partition coefficient, log K_{ow}:
1.82, 1.66, and 1.68 at 25, 35, and 50 °C, respectively (Bhatia and Sandler, 1995)
1.70, 1.69, 1.74, 1.73, 1.78, and 1.75 at 2.2, 6.0, 10.0, 14.1, 18.7, and 24.8 °C, respectively (Dewulf et al., 1999a)
1.79 (Hansch and Leo, 1979)

Solubility in organics:
Miscible with ethanol (U.S. EPA, 1985)

Solubility in water:
6,560, 5.590, 5,500, and 5,400 mg/kg at 0, 10, 20, and 30 °C, respectively (Rex, 1906)
5,060 mg/kg at 25 °C (Gross, 1929)
4,589 mg/L at 30 °C (McNally and Grob, 1984)
4,834.4 mg/L at 30 °C (McNally and Grob, 1983)

Vapor density:
4.04 g/L at 25 °C, 3.42 (air = 1)

Vapor pressure (mmHg):
234 at 25 °C, 270 at 30 °C (Verschueren, 1983)
227 at 25 °C (Howard, 1990)

Environmental fate:
Biological. 1,1-Dichloroethane showed significant degradation with gradual adaptation in a static-culture flask-screening test (settled domestic wastewater inoculum) conducted at 25 °C. At concentrations of 5 and 10 mg/L, percent losses after 4 wk of incubation were 91 and 83, respectively. At a substrate concentration of 5 mg/L, 19% was lost due to volatilization after 10 d (Tabak et al., 1981). Under anoxic conditions, indigenous microbes in uncontaminated sediments produced vinyl chloride (Barrio-Lage et al., 1986).

Surface Water. The following volatilization half-lives were reported for 1,1-dichloroethane: 6-9 d in a pond, 5-8 d in a lake, 24-32 h in a river (Smith et al., 1980).

Photolytic. Titanium dioxide suspended in an aqueous solution and irradiated with UV light (λ = 365 nm) converted 1,1-dichloroethane to carbon dioxide at a significant rate (Matthews, 1986). The initial photodissociation product of 1,1-dichloroethane was reported to be chloro-acetyl chloride (U.S. EPA, 1975). This compound is readily hydrolyzed to hydrochloric acid and chloroacetic acid (Morrison and Boyd, 1971).

The rate constant for the reaction of 1,1-dichloroethane and OH radicals in the atmosphere at 300 K is 1.6 x 10^{11} cm^3/mol·sec (Hendry and Kenley, 1979). At 296 K, a photooxidation rate constant of 2.6 x 10^{-13} cm^3/molecule·sec was reported for the reaction with OH radicals resulting in a half-life of 1.5 months (Howard and Evenson, 1976).

Chemical/Physical. A glass bulb containing air and 1,1-dichloroethane degraded outdoors to carbon dioxide and hydrochloric acid. The half-life for this reaction was 17 wk (Pearson and McConnell, 1975). Hydrolysis of 1,1-dichloroethane under alkaline conditions yielded vinyl chloride, acetaldehyde, and hydrochloric acid (Kollig, 1993). The reported hydrolysis half-life at 25 °C and pH 7 is 61.3 yr (Jeffers et al., 1989).

The evaporation half-life of 1,1-dichloroethane (1 mg/L) from water at 25 °C using a shallow-pitch propeller stirrer at 200 rpm at an average depth of 6.5 cm is 32.2 min (Dilling, 1977).

At influent concentrations of 1.0, 0.1, 0.01, and 0.001 mg/L, the granular activated carbon adsorption capacities at pH 5.3 were 1.8, 0.52, 0.15, and 0.04 mg/g, respectively (Dobbs and Cohen, 1980).

Exposure limits: NIOSH REL: TWA 100 ppm (400 mg/m^3), IDLH 3,000 ppm; OSHA PEL: TWA 100 ppm; ACGIH TLV: TWA 100 ppm (adopted).

Symptoms of exposure: May cause irritation of eyes, nose, throat, and skin (Patnaik, 1992)

Toxicity:

LC$_{50}$ (7-d) for *Poecilia reticulata* 202 mg/L (Könemann, 1981).

LC$_{50}$ (96-h) for bluegill sunfish 550 mg/L, *Menidia beryllina* 480 mg/L (quoted, Verschueren, 1983).

Acute oral LD$_{50}$ for rats 725 mg/kg (quoted, RTECS, 1985).

Drinking water standard: No MCLGs or MCLs have been proposed (U.S. EPA, 1996).

Uses: Extraction solvent; insecticide and fumigant; preparation of vinyl chloride; paint, varnish and finish removers; degreasing and drying metal parts; ore flotation; solvent for plastics, oils, and fats; chemical intermediate for 1,1,1-trichloroethane; in rubber cementing, fabric spreading, and fire extinguishers; formerly used as an anesthetic; organic synthesis.

1,2-DICHLOROETHANE

Synonyms: 1,2-Bichloroethane; Borer sol; Brocide; 1,2-DCA; 1,2-DCE; Destruxol borer-sol; Dichloremulsion; 1,2-Dichlorethane; Dichlormulsion; α,β-Dichloroethane; *sym*-Dichloroethane; Dichloroethylene; Dutch liquid; Dutch oil; EDC; ENT 1656; Ethane dichloride; Ethene dichloride; Ethylene chloride; Ethylene dichloride; 1,2-Ethylene dichloride; Freon 150; Glycol dichloride; NCI-C00511; RCRA waste number U077; UN 1184.

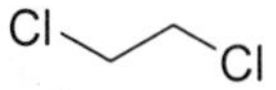

CAS Registry Number: 107-06-2
DOT: 2362
DOT label: Flammable liquid
Molecular formula: $C_2H_4Cl_2$
Formula weight: 98.96
RTECS: KI0525000
Merck reference: 10, 3743

Physical state, color, and odor:
Clear, colorless, oily liquid with a pleasant, chloroform-like odor. Odor threshold concentrations ranged from 6 to 40 ppm (Keith and Walters, 1992).

Melting point (°C):
-35.3 (Weast, 1986)
-36 (Worthing and Hance, 1991)

Boiling point (°C):
83.5 (Weast, 1986)

Density (g/cm³):
1.26000 at 15/4 °C, 1.25280 at 20/4 °C, 1.24530 at 25/4 °C (Standen, 1964)

Diffusivity in water (x 10⁻⁵ cm²/sec):
1.01 at 20 °C using method of Hayduk and Laudie (1974)

Flash point (°C):
13.3 (Fordyce and Meyer, 1940; NIOSH, 1997)
12-15 (Worthing and Hance, 1991)

Lower explosive limit (%):
6.2 (NIOSH, 1997)

Upper explosive limit (%):
16 (NIOSH, 1997)

Heat of fusion (kcal/mol):
2.112 (Riddick et al., 1986)

Henry's law constant (x 10⁻⁴ atm·m³/mol):
11.1 at 25 °C (Warner et al., 1987)
9.8 at 25 °C (Dilling, 1977)

9.09 at 20 °C (Pearson and McConnell, 1975)
11.90 at 25 °C (Hoff et al., 1993)
22.5 at 37 °C (Sato and Nakajima, 1979)
11.5 at 25 °C (Wright et al., 1992)
17.4 at 30 °C (Jeffers et al., 1989)
11.7, 13.0, 14.7, 14.1, and 17.4 at 10, 15, 20, 25, and 30 °C, respectively (Ashworth et al., 1988)
10, 15, 18, and 22 at 20, 30, 35, and 40 °C, respectively (Tse et al., 1992)
Distilled water: 3.43, 4.48, 4.12, 7.47, and 10.09 at 2.0, 6.0, 10.0, 18.2, and 25.0 °C, respectively; natural seawater: 5.52 and 11.3 at 6.0 and 25.0 °C, respectively (Dewulf et al., 1995)
9.47 at 20 °C (Hovorka and Dohnal, 1997)
13.78 at 22.0 °C (mole fraction ratio, Leighton and Calo, 1981)
3.66 and 9.84 at 5.7 and 24.9 °C, respectively (Dewulf et al., 1999a)

Interfacial tension with water (dyn/cm at 25 °C):
28.4 (quoted, Freitas et al., 1997)

Ionization potential (eV):
11.05 (NIOSH, 1997)
11.12 ± 0.05 (Franklin et al., 1969)

Bioconcentration factor, log BCF:
0.30 (bluegill sunfish, Veith et al., 1980)

Soil sorption coefficient, log K_{oc}:
1.279 (Willamette silt loam, Chiou et al., 1979)
1.34 (Jury et al., 1990)
1.88 (Lincoln sand, Wilson et al., 1981)
1.64, 1.65, 1.64, 1.68, 1.70, 1.65, and 1.68 at 2.3, 3.8, 6.2, 8.0, 13.5, 18.6, and 25.0 °C, respectively, for a Leie River (Belgium) clay (Dewulf et al., 1999a)

Octanol/water partition coefficient, log K_{ow}:
1.48 (Konietzko, 1984)
1.45 (Banerjee et al., 1980)
1.51, 1.51, and 1.53 at 25, 35, and 50 °C, respectively (Bhatia and Sandler, 1995)
1.43, 1.42, 1.47, 1.44, 1.50, and 1.46 at 2.2, 6.0, 10.0, 14.1, 18.7, and 24.8 °C, respectively (Dewulf et al., 1999a)

Solubility in organics:
Miscible with chloroform, ethanol, ether (U.S. EPA, 1985), and tetrachloroethylene

Solubility in water:
In g/kg: 9.22 at 0 °C, 8.85 at 10 °C, 8.69 at 20 °C, 8.94 at 30 °C (Rex, 1906)
8,300 mg/L at 25 °C (Warner et al., 1987)
7,986 mg/L at 25 °C (Banerjee et al., 1980)
8,400 mg/kg at 25 °C (McGovern, 1943)
8,650 mg/kg at 25 °C (Gross, 1929)
8,800 mg/L at 20 °C (McConnell et al., 1975)
8,450 mg/L at 20 °C (Chiou et al., 1979)
0.873 wt % at 0 °C (Konietzko, 1984)
In wt %: 0.82 at 0°C, 0.77 at 9.3 °C, 0.72 at 19.7 °C, 0.81 at 29.7 °C, 0.98 at 39.4 °C, 1.06 at

50.3 °C, 1.08 at 61.0 °C, 1.13 at 70.6 °C, 1.06 at 80.7 °C (Stephenson, 1992)
8,720 mg/kg at 15 °C, 9,000 mg/kg at 30 °C (Gross and Saylor, 1931)
8,524 mg/L at 25 °C (Howard, 1990)
8,100 mg/L (Price et al., 1974)
3,506 mg/L at 30 °C (McNally and Grob, 1984)
1.56 x 10^{-3} at 25 °C (mole fraction, Li et al., 1993)
8,690 mg/L at 25 °C (Cowen and Baynes, 1980)

Vapor density:
4.04 g/L at 25 °C, 3.42 (air = 1)

Vapor pressure (mmHg):
87 at 25 °C (ACGIH, 1986)
78.7 at 20 °C (Howard, 1990)
82 at 25 °C (quoted, Nathan, 1978)

Environmental fate:
   Biological. Methanococcus thermolithotrophicus, Methanococcus deltae, and *Methano-bacterium thermoautotrophicum* metabolized 1,2-dichloroethane releasing methane and ethylene (Belay and Daniels, 1987). 1,2-Dichloroethane showed slow to moderate bio-degradative activity with concomitant rate of volatilization in a static-culture flask-screening test (settled domestic wastewater inoculum) conducted at 25 °C. At concentrations of 5 and 10 mg/L, percent losses after 4 wk of incubation were 63 and 53, respectively. At a substrate concentration of 5 mg/L, 27% was lost due to volatilization after 10 d (Tabak et al., 1981).
   Photolytic. Titanium dioxide suspended in an aqueous solution and irradiated with UV light (λ = 365 nm) converted dichloroethane to carbon dioxide at a significant rate (Matthews, 1986).
   The rate constant for the reaction of 1,2-dichloroethane and OH radicals in the atmosphere at 300 K is 1.3 x 10^{11} cm^3/mol·sec (Hendry and Kenley, 1979). At 296 K, a photooxidation rate constant of 2.2 x 10^{-13} cm^3/molecule·sec was reported for the reaction with OH radicals resulting in a half-life of 1.7 months (Howard and Evenson, 1976).
   Chemical/Physical. Anticipated products from the reaction of 1,2-dichloroethane with ozone or OH radicals in the atmosphere are chloroacetaldehyde, chloroacetyl chloride, formaldehyde, and ClHCHO (Cupitt, 1980).
   Hydrolysis of 1,2-dichloroethane under alkaline and neutral conditions yielded vinyl chloride and ethylene glycol, respectively, with 2-chloroethanol and ethylene oxide forming as the intermediates under neutral conditions (Ellington et al., 1988; Jeffers et al., 1989; Kollig, 1993). The reported hydrolysis half-life in distilled water at 25 °C and pH 7 is 72.0 yr (Jeffers et al., 1989), but in a 0.05 M phosphate buffer solution the hydrolysis half-life is 37 yr (Barbash and Reinhard, 1989).
   In an aqueous solution, 1,2-dichloroethane reacted with hydrogen sulfide ions forming 1,2-dithioethane (Barbash and Reinhard, 1989).
   The volatilization half-life of 1,2-dichloroethane (1 mg/L) from water at 25 °C using a shallow-pitch propeller stirrer at 200 rpm at an average depth of 6.5 cm is 28.0 min (Dilling, 1977).
   At 600 °C, 1,2-dichloroethane decomposes to vinyl chloride and hydrogen chloride (NIOSH, 1997).
   At an influent concentration of 1,000 mg/L, treatment with granular activated carbon resulted in an effluent concentration of 189 mg/L. The adsorbability of the carbon used was 163 mg/g carbon (Guisti et al., 1974). Similarly, at influent concentrations of 1.0, 0.1, 0.01, and 0.001 mg/L, the granular activated carbon adsorption capacities at pH 5.3 were 3.6, 0.52, 0.08, and 0.01 mg/g, respectively (Dobbs and Cohen, 1980).

Exposure limits: Potential occupational carcinogen. NIOSH REL: TWA 1 ppm (4 mg/m^3), STEL 2 ppm, IDLH 50 ppm; OSHA PEL: TWA 50 ppm, ceiling 100 ppm, 5-min/3-h peak 200 ppm; ACGIH TLV: TWA 10 ppm (adopted).

Symptoms of exposure: Depression of central nervous system, irritation of eyes, corneal opacity, nausea, vomiting, diarrhea, ulceration, somnolence, cyanosis, pulmonary edema, coma. Ingestion of liquid may cause death (Patnaik, 1992).

Toxicity:

EC_{50} (9-min) for *Crangon crangon* in seawater 1,400 mg/L, *Gobius minutus* in seawater 400 mg/L (quoted, Verschueren, 1983).

LC_{50} (contact) for earthworm (*Eisenia fetida*) 60 μg/cm^2 (Neuhauser et al., 1985).

LC_{50} (7-d) for *Poecilia reticulata* 106 mg/L (Könemann, 1981).

LC_{50} (96-h) for fathead minnows 118 mg/L (Veith et al., 1983), bluegill sunfish 430 mg/L (Spehar et al., 1982), *Crangon crangon* in seawater 65 mg/L, *Gobius minutus* in seawater 185 mg/L (quoted, Verschueren, 1983), *Cyprinodon variegatus* 130-230 ppm using natural seawater (Heitmuller et al., 1981).

LC_{50} (72-h) for *Cyprinodon variegatus* 130-230 ppm (Heitmuller et al., 1981).

LC_{50} (48-h) for *Daphnia magna* 220 mg/L (LeBlanc, 1980), *Cyprinodon variegatus* 130-230 ppm (Heitmuller et al., 1981).

LC_{50} (24-h) for *Daphnia magna* 250 mg/L (LeBlanc, 1980), *Cyprinodon variegatus* 130-230 ppm (Heitmuller et al., 1981).

Acute oral LD_{50} for mice 489 mg/kg, rats 670 mg/kg, rabbits 860 mg/kg (quoted, RTECS, 1985).

Heitmuller et al. (1981) reported a NOEC of 130 ppm.

Drinking water standard (final): MCLG: zero; MCL: 5 μg/L (U.S. EPA, 1996).

Source: Improper use of insecticidal fumigant formulation containing 1,2-dichloropropane and carbon tetrachloride (Granosan)

Uses: Manufacture of acetyl cellulose, vinyl chloride, and ethylenediamine; vinyl chloride solvent; lead scavenger in antiknock unleaded gasoline; paint, varnish, and finish remover; metal degreasers; soap and scouring compounds; wetting and penetrating agents; ore flotation; tobacco flavoring; soil and foodstuff fumigant; solvent for oils, fats, waxes, resins, gums, and rubber.

1,1-DICHLOROETHYLENE

Synonyms: 1,1-DCE; **1,1-Dichloroethene**; *asym*-Dichloroethylene; NCI-C54262; RCRA waste number U078; Sconatex; VDC; Vinylidene chloride; Vinylidene chloride (II); Vinylidene dichloride; Vinylidine chloride.

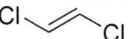

Note: Hydroquinone monomethyl ether (0.02 wt %) is added to prevent polymerization (Gillette et al., 1952).

CAS Registry Number: 75-35-4
DOT: 1303
DOT label: Combustible liquid
Molecular formula: $C_2H_2Cl_2$
Formula weight: 96.94
RTECS: KV9275000
Merck reference: 10, 9798

Physical state, color, and odor:
Colorless liquid or gas with a mild, sweet, chloroform-like odor. Odor threshold concentration is 190 ppm (Amoore and Hautala, 1983).

Melting point (°C):
-122.1 (Weast, 1986)

Boiling point (°C):
31.56 (Boublik et al., 1986)

Density (g/cm³ at 20/4 °C):
1.218 (Weast, 1986)
1.2132 (Riddick et al., 1986)

Diffusivity in water (x 10⁻⁵ cm²/sec):
1.01 at 20 °C using method of Hayduk and Laudie (1974)

Flash point (°C):
-19 (NIOSH, 1997)

Lower explosive limit (%):
6.5 (NFPA, 1984)

Upper explosive limit (%):
15.5 (NFPA, 1984)

Heat of fusion (kcal/mol):
1.557 (Riddick et al., 1986)

Henry's law constant (x 10⁻² atm·m³/mol):
1.5 at 25 °C (Warner et al., 1987)

19 (Pankow and Rosen, 1988)
2.56 at 25 °C (Gossett, 1987)
3.18 at 30 °C (Jeffers et al., 1989)
1.54, 2.03, 2.18, 2.59, and 3.18 at 10, 15, 20, 25, and 30 °C, respectively (Ashworth et al., 1988)
2.29, 3.37, and 4.75 at 20, 30, and 40 °C, respectively (Tse et al., 1992)
3.10 at 24.3 °C (mole fraction ratio, Leighton and Calo, 1981)
0.86, 1.00, 1.27, 1.97, and 2.66 at 2.0, 6.0, 10.0, 18.0, and 25.0 °C, respectively (Dewulf et al., 1999)

Ionization potential (eV):
9.81 ± 0.35 (Franklin et al., 1969)
9.46 (Horvath, 1982)
10.00 (NIOSH, 1997)

Soil sorption coefficient, log K_{oc}:
1.79 (Ellington et al., 1993)

Octanol/water partition coefficient, log K_{ow}:
2.13 (Mabey et al., 1982)
1.48 (HSDB, 1989)

Solubility in organics:
Slightly soluble in ethanol, acetone, benzene, and chloroform (U.S. EPA, 1985)

Solubility in water:
400 mg/L at 20 °C (Pearson and McConnell, 1975)
0.021 wt % at 25 °C (Riddick et al., 1986)
In wt % (°C): 0.24 (15), 0.255 (17), 0.25 (20), 0.225 (25), 0.24 (28.5), 0.255 (29.5), 0.22 (38.5), 0.21 (45), 0.23 (51), 0.24 (60), 0.225 (65), 0.295 (71), 0.25 (74.5), 0.295 (81), 0.37 (85.5), 0.35 (90.5) (DeLassus and Schmidt, 1981)
2,232 mg/L at 30 °C (McNally and Grob, 1984)

Vapor density:
3.96 g/L at 25 °C, 3.35 (air = 1)

Vapor pressure (mmHg):
591 at 25 °C, 720 at 30 °C (Verschueren, 1983)
495 at 20 °C, 760 at 31.8 °C (Standen, 1964)

Environmental fate:
   Biological. 1,1-Dichloroethylene significantly degraded with rapid adaptation in a static-culture flask-screening test (settled domestic wastewater inoculum) conducted at 25 °C. Complete degradation was observed after 14 d. At concentrations of 5 and 10 mg/L, the amount lost due to volatilization at the end of 10 d was 24 and 15%, respectively (Tabak et al., 1981).
   Soil. In a methanogenic aquifer material, 1,1-dichloroethylene biodegraded to vinyl chloride (Wilson et al., 1986). Under anoxic conditions, indigenous microbes in uncontaminated sediments degraded 1,1-dichloroethylene to vinyl chloride (Barrio-Lage et al., 1986).
   Photolytic. Photooxidation of 1,1-dichloroethylene in the presence of nitrogen dioxide and air yielded phosgene, chloroacetyl chloride, formic acid, hydrochloric acid, carbon monoxide, formaldehyde, and ozone (Gay et al., 1976). At 298 K, 1,1-dichloroethylene reacts with ozone

at a rate of 3.7 x 10^{-21} cm^3/molecule·sec (Hull et al., 1973).

Chemical/Physical. Above 0 °C in the presence of oxygen or other catalysts, 1,1-dichloroethylene will polymerize to a plastic (Windholz et al., 1983). The alkaline hydrolysis of 1,1-dichloroethylene yielded chloroacetylene. The reported hydrolysis half-life at 25 °C and pH 7 is 1.2 x 10^8 yr (Jeffers et al., 1989).

The evaporation half-life of 1,1-dichloroethylene (1 mg/L) from water at 25 °C using a shallow-pitch propeller stirrer at 200 rpm at an average depth of 6.5 cm is 27.2 min (Dilling, 1977).

At influent concentrations of 1.0, 0.1, 0.01, and 0.001 mg/L, the granular activated carbon adsorption capacities at pH 5.3 were 4.9, 1.4, 0.41, and 0.12 mg/g, respectively (Dobbs and Cohen, 1980).

Exposure limits: Potential occupational carcinogen. ACGIH TLV: TWA 5 ppm (adopted).

Symptoms of exposure: Irritation of mucous membranes. Narcotic at high concentrations (Patnaik, 1992).

Toxicity:

EC_{10} and EC_{50} concentrations inhibiting the growth of alga *Scenedesmus subspicatus* in 96 h were 240 and 410 mg/L, respectively (Geyer et al., 1985).

LC_{50} (96-h) for bluegill sunfish 74 mg/L, fathead minnows 108 mg/L (Spehar et al., 1982), *Menidia beryllina* 250 mg/L (quoted, Verschueren, 1983), *Cyprinodon variegatus* 250 ppm using natural seawater (Heitmuller et al., 1981).

LC_{50} (72-h) for *Cyprinodon variegatus* 250 ppm (Heitmuller et al., 1981).

LC_{50} (48-h) for *Daphnia magna* 79 mg/L (LeBlanc, 1980), *Cyprinodon variegatus* 250 ppm (Heitmuller et al., 1981).

LC_{50} (24-h) for *Daphnia magna* 98 mg/L (LeBlanc, 1980), *Cyprinodon variegatus* 250 ppm (Heitmuller et al., 1981).

LC_{50} (inhalation) for rats 6,350 ppm/4-h, mice 98 ppm/22-h (quoted, RTECS, 1985).

Acute oral LD_{50} for rats 1,550 mg/kg, male mice 194 mg/kg, female mice 217 mg/kg (Jones and Hathway, 1978); dogs 5,750 mg/kg (Tierney et al., 1979).

Heitmuller et al. (1981) reported a NOEC of 80 ppm.

Drinking water standard (final): MCLG: 7 μg/L; MCL: 7 μg/L (U.S. EPA, 1996).

Uses: Synthetic fibers and adhesives; chemical intermediate in vinylidene fluoride synthesis; co-monomer for food packaging, coating resins, and modacrylic fibers.

trans-1,2-DICHLOROETHYLENE

Synonyms: Acetylene dichloride; *trans*-Acetylene dichloride; 1,2-Dichloroethene; **(*E*)-1,2-Di-chloroethene**; *trans*-Dichloroethylene; 1,2-*trans*-Dichloroethene; 1,2-*trans*-Dichloroethylene; *sym*-Dichloroethylene; Dioform.

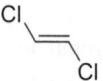

CAS Registry Number: 156-60-5
DOT: 1150 (isomeric mixture)
DOT label: Flammable liquid
Molecular formula: $C_2H_2Cl_2$
Formula weight: 96.94
RTECS: KV9400000
Merck reference: 10, 87

Physical state, color, and odor:
Colorless, viscous liquid with a sweet, pleasant odor. Odor threshold concentration is 17 ppm (Amoore and Hautala, 1983).

Melting point (°C):
-50.0 (McGovern, 1943)

Boiling point (°C):
47.5 (Weast, 1986)

Density (g/cm³):
1.2565 at 20/4 °C (Horvath, 1982)
1.27 at 25/4 °C (isomeric mixture, Weiss, 1986)
1.2631 at 10/4 °C (Standen, 1964)
1.2546 at 25/4 °C (Dean, 1987)

Diffusivity in water (x 10^{-5} cm²/sec):
1.03 at 20 °C using method of Hayduk and Laudie (1974)

Flash point (°C):
2 (Sax, 1984)
4 (Fordyce and Meyer, 1940)

Lower explosive limit (%):
9.7 (Sax, 1984)

Upper explosive limit (%):
12.8 (Sax, 1984)

Heat of fusion (kcal/mol):
2.684 (Riddick et al., 1986)

Henry's law constant (x 10^{-4} atm·m³/mol):
90.9 at 25 °C (Gossett, 1987)

72 (Pankow and Rosen, 1988)
100 at 25 °C (Wright et al., 1992)
121 at 37 °C (Sato and Nakajima, 1979)
57.5 at 30 °C (Jeffers et al., 1989)
5.9, 70.5, 85.7, 94.5, and 121 at 10, 15, 20, 25, and 30 °C, respectively (Ashworth et al., 1988)
79, 118, and 117 at 20, 30, and 40 °C, respectively (Tse et al., 1992)
101, 157, and 206 at 26.2, 35.0, and 46.1 °C, respectively (Hansen et al., 1993, 1995)
76.4 at 20 °C (Hovorka and Dohnal, 1997)

Ionization potential (eV):
9.64 (Franklin et al., 1969)
9.95 (Horvath, 1982)

Soil sorption coefficient, log K_{oc}:
1.58 (Brusseau and Rao, 1991)

Octanol/water partition coefficient, log K_{ow}:
2.09 (Mabey et al., 1982)
2.06 (Hansch and Leo, 1985)

Solubility in organics:
Miscible with acetone, ethanol, and ether and very soluble in benzene and chloroform (U.S. EPA, 1985)

Solubility in water (25 °C):
6.3 g/kg (McGovern, 1943)
6,300 mg/L (Dilling, 1977)
6,260 mg/L (Kamlet et al., 1987)

Vapor density:
3.96 g/L at 25 °C, 3.35 (air = 1)

Vapor pressure (mmHg):
185 at 10 °C, 265 at 20 °C, 410 at 30 °C (Standen, 1964)
340 at 25 °C (Howard, 1990)

Environmental fate:
   Soil. In a methanogenic aquifer material, *trans*-1,2-dichloroethylene biodegraded to vinyl chloride (Wilson et al., 1986). Under anoxic conditions *trans*-1,2-dichloroethylene, when subjected to indigenous microbes in uncontaminated sediments, degraded to vinyl chloride (Barrio-Lage et al., 1986). *trans*-1,2-dichloroethylene showed slow to moderate degradation concomitant with the rate of volatilization in a static-culture flask-screening test (settled domestic wastewater inoculum) conducted at 25 °C. At concentrations of 5 and 10 mg/L, percent losses after 4 wk of incubation were 95 and 93, respectively. The amount lost due to volatilization was 26-33% after 10 d (Tabak et al., 1981).
   Biological. Heukelekian and Rand (1955) reported a 10-d BOD value of 0.05 g/g which is 7.6% of the ThOD value of 0.66 g/g.
   Photolytic. Carbon monoxide, formic and hydrochloric acids were reported to be photooxidation products (Gay et al., 1976).
   The following rate constants were reported for the reaction of *trans*-1,2-dichloroethylene and ozone in the atmosphere: 1.8×10^{-19} cm^3/molecule·sec at 298 K (Atkinson and Carter, 1984) and

3.8 x 10^{-19} cm^3/molecule·sec at 296 K (Blume et al., 1976). Atkinson et al. (1988) reported a rate constant of 1.11 x 10^{-16} cm^3/molecule·sec for the gas-phase reaction with NO$_3$ at 298 K.

The evaporation half-life of *trans*-1,2-dichloroethane (1 mg/L) from water at 25 °C using a shallow-pitch propeller stirrer at 200 rpm at an average depth of 6.5 cm is 24.0 min (Dilling, 1977).

At influent concentrations of 1.0, 0.1, 0.01, and 0.001 mg/L, the granular activated carbon adsorption capacities at pH 6.7 were 3.0, 0.94, 0.29, and 0.09 mg/g, respectively (Dobbs and Cohen, 1980).

Exposure limits (isomeric mixture): OSHA PEL: TWA 200 ppm (790 mg/m^3); ACGIH TLV: TWA 200 ppm (adopted).

Symptoms of exposure: Vapor inhalation may cause somnolence and ataxia. Narcotic at high concentrations (Patnaik, 1992).

Toxicity:

LC$_{50}$ (contact) for earthworm (*Eisenia fetida*) 286 μg/cm^2 (Neuhauser et al., 1985).

LC$_{50}$ (96-h) for bluegill sunfish 140 mg/L (Spehar et al., 1982).

LC$_{50}$ (48-h) for *Daphnia magna* 220 mg/L (LeBlanc, 1980).

LC$_{50}$ (24-h) for *Daphnia magna* 230 mg/L (LeBlanc, 1980).

LD$_{50}$ (gavage administration) for Wistar rats 1,263 mg/kg (Freundt et al., 1977).

Acute oral LD$_{50}$ for mice 2,122 mg/kg (quoted, RTECS, 1985); in Sprague-Dawley rats were 7,902 and 9,939 mg/kg for males and females, respectively (Hayes et al., 1987); for CD-1 mice via gavage administration were 2,212 and 2,391 mg/kg for males and females, respectively (Barnes et al., 1985).

Drinking water standard (final): MCLG: 0.1 mg/L; MCL: 0.1 mg/L (U.S. EPA, 1996).

Uses: A mixture of *cis* and *trans* isomers is used as a solvent for fats, phenols, camphor; ingredient in perfumes; low temperature solvent for sensitive substances such as caffeine; refrigerant; organic synthesis.

DICHLOROFLUOROMETHANE

Synonyms: Algofrene type 5; Arcton 7; Dichloromonofluoromethane; Fluorocarbon 21; Fluorodichloromethane; Freon 21; Genetron 21; Halon 21; Refrigerant 21; UN 1029.

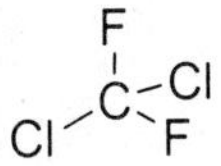

$$\underset{Cl}{\overset{F}{\underset{|}{C}}}\underset{F}{\overset{Cl}{}}$$

CAS Registry Number: 75-43-4
DOT: 1029
Molecular formula: $CHCl_2F$
Formula weight: 120.91
RTECS: PA8400000

Physical state, color, and odor:
Colorless liquid or gas with an ether-like odor

Melting point (°C):
-135 (Horvath, 1982)

Boiling point (°C):
8.92 (Horvath, 1982)

Density (g/cm³):
1.366 at 25/4 °C (Horvath, 1982)

Diffusivity in water (x 10^{-5} cm²/sec):
1.08 at 25 °C using method of Hayduk and Laudie (1974)

Flash point (°C):
Nonflammable gas (NIOSH, 1997)

Ionization potential (eV):
12.39 ± 0.20 (Franklin et al., 1969)

Soil sorption coefficient, log K_{oc}:
1.54 using method of Chiou et al. (1979)

Octanol/water partition coefficient, log K_{ow}:
1.55 (Hansch and Leo, 1979)

Solubility in organics:
Soluble in acetic acid, alcohol, and ether (Weast, 1986)

Solubility in water:
0.7 wt % at 30 °C (NIOSH, 1997)

Vapor density:
4.94 g/L at 25 °C, 4.17 (air = 1)

Vapor pressure (mmHg):
1,216 at 11.8 °C (NIOSH, 1997)

Exposure limits: NIOSH REL: TWA 10 ppm (40 mg/m³), IDLH 5,000 ppm; OSHA PEL: TWA 1,000 ppm (4,200 mg/m³); ACGIH TLV: TWA 10 ppm (adopted).

Toxicity:

LC_{50} (inhalation) for rats 49,900 ppm/4-h (quoted, RTECS, 1985).

Uses: Fire extinguishers; solvent; refrigerant.

sym-DICHLOROMETHYL ETHER

Synonyms: BCME; Bis(chloromethyl) ether; Bis-cme; Chloro(chloromethoxy)methane; Chloromethyl ether; Dimethyl-1,1'-dichloroether; *sym*-Dichlorodimethyl ether; Dichloromethyl ether; **Oxybis(chloromethane)**; RCRA waste number P016; UN 2249.

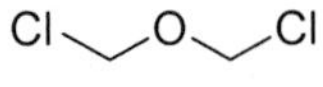

$$Cl\diagdown\diagup O\diagdown\diagup Cl$$

CAS Registry Number: 542-88-1
DOT: 2249
DOT label: Poison and flammable liquid
Molecular formula: $C_2H_4Cl_2O$
Formula weight: 114.96
RTECS: KN1575000

Physical state, color, and odor:
Colorless liquid with a suffocating odor

Melting point (°C):
-41.5 (Weast, 1986)

Boiling point (°C):
104 (Weast, 1986)
106 (Fishbein, 1979)

Density (g/cm³):
1.328 at 15/4 °C (Weast, 1986)
1.315 at 20/4 °C (Fishbein, 1979)

Diffusivity in water:
Not applicable - reacts with water

Flash point (°C):
<19 (Sax and Lewis, 1987)

Henry's law constant (atm·m³/mol):
Not applicable - reacts with water

Soil sorption coefficient, log K_{oc}:
Not applicable - reacts with water

Octanol/water partition coefficient, log K_{ow}:
Not applicable - reacts with water

Solubility in organics:
Soluble in alcohol, ether (Weast, 1986), and benzene (Hawley, 1981)

Solubility in water:
Not applicable - reacts with water (Fishbein, 1979)

Vapor density:
4.70 g/L at 25 °C, 3.97 (air = 1)

Vapor pressure (mmHg):
30 at 22 °C (Dreisbach, 1952)

Environmental fate:
   Chemical/Physical. Reacts rapidly with water forming hydrochloric acid and formaldehyde (Fishbein, 1979; Tou et al., 1974). Tou et al. (1974) reported a hydrolysis half-life of 38 sec for *sym*-dichloromethyl ether at 20 °C.
   Anticipated products from the reaction of *sym*-dichloromethyl ether with ozone or OH radicals in the atmosphere, excluding the decomposition products formaldehyde and hydro-chloric acid, are chloromethyl formate and formyl chloride (Cupitt, 1980).

Exposure limits: Known human carcinogen. ACGIH TLV: TWA 1 ppb (adopted).

Symptoms of exposure: Irritation of eyes, nose, and throat (Patnaik, 1992)

Toxicity:
   The LC_{50} (14-d) inhalation value for rats and hamsters following a single 7-h exposure was 7 ppm (Drew et al., 1975).
   Acute oral LD_{50} for rats is 210 mg/kg (quoted, RTECS, 1985).

Source: *sym*-Dichloromethyl ether may form as an intermediate by-product when form-aldehyde reacts with chloride ions under acidic conditions (Frankel et al., 1974; Tou and Kallos, 1974a; Travenius, 1982). Tou and Kallos (1974) reported that the reactants (formaldehyde and chloride ions) must be in concentrations of mg/L to form *sym*-dichloromethyl ether at concentrations of μg/L.
   Chloromethyl methyl ether may contain 1-8% *sym*-dichloromethyl ether as an impurity (Environment Canada, 1993a).

Uses: Intermediate in anionic-exchange quaternary resins; chloromethylating agent.

2,4-DICHLOROPHENOL

Synonyms: 3-Chloro-4-hydroxychlorobenzene; DCP; 2,4-DCP; 2,4-Dichlorohydroxybenzene; 4,6-Dichlorohydroxybenzene; NCI-C55345; RCRA waste number U081.

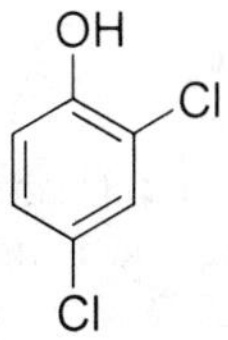

CAS Registry Number: 120-83-2
Molecular formula: $C_6H_4Cl_2O$
Formula weight: 163.00
RTECS: SK8575000

Physical state, color, and odor:
Colorless to yellow crystals with a sweet, musty, or medicinal odor. The reported odor threshold concentration ranged from 5.4 to 210 ppb. The taste threshold concentration in water is 0.98 ppb (Young et al., 1996).

Melting point (°C):
45.0 (Stull, 1947)
43.0 (Renner, 1990)

Boiling point (°C):
210 (Weast, 1986)
216 (Weiss, 1986)

Density (g/cm³):
1.40 at 15/4 °C (Weiss, 1986)

Dissociation constant, pK_a:
7.65 (Keith and Walters, 1992)
7.70 (Xie, 1983)
7.80 (Blackman et al., 1980)
7.85 (Dean, 1973)

Flash point (°C):
113.9, 93.3 (open cup, Weiss, 1986)

Entropy of fusion (cal/mol·K):
15.10 (Poeti et al., 1982)

Henry's law constant (x 10^{-6} atm·m³/mol):
6.66 (calculated, U.S. EPA, 1980a)
3.23 at 25 °C (estimated, Leuenberger et al., 1985)

Bioconcentration factor, log BCF:
2.53 (activated sludge), 2.41 (algae), 2.00 (golden ide) (Freitag et al., 1985)
1.53 (goldfish, Kobayashi, 1979)
1.0 (brown trout, Hattula et al., 1981)

Soil sorption coefficient, log K_{oc}:
3.60 (fine sediments), 3.50 (coarse sediments) (Isaacson and Frink, 1984)
2.10 (Brookstone clay loam, Boyd, 1982)
2.81 (river sediment, Eder and Weber, 1980)
2.17 (loamy sand, Kjeldsen et al., 1990)
2.37 (sandy soil), 2.93 (sand), 2.77 (peaty sand) (Van Gestel and Ma, 1993)
2.22, 2.26, and 2.32 in aerobic, anaerobic, and autoclaved Brookstoneclay loam soil, respectively (Boyd and King, 1984)
3.19, 3.22 (glaciofluvial, sandy aquifer, Nielsen et al., 1996)
2.55, 2.84 (forest soil), >2.68 (agricultural soil) (Seip et al., 1986)

Octanol/water partition coefficient, log K_{ow}:
3.15 (Roberts, 1981)
3.06 (Banerjee et al., 1984; Hansch and Leo, 1979)
3.08 (Krijgsheld and van der Gen, 1986a; Leo et al., 1971)
3.23 (Schellenberg et al., 1984)
3.20 (Kishi and Kobayashi, 1994)

Solubility in organics:
Soluble in ethanol, benzene, ether, chloroform (U.S. EPA, 1985), and carbon tetrachloride (ITII, 1986)

Solubility in water (mg/L):
4,600 at 20 °C, 4,500 at 25 °C (quoted, Verschueren, 1983)
4,500 at 20 °C (Krijgsheld and van der Gen, 1986a)
5,000 at 25 °C (Roberts et al., 1977)
6,194 at 25 °C and pH 5.1 (Blackman, 1955)
15,000 at 25 °C (Caturla, 1988)
3,896, 5,517, 6,075, and 6,501 at 15.3, 25.2, 29.8, and 35.1 °C, respectively (Achard et al., 1996)

Vapor pressure (x 10^{-2} mmHg):
1.5 at 8 °C, 8.9 at 25 °C (quoted, Leuenberger et al., 1985)

Environmental fate:
Biological. In activated sludge, 2.8% mineralized to carbon dioxide after 5 d (Freitag et al., 1985). In freshwater lake sediments, anaerobic reductive dechlorination produced 4-chlorophenol (Kohring et al., 1989). Chloroperoxidase, a fungal enzyme isolated from *Caldariomyces fumago*, converted 9-12% of 2,4-dichlorophenol to 2,4,6-trichlorophenol (Wannstedt et al., 1990). When 2,4-dichlorophenol was statically incubated in the dark at 25 °C with yeast extract and settled domestic wastewater inoculum, significant biodegradation with rapid adaptation was observed. At concentrations of 5 and 10 mg/L, 100 and 99% biodegradation, respectively, were observed after 7 d (Tabak et al., 1981). In activated sludge inoculum, 98.0% COD removal was achieved. The average rate of biodegradation was 10.5 mg COD/g·h (Pitter, 1976).

Surface Water. Hoigné and Bader (1983) reported a 2,4-dichlorophenol reacts with ozone at a rate constant of <1,500/M·sec at the pH range of 1.5 to 3.0.

Groundwater. Nielsen et al. (1996) studied the degradation of 2,4-dichlorophenol in a shallow, glaciofluvial, unconfined sandy aquifer in Jutland, Denmark. As part of the *in situ* microcosm study, a cylinder that was open at the bottom and screened at the top was installed through a cased borehole approximately 5 m below grade. Five liters of water was aerated with atmospheric air to ensure aerobic conditions were maintained. Groundwater was analyzed weekly for approximately 3 months to determine 2,4-dichlorophenol concentrations over time. The experimentally determined first-order biodegradation rate constant and corresponding half-life were 0.20/d and 3.47 d, respectively.

Photolytic. In distilled water, photolysis occurs at a slower rate than in estuarine waters containing humic substances. Photolysis products identified in distilled water were the three isomers of chlorocyclopentadienic acid. The following half-lives were reported for 2,4-dichlorophenol in estuarine water exposed to sunlight and microbes: 0.6 and 2.0 h during summer (24 °C) and winter (10 °C), respectively; in distilled water: 0.8 and 3.0 h during summer and winter, respectively; in poisoned estuarine water: 0.7 and 2.0 h during summer and winter, respectively (Hwang et al., 1986). When titanium dioxide suspended in an aqueous solution was irradiated with UV light ($\lambda = 365$ nm), 2,4-dichlorophenol was converted to carbon dioxide at a significant rate (Matthews, 1986). An aqueous solution containing hydrogen peroxide and irradiated by UV light ($\lambda = 296$ nm) converted 2,4-dichlorophenol to chlorohydroquinone and 1,4-dihydroquinone (Moza et al., 1988). A carbon dioxide yield of 50.4% was achieved when 2,4-dichlorophenol adsorbed on silica gel was irradiated with UV light ($\lambda >290$ nm) for 17 h (Freitag et al., 1985).

Chemical/Physical. 2,4-Dichlorophenol will not hydrolyze to any reasonable extent (Kollig, 1993). Reported second-order rate constants for the reaction of 2,4-dichlorophenol and singlet oxygen in water at 292 K: 7 x 10^6/M·sec at pH 5.5, 2 x 10^6/M·sec at pH 6, 1.0 x 10^5/M·sec at pH 6.65, 1.5 x 10^6/M·sec at pH 7.0, 7.6 x 10^5/M·sec at pH 7.9, 1.20 x 10^4/M·sec at pH 9.0-9.6. At pH 8, the half-life of 2,4-dichlorophenol is 62 h (Scully and Hoigné, 1987). In an aqueous phosphate buffer at 27 °C, 2,4-dichlorophenol reacted with singlet oxygen at a rate of 5.1 x 10^6/M·sec (Tratnyek and Hoigné, 1991). At neutral pH, 2,4-dichlorophenol was completely oxidized when potassium permanganate (2.0 mg/L) after 15 min (quoted, Verschueren, 1983).

Toxicity:

LC$_{50}$ (contact) for earthworm (*Eisenia fetida*) 2.5 μg/cm^2 (Neuhauser et al., 1985).

LC$_{50}$: 10.3-13.5 mg/L (soil porewater concentration) for earthworm (*Eisenia andrei*) and 17-43 mg/L (soil porewater concentration) for earthworm (*Lumbricus rubellus*) (Van Gestel and Ma, 1993).

LC$_{50}$ (8-d) for fathead minnows 6.5 mg/L (Spehar et al., 1982).

LC$_{50}$ (96-h) for bluegill sunfish 2.0 mg/L, fathead minnows 8.2-8.3 mg/L (Spehar, 1982).

LC$_{50}$ (48-h) for fathead minnows 8.4 mg/L (Spehar et al., 1982), *Daphnia magna* 2.6 mg/L (LeBlanc, 1980).

LC$_{50}$ (24-h) for goldfish 4.2 mg/L at pH 4.3 (quoted, Verschueren, 1983), *Daphnia magna* >10 mg/L (LeBlanc, 1980).

Acute oral LD$_{50}$ for mice 1,276 mg/kg, rats 580 mg/kg (quoted, RTECS, 1985).

Drinking water standard: No MCLGs or MCLs have been proposed (U.S. EPA, 1996).

Uses: A chemical intermediate in the manufacture of the pesticide 2,4-dichlorophenoxyacetic acid (2,4-D) and other compounds for use as germicides, antiseptics, and seed disinfectants.

1,2-DICHLOROPROPANE

Synonyms: α,β-Dichloropropane; ENT 15406; NCI-C55141; Propylene chloride; Propylene dichloride; α,β-Propylene dichloride; RCRA waste number U083.

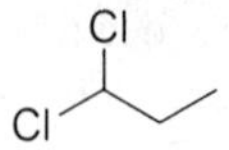

CAS Registry Number: 78-87-5
DOT: 1279
DOT label: Flammable liquid
Molecular formula: $C_3H_6Cl_2$
Formula weight: 112.99
RTECS: TX9625000
Merck reference: 10, 7755

Physical state, color, and odor:
Clear, colorless liquid with a sweet, chloroform-like odor. Odor threshold concentration is 50 ppm (Keith and Walters, 1992).

Melting point (°C):
-100.4 (Dreisbach, 1959)

Boiling point (°C):
96.22 (Boublik et al., 1973)
96.0 (Banerjee et al., 1990)

Density (g/cm³ at 20/4 °C):
1.15597 (Riddick et al., 1986)
1.1560 (Dreisbach, 1959)

Diffusivity in water (x 10⁻⁵ cm²/sec):
0.90 at 20 °C using method of Hayduk and Laudie (1974)

Flash point (°C):
15.6 (NIOSH, 1997)

Lower explosive limit (%):
3.4 (NIOSH, 1997)

Upper explosive limit (%):
14.5 (NIOSH, 1997)

Henry's law constant (x 10⁻³ atm·m³/mol):
2.3 (Pankow and Rosen, 1988)
2.70 at 25 °C (Wright et al., 1992)
4.71 at 37 °C (Sato and Nakajima, 1979)
2.07 at 25 °C (Howard, 1990)
1.22, 1.26, 1.90, 3.57, and 2.86 at 10, 15, 20, 25, and 30 °C, respectively (Ashworth et al., 1988)

2.1, 3.2, and 4.8 at 20, 30, and 40 °C, respectively (Tse et al., 1992)
2.19 at 20 °C (Hovorka and Dohnal, 1997)
2.77 at 24.9 °C (mole fraction ratio, Leighton and Calo, 1981)
0.81, 1.06, 1.32, 2.01, and 2.74 at 2.0, 6.0, 10.0, 18.0, and 25.0 °C, respectively (Dewulf et al., 1999)

Ionization potential (eV):
10.87 (NIOSH, 1997)

Soil sorption coefficient, log K_{oc}:
1.67 (Willamette silt loam, Chiou et al., 1979)

Octanol/water partition coefficient, log K_{ow}:
1.99 at 35 and 50 °C (Bhatia and Sandler, 1995)
2.00 (Hansch and Leo, 1979)

Solubility in organics:
Miscible with organic solvents (U.S. EPA, 1985).

Solubility in water:
2,700 mg/L at 20 °C (Gunther et al., 1968)
2,800 mg/kg at 25 °C (Gross, 1929)
2,700 mg/kg at 25 °C (McGovern, 1943)
2,740 mg/L at 25 °C (Howard, 1990)
2,069 mg/L at 30 °C (McNally and Grob, 1984)
2,420.4 mg/L at 30 °C (McNally and Grob, 1983)
2,096 mg/L at 25 °C (Jones et al., 1977)
In wt %: 0.29 at 0 °C, 0.28 at 9.5 °C, 0.30 at 20.0 °C, 0.29 at 29.7 °C, 0.30 at 40.3 °C, 0.32 at 49.8 °C, 0.35 at 60.0 °C, 0.39 at 70.5 °C, 0.47 at 80.2 °C (Stephenson, 1992)

Vapor density:
4.62 g/L at 25 °C, 3.90 (air = 1)

Vapor pressure (mmHg):
40.0 at 20.5 °C, 59.9 at 28.9 °C, 80.0 at 34.9 °C (Steele et al., 1997)
53.3 at 25 °C (Banerjee et al., 1990)

Environmental fate:
Biological. 1,2-Dichloropropane showed significant degradation with gradual adaptation in a static-culture flask-screening test (settled domestic wastewater inoculum) conducted at 25 °C. At concentrations of 5 and 10 mg/L, percent losses after 4 wk of incubation were 89 and 81, respectively. The amount lost due to volatilization was only 0-3% (Tabak et al., 1981).

Soil. Boesten et al. (1992) investigated the transformation of [^{14}C]1,2-dichloropropane under laboratory conditions of three subsoils collected from the Netherlands (Wassenaar low-humic sand, Kibbelveen peat, Noord-Sleen humic sand podsoil). The groundwater saturated soils were incubated in the dark at 9.5-10.5 °C. In the Wassenaar soil, no transformation of 1,2-dichloropropane was observed after 156 d of incubation. After 608 and 712 d, however, >90% degraded to nonhalogenated volatile compounds, which were detected in the headspace above the soil. These investigators postulated that these compounds can be propylene and propane in a ratio of 8:1. Degradation of 1,2-dichloropropane in the Kibbelveen peat and Noord-Sleen humic sand podsoil was not observed, possibly because the soil redox potentials in both soils (50-180 and 650-670 mV, respectively) were higher than the redox potential in the Wassenaar soil (10-20 mV).

Groundwater. According to the U.S. EPA (1986), 1,2-dichloropropane has a high potential to leach to groundwater.

Photolytic. Distilled water irradiated with UV light (λ = 290 nm) yielded the following photolysis products: 2-chloro-1-propanol, allyl chloride, allyl alcohol, and acetone. The photolysis half-life in distilled water is 50 min, but in distilled water containing hydrogen peroxide, the half-life decreased to less than 30 min (Milano et al., 1988).

Chemical/Physical. Hydrolysis in distilled water at 25 °C produced 1-chloro-2-propanol and hydrochloric acid. The reported half-life for this reaction is 23.6 yr (Milano et al., 1988). The hydrolysis rate constant for 1,2-dichloropropane at pH 7 and 25 °C was determined to be 5 x 10^{-6}/h, resulting in a half-life of 15.8 yr. The half-life is reduced to 24 d at 85 °C and pH 7.15 (Ellington et al., 1987). A volatilization half-life of 50 min was predicted from water stirred in an open container of depth 6.5 cm at 200 rpm (Dilling et al., 1975). Ozonolysis yielded carbon dioxide at low ozone concentrations (Medley and Stover, 1983).

1,2-Dichloropropane (0.12 mM) reacted with OH radicals in water (pH 2.8) at a rate of 3.8 x 10^8/M·sec (Haag and Yao, 1992).

At an influent concentration of 1,000 mg/L, treatment with granular activated carbon resulted in an effluent concentration of 71 mg/L. The adsorbability of the carbon used was 183 mg/g carbon (Guisti et al., 1974). Similarly, at influent concentrations of 1.0, 0.1, 0.01, and 0.001 mg/L, the granular activated carbon adsorption capacities at pH 5.3 were 5.9, 1.5, 0.37, and 0.09 mg/g, respectively (Dobbs and Cohen, 1980).

Emits toxic chloride fumes when heated to decomposition (Lewis, 1990).

Exposure limits: Potential occupational carcinogen. NIOSH REL: IDLH 400 ppm; OSHA PEL: TWA 75 ppm (350 mg/m³); ACGIH TLV: TWA 75 ppm, STEL 110 ppm (adopted).

Toxicity:

LC$_{50}$ (contact) for earthworm (*Eisenia fetida*) 64 μg/cm^2 (Neuhauser et al., 1985).

LC$_{50}$ (7-d) for *Poecilia reticulata* 116 mg/L (Könemann, 1981).

LC$_{50}$ (96-h) for bluegill sunfish 280 mg/L (Spehar et al., 1982) and 320 mg/L, *Menidia beryllina* 240 mg/L (quoted, Verschueren, 1983).

LC$_{50}$ (48-h) for *Daphnia magna* 52 mg/L (LeBlanc, 1980).

LC$_{50}$ (24-h) for *Daphnia magna* 99 mg/L (LeBlanc, 1980).

LC$_{50}$ (6-h inhalation for rats 5,100 mg/m³ (Cottalasso et al., 1994).

Acute oral LD$_{50}$ for guinea pigs 2,000 mg/kg, rats 2,196 mg/kg (quoted, RTECS, 1985).

Drinking water standard (final): MCLG: zero; MCL: 5 μg/L (U.S. EPA, 1996).

Uses: Preparation of tetrachloroethylene and carbon tetrachloride; lead scavenger for antiknock fluids; metal cleanser; soil fumigant for nematodes; solvent for oils, fats, gums, waxes, and resins; spotting agent.

cis-1,3-DICHLOROPROPYLENE

Synonyms: *cis*-1,3-Dichloropropene; *cis*-1,3-Dichloro-1-propene; (*Z*)-1,3-Dichloropropene; (*Z*)-1,3-Dichloro-1-propene; 1,3-Dichloroprop-1-ene; *cis*-1,3-Dichloro-1-propylene.

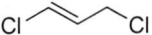

CAS Registry Number: 10061-01-5
DOT: 2047 (isomeric mixture)
DOT label: Flammable liquid
Molecular formula: $C_3H_4Cl_2$
Formula weight: 110.97
RTECS: UC8325000
Merck reference: 10, 3059

Physical state, color, and odor:
Colorless to amber-colored liquid with a pungent, chloroform-like odor

Melting point (°C):
-84 (isomeric mixture, Krijgsheld and van der Gen, 1986)

Boiling point (°C):
104.3 (Horvath, 1982)

Density (g/cm³ at 20/4 °C):
1.224 (Melnikov, 1971)
1.217 (Horvath, 1982)

Diffusivity in water (x 10^{-5} cm²/sec):
0.94 at 20 °C using method of Hayduk and Laudie (1974)

Flash point (isomeric mixture, °C):
35 (NFPA, 1984); 25 (Abel closed cup, Worthing and Hance, 1991)

Lower explosive limit (%):
5.3 (isomeric mixture, NFPA, 1984)

Upper explosive limit (%):
14.5 (isomeric mixture, NFPA, 1984)

Henry's law constant (x 10^{-3} atm·m³/mol):
1.3 (Pankow and Rosen, 1988)

Soil sorption coefficient, log K_{oc}:
1.36 (Kenaga, 1980a)
1.38 (average using 3 soils and computed from vapor-phase sorption) (Leistra, 1970)
1.75 (isomeric mixture, Meylan et al., 1992)

Octanol/water partition coefficient, log K_{ow}:
1.41 (Krijgsheld and van der Gen, 1986)

Solubility in organics:
Miscible with acetone, benzene, carbon tetrachloride, heptane, and methanol (Worthing and Hance, 1991)

Solubility in water (mg/L):
2,700 at 20 °C (Dilling, 1977)
911.2 at 30 °C (McNally and Grob, 1984)
1,071.0 at 30 °C (McNally and Grob, 1983)

Vapor density:
4.54 g/L at 25 °C, 3.83 (air = 1)

Vapor pressure (mmHg):
43 at 25 °C (Verschueren, 1983)
25 at 20 °C (Schwille, 1988)

Environmental fate:
  Biological. cis-1,3-Dichloropropylene was reported to hydrolyze to 3-chloro-2-propen-1-ol and can be biologically oxidized to 3-chloropropenoic acid, which is further oxidized to formylacetic acid. Decarboxylation of this compound yields carbon dioxide (Connors et al., 1990). The isomeric mixture showed significant degradation with gradual adaptation in a static-culture flask-screening test (settled domestic wastewater inoculum) conducted at 25 °C. At concentrations of 5 and 10 mg/L, percent losses after 4 wk of incubation were 85 and 84, respectively. Ten d into the incubation study, 7-19% was lost due to volatilization (Tabak et al., 1981).
  Soil. Hydrolyzes in wet soil forming *cis*-3-chloroallyl alcohol (Castro and Belser, 1966).
  Chemical/Physical. Hydrolysis in distilled water at 25 °C produced *cis*-3-chloro-2-propen-1-ol and hydrochloric acid. The reported half-life for this reaction is 1 d (Milano et al., 1988; Kollig, 1993).
  Chloroacetaldehyde, formyl chloride, and chloroacetic acid were formed from the ozonation of dichloropropylene at approximately 23 °C and 730 mmHg. Chloroacetaldehyde and formyl chloride also formed from the reaction of dichloropropylene and OH radicals (Tuazon et al., 1984).
  The volatilization half-life of *cis*-1,3-dichloropropylene (1 mg/L) from water at 25 °C using a shallow-pitch propeller stirrer at 200 rpm at an average depth of 6.5 cm is 29.6 min (Dilling, 1977).
  Emits chlorinated acids when incinerated. Incomplete combustion may release toxic phosgene (Sittig, 1985).

Exposure limits: ACGIH TLV: TWA 1 ppm for *cis* and *trans* isomers (adopted).

Toxicity:
  LC_{50} (96-h) for bluegill sunfish 6.1 mg/L (Spehar et al., 1982).

Drinking water standard (tentative): MCLG: zero; MCL: none proposed (U.S. EPA, 1996).

Uses: A mixture containing *cis* and *trans* isomers is used as a soil fumigant and a nematocide.

trans-1,3-DICHLOROPROPYLENE

Synonyms: (*E*)-1,3-Dichloropropene; *trans*-1,3-Dichloropropene; **(*E*)-1,3-Dichloro-1-propene**; *trans*-1,3-Dichloro-1-propene; 1,3-Dichloroprop-1-ene; *trans*-1,3-Dichloro-1-propylene.

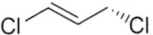

CAS Registry Number: 10061-02-6
DOT: 2047 (isomeric mixture)
DOT label: Flammable liquid
Molecular formula: $C_3H_4Cl_2$
Formula weight: 110.97
RTECS: UC8320000
Merck reference: 10, 3059

Physical state, color, and odor:
Clear, colorless liquid with a chloroform-like odor

Melting point (°C):
-84 (isomeric mixture, Krijgsheld and van der Gen, 1986)

Boiling point (°C):
112.0 (Horvath, 1982)
112.1 (Melnikov, 1971)

Density (g/cm³ at 20/4 °C):
1.224 (Horvath, 1982)
1.217 (Krijgsheld and van der Gen, 1986)

Diffusivity in water (x 10^{-5} cm²/sec):
0.92 at 20 °C using method of Hayduk and Laudie (1974)

Flash point (isomeric mixture, °C):
35 (NFPA, 1984); 25 (Abel closed cup, Worthing and Hance, 1991)

Lower explosive limit (%):
5.3 (isomeric mixture, NFPA, 1984)

Upper explosive limit (%):
14.5 (isomeric mixture, NFPA, 1984)

Henry's law constant (x 10^{-3} atm·m³/mol):
1.3 (Pankow and Rosen, 1988)

Soil sorption coefficient, log K_{oc}:
1.41 (average using 3 soils and computed from vapor-phase sorption) (Leistra, 1970)
1.415 (Kenaga, 1980a)
1.75 (isomeric mixture, Meylan et al., 1992)

Octanol/water partition coefficient, log K_{ow}:
1.41 (Krijgsheld and van der Gen, 1986)

Solubility in organics:
Miscible with acetone, benzene, carbon tetrachloride, heptane, and methanol (Worthing and Hance, 1991)

Solubility in water (mg/L):
2,800 at 20 °C (Dilling, 1977)
1,019.9 at 30 °C (McNally and Grob, 1984)
1,188.1 at 30 °C (McNally and Grob, 1983)

Vapor density:
4.54 g/L at 25 °C, 3.83 (air = 1)

Vapor pressure (mmHg):
34 at 25 °C (Verschueren, 1983)
25 at 20 °C (Schwille, 1988)

Environmental fate:
Biological. The isomeric mixture showed significant degradation with gradual adaptation in a static-culture flask-screening test (settled domestic wastewater inoculum) conducted at 25 °C. At concentrations of 5 and 10 mg/L, percent losses after 4 wk of incubation were 85 and 84, respectively. Ten d into the incubation study, 7-19% was lost due to volatilization (Tabak et al., 1981).

Chemical/Physical. Hydrolysis in distilled water at 25 °C produced *trans*-3-chloro-2-propen-1-ol and hydrochloric acid. The reported half-life for this reaction is only 2 d (Kollig, 1993; Milano et al., 1988). *trans*-1,3-Dichloropropylene was reported to hydrolyze to 3-chloro-2-propen-1-ol and can be biologically oxidized to 3-chloropropenoic acid which is further oxidized to formylacetic acid. Decarboxylation of this compound yields carbon dioxide (Connors et al., 1990). Chloroacetaldehyde, formyl chloride, and chloroacetic acid were formed from the ozonation of dichloropropylene at approximately 23 °C and 730 mmHg. Chloroacet-aldehyde and formyl chloride also formed from the reaction of dichloropropylene and OH radicals (Tuazon et al., 1984).

The volatilization half-life of *trans*-1,3-dichloropropylene (1 mg/L) from water at 25 °C using a shallow-pitch propeller stirrer at 200 rpm at an average depth of 6.5 cm is 24.6 min (Dilling, 1977).

Emits chlorinated acids when incinerated. Incomplete combustion may release toxic phosgene (Sittig, 1985).

Exposure limits: ACGIH TLV: TWA 1 ppm for *cis* and *trans* isomers (adopted).

Toxicity:
LC_{50} (96-h) for bluegill sunfish 6.1 mg/L (Spehar et al., 1982).

Drinking water standard (tentative): MCLG: zero; MCL: none proposed (U.S. EPA, 1996).

Uses: A mixture containing *cis* and *trans* isomers is used as a soil fumigant and a nematocide.

DICHLORVOS

Synonyms: Apavap; Astrobot; Atgard; Atgard C; Atgard V; Bay 19149; Benfos; Bibesol; Brevinyl; Brevinyl E50; Canogard; Cekusan; Chlorvinphos; Cyanophos; Cypona; DDVF; DDVP; Dedevap; Deriban; Derribante; Devikol; Dichlorman; 2,2-Dichloroethenyl dimethyl phosphate; 2,2-Dichloroethenyl phosphoric acid dimethyl ester; Dichlorophos; 2,2-Dichloro-vinyl dimethyl phosphate; 2,2-Dichlorovinyl dimethyl phosphoric acid ester; Dichlorovos; Di-methyl 2,2-dichloroethenyl phosphate; Dimethyl dichlorovinyl phosphate; Dimethyl 2,2-di-chlorovinyl phosphate; *O,O*-Dimethyl *O*-(2,2-dichlorovinyl)phosphate; Divipan; Duo-kill; Duravos; ENT 20738; Equigard; Equigel; Estrosel; Estrosol; Fecama; Fly-die; Fly fighter; Herkal; Herkol; Krecalvin; Lindan; Mafu; Mafu strip; Marvex; Mopari; NA 2783; NCI-C00113; Nerkol; Nogos; Nogos 50; Nogos G; No-pest; No-pest strip; NSC 6738; Nuva; Nuvan; Nuvan 100EC; Oko; OMS 14; **Phosphoric acid 2,2-dichloroethenyl dimethyl ester**; Phos-phoric acid 2,2-dichlorovinyl dimethyl ester; Phosvit; SD 1750; Szklarniak; Tap 9VP; Task; Task tabs; Tenac; Tetravos; UDVF; Unifos; Unifos 50 EC; Vapona; Vaponite; Vapora II; Verdican; Verdipor; Vinyl alcohol 2,2-dichlorodimethyl phosphate; Vinylofos; Vinylophos.

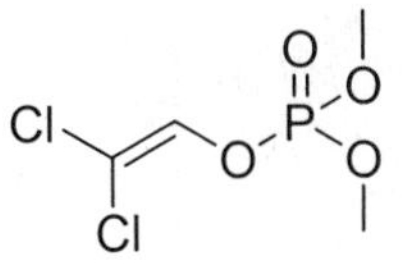

CAS Registry Number: 62-73-7
DOT: 2783
DOT label: Poison
Molecular formula: $C_4H_7Cl_2O_4P$
Formula weight: 220.98
RTECS: TC0350000
Merck reference: 10, 3065

Physical state, color, and odor:
Colorless to yellow liquid with an aromatic odor

Boiling point (°C):
74 at 1 mmHg (Worthing and Hance, 1991)
140 at 20 mmHg (Windholz et al., 1983)

Density (g/cm³):
1.415 at 25/4 °C (Windholz et al., 1983)

Diffusivity in water (x 10^{-5} cm²/sec):
0.78 at 25 °C using method of Hayduk and Laudie (1974)

Flash point (°C):
>80 (NIOSH, 1997)

Henry's law constant (x 10^{-4} atm·m³/mol):
1.23 at 25 °C (Kawamoto and Urano, 1989)

Ionization potential (eV):
9.4 (Lias et al., 1998)

Bioconcentration factor, log BCF:
-0.10 (*Gnathopogon aerulescens*, Devillers et al., 1996)

Soil sorption coefficient, log K_{oc}:
9.57 using method of Saeger et al. (1979)

Octanol/water partition coefficient, log K_{ow}:
1.40 (Leo et al., 1971)
1.16 (Kawamoto and Urano, 1989)

Solubility in organics:
Miscible with alcohol and most non-polar solvents (Windholz et al., 1983)

Solubility in water:
16,000 mg/L at 25 °C (Kawamoto and Urano, 1989)

Vapor density:
9.03 g/L at 25 °C, 7.63 (air = 1)

Vapor pressure (x 10^{-2} mmHg):
1.2 at 20 °C (quoted, Windholz et al., 1983; Kawamoto and Urano, 1989)
5.27 at 25 °C (Kim et al., 1984)

Environmental fate:
   Biological. Dichlorvos incubated with sewage sludge for 1 wk at 29 °C degraded to dichloroethanol, dichloroacetic acid, ethyl dichloroacetate, and an inorganic phosphate. In addition, dimethyl phosphate formed in the presence or absence of microorganisms (Lieberman and Alexander, 1983). Dichlorvos degraded fastest in nonsterile soils and decomposed faster in soils that were sterilized by gamma radiation than in soils that were sterilized by autoclaving. After 1 d of incubation, the percent of dichlorvos degradation that occurred in autoclaved, irradiated and nonsterile soils were 17, 88, and 99, respectively (Getzin and Rosefield, 1968).
   Soil. In a silt loam and sandy loam, reported R_f values were 0.79 and 0.80, respectively (Sharma et al., 1986).
   Plant. Metabolites identified in cotton leaves include dimethyl phosphate, phosphoric acid, methyl phosphate and *O*-dimethyl dichlorvos (Bull and Ridgway, 1969). The estimated half-life of dichlorvos applied to selected ornamental flower crops is 4.8 h (Brouwer et al., 1997).
   Photolytic. Dichlorvos should not undergo direct photolysis because it does not absorb UV light at wavelengths greater than 240 nm (Gore et al., 1971).
   Chemical/Physical. Releases very toxic fumes of chlorides and phosphorous oxides when heated to decomposition (Sax and Lewis, 1987).
   Slowly hydrolyzes in water and in acidic media but is more rapidly hydrolyzed under alkaline conditions to dimethyl hydrogen phosphate and dichloroacetaldehyde (Capel et al., 1988; Hartley and Kidd, 1987; Worthing and Hance, 1991). In the Rhine River (pH 7.4), the hydrolysis half-life of dichlorvos was 6 h (Capel et al., 1988).
   Atkinson and Carter (1984) estimated a half-life of 320 d for the reaction of dichlorvos with ozone in the atmosphere.

Exposure limits (mg/m³): NIOSH REL: TWA 1, IDLH 100; OSHA PEL: TWA 1; ACGIH TLV: TWA 0.9 mg/m3 ppm for *cis* and *trans* isomers (adopted).

Symptoms of exposure: Miosis, eye ache, headache, rhinorrhea, salivation, wheezing, cyanosis, anorexia, vomiting, diarrhea, sweating, muscle fasciculation, paralysis, ataxia, convulsions, low blood pressure (NIOSH, 1997).

Toxicity:

EC_{50} (48-h) for *Daphnia pulex* 0.07 μg/L, *Simocephalus serrulatus* 0.27 μg/L (Mayer and Ellersieck, 1986).

LC_{50} (96-h) for *Mystus vittatus* 400 μg/L, *Ophiocephalus punctatus* 2.04 mg/L (Spehar et al., 1982), bluegill sunfish 869 μg/L *Gammarus lacustris* 0.50 μg/L, *Gammarus faciatus* 0.40 μg/L (quoted, Verschueren, 1983), *Simocephalus serrulatus* 0.26 μg/L, *Daphnia pulex* 0.07 μg/L (Sanders and Cope, 1966), freshwater prawn (*Macrobrachium brevicornis*) 0.78 mg/L (Omkar and Shukla, 1985), marine female copepod (*Tigriopus brevicornis*) 4.6 μg/L (Forget et al., 1998).

LC_{50} (96-h static bioassay) for sand shrimp 4 μg/L, hermit crab 45 μg/L, grass shrimp 15 μg/L (Eisler, 1969), *Pteronarcys californica* 0.10 μg/L (Sanders and Cope, 1968).

LC_{50} (96-h semi-static bioassay) for *Tigriopus brevicornis* females 4.6 μg/L (Forget et al., 1998).

LC_{50} (48-h) for red killifish 81 mg/L (Yoshioka et al., 1986).

LC_{50} (24-h) for bluegill sunfish 1.0 mg/L (Hartley and Kidd, 1987).

Acute oral LD_{50} for wild birds 12 mg/kg, chickens 6.45 mg/kg, ducks 7.8 mg/kg, dogs 1,090 mg/kg, mice 101 mg/kg, pigeons 23.7 mg/kg, pigs 157 mg/kg, quail 23.7 mg/kg, rabbits 10 gm/kg (quoted, RTECS, 1985), female rats 56 mg/kg, male rats 80 mg/kg (Mattson et al., 1955).

Acute percutaneous LD_{50} for rats 300 mg/kg (Worthing and Hance, 1991).

LD_{50} (dermal) for female rats 75 mg/kg, male rats 1070 mg/kg (Mattson et al., 1955).

A NOEL of 10 mg/kg diet was reported for rats in a 2-yr feeding trial (Worthing and Hance, 1991).

Uses: A contact insecticide and fumigant used against Diptera and Culicidae in homes and against Coleoptera, Homoptera, and Lepidoptera in fruit, cotton, and ornamentals.

DIELDRIN

Synonyms: Alvit; Compound 497; Dieldrine; Dieldrite; Dieldrix; ENT 16225; HEOD (>85% active ingredient); Hexachloroepoxyoctahydro-*endo,exo*-dimethanonaphthalene; 1,2,3,4,10,10-Hexachloro-6,7-epoxy-1,4,4a,5,6,7,8,8a-octahydro-1,4-*endo,exo*-5,8-dimethanonaphthalene; **3,4,5,6,9,9-Hexachloro-1a,2,2a,3,6,6a,7,7a-octahydro-2,7:3,6-dimethanonaphth[2,3-*b*]oxirene**; Illoxol; Insecticide 497; NA 2761; NCI-C00124; Octalox; OMS 18; Panoram D-31; Quintox; RCRA waste number P037.

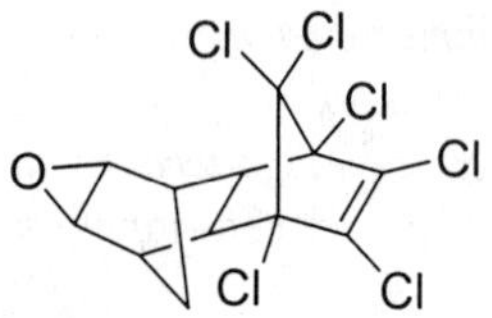

CAS Registry Number: 60-57-1
DOT: 2761
DOT label: Poison
Molecular formula: $C_{12}H_8Cl_6O$
Formula weight: 380.91
RTECS: IO1750000
Merck reference: 10, 3084

Physical state, color, and odor:
White crystals to pale tan flakes with an odorless to mild chemical odor. Odor threshold concentration is 41 μg/L (Keith and Walters, 1992).

Melting point (°C):
175-176 (Weast, 1986)
143-144 (technical grade ≈ 90%, Aldrich, 1988)

Boiling point (°C):
Decomposes (Weast, 1986)

Density (g/cm³):
1.75 at 20/4 °C (Weiss, 1986)

Diffusivity in water (x 10⁻⁵ cm²/sec):
0.44 at 20 °C using method of Hayduk and Laudie (1974)

Flash point (°C):
Nonflammable (Weiss, 1986)

Henry's law constant (x 10⁻⁷ atm·m³/mol):
2 (Eisenreich et al., 1981)
580 at 25 °C (Warner et al., 1987)
290 at 20 °C (Slater and Spedding, 1981)

Bioconcentration factor, log BCF:
2.95 African catfish (*Clarias gariepinus*) (Lamai et al., 1999)
3.53 (*B. subtilis*, Grimes and Morrison, 1975)

4.25 (activated sludge), 3.36 (algae), 3.48 (golden ide) (Freitag et al., 1985)
4.15 (freshwater fish), 3.61 (fish, microcosm) (Garten and Trabalka, 1983)
3.46 *Crassostrea virginica* (quoted, Verschueren, 1983)
3.55 freshwater clam (*Corbicula manilensis*) (Hartley and Johnston, 1983)
3.65 (*Pseudorosbora parva*, Devillers et al., 1996)

Soil sorption coefficient, log K_{oc}:
4.11 (Batcombe silt loam, Briggs, 1981)
4.15 (clay loam, Travis and Arms, 1988)
4.73 (Taichung soil: pH 6.8, % sand: 25, % silt: 40, % clay: 35) (Ding and Wu, 1995)
4.37 (Beverly sandy loam), 4.42 (Plainfield sand), 4.44 (Big Creek sediment) (Sharom et al., 1980a)

Octanol/water partition coefficient, log K_{ow}:
5.16 (Kishi et al., 1990)
6.2 (Briggs, 1981)
5.48 (Mackay, 1982)
4.32 (Geyer et al., 1987)
4.49, 4.51, 4.55, 4.66 (Brooke et al., 1986)
3.692 (Rao and Davidson, 1980)
5.401 (de Bruijn et al., 1989)
5.30 (Hammers et al., 1982)

Solubility in organics:
Soluble in ethanol and benzene (Weast, 1986)

Solubility in water (μg/L):
200 at 20 °C (Weil et al., 1974)
186 at 25-29 °C (Park and Bruce, 1968)
200 at 26.5 °C (Bhavnagary and Jayaram, 1974)
50 at 26 °C (Melnikov, 1971)
90, 195, 400, and 600 at 15, 25, 35, and 45 °C, respectively (particle size ≤5 μ, Biggar and Riggs, 1974)
20-25 °C: 180 (particle size ≤5 μ), 140 (particle size ≤0.04 μ) (Robeck et al., 1965)
50 (Gile and Gillett, 1979)
250 at 20-25 °C (Herzel and Murty, 1984)

Vapor density (ng/L):
54 at 20 °C, 202 at 30 °C, 676 at 40 °C (Spencer and Cliath, 1969)

Vapor pressure (x 10^{-7} mmHg):
31 at 20 °C (Windholz et al., 1983)
28 at 20 °C (Spencer and Cliath, 1969)
100 at 30 °C (Tinsley, 1979)
7.78 at 20.25 °C (Gile and Gillett, 1979)
448 at 25 °C (Bidleman, 1984)
239, 399 at 25 °C (Hinckley et al., 1990)

Environmental fate:
Biological. Identified metabolites of dieldrin from solution cultures containing *Pseudomonas* sp. in soils include aldrin and dihydroxydihydroaldrin. Other unidentified by-products included

a ketone, an aldehyde, and an acid (Matsumura et al., 1968; Kearney and Kaufman, 1976). A pure culture of the marine alga, namely *Dunaliella* sp., degraded dieldrin to photodieldrin and an unknown metabolite at yields of 8.5 and 3.2%, respectively. Photodieldrin and the diol were also identified as metabolites in field-collected samples of marine water, sediments, and associated biological materials (Patil et al., 1972). At least 10 different types of bacteria comprising a mixed anaerobic population degraded dieldrin, via monodechlorination at the methylene bridge carbon, to give *syn-* and *anti-*monodechlorodieldrin. Three isolates, *Clostridium bifermentans*, *Clostridium glycolium* and *Clostridium* sp., were capable of dieldrin dechlorination but the rate was much lower than that of the mixed population (Maule et al., 1987). Using settled domestic wastewater inoculum, dieldrin (5 and 10 mg/L) did not degrade after 28 d of incubation at 25 °C in four successive 7-d incubation periods (Tabak et al., 1981).

Chacko et al. (1966) reported that cultures of six actinomycetes (*Norcardia* sp., *Streptomyces albus*, *Streptomyces antibioticus*, *Streptomyces auerofaciens*, *Streptomyces cinnamoneus*, *Streptomyces viridochromogenes*) and 8 fungi had no effect on the degradation of dieldrin. Matsumura et al. (1970) reported microorganisms isolated from soil and Lake Michigan water converted dieldrin to photodieldrin.

The percentage of dieldrin remaining in a Congaree sandy loam soil after 7 yr was 50% (Nash and Woolson, 1967).

Soil. Dieldrin is very persistent in soil under both aerobic and anaerobic conditions (Castro and Yoshida, 1971; Sanborn and Yu, 1973). Reported half-lives in soil ranged from 175 d to 3 yr (Howard et al., 1991; Jury et al., 1987).

Groundwater. According to the U.S. EPA (1986), dieldrin has a high potential to leach to groundwater.

Surface Water. Mackay and Wolkoff (1973) estimated an evaporation half-life of 723 d from a surface water body that is 25 °C and 1 m deep..

Hargrave et al. (2000) calculated BAFs as the ratio of the compound tissue concentration [wet and lipid weight basis (ng/g)] to the concentration of the compound dissolved in seawater (ng/mL). Average log BAF values for dieldrin in ice algae and phytoplankton collected from the Barrow Strait in the Canadian Archipelago were 4.99 and 5.24, respectively.

Photolytic. Photolysis of a saturated aqueous solution of dieldrin by sunlight for 3 months resulted in a 70% yield of photodieldrin. The direct photolytic half-life under these conditions ranged from 1.8 to 2.1 months (Henderson and Crosby, 1968). A solid film of dieldrin exposed to sunlight for 2 months resulted in a 25% yield of photodieldrin (Benson, 1971). In addition to sunlight, UV light converts dieldrin to photodieldrin (Georgacakis and Khan, 1971). Solid dieldrin exposed to UV light (λ <300 nm) under a stream of oxygen yielded small amounts of photodieldrin (Gäb et al., 1974).

Many other investigators reported photodieldrin as a photolysis product of dieldrin under various conditions (Crosby and Moilanen, 1974; Ivie and Casida, 1970, 1971, 1971a; Rosen and Carey, 1968; Robinson et al., 1976; Rosen et al., 1966). One of the photoproducts identified in addition to photodieldrin was photoaldrin chlorohydrin [1,1,2,3,-3a,5(or 6),7a-heptachloro-6-(or 5)-hydroxydecahydro-2,4,7-metheno-1*H*-cyclopenta[*a*]pentalene] (Lombardo et al., 1972). After a 1 h exposure to sunlight, dieldrin was converted to photodieldrin. Photodecomposition was accelerated by a number of photosensitizing agents (Ivie and Casida, 1971). When an aqueous solution containing dieldrin was photooxidized by UV light at 90-95 °C, 25, 50, and 75% degraded to carbon dioxide after 2.9, 4.8, and 12.5 h, respectively (Knoevenagel and Himmelreich, 1976).

Chemical/Physical. The hydrolysis rate constant for dieldrin at pH 7 and 25 °C was determined to be 7.5×10^{-6}/h, resulting in a half-life of 10.5 yr (Ellington et al., 1987). The epoxide moiety undergoes nucleophilic substitution with water forming dieldrin diol (Kollig, 1993). At higher temperatures, the hydrolysis half-lives decreased significantly. At 69 °C and pH values of 3.13, 7.22, and 10.45, the calculated hydrolysis half-lives were 19.5, 39.5, and

29.2 d, respectively (Ellington et al., 1986).

Products reported from the combustion of dieldrin at 900 °C include carbon monoxide, carbon dioxide, hydrochloric acid, chlorine, and unidentified compounds (Kennedy et al., 1972).

At 33 °C, 35% relative humidity and a 2 mile/h wind speed, the volatility losses of dieldrin as a thick film, droplets on glass, droplets on leaves, and formulation film on glass after 48 h were 78.6, 70.3, 53.5, and 12.5%, respectively (Que Hee et al., 1975).

Exposure limits (mg/m³): Potential occupational carcinogen. NIOSH REL: TWA 0.25, IDLH 50; OSHA PEL: TWA 0.25; ACGIH TLV: TWA 0.25 (adopted).

Symptoms of exposure: Headache, dizziness, nausea, vomiting, malaise, sweating, myoclonic limb jerks, clonic and tonic convulsions, coma, respiratory failure (NIOSH, 1997)

Toxicity:

EC_{50} (48-h) for *Daphnia pulex* 250 μg/L, *Simocephalus serrulatus* 215 μg/L (Mayer and Ellersieck, 1986).

LC_{50} (30-d) for *Pteronarcys californica* 2.0 μg/L, *Acroneuria pacifica* 0.2 μg/L (Jensen and Gaufin, 1966).

LC_{50} (28-d) 8.3 and 53.4 μg/L for tadpoles and adults of the species *Rana pipiens*, respectively (Schuytema et al., 1991).

LC_{50} (24-d) for *Xenopus laevis* tadpoles 5.5 μg/L (Schuytema et al., 1991).

LC_{50} (96-h) for goldfish 37 μg/L (Hartley and Kidd, 1987), bluegill sunfish 8 μg/L, fathead minnows 16 μg/L (Henderson et al., 1959), rainbow trout 10 μg/L, coho salmon 11 μg/L, chinook 6 μg/L (Katz, 1961), pumpkinseed 6.7 μg/L, channel catfish 4.5 μg/L (Verschueren, 1983), juvenile guppy 3-7 μg/L (Spehar et al., 1982), *Pteronarcys californica* 39 μg/L, *Acroneuria pacifica* 24 μg/L (Jensen and Gaufin, 1966), *Classenia sabulosa* 0.58 μg/L, *Pteronarcella badia* 0.5 μg/L (Sanders and Cope, 1968), African clawed frogs (*Xenopus laevis*) 40.4-49.5 μg/L, bullfrogs (*Rana catesbeiana*) 8.7-30.3 μg/L, leopard frogs (*Rana pipiens*) 71.3 μg/L (Schuytema et al., 1991); 11.7 and 4.95 μg/L for 37-day-old fry of African catfish (*Clarias gariepinus*) and *Oreochromis niloticus*, respectively (Lamai et al., 1999).

LC_{50} (96-h static bioassay) for sand shrimp 7 μg/L, hermit crab 18 μg/L, grass shrimp 50 μg/L (Eisler, 1969).

LC_{50} (48-h) for mosquito fish 8 ppb (Verschueren, 1983), red killifish 11 μg/L (Yoshioka et al., 1986), juvenile guppy 5-12 μg/L, adult guppy 35-73 μg/L (Spehar et al., 1982), *Simocephalus serrulatus* 190 μg/L, *Daphnia pulex* 250 μg/L (Sanders and Cope, 1966).

LC_{50} (24-h) for bluegill sunfish 170 ppb and fathead minnows 24 ppb (quoted, Verschueren, 1983).

Acute oral LD_{50} for wild birds 13.3 mg/kg, chickens 20 mg/kg, ducks 381 mg/kg, dogs 65 mg/kg, guinea pigs 49 mg/kg, hamsters 60 mg/kg, monkeys 3 mg/kg, mice 38 mg/kg, pigeons 23,700 mg/kg, pigs 38 mg/kg, quail 10.78 mg/kg, rats 38.3 mg/kg, rabbits 45 mg/kg (quoted, RTECS, 1985).

Drinking water standard: No MCLGs or MCLs have been proposed (U.S. EPA, 1996).

Uses: Insecticide; wool processing industry.

DIETHYLAMINE

Synonyms: Diethamine; *N,N*-Diethylamine; *N*-Ethylethanamine; UN 1154.

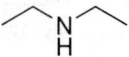

CAS Registry Number: 109-89-7
DOT: 1154
DOT label: Flammable liquid
Molecular formula: $C_4H_{11}N$
Formula weight: 73.14
RTECS: HZ8750000
Merck reference: 10, 3090

Physical state, color, and odor:
Colorless liquid with a fishy, ammonia-like odor. Odor threshold concentration is 140 ppb (Keith and Walters, 1992).

Melting point (°C):
-48 (Weast, 1986)
-50 (Windholz et al., 1983)

Boiling point (°C):
56.3 (Weast, 1986)
55.5 (Windholz et al., 1983)

Density (g/cm³):
0.7056 at 20/4 °C (Weast, 1986)
0.711 at 18/4 °C (Verschueren, 1983)

Diffusivity in water (x 10^{-5} cm²/sec):
1.11 at 25 °C (quoted, Hayduk and Laudie, 1974)

Dissociation constant, pKₐ:
11.090 at 20 °C (Gordon and Ford, 1972)
10.93 at 25 °C (Dean, 1973)

Flash point (°C):
-28 (Aldrich, 1988)

Lower explosive limit (%):
1.8 (NIOSH, 1997)

Upper explosive limit (%):
10.1 (NIOSH, 1997)

Henry's law constant (x 10^{-5} atm·m³/mol):
2.56 at 25 °C (Christie and Crisp, 1967)

Ionization potential (eV):
8.01 ± 0.01 (Franklin et al., 1969)

Soil sorption coefficient, log K_{oc}:
Unavailable because experimental methods for estimation of this parameter for aliphatic amines are lacking in the documented literature. However, its high solubility in water suggests its adsorption to soil will be nominal (Lyman et al., 1982).

Octanol/water partition coefficient, log K_{ow}:
0.43, 0.58 (Sangster, 1989)
0.44 (Collander, 1951)
0.57 (Leo et al., 1971)
0.81 (Eadsforth, 1986)

Solubility in organics:
Miscible with alcohol (Windholz et al., 1983)

Solubility in water:
815,000 mg/L at 14 °C (quoted, Verschueren, 1983)

Vapor density:
2.99 g/L at 25 °C, 2.52 (air = 1)

Vapor pressure (mmHg):
192 at 20 °C (NIOSH, 1997)
200 at 20 °C, 290 at 30 °C (Verschueren, 1983)
233 at 25 °C (Riddick et al., 1986)

Environmental fate:
   Chemical/Physical. Diethylamine reacted with NO_x in the dark forming diethylnitrosamine. In an outdoor chamber, photooxidation by natural sunlight yielded the following products: diethylnitramine, diethylformamide, diethylacetamide, ethylacetamide, ozone, acetaldehyde, and peroxyacetylnitrate (Pitts et al., 1978).
   Reacts with mineral acids forming water-soluble salts (Morrison and Boyd, 1971).

Exposure limits: NIOSH REL: TWA 10 ppm (30 mg/m^3), STEL 25 ppm (75 mg/m^3), IDLH 200 ppm; OSHA PEL: TWA 25 ppm; ACGIH TLV: TWA 5 ppm, STEL 15 ppm (adopted).

Symptoms of exposure: Strong irritant to the eyes, skin, and mucous membranes. Eye contact may cause corneal damage (Patnaik, 1992).

Toxicity:
   LC_{50} (96-h) for rainbow trout 25-182 mg/L (Spehar et al., 1982).
   LC_0 (24-h) and LC_{100} (24-h) for creek chub in Detroit river water were 70 and 100 mg/L, respectively (Gillette et al., 1952).
   Acute oral LD_{50} for mice 500 mg/kg, rats 540 mg/kg (quoted, RTECS, 1985).

Uses: In flotation agents, resins, dyes, resins, pesticides, rubber chemicals, and pharmaceuticals; selective solvent; polymerization and corrosion inhibitors; petroleum chemicals; electroplating; organic synthesis.

2-DIETHYLAMINOETHANOL

Synonyms: DEAE; Diethylaminoethanol; β-Diethylaminoethanol; *N*-Diethylaminoethanol; 2-*N*-Diethylaminoethanol; 2-Diethylaminoethyl alcohol; β-Diethylaminoethyl alcohol; Diethylethanolamine; *N,N*-Diethylethanolamine; Diethyl(2-hydroxyethyl)amine; *N,N*-Diethyl-*N*-(β-hydroxyethyl)amine; 2-Hydroxytriethylamine; UN 2686.

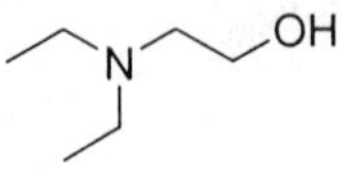

CAS Registry Number: 100-37-8
DOT: 2686
Molecular formula: $C_6H_{15}NO$
Formula weight: 117.19
RTECS: KK5075000
Merck reference: 10, 3092

Physical state, color, and odor:
Colorless, hygroscopic liquid with a nauseating, ammonia-like odor. Odor threshold concentration is 11 ppb (Amoore and Hautala, 1983).

Melting point (°C):
-70 (Dean, 1987)

Boiling point (°C):
163 (Windholz et al., 1983)
161 (Aldrich, 1988)

Density (g/cm³):
0.8800 at 25/4 °C (Windholz et al., 1983)
0.884 at 20/4 °C (Aldrich, 1988)

Diffusivity in water (x 10^{-5} cm²/sec):
0.75 at 20 °C using method of Hayduk and Laudie (1974)

Flash point (°C):
52 (NIOSH, 1997)
60 (open cup, Sax and Lewis, 1987)
48 (Aldrich, 1988)

Ionization potential (eV):
8.58 (Lias et al., 1998)

Soil sorption coefficient, log K_{oc}:
Unavailable because experimental methods for estimation of this parameter for aliphatic alcohols are lacking in the documented literature. However, its high solubility in water suggests its adsorption to soil will be nominal (Lyman et al., 1982).

Solubility in organics:
Soluble in alcohol, benzene, and ether (Windholz et al., 1983)

Solubility in water:
Miscible (NIOSH, 1997)

Vapor density:
4.79 g/L at 25 °C, 4.05 (air = 1)

Vapor pressure (mmHg):
1.4 at 20 °C (Sax and Lewis, 1987)

Exposure limits: NIOSH REL: TWA 10 ppm (50 mg/m^3), IDLH 100 ppm; OSHA PEL: TWA 10 ppm; ACGIH TLV: TWA 2 ppm (adopted).

Toxicity:
   Acute oral LD$_{50}$ for rats 1,300 mg/kg (quoted, RTECS, 1985).

Uses: Water-soluble salts; textile softeners; antirust formulations; fatty acid derivatives; pharmaceuticals; curing agent for resins; emulsifying agents in acid media; organic synthesis.

DIETHYL PHTHALATE

Synonyms: Anozol; **1,2-Benzenedicarboxylic acid diethyl ester**; DEP; Diethyl-*o*-phthalate; Estol 1550; Ethyl phthalate; NCI-C60048; Neantine; Palatinol A; Phthalol; Placidol E; RCRA waste number U088; Solvanol.

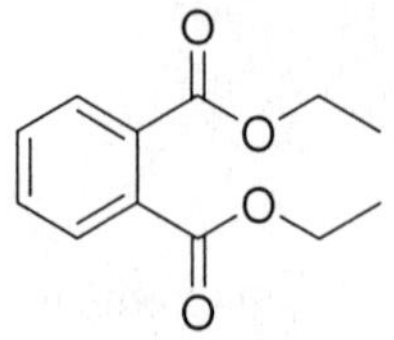

CAS Registry Number: 84-66-2
Molecular formula: $C_{12}H_{14}O_4$
Formula weight: 222.24
RTECS: TI1050000
Merck reference: 10, 7255

Physical state, color, and odor:
Clear, colorless, oily liquid with a mild, chemical odor. Bitter taste.

Melting point (°C):
-40.5 (Verschueren, 1983)

Boiling point (°C):
298 (Weast, 1986)
296 (Standen, 1968)
302 (Sax, 1984)

Density (g/cm³):
1.122 at 14.9/4 °C, 1.118 at 20.1/4 °C, 1.113 at 25.0/4 °C, 1.109 at 30.0/4 °C (De Lorenzi et al., 1997)
1.123 at 25/4 °C (Fishbein and Albro, 1972)

Diffusivity in water (x 10⁻⁵ cm²/sec):
0.59 at 20 °C using method of Hayduk and Laudie (1974)

Flash point (°C):
140 (Windholz et al., 1983)
163 (open cup, Sax, 1984)

Lower explosive limit (%):
0.7 at 186 °C (NFPA, 1984)

Henry's law constant (x 10⁻⁵ atm·m³/mol at 25 °C):
At 25 °C: 5.01, 4.54, 4.78, 4.94, 2.21, and 2.44 at pH values of 2.96, 2.98, 6.18, 6.19, 8.98, and 9.00, respectively (Hakuta et al., 1977).

Interfacial tension with water (dyn/cm at 20.5 °C):
16.27 (Donahue and Bartell, 1952)

Bioconcentration factor, log BCF:
2.07 (bluegill sunfish, Veith et al., 1980)
2.31 (*Chlorella pyrenoidosa*, Yan et al., 1995)

Soil sorption coefficient, log K_{oc}:
1.84 (Broome County, NY composite soil), 1.99 (Conklin, NY sand) (Russell and McDuffie, 1986)

Octanol/water partition coefficient, log K_{ow}:
2.35 (Leyder and Boulanger, 1983)
2.47 (Mabey et al., 1982)
2.42 (Ellington and Floyd, 1996)
2.24 (Howard et al., 1985)
1.40 (Veith et al., 1980)
2.82 (DeKock and Lord, 1987)

Solubility in organics:
Soluble in acetone and benzene; miscible with ethanol, ether, esters, and ketones (U.S. EPA, 1985)

Solubility in water:
928 mg/L at 20 °C (Leyder and Boulanger, 1983)
1,080 mg/L at 25 °C (Howard et al., 1985)
1,200 ppm at 25 °C (Fukano and Obata, 1976)
0.1 wt % at 20 °C (Fishbein and Albro, 1972)
0.15 wt % at 20-25 °C (Fordyce and Meyer, 1940)
0.132, 0.120, and 0.137 at 10.0, 20.0, and 30.0 °C, respectively (Schwarz and Miller, 1980)
680 mg/L at 25 °C (Russell and McDuffie, 1986)

Vapor density:
9.08 g/L at 25 °C, 7.67 (air = 1)

Vapor pressure (x 10^{-3} mmHg):
50 at 70 °C (Fishbein and Albro, 1972)
1.65 at 25 °C (Banerjee et al., 1980; Howard et al., 1985)
2.1 at 25 °C (Hinckley et al., 1990)
3.3 at 25 °C (extrapolated from vapor pressures determined at higher temperatures, Tesconi and Yalkowsky, 1998)

Environmental fate:
Biological. A proposed microbial degradation mechanism is as follows: 4-hydroxy-3-methylbenzyl alcohol to 4-hydroxy-3-methylbenzaldehyde to 3-methyl-4-hydroxybenzoic acid to 4-hydroxyisophthalic acid to protocatechuic acid to β-ketoadipic acid (Chapman, 1972). In anaerobic sludge, diethyl phthalate degraded as follows: monoethyl phthalate to phthalic acid to protocatechuic acid followed by ring cleavage and mineralization (Shelton et al., 1984).

In a static-culture-flask screening test, diethyl phthalate showed significant biodegradation with rapid adaptation. The ester (5 and 10 mg/L) was statically incubated in the dark at 25 °C with yeast extract and settled domestic wastewater inoculum. After 7 d, 100% biodegradation was achieved (Tabak et al., 1981).

Photolytic. An aqueous solution containing titanium dioxide and subjected to UV radiation (λ >290 nm) produced hydroxyphthalates and dihydroxyphthalates as intermediates (Hustert and Moza, 1988).

Chemical/Physical. Under alkaline conditions, diethyl phthalate will initially hydrolyze to ethyl hydrogen phthalate and ethanol. The monoester will undergo further hydrolysis forming *o*-phthalic acid and ethanol (Kollig, 1993). A second-order rate constant of 2.5 x 10^{-2} M/sec was reported for the hydrolysis of diethyl phthalate at 30 °C and pH 8 (Wolfe et al., 1980). At 30 °C, hydrolysis half-lives of 8.8 and 18 yr were reported at pH values 9 and 10-12, respectively (Callahan et al., 1979).

Pyrolysis of diethyl phthalate in a flow reactor at 700 °C yielded the following products: ethanol, ethylene, benzene, naphthalene, phthalic anhydride, and 2-phenylenaphthalene (Bove and Arrigo, 1985).

Exposure limits (mg/m³): NIOSH REL: TWA 5; ACGIH TLV: TWA 5 (adopted).

Symptoms of exposure: Vapor inhalation may cause lacrimation, coughing, and throat irritation (Patnaik, 1992).

Toxicity:
LC_{50} (contact) for earthworm (*Eisenia fetida*) 850 $\mu g/cm^2$ (Neuhauser et al., 1985).
LC_{50} (96-h) for bluegill sunfish 110 mg/L (Spehar et al., 1982), *Cyprinodon variegatus* 30 ppm using natural seawater (Heitmuller et al., 1981).
LC_{50} (72-h) for *Cyprinodon variegatus* 36 ppm (Heitmuller et al., 1981).
LC_{50} (48-h) for red killifish 98 mg/L (Yoshioka et al., 1986), *Daphnia magna* 52 mg/L (LeBlanc, 1980), *Cyprinodon variegatus* 38 ppm (Heitmuller et al., 1981).
LC_{50} (24-h) for *Daphnia magna* 52 mg/L (LeBlanc, 1980), *Cyprinodon variegatus* >69 ppm (Heitmuller et al., 1981).
Acute oral LD_{50} for guinea pigs 8,600 mg/kg, mice 6,172 mg/kg, rats 8,600 mg/kg (quoted, RTECS, 1985).
Heitmuller et al. (1981) reported a NOEC of 22 ppm.

Drinking water standard: No MCLGs or MCLs have been proposed (U.S. EPA, 1996).

Source: Leaching from PVC piping in contact with water (quoted, Verschueren, 1983)

Uses: Plasticizer; plastic manufacturing and processing; denaturant for ethyl alcohol; ingredient in insecticidal sprays and explosives (propellant); dye application agent; wetting agent; perfumery as fixative and solvent; solvent for nitrocellulose and cellulose acetate; camphor substitute.

1,1-DIFLUOROTETRACHLOROETHANE

Synonyms: 1,1-Difluoro-1,2,2,2-tetrachloroethane; 2,2-Difluoro-1,1,1,2-tetrachloroethane; Freon 112a; Halocarbon 112a; Refrigerant 112a; **1,1,1,2-Tetrachloro-2,2-difluoroethane**.

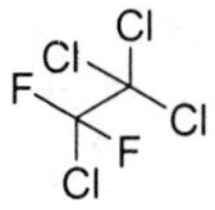

CAS Registry Number: 76-11-9
DOT: 1078
Molecular formula: $C_2Cl_4F_2$
Formula weight: 203.83
RTECS: KI1425000

Physical state, color, and odor:
Colorless solid with a faint ether-like odor

Melting point (°C):
40.6 (Weast, 1986)

Boiling point (°C):
91.5 (Weast, 1986)

Density (g/cm³):
1.65 at 20/4 °C (NIOSH, 1997)

Diffusivity in water (x 10^{-5} cm²/sec):
0.71 at 20 °C using method of Hayduk and Laudie (1974)

Flash point (°C):
Noncombustible solid (NIOSH, 1997)

Solubility in organics:
Soluble in alcohol, ether, and chloroform (Weast, 1986)

Vapor pressure (mmHg):
40 at 20 °C (NIOSH, 1997)

Exposure limits: NIOSH REL: TWA 500 ppm (4,170 mg/m³), IDLH 2,000 ppm; OSHA PEL: TWA 500 ppm; ACGIH TLV: TWA 500 ppm (adopted).

Symptoms of exposure: Eye and skin irritation, drowsiness, central nervous system depression (NIOSH, 1997)

Use: Organic synthesis.

1,2-DIFLUOROTETRACHLOROETHANE

Synonyms: 1,2-Difluoro-1,1,2,2-tetrachloroethane; F 112; Freon 112; Genetron 112; Halo-carbon 112; Refrigerant 112; **1,1,2,2-Tetrachloro-1,2-difluoroethane**.

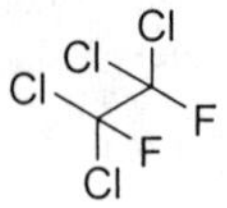

CAS Registry Number: 76-12-0
DOT: 1078
Molecular formula: $C_2Cl_4F_2$
Formula weight: 203.83
RTECS: KI1420000

Physical state, color, and odor:
Colorless liquid or solid with a faint, ether-like odor

Melting point (°C):
25 (Weast, 1986)

Boiling point (°C):
93 (Weast, 1986)
91.58 (Boublik et al., 1984)

Density (g/cm³):
1.6447 at 25/4 °C (Weast, 1986)

Diffusivity in water (x 10⁻⁵ cm²/sec):
0.89 at 25 °C using method of Hayduk and Laudie (1974)

Flash point (°C):
Noncombustible (NIOSH, 1997)

Henry's law constant (atm·m³/mol):
0.102 at 25 °C (approximate - calculated from water solubility and vapor pressure)

Ionization potential (eV):
11.30 (NIOSH, 1997)

Soil sorption coefficient, log K_{oc}:
2.78 using method of Chiou et al. (1979)

Octanol/water partition coefficient, log K_{ow}:
3.39 using method of Hansch et al. (1968)

Solubility in organics:
Soluble in alcohol, chloroform, and ether (Weast, 1986)

Solubility in water:
120 mg/L at 25 °C (Du Pont, 1966)

Vapor density:
8.33 g/L at 25 °C, 7.04 (air = 1)

Vapor pressure (mmHg):
40 at 20 °C (NIOSH, 1997)
45.8 at 25 °C (Boublik et al., 1984)

Exposure limits: NIOSH REL: TWA 500 ppm (4,170 mg/m^3), IDLH 2,000 ppm; OSHA PEL: TWA 500 ppm (adopted).

Toxicity:
   Acute oral LD_{50} for mice 800 mg/kg (quoted, RTECS, 1985).
   LD_{50} (inhalation) for mice 123 gm/m^3/2-h (quoted, RTECS, 1985).

Uses: Organic synthesis.

DIISOBUTYL KETONE

Synonyms: DIBK; *sym*-Diisopropylacetone; 2,6-Dimethylheptan-4-one; **2,6-Dimethyl-4-heptanone**; Isobutyl ketone; Isovalerone; UN 1157; Valerone.

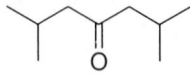

CAS Registry Number: 108-83-8
DOT: 1157
DOT label: Combustible liquid
Molecular formula: $C_9H_{18}O$
Formula weight: 142.24
RTECS: MJ5775000

Physical state, color, and odor:
Clear, colorless liquid with a mild, sweet, ether-like odor. Odor threshold concentration is 0.11 ppm (Amoore and Hautala, 1983).

Melting point (°C):
-46 to -42 (Verschueren, 1983)

Boiling point (°C):
168 (Weast, 1986)

Density (g/cm³):
0.8053 at 20/4 °C (Weast, 1986)

Diffusivity in water (x 10⁻⁵ cm²/sec):
0.63 at 20 °C using method of Hayduk and Laudie (1974)

Flash point (°C):
49 (NIOSH, 1997)

Lower explosive limit (%):
0.8 at 93 °C (NFPA, 1984)

Upper explosive limit (%):
7.1 at 93 °C (NOSH, 1994)

Henry's law constant (x 10⁻⁴ atm·m³/mol):
6.36 at 20 °C (approximate - calculated from water solubility and vapor pressure)

Ionization potential (eV):
9.04 (NIOSH, 1997)

Soil sorption coefficient, log K_{oc}:
Unavailable because experimental methods for estimation of this parameter for aliphatic ketones are lacking in the documented literature

Octanol/water partition coefficient, log K_{ow}:
2.58 using method of Hansch et al. (1968)

Solubility in organics:
Soluble in alcohol and ether (Weast, 1986)

Solubility in water (wt %):
0.069 at 0°C, 0.558 at 9.2 °C, 0.045 at 20.6 °C, 0.040 at 30.8 °C, 0.032 at 40.5 °C, 0.032 at 50.0 °C, 0.031 at 60.7 °C, 0.034 at 70.5 °C, 0.035 at 80.8 °C, 0.037 at 90.7 °C (Stephenson, 1992)

Vapor density:
5.81 g/L at 25 °C, 4.91 (air = 1)

Vapor pressure (mmHg):
1.7 at 20 °C, 2.3 at 30 °C (Verschueren, 1983)

Environmental fate:
Chemical/Physical. Diisobutyl ketone will not hydrolyze because it has no hydrolyzable functional group.

At an influent concentration of 300 mg/L, treatment with granular activated carbon resulted in non-detectable concentrations in the effluent. The adsorbability of the carbon used was 60 mg/g carbon (Guisti et al., 1974).

Exposure limits: NIOSH REL: TWA 25 ppm (150 mg/m^3), IDLH 500 ppm; OSHA PEL: TWA 50 ppm (290 mg/m^3); ACGIH TLV: TWA 25 ppm (adopted).

Symptoms of exposure: Vapor inhalation may cause irritation of eyes, nose, and throat (Patnaik, 1992).

Toxicity:
Acute oral LD$_{50}$ for rats 5,750 mg/kg, mice 1,416 mg/kg (quoted, RTECS, 1985).

Uses: Solvent for nitrocellulose, synthetic resins, rubber, lacquers; coating compositions; inks and stains; organic synthesis.

DIISOPROPYLAMINE

Synonyms: DIPA; *N*-(1-Methylethyl)-2-propanamine; UN 1158.

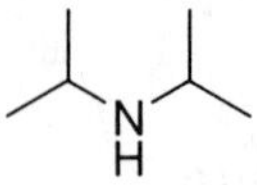

CAS Registry Number: 108-18-9
DOT: 1158
DOT label: Flammable liquid
Molecular formula: $C_6H_{15}N$
Formula weight: 101.19
RTECS: IM4025000
Merck reference: 10, 3182

Physical state, color, and odor:
Colorless liquid with an ammonia-like odor. Odor threshold concentration is 1.8 ppm (Amoore and Hautala, 1983).

Melting point (°C):
-61 (Weast, 1986)
-96.3 (Verschueren, 1983)

Boiling point (°C):
84 (Weast, 1986)

Density (g/cm³):
0.7169 at 20/4 °C (Weast, 1986)
0.722 at 22/4 °C (Windholz et al., 1983)

Diffusivity in water (x 10⁻⁵ cm²/sec):
0.72 at 20 °C using method of Hayduk and Laudie (1974)

Dissociation constant, pK_a:
11.13 at 21 °C (Gordon and Ford, 1972)

Flash point (°C):
-6.7 (NIOSH, 1997)
-1.11 (open cup, Hawley, 1981)

Lower explosive limit (%):
1.1 (NIOSH, 1997)

Upper explosive limit (%):
7.1 (NIOSH, 1997)

Ionization potential (eV):
7.73 ± 0.03 (Franklin et al., 1969)

Soil sorption coefficient, log K_{oc}:
Unavailable because experimental methods for estimation of this parameter for aliphatic amines

are lacking in the documented literature. However, its miscibility in water suggests its adsorption to soil will be nominal (Lyman et al., 1982).

Solubility in organics:
Soluble in acetone, alcohol, benzene, and ether (Weast, 1986)

Solubility in water (wt %):
Miscible at <28.0 °C, 12.39 at 28.0 °C, 10.32 at 30.0 °C, 6.35 at °40.0, 4.57 at 50.0 °C, 2.90 at 60.0 °C, 2.15 at 70.0 °C, 1.76 at 80.0 °C, 1.61 at 84.0 °C (Stephenson, 1993a).

Vapor density:
4.14 g/L at 25 °C, 3.49 (air = 1)

Vapor pressure (mmHg):
70 at 20 °C (NIOSH, 1997)

Environmental fate:
Chemical/Physical. Reacts with acids forming water-soluble salts.

Exposure limits: NIOSH REL: TWA 5 ppm (20 mg/m^3), IDLH 200 ppm; OSHA PEL: TWA 5 ppm; ACGIH TLV: TWA 5 ppm (adopted).

Symptoms of exposure: Severe irritation of eyes, skin, and respiratory tract. Contact with skin causes burns. Visual disturbance and cloudy swelling of cornea accompanied by partial or total loss of vision (Patnaik, 1992).

Toxicity:
LC_{50} (96-h) for rainbow trout 37 mg/L (Spehar et al., 1982).

LC_0 (24-h) and LC_{100} (24-h) for creek chub in Detroit river water were 40 and 60 mg/L, respectively (Gillette et al., 1952).

Acute oral LD_{50} for guinea pigs, 2,800 mg/kg, mice 2,120 mg/kg, rats 770 mg/kg, rabbits 4,700 mg/kg (quoted, RTECS, 1985).

Uses: Intermediate; catalyst; organic synthesis.

N,N-DIMETHYLACETAMIDE

Synonyms: Acetdimethylamide; Acetic acid dimethylamide; Dimethylacetamide; Dimethyl-acetone amide; Dimethylamide acetate; DMA; DMAC; Hallucinogen; NSC 3138; U 5954.

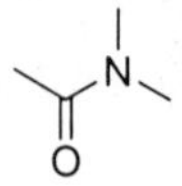

CAS Registry Number: 127-19-5
Molecular formula: C_4H_9NO
Formula weight: 87.12
RTECS: AB7700000
Merck reference: 10, 3218

Physical state, color, and odor:
Clear, colorless liquid with a weak, ammonia-like odor. Odor threshold concentration is 47 ppm (Amoore and Hautala, 1983).

Melting point (°C):
-20 (Weast, 1986)

Boiling point (°C):
165 at 758 mmHg (Weast, 1986)
165.5 (Dean, 1987)

Density (g/cm³):
0.9366 at 25/4 °C (Weast, 1986)

Diffusivity in water (x 10⁻⁵ cm²/sec):
0.89 at 20 °C using method of Hayduk and Laudie (1974)

Flash point (°C):
77.2 (open cup, Sax and Lewis, 1987)
70 (NFPA, 1984)

Lower explosive limit (%):
1.8 at 100 °C (NFPA, 1984)

Upper explosive limit (%):
11.5 at 160 °C (NFPA, 1984)

Ionization potential (eV):
8.81 ± 0.03 (Franklin et al., 1969)
8.60 (HNU, 1986)

Soil sorption coefficient, log K_{oc}:
Unavailable because experimental methods for estimation of this parameter for aliphatic amides are lacking in the documented literature. However, its miscibility in water and low K_{ow} suggest its adsorption to soil will be nominal (Lyman et al., 1982).

Octanol/water partition coefficient, log K_{ow}:
-0.77 (Sangster, 1989)

Solubility in organics:
Miscible with aromatics, esters, ketones, and ethers (Hawley, 1981)

Solubility in water:
Miscible (NIOSH, 1997).

Vapor pressure (mmHg):
2 at 20 °C (NIOSH, 1997)
1.3 at 25 °C, 9 at 60 °C (Verschueren, 1983)

Environmental fate:
   Chemical/Physical. Releases toxic fumes of nitrogen oxides when heated to decomposition (Sax and Lewis, 1987).

Exposure limits: NIOSH REL: TWA 10 ppm (35 mg/m^3), IDLH 300 ppm; OSHA PEL: TWA 10 ppm; ACGIH TLV: TWA 10 ppm (adopted).

Toxicity:
   LC_{50} (1-h inhalation) for rats >2,475ppm (Kennedy and Sherman, 1986).
   Acute oral LD_{50} for male rats 5,809 mg/kg, female rats 4,930 mg/kg (Kennedy and Sherman, 1986).

Uses: Solvent used in organic synthesis; paint removers; solvent for plastics, resins, gums, and electrolytes; intermediate; catalyst.

DIMETHYLAMINE

Synonyms: DMA; *N*-Methylmethanamine; RCRA waste number U092; UN 1032; UN 1160.

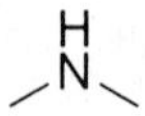

$$\underset{\diagup \, N \, \diagdown}{\overset{\overset{H}{|}}{}}$$

CAS Registry Number: 124-40-3
DOT: 1032 (anhydrous), 1160 (aqueous solution)
DOT label: Flammable gas/flammable liquid (aqueous)
Molecular formula: C_2H_7N
Formula weight: 45.08
RTECS: IP8750000
Merck reference: 10, 3219

Physical state, color, and odor:
Clear, colorless liquid or gas with a strong, ammonia-like odor. Odor threshold concentration is 0.34 ppm (Amoore and Hautala, 1983).

Melting point (°C):
-93 (Weast, 1986)
-96 (Stull, 1947)

Boiling point (°C):
7.4 (Weast, 1986)
6.9 (Dean, 1987)

Density (g/cm³):
0.6804 at 0/4 °C (Weast, 1986)

Diffusivity in water (x 10^{-5} cm²/sec):
1.11 at 20 °C using method of Hayduk and Laudie (1974)

Dissociation constant, pK_a:
10.732 at 25 °C (Gordon and Ford, 1972)

Flash point (°C):
-57, 65.6, 52.8, 39.4, 25.6, -18, -27, -34, and -42 for 0, 1, 2, 5, 10, 40, 50, 60, and 65 wt % aqueous solutions, respectively (Mitchell et al., 1999)
-17.7 (25% solution, Hawley, 1981)

Lower explosive limit (%):
2.8 (NIOSH, 1997)

Upper explosive limit (%):
14.4 (NIOSH, 1997)

Heat of fusion (kcal/mol):
1.420 (Dean, 1987)

Henry's law constant (x 10^{-5} atm·m³/mol):
1.75 at 25 °C (Christie and Crisp, 1967)

Ionization potential (eV):
8.36 (Gibson et al., 1977)

Soil sorption coefficient, log K_{oc}:
2.43 (Melfort loam), 2.49 (Weyburn Oxbow loam), 2.70 (Regina heavy clay), 2.59 (Indian head loam), 2.64 (Asquith sandy loam) (Grover and Smith, 1974)

Octanol/water partition coefficient, log K_{ow}:
-0.38 at pH 13 (Sangster, 1989)

Solubility in organics:
Soluble in alcohol and ether (Weast, 1986)

Solubility in water:
24 wt % at 60 °C (NIOSH, 1997)

Vapor density:
1.84 g/L at 25 °C, 1.56 (air = 1)

Vapor pressure (mmHg):
1,292 at 20 °C (Verschueren, 1983)
1,520 at 25 °C (Howard, 1990)

Environmental fate:
  Photolytic. Dimethylnitramine, nitrous acid, formaldehyde, *N,N*-dimethylformamide and carbon monoxide were reported as photooxidation products of dimethylamine with NO_x. An additional compound was tentatively identified as tetramethylhydrazine (Tuazon et al., 1978). In the atmosphere, dimethylamine reacts with OH radicals forming formaldehyde and/or amides (Atkinson et al., 1978). The rate constant for the reaction of dimethylamine and ozone in the atmosphere is 2.61 x 10^{-18} cm^3/molecule·sec at 296 K (Atkinson and Carter, 1984).
  Soil. After 2 d, degradation yields in an Arkport fine sandy loam (Varna, NY) and sandy soil (Lake George, NY) amended with sewage and nitrite-N were 50 and 20%, respectively. *N*-Nitrosodimethylamine was identified as the major metabolite (Greene et al., 1981). Mills and Alexander (1976) reported that *N*-nitrosodimethylamine also formed in soil, municipal sewage, and lake water supplemented with dimethylamine (ppm) and nitrite-N (100 ppm). They found that nitrosation occurred under nonenzymatic conditions at neutral pHs.
  Chemical/Physical. In an aqueous solution, chloramine reacted with dimethylamine forming *N*-chlorodimethylamine (Isaac and Morris, 1983).
  Reacts with mineral acids forming water soluble ammonium salts and ethanol (Morrison and Boyd, 1971).

Exposure limits: NIOSH REL: TWA 10 ppm (18 mg/m^3), IDLH 500 ppm; OSHA PEL: TWA 10 ppm; ACGIH TLV: TWA 5 ppm, STEL 15 ppm (adopted).

Symptoms of exposure: Strong irritation of eyes, skin, and mucous membranes. Contact with skin may cause necrosis. Eye contact with liquid can cause corneal damage and loss of vision (Patnaik, 1992).

Toxicity:
  LC_{50} (6-h inhalation) for Fischer-344 rats 3,450 ppm (Steinhagen et al., 1982).
  LC_{50} (96-h) for rainbow trout 17 mg/L (Spehar et al., 1982).

Acute oral LD_{50} for guinea pigs 340 mg/kg, mice 316 mg/kg, rats 698 mg/kg, rabbits 240 mg/kg (quoted, RTECS, 1985).

Source: Dimethylamine naturally occurs in soybean seeds (8 ppm), cauliflower (14 ppm), kale leaves (5.5 ppm), barleygrass seeds (1.6 ppm), tobacco leaves, hawthorne leaves, hops flower (1.4 ppm), cabbage leaves (2-2,8 ppm), corn (1-3.5 ppm), celery (5.1 ppm), grapes, grape wine, and grape juice (Duke, 1992).

Uses: Detergent soaps; accelerator for vulcanizing rubber; detection of magnesium; tanning; acid gas absorbent solvent; gasoline stabilizers; textile chemicals; pharmaceuticals; surfactants; manufacture of *N,N*-dimethylformamide and *N,N*-dimethylacetamide; rocket propellants; missile fuels; dehairing agent; electroplating.

p-DIMETHYLAMINOAZOBENZENE

Synonyms: Atul fast yellow R; Benzeneazodimethylaniline; Brilliant fast oil yellow; Brilliant fast spirit yellow; Brilliant fast yellow; Brilliant oil yellow; Butter yellow; Cerasine yellow CG; C.I. 11020; C.I. solvent yellow 2; DAB; Dimethylaminobenzene; 4-Dimethylaminoazobenzene; 4-(*N,N*-Dimethylamino)azobenzene; *N,N*-Dimethyl-4-aminoazobenzene; *N,N*-Dimethyl-*p*-aminoazobenzene; Dimethylaminoazobenzol; 4-Dimethylaminoazobenzol; 4-Dimethylamino-phenylazobenzene; *N,N*-Dimethyl-*p*-azoaniline; *N,N*-Dimethyl-4-(phenylazo)benzamine; *N,N*-Dimethyl-*p*-(phenylazo)benzamine; **N,N-Dimethyl-4-(phenylazo)benzenamine**; *N,N*-Dimethyl-*p*-(phenylazo)benzenamine; Dimethyl yellow; Dimethyl yellow analar; Dimethyl yellow *N,N*-dimethylaniline; DMAB; Enial yellow 2G; Fast oil yellow B; Fast yellow; Fast yellow A; Fast yellow AD OO; Fast yellow ES; Fast yellow ES extra; Fast yellow extra conc.; Fast yellow R; Fast yellow R (8186); Grasal brilliant yellow; Methyl yellow; Oil yellow; Oil yellow 20; Oil yellow 2625; Oil yellow 7463; Oil yellow BB; Oil yellow D; Oil yellow DN; Oil yellow FF; Oil yellow FN; Oil yellow G; Oil yellow G-2; Oil yellow 2G; Oil yellow GG; Oil yellow GR; Oil yellow II; Oil yellow N; Oil yellow PEL; Oleal yellow 2G; Organol yellow ADM; Orient oil yellow GG; PDAB; Petrol yellow WT; RCRA waste number U093; Resinol yellow GR; Resoform yellow GGA; Silotras yellow T2G; Somalia yellow A; Stear yellow JB; Sudan GG; Sudan yellow; Sudan yellow 2G; Sudan yellow 2GA; Toyo oil yellow G; USAF EK-338; Waxoline yellow AD; Waxoline yellow ADS; Yellow G soluble in grease.

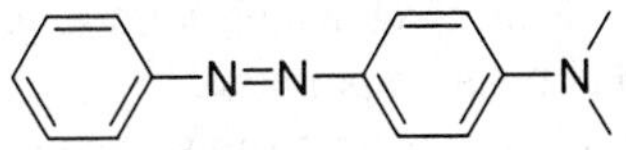

CAS Registry Number: 60-11-7
Molecular formula: $C_{14}H_{15}N_3$
Formula weight: 225.30
RTECS: BX7350000
Merck reference: 10, 3220

Physical state and color:
Yellow leaflets or crystals

Melting point (°C):
114-117 (Windholz et al., 1983)

Boiling point (°C):
Sublimes (Weast, 1986).

Density (g/cm³):
1.212 using method of Lyman et al. (1982)

Diffusivity in water (x 10^{-5} cm²/sec):
0.50 at 20 °C using method of Hayduk and Laudie (1974)

Soil sorption coefficient, log K_{oc}:
3.00 (calculated, Mercer et al., 1990)

Octanol/water partition coefficient, log K_{ow}:
4.58 (quoted, Verschueren, 1983)

Solubility in organics:
Room temperature (g/L): dimethyl sulfoxide (5-10), acetone (50-100), toluene (12-30) (Keith and Walters, 1992).

Solubility in water:
10^{-3} wt % at 20 °C (NIOSH, 1997)
13.6 mg/L at 20-30 °C (Mercer et al., 1990)

Vapor pressure (mmHg):
3 x 10^{-7} (estimated, NIOSH, 1997)

Environmental fate:
   Chemical/Physical. Releases toxic nitrogen oxides when heated to decomposition (Sax and Lewis, 1987).

Exposure limits: Potential occupational carcinogen. Because no standards have been established, NIOSH (1997) recommends the most reliable and protective respirators be used, i.e., a self-contained breathing apparatus that has a full facepiece and is operated under positive-pressure or a supplied-air respirator that has a full facepiece and is operated under pressure-demand or under positive-pressure in combination with a self-contained breathing apparatus operated under pressure-demand or positive-pressure.
   OSHA recommends that worker exposure to this chemical is to be controlled by use of engineering control, proper work practices, and proper selection of personal protective equipment. Specific details of these requirements can be found in CFR 1910.1003-1910.1016.
   ACGIH TLV: TWA 0.5 ppm (adopted).

Toxicity:
   Acute oral LD_{50} for mice 300 mg/kg, rats 200 mg/kg (quoted, RTECS, 1985).

Uses: Not commercially produced in the United States. pH indicator; determines hydrochloric acid in gastric juice; coloring agent; organic research.

DIMETHYLANILINE

Synonyms: Dimethylaminobenzene; *N,N*-Dimethylaniline; ***N,N*-Dimethylbenzenamine**; Dimethylphenylamine; *N,N*-Dimethylphenylamine; NCI-C56428; UN 2253; Versneller NL 63/10.

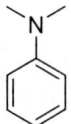

CAS Registry Number: 121-69-7
DOT: 2253
Molecular formula: $C_8H_{11}N$
Formula weight: 121.18
RTECS: BX4725000
Merck reference: 10, 9893

Physical state, color, and odor:
Straw to brown-colored oily liquid with an amine-like odor. Odor threshold concentration is 13 ppb (Amoore and Hautala, 1983).

Melting point (°C):
2.45 (Weast, 1986)

Boiling point (°C):
192-194 (Windholz et al., 1983)
194 (Riddick et al., 1986)

Density (g/cm³):
0.9557 at 20/4 °C (Weast, 1986)

Diffusivity in water (x 10^{-5} cm²/sec):
0.77 at 20 °C using method of Hayduk and Laudie (1974)

Dissociation constant, pK_a:
5.21 at 25 °C (Dean, 1973)

Flash point (°C):
61 (NIOSH, 1997)

Entropy of fusion (cal/mol·K):
9.91 (Tsonopoulos and Prausnitz, 1971)

Heat of fusion (kcal/mol):
2.73 (Tsonopoulos and Prausnitz, 1971)

Henry's law constant (x 10^{-6} atm·m³/mol):
4.98 at 20 °C (approximate - calculated from water solubility and vapor pressure)

Ionization potential (eV):
7.12 (Franklin et al., 1969)

Soil sorption coefficient, log K_{oc}:
2.26 (Meylan et al., 1992)

Octanol/water partition coefficient, log K_{ow}:
2.30 (HPLC, Unger et al., 1978)
2.31 (Leo et al., 1971)
2.62 at pH 7.4 (Rogers and Cammarata, 1969)

Solubility in organics:
Soluble in acetone, alcohol, benzene, chloroform, and ether (Weast, 1986)

Solubility in water:
1,105.2 mg/L at 25 °C (Chiou et al., 1982)

Vapor density:
4.95 g/L at 25 °C, 4.18 (air = 1)

Vapor pressure (mmHg):
0.5 at 20 °C, 1.1 at 30 °C (Verschueren, 1983)
0.52 at 25 °C (Banerjee et al., 1990)

Environmental fate:
   Photolytic. A rate constant of 1.48 x 10^{-10} cm^3/molecule·sec was reported for the reaction of *N,N*-dimethylaniline and OH radicals in air at room temperature (Atkinson et al., 1987).
   Chemical/Physical. Products identified from the gas-phase reaction of ozone with *N,N*-dimethylaniline in synthetic air at 23 °C were: *N*-methylformanilide, formaldehyde, formic acid, hydrogen peroxide, and a nitrated salt having the formula: $[C_6H_6NH(CH_3)_2]^+NO_3^-$ (Atkinson et al., 1987). Reacts with acids forming water-soluble salts.

Exposure limits: NIOSH REL: TWA 5 ppm (25 mg/m^3), STEL 10 ppm (50 mg/m^3), IDLH 100 ppm; OSHA PEL: TWA 5 ppm; ACGIH TLV: TWA 5 ppm, STEL 10 ppm (adopted).

Toxicity:
   LC_{50} (48-h) for red killifish 275 mg/L (Yoshioka et al., 1986).
   Acute oral LD_{50} for rats 1,410 mg/kg (Yoshioka et al., 1986).

Uses: Manufacture of vanillin, Michler's ketone, methyl violet, and other dyes; solvent; reagent for methyl alcohol, hydrogen peroxide, methyl furfural, nitrate, and formaldehyde; chemical intermediate; stabilizer; reagent.

2,2-DIMETHYLBUTANE

Synonyms: Neohexane; UN 1208.

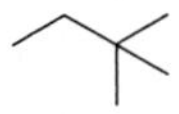

CAS Registry Number: 75-83-2
DOT: 2457
Molecular formula: C_6H_{14}
Formula weight: 86.18
RTECS: EJ9300000

Physical state, color, and odor:
Colorless liquid with a mild gasoline-like odor

Melting point (°C):
-99.9 (Weast, 1986)

Boiling point (°C):
49.7 (Weast, 1986)

Density (g/cm³):
0.6485 at 20/4 °C (Weast, 1986)
0.6570 at 25/4 °C (Hawley, 1981)

Diffusivity in water (x 10^{-5} cm²/sec):
0.75 at 20 °C using method of Hayduk and Laudie (1974)

Dissociation constant, pK_a:
>14 (Schwarzenbach et al., 1993)

Flash point (°C):
-47.8 (Sax and Lewis, 1987)
-34 (Aldrich, 1988)

Lower explosive limit (%):
1.2 (Sax and Lewis, 1987)

Upper explosive limit (%):
7.0 (Sax and Lewis, 1987)

Heat of fusion (kcal/mol):
0.138 (Dean, 1987)

Henry's law constant (atm·m³/mol):
1.69 at 25 °C (Mackay and Shiu, 1981)

Interfacial tension with water (dyn/cm at 25 °C):
49.7 (quoted, Freitas et al., 1997)

Ionization potential (eV):
10.06 (HNU, 1986)

Soil sorption coefficient, log K_{oc}:
Unavailable because experimental methods for estimation of this parameter for aliphatic hydrocarbons are lacking in the documented literature

Octanol/water partition coefficient, log K_{ow}:
3.82 (Hansch and Leo, 1979)

Solubility in organics:
In methanol: 590 and 800 g/L at 5 and 10 °C, respectively. Miscible at higher temperatures (Kiser et al., 1961).

Solubility in water (mg/kg):
21.2 at 25 °C (Price, 1976)
18.4 at 25 °C (McAuliffe, 1963, 1966)
39.4 at 0 °C, 23.8 at 25 °C (Polak and Lu, 1973)

Vapor density:
3.52 g/L at 25 °C, 2.98 (air = 1)

Vapor pressure (mmHg):
325 at 24.47 °C (Willingham et al., 1945)
319.1 at 25 °C (Wilhoit and Zwolinski, 1971)

Environmental fate:
Photolytic. Reported photooxidation rate constants for the reaction of 2,2-dimethylbutane with OH radicals at 297, 299, and 300 K are 2.59 x 10^{-12}, 6.16 x 10^{-12}, and 2.59 x 10^{-12} cm^3/molecule·sec, respectively (Atkinson, 1985, 1990).
Chemical/Physical. Complete combustion in air yields carbon dioxide and water vapor. 2,2-Dimethylbutane will not hydrolyze because it has no hydrolyzable functional group.

Exposure limits: ACGIH TLV: TWA and STEL for all isomers except *n*-hexane are 500 and 1,000 ppm, respectively (adopted).

Source: Comprised 0.1 vol % of total evaporated hydrocarbons from gasoline tank (quoted, Verschueren, 1983).
Schauer et al. (1999) reported 2,2-dimethylbutane in a diesel-powered medium-duty truck exhaust at an emission rate of 310 μg/km.

Use: Intermediate for agricultural chemicals; in high octane fuels.

2,3-DIMETHYLBUTANE

Synonyms: Biisopropyl; Diisopropyl; Isopropyldimethylmethane; UN 2457.

CAS Registry Number: 79-29-8
DOT: 2457
Molecular formula: C_6H_{14}
Formula weight: 86.18
RTECS: EJ9350000

Physical state, color, and odor:
Colorless liquid with a mild gasoline-like odor

Melting point (°C):
-128.5 (Weast, 1986)
-135 (Verschueren, 1983)

Boiling point (°C):
58 (Weast, 1986)

Density (g/cm³):
0.66164 at 20/4 °C, 0.65702 at 25/4 °C (Dreisbach, 1959)

Diffusivity in water (x 10^{-5} cm²/sec):
0.43 at 20 °C using method of Hayduk and Laudie (1974)

Dissociation constant, pK_a:
>14 (Schwarzenbach et al., 1993)

Flash point (°C):
-33 (Aldrich, 1988)

Lower explosive limit (%):
1.2 (NFPA, 1984)

Upper explosive limit (%):
7.0 (NFPA, 1984)

Heat of fusion (kcal/mol):
0.194 (Dean, 1987)

Henry's law constant (atm·m³/mol):
1.28 at 25 °C (Mackay and Shiu, 1981)

Interfacial tension with water (dyn/cm at 25 °C):
49.8 (quoted, Freitas et al., 1997)

Ionization potential (eV):
10.24 (Collin and Lossing, 1959)

Soil sorption coefficient, log K_{oc}:
Unavailable because experimental methods for estimation of this parameter for aliphatic hydrocarbons are lacking in the documented literature

Octanol/water partition coefficient, log K_{ow}:
3.85 (Hansch and Leo, 1979)

Solubility in organics:
In methanol: 495, 593, 760, and 1,700 g/L at 5, 10, 15, and 20 °C, respectively. Miscible at higher temperatures (Kiser et al., 1961).

Solubility in water (mg/kg):
19.1 at 25 °C, 19.2 at 40.1 °C, 23.7 at 55.1 °C, 40.1 at 99.1 °C, 56.8 at 121.3 °C, 97.9 at 137.3
    °C, 171.0 at 149.5 °C (Price, 1976)
32.9 at 0 °C, 22.5 at 25 °C (Polak and Lu, 1973)

Vapor density:
3.52 g/L at 25 °C, 2.98 (air = 1)

Vapor pressure (mmHg):
200 at 20 °C (Verschueren, 1983)
217 at 23.10 °C (Willingham et al., 1945)
234.6 at 25 °C (Wilhoit and Zwolinski, 1971)

Environmental fate:
   Photolytic. Major products reported from the photooxidation of 2,3-dimethylbutane with nitrogen oxides are carbon monoxide and acetone. Minor products included formaldehyde, acetaldehyde and peroxyacyl nitrates (Altshuller, 1983). Synthetic air containing gaseous nitrous acid and exposed to artificial sunlight (λ = 300-450 nm) photooxidized 2,3-dimethyl-butane into acetone, hexyl nitrate, peroxyacetal nitrate, and a nitro aromatic compound tentatively identified as a propyl nitrate (Cox et al., 1980).
   The following rate constants were reported for the reaction of 2,3-dimethylbutane and OH radicals in the atmosphere: 3.1×10^{12} cm³/mol·sec at 300 K (Hendry and Kenley, 1979); 5.16×10^{12} cm³/mol·sec (Greiner, 1970); 6.19×10^{-12} cm³/molecule·sec (Sabljić and Güsten, 1990); 5.67×10^{-12} cm³/molecule·sec at 300 K (Darnall et al., 1978); 6.30×10^{-12} cm³/molecule·sec (Atkinson, 1990). Based on a photooxidation rate constant of 6.3×10^{-12} cm³/molecule·sec for the reaction of 2,3-dimethylbutane and OH radicals in summer sunlight, the atmospheric lifetime is 22 h (Altshuller, 1991). Photooxidation rate constants of 4.06×10^{-16} and 4.06×10^{-16} cm³/molecule·sec were reported for the gas-phase reaction of 2,3-dimethylbutane and NO_3 (Atkinson, 1990; Sabljić and Güsten, 1990).
   Chemical/Physical. Complete combustion in air yields carbon dioxide and water vapor. 2,3-Dimethylbutane will not hydrolyze because it has no hydrolyzable functional group.

Exposure limits: ACGIH TLV: TWA and STEL for all isomers except *n*-hexane are 500 and 1,000 ppm, respectively (adopted).

Source: Comprised 1.6-2.6 vol % of total evaporated hydrocarbons from gasoline tank (quoted, Verschueren, 1983). Schauer et al. (1999) reported 2,3-dimethylbutane in a diesel-powered medium-duty truck exhaust at an emission rate of 570 μg/km.

Uses: Organic synthesis; gasoline component.

cis-1,2-DIMETHYLCYCLOHEXANE

Synonyms: *cis-o*-Dimethylcyclohexane; *cis*-1,2-Hexahydroxylene.

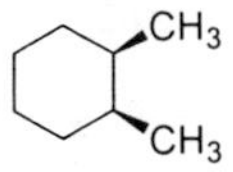

CAS Registry Number: 2207-01-4
DOT: 2263
Molecular formula: C_8H_{16}
Formula weight: 112.22

Physical state and color:
Colorless liquid

Melting point (°C):
-50.1 (Weast, 1986)

Boiling point (°C):
129.7 (Weast, 1986)

Density (g/cm³):
0.79627 at 20/4 °C, 0.79222 at 25/4 °C (Riddick et al., 1986)

Diffusivity in water (x 10⁻⁵ cm²/sec):
0.72 at 20 °C using method of Hayduk and Laudie (1974)

Dissociation constant, pK_a:
>14 (Schwarzenbach et al., 1993)

Flash point (°C):
-12 (Aldrich, 1988)

Heat of fusion (kcal/mol):
0.3932 (Riddick et al., 1986)

Henry's law constant (atm·m³/mol):
0.354 at 25 °C (Hine and Mookerjee, 1975)

Ionization potential (eV):
10.08 ± 0.02 (Franklin et al., 1969)

Soil sorption coefficient, log K_{oc}:
Unavailable because experimental methods for estimation of this parameter for alicyclic hydrocarbons are lacking in the documented literature

Octanol/water partition coefficient, log K_{ow}:
3.26 using method of Hansch et al. (1968)

Solubility in organics:
Soluble in acetone, alcohol, benzene, ether, and ligroin (Weast, 1986)

Solubility in water:
6.0 mg/kg at 25 °C (McAuliffe, 1966)

Vapor density:
4.59 g/L at 25 °C, 3.87 (air = 1)

Vapor pressure (mmHg):
14.5 at 25 °C (Wilhoit and Zwolinski, 1971)

Environmental fate:
  Chemical/Physical. Complete combustion in air yields carbon dioxide and water vapor. *cis*-1,2-Dimethylcyclohexane will not hydrolyze because it has no hydrolyzable functional group.

Source: Component of gasoline (quoted, Verschueren, 1983)

Use: Organic synthesis.

trans-1,4-DIMETHYLCYCLOHEXANE

Synonym: *trans-p*-Dimethylcyclohexane.

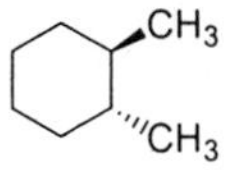

CAS Registry Number: 6876-23-9
DOT: 2263
Molecular formula: C_8H_{16}
Formula weight: 112.22

Physical state and color:
Colorless liquid

Melting point (°C):
-37 (Weast, 1986)

Boiling point (°C):
119.3 (Weast, 1986)

Density (g/cm^3):
0.76255 at 20/4 °C, 0.75835 at 25/4 °C (Dreisbach, 1959)

Diffusivity in water (x 10^{-5} cm^2/sec):
0.70 at 20 °C using method of Hayduk and Laudie (1974)

Dissociation constant, pK$_a$:
>14 (Schwarzenbach et al., 1993)

Flash point (°C):
≈ 10 (isomeric mixture, Hawley, 1981)

Heat of fusion (kcal/mol):
2.73 (Dreisbach, 1955)
2.491-2.508 (Dean, 1987)

Henry's law constant (atm·m^3/mol):
0.91 at 25 °C (Mackay and Shiu, 1981)

Ionization potential (eV):
10.08 ± 0.03 (Franklin et al., 1969)

Soil sorption coefficient, log K$_{oc}$:
Unavailable because experimental methods for estimation of this parameter for alicyclic hydrocarbons are lacking in the documented literature

Octanol/water partition coefficient, log K$_{ow}$:
3.41 using method of Hansch et al. (1968)

Solubility in organics:
Soluble in acetone, alcohol, benzene, ether, and ligroin (Weast, 1986)

Solubility in water:
3.84 mg/kg at 25 °C (Price, 1976)

Vapor density:
4.59 g/L at 25 °C, 3.87 (air = 1)

Vapor pressure (mmHg):
22.65 at 25 °C (Mackay et al., 1982)

Environmental fate:
   Chemical/Physical. Complete combustion in air yields carbon dioxide and water vapor. *trans*-1,2-Dimethylcyclohexane will not hydrolyze because it has no hydrolyzable functional group.

Source: Component of gasoline (quoted, Verschueren, 1983)

Use: Organic synthesis.

N,N-DIMETHYLFORMAMIDE

Synonyms: Dimethylformamide; DMF; DMFA; *N*-Formyldimethylamine; NCI-C60913; NSC 5536; U 4224; UN 2265.

CAS Registry Number: 68-12-2
DOT: 2265
DOT label: Combustible liquid
Molecular formula: C_3H_7NO
Formula weight: 73.09
RTECS: LQ2100000
Merck reference: 10, 3237

Physical state, color, and odor:
Colorless to light yellow, mobile liquid with a faint, ammonia-like odor. Odor threshold concentration is 2.2 ppm (Amoore and Hautala, 1983).

Melting point (°C):
-60.5 (Weast, 1986)

Boiling point (°C):
149-156 (Weast, 1986)

Density (g/cm³):
0.9487 at 20/4 °C (Weast, 1986)
0.9445 at 25/4 °C (Windholz et al., 1983)

Diffusivity in water (x 10^{-5} cm²/sec):
1.03 at 20 °C using method of Hayduk and Laudie (1974)

Flash point (°C):
57.8 (NIOSH, 1997)
67 (open cup, Windholz et al., 1983)

Lower explosive limit (%):
2.2 at 100 °C (NFPA, 1984)

Upper explosive limit (%):
15.2 (NFPA, 1984)

Ionization potential (eV):
9.12 ± 0.02 (Franklin et al., 1969)

Soil sorption coefficient, log K_{oc}:
Unavailable because experimental methods for estimation of this parameter for aliphatic amines are lacking in the documented literature. However, its miscibility in water and low K_{ow} suggest its adsorption to soil will be nominal (Lyman et al., 1982).

Octanol/water partition coefficient, log K_{ow}:
-1.01 (Sangster, 1989)

Solubility in organics:
Miscible with most organic solvents (Windholz et al., 1983)

Solubility in water:
Miscible (NIOSH, 1997). A saturated solution in equilibrium with its own vapor had a concentration of 5,294 g/L at 25 °C (Kamlet et al., 1987).

Vapor density:
2.99 g/L at 25 °C, 2.52 (air = 1)

Vapor pressure (mmHg):
3 at 20 °C (NIOSH, 1997)
3.7 at 25 °C (Sax and Lewis, 1987)

Environmental fate:
   Biological. Incubation of [^{14}C]*N,N*-dimethylformamide (0.1-100 μg/L) in natural seawater resulted in the compound mineralizing to carbon dioxide. The rate of carbon dioxide formation was inversely proportional to the initial concentration (Ursin, 1985).

Exposure limits: NIOSH REL: TWA 10 ppm (30 mg/m^3), IDLH 500 ppm; OSHA PEL: TWA 10 ppm; ACGIH TLV: TWA 10 ppm (adopted).

Toxicity:
   EC_{50} (24-h), EC_{50} (48-h), LC_{50} (24-h), and LC_{50} (48-h) values for *Spirostomum ambiguum* were 9.87, 8.19, 31.7, and 19.7 g/L, respectively (Nałecz-Jawecki and Sawicki, 1999).
   Acute oral LD_{50} for mice 3,750 mg/kg, rats 2,800 mg/kg (quoted, RTECS, 1985).

Uses: Solvent for liquids, gases, and vinyl resins; polyacrylic fibers; gas carrier; catalyst in carboxylation reactions; organic synthesis.

1,1-DIMETHYLHYDRAZINE

Synonyms: UDMH; Dimazine; *asym*-Dimethylhydrazine; *unsym*-Dimethylhydrazine; *N,N*-Dimethylhydrazine.

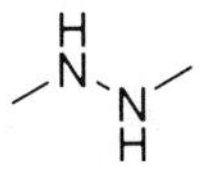

CAS Registry Number: 57-14-7
DOT: 1163 (*asym*), 2382 (*sym*)
DOT label: Flammable liquid and poison
Molecular formula: $C_2H_8N_2$
Formula weight: 60.10
RTECS: MV2450000
Merck reference: 10, 3242

Physical state, color, and odor:
Clear, colorless to yellow, fuming liquid with an amine-like odor. Odor threshold concentration is 1.7 ppm (Amoore and Hautala, 1983).

Melting point (°C):
-57.8 (NIOSH, 1997)

Boiling point (°C):
63.9 (Windholz et al., 1983)

Density (g/cm³):
0.7914 at 22/4 °C (Weast, 1986)

Diffusivity in water (x 10⁻⁵ cm²/sec):
1.04 at 20 °C using method of Hayduk and Laudie (1974)

Flash point (°C):
-15.1 (NIOSH, 1997)
1 (Aldrich, 1988)

Lower explosive limit (%):
2 (NIOSH, 1997)

Upper explosive limit (%):
95 (NIOSH, 1997)

Henry's law constant (x 10⁻⁹ atm·m³/mol):
2.45 at 25 °C (Mercer et al., 1990)

Ionization potential (eV):
7.67 ± 0.05 (Franklin et al., 1969)
8.05 (NIOSH, 1997)

Soil sorption coefficient, log K_{oc}:
-0.70 (calculated, Mercer et al., 1990)

Octanol/water partition coefficient, log K_{ow}:
-2.42 (Mercer et al., 1990)

Solubility in organics:
Miscible with alcohol, *N,N*-dimethylformamide, ethyl ether, and hydrocarbons (Windholz et al., 1983)

Solubility in water:
Miscible (NIOSH, 1997).

Vapor density:
2.46 g/L at 25 °C, 2.07 (air = 1)

Vapor pressure (mmHg):
103 at 20 °C (NIOSH, 1997)
157 at 25 °C (Verschueren, 1983)

Environmental fate:
   Chemical/Physical. Releases toxic nitrogen oxides when heated to decomposition (Sax and Lewis, 1987). Ignites spontaneously in air or in contact with hydrogen peroxide, nitric acid, or other oxidizers (Patnaik, 1992).
   N-Nitrosodimethylamine was the major product of ozonation of 1,1-dimethylhydrazine in the dark. Hydrogen peroxide, methyl hydroperoxide, and methyl diazene were also identified (HSDB, 1989).

Exposure limits: Potential occupational carcinogen. NIOSH REL: 2-h ceiling 0.6 ppm (1.2 mg/m^3), IDLH 15 ppm; OSHA PEL: TWA 0.5 ppm (1 mg/m^3); ACGIH TLV: TWA 0.1 ppm (adopted).

Symptoms of exposure: Irritation of eyes, nose, and throat. May cause diarrhea, stimulation of central nervous system, tremor, and convulsions (Patnaik, 1992).

Toxicity:
   In soft water, LC_{50} static bioassay values for *Lebistes reticulatus* after 24, 48, 72 and 96 h were 82.0, 45.5, 32.4, and 26.5 mg/L, respectively. In hard water, LC_{50} static bioassay values for *Lebistes reticulatus* after 24, 48, 72, and 96 h were 78.4, 29.9, 17.2, and 10.1 mg/L, respectively (Slonium, 1977).
   Acute oral LD_{50} for rats 122 mg/kg, mice 265 mg/kg (quoted, RTECS, 1985).

Uses: Rocket fuel formulations; stabilizer for organic peroxide fuel additives; absorbent for acid gases; plant control agent; photography; in organic synthesis.

2,3-DIMETHYLPENTANE

Synonym: 3,4-Dimethylpentane.

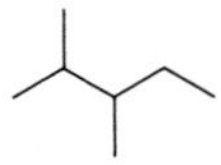

CAS Registry Number: 565-59-3
Molecular formula: C_7H_{16}
Formula weight: 100.20

Physical state and color:
Clear, colorless liquid

Boiling point (°C):
89.8 (Weast, 1986)

Density (g/cm³):
0.6951 at 20/4 °C (Weast, 1986)

Diffusivity in water (x 10^{-5} cm²/sec):
0.71 at 20 °C using method of Hayduk and Laudie (1974)

Dissociation constant, pK_a:
>14 (Schwarzenbach et al., 1993)

Flash point (°C):
-6 (Aldrich, 1988)

Lower explosive limit (%):
1.1 (NFPA, 1984)

Upper explosive limit (%):
6.7 (NFPA, 1984)

Heat of fusion (kcal/mol):
1.389 (Dreisbach, 1959)

Henry's law constant (atm·m³/mol):
1.72 at 25 °C (Mackay and Shiu, 1981)

Interfacial tension with water (dyn/cm at 25 °C):
49.6 (quoted, Freitas et al., 1997)

Soil sorption coefficient, log K_{oc}:
Unavailable because experimental methods for estimation of this parameter for aliphatic hydrocarbons are lacking in the documented literature

Octanol/water partition coefficient, log K_{ow}:
3.26 using method of Hansch et al. (1968)

413

Solubility in organics:
Soluble in acetone, alcohol, benzene, chloroform, and ether (Weast, 1986)

Solubility in water:
5.25 mg/kg at 25 °C (Price, 1976)

Vapor density:
4.10 g/L at 25 °C, 3.46 (air = 1)

Vapor pressure (mmHg):
68.9 at 25 °C (Wilhoit and Zwolinski, 1971)
24.937 at 4.999 °C (Osborn and Douslin, 1974)

Environmental fate:
  Photolytic. A photooxidation rate constant of 3.4 x 10^{-12} cm^3/molecule·sec was reported for the gas-phase reaction of 2,3-dimethylpentane and OH radicals (Atkinson, 1990).
  Chemical/Physical. Complete combustion in air yields carbon dioxide and water vapor. 2,3-Dimethylpentane will not hydrolyze because it has no hydrolyzable functional group.

Source: In diesel engine exhaust at a concentration of 0.9% of emitted hydrocarbons (quoted, Verschueren, 1983). Schauer et al. (1999) reported 2,3-dimethylpentane in a diesel-powered medium-duty truck exhaust at an emission rate of 720 μg/km.

Uses: Organic synthesis; gasoline component.

2,4-DIMETHYLPENTANE

Synonym: Diisopropylmethane.

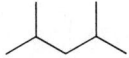

CAS Registry Number: 108-08-7
Molecular formula: C_7H_{16}
Formula weight: 100.20

Physical state and color:
Colorless liquid

Melting point (°C):
-119.2 (Weast, 1986)
-123 (Aldrich, 1988)

Boiling point (°C):
80.5 (Weast, 1986)

Density (g/cm³):
0.6727 at 20/4 °C, 0.66832 at 25/4 °C (Riddick et al., 1986)

Diffusivity in water (x 10⁻⁵ cm²/sec):
0.70 at 20 °C using method of Hayduk and Laudie (1974)

Dissociation constant, pKₐ:
>14 (Schwarzenbach et al., 1993)

Flash point (°C):
-12.1 (Hawley, 1981)

Heat of fusion (kcal/mol):
1.636 (Riddick et al., 1986)

Henry's law constant (atm·m³/mol):
2.94 at 25 °C (Mackay and Shiu, 1981)

Interfacial tension with water (dyn/cm at 25 °C):
50.0 (quoted, Freitas et al., 1997)

Soil sorption coefficient, log K_{oc}:
Unavailable because experimental methods for estimation of this parameter for aliphatic hydrocarbons are lacking in the documented literature

Octanol/water partition coefficient, log K_{ow}:
3.24 using method of Hansch et al. (1968)

Solubility in organics:
Soluble in acetone, alcohol, benzene, chloroform, and ether (Weast, 1986)

Solubility in water (mg/kg):
4.41 at 25 °C (Price, 1976)
3.62 at 25 °C (McAuliffe, 1963)
4.06 at 25 °C (McAuliffe, 1966)
6.50 at 0 °C, 5.50 at 25 °C (Polak and Lu, 1973)

Vapor density:
4.10 g/L at 25 °C, 3.46 (air = 1)

Vapor pressure (mmHg):
98.4 at 25 °C (Wilhoit and Zwolinski, 1971)

Environmental fate:
  Photolytic. Based on a photooxidation rate constant of 5.0 x 10^{-12} cm^3/molecule·sec for the reaction of 2,3-dimethypentane and OH radicals in air, the half-life is 27 h (Alltshuller, 1991).
  Chemical/Physical. Complete combustion in air yields carbon dioxide and water vapor. 2,4-Dimethylpentane will not hydrolyze because it has no hydrolyzable functional group.

Source: In diesel engine exhaust at a concentration of 0.3% of emitted hydrocarbons (quoted, Verschueren, 1983). Schauer et al. (1999) reported 2,4-dimethylpentane in a diesel-powered medium-duty truck exhaust at an emission rate of 410 μg/km.

Uses: Organic synthesis; gasoline component.

3,3-DIMETHYLPENTANE

Synonyms: None.

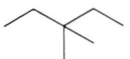

CAS Registry Number: 562-49-2
Molecular formula: C_7H_{16}
Formula weight: 100.20

Physical state and color:
Colorless liquid

Melting point (°C):
-134.4 (Weast, 1986)

Boiling point (°C):
86.1 (Weast, 1986)

Density (g/cm³):
0.69327 at 20/4 °C, 0.68908 at 25/4 °C (Dreisbach, 1959)

Diffusivity in water (x 10⁻⁵ cm²/sec):
0.71 at 20 °C using method of Hayduk and Laudie (1974)

Dissociation constant, pK_a:
>14 (Schwarzenbach et al., 1993)

Flash point (°C):
-6 (Aldrich, 1988)

Heat of fusion (kcal/mol):
1.69 (Dreisbach, 1959)

Henry's law constant (atm·m³/mol):
1.85 at 25 °C (Mackay and Shiu, 1981)

Soil sorption coefficient, log K_{oc}:
Unavailable because experimental methods for estimation of this parameter for aliphatic hydrocarbons are lacking in the documented literature

Octanol/water partition coefficient, log K_{ow}:
3.22 using method of Hansch et al. (1968)

Solubility in organics:
Soluble in acetone, alcohol, benzene, chloroform, and ether (Weast, 1986)

Solubility in water (mg/kg):
5.92 at 25 °C, 6.78 at 40.1 °C, 8.17 at 55.7 °C, 10.3 at 69.7 °C, 15.8 at 99.1 °C, 27.3 at 118.0 °C, 67.3 at 120.4 °C, 86.1 at 150.4 °C (Price, 1976)

Vapor density:
4.10 g/L at 25 °C, 3.46 (air = 1)

Vapor pressure (mmHg):
82.8 at 25 °C (Wilhoit and Zwolinski, 1971)

Environmental fate:
   Chemical/Physical. Complete combustion in air yields carbon dioxide and water vapor. 3,3-Dimethylbutane will not hydrolyze because it has no hydrolyzable functional group.

Uses: Organic synthesis; gasoline component.

2,4-DIMETHYLPHENOL

Synonyms: 4,6-Dimethylphenol; 2,4-DMP; 1-Hydroxy-2,4-dimethylbenzene; 4-Hydroxy-1,3-dimethylbenzene; RCRA waste number U101; 1,3,4-Xylenol; 2,4-Xylenol; *m*-Xylenol.

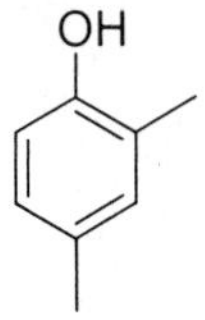

CAS Registry Number: 105-67-9
Molecular formula: $C_8H_{10}O$
Formula weight: 122.17
RTECS: ZE5600000
Merck reference: 10, 9891

Physical state and color:
Colorless solid, slowly turning brown on exposure to air

Melting point (°C):
24.5 (Andon et al., 1960)
27 (Dean, 1987)

Boiling point (°C):
210 (Weast, 1986)
210-212 (Dean, 1987)

Density (g/cm³ at 20/4 °C):
0.9650 (Weast, 1986)
1.02017 (Andon et al., 1960)

Diffusivity in water (x 10^{-5} cm²/sec):
0.77 at 20 °C using method of Hayduk and Laudie (1974)

Dissociation constant, pK_a:
10.63 (Riddick et al., 1986)

Flash point (°C):
>110 (Aldrich, 1988)

Henry's law constant (x 10^{-3} atm·m³/mol):
8.29, 6.74, 10.1, 4.93, and 3.75 at 10, 15, 20, 25, and 30 °C, respectively (Ashworth et al., 1988)

Ionization potential (eV):
8.18 (Lias et al., 1998)

Bioconcentration factor, log BCF:
1.18 (bluegill sunfish, Barrows et al., 1980)
2.18 (bluegill sunfish, Veith et al., 1980)

419

Soil sorption coefficient, log K_{oc}:
2.08 (river sediment), 2.02 (coal wastewater sediment) (Kopinke et al., 1995)
2.19 (activated carbon, Blum et al., 1994)

Octanol/water partition coefficient, log K_{ow}:
2.54 (Sangster, 1989)
2.42 (Veith et al., 1980)
2.30 (Mabey et al., 1982)
2.47 (Garst and Wilson, 1984)
2.34 (Wasik et al., 1981)

Solubility in organics:
Freely soluble in ethanol, chloroform, ether, and benzene (U.S. EPA, 1985)

Solubility in water (mg/L):
4,200 at 20 °C (quoted, Verschueren, 1983)
7,868 at 25 °C (Banerjee et al., 1980)
6,200 at 25 °C (quoted, Leuenberger et al., 1985)
7,819 at 25.0 °C (Wasik et al., 1981)
8,795 at 25 °C and pH 5.1 (Blackman, 1955)
7,888 at 25 °C (Veith et al., 1980)

Vapor pressure (x 10^{-2} mmHg):
6.21 at 20 °C (supercooled liquid, Andon et al., 1960)
9.8 at 25.0 °C (quoted, Leuenberger et al., 1985)

Environmental fate:
Biological. When 2,4-dimethylphenol was statically incubated in the dark at 25 °C with yeast extract and settled domestic wastewater inoculum, significant biodegradation with rapid adaptation was observed. At concentrations of 5 and 10 mg/L, 100 and 99% biodegradation, respectively, were observed after 7 d (Tabak et al., 1981).

Heukelekian and Rand (1955) reported a 5-d BOD value of 1.50 g/g which is 57.2% of the ThOD value of 2.62 g/g. In activated sludge inoculum, 94.5% COD removal was achieved in 5 d. The average rate of biodegradation was 28.2 mg COD/g·h (Pitter, 1976).

Chemical/Physical. Wet oxidation of 2,4-dimethylphenol at 320 °C yielded formic and acetic acids (Randall and Knopp, 1980). 2,4-Dimethylphenol will not hydrolyze because there is no hydrolyzable functional group (Kollig, 1993).

Toxicity:
LC_{50} (contact) for earthworm (*Eisenia fetida*) 2.2 $\mu g/cm^2$ (Neuhauser et al., 1985).

LC_{50} (8-d) for fathead minnows 13-14 mg/L (Spehar et al., 1982).

LC_{50} (96-h) for bluegill sunfish 7.8 mg/L, fathead minnows 17 mg/L (Spehar et al., 1982).

LC_{50} (48-h) for fathead minnows 9.5 mg/L (Spehar et al., 1982), *Daphnia magna* 2.1 mg/L (LeBlanc, 1980).

LC_{50} (24-h) for *Daphnia magna* 8.3 mg/L (LeBlanc, 1980).

Acute oral LD_{50} for mice 809 mg/kg, rats 3,200 mg/kg (quoted, RTECS, 1985).

Source:
Thomas and Delfino (1991) equilibrated contaminant-free groundwater collected from Gainesville, FL with individual fractions of three individual petroleum products at 24-25 °C for 24 h. The aqueous phase was analyzed for organic compounds via U.S. EPA approved test method 625. Average 2,4-dimethylphenol concentrations reported in water-soluble fractions of

unleaded gasoline, kerosene, and diesel fuel were 50, 99, and 108 μg/L, respectively. 2,4-Dichlorophenol may also enter groundwater by leaching from coal tar, asphalt runoff, plastics, and pesticides (quoted, Verschueren, 1983).

Uses: Wetting agent; dyestuffs; preparation of phenolic antioxidants; plastics, resins, solvent, disinfectant, pharmaceuticals, insecticides, fungicides, and rubber chemicals manufacturing; lubricant and gasoline additive; possibly used as a pesticide; plasticizers.

DIMETHYL PHTHALATE

Synonyms: Avolin; **1,2-Benzenedicarboxylic acid dimethyl ester**; Dimethyl-1,2-benzene-dicarboxylate; Dimethylbenzene-*o*-dicarboxylate; DMP; ENT 262; Fermine; Methyl phthalate; Mipax; NTM; Palatinol M; Phthalic acid dimethyl ester; Phthalic acid methyl ester; RCRA waste number U102; Solvanom; Solvarone.

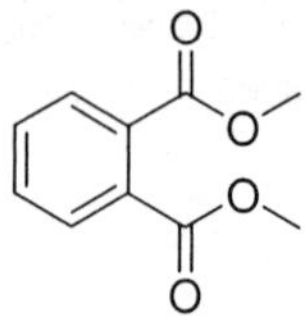

CAS Registry Number: 131-11-3
Molecular formula: $C_{10}H_{10}O_4$
Formula weight: 194.19
RTECS: TI1575000
Merck reference: 10, 3250

Physical state, color, and odor:
Clear, colorless, odorless, moderately viscous, oily liquid

Melting point (°C):
0 (Fishbein and Albro, 1972)
5.5 (U.S. EPA, 1980a)

Boiling point (°C):
283.8 (Weast, 1986)
282-285 (Worthing and Hance, 1991)

Density (g/cm³):
1.194 at 20/20 (Worthing and Hance, 1991)
1.1905 at 20/4 °C (Weast, 1986)
1.189 at 25/4 °C (Fishbein and Albro, 1972)

Diffusivity in water (x 10^{-5} cm²/sec):
0.66 at 20 °C using method of Hayduk and Laudie (1974)

Flash point (°C):
146 (NIOSH, 1997)

Lower explosive limit (%):
0.9 at 180 °C (NFPA, 1984)

Henry's law constant (x 10^{-7} atm·m³/mol at 25 °C):
4.2 (Petrasek et al., 1983)
235, 246, 230, 214, 168, 127, and 118 at pH values of 3.00, 3.37, 5.88, 6.18, 7.62, 8.91, and 8.99, respectively (Hakuta et al., 1977).

Ionization potential (eV):
9.64 (NIOSH, 1997)

Bioconcentration factor, log BCF:
1.76 (bluegill sunfish, Veith et al., 1980)
2.21 (*Chlorella pyrenoidosa*, Yan et al., 1995)
0.77 (sheepshead minnow, Wofford et al., 1981)

Soil sorption coefficient, log K_{oc}:
0.88, 1.63, 1.84 (various Norwegian soils, Seip et al., 1986)
2.28 (Banerjee et al., 1985)
1.56 (Pipestone sand); in Oakville sand: 1.85 and 1.64 for A and B horizons, respectively
    (Maraqa et al., 1998)

Octanol/water partition coefficient, log K_{ow}:
1.53 (Leyder and Boulanger, 1983; Maraqa et al., 1998)
1.56 (Mabey et al., 1982)
1.47 (Howard et al., 1985)
1.62 (Eadsforth, 1986)
1.66 (Renberg et al., 1985)
1.61 (Veith et al., 1980)
1.86 (Eadsforth, 1986)
1.80 (Johnsen et al., 1989)
1.60 (Ellington and Floyd, 1996)

Solubility in organics:
Soluble in ethanol, ether, and benzene (Weast, 1986)

Solubility in water:
4,320 mg/L at 25 °C (Wolfe et al., 1980)
4,290 mg/L at 20 °C (Leyder and Boulanger, 1983)
At 25 °C (mg/L): 4,000 (distilled water), 3,960 (well water), 3,160 (natural seawater) (Howard
    et al., 1985)
4,500 ppm at 25 °C (Fukano and Obata, 1976)
0.5 wt % at 20 °C (Fishbein and Albro, 1972)
0.305 wt % 20-25 °C (Fordyce and Meyer, 1940)

Vapor density:
7.94 g/L at 25 °C, 6.70 (air = 1)

Vapor pressure (x 10^{-3} mmHg at 25 °C):
8.93 (Hinckley et al., 1990)
1.65 (Banerjee et al., 1990; Howard et al., 1985)

Environmental fate:
Biological. In anaerobic sludge, degradation occurred as follows: monomethyl phthalate to
phthalic acid to protocatechuic acid followed by ring cleavage and mineralization (Shelton et
al., 1984). In a static-culture-flask screening test, dimethyl phthalate showed significant
biodegradation with rapid adaptation. The ester (5 and 10 mg/L) was statically incubated in the
dark at 25 °C with yeast extract and settled domestic wastewater inoculum. After 7 d, 100%
biodegradation was achieved (Tabak et al., 1981). Horowitz et al. (1982) reported a
mineralization half-life of approximately 7 d in municipal digested sludge under anaerobic
conditions.
Photolytic. An aqueous solution containing titanium dioxide and subjected to UV radiation

(λ >290 nm) produced hydroxyphthalates and dihydroxyphthalates as intermediates (Hustert and Moza, 1988).

Chemical/Physical. Under alkaline conditions, dimethyl phthalate will initially hydrolyze to methyl hydrogen phthalate and methanol. The monoester will undergo further hydrolysis forming *o*-phthalic acid and methanol (Wolfe et al., 1980; Kollig, 1993). A second-order rate constant of 5.1 x 10^{-2} M/sec was reported for the hydrolysis of dimethyl phthalate at 30 °C and pH 8. Hydrolysis half-lives of 3.2 yr was reported at pH 7 and at pH 9: 11.6 and 25 d at 30 and 18 °C, respectively (Wolfe et al., 1980).

Exposure limits (mg/m^3): NIOSH REL: TWA 5, IDLH 2,000; OSHA PEL: TWA 5; ACGIH TLV: TWA 5 (adopted).

Symptoms of exposure: Irritates nasal passages, upper respiratory system, stomach; eye ache. Ingestion may cause central nervous system depression (NIOSH, 1997).

Toxicity:

LC$_{50}$ (contact) for earthworm (*Eisenia fetida*) 550 μg/cm^2 (Neuhauser et al., 1985).

LC$_{50}$ (8-d) for grass shrimp larvae 100 ppm (Verschueren, 1983).

LC$_{50}$ (96-h) for bluegill sunfish 50 mg/L (Spehar et al., 1982), *Cyprinodon variegatus* 6.7 ppm using natural seawater (Heitmuller et al., 1981).

LC$_{50}$ (72-h) for *Cyprinodon variegatus* 13 ppm (Heitmuller et al., 1981).

LC$_{50}$ (48-h) for *Daphnia magna* 33 mg/L (LeBlanc, 1980), *Cyprinodon variegatus* 58 ppm (Heitmuller et al., 1981).

LC$_{50}$ (24-h) for *Daphnia magna* 150 mg/L (LeBlanc, 1980), *Cyprinodon variegatus* >120 ppm (Heitmuller et al., 1981).

Acute oral LD$_{50}$ for chickens 8,500 mg/kg, guinea pigs 2,400 mg/kg, mice 6,800 mg/kg, rats 6,800 mg/kg, rabbits 4,400 mg/kg (quoted, RTECS, 1985).

Acute percutaneous LD$_{50}$ (9-d) for rats >4,800 mg/kg (Worthing and Hance, 1991).

Heitmuller et al. (1981) reported a NOEC of 21 ppm.

Drinking water standard: No MCLGs or MCLs have been proposed (U.S. EPA, 1996).

Source: May leach from plastic products (e.g., tubing, containers) used in laboratories during chemical analysis of aqueous samples

Uses: Plasticizer for cellulose acetate, nitrocellulose, resins, rubber, elastomers; ingredient in lacquers; coating agents; safety glass; insect repellant; molding powders; perfumes.

2,2-DIMETHYLPROPANE

Synonyms: Neopentaene; Neopentane; *tert*-Pentane; Tetramethylmethane; UN 1265; UN 2044.

CAS Registry Number: 463-82-1
DOT: 2044
Molecular formula: C_5H_{12}
Formula weight: 72.15
RTECS: TY1190000
Merck reference: 10, 6303

Physical state:
Gas

Melting point (°C):
-16.5 (Weast, 1986)
-19.8 (Windholz et al., 1983)

Boiling point (°C):
9.5 (Weast, 1986)

Density (g/cm³):
0.591 at 20/4 °C, 0.5852 at 25/4 °C (Dreisbach, 1959)

Diffusivity in water (x 10⁻⁵ cm²/sec):
0.80 at 20 °C using method of Hayduk and Laudie (1974)

Dissociation constant, pKₐ:
>14 (Schwarzenbach et al., 1993)

Flash point (°C):
-65.0 (Hawley, 1981)

Lower explosive limit (%):
1.4 (NFPA, 1984)

Upper explosive limit (%):
7.5 (NFPA, 1984)

Heat of fusion (kcal/mol):
0.752 (Dean, 1987)

Henry's law constant (atm·m³/mol):
3.70 at 25 °C (Mackay and Shiu, 1981)

Ionization potential (eV):
10.35 (Franklin et al., 1969)

Soil sorption coefficient, log K_{oc}:
Unavailable because experimental methods for estimation of this parameter for aliphatic hydrocarbons are lacking in the documented literature

Octanol/water partition coefficient, log K_{ow}:
3.11 (Hansch and Leo, 1979)

Solubility in organics:
Soluble in alcohol and ether (Weast, 1986).

Solubility in water:
33.2 mg/kg at 25 °C (McAuliffe, 1966)

Vapor density:
2.95 g/L at 25 °C, 2.49 (air = 1)

Vapor pressure (mmHg):
1,287 at 25 °C (Wilhoit and Zwolinski, 1971)
1,074.43 at 19.492 °C, 1,267.75 at 24.560 °C (Osborn and Douslin, 1974)

Environmental fate:
   Photolytic. A rate constant of 6.50×10^{11} cm^3/mol·sec was reported for the reaction of 2-methylpropane with OH radicals in air at 298 (Greiner, 1970). Rate constants of 9.0×10^{-13} and 8.49×10^{-13} cm^3/molecule·sec were reported for the reaction of 2-methylpropane with OH in air (Atkinson et al., 1979; Winer et al., 1979).
   Chemical/Physical. Complete combustion in air yields carbon dioxide and water vapor. 2,2-Dimethylpropane will not hydrolyze because it has no hydrolyzable functional group.

Exposure limits: ACGIH TLV: TWA 600 ppm (adopted).

Toxicity:
   Acute oral LD_{50} (intraperitoneal) for mice 100 mg/kg (quoted, RTECS, 1985).

Uses: Butyl rubber; organic synthesis.

2,7-DIMETHYLQUINOLINE

Synonyms: None.

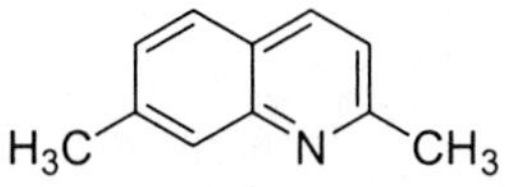

CAS Registry Number: 93-37-8
Molecular formula: $C_{11}H_{11}N$
Formula weight: 157.22

Physical state:
Liquid

Melting point (°C):
61 (Weast, 1986)
57-59 (Verschueren, 1983)

Boiling point (°C):
262-265 (Weast, 1986)

Density (g/cm³):
1.054 using method of Lyman et al. (1982)

Solubility in organics:
Soluble in alcohol, ether, chloroform (Weast, 1986), and benzene (Hawley, 1981)

Solubility in water:
1,795 mg/kg at 25 °C (Price, 1976)

Toxicity:
   LC_{100} (24-h) for *Tetrahymena pyriformis* 200 mg/L (Schultz et al., 1978).

Uses: Organic synthesis; dye intermediate.

DIMETHYL SULFATE

Synonyms: Dimethyl monosulfate; DMS; Methyl sulfate; RCRA waste number U103; **Sulfuric acid dimethyl ester**; UN 1595.

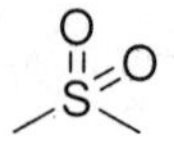

CAS Registry Number: 77-78-1
DOT: 1595
DOT label: Corrosive
Molecular formula: $C_2H_6O_4S$
Formula weight: 126.13
RTECS: WS8225000
Merck reference: 10, 3252

Physical state, color, and odor:
Colorless, oily liquid with an onion-like odor

Melting point (°C):
-31.8 (Du Pont, 1999a)

Boiling point (°C):
188.8 (Du Pont, 1999a)

Density (g/cm³):
1.333 at 15.6 °C (Du Pont, 1999a)
1.3283 at 20/4 °C (Weast, 1986)

Diffusivity in water (x 10^{-5} cm²/sec):
0.91 at 20 °C using method of Hayduk and Laudie (1974)

Flash point (°C):
83 (closed cup), 116 (open cup) (Du Pont, 1999a)

Henry's law constant (x 10^{-6} atm·m³/mol):
2.96 at 20 °C (approximate - calculated from water solubility and vapor pressure)

Soil sorption coefficient, log K_{oc}:
0.61 (calculated, Mercer et al., 1990)

Octanol/water partition coefficient, log K_{ow}:
-1.24 (Mercer et al., 1990)

Solubility in organics:
Soluble in alcohol, benzene, ether (Weast, 1986), 1,4-dioxane, and aromatic hydrocarbons (Windholz et al., 1983)

Solubility in water (g/L):
28 at 18 °C (quoted, Windholz et al., 1983)
28 at 20 °C (Du Pont, 1999a)

Vapor density:
4.35 (air = 1) (Windholz et al., 1983)

Vapor pressure (mmHg):
0.5 at 20 °C (Weast, 1986)
0.7 at 25 °C, 1.1 at 38 °C (Du Pont, 1999a)

Environmental fate:
   Chemical/Physical. Hydrolyzes in water (half-life = 1.2 h) to methyl alcohol and sulfuric acid (Robertson and Sugamori, 1966) via the intermediate methyl sulfuric acid (Du Pont, 1999a)

Exposure limits: Potential occupational carcinogen. NIOSH REL: TWA 0.1 ppm (0.5 mg/m^3), IDLH 7 ppm; OSHA PEL: TWA 1 ppm (5 mg/m^3); ACGIH TLV: TWA 0.1 ppm (adopted).

Toxicity:
   LC_{50} (96-h static bioassay) for bluegill sunfish 7.5 mg/L, *Menidia beryllina* 15 mg/L (quoted, Verschueren, 1983).
   Acute oral LD_{50} for mice 140 mg/kg, rats 205 mg/kg (quoted, RTECS, 1985).

Uses: In organic synthesis as a methylating agent.

1,2-DINITROBENZENE

Synonyms: *o*-Dinitrobenzene; *o*-Dinitrobenzol; UN 1597.

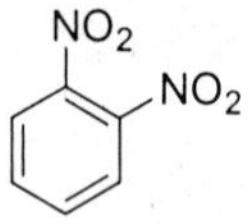

CAS Registry Number: 528-29-0
DOT: 1597
DOT label: Poison
Molecular formula: $C_6H_4N_2O_4$
Formula weight: 168.11
RTECS: CZ7450000
Merck reference: 10, 3273

Physical state and color:
Colorless to yellow needles

Melting point (°C):
118.50 (Martin et al., 1979)

Boiling point (°C):
319 at 775 mmHg (Weast, 1986)

Density (g/cm³):
1.565 at 17/4 °C (Weast, 1986)

Diffusivity in water (x 10^{-5} cm²/sec):
0.79 at 20 °C using method of Hayduk and Laudie (1974)

Flash point (°C):
150 (NIOSH, 1997)

Entropy of fusion (cal/mol·K):
14.0 (Andrews et al., 1926)

Ionization potential (eV):
10.71 (NIOSH, 1997)

Soil sorption coefficient, log K_{oc}:
Unavailable because experimental methods for estimation of this parameter for nitroaromatic
   hydrocarbons are lacking in the documented literature.
K_d = 1.7 mL/g on a Cs^+-kaolinite (Haderlein and Schwarzenbach, 1993).

Octanol/water partition coefficient, log K_{ow}:
1.58 (Leo et al., 1971)

Solubility in organics:
Soluble in alcohol ($\approx$ 16.7 g/L), benzene (50 g/L); freely soluble in chloroform and ethyl
acetate (Windholz et al., 1983).

Solubility in water (mg/L):
151.5 (cold water), 370 at 100 °C (quoted, Windholz et al., 1983)
100 (cold water), 3,800 at 100 °C (quoted, Verschueren, 1983)

Environmental fate:
Biological. Under anaerobic and aerobic conditions using a sewage inoculum, 1,2-dinitro-benzene degraded to nitroaniline (Hallas and Alexander, 1983).

Chemical/Physical. Releases toxic nitrogen oxides when heated to decomposition (Sax and Lewis, 1987). 1,2-Dinitrobenzene will not hydrolyze in water (Kollig, 1993).

Exposure limits (mg/m³): NIOSH REL: TWA 1, IDLH 50; OSHA PEL: TWA 1; ACGIH TLV: TWA 0.15 ppm for all isomers (adopted).

Toxicity:
EC_{50} (15-min) for *Photobacterium phosphoreum* 232 μg/L (Yuan and Lang, 1997).
IC_{50} (24-h) for river bacteria 1.80 mg/L (Yuan and Lang, 1997).

Uses: Organic synthesis; dyes.

1,3-DINITROBENZENE

Synonyms: Binitrobenzene; 2,4-Dinitrobenzene; *m*-Dinitrobenzene; 1,3-Dinitrobenzol; UN 1597.

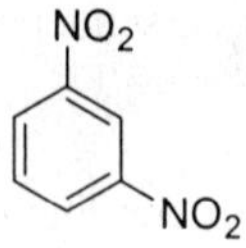

CAS Registry Number: 99-65-0
DOT: 1597
DOT label: Poison
Molecular formula: $C_6H_4N_2O_4$
Formula weight: 168.11
RTECS: CZ7350000
Merck reference: 10, 3273

Physical state and color:
White to yellowish crystals

Melting point (°C):
90 (Weast, 1986)
90.20 (Martin et al., 1979)

Boiling point (°C):
291 at 756 mmHg (Weast, 1986)
300-303 (Windholz et al., 1983)

Density (g/cm³):
1.5751 at 18/4 °C (Weast, 1986)
1.368 at 89/4 °C (Aldrich, 1988)

Diffusivity in water (x 10^{-5} cm²/sec):
0.79 at 20 °C using method of Hayduk and Laudie (1974)

Flash point (°C):
150 (NIOSH, 1997)

Entropy of fusion (cal/mol·K):
11.4 (Andrews et al., 1926)

Henry's law constant (x 10^{-7} atm·m³/mol):
2.75 at 35 °C (approximate - calculated from water solubility and vapor pressure)

Ionization potential (eV):
10.43 (NIOSH, 1997)

Soil sorption coefficient, log K_{oc}:
Unavailable because experimental methods for estimation of this parameter for nitroaromatic hydrocarbons are lacking in the documented literature.
K_d = 1,800 mL/g on a Cs^+-kaolinite (Haderlein and Schwarzenbach, 1993).

Octanol/water partition coefficient, log K_{ow}:
1.49 (Leo et al., 1971)

Solubility in organics:
Soluble in acetone, ether, pyrimidine (Weast, 1986), alcohol (27 g/L), pyridine (3,940 g/kg at 20-25 °C) (Dehn, 1917); freely soluble in benzene, chloroform and ethyl acetate (Windholz et al., 1983).

Solubility in water:
500 mg/L in cold water, 3.13 g/L at 100 °C (quoted, Windholz et al., 1983)
469 mg/L at 15 °C, 3,200 mg/L at 100 °C (quoted, Verschueren, 1983)
654 mg/kg at 30 °C (Gross and Saylor, 1931)
4.67 mM at 35 °C (Hine et al., 1963)
>21.4 g/kg at 20-25 °C (Dehn, 1917)
5.12 mmol/kg at 25.0 °C (Vesala, 1974)

Vapor pressure (mmHg):
8.15 x 10^{-4} at 35 °C (Hine et al., 1963)

Environmental fate:
   Biological. Under anaerobic and aerobic conditions using a sewage inoculum, 1,3-dinitrobenzene degraded to nitroaniline (Hallas and Alexander, 1983). In activated sludge inoculum, following a 20-d adaptation period, no degradation was observed (Pitter, 1976).
   Chemical/Physical. Releases toxic nitrogen oxides when heated to decomposition (Sax and Lewis, 1987). 1,3-Dinitrobenzene will not hydrolyze in water (Kollig, 1993).
   Although no products were identified, 1,3-dinitrobenzene (1.5 x 10^{-5} M) was reduced by iron metal (33.3 g/L acid washed 18-20 mesh) in a carbonate buffer (1.5 x 10^{-2} M) at pH 5.9 and 15 °C. Based on the pseudo-first-order disappearance rate of 0.0339/min, the half-life was 20.4 min (Agrawal and Tratnyek, 1996).

Exposure limits (mg/m³): NIOSH REL: TWA 1, IDLH 50; OSHA PEL: TWA 1; ACGIH TLV: TWA 0.15 ppm for all isomers (adopted).

Toxicity:
   EC_{50} (15-min) for *Photobacterium phosphoreum* 4.03 mg/L (Yuan and Lang, 1997).
   IC_{50} (24-h) for river bacteria 8.62 mg/L (Yuan and Lang, 1997).
   LC_{50} (96-h) for fathead minnows 12.7 mg/L (Spehar et al., 1982).
   Acute oral LD_{50} for rats 83 mg/kg, wild birds 42 mg/kg (quoted, RTECS, 1985).

Drinking water standard: No MCLGs or MCLs have been proposed (U.S. EPA, 1996).

Uses: Organic synthesis; dyes.

1,4-DINITROBENZENE

Synonyms: *p*-Dinitrobenzene; Dithane A-4; UN 1597.

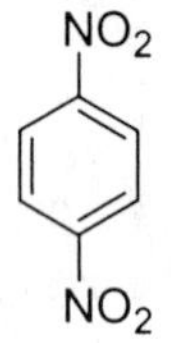

CAS Registry Number: 100-25-4
DOT: 1597
DOT label: Poison
Molecular formula: $C_6H_4N_2O_4$
Formula weight: 168.11
RTECS: CZ7525000
Merck reference: 10, 3273

Physical state and color:
White to yellow crystalline solid

Melting point (°C):
174.00 (Martin et al., 1979)

Boiling point (°C):
299 (NIOSH, 1997)

Density (g/cm³):
1.625 at 18/4 °C (Weast, 1986)

Diffusivity in water (x 10^{-5} cm²/sec):
0.79 at 20 °C using method of Hayduk and Laudie (1974)

Entropy of fusion (cal/mol·K):
15.0 (Andrews et al., 1926)

Henry's law constant (x 10^{-7} atm·m³/mol):
4.79 at 35 °C (approximate - calculated from water solubility and vapor pressure)

Ionization potential (eV):
10.50 (NIOSH, 1997)

Soil sorption coefficient, log K_{oc}:
Unavailable because experimental methods for estimation of this parameter for nitroaromatic hydrocarbons are lacking in the documented literature.
$K_d \approx 4,000$ mL/g on a Cs^+-kaolinite (Haderlein and Schwarzenbach, 1993).

Octanol/water partition coefficient, log K_{ow}:
1.46, 1.49 (Leo et al., 1971)

Solubility in organics:
Soluble in acetone, acetic acid, benzene, toluene (Weast, 1986), and alcohol (3.3 g/L) (Windholz et al., 1983)

Solubility in water:
0.01 wt % at 20 °C (NIOSH, 1997)
80 mg/L in cold water, 1.8 g/L at 100 °C (quoted, Windholz et al., 1983)
0.617 mM at 35 °C (Hine et al., 1963)

Vapor pressure (mmHg):
2.25 x 10^{-4} at 35 °C (Hine et al., 1963)

Environmental fate:
  Biological. In activated sludge inoculum, following a 20-d adaptation period, no biodegradation was observed (Pitter, 1976).
  Chemical/Physical. Releases toxic nitrogen oxides when heated to decomposition (Sax and Lewis, 1987). 1,4-Dinitrobenzene will not hydrolyze in water (Kollig, 1993).

Exposure limits (mg/m³): NIOSH REL: TWA 1, IDLH 50; OSHA PEL: TWA 1 ACGIH TLV: TWA 0.15 ppm for all isomers (adopted).

Toxicity:
  EC_{50} (15-min) for *Photobacterium phosphoreum* 260 μg/L (Yuan and Lang, 1997).
  IC_{50} (24-h) for river bacteria 1.27 mg/L (Yuan and Lang, 1997).

Uses: Organic synthesis; dyes.

4,6-DINITRO-*o*-CRESOL

Synonyms: Antinonin; Antinonnon; Arborol; Capsine; Chemsect DNOC; Degrassan; Dekrysil; Detal; Dinitrocresol; Dinitro-*o*-cresol; 2,4-Dinitro-*o*-cresol; 3,5-Dinitro-*o*-cresol; Dinitrodend-troxal; 3,5-Dinitro-2-hydroxytoluene; Dinitrol; Dinitromethyl cyclohexyltrienol; 2,4-Dinitro-2-methylphenol; 2,4-Dinitro-6-methylphenol; 4,6-Dinitro-2-methylphenol; Dinitrosol; Dinoc; Dinurania; DN; DNC; DN-dry mix no. 2; DNOC; Effusan; Effusan 3436; Elgetol; Elgetol 30; Elipol; ENT 154; Extrar; Hedolit; Hedolite; K III; K IV; Kresamone; Krezotol 50; Lipan; **2-Methyl-4,6-dinitrophenol**; 6-Methyl-2,4-dinitrophenol; Nitrador; Nitrofan; Prokarbol; Rafex; Rafex 35; Raphatox; RCRA waste number P047; Sandolin; Sandolin A; Selinon; Sinox; Trifina; Trifocide; Winterwash.

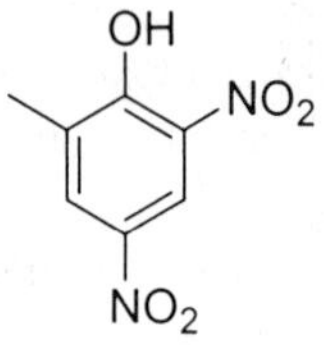

CAS Registry Number: 534-52-1
DOT: 1598
Molecular formula: $C_7H_6N_2O_5$
Formula weight: 198.14
RTECS: GO9625000
Merck reference: 10, 3279

Physical state, color, and odor:
Yellow, odorless crystals

Melting point (°C):
86.5 (Weast, 1986)

Boiling point (°C):
312 (ACGIH, 1986)

Diffusivity in water (x 10^{-5} cm²/sec):
0.69 at 20 °C using method of Hayduk and Laudie (1974)

Dissociation constant, pK_a:
4.35 at 25 °C (Dean, 1973)
4.39, 4.46 (Jafvert et al., 1990)
4.31 at 21.5 °C (Schwarzenbach et al., 1988)

Flash point (°C):
Noncombustible solid (NIOSH, 1997)

Lower explosive limit:
For a dust, 30 mg/m³ is the minimum explosive concentration in air (NIOSH, 1997).

Henry's law constant (x 10^{-6} atm·m³/mol):
1.4 at 25 °C (Warner et al., 1987)

Soil sorption coefficient, log K_{oc}:
2.41 (Meylan et al., 1992)
3.28 (activated carbon, Blum et al., 1994)

Octanol/water partition coefficient, log K_{ow}:
2.14, 2.16 (Jafvert et al., 1990)
2.12 (Schwarzenbach et al., 1988)
2.85 (quoted, Mills et al., 1985)

Solubility in organics:
At 15 °C (mg/L): methanol (7.33), ethanol (9.12), chloroform (37.2), acetone (100.6) (Bailey and White, 1965).

Solubility in water:
0.013 wt % at 15 °C (Berg, 1983)
128 mg/L at 20 °C (quoted, Meites, 1963)
198 mg/L in buffer solution at pH 1.5 and 20 °C (Schwarzenbach et al., 1988)

Vapor pressure (x 10^{-5} mmHg):
5.2 at 25 °C (Melnikov, 1971)
5 at 20 °C (ACGIH, 1986)
32 at 20 °C (Schwarzenbach et al., 1988)

Environmental fate:
Biological. In plants and soils, the nitro groups are reduced to amino groups (Hartley and Kidd, 1987). When 4,6-dinitro-o-cresol was statically incubated in the dark at 25 °C with yeast extract and settled domestic wastewater inoculum, no significant biodegradation and necessary acclimation for optimum biooxidation within the 4-wk incubation period was observed (Tabak et al., 1981). When 3,4-dinitro-o-cresol (100 mg/L) was incubated in a Warburg respirometer for 3 h in the presence of phenol-adapted bacteria, the measured biological oxygen demand was 22.3% (Chambers et al., 1963).
Chemical/Physical. Reacts with alkalies and amines forming water-soluble salts which are indicative of phenols (Morrison and Boyd, 1971).

Exposure limits (mg/m³): NIOSH REL: TWA 0.2, IDLH 5; OSHA PEL: TWA 0.2; ACGIHTLV: TWA 0.2 (adopted).

Symptoms of exposure: Headache, fever, profuse sweating, rapid pulse and respiration, cough, shortness of breath, and coma (Patnaik, 1992)

Toxicity:
    EC_{50} (48-h) for *Daphnia pulex* 145μg/L (Mayer and Ellersieck, 1986).
    LC_{50} (8-d) for fathead minnows 1.3-1.7 mg/L (Spehar et al., 1982).
    LC_{50} (96-h) for fathead minnows 1.9-2.2 mg/L (Spehar et al., 1982).
    LC_{50} (48-h) for fathead minnows 8.6 mg/L (Spehar et al., 1982).
    Acute oral LD_{50} for mice 47 mg/kg, rats 10 mg/kg (quoted, RTECS, 1985); sheep 200 mg/kg (quoted, Worthing and Hance, 1991).
    LD_{50} (skin) for rats 200 mg/kg (quoted, RTECS, 1985).

Uses: Dormant ovicidal spray for fruit trees (highly phototoxic and cannot be used successfully on actively growing plants); selective herbicide and insecticide.

2,4-DINITROPHENOL

Synonyms: Aldifen; Chemox PE; α-Dinitrophenol; DNP; 2,4-DNP; Fenoxyl carbon n; 1-Hydroxy-2,4-dinitrobenzene; Maroxol-50; Nitro kleenup; NSC 1532; RCRA waste number P048; Solfo black B; Solfo black BB; Solfo black 2B supra; Solfo black G; Solfo black SB; Tetrasulphur black PB; Tetrosulphur PBR.

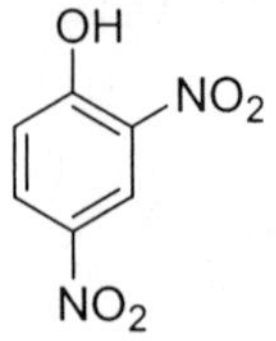

CAS Registry Number: 51-28-5
DOT: 0076
DOT label: Poison
Molecular formula: $C_6H_4N_2O_5$
Formula weight: 184.11
RTECS: SL2800000
Merck reference: 10, 3281

Physical state, color, and odor:
Yellow crystals with a sweet, musty odor

Melting point (°C):
115-116 (Weast, 1986)
106-108 (Aldrich, 1988)
111-113 (Fluka, 1988)

Boiling point (°C):
Sublimes (Weast, 1986)

Density (g/cm³):
1.683 at 24/4 °C (Weast, 1986)
1.68 at 20/4 °C (Weiss, 1986)

Diffusivity in water (x 10⁻⁵ cm²/sec):
0.76 at 20 °C using method of Hayduk and Laudie (1974)

Dissociation constant, pKₐ:
4.09 at 25 °C (Dean, 1973)
3.94 at 21.5 °C (Schwarzenbach et al., 1988)

Entropy of fusion (cal/mol·K):
14.89 (Poeti et al., 1982)

Henry's law constant (x 10⁻⁸ atm·m³/mol):
1.57 at 20 °C (approximate - calculated from water solubility and vapor pressure)

Soil sorption coefficient, log Kₒc:
1.25 (estimated, Montgomery, 1989)

Octanol/water partition coefficient, log K_{ow}:
1.51, 1.54 (Leo et al., 1971)
1.50 (Stockdale and Selwyn, 1971)
1.56 (Korenman et al., 1977)
1.67 (Schwarzenbach et al., 1988)

Solubility in organics:
At 15 °C (wt %): 13.46 in ethyl acetate, 26.42 in acetone, 5.11 in chloroform, 16.72 in pyridine, 0.42 in carbon tetrachloride, 5.98 in toluene (Windholz et al., 1983); 30.5 g/L in alcohol (Meites, 1963).

Solubility in water:
5,600 mg/L at 18 °C, 43,000 mg/L at 100 °C (quoted, Verschueren, 1983)
415, 691 and 975 mg/L at 15.1, 25.0 and 35.0 °C, respectively (Achard et al., 1996)
At pH 1.5 (mg/L): 172 at 5 °C, 207 at 10 °C, 335 at 20 °C, 473 at 30 °C (Schwarzenbach et al., 1988)
3.46 mM at 25 °C (Caturla et al., 1988)
In wt %: 0.137 at 54.5 °C, 0.301 at 75.8 °C, 0.587 at 87.4 °C, 1.22 at 96.2 °C (quoted, Windholz et al., 1983)
202 mg/L at 12.5 °C (quoted, Meites, 1963)

Vapor pressure (x 10^{-5} mmHg):
39 at 20 °C (Schwarzenbach et al., 1988)

Environmental fate:
   Biological. When 2,4-dinitrophenol was statically incubated in the dark at 25 °C with yeast extract and settled domestic wastewater inoculum, significant biodegradation with rapid adaptation was observed. At concentrations of 5 and 10 mg/L, 60 and 68% biodegradation, respectively, were observed after 7 d (Tabak et al., 1981). In activated sludge inoculum, 85.0% COD removal was achieved. The average rate of biodegradation was 6.0 mg COD/g·h (Pitter, 1976).
   Conversely, 2,4-dinitrophenol did not degrade when inoculated with a 1 mL of an enrichment culture isolated from alluvial and pokkali soils (5-day-old cultures of *Flavobacterium* sp. ATCC 27551 and *Pseudomonas* sp. ATCC 29353) (Sudhaker-Barik and Sethunathan, 1978a).
   Photolytic. When an aqueous 2,4-dinitrophenol solution containing titanium dioxide was illuminated by UV light, ammonium and nitrate ions formed as the major products (Low et al., 1991).
   Chemical/Physical. Ozonation of an aqueous solution containing 2,4-dinitrophenol (100 mg/L) yielded formic, acetic, glyoxylic and oxalic acids (Wang, 1990). 2,4-Dinitrophenol will not undergo hydrolysis (Kollig, 1993).

Symptoms of exposure: Heavy sweating, nausea, vomiting, collapse, and death (Patnaik, 1992)

Toxicity:
   LC_{50} (contact) for earthworm (*Eisenia fetida*) 0.6 μg/cm^2 (Neuhauser et al., 1985).
   LC_{50} (8-d) for fathead minnows 16 mg/L (Spehar et al., 1982).
   LC_{50} (96-h) for bluegill sunfish 620 μg/L, fathead minnows 17 mg/L (Spehar et al., 1982), *Cyprinodon variegatus* 29 ppm (Heitmuller et al., 1981).
   LC_{50} (72-h) for *Cyprinodon variegatus* 32 ppm (Heitmuller et al., 1981).

LC_{50} (48-h) for fathead minnows 7.3 mg/L (Spehar et al., 1982); *Daphnia magna* 4.1 mg/L (LeBlanc, 1980), *Cyprinodon variegatus* 32 ppm (Heitmuller et al., 1981).

LC_{50} (24-h) for *Daphnia magna* 4.5 mg/L (LeBlanc, 1980), *Cyprinodon variegatus* 42 ppm (Heitmuller et al., 1981).

Acute oral LD_{50} for mice 45 mg/kg, guinea pigs 81 mg/kg, rats 30 mg/kg, rabbits 30 mg/kg, wild birds 13 mg/kg (quoted, RTECS, 1985).

LD_{50} (subcutaneous) for rats 25 mg/kg (quoted, RTECS, 1985).

Heitmuller et al. (1981) reported a NOEC of 10 ppm.

Uses: Organic synthesis; photographic agent; manufacture of pesticides, herbicides, explosives, and wood preservatives; yellow dyes; preparation of picric acid and diaminophenol (photographic developer); indicator; analytical reagent for potassium and ammonium ions; insecticide.

2,4-DINITROTOLUENE

Synonyms: 2,4-Dinitromethylbenzene; Dinitrotoluol; 2,4-Dinitrotoluol; DNT; 2,4-DNT; **1-Methyl-2,4-dinitrobenzene**; NCI-C01865; RCRA waste number U105.

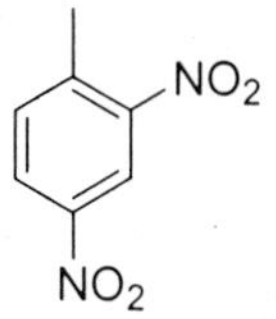

CAS Registry Number: 121-14-2
DOT: 1600 (liquid); 2038 (solid)
Molecular formula: $C_7H_6N_2O_4$
Formula weight: 182.14
RTECS: XT1575000

Physical state, color, and odor:
Yellow to red needles or yellow liquid with a faint, characteristic odor

Melting point (°C):
71.1 (Lenchitz and Velicky, 1970)
67-70 (Aldrich, 1988)

Boiling point (°C):
300 with slight decomposition (Howard, 1989)

Density (g/cm³):
1.521 at 15/4 °C (Sax, 1984)
1.379 at 20/4 °C (Weiss, 1986)
1.3208 at 71/4 °C (Keith and Walters, 1992)

Diffusivity in water (x 10^{-5} cm²/sec):
0.76 at 20 °C using method of Hayduk and Laudie (1974)

Dissociation constant, pK_a:
13.53 (Perrin, 1972)

Flash point (°C):
206.7 (Weiss, 1986)

Entropy of fusion (cal/mol·K):
14.01 (Tsonopoulos and Prausnitz, 1971)

Henry's law constant (x 10^{-7} atm·m³/mol):
8.67 (Howard, 1989)

Soil sorption coefficient, log K_{oc}:
1.79 using method of Karickhoff et al. (1979)

Octanol/water partition coefficient, log K_{ow}:
1.98 (Mabey et al., 1982; Hansch and Leo, 1985)

Solubility in organics:
Soluble in acetone, ethanol, benzene, ether, and pyrimidine (Weast, 1986)

Solubility in water:
270 mg/L at 22 °C (quoted, Verschueren, 1983)

Vapor pressure (x 10^{-4} mmHg):
12.98 at 58.8 °C (Lenchitz and Velicky, 1970)
1.1 at 20 °C (Howard, 1989)

Environmental fate:
Biological. When 2,4-dinitrotoluene was statically incubated in the dark at 25 °C with yeast extract and settled domestic wastewater inoculum, significant biodegradation with gradual acclimation was followed by deadaptive process in subsequent subcultures. At a concentration of 5 mg/L, 77, 61, 50, and 27% losses were observed after 7, 14, 21, and 28-d incubation periods, respectively. At a concentration of 10 mg/L, only 50, 49, 44, and 23% were observed after 7, 14, 21, and 28-d incubation periods, respectively (Tabak et al., 1981).

Razo-Flores et al. (1999) studied the fate of 2,4-dinitrotoluene (120 mg/L) in an upward-flow anaerobic sludge bed reactor containing a mixture of volatile fatty acids and/or glucose as electron donors. 2,4-Dinitrotoluene was transformed to 2,4-diaminotoluene (52% molar yield) in stoichiometric amounts until day 125. Thereafter, the aromatic amine underwent further degradation. Approximately 98.5% of the volatile fatty acids in the reactor was converted to methane during the 202-d experiment.

Chemical/Physical. Wet oxidation of 2,4-dinitrotoluene at 320 °C yielded formic and acetic acids (Randall and Knopp, 1980). 2,4-Dinitrotoluene will not hydrolyze (Kollig, 1993).

Toxicity:
EC_{50} (15-min) for *Photobacterium phosphoreum* 4.17 mg/L (Yuan and Lang, 1997).
IC_{50} (24-h) for river bacteria 35.5 mg/L (Yuan and Lang, 1997).
Acute oral LD_{50} for mice 790 mg/kg, rats 268 mg/kg, guinea pigs 1,300 mg/kg (quoted, RTECS, 1985).

Drinking water standard: No MCLGs or MCLs have been proposed although 2,4-dichloro-toluene has been listed for regulation (U.S. EPA, 1996).

Uses: Organic synthesis; intermediate for toluidine, dyes, and explosives.

2,6-DINITROTOLUENE

Synonyms: 2,6-Dinitromethylbenzene; 2,6-Dinitrotoluol; 2,6-DNT; **2-Methyl-1,3-dinitrobenzene**; RCRA waste number U106.

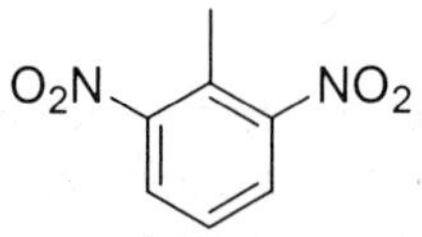

CAS Registry Number: 606-20-2
DOT: 1600 (liquid); 2038 (solid)
Molecular formula: $C_7H_6N_2O_4$
Formula weight: 182.14
RTECS: XT1925000

Physical state and color:
Pale yellow crystals. Odor threshold concentration in water is 100 ppb (Keith and Walters, 1992).

Melting point (°C):
66 (Weast, 1986)
60.5 (Weiss, 1986)

Boiling point (°C):
285 (Maksimov, 1968)

Density (g/cm³):
1.2833 at 111/4 °C (Weast, 1986)

Diffusivity in water (x 10⁻⁵ cm²/sec):
0.76 at 20 °C using method of Hayduk and Laudie (1974)

Flash point (°C):
206.7 (calculated, Weiss, 1986)

Henry's law constant (x 10⁻⁷ atm·m³/mol):
2.17 (Howard, 1989)

Soil sorption coefficient, log K_{oc}:
1.79 using method of Karickhoff et al. (1979)

Octanol/water partition coefficient, log K_{ow}:
1.72 (Hansch and Leo, 1985)
2.00 (quoted, Mills et al., 1985)

Solubility in organics:
Soluble in ethanol (Weast, 1986)

Solubility in water:
≈ 300 mg/L (quoted, Mills et al., 1985)

Vapor pressure (x 10^{-4} mmHg):
3.5 at 20 °C (Howard, 1989)
5.67 at 25 °C (Banerjee et al., 1990)

Environmental fate:

Biological. When 2,6-dinitrotoluene was statically incubated in the dark at 25 °C with yeast extract and settled domestic wastewater inoculum, significant biodegradation with gradual acclimation was followed by deadaptive process in subsequent subcultures. At a concentration of 5 mg/L, 82, 55, 47, and 29% losses were observed after 7, 14, 21, and 28-d incubation periods, respectively. At a concentration of 10 mg/L, only 57, 49, 35, and 13% were observed after 7, 14, 21, and 28-d incubation periods, respectively (Tabak et al., 1981). Under anaerobic and aerobic conditions, a sewage inoculum degraded 2,6-dinitrotoluene to aminonitrotoluene (Hallas and Alexander, 1983).

Photolytic. Simmons and Zepp (1986) estimated the photolytic half-life of 2,6-dinitrotoluene in surface water to range from 2 to 17 h.

Chemical/Physical. 2,6-Dinitrotoluene will not hydrolyze (Kollig, 1993).

Exposure limits (mg/m^3): TWA: 1.5, IDLH: 200 (Weiss, 1986).

Toxicity:

EC_{50} (15-min) for *Photobacterium phosphoreum* 9.78 mg/L (Yuan and Lang, 1997).
IC_{50} (24-h) for river bacteria 63.2 mg/L (Yuan and Lang, 1997).
Acute oral LD_{50} for mice 621 mg/kg, rats 177 mg/kg (quoted, RTECS, 1985).

Drinking water standard: No MCLGs or MCLs have been proposed although 2,6-dinitrotoluene has been listed for regulation (U.S. EPA, 1996).

Uses: Organic synthesis; propellant additive; manufacture of explosives; intermediate in the manufacture of polyurethanes.

DI-*n*-OCTYL PHTHALATE

Synonyms: 1,2-Benzenedicarboxylic acid dioctyl ester; 1,2-Benzenedicarboxylic acid di-*n*-octyl ester; *o*-Benzenedicarboxylic acid dioctyl ester; Celluflex DOP; Dinopol NOP; Di-octyl-*o*-benzenedicarboxylate; Dioctyl phthalate; *n*-Dioctyl phthalate; DNOP; DOP; Octyl phthalate; *n*-Octyl phthalate; Polycizer 162; PX 138; RCRA waste number U107; Vinicizer 85.

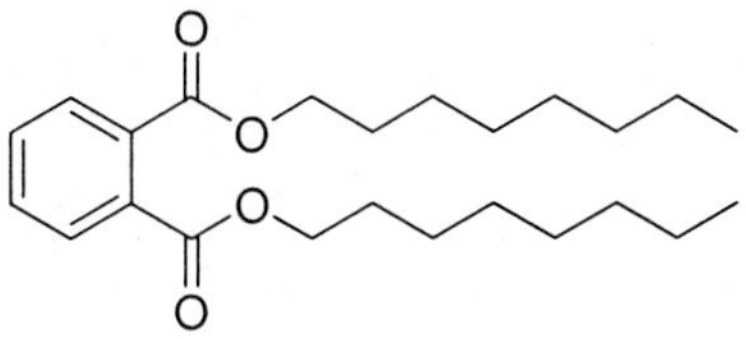

CAS Registry Number: 117-84-0
Molecular formula: $C_{24}H_{38}O_4$
Formula weight: 390.57
RTECS: TI1925000

Physical state, color, and odor:
Clear, light colored, viscous, oily liquid with a slight odor

Melting point (°C):
-30 (Clayton and Clayton, 1981)
-25 (Fishbein and Albro, 1972)

Boiling point (°C):
386 (Weiss, 1986)

Density (g/cm³):
0.978 at 25/4 °C (Standen, 1968)
0.978 at 20/4 °C (Fishbein and Albro, 1972)

Diffusivity in water (x 10^{-5} cm²/sec):
0.39 at 20 °C using method of Hayduk and Laudie (1974)

Flash point (°C):
218.3 (Weiss, 1986)

Lower explosive limit (%):
Not pertinent (Weiss, 1986)

Upper explosive limit (%):
Not pertinent (Weiss, 1986)

Henry's law constant (x 10^{-12} atm·m³/mol):
1.41 at 25 °C (approximate - calculated from water solubility and vapor pressure)

Soil sorption coefficient, log K_{oc}:
8.99 using method of Karickhoff et al. (1979)

Octanol/water partition coefficient, log K_{ow}:
8.18 (Ellington, 1999)

Solubility in water (mg/L):
0.285 at 24 °C (quoted, Verschueren, 1983)
3.0 at 25 °C (Wolfe et al., 1980)
0.020 (DeFoe et al., 1990)
0.00051 at 25 °C (Ellington, 1999)

Vapor density:
16.00 g/L at 25 °C, 13.52 (air = 1)

Vapor pressure (x 10^{-8} mmHg):
5 at 82 °C, 500 at 132 °C (Gross and Colony, 1973)

Environmental fate:
   Biological. o-Phthalic acid was tentatively identified as the major degradation product of di-*n*-octyl phthalate produced by the bacterium *Serratia marcescens* (Mathur and Rouatt, 1975). When di-*n*-octyl phthalate was statically incubated in the dark at 25 °C with yeast extract and settled domestic wastewater inoculum, no degradation was observed after 7 d. Over a 21-d period, however, gradual adaptation did occur, resulting in 94 and 93% losses at concentrations of 5 and 10 mg/L, respectively (Tabak et al., 1981). In the presence of suspended natural populations from unpolluted aquatic systems, the second-order microbial transformation rate constant determined in the laboratory was reported to be $3.7 \pm 0.6 \times 10^{-13}$ L/organism·h (Steen, 1991).
   Chemical/Physical. Under alkaline conditions, di-*n*-octyl phthalate will initially hydrolyze to *n*-octyl hydrogen phthalate and *n*-octanol. The monoester will undergo further hydrolysis forming *o*-phthalic acid and *n*-octanol (Kollig, 1993). The hydrolysis half-life at pH 7 and 25 °C was estimated to be 107 yr (Ellington et al., 1988).

Symptoms of exposure: Ingestion may cause nausea, somnolence, hallucination, and lacrimation (Patnaik, 1992).

Toxicity:
   LC_{50} (contact) for earthworm (*Eisenia fetida*) 3,140 μg/cm^2 (Neuhauser et al., 1985).
   Acute oral LD_{50} for mice 6,513 mg/kg (quoted, RTECS, 1985); rats 13,000 mg/kg (Dogra et al., 1989); mice 1,250-1,954 mg/kg (Etnier, 1987).

Source: Detected in distilled water-soluble fractions of new and used motor oil at concentrations of 1.3-1.4 and 73-78 μg/L, respectively (Chen et al., 1994). May leach from plastic products (e.g., tubing, containers) used in laboratories during chemical analysis of aqueous samples.

Use: Plasticizer to impart flexibility for poly(vinyl chloride) (PVC) and other vinyl polymers.

1,4-DIOXANE

Synonyms: Diethylene dioxide; 1,4-Diethylene dioxide; Diethylene ether; Diethylene oxide; Diokan; 1,4-Dioxacyclohexane; Dioxan; Dioxane-1,4; **1,4-Dioxane**; *p*-Dioxane; Dioxyethylene ether; Glycol ethylene ether; NCI-C03689; RCRA waste number U108; Tetrahydro-1,4-dioxin; Tetrahydro-*p*-dioxin; UN 1165.

CAS Registry Number: 123-91-1
DOT: 1165
DOT label: Flammable liquid
Molecular formula: $C_4H_8O_2$
Formula weight: 88.11
RTECS: JG8225000
Merck reference: 10, 3304

Physical state, color, and odor:
Clear, colorless liquid with a pleasant, ether-like odor. Odor threshold concentration is 24 ppm (Amoore and Hautala, 1983).

Melting point (°C):
11.8 (Weast, 1986)

Boiling point (°C):
101.32 (Riddick et al., 1986)
101.1 (Windholz et al., 1983)

Density (g/cm³):
1.0337 at 20/4 °C (Weast, 1986)
1.0280 at 25.00/4 °C, 1.0221 at 30.00/4 °C, 1.0110 at 40.00/4 °C, 0.9994 at 50.00/4 °C, 0.9886 at 60.00/4 °C, 0.9766 at 70.00/4 °C (Abraham et al., 1971)

Diffusivity in water (x 10^{-5} cm²/sec):
0.97 at 20 °C using method of Hayduk and Laudie (1974)

Flash point (°C):
12.8 (NIOSH, 1997)
18.3 (open cup, Hawley, 1981)

Lower explosive limit (%):
2.0 (NIOSH, 1997)

Upper explosive limit (%):
22 (NIOSH, 1997)

Heat of fusion (kcal/mol):
2.978 (Riddick et al., 1986)

Henry's law constant (x 10^{-6} atm·m³/mol at 25 °C):
7.14 (Friant and Suffet, 1979)

5.00 (Cabani et al., 1971)
4.55 (Rohrschneider, 1973)
9.07 (Amoore and Buttery, 1978)
209 (Hakuta et al., 1977)

Ionization potential (eV):
9.13 ± 0.03 (Franklin et al., 1969)

Soil sorption coefficient, log K_{oc}:
0.54 (calculated, Mercer et al., 1990)

Octanol/water partition coefficient, log K_{ow}:
-0.42 (Collander, 1951)
-0.27 (Hansch and Leo, 1979)

Solubility in organics:
Soluble in acetone, alcohol, benzene, and ether (Weast, 1986). Miscible with most organic solvents (Huntress and Mulliken, 1941).

Solubility in water:
Miscible (Palit, 1947).

Vapor density:
.3.60 g/L at 25 °C, 3.04 (air = 1)

Vapor pressure (mmHg):
29 at 20 °C (NIOSH, 1997)
37 at 25 °C, 50 at 30 °C (Verschueren, 1983)
38.1 at 25 °C (Banerjee et al., 1990)
34.28 at 25.00 °C (Hussam and Carr, 1985)

Environmental fate:
Biological. Heukelekian and Rand (1955) reported a 10-d BOD value of 0.00 g/g which is 0.0% of the ThOD value of 1.89 g/g.

Photolytic. Irradiation of pure 1,4-dioxane through quartz using a 450-W medium-pressure mercury lamp gave meso and racemic forms of 1-hydroxyethyldioxane, a pair of diastereomeric dioxane dimers (Mazzocchi and Bowen, 1975), dioxanone, dioxanol, hydroxymethyldioxane, and hydroxyethylidenedioxane (Houser and Sibbio, 1977). When 1,4-dioxane is subjected to a megawatt ruby laser, 4% was decomposed yielding ethylene, carbon monoxide, hydrogen, and a trace of formaldehyde (Watson and Parrish, 1971).

Chemical/Physical. Anticipated products from the reaction of 1,4-dioxane with ozone or OH radicals in the atmosphere are glyoxylic acid, oxygenated formates, and $OHCOCH_2CH_2OCHO$ (Cupitt, 1980). Storage of 1,4-dioxane in the presence of air resulted in the formation of 1,2-ethanediol monoformate and 1,2-ethane diformate (Jewett and Lawless, 1980).

Schnoor et al. (1997) studied the feasibility and efficacy of phytoremediation of sites contaminated with 1,4-dioxane. Hybrid poplar trees (*Populus deltoides nigra*, DN34, Imperial Carolina) were utilized because of their resistance to contamination and their high growth rates. The poplars were ground in hydroponic solutions for 2 wk in an inorganic nutrient solution. The reactors containing plant, soil and dioxane (200 ppm) were placed in a growth chamber which was maintained at 25 °C. After 7 d, >95% of the dioxane was removed from the hydroponic solution. Losses due to volatilization and sorption to glassware were minor.

1,4-Dioxane will not undergo hydrolysis because it does not contain a hydrolyzable functional group (Kollig, 1993).

Exposure limits: Potential occupational carcinogen. NIOSH REL: 30-min ceiling 1 ppm (3.6 mg/m^3), IDLH 500 ppm; OSHA PEL: TWA 100 ppm; ACGIH TLV: TWA 20 ppm (adopted).

Symptoms of exposure: Ingestion or skin absorption may cause drowsiness, headache, respiratory distress, nausea, and vomiting (Patnaik, 1992).

Toxicity:
LC_{84} values for white rats following 2 and 4 h of inhalation were 52 and 69.5 mg/L, respectively (Pilipiuk et al., 1977).

LC_{50} (96-h static bioassay) for bluegill sunfish >10,000 mg/L, *Menidia beryllina* 6,700 mg/L (quoted, Verschueren, 1983).

LC_{50} (48-h) for red killifish 81,280 mg/L (Yoshioka et al., 1986).

LC_{50} (4-h inhalation) for white rats 46 mg/L (Pilipiuk et al., 1977).

LC_{16} values for white rats following 2 and 4 h of inhalation were 61 and 40 mg/L, respectively (Pilipiuk et al., 1977).

Acute oral LD_{50} for mice 5,700 mg/kg, cats 2,000 mg/kg, guinea pigs 3,150 mg/kg, rats 4,200 mg/kg, rabbits 2,000 mg/kg (quoted, RTECS, 1985).

Drinking water standard: No MCLGs or MCLs have been proposed although 1,4-dioxane has been listed for regulation (U.S. EPA, 1996).

Source: Improper disposal of products listed below may result in 1,4-dioxane leaching into groundwater.

Uses: Solvent for cellulose acetate, benzyl cellulose, ethyl cellulose, waxes, resins, oils, cements, cosmetics, deodorants; fumigants; paint and varnish removers, cleaning and detergent preparations; wetting and dispersing agent in textile processing, dyes baths, stain and printing compositions; polishing compositions; stabilizer for chlorinated solvents; scintillation counter; organic synthesis.

1,2-DIPHENYLHYDRAZINE

Synonyms: *N,N'*-Bianiline; *N,N'*-Diphenylhydrazine; *sym*-Diphenylhydrazine; DPH; Hydrazo-benzene; 1,1'-Hydrazobenzene; Hydrazodibenzene; NCI-C01854; RCRA waste number U109.

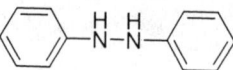

CAS Registry Number: 122-66-7
Molecular formula: $C_{12}H_{12}N_2$
Formula weight: 184.24
RTECS: MW2625000

Physical state and color:
Colorless to pale yellow crystals

Melting point (°C):
131 (Weast, 1986)
126-128 (Fluka, 1988)

Boiling point (°C):
Decomposes near the melting point (U.S. EPA, 1980a)

Density (g/cm³):
1.158 at 16/4 °C (Weast, 1986)

Diffusivity in water (x 10^{-5} cm²/sec):
0.57 at 20 °C using method of Hayduk and Laudie (1974)

Henry's law constant (x 10^{-11} atm·m³/mol):
4.11 at 25 °C (approximate - calculated from water solubility and vapor pressure)

Ionization potential (eV):
7.78 (Lias et al., 1998)

Soil sorption coefficient, log K_{oc}:
2.82 using method of Karickhoff et al. (1979)

Octanol/water partition coefficient, log K_{ow}:
2.94 (Mabey et al., 1982)

Solubility in organics:
Soluble in ethanol (Weast, 1986).

Solubility in water:
221 mg/L at 25 °C (U.S. EPA, 1980a)

Vapor pressure (mmHg):
2.6 x 10^{-5} at 25 °C (Mabey et al., 1982)

Environmental fate:

Biological. When 5 and 10 mg/L of diphenylhydrazine was statically incubated in the dark at 25 °C with yeast extract and settled domestic wastewater inoculum, 80 and 72% biodegradation, respectively, were observed after 7 d (Tabak et al., 1981).

Chemical/Physical. Wet oxidation of 1,2-diphenylhydrazine at 320 °C yielded formic and acetic acids (Randall and Knopp, 1980). 1,2-Diphenylhydrazine will not hydrolyze (Kollig, 1993).

Toxicity:

LC_{50} (96-h) for bluegill sunfish 270 mg/L (Spehar et al., 1982).

LC_{50} (48-h) for *Daphnia magna* 4.1 mg/L (LeBlanc, 1980).

LC_{50} (24-h) for *Daphnia magna* 8.1 mg/L (LeBlanc, 1980).

Acute oral LD_{50} for rats 301 mg/kg (quoted, RTECS, 1985).

Uses: Manufacture of benzidine and starting material for pharmaceutical drugs.

DIURON

Synonyms: AF 101; Cekiuron; Crisuron; Dailon; DCMU; Diater; Dichlorfenidim; 3-(3,4-Dichlorophenol)-1,1-dimethylurea; 3-(3,4-Dichlorophenyl)-1,1-dimethylurea; ***N'*-(3,4-Dichlorophenyl)-*N,N*-dimethylurea**; 1,1-Dimethyl-3-(3,4-dichlorophenyl)urea; Dion; Direx 4L; Diurex; Diurol; DMU; Dynex; Farmco diuron; Herbatox; HW 920; Karmex; Karmex diuron herbicide; Karmex DW; Marmer; NA 2767; Sup'r flo; Telvar; Telvar diuron weed killer; Unidron; Urox D; USAF P-7; USAF XR-42; Vonduron.

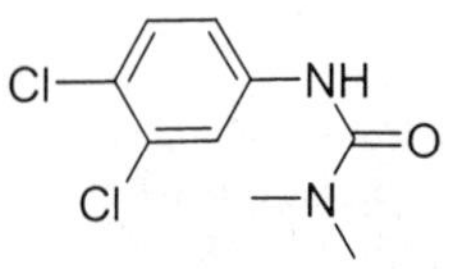

CAS Registry Number: 330-54-1
DOT: 2767
DOT label: Poison
Molecular formula: $C_9H_{10}Cl_2N_2O$
Formula weight: 233.11
RTECS: YS8925000
Merck reference: 10, 3400

Physical state and color:
White, odorless crystalline solid

Melting point (°C):
158-159 (Windholz et al., 1983)
150-155 (Bailey and White, 1965)

Boiling point (°C):
Decomposes at 180 (Hawley, 1981)

Density (g/cm³):
1.385 using method of Lyman et al. (1982).

Diffusivity in water (x 10^{-5} cm²/sec):
0.53 at 20 °C using method of Hayduk and Laudie (1974)

Dissociation constant, pK_a:
-1 to -2 (Bailey and White, 1965)

Flash point (°C):
Noncombustible solid (NIOSH, 1997)

Henry's law constant (x 10^{-9} atm·m³/mol):
1.46 at 25 °C (approximate - calculated from water solubility and vapor pressure)

Soil sorption coefficient, log K_{oc}:
2.51 (Commerce soil), 2.62 (Tracy soil), 2.60 (Catlin soil) (McCall et al., 1981)
2.87 (Webster soil, Nkedi-Kizza et al., 1983)

2.21 (Batcombe silt loam, Briggs, 1981)
2.95 (Eustis fine sand, Wood et al., 1990)

Octanol/water partition coefficient, log K_{ow}:
1.97 (Kenaga and Goring, 1980)
2.68 (Briggs, 1981)
2.60 (Ellgehausen et al., 1981)
2.69, 2.65, and 2.63 at 5, 25, and 40 °C, respectively (Madhun et al., 1986)

Solubility in organics:
In acetone: 5.3 wt % at 27 °C (Meister, 1988).

Solubility in water (mg/L):
40 at 20 °C (Gunther et al., 1968)
42 at 25 °C (Bailey and White, 1965)
22 at 20 °C (Ellgehausen et al., 1981; Fühner and Geiger, 1977)
19.6, 40.1, and 53.4 at 5, 25, and 40 °C, respectively (Madhun et al., 1986)

Vapor pressure (x 10^{-7} mmHg):
31 at 50 °C (Bailey and White, 1965)
2 at 30 °C (Hawley, 1981)

Environmental fate:
   Biological. Degradation of radiolabeled diuron (20 ppm) was not observed after 2 wk of culturing with *Fusarium* and two unidentified microorganisms. After 80 d, only 3.5% of the applied amount evolved as $^{14}CO_2$ (Lopez and Kirkwood, 1974). In 8 wk, <20% of diuron in soil (60 ppm) was detoxified (Corbin and Upchurch, 1967). 3,4-Dichloroaniline was reported as a minor degradation product of diuron in water (Drinking Water Health Advisory, 1989) and soils (Duke et al., 1991).
   Under aerobic conditions, mixed cultures isolated from pond water and sediment degraded diuron (10 µg/mL) to CPDU, 3,4-dichloroaniline, 3-(3,4-dichlorophenyl)-1-methylurea, carbon dioxide, and a monodemethylated product. The extent of biodegradation varied with time, glycerol concentration, and microbial population. The degradation half-life was <70 d at 30 °C (Ellis and Camper, 1982).
   Thom and Agg (1975) reported that diuron is amenable to biological treatment with acclimation.
   Soil. Several degradation pathways were reported. The major products and reaction pathways include formation of 1-methyl-3-(3,4-dichlorophenol) urea and 3-(3,4-dichlorophenyl) urea via *N*-dealkylation, a 6-hydroxy derivative via ring hydroxylation, and formation of 3,4-dichloroaniline, 3,4-dichloronitroaniline, and 3,4-dichloronitrobenzene via hydrolysis and oxidation (Geissbühler et al., 1975).
   Incubation of diuron in soils releases carbon dioxide (Madhun and Freed, 1987). The rate of carbon dioxide formation nearly tripled when the soil temperature was increased from 25 to 35 °C. Reported half-lives in an Adkins loamy sand are 705, 414, and 225 d at 25, 30, and 35 °C, respectively. However, in a Semiahoo mucky peat, the half-lives were considerable higher: 3,991, 2,164, and 1,165 d at 25, 30, and 35 °C, respectively (Madhun and Freed, 1987). Under aerobic conditions, biologically active, organic-rich, diuron-treated pond sediment (40 µg/mL) converted diuron exclusively to 3-(3-chlorophenyl)-1,1-dimethylurea (CPDU) (Attaway et al., 1982, 1982a; Stepp et al., 1985). At 25 and 30 °C, 90% degradation was observed after 55 and 17 d, respectively (Attaway, 1982a).
   The half-lives for diuron in field soils ranged from 133 to 212 d with an average half-life of

328 d (Hill et al., 1955). Hill et al. (1955) studied the degradation of diuron using a Cecil loamy sand (1 ppm) and Brookstone silty clay loam (5 ppm) in the laboratory maintained at 27 °C and 60% relative humidity. In both soils, diuron was applied on four separate occasions over 22 wk. In both instances, the investigators observed 40% of the applied amount degraded in both soils.

In a field application study, diuron did not leach below 5 cm in depth despite repeated applications or water addition (Majka and Lavy, 1977).

Groundwater. According to the U.S. EPA (1986), diuron has a high potential to leach to groundwater.

Photolytic. Tanaka et al. (1985) studied the photolysis of diuron (40 mg/L) in aqueous solution using UV light (λ = 300 nm) or sunlight. After 25 d of exposure to sunlight, diuron degraded to 2,2,3'-trichloro-2,4'-di-*N,N*-dimethylurea biphenyl (yield = 1.3%), and hydrogen chloride (Tanaka et al., 1985). Diuron should undergo direct photolysis because it absorbs UV light at wavelengths greater than 290 nm (Gore et al., 1971).

Chemical/Physical. Diuron decomposes at 180 to 190 °C releasing dimethylamine and 3,4-dichlorophenylisocyanate. Dimethylamine and 3,4-dichloroaniline are produced when hydrolyzed or when acids or bases are added at elevated temperatures (Sittig, 1985). The hydrolysis half-life of diuron in a 0.5 N sodium hydroxide solution at 20 °C is 150 d (El-Dib and Aly, 1976). When diuron was pyrolyzed in a helium atmosphere between 400 and 1,000 °C, the following products were identified: dimethylamine, chlorobenzene, 1,2-dichloro-benzene, benzonitrile, a trichlorobenzene, aniline, 4-chloroaniline, 3,4-dichlorophenyl iso-cyanate, bis(1,3-(3,4-dichlorophenyl)urea), 3,4-dichloroaniline, and monuron [3-(4-chlorophen-yl)-1,1-dimethylurea] (Gomez et al., 1982). Products reported from the combustion of diuron at 900 °C include carbon monoxide, carbon dioxide, chlorine, nitrogen oxides, and hydrochloric acid (Kennedy et al., 1972a).

Diuron is stable in aqueous solutions and does not hydrolyze (Hance, 1967a).

Exposure limits: NIOSH REL: 10 mg/m³; ACGIH TLV: TWA 10 mg/m³ (adopted).

Symptoms of exposure: May irritate eyes, skin, nose, and throat (NIOSH, 1997)

Toxicity:

EC_{50} (48-h) for *Daphnia pulex* 1.4 μg/L, *Simocephalus serrulatus* 2 μg/L (Mayer and Ellersieck, 1986).

LC_{50} (21-d static-renewal) for *Rana catesbeiana* tadpoles 12.7 mg/L (Schuytema and Nebeker, 1998).

LC_{50} (14-d) for Pacific treefrog (*Pseudacris regilla*) 15.2 mg/L, African clawed frog (*Xenoous laevis*) 11.3 mg/L, tadpoles (*Rana aurora*) 22.2 mg/L (Schuytema and Nebeker, 1998).

LC_{50} (10-d) for amphipod (*Hyalella azteca*) 18.4 mg/L, midge (*Chironomus tentans*) 3.3 mg/L, juvenile fathead minnows 27.1 mg/L (Nebeker and Schuytema, 1998).

LC_{50} (7-d) for *Daphnia pulex* 7.1 mg/L, fathead minnow embryos 11.7 mg/L (Nebeker and Schuytema, 1998).

LC_{50} (96-h) for rainbow trout 5.6 mg/L, bluegill sunfish 5.9 mg/L, guppies 25 mg/L (Hartley and Kidd, 1987), *Gammarus fasciatus* 700 μg/L (Sanders, 1970), *Daphnia pulex* 17.9 mg/L, *Hyalella azteca* 19.4 mg/L (Nebeker and Schuytema, 1998).

LC_{50} (48-h) for bluegill sunfish 7.4 ppm, rainbow trout 4.3 ppm, coho salmon 16.0 mg/L (quoted, Verschueren, 1983), *Daphnia pulex* 1,400 μg/L, *Simocephalus serrulatus* 2,000 μg/L (Sanders and Cope, 1966).

Acute oral LD_{50} for rats 3,400 mg/kg (Hartley and Kidd, 1987), 1,017 mg/kg (quoted, RTECS, 1985), 437 mg/kg (Windholz et al., 1983).

NOEL (10-d) for midge (*Chironomus tentans*) 1.9 and 3.4 mg/L, juvenile fathead minnows

<3.4 mg/L, annelid worm (*Lumbriculus variegatus*) 1.8 mg/L, snail (*Physa gyrina*) 13.4 mg/L, Pacific treefrog (*Pseudacris regilla*) embryos 14.5 mg/L (Nebeker and Schuytema, 1998).

Drinking water standard: No MCLGs or MCLs have been proposed although diuron has been listed for regulation (U.S. EPA, 1996).

Uses: Pre-emergence herbicide used in soil to control germinating broadleaf grasses and weeds in crops such as apples, cotton, grapes, pears, pineapples, and alfalfa; sugar cane flowering depressant.

DODECANE

Synonyms: Adakane 12; Bihexyl; Dihexyl; *n*-Dodecane; Duodecane.

CAS Registry Number: 112-40-3
Molecular formula: $C_{12}H_{26}$
Formula weight: 170.33
RTECS: JR2125000

Physical state and color:
Colorless liquid having an odor threshold of 37 mg/m^3 (Lafforts and Dravnieks, 1973)

Melting point (°C):
-9.6 (Weast, 1986)
-12 (Sax and Lewis, 1987)

Boiling point (°C):
216.3 (Weast, 1986)
213 (Hawley, 1981)

Density (g/cm^3):
0.74869 at 20/4 °C, 0.74516 at 25/4 °C (Dreisbach, 1959)

Diffusivity in water (x 10^{-5} cm^2/sec):
0.54 at 20 °C using method of Hayduk and Laudie (1974)

Dissociation constant, pK$_a$:
>14 (Schwarzenbach et al., 1993)

Flash point (°C):
71.1 (Hawley, 1981)
79 (Affens and McLaren, 1972)

Lower explosive limit (%):
0.6 (Sax and Lewis, 1987)

Heat of fusion (kcal/mol):
8.57 (Riddick et al., 1986)

Henry's law constant (atm·m^3/mol at 25 °C):
24.2 (approximate - calculated from water solubility and vapor pressure)

Interfacial tension with water (dyn/cm):
52.8 at 24.5 °C (Demond and Lindner, 1993)
50.48 at 25 °C (Jańczuk et al., 1993)
52.86 at 20 °C (Fowkes, 1980)

Bioconcentration factor, log BCF:
3.80 (*Chlorella fusca*, Freitag et al., 1982; Geyer et al., 1984)
3.11 (activated sludge), 1.70 (golden ide) (Freitag et al., 1985)

Soil sorption coefficient, log K_{oc}:
Unavailable because experimental methods for estimation of this parameter for aliphatic hydrocarbons are lacking in the documented literature

Octanol/water partition coefficient, log K_{ow}:
5.64 (Geyer et al., 1984)
6.10 (Coates et al., 1985)
7.24 (Burkhard et al., 1985a)

Solubility in organics:
Soluble in acetone, alcohol, chloroform, ether (Weast, 1986), and many hydrocarbons

Solubility in water (25 °C):
3.7 μg/kg (distilled water), 2.9 μg/kg (seawater) (Sutton and Calder, 1974)
3.4 μg/L (Mackay and Shiu, 1981)
8.42 μg/kg (Franks, 1966)
5 x 10^{-10} in seawater (mole fraction, Krasnoshchekova and Gubergrits, 1973)

Vapor density:
7.13 g/L at 25 °C, 6.02 (air = 1)

Vapor pressure (mmHg):
0.3 at 20 °C, 1 at 48 °C (Verschueren, 1983)
0.39 at 25 °C (Mackay et al., 1982)

Environmental fate:
   Biological. Dodecane may biodegrade in two ways. The first is the formation of dodecyl hydroperoxide which decomposes to 1-dodecanol. The alcohol is further oxidized forming dodecanoic acid. The other pathway involves dehydrogenation to 1-dodecene, which may react with water, giving 1-dodecanol (Dugan, 1972).
   Estimated half-lives of dodecane (0.3 μg/L) from an experimental marine mesocosm during the spring (8-16 °C), summer (20-22 °C), and winter (3-7 °C) were 1.1, 0.7, and 3.6 d, respectively (Wakeham et al., 1983).
   Chemical/Physical. Complete combustion in air yields carbon dioxide and water. Dodecane will not hydrolyze because it has no hydrolyzable functional group.

Source: Constituent in paraffin fraction of petroleum. Dodecane may be present in stormwater runoff from asphalted roadways and general use of petroleum oils and tars (quoted, Verschueren).
   Schauer et al. (1999) reported anthracene in diesel fuel at a concentration of 15,500 μg/g and in a diesel-powered medium-duty truck exhaust at an emission rate of 503 μg/km.

Uses: Solvent; jet fuel research; rubber industry; manufacturing paraffin products; paper processing industry; standardized hydrocarbon; distillation chaser; gasoline component; organic synthesis.

α-ENDOSULFAN

Synonyms: Benzoepin; Beosit; Bio 5462; Chlorthiepin; Crisulfan; Cyclodan; Endocel; Endosol; Endosulfan; Endosulfan I; Endosulphan; ENT 23979; FMC 5462; 1,2,3,7,7-Hexachlorobicyclo[2.2.1]-2-heptene-5,6-bisoxymethylene sulfite; α,β-1,2,3,7,7-Hexachlorobicyclo[2.2.1]-2-heptene-5,6-bisoxymethylene sulfite; Hexachlorohexahydromethano-2,4,3-benzodioxathiepin-3-oxide; **(3α,5aβ,6α,9α,9aβ)-6,7,8,9,10,10-Hexachloro-1,5,5a,6,9,9a-hexahydro-6,9-methano-2,4,3-benzodioxathiepin-3-oxide**; 1,4,5,6,7,7-Hexachloro-5-norborene-2,3-dimethanol cyclic sulfite; Hildan; HOE 2671; Insectophene; KOP-thiodan; Malix; NCI-C00566; NIA 5462; Niagara 5462; OMS 570; RCRA waste number P050; Thifor; Thimul; Thiodan; Thiofor; Thiomul; Thionex; Thiosulfan; Tionel; Tiovel.

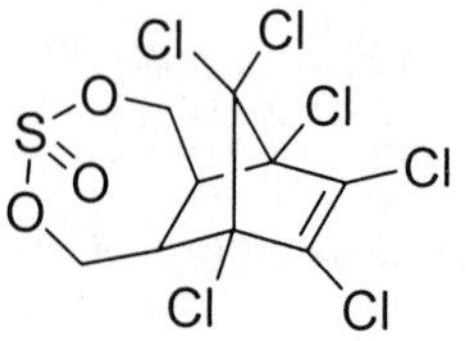

Note: Insecticidal formation contains the two sterioisomers α- and β-endosulfan in an approximate ratio of 70:30 (Goebel, 1983).

CAS Registry Number: 959-98-8
DOT: 2761
DOT label: Poison
Molecular formula: $C_9H_6Cl_6O_3S$
Formula weight: 406.92
RTECS: RB9275000
Merck reference: 10, 3535

Physical state, color, and odor:
Colorless to brown crystals with a sulfur dioxide odor

Melting point (°C):
108-110 (Ali, 1978)

Density (g/cm³):
1.745 at 20/4 °C (Sax, 1984)

Diffusivity in water (x 10⁻⁵ cm²/sec):
0.45 at 20 °C using method of Hayduk and Laudie (1974)

Henry's law constant (x 10⁻⁵ atm·m³/mol at 20 °C):
6.55 and 12.74 in distilled water and 33.3‰ solution of NaCl, respectively. Technical endosulfan [α-Endosulfan (70%) + β-Endosulfan (30%)]: 6.38 and 13.07 in distilled water and 33.3‰ NaCl, respectively (Rice et al., 1997, 1997a).

Bioconcentration factor, log BCF:
3.35 (edible tissue), 3.44 (whole body) (striped mullet, Schimmel et al., 1977b)

Soil sorption coefficient, log K_{oc}:
3.31 using method of Kenaga and Goring (1980)

Octanol/water partition coefficient, log K$_{ow}$:
3.55 (Ali, 1978)

Solubility in water:
530 ppb at 25 °C (Weil et al., 1974)
0.32 mg/L at 22 °C (Worthing and Hance, 1991)
0.51 mg/L at 20 °C (Bowman and Sans, 1983)

Vapor pressure (mmHg):
4.58 x 10^{-5} at 25 °C (Hinckley et al., 1990)

Environmental fate:
Biological. Isolated strains of *Aspergillus niger* degraded endosulfan to endodiol (El Zorghani and Omer, 1974). Biodegradation of α-endosulfan by a bacterial coculture appeared to be at a higher rate than β-endosulfan. The release of chloride ions suggested degradation occurred via dehalogenation. No expected metabolites (e.g., endodiol, endosulfate, endo-hydroxy ether) accumulated, suggesting they were further degraded (Awasthi et al., 1997).

Soil. Metabolites of endosulfan identified in seven soils were: endosulfandiol, endosulfanhydroxy ether, endosulfan lactone, and endosulfan sulfate (Dreher and Podratzki, 1988; Martens, 1977). Endosulfan sulfate was the major biodegradation product in soils under aerobic, anaerobic, and flooded conditions. In flooded soils, endolactone was detected only once, whereas endodiol and endohydroxy ether were identified in all soils under these conditions. Under anaerobic conditions, endodiol formed in low amounts in two soils (Martens, 1977). These compounds, as well as endosulfan ether, were also reported as metabolites identified in aquatic systems (Day, 1991). Endosulfan sulfate was the major biodegradation product in soils under aerobic, anaerobic, and flooded conditions (Martens, 1977). In flooded soils, endolactone was detected only once whereas endodiol and endohydroxy ether were identified in all soils under these conditions. Under anaerobic conditions, endodiol formed in low amounts in two soils (Martens, 1977).

Indigenous microorganisms obtained from a sandy loam degraded endosulfan to endosulfan diol. This diol was converted to endosulfanhydroxy ether and trace amounts of endosulfan ether and both were degraded to endosulfan lactone (Miles and Moy, 1979). Using settled domestic wastewater inoculum, α-endosulfan (5 and 10 mg/L) did not degrade after 28 d of incubation at 25 °C (Tabak et al., 1981).

Awasthi et al. (1997) studied the aerobic degradation of endosulfan (20 mg) in soil (5 g) using a two-membered coculture incubated at 28 °C. The biodegradation half-life of 4 wk was approximately 4 times longer than in the culture medium. The investigators indicated the longer half-life was either due to the adsorption of the endosulfan onto the soil particles or due presence of other carbanaceous materials in the soil (e.g., organic carbon).

Plant. Endosulfan sulfate was formed when endosulfan was translocated from the leaves to roots in both bean and sugar beet plants (Beard and Ware, 1969). In tobacco leaves, α-endosulfan is hydrolyzed to endosulfandiol (Chopra and Mahfouz, 1977). Stewart and Cairns (1974) reported the metabolite endosulfan sulfate was identified in potato peels and pulp at concentrations of 0.3 and 0.03 ppm, respectively. They also reported that the half-life for the conversion of α-endosulfan to β-endosulfan was 60 d.

On apple leaves, direct photolysis of endosulfan by sunlight yielded endosulfan sulfate (Harrison et al., 1967).

In carnation plants, the half-lives of α-endosulfan stored under four different conditions, non-washed and exposed to open air, washed and exposed to open air, non-washed and placed in an enclosed container, and under greenhouse conditions were 6.79, 6.38, 10.45, and 4.22 d, respectively (Cerón et al., 1995).

Surface Water. Endosulfan sulfate was identified as a metabolite in a survey of 11 agricultural watersheds located in southern Ontario, Canada (Frank et al., 1982). When endosulfan (α- and β- isomers, 10 μg/L) was added to Little Miami River water, sealed, and exposed to sunlight and UV light for 1 wk, a degradation yield of 70% was observed. After 2 and 4 wk, 95% and 100% of the applied amount, respectively, degraded. The major degradation product was identified as endosulfan alcohol by IR spectrometry (Eichelberger and Lichtenberg, 1971).

Photolytic. Thin films of endosulfan on glass and irradiated by UV light (λ >300 nm) produced endosulfan diol with minor amounts of endosulfan ether, lactone, α-hydroxyether, and other unidentified compounds (Archer et al., 1972). When an aqueous solution containing endosulfan was photooxidized by UV light at 90-95 °C, 25, 50, and 75% degraded to carbon dioxide after 5.0, 9.5, and 31.0 h, respectively (Knoevenagel and Himmelreich, 1976).

Chemical/Physical. Endosulfan slowly hydrolyzes forming endosulfandiol and endosulfan sulfate (Kollig, 1993; Martens, 1976; Worthing and Hance, 1991). The hydrolysis rate constant for α-endosulfan at pH 7 and 25 °C was determined to be 3.2×10^{-3}/h, resulting in a half-life of 9.0 d (Ellington et al., 1988). The hydrolysis half-lives are reduced significantly at varying pHs and temperature. At temperatures (pH) of 87.0 (3.12), 68.0 (6.89), and 38.0 °C (8.69), the half-lives were 4.3, 0.10, and 0.08 d, respectively (Ellington et al., 1986).

Emits toxic fumes of chlorides and sulfur oxides when heated to decomposition (Lewis, 1990).

Exposure limits: ACGIH TLV: TWA 0.1 mg/m^3 for all isomers (adopted).

Toxicity:

LC$_{50}$ (96-h) for golden orfe 2 μg/L (Hartley and Kidd, 1987), rainbow trout 0.3 μg/L, white sucker 3.0 μg/L (Verschueren, 1983), pink shrimp 0.04 μg/L, grass shrimp 1.3 μg/L, pinfish 0.3 μg/L, striped mullet 0.38 μg/L (Schimmel et al., 1977b), *Pteronarcys californica* 2.3 μg/L (Sanders and Cope, 1968).

Laboratory determined LC$_{50}$ (48-h) values of technical endosulfan (mixed isomers) in river water were 0.6, 1.3, and 0.4 μg/L for early-instar nymphs of *Atalophlebia australis* and *Jappa kutera* and *Cheumatopsyche* sp. larvae, respectively (Leonard et al., 1999).

Ferrando et al. (1992) reported LC$_{50}$ (24-h) values of 0.62 and 5.15 mg/L for *Daphnia magna* and *Brachionus calyciflorus*, respectively.

Use: Insecticide for cotton, tea, and sugar cane, and vegetable crops.

β-ENDOSULFAN

Synonyms: Benzoepin; Beosit; Bio 5462; Chlorthiepin; Crisulfan; Cyclodan; Endocel; Endosol; Endosulfan; Endosulfan II; Endosulphan; ENT 23979; FMC 5462; 1,2,3,7,7-Hexachlorobicyclo[2.2.1]-2-heptene-5,6-bisoxymethylene sulfite; α,β-1,2,3,7,7-Hexachlorobicyclo[2.2.1]-2-heptene-5,6-bisoxymethylene sulfite; Hexachlorohexahydromethano-2,4,3-benzodioxathiepin-3-oxide; **(3α,5aα,6β,9β,9aα)-6,7,8,9,10,10-Hexachloro-1,5,5a,6,9,9a-hexahydro-6,9-methano-2,4,3-benzodioxathiepin-3-oxide**; 1,4,5,6,7,7-Hexachloro-5-norborene-2,3-dimethanol cyclic sulfite; Hildan; HOE 2671; Insectophene; KOP-thiodan; Malix; NCI-C00566; NIA 5,462; Niagara 5462; OMS 570; RCRA waste number P050; Thifor; Thimul; Thiomul; Thiodan; Thiofor; Thionex; Thiosulfan; Tionel; Tiovel.

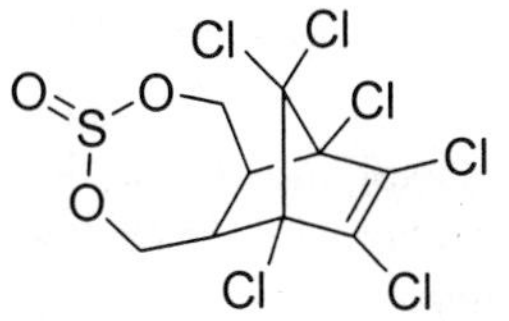

Note: Insecticidal formation contains the two sterioisomers α- and β-endosulfan in an approximate ratio of 70:30 (Goebel, 1983).

CAS Registry Number: 33213-65-9
DOT: 2761
DOT label: Poison
Molecular formula: $C_9H_6Cl_6O_3S$
Formula weight: 406.92
RTECS: RB9275000
Merck reference: 10, 3535

Physical state, color, and odor:
Colorless to brown crystals with a sulfur dioxide odor

Melting point (°C):
207-209 (Ali, 1978)

Density (g/cm³):
1.745 at 20/20 °C (Sax, 1984)

Diffusivity in water (x 10^{-5} cm²/sec):
0.45 at 20 °C using method of Hayduk and Laudie (1974)

Henry's law constant (x 10^{-5} atm·m³/mol at 20 °C):
0.87 and 2.09 in distilled water and 33.3‰ NaCl, respectively. Technical endosulfan [α-endosulfan (70%) + β-endosulfan (30%)]: 6.38 and 13.07 in distilled water and 33.3‰ NaCl, respectively (Rice et al., 1997, 1997a).

Bioconcentration factor, log BCF:
3.35 (edible tissue), 3.44 (whole body) (striped mullet, Schimmel et al., 1977b)

Soil sorption coefficient, log K_{oc}:
3.37 using method of Kenaga and Goring (1980)

Octanol/water partition coefficient, log K_{ow}:
3.62 (Ali, 1978)

Solubility in water ($\mu g/L$):
280 at 25 °C (Weil et al., 1974)
330 at 20 °C (Worthing and Hance, 1991)
450 at 20 °C (Bowman and Sans, 1983)

Vapor pressure (mmHg):
2.40 x 10^{-5} at 25 °C (Hinckley et al., 1990)

Environmental fate:
Biological. Isolated strains of *Aspergillus niger* degraded endosulfan to endodiol (El Zorghani and Omer, 1974). Biodegradation of β-endosulfan by a bacterial coculture appeared to be at a lower rate than α-endosulfan. The release of chloride ions suggested degradation occurred via dehalogenation. No expected metabolites (e.g., endodiol, endosulfate, endohydroxy ether) accumulated, suggesting they were further degraded (Awasthi et al., 1997).

Soil. Metabolites of endosulfan identified in seven soils were: endosulfandiol, endosulfanhydroxy ether, endosulfan lactone, and endosulfan sulfate (Martens, 1977; Dreher and Podratzki, 1988). These compounds, as well as endosulfan ether, were also reported as metabolites identified in aquatic systems (Day, 1991). In soils under aerobic conditions, β-endosulfan is converted to the corresponding alcohol and ether (Perscheid et al., 1973). Endosulfan sulfate was the major biodegradation product in soils under aerobic, anaerobic, and flooded conditions (Martens, 1977). In flooded soils, endolactone was detected only once whereas endodiol and endohydroxy ether were identified in all soils under these conditions. Under anaerobic conditions, endodiol formed in low amounts in two soils (Martens, 1977). Indigenous microorganisms obtained from a sandy loam degraded β-endosulfan to endosulfan diol. This diol was converted to endosulfan α-hydroxy ether and trace amounts of endosulfan ether and both were degraded to endosulfan lactone (Miles and Moy, 1979).

In four successive 7-d incubation periods, β-endosulfan (5 and 10 mg/L) was recalcitrant to degradation in a settled domestic wastewater inoculum (Tabak et al., 1981).

Awasthi et al. (1997) studied the aerobic degradation of endosulfan (20 mg) in soil (5 g) using a two-membered coculture incubated at 28 °C. The biodegradation half-life of 4 wk was approximately 4 times longer than in the culture medium. The investigators indicated the longer half-life was either due to the adsorption of the endosulfan onto the soil particles or due presence of other carbanaceous materials in the soil (e.g., organic carbon).

Plant. Endosulfan sulfate was formed when endosulfan was translocated from the leaves to roots in both bean and sugar beet plants (Beard and Ware, 1969). In tobacco leaves, β-endosulfan hydrolyzed into endosulfandiol (Chopra and Mahfouz, 1977). Stewart and Cairns (1974) reported the metabolite endosulfan sulfate was identified in potato peels and pulp at concentrations of 0.3 and 0.03 ppm, respectively. They also reported that the half-life for the oxidative conversion of β-endosulfan to endosulfan sulfate was 800 d.

In carnation plants, the half-lives of β-endosulfan stored under four different conditions, non-washed and exposed to open air, washed and exposed to open air, non-washed and placed in an enclosed container, and under greenhouse conditions were 23.40, 12.64, 37.42, and 7.62 d, respectively (Cerón et al., 1995).

Surface Water. Endosulfan sulfate was also identified as a metabolite in a survey of 11 agricultural watersheds located in southern Ontario, Canada (Frank et al., 1982). When endosulfan (α- and β- isomers, 10 $\mu g/L$) was added to Little Miami River water, sealed and exposed to sunlight and UV light for 1 wk, a degradation yield of 70% was observed. After 2 and 4 wk, 95% and 100% of the applied amount degraded. The major degradation product identified as endosulfan alcohol by IR spectrometry (Eichelberger and Lichtenberg, 1971).

Photolytic. Thin films of endosulfan on glass and irradiated by UV light (λ >300 nm) produced endosulfan diol with minor amounts of endosulfan ether, lactone, α-hydroxyether and other unidentified compounds (Archer et al., 1972). Gaseous β-endosulfan subjected to UV light (λ >300 nm) produced endosulfan ether, endosulfan diol, endosulfan sulfate, endosulfan lactone, α-endosulfan and a dechlorinated ether (Schumacher et al., 1974). Irradiation of β-endosulfan in *n*-hexane by UV light produced the photoisomer α-endosulfan (Putnam et al., 1975). When an aqueous solution containing endosulfan was photooxidized by UV light at 90-95 °C, 25, 50, and 75% degraded to carbon dioxide after 5.0, 9.5, and 31.0 h, respectively (Knoevenagel and Himmelreich, 1976).

Chemical/Physical. Endosulfan detected in Little Miami River, OH was readily hydrolyzed and tentatively identified as endosulfan diol (Eichelberger and Lichtenberg, 1971). Undergoes slow hydrolysis forming the endosulfan diol and sulfur dioxide (Worthing and Hance, 1991). The hydrolysis half-lives at pH values (temperature) of 3.32 (87.0 °C), 6.89 (68.0 °C), and 8.69 (38.0 °C) were calculated to be 2.7, 0.07, and 0.04 d, respectively (Ellington et al., 1988). Greve and Wit (1971) reported hydrolysis half-lives of β-endosulfan at 20 °C and pH values of 7 and 5.5 were 37 and 187 d, respectively.

Exposure limits: ACGIH TLV: TWA 0.1 mg/m^3 for all isomers (adopted).

Toxicity:

LC$_{50}$ (96-h) for golden orfe 2 μg/L (Hartley and Kidd, 1987), rainbow trout 0.3 μg/L, white sucker 3.0 μg/L (Verschueren, 1983), pink shrimp 0.04 μg/L, grass shrimp 1.3 μg/L, pinfish 0.3 μg/L, striped mullet 0.38 μg/L (Schimmel et al., 1977b), *Pteronarcys californica* 2.3 μg/L (Sanders and Cope, 1968).

Laboratory determined LC$_{50}$ (48-h) values of technical endosulfan (mixed isomers) in river water were 0.6, 1.3, and 0.4 μg/L for early-instar nymphs of *Atalophlebia australis* and *Jappa kutera* and *Cheumatopsyche* sp. larvae, respectively (Leonard et al., 1999).

Ferrando et al. (1992) reported LC$_{50}$ (24-h) values of 0.62 and 5.15 mg/L for *Daphnia magna* and *Brachionus calyciflorus*, respectively.

Use: Insecticide for cotton, tea, and sugar cane, and vegetable crops.

ENDOSULFAN SULFATE

Synonyms: 6,7,8,9,10,10-Hexachloro-1,5,5a,6,9,9a-hexahydro-3,3-dioxide; 6,9-Methano-2,4,3-benzodioxathiepin.

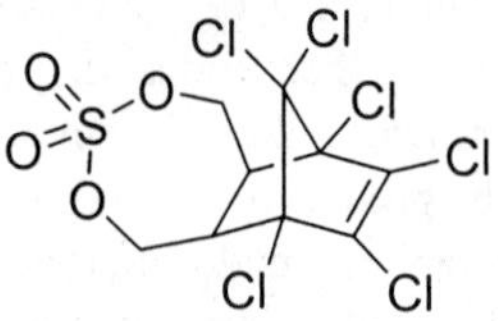

CAS Registry Number: 1031-07-8
DOT: 2761
DOT label: Poison
Molecular formula: $C_9H_6Cl_6O_4S$
Formula weight: 422.92

Physical state:
Solid

Melting point (°C):
181 (Keith and Walters, 1992)
198-201 (Ali, 1978)

Diffusivity in water (x 10^{-5} cm^2/sec):
0.44 at 20 °C using method of Hayduk and Laudie (1974)

Henry's law constant (x 10^{-5} atm·m^3/mol):
4.64 at 25 °C (approximate - calculated from water solubility and vapor pressure)

Soil sorption coefficient, log K_{oc}:
3.37 using method of Kenaga and Goring (1980)

Octanol/water partition coefficient, log K_{ow}:
3.66 (Ali, 1978)

Solubility in water:
117 ppb (Ali, 1978)

Vapor pressure (mmHg):
9.75 x 10^{-6} at 25 °C (Hinckley et al., 1990)

Environmental fate:
  Biological. A mixed culture of soil microorganisms biodegraded endosulfan sulfate to endosulfan ether, endosulfan-α-hydroxy ether and endosulfan lactone (Verschueren, 1983). Indigenous microorganisms obtained from a sandy loam degraded endosulfan sulfate (a metabolite of α- and β-endosulfan) to endosulfan diol. This diol was converted to endosulfan α-hydroxy ether and trace amounts of endosulfan ether and both were degraded to endosulfan lactone (Miles and Moy, 1979). Using settled domestic wastewater inoculum, endosulfan sulfate (5 and 10 mg/L) did not degrade after 28 d of incubation at 25 °C (Tabak et al., 1981).

Plant. In tobacco leaves, endosulfan sulfate was formed to α-endosulfan, which subsequently hydrolyzed to endosulfandiol (Chopra and Mahfouz, 1977).

Chemical/Physical. Eighty-eight percent of endosulfan sulfate was recovered from an aqueous solution after 33 d. Based on this data, a hydrolytic half-life of 178 was reported (Ali, 1978).

Uses: Not known. Based on data obtained from photolysis studies of tobacco, this compound, α- and β-endosulfan and other products were identified in tobacco smoke (Chopra et al., 1977).

ENDRIN

Synonyms: Compound 269; Endrex; Endrine; ENT 17251; Experimental insecticide 269; Hexachloroepoxyoctahydro-*endo,endo*-dimethanonaphthalene; 1,2,3,4,10,10-Hexachloro-6,7-epoxy-1,4,4a,5,6,7,8,8a-octahydro-*endo,endo*-1,4:5,8-dimethanonaphthalene; **3,4,5,6,9,9-Hexa-chloro-1a,2,2a,3,6,6a,7,7a-octahydro-2,7:3,6-dimethanonaphth[2,3-*b*]oxirene**; Hexadrin; Isodrin epoxide; Mendrin; NA 2761; NCI-C00157; Nendrin; OMS 197; RCRA waste number P051.

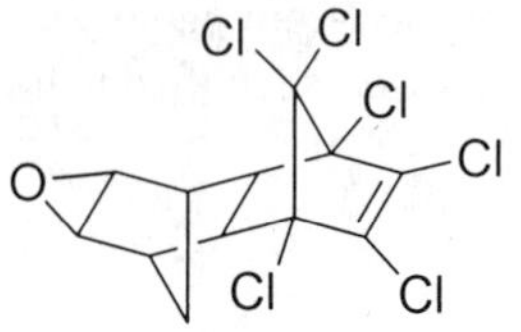

CAS Registry Number: 72-20-8
DOT: 2761
DOT label: Poison
Molecular formula: $C_{12}H_8Cl_6O$
Formula weight: 380.92
RTECS: IO1575000
Merck reference: 10, 5537

Physical state, color, and odor:
White, odorless, crystalline solid when pure; light tan color with faint chemical odor for technical grades. Odor threshold concentrations in air ranged from 18 to 41 ppb (Keith and Walters, 1992).

Melting point (°C):
200 (Caswell et al., 1981)

Boiling point (°C):
Decomposes at 245 (ACGIH, 1986)

Density (g/cm³):
1.65 at 25/4 °C (Weiss, 1986)

Diffusivity in water (x 10^{-5} cm²/sec):
0.44 at 20 °C using method of Hayduk and Laudie (1974)

Flash point (°C):
Noncombustible solid but may be dissolved in flammable liquids (NIOSH, 1997)

Lower explosive limit (%):
1.1 in xylene (Weiss, 1986)

Upper explosive limit (%):
7.0 in xylene (Weiss, 1986)

Henry's law constant (x 10^{-7} atm·m³/mol):
5.0 (U.S. EPA, 1980a)

Bioconcentration factor, log BCF:
2.15-2.35 (algae, Grant, 1976)
3.42 (freshwater fish), 3.13 (fish, microcosm) (Garten and Trabalka, 1983)
3.28 (mussels, Ernst, 1977)
3.17 (rainbow trout) (Neely et al., 1974)

Soil sorption coefficient, log K_{oc}:
3.89 (Beverly sandy loam), 4.15 (Plainfield sand), 4.20 (Big Creek sediment) (Sharom et al., 1980a)

Octanol/water partition coefficient, log K_{ow}:
5.16 (Travis and Arms, 1988)
5.339 (Kenaga and Goring, 1980)
3.209 (Rao and Davidson, 1980)
5.195 (de Bruijn et al., 1989)

Solubility in organics:
At 25 °C (g/L): acetone (170), benzene (138), carbon tetrachloride (33), hexane (71), xylene (183) (Windholz et al., 1983). Soluble in aromatic hydrocarbons, esters, and ketones (ITII, 1986).

Solubility in water (ppb):
230 at 20-25 °C (Geyer et al., 1980)
130 at 15 °C, 250 at 25 °C, 420 at 35 °C, 625 at 45 °C (particle size ≤5 μ, Biggar and Riggs, 1974)
At 20-25 °C: 260 (particle size ≤5 μ), 190 (particle size ≤0.06 μ) (Robeck et al., 1965)

Vapor pressure (mmHg):
2 x 10^{-7} at 25 °C (ACGIH, 1986)

Environmental fate:
Biological. Algae isolated from a stagnant fish pond degraded 24.4% of the applied endrin to ketoendrin (Patil et al., 1972). Microbial degradation of endrin in soil formed several ketones and aldehydes of which *keto*-endrin was the only metabolite identified (Kearney and Kaufman, 1976). In four successive 7-d incubation periods, endrin (5 and 10 mg/L) was recalcitrant to degradation in a settled domestic wastewater inoculum (Tabak et al., 1981).

Soil. In eight Indian rice soils, endrin degraded rapidly to low concentrations after 55 d. Degradation was highest in a pokkali soil and lowest in a sandy soil (Gowda and Sethunathan, 1976). The percentage of endrin remaining in a Congaree sandy loam soil after 14 yr was 41% (Nash and Woolson, 1967).

Under laboratory conditions, endrin degraded to other compounds in a variety of soils maintained at 45 °C. Except for Rutledge sand, endrin disappeared or was transformed in the following soils after 24 h: Lynchburg loamy sand, Magnolia sandy loam, Magnolia sandy clay loam, Greenville sandy clay, and Susquehanna sandy clay. No products were identified (Bowman et al., 1965).

The disappearance half-lives for endrin in field soils under flooded and nonflooded conditions were 130 and 468 d, respectively (Guenzi et al., 1971). The average disappearance half-life in flooded soils under laboratory conditions was 31 d (Gowda and Sethunathan, 1976, 1977).

Groundwater. Endrin has a high potential to leach to groundwater (U.S. EPA, 1986).
Plant. In plants, endrin is converted to the corresponding sulfate (Hartley and Kidd, 1987).

Surface Water. Algae isolated from a stagnant fish pond degraded 24.4% of the applied endrin to ketoendrin (Patil et al., 1972).

Photolytic. Photolysis of thin films of solid endrin using UV light (λ = 254 nm) produced δ-ketoendrin and endrin aldehyde and other compounds (Rosen et al., 1966). When exposed to sunlight for 17 d, endrin completely isomerized to δ-ketoendrin (1,8-*exo*-9,10,11,11-hexachlorocyclo-6.2.1.1^{3,6}.0^{2,7}.0^{4,10}-dodecan-5-one) and minor amounts of endrin aldehyde (Burton and Pollard, 1974). Irradiation of endrin by UV light (λ = 253.7 and 300 nm) or by natural sunlight in cyclohexane and hexane solution resulted in an 80% yield of 1,8-*exo*-9,11,11-pentachloropentacyclo[6.2.1.1^{3,6}.0^{2,7}.0^{4,10}]-dodecan-5-one (Zabik et al., 1971). This compound also formed from the sunlight photolysis of endrin in hexane solution (Fujita et al., 1969).

When an aqueous solution containing endrin was photooxidized by UV light at 90-95 °C, 25, 50, and 75% degraded to carbon dioxide after 15.0, 41.0, and 172.0 h, respectively (Knoevenagel and Himmelreich, 1976). The rate of photolysis of dilute aqueous solutions using UV light at 254 nm was independent over the temperature range of 20 to 40 °C (Bulla and Edgerly (1968).

Chemical/Physical. At 230 °C, endrin isomerizes to an aldehyde and a ketone. When heated to decomposition, hydrogen chloride and phosgene may be released (NIOSH, 1997) but residues containing an aldehyde (15-20%), a ketone (55-60%), a caged alcohol (5%), and other volatile products (15-20%) were reported (Phillips et al., 1962). In water, endrin undergoes nucleophilic attack at the epoxide moiety forming endrin diol (Kollig, 1993).

At 50 °C, endrin was not unaffected by the oxidants chlorine, permanganate, and persulfate (Leigh, 1969). Endrin (0.010 mM) reacted with OH radicals in water (pH 3.4) at a rate of 1.1 x 10^9/M·sec. At pH 2.8 and endrin concentrations of 0.0009 and 0.0005 mM, the reaction rates were 2.7 x 10^8 and 1.3 x 10^9/M·sec, respectively (Haag and Yao, 1992).

Exposure limits (mg/m^3): NIOSH REL: TWA 0.1, IDLH 2; OSHA PEL: TWA 0.1; ACGIH TLV: TWA 0.1 (adopted).

Symptoms of exposure: Epileptiform, convulsions, stupor, headache, dizziness, abdominal discomfort, nausea, vomiting, insomnia, aggressive confusion, lethargy, weakness, anorexia (NIOSH, 1997).

Toxicity:

EC$_{50}$ (48-h) for *Daphnia magna* 39.7 μg/L, *Daphnia pulex* 20 μg/L, *Simocephalus serrulatus* 30 μg/L, *Cypridopsis vidua* 1.8 μg/L (Mayer and Ellersieck, 1986).

LC$_{50}$ (30-d) for *Pteronarcys californica* 1.2 μg/L (Jensen and Gaufin, 1966).

LC$_{50}$ (96-h) for bluegill sunfish 0.6 μg/L, fathead minnows 1.0 μg/L (Henderson et al., 1959), rainbow trout 0.6 μg/L, coho salmon 0.5 μg/L, chinook 1.2 μg/L (Katz, 1961), adult fish (*Cyprinodon variegatus*) 0.36 μg/L (quoted, Reish and Kauwling, 1978), *Claassenia sabulosa* 0.76 μg/L, *Pteronarcella badia* 0.54 μg/L, *Acroneuria pacifica* 0.32 μg/L, *Pteronarcys californica* 0.25 μg/L (Sanders and Cope, 1968), *Pimephales promelas* 0.7 μg/L (Jarvinen et al., 1988), flagfish (*Jordanella floridae*) 0.85 μg/L (Hermanutz, 1978).

LC$_{50}$ (96-h static bioassay) for sand shrimp 1.7 μg/L, hermit crab 12 μg/L, grass shrimp 1.8 μg/L (Eisler, 1969).

LC$_{50}$ (48-h) for red killifish 21 μg/L (Yoshioka et al., 1986), carp 0.8 μg/L, goldfish 1 μg/L, medaka 1.4 μg/L, pond loach 4.9 μg/L (Spehar et al., 1982), *Daphnia pulex* 20 μg/L, *Simocephalus serrulatus* 26 μg/L (Sanders and Cope, 1966).

LC$_{50}$ (28-h) for bullhead fish 0.010 μg/L (Anderson and DeFoe, 1980).

Acute oral LD$_{50}$ for rats 39 mg/kg, wild birds 13.3 mg/kg, chickens 20 mg/kg, ducks 381 mg/kg, dogs 65 mg/kg, guinea pigs 49 mg/kg, hamsters 60 mg/kg, monkeys 3 mg/kg, mice 38 mg/kg, pigeons 23.7 mg/kg, pigs 38 mg/kg, quail 10.78 mg/kg, rats 38.3 mg/kg, rabbits 45 mg/kg (quoted, RTECS, 1985), male rats, 18 mg/kg, female rats 7.5 mg/kg (Windholz et al., 1983).

Drinking water standard (final): MCLG: 2 μg/L; MCL: 2 μg/L (U.S. EPA, 1996).

Use: Insecticide.

ENDRIN ALDEHYDE

Synonym: 2,2a,3,3,4,7-Hexachlorodecahydro-1,2,4-methenocyclopenta[*c,d*]pentalene-5-carboxaldehyde.

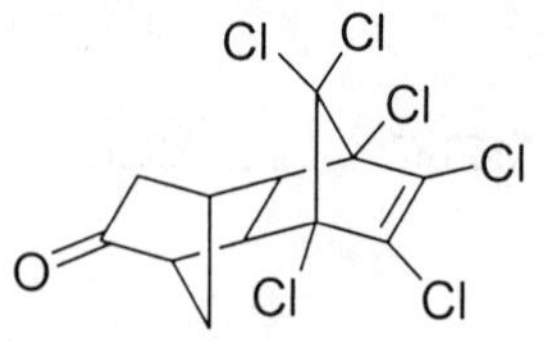

CAS Registry Number: 7421-93-4
DOT: 2761
DOT label: Poison
Molecular formula: $C_{12}H_8Cl_6O$
Formula weight: 380.92

Physical state:
Solid

Melting point (°C):
145-149 (U.S. EPA, 1980a)

Boiling point (°C):
Decomposes at 235 (Callahan et al., 1979)

Diffusivity in water (x 10^{-5} cm²/sec):
0.43 at 20 °C using method of Hayduk and Laudie (1974)

Henry's law constant (x 10^{-7} atm·m³/mol):
3.86 at 25 °C (approximate - calculated from water solubility and vapor pressure)

Soil sorption coefficient, log K_{oc}:
4.43 using method of Kenaga and Goring (1980)

Octanol/water partition coefficient, log K_{ow}:
5.6 (calculated, Neely et al., 1974)

Solubility in water:
260 ppb at 25 °C (Weil et al., 1974)

Vapor pressure (mmHg):
2 x 10^{-7} at 25 °C (Martin, 1972)

Uses: Not known.

EPICHLOROHYDRIN

Synonyms: 1-Chloro-2,3-epoxypropane; 3-Chloro-1,2-epoxypropane; (Chloromethyl)ethylene oxide; **(Chloromethyl)oxirane**; 2-(Chloromethyl)oxirane; 2-Chloropropylene oxide; γ-Chloropropylene oxide; 3-Chloro-1,2-propylene oxide; ECH; α-Epichlorohydrin; (*dl*)-α-Epichlorohydrin; Epichlorophydrin; 1,2-Epoxy-3-chloropropane; 2,3-Epoxypropyl chloride; Glycerol epichlorohydrin; RCRA waste number U041; UN 2023.

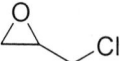

CAS Registry Number: 106-89-8
DOT: 2023
DOT label: Flammable liquid and poison
Molecular formula: C_3H_5ClO
Formula weight: 92.53
RTECS: TX4900000
Merck reference: 10, 3560

Physical state, color, and odor:
Clear, colorless, mobile liquid with a strong, irritating, chloroform-like odor. Odor threshold concentration is 0.93 ppm (Amoore and Hautala, 1983).

Melting point (°C):
-48 (Weast, 1986)
-57 (Verschueren, 1983)

Boiling point (°C):
116.5 (Weast, 1986)
117.9 (Sax and Lewis, 1987)

Density (g/cm³):
1.1801 at 20/4 °C (Weast, 1986)
1.1750 at 25/4 °C (Windholz et al., 1983)

Diffusivity in water (x 10^{-5} cm²/sec):
1.02 at 20 °C using method of Hayduk and Laudie (1974)

Flash point (°C):
40 (open cup, Windholz et al., 1983)
33.9 (NIOSH, 1997)
31 (open cup, NFPA, 1984)

Lower explosive limit (%):
3.8 (NIOSH, 1997)

Upper explosive limit (%):
21.0 (NIOSH, 1997)

Henry's law constant (x 10^{-5} atm·m³/mol):
2.4 at 20 °C (approximate - calculated from water solubility and vapor pressure)

Ionization potential (eV):
10.60 (NIOSH, 1997)

Soil sorption coefficient, log K_{oc}:
1.00 (calculated, Mercer et al., 1990)

Octanol/water partition coefficient, log K_{ow}:
0.30 (Lide, 1990)
0.45 (Deneer et al., 1988)

Solubility in organics:
Soluble in benzene (Weast, 1986). Miscible with alcohol, carbon tetrachloride, chloroform, ether, and tetrachloroethylene (Windholz et al., 1983).

Solubility in water (g/L at 20 °C):
60 (quoted, Verschueren, 1983)
65.8 (Riddick et al., 1986)
66 (Lide, 1990)

Vapor density:
3.78 g/L at 25 °C, 3.19 (air = 1)

Vapor pressure (mmHg):
12.5 at 20 °C (Hawley, 1981)
16.5 at 25 °C (Lide, 1990)

Environmental fate:
Biological. Bridié et al. (1979) reported BOD and COD values 0.03 and 1.16 g/g using filtered effluent from a biological sanitary waste treatment plant. These values were determined using a standard dilution method at 20 °C for a period of 5 d. When a sewage seed was used in a separate screening test, a BOD value of 0.16 g/g was obtained. The ThOD for epichlorohydrin is 1.21 g/g.

Chemical/Physical. Anticipated products from the reaction of epichlorohydrin with ozone or OH radicals in the atmosphere are formaldehyde, glyoxylic acid and $ClCH_2O(O)OHCHO$ (Cupitt, 1980). Haag and Yao (1992) reported a calculated OH radical rate constant in water of 2.9×10^8/M·sec.

Hydrolyzes in water forming 1-chloro-2,3-hydroxypropane which may further hydroylze forming hydrochloric acid and the intermediate 1-hydroxy-2,3-propylene oxide or 1-chloro-2,3-hydroxypropane. The oxide may further hydrolyze forming glycerol (Kollig, 1993). The estimated half-life for this reaction at 20 °C and pH 7 is 8.2 d (Mabey and Mill, 1978).

Emits toxic fumes when heated to decomposition.

Exposure limits: Potential occupational carcinogen. NIOSH REL: IDLH 75 ppm; OSHA PEL: TWA 5 ppm (19 mg/m^3); ACGIH TLV: TWA 0.5 ppm (adopted).

Symptoms of exposure: Irritation of eyes, skin, and respiratory tract (Patnaik, 1992)

Toxicity:
LC_{50} (14-d) for *Poecilia reticulata* 655 μg/L (Hermens et al., 1985).
LC_{50} (96-h static bioassay) for bluegill sunfish 35 mg/L, *Menidia beryllina* 18 mg/L (quoted, Verschueren, 1983).

LC_{50} (48-h) for harlequin fish 36 mg/L (static and flow through).

Acute oral LD_{50} for guinea pigs 280 mg/kg, mice 194 mg/kg, rats 90 mg/kg, rabbits 345 mg/kg (quoted, Verschueren, 1983).

Drinking water standard (final): MCLG: zero; MCL: lowest feasible limit following conventional treatment (U.S. EPA, 1996).

Uses: Solvent for natural and synthetic resins, paints, cellulose esters and ethers, gums, paints, varnishes, lacquers, and nail enamels; manufacturing of glycerol, epoxy resins, surface active agents, pharmaceuticals, insecticides, adhesives, coatings, plasticizers, glycidyl ethers, ion-exchange resins, and fatty acid derivatives; organic synthesis.

EPN

Synonyms: ENT 17298; EPN 300; Ethoxy-4-nitrophenoxy phenylphosphine sulfide; Ethyl *p*-nitrophenyl benzenethionophosphate; Ethyl *p*-nitrophenyl benzenethiophosphonate; Ethyl *p*-nitrophenyl ester; *O*-Ethyl *O*-4-nitrophenyl phenylphosphonothioate; Ethyl *p*-nitrophenyl phenylphosphonothioate; *O*-Ethyl *O*-*p*-nitrophenyl phenylphosphonothioate; Ethyl *p*-nitrophenyl thionobenzenephosphate; Ethyl *p*-nitrophenyl thionobenzenephosphonate; *O*-Ethyl phenyl *p*-nitrophenyl phenylphosphorothioate; Ethyl *p*-nitrophenyl thionobenzenephosphate; *O*-Ethyl phenyl *p*-nitrophenyl thiophosphonate; OMS 219; Phenylphosphonothioic acid *O*-ethyl *O*-*p*-nitrophenyl ester; **Phosphonothioic acid *O,O*-diethyl *O*-(3,5,6-trichloro-2-pyridinyl) ester**; Pin; Santox.

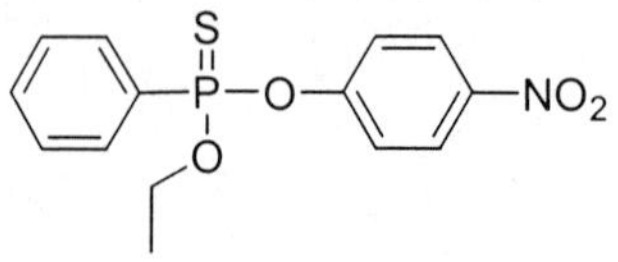

CAS Registry Number: 2104-64-5
Molecular formula: $C_{14}H_{14}NO_4PS$
Formula weight: 323.31
RTECS: TB1925000
Merck reference: 10, 3574

Physical state, color, and odor:
Yellow solid or brown liquid with an aromatic odor

Melting point (°C):
36 (Weast, 1986)

Boiling point (°C):
215 at 5 mmHg (Worthing and Hance, 1991)

Density (g/cm³):
1.27 at 25/4 °C (Weast, 1986)

Flash point (°C):
Noncombustible solid (NIOSH, 1997)

Bioconcentration factor, log BCF:
3.05 (*Oryzias latipes*, Tsuda et al., 1997)

Soil sorption coefficient, log K_{oc}:
3.12 (average of 2 soil types, Kanazawa, 1989)

Octanol/water partition coefficient, log K_{ow}:
3.85 (Kanazawa, 1989)

Solubility in organics:
Miscible with acetone, benzene, methanol, isopropanol, toluene, and xylene (Windholz et al., 1983)

Vapor pressure (x 10⁻⁷ mmHg):

9.4 at 25 °C (Worthing and Hance, 1991)
3,000 at 100 °C (Verschueren, 1983)

Environmental fate:

Biological. From the first-order biotic and abiotic rate constants of EPN in estuarine water and sediment/water systems, the estimated biodegradation half-lives were 6.2 and 9.2 d, respectively (Walker et al., 1988).

Soil. Although no products were reported, the half-life in soil is 15-30 d (Hartley and Kidd, 1987).

Photolytic. EPN may undergo direct photolysis because the insecticide showed some absorption when a 1,4-dioxane was irradiated with UV light (λ >290 nm) (Gore et al., 1971).

Chemical/Physical. On heating, EPN is converted to the *S*-ethyl isomer (Worthing and Hance, 1991). EPN is rapidly hydrolyzed in alkaline solutions to *p*-nitrophenol, alcohol, and benzene thiophosphoric acid (Sittig, 1985).

Releases toxic sulfur, phosphorous, and nitrogen oxides when heated to decomposition (Sax and Lewis, 1987).

Exposure limits (mg/m³): NIOSH REL: TWA 0.5, IDLH 5; OSHA PEL: TWA 0.5; ACGIH TLV: TWA 0.1 mg/m³ (adopted).

Toxicity:

LC_{50} (96-h) for bluegill sunfish 100 μg/L (Sanders and Cope, 1968), fathead minnows 110 mg/L (Solon and Nair, 1970), *Mysidopsis bahia* 3.44 μg/L, *Penaeus duorarum* 0.29 μg/L, *Cyprinidon variegatus* 188.9 μg/L, *Lagodon rhomboides* 18.3 μg/L, *Leiostomus xanthurus* 25.6 μg/L (Schimmel et al., 1979).

LC_{50} (48-h) for killifish (*Oryzias latipes*) 580 μg/L (Tsuda et al., 1997).

Acute oral LD_{50} for wild birds 2.37 mg/kg, chickens 5 mg/kg, ducks 3 mg/kg, dogs 20 mg/kg, mice 14.5 mg/kg, pigeons 4.21 mg/kg, quail 5 mg/kg, rats 7 mg/kg, rabbits 45 mg/kg (quoted, RTECS, 1985), male rats 33-42 mg/kg, female rats 14 mg/kg, bobwhite quail 220 mg/kg, ring-necked pheasants >165 mg/kg (Worthing and Hance, 1991).

Acute percutaneous LD_{50} (24-h) for male white rats 230 mg/kg, female white rats 25 mg/kg (Worthing and Hance, 1991).

The NOEL for dogs in 12-month feeding trials is 2 mg/kg daily (Worthing and Hance, 1991).

Uses: Non-systemic insecticide and acaricide used to control bollworms, *Alabama argillacea* in cotton, and *Chilo* spp. in rice and other leaf-eating larvae in vegetables and fruits (Worthing and Hance, 1991).

ETHANOLAMINE

Synonyms: 2-Aminoethanol; β-Aminoethyl alcohol; Colamine; β-Ethanolamine; Ethylol-amine; Glycinol; 2-Hydroxyethylamine; β-Hydroxyethylamine; MEA; Monoethanolamine; Olamine; Thiofalco M-50; UN 2491; USAF EK-1597.

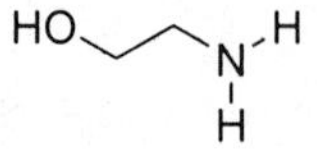

CAS Registry Number: 141-43-5
DOT: 2491
DOT label: Corrosive material
Molecular formula: C_2H_7NO
Formula weight: 61.08
RTECS: KJ5775000
Merck reference: 10, 3675

Physical state, color, and odor:
Colorless, viscous, hygroscopic liquid with an unpleasant, mild, ammonia-like odor. Odor threshold concentration is 2.6 ppm (Amoore and Hautala, 1983).

Melting point (°C):
10.3 (Weast, 1986)

Boiling point (°C):
170 (Weast, 1986)

Density (g/cm³):
1.0180 at 20/4 °C (Weast, 1986)
1.0117 at 25/4 °C (Windholz et al., 1983)

Diffusivity in water (x 10^{-5} cm²/sec):
1.19 at 20 °C using method of Hayduk and Laudie (1974)

Dissociation constant, pK_a:
9.50 at 25 °C (Dean, 1973)

Flash point (°C):
86 (NIOSH, 1997)
90.6 (Windholz et al., 1983)

Lower explosive limit (%):
3.0 at 140 °C (NIOSH, 1997)

Upper explosive limit (%):
23.5 (NIOSH, 1997)

Heat of fusion (kcal/mol):
4.90 (Riddick et al., 1986)

Henry's law constant (x 10^{-10} atm·m³/mol):
1.61 at 20 °C (Bone et al., 1983)

Ionization potential (eV):
8.96 (NIOSH, 1997)

Soil sorption coefficient, log K_{oc}:
Unavailable because experimental methods for estimation of this parameter for amines are lacking in the documented literature. However, its miscibility in water and low K_{ow} suggest its adsorption to soil will be nominal (Lyman et al., 1982).

Octanol/water partition coefficient, log K_{ow}:
-1.31 (Collander, 1951)

Solubility in organics:
Miscible with acetone and methanol; soluble in benzene (1.4 wt %), carbon tetrachloride (0.2 wt %), and ether (2.1 wt %) (Windholz et al., 1983).

Solubility in water:
Miscible (NIOSH, 1997).

Vapor density:
2.50 g/L at 25 °C, 2.11 (air = 1)

Vapor pressure (mmHg):
0.4 at 20 °C, 6 at 60 °C (Verschueren, 1983)
0.48 at 20 °C (Hawley, 1981)

Environmental fate:
   Biological. Bridié et al. (1979) reported BOD and COD values 0.93 and 1.28 g/g using filtered effluent from a biological sanitary waste treatment plant. These values were determined using a standard dilution method at 20 °C for a period of 5 d. Similarly, Heukelekian and Rand (1955) reported a 5-d BOD value of 0.85 g/g which is 65.0% of the ThOD value of 1.31 g/g.
   Chemical/Physical. Aqueous chlorination of ethanolamine at high pH produced *N*-chloroethanolamine, which slowly degraded to unidentified products (Antelo et al., 1981).
   At an influent concentration of 1,012 mg/L, treatment with granular activated carbon resulted in an effluent concentration of 939 mg/L. The adsorbability of the carbon used was 15 mg/g carbon (Guisti et al., 1974).

Exposure limits: NIOSH REL: TWA 3 ppm (8 mg/m^3), STEL 6 ppm (15 mg/m^3), IDLH 30 ppm; OSHA PEL: TWA 3 ppm; ACGIH TLV: TWA 0.3 ppm, STEL 6 ppm (adopted).

Symptoms of exposure: Severe irritation of eyes and moderate irritation of the skin (Patnaik, 1992)

Toxicity:
   LC_{50} (96-h) for goldfish >5,000 mg/L at pH 7 (quoted, Verschueren, 1983).
   LC_{50} (72-h) for goldfish 170 mg/L at pH 10.1 (quoted, Verschueren, 1983).
   LC_{50} (24-h) for goldfish 190 mg/L at pH 10.1 (quoted, Verschueren, 1983).
   Acute oral LD_{50} for guinea pigs 620 mg/kg, mice 700 mg/kg, rats 2,050 mg/kg, rabbits 1,000 mg/kg (Patnaik, 1992).

Uses: Removing carbon dioxide and hydrogen sulfide from natural gas; in emulsifiers, hair waving solutions, polishes; softening agent for hides; agricultural sprays; pharmaceuticals, chemical intermediates; corrosion inhibitor; rubber accelerator; non-ionic detergents used in dry cleaning; wool treatment.

2-ETHOXYETHANOL

Synonyms: Cellosolve; Cellosolve solvent; Dowanol EE; Ektasolve EE; Ethyl cellosolve; Ethylene glycol ethyl ether; Ethylene glycol monoethyl ether; Glycol ether EE; Glycol monoethyl ether; Hydroxy ether; Jeffersol EE; NCI-C54853; Oxitol; Polysolv EE; UN 1171.

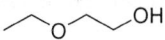

CAS Registry Number: 110-80-5
DOT: 1171
DOT label: Combustible liquid
Molecular formula: $C_4H_{10}O_2$
Formula weight: 90.12
RTECS: KK8050000
Merck reference: 10, 3700

Physical state, color, and odor:
Clear, colorless liquid with a sweetish odor. Odor threshold concentration is 2.7 ppm (Amoore and Hautala, 1983).

Melting point (°C):
-70 (Keith and Walters, 1992)

Boiling point (°C):
135 (Weast, 1986)

Density (g/cm³):
0.9297 at 20/4 °C (Weast, 1986)
0.92515 at 25.00/4 °C (Venkatesulu et al., 1997)

Diffusivity in water (x 10^{-5} cm²/sec):
0.90 at 20 °C using method of Hayduk and Laudie (1974)

Flash point (°C):
43 (NIOSH, 1997)
44 (Windholz et al., 1983)

Lower explosive limit (%):
1.7 at 93 °C (NIOSH, 1997)

Upper explosive limit (%):
15.6 at 93 °C (NIOSH, 1997)

Ionization potential (eV):
9.77 (Lias et al., 1998)

Soil sorption coefficient, log K_{oc}:
Unavailable because experimental methods for estimation of this parameter for aliphatic amines are lacking in the documented literature. However, its miscibility in water and low K_{ow} suggest its adsorption to soil will be nominal (Lyman et al., 1982).

Octanol/water partition coefficient, log K_{ow}:
-0.53 (Collander, 1951)

Solubility in organics:
Miscible with acetone, alcohol, ether, and liquid esters (Windholz et al., 1983)

Solubility in water:
Miscible (Price et al., 1974).

Vapor density:
3.68 g/L at 25 °C, 3.11 (air = 1)

Vapor pressure (mmHg):
3.8 at 20 °C, 7 at 30 °C (Verschueren, 1983)
5.63 at 25 °C (Banerjee et al., 1990)

Environmental fate:
   Biological. Bridié et al. (1979) reported BOD and COD values 1.03 and 1.92 g/g using filtered effluent from a biological sanitary waste treatment plant. These values were determined using a standard dilution method at 20 °C for a period of 5 d. When a sewage seed was used in a separate screening test, a BOD value of 1.27 g/g was obtained. Similarly, Heukelekian and Rand (1955) reported a 5-d BOD value of 1.42 g/g which is 72.4% of the ThOD value of 1.96 g/g.
   Photolytic. Grosjean (1997) reported a rate constants of 1.87 x 10^{-11} cm^3/molecule·sec at 298 for the reaction of 2-ethoxyethanol and OH radicals in the atmosphere. Based on an atmospheric OH radical concentration of 1.0 x 10^6 molecule/cm^3, the reported half-life of methanol is 0.35 d (Grosjean, 1997).
   Chemical/Physical. 2-Ethoxyethanol will not hydrolyze (Kollig, 1993).
   At an influent concentration of 1,024 mg/L, treatment with granular activated carbon resulted in an effluent concentration of 886 mg/L. The adsorbability of the carbon used was 28 mg/g carbon (Guisti et al., 1974).

Exposure limits: NIOSH REL: TWA 0.5 ppm (1.8 mg/m^3), IDLH 500 ppm; OSHA PEL: TWA 200 ppm (740 mg/m^3); ACGIH TLV: TWA 5 ppm (adopted).

Toxicity:
   LC_{50} (7-d) for *Poecilia reticulata* 16,400 mg/L (Könemann, 1981).
   LC_{50} (24-h) for goldfish >5,000 mg/L (quoted, Verschueren, 1983).
   LC_{50} (96-h static bioassay) for bluegill sunfish and *Menidia beryllina* >10,000 mg/L (quoted, Verschueren, 1983).
   Acute oral LD_{50} for guinea pigs 1,400 mg/kg, mice 2,451 mg/kg, rats 3,000 mg/kg, rabbits 3,100 mg/kg (quoted, RTECS, 1985).

Uses: Solvent for lacquers, varnishes, dopes, nitrocellulose, natural and synthetic resins; in cleaning solutions, varnish removers, dye baths; mutual solvent for formation of soluble oils; lacquer thinners; emulsion stabilizer; anti-icing additive for aviation fuels.

2-ETHOXYETHYL ACETATE

Synonyms: Acetic acid 2-ethoxyethyl ester; Cellosolve acetate; CSAC; Ekasolve EE acetate solvent; Ethoxyacetate; **2-Ethoxyethanol acetate**; Ethoxyethyl acetate; β-Ethoxyethyl acetate; 2-Ethoxyethyl ethanoate; Ethylene glycol ethyl ether acetate; Ethylene glycol monoethyl ether acetate; Glycol ether EE acetate; Glycol monoethyl ether acetate; Oxytol acetate; Polysolv EE acetate; UN 1172.

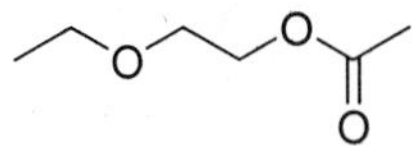

CAS Registry Number: 111-15-9
DOT: 1172
DOT label: Flammable liquid
Molecular formula: $C_6H_{12}O_3$
Formula weight: 132.18
RTECS: KK8225000
Merck reference: 10, 3701

Physical state, color, and odor:
Colorless liquid with a faint, pleasant odor. Odor threshold concentration is 56 ppb (Amoore and Hautala, 1983).

Melting point (°C):
-61.7 (NIOSH, 1997)
-58 (Verschueren, 1983)

Boiling point (°C):
156 (NIOSH, 1997)

Density (g/cm³):
0.975 at 20/20 °C (Windholz et al., 1983)

Diffusivity in water (x 10^{-5} cm²/sec):
0.74 at 20 °C using method of Hayduk and Laudie (1974)

Flash point (°C):
51.5 (NIOSH, 1997)
56 (open cup, Windholz et al., 1983)

Lower explosive limit (%):
1.7 (NIOSH, 1997)

Henry's law constant (x 10^{-7} atm·m³/mol):
9.07 at 25 °C (approximate - calculated from water solubility and vapor pressure)

Soil sorption coefficient, log K_{oc}:
Unavailable because experimental methods for estimation of this parameter for aliphatic amines are lacking in the documented literature. However, its high solubility in water suggests its adsorption to soil will be nominal (Lyman et al., 1982).

Octanol/water partition coefficient, log K_{ow}:
0.50 using method of Hansch et al. (1968)

Solubility in water:
230,000 mg/L at 20 °C (quoted, Verschueren, 1983)

Vapor density:
5.40 g/L at 25 °C, 4.56 (air = 1)

Vapor pressure (mmHg):
2 at 20 °C (NIOSH, 1997)
1.2 at 20 °C, 3.8 at 30 °C (Verschueren, 1983)

Environmental fate:
   Biological. Bridié et al. (1979) reported BOD and COD values 0.74 and 1.76 g/g using filtered effluent from a biological sanitary waste treatment plant. These values were determined using a standard dilution method at 20 °C for a period of 5 d. The ThOD for 2-ethoxyethyl acetate is 1.82 g/g.
   Chemical/Physical.   At an influent concentration of 1,000 mg/L, treatment with granular activated carbon resulted in an effluent concentration of 342 mg/L. The adsorbability of the carbon used was 132 mg/g carbon (Guisti et al., 1974).

Exposure limits: NIOSH REL: TWA 0.5 ppm (2.7 mg/m^3), IDLH 500 ppm; OSHA PEL: TWA 100 ppm (540 mg/m^3); ACGIH TLV: TWA 5 ppm (adopted).

Toxicity:
   Acute oral LD_{50} for guinea pigs 1,910 mg/kg, rats 2,900 mg/kg, rabbits 1,950 mg/kg (quoted, RTECS, 1985).

Uses: Automobile lacquers to reduce evaporation and to impart a high gloss; solvent for nitrocellulose, oils, and resins; varnish removers; wood stains; textiles; leather.

ETHYL ACETATE

Synonyms: Acetic ether; **Acetic acid ethyl ester**; Acetidin; Acetoxyethane; Ethyl acetic ester; Ethyl ethanoate; RCRA waste number U112; UN 1173; Vinegar naphtha.

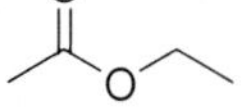

CAS Registry Number: 141-78-6
DOT: 1173
DOT label: Flammable liquid
Molecular formula: $C_4H_8O_2$
Formula weight: 88.11
RTECS: AH5425000
Merck reference: 10, 3706

Physical state, color, and odor:
Clear, colorless, mobile liquid with a pleasant fruity odor. Odor threshold concentration is 3.9 ppm (Amoore and Hautala, 1983).

Melting point (°C):
-83.6 (Weast, 1986)

Boiling point (°C):
77.06 (Weast, 1986)

Density (g/cm³):
0.90062 at 20.00/4 °C (Lee and Tu, 1999)
0.8939 at 25.00/4 °C, 0.8876 at 30.00/4 °C, 0.8755 at 40.00/4 °C, 0.8630 at 50.00/4 °C, 0.8499 at 60.00/4 °C, 0.8364 at 70.00/4 °C (Abraham et al., 1971)
0.8946 at 25.00/4 °C, 0.8885 at 30.00/4 °C, 0.8825 at 35.00/4 °C (Nikam et al., 1998)

Diffusivity in water (x 10⁻⁵ cm²/sec):
1.12 at 25 °C (quoted, Hayduk and Laudie, 1974)

Flash point (°C):
-4.5 (NIOSH, 1997)
7.2 (open cup, Windholz et al., 1983)

Lower explosive limit (%):
2.0 (NFPA, 1984)

Upper explosive limit (%):
11.5 (NFPA, 1984)

Heat of fusion (kcal/mol):
2.505 (Dean, 1987)

Henry's law constant (x 10⁻⁴ atm·m³/mol):
1.69 at 25 °C (Kieckbusch and King, 1979)
1.32 at 25 °C (Butler and Ramchandani, 1935)

1.75 at 28 °C (Nelson and Hoff, 1968)

7.23, 9.86, 9.53, 8.84, and 9.79 at 2.0, 6.0, 10.0, 18.0, and 25.0 °C, respectively (Dewulf et al., 1999)

Interfacial tension with water (dyn/cm):
6.8 at 25 °C (Donahue and Bartell, 1952)

Ionization potential (eV):
10.11 ± 0.02 (Franklin et al., 1969)
10.24 (Gibson et al., 1977)

Bioconcentration factor, log BCF:
4.13 (algae, Geyer et al., 1984)
3.52 (activated sludge), 2.28 (algae), 1.48 (golden ide) (Freitag et al., 1985)

Soil sorption coefficient, log K_{oc}:
Unavailable because experimental methods for estimation of this parameter for aliphatic esters are lacking in the documented literature

Octanol/water partition coefficient, log K_{ow}:
0.66 (Collander, 1951)
0.73 (Hansch and Anderson, 1967)
0.81 (Kamlet et al., 1984)
0.68 (Tewari et al., 1982; Wasik et al., 1981)

Solubility in organics:
Miscible with acetone, alcohol, chloroform, ether (Windholz et al., 1983)

Solubility in water:
80,000 mg/L at 25 °C (Banerjee, 1984)
726 mM at 25 °C (Tewari et al., 1982; Wasik et al., 1981)
85 g/L (Price et al., 1974)
731 mM at 20 °C (Fühner, 1924)
9.50 wt % at 25 °C (Lo et al., 1986)
In wt %: 3.21 at 0 °C, 2.78 at 9.5 °C, 2.26 at 20.0 °C, 1.98 at 30.0 °C, 1.87 at 40.0 °C, 1.72 at 50.0 °C, 1.64 at 60.1 °C, 1.72 at 70.5 °C, 1.66 at 80.0 °C, 1.35 at 90.2 °C (Stephenson and Stuart, 1986)
84.2, 80.4, 77.0, 73.9, and 71.2 g/kg at 20, 25, 30, 35, and 40 °C, respectively (Altshuller and Everson, 1953)
88.0 and 74.0 mL/L at 25 and 60 °C, respectively (Booth and Everson, 1949)

Vapor density:
3.04 (air = 1) (Windholz et al., 1983); 3.60 g/L at 25 °C

Vapor pressure (mmHg):
72.8 at 20 °C, 115 at 30 °C (Verschueren, 1983)
94.5 at 25 °C (Abraham, 1984)

Environmental fate:
Biological. Heukelekian and Rand (1955) reported a 5-d BOD value of 1.00 g/g which is 54.9% of the ThOD value of 1.82 g/g.

Photolytic. Reported rate constants for the reaction of ethyl acetate and OH radicals in the atmosphere (296 K) and aqueous solution are 1.51×10^{-12} and 6.60×10^{-13} cm^3/molecule·sec, respectively (Wallington et al., 1988b).

Chemical/Physical. Hydrolyzes in water forming ethyl alcohol and acetic acid (Kollig, 1993). The estimated hydrolysis half-life at 25 °C and pH 7 is 2.0 yr (Mabey and Mill, 1978).

Exposure limits: NIOSH REL: TWA 400 ppm (1,400 mg/m³), IDLH 2,000 ppm; OSHA PEL: TWA 400 ppm; ACGIH TLV: TWA 400 ppm (adopted).

Symptoms of exposure: Inhalation of vapors may cause irritation of eyes, nose, and throat (Patnaik, 1992)

Toxicity:

LC_{50} (inhalation) for rats 1,600 ppm/8-h (quoted, RTECS, 1985).

Acute oral LD_{50} for mice 4,100 mg/kg, rats 5,620 mg/kg, rabbits 4,935 mg/kg (quoted, RTECS, 1985).

Uses: Manufacture of smokeless powder, photographic film and plates, artificial leather and silk, perfumes; pharmaceuticals; in cleaning textiles; solvent for nitrocellulose, lacquers, varnishes, and airplane dopes.

ETHYL ACRYLATE

Synonyms: Acrylic acid ethyl ester; Ethoxycarbonylethylene; Ethyl propenoate; Ethyl-2-propenoate; NCI-C50384; **2-Propenoic acid ethyl ester**; RCRA waste number U113; UN 1917.

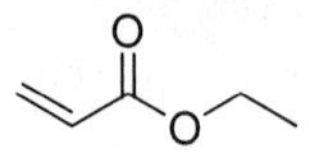

CAS Registry Number: 140-88-5
DOT: 1917
DOT label: Flammable liquid
Molecular formula: $C_5H_8O_2$
Formula weight: 100.12
RTECS: AT0700000
Merck reference: 10, 3708

Physical state, color, and odor:
Colorless liquid with a sharp, penetrating odor. Odor threshold concentration is 1.2 ppb (Amoore and Hautala, 1983).

Melting point (°C):
-71.2 (Weast, 1986)

Boiling point (°C):
99.8 (Weast, 1986)

Density (g/cm³ at 20/4 °C):
0.9234 (Weast, 1986)
0.9405 (Windholz et al., 1983)

Diffusivity in water (x 10^{-5} cm²/sec):
0.84 at 20 °C using method of Hayduk and Laudie (1974)

Flash point (°C):
8.9 (NIOSH, 1997)
15 (open cup, Windholz et al., 1983)
10 (open cup, NFPA, 1984)

Lower explosive limit (%):
1.4 (NIOSH, 1997)

Upper explosive limit (%):
14 (NFPA, 1984)

Henry's law constant (x 10^{-3} atm·m³/mol):
2.25 at 20 °C (approximate - calculated from water solubility and vapor pressure)

Ionization potential (eV):
10.30 (NIOSH, 1997)

Soil sorption coefficient, log K_{oc}:
Unavailable because experimental methods for estimation of this parameter for aliphatic amines are lacking in the documented literature

Octanol/water partition coefficient, log K_{ow}:
1.33 (Tanii et al., 1984)

Solubility in organics:
Soluble in alcohol, chloroform, and ether (Weast, 1986)

Solubility in water:
20 g/L at 20 °C (quoted, Windholz et al., 1983)

Vapor density:
4.09 g/L at 25 °C, 3.45 (air = 1)

Vapor pressure (mmHg):
29 at 20 °C (NIOSH, 1997)
49 at 30 °C (Verschueren, 1983)

Environmental fate:
   Chemical/Physical. Polymerizes on standing and is catalyzed by heat, light, and peroxides (Windholz et al., 1983). Slowly hydrolyzes in water forming ethyl alcohol and acrylic acid.
   At an influent concentration of 1,015 mg/L, treatment with granular activated carbon resulted in an effluent concentration of 226 mg/L. The adsorbability of the carbon used was 157 mg/g carbon (Guisti et al., 1974).

Exposure limits: Potential occupational carcinogen. NIOSH REL: IDLH 300 ppm; OSHA PEL: TWA 25 ppm (100 mg/m^3); ACGIH TLV: TWA 5 ppm, STEL 15 ppm (adopted).

Symptoms of exposure: Strong irritant to eyes, skin, and mucous membranes (Patnaik, 1992)

Toxicity:
   Acute oral LD_{50} for rats is 800 mg/kg, rabbits 400 mg/kg, mice 1,799 mg/kg (quoted, RTECS, 1985).

Uses: Manufacture of water emulsion paints, textile and paper coatings, adhesives, and leather finish resins.

ETHYLAMINE

Synonyms: Aminoethane; 1-Aminoethane; **Ethanamine**; Monoethylamine; UN 1036.

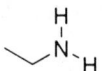

CAS Registry Number: 75-04-7
DOT: 1036
DOT label: Flammable liquid
Molecular formula: C_2H_7N
Formula weight: 45.08
RTECS: KH2100000
Merck reference: 10, 3709

Physical state, color, and odor:
Colorless liquid or gas with a strong ammonia-like odor. Odor threshold concentration is 0.95 ppm (Amoore and Hautala, 1983).

Melting point (°C):
-80.5 (Chao et al., 1990)
-83 (Verschueren, 1983)

Boiling point (°C):
16.6 (Chao et al., 1990)

Density (g/cm³):
0.6829 at 20/4 °C (Weast, 1986)
0.6767 at 25/4 °C (Chao et al., 1990)
0.71 at 0/4 °C (Verschueren, 1983)

Diffusivity in water (x 10^{-5} cm²/sec):
1.13 at 20 °C using method of Hayduk and Laudie (1974)

Dissociation constant, pK_a:
10.807 at 20 °C (Gordon and Ford, 1972)
10.63 at 25 °C (Dean, 1973)

Flash point (°C):
-17.4 (NIOSH, 1997)

Lower explosive limit (%):
3.5 (NIOSH, 1997)

Upper explosive limit (%):
14.0 (NIOSH, 1997)

Henry's law constant (x 10^{-5} atm·m³/mol at 25 °C):
1.00 (Butler and Ramchandani, 1935)
1.23 (Christie and Crisp, 1967)

Ionization potential (eV):
8.86 ± 0.02 (Franklin et al., 1969)
9.19 (Gibson et al., 1977)

Soil sorption coefficient, log K_{oc}:
Unavailable because experimental methods for estimation of this parameter for aliphatic amines are lacking in the documented literature

Octanol/water partition coefficient, log K_{ow}:
-0.13 (Hansch and Leo, 1985)
-0.30 (Sangster, 1989)

Solubility in organics:
Miscible with alcohol and ether (Hawley, 1981)

Solubility in water:
Miscible (NIOSH, 1997). A saturated solution in equilibrium with its own vapor had a concentration of 5,176 g/L at 25 °C (Kamlet et al., 1987).

Vapor density:
1.84 g/L at 25 °C, 1.56 (air = 1)

Vapor pressure (mmHg):
897 at 20 °C, 1,292 at 30 °C (Verschueren, 1983)

Environmental fate:
   Photolytic. The rate constant for the reaction of ethylamine and ozone the atmosphere is 2.76 x 10^{-20} cm^3/molecule·sec at 296 K (Atkinson and Carter, 1984). Atkinson (1985) reported a rate constant of 6.54 x 10^{-11} cm^3/molecule·sec for the vapor-phase reaction of ethylamine and OH radicals at 25.5 °C. The half-life for this reaction is 8.6 h.
   Chemical/Physical. Reacts with OH radicals possibly forming acetaldehyde or acetamide (Atkinson et al., 1978). When ethylamine over kaolin is heated to 600 °C, hydrogen and acetonitrile formed as the major products. Trace amounts of ethylene, ammonia, hydrogen cyanide, and methane were also produced. At 900 °C, however, acetonitrile was not produced (Hurd and Carnahan, 1930).
   Reacts with mineral acids forming water-soluble salts (Morrison and Boyd, 1971).

Exposure limits: NIOSH REL: TWA 10 ppm (18 mg/m^3), IDLH 600 ppm; OSHA PEL: TWA 10 ppm; ACGIH TLV: TWA 5 ppm, STEL 10 ppm (adopted).

Symptoms of exposure: Severe irritant to the eyes, skin, mucous membranes, and respiratory system (Patnaik, 1992; Windholz et al., 1983)

Toxicity:
   LC_{50} (96-h) for goldfish 170 mg/L at pH 10.1 (quoted, Verschueren, 1983).
   LC_{50} (24-h) for goldfish 190 mg/L at pH 10.1, (quoted, Verschueren, 1983).
   LC_0 (24-h) and LC_{100} (24-h) values for creek chub in Detroit river water were 30 and 50 mg/L, respectively (Gillette et al., 1952).
   Acute oral LD_{50} for rats 400 mg/kg (quoted, RTECS, 1985).

Uses: Stabilizer for latex rubber; intermediate for dyestuffs and medicinals; resin and detergent manufacturing; solvent in petroleum and vegetable oil refining; starting material for manufacturing amides; plasticizer; stabilizer for rubber latex; in organic synthesis.

ETHYLBENZENE

Synonyms: EB; Ethylbenzol; NCI-C56393; Phenylethane; UN 1775.

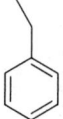

CAS Registry Number: 100-41-4
DOT: 1175
DOT label: Flammable liquid
Molecular formula: C_8H_{10}
Formula weight: 106.17
RTECS: DA0700000
Merck reference: 10, 3714

Physical state, color, and odor:
Clear, colorless liquid with a sweet, gasoline-like odor. The reported odor threshold concentrations in air ranged from 150 ppb to 2.3 ppm (Keith and Walters, 1992; Young et al., 1996). The taste threshold concentration in water is 390 ppb (Young et al., 1996).

Melting point (°C):
-95.0 (Dean, 1973)
-94.4 (Huntress and Mulliken, 1941)

Boiling point (°C):
136.2 (Weast, 1986)

Density (g/cm³):
0.8670 at 20/4 °C (Weast, 1986)
0.86250 at 25/4 °C (Huntress and Mulliken, 1941)

Diffusivity in water (x 10⁻⁵ cm²/sec):
0.81 at 20 °C (Witherspoon and Bonoli, 1969)
0.90 at 25 °C (Bonoli and Witherspoon, 1968)

Dissociation constant, pK_a:
>15 (Christensen et al., 1975)

Flash point (°C):
13 (NIOSH, 1997)
21 (NFPA, 1984)

Lower explosive limit (%):
0.8 (NIOSH, 1997)

Upper explosive limit (%):
6.7 (NIOSH, 1997)

Heat of fusion (kcal/mol):
2.195 (Dean, 1987)

Henry's law constant (x 10^{-3} atm·m^3/mol):
6.6 (Pankow and Rosen, 1988)
6.44 (Valsaraj, 1988)
3.26, 4.51, 6.01, 7.88, and 10.5 at 10, 15, 20, 25, and 30 °C, respectively (Ashworth et al., 1988)
3.02, 4.22, 5.75, 7.84, and 10.3 at 10, 15, 20, 25, and 30 °C, respectively (Perlinger et al., 1993)
Distilled water: 1.93, 2.05, 2.67, 5.02, and 6.62 at 2.0, 6.0, 10.0, 18.2, and 25.0 °C, respectively; natural seawater: 2.98 and 8.27 at 6.0 and 25.0 °C, respectively (Dewulf et al., 1995)
6.67 at 25 °C (Mackay et al., 1979)
7.69 at °C (Robbins et al., 1993)
7.78, 10.17, and 16.4 at 25, 30, and 40 °C, respectively (Robbins et al., 1993)

Interfacial tension with water (dyn/cm):
38.4 at 25 °C (Donahue and Bartell, 1952)
38.70 at 20 °C (Harkins et al., 1920)

Ionization potential (eV):
8.76 ± 0.01 (Franklin et al., 1969)
9.12 (Yoshida et al., 1983)

Bioconcentration factor, log BCF:
1.19 (bluegill sunfish, Ogata et al., 1984)
2.41, 2.52, 2.54, 3.04, 3.20, 3.26, and 5.00 for holly, olive, grass, rosemary, ivy, mock orange, pine, and juniper leaves, respectively (Hiatt, 1998)

Soil sorption coefficient, log K_{oc}:
2.22 (Woodburn silt loam soil, Chiou et al., 1983)
2.27 (St. Clair soil), 2.28 (Oshtemo soil) (Lee et al., 1989)
2.38 (estimated from HPLC capacity factors, Hodson and Williams, 1988)
2.41 (Tamar estuary sediment) (Vowles and Mantoura, 1987)
2.49, 2.73, 2.65, 2.73, 2.77, 2.73, and 2.74 at 2.3, 3.8, 6.2, 8.0, 13.5, 18.6, and 25.0 °C, respectively, for a Leie River (Belgium) clay (Dewulf et al., 1999a)

Octanol/water partition coefficient, log K_{ow}:
3.13 (Wasik et al., 1981, 1983; Yalkowsky et al., 1983a)
3.15 (Campbell and Luthy, 1985; Hansch et al., 1968)

Solubility in organics:
Freely soluble in most solvents (U.S. EPA, 1985)

Solubility in water:
1.76 mM at 25 °C (Wasik et al., 1981, 1983)
187 mg/L at 25 °C (Miller et al., 1985)
152 mg/kg at 25 °C (McAuliffe, 1966)
159 mg/kg at 25 °C (McAuliffe, 1963)
1.31 mM at 25.0 °C (Andrews and Keefer, 1950a)
In mg/L: 219 at 0.4 °C, 213 at 5.2 °C, 207 at 20.7 °C, 207 at 21.2 °C, 208 at 25.0 °C, 209 at 25.6 °C, 211 at 30.2 °C, 221 at 34.9 °C, 231 at 42.8 °C (Bohon and Claussen, 1951)

In mM: 1.85 at 10.0 °C, 1.770 at 20.0 °C, 1.811 at 25.0 °C, 1.777 at 30.0 °C (Owens et al., 1986)

197 mg/kg at 0 °C, 177 mg/kg at 25 °C (Polak and Lu, 1973)

161.2 mg/kg at 25 °C, 111.0 mg/kg in artificial seawater at 25 °C (Sutton and Calder, 1975)

131.0 mg/kg at 25 °C (Price, 1976)

181 mg/L at 20 °C (Burris and MacIntyre, 1986)

77 mg/L in fresh water at 25 °C, 70 mg/L in salt water at 25 °C (Krasnoshchekova and Gubergrits, 1975)

2.00 mM at 25 °C (Ben-Naim and Wilf, 1980)

1.12 mM in 0.5 M NaCl at 25 °C (Wasik et al., 1984)

175 mg/L at 25 °C (Klevens, 1950)

140 mg/kg at 15 °C (Fühner, 1924)

147.7 mg/L at 30 °C (McNally and Grob, 1984)

1.51, 1.59, 1.65 and 1.83 mM at 15, 25, 35 and 45 °C, respectively (Sanemasa et al., 1982)

172 mg/L (Coutant and Keigley, 1988)

1.55 mmol/kg at 25 °C (Morrison and Billett, 1952)

In mg/kg: 196, 192, 186, 187, 181, 183, 180, 184, and 180 at 4.5, 6.3, 7.1, 9.0, 11.8, 12.1, 15.1, 17.9, and 20.1 °C, respectively. In artificial seawater: 140, 133, 129, 125, and 122 at 0.19, 5.32, 10.05, 14.96, and 20.04 °C, respectively (Brown and Wasik, 1974)

1.91 mmol/kg at 25.0 °C (Vesala, 1974)

1.02 mM at 25.00 °C (Sanemasa et al., 1985)

1.37 mM at 25.0 °C (Sanemasa et al., 1987)

0.0196 wt % at 10.0 and 20.0 °C (Schwarz and Miller, 1980)

Vapor density:
4.34 g/L at 25 °C, 3.66 (air = 1)

Vapor pressure (mmHg):
12 at 30 °C (Verschueren, 1983)
7.08 at 20 °C (Burris and MacIntyre, 1986)
9.9 at 25 °C (Mackay et al., 1982)
9.6 at 25 °C (Banerjee et al., 1990)

Environmental fate:

Biological. Phenylacetic acid was reported to be the biooxidation product of ethylbenzene by *Nocardia* sp. in soil using *n*-hexadecane or *n*-octadecane as the substrate. In addition, *Methylosinus trichosporium* OB3b was reported to metabolize ethylbenzene to *o*- and *m*-hydroxybenzaldehyde with methane as the substrate (Keck et al., 1989). Ethylbenzene was oxidized by a strain of *Micrococcus cerificans* to phenacetic acid (Pitter and Chudoba, 1990). A culture of *Nocardia tartaricans* ATCC 31190, growing in a hexadecane medium, oxidized ethylbenzene to 1-phenethanol, which further oxidized to acetophenone (Cox and Goldsmith, 1979). When ethylbenzene (5 mg/L) was statically incubated in the dark at 25 °C with yeast extract and settled domestic wastewater inoculum, complete biodegradation with rapid acclimation was observed after 7 d. At a concentration of 10 mg/L, significant degradation occurred with gradual adaptation. Percent losses of 69, 78, 87, and 100 were obtained after 7, 14, 21, and 28-d incubation periods, respectively (Tabak et al., 1981). Olsen and Davis (1990) reported a first-order degradation rate constant of 0.07/yr and a half-life of 37 d.

Surface Water. The evaporation half-life of ethylbenzene in surface water (1 m depth) at 25 °C is estimated to be from 5 to 6 h (Mackay and Leinonen, 1975). Estimated half-lives of ethylbenzene (3.3 μg/L) from an experimental marine mesocosm during the spring (8-16 °C), summer (20-22 °C), and winter (3-7 °C) were 20, 2.1, and 13 d, respectively (Wakeham et al.,

1983).

Photolytic. Irradiation of ethylbenzene (λ <2537 Å) at low temperatures will form hydrogen, styrene, and free radicals (Calvert and Pitts, 1966).

Chemical/Physical. Complete combustion in air yields carbon dioxide and water vapor. Ethylbenzene will not hydrolyze in water (Kollig, 1993).

At an influent concentration of 115 mg/L, treatment with granular activated carbon resulted in an effluent concentration of 18 mg/L. The adsorbability of the carbon used was 18 mg/g carbon (Guisti et al., 1974). Similarly, at influent concentrations of 10.0, 1.0, 0.1, and 0.01 mg/L, the granular activated carbon adsorption capacities at pH 7.4 were 325, 53, 8.5, and 1.4 mg/g, respectively (Dobbs and Cohen, 1980).

Exposure limits: NIOSH REL: TWA 100 ppm (435 mg/m^3), STEL 125 ppm (545 mg/m^3), IDLH 800 ppm; OSHA PEL: TWA 100 ppm; ACGIH TLV: TWA 100 ppm, STEL 125 ppm (adopted).

Symptoms of exposure: Narcotic at high concentrations. Irritant to the eyes, skin, and nose (Patnaik, 1992).

Toxicity:

EC$_{50}$ (96-h - growth inhibition) for diatoms (*Skeletonema costatum*) 3.6 mg/L, freshwater algae (*Selenastrum capricornutum*) 3.6 mg/L (Masten et al., 1994).

EC$_{50}$ (72-h) for *Selenastrum capricornutum* 4.6 mg/L (Galassi et al., 1988).

LC$_{50}$ (contact) for earthworm (*Eisenia fetida*) 47 μg/cm^2 (Neuhauser et al., 1985).

LC$_{50}$ (96-h) for bluegill sunfish 150 mg/L (Spehar et al., 1982), *Salmo gairdneri* 4.2 mg/L (Galassi et al., 1988), Atlantic silversides (*Menidia menidia*) 5.1 mg/L, mysid shrimp (*Mysidopsis bahia*), 6 mg/L (Masten et al., 1994), *Cyprinodon variegatus* 280 ppm using natural seawater (Heitmuller et al., 1981).

LC$_{50}$ (72-h) for *Cyprinodon variegatus* 320 ppm (Heitmuller et al., 1981).

LC$_{50}$ (48-h) for *Daphnia magna* 75 mg/L (LeBlanc, 1980), *Cyprinodon variegatus* 360 ppm (Heitmuller et al., 1981).

LC$_{50}$ (24-h) for *Daphnia magna* 77 mg/L (LeBlanc, 1980), *Cyprinodon variegatus* 300 ppm (Heitmuller et al., 1981).

Acute oral LD$_{50}$ for rats 3,500 mg/kg (quoted, RTECS, 1985).

Heitmuller et al. (1981) reported a NOEC of 88 ppm.

Drinking water standard (final): MCLG: 0.7 mg/L; MCL: 0.7 mg/L (U.S. EPA, 1996).

Source: Detected in distilled water-soluble fractions of 87 octane gasoline (2.38 mg/L), 94 octane gasoline (7.42 mg/L), Gasohol (3.54 mg/L), No. 2 fuel oil (0.21 mg/L), jet fuel A (0.41 mg/L), diesel fuel (0.17 mg/L), military jet fuel JP-4 (1.57 mg/L) (Potter, 1996), new motor oil (0.15-0.17 μg/L), and used motor oil (117-124 μg/L) (Chen et al., 1994). The average volume percent and estimated mole fraction in American Petroleum Institute PS-6 gasoline are 1.570 and 0.017, respectively (Poulsen et al., 1992).

Thomas and Delfino (1991) equilibrated contaminant-free groundwater collected from Gainesville, FL with individual fractions of three individual petroleum products at 24-25 °C for 24 h. The aqueous phase was analyzed for organic compounds via U.S. EPA approved test method 602. Average ethylbenzene concentrations reported in water-soluble fractions of unleaded gasoline, kerosene, and diesel fuel were 2.025, 0.314, and 0.104 mg/L, respectively. When the authors analyzed the aqueous-phase via U.S. EPA approved test method 610, average ethylbenzene concentrations in water-soluble fractions of unleaded gasoline, kerosene, and diesel fuel were lower, i.e., 1.423, 0.171, and 0.079 mg/L, respectively.

Schauer et al. (1999) reported ethylbenzene in a diesel-powered medium-duty truck exhaust at an emission rate of 470 μg/km.

Detected in 1-yr aged coal tar film and bulk coal tar at concentrations of 350 and 2,100 mg/kg, respectively (Nelson et al., 1996).

Uses: Intermediate in production of styrene, acetophenone, ethylcyclohexane, benzoic acid, 1-bromo-1-phenylethane, 1-chloro-1-phenylethane, 2-chloro-1-phenylethane, *p*-chloroethylbenzene, *p*-chlorostyrene, and many other compounds; solvent; in organic synthesis.

ETHYL BROMIDE

Synonyms: Bromic ether; **Bromoethane**; Halon 2001; Hydrobromic ether; Monobromoethane; NCI-C55481; UN 1891.

$$\diagup\diagdown Br$$

CAS Registry Number: 74-96-4
DOT: 1891
DOT label: Poison
Molecular formula: C_2H_5Br
Formula weight: 108.97
RTECS: KH6475000
Merck reference: 10, 3720

Physical state, color, and odor:
Clear, colorless to yellow, volatile liquid with an ether-like odor. Odor threshold concentration is 3.1 ppm (Amoore and Hautala, 1983).

Melting point (°C):
-119 (Windholz et al., 1983)

Boiling point (°C):
38.4 (Weast, 1986)

Density (g/cm³):
1.4604 at 20/4 °C (Weast, 1986)
1.4505 at 25/4 °C (Riddick et al., 1986)
1.4515 at 25/4 °C (Windholz et al., 1983)
1.4492 at 25/4 °C (Dreisbach, 1959)

Diffusivity in water (x 10⁻⁵ cm²/sec):
1.05 at 20 °C using method of Hayduk and Laudie (1974)

Flash point (°C):
<-20 (NIOSH, 1997)

Lower explosive limit (%):
6.8 (NFPA, 1984)

Upper explosive limit (%):
8.0 (NFPA, 1984)

Heat of fusion (kcal/mol):
1.4 (Dean, 1987)

Henry's law constant (x 10⁻³ atm·m³/mol):
7.56 at 25 °C (Hine and Mookerjee, 1975)

Interfacial tension with water (dyn/cm at 25 °C):
31.20 (Harkins et al., 1920)
31.3 (Donahue and Bartell, 1952)

Ionization potential (eV):
10.29 (Franklin et al., 1969)
10.46 (Gibson et al., 1977)

Soil sorption coefficient, log K_{oc}:
2.67 using method of Chiou et al. (1979)

Octanol/water partition coefficient, log K_{ow}:
1.61 (Hansch et al., 1975)

Solubility in organics:
Miscible with alcohol, chloroform, and ether (Windholz et al., 1983)

Solubility in water:
0.965 wt % at 10 °C (quoted, Windholz et al., 1983)
9,047 mg/L at 20 °C (Fischer and Ehrenberg, 1948)
9,603 mg/L at 17.5 °C (Führer, 1924; Fischer and Ehrenberg, 1948)
1.47×10^{-3} at 25 °C (mole fraction, Li et al., 1993)
9,140 mg/L at 20 °C, 8,960 mg/L at 30 °C (Rex, 1906)

Vapor density:
4.05 g/L at 25 °C, 3.76 (air = 1)

Vapor pressure (mmHg):
386 at 20 °C, 564 at 30 °C (Rex, 1906)

Environmental fate:
   Biological. A strain of *Acinetobacter* sp. isolated from activated sludge degraded ethyl bromide to ethanol and bromide ions (Janssen et al., 1987). When *Methanococcus thermolithotrophicus*, *Methanococcus deltae*, and *Methanobacterium thermoautotrophicum* were grown with H_2-CO_2 in the presence of ethyl bromide, methane and ethane were produced (Belay and Daniels, 1987).
   Groundwater. Groundwater under reducing conditions in the presence of hydrogen sulfide converted ethyl bromide to sulfur-containing products (Schwarzenbach et al., 1985).
   Chemical/Physical. Hydrolyzes in water forming ethyl alcohol and bromide ions. The estimated hydrolysis half-life at 25 °C and pH 7 is 30 d (Mabey and Mill, 1978).

Exposure limits: NIOSH REL: IDLH 2,000 ppm; OSHA PEL: TWA 200 ppm (890 mg/m³); ACGIH TLV: TWA 5 ppm (adopted).

Symptoms of exposure: Irritation of the respiratory system, eyes, pulmonary edema (Patnaik, 1992)

Toxicity:
   Acute oral LD_{50} for rats 1,350 mg/kg (quoted, RTECS, 1985).

Uses: In organic synthesis as an ethylating agent; refrigerant; solvent; grain and fruit fumigant; in medicine as an anesthetic.

ETHYLCYCLOPENTANE

Synonyms: None.

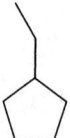

CAS Registry Number: 1640-89-7
Molecular formula: C_7H_{14}
Formula weight: 98.19
RTECS: GY4450000

Physical state and color:
Colorless liquid

Melting point (°C):
-138.4 (Weast, 1986)

Boiling point (°C):
103.5 (Weast, 1986)

Density (g/cm³):
0.7665 at 20/4 °C (Weast, 1986)

Diffusivity in water (x 10^{-5} cm²/sec):
0.76 at 20 °C using method of Hayduk and Laudie (1974)

Dissociation constant, pK$_a$:
>14 (Schwarzenbach et al., 1993)

Flash point (°C):
15 (Aldrich, 1988)

Lower explosive limit (%):
1.1 (Hawley, 1981)

Upper explosive limit (%):
6.7 (Hawley, 1981)

Heat of fusion (kcal/mol):
1.642-1.889 (Dean, 1987)

Henry's law constant (x 10^{-3} atm·m³/mol):
2.10 at 25 °C (approximate - calculated from water solubility and vapor pressure)

Ionization potential (eV):
10.12 (Lias et al., 1998)

Soil sorption coefficient, log K$_{oc}$:
Unavailable because experimental methods for estimation of this parameter for alicyclic hydrocarbons are lacking in the documented literature

Octanol/water partition coefficient, log K_{ow}:
1.90 using method of Hansch et al. (1968)

Solubility in organics:
Soluble in acetone, alcohol, benzene, ether, and petroleum (Weast, 1986)

Solubility in water (mg/kg):
21.9 at 70.5 °C, 52.5 at 113 °C, 224 at 168.5 °C, 759 at 203 °C (Guseva and Parnov, 1964)

Vapor density:
4.01 g/L at 25 °C, 3.39 (air = 1)

Vapor pressure (mmHg):
39.9 at 25 °C (Wilhoit and Zwolinski, 1971)

Environmental fate:
   Chemical/Physical. Complete combustion in air yields carbon dioxide and water vapor. Ethylcyclopentane will not hydrolyze because it has no hydrolyzable functional group.

Uses: Organic research.

ETHYLENE CHLOROHYDRIN

Synonyms: 2-Chloroethanol; δ-Chloroethanol; 2-Chloroethyl alcohol; β-Chloroethyl alcohol; Ethylene chlorhydrin; Glycol chlorohydrin; Glycol monochlorohydrin; 2-Monochloroethanol; NCI-C50135; UN 1135.

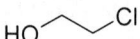

CAS Registry Number: 107-07-3
DOT: 1135
DOT label: Poison and flammable liquid
Molecular formula: C_2H_5ClO
Formula weight: 80.51
RTECS: KK0875000

Physical state, color, and odor:
Colorless liquid with a faint, ether-like odor. Odor threshold concentration is 400 ppb (Keith and Walters, 1992).

Melting point (°C):
-89 (Aldrich, 1988)

Boiling point (°C):
128-130 (Windholz et al., 1983)

Density (g/cm³ at 20/4 °C):
1.2003 (Weast, 1986)
1.121 (Verschueren, 1983)

Diffusivity in water (x 10⁻⁵ cm²/sec):
1.12 at 20 °C using method of Hayduk and Laudie (1974)

Flash point (°C):
61 (NIOSH, 1997)
40 (open cup, Windholz et al., 1983)

Lower explosive limit (%):
4.9 (NIOSH, 1997)

Upper explosive limit (%):
15.9 (NIOSH, 1997)

Ionization potential (eV):
10.90 (NIOSH, 1997)

Soil sorption coefficient, log K_{oc}:
Unavailable because experimental methods for estimation of this parameter for chlorinated aliphatic alcohols are lacking in the documented literature. However, its high solubility in water suggests its adsorption to soil will be nominal (Lyman et al., 1982).

Solubility in organics:
Soluble in alcohol (Weast, 1986)

Solubility in water:
Miscible (Hawley, 1981)

Vapor density:
3.29 g/L at 25 °C, 2.78 (air = 1)

Vapor pressure (mmHg):
4.9 at 20 °C (Hawley, 1981)
8 at 25 °C (quoted, Nathan, 1978)

Environmental fate:
   Biological. Heukelekian and Rand (1955) reported a 10-d BOD value of 0.50 g/g which is 50.0% of the ThOD value of 1.00 g/g.
   Chemical/Physical. Reacts with aqueous sodium bicarbonate solutions at 105 °C producing ethylene glycol (Patnaik, 1992).

Exposure limits: NIOSH REL: ceiling 1 ppm (3 mg/m^3), IDLH 7 ppm; OSHA PEL: TWA 5 ppm (16 mg/m^3); ACGIH TLV: ceiling 1 ppm (adopted).

Symptoms of exposure: Respiratory distress, paralysis, brain damage, nausea, and vomiting (Patnaik, 1992)

Toxicity:
   Acute oral LD$_{50}$ for guinea pigs 110 mg/kg, mice 81 mg/kg, rats 71 mg/kg (quoted, RTECS, 1985).

Uses: Solvent for cellulose acetate, ethylcellulose; manufacturing insecticides, ethylene oxide, and ethylene glycol; treating sweet potatoes before planting; organic synthesis (introduction of the hydroxyethyl group).

ETHYLENEDIAMINE

Synonyms: 1,2-Diaminoethane; Dimethylenediamine; **1,2-Ethanediamine**; 1,2-Ethylenediamine; NCI-C60402; UN 1604.

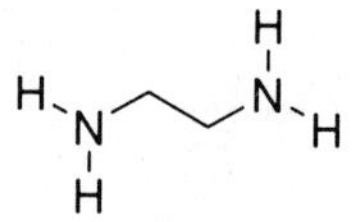

CAS Registry Number: 107-15-3
DOT: 1604
DOT label: Corrosive, flammable liquid
Molecular formula: $C_2H_8N_2$
Formula weight: 60.10
RTECS: KH8575000
Merck reference: 10, 3741

Physical state, color, and odor:
Clear, colorless, volatile, slight viscous, hygroscopic liquid with an ammonia-like odor

Melting point (°C):
8.5 (hydrated, Weast, 1986)
10 (anhydrous, Verschueren, 1983)

Boiling point (°C):
116.5 (Weast, 1986)
118 (hydrated, Verschueren, 1983)

Density (g/cm³):
0.8995 at 20/20 °C (Weast, 1986)
0.8994 at 20/4 °C (anhydrous), 0.963 at 21/4 °C (hydrated) (Verschueren, 1983)

Diffusivity in water (x 10⁻⁵ cm²/sec):
1.12 at 20 °C using method of Hayduk and Laudie (1974)

Dissociation constant (20 °C):
pK_1 = 10.075, pK_2 = 6.985 (Gordon and Ford, 1972)

Flash point (°C):
34.2 (NIOSH, 1997)
43 (Windholz et al., 1983)
66 (open cup, NFPA, 1984)

Lower explosive limit (%):
2.5 at 100 °C (NIOSH, 1997)
4.2 (NFPA, 1984)

Upper explosive limit (%):
12 at 100 °C (NIOSH, 1997)
11.4 (NFPA, 1984)

Henry's law constant (x 10^{-9} atm·m³/mol):
1.69 at 25 °C (Westheimer and Ingraham, 1956)

Ionization potential (eV):
8.60 (NIOSH, 1997)

Bioconcentration factor, log BCF:
3.94 (activated sludge), 2.87 (algae) (hydrochloride salt, Freitag et al., 1985)

Soil sorption coefficient, log K_{oc}:
Unavailable because experimental methods for estimation of this parameter for diamines are lacking in the documented literature. However, its miscibility in water suggests its adsorption to soil will be nominal (Lyman et al., 1982).

Solubility in organics:
Soluble in alcohol; slightly soluble in benzene and ether (Windholz et al., 1983).

Solubility in water:
Miscible (Price et al., 1974).

Vapor density:
2.46 g/L at 25 °C, 2.07 (air = 1)

Vapor pressure (mmHg):
10.7 at 20 °C (Sax and Lewis, 1987)
116 at 20 °C (anhydrous), 9 at 20 °C, 16 at 30 °C (hydrated) (Verschueren, 1983)

Environmental fate:
   Chemical/Physical. Absorbs carbon dioxide forming carbonates (Patnaik, 1992; Windholz et al., 1983).
   At an influent concentration of 1,000 mg/L, treatment with granular activated carbon resulted in an effluent concentration of 893 mg/L. The adsorbability of the carbon used was 21 mg/g carbon (Guisti et al., 1974).

Exposure limits: NIOSH REL: TWA 10 ppm (25 mg/m³), IDLH 1,000 ppm; OSHA PEL: TWA 10 ppm; ACGIH TLV: TWA 10 ppm (adopted).

Symptoms of exposure: Severe skin irritant producing sensitization and blistering of the skin. Liquid splashed in eyes may cause injury (Patnaik, 1992). Inhalation may cause irritation of nose and respiratory system (NIOSH, 1997).

Toxicity:
   LC_{50} (96-h) for fathead minnows 115.7 mg/L (Spehar et al., 1982).
   Acute oral LD_{50} for guinea pigs 470 mg/kg, rats 500 mg/kg (quoted, RTECS, 1985).

Uses: Stabilizing rubber latex; solvent for albumin, casein, shellac, and sulfur; neutralizing oils; in antifreeze as a corrosion inhibitor; emulsifier; adhesives; textile lubricants; fungicides; manufacturing chelating agents such as EDTA (ethylenediaminetetraacetic acid); dimethylol-ethylene-urea resins; organic synthesis.

ETHYLENE DIBROMIDE

Synonyms: Acetylene dibromide; Bromofume; Celmide; DBE; Dibromoethane; **1,2-Dibromoethane**; *sym*-Dibromoethane; α,β-Dibromoethane; Dowfume 40; Dowfume EDB; Dowfume W-8; Dowfume W-85; Dowfume W-90; Dowfume W-100; EDB; EDB-85; E-D-BEE; ENT 15349; Ethylene bromide; Ethylene bromide glycol dibromide; 1,2-Ethylene dibromide; Fumo-gas; Glycol bromide; Glycol dibromide; Iscobrome D; Kopfume; NCI-C00522; Nephis; Pestmaster; Pestmaster EDB-85; RCRA waste number U067; Soilbrom-40; Soilbrom-85; Soilbrom-90; Soilbrom-90EC; Soilbrom-100; Soilfume; UN 1605; Unifume.

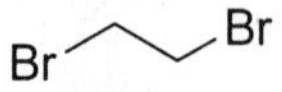

CAS Registry Number: 106-93-4
DOT: 1605
DOT label: Poison
Molecular formula: $C_2H_4Br_2$
Formula weight: 187.86
RTECS: KH9275000
Merck reference: 10, 3742

Physical state and odor:
Colorless liquid with a sweet, chloroform-like odor. Odor threshold concentration is 25 ppb (Keith and Walters, 1992).

Melting point (°C):
9.8 (Weast, 1986)

Boiling point (°C):
131.3 (Weast, 1986)
131.0 (Jones et al., 1977)

Density (g/cm³):
2.1687 at 20/4 °C (Riddick et al., 1986)
2.1688 at 25/4 °C (Dreisbach, 1959)

Diffusivity in water (x 10⁻⁵ cm²/sec):
0.96 at 20 °C using method of Hayduk and Laudie (1974)

Heat of fusion (kcal/mol):
2.616 (Riddick et al., 1986)

Henry's law constant (x 10⁻⁴ atm·m³/mol):
25.0 at 25 °C (Jafvert and Wolfe, 1987)
3.0, 4.8, 6.1, 6.5, and 8.0 at 10, 15, 20, 25, and 30 °C, respectively (Ashworth et al., 1988)
5.21 at 20 °C (Hovorka and Dohnal, 1997)

Interfacial tension with water (dyn/cm at 20 °C):
36.54 (Harkins et al., 1920)

Ionization potential (eV):
9.45 (Franklin et al., 1969)

Soil sorption coefficient, log K_{oc}:
1.64 (Kenaga and Goring, 1980)
1.56-2.21 (soil organic matter content 0.5-21.7%, Mingelgrin and Gerstl, 1983)
1.82 (average of 2 soil adsorbents, Rogers and McFarlane, 1981)
2.13 (Lockwood), 2.11 (Warehouse Point), 2.01 (Broad Brook) (Steinberg et al., 1987)

Octanol/water partition coefficient, log K_{ow}:
1.76 (Rogers and McFarlane, 1981)
1.93 (Steinberg et al., 1987)

Solubility in organics:
Soluble in acetone, alcohol, benzene, and ether (Weast, 1986)

Solubility in water:
4,321 mg/L at 20 °C (Mackay and Yeun, 1983)
3,920 mg/kg at 15 °C, 4,310 mg/kg at 30 °C (Gross and Saylor, 1931)
2,910 mg/L at 25 °C (Jones et al., 1977)
4,200 mg/L at 25 °C (quoted, Dreisbach, 1952)
3,100 g/kg at 25 °C (Tokoro et al., 1986)
At 20 °C and vapor phase concentrations of 10.6, 19.7, 28.4, 38.1, 40.7, and 48.1 mg/m³, the concentrations of ethylene dibromide in water were 468, 816, 1,095, 1,516, 1,653, and 1,930 mg/L, respectively (Call, 1957).
In wt %: 0.395 at 10.1 °C, 0.412 at 19.5 °C, 0.431 at 30.7 °C, 0.444 at 39.6 °C, 0.493 at 50.0 °C, 0.489 at 59.9 °C, 0.542 at 70.2 °C, 0.572 at 80.3 °C, 0.658 at 90.6 °C (Stephenson, 1992)

Vapor density:
7.68 g/L at 25 °C, 6.49 (air = 1)

Vapor pressure (mmHg):
11 at 20 °C (Mackay and Yeun, 1983)
11.4 at 25 °C (Hine and Mookerjee, 1975)
17.4 at 30 °C (Sax and Lewis, 1987)

Environmental fate:
Biological. Complete biodegradation by soil cultures resulted in the formation of ethylene and bromide ions (Castro and Belser, 1968). A mutant of strain *Acinetobacter* sp. GJ70 isolated from activated sludge degraded ethylene dibromide to ethylene glycol and bromide ions (Janssen et al., 1987). When *Methanococcus thermolithotrophicus*, *Methanococcus deltae*, and *Methanobacterium thermoautotrophicum* were grown with H_2-CO_2 in the presence of ethylene dibromide, methane and ethylene were produced (Belay and Daniels, 1987).

In a shallow aquifer material, ethylene dibromide aerobically degraded to carbon dioxide, microbial biomass, and nonvolatile water-soluble compound(s) (Pignatello, 1987).

Soil. In soil and water, chemical and biological mediated reactions can transform ethylene dibromide in the presence of hydrogen sulfides to ethyl mercaptan and other sulfur-containing compounds (Alexander, 1981).

Groundwater. According to the U.S. EPA (1986), ethylene dibromide has a high potential to leach to groundwater.

Chemical/Physical. In an aqueous phosphate buffer solution (0.05 M) containing hydrogen sulfide ion, ethylene dibromide was transformed into 1,2-dithioethane and vinyl bromide. The hydrolysis half-lives for solutions with and without sulfides present ranged from 37-70 d and

0.8-4.6 yr, respectively (Barbash and Reinhard, 1989). Dehydrobromination of ethylene dibromide to vinyl bromide was observed in various aqueous buffer solutions (pH 7-11) over the temperature range of 45 to 90 °C. The estimated half-life for this reaction at 25 °C and pH 7 is 2.5 yr (Vogel and Reinhard, 1986).

Ethylene dibromide may hydrolyze via two pathways. In the first pathway, ethylene dibromide undergoes nucleophilic attack at the carbon-bromine bond by water forming hydrogen bromide and 2-bromoethanol. The alcohol may react further through the formation of ethylene oxide forming ethylene glycol (Kollig, 1993; Leinster et al., 1978). In the second pathway, dehydrobromination of ethylene dibromide to vinyl bromide was observed in various aqueous buffer solutions (pH 7 to 11) over the temperature range of 45 to 90 °C. The estimated hydrolysis half-life for this reaction at 25 °C and pH 7 was 2.5 yr (Vogel and Reinhard, 1986).

The hydrolysis rate constant for ethylene dibromide at pH 7 and 25 °C was determined to be 9.9×10^{-6}/h, resulting in a half-life of 8.0 yr (Ellington et al., 1988). At pH 5 and temperatures of 30, 45, and 60 °C, the hydrolysis half-lives were 180, 29, and 9 d, respectively. When the pH was raised to pH 7, the half-lives increased slightly to 410, 57, and 11 d at temperatures of 30, 45, and 60 °C, respectively. At pH 9, the hydrolysis half-lives were nearly identical to those determined under acidic conditions (Ellington et al., 1986).

Anticipated products from the reaction of ethylene dibromide with ozone or OH radicals in the atmosphere are bromoacetaldehyde, formaldehyde, bromoformaldehyde, and bromide radicals (Cupitt, 1980).

Rajagopal and Burris (1999) studied the degradation reaction of ethylene dibromide in water with zero-valent iron. Ethylene dibromide degraded rapidly following pseudo-first order kinetics. The observed end products were ethylene and bromide ions and were probably formed via a β-elimination pathway. Bromoethane and vinyl bromide were not observed as possible intermediate products.

Emits toxic bromide fumes when heated to decomposition (Lewis, 1990).

Exposure limits: Potential occupational carcinogen. NIOSH REL: TWA 45 ppb, 15-min ceiling 130 ppb, IDLH 100 ppm; OSHA PEL: TWA 20 ppm, ceiling 30 ppm, 5-min peak 50 ppm.

Symptoms of exposure: Irritation of the respiratory system, eyes; dermatitis vesiculation (Patnaik, 1992)

Toxicity:

Adams et al. (1988) reported that the LC_{50} for chronic exposure occurred within 12 h for three laboratory-reared octopuses (*Octopus joubini, Octopus maya, Octopus bimaculoides*).

LC_{50} (96-h) for Japanese medaka (*Oryzias latipes*) 32.1 mg/L (Holcombe et al., 1995).

LC_{50} (48-h) for bluegill sunfish 18 mg/L (Davis and Hardcastle, 1959).

Acute oral LD_{50} for quail 130 mg/kg, rats 108 mg/kg, rabbits 55 mg/kg (quoted, RTECS, 1985).

Drinking water standard (final): MCLG: zero; MCL: 0.05 μg/L (U.S. EPA, 1996).

Uses: Lead scavenger in unleaded gasoline; grain and fruit fumigant; waterproofing preparations; insecticide; medicines; general solvent; organic synthesis.

ETHYLENIMINE

Synonyms: Aminoethylene; Azacyclopropane; Azirane; **Aziridine**; Dihydro-1*H*-azirine; Dihydroazirine; Dimethyleneimine; Dimethylenimine; EI; ENT 50324; Ethyleneimine; Ethylimine; RCRA waste number P054; TL 337; UN 1185.

CAS Registry Number: 151-56-4
DOT: 1185
DOT label: Flammable liquid and poison
Molecular formula: C_2H_5N
Formula weight: 43.07
RTECS: KX5075000
Merck reference: 10, 3748

Physical state, color, and odor:
Colorless liquid with a very strong ammonia odor. Odor threshold concentration is 1.5 ppm (Amoore and Hautala, 1983).

Melting point (°C):
-71.5 (Sax and Lewis, 1987)

Boiling point (°C):
57 (Hawley, 1981)

Density (g/cm³):
0.8321 at 20/4 °C (Weast, 1986)

Diffusivity in water (x 10^{-5} cm²/sec):
1.30 at 20 °C using method of Hayduk and Laudie (1974)

Dissociation constant, pK_a:
8.04 (HSDB, 1989)

Flash point (°C):
-11.21 (NIOSH, 1997)

Lower explosive limit (%):
3.3 (NIOSH, 1997)
3.6 (NFPA, 1984)

Upper explosive limit (%):
54.8 (NIOSH, 1997)

Henry's law constant (x 10^{-7} atm·m³/mol):
1.33 at 25 °C (Mercer et al., 1990)

Ionization potential (eV):
9.20 (NIOSH, 1997)
9.9 (Scala and Salomon, 1976)

Soil sorption coefficient, log K_{oc}:
0.11 (calculated, Mercer et al., 1990)

Octanol/water partition coefficient, log K_{ow}:
-1.01 (Mercer et al., 1990)

Solubility in organics:
Soluble in acetone, alcohol, benzene, and ether (Weast, 1986)

Solubility in water:
Miscible (NIOSH, 1997).

Vapor density:
1.76 g/L at 25 °C, 1.49 (air = 1)

Vapor pressure (mmHg):
160 at 20 °C, 250 at 30 °C (Verschueren, 1983)

Environmental fate:
   Photolytic. The vacuum UV photolysis (λ = 147 nm) and γ radiolysis of ethylenimine resulted in the formation of acetylene, methane, ethane, ethylene, hydrogen cyanide, methyl radicals, and hydrogen (Scala and Salomon, 1976). Photolysis of ethylenimine vapor at krypton and xenon lines yielded ethylene, ethane, methane, acetylene, propane, butane, hydrogen, ammonia, ethylenimino radicals (Iwasaki et al., 1973).
   Chemical/Physical. Polymerizes easily (Windholz et al., 1983). Hydrolyzes in water forming ethanolamine (HSDB, 1989). The estimated hydrolysis half-life in water at 25 °C and pH 7 is 154 d (Mabey and Mill, 1978).

Exposure limits: Potential occupational carcinogen. NIOSH REL: IDLH 100 ppm; ACGIH TLV: TWA 0.5 ppm (1 mg/m³).
   OSHA recommends that worker exposure to this chemical is to be controlled by use of engineering control, proper work practices, and proper selection of personal protective equipment. Specific details of these requirements can be found in CFR 1910.1003-1910.1016.

Symptoms of exposure: Severe irritation of the skin, eyes, and mucous membranes. Eye contact with liquid may cause corneal opacity and loss of vision. Inhalation of vapors may cause eye, nose, throat irritation, and breathing difficulties (Patnaik, 1992). An irritation concentration of 200.00 mg/m³ in air was reported by Ruth (1986).

Toxicity:
   Acute oral LD_{50} for rats 15 mg/kg (quoted, RTECS, 1985).

Uses: Manufacture of triethylenemelamine and other amines; fuel oil and lubricant refining; ion exchange; protective coatings; adhesives; pharmaceuticals; polymer stabilizers; surfactants.

ETHYL ETHER

Synonyms: Aether; Anaesthetic ether; Anesthesia ether; Anesthetic ether; Diethyl ether; Diethyl oxide; Ether; Ethoxyethane; Ethyl oxide; **1,1′-Oxybis(ethane)**; RCRA waste number U117; Solvent ether; Sulfuric ether; UN 1155.

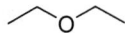

CAS Registry Number: 60-29-7
DOT: 1155
DOT label: Flammable liquid
Molecular formula: $C_4H_{10}O$
Formula weight: 74.12
RTECS: KI5775000
Merck reference: 10, 3751

Physical state, color, and odor:
Colorless, hygroscopic, volatile liquid with a sweet, pungent odor. Odor threshold concentration is 330 ppb (Keith and Walters, 1992).

Melting point (°C):
-116.2 (stable form, Weast, 1986)
-123 (metastable form, Verschueren, 1983)

Boiling point (°C):
34.5 (Weast, 1986)
34.43 (Boublik et al., 1984)

Density (g/cm³):
0.7138 at 20/4 °C (Weast, 1986)
0.79125 at 15/4 °C, 0.70205 at 30/4 °C (Huntress and Mulliken, 1941)
0.71361 at 20/4 °C, 0.70782 at 25/4 °C (Riddick et al., 1986)

Diffusivity in water (x 10⁻⁵ cm²/sec):
0.86 at 20 °C using method of Hayduk and Laudie (1974)

Flash point (°C):
9.5 (NIOSH, 1997)

Lower explosive limit (%):
1.9 (NIOSH, 1997)

Upper explosive limit (%):
36.0 (NIOSH, 1997)

Heat of fusion (kcal/mol):
1.745 (Dean, 1987)

Henry's law constant (x 10⁻⁴ atm·m³/mol at 25 °C):
8.33 (Nielsen et al., 1994)
12.50 (Signer et al., 1969)

Interfacial tension with water (dyn/cm at 20 °C):
10.70 (Harkins et al., 1920)

Ionization potential (eV):
9.53 (NIOSH, 1997)

Soil sorption coefficient, log K_{oc}:
Unavailable because experimental methods for estimation of this parameter for ethers are lacking in the documented literature. However, its high solubility in water suggests its adsorption to soil will be nominal (Lyman et al., 1982).

Octanol/water partition coefficient, log K_{ow}:
0.77 (Leo et al., 1971)
0.89 (Hansch et al., 1975)
0.83 (Collander, 1951)

Solubility in organics:
Soluble in acetone (Weast, 1986). Miscible with lower aliphatic alcohols, benzene, chloroform, petroleum ether, and many oils (Windholz et al., 1983).

Solubility in water:
0.632, 1.010, and 1.2 mol/L at 38, 20, and 25 °C, respectively (Fischer and Ehrenberg, 1948)
6.80 wt % at 20 °C (Palit, 1947)
9.01, 7.95, 6.87, and 6.03 wt % at 10, 15, 20, and 25 °C, respectively (Bennett and Philip, 1928)
In wt %: 11.668 at 0 °C, 9.040 at 10 °C, 7.913 at 15 °C, 6.896 at 20 °C, 6.027 at 25 °C, 5.340 at 30 °C (Hill, 1923)
64 g/kg at 25 °C (Butler and Ramchandani, 1935)
0.80 mol/L at 25 °C (Hine and Weimar, 1965)

Vapor density:
3.03 g/L at 25 °C, 2.55 (air = 1)

Vapor pressure (mmHg):
442 at 20 °C (Verschueren, 1983)
439.8 at 20 °C (Windholz et al., 1983)
537 at 25 °C (Butler and Ramchandani, 1935)

Environmental fate:
Photolytic. The rate constant for the reaction of ethyl ether and OH radicals in the atmosphere at 300 K is 5.4 x 10^{12} cm³/mol·sec (Hendry and Kenley, 1979).
Chemical/Physical. The atmospheric oxidation of ethyl ether by OH radicals in the presence of nitric oxide yielded ethyl formate as the major product. Minor products included formaldehyde and nitrogen dioxide (Wallington and Japar, 1991).
Ethyl ether will not hydrolyze (Kollig, 1993).

Exposure limits: NIOSH REL: IDLH 1,900 ppm; OSHA PEL: TWA 400 ppm (1,200 mg/m³); ACGIH TLV: TWA 400 ppm, STEL 500 ppm (adopted).

Symptoms of exposure: Narcotic at high concentrations and a mild irritant to eyes, nose, and skin (Patnaik, 1992)

Toxicity:

LC_{50} (14-d) for *Poecilia reticulata* 2,138 mg/L (Könemann, 1981).

LC_{50} (96-h static bioassay) for bluegill sunfish and *Menidia beryllina* >10,000 mg/L (quoted, Verschueren, 1983).

Acute oral LD_{50} for rats is 1,215 mg/kg (quoted, RTECS, 1985).

Uses: Solvent for oils, waxes, perfumes, alkaloids, fats, and gums; organic synthesis (Grignard and Wurtz reactions); extractant; manufacture of gun powder, ethylene, and other organic compounds; analytical chemistry; perfumery; alcohol denaturant; primer for gasoline engines; anesthetic.

ETHYL FORMATE

Synonyms: Areginal; Ethyl formic ester; Ethyl methanoate; **Formic acid ethyl ester**; Formic ether; UN 1190.

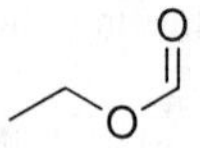

CAS Registry Number: 109-94-4
DOT: 1190
DOT label: Flammable liquid
Molecular formula: $C_3H_6O_2$
Formula weight: 74.08
RTECS: LQ8400000
Merck reference: 10, 3752

Physical state, color, and odor:
Colorless, clear liquid with a pleasant, fruity odor. Odor threshold concentration is 31 ppm (Amoore and Hautala, 1983).

Melting point (°C):
-80.5 (Weast, 1986)
-79 (Verschueren, 1983)

Boiling point (°C):
54.5 (Weast, 1986)
53-54 (Windholz et al., 1983)

Density (g/cm³):
0.9168 at 20/4 °C (Weast, 1986)
0.9208 at 25.00/4 °C (Emmerling et al., 1998)

Diffusivity in water (x 10⁻⁵ cm²/sec):
1.00 at 20 °C using method of Hayduk and Laudie (1974)

Flash point (°C):
-20 (Windholz et al., 1983)

Lower explosive limit (%):
2.8 (NFPA, 1984)

Upper explosive limit (%):
16.0 (NFPA, 1984)

Heat of fusion (kcal/mol):
2.20 (Dean, 1987)

Henry's law constant (x 10⁻³ atm·m³/mol):
5.00 at 25 °C (Hartkopf and Karger, 1973)

Ionization potential (eV):
10.61 ± 0.01 (Franklin et al., 1969)

Soil sorption coefficient, log K_{oc}:
Unavailable because experimental methods for estimation of this parameter for esters are lacking in the documented literature. However, its high solubility in water suggests its adsorption to soil will be nominal (Lyman et al., 1982).

Octanol/water partition coefficient, log K_{ow}:
0.33 (Leo et al., 1971)

Solubility in organics:
Miscible with alcohol, benzene, and ether (Hawley, 1981)

Solubility in water:
9 wt % at 17.9 °C (NIOSH, 1997)
105.0 g/L at 20 °C, 118.0 g/L at 25 °C (quoted, Verschueren, 1983)

Vapor density:
3.03 g/L at 25 °C, 2.56 (air = 1)

Vapor pressure (mmHg):
192 at 20 °C, 300 at 30 °C (Verschueren, 1983)

Environmental fate:
   Photolytic. Reported rate constants for the reaction of ethyl formate and OH radicals in the atmosphere (296 K) and aqueous solution are 1.02×10^{-11} and 6.5×10^{-13} cm^3/molecule·sec, respectively (Wallington et al., 1988b).
   Chemical/Physical. Slowly hydrolyzes in water forming formic acid and ethanol (Windholz et al., 1983).

Exposure limits: NIOSH REL: TWA 100 ppm (300 mg/m^3), IDLH 1,500 ppm; OSHA PEL: TWA 100 ppm; ACGIH TLV: TWA 100 ppm (adopted).

Symptoms of exposure: May irritate eyes and nose (Patnaik, 1992)

Toxicity:
   Acute oral LD_{50} for guinea pigs 1,110 mg/kg, rats 1,850 mg/kg, rabbits 2,075 mg/kg (quoted, RTECS, 1985).

Uses: Solvent for nitrocellulose and cellulose acetate; artificial flavor for lemonades and essences; fungicide and larvacide for cereals, tobacco, dried fruits; acetone substitute; organic synthesis.

ETHYL MERCAPTAN

Synonyms: 2-Aminoethanethiol; **Ethanethiol**; Ethyl hydrosulfide; Ethyl sulfhydrate; Ethyl thioalcohol; LPG ethyl mercaptan 1010; UN 2363; Thioethanol; Thioethyl alcohol.

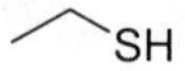

$$\diagup\!\!\!\diagdown\text{SH}$$

CAS Registry Number: 75-08-1
DOT: 2363
DOT label: Flammable liquid
Molecular formula: C_2H_6S
Formula weight: 62.13
RTECS: KI9625000
Merck reference: 10, 3674

Physical state, color, and odor:
Colorless liquid with a strong skunk-like odor. Odor threshold concentration is 0.76 ppb (Amoore and Hautala, 1983).

Melting point (°C):
-144.4 (Weast, 1986)
-147 (Sax and Lewis, 1987)

Boiling point (°C):
35 (Weast, 1986)

Density (g/cm³):
0.8391 at 20/4 °C (Weast, 1986)
0.83147 at 25/4 °C (Windholz et al., 1983)

Diffusivity in water (x 10⁻⁵ cm²/sec):
1.05 at 20 °C using method of Hayduk and Laudie (1974)

Dissociation constant, pK_a:
10.50 at 20 °C (Dean, 1973)

Flash point (°C):
-48.7 (NIOSH, 1997)

Lower explosive limit (%):
2.8 (NIOSH, 1997)

Upper explosive limit (%):
18.0 (NIOSH, 1997)

Heat of fusion (kcal/mol):
1.189 (Dean, 1987)

Henry's law constant (x 10⁻³ atm·m³/mol):
3.57 at 25 °C (Przyjazny et al., 1983)

Interfacial tension with water (dyn/cm at 20 °C):
26.12 (Harkins et al., 1920)

Ionization potential (eV):
9.285 ± 0.005 (Franklin et al., 1969)

Soil sorption coefficient, log K_{oc}:
Unavailable because experimental methods for estimation of this parameter for mercaptans are lacking in the documented literature. However, its high solubility in water suggests its adsorption to soil will be nominal (Lyman et al., 1982).

Octanol/water partition coefficient, log K_{ow}:
1.49 using method of Hansch et al. (1968)

Solubility in organics:
Soluble in acetone, alcohol, and ether (Weast, 1986)

Solubility in water:
6.76 g/L at 20 °C (quoted, Windholz et al., 1983)

Vapor density:
2.54 g/L at 25 °C, 2.14 (air = 1)

Vapor pressure (mmHg):
440 at 20 °C, 640 at 30 °C (Verschueren, 1983)
527.2 at 25 °C (Wilhoit and Zwolinski, 1971)

Environmental fate:
Photolytic. A second-order rate constant of 1.21 x 10^{-12} cm^3/molecule·sec was reported for the reaction of ethyl mercaptan and NO_3 radicals in the atmosphere at 297 K (Atkinson, 1991).
   Chemical/Physical. In the presence of nitric oxide, ethyl mercaptan reacted with OH radical to give ethyl thionitrite (MacLeod et al., 1984).

Exposure limits: NIOSH REL: 15-min ceiling 0.5 ppm (1.3 mg/m^3), IDLH 500 ppm; OSHA PEL: ceiling 10 ppm (25 mg/m^3); ACGIH TLV: TWA 0.5 ppm (adopted).

Symptoms of exposure: May produce irritation of the nose and throat, headache, and fatigue (Patnaik, 1992)

Toxicity:
   Acute oral LD_{50} for rats 682 mg/kg (quoted, RTECS, 1985).

Uses: Odorant for natural gas; manufacturing of plastics, antioxidants, pesticides; adhesive stabilizer; chemical intermediate.

4-ETHYLMORPHOLINE

Synonym: *N*-Ethylmorpholine.

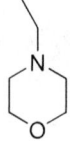

CAS Registry Number: 100-74-3
Molecular formula: $C_6H_{13}NO$
Formula weight: 115.18
RTECS: QE4025000

Physical state, color, and odor:
Colorless liquid with an ammonia-like odor. Odor threshold concentration is 1.4 ppm (Amoore and Hautala, 1983).

Melting point (°C):
-65 to -63 (Verschueren, 1983)

Boiling point (°C):
138-139 at 763 mmHg (Weast, 1986)

Density (g/cm³ at 20/4 °C):
0.9886 (Weast, 1986)
0.905 (Aldrich, 1988)

Diffusivity in water (x 10^{-5} cm²/sec):
0.81 at 20 °C using method of Hayduk and Laudie (1974)

Flash point (°C):
32.5 (open cup, NIOSH, 1997)
27 (Aldrich, 1988)

Soil sorption coefficient, log K_{oc}:
Unavailable because experimental methods for estimation of this parameter for substituted morpholines are lacking in the documented literature. However, its high solubility in water suggests its adsorption to soil will be nominal (Lyman et al., 1982).

Solubility in organics:
Soluble in acetone, alcohol, benzene, and ether (Weast, 1986)

Solubility in water:
Miscible (Hawley, 1981).

Vapor density:
4.71 g/L at 25 °C, 3.98 (air = 1)

Vapor pressure (mmHg at 20 °C):
6 (NIOSH, 1997)
6.1 (Verschueren, 1983)

Environmental fate:

Chemical/Physical. Releases toxic nitrogen oxides when heated to decomposition (Sax and Lewis, 1987).

At an influent concentration of 1,000 mg/L, treatment with granular activated carbon resulted in an effluent concentration of 467 mg/L. The adsorbability of the carbon used was 107 mg/g carbon (Guisti et al., 1974).

Exposure limits: NIOSH REL: TWA 5 ppm (23 mg/m^3), IDLH 100 ppm; OSHA PEL: TWA 20 ppm (94 mg/m^3); ACGIH TLV: TWA 5 ppm (adopted).

Toxicity:

LC_{50} (inhalation) for mice 18,000 mg/m^3/2-h (quoted, RTECS, 1985).

Acute oral LD_{50} for rats 1,780 mg/kg, mice 1,200 mg/kg (quoted, RTECS, 1985).

Uses: Intermediate for pharmaceuticals, dyestuffs, emulsifying agents, and rubber accelerators; solvent for dyes, resins, and oils; catalyst for making polyurethane foams.

2-ETHYLTHIOPHENE

Synonyms: None.

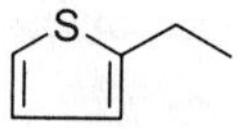

CAS Registry Number: 872-55-9
Molecular formula: C_6H_8S
Formula weight: 112.19

Physical state and odor:
Liquid with a pungent odor

Boiling point (°C):
134 (Wilhoit and Zwolinski, 1971)

Density (g/cm³):
0.9930 at 20/4 °C (Weast, 1986)

Diffusivity in water (x 10⁻⁵ cm²/sec):
0.82 at 20 °C using method of Hayduk and Laudie (1974)

Flash point (°C):
21 (Aldrich, 1988)

Ionization potential (eV):
8.8 ± 0.2 (Franklin et al., 1969)

Soil sorption coefficient, log K_{oc}:
Unavailable because experimental methods for estimation of this parameter for substituted thiophenes are lacking in the documented literature

Octanol/water partition coefficient, log K_{ow}:
2.83 using method of Hansch et al. (1968)

Solubility in organics:
Soluble in alcohol and ether (Weast, 1986)

Solubility in water:
292 mg/kg at 25 °C (Price, 1976)

Vapor density:
4.59 g/L at 25 °C, 3.87 (air = 1)

Vapor pressure (mmHg):
60.9 at 60.3 °C, 92.2 at 70.3 °C, 136 at 80.3 °C, 197 at 90.3 °C, 280 at 100.3 °C (Eon et al., 1971)

Use: Ingredient in crude petroleum.

FLUORANTHENE

Synonyms: 1,2-Benzacenaphthene; Benzo[*jk*]fluorene; Idryl; 1,2-(1,8-Naphthylene)benzene; 1,2-(1,8-Naphthalenediyl)benzene; RCRA waste number U120.

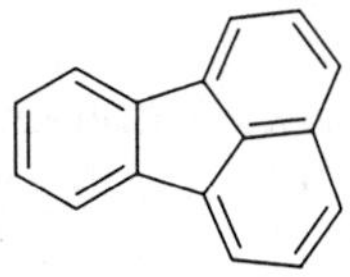

CAS Registry Number: 206-44-0
Molecular formula: $C_{16}H_{10}$
Formula weight: 202.26
RTECS: LL4025000

Physical state and color:
Colorless to light yellow crystals

Melting point (°C):
107 (Verschueren, 1983)
109-110 (Fluka, 1988)

Boiling point (°C):
375 (Weast, 1986)
384 (Aldrich, 1988)
367 (Sax, 1984)

Density (g/cm³):
1.252 at 0/4 °C (Weast, 1986)

Diffusivity in water (x 10^{-5} cm²/sec):
0.56 at 20 °C using method of Hayduk and Laudie (1974)

Dissociation constant, pK_a:
>15 (Christensen et al., 1975)

Entropy of fusion (cal/mol·K):
11.8 (Yalkowsky, 1981)

Henry's law constant (x 10^{-6} atm·m³/mol):
10.90 at 25 °C (de Maagd et al., 1998)
2.57, 6.32, 16.09, 23.49, 57.64, and 61.48 at 10.0, 20.0, 35.0, 40.1, 45.0, and 55.0 °C, respectively (ten Hulscher et al., 1992)
5.53, 8.59, 13.0, 19.3, and 26.8 at 4.1, 11.0, 18.0, 25.0, and 31.0 °C, respectively (Bamford et al., 1998)

Ionization potential (eV):
8.54 (Franklin et al., 1969)

Bioconcentration factor, log BCF:
3.24 (*Daphnia magna*, Newsted and Giesy, 1987)
1.08 (*Polychaete* sp.), 0.76 (*Capitella capitata*) (Bayona et al., 1991)

Apparent values of 4.0 (wet wt) and 5.7 (lipid wt) for freshwater isopods including *Asellus aquaticus* (L.) (van Hattum et al., 1998)

Soil sorption coefficient, log K_{oc}:
4.62, 4.74 (RP-HPLC immobilized humic acids, Szabo et al., 1990)
4.71 (fine aquifer sand, Abdul et al., 1987)
6.38 (average, Kayal and Connell, 1990)
5.26, 5.35, 6.26, 6.60, 7.29 (San Francisco, CA mudflat sediments, Maruya et al., 1996)
6.5 (average value using 8 river bed sediments from the Netherlands, van Hattum et al., 1998)
4.51 (Brown's Lake silty clay), 5.05 (Hamlet City Lake silty sand), 4.16 (Vicksburg, MS silt) (Brannon et al., 1995)
Average K_d values for sorption of fluoranthene to corundum (α-Al_2O_3) and hematite (α-Fe_2O_3) were 0.181 and 0.788 mL/g, respectively (Mader et al., 1997)

Octanol/water partition coefficient, log K_{ow}:
5.22 (Bruggeman et al., 1982)
5.20 (Yoshida et al., 1983)
5.23 at 25 °C (de Maagd et al., 1998)
5.148 (Brooke et al., 1990)
5.155 (Brooke et al., 1990; de Bruijn et al., 1989)

Solubility in organics:
Soluble in acetic acid, benzene, chloroform, carbon disulfide, ethanol, and ether (U.S. EPA, 1985)

Solubility in water:
265 μg/L at 25 °C (Klevens, 1950; Harrison et al., 1975)
260 μg/L at 25 °C (Mackay and Shiu, 1977)
206 and 264 μg/kg at 25 and 29 °C, respectively (May et al., 1978a)
236 μg/L at 25 °C (Schwarz and Wasik, 1976)
240 μg/L at 27 °C (Davis et al., 1942)
133, 275 μg/L at 15 °C, 166 μg/L at 20 °C, 222, 373 μg/L at 25 °C (Kishi and Hashimoto, 1989)
199 μg/L at 25 °C (Walters and Luthy, 1984)
177 μg/L at 25 °C (Vadas et al., 1991)
207 μg/L at 25 °C (de Maagd et al., 1998)
1.4 μmol/L at 25 °C (Akiyoshi et al., 1987)
In mole fraction (x 10^{-8}): 0.7304 at 8.10 °C, 0.9531 at 13.20 °C, 1.321 at 19.70 °C, 1.805 at 24.60 °C, 2.488 at 29.90 °C (May et al., 1983)

Vapor pressure (x 10^{-6} mmHg):
9.23 at 25 °C (Sonnefeld et al., 1983; Wasik et al., 1983)
50 at 25 °C (Bidleman, 1984)
71.6, 116 at 25 °C (Hinckley et al., 1990)

Environmental fate:
Biological. Contaminated soil from a manufactured coal gas plant that had been exposed to crude oil was spiked with fluoranthene (400 mg/kg soil) to which Fenton's reagent (5 mL 2.8 M hydrogen peroxide; 5 mL 0.1 M ferrous sulfate) was added. The treated and nontreated soil samples were incubated at 20 °C for 56 d. Fenton's reagent did not enhance the mineralization of fluoranthene by indigenous microorganisms. The amount of fluoranthene recovered as

carbon dioxide after treatment with and without Fenton's reagent was 20% in both instances. Pretreatment of the soil with a surfactant (10 mM sodium dodecylsulfate) before addition of Fenton's reagent, however, increased the mineralization rate 83% as compared to nontreated soil (Martens and Frankenberger, 1995).

Two pure bacterial strains, *Pasteurella* sp. IFA and *Mycobacterium* sp. PYR-1, degraded fluoranthene in an aqueous medium during 2 wk of incubation at room temperature. Degradation yields of 24 and 46% were observed using *Pasteurella* sp. IFA and *Mycobacterium* sp. PYR-1, respectively. Stable metabolic products identified were 9-fluorenone-1-carboxylic acid, 9-fluorenone, 9-hydroxyfluorene, 9-hydroxy-1-fluorenecarboxylic acid, 2-carboxybenzaldehyde, benzoic acid, and phenyl acetic acid (Sepic et al., 1998).

Soil. The reported half-lives for anthracene in a Kidman sandy loam and McLaurin sandy loam are 377 and 268 d, respectively (Park et al., 1990).

Surface Water. In a 5-m deep surface water body, the calculated half-lives for direct photochemical transformation at 40 °N latitude in the midsummer during midday were 160 and 200 d with and without sediment-water partitioning, respectively (Zepp and Schlotzhauer, 1979).

Photolytic. When an aqueous solution containing fluoranthene was photooxidized by UV light at 90-95 °C, 25, 50, and 75% degraded to carbon dioxide after 75.3, 160.6, and 297.4 h, respectively (Knoevenagel and Himmelreich, 1976).

Behymer and Hites (1985) determined the effect of different substrates on the rate of photooxidation of fluoranthene using a rotary photoreactor. The photolytic half-lives of fluoranthene using silica gel, alumina, and fly ash were 74, 23, and 44 h, respectively. Fluorene reacts with photochemically produced OH radicals in the atmosphere. The atmospheric half-life was estimated to range from 2.02 to 20.2 h (Atkinson, 1987).

Chemical/Physical. 2-Nitrofluoranthene was the principal product formed from the gas-phase reaction of fluoranthene with OH radicals in a NO_x atmosphere. Minor products found include 7- and 8-nitrofluoranthene (Arey et al., 1986). The reaction of fluoranthene with NO_x to form 3-nitrofluoranthene was reported to occur in urban air from St. Louis, MO (Randahl et al., 1982). Chlorination of fluoranthene in polluted, humus-poor lake water gave a large number of mono-, di-, and trichlorofluoranthene derivatives (Johnsen et al., 1989). At pH <4, chlorination of fluoranthene produced 3-chlorofluoranthene as the major product (Oyler et al., 1983). It was suggested that the chlorination of fluoranthene in tap water accounted for the presence of chloro- and dichlorofluoranthenes (Shiraishi et al., 1985).

Complete combustion in air yields carbon dioxide and water vapor. Fluoranthene will not hydrolyze because it has no hydrolyzable functional group (Kollig, 1993).

Exposure limits: Potential occupational carcinogen. No individual standards have been set; however, as a constituent in coal tar pitch volatiles, the following exposure limits have been established (mg/m^3): NIOSH REL: TWA 0.1 (cyclohexane-extractable fraction), IDLH 80; OSHA PEL: TWA 0.2 (benzene-soluble fraction); ACGIH TLV: TWA 0.2 (benzene solubles).

Toxicity:

LC_{50} (contact) for earthworm (*Eisenia fetida*) 2,160 $\mu g/cm^2$ (Neuhauser et al., 1985).

LC_{50} (10-d) for *Rhepoxynius abronius* 3.31 mg/g organic carbon (Swartz et al., 1997).

LC_{50} (96-h) for bluegill sunfish 4.0 mg/L (Spehar et al., 1982), *Cyprinodon variegatus* >560 ppm using natural seawater (Heitmuller et al., 1981).

LC_{50} (72-h) for *Cyprinodon variegatus* >560 ppm (Heitmuller et al., 1981).

LC_{50} (48-h) for *Daphnia magna* 320 mg/L (LeBlanc, 1980), *Cyprinodon variegatus* >560 ppm (Heitmuller et al., 1981).

LC_{50} (24-h) for *Daphnia magna* 1,300 mg/L (LeBlanc, 1980), *Cyprinodon variegatus* >560 ppm (Heitmuller et al., 1981).

Acute oral LD_{50} for rats 2,000 mg/kg (quoted, RTECS, 1985).

Heitmuller et al. (1981) reported a NOEC of 560 ppm.

Source: Detected in groundwater beneath a former coal gasification plant in Seattle, WA at a concentration of 50 μg/L (ASTR, 1995). Detected in 8 diesel fuels at concentrations ranging from 0.060 to 13 mg/L with a mean value of 0.113 mg/L (Westerholm and Li, 1994); in a distilled water-soluble fraction of used motor oil at a concentration range of 1.3 to 1.5 μg/L (Chen et al., 1994). Lee et al. (1992) reported concentration ranges 1.50-125 mg/L and ND-0.5 μg/L in diesel fuel and the corresponding aqueous phase (distilled water), respectively (Lee et al., 1992). Schauer et al. (1999) reported fluoranthene in a diesel-powered medium-duty truck exhaust at an emission rate of 53.0 μg/km. Identified in Kuwait and South Louisiana crude oils at concentrations of 2.9 and 5.0 ppm, respectively (Pancirov and Brown, 1975).

Based on analysis of 7 coal tar samples, fluoranthene concentrations ranged from 1,500 to 13,000 ppm (EPRI, 1990).

Fluoranthene was detected in soot generated from underventilated combustion of natural gas doped with toluene (3 mole %) (Tolocka and Miller, 1995).

Use: Research chemical.

FLUORENE

Synonyms: 2,3-Benzindene; *o*-Biphenylenemethane; *o*-Biphenylmethane; Diphenylenemethane; *o*-Diphenylenemethane; **9H-Fluorene**; 2,2′-Methylenebiphenyl.

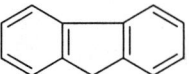

CAS Registry Number: 86-73-7
Molecular formula: $C_{13}H_{10}$
Formula weight: 166.22
RTECS: LL5670000
Merck reference: 10, 4056

Physical state and color:
Small white leaflets or flakes. Fluorescent when impure.

Melting point (°C):
116-117 (Weast, 1986)
114.0 (Pearlman et al., 1984)

Boiling point (°C):
298 (Aldrich, 1988)
294 (Huntress and Mulliken, 1941)

Density (g/cm³):
1.203 at 0/4 °C (Weast, 1986)

Diffusivity in water (x 10^{-5} cm²/sec):
0.61 at 20 °C using method of Hayduk and Laudie (1974)

Dissociation constant, pK_a:
>15 (Christensen et al., 1975)

Entropy of fusion (cal/mol·K):
12.1 (Wauchope and Getzen, 1972)
12.2 (Mackay and Shiu, 1977)

Heat of fusion (kcal/mol):
4.68 (Osborn and Douslin, 1975)
4.67 (Wauchope and Getzen, 1972)

Henry's law constant (x 10^{-5} atm·m³/mol):
21 (Petrasek et al., 1983)
6.3 at 25 °C (Fendinger and Glotfelty, 1990)
6.42 at 25 °C (de Maagd et al., 1998)
9.62 at 25 °C (Shiu and Mackay, 1997)
2.02, 3.49, 5.88, 9.68, and 14.6 at 4.1, 11.0, 18.0, 25.0, and 31.0 °C, respectively (Bamford et al., 1998)

Ionization potential (eV):
8.63 (Franklin et al., 1969)
8.56 (Yoshida et al., 1983)

Bioconcentration factor, log BCF:
2.70 (*Daphnia magna*, Newsted and Giesy, 1987)

Soil sorption coefficient, log K_{oc}:
3.60, 3.66, 3.75 (aquifer material, Abdul et al., 1987)
3.85 (Meylan et al., 1992)
3.95 (humic materials) (Carter and Suffet, 1983)
5.47 (average, Kayal and Connell, 1990)
4.15, 4.21 (RP-HPLC immobilized humic acids, Szabo et al., 1990)
4.16, 4.30, 4.39, 5.28, 5.33 (San Francisco, CA mudflat sediments, Maruya et al., 1996)

Octanol/water partition coefficient, log K_{ow}:
4.12 (Chou and Jurs, 1979)
4.18 (Bruggeman et al., 1982; Hansch et al., 1972; Yalkowsky et al., 1983a)

Solubility in organics:
Soluble in most solvents (U.S. EPA, 1985) including glacial acetic acid, carbon disulfide, ether, and benzene (Windholz et al., 1983)

Solubility in water:
1.685 mg/kg at 25 °C (May et al., 1978a)
1.98 mg/L at 25 °C (Mackay and Shiu, 1977)
1.6622 mg/L at 25 °C (Sahyun, 1966)
1.68 mg/L at 25 °C (Wasik et al., 1983)
In mg/kg: 1.87, 1.88, 1.93 at 24.6 °C, 2.33, 2.34, 2.41 at 29.9 °C, 2.10, 2.23, 2.25 at 30.3 °C, 3.72, 3.73 at 38.4 °C, 3.84, 3.85, 3.88 at 40.1 °C, 5.59, 5.62, 5.68 at 47.5 °C, 6.31, 6.42, 6.54 at 50.1 °C, 6.31, 6.42, 6.54 at 50.1 °C, 6.27 at 50.2 °C, 8.31, 8.41, 8.56 at 54.7 °C, 10.5 at 59.2 °C, 10.7, 11.0, 11.6 at 60.5 °C, 14.1, 14.2, 14.2 at 65.1 °C, 18.5, 18.5, 18.9 at 70.7 °C, 18.8 at 71.9 °C, 21.5 at 73.4 °C (Wauchope and Getzen, 1972)
1.90 mg/L at 25 °C (Walters and Luthy, 1984)
2.23 mg/L at 25 °C (Vadas et al., 1991)
1.96 mg/L at 25 °C (Billington et al., 1988)
12.30 μmol/L at 30 °C (Yalkowsky et al., 1983)
In mole fraction (x 10^{-7}): 0.7786 at 6.60 °C, 1.048 at 13.20 °C, 1.304 at 18.00 °C, 1.751 at 24.00 °C, 2.000 at 27.00 °C, 2.436 at 31.10 °C (May et al., 1983)

Vapor pressure (x 10^{-4} mmHg):
1,130 at 75 °C (Osborn and Douslin, 1975)
1.64 at 33.30 °C, 1.95 at 34.85 °C, 2.5 at 37.20 °C, 2.81 at 38.45 °C, 3.43 at 40.30 °C, 5.43 at 45.00 °C, 7.08 at 47.75 °C, 8.18 at 49.25 °C, 8.33 at 49.55 °C (Bradley and Cleasby, 1953)
6.0 at 25 °C (Sonnefeld et al., 1983; Wasik et al., 1983)
45,100 at 127 °C (Eiceman and Vandiver, 1983)
29 at 25 °C (Bidleman, 1984)
35.5 at 25 °C (Hinckley et al., 1990)

Environmental fate:
Biological. Fluorene was statically incubated in the dark at 25 °C with yeast extract and

settled domestic wastewater inoculum. Significant biodegradation with gradual adaptation was observed. At concentrations of 5 and 10 mg/L, biodegradation yields at the end of 4 wk of incubation were 77 and 45%, respectively (Tabak et al., 1981).

Photolytic. Fluorene reacts with photochemically produced OH radicals in the atmosphere. The atmospheric half-life was estimated to range from 6.81 to 68.1 h (Atkinson, 1987). Behymer and Hites (1985) determined the effect of different substrates on the rate of photooxidation of fluorene (25 μg/g substrate) using a rotary photoreactor. The photolytic half-lives of fluorene using silica gel, alumina, and fly ash were 110, 62, and 37 h, respectively. Gas-phase reaction rate constants for OH radicals, NO_3 radicals and ozone at 24 °C were 1.6 x 10^{-11}, 3.5 x 10^{-15}, and <2 x 10^{-19} in cm^3/molecule·sec, respectively (Kwok et al., 1997).

Chemical/Physical. Oxidation by ozone to fluorenone has been reported (Nikolaou, 1984). Chlorination of fluorene in polluted humus poor lake water gave a chlorinated derivative tentatively identified as 2-chlorofluorene (Johnsen et al., 1989). This compound was also identified as a chlorination product of fluorene at low pH (<4) (Oyler et al., 1983). It was suggested that the chlorination of fluorene in tap water accounted for the presence of chlorofluorene (Shiraishi et al., 1985).

Fluorene will not hydrolyze because it does not contain a hydrolyzable functional group (Kollig, 1993).

Exposure limits: Potential occupational carcinogen. No individual standards have been set; however, as a constituent in coal tar pitch volatiles and asphalt products, the following exposure limits have been established (mg/m^3): NIOSH REL: TWA 0.1 (cyclohexane-extractable fraction), IDLH 80; OSHA PEL: TWA 0.2 (benzene-soluble fraction); ACGIH TLV: TWA 0.2 (benzene solubles).

Toxicity:

EC_{50} (48-h) for *Daphnia magna* 430 μg/L, *Chironomus plumosus* 2.35 mg/L (Mayer and Ellersieck, 1986).

IC_{50} (48-h) for *Daphnia magna* 430 μg/L (Finger et al., 1985).

LC_{50} (contact) for earthworm (*Eisenia fetida*) 171 μg/cm^2 (Neuhauser et al., 1985).

LC_{50} (96-h) for juvenile rainbow trout 820 μg/L (Finger et al., 1985).

Drinking water standard: No MCLGs or MCLs have been proposed (U.S. EPA, 1996).

Source: Fluorene was detected in groundwater beneath a former coal gasification plant in Seattle, WA at a concentration of 140 μg/L (ASTR, 1995). Present in diesel fuel and corresponding aqueous phase (distilled water) at concentrations of 350-900 mg/L and 12-26 μg/L, respectively (Lee et al., 1992). Schauer et al. (1999) reported fluorene in diesel fuel at a concentration of 52 μg/g and in a diesel-powered medium-duty truck exhaust at an emission rate of 34.6 μg/km.

Thomas and Delfino (1991) equilibrated contaminant-free groundwater collected from Gainesville, FL with individual fractions of three individual petroleum products at 24-25 °C for 24 h. The aqueous phase was analyzed for organic compounds via U.S. EPA approved test method 625. Average fluorene concentrations reported in water-soluble fractions of unleaded gasoline, kerosene, and diesel fuel were 1, 3, and 10 μg/L, respectively.

Based on analysis of 7 coal tar samples, fluorene concentrations ranged from 1,100 to 12,000 ppm (EPRI, 1990). Lao et al. (1975) reported a fluorene concentration of 27.39 g/kg in a coal tar sample. Detected in 1-yr aged coal tar film and bulk coal tar at an identical concentration of 4,400 mg/kg (Nelson et al., 1996).

Fluorene was detected in soot generated from underventilated combustion of natural gas doped with toluene (3 mole %) (Tolocka and Miller, 1995).

Uses: Chemical intermediate in numerous applications and in the formation of polyradicals for resins; insecticides and dyestuffs. Derived from industrial and experimental coal gasification operations where the maximum concentrations detected in gas, liquid, and coal tar streams were 9.1, 0.057, and 8.0 mg/m³, respectively (Cleland, 1981).

FORMALDEHYDE

Synonyms: BFV; FA; Fannoform; Formalin; Formalin 40; Formalith; Formic aldehyde; Formol; Fyde; HOCH; Ivalon; Karsan; Lysoform; Methanal; Methyl aldehyde; Methylene glycol; Methylene oxide; Morbicid; NCI-C02799; Oxomethane; Oxymethylene; Paraform; Polyoxymethylene glycols; RCRA waste number U122; Superlysoform; UN 1198; UN 2209.

$$\overset{O}{\underset{\|}{}}$$

CAS Registry Number: 50-00-0
DOT: 1198
DOT label: Combustible liquid (aqueous solutions)
Molecular formula: CH_2O
Formula weight: 30.03
RTECS: LP8925000
Merck reference: 10, 4120

Physical state, color, and odor:
Clear, colorless liquid with a pungent, suffocating odor. Burning taste. Odor threshold concentration is 0.83 ppm (Amoore and Hautala, 1983).

Melting point (°C):
-92 (Weast, 1986)
-118 (Verschueren, 1983)

Boiling point (°C):
-21 (Weast, 1986)
-19.5 (Dean, 1987)

Density (g/cm³):
0.815 at -20/4 °C (Windholz et al., 1983)

Flash point (°C):
50 (37% aqueous solution, NFPA, 1984)

Lower explosive limit (%):
7.0 (NIOSH, 1997)

Upper explosive limit (%):
73 (NIOSH, 1997)

Henry's law constant (x 10^{-7} atm·m³/mol):
3.27 (Dong and Dasgupta, 1986)
34 at 25 °C (Zhou and Mopper, 1990)
1.37 at 15 °C, 3.37 at 25 °C, 6.67 at 35 °C, 1.49 at 45 °C (Betterton and Hoffmann, 1988)

Ionization potential (eV):
10.88 (Franklin et al., 1969)

Soil sorption coefficient, log K_{oc}:
0.56 (calculated, Mercer et al., 1990)

Octanol/water partition coefficient, log K$_{ow}$:
0.35 (Sangster, 1989)

Solubility in organics:
Soluble in acetone, benzene, ether (Weast, 1986), and ethanol (Worthing and Hance, 1991)

Solubility in water:
122 g/L at 25 °C (Dean, 1987)

Vapor density:
1.067 (air = 1) (Windholz et al., 1983)
1.23 g/L at 25 °C

Vapor pressure (mmHg):
3,900 at 25 °C (Lide, 1990)

Environmental fate:
   Biological. Biodegradation products reported include formic acid and ethanol, each of which can further degrade to carbon dioxide (Verschueren, 1983).
   Heukelekian and Rand (1955) reported a 5-d BOD value of 0.74 g/g which is 69.4% of the ThOD value of 1.07 g/g.
   Photolytic. Major products reported from the photooxidation of formaldehyde with nitrogen oxides are carbon monoxide, carbon dioxide, and hydrogen peroxide (Altshuller, 1983). In synthetic air, photolysis of formaldehyde yields hydrogen chloride, and carbon monoxide (Su et al., 1979). Calvert et al. (1972) reported, however, that formaldehyde photodecomposed by direct sunlight in the atmosphere yielding hydrogen, formyl radicals, and carbon monoxide. Photooxidation of formaldehyde in nitrogen oxide-free air using radiation between 2900 and 3500 Å yielded hydrogen peroxide, alkylhydroperoxides, carbon monoxide, and lower molecular weight aldehydes. In the presence of NO$_x$, photooxidation products reported include ozone, hydrogen peroxide, and peroxyacyl nitrates (Kopczynski et al., 1974). Rate constants reported for the reaction of formaldehyde and OH radicals in the atmosphere include 9.0 x 10^{12} cm^3/mol·sec at 300 K (Hendry and Kenley, 1979), 1.4 x 10^{-11} cm^3/molecule·sec (Morris and Niki, 1971), 1.5 x 10^{-11} cm^3/molecule·sec (Niki et al., 1978), 1.11 x 10^{-13} cm^3/molecule·sec at 298 K (Baulch et al., 1984).
   Irradiation of gaseous formaldehyde containing an excess of nitrogen dioxide over chlorine yielded ozone, carbon monoxide, nitrogen pentoxide, nitryl chloride, nitric and hydrochloric acids. Peroxynitric acid was the major photolysis product when chlorine concentration exceeded the nitrogen dioxide concentration (Hanst and Gay, 1977). Formaldehyde also reacts with NO$_3$ in the atmosphere at a rate of 3.2 x 10^{-16} cm^3/molecule·sec (Atkinson and Lloyd, 1984).
   Chemical/Physical. Oxidizes in air to formic acid. Trioxymethylene may precipitate under cold temperatures (Sax, 1984). Polymerizes easily (Windholz et al., 1983) and is a strong reducing agent (Worthing and Hance, 1991).
   Anticipated products from the reaction of formaldehyde with ozone or OH radicals in the atmosphere are carbon monoxide and carbon dioxide (Cupitt, 1980). Major products reported from the photooxidation of formaldehyde with nitrogen oxides are carbon monoxide, carbon dioxide, and hydrogen peroxide (Altshuller, 1983).
   Formaldehyde reacted with hydrogen chloride in moist air to form *sym*-dichloromethyl ether. This compound may also form from an acidic solution containing chloride ions and formaldehyde (Frankel et al., 1974; Travenius, 1982). In an aqueous solution at 25 °C, nearly all the formaldehyde added is hydrated forming a gem-diol (Bell and McDougall, 1960). May

polymerize in an aqueous solution to trioxymethylene (Hartley and Kidd, 1987).

At an influent concentration of 1,000 mg/L, treatment with granular activated carbon resulted in an effluent concentration of 908 mg/L. The adsorbability of the carbon used was 18 mg/g carbon (Guisti et al., 1974).

Exposure limits: Potential occupational carcinogen. NIOSH REL: TWA 16 ppb, 15-min ceiling 100 ppb, IDLH 20 ppm; OSHA PEL: TWA 0.75 ppm, STEL 2 ppm; ACGIH TLV: TWA 1 ppm (adopted) and a proposed ceiling of 0.3 ppm.

Symptoms of exposure: Eye, nose, and throat irritant; coughing, bronchospasm, pulmonary irritation, dermatitis, nausea, vomiting, loss of consciousness (NIOSH, 1997)

Toxicity:

LC_{50} (96-h flow-through) for rainbow trout 118 μL/L, Atlantic salmon 173 μL/L, lake trout 100 μL/L, channel catfish 65.8 μL/L, bluegill sunfish 100 μL/L, black bullhead 62.1 μL/L, lake trout 100 μL/L, green sunfish 173 μL/L, smallmouth bass 136 μL/L, largemouth bass 143 μgL/L (quoted, Verschueren, 1983).

Acute inhalation LD_{50} for rats 820 μg/L (30-min), 480 μg/L (4-h), mice 414 μg/L (4-h) (Worthing and Hance, 1991).

Acute oral LD_{50} for guinea pigs 260 mg/kg, mice 42 mg/kg, rats 800 mg/kg (quoted, RTECS, 1985).

Kennedy and Sherman (1986) reported a NOEL value of 200 ppm for rats over a 90-d feeding period.

Drinking water standard: No MCLGs or MCLs have been proposed (U.S. EPA, 1996).

Source: Formaldehyde naturally occurs in jimsonweed, pears, black currant, horsemint, sago cycas seeds (1,640-2,200 ppm), oats, beets, and wild bergamot (Duke, 1992).

Uses: Manufacture of phenolic, melamine, urea, and acetal resins; polyacetal and phenolic resins; pentaerythritol; fertilizers; dyes; hexamethylenetetramine; ethylene glycol; embalming fluids; textiles; fungicides; air fresheners; cosmetics; in medicine as a disinfectant and germicide; preservative; hardening agent; in oil wells as a corrosion inhibitor; industrial sterilant; reducing agent.

FORMIC ACID

Synonyms: Aminic acid; Formylic acid; Hydrogen carboxylic acid; Methanoic acid; RCRA waste number U123; UN 1779.

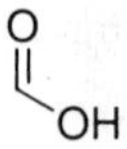

$$\overset{O}{\underset{\displaystyle OH}{\parallel}}$$

CAS Registry Number: 64-18-6
DOT: 1779
DOT label: Corrosive
Molecular formula: CH_2O_2
Formula weight: 46.03
RTECS: LQ4900000
Merck reference: 10, 4123

Physical state, color, and odor:
Colorless, fuming liquid with a pungent, penetrating odor. Odor threshold concentration is 49 ppm (Amoore and Hautala, 1983).

Melting point (°C):
8.1 (Barham and Clark, 1951)

Boiling point (°C):
100.7 (Weast, 1986)

Density (g/cm³):
1.2200 at 20/4 °C (Barham and Clark, 1951)
1.2267 at 15/4 °C (Sax and Lewis, 1987)
1.21045 at 25/4 °C (Huntress and Mulliken, 1941)

Diffusivity in water (x 10⁻⁵ cm²/sec):
1.57 at 20 °C using method of Hayduk and Laudie (1974)

Dissociation constant, pK_a:
3.75 at 20 °C (Gordon and Ford, 1972)

Flash point (°C):
50.4 (90% solution, NIOSH, 1997)
69 (open cup, Hawley, 1981)

Lower explosive limit (%):
18 (90% solution, NIOSH, 1997)

Upper explosive limit (%):
57 (90% solution, NIOSH, 1997)

Heat of fusion (kcal/mol):
3.031 (Riddick et al., 1986)
3.035 (Dean, 1987)

Henry's law constant (x 10⁻⁷ atm·m³/mol):
1.67 at pH 4 (quoted, Gaffney et al., 1987)

1.12 at 25 °C (Johnson et al., 1996)
1.81 at the pH range 1.6 to 1.9 (Khan and Brimblecombe, 1992; Khan et al., 1995)
0.77 at 25 °C (value concentration dependent, Servant et al., 1991)
At 25 °C: 95.2, 75.1, 39.3, 10.7, and 3.17 at pH values of 1.35, 3.09, 4.05, 4.99, and 6.21, respectively (Hakuta et al., 1977).

Ionization potential (eV):
11.05 ± 0.01 (Franklin et al., 1969)
11.33 (Gibson et al., 1977)

Soil sorption coefficient, log K_{oc}:
Unavailable because experimental methods for estimation of this parameter for aliphatic carboxylic acids are lacking in the documented literature. However, its high solubility in water suggests its adsorption to soil will be nominal (Lyman et al., 1982).

Octanol/water partition coefficient, log K_{ow}:
-0.53 (Collander, 1951)
-0.54 (Leo et al., 1971)

Solubility in organics:
Soluble in acetone and benzene (Weast, 1986). Miscible with alcohol, ether, and glycerol (Windholz et al., 1983).

Solubility in water:
Miscible (Price et al., 1974)

Vapor density:
1.88 g/L at 25 °C, 1.59 (air = 1)

Vapor pressure (mmHg):
35 at 20 °C, 54 at 30 °C (Verschueren, 1983)
42.6 at 25 °C (Banerjee et al., 1990)

Environmental fate:
Biological. Near Wilmington, NC, organic wastes containing formic acid (representing 11.4% of total dissolved organic carbon) were injected into an aquifer containing saline water to a depth of about 1,000 feet. The generation of gaseous components (hydrogen, nitrogen, hydrogen sulfide, carbon dioxide, and methane) suggested that formic acid and possibly other waste constituents were anaerobically degraded by microorganisms (Leenheer et al., 1976).

Heukelekian and Rand (1955) reported a 5-d BOD value of 0.20 g/g which is 57.1% of the ThOD value of 0.83 g/g.

Photolytic. Experimentally determined rate constants for the reaction of formic acid with OH radicals in the atmosphere and aqueous solution were 3.7×10^{-13} and 2.2×10^{-13} cm^3/molecule·sec (Dagaut et al., 1988).

Chemical/Physical. Pure formic acid slowly decomposes to carbon monoxide and water. At 20 °C, 0.06 g of water would form in 1 yr by 122 g formic acid. At standard temperature and pressure, this amount of formic acid would produce 0.15 mL of carbon monoxide per hour. Over time, the rate of decomposition will decrease because the water formed acts as a negative catalyst (Barham and Clark, 1951).

Reacts with alkalies forming water-soluble salts. Slowly reacts with alcohols and anhydrides forming formate esters.

At an influent concentration of 1,000 mg/L, treatment with granular activated carbon resulted in an effluent concentration of 765 mg/L. The adsorbability of the carbon used was 47 mg/g carbon (Guisti et al., 1974).

Exposure limits: NIOSH REL: TWA 5 ppm (9 mg/m^3), IDLH 30 ppm; OSHA PEL: TWA 5 ppm; ACGIH TLA: TWA 5 ppm, STEL 10 ppm (adopted).

Symptoms of exposure: Corrosive to skin. Contact with liquid will burn eyes and skin (Patnaik, 1992). An irritation concentration of 27.00 mg/m^3 in air was reported by Ruth (1986).

Toxicity:
Acute oral LD$_{50}$ for mice 700 mg/kg, dogs 4,000 mg/kg, rats 1,100 mg/kg (quoted, RTECS, 1985).

Source: Formic acid naturally occurs in carrots, soybean roots, carob, yarrow, aloe, Levant berries, bearberries, wormwood, ylang-ylang, celandine, jimsonweed, water mint, apples, tomatoes, bay leaves, common juniper, ginkgo, scented boronia, corn mint, European pennyroyal, and bananas (Duke, 1992).

Uses: Chemical analysis; preparation of formate esters and oxalic acid; silvering glass; decalcifier; dehairing and plumping hides; manufacture of fumigants, insecticides, refrigerants, solvents for perfumes, lacquers; reducer in dyeing wool fast colors; ore flotation; electroplating; leather treatment; coagulating rubber latex; vinyl resin plasticizers.

FURFURAL

Synonyms: Artificial ant oil; Artificial oil of ants; Bran oil; Fural; 2-Furaldehyde; Furale; 2-Furanaldehyde; 2-Furancarbonal; **2-Furancarboxaldehyde**; Furfuraldehyde; Furfurol; Furfurole; Furole; α-Furole; 2-Furylmethanal; NCI-C56177; Pyromucic aldehyde; RCRA waste number U125; UN 1199.

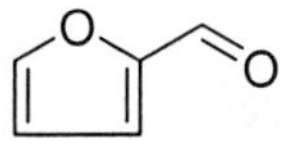

CAS Registry Number: 98-01-1
DOT: 1199
Molecular formula: $C_5H_4O_2$
Formula weight: 96.09
RTECS: LT7000000
Merck reference: 10, 4179

Physical state, color, and odor:
Colorless to yellow liquid with an almond-like odor. Turns reddish brown on exposure to light and air. Odor and taste thresholds are 0.4 and 4 ppm, respectively (Keith and Walters, 1992).

Melting point (°C):
-38.7 (Weast, 1986)
-36.5 (Windholz et al., 1983)

Boiling point (°C):
161.7 (Weast, 1986)

Density (g/cm³):
1.1594 at 20/4 °C (Weast, 1986)
1.1563 at 25/4 °C (Windholz et al., 1983)

Diffusivity in water (x 10^{-5} cm²/sec at 25 °C):
1.12 at 25 °C (quoted, Hayduk and Laudie, 1974)

Flash point (°C):
61 (NIOSH, 1997)
68 (open cup, Windholz et al., 1983)

Lower explosive limit (%):
2.1 (NIOSH, 1997)

Upper explosive limit (%):
19.3 (NIOSH, 1997)

Henry's law constant (x 10^{-6} atm·m³/mol):
1.52 at 20 °C (approximate - calculated from water solubility and vapor pressure)

Interfacial tension with water (dyn/cm):
4.7 at 25 °C (Murphy et al., 1957; Demond and Lindner, 1993)

Ionization potential (eV):
9.21 ± 0.01 (Franklin et al., 1969)

Soil sorption coefficient, log K_{oc}:
Unavailable because experimental methods for estimation of this parameter for aldehydes are lacking in the documented literature. However, its high solubility in water suggests its adsorption to soil will be nominal (Lyman et al., 1982).

Octanol/water partition coefficient, log K_{ow}:
0.41 (Hansch and Leo, 1985)
0.52 (Tewari et al., 1981, 1982)

Solubility in organics:
Soluble in acetone, alcohol, benzene, chloroform, and ether (Weast, 1986)

Solubility in water:
83.0 g/L at 20 °C, 199.0 g/L at 90 °C (quoted, Verschueren, 1983)
77.83 g/L at 25.0 °C (Tewari et al., 1982; Wasik et al., 1981)
92.9 g/L at 26.7 °C (Skrzec and Murphy, 1954)
82.0 g/L at 40 °C (Jones, 1929)
In wt %: 8.22 at 0 °C, 7.86 at 10.0 °C, 7.94 at 20.0 °C, 8.40 at 30 °C, 8.94 at 40.0 °C, 9.50 at 50 °C, 10.28 at 60 °C, 10.97 at 70 °C, 12.56 at 80 °C, 14.74 at 90 °C (Stephenson, 1993b)

Vapor density:
3.93 g/L at 25 °C, 3.32 (air = 1)

Vapor pressure (mmHg):
1 at 20 °C, 3 at 30 °C, 10 at 50 °C (Verschueren, 1983)
2.5 at 20 °C (Riddick et al., 1986)

Environmental fate:
   Biological. Under nitrate-reducing and methanogenic conditions, furfural biodegraded to methane and carbon dioxide (Knight et al., 1990). In activated sludge inoculum, following a 20-d adaptation period, 96.3% COD removal was achieved. The average rate of biodegradation was 37.0 mg COD/g·h (Pitter, 1976).
   Photolytic. Atkinson (1985) reported an estimated photooxidation half-life of 10.5 h for the reaction of furfural with OH radicals in the atmosphere.
   Chemical/Physical. Slowly resinifies at room temperature (Windholz et al., 1983). May polymerize on contact with strong acids or strong alkalies (NIOSH, 1997).

Exposure limits: NIOSH REL: IDLH 100 ppm; OSHA PEL: TWA 5 ppm (20 mg/m³); ACGIH TLV: TWA 2 ppm (adopted).

Toxicity:
   The acute inhalation LC_{50} dose reported for rats is approximately 182 ppm (Gupta et al., 1991).
   Acute oral LD_{50} for dogs 950 mg/kg, guinea pigs 541 mg/kg, mice 400 mg/kg, rats 65 mg/kg (quoted, RTECS, 1985).

Source: Furfural occurs naturally in many plants including rice (90,000-100,000 ppm), lovage roots (2-20 ppm), caraway, strawberry leaves, culantro, java cintronella, cassia, ylang-ylang, sweetflag, Japanese mint, oat husks (100,000 ppm), anise, broad-leaved lavender, myrtle flowers (0-1 ppm), lemon verbena, Karaya gum (123,000 ppm), nutmeg seeds (15,000 ppm), West Indian lemongrass, licorice roots (2 ppm), cinnamon bark (3-12 ppm), Hyssop shoots (1-2 ppm), periwinkle leaves, rockrose leaves, and garden dill (Duke, 1992).

Uses: Synthesizing furan derivatives; adipic acid and adiponitrile; manufacture of furfural-phenol plastics; accelerator for vulcanizing rubber; solvent refining of petroleum oils; solvent for nitrated cotton, nitrocellulose, cellulose acetate, shoe dyes, and gums; insecticide, fungicide and germicide; intermediate for tetrahydrofuran and furfuryl alcohol; weed killer; flavoring; wetting agent in manufacture of abrasive wheels and brake linings; road construction; refining of rare earths and metals; analytical chemistry.

FURFURYL ALCOHOL

Synonyms: 2-Furancarbinol; **2-Furanmethanol**; Furfural alcohol; Furyl alcohol; Furyl-carbinol; 2-Furylcarbinol; α-Furylcarbinol; 2-Furylmethanol; 2-Hydroxymethylfuran; NCI-C56224; UN 2874.

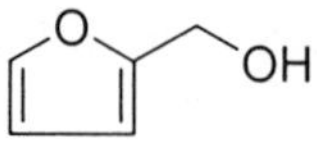

CAS Registry Number: 98-00-0
DOT: 2874
DOT label: Poison
Molecular formula: $C_5H_6O_2$
Formula weight: 98.10
RTECS: LU9100000
Merck reference: 10, 4180

Physical state, color, and odor:
Colorless to yellow, mobile liquid with an irritating odor. Darkens on exposure to air. Odor threshold concentration is 8 ppm (Keith and Walters, 1992).

Melting point (°C):
-14.6 (NIOSH, 1997)
-29 (metastable crystalline form, Verschueren, 1983)

Boiling point (°C):
170 (Windholz et al., 1983)

Density (g/cm³):
1.1296 at 20/4 °C (Weast, 1986)

Diffusivity in water (x 10^{-5} cm²/sec):
0.96 at 20 °C using method of Hayduk and Laudie (1974)

Flash point (°C):
66 (NIOSH, 1997)
75 (open cup, Windholz et al., 1983)

Lower explosive limit (%):
1.8 (NIOSH, 1997)

Upper explosive limit (%):
16.3 (NIOSH, 1997)

Heat of fusion (kcal/mol):
3.12 (Dean, 1987)

Soil sorption coefficient, log K_{oc}:
Unavailable because experimental methods for estimation of this parameter for unsaturated alicyclic alcohols are lacking in the documented literature. However, its high solubility in water suggests its adsorption to soil will be nominal (Lyman et al., 1982).

Solubility in organics:
Soluble in alcohol and ether (Weast, 1986)

Solubility in water:
Miscible (NIOSH, 1997)

Vapor density:
4.01 g/L at 25 °C, 3.39 (air = 1)

Vapor pressure (mmHg):
0.4 at 20 °C (Verschueren, 1983)
0.6 at 25 °C (NIOSH, 1997)

Environmental fate:
   Biological. In activated sludge inoculum, following a 20-d adaptation period, 97.3% COD removal was achieved. The average rate of biodegradation was 41.0 mg COD/g·h (Pitter, 1976).
   Chemical/Physical. Easily resinified by acids (Windholz et al., 1983). Furfuryl alcohol will not hydrolyze because it has no hydrolyzable functional group.

Exposure limits: NIOSH REL: TWA 10 ppm (40 mg/m^3), STEL 15 ppm (60 mg/m^3), IDLH 75 ppm; OSHA PEL: TWA 50 ppm; ACGIH TLV: TWA 10 ppm, STEL 15 ppm (adopted).

Toxicity:
   Acute oral LD$_{50}$ for mice 160 mg/kg, rats 88.3 mg/kg (quoted, RTECS, 1985).

Source: Furfuryl occurs naturally in yarrow, licorice, sesame seeds, clove flowers, and tea leaves (Duke, 1992).

Uses: Solvent for dyes and resins; preparation of furfuryl esters; furan polymers; solvent for textile printing; manufacturing wetting agents and resins; penetrant; flavoring; corrosion-resistant sealants and cements; viscosity reducer for viscous epoxy resins.

GLYCIDOL

Synonyms: Epihydric alcohol; Epihydrin alcohol; 2,3-Epoxypropanol; 2,3-Epoxy-1-propanol; Epoxypropyl alcohol; Glycide; Glycidyl alcohol; 3-Hydroxy-1,2-epoxypropane; Hydroxymethyl ethylene oxide; 2-Hydroxymethyloxiran; 3-Hydroxypropylene oxide; NCI-C55549; **Oxiranemethanol**; Oxiranylmethanol.

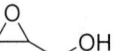

CAS Registry Number: 556-52-5
Molecular formula: $C_3H_6O_2$
Formula weight: 74.08
RTECS: UB4375000
Merck reference: 10, 4353

Physical state, color, and odor:
Clear, colorless, and odorless liquid

Melting point (°C):
-45 (NIOSH, 1997)

Boiling point (°C):
Decomposes at 166-167 (Weast, 1986)
162 (Hawley, 1981)

Density (g/cm³):
1.1143 at 25/4 °C (Weast, 1986)
1.165 at 0/4 °C (Verschueren, 1983)

Diffusivity in water (x 10^{-5} cm²/sec):
0.23 at 25 °C using method of Hayduk and Laudie (1974)

Flash point (°C):
72.8 (NIOSH, 1997)
81 (Aldrich, 1988)

Ionization potential (eV):
10.43 (Lias and Liebman, 1998)

Soil sorption coefficient, log K_{oc}:
Unavailable because experimental methods for estimation of this parameter for epoxy aliphatic alcohols are lacking in the documented literature. However, its high solubility in water and low K_{ow} suggest its adsorption to soil will be nominal (Lyman et al., 1982).

Octanol/water partition coefficient, log K_{ow}:
-0.95 (Deneer et al., 1988)

Solubility in organics:
Soluble in acetone, alcohol, benzene, chloroform, and ether (Weast, 1986)

Solubility in water:
Miscible (NIOSH, 1997)

Vapor density:
3.03 g/L at 25 °C, 2.56 (air = 1)

Vapor pressure (mmHg):
0.9 at 25 °C (Verschueren, 1983)

Environmental fate:
   Chemical/Physical. May hydrolyze in water forming glycerin (Lyman et al., 1982).

Exposure limits: NIOSH REL: TWA 25 ppm (75 mg/m^3), IDLH 150 ppm; OSHA PEL: TWA 50 ppm (150 mg/m^3); ACGIH TLV: TWA 2 ppm (adopted).

Symptoms of exposure: Eye, skin, and lung irritant (Patnaik, 1992)

Toxicity:
   LC_{50} (14-d) for *Poecilia reticulata* >319 μg/L (Deneer et al., 1988).
   Acute oral LD_{50} for mice 431 mg/kg, rats 420 mg/kg (quoted, RTECS, 1985).

Uses: Demulsifier; dye-leveling agent; stabilizer for natural oils and vinyl polymers; organic synthesis.

HEPTACHLOR

Synonyms: Aahepta; Agroceres; Basaklor; 3-Chlorochlordene; Drinox; Drinox H-34; E 3314; ENT 15152; GPKh; H-34; Heptachlorane; 3,4,5,6,7,8,8-Heptachlorodicyclopentadiene; 3,4,5,-6,7,8,8a-Heptachlorodicyclopentadiene; 1(3a),4,5,6,7,8,8-Heptachloro-3a(1),4,7,7a-tetrahydro-4,7-methanoindene; 1,4,5,6,7,8,8-Heptachloro-3a,4,7,7a-tetrahydro-4,7-methanoindene; **1,4,5,-6,7,8,8-Heptachloro-3a,4,7,7a-tetrahydro-4,7-methanol-1*H*-indene**; 1,4,5,6,7,8,8-Hepta-chloro-3a,4,7,7a-tetrahydro-4,7-*endo*-methanoindene; 1,4,5,6,7,8,8a-Heptachloro-3a,4,7,7a-tetrahydro-4,7-methanoindene; 1,4,5,6,7,8,8-Heptachloro-3a,4,7,7a-tetrahydro-4,7-methylene-indene; 1,4,5,6,7,10,10-Heptachloro-4,7,8,9-tetrahydro-4,7-methyleneindene; 1,4,5,6,7,10,10-Heptachloro-4,7,8,9-tetrahydro-4,7-*endo*-methyleneindene; 3,4,5,6,7,8,8a-Heptachloro-α-dicy-clopentadiene; Heptadichlorocyclopentadiene; Heptagran; Heptagranox; Heptamak; Heptamul; Heptasol; Heptox; NA 2761; NCI-C00180; OMS 193; Soleptax; RCRA waste number P059; Rhodiachlor; Velsicol 104; Velsicol heptachlor.

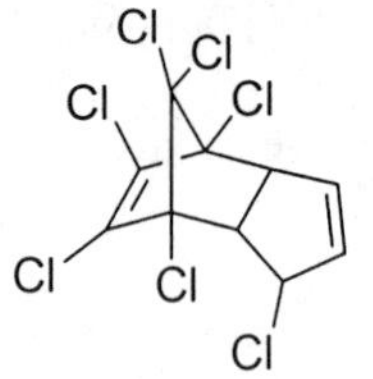

CAS Registry Number: 76-44-8
DOT: 2761
DOT label: Poison
Molecular formula: $C_{10}H_5Cl_7$
Formula weight: 373.32
RTECS: PC0700000
Merck reference: 10, 4547

Physical state, color, and odor:
Crystalline white to light tan, waxy solid with a camphor-like odor

Melting point (°C):
95-96 (Worthing and Hance, 1991)
46-74 (Weiss, 1986)

Boiling point (°C):
Decomposes at 146.2 (NIOSH, 1997)
135-145 at 1-1.5 mmHg (IARC, 1979)

Density (g/cm³):
1.57 at 9/4 °C (Weast, 1986)
1.66 at 20/4 °C (Weiss, 1986)
1.65-1.67 at 25/4 °C (WHO, 1984)

Diffusivity in water (x 10⁻⁵ cm²/sec):
0.46 at 20 °C using method of Hayduk and Laudie (1974)

Flash point (°C):
Noncombustible solid (NIOSH, 1997)

Henry's law constant (x 10^{-3} atm·m^3/mol):
2.3 (Petrasek et al., 1983)

Bioconcentration factor, log BCF:
4.14 (freshwater fish), 3.58 (fish, microcosm) (Garten and Trabalka, 1983)
3.76-3.92 (*Leiostomus xanthurus*, Schimmel et al., 1976a)
4.03 freshwater clam (*Corbicula manilensis*) (Hartley and Johnston, 1983)
3.98 (*Pimephales promelas*, Devillers et al., 1996)

Soil sorption coefficient, log K_{oc}:
4.38 (Jury et al., 1987)
4.90 (Taichung soil: pH 6.8, % sand: 25, % silt: 40, % clay: 35) (Ding and Wu, 1995)

Octanol/water partition coefficient, log K_{ow}:
5.44 (Travis and Arms, 1988)
5.5 (DeKock and Lord, 1987)

Solubility in organics:
At 27 °C (g/L): acetone (750), benzene (1,060), carbon tetrachloride (1,120), cyclohexanone (1,190), alcohol (45), xylene (1,020) (Windholz et al., 1983). Soluble in ether, kerosene and ligroin (U.S. EPA, 1985).

Solubility in water (μg/L):
100 at 15 °C, 180 at 25 °C, 315 at 35 °C, 490 at 45 °C (particle size ≤5 μ, Biggar and Riggs, 1974)
56 at 25-29 °C (Park and Bruce, 1968)

Vapor pressure (x 10^{-4} mmHg):
3 at 20 °C (Sims et al., 1988)
2.33 at 25 °C (Hinckley et al., 1990)

Environmental fate:
Biological. Many soil microorganisms were found to oxidize heptachlor to heptachlor epoxide (Miles et al., 1969). In addition, hydrolysis produced hydroxychlordene with subsequent epoxidation yielding 1-hydroxy-2,3-epoxychlordene (Kearney and Kaufman, 1976). Heptachlor reacted with reduced hematin forming chlordene, which decomposed to hexa-chlorocyclopentadiene and cyclopentadiene (Baxter, 1990). In a model ecosystem containing plankton, *Daphnia magna*, mosquito larva (*Culex pipiens quinquefasciatus*), fish (*Cambusia affinis*), alga (*Oedogonium cardiacum*), and snail (*Physa* sp.), heptachlor degraded to 1-hydroxychlordene, 1-hydroxy-2,3-epoxychlordene, hydroxychlordene epoxide, heptachlor epoxide, and five unidentified compounds (Lu et al., 1975). In four successive 7-d incubation periods, heptachlor (5 and 10 mg/L) was recalcitrant to degradation in a settled domestic wastewater inoculum (Tabak et al., 1981). When heptachlor (10 ppm) in sewage sludge was incubated under anaerobic conditions at 53 °C for 24 h, complete degradation was achieved (Hill and McCarty, 1967).

Heptachlor rapidly degraded when incubated with acclimated, mixed microbial cultures under aerobic conditions. After 4 wk, 95.3% of the applied dosage was removed (Leigh, 1969). In a mixed bacterial culture under aerobic conditions, heptachlor was transformed to chlordene, 1-hydroxychlordene, heptachlor epoxide, and chlordene epoxide in low yields (Miles et al., 1971). Heptachlor rapidly degraded when incubated with acclimated, mixed microbial cultures under aerobic conditions. After 4 wk, 95.3% of the applied dosage was removed (Leigh, 1969).

White rot fungi, namely *Phanerochaete chrysosporium, Pleurotus sajorcaju, Pleurotus florida*, and *Pleurotus eryngii*, biodegraded heptachlor (50 μM) at degradation yields of 92.13-97.30, 59.53-84.76, 66.25-80.00, and 85.85-95.15%, respectively. The experiments were carried out in a culture incubated at 30 °C for 20 d (Arisoy, 1998).

Soil. Harris (1969) studied the mobility of 11 insecticides, including heptachlor, in soil (Hagerstown silty clay and Lakeland sand) using a standard column system. The depth to water in the soil column was decreased. Because the concentration of heptachlor did not change with depth, the author concluded this insecticide is not mobile.

Groundwater. According to the U.S. EPA (1986), heptachlor has a high potential to leach to groundwater.

Plant. On plant surfaces, heptachlor was oxidized to heptachlor epoxide (Gannon and Decker, 1958).

Photolytic. Sunlight and UV light convert heptachlor to photoheptachlor (Georgacakis and Khan, 1971). This is in agreement with Gore et al. (1971) who reported that heptachlor exhibited weakly absorbtion of UV light at wavelengths above 290 nm. Eichelberger and Lichtenberg (1971) reported that heptachlor (10 μg/L) in river water, kept in a sealed jar under sunlight and fluorescent light, was completely converted to 1-hydroxychlordene. Under the same conditions, but in distilled water, 1-hydroxychlordene and heptachlor epoxide formed in yields of 60 and 40%, respectively (Eichelberger and Lichtenberg, 1971). The photolysis of heptachlor in various organic solvents afforded different photoproducts. Photolysis at 253.7 nm in hydrocarbon solvents yielded two olefinic monodechlorination isomers: 1,4,5,7,8,8-hexachloro-3a,4,7,7a-tetrahydro-4,7-methanoindene and 1,4,6,7,8,8-hexachloro-3a,4,7,7a-tetra-hydro-4,7-methanoindene. Irradiation at 300 nm in acetone, 1,2,3,6,9,10,10-heptachloro-pentacyclo[5.3.0.0^{2,5}.0^{3,9}.0^{4,8}]decane is the only product formed. This compound and a C-1 cyclohexyl adduct are formed when heptachlor in a cyclohexane/acetone solvent system is irradiated at 300 nm (McGuire et al., 1972).

Heptachlor reacts with photochemically produced OH radicals in the atmosphere. At a concentration of 5 x 10^5 OH radicals/cm^3, the atmospheric half-life was estimated to be about 6 h (Atkinson, 1987).

Chemical/Physical. Slowly hydrolyzes in water forming hydrogen chloride and 1-hydroxychlordene (Hartley and Kidd, 1987; Kollig, 1993). The hydrolysis half-lives of hepta-chlor in a sterile 1% ethanol/water solution at 25 °C and pH values of 4.5, 5.0, 6.0, 7.0, and 8.0 were 0.77, 0.62, 0.64, 0.64, and 0.43 wk, respectively (Chapman and Cole, 1982). Chemical degradation of heptachlor gave heptachlor epoxide (Newland et al., 1969). Heptachlor degraded in aqueous saturated calcium hypochlorite solution to 1-hydroxychlordene. Although further degradation occurred, no other metabolites were identified (Kaneda et al., 1974).

Heptachlor (1 mM) in ethyl alcohol (30 mL) underwent dechlorination in the presence of nickel boride (generated by the reaction of nickel chloride and sodium borohydride). The catalytic dechlorination of heptachlor by this method yielded a pentachloro derivative as the major product having the empirical formula $C_{10}H_9Cl_5$ (Dennis and Cooper, 1976).

Emits toxic chloride fumes when heated to decomposition (Lewis, 1990).

Exposure limits (mg/m^3): Potential occupational carcinogen. NIOSH REL: TWA 0.5, IDLH 35; OSHA PEL: TWA 0.5; ACGIH TLV: TWA 0.5 (adopted).

Symptoms of exposure: The acceptable daily intake for humans resulting in no appreciable risk is 0.5 μg/kg body weight (Worthing and Hance, 1991).

Toxicity:

EC$_{50}$ (48-h) for *Daphnia pulex* 42 μg/L, *Simocephalus serrulatus* 64 μg/L (Mayer and Ellersieck, 1986).

LC$_{50}$ (181-d) for female mink 10.5 ppm (Crum et al., 1993).

LC$_{50}$ (96-h) for rainbow trout 7 μg/L, bluegill sunfish 26 μg/L, fathead minnows 78-130 μg/L (Hartley and Kidd, 1987), sheapshead minnows 2.7-8.8 μg/L, marine pin perch 0.20-4.4 μg/L (Verschueren, 1983), fish (*Lagodon rhomboides*) 3.77 μg/L, fish (*Leiostomus xanthurus*) 0.85 μg/L, fish (*Cyprinodon variegatus*) 3.68 μg/L (quoted, Reish and Kauwling, 1978), *Lepomis machrochirus* 19 μg/L, *Pimephales promelas* 56 μg/L (Henderson et al., 1959), Claassenia *sabulosa* 29 μg/L, *Pteronarcella badia* 0.98 μg/L, *Pteronarcys californica* 1.1 μg/L (Sanders and Cope, 1968), pink shrimp (*Penaeus duorarum*) 0.03 μg/L, spot (*Leiostomus xanthurus*) 0.86 μg/L (Schimmel et al., 1976).

LC$_{50}$ (48-h) for red killifish 933 μg/L (Yoshioka et al., 1986), *Simocephalus serrulatus* 47 μg/L, *Daphnia pulex* 42 μg/L (Sanders and Cope, 1966); LC$_{50}$ (24-h) for sheapshead minnows 1.22-4.3 μg/L (Verschueren, 1983), northern pike 8 μg/L (Mayer and Ellersieck, 1986).

Acute oral LD$_{50}$ for rats 40 mg/kg, guinea pigs 116 mg/kg, hamsters 100 mg/kg, mice 68 mg/kg (quoted, RTECS, 1985).

Acute percutaneous LD$_{50}$ for rabbits >2,000 mg/kg (Worthing and Hance, 1991).

Dermal LD$_{50}$ for rats is 195-250 mg/kg (Gaines, 1969).

Drinking water standard (final): MCLG: zero; MCL: 0.4 μg/L (U.S. EPA, 1996).

Use: Non-systemic contact and ingested insecticide for termite control.

HEPTACHLOR EPOXIDE

Synonyms: ENT 25584; Epoxy heptachlor; HCE; 1,4,5,6,7,8,8-Heptachloro-2,3-epoxy-2,3,3a,-4,7,7a-hexahydro-4,7-methanoindene; 1,2,3,4,5,6,7,8,8-Heptachloro-2,3-epoxy-3a,4,7,7a-tetra-hydro-4,7-methanoindene; **5a,6,6a-Hexahydro-2,5-methano-2H-indeno[1,2-b]oxirene**; 2,3,-4,5,6,7,7-Heptachloro-1a,1b,5,5a,6,6a-hexahydro-2,5-methano-2H-oxireno[a]indene; Velsicol 53-CS-17.

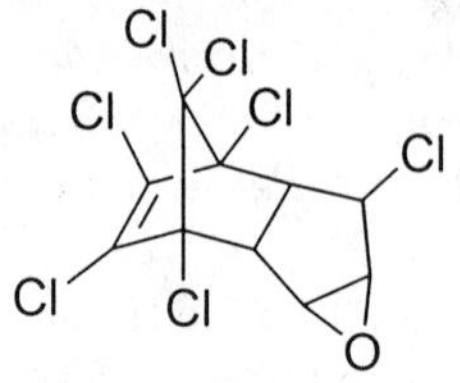

CAS Registry Number: 1024-57-3
DOT: 2761
DOT label: Poison
Molecular formula: $C_{10}H_5Cl_7O$
Formula weight: 389.32
RTECS: PB9450000

Physical state:
Liquid

Melting point (°C):
157-160 (Singh, 1969)

Diffusivity in water (x 10^{-5} cm²/sec):
0.46 at 20 °C using method of Hayduk and Laudie (1974)

Henry's law constant (x 10^{-5} atm·m³/mol):
3.2 at 25 °C (Warner et al., 1987)

Bioconcentration factor, log BCF:
2.90 (*B. subtilis*, Grimes and Morrison, 1975)
4.16 (freshwater fish), 3.69 (fish, microcosm) (Garten and Trabalka, 1983)
3.37 freshwater clam (*Corbicula manilensis*) (Hartley and Johnston, 1983)

Soil sorption coefficient, log K_{oc}:
4.32 (Taichung soil: pH 6.8, % sand: 25, % silt: 40, % clay: 35) (Ding and Wu, 1995)

Octanol/water partition coefficient, log K_{ow}:
5.40 (Travis and Arms, 1988)

Solubility in water (μg/L):
350 at 25-29 °C (Park and Bruce, 1968)
275 at 25 °C (Warner et al., 1987)
110 at 15 °C, 200 at 25 °C, 350 at 35 °C, 600 at 45 °C (particle size ≤5 μ, Biggar and Riggs, 1974)

Vapor pressure (x 10^{-6} mmHg):
2.6 at 20 °C (IARC, 1974)
300 at 30 °C (Nash, 1983)

Environmental fate:

Biological. In a model ecosystem containing plankton, *Daphnia magna*, mosquito larva (*Culex pipiens quinquefasciatus*), fish (*Cambusia affinis*), alga (*Oedogonium cardiacum*) and snail (*Physa* sp.), heptachlor epoxide degraded to hydroxychlordene epoxide (Lu et al., 1975). Using settled domestic wastewater inoculum, heptachlor epoxide (5 and 10 mg/L) did not degrade after 28 d of incubation at 25 °C (Tabak et al., 1981). This is consistent with the findings of Bowman et al. (1965). They reported that heptachlor epoxide was not significantly degraded in 7 air-dried soils that were incubated for 8 d. When heptachlor epoxide was incubated in a sandy loam soil at 28 °C, however, 1-hydroxychlordene formed at yields of 2.8, 5.8, and 12.0% after 4, 8, and 12 wk, respectively (Miles et al., 1971).

Photolytic. Irradiation of heptachlor epoxide by a 450-W high-pressure mercury lamp gave two half-cage isomers, each containing a ketone functional group (Ivie et al., 1972). Benson et al. (1971) reported a degradation yield of 99% when an acetone solution containing heptachlor epoxide was photolyzed at >300 nm for 11 h. An identical degradation yield was achieved in only 60 min when the UV wavelength was reduced to >290 nm.

Graham et al. (1973) reported that when solid heptachlor epoxide was exposed to July sunshine for 23.2 d, 59.3% degradation was achieved. In powdered form, however, only 5 d were required for complete degradation to occur. The products include a semicage ketone and an intermediate that may be converted to an enantiomeric semicage ketone.

Chemical/Physical. Heptachlor epoxide will hydrolyze via nucleophilic attack at the epoxide moiety forming heptachlor diol which may undergo further hydrolysis forming heptachlor triol and hydrogen chloride (Kollig, 1993).

Exposure limits: ACGIH TLV: TWA 0.05 mg/m^3 (adopted).

Toxicity:

LC$_{50}$ (96-h) for pink shrimp (*Penaeus duorarum*) 0.04 μg/L (Schimmel et al., 1976).
Acute oral LD$_{50}$ for rats is 47 mg/kg, mice 39 mg/kg (quoted, RTECS, 1985).

Drinking water standard (final): MCLG: zero; MCL: 0.2 μg/L (U.S. EPA, 1996).

Uses: Not known.

HEPTANE

Synonyms: Dipropylmethane; Gettysolve-C; *n*-Heptane; Heptyl hydride; UN 1206.

CAS Registry Number: 142-82-5
DOT: 1206
DOT label: Flammable liquid
Molecular formula: C_7H_{16}
Formula weight: 100.20
RTECS: MI7700000
Merck reference: 10, 4551

Physical state, color, and odor:
Colorless liquid with a faint, pleasant odor. Odor threshold concentration is 150 ppm (Amoore and Hautala, 1983).

Melting point (°C):
-90.6 (Weast, 1986)

Boiling point (°C):
98.4 (Weast, 1986)

Density (g/cm³):
0.6837 at 20/4 °C (Weast, 1986)
0.6796 at 25.00/4 °C (Aralaguppi et al., 1999)
0.6712 at 35.00/4 °C, 0.6623 at 45.00/4 °C (Sastry et al., 1999)

Diffusivity in water (x 10^{-5} cm²/sec):
0.71 at 20 °C using method of Hayduk and Laudie (1974)

Dissociation constant, pK_a:
>14 (Schwarzenbach et al., 1993)

Flash point (°C):
-3.9 (NIOSH, 1997)
-1 (Affens and McLaren, 1972)

Lower explosive limit (%):
1.05 (NIOSH, 1997)

Upper explosive limit (%):
6.7 (NIOSH, 1997)

Heat of fusion (kcal/mol):
3.355 (Riddick et al., 1986)

Henry's law constant (atm·m³/mol):
0.901, 1.195, and 1.905 at 26.0, 35.8, and 45.0 °C, respectively (Hansen et al., 1995)

Interfacial tension with water (dyn/cm):
50.2 at 25 °C (Donahue and Bartell, 1952)
51.39 at 25 °C (Jańczuk et al., 1993)
51.23 at 20 °C (Fowkes, 1980)

Ionization potential (eV):
9.90 (Franklin et al., 1969)
10.08 (HNU, 1986)

Soil sorption coefficient, log K_{oc}:
Unavailable because experimental methods for estimation of this parameter for aliphatic hydrocarbons are lacking in the documented literature

Octanol/water partition coefficient, log K_{ow}:
4.66 (Tewari et al., 1982; Wasik et al., 1981)

Solubility in organics:
In methanol, g/L: 181 at 5 °C, 200 at 10 °C, 225 at 15 °C, 254 at 20 °C, 287 at 25 °C, 327 at 30 °C, 378 at 35 °C, 450 at 40 °C (Kiser et al., 1961).

Solubility in water:
In mg/kg: 2.24 at 25 °C, 2.63 at 40.1 °C, 3.11 at 55.7 °C (Price, 1976)
2.93 mg/kg at 25 °C (McAuliffe, 1963, 1966)
4.39 mg/kg at 0 °C, 3.37 mg/kg at 25 °C (Polak and Lu, 1973)
35.7 μmol/L at 25.0 °C (Tewari et al., 1982; Wasik et al., 1981)
2.19, 2.66 mg/L at 25 °C (Mackay and Shiu, 1981)
0.07 mL/L at 15.5 °C (Fühner, 1924)
2.9 mg/L (Coutant and Keigley, 1988)
70 mg/kg at 25 °C (Stearns et al., 1947)
Mole fraction x 10^7: 3.51, 3.63, 4.78, 4.07, and 4.32 at 4.3, 13.5, 25.0, 35.0, and 45.0 °C, respectively (Nelson and DeLigny, 1968)
5 x 10^{-7} in seawater at 25 °C (mole fraction, Krasnoshchekova and Gubergrits, 1973)

Vapor density:
4.10 g/L at 25 °C, 3.46 (air = 1)

Vapor pressure (mmHg):
35 at 20 °C, 58 at 30 °C (Verschueren, 1983)
46 at 25 °C (Milligan, 1924)
45.85 at 25 °C (Wilhoit and Zwolinski, 1971)
44.69 at 25.00 °C (Hussam and Carr, 1985)
14.8, 25.0, 45.7, 73.8, and 114.9 at 4.3, 13.5, 25.0, 35.0, and 45.0 °C, respectively (Nelson and DeLigny, 1968)

Environmental fate:
Biological. Heptane may biodegrade in two ways. The first is the formation of heptyl hydroperoxide, which decomposes to 1-heptanol followed by oxidation to heptanoic acid. The other pathway involves dehydrogenation to 1-heptene, which may react with water forming 1-heptanol (Dugan, 1972). Microorganisms can oxidize alkanes under aerobic conditions (Singer and Finnerty, 1984). The most common degradative pathway involves the oxidation of the terminal methyl group forming the corresponding alcohol (1-heptanol). The alcohol may

undergo a series of dehydrogenation steps forming heptanal followed by oxidation forming heptanoic acid. The acid may then be metabolized by β-oxidation to form the mineralization products, carbon dioxide and water (Singer and Finnerty, 1984). Hou (1982) reported hexanoic acid as a degradation product by the microorganism *Pseudomonas aeruginosa*.

Photolytic. The following rate constants were reported for the reaction of hexane and OH radicals in the atmosphere: 7.15 x 10^{-12} cm^3/molecule·sec (Atkinson, 1990). Photooxidation reaction rate constants of 7.19 x 10^{-12} and 1.36 x 10^{-16} cm^3/molecule·sec were reported for the reaction of heptane with OH and NO$_3$, respectively (Sabljić and Güsten, 1990). Based on a photooxidation rate constant 7.15 x 10^{-12} cm^3/molecule·sec for heptane and OH radicals, the estimated atmospheric lifetime is 19 h in summer sunlight (Altshuller, 1991).

Chemical/Physical. Complete combustion in air yields carbon dioxide and water vapor. Heptane will not hydrolyze because it has no hydrolyzable functional group.

Exposure limits: NIOSH REL: TWA 85 ppm (350 mg/m^3), 15-min ceiling 440 ppm (1,800 mg/m^3), IDLH 750 ppm; OSHA PEL: TWA 500 ppm (2,000 mg/m^3); ACGIH TLV: TWA 400 ppm, STEL 500 ppm (adopted).

Symptoms of exposure: May cause nausea and dizziness and may impart a gasoline-like taste (Patnaik, 1992)

Toxicity:
LD$_{50}$ (24-h) for goldfish 4 mg/L (quoted, Verschueren, 1983).
LD$_{50}$ (intravenous) for mice 222 mg/kg (quoted, RTECS, 1985).

Source: Schauer et al. (1999) reported heptane in a diesel-powered medium-duty truck exhaust at an emission rate of 470 μg/km.

Uses: Standard in testing knock of gasoline engines and for octane rating determinations; anesthetic; solvent; organic synthesis.

2-HEPTANONE

Synonyms: Amyl methyl ketone; *n*-Amyl methyl ketone; Methyl amyl ketone; Methyl *n*-amyl ketone; UN 1110.

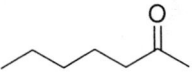

CAS Registry Number: 110-43-0
DOT: 1110
DOT label: Combustible liquid
Molecular formula: $C_7H_{14}O$
Formula weight: 114.19
RTECS: MJ5075000
Merck reference: 10, 4555

Physical state, color, and odor:
Colorless liquid with a banana-like odor. Can be detected at a concentration of 140 μg/kg (Buttery et al., 1969a).

Melting point (°C):
-35.5 (Weast, 1986)
-27 (Verschueren, 1983)

Boiling point (°C):
151.4 (Weast, 1986)
150 (Verschueren, 1983)

Density (g/cm³):
0.8111 at 20/4 °C (Weast, 1986)
0.8197 at 15/4 °C (Sax and Lewis, 1987)
0.8115 at 25/4 °C (Ginnings et al., 1940)
0.81537 at 20/4 °C, 0.81123 at 25/4 °C (Riddick et al., 1986)

Diffusivity in water (x 10^{-5} cm²/sec):
0.72 at 20 °C using method of Hayduk and Laudie (1974)

Flash point (°C):
39.2 (NIOSH, 1997)
47 (Aldrich, 1988)

Lower explosive limit (%):
1.1 at 66.7 °C (NIOSH, 1997)

Upper explosive limit (%):
7.9 at 122.1 °C (NIOSH, 1997)

Henry's law constant (x 10^{-4} atm·m³/mol at 25 °C):
1.44 (Buttery et al., 1969)
1.69 (Shiu and Mackay, 1997)

Interfacial tension with water (dyn/cm):
12.4 at 25 °C (Donahue and Bartell, 1952)

Ionization potential (eV):
9.33 (NIOSH, 1997)

Soil sorption coefficient, log K_{oc}:
Unavailable because experimental methods for estimation of this parameter for aliphatic ketones are lacking in the documented literature

Octanol/water partition coefficient, log K_{ow}:
2.03 (Sangster, 1989)
1.98 (Tewari et al., 1982; Wasik et al., 1981, 1983)

Solubility in organics:
Miscible with organic solvents (Hawley, 1981)

Solubility in water:
4,339 mg/L at 20 °C (Mackay and Yeun, 1983)
In wt %: 0.44 at 20 °C, 0.43 at 25 °C, 0.40 at 30 °C (Ginnings et al., 1940)
35.7 mM at 25.0 °C (Tewari et al., 1982; Wasik et al., 1981, 1983)
In wt %: 0.649 at 0°C, 0.535 at 9.7 °C, 0.436 at 19.8 °C, 0.358 at 30.7 °C, 0.343 at 39.7 °C, 0.336 at 49.8 °C, 0.333 at 60.2 °C, 0.314 at 70.1 °C, 0.348 at 80.2 °C, 0.353 at 90.5 °C (Stephenson, 1992)

Vapor density:
4.67 g/L at 25 °C, 3.94 (air = 1)

Vapor pressure (mmHg at 20 °C):
3 (NIOSH, 1997)
2.6 (Hawley, 1981)

Environmental fate:
   Biological. Heukelekian and Rand (1955) reported a 10-d BOD value of 0.50 g/g which is 17.8% of the ThOD value of 2.81 g/g.
   Chemical/Physical. 2-Heptanone will not hydrolyze because it has no hydrolyzable functional group.

Exposure limits: NIOSH REL: TWA 100 ppm (465 mg/m³), IDLH 800 ppm; OSHA PEL: TWA 100 ppm; ACGIH TLV: TWA 50 ppm (adopted).

Toxicity:
   Acute oral LD_{50} for mice 730 mg/kg, rats 1,670 mg/kg (quoted, RTECS, 1985).

Uses: Ingredient in artificial carnation oils; industrial solvent; synthetic flavoring; solvent for nitrocellulose lacquers; organic synthesis.

3-HEPTANONE

Synonyms: Butyl ethyl ketone; *n*-Butyl ethyl ketone; Ethyl butyl ketone; Heptan-3-one.

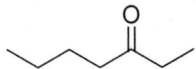

CAS Registry Number: 106-35-4
Molecular formula: $C_7H_{14}O$
Formula weight: 114.19
RTECS: MJ5250000

Physical state, color, and odor:
Colorless liquid with a strong, fruity odor

Melting point (°C):
-39 (Weast, 1986)

Boiling point (°C):
147.8 (Dean, 1987)

Density (g/cm³):
0.8183 at 20/4 °C (Weast, 1986)

Diffusivity in water (x 10⁻⁵ cm²/sec):
0.73 at 20 °C using method of Hayduk and Laudie (1974)

Flash point (°C):
46.5 (open cup, NIOSH, 1997)
41 (Aldrich, 1988)

Henry's law constant (x 10⁻⁵ atm·m³/mol):
4.20 at 20 °C (approximate - calculated from water solubility and vapor pressure)

Ionization potential (eV):
9.02 (NIOSH, 1997)

Soil sorption coefficient, log K_{oc}:
Unavailable because experimental methods for estimation of this parameter for aliphatic ketones are lacking in the documented literature

Octanol/water partition coefficient, log K_{ow}:
1.32 using method of Hansch et al. (1968)

Solubility in organics:
Soluble in alcohol and ether (Weast, 1986)

Solubility in water:
14,300 mg/L at 20 °C (quoted, Verschueren, 1983)
In wt %: 0.717 at 0°C, 0.586 at 9.3 °C, 0.479 at 20.5 °C, 0.431 at 30.7 °C, 0.385 at 39.6 °C, 0.333 at 50.0 °C, 0.309 at 59.8 °C, 0.366 at 70.2 °C, 0.310 at 79.9 °C, 0.309 at 90.1 °C (Stephenson, 1992)

Vapor density:
4.67 g/L at 25 °C, 3.94 (air = 1)

Vapor pressure (mmHg):
4 at 20 °C (NIOSH, 1997)
1.4 at 25 °C (Verschueren, 1983)

Environmental fate:
  Chemical/Physical. 3-Heptanone will not hydrolyze because it has no hydrolyzable functional group.

Exposure limits: NIOSH REL: TWA 50 ppm (230 mg/m^3), IDLH 1,000 ppm; OSHA PEL: TWA 50 ppm; ACGIH TLV: TWA 50 ppm, STEL 75 ppm (adopted).

Symptoms of exposure: May cause irritation of the eyes, skin, and mucous membranes (Patnaik, 1992)

Toxicity:
  Acute oral LD$_{50}$ for rats 2,760 mg/kg (quoted, RTECS, 1985).

Uses: Solvent mixtures for nitrocellulose and polyvinyl resins; in organic synthesis.

cis-2-HEPTENE

Synonyms: (Z)-2-Heptene; *cis*-2-Heptylene.

CAS Registry Number: 6443-92-1
DOT: 2278
Molecular formula: C_7H_{14}
Formula weight: 98.19

Physical state and color:
Colorless liquid

Boiling point (°C):
98.5 (Weast, 1986)
98.41 (Wilhoit and Zwolinski, 1971)

Density (g/cm³):
0.708 at 20/4 °C (Weast, 1986)

Diffusivity in water (x 10^{-5} cm²/sec):
0.73 at 20 °C using method of Hayduk and Laudie (1974)

Dissociation constant, pK_a:
>14 (Schwarzenbach et al., 1993)

Flash point (°C):
-6 (Aldrich, 1988)
-2.2 (commercial grade containing both *cis* and *trans* isomers, Hawley, 1981)

Henry's law constant (atm·m³/mol):
0.413 at 20 °C (approximate - calculated from water solubility and vapor pressure)

Soil sorption coefficient, log K_{oc}:
Unavailable because experimental methods for estimation of this parameter for aliphatic hydrocarbons are lacking in the documented literature

Octanol/water partition coefficient, log K_{ow}:
2.88 using method of Hansch et al. (1968)

Solubility in organics:
Soluble in acetone, alcohol, benzene, chloroform, and ether (Weast, 1986)

Solubility in water:
15 mg/kg at 25 °C (isomeric mixture, McAuliffe, 1966)
15.0 mg/kg at 23.5 °C (isomeric mixture, Schwarz, 1980)
271.6 μmol/L at 25 °C (Natarajan and Venkatachalam, 1972)

Vapor density:
4.01 g/L at 25 °C, 3.39 (air = 1)

Vapor pressure (mmHg):
48 at 25 °C (Wilhoit and Zwolinski, 1971)

Use: Organic synthesis.

trans-2-HEPTENE

Synonyms: (*E*)-2-Heptene; *trans*-2-Heptylene.

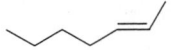

CAS Registry Number: 14686-13-6
DOT: 2278
Molecular formula: C_7H_{14}
Formula weight: 98.19

Physical state and color:
Colorless liquid

Melting point (°C):
-109.5 (Weast, 1986)

Boiling point (°C):
98 (Weast, 1986)

Density (g/cm³):
0.7012 at 20/4 °C (Weast, 1986)

Diffusivity in water (x 10⁻⁵ cm²/sec):
0.72 at 20 °C using method of Hayduk and Laudie (1974)

Dissociation constant, pK_a:
>14 (Schwarzenbach et al., 1993)

Flash point (°C):
-1 (Aldrich, 1988)
-2.2 (commercial grade containing both isomers, Hawley, 1981)

Henry's law constant (atm·m³/mol):
0.422 at 25 °C (approximate - calculated from water solubility and vapor pressure)

Soil sorption coefficient, log K_{oc}:
Unavailable because experimental methods for estimation of this parameter for aliphatic hydrocarbons are lacking in the documented literature

Octanol/water partition coefficient, log K_{ow}:
2.88 using method of Hansch et al. (1968)

Solubility in organics:
Soluble in acetone, alcohol, benzene, chloroform, and ether (Weast, 1986)

Solubility in water:
15 mg/kg at 25 °C (isomeric mixture, McAuliffe, 1966)
15.0 mg/kg at 23.5 °C (isomeric mixture, Schwarz, 1980)
271.6 μmol/L at 25 °C (isomeric mixture, Natarajan and Venkatachalam, 1972)

Vapor density:
4.01 g/L at 25 °C, 3.39 (air = 1)

Vapor pressure (mmHg):
49 at 25 °C (Wilhoit and Zwolinski, 1971)

Uses: Organic synthesis.

HEXACHLOROBENZENE

Synonyms: Amatin; Anticarie; Bunt-cure; Bunt-no-more; Co-op hexa; Granox NM; HCB; Hexa C.B.; Julin's carbon chloride; No bunt; No bunt 40; No bunt 80; No bunt liquid; Penta-chlorophenyl chloride; Perchlorobenzene; Phenyl perchloryl; RCRA waste number U127; Sanocide; Smut-go; Snieciotox; UN 2729.

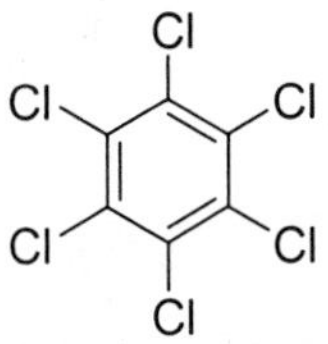

CAS Registry Number: 118-74-1
DOT: 2729
DOT label: Poison
Molecular formula: C_6Cl_6
Formula weight: 284.78
RTECS: DA2975000
Merck reference: 10, 4573

Physical state and color:
Monoclinic, white crystals

Melting point (°C):
230 (Weast, 1986)
227 (Standen, 1964)

Boiling point (°C):
323-326 (Aldrich, 1988)

Density (g/cm³):
1.5691 at 23.6/4 °C (Weast, 1986)
2.049 at 20/4 °C (Melnikov, 1971)

Diffusivity in water (x 10^{-5} cm²/sec):
0.55 at 20 °C using method of Hayduk and Laudie (1974)

Flash point (°C):
242 (Hawley, 1981)

Entropy of fusion (cal/mol·K):
12.0 (Mallard and Linstrom, 1998)
13.7 (Tsonopoulos and Prausnitz, 1971)

Heat of fusion (kcal/mol):
5.354 (Miller et al., 1984)
6.87 (Tsonopoulos and Prausnitz, 1971)

Henry's law constant (x 10^{-4} atm·m³/mol):
17 at 25 °C (Warner et al., 1987)
At 25 °C: 13.1 and 17 at 23 °C in distilled water and seawater, respectively (Atlas et al., 1982)
71 at 20 °C (Oliver, 1985)

557

1.03, 1.97, 2.97, 3.44, 4.15, 4.73, and 4.78 at 14.8, 20.1, 22.1, 24.2, 34.8, 50.5, and 55 °C, respectively (ten Hulscher et al., 1992)
At 25 °C: 13.2 and 17.3 in distilled water and seawater, respectively (Brownawell et al., 1983)
1.0 at 10 °C (Koelmans et al., 1999)

Ionization potential (eV):
9.0 (Lias, 1998)

Bioconcentration factor, log BCF:
4.27 (fathead minnow), 3.73 (rainbow trout), 4.34 (green sunfish) (Veith et al., 1979; Davies and Dobbs, 1984)
4.08-4.30 (rainbow trout, Oliver and Niimi, 1983)
4.54 (activated sludge), 3.37 (golden ide) (Freitag et al., 1985)
4.34 (fathead minnow, Carlson and Kosian, 1987)
4.39 (algae, Geyer et al., 1984)
4.09 (freshwater fish), 3.16 (fish, microcosm) (Garten and Trabalka, 1983)
6.42 (Atlantic croakers), 6.71 (blue crabs), 5.96 (spotted sea trout), 5.98 (blue catfish) (Pereira et al., 1988)
3.09 (Korte et al., 1978)
5.46 (guppy, Könemann and van Leeuwen, 1980)
4.39 (algae), 3.36 (fish), 4.54 (activated sludge) (Klein et al., 1984)
3.54 (*Macoma nasuta*) (Boese et al., 1990)
3.57 (wet weight based), 1.00 (lipid based) (*Gambusia affinis*, Chaisuksant et al., 1997)
5.55 (pond snail, Legierse et al., 1998)

Soil sorption coefficient, log K_{oc}:
3.59 (Kenaga, 1980a)
4.49 (Briggs, 1981)
2.56 (Speyer soil 2.2, pH 5.8), 2.70 (alfisol, pH 7.5) (Rippen et al., 1982)
6.4 (average value using 10 suspended sediment samples collected from the St. Clair and Detroit Rivers, Lau et al., 1989)
3.23 (Captina silt loam), 4.73 (McLaurin sandy loam) (Walton et al., 1992)
K_d = 37.5 mL/g (Barcelona coastal sediments, Bayona et al., 1991)
4.98 (lake sediment, Schrap et al., 1994)
4.66 (fine sand, Enfield et al., 1989)
5.90 (river sediments, Oliver and Charlton, 1984)
5.60 (Detroit River sediment, Jepsen and Lick, 1999)

Octanol/water partition coefficient, log K_{ow}:
5.45 (Travis and Arms, 1988)
5.50 (Chiou et al., 1982; Pereira et al., 1988)
5.66 (Hammers et al., 1982; Isnard and Lambert, 1988)
5.47 (Miller et al., 1984)
5.70, 5.79 (Garst and Wilson, 1984)
5.23 (Mackay, 1982)
5.31 (Watarai et al., 1982)
5.57 (Brooke et al., 1986)
5.20, 5.55 (Geyer et al., 1984)
5.44 (Briggs, 1981)
3.93 (Veith et al., 1980)
5.2 (Platford et al., 1982)

6.06 (Schwarzenbach et al., 1983)
5.68 at 13 °C, 5.70 at 19 °C, 5.58 at 28 °C, 5.17 at 33 °C (Opperhuizen et al., 1988)
6.22 (Rao and Davidson, 1980)
5.00 (Könemann et al., 1979)
6.86 (Burkhard et al., 1985a)
6.42 (DeKock and Lord, 1987)
5.731 (de Bruijn et al., 1989)
5.85 at 25 °C (Paschke et al., 1998)
5.74, 5.60, 5.46, 5.30, and 5.17 at 5, 15, 25, 35, and 45 °C, respectively (Bahadur et al., 1997)
5.55 at 25 °C (Andersson and Schräder, 1999)

Solubility in organics:
Soluble in acetone, benzene, ether, and chloroform (U.S. EPA, 1985)

Solubility in water (μg/L):
110 at 24 °C, 6 at 25 °C (Metcalf et al., 1973a)
4.7 at 25 °C (Miller et al., 1985)
5.0 at 25 °C (Weil et al., 1974; Yalkowsky et al., 1979; Paschkle et al., 1998)
20 (Gile and Gillett, 1979)
40 at 20 °C (Riederer, 1990)
6.2 at 23.5 °C (Farmer et al., 1980)
14 at 22 °C (Schrap et al., 1995)
2.2, 3.5, 5.44, 8.53, and 14 at 5, 15, 25, 35, and 45 °C, respectively (Shiu et al., 1997)
5 at 22 °C (Jepsen and Lick, 1999)

Vapor pressure (x 10^{-6} mmHg):
10.89 at 20 °C (Isensee et al., 1976)
7.5 at 20 °C (Riederer, 1990)
912 at 25 °C (Bidleman, 1984)
18 at 25 °C (Banerjee et al., 1990)
4.2, 8.5, and 17.0 at 15, 20, and 25 °C, respectively (Farmer et al., 1980)
893, 1,190 at 25 °C (Hinckley et al., 1990)

Environmental fate:
   Biological. In activated sludge, only 1.5% of the applied hexachlorobenzene mineralized to carbon dioxide after 5 d (Freitag et al., 1985). Reductive monodechlorination occurred in an anaerobic sewage sludge yielding principally 1,3,5-trichlorobenzene. Other compounds identified include pentachlorobenzene, 1,2,3,5-tetrachlorobenzene, and dichlorobenzenes (Fathepure et al., 1988). In a 5-d experiment, [^{14}C]hexachlorobenzene applied to soil water suspensions under aerobic and anaerobic conditions gave $^{14}CO_2$ yields of 0.4 and 0.2%, respectively (Scheunert et al., 1987).
   When hexachlorobenzene was statically incubated in the dark at 25 °C with yeast extract and settled domestic wastewater inoculum, no significant biodegradation was observed. At a concentration of 5 mg/L, percent losses after 7, 14, 21, and 28-d incubation periods were 56, 30, 8, and 5, respectively. At a concentration of 10 mg/L, only 21 and 3% losses were observed after the 7 and 14-d incubation periods, respectively. The decrease in concentration over time suggests biodegradation followed a deadaptive process (Tabak et al., 1981).
   The half-life of hexachlorobenzene in an anaerobic enrichment culture was 1.4 d (Beurskens et al., 1993).
   Groundwater. Hexachlorobenzene has a high potential to leach to groundwater (U.S. EPA, 1986).

Photolytic. Solid hexachlorobenzene exposed to artificial sunlight for 5 months photolyzed at a very slow rate with no decomposition products identified (Plimmer and Klingebiel, 1976). The sunlight irradiation of hexachlorobenzene (20 g) in a 100-mL borosilicate glass-stoppered Erlenmeyer flask for 56 d yielded 64 ppm pentachlorobiphenyl (Uyeta et al., 1976). A carbon dioxide yield <0.1% was observed when hexachlorobenzene adsorbed on silica gel was irradiated with light (λ >290 nm) for 17 h (Freitag et al., 1985).

Irradiation ($\lambda \geq 285$ nm) of hexachlorobenzene (1.1-1.2 mM/L) in an acetonitrile-water mixture containing acetone (0.553 mM/L) as a sensitizer gave the following products (% yield): pentachlorobenzene (71.0), 1,2,3,4-tetrachlorobenzene (0.6), 1,2,3,5-tetrachlorobenzene (2.2), and 1,2,4,5-tetrachlorobenzene (3.7) (Choudhry and Hutzinger, 1984). Without acetone, the identified photolysis products (% yield) included 1,2,3,4,5-pentachlorobenzene (76.8), 1,2,3,5-tetrachlorobenzene (1.2), 1,2,4,5-tetrachlorobenzene (1.7), 1,2,4-trichlorobenzene (0.2) (Choudhry et al., 1984).

In another study, the irradiation (λ = 290-310 nm) of hexachlorobenzene in an aqueous solution gave only pentachlorobenzene and possibly pentachlorophenol as the transformation products. The photolysis rate increased with the addition of naturally occurring substances (tryptophan and pond proteins) and abiotic sensitizers (diphenylamine and skatole) (Hirsch and Hutzinger, 1989).

When an aqueous solution containing hexachlorobenzene (150 nM) and a nonionic surfactant micelle (0.50 M Brij 58, a polyoxyethylene cetyl ether) was illuminated by a photo-reactor equipped with 253.7-nm monochromatic UV lamps, significant concentrations of penta-chlorobenzene, all tetra-, tri-, and dichlorobenzenes, chlorobenzene, benzene, phenol, hydrogen, and chloride ions were formed. Two compounds, namely 1,2-dichlorobenzene and 1,2,3,4-tetrachlorobenzene, formed in minor amounts (<40 ppb). The half-life for this reaction, based on the first-order photodecomposition rate of 1.44×10^{-2}/sec, is 48 sec (Chu and Jafvert, 1994).

Chemical/Physical. No hydrolysis was observed after 13 d at 85 °C and pH values of 3, 7, and 11 (Ellington et al., 1987).

Exposure limits: ACGIH TLV: TWA 0.002 mg/m^3 (adopted).

Toxicity:
Concentration that reduced the fertility of *Daphnia magna* in 2 wk for 50% (EC$_{50}$) of the population was 0.016 mg/L (Calamari et al., 1983).

EC$_{10}$ and EC$_{50}$ concentrations inhibiting the growth of alga *Scenedesmus subspicatus* in 96 h were both >0.01 mg/L (Geyer et al., 1985).

EC$_{50}$ (96-h) and EC$_{50}$ (3-h) concentrations that inhibit the growth of 50% of *Selenastrum capricornutum* population are <0.03 and 0.030 mg/L, respectively (Calamari et al., 1983).

IC$_{50}$ (24-h) for *Daphnia magna* <0.03 mg/L (Calamari et al., 1983).

LC$_{50}$ (contact) for earthworm (*Eisenia fetida*) >1,000 g/cm^2 (Neuhauser et al., 1985).

LC$_{50}$ (14-d) for *Poecilia reticulata* >319 μg/L (Könemann, 1981).

LC$_{50}$ (48-h) for *Salmo gairdneri* <0.03 mg/L, *Brachydanio rerio* <0.03 mg/L (Calamari et al., 1983).

LC$_{50}$ (inhalation) for mice 4 gm/m^3, rats 3,600 mg/m^3 (Hartley and Kidd, 1987).

LD$_{50}$ for *Larus argentatus* embryos 4.3 mg/kg (Boersma et al., 1986).

Acute oral LD$_{50}$ for cats 1,700 mg/kg, mice 4 gm/kg, rats 10,000 mg/kg, rabbits 2,600 mg/kg (Hartley and Kidd, 1987).

Drinking water standard: MCLG: zero; MCL: 1 μg/L (U.S. EPA, 1996).

Source: Hexachlorobenzene may enter the environment from incomplete combustion of chlorinated compounds including mirex, kepone, chlorobenzenes, pentachlorophenol, PVC, polychlorinated biphenyls, and chlorinated solvents (Ahling et al., 1978; Dellinger et al., 1991). In addition, hexachlorobenzene may enter the environment as a reaction by-product in the production of carbon tetrachloride, dichloroethylene, trichloroethylene, tetrachloroethylene, pentachloronitrobenzene, and vinyl chloride monomer (quoted, Verschueren, 1983).

Uses: Manufacture of pentachlorophenol; seed fungicide; wood preservative.

HEXACHLOROBUTADIENE

Synonyms: Dolen-pur; GP-40-66:120; HCBD; Hexachlorbutadiene; 1,1,2,3,4,4-Hexachloro-butadiene; 1,3-Hexachlorobutadiene; Hexachloro-1,3-butadiene; **1,1,2,3,4,4-Hexachloro-1,3-butadiene**; Perchlorobutadiene; RCRA waste number U128; UN 2279.

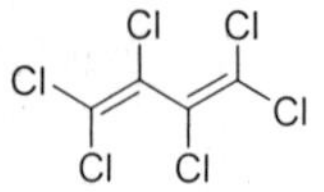

CAS Registry Number: 87-68-3
DOT: 2279
DOT label: Poison
Molecular formula: C_4Cl_6
Formula weight: 260.76
RTECS: EJ0700000

Physical state, color, and odor:
Clear, yellowish-green liquid with a mild to pungent, turpentine-like odor. Odor threshold concentration is 6 ppb (Keith and Walters, 1992).

Melting point (°C):
-21 (Weast, 1986)

Boiling point (°C):
215 (Weast, 1986)

Density (g/cm³):
1.6820 at 20/4 °C (Melnikov, 1971)

Diffusivity in water (x 10^{-5} cm²/sec):
0.65 at 20 °C using method of Hayduk and Laudie (1974)

Henry's law constant (x 10^{-3} atm·m³/mol):
26 (Pankow and Rosen, 1988)
1.02 at 25 °C (Warner et al., 1987)
25.0 at 20 °C (Pearson and McConnell, 1975)
7.8 (Warner et al., 1987)
4.3 at 20 °C (Oliver, 1985)
3.55, 5.87, 6.90, 10.5, and 15.3 at 2.0, 6.0, 10.0, 18.0, and 25.0 °C, respectively (Dewulf et al., 1999)

Bioconcentration factor, log BCF:
3.76, 4.23 (*Oncorhynchus mykiss*, Devillers et al., 1996)
4.50 (Atlantic croakers), 3.97 (blue crabs), 4.06 (spotted sea trout), 4.55 (blue catfish) (Pereira et al., 1988)

Soil sorption coefficient, log K_{oc}:
6.1 (average using 9 suspended sediment samples from the St. Clair and Detroit Rivers, Lau et al., 1989)

Octanol/water partition coefficient, log K_{ow}:
4.78 (Banerjee et al., 1980)
4.90 (Chiou, 1985; Pereira et al., 1988)

Solubility in organics:
Soluble in ethanol and ether (U.S. EPA, 1985)

Solubility in water (mg/L):
3.23 at 25 °C (Banerjee et al., 1980)
4 at 20-25 °C (Geyer et al., 1980)
2.55 at 20 °C (Howard, 1989)

Vapor density:
10.66 g/L at 25 °C, 9.00 (air = 1)

Vapor pressure (mmHg):
0.15 at 20 °C (McConnell et al., 1975)

Environmental fate:
 Chemical/Physical. Hexachlorobutadiene will not hydrolyze to any reasonable extent (Kollig, 1993).

Exposure limits: Potential occupational carcinogen. NIOSH REL: TWA 20 ppb (240 mg/m^3); ACGIH TLV: TWA 0.02 ppm (adopted).

Toxicity:
 LC_{50} (contact) for earthworm (*Eisenia fetida*) 10 μg/cm^2 (Neuhauser et al., 1985).
 LC_{50} (14-d) for *Poecilia reticulata* 394 μg/L (Könemann, 1981).
 LC_{50} (48-h) for zebra fish 1 mg/L (Slooff, 1979).
 LC_{50} (inhalation) for rats 1,600 ppb/4-h; acute oral LD_{50} for rats 113 mg/kg.
 LD_{50} (skin) for rabbits 430 mg/kg (quoted, RTECS, 1985).

Drinking water standard (tentative): MCLG: 1 μg/L; MCL: none proposed (U.S. EPA, 1996).

Source: Hydraulic fluids and rubber (quoted, Verschueren, 1983)

Uses: Solvent for elastomers, natural rubber, synthetic rubber; heat-transfer liquid; transformer and hydraulic fluid; wash liquor for removing C_4 and higher hydrocarbons; sniff gas recovery agent in chlorine plants; chemical intermediate for fluorinated lubricants and rubber compounds; fluid for gyroscopes; fumigant for grapes.

HEXACHLOROCYCLOPENTADIENE

Synonyms: C-56; Graphlox; HCCP; HCCPD; HCPD; Hex; **1,2,3,4,5,5-Hexachloro-1,3-cyclo-pentadiene**; HRS 1655; NCI-C55607; PCL; Perchlorocyclopentadiene; RCRA waste number U130; UN 2646.

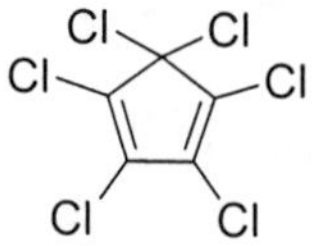

CAS Registry Number: 77-47-4
DOT: 2646
DOT label: Poison
Molecular formula: C_5Cl_6
Formula weight: 272.77
RTECS: GY1225000

Physical state, color, and odor:
Pale yellow to greenish-yellow liquid with a harsh, unpleasant odor. Odor threshold concentrations ranged from 1.4 to 1.6 μg/L (Keith and Walters, 1992).

Melting point (°C):
-9 (Weast, 1986)

Boiling point (°C):
236-238 (Melnikov, 1971)

Density (g/cm³):
1.7019 at 25/4 °C (Weast, 1986)
1.7119 at 20/4 °C (Ungnade and McBee, 1957)

Diffusivity in water (x 10^{-5} cm²/sec):
0.67 at 20 °C using method of Hayduk and Laudie (1974)

Flash point (°C):
Noncombustible liquid (NIOSH, 1997)

Lower explosive limit (%):
Noncombustible liquid (NIOSH, 1997)

Upper explosive limit (%):
Noncombustible liquid (NIOSH, 1997)

Henry's law constant (x 10^{-2} atm·m³/mol):
1.6 (Pankow and Rosen, 1988)

Bioconcentration factor, log BCF:
3.04 (*Chlorella fusca*, Freitag et al., 1982; Geyer et al., 1984)
3.38 (activated sludge), 3.09 (golden ide) (Freitag et al., 1985)
1.47 (fathead minnow, Veith et al., 1979)
2.97 (snail), 3.21 (mosquito) (Lu et al., 1975)

Soil sorption coefficient, log K_oc:
3.63 (Chou and Griffin, 1983)

Octanol/water partition coefficient, log K_ow:
5.00 (McDuffie, 1981)
5.51 (Mackay et al., 1982)
5.04 (Geyer et al., 1984)

Solubility in organics:
Based on structurally similar compounds, hexachlorocyclopentadiene is expected to be soluble in benzene, ethanol, chloroform, methylene chloride, trichloroethylene, and other liquid halogenated solvents.

Solubility in water (mg/L):
0.805 (Lu et al., 1975)
1.8 at 25 °C (Zepp et al., 1979)
At 22 °C: 1.11 (distilled water), 1.14 (deionized water), 1.08 (tap water), 1.08 (Sugar Creek water) (Chou and Griffin, 1983)

Vapor density:
11.15 g/L at 25 °C, 9.42 (air = 1)

Vapor pressure (mmHg):
0.081 at 25 °C (Ungnade and McBee, 1957)
1 at 78-79 °C (Chou and Griffen, 1983)

Environmental fate:
Biological. When hexachlorocyclopentadiene (5 and 10 mg/L) was statically incubated in the dark at 25 °C with yeast extract and settled domestic wastewater inoculum for 7 d, 100% biodegradation with rapid adaptation was observed (Tabak et al., 1981). In a model ecosystem containing plankton, *Daphnia magna*, mosquito larva (*Culex pipiens quinquefasciatus*), fish (*Cambusia affinis*), alga (*Oedogonium cardiacum*), and snail (*Physa* sp.), hexachlorocyclo-pentadiene degraded slightly, but no products were identified (Lu et al., 1975).
Photolytic. The major photolysis and hydrolysis products identified in distilled water are pentachlorocyclopentenone and hexachlorocyclopentenone. In mineralized water, the products identified include *cis-* and *trans*-pentachlorobutadiene, tetrachlorobutenyne, and pentachloro-pentadienoic acid (Chou and Griffin, 1983). In a similar experiment, irradiation of hexachloro-cyclopentadiene in water by mercury-vapor lamps resulted in the formation of 2,3,4,4,5-pentachloro-2-cyclopentenone. This was found to hydrolyze partially to hexachloroindenone (Butz et al., 1982). Other photodegradation products identified include hexachloro-2-cyclo-pentenone and hexachloro-3-cyclopentenone as major products. Secondary photodegradation products reported include pentachloro-*cis*-2,4-pentadienoic acid, *Z-* and *E*-pentachlorobuta-diene, and tetrachlorobutyne (Chou et al., 1987). In natural surface waters, direct photolysis of hexachlorobutadiene via sunlight results in a half-life of 10.7 min (Wolfe et al., 1982).
Anticipated products from the reaction of hexachlorocyclopentadiene with ozone or OH radicals in the atmosphere are phosgene, diacylchlorides, ketones, and chlorine radicals (Cupitt, 1980). Phosgene is hydrolyzed readily to hydrogen chloride and carbon dioxide (Morrison and Boyd, 1971).
Chemical/Physical. Slowly reacts with water forming hydrochloric acid and 1,1-dihy-droxytetrachlorocyclopentadiene (Kollig, 1993; NIOSH, 1997). The diene is unstable and will react forming polymers (Kollig, 1993).

Hexachlorocyclopentadiene (0.020-0.029 mM) reacted with OH radicals in water (pH 2.8) at a rate of 2.2 x 10^9/M·sec. At pH 3.4, the reaction rates at concentrations of 0.0054 and 0.14 mM were 8.8 x 10^8 and 3.4 x 10^9/M·sec, respectively (Haag and Yao, 1992).

Exposure limits: NIOSH REL: 10 ppb (100 mg/m^3); ACGIH TLV: TWA 0.01 ppm (adopted).

Toxicity:
    LC$_{50}$ (30-d) and LC$_{50}$ (96-h) for fathead minnows were 7.0 and 6.7, respectively (Spehar et al., 1979).

Drinking water standard (final): MCLG: 50 μg/L; MCL: 50 μg/L (U.S. EPA, 1996).

Uses: Intermediate in the synthesis of dyes, cyclodiene pesticides, (aldrin, dieldrin, endosulfan), fungicides, and pharmaceuticals; manufacture of chlorendic anhydride and chlorendic acid.

HEXACHLOROETHANE

Synonyms: Avlothane; Carbon hexachloride; Carbon trichloride; Distokal; Distopan; Distopin; Egitol; Ethane hexachloride; Ethylene hexachloride; Falkitol; Fasciolin; HCE; 1,1,1,2,2,2-Hexachloroethane; Hexachloroethylene; Mottenhexe; NA 9037; NCI-C04604; Perchloroethane; Phenohep; RCRA waste number U131.

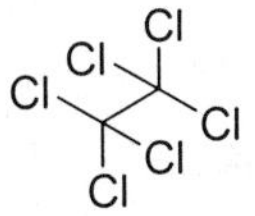

CAS Registry Number: 67-72-1
DOT: 9037
Molecular formula: C_2Cl_6
Formula weight: 236.74
RTECS: KI4025000
Merck reference: 10, 4574

Physical state, color, and odor:
Rhombic, triclinic or cubic, colorless crystals with a camphor-like odor. Odor threshold concentration is 0.15 ppm (Amoore and Hautala, 1983).

Melting point (°C):
186.6 (Weast, 1986)

Boiling point (°C):
187.5 (Dean, 1987)
190-195 (sublimes, Aldrich, 1988)

Density (g/cm³):
2.091 at 20/4 °C (Weast, 1986)

Diffusivity in water (x 10^{-5} cm²/sec):
0.63 at 20 °C using method of Hayduk and Laudie (1974)

Flash point (°C):
Noncombustible solid (NIOSH, 1997)

Heat of fusion (kcal/mol):
0.642 (Dean, 1987)

Henry's law constant (x 10^{-3} atm·m³/mol):
2.5 (Pankow and Rosen, 1988)
10.3 at 30 °C (Jeffers et al., 1989)
5.93, 5.59, 5.91, 8.35, and 10.3 at 10, 15, 20, 25, and 30 °C, respectively (Ashworth et al., 1988)
1.43, 2.81, and 5.31 at 10, 20, and 30 °C, respectively (Munz and Roberts, 1987)

Ionization potential (eV):
11.22 (NIOSH, 1997)
12.11 (Horvath, 1982)

Bioconcentration factor, log BCF:
2.14 (bluegill sunfish, Veith et al., 1980)
2.71 (*Oncorhynchus mykiss*, Devillers et al., 1996)

Soil sorption coefficient, log K_{oc}:
3.34 (Abdul et al., 1987)

Octanol/water partition coefficient, log K_{ow}:
4.62 (Abdul et al., 1987)
3.58 (Könemann et al., 1979)
3.93 (Veith et al., 1980)
4.14 (Chiou, 1985)

Solubility in organics:
Soluble in ethanol, benzene, chloroform, and ether (U.S. EPA, 1985)

Solubility in water:
49.9 mg/L at 22 °C (quoted, Horvath, 1982)
50 mg/kg at 25 °C (McGovern, 1943)
50 ppm at 20 °C (Konietzko, 1984)

Vapor density:
6.3 kg/m^3 at the sublimation point and 750 mmHg (Konietzko, 1984)

Vapor pressure (mmHg):
0.8 at 30 °C (Verschueren, 1983)
0.18 at 20 °C (quoted, Munz and Roberts, 1987)

Environmental fate:
   Biological. Under aerobic conditions or in experimental systems containing mixed cultures, hexachloroethane was reported to degrade to tetrachloroethane (Vogel et al., 1987). In an uninhibited anoxic-sediment water suspension, hexachloroethane degraded to tetrachloroethylene. The reported half-life for this transformation was 19.7 min (Jafvert and Wolfe, 1987). When hexachloroethane (5 and 10 mg/L) was statically incubated in the dark at 25 °C with yeast extract and settled domestic wastewater inoculum for 7 d, 100% biodegradation with rapid adaptation was observed (Tabak et al., 1981).
   Photolytic. When an aqueous solution containing hexachloroethane was photooxidized by UV light at 90-95 °C, 25, 50, and 75% degraded to carbon dioxide after 25.2, 93.7, and 172.0 h, respectively (Knoevenagel and Himmelreich, 1976).
   Chemical/Physical. The reported hydrolysis half-life at 25 °C and pH 7 is 1.8 x 10^9 yr (Jeffers et al., 1989). No hydrolysis was observed after 13 d at 85 °C and pH values of 3, 7, and 11 (Ellington et al., 1987).
   Under laboratory conditions, the evaporation half-life of hexachloroethane (1 mg/L) from water at 25 °C using a shallow-pitch propeller stirrer at 200 rpm at an average depth of 6.5 cm is 40.7 min (Dilling, 1977).

Exposure limits: Potential occupational carcinogen. NIOSH REL: TWA 1 ppm (10 mg/m^3), IDLH 300 ppm; OSHA PEL: TWA 1 ppm; ACGIH TLV: TWA 1 ppm (adopted).

Symptoms of exposure: Vapors may cause irritation to the eyes and mucous membranes (Patnaik, 1992).

Toxicity:

LC$_{50}$ (contact) for earthworm (*Eisenia fetida*) 19 μg/cm^2 (Neuhauser et al., 1985).

LC$_{50}$ (96-h) for fathead minnows 1.5 mg/L (Veith et al., 1983), bluegill sunfish 0.98 mg/L (Spehar et al., 1982), *Cyprinodon variegatus* 2.4 ppm using natural seawater (Heitmuller et al., 1981).

LC$_{50}$ (72-h) for *Cyprinodon variegatus* 2.4 ppm (Heitmuller et al., 1981).

LC$_{50}$ (48-h) for *Daphnia magna* 8.1 mg/L (LeBlanc, 1980), *Cyprinodon variegatus* 2.8 ppm (Heitmuller et al., 1981).

LC$_{50}$ (24-h) for *Daphnia magna* 26 mg/L (LeBlanc, 1980), *Cyprinodon variegatus* 3.1 ppm (Heitmuller et al., 1981).

Acute oral LD$_{50}$ for rats 4,460 mg/kg, guinea pigs 4,970 mg/kg (quoted, RTECS, 1985).

Heitmuller et al. (1981) reported a NOEC of 1.0 ppm.

Drinking water standard: No MCLGs or MCLs have been proposed although hexachloroethane has been listed for regulation (U.S. EPA, 1996).

Uses: Plasticizer for cellulose resins; moth repellant; camphor substitute in cellulose solvent; manufacturing of smoke candles and explosives; rubber vulcanization accelerator; insecticide; refining aluminum alloys.

HEXANE

Synonyms: Gettysolve-B; *n*-Hexane; Hexyl hydride; NCI-C60571; RCRA waste number U114; UN 1208.

CAS Registry Number: 110-54-3
DOT: 1208
DOT label: Flammable liquid
Molecular formula: C_6H_{14}
Formula weight: 86.18
RTECS: MN9275000

Physical state, color, and odor:
Clear, colorless liquid with a faint, gasoline-like odor. Odor threshold concentration is 130 ppm (Amoore and Hautala, 1983).

Melting point (°C):
-95 (Weast, 1986)

Boiling point (°C):
69 (Weast, 1986)
68.74 (Wilhoit and Zwolinski, 1971)

Density (g/cm³):
0.6603 at 20/4 °C (Weast, 1986)
0.65937 at 20/4 °C (Hawley, 1981)
0.6548 at 25.00/4 °C (Aralaguppi et al., 1999)

Diffusivity in water (x 10⁻⁵ cm²/sec):
0.75 at 20 °C using method of Hayduk and Laudie (1974)

Dissociation constant, pK$_a$:
>14 (Schwarzenbach et al., 1993)

Flash point (°C):
15 (Affens and McLaren, 1972)

Lower explosive limit (%):
1.1 (NIOSH, 1997)

Upper explosive limit (%):
7.5 (NIOSH, 1997)

Heat of fusion (kcal/mol):
3.126 (Riddick et al., 1986)

Henry's law constant (atm·m³/mol):
0.238, 0.413, 0.883, 0.768, and 1.56 at 10, 15, 20, 25, and 30 °C, respectively (Ashworth et al., 1988)

Interfacial tension with water (dyn/cm at 25 °C):
51.25 (Harkins et al., 1920)
50.0 (Murphy et al., 1957)
50.80 (20 °C, Fowkes, 1980)
49.7 (Donahue and Bartell, 1952)
49.61 (Jańczuk et al., 1993)

Ionization potential (eV):
10.18 (Franklin et al., 1969)

Soil sorption coefficient, log K_{oc}:
Unavailable because experimental methods for estimation of this parameter for aliphatic
hydrocarbons are lacking in the documented literature

Octanol/water partition coefficient, log K_{ow}:
3.90 (Kamlet et al., 1984)
4.11 (Tewari et al., 1982; Wasik et al., 1981)

Solubility in organics:
Miscible with alcohol, chloroform and ether (Windholz et al., 1983).
In methanol, g/L: 324 at 5 °C, 370 at 10 °C, 427 at 15 °C, 495 at 20 °C, 604 at 25 °C, 830 at
   30 °C. Miscible at higher temperatures (Kiser et al., 1961).

Solubility in water:
In mg/kg: 9.47 at 25 °C, 10.1 at 40.1 °C, 13.2 at 55.7 °C (Price, 1976)
9.5 mg/kg at 25 °C (McAuliffe, 1963, 1966)
16.5 mg/kg at 0 °C, 12.4 mg/kg at 25 °C (Polak and Lu, 1973)
13 mg/L at 20 °C, 75.5 mg/L in salt water at 20 °C (quoted, Verschueren, 1983)
12.3 mg/L at 25 °C. In NaCl solution (salinity, mol/L) at 25 °C, mg/L: 10.48 (0.31), 8.06
   (0.62), 7.54 (1.00), 4.88 (1.50), 3.75 (2.00), 2.55 (2.50) (Aquan-Yuen et al., 1979)
143 μmol/L at 25.0 °C (Tewari et al., 1982; Wasik et al., 1981)
9.52, 18.3 mg/L at 25 °C (Mackay and Shiu, 1981)
0.22 mL/L at 15.5 °C (Führner, 1924)
0.1882 mM at 25 °C (Barone et al., 1967)
At 32 atmHg: 110 and 150 at 20 and 37.8 °C, respectively (Namiot and Beider, 1960)
120 mg/L at 25 °C (McBain and Lissant, 1951)
14 mg/L (Coutant and Keigley, 1988)
24 g/kg at 349.9 °C and 193.7 atmHg (DeLoos et al., 1982)
Mole fraction x 10^{-5}: 3.42, 3.17, 3.83, 2.69, 4.64, and 4.42 at 4.0, 14.0, 25.0, 35.0, 45.0, and
   55.0 °C, respectively (Nelson and DeLigny, 1968)
0.96 mM at 25 °C (Barone et al., 1966)
2.8 and 13 μM at 25 °C in distilled water and seawater, respectively (Krasnoshchekova and
   Gubergrits, 1973)
12.3 mg/L at 25 °C (Leinonen and Mackay, 1973)
0.4 wt % at 37.8 °C (Jones and McCants, 1954)

Vapor density:
3.52 g/L at 25 °C, 2.98 (air = 1)

Vapor pressure (mmHg):
150 at 25 °C (Milligan, 1924)

120 at 20 °C, 190 at 30 °C (Verschueren, 1983)
151.5 at 25 °C (Wilhoit and Zwolinski, 1971)
151.39 at 25.00 °C (Hussam and Carr, 1985)
55.94, 91.8, 151.2, 229.5, 337.9, and 483.3 at 4.0, 14.0, 25.0, 35.0, 45.0, and 55.0 °C, respectively (Nelson and DeLigny, 1968)

Environmental fate:
Biological. n-Hexane may biodegrade in two ways. The first is the formation of hexyl hydroperoxide, which decomposes to 1-hexanol followed by oxidation to hexanoic acid. The other pathway involves dehydrogenation to 1-hexene, which may react with water giving 1-hexanol (Dugan, 1972). Microorganisms can oxidize alkanes under aerobic conditions (Singer and Finnerty, 1984). The most common degradative pathway involves the oxidation of the terminal methyl group forming 1-hexanol. The alcohol may undergo a series of dehydrogenation steps forming a hexanal followed by oxidation to form hexanoic acid. The fatty acid may then be metabolized by β-oxidation to form the mineralization products, carbon dioxide and water (Singer and Finnerty, 1984).

Photolytic. An aqueous solution irradiated by UV light at 50 °C for 1 d resulted in a 50.51% yield of carbon dioxide (Knoevenagel and Himmelreich, 1976). Synthetic air containing gaseous nitrous acid and exposed to artificial sunlight (λ = 300-450 nm) photooxidized hexane into two isomers of hexyl nitrate and peroxyacetal nitrate (Cox et al., 1980).

The following rate constants were reported for the reaction of hexane and OH radicals in the atmosphere: 3.6 x 10^{12} cm³/mol·sec at 300 K (Hendry and Kenley, 1979); 3.8 x 10^9 cm³/mol·sec (Darnall et al., 1976); 5.61 x 10^{-12} cm³/molecule·sec (Atkinson, 1990); 5.19 x 10^{-12} cm³/molecule·sec (DeMore and Bayes, 1999). Photooxidation reaction rate constants of 5.58 x 10^{-12} and 1.05 x 10^{-16} cm³/molecule·sec were reported for the reaction of hexane with OH and NO_3, respectively (Sabljić and Güsten, 1990).

Chemical/Physical. Complete combustion in air yields carbon dioxide and water vapor. Hexane will not hydrolyze because it has no hydrolyzable functional group.

Exposure limits: NIOSH REL: TWA 50 ppm (180 mg/m³), IDLH 1,100 ppm; OSHA PEL: TWA 500 ppm (1,800 mg/m³); ACGIH TLV: TWA 50 ppm (adopted).

Symptoms of exposure: Irritation of respiratory tract. Narcotic at high concentrations (Patnaik, 1992). Dizziness occurs at an air concentration of 5,000 ppm (quoted, Verschueren, 1983). An irritation concentration of 1,800.00 mg/m³ in air was reported by Ruth (1986).

Toxicity:
LD_{50} (24-h) for goldfish 4 mg/L (quoted, Verschueren, 1983).
Acute oral LD_{50} for rats 28,710 mg/kg (quoted, RTECS, 1985).

Drinking water standard: No MCLGs or MCLs have been proposed (U.S. EPA, 1996).

Source: In diesel engine exhaust at a concentration of 1.2% of emitted hydrocarbons (quoted, Verschueren, 1983).

Uses: Determining refractive index of minerals; paint diluent; dyed hexane is used in thermometers instead of mercury; polymerization reaction medium; calibrations; solvent for vegetable oils; alcohol denaturant; chief constituent of petroleum ether, rubber solvent, and gasoline; in organic synthesis.

2-HEXANONE

Synonyms: Butyl ketone; Butyl methyl ketone; *n*-Butyl methyl ketone; Hexanone-2; MBK; Methyl *n*-butyl ketone; MNBK; Propylacetone.

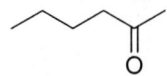

CAS Registry Number: 591-78-6
Molecular formula: $C_6H_{12}O$
Formula weight: 100.16
RTECS: MP1400000
Merck reference: 10, 5909

Physical state, color, and odor:
Clear, colorless liquid with an odor resembling acetone

Melting point (°C):
-56.9 (Dean, 1973)
-55.8 (Riddick et al., 1986)

Boiling point (°C):
128 (Weast, 1986)

Density (g/cm³ at 20/4 °C):
0.8113 (Weast, 1986)
0.8300 (Standen, 1967)

Diffusivity in water (x 10⁻⁵ cm²/sec):
0.78 at 20 °C using method of Hayduk and Laudie (1974)

Dissociation constant, pK$_a$:
-8.30 (Riddick et al., 1986)

Flash point (°C):
-9 (NFPA, 1984)

Upper explosive limit (%):
8.0 (NIOSH, 1997)

Heat of fusion (kcal/mol):
3.56 (Riddick et al., 1986)

Henry's law constant (x 10⁻³ atm·m³/mol):
1.75 at 25 °C (approximate - calculated from water solubility and vapor pressure)

Interfacial tension with water (dyn/cm at 20 °C):
9.73 (Harkins et al., 1920)

Ionization potential (eV):
9.35 (Franklin et al., 1969)

Soil sorption coefficient, log K_{oc}:
2.13 using method of Kenaga and Goring (1980)

Octanol/water partition coefficient, log K_{ow}:
1.19 (Sangster, 1989)
1.38 (Leo et al., 1971)

Solubility in organics:
Soluble in acetone, ethanol, and ether (Weast, 1986)

Solubility in water (wt %):
1.75 at 20 °C, 1.64 at 25 °C, 1.52 at 30 °C (Ginnings et al., 1940)
2.46 at 0 °C, 1.91 at 9.6 °C, 1.51 at 19.8 °C, 1.37 at 29.7 °C, 1.24 at 39.6 °C, 1.16 at 50.0 °C, 1.12 at 60.5 °C, 1.12 at 70.3 °C, 1.15 at 80.7 °C, 1.19 at 91.5 °C (Stephenson, 1992)

Vapor density:
4.09 g/L at 25 °C, 3.46 (air = 1)

Vapor pressure (mmHg):
2 at 20 °C (Verschueren, 1983)
3.8 at 25 °C (ACGIH, 1986)

Exposure limits: NIOSH REL: TWA 1 ppm (4 mg/m³), IDLH 1,600 ppm; OSHA PEL: TWA 100 ppm (410 mg/m³); ACGIH TLV: TWA 5 ppm, STEL 10 ppm (adopted).

Symptoms of exposure: Vapor inhalation may cause muscle weakness, weakness in ankles and hand, and difficulty in grasping objects (Patnaik, 1992).

Toxicity:
Acute oral LD_{50} for rats 2,590 mg/kg, mice 2,430 mg/kg (quoted, RTECS, 1985).

Environmental fate:
Photolytic. A second-order photooxidation rate constant of 8.97 x 10⁻¹² cm³/molecule·sec for the reaction of 2-hexanone and OH radicals in the atmosphere at 299 K was reported by Atkinson (1985).
Chemical/Physical. 2-Hexanone will not hydrolyze because it has no hydrolyzable functional group (Kollig, 1995).

Uses: Solvent for paints, varnishes, nitrocellulose lacquers, oils, fats, and waxes; denaturant for ethyl alcohol; in organic synthesis.

1-HEXENE

Synonyms: Butylethylene; Hex-1-ene; Hexylene; UN 2370.

CAS Registry Number: 592-41-6
DOT: 2370
Molecular formula: C_6H_{12}
Formula weight: 84.16
RTECS: MP6600100

Physical state and color:
Colorless liquid

Melting point (°C):
-139.8 (Weast, 1986)

Boiling point (°C):
63.3 (Weast, 1986)

Density (g/cm³):
0.67317 at 20/4 °C, 0.66848 at 25/4 °C (Dreisbach, 1959)

Diffusivity in water (x 10⁻⁵ cm²/sec):
0.77 at 20 °C using method of Hayduk and Laudie (1974)

Dissociation constant, pK_a:
>14 (Schwarzenbach et al., 1993)

Flash point (°C):
-26.1 (Hawley, 1981)

Heat of fusion (kcal/mol):
2.2341 (Riddick et al., 1986)

Henry's law constant (atm·m³/mol):
0.435 at 25 °C (Hine and Mookerjee, 1975)

Ionization potential (eV):
9.45 ± 0.02 (Franklin et al., 1969)

Soil sorption coefficient, log K_{oc}:
Unavailable because experimental methods for estimation of this parameter for aliphatic hydrocarbons are lacking in the documented literature

Octanol/water partition coefficient, log K_{ow}:
2.25 (Coates et al., 1985)
2.70 (Hansch and Leo, 1979)
3.39 (Schantz and Martire, 1987)
3.47 (Wasik et al., 1981)

Solubility in organics:
Soluble in alcohol, benzene, chloroform, ether, and petroleum (Weast, 1986)

Solubility in water:
49 mg/L at 23 °C (Coates et al., 1985)
50 mg/kg at 25 °C (McAuliffe, 1966)
778, 643, and 501 μM in 1 mM nitric acid at 20, 25, and 30 °C, respectively (Natarajan and Venkatachalam, 1972)
82.8 mM at 25.0 °C (Wasik et al., 1981)
55.4 mg/L at 25 °C (Leinonen and Mackay, 1973)

Vapor density:
3.44 g/L at 25 °C, 2.91 (air = 1)

Vapor pressure (mmHg):
176 at 23.7 °C (Forziati et al., 1950)
186.0 at 25 °C (Wilhoit and Zwolinski, 1971)

Environmental fate:
   Biological. Biooxidation of 1-hexene may occur yielding 5-hexen-1-ol, which may further oxidize to give 5-hexenoic acid (Dugan, 1972). Washed cell suspensions of bacteria belonging to the genera *Mycobacterium, Nocardia, Xanthobacter,* and *Pseudomonas* and growing on selected alkenes metabolized 1-hexene to 1,2-epoxyhexane (Van Ginkel et al., 1987).
   Photolytic. The following rate constants were reported for the reaction of 1-hexene and OH radicals in the atmosphere: 1.9×10^{12} cm^3/mol·sec at 300 K (Hendry and Kenley, 1979); 3.75×10^{-11} cm^3/molecule·sec at 295 K (Atkinson and Carter, 1984); 3.18×10^{-11} cm^3/molecule·sec (Atkinson, 1990). The following rate constants were reported for the reaction of 1-hexene and ozone in the atmosphere: 1.10×10^{-17} cm^3/molecule·sec (Bufalini and Altshuller, 1985); 9.0×10^{-17} cm^3/molecule·sec (Cadle and Schadt, 1952); 1.40×10^{-17} (Cox and Penkett, 1972); 1.08×10^{-17} at 294 K (Adeniji et al., 1981).

Exposure limits: ACGIH TLV: TWA 30 ppm (adopted).

Uses: Synthesis of perfumes, flavors, dyes, and resins; polymer modifier; organic synthesis.

sec-HEXYL ACETATE

Synonyms: Acetic acid 1,3-dimethylbutyl ester; 1,3-Dimethylbutyl acetate; *sec*-Hexyl acetate; *sec*-Hexyl ethanoate; MAAC; Methylamyl acetate; Methylisoamyl acetate; Methylisobutyl-carbinol acetate; **4-Methyl-2-pentanol acetate**; 4-Methyl-2-pentyl acetate; UN 1233.

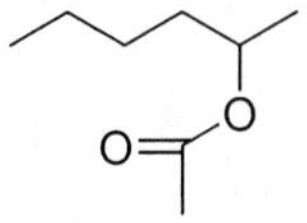

CAS Registry Number: 108-84-9
DOT: 1233
DOT label: Combustible liquid
Molecular formula: $C_8H_{16}O_2$
Formula weight: 144.21
RTECS: SA7525000

Physical state, color, and odor:
Colorless liquid with a fruity odor

Melting point (°C):
-64.4 (NIOSH, 1997)

Boiling point (°C):
158 (Weast, 1986)
146.3 (Hawley, 1981)

Density (g/cm³):
0.8658 at 20/4 °C (Weast, 1986)

Diffusivity in water (x 10⁻⁵ cm²/sec):
0.65 at 20 °C using method of Hayduk and Laudie (1974)

Flash point (°C):
45 (NIOSH, 1997)

Henry's law constant (x 10⁻³ atm·m³/mol):
9.5 at 20 °C (approximate - calculated from water solubility and vapor pressure)

Soil sorption coefficient, log K_{oc}:
Unavailable because experimental methods for estimation of this parameter for aliphatic esters are lacking in the documented literature

Octanol/water partition coefficient, log K_{ow}:
3.37 using method of Hansch et al. (1968)

Solubility in organics:
Soluble in alcohol and ether (Weast, 1986)

Solubility in water:
0.008 wt % at 20 °C (NIOSH, 1997)

Vapor density:
5.89 g/L at 25 °C, 4.98 (air = 1)

Vapor pressure (mmHg at 20 °C):
4 (NIOSH, 1997)
3 (Hawley, 1981)

Environmental fate:
   Chemical/Physical. Slowly hydrolyzes in water forming 4-methyl-2-pentanol and acetic acid.

Exposure limits: NIOSH REL: TWA 50 ppm (300 mg/m^3), IDLH 500 ppm; OSHA PEL: TWA 50 ppm; ACGIH TLV: TWA 50 ppm (adopted).

Symptoms of exposure: An irritation concentration of 600.00 mg/m^3 in air was reported by Ruth (1986).

Toxicity:
   Acute oral LD$_{50}$ for rats 6,160 mg/kg (quoted, RTECS, 1985).

Uses: Solvent for nitrocellulose and other lacquers.

HYDROQUINONE

Synonyms: Arctuvin; *p*-Benzenediol; **1,4-Benzenediol**; Benzohydroquinone; Benzoquinol; Black and white bleaching cream; Dihydroxybenzene; *p*-Dihydroxybenzene; 1,4-Dihydroxy-benzene; *p*-Dioxobenzene; Eldopaque; Eldoquin; Hydroquinol; Hydroquinole; α-Hydro-quinone; *p*-Hydroquinone; 4-Hydroxyphenol; *p*-Hydroxyphenol; NCI-C55834; Quinol; β-Quinol; Quinone; Tecquinol; Tenox HQ; Tequinol; UN 2662; USAF EK-356.

CAS Registry Number: 123-31-9
DOT: 2662
Molecular formula: $C_6H_6O_2$
Formula weight: 110.11
RTECS: MX3500000
Merck reference: 10, 4719

Physical state, color, and odor:
Colorless to pale brown, odorless, hexagonal crystals

Melting point (°C):
173-174 (Weast, 1986)
170-171 (Windholz et al., 1983)
174.00 (Martin et al., 1979)

Boiling point (°C):
285 at 750 mmHg (Weast, 1986)
285-287 (Windholz et al., 1983)

Density (g/cm³):
1.328 at 15/4 °C (Weast, 1986)
1.358 at 20/4 °C (Sax and Lewis, 1987)

Diffusivity in water (x 10⁻⁵ cm²/sec):
0.83 at 20 °C using method of Hayduk and Laudie (1974)

Dissociation constant (at 25 °C):
$pK_1 = 10.0$, $pK_2 = 12.0$ (Dean, 1973)

Flash point (°C):
166 (molten, NIOSH, 1997)

Entropy of fusion (cal/mol·K):
14.6 (Andrews et al., 1926)

Heat of fusion (kcal/mol):
6.48 (Tsonopoulos and Prausnitz, 1971)

Henry's law constant (x 10^{-9} atm·m³/mol):
<2.07 at 20 °C (approximate - calculated from water solubility and vapor pressure)

Ionization potential (eV):
7.95 (NIOSH, 1997)

Bioconcentration factor, log BCF:
1.54 (algae, Geyer et al., 1984)
2.94 (activated sludge), 1.60 (algae), 1.60 (golden ide) (Freitag et al., 1985)

Soil sorption coefficient, log K_{oc}:
0.98 using method of Kenaga and Goring (1980)

Octanol/water partition coefficient, log K_{ow}:
0.59 (Leo et al., 1971)
0.50 at pH 5.62 (Umeyama et al., 1971)
0.55 (Geyer et al., 1984)
0.54 (Nahum and Horvath, 1980)
0.46 (Janini and Attari, 1983)

Solubility in organics:
Soluble in acetone, alcohol, and ether (Weast, 1986). Slightly soluble in benzene (Windholz et al., 1983).

Solubility in water (mg/L):
80,135 at 25 °C (Korman and La Mer, 1936)
59,000 at 15 °C, 70,000 at 25 °C, 94,000 at 28 °C (quoted, Verschueren, 1983)

Vapor pressure (mmHg):
<10^{-3} at 20 °C (Sax and Lewis, 1987)
4 at 150 °C (Verschueren, 1983)

Environmental fate:
   Biological. In activated sludge, 7.5% mineralized to carbon dioxide after 5 d (Freitag et al., 1985). Under methanogenic conditions, inocula from a municipal sewage treatment plant digester degraded hydroquinone to phenol prior to being mineralized to carbon dioxide and methane (Young and Rivera, 1985). In various pure cultures, hydroquinone degraded to the following intermediates: benzoquinone, 2-hydroxy-1,4-benzoquinone, and β-ketoadipic acid. Hydroquinone also degraded in activated sludge but no products were identified (Harbison and Belly, 1982).
   Heukelekian and Rand (1955) reported a 5-d BOD value of 0.74 g/g which is 39.2% of the ThOD value of 1.89 g/g. In activated sludge inoculum, following a 20-d adaptation period, 90.0% COD removal was achieved. The average rate of biodegradation was 54.2 mg COD/g·h (Pitter, 1976).
   Photolytic. A carbon dioxide yield of 53.7% was achieved when hydroquinone adsorbed on silica gel was irradiated with light (λ >290 nm) for 17 h (Freitag et al., 1985).
   Chemical/Physical. Ozonolysis products reported are *p*-quinone and dibasic acids (Verschueren, 1983). Moussavi (1979) studied the autoxidation of hydroquinone in slightly alkaline (pH 7-9) aqueous solutions at room temperature. The oxidation of hydroquinone by oxygen followed first-order kinetics that yielded hydrogen peroxide and *p*-quinone as products. At pH values of 7.0, 8.0, and 9.0, the calculated half-lives of this reaction were 111, 41, and

0.84 h, respectively (Moussavi, 1979).

Chlorine dioxide reacted with hydroquinone in an aqueous solution forming *p*-benzoquinone (Wajon et al., 1982). Kanno et al. (1982) studied the aqueous reaction of hydroquinone and other substituted aromatic hydrocarbons (aniline, toluidine, 1- and 2-naphthylamine, phenol, cresol, pyrocatechol, resorcinol, and 1-naphthol) with hypochlorous acid in the presence of ammonium ion. They reported that the aromatic ring was not chlorinated as expected but was cleaved by chloramine forming cyanogen chloride. As the pH was lowered, the amount of cyanogen chloride formed increased (Kanno et al., 1982).

Exposure limits (mg/m^3): NIOSH REL: 15-min ceiling 2, IDLH 50; OSHA PEL: TWA 2; ACGIH TLV: TWA 2 (adopted).

Toxicity:

Acute oral LD$_{50}$ for cats 70 mg/kg, dogs 200 mg/kg, guinea pigs 550 mg/kg, mice 245 mg/kg, pigeons 300 mg/kg, rats 320 mg/kg (quoted, RTECS, 1985).

Source: Hydroquinone occurs naturally in strawberry tree leaves, pears, blackberries, Chinese alpenrose, bilberries, blackberries, hyacinth flowers, anise, cowberries, and lingonberries (Duke, 1992).

Uses: Antioxidant; photographic reducer and developer for black and white film; determination of phosphate; dye intermediate; medicine; in monomeric liquids to prevent polymerization; stabilizer in paints and varnishes; motor fuels and oils.

INDAN

Synonyms: 2,3-Dihydroindene; **2,3-Dihydro-1*H*-indene**; Hydrindene.

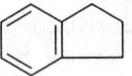

CAS Registry Number: 496-11-7
Molecular formula: C_9H_{10}
Formula weight: 118.18
RTECS: NK3750000
Merck reference: 10, 4825

Physical state:
Liquid

Melting point (°C):
-51.4 (Weast, 1986)
-51.0 (Bjørrseth, 1983)

Boiling point (°C):
178 (Weast, 1986)
176.5 (Hawley, 1981)

Density (g/cm³):
0.9639 at 20/4 °C (Weast, 1986)

Diffusivity in water (x 10⁻⁵ cm²/sec):
0.71 at 20 °C using method of Hayduk and Laudie (1974)

Dissociation constant, pK$_a$:
>14 (Schwarzenbach et al., 1993)

Flash point (°C):
50 (Aldrich, 1988)

Entropy of fusion (cal/mol·K):
9.266 (Mallard and Linstrom, 1998)

Henry's law constant (x 10⁻³ atm·m³/mol):
2.14 at 25 °C (approximate - calculated from water solubility and vapor pressure)

Soil sorption coefficient, log K$_{oc}$:
2.48 using method of Karickhoff et al. (1979)

Octanol/water partition coefficient, log K$_{ow}$:
3.33 (Hansch and Leo, 1985)

Solubility in organics:
Soluble in alcohol and ether (Weast, 1986)

Solubility in water:

88.9 mg/kg at 25 °C (Price, 1976)

109.1 mg/kg at 25 °C (Mackay and Shiu, 1977)

6.93 mg/L (water soluble fraction of a 15-component simulated jet fuel mixture (JP-8) containing 7.5 wt % indan, MacIntyre and deFur, 1985)

Vapor density:

4.83 g/L at 25 °C, 4.08 (air = 1)

Vapor pressure (mmHg):

1.5 at 25 °C (extrapolated, Ambrose and Sprake, 1975)

Environmental fate:

Photolytic. Gas-phase reaction rate constants for OH radicals, NO_3 radicals, and ozone at 24 °C were 1.9×10^{-11}, 6.6×10^{-15}, and $<3 \times 10^{-19}$ cm^3/molecule·sec, respectively (Kwok et al., 1997).

Source: Detected in distilled water-soluble fractions of 87 octane gasoline (0.40 mg/L), 94 octane gasoline (0.23 mg/L), Gasohol (0.50 mg/L), No. 2 fuel oil (0.05 mg/L), jet fuel A (0.15 mg/L), and diesel fuel (0.06 mg/L) (Potter, 1996). Based on analysis of 7 coal tar samples, indan concentrations ranged from ND to 3,800 ppm (EPRI, 1970).

Use: Organic synthesis.

INDENO[1,2,3-*cd*]PYRENE

Synonyms: Indenopyren; IP; 1,10-(*o*-Phenylene)pyrene; 2,3-Phenylenepyrene; 2,3-*o*-Phenylenepyrene; 3,4-(*o*-Phenylene)pyrene; 2,3-Phenylene-*o*-pyrene; 1,10-(1,2-Phenylene) pyrene; *o*-Phenylpyrene; RCRA waste number U137.

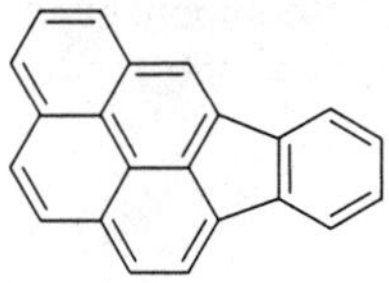

CAS Registry Number: 193-39-5
Molecular formula: $C_{22}H_{12}$
Formula weight: 276.34
RTECS: NK9300000

Physical state:
Solid

Melting point (°C):
160-163 (Verschueren, 1983)

Boiling point (°C):
536 (Verschueren, 1983)

Diffusivity in water (x 10^{-5} cm²/sec):
0.48 at 20 °C using method of Hayduk and Laudie (1974)

Dissociation constant, pK_a:
>15 (Christensen et al., 1975)

Henry's law constant (x 10^{-7} atm·m³/mol):
1.78, 2.86, 5.63, 6.02, 7.60, and 10.36 at 10.0, 20.0, 35.0, 40.1, 45.0, and 55.0 °C, respectively (ten Hulscher et al., 1992)

Bioconcentration factor, log BCF:
Apparent values of 3.4 (wet wt) and 5.1 (lipid wt) for freshwater isopods including *Asellus aquaticus* (L.) (van Hattum et al., 1998)

Soil sorption coefficient, log K_{oc}:
6.93 (San Francisco, CA mudflat sediments, Maruya et al., 1996)
6.3 (average value using 8 river bed sediments from the Netherlands, van Hattum et al., 1998)

Octanol/water partition coefficient, log K_{ow}:
5.97 (Sims et al., 1988)

Solubility in organics:
Soluble in most solvents (U.S. EPA, 1985)

Solubility in water:
62 μg/L (Sims et al., 1988)

Vapor pressure (mmHg):
1.01 x 10^{-10} at 25 °C (McVeety and Hites, 1988)

Environmental fate:
   Chemical/Physical. Indeno[1,2,3-*cd*] will not hydrolyze (Kollig, 1993).

Exposure limits: Potential occupational carcinogen. No individual standards have been set; however, as a constituent in coal tar pitch volatiles, the following exposure limits have been established (mg/m^3): NIOSH REL: TWA 0.1 (cyclohexane-extractable fraction), IDLH 80; OSHA PEL: TWA 0.2 (benzene-soluble fraction); ACGIH TLV: TWA 0.2 (benzene solubles).

Drinking water standard: No MCLGs or MCLs have been proposed (U.S. EPA, 1996).

Source: Detected in 8 diesel fuels at concentrations ranging from 0.056 to 0.85 mg/L with a mean value of 0.143 mg/L (Westerholm and Li, 1994). Indeno[1,2,3-*cd*]pyrene was also reported in gasoline (59 μg/L), fresh motor oil (30 μg/L), and used motor oil (34.0-83.2 mg/kg) (quoted, Verschueren, 1983).
   Based on analysis of 7 coal tar samples, indeno[1,2,3-*cd*]pyrene concentrations ranged from ND to 1,400 ppm (EPRI, 1990).

Uses: Produced primarily for research purposes. Derived from industrial and experimental coal gasification operations where the maximum concentration detected in coal tar streams was 1.7 mg/m^3 (Cleland, 1981).

INDOLE

Synonyms: 1-Azaindene; 1-Benzazole; Benzopyrrole; Benzo[*b*]pyrrole; 1-Benzo[*b*]pyrrole; 2,3-Benzopyrrole; **1*H*-Indole**; Ketole.

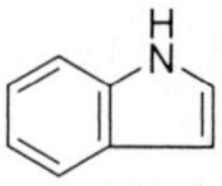

CAS Registry Number: 120-72-9
Molecular formula: C_8H_7N
Formula weight: 117.15
RTECS: NL2450000
Merck reference: 10, 4847

Physical state, color, and odor:
Colorless to yellow scales with an unpleasant odor. Turns red on exposure to light and air. An odor threshold of 0.3 mg/L was reported (quoted, Verschueren, 1983).

Melting point (°C):
52.5 (Weast, 1986)

Boiling point (°C):
254 (Weast, 1986)

Density (g/cm³):
1.22 (Weast, 1986)
1.643 (Dean, 1987)

Diffusivity in water (x 10⁻⁵ cm²/sec):
0.76 at 20 °C using method of Hayduk and Laudie (1974)

Dissociation constant, pK_a:
3.17 (Sangster, 1989)

Flash point (°C):
>110 (Aldrich, 1988)

Ionization potential (eV):
7.76 (Lias, 1998)

Soil sorption coefficient, log K_{oc}:
1.69 using method of Kenaga and Goring (1980)

Octanol/water partition coefficient, log K_{ow}:
2.00, 2.25 (Leo et al., 1971)
2.14 (Hansch and Anderson, 1967)
1.81 (Eadsforth, 1986)
2.16 (Garst and Wilson, 1984)
2.28 (Rogers and Cammarata, 1969)

Solubility in organics:
Soluble in alcohol, benzene, ether, and ligroin (Weast, 1986)

Solubility in water:
3,558 mg/kg at 25 °C (Price, 1976)

Environmental fate:
Biological. In 9% anaerobic municipal sludge, indole degraded to 1,3-dihydro-2H-indol-2-one (oxindole), which further degraded to methane and carbon dioxide (Berry et al., 1987). Heukelekian and Rand (1955) reported a 5-d BOD value of 1.70 g/g which is 65.4% of the ThOD value of 2.48 g/g.

Chemical/Physical. The aqueous chlorination of indole by hypochlorite/hypochlorous acid, chlorine dioxide, and chloramines produced oxindole, isatin, and possibly 3-chloroindole (Lin and Carlson, 1984).

Toxicity:
Acute oral LD_{50} for rats 1,000 mg/kg (quoted, RTECS, 1985).

Source: Indole was detected in jasmine flowers (*Jasminum officinale*), licorice (*Glycyrrhiza glabra*), kohlrabi stems (*Brassica oleracea var. gongylodes*), and hyacinth flowers (*Hyacinthus orientalis*) at concentrations of 42-95, 2, 1.33, and 0.24-3.45 ppm, respectively. Indole also occurs in tea leaves, black locust flowers, corn leaves, petitgrain, and yellow elder (Duke, 1992).

Uses: Chemical reagent; medicine; flavoring agent; perfumery; constituent of coal tar.

INDOLINE

Synonyms: 2,3-Dihydroindole; **2,3-Dihydro-1*H*-indole**.

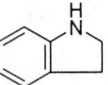

CAS Registry Number: 496-15-1
Molecular formula: C_8H_9N
Formula weight: 119.17

Physical state and color:
Dark brown liquid

Boiling point (°C):
228-230 (Weast, 1986)
220-221 (Aldrich, 1988)

Density (g/cm³):
1.069 at 20/4 °C (Weast, 1986)

Diffusivity in water (x 10⁻⁵ cm²/sec):
0.83 at 20 °C using method of Hayduk and Laudie (1974)

Ionization potential (eV):
7.15 (Lias, 1998)

Soil sorption coefficient, log K_{oc}:
1.42 using method of Kenaga and Goring (1980)

Octanol/water partition coefficient, log K_{ow}:
0.16 using method of Kenaga and Goring (1980)

Solubility in organics:
Soluble in acetone, benzene, and ether (Weast, 1986)

Solubility in water:
10,800 mg/kg at 25 °C (Price, 1976)

Uses: Organic synthesis.

1-IODOPROPANE

Synonyms: Propyl iodide; *n*-Propyl iodide.

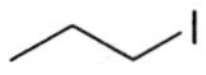

$$\text{/\!\!\!\!\!\!\!\backslash\!\!\!\!\!/\!\!\!\!\!I}$$

CAS Registry Number: 107-08-4
DOT: 2392
DOT label: Combustible liquid
Molecular formula: C_3H_7I
Formula weight: 169.99
RTECS: TZ4100000
Merck reference: 10, 7763

Physical state and color:
Colorless liquid

Melting point (°C):
-101 (Weast, 1986)
-98 (Windholz et al., 1983)

Boiling point (°C):
102.4 (Weast, 1986)

Density (g/cm^3):
1.7489 at 20/4 °C (Weast, 1986)

Diffusivity in water (x 10^{-5} cm^2/sec):
0.90 at 20 °C using method of Hayduk and Laudie (1974)

Flash point (°C):
None (Dean, 1987).

Henry's law constant (x 10^{-3} atm·m^3/mol):
9.09 at 25 °C (Hine and Mookerjee, 1975)

Ionization potential (eV):
9.26 ± 0.01 (Franklin et al., 1969)

Soil sorption coefficient, log K_{oc}:
2.16 using method of Chiou et al. (1979)

Octanol/water partition coefficient, log K_{ow}:
2.49 using method of Hansch et al. (1968)

Solubility in organics:
Soluble in benzene and chloroform (Weast, 1986). Miscible with alcohol and ether (Windholz et al., 1983).

Solubility in water:
0.1065 wt % at 23.5 °C (Schwarz, 1980)

1,040 mg/kg at 30 °C (Gross and Saylor, 1931)
5.1 mM at 20 °C (Fühner, 1924)
1.17 x 10^{-4} at 25 °C (mole fraction, Li et al., 1993)
1,070 mg/kg at 25 °C, 1,030 mg/kg at 30 °C (Rex, 1906)

Vapor density:
6.95 g/L at 25 °C, 5.87 (air = 1)

Vapor pressure (mmHg):
43.1 at 25 °C (Abraham, 1984)
35.1 at 20 °C, 54.8 at 30 °C (Rex, 1906)

Environmental fate:
Biological. A strain of *Acinetobacter* sp. isolated from activated sludge degraded 1-iodopropane to 1-propanol and iodide ions (Janssen et al., 1987).
Chemical/Physical. Slowly hydrolyzes in water forming 1-propanol and hydroiodic acid.

Toxicity:
LD_{50} (inhalation) for rats 73,000 mg/m^3/30-min (quoted, RTECS, 1985).

Uses: Organic synthesis.

ISOAMYL ACETATE

Synonyms: Acetic acid isopentyl ester; Acetic acid 3-methylbutyl ester; Banana oil; Isoamyl ethanoate; Isopentyl acetate; Isopentyl alcohol acetate; **3-Methyl-1-butanol acetate**; 3-Methyl-butyl acetate; 3-Methyl-1-butyl acetate; 3-Methylbutyl ethanoate; Pear oil.

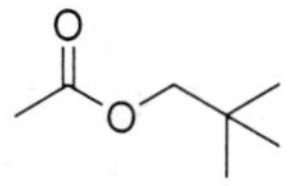

CAS Registry Number: 123-92-2
DOT: 1104
DOT label: Flammable liquid
Molecular formula: $C_7H_{14}O_2$
Formula weight: 130.19
RTECS: NS9800000
Merck reference: 10, 4957

Physical state, color, and odor:
Clear, colorless liquid with a banana or pear-like odor. Odor threshold concentration is 7 ppm (Keith and Walters, 1992).

Melting point (°C):
-78.5 (Weast, 1986)

Boiling point (°C):
142 (Weast, 1986)

Density (g/cm³):
0.8670 at 20/4 °C (Weast, 1986)
0.876 at 15/4 °C (Hawley, 1981)

Diffusivity in water (x 10⁻⁵ cm²/sec):
0.70 at 20 °C using method of Hayduk and Laudie (1974)

Flash point (°C):
25 (NIOSH, 1997)
33, 38 (open cup, Windholz et al., 1983)

Lower explosive limit (%):
1.0 at 100 °C (NIOSH, 1997)

Upper explosive limit (%):
7.5 (NIOSH, 1997)

Henry's law constant (x 10⁻² atm·m³/mol):
5.87 at 25 °C (Hine and Mookerjee, 1975)

Soil sorption coefficient, log K_{oc}:
1.95 using method of Chiou et al. (1979)

Octanol/water partition coefficient, log K_{ow}:
2.30 using method of Hansch et al. (1968)

Solubility in organics:
Miscible with alcohol, amyl alcohol, ether, and ethyl acetate (Windholz et al., 1983)

Solubility in water (wt %):
0.340 at 0 °C, 0.265 at 9.1 °C, 0.212 at 19.4 °C, 0.208 at 30.3 °C, 0.184 at 39.7 °C, 0.174 at 50.0 °C, 0.152 at 60.1 °C, 0.203 at 70.2 °C, 0.182 at 80.3 °C, 0.205 at 90.7 °C (Stephenson and Stuart, 1986)

Vapor density:
5.32 g/L at 25 °C, 4.49 (air = 1)

Vapor pressure (mmHg):
4 at 20 °C (NIOSH, 1997)

Environmental fate:
Chemical/Physical. Slowly hydrolyzes in water forming 3-methyl-1-butanol and acetic acid.

Exposure limits: NIOSH REL: TWA 100 ppm (525 mg/m^3), IDLH 1,000 ppm; OSHA PEL: TWA 100 ppm; ACGIH TLV: proposed TWA and STEL values for all isomers are 50 and 100 ppm, respectively.

Symptoms of exposure: Vapors may cause irritation to the eyes, nose, and throat; fatigue, increased pulse rate, and narcosis (Patnaik, 1992).

Toxicity:
Acute oral LD$_{50}$ for rabbits 7.422 g/kg, rats 16.6 g/kg (quoted, RTECS, 1985).

Uses: Artificial pear flavor in mineral waters and syrups; dyeing and finishing textiles; solvent for tannins, lacquers, nitrocellulose, camphor, oil colors, and celluloid; manufacturing of artificial leather, pearls or silk, photographic films, swelling bath sponges; celluloid cements, waterproof varnishes; bronzing liquids; metallic paints; perfumery; masking undesirable odors.

ISOAMYL ALCOHOL

Synonyms: Fermentation amyl alcohol; Fusel oil; Isoamylol; Isobutyl carbinol; Isopentanol; Isopentyl alcohol; 2-Methyl-4-butanol; 3-Methylbutanol; 3-Methylbutan-1-ol; **3-Methyl-1-butanol**; Primary isoamyl alcohol; Primary isobutyl alcohol; UN 1105.

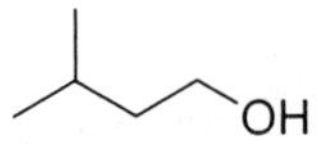

CAS Registry Number: 123-51-3
DOT: 1105
DOT label: Combustible liquid
Molecular formula: $C_5H_{12}O$
Formula weight: 88.15
RTECS: EL5425000
Merck reference: 10, 5042

Physical state, color, and odor:
Clear, colorless liquid with a pungent odor. Odor threshold concentration is 22 ppm (Amoore and Hautala, 1983).

Melting point (°C):
-118.2 (NIOSH, 1997)

Boiling point (°C):
128.5 at 750 mmHg (Weast, 1986)
132 (Windholz et al., 1983)

Density (g/cm³):
0.8092 at 20/4 °C (Weast, 1986)
0.813 at 15/4 °C (Hawley, 1981)
0.8088 at 25.00/4 °C, 0.8046 at 30.00/4 °C, 0.7968 at 40.00/4 °C, 0.7892 at 50.00/4 °C, 0.7814 at 60.00/4 °C, 0.7736 at 70.00/4 °C (Abraham et al., 1971)

Diffusivity in water (x 10⁻⁵ cm²/sec):
0.84 at 20 °C using method of Hayduk and Laudie (1974)

Dissociation constant, pKₐ:
>14 (Schwarzenbach et al., 1993)

Flash point (°C):
43.1 (NIOSH, 1997)
45, 55 (open cup, Windholz et al., 1983)

Lower explosive limit (%):
1.2 (NIOSH, 1997)

Upper explosive limit (%):
9.0 at 100 °C (NIOSH, 1997)

Henry's law constant (x 10⁻⁴ atm·m³/mol):
5.16 at 25 °C (Hakuta et al., 1977)

Interfacial tension with water (dyn/cm):
4.8 at 25 °C (Donahue and Bartell, 1952)
4.4 at 20 °C (Harkins et al., 1920)

Soil sorption coefficient, log K_{oc}:
Unavailable because experimental methods for estimation of this parameter for aliphatic esters are lacking in the documented literature

Octanol/water partition coefficient, log K_{ow}:
1.16 (Leo et al., 1971)
1.42 (Sangster, 1989)

Solubility in organics:
Soluble in acetone (Weast, 1986). Miscible with alcohol, benzene, chloroform, ether, glacial acetic acid, oils, and petroleum ether (Windholz et al., 1983).

Solubility in water:
20 g/L at 14 °C (quoted, Windholz et al., 1983)
35 mL/L at 25.0 °C (Booth and Everson, 1948)
28.5, 26.7, and 25.3 g/kg at 20, 25, and 30 °C, respectively (Ginnings and Baum, 1937)
27 g/L at 25 °C (Crittenden and Hixon, 1948)
3.18 wt % at 20 °C (Palit, 1947)
0.313 M at 18 °C (Fühner, 1924)
In wt %: 3.73 at 0 °C, 3.14 at 10.1 °C, 2.64 at 19.8 °C, 2.29 at 30.2 °C, 2.18 at 40.0 °C, 2.03 at
    49.9 °C, 2.19 at 59.8 °C, 2.11 at 70.0 °C, 2.20 at 80.0 °C, 2.27 at 90.0 °C (Stephenson et al.,
    1984)

Vapor density:
3.60 g/L at 25 °C, 3.04 (air = 1)

Vapor pressure (mmHg):
28 at 20 °C (NIOSH, 1997)
2.3 at 20 °C, 4.8 at 30 °C (Verschueren, 1983)

Environmental fate:
   Chemical/Physical. Isoamyl alcohol will not hydrolyze because it has no hydrolyzable functional group (Kollig, 1993).

Exposure limits: NIOSH REL: TWA 100 ppm (360 mg/m³), IDLH 500 ppm; OSHA PEL: TWA 100 ppm; ACGIH TLV: TWA 100 ppm, STEL 125 ppm (adopted).

Symptoms of exposure: An irritation concentration of 360.00 mg/m³ in air was reported by Ruth (1986).

Uses: Determining fat content in milk; solvent for alkaloids, fats, oils; manufacturing isovaleric acid, isoamyl or amyl compounds, esters, mercury fulminate, artificial silk, smokeless powders, lacquers, pyroxylin; photographic chemicals; pharmaceutical products; microscopy; in organic synthesis.

ISOBUTYL ACETATE

Synonyms: Acetic acid isobutyl ester; **Acetic acid 2-methylpropyl ester**; Isobutyl ethanoate; 2-Methylpropyl acetate; 2-Methyl-1-propyl acetate; β-Methylpropyl ethanoate; UN 1213.

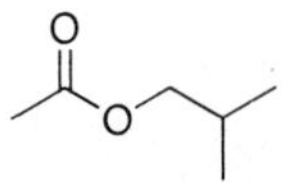

CAS Registry Number: 110-19-0
DOT: 1213
DOT label: Combustible liquid
Molecular formula: $C_6H_{12}O_2$
Formula weight: 116.16
RTECS: AI4025000
Merck reference: 10, 4977

Physical state, color, and odor:
Colorless liquid with a fruity odor. Odor threshold concentration is 0.64 ppm (Amoore and Hautala, 1983).

Melting point (°C):
-98.58 (Weast, 1986)

Boiling point (°C):
117.2 (Weast, 1986)

Density (g/cm³):
0.8712 at 20/4 °C (Weast, 1986)

Diffusivity in water (x 10^{-5} cm²/sec):
0.75 at 20 °C using method of Hayduk and Laudie (1974)

Flash point (°C):
17.9 (NIOSH, 1997)

Lower explosive limit (%):
1.3 (NFPA, 1984)

Upper explosive limit (%):
10.5 (NFPA, 1984)

Henry's law constant (x 10^{-4} atm·m³/mol):
4.85 at 25 °C (approximate - calculated from water solubility and vapor pressure)

Interfacial tension with water (dyn/cm at 25 °C):
13.2 (quoted, Freitas et al., 1997)

Ionization potential (eV):
9.97 (Franklin et al., 1969)

Soil sorption coefficient, log K_{oc}:
Unavailable because experimental methods for estimation of this parameter for aliphatic esters are lacking in the documented literature

Octanol/water partition coefficient, log K_{ow}:
1.76 using method of Hansch et al. (1968)

Solubility in organics:
Miscible with most organic solvents (Patnaik, 1992)

Solubility in water:
6,300 mg/L at 25 °C (quoted, Verschueren, 1983)
658 mM at 20 °C (Fühner, 1924)
In wt % (°C): 1.03 (0), 0.83 (10.0), 0.66 (19.7), 0.61 (29.9), 0.54 (39.7), 0.49 (50.0), 0.57 (60.5), 0.53 (70.1), 0.55 (80.2) (Stephenson and Stuart, 1986)

Vapor density:
4.75 g/L at 25 °C, 4.01 (air = 1)

Vapor pressure (mmHg):
13 at 20 °C (NIOSH, 1997)
10 at 16 °C, 20 at 25 °C (Verschueren, 1983)

Environmental fate:
   Chemical/Physical. Slowly hydrolyzes in water forming 2-methylpropanol and acetic acid.
   At an influent concentration of 1,000 mg/L, treatment with granular activated carbon resulted in an effluent concentration of 180 mg/L. The adsorbability of the carbon used was 164 mg/g carbon (Guisti et al., 1974).

Exposure limits: NIOSH REL: TWA 150 ppm (700 mg/m³), IDLH 1,300 ppm; OSHA PEL: TWA 150 ppm; ACGIH TLV: TWA 150 ppm (adopted).

Symptoms of exposure: Headache, drowsiness, and irritation of upper respiratory tract (Patnaik, 1992). An irritation concentration of 1,350.00 mg/m³ in air was reported by Ruth (1986).

Toxicity:
   Acute oral LD_{50} for rabbits 4,763 mg/kg, rats 13,400 mg/kg (quoted, RTECS, 1985).

Source: A product of whiskey fermentation (quoted, Verschueren, 1983)

Uses: Solvent for nitrocellulose; in thinners, sealers, and topcoat lacquers; flavoring agent; perfumery.

ISOBUTYL ALCOHOL

Synonyms: Fermentation butyl alcohol; 1-Hydroxymethylpropane; IBA; Isobutanol; Isopropylcarbinol; 2-Methylpropanol; 2-Methylpropanol-1; **2-Methyl-1-propanol**; 2-Methyl-1-propan-1-ol; 2-Methylpropyl alcohol; RCRA waste number U140; UN 1212.

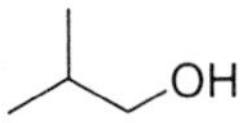

CAS Registry Number: 78-83-1
DOT: 1212
DOT label: Combustible liquid
Molecular formula: $C_4H_{10}O$
Formula weight: 74.12
RTECS: NP9625000
Merck reference: 10, 4978

Physical state, color, and odor:
Colorless, oily liquid with a sweet, musty odor. Burning taste. Odor threshold concentration is 1.6 ppm (Amoore and Hautala, 1983).

Melting point (°C):
-72.8 (NIOSH, 1997)

Boiling point (°C):
108.1 (Weast, 1986)
107 (Hawley, 1981)

Density (g/cm³):
0.8018 at 20/4 °C (Weast, 1986)
0.798 at 25/4 °C (Verschueren, 1983)
0.806 at 15/4 °C (Hawley, 1981)

Diffusivity in water (x 10⁻⁵ cm²/sec at 25 °C):
1.08 (quoted, Hayduk and Laudie, 1974)
0.95 (Hao and Leist, 1996)

Dissociation constant, pK$_a$:
>14 (Schwarzenbach et al., 1993)

Flash point (°C):
28 (NIOSH, 1997)
37 (open cup, Hawley, 1981)

Lower explosive limit (%):
1.7 at 51 °C (NIOSH, 1997)

Upper explosive limit (%):
10.6 at 95 °C (NIOSH, 1997)

Heat of fusion (kcal/mol):
1.511 (Riddick et al., 1986)

Henry's law constant (x 10^{-6} atm·m^3/mol at 25 °C):
9.79 (Snider and Dawson, 1985)
440 (Hakuta et al., 1977)
11.8 (Butler et al., 1935)
26.95 (Shiu and Mackay, 1997)

Interfacial tension with water (dyn/cm):
2.0 at 25 °C (Donahue and Bartell, 1952)
1.9 at 25 °C (Murphy et al., 1957)
1.8 at 20 °C (Harkins et al., 1920)

Ionization potential (eV):
10.02 (Lias, 1998)

Soil sorption coefficient, log K_{oc}:
Unavailable because experimental methods for estimation of this parameter for aliphatic alcohols are lacking in the documented literature

Octanol/water partition coefficient, log K_{ow}:
0.65 (Sangster, 1989)
0.76 (Hansch and Leo, 1985)
0.83 (Collander, 1951)

Solubility in organics:
Miscible with alcohol and ether (Windholz et al., 1983)

Solubility in water:
95,000 mg/L at 18 °C (quoted, Verschueren, 1983)
111 mL/L at 25 °C (Booth and Everson, 1948)
94.87 g/L at 20 °C (Mackay and Yeun, 1983)
85 g/L (Price et al., 1974)
75.6 g/kg at 25 °C (Donahue and Bartell, 1952)
94 g/kg at 25 °C (De Santis et al., 1976)
1.351 M at 18 °C (Fühner, 1924)
In wt %: 11.60 at 0 °C, 10.05 at 9.8 °C, 8.84 at 19.7 °C, 7.87 at 30.6 °C, 7.30 at 40.4 °C, 7.08 at 50.1 °C, 7.05 at 60.2 °C, 6.93 at 70.3 °C, 7.31 at 80.5 °C, 7.71 at 90.7 °C (Stephenson and Stuart, 1986)

Vapor density:
3.03 g/L at 25 °C, 2.56 (air = 1)

Vapor pressure (mmHg at 20 °C):
9 (NIOSH, 1997)
10.0 (Mackay and Yeun, 1983)

Environmental fate:
Biological. Bridié et al. (1979) reported BOD and COD values 0.41 and 2.46 g/g using filtered effluent from a biological sanitary waste treatment plant. These values were determined using a standard dilution method at 20 °C for a period of 5 d. When a sewage seed was used in a separate screening test, a BOD value of 1.63 g/g was obtained. Heukelekian and Rand (1955) reported a 5-d BOD value of 1.66 g/g which is 64.0% of the ThOD value of 2.59 g/g.

Chemical/Physical. Isobutyl alcohol will not hydrolyze because it does not have a hydrolyzable functional group (Kollig, 1993).

At an influent concentration of 1,000 mg/L, treatment with granular activated carbon resulted in an effluent concentration of 581 mg/L. The adsorbability of the carbon used was 84 mg/g carbon (Guisti et al., 1974).

Exposure limits: NIOSH REL: TWA 50 ppm (150 mg/m^3), IDLH 1,600 ppm; OSHA PEL: TWA 100 ppm (300 mg/m^3); ACGIH TLV: TWA 50 ppm (adopted).

Symptoms of exposure: Inhalation of vapors may cause eye and throat irritation and headache. Contact with skin may cause cracking (Patnaik, 1992). An irritation concentration of 300.00 mg/m^3 in air was reported by Ruth (1986).

Toxicity:
LC$_{50}$ (96-h) for fathead minnows 1,430 mg/L (Veith et al., 1983).
Acute oral LD$_{50}$ for rats 2,460 mg/kg (quoted, RTECS, 1985).

Source: A product of whiskey fermentation (quoted, Verschueren, 1983). Isobutyl alcohol also occurs in tea leaves and java cintronella plants (Duke, 1992).

Uses: Preparation of esters for the flavoring industry; solvent for plastics, textiles, oils, perfumes, paint, and varnish removers; intermediate for amino coating resins; liquid chromatography; fluorometric determinations; in organic synthesis.

ISOBUTYLBENZENE

Synonyms: (2-Methylpropyl)benzene; 2-Methyl-1-phenylpropane.

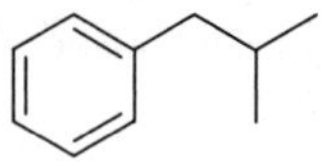

CAS Registry Number: 538-93-2
Molecular formula: $C_{10}H_{14}$
Formula weight: 134.22
RTECS: DA3550000
Merck reference: 10, 4981

Physical state and color:
Colorless liquid. An odor threshold concentration of 80 μg/kg was reported (quoted, Verschueren, 1983)

Melting point (°C):
-51.5 (Weast, 1986)

Boiling point (°C):
172.8 (Weast, 1986)
170.5 (Windholz et al., 1983)

Density (g/cm³ at 20/4 °C):
0.8532 (Weast, 1986)
0.8673 (Windholz et al., 1983)

Diffusivity in water (x 10⁻⁵ cm²/sec):
0.68 at 20 °C using method of Hayduk and Laudie (1974)

Dissociation constant, pK$_a$:
>14 (Schwarzenbach et al., 1993)

Flash point (°C):
60 (Hawley, 1981)
55 (Aldrich, 1988)

Lower explosive limit (%):
0.8 (NFPA, 1984)

Upper explosive limit (%):
6.0 (NFPA, 1984)

Henry's law constant (x 10⁻² atm·m³/mol):
1.09 at 25 °C (Hine and Mookerjee, 1975)

Ionization potential (eV):
8.68 (Lias, 1998)

Soil sorption coefficient, log K_{oc}:
3.90 using method of Karickhoff et al. (1979)

Octanol/water partition coefficient, log K_{ow}:
4.11 (Chiou et al., 1982)

Solubility in organics:
Soluble in acetone, alcohol, benzene, ether, and petroleum hydrocarbons (Weast, 1986)

Solubility in water:
10.1 mg/kg at 25 °C (Price, 1976)
33.71 mg/L at 25 °C (Chiou et al., 1982)
70.2 μmol/L in 0.5 M NaCl at 25 °C (Wasik et al., 1984)

Vapor density:
5.49 g/L at 25 °C, 4.63 (air = 1)

Vapor pressure (mmHg):
2.06 at 25 °C (Mackay et al., 1982)

Environmental fate:
   Biological. Oxidation of isobutylbenzene by *Pseudomonas desmolytica* S44B1 and *Pseudomonas convexa* S107B1 yielded 3-isobutylcatechol and (+)-2-hydroxy-8-methyl-6-oxo-nonanoic acid (Jigami et al., 1975).
   Chemical/Physical. Complete combustion in air yields carbon dioxide and water vapor. Isobutylbenzene will not hydrolyze because it has no hydrolyzable functional group.

Uses: Perfume synthesis; flavoring; pharmaceutical intermediate.

ISOPHORONE

Synonyms: Isoacetophorone; Isoforon; Isoforone; Isooctaphenone; Isophoron; NCI-C55618; 1,1,3-Trimethyl-3-cyclohexene-5-one; Trimethylcyclohexenone; **3,5,5-Trimethyl-2-cyclo-hexen-1-one**.

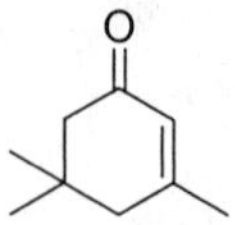

CAS Registry Number: 78-59-1
DOT: 1224
DOT label: Combustible liquid
Molecular formula: $C_9H_{14}O$
Formula weight: 138.21
RTECS: GW7700000

Physical state, color, and odor:
Clear, colorless liquid with a sharp peppermint or camphor-like odor. Odor threshold concentration is 0.20 ppm (Amoore and Hautala, 1983).

Melting point (°C):
-8.1 (Dean, 1973)

Boiling point (°C):
215.2 (Sax, 1984)
208-212 (Fluka, 1988)

Density (g/cm³):
0.9229 at 20/4 °C (Weast, 1986)
0.921 at 25/4 °C (Weiss, 1986)

Diffusivity in water (x 10⁻⁵ cm²/sec):
0.70 at 20 °C using method of Hayduk and Laudie (1974)

Flash point (°C):
85.1 (NIOSH, 1997)

Lower explosive limit (%):
0.8 (NIOSH, 1997)

Upper explosive limit (%):
3.8 (NIOSH, 1997)

Henry's law constant (x 10⁻⁶ atm·m³/mol):
5.8 (calculated, U.S. EPA, 1980a)

Ionization potential (eV):
9.07 (NIOSH, 1997)

Bioconcentration factor, log BCF:
0.85 (bluegill sunfish, Veith et al., 1980)

Soil sorption coefficient, log K_{oc}:
1.49 using method of Karickhoff et al. (1979)

Octanol/water partition coefficient, log K_{ow}:
1.67 (Veith et al., 1980)

Solubility in organics:
Soluble in acetone, ethanol, and ether (Weast, 1986)

Solubility in water:
12,000 mg/L at 25 °C (quoted, Amoore and Hautala, 1983)
In wt %: 2.40 at 0 °C, 1.80 at 9.3 °C, 1.57 at 19.8 °C, 1.32 at 30.9 °C, 1.24 at 40.5 °C, 1.19 at
50.4 °C, 1.18 at 60.2 °C, 1.14 at 70.0 °C, 1.27 at 80.5 °C, 1.35 at 91.1 °C (Stephenson, 1992)

Vapor density:
5.65 g/L at 25 °C, 4.77 (air = 1)

Vapor pressure (mmHg):
0.2 at 20 °C (ACGIH, 1986)
0.40 at 25 °C (Banerjee et al., 1990)

Environmental fate:
Biological. The pure culture *Aspergillus niger* biodegraded isophorone to 3,5,5-trimethyl-2-cyclohexene-1,4-dione, 3,5,5-trimethylcyclohexane-1,4-dione, (*S*)-4-hydroxy-3,5,5-trimethyl-2-cyclohex-1-one, and 3-hydroxymethyl-5,5-dimethyl-2-cyclohexen-1-one (Mikami et al., 1981).

Chemical/Physical. At an influent concentration of 1,000 mg/L, treatment with granular activated carbon resulted in an effluent concentration of 34 mg/L. The adsorbability of the carbon used was 193 mg/g carbon (Guisti et al., 1974).

Isophorone will not hydrolyze in water because it does not contain a hydrolyzable functional group.

Exposure limits: NIOSH REL: TWA 4 ppm (23 mg/m^3), IDLH 200 ppm; OSHA PEL: TWA 25 ppm (140 mg/m^3); ACGIH TLV: TWA 5 ppm (adopted).

Symptoms of exposure: Vapors may cause mild irritation to the eyes, nose, and throat (NIOSH, 1997; Patnaik, 1992), headache, nausea, dizziness, fatigue, discomfort, and dermatitis. May be narcotic at high concentrations (NIOSH, 1997). An irritation concentration of 50.00 mg/m^3 in air was reported by Ruth (1986).

Toxicity:
LC_{50} (96-h) for bluegill sunfish 220 mg/L (Spehar et al., 1982), *Cyprinodon variegatus* 170-300 ppm using natural seawater (Heitmuller et al., 1981).

LC_{50} (72-h) for *Cyprinodon variegatus* 170-300 ppm (Heitmuller et al., 1981).

LC_{50} (48-h) for *Daphnia magna* 120 mg/L (LeBlanc, 1980), *Cyprinodon variegatus* 170-300 ppm (Heitmuller et al., 1981).

LC_{50} (24-h) for *Daphnia magna* 430 mg/L (LeBlanc, 1980), *Cyprinodon variegatus* 170-300 ppm (Heitmuller et al., 1981).

LD_{50} (skin) for rabbits 1,500 mg/kg (quoted, RTECS, 1985).

Acute oral LD_{50} for rats 2,330 mg/kg (quoted, RTECS, 1985).

Heitmuller et al. (1981) reported a NOEC of 170 ppm.

Drinking water standard: No MCLGs or MCLs have been proposed although isophorone has been listed for regulation (U.S. EPA, 1996).

Uses: Solvent for paints, tin coatings, agricultural chemicals, and synthetic resins; excellent solvent for vinyl resins, cellulose esters, and ethers; pesticides; storing lacquers; pesticide manufacturing; intermediate in the manufacture of 3,5-xylenol, 2,3,5-trimethylcyclohexanol, and 3,5-dimethylaniline.

ISOPROPYL ACETATE

Synonyms: Acetic acid isopropyl ester; **Acetic acid 1-methylethyl ester**; 2-Acetoxypropane; Isopropyl ethanoate; 2-Propyl acetate; UN 1220.

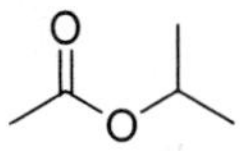

CAS Registry Number: 108-21-4
DOT: 1993
DOT label: Combustible liquid
Molecular formula: $C_5H_{10}O_2$
Formula weight: 102.13
RTECS: AI4930000
Merck reference: 10, 5054

Physical state, color, and odor:
Colorless liquid with an aromatic odor. Odor threshold concentration is 2.7 ppm (Amoore and Hautala, 1983).

Melting point (°C):
-73.4 (Weast, 1986)
-69.3 (Sax and Lewis, 1987)

Boiling point (°C):
90 (Weast, 1986)

Density (g/cm³):
0.8718 at 20/4 °C (Weast, 1986)
0.877 at 16/4 °C (Verschueren, 1983)
0.8690 at 25/4 °C (Hawley, 1981)

Diffusivity in water (x 10^{-5} cm²/sec):
0.80 at 20 °C using method of Hayduk and Laudie (1974)

Flash point (°C):
2.2 (NIOSH, 1997)
2-4 (open cup, Windholz et al., 1983)

Lower explosive limit (%):
1.8 at 38.1 °C (NIOSH, 1997)

Upper explosive limit (%):
8.0 (NIOSH, 1997)

Henry's law constant (x 10^{-4} atm·m³/mol):
2.81 at 25 °C (Hine and Mookerjee, 1975)

Interfacial tension with water (dyn/cm at 25 °C):
10.0 (quoted, Freitas et al., 1997)

Ionization potential (eV):
9.98 ± 0.02 (Franklin et al., 1969)
10.08 (Gibson et al., 1977)

Soil sorption coefficient, log K_{oc}:
Unavailable because experimental methods for estimation of this parameter for aliphatic esters are lacking in the documented literature

Octanol/water partition coefficient, log K_{ow}:
1.03 using method of Hansch et al. (1968)

Solubility in organics:
Miscible with alcohol and ether (Windholz et al., 1983)

Solubility in water:
18,000 mg/L at 20 °C, 30,900 mg/L at 20 °C (quoted, Verschueren, 1983)
29 g/L (Price et al., 1974)
0.303 M at 20 °C (Führer, 1924)
In wt %: 4.08 at 0 °C, 3.46 at 9.0 °C, 2.79 at 19.9 °C, 2.44 at 29.7 °C, 2.19 at 39.8 °C, 2.07 at
    50.0 °C, 1.92 at 62.2 °C, 1.80 at 74.6 °C (Stephenson and Stuart, 1986)

Vapor density:
4.17 g/L at 25 °C, 3.53 (air = 1)

Vapor pressure (mmHg):
47.5 at 20 °C, 73 at 25 °C (Verschueren, 1983)

Environmental fate:
    Chemical/Physical. Hydrolyzes in water forming isopropyl alcohol and acetic acid (Morrison and Boyd, 1971). The estimated hydrolysis half-life at 25 °C and pH 7 is 8.4 yr (Mabey and Mill, 1978).
    At an influent concentration of 1,000 mg/L, treatment with granular activated carbon resulted in an effluent concentration of 319 mg/L. The adsorbability of the carbon used was 137 mg/g carbon (Guisti et al., 1974).

Exposure limits: NIOSH REL: IDLH 1,800 ppm; OSHA PEL: TWA 250 ppm (950 mg/m³); ACGIH TLV: TWA 250 ppm, STEL 310 ppm (adopted) and proposed TWA and STEl values of 100 and 200 ppm, respectively.

Symptoms of exposure: May irritate eyes, nose, and throat (Patnaik, 1992)

Toxicity:
    Acute oral LD_{50} for rats 3,000 mg/kg, rabbits 6,949 mg/kg (quoted, RTECS, 1985).
    Toxicity threshold for *Pseudomonas putida* is 190 mg/L (Bringmann and Kühn, 1980).

Uses: Solvent for plastics, oils, fats, and cellulose derivatives; perfumes; paints, lacquers, and printing inks; odorant and flavoring agent; in organic synthesis.

ISOPROPYLAMINE

Synonyms: 2-Aminopropane; 1-Methylethylamine; Monoisopropylamine; **2-Propanamine**; *sec*-Propylamine; 2-Propylamine; UN 1221.

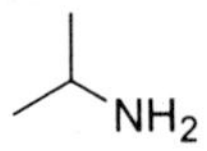

CAS Registry Number: 75-31-0
DOT: 1221
DOT label: Flammable/combustible liquid
Molecular formula: C_3H_9N
Formula weight: 59.11
RTECS: NT8400000
Merck reference: 10, 5058

Physical state, color, and odor:
Colorless liquid with a penetrating, ammonia-like odor. Odor threshold concentration is 1.2 ppm (Amoore and Hautala, 1983).

Melting point (°C):
-95.2 (Weast, 1986)
-101 (Windholz et al., 1983)

Boiling point (°C):
32.4 (Weast, 1986)
33-34 (Windholz et al., 1983)

Density (g/cm³):
0.6891 at 20/4 °C (Weast, 1986)
0.686 at 25/4 °C (Dean, 1987)
0.694 at 15/4 °C (Sax and Lewis, 1987)

Diffusivity in water (x 10^{-5} cm²/sec):
0.97 at 20 °C using method of Hayduk and Laudie (1974)

Dissociation constant, pK_a:
10.53 at 25 °C (Dean, 1973)

Flash point (°C):
-37.5 (open cup, NIOSH, 1997)
-26 (open cup, Windholz et al., 1983)

Ionization potential (eV):
8.72 ± 0.03 (Franklin et al., 1969)
8.86 (Gibson et al., 1977)

Soil sorption coefficient, log K_{oc}:
Unavailable because experimental methods for estimation of this parameter for aliphatic amines are lacking in the documented literature. However, its high solubility in water and low K_{ow} suggest its adsorption to soil will be nominal (Lyman et al., 1982).

Octanol/water partition coefficient, log K_{ow}:
0.26 (Sangster, 1989)
-0.03 (Leo et al., 1971)

Solubility in organics:
Miscible with alcohol and ether (Windholz et al., 1983)

Solubility in water:
Miscible (NIOSH, 1997).

Vapor density:
2.42 g/L at 25 °C, 2.04 (air = 1)

Vapor pressure (mmHg):
460 at 20 °C (Verschueren, 1983)

Environmental fate:
   Chemical/Physical. Releases toxic nitrogen oxides when heated to decomposition (Sax and Lewis, 1987). Forms water-soluble salts with acids.

Exposure limits: NIOSH REL: IDLH 750 ppm; OSHA PEL: TWA 5 ppm (12 mg/m^3); ACGIH TLV: TWA 5 ppm, STEL 10 ppm (adopted).

Symptoms of exposure: Strong irritation to the eyes, skin, throat, and respiratory system; pulmonary edema. Skin contact may cause dermatitis and skin burns (Patnaik, 1992).

Toxicity:
   Acute oral LD_{50} for guinea pigs 2,700 mg/kg, mice 2,200 mg/kg, rats 820 mg/kg, rabbits 3,200 mg/kg (quoted, RTECS, 1985).

Uses: Intermediate in the synthesis of rubber accelerators, dyes, pharmaceuticals, insecticides, bactericides, textiles, and surface-active agents; solvent; dehairing agent; solubilizer for 2,4-D.

ISOPROPYLBENZENE

Synonyms: Cumene; Cumol; Isopropylbenzol; **(1-Methylethyl)benzene**; 2-Phenylpropane; RCRA waste number U055; UN 1221.

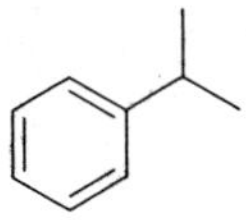

CAS Registry Number: 98-82-8
DOT: 1918
Molecular formula: C_9H_{12}
Formula weight: 120.19
RTECS: GR8575000
Merck reference: 10, 2605

Physical state, color, and odor:
Colorless liquid with an aromatic odor. The reported odor threshold concentrations in air ranged from 10 to 88 ppb (Keith and Walters, 1992; Young et al., 1996). The taste threshold concentration in water is 60 ppb (Young et al., 1996).

Melting point (°C):
-96 (Weast, 1986)
-96.8 (Mackay et al., 1982)

Boiling point (°C):
152.4 (Weast, 1986)
164.6 (Mackay et al., 1982)

Density (g/cm³):
0.8618 at 20/4 °C (Weast, 1986)

Diffusivity in water (x 10^{-5} cm²/sec):
0.73 at 20 °C using method of Hayduk and Laudie (1974)

Dissociation constant, pK$_a$:
>14 (Schwarzenbach et al., 1993)

Flash point (°C):
39 (Windholz et al., 1983)
36 (NIOSH, 1997)

Lower explosive limit (%):
0.9 (NIOSH, 1997)

Upper explosive limit (%):
6.5 (NIOSH, 1997)
8.8 (Butler and Webb, 1957)

Heat of fusion (kcal/mol):
1.86 (Dean, 1987)

Isopropylbenzene

Henry's law constant (x 10^{-3} atm·m³/mol):

3.51, 4.20, 5.03, 5.58, and 7.70 at 10, 15, 20, 25, and 30 °C, respectively (Ashworth et al., 1988)

13.0, 15.3, and 23.9 at 28.0, 35.0, and 46.1 °C, respectively (Hansen et al., 1993)

Ionization potential (eV):

9.13 (Yoshida et al., 1983)

8.75 (NIOSH, 1997)

Bioconcentration factor, log BCF:

1.55 (goldfish, Ogata et al., 1984)

Soil sorption coefficient, log K_{oc}:

3.45 using method of Karickhoff et al. (1979)

3.40 (estimated, Ellington et al., 1993)

Octanol/water partition coefficient, log K_{ow}:

3.63 (Chiou et al., 1977, 1982)

3.66 (Leo et al., 1971)

3.51 (quoted, Galassi et al., 1988)

Solubility in organics:

Soluble in acetone, alcohol, benzene, and ether (Weast, 1986)

Solubility in water:

48.3 mg/kg at 25 °C (Price, 1976)

50.5 mg/kg at 25 °C (McAuliffe, 1966)

53 mg/kg at 25 °C (McAuliffe, 1963)

65.3 mg/kg at 25 °C, 42.5 mg/kg in seawater at 25 °C (Sutton and Calder, 1975).

0.477 mM at 25.0 °C (Andrews and Keefer, 1950a)

0.299 mM in 0.5 M NaCl at 25 °C (Wasik et al., 1984)

495, 510, 568, and 638 μM at 15, 25, 35, and 45 °C, respectively (Sanemasa et al., 1982)

500 mg/L in artificial seawater at 25 °C (Price et al., 1974)

Mole fraction x 10^5 (°C): 1.2050 (24.94), 1.2416 (29.98), 1.2825 (34.92), 1.3446 (39.96), 1.4162 (44.91), 1.5037 (49.90), 1.6011 (54.92), 1.7221 (59.98), 1.8624 (65.17), 2.0302 (70.32), 2.2064 (75.10), 2.4212 (80.21) (Glew and Robertson, 1956)

170 mg/kg at 25 °C (Stearns et al., 1947)

9.80 x 10^{-6} at 25 °C (mole fraction, Li et al., 1993)

Vapor density:

4.91 g/L at 25 °C, 4.15 (air = 1)

Vapor pressure (mmHg):

3.2 at 20 °C (Verschueren, 1983)

4.6 at 25 °C (Mackay et al., 1982)

Environmental fate:

Biological. When isopropylbenzene was incubated with *Pseudomonas putida*, the substrate was converted to *ortho*-dihydroxy compounds in which the isopropyl part of the compound remained intact (Gibson, 1968). Oxidation of isopropylbenzene by *Pseudomonas desmolytica* S44B1 and *Pseudomonas convexa* S107B1 yielded 3-isopropylcatechol and a ring fission product, (+)-2-hydroxy-7-methyl-6-oxooctanoic acid (Jigami et al., 1975).

Surface Water. Mackay and Wolkoff (1973) estimated an evaporation half-life of 14.2 min from a surface water body that is 25 °C and 1 m deep.

Photolytic. Major products reported from the photooxidation of isopropylbenzene with nitrogen oxides include nitric acid and benzaldehyde (Altshuller, 1983). A *n*-hexane solution containing isopropylbenzene and spread as a thin film (4 mm) on cold water (10 °C) was irradiated by a mercury medium pressure lamp. In 3 h, 22% of the applied isopropylbenzene photooxidized into α,α-dimethylbenzyl alcohol, 2-phenylpropionaldehyde, and allylbenzene (Moza and Feicht, 1989).

A rate constant of 3.7×10^9 L/mol·sec was reported for the reaction of isopropylbenzene with OH radicals in the gas phase (Darnall et al., 1976). Similarly, a room temperature rate constant of 6.6×10^{-12} cm^3/molecule·sec was reported for the vapor-phase reaction of isopropylbenzene with OH radicals (Atkinson, 1985). At 25 °C, a rate constant of 6.25×10^{-12} cm^3/molecule·sec was reported for the same reaction (Ohta and Ohyama, 1985).

Chemical/Physical. Complete combustion in air yields carbon dioxide and water vapor.

Isopropylbenzene will not hydrolyze because it has no hydrolyzable functional group.

The calculated evaporation half-life of isopropylbenzene from surface water 1 m deep at 25 °C is 5.79 h (Mackay and Leinonen, 1975).

Exposure limits: NIOSH REL: TWA 50 ppm (245 mg/m^3), IDLH 900 ppm; OSHA PEL: TWA 50 ppm; ACGIH TLV: TWA 50 ppm (adopted).

Symptoms of exposure: Vapors may cause irritation to the eyes, skin, and upper respiratory system. Narcotic at high concentrations (Patnaik, 1992).

Toxicity:

EC$_{50}$ (72-h) for *Selenastrum capricornutum* 2.6 mg/L (Galassi et al., 1988).

LC$_{50}$ (96-h) for *Salmo gairdneri* 2.7 mg/L, *Poecilia reticulata* 5.1 mg/L (Galassi et al., 1988).

Acute oral LD$_{50}$ for rats 1,400 mg/kg (quoted, RTECS, 1985).

Drinking water standard: As of October 1995, no MCLGs or MCLs have been proposed although isopropylbenzene has been listed for regulation (U.S. EPA, 1996).

Source: Detected in distilled water-soluble fractions of 94 octane gasoline and Gasohol at concentrations of 0.14 and 0.15 mg/L, respectively (Potter, 1996).

Thomas and Delfino (1991) equilibrated contaminant-free groundwater collected from Gainesville, FL with individual fractions of three individual petroleum products at 24-25 °C for 24 h. The aqueous phase was analyzed for organic compounds via U.S. EPA approved test method 602. Average isopropylbenzene concentrations reported in water-soluble fractions of unleaded gasoline and kerosene were 235 and 28 μg/L, respectively. When the authors analyzed the aqueous-phase via U.S. EPA approved test method 610, average isopropylbenzene concentrations in water-soluble fractions of unleaded gasoline and kerosene were lower, i.e., 206 and 22 μg/L, respectively. Isopropylbenzene was detected in both water-soluble fractions of diesel fuel but were not quantified.

Isopropylbenzene naturally occurs in Ceylon cinnamon, cumin, and ginger (1 ppm in rhizome) (Duke, 1992).

Uses: Manufacture of acetone, acetophenone, diisopropylbenzene, α-methylstyrene and phenol, polymerization catalysts; constituent of motor fuel, asphalt, and naphtha; catalyst for acrylic and polyester-type resins; octane booster for gasoline; solvent.

ISOPROPYL ETHER

Synonyms: Diisopropyl ether; Diisopropyl oxide; DIPE; IPE; 2-Isopropoxypropane; **2,2'-Oxy-bis(propane)**; UN 1159.

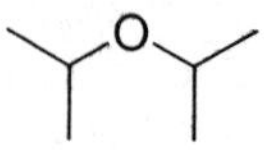

CAS Registry Number: 108-20-3
DOT: 1159
DOT label: Flammable liquid
Molecular formula: $C_6H_{14}O$
Formula weight: 102.18
RTECS: TZ5425000
Merck reference: 10, 5061

Physical state, color, and odor:
Colorless liquid with a penetrating, sweet, ether-like odor. Odor threshold concentration is 17 ppb (Amoore and Hautala, 1983).

Melting point (°C):
-85.9 (Weast, 1986)
-60 (Verschueren, 1983)
-85.5 (Riddick et al., 1986)

Boiling point (°C):
68 (Weast, 1986)
68.34 (Boublik et al., 1984)

Density (g/cm³):
0.7241 at 20/4 °C (Weast, 1986)
0.71854 at 25/4 °C (Riddick et al., 1986)

Diffusivity in water (x 10⁻⁵ cm²/sec):
0.72 at 20 °C using method of Hayduk and Laudie (1974)

Flash point (°C):
-28 (NIOSH, 1997)
-9 (open cup, Windholz et al., 1983)

Lower explosive limit (%):
1.4 (NIOSH, 1997)

Upper explosive limit (%):
7.9 (NIOSH, 1997)

Heat of fusion (kcal/mol):
2.876 (Riddick et al., 1986)

Henry's law constant (x 10⁻³ atm·m³/mol):
2.04 at 25 °C (Nielsen et al., 1994)
2.28 at 23 °C (Miller and Stuart, 2000)

Interfacial tension with water (dyn/cm):
17.9 at 20 °C (Donahue and Bartell, 1952)

Ionization potential (eV):
9.20 (Franklin et al., 1969)

Soil sorption coefficient, log K_{oc}:
Unavailable because experimental methods for estimation of this parameter for aliphatic ethers are lacking in the documented literature

Octanol/water partition coefficient, log K_{ow}:
1.52 (Funasaki et al., 1985)

Solubility in organics:
Miscible with alcohol and ether (Windholz et al., 1983)

Solubility in water:
9,000 mg/L at 20 °C (quoted, Verschueren, 1983)
0.02 mol/L at 25 °C (Hine and Weimar, 1965)
1.59×10^{-3} at 25 °C (mole fraction, Li et al., 1993)
In wt %: 2.28 at 0 °C, 1.02 at 9.7 °C, 0.79 at 20.0 °C, 0.54 at 31.0 °C, 0.41 at 40.8 °C, 0.28 at
    50.7 °C, 0.22 at 61.0 °C (Stephenson, 1992)

Vapor density:
4.18 g/L at 25 °C, 3.53 (air = 1)

Vapor pressure (mmHg):
119 at 20 °C (NIOSH, 1997)

Environmental fate:
  Biological. Bridié et al. (1979) reported BOD and COD values 0.19 and 1.75 g/g using filtered effluent from a biological sanitary waste treatment plant. These values were determined using a standard dilution method at 20 °C and stirred for a period of 5 d. The ThOD for isopropyl ether is 2.82 g/g.
  Chemical/Physical. May form explosive peroxides on standing with air (NIOSH, 1997).
  At an influent concentration of 1,018 mg/L, treatment with granular activated carbon resulted in an effluent concentration of 203 mg/L. The adsorbability of the carbon used was 162 mg/g carbon (Guisti et al., 1974).

Exposure limits: NIOSH REL: TWA 500 ppm (2,100 mg/m^3), IDLH 1,400 ppm; OSHA PEL: TWA 250 ppm.

Symptoms of exposure: Narcotic and irritant to skin and mucous membranes. Exposure to high concentrations may cause intoxication, respiratory arrest, and death (Patnaik, 1992).

Toxicity:
  LC_{50} (96-h) for fathead minnows 91.7 mg/L (Veith et al., 1983).
  Acute oral LD_{50} for rats 8,470 mg/kg; LD_{50} (inhalation) for mice 131 gm/m^3, rats 162 gm/m^3 (quoted, RTECS, 1985).

Uses: Solvent for waxes, resins, dyes, animal, vegetable and mineral oils; paint and varnish removers; rubber cements; extracting acetic acid from solution; spotting compositions; rubber cements; oxygenate in gasoline; organic synthesis.

KEPONE

Synonyms: Chlordecone; CIBA 8514; Compound 1189; 1,2,3,5,6,7,8,9,10,10-Decachloro-[5.2.1.0^{2,6}.0^{3,9}.0^{5,8}]decano-4-one; Decachloroketone; Decachloro-1,3,4-metheno-2H-cyclobuta-[cd]pentalen-2-one; Decachlorooctahydrokepone-2-one; Decachlorooctahydro-1,3,4-metheno-2H-cyclobuta[cd]pentalen-2-one; **1,1a,3,3a,4,5,5a,5b,6-Decachlorooctahydro-1,3,4-metheno-2H-cyclo-buta[cd]pentalen-2-one**; Decachloropentacyclo[5.2.1.0^{2,6}.0^{3,9}.0^{5,8}]decan-3-one; Decachloropentacyclo[5.2.1.0^{2,6}.0^{4,10}.0^{5,9}]decan-3-one; Decachlorotetracyclodecanone; Decachloro-tetrahydro-4,7-methanoindeneone; ENT 16391; GC-1189; General chemicals 1189; Merex; NA 2761; NCI-C00191; RCRA waste number U142.

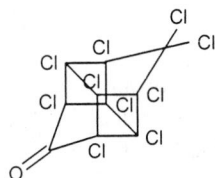

CAS Registry Number: 143-50-0
DOT: 2761
DOT label: Poison
Molecular formula: $C_{10}Cl_{10}O$
Formula weight: 490.68
RTECS: PC8575000
Merck reference: 10, 2048

Physical state, color, and odor:
Colorless to tan, odorless, crystalline solid

Melting point (°C):
Sublimes at 352.8 (NIOSH, 1997)

Boiling point (°C):
Decomposes at 350 (Windholz et al., 1983)

Flash point (°C):
Noncombustible solid (NIOSH, 1997)

Henry's law constant (x 10^{-2} atm·m^3/mol):
3.11 at 25 °C (approximate - calculated from water solubility and vapor pressure)

Bioconcentration factor, log BCF:
4.00 (activated sludge), 2.65 (algae), 2.76 (golden ide) (Freitag et al., 1985)
2.84 (grass shrimp, *Palaemonetes pugio*), 0.91 (blue crab, *Callinectes sapidus*), 3.19 (sheeps-head minnow, *Cyprinodon variegatus*), 3.09 (spot, *Leiostomus xanthurus*) (Schimmel and Wilson, 1977)

Soil sorption coefficient, log K$_{oc}$:
4.74 (calculated, Mercer et al., 1990)

Octanol/water partition coefficient, log K$_{ow}$:
4.07 using method of Kenaga and Goring (1980)

Solubility in organics:
Soluble in acetic acid, alcohols, and ketones (Windholz et al., 1983)

Solubility in water (mg/L):
7.600 at 24 °C (Hollifield, 1979)
2.7 at 20-25 °C (Kilzer et al., 1979)

Vapor pressure (mmHg):
2.25×10^{-7} at 25 °C (Kilzer et al., 1979)

Environmental fate:
   Photolytic. Kepone-contaminated soils obtained from a site in Hopewell, VA were analyzed by GC/MS. 8-Chloro and 9-chloro homologs identified suggested these were photodegradation products of kepone (Borsetti and Roach, 1978).
   Products identified from the photolysis of kepone in cyclohexane were 1,2,3,4,6,7,9,10,10-nonachloro-5,5-dihydroxypentacyclo[$5.3.0.0^{2.6}.0^{3.9}.0^{4.8}$]decane for the hydrate and 1,2,3,4,6,7,-9,10,10-nonachloro-5,5-dimethoxypentacyclo[$5.3.0.0^{2.6}.0^{3.9}.0^{4.8}$]decane (Alley et al., 1974).
   Chemical/Physical. Readily reacts with moisture forming hydrates (Hollifield, 1979). Decomposes at 350 °C (Windholz et al., 1983), probably emitting toxic chlorine fumes.
   Kepone will not hydrolyze to any reasonable extent (Kollig, 1993).

Exposure limits: Potential occupational carcinogen. NIOSH REL: TWA 1 μg/m^3. NIOSH (1997) recommends the most reliable and protective respirators be used, i.e., a self-contained breathing apparatus that has a full facepiece and is operated under positive-pressure or a supplied-air respirator that has a full facepiece and is operated under pressure-demand or under positive-pressure in combination with a self-contained breathing apparatus operated under pressure-demand or positive-pressure.

Symptoms of exposure: May cause tremors (NIOSH, 1997)

Toxicity:
   EC_{50} (48-h) for *Chironomus plumosus* 320 μg/L, *Daphnia magna* 260 μg/L (Mayer and Ellersieck, 1986), *Chironomus plumosus* 350 μg/L (Sanders et al., 1981).
   LC_{50} (26 to 30-d) for adult fathead minnows 31 μg/L (Spehar et al., 1982).
   LC_{50} (96-h) for trout 0.02 ppm, bluegill sunfish 0.051, rainbow trout 0.036 ppm (Verschueren, 1983), *Cyprinodon variegatus* 69.5 μg/L, *Leiostomus xanthurus* 6.6 μg/L, decapod *Palaemonete pugio* 120.9 μg/L (Schimmel and Wilson, 1977), decapod (*Callinectes sapidus*) >210 μg/L (quoted, Reish and Kauwling, 1978), *Salmo gairdneri* 29.5 μg/L (Mayer and Ellersieck, 1986), amphipod (*Gammarus pseudolimnaeus*) 180 μg/L (Sanders et al., 1981).
   LC_{50} (48-h) for sunfish 0.27 ppm, trout 38 ppb (Verschueren, 1983).
   LC_{50} (24-h) for bluegill sunfish 257 ppb, trout 66 ppb, rainbow trout 156 ppb (quoted, Verschueren, 1983), *Salmo gairdneri* 82 μg/L (Mayer and Ellersieck, 1986).
   Acute oral LD_{50} for dogs 250 mg/kg, quail 237 mg/kg, rats 95 mg/kg, rabbits 65 mg/kg (quoted, RTECS, 1985).

Uses: Not commercially produced in the United States. Insecticide; fungicide.

LINDANE

Synonyms: Aalindan; Aficide; Agrisol G-20; Agrocide; Agrocide 2; Agrocide 6G; Agrocide 7; Agrocide III; Agrocide WP; Agronexit; Ameisenatod; Ameisenmittel merck; Aparasin; Aphtiria; Aplidal; Arbitex; BBX; Benhex; Bentox 10; Benzene hexachloride; Benzene-γ-hexachloride; γ-Benzene hexachloride; Bexol; BHC; γ-BHC; Celanex; Chloran; Chloresene; Codechine; DBH; Detmol-extrakt; Detox 25; Devoran; Dol granule; ENT 7796; Entomoxan; Exagama; Forlin; Gallogama; Gamacid; Gamaphex; Gamene; Gamiso; Gammahexa; Gammalin; Gammexene; Gammopaz; Gexane; HCCH; HCH; γ-HCH; Heclotox; Hexa; γ-Hexachlor; Hexachloran; γ-Hexachloran; Hexachlorane; γ-Hexachlorane; γ-Hexachlorobenzene; 1,2,3,4,5,6-Hexachlorocyclohexane; **1α,2α,3β,4α,5α,6β-Hexachlorocyclohexane**; 1,2,3,4,5,6-Hexachloro-γ-cyclohexane; γ-Hexachlorocyclohexane; γ-1,2,3,4,5,6-Hexachlorocyclohexane; Hexatox; Hexaverm; Hexicide; Hexyclan; HGI; Hortex; Inexit; γ-Isomer; Isotox; Jacutin; Kokotine; Kwell; Lendine; Lentox; Lidenal; Lindafor; Lindagam; Lindagrain; Lindagranox; γ-Lindane; Lindapoudre; Lindatox; Lindosep; Lintox; Lorexane; Milbol 49; Mszycol; NA 2761; NCI-C00204; Neo-scabicidol; Nexen FB; Nexit; Nexit-stark; Nexol-E; Nicochloran; Novigam; Omnitox; Ovadziak; Owadziak; Pedraczak; Pflanzol; Quellada; RCRA waste number U129; Silvanol; Spritz-rapidin; Spruehpflanzol; Streunex; Tap 85; TBH; Tri-6; Viton.

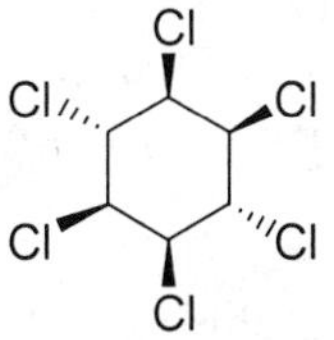

CAS Registry Number: 58-89-9
DOT: 2761
DOT label: Poison
Molecular formula: $C_6H_6Cl_6$
Formula weight: 290.83
RTECS: GV4900000
Merck reference: 10, 5329

Physical state, color, and odor:
Colorless to yellow crystalline solid with a slight, musty odor. Odor threshold concentration is 12.0 mg/kg (Sigworth, 1964).

Melting point (°C):
112.5 (Weast, 1986)
112.9 (Sunshine, 1969)

Boiling point (°C):
323.4 (Weast, 1986)

Density (g/cm³):
1.5691 at 23.6/4 °C (Weast, 1986)

Diffusivity in water (x 10⁻⁵ cm²/sec):
0.59 at 20 °C using method of Hayduk and Laudie (1974)

Flash point (°C):
Noncombustible solid (NIOSH, 1997)

Henry's law constant (x 10^{-7} atm·m^3/mol):
4.8 (Eisenreich et al., 1981)
245 (Spencer and Cliath, 1988)
2.43 at 23 °C (Fendinger et al., 1989)
11.9 at 23 °C (Fendinger and Glotfelty, 1988)

Bioconcentration factor, log BCF:
3.08-3.30 (fish tank), 3.00 (Lake Ontario) (rainbow trout, Oliver and Niimi, 1985)
2.00 (*B. subtilis*, Grimes and Morrison, 1975)
2.65 (brown trout, Sugiura et al., 1979)
2.38 (algae, Geyer et al., 1984)
2.91 (activated sludge), 2.57 (golden ide) (Freitag et al., 1985)
2.65 (freshwater fish), 2.63 (fish, microcosm) (Garten and Trabalka, 1983)
1.92 (pink shrimp), 2.34 (pin fish), 1.80 (grass shrimp), 2.69 (sheepshead minnow) (Schimmel et al., 1977a)
3.42 freshwater clam (*Corbicula manilensis*) (Hartley and Johnston, 1983)
2.08 mussel (*Mytilus edulis*) (Renberg et al., 1985)
2.26 (*Pimephales promelas*, Devillers et al., 1996)
1.57-1.75 (*Lymnaea palustris,* Thybaud and Caquet, 1991)

Soil sorption coefficient, log K_{oc}:
2.86, 2.94 (Svea silt loam), 2.88 (Hegne silty clay), 2.89 (Fargo silty clay), 2.91 (sandy loam, Ontonagon clay), 2.92 (Milaca silty loam), 2.93 (Kranzburg silty clay), 2.99 (Fayette silty loam), 3.00 (Canisteo clay), 3.03 (Lester fine silty loam), 3.05 (Zimmerman soil), 3.10 (Brainerd fine silty loam), 3.11 (Bearden silt), 3.26, 3.36 (Blue Earth silt), 3.51 (Ulen silty loam) (Adams and Li, 1971)
2.88 (Gila silt loam, Huggenberger et al., 1972)
3.03 (Wahid and Sethunathan, 1979)
2.87 (log K_{om}, Chiou et al., 1985)
3.11 (Spencer and Cliath, 1988)
3.52 (Ramamoorthy, 1985)
2.93 (log K_{om}, Sharom et al., 1980a)
3.42 (Chin et al., 1988)
3.238 (silt loam, Chiou et al., 1979)
3.2885 (Reinbold et al., 1979)
2.99, 3.01, and 3.09 for (muck, Fox loamy sand, and Brookstone sandy loam, respectively (Kay and Elrick, 1971)
3.00 (Alfisol, pH 7.5) (Rippen et al., 1982)
2.81 (Rao and Davidson, 1980)
2.76 (sandy clay loam), 3.04 (clay) (Albanis et al., 1989)
2.88 (Commerce soil), 2.96 (Tracy soil), 2.87 (Catlin soil) (McCall et al., 1981)
2.38 (average of 2 soil types, Kanazawa, 1989)
2.84-3.11 (5 soils, Kishi et al., 1990)
Ca-Staten peaty muck: 3.47, 3.41, 3.32, 3.19 at 10, 20, 30, and 40 °C, respectively; Ca-Venanado clay: 3.19, 3.12, 3.07, 3.02 at 10, 20, 30, and 40 °C, respectively (Mills and Biggar, 1969)
3.27 (Powerville sediment (Caron et al., 1985)

Octanol/water partition coefficient, log K_{ow}:
3.89 (DeKock and Lord, 1987; Rippen et al., 1982)
3.70 (Chiou et al., 1986)

3.66 (Travis and Arms, 1988)
3.72 (Kurihara et al., 1973)
3.30 (Geyer et al., 1987)
3.20 (Geyer et al., 1984)
3.57 (Kishi and Hashimoto, 1989)
3.688 (de Bruijn et al., 1989)
3.69 at 25 °C (Paschke et al., 1998)

Solubility in organics (wt % at 20 °C):
Acetone (30.31), benzene (22.42), chloroform (19.35), ether (17.22), ethanol (6.02) (Windholz et al., 1983); 9.76 and 13.97 g/L in hexane at 10 and 20 °C, respectively (Mills and Biggar, 1969).

Solubility in water:
7.52 ppm at 25 °C (Masterton and Lee, 1972)
7.8 ppm at 25 °C (Weil et al., 1974)
7.3 ppm at 25 °C, 12 ppm at 35 °C, 14 ppm at 45 °C (Berg, 1983)
6.98 ppm (Caron et al., 1975)
7.5 g/m^3 at 25 °C (Spencer and Cliath, 1988)
7.87 mg/L at 24 °C (Chiou et al., 1986)
17.0 mg/L at 24 °C (Hollifield, 1979)
7.8 mg/L at 25 °C (Chiou et al., 1979)
5.75-7.40 ppm at 28 °C (Kurihara et al., 1973)
12 ppm at 26.5 °C (Bhavnagary and Jayaram, 1974)
In ppb: 2,150 at 15 °C, 6,800 at 25 °C, 11,400 at 35 °C, 15,200 at 45 °C (particle size ≤5 μ, Biggar and Riggs, 1974)
At 20-25 °C: 6,600 ppb (particle size ≤5 μ), 500 ppb (particle size ≤0.04 μ) (Robeck et al., 1975)
9.2 mg/L at 25 °C (Saleh et al., 1982)
8.5 mg/L at 20 °C (Mills and Biggar, 1969)
7.2 mg/L at 25 °C (Paschke et al., 1998)

Vapor density (ng/L):
518 at 20 °C, 1,971 at 30 °C, 6,784 at 40 °C (Spencer and Cliath, 1970)

Vapor pressure (x 10^{-6} mmHg):
802.5 at 25 °C (Hinckley et al., 1990)
67 at 25 °C (Spencer and Cliath, 1988)
35.4 at 15 °C, 75.8 at 20 °C, 117.8 at at 25 °C, 157.5 at 25 °C, 634.5 at 25 °C, 2,348 at 45 °C (Wania et al., 1994)
32.25 at 20 °C (Dobbs and Cull, 1982)
128 at 30 °C (Nash, 1983; Tinsley, 1979)
14.3 at 25 °C (Klöpffer et al., 1988)
491 at 25 °C (Bidleman, 1984)
28.0, 55.3, 87.0, 168.8, 285.8, 297.0, and 538.5 at 19.58, 24.95, 28.42, 33.58, 37.82, 37.86, and 43.32 °C, respectively (Boehncke et al., 1996)
653 at 25 °C (extrapolated from experimental vapor pressures. Tesconi and Yalkowsky, 1998)

Environmental fate:
Biological. In a laboratory experiment, a strain of *Pseudomonas putida* culture transformed lindane to γ-3,4,5,6-tetrachlorocyclohexane (γ-TCCH), γ-pentachlorocyclohexane (γ-PCCH),

and α-BHC (Benezet and Matsumura, 1973). γ-TCCH was also reported as a product of lindane degradation by *Clostridium sphenoides* (MacRae et al., 1969; Heritage and MacRae, 1977, 1977a), an anaerobic bacterium isolated from flooded soils (MacRae et al., 1969; Sethunathan and Yoshida, 1973a). Lindane degradation by *Escherichia coli* also yielded γ-PCCH (Francis et al., 1975). Evidence suggests that degradation of lindane in anaerobic cultures or flooded soils amended with lindane occurs via reductive dehalogenation producing chlorine-free volatile metabolites (Sethunathan and Yoshida, 1973a).

After a 30-d incubation period, the white rot fungus *Phanerochaete chrysosporium* converted lindane to carbon dioxide. Mineralization began between the third and sixth day of incubation. The production of carbon dioxide was highest between 3-18 d of incubation, after which the rate of carbon dioxide produced decreased until the 30th day. It was suggested that the metabolism of lindane and other compounds, including *p,p'*-DDT, TCDD, and benzo[*a*]pyrene, was dependent on the extracellular lignin-degrading enzyme system of this fungus (Bumpus et al., 1985). White rot fungi, namely *Phanerochaete chrysosporium*, *Pleurotus sajorcaju*, *Pleurotus florida*, and *Pleurotus eryngii*, biodegraded lindane (50 μM) at degradation yields of 10.57-34.34, 20.68-89.11, 17.76-28.90, and 67.13-82.40%, respectively. The experiments were carried out in a culture incubated at 30 °C for 20 d (Arisoy, 1998).

Beland et al. (1976) studied the degradation of lindane in sewage sludge under anaerobic conditions. Lindane underwent reductive hydrodechlorination forming 3,4,5,6-tetrachloro-cyclohex-1-ene (γ-BTC). The amount of γ-BTC that formed reached a maximum concentration of 5% after 2 wk. Further incubation with sewage sludge resulted in decreased concentrations. The evidence suggested that γ-BTC underwent further reduction affording benzene (Beland et al., 1976). Hill and McCarty (1967) reported that lindane (1 ppm) in sewage sludge completely degraded in 24 h. Under anaerobic conditions, the rate of degradation was higher.

When lindane was incubated in aerobic and anaerobic soil suspensions for 3 wk, 0 and 63.8% was lost, respectively (MacRae et al., 1984). Using settled domestic wastewater inoculum, lindane (5 and 10 mg/L) did not degrade after 28 d of incubation at 25 °C (Tabak et al., 1981). When lindane was incubated river water samples and sediments for 3 wk, 80% of the applied amount had degraded. Under sterilized conditions, >95% was recovered after 12 wk. Under unsterile and sterile conditions, 20 and 80% of the recovered lindane was bound to sediments (Oloffs et al., 1973; Oloffs and Albright, 1974).

Soil. Lindane degraded rapidly in flooded rice soils (Raghu and MacRae, 1966). In moist soils, lindane biodegraded to γ-pentachlorocyclohexene (Elsner et al., 1972; Kearney and Kaufman, 1976; Fuhremann and Lichtenstein, 1980). Under anaerobic conditions, degradation by soil bacteria yielded γ-BTC and α-BHC (Kobayashi and Rittman, 1982). Other reported biodegradation products include penta- and tetrachloro-1-cyclohexanes and penta- and tetra-chlorobenzenes (Moore and Ramamoorthy, 1984). Incubation of lindane for 6 wk in a sandy loam soil under flooded conditions yielded γ-TCCH, γ-2,3,4,5,6-pentachlorocyclohex-1-ene, and small amounts of 1,2,4-trichlorobenzene, 1,2,3,4-tetrachlorobenzene, 1,2,3,5-, and/or 1,2,4,5-tetrachlorobenzene (Mathur and Saha, 1975). Incubation of lindane in moist soil for 8 wk yielded the following metabolites: γ-BTC, γ-1,2,3,4,5-pentachlorocyclohex-1-ene, penta-chlorobenzene, 1,2,3,4-, 1,2,3,5-, and/or 1,2,4,5-tetrachlorobenzene, 1,2,4-trichlorobenzene, 1,3,5-trichlorobenzene, *m*- and/or *p*-dichlorobenzene (Mathur and Saha, 1977).

Microorganisms isolated from a loamy sand soil degraded lindane and some of the metabolites identified were pentachlorobenzene, 1,2,4,5-tetrachlorobenzene, 1,2,3,5-tetra-chlorobenzene, γ-PCCH, γ-TCCH and β-3,4,5,6-tetrachloro-1-cyclohexane (β-TCCH) (Tu, 1976). γ-PCCH was also reported as a metabolite of lindane in an Ontario soil that was pretreated with *p,p'*-DDT, dieldrin, lindane, and heptachlor (Yule et al., 1967). The reported half-life in soil is 266 d (Jury et al., 1987).

Indigenous microbes in soil partially degraded lindane to carbon dioxide (MacRae et al., 1967). In a 42-d experiment, ^{14}C-labeled lindane applied to soil-water suspensions under aerobic and anaerobic conditions gave $^{14}CO_2$ yields of 1.9 and 3.0%, respectively (Scheunert et al., 1987).

In a moist Hatboro silt loam, volatilization yields of 50 and 90% were found after 6 h and 6 d, respectively. In a dry Norfolk sand loam, 12% volatilization was reported after 50 h (Glotfelty et al., 1984). The average half-lives for lindane in aerobic and flooded soils under laboratory conditions were 276 and 114 d, respectively (Mathur and Saha, 1977). In field soils, the half-lives ranged from 88 d to 3.2 yr with an average half-life of 426 d (Lichtenstein and Schulz, 1959, 1959a; Lichtenstein et al., 1971; Voerman and Besemer, 1975; Mathur and Saha, 1977).

Surface Water. Lindane degraded in simulated lake impoundments under both aerobic (15%) and anaerobic (90%) conditions. Lindane degraded primarily to α-BHC with trace amounts of δ-BHC (Newland et al., 1969).

Mackay and Wolkoff (1973) estimated an evaporation half-life of 289 d from a surface water body that is 25 °C and 1 m deep.

Surface Water. Hargrave et al. (2000) calculated BAFs as the ratio of the compound tissue concentration [wet and lipid weight basis (ng/g)] to the concentration of the compound dissolved in seawater (ng/mL). Average log BAF values for lindane in ice algae and phytoplankton collected from the Barrow Strait in the Canadian Archipelago were 5.68 and 5.49, respectively.

Plant. Lindane appeared to be metabolized by several grasses to hexachlorobenzene and α-BHC, the latter isomerizing to β-BHC (Steinwandter, 1978; Steinwandter and Schluter, 1978). Oat plants were grown in two soils treated with [^{14}C]lindane. 2,4,5-Trichlorophenol and possibly γ-PCCH were identified in soils but no other compounds other than lindane were identified in the oat roots or tops (Fuhremann and Lichtenstein, 1980). The half-life of lindane in alfalfa was 3.3 d (Treece and Ware, 1965).

Photolytic. Photolysis of lindane in aqueous solutions gives β-BHC (U.S. Department of Health and Human Services, 1989). When an aqueous solution containing lindane was photooxidized by UV light at 90-95 °C, 25, 50, and 75% degraded to carbon dioxide after 3.0, 17.4, and 45.8 h, respectively (Knoevenagel and Himmelreich, 1976).

Chemical/Physical. In basic aqueous solutions, lindane dehydrochlorinates to form pentachlorocyclohexene, then to trichlorobenzenes. In a buffered aqueous solution at pH 8 and 5 °C, the calculated hydrolysis half-life was determined to be 42 yr (Ngabe et al., 1993). The hydrolysis rate constant for lindane at pH 7 and 25 °C was determined to be 1.2×10^{-4}/h, resulting in a half-life of 241 d (Ellington et al., 1987). The hydrolysis half-lives decreases as the temperature and pH increases. Hydrolysis half-lives of 4.4 d, 2.4 d, 9.6 min, and 1.8 min were reported at 65.5 °C (pH 6.85), 65.5 °C (pH 7.24), 46.0 °C (pH 10.98), and 37.0 °C (pH 11.29), respectively (Ellington et al., 1987).

In weakly basic media, lindane undergoes *trans*-dehydrochlorination of the axial chlorines to give the intermediate 1,3,4,5,6-pentachlorocyclohexane. This compound will further react with water to give 1,2,4-trichlorobenzene, 1,2,3 trichlorobenzene, and hydrochloric acid. Three molecules of the acid are produced for every molecule of lindane that reacts (Cremlyn, 1991; Kollig, 1993).

When lindane in hydrogen-saturated deionized water was exposed to a slurry of palladium catalyst (1%) at room temperature, benzene and chloride ions formed as the final products (Schüth and Reinhard, 1997).

Lindane (0.004-0.005 mM) reacted with OH radicals generated from Fenton's reagent in water (pH 2.8) at a rate of 1.1×10^9/M·sec. At pH 2.9 and concentrations of 0.007 and 0.004 mM, the reaction rates were 5.8×10^8 and 5.2×10^8/M·sec, respectively (Haag and Yao, 1992).

Exposure limits (mg/m^3): NIOSH REL: TWA 0.5, IDLH 50; OSHA PEL: TWA 0.5; ACGIH TLV: TWA 0.5 (adopted).

Symptoms of exposure: Irritates eyes, nose, throat, and skin; headache, nausea, clonic convulsions, respiratory problems, cyanosis, aplastic anemia, and muscle spasms (NIOSH, 1997).

Toxicity:

EC_{10} and EC_{50} concentrations inhibiting the growth of alga *Scenedesmus subspicatus* in 96 h were 0.5 and 2.5 mg/L, respectively (Geyer et al., 1985).

EC_{50} (48-h) for *Daphnia pulex* 460 µg/L, *Simocephalus serrulatus* 700 µg/L (Mayer and Ellersieck, 1986).

EC_{50} (24-h) for *Daphnia magna* 14.5 mg/L (Lilius et al., 1995).

LC_{50} (10-d) for *Gammarus pulex* juvenile 7 µg/L, *Chironomus riparius* second larval instar 13 µg/L (Taylor et al., 1991), *Hyalella asteca* 9.8 µg/L (Blockwell et al., 1998).

LC_{50} (5-d) *Chironomus riparius* second larval instar 27 µg/L (Taylor et al., 1991).

LC_{50} (24, 38, 72, and 96-h) values, which were identical regardless of exposure time, were 0.36, 0.16, and 0.14 mg/L for *Poecilia reticulata*, *Brachydanio rerio*, and *Paracheirodon axelrodi*, respectively (Oliveira-Filho and Paumgartten, 1997).

LC_{50} (96-h) for bullhead 64 µg/L, carp 90 µg/L, catfish 44 µg/L, coho salmon 41 µg/L, goldfish 131 µg/L, perch 68 µg/L, bass 32 µg/L, minnows 87 µg/L, bluegill sunfish 68 µg/L, rainbow trout 27 µg/L (Macek and McAllister, 1970), *Gammarus italicus* 26 µg/L, *Echinogammarus tibaldii* 5.1 µg/L (Pantani et al., 1997), *Mysidopsis bahia* 6.3 µg/L, *Penaeus duorarum* 0.17 µg/L, *Palaemonetes pugio* 4.4 µg/L, *Cyprinodon variegatus* 104 µg/L, *Lagodon rhomboides* 30.6 µg/L (Schimmel et al., 1977a), 12.9 and 42.8 µg/L for neonate and adult stage *Hyalella azteca*, respectively (Blockwell et al., 1998), bluegill sunfish 57 µg/L (technical), 138 µg/L (1% dust), *Daphnia magna* 516 µg/L (technical), 6,442 µg/L (1% dust) (Randall et al., 1979), 4.0 and 3.9 µg/L for fourth instar *Chaoborus flavicans* and fourth or fifth instar *Sigara striata*, respectively (Maund et al., 1992), *Gammarus pulex* juvenile 79 µg/L, *Chironomus riparius* second larval instar 34 µg/L (Taylor et al., 1991), *Gammarus pulex* adult 34 µg/L (Abel, 1980) and 22.5 µg/L (Green et al., 1986), *Chironomus riparius* fourth instar 23.5 µg/L, *Baetis rhodani* larva 5.4 µg/L (Green et al., 1986).

LC_{50} (72-h) 13.2 and 45.4 µg/L for neonate and adult stage *Hyalella azteca*, respectively (Blockwell et al., 1998); for freshwater insect (*Chironomus riparius*) fourth instar 6.5 µg/L (Maund et al., 1992).

LC_{50} (48-h) for guppies 0.16-0.3 mg/L (Hartley and Kidd, 1987), 14.8 and 47.6 µg/L for neonate and adult stage *Hyalella azteca*, respectively (Blockwell et al., 1998), *Simocephalus serrulatus* 520 µg/L, *Daphnia pulex* 460 µg/L (Sanders and Cope, 1966), zebrafish 120 µg/L (Slooff, 1979), *Chironomus riparius* second larval instar 55 µg/L (Taylor et al., 1991), three spined stickleback (*Gasterosteus aculeatus*), 45 and 51 µg/L, carp (*Cyprinus carpio*) 94 µg/L, goldfish (*Carassius auratus*) 230 µg/L (Geyer et al., 1994), *Gammarus fasciatus* 39 µg/L, *Salvelinus fontinalis* 44 µg/L, *Pimephales promelas* >100 µg/L, *Chironomus tentans* 1st instar 207 µg/L, *Daphnia magna* 485 µg/L (Macek et al., 1976a).

LC_{50} (24-h) for neonate *Hyalella azteca* 29.5 µg/L (Blockwell et al., 1998), *Brachionus calyciflorus* 22.50 µg/L, *Brachionus plicatilis* 35.89 µg/L (Ferrando and Andreu-Moliner, 1991), *Daphnia magna* 1.64 mg/L, *Brachionus calyciflorus* 22.5 mg/L (Ferrando et al., 1992), for freshwater insect (*Chironomus riparius*) second instar 2.0 µg/L (Maund et al., 1992), *Chironomus riparius* second larval instar 61 µg/L (Taylor et al., 1991), Protozoan (*Spirostomum teres*) 25.26 mg/L (Twagilimana et al., 1998).

LC_{50} (30-min) for Protozoan (*Spirostomum teres*) 11.02 mg/L (Twagilimana et al., 1998).

Acute oral LD_{50} for rats 88-91 mg/kg (Reuber, 1979), cats 25 mg/kg, dogs 40 mg/kg, guinea pigs 127 mg/kg, hamsters 360 mg/kg, rabbits 60 mg/kg (RTECS, 1985); mice 59-246 mg/kg (Worthing and Hance, 1991).

LD_{50} (skin) for rats 500 mg/kg, rabbits 50 mg/kg (quoted, RTECS, 1985).

NOELs of 25 and 50 mg/kg diet were reported for rats and dogs during 2-yr feeding trials (Worthing and Hance, 1991).

Drinking water standard (final): MCLG: 0.2 μg/L; MCL: 0.2 μg/L (U.S. EPA, 1996).

Uses: Ingested insecticide and by contact. Used against soil-dwelling and phytophagous insects including Aphididae and larvae of Coleoptera, Curculiodidae, Diplopoda, Diptera, Lepidoptera, Symphyla, and Thysanoptera (Worthing and Hance, 1991).

MALATHION

Synonyms: American Cyanamid 4049; *S*-1,2-Bis(carbethoxy)ethyl-*O,O*-dimethyl dithiophosphate; *S*-1,2-Bis(ethoxycarbonyl)ethyl-*O,O*-dimethyl phosphorodithioate; *S*-1,2-Bis(ethoxycarbonyl)ethyl-*O,O*-dimethyl thiophosphate; Calmathion; Carbethoxy malathion; Carbetovur; Carbetox; Carbofos; Carbophos; Celthion; Chemathion; Cimexan; Compound 4049; Cythion; Detmol MA; Detmol MA 96%; *S*-1,2-Dicarbethoxyethyl-*O,O*-dimethyl dithiophosphate; Dicarboethoxyethyl-*O,O*-dimethyl phosphorodithioate; 1,2-Di(ethoxycarbonyl)ethyl-*O,O*-dimethyl phosphorodithioate; *S*-1,2-Di(ethoxycarbonyl)ethyl dimethyl phosphorothiolothionate; Diethyl (dimethoxyphosphinothioylthio) butanedioate; Diethyl (dimethoxyphosphinothioylthio) succinate; Diethyl mercaptosuccinate, *O,O*-dimethyl phosphorodithioate; Diethyl mercaptosuccinate, *O,O*-dimethyl thiophosphate; Diethylmercaptosuccinic acid *O,O*-dimethyl phosphorodithioate; **[(Dimethoxyphosphinothioyl)thio]butanedioic acid diethyl ester**; *O,O*-Dimethyl-*S*-1,2-bis(ethoxycarbonyl)ethyldithiophosphate; *O,O*-Dimethyl-*S*-(1,2-dicarbethoxyethyl)dithiophosphate; *O,O*-Dimethyl-*S*-(1,2-dicarbethoxyethyl)phosphorodithioate; *O,O*-Dimethyl-*S*-(1,2-dicarbethoxyethyl)thiothionophosphate; *O,O*-Dimethyl-*S*-1,2-di(ethoxycarbamyl)ethyl phosphorodithioate; *O,O*-Dimethyldithiophosphate dimethylmercaptosuccinate; EL 4049; Emmatos; Emmatos extra; ENT 17034; Ethiolacar; Etiol; Experimental insecticide 4049; Extermathion; Formal; Forthion; Fosfothion; Fosfotion; Four thousand forty-nine; Fyfanon; Hilthion; Hilthion 25WDP; Insecticide 4049; Karbofos; Kopthion; Kypfos; Malacide; Malafor; Malakill; Malagran; Malamar; Malamar 50; Malaphele; Malaphos; Malasol; Malaspray; Malathion E50; Malathion LV concentrate; Malathion ULV concentrate; Malathiozoo; Malathon; Malathyl LV concentrate & ULV concentrate; Malatol; Malatox; Maldison; Malmed; Malphos; Maltox; Maltox MLT; Mercaptosuccinic acid diethyl ester; Mercaptothion; MLT; Moscardia; NA 2783; NCI-C00215; Oleophosphothion; OMS 1; Orthomalathion; Phosphothion; Prioderm; Sadofos; Sadophos; SF 60; Siptox I; Sumitox; Tak; TM-4049; Vegfru malatox; Vetiol; Zithiol.

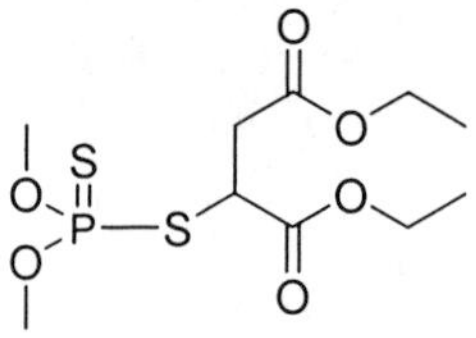

CAS Registry Number: 121-75-5
DOT: 2783
DOT label: Poison
Molecular formula: $C_{10}H_{19}O_6PS_2$
Formula weight: 330.36
RTECS: WM8400000
Merck reference: 10, 5522

Physical state, color, and odor:
Clear yellow to brown liquid with a garlic odor

Melting point (°C):
2.9 (Windholz et al., 1983)

Boiling point (°C):
156-157 at 0.7 mmHg (Windholz et al., 1983)
120 °C at 0.2 mmHg (Freed et al., 1977)

Density (g/cm³):
1.23 at 25/4 °C (Windholz et al., 1983)

Diffusivity in water (x 10⁻⁵ cm²/sec):
0.44 at 20 °C using method of Hayduk and Laudie (1974)

Flash point (°C):
>162.8 (open cup, Meister, 1988)

Henry's law constant (x 10⁻⁹ atm·m³/mol):
4.89 at 25 °C (Fendinger and Glotfelty, 1990)

Bioconcentration factor, log BCF:
0.00 (fish, microcosm) (Garten and Trabalka, 1983)
1.04 (*Oryzias latipes*, Tsuda et al., 1997)
Values reported in the whole body of the freshwater fish, willow shiner (*Gnathopogon caeru-
lescens*) after 0.98, 1.39, 1.56, 1.48, 1.51, and 1.58 following exposure times of 6, 12, 24, 48,
72, and 168 h, respectively (Tsuda et al., 1989).

Soil sorption coefficient, log K_{oc}:
2.61 (Lihue silty clay soil, Miles and Takashima, 1991)

Octanol/water partition coefficient, log K_{ow}:
2.89 (Chiou et al., 1977; Freed et al., 1979)
2.84 (Bowman and Sans, 1983a)

Solubility in organics:
Miscible in most organic solvents (Meister, 1988)

Solubility in water (mg/L):
300 at 30 °C (Freed et al., 1977)
141 at 10 °C, 145 at 20 °C, 164 at 30 °C (Bowman and Sans, 1985)
143 at 20 °C (Bowman and Sans, 1983; Miles and Takashima, 1991)
145 at 20 °C (Fühner and Geiger, 1977)

Vapor density:
13.50 g/L at 25 °C, 11.40 (air = 1)

Vapor pressure (x 10⁻⁶ mmHg):
35.3 at 25 °C (Hinckley et al., 1990)
1.25 at 20 °C (Freed et al., 1977)
40 at 20 °C (Kenaga and Goring, 1980)
7.95 at 25 °C (Kim et al., 1984)

Environmental fate:
Biological. Walker (1976) reported that 97% of malathion added to both sterile and
nonsterile estuarine water was degraded after incubation in the dark for 18 d. Complete
degradation was obtained after 25 d. Malathion degraded fastest in nonsterile soils and
decomposed faster in soils that were sterilized by gamma radiation than in soils that were
sterilized by autoclaving. After 1 d of incubation, the percents of malathion degradation that
occurred in autoclaved, irradiated, and nonsterile soils were 7, 90, and 97, respectively (Getzin

and Rosefield, 1968). Degradation of malathion in organic-rich soils was 3 to 6 times higher than in soils not containing organic matter. The half-life in an organic-rich soil was about 1 d (Gibson and Burns, 1977). Malathion was degraded by soil microcosms isolated from an agricultural area on Kauai, HI. Degradation half-lives in laboratory and field experiments were 8.2 and 2 h, respectively. Dimethyl phosphorodithioic acid and diethyl fumarate were identified as degradation products (Miles and Takashima, 1991). Mostafa et al. (1972) found that the soil fungi *Penicillium notatum*, *Aspergillus niger*, *Rhizoctonia solani*, *Rhizobium trifolii*, and *Rhizobium leguminosarum* converted malathion to the following metabolites: malathion diacid, dimethyl phosphorothioate, dimethyl phosphorodithioate, dimethyl phosphate, monomethyl phosphate, and thiophosphates. Malathion also degraded in groundwater and seawater, but at a slower rate (half-life = 4.7 d). Microorganisms isolated from papermill effluents were responsible for the formation of malathion monocarboxylic acid (Singh and Seth, 1989).

Paris et al. (1975) isolated a heterogenous bacterial population that was capable of degrading low concentrations of malathion to β-malathion monoacid. About 1% of the original malathion concentration degraded to malathion dicarboxylic acid, *O,O*-dimethyl phosphorodithioic acid and diethyl maleate. The major metabolite in soil is β-malathion monoacid (Paris et al., 1981). Rosenberg and Alexander (1979) demonstrated that two strains of *Pseudomonas* used malathion as the sole source of phosphorus. It was suggested that degradation of malathion resulted from an induced enzyme or enzyme system that catalytically hydrolyzed the aryl P-O bond, forming dimethyl phosphorothioate as the major product.

Matsumura and Bousch (1966) isolated carboxylesterase(s) enzymes from the soil fungus *Trichoderma viride* and a bacterium *Pseudomonas* sp., obtained from Ohio soil samples, that were capable of degrading malathion. Compounds identified included diethyl maleate, desmethyl malathion, carboxylesterase products, other hydrolysis products, and unidentified metabolites. The authors found that these microbial populations did not have the capability to oxidize malathion due to the absence of malaoxon. However, the major degradative pathway appeared to be desmethylation and the formation of carboxylic acid derivatives.

Soil. In soil, malathion was degraded by *Arthrobacter* sp. to malathion monoacid, malathion dicarboxylic acid, potassium dimethyl phosphorothioate, and potassium dimethyl phosphorodithioate. After 10 d, degradation yields in sterile and non-sterile soils were 8, 5, and 19% and 92, 94, and 81%, respectively (Walker and Stojanovic, 1974). Chen et al. (1969) reported that the microbial conversion of malathion to malathion monoacid was a result of demethylation of the *O*-methyl group. Malathion was converted by unidentified microorganisms in soil to thiomalic acid, dimethyl thiophosphoric acid, and diethylthiomaleate (Konrad et al., 1969).

The half-lives for malathion in soil incubated in the laboratory under aerobic conditions ranged from 0.2 to 2.1 d with an average of 0.8 d (Konrad et al., 1969; Walker and Stojanovic, 1973; Gibson and Burns, 1977).

In a silt loam and sandy loam, reported R_f values were 0.88 and 0.90, respectively (Sharma et al., 1986).

Plant. When malathion on ladino clover seeds (10.9 ppm) was exposed to UV light (2537 Å) for 168 h, malathion was the only residue detected. It was reported that 66.1% of the applied amount was lost due to volatilization (Archer, 1971).

Residues identified on field-treated kale other than malathion included the oxygen analog and the impurity identified by nuclear magnetic resonance and mass spectrometry as ethyl butyl mercaptosuccinate, *S*-ester with *O,O*-dimethyl phosphorodithioate. This compound did not form as an alteration product of malathion but was present in the 50% emulsifiable concentrate (Gardner et al., 1969).

Dogger and Bowery (1958) reported a half-life of malathion in alfalfa of 4.1 d. Foliar half-life of 0.3-4.9 d were reported by many investigators (Brett and Bowery, 1958; Dogger and Bowery, 1958; Gardner et al., 1969; Nigg et al., 1981; Polles and Vinson, 1969; Saini and Dorough, 1970; Smith et al., 1960; Waites and Van Middelem, 1958; Wheeler et al., 1967).

Surface Water. In raw river water (pH 7.3-8.0), 90% degraded within 2 wk, presumably by biological activity (Eichelberger and Lichtenberg, 1971). In estuarine water, the half-life of malathion ranged from 4.4 to 4.9 d (Lacorte et al., 1995).

Photolytic. Malathion absorbs UV light at wavelengths more than 290 nm indicating direct photolysis should occur (Gore et al., 1971). When malathion was exposed to UV light, malathion monoacid, malathion diacid, *O,O*-diethyl phosphorothioic acid, dimethyl phosphate, and phosphoric acid were formed (Mosher and Kadoum, 1972).

In sterile water and river water, photolytic half-lives of 41.25 d and 16 h were reported, respectively (Archer, 1971). Zepp and Schlotzhauer (1983) found photolysis of malathion in water containing algae to occur at a rate more than 25 times faster than in distilled water.

Chemical/Physical. Hydrolyzes in water forming *cis*-diethyl fumarate, *trans*-diethyl fumarate (Suffet et al., 1967), thiomalic acid, and dimethyl thiophosphate (Mulla et al., 1981). The reported hydrolysis half-lives at pH 7.4 and temperatures of 20 and 37.5 °C were 10.5 and 1.3 d, respectively (Freed et al., 1977). In a preliminary study, Librando and Lane (1997) concluded that the hydrolysis of malathion is very sensitive to pH. At pH 8.5, <5% of the malathion remains after 2 d, whereas at pH 5.7, >90% remains after 20 d.

Day (1991) reported that the hydrolysis products are dependent upon pH. In basic solutions, malathion hydrolyzes to diethyl fumarate and dimethyl phosphorodithioic acid (Bender, 1969; Day, 1991). Dimethyl phosphorothionic acid and 2-mercaptodiethyl succinate formed in acidic solutions (Day, 1991). The hydrolysis half-lives of malathion in a sterile 1% ethanol/water solution at 25 °C and pH values of 4.5, 6.0, 7.0, and 8.0, were 18, 5.8, 1.7, and 0.53 wk, respectively (Chapman and Cole, 1982). The reported hydrolysis half-lives at pH 7.4 at 20 and 37.5 °C were 10.5 and 1.3 d, respectively. At 20 °C and pH 6.1, the hydrolysis half-life is 120 d (Freed et al., 1979). Konrad et al. (1969) reported that after 7 d at pH values of 9.0 and 11.0, 25 and 100% of the malathion was hydrolyzed. Hydrolysis of malathion in acidic and alkaline (0.5 M sodium hydroxide) conditions gives $(CH_3O)_2P(S)Na$ and $(CH_3O)_2P(S)OH$ (Sittig, 1985).

Malaoxon and phosphoric acid were reported as ozonation products of malathion in drinking water (Richard and Bréner, 1984).

At 87 °C and pH 2.5, malathion degraded in water to malathion α-monoacid and malathion β-monoacid. From the extrapolated acid degradation constant at 27 °C, the half-life was calculated to be >4 yr (Wolfe et al., 1977a). Under alkaline conditions (pH 8 and 27 °C), malathion degraded in water to malathion monoacid, diethyl fumarate, ethyl hydrogen fumarate, and *O,O*-dimethyl phosphorodithioic acid. At pH 8, the reported half-lives at 0, 27, and 40 °C are 40 d, 36 h, and 1 h, respectively (Wolfe et al., 1977a). However, under acidic conditions, it was reported that malathion degraded into diethyl thiomalate and *O,O*-dimethyl phosphorothionic acid (Wolfe et al., 1977a).

When applied as an aerial spray, malathion was converted to malaoxon and diethyl fumarate via oxidation and hydrolysis, respectively (Brown et al., 1993).

Emits toxic fumes of nitrogen and phosphorus oxides when heated to decomposition (Lewis, 1990). Products reported from the combustion of malathion at 900 °C include carbon monoxide, carbon dioxide, chlorine, sulfur oxides, nitrogen oxides, hydrogen sulfide, and oxygen (Kennedy et al., 1972).

Exposure limits (mg/m³): NIOSH REL: TWA 10, IDLH 250; OSHA PEL: 15; ACGIH TLV: TWA 10 (adopted).

Symptoms of exposure: Miosis; eye and skin irritation; rhinorrhea; headache; tight chest, wheezing, laryngeal spasm; salivation; anorexia, nausea, vomiting, abdominal cramps; diarrhea, ataxia (NIOSH, 1997).

The acceptable daily intake for humans is 0.02 mg/kg body weight (Worthing and Hance, 1991).

Toxicity:

EC_{50} (48-h) for *Daphnia magna* 1 μg/L, *Daphnia pulex* 1.8 μg/L, *Simocephalus serrulatus* 3.4 μg/L (Mayer and Ellersieck, 1986).

EC_{50} (24-h) for *Daphnia magna* 353 μg/L, *Daphnia pulex* 6.6 μg/L (Lilius et al., 1995), brine shrimp (*Artemia* sp.) >140 mg/L, for estuarine rotifer (*Brachionus plicatilis*) 74 mg/L (Guzzella et al., 1997).

LC_{50} (96-h) for bluegill sunfish 100 μg/L (Hartley and Kidd, 1987), largemouth bass 285 μg/L (Worthing and Hance, 1991), coho salmon 100 μg/L, brown trout 200 μg/L, channel catfish 9.0 mg/L, channel black bullhead 12.9 mg/L, fathead minnows 8.7 mg/L, rainbow trout 170 μg/L, perch 260 μg/L (Macek and McAllister, 1970), green sunfish 120 μg/L (Verschueren, 1983), fish (*Cyprinodon variegatus*) 51 μg/L (quoted, Reish and Kauwling, 1978), flagfish (*Jordanella floridae*) 349 μg/L (Hermanutz, 1978), 4.51 and 3.89 mg/L for technical and commercial formulations to freshwater fish (*Channa punctatus*), respectively (Haider and Inbaraj, 1986).

LC_{50} (96-h semi-static bioassay) for *Tigriopus brevicornis* females 24.3 μg/L (Forget et al., 1998), Indian catfish (*Heteropneustes fossilis*) 11.676 mg/L (Munshi et al., 1999).

LC_{50} (48-h) for red killifish 3 mg/L (Yoshioka et al., 1986), *Tilapia mossambica* 5.6 mg/L (Spehar et al., 1982), killifish (*Oryzias latipes*) 1.8 mg/L (Tsuda et al., 1997).

LC_{50} (24-h) for bluegill sunfish 120 ppb, rainbow trout 100 ppb (Verschueren, 1983), *Brachionus calyciflorus* 33.7 mg/L (Ferrando and Andreu-Moliner, 1991), *Brachionus rubens* 35.3 mg/L (Snell and Persoone, 1989), first instar *Toxorhynchites splendens* 69.1 ppb (Tietze et al., 1993).

LC_5, LC_{50}, and LC_{95} values for *Schistosoma mansoni miracidia* after 2 h of exposure were 83.38, 153.11, and 245.85 ppm, respectively. After 4 h of exposure, the LC_5, LC_{50}, and LC_{95} values 76.86, 116.48, and 172.04 ppm, respectively (Tchounwou et al., 1991a).

Acute oral LD_{50} for wild birds 400 mg/kg, chickens 600 mg/kg, cattle 53 mg/kg, ducks 1.485 mg/kg, guinea pigs 570 mg/kg, rats 370 mg/kg (quoted, RTECS, 1985), 1,680 mg/kg (Berteau and Deen, 1978).

Acute percutaneous LD_{50} (24-h) for rabbits 4,100 mg/kg (Worthing and Hance, 1991).

LD_{50} (inhalation) for mice >759 mg/kg (Berteau and Deen, 1978).

Drinking water standard: No MCLGs or MCLs have been proposed (U.S. EPA, 1996).

Use: Non-systemic insecticide and acaricide for control of sucking and chewing insects and spider mites on vegetables, fruits, ornamentals, field crops, greenhouses, gardens, and forestry.

MALEIC ANHYDRIDE

Synonyms: *cis*-Butenedioic anhydride; **2,5-Furanedione**; Maleic acid anhydride; RCRA waste number U147; Toxilic anhydride; UN 2215.

CAS Registry Number: 108-31-6
DOT: 2215
Molecular formula: $C_4H_2O_3$
Formula weight: 98.06
RTECS: ON3675000
Merck reference: 10, 5526

Physical state and color:
White crystals. Odor threshold concentration is 0.32 ppm (Amoore and Hautala, 1983).

Melting point (°C):
60 (Weast, 1986)
52.8 (Windholz et al., 1983)

Boiling point (°C):
197-199 (Weast, 1986)
202 (Windholz et al., 1983)

Density (g/cm³):
1.314 at 60/4 °C (Weast, 1986)
1.48 at 20/4 °C (Sax and Lewis, 1987)

Diffusivity in water:
Not applicable - reacts with water

Flash point (°C):
104.2 (NIOSH, 1997)
103 (Dean, 1987)

Lower explosive limit (%):
1.4 (NFPA, 1984)

Upper explosive limit (%):
7.1 (NIOSH, 1997)

Entropy of fusion (cal/mol·K):
9.94 (Domalski and Hearing, 1998)

Henry's law constant (atm·m³/mol):
Not applicable - reacts with water

Ionization potential (eV):
11.07-11.45 (Lias et al., 1998)

Soil sorption coefficient, log K$_{oc}$:
Not applicable - reacts with water

Octanol/water partition coefficient, log K$_{ow}$:
Not applicable - reacts with water

Solubility in organics (wt % at 25 °C):
Acetone (227), benzene (50), carbon tetrachloride (0.60), chloroform (52.5), ethyl acetate (112), ligroin (0.25), toluene (23.4), *o*-xylene (19.4) (Windholz et al., 1983). Soluble in alcohol and 1,4-dioxane (Hawley, 1981).

Vapor pressure (x 10^{-5} mmHg):
5 at 20 °C, 200 at 30 °C (Verschueren, 1983)

Environmental fate:
   Chemical/Physical. Reacts with water forming maleic acid (Bunton et al., 1963). Anticipated products from the reaction of maleic anhydride with ozone or OH radicals in the atmosphere are carbon monoxide, carbon dioxide, aldehydes, and esters (Cupitt, 1980).

Exposure limits: NIOSH REL: TWA 1 ppm (0.25 mg/m^3), IDLH 10 ppm; OSHA PEL: TWA 0.25 ppm; ACGIH TLV: TWA 0.25 ppm with an intended change of 0.1 ppm.

Toxicity:
   Acute oral LD$_{50}$ for rats 400 mg/kg, guinea pigs 390 mg/kg, rabbits 875 mg/kg, mice 465 mg/kg (quoted, RTECS, 1985).

Uses: In organic synthesis (Diels-Alder reactions); manufacturing of agricultural chemicals, dye intermediates, pharmaceuticals, and alkyd-type resins; manufacture of fumaric and tartaric acids; pesticides; preservative for oils and fats.

MESITYL OXIDE

Synonyms: Isobutenyl methyl ketone; Isopropylidene acetone; Methyl isobutenyl ketone; 2-Methyl-2-penten-4-one; **4-Methyl-2-penten-2-one**; UN 1229.

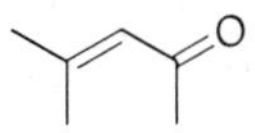

CAS Registry Number: 141-79-7
DOT: 1229
DOT label: Flammable liquid
Molecular formula: $C_6H_{10}O$
Formula weight: 98.14
RTECS: SB4200000
Merck reference: 10, 5753

Physical state, color, and odor:
Clear, pale yellow liquid with a strong, peppermint, or honey-like odor. Odor threshold concentration is 0.45 ppm (Amoore and Hautala, 1983).

Melting point (°C):
-52.9 (Weast, 1986)
-41.5 (Windholz et al., 1983)

Boiling point (°C):
129.7 (Weast, 1986)

Density (g/cm³):
0.8653 at 20/4 °C (Weast, 1986)
0.8592 at 15/4 °C (Windholz et al., 1983)

Diffusivity in water (x 10⁻⁵ cm²/sec):
0.82 at 20 °C using method of Hayduk and Laudie (1974)

Flash point (°C):
30.8 (NIOSH, 1997)
32.2 (Hawley, 1981)

Lower explosive limit (%):
1.4 (NFPA, 1984)

Upper explosive limit (%):
7.2 (NFPA, 1984)

Henry's law constant (x 10⁻⁶ atm·m³/mol):
4.01 at 20 °C (approximate - calculated from water solubility and vapor pressure)

Ionization potential (eV):
9.08 ± 0.03 (Franklin et al., 1969)

Soil sorption coefficient, log K_{oc}:
Unavailable because experimental methods for estimation of this parameter for ketones are lacking in the documented literature

Octanol/water partition coefficient, log K_{ow}:
1.25 using method of Hansch et al. (1968)

Solubility in organics:
Soluble in acetone, alcohol, and ether (Weast, 1986)

Solubility in water (20 °C):
3 wt % (NIOSH, 1997)
28 g/L (quoted, Verschueren, 1983)

Vapor density:
4.01 g/L at 25 °C, 3.39 (air = 1)

Vapor pressure (mmHg):
8.7 at 20 °C (Verschueren, 1983)

Exposure limits: NIOSH REL: TWA 10 ppm (40 mg/m^3), IDLH 1,400 ppm; OSHA PEL: TWA 25 ppm (100 mg/m^3); ACGIH TLV: TWA 15 ppm, STEL 25 ppm (adopted).

Symptoms of exposure: Irritant to the eyes, skin, and mucous membranes. Narcotic at high concentrations (Patnaik, 1992).

Toxicity:
Acute oral LD$_{50}$ for rats 1,120 mg/kg, mice 710 mg/kg (quoted, RTECS, 1985).

Uses: Solvent for nitrocellulose, gums, and resins; roll-coating inks, varnishes, lacquers, stains, and enamels; starting material for synthesizing methyl isobutyl ketone; insect repellent; ore flotation.

METHANOL

Synonyms: Carbinol; Colonial spirit; Columbian spirits; Columbian spirits (wood alcohol); Methyl alcohol; Methyl hydroxide; Methylol; Monohydroxymethane; NA 1230; Pyroxylic spirit; RCRA waste number U154; Wood alcohol; Wood naphtha; Wood spirit.

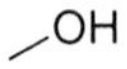

$\diagdown$OH

CAS Registry Number: 67-56-1
DOT: 1230
Molecular formula: CH_4O
Formula weight: 32.04
RTECS: PC1400000
Merck reference: 10, 5816

Physical state, color, and odor:
Clear, colorless liquid with a characteristic odor. Odor threshold concentration is 4.3 ppm (Keith and Walters, 1992).

Melting point (°C):
-93.9 (Weast, 1986)
-97.8 (Windholz et al., 1983)

Boiling point (°C):
65 (Weast, 1986)

Density (g/cm³):
0.796 at 15/4 °C (Verschueren, 1983)
0.7914 at 20/4 °C (Weast, 1986)
0.7869 at 25.00/4 °C (Aralaguppi et al., 1999a)
0.78162 at 30.00/4 °C (Nikam et al., 2000)

Diffusivity in water (x 10⁻⁵ cm²/sec):
1.54 at 25 °C (Hao and Leist, 1996)
At 5 °C (mole fraction): 0.772 (0.0373), 0.710 (0.0763), 0.590 (0.1776), 0.522 (0.2692), 0.586 (0.4814), 0.827 (0.6867), 1.212 (0.8943); at 25 °C: 1.387 (0.04402), 1.281 (0.08502), 1.014 (0.27789), 1.038 (0.49099), 1.362 (0.68519), 1.735 (0.83484) (Derlacki et al., 1985)

Dissociation constant, pK_a:
≈ 16 (Gordon and Ford, 1972)

Flash point (°C):
16.9 (NIOSH, 1997)
12.2 (open cup, Hawley, 1981)

Lower explosive limit (%):
6.0 (NIOSH, 1997)

Upper explosive limit (%):
36 (NIOSH, 1997)

Heat of fusion (kcal/mol):
0.768 (Dean, 1987)

Henry's law constant (x 10^{-6} atm·m^3/mol):
4.44 at 25 °C (Snider and Dawson, 1985)
4.35 at 25 °C (Butler et al., 1935; Burnett, 1963)

Ionization potential (eV):
10.84 (Franklin et al., 1969)

Bioconcentration factor, log BCF:
4.45 (algae, Geyer et al., 1984)
2.67 (activated sludge, Freitag et al., 1985)

Soil sorption coefficient, log K_{oc}:
0.44 (Gerstl and Helling, 1987)

Octanol/water partition coefficient, log K_{ow}:
-0.32, -0.52, -0.68, -0.71, -0.77, -0.82 (Sangster, 1989)
-0.66 (Hansch and Anderson, 1967)
-0.81 (Collander, 1951)
-0.70 (Geyer et al., 1984)

Solubility in organics:
Miscible with benzene, ethanol, ether, ketone, and many other organic solvents (Windholz et al., 1983)

Solubility in water:
Miscible (Palit, 1947). A saturated solution in equilibrium with its own vapor had a concentration of 1,163 g/L at 25 °C (Kamlet et al., 1987).

Vapor density:
1.31 g/L at 25 °C, 1.11 (air = 1)

Vapor pressure (mmHg):
74.1 at 15 °C, 97.6 at 20 °C, 127.2 at 25 °C, 164.2 at 30 °C (Gibbard and Creek, 1974)

Environmental fate:
Biological. In a 5-d experiment, [^{14}C]methanol applied to soil water suspensions under aerobic and anaerobic conditions gave $^{14}CO_2$ yields of 53.4 and 46.3%, respectively (Scheunert et al., 1987). Heukelekian and Rand (1955) reported a 5-d BOD value of 0.85 g/g which is 56.7% of the ThOD value of 1.50 g/g.

Photolytic. Photooxidation of methanol in an oxygen-rich atmosphere (20%) yielded formaldehyde and hydroxy-peroxyl radicals. With chlorine, formaldehyde, carbon monoxide, hydrogen peroxide, and formic acid were detected (Whitbeck, 1983). Reported rate constants for the reaction of methanol and OH radicals in the atmosphere: 5.7 x 10^{11} cm^3/mol·sec at 300 K (Hendry and Kenley, 1979), 5.7 x 10^8 L/mol·sec (second-order) at 292 K (Campbell et al., 1976), 1.00 x 10^{-12} cm^3/molecule·sec at 292 K (Meier et al., 1985), 7.6 x 10^{-13} cm^3/molecule·sec at 298 K (Ravishankara and Davis, 1978), 6.61 x 10^{-13} cm^3/molecule·sec at room temperature (Wallington et al., 1988a). Based on an atmospheric OH concentration of 1.0 x 10^6 molecule/cm^3, the reported half-life of methanol is 8.6 d (Grosjean, 1997).

Chemical/Physical. In a smog chamber, methanol reacted with nitrogen dioxide to give methyl nitrite and nitric acid (Takagi et al., 1986). The formation of these products was facilitated when this experiment was accompanied by UV light (Akimoto and Takagi, 1986).

Methanol will not hydrolyze because it has does not have a hydrolyzable functional group (Kollig, 1993).

At an influent concentration of 1,000 mg/L, treatment with granular activated carbon resulted in an effluent concentration of 964 mg/L. The adsorbability of the carbon used was 7 mg/g carbon (Guisti et al., 1974).

Hydroxyl radicals reacts with methanol in aqueous solution at a reaction rate of 1.60×10^{-12} cm^3/molecule·sec (Wallington et al., 1988).

Exposure limits: NIOSH REL: TWA 200 ppm (260 mg/m^3), STEL 250 ppm (325 mg/m^3), IDLH 6,000 ppm; OSHA PEL: TWA 200 ppm; ACGIH TLV: TWA 200 ppm, STEL 250 ppm (adopted).

Symptoms of exposure: Ingestion may cause acidosis and blindness. Symptoms of poisoning include nausea, abdominal pain, headache, blurred vision, shortness of breath, and dizziness (Patnaik, 1992). An irritation concentration of 22.875 g/m^3 in air was reported by Ruth (1986).

Toxicity:

EC$_{50}$ (48-h) for *Spirostomum ambiguum* were 17.6g/L (Nałecz-Jawecki and Sawicki, 1999).

EC$_{50}$ (24-h) for *Daphnia magna* 20.8 g/L, *Daphnia pulex* 27.5 g/L (Lilius et al., 1995), *Spirostomum ambiguum* 23.5 g/L (Nałecz-Jawecki and Sawicki, 1999).

LC$_{50}$ (96-h) for fathead minnows 28,100 mg/L (Veith et al., 1983).

LC$_{50}$ (48-h) and LC$_{50}$ (24-h) values for *Spirostomum ambiguum* were 36.8 and 37.4 g/L, respectively (Nałecz-Jawecki and Sawicki, 1999).

LC$_{50}$ (inhalation) for rats 64,000 ppm/4-h (quoted, RTECS, 1985).

Source: Methanol occurs naturally in small-flowered oregano (5-45 ppm) (Baser et al., 1991), Guveyoto shoots (700 ppb) (Baser et al., 1992), orange juice (0.8-80 ppm), onion bulbs, pineapples, black currant, spearmint, apples, jimsonweed leaves, soybean plants, wild parsnip, blackwood, soursop, cauliflower, caraway, petitgrain, bay leaves, tomatoes, parsley leaves, and geraniums (Duke, 1992).

Methanol may enter the environment from methanol spills because it is used in formaldehyde solutions to prevent polymerization (Worthing and Hance, 1991).

Uses: Solvent for nitrocellulose, ethyl cellulose, polyvinyl butyral, rosin, shellac, manila resin, dyes; fuel for utility plants; home heating oil extender; preparation of methyl esters, form-aldehyde, methacrylates, methylamines, dimethyl terephthalate, polyformaldehydes; methyl halides, ethylene glycol; in gasoline and diesel oil antifreezes; octane booster in gasoline; source of hydrocarbon for fuel cells; extractant for animal and vegetable oils; denaturant for ethanol; in formaldehyde solutions to inhibit polymerization; softening agent for certain plastics; dehydrator for natural gas.

METHOXYCHLOR

Synonyms: 2,2-Bis(*p*-anisyl)-1,1,1-trichloroethane; 1,1-Bis(*p*-methoxyphenyl)-2,2,2-trichloro-ethane; 2,2-Bis(*p*-methoxyphenyl)-1,1,1-trichloroethane; Chemform; 2,2-Di-*p*-anisyl-1,1,1-tri-chloroethane; Dimethoxy-DDT; *p,p'*-Dimethoxydiphenyltrichloroethane; Dimethoxy-DT; 2,2-Di(*p*-methoxyphenyl)-1,1,1-trichloroethane; Di(*p*-methoxyphenyl)trichloromethylmethane; DMDT; 4,4'-DMDT; *p,p'*-DMDT; DMTD; ENT 1716; Maralate; Marlate; Marlate 50; Meth-oxide; Methoxo; 4,4'-Methoxychlor; *p,p'*-Methoxychlor; Methoxy-DDT; Metox; Moxie; NCI-C00497; OMS 466; RCRA waste number U247; 1,1,1-Trichloro-2,2-bis(*p*-anisyl)ethane; 1,1,1-Trichloro-2,2-bis(*p*-methoxyphenol)ethanol; 1,1,1-Trichlor-*o*-2,2-bis(*p*-methoxyphenyl)ethane; 1,1,1-Trichloro-2,2-di(4-methoxyphenyl)ethane; **1,1'-(2,2,2-Trichloroethylidene)bis-(4-meth-oxybenzene)**.

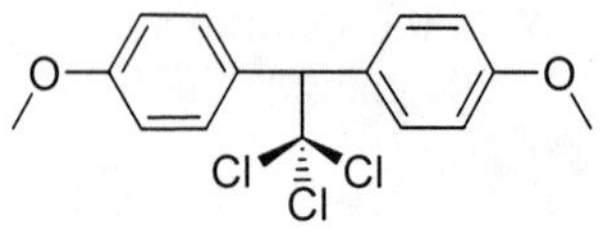

Note: Technical grades are grayish and contain ≥88% methoxychlor and ≤12 related isomers (Worthing and Hance, 1991).

CAS Registry Number: 72-43-5
DOT: 2761
DOT label: Poison
Molecular formula: $C_{16}H_{15}Cl_3O_2$
Formula weight: 345.66
RTECS: KJ3675000
Merck reference: 10, 5868

Physical state, color, and odor:
White, gray, or pale yellow crystals or powder. May be dissolved in an organic solvent or petroleum distillate for application. Pungent to mild, fruity odor. Odor threshold concentration in water is 4.7 mg/kg (Keith and Walters, 1992).

Melting point (°C):
98 (Verschueren, 1983)
92 (Kapoor et al., 1970)
89 (Worthing and Hance, 1991)
77 (technical grade, Sunshine, 1969)

Boiling point (°C):
Decomposes (Weast, 1986)

Density (g/cm³):
1.41 at 25/4 °C (Verschueren, 1983)

Diffusivity in water (x 10^{-5} cm²/sec):
0.44 at 20 °C using method of Hayduk and Laudie (1974)

Flash point (°C):
Burns only at high temperatures (Weiss, 1986)

Lower explosive limit (%):
Not pertinent (Weiss, 1986)

Upper explosive limit (%):
Not pertinent (Weiss, 1986)

Henry's law constant (x 10^{-5} atm·m^3/mol):
1.58 at 25 °C (estimated, Howard, 1991)

Bioconcentration factor, log BCF:
3.92 (freshwater fish), 3.19 (fish, microcosm) (Garten and Trabalka, 1983)
4.08 mussel (*Mytilus edulis*) (Renberg et al., 1985)

Soil sorption coefficient, log K_{oc}:
4.90 (Kenaga, 1980a)
4.95 (clay, Karickhoff et al., 1979)

Octanol/water partition coefficient, log K_{ow}:
4.30 (Mackay, 1982)
4.68 (Kenaga, 1980)
4.40 (Garten and Trabalka, 1983)
3.40 (Wolfe et al., 1977)

Solubility in organics:
Soluble in ethanol (Windholz et al., 1983), chloroform (440 g/kg), xylene (440 g/kg), and methanol (50 g/kg) (Worthing and Hance, 1991)

Solubility in water:
40 μg/L at 24 °C (Hollifield, 1979)
100 μg/L at 25 °C (quoted, IARC, 1979)
620 ppb (Kapoor et al., 1970)
In ppb: 20 at 15 °C, 45 at 25 °C, 95 at 35 °C, 185 at 45 °C (particle size ≤5 μ) (Biggar and
   Riggs, 1974)
120 ppb at 25 °C (Zepp et al., 1976)

Environmental fate:
   Biological. Degradation by *Aerobacter aerogenes* under aerobic or anaerobic conditions
yielded 1,1-dichloro-2,2-bis(*p*-methoxyphenyl)ethylene and 1,1-dichloro-2,2-bis(*p*-methoxy-
phenyl)ethane (Mendel and Walton, 1966; Kobayashi and Rittman, 1982). Methoxychlor
degrades at a faster rate in flooded/anaerobic soils than in nonflodded/aerobic soils (Fogel et al.,
1982; Golovleva et al., 1984). In anaerobic soil, 90% of the applied dosage was lost after 3
months. In aerobic soil, only 0.3% was lost as carbon dioxide after 410 d (Fogel et al., 1982).
   In a model aquatic ecosystem, methoxychlor degraded to ethanol, dihydroxyethane, dihy-
droxyethylene, and unidentified polar metabolites (Metcalf et al., 1971). Kapoor et al. (1970)
also studied the biodegradation of methoxychlor in a model ecosystem containing snails,
plankton, mosquito larvae, *Daphnia magna*, and mosquito fish (*Gambusia affinis*). The following metabolites were identified: 2-(*p*-methoxyphenyl)-2-(*p*-hydroxyphenyl)-1,1,1-trichloro-
ethane, 2,2-bis(*p*-hydroxyphenyl)-1,1,1-trichloroethane, 2,2-bis(*p*-hydroxyphenyl)-1,1,1-tri-
chloroethylene, and polar metabolites (Kapoor et al., 1970).
   From the first-order biotic and abiotic rate constants of methoxychlor in estuarine water and
sediment/water systems, the estimated biodegradation half-lives were 208-8,837 and 12-45 d,

respectively (Walker et al., 1988).

Groundwater. Methoxychlor has a high potential to leach to groundwater (U.S. EPA, 1986).

Plant. Brett and Bowery (1958) and Johansen (1954) reported foliar half-lives of 1.8 and 6.3 d in collards and cherries, respectively.

Photolytic. In air-saturated distilled water, direct photolysis of methoxychlor by >280 nm light produced 1,1-bis(*p*-methoxyphenyl)-2,2-dichloroethylene (DMDE), which photolyzed to *p*-methoxybenzaldehyde. The photolysis half-life was estimated to be 4.5 months (Zepp et al., 1976).

Methoxychlor-DDE and *p,p*-dimethoxybenzophenone were formed when methoxychlor in water was irradiated by UV light (Paris and Lewis, 1973). Compounds reported from the photolysis of methoxychlor in aqueous, alcoholic solutions were *p,p*-dimethoxybenzophenone, *p*-methoxybenzoic acid, and 4-methoxyphenol (Wolfe et al., 1976). However, when methoxychlor in milk was irradiated by UV light (λ = 220 and 330 nm), 4-methoxyphenol, methoxychlor-DDE, *p,p*-dimethoxybenzophenone, and 1,1,4,4-tetrakis(*p*-methoxyphenyl)-1,2,3-butatriene were formed (Li and Bradley, 1969).

Chemical/Physical. Hydrolysis at common aquatic pHs produced anisoin, anisil, hydrochloric acid, and 2,2-bis(*p*-methoxyphenyl)-1,1-dichloroethylene (estimated half-life = 270 d at 25 °C and pH 7.1) (Wolfe et al., 1977). Above pH 10, 2,2-bis(*p*-methoxyphenyl)-1,1-dichloroethylene is the only reported product. Below pH 10, anisoin is formed (Kollig, 1993). The estimated hydrolysis half-life in water at 25 °C and pH 7.1 is 270 d (Mabey and Mill, 1978).

Emits toxic chloride fumes when heated to decomposition (Lewis, 1990).

Exposure limits (mg/m³): Potential occupational carcinogen. NIOSH REL: IDLH 5,000; OSHA PEL: TWA 15; ACGIH TLV: TWA 10 (adopted).

Symptoms of exposure: Slightly irritating to skin (NIOSH, 1997). An acceptable daily intake reported for humans is 0.1 mg/kg body weight (Worthing and Hance, 1991).

Toxicity:

EC_{50} (48-h) for *Daphnia pulex* 0.78 μg/L, *Simocephalus serrulatus* 5.3 μg/L, *Cypridopsis vidua* 32 μg/L (Mayer and Ellersieck, 1986).

LC_{50} (96-h) for fathead minnows 7.5 μg/L, bluegill sunfish 62.0 μg/L, rainbow trout 62.6 μg/L, coho salmon 66.2 μg/L, chinook 27.9 μg/L, perch 20.0 μg/L (Verschueren, 1983), adult decapod (*Cancer magister*) 1.30 μg/L, fish (*Cyprinodon variegatus*) 49 μg/L (quoted, Reish and Kauwling, 1978), *Gammarus italicus* 5.8 μg/L, *Echinogammarus tibaldii* 1.6 μg/L (Pantani et al., 1997).

LC50 (48-h) for *Daphnia magna* 0.78 μg/L (Worthing and Hance, 1991).

LC_{50} (24-h) for bluegill sunfish 67 μg/L, rainbow trout 52 μg/L (Worthing and Hance, 1991), isopod (*Asellus communis*) 0.42 μg/L (Anderson and DeFoe, 1980), northern pike 13.5 μg/L (Mayer and Ellersieck, 1986).

Acute oral LD_{50} for mice 1,850 mg/kg, rats 5,000 mg/kg (quoted, RTECS, 1985), 10.0 μg/roach and 15.2 μg/roach for male and female, respectively (Gardner and Vincent, 1978); mallard ducks >2,000 mg (technical grade)/kg (Worthing and Hance, 1991).

NOELs of 200 and 300 mg/kg diet were reported for rats (2-yr feeding trial) and dogs (12-month feeding trial), respectively (Worthing and Hance, 1991).

Drinking water standard (final): MCLG: 40 μg/L; MCL: 40 μg/L (U.S. EPA, 1996).

Uses: Contact and stomach insecticide used to control mosquito larvae, house flies, ectoparasites on cattle, sheep, and goats; recommended for use in dairy barns. Effective against a wide variety of pests in field, forage, fruit, and vegetable crops. Methoxychlor is also used to control household and industrial pests (Worthing and Hance, 1991).

METHYL ACETATE

Synonyms: Devoton; **Acetic acid methyl ester**; Methyl ethanoate; Tereton; UN 1231.

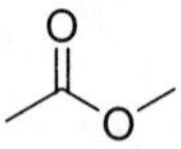

CAS Registry Number: 79-20-9
DOT: 1231
DOT label: Flammable liquid
Molecular formula: $C_3H_6O_2$
Formula weight: 74.08
RTECS: AI9100000
Merck reference: 10, 5886

Physical state, color, and odor:
Colorless liquid with a pleasant odor. Odor threshold concentration is 4.6 ppm (Amoore and Hautala, 1983).

Melting point (°C):
-98.1 (Weast, 1986)
-99 (Verschueren, 1983)

Boiling point (°C):
57 (Weast, 1986)
54.05 (Hawley, 1981)

Density (g/cm³):
0.93364 at 20.00/4 °C (Lee and Tu, 1999)
0.9342 at 20/4 °C, 0.9279 at 25/4 °C (Windholz et al., 1983)

Diffusivity in water (x 10^{-5} cm²/sec):
1.01 at 20 °C using method of Hayduk and Laudie (1974)

Flash point (°C):
-5.6 (NIOSH, 1997)

Lower explosive limit (%):
3.1 (NIOSH, 1997)

Upper explosive limit (%):
16 (NIOSH, 1997)

Henry's law constant (x 10^{-4} atm·m³/mol at 25 °C):
1.28 (Kieckbusch and King, 1979)
0.91 (Butler and Ramchandani, 1935)
1.15 (Buttery et al., 1969)

Interfacial tension with water (dyn/cm at 25 °C):
1.6 (quoted, Freitas et al., 1997)

Ionization potential (eV):
10.27 ± 0.02 (Franklin et al., 1969)

Soil sorption coefficient, log K_{oc}:
Unavailable because experimental methods for estimation of this parameter for aliphatic esters are lacking in the documented literature

Octanol/water partition coefficient, log K_{ow}:
0.17 (Collander, 1951)

Solubility in organics:
Soluble in acetone, benzene, and chloroform (Weast, 1986). Miscible with alcohol and ether (Sax and Lewis, 1987).

Solubility in water:
245,000 mg/L at 20 °C (Riddick et al., 1986)
319,000 mg/L, 240,000 mg/L at 20 °C (quoted, Verschueren, 1983)
3.29 M at 20 °C (Fühner, 1924)

Vapor density:
3.03 g/L at 25 °C, 2.56 (air = 1)

Vapor pressure (mmHg):
173 at 20 °C (NIOSH, 1997)
170 at 20 °C, 235 at 25 °C, 255 at 30 °C (Verschueren, 1983)
216 at 25 °C (Abraham, 1984)

Environmental fate:
   Photolytic. A rate constant of 2.00 x 10^{-13} cm³/molecule·sec was reported for the reaction of methyl acetate and OH radicals in aqueous solution (Wallington et al., 1988b).
   Chemical/Physical. Slowly hydrolyzes in water forming methyl alcohol and acetic acid (NIOSH, 1997).
   At an influent concentration of 1,030 mg/L, treatment with granular activated carbon resulted in an effluent concentration of 760 mg/L. The adsorbability of the carbon used was 54 mg/g carbon (Guisti et al., 1974).

Exposure limits: NIOSH REL: TWA 200 ppm (610 mg/m³), STEL 250 ppm (760 mg/m³), IDLH 3,100 ppm; OSHA PEL: TWA 200 ppm; ACGIH TLV: TWA 200 ppm, STEL, 250 ppm (adopted).

Symptoms of exposure: Inflammation of the eyes, visual and nervous disturbances, tightness of the chest, drowsiness, and narcosis (Patnaik, 1992). An irritation concentration of 30.5 g/m³ in air was reported by Ruth (1986).

Toxicity:
   Acute oral LD_{50} for rabbits 3,705 mg/kg, rats 5,450 mg/kg (quoted, RTECS, 1985).

Uses: Solvent for resins, lacquers, oils, acetylcellulose, nitrocellulose; paint removers; synthetic flavoring.

METHYL ACRYLATE

Synonyms: Acrylic acid methyl ester; Curithane 103; Methoxycarbonylethylene; Methyl propenate; Methyl propenoate; Methyl-2-propenoate; Propenoic acid methyl ester; **2-Propenoic acid methyl ester**; UN 1919.

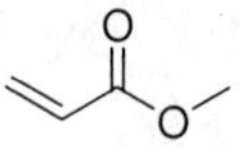

CAS Registry Number: 96-33-3
DOT: 1919
DOT label: Flammable liquid
Molecular formula: $C_4H_6O_2$
Formula weight: 86.09
RTECS: AT2800000
Merck reference: 10, 5889

Physical state, color, and odor:
Clear, colorless liquid with a heavy, sweet odor. Odor threshold concentration is 17 $\mu g/m^3$ (Keith and Walters, 1992).

Melting point (°C):
-77 (NIOSH, 1997)

Boiling point (°C):
80.5 (Weast, 1986)

Density (g/cm³):
0.9561 at 20/4 °C (Windholz et al., 1983)

Diffusivity in water (x 10^{-5} cm²/sec):
0.94 at 20 °C using method of Hayduk and Laudie (1974)

Flash point (°C):
-2.8 (NIOSH, 1997)
-3.8 (open cup, Hawley, 1981)

Lower explosive limit (%):
2.8 (NIOSH, 1997)

Upper explosive limit (%):
25.0 (NIOSH, 1997)

Henry's law constant (x 10^{-4} atm·m³/mol):
1.3 at 20 °C (approximate - calculated from water solubility and vapor pressure)

Ionization potential (eV):
9.90 (NIOSH, 1997)
9.19 ± 0.05 (Franklin et al., 1969)

Soil sorption coefficient, log K_{oc}:
Unavailable because experimental methods for estimation of this parameter for aliphatic esters are lacking in the documented literature

Octanol/water partition coefficient, log K_{ow}:
0.80 (Tanii and Hashimoto, 1982)

Solubility in organics:
Soluble in acetone, alcohol, benzene, and ether (Weast, 1986)

Solubility in water (g/L):
60 at 20 °C, 50 at 40 °C (quoted, Windholz et al., 1983)
52.0 g/L (quoted, Verschueren, 1983)

Vapor density:
3.52 g/L at 25 °C, 2.97 (air = 1)

Vapor pressure (mmHg):
70 at 20 °C, 110 at 30 °C (Verschueren, 1983)

Environmental fate:
   Photolytic. Polymerizes on standing and is accelerated by heat, light, and peroxides (Windholz et al., 1983). Methyl acrylate reacts with OH radicals in the atmosphere (296 K) and aqueous solution at rates of 3.04 x 10^{-12} and 2.80 x 10^{-12} cm^3/molecule·sec, respectively (Wallington et al., 1988b).
   Chemical/Physical. Begins to polymerize at 80.2 °C (Weast, 1986). Slowly hydrolyzes in water forming methanol and acrylic acid (Morrison and Boyd, 1971). Based on a hydrolysis rate constant of 0.0779/M·h at pH 9 at 25 °C, an estimated half-life of 2.8 yr at pH 7 was reported (Roy, 1972).

Exposure limits: NIOSH REL: TWA 10 ppm (35 mg/m^3), IDLH 250 ppm; OSHA PEL: TWA 10 ppm; ACGIH TLV: TWA 2 ppm (adopted).

Symptoms of exposure: Lacrimation, irritation of respiratory tract, lethargy, and convulsions (Patnaik, 1992). An irritation concentration of 262.5 mg/m^3 in air was reported by Ruth (1986).

Toxicity:
   Acute oral LD$_{50}$ for rats 277 mg/kg, mice 827 mg/kg (quoted, RTECS, 1985).

Uses: Manufacturing plastic films, leather finish resins, textile and paper coatings; amphoteric surfactants; chemical intermediate.

METHYLAL

Synonyms: Anesthenyl; **Dimethoxymethane**; Formal; Formaldehyde dimethylacetal; Methyl formal; Methylene dimethyl ether; UN 1234.

CAS Registry Number: 109-87-5
DOT: 1234
Molecular formula: $C_3H_8O_2$
Formula weight: 76.10
RTECS: PA8750000
Merck reference: 10, 5890

Physical state, color, and odor:
Colorless liquid with a pungent, chloroform-like odor

Melting point (°C):
-104.8 (Weast, 1986)

Boiling point (°C):
45.5 (Weast, 1986)
42.3 (Sax and Lewis, 1987)

Density (g/cm³):
0.8593 at 20/4 °C (Weast, 1986)
0.86645 at 15/4 °C (Huntress and Mulliken, 1941)

Diffusivity in water (x 10^{-5} cm²/sec):
0.95 at 20 °C using method of Hayduk and Laudie (1974)

Flash point (°C):
-17 (Dean, 1987)
-18 (Windholz et al., 1983)
-32 (open cup, NFPA, 1984)

Lower explosive limit (%):
2.2 (NFPA, 1984)

Upper explosive limit (%):
13.8 (NFPA, 1984)

Henry's law constant (x 10^{-4} atm·m³/mol):
1.73 at 25 °C (Hine and Mookerjee, 1975)

Ionization potential (eV):
10.00 (NIOSH, 1997)

Soil sorption coefficient, log K_{oc}:
Unavailable because experimental methods for estimation of this parameter for aliphatic ethers

are lacking in the documented literature. However, its high solubility in water and low K_{ow} suggest its adsorption to soil will be nominal (Lyman et al., 1982).

Octanol/water partition coefficient, log K_{ow}:
-0.01 (Collander, 1951)

Solubility in organics:
Soluble in acetone and benzene (Weast, 1986). Miscible with alcohol, ether, and oils (Windholz et al., 1983).

Solubility in water:
330,000 mg/L (quoted, Verschueren, 1983)

Vapor density:
3.11 g/L at 25 °C, 2.63 (air = 1)

Vapor pressure (mmHg):
330 at 20 °C, 400 at 25 °C (Verschueren, 1983)

Exposure limits: NIOSH REL: TWA 1,000 ppm (3,100 mg/m³), IDLH 2,200 ppm; OSHA PEL: TWA 1,000 ppm; ACGIH TLV: TWA 1,000 ppm (adopted).

Toxicity:
   LC_{50} (inhalation) for rats 15,000 ppm (quoted, RTECS, 1985).
   Acute oral LD_{50} for rats 5,708 mg/kg (quoted, RTECS, 1985).

Uses: Artificial resins; perfumery; solvent; adhesives; protective coatings; special fuel; organic synthesis (Grignard and Reppe reactions).

METHYLAMINE

Synonyms: Aminomethane; Carbinamine; Mercurialin; **Methanamine**; Monomethylamine; UN 1061; UN 1235.

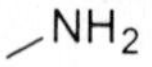

$$\diagup NH_2$$

CAS Registry Number: 74-89-5
DOT: 1061
DOT label: Flammable gas (anhydrous)/flammable liquid
Molecular formula: CH_5N
Formula weight: 31.06
RTECS: PF6300000
Merck reference: 10, 5891

Physical state, color, and odor:
Colorless gas with a strong ammonia-like odor. Odor threshold concentration is 3.2 ppm (Amoore and Hautala, 1983).

Melting point (°C):
-93.5 (Weast, 1986)

Boiling point (°C):
-6.3 (Weast, 1986)

Density (g/cm³):
0.6628 at 20/4 °C (Weast, 1986)
0.769 at -70/4 °C (Verschueren, 1983)

Diffusivity in water (x 10^{-5} cm²/sec):
1.38 at 20 °C using method of Hayduk and Laudie (1974)

Dissociation constant, pK_a:
10.657 at 25 °C (Gordon and Ford, 1972)

Flash point (°C):
-62, -4, -11, and -23 at 0, 25, 40, and 50 wt % aqueous solutions, respectively (Mitchell et al., 1999)

Lower explosive limit (%):
4.9 (NFPA, 1984)

Upper explosive limit (%):
20.7 (NFPA, 1984)

Heat of fusion (kcal/mol):
1.466 (Dean, 1987)

Henry's law constant (x 10^{-5} atm·m³/mol):
1.11 at 25 °C (Christie and Crisp, 1967)

Ionization potential (eV):
8.97 (Franklin et al., 1969)
9.18 (Gibson et al., 1977)

Soil sorption coefficient, log K_{oc}:
Unavailable because experimental methods for estimation of this parameter for aliphatic amines are lacking in the documented literature. However, its high solubility in water and low K_{ow} suggest its adsorption to soil will be nominal (Lyman et al., 1982).

Octanol/water partition coefficient, log K_{ow}:
-0.57 (Collander, 1951)

Solubility in organics:
Soluble in benzene (105 g/L at 25 °C) and miscible with ether (Windholz et al., 1983)

Solubility in water:
1,154 volumes at 12.5 °C, 959 volumes at 25 °C (quoted, Windholz et al., 1983)

Vapor density:
1.27 g/L at 25 °C, 1.07 (air = 1)

Vapor pressure (mmHg):
2,356 at 20 °C, 3,268 at 30 °C (Verschueren, 1983)

Environmental fate:
Photolytic. The rate constant for the reaction of methylamine and OH radicals in the atmosphere at 300 K is 1.3 x 10^{13} cm^3/mol·sec (Hendry and Kenley, 1979).
Chemical/Physical. In an aqueous solution, chloramine reacted with methylamine to form *N*-chloromethylamine (Isaac and Morris, 1983).
Reacts with acids forming water-soluble salts.

Exposure limits: NIOSH REL: TWA 10 ppm (12 mg/m^3), IDLH 100 ppm; OSHA PEL: TWA 10 ppm; ACGIH TLV: TWA 5 ppm, STEL 15 ppm (adopted).

Symptoms of exposure: Severe irritant to eyes, skin, and respiratory tract (Patnaik, 1992). An irritation concentration of 30.00 mg/m^3 in air was reported by Ruth (1986).

Toxicity:
LC$_{50}$ (inhalation) for mice 2,400 mg/kg/2-h (quoted, RTECS, 1985).

Source: Methylamine was detected in cauliflower (65 ppm), carrots (3,970 ppm), tea leaves (50 ppm), red and white cabbage (3.4-22.7 ppm), corn (27 ppm), kale leaves (16.6 ppm), barley seeds (4.5 ppm), epidermis of apples (4.5 ppm), celery (6.4 ppm), sweetflag, celandine, and tobacco leaves (Duke, 1992).

Uses: Tanning; intermediate for accelerators, dyes, pharmaceuticals, insecticides, fungicides, tanning, surface active agents, fuel additive, dyeing of acetate textiles; polymerization inhibitor; ingredient in paint removers; photographic developer; solvent; rocket propellant; fuel additive; solvent; in organic synthesis.

METHYLANILINE

Synonyms: Anilinomethane; MA; (Methylamino)benzene; *N*-Methylaminobenzene; *N*-Methyl-aniline; **N-Methylbenzenamine**; Methylphenylamine; *N*-Methylphenylamine; Monomethyl-aniline; *N*-Monomethylaniline; *N*-Phenylmethylamine; UN 2294.

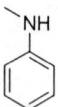

CAS Registry Number: 100-61-8
DOT: 2294
Molecular formula: C_7H_9N
Formula weight: 107.16
RTECS: BY4550000
Merck reference: 10, 5894

Physical state, color, and odor:
Colorless to yellow to pale brown liquid with a faint, ammonia-like odor. Odor threshold concentration is 1.7 ppm (Amoore and Hautala, 1983).

Melting point (°C):
-57 (Weast, 1986)

Boiling point (°C):
196.25 (Weast, 1986)
190-191 (Hawley, 1981)

Density (g/cm³):
0.9891 at 20/4 °C (Weast, 1986)

Diffusivity in water (x 10^{-5} cm²/sec):
0.84 at 20 °C using method of Hayduk and Laudie (1974)

Dissociation constant, pK_a:
4.848 at 25 °C (Gordon and Ford, 1972)

Flash point (°C):
80.1 (NIOSH, 1997)

Henry's law constant (x 10^{-5} atm·m³/mol):
1.19 at 25 °C (approximate - calculated from water solubility and vapor pressure)

Ionization potential (eV):
7.32 (Franklin et al., 1969)

Soil sorption coefficient, log K_{oc}:
2.28 (Meylan et al., 1992)

Octanol/water partition coefficient, log K_{ow}:
1.66, 1.82 (Leo et al., 1971)
1.40 (Johnson and Westall, 1990)

Solubility in organics:
Soluble in alcohol (Weast, 1986) and ether (Windholz et al., 1983)

Solubility in water:
5.624 g/L at 25 °C (Chiou et al., 1982)

Vapor density:
4.38 g/L at 25 °C, 3.70 (air = 1)

Vapor pressure (mmHg):
0.3 at 20 °C, 0.65 at 30 °C (Verschueren, 1983)

Environmental fate:
Soil. Reacts slowly with humic acids or humates forming quinoidal structures (Parris, 1980).

Exposure limits: NIOSH REL: TWA 0.5 ppm (2 mg/m^3), IDLH 100 ppm; OSHA PEL: TWA 2 ppm (9 mg/m^3); ACGIH TLV: TWA 0.5 ppm (adopted).

Toxicity:
LC$_{50}$ (48-h) for red killifish 355 mg/L (Yoshioka et al., 1986)

Uses: Solvent; acid acceptor; organic synthesis.

2-METHYLANTHRACENE

Synonym: β-Methylanthracene.

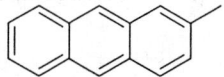

CAS Registry Number: 613-12-7
Molecular formula: $C_{15}H_{12}$
Formula weight: 192.96
RTECS: CB0680000

Physical state:
Solid

Melting point (°C):
209 (Weast, 1986)

Boiling point (°C):
Sublimes (Weast, 1986)
358.5 (Wilhoit and Zwolinski, 1971)

Density (g/cm³):
1.165 using method of Lyman et al. (1982)

Diffusivity in water (x 10⁻⁵ cm²/sec):
0.56 at 20 °C using method of Hayduk and Laudie (1974)

Dissociation constant, pK_a:
>14 (Schwarzenbach et al., 1993)

Ionization potential (eV):
7.37 (Rosenstock et al., 1998)

Soil sorption coefficient, log K_{oc}:
5.12 using method of Karickhoff et al. (1979)

Octanol/water partition coefficient, log K_{ow}:
5.14 (Yalkowsky et al., 1983a)

Solubility in organics:
Soluble in benzene and chloroform (Weast, 1986)

Solubility in water:
In μg/kg: 7.06 at 6.3 °C, 8.48 at 9.1 °C, 9.43 at 10.8 °C, 11.1 at 13.9 °C, 14.5 at 18.3 °C, 19.1 at 23.1 °C, 24.2 at 27.0 °C, 32.1 at 31.1 °C (May et al., 1978)
21.3 μg/kg at 25 °C (May et al., 1978a)
39 μg/L at 25 °C (Mackay and Shiu, 1977)
In nmol/L: 39.2, 50.4, 63.9, 84.0, and 117 at 8.8, 12.9, 17.0, 21.1, and 25.3 °C, respectively. In seawater (salinity = 36.5 g/kg): 15.6, 19.5, 26.2, 35.7, 49.4, and 70.8 at 4.6, 8.8, 12.9, 17.0, 21.1, and 25.3 °C, respectively (Whitehouse, 1984)

In mole fraction (x 10^{-9}): 0.6615 at 6.30 °C, 0.7946 at 9.38 °C, 0.8836 at 10.80 °C, 1.040 at 13.90 °C, 1.359 at 18.30 °C, 1.790 at 23.10 °C, 2.268 at 27.00 °C, 3.008 at 31.10 °C (May et al., 1983)

Environmental fate:
  Chemical/Physical. 2-Methylanthracene will not hydrolyze because it has no hydrolyzable functional group.

Source: Detected in 8 diesel fuels at concentrations ranging from 0.18 to 160 mg/L with a mean value of 0.46.16 mg/L (Westerholm and Li, 1994). Schauer et al. (1999) reported 2-methylanthracene in diesel fuel at a concentration of 6 μg/g and in a diesel-powered medium-duty truck exhaust at an emission rate of 10.4 μg/km.

Uses: Organic synthesis.

METHYL BROMIDE

Synonyms: Brom-o-gas; Brom-o-gaz; **Bromomethane**; Celfume; Dawson 100; Dowfume; Dowfume MC-2; Dowfume MC-2 soil fumigant; Dowfume MC-33; Edco; Embafume; Fumigant-1; Halon 1001; Iscobrome; Kayafume; MB; M-B-C Fumigant; MBX; MEBR; Metafume; Methogas; Monobromomethane; Pestmaster; Profume; R 40B1; RCRA waste number U029; Rotox; Terabol; Terr-o-gas 100; UN 1062; Zytox.

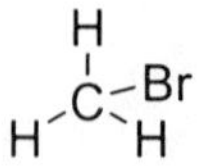

$$\begin{array}{c} H \\ | \\ H\!-\!\overset{\displaystyle}{C}\!-\!Br \\ | \\ H \end{array}$$

CAS Registry Number: 74-83-9
DOT: 1062
DOT label: Poison
Molecular formula: CH_3Br
Formula weight: 94.94
RTECS: PA4900000
Merck reference: 10, 5905

Physical state, color, and odor:
Colorless liquid or gas with an odor similar to chloroform at high concentrations

Melting point (°C):
-93.6 (Weast, 1986)

Boiling point (°C):
3.55 (Kudchadker et al., 1979)
4.5 (Worthing and Hance, 1991)

Density (g/cm³):
1.6755 at 20/4 °C (Weast, 1986)
1.732 at 0/0 °C (Sax, 1984)

Diffusivity in water (x 10^{-5} cm²/sec):
1.23 at 20 °C using method of Hayduk and Laudie (1974)

Flash point (°C):
Practically nonflammable (NFPA, 1984)

Lower explosive limit (%):
10 (NFPA, 1984)

Upper explosive limit (%):
16 (NFPA, 1984)

Heat of fusion (kcal/mol):
1.429 (Dean, 1987)

Henry's law constant (x 10^{-2} atm·m³/mol):
62.3 (Glew and Moelwyn-Hughes, 1953)

652

3.18 (low ionic strength, Jury et al., 1984)
0.72 at 21 °C (Gan and Yates, 1996)

Ionization potential (eV):
10.54 (NIOSH, 1997)
10.53 (Franklin et al., 1969)
10.69 (Gibson, 1982)

Soil sorption coefficient, log K_{oc}:
2.236, 2.241, and 2.215 for Naaldwijk loamy sand, Aalsmeer loam, and Boskoop peaty clay, respectively (Howard, 1989)
1.12 (Greenfield sandy loam), 0.79 (Carsetas loamy sand), 0.76 (Linne clay loam) (Gan and Yates, 1996)

Octanol/water partition coefficient, log K_{ow}:
1.00 (quoted, Mills et al., 1985)
1.19 (Hansch and Leo, 1979; Leo et al., 1975)

Solubility in organics:
Soluble in ethanol, ether (Weast, 1986), chloroform, carbon disulfide, carbon tetrachloride, and benzene (ITII, 1986)

Solubility in water (mg/L):
20,700 at 20.01 °C, 24,137 at 24.98 °C (Glew and Moelwyn-Hughes, 1953)
17,500 at 20 °C under 748 mmHg atmosphere consisting of methyl bromide and water vapor (Standen, 1964)
26.79, 18.30, 13.41, and 11.49 g/kg at 10, 17, 25, and 32 °C, respectively (Haight, 1951)

Vapor density:
3.88 g/L at 25 °C, 3.28 (air = 1)

Vapor pressure (mmHg):
1,633 at 25 °C (Howard, 1989)
1,420 at 20 °C (U.S. EPA, 1976)

Environmental fate:
Photolytic. When methyl bromide and bromine gas (concentration = 3%) were irradiated at 1850 Å, methane was produced (Kobrinsky and Martin, 1968).

Chemical/Physical. Hydrolyzes in water forming methanol and hydrobromic acid. The estimated hydrolysis half-life in water at 25 °C and pH 7 is 20 d (Mabey and Mill, 1978). Castro and Belser (1981) reported a hydrolysis rate constant of 3×10^{-7}/sec or a half-life of 26.7 d. Forms a voluminous crystalline hydrate at 0-5 °C (Keith and Walters, 1992).

When methyl bromide was heated to 550 °C in the absence of oxygen, methane, hydrobromic acid, hydrogen, bromine, ethyl bromide, anthracene, pyrene, and free radicals were produced (Chaigneau et al., 1966).

Exposure limits: Potential occupational carcinogen. NIOSH REL: IDLH 250 ppm; OSHA PEL: ceiling 20 ppm (80 mg/m³); ACGIH TLV: TWA 1 ppm (adopted).

Symptoms of exposure: Inhalation may cause headache, visual disturbance, vertigo, nausea, vomiting, malaise, hand tremor, convulsions, eye and skin irritation (NIOSH, 1997).

Toxicity:
   LC_{50} (8-h inhalation) for rats 302 ppm (Honma et al., 1985).
   LC_{50} (2-h inhalation) for mice 1,540 mg/m^3 (quoted, RTECS, 1985).
   LC_{50} (60-min inhalation) for mice 4.69 mg/L or approximately 1,200 ppm (Alexeeff et al., 1985.
   Acute oral LD_{50} for rats is 100 mg/kg (Ashton and Monaco, 1991).

Drinking water standard: No MCLGs or MCLs have been proposed (U.S. EPA, 1996).

Uses: Soil, space, and food fumigant; organic synthesis; fire extinguishing agent; refrigerant; disinfestation of potatoes, tomatoes, and other crops; solvent for extracting vegetable oils.

2-METHYL-1,3-BUTADIENE

Synonyms: Hemiterpene; Isoprene; β-Methylbivinyl; 2-Methylbutadiene; UN 1218.

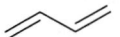

CAS Registry Number: 78-79-5
DOT: 1218 (inhibited)
Molecular formula: C_5H_8
Formula weight: 68.12
RTECS: NT4037000
Merck reference: 10, 5048

Physical state and color:
Colorless, volatile liquid

Melting point (°C):
-146 (Weast, 1986)

Boiling point (°C):
34 (Weast, 1986)

Density (g/cm³):
0.6810 at 20/4 °C (Weast, 1986)

Diffusivity in water (x 10^{-5} cm²/sec):
0.88 at 20 °C using method of Hayduk and Laudie (1974)

Dissociation constant, pK_a:
>14 (Schwarzenbach et al., 1993)

Flash point (°C):
-53.9 (Sax and Lewis, 1987)

Lower explosive limit (%):
1.5 (NFPA, 1984)

Upper explosive limit (%):
8.9 (NFPA, 1984)

Heat of fusion (kcal/mol):
1.115 (Dean, 1987)

Henry's law constant (x 10^{-2} atm·m³/mol):
7.7 at 25 °C (Hine and Mookerjee, 1975)

Ionization potential (eV):
8.845 ± 0.005 (Franklin et al., 1969)

Soil sorption coefficient, log K_{oc}:
Unavailable because experimental methods for estimation of this parameter for aliphatic hydrocarbons are lacking in the documented literature

Octanol/water partition coefficient, log K_{ow}:
1.76 using method of Hansch et al. (1968)

Solubility in organics:
Miscible with alcohol and ether (Windholz et al., 1983)

Solubility in water:
642 mg/kg at 25 °C (McAuliffe, 1966)

Vapor density:
2.78 g/L at 25 °C, 2.35 (air = 1)

Vapor pressure (mmHg):
493 at 20 °C, 700 at 30 °C (Verschueren, 1983)
550.1 at 25 °C (Wilhoit and Zwolinski, 1971)

Environmental fate:
Photolytic. Methyl vinyl ketone and methacrolein were reported as major photooxidation products for the reaction of 2-methyl-1,3-butadiene with OH radicals. Formaldehyde, nitrogen dioxide, nitric oxide, and HO_2 were reported as minor products (Lloyd et al., 1983). Synthetic air containing gaseous nitrous acid and exposed to artificial sunlight (λ = 300-450 nm) photo-oxidized 2-methyl-1,3-butadiene into formaldehyde, methyl nitrate, peroxyacetal nitrate, and a compound tentatively identified as methyl vinyl ketone (Cox et al., 1980).

The following rate constants were reported for the reaction of 2-methyl-1,3-butadiene with OH radicals in the atmosphere: 9.26-9.98 x 10^{-11} cm³/molecule·sec (Atkinson et al., 1985), 5.91 x 10^{-11} cm³/molecule·sec at 298 K (Sabljić and Güsten, 1990), 1.10 x 10^{-10} cm³/molecule·sec (Atkinson et al., 1990), 4.7 x 10^{13} cm³/mol·sec at 300 K (Hendry and Kenley, 1979) and 5.94 x 10^{-13} cm³/molecule·sec at 298 K (Atkinson, 1991); with ozone in the gas-phase: 1.65 x 10^{-17} cm³/molecule·sec at 294 K (Adeniji et al., 1981), 5.80 x 10^{-18} to 1.25 x 10^{-17} cm³/molecule·sec (Atkinson and Carter, 1984), 1.43 x 10^{-17} cm³/molecule·sec at 298 K (Atkinson, 1990); with NO_3 in the atmosphere: 1.3 x 10^{-12} cm³/molecule·sec (Benter and Schindler, 1988) and 6.52 x 10^{-13} cm³/molecule·sec at 297 K (Atkinson, 1991). The estimated atmospheric lifetimes for the reaction of 2-methyl-1,3-butadiene with ozone, OH, and NO_3 radicals are 28.3, 2.9, and 0.083 h, respectively (Atkinson and Carter, 1984).

Chemical/Physical. Slowly oxidizes and polymerizes in air (Huntress and Mulliken, 1941).

Toxicity:
LC_{50} (inhalation) for mice 139 g/m³/2-h, rats 180 g/m³/4-h (quoted, RTECS, 1985).

Uses: Manufacture of butyl and synthetic rubber; gasoline component; organic synthesis.

2-METHYLBUTANE

Synonyms: Ethyldimethylmethane; Isoamylhydride; Isopentane; UN 1265.

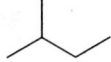

CAS Registry Number: 78-78-4
DOT: 1265
Molecular formula: C_5H_{12}
Formula weight: 72.15
RTECS: EK4430000

Physical state, color, and odor:
Colorless liquid with a pleasant odor

Melting point (°C):
-159.9 (Weast, 1986)

Boiling point (°C):
27.8 (Weast, 1986)
28.88 (Wilhoit and Zwolinski, 1971)

Density (g/cm³):
0.61067 at 20/4 °C, 0.61462 at 25/4 °C (Dreisbach, 1959)
0.6201 at 20/4 °C (Weast, 1986)

Diffusivity in water (x 10^{-5} cm²/sec):
0.81 at 20 °C using method of Hayduk and Laudie (1974)

Dissociation constant, pK$_a$:
>14 (Schwarzenbach et al., 1993)

Flash point (°C):
-57 (Hawley, 1981)

Lower explosive limit (%):
1.4 (Sax and Lewis, 1987)

Upper explosive limit (%):
7.6 (Sax and Lewis, 1987)

Heat of fusion (kcal/mol):
1.231 (Riddick et al., 1986)

Henry's law constant (atm·m³/mol):
1.38 at 25 °C (calculated, Mackay et al., 1979)

Interfacial tension with water (dyn/cm at 20 °C):
49.64 (Harkins et al., 1920)

Ionization potential (eV):
10.32 (Franklin et al., 1969)
10.60 (Collin and Lossing, 1959)

Soil sorption coefficient, log K_{oc}:
Unavailable because experimental methods for estimation of this parameter for aliphatic hydrocarbons are lacking in the documented literature

Octanol/water partition coefficient, log K_{ow}:
2.23 (Coates et al., 1985)

Solubility in organics:
Soluble in alcohol, ether (Weast, 1986), hydrocarbons, and oils (Hawley, 1981)

Solubility in water (mg/kg):
51.8 at 23 °C (Coates et al., 1985)
48 at 25 °C (Price, 1976)
47.8 at 25 °C (McAuliffe, 1963, 1966)
72.4 at 0 °C, 49.6 at 25 °C (Polak and Lu, 1973)

Vapor density:
2.95 g/L at 25 °C, 2.49 (air = 1)

Vapor pressure (mmHg):
621 at 22.04 °C (Schumann et al., 1942)
628 at 22.44 °C (Willingham et al., 1945)

Environmental fate:
   Photolytic. When synthetic air containing gaseous nitrous acid and 2-methylbutane was exposed to artificial sunlight (λ = 300-450 nm), acetone, acetaldehyde, methyl nitrate, peroxyacetal nitrate, propyl nitrate, and pentyl nitrate were formed (Cox et al., 1980).
   Based upon a photooxidation rate constant of 3.90 x 10^{-12} cm^3/molecule·sec with OH radicals in summer daylight, the atmospheric lifetime is 36 h (Altshuller, 1991). At atmospheric pressure and 300 K, Darnall et al. (1978) reported a rate constant of 3.78 x 10^{-12} cm^3/molecule·sec for the same reaction.
   Chemical/Physical. Complete combustion in air gives carbon dioxide and water vapor. 2-Methylbutane will not hydrolyze because it does not contain a hydrolyzable functional group.

Exposure limits: ACGIH TLV: TWA 600 ppm (adopted).

Symptoms of exposure: May be narcotic at high concentrations (NIOSH, 1997; Patnaik, 1992)

Uses: Solvent; blowing agent for polystyrene; manufacturing chlorinated derivatives.

3-METHYL-1-BUTENE

Synonyms: Isopentene; Isopropylethylene; α-Isoamylene; UN 2561.

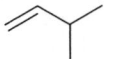

CAS Registry Number: 563-45-1
DOT: 2561
DOT label: Combustible liquid
Molecular formula: C_5H_{10}
Formula weight: 70.13
RTECS: EM7600000

Physical state, color, and odor:
Colorless liquid with a disagreeable odor

Melting point (°C):
-168.5 (Weast, 1986)

Boiling point (°C):
20 (Weast, 1986)
21, 25 (Verschueren, 1983)

Density (g/cm³):
0.6272 at 20/4 °C, 0.6219 at 25/4 °C (Dreisbach, 1959)

Diffusivity in water (x 10^{-5} cm²/sec):
0.83 at 20 °C using method of Hayduk and Laudie (1974)

Dissociation constant, pK_a:
>14 (Schwarzenbach et al., 1993)

Flash point (°C):
-57 (Hawley, 1981)

Lower explosive limit (%):
1.5 (NFPA, 1984)

Upper explosive limit (%):
9.1 (NFPA, 1984)

Heat of fusion (kcal/mol):
1.281 (Dean, 1987)

Henry's law constant (atm·m³/mol):
0.535 at 25 °C (Hine and Mookerjee, 1975)

Ionization potential (eV):
9.51 (Franklin et al., 1969)
9.60 (Collin and Lossing, 1959)

Soil sorption coefficient, log K_{oc}:
Unavailable because experimental methods for estimation of this parameter for aliphatic hydrocarbons are lacking in the documented literature

Octanol/water partition coefficient, log K_{ow}:
2.30 using method of Hansch et al. (1968)

Solubility in organics:
Soluble in alcohol, benzene, and ether (Weast, 1986)

Solubility in water:
130 mg/kg at 25 °C (McAuliffe, 1966)

Vapor density:
2.87 g/L at 25 °C, 2.42 (air = 1)

Vapor pressure (mmHg):
902.1 at 25 °C (Wilhoit and Zwolinski, 1971)

Environmental fate:
  Photolytic. The following rate constants were reported for the reaction of 3-methyl-1-butene and OH radicals in the atmosphere: 3.0 x 10^{-11} cm^3/molecule·sec (Atkinson et al., 1979); 6.07-9.01 x 10^{-11} cm^3/molecule·sec (Atkinson, 1985); 3.18 x 10^{-11} cm ^{3}/molecule·sec (Atkinson, 1990).

Source: Schauer et al. (1999) reported 3-methyl-1-butene in a diesel-powered medium-duty truck exhaust at an emission rate of 160 μg/km.

Uses: In high octane fuels; organic synthesis.

METHYL CELLOSOLVE

Synonyms: Dowanol EM; EGM; EGME; Ektasolve; Ethylene glycol methyl ether; Ethylene glycol monomethyl ether; Glycol ether EM; Glycol methyl ether; Glycol monomethyl ether; Jeffersol EM; MECS; **2-Methoxyethanol**; Methoxyhydroxyethane; Methyl ethoxol; Methyl glycol; Methyl oxitol; Polysolv EM; Prist; UN 1188.

CAS Registry Number: 109-86-4
DOT: 1188
DOT label: Flammable liquid
Molecular formula: $C_3H_8O_2$
Formula weight: 76.10
RTECS: KL5775000
Merck reference: 10, 5915

Physical state, color, and odor:
Colorless liquid with a mild, ether-like odor. Odor threshold concentration is 900 ppb (Keith and Walters, 1992).

Melting point (°C):
-85.1 (Weast, 1986)

Boiling point (°C):
124.6 (Dean, 1987)

Density (g/cm³):
0.9647 at 20/4 °C (Weast, 1986)
0.96020 at 25.00/4 °C (Venkatesulu et al., 1997)

Diffusivity in water (x 10^{-5} cm²/sec):
1.02 at 20 °C using method of Hayduk and Laudie (1974)

Flash point (°C):
46.1 (open cup, Windholz et al., 1983)
39 (NFPA, 1984)

Lower explosive limit (%):
1.8 (NIOSH, 1997)
1.8 at 0 °C (NFPA, 1984)

Upper explosive limit (%):
14 (NIOSH, 1997)
14 at 0 °C (NFPA, 1984)

Henry's law constant (x 10^{-2} atm·m³/mol):
4.41, 3.63, 11.6, 3.09, and 3.813 at 10, 15, 20, 25, and 30 °C, respectively (Ashworth et al., 1988)

Ionization potential (eV):
10.13 (Lias et al., 1998)

Soil sorption coefficient, log K$_{oc}$:
Unavailable because experimental methods for estimation of this parameter for methoxy alcohols are lacking in the documented literature. However, its miscibility in water suggests its adsorption to soil will be nominal (Lyman et al., 1982).

Octanol/water partition coefficient, log K$_{ow}$:
Unavailable because experimental methods for estimation of this parameter for methoxy alcohols are lacking in the documented literature

Solubility in organics:
Very soluble in acetone, dimethylsulfoxide, and 95% ethanol (Keith and Walters, 1992). Miscible with *N,N*-dimethylformamide, ether, and glycerol (Windholz et al., 1983).

Solubility in water:
Miscible (Price et al., 1974)

Vapor density:
3.11 g/L at 25 °C, 2.63 (air = 1)

Vapor pressure (mmHg):
6.2 at 20 °C, 14 at 30 °C (Verschueren, 1983)
9.4 at 25 °C (Banerjee et al., 1990)

Environmental fate:
Photolytic. Grosjean (1997) reported an atmospheric rate constant of 1.25 x 10^{-11} cm^3/molecule·sec at 298 K for the reaction of methyl cellosolve and OH radicals. Based on an atmospheric OH concentration of 1.0 x 10^6 molecule/cm^3, the reported half-life of methanol is 0.64 d (Grosjean, 1997).

Chemical/Physical. At an influent concentration of 1,000 mg/L, treatment with granular activated carbon resulted in an effluent concentration of 342 mg/L. The adsorbability of the carbon used was 132 mg/g carbon (Guisti et al., 1974).

Exposure limits: NIOSH REL: TWA 0.1 ppm (0.3 mg/m^3), IDLH 200 ppm; OSHA PEL: TWA 25 ppm (80 mg/m^3); ACGIH TLV: TWA 5 ppm (16 mg/m^3).

Symptoms of exposure: Inhalation of vapors may cause headache, weakness, eye irritation, ataxia, and tremor (Patnaik, 1992). An irritation concentration of 368.00 mg/m^3 in air was reported by Ruth (1986).

Toxicity:
Acute oral LD$_{50}$ for guinea pigs 950 mg/kg, rats 2,460 mg/kg, rabbits 890 mg/kg (quoted, RTECS, 1985).

Uses: Solvent for natural and synthetic resins, cellulose acetate, nitrocellulose, and some dyes; nail polishes; dyeing leather; sealing moisture-proof cellophane; lacquers, varnishes, enamels, wood stains; in solvent mixtures; perfume fixative; jet fuel de-icing additive.

METHYL CELLOSOLVE ACETATE

Synonyms: Acetic acid 2-methoxyethyl ester; Ethylene glycol methyl ether acetate; Ethylene glycol monomethyl ether acetate; Glycol ether EM acetate; Glycol monomethyl ether acetate; **2-Methoxyethanol acetate**; 2-Methoxyethyl acetate; Methyl cellosolve ethanoate; Methyl glycol acetate; Methyl glycol monoacetate; UN 1189.

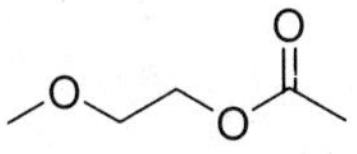

CAS Registry Number: 110-49-6
DOT: 1189
DOT label: Combustible liquid
Molecular formula: $C_5H_{10}O_3$
Formula weight: 118.13
RTECS: KL5950000
Merck reference: 10, 5916

Physical state, color, and odor:
Colorless liquid with a mild, ether-like odor

Melting point (°C):
-65 (NIOSH, 1997)
-70 (Windholz et al., 1983)

Boiling point (°C):
144-145 (Weast, 1986)

Density (g/cm³):
1.0090 at 19/19 °C (Weast, 1986)

Diffusivity in water (x 10⁻⁵ cm²/sec):
0.81 at 19 °C using method of Hayduk and Laudie (1974)

Flash point (°C):
48.9 (NIOSH, 1987)
54 (Windholz et al., 1983)

Lower explosive limit (%):
1.7 (NFPA, 1984)

Upper explosive limit (%):
8.2 (NFPA, 1984)

Soil sorption coefficient, log K_{oc}:
Unavailable because experimental methods for estimation of this parameter for cellosolve esters are lacking in the documented literature. However, its miscibility in water suggests its adsorption to soil will be nominal (Lyman et al., 1982).

Octanol/water partition coefficient, log K_{ow}:
Unavailable because experimental methods for estimation of this parameter for cellosolve esters are lacking in the documented literature

Solubility in organics:
Soluble in alcohol and ether (Weast, 1986)

Solubility in water:
Miscible (Lyman et al., 1982)

Vapor density:
4.83 g/L at 25 °C, 4.08 (air = 1)

Vapor pressure (mmHg at 20 °C):
2 (NIOSH, 1997)
7 (Verschueren, 1983)

Environmental fate:
   Chemical/Physical. Hydrolyzes in water forming methyl cellosolve and acetic acid.
   At an influent concentration of 1,024 mg/L, treatment with granular activated carbon resulted in an effluent concentration of 886 mg/L. The adsorbability of the carbon used was 28 mg/g carbon (Guisti et al., 1974).

Exposure limits: NIOSH REL: TWA 0.1 ppm (0.5 mg/m^3), IDLH 200 ppm; OSHA PEL: TWA 25 ppm (120 mg/m^3); ACGIH TLV: TWA 5 ppm (24 mg/m^3).

Toxicity:
   Acute oral LD$_{50}$ for guinea pigs 1,250 mg/kg, mice 3,390 mg/kg (quoted, RTECS, 1985).

Uses: Solvent for cellulose acetate, nitrocellulose, various gums, resins, waxes, oils; textile printing; lacquers; dopes; textile printing; photographic film.

METHYL CHLORIDE

Synonyms: Artic; **Chloromethane**; Monochloromethane; RCRA waste number U045; UN 1063.

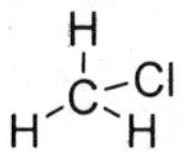

$$\begin{array}{c} H \\ | \\ H-C-Cl \\ | \\ H \end{array}$$

CAS Registry Number: 74-87-3
DOT: 1063
DOT label: Flammable gas
Molecular formula: CH_3Cl
Formula weight: 50.48
RTECS: PA6300000
Merck reference: 10, 5918

Physical state, color, and odor:
Liquefied compressed gas, colorless, odorless or sweet, ethereal odor

Melting point (°C):
-97.1 (Weast, 1986)
-97.6 (McGovern, 1943)

Boiling point (°C):
-24.22 (Dreisbach, 1959)
-23.76 (McGovern, 1943)

Density (g/cm³ at 20/4 °C):
0.9159 (Weast, 1986)
0.9214 (Riddick et al., 1986)

Diffusivity in water (x 10⁻⁵ cm²/sec):
1.49 at 25 °C (quoted, Hayduk and Laudie, 1974)

Flash point (°C):
-50 (NFPA, 1984)

Lower explosive limit (%):
8.1 (NIOSH, 1997)

Upper explosive limit (%):
17.4 (NIOSH, 1997)

Heat of fusion (kcal/mol):
1.537 (Riddick et al., 1986)

Henry's law constant (x 10⁻³ atm·m³/mol):
7.69 at 20 °C (McConnell et al., 1975; Pearson and McConnell, 1975)
8.33 at 25 °C (Gossett, 1987)
9.41 (Glew and Moelwyn-Hughes, 1953)

6.6 (low ionic strength, Pankow and Rosen, 1988)
In seawater: 3.9, 4.6, and 5.3 at 0, 3, and 6 °C, respectively; in water: 10.64 at 25 °C (Moore et al., 1995)

Interfacial tension with water (dyn/cm at 20 °C):
28.3 (quoted, Freitas et al., 1997)

Ionization potential (eV):
11.26, 11.28 (Horvath, 1982)
11.33 (Gibson, 1977; Yoshida et al., 1983a)
11.3 (Franklin et al., 1969)

Soil sorption coefficient, log K_{oc}:
0.78 (Jury et al., 1990)

Octanol/water partition coefficient, log K_{ow}:
0.91 (Hansch et al., 1975)

Solubility in organics:
Miscible with chloroform, ether, glacial acetic acid (U.S. EPA, 1985)

Solubility in water:
7.4 g/kg at 30 °C (McGovern, 1943)
6,450-7,250 mg/L at 20 °C (Pearson and McConnell, 1975)
5,350 mg/L at 25.01 °C (Glew and Moelwyn-Hughes, 1953)
4,800 mg/L at 25 °C (Standen, 1964)
In mmol/atm: 167 at 10.2 °C, 97.9 at 23.5 °C, 71.4 at 36.9 °C, 69.1 at 37.4 °C, 45.2 at 59.2 °C (Boggs and Buck, 1958)

Vapor density:
2.06 g/L at 25 °C, 1.74 (air = 1)

Vapor pressure (mmHg):
3,756 at 20 °C (McConnell et al., 1975)
4,962 at 30 °C, 7,313 at 50 °C (Hsu et al., 1964)
4,309.7 at 30 °C (Howard, 1989)
4,028 at 25 °C (quoted, Nathan, 1978)

Environmental fate:
   Biological. Enzymatic degradation of methyl chloride yielded formaldehyde (Vogel et al., 1987).
   Photolytic. Reported photooxidation products via OH radicals include formyl chloride, carbon monoxide, hydrogen chloride, and phosgene (Spence et al., 1976). In the presence of water, formyl chloride hydrolyzes to hydrochloric acid and carbon monoxide, whereas phosgene hydrolyzes to hydrogen chloride and carbon monoxide (Morrison and Boyd, 1971).
   Methyl chloride reacts with OH radicals in the atmosphere at a rate of 8.5 x 10^{-14} cm³/sec with a lifetime of 135 d (Cox et al., 1976).
   Chemical/Physical. The estimated hydrolysis in water at 25 °C and pH 7 is 0.93 yr (Mabey and Mill, 1978).
   The evaporation half-life of methyl chloride (1 mg/L) from water at 25 °C using a shallow-pitch propeller stirrer at 200 rpm at an average depth of 6.5 cm is 27.6 min (Dilling, 1977).

Exposure limits: Potential occupational carcinogen. NIOSH REL: IDLH 2,000 ppm; OSHA PEL: TWA 100 ppm, ceiling 200 ppm, 5-min/3-h peak 300 ppm; ACGIH TLV: TWA 50 ppm, STEL 100 ppm (adopted).

Symptoms of exposure: Inhalation of vapors may cause headache, dizziness, drowsiness, nausea, vomiting, convulsions, coma, and respiratory failure (Patnaik, 1992). An irritation concentration of 1,050.00 mg/m^3 in air was reported by Ruth (1986).

Toxicity:

LC_{50} (14-d) for *Poecilia reticulata* 175 mg/L (Könemann, 1981).

LC_{50} (inhalation) for mice 3,146 ppm/7-h, rats 152,000 mg/m^3/30-min (quoted, RTECS, 1985).

Drinking water standard: No MCLGs or MCLs have been proposed although methyl chloride has been listed for regulation (U.S. EPA, 1996).

Uses: Coolant and refrigerant; herbicide and fumigant; organic synthesis-methylating agent; manufacturing of silicone polymers, pharmaceuticals, tetraethyl lead, synthetic rubber, methyl cellulose, agricultural chemicals and non-flammable films; preparation of methylene chloride, carbon tetrachloride, chloroform; low temperature solvent and extractant; catalytic carrier for butyl rubber polymerization; topical anesthetic; fluid for thermometric and thermostatic equipment.

METHYLCYCLOHEXANE

Synonyms: Cyclohexylmethane; Hexahydrotoluene; Sextone B; Toluene hexahydride; UN 2296.

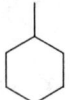

CAS Registry Number: 108-87-2
DOT: 2296
Molecular formula: C_7H_{14}
Formula weight: 98.19
RTECS: GV6125000

Physical state and color:
Colorless liquid. Odor threshold concentration is 630 ppm (Amoore and Hautala, 1983).

Melting point (°C):
-126.6 (Weast, 1986)

Boiling point (°C):
100.9 (Weast, 1986)

Density (g/cm³):
0.7694 at 20/4 °C, 0.76506 at 25/4 °C (Dreisbach, 1959)
0.7864 at 0/4 °C (Sax and Lewis, 1987)

Diffusivity in water (x 10^{-5} cm²/sec):
0.77 at 20 °C using method of Hayduk and Laudie (1974)

Dissociation constant, pK_a:
>14 (Schwarzenbach et al., 1993)

Flash point (°C):
-3.9 (NIOSH, 1997)

Lower explosive limit (%):
1.2 (NIOSH, 1997)

Upper explosive limit (%):
6.7 (NIOSH, 1997)

Heat of fusion (kcal/mol):
1.615 (Riddick et al., 1986)

Henry's law constant (atm·m³/mol):
0.125, 0.342, and 0.715 at 27.3, 35.8, and 45.0 °C, respectively (Hansen et al., 1995)
0.0678 at 25.0 °C (Ramachandran et al., 1996)

Ionization potential (eV):
9.85 ± 0.03 (Franklin et al., 1969)

Interfacial tension with water (dyn/cm at 25 °C):
41.9 (quoted, Freitas et al., 1997)

Soil sorption coefficient, log K_{oc}:
Unavailable because experimental methods for estimation of this parameter for alicyclic hydrocarbons are lacking in the documented literature

Octanol/water partition coefficient, log K_{ow}:
2.82 (Hansch and Leo, 1979)

Solubility in organics:
In methanol, g/L: 269 at 5 °C, 298 at 10 °C, 332 at 15 °C, 372 at 20 °C, 422 at 25 °C, 488 at 30 °C, 575 at 35 °C, 709 at 40 °C (Kiser et al., 1961).

Solubility in water:
In mg/kg: 16.0 at 25 °C, 18.0 at 40.1 °C, 18.9 at 55.7 °C, 33.8 at 99.1 °C, 79.5 at 120.0 °C, 139.0 at 137.3 °C, 244.0 at 149.5 °C (Price, 1976)
14.0 mg/kg at 25 °C (McAuliffe, 1963, 1966)
15.2 mg/L at 20 °C (Burris and MacIntyre, 1986)
At 25 °C: 16.7 mg/L (distilled water), 11.5 mg/L (3.3% NaCl) (Groves, 1988)

Vapor density:
4.01 g/L at 25 °C, 3.39 (air = 1)

Vapor pressure (mmHg):
47.7 at 24.46 °C (Willingham et al., 1945)
36.14 at 20 °C (Burris and MacIntyre, 1986)

Environmental fate:
   Biological. May be oxidized by microbes to 4-methylcyclohexanol, which may further oxidize to give 4-methylcycloheptanone (Dugan, 1972).
   Photolytic. Based on a photooxidation rate constant 1.04×10^{-11} cm³/molecule·sec for the reaction of cyclohexane and OH radicals in the atmosphere at 298, the estimated lifetime is 13 h (Altshuller, 1991).
   Chemical/Physical. Complete combustion in air gives carbon dioxide and water vapor. Methycyclohexane will not hydrolyze in water (Kollig, 1993).

Exposure limits: NIOSH REL: TWA 400 ppm (1,600 mg/m³), IDLH 1,200 ppm; OSHA PEL: TWA 500 ppm (2,000 mg/m³); ACGIH TLV: TWA 400 ppm (adopted).

Symptoms of exposure: Vapors may irritate mucous membranes (NIOSH, 1997; Patnaik, 1992).

Toxicity:
   Acute oral LD_{50} for mice 2,250 mg/kg (quoted, RTECS, 1985).
   LC_{50} (inhalation) for mice 41,500 mg/m³/2-h (quoted, RTECS, 1985).

Source: Schauer et al. (1999) reported methylcyclohexane in a diesel-powered medium-duty truck exhaust at an emission rate of 620 μg/km.

Uses: Solvent for cellulose ethers and other organics; gasoline component; organic synthesis.

o-METHYLCYCLOHEXANONE

Synonyms: 2-Methylcyclohexanone; Tetrahydro-*o*-cresol.

CAS Registry Number: 583-60-8
DOT: 2297
DOT label: Combustible liquid
Molecular formula: $C_7H_{12}O$
Formula weight: 112.17
RTECS: GW1750000

Physical state, color, and odor:
Colorless liquid with a weak, peppermint-like odor

Melting point (°C):
-13.9 (Weast, 1986)
-19 (Verschueren, 1983)

Boiling point (°C):
165 at 757 mmHg (Weast, 1986)

Density (g/cm³):
0.9250 at 20/4 °C (Weast, 1986)

Diffusivity in water (x 10⁻⁵ cm²/sec):
0.79 at 20 °C using method of Hayduk and Laudie (1974)

Flash point (°C):
48.2 (NIOSH, 1997)

Ionization potential (eV):
9.05 (Lias et al., 1998)

Solubility in organics:
Soluble in alcohol and ether (Weast, 1986)

Solubility in water (wt %):
2.94 at 0 °C, 2.41 at 9.0 °C, 1.98 at 19.5 °C, 1.68 at 31.0 °C, 1.56 at 40.8 °C, 1.47 at 50.2 °C, 1.45 at 60.3 °C, 1.47 at 71.0 °C, 1.46 at 81.3 °C, 1.54 at 90.3 °C (Stephenson, 1992)

Vapor density:
4.58 g/L at 25 °C, 3.87 (air = 1)

Vapor pressure (mmHg):
1 at 20 °C (NIOSH, 1997)

Exposure limits: NIOSH REL: TWA 50 ppm (230 mg/m^3), STEL 75 ppm (345 mg/m^3), IDLH 600 ppm; OSHA PEL: TWA 100 ppm (460 mg/m^3); ACGIH TLV: TWA 50 ppm, STEL 75 ppm (adopted).

Toxicity:

Acute oral LD$_{50}$ for rats 2,140 mg/kg, rabbits 1 g/kg (quoted, RTECS, 1985).

Uses: Solvent; lacquers; organic synthesis.

1-METHYLCYCLOHEXENE

Synonyms: 1-Methyl-1-cyclohexene; 2,3,4,5-Tetrahydrotoluene.

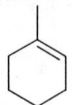

CAS Registry Number: 591-49-1
Molecular formula: C_7H_{12}
Formula weight: 96.17

Physical state and color:
Colorless liquid

Melting point (°C):
-121 (Weast, 1986)

Boiling point (°C):
110 (Weast, 1986)

Density (g/cm³):
0.8102 at 20/4 °C, 0.8058 at 25/4 °C (Dreisbach, 1959)

Diffusivity in water (x 10^{-5} cm²/sec):
0.80 at 20 °C using method of Hayduk and Laudie (1974)

Dissociation constant, pK_a:
>14 (Schwarzenbach et al., 1993)

Flash point (°C):
-3 (Dean, 1987)

Henry's law constant (x 10^{-2} atm·m³/mol):
7.45 at 25 °C (approximate - calculated from water solubility and vapor pressure)

Ionization potential (eV):
8.67 (Lias et al., 1998)

Soil sorption coefficient, log K_{oc}:
Unavailable because experimental methods for estimation of this parameter for alicyclic hydrocarbons are lacking in the documented literature

Octanol/water partition coefficient, log K_{ow}:
2.44 using method of Hansch et al. (1968)

Solubility in organics:
Soluble in benzene and ether (Weast, 1986)

Solubility in water:
52 mg/kg at 25 °C (McAuliffe, 1966)

Vapor density:
3.93 g/L at 25 °C, 3.32 (air = 1)

Vapor pressure (mmHg):
30.6 at 25 °C (estimated using Antoine equation, Dreisbach, 1959)

Environmental fate:
Photolytic. Atkinson (1985) reported a rate constant of 9.45 x 10^{-11} cm^3/molecule·sec for the reaction of 1-methylcyclohexene with OH radicals in the atmosphere.

Uses: Organic synthesis.

METHYLCYCLOPENTANE

Synonym: UN 2298.

CAS Registry Number: 96-37-7
DOT: 2298
DOT label: Combustible liquid
Molecular formula: C_6H_{12}
Formula weight: 84.16
RTECS: GY4640000

Physical state, color, and odor:
Colorless liquid with a sweetish odor

Melting point (°C):
-142.4 (Weast, 1986)

Boiling point (°C):
71.8 (Weast, 1986)

Density (g/cm³):
0.74864 at 20/4 °C, 0.74394 at 25/4 °C (Dreisbach, 1959)

Diffusivity in water (x 10⁻⁵ cm²/sec):
0.85 at 20 °C (Witherspoon and Bonoli, 1969)
0.90 at 25 °C (Bonoli and Witherspoon, 1968)

Dissociation constant, pK_a:
>14 (Schwarzenbach et al., 1993)

Flash point (°C):
<-7 (NFPA, 1984)

Lower explosive limit (%):
1.0 (NFPA, 1984)

Upper explosive limit (%):
8.35 (NFPA, 1984)

Entropy of fusion (cal/mol·K):
12.67 (Domalski and Hearing, 1998)

Heat of fusion (kcal/mol):
1.656 (Riddick et al., 1986)

Henry's law constant (atm·m³/mol):
0.362 at 25 °C (Hine and Mookerjee, 1975)

Ionization potential (eV):
9.7 (Lias, 1998)

Soil sorption coefficient, log K$_{oc}$:
Unavailable because experimental methods for estimation of this parameter for alicyclic hydrocarbons are lacking in the documented literature

Octanol/water partition coefficient, log K$_{ow}$:
3.37 (Sangster, 1989)

Solubility in organics:
In methanol, g/L: 380 at 5 °C, 415 at 10 °C, 500 at 15 °C, 595 at 20 °C, 740 at 25 °C, 1,100 at 30 °C. Miscible at higher temperatures (Kiser et al., 1961).

Solubility in water (mg/kg):
41.8 at 25 °C, 18.0 at 40.1 °C, 18.9 at 55.7 °C, 33.8 at 99.1 °C, 79.5 at 120.0 °C, 139.0 at 137.3 °C, 244.0 at 149.5 °C. In NaCl solution at 25 °C (salinity, g/kg): 38 (1.002), 36.3 (10.000), 29.2 (34.472), 27.0 (50.030), 12.7 (125.100), 5.72 (199.900), 3.36 (279.800), 1.89 (358.700) (Price, 1976)
42 at 25 °C (McAuliffe, 1966)
42.6 at 25 °C (McAuliffe, 1963)

Vapor density:
3.44 g/L at 25 °C, 2.91 (air = 1)

Vapor pressure (mmHg):
125.1 at 24.75 °C (Willingham et al., 1945)

Environmental fate:
Photolytic. A photooxidation rate constant of 7.0 x 10^{-12} cm^3/molecule·sec was reported for the reaction of methylcyclopentane and OH radicals in the atmosphere (Atkinson, 1990).
Chemical/Physical. Complete combustion in air gives carbon dioxide and water vapor. Methylcyclopentane will not hydrolyze because it does not contain a hydrolyzable functional group.
At elevated temperatures, rupture of the ring occurs and 1-propene is produced in a 40% yield. Other products include hydrogen and cyclic mono- and diolefins (Rice and Murphy, 1942).

Symptoms of exposure: Vapors may irritate respiratory tract (Patnaik, 1992).

Uses: Extractive solvent; azeotropic distillation agent; organic synthesis.

METHYLENE CHLORIDE

Synonyms: Aerothene MM; DCM; **Dichloromethane**; Freon 30; Methane dichloride; Methylene bichloride; Methylene dichloride; Narcotil; NCI-C50102; RCRA waste number U080; Solaesthin; Solmethine; UN 1593.

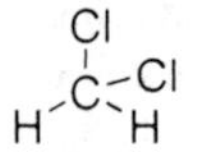

$$H-\overset{\displaystyle \overset{Cl}{|}}{\underset{\displaystyle H}{C}}-Cl$$

CAS Registry Number: 75-09-2
DOT: 1593
Molecular formula: CH_2Cl_2
Formula weight: 84.93
RTECS: PA8050000
Merck reference: 10, 5936

Physical state, color, and odor:
Clear, colorless liquid with a sweet, penetrating, ethereal odor. Odor threshold concentrations ranged from 205 to 307 ppm (Keith and Walters, 1992).

Melting point (°C):
-95.14 (Dreisbach, 1959)
-96.7 (Carlisle and Levine, 1932)
-94.92 (Riddick et al., 1986)

Boiling point (°C):
39.75 (Dreisbach, 1959)
40.1 (McConnell et al., 1975)

Density (g/cm³):
1.3361 at 20/4 °C (Carlisle and Levine, 1932)
1.3266 at 20/4 °C (Horvath, 1982)
1.3163 at 25/4 °C (Dreisbach, 1959)

Diffusivity in water (x 10^{-5} cm²/sec):
1.15 at 20 °C using method of Hayduk and Laudie (1974)

Flash point (°C):
≥30 (Kuchta et al., 1968)

Lower explosive limit (%):
13 (NIOSH, 1997)
15.1 at 103 °C (Coffee et al., 1972)

Upper explosive limit (%):
23 (NIOSH, 1997)
15.1 at 103 °C (Coffee et al., 1972)

Heat of fusion (kcal/mol):
1.1 (Dean, 1987)

Henry's law constant (x 10^{-3} atm·m³/mol):
2.0 at 25 °C (Pankow and Rosen, 1988)
3.19 at 25 °C (Warner et al., 1987)
2.69 at 25 °C (Dilling, 1977)
3.53 at 37 °C (Sato and Nakajima, 1979)
1.40, 1.69, 2.44, 2.96, and 3.61 at 10, 15, 20, 25, and 30 °C, respectively (Ashworth et al., 1988)
2.1, 3.1, 37, and 4.5 at 20, 30, 35, and 40 °C, respectively (Tse et al., 1992)
1.94 at 20 °C (Hovorka and Dohnal, 1997)
3.03 at 20 °C (Pearson and McConnell, 1975)
2.13 at 25 °C (Gossett, 1987)
2.44 at 25 °C (Hoff et al., 1993)
2.50 at 25 ° (Wright et al., 1992)
2.93 at 24.9 °C (mole fraction ratio, Leighton and Calo, 1981)

Interfacial tension with water (dyn/cm at 20 °C):
28.31 (Harkins et al., 1920)

Ionization potential (eV):
11.35 ± 0.01 (Franklin et al., 1969)
11.33 (Horvath, 1982)

Soil sorption coefficient, log K_{oc}:
1.00 (Daniels et al., 1985)
1.44 (log K_{om}, Sabljić, 1984)

Octanol/water partition coefficient, log K_{ow}:
1.35, 1.34, and 1.37 at 25, 30, and 50 °C, respectively (Bhatia and Sandler, 1995)
1.25 (Hansch et al., 1975)

Solubility in organics:
Miscible with ethanol, ether (U.S. EPA, 1985), and many chlorinated solvents including carbon tetrachloride and chloroform

Solubility in water:
In g/kg H_2O: 23.63 at 0 °C, 21.22 at 10 °C, 20.00 at 20 °C, 19.69 at 30 °C (Rex, 1906)
20,000 mg/kg (Carlisle and Levine, 1932)
13,000 mg/L at 25 °C (Fluka, 1988; Haque, 1990)
19,400 mg/L at 25 °C (Dilling, 1977)
13,200 mg/L at 25 °C (Pearson and McConnell, 1975)
2.03, 1.92, 1.80, 1.72, and 1.77 at 0, 9.2, 17.3, 26.8, and 35.7 °C, respectively (Stephenson, 1992)
22,700 mg/L at 1.5 °C (Suntio et al., 1988)
3.95 x 10^{-3} at 25 °C (mole fraction, Li et al., 1993)

Vapor density:
3.47 g/L at 25 °C, 2.93 (air = 1)

Vapor pressure (mmHg):
440 at 25 °C (ACGIH, 1986)
362.4 at 20 °C (McConnell et al., 1975)

455 at 25 °C (Valsaraj, 1988)
380 at 22 °C (Sax, 1984)
420 at 25 °C (quoted, Nathan, 1978)

Environmental fate:

Biological. Complete microbial degradation to carbon dioxide was reported under anaerobic conditions by mixed or pure cultures. Under enzymatic conditions formaldehyde was the only product reported (Vogel et al., 1987). In a static-culture-flask screening test, methylene chloride (5 and 10 mg/L) was statically incubated in the dark at 25 °C with yeast extract and settled domestic wastewater inoculum. After 7 d, 100% biodegradation with rapid adaptation was observed (Tabak et al., 1981).

Under aerobic conditions with sewage seed or activated sludge, complete biodegradation was observed between 6 h to 1 wk (Rittman and McCarty, 1980).

Soil. Methylene chloride undergoes biodegradation in soil under aerobic and anaerobic conditions. Under aerobic conditions, the following half-lives were reported: 54.8 d in sand (500 ppb); 1.3, 9.4 and 191.4 d at concentrations of 160, 500, and 5,000 ppb, respectively, in sandy loam soil; 12.7 d (500 ppb) in sandy clay loam soil; 7.2 d (500 ppb) following a 50-d lag time. Under anaerobic conditions, the half-life of methylene chloride in clay following a 70-d lag time is 21.5 d (Davis and Madsen, 1991). The estimated volatilization half-life of methylene chloride in soil is 100 d (Jury et al., 1990).

Photolytic. Reported photooxidation products via OH radicals include carbon dioxide, carbon monoxide, formyl chloride, and phosgene (Spence et al., 1976). In the presence of water, phosgene hydrolyzes to hydrochloric acid and carbon dioxide, whereas formyl chloride hydrolyzes to hydrogen chloride and carbon monoxide (Morrison and Boyd, 1971).

The rate constant for the reaction of methylene chloride and OH radicals in the atmosphere is 9×10^7 cm³/mol·sec (Hendry and Kenley, 1979). Methylene chloride reacts with OH radicals in water (pH 8.5) at a rate of 9.6×10^7/M·sec (Haag and Yao, 1992).

Chemical/Physical. Under laboratory conditions, methylene chloride hydrolyzed with subsequent oxidation and reduction to produce methyl chloride, methanol, formic acid, and formaldehyde (Smith and Dragun, 1984). The experimental half-life for hydrolysis in water at 25 °C is approximately 18 months (Dilling et al., 1975).

The evaporation half-life of methylene chloride (1 mg/L) from water at 25 °C using a shallow-pitch propeller stirrer at 200 rpm at an average depth of 6.5 cm is 20.2 min (Dilling, 1977).

At influent concentrations of 10.0, 1.0, 0.1, and 0.01 mg/L, the granular activated carbon adsorption capacities at pH 5.8 were 19.0, 1.3, 0.09, and 0.006 mg/g, respectively (Dobbs and Cohen, 1980).

Methylene chloride is stable up to 120 °C (Carlisle and Levine, 1932).

Exposure limits: Potential occupational carcinogen. NIOSH REL: IDLH 2,300 ppm; OSHA PEL: TWA 500 ppm, ceiling 1,000 ppm, 5-min/2-h peak 2,000 ppm; ACGIH TLV: TWA 50 ppm (adopted).

Symptoms of exposure: May produce fatigue, weakness, headache, lightheadedness, euphoria, nausea, and sleep. May be narcotic at high concentrations (Patnaik, 1992). An irritation concentration of 8,280.00 mg/m³ in air was reported by Ruth (1986).

Toxicity:

EC_{50} (24-h) for *Daphnia magna* 1,942 mg/L (Lilius et al., 1995).

LC_{50} (contact) for earthworm (*Eisenia fetida*) 304 μg/cm² (Neuhauser et al., 1985).

LC_{50} (96-h) for bluegill sunfish 220 mg/L (Spehar et al., 1982), fathead minnows 193 mg/L (flow-through), 310 mg/L (static bioassay) (Alexander et al., 1978), *Cyprinodon variegatus* 330 ppm using natural seawater (Heitmuller et al., 1981).

LC_{50} (72-h) for *Cyprinodon variegatus* 360 ppm (Heitmuller et al., 1981).

LC_{50} (48-h) for *Daphnia magna* 220 mg/L (LeBlanc, 1980), *Cyprinodon variegatus* 360 ppm (Heitmuller et al., 1981).

LC_{50} (24-h) for *Daphnia magna* 310 mg/L (LeBlanc, 1980), *Cyprinodon variegatus* 370 ppm (Heitmuller et al., 1981).

LC_{50} (6-h) for B6C3F1 mice 2,200 ppm (Chellman et al., 1986).

LC_{50} (inhalation) for mice 14,400 ppm/7-h, rats 88,000 mg/m^3/30-min (quoted, RTECS, 1985).

Acute oral LD_{50} for rats 2,136 mg/kg (quoted, RTECS, 1985).

LD_{50} (subcutaneous) for mice 6,452 mg/kg (Klaassen and Plaa, 1966).

Heitmuller et al. (1981) reported a NOEC of 130 ppm.

Drinking water standard (final): MCLG: zero; MCL: 5 μg/L (U.S. EPA, 1996).

Uses: Low temperature solvent; ingredient in paint and varnish removers; cleaning, degreasing, and drying metal parts; fumigant; manufacturing of aerosols; refrigerant; dewaxing; blowing agent in foams; solvent for cellulose acetate; organic synthesis.

METHYL FORMATE

Synonyms: Formic acid methyl ester; Methyl methanoate; UN 1243.

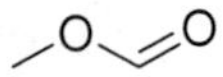

CAS Registry Number: 107-31-3
DOT: 1243
DOT label: Flammable liquid
Molecular formula: $C_2H_4O_2$
Formula weight: 60.05
RTECS: LQ8925000
Merck reference: 10, 5947

Physical state, color, and odor:
Clear, colorless, mobile liquid with a pleasant odor. Odor threshold concentration is 600 ppm (Amoore and Hautala, 1983).

Melting point (°C):
-99 (Weast, 1986)

Boiling point (°C):
31.5 (Weast, 1986)

Density (g/cm³):
0.9742 at 20/4 °C (Weast, 1986)
0.9713 at 25.00/4 °C (Emmerling et al., 1998)

Diffusivity in water (x 10^{-5} cm²/sec):
1.17 at 20 °C using method of Hayduk and Laudie (1974)

Flash point (°C):
-19 (NIOSH, 1997)

Lower explosive limit (%):
4.5 (NFPA, 1984)

Upper explosive limit (%):
23 (NFPA, 1984)

Heat of fusion (kcal/mol):
1.800 (Dean, 1987)

Henry's law constant (x 10^{-4} atm·m³/mol):
2.44 at 25 °C (Hoff et al., 1993)

Ionization potential (eV):
10.815 ± 0.005 (Franklin et al., 1969)

Soil sorption coefficient, log K_{oc}:
Unavailable because experimental methods for estimation of this parameter for aliphatic esters

are lacking in the documented literature. However, its high solubility in water suggests its adsorption to soil will not be nominal (Lyman et al., 1982).

Octanol/water partition coefficient, log K_{ow}:
-0.18 using method of Hansch et al. (1968)

Solubility in organics:
Soluble in ether (Weast, 1986). Miscible with alcohol (Windholz et al., 1983).

Solubility in water (g/L):
304 at 20 °C (quoted, Verschueren, 1983)
203 at 25 °C (Riddick et al., 1986)

Vapor density:
2.45 g/L at 25 °C, 2.07 (air = 1)

Vapor pressure (mmHg):
476 at 20 °C (NIOSH, 1997)
625 at 25 °C (Abraham, 1984)

Environmental fate:
   Photolytic. Methyl formate, formed from the irradiation of dimethyl ether in the presence of chlorine, degraded to carbon dioxide, water, and small amounts of formic acid. Continued irradiation degraded formic acid to carbon dioxide, water, and hydrogen chloride (Kallos and Tou, 1977; Good et al., 1999).
   A rate constant of 2.27 x 10^{-12} cm^3/molecule·sec was reported for the reaction of methyl formate and OH radicals in the atmosphere (Atkinson, 1989).
   Chemical/Physical. Hydrolyzes slowly in water forming methyl alcohol and formic acid (NIOSH, 1997). Hydrolysis half-lives reported at 25 °C: 0.91 h at pH 9, 9.1 h at pH 8, 2.19 d at pH 7, and 21.9 d at pH 6 (Mabey and Mill, 1978).

Exposure limits: NIOSH REL: TWA 100 ppm (250 mg/m³), STEL 150 ppm (375 mg/m³), IDLH 4,500 ppm; OSHA PEL: TWA 100 ppm; ACGIH TLV: TWA 100 ppm, STEL 150 ppm (adopted).

Symptoms of exposure: May irritate eyes, nose, and throat; inhalation may produce visual disturbances, narcotic effects, and respiratory distress (Patnaik, 1992). An irritation concentration of 8,750.00 mg/m³ in air was reported by Ruth (1986).

Toxicity:
   Acute oral LD_{50} for rabbits 1,622 mg/kg (quoted, RTECS, 1985).

Uses: Fumigant and larvacide for tobacco, cereals, dried fruits; cellulose acetate solvent; military poison gases; organic synthesis.

3-METHYLHEPTANE

Synonym: 5-Methylheptane.

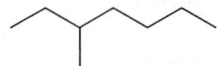

CAS Registry Number: 589-81-1
Molecular formula: C_8H_{18}
Formula weight: 114.23

Physical state and color:
Colorless liquid

Melting point (°C):
-120.5 (Weast, 1986)

Boiling point (°C):
119 (Weast, 1986)

Density (g/cm³):
0.70582 at 20/4 °C, 0.70175 at 25/4 °C (Dreisbach, 1959)

Diffusivity in water (x 10⁻⁵ cm²/sec):
0.67 at 20 °C using method of Hayduk and Laudie (1974)

Dissociation constant, pK_a:
>14 (Schwarzenbach et al., 1993)

Heat of fusion (kcal/mol):
2.779 (Dean, 1987)

Henry's law constant (atm·m³/mol):
3.70 at 25 °C (approximate - calculated from water solubility and vapor pressure)

Soil sorption coefficient, log K_{oc}:
Unavailable because experimental methods for estimation of this parameter for aliphatic hydrocarbons are lacking in the documented literature

Octanol/water partition coefficient, log K_{ow}:
3.97 using method of Hansch et al. (1968)

Solubility in organics:
In methanol, g/L: 154 at 5 °C, 170 at 10 °C, 190 at 15 °C, 212 at 20 °C, 242 at 25 °C, 274 at 30 °C, 314 at 35 °C, 365 at 40 °C (Kiser et al., 1961)

Solubility in water:
792 μg/kg at 25 °C (Price, 1976)

Vapor density:
4.67 g/L at 25 °C, 3.94 (air = 1)

Vapor pressure (mmHg):
19.5 at 25 °C (Wilhoit and Zwolinski, 1971)
8.260 at 10.000 °C, 11.146 at 15.000 °C (Osborn and Douslin, 1974)

Environmental fate:
   Photolytic. Based on a photooxidation rate constant of 8.90 x 5.61 x 10^{-12} cm^3/molecule·sec for the reaction of 3-methylheptane and OH radicals, the estimated lifetime is 16 h during summer sunlight (Altshuller, 1991).
   Chemical/Physical. Complete combustion in air gives carbon dioxide and water vapor. 3-Methylheptane will not hydrolyze because it does not contain a hydrolyzable functional group.

Uses: Calibration; gasoline component; organic synthesis.

5-METHYL-3-HEPTANONE

Synonyms: Amyl ethyl ketone; EAK; Ethyl amyl ketone; Ethyl *sec*-amyl ketone; 3-Methyl-5-heptanone.

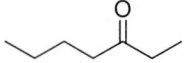

CAS Registry Number: 541-85-5
DOT: 2271
DOT label: Combustible liquid
Molecular formula: $C_8H_{16}O$
Formula weight: 128.21
RTECS: MJ7350000
Merck reference: 10, 3712

Physical state, color, and odor:
Colorless liquid with a fruity odor

Melting point (°C):
-57.1 (NIOSH, 1997)

Boiling point (°C):
158.5 (NIOSH, 1987)
160, 162 (Verschueren, 1983)

Density (g/cm³):
0.820-0.824 at 20/20 °C (Windholz et al., 1983)
0.85 at 0/4 °C (Verschueren, 1983)

Diffusivity in water (x 10⁻⁵ cm²/sec):
0.68 at 20 °C using method of Hayduk and Laudie (1974)

Flash point (°C):
59 (NIOSH, 1997)

Henry's law constant (x 10⁻⁴ atm·m³/mol):
1.30 at 20 °C (approximate - calculated from water solubility and vapor pressure)

Soil sorption coefficient, log K_{oc}:
Unavailable because experimental methods for estimation of this parameter for aliphatic hydrocarbons are lacking in the documented literature

Octanol/water partition coefficient, log K_{ow}:
1.96 using method of Hansch et al. (1968)

Solubility in organics:
Mixes readily with alcohols, ether, ketones, and other organic solvents (Patnaik, 1992)

Solubility in water (wt %):
0.275 at 0°C, 0.227 at 9.3 °C, 0.192 at 19.9 °C, 0.151 at 30.9 °C, 0.151 at 40.8 °C, 0.142 at

50.2 °C, 0.124 at 59.8 °C, 0.132 at 70.5 °C, 0.128 at 80.2 °C, 0.131 at 90.5 °C (Stephenson, 1992)

Vapor density:
5.24 g/L at 25 °C, 4.43 (air = 1)

Vapor pressure (mmHg):
2 at 20 °C (NIOSH, 1987)
2 at 25 °C (Verschueren, 1983)

Environmental fate:
   Chemical/Physical. 5-Methyl-3-heptanone will not hydrolyze because it does not contain a hydrolyzable functional group.

Exposure limits: NIOSH REL: TWA 25 ppm (130 mg/m^3), IDLH 100 ppm; OSHA PEL: TWA 25 ppm.

Symptoms of exposure: May irritate eyes, nose, and throat. At high concentrations, ataxia, prostration, respiratory, pain and narcosis may occur (Patnaik, 1992).

Toxicity:
   Acute oral LD$_{50}$ for guinea pigs 2,500 mg/kg, mice 3,800 mg/kg, rats 3,500 mg/kg (quoted, RTECS, 1985).

Uses: Solvent for vinyl resins, nitrocellulose-alkyd and nitrocellulose-maleic acid resins.

2-METHYLHEXANE

Synonyms: Ethylisobutylmethane; Isoheptane.

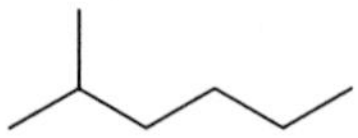

CAS Registry Number: 591-76-4
Molecular formula: C_7H_{16}
Formula weight: 100.20

Physical state and color:
Colorless liquid

Melting point (°C):
-118.3 (Weast, 1986)

Boiling point (°C):
90 (Weast, 1986)

Density (g/cm³):
0.67859 at 20/4 °C, 0.67439 at 25/4 °C (Dreisbach, 1959)

Diffusivity in water (x 10⁻⁵ cm²/sec):
0.70 at 20 °C using method of Hayduk and Laudie (1974)

Dissociation constant, pK$_a$:
>14 (Schwarzenbach et al., 1993)

Flash point (°C):
<-17.7 (Hawley, 1981)

Heat of fusion (kcal/mol):
2.195 (Riddick et al., 1986)

Henry's law constant (atm·m³/mol):
0.512, 0.311, and 0.256 at 26.9, 35.0, and 45.0 °C, respectively (Hansen et al., 1995)

Soil sorption coefficient, log K$_{oc}$:
Unavailable because experimental methods for estimation of this parameter for aliphatic hydrocarbons are lacking in the documented literature

Octanol/water partition coefficient, log K$_{ow}$:
3.30 (Coates et al., 1985)

Solubility in organics:
Soluble in acetone, alcohol, benzene, chloroform, ligroin, and ether (Weast, 1986)

Solubility in water:
3.8 mg/L at 23 °C (Coates et al., 1985)
2.54 mg/kg at 25 °C (Price, 1976)

Vapor density:
4.10 g/L at 25 °C, 3.46 (air = 1)

Vapor pressure (mmHg):
65.9 at 25 °C (Wilhoit and Zwolinski, 1971)

Environmental fate:

Biological. Riser-Roberts (1992) reported 2- and 5-methylhexanoic acids as metabolites by the microorganism *Pseudomonas aeruginosa*.

Photolytic. Based on a reported photooxidation reaction rate constant of 6.80 x 10^{-12} cm³/molecule·sec with OH radicals, the half-life of 2-methylhexane is 25 h (Altshuller, 1990).

Chemical/Physical. Complete combustion in air gives carbon dioxide and water vapor. 2-Methylhexane will not hydrolyze because it does not contain a hydrolyzable functional group.

Source: Schauer et al. (1999) reported 2-methylhexane in a diesel-powered medium-duty truck exhaust at an emission rate of 570 μg/km.

Use: Organic synthesis. Component of gasoline.

3-METHYLHEXANE

Synonym: 4-Methylhexane.

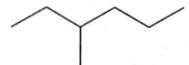

CAS Registry Number: 589-34-4
Molecular formula: C_7H_{16}
Formula weight: 100.20

Physical state and color:
Colorless liquid

Melting point (°C):
-119 (Weast, 1986)

Boiling point (°C):
91.85 (Wilhoit and Zwolinski, 1971)

Density (g/cm³):
0.68713 at 20/4 °C, 0.68295 at 25/4 °C (Dreisbach, 1959)

Diffusivity in water (x 10^{-5} cm²/sec):
0.71 at 20 °C using method of Hayduk and Laudie (1974)

Dissociation constant, pK_a:
>14 (Schwarzenbach et al., 1993)

Flash point (°C):
-3.9 (Hawley, 1981)

Henry's law constant (atm·m³/mol):
2.38 at 25 °C (Mackay and Shiu, 1981)

Interfacial tension with water (dyn/cm at 25 °C):
50.4 (quoted, Freitas et al., 1997)

Soil sorption coefficient, log K_{oc}:
Unavailable because experimental methods for estimation of this parameter for aliphatic hydrocarbons are lacking in the documented literature

Octanol/water partition coefficient, log K_{ow}:
3.41 (Coates et al., 1985)

Solubility in organics:
Soluble in acetone, alcohol, benzene, chloroform, ligroin, and ether (Weast, 1986)

Solubility in water:
2.90 mg/L at 23 °C (Coates et al., 1985)

2.64 mg/kg at 25 °C (Price, 1976)
5.24 mg/kg at 0 °C, 4.95 mg/kg at 25 °C (Polak and Lu, 1973)

Vapor density:
4.10 g/L at 25 °C, 3.46 (air = 1)

Vapor pressure (mmHg):
61.6 at 25 °C (Wilhoit and Zwolinski, 1971)

Environmental fate:
   Photolytic. Based on a reported photooxidation reaction rate constant of 7.20 x 10^{-12} cm^3/molecule·sec with OH radicals, the half-life of 3-methylhexane is 20 h (Altshuller, 1990).
   Chemical/Physical. Complete combustion in air gives carbon dioxide and water vapor. 3-Methylhexane will not hydrolyze because it does not contain a hydrolyzable functional group.

Source: Schauer et al. (1999) reported 3-methylhexane in a diesel-powered medium-duty truck exhaust at an emission rate of 310 μg/km.

Uses: Oil extender solvent; gasoline component; organic synthesis.

METHYLHYDRAZINE

Synonyms: Hydrazomethane; 1-Methylhydrazine; MMH; Monomethylhydrazine.

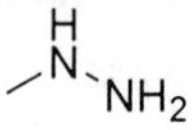

CAS Registry Number: 60-34-4
DOT: 1244
DOT label: Flammable liquid
Molecular formula: CH_6N_2
Formula weight: 46.07
RTECS:,MV5600000
Merck reference: 10, 5957

Physical state, color, and odor:
Fuming, clear, colorless liquid with an ammonia-like odor. Odor threshold concentrations ranged from 1 to 3 ppm (Keith and Walters, 1992).

Melting point (°C):
-52.4 (Weast, 1986)
-20.9 (Sax and Lewis, 1987)

Boiling point (°C):
87.5 (Weast, 1986)

Density (g/cm³):
0.874 at 25/4 °C (Weast, 1986)

Diffusivity in water (x 10^{-5} cm²/sec):
1.47 at 25 °C using method of Hayduk and Laudie (1974)

Flash point (°C):
-8.4 (NIOSH, 1997)

Lower explosive limit (%):
2.5 (NIOSH, 1997)

Upper explosive limit (%):
92 (NFPA, 1984)

Ionization potential (eV):
8.00 ± 0.06 (Franklin et al., 1969)

Soil sorption coefficient, log K_{oc}:
Unavailable because experimental methods for estimation of this parameter for hydrazines are lacking in the documented literature. However, its miscibility in water suggests its adsorption to soil will be nominal (Lyman et al., 1982).

Octanol/water partition coefficient, log K_{ow}:
Unavailable because experimental methods for estimation of this parameter for hydrazines are lacking in the documented literature

Solubility in organics:
Soluble in alcohol and ether (Weast, 1986)

Solubility in water:
Miscible

Vapor density:
1.88 g/L at 25 °C, 1.59 (air = 1)

Vapor pressure (mmHg):
38 at 20 °C (NIOSH, 1997)
49.6 at 25 °C (Sax, 1985)

Environmental fate:
  Biological. It was suggested that the rapid disappearance of methylhydrazine in sterile and nonsterile soil (Arrendondo fine sand) under aerobic conditions was due to chemical oxidation. Although the oxidation product was not identified, it biodegraded to carbon dioxide in the nonsterile soil. The oxidation product did not degrade further in the sterile soil (Ou and Street, 1988).

Exposure limits: Potential occupational carcinogen. NIOSH REL: 2-h ceiling 0.04 ppm (0.08 mg/m^3), IDLH 20 ppm; OSHA PEL: ceiling 0.2 ppm (0.35 mg/m^3); ACGIH TLV: TWA 0.01 ppm (adopted).

Toxicity:
  Acute oral LD$_{50}$ for hamsters 22 mg/kg, mice 29 mg/kg, rats 32 mg/kg (quoted, RTECS, 1985).

Uses: Rocket fuel; solvent; intermediate; organic synthesis.

METHYL IODIDE

Synonyms: Halon 10001; **Iodomethane**; RCRA waste number U138; UN 2644.

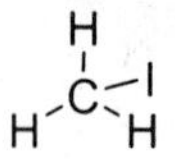

$$\begin{array}{c} H \\ | \\ H-C-I \\ | \\ H \end{array}$$

CAS Registry Number: 74-88-4
DOT: 2644
DOT label: Poison
Molecular formula: CH_3I
Formula weight: 141.94
RTECS: PA9450000
Merck reference: 10, 5958

Physical state and color:
Clear, colorless liquid which may become yellow, red, or brown on exposure to light and moisture

Melting point (°C):
-64.4 (Stull, 1947)

Boiling point (°C):
42.4 (Weast, 1986)

Density (g/cm³):
2.279 at 20/4 °C (Weast, 1986)

Diffusivity in water (x 10^{-5} cm²/sec):
1.17 at 20 °C using method of Hayduk and Laudie (1974)

Flash point (°C):
Noncombustible liquid (NIOSH, 1997).

Henry's law constant (x 10^{-3} atm·m³/mol):
5.87 at 25 °C (Liss and Slater, 1974)
5.26 at 25 °C (Hunter-Smith et al., 1983)
In seawater (salinity 30.4‰): 1.71, 3.16, and 5.40 at 0, 10, and 20 °C, respectively (Moore et al., 1995)
5.06 at 21 °C (Gan and Yates, 1996)

Ionization potential (eV):
9.54 (Franklin et al., 1969)
9.1 (Horvath, 1982)
9.86 (Gibson et al., 1977)

Soil sorption coefficient, log K_{oc}:
1.23 (Greenfield sandy loam), 1.04 (Carsetas loamy sand), 0.94 (Linne clay loam) (Gan and Yates, 1996)

Octanol/water partition coefficient, log K_{ow}:
1.69 at 19 °C (Collander, 1951)

1.51 (Hansch et al., 1975; Hansch and Leo, 1979)
1.69 (Leo et al., 1971)

Solubility in organics:
Soluble in acetone and benzene (Weast, 1986). Miscible with alcohol and ether (Windholz et al., 1983).

Solubility in water:
In g/kg H_2O: 15.65 at 0 °C, 14.46 at 10 °C, 14.19 at 20 °C, 14.29 at 30 °C (Rex, 1906)
18.217 mM at 40.34 °C (Swain and Thornton, 1962)
95.9 mM at 22 °C (Führner, 1924)

Vapor density:
5.80 g/L at 25 °C, 4.90 (air = 1)

Vapor pressure (mmHg):
331 at 20 °C, 483 at 30 °C (Rex, 1906)
405 at 25 °C (calculated, Kudchadker et al., 1979)

Environmental fate:
  Chemical/Physical. Anticipated products from the reaction of methyl iodide with ozone or OH radicals in the atmosphere are formaldehyde, iodoformaldehyde, carbon monoxide, and iodine radicals (Cupitt, 1980). With OH radicals, CH_2, methyl radical, HOI and water are possible reaction products (Brown et al., 1990). The estimated half-life of methyl iodide in the atmosphere, based on a measured rate constant for the vapor phase reaction with OH radicals, ranges from 535 h to 32 wk (Garraway and Donovan, 1979).
  Hydrolyzes in water forming methyl alcohol and hydriodic acid. The estimated half-life in water at 25 °C and pH 7 is 110 d (Mabey and Mill, 1978). May react with chlorides in seawater to form methyl chloride (Zafiriou, 1975).

Exposure limits: Potential occupational carcinogen. NIOSH REL: TWA 2 ppm (10 mg/m³), IDLH 100 ppm; OSHA PEL: TWA 5 ppm (28 mg/m³); ACGIH TLV: TWA 2 ppm (adopted).

Symptoms of exposure: An irritation concentration of 21.50 g/m³ in air was reported by Ruth (1986).

Toxicity:
  LD_{50} (intraperitoneal) for guinea pigs 51 mg/kg, mice 172 mg/kg, rats 101 mg/kg (quoted, RTECS, 1985).
  LD_{50} (subcutaneous) for mice 110 mg/kg (quoted, RTECS, 1985).

Uses: Microscopy; medicine; testing for pyridine; methylating agent.

METHYL ISOCYANATE

Synonyms: Isocyanic acid methyl ester; **Isocyanatomethane**; RCRA waste number P064; TL 1450; UN 2480.

$$\text{\_N} \diagdown \text{O}$$

CAS Registry Number: 624-83-9
DOT: 2480
DOT label: Flammable liquid and poison
Molecular formula: C_2H_3NO
Formula weight: 57.05
RTECS: NQ9450000

Physical state, color, and odor:
Colorless liquid with a sharp, penetrating odor. Odor threshold concentration is 2.1 ppm (Amoore and Hautala, 1983).

Melting point (°C):
-45 (Weast, 1986)

Boiling point (°C):
39.1-40.1 (Weast, 1986)

Density (g/cm³):
0.9230 at 27/4 °C (Weast, 1986)

Diffusivity in water (x 10^{-5} cm²/sec):
1.34 at 25 °C using method of Hayduk and Laudie (1974)

Flash point (°C):
-7 (NFPA, 1984)

Lower explosive limit (%):
5.3 (NIOSH, 1997)

Upper explosive limit (%):
26 (NIOSH, 1997)

Ionization potential (eV):
10.67 (Lias et al., 1998)

Soil sorption coefficient, log K_{oc}:
Unavailable because experimental methods for estimation of this parameter for isocyanates are lacking in the documented literature

Octanol/water partition coefficient, log K_{ow}:
Unavailable because experimental methods for estimation of this parameter for isocyanates are lacking in the documented literature

Solubility in water:
10 wt % at 15 °C (NIOSH, 1997)

Vapor density:
2.33 g/L at 25 °C, 1.97 (air = 1)

Vapor pressure (mmHg):
348 at 20 °C (NIOSH, 1997)

Exposure limits: NIOSH REL: TWA 0.02 ppm (0.05 mg/m^3), IDLH 3 ppm; OSHA PEL: TWA 0.02 ppm; ACGIH TLV: TWA 0.02 ppm (adopted).

Symptoms of exposure: Exposure to vapors may cause lacrimation, nose, and throat irritation and respiratory difficulty. Oral intake or absorption through skin may produce asthma, chest pain, dyspnea, pulmonary edema, breathing difficulty, and death (Patnaik, 1992). An irritation concentration of 5.00 mg/m^3 in air was reported by Ruth (1986).

Toxicity:
   LC$_{50}$ (6-h inhalation) for Fischer 344 rats 6.1 ppm, B6C3F1 mice 12.2 ppm, Hartley guinea pigs 5.4 ppm (Dodd et al., 1986).
   LC$_{50}$ (15-min inhalation) for rats 171 ppm, guinea pigs 112 ppm (Dodd et al., 1987).
   In mice, the oral LD$_{50}$ is 69 mg/kg (Patnaik, 1992).

Uses: Manufacture of pesticides, e.g., aldicarb and carbaryl; chemical intermediate in production of plastics and polyurethane foams.

METHYL MERCAPTAN

Synonyms: Mercaptomethane; **Methanethiol**; Methyl sulfhydrate; Methyl thioalcohol; RCRA waste number U153; Thiomethanol; Thiomethyl alcohol; UN 1064.

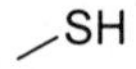

SH

CAS Registry Number: 74-93-1
DOT: 1064
DOT label: Flammable gas
Molecular formula: CH_4S
Formula weight: 48.10
RTECS: PB4375000
Merck reference: 10, 5815

Physical state, color, and odor:
Colorless gas with a garlic-like or rotten cabbage odor. Odor threshold concentration is 1.6 ppb (Amoore and Hautala, 1983).

Melting point (°C):
-123 (Weast, 1986)
-121 (Hawley, 1981)

Boiling point (°C):
6.2 (Weast, 1986)
6-7.6 (Verschueren, 1983)
5.956 (Wilhoit and Zwolinski, 1971)

Density (g/cm³):
0.8665 at 20/4 °C (Weast, 1986)

Diffusivity in water (x 10^{-5} cm²/sec):
1.25 at 20 °C using method of Hayduk and Laudie (1974)

Dissociation constant, pK_a:
10.70 at 25 °C (Dean, 1973)

Flash point (°C):
-17.9 (open cup, NIOSH, 1997)

Lower explosive limit (%):
3.9 (NIOSH, 1997)

Upper explosive limit (%):
21.8 (NIOSH, 1997)

Henry's law constant (x 10^{-3} atm·m³/mol at 25 °C):
2.56 (Przyjazny et al., 1983)
5.00 (De Bruyn et al., 1994)
3.03 (Hine and Weimar, 1965)

Ionization potential (eV):
9.440 ± 0.005 (Franklin et al., 1969)

Soil sorption coefficient, log K_{oc}:
Unavailable because experimental methods for estimation of this parameter for mercaptans are lacking in the documented literature

Octanol/water partition coefficient, log K_{ow}:
Unavailable because experimental methods for estimation of this parameter for mercaptans are lacking in the documented literature

Solubility in organics:
Soluble in alcohol, ether (Weast, 1986), and petroleum naphtha (Hawley, 1981)

Solubility in water:
23.30 g/L at 20 °C (quoted, Windholz et al., 1983)
0.330 mol/L at 25 °C (Hine and Weimar, 1965)

Vapor density:
1.97 g/L at 25 °C, 1.66 (air = 1)

Vapor pressure (mmHg):
1,292 at 20 °C (NIOSH, 1997)
1,516 at 25 °C (Wilhoit and Zwolinski, 1971)

Environmental fate:
  Photolytic. Sunlight irradiation of a methyl mercaptan-nitrogen oxide mixture in an outdoor chamber yielded formaldehyde, sulfur dioxide, nitric acid, methyl nitrate, methanesulfonic acid, and an inorganic sulfate (Grosjean, 1984a).
  Chemical/Physical. In the presence of nitric oxide, gaseous methyl mercaptan reacted with OH radicals to give methyl sulfenic acid and methyl thionitrite (MacLeod et al., 1984). Forms a crystalline hydrate with water (Patnaik, 1992).

Exposure limits: NIOSH REL: 15-min ceiling 0.5 ppm (1 mg/m³), IDLH 150 ppm; OSHA PEL: TWA 10 ppm (20 mg/m³); ACGIH TLV: TWA 0.5 ppm (adopted).

Symptoms of exposure: Inhalation of vapors may cause headache, narcosis, nausea, pulmonary irritation, and convulsions (Patnaik, 1992). Exposure of skin to liquid may cause frostbite (NIOSH, 1997).

Toxicity:
  LC_{50} (inhalation) for mice 6,530 μg/m³/2-h, rats 675 ppm (quoted, RTECS, 1985).

Source: Occurs naturally in kohlrabi stems (*Brassica oleracea var. gongylodes*) and potato plants (Duke, 1992)

Uses: Synthesis of methionine; intermediate in the manufacture of pesticides, fungicides, jet fuels, plastics; catalyst; added to natural gas to give odor.

METHYL METHACRYLATE

Synonyms: Diakon; Methacrylic acid methyl ester; Methyl-α-methylacrylate; Methyl methacrylate monomer; Methyl-2-methyl-2-propenoate; MME; Monocite methacrylate monomer; NA 1247; NCI-C50680; **2-Propenoic acid 2-methyl methyl ester**; RCRA waste number U162; UN 1247.

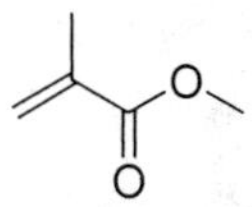

CAS Registry Number: 80-62-6
DOT: 1247
DOT label: Combustible liquid
Molecular formula: $C_5H_8O_2$
Formula weight: 100.12
RTECS: OZ5075000
Merck reference: 10, 5796

Physical state, color, and odor:
Clear, colorless liquid with a penetrating, fruity odor. Odor threshold concentration is 83 ppb (Amoore and Hautala, 1983).

Melting point (°C):
-48 (Weast, 1986)
-50 (Verschueren, 1983)

Boiling point (°C):
100-101 (Weast, 1986)

Density (g/cm³ at 20/4 °C):
0.9440 (Weast, 1986)
0.936 (Sax and Lewis, 1987)

Diffusivity in water (x 10⁻⁵ cm²/sec):
0.85 at 20 °C using method of Hayduk and Laudie (1974)

Flash point (°C):
10 (open cup, NIOSH, 1997)

Lower explosive limit (%):
1.7 (NIOSH, 1997)
2.1 (Sax and Lewis, 1987)

Upper explosive limit (%):
8.2 (NIOSH, 1997)
12.5 (Sax and Lewis, 1987)

Entropy of fusion (cal/mol·K):
14.26 (Karabaev et al., 1985)
15.3 (Lebedev and Rabinovich, 1971)

Henry's law constant (x 10^{-4} atm·m^3/mol):
2.46 at 20 °C (approximate - calculated from water solubility and vapor pressure)

Ionization potential (eV):
9.7, 10.06, 10.28 (Lias et al., 1998)

Soil sorption coefficient, log K_{oc}:
Unavailable because experimental methods for estimation of this parameter for unsaturated esters are lacking in the documented literature

Octanol/water partition coefficient, log K_{ow}:
0.67, 0.70 (Fujisawa and Masuhara, 1981)

Solubility in organics:
Soluble in acetone, alcohol, ether (Weast, 1986), methyl ethyl ketone, tetrahydrofuran, esters, aromatic and chlorinated hydrocarbons (Windholz et al., 1983)

Solubility in water:
15,600 mg/L at 20 °C (Riddick et al., 1986)

Vapor density:
4.09 g/L at 25 °C, 3.46 (air = 1)

Vapor pressure (mmHg):
28 at 20 °C, 40 at 26 °C, 49 at 30 °C (Verschueren, 1983)

Environmental fate:
   Chemical/Physical. Polymerizes easily (Windholz et al., 1983). Methyl methacrylate under-goes nucleophilic attack by hydroxyl ions in water (hydrolysis) resulting in the formation of methacrylic acid and methanol (Kollig, 1993). Hydrolysis occurs at a rate of 171/M·h at 25 °C (Sharma and Sharma, 1970). No measurable hydrolysis was observed at 85.0 °C (pH 7) and 25 °C (pH 7.07). Hydrolysis half-lives of 9 and 134 min were observed at 66.0 °C (pH 9.86) and 25.0 °C (pH 11.3), respectively (Ellington et al., 1987).

Exposure limits: NIOSH REL: TWA 100 ppm (410 mg/m^3), IDLH 1,000 ppm; OSHA PEL: TWA 100 ppm; ACGIH TLV: TWA 100 ppm with intended TWA and STEL values of 50 and 100 ppm, respectively.

Symptoms of exposure: An irritation concentration of 697.00 mg/m^3 in air was reported by Ruth (1986).

Toxicity:
   LC_{50} (96-h flow-through assay) for juvenile bluegill sunfish 191 mg/L (Bailey et al., 1985).
   LC_{50} (96-h) for *Selenastrum capricornutum* 170 mg/L (Forbis, 1990).
   Acute oral LD_{50} for guinea pigs 6,300 mg/kg, mice 5,204 mg/kg, rats 7,872 mg/kg (quoted, RTECS, 1985).

Uses: Manufacturing methacrylate resins and plastics (e.g., Plexiglas and Lucite); impregnation of concrete.

2-METHYLNAPHTHALENE

Synonym: β-Methylnaphthalene.

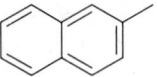

CAS Registry Number: 91-57-6
Molecular formula: $C_{11}H_{10}$
Formula weight: 142.20
RTECS: QJ9635000

Physical state:
Solid

Melting point (°C):
34.6 (Weast, 1986)

Boiling point (°C):
241.052 (Wilhoit and Zwolinski, 1971)

Density (g/cm³):
1.0058 at 20/4 °C (Weast, 1986)

Diffusivity in water (x 10^{-5} cm²/sec):
0.72 at 20 °C using method of Hayduk and Laudie (1974)

Dissociation constant, pK_a:
>15 (Christensen et al., 1975)

Flash point (°C):
97 (Aldrich, 1988)

Entropy of fusion (cal/mol·K):
9.38 (Tsonopoulos and Prausnitz, 1971)

Heat of fusion (kcal/mol):
2.808 (Dean, 1987)
2.85 (Tsonopoulos and Prausnitz, 1971)

Henry's law constant (x 10^{-4} atm·m³/mol at 25 °C):
3.18 (Fendinger and Glotfelty, 1990)
4.54 (de Maagd et al., 1998)
2,000, 2,260, and 2,590 at 26.0, 35.8, and 46.0 °C, respectively (Hansen et al., 1995)
1.28, 2.07, 3.28, 5.06, and 7.23 at 4.1, 11.0, 18.0, 25.0, and 31.0 °C, respectively (Bamford et al., 1998)

Ionization potential (eV):
7.91 (Lias, 1998)
8.48 (Yoshida et al., 1983a)

Bioconcentration factor, log BCF:
3.79 (Davies and Dobbs, 1984)

Soil sorption coefficient, log K_{oc}:
3.93 (Abdul et al., 1987)
3.40 (estuarine sediment, Vowles and Mantoura, 1987)
3.66 (Tinker), 3.23 (Carswell), 2.96 (Barksdale), 3.29 (Blytheville), 3.83 (Traverse City), 3.51
(Borden), 3.40 (Lula) (Stauffer et al., 1989).

Octanol/water partition coefficient, log K_{ow}:
4.11 (Abdul et al., 1987)
3.86 (Hansch and Leo, 1979; Yoshida et al., 1983)
3.864 (Krishnamurthy and Wasik, 1978)

Solubility in organics:
Soluble in most solvents (U.S. EPA, 1985).

Solubility in water:
24.6 mg/kg at 25 °C (Eganhouse and Calder, 1976)
25.4 mg/L at 25 °C (Mackay and Shiu, 1977)
1.53 mg/L (water soluble fraction of a 15-component simulated jet fuel mixture (JP-8) containing 6.3 wt % 2-methylnaphthalene (MacIntyre and deFur, 1985)
27.3 mg/L at 25 °C (Vadas et al., 1991)
20.0 mg/L at 25 °C (Vozňáková et al., 1978)

Vapor pressure (x 10^{-2} mmHg at 25 °C):
5.1 (calculated, Gherini et al., 1988)
5.4 (extrapolated, Mackay et al., 1982)

Environmental fate:
   Biological. 2-Naphthoic acid was reported as the biooxidation product of 2-methyl-
naphthalene by *Nocardia* sp. in soil using *n*-hexadecane as the substrate (Keck et al., 1989).
   Estimated half-lives of 2-methylnaphthalene (0.6 μg/L) from an experimental marine mesocosm during the spring (8-16 °C), summer (20-22 °C), and winter (3-7 °C) were 11, 1.0, and
13 d, respectively (Wakeham et al., 1983).
   Photolytic. Based upon a rate constant of 0.042/h, the photolytic half-life of 2-methyl-
naphthalene in water is 16.4 h (Fukuda et al., 1988).
   Chemical/Physical. An aqueous solution containing chlorine dioxide in the dark for 3.5 d at
room temperature oxidized 2-methylnaphthalene into the following: 1-chloro-2-methyl-
naphthalene, 3-chloro-2-methylnaphthalene, 1,3-dichloro-2-methylnaphthalene, 3-hydroxy-
methylnaphthalene, 2-naphthaldehyde, 2-naphthoic acid, and 2-methyl-1,4-naphthoquinone
(Taymaz et al., 1979).

Toxicity:
   Acute oral LD_{50} for rats 1,630 mg/kg (quoted, RTECS, 1985).

Source: Detected in distilled water-soluble fractions of No. 2 fuel oil (0.42 mg/L), jet fuel A
(0.17 mg/L), diesel fuel (0.27 mg/L), military jet fuel JP-4 (0.07 mg/L) (Potter, 1996), new
motor oil (0.42-0.66 μg/L), and used motor oil (46-54 μg/L) (Chen et al., 1994). Present in
diesel fuel and corresponding aqueous phase (distilled water) at concentrations of 3.5-9.0 g/L
and 180-340 μg/L, respectively (Lee et al., 1992). Schauer et al. (1999) reported 2-methyl-

naphthalene in diesel fuel at a concentration of 980 μg/g and in a diesel-powered medium-duty truck exhaust at an emission rate of 511 μg/km.

Thomas and Delfino (1991) equilibrated contaminant-free groundwater collected from Gainesville, FL with individual fractions of three individual petroleum products at 24-25 °C for 24 h. The aqueous phase was analyzed for organic compounds via U.S. EPA approved test method 625. Average 2-methylnaphthalene concentrations reported in water-soluble fractions of unleaded gasoline, kerosene, and diesel fuel were 256, 354, and 267 μg/L, respectively.

Based on analysis of 7 coal tar samples, 2-methylnaphthalene concentrations ranged from 680 to 42,000 ppm (EPRI, 1990). Detected in 1-yr aged coal tar film and bulk coal tar at concentrations of 25,000 and 26,000 mg/kg, respectively (Nelson et al., 1996).

Uses: Organic synthesis; insecticides; jet fuel component. Derived from industrial and experimental coal gasification operations where the maximum concentrations detected in gas, liquid, and coal tar streams were 2.1, 0.22, and 10 mg/m^3, respectively (Cleland, 1981).

4-METHYLOCTANE

Synonym: 5-Methyloctane.

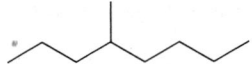

CAS Registry Number: 2216-34-4
Molecular formula: C_9H_{20}
Formula weight: 128.26

Physical state:
Liquid

Melting point (°C):
-113.2 (Weast, 1986)

Boiling point (°C):
142.4 (Weast, 1986)

Density (g/cm³):
0.7199 at 20/4 °C, 0.7169 at 25/4 °C (Dreisbach, 1959)

Diffusivity in water (x 10^{-5} cm²/sec):
0.63 at 20 °C using method of Hayduk and Laudie (1974)

Dissociation constant, pK_a:
>14 (Schwarzenbach et al., 1993)

Henry's law constant (atm·m³/mol):
10 at 25 °C (Mackay and Shiu, 1981)

Soil sorption coefficient, log K_{oc}:
Unavailable because experimental methods for estimation of this parameter for aliphatic hydrocarbons are lacking in the documented literature

Octanol/water partition coefficient, log K_{ow}:
4.69 using method of Hansch et al. (1968)

Solubility in organics:
Soluble in acetone, alcohol, benzene, and ether (Weast, 1986)

Solubility in water:
115 μg/kg at 25 °C (Price, 1976)

Vapor density:
5.24 g/L at 25 °C, 4.43 (air = 1)

Vapor pressure (mmHg):
7 at 25 °C (Wilhoit and Zwolinski, 1971)

Environmental fate:

Chemical/Physical. Complete combustion in air gives carbon dioxide and water vapor. 2-Methyloctane will not hydrolyze because it does not contain a hydrolyzable functional group.

Exposure limits: ACGIH TLV: TWA for all isomers 200 ppm (adopted).

Uses: Gasoline component; organic synthesis.

2-METHYLPENTANE

Synonyms: Dimethylpropylmethane; Isohexane; UN 2462.

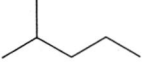

CAS Registry Number: 107-83-5
DOT: 2462
DOT label: Flammable liquid
Molecular formula: C_6H_{14}
Formula weight: 86.18
RTECS: SA2995000

Physical state and color:
Colorless liquid

Melting point (°C):
-153.7 (Weast, 1986)

Boiling point (°C):
60.3 (Weast, 1986)

Density (g/cm³):
0.63215 at 20/4 °C, 0.64852 at 25/4 °C (Dreisbach, 1959)

Diffusivity in water (x 10^{-5} cm²/sec):
0.75 at 20 °C using method of Hayduk and Laudie (1974)

Dissociation constant, pK_a:
>14 (Schwarzenbach et al., 1993)

Flash point (°C):
-23.3 (Hawley, 1981)
-7 (NFPA, 1984)

Lower explosive limit (%):
1.2 (NFPA, 1984)

Upper explosive limit (%):
7.0 (NFPA, 1984)

Heat of fusion (kcal/mol):
1.498 (Riddick et al., 1986)

Henry's law constant (atm·m³/mol):
0.697, 0.694, 0.633, 0.825, and 0.848 at 10, 15, 20, 25, and 30 °C, respectively (Ashworth et al., 1988)

Interfacial tension with water (dyn/cm at 25 °C):
48.9 (quoted, Freitas et al., 1997)

Ionization potential (eV):
10.04 (Lias and Liebman, 1998)
10.34 (Collin and Lossing, 1959)

Soil sorption coefficient, log K_{oc}:
Unavailable because experimental methods for estimation of this parameter for aliphatic hydrocarbons are lacking in the documented literature

Octanol/water partition coefficient, log K_{ow}:
2.77 (Coates et al., 1985)

Solubility in organics:
Soluble in acetone, alcohol, benzene, chloroform, and ether (Weast, 1986)

Solubility in water:
14.0 mg/L at 23 °C (Coates et al., 1985)
In mg/kg: 13.0 at 25 °C, 13.8 at 40.1 °C, 15.7 at 55.7 °C, 27.1 at 99.1 °C, 44.9 at 118.0 °C, 86.8 at 137.3 °C, 113.0 at 149.5 °C (Price, 1976)
13.8 mg/kg at 25 °C (McAuliffe, 1963, 1966)
19.45 mg/kg at 0 °C, 15.7 mg/kg at 25 °C (Polak and Lu, 1973)
14.2 mg/L at 25 °C (Leinonen and Mackay, 1973)
16.21 mg/L at 25 °C (Barone et al., 1966)

Vapor density:
3.52 g/L at 25 °C, 2.98 (air = 1)

Vapor pressure (mmHg):
211.8 at 25 °C (Wilhoit and Zwolinski, 1971)

Environmental fate:
Photolytic. When synthetic air containing gaseous nitrous acid and 2-methylpentane was exposed to artificial sunlight (λ = 300-450 nm), acetone, propionaldehyde, peroxyacetal nitrate, peroxypropionyl nitrate, and possibly two isomers of hexyl nitrate and propyl nitrate formed as products (Cox et al., 1980).

Based on a photooxidation rate constant of 5.6 x 10^{-12} cm^3/molecule·sec for the reaction of 2,3-dimethylbutane and OH radicals, the atmospheric lifetime is 25 h (Altshuller, 1991).

Chemical/Physical: Complete combustion is air yields carbon dioxide and water vapor. 2-Methylpentane will not hydrolyze because it does not contain a hydrolyzable functional group.

Exposure limits: ACGIH TLV: TWA and STEL for all isomers except *n*-hexane are 500 and 1,000 ppm, respectively (adopted).

Symptoms of exposure: Inhalation of vapors may cause irritation of respiratory tract (Patnaik, 1992)

Source: Schauer et al. (1999) reported 2-methylpentane in a diesel-powered medium-duty truck exhaust at an emission rate of 930 μg/km.

Uses: Solvent; gasoline component; organic synthesis.

3-METHYLPENTANE

Synonyms: Diethylmethylmethane; UN 2462.

CAS Registry Number: 96-14-0
DOT: 2462
DOT label: Combustible liquid
Molecular formula: C_6H_{14}
Formula weight: 86.18
RTECS: SA2995500

Physical state and color:
Colorless liquid

Melting point (°C):
-117.8 (Exxon Corp., 1985)

Boiling point (°C):
63.3 (Weast, 1986)

Density (g/cm³):
0.66431 at 20/4 °C, 0.65976 at 25/4 °C (Dreisbach, 1959)

Diffusivity in water (x 10^{-5} cm²/sec):
0.76 at 20 °C using method of Hayduk and Laudie (1974)

Dissociation constant, pK_a:
>14 (Schwarzenbach et al., 1993)

Flash point (°C):
<-6.6 (Hawley, 1981)

Lower explosive limit (%):
1.2 (NFPA, 1984)

Upper explosive limit (%):
7.0 (NFPA, 1984)

Entropy of fusion (cal/mol·K):
12.53 (Douslin and Huffman, 1946)

Heat of fusion (kcal/mol):
1.2675 (Riddick et al., 1986)

Henry's law constant (atm·m³/mol):
1.693 at 25 °C (Hine and Mookerjee, 1975)

Interfacial tension with water (dyn/cm at 25 °C):
49.9 (quoted, Freitas et al., 1997)

Ionization potential (eV):
10.04 (Lias and Liebman, 1998)
10.30 (Collin and Lossing, 1959)

Soil sorption coefficient, log K_{oc}:
Unavailable because experimental methods for estimation of this parameter for aliphatic hydrocarbons are lacking in the documented literature

Octanol/water partition coefficient, log K_{ow}:
2.88 (Coates et al., 1985)

Solubility in organics:
In methanol, g/L: 389 at 5 °C, 450 at 10 °C, 530 at 15 °C, 650 at 20 °C, 910 at 25 °C. Miscible at higher temperatures (Kiser et al., 1961).

Solubility in water:
10.5 mg/L at 23 °C (Coates et al., 1985)
13.1 mg/kg at 25 °C (Price, 1976)
12.8 mg/kg at 25 °C (McAuliffe, 1966)
21.5 mg/kg at 0 °C, 17.9 mg/kg at 25 °C (Polak and Lu, 1973)

Vapor density:
3.52 g/L at 25 °C, 2.98 (air = 1)

Vapor pressure (mmHg):
217.8 at 23.2 °C (Willingham et al., 1945)

Environmental fate:
   Chemical/Physical. Complete combustion in air gives carbon dioxide and water vapor. 3-Methylpentane will not hydrolyze because it does not contain a hydrolyzable functional group.
   Photolytic. The following rate constants were reported for the reaction of 3-methylpentane and OH radicals in the atmosphere: 4.30×10^{-9} cm³/molecule·sec at 300 K (Darnall et al., 1976); 6.8×10^{-12} cm³/molecule·sec at 305 K (Darnall et al., 1978); 5.7×10^{-12} cm³/molecule·sec (Altshuller, 1991).

Exposure limits: ACGIH TLV: TWA and STEL for all isomers except *n*-hexane are 500 and 1,000 ppm, respectively (adopted).

Symptoms of exposure: Inhalation of vapors may cause irritation to respiratory tract. Narcotic at high concentrations (Patnaik, 1992).

Source: Schauer et al. (1999) reported 3-methylpentane in a diesel-powered medium-duty truck exhaust at an emission rate of 670 μg/km.

Uses: Solvent; gasoline component; organic synthesis.

4-METHYL-2-PENTANONE

Synonyms: Hexanone; Hexone; Isobutyl methyl ketone; Isopropylacetone; Methyl isobutyl ketone; 2-Methyl-4-pentanone; MIBK; MIK; RCRA waste number U161; Shell MIBK; UN 1245.

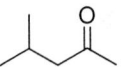

CAS Registry Number: 108-10-1
DOT: 1245
DOT label: Flammable liquid
Molecular formula: $C_6H_{12}O$
Formula weight: 100.16
RTECS: SA9275000
Merck reference: 10, 5056

Physical state, color, and odor:
Clear, colorless, watery liquid with a mild, pleasant odor. Odor threshold concentration is 100 ppb (Keith and Walters, 1992).

Melting point (°C):
-84.7 (Weast, 1986)

Boiling point (°C):
116.8 (Weast, 1986)

Density (g/cm³ at 20/4 °C):
0.7978 (Weast, 1986)
0.8008 (Huntress and Mulliken, 1941)

Diffusivity in water (x 10⁻⁵ cm²/sec):
0.77 at 20 °C using method of Hayduk and Laudie (1974)

Flash point (°C):
17.9 (NIOSH, 1997)

Lower explosive limit (%):
1.2 at 94 °C (NIOSH, 1997)

Upper explosive limit (%):
8.0 at 94 °C (NIOSH, 1997)

Henry's law constant (x 10⁻⁴ atm·m³/mol):
6.6, 3.7, 2.9, 3.9, and 6.8 at 10, 15, 20, 25, and 30 °C, respectively (Ashworth et al., 1988)

Interfacial tension with water (dyn/cm):
10.1 at 25 °C (Donahue and Bartell, 1952)

Ionization potential (eV):
9.30 (Franklin et al., 1969)

Soil sorption coefficient, log K_{oc}:
0.79 (estimated, Montgomery, 1989)

Octanol/water partition coefficient, log K_{ow}:
1.31 (Sangster, 1989)
1.09 (Hansch et al., 1968)

Solubility in organics:
Soluble in acetone, ethanol, benzene, chloroform, ether, and many other solvents (U.S. EPA, 1985)

Solubility in water:
0.097 at 25 °C (mole fraction, Amidon et al., 1975)
In wt %: 2.04 at 20 °C, 1.91 at 25 °C, 1.78 at 30 °C (Ginnings et al., 1940)
1.77 wt % at 25 °C (Lo et al., 1986)
In wt %: 2.92 at 0 °C, 2.21 at 9.5 °C, 1.92 at 19.4 °C, 1.66 at 30.8 °C, 1.47 at 39.6 °C, 1.38 at 50.1 °C, 1.29 at 60.4 °C, 1.24 at 70.2 °C, 1.18 at 80.1 °C, 1.22 at 90.4 °C (Stephenson, 1992)

Vapor density:
4.09 g/L at 25 °C, 3.46 (air = 1)

Vapor pressure (mmHg):
14.5 at 20 °C (Howard, 1990)
19.9 at 25 °C (Banerjee et al., 1990)

Environmental fate:
Biological. Bridié et al. (1979) reported BOD and COD values 2.06 and 2.16 g/g using filtered effluent from a biological sanitary waste treatment plant. These values were determined using a standard dilution method at 20 °C and stirred for a period of 5 d. Heukelekian and Rand (1955) reported a 5-d BOD value of 1.51 g/g which is 55.5% of the ThOD value of 2.72 g/g.

Photolytic. When synthetic air containing gaseous nitrous acid and 4-methyl-2-pentanone was exposed to artificial sunlight (λ = 300-450 nm), photooxidation products identified were acetone, peroxyacetal nitrate, and methyl nitrate (Cox et al., 1980). In a subsequent experiment, the hydroxyl-initiated photooxidation of 4-methyl-2-pentanone in a smog chamber produced acetone (90% yield) and peroxyacetal nitrate (Cox et al., 1981). Irradiation at 3130 Å resulted in the formation of acetone, propyldiene, and free radicals (Calvert and Pitts, 1966).

Second-order photooxidation rate constants for the reaction of 4-methyl-2-butanone and OH radicals in the atmosphere are 1.4×10^{-10}, 1.42×10^{-10}, and 1.32×10^{-10} cm^3/molecule·sec at 295, 299, and 300 K, respectively (Atkinson, 1985). The atmospheric lifetime was estimated to be 1 to 5 d (Kelly et al., 1994).

Chemical/Physical. 4-Methyl-2-pentanone will not hydrolyze in water because it does not contain a hydrolyzable functional group (Kollig, 1993).

At an influent concentration of 1,000 mg/L, treatment with granular activated carbon resulted in an effluent concentration of 152 mg/L. The adsorbability of the carbon used was 169 mg/g carbon (Guisti et al., 1974).

Complete combustion in air yields carbon dioxide and water vapor.

Exposure limits: NIOSH REL: TWA 50 ppm (205 mg/m^3), STEL 75 ppm (300 mg/m^3), IDLH 500 ppm; OSHA PEL: TWA 100 ppm (410 mg/m^3); ACGIH TLV: TWA 50 ppm, STEL 75 ppm (adopted).

Symptoms of exposure: Mild irritant and strong narcotic (Patnaik, 1992). An irritation concentration of 410.00 mg/m^3 in air was reported by Ruth (1986).

Toxicity:

EC_{50} (48-h) and EC_{50} (24-h) values for *Spirostomum ambiguum* were 9,830 and 1,482 mg/L, respectively (Nałecz-Jawecki and Sawicki, 1999).

LC_{50} (96-h) for fathead minnows 505 mg/L (Veith et al., 1983).

LC_{50} (48-h) and LC_{50} (24-h) values for *Spirostomum ambiguum* were 3,602 and 3,621 mg/L, respectively (Nałecz-Jawecki and Sawicki, 1999).

LC_{50} (inhalation) for rats 8,000 ppm/4-h (quoted, RTECS, 1985).

Acute oral LD_{50} for mice 2,671 mg/kg, rats 2,080 mg/kg (quoted, RTECS, 1985).

Uses: Denaturant for ethyl alcohol; solvent for paints, varnishes, cellulose acetate, nitrocellulose lacquers, resins, fats, oils, and waxes; preparation of methyl amyl alcohol; in hydraulic fluids and antifreeze; extraction of uranium from fission products; organic synthesis.

2-METHYL-1-PENTENE

Synonyms: 2-Methylpentene; 1-Methyl-1-propylethene; 1-Methyl-1-propylethylene.

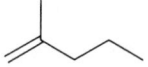

CAS Registry Number: 763-29-1
DOT: 2288
Molecular formula: C_6H_{12}
Formula weight: 84.16
RTECS: SB2230000

Physical state and color:
Colorless liquid

Melting point (°C):
-135.7 (Weast, 1986)

Boiling point (°C):
60.7 (Weast, 1986)
61.5 (Verschueren, 1983)
62.6 (Hawley, 1981)
62.11 (Wilhoit and Zwolinski, 1971)

Density (g/cm³):
0.6799 at 20/4 °C, 0.6751 at 25/4 °C (Dreisbach, 1959)

Diffusivity in water (x 10^{-5} cm²/sec):
0.78 at 20 °C using method of Hayduk and Laudie (1974)

Dissociation constant, pK_a:
>14 (Schwarzenbach et al., 1993)

Flash point (°C):
-26.1 (Dean, 1987)

Henry's law constant (atm·m³/mol):
0.28 at 25 °C (Mackay and Shiu, 1981)

Ionization potential (eV):
9.08 (Lias, 1998)

Soil sorption coefficient, log K_{oc}:
Unavailable because experimental methods for estimation of this parameter for aliphatic hydrocarbons are lacking in the documented literature

Octanol/water partition coefficient, log K_{ow}:
2.54 using method of Hansch et al. (1968)

Solubility in organics:
Soluble in alcohol, benzene, chloroform, and petroleum (Weast, 1986)

Solubility in water:
78 mg/kg at 25 °C (McAuliffe, 1966)

Vapor density:
3.44 g/L at 25 °C, 2.91 (air = 1)

Vapor pressure (mmHg):
195.4 at 25 °C (Wilhoit and Zwolinski, 1971)

Environmental fate:
   Photolytic. The reported reaction rate constants for the reaction of 2-methyl-1-pentene with OH radicals and ozone in the atmosphere are 1.05 x 10^{-17} and 6.26 x 10^{-11} cm^3/molecule·sec, respectively (Atkinson and Carter, 1984; Atkinson, 1985).

Toxicity:
   LC_{50} (inhalation) for mice 127 g/m^3/2-h, rats 115 g/^{3}/4-h (quoted, RTECS, 1985).

Uses: Flavors; perfumes; medicines; dyes; oils; resins; organic synthesis.

4-METHYL-1-PENTENE

Synonyms: 1-Isopropyl-2-methylethene; 1-Isopropyl-2-methylethylene.

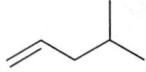

CAS Registry Number: 691-37-2
Molecular formula: C_6H_{12}
Formula weight: 84.16

Physical state and color:
Colorless liquid

Melting point (°C):
-153.6 (Weast, 1986)

Boiling point (°C):
53.9 (Weast, 1986)

Density (g/cm^3):
0.6642 at 20/4 °C, 0.6594 at 25/4 °C (Dreisbach, 1959)

Diffusivity in water (x 10^{-5} cm^2/sec):
0.77 at 20 °C using method of Hayduk and Laudie (1974)

Dissociation constant, pK_a:
>14 (Schwarzenbach et al., 1993)

Flash point (°C):
-31.6 (Hawley, 1981)

Henry's law constant (atm·m^3/mol):
0.615 at 25 °C (Hine and Mookerjee, 1975)

Ionization potential (eV):
9.45 (Lias, 1998)

Soil sorption coefficient, log K_{oc}:
Unavailable because experimental methods for estimation of this parameter for aliphatic hydrocarbons are lacking in the documented literature

Octanol/water partition coefficient, log K_{ow}:
2.70 using method of Hansch et al. (1968)

Solubility in organics:
Soluble in alcohol, benzene, chloroform, and petroleum (Weast, 1986)

Solubility in water:
48 mg/kg at 25 °C (McAuliffe, 1966)

Vapor density:
3.44 g/L at 25 °C, 2.91 (air = 1)

Vapor pressure (mmHg):
270.8 at 25 °C (Wilhoit and Zwolinski, 1971)

Environmental fate:
  Photolytic. Atkinson and Carter (1984) reported a rate constant of 1.06 x 10^{-16} cm³/molecule·sec for the reaction of 4-methyl-1-pentene in the atmosphere.

Uses: Manufacture of plastics used in automobiles, laboratory ware, and electronic components; organic synthesis.

1-METHYLPHENANTHRENE

Synonym: α-Methylphenanthrene.

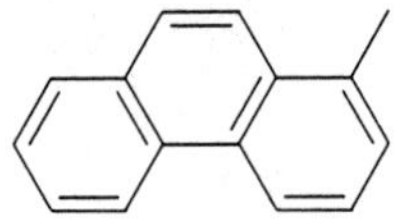

CAS Registry Number: 832-69-9
Molecular formula: $C_{15}H_{12}$
Formula weight: 192.26
RTECS: SF7810000

Physical state and color:
White powder or solid.

Melting point (°C):
123 (Weast, 1986)

Boiling point (°C):
358.6 (Wilhoit and Zwolinski, 1971)

Density (g/cm³):
1.161 using method of Lyman et al. (1982)

Diffusivity in water (x 10⁻⁵ cm²/sec):
0.55 at 20 °C using method of Hayduk and Laudie (1974)

Dissociation constant, pK_a:
>14 (Schwarzenbach et al., 1993)

Ionization potential (eV):
7.70 (Rosenstock et al., 1998)

Henry's law constant (x 10⁻⁵ atm·m³/mol):
1.56, 2.33, 3.42, 4.93, and 6.68 at 4.1, 11.0, 18.0, 25.0, and 31.0 °C, respectively (Bamford et al., 1998)

Soil sorption coefficient, log K_{oc}:
4.56 using method of Karickhoff et al. (1979)

Octanol/water partition coefficient, log K_{ow}:
5.27 using method of Yalkowsky and Valvani (1979)

Solubility in organics:
Soluble in alcohol (Weast, 1986)

Solubility in water:
95.2 μg/L at 6.6 °C, 114 μg/L at 8.9 °C, 147 μg/L at 14.0 °C, 193 μg/L at 19.2 °C, 255 μg/L at 24.1 °C, 304 μg/L at 26.9 °C, 355 μg/L at 29.9 °C (May et al., 1978)
269 μg/L at 25 °C (May et al., 1978a)

173 μg/L at 25 °C, 300 μg/L in seawater at 22 °C (quoted, Verschueren, 1983)

In mole fraction (x 10^{-8}): 0.8921 at 6.60 °C, 1.068 at 8.90 °C, 1.377 at 14.00 °C, 1.808 at 19.20 °C, 2.389 at 24.10 °C, 2.849 at 26.90 °C, 3.326 at 29.90 °C (May et al., 1983)

Source: Detected in 8 diesel fuels at concentrations ranging from 0.10 to 210 mg/L with a mean value of 44.33 mg/L (Westerholm and Li, 1994). Identified in a South Louisiana crude oil at a concentrations of 111 ppm (Pancirov and Brown, 1975). Schauer et al. (1999) reported 1-methylphenanthrene in diesel fuel at a concentration of 28 μg/g and in a diesel-powered medium-duty truck exhaust at an emission rate of 17.0 μg/km.

Uses: Chemical research; organic synthesis.

2-METHYLPHENOL

Synonyms: 2-Cresol; *o*-Cresol; *o*-Cresylic acid; 1-Hydroxy-2-methylbenzene; 2-Hydroxy-toluene; *o*-Hydroxytoluene; 2-Methylhydroxybenzene; *o*-Methylhydroxybenzene; *o*-Methyl-phenol; *o*-Methylphenylol; Orthocresol; *o*-Oxytoluene; RCRA waste number U052; 2-Toluol; *o*-Toluol; UN 2076.

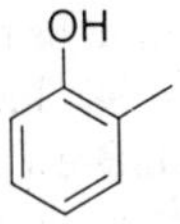

CAS Registry Number: 95-48-7
DOT: 2076
DOT label: Corrosive material, poison
Molecular formula: C_7H_8O
Formula weight: 108.14
RTECS: GO6300000
Merck reference: 10, 2566

Physical state, color, and odor:
Colorless solid or liquid with a phenolic odor; darkens on exposure to air

Melting point (°C):
30.9 (Weast, 1986)

Boiling point (°C):
191.0 (Dean, 1973)

Density (g/cm³ at 20/4 °C):
1.0273 (Weast, 1986)
1.0465 (Standen, 1965)

Diffusivity in water (x 10^{-5} cm²/sec):
0.77 at 20 °C using method of Hayduk and Laudie (1974)

Dissociation constant, pK$_a$:
10.29 (Riddick et al., 1986)

Flash point (°C):
82 (NIOSH, 1997)

Lower explosive limit (%):
1.4 at 150 °C (NIOSH, 1997)

Entropy of fusion (cal/mol·K):
10.99 (Poeti et al., 1982)
12.43 (Andon et al., 1967)

Heat of fusion (kcal/mol):
3.33 (Poeti et al., 1982)
3.78 (Andon et al., 1967)

Henry's law constant (x 10^{-6} atm·m^3/mol):
1.20 at 25 °C (Parsons et al., 1972)

Ionization potential (eV):
8.14 (Mallard and Linstrom, 1998)

Bioconcentration factor, log BCF:
1.03 (*Brachydanio rerio*, Devillers et al., 1996)

Soil sorption coefficient, log K_{oc}:
1.70 (river sediment), 1.75 (coal wastewater sediment) (Kopinke et al., 1995)
1.34 at pH 5.7 (Brookstone clay loam, Boyd, 1982)

Octanol/water partition coefficient, log K_{ow}:
1.93 (Howard, 1989)
1.95 (Leo et al., 1971)
1.99 (Dearden, 1985)
1.96 (Wasik et al., 1981)

Solubility in organics:
Miscible with ethanol, benzene, ether, glycerol (U.S. EPA, 1985)

Solubility in water:
31,000 mg/L at 40 °C, 56,000 mg/L at 100 °C (quoted, Verschueren, 1983)
13.0 g/L at 0 °C, 29.0 g/L at 46.2 °C (Standen, 1965)
30.8 g/L at 40 °C (Howard, 1989)
23 g/L at 8 °C, 26 g/L at 25 °C (quoted, Leuenberger et al., 1985)
23,000 mg/L at 23 °C (Pinal et al., 1990)
25.2 mM at 25.0 °C (Wasik et al., 1981)
22.0 and 30.8 mL/L at 25 and 60 °C, respectively (Booth and Everson, 1949)

Vapor pressure (mmHg):
0.31 at 25 °C (Howard, 1989)
0.045 at 8 °C, 0.29 at 25 °C (quoted, Leuenberger et al., 1985)

Environmental fate:
Biological. Bacterial degradation of 2-methylphenol may introduce a hydroxyl group to produce *m*-methylcatechol (Chapman, 1972). In phenol-acclimated activated sludge, metabolites identified include 3-methylcatechol, 4-methylresorcinol, methylhydroquinone, α-ketobutyric acid, dihydroxybenzaldehyde, and trihydroxytoluene (Masunaga et al., 1986).

Chloroperoxidase, a fungal enzyme isolated from *Caldariomyces fumago*, reacted with 2-methylphenol forming 2-methyl-4-chlorophenol (38% yield) and 2-methyl-6-chlorophenol (Wannstedt et al., 1990).

Heukelekian and Rand (1955) reported a 5-d BOD value of 1.70 g/g which is 67.5% of the ThOD value of 2.72 g/g. In activated sludge inoculum, 95.0% COD removal was achieved. The average rate of biodegradation was 54.0 mg COD/g·h (Pitter, 1976).

Soil. In laboratory microcosm experiments kept under aerobic conditions, half-lives of 5.1 and 1.6 d were reported for 2-methylphenol in an acidic clay soil (<1% organic matter) and slightly basic sandy loam soil (3.25% organic matter) (Loehr and Matthews, 1992).

Surface Water. In river water, the half-life of 2-methylphenol was 2 and 4 d at 20 and 4 °C, respectively (Ludzack and Ettinger, 1960).

Groundwater. Nielsen et al. (1996) studied the degradation of 2-methylphenol in a shallow, glaciofluvial, unconfined sandy aquifer in Jutland, Denmark. As part of the *in situ* microcosm study, a cylinder that was open at the bottom and screened at the top was installed through a cased borehole approximately 5 m below grade. Five liters of water was aerated with atmospheric air to ensure aerobic conditions were maintained. Groundwater was analyzed weekly for approximately 3 months to determine 2-methylphenol concentrations over time. The experimentally determined first-order biodegradation rate constant and corresponding half-life were 0.2/d and 3.5 d, respectively. Groundwater contaminated with phenol and other phenols degraded in a methanogenic aquifer to methane and carbon dioxide. These results could not be duplicated in the laboratory utilizing an anaerobic digester (Godsy et al., 1983).

Photolytic. Sunlight irradiation of 2-methylphenol and nitrogen oxides in air yielded the following gas-phase products: acetaldehyde, formaldehyde, pyruvic acid, peroxyacetylnitrate, nitrocresols, and trace levels of nitric acid and methyl nitrate. Particulate phase products were also identified and these include 2-hydroxy-3-nitrotoluene, 2-hydroxy-5-nitrotoluene, 2-hydroxy-3,5-dinitrotoluene, and tentatively identified nitrocresol isomers (Grosjean, 1984).

Reported rate constants for the reaction of 2-methylphenol and OH radicals in the atmosphere: 2.0×10^{13} cm^3/mol·sec at 300 K (Hendry and Kenley, 1979), 3.7×10^{-11} cm^3/molecule·sec at room temperature (Atkinson, 1985); with ozone in the atmosphere: 2.6×10^{-19} cm^3/molecule·sec at 296 K (Atkinson et al., 1982); with NO$_3$ radicals in the atmosphere: 1×10^{-11} cm^3/molecule·sec (Atkinson and Lloyd, 1984).

Chemical/Physical. Ozonation of an aqueous solution containing 2-methylphenol (200-600 mg/L) yielded formic, acetic, propionic, glyoxylic, oxalic, and salicylic acids (Wang, 1990). In a different experiment, however, an aqueous solution containing 2-methylphenol (1 mM) reacted with ozone (11.7 mg/min) forming 2-methylmuconic acid and hydrogen peroxide as end products. The proposed pathway of degradation involved electrophilic aromatic substitution by the first ozone molecule followed by a 1,3-dipolar addition of the second ozone molecule to the cleaved ring (Beltran et al., 1990).

In a smog chamber experiment, 2-methylphenol reacted with nitrogen oxides to form nitro-cresols, dinitrocresols, and hydroxynitrocresols (McMurry and Grosjean, 1985). Anticipated products from the reaction of 2-methylphenol with ozone or OH radicals in the atmosphere are hydroxynitrotoluenes and ring cleavage compounds (Cupitt, 1980).

Kanno et al. (1982) studied the aqueous reaction of 2-methylphenol (*o*-cresol) and other sub-stituted aromatic hydrocarbons (toluidine, 1-naphthylamine, phenol, *m*- and *p*-cresol, pyro-catechol, resorcinol, hydroquinone, and 1-naphthol) with hypochlorous acid in the presence of ammonium ion. They reported that the aromatic ring was not chlorinated as expected but was cleaved by chloramine forming cyanogen chloride. The amount of cyanogen chloride formed was increased as the pH was lowered (Kanno et al., 1982).

In aqueous solution, 2-methylphenol was degraded by ozone at a reaction rate of 1.4×10^4/M·sec at pH range of 1.5 to 2 (Hoigné and Bader, 1983).

Exposure limits: NIOSH REL: TWA 2.3 ppm (10 mg/m^3), IDLH 250 ppm; OSHA PEL: TWA 5 ppm (22 mg/m^3); ACGIH TLV: TWA for all isomers 5 ppm (adopted).

Symptoms of exposure: May cause weakness, confusion, depression of central nervous system, dyspnea, and respiratory failure. Eye and skin irritant. Contact with skin may cause burns and dermatitis. Chronic effects may include gastrointestinal disorders, nervous disorders, tremor, confusion, skin eruptions, oliguria, jaundice, and liver damage (Patnaik, 1992).

Toxicity:
   Acute oral LD$_{50}$ for rats 121 mg/kg, mice 344 mg/kg (quoted, RTECS, 1985).
   LD$_{50}$ (skin) 620 mg/kg, rats 620 mg/kg, rabbits 890 mg/kg (quoted, RTECS, 1985).

Source: Detected in distilled water-soluble fractions of 87 octane gasoline (6.61 mg/L), 94 octane gasoline (0.57 mg/L), Gasohol (1.17 mg/L), No. 2 fuel oil (2.64 mg/L), jet fuel A (0.72 mg/L), diesel fuel (1.36 mg/L), and military jet fuel JP-4 (1.51 mg/L) (Potter, 1996).

Occurs naturally in white sandlewood, sour cherries, peppermint leaves (1-10 ppb), tarragon, asparagus shoots, tea leaves, coffee beans, Japanese privet, tomatoes, licorice roots, and African palm oil (Duke, 1992).

Uses: Disinfectant; phenolic resins; tricresyl phosphate; ore flotation; textile scouring agent; organic intermediate; manufacturing salicylaldehyde, coumarin, and herbicides; surfactant; synthetic food flavors (*para* isomer only); food antioxidant; dye, perfume, plastics, and resins manufacturing.

4-METHYLPHENOL

Synonyms: 4-Cresol; *p*-Cresol; *p*-Cresylic acid; 1-Hydroxy-4-methylbenzene; *p*-Hydroxy-toluene; 4-Hydroxytoluene; *p*-Kresol; 1-Methyl-4-hydroxybenzene; 4-Methylhydroxybenzene; *p*-Methylhydroxybenzene; *p*-Methylphenol; 4-Oxytoluene; *p*-Oxytoluene; Paracresol; Para-methylphenol; RCRA waste number U052; 4-Toluol; *p*-Toluol; *p*-Tolyl alcohol; UN 2076.

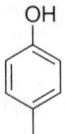

CAS Registry Number: 106-44-5
DOT: 2076
DOT label: Corrosive material, poison
Molecular formula: C_7H_8O
Formula weight: 108.14
RTECS: GO6475000
Merck reference: 10, 2567

Physical state, color, and odor:
Colorless to pink crystals with a phenolic odor. Odor threshold concentration is <1 ppm (Keith and Walters, 1992).

Melting point (°C):
34.8 (Weast, 1986)
36 (Huntress and Mulliken, 1941)

Boiling point (°C):
201.9 (Weast, 1986)

Density (g/cm³ at 20/4 °C):
1.0178 (Weast, 1986)
1.0341 (Standen, 1965)

Diffusivity in water (x 10⁻⁵ cm²/sec):
0.77 at 20 °C using method of Hayduk and Laudie (1974)

Dissociation constant, pK$_a$:
10.26 at 25 °C (Dean, 1973)
10.20 (Hoigné and Bader, 1983)

Flash point (°C):
87 (NIOSH, 1997)

Lower explosive limit (%):
1.1 at 151 °C (NIOSH, 1997)

Entropy of fusion (cal/mol·K):
9.1 (Meva'a and Lichanot, 1990)
9.195 (Poeti et al., 1982)
9.861 (Andon et al., 1967)

Heat of fusion (kcal/mol):
2.8202 (Meva'a and Lichanot, 1990)
2.8410 (Peoti et al., 1982)
3.0370 (Andon et al., 1967)

Henry's law constant (x 10^{-7} atm·m³/mol):
2.20 at 8 °C (estimated, Leuenberger et al., 1985)
7.69 at 25 °C (Parsons et al., 1972)

Ionization potential (eV):
8.34 (Lias, 1998)

Soil sorption coefficient, log K_{oc}:
1.69 (Brookstone clay loam, Boyd, 1982)
3.53 (Apison and Fullerton soils), 2.06 (Dormont soil) (Southworth and Keller, 1986)
2.81 (Coyote Creek sediments, Smith et al., 1978)
2.70 (Meylan et al., 1992)
K_d = 0.9 mL/g on a Cs^+-kaolinite (Haderlein and Schwarzenbach, 1993)

Octanol/water partition coefficient, log K_{ow}:
1.91, 1.95, 1.99, 2.10 (Sangster, 1989)
1.67 (Neely and Blau, 1985)
3.01 (Mackay and Paterson, 1981)
1.92 (Leo et al., 1971)
1.94 (Campbell and Luthy, 1985; Fujita et al., 1964)
1.98 (Garst and Wilson, 1984)

Solubility in organics:
Miscible with ethanol, benzene, ethyl ether, glycerol (U.S. EPA, 1985), ethylene glycol, and toluene.

Solubility in water:
24,000 and 53,000 mg/L at 40 and 100 °C, respectively (quoted, Verschueren, 1983)
22,100, 54,000, and 164,000 mg/L at 30, 105, and 138 °C, respectively (Standen, 1965)
13,000 and 18,000 mg/L at 8 and 25 °C, respectively (quoted, Leuenberger et al., 1985)
230 mM at 25 °C (Southworth and Keller, 1986)

Vapor pressure (x 10^{-2} mmHg):
4 at 20 °C (Verschueren, 1983)
8 at 25 °C (Valsaraj, 1988)
13 at 25 °C (Howard, 1989)
2.0 at 8 °C, 12 at 25 °C (quoted, Leuenberger et al., 1985)

Environmental fate:
   Biological. Protocatechuic acid (3,4-dihydroxybenzoic acid) is the central metabolite in the bacterial degradation of 4-methylphenol. Intermediate by-products include 4-hydroxybenzyl alcohol, 4-hydroxybenzaldehyde, and 4-hydroxybenzoic acid. In addition, 4-methylphenol may undergo hydroxylation to form 4-methylcatechol (Chapman, 1972). Chloroperoxidase, a fungal enzyme isolated from *Caldariomyces fumago*, reacted with 4-methylphenol forming 4-methyl-2-chlorophenol (Wannstedt et al., 1990). Under methanogenic conditions, inocula from a municipal sewage treatment plant digester degraded 4-methylphenol to phenol prior to being

mineralized to carbon dioxide and methane (Young and Rivera, 1985).

A species of *Pseudomonas*, isolated from creosote-contaminated soil, degraded 4-methylphenol into 4-hydroxybenzaldehyde and 4-hydroxybenzoate. Both metabolites were then converted into protocatechuate (O'Reilly and Crawford, 1989). In the presence of suspended natural populations from unpolluted aquatic systems, the second-order microbial transformation rate constant determined in the laboratory was reported to be $2.7 \pm 1.3 \times 10^{-10}$ L/organism·h (Steen, 1991).

Heukelekian and Rand (1955) reported a 5-d BOD value of 1.44 g/g which is 57.2% of the ThOD value of 2.52 g/g. In activated sludge inoculum, 96.0% COD removal was achieved. The average rate of biodegradation was 55.0 mg COD/g·h (Pitter, 1976).

Photolytic. Photooxidation products reported include 2,2'-dihydroxy-4,4'-dimethylbiphenyl, 2-hydroxy-3,4'-dimethylbiphenyl ether, and 4-methyl catechol (Smith et al., 1978). Anticipated products from the reaction of 4-methylphenol with ozone or OH radicals in the atmosphere are hydroxynitrotoluene and ring cleavage compounds (Cupitt, 1980).

Reaction rate constants for the reaction of 4-methylphenol and NO_3 in the atmosphere: 1.3×10^{-11} cm^3/molecule·sec at 300 K (Japar and Niki, 1975), 1×10^{-11} cm^3/molecule·sec at 300 K (Atkinson and Lloyd, 1984), 2.19×10^{-12} cm^3/molecule·sec at 298 K (Atkinson et al., 1988), 1.07×10^{-11} cm^3/molecule·sec at 296 K (Atkinson et al., 1992); with OH radicals in the atmosphere: 3.8×10^{-11} cm^3/molecule·sec (half-life 10 h) at 300 K (Atkinson et al., 1979), 4.50×10^{-11} cm^3/molecule·sec at 298 K (Atkinson and Lloyd, 1984; Atkinson, 1985); with ozone in the atmosphere: 4.71×10^{-19} cm^3/molecule·sec at 296 K (Atkinson et al., 1982).

Chemical/Physical. Kanno et al. (1982) studied the aqueous reaction of 4-methylphenol and other substituted aromatic hydrocarbons (toluidine, 1-naphthylamine, phenol, 2- and 3-methylphenol, pyrocatechol, resorcinol, hydroquinone, and 1-naphthol) with hypochlorous acid in the presence of ammonium ion. They reported that the aromatic ring was not chlorinated as expected but was cleaved by chloramine forming cyanogen chloride. The amount of cyanogen chloride formed was increased as the pH was lowered (Kanno et al., 1982).

In aqueous solution, 4-methylphenol was degraded by ozone at a reaction rate of 3.0×10^4/M·sec at pH range of 1.5 to 2 (Hoigné and Bader, 1983). Reported rate constants for the reaction of 4-methylphenol and singlet oxygen in water at 292 K: 1.1×10^7/M·sec at pH 8.3, 2.4×10^7/M·sec at pH 8.8, 1.6×10^8/M·sec at pH 10, 3.5×10^8/M·sec at pH 11.5 (Scully and Hoigné, 1987), and 9.6×10^6/M·sec for an aqueous phosphate buffer solution at 27 °C (Tratnyek and Hoigné, 1991).

Exposure limits: NIOSH REL: TWA 2.3 ppm (10 mg/m^3), IDLH 250 ppm; OSHA PEL: TWA 5 ppm (22 mg/m^3); ACGIH TLV: TWA for all isomers 5 ppm (adopted).

Symptoms of exposure: May cause weakness, confusion, depression of central nervous system, dyspnea, weak pulse, and respiratory failure. May irritate eyes and mucous membranes. Contact with skin may cause burns and dermatitis. Chronic effects may include gastrointestinal disorders, nervous disorders, tremor, confusion, skin eruptions, oliguria, jaundice, and liver damage (NIOSH, 1997; Patnaik, 1992).

Toxicity:
Acute oral LD_{50} for rats 207 mg/kg, mice 344 mg/kg (quoted, RTECS, 1985).
LD_{50} (skin) for rats 750 mg/kg, rabbits 301 mg/kg (quoted, RTECS, 1985).

Source: As 3+4-methylphenol, detected in distilled water-soluble fractions of 87 octane gasoline (6.03 mg/L), 94 octane gasoline (0.60 mg/L), Gasohol (1.76 mg/L), No. 2 fuel oil (1.84 mg/L), jet fuel A (0.43 mg/L), diesel fuel (1.318 mg/L), and military jet fuel JP-4 (0.92 mg/L) (Potter, 1996).

Occurs naturally in brown juniper, Spanish cedar, peppermint (2-20 ppb), tarragon, asparagus shoots, ylang-ylang, jasmine, tea leaves, coffee beans, Japanese privet, white mulberries, raspberries, vanilla, blueberries, sour cherries, anise, and tamarind (Duke, 1992).

Uses: Disinfectant; phenolic resins; tricresyl phosphate; ore flotation; textile scouring agent; organic intermediate; manufacturing of salicylaldehyde, coumarin, and herbicides; surfactant; synthetic food flavors.

2-METHYLPROPANE

Synonyms: Isobutane; Liquefied petroleum gas; Trimethylmethane; UN 1075; UN 1969.

CAS Registry Number: 75-28-5
DOT: 1011
DOT label: Flammable gas
Molecular formula: C_4H_{10}
Formula weight: 58.12
RTECS: TZ4300000

Physical state, color, and odor:
Colorless gas with a faint odor

Melting point (°C):
-159.4 (Weast, 1986)
-145 (Verschueren, 1983)

Boiling point (°C):
-11.633 (Weast, 1986)
-11.83 (Wilhoit and Zwolinski, 1971)

Density (g/cm³):
0.5572 at 20/4 °C, 0.5510 at 25/4 °C (Dreisbach, 1959)
0.549 at 20/4 °C (Weast, 1986)

Diffusivity in water (x 10⁻⁵ cm²/sec):
0.85 at 20 °C using method of Hayduk and Laudie (1974)

Dissociation constant, pK$_a$:
>14 (Schwarzenbach et al., 1993)

Flash point (°C):
-83 (Hawley, 1981)

Lower explosive limit (%):
1.6 (NIOSH, 1997)

Upper explosive limit (%):
8.4 (NFPA, 1984)

Entropy of fusion (cal/mol·K):
9.541 (Aston et al., 1940)
9.496 (Parks et al., 1937)

Heat of fusion (kcal/mol):
1.085 (Aston et al., 1940)
1.075 (Parks et al., 1937)

Henry's law constant (atm·m³/mol):
1.171 at 25 °C (Hine and Mookerjee, 1975)

Ionization potential (eV):
10.68 (Lias, 1998)
10.79 (Collin and Lossing, 1959)

Soil sorption coefficient, log K_{oc}:
Unavailable because experimental methods for estimation of this parameter for aliphatic hydrocarbons are lacking in the documented literature

Octanol/water partition coefficient, log K_{ow}:
2.76 (Leo et al., 1975)

Solubility in organics (mole fraction):
In 1-butanol: 0.0897, 0.0491, and 0.0308 at 25, 30, and 70 °C, respectively; chlorobenzene: 0.157, 0.0837, and 0.0542 at 25, 30, and 70 °C, respectively and octane: 0.301, 0.161, and 0.101 at 25, 30, and 70 °C, respectively (Hayduk et al., 1988).
In 1-butanol: 0.0889 and 0.0486 at 25 and 70 °C, respectively; in chlorobenzene: 0.162 and 0.0853 at 25 and 70 °C, respectively, and in carbon tetrachloride: 0.238 and 0.132 at 25 and 70 °C, respectively (Blais and Hayduk, 1983).

Solubility in water:
48.9 mg/kg at 25 °C (McAuliffe, 1963, 1966)

Vapor density:
2.38 g/L at 25 °C, 2.01 (air = 1)

Vapor pressure (mmHg):
2,611 at 25 °C (Riddick et al., 1986)

Environmental fate:
Photolytic. Based upon a photooxidation rate constant of 2.34 x 10^{-12} cm³/molecule·sec with OH radicals in summer daylight, the atmospheric lifetime is 59 h (Altshuller, 1991). At atmospheric pressure and 300 K, Darnall et al. (1978) reported a rate constant of 2.52 x 10^{-12} cm³/molecule·sec for the same reaction. Rate constants of 1.28 x 10^9 and 6.03 x 10^{12} L/mol·sec were reported for the reaction of 2-methylpropane with OH radicals in air at 300 and 296 K, respectively (Greiner, 1967, 1970). Rate constants of 7.38 x 10^{-13} and 6.50 x 10^{-17} cm³/molecule·sec were reported for the reaction of 2-methylpropane with OH and NO₃, respectively (Sabljić and Güsten, 1990).

Chemical/Physical. Complete combustion in air gives carbon dioxide and water vapor. 2-Methylpropane will not hydrolyze because it does not contain a hydrolyzable functional group.

Exposure limits: NIOSH REL: TWA 800 ppm (1,900 mg/m³).

Symptoms of exposure: An asphyxiate. Inhalation of concentrations at 1% may cause narcosis and drowsiness (Patnaik, 1992).

Uses: Gasoline component; in liquefied petroleum gas; organic synthesis.

2-METHYLPROPENE

Synonyms: γ-Butylene; *unsym*-Dimethylethylene; Isobutene; Isobutylene; Methylpropene; **2-Methyl-1-propene**; 2-Methylpropylene.

CAS Registry Number: 115-11-7
DOT: 1055
DOT label: Flammable gas
Molecular formula: C_4H_8
Formula weight: 56.11
RTECS: UD0890000
Merck reference: 10, 4987

Physical state, color, and odor:
Volatile liquid or colorless gas with a coal gas odor

Melting point (°C):
-140.3 (Sax and Lewis, 1987)
-146.8 (McAuliffe, 1966)

Boiling point (°C):
-6.900 (Windholz et al., 1983)

Density (g/cm³):
0.5942 at 20/4 °C, 0.5879 at 25/4 °C, 0.5815 at 30/4 °C (Windholz et al., 1983)

Diffusivity in water (x 10^{-5} cm²/sec):
0.91 at 20 °C using method of Hayduk and Laudie (1974)

Dissociation constant, pK_a:
>14 (Schwarzenbach et al., 1993)

Flash point (°C):
-76 (Hawley, 1981)

Lower explosive limit (%):
1.8 (Sax and Lewis, 1987)

Upper explosive limit (%):
9.6 (Sax and Lewis, 1987)

Heat of fusion (kcal/mol):
1.418 (Dean, 1987)

Henry's law constant (atm·m³/mol):
0.21 at 25 °C (Hine and Mookerjee, 1975)

Ionization potential (eV):
9.23 (HNU, 1986)
9.26 (Collin and Lossing, 1959)

Soil sorption coefficient, log K_{oc}:
Unavailable because experimental methods for estimation of this parameter for aliphatic hydrocarbons are lacking in the documented literature

Octanol/water partition coefficient, log K_{ow}:
2.34, 2.40 (Sangster, 1989)

Solubility in organics (mole fraction):
In 1-butanol: 0.131, 0.0695, and 0.0458 at 25, 30, and 70 °C, respectively; chlorobenzene: 0.234, 0.132, and 0.0796 at 25, 30, and 70 °C, respectively; octane: 0.333, 0.184, and 0.119 at 25, 30, and 70 °C, respectively (Hayduk et al., 1988).

Solubility in water:
263 mg/kg at 25 °C (McAuliffe, 1966)

Vapor density:
2.29 g/L at 25 °C, 1.94 (air = 1)

Vapor pressure (mmHg):
1,976 at 20 °C, 2,736 at 30 °C (Verschueren, 1983)
2,270 at 25 °C (Wilhoit and Zwolinski, 1971)

Environmental fate:
   Photolytic. Products identified from the photoirradiation of 2-methylpropene with nitrogen dioxide in air are 2-butanone, 2-methylpropanal, acetone, carbon monoxide, carbon dioxide, methanol, methyl nitrate, and nitric acid (Takeuchi et al., 1983).
   The following rate constants were reported for the reaction of 2-methylpropene and OH radicals in the atmosphere: 3.0×10^{13} cm³/mol·sec at 300 K (Hendry and Kenley, 1979); 5.40×10^{-11} cm³/molecule·sec (Atkinson et al., 1979); 5.14×10^{-11} at 298 K (Atkinson, 1990). Reported reaction rate constants for 2-methylpropene and ozone in the atmosphere include 2.3×10^{-19} cm³/mol·sec (Bufalini and Altshuller, 1965); 1.17×10^{-19} cm³/mol·sec at 300 K (Adeniji and et al., 1965); 1.21×10^{-17} cm³/molecule·sec at 298 K (Atkinson, 1990).

Symptoms of exposure: An asphyxiant (Patnaik, 1992)

Toxicity:
   LC_{50} (inhalation) for mice 415 g/m³/2-h, rats 620 g/m³/4-h (quoted, RTECS, 1985).

Uses: Production of isooctane, butyl rubber, polyisobutene resins, high octane aviation fuels, *t*-butyl chloride, *t*-butyl methacrylates; copolymer resins with acrylonitrile, butadiene, and other unsaturated hydrocarbons; organic synthesis.

α-METHYLSTYRENE

Synonyms: AMS; Isopropenylbenzene; **(1-Methylethenyl)benzene**; 1-Methyl-1-phenylethylene; 2-Phenylpropene; β-Phenylpropene; 2-Phenylpropylene; β-Phenylpropylene.

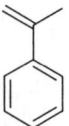

CAS Registry Number: 98-83-9
DOT: 2303
Molecular formula: C_9H_{10}
Formula weight: 118.18
RTECS: WL5250000

Physical state, color, and odor:
Colorless liquid with a sharp aromatic odor. Odor threshold concentration is 290 ppb (Amoore and Hautala, 1983).

Melting point (°C):
24.3 (Weast, 1986)
-96 (Sax and Lewis, 1987)
-23.21 (Hawley, 1981)

Boiling point (°C):
163-164 (Weast, 1986)
152.4 (Sax and Lewis, 1987)
165.38 °C (Hawley, 1981)

Density (g/cm³ at 20/4 °C):
0.9082 (Weast, 1986)
0.862 (Sax and Lewis, 1987)

Diffusivity in water (x 10^{-5} cm²/sec):
0.76 at 20 °C using method of Hayduk and Laudie (1974)

Dissociation constant, pK_a:
>14 (Schwarzenbach et al., 1993)

Flash point (°C):
54.3 (NIOSH, 1997)

Lower explosive limit (%):
1.9 (NIOSH, 1997)

Upper explosive limit (%):
6.1 (NIOSH, 1997)

Entropy of fusion (cal/mol·K):
11.36 (Lebedev and Rabinovich, 1971a)

Heat of fusion (kcal/mol):
2.8499 (Lebedev and Rabinovich, 1971)

Ionization potential (eV):
8.35 ± 0.01 (Franklin et al., 1969)

Solubility in organics:
Soluble in benzene and chloroform (Weast, 1986). Miscible with alcohol and ether (Sax and Lewis, 1987).

Solubility in water:
115.5 mg/L at 25 °C (Deno and Berkheimer, 1960)

Vapor density:
4.83 g/L at 25 °C, 4.08 (air = 1)

Vapor pressure (mmHg):
2 at 20 °C (NIOSH, 1997)

Environmental fate:
   Chemical/Physical. Polymerizes in the presence of heat or catalysts (Hawley, 1981).

Exposure limits: NIOSH REL: TWA 50 ppm (240 mg/m^3), STEL 100 ppm (485 mg/m^3), IDLH 700 ppm; OSHA PEL: ceiling 100 ppm; ACGIH TLV: TWA 50 ppm, STEL 100 ppm (adopted).

Symptoms of exposure: An irritation concentration of 960.00 mg/m^3 in air was reported by Ruth (1986).

Toxicity:
   LC$_{50}$ (inhalation) for mice 3,020 mg/m^3 (quoted, RTECS, 1985).
   Acute oral LD$_{50}$ for mice 3,160 mg/kg, rats 4 g/kg (RTECS, 1985).

Uses: Manufacture of polyesters.

MEVINPHOS

Synonyms: Apavinphos; 2-Butenoic acid 3-[(dimethoxyphosphinyl)oxy]methyl ester; 2-Carbo-methoxy-1-methylvinyl dimethyl phosphate; α-Carbomethoxy-1-methylvinyl dimethyl phosphate; 2-Carbomethoxy-1-propen-2-yl dimethyl phosphate; CMDP; Compound 2046; **3-[(Di-methoxyphosphinyl)oxy]-2-butenoic acid methyl ester**; *O,O*-Dimethyl-*O*-(2-carbomethoxy-1-methylvinyl)phosphate; Dimethyl-1-carbomethoxy-1-propen-2-yl phosphate; *O,O*-Dimethyl 1-carbomethoxy-1-propen-2-yl phosphate; Dimethyl 2-methoxycarbonyl-1-methylvinyl phosphate; Dimethyl methoxycarbonylpropenyl phosphate; Dimethyl (1-methoxycarboxypropen-2-yl)phosphate; *O,O*-Dimethyl *O*-(1-methyl-2-carboxyvinyl)phosphate; Dimethyl phosphate of methyl-3-hydroxy-*cis*-crotonate; Duraphos; ENT 22324; Fosdrin; Gesfid; Gestid; 3-Hydroxy-crotonic acid methyl ester dimethyl phosphate; Meniphos; Menite; 2-Methoxycarbonyl-1-methylvinyl dimethyl phosphate; *cis*-2-Methoxycarbonyl-1-methylvinyl dimethyl phosphate; 1-Methoxycarbonyl-1-propen-2-yl dimethyl phosphate; Methyl 3-(dimethoxyphosphinyloxy)-crotonate; NA 2783; OS 2046; PD 5; Phosdrin; *cis*-Phosdrin; Phosfene; Phosphoric acid (1-methoxycarboxypropen-2-yl) dimethyl ester.

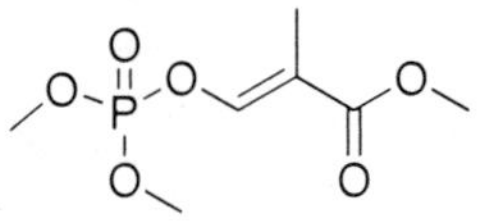

CAS Registry Number: 7786-34-7
DOT: 2783
DOT label: Poison
Molecular formula: $C_7H_{13}O_6P$
Formula weight: 224.16
RTECS: GQ5250000
Merck reference: 10, 6043

Physical state, color, and odor:
Colorless to pale yellow liquid with a weak odor

Melting point (°C):
6.7 (*trans*), 21.3 (*cis*), (NIOSH, 1997)

Boiling point (°C):
106-107.5 at 1 mmHg (Windholz et al., 1983)
99-103 at 0.03 mmHg (Verschueren, 1983)

Density (g/cm³):
1.25 at 20/4 °C (Windholz et al., 1983)

Diffusivity in water (x 10⁻⁵ cm²/sec):
0.63 at 20 °C using method of Hayduk and Laudie (1974)

Flash point (°C):
176 (open cup, NIOSH, 1997)

Soil sorption coefficient, log K_{oc}:
Unavailable because experimental methods for estimation of this parameter for miscible insecticides are lacking in the documented literature. However, its miscibility in water suggests its adsorption to soil will be nominal (Lyman et al., 1982).

Octanol/water partition coefficient, log K_{ow}:
Unavailable because experimental methods for estimation of this parameter for miscible insecticides are lacking in the documented literature

Solubility in organics:
Miscible with acetone, benzene, carbon tetrachloride, chloroform, ethanol, isopropanol, ketones, toluene, and xylene. Soluble in carbon disulfide and kerosene (50 g/L) (Windholz et al., 1983; Worthing and Hance, 1991).

Solubility in water:
Miscible (Gunther et al., 1968)

Vapor density:
9.16 g/L at 25 °C, 7.74 (air = 1)

Vapor pressure (x 10^{-3} mmHg):
2.2 at 20 °C, 5.7 at 29 °C (Freed et al., 1977)

Environmental fate:
   Plant. In plants, mevinphos is hydrolyzed to phosphoric acid dimethyl ester, phosphoric acid, and other less toxic compounds (Hartley and Kidd, 1987). In one day, the compound is almost completely degraded in plants (Cremlyn, 1991). Casida et al. (1956) proposed two degradative pathways of mevinphos in bean plants and cabbage. In the first degradative pathway, cleavage of the vinyl phosphate bond affords methylacetoacetate and acetoacetic acid, which may be precursors to the formation of the end products dimethyl phosphoric acid, methanol, acetone, and carbon dioxide. In the other degradative pathway, direct hydrolysis of the carboxylic ester would yield vinyl phosphates as intermediates. The half-life of mevinphos in bean plants was 0.5 d (Casida et al., 1956). In alfalfa, the half-life was 17 h (Huddelston and Gyrisco, 1961).
   Chemical/Physical. The reported hydrolysis half-lives of *cis*-mevinphos and *trans*-mevinphos at pH 11.6 were 1.8 and 3.0 h, respectively. The volatility half-lives for the *cis* and *trans* forms at 28 °C were 21 and 24 h, respectively (Casida et al., 1956). Worthing and Hance (1991) reported that at pH values of 6, 7, 9, and 11, the hydrolysis half-lives were 120 d, 35 d, 3.0 d, and 1.4 h, respectively (Worthing and Hance, 1991).
   Emits toxic phosphorus oxide fumes when heated to decomposition (Lewis, 1990).

Exposure limits: NIOSH REL: TWA 0.01 ppm (0.1 mg/m³), STEL 0.03 ppm (0.3 mg/m³), IDLH 4 ppm; OSHA PEL: TWA 0.1 mg/m³; ACGIH TLV: TWA 0.9 mg/m³, STEL 0.27 mg/m³ (adopted).
   The acceptable daily intake for humans is 1.5 μg/kg body weight (Worthing and Hance, 1991).

Toxicity:
   EC_{50} (48-h) for *Daphnia pulex* 0.18 μg/L, *Simocephalus serrulatus* 0.46 μg/L (Mayer and Ellersieck, 1986).
   EC_{50} (48-h) for *Daphnia pulex* 0.23 μg/L, *Simocephalus serrulatus* 0.87 μg/L (Mayer and Ellersieck, 1986).
   LC_{50} for fish 0.02-31.6 mg/L (Worthing and Hance, 1991).
   Acute oral LD_{50} for wild birds 1.78 mg/kg, ducks 4.6 mg/kg, mice 4 mg/kg, pigeons 4.21 mg/kg, quail 23.7 mg/kg, rats 3 mg/kg (quoted, RTECS, 1985).
   Acute percutaneous LD_{50} for rats 4-90 mg/kg, rabbits 16-33 mg/kg (Worthing and Hance, 1991).

In 2-yr feeding trials, NOELs were observed in rats and dogs receiving 4 and 5 mg/kg diet, respectively (Worthing and Hance, 1991).

Uses: Contact and systemic insecticide and acaricide used on hops, vegetables, and tobacco for controlling Acarina and Coleoptera (Worthing and Hance, 1991).

MORPHOLINE

Synonyms: Diethyleneimide oxide; Diethyleneimid oxide; Diethylene oximide; Diethylenimide oxide; 1-Oxa-4-azacyclohexane; Tetrahydro-1,4-isoxazine; Tetrahydro-1,4-oxazine; Tetrahydro-2*H*-1,4-oxazine; UN 2054.

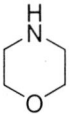

CAS Registry Number: 110-91-8
DOT: 1760 (aqueous), 2054
DOT label: Flammable liquid
Molecular formula: C_4H_9NO
Formula weight: 87.12
RTECS: QD6475000
Merck reference: 10, 6137

Physical state, color, and odor:
Colorless liquid with a weak ammonia-like odor. Hygroscopic. Odor threshold concentration is 10 ppb (Amoore and Hautala, 1983).

Melting point (°C):
-4.7 (Weast, 1986)

Boiling point (°C):
128.3 (Weast, 1986)

Density (g/cm³):
1.0005 at 20/4 °C (Weast, 1986)

Diffusivity in water (x 10⁻⁵ cm²/sec):
0.96 at 20 °C using method of Hayduk and Laudie (1974)

Dissociation constant, pK$_a$:
8.33 at 25 °C (Gordon and Ford, 1972)

Flash point (°C):
37 (NIOSH, 1997)
38 (open cup, Windholz et al., 1983)

Lower explosive limit (%):
1.4 (NFPA, 1984)

Upper explosive limit (%):
11.2 (NFPA, 1984)

Ionization potential (eV):
8.88 (Lias et al., 1998)

Soil sorption coefficient, log K$_{oc}$:
Unavailable because experimental methods for estimation of this parameter for morpholines are

lacking in the documented literature. However, its miscibility in water suggest its adsorption to soil will be nominal (Lyman et al., 1982).

Octanol/water partition coefficient, log K_{ow}:
-1.08 (Leo et al., 1971)

Solubility in organics:
Miscible with acetone, benzene, castor oil, ethanol, ether, ethylene glycol, 2-hexanone, linseed oil, methanol, pine oil, and turpentine (Windholz et al., 1983)

Solubility in water:
Miscible (NIOSH, 1997)

Vapor density:
3.56 g/L at 25 °C, 3.01 (air = 1)

Vapor pressure (mmHg):
6 at 20 °C (NIOSH, 1997)
4.3 at 10 °C, 8.0 at 20 °C, 13.4 at 25 °C (Verschueren, 1983)

Environmental fate:
Biological. Heukelekian and Rand (1955) reported a 5-d BOD value of 0.0 g/g which is 0.0% of the ThOD value of 1.84 g/g.

Chemical/Physical. In an aqueous solution, chloramine reacted with morpholine to form N-chloromorpholine (Isaac and Morris, 1983). The aqueous reaction of nitrogen dioxide (1-99 ppm) and morpholine yielded N-nitromorpholine and N-nitromorpholine (Cooney et al., 1987).

Exposure limits: NIOSH REL: TWA 20 ppm (70 mg/m³), STEL 30 ppm (105 mg/m³), IDLH 1,400 ppm; OSHA PEL: TWA 20 ppm; ACGIH TLV: TWA 20 ppm (adopted).

Symptoms of exposure: Inhalation of vapors may cause visual disturbance, nasal irritation, coughing, and at high concentrations, respiratory distress (Patnaik, 1992).

Toxicity:
LC_{50} (96-h) for rainbow trout 180 mg/L (Spehar et al., 1982).
LC_{50} (inhalation) for mice 1,320 mg/m³/2-h, rats 8,000 ppm/8-h (quoted, RTECS, 1985).
Acute LD_{50} for mice 525 mg/kg, rats 1,050 mg/kg (quoted, RTECS, 1985).
LD_{50} (skin) for rabbits 500 mg/kg (quoted, RTECS, 1985).

Uses: Solvent for waxes, casein, dyes, and resins; rubber accelerator; solvent; optical brightener for detergents; corrosion inhibitor; additive to boiler water; preservation of book paper; organic synthesis.

NALED

Synonyms: Arthodibrom; Bromchlophos; Bromex; Dibrom; 1,2-Dibromo-2,2-dichloroethyldimethyl phosphate; Dimethyl 1,2-dibromo-2,2-dichloroethyl phosphate; *O,O*-Dimethyl-*O*-(1,2-dibromo-2,2-dichloroethyl)phosphate; *O,O*-Dimethyl *O*-(2,2-dichloro-1,2-dibromoethyl)phosphate; ENT 24988; Hibrom; NA 2783; Ortho 4355; Orthodibrom; Orthodibromo; **Phosphoric acid 1,2-dibromo-2,2-dichloroethyl dimethyl ester**; RE-4355.

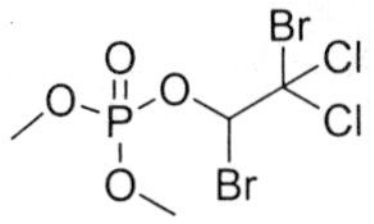

CAS Registry Number: 300-76-5
DOT: 2783
DOT label: Poison
Molecular formula: $C_4H_7Br_2Cl_2O_4P$
Formula weight: 380.79
RTECS: TB9450000
Merck reference: 10, 6204

Physical state, color, and odor:
Colorless to pale yellow liquid or solid with a pungent odor

Melting point (°C):
26.5-27.5 (Windholz et al., 1983)

Boiling point (°C):
110 at 0.5 mmHg (Windholz et al., 1983)

Density (g/cm³):
1.96 at 25/4 °C (Windholz et al., 1983)

Diffusivity in water (x 10^{-5} cm²/sec):
0.68 at 20 °C using method of Hayduk and Laudie (1974)

Flash point (°C):
Noncombustible solid (NIOSH, 1997)

Soil sorption coefficient, log K_{oc}:
Not applicable - reacts with water

Octanol/water partition coefficient, log K_{ow}:
Not applicable - reacts with water

Solubility in organics:
Freely soluble in ketone, alcohols, aromatic and chlorinated hydrocarbons but sparingly soluble in petroleum solvents and mineral oils (Windholz et al., 1983)

Vapor pressure (x 10^{-3} mmHg at 20 °C):
0.2 (NIOSH, 1997)
2 (Verschueren, 1983)

Environmental fate:

Chemical/Physical. Completely hydrolyzed in water within 2 d (Windholz et al., 1983). In the presence of metals or reducing agents, dichlorvos is formed (Worthing and Hance, 1991).

Emits toxic fumes of bromides, chlorides, and phosphorus oxides when heated to decomposition (Lewis, 1990).

Exposure limits (mg/m³): NIOSH REL: TWA 3, IDLH 200; OSHA PEL: TWA 3; ACGIH TLV: TWA 3 (adopted).

Toxicity:

EC_{50} (48-h) for *Daphnia pulex* 0.4 μg/L, *Simocephalus serrulatus* 1.1 μg/L (Mayer and Ellersieck, 1986).

LC_{50} (24-h) for first instar *Toxorhynchites splendens* 623 ppb (Tietze et al., 1993), mosquitofish (*Gambusia affinis*) 3.50 ppm (Tietze et al., 1991).

LC_{50} (inhalation) for mice 156 mg/kg, rats 7.70 mg/kg (quoted, RTECS, 1985).

Acute oral LD_{50} for ducks 52 mg/kg, mice 330 mg/kg, rats 250 mg/kg (quoted, RTECS, 1985).

Uses: Not produced commercially in the United States. Insecticide; acaricide.

NAPHTHALENE

Synonyms: Camphor tar; Mighty 150; Mighty RD1; Moth balls; Moth flakes; Naphthalin; Naphthaline; Naphthene; NCI-C52904; RCRA waste number U165; Tar camphor; UN 1334; White tar.

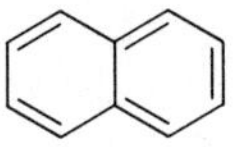

CAS Registry Number: 91-20-3
DOT: 1334 (crude/refined), 2304 (molten)
DOT label: Flammable solid
Molecular formula: $C_{10}H_8$
Formula weight: 128.18
RTECS: QJ0525000
Merck reference: 10, 2830

Physical state, color, and odor:
White, crystalline flakes, or powder with a strong aromatic odor resembling coal-tar or moth balls. The reported odor threshold concentrations ranged from 2.5 to 3 ppb (Keith and Walters, 1992; Young et al., 1996). The taste threshold concentration in water is 25 ppb (Young et al., 1996).

Melting point (°C):
80.5 (Weast, 1986)
80.28 (Fowler et al., 1968)

Boiling point (°C):
217.942 (Wilhoit and Zwolinski, 1971)

Density (g/cm³):
0.9625 at 100/4 °C (Weast, 1986)
1.145 at 20/4 °C (Weiss, 1986)
1.01813 at 30/4 °C, 0.9752 at 85/4 °C (Standen, 1967)

Diffusivity in water (x 10^{-5} cm²/sec):
0.70 at 20 °C using method of Hayduk and Laudie (1974)

Dissociation constant, pK_a:
>15 (Christensen et al., 1975)

Flash point (°C):
79.5 (NIOSH, 1997)

Lower explosive limit (%):
0.9 (NIOSH, 1997)

Upper explosive limit (%):
5.9 (NIOSH, 1997)

Entropy of fusion (cal/mol·K):
13.0 (Spaght et al., 1932)

12.7 (Ueberreiter and Orthmann, 1950)
12.33 (McCullough et al., 1957a)
12.9 (Andrews et al., 1926; David, 1964; Rastogi and Bassi, 1964)
12.84 (Tsonopoulos and Prausnitz, 1971)
12.8 (Syunyaev et al., 1984)

Heat of fusion (kcal/mol):
4.541 (Andrews et al., 1926)
4.589 (Spaght et al., 1932)
4.490 (Ueberreiter and Orthmann, 1950)
4.496 (Mastrangelo, 1957)
4.356 (McCullough et al., 1957a)
4.601 (David, 1964)
4.565 (Rastogi and Bassi, 1964)
4.56 (Wauchope and Getzen, 1972)
4.546 (Syunyaev et al., 1984)

Henry's law constant (x 10^{-4} atm·m^3/mol):
4.6 (Pankow and Rosen, 1988)
4.8 (Valsaraj, 1988)
4.76 at 25 °C (Mackay et al., 1979)
3.6 (Petrasek et al., 1983)
4.44 at 25 °C (de Maagd et al., 1998)
5.53 (Southworth, 1979)
7.34 at 25 °C (Fendinger and Glotfelty, 1990)
4.40 at 25 °C (Shiu and Mackay, 1997)
4.19 at 25 °C (Alaee et al., 1996)
2.94, 2.75, 4.19, 5.74, and 7.59 at 2.0, 6.0, 10.0, 18.0, and 25.0 °C, respectively (Dewulf et al., 1999)

Ionization potential (eV):
8.26 (Krishna and Gupta, 1970; Yoshida et al., 1983a)
8.144 (Lias, 1998)

Bioconcentration factor, log BCF:
2.12 (*Daphnia pulex*, Southworth et al., 1978)
2.63 (fish, Veith et al., 1979)
2.11 (*Chlorella fusca,* Geyer et al., 1984)
1.64 (mussel, Lee et al., 1972)
3.00 (activated sludge), 1.48 (golden ide) (Freitag et al., 1985)
2.50 (bluegill sunfish, McCarthy and Jimenez, 1985)
2.11 (algae, Geyer et al., 1984)
4.10 green alga, *Selenastrum capricornutum* (Casserly et al., 1983)
Apparent values of 3.4 (wet wt) and 5.1 (lipid wt) for freshwater isopods including *Asellus aquaticus* (L.) (van Hattum et al., 1998)
3.15, 3.46, 3.76, 3.90, 4.11, 4.59, 4.82, and 5.45 for olive, holly, grass, ivy, pine, mock orange, rosemary, and pine leaves, respectively (Hiatt, 1998)

Soil sorption coefficient, log K_{oc}:
2.74 (aquifer sands, Abdul et al., 1987)
3.11 (Karickhoff et al., 1979)

3.11, 3.52 (Chin et al., 1988)

2.96 (quoted, Hodson and Williams, 1988)

3.11 (Alfisol, pH 7.5), 3.16 (Entisol, pH 7.9), 3.21 (Speyer soil 2.3, pH 7.1), 3.50 (Speyer soil 2.1, pH 7.0), 4.43 (Speyer soil 2.2, pH 5.8) (Rippen et al., 1982)

2.72-3.32 (average = 3.04 for 10 Danish soils, Løkke, 1984)

2.62 (average for 5 soils, Briggs, 1981)

3.15 (Menlo Park soil), 2.76 (Eustis sand) (Podoll et al., 1989)

2.77 (Appalachee soil, Stauffer and MacIntyre, 1986)

2.93 (estuarine sediment, Vowles and Mantoura, 1987)

2.67 (average for 5 soils, Kishi et al., 1990)

3.02, 3.34 (Allerod), 3.10 (Borris), 3.67, 3.83 (Brande), 3.12 (Finderup), 3.12, 3.44 (Gunderup), 3.87 (Herborg), 3.71 (Rabis), 3.07, 3.21 (Tirstrup), 3.10 (Tylstrup), 3.10 (Vasby), 2.53, 3.56 (Vejen), 3.32, 3.69 (Vorbasse) (Larsen et al., 1992)

3.91 (Tinker), 3.24 (Carswell), 2.73 (Barksdale), 3.74 (Traverse City), 3.24 (Borden), 3.13 (Lula, Stauffer et al., 1989)

3.29 (Eustis fine sand, Wood et al., 1990)

5.00 (average, Kayal and Connell, 1990)

3.18 (Oshtemo, Sun and Boyd, 1993)

3.00 (river sediment), 3.08 (coal wastewater sediment) (Kopinke et al., 1995)

2.92 (light clay, Kishi et al., 1990)

3.32, 3.36, 3.55 (glaciofluvial, sandy aquifer, Nielsen et al., 1996)

4.8 (average value using 8 river bed sediments from the Netherlands, van Hattum et al., 1998)

3.48 (Mt. Lemmon soil, Hu et al., 1995)

Average K_d values for sorption of naphthalene to corundum (α-Al_2O_3) and hematite (α-Fe_2O_3) were 0.00899 and 0.00419 mL/g, respectively (Mader et al., 1997)

Octanol/water partition coefficient, log K_{ow}:

3.40 at 25 °C (Andersson and Schräder, 1999)

3.36 (Karickhoff et al., 1979; Briggs, 1981)

3.59 (Mackay, 1982)

3.23, 3.24, 3.26, 3.28 (Brooke et al., 1986)

3.30 (Campbell and Luthy, 1985; Geyer et al., 1984)

3.31 (Kenaga and Goring, 1980)

3.37 (Hansch and Fujita, 1964)

3.35 (Bruggeman et al., 1982; Wasik et al., 1981, 1983)

3.29 (DeKock and Lord, 1987)

3.43 (Garst and Wilson, 1984)

3.395 (Krishnamurthy and Wasik, 1978)

3.33 at 25 °C (de Maagd et al., 1998)

Solubility in organics:

Soluble in methanol (1 g/13 mL), benzene or toluene (285 g/L), olive oil or turpentine (125 g/L), chloroform, carbon disulfide, or carbon tetrachloride (500 g/L) (Windholz et al., 1983).

66.2, 95.2, 97.8, and 334.0 g/L in methanol (23 °C), ethanol, acetone, and propanol, respectively (Fu et al., 1986).

At 25.0 °C (mol/L): glycerol (0.01052), formamide (0.0539), ethylene glycol (0.0896), methanol (0.579), acetic acid (0.882), acetonitrile (1.715), dimethyl sulfoxide (2.02), acetone (2.75), N,N-dimethylformamide (3.12) (Van Meter and Neumann, 1976).

Solubility in water:

0.239 mM at 25 °C (Wasik et al., 1981, 1983)

30 μg/L at 23 °C (Pinal et al., 1991)

In mg/L: 1.37 at 0 °C, 1.37 at 0.4 °C, 1.38 at 0.5 °C, 1.46 at 0.9 °C, 1.50 at 19 °C, 19.6 at 9.4 °C, 19.4 at 10.0 °C, 23.4 at 14.9 °C, 24.6 at 15.9 °C, 28.0 at 19.3 °C, 34.4 at 25 °C, 35.8 at 25.6 °C, 43.0 at 30.1 °C, 43.9 at 30.2 °C, 54.5 at 35.2 °C, 54.8 at 36.0 °C, 73.5 at 42.8 °C (Bohon and Claussen, 1951)

31.7 mg/L at 25 °C (Mackay and Shiu, 1977)

33.9 mg/L at 22 °C (Coyle et al., 1997)

31.5 mg/kg at 25 °C (Andrews and Keefer, 1949)

22 mg/L at 25 °C (Schwarz and Wasik, 1976)

33.6 mg/L at 25 °C. In 35% NaCl solution: 23.6 mg/kg at 25 °C (Gordon and Thorne, 1967)

31.3 mg/kg at 25 °C. In artificial seawater (salinity = 35 g/kg): 22.0 mg/kg at 25 °C (Eganhouse and Calder, 1976)

In mg/kg: 28.8, 28.8, 29.1 at 22.2 °C, 30.1, 30.7, 30.8 at 24.5 °C, 38.1, 38.2, 38.3 at 29.9 °C, 37.6, 37.6, 38.1 at 30.3 °C, 43.8, 44.6 at 34.5 °C, 52.6, 52.8 at 39.2 °C, 54.8 at 40.1 °C, 65.3, 65.5, 66.0 at 44.7 °C, 78.6 at 50.2 °C, 106 at 55.6 °C, 151, 157, 166 at 64.5 °C, 240, 244, 247 at 73.4 °C (Wauchope and Getzen, 1972)

37.7 mg/L at 20-25 °C (Geyer et al., 1982)

20.315 mg/L at 25 °C (Sahyun, 1966)

0.22 mM at 21 °C (Almgren et al., 1979)

157, 190 and 234 μmol/L at 12, 18 and 25 °C, respectively (Schwarz and Wasik, 1977)

In μmol/L: 140 at 8.4 °C, 149 at 11.1 °C, 166 at 14.0 °C, 188 at 17.5 °C, 207 at 20.2 °C, 222 at 23.2 °C, 236 at 25.0 °C, 248 at 26.3 °C, 268 at 29.2 °C, 283 at 31.8 °C. In 0.5 M NaCl: 84 at 8.4 °C, 92 at 11.1 °C, 109 at 14.0 °C, 123 at 17.1 °C, 137 at 20.0 °C, 158 at 23.0 °C, 173 at 25.0 °C, 222 at 31.8 °C (Schwarz, 1977)

In mg/L: 13.66 at 1 °C, 29.41 at 23 °C, 53.90 at 40 °C (Klöpffer et al., 1988)

In mg/L: 12.07 at 1.9 °C, 17.19 at 10.7 °C, 21.64 at 15.4 °C, 26.72 at 21.7 °C, 30.72 at 25.2 °C, 40.09 at 30.7 °C, 46.31 at 35.1 °C, 54.80 at 39.3 °C, 68.90 at 44.9 °C (Bennett and Canady, 1984)

30 mg/L at 23 °C (Pinal et al., 1990)

12.5 mg/L at 25 °C (Klevens, 1950)

32.9 mg/L at 25 °C (Walters and Luthy, 1984)

300 μmol/L at 25 °C (Edwards et al., 1991)

At 20 °C: 190, 134, 120, and 146 μmol/L in distilled water, Pacific seawater, artificial seawater, and NaCl solution (35 wt %), respectively (Hashimoto et al., 1984)

In NaCl (g/kg) at 25 °C, mg/kg: 30.1 (12.40), 25.2 (25.31), 25.3 (30.59), 20.9 (43.70), 16.9 (61.63) (Paul, 1952)

31.69 mg/kg at 25 °C (May et al., 1978a)

30.6 mg/L at 25 °C (Vadas et al., 1991)

31.3 and 31.9 mg/L at 25 °C (Billington et al., 1988)

235 μmol/L at 25 °C (Akiyoshi et al., 1987)

251 μmol/kg at 25.0 °C (Vesala, 1974)

30.6 mg/L at 23 °C (Fu et al., 1986)

In mM: 0.135 at 5.0 °C, 0.158 at 10.0 °C, 0.190 at 15.0 °C, 0.224 at 20.0 °C, 0.263 at 25.0 °C, 0.324 at 30.0 °C, 0.371 at 35.0 °C, 0.436 at 40.00 °C (Pérez-Tejeda et al., 1990)

254 μmol/L at 25.0 °C (Vesala and Lönnberg, 1980)

234 μmol/L at 25.0 °C (Van Meter and Neumann, 1976)

34,800 μg/L at 25 °C (de Maagd et al., 1998)

29.9 mg/L at 25 °C (Etzweiler et al., 1995)

In mole fraction (x 10^{-5}): 0.2376 at 8.2 °C, 0.2703 at 11.5 °C, 0.2863 at 13.4 °C, 0.3019 at 15.1 °C, 0.3624 at 19.3 °C, 0.4142 at 23.4 °C, 0.4485 at 25.0 °C, 0.4799 at 27.0 °C (May et al., 1983)

Vapor pressure (x 10^{-2} mmHg):

23 at 25 °C (quoted, Mackay and Wolkoff, 1973)

7.8 at 25 °C (Wasik et al., 1983)

17 at 25 °C (Hinckley et al., 1990)

2.32, 4.19, 4.45, and 9.44 at 12.80, 18.40, 18.85, and 26.40 °C, respectively (Macknick and Prausnitz, 1979)

0.122 at 6.70 °C, 0.141 at 8.10 °C, 0.222 at 12.30 °C, 0.235 at 12.70 °C, 0.263 at 13.85 °C, 0.320 at 15.65 °C, 0.350 at 16.85 °C, 0.382 at 17.35 °C, 0.383 at 17.55 °C, 0.438 at 18.70 °C, 0.534 at 20.70 °C (Bradley and Cleasby, 1953)

0.62 at 1 °C, 7.1 at 23 °C, 34 at 40 °C (Klöppfer et al., 1982)

35.4 at 40.33 °C (Fowler et al., 1968)

7.8 at 25 °C (Sonnefeld et al., 1983)

8.2 at 25 °C (Mackay et al., 1982)

21 at 25 °C (Bidleman, 1984)

8.5 at 25 °C (de Kruif, 1980)

10.4 at 25 °C (extrapolated, Tesconi and Yalkowsky, 1998)

Environmental fate:

Biological. In activated sludge, 9.0% of the applied amount mineralized to carbon dioxide after 5 d (Freitag et al., 1985). Under certain conditions, *Pseudomonas* sp. oxidized naphthalene to *cis*-1,2-dihydro-1,2-dihydroxynaphthalene (Dagley, 1972). This metabolite may be further oxidized by *Pseudomonas putida* to carbon dioxide and water (Jerina et al., 1971). Under aerobic conditions, *Cunninghamella elegans* degraded naphthalene to α-naphthol, β-naphthol, *trans*-1,2-dihydroxy-1,2-dihydronaphthalene, 4-hydroxy-1-tetralone, and 1,4-naphthoquinone. Under aerobic conditions, *Agnenellum, Oscillatoria*, and *Anabaena* degraded naphthalene to 1-naphthol, *cis*-1,2-dihydroxyl-1,2-dihydronaphthalene, and 4-hydroxy-1-tetralone (Kobayashi and Rittman, 1982; Riser-Roberts, 1992). *Candida lipolytica, Candida elegans* and species of *Cunninghamella, Syncephalastrum* and *Mucor* oxidized naphthalene to α-naphthol, β-naphthol, *trans*-1,2-dihydroxy-1,2-dihydronaphthalene, 4-hydroxy-1-tetralone, 1,2-naphthoquinone, and 1,4-naphthoquinone (Cerniglia et al., 1978, 1980; Dodge and Gibson, 1980).

Cultures of *Bacillus* sp. oxidized naphthalene to (+)-*trans*-1,2-dihydro-1,2-dihydroxy-naphthalene. In the presence of reduced nicotinamide adeninedinucleotide phosphate (NADPH$_2$) and ferrous ions, a cell extract oxidized naphthalene to *trans*-naphthalenediol (Gibson, 1968). Hydroxylation by pure microbial cultures yielded an unidentified phenol, 1- and 2-hydroxynaphthalene (Smith and Rosazza, 1974).

In a soil-water system under anaerobic conditions, no significant degradation of naphthalene was observed after 45 d. Under denitrification conditions, naphthalene (water concentration 700 µg/L) degraded to non-detectable levels in 45 d. In both studies, the acclimation period was approximately 2 d (Mihelcic and Luthy, 1988).

In a static-culture-flask screening test, naphthalene (5 and 10 mg/L) was statically incubated in the dark at 25 °C with yeast extract and settled domestic wastewater inoculum. After 7 d, 100% biodegradation with rapid adaptation was observed (Tabak et al., 1981). In freshwater sediments, naphthalene biodegraded to *cis*-1,2-dihydroxy-1,2-dihydronaphthalene, α-naphthol, salicylic acid, and catechol.

Contaminated soil from a manufactured coal gas plant that had been exposed to crude oil was spiked with naphthalene (400 mg/kg soil) to which Fenton's reagent (5 mL 2.8 M hydrogen peroxide; 5 mL 0.1 M ferrous sulfate) was added. The treated and nontreated soil samples were incubated at 20 °C for 56 d. Fenton's reagent enhanced the mineralization of naphthalene by indigenous microorganisms. The amounts of naphthalene recovered as carbon dioxide after treatment with and without Fenton's reagent were 62 and 53%, respectively

(Martens and Frankenberger, 1995).

The estimated half-life of naphthalene from an experimental marine mesocosm during the winter (3-7 °C) was 12 d (Wakeham et al., 1983).

Soil. The half-lives of naphthalene in pristine and oil-contaminated sediments are >88 d and 4.9 h, respectively (Herbes and Schwall, 1978). Reported half-lives for naphthalene in a Kidman sandy loam and McLaurin sandy loam are 2.1 and 2.2 d, respectively (Park et al., 1990).

Surface Water. The volatilization half-life of naphthalene from surface water (1 m deep, water velocity 0.5 m/sec, wind velocity 22.5 m/sec) using experimentally determined Henry's law constants is estimated to be 16 h (Southworth, 1979). The reported half-lives of naphthalene in an oil-contaminated estuarine stream, clean estuarine stream, coastal waters, and in the Gulf stream are 7, 24, 63, and 1,700 d, respectively (Lee, 1977). Mackay and Wolkoff (1973) estimated an evaporation half-life of 2.9 h from a surface water body that is 25 °C and 1 m deep.

Groundwater. The estimated half-life of naphthalene in groundwater in the Netherlands was 6 months (Zoeteman et al., 1981). Nielsen et al. (1996) studied the degradation of naphthalene in a shallow, glaciofluvial, unconfined sandy aquifer in Jutland, Denmark. As part of the *in situ* microcosm study, a cylinder that was open at the bottom and screened at the top was installed through a cased borehole approximately 5 m below grade. Five liters of water was aerated with atmospheric air to ensure aerobic conditions were maintained. Groundwater was analyzed weekly for approximately 3 months to determine naphthalene concentrations over time. The experimentally determined first-order biodegradation rate constant and corresponding half-life following a 6-d lag phase were 0.8/d and 20.8 h, respectively.

Photolytic. Irradiation of naphthalene and nitrogen dioxide using a high pressure mercury lamp (λ >290 nm) yielded the following principal products: 1- and 2-hydroxynaphthalene, 1-hydroxy-2-nitronaphthalene, 1-nitronaphthalene, 2,3-dinitronaphthalene, phthalic anhydride, 1,3-, 1,5- and 1,8-dinitronaphthalene (Barlas and Parlar, 1987). In a similar experiment, naphthalene crystals was heated to 50 °C and exposed to pure air containing NO and OH radicals. Photodecomposition followed first-order kinetics indicating the concentration of OH radicals remained constant throughout the reaction. Degradation products identified by GC/MS included: 1-anphthol, 2-naphthol, 1-nitronaphthalene, 2-nitronaphthalene, 1,4-naphthoquinone, 1,4-naphthoquinone-2,3-epoxide, 3-nitrophthalic anhydride, phthalic anhydride, 4-methyl-2*H*-1-benzopyran-2-one, 1(3*H*)-isobenzofuranone, 1,2-benzenecarboxaldehyde, (*E*)-2-formyl-cinnamaldehyde, (*Z*)-2-formylcinnamaldehyde, and phthalide. The following were tentatively identified: 2,7-naphthalenediol, 2-nitro-1-naphthol, 4-nitro-1-naphthol, and 2,4-dinitro-1-naphthol. Photoproducts identified by HPLC included: benzoic acid, cinnamic acid, 2,4-dinitro-1-naphthol, 2-formylcinnamic acid, (*E*)-2-formylcinnamaldehyde, (*Z*)-2-formylcinnamalde-hyde, 1-nitronaphthalene, 2-nitronaphthalene, 1-naphthol, 2-naphthol, 1,4-naphthoquinone, 1,4-naphthoquinone-2,3-epoxide, 3-nitrophthalic anhydride, oxalic acid, phthalic acid, phthal-aldehyde, and phthalide (Lane et al., 1997).

A carbon dioxide yield of 30.0% was achieved when naphthalene adsorbed on silica gel was irradiated with light (λ >290 nm) for 17 h (Freitag et al., 1985).

Fukuda et al. (1988) studied the photolysis of naphthalene in distilled water using a high pressure mercury lamp. After 96 h of irradiation, a rate constant of 0.028/h with a half-life of 25 h was determined. When the experiment was replicated in the presence of various sodium chloride concentrations, they found that the rate of photolysis increased proportionately to the concentration of sodium chloride. The photolysis rates of naphthalene in sodium chloride concentrations of 0.2, 0.3, 0.4, and 0.5 M following 3 h of irradiation were 33.3, 50.6, 91.6, and 99.2%, respectively.

Tuhkanen and Beltrán (1995) studied the decomposition of naphthalene in water using hydrogen peroxide and UV light via a low pressure mercury lamp (λ = 254 nm). Hydrogen peroxide alone did not cause any decrease in naphthalene concentration. However, UV light or UV light/hydrogen peroxide causes photolytic degradation of naphthalene. Intermediates identified from the direct photolysis of naphthalene in solution include 2,4-dimethyl-1,3-penta-diene, bicyclo[4.2.0]octa-1,3,5-triene, benzaldehyde, phenol, 2-hydroxybenzaldehyde, 2,3-di-hydrobenzofuran, 2'-hydroxyacetophenone, and 1(3H)isobenzofuranone. During the oxidation of naphthalene using UV/hydrogen peroxide the following intermediates formed: 2,4-dimethyl-1,3-pentadiene, bicyclo[4.2.0]octa-1,3,5-triene, p-xylene, phenol, and 2-hydroxybenzaldehyde. The researchers concluded that naphthalene was first oxidized to naphthol and then to naphthoquinone, benzaldehyde, phthalic and benzoic acids. Continued irradiation with or without the presence of OH radicals would eventually result in the complete mineralization of naphthalene and its intermediates, i.e., carbon dioxide and water.

Naphthalene (50.0 mg/L) in a methanol-water solution (3:7 v/v) was subjected to a high pressure mercury lamp or sunlight. Based on a rate constant of 6.0 x 10^{-4}/min, the corresponding half-life is 19.18 h (Wang et al., 1991). The estimated photooxidation half-life of naphthalene in the atmosphere via OH radicals is <24 h (Atkinson, 1987).

Chemical/Physical. An aqueous solution containing chlorine dioxide in the dark for 3.5 d oxidized naphthalene to chloronaphthalene, 1,4-dichloronaphthalene, and methyl esters of phthalic acid (Taymaz et al., 1979). In the presence of bromide ions and a chlorinating agent (sodium hypochlorite), major products identified at various reaction times and pHs include 1-bromonaphthalene, dibromonaphthalene, and 2-bromo-1,4-naphthoquinone. Minor products identified include chloronaphthalene, dibromonaphthalene, bromochloronaphthalene, bromo-naphthol, dibromonaphthol, 2-bromonaphthoquinone, dichloronaphthalene, and chlorodibromo-naphthalene (Lin et al., 1984).

The gas-phase reaction of N_2O_5 and naphthalene in an environmental chamber at room temperature resulted in the formation of 1- and 2-nitronaphthalene with approximate yields of 18 and 7.5%, respectively (Pitts et al., 1985). The reaction of naphthalene with NO_x to form nitronaphthalene was reported to occur in urban air from St. Louis, MO (Randahl et al., 1982).

It was suggested that the chlorination of naphthalene in tap water accounted for the presence of chloro- and dichloronaphthalenes (Shiraishi et al., 1985). Kanno et al. (1982) studied the aqueous reaction of naphthalene and other aromatic hydrocarbons (benzene, toluene, o-, m-, and p-xylene) with hypochlorous acid in the presence of ammonium ion. They reported that the aromatic ring was not chlorinated as expected but was cleaved by chloramine forming cyanogen chloride. The amount of cyanogen chloride increased at lower pHs (Kanno et al., 1982).

When naphthalene (2 mg/L) in hydrogen-saturated deionized water was exposed to a slurry of palladium catalyst (1 g/L Pd on alumina) at room temperature for 35 min, 95% was converted to tetrahydronaphthalene (Schüth and Reinhard, 1997).

At influent concentrations of 1.0, 0.1, 0.01, and 0.001 mg/L, the granular activated carbon adsorption capacities at pH 5.6 were 132, 50, 19, and 7.3 mg/g, respectively (Dobbs and Cohen, 1980).

Exposure limits: NIOSH REL: TWA 10 ppm (50 mg/m^3), STEL 15 ppm (75 mg/m^3), IDLH 250 ppm; OSHA PEL: TWA 10 ppm; ACGIH TLV: TWA 10 ppm, STEL 15 ppm (adopted).

Symptoms of exposure: Inhalation of vapors may cause irritation of eyes, skin, respiratory tract, headache, nausea, confusion, and excitement. Ingestion may cause gastrointestinal pain and kidney damage (Patnaik, 1992). An irritation concentration of 75.00 mg/m^3 in air was reported by Ruth (1986).

Toxicity:

EC$_{50}$ (48-h) and EC$_{50}$ (24-h) values for *Spirostomum ambiguum* were 36.4 and 37.6 mg/L, respectively (Nałecz-Jawecki and Sawicki, 1999).

LC$_{50}$ (contact) for earthworm (*Eisenia fetida*) 4,670 μg/cm^2 (Neuhauser et al., 1985).

LC$_{50}$ (96-h) for *Pimephales promelas* 7,900 μg/L (DeGraeve et al., 1982).

LC$_{50}$ (72-h) for embryos of *Micropterus salmoides* 240 μg/L (Black et al., 1983).

LC$_{50}$ (48-h) for *Daphnia magna* 8.6 mg/L (LeBlanc, 1980), *Spirostomum ambiguum* 40.1 mg/L (Nałecz-Jawecki and Sawicki, 1999).

LC$_{50}$ (24-h) for *Daphnia magna* 17 mg/L (LeBlanc, 1980), *Spirostomum ambiguum* 43.7 mg/L (Nałecz-Jawecki and Sawicki, 1999).

Acute oral LD$_{50}$ for guinea pigs 1,200 mg/kg, mice 533 mg/kg, rats 1,250 mg/kg (quoted, RTECS, 1985).

Drinking water standard: No MCLGs or MCLs have been proposed (U.S. EPA, 1996).

Source: Schauer et al. (1999) reported naphthalene in diesel fuel at a concentration of 600 μg/g and in a diesel-powered medium-duty truck exhaust at an emission rate of 617 μg/km. Detected in distilled water-soluble fractions of 87 octane gasoline (0.24 mg/L), 94 octane gasoline (0.21 mg/L), Gasohol (0.29 mg/L), No. 2 fuel oil (0.60 mg/L), jet fuel A (0.34 mg/L), diesel fuel (0.25 mg/L), military jet fuel JP-4 (0.18 mg/L) (Potter, 1996), and used motor oil (116-117 μg/L) (Chen et al., 1994). Lee et al. (1992) investigated the partitioning of aromatic hydrocarbons into water. They reported concentration ranges 350-1,500 mg/L and 80-300 μg/L in diesel fuel and the corresponding aqueous phase (distilled water), respectively.

Thomas and Delfino (1991) equilibrated contaminant-free groundwater collected from Gainesville, FL with individual fractions of three individual petroleum products at 24-25 °C for 24 h. The aqueous phase was analyzed for organic compounds via U.S. EPA approved test method 625. Average naphthalene concentrations reported in water-soluble fractions of unleaded gasoline, kerosene, and diesel fuel were 989, 644, and 167 μg/L, respectively.

Based on analysis of 7 coal tar samples, naphthalene concentrations ranged from 940 to 71,000 ppm (EPRI, 1990). Detected in 1-yr aged coal tar film and bulk coal tar at concentrations of 26,000 and 29,000 mg/kg, respectively (Nelson et al., 1996).

Naphthalene was detected in soot generated from underventilated combustion of natural gas doped with toluene (3 mole %) (Tolocka and Miller, 1995).

Uses: Intermediate for phthalic anhydride, naphthol, 1,4-naphthoquinone, 1,4-dihydronaphthalene, 1,2,3,4-tetrahydronaphthalene (tetralin), decahydronaphthalene (decalin), 1-nitronaphthalene, halogenated naphthalenes, naphthyl, naphthol derivatives, dyes, explosives; mothballs manufacturing; preparation of pesticides, fungicides, detergents and wetting agents, synthetic resins, celluloids, and lubricants; synthetic tanning; preservative; textile chemicals; emulsion breakers; scintillation counters; smokeless powders.

1-NAPHTHYLAMINE

Synonyms: 1-Aminonaphthalene; C.I. azoic diazo component 114; Fast garnet B base; Fast garnet base B; **1-Naphthalenamine**; Naphthalidam; Naphthalidine; α-Naphthylamine; RCRA waste number U167; UN 2077.

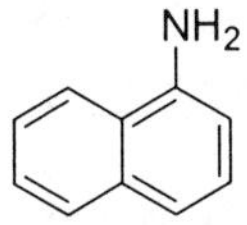

CAS Registry Number: 134-32-7
DOT: 2077
DOT label: Poison
Molecular formula: $C_{10}H_9N$
Formula weight: 143.19
RTECS: QM1400000
Merck reference: 10, 6250

Physical state, color, and odor:
Colorless crystals or yellow, rhombic needles with an unpleasant odor. Becomes purplish-red in color on exposure to air. Odor threshold concentrations ranged from 140 to 290 $\mu g/m^3$ (Keith and Walters, 1992).

Melting point (°C):
50 (Weast, 1986)

Boiling point (°C):
Sublimes at 300.8 (Weast, 1986).

Density (g/cm³):
1.1229 at 25/25 °C (Weast, 1986)

Diffusivity in water (x 10⁻⁵ cm²/sec):
0.67 at 20 °C using method of Hayduk and Laudie (1974)

Dissociation constant, pK_a:
3.92 at 25 °C (Dean, 1973)

Flash point (°C):
158.5 (NIOSH, 1997)

Henry's law constant (x 10⁻¹⁰ atm·m³/mol):
1.27 at 25 °C (Mercer et al., 1990)

Ionization potential (eV):
7.09-7.39 (Mallard and Linstrom, 1998)

Soil sorption coefficient, log K_{oc}:
3.51 (average of 3 soils, Graveel et al., 1986)

Octanol/water partition coefficient, log K_{ow}:
2.27 (Sangster, 1989)

Solubility in organics:
Soluble in alcohol and ether (Weast, 1986)

Solubility in water:
1.6 g/L at 20 °C (Patnaik, 1992)
1,700 mg/L (quoted, Verschueren, 1983)

Vapor pressure (x 10^{-5} mmHg):
100,000 at 158.5 °C (NIOSH, 1997)
6.5 at 20-30 °C (Mercer et al., 1990)

Environmental fate:
   Biological. 1-Naphthylamine added to three different soils was incubated in the dark at 23 °C under a carbon dioxide-free atmosphere. After 308 d, 16.6 to 30.7% of the 1-naphthylamine added to soil biodegraded to carbon dioxide (Graveel et al., 1986).
   Heukelekian and Rand (1955) reported a 5-d BOD value of 0.89 g/g which is 34.6% of the ThOD value of 2.57 g/g. In activated sludge inoculum, following a 20-d adaptation period, no degradation was observed (Pitter, 1976).
   Chemical/Physical. Kanno et al. (1982) studied the aqueous reaction of 1-naphthylamine and other substituted aromatic hydrocarbons (aniline, toluidine, 2-naphthylamine, phenol, cresol, pyrocatechol, resorcinol, hydroquinone, and 1-naphthol) with hypochlorous acid in the presence of ammonium ion. They reported that the aromatic ring was not chlorinated as expected but was cleaved by chloramine forming cyanogen chloride. The amount of cyanogen chloride that formed increased as the pH was lowered (Kanno et al., 1982).
   1-Methylnaphthylamine will not hydrolyze because it does not contain a hydrolyzable functional group (Kollig, 1993).

Exposure limits: Potential occupational carcinogen. Because no standards have been established, NIOSH (1997) recommends the most reliable and protective respirators be used, i.e., a self-contained breathing apparatus that has a full facepiece and is operated under positive-pressure or a supplied-air respirator that has a full facepiece and is operated under pressure-demand or under positive-pressure in combination with a self-contained breathing apparatus operated under pressure-demand or positive-pressure.
   OSHA recommends that worker exposure to this chemical is to be controlled by use of engineering control, proper work practices, and proper selection of personal protective equipment. Specific details of these requirements can be found in CFR 1910.1003-1910.1016.

Symptoms of exposure: Ingestion, skin contact, or inhalation of vapors can cause acute hemorrhagic cystitis, dyspnea, ataxia, dysuria, and hematuria (Patnaik, 1992)

Toxicity:
   LC_{50} (48-h) for red killifish 49 mg/L (Yoshioka et al., 1986).
   Acute oral LD_{50} for rats 779 mg/kg (quoted, RTECS, 1985).
   LD_{50} (intraperitoneal) for mice 96 mg/kg (quoted, RTECS, 1985).

Uses: Manufacture of dyes and dye intermediates; agricultural chemicals.

2-NAPHTHYLAMINE

Synonyms: 2-Aminonaphthalene; C.I. 37270; Fast scarlet base B; 2-Naphthalamine; **2-Naphthalenamine**; β-Naphthylamine; 6-Naphthylamine; 2-Naphthylamine mustard; RCRA waste number U168; UN 1650; USAF CB-22.

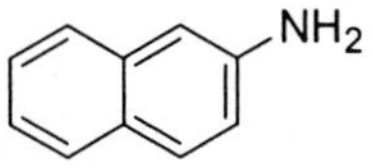

CAS Registry Number: 91-59-8
DOT: 1650
DOT label: Poison
Molecular formula: $C_{10}H_9N$
Formula weight: 143.19
RTECS: QM2100000
Merck reference: 10, 6251

Physical state and color:
Colorless to white crystals. Becomes purplish-red in color on exposure to air. Odor threshold concentrations ranged from 1.4 to 1.9 mg/m³ (Keith and Walters, 1992).

Melting point (°C):
113 (Weast, 1986)
110.2 (Verschueren, 1983)

Boiling point (°C):
306.1 (Weast, 1986)

Density (g/cm³):
1.0614 at 98/4 °C (Weast, 1986)

Diffusivity in water (x 10^{-5} cm²/sec):
0.67 at 20 °C using method of Hayduk and Laudie (1974)

Dissociation constant, pK_a:
4.11 (Dean, 1973)

Flash point (°C):
158.5 (NIOSH, 1997)

Henry's law constant (x 10^{-9} atm·m³/mol):
2.01 at 25 °C (Mercer et al., 1990)

Ionization potential (eV):
7.1 (Lias, 1998)

Soil sorption coefficient, log K_{oc}:
2.11 (calculated, Mercer et al., 1990)

Octanol/water partition coefficient, log K_{ow}:
2.40 (Sangster, 1989)

Solubility in organics:
Soluble in alcohol and ether (Weast, 1986)

Solubility in water:
0.002 wt % at 20 °C (NIOSH, 1997)

Vapor pressure (x 10^{-4} mmHg):
10,000 at 108.6 °C (NIOSH, 1997)
2.56 at 20-30 °C (Mercer et al., 1990)

Environmental fate:
   Chemical/Physical. Kanno et al. (1982) studied the aqueous reaction of 1-naphthylamine and other substituted aromatic hydrocarbons (aniline, toluidine, 2-naphthylamine, phenol, cresol, pyrocatechol, resorcinol, hydroquinone, and 1-naphthol) with hypochlorous acid in the presence of ammonium ion. They reported that the aromatic ring was not chlorinated as expected but was cleaved by chloramine forming cyanogen chloride. At lower pHs, the amount of cyanogen chloride formed increased (Kanno et al., 1982).
   1-Methylnaphthylamine will not hydrolyze because it does not contain a hydrolyzable functional group (Kollig, 1993).

Exposure limits: Potential occupational carcinogen. Because no standards have been established, NIOSH (1997) recommends the most reliable and protective respirators be used, i.e., a self-contained breathing apparatus that has a full facepiece and is operated under positive-pressure or a supplied-air respirator that has a full facepiece and is operated under pressure-demand or under positive-pressure in combination with a self-contained breathing apparatus operated under pressure-demand or positive-pressure.
   OSHA recommends that worker exposure to this chemical is to be controlled by use of engineering control, proper work practices, and proper selection of personal protective equipment. Specific details of these requirements can be found in CFR 1910.1003-1910.1016.

Symptoms of exposure: Ingestion, skin contact, or inhalation of vapors can cause acute hemorrhagic cystitis, respiratory stress, and hematuria (Patnaik, 1992).

Toxicity:
   Acute oral LD$_{50}$ for rats 727 mg/kg (quoted, RTECS, 1985).
   LD$_{50}$ (intraperitoneal) for mice 200 mg/kg (quoted, RTECS, 1985).

Uses: Manufacture of dyes and in rubber.

NITRAPYRIN

Synonyms: 2-Chloro-6-(trichloromethyl)pyridine; Dowco-163; N-serve; N-serve nitrogen stabilizer; Nitrapyrine.

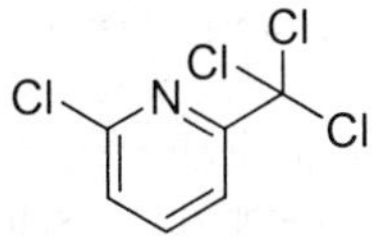

CAS Registry Number: 1929-82-4
Molecular formula: $C_6H_3Cl_4N$
Formula weight: 230.90
RTECS: US7525000

Physical state, color, and odor:
Colorless to white crystalline solid with a mild, sweet odor

Melting point (°C):
62-63 (Verschueren, 1983)

Density (g/cm³):
1.744 using method of Lyman et al. (1982)

Entropy of fusion (cal/mol·K):
14.39 (Tan et al., 1989)

Heat of fusion (kcal/mol):
4.85 (Tan et al., 1989)

Henry's law constant (x 10^{-3} atm·m³/mol):
2.13 (approximate - calculated from water solubility and vapor pressure)

Soil sorption coefficient, log K_{oc}:
2.64 (Commerce soil), 2.68 (Tracy soil), 2.66 (Tracy soil) (McCall et al., 1981)
2.62 (Kenaga and Goring, 1980)
2.24 (Cottenham sandy loam, Briggs, 1981)

Octanol/water partition coefficient, log K_{ow}:
3.41 (Kenaga and Goring, 1980)
3.02 (Briggs, 1981)

Solubility in organics:
At 20 °C: acetone (1.98 kg/kg), methylene chloride (1.85 kg/kg); 22 °C: anhydrous ammonia (540 g/kg), ethanol (300 g/kg); 26 °C: xylenes (1.04 kg/kg) (Worthing and Hance, 1991)

Solubility in water:
40 mg/L at 22 °C (Worthing and Hance, 1991)

Vapor pressure (mmHg):
2.8 x 10^{-3} at 20 °C (Verschueren, 1983)

Environmental fate:

Biological. 6-Chloropicolinic acid and carbon dioxide were reported as biodegradation products (Verschueren, 1983).

Soil. Hydrolyzes in soil to 6-chloropyridine-2-carboxylic acid which is readily absorbed by plants (Worthing and Hance, 1991).

Photolytic. Photolysis of nitrapyrin in water yielded 6-chloropicolinic acid, 6-hydroxy-picolinic acid, and an unidentified polar material (Verschueren, 1983).

Chemical/Physical. Emits toxic nitrogen oxides and chloride fumes when heated to decomposition (Lewis, 1990).

Exposure limits (mg/m³): NIOSH REL: TWA 10, STEL 20, IDLH 250; OSHA PEL: TWA 15; ACGIH TLV: TWA 10, STEL, 20 (adopted).

Toxicity:

LC_{50} for channel catfish 5.8 mg/L (Worthing and Hance, 1991).

LC_{50} (8-d) for mallard ducks 1,466 mg/kg diet, Japanese quail 820 mg/kg diet (Worthing and Hance, 1991).

Acute oral LD_{50} for chickens 235 mg/kg, mice 710 mg/kg, rats 940 mg/kg, rabbits 500 mg/kg (quoted, RTECS, 1985).

Acute percutaneous LD_{50} (24-h) for rabbits 2,830 mg/kg (Worthing and Hance, 1991).

In 94-d feeding trials using rats and dogs, the NOELS were 300 and 600 mg/kg diet, respectively (Worthing and Hance, 1991).

Uses: Not commercially produced in the United States. Nitrification inhibitor in ammonium fertilizers.

2-NITROANILINE

Synonyms: 1-Amino-2-nitrobenzene; Azoene fast orange GR base; Azoene fast orange GR salt; Azofix orange GR salt; Azogene fast orange GR; Azoic diazo component 6; Brentamine fast orange GR base; Brentamine fast orange GR salt; C.I. 37025; C.I. azoic diazo component 6; Devol orange B; Devol orange salt B; Diazo fast orange GR; Fast orange base GR; Fast orange base GR salt; Fast orange base JR; Fast orange GR base; Fast orange O base; Fast orange O salt; Fast orange salt JR; Hiltonil fast orange GR base; Hiltosal fast orange GR salt; Hindasol orange GR salt; Natasol fast orange GR salt; *o*-Nitraniline; *o*-Nitroaniline; **2-Nitrobenzenamine**; ONA; Orange base CIBA II; Orange base IRGA II; Orange GRS salt; Orange salt CIBA II; Orange salt IRGA II; Orthonitroaniline; UN 1661.

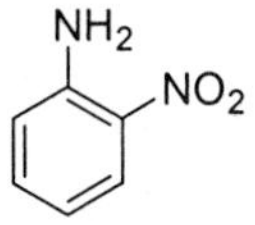

CAS Registry Number: 88-74-4
DOT: 1661
DOT label: Poison
Molecular formula: $C_6H_6N_2O_2$
Formula weight: 138.13
RTECS: BY6650000
Merck reference: 10, 6429

Physical state, color, and odor:
Orange-yellow crystals with a musty odor

Melting point (°C):
71.5 (Weast, 1986)
69.3 (Collett and Johnston, 1926)
69-70 (Dean, 1987)

Boiling point (°C):
284.1 (Dean, 1973)

Density (g/cm³):
1.442 at 15/4 °C (Weast, 1986)
0.9015 at 25/4 °C (Sax, 1984)
1.44 at 20/4 °C (Weiss, 1986)

Diffusivity in water (x 10⁻⁵ cm²/sec):
0.78 at 20 °C using method of Hayduk and Laudie (1974)

Dissociation constant, pK_a:
13.25 (Keith and Walters, 1992)

Flash point (°C):
168 (Hawley, 1981)

Lower explosive limit (%):
Not pertinent (Weiss, 1986)

Entropy of fusion (cal/mol·K):
11.2 (Domalski and Hearing, 1998)

Heat of fusion (kcal/mol):
3.85 (Tsonopoulos and Prausnitz, 1971)

Henry's law constant (x 10^{-5} atm·m³/mol):
9.72 at 25 °C (approximate - calculated from water solubility and vapor pressure)

Bioconcentration factor, log BCF:
0.91 (*Brachydanio rerio*, Kalsch et al., 1991)

Ionization potential (eV):
8.27, 8.43, 8.66 (Mallard and Linstrom, 1998)

Soil sorption coefficient, log K_{oc}:
1.23-1.62 using method of Karickhoff et al. (1979)

Octanol/water partition coefficient, log K_{ow}:
1.44 (Fujita et al., 1964)
1.79 (Leo et al., 1971)
1.73 (Kramer and Henze, 1990)
1.83 (Hansch and Anderson, 1967)

Solubility in organics:
In g/kg: benzene (208 at 25 °C), chloroform (11.7 at 0 °C), and ethanol (278.7 at 25 °C) (Collett and Johnston, 1926)

Solubility in water:
1.0 g/L at 30 °C (Phatak and Gaikar, 1996)
1.47 g/L at 30 °C (Gross et al., 1933)
1.212 and 2.423 g/kg at 25 and 40 °C, respectively (Collett and Johnston, 1926)

Vapor pressure (mmHg):
8.1 at 25 °C (Mabey et al., 1982)

Environmental fate:
   Biological. Under aerobic and anaerobic conditions using a sewage inoculum, 2-nitroaniline degraded to 2-methylbenzimidazole and 2-nitroacetanilide (Hallas and Alexander, 1983). A *Pseudomonas* sp. strain P6, isolated from a Matapeake silt loam, did not grow on 2-nitroaniline as the sole source of carbon. However, in the presence of 4-nitroaniline, approximately 50% of the applied 2-nitroaniline metabolized to nonvolatile products which could not be identified by HPLC (Zeyer and Kearney, 1983). In activated sludge inoculum, following a 20-d adaptation period, no degradation was observed (Pitter, 1976).
   Plant. 2-Nitroaniline was degraded by tomato cell suspension cultures (*Lycopericon lycopersicum*). Transformation products identified included 2-nitroanilino-β-D-glucopyranoside, β-(2-amino-3-nitrophenyl)glucopyranoside, and β-(4-amino-3-nitrophenyl)glucopyranoside (Pogány et al., 1990).

Toxicity:
   EC_{50} (15-min) for *Photobacterium phosphoreum* 26.9 mg/L (Yuan and Lang, 1997).

IC_{50} (24-h) for river bacteria 34.7 mg/L (Yuan and Lang, 1997).

Acute LD_{50} for wild birds 750 mg/kg, guinea pigs 2,350 mg/kg, mice 1,070 mg/kg, quail 750 mg/kg, rats 1,600 mg/kg (quoted, RTECS, 1985).

Use: Organic synthesis.

3-NITROANILINE

Synonyms: Amarthol fast orange R base; 1-Amino-3-nitrobenzene; 3-Aminonitrobenzene; *m*-Aminonitrobenzene; Azobase MNA; C.I. 37030; C.I. azoic diazo component 7; Daito orange base R; Devol orange R; Diazo fast orange R; Fast orange base R; Fast orange M base; Fast orange MM base; Fast orange R base; Fast orange R salt; Hiltonil fast orange R base; MNA; Naphtolean orange R base; Nitranilin; *m*-Nitraniline; 3-Nitroaminobenzene; *m*-Nitroamino-benzene; *m*-Nitroaniline; **3-Nitrobenzenamine**; *m*-Nitrobenzenamine; *m*-Nitrophenylamine; Orange base IRGA I; UN 1661.

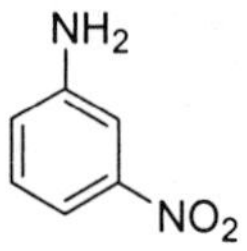

CAS Registry Number: 99-09-2
DOT: 1661
DOT label: Poison
Molecular formula: $C_6H_6N_2O_2$
Formula weight: 138.13
RTECS: BY6825000
Merck reference: 10, 6428

Physical state and color:
Yellow, rhombic crystals

Melting point (°C):
114 (Weast, 1986)

Boiling point (°C):
306.4 (Dean, 1973)

Density (g/cm³):
0.9011 at 25/4 °C (Sax, 1984)

Diffusivity in water (x 10^{-5} cm²/sec):
0.78 at 20 °C using method of Hayduk and Laudie (1974)

Dissociation constant, pK$_a$:
2.46 at 25 °C (Dean, 1973)

Entropy of fusion (cal/mol·K):
14.7 (Andrews et al., 1926)

Heat of fusion (kcal/mol):
5.66 (Andrews et al., 1926; Singh et al., 1990)

Henry's law constant (x 10^{-5} atm·m³/mol):
1.93 at 25 °C (approximate - calculated from water solubility and vapor pressure)

Ionization potential (eV):
8.80 (Franklin et al., 1969)

Bioconcentration factor, log BCF:
0.92 (*Brachydanio rerio*, Kalsch et al., 1991)

Soil sorption coefficient, log K_{oc}:
1.26 using method of Karickhoff et al. (1979)

Octanol/water partition coefficient, log K_{ow}:
1.37 (Fujita et al., 1964)

Solubility in organics:
In g/kg at 25 °C: benzene (27.18), chloroform (32.16), and ethanol (77.78) (Collett and Johnston, 1926)

Solubility in water:
890 mg/L at 25 °C (quoted, Verschueren, 1983)
0.910 and 1.785 g/kg at 25 and 40.1 °C, respectively (Collett and Johnston, 1926)
1,100 mg/L at 20 °C (Dean, 1973)
1.21 g/L at 30 °C (Gross et al., 1933)

Vapor pressure (mmHg):
9.56×10^{-5} at 25 °C (Banerjee et al., 1990)

Environmental fate:
Biological. A bacterial culture isolated from the Oconee River in North Georgia degraded 3-nitroaniline to the intermediate 4-nitrocatechol (Paris and Wolfe, 1987). A *Pseudomonas* sp. strain P6, isolated from a Matapeake silt loam, did not grow on 3-nitroaniline as the sole source of carbon. However, in the presence of 4-nitroaniline, all of the applied 3-nitroaniline metabolized completely to carbon dioxide (Zeyer and Kearney, 1983). In the presence of suspended natural populations from unpolluted aquatic systems, the second-order microbial transformation rate constant determined in the laboratory was reported to be $4.6 \pm 0.1 \times 10^{-13}$ L/organism·h (Steen, 1991).

In activated sludge inoculum, following a 20-d adaptation period, no degradation was observed (Pitter, 1976).

Chemical/Physical. Will react with acids forming water soluble salts.

Toxicity:
EC_{50} (15-min) for *Photobacterium phosphoreum* 85.2 mg/L (Yuan and Lang, 1997).
IC_{50} (24-h) for river bacteria 45.7 mg/L (Yuan and Lang, 1997).
Acute LD_{50} for guinea pigs 450 mg/kg, mice 308 mg/kg, quail 562 mg/kg, rats 535 mg/kg (quoted, RTECS, 1985).

Use: Organic synthesis.

4-NITROANILINE

Synonyms: 1-Amino-4-nitrobenzene; 4-Aminonitrobenzene; *p*-Aminonitrobenzene; Azoamine red ZH; Azofix Red GG salt; Azoic diazo component 37; C.I. 37035; C.I. azoic diazo component 37; C.I. developer 17; Developer P; Devol red GG; Diazo fast red GG; Fast red base GG; Fast red base 2J; Fast red 2G base; Fast red 2G salt; Fast red GG base; Fast red GG salt; Fast red MP base; Fast red P base; Fast red P salt; Fast red salt GG; Fast red salt 2J; IG base; Naphtolean red GG base; NCI-C60786; 4-Nitraniline; *p*-Nitraniline; Nitrazol 2F extra; *p*-Nitroaniline; **4-Nitrobenzenamine**; *p*-Nitrobenzenamine; *p*-Nitrophenylamine; PNA; RCRA waste number P077; Red 2G base; Shinnippon fast red GG base; UN 1661.

CAS Registry Number: 100-01-6
DOT: 1661
DOT label: Poison
Molecular formula: $C_6H_6N_2O_2$
Formula weight: 138.13
RTECS: BY7000000
Merck reference: 10, 6430

Physical state, color, and odor:
Bright yellow crystalline powder with a faint, ammonia-like, slightly pungent odor

Melting point (°C):
146 (Dean, 1987)
148-149 (Weast, 1986)

Boiling point (°C):
331.7 (Weast, 1986)
336 (Weiss, 1986)

Density (g/cm³):
1.424 at 20/4 °C (Weast, 1986)

Diffusivity in water (x 10^{-5} cm²/sec):
0.78 at 20 °C using method of Hayduk and Laudie (1974)

Dissociation constant, pK_a:
0.99 at 25 °C (Dean, 1973)

Flash point (°C):
165 (Aldrich, 1988)
200.5 (NIOSH, 1997)

Entropy of fusion (cal/mol·K):
12.0 (Andrews et al., 1926)

Heat of fusion (kcal/mol):
5.04 (Andrews et al., 1926)

Henry's law constant (x 10^{-8} atm·m^3/mol):
1.14 at 25 °C (approximate - calculated from water solubility and vapor pressure)

Ionization potential (eV):
8.85 (NIOSH, 1997)

Bioconcentration factor, log BCF:
0.64 (*Brachydanio rerio*, Kalsch et al., 1991)

Soil sorption coefficient, log K_{oc}:
1.88 (Batcombe silt loam, Briggs, 1981)

Octanol/water partition coefficient, log K_{ow}:
1.39 (Fujita et al., 1964)
1.40 (Campbell and Luthy, 1985)
1.51 (Kramer and Henze, 1990)

Solubility in organics:
In g/kg at 25 °C: benzene (5.794), chloroform (9.290), and ethanol (60.48) (Collett and Johnston, 1926)

Solubility in water:
0.8 g/L at 30 °C (Phatak and Gaikar, 1996)
0.568 and 1.157 g/kg at 25 and 40 °C, respectively (Collett and Johnston, 1926).
728 mg/kg at 30 °C (Gross and Saylor, 1931)
380, 390, 400 at 20 °C (Hashimoto et al., 1982)

Vapor pressure (x 10^{-3} mmHg):
1.5 at 20 °C, 7 at 30 °C (Verschueren, 1983)

Environmental fate:
Biological. A *Pseudomonas* sp. strain P6, isolated from a Matapeake silt loam, was grown using a yeast extract. After 8 d, 4-nitroaniline degraded completely to carbon dioxide (Zeyer and Kearney, 1983). In activated sludge inoculum, following a 20-d adaptation period, no degradation was observed (Pitter, 1976).

Chemical/Physical: Spacek et al. (1995) investigated the photodegradation of 4-nitroaniline using titanium dioxide-UV light and Fenton's reagent (hydrogen peroxide:substance - 10:1; Fe^{2+} 2.5 x 10^{-4} mol/L). Both experiments were carried out at 25 °C. The decomposition rate of 4-nitroaniline was very high by the photo-Fenton reaction in comparison to titanium dioxide-UV light (λ = 365 nm). Decomposition products identified in both reactions were nitrobenzene, *p*-benzoquinone, hydroquinone, oxalic acid, and resorcinol. Oxalic acid, hydroquinone, and *p*-benzoquinone were identified as intermediate products using HPLC.

Reacts with mineral acids forming water soluble salts.

Exposure limits (mg/m^3): NIOSH REL: TWA 3; IDLH 300; OSHA PEL: TWA 6; ACGIH TLV: TWA 3 (adopted).

Toxicity:
EC_{50} (15-min) for *Photobacterium phosphoreum* 14.8 mg/L (Yuan and Lang, 1997).
IC_{50} (24-h) for river bacteria 28.2 mg/L (Yuan and Lang, 1997).
LC_{50} (96-h) for fathead minnows 106 mg/L (Spehar et al., 1982).

LC_{50} (48-h) for red killifish 363 mg/L (Yoshioka et al., 1986).

LD_{50} (intraperitoneal) for mice 250 mg/kg (quoted, RTECS, 1985).

Acute LD_{50} for wild birds 75 mg/kg, guinea pigs 450 mg/kg, mice 810 mg/kg, quail 1,000 mg/kg, rats 750 mg/kg (quoted, RTECS, 1985).

Uses: Intermediate for dyes and antioxidants; inhibits gum formation in gasoline; corrosion inhibiter; organic synthesis (preparation of *p*-phenylenediamine).

NITROBENZENE

Synonyms: Essence of mirbane; Essence of myrbane; Mirbane oil; NCI-C60082; Nitrobenzol; Oil of bitter almonds; Oil of mirbane; Oil of myrbane; RCRA waste number U169; UN 1662.

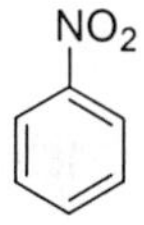

CAS Registry Number: 98-95-3
DOT: 1662
DOT label: Poison
Molecular formula: $C_6H_5NO_2$
Formula weight: 123.11
RTECS: DA6475000
Merck reference: 10, 6434

Physical state, color, and odor:
Clear, light yellow to brown, oily liquid with an almond-like or shoe polish odor. May darken on exposure to air. Odor threshold concentration is 18 ppb (Amoore and Hautala, 1983).

Melting point (°C):
5.7 (Weast, 1986)

Boiling point (°C):
210.8 (Weast, 1986)

Density (g/cm³):
1.20329 at 20.00/4 °C (Tsierkezos et al., 2000)
1.2125 at 10/4 °C, 1.205 at 18/4 °C, 1.1986 at 25/4 °C (Standen, 1967)

Diffusivity in water (x 10⁻⁵ cm²/sec):
0.87 at 20 °C using method of Hayduk and Laudie (1974)

Dissociation constant, pK$_a$:
>15 (Christensen et al., 1975)

Flash point (°C):
88.5 (NIOSH, 1997)

Lower explosive limit (%):
1.8 at 100 °C (NIOSH, 1997)

Entropy of fusion (cal/mol·K):
9.27 (Pacor, 1967)
10.39 (Parks et al., 1936)

Heat of fusion (kcal/mol):
2.58 (Pacor, 1967)
2.90 (Parks et al., 1936)

Henry's law constant (x 10^{-5} atm·m³/mol):
2.45 at 25 °C (Warner et al., 1987)
0.95, 5.27, 8.82, 16.5, and 31.8 at 2.0, 6.0, 10.0, 18.0, and 25.0 °C, respectively (Dewulf et al., 1999)

Interfacial tension with water (dyn/cm at 20 °C):
25.66 (Harkins et al., 1920)

Ionization potential (eV):
9.92 (Franklin et al., 1969)

Bioconcentration factor, log BCF:
2.41 (algae, Geyer et al., 1984)
. 1.60 (activated sludge), 1.30 (algae) (Freitag et al., 1985)
1.18 (fathead minnow, Veith et al., 1979)

Soil sorption coefficient, log K_{oc}:
1.85 (river sediment), 1.95 (coal wastewater sediment) (Kopinke et al., 1995)
2.36 (Løkke, 1984)
1.49, 1.95, >2.01 (various Norwegian soils, Seip et al., 1986)
1.94 (Batcombe silt loam, Briggs, 1981)
2.28 (Lincoln sand, Wilson et al., 1981)
2.15 (Delta soil, Miller and Weber, 1986)
1.95 (Captina silt loam), 2.02 (McLaurin sandy loam) (Walton et al., 1992)
K_d = 3.5 mL/g on a Cs^+-kaolinite (Haderlein and Schwarzenbach, 1993)

Octanol/water partition coefficient, log K_{ow}:
1.85 (Briggs, 1981, Campbell and Luthy, 1985; Fujita et al., 1964; Walton et al., 1992; Wasik et al., 1981)
1.83 (Banerjee et al., 1980; Brooke et al., 1990; Garst and Wilson, 1984; de Bruijn et al., 1989)
1.84 (Geyer et al., 1984; Brooke et al., 1990)
1.792 (Lu and Metcalf, 1975)
1.88 (Leo et al., 1971; Andersson and Schräder, 1999)
1.70 (Hammers et al., 1982)

Solubility in organics:
Soluble in acetone, ethanol, benzene, ether (Weast, 1986), and many other hydrocarbons including toluene and ethylbenzene

Solubility in water:
1,900 mg/L at 20 °C, 8,000 mg/L at 80 °C (quoted, Verschueren, 1983)
2,000 mg/L at 25 °C (Warner et al., 1987)
2,090 mg/L at 25 °C (Banerjee et al., 1980)
2,043 mg/L at 25 °C (Chiou, 1985)
1,780 mg/kg at 15 °C, 2,050 mg/kg at 30 °C (Gross and Saylor, 1931)
1,930 mg/L at 25 °C (Andrews and Keefer, 1950)
18.35 mM at 35 °C (Hine et al., 1963)
31.1 mM at 25 °C (Tewari et al., 1982; Wasik et al., 1981)

Vapor density:
5.03 g/L at 25 °C, 4.25 (air = 1)

Vapor pressure (mmHg):
0.15 at 20.0 °C, 0.35 at 30.0 °C (Verschueren, 1983)
0.28 at 25 °C (Warner et al., 1987)
0.600 at 35 °C (Hine et al., 1963)

Environmental fate:

Biological. In activated sludge, 0.4% of the applied nitrobenzene mineralized to carbon dioxide after 5 d (Freitag et al., 1985). Under anaerobic conditions using a sewage inoculum, nitrobenzene degraded to aniline (Hallas and Alexander, 1983). When nitrobenzene (5 and 10 mg/L) was statically incubated in the dark at 25 °C with yeast extract and settled domestic wastewater inoculum, complete biodegradation with rapid acclimation was observed after 7-14 d (Tabak et al., 1981). In activated sludge inoculum, 98.0% COD removal was achieved in 5 d. The average rate of biodegradation was 14.0 mg COD/g·h (Pitter, 1976).

Razo-Flores et al. (1999) studied the fate of nitrobenzene (50 mg/L) in an upward-flow anaerobic sludge bed reactor containing a mixture of volatile fatty acids and/or glucose as electron donors. The nitrobenzene loading rate and hydraulic retention time for this experiment were 43 mg/L·d and 28 h, respectively. Nitrobenzene was effectively reduced (>99.9%) to aniline (92% molar yield) in stoichiometric amounts for the 100-d experiment.

Photolytic. Irradiation of nitrobenzene in the vapor phase produced nitrosobenzene and 4-nitrophenol (HSDB, 1989). Titanium dioxide suspended in an aqueous solution and irradiated with UV light (λ = 365 nm) converted nitrobenzene to carbon dioxide at a significant rate (Matthews, 1986). A carbon dioxide yield of 6.7% was achieved when nitrobenzene adsorbed on silica gel was irradiated with light (λ >290 nm) for 17 h (Freitag et al., 1985).

An aqueous solution containing nitrobenzene (500 μM) and hydrogen peroxide (100 μM) was irradiated with UV light (λ = 285-360 nm). After 18 h, 2% of the substrate was converted into *o*-, *m*-, and *p*-nitrophenols having an isomer distribution of 50, 29.5, and 20.5%, respectively (Draper and Crosby, 1984).

When nitrobenzene, with nitrogen as a carrier gas, was passed through a quartz cell and irradiated by two 220-volt arcs, nitrosobenzene and *p*-nitrophenol formed as the major products (Hastings and Matsen, 1948). A rate constant of 1.4 x 10^{-13} cm^3/molecule·sec was reported for the gas-phase reaction of nitrobenzene and OH radicals in air (Witte et al., 1986).

Chemical/Physical. In an aqueous solution, nitrobenzene (100 μM) reacted with Fenton's reagent (35 μM). After 15 min, 2-, 3-, and 4-nitrophenol were identified as products. After 6 h, about 50% of the nitrobenzene was destroyed (Lipczynska-Kochany, 1991). Under anaerobic conditions, nitrobenzene in distilled water was reduced in the presence of iron metal (33.3 g/L acid washed 18-20 mesh). Aniline formed with nitrosobenzene as an intermediate product. In about 3 h, >98% of nitrobenzene (1.5 x 10^{-5} M) had reacted. Based on the pseudo-first-order disappearance rate of 0.035/min, the half-life was 19.7 min (Agrawal and Tratnyek, 1996).

Yu and Bailey (1992) studied the reaction of nitrobenzene with four sulfide minerals under anaerobic conditions. Observed half-lives of nitrobenzene were 7.5, 40, 105, and 360 h for the reaction with sodium sulfide, alabandite (manganese sulfide), sphalerite (zinc sulfide), and molybdenite (molybdenum sulfide), respectively. Aniline and elemental sulfur were found as reduction products of nitrobenzene-manganese sulfide reaction. Aniline was also a reduction product in the nitrobenzene-molybdenum sulfide and nitrobenzene-sodium sulfide reactions. Several unidentified products formed in the reaction of nitrobenzene and sphalerite (Yu and Bailey, 1992).

At an influent concentration of 1,023 mg/L, treatment with granular activated carbon resulted in an effluent concentration of 44 mg/L. The adsorbability of the carbon used was 196 mg/g carbon (Guisti et al., 1974).

Nitrobenzene will not hydrolyze in water because it does not contain a hydrolyzable group (Kollig, 1993).

Exposure limits: NIOSH REL: TWA 1 ppm (5 mg/m³), IDLH 200 ppm; OSHA PEL: TWA 1 ppm; ACGIH TLV: TWA 1 ppm (adopted).

Symptoms of exposure: Chronic exposure may cause anemia. Acute effects include headache, dizziness, nausea, vomiting, dyspnea (Patnaik, 1992), drowsiness, methemoglobinemia with cyanosis (Windholz et al., 1983). An irritation concentration of 230.00 mg/m³ in air was reported by Ruth (1986).

Toxicity:

EC_{50} (15-min) for *Photobacterium phosphoreum* 67.7 mg/L (Yuan and Lang, 1997).

IC_{50} (24-h) for river bacteria 79.5 mg/L (Yuan and Lang, 1997).

LC_{50} (contact) for earthworm (*Eisenia fetida*) 16 μg/cm² (Neuhauser et al., 1985).

LC_{50} (96-h) for bluegill sunfish 43 mg/L (Spehar et al., 1982), *Cyprinodon variegatus* 59 ppm using natural seawater (Heitmuller et al., 1981).

LC_{50} (72-h) for *Cyprinodon variegatus* >120 ppm (Heitmuller et al., 1981).

LC_{50} (48-h) for red killifish 275 mg/L (Yoshioka et al., 1986), *Daphnia magna* 27 mg/L (LeBlanc, 1980), *Cyprinodon variegatus* >120 ppm (Heitmuller et al., 1981).

LC_{50} (24-h) for *Daphnia magna* 24 mg/L (LeBlanc, 1980), *Cyprinodon variegatus* >120 ppm (Heitmuller et al., 1981).

Acute oral LD_{50} for mice 590 mg/kg, rats 489 mg/kg (Yoshioka et al., 1986).

Heitmuller et al. (1981) reported a NOEC of 22 ppm.

Uses: Solvent for cellulose ethers; modifying esterification of cellulose acetate; ingredient of metal polishes and shoe polishes; manufacture of aniline, benzidine, quinoline, azobenzene, drugs, photographic chemicals.

4-NITROBIPHENYL

Synonyms: 4-Nitro-1,1′-biphenyl; 4-Nitrodiphenyl; *p*-Nitrobiphenyl; 4-Phenylnitrobenzene; *p*-Phenylnitrobenzene; PNB.

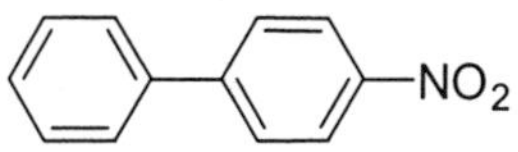

CAS Registry Number: 92-93-3
Molecular formula: $C_{12}H_9NO_2$
Formula weight: 199.21
RTECS: DV5600000
Merck reference: 10, 6439

Physical state, color, and odor:
White to yellow crystals with a sweetish odor

Melting point (°C):
114 (Weast, 1986)

Boiling point (°C):
340 (Weast, 1986)

Diffusivity in water (x 10^{-5} cm^2/sec):
0.59 at 20 °C using method of Hayduk and Laudie (1974)

Flash point (°C):
144.5 (NIOSH, 1997)

Lower explosive limit (%):
Not applicable (NFPA, 1984)

Upper explosive limit (%):
Not applicable (NFPA, 1984)

Solubility in organics:
Soluble in acetic acid, benzene, chloroform, and ether (Weast, 1986)

Exposure limits: Potential occupational carcinogen. Because no standards have been established, NIOSH (1997) recommends the most reliable and protective respirators be used, i.e., a self-contained breathing apparatus that has a full facepiece and is operated under positive-pressure or a supplied-air respirator that has a full facepiece and is operated under pressure-demand or under positive-pressure in combination with a self-contained breathing apparatus operated under pressure-demand or positive-pressure.

OSHA recommends that worker exposure to this chemical is to be controlled by use of engineering control, proper work practices, and proper selection of personal protective equipment. Specific details of these requirements can be found in CFR 1910.1003-1910.1016.

Toxicity:
Acute oral LD_{50} for rats 2,230 mg/kg, rabbits 1,970 mg/kg (quoted, RTECS, 1985).

Uses: Organic synthesis.

NITROETHANE

Synonym: UN 2842.

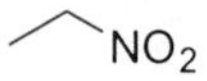

$$\diagup\!\!\!\diagdown NO_2$$

CAS Registry Number: 79-24-3
DOT: 2842
DOT label: Combustible liquid
Molecular formula: $C_2H_5NO_2$
Formula weight: 75.07
RTECS: KI5600000
Merck reference: 10, 6443

Physical state, color, and odor:
Colorless liquid with a fruity odor. Odor threshold concentration is 2.1 ppm (Amoore and Hautala, 1983).

Melting point (°C):
-50 (Weast, 1986)

Boiling point (°C):
114.1 (Dean, 1987)
115 (Weast, 1986)

Density (g/cm³):
1.0448 at 25/4 °C (Weast, 1986)

Diffusivity in water (x 10^{-5} cm²/sec):
1.23 at 25 °C using method of Hayduk and Laudie (1974)

Dissociation constant, pK_a:
8.5 (Gordon and Ford, 1972)
8.44 at 25 °C (Dean, 1973)

Flash point (°C):
28.0 (NIOSH, 1997)
41.11 (open cup, Windholz et al., 1983)

Lower explosive limit (%):
3.4 (NIOSH, 1997)

Entropy of fusion (cal/mol·K):
12.82 (Liu and Ziegler, 1966)

Heat of fusion (kcal/mol):
2.36 (Liu and Ziegler, 1966)

Henry's law constant (x 10^{-6} atm·m³/mol):
7.14 at 25 °C (Friant and Suffet, 1979)

Ionization potential (eV):
10.88 ± 0.05 (Franklin et al., 1969)

Soil sorption coefficient, log K_{oc}:
Unavailable because experimental methods for estimation of this parameter for nitroaliphatics are lacking in the documented literature. However, its moderate solubility in water and low K_{ow} suggest its adsorption to soil will be low (Lyman et al., 1982).

Octanol/water partition coefficient, log K_{ow}:
0.18 (Hansch and Anderson, 1967)

Solubility in organics:
Soluble in acetone (Weast, 1986). Miscible with alcohol, chloroform, and ether (Sax and Lewis, 1987).

Solubility in water (wt %):
4.32 at 0 °C, 4.22 at 9.2 °C, 4.38 at 19.6 °C, 4.44 at 31.0 °C, 4.60 at 40.6 °C, 5.19 at 50.5 °C, 5.40 at 60.8 °C, 5.82 at 70.4 °C, 6.27 at 80.4 °C, 6.55 at 90.1 °C (Stephenson, 1992)

Vapor density:
3.07 g/L at 25 °C, 2.59 (air = 1)

Vapor pressure (mmHg):
21 at 25.2 °C (NIOSH, 1997)

Environmental fate:
   Chemical/Physical. 2-Nitroethane will not hydrolyze because it does not contain a hydrolyzable functional group.

Exposure limits: NIOSH REL: TWA 100 ppm (310 mg/m^3), IDLH 1,000 ppm; OSHA PEL: TWA 100 ppm; ACGIH TLV: TWA 100 ppm (adopted).

Symptons of exposure: An irritation concentration of 310.00 mg/m^3 in air was reported by Ruth (1986).

Toxicity:
   Acute oral LD_{50} for mice 860 mg/kg, rats 1,100 mg/kg (quoted, RTECS, 1985).

Uses: Solvent for nitrocellulose; cellulose acetate; cellulose acetobutyrate; cellulose aceto-propionate, waxes, fats, dyestuffs, vinyl, and alkyd resins; experimental propellant; fuel additive; organic synthesis (Friedel-Crafts reactions).

NITROMETHANE

Synonyms: Nitrocarbol; UN 1261.

$$\diagdown NO_2$$

CAS Registry Number: 75-52-5
DOT: 1261
DOT label: Combustible liquid
Molecular formula: CH_3NO_2
Formula weight: 61.04
RTECS: PA9800000
Merck reference: 10, 6457

Physical state, color, and odor:
Colorless liquid with a strong, disagreeable odor. Odor threshold concentration is 3.5 ppm (Amoore and Hautala, 1983).

Melting point (°C):
-17 (Weast, 1986)
-29 (Windholz et al., 1983)

Boiling point (°C):
100.8 (Weast, 1986)

Density (g/cm³):
1.13754 at 20.00/4 °C (Lee and Tu, 1999)
1.1322 at 25/4 °C (Windholz et al., 1983)

Diffusivity in water (x 10^{-5} cm²/sec):
1.27 at 20 °C using method of Hayduk and Laudie (1974)

Dissociation constant, pK_a:
10.21 at 25 °C (Dean, 1973)

Flash point (°C):
35 (NIOSH, 1997)
44.4 (Windholz et al., 1983)

Lower explosive limit (%):
7.3 (NIOSH, 1997)

Entropy of fusion (cal/mol·K):
9.474 (Jones and Giauque, 1947)

Heat of fusion (kcal/mol):
2.319 (Jones and Giauque, 1947)

Henry's law constant (x 10^{-5} atm·m³/mol):
2.22 at 25 °C (Rohrschneider, 1973)

Interfacial tension with water (dyn/cm):
5.7 at 25 °C (Murphy et al., 1957)
9.5 at 25 °C (Donahue and Bartell, 1952)
9.66 at 20 °C (Harkins et al., 1920)

Ionization potential (eV):
11.08 (Lias, 1998)

Soil sorption coefficient, log K_{oc}:
Unavailable because experimental methods for estimation of this parameter for nitroaliphatics are lacking in the documented literature. However, its high solubility in water and low K_{ow} suggest its adsorption to soil will be low (Lyman et al., 1982).

Octanol/water partition coefficient, log K_{ow}:
-0.33 (Hansch and Anderson, 1967)
-0.35 (Hansch and Leo, 1979)

Solubility in organics:
Soluble in acetone, alcohol, ether (Weast, 1986), and *N,N*-dimethylformamide (Windholz et al., 1983).

Solubility in water (wt %):
9.0 at 0 °C, 9.7 at 9.5 °C, 10.4 at 19.7 °C, 11.7 at 31.0 °C, 12.8 at 40.4 °C, 14.8 at 50.0 °C, 15.1 at 60.5 °C, 17.1 at 70.5 °C, 19.6 at 80.2 °C, 20.8 at 89.8 °C (Stephenson, 1992)

Vapor density:
2.49 g/L at 25 °C, 2.11 (air = 1)

Vapor pressure (mmHg):
27.8 at 20 °C, 46 at 30 °C (Verschueren, 1983)
35.26 at 25.00 °C (Hussam and Carr, 1985)

Environmental fate:
Chemical/Physical. Nitromethane will not hydrolyze because it does not contain a hydrolyzable functional group.

Exposure limits: NIOSH REL: IDLH 750 ppm; OSHA PEL: TWA 100 ppm (250 mg/m³); ACGIH TLV: TWA 20 ppm (adopted).

Symptoms of exposure: An irritation concentration of 500.00 mg/m³ in air was reported by Ruth (1986).

Toxicity:
LC_{50} (96-h) for fathead minnows >650 mg/L (Spehar et al., 1982).
Acute oral LD_{50} for mice 950 mg/kg, rats 940 mg/kg (quoted, RTECS, 1985).
LD_{50} (intraperitoneal) for mice 110 mg/kg (quoted, RTECS, 1985).

Uses: Rocket fuel; coatings industry; solvent for cellulosic compounds, polymers, waxes, fats; gasoline additive; organic synthesis.

2-NITROPHENOL

Synonyms: 2-Hydroxynitrobenzene; *o*-Hydroxynitrobenzene; 2-Nitro-1-hydroxybenzene; *o*-Nitrophenol; ONP; UN 1663.

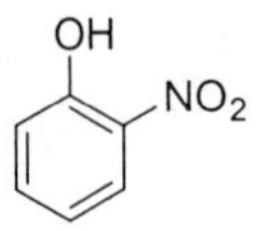

CAS Registry Number: 88-75-5
DOT: 1663
DOT label: Poison
Molecular formula: $C_6H_5NO_3$
Formula weight: 139.11
RTECS: SM2100000
Merck reference: 10, 6466

Physical state, color, and odor:
Pale yellow crystals with an aromatic odor

Melting point (°C):
45-46 (Weast, 1986)
44-45 (Standen, 1967)

Boiling point (°C):
216 (Weast, 1986)

Density (g/cm³):
1.485 at 14/4 °C (Weast, 1986)
1.495 at 20/4 °C (Sax, 1984)
1.2942 at 40/4 °C, 1.2712 at 60/4 °C, 1.2482 at 80/4 °C (Standen, 1967)

Diffusivity in water (x 10⁻⁵ cm²/sec):
0.81 at 20 °C using method of Hayduk and Laudie (1974)

Dissociation constant, pK$_a$:
7.23 at 25 °C (Dean, 1973)

Flash point (°C):
73.5 (Sax, 1985)

Lower explosive limit (%):
Not pertinent (Weiss, 1986)

Upper explosive limit (%):
Not pertinent (Weiss, 1986)

Entropy of fusion (cal/mol·K):
13.11 (Poeti et al., 1982)

Heat of fusion (kcal/mol):
4.17 (Poeti et al., 1982)
4.20 (Tsonopoulos and Prausnitz, 1971)

Henry's law constant (x 10^{-6} atm·m^3/mol):
3.5 (Howard, 1989)

Ionization potential (eV):
9.29 (Mallard and Linstrom, 1998)

Bioconcentration factor, log BCF:
1.48 (activated sludge), 1.48 (algae), 1.60 (golden ide) (Freitag et al., 1985)

Soil sorption coefficient, log K_{oc}:
1.79 (river sediment), 2.32 (coal wastewater sediment) (Kopinke et al., 1995)
2.06 (Brookstone clay loam, Boyd, 1982)
2.21 (coarse sand), 1.75 (loamy sand) (Kjeldsen et al., 1990)
$K_d = 35$ mL/g on a Cs^+-kaolinite (Haderlein and Schwarzenbach, 1993)
2.96 (glaciofluvial, sandy aquifer, Nielsen et al., 1996)

Octanol/water partition coefficient, log K_{ow}:
1.62 at pH 7 (HPLC, Unger et al., 1978)
1.73 (Leo et al., 1971)
1.79 (Fujita et al., 1964)
1.89 in buffer solution at pH 1.5 (Schwarzenbach et al., 1988)

Solubility in organics:
In g/kg at 25 °C: benzene (1,472; 3,597 at 30 °C) and ethanol (460) (Windholz et al., 1983)
0.33 and 1,270 mM at 25 °C in isooctane and butyl ether, respectively (Anderson et al., 1980)

Solubility in water:
3.30 g/kg at 40 °C (Palit, 1947)
2,100 mg/L at 20 °C, 10,800 mg/L at 100 °C (quoted, Verschueren, 1983)
1.076, 1.697, and 2.935 g/L at 15.6, 24.8, and 34.7 °C, respectively (Achard et al., 1996)
3.89 g/L at 48 °C (quoted, Stephen and Stephen, 1963)
3,200 mg/L at 38 °C (quoted, Standen, 1967)
1,060 mg/L at 20 °C, 2,500 mg/L at 25 °C (Howard, 1989)
1.4 g/L at 20 °C (quoted, Leuenberger et al., 1985)
1,300 mg/L at 25 °C (Riederer, 1990)
1,079 mg/L at 20 °C (buffer solution at pH 1.5, Schwarzenbach et al., 1988)
10.0, 8.34, 8.70, and 8.55 mM in doubly distilled water, Pacific seawater, artificial seawater, and 35% NaCl at 20 °C, respectively (Hashimoto et al., 1984)

Vapor pressure (x 10^{-2} mmHg):
1.9 at 8 °C, 12 at 25 °C (quoted, Leuenberger et al., 1985)
9.3 at 25 °C (Sonnefeld et al., 1983)
8.9 at 25 °C (Riederer, 1990)

Environmental fate:
Biological. A microorganism, *Pseudomonas putida*, isolated from soil degraded 2-nitrophenol to nitrite. Degradation by enzymatic mechanisms produced nitrite and catechol.

Catechol subsequently degraded to β-ketoadipic acid (Zeyer and Kearney, 1984). When 2-nitrophenol was statically incubated in the dark at 25 °C with yeast extract and settled domestic wastewater inoculum, 100% biodegradation with rapid adaptation was achieved after 7 d (Tabak et al., 1981). In a similar study, 2-nitrophenol degraded rapidly from flooded alluvial and pokkali (organic matter-rich acid sulfate) soils that were inoculated with parathion-emrichment culture containing 5-day-old cultures of *Flavobacterium* sp. ATCC 27551 and *Pseudomonas* sp. ATCC 29353 (Sudhaker-Barik and Sethunathan, 1978a). 2-Nitrophenol disappeared completely with the formation of nitrite, particularly in the inoculated soils rather than in the uninoculated soils.

In activated sludge inoculum, 97.0% COD removal was achieved. The average rate of biodegradation was 14.0 mg COD/g·h (Pitter, 1976).

Groundwater. Nielsen et al. (1996) studied the degradation of 2-nitrophenol in a shallow, glaciofluvial, unconfined sandy aquifer in Jutland, Denmark. As part of the *in situ* microcosm study, a cylinder that was open at the bottom and screened at the top was installed through a cased borehole approximately 5 m below grade. Five liters of water was aerated with atmospheric air to ensure aerobic conditions were maintained. Groundwater was analyzed weekly for approximately 3 months to determine 2-nitrophenol concentrations over time. The experimentally determined first-order biodegradation rate constant and corresponding half-life were 0.05/d and 13.86 d, respectively.

Photolytic. A second-order reaction rate constant of 9×10^{-13} cm^3/molecule·sec was reported for the reaction of 2-nitrophenol and OH radicals in the atmosphere (Atkinson, 1985).

Chemical/Physical. Oxidation by Fenton's reagent (hydrogen peroxide and Fe^{3+}) produced nitrohydroquinone and 3-nitrocatechol (Andersson et al., 1986).

Toxicity:

EC_{50} (15-min) for *Photobacterium phosphoreum* 41.1 mg/L (Yuan and Lang, 1997).

IC_{50} (24-h) for river bacteria 43.0 mg/L (Yuan and Lang, 1997).

LC_{50} (contact) for earthworm (*Eisenia fetida*) 5.9 μg/cm^2 (Neuhauser et al., 1985).

LC_{50} (48-h) for red killifish 275 mg/L (Yoshioka et al., 1986).

Acute oral LD_{50} for mice 1,300 mg/kg, rats 334 mg/kg (quoted, RTECS, 1985).

Uses: Indicator; preparation of *o*-nitroanisole and other organic compounds.

4-NITROPHENOL

Synonyms: 4-Hydroxynitrobenzene; *p*-Hydroxynitrobenzene; NCI-C55992; 4-Nitro-1-hy-droxybenzene; *p*-Nitrophenol; PNP; RCRA waste number U170; UN 1663.

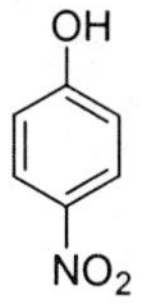

CAS Registry Number: 100-02-7
DOT: 1663
DOT label: Poison
Molecular formula: $C_6H_5NO_3$
Formula weight: 139.11
RTECS: SM2275000
Merck reference: 10, 6467

Physical state, color, and odor:
Colorless to pale yellow, odorless crystals

Melting point (°C):
112-114 (Dean, 1987)
114 (Standen, 1967)

Boiling point (°C):
279 (Aldrich, 1990)

Density (g/cm³):
1.479 at 20/4 °C, 1.270 at 120/4 °C (Standen, 1967)

Diffusivity in water (x 10^{-5} cm²/sec):
0.81 at 20 °C using method of Hayduk and Laudie (1974)

Dissociation constant, pK_a:
7.15 at 25 °C (Dean, 1973)
7.08 at 21.5 °C (Schwarzenbach et al., 1988)

Flash point (°C):
Not pertinent (combustible solid, Weiss, 1986)

Entropy of fusion (cal/mol·K):
15.0 (Campbell and Campbell, 1941)
11.27 (Poeti et al., 1982)
18.7 (Singh and Kumar, 1986)

Heat of fusion (kcal/mol):
3.80 (Tsonopoulos and Prausnitz, 1971)
5.80 (Campbell and Campbell, 1941)
4.36 (Poeti et al., 1982)
7.20 (Singh and Kumar, 1986)

Henry's law constant (x 10^{-10} atm·m^3/mol):
3.85 at 25 °C (Parsons et al.,1971)

Ionization potential (eV):
9.52 (Gordon and Ford, 1972)

Bioconcentration factor, log BCF:
1.48 (algae, Geyer et al., 1984)
1.48 (activated sludge), 1.60 (golden ide) (Freitag et al., 1985)
1.90, 2.44 (fathead minnow, Call et al., 1980)
2.10 (fathead minnow, Veith et al., 1979)

Soil sorption coefficient, log K_{oc}:
1.75-2.73 (average = 2.33 for 10 Danish soils, Løkke, 1984)
1.74 (Brookstone clay loam, Boyd, 1982)
2.37 (Meylan et al., 1992)
1.94 (loamy sand, Kjeldsen et al., 1990)

Octanol/water partition coefficient, log K_{ow}:
1.68 (HPLC, Unger et al., 1978)
1.91 (Leo et al., 1971; Campbell and Luthy, 1985)
1.85, 1.92 (Geyer et al., 1984)
1.93 (Beltrame et al., 1993)
2.04 (Schwarzenbach et al., 1988)
1.96 (Fujita et al., 1964; Garst and Wilson, 1984)

Solubility in organics:
Soluble in benzene (9.2 g/kg at 20 °C), ethanol (1,895 g/kg at 25 °C) and toluene (227 g/kg at 70 °C) (Palit, 1947). Also soluble in isooctane and *n*-butyl ether at 0.046 and 176.7 mg/L at 25 °C, respectively (Anderson et al., 1971)

Solubility in water:
32.8 g/kg at 40 °C (Palit, 1947)
16,000 mg/L at 25 °C, 269,000 mg/L at 90 °C (Smith et al., 1976)
10.162, 15.599, 19.600, and 26.846 g/L at 15.3, 25.0, 30.3, and 34.9 °C, respectively (Achard et al., 1996)
8.04, 16, and 29.1 g/L at 15, 25, and 90 °C, respectively (Standen, 1967)
11,300 mg/L at 20 °C, 25,000 mg/L at 25 °C (Howard, 1989)
11.57 g/L at 20 °C in a buffered solution (pH 1.5, Schwarzenbach et al., 1988)
0.1 M at 25 °C (Caturla et al., 1988)
14,746 mg/L at 25 °C (Riederer, 1990)
At 20 °C: 97, 77.6, 78.4, and 77.6 mM in distilled water, Pacific seawater, artificial seawater, and 35 wt % NaCl solution, respectively (Hashimoto et al., 1984)

Vapor pressure (x 10^{-5} mmHg):
10 at 20 °C (Schwarzenbach et al., 1988)
30 at 30 °C (extrapolated, McCrady et al., 1985)
4.05 at 25 °C (Riederer, 1990)

Environmental fate:
   Biological. Under anaerobic conditions, 4-nitrophenol may undergo nitroreduction producing

4-aminophenol (Kobayashi and Rittman, 1982). Estuarine sediment samples collected from the Mississippi River near Leeville, LA were used to study the mineralization of 4-nitrophenol under aerobic and anaerobic conditions. The rate of mineralization to carbon dioxide was found to be faster under aerobic conditions (1.04 x 10^{-3} μg/day/g dry sediment) than under anaerobic conditions (2.95 x 10^{-5} μg/day/g dry sediment) (Siragusa and DeLaune, 1986). In lake water samples collected from Beebe and Cayuga Lakes, Ithaca, NY, 4-nitrophenol at 50, 75, and 100 μg/L was not mineralized after 7 d. When the lake water samples were inoculated with the microorganism *Corynebacterium* sp., extensive mineralization was observed. However, at a concentration of 26 μg/L, the extent of mineralization was much lower than at higher concentrations. The presence of a eucaryotic inhibitor (cycloheximide) also inhibited mineralization at the lower concentration but did not affect mineralization at the higher concentrations (Zaidi et al., 1989).

4-Nitrophenol degraded rapidly from flooded alluvial and pokkali (organic matter-rich acid sulfate) soils that were inoculated with parathion-enrichment culture containing 5-day-old cultures of *Flavobacterium* sp. ATCC 27551 and *Pseudomonas* sp. ATCC 29353 (Sudhaker-Barik and Sethunathan, 1978a). 4-Nitrophenol disappeared completely with the formation of nitrite, particularly in the inoculated soils rather than in the uninoculated soils.

In activated sludge, 0.5% mineralized to carbon dioxide after 5 d (Freitag et al., 1985). Intermediate products include 4-nitrophenol, which further degraded to hydroquinone with lesser quantities of oxyhydroquinone (Nyholm et al., 1984). When 4-nitrophenol was statically incubated in the dark at 25 °C with yeast extract and settled domestic wastewater inoculum, 100% biodegradation with rapid adaptation was achieved after 7 d (Tabak et al., 1981). In the presence of suspended natural populations from unpolluted aquatic systems, the second-order microbial transformation rate constant determined in the laboratory was reported to be 3.8 ± 1.4 x 10^{-11} L/organism·h (Steen, 1991). In activated sludge inoculum, 95.0% COD removal was achieved. The average rate of biodegradation was 17.5 mg COD/g·h (Pitter, 1976).

Surface Water. Photodegration half-lives of 5.7, 6.7, and 13.7 d were reported at pH values of 5, 7, and 9, respectively (Hustert et al., 1981).

Groundwater. Nielsen et al. (1996) studied the degradation of 4-nitrophenol in a shallow, glaciofluvial, unconfined sandy aquifer in Jutland, Denmark. As part of the *in situ* microcosm study, a cylinder that was open at the bottom and screened at the top was installed through a cased borehole approximately 5 m below grade. Five liters of water was aerated with atmospheric air to ensure aerobic conditions were maintained. Groundwater was analyzed weekly for approximately 3 months to determine 4-nitrophenol concentrations over time. The experimentally determined first-order biodegradation rate constant and corresponding half-life were 0.2/d and 3.47 d, respectively.

Photolytic. An aqueous solution containing 200 ppm 4-nitrophenol exposed to sunlight for 1-2 months yielded hydroquinone, 4-nitrocatechol and an unidentified polymeric substance (Callahan et al., 1979). Under artificial sunlight, river water containing 2-5 ppm 4-nitrophenol photodegraded to produce trace amounts of 4-aminophenol (Mansour et al., 1989). A carbon dioxide yield of 39.5% was achieved when 4-nitrophenol adsorbed on silica gel was irradiated with light (λ >290 nm) for 17 h (Freitag et al., 1985).

Chemical/Physical. Wet oxidation of 4-nitrophenol at 320 °C yielded formic and acetic acids (Randall and Knopp, 1980). Wet oxidation of 4-nitrophenol at an elevated pressure and temperature gave the following products: acetone, acetaldehyde, formic, acetic, maleic, oxalic and succinic acids (Keen and Baillod, 1985).

In an aqueous solution, 4-nitrophenol (100 μM) reacted with Fenton's reagent (35 μM). After 15 min into the reaction, the following products were identified: 1,2,4-trihydroxybenzene, hydroquinone, hydroxy-*p*-benzoquinone, *p*-benzoquinone, and 4-nitrocatechol. After 3.5 h, 90% of the 4-nitrophenol was destroyed (Lipczynska-Kochany, 1991). In a dilute aqueous solution at pH 6.0, 4-nitrophenol reacted with excess hypochlorous acid forming 2,6-dichlorobenzoquinone, 2,6-dichloro-4-nitrophenol, and 2,3,4,6-tetrachlorophenol at yields of 20, 1, and 0.3%, respectively (Smith et al., 1976).

Toxicity:

EC_{10} and EC_{50} concentrations inhibiting the growth of alga *Scenedesmus subspicatus* in 96 h were 8.0 and 32.0 mg/L, respectively (Geyer et al., 1985).

EC_{50} (48-h) for *Daphnia magna* 7.68 mg/L (Keen and Baillod, 1985).

EC_{50} (15-min) for *Photobacterium phosphoreum* 27.8 mg/L (Yuan and Lang, 1997).

IC_{50} (24-h) for river bacteria 30.4 mg/L (Yuan and Lang, 1997).

LC_{50} (contact) for earthworm (*Eisenia fetida*) 0.7 $\mu g/cm^2$ (Neuhauser et al., 1985).

LC_{50} (8-d) for fathead minnows 49-40 mg/L (Spehar et al., 1982).

LC_{50} (96-h) for bluegill sunfish 8.3 mg/L, fathead minnows 59-62 mg/L (Spehar et al., 1982), *Cyprinodon variegatus* 27 ppm using natural seawater (Heitmuller et al., 1981).

LC_{50} (72-h) for *Cyprinodon variegatus* 27 ppm (Heitmuller et al., 1981).

LC_{50} (48-h) for red killifish 100 mg/L (Yoshioka et al., 1986), fathead minnows 10 mg/L (quoted, RTECS, 1985), *Daphnia magna* 22 mg/L (LeBlanc, 1980), *Cyprinodon variegatus* 28 ppm (Heitmuller et al., 1981).

LC_{50} (24-h) for *Daphnia magna* 24 mg/L (LeBlanc, 1980), *Cyprinodon variegatus* 28 ppm (Heitmuller et al., 1981).

Acute oral LD_{50} for mice 380 mg/kg, rats 250 mg/kg (quoted, RTECS, 1985).

Heitmuller et al. (1981) reported a NOEC of 24 ppm.

Drinking water standard: No MCLGs or MCLs have been proposed (U.S. EPA, 1996).

Uses: Fungicide for leather; production of parathion; preparation of *p*-nitrophenyl acetate and other organic compounds.

1-NITROPROPANE

Synonym: UN 2608.

$$\diagdown\diagup\diagdown\diagup\text{NO}_2$$

CAS Registry Number: 108-03-2
DOT: 2608
DOT label: Combustible liquid
Molecular formula: $C_3H_7NO_2$
Formula weight: 89.09
RTECS: TZ5075000
Merck reference: 10, 6473

Physical state, color, and odor:
Colorless, oily liquid with a mild, fruity odor. Odor threshold concentration is 11 ppm (Amoore and Hautala, 1983).

Melting point (°C):
-108 (Weast, 1986)

Boiling point (°C):
130-131 (Weast, 1986)

Density (g/cm³):
1.0081 at 24/4 °C (Weast, 1986)
0.9934 at 25/4 °C (Windholz et al., 1983)

Diffusivity in water (x 10⁻⁵ cm²/sec):
1.08 at 25 °C using method of Hayduk and Laudie (1974)

Dissociation constant, pKₐ:
8.98 at 25 °C (Dean, 1987)

Flash point (°C):
35.8 (NIOSH, 1997)
34 (Windholz et al., 1983)

Lower explosive limit (%):
2.2 (NIOSH, 1997)

Henry's law constant (x 10⁻⁵ atm·m³/mol):
8.68 at 25 °C (Hine and Mookerjee, 1975)

Ionization potential (eV):
10.78 (Lias, 1998)

Soil sorption coefficient, log K_{oc}:
Unavailable because experimental methods for estimation of this parameter for nitroaliphatics are lacking in the documented literature. However, its moderate solubility in water suggests its adsorption to soil will be low (Lyman et al., 1982).

Octanol/water partition coefficient, log K_{ow}:
0.65 (Hansch and Anderson, 1967)
0.87 (Hansch and Leo, 1979)

Solubility in organics:
Soluble in alcohol, chloroform, and ether (Weast, 1986). Miscible with many organic solvents (Windholz et al., 1983).

Solubility in water (wtv%):
1.78 at 0 °C, 1.65 at 9.5 °C, 1.58 at 20.0 °C, 1.70 at 31.0 °C, 1.73 at 41.0 °C, 1.78 at 50.2 °C, 1.91 at 60.2 °C, 2.05 at 70.1 °C, 2.17 at 81.2 °C, 2.29 at 90.5 °C (Stephenson, 1992)

Vapor density:
3.64 g/L at 25 °C, 3.08 (air = 1)

Vapor pressure (mmHg):
8 at 20 °C (NIOSH, 1997)

Exposure limits: NIOSH REL: TWA 25 ppm (90 mg/m^3), IDLH 1,000 ppm; OSHA PEL: TWA 25 ppm; ACGIH TLV: TWA 25 ppm (adopted).

Symptoms of exposure: An irritation concentration of 360.00 mg/m^3 in air was reported by Ruth (1986).

Toxicity:
Acute oral LD$_{50}$ for mice 800 mg/kg, rats 455 mg/kg (quoted, RTECS, 1985).

Uses: Solvent for cellulose acetate, lacquers, vinyl resins, fats, oils, dyes, synthetic rubbers; chemical intermediate; propellant; gasoline additive.

2-NITROPROPANE

Synonyms: Dimethylnitromethane; Isonitropropane; Nipar S-20 solvent; Nipar S-30 solvent; Nitroisopropane; 2-NP; RCRA waste number U171; UN 2608.

CAS Registry Number: 79-46-9
DOT: 2608
DOT label: Combustible liquid
Molecular formula: $C_3H_7NO_2$
Formula weight: 89.09
RTECS: TZ5250000
Merck reference: 10, 6474

Physical state, color, and odor:
Colorless liquid with a mild, fruity odor. Odor threshold concentration is 70 ppm (Amoore and Hautala, 1983).

Melting point (°C):
-93 (Weast, 1986)

Boiling point (°C):
120 (Weast, 1986)

Density (g/cm³):
0.9876 at 20/4 °C (Weast, 1986)
0.9821 at 25/4 °C (Windholz et al., 1983)

Diffusivity in water (x 10⁻⁵ cm²/sec):
0.94 at 20 °C using method of Hayduk and Laudie (1974)

Dissociation constant, pK$_a$:
7.675 at 25 °C (Dean, 1987)

Flash point (°C):
24.1 (NIOSH, 1997)

Lower explosive limit (%):
2.6 (NIOSH, 1997)

Upper explosive limit (%):
11.0 (NIOSH, 1997)

Henry's law constant (x 10⁻⁴ atm·m³/mol):
1.23 at 25 °C (Hine and Mookerjee, 1975)

Ionization potential (eV):
10.74 (Lias, 1998)

Bioconcentration factor, log BCF:
1.85 (activated sludge), 1.30 (algae) (Freitag et al., 1985)

Soil sorption coefficient, log K_{oc}:
Unavailable because experimental methods for estimation of this parameter for nitroaliphatics are lacking in the documented literature. However, its moderate solubility in water suggests its adsorption to soil will be low (Lyman et al., 1982).

Octanol/water partition coefficient, log K_{ow}:
Unavailable because experimental methods for estimation of this parameter for nitroaliphatics are lacking in the documented literature

Solubility in organics:
Miscible with many organic solvents (Windholz et al., 1983)

Solubility in water (wt %):
3.07 at 0 °C, 1.80 at 19.7 °C, 1.70 at 30.9 °C, 1.78 at 40.0 °C, 1.79 at 50.5 °C, 2.07 at 61.1 °C, 2.09 at 70.6 °C, 2.26 at 81.0 °C, 2.36 at 90.2 °C (Stephenson, 1992)

Vapor density:
3.64 g/L at 25 °C, 3.08 (air = 1)

Vapor pressure (mmHg):
13 at 20 °C (NIOSH, 1997)

Environmental fate:
Photolytic. Anticipated products from the reaction of 2-nitropropane with ozone or OH radicals in the atmosphere are formaldehyde and acetaldehyde (Cupitt, 1980).

Exposure limits: Potential occupational carcinogen. NIOSH REL: IDLH 100 ppm; OSHA PEL: TWA 25 ppm (90 mg/m^3); ACGIH TLV: TWA 10 ppm (adopted).

Toxicity:
Acute oral LD_{50} for rats 720 mg/kg (quoted, RTECS, 1985).

Uses: Solvent for cellulose acetate, lacquers, vinyl resins, fats, oils, dyes, synthetic rubbers; chemical intermediate; propellant; gasoline additive.

N-NITROSODIMETHYLAMINE

Synonyms: Dimethylnitrosamine; *N*-Dimethylnitrosamine; *N,N*-Dimethylnitrosamine; Dimethylnitrosomine; DMN; DMNA; **N-Methyl-*N*-nitrosomethanamine**; NDMA; Nitrous dimethylamide; RCRA waste number P082.

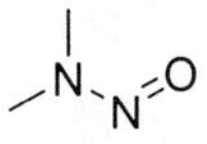

CAS Registry Number: 62-75-9
DOT: 1955
DOT label: Poison
Molecular formula: $C_2H_6N_2O$
Formula weight: 74.09
RTECS: IQ0525000
Merck reference: 10, 6483

Physical state, color, and odor:
Yellow, oily liquid with a faint, characteristic odor

Boiling point (°C):
151-153 (Dean, 1987)
154 (Weast, 1986)

Density (g/cm³):
1.0059 at 20/4 °C (Weast, 1986)
1.0049 at 18/4 °C (Weast and Astle, 1986)

Diffusivity in water (x 10^{-5} cm²/sec):
1.06 at 20 °C using method of Hayduk and Laudie (1974)

Flash point (°C):
61 (Aldrich, 1988)

Henry's law constant (atm·m³/mol):
0.143 at 25 °C (estimated using a solubility of 1,000 g/L)

Ionization potential (eV):
8.69 (Lias et al., 1998)

Soil sorption coefficient, log K_{oc}:
1.41 using method of Kenaga and Goring (1980)

Octanol/water partition coefficient, log K_{ow}:
0.06 (Radding et al., 1976)

Solubility in organics:
Soluble in solvents (U.S. EPA, 1985), including ethanol and ether (Weast, 1986)

Solubility in water:
Miscible (Mirvish et al., 1976)

Vapor pressure (mmHg):
8.1 at 25 °C (Mabey et al., 1982)
2.7 at 20 °C (Klein, 1982)

Environmental fate:
Biological. Two of seven microorganisms, *Escherichia coli* and *Pseudomonas fluorescens*, were capable of slowly degrading *N*-nitrosodimethylamine to dimethylamine (Mallik and Tesfai, 1981).

Photolytic. A Teflon bag containing air and *N*-nitrosodimethylamine was subjected to sunlight on two different d. On a cloudy day, half of the *N*-nitrosodimethylamine was photolyzed in 60 min. On a sunny day, half of the *N*-nitrosodimethylamine was photolyzed in 30 min. Photolysis products include nitric oxide, carbon monoxide, formaldehyde, and an unidentified compound (Hanst et al., 1977). In a separate experiment, Tuazon et al. (1984a) irradiated an ozone-rich atmosphere containing *N*-nitrosodimethylamine. Photolysis products identified include dimethylnitramine, nitromethane, formaldehyde, carbon monoxide, nitrogen dioxide, nitrogen pentoxide, and nitric acid.

Chemical/Physical. *N*-Nitrosodimethylamine will not hydrolyze because it does not contain a hydrolyzable functional group (Kollig, 1993).

Exposure limits: Potential occupational carcinogen. Because no standards have been established, NIOSH (1997) recommends the most reliable and protective respirators be used, i.e., a self-contained breathing apparatus that has a full facepiece and is operated under positive-pressure or a supplied-air respirator that has a full facepiece and is operated under pressure-demand or under positive-pressure in combination with a self-contained breathing apparatus operated under pressure-demand or positive-pressure.

OSHA recommends that worker exposure to this chemical is to be controlled by use of engineering control, proper work practices, and proper selection of personal protective equipment. Specific details of these requirements can be found in CFR 1910.1003-1910.1016.

Toxicity:
LC_{50} (inhalation) for mice 57 ppm/4-h, rats 78 ppm/4-h (quoted, RTECS, 1985).
Acute oral LD_{50} for hamsters 28 mg/kg, rats 45 mg/kg (quoted, RTECS, 1985).

Source: After 2 d, *N*-nitrodimethylamine was identified as a major metabolite of dimethylamine in an Arkport fine sandy loam (Varna, NY) and sandy soil (Lake George, NY) amended with sewage and nitrite-N. Mills and Alexander (1976) reported that *N*-nitrosodimethylamine also formed in soil, municipal sewage, and lake water supplemented with dimethylamine (ppm) and nitrite-N (100 ppm). They found that nitrosation occurred under nonenzymatic conditions at neutral pHs.

Uses: Rubber accelerator; solvent in fiber and plastic industry; rocket fuels; lubricants; condensers to increase dielectric constant; industrial solvent; antioxidant; nematocide; softener of copolymers; research chemical; plasticizer in acrylonitrile polymers; inhibit nitrification in soil; chemical intermediate for 1,1-dimethylhydrazine.

N-NITROSODIPHENYLAMINE

Synonyms: Benzenamine; Curetard A; Delac J; Diphenylnitrosamine; Diphenyl-*N*-nitrosamine; *N*,*N*-Diphenylnitrosamine; Naugard TJB; NCI-C02880; NDPA; NDPhA; Nitrosodiphenylamine; *N*-Nitroso-*n*-phenylamine; *N*-Nitroso-*n*-phenylbenzenamine; Nitrous diphenylamide; Redax; Retarder J; TJB; Vulcalent A; Vulcatard; Vulcatard A; Vultrol.

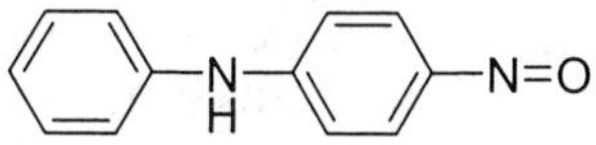

CAS Registry Number: 86-30-6
Molecular formula: $C_{12}H_{10}N_2O$
Formula weight: 198.22
RTECS: JJ9800000
Merck reference: 10, 6485

Physical state and color:
Yellow to brown to orange power or flakes

Melting point (°C):
66.5 (Weast, 1986)
65-66 (Fluka, 1988)

Diffusivity in water (x 10^{-5} cm²/sec):
0.58 at 20 °C using method of Hayduk and Laudie (1974)

Henry's law constant (x 10^{-8} atm·m³/mol):
2.33 at 25 °C (approximate - calculated from water solubility and vapor pressure)

Bioconcentration factor, log BCF:
2.34 (bluegill sunfish, Veith et al., 1980)

Soil sorption coefficient, log K_{oc}:
2.76 (estimated, Montgomery, 1989)

Octanol/water partition coefficient, log K_{ow}:
3.13 (Veith et al., 1980)
3.18 (Garst and Wilson, 1984)

Solubility in organics:
Soluble in ethanol, benzene, ethylene dichloride, gasoline, (Keith and Walters, 1992), ether, chloroform, and slightly soluble in petroleum ether (Windholz et al., 1983)

Solubility in water:
35.1 mg/L at 25 °C (Banerjee et al., 1980)

Vapor pressure (mmHg):
0.1 at 25 °C (assigned by analogy, Mabey et al., 1982)

Environmental fate:
Chemical/Physical. Above 85 °C, technical grades may decompose to nitrogen oxides

(IARC, 1978). *N*-Nitrosodiphenylamine will not hydrolyze because it does not contain a hydrolyzable functional group (Kollig, 1993).

Toxicity:

LC$_{50}$ (contact) for earthworm (*Eisenia fetida*) 2.4 μg/cm^2 (Neuhauser et al., 1985).

LC$_{50}$ (96-h) for bluegill sunfish 5.8 mg/L (Spehar et al., 1982).

LC$_{50}$ (48-h) for *Daphnia magna* 7.8 mg/L (LeBlanc, 1980).

LC$_{50}$ (24-h) for *Daphnia magna* >46 mg/L (LeBlanc, 1980).

Acute oral LD$_{50}$ for mice 3,850 mg/kg, rats 1,650 mg/kg (quoted, RTECS, 1985).

Uses: Chemical intermediate for *N*-phenyl-*p*-phenylenediamine; rubber processing (vulcanization retarder).

N-NITROSODI-*n*-PROPYLAMINE

Synonyms: Dipropylnitrosamine; Di-*n*-propylnitrosamine; DPN; DPNA; NDPA; *N*-Nitrosodi-propylamine; **N-Nitroso-*n*-propyl-1-propanamine**; RCRA waste number U111.

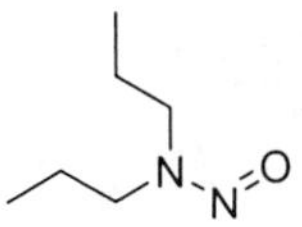

CAS Registry Number: 621-64-7
Molecular formula: $C_6H_{14}N_2O$
Formula weight: 130.19
RTECS: JL9700000

Physical state and color:
Yellow to gold colored liquid

Boiling point (°C):
205.9 (Dean, 1973)

Density (g/cm³):
0.9160 at 20/4 °C (IARC, 1978)

Diffusivity in water (x 10^{-5} cm²/sec):
0.72 at 20 °C using method of Hayduk and Laudie (1974)

Soil sorption coefficient, log K_{oc}:
1.01 (estimated, Montgomery, 1989)

Octanol/water partition coefficient, log K_{ow}:
1.31 (calculated, U.S. EPA, 1980a)

Solubility in organics:
Very soluble in ethanol and ether (Keith and Walters, 1992)

Solubility in water:
9,900 mg/L at 25 °C (Mirvish et al., 1976)

Environmental fate:
   Chemical/Physical. *N*-Nitroso-*n*-propylamine will not hydrolyze because it does not contain a hydrolyzable functional group (Kollig, 1993).

Toxicity:
   LC_{50} (contact) for earthworm (*Eisenia fetida*) 11.0 μg/cm² (Neuhauser et al., 1985).
   Acute oral LD_{50} for rats 480 mg/kg (quoted, RTECS, 1985).

Use: Research chemical.

2-NITROTOLUENE

Synonyms: 1-Methyl-2-nitrobenzene; 2-Methylnitrobenzene; *o*-Methylnitrobenzene; *o*-Nitro-toluene; ONT; UN 1664.

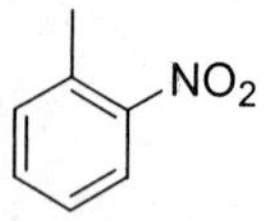

CAS Registry Number: 88-72-2
DOT: 1664
DOT label: Poison
Molecular formula: $C_7H_7NO_2$
Formula weight: 137.14
RTECS: XT3150000
Merck reference: 10, 6497

Physical state, color, and odor:
Yellowish liquid with a faint, aromatic odor

Melting point (°C):
-9.5 (needles), -2.5 °C (crystals) (Weast, 1986)
-4.1 (Stull, 1947)

Boiling point (°C):
221.7 (Weast, 1986)
225 (Gross et al., 1933)
220.4 (Hawley, 1981)

Density (g/cm³):
1.1629 at 20/4 °C (Weast, 1986)

Diffusivity in water (x 10⁻⁵ cm²/sec):
0.80 at 20 °C using method of Hayduk and Laudie (1974)

Flash point (°C):
107 (NIOSH, 1997)

Lower explosive limit (%):
2.2 (NIOSH, 1997)

Henry's law constant (x 10⁻⁵ atm·m³/mol):
4.51 at 20 °C (approximate - calculated from water solubility and vapor pressure)

Interfacial tension with water (dyn/cm at 20 °C):
27.19 (Harkins et al., 1920)

Ionization potential (eV):
9.24 (Lias, 1998)

Soil sorption coefficient, log K_{oc}:
Unavailable because experimental methods for estimation of this parameter for nitroaromatics are lacking in the documented literature

Octanol/water partition coefficient, log K_{ow}:
2.30 (Leo et al., 1971)

Solubility in organics:
Soluble in alcohol, ether (Weast, 1986), benzene, and petroleum ether (Windholz et al., 1983)

Solubility in water:
652 mg/kg at 30 °C (Gross et al., 1933)

Vapor density:
5.61 g/L at 25 °C, 4.73 (air = 1)

Vapor pressure (mmHg):
0.1 at 20 °C (NIOSH, 1987)
0.25 at 30 °C (Verschueren, 1983)

Exposure limits: NIOSH REL: TWA 2 ppm (11 mg/m^3), IDLH 200 ppm; OSHA PEL: TWA 5 ppm (30 mg/m^3); ACGIH TLV: TWA 2 ppm (adopted).

Toxicity:
   LC_{50} (48-h) for red killifish 245 mg/L (Yoshioka et al., 1986).
   Acute oral LD_{50} for mice 970 mg/kg, rats 891 mg/kg (quoted, RTECS, 1985).

Uses: Manufacture of dyes, nitrobenzoic acids, toluidines, etc.

3-NITROTOLUENE

Synonyms: 1-Methyl-3-nitrobenzene; 3-Methylnitrobenzene; *m*-Methylnitrobenzene; MNT; *m*-Nitrotoluene; 3-Nitrotoluol; UN 1664.

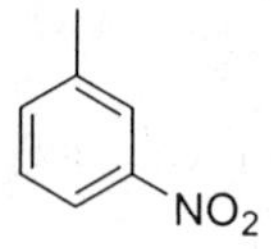

CAS Registry Number: 99-08-1
DOT: 1664
DOT label: Poison
Molecular formula: $C_7H_7NO_2$
Formula weight: 137.14
RTECS: XT2975000
Merck reference: 10, 6497

Physical state, color, and odor:
Clear, yellowish liquid with an aromatic odor. Odor threshold concentration is 600 ppm (Amoore and Hautala, 1983).

Melting point (°C):
16 (Weast, 1986)

Boiling point (°C):
232.6 (Weast, 1986)

Density (g/cm³):
1.1571 at 20/4 °C (Weast, 1986)

Diffusivity in water (x 10^{-5} cm²/sec):
0.80 at 20 °C using method of Hayduk and Laudie (1974)

Flash point (°C):
107 (NIOSH, 1997)

Lower explosive limit (%):
1.6 (NIOSH, 1997)

Henry's law constant (x 10^{-5} atm·m³/mol):
5.41 at 20 °C (approximate - calculated from water solubility and vapor pressure)

Interfacial tension with water (dyn/cm at 20 °C):
27.68 (Harkins et al., 1920)

Ionization potential (eV):
9.45 (Lias, 1998)
9.65 ± 0.05 (Franklin et al., 1969)

Soil sorption coefficient, log K_{oc}:
Unavailable because experimental methods for estimation of this parameter for nitroaromatics are lacking in the documented literature

Octanol/water partition coefficient, log K_{ow}:
2.40, 2.45 (Leo et al., 1971)
2.42 (Fujita et al., 1964)

Solubility in organics:
Soluble in alcohol, benzene, and ether (Weast, 1986)

Solubility in water:
498 mg/kg at 30 °C (Gross et al., 1933)

Vapor density:
5.61 g/L at 25 °C, 4.73 (air = 1)

Vapor pressure (mmHg):
0.1 at 20 °C (NIOSH, 1987)
0.25 at 25 °C (Verschueren, 1983)

Environmental fate:
Biological. Under anaerobic conditions using a sewage inoculum, 3-nitrotoluene, and 4-nitrotoluene both degraded to toluidine (Hallas and Alexander, 1983).

Exposure limits: NIOSH REL: TWA 2 ppm (11 mg/m^3), IDLH 200 ppm; OSHA PEL: TWA 5 ppm (30 mg/m^3); ACGIH TLV: TWA 2 ppm (adopted).

Toxicity:
Acute oral LD$_{50}$ for guinea pigs 3,600 mg/kg, mice 330 mg/kg, rats 1,072 mg/kg, rabbits 2,400 (quoted, RTECS, 1985).

Uses: Manufacture of dyes, nitrobenzoic acids, toluidines; organic synthesis.

4-NITROTOLUENE

Synonyms: 1-Methyl-4-nitrobenzene; 4-Methylnitrobenzene; *p*-Methylnitrobenzene; NCI-C60537; *p*-Nitrotoluene; 4-Nitrotoluol; PNT; UN 1664.

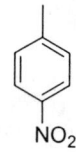

CAS Registry Number: 99-99-0
DOT: 1664
DOT label: Poison
Molecular formula: $C_7H_7NO_2$
Formula weight: 137.14
RTECS: XT3325000
Merck reference: 10, 6497

Physical state, color, and odor:
Yellowish crystals with a weak, aromatic odor

Melting point (°C):
54.5 (Weast, 1986)
51.3 (Verschueren, 1983)
51.9 (Stull, 1947)

Boiling point (°C):
238.3 (Weast, 1986)

Density (g/cm³):
1.1038 at 75/4 °C (Weast, 1986)
1.286 at 20 °C (Verschueren, 1983)

Diffusivity in water (x 10^{-5} cm²/sec):
0.81 at 20 °C using method of Hayduk and Laudie (1974)

Dissociation constant, pK_a:
11.27 (Perrin, 1972)

Flash point (°C):
107 (NIOSH, 1997)

Lower explosive limit (%):
1.6 (NIOSH, 1997)

Entropy of fusion (cal/mol·K):
12.6 (Tsonopoulos and Prausnitz, 1971)

Heat of fusion (kcal/mol):
4.10 (Tsonopoulos and Prausnitz, 1971)

Henry's law constant (x 10^{-5} atm·m³/mol):
5.0 at 25 °C using method of Hine and Mookerjee (1975)

Ionization potential (eV):
9.46 (Lias, 1998)
9.82 (Franklin et al., 1969)

Soil sorption coefficient, log K_{oc}:
Unavailable because experimental methods for estimation of this parameter for nitroaromatics are lacking in the documented literature

Octanol/water partition coefficient, log K_{ow}:
2.37 (Fujita et al., 1964)
2.42 (Leo et al., 1971)

Solubility in organics:
Soluble in acetone, alcohol, benzene, ether (Weast, 1986), and chloroform (Windholz et al., 1983)

Solubility in water:
442 mg/kg at 30 °C (Gross et al., 1933)
At 20 °C: 210, 183, 187, and 177 μmol/L in distilled water, Pacific seawater, artificial sea-
    water, and 35 wt % NaCl, respectively (Hashimoto et al., 1984)

Vapor pressure (mmHg):
0.1 at 20 °C (NIOSH, 1997)
0.22 at 30 °C (Verschueren, 1983)
0.004633 at 23.9 °C, 0.005484 at 26.0 °C (Lenchitz and Velicky, 1970)

Environmental fate:
    Biological. Under anaerobic conditions using a sewage inoculum, 3- and 4-nitrotoluene both degraded to toluidine (Hallas and Alexander, 1983).
    Chemical. Though no products were identified, 4-nitrotoluene (1.5 x 10^{-5} M) was reduced by iron metal (33.3 g/L acid washed 18-20 mesh) in a carbonate buffer (1.5 x 10^{-2} M) at pH 5.9 and 15 °C. Based on the pseudo-first-order disappearance rate of 0.0335/min, the half-life was 20.7 min (Agrawal and Tratnyek, 1996).

Exposure limits: NIOSH REL: TWA 2 ppm (11 mg/m^3), IDLH 200 ppm; OSHA PEL: TWA 5 ppm (30 mg/m^3); ACGIH TLV: TWA 2 ppm (adopted).

Toxicity:
    LC_{50} (48-h) for red killifish 537 mg/L (Yoshioka et al., 1986).
    Acute oral LD_{50} for mice 1,231 mg/kg, rats 1,960 mg/kg (quoted, RTECS, 1985).

Uses: Manufacture of dyes, nitrobenzoic acids, toluidines, etc.

NONANE

Synonyms: *n*-Nonane; Nonyl hydride; Shellsol 140; UN 1920.

CAS Registry Number: 111-84-2
DOT: 1920
Molecular formula: C_9H_{20}
Formula weight: 128.26
RTECS: RA6115000

Physical state, color, and odor:
Clear, colorless liquid with a gasoline-like odor. Odor threshold concentration is 47 ppm (Amoore and Hautala, 1983).

Melting point (°C):
-53.7 (Stull, 1947)

Boiling point (°C):
150.8 (Weast, 1986)

Density (g/cm³):
0.71763 at 20/4 °C, 0.71381 at 25/4 °C (Dreisbach, 1959)

Diffusivity in water (x 10⁻⁵ cm²/sec):
0.63 at 20 °C using method of Hayduk and Laudie (1974)

Dissociation constant, pK_a:
>14 (Schwarzenbach et al., 1993)

Flash point (°C):
31.4 (NIOSH, 1997)
33 (Affens and McLaren, 1972)

Lower explosive limit (%):
0.8 (NIOSH, 1997)

Upper explosive limit (%):
2.9 (NIOSH, 1997)

Heat of fusion (kcal/mol):
3.697 (Riddick et al., 1986)

Henry's law constant (atm·m³/mol):
0.400, 0.496, 0.332, 0.414, and 0.465 at 10, 15, 20, 25, and 30 °C, respectively (Ashworth et al., 1988).
2.32 at 25 °C (Jönsson et al., 1982)

Interfacial tension with water (dyn/cm):
51.96 at 20 °C (Fowkes, 1980)
49.86 at 25 °C (Jańczuk et al., 1993)

Ionization potential (eV):
9.71 (Lias, 1998)

Soil sorption coefficient, log K_{oc}:
Unavailable because experimental methods for estimation of this parameter for aliphatic hydrocarbons are lacking in the documented literature

Octanol/water partition coefficient, log K_{ow}:
4.67 using method of Hansch et al. (1968)

Solubility in organics:
In methanol, g/L: 84 at 5 °C, 95 at 10 °C, 105 at 15 °C, 116 at 20 °C, 129 at 25 °C, 142 at 30 °C, 155 at 35 °C, 170 at 40 °C (Kiser et al., 1961)

Solubility in water:
In mg/kg: 0.122 at 25 °C, 0.309 at 69.7 °C, 0.420 at 99.1 °C, 1.70 at 121.3 °C, 5.07 at 136.6 °C (Price, 1976)
0.07 mg/L at 20 °C, 0.43 mg/L in seawater at 20 °C (quoted, Verschueren, 1983)
220 μg/kg at 25 °C (McAuliffe, 1969)
1 x 10^{-8} at 25 °C (mole fraction, Krasnoshchekova and Gubergrits, 1973)
272 mg/L at 25 °C (Jönsson et al., 1982)

Vapor density:
5.24 g/L at 25 °C, 4.43 (air = 1)

Vapor pressure (mmHg):
3.22 at 20 °C (Verschueren, 1983)
4.35 at 25 °C (Wilhoit and Zwolinski, 1971)

Environmental fate:
Biological. Nonane may biodegrade in two ways. This first is the formation of nonyl hydroperoxide, which decomposes to 1-nonanol followed by oxidation to nonanoic acid. The other pathway involves dehydrogenation to 1-nonene, which may react with water giving 1-nonanol (Dugan, 1972). Microorganisms can oxidize alkanes under aerobic conditions. The most common degradative pathway involves the oxidation of the terminal methyl group forming the 1-nonanol. The alcohol may undergo a series of dehydrogenation steps forming nonanal then a fatty acid (nonanoic acid). The fatty acid may then be metabolized by β-oxidation to form the mineralization products, carbon dioxide, and water (Singer and Finnerty, 1984).

Photolytic. Atkinson reported a photooxidation rate constant of 1.02 x 10^{-11} cm^3/molecule·sec for the reaction of nonane and OH radicals in the atmosphere. Photooxidation reaction rate constants of 1.0 x 10^{-11} and 2.30 x 10^{-16} cm^3/molecule·sec were reported for the reaction of nonane with OH and NO$_3$, respectively (Sabljić and Güsten, 1990).

Chemical/Physical. Complete combustion in air yields carbon dioxide and water.

Exposure limits: NIOSH REL: TWA 200 ppm (1,050 mg/m^3); ACGIH TLV: TWA 200 ppm (adopted).

Toxicity:
LC$_{50}$ (8-h inhalation) for Sprague-Dawley rats was estimated to be 4,467 ppm (Nilsen et al., 1988).
LC$_{50}$ (inhalation) for rats 3,200 ppm/4-h (quoted, RTECS, 1985).

Source: Schauer et al. (1999) reported nonane in a diesel-powered medium-duty truck exhaust at an emission rate of 160 μg/km.

Uses: Solvent; standardized hydrocarbon; manufacturing paraffin products; biodegradable detergents; jet fuel research; rubber industry; paper processing industry; distillation chaser; component of gasoline and similar fuels; organic synthesis.

OCTACHLORONAPHTHALENE

Synonym: Halowax 1051.

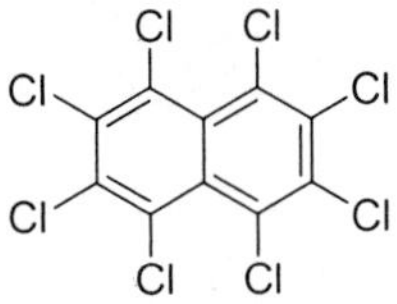

CAS Registry Number: 2234-13-1
Molecular formula: $C_{10}Cl_8$
Formula weight: 403.73
RTECS: QK0250000

Physical state, color, and odor:
Waxy, light yellow solid with an aromatic odor

Melting point (°C):
197-198 (Weast, 1986)

Boiling point (°C):
413 (NIOSH, 1997)

Density (g/cm³):
2.00 at 20/4 °C (NIOSH, 1997)

Diffusivity in water (x 10^{-5} cm²/sec):
0.45 at 20 °C using method of Hayduk and Laudie (1974)

Flash point (°C):
Noncombustible solid (NIOSH, 1997)

Bioconcentration factor, log BCF:
2.52 (rainbow trout, Oliver and Niimi, 1985)

Solubility in organics:
Soluble in benzene, chloroform, and ligroin (Weast, 1986)

Vapor pressure (mmHg):
4.20 x 10^{-7} at 25.00 °C (Lei et al., 1999)

Exposure limits (mg/m³): NIOSH REL: TWA 0.1, STEL 0.3; OSHA PEL: TWA 0.1; ACGIH TLV: TWA 0.1, STEL, 0,3 (adopted).

Toxicity:
   LC_{50} (96-h) for *Cyprinodon variegatus* >560 ppm using natural seawater (Heitmuller et al., 1981).
   LC_{50} (48-h) and LC_{50} (24-h) values for *Daphnia magna* >525 mg/L (Mayer and Ellersieck, 1986), *Cyprinodon variegatus* >560 ppm (Heitmuller et al., 1981).
   Heitmuller et al. (1981) reported a NOEC of 560 ppm.

Uses: Chemical research; organic synthesis.

OCTANE

Synonyms: *n*-Octane; UN 1262.

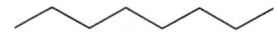

CAS Registry Number: 111-65-9
DOT: 1262
DOT label: Flammable liquid
Molecular formula: C_8H_{18}
Formula weight: 114.23
RTECS: RG8400000
Merck reference: 10, 6592

Physical state, color, and odor:
Colorless liquid with a gasoline-like odor. Odor threshold concentration is 48 ppm (Amoore and Hautala, 1983).

Melting point (°C):
-56.8 (Weast, 1986)

Boiling point (°C):
125.67 (Dreisbach, 1959)
125.66 (Stephenson and Malanowski, 1987)

Density (g/cm³):
0.70252 at 20/4 °C, 0.69849 at 25/4 °C (Dreisbach, 1959)
0.70267 at 20/4 °C, 0.68862 at 25/4 °C (Riddick et al., 1986)
0.6984 at 25.00/4 °C (Aralaguppi et al., 1999)

Diffusivity in water (x 10^{-5} cm²/sec):
0.66 at 20 °C using method of Hayduk and Laudie (1974)

Dissociation constant, pK_a:
>14 (Schwarzenbach et al., 1993)

Flash point (°C):
13.4 (NIOSH, 1997)
48 (Affens and McLaren, 1972)

Lower explosive limit (%):
1.0 (NIOSH, 1997)

Upper explosive limit (%):
6.5 (NIOSH, 1997)
4.7 (Sax and Lewis, 1987)

Entropy of fusion (cal/mol·K):
22.27 (Parks et al., 1930)
22.9 (Huffman et al., 1931)
22.91 (Finke et al., 1954)

Heat of fusion (kcal/mol):
4.802 (Parks et al., 1930)
4.936 (Huffman et al., 1931)
4.957 (Finke et al., 1954)

Henry's law constant (atm·m³/mol):
4.45 at 25 °C (Jönsson et al., 1982)

Interfacial tension with water (dyn/cm):
50.2 at 25 °C (Donahue and Bartell, 1952)
51.68 at 20 °C (Fowkes, 1980)
50.81 at 20 °C (Harkins et al., 1920)
51.09 at 25 °C (Jańczuk et al., 1993)

Ionization potential (eV):
9.80 (Lias, 1998)

Soil sorption coefficient, log K_{oc}:
Unavailable because experimental methods for estimation of this parameter for aliphatic hydrocarbons are lacking in the documented literature

Octanol/water partition coefficient, log K_{ow}:
4.00 (Coates et al., 1985)
5.18 (Tewari et al., 1982; Wasik et al., 1981)

Solubility in organics:
In methanol, g/L: 122 at 5 °C, 136 at 10 °C, 152 at 15 °C, 167 at 20 °C, 184 at 25 °C, 206 at 30 °C, 230 at 35 °C, 260 at 40 °C (Kiser et al., 1961)

Solubility in water:
In mg/kg: 0.431 at 25 °C, 0.524 at 40.1 °C, 0.907 at 69.7 °C, 1.12 at 99.1 °C, 4.62 at 121.3 °C, 8.52 at 136.6 °C, 11.8 at 149.5 °C (Price, 1976)
0.66 mg/kg at 25 °C (Coates et al., 1985; McAuliffe, 1963, 1966)
1.35 mg/kg at 0 °C, 0.85 mg/kg at 25 °C (Polak and Lu, 1973)
0.884 mg/L at 20 °C, 1.66 mg/L at 70 °C (Burris and MacIntyre, 1986)
9.66 μmol/L at 25.0 °C (Tewari et al., 1982; Wasik et al., 1981)
0.493 and 0.88 mg/L at 25 °C (Mackay and Shiu, 1981)
0.02 mL/L at 16 °C (Führer, 1924)
0.090 mL/L (Baker, 1980)
1.25 mg/L (Coutant and Keigley, 1988)
Mole fraction x 10^7: 2.6, 1.4, and 2.9 at 5.0, 25.0, and 45.0 °C, respectively (Nelson and De-Ligny, 1968)
10^{-7} at 25 °C (mole fraction, Krasnoshchekova and Gubergrits, 1973)
0.615 mg/L at 25 °C (Jönsson et al., 1982)
1.25 mg/L at 25 °C (Coutant and Keigley, 1988)

Vapor density:
4.67 g/L at 25 °C, 3.94 (air = 1)

Vapor pressure (mmHg):
11 at 20 °C, 18 at 30 °C (Verschueren, 1983)

14.14 at 25 °C (Wilhoit and Zwolinski, 1971)
10.37 at 20 °C, 118.4 at 70 °C (Burris and MacIntyre, 1986)
13.60 at 25.00 °C (Hussam and Carr, 1985)
4.2, 14.0, and 39.8 at 5.0, 25.0, and 45.0 °C, respectively (Nelson and DeLigny, 1968)

Environmental fate:
  Biological. *n*-Octane may biodegrade in two ways. This first is the formation of octyl hydroperoxide, which decomposes to 1-octanol followed by oxidation to octanoic acid. The other pathway involves dehydrogenation to 1-octene, which may react with water giving 1-octanol (Dugan, 1972). 1-Octanol was reported as the biodegradation product of octane by a *Pseudomonas* sp. (Riser-Roberts, 1992). Microorganisms can oxidize alkanes under aerobic conditions (Singer and Finnerty, 1984). The most common degradative pathway involves the oxidation of the terminal methyl group forming the corresponding alcohol (1-octanol). The alcohol may undergo a series of dehydrogenation steps forming an aldehyde (octanal) then a fatty acid (octanoic acid). The fatty acid may then be metabolized by β-oxidation to form the mineralization products, carbon dioxide and water (Singer and Finnerty, 1984).
  Photolytic. The following rate constants were reported for the reaction of octane and OH radicals in the atmosphere: 5.1 x 10^{12} cm^3/mol·sec at 300 K (Hendry and Kenley, 1979); 1.34 x 10^{-12} cm^3/molecule·sec (Greiner, 1970); 8.40 x 10^{-12} cm^3/molecule·sec (Atkinson et al., 1979), 8.42 x 10^{-12} cm^3/molecule·sec at 295 K (Darnall et al., 1978). Photooxidation reaction rate constants of 8.71 x 10^{-12} and 1.81 x 10^{-18} cm^3/molecule·sec were reported for the reaction of octane with OH and NO_3, respectively (Sabljić and Güsten, 1990).
  Surface Water. Mackay and Wolkoff (1973) estimated an evaporation half-life of 3.8 sec from a surface water body that is 25 °C and 1 m deep.
  Chemical/Physical. Complete combustion in air gives carbon dioxide and water vapor. Octane will not hydrolyze because it does not contain a hydrolyzable functional group.

Exposure limits: NIOSH REL: TWA 75 ppm (350 mg/m³), 15-min ceiling 385 ppm (1,800 mg/m³), IDLH 1,000 ppm; OSHA PEL: TWA 500 ppm (2,350 mg/m³); ACGIH TLV: TWA 300 ppm (adopted).

Symptoms of exposure: Irritates mucous membranes. Narcotic at high concentrations (Patnaik, 1992). An irritation concentration of 1,450.00 mg/m³ in air was reported by Ruth (1986).

Source: Schauer et al. (1999) reported octane in a diesel-powered medium-duty truck exhaust at an emission rate of 260 µg/km.

Uses: Solvent; rubber and paper industries; calibrations; azeotropic distillations; occurs in gasoline and petroleum naphtha; organic synthesis.

1-OCTENE

Synonyms: 1-Caprylene; 1-Octylene.

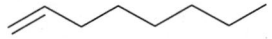

CAS Registry Number: 111-66-0
Molecular formula: C_8H_{16}
Formula weight: 112.22
Merck reference: 10, 1738

Physical state and color:
Colorless liquid

Melting point (°C):
-101.7 (Weast, 1986)

Boiling point (°C):
121.3 (Weast, 1986)

Density (g/cm³):
0.71492 at 20/4 °C, 0.71085 at 25/4 °C (Dreisbach, 1959)

Diffusivity in water (x 10⁻⁵ cm²/sec):
0.68 at 20 °C using method of Hayduk and Laudie (1974)

Dissociation constant, pK$_a$:
>14 (Schwarzenbach et al., 1993)

Flash point (°C):
21 (open cup, NFPA, 1984)

Entropy of fusion (cal/mol·K):
21.35 (McCullough et al., 1957)

Heat of fusion (kcal/mol):
3.660 (McCullough et al., 1957)
3.721 (Riddick et al., 1986)

Henry's law constant (atm·m³/mol):
0.952 at 25 °C (Hine and Mookerjee, 1975)

Ionization potential (eV):
9.43 (Lias, 1998)

Soil sorption coefficient, log K$_{oc}$:
Unavailable because experimental methods for estimation of this parameter for aliphatic hydrocarbons are lacking in the documented literature

Octanol/water partition coefficient, log K$_{ow}$:
2.79 (Coates et al., 1985)

4.56 (Shantz and Martire, 1987)
4.88 (Wasik et al., 1981)

Solubility in organics:
Soluble in acetone, benzene, and chloroform (Weast, 1986). Miscible with alcohol and ether (Windholz et al., 1983).

Solubility in water:
3.6 mg/L at 23 °C (Coates et al., 1985)
2.7 mg/kg at 25 °C (McAuliffe, 1966)
36.5 μmol/L at 25.0 °C (Wasik et al., 1981)
In 10^{-3} M nitric acid: 0.239, 0.197, and 0.153 mM at 20, 25, and 30 °C, respectively (Natarajan and Venkatachalam, 1972)

Vapor density:
4.59 g/L at 25 °C, 3.87 (air = 1)

Vapor pressure (mmHg):
17.4 at 25 °C (Wilhoit and Zwolinski, 1971)

Environmental fate:
Biological. Biooxidation of 1-octene may occur yielding 7-octen-1-ol, which may further oxidize to 7-octenoic acid (Dugan, 1972).

Photolytic. Atkinson and Carter (1984) reported a rate constant of 8.1 x 10^{-18} cm^3/molecule·sec for the reaction of 1-octene and OH radicals in the atmosphere.

Chemical/Physical. The reaction of ozone and OH radicals with 1-octene was studied in a flexible outdoor Teflon chamber (Paulson and Seinfeld, 1992). 1-Octene reacted with ozone producing heptanal, a thermally stabilized C$_7$ biradical, and hexane at yields of 80, 10, and 1%, respectively. With OH radicals, only 15% of 1-octene was converted to heptanal. In both reactions, the remaining compounds were tentatively identified as alkyl nitrates (Paulson and Seinfeld, 1977).

Uses: Plasticizer; surfactants; organic synthesis.

OXALIC ACID

Synonyms: Ethanedioic acid; Ethanedionic acid; NCI-C55209.

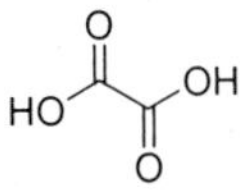

CAS Registry Number: 144-62-7
DOT: 2449
Molecular formula: $C_2H_2O_4$
Formula weight: 90.04
RTECS: RO2450000
Merck reference: 10, 6784

Physical state, color, and odor:
Colorless and odorless rhombic crystals. Hygroscopic.

Melting point (°C):
182-189.5 (anhydrous), 101.5 °C (hydrated) (Weast, 1986)

Boiling point (°C):
Sublimes at 157 (Weast, 1986)
Sublimes at 150 (Verschueren, 1983)

Density (g/cm³):
1.895-1.900 at 17/4 °C (Weast, 1986)
1.653 at 18/4 °C (Windholz et al., 1983)

Diffusivity in water (x 10^{-5} cm²/sec):
0.97 at 20 °C using method of Hayduk and Laudie (1974)

Dissociation constant (at 25 °C):
pK_1 = 1.27, pK_2 = 4.27 (Windholz et al., 1983)

Henry's law constant (x 10^{-10} atm·m³/mol):
1.43 at pH 4 (quoted, Gaffney et al., 1987)

Ionization potential (eV):
11.20 (Lias et al., 1998)

Soil sorption coefficient, log K_{oc}:
0.89 using method of Kenaga and Goring (1980)

Octanol/water partition coefficient, log K_{ow}:
-0.43, -0.81 (quoted, Verschueren, 1983)

Solubility in organics (g/L):
Alcohol (40), ether (10), glycerol (181.8) (Windholz et al., 1983)

Solubility in water:
6.71 wt % at 15 °C, 8.34 wt % at 20 °C, 9.81 wt % at 25 °C (quoted, Windholz et al., 1983)
95,000 mg/L at 15 °C, 1,290 g/L at 90 °C (quoted, Verschueren, 1983)

Vapor pressure (mmHg):
3 x 10^{-4} at 30 °C (Verschueren, 1983)

Environmental fate:
   Biological. Heukelekian and Rand (1955) reported a 5-d BOD value of 0.12 g/g which is 66.7% of the ThOD value of 0.18 g/g.
   Chemical/Physical. Above 189.5 °C, decomposes to carbon dioxide, carbon monoxide, formic acid, and water (Windholz et al., 1983). Ozonolysis of oxalic acid in distilled water at 25 °C under acidic conditions (pH 6.3) yielded carbon dioxide (Kuo et al., 1977).
   Absorbs moisture in air forming the dihydrate (Huntress and Mulliken, 1941).

Exposure limits (mg/m³): NIOSH REL: TWA 1, STEL 2, IDLH 500; OSHA PEL: TWA 1; ACGIH TLV: TWA 1, STEL 2 (adopted).

Symptoms of exposure: Ingestion may cause vomiting, diarrhea, severe gastrointestinal disorder, renal damage, shock, convulsions, and coma (Patnaik, 1992).

Toxicity:
   Acute oral LD_{50} in rats 375 mg/kg (quoted, RTECS, 1985).

Source: Oxalic acid occurs naturally in many plants including buckwheat leaves (111,000 ppm), lambsquarter (140,000-300,000 ppm), black pepper (4,000-34,000 ppm), star fruit (50,000-95,800 ppm), purslane (1,679-16,790 ppm), nance bark (27,300 ppm), rhubarb 4,400-13,360 ppm), tea leaves (2,192-10,000 ppm), bitter lettuce (10,000 ppm), spinach (6,580 ppm), cacao (1,520-5,000 ppm), bananas (22-5,240 ppm), ginger (5,000 ppm in rhizome), cashews (3,184 ppm), almonds (4,073 ppm), taro roots (1,334 ppm), tamarind (1,960 ppm), garden sorrel (3,000 ppm), mustard green leaves (1,287 ppm), peppers (257-1,171 ppm), sweet potato roots (1,000 ppm), pumpkins, oats (400 ppm), beets tomatillo (109-536 ppm), various cabbage leaves (59-350 ppm), and horseradish (Duke, 1992).

Uses: Calico printing and dyeing, analytical and laboratory reagent; bleaching agent; removing ink or rust stains, paint or varnish; stripping agent for permanent press resins; reducing agent; metal polishes; ceramics and pigments; cleaning wood; purifying methanol; cleanser in the metallurgical industry; in paper industry, photography, and process engraving; producing glucose from starch; leather tanning; rubber manufacturing industry; automobile radiator cleanser; purifying agent and intermediate for many compounds; catalyst; rare earth processing.

PARATHION

Synonyms: AAT; AATP; AC 3422; ACC 3422; Alkron; Alleron; American Cyanamid 3422; Aphamite; B 404; Bay E-605; Bladan; Bladan F; Compound 3422; Corothion; Corthion; Corthione; Danthion; DDP; *O,O*-Diethyl-*O*-4-nitrophenyl phosphorothioate; *O,O*-Diethyl *O-p*-nitrophenyl phosphorothioate; Diethyl-4-nitrophenyl phosphorothionate; Diethyl-*p*-nitrophenyl thionophosphate; *O,O*-Diethyl-*O*-4-nitrophenyl thionophosphate; *O,O*-Diethyl-*O-p*-nitrophenyl thionophosphate; Diethyl-*p*-nitrophenyl thiophosphate; *O,O*-Diethyl-*O-p*-nitrophenyl thiophosphate; Diethylparathion; DNTP; DPP; Drexel parathion 8E; E 605; Ecatox; Ekatin WF & WF ULV; Ekatox; ENT 15108; Ethlon; Ethyl parathion; Etilon; Folidol; Folidol E605; Folidol E & E 605; Fosfermo; Fosferno; Fosfex; Fosfive; Fosova; Fostern; Fostox; Gearphos; Genithion; Kolphos; Kypthion; Lethalaire G-54; Lirothion; Murfos; NA 2783; NCI-C00226; Niran; Niran E-4; Nitrostigmine; Nitrostygmine; Niuif-100; Nourithion; Oleofos 20; Oleoparaphene; Oleoparathion; Orthophos; Pac; Panthion; Paradust; Paraflow; Paramar; Paramar 50; Paraphos; Paraspray; Parathene; Parathion-ethyl; Parawet; Penphos; Pestox plus; Pethion; Phoskil; Phosphemol; Phosphenol; **Phosphorothioic acid *O,O*-diethyl *O*-(4-nitrophenyl) ester**; Phosphostigmine; RB; RCRA waste number P089; Rhodiasol; Rhodiatox; Rhodiatrox; Selephos; Sixty-three special E.C. insecticide; SNP; Soprathion; Stabilized ethyl parathion; Stathion; Strathion; Sulphos; Super rodiatox; T-47; Thiofos; Tiophos; Thiophos 3422; Tox 47; Vapophos; Vitrex.

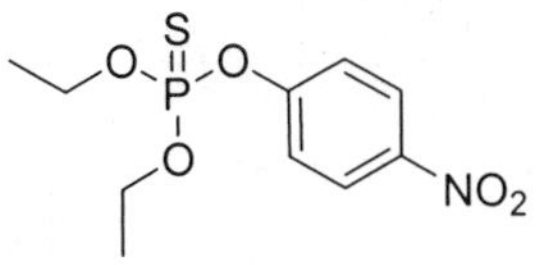

CAS Registry Number: 56-38-2
DOT: 2783 (liquid/dry), 2784 (flammable liquid)
DOT label: Poison
Molecular formula: $C_{10}H_{14}NO_5PS$
Formula weight: 291.27
RTECS: TF4550000
Merck reference: 10, 6897

Physical state, color, and odor:
Pale yellow to dark brown liquid with a garlic-like odor

Melting point (°C):
6.1 (Weast, 1986)

Boiling point (°C):
375 (Weast, 1986)

Density (g/cm³):
1.2704 at 20/20 °C (Weast, 1986)
1.2656 at 25/4 °C (Williams, 1951)

Diffusivity in water (x 10^{-5} cm²/sec):
0.51 at 20 °C using method of Hayduk and Laudie (1974)

Flash point (°C):
174 (Meister, 1988)
202 (open cup, NIOSH, 1997)

Henry's law constant (x 10^{-8} atm·m³/mol):
8.56 at 25 °C (Fendinger and Glotfelty, 1990)

Soil sorption coefficient, log K_{oc}:
3.68 (Kenaga and Goring, 1980)
3.02 (Batcombe silt loam, Briggs, 1981)
2.50-4.20 (organic matter content 0.2-6.1%, Mingelgrin and Gerstl, 1983)
3.25 (Mediterranean soil), 3.43 (Terra rossa) and 3.79 (Dark rendzina) (Saltzman et al., 1972)
3.02-3.14 (average = 3.07 for 5 sterilized Iowa soils, Felsot and Dahm, 1979)
2.43 (Netanya soil), 2.25 (Mivtahim soil), 2.25 (Golan soil), 2.39 (Gilat soil), 2.31 (Shefer soil), 2.32 (Bet Dagan soil), 2.40 (Neve Yaar soil), 2.45 (Malkiya soil), 2.75 (Kinneret sediment), 2.67 (Kinneret-A sediment), 2.70 (Kinneret-F sediment), 2.72 (Kinneret-G sediment, Gerstl and Mingelgrin, 1984)

Octanol/water partition coefficient, log K_{ow}:
3.81 (Chiou et al., 1977; Freed et al., 1979a)
3.91 (quoted, Verschueren, 1983)
2.15 (Leo et al., 1971)
3.93 (Briggs, 1981)
3.76 (Bowman and Sans, 1983a)

Solubility in organics:
2,900 and 2,700 g/kg in petroleum ether and heptane, respectively (Williams, 1951)

Solubility in water:
6.54 ppm at 24 °C (Felsot and Dahm, 1979)
24 mg/L at 25 °C (Williams, 1951)
11.9 and 11 ppm at 20 and 40 °C, respectively (Freed et al., 1977)
10.3 and 15.2 mg/L at 10 and 30 °C, respectively (Bowman and Sans, 1985)
11 mg/L at 20 °C (Worthing and Hance, 1991)
12.9 mg/L at 20 °C (Bowman and Sans, 1977, 1985)
4.3 mg/L at 10 °C (Yaron and Saltzman, 1972)
12.4 mg/L at 20.0 °C (Bowman and Sans, 1979)

Vapor density:
11.91 g/L at 25 °C, 10.06 (air = 1)

Vapor pressure (x 10^{-5} mmHg):
0.98 at 25 °C (Kim et al., 1984)
3.78, 6.95, 17.5, 41.6, and 93 at 20.0, 26.8, 37.8, 49.0, and 60.0 °C, respectively (Bright et al., 1950)

Environmental fate:
Biological. Parathion was reported to biologically hydrolyze to *p*-nitrophenol in different soils under flooded conditions (Sudhakar and Sethunathan, 1978). *p*-Nitrophenol also was identified as a hydrolysis product (Suffet et al., 1967).

When equilibrated with a prereduced pokkali soil, parathion instantaneously degraded to aminoparathion. The quick rate of reaction was reportedly due to soil enzymes and/or other heat labile substances (Wahid et al., 1980). Aminoparathion also was formed when parathion (500 ppm) was incubated in a flooded alluvial soil. The amount of parathion remaining after 6 and 12 d was 43.0 and 0.09%, respectively (Freed et al., 1979).

Initial hydrolysis products include diethyl-*O*-thiophosphoric acid, *p*-nitrophenol (Munnecke and Hsieh, 1976; Sethunathan, 1973, 1973a; Sethunathan et al., 1977; Verschueren, 1983), and the biodegradation products *p*-aminoparathion and *p*-aminophenol (Sethunathan, 1973). Mixed bacterial cultures were capable of growing on technical parathion as the sole carbon and energy source (Munnecke and Hsieh, 1976). Three oxidative pathways were reported. The primary degradative pathway is initial hydrolysis to yield *p*-nitrophenol and diethylthiophosphoric acid. The secondary pathway involves the formation of paraoxon (diethyl *p*-nitrophenyl phosphate), which subsequently undergoes hydrolysis to yield *p*-nitrophenol and diethylphosphoric acid. The third degradative pathway involved reduction of parathion under low oxygen conditions to yield *p*-aminoparathion followed by hydrolysis to *p*-aminophenol and diethylphosphoric acid. Other potential degradation products include hydroquinone, 2-hydroxyhydroquinone, ammonia, and polymeric substances (Munnecke and Hsieh, 1976). The reported half-life in soil is 18 d (Jury et al., 1987).

A *Flavobacterium* sp. (ATCC 27551), isolated from rice paddy water, degraded parathion to *p*-nitrophenol. The microbial hydrolysis half-life of this reaction was <1 h (Sethunathan and Yoshida, 1973; Forrest, 1981). When parathion (40 μg) was incubated in a mineral salts medium containing 5-day-old cultures of *Flavobacterium* sp. ATCC 27551, complete hydrolysis occurred in 72 h. The major degradation product was 4-nitrophenol (18.6 μg) (Sudhaker-Barik and Sethunathan, 1978a).

Sharmila et al. (1989) isolated a *Bacillus* sp. from a laterite soil which degraded parathion in the presence yeast extracts. At yeast concentrations of 0.05, 0.1, and 0.25%, parathion degraded via hydrolysis, hydrolysis and nitro group reduction, and exclusively by nitro group reduction, respectively. Aminoparathion and *p*-nitrophenol were the identified metabolites under these conditions. Rosenberg and Alexander (1979) demonstrated that two strains of *Pseudomonas* used parathion as the sole source of phosphorus. It was suggested that degradation of parathion resulted from an induced enzyme or enzyme system that catalytically hydrolyzed the aryl P-O bond, forming dimethyl phosphorothioate as the major product.

In both soils and water, chemical and biological mediated reactions transform parathion to paraoxon (Alexander, 1981). Parathion was reported to biologically hydrolyze to *p*-nitrophenol in different soils under flooded conditions (Ferris and Lichtenstein, 1980; Sudhakar-Barik and Sethunathan, 1978).

p-Nitrophenol, paraoxon and three unidentified metabolites were identified in a model ecosystem containing algae, *Daphnia magna*, fish, mosquitoes, and snails (Yu and Sanborn, 1975).

Soil. A *Pseudomonas* sp. (ATCC 29354), isolated from parathion-amended treated soil, degraded *p*-nitrophenol to *p*-nitrocatechol, which was recalcitrant to further degradation. In an unsterilized soil, however, *p*-nitrocatechol was further degraded to nitrites and other unidentified compounds (Sudhakar-Barik et al., 1978a). *Pseudomonas* sp. and *Bacillus* sp., isolated from a parathion-amended flooded soil, degraded *p*-nitrophenol (parathion hydrolysis product) to nitrite ions (Siddaramappa et al., 1973; Sudhakar-Barik et al., 1976) and carbon dioxide (Sudhakar-Barik et al., 1976).

When parathion was equilibrated with aerobic soils, virtually no degradation was observed (Adhya et al., 1981a). However, in flooded (anaerobic) acid sulfate soils or low sulfate soils, aminoparathion and desethyl aminoparathion formed as the major metabolites (Adhya et al., 1981). However, under flooded acid conditions, parathion hydrolyzed to form *p*-nitrophenol (Sethunathan, 1973a). The rate of hydrolysis was found to be higher in soils containing higher organic matter content. The bacterium *Bacillus* sp., isolated from parathion-amended flooded alluvial soil, decomposed *p*-nitrophenol as a sole carbon source (Sethunathan, 1973a). In a similar study, Rajaram and Sethunathan (1975) were able to increase the rate of hydrolysis of parathion in soil by a variety of organic materials. They found that the rate of hydrolysis was as follows: glucose >rice straw >algal crust >farmyard manure >unamended soil. However, these organic materials inhibited the rate of hydrolysis when the soils were inoculated with a parathion-hydrolyzing bacterial culture. Gerztl et al. (1979) studied the degradation of parathion in three soils ranging in concentration from 1.4 to 28 mg/kg. The rate of degradation increased as the parathion concentration and soil moisture increased. At low-parathion concentrations and moisture content, only 20% of the applied amount disappeared after 11 d, whereas at high parathion concentrations and moisture content, 96% of the applied amount had disappeared.

p-Nitrophenol was identified as a hydrolysis product in soil (Camper, 1991; Miles et al., 1979; Somasundaram et al., 1991; Suffet et al., 1967). The rate of hydrolysis in soil is accelerated following repeated applications of parathion (Ferris and Lichtenstein, 1980) or the surface catalysis of clays (Mingelgrin et al., 1977).

The reported hydrolysis half-lives at pH 7.4 and 20 and 37.5 °C were 130 and 26.8 d, respectively. At pH 6.1 and 20 °C, the hydrolysis half-life is 170 d (Freed et al., 1979). When equilibrated with a prereduced pokkali soil (acid sulfate), parathion instantaneously degraded to aminoparathion. The quick rate of reaction was reportedly due to soil enzymes and/or other heat labile substances. Desethyl aminoparathion was also identified as a metabolite in two separate studies (Wahid and Sethunathan, 1979; Wahid et al., 1980).

Aminoparathion also was formed when parathion (500 ppm) was incubated in a flooded alluvial soil. The amounts of parathion remaining after 6 and 12 d were 43.0 and 0.09%, respectively (Freed et al., 1979). It was observed that the degradation of parathion in flooded soil was more rapid in the presence of ferrous sulfate. Parathion degraded to aminoparathion and four unknown products (Rao and Sethunathan, 1979). In flooded alluvial soils, parathion degraded to aminoparathion via nitro group reduction. The rate of degradation remained constant despite variations in the redox potential of the soils (Adhya et al., 1981a). In soil, parathion may degrade via two oxidative pathways. The primary pathway is hydrolysis to *p*-nitrophenol (Sudhakar-Barik et al., 1979) and diethylthiophosphoric acid (Miles et al., 1979). The other pathway involves oxidation to paraoxon, but aminoparathion is formed under anaerobic (Miles et al., 1979) and flooded conditions (Sudhakar-Barik et al., 1979). The degradation pathway as well as the rate of degradation of parathion in a flooded soil changed following each successive application of parathion (Sudhakar-Barik et al., 1979). After the first application, nitro group reduction gave aminoparathion as the major product. After the second application, both the hydrolysis product (*p*-nitrophenol) and aminoparathion were found. Following the third addition of parathion, *p*-nitrophenol was the only product detected. It was reported that the change from nitro group reduction to hydrolysis occurred as a result of rapid proliferation of parathion-hydrolyzing microorganisms that utilized *p*-nitrophenol as the carbon source (Sudhakar-Barik et al., 1979).

In a cranberry soil pretreated with *p*-nitrophenol, parathion was rapidly mineralized to carbon dioxide by indigenous microorganisms (Ferris and Lichtenstein, 1980). The half-lives of parathion (10 ppm) in a nonsterile sandy loam and a nonsterile organic soil were <1 and 1.5 wk, respectively (Miles et al., 1979). Walker (1976) reported that 16-23% of parathion added to both sterile and nonsterile estuarine water was degraded after incubation in the dark for 40 d.

The percentage of the initial dosage (1 ppm) of parathion remaining after 8 wk of incubation in an organic and mineral soil were 6 and <2%, respectively, while in sterilized controls 95 and 80% remained, respectively (Chapman et al., 1981).

Six yr after applying parathion to soil at concentrations of 30,000-95,000 ppm, very small concentrations were detected 9 cm below the surface (Wolfe et al., 1973).

Plant. Oat plants were grown in two soils treated with [^{14}C]parathion. Less than 2% of the applied [^{14}C]parathion was translocated to the oat plant. Metabolites identified in both soils and leaves were paraoxon, aminoparaoxon, aminoparathion, *p*-nitrophenol, and an aminophenol (Fuhremann and Lichtenstein, 1980).

The following metabolites were identified in a soil-oat system: paraoxon, aminoparathion, *p*-nitrophenol, and *p*-aminophenol (Lichtenstein, 1980; Lichtenstein et al., 1982). Mick and Dahm

(1970) reported that *Rhizobium* sp. converted 85% [^{14}C]parathion to aminoparathion and 10% diethylphosphorothioic acid in 1 d.

One month after application of [^{14}C]parathion to cotton plants, 6.5-10.5% of the total reactivity was found to be unreacted [^{14}C]parathion. Photoalteration products identified included *S*-ethyl parathion, *S*-phenyl parathion, paraoxon, and *p*-nitrophenol (Joiner and Baetcke, 1973). Reddy and Sethunathan (1983) studied the mineralization of ring-labeled [2,6-^{14}C]parathion in the rhizosphere of rice seedlings under flooded and nonflooded soil conditions. In unplanted soil, only 5.5% of the ^{14}C in the parathion was evolved as $^{14}CO_2$ in 15 d under flooded and non-flooded conditions. However, in soils planted with rice, 9.2 and 22.6% of the ^{14}C in the parathion evolved as $^{14}CO_2$ under nonflooded and flooded conditions, respectively. In an earlier study, the presence of rice straw in a flooded alluvial soil inoculated with an enrichment culture greatly inhibited the hydrolysis of parathion to *p*-nitrophenol and *O,O*-diethylphosphorothioic acid. In uninoculated soils, however, rice straw enhanced the degradation of parathion via nitro group reduction to *p*-aminoparathion and a compound possessing a P=S bond (Sethunathan, 1973).

Photolytic. *p*-Nitrophenol and paraoxon were formed from the irradiation of parathion in water, aqueous methanol, and aqueous *n*-propyl alcohol solutions by a low-pressure mercury lamp. Degradation was more rapid in water than in organic solvent/water mixtures with *p*-nitrophenol forming as the major product (Mansour et al., 1983). When parathion in aqueous tetrahydrofuran or ethanol solutions (80%) was irradiated at 2537 Å, *O,O,S*-triethylthiophosphate formed as the major product (Grunwell and Erickson, 1973). Minor photoproducts included *O,O,O*-triethylthiophosphate, triethylphosphate, paraoxon, and traces of ethanethiol and *p*-nitrophenol. Zepp and Schlotzhauer (1983) found that photolysis of parathion in water containing algae occurred at a rate more than 27 times faster than in distilled water.

Parathion degraded on both glass surfaces and on bean plant leaves. Metabolites reported were paraoxon, *p*-nitrophenol, and a compound tentatively identified as *s*-ethyl parathion (El-Refai and Hopkins, 1966). Upon exposure to high intensity UV light, parathion was altered to the following photoproducts: paraoxon, *O,S*-diethyl *O*-4-nitrophenyl phosphorothioate, *O,O*-diethyl *S*-4-nitrophenyl phosphorothioate, *O,O*-bis(4-nitrophenyl) *O*-ethyl phosphorothioate, *O,O*-bis(4-nitrophenyl) *O*-ethyl phosphate, *O,O*-diethyl *O*-phenyl phosphorothioate, and *O,O*-diethyl *O*-phenyl phosphate (Joiner et al., 1971).

When parathion was released in the atmosphere on a sunny day, it was rapidly converted to the photochemical paraoxon. The estimated photolytic half-life is 2 min (Woodrow et al., 1978). The reaction involving the oxidation of parathion to paraoxon is catalyzed in the presence of UV light, ozone, soil dust, or clay minerals (Spencer et al., 1980, 1980a).

Ozone and UV light alone cannot oxide parathion. However, in the presence of ozone (300 ppb) and UV light, dry kaolinite clays were more effective than the montmorillonite clays in the oxidation of parathion. The Cu-saturated clays were most effective oxidizing parathion to paraoxon and the Ca-saturated clays were the least effective (Spencer et al., 1980).

When applied as thin films on leaf surfaces, parathion was converted to paraoxon and *p*-nitrophenol. The photodegradation half-life was reported to be 88 h (HSDB, 1989).

When an aqueous solution containing parathion was photooxidized by UV light at 90-95 °C, 25, 50, and 75% degraded to carbon dioxide after 15.1, 57.6, and 148.8 h, respectively (Knoevenagel and Himmelreich, 1976). *p*-Nitrophenol and paraoxon were the major products identified following the sunlight irradiation of parathion in distilled water and river water. The photolysis half-life of parathion in river water was 15 h (Mansour et al., 1989). Kotronarou et al. (1992) studied the degradation of parathion-saturated deionized water solution at 30 °C by ultrasonic irradiation (sonolysis). After 2 h of sonolysis, all the parathion degraded to the following final end products: sulfate ions, phosphate ions, nitrate ions, hydrogen ions, and carbon dioxide. Precursors/intermediate compounds to the final end products included *p*-nitrophenol, diethylmonothiophosphoric acid, hydroquinone, benzoquinone, formic acid, oxalic acid, 4-nitrocatechol, nitrite ions, and ethanol.

Chemical/Physical. The reported hydrolysis half-lives at pH 7.4 at 20 and 37.5 °C were 130 and 26.8 d, respectively (Freed et al., 1977). The hydrolysis half-lives of parathion in a sterile 1% ethanol/water solution at 25 °C and pH values of 4.5, 5.0, 6.0, 7.0, and 8.0, were 39, 43, 33, 24, and 15 wk, respectively (Chapman and Cole, 1982). Kollig (1993) reported the following products of hydrolysis: *O*-ethyl-*O*-(*p*-nitrophenyl)phosphorothioic acid, *O*-(*p*-nitrophenyl)-phosphorothioic acid, phosphorothioic acid, phosphoric acid, *p*-nitrophenol, *O,O*-diethylphosphorothioic acid, *O*-ethylphosphorothioic acid, and ethanol. Low concentrations of Ca^{2+} and Cu^{2+} will catalyze the hydrolysis of parathion in water (Plastourgou and Hoffman, 1984).

Reported ozonation products of parathion in drinking water include sulfuric acid (Richard and Bréner, 1984), paraoxon, 2,4-dinitrophenol, picric, and phosphoric acids (Laplanche et al., 1984).

Paraoxon was also found in fogwater collected near Parlier, CA (Glotfelty et al., 1990). It was suggested that parathion was oxidized in the atmosphere during daylight hours prior to its partitioning in the fog. On January 12, 1986, the distributions of parathion (9.4 ng/m^3) in the vapor phase, dissolved phase, air particles, and water particles were 78, 10, 11, and 0.6%, respectively. For paraoxon (2.3 ng/m^3), the distribution in the vapor phase, dissolved phase, air particles, and water particles were 7.8, 35.5, 56.7, and 0.09%, respectively (Glotfelty et al., 1990).

At 130 °C, parathion isomerizes to *O,S*-diethyl *O-p*-nitrophenyl phosphorothioate (Hartley and Kidd, 1987; Worthing and Hance, 1991). An 85% yield of this compound was reported when parathion was heated at 150 °C for 24 h (Wolfe et al., 1976). Emits toxic oxides of nitrogen, sulfur, and phosphorus when heated to decomposition (Lewis, 1990).

Although no products were identified, parathion (1.5 x 10^{-5} M) was reduced by iron metal (33.3 g/L acid washed 18-20 mesh) in a carbonate buffer (1.5 x 10^{-2} M) at pH 5.9 and 15 °C. Based on the pseudo-first-order disappearance rate of 0.0250/min, the half-life was 27.7 min (Agrawal and Tratnyek, 1996).

Exposure limits (mg/m^3): NIOSH REL: TWA 0.05, IDLH 10; OSHA PEL: TWA 0.1; ACGIH TLV: TWA 0.1 (adopted).

Toxicity:

EC_{50} (48-h) for *Daphnia pulex* 0.6 $\mu g/L$, *Simocephalus serrulatus* 0.42 $\mu g/L$ (Mayer and Ellersieck, 1986).

EC_{50} (24-h) for brine shrimp (*Artemia* sp.) >25 mg/L, estuarine rotifer (*Brachionus plicatilis*) >25 mg/L (Guzzella et al., 1997), *Daphnia magna* 3.25 $\mu g/L$, *Simocephalus serrulatus* 1.28 $\mu g/L$ (Mayer and Ellersieck, 1986).

EC_{50} (5-min) for *Photobacterium phosphoreum* 8.5 mg/L (Somasundaram et al., 1990).

LC_{50} (96-h) for rainbow trout 1.5 mg/L, golden orfe 570 $\mu g/L$, fathead minnow 1.4-2.7 mg/L (Worthing and Hance, 1991).

LC_{50} (48-h) for red killifish 10 mg/L (Yoshioka et al., 1986), carp 4.5 mg/L, goldfish 1.7 mg/L, medaka 2.9 mg/L, pond loach 1.4 mg/L (Spehar et al., 1982).

LC_{50} (24-h) for Protozoan (*Spirostomum teres*) 19.71 mg/L (Twagilimana et al., 1998).

LC_{50} (30-min) for Protozoan (*Spirostomum teres*) 5.99 mg/L (Twagilimana et al., 1998).

Acute oral LD_{50} for wild birds 1,330 mg/kg, cats 930 $\mu g/kg$, dogs 3 mg/kg, guinea pigs 8 mg/kg, mice 6 mg/kg, pigeons 1.33 mg/kg, quail 4.04 mg/kg, rats 2 mg/kg, rabbits 10 mg/kg (RTECS, 1985), pheasant 12.4 mg/kg, mallard ducks 1.9-2.1 mg/L (Worthing and Hance, 1991).

LC_{50} (inhalation) for rats 84 mg/m^3/4-h (quoted, RTECS, 1985).

A NOEL of 2 mg/kg diet was reported in a 2-yr feeding trial using rats (Worthing and Hance, 1991).

Uses: Non-systemic insecticide and acaricide for controlling soil-swelling insects in cotton, fruits, grapes, and vegetables (Worthing and Hance, 1991).

PCB-1016

Synonyms: Arochlor 1016; **Aroclor 1016**; Chlorodiphenyl (41% Cl).

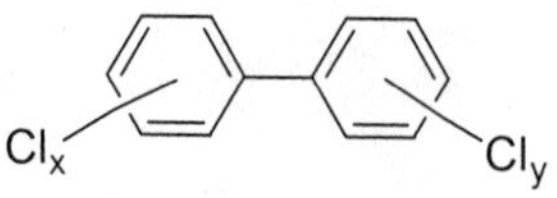

Cl_x ⬡—⬡ Cl_y

x + y = 0 thru 6

CAS Registry Number: 12674-11-2
DOT: 2315
Molecular formula: Not definitive. PCB-1016 is a mixture of many biphenyls with varying degrees of chlorination. According to Hutzinger et al. (1974), the approximate composition of PCB-1016 by weight is as follows: biphenyls (<0.1%), chlorobiphenyls (1%), dichlorobiphenyls (20%), trichlorobiphenyls (57%), tetrachlorobiphenyls (21%), pentachlorobiphenyls (1%), hexachlorobiphenyls (<0.1%) and heptachlorobiphenyls (0%). Lee et al. (1979) reported the following composition of PCB-1016 based on quantification of water-soluble isomers: chlorobiphenyls (12%), dichlorobiphenyls (34%), trichlorobiphenyls (35%), and tetrachlorobiphenyls (19%).
Formula weight: 257.9 (average, Hutzinger et al., 1974)
RTECS: TQ1351000

Physical state, color, and odor:
Viscous, oily, light yellow liquid or white powder with a weak odor

Boiling point (°C):
Distills at 325-356 (Monsanto, 1974)

Density (g/cm³):
1.33 at 25/4 °C (Monsanto, 1974)

Diffusivity in water (x 10⁻⁵ cm²/sec):
0.68 at 25 °C using method of Hayduk and Laudie (1974)

Flash point (°C):
Nonflammable (Sittig, 1985)

Henry's law constant (atm/mol fraction):
750 (Paris et al., 1978)

Bioconcentration factor, log BCF:
4.63 (freshwater fish, Garten and Trabalka, 1983)
3.81 (Doe Run bacteria), 3.73 (Hickory Hills pond bacteria) (Paris et al., 1978)
Creek chubsucker (*Erimyzon oblongus*) 4.34, yellow perch (*Perca flavescens*) 4.26, pumpkinseed (*Lepomis gibbosus*) 4.40, brown bullhead (*Ictalurus nebulosus*) 4.45-4.58 (Skea et al., 1979)

Soil sorption coefficient, log K_{oc}:
5.51 (Oconee River sediment), 5.21 (USDA pond sediment), 4.99 (Doe Run pond sediment), 4.41 (Hickory Hill pond sediment, Steen et al., 1978)

Octanol/water partition coefficient, log K_{ow}:
4.38 (Paris et al., 1978)
5.88 (Mackay, 1982)

Solubility in organics:
Soluble in most solvents (U.S. EPA, 1985).

Solubility in water (μg/L):
420 (Paris et al., 1978)
49 at 24 °C (Hollifield, 1979)
906 at 23 °C (Lee et al., 1979)

Vapor pressure (mmHg):
9.03 x 10^{-4} at 25 °C (Foreman and Bidleman, 1985)

Environmental fate:
Biological. Reported degradation products by the microorganism *Alcaligenes* BM-2 for a mixture of polychlorinated biphenyls include monohydroxychlorobiphenyl, 2-hydroxy-6-oxochlorophenylhexa-2,4-dieonic acid, chlorobenzoic acid, chlorobenzoylpropionic acid, chlorophenylacetic acid, and 3-chlorophenyl-2-chloropropenic acid (Yagi and Sudo, 1980). When PCB-1016 was statically incubated in the dark at 25 °C with yeast extract and settled domestic wastewater inoculum, no significant biodegradation was observed. At a concentration of 5 mg/L, percent losses after 7, 14, 21, and 28-d incubation periods were 44, 47, 46, and 48, respectively. At a concentration of 10 mg/L, only 22, 46, 20, and 13% losses were observed after the 7, 14, 21, and 28-d incubation periods, respectively (Tabak et al., 1981).
Chemical/Physical. PCB-1016 will not hydrolyze to any reasonable extent (Kollig, 1993).

Exposure limits: Potential occupational carcinogen. NIOSH REL: TWA 1.0 μg/m^3, IDLH 5 mg/m^3.

Toxicity:
A teratogen having a low toxicity (Patnaik, 1992).

Drinking water standard (final): For all PCBs, the MCLG and MCL are zero and 0.5 μg/L, respectively (U.S. EPA, 1996).

Uses: Insulator fluid for electric condensers; additive in high pressure lubricants.

PCB-1221

Synonyms: Arochlor 1221; **Aroclor 1221**; Chlorodiphenyl (21% Cl).

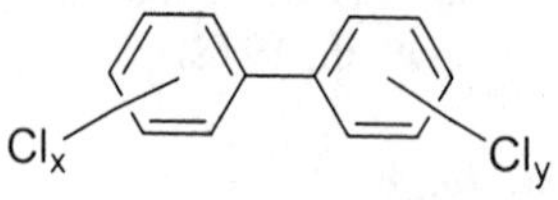

x + y = 0 thru 5

CAS Registry Number: 11104-28-2
DOT: 2315
Molecular formula: Not definitive. PCB-1221 is a mixture of many biphenyls with varying degrees of chlorination. According to Hutzinger et al. (1974), the approximate composition of PCB-1221 by weight is as follows: biphenyls (11%), chlorobiphenyls (51%), dichlorobiphenyls (32%), trichlorobiphenyls (4%), tetrachlorobiphenyls (2%), pentachlorobiphenyls (<0.5%), hexachlorobiphenyls (0%) and heptachlorobiphenyls (0%). Lee et al. (1979) reported the following composition of PCB-1221 based on quantification of water-soluble isomers: chlorobiphenyls (92%), dichlorobiphenyls (7%), trichlorobiphenyls (1%), and tetrachlorobiphenyls (trace amounts).
Formula weight: 192 (average, Hutzinger et al., 1974)
RTECS: TQ1352000

Physical state, color, and odor:
Viscous, oily, colorless to light yellow, mobile liquid with a faint odor

Melting point (°C):
Crystals form at 1 °C (Broadhurst, 1972)

Boiling point (°C):
Distills at 275-320 (Monsanto, 1974)

Density (g/cm³):
1.15 at 25/4 °C (Hutzinger et al., 1974)

Diffusivity in water (x 10^{-5} cm²/sec):
0.75 at 25 °C using method of Hayduk and Laudie (1974)

Flash point (°C):
141-150 (Broadhurst, 1972; Hutzinger et al., 1974)

Henry's law constant (x 10^{-4} atm·m³/mol):
0.286, 0.452, 0.697, 1.06, 1.57, 2.28, 3.26, 4.83, and 6.68 at 0.0, 5.0, 10.0, 15.0, 20.0, 25.0, 30.0, 35.5, and 40.0 °C, respectively (predicted, Burkhard et al., 1985)

Soil sorption coefficient, log K_{oc}:
2.44 (estimated, Montgomery, 1989)

Octanol/water partition coefficient, log K_{ow}:
4.08 (Callahan et al., 1979)
2.81 (Pal et al., 1980)

Solubility in organics:
Soluble in most solvents (U.S. EPA, 1985)

Solubility in water (mg/L):
0.590 at 24 °C (Hollifield, 1979)
3.5 at 23 °C (Lee et al., 1979)
5.0 (Zitko, 1970)

Vapor pressure (mmHg):
7 x 10^{-3} at 25 °C (Pal et al., 1980)

Environmental fate:
Biological. Reported degradation products by the microorganism *Alcaligenes* BM-2 for a mixture of polychlorinated biphenyls include monohydroxychlorobiphenyl, 2-hydroxy-6-oxochlorophenylhexa-2,4-dieonic acid, chlorobenzoic acid, chlorobenzoylpropionic acid, chlorophenylacetic acid, and 3-chlorophenyl-2-chloropropenic acid (Yagi and Sudo, 1980).

In sewage wastewater, *Pseudomonas* sp. 7509 degraded PCB-1221 into a yellow compound tentatively identified as a chlorinated derivative of α-hydroxymuconic acid (Liu, 1981). When PCB-1221 was statically incubated in the dark at 25 °C with yeast extract and settled domestic wastewater inoculum for 7 d, significant biodegradation with rapid adaptation was observed (Tabak et al., 1981).

In activated sludge, 80.6% degraded after a 47-h time period (Pal et al., 1980).

Chemical/Physical. Zhang and Rusling (1993) evaluated the bicontinuous microemulsion of surfactant/oil/water as a medium for the dechlorination of polychlorinated biphenyls by electrochemical catalytic reduction. The microemulsion (20 mL) contained didodecyldimethylammonium bromide, dodecane, and water at 21, 57, and 22 wt %, respectively. The catalyst used was zinc phthalocyanine (2.5 nM). When PCB-1221 (72 mg), the emulsion and catalyst were subjected to a current of mA/cm^2 on 11.2 cm^2 lead electrode for 10 h, a dechlorination yield of 99% was achieved. Reaction products included monochlorobiphenyl (0.9 mg) and biphenyl and reduced alkylbenzene derivatives.

PCB-1221 will not hydrolyze to any reasonable extent (Kollig, 1993).

Exposure limits: Potential occupational carcinogen. NIOSH REL: TWA 1.0 $\mu g/m^3$, IDLH 5 mg/m^3.

Toxicity:
A teratogen and suspected human carcinogen (Patnaik, 1992).

Drinking water standard (final): For all PCBs, the MCLG and MCL are zero and 0.5 $\mu g/L$, respectively (U.S. EPA, 1996).

Uses: In polyvinyl acetate to improve fiber-tear properties; plasticizer for polystyrene; in epoxy resins to improve adhesion and resistance to chemical attack; as an insulator fluid for electric condensers and as an additive in very high pressure lubricants.

PCB-1232

Synonyms: Arochlor 1232; **Aroclor 1232**; Chlorodiphenyl (32% Cl).

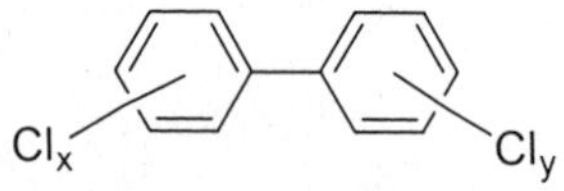

$$x + y = 0 \text{ thru } 6$$

CAS Registry Number: 11141-16-5
DOT: 2315
Molecular formula: Not definitive. PCB-1232 is a mixture of many biphenyls with varying degrees of chlorination. According to Hutzinger et al. (1974), the approximate composition of PCB-1232 by weight is as follows: biphenyls (<0.1%), chlorobiphenyls (31%), dichloro-biphenyls (24%), trichlorobiphenyls (28%), tetrachlorobiphenyls (12%), pentachlorobiphenyls (4%), hexachlorobiphenyls (<0.1%), and heptachlorobiphenyls (0%).
Formula weight: 221 (average, Hutzinger et al., 1974)
RTECS: TQ1354000

Physical state, color, and odor:
Viscous, oily, almost colorless to light yellow liquid with a faint odor

Melting point (°C):
-35.5 (pour point, Broadhurst, 1972)

Boiling point (°C):
Distills at 290-325 (Monsanto, 1974)

Density (g/cm³):
1.24 at 25/4 °C (Monsanto, 1974)
1.270-1.280 at 25/15.5 °C (Standen, 1964)

Diffusivity in water (x 10^{-5} cm²/sec):
0.72 at 20 °C using method of Hayduk and Laudie (1974)

Flash point (°C):
152-154 (Broadhurst, 1972; Hutzinger et al., 1974)

Henry's law constant (x 10^{-4} atm·m³/mol):
8.64 (calculated, U.S. EPA, 1980a)

Soil sorption coefficient, log K_{oc}:
2.83 (estimated, Montgomery, 1989)

Octanol/water partition coefficient, log K_{ow}:
3.23 (Pal et al., 1980)

Solubility in water:
1.45 mg/L at 25 °C (Pal et al., 1980)

Vapor density:
9.03 g/L at 25 °C, 7.63 (air = 1)

Vapor pressure (mmHg):
4.6 x 10^{-3} at 25 °C (estimated, U.S. EPA, 1980a)

Environmental fate:
   Biological. Reported degradation products by the microorganism *Alcaligenes* BM-2 for a mixture of polychlorinated biphenyls include monohydroxychlorobiphenyl, 2-hydroxy-6-oxo-chlorophenylhexa-2,4-dieonic acid, chlorobenzoic acid, chlorobenzoylpropionic acid, chloro-phenylacetic acid, and 3-chlorophenyl-2-chloropropenic acid (Yagi and Sudo, 1980).
   When PCB-1232 was statically incubated in the dark at 25 °C with yeast extract and settled domestic wastewater inoculum for 7 d, significant biodegradation with rapid adaptation was observed (Tabak et al., 1981).
   Photolytic. PCB-1232 in a 90% acetonitrile/water solution containing 0.2-0.3 M sodium borohydride and irradiated with UV light (λ = 254 nm) reacted to yield dechlorinated biphenyls. Without sodium borohydride, the reaction proceeded more slowly (Epling et al., 1988).
   Chemical/Physical. Zhang and Rusling (1993) evaluated the bicontinuous microemulsion of surfactant/oil/water as a medium for the dechlorination of polychlorinated biphenyls by electro-chemical catalytic reduction. The microemulsion (20 mL) contained didodecyldimethyl-ammonium bromide, dodecane, and water at concentrations of 21, 57, and 22 wt %, respectively. The catalyst used was zinc phthalocyanine (3.5 nM). When PCB-1232 (69 mg), the microemulsion and catalyst were subjected to an electrical current of mA/cm^2 on 11.2 cm^2 lead electrode for 12 h, a dechlorination yield of >99.8% was achieved. Reaction products included minor amounts of mono- and dichloropbiphenyls (0.01 mg), biphenyl and reduced alkylbenzene derivatives.
   PCB-1232 will not hydrolyze to any reasonable extent (Kollig, 1993).

Exposure limits: NIOSH REL: TWA 1.0 μg/m^3, IDLH 5 mg/m^3.

Toxicity:
   A teratogen and suspected human carcinogen having a low toxicity (Patnaik, 1992).
   Acute oral LD$_{50}$ for rats 4,470 mg/kg (quoted, RTECS, 1985).

Drinking water standard (final): For all PCBs, the MCLG and MCL are zero and 0.5 μg/L, respectively (U.S. EPA, 1996).

Uses: In polyvinyl acetate to improve fiber-tear properties; as an insulator fluid for electric condensers and as an additive in very high pressure lubricants.

PCB-1242

Synonyms: Arochlor 1242; **Aroclor 1242**; Chlorodiphenyl (42% Cl).

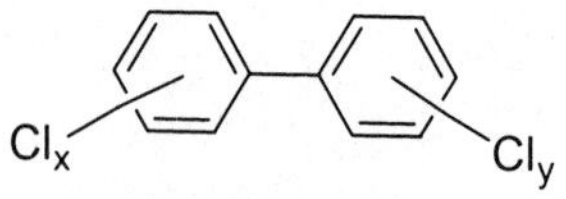

$x + y = 0$ thru 7

CAS Registry Number: 53469-21-9
DOT: 2315
Molecular formula: Not definitive. PCB-1242 is a mixture of many biphenyls with varying degrees of chlorination. According to Hutzinger et al. (1974), the approximate composition of PCB-1242 by weight is as follows: biphenyls (<0.1%), chlorobiphenyls (1%), dichloro-biphenyls (16%), trichlorobiphenyls (49%), tetrachlorobiphenyls (25%), pentachlorobiphenyls (8%), hexachlorobiphenyls (1%), and heptachlorobiphenyls (<0.1%). Lee et al. (1979) reported the following composition of PCB-1242 based on quantification of water-soluble isomers: chlorobiphenyls (19%), dichlorobiphenyls (35%), trichlorobiphenyls (31%), and tetrachlorobi-phenyls (14%), pentachlorobiphenyls (1%).
Formula weight: ranges from 154 to 358 (Nisbet and Sarofim, 1972) with an average value of 261 (Hutzinger et al., 1974)
RTECS: TQ1356000
Merck reference: 10, 7437

Physical state, color, and odor:
Colorless to light yellow, viscous, oily liquid with a weak, hydrocarbon odor

Melting point (°C):
-19 (pour point, Broadhurst, 1972)

Boiling point (°C):
Distills at 325-366 (Broadhurst, 1972; Monsanto, 1974)

Density (g/cm³):
1.392 at 15/4 °C, 1.381-1.392 at 25/15.5 °C (Standen, 1964)

Diffusivity in water (x 10^{-5} cm²/sec):
0.61 at 20 °C using method of Hayduk and Laudie (1974)

Flash point (°C):
176-180 (Hutzinger et al., 1974)

Henry's law constant (x 10^{-4} atm·m³/mol):
5.6 (Eisenreich et al., 1981)
2.8 at 20 °C (Murphy et al., 1987)
7.78 and 29.2 at 23 °C in distilled water and seawater, respectively (Atlas et al., 1972)
0.347, 0.575, 0.928, 1.47, 2.27, 3.43, 5.09, 7.60, and 10.8 at 0.0, 5.0, 10.0, 15.0, 20.0, 25.0, 30.0, 35.5, and 40.0 °C, respectively (predicted, Burkhard et al., 1985)

Bioconcentration factor, log BCF:
4.51-5.44 for fathead minnows (quoted, Skea et al., 1979)

Soil sorption coefficient, log K_{oc}:
5.50 (Oconee River sediment), 5.13 (USDA Pond sediment), 4.94 (Doe Run Pond sediment),
   4.35 (Hickory Hill Pond sediment) (Steen et al., 1978)
3.68, 4.37, 4.48, 4.51, 4.53, 4.75 (Paya-Perez et al., 1991)

Octanol/water partition coefficient, log K_{ow}:
4.11 (Paris et al., 1978)

Solubility in organics:
Soluble in most solvents (U.S. EPA, 1985)

Solubility in water:
100 μg/L at 24 °C (Hollifield, 1979)
240 μg/L at 25 °C (Monsanto, 1974; U.S. EPA, 1980a)
340 μg/L (Paris et al., 1978)
200 μg/L at 20 °C (Nisbet and Sarofim, 1972)
132.9 and 18.6 ppb at 11.5 °C in distilled water and artificial seawater, respectively (Dexter and
   Pavlou, 1978)
277 μg/L at 20 °C (Murphy et al., 1987)
16 ppb (Watanabe et al., 1985)
703 μg/L at 23 °C (Lee et al., 1979)

Vapor density:
10.67 g/L at 25 °C, 9.01 (air = 1)

Vapor pressure (x 10^{-4} mmHg):
4.06 at 25 °C (quoted, Mackay and Wolkoff, 1973)
10 at 38 °C (Nisbet and Sarofim, 1972)
2.5 at 20 °C (Murphy et al., 1987)
5.74 at 25 °C (Foreman and Bidleman, 1985)

Environmental fate:
   Biological. A strain of *Alcaligenes eutrophus* degraded 81% of the congeners by
dechlorination under anaerobic conditions (Bedard et al., 1987). A bacterial culture isolated
from Hamilton Harbour, Ontario was capable of degrading a commercial mixture of PCB-1242.
The metabolites identified by GC/MS included isohexane, isooctane, ethylbenzene, isoheptane,
isopropylbenzene, *n*-propylbenzene, isobutylbenzene, *n*-butylbenzene, and isononane (Kaiser
and Wong, 1974). A strain of *Pseudomonas*, isolated from activated sludge and grown with
biphenyl as the sole carbon source, degraded 2,4'-dichlorobiphenyl yielding the following
compounds: monochlorobenzoic acids, two monohydroxydichlorobiphenyls, and the yellow
compound hydroxyoxo(chlorophenyl)chlorohexadienoic acid. Irradiation of the mixture
containing these compounds led to the formation of two monochloroacetophenones and the
disappearance of the yellow compound. Similar compounds were found when 2,4'-dichloro-
biphenyl was replaced with a PCB-1242 mixture (Baxter and Sutherland, 1984).
   Reported degradation products by the microorganism *Alcaligenes* BM-2 for a mixture of
polychlorinated biphenyls include monohydroxychlorobiphenyl, 2-hydroxy-6-oxochloro-
phenylhexa-2,4-dieonic acid, chlorobenzoic acid, chlorobenzoylpropionic acid, chlorophenylacetic acid, and 3-chlorophenyl-2-chloropropenic acid (Yagi and Sudo, 1980).
   When PCB-1242 was statically incubated in the dark at 25 °C with yeast extract and settled
domestic wastewater inoculum, no significant biodegradation was observed. At a concentration
of 5 mg/L, percent losses after 7, 14, 21, and 28-d incubation periods were 37, 41, 47, and 66,

respectively. At a concentration of 10 mg/L, only 34, 33, 15, and 0% losses were observed after the 7, 14, 21, and 28-d incubation periods, respectively (Tabak et al., 1981).

Soil. In Hudson River, NY sediments, the presence of adsorbed PCB-1242 in core samples suggests it is very persistent in this environment. The estimated half-life ranged from 1.3 to 2.1 yr (Bopp et al., 1982).

Photolytic. PCB-1242 in a 90% acetonitrile/water solution containing 0.2-0.3 M sodium borohydride and irradiated with UV light (λ = 254 nm) reacted to yield dechlorinated biphenyls. Without sodium borohydride, the reaction proceeded much more slowly (Epling et al., 1988).

Chemical/Physical. When PCB-1242-contaminated sand was treated with a poly(ethylene glycol)/potassium hydroxide mixture at room temperature, 27% reacted after 2 wk forming aryl poly(ethylene glycols) (Brunelle and Singleton, 1985).

Trstenjak and Perdih (1999) reported that the fungus *Phanerochaete chrysosporium* removed the majority of PCB-1242 from a saturated water solution at room temperature up to a volumetric ratio of one to several throusand. In static column tests containing pelletized *Phanerochaete chrysosporium*, the percentage of PCB-1242 not removed ranged from 2.4 to 12% at pellets to contaminated water ratios of 1:5,000 and 1:500,000, respectively.

When exposed to fire, black soot containing PCBs, polychlorinated dibenzofurans and chlorinated dibenzo-*p*-dioxins is formed (NIOSH, 1997).

PCB-1242 will not hydrolyze to any reasonable extent (Kollig, 1993).

Mackay and Wolkoff (1973) estimated an evaporation half-life of 5.96 h from a surface water body that is 25 °C and 1 m deep.

Exposure limits: Potential occupational carcinogen. NIOSH REL: TWA 1.0 $\mu g/m^3$, IDLH 5 mg/m^3; OSHA PEL: TWA 1 mg/m^3; ACGIH TLV: TWA 1 mg/m^3 (adopted).

Toxicity:
Acute oral LD$_{50}$ for rats 4,250 mg/kg (quoted, RTECS, 1985).

Drinking water standard (final): For all PCBs, the MCLG and MCL are zero and 0.5 $\mu g/L$, respectively (U.S. EPA, 1996).

Uses: Dielectric liquids; heat-transfer liquid widely used in transformers; swelling agents for transmission seals; ingredient in lubricants, oils, and greases; plasticizers for cellulosics, vinyl, and chlorinated rubbers; in polyvinyl acetate to improve fiber-tear properties.

PCB-1248

Synonyms: Arochlor 1248; **Aroclor 1248**; Chlorodiphenyl (48% Cl).

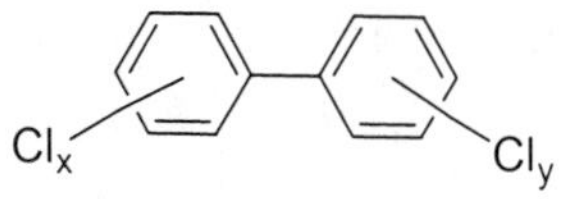

x + y = 2 thru 6

CAS Registry Number: 12672-29-6
DOT: 2315
Molecular formula: Not definitive. PCB-1248 is a mixture of many biphenyls with varying degrees of chlorination. According to Hutzinger et al. (1974), the approximate composition of PCB-1248 by weight is as follows: biphenyls (0%), chlorobiphenyls (0%), dichlorobiphenyls (2%), trichlorobiphenyls (18%), tetrachlorobiphenyls (40%), pentachlorobiphenyls (36%), hexachlorobiphenyls (4%), and heptachlorobiphenyls (0%).
Formula weight: ranges from 222 to 358 (Nisbet and Sarofim, 1972) with an average value of 288 (Hutzinger et al., 1974)
RTECS: TQ1358000

Physical state, color, and odor:
Viscous, oily, light yellow to yellow-green, mobile liquid with a faint odor

Melting point (°C):
-7 (pour point, Broadhurst, 1972)

Boiling point (°C):
Distills at 340-375 (Monsanto, 1974)

Density (g/cm^3):
1.41 at 25/4 °C (Monsanto, 1974)

Diffusivity in water (x 10^{-5} cm^2/sec):
0.66 at 20 °C using method of Hayduk and Laudie (1974)

Flash point (°C):
193-196 (Hutzinger et al., 1974)

Henry's law constant (x 10^{-3} atm·m^3/mol):
0.413, 0.696, 1.14, 1.83, 2.88, 4.40, 6.62, 9.91, and 14.3 at 0.0, 5.0, 10.0, 15.0, 20.0, 25.0, 30.0, 35.5, and 40.0 °C, respectively (predicted, Burkhard et al., 1985)
3.67 at 25 °C (Slinn et al., 1978)

Bioconcentration factor, log BCF:
4.85 (freshwater fish, Garten and Trabalka, 1983)
4.42 (bluegill sunfish, Stalling and Meyer, 1972)
5.08 (fathead minnow, DeFoe et al., 1978)
4.75 for channel catfish (quoted, Skea et al., 1979)

Soil sorption coefficient, log K$_{oc}$:
5.64 (estimated, Montgomery, 1989)

Octanol/water partition coefficient, log K$_{ow}$:
6.110 (for 2,2′,4,4′- and 2,2′,5,5′-tetrachlorobiphenyl, Kenaga and Goring, 1980)
6.0 (Mills et al., 1982)
6.11 (Chiou et al., 1977)

Solubility in organics:
Soluble in most solvents (U.S. EPA, 1985)

Solubility in water (μg/L):
54 (Monsanto, 1974)
50 at 20 °C (Nisbet and Sarofim, 1972)
60 at 24 °C (Hollifield, 1979)

Vapor pressure (x 10^{-4} mmHg):
4.94 at 25 °C (quoted, Mackay and Wolkoff, 1973)
0.6 at 38 °C (Nisbet and Sarofim, 1972)
1.835 at 25 °C (Foreman and Bidleman, 1985)

Environmental fate:
Biological. Reported degradation products by the microorganism *Alcaligenes* BM-2 for a mixture of polychlorinated biphenyls include monohydroxychlorobi-phenyl, 2-hydroxy-6-oxochlorophenylhexa-2,4-dieonic acid, chlorobenzoic acid, chlorobenzoylpropionic acid, chlorophenylacetic acid, and 3-chlorophenyl-2-chloropropenic acid (Yagi and Sudo, 1980). When PCB-1248 (5 and 10 mg/) was statically incubated in the dark at 25 °C with yeast extract and settled domestic wastewater inoculum, no biodegradation was observed (Tabak et al., 1981).

Chemical/Physical. Heating PCB-1248 in oxygen at 270-300 °C for 1 wk resulted in the formation of mono-, di-, tri-, tetra-, and pentachlorodibenzofurans (PCDFs). Above 330 °C, PCDFs decompose (Morita et al., 1978).

PCB-1248 will not hydrolyze to any reasonable extent (Kollig, 1993).

Mackay and Wolkoff (1973) estimated an evaporation half-life of 58.3 min from a surface water body that is 25 °C and 1 m deep.

Exposure limits: Potential occupational carcinogen. NIOSH REL: TWA 1.0 μg/m^3, IDLH 5 mg/m^3.

Toxicity:
Acute oral LD$_{50}$ for rats 11 g/kg (quoted, RTECS, 1985).

Drinking water standard (final): For all PCBs, the MCLG and MCL are zero and 0.5 μg/L, respectively (U.S. EPA, 1996).

Uses: In epoxy resins to improve adhesion and resistance to chemical attack; as an insulator fluid for electric condensers and as an additive in very high pressure lubricants.

PCB-1254

Synonyms: Arochlor 1254; **Aroclor 1254**; Chlorodiphenyl (54% Cl); NCI-C02664.

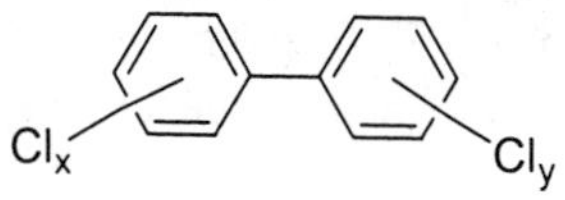

$$x + y = 0 \text{ thru } 7$$

CAS Registry Number: 11097-69-1
DOT: 2315
Molecular formula: Not definitive. PCB-1254 is a mixture of many biphenyls with varying degrees of chlorination. According to Hutzinger et al. (1974), the approximate composition of PCB-1254 by weight is as follows: biphenyls (<0.1%), chlorobiphenyls (<0.1%), dichlorobiphenyls (0.5%), trichlorobiphenyls (1%), tetrachlorobiphenyls (21%), pentachlorobiphenyls (48%), hexachlorobiphenyls (23%), and heptachlorobiphenyls (6%).
Formula weight: 327 (average, Hutzinger et al., 1974)
RTECS: TQ1360000
Merck reference: 10, 7437

Physical state, color, and odor:
Light yellow, viscous, oily liquid with a faint odor

Melting point (°C):
10 (pour point, Broadhurst, 1972)

Boiling point (°C):
Distills at 365-390 (Monsanto, 1974)

Density (g/cm³):
1.505 at 15.5/4 °C (Standen, 1964)
1.38 at 25/4 °C (NIOSH, 1997)

Diffusivity in water (x 10^{-5} cm²/sec):
0.56 at 20 °C using method of Hayduk and Laudie (1974)

Henry's law constant (x 10^{-4} atm·m³/mol):
27 (Eisenreich et al., 1981)
26.9 (Slinn et al., 1978)
23 (Petrasek et al., 1983)
1.9 at 20 °C (Murphy et al., 1987)
0.226, 0.394, 0.670, 1.11, 1.80, 2.83, 4.37, 6.65, and 9.81 at 0.0, 5.0, 10.0, 15.0, 20.0, 25.0, 30.0, 35.5, and 40.0 °C, respectively (predicted, Burkhard et al., 1985)

Bioconcentration factor, log BCF:
3.34 (guppies, Gooch and Hamdy, 1983)
4.70 (freshwater fish, Garten and Trabalka, 1983)
4.79 for channel catfish (quoted, Skea et al., 1979)

Soil sorption coefficient, log K_{oc}:
5.61 (Lake Michigan sediments, Voice and Weber, 1985)

4.628 (2,2′,4,5,5′-pentachlorobiphenyl, Kenaga and Goring, 1980)
4.88 (freshly amended Glendale soil), 4.73 (preconditioned Glendale soil), 4.47, 4.50 (Harvey soils), 4.40, 4.55 (Lea soils) (Fairbanks and O'Connor, 1984)
6.72 (montmorillonite), 5.94 (blue clay), 8.21 (Saginaw Bay sand), 6.28 (Huron River suspended sediments) (Weber et al., 1983)
3.16 (Woodburn silt loam, Haque et al., 1974)

Octanol/water partition coefficient, log K_{ow}:
6.47 (Travis and Arms, 1988)
5.61 (Voice and Weber, 1985)

Solubility in organics:
Soluble in most solvents (U.S. EPA, 1985)

Solubility in water (μg/L):
57 at 24 °C (Hollifield, 1979)
50 at 20 °C, ≈ 56 at 26 °C (Haque et al., 1974)
24.2 and 4.34 at 11.5 °C in distilled water and artificial seawater, respectively (Dexter and Pavlou, 1978)
43 at 20 °C (Murphy et al., 1987)
28.1 and 24.7 at 16.5 °C for membrane and carbon-filtered seawater, respectively. The following congeners and their respective concentrations (μg/L) for carbon filtered seawater were reported: 2,2′,5,5′-tetrachlorobiphenyl (4.52), 2,2′,3,5′-tetrachlorobiphenyl (4.52), 2,2′,3,5′,6-pentachlorobiphenyl (5.48), 2,2′,4,5,5′-pentachlorobiphenyl (3.19), 2,2′,3′,4,5-pentachlorobiphenyl (4.54), 2,3′,4,4′,5-pentachlorobiphenyl (1.28), 2,2′,4,4′,5,5′-hexachlorobiphenyl (0.79), 2,2′,3,4,4′,6-hexachlorobiphenyl (0.42) (Wiese and Griffin, 1978)
≈ 70 at 23 °C (Lee et al., 1979)

Vapor density:
13.36 g/L at 25 °C, 11.29 (air = 1)

Vapor pressure (x 10^{-5} mmHg):
7.71 at 25 °C (quoted, Mackay and Wolkoff, 1973)
2.2 at 20 °C (Murphy et al., 1987)
3.22 at 25 °C (Foreman and Bidleman, 1985)

Environmental fate:
Biological. A strain of *Alcaligenes eutrophus* degraded 35% of the congeners by dechlorination under anaerobic conditions (Bedard et al., 1987). Indigenous microbes in the Center Hill Reservoir, TN oxidized 2-chlorobiphenyl (a congener present in trace quantities) into chlorobenzoic acid and chlorobenzoylformic acid. Biooxidation of the PCB mixture containing 54 wt % chlorine was not observed (Shiaris and Sayler, 1982).

When PCB-1254 was statically incubated in the dark at 25 °C with yeast extract and settled domestic wastewater inoculum, no significant biodegradation was observed. At a concentration of 5 mg/L, percent losses after 7, 14, 21, and 28-d incubation periods were 11, 42, 15, and 0, respectively. At a concentration of 10 mg/L, only 10 and 26% losses were observed after the 7 and 14-d incubation periods, respectively (Tabak et al., 1981).

Reported degradation products by the microorganism *Alcaligenes* BM-2 for a mixture of polychlorinated biphenyls include monohydroxychlorobiphenyl, 2-hydroxy-6-oxochlorophenylhexa-2,4-dieonic acid, chlorobenzoic acid, chlorobenzoylpropionic acid, chlorophenylacetic acid, and 3-chlorophenyl-2-chloropropenic acid (Yagi and Sudo, 1980).

Soil. In two California soils (organic carbon content 0.3 and 11.3%), 95% of the applied amount was recovered after 1 yr with no change in the congener composition. In the four remaining soils containing less organic carbon (0.06 to 1.9%), degradation resulted in preferential loss of congeners with the lowest molecular weight (Iwata et al., 1973).

Photolytic. PCB-1254 in a 90% acetonitrile/water solution containing 0.2-0.3 M sodium borohydride and irradiated with UV light (λ = 254 nm) reacted to yield dechlorinated biphenyls. After 16 h, no chlorinated biphenyls were detected. Without sodium borohydride, only 25% of PCB-1254 were destroyed after 16 h (Epling et al., 1988). In a similar experiment, PCB-1254 (1,000 mg/L) in an alkaline 2-propanol solution was exposed to UV light (λ = 254 nm). After 30 min, all of the PCB-1254 isomers were converted to biphenyl. When the radiation source was sunlight, only 25% was degraded after a 20-h exposure. But when the sensitizer phenothiazine (5 mM) was added to the solution, photodechlorination of PCB-1254 was complete after 1 h at 350 nm. In addition, when PCB-1254 contaminated soil was heated at about 80 °C in the presence of di-*t*-butyl peroxide, complete dechlorination to biphenyl was observed (Hawari et al., 1992).

When PCB-1254 in a methanol/water solution containing sodium methyl siliconate was irradiated with UV light (λ = 300 nm) for 5 h, most of the congeners dechlorinated to biphenyl (Hawari et al., 1991). In a similar experiment, a dilute solution of PCB-1254 in cyclohexane was exposed to a Montreal sunlight in December and January for 55 d at 10 °C. Dechlorination of the higher chlorinated congeners at the *ortho* position was observed. In addition, two cyclohexyl adducts of polychlorinated biphenyls containing three and four chlorine atoms were also formed in small amounts (0.6% of the original PCB-1254 added) (Lépine et al., 1992). Photoproducts resulting from the irradiation of PCB-1254 in a dioxane-water solution include dechlorinated, hydroxylated, and hydrated compounds. The photolysis rate decreases as the oxygen content increases (Safe et al., 1976).

Chemical/Physical. When PCB-1254-contaminated sand was treated with a poly(ethylene glycol)/potassium hydroxide mixture at room temperature, 81% reacted after 6 d forming aryl poly(ethylene glycols) (Brunelle and Singleton, 1985).

Using methanol, ethanol, or 2-propanol in the presence of nickel chloride and sodium borohydride, dechlorination resulted in the formation of biphenyls with smaller quantities of mono- and dichlorobiphenyls (Dennis et al., 1979).

Zhang and Wang (1997) studied the reaction of zero-valent iron powder and palladium-coated iron particles with trichloroethylene and PCBs. In the batch scale experiments, 50 μL of 200 μg/mL PCB-1254 in methanol was mixed with 1 mL ethanol/water solution (volume ratio = 1/9) and 0.1 g of wet iron or palladium/iron powder in a 2-mL vial. The vial was placed on a rotary shaker (30 rpm) at room temperature for 17 h. Trichloroethylene was completely dechlorinated by the nanoscale palladium/iron powders within the 17-h time period. Only partial dechlorination was observed when wet iron powder was used.

When exposed to fire, black soot containing PCBs, polychlorinated dibenzofurans and chlorinated dibenzo-*p*-dioxins is formed (NIOSH, 1997).

PCB-1254 will not hydrolyze to any reasonable extent (Kollig, 1993).

Mackay and Wolkoff (1973) estimated an evaporation half-life of 1.2 min from a surface water body that is 25 °C and 1 m deep.

Exposure limits: Potential occupational carcinogen. NIOSH REL: TWA 1 μg/m^3, IDLH 5 mg/m^3; OSHA PEL: TWA 0.5 mg/m^3; ACGIH TLV: TWA 1 mg/m^3 (adopted).

Toxicity:

LC$_{50}$ for juvenile shrimp, *Palaemonetes pugio* ranged from 6.1 (salinity 1 and 7‰) to 7.8 μg/L (salinity 14-35‰), respectively (Roesijadi et al., 1976).

Acute oral LD$_{50}$ for rats 1,010 mg/kg (quoted, RTECS, 1985).

Drinking water standard (final): For all PCBs, the MCLG and MCL are zero and 0.5 μg/L, respectively (U.S. EPA, 1996).

Uses: Secondary plasticizer for polyvinyl chloride; co-polymers of styrene-butadiene and chlorinated rubber to improve chemical resistance to attack.

PCB-1260

Synonyms: Arochlor 1260; **Aroclor 1260**; Chlorodiphenyl (60% Cl); Clophen A60; Kanechlor; Phenoclor DP6.

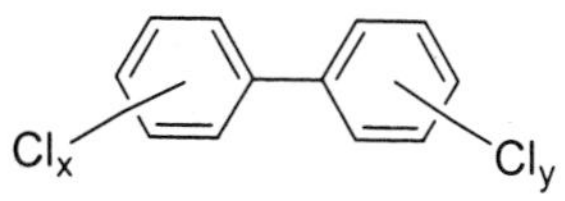

x + y = 4 thru 7

CAS Registry Number: 11096-82-5
DOT: 2315
Molecular formula: Not definitive. PCB-1260 is a mixture of many biphenyls with varying degrees of chlorination. According to Hutzinger et al. (1974), the approximate composition of PCB-1260 by weight is as follows: biphenyls (0%), chlorobiphenyls (0%), dichlorobiphenyls (0%), trichlorobiphenyls (0%), tetrachlorobiphenyls (1%), pentachlorobiphenyls (12%), hexachlorobiphenyls (38%), and heptachlorobiphenyls (41%).
Formula weight: ranges from 324 to 460 (Nisbet and Sarofim, 1972) with an average value of 370 (Hutzinger et al., 1974)
RTECS: TQ1362000
Merck reference: 10, 7437

Physical state, color, and odor:
Light yellow sticky, soft resin with a faint odor

Melting point (°C):
31 (pour point, Broadhurst, 1972)

Boiling point (°C):
Distills at 385-420 (Monsanto, 1974)

Density (g/cm³):
1.566 at 15.5/4 °C, 1.555-1.566 at 90/15.5 °C (Standen, 1964)
1.58 (Mills et al., 1982)

Diffusivity in water (x 10^{-5} cm²/sec):
0.53 at 20 °C using method of Hayduk and Laudie (1974)

Flash point (°C):
None to the boiling point (Hutzinger et al., 1974).

Bioconcentration factor, log BCF:
5.29 (fathead minnow, Veith et al., 1979a)

Henry's law constant (x 10^{-4} atm·m³/mol):
1.7 at 20 °C (Murphy et al., 1987)
0.244, 0.435, 0.754, 1.27, 2.10, 3.36, 5.27, 8.09, and 12.1 at 0.0, 5.0, 10.0, 15.0, 20.0, 25.0, 30.0, 35.5, and 40.0 °C, respectively (predicted, Burkhard et al., 1985)

Soil sorption coefficient, log K_{oc}:
6.42 (estimated, Montgomery, 1989)

Octanol/water partition coefficient, log K$_{ow}$:
6.91 (Mackay, 1982)
6.11 (Chiou et al., 1977)

Solubility in organics:
Soluble in most solvents (U.S. EPA, 1985)

Solubility in water (μg/L):
80 at 24 °C (Hollifield, 1979)
2.7 (Monsanto, 1974)
14.4 at 20 °C (Murphy et al., 1987)

Vapor pressure (x 10^{-7} mmHg):
405 at 25 °C (quoted, Mackay and Wolkoff, 1973)
2 at 38 °C (Nisbet and Sarofim, 1972)
63.1 at 20 °C (Murphy et al., 1987)
130.5 at 25 °C (Foreman and Bidleman, 1985)

Environmental fate:

Biological. Reported degradation products by the microorganism *Alcaligenes* BM-2 for a mixture of polychlorinated biphenyls include monohydroxychlorobiphenyl, 2-hydroxy-6-oxo-chlorophenylhexa-2,4-dieonic acid, chlorobenzoic acid, chlorobenzoylpropionic acid, chloro-phenylacetic acid, and 3-chlorophenyl-2-chloropropenic acid (Yagi and Sudo, 1980). When PCB-1260 was statically incubated in the dark at 25 °C with yeast extract and settled domestic wastewater inoculum, no significant biodegradation was observed (Tabak et al., 1981).

Photolytic. PCB-1260 in a 90% acetonitrile/water solution containing 0.2-0.3 M sodium borohydride and irradiated with UV light (λ = 254 nm) reacted to yield dechlorinated biphenyls. After 2 h, about 75% of the congeners were destroyed. Without sodium borohydride, only 10% of the congeners had reacted. Products identified by GC include biphenyl, 2-, 3- and 4-chlorobiphenyl, six dichlorobiphenyls, three trichlorobiphenyls, 1-phenyl-1,4-cyclohexadiene, and 1-phenyl-3-cyclohexene (Epling et al., 1988).

Chemical/Physical. Zhang and Rusling (1993) evaluated the bicontinuous microemulsion of surfactant/oil/water as a medium for the dechlorination of polychlorinated biphenyls by electrochemical catalytic reduction. The microemulsion (20 mL) contained didodecyldimethyl-ammonium bromide, dodecane, and water at 21, 57, and 22 wt %, respectively. The catalyst used was zinc phthalocyanine (4.5 nM). When PCB-1260 (100 mg), the emulsion and catalyst were subjected to a current of mA/cm^2 on 11.2 cm^2 lead electrode for 18 h, a dechlorination yield of >99.8 % was achieved. Reaction products included minor amounts of mono- and dichloropbiphenyls (0.02 mg), biphenyl and reduced alkylbenzene derivatives.

When PCB-1260-contaminated sand was treated with a poly(ethylene glycol)/potassium hydroxide mixture at room temperature, more than 99% reacted after 2 d forming aryl poly(ethylene glycols) (Brunelle and Singleton, 1985).

PCB-1260 will not hydrolyze to any reasonable extent (Kollig, 1993).

Mackay and Wolkoff (1973) estimated an evaporation half-life of 28.8 min from a surface water body that is 25 °C and 1 m deep.

Toxicity:

Suspected human carcinogen. Acute oral LD$_{50}$ for rats 1,315 mg/kg (quoted, RTECS, 1985).

Drinking water standard (final): For all PCBs, the MCLG and MCL are zero and 0.5 μg/L, respectively (U.S. EPA, 1996).

Uses: Secondary plasticizer for polyvinyl chloride; in polyester resins to increase strength of fiberglass; varnish formulations to improve water and alkali resistance; as an insulator fluid for electric condensers and as an additive in very high pressure lubricants.

PENTACHLOROBENZENE

Synonyms: QCB; RCRA waste number U183.

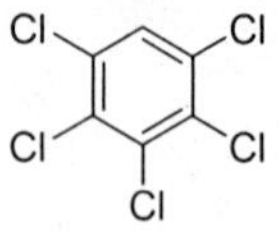

CAS Registry Number: 608-93-5
Molecular formula: C_6HCl_5
Formula weight: 250.34
RTECS: DA6640000

Physical state and color:
White needles

Melting point (°C):
86 (Weast, 1986)
82-85 (Verschueren, 1983)

Boiling point (°C):
277 (Weast, 1986)

Density (g/cm³):
1.8342 at 16.5/4 °C (Weast, 1986)

Diffusivity in water (x 10^{-5} cm²/sec):
0.59 at 20 °C using method of Hayduk and Laudie (1974)

Flash point (°C):
None (Dean, 1987)

Entropy of fusion (cal/mol·K):
13.5 (Domalski and Hearing, 1998)
13.8 (Miller et al., 1984)

Heat of fusion (kcal/mol):
4.92 (Miller et al., 1984)
4.80 (Domalski and Hearing, 1998)

Henry's law constant (x 10^{-4} atm·m³/mol):
7.1 at 20 °C (Oliver, 1985)
3.69, 4.88, 6.72, 6.58, 12.25, and 27.26 at 14.8, 20.1, 22.1, 24.2, 34.8, and 50.5 °C, respectively
  (ten Hulscher et al., 1992)

Ionization potential (eV):
8.8, 9.11, 9.21 (Lias et al., 1998)

Bioconcentration factor, log BCF:
3.53 (bluegill sunfish, Veith et al., 1980)
3.60 (algae, Geyer et al., 1984)

4.16 (activated sludge), 3.48 (golden ide) (Freitag et al., 1985)

3.84 (freshwater fish, Garten and Trabalka, 1983)

3.63 mussel (*Mytilus edulis*) (Renberg et al., 1985)

5.93 (Atlantic croakers), 6.12 (blue crabs), 4.96 (spotted sea trout), 5.57 (blue catfish) (Pereira et al., 1988)

5.41 (guppy, Könemann and van Leeuwen, 1980)

3.71 (bluegill sunfish), 3,65 (rainbow trout), 3.86 (guppy) (Banerjee et al., 1984)

3.92 (fathead minnow, Carlson and Kosian, 1987)

3.67 (guppy, Van Hoogan and Opperhuizen, 1988)

5.22 (pond snail, Legierse et al., 1998)

Soil sorption coefficient, log K_{oc}:

6.3 (average of 7 suspended sediment samples from the St. Clair and Detroit Rivers, Lau et al., 1989)

4.36, 4.52, 5.57 (Paya-Perez et al., 1991)

4.68 (lake sediment, Schrap et al., 1994)

4.06, 4.55, and 4.74 in Iowa soil, North River, and Charles River sediments, respectively (Wu and Gschwend, 1986)

4.60 (Georgia sediments, Karickhoff and Morris, 1985a)

4.89 (Oostvaardersplassen), 4.39 (Ransdorperdie), 4.63 (Noordzeekanaal) (Netherlands sediments, Sijm et al., 1997)

5.38 (sandy soil, Van Gestel and Ma, 1993)

Octanol/water partition coefficient, log K_{ow}:

5.17 (Banerjee, 1984; Watarai et al., 1982)

4.94 (Banerjee et al., 1980; Veith et al., 1980)

5.19 (Kenaga and Goring, 1980)

4.88 (Geyer et al., 1984; Könemann et al., 1979)

5.75 (DeKock and Lord, 1987)

5.20 at 13 °C, 5.05 at 19 °C, 4.70 at 28 °C, 4.66 at 33 °C (Opperhuizen et al., 1988)

5.183 (de Bruijn et al., 1989)

5.06 (Hammers et al., 1982)

5.03 (Miller et al., 1984)

5.69 (Bruggeman et al., 1982)

5.20 (Pereira et al., 1988)

5.20, 5.06, 4.94, 4.79, and 4.66 at 5, 15, 25, 35, and 45 °C, respectively (Bahadur et al., 1997)

Solubility in organics:

Very soluble in ether (Sax and Lewis, 1987).

Solubility in water:

5.37 μmol/L at 20 °C (Veith et al., 1980)

5.32 μmol/L at 25 °C (Banerjee et al., 1980)

240 ppb at 22 °C (quoted, Verschueren, 1983)

3.32 μmol/L at 25 °C (Miller et al., 1984)

2.24 μmol/L at 25 °C (Yalkowsky et al., 1979)

145, 254, 419, 447, 618, and 856 μg/L at 5, 15, 25, 27, 35, and 45 °C, respectively (Shiu et al., 1997)

Vapor pressure (mmHg):

6 x 10^{-3} at 20-30 °C (Mercer et al., 1990)

Environmental fate:

Biological. In activated sludge, <0.1% mineralized to carbon dioxide after 5 d (Freitag et al., 1985). The half-life of pentachlorobenzene in an anaerobic enrichment culture was 24 h (Beurskens et al., 1993).

Soil. An estimated degradation half-life of 270 d was reported for pentachlorobenzene in soil (Beck and Hansen, 1974).

Photolytic. UV irradiation ($\lambda = 2537$ Å) of pentachlorobenzene in *n*-hexane solution for 3 h produced a 50% yield of 1,2,4,5-tetrachlorobenzene and a 13% yield of 1,2,3,5-tetrachlorobenzene (Crosby and Hamadmad, 1971). Irradiation ($\lambda \geq 285$ nm) of pentachlorobenzene (1.1-1.2 mM/L) in an acetonitrile-water mixture containing acetone (0.553 mM/L) as a sensitizer gave the following products (% yield): 1,2,3,4-tetrachlorobenzene (6.6), 1,2,3,5-tetrachlorobenzene (52.8), 1,2,4,5-tetrachlorobenzene (15.1), 1,2,4-trichlorobenzene (1.9), 1,3,5-trichlorobenzene (5.3), 1,3-dichlorobenzene (0.9), 2,2′,3,3′,4,4′,5,6,6′-nonachlorobiphenyl (2.08), 2,2′,3,3′,4,4′,5,5′,6-nonachlorobiphenyl (0.34), 2,2′,3,3′,4,5,5′,6,6′-nonachlorobiphenyl (trace), one octachlorobiphenyl (0.53), and one heptachlorobiphenyl (0.49) (Choudhry and Hutzinger, 1984). Without acetone, the identified photolysis products (% yield) included 1,2,3,4-tetrachlorobenzene (3.7), 1,2,3,5-tetrachlorobenzene (13.5), 1,2,4,5-tetrachlorobenzene (2.8), 1,2,4-trichlorobenzene (12.7), 1,3,5-trichlorobenzene (1.0) and 1,4-dichlorobenzene (6.7) (Choudhry and Hutzinger, 1984).

A carbon dioxide yield of 2.0% was achieved when pentachlorobenzene adsorbed on silica gel was irradiated with light ($\lambda > 290$ nm) for 17 h (Freitag et al., 1985).

The experimental first-order decay rate for pentachlorobenzene in an aqueous solution containing a nonionic surfactant micelle (Brij 58, a polyoxyethylene cetyl ether) and illuminated by a photoreactor equipped with 253.7-nm monochromatic UV lamps, is 1.47×10^{-2}/sec. The corresponding half-life is 47 sec. Photoproducts reported include, all tetra-, tri- and dichlorobenzenes, chlorobenzene, benzene, phenol, hydrogen, and chloride ions (Chu and Jafvert, 1994).

Chemical/Physical. Emits toxic chlorinated acids and phosphene when incinerated (Sittig, 1985).

Based on an assumed base-mediated 1% disappearance after 16 d at 85 °C and pH 9.70 (pH 11.26 at 25 °C), the hydrolysis half-life was estimated to be >900 yr (Ellington et al., 1988).

Toxicity:

EC_{50} (48-h) for *Daphnia magna* 122 μg/L Hermens et al., 1984).

EC_{50} (growth reduction) for mosquito fish (*Gambusia affinis*) 37 nmol/L (Chaisuksant et al., 1998).

LC_{50} (14-d) for *Poecilia reticulata* 177 μg/L (Könemann, 1981).

LC_{50} (96-h) for bluegill sunfish 250 μg/L (Spehar et al., 1982), mosquito fish (*Gambusia affinis*) 0.80 μmol/L (Chaisuksant et al., 1998); for *Poecilia reticulata* 135 μg/L (van Hoogen and Opperhuizen, 1988), *Cyprinodon variegatus* 800 ppb using natural seawater (Heitmuller et al., 1981).

LC_{50} (72-h) for *Cyprinodon variegatus* 3.2-10 ppm (Heitmuller et al., 1981).

LC_{50} (48-h) for *Daphnia magna* 5.3 mg/L (LeBlanc, 1980), *Cyprinodon variegatus* 9.6 ppm (Heitmuller et al., 1981).

LC_{50} (24-h) for *Daphnia magna* 17 mg/L (LeBlanc, 1980), *Cyprinodon variegatus* >32 ppm (Heitmuller et al., 1981).

LC_{50} 11.8 and 62.5 μg/L (soil porewater concentration) for earthworm (*Eisenia andrei*) and 55 and 108 μg/L (soil porewater concentration) for earthworm (*Lumbricus rubellus*) (Van Gestel and Ma, 1993).

Acute oral LD_{50} for rats 1,080 mg/kg, mice 1,175 mg/kg (quoted, RTECS, 1985).

Heitmuller et al. (1981) reported a NOEC of 300 ppb.

Source: Pentachlorobenzene may enter the environment from leaking dielectric fluids containing this compound. Pentachlorobenzene may be present as an undesirable by-product in the chemical manufacture of hexachlorobenzene, pentachloronitrobenzene, tetrachlorobenzenes, tetrachloroethylene, trichloroethylene, and 1,2-dichloroethane (U.S. EPA, 1980).

Uses: No commercial use for this compound except for chemical research and organic synthesis.

PENTACHLOROETHANE

Synonyms: Ethane pentachloride; NCI-C53894; Pentalin; RCRA waste number U184; UN 1669.

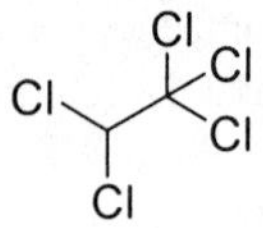

$$Cl_3C-CHCl_2$$

CAS Registry Number: 76-01-7
DOT: 1669
DOT label: Poison
Molecular formula: C_2HCl_5
Formula weight: 202.28
RTECS: KI6300000
Merck reference: 10, 6969

Physical state, color, and odor:
Clear, colorless liquid with a sweetish, chloroform-like odor

Melting point (°C):
-29.0 (Horvath, 1982)

Boiling point (°C):
162 (Weast, 1986)
159.72 (Boublik et al., 1973)
160.5 (Dean, 1987)

Density (g/cm³):
1.6796 at 20/4 °C (Weast, 1986)
1.6808 at 20/4 °C, 1.6732 at 25/4 °C (Riddick et al., 1986)

Diffusivity in water (x 10⁻⁵ cm²/sec):
0.79 at 20 °C using method of Hayduk and Laudie (1974)

Heat of fusion (kcal/mol):
2.7 (Dean, 1987)

Henry's law constant (x 10⁻³ atm·m³/mol):
2.45 at 25 °C (Hine and Mookerjee, 1975)

Ionization potential (eV):
11.0, 11.28 (Lias et al., 1998)

Bioconcentration factor, log BCF:
1.83 (bluegill sunfish, Veith et al., 1980)

Soil sorption coefficient, log K_{oc}:
3.28 (Mercer et al., 1990)

Octanol/water partition coefficient, log K_{ow}:
2.89 (Veith et al., 1980)

Solubility in organics:
Miscible with alcohol and ether (Windholz et al., 1983)

Solubility in water:
3.8 mM at 20 °C (Veith et al., 1980)
470 mg/kg H_2O at 25 °C (O'Connell, 1963)
500 mg/kg at 25 °C (McGovern, 1943)
500 and 769 mg/L at 20 and 25 °C, respectively (Mackay and Shiu, 1981)

Vapor density:
8.27 g/L at 25 °C, 6.98 (air = 1)

Vapor pressure (mmHg):
3.4 at 20 °C, 6 at 30 °C (Verschueren, 1983)
4.5 at 25 °C (Mackay and Shiu, 1981; Mackay et al., 1982)

Environmental fate:
   Chemical/Physical. At various pHs, pentachloroethane hydrolyzed to tetrachloroethylene (Jeffers et al., 1989; Roberts and Gschwend, 1991). Dichloroacetic acid was also reported as a hydrolysis product. Reacts with alkalies and metals producing explosive chloroacetylenes (NIOSH, 1997). The reported hydrolysis half-life at 25 °C and pH 7 is 3.6 d (Jeffers et al., 1989).
   The evaporation half-life of pentachloroethane (1 mg/L) from water at 25 °C using a shallow-pitch propeller stirrer at 200 rpm at an average depth of 6.5 cm is 46.5 min (Dilling, 1977).

Toxicity:
   LC_{50} (7-d) for *Poecilia reticulata* 15 mg/L (Könemann, 1981).
   LC_{50} (96-h) for fathead minnows 7.3 mg/L (Veith et al., 1983), bluegill sunfish 7.2 mg/L (Spehar et al., 1982).
   LC_{50} (48 and 24-h) for *Daphnia magna* 63 mg/L (LeBlanc, 1980).

Drinking water standard: No MCLGs or MCLs have been proposed (U.S. EPA, 1996).

Uses: Solvent for oil and grease in metal cleaning.

PENTACHLOROPHENOL

Synonyms: Acutox; Chempenta; Chemtol; Chlorophen; Cryptogil OL; Dowcide 7; Dowicide 7; Dowicide EC-7; Dowicide G; Dow pentachlorophenol DP-2 antimicrobial; Durotox; EP 30; Fungifen; Fungol; Glazd penta; Grundier arbezol; Lauxtol; Lauxtol A; Liroprem; Monsanto penta; Moosuran; NCI-C54933; NCI-C55378; NCI-C56655; PCP; Penchlorol; Penta; Penta-chlorofenol; Pentachlorofenolo; Pentachlorophenate; Pentachlorphenol; 2,3,4,5,6-Pentachloro-phenol; Pentacon; Penta-kil; Pentasol; Penwar; Peratox; Permacide; Permaguard; Permasan; Permatox DP-2; Permatox Penta; Permite; Priltox; RCRA waste number U242; Santobrite; Santophen; Santophen 20; Sinituho; Term-i-trol; Thompson's wood fix; Weedone; Witophen P.

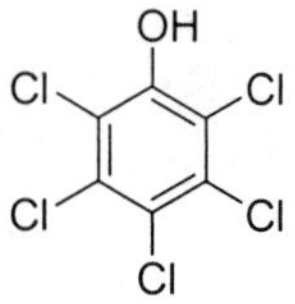

CAS Registry Number: 87-86-5
DOT: 2020
Molecular formula: C_6HCl_5O
Formula weight: 266.34
RTECS: SM6300000
Merck reference: 10, 6970

Physical state, color, and odor:
White to dark-colored flakes or crystalline solid with a phenolic odor. The reported odor threshold concentrations in air ranged from 9.3 to 857 ppb (Keith and Walters, 1992; Young et al., 1996). The taste threshold concentration in water is 8 ppb (Young et al., 1996).

Melting point (°C):
191 (Weast, 1986)
174 (IARC, 1979)
188 (Weiss, 1986)

Boiling point (°C):
310 (Melnikov, 1971)
293 (Bailey and White, 1965)

Density (g/cm³):
1.978 at 22/4 °C (Weast, 1986)

Diffusivity in water (x 10^{-5} cm²/sec):
0.49 at 20 °C using method of Hayduk and Laudie (1974)

Dissociation constant, pK_a:
4.74 (quoted, Mills et al., 1985)
4.80 (Blackman et al., 1955)
5.3 (Eder and Weber, 1980)

Flash point (°C):
Noncombustible solid (NIOSH, 1997)

Henry's law constant (x 10^{-7} atm·m^3/mol):
2.8 (Eisenreich et al., 1981)
21 (Petrasek et al., 1983)

Bioconcentration factor, log BCF:
3.10 (algae, Geyer et al., 1984)
1.60 (killifish, Trujillo et al., 1982)
1.90-2.08 and 1.79-1.93 for freshwater mussels *Anodanta anatina* and *Pseudanodonta complanata*, respectively (Makela and Oikari, 1995)
1.72, 2.11, and 2.78 pHs of 9, 8, and 7, respectively (Stehly and Hayton, 1990)
2.63, 2.80 (earthworm, Van Gestel and Ma, 1988)
3.04 (activated sludge), 2.41 (golden ide) (Freitag et al., 1985)
2.54 (mussel, Geyer et al., 1982)
2.89 (freshwater fish), 2.11 (fish, microcosm) (Garten and Trabalka, 1983)
2.00 (trout, Hattula et al., 1981)
2.33 (*Jordanella floridae*, Devillers et al., 1996)
1.81 (*Crassostrea virginica*, Schimmel and Garnas, 1981)

Soil sorption coefficient, log K_{oc}:
2.95 (Kenaga, 1980a)
2.96, 2.47, >2.68 (various Norwegian soils, Seip et al., 1986)
3.41 (Detroit River sediment, Jepsen and Lick, 1999)
4.16 (Jury et al., 1987)
3.10-3.26 and 4.40 calculated for dissociated and undissociated species, respectively (Lagas, 1988)
K_d = 1.73 mL/g (Eder and Weber, 1980)
Bluepoint soil: 5.27 and 5.71 at pHs 7.8 and 7.4, respectively; Glendale soil: 5.58 and 5.52 at pHs 7.3 and 4.3, respectively (Bellin et al., 1990)
2.76 (coarse sand), 2.48 (loamy sand) (Kjeldsen et al., 1990)
3.89 at pH 5.2 (Humaquept sand, Lafrance et al., 1994)
3.75 (sandy soil), 3.55 (sand), 3.90 (peaty sand) (Van Gestel and Ma, 1993)

Octanol/water partition coefficient, log K_{ow}:
5.01 (Leo et al., 1971)
5.86 (Banerjee et al., 1984)
4.84 at pH 1.2, 4.72 at pH 2.4, 4.62 at pH 3.4, 4.57 at pH 4.7, 3.72 at pH 5.9, 3.56 at pH 6.5, 3.32 at pH 7.2, 3.20 at pH 7.8, 3.10 at pH 8.4, 2.75 at pH 8.9, 1.45 at pH 9.3, 1.36 at pH 9.8, 1.30 at pH 10.5, 2.42 at pH 10.5, 1.30 at pH 11.5, 1.67 at pH 12.5, 3.86 at pH 13.5 (Kaiser and Valdmanis, 1982)
3.69 (Geyer et al., 1982)
5.24 (Schellenberg et al., 1984)
3.807 (Lu and Metcalf, 1975)
5.0 (Van Gestel and Ma, 1988)
4.16 (Rao and Davidson, 1980)
4.07 (Riederer, 1990)
5.11 (Garst and Wilson, 1984)
5.02 (Kishino and Kobayashi, 1994)

Solubility in organics:
At 20 °C (g/100 g solution): methanol (57.0), anhydrous ethanol (53.0), 95% ethanol (47.5) diethylene glycol monomethyl ether (48.0), pine oil (32.0), diethylene glycol monoethyl ether

(30.0), diethylene glycol (27.5), 2-ethoxyethanol (27.0), dioxane (11.5), benzene (11.0), ethylene glycol (6.0), diesel oil (3.1), fuel oil (2.6) (Carswell and Nason, 1938).

Solubility in water:
37.7 mM at 22.7 °C and pH 9.46 (Arcand et al., 1995)
16 mg/L (Mills and Hoffman, 1993)
35 mg/L at 50 °C, 85 mg/L at 70 °C (quoted, Verschueren, 1983)
20 mg/L at 20 °C (Kearney and Kaufman, 1976)
In mg/kg solution: 5, 18, 35, 58, and 85 at 0, 27, 50, 62, and 70 °C, respectively (Carswell and Nason, 1938)
14 ppm at 20 °C, 19 ppm at 30 °C (Bevenue and Beckman, 1967)
14 mg/L at 20 °C, 20 mg/L at 30 °C (Gunther et al., 1968)
2.0 mg/L (Gile and Gillett, 1979)
12 mg/L at 20 °C (Riederer, 1990)
21.4 mg/L at 25.1 °C (Achard et al., 1996)
9.59 mg/L at 25 °C and pH 5.1 (Blackman et al., 1955)
10.8 mg/L at 27 °C (pH 5.0, Toyota and Kuwahara, 1967)
11 mg/L at 22 °C (Jepsen and Lick, 1999)

Vapor pressure (x 10^{-5} mmHg):
1.7 at 0 °C, 17 at 20 °C, 310 at 50 °C, 2,400 at 75 °C, 140,000 at 100 °C (Carswell and Nason, 1938)
2.14 at 20 °C (Dobbs and Cull, 1982)
11 at 20 °C, 1.7 at 23 °C, 0.76 at 30 °C, 26 at 40 °C (Klöpffer et al., 1988)
11 at 20 °C (Bevenue and Beckman, 1967)
1.7 at 20.25 °C (Gile and Gillett, 1979)
7.5 at 25 °C (extrapolated from vapor pressures determined at higher temperatures, Tesconi and Yalkowsky, 1998)

Environmental fate:
Biological. Under aerobic conditions, microbes in estuarine water partially dechlorinated pentachlorophenol to trichlorophenol (Hwang et al., 1986). The disappearance of pentachlorophenol was studied in four aquaria, with and without mud, under aerobic and anaerobic conditions. Potential biological and/or chemical products identified include pentachloroanisole, 2,3,4,5-, 2,3,4,6-, and 2,3,5,6-tetrachlorophenol (Boyle, 1980).

Pentachlorophenol degraded in anaerobic sludge to 3,4,5-trichlorophenol, which was further reduced to 3,5-dichlorophenol (Mikesell and Boyd, 1985). In activated sludge, only 0.2% of the applied amount was mineralized to carbon dioxide after 5 d (Freitag et al., 1985).

Pentachlorophenol was statically incubated in the dark at 25 °C with yeast extract and settled domestic wastewater inoculum. Significant biooxidation was observed but with a gradual adaptation over a 14-d period to achieve complete degradation at 5 mg/L substrate cultures. At a concentration of 10 mg/L, it took 28 d for pentachlorophenol to degrade completely (Tabak et al., 1981).

Pentachlorophenol was also subject to methylation by a culture medium containing *Trichoderma viride* affording pentachloroanisole (Cserjesi and Johnson, 1972).

Melcer and Bedford (1988) studied the fate of pentachlorophenol in municipal activated sludge reactor systems that were operated at solids retention times of 10 to 20 d and hydraulic retention times of 120 d. Under these conditions, pentachlorophenol concentrations degreased from 0.1 and 12 mg/L to <10 μg/L. At solids retentions times of 5 d or less, pentrachlorophenol degradation was incomplete.

Soil. Under anaerobic conditions, pentachlorophenol may undergo sequential dehalogenation to produce tetra-, tri-, di-, and *m*-chlorophenol (Kobayashi and Rittman, 1982). In aerobic and anaerobic soils, pentachloroanisole was the major metabolite, with minor quantities of 2,3,6-trichlorophenol, 2,3,4,5- and 2,3,5,6-tetrachlorophenol (Murthy et al., 1979). Knowlton and Huckins (1983) reported 2,3,5,6-tetrachlorophenol and carbon dioxide as major metabolites in soil. Degradation was rapid in estuarine sediments having pH values of 6.5 and 8.0. Following a 17-d lag period, 70% of pentachlorophenol degraded (DeLaune, 1983). After 160 d in aerobic soil, only 6% biodegradation was observed (Baker and Mayfield, 1980).

Weiss et al. (1982) studied the fate of [^{14}C]pentachlorophenol added to flooded rice soil in a plant growth chamber. After one growing season, the following residues were observed (% of applied radioactivity): unidentified unextractable/bound compounds (28.61%), pentachlorophenol (0.51%), conjugated pentachlorophenol (0.61%), 2,3,6-, 2,4,5-, 2,4,6-, 2,3,4-, 2,3,5-, and 3,4,5-trichlorophenols (1.27%), 2,3,4,5-tetrachlorophenol (0.38%), 2,3,4-, 2,3,6-, 2,4,6-, and 3,4,5-trichloroanisoles (0.08%), 2,3,4,5-tetrachloroanisole (0.02%), pentachloroanisole (0.02%), and unidentified conversion products (including highly polar hydrolyzable and nonhydrolyzable compounds) (4.74%).

Metabolites identified in soil beneath a sawmill environment where pentachlorophenol was used as a wood preservative include pentachloroanisole, 2,3,4,6-tetrachloroanisole, tetrachlorocatechol, tetrachlorohydroquinone, 3,4,5-trichlorocatechol, 2,3,6-trichlorohydroquinone, 3,4,6-trichlorocatechol, and 2,3,4,6-tetrachlorophenol (Knuutinen et al., 1990).

Surface Water. Crossland and Wolff (1985) reported that the phototransformation half-life of pentachlorophenol in surface water ranged from 1.5 to 3.0 d.

Groundwater. According to the U.S. EPA (1986), pentachlorophenol has a high potential to leach to groundwater.

Photolytic. When pentachlorophenol in distilled water is exposed to UV irradiation, it is photolyzed to give tetrachlorophenols, trichlorophenols, chlorinated dihydroxybenzenes, and dichloromaleic acid. Phototransformation rates of 0.6/h (half-life 1 h) and 0.37/h (half-life 2 h) were reported at temperatures of 25 and 11 °C, respectively (Hwang et al., 1986). Wood treated with pure pentachlorophenol did not photolyze under natural sunlight or laboratory-induced UV radiation. However, in the presence of an antimicrobial (Dowcide EC-7), pure pentachlorophenol degraded to chlorinated dibenzo-*p*-dioxin. Wood containing composited technical grade pentachlorophenol yielded similar results (Lamparski et al., 1980).

Photodecomposition of pentachlorophenol was observed when an aqueous solution was exposed to sunlight for 10 d. The violet-colored solution contained 3,4,5-trichloro-6-(2'-hydroxy-3',4',5',6'-tetrachlorophenoxy)-*o*-benzoquinone as the major product. Minor photodecomposition products (% yield) included tetrachlororesorcinol (0.10%), 2,5-dichloro-3-hydroxy-6-pentachlorophenoxy-*p*-benzoquinone (0.16%), and 3,5-dichloro-2-hydroxy-5-2',4',5',6'-tetrachloro-3-hydroxyphenoxy-*p*-benzoquinone (0.08%) (Plimmer, 1970).

An aqueous solution containing pentachlorophenol and exposed to sunlight or laboratory UV light yielded tetrachlorocatechol, tetrachlororesorcinol, and tetrachlorohydroquinone. These compounds were air-oxidized to chloranil, hydroxyquinones, and 2,3-dichloromaleic acid. Other compounds identified include a cyclic dichlorodiketone, 2,3,5,6- and 2,3,4,6-tetrachlorophenol, and trichlorophenols (Munakata and Kuwahara, 1969; Wong and Crosby, 1981). Degradation was more rapid at pH 7.3 and less rapid at pH 3.3 (Wong and Crosby, 1981).

UV irradiation (λ = 2537 Å) of pentachlorophenol in hexane solution for 32 h produced a 30% yield of 2,3,5,6-tetrachlorophenol and about a 10% yield of a compound tentatively identified as an isomeric tetrachlorophenol (Crosby and Hamadmad, 1971).

A carbon dioxide yield of 50.0% was achieved when pentachlorophenol adsorbed on silica gel was irradiated with light (λ >290 nm) for 17 h (Freitag et al., 1985). In soil, photodegradation was not significant (Baker et al., 1980).

When an aqueous solution containing pentachlorophenol (45 μM) and a suspension of titanium dioxide (2 g/L) was irradiated with UV light, carbon dioxide, and hydrochloric acid formed in quantitative amounts. The half-life for this reaction at 45-50 °C is 8 min (Barbeni et al., 1985). When an aqueous solution containing pentachlorophenol was photooxidized by UV light at 90-95 °C, 25, 50, and 75% degraded to carbon dioxide after 31.7, 66.0, and 180.7 h, respectively (Knoevenagel and Himmelreich, 1976). The photolysis half-lives of pentachloro-phenol under sunlight irradiation in distilled water and river water were 27 and 53 h, respectively (Mansour et al., 1989).

A rate constant of 2 x 10^7/M·sec was reported for the reaction of pentachlorophenol and singlet oxygen in an aqueous phosphate buffer solution at 27 °C (Tratnyek and Hoigné, 1991).

Pentachlorophenol (47 μM) in an air-saturated solution containing titanium dioxide suspension was irradiated with UV light (λ = 330-370 nm). Major chemical intermediates included p-choranil, tetrachlorohydroquinone, hydrogen peroxide, and o-chloranil. The inter-mediate compounds were attacked by OH radicals during the latter stages of irradiation forming HCO_2^-, acetate and formate ions, carbon dioxide, and hydrochloric acid (Mills and Hoffman, 1993).

An atmospheric half-life of 216 h was reported for the reaction of pentachlorophenol and OH radicals in January (Bunce et al., 1991).

The following half-lives were reported for pentachlorophenol in estuarine water exposed to sunlight and microbes: 10 and 7 h during summer (24 °C) and winter (10 °C), respectively; in poisoned estuarine water: 6 and 10 h during summer and winter, respectively (Hwang et al., 1986).

Chemical/Physical. Wet oxidation of pentachlorophenol at 320 °C yielded formic and acetic acids (Randall and Knopp, 1980). In a dilute aqueous solution at pH 6.0, pentachlorophenol reacted with excess of hypochlorous acid forming 2,3,5,6-tetrachlorobenzoquinone (3% yield), 2,3,4,4,5,6-hexachlorobenzoquinone (20% yield), and two other chlorinated compounds (Smith et al., 1976).

Petrier et al. (1992) studied the sonochemical degradation of pentachlorophenol in aqueous solutions saturated with different gases at 24 °C. Ultrasonic irradiation of solutions saturated with air or oxygen resulted in the liberation of chloride ions and mineralization of the parent compound to carbon dioxide. When the solution was saturated with argon, pentachlorophenol completely degraded to carbon monoxide and chloride ions. In aqueous solution, pentachlorophenol was degraded by ozone at a reaction rate of >3.0 x 10^5/M·sec at pH 2.0 (Hoigné and Bader, 1983).

Reacts with amines and alkali metals forming water-soluble salts (Sanborn et al., 1977).

Hexachlorobenzene, octachlorodiphenylene dioxide, and other polymeric compounds were formed when pentachlorophenol was heated to 300 °C for 24 h (Sanderman et al., 1957).

Pentachlorophenol will not hydrolyze to any reasonable extent (Kollig, 1993).

Exposure limits (mg/m^3): NIOSH REL: TWA 0.5, IDLH 2.5; OSHA PEL: TWA 0.5; ACGIH TLV: TWA 0.5 (adopted).

Symptoms of exposure: Ingestion may cause fluctuation in blood pressure, respiration, fever, urinary output, weakness, convulsions, and possibly death (NIOSH, 1997). An irritation concentration of 10.90 mg/m^3 in air was reported by Ruth (1986).

Toxicity:

EC$_{10}$ and EC$_{50}$ concentrations inhibiting the growth of alga *Scenedesmus subspicatus* in 96 h were 30 and 90 μg/L, respectively (Geyer et al., 1985).

EC$_{50}$ (48-h) for *Daphnia magna* 320 μg/L (Mayer and Ellersieck, 1986).

EC$_{50}$ (24-h) for *Daphnia magna* 858 μg/L (Lilius et al., 1995).

LC$_{50}$ (contact) for earthworm (*Eisenia fetida*) 2.4 μg/cm^2 (Neuhauser et al., 1985).

LC$_{50}$ (96-h) for juvenile guppy 720-880 μg/L, fathead minnows 220-230 μg/L (Spehar et al., 1982).

LC$_{50}$ (72-h) *Prionchulus punctatus* 0.5 μmol/L, *Dorylaimus stagnalis* 3.6 μmol/L, *Aporcelaimellus obtusicaudatus* 3.6 μmol/L, *Tobrilus gracilis* 1.9 μmol/L, *Plectus acuminatus* 18.7 μmol/L, *Cephalobus persegnis* 9.6 μmol/L, *Rhabditis* sp. 9.1 μmol/L, *Diplogasteritus* sp. 25.4 μmol/L, *Tylenchus elegans* 4.5 μmol/L (Kammenga et al., 1994).

LC$_{50}$ (48-h) for rainbow trout 0.17 mg/L (sodium salt) (Hartley and Kidd, 1987), juvenile guppy 1.05 mg/L, fathead minnows 7.9 mg/L (Spehar et al., 1982), zebra fish 400 μg/L (Slooff, 1979); 1.78, 4.59, and 0.51 mg/L for *Daphnia magna*, *Daphnia pulex*, and *Daphnia galeata*, adults, respectively (Stephenson et al., 1991), *Prionchulus punctatus* 1.1 μmol/L, *Dorylaimus stagnalis* 3.6 μmol/L, *Aporcelaimellus obtusicaudatus* 3.8 μmol/L, *Tobrilus gracilis* 2.6 μmol/L, *Plectus acuminatus* 19.1 μmol/L, *Cephalobus persegnis* >34.5 μmol/L, *Rhabditis* sp. >34.5 μmol/L, *Diplogasteritus* sp. 26.4 μmol/L, *Tylenchus elegans* 6.4 μmol/L (Kammenga et al., 1994).

LC$_{50}$ (24-h) for goldfish 270 ppb (Verschueren, 1983), *Daphnia magna* 0.39 mg/L, *Brachionus calyciflorus* 2.16 mg/L (Ferrando et al., 1992), *Prionchulus punctatus* 6.4 μmol/L, *Dorylaimus stagnalis* 5.0 μmol/L, *Aporcelaimellus obtusicaudatus* 7.5 μmol/L, *Tobrilus gracilis* 2.8 μmol/L, *Plectus acuminatus* 19.5 μmol/L, *Cephalobus persegnis* >34.5 μmol/L, *Rhabditis* sp. >34.5 μmol/L, *Diplogasteritus* sp. 26.6 μmol/L, *Tylenchus elegans* 13.9 μmol/L (Kammenga et al., 1994).

LC$_{50}$ (6-h) for red abalone (*Haliotis rufescens*) 1.6 mg/L (Tjeerdema et al., 1991).

LC$_{50}$ 639-1,518 μg/L (soil porewater concentration) for earthworm (*Eisenia andrei*) and 3.5-15.7 mg/L (soil porewater concentration) for earthworm (*Lumbricus rubellus*) (Van Gestel and Ma, 1993).

LC$_{50}$ (45-min inhalation) for rates exposed to sodium pentachlorophenate was 14 mg/m^3 (Hoben et al., 1976).

Acute oral LD$_{50}$ for ducks 380 mg/kg, hamsters 168 mg/kg, mice 117 mg/kg, rats 27 mg/kg (quoted, RTECS, 1985).

LD$_{50}$ (gavage administration) for male and female mice were 129 and 134 mg/kg, respectively (Renner et al., 1986).

LC$_{50}$ (inhalation) for mice 225 mg/m^3, rats 355 mg/m^3 (quoted, RTECS, 1985).

Drinking water standard (final): MCLG: zero; MCL: 1 μg/L (U.S. EPA, 1996)

Uses: Manufacture of insecticides (termite control), algicides, herbicides, fungicides, and bactericides; wood preservative.

1,4-PENTADIENE

Synonyms: None.

CAS Registry Number: 591-93-5
Molecular formula: C_5H_8
Formula weight: 68.12

Physical state:
Liquid or gas

Melting point (°C):
-148.3 (Weast, 1986)

Boiling point (°C):
26 (Weast, 1986)

Density (g/cm³):
0.66076 at 20/4 °C, 0.65571 at 25/4 °C (Dreisbach, 1959)

Diffusivity in water (x 10^{-5} cm²/sec):
0.87 at 20 °C using method of Hayduk and Laudie (1974)

Dissociation constant, pK_a:
>14 (Schwarzenbach et al., 1993)

Flash point (°C):
<0 (Sax and Lewis, 1987)

Entropy of fusion (cal/mol·K):
11.62 (Messerly et al., 1970)
11.81 (Parks et al., 1936a)

Heat of fusion (kcal/mol):
1.452 (Messerly et al., 1970)
1.468 (Parks et al., 1936a)

Henry's law constant (atm·m³/mol):
0.120 at 25 °C (Hine and Mookerjee, 1975)

Ionization potential (eV):
9.58 (Collin and Lossing, 1959)

Soil sorption coefficient, log K_{oc}:
Unavailable because experimental methods for estimation of this parameter for aliphatic hydrocarbons are lacking in the documented literature

Octanol/water partition coefficient, log K_{ow}:
1.48 (Hansch and Leo, 1979)

Solubility in organics:
Soluble in acetone, alcohol, benzene, and ether (Weast, 1986)

Solubility in water:
558 mg/kg at 25 °C (McAuliffe, 1966)

Vapor density:
2.78 g/L at 25 °C, 2.35 (air = 1)

Vapor pressure (mmHg):
734.6 at 25 °C (Wilhoit and Zwolinski, 1971)

Uses: Chemical research; organic synthesis.

PENTANE

Synonyms: Amyl hydride; Dimethylmethane; *n*-Pentane; UN 1265.

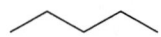

CAS Registry Number: 109-66-0
DOT: 1265
DOT label: Flammable liquid
Molecular formula: C_5H_{12}
Formula weight: 72.15
RTECS: RZ9450000
Merck reference: 10, 6983

Physical state, color, and odor:
Clear, colorless, volatile liquid with an odor resembling gasoline. Odor threshold concentration is 10 ppm (Keith and Walters, 1992).

Melting point (°C):
-130 (Weast, 1986)

Boiling point (°C):
36.1 (Weast, 1986)

Density (g/cm³):
0.62624 at 20/4 °C, 0.62139 at 25/4 °C (Dreisbach, 1959)
0.6290 at 17.2/4 °C, 0.62139 at 25.00/4 °C (Curtice et al., 1972)

Diffusivity in water (x 10^{-5} cm²/sec):
0.84 at 20 °C (Witherspoon and Bonoli, 1969)
0.97 at 25 °C (quoted, Hayduk and Laudie, 1974)

Dissociation constant, pK_a:
>14 (Schwarzenbach et al., 1993)

Flash point (°C):
-49.8 (NIOSH, 1997)
-40 (Windholz et al., 1983)

Lower explosive limit (%):
1.5 (NIOSH, 1997)

Upper explosive limit (%):
7.8 (NIOSH, 1997)

Entropy of fusion (cal/mol·K):
14.00 (Messerly et al., 1967)
14.02 (Messerly and Kennedy, 1940)
13.96 (Parks and Huffman, 1930)

Heat of fusion (kcal/mol):
2.008 (Messerly et al., 1967)
2.011 (Messerly and Kennedy, 1940)
2.002 (Parks and Huffman, 1930)

Henry's law constant (atm·m³/mol):
1.20 at 25 °C (Jönsson et al., 1982)

Interfacial tension with water (dyn/cm at 25 °C):
49.0 at 25 °C (Donahue and Bartell, 1952)

Ionization potential (eV):
10.28 (Lias, 1998)

Soil sorption coefficient, log K_{oc}:
Unavailable because experimental methods for estimation of this parameter for aliphatic hydrocarbons are lacking in the documented literature

Octanol/water partition coefficient, log K_{ow}:
3.23, 3.39 (Hansch and Leo, 1979)
3.62 (Schantz and Martire, 1987; Tewari et al., 1982; Wasik et al., 1981)

Solubility in organics:
In methanol: 620 and 810 g/L at 5 and 10 °C, respectively. Miscible at higher temperatures (Kiser et al., 1961).

Solubility in water:
In mg/kg: 39.5 at 25 °C, 39.8 at 40.1 °C, 41.8 at 55.7 °C, 69.4 at 99.1 °C. In NaCl solution at 25 °C (salinity, g/kg): 36.8 (1.002), 34.5 (10.000), 27.6 (34.472), 22.6 (50.03), 10.9 (125.10), 5.91 (199.90), 2.64 (279.80), 2.01 (358.70) (Price, 1976)
38.5 mg/kg at 25 °C (McAuliffe, 1963, 1966)
65.7 mg/kg at 0 °C, 47.6 mg/kg at 25 °C (Polak and Lu, 1973)
39 mg/kg at 25 °C (Krzyzanowska and Szeliga, 1978)
360 mg/L at 16 °C (Fischer and Ehrenberg, 1948)
565 μmol/L at 25.0 °C (Tewari et al., 1982; Wasik et al., 1981)
40.0, 40.4 and 47.6 mg/L at 25 °C (Mackay and Shiu, 1981)
0.60 mL/L at 16 °C (Fühner, 1924)
0.11 g/kg at 20 °C and 32 atmHg (Namiot and Beider, 1960)
700 mg/L at 20 °C (Korenman and Aref'eva, 1977)
Mole fraction x 10^5: 1.02, 1.07, 0.98, 1.01, and 1.01 at 4.0, 10.0, 20.0, 25.0, and 30.0 °C, respectively (Nelson and DeLigny, 1968)
1.03 mM at 25 °C (Barone et al., 1966)
39.0 mg/L at 25 °C (Kryzanowska and Szeliga, 1978)
40.6 mg/L at 25 °C (Jönsson et al., 1982)

Vapor density:
2.95 g/L at 25 °C, 2.49 (air = 1)

Vapor pressure (mmHg):
512.8 at 25 °C (Wilhoit and Zwolinski, 1971)
433.67 at 20.572 °C, 525.95 at 25.698 °C (Osborn and Douslin, 1974)

516.65 at 25.00 °C (Hussam and Carr, 1985)

219.3, 283.7, 424.1, 511.3, and 614.8 at 4.0, 10.0, 20.0, 25.0, and 30.0 °C, respectively (Nelson and DeLigny, 1968)

Environmental fate:

Biological. n-Pentane may biodegrade in two ways. The first is the formation of pentyl hydroperoxide, which decomposes to 1-pentanol followed by oxidation to pentanoic acid. The other pathway involves dehydrogenation to 1-pentene, which may react with water giving 1-pentanol (Dugan, 1972). Microorganisms can oxidize alkanes under aerobic conditions (Singer and Finnerty, 1984). The most common degradative pathway involves the oxidation of the terminal methyl group forming 1-pentanol. The alcohol may undergo a series of dehydrogenation steps forming an aldehyde (valeraldehyde) then a fatty acid (valeric acid). The fatty acid may then be metabolized by β-oxidation to form the mineralization products, carbon dioxide and water (Singer and Finnerty, 1984). *Mycobacterium smegnatis* was capable of degrading pentane to 2-pentanone (Riser-Roberts, 1992).

Photolytic. When synthetic air containing gaseous nitrous acid and pentane was exposed to artificial sunlight (λ = 300-450 nm) methyl nitrate, pentyl nitrate, peroxyacetal nitrate, and peroxypropionyl nitrate formed as products (Cox et al., 1980).

The following rate constants were reported for the reaction of pentane and OH radicals in the atmosphere: 3.9×10^{12} cm^3/mol·sec (Hendry and Kenley, 1979); 4.06×10^{-12} cm^3/molecule·sec (Sabljić and Güsten, 1990); 3.74×10^{-12} cm^3/molecule·sec at 300 K (Darnall et al., 1978); 3.94×10^{-12} cm^3/molecule·sec at 297 K (Atkinson, 1990). A photooxidation rate constant of 8.1×10^{-17} cm^3/molecule·sec was reported for the reaction of pentane and NO_3 (Altshuller, 1991).

Exposure limits: NIOSH REL: TWA 120 ppm (350 mg/m^3), ceiling 610 ppm (1,800 mg/m^3), IDLH 1,500 ppm; OSHA PEL: TWA 1,000 ppm (2,950 mg/m^3); ACGIH TLV: TWA 600 ppm (adopted).

Symptoms of exposure: Inhalation may cause narcosis and irritation of respiratory tract (Patnaik, 1992).

Toxicity:

LD$_{50}$ (intravenous) for mice 466 mg/kg (quoted, RTECS, 1985).

Source: Schauer et al. (1999) reported pentane in a diesel-powered medium-duty truck exhaust at an emission rate of 1,860 μg/km.

Uses: Solvent recovery and extraction; blowing agent for plastic foams; low temperature thermometers; natural gas processing plants; production of olefin, hydrogen, ammonia; fuel production; pesticide; manufacture of artificial ice; organic synthesis.

2-PENTANONE

Synonyms: Ethyl acetone; Methyl propyl ketone; Methyl *n*-propyl ketone; MPK; Propyl methyl ketone; UN 1249.

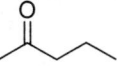

CAS Registry Number: 107-87-9
DOT: 1249
DOT label: Flammable liquid
Molecular formula: $C_5H_{10}O$
Formula weight: 86.13
RTECS: SA7875000
Merck reference: 10, 5988

Physical state, color, and odor:
Colorless liquid with a characteristic, pungent odor. Odor threshold concentration is 8 ppm (Keith and Walters, 1992).

Melting point (°C):
-77.8 (Weast, 1986)

Boiling point (°C):
102 (Weast, 1986)
103.3 (Stull, 1947)

Density (g/cm³):
0.8089 at 20/4 °C (Weast, 1986)
0.8018 at 25/4 °C (Ginnings et al., 1940)

Diffusivity in water (x 10^{-5} cm²/sec):
0.85 at 20 °C using method of Hayduk and Laudie (1974)

Flash point (°C):
7.3 (NIOSH, 1997)

Lower explosive limit (%):
1.5 (NIOSH, 1997)

Upper explosive limit (%):
8.2 (NIOSH, 1997)

Entropy of fusion (cal/mol·K):
12.93 (Oetting, 1965)

Heat of fusion (kcal/mol):
2.54 (Oetting, 1965)

Henry's law constant (x 10^{-5} atm·m³/mol at 25 °C):
6.36 (Buttery et al., 1969)

10.8 (28 °C, Nelson and Hoff, 1968)
11 (Hawthorne et al., 1985)
8.36 (Shiu and Mackay, 1997)

Ionization potential (eV):
9.37 ± 0.02 (Franklin et al., 1969)

Soil sorption coefficient, log K_{oc}:
Unavailable because experimental methods for estimation of this parameter for ketones are lacking in the documented literature

Octanol/water partition coefficient, log K_{ow}:
0.78 (Sangster, 1989)
0.91 (HPLC, Unger et al., 1978)

Solubility in organics:
Miscible with alcohol and ether (Windholz et al., 1983).

Solubility in water:
43,065 mg/L at 20 °C (Mackay and Yeun, 1983)
In wt %: 5.95 at 20 °C, 5.51 at 25 °C, 5.18 at 30 °C (Ginnings et al., 1940)
0.630 and 0.515 mol % at 30 and 50 °C, respectively (Palit, 1947)

Vapor density:
3.52 g/L at 25 °C, 2.97 (air = 1)

Vapor pressure (mmHg):
27 at 20 °C (NIOSH, 1997)
12 at 20 °C, 16 at 25 °C, 21 at 30 °C (Verschueren, 1983)

Environmental fate:
  Chemical/Physical. At an influent concentration of 1,000 mg/L, treatment with granular activated carbon resulted in an effluent concentration of 305 mg/L. The adsorbability of the carbon used was 139 mg/g carbon (Guisti et al., 1974).

Exposure limits: NIOSH REL: TWA 150 ppm (530 mg/m³), IDLH 1,500 ppm; OSHA PEL: TWA 200 (700 mg/m³); ACGIH TLV: TWA 200 ppm, STEL 250 ppm (adopted).

Symptoms of exposure: Inhalation of vapors may cause narcosis and irritation of eyes and respiratory tract (Patnaik, 1992).

Toxicity:
  Acute oral LD_{50} for mice 2,205 mg/kg, rats 3,730 mg/kg (quoted, RTECS, 1985).

Uses: Solvent; substitute for 3-pentanone; flavoring.

1-PENTENE

Synonyms: α-*n*-Amylene; Propylethylene.

CAS Registry Number: 109-67-1
DOT: 1108
DOT label: Combustible liquid
Molecular formula: C_5H_{10}
Formula weight: 70.13
Merck reference: 10, 6991

Physical state and color:
Colorless liquid

Melting point (°C):
-165.219 (Dreisbach, 1959)

Boiling point (°C):
30 (Weast, 1986)

Density (g/cm³):
0.64050 at 20/4 °C, 0.63533 at 25/4 °C (Dreisbach, 1959)

Diffusivity in water (x 10^{-5} cm²/sec):
0.84 at 20 °C using method of Hayduk and Laudie (1974)

Dissociation constant, pK_a:
>14 (Schwarzenbach et al., 1993)

Flash point (°C):
-17.7 (open cup, NFPA, 1984)

Lower explosive limit (%):
1.5 (Sax and Lewis, 1987)

Upper explosive limit (%):
8.7 (Sax and Lewis, 1987)

Entropy of fusion (cal/mol·K):
12.86 (Chao et al., 1983; Todd et al., 1947)
13.14 (Messerly et al., 1990)

Heat of fusion (kcal/mol):
1.388 (Chao et al., 1983; Todd et al., 1947)
1.419 (Messerly et al., 1990)

Henry's law constant (atm·m³/mol):
0.406 at 25 °C (Hine and Mookerjee, 1975)

Ionization potential (eV):
9.49 (Lias, 1998)
9.67 (Collin and Lossing, 1959)

Soil sorption coefficient, log K_{oc}:
Unavailable because experimental methods for estimation of this parameter for aliphatic hydrocarbons are lacking in the documented literature

Octanol/water partition coefficient, log K_{ow}:
2.26 using method of Hansch et al. (1968)

Solubility in organics:
Miscible with alcohol, benzene, and ether (Windholz et al., 1983)

Solubility in water:
148 mg/kg at 25 °C (McAuliffe, 1966)

Vapor density:
2.87 g/L at 25 °C, 2.42 (air = 1)

Vapor pressure (mmHg):
628.2 at 24.6 °C (Forziati, 1950)
637.7 at 25 °C (Wilhoit and Zwolinski, 1971)

Environmental fate:
   Biological. Biooxidation of 1-pentene may occur yielding 4-penten-1-ol, which may further oxidize to give 4-pentenoic acid (Dugan, 1972). Washed cell suspensions of bacteria belonging to the genera *Mycobacterium*, *Nocardia*, *Xanthobacter*, and *Pseudomonas* and growing on selected alkenes metabolized 1-pentene to 1,2-epoxypentane. *Mycobacterium* sp., growing on ethene, hydrolyzed 1,2-epoxypropane to 1,2-propanediol (Van Ginkel et al., 1987).
   Photolytic. The following rate constants were reported for the reaction of 1-pentene and OH radicals in the atmosphere: 1.8×10^{13} cm^3/mol·sec at 300 K (Hendry and Kenley, 1979); 3.14×10^{-11} cm^3/molecule·sec (Atkinson, 1990). Atkinson (1990) also reported a photooxidation rate constant of 1.10×10^{-17} cm^3/molecule·sec for the reaction of 1-pentene and ozone.

Uses: Blending agent for high octane motor fuel; organic synthesis.

cis-2-PENTENE

Synonyms: *cis*-Pentene-2; **(Z)-2-Pentene**.

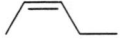

CAS Registry Number: 627-20-3
Molecular formula: C_5H_{10}
Formula weight: 70.13
Merck reference: 10, 6992

Physical state and color:
Colorless liquid

Melting point (°C):
-151.4 (Weast, 1986)

Boiling point (°C):
36.9 (Weast, 1986)

Density (g/cm³):
0.6556 at 20/4 °C, 0.6504 at 25/4 °C (Dreisbach, 1959)
0.6614 at 16.50/4 °C, 0.65152 at 25.00/4 °C (Curtice et al., 1972)
0.6503 at 20/4 °C (Dean, 1987)

Diffusivity in water (x 10^{-5} cm²/sec):
0.85 at 20 °C using method of Hayduk and Laudie (1974)

Dissociation constant, pK_a:
>14 (Schwarzenbach et al., 1993)

Flash point (°C):
-18 (Sax and Lewis, 1987)

Entropy of fusion (cal/mol·K):
13.96 (Chao et al., 1983; Todd et al., 1947)

Heat of fusion (kcal/mol):
1.700 (Chao et al., 1983; Todd et al., 1947)

Henry's law constant (atm·m³/mol):
0.225 at 25 °C (approximate - calculated from water solubility and vapor pressure)

Ionization potential (eV):
9.01 (Lias, 1998)
9.11 (Collin and Lossing, 1959)

Soil sorption coefficient, log K_{oc}:
Unavailable because experimental methods for estimation of this parameter for aliphatic hydrocarbons are lacking in the documented literature

Octanol/water partition coefficient, log K_{ow}:
2.15 using method of Hansch et al. (1968)

Solubility in organics:
Soluble in alcohol, benzene, and ether (Weast, 1986)

Solubility in water:
203 mg/kg at 25 °C (McAuliffe, 1966)

Vapor density:
2.87 g/L at 25 °C, 2.42 (air = 1)

Vapor pressure (mmHg):
494.6 at 25 °C (Wilhoit and Zwolinski, 1971)

Environmental fate:
   Photolytic. The rate constant for the reaction of *cis*-2-pentene and OH radicals in the atmosphere at 300 K is 3.9 x 10^{13} cm^3/mol·sec (Hendry and Kenley, 1979). Atkinson (1990) reported a rate constant of 6.50 x 10^{11} cm^3/mol·sec for the same reaction.

Uses: Polymerization inhibitor; organic synthesis.

trans-2-PENTENE

Synonyms: (*E*)-2-Pentene; *trans*-Pentene-2.

CAS Registry Number: 646-04-8
Molecular formula: C_5H_{10}
Formula weight: 70.13
Merck reference: 10, 6992

Physical state and color:
Colorless liquid

Melting point (°C):
-136 (Weast, 1986)
-140.2 (Dean, 1987)

Boiling point (°C):
36.3 (Weast, 1986)

Density (g/cm³):
0.6482 at 20/4 °C (Weast, 1986)

Diffusivity in water (x 10^{-5} cm²/sec):
0.84 at 20 °C using method of Hayduk and Laudie (1974)

Dissociation constant, pK_a:
>14 (Schwarzenbach et al., 1993)

Flash point (°C):
<-20 (NFPA, 1984)

Entropy of fusion (cal/mol·K):
15.02 (Chao et al., 1983)
15.01 (Todd et al., 1947)

Heat of fusion (kcal/mol):
1.996 (Chao et al., 1983; Todd et al., 1947)

Henry's law constant (atm·m³/mol):
0.234 at 25 °C (Hine and Mookerjee, 1975)

Ionization potential (eV):
9.06 (Collin and Lossing, 1959)
9.04 (Lias, 1998)

Soil sorption coefficient, log K_{oc}:
Unavailable because experimental methods for estimation of this parameter for aliphatic hydrocarbons are lacking in the documented literature

Octanol/water partition coefficient, log K$_{ow}$:
2.15 using method of Hansch et al. (1968)

Solubility in organics:
Soluble in alcohol, benzene, and ether (Weast, 1986)

Solubility in water:
203 mg/kg at 25 °C (McAuliffe, 1966)

Vapor density:
2.87 g/L at 25 °C, 2.42 (air = 1)

Vapor pressure (mmHg):
505.5 at 25 °C (Wilhoit and Zwolinski, 1971)

Source: Schauer et al. (1999) reported *trans*-2-pentene in a diesel-powered medium-duty truck exhaust at an emission rate of 50 μg/km.

Uses: Polymerization inhibitor; organic synthesis.

PENTYLCYCLOPENTANE

Synonym: 1-Cyclopentylpentane.

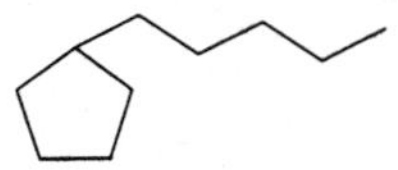

CAS Registry Number: 3741-00-2
Molecular formula: $C_{10}H_{20}$
Formula weight: 140.28

Physical state:
Liquid

Melting point (°C):
-83 (Dreisbach, 1959)

Boiling point (°C):
180.6 (Wilhoit and Zwolinski, 1971)

Density (g/cm³):
0.7912 and 0.7874 at 20/4 and 20/5 °C, respectively (Dreisbach, 1959)

Diffusivity in water (x 10^{-5} cm²/sec):
0.57 at 20 °C using method of Hayduk and Laudie (1974)

Dissociation constant, pK_a:
>14 (Schwarzenbach et al., 1993)

Henry's law constant (atm·m³/mol):
1.82 at 25 °C (Mackay and Shiu, 1981)

Ionization potential (eV):
9.91 (Lias et al., 1998)

Soil sorption coefficient, log K_{oc}:
Unavailable because experimental methods for estimation of this parameter for alicyclic hydrocarbons are lacking in the documented literature

Octanol/water partition coefficient, log K_{ow}:
4.90 using method of Hansch et al. (1968)

Solubility in water:
115 μg/kg at 25 °C (Price, 1976)

Vapor density:
5.73 g/L at 25 °C, 4.84 (air = 1)

Uses: Organic synthesis; gasoline component.

PHENANTHRENE

Synonyms: Phenanthren; Phenantrin.

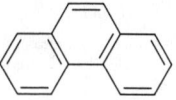

CAS Registry Number: 85-01-8
Molecular formula: $C_{14}H_{10}$
Formula weight: 178.24
RTECS: SF7175000
Merck reference: 10, 7075

Physical state, color, and odor:
Colorless, monoclinic crystals with a faint, aromatic odor

Melting point (°C):
100.5 (Dean, 1973)
98-100 (Fluka, 1988)

Boiling point (°C):
341.2 (Wilhoit and Zwolinski, 1971)

Density (g/cm³):
1.179 at 25/4 °C (Dean, 1987)

Diffusivity in water (x 10^{-5} cm²/sec):
0.59 at 20 °C using method of Hayduk and Laudie (1974)

Dissociation constant, pK_a:
>15 (Christensen et al., 1975)

Flash point (°C):
171 (open cup, NFPA, 1984)

Entropy of fusion (cal/mol·K):
11.9 (Rai et al., 1987)
11.5 (Rastogi and Bassi, 1964)

Heat of fusion (kcal/mol):
3.757 (Sabbah and El Watik, 1992)
4.452 (Rai et al., 1987)
4.302 (Rastogi and Bassi, 1964)

Henry's law constant (x 10^{-5} atm·m³/mol):
3.9 at 25 °C (Mackay et al., 1979)
2.35 at 25 °C (Fendinher and Glotfelty, 1990; Alaee et al., 1996)
13 (Petrasek et al., 1983)

2.86 at 25 °C (de Maagd et al., 1998)

5.48 (Southworth, 1979)

3.56 at 25 °C (Shiu and Mackay, 1997)

0.93, 1.58, 2.62, 4.23, and 6.30 at 4.1, 11.0, 18.0, 25.0, and 31.0 °C, respectively (Bamford et al., 1998)

Ionization potential (eV):

7.891 (Lias, 1998)

8.19 (Cavalieri and Rogan, 1985)

8.03 (Krishna and Gupta, 1970)

Bioconcentration factor, log BCF:

2.51 (*Daphnia pulex*, Southworth et al., 1978)

3.42 (fathead minnow, Veith et al., 1979)

2.51 (*Daphnia magna*, Newsted and Giesy, 1987)

0.77 (*Polychaete* sp.), 1.49 (*Capitella capitata*) (Bayona et al., 1991)

3.25 (algae, Geyer et al., 1984)

2.97 (activated sludge), 3.25 (golden ide) (Freitag et al., 1985)

4.38 (*Selenastrum capricornutum*, Casserly et al., 1983)

Apparent values of 3.6 (wet wt) and 5.4 (lipid wt) for freshwater isopods including *Asellus aquaticus* (L.) (van Hattum et al., 1998)

Soil sorption coefficient, log K_{oc}:

3.72 (aquifer sands, Abdul et al., 1987)

4.36 (Karickhoff et al., 1979)

4.59 (Socha and Carpenter, 1987)

4.70 (Gauthier et al., 1986)

4.28 (estuarine sediment, Vowles and Mantoura, 1987)

4.07 (dark sand from Newfield, NY, Magee et al., 1991)

3.77, 3.15 and 3.76 in Apison, Fullerton, and Dormont soils, respectively (Southworth and Keller, 1986)

4.42 (Eustis fine sand, Wood et al., 1990)

6.12 (average, Kayal and Connell, 1990)

4.12 (Oshtemo, Sun and Boyd, 1993)

4.60 (humic acid, Gauthier et al., 1986)

4.46 (Rotterdam Harbor sediment, Hegeman et al., 1995)

4.18, 4.39, 4.91, 5.28, 5.91 (San Francisco, CA mudflat sediments, Maruya et al., 1996)

5.9 (average value using 8 river bed sediments from the Netherlands, van Hattum et al., 1998)

K_d value = 0.130 mL/g for sorption to hematite (α-Fe_2O_3) (Mader et al., 1997)

Octanol/water partition coefficient, log K_{ow}:

4.57 at 25 °C (Karickhoff et al., 1979; de Maagd et al., 1998)

4.60 at 25 °C (Andersson and Schräder, 1999)

4.46 (Hansch and Fujita, 1964)

4.52 (Kenaga and Goring, 1980; Yoshida et al., 1983)

4.16 (Landrum et al., 1984)

4.562 (Brooke et al., 1990; de Bruijn et al., 1989)

4.374 (Brooke et al., 1990)

4.54 (Hammers et al., 1982)

4.63 (Bruggeman et al., 1982)

4.45 (Wang et al., 1992)

Solubility in organics:
Soluble in carbon tetrachloride (417 g/L) and ethyl ether (303 g/L) (Windholz et al., 1983)
24.5 and 31.7 g/L at 26 °C in methanol and ethanol, respectively (Fu et al., 1986).

Solubility in water:
At 25 °C: 0.85, 0.67, 0.48, and 0.33 mg/L in 0.50, 1.00, 1.50, and 2.00 M NaCl, respectively
   (Aquan-Yuen et al., 1979)
In mg/kg: 1.11-1.12 at 24.6 °C, 1.49 at 29.9 °C, 1.47-1.48 at 30.3 °C, 2.44-2.45 at 38.4 °C,
   2.25-2.28 at 40.1 °C, 3.81-3.88 at 47.5 °C, 4.30-4.38 at 50.1 °C, 4.04-4.11 at 50.2 °C,
   5.63-5.66 at 54.7 °C, 7.17-7.21 at 59.2 °C, 7.2-7.6 at 60.5 °C, 9.7-9.8 at 65.1 °C, 12.4-12.6
   at 70.7 °C (Wauchope and Getzen, 1972)
1.29 mg/L at 25 °C (Mackay and Shiu, 1977; Walters and Luthy, 1984)
994 μg/L at 25 °C (Andrews and Keefer, 1949)
1.002 and 1.220 mg/kg at 25 and 29 °C, respectively (May et al., 1978a)
At 25.0 °C: 1.07 mg/kg (distilled water), 0.71 mg/kg in artificial seawater (salinity = 35 g/kg)
   (Eganhouse and Calder, 1976)
1.55-1.65 mg/L at 27 °C (Davis et al., 1942)
1.650 mg/L at 27 °C (Davis and Parke, 1942)
16 ng/L at 25 °C (Klevens, 1950)
710 μg/L at 25 °C (Sahyun, 1966)
In μg/kg: 423 at 8.5 °C, 468 at 10.0 °C, 512 at 12.5 °C, 601 at 15 °C, 816 at 21.0 °C, 995 at
   24.3 °C, 1,277 at 29.9 °C (May et al., 1978)
In μmol/L: 2.81 at 8.4 °C, 3.09 at 11.1 °C, 3.59 at 14.0 °C, 4.40 at 17.5 °C, 4.94 at 20.2 °C,
   6.09 at 23.3 °C, 6.46 at 25.0 °C, 7.7 at 29.3 °C, 9.13 at 31.8 °C (Schwarz, 1977)
235 ppb at 20-25 °C in Narragansett Bay water (pH 7.7, dissolved organic carbon 2.9 mg/L,
   salinity 27 wt %) (Boehm and Quinn, 1973)
5.61 μmol/L at 25 °C (Wasik et al., 1983)
9.5 μmol/L at 29 °C (Stucki and Alexander, 1987)
9 μmol/L at 25 °C (Edwards et al., 1991)
In μmol/L: 2.01, 2.45, 3.12, 4.04, 4.94, and 6.16 at 4.6, 8.8, 12.9, 17.0, 21.1, and 25.3 °C, respectively. In seawater (salinity 36.5 g/kg): 1.36, 1.75, 2.21, 2.81, 3.59, and 4.54 at 4.6, 8.8,
   12.9, 17.0, 21.1, and 25.3 °C, respectively (Whitehouse, 1984)
At 20 °C: 6.2, 4.15, 4.01, and 4.14 μmol/L in distilled water, Pacific seawater, artificial seawater, and 35% NaCl solution, respectively (Hashimoto et al., 1984)
1.00 mg/L at 25 °C (Vadas et al., 1991)
1.03 mg/L at 25 °C (Etzweiler et al., 1995)
1.08 mg/L at 25 °C (Billington et al., 1988)
6.77 μmol/kg at 25.0 °C (Vesala, 1974)
823 μg/L at 25 °C (de Maagd et al., 1998)
In mole fraction (x 10^{-7}): 0.3649 at 4.00 °C, 0.4275 at 8.50 °C, 0.4730 at 10.00 °C, 0.5175 at
   12.50 °C, 0.6075 at 15.00 °C, 0.7955 at 20.00 °C, 0.8248 at 21.00 °C, 0.9653 at 24.30 °C,
   1.240 at 29.90 °C (May et al., 1983)

Vapor pressure (x 10^{-4} mmHg):
1.21 at 25 °C (Sonnefeld et al., 1983; Wasik et al., 1983)
6.80 at 25 °C (Radding et al., 1976)
4.2, 8.3 at 25 °C (Hinckley et al., 1990)
0.64 at 36.70 °C, 0.81 at 39.15 °C, 0.89 at 39.85 °C, 1.11 at 42.10 °C, 8.1 at 42.60 °C, 1.45 at
   44.62 °C, 1.85 at 46.70 °C (Bradley and Cleasby, 1953)
0.95 at 25 °C (McVeety and Hites, 1988)
15,500 at 127 °C (Eiceman and Vandiver, 1983)

5.16 at 25 °C (Bidleman, 1984)

2.68, 6.82, 10.50, 17.03, 24.23, and 74.85 at 30.34, 37.22, 40.77, 44.79, 48.10, and 59.78 °C, respectively (Oja and Suuberg, 1998)

3.1 at 25 °C (extrapolated from vapor pressures determined at higher temperatures, Tesconi and Yalkowsky, 1998)

Environmental fate:

Biological. Catechol is the central metabolite in the bacterial degradation of phenanthrene. Intermediate by-products include 1-hydroxy-2-naphthoic acid, 1,2-dihydroxynaphthalene, and salicylic acid (Chapman, 1972; Hou, 1982). It was reported that *Beijerinckia*, under aerobic conditions, degraded phenanthrene to *cis*-3,4-dihydroxy-3,4-dihydrophenanthracene (Kobayashi and Rittman, 1982).

In activated sludge, 39.6% mineralized to carbon dioxide (Freitag et al., 1985). When phenanthrene (5 and 10 mg/L) was statically incubated in the dark at 25 °C with yeast extract and settled domestic wastewater inoculum for 7 d, 100% biodegradation with rapid adaptation was observed (Tabak et al., 1981).

Contaminated soil from a manufactured coal gas plant that had been exposed to crude oil was spiked with phenanthrene (400 mg/kg soil) to which Fenton's reagent (5 mL 2.8 M hydrogen peroxide; 5 mL 0.1 M ferrous sulfate) was added. The treated and nontreated soil samples were incubated at 20 °C for 56 d. Fenton's reagent greatly enhanced the mineralization of phenanthrene by indigenous microorganisms. The amounts of phenanthrene recovered as carbon dioxide after treatment with and without Fenton's reagent were 29 and 3%, respectively. Pretreatment of the soil with a surfactant (10 mM sodium dodecylsulfate) before addition of Fenton's reagent, increased the mineralization rate 84% as compared to nontreated soil (Martens and Frankenberger, 1995).

Soil. The reported half-lives for phenanthrene in a Kidman sandy loam and McLaurin sandy loam are 16 and 35 d, respectively (Park et al., 1990). Manilal and Alexander (1991) reported a half-life of 11 d in a Kendaia soil.

Surface Water. In a 5-m deep surface water body, the calculated half-lives for direct photochemical transformation at 40 °N latitude, in the midsummer during midday were 59 and 69 d with and without sediment-water partitioning, respectively (Zepp and Schlotzhauer, 1979).

Photolytic. A carbon dioxide yield of 24.2% was achieved when phenanthrene adsorbed on silica gel was irradiated with light (λ >290 nm) for 17 h (Freitag et al., 1985). In a 2-wk experiment, [^{14}C]phenanthrene applied to soil-water suspensions under aerobic and anaerobic conditions gave $^{14}CO_2$ yields of 7.2 and 6.3%, respectively (Scheunert et al., 1987).

Wang et al. (1995) investigated the photodegradation of phenanthrene in water using artificial light (λ >290 nm) in the presence of fulvic acids as sensitizers and hydrogen peroxide as an oxidant. The major photoproducts identified were 9,10-phenanthrenequinone, 1,3,4-trihydroxyphenanthrene, 9-hydroxyphenanthrene, 2,2′-biphenyldialdehyde, 2,2′-biphenyldicarbonic acid, and 2-phenylbenzaldehyde. The presence of fulvic acids/humic substances produced mixed results. The rate of photodegradation was retarded or accelerated depending upon the origins of the humic substances. Experimentally determined photolysis half-lives of phenanthrene in water using fulvic acids obtained from six different locations ranged from 1.30 to 5.78 h. It was suggested that the formation of the photoproducts involved oxidation of phenanthrene via OH radicals generated from hydrogen peroxide.

Behymer and Hites (1985) determined the effect of different substrates on the rate of photooxidation of phenanthrene using a rotary photoreactor. The photolytic half-lives of phenanthrene using silica gel, alumina, and fly ash were 150, 45, and 49 h, respectively.

Phenanthrene (5.0 mg/L) in a methanol-water solution (2:3 v/v) was subjected to a high pressure mercury lamp or sunlight. Based on a rate constant of 6.53 x 10^{-3}/min, the corresponding half-life is 1.78 h (Wang et al., 1991).

Chemical/Physical. The aqueous chlorination of phenanthrene at pH <4 produced phenanthrene-9,10-dione and 9-chlorophenanthrene. At high pH (>8.8), phenanthrene-9,10-oxide, phenanthrene-9,10-dione, and 9,10-dihydrophenanthrenediol were identified as major products (Oyler et al., 1983). It was suggested that the chlorination of phenanthrene in tap water accounted for the presence of chloro- and dichlorophenanthrenes (Shiraishi et al., 1985).

When phenanthrene (0.65 mg/L) in hydrogen-saturated deionized water was exposed to a slurry of palladium catalyst (1%) at room temperature for approximately 2 h, 1,2,3,4,5,6,7,8-octahydrophenanthrene and 1,2,3,4,4a,9,10,10a-octahydrophenanthrene formed as products via the intermediates 1,2,3,4-tetrahydrophenanthrene and 9,10-dihydrophenanthrene, respectively (Schüth and Reinhard, 1997).

Exposure limits: Potential occupational carcinogen. No individual standards have been set; however, as a constituent in coal tar pitch volatiles, the following exposure limits have been established (mg/m^3): NIOSH REL: TWA 0.1 (cyclohexane-extractable fraction), IDLH 80; OSHA PEL: TWA 0.2 (benzene-soluble fraction); ACGIH TLV: TWA 0.2 (benzene solubles).

Toxicity:
LC$_{50}$ (27-d) for *Oncorhynchus mykiss* embryos 30 μg/L (Call et al., 1976).
LC$_{50}$ (10-d) for *Rhepoxynius abronius* 2.22 mg/g organic carbon (Swartz et al., 1997).
LC$_{50}$ (96-h) for juvenile *Oncorhynchus mykiss* 375 μg/L (Call et al., 1976).
LD$_{50}$ (intraperitoneal) for mice 700 mg/kg (Salamone, 1981).
Acute oral LD$_{50}$ for mice 700 mg/kg (Simmon et al., 1979).

Drinking water standard: No MCLGs or MCLs have been proposed (U.S. EPA, 1996).

Source: Detected in groundwater beneath a former coal gasification plant in Seattle, WA at a concentration of 130 μg/L (ASTR, 1995). Detected in 8 diesel fuels at concentrations ranging from 0.17 to 110 mg/L with a mean value of 41.43 mg/L (Westerholm and Li, 1994) and in distilled water-soluble fractions of new and used motor oil at concentrations of 1.9-2.1 and 2.1-2.2 μg/L, respectively (Chen et al., 1994). Lee et al. (1992) reported concentration ranges 100-300 mg/L and 15-25 μg/L in diesel fuel and the corresponding aqueous phase (distilled water), respectively. Schauer et al. (1999) reported phenanthrene in diesel fuel at a concentration of 57 μg/g and in a diesel-powered medium-duty truck exhaust at an emission rate of 93.1 μg/km. Identified in Kuwait and South Louisiana crude oils at concentrations of 26 and 70 ppm, respectively (Pancirov and Brown, 1975).

Thomas and Delfino (1991) equilibrated contaminant-free groundwater collected from Gainesville, FL with individual fractions of three individual petroleum products at 24-25 °C for 24 h. The aqueous phase was analyzed for organic compounds via U.S. EPA approved test method 625. Phenanthrene was only detected in the water-soluble fraction of diesel fuel at an average concentration of 17 μg/L.

Based on analysis of 7 coal tar samples, phenanthrene concentrations ranged from 3,100 to 35,000 ppm (EPRI, 1990). Detected in 1-yr aged coal tar film and bulk coal tar at an identical concentration of 10,000 mg/kg (Nelson et al., 1996).

Uses: Explosives; dyestuffs; biochemical research; synthesis of drugs; preparation of 9,10-phenanthrenequinone, 9,10-dihydrophenanthrene, 9-bromophenanthrene, 9,10-dibromo-9,10-di-hydrophenanthrene, and many other organic compounds.

PHENOL

Synonyms: Baker's P and S liquid and ointment; Benzenol; Carbolic acid; Hydroxybenzene; Monohydroxybenzene; NA 2821; NCI-C50124; Oxybenzene; Phenic acid; Phenyl hydrate; Phenyl hydroxide; Phenylic acid; Phenylic alcohol; RCRA waste number U188; UN 1671; UN 2312; UN 2821.

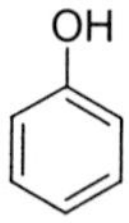

CAS Registry Number: 108-95-2
DOT: 1671 (solid); 2312 (molten); 2821 (solution)
DOT label: Poison
Molecular formula: C_6H_6O
Formula weight: 94.11
RTECS: SJ3325000
Merck reference: 10, 7115

Physical state, color, and odor:
White crystals or light pink liquid which slowly turns brown on exposure to air. Sweet, tarry odor. The reported odor threshold concentrations in air ranged from 9.5 to 16 ppb (Keith and Walters, 1992; Young et al., 1996). The taste threshold concentration in water is <2 ppb (Young et al., 1996).

Melting point (°C):
43 (Weast, 1986)
40.9 (Weiss, 1986)

Boiling point (°C):
181.7 (Weast, 1986)
183 (Huntress and Mulliken, 1941)

Density (g/cm³):
1.0576 at 20/4 °C (Weast, 1986)
1.05760 at 41/4 °C (Standen, 1968)

Diffusivity in water (x 10^{-5} cm²/sec):
0.87 at 20 °C using method of Hayduk and Laudie (1974)

Dissociation constant, pK_a:
9.99 at 25 °C (Dean, 1973)
9.90 (Blackman et al., 1955)

Flash point (°C):
80 (NIOSH, 1997)

Lower explosive limit (%):
1.8 (NIOSH, 1997)

Upper explosive limit (%):
8.6 (NIOSH, 1997)

Entropy of fusion (cal/mol·K):
8.762 (Andon et al., 1963)
8.755 (Tsomopoulos and Prausnitz, 1971)

Heat of fusion (kcal/mol):
2.752 (Andon et al., 1963)
2.898 (Mastrangelo, 1957)

Henry's law constant (x 10^{-7} atm·m^3/mol):
2.7 (Petrasek et al., 1983)
3.937 at 27.0 °C (Abd-El-Bary et al., 1986)
3.45 (Parsons et al., 1971)

Interfacial tension with water (dyn/cm at 20 °C):
0.8 (quoted, Freitas et al., 1997)

Ionization potential (eV):
8.51 (Franklin et al., 1969)

Bioconcentration factor, log BCF:
1.24 (*Brachydanio rerio*, Devillers et al., 1996)
1.28 (*Daphnia magna*, Dauble et al., 1986)
4.20 (fathead minnow, Call et al., 1980)
0.54 (algae, Hardy et al., 1985)
1.30 (golden ide), 2.30 (algae), 3.34 (activated sludge) (Freitag et al., 1985)

Soil sorption coefficient, log K_{oc}:
1.72 (Batcombe silt loam, Briggs, 1981)
3.46 (fine sediments), 3.49 (coarse sediments) (Isaacson and Frink, 1984)
1.21 (Brookstone clay loam soil, Campbell and Luthy, 1985)
2.40 (Meylan et al., 1992)
1.74 (Apison soil), 2.85 (Fullerton soil), 0.845 (Dormont soil) (Southworth and Keller, 1986)
1.24 (Boyd, 1982)
1.57, 1.96 (silt loam, Scott et al., 1983)
1.4 (river sediment), 1.6 (coal wastewater sediment) (Kopinke et al., 1995)
2.70 (glaciofluvial, sandy aquifer, Nielsen et al., 1996)

Octanol/water partition coefficient, log K_{ow}:
1.48 (Leo et al., 1971)
1.46 (Fujita et al., 1964; Leo et al., 1971; Briggs, 1981; Berthod et al., 1988)
1.39 (Riederer, 1990)
1.31-1.43 (Kubáň, 1991)
1.57 (Kishino and Kobayashi, 1994)
1.55 (Garst and Wilson, 1984)
1.51 (Umeyama et al., 1977)
1.45 (Mirrlees et al., 1976; Wasik et al., 1981)
1.47 (Leo et al., 1971)

Solubility in organics:
Soluble in carbon disulfide and chloroform; very soluble in ether; miscible with carbon tetrachloride, hot benzene (U.S. EPA, 1985), and alcohol (Meites, 1963)

Solubility in water:
82,000 mg/L at 15 °C (quoted, Verschueren, 1983)
67,000 mg/L at 25 °C (Warner et al., 1987)
0.866 wt % at 25 °C (Riddick et al., 1986)
84,120 mg/L (Reinbold et al., 1979)
0.90 M at 25 °C (Caturla et al., 1988)
74 g/L at 8 °C, 82 g/L at 25 °C (quoted, Leuenberger et al., 1985)
670 mg/L at 16 °C (quoted, Meites, 1963)
84 g/L (Price et al., 1974)
0.813 M at 25.0 °C (Wasik et al., 1981)
At 20 °C: 1.6, 1.35, 1.39, and 1.33 M in distilled water, Pacific seawater, artificial seawater, and 35 wt % NaCl, respectively (Hashimoto et al., 1984)
76.044, 84.045, and 93.098 g/L at 15.1, 25.0, and 35.0 °C, respectively (Achard et al., 1996)
0.88 M at 25 °C (Southworth and Keller, 1986)

Vapor pressure (x 10^{-2} mmHg):
20 at 20 °C, 100 at 40 °C (Verschueren, 1983)
35 at 25 °C (ACGIH, 1986)
6.4 at 8 °C, 34 at 25 °C (quoted, Leuenberger et al., 1985)
62 at 25 °C (Riederer, 1990)

Environmental fate:
Biological. Under methanogenic conditions, inocula from a municipal sewage treatment plant digester degraded phenol to carbon dioxide and methane (Young and Rivera, 1985). In a methanogenic enrichment culture, phenol anaerobically biodegraded to carbon dioxide and methane (yield 70%) (Healy and Young, 1979). Chloroperoxidase, a fungal enzyme isolated from *Caldariomyces fumago*, reacted with phenol forming 2- and 4-chlorophenol, the latter in a 25% yield (Wannstedt et al., 1990). In activated sludge, 41.4% mineralized to carbon dioxide after 5 d (Freitag et al., 1985). When phenol was statically incubated in the dark at 25 °C with yeast extract and settled domestic wastewater inoculum, significant biodegradation with rapid adaptation was observed. At concentrations of 5 and 10 mg/L, 96 and 97% biodegradation, respectively, were observed after 7 d (Tabak et al., 1981). Phenol is rapidly degraded in aerobically incubated soil but is much slower under anaerobic conditions (Baker and Mayfield, 1980).

In a continuous stirred reactor maintained at 20 °C, phenol degraded at rates of 0.094 and 0.007/h at feed concentrations of 180 and 360 mg/L, respectively (Beltrame et al., 1984). In the presence of suspended natural populations from unpolluted aquatic systems, the second-order microbial transformation rate constant determined in the laboratory was reported to be 3.3±1.2 x 10^{-10} L/organism·h (Steen, 1991).

Bridié et al. (1979) reported BOD and COD values 1.68 and 2.33 g/g using filtered effluent from a biological sanitary waste treatment plant. These values were determined using a standard dilution method at 20 °C of 5 d. Similarly, Heukelekian and Rand (1955) reported a 5-d BOD value of 1.81 g/g which is 76.0% of the ThOD value of 2.38 g/g. In activated sludge inoculum, 98.5% COD removal was achieved. The average rate of biodegradation was 80.0 mg COD/g·h (Pitter, 1976).

Soil. Loehr and Matthews (1992) studied the degradation of phenol in different soils under aerobic conditions. In a slightly basic basic sandy loam (3.25% organic matter) and in acidic clay soil (<1.0% organic matter), the resultant degradation half-lives were 4.1 and 23 d, respectively.

Soil sorption distribution coefficients (K_d) were determined from centrifuge column tests using kaolinite as the absorbent (Celorie et al., 1989). Values for K_d ranged from 0.010 to 0.054

L/g.

Surface Water. Vaishnav and Babeu (1987) reported a half-life of 11 d in river waters and 3 d in harbor waters.

Groundwater. Nielsen et al. (1996) studied the degradation of phenol in a shallow, glaciofluvial, unconfined sandy aquifer in Jutland, Denmark. As part of the *in situ* microcosm study, a cylinder that was open at the bottom and screened at the top was installed through a cased borehole approximately 5 m below grade. Five liters of water was aerated with atmospheric air to ensure aerobic conditions were maintained. Groundwater was analyzed weekly for approximately 3 months to determine phenol concentrations over time. The experimentally determined first-order biodegradation rate constant and corresponding half-life were 0.5/d and 33.4 h, respectively. Vaishnav and Babeu (1987) reported a biodegradation rate constant of 0.035/d and a half-life of 20 d in groundwater.

Photolytic. In an aqueous, oxygenated solution exposed to artificial light (λ = 234 nm), phenol was photolyzed to hydroquinone, catechol, 2,2'-, 2,4'- and 4,4'-dihydroxybiphenyl (Callahan et al., 1979). When an aqueous solution containing potassium nitrate (10 mM) and phenol (1 mM) was irradiated with UV light (λ = 290-350 nm) up to a conversion of 10%, the following products formed: hydroxyhydroquinone, hydroquinone, resorcinol, hydroxybenzo-quinone, benzoquinone, catechol, nitrosophenol, 4-nitrocatechol, nitrohydroquinone, 2- and 4-nitrophenol (Niessen et al., 1988). Titanium dioxide suspended in an aqueous solution and irradiated with UV light (λ = 365 nm) converted phenol to carbon dioxide at a significant rate (Matthews, 1986).

Irradiation of phenol with UV light (λ = 254 nm) in the presence of oxygen yielded substituted biphenyls, hydroquinone, and catechol (Joschek and Miller, 1966). A carbon dioxide yield of 32.5% was achieved when phenol adsorbed on silica gel was irradiated with light (λ >290 nm) for 17 h (Freitag et al., 1985). When an aqueous solution containing phenol was photooxidized by UV light at 50 °C, 10.96% degraded to carbon dioxide after 24 h (Knoevenagel and Himmelreich, 1976). The dye-sensitized photodegradation of phenol in aqueous solution was studied by Okamoto et al. (1982). They reported a first-order rate constant of 6.5 x 10^3 /sec. Second-order rate constants of 2.66 x 10^8, 6.16 x 10^7, and 1.95 x 10^7 L/mol·sec were determined at pH values 10.3, 9, and 8, respectively.

The following half-lives were reported for phenol in estuarine water exposed to sunlight and microbes: 39 and 94 h during summer (24 °C) and winter (10 °C), respectively; in distilled water: 46 and 173 h during summer and winter, respectively; in poisoned estuarine water: 43 and 118 h during summer and winter, respectively (Hwang et al., 1986).

Anticipated products from the reaction of phenol with ozone or OH radicals in the atmosphere are dihydroxybenzenes, nitrophenols, and ring cleavage products (Cupitt, 1980). Reported rate constants for the reaction of phenol and OH radicals in the atmosphere: 2.8 x 10^{-11} cm³/molecule·sec at room temperature (Atkinson, 1985) and with NO_3 in the atmosphere: 2.1 x 10^{-12} cm³/molecule·sec at 296 K (Atkinson et al., 1984).

Chemical/Physical. In an environmental chamber, nitrogen trioxide (10,000 ppb) reacted quickly with phenol (concentration 200-1400 ppb) to form phenoxy radicals and nitric acid (Carter et al., 1981). The phenoxy radicals may react with oxygen and nitrogen dioxide to form quinones and nitrohydroxy derivatives, respectively (Nielsen et al., 1983).

Reported rate constants for the reaction of phenol and singlet oxygen in water at 292 K: 2.5 x 10^6/M·sec at pH 8, 1.9 x 10^7/M·sec at pH 9, 4.6 x 10^7/M·sec at pH 9.5, 9.0 x 10^7/M·sec at pH 10 and 1.8 x 10^8/M·sec at pH 11.5 (Scully and Hoigné, 1987).

Groundwater contaminated with various phenols degraded in a methanogenic aquifer. Similar results were obtained in the laboratory utilizing an anaerobic digester. Methane and carbon dioxide were reported as degradation products (Godsy et al., 1983).

Ozonization of phenol in water resulted in the formation of many oxidation products. The identified products in the order of degradation are catechol, hydroquinone, *o*-quinone, *cis,cis-*

muconic acid, maleic (or fumaric) and oxalic acids (Eisenhauer, 1968). In addition, glyoxylic, formic, and acetic acids also were reported as ozonization products prior to further oxidation to carbon dioxide (Kuo et al., 1977). Ozonation of an aqueous solution of phenol subjected to UV light (120 watt low pressure mercury lamp) gave glyoxal, glyoxylic, oxalic, and formic acids as major products. Minor products included catechol, hydroquinone, muconic, fumaric, and maleic acids (Takahashi, 1990). Wet oxidation of phenol at 320 °C yielded formic and acetic acids (Randall and Knopp, 1980).

Chlorination of water containing bromide ions converted phenol to 2,4,6-tribromophenol. Bromodichlorophenol, dibromochlorophenol, and tribromophenol have also been reported to form from the chlorination of natural water under simulated conditions (Watanabe et al., 1984).

Wet oxidation of phenol at elevated pressure and temperature gave the following products: acetone, acetaldehyde, formic, acetic, maleic, oxalic, and succinic acids (Keen and Baillod, 1985). Chlorine dioxide reacted with phenol in an aqueous solution forming p-benzoquinone and hypochlorous acid (Wajon et al., 1982).

Kanno et al. (1982) studied the aqueous reaction of phenol and other substituted aromatic hydrocarbons (aniline, toluidine, 1-naphthylamine, cresol, pyrocatechol, resorcinol, hydroquinone, and 1-naphthol) with hypochlorous acid in the presence of ammonium ion. They reported that the aromatic ring was not chlorinated as expected but was cleaved by chloramine forming cyanogen chloride (Kanno et al., 1982). The amount of cyanogen chloride formed increased at lower pHs. At pH 6, the greatest amount of cyanogen chloride was formed when the reaction mixture contained ammonium ion and hypochlorous acid at a ratio of 2:3 (Kanno et al., 1982).

Spacek et al. (1995) investigated the photodegradation of phenol using titanium dioxide-UV light and Fenton's reagent (hydrogen peroxide:substance - 10:1; Fe^{2+} 2.5 x 10^{-4} mol/L) at 25 °C. The decomposition rate of 4-nitroaniline was very high by the photo-Fenton reaction in comparison to titanium dioxide-UV light (λ = 365 nm). Decomposition products identified in both reactions were p-benzoquinone, hydroquinone, and oxalic acid.

Phenol will not hydrolyze in water (Kollig, 1993).

Reacts with sodium and potassium hydroxide forming sodium and potassium phenolate, respectively (Morrison and Boyd, 1971).

At an influent concentration of 1,000 mg/L, treatment with granular activated carbon resulted in an effluent concentration of 194 mg/L. The adsorbability of the carbon used was 161 mg/g carbon (Guisti et al., 1974).

Exposure limits: NIOSH REL: TWA 5 ppm (19 mg/m^3), 15-min ceiling 15.6 ppm (60 mg/m^3), IDLH 250 ppm; OSHA PEL: 5 ppm; ACGIH TLV: TWA 5 ppm (adopted).

Symptoms of exposure: Ingestion of 5-10 mg may cause death. Toxic symptoms include nausea, vomiting, weakness, cyanosis, tremor, convulsions, kidney and liver damage. Eye, nose, and throat irritant. Burns skin on contact and may cause dermatitis (Patnaik, 1992). An irritation concentration of 182.40 mg/m^3 in air was reported by Ruth (1986).

Toxicity:

EC_{50} (48-h) for *Daphnia magna* 6.60 mg/L (Keen and Baillod, 1985).

EC_{50} (24-h) for *Daphnia magna* 9.1 mg/L, *Daphnia pulex* 17.9 mg/L (Lilius et al., 1995).

LC_{50} (contact) for earthworm (*Eisenia fetida*) 5.0 $\mu g/cm^2$ (Neuhauser et al., 1985).

LC_{50} (8-d) for fathead minnows 22-23 mg/L (Spehar et al., 1982).

LC_{50} (96-h) for fathead minnows 28-29 mg/L (Spehar et al., 1982), mud crab (*Panopeus herbstii*) 1.06 mg/L (Key and Scott, 1986), Japanese medaka (*Oryzias latipes*) 38.3 mg/L (Holcombe et al., 1995).

LC_{50} (48-h) for fathead minnows 8.3 mg/L (Spehar et al., 1982), zebra fish 60 mg/L (Slooff, 1979), *Daphnia magna* 12 mg/L (LeBlanc, 1980).

LC_{50} (24-h) for *Daphnia magna* 29 mg/L (LeBlanc, 1980).
Acute oral LD_{50} for mice 270 mg/kg, rats 317 mg/kg (quoted, RTECS, 1985).
LC_{50} (inhalation) for mice 177 mg/m^3 (quoted, RTECS, 1985).

Drinking water standard: No MCLGs or MCLs have been proposed (U.S. EPA, 1996).

Source: Detected in distilled water-soluble fractions of 87 octane gasoline (1.53 mg/L), 94 octane gasoline (0.19 mg/L), Gasohol (0.33 mg/L), No. 2 fuel oil (0.09 mg/L), jet fuel A (0.09 mg/L), diesel fuel (0.07 mg/L), and military jet fuel JP-4 (0.22 mg/L) (Potter, 1996).

Thomas and Delfino (1991) equilibrated contaminant-free groundwater collected from Gainesville, FL with individual fractions of three individual petroleum products at 24-25 °C for 24 h. The aqueous phase was analyzed for organic compounds via U.S. EPA approved test method 625. Average phenol concentrations reported in water-soluble fractions of unleaded gasoline, kerosene, and diesel fuel were 20, 8, and 19 μg/L, respectively.

Phenol occurs naturally in many plants including blueberries (10-60 ppb), marjoram (1,431-8,204 ppm), sweetflag, safflower buds (40 ppb), mud plantain, capillary wormwood, asparagus shoots, tea leaves, petitgrain, cinnamon, cassia, licorice, witch hazel, Japanese privet, St. John's wort, European pennyroyal, tomatoes, white mulberries, tobacco leaves, benneseed, sesame seeds, tamarind, white sandlewood, patchouli leaves, rue, slash pine, bayberries, Scotch pine, and tarragon (Duke, 1992).

Uses: Antiseptic and disinfectant; pharmaceuticals; dyes; indicators; slimicide; phenolic resins; epoxy resins (bisphenol-A); nylon-6 (caprolactum); 2,4-D; solvent for refining lubricating oils; preparation of adipic acid, salicylic acid, phenolphthalein, pentachlorophenol, acetophenetidin, picric acid, anisole, phenoxyacetic acid, phenyl benzoate, 2-phenolsulfonic acid, 4-phenolsulfonic acid, 2-nitrophenol, 4-nitrophenol, 2,4,6-tribromophenol, 4-bromophenol, 4-*tert*-butylphenol, salicylaldehyde, and many other organic compounds; germicidal paints; laboratory reagent.

p-PHENYLENEDIAMINE

Synonyms: 4-Aminoaniline; *p*-Aminoaniline; BASF ursol D; **1,4-Benzenediamine**; *p*-Benzenediamine; Benzofur D; C.I. 76060; C.I. developer 13; C.I. oxidation base 10; Developer 13; Developer PF; 1,4-Diaminobenzene; *p*-Diaminobenzene; Durafur black R; Fouramine D; Fourrine 1; Fourrine D; Fur black 41867; Fur brown 41866; Furro D; Fur yellow; Futramine D; Nako H; Orsin; Oxidation base 10; Para; Pelagol D; Pelagol DR; Pelagol grey D; Peltol D; 1,4-Phenylenediamine; PPD; Renal PF; Santoflex IC; Tertral D; UN 1673; Ursol D; USAF EK-394; Vulkanox 4020; Zoba black D.

CAS Registry Number: 106-50-3
DOT: 1673
DOT label: Poison
Molecular formula: $C_6H_8N_2$
Formula weight: 108.14
RTECS: SS8050000
Merck reference: 10, 7166

Physical state and color:
White, red, or brown crystals. May darken on exposure to air.

Melting point (°C):
140 (Weast, 1986)
145-147 (Windholz et al., 1983)

Boiling point (°C):
267 (Weast, 1986)
271 (Du Pont, 1999h)

Diffusivity in water (x 10^{-5} cm²/sec):
0.78 at 20 °C using method of Hayduk and Laudie (1974)

Dissociation constant (at 25 °C):
$pK_1 = 3.29$, $pK_2 = 6.08$ (Dean, 1973)

Flash point (°C):
156.8 (NIOSH, 1997)
154 (Du Pont, 1999h)

Entropy of fusion (cal/mol·K):
14.3 (Rai and Mandal, 1990)

Heat of fusion (kcal/mol):
5.951 (Rai and Mandal, 1990)

Ionization potential (eV):
6.87 (Lias, 1998)

Bioconcentration factor, log BCF:
2.66 (activated sludge), 2.65 (algae) (hydrochloride, Freitag et al., 1985)

Soil sorption coefficient, log K_{oc}:
Unavailable because experimental methods for estimation of this parameter for aromatic amines are lacking in the documented literature

Octanol/water partition coefficient, log K_{ow}:
Unavailable because experimental methods for estimation of this parameter for aromatic amines are lacking in the documented literature

Solubility in organics:
Soluble in alcohol ether (Weast, 1986) and slight soluble in chloroform (Du Pont, 1999h)

Solubility in water:
4.7 wt % at 20 °C (NIOSH, 1987)
38,000 mg/L at 24 °C, 6,690 g/L at 107 °C (quoted, Verschueren, 1983)
10.0 wt % at 40 °C (Du Pont, 1999h)

Vapor pressure (mmHg):
0.00091 and 1.0797 at 39 and 100.0 °C, respectively (Du Pont, 1999h)

Environmental fate:
Biological. In activated sludge, 3.8% mineralized to carbon dioxide after 5 d (Freitag et al., 1985). In activated sludge inoculum, following a 20-d adaptation period, 80.0% COD removal was achieved (Pitter, 1976).

Photolytic. A carbon dioxide yield of 53.7% was achieved when phenylenediamine (presumably an isomeric mixture) adsorbed on silica gel was irradiated with light (λ >290 nm) for 17 h (Freitag et al., 1985).

Chemical/Physical. *p*-Phenylenediamine will not hydrolyze because it does not contain a hydrolyzable functional group (Kollig, 1993).

Exposure limits (mg/m³): NIOSH REL: TWA 0.1, IDLH 25; OSHA PEL: TWA 0.1; ACGIH TLV: TWA 0.1 (adopted).

Symptoms of exposure: May include vertigo, gastritis, jaundice, allergic asthma, dermatitis, cornea ulcer, eye burn (Patnaik, 1992)

Toxicity:
LC_{50} (48-h) for red killifish 186 mg/L (Yoshioka et al., 1986).
Acute oral LD_{50} for wild birds 100 mg/kg, quail 100 mg/kg, rats 80 mg/kg (quoted, RTECS, 1985).

Source: Bulk quantitities may contain *m*- and *o*-phenylenediamine and aniline as impurities.

Uses: Manufacturing azo dyes, intermediates for antioxidants and accelerators for rubber; photochemical measurements; laboratory reagent; dyeing hair and fur.

PHENYL ETHER

Synonyms: Biphenyl ether; Biphenyl oxide; Diphenyl ether; Diphenyl oxide; Geranium crystals; **1,1-Oxybisbenzene**; Phenoxybenzene.

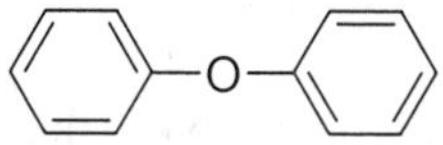

CAS Registry Number: 101-84-8
Molecular formula: $C_{12}H_{10}O$
Formula weight: 170.21
RTECS: KN8970000
Merck reference: 10, 7169

Physical state, color, and odor:
Colorless solid or liquid with a geranium-like odor. Odor threshold concentration is 1.2 ppb (Amoore and Hautala, 1983).

Melting point (°C):
28 (Verschueren, 1983)
26.87 (Riddick et al., 1986)

Boiling point (°C):
257.9 (Weast, 1986)

Density (g/cm³):
1.0748 at 20/4 °C (Weast, 1986)

Diffusivity in water (x 10^{-5} cm²/sec):
0.67 at 20 °C using method of Hayduk and Laudie (1974)

Flash point (°C):
116 (NIOSH, 1997)

Lower explosive limit (%):
0.7 (NIOSH, 1997)

Upper explosive limit (%):
6.0 (NIOSH, 1997)

Entropy of fusion (cal/mol·K):
13.71 (Ginnings and Furukawa, 1953)

Heat of fusion (kcal/mol):
4.115 (Riddick et al., 1986)

Henry's law constant (x 10^{-4} atm·m³/mol):
2.13 at 20 °C (approximate - calculated from water solubility and vapor pressure)

Ionization potential (eV):
8.82 ± 0.05 (Franklin et al., 1969)
8.09 (Lias, 1998)

Bioconcentration factor, log BCF:
2.29 (fish, Mackay, 1982)

Soil sorption coefficient, log K_{oc}:
Unavailable because experimental methods for estimation of this parameter for aromatic ethers are lacking in the documented literature

Octanol/water partition coefficient, log K_{ow}:
4.21, 4.36 (Leo et al., 1971)
3.79 (Burkhard et al., 1985a)
4.08 (Banerjee et al., 1980)
4.20 (Chiou et al., 1977)

Solubility in organics:
Soluble in acetic acid, alcohol, benzene, and ether (Weast, 1986)

Solubility in water (mg/L at 25 °C):
18.04 (Banerjee et al., 1980)
21 (Chiou et al., 1977)

Vapor pressure (mmHg):
0.02 at 20 °C, 0.12 at 30 °C (Verschueren, 1983)

Exposure limits: NIOSH REL: TWA 1 ppm (7 mg/m³), IDLH 100 ppm; OSHA PEL: TWA 1 ppm; ACGIH TLV: TWA 0.1, STEL 2 ppm (adopted).

Symptoms of exposure: Mild skin irritant. Ingestion may cause liver and kidney damage (Patnaik, 1992).

Toxicity:
   LC_{50} (96-h) for fathead minnows 4.0 mg/L (Veith et al., 1983).
   LC_{50} (48-h) for *Daphnia magna* 670 μg/L (LeBlanc, 1980).
   LC_{50} (24-h) for *Daphnia magna* 1.4 mg/L (LeBlanc, 1980).
   Acute oral LD_{50} for rats 3,370 mg/kg (quoted, RTECS, 1985).

Uses: Heat transfer liquid; perfuming soaps; resins for laminated electrical insulation; organic synthesis.

PHENYLHYDRAZINE

Synonyms: Hydrazine-benzene; Hydrazinobenzene; UN 2562.

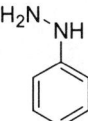

CAS Registry Number: 100-63-0
DOT: 2572
DOT label: Poison
Molecular formula: $C_6H_8N_2$
Formula weight: 108.14
RTECS: MV8925000
Merck reference: 10, 7174

Physical state, color, and odor:
Yellow monoclinic crystals or oil with a faint, aromatic odor. Turns reddish-brown on exposure to air.

Melting point (°C):
19.8 (Weast, 1986)

Boiling point (°C):
243 (Weast, 1986) with decomposition (Windholz et al., 1983)

Density (g/cm³):
1.0986 at 20/4 °C (Weast, 1986)

Diffusivity in water (x 10^{-5} cm²/sec):
0.89 at 20 °C using method of Hayduk and Laudie (1974)

Dissociation constant, pK_a:
8.79 at 15 °C (Windholz et al., 1983)
5.20 at 25 °C (Dean, 1973)

Flash point (°C):
88.5 (NIOSH, 1997)

Ionization potential (eV):
7.64, 7.74 (Rosenstock et al., 1998)

Soil sorption coefficient, log K_{oc}:
Unavailable because experimental methods for estimation of this parameter for hydrazines are lacking in the documented literature

Octanol/water partition coefficient, log K_{ow}:
1.25 (Leo et al., 1971)

Solubility in organics:
Soluble in acetone (Weast, 1986). Miscible with alcohol, benzene, chloroform, and ether (Windholz et al., 1983).

Vapor density:
4.42 g/L at 25 °C, 3.73 (air = 1)

Vapor pressure (mmHg):
0.04 at 25 °C (NIOSH, 1997)

Exposure limits: Potential occupational carcinogen. NIOSH REL: 2-h ceiling 0.14 ppm (0.6 mg/m^3), IDLH 15 ppm; OSHA PEL: TWA 5 ppm (22 mg/m^3); ACGIH TLV: TWA 0.1 ppm (adopted).

Symptoms of exposure: Acute toxic symptoms include hematuria, vomiting, convulsions, and respiratory arrest (Patnaik, 1992).

Toxicity:
    Acute oral LD$_{50}$ for guinea pigs 80 mg/kg, rats 188 mg/kg, rabbits 80 mg/kg (quoted, RTECS, 1985).

Uses: Reagent for aldehydes, ketones, sugars; manufacturing dyes, and antipyrine; organic synthesis.

PHTHALIC ANHYDRIDE

Synonyms: 1,2-Benzenedicarboxylic acid anhydride; 1,3-Dioxophthalan; ESEN; 1,3-Dihydro-1,3-dioxoisobenzofuran; **1,3-Isobenzofurandione**; NCI-C03601; Phthalandione; 1,3-Phthalandione; Phthalic acid anhydride; RCRA waste number U190; Retarder AK; Retarder ESEN; Retarder PD; UN 2214.

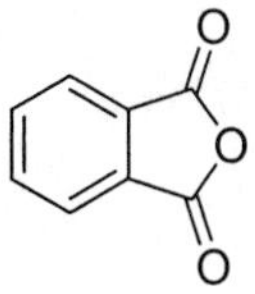

CAS Registry Number: 85-44-9
DOT: 2214
Molecular formula: $C_8H_4O_3$
Formula weight: 148.12
RTECS: TI3150000
Merck reference: 10, 7256

Physical state, color, and odor:
White to pale cream crystals with a characteristic, choking odor. Odor threshold concentration is 53 ppb (Amoore and Hautala, 1983).

Melting point (°C):
131.6 (Weast, 1986)
130.8 (Windholz et al., 1983)

Boiling point (°C):
Sublimes at 295 (Weast, 1986)

Density (g/cm³):
1.527 at 4/4 °C (Verschueren, 1983)

Flash point (°C):
152.9 (NIOSH, 1997)

Lower explosive limit (%):
1.7 (NIOSH, 1997)

Upper explosive limit (%):
10.5 (NIOSH, 1997)

Henry's law constant (x 10^{-9} atm·m³/mol):
6.29 at 20 °C (approximate - calculated from water solubility and vapor pressure)

Ionization potential (eV):
10.1 (Lias et al., 1998)

Soil sorption coefficient, log K_{oc}:
Not applicable - reacts with water

Octanol/water partition coefficient, log K_{ow}:
-0.62 (Kenaga and Goring, 1980)

Solubility in organics:
One part in 125 parts carbon disulfide (Windholz et al., 1983).
At 25 °C (mol/L): acetic acid (0.27), acetone (1.15), *t*-butyl alcohol (0.04), chloroform (0.78), cyclohexane (0.0064), decane (0.0049), di-*n*-propyl ether (0.04), dodecane (0.0048), ethyl ether (0.116), heptane (0.0049), hexane (0.0050), hexadecane (0.0047), isooctane (0.00422), pyridine (2.9), and tetrahydrofuran (1.2) (Fung and Higuchi, 1971)

Solubility in water:
0.6 wt % at 20 °C (NIOSH, 1997)

Vapor pressure (x 10^{-4} mmHg):
2 at 20 °C, 10 at 30 °C (Verschueren, 1983)

Environmental fate:
   Chemical/Physical. Reacts with water to form *o*-phthalic acid (Kollig, 1993; Windholz et al., 1983). Based on an observed rate constant of 7.9 x 10^{-9}/sec, the hydrolysis half-life is 88 sec (Hawkins, 1975).
   Pyrolysis of phthalic anhydride in the presence of polyvinyl chloride at 600 °C gave the following compounds: biphenyl, fluorene, benzophenone, 9-fluorenone, *o*-terphenyl, 9-phenylfluorene, and three unidentified compounds (Bove and Dalven, 1984).

Exposure limits: NIOSH REL: TWA 6 mg/m^3 (1 ppm), IDLH 60 mg/m^3; OSHA PEL: TWA 12 mg/m^3 (2 ppm); ACGIH TLV: TWA 1 ppm (adopted).

Symptoms of exposure: An irritation concentration of 30.00 mg/m^3 in air was reported by Ruth (1986).

Toxicity:
   Acute oral LD$_{50}$ for mice 2 g/kg, rats 4,020 mg/kg (quoted, RTECS, 1985).

Uses: Manufacturing phthalates, phthaleins, benzoic acid, synthetic indigo, pharmaceuticals, insecticides, chlorinated products, and artificial resins.

PICRIC ACID

Synonyms: Carbazotic acid; C.I. 10305; 2-Hydroxy-1,3,5-trinitrobenzene; Lyddite; Melinite; Nitroxanthic acid; Pertite; Phenol trinitrate; Picronitric acid; Shimose; 1,3,5-Trinitrophenol; **2,4,6-Trinitrophenol**; UN 0154.

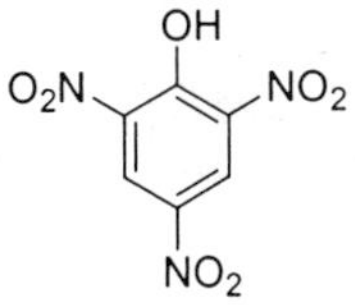

CAS Registry Number: 88-89-1
DOT: 1344
DOT label: Class A explosive
Molecular formula: $C_6H_3N_3O_7$
Formula weight: 229.11
RTECS: TJ7875000
Merck reference: 10, 7288

Physical state, color, and odor:
Colorless to yellow, odorless liquid or crystals

Melting point (°C):
122-123 (Weast, 1986)

Boiling point (°C):
Sublimes and explodes >300 °C (Weast, 1986).

Density (g/cm³):
1.763 (Windholz et al., 1983)

Diffusivity in water (x 10^{-5} cm²/sec):
0.72 at 20 °C using method of Hayduk and Laudie (1974)

Dissociation constant, pK_a:
0.29 at 25 °C (Dean, 1973)

Flash point (°C):
151 (NIOSH, 1997)

Entropy of fusion (cal/mol·K):
10.4 (Farrell, et al., 1979)

Heat of fusion (kcal/mol):
4.087 (Farrell et al., 1979)

Octanol/water partition coefficient, log K_{ow}:
2.03 (quoted, Verschueren, 1983)
1.34 (Hansch and Anderson, 1967)

Solubility in organics:
In g/L: alcohol (83.3), benzene (100), chloroform (28.57), and ether (15.38) (Windholz et al., 1983)

At 25 °C (mM): cyclohexane (0.41), decane (0.42), dodecane (0.33), heptane (0.33), hexane (0.30), hexadecane (0.39), and isooctane (0.25) (Fung and Higuchi, 1971)

Solubility in water (mg/L):
66,670 at 100 °C (quoted, Windholz et al., 1983)
14,000 at 20 °C, 68,000 at 100 °C (quoted, Verschueren, 1983)

Vapor pressure (mmHg):
1 at 197 °C (NIOSH, 1997)

Environmental fate:
   Chemical/Physical. Picric acid explodes when heated >300 °C (Weast, 1986). Shock sensitive! (Keith and Walters, 1992).

Exposure limits (mg/m³): NIOSH REL: TWA 0.1, STEL 0.3, IDLH 75; OSHA PEL: TWA 0.1; ACGIH TLV: TWA 0.1 (adopted).

Symptoms of exposure: Contact with skin may cause sensitization dermatitis. Ingestion may cause severe poisoning. Toxic symptoms include headache, nausea, vomiting, abdominal pain, and yellow coloration of skin (Patnaik, 1992).

Toxicity:
   In rabbits, the lethal dose is 120 mg/kg (quoted, RTECS, 1985).
   LC_{50} (48-h) for red killifish 513 mg/L (Yoshioka et al., 1986).

Uses: Explosives, matches; electric batteries; in leather industry; manufacturing colored glass; etching copper; textile mordant; reagent.

PINDONE

Synonyms: Chemrat; **2-(2,2-Dimethyl-1-oxopropyl)-1*H*-indene-1,3(2*H*)-dione**; Pivacin; Pival; Pivaldione; 2-Pivaloylindane-1,3-dione; 2-Pivaloyl-1,3-indanedione; Pivalyl; Pivalyl indandione; 2-Pivalyl-1,3-indandione; Pivalyl valone; Tri-Ban; UN 2472.

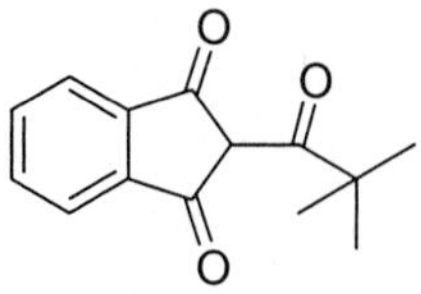

CAS Registry Number: 83-26-1
DOT: 2472
Molecular formula: $C_{14}H_{14}O_3$
Formula weight: 230.25
RTECS: NK6300000
Merck reference: 10, 7318

Physical state and color:
Bright yellow crystals

Melting point (°C):
108-110 (Windholz et al., 1983)

Diffusivity in water (x 10^{-5} cm²/sec):
0.51 at 20 °C using method of Hayduk and Laudie (1974)

Soil sorption coefficient, log K_{oc}:
2.95 using method of Kenaga and Goring (1980)

Octanol/water partition coefficient, log K_{ow}:
3.18 using method of Kenaga and Goring (1980)

Solubility in organics:
Soluble in most organic solvents (Worthing and Hance, 1991)

Solubility in water:
18 ppm at 25 °C (Gunther et al., 1968)

Exposure limits (mg/m³): NIOSH REL: TWA 0.1, IDLH 100; OSHA PEL: TWA 0.1; ACGIH TLV: TWA 0.1 (adopted).

Toxicity:
Acute oral LD_{50} for dogs 75 mg/kg, rats 280 mg/kg (quoted, RTECS, 1985).

Uses: Insecticide; rodenticide; intermediate in manufacturing pharmaceuticals.

PROPANE

Synonyms: Dimethylmethane; Liquefied petroleum gas; LPG; Propyl hydride; UN 1075; UN 1978.

CAS Registry Number: 74-98-6
DOT: 1978
DOT label: Flammable gas
Molecular formula: C_3H_8
Formula weight: 44.10
RTECS: TX2275000
Merck reference: 10, 7701

Physical state, color, and odor:
Colorless gas. Odor threshold concentration is 16,000 ppm (Amoore and Hautala, 1983).

Melting point (°C):
-189.7 (Weast, 1986)
-187.7 (Dean, 1987)

Boiling point (°C):
-42.1 (Weast, 1986)

Density (g/cm³):
0.5843 at -45/4 °C (Weast, 1986)

Diffusivity in water (x 10^{-5} cm²/sec):
1.16 at 25 °C (quoted, Hayduk and Laudie, 1974)

Dissociation constant, pK_a:
≈ 44 (Gordon and Ford, 1972)

Flash point (°C):
-105 (Hawley, 1981)

Lower explosive limit (%):
2.1 (NIOSH, 1997)

Upper explosive limit (%):
9.5 (NIOSH, 1997)

Heat of fusion (kcal/mol):
0.842 (Dean, 1987)

Henry's law constant (atm·m³/mol):
0.706 at 25 °C (Hine and Mookerjee, 1975)

Ionization potential (eV):
10.94 (Lias, 1998)
11.12 (Svec and Junk, 1967)

Soil sorption coefficient, log K$_{oc}$:
Unavailable because experimental methods for estimation of this parameter for aliphatic hydrocarbons are lacking in the documented literature

Octanol/water partition coefficient, log K$_{ow}$:
2.36 (Hansch et al., 1975)

Solubility in organics (vol %):
Alcohol (790 at 16.6 °C and 754 mmHg), benzene (1,452 at 21.5 °C and 757 mmHg), chloroform (1,299 at 21.6 °C and 757 mmHg), ether (926 at 16.6 °C and 757 mmHg), and turpentine (1,587 at 17.7 °C and 757 mmHg) (Windholz et al., 1983).

Solubility in water:
62.4 mg/kg at 25 °C (McAuliffe, 1963, 1966)
6.5 vol % at 17.8 °C and 753 mmHg (quoted, Windholz et al., 1983)
At 0 °C, 0.0394 volumes dissolved in a unit volume of water at 19.8 °C (Claussen and Pol-
    glase, 1952)
1.50 mM at 25 °C (Barone et al., 1966)
3.46, 1.53, and 0.84 mM at 4, 25, and 50 °C, respectively (Kresheck et al., 1965)

Vapor density:
2.0200 g/L at 0 °C, 1.8324 g/L at 25 °C (Windholz et al., 1983)
1.52 (air = 1)

Vapor pressure (mmHg):
7,904 at 27.6 °C (Francis and Robbins, 1933)
6,460 at 20 °C, 8,360 at 30 °C (Verschueren, 1983)

Environmental fate:
Biological. In the presence of methane, *Pseudomonas methanica* degraded propane to 1-propanol, propionic acid, and acetone (Leadbetter and Foster, 1959). The presence of carbon dioxide was required for "*Nocardia paraffinicum*" to degrade propane to propionic acid (MacMichael and Brown, 1987). Propane may biodegrade in two pathways. The first is the formation of propyl hydroperoxide, which decomposes to propanol followed by oxidation to propanoic acid. The other pathway involves dehydrogenation to 1-propene, which may react with water giving propanol (Dugan, 1972). Microorganisms can oxidize alkanes under aerobic conditions (Singer and Finnerty, 1984). The most common degradative pathway involves the oxidation of the terminal methyl group forming the corresponding alcohol (propyl alcohol). The alcohol may undergo a series of dehydrogenation steps forming an aldehyde (propionaldehyde), then a fatty acid (propionic acid). The fatty acid may then be metabolized by β-oxidation to form the mineralization products carbon dioxide and water (Singer and Finnerty, 1984).
Photolytic. When synthetic air containing propane and nitrous acid was exposed to artificial sunlight (λ = 300-450 nm), propane photooxidized to acetone with a yield of 56% (Cox et al., 1980). The rate constants for the reaction of propane and OH radicals in the atmosphere at 298 and 300 K were 1.11 x 10^{-12} cm^3/molecule·sec (DeMore and Bayes, 1999) and 1.3 x 10^{12} cm^3/mol·sec (Hendry and Kenley, 1979).
Chemical/Physical. Incomplete combustion of propane in the presence of excess hydrogen chloride resulted in a high number of different chlorinated compounds including, but not limited to alkanes, alkenes, monoaromatics, alicyclic hydrocarbons, and polynuclear aromatic hydrocarbons. Without hydrogen chloride, 13 non-chlorinated polynuclear aromatic hydrocarbons were formed (Eklund et al., 1987).

Exposure limits: NIOSH REL: TWA 1,000 ppm (1,800 mg/m^3), IDLH 2,100 ppm; OSHA PEL: TWA 1,000 ppm (1,800 mg/m^3); ACGIH TLV: TWA 2,500 ppm (adopted).

Symptoms of exposure: An asphyxiate. Narcotic at high concentrations (Patnaik, 1992).

Uses: Organic synthesis; refrigerant; fuel gas; manufacture of ethylene; solvent; extractant; aerosol propellant; mixture for bubble chambers.

1-PROPANOL

Synonyms: Ethyl carbinol; 1-Hydroxypropane; Optal; Osmosol extra; Propanol; 1-Propanol; *n*-Propanol; Propyl alcohol; 1-Propyl alcohol; *n*-Propyl alcohol; Propylic alcohol; UN 1274.

CAS Registry Number: 71-23-8
DOT: 1274
DOT label: Combustible liquid
Molecular formula: C_3H_8O
Formula weight: 60.10
RTECS: UH8225000
Merck reference: 10, 7742

Physical state, color, and odor:
Colorless liquid with a mild, alcohol-like odor. Odor threshold concentration is 2.6 ppm (Amoore and Hautala, 1983).

Melting point (°C):
-126.5 (Weast, 1986)

Boiling point (°C):
97.4 (Weast, 1986)

Density (g/cm³):
0.8035 at 20/4 °C (Weast, 1986)
0.8036 at 20.00/4 °C, 0.7995 at 25.00/4 °C (Shan and Asfour, 1999)

Diffusivity in water (x 10⁻⁵ cm²/sec at 25 °C):
1.12 (quoted, Hayduk and Laudie, 1974)
1.05 (Hao and Leaist, 1996)

Dissociation constant, pK_a:
>14 (Schwarzenbach et al., 1993)

Flash point (°C):
22 (NIOSH, 1997)
25 (open cup, Hawley, 1981)

Lower explosive limit (%):
2.2 (NIOSH, 1997)

Upper explosive limit (%):
13.7 (NIOSH, 1997)

Heat of fusion (kcal/mol):
1.294 (Riddick et al., 1986)

Henry's law constant (x 10⁻⁶ atm·m³/mol):
6.12 at 25 °C (Burnett, 1963)

6.85 at 25 °C (Butler et al., 1935)
7.41 at 25 °C (Snider and Dawson, 1985)
80.30 at 60 °C (Chai and Zhu, 1998)

Ionization potential (eV):
10.1 (Franklin et al., 1969)
10.15 (NIOSH, 1997)

Soil sorption coefficient, log K_{oc}:
0.48 (Gerstl and Helling, 1987)

Octanol/water partition coefficient, log K_{ow}:
0.34 (Hansch and Anderson, 1967)
0.25 (Hansch and Leo, 1979)

Solubility in organics:
Soluble in acetone and benzene (Weast, 1986). Miscible with alcohol and ether (Windholz et al., 1983).

Solubility in water:
Miscible (Palit, 1947). A saturated solution in equilibrium with its own vapor had a concentration of 250.4 g/L at 25 °C (Kamlet et al., 1987).

Vapor density:
2.46 g/L at 25 °C, 2.07 (air = 1)

Vapor pressure (mmHg):
14.5 at 20 °C, 20.8 at 25 °C, 27 at 30 °C (Verschueren, 1983)

Environmental fate:
Biological. In activated sludge inoculum, following a 20-d adaptation period, 98.8% COD removal was achieved. The average rate of biodegradation was 71.0 mg COD/g·h (Pitter, 1976).

Photolytic. Reported rate constants for the reaction of methanol and OH radicals in the atmosphere: 2.3×10^{12} cm^3/mol·sec at 300 K (Hendry and Kenley, 1979); 2.3×10^9 L/mol·sec (second-order) at 292 K (Campbell et al., 1976), 5.33×10^{-12} cm^3/molecule·sec at 296 K (Overend and Paraskevopoulos, 1978). Based on an atmospheric OH concentration of 1.0×10^6 molecule/cm^3, the reported half-life of 1-propanol is 1.5 d (Grosjean, 1997).

Chemical/Physical. At an influent concentration of 1,000 mg/L, treatment with granular activated carbon resulted in an effluent concentration of 811 mg/L. The adsorbability of the carbon used was 38 mg/g carbon (Guisti et al., 1974).

Exposure limits: NIOSH REL: TWA 200 ppm (500 mg/m^3), STEL 250 ppm (625 mg/m^3), IDLH 800 ppm; OSHA PEL: TWA 200 ppm; ACGIH TLV: TWA 200, STEL 250 ppm (adopted).

Symptoms of exposure: Ingestion causes headache, drowsiness, abdominal cramps, gastrointestinal pain, nausea, and diarrhea. May irritate eyes on contact (Patnaik, 1992).

Toxicity:
EC_{50} (48-h) and EC_{50} (24-h) values for *Spirostomum ambiguum* were 5.95 and 7.99 g/L,

respectively (Nałecz-Jawecki and Sawicki, 1999).

LC_{50} (48-h) and LC_{50} (24-h) values for *Spirostomum ambiguum* were 12.4 and 12.5 mg/L, respectively (Nałecz-Jawecki and Sawicki, 1999).

LC_{50} for red killifish 83.2 g/L (Yoshioka et al., 1980).

Acute oral LD_{50} for mice 6,800 mg/kg, rats 1,870 mg/kg (quoted, RTECS, 1985).

Uses: Solvent for cellulose esters and resins; in manufacturing of printing inks, nail polishes, polymerization and spinning of acrylonitrile, dyeing wool, polyvinyl chloride adhesives, esters, waxes, vegetable oils; brake fluids; solvent degreasing; antiseptic; organic synthesis.

β-PROPIOLACTONE

Synonyms: Betaprone; BPL; Hydracrylic acid β-lactone; 3-Hydroxypropionic acid lactone; β-Lactone; NSC 21626; **2-Oxetanone**; Propanolide; Propiolactone; 1,3-Propiolactone; 3-Propiolactone; β-Propionolactone; 3-Propionolactone; β-Proprolactone.

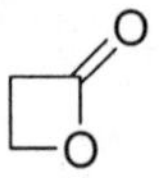

CAS Registry Number: 57-57-8
Molecular formula: $C_3H_4O_2$
Formula weight: 72.06
RTECS: RQ7350000
Merck reference: 10, 7721

Physical state, color, and odor:
Colorless liquid with a sweet but irritating odor

Melting point (°C):
-33.4 (Weast, 1986)

Boiling point (°C):
Decomposes at 162 (Weast, 1986)
155 (commercial grade, James and Wellington, 1969)

Density (g/cm³):
1.1460 at 20/5 °C (Weast, 1986)
1.1425 at 25/4 °C (Windholz et al., 1983)

Diffusivity in water (x 10⁻⁵ cm²/sec):
1.16 at 20 °C using method of Hayduk and Laudie (1974)

Flash point (°C):
75 (NIOSH, 1997)
70 (Dean, 1987)

Lower explosive limit (%):
2.9 (NIOSH, 1997)

Entropy of fusion (cal/mol·K):
9.376 (Lebedev and Yevstropov, 1983)

Heat of fusion (kcal/mol):
2.249 (Lebedev and Yevstropov, 1983)

Henry's law constant (x 10⁻⁷ atm·m³/mol):
7.6 at 25 °C (approximate - calculated from water solubility and vapor pressure)

Ionization potential (eV):
9.70 (Mallard and Linstrom, 1998)

Soil sorption coefficient, log K_{oc}:
Unavailable because experimental methods for estimation of this parameter for lactones are lacking in the documented literature

Octanol/water partition coefficient, log K_{ow}:
Unavailable because experimental methods for estimation of this parameter for lactones are lacking in the documented literature

Solubility in organics:
Miscible with acetone, alcohol, chloroform, and ether (Windholz et al., 1983)

Solubility in water:
37 wt % at 25 °C (NIOSH, 1997)

Vapor density:
2.95 g/L at 25 °C, 2.49 (air = 1)

Vapor pressure (mmHg):
3 at 25 °C (NIOSH, 1997)

Environmental fate:
   Chemical/Physical. Slowly hydrolyzes to hyracrylic acid (Windholz et al., 1983). In a reactor heated to 250 °C and a pressure of 12 mmHg, β-propiolactone decomposed to give equal amounts of ethylene and carbon dioxide (James and Wellington, 1969).

Exposure limits: Potential occupational carcinogen. ACGIH TLV: TWA 0.5 ppm.
   OSHA recommends that worker exposure to this chemical is to be controlled by use of engineering control, proper work practices, and proper selection of personal protective equipment. Specific details of these requirements can be found in CFR 1910.1003-1910.1016.
   ACGIH TLV: TWA 0.5 ppm (adopted).

Toxicity:
   LC_{50} (inhalation) for rats 25 ppm/6-h (quoted, RTECS, 1985).

Uses: Organic synthesis; disinfectant; vapor sterilant.

n-PROPYL ACETATE

Synonyms: Acetic acid propyl ester; Acetic acid *n*-propyl ester; 1-Acetoxypropane; Propyl acetate; 1-Propyl acetate; *n*-Propyl ethanoate; UN 1276.

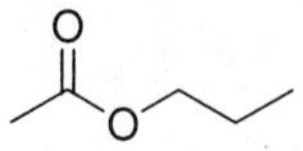

CAS Registry Number: 109-60-4
DOT: 1276
DOT label: Flammable liquid
Molecular formula: $C_5H_{10}O_2$
Formula weight: 102.12
RTECS: AJ3675000
Merck reference: 10, 7741

Physical state, color, and odor:
Clear, colorless liquid with a pleasant, pear-like odor. Odor threshold concentration is 670 ppb (Amoore and Hautala, 1983).

Melting point (°C):
-95 (Weast, 1986)
-92 (Verschueren, 1983)

Boiling point (°C):
101.6 (Weast, 1986)

Density (g/cm³ at 20/4 °C):
0.88770 (Lee and Tu, 1999)
0.836 (Windholz et al., 1983)

Diffusivity in water (x 10⁻⁵ cm²/sec):
0.81 at 20 °C using method of Hayduk and Laudie (1974)

Flash point (°C):
13 (NFPA, 1984)

Lower explosive limit (%):
1.7 at 38 °C (NFPA, 1984)

Upper explosive limit (%):
8 (NFPA, 1984)

Henry's law constant (x 10⁻⁴ atm·m³/mol):
2.17 at 25 °C (Kieckbusch and King, 1979)

Ionization potential (eV):
10.04 ± 0.03 (Franklin et al., 1969)

Soil sorption coefficient, log K_{oc}:
Unavailable because experimental methods for estimation of this parameter for aliphatic esters are lacking in the documented literature

Octanol/water partition coefficient, log K_{ow}:
1.24 (Tewari et al., 1982; Wasik et al., 1981)

Solubility in organics:
Miscible with alcohol, ether (Windholz et al., 1983), hydrocarbons, and ketones (Hawley, 1981)

Solubility in water:
18,900 mg/L at 20 °C, 26,000 mg/L at 20 °C (quoted, Verschueren, 1983)
16 g/L at 16 °C (quoted, Windholz et al., 1983)
200 mM at 25.0 °C (Tewari et al., 1982; Wasik et al., 1981)
10.3 g/L in artificial seawater at 25 °C (Price et al., 1974)
185 mM at 20 °C (Fühner, 1924)
In wt % (°C): 3.21 (0), 2.78 (9.5), 2.26 (20.0), 1.98 (30.0), 1.87 (40.0), 1.72 (50.0), 1.66 (80.0), 1.35 (90.2) (Stephenson and Stuart, 1986)
22.6 g/kg at 25 °C (Butler and Ramchandani, 1935)

Vapor density:
4.17 g/L at 25 °C, 3.53 (air = 1)

Vapor pressure (mmHg):
25 at 20 °C, 35 at 25 °C, 42 at 30 °C (Verschueren, 1983)
33 at 25 °C (Butler and Ramchandani, 1935)

Environmental fate:
Photolytic. Reported rate constants for the reaction of *n*-propyl acetate and OH radicals in the atmosphere and aqueous solution are 2.7 x 10^{12} cm^3/mol·sec (Hendry and Kenley, 1979) and 2.30 x 10^{-13} cm^3/molecule·sec (Wallington et al., 1988b).
Chemical/Physical. Slowly hydrolyzes in water forming acetic acid and propyl alcohol.
At an influent concentration of 1,000 mg/L, treatment with granular activated carbon resulted in an effluent concentration of 248 mg/L. The adsorbability of the carbon used was 149 mg/g carbon (Guisti et al., 1974).

Exposure limits: NIOSH REL: TWA 200 ppm (840 mg/m^3), STEL 250 ppm (1,050 mg/m^3), IDLH 1,700 ppm; OSHA PEL: TWA 200 ppm; ACGIH TLV: TWA 200 ppm, STEL 250 ppm (adopted).

Symptoms of exposure: Vapors may irritate eyes, nose, and throat and cause narcotic effects. Ingestion may cause narcotic action. At high concentrations death may occur (Patnaik, 1992).

Toxicity:
Acute oral LD_{50} in mice 8,300 mg/kg, rats 9,370, rabbits 6,640 mg/kg (Patnaik, 1992).

Uses: Manufacture of flavors and perfumes; solvent for plastics, cellulose products, and resins; lacquers, paints; natural and synthetic resins; lab reagent; organic synthesis.

n-PROPYLBENZENE

Synonyms: Isocumene; 1-Phenylpropane; **Propylbenzene**; UN 2364.

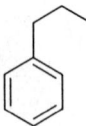

CAS Registry Number: 103-65-1
DOT: 2364
DOT label: Combustible liquid
Molecular formula: C_9H_{12}
Formula weight: 120.19
RTECS: DA8750000
Merck reference: 10, 7744

Physical state and color:
Colorless liquid

Melting point (°C):
-99.5 (Weast, 1986)
-101.75 (Mackay et al., 1982)

Boiling point (°C):
159.2 (Weast, 1986)

Density (g/cm³):
0.8620 at 20/4 °C (Weast, 1986)

Diffusivity in water (x 10^{-5} cm²/sec):
0.7368 at 20 °C using method of Hayduk and Laudie (1974)

Dissociation constant, pK_a:
>14 (Schwarzenbach et al., 1993)

Flash point (°C):
30 (Hawley, 1981)

Lower explosive limit (%):
0.8 (Sax and Lewis, 1987)

Upper explosive limit (%):
6 (Sax and Lewis, 1987)

Heat of fusion (kcal/mol):
2.03-2.215 (Dean, 1987)

Henry's law constant (x 10^{-3} atm·m³/mol):
5.68, 7.31, 8.81, 10.8, and 13.7 at 10, 15, 20, 25, and 30 °C, respectively (Ashworth et al., 1988)
4.35, 6.21, 8.37, 11.6, and 15.3 at 10, 15, 20, 25, and 30 °C, respectively (Perlinger et al., 1993)

886

Interfacial tension with water (dyn/cm at 20 °C):
38.5 (Demond and Lindner, 1993)

Ionization potential (eV):
9.14 (Yoshida et al., 1983a)
8.713 (Lias, 1998)

Soil sorption coefficient, log K$_{oc}$:
2.87 (estuarine sediment, Vowles and Mantoura, 1987)

Octanol/water partition coefficient, log K$_{ow}$:
3.57 (Leo et al., 1971)
3.68 (Hansch et al., 1968)
3.60 (quoted, Galassi et al., 1988)
3.69 (Schantz and Martire, 1987; Tewari et al., 1982; Wasik et al., 1981, 1983)
3.72 (DeVoe et al., 1981; Camilleri et al., 1988)
3.44 (Nahum and Horvath, 1980)
3.71, 3.72, 3.73 at 23 °C (Wasik et al., 1981)

Solubility in organics:
Miscible with alcohol and ether (Sax and Lewis, 1987)

Solubility in water:
360 μmol/L at 25 °C (Andrews and Keefer, 1950a)
60.24 mg/L at 25 °C (Chiou et al., 1982)
447, 435, 452, 443, and 437 μmol/L at 10.0, 15.0, 20.0, 25.0, and 30.0 °C, respectively (Owens et al., 1986)
434 μmol/L at 25.0 °C (Tewari et al., 1982; Wasik et al., 1981, 1983)
282 μmol/L in 0.5 M NaCl at 25 °C (Wasik et al., 1984)
388, 423, 455, and 528 μmol/L at 15.0, 25.0, 35.0, and 45.0 °C, respectively (Sanemasa et al., 1982)
426, 425, 427, 432, and 445 μmol/L at 15, 20, 23, 25, and 30 °C, respectively (Wasik et al., 1981)
70 mg/L at 25 °C (Krasnoshchekova and Gubergrits, 1975)
120 mg/L at 25 °C (Klevens, 1950)
60 mg/kg at 15 °C (Fühner, 1924)
432 μmol/L at 25 °C (DeVoe et al., 1981)
110 mg/kg at 25 °C (Stearns et al., 1947)
239 μmol/L at 25.00 °C (Sanemasa et al., 1985)
415 μmol/L at 25.0 °C (Sanemasa et al., 1987)
7.35 x 10^{-6} at 25 °C (mole fraction, Li et al., 1993)

Vapor density:
4.91 g/L at 25 °C, 4.15 (air = 1)

Vapor pressure (mmHg):
2.5 at 20 °C (Verschueren, 1983)
3.43 at 25 °C (Mackay et al., 1982)

Environmental fate:
   Biological. A *Nocardia* sp., growing on *n*-octadecane, biodegraded *n*-propylbenzene to

phenyl acetic acid (Davis and Raymond, 1981). *n*-Propylbenzene was cometabolized by a strain of *Micrococcus cerificans* to cinnamic acid (Pitter and Chudoba, 1990).

Estimated half-lives of *n*-propylbenzene (0.8 μg/L) from an experimental marine mesocosm during the spring (8-16 °C), summer (20-22 °C), and winter (3-7 °C) were 19, 1.3, and 11 d, respectively (Wakeham et al., 1983).

Photolytic. A rate constant of 3.7 x 10^9 L/mol·sec was reported for the reaction of *n*-propylbenzene with OH radicals in the gas phase (Darnall et al., 1976). Similarly, a room temperature rate constant of 5.7 x 10^{-12} cm^3/molecule·sec was reported for the vapor-phase reaction of *n*-propylbenzene with OH radicals (Atkinson, 1985). At 25 °C, a rate constant of 6.58 x 10^{-12} cm^3/molecule·sec was reported for the same reaction (Ohta and Ohyama, 1985).

Chemical/Physical. n-Propylbenzene will not hydrolyze because it does not contain a hydrolyzable functional group (Kollig, 1993).

Toxicity:

EC_{50} (72-h) for *Selenastrum capricornutum* 1.8 mg/L (Galassi et al., 1988).

LC50 (96-h) for *Salmo gairdneri* 1.55 mg/L (Galassi et al., 1988).

Acute oral LD_{50} for rats 6,040 mg/kg (quoted, RTECS, 1985).

Drinking water standard: No MCLGs or MCLs have been proposed (U.S. EPA, 1996).

Source: Thomas and Delfino (1991) equilibrated contaminant-free groundwater collected from Gainesville, FL with individual fractions of three individual petroleum products at 24-25 °C for 24 h. The aqueous phase was analyzed for organic compounds via U.S. EPA approved test method 602. Average *n*-propylbenzene concentrations reported in water-soluble fractions of unleaded gasoline, kerosene, and diesel fuel were 246, 82, and 23 μg/L, respectively. When the authors analyzed the aqueous-phase via U.S. EPA approved test method 610, average *n*-propyl-benzene concentrations in water-soluble fractions of unleaded gasoline, kerosene, and diesel fuel were generally lower, i.e., 210, 57, and 25 μg/L, respectively.

Schauer et al. (1999) reported *n*-propylbenzene in a diesel-powered medium-duty truck exhaust at an emission rate of 100 μg/km.

Uses: In textile dyeing and printing; solvent for cellulose acetate; manufacturing methylstyrene.

PROPYLCYCLOPENTANE

Synonym: 1-Cyclopentylpropane.

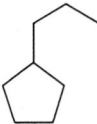

CAS Registry Number: 2040-96-2
Molecular formula: C_8H_{16}
Formula weight: 112.22
RTECS: GY4700000

Physical state, color, and odor:
Colorless liquid with an ether-like odor

Melting point (°C):
-117.3 (Weast, 1986)

Boiling point (°C):
131 (Weast, 1986)

Density (g/cm³):
0.77633 at 20/4 °C, 0.77229 at 25/4 °C (Dreisbach, 1959)

Diffusivity in water (x 10^{-5} cm²/sec):
0.71 at 20 °C using method of Hayduk and Laudie (1974)

Dissociation constant, pK_a:
>14 (Schwarzenbach et al., 1993)

Entropy of fusion (cal/mol·K):
15.40 (Messerly et al., 1965)

Heat of fusion (kcal/mol):
2.398 (Messerly et al., 1965)

Henry's law constant (atm·m³/mol):
1.1 at 25 °C (Mackay and Shiu, 1981)

Ionization potential (eV):
9.34 (Lias and Liebman, 1998)

Soil sorption coefficient, log K_{oc}:
Unavailable because experimental methods for estimation of this parameter for alicyclic hydrocarbons are lacking in the documented literature

Octanol/water partition coefficient, log K_{ow}:
3.63 using method of Hansch et al. (1968)

Solubility in organics:
Soluble in acetone, alcohol, benzene, and ether (Weast, 1986)

Solubility in water (at 25 °C):
2.04 mg/kg (Price, 1976)
1.77 mg/L (Kryzanowska and Szeliga, 1978)

Vapor density:
4.59 g/L at 25 °C, 3.87 (air = 1)

Vapor pressure (mmHg):
12.3 at 25 °C (Wilhoit and Zwolinski, 1971)

Environmental fate:
   Chemical/Physical. Complete combustion in air gives carbon dioxide and water vapor. Propylcyclopentane will not hydrolyze because it does not contain a hydrolyzable functional group (Kollig, 1993).

Uses: Organic synthesis; gasoline component.

PROPYLENE OXIDE

Synonyms: Epoxypropane; 1,2-Epoxypropane; Methyl ethylene oxide; **Methyloxirane**; NCI-C50099; Propene oxide; 1,2-Propylene oxide; UN 1280.

CAS Registry Number: 75-56-9
DOT: 1280
DOT label: Flammable liquid
Molecular formula: C_3H_6O
Formula weight: 58.08
RTECS: TZ2975000
Merck reference: 10, 7757

Physical state, color, and odor:
Clear, colorless liquid with an agreeable, ether-like odor. Odor threshold concentration is 44 ppm (Amoore and Hautala, 1983).

Melting point (°C):
-113 (NIOSH, 1997)
-104.4 (Verschueren, 1983)

Boiling point (°C):
34.3 (Weast, 1986)

Density (g/cm³):
0.859 at 0/4 °C (Weast, 1986)

Flash point (°C):
-37.5 (NIOSH, 1997)
-35 (Windholz et al., 1983)

Lower explosive limit (%):
2.8 (Sax and Lewis, 1987)
2.3 (NFPA, 1984)

Upper explosive limit (%):
36 (NFPA, 1984)

Entropy of fusion (cal/mol·K):
9.685 (Oetting, 1964)
9.737 (Beaumont et al., 1966)

Heat of fusion (kcal/mol):
1.561 (Oetting, 1964)
1.570 (Beaumont et al., 1966)

Henry's law constant (atm·m³/mol):
Not an environmentally important parameter because propylene oxide reacts rapidly with water

Ionization potential (eV):
10.22 (Lias, 1998)

Soil sorption coefficient, log K_{oc}:
Not applicable - reacts with water

Octanol/water partition coefficient, log K_{ow}:
0.08 (Deneer et al., 1988)

Solubility in organics:
Miscible with alcohol and ether (Weast, 1986)

Solubility in water:
405,000 mg/L at 20 °C, 650,000 mg/L at 30 °C (quoted, Verschueren, 1983)
40 wt % at 20 °C (Gunther et al., 1968)

Vapor density:
2.37 g/L at 25 °C, 2.01 (air = 1)

Vapor pressure (mmHg):
445 at 20 °C (NIOSH, 1987)

Environmental fate:
Biological. Bridié et al. (1979) reported BOD and COD values 0.17 and 1.77 g/g using filtered effluent from a biological sanitary waste treatment plant. These values were determined using a standard dilution method at 20 °C for a period of 5 d. When a sewage seed was used in a separate screening test, a BOD value of 0.20 g/g was obtained. The ThOD for propylene oxide is 2.21 g/g.

Photolytic. Anticipated products from the reaction of propylene oxide with ozone or OH radicals in the atmosphere are formaldehyde, pyruvic acid, $CH_3C(O)OCHO$, and $HC(O)OCHO$ (Cupitt, 1980). An experimentally determined reaction rate constant of 5.2 x 10^{-13} cm^3/molecule·sec was reported for the gas phase reaction of propylene oxide with OH radicals (Güsten et al., 1981).

Chemical/Physical. The reported hydrolysis half-life for the conversion of propylene oxide to 1,2-propanediol in water at 25 °C and pH 7 is 14.6 d (Mabey and Mill, 1978).

May polymerize at high temperatures or on contact with alkalies, aqueous acids, amines, and acid alcohols (NIOSH, 1997).

At an influent concentration of 1,000 mg/L, treatment with granular activated carbon resulted in an effluent concentration of 739 mg/L. The adsorbability of the carbon used was 52 mg/g carbon (Guisti et al., 1974).

Exposure limits: Potential occupational carcinogen. NIOSH REL: IDLH 400 ppm; OSHA PEL: TWA 100 ppm (240 mg/m³); ACGIH TLV: TWA 20 ppm (adopted) with an intended TWA value 5 ppm.

Symptoms of exposure: Vapors may irritate eyes, mucous membranes, and skin. Inhalation may cause weakness and drowsiness (Patnaik, 1992). May cause blisters or burns (NIOSH, 1997). An irritation concentration of 1,125.00 mg/m³ in air was reported by Ruth (1986).

Toxicity:
LC_{50} (48-h) for *Poecilia reticulata* 32 mg/L (Deneer et al., 1988).

LC_{50} (inhalation) for mice 1,740 ppm/4-h (quoted, RTECS, 1985).

Acute oral LD_{50} for guinea pigs 660 mg/kg, mice 630 mg/kg, rats 520 mg/kg (quoted, RTECS, 1985).

LD_{50} (skin) for rabbits 1,245 mg/kg (quoted, RTECS, 1985).

Uses: Preparation of propylene and dipropylene glycols, poly(propylene oxide), lubricants, oil demulsifiers, surfactants, isopropanol amines, polyols for urethane foams; solvent; soil sterilant; fumigant.

n-PROPYL NITRATE

Synonyms: Nitric acid propyl ester; Propyl nitrate; UN 1865.

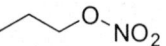

CAS Registry Number: 627-13-4
DOT: 1865
Molecular formula: $C_3H_7NO_3$
Formula weight: 105.09
RTECS: UK0350000
Merck reference: 10, 7765

Physical state, color, and odor:
Colorless to light yellow liquid with an ether-like odor. Odor threshold concentration is 50 ppm (Amoore and Hautala, 1983).

Melting point (°C):
-101 (NIOSH, 1997)

Boiling point (°C):
110 at 762 mmHg (Weast, 1986)

Density (g/cm³ at 20/4 °C):
1.0538 (Weast, 1986)
1.07 (Hawley, 1981)

Diffusivity in water (x 10^{-5} cm²/sec):
0.88 at 20 °C using method of Hayduk and Laudie (1974)

Flash point (°C):
20 (NIOSH, 1997)

Lower explosive limit (%):
2 (NIOSH, 1997)

Upper explosive limit (%):
100 (NIOSH, 1997)

Henry's law constant (x 10^{-4} atm·m³/mol):
9.09 at 25 °C (Kames and Schurath, 1992)

Ionization potential (eV):
11.07 (Rosenstock et al., 1998)

Soil sorption coefficient, log K_{oc}:
Unavailable because experimental methods for estimation of this parameter for aliphatic nitrates are lacking in the documented literature

Octanol/water partition coefficient, log K$_{ow}$:
Unavailable because experimental methods for estimation of this parameter for aliphatic nitrates are lacking in the documented literature

Solubility in organics:
Soluble in alcohol and ether (Weast, 1986)

Vapor density:
4.30 g/L at 25 °C, 3.63 (air = 1)

Vapor pressure (mmHg):
18 at 20 °C (NIOSH, 1997)

Exposure limits: NIOSH REL: TWA 25 ppm (109 mg/m^3), STEL 40 ppm (170 mg/m^3), IDLH 500 ppm; OSHA PEL: TWA 25 ppm; ACGIH TLV: TWA 25 ppm, STEL 40 ppm (adopted).

Uses: Rocket fuel formulations; organic synthesis.

PROPYNE

Synonyms: Allylene; Methyl acetylene; Propine; **1-Propyne**.

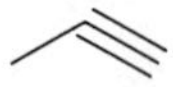

CAS Registry Number: 74-99-7
Molecular formula: C_3H_4
Formula weight: 40.06
RTECS: UK4250000

Physical state, color, and odor:
Colorless gas with a sweet odor

Melting point (°C):
-101.5 (Weast, 1986)

Boiling point (°C):
-23.2 (Weast, 1986)
-27.5 (Verschueren, 1983)

Density (g/cm³):
0.7062 at -50/4 °C (Weast, 1986)
0.678 at -27/4 °C (Verschueren, 1983)

Lower explosive limit (%):
1.7 (NIOSH, 1997)

Henry's law constant (atm·m³/mol):
0.11 at 25 °C (Hine and Mookerjee, 1975)

Ionization potential (eV):
10.36 ± 0.01 (Franklin et al., 1969)

Soil sorption coefficient, log K_{oc}:
Unavailable because experimental methods for estimation of this parameter for aliphatic hydrocarbons are lacking in the documented literature

Octanol/water partition coefficient, log K_{ow}:
1.61 using method of Hansch et al. (1968)

Solubility in organics:
Soluble in alcohol, benzene, and chloroform (Weast, 1986)

Solubility in water:
3,640 mg/L at 20 °C (quoted, Verschueren, 1983)

Vapor density:
1.64 g/L at 25 °C, 1.38 (air = 1)

Vapor pressure (mmHg):
3,952 at 20 °C, 5,244 at 30 °C (Verschueren, 1983)
4,310 at 25 °C (Wilhoit and Zwolinski, 1971)

Environmental fate:
   Chemical/Physical. When passed through a cold solution containing hydrobromite ions, 1-bromo-1-propyne was formed (Hatch and Kidwell, 1954).

Exposure limits: NIOSH REL: TWA 1,000 ppm (1,650 mg/m³), IDLH 1,700 ppm; OSHA PEL: TWA 1,000 ppm; ACGIH TLV: TWA 1,000 ppm (adopted).

Symptoms of exposure: An asphyxiant. Toxic at high concentrations (Patnaik, 1992).

Uses: Chemical intermediate; specialty fuel.

PYRENE

Synonyms: Benzo[*def*]phenanthrene; β-Pyrene; β-Pyrine.

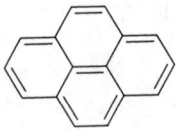

CAS Registry Number: 129-00-0
Molecular formula: $C_{16}H_{10}$
Formula weight: 202.26
RTECS: UR2450000
Merck reference: 10, 7864

Physical state and color:
Colorless solid (tetracene impurities impart a yellow color) or monoclinic prisms crystallized from alcohol. Solutions have a slight blue fluorescence.

Melting point (°C):
156 (Weast, 1986)
151 (Aldrich, 1988)
149 (Sims et al., 1988)

Boiling point (°C):
393 (Weast, 1986)
388 (Pearlman et al., 1984)
404 (Sax, 1984)

Density (g/cm³):
1.271 at 23/4 °C (Weast, 1986)

Diffusivity in water (x 10^{-5} cm²/sec):
0.56 at 20 °C using method of Hayduk and Laudie (1974)

Dissociation constant, pK_a:
>15 (Christensen et al., 1975)

Entropy of fusion (cal/mol·K):
8.6 (Wauchope and Getzen, 1972)

Heat of fusion (kcal/mol):
3.66 (Wauchope and Getzen, 1972)

Henry's law constant (x 10^{-5} atm·m³/mol):
1.09 (Mackay and Shiu, 1981)
1.97 at 25 °C (de Maagd et al., 1998)
1.87 (Southworth, 1979)
1.19 at 25 °C (Shiu and Mackay, 1997)
0.42, 0.68, 1.09, 1.69, and 2.42 at 4.1, 11.0, 18.0, 25.0, and 31.0 °C, respectively (Bamford et al., 1998)

Ionization potential (eV):
7.55 (Yoshida et al., 1983a)
7.50 (Cavalieri and Rogan, 1985)
7.43 (Lias, 1998)

Bioconcentration factor, log BCF:
3.43 (*Daphnia pulex*, Southworth et al., 1978)
3.43 (*Daphnia magna*, Newsted and Giesy, 1987)
2.66 (goldfish, Ogata et al., 1984)
0.72 (*Polychaete* sp.), 1.12 (*Capitella capitata*) (Bayona et al., 1991)
4.56 alga, *Selenastrum capricornutum* (Casserly et al., 1983)
Apparent values of 4.2 (wet wt) and 6.0 (lipid wt) for freshwater isopods including *Asellus aquaticus* (L.) (van Hattum et al., 1998)

Soil sorption coefficient, log K_{oc}:
4.66 (aquifer material, Abdul et al., 1987)
4.92 (Schwarzenbach and Westall, 1981)
4.88 (Chin et al., 1988)
4.81 (Means et al., 1979)
4.74 (Webster soil, Woodburn et al., 1989)
4.80 (Hassett et al., 1980)
4.79 (Flint aquifer sample, Abdul and Gibson, 1986)
5.13 (Vowles and Mantoura, 1987)
4.67 (Socha and Carpenter, 1987)
5.23 (Gauthier et al., 1987)
5.18 (Karickhoff et al., 1979)
6.51 (average, Kayal and Connell, 1990)
4.88 (fine sand, Enfield et al., 1989)
4.77, 4.82 (RP-HPLC immobilized humic acids, Szabo et al., 1990)
4.94 (Mississippi sediment), 4.98 (Ohio River sediment) (Karickhoff and Morris, 1985)
4.83 (Karickhoff, 1981a)
4.64, 4.68, 4.77, 4.78, 4.83 (Illinois sediment), 4.70, 4.71 (North Dakota sediment), 4.76 (West Virginia soil), 4.92 (Indiana sediment), 4.80 (Georgia sediment), 4.81, 4.83 (Iowa loess), 4.88 (Kentucky sediment), 4.93 (South Dakota sediment) (Means et al., 1980)
4.97, 5.11, 6.02, 6.11, 6.53 (San Francisco, CA mudflat sediments, Maruya et al., 1996)
6.6 (average value using 8 river bed sediments from the Netherlands, van Hattum et al., 1998)
Average K_d values for sorption of pyrene to corundum (α-Al$_2$O$_3$) and hematite (α-Fe$_2$O$_3$) were 0.231 and 0.983 mL/g, respectively (Mader et al., 1997)

Octanol/water partition coefficient, log K_{ow}:
4.77 at 25 °C (Andersson and Schräder, 1999)
4.88 (Chou and Jurs, 1979)
5.09 (Means et al., 1979)
5.52 (Burkhard et al., 1985a)
5.22 (Bruggeman et al., 1982)

Solubility in organics:
Soluble in most solvents (U.S. EPA, 1985)

Solubility in water:
160 μg/L at 26 °C, 32 μg/L at 24 °C (practical grade, Verschueren, 1983)

13 mg/kg at 25 °C. In seawater (salinity = 35 g/kg): 56, 71, and 89 μg/kg at 15, 20, and 25 °C, respectively (Rossi and Thomas, 1981)

13 μg/L at 25 °C (Miller et al., 1985)

135 μg/L at 25 °C (Mackay and Shiu, 1977)

135 μg/L at 24 °C (Means et al., 1979)

132 μg/kg at 25 °C, 162 μg/kg at 29 °C (May et al., 1978a)

171 μg/L at 25 °C (Schwarz and Wasik, 1976)

In mg/kg: 0.124, 0.128, 0.129 at 22.2 °C, 0.228, 0.235 at 34.5 °C, 0.395, 0.397, 0.405 at 44.7 °C, 0.556, 0.558, 0.576 at 50.1 °C, 0.75, 0.75, 0.77 at 55.6 °C, 0.74 at 56.0 °C, 0.90, 0.95, 0.96 at 60.7 °C, 1.27, 1.29 at 65.2 °C, 1.83, 1.86, 1.89 at 71.9 °C, 2.21 at 74.7 °C (Wauchope and Getzen, 1972)

160, 165 μg/L at 27 °C (Davis et al., 1942)

In nmol/L: 270 at 12.2 °C, 339 at 15.5 °C, 391 at 17.4 °C, 457 at 20.3 °C, 578 at 23.0 °C, 582 at 23.3 °C, 640 at 25.0 °C, 713 at 26.2 °C, 7.18 at 26.7 °C, 809 at 28.5 °C, 930 at 31.3 °C (Schwarz, 1977)

175 μg/L at 25 °C (Klevens, 1950)

133 μg/L at 25 °C (Walters and Luthy, 1984)

150 μg/L in Lake Michigan water at $\approx$ 25 °C (Eadie et al., 1990; Fatiadi, 1967)

1 μmol/L at 25 °C (Edwards et al., 1991)

At 20 °C: 470, 322, 318, and 311 nmol/L in doubly distilled water, Pacific seawater, artificial seawater, and 35% NaCl, respectively (Hashimoto et al., 1984)

107 μg/L at 25 °C (Vadas et al., 1991)

118 μg/L at 25 °C (Billington et al., 1988)

In mole fraction (x 10^{-8}): 4.832 at 4.70 °C, 5.211 at 9.50 °C, 6.413 at 14.30 °C, 8.310 at 18.70 °C, 9.709 at 21.20 °C, 12.11 at 25.50 °C, 15.14 at 29.90 °C (May et al., 1983)

Vapor pressure (x 10^{-7} mmHg):

6.85 at 25 °C (Radding et al., 1976)

25 at 25 °C (Mabey et al., 1982)

379 at 68.90 °C, 1,080 at 71.75 °C, 1,410 at 74.15 °C, 1,670 at 75.85 °C, 2,060 at 78.20 °C, 2,160 at 78.90 °C, 2,920 at 81.70 °C, 3,070 at 82.65 °C, 3,040 at 82.70 °C, 3,790 at 85.00 °C, 3,800 at 85.25 °C (Bradley and Cleasby, 1953)

45 at 25 °C (Sonnefeld et al., 1983; Wasik et al., 1983)

367 at 25 °C (Bidleman, 1984)

564, 848 at 25 °C (Hinckley et al., 1990)

647, 1,823, and 1,568 at 46.95, 56.99, and 57.98 °C, respectively (Oja and Suuberg, 1998)

180 at 25 °C (extrapolated from vapor pressures determined at higher temperatures, Tesconi and Yalkowsky, 1998)

Environmental fate:

Biological. When pyrene was statically incubated in the dark at 25 °C with yeast extract and settled domestic wastewater inoculum, complete degradation was demonstrated at the 5 mg/L substrate concentration after 2 wk. At a substrate concentration of 10 mg/L, however, only 11 and 2% losses were observed after 7 and 14 d, respectively (Tabak et al., 1981).

Contaminated soil from a manufactured coal gas plant that had been exposed to crude oil was spiked with pyrene (400 mg/kg soil) to which Fenton's reagent (5 mL 2.8 M hydrogen peroxide; 5 mL 0.1 M ferrous sulfate) was added. The treated and nontreated soil samples were incubated at 20 °C for 56 d. Fenton's reagent greatly enhanced the mineralization of pyrene by indigenous microorganisms. The amounts of pyrene recovered as carbon dioxide after treatment with and without Fenton's reagent were 16 and 5%, respectively. Pretreatment of the soil with a surfactant (10 mM sodium dodecylsulfate) before addition of Fenton's reagent, increased the mineralization rate 55% as compared to nontreated soil (Martens and Frankenberger, 1995).

Soil. The reported half-lives for pyrene in a Kidman sandy loam and McLaurin sandy loam are 260 and 199 d, respectively (Park et al., 1990).

Plant. Hückelhoven et al. (1997) studied the metabolism of pyrene by suspended plant cell cultures of soybean, wheat, jimsonweed, and purple foxglove. Soluble metabolites were only detected in foxglove and wheat. Approximately 90% of pyrene was transformed in wheat. In foxglove, 1-hydroxypyrene methyl ether was identified as the main metabolite but in wheat, the metabolites were identified as conjugates of 1-hydroxypyrene.

Photolytic. Adsorption onto garden soil for 10 d at 32 °C and irradiated with UV light produced 1,1'-bipyrene, 1,6-pyrenedione, 1,8-pyrenedione, and three unidentified compounds (Fatiadi, 1967). Microbial degradation by *Mycobacterium* sp. yielded the following ring-fission products: 4-phenanthroic acid, 4-hydroxyperinaphthenone, cinnamic acid, and phthalic acids. The compounds pyrenol and the *cis-* and *trans*-4,5-dihydrodiols of pyrene were identified as ring-oxidation products (Heitkamp et al., 1988).

Silica gel coated with pyrene and suspended in an aqueous solution containing nitrite ion and subjected to UV radiation yielded the tentatively identified product 1-nitropyrene (Suzuki et al., 1987). 1-Nitropyrene coated on glass surfaces and exposed to natural sunlight resulted in the formation of hydroxypyrene, possibly pyrene dione and dihydroxy pyrene and other unidentified compounds (Benson et al., 1985).

1-Nitropyrene also formed when pyrene deposited on glass filter paper containing sodium nitrite was irradiated with UV light at room temperature (Ohe, 1984). This compound was reported to have formed from the reaction of pyrene with NO_x in urban air from St. Louis, MO (Randahl et al., 1982). Behymer and Hites (1985) determined the effect of different substrates on the rate of photooxidation of pyrene using a rotary photoreactor. The photolytic half-lives of pyrene using silica gel, alumina, and fly ash were 21, 31, and 46 h, respectively.

In a 5-m deep surface water body, the calculated half-lives for direct photochemical transformation at 40 °N latitude, in the midsummer during midday were 5.9 and 4.2 d with and without sediment-water partitioning, respectively (Zepp and Schlotzhauer, 1979).

Chemical/Physical. At room temperature, concentrated sulfuric acid will react with pyrene to form a mixture of disulfonic acids. In addition, an atmosphere containing 10% sulfur dioxide transformed pyrene into many sulfur compounds, including pyrene-1-sulfonic acid and pyrenedisulfonic acid (Nielsen et al., 1983).

2-Nitropyrene was the sole product formed from the gas-phase reaction of pyrene with OH radicals in a NO_x atmosphere (Arey et al., 1986). Pyrene adsorbed on glass fiber filters reacted rapidly with N_2O_5 to form 1-nitropyrene with a 60-70% yield (Pitts et al., 1985).

When pyrene, adsorbed from the vapor phase onto coal fly ash, silica, and alumina, was exposed to nitrogen dioxide, no reaction occurred. However, in the presence of nitric acid, nitrated compounds were produced (Yokley et al., 1985). Ozonation of water containing pyrene (10-200 μg/L) yielded short-chain aliphatic compounds as the major products (Corless et al., 1990). A monochlorinated pyrene was the major product formed during the chlorination of pyrene in aqueous solutions. At pH 4, the reported half-lives at chlorine concentrations of 0.6 and 10 mg/L were 8.8 and <0.2 h, respectively (Mori et al., 1991).

A volatilization rate constant of 1.1 x 10^{-4}/sec was determined when pyrene on a glass surface was subjected to an air flow rate of 3 L/min at 24 °C (Cope and Kalkwarf, 1987).

Pyrene will not hydrolyze because it does not contain a hydrolyzable functional group (Kollig, 1993).

Exposure limits: Potential occupational carcinogen. No individual standards have been set; however, as a constituent in coal tar pitch volatiles, the following exposure limits have been established (mg/m³): NIOSH REL: TWA 0.1 (cyclohexane-extractable fraction), IDLH 80; OSHA PEL: TWA 0.2 (benzene-soluble fraction); ACGIH TLV: TWA 0.2 (benzene solubles).

Toxicity:

LC_{50} (10-d) for *Rhepoxynius abronius* 2.81 mg/g organic carbon (Swartz et al., 1997).

LC_{50} (inhalation) for rats 170 mg/m³ (quoted, RTECS, 1985).

LD_{50} (4-d intraperitoneal) and LD_{50} (7-d intraperitoneal) values for mice were 514 and 678 mg/kg, respectively (Salamone, 1981).

Acute oral LD_{50} for mice 800 mg/kg, rats 2,700 mg/kg (quoted, RTECS, 1985).

Drinking water standard: No MCLGs or MCLs have been proposed (U.S. EPA, 1996).

Source: Detected in groundwater beneath a former coal gasification plant in Seattle, WA at a concentration of 180 μg/L (ASTR, 1995). Detected in 8 diesel fuels at concentrations ranging from 0.16 to 24 mg/L with a mean value of 5.54 mg/L (Westerholm and Li, 1994). Identified in Kuwait and South Louisiana crude oils at concentrations of 4.5 and 3.5 ppm, respectively (Pancirov and Brown, 1975).

Schauer et al. (1999) reported pyrene in diesel fuel at a concentration of 64 μg/g and in a diesel-powered medium-duty truck exhaust at an emission rate of 71.9 μg/km.

Based on analysis of 7 coal tar samples, pyrene concentrations ranged from 900 to 18,000 ppm (EPRI, 1990). Detected in 1-yr aged coal tar film and bulk coal tar at concentrations of 2,700 and 2,900 mg/kg, respectively (Nelson et al., 1996).

Use: Research chemical. Derived from industrial and experimental coal gasification operations where the maximum concentrations detected in gas and coal tar streams were 9.2 and 24 mg/m³, respectively (Cleland, 1981).

PYRIDINE

Synonyms: Azabenzene; Azine; NCI-C55301; RCRA waste number U196; UN 1282.

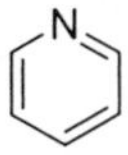

CAS Registry Number: 110-86-1
DOT: 1282
DOT label: Flammable liquid
Molecular formula: C_5H_5N
Formula weight: 79.10
RTECS: UR8400000
Merck reference: 10, 7869

Physical state, color, and odor:
Colorless liquid with a sharp, penetrating, fish-like odor. Odor threshold concentration is 21 ppb (Keith and Walters, 1992).

Melting point (°C):
-42 (Weast, 1986)

Boiling point (°C):
115.5 (Weast, 1986)

Density (g/cm³):
0.9819 at 20/4 °C (Weast, 1986)
0.9780 at 25/4 °C (Windholz et al., 1983)

Diffusivity in water (x 10^{-5} cm²/sec):
1.00 at 20 °C using method of Hayduk and Laudie (1974)

Dissociation constant, pK_a:
5.19 (Windholz et al., 1983)
5.21 at 25 °C (Gordon and Ford, 1972)

Flash point (°C):
20 (NIOSH, 1997)

Lower explosive limit (%):
1.8 (NIOSH, 1997)

Upper explosive limit (%):
12.4 (NIOSH, 1997)

Entropy of fusion (cal/mol·K):
8.554 (Parks et al., 1936)

Heat of fusion (kcal/mol):
1.977 (Parks et al., 1936)

Henry's law constant (x 10^{-6} atm·m³/mol):
12 at 25 °C (Hawthorne et al., 1985)
8.83 at 25 °C (Andon et al., 1954)
14.0 at 25 °C (Amoore and Buttery, 1978)
18.13 at 25 °C (Chaintreau et al., 1995)
361, 413, 349, 263, and 210 at 2.0, 6.0, 10.0, 18.0, and 25.0 °C, respectively (Dewulf et al., 1999)

Ionization potential (eV):
9.32 (Franklin et al., 1969)
9.26 (Lias, 1998)

Soil sorption coefficient, log K_{oc}:
Unavailable because experimental methods for estimation of this parameter for pyridines are lacking in the documented literature. However, its miscibility in water and low K_{ow} suggest its adsorption to soil will be nominal (Lyman et al., 1982).

Octanol/water partition coefficient, log K_{ow}:
0.65 (Leo et al., 1971; Campbell and Luthy, 1985)
0.70 (Berthod et al., 1988)
1.28 (Eadsforth, 1986)
0.63, 0.68, 0.72 (Garst and Wilson, 1984)
0.66 (Mirrlees et al., 1976)

Solubility in organics:
Soluble in acetone, alcohol, benzene, and ether (Weast, 1986). Miscible with alcohol, ether, petroleum ether, oils, and many other organic liquids (Windholz et al., 1983).

Solubility in water:
Miscible (Fischer and Ehrenberg, 1948; Amoore and Buttery, 1978). A saturated solution in equilibrium with its own vapor had a concentration of 233.4 g/L at 25 °C (Kamlet et al., 1987).

Vapor density:
3.23 g/L at 25 °C, 2.73 (air = 1)

Vapor pressure (mmHg):
18 at 20 °C (NIOSH, 1987)
14 at 20 °C, 20 at 25 °C, 26 at 30 °C (Verschueren, 1983)

Environmental fate:
   Biological. Heukelekian and Rand (1955) reported a 5-d BOD value of 1.31 g/g which is 58.7% of the ThOD value of 2.23 g/g.
   Photolytic. Irradiation of an aqueous solution at 50 °C for 24 h resulted in a 23.06% yield of carbon dioxide (Knoevenagel and Himmelreich, 1976). When an aqueous pyridine solution in the presence of titanium dioxide was illuminated by UV light, ammonium and nitrate ions formed as the major products (Low et al., 1991).
   A rate constant of 4.9 x 10^{-13} cm³/molecule·sec was reported for the reaction of pyridine and OH radicals in air at room temperature (Atkinson et al., 1987).
   Chemical/Physical. The gas-phase reaction of ozone with pyridine in synthetic air at 23 °C yielded a nitrated salt having the formula: $[C_6H_5NH]^+NO_3^-$ (Atkinson et al., 1987).
   Ozonation of pyridine in aqueous solutions at 25 °C were studied with and without the

addition of *t*-butyl alcohol (20 mM) as a radical scavenger. With *t*-butyl alcohol, ozonation of pyridine yielded mainly pyridine *N*-oxide (80% yield), which was very stable towards ozone. Without *t*-butyl alcohol, the heterocyclic ring is rapidly cleaved forming ammonia, nitrate, and the amidic compound *N*-formyl oxamic acid (Andreozzi et al., 1991).

Forms water-soluble salts with strong acids.

Pyridine will not hydrolyze because it does not contain a hydrolyzable functional group (Kollig, 1993).

At an influent concentration of 1,000 mg/L, treatment with granular activated carbon resulted in an effluent concentration of 527 mg/L. The adsorbability of the carbon used was 95 mg/g carbon (Guisti et al., 1974).

Exposure limits: NIOSH REL: TWA 5 ppm (15 mg/m^3), IDLH 1,000 ppm; OSHA PEL: TWA 5 ppm; ACGIH TLV: TWA 5 ppm (adopted).

Symptoms of exposure: Headache, dizziness, nervousness, nausea, insomnia, frequent urination, and abdominal pain (Patnaik, 1992). An irritation concentration of 90.00 mg/m^3 in air was reported by Ruth (1986).

Toxicity:

EC_{50} (48-h) and EC_{50} (24-h) values for *Spirostomum ambiguum* were 989 and 1,226 mg/L, respectively (Nałecz-Jawecki and Sawicki, 1999).

LC_{50} (48-h) and LC_{50} (24-h) values for *Spirostomum ambiguum* were 2,547 and 3,117 mg/L, respectively (Nałecz-Jawecki and Sawicki, 1999).

LC_{50} for red killifish 9,330 mg/L (Yoshioka et al., 1986).

Acute oral LD_{50} for mice 1,500 mg/kg, rats 891 mg/kg (quoted, RTECS, 1985).

Source: Pyridine occurs naturally in potatoes, anabasis, henbane leaves, peppermint (0-1 ppb), tea leaves, and tobacco leaves (Duke, 1992).

Uses: Organic synthesis (vitamins and drugs); analytical chemistry (cyanide analysis); solvent for anhydrous mineral salts; denaturant for alcohol; antifreeze mixtures; textile dyeing; waterproofing; fungicides; rubber chemicals.

p-QUINONE

Synonyms: 1,4-Benzoquine; Benzoquinone; 1,4-Benzoquinone; *p*-Benzoquinone; Chinone; Cyclohexadienedione; 1,4-Cyclohexadienedione; **2,5-Cyclohexadiene-1,4-dione**; 1,4-Cyclohexadiene dioxide; 1,4-Dioxybenzene; NCI-C55845; Quinone; 4-Quinone; RCRA waste number U197; UN 2587; USAF P-220.

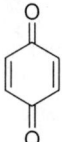

CAS Registry Number: 106-51-4
DOT: 2587
DOT label: Poison
Molecular formula: $C_6H_4O_2$
Formula weight: 108.10
RTECS: DK2625000
Merck reference: 10, 7998

Physical state, color, and odor:
Light yellow crystals with an acrid odor resembling chlorine. Odor threshold concentration is 84 ppb (Amoore and Hautala, 1983).

Melting point (°C):
115-117 (Weast, 1986)

Boiling point (°C):
Sublimes (Weast, 1986)

Density (g/cm³):
1.318 at 20/4 °C (Weast, 1986)

Diffusivity in water (x 10^{-5} cm²/sec):
0.87 at 20 °C using method of Hayduk and Laudie (1974)

Flash point (°C):
38-94 (NIOSH, 1997)

Entropy of fusion (cal/mol·K):
11.4 (Andrews et al., 1926)

Heat of fusion (kcal/mol):
4.4096 (Andrews et al., 1926)

Ionization potential (eV):
9.67 ± 0.02 (Franklin et al., 1969)
10.0 (Lias, 1998)

Soil sorption coefficient, log K_{oc}:
Unavailable because experimental methods for estimation of this parameter for quinones are lacking in the documented literature

Octanol/water partition coefficient, log K_{ow}:
0.20 (Leo et al., 1971)

Solubility in organics:
Soluble in alcohol, ether (Weast, 1986), and hot petroleum ether (Windholz et al., 1983)

Vapor pressure (mmHg):
0.1 at 20 °C (NIOSH, 1997)

Exposure limits: NIOSH REL: TWA 0.4 mg/m^3 (0.1 ppm), IDLH 100 mg/m^3; OSHA PEL: TWA 0.4 mg/m^3; ACGIH TLV: TWA 0.1 ppm (adopted).

Symptoms of exposure: Prolonged exposure may cause eye irritation. May damage cornea on contact. Contact with skin may cause irritation, ulceration, and necrosis (Patnaik, 1992). An irritation concentration of 2.00 mg/m^3 in air was reported by Ruth (1986).

Toxicity:
Acute oral LD$_{50}$ for rats 130 mg/kg (quoted, RTECS, 1985).

Uses: Oxidizing agent; manufacturing hydroquinone and dyes; in photography; strengthening animal fibers; tanning hides; analytical reagent; making gelatin insoluble; fungicides.

RONNEL

Synonyms: Dermaphos; Dimethyl trichlorophenyl thiophosphate; *O,O*-Dimethyl-*O*-2,4,5-trichlorophenyl phosphorothioate; *O,O*-Dimethyl *O*-(2,4,5-trichlorophenyl)thiophosphate; Dow ET 14; Dow ET 57; Ectoral; ENT 23284; ET 14; ET 57; Etrolene; Fenchlorfos; Fenchlorophos; Fenchchlorphos; Karlan; Korlan; Korlane; Nanchor; Nanker; Nankor; OMS 123; **Phosphorothioic acid *O,O*-dimethyl *O*-(2,4,5-trichlorophenyl) ester**; Trichlorometafos; Trolen; Trolene; Viozene.

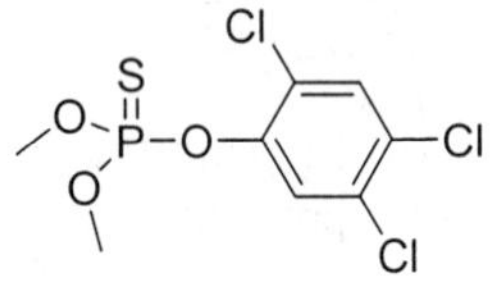

CAS Registry Number: 299-84-3
DOT: 2922
DOT label: Corrosive, poisonous
Molecular formula: $C_8H_8Cl_3O_3PS$
Formula weight: 321.57
RTECS: TG0525000
Merck reference: 10, 8129

Physical state and color:
White to light brown, waxy solid

Melting point (°C):
41 (Windholz et al., 1983)
35-37 (Freed et al., 1977)

Boiling point (°C):
Decomposes (NIOSH, 1997)
97 at 0.01 mmHg (Freed et al., 1977)

Density (g/cm³):
1.48 at 25/4 °C (Verschueren, 1983)

Diffusivity in water (x 10⁻⁵ cm²/sec):
0.51 at 20 °C using method of Hayduk and Laudie (1974)

Flash point (°C):
Noncombustible solid (NIOSH, 1997)

Henry's law constant (x 10⁻⁶ atm·m³/mol):
8.46 at 25 °C (approximate - calculated from water solubility and vapor pressure)

Bioconcentration factor, log BCF:
4.64 (*Poecilia reticulata* - express on a lipid content basis, Devillers et al., 1996)

Soil sorption coefficient, log K_{oc}:
2.76 using method of Kenaga and Goring (1980)

Octanol/water partition coefficient, log K_{ow}:
4.88 (Chiou et al., 1977; Freed et al., 1979a)
4.67 (Kenaga and Goring, 1980)
5.068 (de Bruijn et al., 1989)
4.81 (Bowman and Sans, 1983a)

Solubility in organics:
Soluble in acetone, carbon tetrachloride, ether, kerosene, methylene chloride, and toluene (Windholz et al., 1983)

Solubility in water (mg/L):
1.08 at 20 °C (Chiou et al., 1977; Freed et al., 1977, 1979a)
40 at 25 °C (quoted, Windholz et al., 1983)
0.6 at 20.0 °C (Bowman and Sans, 1979)

Vapor pressure (x 10^{-5} mmHg):
80 at 25 °C (Windholz et al., 1983)
5.29 at 20 °C, 18.6 at 30 °C (Freed et al., 1977)

Environmental fate:
   Chemical/Physical. Although no products were identified, the reported hydrolysis half-life at pH 7.4 and 70 °C using a 1:4 ethanol/water mixture is 10.2-10.4 h (Freed et al., 1977). Ronnel decomposed at elevated temperatures on five clay surfaces, each treated with hydrogen, calcium, magnesium, aluminum, and iron ions. At temperatures <950 °C (125, 300, and 750 °C), bentonite clays impregnated with technical ronnel (18.6 wt %) decomposed to 2,4,5-trichlorophenol and a rearrangement product tentatively identified as *O*-methyl *S*-methyl-*O*-(2,4,5-trichlorophenyl) phosphorothioate (Rosenfield and Van Valkenburg, 1965). At 950 °C, only the latter product formed. It was postulated that this compound resulted from an acid-catalyzed molecular rearrangement reaction. Ronnel also undergoes base-catalyzed hydrolysis at elevated temperatures. Products include methanol and a new compound that is formed via cleavage of a methyl group from one of the methoxy groups, which is then bonded to the sulfur atom (Rosenfield and Van Valkenburg, 1965).
   Ronnel is stable to hydrolysis over the pH range of 5-6 (Mortland and Raman, 1967). However, in the presence of a Cu(II) salt (as cupric chloride) or when present as the exchangeable Cu(II) cation in montmorillonite clays, ronnel is completely hydrolyzed via first-order kinetics in <24 h at 20 °C. The calculated half-life of ronnel at 20 °C for this reaction is 6.0 h. It was suggested that decomposition in the presence of Cu(II) was a result of coordination of the copper atom through the oxygen or sulfur on the phosphorus atom, resulting in the cleavage of the side chain containing the phosphorus atom, forming *O,O*-ethyl-*O*-phosphorothioate and 1,2,4-trichlorobenzene (Mortland and Raman, 1967).
   Emits toxic fumes of chlorides, sulfur, and phosphorus oxides when heated to decomposition (Lewis, 1990).

Exposure limits (mg/m³): NIOSH REL: TWA 10, IDLH 300; OSHA PEL: TWA 15; ACGIH TLV: TWA 10 (adopted).

Symptoms of exposure: In animals: cholinesterase inhibition; irritates eyes; liver and kidney damage.

Toxicity:
   LC_{50} (96-h) for fathead minnows 305 μg/L (Verschueren, 1983).

Acute oral LD_{50} for wild birds 80 mg/kg, chickens 6,375 mg/kg, ducks 3,500 mg/kg, dogs 500 mg/kg, guinea pigs 1,400 mg/kg, mice 2,000 mg/kg, rats 625 mg/kg, rabbits 420 mg/kg, turkeys 500 mg/kg (quoted, RTECS, 1985).

Uses: Not commercially produced in the United States. Systemic insecticide.

STRYCHNINE

Synonyms: Certox; Dolco mouse cereal; Kwik-kil; Mole death; Mole-nots; Mouse-rid; Mouse-tox; Pied piper mouse seed; RCRA waste number P108; Rodex; Sana-seed; **Strychnidin-10-one**; Strychnos; UN 1692.

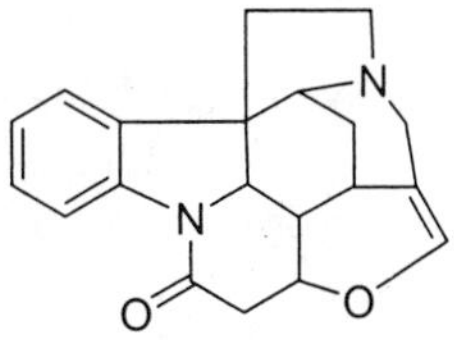

CAS Registry Number: 57-24-9
DOT: 1692
DOT label: Poison
Molecular formula: $C_{21}H_{22}N_2O_2$
Formula weight: 334.42
RTECS: WL2275000
Merck reference: 10, 8724

Physical state, color, and odor:
Colorless to white, odorless crystals. Bitter taste.

Melting point (°C):
286-288 (Weast, 1986)
268 (Sax and Lewis, 1987)
284-286 (Dean, 1987)

Boiling point (°C):
270 at 5 mmHg (Weast, 1986)

Density (g/cm³):
1.36 at 20/4 °C (Weast, 1986)

Dissociation constant (20 °C):
$pK_1 = 6.0$, $pK_2 = 11.7$ (Windholz et al., 1983)

Soil sorption coefficient, log K_{oc}:
4.20 (Valent loamy sand, pH 6.7), 4.14 (Ascalon sandy loam, pH 8.3), 3.97 (Table Mountain sandy clay loam, pH 7.7), 4.20 (Kim loam, pH 8.0) (Starr et al., 1996).
3.24 (Orovada fine sandy loam), 4.24 (Rose Creek Variant loamy fine sand), 3.86 (Truckee silt loam) (Miller et al., 1983)

Octanol/water partition coefficient, log K_{ow}:
1.93 (Mercer et al., 1990)

Solubility in organics (g/L):
Alcohol (6.67), amyl alcohol (4.55), benzene (5.56), chloroform (200), glycerol (3.13), methanol (3.85), toluene (≈ 5) (Windholz et al., 1983), pyridine (>12.3 g/kg at 20-25 °C) (Dehn, 1917)

Solubility in water:
156.25 mg/L, 322.6 mg/L at 100 °C (quoted, Windholz et al., 1983)

143 ppm at 20-25 °C (quoted, Verschueren, 1983)
>0.2 g/kg at 20-25 °C (Dehn, 1917)

Environmental fate:
  Biological. Strychnine degraded in sandy loam and sandy clay loam soils. The disappearance half-life ranged from 24 to 27 d (Starr et al., 1996). Initial degradation products (i.e., strychnine *N*-oxide, 16-hydroxystrychnine, and/or 2-hydroxystrychnin) and unidentified polar compounds were identified. The polar compounds were likely adsorbed strongly onto soil colloidal fractions.
  Chemical/Physical. Strychnine will not hydrolyze to any reasonable extent (Kollig, 1993). Emits toxic nitrogen oxides when heated to decomposition (Lewis, 1990).
  Forms water-soluble salts with acids (Worthing and Hance, 1991).

Exposure limits (mg/m³): NIOSH REL: TWA 0.15, IDLH 3; OSHA PEL: TWA 0.15; ACGIH TLV: TWA 0.15 (adopted).

Symptoms of exposure: Powerful convulsant and toxic. Ingestion of 30-60 mg/kg may be fatal to humans (Patnaik, 1992).

Toxicity:
  Acute oral LD_{50} for wild birds 4 mg/kg, cats 500 μg/kg, ducks 3 mg/kg, dogs 500 μg/kg, mice 2 mg/kg, quail 23 mg/kg, rats 16 mg/kg (quoted, RTECS, 1985).

Source: Strychnine silver morning glory, wood rose, and in *Strychnos nusvomica* L.: 15,800, 400-12,000, 8,000, and 7,030 ppm in bark, seeds, leaves, and roots, respectively (Duke, 1992).

Uses: Destroying rodents (e.g., moles) and predatory animals; trapping fur-bearing animals; medicine.

STYRENE

Synonyms: Cinnamene; Cinnamenol; Cinnamol; Diarex HF77; **Ethyenyl benzene**; NCI-C02200; Phenethylene; Phenylethene; Phenylethylene; Styrene monomer; Styrol; Styrolene; Styron; Styropol; Styropor; UN 2055; Vinyl benzene; Vinyl benzol.

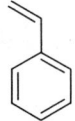

CAS Registry Number: 100-42-5
DOT: 2055
DOT label: Flammable or combustible liquid
Molecular formula: C_8H_8
Formula weight: 104.15
RTECS: WL3675000
Merck reference: 10, 8732

Physical state, color, and odor:
Colorless to light yellow, oily liquid with a sweet, penetrating odor. The reported odor threshold concentrations in air ranged from 37 to 148 ppb (Keith and Walters, 1992; Young et al., 1996). The taste threshold concentration in water is 94 ppb (Young et al., 1996).

Melting point (°C):
-30.6 (Weast, 1986)
-33 (Huntress and Mulliken, 1941)

Boiling point (°C):
145.2 (Weast, 1986)

Density (g/cm³):
0.9237 at 0/4 °C, 0.9148 at 10/4 °C, 0.9059 at 20/4 °C, 0.8970 at 30/4 °C, 0.8880 at 40/4 °C, 0.8702 at 60/4 °C (Standen, 1969)

Diffusivity in water (x 10^{-5} cm²/sec):
0.81 at 20 °C using method of Hayduk and Laudie (1974)

Dissociation constant, pK_a:
>14 (Schwarzenbach et al., 1993)

Flash point (°C):
31 (NIOSH, 1997)

Lower explosive limit (%):
0.9 (NIOSH, 1997)

Upper explosive limit (%):
6.8 (NIOSH, 1997)

Entropy of fusion (cal/mol·K):
10.79 (Guttman et al., 1943; Pitzer et al., 1946)

10.8 (Lebedev et al., 1985)
10.83 (Warfield and Petree, 1961)

Heat of fusion (kcal/mol):
2.617 (Guttman et al., 1943; Pitzer et al., 1946; Lebedev et al., 1985)
2.620 (Warfield and Petree, 1961)

Henry's law constant (x 10^{-3} atm·m^3/mol):
2.61 (calculated, U.S. EPA, 1980a)

Interfacial tension with water (dyn/cm at 19 °C):
35.48 (Demond and Lindner, 1993)

Ionization potential (eV):
8.47 (Franklin et al., 1969)
8.71 (Rav-Acha and Choshen, 1987)
8.86 (Krishna and Gupta, 1970)

Bioconcentration factor, log BCF:
1.13 (goldfish, Ogata et al., 1984)

Soil sorption coefficient, log K_{oc}:
2.96 (Meylan et al., 1992)
5.06 (aquifer sand), 4.25 (Lima loam), 3.65 (Edwards muck) (Fu and Alexander, 1992)

Octanol/water partition coefficient, log K_{ow}:
2.76 (Fujisawa and Masuhara, 1981)
2.95 (Chou and Jurs, 1979)
3.16 (Banerjee et al., 1980)

Solubility in organics:
Soluble in acetone, ethanol, benzene, ether, carbon disulfide (Weast, 1986), and many aromatic
solvents

Solubility in water:
1.54 mM at 25.0 °C (Banerjee et al., 1980)
2.22 mM at 25 °C (Andrews and Keefer, 1950a)
9.6 g/kg at 60.3 °C (Fordyce and Chapin, 1947)
In wt % (°C): 0.032 (6), 0.066 (25), 0.084 (31), 0.101 (40), 0.123 (51) (Lane, 1946)
3.19 x 10^{-3} at 25 °C (mole fraction, Li et al., 1993)

Vapor density:
4.26 g/L at 25 °C, 3.60 (air = 1)

Vapor pressure (mmHg):
5 at 20 °C, 9.5 at 30 °C (Verschueren, 1983)
4.3 at 15 °C (ACGIH, 1987)
6.45 at 25 °C (WHO, 1983a)

Environmental fate:
   Biological. Fu and Alexander (1992) observed that despite the high degree of adsorption

onto soils, styrene was mineralized to carbon dioxide under aerobic conditions. Rates of mineralization from highest to lowest were sewage sludge, Lima soil (pH 7.23, 7.5% organic matter), groundwater (pH 8.25, 30.5 mg/L organic matter), Beebe Lake water from Ithaca, NY (pH 7.5, 50-60 mg/L organic matter), aquifer sand (pH 6.95, 0.4% organic matter), Erie silt loam (pH 4.87, 5.74% organic matter). Styrene did not mineralize in sterile environmental samples.

Bridié et al. (1979) reported BOD and COD values 1.29 and 2.80 g/g using filtered effluent from a biological sanitary waste treatment plant. These values were determined using a standard dilution method at 20 °C and stirred for a period of 5 d. When a sewage seed was used in a separate screening test, a BOD value of 2.45 g/g was obtained. The ThOD for styrene is 3.08 g/g.

Photolytic. Irradiation of styrene in solution forms polystyrene. In a benzene solution, irradiation of polystyrene will result in depolymerization to presumably styrene (Calvert and Pitts, 1966).

Atkinson (1985) reported a photooxidation reaction rate of 5.25 x 10^{-11} cm^3/molecule·sec for styrene and OH radicals in the atmosphere. A reaction rate of 1.8 x 10^4 L/mol·sec at 303 K was reported for the reaction of styrene and ozone in the vapor phase (Bufalini and Altshuller, 1965).

Chemical/Physical. In the dark, styrene reacted with ozone forming benzaldehyde, formaldehyde, benzoic acid, and trace amounts of formic acid (Grosjean, 1985). Polymerizes readily in the presence of heat, light, or a peroxide catalyst. Polymerization is exothermic and may become explosive (NIOSH, 1997).

Styrene will not hydrolyze because it does not contain a hydrolyzable functional group (Kollig, 1993).

At an influent concentration of 180 mg/L, treatment with granular activated carbon resulted in an effluent concentration of 18 mg/L. The adsorbability of the carbon used was 28 mg/g carbon (Guisti et al., 1974).

Exposure limits: NIOSH REL: TWA 50 ppm (215 mg/m^3), STEL 100 ppm (425 mg/m^3), IDLH 700 ppm; OSHA PEL: TWA 100 ppm, ceiling 200 ppm, 5-min/3-h peak 600 ppm; ACGIH TLV: TWA 20 ppm, STEL 40 ppm (adopted).

Symptoms of exposure: Irritates, eye, skin, and mucous membranes. Narcotic at high concentrations (Patnaik, 1992). An irritation concentration of 430.00 mg/m^3 in air was reported for uninhibited styrene (Ruth, 1986).

Toxicity:

EC_{50} (48-h) for *Daphnia magna* 4.7 mg/L (Cushman et al., 1997).

LC_{50} (14-d) for *Eisenia fostida* 120 mg/kg (Cushman et al., 1997).

LC_{50} (96-h) for *Pimephales promelas* 10 mg/L, *Hyalella azteca* 9.5 mg/L, *Selenastrum capricornutum* 0.72 mg/L (Cushman et al., 1997).

LC_{50} (48-h) for *Daphnia magna* 23 mg/L (LeBlanc, 1980).

LC_{50} (24-h) for *Daphnia magna* 27 mg/L (LeBlanc, 1980); *Oncorhynchus mykiss* 2.5 mg/L (Qureshi et al., 1982), *Cyprinodon variegatus* 9.1 mg/L (Heitmuller et al., 1984).

LC_{50} (inhalation) for mice 21,600 mg/m^3/2-h, rats 24 g/m^3/4-h (quoted, RTECS, 1985).

Acute oral LD_{50} for mice 316 mg/kg, rats 5,000 mg/kg (quoted, RTECS, 1985).

Drinking water standard (final): MCLG: 0.1 mg/L; MCL: 0.1 mg/L (U.S. EPA, 1996).

Source: Based on analysis of 7 coal tar samples, styrene concentrations ranged from ND to 2,500 ppm (EPRI, 1990).

Styrene occurs naturally in benzoin, rosemary, sweetgum, cassia, Oriental styrax, and Peru balsam (Duke, 1992).

Uses: Preparation of polystyrene, styrene oxide, ethylbenzene, ethylcyclohexane, benzoic acid, synthetic rubber, resins, protective coatings, and insulators.

SULFOTEPP

Synonyms: ASP 47; Bay E-393; Bayer-E 393; Bis-*O,O*-diethylphosphorothionic anhydride; Bladafum; Bladafume; Bladafun; Dithio; Dithione; Dithiophos; Dithiophosphoric acid tetraethyl ester; Dithiotep; E 393; ENT 16273; Ethyl thiopyrophosphate; Lethalaire G-57; Pirofos; Plant dithio aerosol; Plantfume 103 smoke generator; Pyrophosphorodithioic acid tetraethyl ester; Pyrophosphorodithioic acid *O,O,O,O*-tetraethyl dithionopyrophosphate; RCRA waste number P109; Sulfatep; Sulfotep; TEDP; TEDTP; Tetraethyl dithionopyrophosphate; Tetraethyl dithiopyrophosphate; *O,O,O,O*-Tetraethyl dithiopyrophosphate; Thiodiphosphoric acid tetraethyl ester; **Thiophosphoric acid tetraethyl ester**; Thiopyrophosphoric acid tetraethyl ester; Thiotepp; UN 1704.

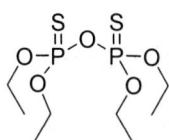

CAS Registry Number: 3689-24-5
DOT: 1704
DOT label: Poison
Molecular formula: $C_8H_{20}O_5P_2S_2$
Formula weight: 322.30
RTECS: XN4375000
Merck reference: 10, 8846

Physical state, color, and odor:
Pale yellow mobile liquid with a garlic-like odor

Boiling point (°C):
136-139 at 2 mmHg (Windholz et al., 1983)
Decomposes (NIOSH, 1997)

Density (g/cm³):
1.196 at 25/4 °C (Windholz et al., 1983)
1.796 at 20/4 °C (Worthing and Hance, 1991)

Diffusivity in water (x 10^{-5} cm²/sec):
0.63 at 20 °C using method of Hayduk and Laudie (1974)

Henry's law constant (x 10^{-6} atm·m³/mol):
2.88 at 20 °C (approximate - calculated from water solubility and vapor pressure)

Soil sorption coefficient, log K_{oc}:
2.87 using method of Kenaga and Goring (1980)

Octanol/water partition coefficient, log K_{ow}:
3.02 using method of Kenaga and Goring (1980)

Solubility in organics:
Miscible with methyl chloride and most organic solvents (Worthing and Hance, 1991)

Solubility in water (mg/L):
25 (Windholz et al., 1983)
30 at 20 °C (Worthing and Hance, 1991)

Vapor density:
13.17 g/L at 25 °C, 11.13 (air = 1)

Vapor pressure (mmHg):
1.7 x 10^{-4} at 20 °C (Windholz et al., 1983)

Environmental fate:
Soil. Cleavage of the molecule yields diethyl phosphate, monoethyl phosphate, and phosphoric acid (Hartley and Kidd, 1987).
Chemical/Physical. Emits toxic oxides of sulfur and phosphorus when heated to decomposition (Lewis, 1990).

Exposure limits (mg/m^3): NIOSH REL: TWA 0.2, IDLH 10; OSHA PEL: TWA 0.2; ACGIH TLV: TWA 0.2 (adopted).

Toxicity:
LC_{50} (96-h) for fathead minnows 178 μg/L, bluegill sunfish 1.6 μg/L and rainbow trout 18 μg/L (Verschueren, 1983).
Acute inhalation LC_{50} (4-h) for rats ≈0.05 mg/L air (Worthing and Hance, 1991).
Acute oral LD_{50} for wild birds 100 mg/kg, cats 3 mg/kg, chickens 25 mg/kg, dogs 5 mg/kg, mice 22 mg/kg, rats 5 mg/kg, rabbits 25 mg/kg (quoted, RTECS, 1985).
Acute percutaneous LD_{50} (7-d) for rats 65 mg/kg (Worthing and Hance, 1991).
In 2-yr feeding trials, the NOEL for rats was 10 mg/kg diet (Worthing and Hance, 1991).

Uses: Not commercially produced in the United States. Non-systemic insecticide for controlling pests in vegetables and greenhouse ornamentals.

2,4,5-T

Synonyms: Amine 2,4,5-T for rice; BCF-bushkiller; Brush-off 445 low volatile brush killer; Brush-rhap; Brushtox; Dacamine; Debroussaillant concentre; Debroussaillant super concentre; Decamine 4T; Ded-weed brush killer; Ded-weed LV-6 brush-kil and T-5 brush-kil; Dinoxol; Envert-T; Estercide T-2 and T-245; Esteron; Esterone 245; Esteron 245 BE; Esteron brush killer; Farmco fence rider; Fence rider; Forron; Forst U 46; Fortex; Fruitone A; Inverton 245; Line rider; NA 2765; Phortox; RCRA waste number U232; Reddon; Reddox; Spontox; Super D weedone; Tippon; Tormona; Transamine; Tributon; **(2,4,5-Trichlorophenoxy)acetic acid**; Trinoxol; Trioxon; Trioxone; U 46; Veon; Veon 245; Verton 2T; Visko-rhap low volatile ester; Weddar; Weedone; Weedone 2,4,5-T.

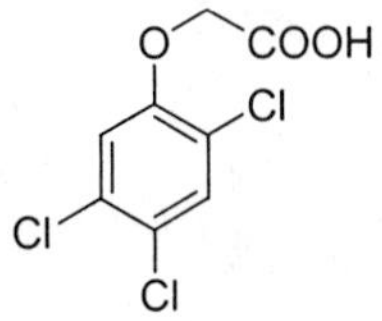

CAS Registry Number: 93-76-5
DOT: 2765
DOT label: Poison
Molecular formula: $C_8H_5Cl_3O_3$
Formula weight: 255.48
RTECS: AJ8400000
Merck reference: 10, 8902

Physical state and color:
Colorless to pale brown crystals. Odor threshold from water is 2.92 mg/kg (Keith and Walters, 1992). Metallic taste.

Melting point (°C):
153 (Windholz et al., 1983)

Density (g/cm³):
1.80 at 20/20 °C (Windholz et al., 1983)

Diffusivity in water (x 10⁻⁵ cm²/sec):
0.54 at 20 °C using method of Hayduk and Laudie (1974)

Dissociation constant, pK$_a$:
2.80, 2.83 (Jafvert et al., 1990)
2.88 (Nelson and Faust, 1969)

Henry's law constant (x 10⁻⁸ atm·m³/mol):
4.87 at 25 °C (approximate - calculated from water solubility and vapor pressure)

Bioconcentration factor, log BCF:
1.36 (fish, microcosm) (Garten and Trabalka, 1983)

Soil sorption coefficient, log K$_{oc}$:
1.72 (Kenaga and Goring, 1980)
2.27 (Webster soil, Nkedi-Kizza et al., 1983)

Octanol/water partition coefficient, log K_{ow}:
3.31, 3.38 (Jafvert et al., 1990)
3.40 (Riederer, 1990)

Solubility in organics (°C):
Soluble in ethanol (548.2 mg/L), ether (243.2 mg/L), heptane (400 mg/L), xylenes (6.8 g/L), methanol (496 g/L), toluene (7.32 g/L) (Keith and Walters, 1992).

Solubility in water:
278 ppm at 25 °C (quoted, Verschueren, 1983)
240 mg/L at 25 °C (Klöpffer et al., 1982)
220 mg/L at 20 °C (Riederer, 1990)
1.05 mM at 25 °C (Gunther et al., 1968)
150 g/L at 25 °C (Lewis, 1989)

Vapor pressure (x 10^{-6} mmHg):
37.5 at 20 °C (Riederer, 1990)
6.46 at 25 °C (Lewis, 1989)

Environmental fate:
  Biological. 2,4,5-T degraded in anaerobic sludge by reductive dechlorination to 2,4,5-trichlorophenol, 3,4-dichlorophenol, and 4-chlorophenol (Mikesell and Boyd, 1985). An anaerobic methanogenic consortium, growing on 3-chlorobenzoate, metabolized 2,4,5-T to (2,5-dichlorophenoxy)acetic acid at a rate of 1.02×10^{-7} M/h. The half-life was reported to be 2 d at 37 °C (Suflita et al., 1984). Under aerobic conditions, 2,4,5-T degraded to 2,4,5-trichlorophenol and 3,5-dichlorocatechol, which may further degrade to 4-chlorocatechol or *cis,cis*-2,4-dichloromuconic acid, 2-chloro-4-carboxymethylenebut-2-enolide, chlorosuccinic acid, and succinic acid (Byast and Hance, 1975). The cometabolic oxidation of 2,4,5-T by *Brevibacterium* sp. yielded a product tentatively identified as 3,5-dichlorocatechol (Horvath, 1970). The cometabolism of this compound by *Achromobacter* sp. gave 3,5-dichloro-2-hydroxymuconic semialdehyde (Horvath, 1970a). Rosenberg and Alexander (1980) reported that 2,4,5-trichlorophenol, the principal degradation product of 2,4,5-T by microbes, was further metabolized to 3,5-dichlorocatechol, 4-chlorocatechol, succinate, *cis,cis*-2,4-dichloromuconate, 2-chloro-4-(carboxymethylene)but-2-enolide, and chlorosuccinate.
  Soil. 2,4,5-Trichlorophenol and 2,4,5-trichloroanisole were formed when 2,4,5-T was incubated in soil at 25 °C under aerobic conditions. The half-life under these conditions was 14 d (McCall et al., 1981). When 2,4,5-T (10 μg), in unsterilized tropical clay and silty clay soils, was incubated for 4 months, 5-35% degradation yields were observed (Rosenberg and Alexander, 1980). Hydrolyzes in soil to 2,4,5-trichlorophenol (Somasundaram et al., 1989, 1991) and 2,4,5-trichloroanisole (Somasundaram et al., 1989). The rate of 2,4,5-T degradation in soil remained unchanged in a soil pretreated with its hydrolysis metabolite (2,4,5-trichlorophenol) (Somasundaram et al., 1989).
  The half-lives of 2,4,5-T in soil incubated in the laboratory under aerobic conditions ranged from 14 to 64 d with an average of 33 d (Altom and Stritzke, 1973; Foster and McKercher, 1973; Yoshida and Castro, 1975). In field soils, the disappearance half-lives were lower and ranged from 8 to 54 d with an average of 16 d (Radosevich and Winterlin, 1977; Stewart and Gaul, 1977).
  Groundwater. According to the U.S. EPA (1986), 2,4,5-T has a high potential to leach to groundwater.
  Photolytic. When 2,4,5-T (100 μM), in oxygenated water containing titanium dioxide (2 g/L) suspension, was irradiated by sunlight ($\lambda \geq 340$ nm), 2,4,5-trichlorophenol, 2,4,5-trichlorophenyl

formate, and nine chlorinated aromatic hydrocarbons formed as major intermediates. Complete mineralization yielded hydrochloric acid, carbon dioxide, and water (Barbeni et al., 1987).

Crosby and Wong (1973) studied the photolysis of 2,4,5-T in aqueous solutions (100 mg/L) under alkaline conditions (pH 8) using both outdoor sunlight and indoor irradiation (λ = 300-450 nm). 2,4,5-Trichlorophenol and 2-hydroxy-4,5-dichlorophenoxyacetic acid formed as major products. Minor photodecomposition products included 4,6-dichlororesorcinol, 4-chloro-resorcinol, 2,5-dichlorophenol, and a dark polymeric substance. The rate of photolysis increased 11-fold in the presence of sensitizers (acetone or riboflavin) (Crosby and Wong, 1973). The rate of photolysis of 2,4,5-T was also higher in natural waters containing fulvic acids when compared to distilled water. The major photoproduct found in the humic acid-induced reaction was 2,4,5-trichlorophenol. In addition, the presence of ferric ions and/or hydrogen peroxides may contribute to the sunlight-induced photolysis of 2,4,5-T in acidic, weakly absorbing natural waters (Skurlatov et al., 1983).

Chemical/Physical. Carbon dioxide, chloride, dichloromaleic, oxalic and glycolic acids, were reported as ozonation products of 2,4,5-T in water at pH 8 (Struif et al., 1978). Reacts with alkali metals and amines forming water-soluble salts (Worthing and Hance, 1991). When 2,4,5-T was heated at 900 °C, carbon monoxide, carbon dioxide, chlorine, hydrochloric acid, and oxygen were produced (Kennedy et al., 1972, 1972a).

2,3,5-T will not hydrolyze to any reasonable extent (Kollig, 1993).

Exposure limits (mg/m³): NIOSH REL: TWA 10, IDLH 250; OSHA PEL: TWA 10; ACGIH TLV: TWA 10 (adopted).

Symptoms of exposure: Skin irritation. May also cause eye, nose, and throat irritation (NIOSH, 1987). An acceptable daily intake reported for humans is 0.03 mg/kg body weight provided the product contains ≤0.01 mg TCDD/kg 2,4,5-T (Worthing and Hance, 1991).

Toxicity:
EC_{50} (5-min) for *Photobacterium phosphoreum* 51.7 mg/L (Somasundaram et al., 1990).
LC_{50} (96-h) for rainbow trout 350 mg/L and carp 355 mg/L (Hartley and Kidd, 1987).
Acute oral LD_{50} for chickens 310 mg/kg, dogs 100 mg/kg, guinea pigs 381 mg/kg, hamsters 425 mg/kg, mice 389 mg/kg, rats 300 mg/kg (quoted, RTECS, 1985).
Acute oral LD_{50} for rats >5,000 mg/kg (Worthing and Hance, 1991).
A NOEL of 30 mg/kg diet was observed for rats during 2-yr feeding trials (Worthing and Hance, 1991).

Drinking water standard: No MCLGs or MCLs have been proposed although 2,4,5-T has been listed for regulation (U.S. EPA, 1996).

Uses: Plant hormone; defoliant. Formerly used as a herbicide. Banned by the U.S. EPA.

TCDD

Synonyms: Dioxin; Dioxin (herbicide contaminant); Dioxine; NCI-C03714; TCDBD; 2,3,7,8-TCDD; 2,3,7,8-Tetrachlorodibenzodioxin; **2,3,7,8-Tetrachlorodibenzo[*b,e*][1,4]dioxin**; 2,3,7,-8-Tetrachlorodibenzo-1,4-dioxin; 2,3,7,8-Tetrachlorodibenzo-*p*-dioxin; Tetradioxin.

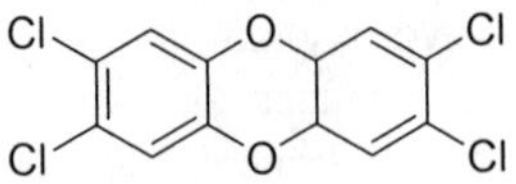

CAS Registry Number: 1746-01-6
Molecular formula: $C_{12}H_4Cl_4O_2$
Formula weight: 321.98
RTECS: HP3500000
Merck reference: 10, 8957

Physical state and color:
Colorless to white needles

Melting point (°C):
305.0 (Schroy et al., 1985)
303-305 (Crummett and Stehl, 1973)

Boiling point (°C):
421.2 (estimated, Schroy et al., 1985)
Begins to decompose at 500 (U.S. EPA, 1985)

Density (g/cm³):
1.827 at 25 °C (estimated, Schroy et al., 1985)

Diffusivity in water (x 10^{-5} cm²/sec):
0.49 at 20 °C using method of Hayduk and Laudie (1974)

Entropy of fusion (cal/mol·K):
16.49 (Rordorf, 1989)

Heat of fusion (kcal/mol):
9.30 (Shroy et al., 1985)

Henry's law constant (x 10^{-23} atm·m³/mol):
5.40 at 20 °C (approximate - calculated from water solubility and vapor pressure)

Ionization potential (eV):
9.148 (calculated, Koester and Hites, 1988)

Bioconcentration factor, log BCF:
3.73 (fish, Kenega and Goring, 1980)
3.90 (fathead minnow, Adams et al., 1986)
3.97 (rainbow trout, Branson et al., 1985)
4.43 (rainbow trout, Mehrle et al., 1988)
3.02 (Garten and Trabalka, 1983)

3.30-4.27 (algae), 3.30-4.45 (catfish), 3.89-4.68 (daphnids), 3.08-3.70 (duckweed), 3.00-4.80 (mosquito fish), 3.15-4.67 (snails) (Isensee and Jones, 1975)
3.02 (fish, microcosm) (Garten and Trabalka, 1983)
4.56 (*Oncorhynchus mykiss*, Devillers et al., 1996)
1.69 (daphnids), 2.34 (ostracod), 2.08 (brine shrimp) (Matsumura and Benezet, 1973)
5.80-5.90 (goldfish, Sijm et al., 1989)
1.73 (silversides), 3.20 (brine shrimp) (Callahan et al., 1979)

Soil sorption coefficient, log K_{oc}:
6.6 (Times Beach, MO soil, Walters and Guiseppi-Elie, 1988)
6.44 (soil 96 - pH 6.8; CEC 5.4 meq/100 g; 44% sand, 42% silt, 14% clay), 6.66 (soil 91: pH 5.8; CEC 15.3 meq/100 g; 38% sand, 40% silt, 22% clay) (Walters et al., 1989)
6.30 (lake Ontario sediment), 7.25 (organic carbon) (Lodge and Cook, 1989)

Octanol/water partition coefficient, log K_{ow}:
6.15 (Schroy et al., 1985; Travis and Arms, 1988)
6.64 (Marple et al., 1986)
6.20 (Doucette and Andren, 1988)
7.02 (Burkhard and Kuehl, 1986)
5.38 (Rappe et al., 1987)

Solubility in organics (mg/L):
Acetone (110), benzene (570), chlorobenzene (720) chloroform (370), *o*-dichlorobenzene (1,400), methanol (10) and octanol (48-50) (Crummett and Stehl, 1973; Arthur and Frea, 1989); benzene (570), tetrachloroethylene (680), lard oil (40), hexane (280) (Keith and Walters, 1992).

Solubility in water:
0.317 μg/L at 25 °C (Schroy et al., 1985)
0.2 ppb (Crummett and Stehl, 1973)
0.0193 ppb at 22 °C (Marple et al., 1986)
0.32 ppb at 25 °C (Rappe et al., 1987)
0.0129 and 0.483 ppb at 0.2 and 17.3 °C, respectively (Lodge, 1989)
7.91 ± 2.7 ng/L at 20-22 °C (Adams and Blaine, 1986)

Vapor pressure (x 10^{-10} mmHg):
34.6 at 30.1 °C (Rodorf, 1986)
2.7 at 15 °C, 6.4 at 20 °C, 14 at 25 °C, 35 at 30 °C, 163 at 40 °C (Rappe et al., 1987)
7.2 at 25 °C (Podoll et al., 1986)
340 at 25 °C (Rodorf, 1985)

Environmental fate:
Biological. After a 30-d incubation period, the white rot fungus *Phanerochaete chrysosporium* was capable of oxidizing TCDD to carbon dioxide. Mineralization began between the third and sixth day of incubation. The production of carbon dioxide was highest between 3-18 d of incubation, after which the rate of a carbon dioxide produced decreased until the 30th day. It was suggested that the metabolism of TCDD and other compounds, including *p,p'*-DDT, benzo[*a*]pyrene and lindane, was dependent on the extracellular lignin-degrading enzyme system of this fungus (Bumpus et al., 1985).

A half-life of 418 d was calculated based on die away test data (Kearney et al., 1971).

Soil. In a laboratory sediment-water system incubated under anaerobic conditions, the half-life of TCDD was determined to range 500 to 600 d (Ward and Matsumura, 1978). In shallow and deep soils, reported half-lives were 10 and 100 yr, respectively (Nauman and Schaum, 1987). Due to its low aqueous solubility, TCDD will not undergo significant leaching by runoff (Helling et al., 1973).

Surface Water. Plimmer et al. (1973) reported that the photolysis half-life of TCDD in a methanol solution exposed to sunlight was 3 h. Volatilization half-lives of 32 and 16 d were reported for lakes and rivers, respectively (Podoll et al., 1986).

Photolytic. Pure TCDD did not photolyze under UV light. However, in aqueous solutions containing cationic (1-hexadecylpyridinium chloride), anionic (sodium dodecyl sulfate) and nonionic (methanol) surfactants, TCDD decomposed into the end product tentatively identified as 2-phenoxyphenol. The times required for total TCDD decomposition using the cationic, anionic, and nonionic solutions were 4, 8, and 16 h, respectively (Botré et al., 1978). TCDD photodegrades rapidly in alcoholic solutions by reductive dechlorination. In water, however, the reaction was very slow (Crosby et al., 1973). In an earlier study, Crosby et al. (1971) reported a photolytic half-life of 14 d when TCDD in distilled water was exposed to sunlight. The major photodegradative pathway of TCDD involves a replacement of the chlorine atom by a hydrogen atom. The proposed degradative pathway is TCDD to 2,7,8-trichlorodi-benzo[*b,e*][1,4]dioxin to 2,7-dichlorodibenzo[*b,e*][1,4]dioxin to 2-chlorodibenzo[*b,e*][1,4]di-oxin to dibenzo[*b,e*][1,4]dioxin to 2-hydroxydiphenyl ether, which undergoes polymerization (Makino et al., 1992).

[^{14}C]TCDD on a silica plate was exposed to summer sunlight at Beltsville, MD for 20 h. A polar product was formed which was not identified. An identical experiment using soil demon-strated that photodegradation occurred but to a smaller extent. No photoproducts were identified (Plimmer, 1978). TCDD on a silica plate was exposed to a GE Model RS sunlamp at a distance of 1.0 m. The estimated photolytic half-life for this reaction is 140 h (Nestrick et al., 1980).

When TCDD in isooctane (3.1 μM) was irradiated by UV light ($\lambda \leq$ 310 nm) at 31-33 °C, $\approx$ 10% was converted to 2,3,7-trichlorodibenzo-*p*-dioxin. In addition, 4,4',5,5'-tetrachloro-2,2'-di-hydroxybiphenyl was identified as a new photoproduct formed by the reductive rearrangement of TCDD (Kieatiwong et al., 1990).

Hilarides and others (1994) investigated the destruction of TCDD on artificially contaminated soils using ^{60}Co γ radiation. It appeared that TCDD underwent stepwise reduction dechlorination from tetra- to tri-, then di- to monochlordioxin, and then to presumably nonchlorinated dioxins and phenols. The investigators discovered that the greatest amount of TCDD destruction (92%) occurred when soils were amended with 25% water and 2% nonionic surfactant [alkoxylated fatty alcohol (Plurafac RA-40)]. Replicate experiments conducted without the surfactant lowered the rate of TCDD destruction.

The estimated photooxidation half-life of TCDD in the atmosphere via OH radicals ranged from 22.3 to 223 h (Atkinson, 1987a). An atmospheric half-life of 58 min was reported for TCDD exposed to summer sunlight at 40 ° N latitude (Buser, 1988).

Plimmer et al. (1973) reported that the photolysis half-life of TCDD in a methanol solution exposed to sunlight was 3 h. In sunlight at 40 ° N latitude, estimated photolytic half-lives of TCDD in surface water in winter, spring, summer, and fall were 118, 27, 21, and 51 h, respectively (Podoll et al., 1986). In shallow surface water bodies, a photolytic half-life of 40 h was reported (Travis and Hattemer-Frey, 1987).

TCDD in a water-acetonitrile mixture (90:10) was exposed to summer sunlight. Based on the measured rate constant for this reaction, the estimated half-life is 27 h (Dulin et al., 1986).

Chemical/Physical. TCDD was dehalogenated by a solution composed of poly(ethylene glycol), potassium carbonate and sodium peroxide. After 2 h at 85 °C, >99.9% of the applied TCDD decomposed. Chemical intermediates identified include tri-, di-, and mono-chloro[*b,e*]dibenzo[1,4]dioxin, dibenzodioxin, hydrogen, carbon monoxide, methane, ethylene, and acetylene (Tundo et al., 1985).

TCDD will not hydrolyze to any reasonable extent (Kollig, 1993).

Exposure limits: An IDLH of 1 ppb was recommended by Schroy et al. (1985).

Symptoms of exposure: Extremely toxic. Causes chromosome damage. May cause miscarriage, birth defects, and fetotoxicity. Fatigue, headache, irritability, abdominal pain, blurred vision, and ataxia have been observed. Contact with skin may cause dermatitis and allergic reactions (Patnaik, 1992).

Toxicity:
LC_{10} (eggs) for fathead minnows 292 pg/g, channel catfish 429 pg/g, lake herring 509 pg/g, medaka 656 pg/g, white sucker 1,590 pg/g, northern pike 1,530 pg/g and zebrafish 1,610 pg/g (Elonen et al., 1997).

LC_{50} (3-d post hatch) for Japanese medaka (*Oryzias latipes*) ranged from 9 to 13 ng/L (Wisk and Cooper, 1990, 1990a).

LD_{50} (125-d) and LD_{50} (28-d) values for adult female mink (*Mustela vison*) were 0.047 and 0.264 μg/kg/day, respectively (Hochstein et al., 1998).

Acute oral LD_{50} for dogs 1 μg/kg, frogs mg/kg, guinea pigs 500 ng/kg, hamsters 1.157 mg/kg, monkeys 2 μg/kg, mice 114 μg/kg, rats 20 μg/kg, guinea pigs 2 μg/kg, male rats 22 μg/kg, rabbits 115 μg/kg and mice 284 μg/kg (Kriebel, 1981).

LD_{50} (skin) for rabbits 275 μg/kg (quoted, RTECS, 1985).

LD_{50} (10-d post-fertilization) for zebrafish 2.5 ng/g egg (Henry et al., 1997).

LOECs (pg/g wet weight) for fathead minnows 435, channel catfish 855, lake herring 270, medaka 949, white sucker 1,220, northern pike 1,800, and zebrafish 2,000 (Elonen et al., 1997).

NOECs (pg/g wet weight) for fathead minnows 235, channel catfish 385, lake herring 175, medaka 455, white sucker 848, northern pike 1,190, and zebrafish 424 (Elonen et al., 1997).

Drinking water standard (final): MCLG: zero; MCL: 3×10^{-5} μg/L (U.S. EPA, 1996). In Canada, the Ontario Ministry of Environment has established an Interim Drinking Water Objective of 10 parts per quadrillion (Boddington, 1990).

Source: Although not produced commercially, TCDD is formed as a by-product in the synthesis of 2,4,5-trichlorophenol. TCDD was found in 85% of soil samples obtained from a trichlorophenol manufacturing site. Concentrations ranged from approximately 20 ng/kg to 600 μg/kg (Van Ness et al., 1980). TCDD may be present in the herbicide 2,4-D which contains a mixture of dichloro-, trichloro-, and tetrachlorodioxins. TCDD is commonly found as a contaminant associated with pulp and paper mills (Boddington, 1990). In addition, during the manufacture of 2,4,5-T and silvex from trichlorophenol, TCDD was found at concentrations averaging 20 parts per billion (Newton and Snyder, 1978).

Uses: Toxicity studies.

1,2,4,5-TETRABROMOBENZENE

Synonym: *sym*-Tetrabromobenzene.

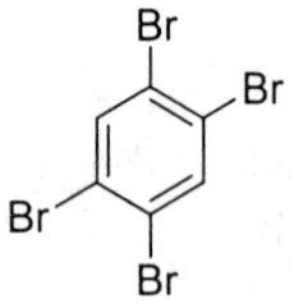

CAS Registry Number: 636-28-2
Molecular formula: $C_6H_2Br_4$
Formula weight: 393.70

Physical state:
Solid

Melting point (°C):
189 (Weast, 1986)

Boiling point (°C):
329 (Weast, 1986)

Diffusivity in water (x 10^{-5} cm²/sec):
0.61 at 20 °C using method of Hayduk and Laudie (1974)

Bioconcentration factor, log BCF:
3.57-3.81 (rainbow trout, Oliver and Niimi, 1985)
3.45 (wet weight based), 0.92 (lipid based) (*Gambusia affinis*, Chaisuksant et al., 1997)

Ionization potential (eV):
8.65, 8.89 (Mallard and Linstrom, 1998)

Soil sorption coefficient, log K_{oc}:
4.82 using method of Chiou et al. (1979)

Octanol/water partition coefficient, log K_{ow}:
5.13 (Watarai et al., 1982)

Solubility in organics:
Soluble in alcohol, benzene, and ether (Weast, 1986)

Solubility in water:
40 μg/L at 20-25 °C (Kim and Saleh, 1990)
104.7 nmol/L at 25 °C (Yalkowsky et al., 1979)

Environmental fate:
   Chemical/Physical. 1,2,4,5-Tetrabromobenzene (150 mg in 250-mL distilled water) was stirred in the dark for 1 day. GC analysis of the solution showed 1,2,4-tribromobenzene as a major transformation product (Kim and Saleh, 1990).

Uses: Organic synthesis.

1,1,2,2-TETRABROMOETHANE

Synonyms: Acetylene tetrabromide; Muthmann's liquid; TBE; Tetrabromoacetylene; Tetrabromoethane; *sym*-Tetrabromoethane; UN 2504.

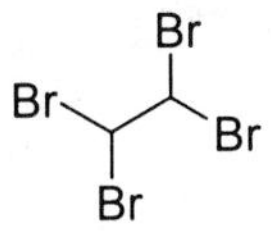

CAS Registry Number: 79-27-6
DOT: 2504
Molecular formula: $C_2H_2Br_4$
Formula weight: 345.65
RTECS: KI8225000
Merck reference: 10, 9012

Physical state, color, and odor:
Colorless to pale yellow liquid with a pungent odor resembling camphor and iodoform

Melting point (°C):
0 (Weast, 1986)

Boiling point (°C):
243.5 (Weast, 1986)
239-242 (Verschueren, 1983)

Density (g/cm³ at 20/4 °C):
2.8748 (Weast, 1986)
2.964 (Verschueren, 1983)
2.9529 (25/4 °C, Dean, 1987)

Diffusivity in water (x 10^{-5} cm²/sec):
0.79 at 20 °C using method of Hayduk and Laudie (1974)

Flash point (°C):
Noncombustible liquid (NIOSH, 1997)

Henry's law constant (x 10^{-5} atm·m³/mol):
6.40 at 20 °C (approximate - calculated from water solubility and vapor pressure)

Interfacial tension with water (dyn/cm at 20 °C):
38.82 (Harkins et al., 1920)

Soil sorption coefficient, log K_{oc}:
2.45 using method of Chiou et al. (1979)

Octanol/water partition coefficient, log K_{ow}:
2.91 using method of Hansch et al. (1968)

Solubility in organics:
Soluble in acetone and benzene (Weast, 1986). Miscible with acetic acid, alcohol, aniline, chloroform, and ether (Windholz et al., 1983).

Solubility in water:
0.07 wt % at 20 °C (NIOSH, 1997)
651 mg/L at 30 °C (quoted, Verschueren, 1983)

Vapor density:
14.13 g/L at 25 °C, 11.93 (air = 1)

Vapor pressure (mmHg):
0.1 at 20 °C (Verschueren, 1983)

Exposure limits: NIOSH REL: IDLH 8 ppm; OSHA PEL: 1 ppm (14 mg/m^3).

Toxicity:
   Acute oral LD_{50} for guinea pigs 400 mg/kg, mice 269 mg/kg, rats 1,100 mg/kg, rabbits 400 mg/kg (quoted, RTECS, 1985).

Uses: Solvent; in microscopy; separating minerals by density.

1,2,3,4-TETRACHLOROBENZENE

Synonym: 1,2,3,4-TCB.

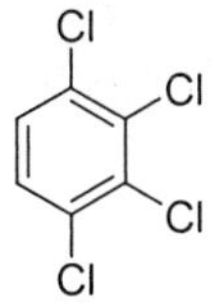

CAS Registry Number: 634-66-2
Molecular formula: $C_6H_2Cl_4$
Formula weight: 215.89
RTECS: DB9440000

Physical state and color:
White crystals or needles

Melting point (°C):
47.5 (Weast, 1986)
46.6 (Hawley, 1981)

Boiling point (°C):
254 (Weast, 1986)

Diffusivity in water (x 10^{-5} cm²/sec):
0.63 at 20 °C using method of Hayduk and Laudie (1974)

Flash point (°C):
112 (Dean, 1987)

Entropy of fusion (cal/mol·K):
12.7 (Domalski and Hearing, 1998)

Heat of fusion (kcal/mol):
4.0534 (Domalski and Hearing, 1998)

Henry's law constant (x 10^{-4} atm·m³/mol):
6.9 at 20 °C (Oliver, 1985)
4.79, 5.13, 6.72, 7.00, 12.62, and 27.26 at 14.8, 20.1, 22.1, 24.2, 34.8, and 50.5 °C, respectively
    (ten Hulscher et al., 1992)
14.2 at 25 °C (Shiu and Mackay, 1997)
3.6 at 10 °C (Koelmans et al., 1999)

Ionization potential (eV):
8.9, 9.11, 9.23 (Lias et l., 1998)

Bioconcentration factor, log BCF:
3.89 (rainbow trout, Oliver and Niimi, 1985)
5.46 (Atlantic croakers), 5.70 (blue crabs), 4.68 (spotted sea trout), 5.30 (blue catfish) (Pereira
    et al., 1988)
3.72 (*Oncorchynchus mykiss*, Devillers et al., 1996)
4.86 (guppy, Könemann and van Leeuwen, 1980)

3.70 (rainbow trout, Banerjee et al., 1984)
3.38 (fathead minnow, Carlson and Kosian, 1987)
3.36 (guppy, Van Hoogan and Opperhuizen, 1988)
4.28 (pond snail, Legierse et al., 1998)

Soil sorption coefficient, log K_{oc}:
5.4 (average of 5 suspended sediment samples from the St. Clair and Detroit Rivers, Lau et al., 1989)
3.47 (average for 5 soils, Kishi et al., 1990)
K_d = 5.4 mL/g on a Cs^+-kaolinite (Haderlein and Schwarzenbach, 1993)
4.27, 4.35, 5.26 (Paya-Perez et al., 1991)
4.26 (lake sediment, Schrap et al., 1994)
5.00 (river sediments, Oliver and Charlton, 1984)
4.46 (Oostvaardersplassen), 4.00 (Ransdorperdie), 4.08 (Noordzeekanaal) (Netherlands sediments, Sijm et al., 1997)
4.07 (sandy soil, Van Gestel and Ma, 1993)
Average K_d values for sorption of 1,2,3,4-tetrachlorobenzene to corundum (α-Al_2O_3) and hematite (α-Fe_2O_3) were 0.0473 and 0.0527 mL/g, respectively (Mader et al., 1997)

Octanol/water partition coefficient, log K_{ow}:
4.37 (Watarai et al., 1982)
4.46 (Könemann et al., 1979)
4.83 at 13 °C, 4.61 at 19 °C, 4.37 at 28 °C, 4.25 at 33 °C (Opperhuizen et al., 1988)
4.635 (de Bruijn et al., 1989)
4.41 (Hammers et al., 1982)
4.55 (Miller et al., 1984)
4.60 (Chiou, 1985; Pereira et al., 1988)
4.94 (Bruggeman et al., 1982)
4.65, 4.53, 4.41, 4.28, and 4.15 at 5, 15, 25, 35, and 45 °C, respectively (Bahadur et al., 1997)

Solubility in organics:
Soluble in acetic acid, ether, and ligroin (Weast, 1986)

Solubility in water:
3.5 ppm at 22 °C (quoted, Verschueren, 1983)
3.27 mg/L at 20-25 °C (Kim and Saleh, 1990)
5.92 mg/L at 25 °C (Banerjee, 1984)
56.5 μmol/L at 25 °C (Miller et al., 1984)
20.0 μmol/L at 25 °C (Yalkowsky et al., 1979)

Vapor pressure (x 10^{-2} mmHg at 25 °C):
2.6 (Bidleman, 1984)
3.0 (Hinckley et al., 1990)
1.5 at 25 °C (extrapolated from vapor pressures determined at higher temperatures, Tesconi and Yalkowsky, 1998)

Environmental fate:
Biological. A mixed culture of soil bacteria or a *Pseudomonas* sp. transformed 1,2,3,4-tetrachlorobenzene to 2,3,4,5-tetrachlorophenol (Ballschiter and Scholz, 1980). After incubation in sewage sludge for 32 d under anaerobic conditions, 1,2,3,4-tetrachlorobenzene did not biodegrade (Kirk et al., 1989). The half-life of 1,2,3,4-tetrachlorobenzene in an anaerobic

enrichment culture was 26.4 h (Beurskens et al., 1993).

Photolytic. Irradiation ($\lambda \geq 285$ nm) of 1,2,3,4-tetrachlorobenzene (1.1-1.2 mM/L) in an acetonitrile-water mixture containing acetone (0.553 mM/L) as a sensitizer gave the following products (% yield): 1,2,3-trichlorobenzene (9.2), 1,2,4-trichlorobenzene (32.6), 1,3-dichloro-benzene (5.2), 1,4-dichlorobenzene (1.5), 2,2',3,3',4,4',5-heptachlorobiphenyl (2.52), 2,2',3,3',-4,5,6'-heptachlorobiphenyl (1.22), 10 hexachlorobiphenyls (3.50), five pentachlorobiphenyls (0.87), dichlorophenyl cyanide, two trichloroacetophenones, trichlorocyanophenol, (trichloro-phenyl)acetonitriles, and 1-(trichlorophenyl)-2-propanone (Choudhry and Hutzinger, 1984). Without acetone, the identified photolysis products (% yield) included 1,2,3-trichlorobenzene (7.8), 1,2,4-trichlorobenzene (26.8), 1,2-dichlorobenzene (0.5), 1,3-dichlorobenzene (0.7), 1,4-dichlorobenzene (30.4), 1,2,3,5-tetrachlorobenzene (2.26), 1,2,4,5-tetrachlorobenzene (0.72), 2,2',3,3',4,4',5-heptachlorobiphenyl (<0.01), and 2,2',3,3',4,5,6'-heptachlorobiphenyl (<0.01) (Choudhry and Hutzinger, 1984). The sunlight irradiation of 1,2,3,4-tetrachlorobenzene (20 g) in a 100-mL borosilicate glass-stoppered Erlenmeyer flask for 56 d yielded 4,280 ppm heptachlorobiphenyl (Uyeta et al., 1976).

When an aqueous solution containing 1,2,3,4-tetrachlorobenzene and a nonionic surfactant micelle (Brij 58, a polyoxyethylene cetyl ether) was illuminated by a photoreactor equipped with 253.7-nm monochromatic UV lamps, no photoisomerization was observed. However, based on photodechlorination of other polychlorobenzenes under similar conditions, it was suggested that tri- and dichlorobenzenes, chlorobenzene, benzene, phenol, hydrogen, and chloride ions are likely to form. The photodecomposition half-life for this reaction, based on the first-order photodecomposition rate of 5.24×10^{-3}/sec, is 2.2 min (Chu and Jafvert, 1994).

Toxicity:

LC_{50} (chronic 28-d) for *Brachydanio rerio* 410 μg/L (van Leeuwen et al., 1990).

LC_{50} (14-d) for *Poecilia reticulata* 803 μg/L (Könemann, 1981).

LC_{50} (96-h) for fathead minnows 1.1 mg/L (Veith et al., 1983), *Poecilia reticulata* 365 μg/L (van Hoogen and Opperhuizen, 1988).

LC_{50} 259 and 345 mg/L (soil porewater concentration) for earthworm (*Eisenia andrei*) and 237 and 497 mg/L (soil porewater concentration) for earthworm (*Lumbricus rubellus*) (Van Gestel and Ma, 1993).

Acute oral LD_{50} for rats 1,167 mg/kg (quoted, RTECS, 1985).

Uses: Dielectric fluids; organic synthesis.

1,2,3,5-TETRACHLOROBENZENE

Synonym: 1,2,3,5-TCB.

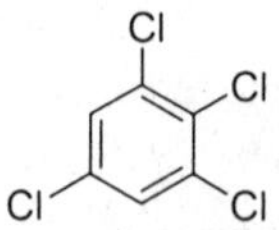

CAS Registry Number: 634-90-2
Molecular formula: $C_6H_2Cl_4$
Formula weight: 215.89
RTECS: DB9445000

Physical state:
Solid

Melting point (°C):
54.5 (Weast, 1986)
50-52 (Verschueren, 1983)

Boiling point (°C):
246 (Weast, 1986)

Diffusivity in water (x 10^{-5} cm²/sec):
0.63 at 20 °C using method of Hayduk and Laudie (1974)

Entropy of fusion (cal/mol·K):
13.5 (Domalski and Hearing, 1998)

Heat of fusion (kcal/mol):
4.379 (Domalski and Hearing, 1998)
4.54 (Miller et al., 1984)

Henry's law constant (x 10^{-4} atm·m³/mol):
9.77 at 20.0 °C (ten Hulscher et al., 1992)
58.0 at 25 °C (Shiu and Mackay, 1997)

Ionization potential (eV):
9.02, 9.16, 9.26 (Lias et al., 1998)

Bioconcentration factor, log BCF:
3.26 (bluegill sunfish, Veith et al., 1980)
5.05 (Atlantic croakers), 5.20 (blue crabs), 4.27 (spotted sea trout), 4.90 (blue catfish) (Pereira et al., 1988)
3.12 (wet weight based), 0.63 (lipid based) (*Gambusia affinis*, Chaisuksant et al., 1997)
3.64 (*Poecilia reticulata*, Devillers et al., 1996)
3.46 (bluegill sunfish, Banerjee et al., 1984)
4.27 (pond snail, Legierse et al., 1998)

Soil sorption coefficient, log K_{oc}:
6.0 (average using 2 suspended sediment samples from the St. Clair and Detroit Rivers, Lau et

al., 1989)

3.49 (Rothamsted Farm soil, Lord et al., 1980)

K_d = 7.2 mL/g on a Cs^+-kaolinite (Haderlein and Schwarzenbach, 1993)

4.23, 4.27, 5.19 (Paya-Perez et al., 1991)

Octanol/water partition coefficient, log K_{ow}:

4.46 (Veith et al., 1980)

4.658 (de Bruijn et al., 1989)

4.53 (Hammers et al., 1982)

4.65 (Miller et al., 1984)

4.50 (Könemann et al., 1979)

4.59 (Chiou, 1985; Pereira et al., 1988)

4.80, 4.67, 4.55, 4.42, and 4.33 at 5, 15, 25, 35, and 45 °C, respectively (Bahadur et al., 1997)

Solubility in organics:

Soluble in alcohol, benzene, ether, and ligroin (Weast, 1986)

Solubility in water:

19.055 μmol/L at 20 °C (Veith et al., 1980)

2.4 ppm at 22 °C (quoted, Verschueren, 1983)

2.79 mg/L at 20-25 °C (Kim and Saleh, 1990)

5.19 mg/L at 25 °C (Banerjee, 1984)

13.4 μmol/L at 25 °C (Miller et al., 1984)

16.2 μmol/L at 25 °C (Yalkowsky et al., 1979)

1.70, 2.43, 3.44, 3.79, 5.08, and 7.03 μg/L at 5, 15, 25, 27, 35, and 45 °C, respectively (Shiu et al., 1997)

Vapor pressure (mmHg):

0.06 at 25 °C (extrapolated from vapor pressures determined at higher temperatures, Tesconi and Yalkowsky, 1998)

0.07 at 25 °C (extrapolated, Mackay et al., 1982)

Environmental fate:

Biological. A mixed culture of soil bacteria or a *Pseudomonas* sp. transformed 1,2,3,5-tetrachlorobenzene to 2,3,4,6-tetrachlorophenol (Ballschiter and Scholz, 1980). The half-life of 1,2,3,5-tetrachlorobenzene in an anaerobic enrichment culture was 1.8 d (Beurskens et al., 1993).

Photolytic. Irradiation (λ $\geq$285 nm) of 1,2,3,5-tetrachlorobenzene (1.1-1.2 mM/L) in an acetonitrile-water mixture containing acetone (0.553 mM/L) as a sensitizer gave the following products (% yield): 1,2,3-trichlorobenzene (5.3), 1,2,4-trichlorobenzene (4.9), 1,3,5-trichlorobenzene (49.3), 1,3-dichlorobenzene (1.8), 2,3,4,4',5,5',6-heptachlorobiphenyl (1.41), 2,2',3,4,4',6,6'-heptachlorobiphenyl (1.10), 2,2',3,3',4,5',6-heptachlorobiphenyl (4.50), four hexachlorobiphenyls (4.69), one pentachlorobiphenyl (0.64), trichloroacetophenone, 1-(trichlorophenyl)-2-propanone, and (trichlorophenyl)acetonitrile (Choudhry and Hutzinger, 1984). Without acetone, the identified photolysis products (% yield) included 1,2,3-trichlorobenzene (trace), 1,2,4-trichlorobenzene (24.3), 1,3,5-trichlorobenzene (11.7), 1,3-dichlorobenzene (0.5), 1,4-dichlorobenzene (3.3), 1,2,3,4,5-pentachlorobenzene (1.43), 1,2,3,4-tetrachlorobenzene (5.99), two heptachlorobiphenyls (1.40), two hexachlorobiphenyls (<0.01), and one pentachlorobiphenyl (0.75) (Choudhry and Hutzinger, 1984).

When an aqueous solution containing 1,2,3,5-tetrachlorobenzene and a nonionic surfactant micelle (Brij 58, a polyoxyethylene cetyl ether) was illuminated by a photoreactor equipped with 253.7-nm monochromatic UV lamps, no photoisomerization was observed. However, based on photodechlorination of other polychlorobenzenes under similar conditions, it was suggested that tri- and dichlorobenzenes, chlorobenzene, benzene, phenol, hydrogen, and chloride ions are likely to form. The photodecomposition half-life for this reaction, based on the first-order photodecomposition rate of 2.67×10^{-3}/sec, is 4.3 min (Chu and Jafvert, 1994).

Toxicity:

LC_{50} (14-d) for *Poecilia reticulata* 803 μg/L (Könemann, 1981).

LC_{50} (96-h) for bluegill sunfish 6.4 mg/L (Spehar et al., 1982), *Cyprinodon variegatus* 3.7 ppm using natural seawater (Heitmuller et al., 1981).

LC_{50} (72-h) for *Cyprinodon variegatus* 4.7 ppm (Heitmuller et al., 1981).

LC_{50} (48-h) for *Daphnia magna* 9.7 mg/L (LeBlanc, 1980), *Cyprinodon variegatus* 5.6 ppm (Heitmuller et al., 1981).

LC_{50} (24-h) for *Daphnia magna* 18 mg/L (LeBlanc, 1980), *Cyprinodon variegatus* >7.5 ppm (Heitmuller et al., 1981).

Acute oral LD_{50} for rats 1,727 mg/kg (quoted, RTECS, 1985).

Heitmuller et al. (1981) reported a NOEC of 1.0 ppm.

Uses: Organic synthesis.

1,2,4,5-TETRACHLOROBENZENE

Synonyms: *sym*-Tetrachlorobenzene; RCRA waste number U207.

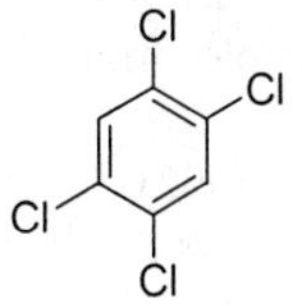

CAS Registry Number: 95-94-3
Molecular formula: $C_6H_2Cl_4$
Formula weight: 215.89
RTECS: DB9450000

Physical state, color, and odor:
White, odorless flakes or needles

Melting point (°C):
139-140 (Weast, 1986)

Boiling point (°C):
243-246 (Weast, 1986)

Density (g/cm³):
1.858 at 21/4 °C (Verschueren, 1983)
1.734 at 10/4 °C (Bailey and White, 1965)

Diffusivity in water (x 10^{-5} cm²/sec):
0.63 at 20 °C using method of Hayduk and Laudie (1974)

Flash point (°C):
155 (Hawley, 1981)

Entropy of fusion (cal/mol·K):
14.0 (Miller et al., 1984)
14.4 (Domalski and Hearing, 1998)

Heat of fusion (kcal/mol):
5.9607 (Domalski and Hearing, 1998)
5.76 (Miller et al., 1984)

Henry's law constant (x 10^{-2} atm·m³/mol):
0.10 at 20 °C (Oliver, 1985)

Ionization potential (eV):
9.0 (Lias, 1998)

Bioconcentration factor, log BCF:
3.61 (*Jordanella floridae*, Devillers et al., 1996)
5.05 (Atlantic croakers), 5.20 (blue crabs), 4.27 (spotted sea trout), 4.90 (blue catfish) (Pereira et al., 1988)

Soil sorption coefficient, log K_{oc}:
3.72, 3.89, 3.91 (Schwarzenbach and Westall, 1981)
6.1 (average of 5 suspended sediment samples from the St. Clair and Detroit Rivers, Lau et al., 1989)
2.79 (McLaurin sandy loam, Walton et al., 1992)
5.10 (river sediments, Oliver and Charlton, 1984)

Octanol/water partition coefficient, log K_{ow}:
4.56 (Watarai et al., 1982)
4.67 (Kenaga and Goring, 1980)
4.52 (Hammers et al., 1982; Könemann et al., 1979)
4.604 (de Bruijn et al., 1989)
4.51 (Miller et al., 1984)
4.70 (Chiou, 1985; Pereira et al., 1988)

Solubility in organics:
Soluble in benzene, chloroform, and ether (Weast, 1986)

Solubility in water:
10.9 μmol/L at 25 °C (Miller et al., 1984)
0.3 ppm at 22 °C (quoted, Verschueren, 1983)
0.56 mg/L at 20-25 °C (Kim and Saleh, 1990)
0.6 mg/L (Geyer et al., 1980)
0.465 mg/L at 25 °C (Banerjee, 1984)
2.75 μmol/L at 25 °C (Yalkowsky et al., 1979)
208, 322, 528, 543, 739, and 1,127 μg/L at 5, 15, 25, 27, 35, and 45 °C, respectively (Shiu et al., 1997)

Vapor pressure (x 10^{-3} mmHg):
5 at 25 °C (extrapolated, Mackay et al., 1982)

Environmental fate:
Biological. A mixed culture of soil bacteria or a *Pseudomonas* sp. transformed 1,2,4,5-tetrachlorobenzene to 2,3,5,6-tetrachlorophenol (Ballschiter and Scholz, 1980). After incubation in sewage sludge for 32 d under anaerobic conditions, 1,2,4,5-tetrachlorobenzene did not biodegrade (Kirk et al., 1989).

Photolytic. Irradiation ($\lambda \geq 285$ nm) of 1,2,4,5-tetrachlorobenzene (1.1-1.2 mM/L) in an acetonitrile-water mixture containing acetone (0.553 mM) as a sensitizer gave the following products (% yield): 1,2,4-trichlorobenzene (25.3), 1,3-dichlorobenzene (8.1), 1,4-dichloro-benzene (3.6), 2,2′,3,4′,5,5′,6-heptachlorobiphenyl (4.19), four hexachlorobiphenyls (6.78), four pentachlorobiphenyls (2.33), one tetrachlorobiphenyl (0.32), 2,4,5-trichloroacetophenone, and (2,4,5-trichlorophenyl)acetonitrile (Choudhry and Hutzinger, 1984). Without acetone, the identified photolysis products (% yield) included 1,2,4-trichlorobenzene (27.7), 1,3-dichlorobenzene (0.3), 1,4-dichlorobenzene (8.5), 1,2,3,4,5-pentachlorobenzene (trace), 1,2,3,4-tetrachlorobenzene (0.45), 1,2,3,5-tetrachlorobenzene (1.11), 2,2′,3,4′,5,5′,6-heptahlorobi-phenyl (1.24), three hexachlorobiphenyls (1.19), and four pentachlorobiphenyls (0.56) (Choudhry and Hutzinger, 1984). The sunlight irradiation of 1,2,4,5-tetrachlorobenzene (20 g) in a 100-mL borosilicate glass-stoppered Erlenmeyer flask for 28 d yielded 26 ppm heptachlorobiphenyl (Uyeta et al., 1976).

When an aqueous solution containing 1,2,4,5-tetrachlorobenzene and a nonionic surfactant micelle (Brij 58, a polyoxyethylene cetyl ether) was illuminated by a photoreactor equipped with 253.7-nm monochromatic UV lamps, no photoisomerization was observed. However, based on photodechlorination of other polychlorobenzenes under similar conditions, it was suggested that tri- and dichlorobenzenes, chlorobenzene, benzene, phenol, hydrogen, and chloride ions are likely to form. The photodecomposition half-life for this reaction, based on the first-order photodecomposition rate of 1.88×10^{-3}/sec, is 6.1 min (Chu and Jafvert, 1994).

Chemical/Physical. Based on an assumed base-mediated 1% disappearance after 16 d at 85 °C and pH 9.70 (pH 11.26 at °C), the hydrolysis half-life was estimated to be >900 yr (Ellington et al., 1988).

Toxicity:

LC_{50} (14-d) for *Poecilia reticulata* 304 μg/L (Könemann, 1981).

LC_{50} (96-h) for bluegill sunfish 1.6 mg/L (Spehar et al., 1982), *Cyprinodon variegatus* 900 ppb using natural seawater (Heitmuller et al., 1981).

LC_{50} (72-h) for *Cyprinodon variegatus* 800 ppb (Heitmuller et al., 1981).

LC_{50} (48-h) for *Daphnia magna* >530 mg/L (LeBlanc, 1980), *Cyprinodon variegatus* 900 ppb (Heitmuller et al., 1981).

LC_{50} (24-h) for *Daphnia magna* >530 mg/L (LeBlanc, 1980), *Cyprinodon variegatus* >1.8 ppm (Heitmuller et al., 1981).

Acute oral LD_{50} for rats 1,500 mg/kg, mice 1,035 mg/kg (quoted, RTECS, 1985).

Heitmuller et al. (1981) reported a NOEC of 300 ppb.

Uses: Insecticides; intermediate for herbicides and defoliants; electrical insulation; impregnant for moisture resistance.

1,1,2,2-TETRACHLOROETHANE

Synonyms: Acetosol; Acetylene tetrachloride; Bonoform; Cellon; 1,1-Dichloro-2,2-dichloro-ethane; Ethane tetrachloride; NCI-C03554; RCRA waste number U208; TCE; Tetrachlor-ethane; Tetrachloroethane; *sym*-Tetrachloroethane; UN 1702; Westron.

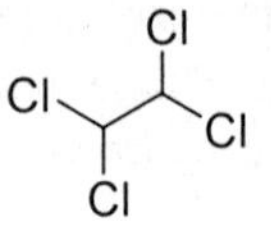

CAS Registry Number: 79-34-5
DOT: 1702
DOT label: Poison
Molecular formula: $C_2H_2Cl_4$
Formula weight: 167.85
RTECS: KI8575000
Merck reference: 10, 9016

Physical state, color, and odor:
Colorless to pale yellow liquid with a sweet, chloroform-like odor. Odor threshold concentration is 500 ppb (Keith and Walters, 1992).

Melting point (°C):
-36 (Weast, 1986)
-42.5 (Standen, 1964)

Boiling point (°C):
146.2 (Weast, 1986)
145.1 (Riddick et al., 1986)

Density (g/cm³):
1.5953 at 20/4 °C (Weast, 1986)
1.60255 at 15/4 °C, 1.5869 at 25/4 °C, 1.57860 at 30/4 °C (Standen, 1964)
1.5745 at 35.00/4 °C, 1.5642 at 45.00/4 °C (Sastry et al., 1999)

Diffusivity in water (x 10⁻⁵ cm²/sec):
0.86 at 20 °C using method of Hayduk and Laudie (1974)

Flash point (°C):
Noncombustible liquid (NIOSH, 1997)

Henry's law constant (x 10⁻⁴ atm·m³/mol):
3.8 (Pankow and Rosen, 1988)
5.00 at 25 °C (Wright et al., 1992)
7.1 at 37 °C (Sato and Nakajima, 1979)
3.61 at 22.0 °C (mole fraction ratio, Leighton and Calo, 1981)
7 at 30 °C (Jeffers et al., 1989)
3.3, 2.0, 7.3, 2.5, and 7.0 at 10, 15, 20, 25, and 30 °C, respectively (Ashworth et al., 1988)
3, 5, 6, and 9 at 20, 30, 35, and 40 °C, respectively (Tse et al., 1992)
3.40 at 20 °C (Hovorka and Dohnal, 1997)

Ionization potential (eV):
11.10 (Rosenstock et al., 1998)

Bioconcentration factor, log BCF:
0.90 (bluegill sunfish, Veith et al., 1980)

Soil sorption coefficient, log K_{oc}:
1.90 (Willamette silt loam, Chiou et al., 1979)

Octanol/water partition coefficient, log K_{ow}:
2.56 (quoted, Mills et al., 1985)
2.39 (Banerjee et al., 1980; Yoshida et al., 1983)

Solubility in organics:
Soluble in acetone, ethanol, ether, benzene, carbon tetrachloride, petroleum ether, carbon disulfide, *N,N*-dimethylformamide, and oils (U.S. EPA, 1985). Miscible with alcohol and chloroform (Meites, 1963).

Solubility in water:
2,970 mg/L at 25 °C (Banerjee et al., 1980)
3,230 mg/L at 20 °C (Chiou et al., 1979)
2,900 mg/kg at 25 °C (McGovern, 1943)
0.29 mass % at 20 °C (Konietzko, 1984)
0.296 wt % at 23.5 °C (Schwarz, 1980)
0.372, 0.385, and 0.367 wt % at 10.0, 20.0, and 30.0 °C, respectively (Schwarz and Miller, 1980)
2,962 mg/L at 25 °C (Howard, 1990)
2,915 mg/L at 30 °C (McNally and Grob, 1984)
In wt %: 0.317 at 0 °C, 0.290 at 9.5 °C, 0.291 at 20.0 °C, 0.292 at 29.7 °C, 0.301 at 39.6 °C, 0.316 at 50.1 °C, 0.357 at 61.0 °C, 0.385 at 70.5 °C, 0.425 at 80.6 °C, 0.474 at 90.8 °C (Stephenson, 1992)

Vapor density:
5 kg/m^3 at the boiling point (Konietzko, 1984)
6.86 g/L at 25 °C, 5.79 (air = 1)

Vapor pressure (mmHg):
5 at 20 °C, 8.5 at 30 °C (Verschueren, 1983)
6.5 at 25 °C (Mackay et al., 1982)

Environmental fate:
Biological. Monodechlorination by microbes under laboratory conditions produced 1,1,2-trichloroethane (Smith and Dragun, 1984). In a static-culture-flask screening test, 1,1,2,2-tetrachloroethane (5 and 10 mg/L) was statically incubated in the dark at 25 °C with yeast extract and settled domestic wastewater inoculum. No significant degradation was observed after 28 d of incubation (Tabak et al., 1981).
Chemical/Physical. In an aqueous solution containing 0.100 M phosphate-buffered distilled water, 1,1,2,2-tetrachloroethane was abiotically transformed to 1,1,2-trichloroethane. This reaction was investigated over a temperature range of 30-95 °C at various pHs (5-9) (Cooper et al., 1987). Abiotic dehydrohalogenation of 1,1,2,2-tetrachloroethane yielded 1,1,1-trichloroethylene (Vogel et al., 1987) and hydrochloric acid (Kollig, 1993). The half-life for this reaction

at 20 °C was reported to be 0.8 yr (Vogel et al., 1987). Under alkaline conditions, 1,1,2,2-tetrachloroethane dehydrohalogenated to trichloroethylene. The reported hydrolysis half-life of 1,1,2,2-tetrachloroethane in water at 25 °C and pH 7 is 146 d (Jeffers et al., 1989).

The evaporation half-life of 1,1,2,2-tetrachloroethane (1 mg/L) from water at 25 °C using a shallow-pitch propeller stirrer at 200 rpm at an average depth of 6.5 cm is 55.2 min (Dilling, 1977).

At influent concentrations of 1.0, 0.1, 0.01, and 0.001 mg/L, the granular activated carbon adsorption capacities at pH 5.3 were 11, 4.5, 1.9, and 0.8 mg/g, respectively (Dobbs and Cohen, 1980).

Exposure limits: Potential occupational carcinogen. NIOSH REL: TWA 1 ppm (7 mg/m^3), IDLH 100 ppm; OSHA PEL: TWA 5 ppm (35 mg/m^3); ACGIH TLV: TWA 1 ppm (adopted).

Symptoms of exposure: An irritation concentration of 1,302.00 mg/m^3 in air was reported by Ruth (1986).

Toxicity:

LC_{50} (contact) for earthworm (*Eisenia fetida*) 14 μg/cm^2 (Neuhauser et al., 1985).

LC_{50} (7-d) for *Poecilia reticulata* 36.7 mg/L (Könemann, 1981).

LC_{50} (96-h) for fathead minnows 20.3 mg/L (Veith et al., 1983), bluegill sunfish 21 mg/L (Spehar et al., 1982), *Cyprinodon variegatus* 12 ppm using natural seawater (Heitmuller et al., 1981).

LC_{50} (72-h) for *Cyprinodon variegatus* 13 ppm (Heitmuller et al., 1981).

LC_{50} (48-h) values of 23 and 25 mg/L for unfed and fed *Daphnia magna*, respectively (Richter et al., 1983); *Daphnia magna* 9.3 mg/L (LeBlanc, 1980), *Cyprinodon variegatus* 16 ppm (Heitmuller et al., 1981).

LC_{50} (24-h) for *Daphnia magna* 18 mg/L (LeBlanc, 1980), *Cyprinodon variegatus* 19 ppm (Heitmuller et al., 1981).

Acute oral LD_{50} for rats 800 mg/kg (quoted, RTECS, 1985).

Heitmuller et al. (1981) reported a NOEC of <8.8 ppm.

Drinking water standard: No MCLGs or MCLs have been proposed although 1,1,2,2-tetrachloroethane has been listed for regulation (U.S. EPA, 1996).

Uses: Solvent for chlorinated rubber; insecticide and bleach manufacturing; paint, varnish and rust remover manufacturing; degreasing, cleansing, and drying of metals; denaturant for ethyl alcohol; preparation of 1,1-dichloroethylene; extractant and solvent for oils and fats; insecticides; weed killer; fumigant; intermediate in the manufacturing of other chlorinated hydrocarbons; herbicide.

TETRACHLOROETHYLENE

Synonyms: Ankilostin; Antisol 1; Carbon bichloride; Carbon dichloride; Dee-Solv; Didakene; Dow-per; ENT 1860; Ethylene tetrachloride; Fedal-UN; NCI-C04580; Nema; PCE; Perawin; PERC; Perchlor; Perchlorethylene; Perchloroethylene; Perclene; Perclene D; Percosolv; Perk; Perklone; Persec; RCRA waste number U210; Tetlen; Tetracap; Tetrachlorethylene; **Tetrachloroethene**; 1,1,2,2-Tetrachloroethylene; Tetraleno; Tetralex; Tetravec; Tetroguer; Tetropil; UN 1897.

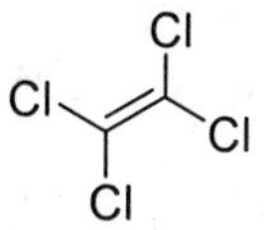

CAS Registry Number: 127-18-4
DOT: 1897
Molecular formula: C_2Cl_4
Formula weight: 165.83
RTECS: KX3850000
Merck reference: 10, 9017

Physical state, color, and odor:
Colorless liquid with a chloroform or sweet, ethereal odor. Odor threshold concentration is 27 (Amoore and Hautala, 1983).

Melting point (°C):
-19 (Weast, 1986)
-22.35 (McGovern, 1943)

Boiling point (°C):
121.2 (Dean, 1973)

Density (g/cm³):
1.623 at 20/4 °C (McGovern, 1943)
1.63109 at 15/4 °C, 1.62260 at 20/4 °C, 1.60640 at 30/4 °C (Standen, 1964)

Diffusivity in water (x 10⁻⁵ cm²/sec):
0.87 at 20 °C using method of Hayduk and Laudie (1974)

Entropy of fusion (cal/mol·K):
10.37 (Novoselova et al., 1986)

Heat of fusion (kcal/mol):
2.6003 (Novoselova et al., 1986)

Henry's law constant (x 10⁻³ atm·m³/mol):
15.3 (Pankow and Rosen, 1988)
2.87 at 25 °C (Warner et al., 1987)
20.0 at 20 °C (Pearson and McConnell, 1975)
59.2 at 37 °C (Sato and Nakajima, 1979)
17.5 at 25 °C (Gossett, 1987)

15.6 at 25 °C (Hoff et al., 1993)

17.0, 23.1, 30.3, and 38.1 at 25, 30, 40, and 45 °C, respectively (Robbins et al., 1993)

14.6 at 20 °C (Roberts et al., 1985)

20.8 at 25 °C (Tancréde and Yanagisawa, 1990)

24.5 at 30 °C (Jeffers et al., 1989)

8.46, 11.1, 14.1, 17.1, and 24.5 at 10, 15, 20, 25, and 30 °C, respectively (Ashworth et al., 1988)

Distilled water: 3.85, 5.19, 6.27, 10.07, and 14.72 at 2.0, 6.0, 10.0, 18.2, and 25.0 °C, respectively; natural seawater: 7.07 and 17.8 at 6.0 and 25.0 °C, respectively (Dewulf et al., 1995)

11.9 at 20 °C (Hovorka and Dohnal, 1997)

16.4 at 25.3 °C (mole fraction ratio, Leighton and Calo, 1981)

At 25 °C (NaCl concentration, mol/L): 12.66 (0), 13.44 (0.1), 14.79 (0.3), 16.45 (0.5), 18.04 (0.7), 20.93 (1.0); salinity, NaCl = 37.34 g/L (°C): 13.32 (15), 17.91 (20), 21.61 (25), 26.03 (30), 31.12 (35), 41.80 (40), 51.27 (45) (Peng and Wan, 1998).

8.36, 1.43, and 2.35 at 10, 20, and 30 °C, respectively (Munz and Roberts, 1987)

19.91, 33.50, 47.04, and 68.13 at 30, 40, 50, and 60 °C, respectively (Vane and Giroux, 2000)

Interfacial tension with water (dyn/cm at 20 °C):
47.48 (Demond and Lindner, 1993)

Ionization potential (eV):
9.326 (Lias, 1998)
9.71 (Yoshida et al., 1983a)

Bioconcentration factor, log BCF:
1.69 (bluegill sunfish, Veith et al., 1980)
1.54 (trout, Neely et al., 1974)
1.79 (rainbow trout, Saito et al., 1992)
2.54, 2.34, 2.32, 2.54, 2.64, 3.73, and 2.62 for olive, grass, holly, mock orange, pine, rosemary, and juniper leaves, respectively (Hiatt, 1998)

Soil sorption coefficient, log K_{oc}:
2.35 (Piwoni and Banerjee, 1989)
2.35 (Lincoln sand, Wilson et al., 1981)
2.38 (Friesel et al., 1984)
2.42 (Abdul et al., 1987)
2.43 (Agawam fine sandy loam, Pignatello, 1990)
2.43 (Tampa sandy aquifer material, Brusseau and Rao, 1991)
2.56 (Willamette silt loam, Chiou et al., 1979)
2.25, 2.31, 2.54 (various Norwegian soils, Seip et al., 1986)
2.63 (Catlin soil, Roy et al., 1985)
2.65 (silty clay), 2.55, 2.99 (coarse sand) (Pavlostathis and Mathavan, 1992)
2.57 (Schwarzenbach and Westall, 1981)
3.04 (Detroit River sediment, Jepsen and Lick, 1999)
2.25 (Mt. Lemmon soil, Hu et al., 1995)
2.62, 2.79, 2.74, 2.80, 2.85, 2.78, and 2.83 at 2.3, 3.8, 6.2, 8.0, 13.5, 18.6, and 25.0 °C, respectively, for a Leie River (Belgium) clay (Dewulf et al., 1999a)
2.28 (Eustis sand, Brusseau et al., 1990)

Octanol/water partition coefficient, log K_{ow}:
2.10 (Banerjee et al., 1980)

2.53 (Veith et al., 1980)
2.60 (Hansch and Leo, 1979)

Solubility in organics:
Miscible with many organic chlorinated solvents including methylene chloride, carbon tetra-chloride, chloroform, trichloroethylene, 1,1,1-trichloroethane, 1,1,2-trichloroethane, and 1,1,2,2-tetrachloroethane

Solubility in water:
150 mg/L at 20 °C (Pearson and McConnell, 1975)
150 mg/L at 22 °C (Jepsen and Lick, 1999)
2,200 mg/L at 20 °C (Chiou et al., 1977)
150 mg/kg at 25 °C (McGovern, 1943)
485 mg/L at 25 °C (Banerjee et al., 1980)
240 mg/L at 23-24 °C (Broholm and Feenstra, 1995)
2.78×10^{-5} at 25 °C (mole fraction, Li et al., 1993)
In wt %: 0.0273 at 0 °C, 0.0270 at 9.5 °C, 0.0286 at 19.5 °C, 0.0221 at 31.1 °C, 0.0213 at 40.0 °C, 0.0273 at 50.1 °C, 0.0304 at 61.3 °C, 0.0377 at 71.0 °C, 0.0380 at 80.2 °C, 0.0523 at 91.8 °C (Stephenson, 1992)

Vapor density:
6.78 g/L at 25 °C, 5.72 (air = 1)

Vapor pressure (mmHg):
24 at 30 °C (Verschueren, 1983)
18.6 at 25 °C (Mackay et al., 1982)
20 at 25 °C (Valsaraj, 1988)
14 at 20 °C (McConnell et al., 1975)

Environmental fate:
Biological. Sequential dehalogenation by microbes under laboratory conditions produced trichloroethylene, *cis*-1,2-dichloroethylene, *trans*-1,2-dichloroethylene, and vinyl chloride (Smith and Dragun, 1984). A microcosm composed of aquifer water and sediment collected from uncontaminated sites in the Everglades biotransformed tetrachloroethylene to *cis*- and *trans*-1,2-dichloroethylene (Parsons and Lage, 1985). Microbial degradation to trichloro-ethylene under anaerobic conditions or using mixed cultures was also reported (Vogel et al., 1987).

In a continuous-flow mixed-film methanogenic column study, tetrachloroethylene degraded to trichloroethylene with traces of vinyl chloride, dichloroethylene isomers, and carbon dioxide (Vogel and McCarty, 1985). In a static-culture-flask screening test, tetrachloroethylene (5 and 10 mg/L) was statically incubated in the dark at 25 °C with yeast extract and settled domestic wastewater inoculum. Significant degradation with gradual adaptation was observed after 28 d of incubation. The amount lost due to volatilization after 10 d was 16-23% (Tabak et al., 1981).

Surface Water. Estimated half-lives of tetrachloroethylene (3.2 μg/L) from an experimental marine mesocosm during the spring (8-16 °C), summer (20-22 °C), and winter (3-7 °C) were 28, 13, and 15 d, respectively (Wakeham et al., 1983).

Photolytic. Photolysis in the presence of nitrogen oxides yielded phosgene (carbonyl chloride) with minor amounts of carbon tetrachloride, dichloroacetyl chloride, and trichloroacetyl chloride (Howard, 1990). In sunlight, photolysis products reported include chlorine, hydrogen chloride, and trichloroacetic acid. Tetrachloroethylene reacts with ozone to produce a mixture of phosgene and trichloroacetyl chloride with a reported half-life of 8 d

(Fuller, 1976). Reported photooxidation products in the troposphere include trichloroacetyl chloride and phosgene (Andersson et al., 1975; Gay et al., 1976; U.S. EPA, 1975). Phosgene is hydrolyzed readily to hydrogen chloride and carbon dioxide (Morrison and Boyd, 1971).

Reported rate constants for the reaction of tetrachloroethylene and OH radicals in the atmosphere are: 1.0×10^{11} cm^3/mol·sec at 300 K (Hendry and Kenley, 1979), 1.70×10^{-13} cm^3/molecule·sec (Atkinson, 1985); with ozone: $<2 \times 10^{-23}$ cm^3/molecule·sec (Mathias et al., 1974); with NO_3: $>6.19 \times 10^{-17}$ cm^3/molecule·sec (Atkinson et al., 1988).

Chemical/Physical. The experimental half-life for hydrolysis of tetrachloroethylene in water at 25 °C is 8.8 months (Dilling et al., 1975).

The volatilization half-life of tetrachloroethylene (1 mg/L) from water at 25 °C using a shallow-pitch propeller stirrer at 200 rpm at an average depth of 6.5 cm is 25.4 min (Dilling, 1977).

At influent concentrations of 1.0, 0.1, 0.01, and 0.001 mg/L, the granular activated carbon adsorption capacities at pH 5.3 were 51, 14.0, 3.9, and 1.1 mg/g, respectively (Dobbs and Cohen, 1980).

At elevated temperatures, tetrachloroethylene decomposes to hydrogen chloride and phosgene (NIOSH, 1997).

Exposure limits: Potential occupational carcinogen. NIOSH REL: IDLH 150 ppm; OSHA PEL: TWA 100 ppm, ceiling 200 ppm, 5-min/3-h peak 300 ppm; ACGIH TLV: TWA 25 ppm, STEL 100 ppm (adopted).

Symptoms of exposure: May cause headache, dizziness, drowsiness, incoordination, irritation of eyes, nose, and throat. Narcotic at high concentrations (Patnaik, 1992). An irritation concentration of 710.20 mg/m^3 in air was reported by Ruth (1986).

Toxicity:

LC_{50} (7-d) for *Poecilia reticulata* 69 mg/L (Könemann, 1981).

LC_{50} (96-h) for fathead minnows 13.5 mg/L (Veith et al., 1983), 18.4 mg/L (flow-through), 21.4 mg/L (static bioassay) (Alexander et al., 1978), bluegill sunfish 13 mg/L (Spehar et al., 1982), *Cyprinodon variegatus* 29-52 ppm using natural seawater (Heitmuller et al., 1981).

LC_{50} (48-h) for *Daphnia magna* 18 mg/L (Yoshioka et al., 1986), *Cyprinodon variegatus* >52 ppm (Heitmuller et al., 1981).

LC_{50} (24-h) for *Daphnia magna* 18 mg/L (Yoshioka et al., 1986), *Cyprinodon variegatus* >52 ppm (Heitmuller et al., 1981).

LC_{50} (4-h inhalation) for mice 5,200 ppm (Friberg et al., 1953).

LC_{50} for red killifish 263 mg/L (Yoshioka et al., 1986).

Acute oral LD_{50} for mice 8,100 mg/kg, rats 3,005 mg/kg (quoted, RTECS, 1985).

Heitmuller et al. (1981) reported a NOEC of 29 ppm.

Drinking water standard (final): MCLG: zero; MCL: 5 μg/L (U.S. EPA, 1996).

Uses: Dry cleaning fluid; degreasing and drying metals and other solids; solvent for waxes, greases, fats, oils, gums; manufacturing printing inks and paint removers; preparation of fluorocarbons and trichloroacetic acid; vermifuge; heat-transfer medium; organic synthesis.

TETRAETHYL PYROPHOSPHATE

Synonyms: Bis-*O,O*-diethylphosphoric anhydride; Bladan; **Diphosphoric acid tetraethyl ester**; ENT 18771; Ethyl pyrophosphate; Fosvex; Grisol; Hept; Hexamite; Killax; Kilmite 40; Lethalaire G-52; Lirohex; Mortopal; NA 2783; Nifos; Nifos T; Nifost; Pyrophosphoric acid tetraethyl ester; RCRA waste number P111; TEP; TEPP; Tetraethyl diphosphate; Tetraethyl pyrofosfaat; Tetrastigmine; Tetron; Tetron-100; Vapotone.

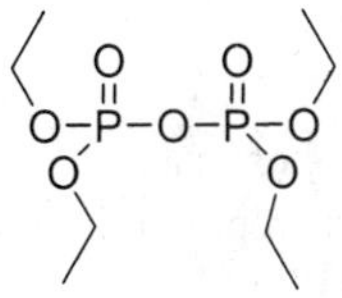

CAS Registry Number: 107-49-3
DOT: 2784
Molecular formula: $C_8H_{10}O_7P_2$
Formula weight: 290.20
RTECS: UX6825000
Merck reference: 10, 9030

Physical state, color, and odor:
Colorless to amber liquid with an agreeable, fruity odor

Melting point (°C):
0 (NIOSH, 1997)

Boiling point (°C):
Decomposes at 170-213 °C releasing ethylene (Windholz et al., 1983).
135-138 at 1 mmHg (Verschueren, 1983)

Density (g/cm³):
1.185 at 20/4 °C (Windholz et al., 1983)

Diffusivity in water:
Not applicable - reacts with water

Flash point (°C):
Noncombustible liquid (NIOSH, 1997)

Soil sorption coefficient, log K_{oc}:
Not applicable - reacts with water

Octanol/water partition coefficient, log K_{ow}:
Not applicable - reacts with water

Solubility in organics:
Miscible with acetone, benzene, carbon tetrachloride, chloroform, ethanol, ethylene glycol, glycerol, methanol, propylene glycol, toluene, and xylene (Windholz et al., 1983)

Solubility in water:
Miscible

Vapor density:
11.86 g/L at 25 °C, 10.02 (air = 1)

Vapor pressure (x 10⁻⁴ mmHg):
4.7 at 30 °C (Windholz et al., 1983)
1.55 at 20 °C (Verschueren, 1983)

Environmental fate:
   Chemical/Physical. Tetraethyl pyrophosphate is quickly hydrolyzed by water forming pyrophosphoric acid (NIOSH, 1997). The reported hydrolysis half-life at 25 °C and pH 7 is 7.5 h (Ketelaar and Bloksma, 1948; Coates, 1949).
   Decomposes at 170-213 °C releasing large amounts of ethylene (Hartley and Kidd, 1987; Keith and Walters, 1992).

Exposure limits (mg/m³): NIOSH REL: TWA 0.05, IDLH 5; OSHA PEL: TWA 0.05.

Toxicity:
   LC_{50} (48-h) for red killifish 1,020 mg/L (Yoshioka et al., 1986).
   Acute oral LD_{50} for wild birds 1.30 mg/kg, ducks 3.56 mg/kg, guinea pigs 2.3 mg/kg, mice 7 mg/kg, rats 500 μg/kg (quoted, RTECS, 1985).

Uses: Insecticide for mites and aphids; rodenticide.

TETRAHYDROFURAN

Synonyms: Butylene oxide; Cyclotetramethylene oxide; Diethylene oxide; Furanidine; Hydrofuran; NCI-C60560; Oxacyclopentane; Oxolane; RCRA waste number U213; Tetramethylene oxide; THF; UN 2506.

CAS Registry Number: 109-99-9
DOT: 2056
DOT label: Flammable liquid
Molecular formula: C_4H_8O
Formula weight: 72.11
RTECS: LU5950000
Merck reference: 10, 9036

Physical state, color, and odor:
Colorless liquid with an ether-like odor. Odor threshold concentration is 2 ppm (Amoore and Hautala, 1983).

Melting point (°C):
-108 (Weast, 1986)
-65 (Hawley, 1981)

Boiling point (°C):
67 (Weast, 1986)
65.4 (Sax and Lewis, 1987)

Density (g/cm³):
0.8892 at 20/4 °C (Weast, 1986)

Diffusivity in water (x 10^{-5} cm²/sec):
1.00 at 20 °C using method of Hayduk and Laudie (1974)

Flash point (°C):
-14.6 (NIOSH, 1997)
-17.2 (Windholz et al., 1983)

Lower explosive limit (%):
2 (NIOSH, 1997)
1.8 (Sax and Lewis, 1987)

Upper explosive limit (%):
11.8 (NIOSH, 1997)

Entropy of fusion (cal/mol·K):
12.4 (Lebedev et al., 1978)
12.39 (Lebedev and Lityagov, 1977)

Heat of fusion (kcal/mol):
2.041 (Lebedev and Lityagov, 1976; Lebedev et al., 1978)

Henry's law constant (x 10^{-5} atm·m^3/mol at 25 °C):
7.14 (Cabani et al., 1971)
4.55 (Signer et al., 1969)

Ionization potential (eV):
9.40 (Lias, 1998)

Soil sorption coefficient, log K_{oc}:
1.37 and 1.26 for Captina silt loam and McLaurin sandy loam, respectively (Walton et al., 1992)

Octanol/water partition coefficient, log K_{ow}:
0.46 (Hansch and Leo, 1979)

Solubility in organics:
Soluble in alcohols, ketones, esters, ethers, and hydrocarbons (Windholz et al., 1983)

Solubility in water:
Miscible (Fischer and Ehrenberg, 1948; Palit, 1947)
4.2 M at 25 °C (Fischer and Ehrenberg, 1948)

Vapor density:
2.95 g/L at 25 °C, 2.49 (air = 1)

Vapor pressure (mmHg):
83.6 at 10 °C, 131.5 at 20 °C, 197.6 at 30 °C (Verschueren, 1983)
114 at 15 °C (Sax and Lewis, 1987)

Environmental fate:
Photolytic. The rate constant for the reaction of tetrahydrofuran and OH radicals in the atmosphere at 300 K is 8.8 x 10^{12} cm^3/mol·sec (Hendry and Kenley, 1979). Atkinson et al. (1988) reported a rate constant of 4.875 x 10^{-15} cm^3/molecule·sec for the reaction with NO$_3$ radicals in air.

Exposure limits: NIOSH REL: TWA 200 ppm (590 mg/m^3), STEL 250 ppm (735 mg/m^3), IDLH 2,000 ppm; OSHA PEL: TWA 200 ppm; ACGIH TLV: TWA 200 ppm, STEL 250 ppm (adopted).

Symptoms of exposure: Vapors may irritate respiratory tract and eyes. An anesthetic at high concentrations (Patnaik, 1992).

Toxicity:
LC$_{50}$ (96-h) for fathead minnows 2,160 mg/L (Veith et al., 1983).
Acute oral LD$_{50}$ for rats 2,816 mg/kg (quoted, RTECS, 1985).

Source: Leaches from PVC cement used to join tubing (Wang and Bricker, 1979)

Uses: Solvent for uncured rubber and polyvinyl chlorides, vinyl chloride copolymers, vinylidene chloride copolymers, natural resins; topcoating solutions; cellophane; magnetic tapes; adhesives; printing inks; organic synthesis.

1,2,4,5-TETRAMETHYLBENZENE

Synonyms: Durene; Durol; *sym*-1,2,4,5-Tetramethylbenzene.

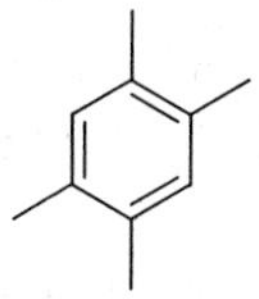

CAS Registry Number: 95-93-2
Molecular formula: $C_{10}H_{14}$
Formula weight: 134.22
RTECS: DC0500000
Merck reference: 10, 3455

Physical state, color, and odor:
Colorless crystals or scales with a camphor-like odor

Melting point (°C):
79.2 (Weast, 1986)
77 (Verschueren, 1983)

Boiling point (°C):
196.8 (Weast, 1986)
191-193 (Windholz et al., 1983)

Density (g/cm³):
0.8380 at 81/4 °C (Weast, 1986)

Diffusivity in water (x 10^{-5} cm²/sec):
0.62 at 20 °C using method of Hayduk and Laudie (1974)

Dissociation constant, pK$_a$:
>14 (Schwarzenbach et al., 1993)

Flash point (°C):
54 (NFPA, 1984)
73 (Dean, 1987)

Entropy of fusion (cal/mol·K):
14.2 (Domalski and Hearing, 1998)

Heat of fusion (kcal/mol):
4.9903 (Domalski and Hearing, 1998)

Henry's law constant (x 10^{-2} atm·m³/mol):
2.49 at 25 °C (approximate - calculated from water solubility and vapor pressure)

Ionization potential (eV):
8.06 (Lias, 1998)

Soil sorption coefficient, log K$_{oc}$:
3.12 (Schwarzenbach and Westall, 1981)

Octanol/water partition coefficient, log K_{ow}:
3.84 (Garst, 1984)
4.00 (Camilleri et al., 1988)

Solubility in organics:
Soluble in acetone, alcohol, benzene, and ether (Weast, 1986)

Solubility in water (at 25 °C):
3.48 mg/kg (Price, 1976)
19.4 mg/L (Deno and Berkheimer, 1960)

Vapor pressure (mmHg):
0.49 at 25 °C (Mackay et al., 1982)

Toxicity:
  Acute oral LD_{50} for rats 6,989 mg/kg (quoted, RTECS, 1985).

Uses: Plasticizers; polymers; fibers; organic synthesis.

TETRANITROMETHANE

Synonyms: NCI-C55947; RCRA waste number P112; Tetan; TNM; UN 1510.

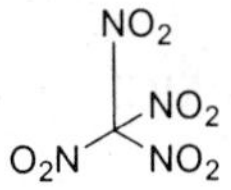

$$NO_2$$
$$O_2N - C(NO_2) - NO_2$$

CAS Registry Number: 509-14-8
DOT: 1510
Molecular formula: CN_4O_8
Formula weight: 196.03
RTECS: PB4025000
Merck reference: 10, 9057

Physical state, color, and odor:
Colorless to pale yellow liquid or solid with a pungent odor

Melting point (°C):
14.2 (Weast, 1986)
12.5 (Hawley, 1981)

Boiling point (°C):
126 (Weast, 1986)

Density (g/cm³):
1.6380 at 20/4 °C (Weast, 1986)
1.6229 at 25/4 °C (Windholz et al., 1983)
1.650 at 13/4 °C (Hawley, 1981)

Diffusivity in water (x 10^{-5} cm²/sec):
0.79 at 20 °C using method of Hayduk and Laudie (1974)

Flash point (°C):
>112 (Dean, 1987)

Ionization potential (eV):
12.55 (Lias and Liebman, 1998)

Solubility in organics:
Miscible with alcohol and ether (Hawley, 1981)

Vapor density:
8.01 g/L at 25 °C, 6.77 (air = 1)

Vapor pressure (mmHg):
8 at 20 °C, 13 at 25 °C, 15 at 30 °C (Verschueren, 1983)

Exposure limits: NIOSH REL: TWA 1 ppm (8 mg/m³), IDLH 4 ppm; OSHA PEL: TWA 5 ppb (adopted).

Toxicity:

Acute oral LD_{50} for rats 130 mg/kg, mice 375 mg/kg (quoted, RTECS, 1985).

LC_{50} (inhalation) for mice 54 ppm/4-h, rats 18 ppm/4-h (quoted, RTECS, 1985).

Drinking water standard: No MCLGs or MCLs have been proposed (U.S. EPA, 1996).

Uses: Laboratory reagent for detecting double bonds in organic compounds; oxidizer in rocket propellants; diesel fuel booster.

TETRYL

Synonyms: *N*-Methyl-*N*,2,4,6-tetranitroaniline; **N-Methyl-N,2,4,6-tetranitrobenzenamine**; Nitramine; Picrylmethylnitramine; Picrylnitromethylamine; Tetralit; Tetralite; Tetril; 2,4,6-Tetryl; Trinitrophenylmethylnitramine; 2,4,6-Trinitrophenylmethylnitramine; 2,4,6-Trinitrophenyl-*N*-methylnitramine; UN 0208.

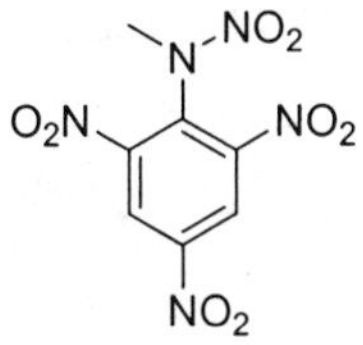

CAS Registry Number: 479-45-8
DOT: 0208
DOT label: Class A explosive
Molecular formula: $C_7H_5N_5O_8$
Formula weight: 287.15
RTECS: BY6300000
Merck reference: 10, 6416

Physical state and color:
Colorless to pale yellow crystals

Melting point (°C):
131-132 (Weast, 1986)

Boiling point (°C):
Explodes at 187 (Weast, 1986)

Density (g/cm³):
1.57 at 10/4 °C (Weast, 1986)
1.57 at 19/4 °C (Verschueren, 1983)

Diffusivity in water (x 10⁻⁵ cm²/sec):
0.64 at 20 °C using method of Hayduk and Laudie (1974)

Flash point (°C):
Explodes (NIOSH, 1997)

Entropy of fusion (cal/mol·K):
13.6 (Krien et al., 1973)

Heat of fusion (kcal/mol):
5.4803 (Krien et al., 1973)
6.1798 (Hall, 1971)

Henry's law constant (x 10⁻³ atm·m³/mol):
<1.89 at 20 °C (approximate - calculated from water solubility and vapor pressure)

Soil sorption coefficient, log K_{oc}:
2.37 using method of Kenaga and Goring (1980)

Octanol/water partition coefficient, log K$_{ow}$:
2.04 using method of Kenaga and Goring (1980)

Solubility in organics:
Soluble in acetone, benzene (Weast, 1986), glacial acetic acid, and ether (Windholz et al., 1983)

Solubility in water:
0.02 wt % at 20 °C (NIOSH, 1997)

Vapor pressure (mmHg):
<1 at 20 °C (NIOSH, 1997)

Environmental fate:
   Chemical/Physical. Produces highly toxic nitrogen oxides on decomposition (Lewis, 1990).

Exposure limits (mg/m^3): NIOSH REL: 1.5, IDLH 750; OSHA PEL: TWA 1.5; ACGIH TLV: TWA 1.5 (adopted).

Uses: As an indicator in analytical chemistry; as a booster in artillery explosives.

THIOPHENE

Synonyms: CP 34; Divinylene sulfide; Huile H50; Huile HSO; Thiacyclopentadiene; Thiaphene; Thiofuran; Thiofuram; Thiofurfuran; Thiole; Thiophen; Thiotetrole; UN 2414; USAF EK-1860.

CAS Registry Number: 110-02-1
DOT: 2414
Molecular formula: C_4H_4S
Formula weight: 84.14
RTECS: XM7350000
Merck reference: 10, 9195

Physical state, color, and odor:
Clear, colorless liquid with an aromatic odor resembling benzene

Melting point (°C):
-38.2 (Weast, 1986)
-30 (Verschueren, 1983)

Boiling point (°C):
84.2 (Weast, 1986)

Density (g/cm³):
1.0649 at 20/4 °C (Weast, 1986)
1.0873 at 0/4 °C, 1.0573 at 25/4 °C (Windholz et al., 1983)

Diffusivity in water (x 10⁻⁵ cm²/sec):
1.01 at 20 °C using method of Hayduk and Laudie (1974)

Flash point (°C):
-1.1 (Hawley, 1981)

Heat of fusion (kcal/mol):
1.216 (Riddick et al., 1986)

Henry's law constant (x 10⁻³ atm·m³/mol at 25 °C):
2.33 and 2.70 in distilled water and seawater, respectively (Przyjazny et al., 1983)

Ionization potential (eV):
8.860 (Lias, 1998)

Soil sorption coefficient, log K_{oc}:
1.73 using method of Kenaga and Goring (1980)

Octanol/water partition coefficient, log K_{ow}:
1.81 (Leo et al., 1971)
1.82 (Sangster, 1989)

Solubility in organics:
Miscible with carbon tetrachloride, heptane, pyrimidine, dioxane, toluene, and many organic solvents (Keith and Walters, 1992)

Solubility in water:
3,015 mg/kg at 25 °C (Price, 1976)
3,600 mg/L at 18 °C (quoted, Verschueren, 1983)
2-5 mM at 20 °C (Fischer and Ehrenberg, 1948)

Vapor density:
3.44 g/L at 25 °C, 2.90 (air = 1)

Vapor pressure (mmHg):
60 at 20 °C (Verschueren, 1983)
79.7 at 25 °C (Wilhoit and Zwolinski, 1971)
337 at 60.3 °C, 486 at 70.3 °C, 686 at 80.3 °C, 951 at 90.3 °C (Eon et al., 1971)

Environmental fate:
  Photolytic. A rate constant 9.70×10^{-12} cm^3/molecule·sec was reported for the reaction of thiophene and OH radicals in the atmosphere at room temperature (Atkinson, 1985). Thiophene also reacts with NO$_3$ radicals in the atmosphere at rate constants ranging from 3.2×10^{-14} (Atkinson et al., 1985) to 3.93×10^{-14} cm^3/molecule·sec (Atkinson, 1991).

Toxicity:
  LD$_{50}$ (intraperitoneal) for mice 100 mg/kg (quoted, RTECS, 1985).

Uses: Solvent; manufacturing resins, dyes, and pharmaceuticals.

THIRAM

Synonyms: Aatack; Accelerator thiuram; Aceto TETD; Arasan; Arasan 70; Arasan 75; Arasan-M; Arasan 42-S; Arasan-SF; Arasan-SF-X; Aules; Bis(dimethylamino)carbonothioyl disulfide; Bis(dimethylthiocarbamoyl) disulfide; Bis(dimethylthiocarbamyl) disulfide; Chipco thiram 75; Cyuram DS; α,α'-Dithiobis(dimethylthio) formamide; *N,N'*-(Dithiodicarbono-thioyl)bis(*N*-methylmethanamine); Ekagom TB; ENT 987; Falitram; Fermide; Fernacol; Fernasan; Fernasan A; Fernide; Flo pro T seed protectant; Hermal; Hermat TMT; Heryl; Hexathir; Kregasan; Mercuram; Methyl thiram; Methyl thiuramdisulfide; Methyl tuads; NA 2771; Nobecutan; Nomersan; Normersan; NSC 1771; Panoram 75; Polyram ultra; Pomarsol; Pomersol forte; Pomasol; Puralin; RCRA waste number U244; Rezifilm; Royal TMTD; Sadoplon; Spotrete; Spotrete-F; Spotrete 75WDG; SQ 1489; Tersan; Tersan 75; Tetramethyldiurane sulphite; Tetramethylthiuram bisulfide; Tetramethylthiuram bisulphide; Tetramethylenethiuram disulfide; **Tetramethylthioperoxydicarbonic diamide**; Tetramethyl-thiocarbamoyl disulfide; Tetramethylthioperoxydicarbonic diamide; Tetramethylthiuram di-sulfide; Tetramethylthiuram disulphide; *N,N*-Tetramethylthiuram disulfide; *N,N,N',N'*-Tetramethylthiuram disulfide; Tetramethylthiuran disulfide; Tetramethylthiurane disulphide; Tetramethylthiurum disulfide; Tetramethylthiurum disulphide; Tetrapom; Tetrasipton; Tetrathiuram disulfide; Tetrathiuram disulphide; Thillate; Thimer; Thiosan; Thiotex; Thiotox; Thiram 75; Thiramad; Thiram B; Thirame; Thirasan; Thiulix; Thiurad; Thiuram; Thiuram D; Thiuramin; Thiuram M; Thiuram M rubber accelerator; Thiuramyl; Thylate; Tirampa; Tiuramyl; TMTD; TMTDS; Trametan; Tridipam; Tripomol; TTD; Tuads; Tuex; Tulisan; USAF B-30; USAF EK-2089; USAF P-5; Vancida TM-95; Vancide TM; Vulcafor TMTD; Vulkacit MTIC.

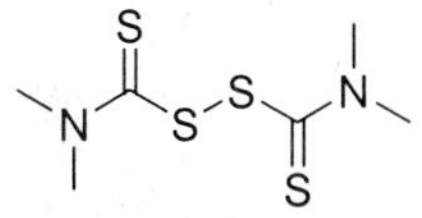

CAS Registry Number: 137-26-8
DOT: 2771
DOT label: Poison
Molecular formula: $C_6H_{12}N_2S_4$
Formula weight: 240.44
RTECS: JO1400000
Merck reference: 10, 9216

Physical state and color:
Colorless to white to cream-colored crystals. May darken on exposure to air or light.

Melting point (°C):
155.6 (Weast, 1986)
146 (Worthing and Hance, 1991)

Boiling point (°C):
310-315 at 15 mmHg (Weast, 1986)
129 at 20 mmHg (Sax and Lewis, 1987)

Density (g/cm³):
1.29 at 20/4 °C (Hawley, 1981)

Solubility in organics:
Soluble in acetone (1.2 wt %), alcohol (<0.2 wt %), benzene (2.5 wt %), ether (<0.2 wt %) (Windholz et al., 1983), chloroform (230 g/L) (Worthing and Hance, 1991)

Solubility in water:
30 mg/L at room temperature (Worthing and Hance, 1991)

Vapor pressure (mmHg):
8 x 10^{-6} at 20 °C (NIOSH, 1997)

Environmental fate:
   Biological. Odeyemi and Alexander (1977) isolated three strains of *Rhizobium* sp. that degraded thiram. One of these strains, *Rhizobium meliloti*, metabolized thiram to yield dimethylamine (DMA) and carbon disulfide, which formed spontaneously from dimethyldithiocarbamate (DMDT). The conversion of DMDT to DMA and carbon disulfide occurred via enzymatic and nonenzymatic mechanisms.
   When thiram (100 ppm) was inoculated with activated sludge (30 ppm) at 25 °C and pH 7.0 for 2 wk, 30% degraded. Metabolites included methionine, elemental sulfur, formaldehyde, dimethyldithiocarbamate-α-aminobutyric acid, and the corresponding keto acid (Kawasaki, 1980).
   Soil. In both soils and water, chemical and biological mediated reactions can transform thiram to compounds containing the mercaptan group (Alexander, 1981). Decomposes in soils to carbon disulfide and dimethylamine (Sisler and Cox, 1954; Kaars Sijpesteijn et al., 1977). When a spodosol (pH 3.8) pretreated with thiram was incubated for 24 d at 30 °C and relative humidity of 60-90%, dimethylamine formed as the major product. Minor degradative products included nitrite ions (nitration reduction) and dimethylnitrosamine (Ayanaba et al., 1973).
   Plant. Major plant metabolites are ethylene thiourea, thiram monosulfide, ethylene thiram disulfide, and sulfur (Hartley and Kidd, 1987).
   Photolytic. In methanol, thiram absorbed UV light at wavelengths >290 nm (Gore et al., 1971).
   Chemical/Physical. Thiram is not flammable and is stable at room temperature. Emits nitrogen and sulfur oxides when heated to decomposition (Turf & Ornamental Chemicals Reference, 1991).
   Although no products were reported, the calculated hydrolysis half-life at 25 °C and pH 7 is 5.3 d (Ellington et al., 1988). In an earlier study, Ellington et al. (1987) reported the following hydrolysis half-lives: 11.8 h at 85 °C (pH 3.22), 2.6 h at 65 °C (pH 6.97), and 23 min at 24 °C (pH 10.49).

Exposure limits (mg/m³): NIOSH REL: TWA 5, IDLH 100; OSHA PEL: TWA 5; ACGIH TLV: TWA 1 (adopted).

Symptoms of exposure: Irritates mucous membranes; dermatitis; with ethanol consumption: flush, erythema, pruritus, urticaria, headache, nausea, vomiting, diarrhea, weakness, dizziness, difficulty in breathing. Contact with skin may cause allergic reaction (NIOSH, 1997).

Toxicity:
   LC_{50} (96-h) for rainbow trout 0.13 mg/L, bluegill sunfish 0.23 mg/L, carp 4.0 mg/L (Hartley and Kidd, 1987), crawfish 4.3 ppm (Turf & Ornamental Chemicals Reference, 1991).
   LC_{50} (48-h) for bluegill sunfish 230 ppb, channel catfish 630 ppb, rainbow trout 130 ppb (Turf & Ornamental Chemicals Reference, 1991).
   Acute inhalation LC_{50} (4-h) for rats >0.5 mg/L air (Worthing and Hance, 1991).

Acute oral LD$_{50}$ for wild birds 300 mg/kg, mice 1,350 mg/kg, rats 560 mg/kg, rabbits 210 mg/kg (quoted, RTECS, 1985).

Reported NOELs for rats and dogs were <250 and 200 mg/kg diet, respectively (Worthing and Hance, 1991).

Uses: Vulcanizer; seed disinfectant; rubber accelerator; rabbit, deer and rodent repellant; bacteriostat in soap. Protective fungcide applied to foliage to control *Botrytis* spp. On ornamentals, lettuce, soft fruit, and vegetables; *Venturia pirina* on pears. Also used in fields and orchards to control birds, rodents, and deer (Worthing and Hance, 1991). Prevents infestation common turf diseases such as Dollar Spot and Brown Patch and controls Snow Mold (Turf & Ornamental Chemicals Reference, 1991).

TOLUENE

Synonyms: Antisal 1a; Methacide; **Methylbenzene**; Methylbenzol; NCI-C07272; Phenyl-methane; RCRA waste number U220; Toluol; Tolusol; UN 1294.

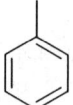

CAS Registry Number: 108-88-3
DOT: 1294
DOT label: Flammable liquid
Molecular formula: C_7H_8
Formula weight: 92.14
RTECS: XS5250000
Merck reference: 10, 9357

Physical state, color, and odor:
Colorless, water-white liquid with a pleasant, sweet-like odor similar to benzene. The reported odor threshold concentrations ranged from 170 to 960 ppb (Keith and Walters, 1992; Young et al., 1996). The taste threshold concentration in water is 960 ppb (Young et al., 1996).

Melting point (°C):
-95 (Weast, 1986)

Boiling point (°C):
110.6 (Weast, 1986)

Density (g/cm³):
0.86689 at 20.00/4 °C (Tsierkezos et al., 2000)
0.8614 at 25.00/4 °C (Aminabhavi and Banerjee, 1999)
0.8666 at 20/4 °C, 0.8573 at 30/4 °C, 0.8480 at 40/4 °C (Sumer et al., 1968)

Diffusivity in water (x 10^{-5} cm²/sec):
0.85 at 20 °C (Witherspoon and Bonoli, 1969)
0.95 at 25 °C (Bonoli and Witherspoon, 1968)
0.93 at 25 °C (mole fraction = 10^{-5}) (Gabler et al., 1996)

Dissociation constant, pK_a:
≈ 35 (Gordon and Ford, 1972)

Flash point (°C):
4.5 (NIOSH, 1997)

Lower explosive limit (%):
1.1 (NIOSH, 1997)

Upper explosive limit (%):
7.1 (NIOSH, 1997)

Entropy of fusion (cal/mol·K):
8.891 (Kelley, 1929)
8.903 (Scott et al., 1962)
8.793 (Ziegler and Andrews, 1942)

Heat of fusion (kcal/mol):
1.582 (Kelley, 1929)
1.586 (Scott et al., 1962)
1.565 (Ziegler and Andrews, 1942)

Henry's law constant (x 10^{-3} atm·m^3/mol):
6.66 at 23 °C (Anderson, 1992)
6.46 at 25.0 °C (Ramachandran et al., 1996)
6.7 (Pankow and Rosen, 1988)
7.69 (Hoff et al., 1993)
3.81, 4.92, 5.55, 6.42, and 8.08 at 10, 15, 20, 25, and 30 °C, respectively (Ashworth et al., 1988)
2.89, 3.85, 4.92, 6.51, and 8.27 at 10, 15, 20, 25, and 30 °C, respectively (Perlinger et al., 1993)
Distilled water: 1.88, 2.14, 2.60, 4.29, and 5.49 at 2.0, 6.0, 10.0, 18.2, and 25.0 °C, respectively; natural seawater: 2.77 and 6.90 at 6.0 and 25.0 °C, respectively (Dewulf et al., 1995)
6.67 at 25 °C (Mackay et al., 1979)
5.88 (Wasik and Tsang, 1970)
5.26 at 25 °C (Vitenberg et al., 1975)
5.43, 8.38, 11.2, 14.0, and 15.5 at 25, 30, 40, 45, and 50 °C, respectively (Robbins et al., 1993)
6.25 at 25 °C (Nielsen et al., 1994)
6.38 at 25.0 °C (Ettre et al., 1993)
6.36 at 25.0 °C (Kolb et al., 1992)
6.65 at 25.0 °C (McAuliffe, 1971)
6.19 at 23.0 °C (mole fraction ratio, Leighton and Calo, 1981)
2.01 and 5.50 at 5.7 and 24.9 °C, respectively (Dewulf et al., 1999a)
At 25 °C (NaCl concentration, mol/L): 4.80 (0), 5.04 (0.1), 5.65 (0.3), 6.24 (0.5), 7.05 (0.7), 8.32 (1.0); salinity, NaCl = 37.34 g/L (°C): 4.92 (15), 6.62 (20), 7.98 (25), 9.48 (30), 11.74 (35), 14.29 (40), 16.74 (45) (Peng and Wan, 1998).
1.85, 2.20, 2.83, 4.71, and 6.57 at 2.0, 6.0, 10.0, 18.0, and 25.0 °C, respectively (Dewulf et al., 1999)
6.25, 10.80, 14.80, and 21.17 at 30, 40, 50, and 60 °C, respectively (Vane and Giroux, 2000)

Interfacial tension with water (dyn/cm at 25 °C):
27.2 (Murphy et al., 1957)
36.06 (Harkins et al., 1920)

Ionization potential (eV):
8.82 ± 0.01 (Franklin et al., 1969)

Bioconcentration factor, log BCF:
0.92 (goldfish, Ogata et al., 1984)
1.12 (eels, Ogata and Miyake, 1978)
0.62 (mussels, Geyer et al., 1982)
2.58 (algae, Geyer et al., 1984)
3.28 (activated sludge), 1.95 (golden ide) (Freitag et al., 1985)
3.18 green alga, *Selenastrum capricornutum* (Casserly et al., 1983)

2.26, 2.30, 2.34, 2.28, 2.77, 3.36, 4.15, and 2.95 for olive, grass, holly, ivy, mock orange, pine, rosemary, and juniper leaves, respectively (Hiatt, 1998)

Soil sorption coefficient, log K_{oc}:
2.06 (Abdul et al., 1987)
2.18 (Sapsucker woods soil, Garbarini and Lion, 1986)
1.74, 1.97, 2.13 (various Norwegian soils, Seip et al., 1986)
1.66 (Vandreil sandy loam), 2.20 (Grimsby silt loam), 1.57 (Wendover silty loam) (Nathwani and Phillips, 1977)
2.25 (sandy soil, Wilson et al., 1981)
2.00 (estuarine sediment, Vowles and Mantoura, 1987)
1.99-3.05 (silty clay), 3.03, 3.39 (coarse sand) (Pavlostathis and Mathavan, 1992)
2.22 and 2.16 in Captina silt loam and McLaurin sandy loam, respectively (Walton et al., 1992)
2.08 (river sediment), 2.03 (coal wastewater sediment) (Kopinke et al., 1995)
2.26 (RP-HPLC immobilized humic acids, Szabo et al., 1990)
2.39 (Schwarzenbach and Westall, 1981)
2.96, 2.97, 3.03 (glaciofluvial, sandy aquifer, Nielsen et al., 1996)
1.65 (Mt. Lemmon soil, Hu et al., 1995)
2.74, 2.87, 3.41 (Cohansey sand), 2.92, 3.33, 4.04 (Potomac-Raritan-Magothy sandy loam) (Uchrin and Mangels, 1987)
2.23, 2.32, 2.33, 2.34, 2.40, 2.31, and 2.34 at 2.3, 3.8, 6.2, 8.0, 13.5, 18.6, and 25.0 °C, respectively, for a Leie River (Belgium) clay (Dewulf et al., 1999a)

Octanol/water partition coefficient, log K_{ow}:
2.65 (Tewari et al., 1982; Wasik et al., 1981, 1983)
2.69 (Hansch et al., 1968)
2.21 (Veith et al., 1980)
2.63 (Brooke et al., 1990; Yalkowsky et al., 1983a)
2.50 (Walton et al., 1989)
2.11, 2.80 (Leo et al., 1971)
2.73 (Campbell and Luthy, 1985; Leo et al., 1971)
2.78 (Burkhard et al., 1985a)
2.79 (Fujita et al., 1964; Brooke et al., 1990; de Bruijn et al., 1989)
2.68 (Nahum and Horvath, 1980)
2.72 (Garst and Wilson, 1984)
2.32, 2.40, 2.46, 2.38, 2.41, and 2.32 at 2.2, 6.0, 10.0, 14.1, 18.7, and 24.8 °C, respectively (Dewulf et al., 1999a)

Solubility in organics:
Soluble in acetone, carbon disulfide, and ligroin; miscible with acetic acid, ethanol, benzene, ether, chloroform (U.S. EPA, 1985), and other organic solvents including xylene and ethyl-benzene.

Solubility in water:
5.58, 5.71, 5.88, and 6.28 mM at 15.0, 25.0, 35.0, and 45.0 °C, respectively (Sanemasa et al., 1982)
519.5 mg/L at 25 °C (Mackay and Shiu, 1975)
563.3 μL/L at 25 °C (Sada et al., 1975)
500 mg/L at 25 °C (Klevens, 1950)
16.8 mM at 25 °C (Banerjee et al., 1980)
524 mg/L at 25 °C (Banerjee, 1984)

506.7 mg/kg at 25 °C. In natural seawater: 410, 410, and 418.5 mg/kg at 15, 20, and 25 °C, respectively (Rossi and Thomas, 1981)

515 mg/kg at 25 °C (McAuliffe, 1966)

In mg/L: 658 at 0.4 °C, 646 at 3.6 °C, 628 at 10.0 °C, 624 at 11.2 °C, 623 at 14.9 °C, 621 at 15.9 °C, 627 at 25.0 °C, 625 at 25.6 °C, 640 at 30.0 °C, 642 at 30.2 °C, 657 at 35.2 °C, 701 at 42.8 °C, 717 at 45.3 °C (Bohon and Claussen, 1951).

530 mg/L at 25 °C (Andrews and Keefer, 1949)

570 mg/kg at 30 °C (Gross and Saylor, 1931)

724 mg/L at 0 °C (Brookman et al., 1985)

724 mg/kg at 0 °C, 573 mg/kg at 25 °C (Polak and Lu, 1973)

534.8 mg/kg at 25 °C, 379.3 mg/kg in artificial seawater at 25 °C (Sutton and Calder, 1975)

554.0 mg/kg at 25 °C. In NaCl solution at 25 °C (salinity, g/kg), mg/kg: 526 (1.002), 490 (10.000), 402.0 (34.472), 359 (50.030), 182 (125.100), 106 (199.900), 53.8 (279.800), 37.2 (358.700) (Price, 1976)

6.28 mM at 25 °C (Tewari et al., 1982; Wasik et al., 1981, 1983)

6.29 mM at 25.00 °C (Keeley et al., 1988)

220 mg/L in fresh water at 25 °C, 230 mg/L in salt water at 25 °C (Krasnoshchekova and Gubergrits, 1975)

660, 670 mg/kg at 23.5 °C (Schwarz, 1980)

6.69 mM at 25 °C (Ben-Naim and Wilf, 1980)

392 mg/L in seawater at 25 °C (Bobra et al., 1979)

4.19 mM in 0.5 M NaCl at 25 °C (Wasik et al., 1984)

7.13 mM at 35 °C (Hine et al., 1963)

479 mg/kg at 25 °C (Chey and Calder, 1972)

5.1 mM at 16 °C (Fühner, 1924)

26 mM at 25 °C (Hogfeldt and Bolander, 1963)

466.9 mg/L at 30 °C (McNally and Grob, 1984)

0.052 wt % at 25 °C (Lo et al., 1986)

538 mg/L (Coutant and Keigley, 1988)

5.82 mmol/kg at 25 °C (Morrison and Billett, 1952)

In mg/kg: 612, 601, 586, 587, 573, 575, 569, 577, and 566 at 4.5, 6.3, 7.1, 9.0, 11.8, 12.1, 15.1, 17.9, and 20.1 °C, respectively. In artificial seawater: 449, 429, 416, 405, and 397 at 0.19, 5.32, 10.05, 14.96, and 20.04 °C, respectively (Brown and Wasik, 1974)

In wt % (°C): 0.823 (114), 1.640 (147), 2.387 (169), 2.790 (183), 4.113 (207), 5.072 (224) (Guseva and Parnov, 1963)

6.81 mmol/kg at 25.0 °C (Vesala, 1974)

3.43 mM at 25.00 °C (Sanemasa et al., 1985)

5.65 mM at 25.0 °C (Sanemasa et al., 1987)

547 and 12,600 mg/L at 25 and 200 °C, respectively (Yang et al., 1997)

In wt % (°C): 0.12 (0), 0.09 (9.5), 0.08 (19.8), 0.08 (29.7), 0.10 (39.6), 0.09 (50.0), 0.10 (60.1), 0.09 (70.4), 0.13 (81.0), 0.12 (90.2) (Stephenson, 1992)

0.0768, 0.0735, and 0.0837 wt % at 10.0, 20.0, and 30.0 °C, respectively (Schwarz and Miller, 1980)

At 25 °C: 562.9 mg/L in distilled water; 347.9, 216.2, and 144.4 mg/L in 1.00, 2.00, and 3.00 molal NaCl solutions, respectively (Poulson et al., 1999)

Vapor density:
3.77 g/L at 25 °C, 3.18 (air = 1)

Vapor pressure (mmHg):
22 at 20 °C (Verschueren, 1983)

36.7 at 30 °C (Sax, 1984)
28.4 at 25 °C (Howard, 1990)
28.1 at 25 °C (Mackay et al., 1982)
46.7 at 35 °C (Chey and Calder, 1972)
27.82 at 25.00 °C (Hussam and Carr, 1985)

Environmental fate:

Biological. Toluene can undergo two types of microbial attack. The first type proceeds via immediate hydroxylation of the benzene ring, followed by ring cleavage. The second type of attack proceeds via oxidation of the methyl group followed by hydroxylation and ring cleavage (Fewson, 1981). A mutant of *Pseudomonas putida* oxidized toluene to (+)-*cis*-2,3-dihydroxy-1-methylcyclohexa-1,4-diene (Dagley, 1972). Claus and Waker (1964) reported that *Pseudomonas* sp. and an *Achromobacter* sp. oxidized toluene to 3-methyl catechol. Other metabolites identified in the microbial degradation of toluene include *cis*-2,3-dihydroxy-2,3-dihydrotoluene, 3-methyl catechol, benzyl alcohol, benzaldehyde, benzoic acid, catechol (Verschueren, 1983), and 1-hydroxy-2-naphthoic acid (Claus and Walker, 1964). In a methanogenic aquifer material, toluene degraded completely to carbon dioxide (Wilson et al., 1986). In activated sludge, 26.3% of the applied toluene mineralized to carbon dioxide after 5 d (Freitag et al., 1985). Based on a first-order degradation rate constant of 0.07/yr, the half-life of toluene is 39 d (Zoeteman et al., 1981).

In anoxic groundwater near Bemidji, MI, toluene anaerobically biodegraded to the intermediate benzoic acid (Cozzarelli et al., 1990). Methylmuconic acid was reported to be the biooxidation product of toluene by *Nocardia corallina* V-49, using *n*-hexadecane as the substrate. With methane as the substrate and *Methylosinus trichosporium* OB3b as the microorganism, *p*-hydroxytoluene and benzoic acid are the products of biooxidation. In addition, *Methyloccus capsulatus* was reported to bioxidize toluene to benzyl alcohol and cresol (Keck et al., 1989). When toluene (5 and 10 mg/L) was statically incubated in the dark at 25 °C with yeast extract and settled domestic wastewater inoculum for 7 d, 100% biodegradation with rapid adaptation was observed (Tabak et al., 1981). Pure microbial cultures isolated from soil hydroxylated toluene to 2- and 4-hydroxytoluene (Smith and Rosazza, 1974). When toluene (5 and 10 mg/L) was statically incubated in the dark at 25 °C with yeast extract and settled domestic wastewater inoculum for 7 d, complete biodegradation with rapid acclimation was observed (Tabak et al., 1981).

Bridié et al. (1979) reported BOD and COD values of 2.15 and 2.52 g/g, respectively, using filtered effluent from a biological sanitary waste treatment plant. These values were determined using a standard dilution method at 20 °C and stirred for a period of 5 d. The ThOD for toluene is 3.13 g/g.

Estimated half-lives of toluene (3.6 μg/L) from an experimental marine mesocosm during the spring (8-16 °C), summer (20-22 °C), and winter (3-7 °C) were 16, 1.5, and 13 d, respectively (Wakeham et al., 1983).

Surface Water. Mackay and Wolkoff (1973) estimated an evaporation half-life of 30.6 min from a surface water body that is 25 °C and 1 m deep.

Groundwater. Nielsen et al. (1996) studied the degradation of toluene in a shallow, glaciofluvial, unconfined sandy aquifer in Jutland, Denmark. As part of the *in situ* microcosm study, a cylinder that was open at the bottom and screened at the top was installed through a cased borehole approximately 5 m below grade. Five liters of water was aerated with atmospheric air to ensure aerobic conditions were maintained. Groundwater was analyzed weekly for approximately 3 months to determine toluene concentrations over time. The experimentally determined first-order biodegradation rate constant and corresponding half-life following a 5-d lag phase were 0.4/d and 1.73 d, respectively.

Photolytic. Synthetic air containing gaseous nitrous acid and toluene exposed to artificial sunlight (λ = 300-450 nm) yielded methyl nitrate, peroxyacetal nitrate, and a nitro aromatic compound tentatively identified as a nitrophenol or nitrocresol (Cox et al., 1980). An *n*-hexane solution containing toluene and spread as a thin film (4 mm) on cold water (10 °C) was irradiated by a mercury medium pressure lamp. In 3 h, 26% of the toluene photooxidized into benzaldehyde, benzyl alcohol, benzoic acid, and *m*-cresol (Moza and Feicht, 1989). Methane and ethane were reported as products of the gas-phase photolysis of toluene at 2537 Å (Calvert and Pitts, 1966).

Irradiation of toluene (80 ppm) by UV light (λ = 200-300 nm) on titanium dioxide in the presence of oxygen (20%) and moisture resulted in the formation of benzaldehyde and carbon dioxide. Carbon dioxide concentrations increased linearly with the increase in relative humidity. However, the concentration of benzaldehyde decreased with an increase in relative humidity. An identical experiment, but without moisture, resulted in the formation of benzaldehyde, carbon dioxide, hydrogen cyanide, and nitrotoluenes. In an atmosphere containing moisture and nitrogen dioxide (80 ppm), cresols, benzaldehyde, carbon dioxide, and nitrotoluenes were the photoirradiation products (Ibusuki and Takeuchi, 1986).

Irradiation of toluene in the presence of chlorine yielded benzyl hydroperoxide, benzaldehyde, peroxybenzoic acid, carbon monoxide, carbon dioxide, and other unidentified products (Hanst and Gay, 1983). The photooxidation of toluene in the presence of nitrogen oxides (NO and NO_2) yielded small amounts of formaldehyde and traces of acetaldehyde or other low molecular weight carbonyls (Altshuller et al., 1970). Other photooxidation products not previously mentioned include phenol, phthalaldehydes, and benzoyl alcohol (Altshuller, 1983). A carbon dioxide yield of 8.4% was achieved when toluene adsorbed on silica gel was irradiated with light (λ >290 nm) for 17 h (Freitag et al., 1985).

Chemical/Physical. Products identified from the reaction of toluene with nitric oxide and OH radicals include benzaldehyde, benzyl alcohol, *m*-nitrotoluene, *p*-methylbenzoquinone, and *o*-, *m*-, and *p*-cresol (Kenley et al., 1978). Gaseous toluene reacted with nitrate radicals in purified air forming the following products: benzaldehyde, benzyl alcohol, benzyl nitrate, and 2-, 3- and 4-nitro-toluene (Chiodini et al., 1993). Under atmospheric conditions, the gas-phase reaction with OH radicals and nitrogen oxides resulted in the formation of benzaldehyde, benzyl nitrate, *m*-nitrotoluene, and *o*-, *m*-, and *p*-cresol (Finlayson-Pitts and Pitts, 1986; Atkinson, 1990).

Kanno et al. (1982) studied the aqueous reaction of toluene and other aromatic hydrocarbons (benzene, *o*-, *m*-, and *p*-xylene, and naphthalene) with hypochlorous acid in the presence of ammonium ion. Although chlorination of the aromatic was observed with the formation of 2- and 4-chlorotoluene, the major degradative pathway was cleavage of the aromatic ring by chloramine, forming cyanogen chloride (Kanno et al., 1982). The amount of cyanogen chloride formed was inversely proportional to the pH of the solution. At pH 6, the greatest amount of cyanogen chloride was formed when the reaction mixture contained ammonium ion and hypochlorous acid at a ratio of 2:3 (Kanno et al., 1982).

Toluene will not hydrolyze because it does not contain a hydrolyzable functional group (Kollig, 1993).

At an influent concentration of 317 mg/L, treatment with granular activated carbon resulted in an effluent concentration of 66 mg/L. The adsorbability of the carbon used was 50 mg/g carbon (Guisti et al., 1974). Similarly, at influent concentrations of 1.0, 0.1, 0.01, and 0.001 mg/L, the granular activated carbon adsorption capacities at pH 5.6 were 26, 9.4, 3.4, and 1.2 mg/g, respectively (Dobbs and Cohen, 1980).

In a marine ecosystem, the volatilization half-life of toluene at the temperature range of 2 to 10 °C was 6 d (Wakeham et al., 1985).

Exposure limits: NIOSH REL: TWA 100 ppm (375 mg/m³), STEL 150 ppm (560 mg/m³), IDLH 500 ppm; OSHA PEL: TWA 200 ppm, ceiling 300 ppm, ceiling 10-min peak 500 ppm; ACGIH TLV: TWA 50 ppm (adopted).

Symptoms of exposure: May cause headache, dizziness, excitement, euphoria, hallucination, distorted perceptions, and confusion. Narcotic at high concentrations (Patnaik, 1992).

Toxicity:

EC_{50} (8-d) for *Selenastrum capricornutum* 9.4 mg/L (Herman et al., 1990).

EC_{50} (72-h) for *Selenastrum capricornutum* 12.5 mg/L (Galassi et al., 1988).

LC_{50} (contact) for earthworm (*Eisenia fetida*) 75 $\mu g/cm^2$ (Neuhauser et al., 1985).

LC_{50} (14-d) for *Poecilia reticulata* 68.3 mg/L (Könemann, 1981).

LC_{50} (96-h) for *Salmo gairdneri* 5.8 mg/L, *Poecilia reticulata* 28.2 mg/L (Galassi et al., 1988), bluegill sunfish 13 mg/L (Spehar et al., 1982), *Oncorhynchus kisutch* fry 5.5 mg/L (Moles et al., 1981), *Cancer magister* larvae 28 mg/L (Caldwell et al., 1976), *Cyprinodon variegatus* 280-480 ppm using natural seawater (Heitmuller et al., 1981).

LC_{50} (72-h) for *Cyprinodon variegatus* 280-480 ppm (Heitmuller et al., 1981).

LC_{50} (48-h) for zebra fish 60 mg/L (Slooff, 1979), *Daphnia magna* 11.5 mg/L (Bobra et al., 1983) and 310 mg/L (LeBlanc, 1980), *Cyprinodon variegatus* 280-480 ppm (Heitmuller et al., 1981).

LC_{50} (24-h) for *Palaemonetes pugio* larvae 25.8 mg/g, adults 17.2 mg/L (Potera, 1975), *Daphnia magna* 310 mg/L (LeBlanc, 1980), grass shrimp (*Palaemonetes pugio*) larvae and adults were 25.8 and 17.2 mg/L, respectively (Environment Canada, 1982), *Cyprinodon variegatus* 280-480 ppm (Heitmuller et al., 1981).

Acute oral LD_{50} for rats 5,000 mg/kg (quoted, RTECS, 1985).

LC_{50} (inhalation) for mice 5,320 ppm/8-h (quoted, RTECS, 1985).

Heitmuller et al. (1981) reported a NOEC of 280 ppm.

Drinking water standard (final): MCLG: 1 mg/L; MCL: 1 mg/L (U.S. EPA, 1996).

Source: Detected in distilled water-soluble fractions of 87 octane gasoline (25.9 mg/L), 94 octane gasoline (86.9 mg/L), Gasohol (60.8 mg/L), No. 2 fuel oil (1.54 mg/L), jet fuel A (1.05 mg/L), diesel fuel (0.86 mg/L), military jet fuel JP-4 (32.0 mg/L) (Potter, 1996), new motor oil (16.3-16.9 μg/L), and used motor oil (781-814 μg/L) (Chen et al., 1994). The average volume percent and estimated mole fraction in American Petroleum Institute PS-6 gasoline are 3.519 and 0.04392, respectively (Poulsen et al., 1992). Schauer et al. (1999) reported toluene in a diesel-powered medium-duty truck exhaust at an emission rate of 3,980 μg/km.

Thomas and Delfino (1991) equilibrated contaminant-free groundwater collected from Gainesville, FL with individual fractions of three individual petroleum products at 24-25 °C for 24 h. The aqueous phase was analyzed for organic compounds via U.S. EPA approved test method 602. Average toluene concentrations reported in water-soluble fractions of unleaded gasoline, kerosene, and diesel fuel were 23.676, 1.065, and 0.552 mg/L, respectively. When the authors analyzed the aqueous-phase via U.S. EPA approved test method 610, average toluene concentrations in water-soluble fractions of unleaded gasoline, kerosene, and diesel fuel were lower, i.e., 12.969, 0.448, and 0.030 mg/L, respectively.

In 7 coal tar samples, toluene concentrations ranged from ND to 7,000 ppm (EPRI, 1990). Detected in 1-yr aged coal tar film and bulk coal tar at concentrations of <75 and 220 mg/kg, respectively (Nelson et al., 1996).

Uses: Manufacture of caprolactum, medicines, dyes, perfumes, benzoic acid, trinitrotoluene, nitrotoluenes, *o*- and *p*-toluenesulfonic acid, *o*- and *p*-xylene, benzyl chloride, benzal chloride, benzotrichloride, halogenated toluenes; solvent for paints and coatings, gums, resins, rubber, oils, and vinyl compounds; adhesive solvent in plastic toys and model airplanes; diluent and thinner for nitrocellulose lacquers; detergent manufacturing; aviation gasoline and high-octane blending stock; preparation of toluene diisocyanate for polyurethane resins.

2,4-TOLUENE DIISOCYANATE

Synonyms: Desmodur T80; **2,4-Diisocyanato-1-methylbenzene**; Diisocyanatoluene; 2,4-Diisocyanotoluene; Isocyanic acid methylphenylene ester; Isocyanic acid 4-methyl-*m*-phenylene ester; Hylene T; Hylene TCPA; Hylene TLC; Hylene TM; Hylene TM-65; Hylene TRF; 4-Methylphenylene diisocyanate; 4-Methylphenylene isocyanate; Mondur TD; Mondur TD-80; Mondur TDS; Nacconate 100; NCI-C50533; Niax TDI; Niax TDI-P; RCRA waste number U223; Rubinate TDI 80/20; TDI; 2,4-TDI; TDI-80; Toluene 2,4-diisocyanate; Tolyene-2,4-diisocyanate; Tolylene-2,4-diisocyanate; 2,4-Tolylene diisocyanate; *m*-Tolylene diisocyanate.

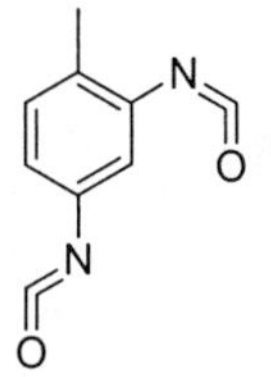

CAS Registry Number: 584-84-9
DOT: 2078
DOT label: Poison
Molecular formula: $C_9H_6N_2O_2$
Formula weight: 174.15
RTECS: CZ6300000
Merck reference: 10, 9358

Physical state, color, and odor:
Clear, colorless to light yellow liquid with a pungent, fruity odor. Odor threshold concentrations ranged from 0.4 to 2.14 ppm (Keith and Walters, 1992).

Melting point (°C):
19.5-21.5 (Windholz et al., 1983)
20-21 (Dean, 1987)

Boiling point (°C):
251 (Windholz et al., 1983)

Density (g/cm³):
1.2244 at 20/4 °C (Windholz et al., 1983)

Flash point (°C):
127 (NFPA, 1984)

Lower explosive limit (%):
0.9 (NFPA, 1984)

Upper explosive limit (%):
9.5 (NIOSH, 1997)

Soil sorption coefficient, log K_{oc}:
Not applicable - reacts with water

Octanol/water partition coefficient, log K_{ow}:
Not applicable - reacts with water

Solubility in organics:
Miscible with acetone, alcohol (decomposes), benzene, carbon tetrachloride, diglycol monomethyl ether, ether, kerosene, olive oil (Windholz et al., 1983), chlorobenzene, and kerosene (Keith and Walters, 1992)

Solubility in water:
Not applicable - reacts with water

Vapor density:
7.12 g/L at 25 °C, 6.01 (air = 1)

Vapor pressure (mmHg):
0.01 at 20 °C, 1 at 80 °C (Verschueren, 1983)

Environmental fate:
  Chemical/Physical. Slowly reacts with water forming carbon dioxide and polyureas (NIOSH, 1997; Windholz et al., 1983).

Exposure limits: NIOSH REL: IDLH 2.5 ppm; OSHA PEL: ceiling 0.02 ppm (0.14 mg/m^3); ACGIH TLV: TWA 5 ppb, STEL 20 ppb (adopted).

Symptoms of exposure: Vapors may cause bronchitis, headache, sleeplessness, pulmonary edema, wheezing, shortness of breath, and chest congestion. Ingestion may cause coughing, vomiting, and gastrointestinal pain. Contact with skin may cause nausea, vomiting, abdominal pain, dermatitis, and skin sensitization (Patnaik, 1992). An irritation concentration of 4.00 mg/m^3 in air was reported by Ruth (1986).

Toxicity:
  Acute oral LD$_{50}$ for rats 5,800 mg/kg, wild birds 100 mg/kg (quoted, RTECS, 1985).
  LC$_{50}$ values reported for fathead minnows at exposure time of 24, 72, and 96 h were 194.9, 172.1, and 164.5 mg/L, respectively (Curtis et al., 1978).
  LC$_{50}$ (inhalation) for guinea pigs 13 ppm/4-h, mice 10 ppm/4-h, rats 14 ppm/4-h (quoted, RTECS, 1985).

Uses: Manufacturing of polyurethane foams and other plastics; cross-linking agent for nylon 6.

o-TOLUIDINE

Synonyms: 1-Amino-2-methylbenzene; 2-Amino-1-methylbenzene; 2-Aminotoluene; *o*-Aminotoluene; C.I. 37077; 1-Methyl-2-aminobenzene; 2-Methyl-1-aminobenzene; 2-Methylaniline; *o*-Methylaniline; **2-Methylbenzenamine**; *o*-Methylbenzenamine; 2-Toluidine; *o*-Tolylamine; UN 1708.

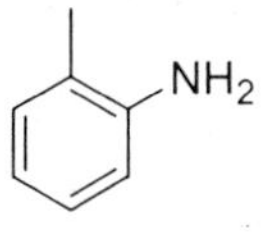

CAS Registry Number: 95-53-4
DOT: 1708
DOT label: Poison
Molecular formula: C_7H_9N
Formula weight: 107.16
RTECS: XU2975000
Merck reference: 10, 9365

Physical state, color, and odor:
Colorless to pale yellow liquid with an aromatic, aniline-like odor. Becomes reddish-brown on exposure to air and light. Odor threshold concentration is 250 ppb (Amoore and Hautala, 1983).

Melting point (°C):
-14.7 (Weast, 1986)

Boiling point (°C):
200.2 (Weast, 1986)

Density (g/cm³ at 20/4 °C):
0.9984 (Weast, 1986)
1.004 (Verschueren, 1983)

Diffusivity in water (x 10^{-5} cm²/sec):
0.85 at 20 °C using method of Hayduk and Laudie (1974)

Dissociation constant, pK_a:
4.45 at 25 °C (Gordon and Ford, 1972)

Flash point (°C):
86 (NIOSH, 1997)

Entropy of fusion (cal/mol·K):
7.6 (Meva'a and Lichanot, 1990)

Heat of fusion (kcal/mol):
1.936 (Meva'a and Lichanot, 1990)

Henry's law constant (x 10^{-3} atm·m³/mol):
3.01 at 25 °C (approximate - calculated from water solubility and vapor pressure)

Ionization potential (eV):
7.47 (Lias, 1998)

Soil sorption coefficient, log K_{oc}:
2.61 (calculated, Mercer et al., 1990)

Octanol/water partition coefficient, log K_{ow}:
1.29 (Leo et al., 1971)
1.32 (Carlson et al., 1975)

Solubility in organics:
Soluble in alcohol and ether (Weast, 1986).

Solubility in water (g/L at 25 °C):
15 (quoted, Verschueren, 1983)
16.33 (Chiou et al., 1982)

Vapor density:
4.38 g/L at 25 °C, 3.70 (air = 1)

Vapor pressure (mmHg):
0.1 at 20 °C, 0.3 at 30 °C (Verschueren, 1983)
0.32 at 25 °C (Banerjee et al., 1990)

Environmental fate:
Biological. Heukelekian and Rand (1955) reported a 5-d BOD value of 1.40 g/g which is 55.1% of the ThOD value of 2.54 g/g.

Chemical/Physical. Kanno et al. (1982) studied the aqueous reaction of *o*-toluidine and other substituted aromatic hydrocarbons (aniline, toluidine, 1- and 2-naphthylamine, phenol, cresol, pyrocatechol, resorcinol, hydroquinone, and 1-naphthol) with hypochlorous acid in the presence of ammonium ion. They reported that the aromatic ring was not chlorinated as expected but was cleaved by chloramine forming cyanogen chloride. As the pH was lowered, the amount of cyanogen chloride formed increased (Kanno et al., 1982).

o-Toluidine will not hydrolyze because it does not contain a hydrolyzable functional group (Kollig, 1993).

Exposure limits: Potential occupational carcinogen. NIOSH REL: IDLH 50 ppm; OSHA PEL: TWA 5 ppm (22 mg/m^3); ACGIH TLV: TWA 2 ppm (adopted).

Symptoms of exposure: May cause methemoglobinemia, anemia, and reticulocytosis. Contact with skin may cause irritation and dermatitis (Patnaik, 1992).

Toxicity:
EC_{50} (48-h) and EC_{50} (24-h) values for *Spirostomum ambiguum* were 3,708 mg/L (Nałecz-Jawecki and Sawicki, 1999).

LC_{50} (48-h) and LC_{50} (24-h) values for *Spirostomum ambiguum* were 5,412 mg/L (Nałecz-Jawecki and Sawicki, 1999).

Acute oral LD_{50} in mice 520 mg/kg, rats 670 mg/kg, rabbits 840 mg/kg (quoted, RTECS, 1985).

Source: As *o*+*p*-toluidine, detected in distilled water-soluble fractions of 87 octane gasoline and Gasohol at concentrations of 0.80 and 0.19 mg/L, respectively (Potter, 1996).

Uses: Manufacture of dyes; vulcanization accelerator; organic synthesis.

TOXAPHENE

Synonyms: Agricide maggot killer (F); Alltex; Alltox; Attac 4-2; Attac 4-4; Attac 6; Attac 6-3; Attac 8; Camphechlor; Camphochlor; Camphoclor; Chemphene M5055; Chlorinated camphene; Chlorocamphene; Chlorter; Clorchem T-590; Compound 3956; Crestoxo; Crestoxo 90; ENT 9735; Estonox; Fasco-terpene; Geniphene; Gyphene; Hercules 3956; Hercules toxaphene; Huilex; Kamfochlor; M 5055; Melipax; Motox; NA 2761; NCI-C00259; Octachlorocamphene; PCC; Penphene; Phenacide; Phenatox; Phenphane; Polychlorcamphene; Polychlorinated camphenes; Polychlorocamphene; RCRA waste number P123; Strobane-T; Strobane T-90; Synthetic 3956; Texadust; Toxakil; Toxadust; Toxadust 10; Toxon 63; Toxyphen; Vertac 90%; Vertac toxaphene 90.

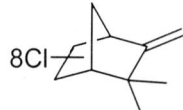

Note: No definitive structure can be illustrated. Toxaphene is a complex mixture of at least 175 chlorinated camphenes; of this number, less than 10 structures are known (Casida et al., 1974). According to Pollock and Kilgore (1978), the average chlorine content is 67-69%.

CAS Registry Number: 8001-35-2
DOT: 2761
DOT label: Poison
Molecular formula: $C_{10}H_{10}Cl_8$
Formula weight: 413.82
RTECS: XW5250000
Merck reference: 10, 9382

Physical state, color, and odor:
Yellow, waxy solid with a chlorine or terpene-like odor. Odor threshold concentration from water is 140 μg/L (Keith and Walters, 1992).

Melting point (°C):
65-90 (IARC, 1979)
85 (Sims et al., 1988)

Boiling point (°C):
Decomposes at 120 (U.S. EPA, 1980a)

Density (g/cm³):
1.6 at 20/4 °C (Melnikov, 1971)
1.519-1.567 at 25/25 °C (Berg, 1983)
1.6 at 15/4 °C (Weiss, 1986)

Flash point (°C):
Nonflammable solid (NIOSH, 1997)
28.9 in 10% xylene (Keith and Walters, 1992)

Henry's law constant (x 10^{-2} atm·m³/mol):
6.3 (Petrasek et al., 1983)

Bioconcentration factor, log BCF:
3.53 (*B. subtilis*, Paris et al., 1977)
3.81 (freshwater fish), 3.72 (fish, microcosm) (Garten and Trabalka, 1983)
6.30 (Arctic cod, Kucklick et al., 1994)

Soil sorption coefficient, log K_{oc}:
3.18 using method of Kenaga and Goring (1980)

Octanol/water partition coefficient, log K_{ow}:
3.30 (Paris et al., 1977)
5.50 (Travis and Arms, 1988)
3.23 (Rao and Davidson, 1980)

Solubility in organics:
120 g/L in alcohol at 25-30 °C (quoted, Meites, 1963)

Solubility in water:
≈ 3 ppm at 25 °C (Brooks, 1974)
740 ppb at 25 °C (Weil et al., 1974)
1.75 mg/L at 25 °C (Warner et al., 1987)
400 ppb at 20-25 °C (Weber, 1972)
550 μg/L at 20 °C (Murphy et al., 1987)

Vapor pressure (x 10^{-6} mmHg):
33 at 20-25 °C (quoted, WHO, 1984a)
1 (Sims et al., 1988)
15.8 at 25 °C (2,2,5-*endo*-6-*exo*-8,9,10-heptachlorobornane, toxaphene component, Hinckley et al., 1990)

Environmental fate:
Soil. Under reduced soil conditions, about 50% of the C-Cl bonds were cleaved (dechlorinated) by Fe^{2+} porphyrins forming two major toxicants having the molecular formulas $C_{10}H_{10}Cl_8$ (Toxicant A) and $C_{10}H_{11}Cl_7$ (Toxicant B). Toxicant A reacted with reduced hematin yielding two reductive dechlorination products ($C_{10}H_{11}Cl_7$), two dehydrodechlorination products ($C_{10}H_9Cl_7$), and two other products ($C_{10}H_{10}Cl_6$). Similarly, products formed from the reaction of Toxicant B with reduced hematin included two reductive dechlorination products ($C_{10}H_{12}Cl_6$), one dehydrochlorination product ($C_{10}H_{10}Cl_6$), and two products having the molecular formula $C_{10}H_{11}Cl_5$ (Khalifa et al., 1976). The reported dissipation rate of toxaphene from soil is 0.010/d (Seiber et al., 1979).

Surface Water. Hargrave et al. (2000) calculated BAFs as the ratio of the compound tissue concentration [wet and lipid weight basis (ng/g)] to the concentration of the compound dissolved in seawater (ng/mL). Average log BAF values for toxaphene in ice algae and phytoplankton collected from the Barrow Strait in the Canadian Archipelago were <5.38 and <5.46, respectively.

Photolytic. Dehydrochlorination will occur after prolonged exposure to sunlight, releasing hydrochloric acid (HSDB, 1989). Two compounds isolated from toxaphene, 2-*exo*, 3-*exo*, 5,5,6-*endo*, 8,9,10,10-nonachloroborane, and 2-*exo*, 3-*exo*, 5,5,6-*endo*, 8,10,10-octachloroborane were irradiated with UV light (λ >290 nm) in a neutral aqueous solution and on a silica gel surface. Both compounds underwent reductive dechlorination, dehydrochlorination, and/or oxidation to yield numerous products including bicyclo[2.1.1]hexane derivatives (Parlar, 1988).

Chemical/Physical. Saleh and Casida (1978) demonstrated that toxicant B (2,2,5-*endo*,6-

exo,8,9,10-heptachlorobornane), the most active component of toxaphene, underwent reductive dechlorination at the geminal dichloro position, yielding 2-*endo*,5-*endo*,6-*exo*,8,9,10-hexachlorobornane, and 2-*exo*,5-*endo*,6-*exo*,8,9,10-hexachlorobornane in various chemical, photochemical, and metabolic systems.

Toxaphene will slowly undergo hydrolysis resulting in the loss of chlorine atoms and the formation of hydrochloric acid (Kollig, 1993). The hydrolysis rate constant for toxaphene at pH 7 and 25 °C was determined to be 8×10^{-6}/h, resulting in a half-life of 9.9 yr. At 85 °C, experimentally determined hydrolysis half-lives were 9.8, 5.2, and 1.6 d at pH values of 3.24, 7.20, and 9.63, respectively (Ellington et al., 1987).

Emits toxic fumes of chlorides when heated to decomposition (Lewis, 1990).

Exposure limits (mg/m³): Potential occupational carcinogen. NIOSH REL: IDLH 200; OSHA PEL: TWA 0.5; ACGIH TLV: TWA 0.5, STEL 1 (adopted).

Symptoms of exposure: Nausea, confusion, agitation, tremors, convulsions, unconsciousness, dry red skin (NIOSH, 1987)

Toxicity:

EC_{50} (48-h) for *Daphnia magna* 10 μg/L, *Daphnia pulex* 14.2 μg/L, *Simocephalus serrulatus* 15 μg/L (Mayer and Ellersieck, 1986).

EC_{50} (24-h) for *Daphnia pulex* 23 μg/L, *Simocephalus serrulatus* 76 μg/L (Mayer and Ellersieck, 1986).

LC_{50} (34-d) for bluegill sunfish 0.7 μg/L; LC_{50} (7-d) for bluegill sunfish 1.4 μg/L (Spehar et al., 1982).

LC_{50} (96-h) for fish (*Cyprinodon variegatus*) 1.1 μg/L, fish (*Leiostomus xanthurus*) 0.5 μg/L, decapod (*Palaemonete pugio*) 4.4 μg/L, decapod (*Penaeus duorarum*) 1.4 μg/L (quoted, Reish and Kauwling, 1978), bluegill sunfish 2.4-4.7 μg/L, fathead minnows 7.0-23 μg/L (Spehar et al., 1982); freshwater mussels (*Anodonta imbecilis*) 740 μg/L (Keller, 1993).

Acute oral LD_{50} for ducks 31 mg/kg, dogs 15 mg/kg, guinea pigs 250 mg/kg, mice 112 mg/kg, rats 55 mg/kg (quoted, RTECS, 1985).

Drinking water standard (final): MCLG: zero; MCL: 3 μg/L (U.S. EPA, 1996).

Uses: Pesticide used primarily on cotton, lettuce, tomatoes, corn, peanuts, wheat, and soybean. Its use was banned by the U.S. EPA in 1982.

1,3,5-TRIBROMOBENZENE

Synonym: *sym*-Tribromobenzene.

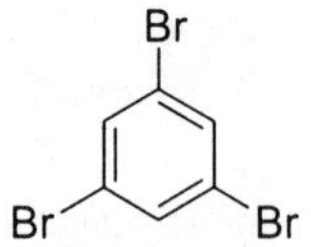

CAS Registry Number: 626-39-1
Molecular formula: $C_6H_3Br_3$
Formula weight: 314.80

Physical state:
Solid

Melting point (°C):
121-122 (Weast, 1986)

Boiling point (°C):
271 at 765 mmHg (Weast, 1986)

Diffusivity in water (x 10^{-5} cm²/sec):
0.62 at 20 °C using method of Hayduk and Laudie (1974)

Bioconcentration factor, log BCF:
3.97-4.08 (rainbow trout, Oliver and Niimi, 1985)

Ionization potential (eV):
8.91-9.21 (Lias et al., 1998)

Soil sorption coefficient, log K_{oc}:
4.05 using method of Chiou et al. (1979)

Octanol/water partition coefficient, log K_{ow}:
4.51 (Watarai et al., 1982)
5.26 (Gobas et al., 1978)

Solubility in organics:
Soluble in benzene, chloroform, and ether (Weast, 1986)

Solubility in water:
2.51 μmol/L at 25 °C (Yalkowsky et al., 1979)

Uses: Organic synthesis.

TRIBUTYL PHOSPHATE

Synonyms: Celluphos 4; **Phosphoric acid tributyl ester**; TBP; Tri-*n*-butyl phosphate.

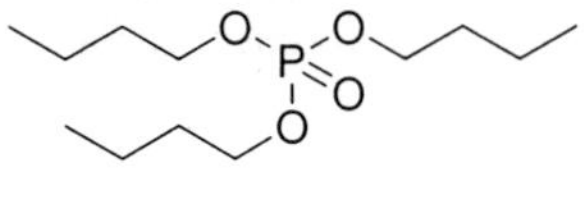

CAS Registry Number: 126-73-8
Molecular formula: $C_{12}H_{27}O_4P$
Formula weight: 266.32
RTECS: TC7700000
Merck reference: 10, 9431

Physical state, color, and odor:
Colorless to pale yellow, odorless liquid

Melting point (°C):
-81 (NIOSH, 1997)

Boiling point (°C):
289 (Weast, 1986) with decomposition (Windholz et al., 1983)

Density (g/cm³):
0.981 at 14.9/4 °C, 0.976 at 20.1/4 °C, 0.972 at 25.0/4 °C, 0.968 at 30.0/4 °C (De Lorenzi et al., 1997)

Diffusivity in water (x 10^{-5} cm²/sec):
0.49 at 20 °C using method of Hayduk and Laudie (1974)

Flash point (°C):
146 (open cup, NFPA, 1984)

Soil sorption coefficient, log K_{oc}:
2.29 (commercial mixture) using method of Kenaga and Goring (1980)

Octanol/water partition coefficient, log K_{ow}:
4.00 (commercial mixture, Saeger et al., 1979)

Solubility in organics:
Miscible with alcohol, ether (Sax and Lewis, 1987), and many other organic solvents (Dean, 1987)

Solubility in water:
≈ 0.61 vol % (quoted, Windholz et al., 1983)
280 ppm at 20-25 °C (commercial mixture, Saeger et al., 1979)

Vapor density:
10.89 g/L at 25 °C, 9.19 (air = 1)

Vapor pressure (x 10^{-3} mmHg):
4 at 25 °C (NIOSH, 1997)

Environmental fate:

Biological. Indigenous microbes in Mississippi River water degraded tributyl phosphate to carbon dioxide. After 4 wk, 90.8% of the theoretical carbon dioxide had evolved (Saeger et al., 1979).

Exposure limits: NIOSH REL: TWA 0.2 ppm (2.5 mg/m^3), IDLH 30 mg/m^3; OSHA PEL: TWA 5 mg/m^3; ACGIH TLV: TWA 0.2 ppm (adopted).

Symptoms of exposure: Depression of central nervous system and irritation of skin, eyes, and respiratory tract (Patnaik, 1992)

Toxicity:

LC$_{50}$ (48-h) for red killifish 68 mg/L (Yoshioka et al., 1986).

Acute oral LD$_{50}$ for mice 1,189 mg/kg, rats 3,000 mg/kg (quoted, RTECS, 1985).

Uses: Plasticizer for lacquers, plastics, cellulose esters, and vinyl resins; heat-exchange liquid; solvent extraction of metal ions from solution of reactor products; pigment grinding assistant; antifoaming agent; solvent for nitrocellulose and cellulose acetate.

1,2,3-TRICHLOROBENZENE

Synonyms: 1,2,3-TCB; UN 2321.

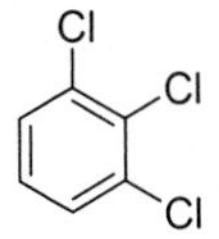

CAS Registry Number: 87-61-6
DOT: 2321 (liquid)
Molecular formula: $C_6H_3Cl_3$
Formula weight: 181.45
RTECS: DC2095000
Merck reference: 10, 9442

Physical state and color:
White crystals or platelets

Melting point (°C):
53-54 (Weast, 1986)

Boiling point (°C):
218-219 (Weast, 1986)

Density (g/cm³):
1.69 (Windholz et al., 1983)

Diffusivity in water (x 10^{-5} cm²/sec):
0.67 at 20 °C using method of Hayduk and Laudie (1974)

Flash point (°C):
113 (Windholz et al., 1983)

Entropy of fusion (cal/mol·K):
13.6 (Miller et al., 1984)

Heat of fusion (kcal/mol):
4.15 (Tsonopoulos and Prausnitz, 1971)
4.30 (Miller et al., 1984)

Henry's law constant (x 10^{-4} atm·m³/mol):
8.9 at 20 °C (Oliver, 1985)
7.10 at 20 °C (ten Hulscher et al., 1992)
12.5 at 25 °C (Shiu and Mackay, 1997)
8.36, 10.3, 13.2, 20.6, and 26.5 at 2.0, 6.0, 10.0, 18.0, and 25.0 °C, respectively (Dewulf et al., 1999)

Ionization potential (eV):
9.18, 9.22 (Lias et al., 1998)

Bioconcentration factor, log BCF:
2.90, 3.28 (*Poecilia reticulata*), 3.08 (*Oncorhynchus mykiss*) (Devillers et al., 1996)
4.11 (guppy, Könemann and van Leeuwen, 1980)
4.54 (Atlantic croakers), 4.77 (blue crabs), 3.13 (spotted sea trout), 4.49 (blue catfish) (Pereira et al., 1988)
2.63 (wet weight based), 0.08 (lipid based) (*Gambusia affinis*, Chaisuksant et al., 1997)
2.94 (pond snail, Legierse et al., 1998)

Soil sorption coefficient, log K_{oc}:
3.24 (average for 5 soils, Kishi et al., 1990)
3.97, 4.19, 5.09 (Paya-Perez et al., 1991)
3.81 (lake sediment, Schrap et al., 1994)
4.07 (Oostvaardersplassen), 3.54 (Ransdorperdie), 3.62 (Noordzeekanaal) (Netherlands sediments, Sijm et al., 1997)
3.35 (sandy soil), 3.38 (sand), 3.36 (peaty sand) (Van Gestel and Ma, 1993)

Octanol/water partition coefficient, log K_{ow}:
3.97 (Watarai et al., 1982)
4.05 (Watarai et al., 1982)
4.02 (McDuffie, 1981)
4.11 (Könemann et al., 1979)
4.04 (Miller et al., 1984)
4.14 (Chiou, 1985; Pereira et al., 1988)
4.27 (Leo et al., 1981)

Solubility in organics:
Soluble in benzene and ether (Weast, 1986)

Solubility in water:
67.6 μmol/L at 25 °C (Miller et al., 1984)
12 ppm at 22 °C (quoted, Verschueren, 1983)
18.0 mg/L at 25 °C (Banerjee, 1984; Chiou et al., 1986)
174 μmol/L at 25 °C (Yalkowsky et al., 1979)
7.66, 19.31, and 45.61 mg/L at 4, 25, and 50 °C, respectively (Shiu et al., 1997)

Vapor pressure (mmHg):
2.1 at 25 °C (Banerjee et al., 1990)
0.14 at 25 °C (extrapolated from vapor pressures determined at higher temperatures, Tesconi and Yalkowsky, 1998)

Environmental fate:
Biological. Under aerobic conditions, soil microbes are capable of degrading 1,2,3-trichlorobenzene to 1,3- and 2,3-dichlorobenzene and carbon dioxide (Kobayashi and Rittman, 1982). A mixed culture of soil bacteria or a *Pseudomonas* sp. transformed 1,2,3-trichlorobenzene to 2,3,4-, 3,4,5-, and 2,3,6-trichlorophenol (Ballschiter and Scholz, 1980).
Photolytic. The sunlight irradiation of 1,2,3-trichlorobenzene (20 g) in a 100-mL borosilicate glass-stoppered Erlenmeyer flask for 56 d yielded 32 ppm pentachlorobiphenyl (Uyeta et al., 1976).
When an aqueous solution containing 1,2,3-trichlorobenzene and a nonionic surfactant micelle (Brij 58, a polyoxyethylene cetyl ether) was illuminated by a photoreactor equipped with 253.7-nm monochromatic UV lamps, 1,2,4- and 1,3,5-trichlorobenzene formed as photo-

isomerization products. Continued irradiation of the solution would yield all dichlorobenzenes, chlorobenzene, benzene, phenol, hydrogen, and chloride ions. The photodecomposition half-life for this reaction, based on the first-order photodecomposition rate of 1.10×10^{-3}/sec, is 10.5 min (Chu and Jafvert, 1994).

Chemical/Physical. At 70.0 °C and pH values of 3.07, 7.13, and 9.80, the hydrolysis half-lives were calculated to be 19.2, 15.0, and 34.4 d, respectively (Ellington et al., 1986).

Emits toxic chloride fumes when heated to decomposition.

Toxicity:

Concentrations that reduce the fertility of *Daphnia magna* in 2 wk for 50% (EC_{50}) and 16% (EC_{16}) of the population are 0.20 and 0.08 mg/L, respectively (Calamari et al., 1983). An EC_{50} value 540 nmol/L for growth rate reduction was determined for mosquito fish (Chaisuksant et al., 1994).

EC_{50} (96-h) and EC_{50} (3-h) concentrations that inhibit the growth of 50% of *Selenastrum capricornutum* population are 0.9 and 2.2 mg/L, respectively (Calamari et al., 1983).

IC_{50} (24-h) for *Daphnia magna* 0.35 mg/L (Calamari et al., 1983).

LC_{50} (14-d) for *Poecilia reticulata* 2.3 mg/L (Könemann, 1981).

LC_{50} (96-h) for mosquito fish (*Gambusia affinis*) 12.1 μmol/L (Chaisuksant et al., 1998).

LC_{50} (48-h) for *Salmo gairdneri* 0.71 mg/L, *Brachydanio rerio* 3.1 mg/L (Calamari et al., 1983).

LC_{50} 2,177-3,084 μg/L (soil porewater concentration) for earthworm (*Eisenia andrei*) and 2,540-3,084 μg/L (soil porewater concentration) for earthworm (*Lumbricus rubellus*) (Van Gestel and Ma, 1993).

Uses: The isomeric mixture is used to control termites; organic synthesis.

1,2,4-TRICHLOROBENZENE

Synonyms: 1,2,4-TCB; *unsym*-Trichlorobenzene; UN 2321.

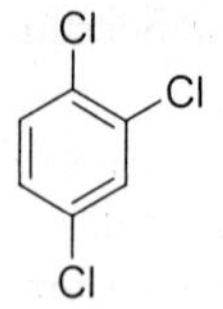

CAS Registry Number: 120-82-1
DOT: 2321
Molecular formula: $C_6H_3Cl_3$
Formula weight: 181.45
RTECS: DC2100000
Merck reference: 10, 9443

Physical state, color, and odor:
Colorless liquid with an odor similar to *o*-dichlorobenzene. Odor threshold concentration is 1.4 (Amoore and Hautala, 1983).

Melting point (°C):
17 (Weast, 1986)

Boiling point (°C):
213.5 (Weast, 1986)
210 (Standen, 1964)

Density (g/cm³):
1.4542 at 20/4 °C (Weast, 1986)
1.4460 at 25/4 °C (Standen, 1964)

Diffusivity in water (x 10^{-5} cm²/sec):
0.78 at 20 °C using method of Hayduk and Laudie (1974)

Flash point (°C):
105 (NFPA, 1984)

Lower explosive limit (%):
2.5 at 150 °C (NFPA, 1984)

Upper explosive limit (%):
6.6 at 150 °C (NFPA, 1984)

Entropy of fusion (cal/mol·K):
12.8 (Tsonopoulos and Prausnitz, 1971)

Heat of fusion (kcal/mol):
3.70 (Tsonopoulos and Prausnitz, 1971)

Henry's law constant (x 10^{-3} atm·m³/mol):
2.32 (Valsaraj, 1988)

1.2 (Oliver, 1985)

1.42 at 25 °C (Warner et al., 1987)

1.29, 1.05, 1.83, 19.2, and 29.7 at 10, 15, 20, 25, and 30 °C, respectively (Ashworth et al., 1988)

0.997 at 20.0 °C (ten Hulscher et al., 1992)

1.24, 2.27, 2.58, 3.06, and 3.90 at 2.0, 6.0, 10.0, 18.0, and 25.0 °C, respectively (Dewulf et al., 1999)

Ionization potential (eV):
9.04 (Lias et al., 1998)

Bioconcentration factor, log BCF:
3.36-3.57 (fish tank), 3.08 (Lake Ontario) (rainbow trout, Oliver and Niimi, 1985)

2.40 (algae, Geyer et al., 1984)

2.23 (*Crassostrea virginica*, Schimmel and Garnas, 1981)

3.15 (activated sludge), 2.69 (golden ide) (Freitag et al., 1985)

4.76 (Atlantic croakers), 4.90 (blue crabs), 3.54 (spotted sea trout), 4.68 (blue catfish) (Pereira et al., 1988)

3.45 (fathead minnow), 3.37 (green sunfish), 2.95 (rainbow trout) (Veith et al., 1979)

3.11 (rainbow trout, Oliver and Niimi, 1985)

2.61 (fathead minnow, Carlson and Kosian, 1987)

In fingerling rainbow trout (*Salmo gairdneri*), values of 1.71, 2.01, and 2.38 were reported for muscle, liver, and bile, respectively, after an 8-h exposure; after a 35-d exposure, values of 1.95, 2.59, and 3.15 were reported for muscle, liver, and bile, respectively (Melancon and Lech, 1980)

Soil sorption coefficient, log K_{oc}:
2.94 (Woodburn silt loam soil, Chiou et al., 1983)

3.01 (Scheunert et al., 1994)

3.14 (Wilson et al., 1981)

3.16 (peaty soil, Friesel et al., 1984)

3.98, 4.61 (lacustrine sediments, Chin et al., 1988)

3.09, 3.16 (Banerjee et al., 1985)

3.19, 3.27 (Marlette soil, Lee et al., 1989)

3.32 (Apison soil), 3.11 (Fullerton soil), 2.95 (Dormont soil) (Southworth and Keller, 1986)

3.49 (Charles River sediment, Wu and Gschwend, 1986)

4.08, 4.41, 5.11 (Paya-Perez et al., 1991)

2.73 (muck), 2.89 (Eustis sand) (Brusseau et al., 1990)

Average K_d values for sorption of 1,2,4-trichlorobenzene to corundum (α-Al_2O_3) and hematite (α-Fe_2O_3) were 0.0110 and 0.0232 mL/g, respectively (Mader et al., 1997)

Octanol/water partition coefficient, log K_{ow}:
4.02 (Chiou, 1985; Pereira et al., 1988)

3.98 (Miller et al., 1984; Chin et al., 1986)

4.23 (Mackay, 1982)

4.11 (Hawker and Connell, 1988)

3.93 (Könemann et al., 1979)

4.176 (Kenaga and Goring, 1980)

4.12 (Anliker and Moser, 1987)

4.07 (DeKock and Lord, 1987)

4.05 (de Bruijn et al., 1989; Leo et al., 1971)

4.22 (DeKock and Lord, 1987)
3.96 (Hammers et al., 1982)
3.63 (Wasik et al., 1981)

Solubility in organics:
Soluble in ether (Weast, 1986), and in other organic solvents and oils (ITII, 1986)

Solubility in water:
31.3 mg/L at 25 °C (Banerjee, 1984)
48.8 mg/L at 25 °C (Neely and Blau, 1985)
34.57 mg/L at 25 °C (Yalkowsky et al., 1979)
0.254 mM at 25 °C (Miller et al., 1984)
64.5 mg/L at 30 °C (McNally and Grob, 1983, 1984)
28.6 mg/L 20 °C (Wilson et al., 1981)

Vapor density:
7.42 g/L at 25 °C, 6.26 (air = 1)

Vapor pressure (mmHg):
0.4 at 25 °C (Mackay et al., 1982; Neely and Blau, 1985)
0.29 at 25 °C (Warner et al., 1987)

Environmental fate:
Biological. Under aerobic conditions, biodegradation products may include 2,3-di-chlorobenzene, 2,4-dichlorobenzene, 2,5-dichlorobenzene, 2,6-dichlorobenzene, and carbon dioxide (Kobayashi and Rittman, 1982). A mixed culture of soil bacteria or a *Pseudomonas* sp. transformed 1,2,4-trichlorobenzene to 2,4,5- and 2,4,6-trichlorophenol (Ballschiter and Scholz, 1980). When 1,2,4-trichlorobenzene was statically incubated in the dark at 25 °C with yeast extract and settled domestic wastewater inoculum, significant biodegradation occurred, with gradual acclimation followed by a deadaptive process in subsequent subcultures. At a concentration of 5 mg/L, 54, 70, 59, and 24% losses were observed after 7, 14, 21, and 28-d incubation periods, respectively. At a concentration of 10 mg/L, only 43, 54, 14, and 0% were observed after 7, 14, 21, and 28-d incubation periods, respectively (Tabak et al., 1981). In activated sludge, <0.1% mineralized to carbon dioxide after 5 d (Freitag et al., 1985).
Surface Water. Estimated half-lives of 1,2,4-trichlorobenzene (0.5 μg/L) from an experimental marine mesocosm during the spring (8-16 °C), summer (20-22 °C), and winter (3-7 °C) were 22, 11, and 12 d, respectively (Wakeham et al., 1983).
Photolytic. A carbon dioxide yield of 9.8% was achieved when 1,2,4-trichlorobenzene adsorbed on silica gel was irradiated with light (λ >290 nm) for 17 h (Freitag et al., 1985).
The sunlight irradiation of 1,2,4-trichlorobenzene (20 g) in a 100-mL borosilicate glass-stop-pered Erlenmeyer flask for 56 d yielded 9,770 ppm 2,4,5,2′,5′-pentachlorobiphenyl (Uyeta et al., 1976).
When an aqueous solution containing 1,2,4-trichlorobenzene (45 μM) and a nonionic surfactant micelle (Brij 58, a polyoxyethylene cetyl ether) was illuminated by a photoreactor equipped with 253.7-nm monochromatic UV lamps for 48 min, the chloride ion concentration increased from 9.4 x 10^{-7} to 1.1 x 10^{-4} M and the pH decreased from 6.9 to 4.0. Intermediate products identified during this reaction included all dichlorobenzenes and the photoisomerization products 1,2,3- and 1,3,5-trichlorobenzene. The photodecomposition half-life for this reaction, based on the first-order photodecomposition rate of 1.21 x 10^{-3}/sec, is 9.6 min (Chu and Jafvert, 1994). A room temperature rate constant of 5.32 x 10^{-13} cm^3/molecule·sec was reported for the vapor-phase reaction of 1,2,4-trichlorobenzene with OH

radicals (Atkinson, 1985).

Chemical/Physical. The hydrolysis half-life was estimated to be >900 yr (Ellington et al., 1988). At 70.0 °C and pH values of 3.10, 7.11, and 9.77, the hydrolysis half-lives were calculated to be 18.4, 6.6, and 5.9 d, respectively (Ellington et al., 1986).

At influent concentrations of 1.0, 0.1, 0.01, and 0.001 mg/L, the granular activated carbon adsorption capacities at pH 5.3 were 157, 77.6, 38.4, and 19.0 mg/g, respectively (Dobbs and Cohen, 1980).

Exposure limits: NIOSH REL: TWA ceiling 5 ppm (40 mg/m^3); ACGIH TLV: ceiling 5 ppm (adopted).

Symptoms of exposure: An irritation concentration of 40.00 mg/m^3 in air was reported by Ruth (1986).

Toxicity:

EC_{10} and EC_{50} concentrations inhibiting the growth of alga *Scenedesmus subspicatus* in 96 h were 3.0 and 8.4 mg/L, respectively (Geyer et al., 1985).

EC_{50} (96-h) and EC_{50} (3-h) concentrations that inhibit the growth of 50% of *Selenastrum capricornutum* population 1.4 and 3.9 mg/L, respectively (Calamari et al., 1983).

Concentrations that reduce the fertility of *Daphnia magna* in 2 wk for 50% (EC_{50}) and 16% (EC_{16}) of the population are 0.45 and 0.32 mg/L, respectively (Calamari et al., 1983).

IC_{50} (24-h) for *Daphnia magna* 1.2 mg/L (Calamari et al., 1983).

LC_{50} (contact) for earthworm (*Eisenia fetida*) 27 μg/cm^2 (Neuhauser et al., 1985).

LC_{50} for red killifish 65 mg/L (Yoshioka et al., 1986).

LC_{50} (14-d) for *Poecilia reticulata* 2.4 mg/L (Könemann, 1981).

LC_{50} (96-h) for fathead minnows 2.9 mg/L (Veith et al., 1983), bluegill sunfish 3.4 mg/L (Spehar et al., 1982), *Cyprinodon variegatus* 21 ppm using natural seawater (Heitmuller et al., 1981).

LC_{50} (72-h) for *Cyprinodon variegatus* 47 ppm (Heitmuller et al., 1981).

LC_{50} (48-h) for *Daphnia magna* 50 mg/L (LeBlanc, 1980), *Salmo gairdneri* 1.95 mg/L, *Brachydanio rerio* 6.3 mg/L (Calamari et al., 1983), *Cyprinodon variegatus* >47 ppm (Heitmuller et al., 1981).

LC_{50} (24-h) for *Daphnia magna* 110 mg/L (LeBlanc, 1980), *Cyprinodon variegatus* >47 ppm (Heitmuller et al., 1981).

Acute oral LD_{50} for mice 300 mg/kg, rats 756 mg/kg (quoted, RTECS, 1985).

Heitmuller et al. (1981) reported a NOEC of 15 ppm.

Drinking water standard (final): MCLG: 70 μg/L; MCL: 70 μg/L (U.S. EPA, 1996).

Uses: Solvent in chemical manufacturing; dyes and intermediates; dielectric fluid; synthetic transformer oils; lubricants; heat-transfer medium; insecticides; organic synthesis.

1,3,5-TRICHLOROBENZENE

Synonyms: 1,3,5-TCB; *sym*-Trichlorobenzene; UN 2321.

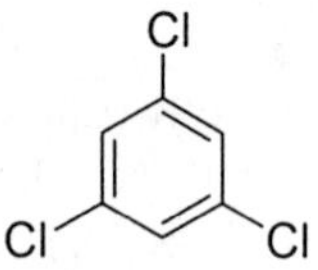

CAS Registry Number: 108-70-3
DOT: 2321 (liquid)
Molecular formula: $C_6H_3Cl_3$
Formula weight: 181.45
RTECS: DC2100100
Merck reference: 10, 9444

Physical state:
Crystals

Melting point (°C):
63-64 (Weast, 1986)

Boiling point (°C):
208 at 763 mmHg (Weast, 1986)

Diffusivity in water (x 10^{-5} cm^2/sec):
0.67 at 20 °C using method of Hayduk and Laudie (1974)

Flash point (°C):
107 (Windholz et al., 1983)

Entropy of fusion (cal/mol·K):
13.4 (Miller et al., 1984)
13.9 (Tsonopoulos and Prausnitz, 1971)

Heat of fusion (kcal/mol):
4.49 (Miller et al., 1984)

Henry's law constant (x 10^{-3} atm·m^3/mol):
1.9 at 20 °C (Oliver, 1985)
1.89 at 20 °C (ten Hulscher et al., 1992)
1.83, 2.04, 2.64, 4.49, and 5.24 at 2.0, 6.0, 10.0, 18.0, and 25.0 °C, respectively (Dewulf et al., 1999)

Ionization potential (eV):
9.30 (Lias, 1998)

Bioconcentration factor, log BCF:
3.26 (*Oncorhynchus mykiss*), 3.48 (*Poecilia reticulata*) (Devillers et al., 1996)
4.40 (Atlantic croakers), 4.45 (blue crabs), 3.51 (spotted sea trout), 4.22 (blue catfish) (Pereira et al., 1988)

4.15 (guppy, Könemann and van Leeuwen, 1980)

2.68 (pond snail, Legierse et al., 1998)

Soil sorption coefficient, log K_{oc}:

>2.55, 2.85 (forest soil), >2.61 (agricultural soil) (Seip et al., 1986)

5.7 (average of 5 suspended sediment samples from the St. Clair and Detroit Rivers, Lau et al., 1989)

K_d = 1.3 mL/g on a Cs^+-kaolinite (Haderlein and Schwarzenbach, 1993)

4.18, 4.39, 5.23 (Paya-Perez et al., 1991)

3.96 (lake sediment, Schrap et al., 1994)

5.10 (Oliver and Charlton, 1984)

Octanol/water partition coefficient, log K_{ow}:

4.19 (Watarai et al., 1982)

4.15 (Könemann et al., 1979)

4.40 at 13 °C, 4.32 at 19 °C, 4.04 at 28 °C, 3.93 at 33 °C (Opperhuizen et al., 1988)

4.02 (Leo et al., 1971; Miller et al., 1984)

4.31 (Chiou, 1985; Pereira et al. 1988)

4.18 (Garst, 1984)

4.52, 4.43, 4.34, 4.21, and 4.09 at 5, 15, 25, 35, and 45 °C, respectively (Bahadur et al., 1997)

Solubility in organics:

Soluble in benzene, ether, ligroin (Weast, 1986), glacial acetic acid, carbon disulfide, and petroleum ether (Windholz et al., 1983)

Solubility in water:

2.7 μmol/L at 25 °C (Miller et al., 1984)

5.8 ppm at 20 °C (quoted, Verschueren, 1983)

6.01 mg/L at 25 °C (Banerjee, 1984)

8.46 mg/L at 25 °C (Shiu et al., 1997)

Vapor pressure (mmHg):

0.58 at 25 °C (Mackay et al., 1982)

Environmental fate:

Biological. Under aerobic conditions, soil microbes degraded 1,3,5-trichlorobenzene to 1,4- and 2,4-dichlorobenzene and carbon dioxide (Kobayashi and Rittman, 1982). A mixed culture of soil bacteria or a *Pseudomonas* sp. transformed 1,3,5-trichlorobenzene to 2,4,6-trichlorophenol (Ballschiter and Scholz, 1980).

Photolytic. The sunlight irradiation of 1,3,5-trichlorobenzene (20 g) in a 100-mL borosilicate glass-stoppered Erlenmeyer flask for 56 d yielded 160 ppm pentachlorobiphenyl (Uyeta et al., 1976). A photooxidation half-life of 6.17 months was reported for the vapor-phase reaction of 1,3,5-trichlorobenzene with OH radicals (Atkinson, 1985).

When an aqueous solution, containing 1,3,5-trichlorobenzene and a nonionic surfactant micelle (Brij 58, a polyoxyethylene cetyl ether), was illuminated by a photoreactor equipped with 253.7-nm monochromatic UV lamps, 1,2,4-trichlorobenzene formed as a result of photo-isomerization. Based on photodechlorination of other polychlorobenzenes under similar conditions, it was suggested that dichlorobenzenes, chlorobenzene, benzene, phenol, hydrogen, and chloride ions would form as photodegradation products. The photodecomposition half-life for this reaction, based on the first-order photodecomposition rate of 1.07×10^{-3}/sec, is 10.8 min (Chu and Jafvert, 1994).

Toxicity:
  LC_{50} (14-d) for *Poecilia reticulata* 7.4 μg/L (Könemann, 1981).

Drinking water standard: No MCLGs or MCLs have been proposed (U.S. EPA, 1996).

Uses: Organic synthesis.

1,1,1-TRICHLOROETHANE

Synonyms: Aerothene; Aerothene TT; Baltana; Chloroethene; Chloroethene NU; Chlorothane NU; Chlorothene; Chlorothene NU; Chlorothene VG; Chlorten; Genklene; Inhibisol; Methyl chloroform; Methyltrichloromethane; NCI-C04626; RCRA waste number U226; Solvent III; α-T; 1,1,1-TCA; 1,1,1-TCE; α-Trichloroethane; Triethane; UN 2831.

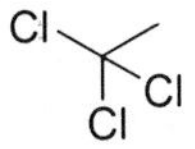

CAS Registry Number: 71-55-6
DOT: 2831
Molecular formula: $C_2H_3Cl_3$
Formula weight: 133.40
RTECS: KJ2975000
Merck reference: 10, 9449

Physical state, color, and odor:
Colorless, watery liquid with an odor similar to chloroform. The reported odor threshold concentrations in air ranged from 3.2 to 100 ppm (Keith and Walters, 1992; Young et al., 1996). The taste threshold concentration in water is 1.5 ppm (Young et al., 1996).

Melting point (°C):
-30.6 (Stull, 1947)
-32.62 (Standen, 1964)

Boiling point (°C):
74.1 (Dreisbach, 1959)

Density (g/cm³):
1.3390 at 20/4 °C (Weast, 1986)
1.37068 at 0/4 °C, 1.34587 at 15/4 °C, 1.3296 at 30/4 °C (Standen, 1964)

Diffusivity in water (x 10^{-5} cm²/sec):
0.89 at 20 °C using method of Hayduk and Laudie (1974)

Flash point (°C):
None (NFPA, 1984)
≤25 (Kuchta et al., 1968)

Lower explosive limit (%):
7.5 (NFPA, 1984)

Upper explosive limit (%):
12.5 (NFPA, 1984)

Heat of fusion (kcal/mol):
0.562 (Riddick et al., 1986)

Henry's law constant (x 10^{-2} atm·m³/mol):
3.45 at 25 °C (Pearson and McConnell, 1975)
1.5 at 20 °C (Roberts and Dändliker, 1983)

2.74 at 37 °C (Sato and Nakajima, 1979)

2.1 at 30 °C (Jeffers et al., 1989)

0.561 and 1.36 at 5.7 and 24.9 °C, respectively (Dewulf et al., 1999a)

0.965, 1.15, 1.46, 1.74, and 2.11 at 10, 15, 20, 25, and 30 °C, respectively (Ashworth et al., 1988)

1.26, 2.00, 2.35, and 2.81 at 20, 30, 35, and 40 °C, respectively (Tse et al., 1992)

At 25 °C: 1.30 and 2.30 in distilled water and seawater, respectively (Hunter-Smith et al., 1983)

Distilled water: 0.480, 0.635, 0.696, 1.127, and 1.490 at 2.0, 6.0, 10.0, 18.2, and 25.0 °C, respectively; natural seawater: 7.90 and 18.6 at 6.0 and 25.0 °C, respectively (Dewulf et al., 1995)

1.74, 2.38, and 3.19 at 26.3, 35.0, and 44.8 °C, respectively (Hansen et al., 1993, 1995)

1.69 at 25 °C (Gossett, 1987)

1.98 at 25.3 °C (mole fraction ratio, Leighton and Calo, 1981)

1.85 at 25 °C (Hoff et al., 1993)

1.64 at 25 °C (Wright et al., 1992)

1.76, 2.18, 2.64, 3.55, and 4.11 at 25, 30, 40, 45, and 50 °C, respectively (Robbins et al., 1993)

1.27 at 20 °C (Hovorka and Dohnal, 1997)

0.823, 1.34, and 2.12 at 10, 20, and 30 °C, respectively (Munz and Roberts, 1987)

1.87, 2.95, 3.89, and 5.33 at 30, 40, 50, and 60 °C, respectively (Vane and Giroux, 2000)

Interfacial tension with water (dyn/cm at 25 °C):
36.6 (quoted, Freitas et al., 1997)

Ionization potential (eV):
10.82 (Horvath, 1982)

Bioconcentration factor, log BCF:
0.95 (bluegill sunfish, Veith et al., 1980)

Soil sorption coefficient, log K_{oc}:
2.25 (Willamette silt loam, Chiou et al., 1979)

2.02 (Eerd soil), 2.24 (Podzol soil) (Loch et al., 1986)

2.03 (Friesel et al., 1984)

2.15, 2.50 (Allerod), 2.40 (Borris), 3.18, 3.19 (Brande), 2.15, 2.85 (Finderup), 2.79 (Gunderup), 2.41 (Herborg), 3.01 (Rabis), 2.60, 2.63 (Tirstrup), 2.26 (Tylstrup), 3.24 (Vasby), 2.68, 2.80 (Vejen), 2.52, 3.29, 3.40 (Vorbasse) (Larsen et al., 1992)

1.95, 1.98, 1.98, 1.99, 2.01, 1.98, and 2.03 at 2.3, 3.8, 6.2, 8.0, 13.5, 18.6, and 25.0 °C, respectively, for a Leie River (Belgium) clay (Dewulf et al., 1999a)

Octanol/water partition coefficient, log K_{ow}:
2.48 (Neely and Blau, 1985)

2.49 (Hansch and Leo, 1979)

2.47 (Veith et al., 1980)

2.17 (Schwarzenbach et al., 1983)

2.47, 2.50, and 2.52 at 25, 35, and 50 °C, respectively (Bhatia and Sandler, 1995)

2.18, 2.18, 2.28, 2.24, 2.29, and 2.20 at 2.2, 6.0, 10.0, 14.1, 18.7, and 24.8 °C, respectively (Dewulf et al., 1999a)

Solubility in organics:
Sparingly soluble in ethyl alcohol; freely soluble in carbon disulfide, benzene, ethyl ether, methanol, carbon tetrachloride (U.S. EPA, 1985), and many other organic solvents

Solubility in water:
480 mg/L at 20 °C (Pearson and McConnell, 1975)
1,334 mg/L at 25 °C (Neely and Blau, 1985)
730 mg/L at 20 °C (Mackay and Shiu, 1981)
1,360 mg/L at 20 °C (Chiou et al., 1979)
0.44 mass % at 20 °C (Konietzko, 1984)
0.1175 wt % 23.5 °C (Schwarz, 1980)
347 mg/L at 25 °C (Howard, 1990)
479.8 mg/L at 30 °C (McNally and Grob, 1984)
1,250 mg/L at 23-24 °C (Broholm and Feenstra, 1995)
1.69×10^{-4} at 25 °C (mole fraction, Li et al., 1993)
In wt %: 0.147 at 0 °C, 0.070 at 20.2 °C, 0.076 at 31.6 °C, 0.101 at 41.1 °C, 0.106 at 51.3 °C,
    0.103 at 61.5 °C, 0.114 at 71.5 °C (Stephenson, 1992)
    0.180, 0.185, and 0.159 wt % at 10.0, 20.0, and 30.0 °C, respectively (Schwarz and Miller,
    1980)

Vapor density:
5.45 g/L at 25 °C, 4.60 (air = 1)

Vapor pressure (mmHg):
62 at 10 °C, 100 at 20 °C, 150 at 30 °C (Anliker and Moser, 1987)
124 at 25 °C (Neely and Blau, 1985)
96 at 20 °C (U.S. EPA, 1980a)

Environmental fate:
   Biological. Microbial degradation by sequential dehalogenation under laboratory conditions
produced 1,1-dichloroethane, *cis-* and *trans*-1,2-dichloroethylene, chloroethane, and vinyl
chloride. Hydrolysis products via dehydrohalogenation included acetic acid, 1,1-dichloroeth-
ylene (Dilling et al., 1975; Smith and Dragun, 1984) and hydrochloric acid (Dilling et al.,
1975). The reported half-lives for this reaction at 20 and 25 °C are 0.5-2.5 and 1.1 yr,
respectively (Vogel et al., 1987; ten Hulscher et al., 1992).
   In an anoxic aquifer beneath a landfill in Ottawa, Ontario, Canada, there was evidence that
1,1,1-trichloroethane was biotransformed to 1,1-dichloroethane, 1,1-dichloroethylene, and vinyl
chloride (Lesage et al., 1990). In a similar study, 1,1,1-trichloroethane rapidly degraded in
samples from an alluvial aquifer (Norman, OK) under both methanogenic and sulfate reducing
conditions (Klečka et al., 1990). 1,1-Dichloroethane, ethyl chloride, possibly acetic acid and
carbon dioxide (mineralization of acetic acid) were formed by biological processes, whereas
1,1-dichloroethylene and acetic acid were formed under abiotic conditions. The reported
biological and abiotic half-lives were 46-206 and 1,155-1,390 d, respectively. Under aerobic or
denitrifying conditions, 1,1,1-trichloroethane was recalcitrant to biodegradation (Klečka et al.,
1990).
   An anaerobic species of *Clostridium* biotransformed 1,1,1-trichloroethane to 1,1-dichloro-
ethane, acetic acid, and unidentified products (Gälli and McCarty, 1989). A microcosm
composed of aquifer water and sediment collected from uncontaminated sites in the Everglades
biotransformed 1,1,1-trichloroethane to 1,1-dichloroethylene (Parsons and Lage, 1985). In a
static-culture-flask screening test, 1,1,1-trichloroethane was statically incubated in the dark at
25 °C with yeast extract and settled domestic wastewater inoculum. Percent degradation was 83
and 75% after 28 d at substrate concentrations of 5 and 10 mg/L, respectively. The amount lost
due to volatilization was 7-27%, respectively (Tabak et al., 1981).
   Surface Water. Estimated half-lives of 1,1,1-trichloroethane (4.3 μg/L) from an experimental
marine mesocosm during the spring (8-16 °C), summer (20-22 °C), and winter (3-7 °C) were

24, 12, and 11 d, respectively (Wakeham et al., 1983).

Photolytic. Reported photooxidation products include phosgene, chlorine, hydrochloric acid, and carbon dioxide (McNally and Grob, 1984). Acetyl chloride (Christiansen et al., 1972) and trichloroacetaldehyde (U.S. EPA, 1975) have also been reported as photooxidation products. 1,1,1-Trichloroethane may react with OH radicals in the atmosphere producing chlorine atoms and chlorine oxides (McConnell and Schiff, 1978). The rate constant for this reaction at 300 K is 9.0×10^9 cm^3/mol·sec (Hendry and Kenley, 1979).

Chemical/Physical. The evaporation half-life of 1,1,1-trichloroethane (1 mg/L) from water at 25 °C using a shallow-pitch propeller stirrer at 200 rpm at an average depth of 6.5 cm is 18.7 min (Dilling, 1977).

The experimental half-life for hydrolysis of 1,1,1-trichloroetane in water at 25 °C is 6 months (Dilling et al., 1975). Products of hydrolysis include acetic acid, 1,1-dichloroethylene, and hydrochloric acid (Kollig, 1993).

At influent concentrations of 1.0, 0.1, 0.01, and 0.001 mg/L, the granular activated carbon adsorption capacities at pH 5.3 were 2.5, 1.1, 0.51, and 0.23 mg/g, respectively (Dobbs and Cohen, 1980).

Slowly reacts with water forming hydrochloric acid (NIOSH, 1997).

Exposure limits: NIOSH REL: 15-min ceiling 350 ppm (1,900 mg/m^3), IDLH 700 ppm; OSHA PEL: TWA 350 ppm; ACGIH TLV: TWA 350 ppm, STEL 450 ppm (adopted).

Symptoms of exposure: Vapors are irritating to eyes and mucous membranes (Patnaik, 1992). An irritation concentration of 5,428.57 mg/m^3 in air was reported by Ruth (1986).

Toxicity:

LC$_{50}$ (contact) for earthworm (*Eisenia fetida*) 83 μg/cm^2 (Neuhauser et al., 1985).

LC$_{50}$ (7-d) for *Poecilia reticulata* 133 mg/L (Könemann, 1981).

LC$_{50}$ (96-h) for bluegill sunfish 72 μg/L (Spehar et al., 1982), for fathead minnows 52.8 mg/L (flow-through), 105 mg/L (static bioassay) (Alexander et al., 1978), *Cyprinodon variegatus* 71 ppm using natural seawater (Heitmuller et al., 1981).

LC$_{50}$ (72-h) for *Cyprinodon variegatus* 71 ppm (Heitmuller et al., 1981).

LC$_{50}$ (48-h) for *Daphnia magna* >530 mg/L (LeBlanc, 1980), *Cyprinodon variegatus* 71 ppm (Heitmuller et al., 1981).

LC$_{50}$ (24-h) for *Daphnia magna* >530 mg/L (LeBlanc, 1980), *Cyprinodon variegatus* 68 ppm (Heitmuller et al., 1981).

Acute oral LD$_{50}$ for dogs 750 mg/kg, guinea pigs 9,470 mg/kg, mice 11,240 mg/kg, rats 10,300 rabbits 5,660 mg/kg (quoted, RTECS, 1985).

Heitmuller et al. (1981) reported a NOEC of 43 ppm.

Drinking water standard (final): MCLG: 0.2 mg/L; MCL: 0.2 mg/L (U.S. EPA, 1996).

Uses: Organic synthesis; solvent for metal cleaning of precision instruments; textile processing; aerosol propellants; pesticide.

1,1,2-TRICHLOROETHANE

Synonyms: Ethane trichloride; NCI-C04579; RCRA waste number U227; β-T; 1,1,2-TCA; 1,2,2-Trichloroethane; β-Trichloroethane; Vinyl trichloride.

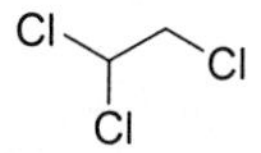

CAS Registry Number: 79-00-5
DOT: 2831
Molecular formula: $C_2H_3Cl_3$
Formula weight: 133.40
RTECS: KJ3150000
Merck reference: 10, 9450

Physical state, color, and odor:
Colorless liquid with a pleasant, sweet, chloroform-like odor

Melting point (°C):
-36.5 (Weast, 1986)
-37.0 (Standen, 1964)

Boiling point (°C):
113.8 (Weast, 1986)
111-114 (Fluka, 1988)

Density (g/cm³ at 20/4 °C):
1.4397 (Weast, 1986)
1.434 (Fluka, 1988)
1.4410 (Standen, 1964)

Diffusivity in water (x 10^{-5} cm²/sec):
0.92 at 20 °C using method of Hayduk and Laudie (1974)

Flash point (°C):
None (Dean, 1973)

Lower explosive limit (%):
6 (NIOSH, 1997)

Upper explosive limit (%):
15.5 (NIOSH, 1997)

Entropy of fusion (cal/mol·K):
10.9 (Golovanova and Kolesov, 1984)
11.5 (Crowe and Smyth, 1950)

Heat of fusion (kcal/mol):
2.6003 (Golovanova and Kolesov, 1984)
2.7198 (Crowe and Smyth, 1950)

Henry's law constant (x 10^{-4} atm·m^3/mol):
7.4 (Pankow and Rosen, 1988)
8.33 at 25 °C (Wright et al., 1992)
14.9 at 37 °C (Sato and Nakajima, 1979)
13.3 at 30 °C (Jeffers et al., 1989)
3.9, 6.3, 7.4, 9.1, and 13.3 at 10, 15, 20, 25, and 30 °C, respectively (Ashworth et al., 1988)
7, 11, and 17 at 20, 30, and 40 °C, respectively (Tse et al., 1992)
8.09, 18.2, and 25.3 at 26.2, 35.0, and 44.8 °C, respectively (Hansen et al., 1993, 1995)
6.60 at 20 °C (Hovorka and Dohnal, 1997)
7.95 at 24.3 °C (mole fraction ratio, Leighton and Calo, 1981)
2.49, 3.21, 4.18, 6.45, and 9.07 at 2.0, 6.0, 10.0, 18.0, and 25.0 °C, respectively (Dewulf et al., 1999)

Interfacial tension with water (dyn/cm at 25 °C):
27.1 (Murphy et al., 1957)
29.6 (Demond and Lindner, 1993)

Ionization potential (eV):
11.00 (NIOSH, 1997)

Soil sorption coefficient, log K_{oc}:
1.78, 2.03 (forest soil), 1.80 (agricultural soil) (Seip et al., 1986)
1.88 (Lincoln sand, Wilson et al., 1981)

Octanol/water partition coefficient, log K_{ow}:
1.98, 1.93, and 1.94 at 25, 35, and 50 °C, respectively (Bhatia and Sandler, 1995)

Solubility in organics:
Soluble in ethanol and chloroform (U.S. EPA, 1985)

Solubility in water:
4,400 mg/kg at 25 °C (McGovern, 1943)
3,704 mg/L at 25 °C (Van Arkel and Vles, 1936)
4,500 mg/L at 19.6 °C (Gladis, 1960)
0.45 wt % at 20 °C (Konietzko, 1984)
4,365.3 mg/L at 30 °C (McNally and Grob, 1984)
In wt %: 0.464 at 0 °C, 0.439 at 9.2 °C, 0.458 at 31.3 °C, 0.483 at 41.0 °C, 0.518 at 50.6 °C, 0.497 at 60.5 °C, 0.555 at 71.0 °C, 0.658 at 81.7 °C, 0.703 at 90.8 °C (Stephenson, 1992)

Vapor density:
5.45 g/L at 25 °C, 4.60 (air = 1); 4 kg/m^3 at the boiling point (Konietzko, 1984)

Vapor pressure (mmHg):
19 at 20 °C, 32 at 30 °C (Verschueren, 1983)
30.3 at 25 °C (Mackay et al., 1982)

Environmental fate:
Biological. Vinyl chloride was reported to be a biodegradation product from an anaerobic digester at a wastewater treatment facility (Howard, 1990). Under aerobic conditions, *Pseudomonas putida* oxidized 1,1,2-trichloroethane to chloroacetic and glyoxylic acids. Simultaneously, 1,1,2-trichloroethane is reduced to vinyl chloride exclusively (Castro and

Belser, 1990). In a static-culture-flask screening test, 1,1,2-trichloroethane was statically incubated in the dark at 25 °C with yeast extract and settled domestic wastewater inoculum. Biodegradative activity was slow to moderate, concomitant with a significant rate of volatilization (Tabak et al., 1981).

Chemical/Physical. Products of hydrolysis include chloroacetaldehyde, 1,1-dichloroethylene, and hydrochloric acid. The aldehyde is subject to further hydrolysis breaking down to hydroxyacetaldehyde and hydrochloric acid (Kollig, 1993). The reported half-life for this reaction at 20 °C is 170 yr (Vogel et al., 1987). Under alkaline conditions, 1,1,2-trichloroethane hydrolyzed to 1,2-dichloroethylene. The reported hydrolysis half-life in water at 25 °C and pH 7 is 139.2 yr (Sata and Nakajima, 1979).

1,1,2-Trichloroethane (0.22 mM) reacted with OH radicals in water (pH 2.8) at a rate of 1.3 x 10^8/M·sec (Haag and Yao, 1992).

The volatilization half-life of 1,1,2-trichloroethane (1 mg/L) from water at 25 °C using a shallow-pitch propeller stirrer at 200 rpm at an average depth of 6.5 cm is 35.1 min (Dilling, 1977).

At influent concentrations of 1.0, 0.1, 0.01, and 0.001 mg/L, the granular activated carbon adsorption capacities were 5.8, 1.4, 0.36, and 0.09 mg/g, respectively (Dobbs and Cohen, 1980).

Exposure limits: Potential occupational carcinogen. NIOSH REL: TWA 10 ppm (45 mg/m³), IDLH 100 ppm; OSHA PEL: TWA 10 ppm (adopted).

Symptoms of exposure: Irritant to eyes and mucous membranes. Ingestion may cause somnolence, nausea, vomiting, ulceration, hepatitis, and necrosis (Patnaik, 1992).

Toxicity:

LC_{50} (contact) for earthworm (*Eisenia fetida*) 42 μg/cm² (Neuhauser et al., 1985).

LC_{50} (7-d) for juvenile guppy 70 mg/L, adult guppy 75 mg/L (Spehar et al., 1982), *Poecilia reticulata* 94 mg/L (Könemann, 1981).

LC_{50} (96-h) for fathead minnows 81.7 mg/L (Veith et al., 1983), bluegill sunfish 40 mg/L (Spehar et al., 1982).

LC_{50} (48-h) for *Daphnia magna* 18 mg/L (LeBlanc, 1980).

LC_{50} (24-h) for *Daphnia magna* 140 mg/L, juvenile guppy 72 mg/L, adult guppy 85 mg/L (Spehar et al., 1982), *Daphnia magna* 19 mg/L (LeBlanc, 1980).

Acute oral LD_{50} for mice 378 mg/kg, rats 580 mg/kg (quoted, RTECS, 1985).

Drinking water standard (final): MCLG: 3 μg/L; MCL: 5 μg/L (U.S. EPA, 1996).

Uses: Solvent for fats, oils, resins, waxes, resins, and other products; organic synthesis.

TRICHLOROETHYLENE

Synonyms: Acetylene trichloride; Algylen; Anamenth; Benzinol; Blacosolv; Blancosolv; Cecolene; Chlorilen; 1-Chloro-2,2-dichloroethylene; Chlorylea; Chlorylen; Circosolv; Crawhaspol; Densinfluat; 1,1-Dichloro-2-chloroethylene; Dow-tri; Dukeron; Ethinyl trichloride; Ethylene trichloride; Fleck-flip; Flock-flip; Fluate; Gemalgene; Germalgene; Lanadin; Lethurin; Narcogen; Narkogen; Narkosoid; NCI-C04546; Nialk; Perm-a-chlor; Perm-a-clor; Petzinol; Philex; RCRA waste number U228; TCE; Threthylen; Threthylene; Trethylene; Tri; Triad; Trial; Triasol; Trichloran; Trichloren; **Trichloroethene**; 1,1,2-Trichloroethene; 1,2,2-Trichloroethene; 1,1,2-Trichloroethylene; 1,2,2-Trichloroethylene; Triclene; Trielene; Trieline; Triklone; Trilen; Trilene; Triline; Trimar; Triol; Triplus; Triplus M; UN 1710; Vestrol; Vitran; Westrosol.

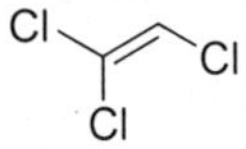

CAS Registry Number: 79-01-6
DOT: 1710
Molecular formula: C_2HCl_3
Formula weight: 131.39
RTECS: KX4550000
Merck reference: 10, 9452

Physical state, color, and odor:
Clear, colorless, watery-liquid with a chloroform-like odor. Odor threshold concentration is 21.4 ppm (Keith and Walters, 1992).

Melting point (°C):
-86.4 (McGovern, 1943)
-87.1 (Standen, 1964)

Boiling point (°C):
87.2 (Dean, 1973)
86.7 (Standen, 1964)

Density (g/cm³):
1.464 at 20/4 °C (McGovern, 1943)
1.461 at 20/4 °C (Fluka, 1988)
1.4996 at 0/4 °C, 1.4762 at 15/4 °C, 1.4514 at 30/4 °C (Carlisle and Levine, 1932a)

Diffusivity in water (x 10⁻⁵ cm²/sec):
0.94 at 20 °C using method of Hayduk and Laudie (1974)

Flash point (°C):
32.2 (Weiss, 1986)

Lower explosive limit (%):
8 at 25 °C, 7.8 at 100 °C (NFPA, 1984)

Upper explosive limit (%):
10.5 at 25 °C, 52 at 100 °C (NFPA, 1984)

Entropy of fusion (cal/mol·K):
10.7 (Golovanova and Kolesov, 1984)

Heat of fusion (kcal/mol):
2.020 (Golovanova and Kolesov, 1984)

Henry's law constant (x 10^{-3} atm·m^3/mol):
9.1 (Pankow and Rosen, 1988)
9.9 at 20 °C (Roberts and Dändliker, 1983)
19.6 at 37 °C (Sato and Nakajima, 1979)
10.1 at 20 °C (Roberts et al., 1985)
7.69 at 25 °C (Nielsen et al., 1994)
9.09 at 20 °C (Pearson and McConnell, 1975)
9.09 at 25 °C (Gossett, 1987; Wright et al., 1992; Hodd et al., 1993)
12.8 at 30 °C (Jeffers et al., 1989)
10.3, 13.1, 16.5, 22.4, and 26.2 at 25, 30, 40, 45, and 50 °C, respectively (Robbins et al., 1993)
14.5 at 25 °C (Tancréde and Yanagisawa, 1990)
5.38, 6.67, 8.42, 10.2, and 12.8 at 10, 15, 20, 25, and 30 °C, respectively (Ashworth et al., 1988)
7.0, 11.4, and 17.3 at 20, 30, and 40 °C, respectively (Tse et al., 1992)
Distilled water: 2.47, 3.06, 3.41, 6.22, and 8.60 at 2.0, 6.0, 10.0, 18.2, and 25.0 °C, respectively; natural seawater: 3.75 and 10.6 at 6.0 and 25.0 °C, respectively (Dewulf et al., 1995)
7.42 at 20 °C (Hovorka and Dohnal, 1997)
9.86 at 25.3 °C (mole fraction ratio, Leighton and Calo, 1981)
At 25 °C (NaCl concentration, mol/L): 7.12 (0), 7.39 (0.1), 8.25 (0.3), 9.25 (0.5), 10.21 (0.7), 11.80 (1.0); salinity, NaCl = 37.34 g/L (°C): 7.22 (15), 9.29 (20), 11.80 (25), 13.76 (30), 17.38 (35), 21.72 (40), 26.72 (45) (Peng and Wan, 1998).
4.47, 7.86, and 13.3 at 10, 20, and 30 °C, respectively (Munz and Roberts, 1987)
11.80 at 25.0 °C (Ramachandran et al., 1996)
2.71, 3.30, 4.32, 7.22, and 10.2 at 2.0, 6.0, 10.0, 18.0, and 25.0 °C, respectively (Dewulf et al., 1999)
0.32 and 7.95 at 5.7 and 24.9 °C, respectively (Dewulf et al., 1999a)
10.63, 17.82, 24.46, and 34.82 at 30, 40, 50, and 60 °C, respectively (Vane and Giroux, 2000)

Interfacial tension with water (dyn/cm at 24 °C):
34.5 (Demond and Lindner, 1993)

Ionization potential (eV):
9.94 (Yoshida et al., 1983a)
9.45 (Franklin et al., 1969)

Bioconcentration factor, log BCF:
1.23 (bluegill sunfish, Veith et al., 1980)
1.59 (rainbow trout, Neely et al., 1974)
3.06 (algae, Geyer et al., 1984)
3.00 (activated sludge), 1.95 (golden ide) (Freitag et al., 1985)

Soil sorption coefficient, log K_{oc}:
1.81 (Abdul et al., 1987)
2.025 (Garbarini and Lion, 1986)
1.86, 1.98, 2.15 (various Norwegian soils, Seip et al., 1986)

1.79 (humic-coated alumina, Peterson et al., 1988)
1.66, 2.64, 2.83 (silty clay, Pavlostathis and Mathavan, 1992)
1.76 (peat, Rutherford and Chiou, 1992)
1.80 (Smith et al., 1990)
1.84 (Tampa sandy aquifer material, Brusseau and Rao, 1991)
1.92 (Appalachee soil, Stauffer and MacIntyre, 1986)
1.93 (aquifer material, Piwoni and Banerjee, 1989)
1.94 (Overton silty clay loam), 2.17 (Hastings silty clay loam) (Rogers and McFarlane, 1981)
1.96 (Lincoln sand, Wilson et al., 1981)
2.00 (Friesel et al., 1984)
2.01 (Eerd soil, Loch et al., 1986)
1.70 (Mt. Lemmon soil, Hu et al., 1995)
2.23, 2.32, 2.35, 2.34, 2.34, 2.36, and 2.41 at 2.3, 3.8, 6.2, 8.0, 13.5, 18.6, and 25.0 °C, respectively, for a Leie River (Belgium) clay (Dewulf et al., 1999a)
2.36 (muck), 1.77 (Eustis sand), 2.37 (Grayling sand), 2.14 (Keewenaw silty sand) (Brusseau et al., 1990)

Octanol/water partition coefficient, log K_{ow}:
2.53 (Miller et al., 1985; Wasik et al., 1981)
2.29 (Leo et al., 1971; Schwarzenbach et al., 1983)
2.42 (Banerjee et al., 1980)
2.60 (Hawker and Connell, 1988)
3.24, 3.30 (Geyer et al., 1984)
2.37 (Green et al., 1983)
3.03 (Tewari et al., 1982)
2.25, 2.27, 2.38, 2.32, 2.36, and 2.27 at 2.2, 6.0, 10.0, 14.1, 18.7, and 24.8 °C, respectively (Dewulf et al., 1999a)

Solubility in organics:
Soluble in acetone, ethanol, chloroform, ether (U.S. EPA, 1985), and other organic solvents including bromoform, carbon tetrachloride, methylene chloride, trichloroethylene, and tetrachloroethylene.

Solubility in water:
1,100 mg/L at 20 °C (Pearson and McConnell, 1975)
1.1 g/kg at 25 °C (McGovern, 1943)
10.4 mM at 25 °C (Tewari et al., 1982; Wasik et al., 1981)
1,250 mg/L at 60 °C (quoted, Standen, 1964)
743.1 mg/L at 30 °C (McNally and Grob, 1984)
1,400 mg/L at 23-24 °C (Broholm and Feenstra, 1995)
1.14×10^{-4} at 25 °C (mole fraction, Li et al., 1993)
1,300 and 1,500 mg/L at 9 and 71 °C, respectively (Heron et al., 1998)

Vapor density:
5.37 g/L at 25 °C, 4.54 (air = 1)

Vapor pressure (mmHg):
74 at 25 °C (Mackay and Shiu, 1981)
100 at 32 °C (Sax, 1984)
56.8 at 20 °C, 72.6 at 25 °C, 91.5 at 30 °C (Klöpffer et al., 1988)
69 at 25 °C (Howard, 1990)

Environmental fate:

Biological. Microbial degradation of trichloroethylene via sequential dehalogenation produced *cis-* and *trans*-1,2-dichloroethylene and vinyl chloride (Smith and Dragun, 1984). Anoxic microcosms in sediment and water degraded trichloroethylene to 1,2-dichloroethylene and then to vinyl chloride (Barrio-Lage et al., 1986). Trichloroethylene in soil samples collected from Des Moines, IA anaerobically degraded to 1,2-dichloroethylene. The production of 1,1-dichloroethylene was not observed in this study (Kleopfer, 1985).

In a methanogenic aquifer, trichloroethylene biodegraded to 1,2-dichloroethylene and vinyl chloride (Wilson et al., 1986). Dichloroethylene was reported as a biotransformation product under anaerobic conditions and in experimental systems using mixed or pure cultures. Under aerobic conditions, carbon dioxide was the principal degradation product in experiments containing pure and mixed cultures (Vogel et al., 1987). A microcosm composed of aquifer water and sediment collected from uncontaminated sites in the Everglades biotransformed trichloroethylene to *cis-* and *trans*-1,2-dichloroethylene (Parsons and Lage, 1985). Trichloroethylene biodegraded to dichloroethylene in both anaerobic and aerobic soils of different soil types (Barrio-Lage et al., 1987). Under anaerobic conditions, nonmethanogenic fermenters, and methanogens degraded trichloroethylene to chloroethane, methane, 1,1-dichloroethylene, 1,2-dichloroethylene, and vinyl chloride (Baek and Jaffé, 1989). Titanium dioxide suspended in an aqueous solution and irradiated with UV light (λ = 365 nm) converted trichloroethylene to carbon dioxide at a significant rate (Matthews, 1986). In a static-culture-flask screening test, trichloroethylene was statically incubated in the dark at 25 °C with yeast extract and settled domestic wastewater inoculum. Percent degradation was 87 and 84% after 28 d at substrate concentrations of 5 and 10 mg/L, respectively. However, losses due to volatilization were 22-29% after 10 d (Tabak et al., 1981).

Surface Water. Estimated half-lives of trichloroethylene (3.2 μg/L) from an experimental marine mesocosm during the spring (8-16 °C), summer (20-22 °C), and winter (3-7 °C) were 28, 13, and 15 d, respectively (Wakeham et al., 1983).

Photolytic. Under smog conditions, indirect photolysis via OH radicals yielded phosgene, dichloroacetyl chloride, and formyl chloride (Howard, 1990). These compounds are readily hydrolyzed to hydrochloric acid, carbon monoxide, carbon dioxide, and dichloroacetic acid (Morrison and Boyd, 1971). Dichloroacetic acid and hydrogen chloride were reported to be aqueous photodecomposition products (Dilling et al., 1975). Reported rate constants for the reaction of trichloroethylene and OH radicals in the atmosphere: 1.2 x 10^{12} cm^3/mol·sec at 300 K (Hendry and Kenley, 1979), 2 x 10^{-12} cm^3/molecule·sec (Howard, 1976), 2.36 x 10^{-12} cm^3/molecule·sec at 298 K (Atkinson, 1985), 2.86 x 10^{-12} cm^3/molecule·sec at 296 K (Edney et al., 1986); with NO$_3$: 2.96 x 10^{-16} cm^3/molecule·sec at 296 K (Atkinson et al., 1988).

An aqueous solution containing 300 ng/μL trichloroethylene and colloidal platinum catalyst was irradiated with UV light. After 12 h, 7.4 ng/μL trichloroethylene and 223.9 ng/μL ethane were detected. A duplicate experiment was performed but 1 g zinc was added to the system. After 5 h, 259.9 ng/μL ethane was formed and trichloroethylene was non-detectable (Wang and Tan, 1988). Major products identified from the pyrolysis of trichloroethylene between 300-800 °C were carbon tetrachloride, tetrachloroethylene, hexachloroethane, hexachlorobutadiene, and hexachlorobenzene (Yasuhara and Morita, 1990).

Jacoby et al. (1994) studied the photocatalytic reaction of gaseous trichloroethylene in air in contact with UV-irradiated titanium dioxide catalyst. The UV radiation was kept below the maximum wavelength so that the catalyst could be excited by photons, i.e., λ <356 nm. Two reaction pathways were proposed. The first pathway includes the formation of the intermediate dichloroacetyl chloride. This compound has a very short residence time and is quickly converted to the following compounds: phosgene, carbon dioxide, carbon monoxide, carbon dioxide, and hydrogen chloride. The second pathway involves the formation of the final products without the formation of the intermediate.

Chemical/Physical. The evaporation half-life of trichloroethylene (1 mg/L) from water at 25 °C using a shallow-pitch propeller stirrer at 200 rpm at an average depth of 6.5 cm is 18.5 min (Dilling, 1977).

The experimental hydrolysis half-life of trichloroethylene in water at 25 °C and pH 7 is 10.7 months (Dilling et al., 1975). The reported hydrolysis half-life at 25 °C and pH 7 is 1.3×10^{-6} yr (Jeffers et al., 1989). In the moisture, hydrochloric acid is formed (Windholz et al., 1983).

In the laboratory, simulated groundwater containing various concentrations of trichloroethylene (4.7, 10.2 and 61.0 mg/L) was pumped through a silica sand column containing iron powder. The residence time for 1 pore volume was 37.8 h. Under these conditions, the average degradation half-life was 3.25 h. The major degradation products observed in solution, in order of decreasing concentration, were *cis*-1,2-dichloroethylene, 1,1-dichloroethylene, vinyl chloride, and *trans*-1,2-dichloroethylene. At trichloroethylene concentrations between 4.69 and 61.0 mg/L, the major gases formed were ethylene and ethane followed by methane, propylene, propane, 1-butene, and butane. Under steady-state conditions, the amount of trichloroethylene that degraded at dissolved concentrations of 4.7, 10.2, and 61.0 mg/L were 3.5, 0.8, and 3.0%, respectively (Orth and Gillham, 1996).

In a similar study, Zhang and Wang (1997) studied the reaction of zero-valent iron powder and palladium-coated iron particles with trichloroethylene and PCBs. In the batch scale experiments, 50 mL of 20 mg/L trichloroethylene solution and 1.0 g of iron or palladium-coated iron were placed into a 50 mL vial. The vial was placed on a rotary shaker (30 rpm) at room temperature. Trichloroethylene was completely degraded by palladium/commercial iron powders (<2 h), by nanoscale iron powder (<1.7 h), and nanoscale palladium/iron bimetallic powders (<30 min). Degradation products included ethane, ethene, propene, propane, propene, butane, butene, and pentane. The investigators concluded that nanoscale iron powder was more reactive than commercial iron powders due to the high specific surface area and less surface area of the iron oxide layer. In addition, air-dried nanoscale iron powder was not effective in the dechlorination process because of the formation of iron oxide.

At influent concentrations of 1.0, 0.1, 0.01, and 0.001 mg/L, granular activated carbon adsorption capacities at pH 5.3 were 28, 6.7, 1.6, and 0.38 mg/g, respectively (Dobbs and Cohen, 1980).

Exposure limits: Potential occupational carcinogen. NIOSH recommends a REL of 2 ppm (1 h ceiling) when trichloroethylene is used as an anesthetic agent and a 10-h TWA of 25 ppm during all other exposures. OSHA PEL: TWA 100 ppm, ceiling 200 ppm, 5-min/2-h peak 300 ppm; ACGIH TLV: TWA 50 ppm, STEL 200 ppm (adopted).

Symptoms of exposure: Inhalation of vapors may cause dizziness, headache, fatigue, and visual disturbances. Narcotic at high concentrations. Ingestion may cause nausea, vomiting, diarrhea, and gastric disturbances (Patnaik, 1992). An irritation concentration of 864.00 mg/m^3 in air was reported by Ruth (1986).

Toxicity:

EC_{10} and EC_{50} concentrations inhibiting the growth of alga *Scenedesmus subspicatus* in 96 h were 300 and 450 mg/L, respectively (Geyer et al., 1985).

LC_{50} (contact) for earthworm (*Eisenia fetida*) 105 μg/cm^2 (Neuhauser et al., 1985).

LC_{50} (7-d) for *Poecilia reticulata* 54.8 mg/L (Könemann, 1981).

LC_{50} (96-h) for fathead minnows 44.1 mg/L (Veith et al., 1983), 40.7 mg/L (flow-through), 60.8 mg/L (static bioassay) (Alexander et al., 1978), bluegill sunfish 45 mg/L (Spehar et al., 1982).

LC_{50} (48-h) for zebra fish 60 mg/L (Slooff, 1979), *Daphnia magna* 18 mg/L (LeBlanc, 1980).

LC_{50} (24-h) for *Daphnia magna* 22 mg/L (LeBlanc, 1980).
LC_{50} for red killifish 630 mg/L (Yoshioka et al., 1986).
Acute oral LD_{50} for mice 2,402 mg/kg, rats 3,670 mg/kg (quoted, RTECS, 1985).

Drinking water standard (final): MCLG: zero; MCL: 5 μg/L (U.S. EPA, 1996).

Uses: Dry cleaning fluid; degreasing and drying metals and electronic parts; extraction solvent for oils, waxes, and fats; solvent for cellulose esters and ethers; removing caffeine from coffee; refrigerant and heat exchange liquid; fumigant; diluent in paints and adhesives; textile processing; aerospace operations (flushing liquid oxygen); anesthetic; organic synthesis.

TRICHLOROFLUOROMETHANE

Synonyms: Algofrene type 1; Arcton 9; Electro-CF 11; Eskimon 11; F 11; FC 11; Fluoro-carbon 11; Fluorotrichloromethane; Freon 11; Freon 11A; Freon 11B; Freon HE; Freon MF; Frigen 11; Genetron 11; Halocarbon 11; Isceon 11; Isotron 11; Ledon 11; Monofluorotrichloro-methane; NCI-C04637; RCRA waste number U121; Refrigerant 11; Trichloromonofluorometh-ane; Ucon 11; Ucon fluorocarbon 11; Ucon refrigerant 11.

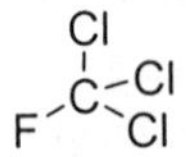

CAS Registry Number: 75-69-4
DOT: 1078
Molecular formula: CCl_3F
Formula weight: 137.37
RTECS: PB6125000
Merck reference: 10, 9453

Physical state, color, and odor:
Colorless, odorless liquid

Melting point (°C):
-111 (Windholz et al., 1983)

Boiling point (°C):
23.63 (Kudchadker et al., 1979)

Density (g/cm^3):
1.476 at 25/4 °C (Neely and Blau, 1985)
1.484 at 17.2/4 °C (Sax, 1984)
1.487 at 20/4 °C (Fluka, 1988)
1.467 at 25/4 °C (Horvath, 1982)

Diffusivity in water (x 10^{-5} cm^2/sec):
0.93 at 20 °C using method of Hayduk and Laudie (1974)

Flash point (°C):
Nonflammable gas (NIOSH, 1997)

Entropy of fusion (cal/mol·K):
20.75 (Osborne et al., 1941)

Heat of fusion (kcal/mol):
1.6476 (Osborne et al., 1941)

Henry's law constant (x 10^{-3} atm·m^3/mol):
110 at 20-25 °C (Pankow and Rosen, 1988)
5.83 at 25 °C (Warner et al., 1987)
833 at 20 °C (Pearson and McConnell, 1975)
120 at 25 °C (Liss and Slater, 1974)

5.38, 6.67, 8.42, 10.2, and 12.8 at 10, 15, 20, 25, and 30 °C, respectively (Ashworth et al., 1988)
At 25 °C: 88.2 and 123 in distilled water and seawater, respectively (Hunter-Smith et al., 1983)

Ionization potential (eV):
11.68 (Lias, 1998)
11.77 ± 0.02 (Franklin et al., 1969)

Soil sorption coefficient, log K_{oc}:
2.20 (Schwille, 1988)
2.13 (Neely and Blau, 1985)

Octanol/water partition coefficient, log K_{ow}:
2.53 (Hansch et al., 1975)

Solubility in organics:
Soluble in ethanol, ether, and other solvents (U.S. EPA, 1985)

Solubility in water (mg/L):
1,240 at 25 °C (Neely and Blau, 1985)
1,100 at 20 °C (Du Pont, 1966; Pearson and McConnell, 1975)
1,080 at 30 °C (Horvath, 1982)

Vapor density:
5.85 g/L at 23.77 °C (Braker and Mossman, 1971)
5.61 g/L at 25 °C, 4.74 (air = 1)

Vapor pressure (mmHg):
687 at 20 °C, 980 at 30 °C (Verschueren, 1983)
792 at 25 °C (ACGIH, 1986)
667.4 at 20 °C (McConnell et al., 1975)
802.8 at 25 °C (Howard, 1990)

Environmental fate:
Biological. In a static-culture-flask screening test, trichlorofluoromethane was statically incubated in the dark at 25 °C with yeast extract and settled domestic wastewater inoculum. No significant degradation was observed after 28 d of incubation. At substrate concentrations of 5 and 10 mg/L, percent losses due to volatilization were 58 and 37% after 10 d (Tabak et al., 1981).

Chemical/Physical. When trichlorofluoromethane (50 μg/L) in an ultrasonicator was exposed to 20-kHz ultrasound at 5 °C, nearly 100% degradation was achieved after 6 min. During sonication, the pH of the aqueous solution decreased, which is consistent with the formation of hydrochloric acid, hydrofluoric acid, and acidic species from fluorine and chlorine. In this experiment <5% of trichlorofluoroethane was lost to volatilization (Cheung and Kurup, 1994).

At influent concentrations of 1.0, 0.1, 0.01, and 0.001 mg/L, the granular activated carbon adsorption capacities at pH 5.3 were 5.6, 3.2, 1.9, and 1.1 mg/g, respectively (Dobbs and Cohen, 1980).

Trichlorofluoromethane is not expected to hydrolyze to any reasonable extent (Kollig, 1993).

Exposure limits: NIOSH REL: ceiling 1,000 ppm (5,600 mg/m^3), IDLH 2,000 ppm; OSHA PEL: TWA 1,000 ppm; ACGIH TLV: ceiling 1,000 ppm (adopted).

Toxicity:

LC$_{50}$ (inhalation) for guinea pigs 25 pph/30-min, mice 10 pph/30-min, rabbits 25 pph/30-min (quoted, RTECS, 1985).

Drinking water standard: No MCLGs or MCLs have been proposed although trichlorofluoromethane has been listed for regulation (U.S. EPA, 1996).

Uses: Aerosol propellant; refrigerant; solvent; blowing agent for polyurethane foams; fire extinguishing; chemical intermediate; organic synthesis.

2,4,5-TRICHLOROPHENOL

Synonyms: Collunosol; Dowicide 2; Dowicide B; NCI-C61187; Nurelle; Phenachlor; Preventol I; RCRA waste number U230; 2,4,5-TCP; 2,4,5-TCP-Dowicide 2.

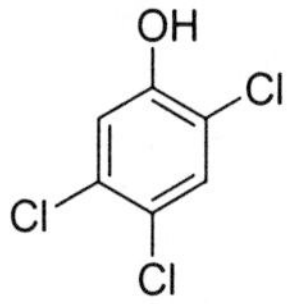

CAS Registry Number: 95-95-4
DOT: 2020
Molecular formula: $C_6H_3Cl_3O$
Formula weight: 197.45
RTECS: SN1400000
Merck reference: 10, 9455

Physical state, color, and odor:
Colorless crystals or yellow to gray flakes with a strong, disinfectant or phenolic odor. The reported odor threshold concentration in air and the taste threshold concentration in water are 63 and 190 ppb, respectively (Young et al., 1996).

Melting point (°C):
68-70 (Weast, 1986)
61-63 (Sax, 1984)
57 (Weiss, 1986)

Boiling point (°C):
252 (Weast, 1986)

Density (g/cm³):
1.5 at 75/4 °C (Verschueren, 1983)
1.678 at 25/4 °C (Sax, 1984)

Dissociation constant, pK$_a$:
6.90 (Hoigné and Bader, 1983)
7.00 (Blackman et al., 1955)
7.37 at 25 °C (Dean, 1973)

Diffusivity in water (x 10^{-5} cm²/sec):
0.62 at 20 °C using method of Hayduk and Laudie (1974)

Flash point (°C):
Nonflammable (Weiss, 1986)

Lower explosive limit (%):
Nonflammable (Weiss, 1986)

Upper explosive limit (%):
Nonflammable (Weiss, 1986)

Henry's law constant (x 10^{-7} atm·m^3/mol):
1.76 at 25 °C (estimated, Leuenberger et al., 1985)

Bioconcentration factor, log BCF:
3.27 (fathead minnows, Call et al., 1980)

Soil sorption coefficient, log K_{oc}:
2.56 (Brookstoneclay loam, Boyd, 1982)
2.45 (river sediment, Eder and Weber, 1980)
2.55 (loamy sand, Kjeldsen et al., 1990)
3.30 (sandy soil), 3.34 (sand), 3.38 (peaty sand) (Van Gestel, 1993)

Octanol/water partition coefficient, log K_{ow}:
3.72 (Leo et al., 1971)
4.10 (Xie et al., 1984)
4.19 (Schellenberg et al., 1984)
3.80 (Saarikoski and Viluksela, 1982)
3.85 (Schultz et al., 1989)
4.02 (Kishino and Kobayashi, 1994)

Solubility in organics:
Soluble in ethanol and ligroin (U.S. EPA, 1985)

Solubility in water:
1,190 mg/kg at 25 °C (quoted, Verschueren, 1983)
948 mg/L at pH 5.1 (Blackman et al., 1955)
1.2 g/L at 25 °C (quoted, Leuenberger et al., 1985)
649 mg/L at 25 °C and pH 4.9 (Ma et al., 1993)

Vapor pressure (x 10^{-3} mmHg):
3.5 at 8 °C, 22 at 25 °C (quoted, Leuenberger et al., 1985)

Environmental fate:
   Biological. Chloroperoxidase, a fungal enzyme isolated from *Caldariomyces fumago*, chlorinated 2,4,5-trichlorophenol to give 2,3,4,6-tetrachlorophenol (Wannstedt et al., 1990).
   Photolytic. When 2,4,5-trichlorophenol (100 μM) in an oxygenated, titanium dioxide (2 g/L) suspension was irradiated by sunlight ($\lambda \geq 340$ nm), complete mineralization to carbon dioxide and water and chloride ions was observed (Barbeni et al., 1987).
   The following phototransformation half-lives were reported for 2,4,5-trichlorophenol in estuarine water exposed to sunlight and microbes: 1 h during winter; in distilled water: 0.6 and 1 h during summer and winter, respectively; in poisoned estuarine water: 14 and 24 h during summer and winter, respectively (Hwang et al., 1986).
   A photooxidation rate constant of <3,000/M·sec was reported for the reaction of 2,4,5-trichlorophenol and ozone in water at a pH range of 1.2 to 1.5 (Hoigné and Bader, 1983).
   Chemical/Physical. 2,4,5-Trichlorophenol will not hydrolyze because it does not contain a hydrolyzable functional group (Kollig, 1993).
   During the manufacture/synthesis of 2,4,5-T using alkalies at high temperatures, some TCDD may form (Worthing and Hance, 1991).

Toxicity:
   EC$_{50}$ (48-h) for marine polychaete *Platynereis dumerilii* 2.55 mg/L (Palau-Casellas and

Hutchinson, 1998).

LC$_{50}$ 750-1,106 μg/L (soil porewater concentration) for earthworm (*Eisenia andrei*) and 4.1 and 6.9 mg/L (soil porewater concentration) for earthworm (*Lumbricus rubellus*) (Van Gestel and Ma, 1993).

LC$_{50}$ (96-h) for bluegill sunfish 450 μg/L (Spehar et al., 1982), *Cyprinodon variegatus* 1.7 ppm using natural seawater (Heitmuller et al., 1981).

LC$_{50}$ (72-h) for *Cyprinodon variegatus* 1.7 ppm (Heitmuller et al., 1981).

LC$_{50}$ (48-h) for red killifish 12 mg/L (Yoshioka et al., 1986), *Daphnia magna* 2.7 mg/L (LeBlanc, 1980), marine polychaete *Platynereis dumerilii* 4.24 mg/L (Palau-Casellas and Hutchinson, 1998), *Cyprinodon variegatus* 1.7 ppm (Heitmuller et al., 1981).

LC$_{50}$ (24-h) for *Daphnia magna* 3.8 mg/L (LeBlanc, 1980), Protozoan (*Spirostomum teres*) 2.00 mg/L (Twagilimana et al., 1998), *Cyprinodon variegatus* 2.4 ppm (Heitmuller et al., 1981).

LC$_{50}$ (30-min) for Protozoan (*Spirostomum teres*) 12.72 mg/L (Twagilimana et al., 1998).

Acute oral LD$_{50}$ for guinea pigs 1,000 mg/kg, mice 600 mg/kg, rats 820 mg/kg (quoted, RTECS, 1985).

Heitmuller et al. (1981) reported a NOEC of 1.0 ppm.

Uses: Fungicide; bactericide; organic synthesis.

2,4,6-TRICHLOROPHENOL

Synonyms: Dowicide 2S; NCI-C02904; Omal; Phenachlor; RCRA waste number F027; 2,4,6-TCP; 2,4,6-TCP-Dowicide 25.

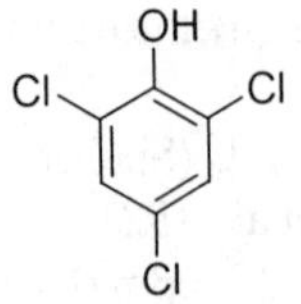

CAS Registry Number: 88-06-2
DOT: 2020
Molecular formula: $C_6H_3Cl_3O$
Formula weight: 197.45
RTECS: SN1575000
Merck reference: 10, 9456

Physical state, color, and odor:
Colorless needles or yellow solid with a strong, phenolic odor. The reported odor threshold concentration in air and the taste threshold concentration in water are 380 and >12 ppb, respectively (Young et al., 1996).

Melting point (°C):
69.5 (Weast, 1986)

Boiling point (°C):
246 (Weast, 1986)

Density (g/cm³):
1.4901 at 75/4 °C (Weast, 1986)

Diffusivity in water (x 10^{-5} cm²/sec):
0.62 at 20 °C using method of Hayduk and Laudie (1974)

Dissociation constant, pK_a:
7.42 at 25 °C (Howard, 1989)
6.10 (Blackman et al., 1955)
6.0 (Eder and Weber, 1980)

Henry's law constant (x 10^{-8} atm·m³/mol):
9.07 at 25 °C (estimated, Leuenberger et al., 1985)

Bioconcentration factor, log BCF:
1.71 (algae, Geyer et al., 1984)
2.75, 2.84 (Atlantic salmon, Carlberg et al., 1986)
3.48 (snail, Virtanen and Hattula, 1982)
1.78 (activated sludge), 2.49 (golden ide) (Freitag et al., 1985)
1.94 (*Jordanella floridae*, Devillers et al., 1996)
1.60 (mussel, Geyer et al., 1982)

Soil sorption coefficient, log K_{oc}:
3.03 (Howard, 1989)

3.05, 3.27 (Robinson and Novak, 1994)
1.72 (river sediment, Eder and Weber, 1980)
>2.68, 2.74 (forest soil), 1.96 (agricultural soil) (Seip et al., 1986)

Octanol/water partition coefficient, log K_{ow}:
2.80 (Geyer et al., 1982)
3.72 (Schellenberg et al., 1984)
3.06, 3.69 (Leo et al., 1971)
3.67 (Kishino and Kobayashi, 1994)

Solubility in organics:
Soluble in ethanol and ether (Weast, 1986)

Solubility in water (mg/L):
800 at 25 °C, 2,430 at 96 °C (quoted, Verschueren, 1983)
420 at 20-25 °C (Geyer et al., 1982)
900 at 20-25 °C (Kilzer et al., 1979)
410, 427, and 692 at 19.5, 20.1, and 24.9 °C, respectively (Achard et al., 1996)
434 at 20 °C (Blackman et al., 1955)

Vapor pressure (x 10^{-3} mmHg):
8.4 at 24 °C (Howard, 1989)
2.5 at 8 °C, 17 at 25 °C (quoted, Leuenberger et al., 1985)

Environmental fate:
   Biological. In activated sludge, only 0.3% mineralized to carbon dioxide after 5 d (Freitag et al., 1985). In anaerobic sludge, 2,4,6-trichlorophenol degraded to *p*-chlorophenol (Mikesell and Boyd, 1985). When 2,4,6-trichlorophenol was statically incubated in the dark at 25 °C with yeast extract and settled domestic wastewater inoculum, significant biodegradation with rapid adaptation was observed. At concentrations of 5 and 10 mg/L, 96 and 97% biodegradation, respectively, were observed after 7 d (Tabak et al., 1981).
   Photolytic. Titanium dioxide suspended in an aqueous solution and irradiated with UV light (λ = 365 nm) converted 2,4,6-trinitrophenol to carbon dioxide at a significant rate (Matthews, 1986). A carbon dioxide yield of 65.8% was achieved when 2,4,6-trichlorophenol adsorbed on silica gel was irradiated with light (λ >290 nm) for 17 h (Freitag et al., 1985).
   Scully and Hoigné (1987) reported a half-life of 62 h for the reaction of 2,4,6-trichlorophenol and singlet oxygen in water at 19 °C and pH 8. They reported the following rate constants: 2 x 10^6/M·sec at pH 4.2, 6 x 10^8/M·sec at pH 4.8, 1.5 x 10^7/M·sec at pH 5.2, 2.6 x 10^7/M·sec at pH 5.5, 5.5 x 10^7/M·sec at pH 6.0, 9.50 x 10^7/M·sec at pH 7, and 1.2 x 10^8/M·sec at pH 9. Tratnyek and Hoigné (1991) reported a rate constant of 1.7 x 10^7/M·sec for the same reaction in an aqueous phosphate buffer solution at 27 °C. A photooxidation rate constant of <10,000/M·sec was reported for the reaction of 2,4,6-trichlorophenol and ozone in water at a pH range of 1.3 to 1.5 (Hoigné and Bader, 1983).
   Chemical/Physical. An aqueous solution containing chloramine reacted with 2,4,6-tri-chlorophenol to yield the following intermediate products after 2 h at 25 °C: 2,6-dichloro-1,4-benzoquinone-4-(*N*-chloro)imine   and   4,6-dichloro-1,2-benzoquinone-2-(*N*-chloro)imine (Maeda et al., 1987).
   In a dilute aqueous solution at pH 6.0, 2,4,6-trichlorophenol reacted with an excess of hypochlorous acid forming 2,6-dichlorobenzoquinone (18% yield) and other chlorinated compounds (Smith et al., 1976). Based on an assumed 5% disappearance after 330 h at 85 °C, the hydrolysis half-life was estimated to be >40 yr (Ellington et al., 1988).

Toxicity:

EC_{10} and EC_{50} concentrations inhibiting the growth of alga *Scenedesmus subspicatus* in 96 h were 1.1 and 5.6 mg/L, respectively (Geyer et al., 1985).

LC_{50} (contact) for earthworm (*Eisenia fetida*) 5.0 μg/cm^2 (Neuhauser et al., 1985).

LC_{50} (96-h) for bluegill sunfish 320 μg/L (Spehar et al., 1982), *Cyprinodon variegatus* 130 ppm (Heitmuller et al., 1981).

LC_{50} (72-h) *Cyprinodon variegatus* 130 ppm (Heitmuller et al., 1981).

LC_{50} (48-h) for *Daphnia magna* 85 mg/L (LeBlanc, 1980), *Cyprinodon variegatus* 130 ppm (Heitmuller et al., 1981).

LC_{50} (24-h) for *Daphnia magna* >220 mg/L (LeBlanc, 1980), *Cyprinodon variegatus* 130 ppm (Heitmuller et al., 1981).

LC_{50} for red killifish 7.6 mg/L (Yoshioka et al., 1986).

Acute oral LD_{50} for rats 820 mg/kg (quoted, RTECS, 1985).

Heitmuller et al. (1981) reported a NOEC of 100 ppm.

Drinking water standard: No MCLGs or MCLs have been proposed although 2,4,6-trichlorophenol has been listed for regulation (U.S. EPA, 1996).

Uses: Manufacture of fungicides, bactericides, antiseptics, germicides; wood and glue preservatives; in textiles to prevent mildew; defoliant; disinfectant; organic synthesis.

1,2,3-TRICHLOROPROPANE

Synonyms: Allyl trichloride; Glycerin trichlorohydrin; Glycerol trichlorohydrin; Glyceryl trichlorohydrin; NCI-C60220; Trichlorohydrin.

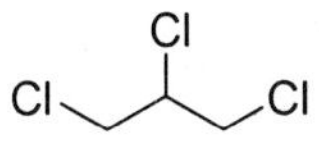

CAS Registry Number: 96-18-4
Molecular formula: $C_3H_5Cl_3$
Formula weight: 147.43
RTECS: TZ9275000

Physical state, color, and odor:
Clear, colorless liquid with a strong, chloroform-like odor

Melting point (°C):
-14.7 (Weast, 1986)

Boiling point (°C):
156.8 (Weast, 1986)

Density (g/cm³):
1.3889 at 20/4 °C (Horvath, 1982)
1.417 at 15/4 °C (Verschueren, 1983)

Diffusivity in water (x 10⁻⁵ cm²/sec):
0.85 at 20 °C using method of Hayduk and Laudie (1974)

Flash point (°C):
71.7 (NIOSH, 1997)
82.2 (open cup, Hawley, 1981)

Lower explosive limit (%):
3.2 at 121 °C (NIOSH, 1997)

Upper explosive limit (%):
12.6 at 151 °C (NIOSH, 1997)

Henry's law constant (x 10⁻⁵ atm·m³/mol):
9.79 at 25 °C (Tancréde and Yanagisawa, 1990)
35.8 at 24.9 °C (mole fraction ratio, Leighton and Calo, 1981)

Interfacial tension with water (dyn/cm at 20 °C):
38.50 (Demond and Lindner, 1993)

Solubility in organics:
Soluble in alcohol, chloroform, and ether (Weast, 1986)

Solubility in water:
1,900 mg/L (Afghan and Mackay, 1980; Dilling, 1977)

Vapor density:
6.03 g/L at 25 °C, 5.09 (air = 1)

Vapor pressure (mmHg):
2 at 20 °C, 4 at 30 °C (Verschueren, 1983)
3.1 at 25 °C (Banerjee et al., 1990)

Environmental fate:
Chemical/Physical. The hydrolysis rate constant for 1,2,3-trichloropropane at pH 7 and 25 °C was determined to be 1.8 x 10^{-6}/h, resulting in a half-life of 43.9 yr (Ellington et al., 1988). The hydrolysis half-lives decrease at varying pHs and temperature. At 87 °C, the hydrolysis half-lives at pH values of 3.07, 7.12, and 9.71 were 21.1, 11.6, and 0.03 d, respectively (Ellington et al., 1986). By analogy to 1,2-dibromo-2-chloropropane, the following hydrolysis products would be formed: 2,3-dichloro-1-propanol, 2,3-dichloropropene, epichlorohydrin, 1-chloro-2,3-dihydroxypropane, glycerol, 1-hydroxy-2,3-propylene oxide, 2-chloro-3-hydroxy-propene, and hydrochloric acid (Kollig, 1993).

The volatilization half-life of 1,2,3-trichloropropane (1 mg/L) from water at 25 °C using a shallow-pitch propeller stirrer at 200 rpm at an average depth of 6.5 cm is 56.1 min (Dilling, 1977).

Exposure limits: Potential occupational carcinogen. NIOSH REL: TWA 10 ppm (60 mg/m³), IDLH 100 ppm; OSHA PEL: TWA 50 ppm (300 mg/m³); ACGIH TLV: TWA 10 ppm (adopted).

Symptoms of exposure: Inhalation of vapors may cause depression of central nervous system, narcosis, and convulsions (Patnaik, 1992). An irritation concentration of 300.00 mg/m³ in air was reported by Ruth (1986).

Toxicity:
   LC_{50} (7-d) for *Poecilia reticulata* 41.6 mg/L (Könemann, 1981).
   Acute oral LD_{50} for rats 320 mg/kg (quoted, RTECS, 1985).
   LC_{50} (inhalation) for mice 3,400 mg/m³/2-h (quoted, RTECS, 1985).

Drinking water standard: No MCLGs or MCLs have been proposed although 1,2,3-trichloropropane has been listed for regulation (U.S. EPA, 1996).

Uses: Solvent; degreaser; paint and varnish removers.

1,1,2-TRICHLOROTRIFLUOROETHANE

Synonyms: Fluorocarbon 113; Freon 113; Frigen 113 TR-T; Halocarbon 113; Kaiser chemicals 11; R 113; Refrigerant 113; 1,1,2-Trichloro-1,2,2-trifluoroethane; Trichlorotrifluoro-ethane; 1,1,2-Trifluoro-1,2,2-trichloroethane; TTE; Ucon 113; Ucon fluorocarbon 113; Ucon 113/halocarbon 113.

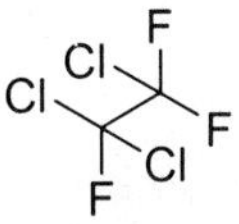

CAS Registry Number: 76-13-1
DOT: 1078
Molecular formula: $C_2Cl_3F_3$
Formula weight: 187.38
RTECS: KJ4000000

Physical state, color, and odor:
Colorless liquid with a carbon tetrachloride-like odor at high concentrations

Melting point (°C):
-36.4 (Weast, 1986)

Boiling point (°C):
47.7 (Weast, 1986)

Density (g/cm³):
1.5635 at 20/4 °C (Weast, 1986)
1.42 at 25/4 °C (Hawley, 1981)

Diffusivity in water (x 10^{-5} cm²/sec):
0.79 at 20 °C using method of Hayduk and Laudie (1974)

Flash point (°C):
Noncombustible at ordinary temperatures (NIOSH, 1997)

Entropy of fusion (cal/mol·K):
2.03 (Golovanova and Kolesov, 1984)

Heat of fusion (kcal/mol):
0.556 (Golovanova and Kolesov, 1984)

Henry's law constant (atm·m³/mol):
0.154, 0.215, 0.245, 0.319, and 0.321 at 10, 15, 20, 25, and 30 °C, respectively (Ashworth et al., 1988)

Ionization potential (eV):
11.99 ± 0.02 (Franklin et al., 1969)

Soil sorption coefficient, log K_{oc}:
2.59 using method of Chiou et al. (1979)

Octanol/water partition coefficient, log K_{ow}:
2.57 (Lesage et al., 1990)

Solubility in organics:
Soluble in alcohol, benzene, and ether (Weast, 1986)

Solubility in water (mg/L):
136 at 10 °C (Lesage et al., 1990)
170 at 25 °C (Du Pont, 1966; Jones et al., 1977)

Vapor density:
7.66 g/L at 25 °C, 6.47 (air = 1)

Vapor pressure (mmHg at 20 °C):
285 (NIOSH, 1997)
270 (Verschueren, 1983)

Environmental fate:
  Biological. In an anoxic aquifer beneath a landfill in Ottawa, Ontario, Canada, there was evidence to suggest that 1,1,2-trichlorotrifluoroethane underwent reductive dehalogenation to give 1,2-difluoro-1,1,2-trichloroethylene and 1,2-dichloro-1,1,2-trifluoroethane. It was proposed that the latter compound was further degraded via dehydrodehalogenation to give 1-chloro-1,1,2-trifluoroethylene (Lesage et al., 1990).
  Chemical/Physical. 1,1,2-Trichlorotrifluoroethane will not hydrolyze to any reasonable extent (Kollig, 1993).

Exposure limits: NIOSH REL: TWA 1,000 ppm (7,600 mg/m³), STEL 1,250 ppm (9,500 mg/m³), IDLH 2,000 ppm; OSHA PEL: TWA 1,000 ppm, STEL 1,250 ppm (adopted).

Symptoms of exposure: May produce a weak narcotic effect, cardiac sensitization, and irritation of respiratory passage (Patnaik, 1992)

Toxicity:
  Acute oral LD_{50} for rats 43 mg/kg (quoted, RTECS, 1985).

Drinking water standard: No MCLGs or MCLs have been proposed although 1,1,2-trichloro-trifluoroethane has been listed for regulation (U.S. EPA, 1996).

Uses: Fire extinguishers; dry-cleaning solvent; manufacture of chlorotrifluoroethylene; refrigerant; polymer intermediate; blowing agent; drying electronic parts; extraction solvent for analyzing hydrocarbons, oils, and greases.

TRI-*o*-CRESYL PHOSPHATE

Synonyms: *o*-Cresyl phosphate; Phosflex 179-C; **Phosphoric acid tris(2-methylphenyl) ester**; Phosphoric acid tri-2-tolyl ester; Phosphoric acid tri-*o*-cresyl ester; TCP; TOCP; TOFK; *o*-Tolyl phosphate; TOTP; Tricresyl phosphate; Triorthocresyl phosphate; Tri-2-methylphenyl phosphate; Tris(*o*-cresyl)phosphate; Tris(*o*-methylphenyl)phosphate; Tris(*o*-tolyl)phosphate; Tri-2-tolyl phosphate; Tri-*o*-tolyl phosphate.

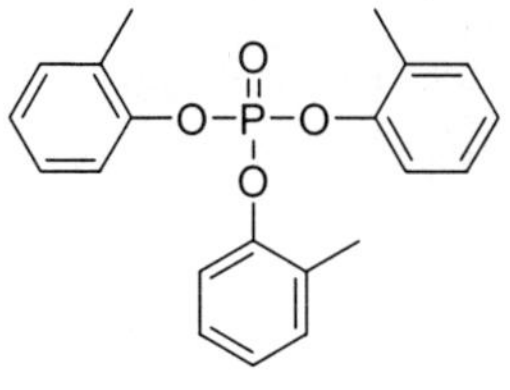

CAS Registry Number: 78-30-8
DOT: 2574
DOT label: Poison
Molecular formula: $C_{21}H_{21}O_4P$
Formula weight: 368.37
RTECS: TD0350000
Merck reference: 10, 9567

Physical state, color, and odor:
Colorless to pale yellow, odorless liquid

Melting point (°C):
11 (Weast, 1986)
-25, -30 (Verschueren, 1983)

Boiling point (°C):
410 (Weast, 1986)
420 (Verschueren, 1983)

Density (g/cm³):
1.180 at 26.5/4 °C, 1.179 at 30.1/4 °C, 1.178 at 34.9/4 °C, 1.177 at 39.7/4 °C, 1.176 at 43.3/4 °C, 1.175 at 48.1/4 °C (Kannan and Kishore, 1999)

Diffusivity in water (x 10^{-5} cm²/sec):
0.61 at 20 °C using method of Hayduk and Laudie (1974)

Flash point (°C):
227 (NIOSH, 1997)

Bioconcentration factor, log BCF:
2.22 (*Pimephales promelas*), 2.90 (*Alburnus alburnus*) (Devillers et al., 1996)

Soil sorption coefficient, log K_{oc}:
3.37 using method of Kenaga and Goring (1980)

Octanol/water partition coefficient, log K_{ow}:
5.11 (commercial mixture containing tricresyl phosphates, Saeger et al., 1979)

Solubility in organics:
Soluble in acetic acid, alcohol, ether, and toluene (Weast, 1986)

Solubility in water (ppm):
0.36 and 0.34 at 20-25 and 25 °C, respectively (mixture containing tricresyl phosphates, Ofstad and Sletten, 1985; Saeger et al., 1979)
3.1 at 25 °C ("Pliabrac 521," Ofstad and Sletten, 1985)
0.008 wt % 20-25 °C (Fordyce and Meyer, 1940)

Vapor density:
15.06 g/L at 25 °C, 12.72 (air = 1)

Vapor pressure (x 10^{-5} mmHg):
2 at 25 °C (NIOSH, 1997)

Environmental fate:
Biological. A commercial mixture containing tricresyl phosphates was completely degraded by indigenous microbes in Mississippi River water to carbon dioxide. After 4 wk, 82.1% of the theoretical carbon dioxide had evolved (Saeger et al., 1979).

Chemical/Physical. Tri-*o*-cresyl phosphate hydrolyzed rapidly in Lake Ontario water, presumably to di-*o*-cresyl phosphate (Howard and Doe, 1979). When an aqueous solution containing a mixture of isomers (0.1 mg/L) and chlorine (3-1,000 mg/L) was stirred in the dark at 20 °C for 24 h, the benzene ring was substituted with 1-3 chlorine atoms (Ishikawa and Baba, 1988).

Exposure limits (mg/m^3): NIOSH REL: TWA 0.1, IDLH 40; OSHA PEL: TWA 0.1.

Symptoms of exposure: Ingestion may cause gastrointestinal pain, diarrhea, weakness, muscle pain, kidney damage, paralysis, and death (Patnaik, 1992).

Toxicity:
Acute oral LD_{50} for rats 160 mg/kg (quoted, RTECS, 1985).

Uses: Plasticizer in lacquers, varnishes, polyvinyl chloride, polystyrene, nitrocellulose; waterproofing agent; hydraulic fluid and heat exchange medium; fire retardant for plastics; solvent mixtures; synthetic lubricant; gasoline additive to prevent pre-ignition.

TRIETHYLAMINE

Synonyms: Diethylaminoethane; *N,N*-**Diethylethanamine**; TEN; UN 1296.

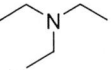

CAS Registry Number: 121-44-8
DOT: 1296
DOT label: Flammable liquid
Molecular formula: $C_6H_{15}N$
Formula weight: 101.19
RTECS: YE0175000
Merck reference: 10, 9477

Physical state, color, and odor:
Colorless liquid with a strong, ammonia-like odor. Odor threshold concentration is 480 ppb (Amoore and Hautala, 1983).

Melting point (°C):
-114.7 (Weast, 1986)

Boiling point (°C):
89.3 (Weast, 1986)

Density (g/cm³):
0.7275 at 20/4 °C (Weast, 1986)
0.7255 at 25/4 °C (Windholz et al., 1983)

Diffusivity in water (x 10^{-5} cm²/sec):
0.73 at 20 °C using method of Hayduk and Laudie (1974)

Dissociation constant, pK_a:
11.01 at 18 °C (Gordon and Ford, 1972)
10.72 at 25 °C (Dean, 1973)

Flash point (°C):
-6.7 (NIOSH, 1997)

Lower explosive limit (%):
1.2 (NIOSH, 1997)

Upper explosive limit (%):
8.0 (NIOSH, 1997)

Henry's law constant (x 10^{-4} atm·m³/mol):
1.79 at 25 °C (Christie and Crisp, 1967)

Interfacial tension with water (dyn/cm at 20 °C):
0.1 (quoted, Freitas et al., 1997)

Ionization potential (eV):
7.50 ± 0.02 (Franklin et al., 1969)

Soil sorption coefficient, log K_{oc}:
Unavailable because experimental methods for estimation of this parameter for aliphatic amines are lacking in the documented literature

Octanol/water partition coefficient, log K_{ow}:
1.15 (pH 9.4), 1.44, 1.45 (pH 13) (Sangster, 1989)

Solubility in organics:
Miscible with alcohol and ether (Windholz et al., 1983)

Solubility in water:
15,000 mg/L at 20 °C, 19,700 mg/L at 95 °C (quoted, Verschueren, 1983)
1.41 M at 20 °C (Fischer and Ehrenberg, 1948)
728 mmol/kg at 25.0 °C (Vesala, 1974)
In wt %: miscible at <18.0 °C, 14.38 at 18.0 °C, 12.38 at 19.0 °C, 11.24 at 20.0 °C, 10.23 at
   21.0 °C, 10.29 at 22.0 °C, 8.91 at 24.0 °C, 8.00 at 26.0 °C, 6.51 at 28.0 °C, 3.93 at 30.0 °C,
   3.60 at 40.0 °C, 2.59 at 50.0 °C, 1.96 at 60.0 °C, 1.60 at 70.0 °C, 1.38 at 80.0 °C
   (Stephenson, 1993a)

Vapor density:
4.14 g/L at 25 °C, 3.49 (air = 1)

Vapor pressure (mmHg at 20 °C):
53 (NIOSH, 1997)
50 (Verschueren, 1983)

Environmental fate:
   Chemical/Physical. Triethylamine reacted with NO_x in the dark to form diethylnitrosamine. In an outdoor chamber, photooxidation by natural sunlight yielded the following products: diethylnitramine, diethylformamide, diethylacetamide, ethylacetamide, diethylhydroxylamine, ozone, acetaldehyde, and peroxyacetylnitrate (Pitts et al., 1978).

Exposure limits: NIOSH REL: IDLH 200 ppm; OSHA PEL: TWA 25 ppm (100 mg/m³);
ACGIH TLV: TWA 1 ppm, STEL 3 ppm (adopted).

Symptoms of exposure: Irritates eyes and mucous membranes. An irritation concentration of 200.00 mg/m³ in air was reported by Ruth (1986).

Toxicity:
   Acute oral LD_{50} for mice 546 mg/kg, rats 460 mg/kg (quoted, RTECS, 1985).

Uses: Curing and hardening polymers; catalytic solvent in chemical synthesis; accelerator activators for rubber; wetting, penetrating, and waterproofing agents of quaternary ammonium types; corrosion inhibitor; propellant.

TRIFLURALIN

Synonyms: Agreflan; Agriflan 24; Crisalin; Digermin; 2,6-Dinitro-*N,N*-dipropyl-4-trifluoro-methylaniline; **2,6-Dinitro-*N,N*-dipropyl-4-(trifluoromethyl)benzenamine**; 2,6-Dinitro-*N,N*-di-*n*-propyl-α,α,α-trifluoro-*p*-toluidine; 4-(Di-*n*-propylamino)-3,5-dinitro-1-trifluoromethyl-benzene; *N,N*-Di-*n*-propyl-2,6-dinitro-4-trifluoromethylaniline; *N,N*-Dipropyl-4-trifluoromethyl-2,6-dinitroaniline; Elancolan; L-36352; Lilly 36352; NCI-C00442; Nitran; Olitref; Trefano-cide; Treficon; Treflam; Treflan; Treflanocide elancolan; Trifluoralin; α,α,α-Trifluoro-2,6-dinitro-*N,N*-dipropyl-*p*-toluidine; Trifluraline; Triflurex; Trikepin; Trim.

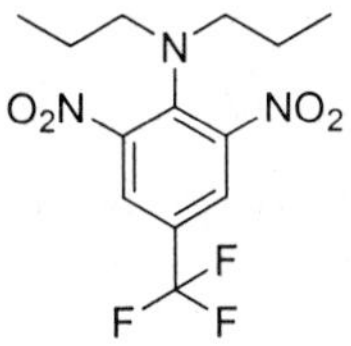

CAS Registry Number: 1582-09-8
DOT: 1609
Molecular formula: $C_{13}H_{16}F_3N_3O_4$
Formula weight: 335.29
RTECS: XU9275000
Merck reference: 10, 9493

Physical state and color:
Yellow to orange crystals

Melting point (°C):
46-47 (Windholz et al., 1983)
48.5-49 (Verschueren, 1983)
43 (Bailey and White, 1965)

Boiling point (°C):
139-140 at 4.2 mmHg (Windholz et al., 1983)
96-97 at 0.18 mmHg (Probst et al., 1967)

Density (g/cm³):
1.294 at 25/4 °C (Keith and Walters, 1992)

Diffusivity in water (x 10⁻⁵ cm²/sec):
0.47 at 20 °C using method of Hayduk and Laudie (1974)

Henry's law constant (x 10⁻⁵ atm·m³/mol):
5.83 at 23 °C (Fendinger et al., 1989)
10.19 and 14.89 in distilled water and 33.3‰ NaCl at 20 °C, respectively (Rice et al., 1997a)

Bioconcentration factor, log BCF:
3.03 (freshwater fish), 2.45 (fish, microcosm) (Garten and Trabalka, 1983)

Soil sorption coefficient, log K_{oc}:
3.73 (Commerce soil), 3.67 (Tracy soil), 3.64 (Catlin soil) (McCall et al., 1981)
4.14 (Kenaga and Goring, 1980)

4.49 (Hickory Hill sediment, Brown and Flagg, 1981)

2.94 (average of 2 soil types, Kanazawa, 1989)

1.705 (Cecil sandy loam), 1.64 (Eustis fine sand), 2.25 (Glendale sandy clay loam), 1.88 (Webster silty clay loam, Davidson et al., 1980)

3.70 (Mivtahim soil), 3.44 (Gilat soil), 3.46 (Neve Yaar soil), 3.63 (Malkiya soil), 4.00 (Kinneret-A sediment), 3.97 (Kinneret-G sediment) (Gerstl and Mingelgrin, 1984)

Octanol/water partition coefficient, log K_{ow}:
5.34 (Kenaga and Goring, 1980)
5.28 (Brown and Flagg, 1981)
5.07 at pH 7 and 25 °C (Worthing and Hance, 1991)

Solubility in organics:
Freely soluble in Stoddard solvent (Windholz et al., 1983), chloroform, methanol (Probst et al., 1967), acetone (400 g/L), xylene (580 g/L) (Worthing and Hance, 1991), ether, and ethanol (Bailey and White, 1965)

Solubility in water:
240 mg/L (Bailey and White, 1965)
4 ppm at 27 °C (quoted, Verschueren, 1983)
3 μmol/L at 25 °C (LaFleur, 1979)
4 mg/L at 20 °C (Fühner and Geiger, 1977)

Vapor pressure (x 10^{-4} mmHg):
1.99 at 29.5 °C (Verschueren, 1983)
1.1 at 25 °C (Lewis, 1989)
2.42 at 30 °C (Nash, 1983)

Environmental fate:
Biological. Laanio et al. (1973) incubated [$^{14}CF_3$]trifluralin with *Paecilomyces*, *Fusarium oxysporum*, or *Aspergillus fumigatus* and reported that <1% was converted to $^{14}CO_2$. From the first-order biotic and abiotic rate constants of trifluralin in estuarine water and sediment/water systems, the estimated biodegradation half-lives were 1.2-9.7 and 2.4-7.1 d, respectively (Walker et al., 1988).

Soil. Anaerobic degradation in a Crowley silt loam yielded α,α,α-trifluoro-N^4,N^4-dipropyl-5-nitrotoluene-3,4-diamine and α,α,α-trifluoro-N^4,N^4-dipropyltoluene-3,4,5-triamine (Parr and Smith, 1973). Probst and Tepe (1969) reported that trifluralin degradation in many soils was probably by chemical reduction of the nitro groups into amino groups. Reported degradation products in aerobic soils include α,α,α-trifluoro-2,6-dinitro-*N*-propyl-*p*-toluidine, α,α,α-trifluoro-2,6-dinitro-*p*-toluidine, α,α,α-trifluoro-5-nitrotoluene-3,4-diamine, and α,α,α-trifluoro-*N*,*N*-dipropyl-5-nitrotoluene-3,4-diamine. Anaerobic degradation products identified include α,α,α-trifluoro-*N*,*N*-dipropyl-5-nitrotoluene-3,4-diamine, α,α,α-trifluoro-*N*,*N*-dipropyltoluene-3,4,5-triamine, α,α,α-trifluorotoluene-3,4,5-triamine, and α,α,α-trifluoro-*N*-propyltoluene-3,4,5-triamine. α,α,α-Trifluoro-5-nitro-*N*-propyltoluene-3,4-diamine was identified in both aerobic and anaerobic soils (Probst et al., 1967). The following compounds were reported as major soil metabolites: α,α,α-trifluoro-2,6-dinitro-*N*-propyl-*p*-toluidine, α,α,α-trifluoro-2,6-dinitro-*p*-toluidine, α,α,α-trifluoro-5-nitrotoluene-3,4-diamine, α,α,α-trifluorotoluene-3,4,5-triamine, 2-ethyl-7-nitro-1-propyl-5-(trifluoromethyl)benzimidazole, 2-ethyl-7-nitro-5-(trifluoromethyl)benzimidazole, 7-nitro-1-propyl-5-(trifluoromethyl)benzimidazole, 4-(dipropylamino)-3,5-dinitrobenzoic acid, 2,2′-azoxybis(α,α,α-trifluoro-6-nitro-*N*-propyl-*p*-toluidine), 2,2′-azo-bis(α,α,α-trifluoro-6-nitro-*N*-propyl-*p*-toluidine), 2,6-dinitro-*N*,*N*-dipropyl-4-(trifluoromethyl)-

m-anisidine, and α,α,α-trifluoro-2′,6′-dinitro-*N*-propyl-*p*-propionotoluidine (Koskinen et al., 1984, 1985).

Golab et al. (1979) studied the degradation of trifluralin in soil over a 3-yr period. They found that the herbicide undergoes *N*-dealkylation, reduction of nitro substituents, followed by the formation cyclized products. Of the 28 transformations products identified, none exceeded 3% of the applied amount. These compounds were: α,α,α-trifluoro-2,6-dinitro-*N*-propyl-*p*-toluidine, α,α,α-trifluoro-2,6-dinitro-*p*-toluidine, α,α,α-trifluoro-5-nitro-*N*⁴,*N*⁴-dipropyltoluene-3,4-diamine, α,α,α-trifluoro-5-nitro-*N*⁴-propyltoluene-3,4-diamine, α,α,α-trifluoro-5-nitrotoluene-3,4-diamine, α,α,α-trifluoro-*N*⁴,*N*⁴-dipropyltoluene-3,4,5-triamine, α,α,α-trifluoro-*N*⁴-propyltoluene-3,4,5-triamine, α,α,α-trifluorotoluene-3,4,5-triamine, α,α,α-trifluoro-2′-hydroxyamino-6′-nitro-*N*-propyl-*p*-propionotoluidide, 2-ethyl-7-nitro-1-propyl-5-(trifluoromethyl)benzimidazole 3-oxide, 2-ethyl-7-nitro-5-(trifluoromethyl)benzimidazole 3-oxide, 2-ethyl-7-nitro-1-propyl-5-(trifluoromethyl)benzimidazole, 7-amino-2-ethyl-1-propyl-5-(trifluoromethyl)benzimidazole, 2-ethyl-7-nitro-5-(trifluoromethyl)benzimidazole, 7-amino-2-ethyl-5-(trifluoromethyl)benzimidazole, 7-nitro-1-propyl-5-(trifluoromethyl)benzimidazole, 7-nitro-5-(trifluoromethyl)benzimidazole, 7-amino-5-(trifluoromethyl)benzimidazole, α,α,α-trifluoro-2,6-dinitro-*p*-cresol, 4-(dipropylamino)-3,5-dinitrobenzoic acid, 2,2′-azoxybis(α,α,α-trifluoro-6-nitro-*N*-propyl-*p*-toluidine), *N*-propyl-2,2′-azoxybis(α,α,α-trifluoro-6-nitro-*p*-toluidine), 2,2′-azoxybis(α,α,α-trifluoro-6-nitro-*p*-toluidine), 2,2′-azobis(α,α,α-trifluoro-6-nitro-*N*-propyl-*p*-toluidine), α,α,α-trifluoro-4,6-dinitro-5-(dipropylamino)-*o*-cresol, α,α,α-trifluoro-2-hydroxyamino-6-nitro-*N*,*N*-dipropyl-*p*-toluidine, α,α,α-trifluoro-2′,6′-dinitro-*N*-propyl-*p*-propionotoluidide, and α,α,α-trifluoro-2,6-dinitro-*N*-(propan-2-ol)-*N*-propyl-*p*-toluidine (Golab et al., 1979).

Zayed et al. (1983) studied the degradation of trifluralin by the microbes *Aspergillus carneus*, *Fusarium oxysporum*, and *Trichoderma viride*. Following an inoculation and incubation period of 10 d in the dark at 25 °C, the following metabolites were identified: α,α,α-trifluoro-2,6-dinitro-*N*-propyl-*p*-toluidine, α,α,α-trifluoro-2,6-dinitro-*p*-toluidine, 2-amino-6-nitro-α,α,α-trifluoro-*p*-toluidine and 2,6-dinitro-4-trifluoromethyl phenol. The reported half-life in soil is 132 d (Jury et al., 1987).

Groundwater. According to the U.S. EPA (1986), trifluralin has a high potential to leach to groundwater.

Plant. Trifluralin was absorbed by carrot roots in greenhouse soils pretreated with the herbicide (0.75 lb/acre). The major metabolite formed was α,α,α-trifluoro-2,6-dinitro-*N*-(*n*-propyltoluene)-*p*-toluidine (Golab et al., 1967). Two metabolites of trifluralin that were reported in goosegrass (*Eleucine indica*) were 3-methoxy-2,6-dinitro-*N*,*N*′-dipropyl-4-(trifluoromethyl)benzenamine and *N*-(2,6-dinitro-4-(trifluoromethyl)phenyl)-*N*-propylpropanamide (Duke et al., 1991).

Photolytic. Irradiation of trifluralin in hexane by laboratory light produced α,α,α-trifluoro-2,6-dinitro-*N*-propyl-*p*-toluidine and α,α,α-trifluoro-2,6-dinitro-*p*-toluidine. The sunlight irradiation of trifluralin in water yielded α,α,α-trifluoro-*N*⁴,*N*⁴-dipropyl-5-nitrotoluene-3,4-diamine, α,α,α-trifluoro-*N*⁴,*N*⁴-dipropyltoluene-3,4,5-triamine, 2-ethyl-7-nitro-5-trifluoromethylbenzimidazole, 2,3-dihydroxy-2-ethyl-7-nitro-1-propyl-5-trifluoromethylbenzimidazoline, and 2-ethyl-7-nitro-5-trifluoromethylbenzimidazole. 2-Amino-6-nitro-α,α,α-trifluoro-*p*-toluidine and 2-ethyl-5-nitro-7-trifluoromethylbenzimidazole also were reported as major products under acidic and basic conditions, respectively (Crosby and Leitis, 1973). In a later study, Leitis and Crosby (1974) reported that trifluralin in aqueous solutions was very unstable to sunlight, especially in the presence of methanol. The photodecomposition of trifluralin involved oxidative *N*-dealkylation, nitro reduction, and reductive cyclization. The principal photodecomposition products of trifluralin were 2-amino-6-nitro-α,α,α-trifluoro-*p*-toluidine, 2-ethyl-7-nitro-5-trifluoromethylbenzimidazole 3-oxide, 2,3-dihydroxy-2-ethyl-7-nitro-1-propyl-5-trifluoromethylbenzimidazole, and two azoxybenzenes. Under alkaline conditions, the principal photodecomposition product was 2-ethyl-7-nitro-5-trifluoromethylbenzimidazole (Leitis and

Crosby, 1974).

When trifluralin was released in the atmosphere on a sunny day, it was rapidly converted to the photochemical 2,6-dinitro-N-propyl-α,α,α-trifluoro-p-toluidine. The estimated half-life is 20 min (Woodrow et al., 1978). The vapor-phase photolysis of trifluralin was studied in the laboratory using a photoreactor, which simulated sunlight conditions (Soderquist et al., 1975). Vapor-phase photoproducts of trifluralin were identified as 2,6-dinitro-N-propyl-α,α,α-trifluoro-p-toluidine, 2,6-dinitro-α,α,α-trifluoro-p-toluidine, 2-ethyl-7-nitro-1-propyl-5-trifluoromethyl-benzimidazole, 2-ethyl-7-nitro-5-trifluoromethylbenzimidazole, and four benzimidazole precursors, reported by Leitis and Crosby (1974). Similar photoproducts were also identified in air above both bare surface treated soil and soil incorporated fields (Soderquist et al., 1975).

Sullivan et al. (1980) studied the UV photolysis of trifluralin in anaerobic benzene solutions. Products identified included the three azoxybenzene derivatives N-propyl-2,2'-azoxybis(α,α,α-trifluoro-6-nitro-p-toluidine), 2,2'-azoxybis(α,α,α-trifluoro-6-nitro-N-propyl-p-toluidine), and 2,2'-azoxybis(α,α,α-trifluoro-6-nitro-p-toluidine), and two azobenzene derivatives, N-propyl-2,2'-azobis(α,α,α-trifluoro-6-nitro-p-toluidine) and 2,2'-azobis(α,α,α-trifluoro-6-nitro-N-propyl-p-toluidine) (Sullivan et al., 1980). [^{14}C]Trifluralin on a silica plate was exposed to summer sunlight for 7.5 h. Although 52% of trifluralin was recovered, no photodegradation products were identified. In soil, no significant photodegradation of trifluralin was observed after 9 h of irradiation (Plimmer, 1978). When trifluralin on glass plates was irradiated for 4-6 h, it photodegraded to 2,3-dihydroxy-2-ethyl-7-nitro-1-propyl-5-trifluoromethylbenzimidazoline and 2-ethyl-7-nitro-5-trifluoromethylbenzimidazole-3-oxide (Wright and Warren, 1965).

Chemical/Physical. Releases carbon monoxide, carbon dioxide, and ammonia when heated to 900 °C (Kennedy, 1972, 1972a). Incineration may also release hydrofluoric acid in the off-gases (Sittig, 1985).

Toxicity:

EC_{50} (48-h) for *Daphnia magna* 560 μg/L, *Daphnia pulex* 625 μg/L, *Simocephalus serrulatus* 900 μg/L (Mayer and Ellersieck, 1986).

LC_{50} (96-h) for young rainbow trout 10-40 μg/L, young bluegill sunfish 20-90 μg/L (Hartley and Kidd, 1987), carp 45 μg/L (Poleksic and Karan, 1999).

LC_{50} (48-h) for bluegill sunfish 19 ppb, rainbow trout 11 ppb (Verschueren, 1983), carp 1.0 mg/L, goldfish 850 μg/L, medaka 430 μg/L, pond loach 350 μg/L (Spehar et al., 1982).

Acute oral LD_{50} for mice 5,000 mg/kg (quoted, RTECS, 1985), mice 500 mg/kg, dogs and rabbits >2,000 mg/kg (Worthing and Hance, 1991).

In 2-yr feeding trials, rats and dogs receiving 2,000 and 1,000 mg/kg, respectively, suffered no ill effects (Worthing and Hance, 1991).

Drinking water standard: No MCLGs or MCLs have been proposed although trifluralin has been listed for regulation (U.S. EPA, 1996).

Use: Pre-emergence herbicide used to control annual grasses and broad-leaved weeds in beans, brassicas, cotton, groundnuts, forage legumes, orchards, ornamentals, transplanted peppers, soyabeans, sugar beets, sunflowers, tomatoes, and vineyards (Worthing and Hance, 1991).

1,2,3-TRIMETHYLBENZENE

Synonym: Hemimellitene.

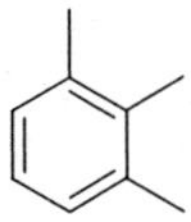

CAS Registry Number: 526-73-8
DOT: 2325
DOT label: Combustible liquid
Molecular formula: C_9H_{12}
Formula weight: 120.19
RTECS: DC3300000

Physical state, color, and odor:
Colorless liquid with an aromatic odor

Melting point (°C):
-25.4 (Weast, 1986)

Boiling point (°C):
176.1 (Weast, 1986)

Density (g/cm³):
0.8944 at 20/4 °C (Weast, 1986)

Diffusivity in water (x 10^{-5} cm²/sec):
0.74 at 20 °C using method of Hayduk and Laudie (1974)

Dissociation constant, pK_a:
>14 (Schwarzenbach et al., 1993)

Flash point (°C):
48 (Dean, 1987)
53 (90.5% solution, NFPA, 1984)

Lower explosive limit (%):
0.8 (NIOSH, 1997)

Upper explosive limit (%):
6.6 (NIOSH, 1997)

Heat of fusion (kcal/mol):
1.955 (Dean, 1987)

Henry's law constant (x 10^{-3} atm·m³/mol):
3.18 at 25 °C (approximate - calculated from water solubility and vapor pressure)

Ionization potential (eV):
8.42 (Lias, 1998)

Soil sorption coefficient, log K_{oc}:
2.80 (Schwarzenbach and Westall, 1981)

Octanol/water partition coefficient, log K_{ow}:
3.55 (Tewari et al., 1982; Wasik et al., 1981)
3.66 (Camilleri et al., 1988; Hammers et al., 1982; Hansch and Leo, 1979)

Solubility in organics:
Soluble in acetone, alcohol, benzene, and ether (Weast, 1986)

Solubility in water:
75.2 mg/kg at 25 °C, 48.6 mg/kg in artificial seawater (salinity = 34.5 g/kg) at 25.0 °C (Sutton and Calder, 1975)
545 μmol/L at 25.0 °C (Tewari et al., 1982; Wasik et al., 1981)
498, 520, 597, and 702 μmol/L at 15.0, 25.0, 35.0, and 45.0 °C, respectively (Sanemasa et al., 1982)

Vapor density:
4.91 g/L at 25 °C, 4.15 (air = 1)

Vapor pressure (mmHg):
1.51 at 25 °C (Mackay et al., 1982)

Environmental fate:
Photolytic. Glyoxal, methylglyoxal, and biacetyl were produced from the photooxidation of 1,2,3-trimethylbenzene by OH radicals in air at 25 °C (Tuazon et al., 1986a). The rate constant for the reaction of 1,2,3-trimethylbenzene and OH radicals at room temperature was 1.53 x 10^{-11} cm^3/molecule·sec (Hansen et al., 1975). A rate constant of 1.49 x 10^8 L/mol·sec was reported for the reaction of 1,2,3-trimethylbenzene with OH radicals in the gas phase (Darnall et al., 1976). Similarly, a room temperature rate constant of 3.16 x 10^{-11} cm^3/molecule·sec was reported for the vapor-phase reaction of 1,2,3-trimethylbenzene with OH radicals (Atkinson, 1985). At 25 °C, a rate constant of 2.69 x 10^{-11} cm^3/molecule·sec was reported for the same reaction (Ohta and Ohyama, 1985).
Chemical/Physical. 1,2,3-Trimethylbenzene will not hydrolyze (Kollig, 1993).

Exposure limits: NIOSH REL: TWA 25 ppm (125 mg/m^3); ACGIH TLV: TWA for mixed isomers 25 ppm (adopted).

Source: Detected in distilled water-soluble fractions of 87 octane gasoline (0.30 mg/L), 94 octane gasoline (0.81 mg/L), Gasohol (0.80 mg/L), No. 2 fuel oil (0.22 mg/L), diesel fuel (0.09 mg/L), and military jet fuel JP-4 (0.19 mg/L) (Potter, 1996).
Thomas and Delfino (1991) equilibrated contaminant-free groundwater collected from Gainesville, FL with individual fractions of three individual petroleum products at 24-25 °C for 24 h. The aqueous phase was analyzed for organic compounds via U.S. EPA approved test method 602. Average 1,2,3-trimethylbenzene concentrations detected in water-soluble fractions of unleaded gasoline, kerosene, and diesel fuel were 1.219, 0.405, and 0.118 mg/L, respectively. When the authors analyzed the aqueous-phase via U.S. EPA approved test method 610, average 1,2,3-trimethylbenzene concentrations in water-soluble fractions of unleaded gasoline, kerosene, and diesel fuel were lower, i.e., 742, 291, and 105 μg/L, respectively.

Uses: Organic synthesis.

1,2,4-TRIMETHYLBENZENE

Synonyms: *asym*-Trimethylbenzene; Pseudocumene; Pseudocumol.

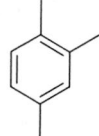

CAS Registry Number: 95-63-6
DOT: 2325
DOT label: Combustible liquid
Molecular formula: C_9H_{12}
Formula weight: 120.19
RTECS: DC3325000
Merck reference: 10, 7816

Physical state, color, and odor:
Colorless liquid with a slight aromatic odor

Melting point (°C):
-43.8 (Weast, 1986)

Boiling point (°C):
169.3 (Weast, 1986)

Density (g/cm³):
0.8758 at 20/4 °C (Weast, 1986)
0.888 at 4/4 °C (Sax and Lewis, 1987)

Diffusivity in water (x 10^{-5} cm²/sec):
0.73 at 20 °C using method of Hayduk and Laudie (1974)

Dissociation constant, pK$_a$:
>14 (Schwarzenbach et al., 1993)

Flash point (°C):
54.4 (Sax and Lewis, 1987)
44 (NFPA, 1984)

Lower explosive limit (%):
0.9 (NFPA, 1984)

Upper explosive limit (%):
6.4 (NFPA, 1984)

Entropy of fusion (cal/mol·K):
13.2 (Huffman et al., 1931)

Heat of fusion (kcal/mol):
3.023 (Huffman et al., 1931)

Henry's law constant (x 10^{-3} atm·m³/mol):
6.946, 11.202, and 15.702 at 27.0, 35.0, and 45.0 °C, respectively (Hansen et al., 1995)

Ionization potential (eV):
8.27 (Lias, 1998)

Bioconcentration factor, log BCF:
2.90, 3.26, 3.18, 3.95, 5.04, and 4.72 for olive, grass, holly, ivy, mock orange, pine, and rosemary, respectively (Hiatt, 1998)

Soil sorption coefficient, log K_{oc}:
3.57 using method of Karickhoff et al. (1979)

Octanol/water partition coefficient, log K_{ow}:
3.65 (Mackay et al., 1980)
3.78 (Camilleri et al., 1988; Hammers et al., 1982)

Solubility in organics:
Soluble in acetone, alcohol, benzene, and ether (Weast, 1986)

Solubility in water:
51.9 mg/kg at 25 °C (Price, 1976)
57 mg/kg at 25 °C (McAuliffe, 1966)
59.0 mg/kg at 25 °C (Sutton and Calder, 1975)
0.313 mM in 0.5 M NaCl at 25 °C (Wasik et al., 1984)
435, 469, 514, and 571 μM at 15.0, 25.0, 35.0, and 45.0 °C, respectively (Sanemasa et al., 1982)
3.47 mg/L (water soluble fraction of a 15-component simulated jet fuel mixture (JP-8) containing 6.9 wt % 1,2,4-trimethylbenzene, MacIntyre and deFur, 1985)

Vapor density:
4.91 g/L at 25 °C, 4.15 (air = 1)

Vapor pressure (mmHg):
2.03 at 25 °C (Mackay et al., 1982)

Environmental fate:
Biological. In anoxic groundwater near Bemidji, MI, 1,2,4-trimethylbenzene anaerobically biodegraded to the intermediate 3,4-dimethylbenzoic acid and the tentatively identified compounds 2,4- and/or 2,5-dimethylbenzoic acid (Cozzarelli et al., 1990).

Photolytic. Glyoxal, methylglyoxal, and biacetyl were produced from the photooxidation of 1,2,4-trimethylbenzene by OH radicals in air at 25 °C (Tuazon et al., 1986a). A rate constant of 2.0 x 10^8 L/mol·sec was reported for the reaction of 1,2,4-trimethylbenzene with OH radicals in the gas phase (Darnall et al., 1976). Similarly, the rate constants for the reaction of 1,2,4-trimethylbenzene and OH radicals at room temperature were 3.35 x 10^{-11} (Hansen et al., 1975) and 3.84 x 10^{-11} cm³/molecule·sec (Atkinson, 1985). At 25 °C, a rate constant of 3.15 x 10^{-11} cm³/molecule·sec was reported for the same reaction (Ohta and Ohyama, 1985).

Chemical/Physical. 1,2,4-Trimethylbenzene will not hydrolyze in water (Kollig, 1993).

Exposure limits: NIOSH REL: TWA 25 ppm (125 mg/m³); ACGIH TLV: TWA for mixed isomers 25 ppm (adopted).

Toxicity:
LC_{50} (inhalation) for rats 18 g/m^3/4-h (quoted, RTECS, 1985).

Drinking water standard: No MCLGs or MCLs have been proposed (U.S. EPA, 1996).

Source: Detected in distilled water-soluble fractions of 87 octane gasoline (1.11 mg/L), 94 octane gasoline (3.11 mg/L), Gasohol (2.90 mg/L), No. 2 fuel oil (0.51 mg/L), jet fuel A (0.44 mg/L), diesel fuel (0.39 mg/L), and military jet fuel JP-4 (0.39 mg/L) (Potter, 1996). Schauer et al. (1999) reported 1,2,4-trimethylbenzene in a diesel-powered medium-duty truck exhaust at an emission rate of 880 μg/km.

Thomas and Delfino (1991) equilibrated contaminant-free groundwater collected from Gainesville, FL with individual fractions of three individual petroleum products at 24-25 °C for 24 h. The aqueous phase was analyzed for organic compounds via U.S. EPA approved test method 602. Average 1,2,4-trimethylbenzene concentrations reported in water-soluble fractions of unleaded gasoline, kerosene, and diesel fuel were 1.952, 0.478, and 0.130 mg/L, respectively. When the authors analyzed the aqueous-phase via U.S. EPA approved test method 610, average 1,2,4-trimethylbenzene concentrations in water-soluble fractions of unleaded gasoline, kerosene, and diesel fuel were generally greater, i.e., 1.968, 0.401, and 0.146 mg/L, respectively.

Uses: Manufacture of dyes, resins, perfumes, trimellitic anhydride, pseudo-cumidine.

1,3,5-TRIMETHYLBENZENE

Synonyms: Mesitylene; Fleet-X; TMB; *sym*-Trimethylbenzene; Trimethylbenzol; UN 2325.

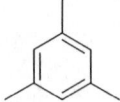

CAS Registry Number: 108-67-8
DOT: 2325
DOT label: Combustible liquid
Molecular formula: C_9H_{12}
Formula weight: 120.19
RTECS: OX6825000
Merck reference: 10, 5752

Physical state, color, and odor:
Colorless liquid with a peculiar odor. Odor threshold concentration is 550 ppb (Amoore and Hautala, 1983).

Melting point (°C):
-44.7 (Weast, 1986)
-52.7 (Verschueren, 1983)

Boiling point (°C):
164.7 (Weast, 1986)

Density (g/cm³):
0.8652 at 20/4 °C (Weast, 1986)
0.8606 at 25.00/4 °C (Aminabhavi and Banerjee, 1999)

Diffusivity in water (x 10⁻⁵ cm²/sec):
0.73 at 20 °C using method of Hayduk and Laudie (1974)

Dissociation constant, pK$_a$:
>14 (Schwarzenbach et al., 1993)

Flash point (°C):
44 (Dean, 1987)
50 (NFPA, 1984)

Heat of fusion (kcal/mol):
1.892-2.274 (Dean, 1987)

Henry's law constant (x 10⁻³ atm·m³/mol):
4.03, 4.60, 5.71, 6.73, and 9.63 at 10, 15, 20, 25, and 30 °C, respectively (Ashworth et al., 1988)

Interfacial tension with water (dyn/cm at 20 °C):
38.70 (Demond and Lindner, 1993)

Ionization potential (eV):
8.40 (Lias, 1998)

Soil sorption coefficient, log K_{oc}:
2.82 (Schwarzenbach and Westall, 1981)

Octanol/water partition coefficient, log K_{ow}:
3.41 (Campbell and Luthy, 1985)
3.42 (Chiou et al., 1982; Hansch and Leo, 1979)
3.78 (Hammers et al., 1978)

Solubility in organics:
Miscible with alcohol, benzene, and ether (Windholz et al., 1983)

Solubility in water:
48.2 mg/kg at 25 °C (Sutton and Calder, 1975)
640 μmol/L at 25.0 °C (Andrews and Keefer, 1950a)
97.7 mg/L at 25 °C (Chiou et al., 1982)
282 μmol/L in 0.5 M NaCl at 25 °C (Wasik et al., 1984)
383, 415, 455, and 485 μmol/L at 15.0, 25.0, 35.0, and 45.0 °C, respectively (Sanemasa et al., 1982)
<200 mg/L at 25 °C (Booth and Everson, 1948)
328 μmol/kg at 25.0 °C (Vesala, 1974)
8.55 x 10^{-6} at 25 °C (mole fraction, Li et al., 1993)

Vapor density:
4.91 g/L at 25 °C, 4.15 (air = 1)

Vapor pressure (mmHg):
2 at 20 °C (NIOSH, 1997)
2.42 at 25 °C (Mackay et al., 1982)

Environmental fate:
 Biological. In anoxic groundwater near Bemidji, MI, 1,3,5-trimethylbenzene anaerobically biodegraded to the intermediate tentatively identified as 3,5-dimethylbenzoic acid (Cozzarelli et al., 1990).
 Photolytic. Glyoxal, methylglyoxal, and biacetyl were produced from the photooxidation of 1,3,5-trimethylbenzene by OH radicals in air at 25 °C (Tuazon et al., 1986a). The rate constant for the reaction of 1,3,5-trimethylbenzene and OH radicals at room temperature was 4.72 x 10^{-11} cm^3/molecule·sec (Hansen et al., 1975). A rate constant of 2.97 x 10^8 L/mol·sec was reported for the reaction of 1,3,5-trimethylbenzene with OH radicals in the gas phase (Darnall et al., 1976). Similarly, a room temperature rate constant of 6.05 x 10^{-11} cm^3/molecule·sec was reported for the vapor-phase reaction of 1,3,5-trimethylbenzene with OH radicals (Atkinson, 1985). At 25 °C, a rate constant of 3.87 x 10^{-11} cm^3/molecule·sec was reported for the same reaction (Ohta and Ohyama, 1985).
 Chemical/Physical. 1,3,5-Trimethylbenzene will not hydrolyze because it does not contain a hydrolyzable functional group (Kollig, 1993).
 Complete combustion in air yields carbon dioxide and water vapor.

Exposure limits: NIOSH REL: TWA 25 ppm (125 mg/m^3); ACGIH TLV: TWA for mixed isomers 25 ppm (adopted).

Toxicity:

LD$_{50}$ (inhalation) for rats 24 g/m^3/4-h (quoted, RTECS, 1985).

Drinking water standard: No MCLGs or MCLs have been proposed (U.S. EPA, 1996).

Source: Detected in distilled water-soluble fractions of 87 octane gasoline (0.34 mg/L), 94 octane gasoline (1.29 mg/L), Gasohol (0.48 mg/L), No. 2 fuel oil (0.08 mg/L), jet fuel A (0.09 mg/L), diesel fuel (0.03 mg/L), and military jet fuel JP-4 (0.09 mg/L) (Potter, 1996). Schauer et al. (1999) reported 1,3,5-trimethylbenzene in a diesel-powered medium-duty truck exhaust at an emission rate of 260 μg/km.

Thomas and Delfino (1991) equilibrated contaminant-free groundwater collected from Gainesville, FL with individual fractions of three individual petroleum products at 24-25 °C for 24 h. The aqueous phase was analyzed for organic compounds via U.S. EPA approved test method 602. Average 1,3,5-trimethylbenzene concentrations reported in water-soluble fractions of unleaded gasoline, kerosene, and diesel fuel were 333, 86, and 13 μg/L, respectively. When the authors analyzed the aqueous-phase via U.S. EPA approved test method 610, average 1,3,5-trimethylbenzene concentrations in water-soluble fractions of unleaded gasoline, kerosene, and diesel fuel were greater, i.e., 441, 91, and 27 μg/L, respectively.

Uses: UV oxidation stabilizer for plastics; manufacturing anthraquinone dyes.

1,1,3-TRIMETHYLCYCLOHEXANE

Synonym: 1,3,3-Trimethylcyclohexane.

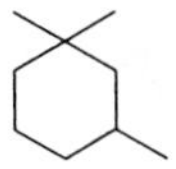

CAS Registry Number: 3073-66-3
Molecular formula: C_9H_{18}
Formula weight: 126.24
RTECS: GV7650000

Physical state:
Liquid

Melting point (°C):
-65.7 (Dreisbach, 1959)

Boiling point (°C):
136.6 (Dean, 1987)
138-139 (Weast, 1986)

Density (g/cm³):
0.7664 at 20/4 °C (Weast, 1986)

Diffusivity in water (x 10^{-5} cm²/sec):
0.66 at 20 °C using method of Hayduk and Laudie (1974)

Dissociation constant, pK_a:
>14 (Schwarzenbach et al., 1993)

Henry's law constant (atm·m³/mol):
1.1 at 25 °C (Mackay and Shiu, 1981)

Ionization potential (eV):
9.39 (Mallard and Linstrom, 1998)

Soil sorption coefficient, log K_{oc}:
Unavailable because experimental methods for estimation of this parameter for alicyclic hydrocarbons are lacking in the documented literature

Octanol/water partition coefficient, log K_{ow}:
Unavailable because experimental methods for estimation of this parameter for alicyclic hydrocarbons are lacking in the documented literature

Solubility in water:
1.77 mg/kg at 25 °C (Price, 1976)

Vapor density:
4.36 g/L at 25 °C, 3.68 (air = 1)

Vapor pressure (mmHg):
11.1 (estimated using Antoine equation, Dean, 1987)

Environmental fate:
   Chemical/Physical. Complete combustion in air gives carbon dioxide and water vapor. 1,1,3-Trimethylcyclohexane will not hydrolyze because it does not contain a hydrolyzable functional group.

Uses: Organic synthesis; gasoline component.

1,1,3-TRIMETHYLCYCLOPENTANE

Synonym: 1,3,3-Trimethylcyclopentane.

CAS Registry Number: 4516-69-2
Molecular formula: C_8H_{16}
Formula weight: 112.22

Physical state and color:
Colorless liquid

Melting point (°C):
-14.4 (Mackay et al., 1982)

Boiling point (°C):
104.89 °C (Wilhoit and Zwolinski, 1971)

Density (g/cm³):
0.7703 at 20/4 °C (Weast, 1986)

Diffusivity in water (x 10^{-5} cm²/sec):
0.71 at 20 °C using method of Hayduk and Laudie (1974)

Dissociation constant, pK_a:
>14 (Schwarzenbach et al., 1993)

Henry's law constant (atm·m³/mol):
1.56 at 25 °C (Mackay and Shiu, 1981)

Soil sorption coefficient, log K_{oc}:
Unavailable because experimental methods for estimation of this parameter for alicyclic hydrocarbons are lacking in the documented literature

Octanol/water partition coefficient, log K_{ow}:
3.34 (calculated, Wang et al., 1992)

Solubility in water:
3.73 mg/kg at 25 °C (Price, 1976)

Vapor density:
4.59 g/L at 25 °C, 3.87 (air = 1)

Vapor pressure (mmHg):
39.7 at 25 °C (Wilhoit and Zwolinski, 1971)

Environmental fate:
Chemical/Physical. 1,1,3-Trimethylcyclopentane will not hydrolyze because it does not

contain a hydrolyzable functional group (Kollig, 1993).

Complete combustion in air yields carbon dioxide and water vapor.

Uses: Organic synthesis; gasoline component.

2,2,5-TRIMETHYLHEXANE

Synonym: 2,5,5-Trimethylhexane.

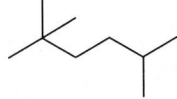

CAS Registry Number: 3522-94-9
Molecular formula: C_9H_{20}
Formula weight: 128.26

Physical state and color:
Colorless liquid

Melting point (°C):
-105.8 (Weast, 1986)

Boiling point (°C):
124 (Weast, 1986)

Density (g/cm³):
0.70721 at 20/4 °C, 0.70322 at 25/4 °C (Dreisbach, 1959)

Diffusivity in water (x 10⁻⁵ cm²/sec):
0.62 at 20 °C using method of Hayduk and Laudie (1974)

Dissociation constant, pK_a:
>14 (Schwarzenbach et al., 1993)

Flash point (°C):
12.7 (Hawley, 1981)

Heat of fusion (kcal/mol):
1.48 (Riddick et al., 1986)

Henry's law constant (atm·m³/mol):
3.45 at 25 °C (Mackay and Shiu, 1981)

Soil sorption coefficient, log K_{oc}:
Unavailable because experimental methods for estimation of this parameter for aliphatic hydrocarbons are lacking in the documented literature

Octanol/water partition coefficient, log K_{ow}:
3.88 using method of Hansch et al. (1968)

Solubility in organics:
In methanol, g/L: 162 at 5 °C, 179 at 10 °C, 200 at 15 °C, 221 at 20 °C, 247 at 25 °C, 280 at 30 °C, 316 at 35 °C, 360 at 40 °C (Kiser et al., 1961)

Solubility in water (mg/kg):
0.79 at 0 °C, 0.54 at 25 °C (Polak and Lu, 1973)
1.15 at 25 °C (McAuliffe, 1966)

Vapor density:
5.24 g/L at 25 °C, 4.43 (air = 1)

Vapor pressure (mmHg):
16.5 at 25 °C (Wilhoit and Zwolinski, 1971)

Environmental fate:
   Chemical/Physical. Complete combustion in air gives carbon dioxide and water vapor. 2,2,5-Trimethylhexane will not hydrolyze because it does not contain a hydrolyzable functional group.

Exposure limits: ACGIH TLV: TWA for all isomers 200 ppm (adopted).

Uses: Motor fuel additive; organic synthesis.

2,2,4-TRIMETHYLPENTANE

Synonyms: Isobutyltrimethylmethane; Isooctane; UN 1262.

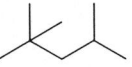

CAS Registry Number: 540-84-1
DOT: 1262
DOT label: Flammable liquid
Molecular formula: C_8H_{18}
Formula weight: 114.23
RTECS: SA3320000
Merck reference: 10, 5040

Physical state, color, and odor:
Colorless liquid with a gasoline-like odor

Melting point (°C):
-107.4 (Weast, 1986)
-116 (Sax and Lewis, 1987)

Boiling point (°C):
99.2 (Weast, 1986)

Density (g/cm³):
0.69192 at 20/4 °C, 0.68777 at 25/4 °C (Dreisbach, 1959)
0.6879 at 25.00/4 °C (Aralaguppi et al., 1999)

Diffusivity in water (x 10⁻⁵ cm²/sec):
0.66 at 20 °C using method of Hayduk and Laudie (1974)

Dissociation constant, pK_a:
>14 (Schwarzenbach et al., 1993)

Flash point (°C):
-12 (Windholz et al., 1983)

Lower explosive limit (%):
1.1 (Sax and Lewis, 1987)

Upper explosive limit (%):
6.0 (Sax and Lewis, 1987)

Entropy of fusion (cal/mol·K):
13.07 (Parks et al., 1930)

Heat of fusion (kcal/mol):
2.161 (Parks et al., 1930)
2.198 (Riddick et al., 1986)

Henry's law constant (atm·m³/mol):
3.23 at 25 °C (Mackay and Shiu, 1981)

Interfacial tension with water (dyn/cm):
50.1 at 25 °C (Demond and Lindner, 1993)

Ionization potential (eV):
9.89 (Lias, 1998)

Soil sorption coefficient, log K_{oc}:
Unavailable because experimental methods for estimation of this parameter for aliphatic hydrocarbons are lacking in the documented literature

Octanol/water partition coefficient, log K_{ow}:
5.83 (Burkhard et al., 1985a)

Solubility in organics:
In methanol, g/L: 249 at 5 °C, 279 at 10 °C, 314 at 15 °C, 353 at 20 °C, 402 at 25 °C, 460 at 30 °C, 560 at 35 °C, 760 at 40 °C (Kiser et al., 1961)

Solubility in water:
1.14 mg/kg at 25 °C (Price, 1976)
2.46 mg/kg at 0 °C, 2.05 mg/kg at 25 °C (Polak and Lu, 1973)
0.56 mg/L at 25 °C (quoted, Verschueren, 1983)
2.44 mg/kg at 25 °C (McAuliffe, 1963, 1966)
0.900 and 1.296 mL/L in water and 0.1 wt % aqueous sodium naphthenate, respectively (Baker, 1980)

Vapor density:
4.67 g/L at 25 °C; 3.94 (air = 1)

Vapor pressure (mmHg):
47.8 at 24.4 °C (Willingham et al., 1945)
49.3 at 25 °C (Wilhoit and Zwolinski, 1971)

Environmental fate:
   Surface Water. Mackay and Wolkoff (1973) estimated an evaporation half-life of 4.1 sec from a surface water body that is 25 °C and 1 m deep.
   Photolytic. The following rate constants were reported for the reaction of 2,2,4-trimethyl-pentane and OH radicals in the atmosphere: 2.3 x 10^{12} cm³/mol·sec at 300 K (Hendry and Kenley, 1979); 2.83 x 10^{-12} cm³/mol·sec at 298 K (Greiner, 1970); 3.73 x 10^{-12} cm³/molecule·sec at 298-305 K (Darnall et al., 1978); 3.7 x 10^{-12} cm³/molecule·sec (Atkinson et al., 1979); 3.90 x 10^{-12} cm³/molecule·sec at 298 K (Atkinson, 1985). Based on a photooxidation rate constant of 3.68 x 10^{-12} cm³/molecule·sec for the reaction of 2,2,4-trimethylpentane and OH radicals in summer sunlight, the lifetime is 16 h (Altshuller, 1991).
   Chemical/Physical. Complete combustion in air gives carbon dioxide and water vapor.

Exposure limits: ACGIH TLV: TWA for all isomers 300 ppm (adopted).

Symptoms of exposure: High concentrations may cause irritation of respiratory tract (Patnaik, 1992).

Source: Schauer et al. (1999) reported 2,2,4-trimethylpentane in a diesel-powered medium-duty truck exhaust at an emission rate of 1,240 μg/km.

Uses: Determining octane numbers of fuels; solvent and thinner; in spectrophotometric analysis.

2,3,4-TRIMETHYLPENTANE

Synonyms: None.

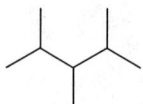

CAS Registry Number: 565-75-3
Molecular formula: C_8H_{18}
Formula weight: 114.23

Physical state and color:
Colorless liquid

Melting point (°C):
-109.2 (Weast, 1986)

Boiling point (°C):
113.4 (Weast, 1986)

Density (g/cm³):
0.71906 at 20/4 °C, 0.71503 at 25/4 °C (Dreisbach, 1959)

Diffusivity in water (x 10^{-5} cm²/sec):
0.67 at 20 °C using method of Hayduk and Laudie (1974)

Dissociation constant, pK_a:
>14 (Schwarzenbach et al., 1993)

Entropy of fusion (cal/mol·K):
13.54 (Pitzer and Scott, 1941)

Heat of fusion (kcal/mol):
2.215 (Pitzer and Scott, 1941)

Henry's law constant (atm·m³/mol):
1.89 at 25 °C (Mackay and Shiu, 1981)

Soil sorption coefficient, log K_{oc}:
Unavailable because experimental methods for estimation of this parameter for aliphatic hydrocarbons are lacking in the documented literature

Octanol/water partition coefficient, log K_{ow}:
3.78 using method of Hansch et al. (1968)

Solubility in organics:
Soluble in acetone, alcohol, benzene, chloroform, and ether (Weast, 1986)

Solubility in water (mg/kg):
1.36 at 25 °C (Price, 1976)
2.34 at 0 °C, 2.30 at 25 °C (Polak and Lu, 1973)

Vapor density:
4.67 g/L at 25 °C, 3.94 (air = 1)

Vapor pressure (mmHg):
27.0 at 25 °C (Wilhoit and Zwolinski, 1971)

Environmental fate:

Photolytic. Atkinson (1990) reported a rate constant of 7.0 x 10^{-12} cm³/molecule·sec for the reaction of 2,3,4-trimethylpentane and OH radicals in the atmosphere at 298 K. Based on this reaction rate constant, the estimated lifetime is 20 h (Altshuller, 1991).

Chemical/Physical. Complete combustion in air gives carbon dioxide and water vapor. 2,3,4-Trimethylpentane will not hydrolyze because it does not contain a hydrolyzable functional group.

Exposure limits: ACGIH TLV: TWA for all isomers 300 ppm (adopted).

Source: Schauer et al. (1999) reported 2,3,4-trimethylpentane in a diesel-powered medium-duty truck exhaust at an emission rate of 310 μg/km.

Uses: Organic synthesis.

2,4,6-TRINITROTOLUENE

Synonyms: Entsufon; 1-Methyl-2,4,6-trinitrobenzene; **2-Methyl-1,3,5-trinitrobenzene**; NCI-C56155; TNT; α-TNT; TNT-tolite; Tolit; Tolite; Trilit; Trinitrotoluene; *sym*-Trinitrotoluene; Trinitrotoluol; α-Trinitrotoluol; *sym*-Trinitrotoluol; Tritol; Triton; Trotyl; Trotyl oil; UN 0209.

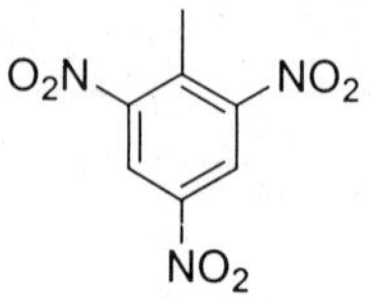

CAS Registry Number: 118-96-7
DOT: 1356
DOT label: Class A explosive
Molecular formula: $C_7H_5N_3O_6$
Formula weight: 227.13
RTECS: XU0175000
Merck reference: 10, 9534

Physical state, color, and odor:
Colorless to light yellow, odorless monoclinic crystals

Melting point (°C):
82 (Weast, 1986)
80.1 (Windholz et al., 1983)

Boiling point (°C):
Explodes at 240 (Weast, 1986)

Density (g/cm³):
1.654 at 20/4 °C (Weast, 1986)

Diffusivity in water (x 10^{-5} cm²/sec):
0.68 at 20 °C using method of Hayduk and Laudie (1974)

Flash point (°C):
Explodes (NIOSH, 1997)

Ionization potential (eV):
10.59 (Lias et al., 1998)

Soil sorption coefficient, log K_{oc}:
2.48 using method of Kenaga and Goring (1980)

Octanol/water partition coefficient, log K_{ow}:
2.25 using method of Kenaga and Goring (1980)

Solubility in organics:
Soluble in acetone, benzene, ether, and pyrimidine (Weast, 1986)

Solubility in water:
0.013 wt % at 20 °C (NIOSH, 1987)

1.42 g/L at 100 °C (quoted, Windholz et al., 1983)
200 mg/L at 15 °C (quoted, Verschueren, 1983)

Vapor pressure (x 10⁻³ mmHg):
0.2 at 20 °C (NIOSH, 1997)
4.26 at 54.8 °C, 2.557 at 72.5 °C, 4.347 at 76.1 °C (Lenchitz and Velicky, 1970)

Environmental fate:
   Biological. 4-Amino-2,6-dinitrotoluene and 2-amino-4,6-dinitrotoluene, detected in contaminated groundwater beneath the Hawthorne Naval Ammunition Depot, NV, were reported to have formed from the microbial degradation of 2,4,6-trinitrotoluene (Pereira et al., 1979).
   Chemical. Although no products were identified, 2,4,6-trinitrotoluene (1.5 x 10⁻⁵ M) was reduced by iron metal (33.3 g/L acid washed 18-20 mesh) in a carbonate buffer (1.5 x 10⁻² M) at pH 5.9 and 15 °C. Based on the pseudo-first-order disappearance rate of 0.0330/min, the half-life was 21.0 min (Agrawal and Tratnyek, 1996).
   Will detonate upon heating (NIOSH, 1997).

Exposure limits (mg/m³): NIOSH 0.5, IDLH 500; OSHA PEL: TWA 1.5; ACGIH TLV: TWA 0.1 (adopted).

Symptoms of exposure: May cause dermatitis, cyanosis, gastritis, sneezing, sore throat, muscular pain, somnolence, tremor, convulsions, and asplastic anemia (Patnaik, 1992)

Toxicity:
   Acute oral LD₅₀ in mice 660 mg/kg, rats 795 (quoted, RTECS, 1985).

Drinking water standard: No MCLGs or MCLs have been proposed (U.S. EPA, 1996).

Uses: High explosive; intermediate in dyestuffs and photographic chemicals.

TRIPHENYL PHOSPHATE

Synonyms: Celluflex TPP; **Phosphoric acid triphenyl ester**; Phenyl phosphate; TPP.

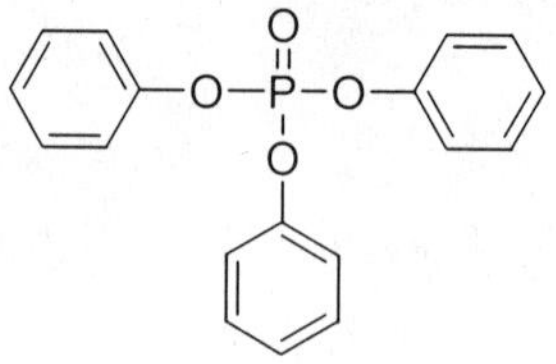

CAS Registry Number: 115-86-6
Molecular formula: $C_{18}H_{15}O_4P$
Formula weight: 326.29
RTECS: TC8400000
Merck reference: 10, 9547

Physical state, color, and odor:
Colorless, solid with a faint, phenol-like odor

Melting point (°C):
48.5 (Fordyce and Meyer, 1940)
50-51 (Weast, 1986)

Boiling point (°C):
245 at 11 mmHg (Weast, 1986)

Density (g/cm³):
1.2055 at 50/4 °C (Weast, 1986)
1.268 at 60/4 °C (Sax and Lewis, 1987)

Diffusivity in water (x 10^{-5} cm²/sec):
0.46 at 20 °C using method of Hayduk and Laudie (1974)

Flash point (°C):
222 (NIOSH, 1997)
223 (Dean, 1987)

Entropy of fusion (cal/mol·K):
21.95 (Rabinovich et al., 1986)

Heat of fusion (kcal/mol):
7.077 (Rabinovich et al., 1986)

Henry's law constant (x 10^{-2} atm·m³/mol):
5.88 at 20 °C (approximate - calculated from water solubility and vapor pressure)

Bioconcentration factor, log BCF:
2.60 (*Alburnus alburnus*, Devillers et al., 1996)

Soil sorption coefficient, log K_{oc}:
3.72 using method of Kenaga and Goring (1980)

Octanol/water partition coefficient, log K_{ow}:
5.27 using method of Saeger et al. (1979)

Solubility in organics:
Soluble in alcohol, benzene, chloroform, and ether (Weast, 1986)

Solubility in water:
0.002 wt % at 54 °C (NIOSH, 1997)
0.73 mg/L at 24 °C (practical grade, Verschueren, 1983)

Vapor pressure (mmHg):
1 at 195 °C (NIOSH, 1997)
<0.1 at 20 °C (Verschueren, 1983)

Environmental fate:
   Chemical/Physical. When an aqueous solution containing triphenyl phosphate (0.1 mg/L) and chlorine (3-1,000 mg/L) was stirred in the dark at 20 °C for 24 h, the benzene ring was substituted with 1-3 chlorine atoms (Ishikawa and Baba, 1988). The reported hydrolysis half-lives at pH values of 8.2 and 9.5 were 7.5 and 1.3 d, respectively (Howard and Doe, 1979).

Exposure limits (mg/m³): NIOSH REL: TWA 3, IDLH 1,000; OSHA PEL: TWA 3; ACGIH TLV: TWA 3 (adopted).

Symptoms of exposure: May cause depression of central nervous system and irritation of eyes, skin, and respiratory tract (Patnaik, 1992)

Toxicity:
   EC_{50} (96-h) for scud 0.25 mg/L (Huckins et al., 1991).
   EC_{50} (48-h) for midge 0.36 mg/L (Huckins et al., 1991).
   LC_{50} (96-h) for blugill sunfish 0.78 mg/L (Huckins et al., 1991).
   Acute oral LD_{50} for mice 1,320 mg/kg, rats 3,500 mg/kg (quoted, RTECS, 1985).

Uses: Camphor substitute in celluloid; impregnating roofing paper; plasticizer in lacquers and varnishes; renders acetylcellulose, airplane "dope," nitrocellulose, stable and fireproof; gasoline additives; insecticides; floatation agents; antioxidants, stabilizers, and surfactants.

VINYL ACETATE

Synonyms: Acetic acid ethenyl ester; Acetic acid ethylene ester; Acetic acid vinyl ester; 1-Acetoxyethylene; Ethenyl acetate; Ethenyl ethanoate; UN 1301; VAC; VAM; Vinyl acetate H.Q.; Vinyl ethanoate.

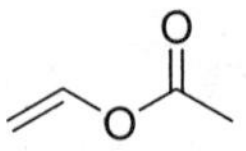

CAS Registry Number: 108-05-4
DOT: 1301
DOT label: Combustible liquid
Molecular formula: $C_4H_6O_2$
Formula weight: 86.09
RTECS: AK0875000
Merck reference: 10, 9794

Physical state, color, and odor:
Colorless, watery liquid with a pleasant, fruity odor. Odor threshold concentration is 500 ppb (Amoore and Hautala, 1983).

Melting point (°C):
-93.2 (Weast, 1986)

Boiling point (°C):
72.2 (Weast, 1986)

Density (g/cm³):
0.9317 at 20/4 °C (Weast, 1986)

Diffusivity in water (x 10⁻⁵ cm²/sec):
0.93 at 20 °C using method of Hayduk and Laudie (1974)

Flash point (°C):
-8 (NIOSH, 1997)

Lower explosive limit (%):
2.6 (NIOSH, 1997)

Upper explosive limit (%):
13.4 (NIOSH, 1997)

Henry's law constant (x 10⁻⁴ atm·m³/mol):
4.81 (calculated, Howard, 1989)

Ionization potential (eV):
9.19 ± 0.05 (Franklin et al., 1969)

Soil sorption coefficient, log K_{oc}:
0.45 (estimated, Montgomery, 1989)

Octanol/water partition coefficient, log K_{ow}:
0.60 (Fujisawa and Masuhara, 1980)
0.73 (Sangster, 1989)

Solubility in organics:
Soluble in acetone, ethanol, benzene, chloroform, and ether (Weast, 1986)

Solubility in water (mg/L):
25,000 at 25 °C (Amoore and Hautala, 1983)
20,000 at 20 °C (Dean, 1973)

Vapor density:
3.52 g/L at 25 °C, 2.97 (air = 1)

Vapor pressure (mmHg):
83 at 20 °C, 115 at 25 °C, 140 at 30 °C (Verschueren, 1983)

Environmental fate:
   Chemical/Physical. Anticipated hydrolysis products would include acetic acid and vinyl alcohol. Slowly polymerizes in light to a colorless, transparent mass.
   At an influent concentration of 1,000 mg/L, treatment with granular activated carbon resulted in an effluent concentration of 357 mg/L. The adsorbability of the carbon used was 129 mg/g carbon (Guisti et al., 1974).

Exposure limits: NIOSH REL: 15-min ceiling 4 ppm (15 mg/m^3); ACGIH TLV: TWA 10 ppm, STEL 15 ppm (adopted).

Toxicity:
   LC_{50} (inhalation) for mice 1,550 ppm/4-h, rats 4,000 ppm/2-h, rabbits 2,500 ppm/4-h (quoted, RTECS, 1985).
   Acute oral LD_{50} for mice 1,613 mg/kg, rats 2,920 mg/kg (quoted, RTECS, 1985).

Uses: Manufacture of polyvinyl acetate, polyvinyl alcohol, polyvinyl chloride-acetate resins; used particularly in latex paint; paper coatings; adhesives; textile finishing; safety glass interlayers.

VINYL CHLORIDE

Synonyms: Chlorethene; Chlorethylene; Chloroethene; 1-Chloroethene; Chloroethylene; 1-Chloroethylene; Ethylene monochloride; Monochloroethene; Monochloroethylene; MVC; RCRA waste number U043; Trovidur; UN 1086; VC; VCM; Vinyl C monomer; Vinyl chloride monomer.

CAS Registry Number: 75-01-4
DOT: 1086
DOT label: Flammable gas
Molecular formula: C_2H_3Cl
Formula weight: 62.50
RTECS: KU9625000
Merck reference: 10, 9796

Physical state, color, and odor:
Colorless, liquefied compressed gas with a faint, sweetish odor

Melting point (°C):
-153.8 (Weast, 1986)

Boiling point (°C):
-13.4 (Weast, 1986)

Density (g/cm³):
0.9106 at 20/4 °C (Weast, 1986)
0.94 at 13.9/4 °C, 0.9121 at 20/4 °C (Standen, 1964)

Diffusivity in water (x 10^{-5} cm²/sec):
1.34 at 25 °C (quoted, Hayduk and Laudie, 1974)

Lower explosive limit (%):
3.6 (NIOSH, 1997)

Upper explosive limit (%):
33 (NIOSH, 1997)

Entropy of fusion (cal/mol·K):
9.85 (Lebedev et al., 1967)

Heat of fusion (kcal/mol):
1.175 (Lebedev et al., 1967)

Henry's law constant (x 10^{-2} atm·m³/mol):
278 (Gossett, 1987)
2.2 (Pankow and Rosen, 1988)
122 at 10 °C (Dilling, 1977)
121.9 at 20 °C (Pearson and McConnell, 1975)
1.50, 1.68, 2.17, 2.65, and 2.8 at 10, 15, 20, 25, and 30 °C, respectively (Ashworth et al., 1988)

Ionization potential (eV):
9.99 (Horvath, 1982; Lias, 1998)

Bioconcentration factor, log BCF:
3.04 (activated sludge), 1.60 (algae) (Freitag et al., 1985)

Soil sorption coefficient, log K_{oc}:
0.39 using method of Karickhoff et al. (1979)

Octanol/water partition coefficient, log K_{ow}:
0.60 (Radding et al., 1976)

Solubility in organics:
Soluble in ethanol, carbon tetrachloride, ether (U.S. EPA, 1985), chloroform, tetrachloro-ethylene, and trichloroethylene

Solubility in water:
1,100 mg/L at 25 °C (quoted, Verschueren, 1983)
60 mg/L at 10 °C (Pearson and McConnell, 1975)
2,700 mg/L at 25 °C (Dilling, 1977)
In wt %: 0.95 at 15 °C, 0.995 at 16 °C, 0.915 at 20.5 °C, 0.88 at 26 °C, 0.89 at 29.5 °C, 0.94 at 35 °C, 0.89 at 41 °C, 0.88 at 46.5 °C, 0.95 at 55 °C, 0.92 at 65 °C, 0.98 at 72.5 °C, 1.00 at 80 °C, 1.12 at 85 °C (DeLassus and Schmidt, 1981)
14 mM at 20 °C (Fischer and Ehrenberg, 1948)

Vapor density:
2.86 g/L at 0 °C (Hayduk and Laudie, 1974a)
2.55 g/L at 25 °C, 2.16 (air = 1)

Vapor pressure (mmHg):
2,531 at 20 °C, 3,428 at 30 °C (Standen, 1964)
2,320 at 20 °C (McConnell et al., 1975)
2,660 at 25 °C (quoted, Nathan, 1978)

Environmental fate:
   Biological. Under anaerobic or aerobic conditions, degradation to carbon dioxide was reported in experimental systems containing mixed or pure cultures (Vogel et al., 1987). The anaerobic degradation of vinyl chloride dissolved in groundwater by static microcosms was enhanced by the presence of nutrients (methane, methanol, ammonium phosphate, phenol). Methane and ethylene were reported as the biodegradation end products (Barrio-Lage et al., 1990). When vinyl chloride (1 mM) was incubated with resting cells of *Pseudomonas* sp (0.1 g/L) in a 0.1 M phosphate buffer at pH 7.4, hydroxylation of the C-Cl bond occurred yielding acetaldehyde and chloride ions. Oxidation at both the methyl and carbonyl carbons produced acetic acid and hydroxyacetaldehyde, which underwent further oxidation to give glycolic acid (hydroxyacetic acid). The acid was further oxidized to carbon dioxide (Castro et al., 1992).
   Under methanogenic conditions, microorganisms isolated from bed sediments collected from a black-water stream at the Naval Air Station Cecil Field, FL rapidly degraded vinyl chloride to methane and carbon dioxide via the acetate as an intermediate product (Bradley and Chapelle, 1999).
   Surface Water. In natural surface waters, vinyl chloride was resistant to biological and chemical degradation (Hill et al., 1976).

Groundwater. Under aerobic conditions, >99% vinyl chloride degraded in shallow groundwater after 108 d and 65% was completely mineralized (Davis and Carpenter, 1990).

Photolytic. Irradiation of vinyl chloride in the presence of nitrogen dioxide for 160 min produced formic acid, hydrochloric acid, carbon monoxide, formaldehyde, ozone, and trace amounts of formyl chloride and nitric acid. In the presence of ozone, however, vinyl chloride photooxidized to carbon monoxide, formaldehyde, formic acid, and small amounts of hydrochloric acid (Gay et al., 1976). Reported photooxidation products in the troposphere include hydrogen chloride and/or formyl chloride (U.S. EPA, 1985). In the presence of moisture, formyl chloride will decompose to carbon monoxide and hydrochloric acid (Morrison and Boyd, 1971). Vinyl chloride reacts rapidly with OH radicals in the atmosphere. Based on a reaction rate of 6.6×10^{-12} cm^3/molecule·sec, the estimated half-life for this reaction at 299 K is 1.5 d (Perry et al., 1977). Vinyl chloride reacts also with ozone and NO$_3$ in the gas-phase. Sanhueza et al. (1976) reported a rate constant of 6.5×10^{-21} cm^3/molecule·sec for the reaction with OH radicals in air at 295 K. Atkinson et al. (1988) reported a rate constant of 4.45×10^{-16} cm^3/molecule·sec for the reaction with NO$_3$ radicals in air at 298 K.

Chemical/Physical. In a laboratory experiment, it was observed that the leaching of a vinyl chloride monomer from a polyvinyl chloride pipe into water reacted with chlorine to form chloroacetaldehyde, chloroacetic acid, and other unidentified compounds (Ando and Sayato, 1984).

Deng et al. (1997) studied the reaction of metallic iron powder (5 g 40 mesh) and vinyl chloride (15.0 mL) under anaerobic conditions at various temperatures. In the experiments, the vials containing the iron and vinyl chloride were placed on a roller drum set at 8 rpm. Separate reactions were performed at 4, 20, 32, and 45 °C. The major degradate produced was ethylene. Degradation followed pseudo-first-order kinetics. The rate of degradation increased as the temperature increased. Based on the estimated activation energy for vinyl chloride reduction of 40 kilojoules/mol, the investigators concluded that the overall rate of reaction was controlled at the surface rather than the solution.

The evaporation half-life of vinyl chloride (1 mg/L) from water at 25 °C using a shallow-pitch propeller stirrer at 200 rpm at an average depth of 6.5 cm is 27.6 min (Dilling, 1977).

Vinyl chloride will not hydrolyze to any reasonable extent (Kollig, 1993).

Exposure limits: Known human carcinogen. OSHA PEL: TWA 1 ppm, 15-min ceiling 5 ppm; ACGIH TLV: TWA 1 ppm (adopted).

Symptoms of exposure: Narcotic at high concentrations (Patnaik, 1992)

Toxicity:
Acute oral LD$_{50}$ for rats 500 mg/kg (quoted, RTECS, 1985).

Drinking water standard (final): MCLG: zero; MCL: 2 μg/L (U.S. EPA, 1996).

Source: Vinyl chloride in soil and/or groundwater may form from the biotransformation of 1,1,1-trichloroethane (Lesage et al., 1990), trichloroethylene, 1,2-dichloroethylene (Smith and Dragun, 1984; Wilson et al., 1986), and from the chemical reduction of trichloroethylene by zero-valent iron (Orth and Gillham, 1996).

Uses: Manufacture of polyvinyl chloride and copolymers; adhesives for plastics; refrigerant; extraction solvent; organic synthesis.

WARFARIN

Synonyms: 3-(Acetonylbenzyl)-4-hydroxycoumarin; 3-(α-Acetonylbenzyl)-4-hydroxycoumarin; Athrombine-K; Athrombin-K; Brumolin; Compound 42; Corax; Coumadin; Coumafen; Coumafene; Cov-r-tox; D-con; Dethmor; Dethnel; Eastern states duocide; Fasco fascrat powder; 1-(4'-Hydroxy-3'-coumarinyl)-1-phenyl-3-butanone; **4-Hydroxy-3-(3-oxo-1-phenylbutyl)-2*H*-1-benzopyran-2-one**; 4-Hydroxy-3-(1-phenyl-3-oxobutyl)coumarin; Kumader; Kumadu; Kypfarin; Liquatox; Marfrin; Martin's marfrin; Maveran; Mouse pak; 3-(1'-Phenyl-2'-acetylethyl)-4-hydroxycoumarin; 3-α-Phenyl-β-acetylethyl-4-hydroxycoumarin; Prothromadin; Rat-a-way; Rat-b-gon; Rat-gard; Rat-kill; Rat & mice bait; Rat-mix; Rat-o-cide #2; Ratola; Ratorex; Ratox; Ratoxin; Ratron; Ratron G; Rats-no-more; Rattrol; Rattunal; Rax; RCRA waste number P001; Rodafarin; Rodeth; Rodex; Rodex blox; Rosex; Rough & ready mouse mix; Solfarin; Spraytrol brand rodentrol; Temus W; Toxhid; Twin light rat away; Vampirinip II; Vampirin III; Waran; WARF. 42; Warfarat; Warfarine; Warfarin plus; Warfarin Q; Warf compound 42; Warficide.

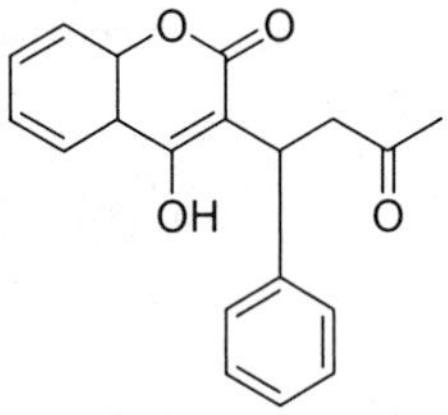

CAS Registry Number: 81-81-2
DOT: 3027
DOT label: Poison
Molecular formula: $C_{19}H_{16}O_4$
Formula weight: 308.33
RTECS: GN4550000
Merck reference: 10, 9852

Physical state, color, and odor:
Colorless, odorless crystals

Melting point (°C):
161 (Weast, 1986)
159-161 (Worthing and Hance, 1991)
159-165 (Berg, 1983)

Boiling point (°C):
Decomposes (NIOSH, 1997)

Diffusivity in water (x 10^{-5} cm²/sec):
0.44 at 20 °C using method of Hayduk and Laudie (1974)

Soil sorption coefficient, log K_{oc}:
2.96 using method of Kenaga and Goring (1980)

Octanol/water partition coefficient, log K_{ow}:
3.20 using method of Kenaga and Goring (1980)

Solubility in organics:
Soluble in alcohol, benzene, 1,4-dioxane (Weast, 1986), and acetone (Sax and Lewis, 1987). Moderately soluble in methanol, ethanol, isopropanol, and some oils (Windholz et al., 1983).

Solubility in water (mg/L):
17 at 20 °C, 72 at 100 °C (Gunther et al., 1968)

Vapor pressure (mmHg):
0.09 at 22 °C (NIOSH, 1997)

Environmental fate:
   Photolytic. Warfarin may undergo direct photolysis because the pesticide showed an absorption maximum of 330 nm (Gore et al., 1971).
   Chemical/Physical. The hydrolysis half-lives at 68.0 °C and pH values of 3.09, 7.11, and 10.18 were calculated to be 12.9, 57.4, and 23.9 d, respectively. At 25 °C and pH 7, the half-life was estimated to be 16 yr (Ellington et al., 1986).

Exposure limits (mg/m³): NIOSH REL: TWA 0.1, IDLH 100; OSHA PEL: TWA 0.1; ACGIH TLV: TWA 0.1 (adopted).

Symptoms of exposure: Hematuria, back pain, hematoma in arms, legs; epistaxis, bleeding lips, mucous membrane hemorrhage, abdominal pain, vomiting, fecal blood; petechial rash; abnormal hematology (NIOSH, 1987)

Toxicity:
   Acute oral LD_{50} for dogs 200 mg/kg, guinea pigs 182 mg/kg, mice 331 mg/kg, rats 3 mg/kg (quoted, RTECS, 1985).

Use: Rodenticide.

o-XYLENE

Synonyms: 1,2-Dimethylbenzene; *o*-Dimethylbenzene; *o*-Methyltoluene; UN 1307; 1,2-Xylene; *o*-Xylol.

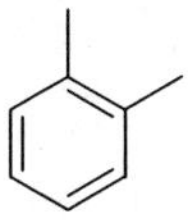

CAS Registry Number: 95-47-6
DOT: 1307
DOT label: Flammable liquid
Molecular formula: C_8H_{10}
Formula weight: 106.17
RTECS: ZE2450000
Merck reference: 10, 9890

Physical state, color, and odor:
Clear, colorless liquid with an aromatic odor

Melting point (°C):
-25.2 (Weast, 1986)

Boiling point (°C):
144.4 (Weast, 1986)

Density (g/cm³):
0.8802 at 20/4 °C (Weast, 1986)

Diffusivity in water (x 10⁻⁵ cm²/sec):
0.79 at 20 °C using method of Hayduk and Laudie (1974)

Dissociation constant, pK_a:
>15 (Christensen et al., 1975)

Flash point (°C):
33 (NIOSH, 1997)
17 (Windholz et al., 1976)

Lower explosive limit (%):
0.9 (NIOSH, 1997)

Upper explosive limit (%):
6.7 (NIOSH, 1997)

Entropy of fusion (cal/mol·K):
13.11 (Pitzer and Scott, 1943)

Heat of fusion (kcal/mol):
3.250 (Pitzer and Scott, 1941)

Henry's law constant (x 10⁻³ atm·m³/mol):
5.0 (Pankow and Rosen, 1988)

3.45 (Wasik and Tsang, 1970)

2.85, 3.61, 4.74, 4.87, and 6.26 at 10, 15, 20, 25, and 30 °C, respectively (Ashworth et al., 1988)

Distilled water: 1.47, 1.24, 1.62, 3.28, and 4.24 at 2.0, 6.0, 10.0, 18.2, and 25.0 °C, respectively; natural seawater: 1.86 and 4.89 at 6.0 and 25.0 °C, respectively (Dewulf et al., 1995)

4.99, 6.29, 10.9, 10.6, and 11.6 at 25, 30, 40, 45, and 50 °C, respectively (Robbins et al., 1993)

5.81 at 23 °C (Anderson, 1992)

Interfacial tension with water (dyn/cm):
37.2 at 25 °C (Donahue and Bartell, 1952)

36.06 at 20 °C (Harkins et al., 1920)

Ionization potential (eV):
8.56 (Franklin et al., 1969)

9.04 (Yoshida et al., 1983a)

Bioconcentration factor, log BCF:
0.70 (clams, Nunes and Benville, 1979)

1.33 (eels, Ogata and Miyake, 1978)

1.15 (goldfish, Ogata et al., 1984)

2.34 (*Selenastrum capricornutum*, Herman et al., 1991)

Soil sorption coefficient, log K_{oc}:
2.11 (Abdul et al., 1987)

2.41 (Catlin soil, Roy et al., 1985, 1987)

2.35 (estuarine sediment, Vowles and Mantoura, 1987)

2.25 (Meylan et al., 1992)

2.03 (Grimsby silt loam), 1.92 (Vaudreil sandy loam), 2.07 (Wendover silty clay) (Nathwani and Phillips, 1977)

2.32 (river sediment), 2.40 (coal wastewater sediment) (Kopinke et al., 1995)

2.37, 2.40 (RP-HPLC immobilized humic acids, Szabo et al., 1990)

3.07, 3.12, 3.30 (glaciofluvial, sandy aquifer, Nielsen et al., 1996)

1.68-1.83 (Nathwani and Phillips, 1977)

2.40, 2.70, 2.58, 2.68, 2.73, 2.69, and 2.68 at 2.3, 3.8, 6.2, 8.0, 13.5, 18.6, and 25.0 °C, respectively, for a Leie River (Belgium) clay (Dewulf et al., 1999a)

Octanol/water partition coefficient, log K_{ow}:
3.13 (Tewari et al., 1982; Wasik et al., 1981, 1983)

2.73 (Banerjee et al., 1980)

2.77 (Leo et al., 1971)

3.08 (quoted, Galassi et al., 1988)

3.12 (Tuazon et al., 1986a)

3.18 (Garst and Wilson, 1984)

3.19 (Hammers et al., 1982)

Solubility in organics:
Soluble in acetone, ethanol, benzene, and ether (Weast, 1986) and many other organic solvents

Solubility in water:
152 mg/L at 20 °C (Mackay and Shiu, 1981)

204 mg/L at 25 °C (Andrews and Keefer, 1949)

175 mg/kg at 25 °C (McAuliffe, 1963)
2.08 mM at 25 °C (Tewari et al., 1982; Wasik et al., 1981, 1983)
142 mg/L at 0 °C, 167 mg/L at 25 °C (Brookman et al., 1985)
142 mg/kg at 0 °C, 213 mg/kg at 25 °C (Polak and Lu, 1973)
170.5 mg/kg at 25 °C, 129.6 mg/kg in artificial seawater at 25 °C (Sutton and Calder, 1975)
176.2 mg/L at 25 °C (Hermann, 1972)
1.742 mM at 20 °C (quoted, Galassi et al., 1988)
179 mg/L in seawater at 25 °C (Bobra et al., 1979)
1.58, 1.68, 1.85, and 2.00 mM at 15, 25, 35, and 45 °C, respectively (Sanemasa et al., 1982)
167.0 mg/kg at 25 °C (Price, 1976)
135 ppm (Brooker and Ellison, 1974)
176 mg/L (Coutant and Keigley, 1988)
In wt % (°C): 0.047 (139), 0.093 (162), 0.407 (207), 0.960 (251) (Guseva and Parnov, 1963)
1.68 mM at 25.0 °C (Sanemasa et al., 1987)
3.28 x 10^{-5} at 25 °C (mole fraction, Li et al., 1993)

Vapor density:
4.34 g/L at 25 °C, 3.66 (air = 1)

Vapor pressure (mmHg):
6.6 at 25 °C (quoted, Mackay and Wolkoff, 1973)

Environmental fate:
   Biological. Reported biodegradation products of the commercial product containing xylene include α-hydroxy-*p*-toluic acid, *p*-methylbenzyl alcohol, benzyl alcohol, 4-methylcatechol, *m*- and *p*-toluic acids (Fishbein, 1985). *o*-Xylene was also cometabolized resulting in the formation of *o*-toluic acid (Pitter and Chudoba, 1990). In anoxic groundwater near Bemidji, MI, *o*-xylene anaerobically biodegraded to the intermediate *o*-toluic acid (Cozzarelli et al., 1990).
   Bridié et al. (1979) reported BOD and COD values 1.64 and 2.91 g/g using filtered effluent from a biological sanitary waste treatment plant. These values were determined using a standard dilution method at 20 °C and stirred for a period of 5 d. When a sewage seed was used in a separate screening test, a BOD value of 1.80 g/g was obtained. The ThOD for *o*-xylene is 3.17 g/g.
   Surface Water. The evaporation half-life of *o*-xylene in surface water (1 m depth) at 25 °C is estimated to be 5.18 h (Mackay and Leinonen, 1975).
   Groundwater. Nielsen et al. (1996) studied the degradation of *o*-xylene in a shallow, glaciofluvial, unconfined sandy aquifer in Jutland, Denmark. As part of the *in situ* microcosm study, a cylinder that was open at the bottom and screened at the top was installed through a cased borehole approximately 5 m below grade. Five liters of water was aerated with atmospheric air to ensure aerobic conditions were maintained. Groundwater was analyzed weekly for approximately 3 months to determine *o*-xylene concentrations over time. The experimentally determined first-order biodegradation rate constant and corresponding half-life following a 7-d lag phase were 0.1/d and 6.93 d, respectively.
   Photolytic. When synthetic air containing gaseous nitrous acid and *o*-xylene was exposed to artificial sunlight (λ = 300-450 nm) biacetyl, peroxyacetal nitrate, and methyl nitrate formed as products (Cox et al., 1980). An *n*-hexane solution containing *o*-xylene and spread as a thin film (4 mm) on cold water (10 °C) was irradiated by a mercury medium pressure lamp. In 3 h, 13.6% of the *o*-xylene photooxidized into *o*-methylbenzaldehyde, *o*-benzyl alcohol, *o*-benzoic acid, and *o*-methylacetophenone (Moza and Feicht, 1989). Irradiation of *o*-xylene at ≈ 2537 Å at 35 °C and 6 mmHg isomerizes to *m*-xylene (Calvert and Pitts, 1966). Glyoxal, methyl-glyoxal, and biacetyl were produced from the photooxidation of *o*-xylene by OH radicals in air

at 25 °C (Tuazon et al., 1986a).

Major products reported from the photooxidation of *o*-xylene with nitrogen oxides not previously mentioned include formaldehyde, acetaldehyde, peroxyacetyl nitrate, glyoxal, and methylglyoxal (Altshuller, 1983). The rate constant for the reaction of *o*-xylene and OH radicals at room temperature was 1.53×10^{-11} cm^3/molecule·sec (Hansen et al., 1975). A rate constant of 8.4×10^9 L/mol·sec was reported for the reaction of *o*-xylene with OH radicals in the gas phase (Darnall et al., 1976). Similarly, a room temperature rate constant of 1.34×10^{-11} cm^3/molecule·sec was reported for the vapor-phase reaction of *o*-xylene with OH radicals (Atkinson, 1985). At 25 °C, a rate constant of 1.25×10^{-11} cm^3/molecule·sec was reported for the same reaction (Ohta and Ohyama, 1985).

Chemical/Physical. Under atmospheric conditions, the gas-phase reaction of *o*-xylene with OH radicals and nitrogen oxides resulted in the formation of *o*-tolualdehyde, *o*-methylbenzyl nitrate, nitro-*o*-xylenes, 2,3-and 3,4-dimethylphenol (Atkinson, 1990). Kanno et al. (1982) studied the aqueous reaction of *o*-xylene and other aromatic hydrocarbons (benzene, toluene, *m*- and *p*-xylene, and naphthalene) with hypochlorous acid in the presence of ammonium ion. They reported that the aromatic ring was not chlorinated as expected but was cleaved by chloramine forming cyanogen chloride. The amount of cyanogen chloride formed increased at lower pHs (Kanno et al., 1982). In the gas phase, *o*-xylene reacted with nitrate radicals in purified air forming the following products: 5-nitro-2-methyltoluene and 6-nitro-2-methyltoluene, 2-methylbenzaldehyde and an aryl nitrate (Chiodini et al., 1993).

o-Xylene will not hydrolyze because it does not contain a hydrolyzable functional group (Kollig, 1993).

Exposure limits: NIOSH REL: 100 ppm (435 mg/m^3), STEL 150 ppm (655 mg/m^3), IDLH 900 ppm; OSHA PEL: TWA 100 ppm; ACGIH TLV: TWA 100 ppm, STEL 150 ppm (adopted).

Symptoms of exposure: May cause irritation of eyes, nose, and throat, headache, dizziness, excitement, drowsiness, nausea, vomiting, abdominal pain, and dermatitis (Patnaik, 1992)

Toxicity:

EC_{50} (72-h) for *Selenastrum capricornutum* 4.7 mg/L (Galassi et al., 1988).

LC_{50} (7-d) for *Poecilia reticulata* 35 mg/L (Könemann, 1981).

LC_{50} (96-h) for *Salmo gairdneri* 7.6 mg/L, *Poecilia reticulata* 12.0 mg/L (Galassi et al., 1988).

Acute oral LD_{50} in rats is approximately 5 g/kg (quoted, RTECS, 1992).

Drinking water standard (final): For all xylenes, the MCLG and MCL are both 10 mg/L (U.S. EPA, 1996).

Source: Detected in distilled water-soluble fractions of 87 octane gasoline (3.83 mg/L), 94 octane gasoline (11.4 mg/L), Gasohol (8.49 mg/L), No. 2 fuel oil (1.73 mg/L), jet fuel A (0.87 mg/L), diesel fuel (1.75 mg/L), military jet fuel JP-4 (1.99 mg/L) (Potter, 1996), new motor oil (16.2-17.5 μg/L), and used motor oil (294-308 μg/L) (Chen et al., 1994). The average volume percent and estimated mole fraction in American Petroleum Institute PS-6 gasoline are 2.088 and 0.01959, respectively (Poulsen et al., 1992). Schauer et al. (1999) reported *o*-xylene in a diesel-powered medium-duty truck exhaust at an emission rate of 830 μg/km.

Thomas and Delfino (1991) equilibrated contaminant-free groundwater collected from Gainesville, FL with individual fractions of three individual petroleum products at 24-25 °C for 24 h. The aqueous phase was analyzed for organic compounds via U.S. EPA approved test method 602. Average *o*-xylene concentrations reported in water-soluble fractions of unleaded gasoline, kerosene, and diesel fuel were 3.830, 0.382, and 0.172 mg/L, respectively. When the authors analyzed the aqueous-phase via U.S. EPA approved test method 610, average *o*-xylene concentrations in water-soluble fractions of unleaded gasoline, kerosene, and diesel fuel were lower, i.e., 2.931, 0.232, and 0.138 mg/L, respectively.

Based on analysis of 7 coal tar samples, *o*-xylene concentrations ranged from 2 to 2,000 ppm (EPRI, 1990).

Uses: Preparation of phthalic acid, phthalic anhydride, terephthalic acid, isophthalic acid; solvent for alkyd resins, lacquers, enamels, rubber cements; manufacture of dyes, pharmaceuticals, and insecticides; motor fuels.

m-XYLENE

Synonyms: 1,3-Dimethylbenzene; *m*-Dimethylbenzene; *m*-Methyltoluene; UN 1307; 1,3-Xylene; *m*-Xylol.

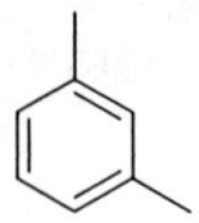

CAS Registry Number: 108-38-3
DOT: 1307
DOT label: Flammable liquid
Molecular formula: C_8H_{10}
Formula weight: 106.17
RTECS: ZE2275000
Merck reference: 10, 9890

Physical state, color, and odor:
Clear, colorless, watery liquid with a sweet, aromatic odor. Odor threshold concentration is 1.1 ppm (Amoore and Hautala, 1983).

Melting point (°C):
-47.9 (Weast, 1986)
-47.40 (Martin et al., 1979)

Boiling point (°C):
139.1 (Weast, 1986)

Density (g/cm³):
0.8642 at 20/4 °C (Weast, 1986)
0.8684 at 25/4 °C (Hawley, 1981)
0.86407 at 20/4 °C, 0.85979 at 25/4 °C (Huntress and Mulliken, 1941)

Diffusivity in water (x 10^{-5} cm²/sec):
0.78 at 20 °C using method of Hayduk and Laudie (1974)

Dissociation constant, pK_a:
>15 (Christensen et al., 1975)

Flash point (°C):
28 (NIOSH, 1997)

Lower explosive limit (%):
1.1 (NIOSH, 1997)

Upper explosive limit (%):
7.0 (NIOSH, 1997)

Entropy of fusion (cal/mol·K):
12.28 (Pitzer and Scott, 1941)

Heat of fusion (kcal/mol):
2.765 (Pitzer and Scott, 1941)

Henry's law constant (x 10^{-3} atm·m³/mol):
7.0 (Pankow and Rosen, 1988)
7.68 (Tuazon et al., 1986a)
4.11, 4.96, 5.98, 7.44, and 8.87 at 10, 15, 20, 25, and 30 °C, respectively (Ashworth et al., 1988)
Distilled water: 2.24, 2.15, 2.74, 4.78, and 6.08 at 2.0, 6.0, 10.0, 18.2, and 25.0 °C, respectively; natural seawater: 2.93 and 7.58 at 6.0 and 25.0 °C, respectively (Dewulf et al., 1995)

Interfacial tension with water (dyn/cm at 20 °C):
37.89 (Harkins et al., 1920)

Ionization potential (eV):
8.58 (Franklin et al., 1969)
9.05 (Yoshida et al., 1983a)

Soil sorption coefficient, log K_{oc}:
2.22 (Abdul et al., 1987)
2.11, 2.46 (forest soil), 2.20 (agricultural soil) (Seip et al., 1986)

Bioconcentration factor, log BCF:
0.78 (clams, Nunes and Benville, 1979)
1.37 (eels, Ogata and Miyake, 1978)
2.40 (*Selenastrum capricornutum*, Herman et al., 1991)

Octanol/water partition coefficient, log K_{ow}:
3.20 (Leo et al., 1971)
3.13 (Wasik et al., 1981, 1983)
3.28 (Garst and Wilson, 1984)

Solubility in organics:
Soluble in acetone, ethanol, benzene, and ether (Weast, 1986)

Solubility in water:
In mg/L: 209 at 0.4 °C, 201 at 5.2 °C, 192 at 14.9 °C, 196 at 21.0 °C, 196 at 25.0 °C, 196 at 25.6 °C, 198 at 30.3 °C, 203 at 34.9 °C, 218 at 39.6 °C (Bohon and Claussen, 1951)
173 mg/L at 25 °C (Andrews and Keefer, 1949)
1.51 mM at 25 °C (Tewari et al., 1982; Wasik et al., 1981, 1983)
196 mg/kg at 0 °C, 162 mg/kg at 25 °C (Polak and Lu, 1973)
146.0 mg/kg at 25 °C, 106.0 mg/kg in artificial seawater at 25 °C (Sutton and Calder, 1975)
157.0 mg/L at 25 °C (Hermann, 1972)
1.525 mM at 20 °C (quoted, Galassi et al., 1988)
1.92 mM at 35 °C (Hine et al., 1963)
1.49, 1.52, 1.57, and 1.73 mM at 15, 25, 35, and 45 °C, respectively (Sanemasa et al., 1982)
134.0 mg/kg at 25 °C (Price, 1976)
In wt % (°C): 0.031 (127), 0.072 (149), 0.168 (187), 0.648 (239) (Guseva and Parnov, 1963)
1.94 mmol/kg at 25.0 °C (Vesala, 1974)
1.33 mM at 25.0 °C (Sanemasa et al., 1987)
3.01 x 10^{-5} at 25 °C (mole fraction, Li et al., 1993)

Vapor density:
4.34 g/L at 25 °C, 3.66 (air = 1)

Vapor pressure (mmHg):
8.3 at 25 °C (Mackay et al., 1982)
15.2 at 35 °C (Hine et al., 1963)

Environmental fate:
Biological. Microbial degradation produced 3-methylbenzyl alcohol, 3-methylbenzaldehyde, *m*-toluic acid, and 3-methyl catechol (Verschueren, 1983). *m*-Toluic acid was reported to be the biooxidation product of *m*-xylene by *Nocardia corallina* V-49 using *n*-hexadecane as the substrate (Keck et al., 1989). Reported biodegradation products of the commercial product containing xylene include α-hydroxy-*p*-toluic acid, *p*-methylbenzyl alcohol, benzyl alcohol, 4-methylcatechol, *m*- and *p*-toluic acids (Fishbein, 1985). In anoxic groundwater near Bemidji, MI, *m*-xylene anaerobically biodegraded to the intermediate *m*-toluic acid (Cozzarelli et al., 1990).
Bridié et al. (1979) reported BOD and COD values 2.53 and 2.62 g/g using filtered effluent from a biological sanitary waste treatment plant. These values were determined using a standard dilution method at 20 °C and stirred for a period of 5 d. The ThOD for *m*-xylene is 3.17 g/g.
Photolytic. When synthetic air containing gaseous nitrous acid and *m*-xylene was exposed to artificial sunlight (λ = 300-450 nm) biacetyl, peroxyacetal nitrate, and methyl nitrate were formed (Cox et al., 1980). An *n*-hexane solution containing *m*-xylene and spread as a thin film (4 mm) on cold water (10 °C) was irradiated by a mercury medium pressure lamp. In 3 h, 25% of the *m*-xylene photooxidized into *m*-methylbenzaldehyde, *m*-benzyl alcohol, *m*-benzoic acid, and *m*-methylacetophenone (Moza and Feicht, 1989).
Irradiation of *m*-xylene isomerizes to *p*-xylene (Calvert and Pitts, 1966). Glyoxal, methylglyoxal, and biacetyl were produced from the photooxidation of *m*-xylene by OH radicals in air at 25 °C (Tuazon et al., 1986a). The photooxidation of *m*-xylene in the presence of nitrogen oxides (NO and NO_2) yielded small amounts of formaldehyde and a trace of acetaldehyde (Altshuller et al., 1970). *m*-Tolualdehyde and nitric acid also were identified as photooxidation products of *m*-xylene with nitrogen oxides (Altshuller, 1983). The rate constant for the reaction of *m*-xylene and OH radicals at room temperature was 2.36 x 10^{-11} cm^3/molecule·sec (Hansen et al., 1975). A rate constant of 1.41 x 10^8 L/mol·sec was reported for the reaction of *m*-xylene with OH radicals in the gas phase (Darnall et al., 1976). Similarly, a room temperature rate constant of 2.35 x 10^{-11} cm^3/molecule·sec was reported for the vapor-phase reaction of *m*-xylene with OH radicals (Atkinson, 1985). At 25 °C, a rate constant of 2.22 x 10^{-11} cm^3/molecule·sec was reported for the same reaction (Ohta and Ohyama, 1985).
Chemical/Physical. Under atmospheric conditions, the gas-phase reaction with OH radicals and nitrogen oxides resulted in the formation of *m*-tolualdehyde, *m*-methylbenzyl nitrate, nitro-*m*-xylenes, 2,4- and 2,6-dimethylphenol (Atkinson, 1990). Kanno et al. (1982) studied the aqueous reaction of *m*-xylene and other aromatic hydrocarbons (benzene, toluene, *o*- and *p*-xylene, and naphthalene) with hypochlorous acid in the presence of ammonium ion. They reported that the aromatic ring was not chlorinated as expected but was cleaved by chloramine forming cyanogen chloride. The amount of cyanogen chloride formed increased at lower pHs (Kanno et al., 1982). In the gas phase, *m*-xylene reacted with nitrate radicals in purified air forming 3-methylbenzaldehyde, an aryl nitrate and trace amounts of 2,6-dimethylnitrobenzene, 2,4-dimethylnitrobenzene, and 3,5-dimethylnitrobenzene (Chiodini et al., 1993).
m-Xylene will not hydrolyze (Kollig, 1993).

Exposure limits: NIOSH REL: 100 ppm (435 mg/m³), STEL 150 ppm (655 mg/m³), IDLH 900 ppm; OSHA PEL: TWA 100 ppm; ACGIH TLV: TWA 100 ppm, STEL 150 ppm (adopted).

Symptoms of exposure: May cause irritation of eyes, nose, and throat, headache, dizziness, excitement, drowsiness, nausea, vomiting, abdominal pain, and dermatitis (Patnaik, 1992)

Toxicity:
EC$_{50}$ (72-h) for *Selenastrum capricornutum* 4.9 mg/L (Galassi et al., 1988).
LC$_{50}$ (14-d) for *Poecilia reticulata* 37.7 mg/L (Könemann, 1981).
LC$_{50}$ (96-h) for *Salmo gairdneri* 8.4 mg/L, *Poecilia reticulata* 12.9 mg/L (Galassi et al., 1988).
Acute oral LD$_{50}$ in rats 5 g/kg (quoted, RTECS, 1985).

Drinking water standard (final): For all xylenes, the MCLG and MCL are both 10 mg/L (U.S. EPA, 1996).

Source: As *m+p*-xylene, detected in distilled water-soluble fractions of 87 octane gasoline, 94 octane gasoline, and Gasohol at concentrations of 7.00, 20.1, and 14.6 mg/L, respectively (Potter, 1996); in distilled water-soluble fractions of new and used motor oil at concentrations of 0.26-0.29 and 302-339 µg/L, respectively (Chen et al., 1994). The average volume percent and estimated mole fraction in American Petroleum Institute PS-6 gasoline are 4.072 and 0.04406, respectively (Poulsen et al., 1992).

Thomas and Delfino (1991) equilibrated contaminant-free groundwater collected from Gainesville, FL with individual fractions of three individual petroleum products at 24-25 °C for 24 h. The aqueous phase was analyzed for organic compounds via U.S. EPA approved test method 602. Average *m+p*-xylene concentrations reported in water-soluble fractions of unleaded gasoline, kerosene, and diesel fuel were 8.611, 0.658, and 0.228 mg/L, respectively. When the authors analyzed the aqueous-phase via U.S. EPA approved test method 610, average m+p-xylene concentrations in water-soluble fractions of unleaded gasoline, kerosene, and diesel fuel were lower, i.e., 6.068, 0.360, and 0.222 mg/L, respectively.

Based on analysis of 7 coal tar samples, *m+p*-xylene concentrations ranged from ND to 6,000 ppm (EPRI, 1990).

Uses: Solvent; preparation of isophthalic acid, intermediate for dyes; insecticides; aviation fuel.

p-XYLENE

Synonyms: Chromar; 1,4-Dimethylbenzene; *p*-Dimethylbenzene; *p*-Methyltoluene; Scintillar; UN 1307; 1,4-Xylene; *p*-Xylol.

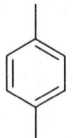

CAS Registry Number: 106-42-3
DOT: 1307
DOT label: Flammable liquid
Molecular formula: C_8H_{10}
Formula weight: 106.17
RTECS: ZE2625000
Merck reference: 10, 9890

Physical state, color, and odor:
Clear, colorless, watery liquid with a sweet odor

Melting point (°C):
13.3 (Weast, 1986)
13.50 (Martin et al., 1979)

Boiling point (°C):
138.3 (Weast, 1986)

Density (g/cm³):
0.86314 at 20.00/4 °C (Tsierkezos et al., 2000)
0.8566 at 25.00/4 °C (Aminabhavi and Banerjee, 1999)
0.85655 at 25/4 °C (Kirchnerová and Cave, 1976)

Diffusivity in water (x 10⁻⁵ cm²/sec):
0.79 at 20 °C using method of Hayduk and Laudie (1974)

Dissociation constant, pK_a:
>15 (Christensen et al., 1975)

Flash point (°C):
27 (NIOSH, 1997)

Lower explosive limit (%):
1.1 (NIOSH, 1997)

Upper explosive limit (%):
7.0 (NIOSH, 1997)

Entropy of fusion (cal/mol·K):
14.28 (Pitzer and Scott, 1941; Corruccini and Ginnings, 1947)
14.29 (Messerly et al., 1988)
14.13 (Huffman et al., 1930)

Heat of fusion (kcal/mol):
4.087 (Corruccini and Ginnings, 1947)
4.091 (Messerly et al., 1988)
4.090 (Pitzer and Scott, 1941)
4.047 (Huffman et al., 1930)

Henry's law constant (x 10⁻³ atm·m³/mol):
7.1 (Pankow and Rosen, 1988)
7.68 (Tuazon et al., 1986a)
4.20, 4.83, 6.45, 7.44, and 9.45 at 10, 15, 20, 25, and 30 °C, respectively (Ashworth et al., 1988)
4.35 (Wasik and Tsang, 1970)
Distilled water: 1.89, 1.67, 2.62, 4.73, and 5.68 at 2.0, 6.0, 10.0, 18.2, and 25.0 °C, respectively; natural seawater: 3.13 and 7.53 at 6.0 and 25.0 °C, respectively (Dewulf et al., 1995)

Interfacial tension with water (dyn/cm at 20 °C):
37.77 (Harkins et al., 1920)

Ionization potential (eV):
8.44 (Franklin et al., 1969)
8.99 (Yoshida et al., 1983a)

Soil sorption coefficient, log K_{oc}:
2.31 (Abdul et al., 1987)
2.42 (estuarine sediment, Vowles and Mantoura, 1987)
2.72 (Captina silt loam), 2.87 (McLaurin sandy loam) (Walton et al., 1992)
2.52 (Schwarzenbach and Westall, 1984)
2.49, 2.75, 2.65, 2.76, 2.79, 2.77, and 2.78 at 2.3, 3.8, 6.2, 8.0, 13.5, 18.6, and 25.0 °C, respectively, for a Leie River (Belgium) clay (Dewulf et al., 1999a)

Bioconcentration factor, log BCF:
1.37 (eels, Ogata and Miyake, 1978)
1.17 (goldfish, Ogata et al., 1984)
2.41 (*Selenastrum capricornutum*, Herman et al., 1991)

Octanol/water partition coefficient, log K_{ow}:
3.18 (Tewari et al., 1982; Wasik et al., 1981, 1983)
3.15 (Campbell and Luthy, 1985; Leo et al., 1971)

Solubility in organics:
Soluble in acetone, ethanol, and benzene (Weast, 1986)

Solubility in water:
200 mg/L at 25 °C (Andrews and Keefer, 1949)
2.02 mM at 25 °C (Tewari et al., 1982; Wasik et al., 1981, 1983)
In mg/L: 156 at 0.4 °C, 188 at 10.0 °C, 195 at 14.9 °C, 197 at 21.0 °C, 198 at 25.0 °C, 199 at 25.6 °C, 201 at 30.2 °C, 204 at 30.3 °C, 207 at 34.9 °C, 207 at 35.2 °C, 222 at 42.8 °C (Bohon and Claussen, 1951)
180 mg/L at 25 °C (Banerjee, 1984)
0.0156 wt % at 20 °C (Riddick et al., 1986)
185 mg/kg at 25 °C (Polak and Lu, 1973)

156.0 mg/kg at 25 °C, 110.9 mg/kg in artificial seawater at 25 °C (Sutton and Calder, 1975)

163.3 mg/L at 25 °C (Hermann, 1972)

1.94 mM at 35 °C (Hine et al., 1963)

1.48, 1.53, 1.61, and 1.66 mM at 15, 25, 35, and 45 °C, respectively (Sanemasa et al., 1982)

0.019 wt % at 25 °C (Lo et al., 1986)

157.0 mg/kg at 25 °C (Price, 1976)

In wt % (°C): 0.049 (141), 0.096 (169), 0.231 (194), 0.607 (231), 1.283 (258) (Guseva and Parnov, 1963)

1.51 mM at 25.0 °C (Sanemasa et al., 1987)

3.00 x 10^{-5} at 25 °C (mole fraction, Li et al., 1993)

Vapor density:
4.34 g/L at 25 °C, 3.66 (air = 1)

Vapor pressure (mmHg):
8.8 at 25 °C (Mackay et al., 1982)
15.8 at 35 °C (Hine et al., 1963)

Environmental fate:

Biological. Microbial degradation of *p*-xylene produced 4-methylbenzyl alcohol, 4-methylbenzaldehyde, *p*-toluic acid, and 4-methyl catechol (Verschueren, 1983). Dimethyl-*cis,cis*-muconic acid, and 2,3-dihydroxy-*p*-toluic acid were reported to be biooxidation products of *p*-xylene by *Nocardia corallina* V-49 using *n*-hexadecane as the substrate (Keck et al., 1989). Reported biodegradation products of the commercial product containing xylene include α-hydroxy-*p*-toluic acid, *p*-methylbenzyl alcohol, benzyl alcohol, 4-methylcatechol, *m*- and *p*-toluic acids (Fishbein, 1985). It was reported that *p*-xylene was cometabolized resulting in the formation of *p*-toluic and 2,3-dihydroxy-*o*-toluic acids (Pitter and Chudoba, 1990). In anoxic groundwater near Bemidji, MI, *p*-xylene anaerobically biodegraded to the intermediate *p*-toluic acid (Cozzarelli et al., 1990).

Bridié et al. (1979) reported BOD and COD values 1.40 and 2.56 g/g using filtered effluent from a biological sanitary waste treatment plant. These values were determined using a standard dilution method at 20 °C and stirred for a period of 5 d. When a sewage seed was used in a separate screening test, a BOD value of 2.35 g/g was obtained. The ThOD for *p*-xylene is 3.17 g/g.

Photolytic. An *n*-hexane solution containing *m*-xylene and spread as a thin film (4 mm) on cold water (10 °C) was irradiated by a mercury medium pressure lamp. In 3 h, 18.5% of the *p*-xylene photooxidized into *p*-methylbenzaldehyde, *p*-benzyl alcohol, *p*-benzoic acid, and *p*-methylacetophenone (Moza and Feicht, 1989). Glyoxal and methylglyoxal were produced from the photooxidation of *p*-xylene by OH radicals in air at 25 °C (Tuazon et al., 1986a). The rate constant for the reaction of *p*-xylene and OH radicals at room temperature was 1.22 x 10^{-11} cm³/molecule·sec (Hansen et al., 1975). A rate constant of 7.45 x 10^9 L/mol·sec was reported for the reaction of *p*-xylene with OH radicals in the gas phase (Darnall et al., 1976). Similarly, a room temperature rate constant of 1.41 x 10^{-11} cm³/molecule·sec was reported for the vapor-phase reaction of *p*-xylene with OH radicals (Atkinson, 1985). At 25 °C, a rate constant of 1.29 x 10^{-11} cm³/molecule·sec was reported for the same reaction (Ohta and Ohyama, 1985).

Chemical/Physical. Under atmospheric conditions, the gas-phase reaction with OH radicals and nitrogen oxides resulted in the formation of *p*-tolualdehyde (Atkinson, 1990). Kanno et al. (1982) studied the aqueous reaction of *p*-xylene and other aromatic hydrocarbons (benzene, toluene, *o*- and *m*-xylene, and naphthalene) with hypochlorous acid in the presence of ammonium ion. They reported that the aromatic ring was not chlorinated as expected but was cleaved by chloramine forming cyanogen chloride. The amount of cyanogen chloride formed increased at lower pHs (Kanno et al., 1982).

In the gas phase, *p*-xylene reacted with nitrate radicals in purified air forming 3,6-dimethylnitrobenzene, 4-methylbenzaldehyde, and an aryl nitrate (Chiodini et al., 1993).

At influent concentrations of 10, 1.0, 0.1, and 0.01 mg/L, the granular activated carbon adsorption capacities at pH 7.3 were 130, 85, 54, and 35 mg/g, respectively (Dobbs and Cohen, 1980).

p-Xylene will not hydrolyze because it does not contain a hydrolyzable functional group (Kollig, 1993).

Exposure limits: NIOSH REL: 100 ppm (435 mg/m^3), STEL 150 ppm (655 mg/m^3), IDLH 900 ppm; OSHA PEL: TWA 100 ppm; ACGIH TLV: TWA 100 ppm, STEL 150 ppm (adopted).

Symptoms of exposure: May cause irritation of eyes, nose, and throat, headache, dizziness, excitement, drowsiness, nausea, vomiting, abdominal pain, and dermatitis (Patnaik, 1992)

Toxicity:

EC_{50} (72-h) for *Selenastrum capricornutum* 3.2 mg/L (Galassi et al., 1988).

LC_{50} (7-d) for *Poecilia reticulata* 35 mg/L (Könemann, 1981).

LC_{50} (96-h) for *Salmo gairdneri* 2.6 mg/L, *Poecilia reticulata* 8.8 mg/L (Galassi et al., 1988).

Acute oral LD_{50} in rats 5 g/kg; LC_{50} (inhalation) for rats 4,550 ppm/4-h (quoted, RTECS, 1985).

Drinking water standard (final): For all xylenes, the MCLG and MCL are both 10 mg/L (U.S. EPA, 1996).

Source: Detected in distilled water-soluble fractions No. 2 fuel oil (1.11 mg/L), jet fuel A (1.23 mg/L), diesel fuel (0.56 mg/L), and military jet fuel JP-4 (5.48 mg/L) (Potter, 1996); in new and used motor oil at concentrations of 0.26-0.29 and 302-339 μg/L, respectively (Chen et al., 1994). The average volume percent and estimated mole fraction in American Petroleum Institute PS-6 gasoline are 1.809 and 0.02263, respectively (Poulsen et al., 1992).

Thomas and Delfino (1991) equilibrated contaminant-free groundwater collected from Gainesville, FL with individual fractions of three individual petroleum products at 24-25 °C for 24 h. The aqueous phase was analyzed for organic compounds via U.S. EPA approved test method 602. Average *m+p*-xylene concentrations reported in water-soluble fractions of unleaded gasoline, kerosene, and diesel fuel were 8.611, 0.658, and 0.228 mg/L, respectively. When the authors analyzed the aqueous-phase via U.S. EPA approved test method 610, average *m+p*-xylene concentrations in water-soluble fractions of unleaded gasoline, kerosene, and diesel fuel were lower, i.e., 6.068, 0.360, and 0.222 mg/L, respectively.

Based on analysis of 7 coal tar samples, *m+p*-xylene concentrations ranged from ND to 6,000 ppm (EPRI, 1990). Detected in 1-yr aged coal tar film and bulk coal tar at concentrations of 260 and 830 mg/kg, respectively (Nelson et al., 1996).

Uses: Preparation of terephthalic acid for polyester resins and fibers (Dacron, Mylar, and Terylene), vitamins, pharmaceuticals, and insecticides. Major constituent in gasoline.

Table 1. Conversion Factors for Various Concentration Units

To obtain	From	Compute
$atm \cdot m^3/mol$	M/atm	$K_H = 1/[(M/atm)(1000)]$
g/L of solution	molality	$g/L = 1000dm(MW_1)/[1000+m(MW_1)]$
g/L of solution	molarity	$g/L = M(MW_1)$
g/L of solution	mole fraction	$g/L = 1000dx(MW_1)/[x(MW_1)+(1-x)MW_2]$
g/L of solution	wt %	$g/L = 10d(wt\ \%)$
$atm \cdot m^3/mol$	$kPa \cdot m^3/mol$	$K_H = (kPa \cdot m^3/mol)/101.325$
$atm \cdot m^3/mol$	$Pa \cdot m^3/mol$	$K_H = (Pa \cdot m^3/mol)/101,325$
M/atm	$atm \cdot m^3/mol$	$H^* = 1/[(1000)(atm \cdot m^3/mol)]$
molality	molarity	$m = 1000M/[1000d-M(MW_1)]$
molality	mole fraction	$m = 1000x/[MW_2-x(MW_2)]$
molality	g/L of solution	$m = 1000G/[MW_1(1000d-G)]$
molality	wt % of solute	$m = 1000(wt\ \%)/[MW_1(100-wt\ \%)]$
molarity	g/L of solution	$M = G/MW_1$
molarity	molality	$M = 1000dm/[1000+m(MW_1)]$
molarity	mole fraction	$M = 1000dx/[x(MW_1)+(1-x)MW_2]$
molarity	wt % of solute	$M = 10d(wt\ \%)/MW_1$
mole fraction	g/L of solution	$x = G(MW_2)/[G(MW_2-MW_1)+1000d(MW_1)]$
mole fraction	molality	$x = m(MW_2)/[m(MW_2)+1000]$
mole fraction	molarity	$x = M(MW_2)/[M(MW_2-MW_1)+1000d]$
mole fraction	wt % of solute	$x = [(wt\ \%)/MW_1]/[(wt\ \%/MW_1)+(100-wt\ \%)MW_2]$
wt % of solute	g/L solution	$wt\ \% = G/10d$
wt % of solute	molality	$wt\ \% = 100m(MW_1)/[1000+m(MW_1)]$
wt % of solute	molarity	$wt\ \% = M(MW_1)/10d$
wt % of solute	mole fraction	$wt\ \% = 100x(MW_1)/[x(MW_1)+(1-x)MW_2]$

d = density of solution (g/L); G = g of solute/L of solution; H^* = effective Henry's law constant; K_H = Henry's law constant; m = molality; M = molarity; MW_1 = formula weight of solute; MW_2 = formula weight of solvent; x = mole fraction; wt % = weight percent of solute

Table 2. Conversion Factors

To convert	Into	Multiply by
	A	
acre-feet	feet^3	4.356×10^4
acre-feet	gallons (U.S.)	3.529×10^5
acre-feet	inches^3	7.527×10^7
acre-feet	liters	1.233×10^6
acre-feet	meters^3	1,233
acre-feet	yards^3	1,613
acre-feet/day	feet^3/second	5.042×10^{-1}
acre-feet/day	gallons (U.S.)/minute	226.3
acre-feet/day	liters^3/second	14.28
acre-feet/day	meters^3/day	1,234
acre-feet/day	meters^3/second	1.428×10^{-2}
acres	feet^2	43,560
acres	hectares	4.047×10^{-1}
acres	inches^2	6.273×10^6
acres	kilometers^2	4.047×10^{-3}
acres	meters^2	4,047
acres	miles^2	1.563×10^{-3}
angstroms	inches	3.937×10^{-6}
angstroms	meters	10^{-10}
angstroms	microns	10^{-4}
atmospheres	bars	1.01325
atmospheres	feet of water (at 4 °C)	33.90
atmospheres	inches of Hg (at 0 °C)	29.92126
atmospheres	millibars	1013.25
atmospheres	millimeters of Hg	760
atmospheres	millimeters of water	1.033227×10^4
atmospheres	pascals	1.01325×10^5
atmospheres	pounds/inch^2	14.70
atmospheres	tons/inch^2	7.348×10^{-3}
atmospheres	torrs	760
	B	
barrels (petroleum)	liters	159.0
bars	atmospheres	9.869×10^{-1}
bars	pounds/inch^2	14.50
British thermal unit (Btu)	joules	1.055×10^{-3}
British thermal unit (Btu)	kilowatt-hour	2.928×10^{-4}
British thermal unit (Btu)	watts	2.931×10^{-1}
Btu/hour	watts	2.931×10^{-1}
bushels	feet^3	1.2445
bushels	pecks	4.0
bushels	pints (dry)	64.0
bushels	quarts (dry)	32.0
	C	
calories	joules	4.187
Celsius (°C)	Fahrenheit (°F)	(1.8 x °C)+32
Celsius (°C)	Kelvin (K)	°C+273.15
centiliters	drams	2.705
centiliters	inches^3	6.103×10^{-2}
centiliters	liters	10^{-2}
centimeters	feet	3.291×10^{-2}
centimeters	inches	3.937×10^{-1}
centimeters	miles	6.214×10^{-6}
centimeters	millimeters	10
centimeters	mils	393.7
centimeters	yards	1.094×10^{-2}
centimeters^2	feet^2	1.076×10^{-3}
centimeters^2	inches^2	1.55×10^{-1}
centimeters^2	meters^2	1×10^{-4}
centimeters^2	yards^2	1.196×10^{-4}
centimeters^3	fluid ounces	3.381×10^{-2}
centimeters^3	feet^3	3.5314×10^{-5}
centimeters^3	inches^3	6.102×10^{-2}
centimeters^3	liters	1×10^{-3}

To convert	Into	Multiply by
centimeters^3	ounces (U.S., fluid)	3.381×10^{-2}
centimeters/second	feet/day	2,835
centimeters/second	feet/minute	1.968
centimeters/second	feet/second	3.281×10^{-2}
centimeters/second	kilometers/hour	3.6×10^{-2}
centimeters/second	liters/meter/second	9.985
centimeters/second	meters/minute	6×10^{-1}
centimeters/second	miles/hour	2.237×10^{-2}
centimeters/second	miles/minute	3.728×10^{-4}
chain	inches	792.00
chain	meters	20.12
circumference	radians	6.283
cords	cord feet	8
coulombs	faradays	1.036×10^{-5}

D

To convert	Into	Multiply by
Darcy	centimeters/second	9.66×10^{-4}
Darcy	feet/second	3.173×10^{-5}
Darcy	liters/meter/second	8.58×10^{-3}
days	seconds	86,400
deciliters	liters	10^{-1}
decimeters	meters	10^{-1}
dekagrams	grams	10.0
dekaliters	liters	10.0
dekameters	meters	10.0
degrees (angle)	quadrants	1.111×10^{-2}
degrees (angle)	seconds	3,600.0
dynes	grams	1.020×10^{-3}
dynes	kilograms	1.020×10^{-6}
dynes/centimeter^2	atmospheres	9.869×10^{-7}
dynes/centimeter^2	inches of Hg at 0 °C	2.953×10^{-5}
dynes/centimeter^2	inches of water at 4 °C	4.015×10^{-4}

E

To convert	Into	Multiply by
ell	centimeters	114.30
ergs	Btu	9.480×10^{-11}
ergs	dynes-centimeters	1.0
ergs	foot-pound	7.367×10^{-8}
ergs	kilowatt-hour	2.778×10^{-14}
ergs/second	kilowatts	10^{-10}

F

To convert	Into	Multiply by
Fahrenheit (°F)	Celsius (°C)	5(°F-32)/9
Fahrenheit (°F)	Kelvin (K)	5(°F+459.67)/9
Fahrenheit (°F)	Rankine (°R)	°F+459.67
fathoms	feet	6.0
feet	centimeters	30.48
feet	inches	12
feet	kilometers	3.048×10^{-4}
feet	meters	3.048×10^{-1}
feet	miles	1.894×10^{-4}
feet	millimeters	304.8
feet	yards	3.33×10^{-1}
feet of water	inches of Hg	8.826×10^{-1}
feet/day	feet/second	1.157×10^{-5}
feet/day	meters/second	3.528×10^{-6}
feet^2	acres	2.296×10^{-5}
feet^2	hectares	9.29×10^{-9}
feet^2	inches^2	144
feet^2	kilometers^2	9.29×10^{-8}
feet^2	meters^2	9.29×10^{-2}
feet^2	miles^2	3.587×10^{-8}
feet^2	yards^2	1.111×10^{-1}
feet^2/day	meters^2/day	9.290×10^{-2}
feet^2/second	meters^2/day	8,027

To convert	Into	Multiply by
feet^3	acre-feet	2.296 x 10^{-5}
feet^3	gallons (U.S.)	7.481
feet^3	inches^3	1,728
feet^3	liters	28.32
feet^3	meters^3	2.832 x 10^{-2}
feet^3	yards^3	3.704 x 10^{-2}
feet^3/foot/day	gallons/foot/day	7.48052
feet^3/foot/day	liters/meter/day	92.903
feet^3/foot/day	meters^3/meter/day	9.29 x 10^{-2}
feet^3/foot/day	feet^3/foot^2/minute	6.944 x 10^{-4}
feet^3/foot/day	gallons/foot^2/day	7.4805
feet^3/foot/day	inches^3/inch^2/hour	5 x 10^{-1}
feet^3/foot/day	liters/meter^2/day	304.8
feet^3/foot/day	meters^3/meter^2/day	3.048 x 10^{-1}
feet^3/foot/day	millimeters^3/inch^2/hour	5 x 10^{-1}
feet^3/foot^2/day	millimeters^3/millimeter^2/hour	25.4
feet/mile	meters/kilometer	1.894 x 10^{-1}
feet/second	centimeters/second	5.080 x 10^{-1}
feet/second	feet/day	86,400
feet/second	feet/hour	3,600
feet/second	gallons (U.S.)/foot^2/day	5.737 x 10^5
feet/second	kilometers/hour	1.097
feet/second	meters/second	3.048 x 10^{-1}
feet/second	miles/hour	6.818 x 10^{-1}
feet/year	centimeters/second	9.665 x 10^{-7}
feet^3/second	acre-feet/day	1.983
feet^3/second	feet^3/minute	60.0
feet^3/second	gallons (U.S.)/minute	448.8
feet^3/second	liters/second	28.32
feet^3/second	meters^3/second	2.832 x 10^{-2}
feet^3/second	meters^3/day	2,447
furlongs	feet	660.0

G

To convert	Into	Multiply by
gallons (U.K.)	gallons (U.S.)	1.200
gallons (U.S.)	acre-feet	3.068 x 10^{-6}
gallons (U.S.)	feet^3	1.337 x 10^{-1}
gallons (U.S.)	fluid ounces	128.0
gallons (U.S.)	liters	3.785
gallons (U.S.)	meters^3	3.785 x 10^{-3}
gallons (U.S.)	yards^3	4.951 x 10^{-3}
gallons/day	acre-feet/year	1.12 x 10^{-3}
gallons/foot/day	feet^3/foot/day	1.3368 x 10^{-1}
gallons/foot/day	liters/meter/day	12.42
gallons/foot/day	meters^3/meter/day	1.242 x 10^{-2}
gallons (U.S.)/foot^2/day	centimeters/second	4.717 x 10^{-5}
gallons (U.S.)/foot^2/day	Darcy	5.494 x 10^{-2}
gallons (U.S.)/foot^2/day	feet/day	1.3368 x 10^{-1}
gallons (U.S.)/foot^2/day	feet/second	1.547 x 10^{-6}
gallons (U.S.)/foot^2/day	gallons/foot^2/minute	6.944 x 10^{-4}
gallons (U.S.)/foot^2/day	liters/meter^2/day	40.7458
gallons (U.S.)/foot^2/day	meters/day	4.07458 x 10^{-2}
gallons (U.S.)/foot^2/day	meters/minute	2.83 x 10^{-5}
gallons (U.S.)/foot^2/day	meters/second	4.716 x 10^{-7}
gallons (U.S.)/foot^2/minute	meters/day	58.67
gallons (U.S.)/foot^2/minute	meters/second	6.791 x 10^{-2}
gallons (U.S.)/minute	acre-feet/day	4.419 x 10^{-3}
gallons (U.S.)/minute	feet^3/second	2.228 x 10^{-3}
gallons (U.S.)/minute	feet^3/hour	8.0208
gallons (U.S.)/minute	liters/second	6.309 x 10^{-2}
gallons (U.S.)/minute	meters^3/day	5.30
gallons (U.S.)/minute	meters^3/second	6.309 x 10^{-5}
gills	centimeter^3	118.2941
gills	gallons (U.S.)	3.125 x 10^{-2}
gills	milliliters	118.2941

To convert	Into	Multiply by
gills	ounces (fluid)	4.0
gills	pints (liquid)	2.5×10^{-1}
grams/centimeter^3	kilograms/meter^3	1,000
grams/centimeter^3	pounds/feet^3	62.428
grams/centimeter^3	pounds/gallon (U.S.)	8.345
grams	kilograms	1×10^{-3}
grams	ounces (avoirdupois)	3.527×10^{-2}
grams	pounds	2.2046×10^{-3}
grams/liter	grains/gallon (U.S.)	58.4178
grams/liter	grams/centimeter^3	1×10^{-3}
grams/liter	kilograms/meter^3	1
grams/liter	pounds/feet^3	6.24×10^{-2}
grams/liter	pounds/inch^3	3.61×10^{-5}
grams/liter	pounds/gallon (U.S.)	8.35×10^{-3}
grams/meter^3	grains/feet^3	4.37×10^{-1}
grams/meter^3	milligrams/liter	1.0
grams/meter^3	pounds/gallon (U.S.)	8.345×10^{-5}
grams/meter^3	pounds/inch^3	7.433×10^{-3}

H

To convert	Into	Multiply by
hectares	acres	2.471
hectares	feet^2	1.076×10^{5}
hectares	inches^2	1.55×10^{7}
hectares	kilometers^2	1×10^{-2}
hectares	meters^2	10,000
hectares	miles^2	3.861×10^{-3}
hectares	yards^2	11,959.90
horsepower	kilowatts/hour	7.457×10^{-1}

I

To convert	Into	Multiply by
inches	centimeters	2.540
inches	feet	8.333×10^{-1}
inches	kilometers	2.54×10^{-5}
inches	meters	2.54×10^{-2}
inches	miles	1.578×10^{-5}
inches	millimeters	25.4
inches	yards	2.778×10^{-2}
inches of Hg	pascals	3,386
inches^2	acres	1.594×10^{-8}
inches^2	centimeters^2	6.4516
inches^2	feet^2	6.944×10^{-3}
inches^2	hectares	6.452×10^{-8}
inches^2	kilometers^2	6.452×10^{-10}
inches^2	meters^2	6.452×10^{-4}
inches^2	millimeters^2	645.16
inches^3	acre-feet	1.329×10^{-8}
inches^3	centimeters^3	16.39
inches^3	feet^3	5.787×10^{-4}
inches^3	gallons (U.S.)	4.329×10^{-3}
inches^3	liters	1.639×10^{-2}
inches^3	meters^3	1.639×10^{-5}
inches^3	milliliters	16.387
inches^3	yards^3	2.143×10^{-5}

J

To convert	Into	Multiply by
joules	Btu	9.480×10^{-4}
joules	ergs	7.376×10^{-1}
joules	foot-pound	10^{7}
joules/centimeter	pounds	22.48

K

To convert	Into	Multiply by
Kelvin (K)	Celsius	K - 273.15
Kelvin (K)	Fahrenheit	1.8(K) - 459.67
kilograms	grams	1,000
kilograms	milligrams	1×10^{6}

To convert	Into	Multiply by
kilograms	ounces (avoirdupois)	35.28
kilograms	pounds	2.205
kilograms	tons (long)	9.842×10^{-4}
kilograms	tons (metric)	1×10^{-3}
kilograms/meter^3	grams/centimeter^3	1×10^{-3}
kilograms/meter^3	grams/liter	1.0
kilograms/meter^3	pounds/inch^3	3.613×10^{-5}
kilograms/meter^3	pounds/feet^3	0.0624
kilometers	feet	3,281
kilometers	inches	39,370
kilometers	meters	1,000
kilometers	miles	6.214×10^{-1}
kilometers	millimeters	1×10^{6}
kilometers^2	acres	247.1
kilometers^2	hectares	100
kilometers^2	inches^2	1.55×10^{9}
kilometers^2	meters^2	1×10^{6}
kilometers^2	miles^2	0.3861
kilometers/hour	feet/day	78,740
kilometers/hour	feet/second	9.113×10^{-1}
kilometers/hour	meters/second	2.778×10^{-1}
kilometers/hour	miles/hour	6.214×10^{-1}

L

To convert	Into	Multiply by
liters	acre-feet	8.106×10^{-7}
liters	feet^3	3.531×10^{-2}
liters	fluid ounces	33.814
liters	gallons (U.S.)	2.642×10^{-1}
liters	inches^3	61.02
liters	meters^3	10^{-3}
liters	yards^3	1.308×10^{-3}
liters/second	acre-feet/day	7.005×10^{-2}
liters/second	feet^3/second	3.531×10^{-2}
liters/second	gallons (U.S.)/minute	15.85
liters/second	meters^3/day	86.4

M

To convert	Into	Multiply by
meters	feet	3.28084
meters	inches	39.3701
meters	kilometers	1×10^{-3}
meters	miles	6.214×10^{-4}
meters	millimeters	1,000
meters	yards	1.0936
meters^2	acres	2.471×10^{-4}
meters^2	feet^2	10.76
meters^2	hectares	1×10^{-4}
meters^2	inches^2	1.550
meters^2	kilometers	1×10^{-6}
meters^2	miles^2	3.861×10^{-7}
meters^2	yards^2	1.196
meters^3	acre-feet	8.106×10^{-4}
meters^3	feet^3	35.31
meters^3	gallons (U.S.)	264.2
meters^3	inches^3	6.102×10^{4}
meters^3	liters	1,000
meters^3	yards^3	1.308
meters/second	feet/minute	196.8
meters/second	feet/day	283,447
meters/second	feet/second	3.281
meters/second	kilometers/hour	3.6
meters/second	miles/hour	2.237
meters^3/day	acre-feet/day	6.051×10^{6}
meters^3/day	feet^3/second	3.051×10^{6}
meters^3/day	gallons (U.S.)/minute	1.369×10^{9}
meters^3/day	liters/second	8.64×10^{7}

To convert	Into	Multiply by
miles	feet	5,280
miles	kilometers	1.609
miles	meters	1,609
miles	millimeters	1.609×10^6
miles^2	acres	640
miles^2	feet^2	2.778×10^7
miles^2	hectares	259
miles^2	inches^2	4.014×10^9
miles^2	kilometers^2	2.590
miles^2	meters^2	2.59×10^6
miles/hour	feet/day	1.267×10^5
miles/hour	kilometers/hour	1.609
miles/hour	meters/second	4.47×10^{-1}
millibars	pascals	100.0
milligrams/liter	grams/meter^3	1.0
milligrams/liter	parts/million	1.0
milligrams/liter	pounds/feet^3	6.2428×10^{-5}
milliliters	centimeters^3	1.0
milliliters	liters	1×10^{-3}
millimeters	centimeters	1×10^{-1}
millimeters	feet	3.281×10^{-3}
millimeters	inches	0.03937
millimeters	kilometers	1.0×10^{-6}
millimeters	meters	1.0×10^{-3}
millimeters	miles	6.214×10^{-7}
millimeters^2	centimeters^2	1×10^{-2}
millimeters^2	inches^2	1.55×10^{-3}
millimeters^3	centimeters^3	1×10^{-3}
millimeters^3	inches^3	6.102×10^{-5}
millimeters^3	liters	10^{-6}
millimeters of Hg	atmospheres	1.316×10^{-3}
millimeters of Hg	pascals	133.3224
millimeters of Hg	torrs	1.0

O

To convert	Into	Multiply by
ounces (avoirdupois)	grams	28.35
ounces (avoirdupois)	kilograms	2.8355×10^{-2}
ounces (avoirdupois)	pounds	6.25×10^{-2}

P

To convert	Into	Multiply by
pascals	atmospheres	9.869×10^{-6}
pascals	millimeters of Hg	7.501×10^{-3}
pounds (avoirdupois)	kilograms	4.535×10^{-1}
pounds (avoirdupois)	ounces (avoirdupois)	16
pounds/centimeter^3	pounds/inch^3	3.61×10^2
pounds/inch^2	pascals	6,895
pounds/inch^2	kilograms/centimeter^2	7.31×10^2
pounds/feet^3	grams/centimeter^3	1.6×10^2
pounds/feet^3	grams/liter	27,680
pounds/feet^3	pounds/inch^3	5.787×10^{-4}
pounds/feet^3	pounds/gallon (U.S.)	1.337×10^{-1}
pounds/foot^2	pascals	47.88
pounds/inch^3	pounds/centimeter^3	27.68
pounds/inch^3	pounds/feet^3	1,728
pounds/inch^3	pounds/gallon (U.S.)	231
pounds/gallon (U.S.)	grams/centimeter^3	1.198×10^{-1}
pounds/gallon (U.S.)	grams/liter	119.8
pounds/gallon (U.S.)	pounds/feet^3	7.481
pounds/gallon (U.S.)	pounds/inch^3	4.329×10^{-3}

Q

To convert	Into	Multiply by
quadrants (angle)	radians	1.571
quarts (U.S.)	gallons	0.25
quarts (liquid)	feet^3	3.342×10^{-2}
quarts (liquid)	inches^3	57.75

To convert	Into	Multiply by
quarts (liquid)	liters	9.464×10^{-1}
quarts	pints	2.0
quarts (liquid)	yards^3	1.238×10^{-3}
	R	
radians	degrees	57.30
radians	minutes	3,438
radians	quadrants	6.366×10^{-1}
radians	seconds	2.063×10^5
revolutions	degrees	360.0
revolutions	quadrants	4.0
revolutions	radians	6.283
revolutions/minute	degrees/second	6.0
revolutions/second	degrees/second	360.0
revolutions/second	radians/second	6.283
revolutions/second	revolutions/minute	60.0
rods	feet	16.5
rods	meters	5.029
rods (surveyors'measure)	yards	5.5
	S	
scruples	grains	20.0
scruples	grams	1.296
scruples	milligrams	1,295.97
scruples	pennyweights	8.333×10^{-1}
slugs	grams	1.459×10^4
slugs	kilograms	14.59
	T	
tons (short)	kilograms	907.1847
tons (short)	pounds	2,000.0
torrs	atmospheres	1.316×10^{-3}
torrs	millimeters of Hg	1.0
torrs	pascals	133.322
	W	
watt-hours	Btu	3.413
watt-hours	foot-pound	2,656
watts	Btu/hour	3.4129
watts	Btu/minute	5.688×10^{-2}
watts	horsepower	1.341×10^{-3}
watts	horsepower (metric)	1.360×10^{-3}
watts	kilowatts	10^{-3}
webers	maxwells	10^8
	Y	
yards	centimeters	91.44
yards	fathoms	5×10^{-1}
yards	feet	3.0
yards	inches	36.0
yards	kilometers	9.144×10^{-4}
yards	meters	9.144×10^{-1}
yards	millimeters	914.4
yards	miles	5.682×10^{-4}
yards	feet^3	27
yards	gallons (U.S.)	202
yards	inches^3	4.666×10^4
yards	liters	764.6
yards	meters^3	7.646×10^{-1}
yards^2	meters^2	8.361×10^{-1}
yards^3	meters^3	7.646×10^{-1}
years (normal calendar)	hours	8,760
years (normal calendar)	minutes	5.256×10^5
years (normal calendar)	seconds	3.1536×10^7
years (normal calendar)	weeks	52.1428

Table 3. U.S. EPA Approved Test Methods

The numbers that follow each compound name below are the U.S. EPA-approved test methods based on information obtained from the *Guide to Environmental Analytical Methods* (Schenectady, NY: Genium Publishing, 1992) and the U.S. EPA's Sampling and Analysis Methods Database. The database information is available in hard copy from Keith, L.H., *Compilation of EPA's Sampling and Analysis Methods* (Chelsea, MI: Lewis Publishers, 1991), 803 p.

Chemical Name	Test Methods
Acrolein	8030, 8240
Acrylonitrile	8030, 8240
Alachlor	525
Aldrin	508, 525, 608, 625, SM-6410, SM-6630, 8080, 8270
Anilazine	8270
Atrazine	525
Azinphos-methyl	8140, 8270
α-BHC	508, 608, 625, SM-6410, SM-6630, 8080, 8270
β-BHC	508, 608, 625, SM-6410, SM-6630, 8080, 8270
δ-BHC	508, 608, 625, SM-6410, SM-6630, 8080, 8270
Bis(2-chloroethyl) ether	625, SM-6410, 8270
Bis(2-chloroisopropyl) ether	625, SM-6410, 8270
Bromoxynil	8270
Carbaryl	8270
Carbofuran	8270
Carbon disulfide	8240
Carbon tetrachloride	502.1, 502.2, 601, 524.1, 524.2, 624, SM-6210, SM-6230, 8010, 8240
Carbophenothion	8270
Chloramben	515.1
Chlordane (technical)	508, 608, SM-6630, 8080, 8270
Chlorobenzilate	508, 8270
Chlorothalonil	508
Chlorpyrifos	508, 8140
Crotoxyphos	8270
2,4-D	515.1, 8150
Dalapon	8150
p,p'-DDD	508, 608, SM-6630, 8080, 8270
p,p'-DDE	508, 608, SM-6630, 8080, 8270
p,p'-DDT	508, 608, SM-6630, 8080, 8270
Diallate	8270
Diazinon	8140
1,2-Dibromo-3-chloropropane	502.2, 504, 524.1, 524.2, SM-6210, 8240
Di-*n*-butyl phthalate	525, 625, SM-6410, 8060, 8270
Dicamba	515.1, 8150
Dichlone	8270
1,2-Dichloropropane	502.1, 502.2, 524.1, 524.2, 601, 624, SM-6210, SM-6230, 8010, 8240
cis-1,3-Dichloropropylene	502.1, 502.2, 524.1, 601, 624, SM-6210, SM-6230, 8010, 8240
trans-1,3-Dichloropropylene	502.1, 502.2, 524.1, 601, 624, SM-6210, SM-6230, 8010, 8240
Dichlorvos	8140, 8270
Dicrotophos	8270
Dieldrin	508, 608, SM-6630, 8080, 8270
Dimethoate	8270
Dimethyl phthalate	525, 625, SM-6410, 8060, 8270
4,6-Dinitro-*o*-cresol	8040
Dinoseb	515.1, 8150, 8270
Disulfoton	8140, 8270
α-Endosulfan	508, 608, 625, SM-6410, SM-6630, 8080, 8270
β-Endosulfan	508, 608, 625, SM-6410, SM-6630, 8080, 8270
Endosulfan sulfate	508, 608, 625, SM-6410, SM-6630, 8080, 8270
Endrin	508, 525, 608, 625, SM-6410, SM-6630, 8080, 8270
Endrin aldehyde	508, 608, 625, SM-6410, SM-6630, 8080, 8270
EPN	8270
Ethion	8270
Ethoprop	8140
Ethylene dibromide	502.1, 502.2, 504, 524.1, 524.2, SM-6210, 8240
Fensulfothion	8140, 8270
Fenthion	8140, 8270
Heptachlor	508, 525, 608, 625, SM-6410, SM-6630, 8080, 8270
Heptachlor epoxide	508, 525, 608, 625, SM-6410, SM-6630, 8080, 8270
Hexachlorobenzene	508, 525, 625, SM-6410, 8120, 8270
Kepone	8270

Chemical Name	Test Methods
Lindane	508, 525, 608, SM-6630, 8080, 8270
Malathion	8270
MCPA	8150
Methoxychlor	508, 525, SM-6630, 8080, 8270
Methyl bromide	502.1, 502.2, 524.1, 524.2, 601, 624, SM-6210, SM-6230, 8010, 8240
Mevinphos	8140, 8270
Monocrotophos	8270
Naled	8140, 8270
Parathion	8270
Pentachlorobenzene	8270
Pentachlorophenol	525, 625, SM-6410, 8040, 8270
Phorate	8140, 8270
Phosalone	8270
Phosmet	8270
Phosphamidon	8270
Picloram	515.1
Propachlor	508
Propyzamide	8270
Ronnel	8140
Simazine	525
Strychnine	525
Sulprofos	8140
2,4,5-T	515.1, 8150, SM-6640
Terbufos	8270
Tetraethyl pyrophosphate	8270
Toxaphene	508, 525, 608, 625, SM-6630, 8080, 8270
Trifluralin	508, 8270

Note: Methods beginning with the prefix "SM-" refer to standard methods found in the 17th edition of the *Standard Methods for the Examination of Water and Waste Water*, American Water Works Association, 1989.

Table 4. Typical Bulk Density Values of Selected Soils and Rocks

Soil/Rock Type	Bulk Density, B
Silt	1.38
Clay	1.49
Loess	1.45
Sand, dune	1.58
Sand, fine	1.55
Sand, medium	1.69
Sand, coarse	1.73
Gravel, fine	1.76
Gravel, medium	1.85
Gravel, coarse	1.93
Till, predominantly silt	1.78
Till, predominantly sand	1.88
Till, predominantly gravel	1.91
Glacial drift, predominantly silt	1.38
Glacial drift, predominantly sand	1.55
Glacial drift, predominantly gravel	1.60
Siltstone	1.61
Claystone	1.51
Sandstone, fine-grained	1.76
Sandstone, medium-grained	1.68
Limestone	1.94
Dolomite	2.02
Schist	1.76
Basalt	2.53
Shale	2.53
Gabbro, weathered	1.73
Granite, weathered	1.50

Reference: Morris and Johnson (1967)

Table 5. Ranges of Porosity Values of Selected Soils and Rocks

Soil/Rock Type	Porosity, n
Peat	0.60-0.80
Silt	0.34-0.61
Clay	0.34-0.57
Loess	0.40-0.57
Sand, dune	0.35-0.51
Sand, fine	0.25-0.55
Sand, medium	0.28-0.49
Sand, coarse	0.30-0.46
Gravel, fine	0.25-0.40
Gravel, medium	0.24-0.44
Gravel, coarse	0.25-0.35
Sand + gravel	0.20-0.35
Till, predominantly silt	0.30-0.41
Till, predominantly sand	0.22-0.37
Till, predominantly gravel	0.22-0.30
Till, clay-loam	0.30-0.35
Glacial drift, predominantly silt	0.38-0.59
Glacial drift, predominantly sand	0.36-0.48
Glacial drift, predominantly gravel	0.35-0.42
Siltstone	0.20-0.41
Claystone	0.41-0.45
Sandstone, fine-grained	0.14-0.49
Sandstone, medium-grained	0.30-0.44
Volcanic, dense	0.01-0.10
Volanic, pumice	0.80-0.90
Volcanic, vesicular	0.10-0.50
Volcanic, tuff	0.10-0.40
Limestone	0.05-0.56
Dolomite	0.19-0.33
Schist	0.05-0.55
Basalt	0.03-0.35
Shale	0.01-0.10
Igneous, dense metamorphic and plutonic	0.01-0.05
Igneous, weathered metamorphic and plutonic	0.34-0.55

References: Davis and DeWiest (1966); Grisak et al. (1980); Morris and Johnson (1967)

Table 6. Aqueous Solubility Data of Miscellaneous Organic Compounds

Compound [CAS No.]	Temp. °C	Log S mol/L	Reference
Acetal [105-57-7]	25.0	-0.12	Tewari et al., 1982
Acetochlor [34256-82-1]	25	-3.08	Humburg et al., 1989
Acetophenone [98-86-2]	19.2	-1.26	Stephenson, 1992*
	25.0	-1.34	Andrews and Keefer, 1950
	25	-1.29	Southworth and Keller, 1986
	29.5	-1.23	Stephenson, 1992*
	39.5	-1.16	Stephenson, 1992*
	49.8	-1.17	Stephenson, 1992*
	60.1	-1.13	Stephenson, 1992*
	70.2	-1.08	Stephenson, 1992*
	80.2	-1.00	Stephenson, 1992*
9-Acetylanthracene [784-04-3]	25	-4.85	Southworth and Keller, 1986
4-Acetylbiphenyl [92-91-1]	25	-4.21	Southworth and Keller, 1986
2-Acetylnaphthalene [93-08-3]	25	-3.09	Southworth and Keller, 1986
Acetylpyridine [1122-62-9]	9	0.46	Stephenson, 1993*
	12.0	0.38	Stephenson, 1993*
	20.0	0.25	Stephenson, 1993*
	25.0	0.11	Stephenson, 1993*
	30.0	0.04	Stephenson, 1993*
	40.0	-0.05	Stephenson, 1993*
	50.0	-0.10	Stephenson, 1993*
	60.0	-0.09	Stephenson, 1993*
	70.0	-0.13	Stephenson, 1993*
	80.0	-0.13	Stephenson, 1993*
	90.0	-0.12	Stephenson, 1993*
Aclonifen [74070-46-5]	20	-5.02	Worthing and Hance, 1991
Acridine [260-94-6]	25	-3.67	Banwart et al., 1982
Adiponitrile [111-69-3]	20	-0.08	Du Pont, 1999*
	102	0.67	Du Pont, 1999*
Alachlor [15972-60-8]	25	-3.05	Humburg et al., 1989
Allethrin [584-79-2]	20	-4.82	Worthing and Hance, 1991
Allidochlor [93-71-0]	25	-0.95	Bailey and White, 1965
Alloxydim-sodium [66003-55-2]	30	>0.76	Worthing and Hance, 1991
Allyl bromide [106-95-6]	25.0	-1.50	Tewari et al., 1982
Ametryn [834-12-8]	20	-3.08	Fühner and Geiger, 1977
	25	-3.09	Humburg et al., 1989
	26.0	-1.75	pH 2.0, Ward and Weber, 1968
	26.0	-2.75	pH 3.0, Ward and Weber, 1968
	26.0	-3.07	pH 5.0, Ward and Weber, 1968
	26.0	-3.07	pH 7.0, Ward and Weber, 1968
	26.0	-3.07	pH 10.0, Ward and Weber, 1968
Amitrole [61-82-5]	25	0.52	Humburg et al., 1989
Ammonium sulfamate [773-06-0]	25	1.28	Humburg et al., 1989
Ancymidol [12771-68-5]	25	-2.60	Worthing and Hance, 1991
Androst-16-en-3-ol [1153-51-1]	25	-6.08	Amoore and Buttery, 1978
Anilazine [101-05-3]	20	-4.54	Worthing and Hance, 1991
Anilofos [64249-01-0]	20	-4.43	Worthing and Hance, 1991
Anisole [100-66-3]	10.2	-1.66	Stephenson, 1992*
	20.0	-1.73	Stephenson, 1992*
	25.0	-1.85	Vesala, 1974
	29.7	-1.76	Stephenson, 1992*
	39.9	-1.77	Stephenson, 1992*
	50.2	-1.73	Stephenson, 1992*
	60.2	-1.63	Stephenson, 1992*
	70.2	-1.63	Stephenson, 1992*
	81.2	-1.57	Stephenson, 1992*
	90.7	-1.49	Stephenson, 1992*
	25.0	-1.89	Andrews and Keefer, 1950
9-Anthracenemethanol [1468-95-7]	25	-4.72	Southworth and Keller, 1986
Anthraquinone [84-65-1]	20	-5.54	Worthing and Hance, 1991
Arsenous oxide [1327-53-3]	16	-1.07	Worthing and Hance, 1991
Atratone [1610-17-9]	26.0	-2.04	pH 3.0, Ward and Weber, 1968
	26.0	-2.12	pH 7.0, Ward and Weber, 1968
	26.0	-2.11	pH 10.0, Ward and Weber, 1968
Asulam [3337-71-1]	25	0.34	Humburg et al., 1989

Compound [CAS No.]	Temp. °C	Log S mol/L	Reference
Atrazine [1912-24-9]	1	-3.86	pH 6, Gaynor and Van Volk, 1981
	8	-3.85	pH 6, Gaynor and Van Volk, 1981
	20	-3.80	pH 6, Gaynor and Van Volk, 1981
	20	-3.82	Ellgehausen et al., 1980
	20	-3.86	Ellgehausen et al., 1981
	20	-3.86	Fühner and Geiger, 1977
	22	-3.80	Mills and Thurman, 1994
	26.0	-3.84	pH 3.0, Ward and Weber, 1968
	26.0	-3.79	pH 7.0, Ward and Weber, 1968
	26.0	-3.77	pH 10.0, Ward and Weber, 1968
	29	-3.71	pH 6, Gaynor and Van Volk, 1981
Azaconazole [60207-31-0]	20	-3.00	Worthing and Hance, 1991
Azamethiphos [35575-96-3]	20	-2.47	Worthing and Hance, 1991
Azinphos-methyl [86-50-0]	20	-4.05	Worthing and Hance, 1991
Aziprotryne [4658-28-0]	20	-3.61	Worthing and Hance, 1991
Barban [101-27-9]	25	-4.47	Humburg et al., 1989
Benalaxyl [71626-11-4]	25	-3.94	Worthing and Hance, 1991
Benazolin [3813-05-6]	20	-2.61	Humburg et al., 1989
Benazolin-ethyl [25059-80-7]	25	-3.76	Worthing and Hance, 1991
Bendiocarb [22781-23-3]	25	-0.93	Worthing and Hance, 1991
Benfluralin [1861-40-1]	25	-6.53	Humburg et al., 1989
Benfuracarb [82560-54-1]	20	-4.71	Worthing and Hance, 1991
Benfuresate [68505-69-1]	25	-3.13	Worthing and Hance, 1991
Benomyl [17804-35-2]	25	-1.20	pH 1, Singh and Chiba, 1985
	25	-1.86	pH 3, Singh and Chiba, 1985
	25	-1.91	pH 5, Singh and Chiba, 1985
	25	-2.02	pH 7, Singh and Chiba, 1985
	25	-1.99	pH 8, Singh and Chiba, 1985
	25	-2.18	pH 9, Singh and Chiba, 1985
	25	-2.21	pH 10, Singh and Chiba, 1985
	25	-1.52	pH 11, Singh and Chiba, 1985
	25	-1.81	pH 12, Singh and Chiba, 1985
Benoxacor [98730-04-2]	20	-4.11	Worthing and Hance, 1991
Bensulfuron-methyl [83055-99-6]	25	-5.14	pH 5, Humburg et al., 1989
	25	-3.53	pH 7, Humburg et al., 1989
Bensulfide [741-58-2]	20	-4.20	Humburg et al., 1989
Bensultap [17606-31-4]	25	-4.76	Worthing and Hance, 1991
Bentazon [25057-89-0]	20	-2.68	Humburg et al., 1989
Benzaldehyde [100-52-7]	0	-1.10	Stephenson, 1993b*
	20.0	-1.17	Stephenson, 1993b*
	30.0	-1.16	Stephenson, 1993b*
	40.0	-1.13	Stephenson, 1993b*
	50.0	-1.11	Stephenson, 1993b*
	60.0	-1.06	Stephenson, 1993b*
	70.0	-1.02	Stephenson, 1993b*
	80.0	-0.93	Stephenson, 1993b*
	90.0	-0.88	Stephenson, 1993b*
2,3-Benzanthracene [92-24-0]	25	-8.60	Mackay and Shiu, 1977
	27	-8.36	Davis et al., 1942
Benziodarone [68-90-6]	20	-4.95	Hafkenscheid and Tomlinson, 1983
1,2-Benzofluorene [238-84-6]	25	-6.68	Mackay and Shiu, 1977
2,3-Benzofluorene [243-17-4]	25	-7.73	Billington et al., 1988
	25	-8.03	Mackay and Shiu, 1977
Benzonitrile [100-47-0]	25	-1.38	McGowan et al., 1966
Benzophenone [119-61-9]	20	-3.39	Hafkenscheid and Tomlinson, 1983
	25	-2.96	Southworth and Keller, 1986
Benzoximate [29104-30-1]	25	-4.08	Worthing and Hance, 1991
Benzthiazuron [929-88-0]	20	-4.14	Worthing and Hance, 1991
Benzyl acetate [140-11-4]	25.0	-1.77	Stephenson and Stuart, 1986*
Benzyl formate [104-57-4]	0	-1.05	Stephenson, 1992*
	9.8	-1.06	Stephenson, 1992*
	19.6	-1.10	Stephenson, 1992*
	29.8	-1.10	Stephenson, 1992*
	39.7	-1.15	Stephenson, 1992*

Compound [CAS No.]	Temp. °C	Log S mol/L	Reference
	49.7	-1.11	Stephenson, 1992*
	60.0	-1.09	Stephenson, 1992*
	70.3	-1.04	Stephenson, 1992*
	80.1	-0.98	Stephenson, 1992*
	90.5	-0.69	Stephenson, 1992*
Bibenzyl [103-29-7]	20	-5.62	Swann et al., 1983
	25.0	-4.80	Andrews and Keefer, 1950a
Bifenox [42576-02-3]	25	-5.99	Humburg et al., 1989
Bifenthrin [82657-04-3]	20-25	-6.63	Worthing and Hance, 1991
Bioallethrin ((S)-cyclopentenyl isomer) [28434-00-6]	20	-4.82	Worthing and Hance, 1991
4-Biphenylmethanol [3597-91-9]	25	-3.39	Southworth and Keller, 1986
2,2'-Biquinoline [119-91-5]	25	-5.40	Banwart et al., 1982
Bis(2-ethylhexyl) isophthalate [137-89-3]	24	-7.55	Hollifield, 1979
Bolasterone [1605-89-6]	37	-3.74	Hamlin et al., 1965
Borax-anhydrous [7775-19-1]	20	-1.17	Humburg et al., 1989
Borneol [507-70-0]	25	-2.32	Mitchell, 1926
1S-endo-(–)-Borneol [464-45-9]	25	-2.52	Fichan et al., 1999
Bromacil [314-40-9]	4	-2.62	Madhun et al., 1986
	25	-2.53	Madhun et al., 1986
	25	-2.51	Humburg et al., 1989
	25	-2.39	Gerstl and Yaron, 1983
	40	-2.40	Madhun et al., 1986
Bromadiolone [28772-56-7]	20	-4.44	Worthing and Hance, 1991
4-Bromobiphenyl [92-66-0]	4.0	-6.00	Doucette and Andren, 1988a
	25.0	-5.55	Doucette and Andren, 1988a
	40.0	-5.43	Doucette and Andren, 1988a
1-Bromobutane [109-65-9]	25.0	-2.20	Tewari et al., 1982
	30	-2.35	Gross and Saylor, 1931
4-Bromo-1-butene [5162-44-7]	25.0	-2.25	Tewari et al., 1982
Bromobutide [74712-19-9]	25	-4.95	Worthing and Hance, 1991
2-Bromochlorobenzene [694-80-4]	25	-3.19	Yalkowsky et al., 1979
3-Bromochlorobenzene [108-37-2]	25	-3.21	Yalkowsky et al., 1979
4-Bromochlorobenzene [106-39-8]	25	-3.63	Yalkowsky et al., 1979
1-Bromo-2-chloroethane [107-04-0]	30.00	-1.32	Gross et al., 1933
1-Bromo-3-chloropropane [109-70-6]	25.0	-1.85	Tewari et al., 1982
2-Bromoethylacetate [927-68-4]	25.0	-0.67	Tewari et al., 1982
Bromofenoxim [13181-17-4]	20	-6.66	Worthing and Hance, 1991
1-Bromoheptane [629-04-9]	25.0	-4.43	Tewari et al., 1982
1-Bromohexane [111-25-1]	25.0	-3.81	Tewari et al., 1982
4-Bromoiodobenzene [589-87-7]	25	-4.56	Yalkowsky et al., 1979
1-Bromooctane [111-83-1]	25.0	-5.06	Tewari et al., 1982
1-Bromopentane [110-53-2]	25.0	-3.08	Tewari et al., 1982
Bromophos [2104-96-3]	20	-5.56	Fühner and Geiger, 1977
Bromophos-ethyl [4824-78-6]	20	-5.12	Fühner and Geiger, 1977
Bromopol [52-51-7]	22-24	0.10	Worthing and Hance, 1991
Bromopropylate [18181-80-1]	20	<-5.93	Fühner and Geiger, 1977
2-Bromopyridine [109-04-6]	0	-0.76	Stephenson, 1993*
	10.0	-0.80	Stephenson, 1993*
	20.0	-0.84	Stephenson, 1993*
	30.0	-0.84	Stephenson, 1993*
	40.0	-0.87	Stephenson, 1993*
	50.0	-0.86	Stephenson, 1993*
	60.0	-0.84	Stephenson, 1993*
	70.0	-0.81	Stephenson, 1993*
	80.0	-0.80	Stephenson, 1993*
	90.0	-0.74	Stephenson, 1993*
Bromoxynil [1689-84-5]	25	-3.33	Humburg et al., 1989
Buprofezin [69327-76-0]	25	-5.53	Worthing and Hance, 1991
Butachlor [23184-66-9]	24	-4.13	Humburg et al., 1989
Butamifos [36335-67-8]	20	-4.81	Worthing and Hance, 1991
1,3-Butane dinitrate [6423-44-5]	25	-1.66	Fischer and Ballschmiter, 1998, 1998a
1,4-Butane dinitrate [3457-91-8]	25	-1.68	Fischer and Ballschmiter, 1998, 1998a
2,3-Butanedione [431-03-8]	0	0.60	Stephenson, 1992*
	9.2	0.59	Stephenson, 1992*

Compound [CAS No.]	Temp. °C	Log S mol/L	Reference
	19.2	0.57	Stephenson, 1992*
	31.0	0.54	Stephenson, 1992*
	40.0	0.52	Stephenson, 1992*
	50.7	0.48	Stephenson, 1992*
	61.1	0.45	Stephenson, 1992*
	70.3	0.42	Stephenson, 1992*
	80.3	0.40	Stephenson, 1992*
Butenachlor [87310-56-3]	27	-4.03	Worthing and Hance, 1991
Butoxycarboxim [34681-23-7]	25	-0.03	Worthing and Hance, 1991
Butralin [33629-47-9]	24-26	-5.47	Worthing and Hance, 1991
2-Butylacrolein [1070-66-2]	0	-2.10	Stephenson, 1993*
	10.0	-2.20	Stephenson, 1993*
	20.0	-2.20	Stephenson, 1993*
	40.0	-2.01	Stephenson, 1993*
	50.0	-2.05	Stephenson, 1993*
	60.0	-2.05	Stephenson, 1993*
	70.0	-1.82	Stephenson, 1993*
	80.0	-1.87	Stephenson, 1993*
	90.0	-1.94	Stephenson, 1993*
Butylate [2008-41-5]	20	-3.67	Worthing and Hance, 1991
	22	-3.68	Humburg et al., 1989
Butyl butyrate [109-21-7]	25.0	-2.37	Stephenson and Stuart, 1986*
Butyl ether [142-96-1]	0	-2.51	Stephenson, 1992*
	9.3	-2.61	Stephenson, 1992*
	19.9	-2.75	Stephenson, 1992*
	30.9	-2.75	Stephenson, 1992*
	40.3	-2.81	Stephenson, 1992*
	50.5	-2.77	Stephenson, 1992*
	61.3	-3.04	Stephenson, 1992*
	70.5	-2.94	Stephenson, 1992*
	80.7	-3.16	Stephenson, 1992*
	90.5	-3.12	Stephenson, 1992*
Butyl ethyl ether [628-81-9]	0	-0.97	Stephenson, 1992*
	9.3	-1.09	Stephenson, 1992*
	20.0	-1.29	Stephenson, 1992*
	31.2	-1.29	Stephenson, 1992*
	39.7	-1.25	Stephenson, 1992*
	50.8	-1.35	Stephenson, 1992*
	60.2	-1.30	Stephenson, 1992*
	70.2	-1.42	Stephenson, 1992*
	80.2	-1.38	Stephenson, 1992*
	90.7	-1.41	Stephenson, 1992*
Butyl formate [592-84-7]	25.0	-0.97	Stephenson and Stuart, 1986*
Butyl isobutyrate [97-87-0]	25.0	-2.27	Stephenson and Stuart, 1986*
Butyl lactate [138-22-7]	25	-0.53	Rehberg and Dixon, 1950
5-Butyl-2-methylpyridine [702-16-9]	0	-1.55	Stephenson, 1993*
	10.0	-1.78	Stephenson, 1993*
	20.0	-1.92	Stephenson, 1993*
	30.0	-1.90	Stephenson, 1993*
	40.0	-1.81	Stephenson, 1993*
	50.0	-1.68	Stephenson, 1993*
	60.0	-1.85	Stephenson, 1993*
	70.0	-1.79	Stephenson, 1993*
	80.0	-1.87	Stephenson, 1993*
	90.0	-1.94	Stephenson, 1993*
1-Butyl nitrate [928-15-0]	25	-2.03	Hauff et al., 1998
4-sec-Butyl-2-nitrophenol [3555-18-8]	20.0	-3.84	Schwarzenbach et al., 1988
Butyl propionate [590-01-2]	25.0	-1.85	Stephenson and Stuart, 1986*
4-tert-Butylpyridine [3978-81-2]	0	-1.38	Stephenson, 1993*
	10.0	-1.49	Stephenson, 1993*
	20.0	-1.50	Stephenson, 1993*
	30.0	-1.53	Stephenson, 1993*
	40.0	-1.56	Stephenson, 1993*
	50.0	-1.59	Stephenson, 1993*

Compound [CAS No.]	Temp. °C	Log S mol/L	Reference
	60.0	-1.55	Stephenson, 1993*
	70.0	-1.54	Stephenson, 1993*
	80.0	-1.48	Stephenson, 1993*
	90.0	-1.46	Stephenson, 1993*
Butyraldehde [123-72-8]	0	0.22	Stephenson, 1993b*
	10.0	0.10	Stephenson, 1993b*
	20.0	0.01	Stephenson, 1993b*
	30.0	-0.12	Stephenson, 1993b*
	40.0	- 0.19	Stephenson, 1993b*
	50.0	- 0.23	Stephenson, 1993b*
	60.0	-0.24	Stephenson, 1993b*
	70.0	-0.27	Stephenson, 1993b*
Captan [133-06-2]	20	<-5.78	Führer and Geiger, 1977
	25	-4.96	Worthing and Hance, 1991
Carbazole [86-74-8]	20	-5.14	Hashimoto et al., 1982
Carbendazim [10605-21-7]	20	-3.83	Worthing and Hance, 1991
Carbon tetrabromide [558-13-4]	30	-3.14	Gross and Saylor, 1931
	30.00	-2.71	Gross et al., 1933
Carbosulfan [55285-14-8]	25	-7.10	Worthing and Hance, 1991
Carboxin [5234-68-4]	25	-3.07	Worthing and Hance, 1991
(−)-Carveol [99-48-9]	25	-1.72	Fichan et al., 1999
(S)-(+)-Carvone [2244-16-8]	25	-2.06	Fichan et al., 1999
Chinomethionat [2439-01-2]	20	-5.37	Worthing and Hance, 1991
Chlomethoxyfen [32861-85-1]	15	-6.02	Worthing and Hance, 1991
Chloralose [15879-93-3]	15	-1.84	Worthing and Hance, 1991
Chloramben [133-90-4]	20	-2.47	Humburg et al., 1989
Chlorazine [580-48-3]	26.0	-4.04	pH 3.0, Ward and Weber, 1968
	26.0	-4.07	pH 7.0, Ward and Weber, 1968
	26.0	-4.08	pH 10.0, Ward and Weber, 1968
Chlorbromuron [13360-45-7]	20	-3.92	Worthing and Hance, 1991
Chlorbufam [1967-16-4]	20	-2.62	Worthing and Hance, 1991
Chlorfenvinphos [18708-87-7]	20	-3.41	Führer and Geiger, 1977
Chlorflurenol [2464-31-4]	20	-4.16	Humburg et al., 1989
Chlorimuron [99283-00-8]	25	-4.55	pH 5, Humburg et al., 1989
	25	-2.51	pH 7, Humburg et al., 1989
Chlormephos [24934-91-6]	20	-3.59	Worthing and Hance, 1991
Chlornitrofen [1836-77-7]	25	-6.11	Worthing and Hance, 1991
2-Chloroallyl N,N-diethyldithiocarbamate [95-06-7]	25	-3.43	Bailey and White, 1965
2-Chloroanisole [766-51-8]	25	-2.46	Lun et al., 1995
3-Chloroanisole [2845-89-8]	25	-2.78	Lun et al., 1995
4-Chloroanisole [623-12-1]	25	-2.78	Lun et al., 1995
Chlorobenzilate [510-15-6]	20	-4.40	Führer and Geiger, 1977
2-Chlorobenzoic acid [118-91-2]	30	-1.87	Phatak and Gaikar, 1996
4-Chlorobenzoic acid [74-11-3]	30	-3.29	Phatak and Gaikar, 1996
2-Chlorobiphenyl [2051-60-7]	25	-4.57	Miller et al., 1984
4-Chlorobiphenyl [2051-62-9]	25	-5.15	Li and Andren, 1994
	25	-5.15	Li and Doucette, 1993
1-Chlorobutane [109-69-3]	25.0	-2.03	Tewari et al., 1982
3-Chloro-2-butanone [4091-39-8]	0	-0.51	Stephenson, 1992*
	9.7	-0.56	Stephenson, 1992*
	19.1	-0.58	Stephenson, 1992*
	41.1	-0.60	Stephenson, 1992*
	51.1	-0.61	Stephenson, 1992*
	61.1	-0.59	Stephenson, 1992*
	71.5	-0.57	Stephenson, 1992*
	91.8	-0.50	Stephenson, 1992*
1-Chlorodibenzo-p-dioxin [39227-53-7]	5	-6.21	Shiu et al., 1988
	15	-5.97	Shiu et al., 1988
	25	-5.72	Shiu et al., 1988
	35	-5.48	Shiu et al., 1988
	45	-5.25	Shiu et al., 1988
2-Chlorodibenzo-p-dioxin [39227-54-8]	3.9	-6.21	Doucette and Andren, 1988a
	5	-6.54	Shiu et al., 1988
	15	-6.20	Shiu et al., 1988

Compound [CAS No.]	Temp. °C	Log S mol/L	Reference
	25	-5.90	Shiu et al., 1988
	35	-5.52	Shiu et al., 1988
	45	-5.29	Shiu et al., 1988
	25.0	-5.84	Doucette and Andren, 1988a
	39.0	-5.46	Doucette and Andren, 1988a
4-Chloro-2,5-dimethylphenol [1124-06-7]	25	-1.20	pH 5.1, Blackman et al., 1955
4-Chloro-2,6-dimethylphenol [1123-63-3]	25	-2.44	pH 5.1, Blackman et al., 1955
4-Chloro-3,5-dimethylphenol [88-04-0]	25	-1.61	pH 5.1, Blackman et al., 1955
4-Chloro-3,5-dimethyl-2-phenylmethylphenol [11867-85-2]	25	-4.27	pH 5.1, Blackman et al., 1955
2-Chloroethyl methyl ether [627-42-9]	0	0.01	Stephenson, 1992*
	9.8	-0.05	Stephenson, 1992*
	20.2	-0.08	Stephenson, 1992*
	29.7	-0.10	Stephenson, 1992*
	39.7	-0.14	Stephenson, 1992*
	50.0	-0.18	Stephenson, 1992*
	70.2	-0.18	Stephenson, 1992*
	80.8	-0.18	Stephenson, 1992*
4-Chloroguaiacol [16766-30-6]	10.0	-1.56	Larachi et al., 2000
	15.0	-1.54	Larachi et al., 2000
	19.0	-1.51	Larachi et al., 2000
	25	-1.45	Tam et al., 1994
	36.3	-1.47	Larachi et al., 2000
1-Chloroheptane [629-06-1]	25.0	-4.00	Tewari et al., 1982
2-Chloroiodobenzene [615-41-8]	25	-3.54	Yalkowsky et al., 1979
3-Chloroiodobenzene [625-99-0]	25	-3.55	Yalkowsky et al., 1979
4-Chloroiodobenzene [637-87-6]	25	-4.03	Yalkowsky et al., 1979
Chloromethiuron [28217-97-2]	20	-3.66	Worthing and Hance, 1991
4-Chloro-3-methyl-5-ethylphenol [1125-66-2]	25	-2.64	pH 5.1, Blackman et al., 1955
2-Chloro-6-methylphenol [87-64-9]	25	-1.60	pH 5.1, Blackman et al., 1955
4-Chloro-2-methylphenol [1570-64-5]	25	-1.32	pH 5.1, Blackman et al., 1955
4-Chloro-3-methylphenol [59-50-7]	25	-1.55	pH 5.1, Blackman et al., 1955
Chloroneb [2675-77-6]	25	-4.41	Worthing and Hance, 1991
2-Chloronitrobenzene [88-73-3]	20	-2.55	Eckert, 1962
3-Chloronitrobenzene [121-73-3]	20	-2.76	Eckert, 1962
4-Chloronitrobenzene [100-00-5]	20	-2.38	Hafkenscheid and Tomlinson, 1983
	20	-2.54	Eckert, 1962
4-Chloro-2-nitrophenol [89-64-5]	20.0	-3.09	Schwarzenbach et al., 1988
6-Chloro-3-nitrotoluene [7147-89-9]	20.0	-4.39	Schwarzenbach et al., 1988
5-Chloro-2-pentanone [5891-21-4]	0	-0.11	Stephenson, 1992*
	22.3	-0.41	Stephenson, 1992*
	30.8	-0.47	Stephenson, 1992*
	50.1	-0.44	Stephenson, 1992*
	60.7	-0.25	Stephenson, 1992*
	70.7	0.05	Stephenson, 1992*
3-Chlorophenol [7732-18-5]	20	-0.76	Mulley and Metcalf, 1966
4-Chlorophenol [106-48-9]	15.1	-0.74	Achard et al., 1996
	20	-0.69	Mulley and Metcalf, 1966
	20	-0.99	Hafkenscheid and Tomlinson, 1983
	25	-0.68	pH 5.1, Blackman et al., 1955
	25.2	-0.70	Achard et al., 1996
4-Chlorophenoxyacetic acid [122-48-3]	25	-2.69	Bailey and White, 1965
Chloropropylate [5836-10-2]	20	-5.35	Fühner and Geiger, 1977
2-Chloropyridine [109-09-1]	0	-0.50	Stephenson, 1993*
	10.0	-0.57	Stephenson, 1993*
	20.0	-0.61	Stephenson, 1993*
	30.0	-0.65	Stephenson, 1993*
	40.0	-0.66	Stephenson, 1993*
	50.0	-0.67	Stephenson, 1993*
	60.0	-0.66	Stephenson, 1993*
	70.0	-0.65	Stephenson, 1993*
	80.0	-0.64	Stephenson, 1993*
	90.0	-0.60	Stephenson, 1993*
Chlorothalonil [1897-45-6]	25	-5.65	Worthing and Hance, 1991

Compound [CAS No.]	Temp. °C	Log S mol/L	Reference
4-Chlorotoluene [106-43-4]	20	-2.97	Hafkenscheid and Tomlinson, 1983
	20	-3.08	Hafkenscheid and Tomlinson, 1981
Chlorotoluron [15545-48-9]	5	-3.58	Madhun et al., 1986
	20	-3.48	Worthing and Hance, 1991
	25	-3.42	Madhun et al., 1986
	40	-3.33	Madhun et al., 1986
Chloroxuron [1982-47-4]	20	-4.86	Worthing and Hance, 1991
Chlorpropham [101-21-3]	25	-3.39	Humburg et al., 1989
Chlorprothixene [113-59-7]	20	-4.41	Hafkenscheid and Tomlinson, 1983
Chlorpyrifos-methyl [5598-13-0]	24	-4.91	Worthing and Hance, 1991
Chlorsulfuron [64902-72-3]	25	-3.08	pH 5, Humburg et al., 1989
	25	-1.11	pH 7, Humburg et al., 1989
Chlorthal-dimethyl [1861-32-1]	25	-5.82	Humburg et al., 1989
Cholanthrene [479-23-2]	27	-7.86	Davis et al., 1942
Chlozolinate [72391-46-9]	25	-4.02	Worthing and Hance, 1991
Cinmethylin [87818-31-3]	20	-3.64	Humburg et al., 1989
Cinnamic acid [621-82-9]	30.00	-2.39	Gross et al., 1933
Cinosulfuron [94593-91-6]	20	-4.36	Worthing and Hance, 1991
Cloethocarb [51487-69-5]	20	-2.30	Worthing and Hance, 1991
Clomazone [81777-89-1]	25	-2.34	Humburg et al., 1989
Clomeprop [84496-56-0]	25	-7.01	Worthing and Hance, 1991
Clopyralid [1702-17-6]	25	-2.28	Humburg et al., 1989
Coronene [191-07-1]	25	-9.48	Billington et al., 1988
	25	-9.33	Mackay and Shiu, 1977
Coumachlor [81-82-3]	20	-5.84	Worthing and Hance, 1991
Coumatetralyl [5836-29-3]	20	-4.86	Worthing and Hance, 1991
Crotonaldehyde [123-73-9]	0	0.43	Stephenson, 1993b*
	10.0	0.41	Stephenson, 1993b*
	20.0	0.40	Stephenson, 1993b*
	30.0	0.34	Stephenson, 1993b*
	40.0	0.31	Stephenson, 1993b*
	50.0	0.32	Stephenson, 1993b*
	60.0	0.32	Stephenson, 1993b*
	70.0	0.32	Stephenson, 1993b*
	80.0	0.31	Stephenson, 1993b*
	90.0	0.36	Stephenson, 1993b*
Cupric hydroxide [20427-59-2]	25	-4.53	Worthing and Hance, 1991
Cyanazine [21725-46-2]	23	-3.18	Humburg et al., 1989
	25	-3.15	Humburg et al., 1989
Cyanophos [2636-26-2]	30	-3.72	Worthing and Hance, 1991
Cycloate [1134-23-2]	22	-3.40	Humburg et al., 1989
Cycloheptanol [502-41-0]	25.0	-0.89	Stephenson and Stuart, 1986*
Cycloheptanone [502-42-1]	0	-0.33	Stephenson, 1992*
	9.5	-0.41	Stephenson, 1992*
	19.8	-0.49	Stephenson, 1992*
	31.0	-0.56	Stephenson, 1992*
	39.8	-0.62	Stephenson, 1992*
	50.2	-0.62	Stephenson, 1992*
	60.8	-0.62	Stephenson, 1992*
	70.7	-0.63	Stephenson, 1992*
	81.3	-0.61	Stephenson, 1992*
	91.9	-0.60	Stephenson, 1992*
Cycloheptatriene [544-25-2]	25	-2.17	McAuliffe, 1966
Cycloheptene [628-92-2]	25	-3.16	McAuliffe, 1966
1,4-Cyclohexadiene [628-41-1]	25	-2.06	McAuliffe, 1966
Cyclohexyl acetate [622-45-7]	25.0	-1.71	Stephenson and Stuart, 1986*
Cyclohexyl butyrate [1551-44-6]	0	-2.23	Stephenson, 1992*
	9.8	-1.98	Stephenson, 1992*
	19.7	-2.19	Stephenson, 1992*
	29.8	-1.95	Stephenson, 1992*
	39.8	-1.87	Stephenson, 1992*
	50.1	-2.09	Stephenson, 1992*
	60.2	-2.19	Stephenson, 1992*
	70.4	-2.12	Stephenson, 1992*

Compound [CAS No.]	Temp. °C	Log S mol/L	Reference
	80.3	-2.33	Stephenson, 1992*
	90.5	-2.28	Stephenson, 1992*
Cyclohexyl formate [4351-54-6]	25.0	-1.19	Stephenson and Stuart, 1986*
Cyclohexylmethanol [100-49-2]	25.0	-1.14	Stephenson and Stuart, 1986*
Cyclohexyl propionate [6222-35-1]	25.0	-2.31	Stephenson and Stuart, 1986*
Cyclooctane [292-64-8]	25	-4.15	McAuliffe, 1966
Cyclooctanol [696-71-9]	25.0	-1.30	Stephenson and Stuart, 1986*
Cyclopentanol [96-41-3]	25.0	0.07	Stephenson and Stuart, 1986*
Cyclopentanone [120-92-3]	0	0.65	Stephenson, 1992*
	10.0	0.61	Stephenson, 1992*
	20.1	0.57	Stephenson, 1992*
	30.0	0.52	Stephenson, 1992*
	40.2	0.49	Stephenson, 1992*
	50.0	0.46	Stephenson, 1992*
	60.6	0.45	Stephenson, 1992*
	70.5	0.45	Stephenson, 1992*
	80.0	0.47	Stephenson, 1992*
	90.7	0.49	Stephenson, 1992*
Cycloprothrin [63935-38-6]	25	-6.72	Worthing and Hance, 1991
Cycloxydim [101205-02-1]	20	-3.58	Humburg et al., 1989
β-Cyfluthrin [68359-37-5]	20	-8.34	Worthing and Hance, 1991
Cyhalothrin [68085-85-8]	20	-8.18	Worthing and Hance, 1991
Cymoxanil [57966-95-7]	25	-2.30	Worthing and Hance, 1991
Cyproconazole [113096-99-4]	25	-3.32	Worthing and Hance, 1991
Cyromazine [66215-27-8]	20	-1.18	Worthing and Hance, 1991
L-Cystine [58-89-3]	25	-3.16	pH 7.0, Carta and Tola, 1996
Dalapon-sodium [127-20-8]	25	0.75	Worthing and Hance, 1991
Daminozide [1596-84-5]	25	-0.20	Worthing and Hance, 1991
Dazomet [533-74-4]	25	-1.73	Worthing and Hance, 1991
2,4-DB [10433-59-7]	25	-3.73	Worthing and Hance, 1991
o,p'-DDE [3424-92-6]	25	-6.36	Biggar and Riggs, 1974
o,p'-DDT [789-02-6]	15	-6.85	Biggar and Riggs, 1974
	25	-6.62	Biggar and Riggs, 1974
	35	-6.42	Biggar and Riggs, 1974
	45	-6.25	Biggar and Riggs, 1974
Decachlorobiphenyl [2051-24-3]	22.0	-10.38	Opperhuizen et al., 1988
	25	-11.89	Dickhut et al., 1986
	25	-10.83	Miller et al., 1984
	25	-10.80	Paschkle et al., 1998
Decanal [112-31-2]	0	-2.24	Stephenson, 1993b*
	10.0	-2.72	Stephenson, 1993b*
	20.0	-2.72	Stephenson, 1993b*
	30.0	-2.89	Stephenson, 1993b*
	40.0	-2.89	Stephenson, 1993b*
	50.0	-2.89	Stephenson, 1993b*
	60.0	-2.89	Stephenson, 1993b*
	70.0	-2.72	Stephenson, 1993b*
	80.0	-2.72	Stephenson, 1993b*
	90.0	-2.59	Stephenson, 1993b*
	90.0	-2.59	Stephenson, 1993b*
Decanoic acid [334-48-5]	20.0	-3.06	Ralston and Hoerr, 1942
1-Decanol [112-30-1]	20.0	-3.57	Hommelen, 1959
	25	-3.50	Stearns et al., 1947
	25.0	-3.64	Etzweiler et al., 1995
	29.6	-2.88	Stephenson and Stuart, 1986*
2-Decanone [693-54-9]	25.0	-3.30	Tewari et al., 1982
4-Decanone [624-16-8]	0	-2.60	Stephenson, 1992*
	9.2	-2.72	Stephenson, 1992*
	19.7	-2.82	Stephenson, 1992*
	30.2	-2.94	Stephenson, 1992*
	39.6	-3.00	Stephenson, 1992*
	50.2	-3.20	Stephenson, 1992*
	70.6	-3.25	Stephenson, 1992*
	80.2	-3.39	Stephenson, 1992*

Compound [CAS No.]	Temp. °C	Log S mol/L	Reference
	91.5	-3.10	Stephenson, 1992*
1-Decene [872-05-9]	15	-4.09[a]	Natarajan and Venkatachalam, 1972
	20	-4.21[a]	Natarajan and Venkatachalam, 1972
	25	-4.39[a]	Natarajan and Venkatachalam, 1972
Decylamine [2016-57-1]	20.0	-1.56	Stephenson, 1993a*
Decylbenzene [104-72-3]	25.0	-7.94	Sherblom et al., 1992
Deltamethrin [52918-63-5]	20	-8.40	Worthing and Hance, 1991
Demeton-S-methyl [867-27-6]	20	-1.84	Worthing and Hance, 1991
Desmedipham [13684-56-5]	25	-4.63	Humburg et al., 1989
Desmetryne [1014-69-3]	20	-2.55	Führner and Geiger, 1977
Diafenthiuron [80060-09-9]	20	-6.89	Worthing and Hance, 1991
Diallate [2303-16-4]	25	-4.29	Humburg et al., 1989
Diallylamine [124-02-7]	6.0	0.38	Stephenson, 1993a*
	8.0	0.33	Stephenson, 1993a*
	10.0	0.22	Stephenson, 1993a*
	20.0	0.08	Stephenson, 1993a*
	30.0	0.01	Stephenson, 1993a*
	40.0	-0.19	Stephenson, 1993a*
	50.0	-0.29	Stephenson, 1993a*
	60.0	-0.33	Stephenson, 1993a*
	70.0	-0.39	Stephenson, 1993a*
	80.0	-0.42	Stephenson, 1993a*
	90.0	-0.44	Stephenson, 1993a*
Dially phthalate [131-17-9]	20	-3.13	Leyder and Boulanger, 1983
Diazinon [333-41-5]	25	-2.47	Somasundaram et al., 1991
Dicamba [1918-00-9]	25	-1.53	Worthing and Hance, 1991
Dicamba-dimethylammonium [2300-66-5]	25	-0.57	Humburg et al., 1989
1,2,3,4-Dibenzanthracene [215-58-7]	25	-8.24	Billington et al., 1988
1,2,7,8-Dibenzocarbazole [239-64-5]	25	-7.41	Banwart et al., 1982
Dibenzo-p-dioxin [262-12-4]	4.1	-5.82	Doucette and Andren, 1988a
	25.0	-5.31	Doucette and Andren, 1988a
	40.0	-4.89	Doucette and Andren, 1988a
	25	-5.34	Shiu et al., 1988
1,2-Dibromobenzene [583-53-9]	25	-3.50	Yalkowsky et al., 1979
1,3-Dibromobenzene [108-36-1]	35.0	-3.54	Hine et al., 1963
	25	-3.38	Yalkowsky et al., 1979
Dibromomethane [74-95-3]	0	-1.17	Stephenson, 1992*
	9.7	-1.19	Stephenson, 1992*
	19.3	-1.13	Stephenson, 1992*
	29.5	-1.18	Stephenson, 1992*
	30	-1.16	Gross and Saylor, 1931
	39.5	-1.16	Stephenson, 1992*
	49.5	-1.14	Stephenson, 1992*
	59.9	-1.11	Stephenson, 1992*
	69.9	-1.11	Stephenson, 1992*
	79.8	-1.03	Stephenson, 1992*
	90.1	-1.06	Stephenson, 1992*
1,3-Dibromopropane [109-64-8]	30.00	-2.08	Gross et al., 1933
Dibutylamine [111-92-2]	0	-1.03	Stephenson, 1993a*
	10.0	-1.20	Stephenson, 1993a*
	20.0	-1.36	Stephenson, 1993a*
	30.0	-1.48	Stephenson, 1993a*
	40.0	-1.57	Stephenson, 1993a*
	50.0	-1.61	Stephenson, 1993a*
	60.0	-1.75	Stephenson, 1993a*
	70.0	-1.73	Stephenson, 1993a*
	80.0	-1.77	Stephenson, 1993a*
	90.0	-1.77	Stephenson, 1993a*
Di-sec-butylamine [626-23-3]	0	-0.85	Stephenson, 1993a*
	10.0	-1.10	Stephenson, 1993a*
	20.0	-1.32	Stephenson, 1993a*
	30.0	-1.51	Stephenson, 1993a*
	40.0	-1.67	Stephenson, 1993a*
	50.0	-1.79	Stephenson, 1993a*

Compound [CAS No.]	Temp. °C	Log S mol/L	Reference
	60.0	-1.89	Stephenson, 1993a*
	70.0	-1.92	Stephenson, 1993a*
	80.0	-2.16	Stephenson, 1993a*
	90.0	-2.61	Stephenson, 1993a*
Dicapthon [2463-84-5]	20	-4.68	Chiou et al., 1977
Dichlobenil [1194-65-6]	20	-3.98	Humburg et al., 1989
Dichlofluanid [1085-98-9]	20	-5.41	Worthing and Hance, 1991
Dichlone [117-80-6]	25	-6.36	Worthing and Hance, 1991
Dichlormid [37764-25-3]	20	-1.62	Humburg et al., 1989
3,5-Dichloroaniline [626-43-7]	23	-2.37	Fu and Luthy, 1986
2,3-Dichloroanisole [1984-59-4]	25	-3.31	Lun et al., 1995
2,6-Dichloroanisole [1984-65-2]	25	-3.10	Lun et al., 1995
2,2'-Dichlorobiphenyl [13029-08-8]	22	-6.61	Jepsen and Lick, 1999
	25	-5.27	Dunnivant and Elzerman, 1988
	25	-5.45	Dunnivant and Elzerman, 1988
2,4-Dichlorobiphenyl [33284-50-3]	25	-5.29	Dunnivant and Elzerman, 1988
2,5-Dichlorobiphenyl [34883-39-1]	25	-5.30	Dunnivant and Elzerman, 1988
	25	-5.06	Miller et al., 1984
	25	-5.59	Dunnivant and Elzerman, 1988
2,6-Dichlorobiphenyl [33146-45-1]	22.0	-5.62	Opperhuizen et al., 1988
	25	-4.97	Dunnivant and Elzerman, 1988
	25	-5.21	Miller et al., 1984
3,3'-Dichlorobiphenyl [2050-67-1]	25	-5.80	Dunnivant and Elzerman, 1988
3,4-Dichlorobiphenyl [2974-92-7]	25	-7.45	Dunnivant and Elzerman, 1988
3,5-Dichlorobiphenyl [31883-41-5]	25	-6.37	Dulfer et al., 1995
4,4'-Dichlorobiphenyl [2050-08-2]	20	-6.56	Freed et al., 1979
	25	-6.56	Chiou et al., 1977
	25	-6.32	Dulfer et al., 1995
2,4-Dichloro-6-butylphenol [91399-13-2]	25	-3.62	pH 5.1, Blackman et al., 1955
2,3-Dichlorodibenzo-p-dioxin [29446-15-9]	5	-7.84	Shiu et al., 1988
	15	-7.55	Shiu et al., 1988
	25	-7.23	Shiu et al., 1988
	35	-6.92	Shiu et al., 1988
	45	-6.64	Shiu et al., 1988
2,7-Dichlorodibenzo-p-dioxin [33857-26-0]	5	-8.37	Shiu et al., 1988
	15	-8.10	Shiu et al., 1988
	25	-8.05	Santl et al., 1994
	25	-7.83	Shiu et al., 1988
	35	-7.54	Shiu et al., 1988
	45	-7.28	Shiu et al., 1988
2,8-Dichlorodibenzofuran [5409-83-6]	4.5	-7.79	Doucette and Andren, 1988a
	5	-7.76	Shiu et al., 1988
	15	-7.47	Shiu et al., 1988
	25	-7.21	Doucette and Andren, 1988a
	25	-7.18	Shiu et al., 1988
	35	-6.96	Shiu et al., 1988
	39.5	-6.84	Doucette and Andren, 1988a
	45	-6.69	Shiu et al., 1988
cis-1,2-Dichloroethylene [156-59-2]	25	-1.44	McGovern, 1943
2,4-Dichloro-6-ethylphenol [24539-94-4]	25	-2.89	pH 5.1, Blackman et al., 1955
2,4-Dichlroo-6-methylphenol [1570-65-6]	25	-2.80	pH 5.1, Blackman et al., 1955
2,6-Dichloro-4-methylphenol [2432-12-4]	25	-2.42	pH 5.1, Blackman et al., 1955
2,3-Dichloronitrobenzene [3209-22-1]	20	-3.49	Eckert, 1962
2,5-Dichloronitrobenzene [89-61-2]	20	-3.32	Eckert, 1962
	25	-6.79	Dunnivant and Elzerman, 1988
	25	-6.60	Dunnivant and Elzerman, 1988
3,4-Dichloronitrobenzene [99-54-7]	20	-3.20	Eckert, 1962
Dichlorophen [97-23-4]	25	-3.95	Worthing and Hance, 1991
2,4-Dichloro-6-phenylmethylphenol [19578-81-5]	25	-4.64	pH 5.1, Blackman et al., 1955
2,4-Dichloro-6-propylphenol [91399-12-1]	25	-3.31	pH 5.1, Blackman et al., 1955
3,6-Dichloropyridazine [141-30-0]	25	1.17	Liao et al., 1996
3,6-Dichloropyrimidine [3934-20-1]	25	1.66	Liao et al., 1996
4,5-Dichloroveratrole [2772-46-5]	25	-3.46	Lun et al., 1995
Dichlorprop [120-36-5]	20	-2.83	Worthing and Hance, 1991

Compound [CAS No.]	Temp. °C	Log S mol/L	Reference
Dichlorprop-P [15165-67-0]	20	-2.60	Worthing and Hance, 1991
Diclofop-methyl [51338-27-3]	22	-2.04	Humburg et al., 1989
Diclomezine [62865-36-5]	25	-5.54	Worthing and Hance, 1991
Dicloran [99-30-9]	20	-4.52	Worthing and Hance, 1991
Dicofol [115-32-2]	20	-5.67	Walsh and Hites, 1979
Dicyclohexyl phthalate [84-61-7]	24	-4.92	Hollifield, 1979
Di-n-decyl phthalate [2432-90-8]	24	-6.18	Hollifield, 1979
Diethatyl-ethyl [58727-55-8]	25	-3.47	Humburg et al., 1989
Diethofencarb [87130-20-9]	20	-4.00	Worthing and Hance, 1991
Diethyl carbonate [105-58-8]	25.0	-0.70	Stephenson and Stuart, 1986*
Diethyl glutarate [818-38-2]	0	-1.09	Stephenson, 1992*
	9.1	-1.19	Stephenson, 1992*
	20.3	-1.27	Stephenson, 1992*
	29.7	-1.20	Stephenson, 1992*
	39.7	-1.17	Stephenson, 1992*
	49.9	-1.30	Stephenson, 1992*
	70.6	-1.35	Stephenson, 1992*
	80.7	-1.32	Stephenson, 1992*
	90.6	-1.32	Stephenson, 1992*
Diethyl maleate [141-05-9]	0	-0.95	Stephenson, 1992*
	9.1	-0.98	Stephenson, 1992*
	19.5	-1.04	Stephenson, 1992*
	29.5	-1.08	Stephenson, 1992*
	40.8	-1.09	Stephenson, 1992*
	50.5	- 1.09	Stephenson, 1992*
	60.7	-1.06	Stephenson, 1992*
	71.1	-1.04	Stephenson, 1992*
	80.8	-1.01	Stephenson, 1992*
	90.9	-0.99	Stephenson, 1992*
N,N-Diethylmethylamine [616-39-7]	52.0	0.34	Stephenson, 1993a*
	54.0	0.23	Stephenson, 1993a*
	56.0	0.17	Stephenson, 1993a*
	58.0	0.11	Stephenson, 1993a*
	60.0	0.08	Stephenson, 1993a*
	62.0	0.02	Stephenson, 1993a*
	64.0	-0.04	Stephenson, 1993a*
Diethyl oxalate [95-92-1]	25.0	-0.05	Stephenson and Stuart, 1986*
Difenoconazole [119446-68-3]	20	-5.09	Worthing and Hance, 1991
Difenoxuron [14214-32-5]	20	-4.16	Worthing and Hance, 1991
Difenzoquat methyl sulfate [43222-48-6]	23	0.33	Humburg et al., 1989
Diflubenzuron [35367-38-5]	20	-6.49	Worthing and Hance, 1991
1,2-Difluorobenzene [367-11-3]	25	-2.00	Yalkowsky et al., 1979
1,3-Difluorobenzene [372-18-9]	25	-2.00	Yalkowsky et al., 1979
1,4-Difluorobenzene [540-36-3]	25	-1.97	Yalkowsky et al., 1979
3,6-Dihydroxylpyrimidine [1193-24-4]	25	1.63	Liao et al., 1996
3,4-Dihydropyran [110-87-2]	0	-0.93	Stephenson, 1992*
	9.5	-0.86	Stephenson, 1992*
	19.7	-0.91	Stephenson, 1992*
	31.2	-0.92	Stephenson, 1992*
	41.1	-0.92	Stephenson, 1992*
	51.1	-0.84	Stephenson, 1992*
	61.9	-0.80	Stephenson, 1992*
	71.5	-0.75	Stephenson, 1992*
	81.6	-0.57	Stephenson, 1992*
2,5-Dihydroxybenzaldehyde [1194-98-5]	25	-1.00	Jin et al., 1998
3,4-Dihydroxybenzaldehyde [139-85-5]	25	-1.34	Jin et al., 1998
1,2-Diiodobenzene [615-42-9]	25	-4.24	Yalkowsky et al., 1979
1,4-Diiodobenzene [624-38-4]	25	-5.25	Yalkowsky et al., 1979
	25.0	-5.50	Andrews and Keefer, 1950
Diiodomethane [75-11-6]	30	-2.33	Gross and Saylor, 1931
Diisobutylamine [110-96-3]	0	-1.22	Stephenson, 1993a*
	10.0	-1.50	Stephenson, 1993a*
	20.0	-1.68	Stephenson, 1993a*
	30.0	-1.81	Stephenson, 1993a*

Compound [CAS No.]	Temp. °C	Log S mol/L	Reference
	40.0	-2.00	Stephenson, 1993a*
	50.0	-1.91	Stephenson, 1993a*
	60.0	-1.94	Stephenson, 1993a*
	70.0	-2.03	Stephenson, 1993a*
	80.0	-2.00	Stephenson, 1993a*
	90.0	-1.94	Stephenson, 1993a*
Diisobutyl phthalate [84-69-5]	20	-4.14	Leyder and Boulanger, 1983
	24	-4.65	Hollifield, 1979
N,N-Diisopropylethylamine [7087-68-5]	0	-1.17	Stephenson, 1993a*
	20.0	-1.56	Stephenson, 1993a*
	30.0	-1.77	Stephenson, 1993a*
	40.0	-1.88	Stephenson, 1993a*
	50.0	-1.91	Stephenson, 1993a*
	60.0	-1.96	Stephenson, 1993a*
	70.0	-2.03	Stephenson, 1993a*
	80.0	-2.00	Stephenson, 1993a*
	90.0	-2.21	Stephenson, 1993a*
Diisopropyl phthalate [605-45-8]	20	-2.88	Leyder and Boulanger, 1983
Dikegulac [18467-77-1]	25	0.33	Worthing and Hance, 1991
Dimepiperate [61432-55-1]	25	-4.12	Worthing and Hance, 1991
Dimethachlor [50563-36-5]	20	-2.09	Worthing and Hance, 1991
Dimethametryn [22936-75-0]	20	-3.71	Worthing and Hance, 1991
Dimethipin [55290-64-7]	25	-1.85	Worthing and Hance, 1991
Dimethirimol [5221-53-4]	25	-2.24	Worthing and Hance, 1991
Dimethoate [60-51-5]	21	-0.96	Worthing and Hance, 1991
2,5-Dimethoxybenzaldehyde [93-02-7]	25	-2.32	Jin et al., 1998
3,4-Dimethoxybenzaldehyde [120-14-9]	25	-1.42	Jin et al., 1998
1,2-Dimethoxybenzene [91-16-7]	19.9	-1.29	Stephenson, 1992*
	25	-1.31	Lun et al., 1995
	31.0	-1.29	Stephenson, 1992*
	41.1	-1.28	Stephenson, 1992*
	50.6	-1.26	Stephenson, 1992*
	60.2	-1.24	Stephenson, 1992*
	70.6	-1.20	Stephenson, 1992*
	80.7	-1.15	Stephenson, 1992*
	91.8	-1.11	Stephenson, 1992*
1,3-Dimethoxybenzene [151-10-1]	25	-2.06	McGowan et al., 1966
2,5-Dimethoxytetrahydrofuran [696-59-3]	0	0.42	Stephenson, 1992*
	9.3	0.40	Stephenson, 1992*
	21.0	0.38	Stephenson, 1992*
	31.0	0.36	Stephenson, 1992*
	40.7	0.31	Stephenson, 1992*
	50.1	0.28	Stephenson, 1992*
	61.1	0.24	Stephenson, 1992*
	71.3	0.20	Stephenson, 1992*
	81.0	0.18	Stephenson, 1992*
	90.2	0.16	Stephenson, 1992*
Dimethyl adipate [627-93-0]	25.0	-0.77	Stephenson and Stuart, 1986*
9,10-Dimethylanthracene [781-43-1]	25	-6.57	Sutton and Calder, 1975
	25	-5.87	Mackay and Shiu, 1977
Dimethylarsinic acid [75-60-5]	20	0.68	Humburg et al., 1989
	25	1.16	Worthing and Hance, 1991
7,12-Dimethylbenz[a]anthracene [57-97-6]	25	-7.02	Means et al., 1980
	25	-6.62	Mackay and Shiu, 1977
9,10-Dimethylbenz[a]anthracene [58429-99-5]	24	-6.67	Hollifield, 1979
	27	-6.78	Davis et al., 1942
4,4'-Dimethylbiphenyl [613-33-2]	4.0	-6.42	Doucette and Andren, 1988a
	25.0	-6.02	Doucette and Andren, 1988a
	40.0	-5.62	Doucette and Andren, 1988a
2,2-Dimethyl-1-butanol [1185-33-7]	20	-1.09	Ginnings and Webb, 1938
	25	-1.13	Ginnings and Webb, 1938
	30	-1.16	Ginnings and Webb, 1938
2,3-Dimethyl-2-butanol [594-60-5]	20	-0.34	Ginnings and Webb, 1938
	25	-0.39	Ginnings and Webb, 1938

Compound [CAS No.]	Temp. °C	Log S mol/L	Reference
	30	-0.43	Ginnings and Webb, 1938
3,3-Dimethyl-2-butanol [464-07-3]	20	-0.59	Ginnings and Webb, 1938
	25	-0.62	Ginnings and Webb, 1938
	25.0	-0.68	Stephenson and Stuart, 1986*
	30	-0.65	Ginnings and Webb, 1938
3,3-Dimethyl-2-butanone [75-97-8]	0	-0.53	Stephenson, 1992*
	9.5	-0.62	Stephenson, 1992*
	19.2	-0.71	Stephenson, 1992*
	25	-0.72	Ginnings et al., 1940
	31.0	-0.77	Stephenson, 1992*
	39.5	-0.81	Stephenson, 1992*
	50.2	-0.83	Stephenson, 1992*
	59.8	-0.86	Stephenson, 1992*
	70.4	-0.86	Stephenson, 1992*
	80.1	-0.88	Stephenson, 1992*
	90.2	-0.94	Stephenson, 1992*
N,N-Dimethylbutylamine [927-62-8]	3.8	-0.07	Stephenson, 1993a*
	6.0	-0.30	Stephenson, 1993a*
	10.0	-0.31	Stephenson, 1993a*
	20.0	-0.46	Stephenson, 1993a*
	30.0	-0.61	Stephenson, 1993a*
	40.0	-0.83	Stephenson, 1993a*
	50.0	-0.96	Stephenson, 1993a*
	60.0	-1.08	Stephenson, 1993a*
	70.0	-1.11	Stephenson, 1993a*
	80.0	-1.26	Stephenson, 1993a*
	90.0	-1.27	Stephenson, 1993a*
Dimethyl carbonate [616-38-6]	25.0	0.03	Stephenson and Stuart, 1986*
5,6-Dimethylchrysene [3697-27-6]	27	-7.01	Davis et al., 1942
2,6-Dimethylcyclohexanol [5337-72-4]	25.0	-1.46	Stephenson and Stuart, 1986*
Dimethyl glutarate [1119-40-0]	25.0	-0.43	Stephenson and Stuart, 1986*
2,6-Dimethyl-4-heptanol [108-82-7]	25.0	-2.35	Stephenson and Stuart, 1986*
	25	-2.16	Crittenden and Hixon, 1954
(1,5-Dimethylhexyl)amine [543-82-8]	0	-1.50	Stephenson, 1993a*
	20.0	-1.68	Stephenson, 1993a*
	30.0	-1.71	Stephenson, 1993a*
	40.0	-1.68	Stephenson, 1993a*
	50.0	-1.77	Stephenson, 1993a*
	60.0	-1.86	Stephenson, 1993a*
	70.0	-1.86	Stephenson, 1993a*
	80.0	-1.83	Stephenson, 1993a*
	90.0	-1.88	Stephenson, 1993a*
Dimethyl maleate [624-48-6]	25.0	-0.27	Stephenson and Stuart, 1986*
Dimethyl malonate [108-59-8]	0	0.02	Stephenson, 1992*
	9.2	0.05	Stephenson, 1992*
	19.4	0.05	Stephenson, 1992*
	29.7	0.09	Stephenson, 1992*
	39.5	0.09	Stephenson, 1992*
	50.2	0.13	Stephenson, 1992*
	61.0	0.18	Stephenson, 1992*
	70.3	0.25	Stephenson, 1992*
	79.8	0.25	Stephenson, 1992*
	90.2	0.35	Stephenson, 1992*
1,3-Dimethylnaphthalene [575-41-7]	25	-3.79	Mackay and Shiu, 1977
1,4-Dimethylnaphthalene [571-58-4]	25	-4.29	Mackay and Shiu, 1977
1,5-Dimethylnaphthalene [571-61-9]	25.0	-4.74	Eganhouse and Calder, 1976
	25	-4.14	Mackay and Shiu, 1977
N,N-Dimethyloctylamine [7378-99-6]	0	-2.40	Stephenson, 1993a*
	10.0	-2.55	Stephenson, 1993a*
	20.0	-2.64	Stephenson, 1993a*
2,2-Dimethylpentane [590-35-2]	25	-4.36	Price, 1976
2,2-Dimethyl-3-pentanol [3970-62-5]	20	-1.12	Ginnings and Hauser, 1938
	25	-1.15	Ginnings and Hauser, 1938
	30	-1.17	Ginnings and Hauser, 1938

Compound [CAS No.]	Temp. °C	Log S mol/L	Reference
2,3-Dimethyl-2-pentanol [4911-70-0]	20	-0.84	Ginnings and Hauser, 1938
	25	-0.88	Ginnings and Hauser, 1938
	30	-0.92	Ginnings and Hauser, 1938
2,3-Dimethyl-3-pentanol [595-41-5]	20	-0.79	Ginnings and Hauser, 1938
	25	-0.85	Ginnings and Hauser, 1938
	30	-0.91	Ginnings and Hauser, 1938
2,4-Dimethyl-2-pentanol [625-06-9]	20	-0.89	Ginnings and Hauser, 1938
	25	-0.94	Ginnings and Hauser, 1938
	30	-0.98	Ginnings and Hauser, 1938
2,4-Dimethyl-3-pentanol [600-36-2]	0	-0.99	Stephenson et al., 1984*
	10.0	-1.04	Stephenson et al., 1984*
	20	-1.17	Ginnings and Hauser, 1938
	20.2	-1.21	Stephenson et al., 1984*
	25	-1.22	Ginnings and Hauser, 1938
	30	-1.24	Ginnings and Hauser, 1938
	30.6	-1.26	Stephenson et al., 1984*
	39.5	-1.34	Stephenson et al., 1984*
	49.7	-1.36	Stephenson et al., 1984*
	60.3	-1.40	Stephenson et al., 1984*
	70.2	-1.42	Stephenson et al., 1984*
	80.2	-1.43	Stephenson et al., 1984*
	90.6	-1.41	Stephenson et al., 1984*
2,4-Dimethyl-3-pentanone [565-80-0]	0	-1.17	Stephenson, 1992*
	9.3	-1.31	Stephenson, 1992*
	20.0	-1.34	Stephenson, 1992*
	25	-1.30	Ginnings et al., 1940
	29.7	-1.44	Stephenson, 1992*
	39.6	-1.50	Stephenson, 1992*
	50.5	-1.51	Stephenson, 1992*
	60.6	-1.54	Stephenson, 1992*
	70.2	-1.55	Stephenson, 1992*
	80.2	-1.55	Stephenson, 1992*
	90.3	-1.58	Stephenson, 1992*
3,5-Dimethylphenol [106-68-9]	25	-1.37	Southworth and Keller, 1986
2,6-Dimethylpiperidine [504-03-0]	27.0	0.02	Stephenson, 1993*
	28.0	-0.09	Stephenson, 1993*
	29.0	-0.11	Stephenson, 1993*
	30.0	-0.14	Stephenson, 1993*
	31.0	-0.20	Stephenson, 1993*
	32.0	-0.22	Stephenson, 1993*
	33.0	-0.24	Stephenson, 1993*
	34.0	-0.26	Stephenson, 1993*
	35.0	-0.29	Stephenson, 1993*
	36.0	-0.33	Stephenson, 1993*
	38.0	-0.31	Stephenson, 1993*
	40.0	-0.35	Stephenson, 1993*
	45.0	-0.45	Stephenson, 1993*
	50.0	-0.51	Stephenson, 1993*
	55.0	-0.57	Stephenson, 1993*
	60.0	-0.58	Stephenson, 1993*
	65.0	-0.63	Stephenson, 1993*
	70.0	-0.67	Stephenson, 1993*
	75.0	-0.72	Stephenson, 1993*
	80.0	-0.73	Stephenson, 1993*
	85.0	-0.76	Stephenson, 1993*
	92.0	-0.77	Stephenson, 1993*
3,3-Dimethylpiperidine [1193-12-0]	0	-0.21	Stephenson, 1993*
	10.0	-0.37	Stephenson, 1993*
	20.0	-0.46	Stephenson, 1993*
	30.0	-0.49	Stephenson, 1993*
	40.0	-0.58	Stephenson, 1993*
	50.0	-0.62	Stephenson, 1993*
	60.0	-0.68	Stephenson, 1993*
	70.0	-0.74	Stephenson, 1993*

Compound [CAS No.]	Temp. °C	Log S mol/L	Reference
	80.0	-0.77	Stephenson, 1993*
	90.0	-0.75	Stephenson, 1993*
3,5-Dimethylpiperidine [35794-11-7]	0	-0.47	Stephenson, 1993*
	10.0	-0.55	Stephenson, 1993*
	20.0	-0.66	Stephenson, 1993*
	30.0	-0.74	Stephenson, 1993*
	40.0	-0.82	Stephenson, 1993*
	50.0	-0.85	Stephenson, 1993*
	60.0	-0.88	Stephenson, 1993*
	70.0	-0.90	Stephenson, 1993*
	80.0	-0.90	Stephenson, 1993*
	90.0	-0.90	Stephenson, 1993*
2,2-Dimethyl-1-propanol [75-84-3]	12.0	-0.38	Stephenson et al., 1984*
	18.8	-0.40	Stephenson et al., 1984*
	20	-0.37	Ginnings and Baum, 1937
	25	-0.41	Ginnings and Baum, 1937
	30	-0.43	Ginnings and Baum, 1937
	30.0	-0.48	Stephenson et al., 1984*
	40.0	-0.51	Stephenson et al., 1984*
	50.0	-0.57	Stephenson et al., 1984*
	60.0	-0.55	Stephenson et al., 1984*
	70.2	-0.61	Stephenson et al., 1984*
	80.0	-0.53	Stephenson et al., 1984*
	90.0	-0.57	Stephenson et al., 1984*
2,3-Dimethylpyridine [583-61-9]	16.0	0.20	Stephenson, 1993*
	18.0	0.06	Stephenson, 1993*
	20.0	-0.01	Stephenson, 1993*
	25.0	-0.10	Stephenson, 1993*
	30.0	-0.19	Stephenson, 1993*
	35.0	-0.25	Stephenson, 1993*
	40.0	-0.30	Stephenson, 1993*
	50.0	-0.35	Stephenson, 1993*
	60.0	-0.39	Stephenson, 1993*
	70.0	-0.40	Stephenson, 1993*
	80.0	-0.39	Stephenson, 1993*
	90.0	-0.39	Stephenson, 1993*
2,4-Dimethylpyridine [108-47-4]	24.0	0.16	Stephenson, 1993*
	26.0	0.04	Stephenson, 1993*
	28.0	-0.02	Stephenson, 1993*
	30.0	-0.09	Stephenson, 1993*
	35.0	-0.19	Stephenson, 1993*
	40.0	-0.26	Stephenson, 1993*
	45.0	-0.32	Stephenson, 1993*
	50.0	-0.35	Stephenson, 1993*
	60.0	-0.40	Stephenson, 1993*
	70.0	-0.41	Stephenson, 1993*
	80.0	-0.42	Stephenson, 1993*
	90.0	-0.41	Stephenson, 1993*
2,5-Dimethylpyridine [589-93-5]	13.0	0.25	Stephenson, 1993*
	14.0	0.15	Stephenson, 1993*
	16.0	0.02	Stephenson, 1993*
	18.0	-0.04	Stephenson, 1993*
	20.0	-0.12	Stephenson, 1993*
	25.0	-0.21	Stephenson, 1993*
	30.0	-0.27	Stephenson, 1993*
	40.0	-0.34	Stephenson, 1993*
	50.0	-0.47	Stephenson, 1993*
	60.0	-0.48	Stephenson, 1993*
	70.0	-0.46	Stephenson, 1993*
	80.0	-0.49	Stephenson, 1993*
	90.0	-0.48	Stephenson, 1993*
2,6-Dimethylpyridine [108-48-5]	34.0	0.33	Stephenson, 1993*
	35.0	0.11	Stephenson, 1993*
	36.0	0.08	Stephenson, 1993*

Compound [CAS No.]	Temp. °C	Log S mol/L	Reference
	37.0	0.03	Stephenson, 1993*
	38.0	-0.03	Stephenson, 1993*
	39.0	-0.06	Stephenson, 1993*
	40.0	-0.07	Stephenson, 1993*
	41.0	-0.12	Stephenson, 1993*
	43.0	-0.16	Stephenson, 1993*
	45.0	-0.18	Stephenson, 1993*
	50.0	-0.29	Stephenson, 1993*
	60.0	-0.39	Stephenson, 1993*
	70.0	-0.44	Stephenson, 1993*
	80.0	-0.46	Stephenson, 1993*
	90.0	-0.47	Stephenson, 1993*
3,4-Dimethylpyridine [583-58-4]	0	-0.11	Stephenson, 1993*
	10.0	-0.24	Stephenson, 1993*
	20.0	-0.33	Stephenson, 1993*
	30.0	-0.40	Stephenson, 1993*
	40.0	-0.43	Stephenson, 1993*
	50.0	-0.45	Stephenson, 1993*
	60.0	-0.44	Stephenson, 1993*
	70.0	-0.43	Stephenson, 1993*
	80.0	-0.42	Stephenson, 1993*
	90.0	-0.39	Stephenson, 1993*
3,5-Dimethylpyridine [591-22-0]	0	-0.41	Stephenson, 1993*
	10.0	-0.49	Stephenson, 1993*
	20.0	-0.57	Stephenson, 1993*
	30.0	-0.63	Stephenson, 1993*
	40.0	-0.65	Stephenson, 1993*
	50.0	-0.67	Stephenson, 1993*
	60.0	-0.67	Stephenson, 1993*
	70.0	-0.65	Stephenson, 1993*
	80.0	-0.67	Stephenson, 1993*
	90.0	-0.65	Stephenson, 1993*
2,7-Dimethylquinoline [93-37-8]	25	-1.94	Price, 1976
Dimethyl succinate [106-65-0]	20.6	-0.07	Stephenson, 1992*
	30.1	-0.08	Stephenson, 1992*
	39.6	-0.05	Stephenson, 1992*
	50.0	-0.05	Stephenson, 1992*
	60.8	-0.03	Stephenson, 1992*
	70.5	-0.01	Stephenson, 1992*
	80.0	0.02	Stephenson, 1992*
	91.6	0.07	Stephenson, 1992*
Dimethyl sulfide [75-18-3]	25	-0.45	Hine and Weimar, 1965
2,5-Dimethyltetrahydrofuran [1003-38-9]	0	0.03	Stephenson, 1992*
	9.2	-0.10	Stephenson, 1992*
	19.5	-0.25	Stephenson, 1992*
	31.0	-0.39	Stephenson, 1992*
	40.8	-0.50	Stephenson, 1992*
	50.7	-0.59	Stephenson, 1992*
	61.5	-0.66	Stephenson, 1992*
	71.8	-0.72	Stephenson, 1992*
	83.0	-0.81	Stephenson, 1992*
Diniconazole [83657-24-3]	25	-4.91	Worthing and Hance, 1991
Dinitramine [29091-05-2]	25	-5.47	Worthing and Hance, 1991
2,5-Dinitrophenol [329-71-5]	20.0	-2.68	Schwarzenbach et al., 1988
Dinoseb [88-85-7]	25	-3.66	Humburg et al., 1989
Dioxabenzofos [3811-49-2]	20	-3.70	Worthing and Hance, 1991
Dioxacarb [6988-21-2]	20	-1.57	Worthing and Hance, 1991
Di-n-pentylamine [2050-92-2]	10.0	-2.18	Stephenson, 1993a*
	20.0	-2.29	Stephenson, 1993a*
	30.0	-2.60	Stephenson, 1993a*
Di-n-pentyl phthalate [131-18-0]	20	-5.84	Leyder and Boulanger, 1983
Diphenamide [957-51-7]	27	-2.96	Humburg et al., 1989
1,1-Diphenylethylene [530-48-3]	25.0	-4.52	Andrews and Keefer, 1950a
Diphenyl phosphate [838-85-7]	24	-2.97	Hollifield, 1979

Compound [CAS No.]	Temp. °C	Log S mol/L	Reference
Diphenyl phthalate [84-62-8]	24	-6.59	Hollifield, 1979
Diphenylamine [122-39-4]	20	-3.55	Hashimoto et al., 1982
Diphenylmethane [101-81-5]	24	-4.75	Hollifield, 1979
	25.0	-5.08	Andrews and Keefer, 1949
Diphenylmethyl phosphate [115-89-9]	24	-5.44	Hollifield, 1979
Dipropetryn [4147-51-7]	20	-4.20	Humburg et al., 1989
Dipropylamine [142-84-7]	0	0.11	Stephenson, 1993a*
	10.0	-0.12	Stephenson, 1993a*
	20.0	-0.28	Stephenson, 1993a*
	30.0	-0.48	Stephenson, 1993a*
	40.0	-0.58	Stephenson, 1993a*
	50.0	-0.71	Stephenson, 1993a*
	60.0	-0.79	Stephenson, 1993a*
	70.0	-0.84	Stephenson, 1993a*
	80.0	-0.94	Stephenson, 1993a*
	90.0	-0.99	Stephenson, 1993a*
Di-n-propyl phthalate [131-16-8]	20	-3.36	Leyder and Boulanger, 1983
Diquat dibromide [85-00-7]	20	0.31	Humburg et al., 1989
Disulfoton [298-04-4]	22	-4.36	Worthing and Hance, 1991
Ditridecyl phthalate [119-06-2]	24	-6.19	Hollifield, 1979
	25	-0.40	Ginnings and Baum, 1937
Diuron [330-54-1]	25	-3.74	Humburg et al., 1989
Dodecanedioic acid [693-23-2]	60	-3.28	Du Pont, 1999b*
	100	-1.76	Du Pont, 1999b*
Dodecanoic acid [143-07-7]	20.0	-3.56	Ralston and Hoerr, 1942
	25	-4.58	pH 5.7, Robb, 1966
1-Dodecanol [112-53-8]	25	-4.64	Robb, 1966
	29.5	-2.67	Stephenson and Stuart, 1986*
Dodecylamine [124-22-1]	40.0	-2.46	Stephenson, 1993a*
	70.0	-2.34	Stephenson, 1993a*
	80.0	-2.31	Stephenson, 1993a*
	90.0	-1.06	Stephenson, 1993a*
Dodine [2439-10-3]	25	-2.66	Worthing and Hance, 1991
Eglinazine [68228-19-3]	25	-2.89	Worthing and Hance, 1991
Eicosane [112-95-8]	25.0	-8.18	Sutton and Calder, 1974
Empenthrin [54406-48-3]	25	-5.06	Worthing and Hance, 1991
Endothal [145-73-3]	20	-0.27	Humburg et al., 1989
EPTC [759-94-4]	20	-2.71	Humburg et al., 1989
Esfenvalerate [66230-04-4]	25	-6.15	Worthing and Hance, 1991
Esprocarb [85785-20-2]	20	-4.73	Worthing and Hance, 1991
Etacelasil [37894-46-5]	20	-1.10	Worthing and Hance, 1991
Ethafluralin [55283-68-6]	25	-5.05	Humburg et al., 1989
1,2-Ethane dinitrate [628-96-6]	25	-0.82	Fischer and Ballschmiter, 1998, 1998a
Ethidimuron [30043-49-3]	20	-1.94	Worthing and Hance, 1991
Ethiofencarb [29973-13-5]	20	-2.10	Worthing and Hance, 1991
Ethirimol [23947-60-6]	20	-2.92	Worthing and Hance, 1991
Ethofumesate [26225-79-6]	25	-3.76	Worthing and Hance, 1991
3-Ethoxy-4-hydroxybenzaldehyde [121-32-4]	25	-1.77	Jin et al., 1998
Ethyl adipate [141-28-6]	20	-1.68	Sobotka and Kahn, 1931
	30	-1.68	Gross et al., 1933
2-Ethylanthracene [52251-71-5]	25.0	-6.78	Vadas et al., 1991
	25.3	-6.89	Whitehouse, 1984
10-Ethylbenz[a]anthracene [3697-30-1]	27	-6.81	Davis and Parke, 1942
Ethyl benzoate [93-89-0]	25.0	-2.40	Andrews and Keefer, 1950
	25.0	-2.26	Stephenson and Stuart, 1986*
2-Ethyl-1-butanol [97-95-0]	25.0	-1.06	Stephenson and Stuart, 1986*
	25	-1.41	Crittenden and Hixon, 1954
N-Ethylbutylamine [13360-63-9]	17.4	-0.23	Stephenson, 1993a*
	20.0	-0.30	Stephenson, 1993a*
	30.0	-0.38	Stephenson, 1993a*
	40.0	-0.60	Stephenson, 1993a*
	50.0	-0.75	Stephenson, 1993a*
	60.0	-0.86	Stephenson, 1993a*
	70.0	-0.90	Stephenson, 1993a*

Compound [CAS No.]	Temp. °C	Log S mol/L	Reference
	80.0	-0.96	Stephenson, 1993a*
	90.0	-0.99	Stephenson, 1993a*
2-Ethylbutyraldehyde [97-96-1]	0	-1.18	Stephenson, 1993b*
	10.0	-1.33	Stephenson, 1993b*
	20.0	-1.34	Stephenson, 1993b*
	30.0	-1.40	Stephenson, 1993b*
	40.0	-1.44	Stephenson, 1993b*
	50.0	-1.48	Stephenson, 1993b*
	60.0	-1.50	Stephenson, 1993b*
	70.0	-1.44	Stephenson, 1993b*
	80.0	-1.55	Stephenson, 1993b*
	90.0	-1.57	Stephenson, 1993b*
Ethyl butyrate [105-54-4]	25.0	-1.26	Stephenson and Stuart, 1986*
Ethyl chloroacetate [105-39-5]	25.0	-0.80	Stephenson and Stuart, 1986*
2-Ethylcrotonaldehyde [19780-25-7]	0	-0.94	Stephenson, 1993*
	10.0	-1.01	Stephenson, 1993*
	20.0	-1.07	Stephenson, 1993*
	30.0	-1.11	Stephenson, 1993*
	40.0	-1.13	Stephenson, 1993*
	50.0	-1.13	Stephenson, 1993*
	60.0	-1.16	Stephenson, 1993*
	70.0	-1.10	Stephenson, 1993*
	80.0	-1.08	Stephenson, 1993*
	90.0	-1.05	Stephenson, 1993*
Ethyl decanoate [110-38-3]	20	-4.13	Sobotka and Kahn, 1931
Ethylene glycol diacetate [111-55-7]	20-25	-0.91	Fordyce and Meyer, 1940
Ethyl heptanoate [106-30-9]	20	-2.74	Sobotka and Kahn, 1931
2-Ethylhexanal [123-05-7]	0	-2.60	Stephenson, 1993b*
	10.0	-2.39	Stephenson, 1993b*
	20.0	-2.65	Stephenson, 1993b*
	30.0	-2.42	Stephenson, 1993b*
	40.0	-2.55	Stephenson, 1993b*
	50.0	-2.47	Stephenson, 1993b*
	60.0	-2.33	Stephenson, 1993b*
	70.0	-2.44	Stephenson, 1993b*
	80.0	-2.39	Stephenson, 1993b*
	90.0	-2.39	Stephenson, 1993b*
2-Ethyl-1,3-hexanediol [94-96-2]	25.0	-2.81	Tewari et al., 1982
Ethyl hexanoate [123-66-0]	20	-2.36	Sobotka and Kahn, 1931
	25.0	-2.26	Stephenson and Stuart, 1986*
2-Ethyl-1-hexanol [104-76-7]	10.2	-1.99	Stephenson et al., 1984*
	19.8	-2.02	Stephenson et al., 1984*
	20.0	-2.17	Hommelen, 1959
	29.6	-2.13	Stephenson et al., 1984*
	40.1	-2.07	Stephenson et al., 1984*
	50.2	-2.25	Stephenson et al., 1984*
	60.3	-2.18	Stephenson et al., 1984*
	70.1	-2.12	Stephenson et al., 1984*
	80.1	-2.08	Stephenson et al., 1984*
	90.3	-2.05	Stephenson et al., 1984*
2-Ethyl-2-hexenal [645-62-5]	0	-1.86	Stephenson, 1993b*
	10.0	-2.02	Stephenson, 1993b*
	20.0	-2.24	Stephenson, 1993b*
	30.0	-2.07	Stephenson, 1993b*
	40.0	-2.00	Stephenson, 1993b*
	50.0	-1.96	Stephenson, 1993b*
	60.0	-2.03	Stephenson, 1993b*
	70.0	-1.93	Stephenson, 1993b*
	80.0	-1.97	Stephenson, 1993b*
	90.0	-1.99	Stephenson, 1993b*
(2-Ethylhexyl)amine [104-75-6]	0	-1.38	Stephenson, 1993a*
	10.0	-1.44	Stephenson, 1993a*
	20.0	-1.57	Stephenson, 1993a*
	30.0	-1.69	Stephenson, 1993a*

Compound [CAS No.]	Temp. °C	Log S mol/L	Reference
	40.0	-1.74	Stephenson, 1993a*
	50.0	-1.71	Stephenson, 1993a*
	60.0	-1.73	Stephenson, 1993a*
	70.0	-1.74	Stephenson, 1993a*
	80.0	-1.78	Stephenson, 1993a*
	90.0	-1.69	Stephenson, 1993a*
Ethyl hydrocinnamate [2021-28-5]	25.0	-3.01	Andrews and Keefer, 1950a
Ethyl iodide [75-03-6]	30	-1.59	Gross and Saylor, 1931
Ethyl isobutyrate [97-62-1]	25.0	-1.27	Stephenson and Stuart, 1986*
Ethyl isopentanoate [108-64-5]	25.0	-1.87	Stephenson and Stuart, 1986*
Ethyl malonate [105-53-3]	0	-0.71	Stephenson, 1992*
	9.1	-0.76	Stephenson, 1992*
	20.0	-0.85	Stephenson, 1992*
	20	-0.89	Sobotka and Kahn, 1931
	31.0	-0.88	Stephenson, 1992*
	39.6	-0.88	Stephenson, 1992*
	49.9	-0.89	Stephenson, 1992*
	60.2	-0.88	Stephenson, 1992*
	70.5	-0.87	Stephenson, 1992*
	81.0	-0.84	Stephenson, 1992*
	90.6	-0.81	Stephenson, 1992*
N-Ethyl-2-methylallylamine [18328-90-0]	18.6	0.09	Stephenson, 1993a*
	20.0	0.06	Stephenson, 1993a*
	30.0	-0.16	Stephenson, 1993a*
	40.0	-0.24	Stephenson, 1993a*
	50.0	-0.42	Stephenson, 1993a*
	60.0	-0.56	Stephenson, 1993a*
	70.0	-0.70	Stephenson, 1993a*
	80.0	-0.71	Stephenson, 1993a*
	90.0	-0.79	Stephenson, 1993a*
Ethyl 2-methyl butyrate [7452-79-1]	0	-1.51	Stephenson, 1992*
	9.0	-1.61	Stephenson, 1992*
	18.7	-1.71	Stephenson, 1992*
	31.1	-1.79	Stephenson, 1992*
	40.7	-1.84	Stephenson, 1992*
	50.5	-1.86	Stephenson, 1992*
	60.8	-1.91	Stephenson, 1992*
	70.0	-1.90	Stephenson, 1992*
	79.7	-1.91	Stephenson, 1992*
	91.3	-1.94	Stephenson, 1992*
5-Ethyl-2-methylpyridine [104-90-5]	0.0	-0.76	Stephenson, 1993*
	10.0	-0.90	Stephenson, 1993*
	20.0	-1.00	Stephenson, 1993*
	30.0	-1.06	Stephenson, 1993*
	40.0	-1.14	Stephenson, 1993*
	50.0	-1.15	Stephenson, 1993*
	60.0	-1.12	Stephenson, 1993*
	70.0	-1.12	Stephenson, 1993*
	80.0	-1.14	Stephenson, 1993*
	90.0	-1.13	Stephenson, 1993*
1-Ethylnaphthalene [1127-26-0]	8.1	-4.28	Schwarz, 1977
	10	-4.28	Schwarz and Wasik, 1977
	11.1	-4.28	Schwarz, 1977
	14	-4.28	Schwarz and Wasik, 1977
	14.0	-4.28	Schwarz, 1977
	17.1	-4.26	Schwarz, 1977
	20	-4.19	Schwarz and Wasik, 1977
	20.0	-4.27	Schwarz, 1977
	23.0	-4.26	Schwarz, 1977
	25	-4.19	Schwarz and Wasik, 1977
	25.0	-4.19	Schwarz, 1977
	25	-4.16	Mackay and Shiu, 1977
	26.1	-4.20	Schwarz, 1977
	31.7	-4.12	Schwarz, 1977

Compound [CAS No.]	Temp. °C	Log S mol/L	Reference
2-Ethylnaphthalene [939-27-5]	25.0	-4.29	Eganhouse and Calder, 1976
Ethyl octanoate [123-29-5]	20	-3.79	Sobotka and Kahn, 1931
3-Ethyl-3-pentanol [597-49-9]	20	-0.78	Ginnings and Hauser, 1938
	25	-0.84	Ginnings and Hauser, 1938
	30	-0.89	Ginnings and Hauser, 1938
	40	-0.96	Ginnings and Hauser, 1938
1-Ethylpiperidine [766-09-6]	9.0	-0.16	Stephenson, 1993*
	10.0	-0.10	Stephenson, 1993*
	11.0	-0.12	Stephenson, 1993*
	12.0	-0.18	Stephenson, 1993*
	14.0	-0.22	Stephenson, 1993*
	16.0	-0.23	Stephenson, 1993*
	17.0	-0.27	Stephenson, 1993*
	18.0	-0.25	Stephenson, 1993*
	19.0	-0.31	Stephenson, 1993*
	20.0	-0.33	Stephenson, 1993*
	25.0	-0.38	Stephenson, 1993*
	30.0	-0.52	Stephenson, 1993*
	40.0	-0.69	Stephenson, 1993*
	50.0	-0.82	Stephenson, 1993*
	60.0	-0.91	Stephenson, 1993*
	70.0	-0.97	Stephenson, 1993*
	80.0	-1.05	Stephenson, 1993*
	90.0	-1.10	Stephenson, 1993*
2-Ethylpiperidine [1484-80-6]	8.0	0.12	Stephenson, 1993*
	10.0	-0.01	Stephenson, 1993*
	15.0	-0.19	Stephenson, 1993*
	20.0	-0.27	Stephenson, 1993*
	30.0	-0.37	Stephenson, 1993*
	40.0	-0.48	Stephenson, 1993*
	50.0	-0.56	Stephenson, 1993*
	60.0	-0.64	Stephenson, 1993*
	70.0	-0.68	Stephenson, 1993*
	80.0	-0.73	Stephenson, 1993*
	90.0	-0.77	Stephenson, 1993*
Ethyl propionate [105-37-3]	25.0	-0.83	Tewari et al., 1982
	25.0	-0.69	Stephenson and Stuart, 1986*
N-Ethylpropylamine [20193-20-8]	49.0	0.39	Stephenson, 1993a*
	50.0	0.29	Stephenson, 1993a*
	52.0	0.15	Stephenson, 1993a*
	60.0	-0.06	Stephenson, 1993a*
	70.0	-0.19	Stephenson, 1993a*
	79.0	-0.25	Stephenson, 1993a*
(1-Ethylpropyl)amine [616-24-0]	64.0	0.34	Stephenson, 1993a*
	66.0	0.12	Stephenson, 1993a*
	68.0	0.08	Stephenson, 1993a*
	70.0	0.00	Stephenson, 1993a*
	80.0	-0.11	Stephenson, 1993a*
	90.0	-0.16	Stephenson, 1993a*
2-Ethylpyridine [100-71-0]	0	-0.04	Stephenson, 1993*
	10.0	-0.27	Stephenson, 1993*
	20.0	-0.39	Stephenson, 1993*
	30.0	-0.47	Stephenson, 1993*
	40.0	-0.53	Stephenson, 1993*
	50.0	-0.60	Stephenson, 1993*
	60.0	-0.62	Stephenson, 1993*
	70.0	-0.64	Stephenson, 1993*
	80.0	-0.63	Stephenson, 1993*
	90.0	-0.62	Stephenson, 1993*
3-Ethylpyridine [536-78-7]	0	-0.42	Stephenson, 1993*
	10.0	-0.53	Stephenson, 1993*
	20.0	-0.59	Stephenson, 1993*
	30.0	-0.64	Stephenson, 1993*
	40.0	-0.64	Stephenson, 1993*

Compound [CAS No.]	Temp. °C	Log S mol/L	Reference
	50.0	-0.64	Stephenson, 1993*
	60.0	-0.67	Stephenson, 1993*
	70.0	-0.66	Stephenson, 1993*
	80.0	-0.67	Stephenson, 1993*
	90.0	-0.64	Stephenson, 1993*
4-Ethylpyridine [536-75-4]	0	-0.34	Stephenson, 1993*
	10.0	-0.43	Stephenson, 1993*
	20.0	-0.51	Stephenson, 1993*
	30.0	-0.52	Stephenson, 1993*
	40.0	-0.57	Stephenson, 1993*
	50.0	-0.58	Stephenson, 1993*
	60.0	-0.59	Stephenson, 1993*
	70.0	-0.57	Stephenson, 1993*
	80.0	-0.58	Stephenson, 1993*
	90.0	-0.57	Stephenson, 1993*
Ethyl salicylate [119-61-6]	25.0	-2.65	Stephenson and Stuart, 1986*
Ethyl succinate [123-25-1]	20	-0.96	Sobotka and Kahn, 1931
	25.0	-0.93	Stephenson and Stuart, 1986*
2-Ethylthiophene [872-55-9]	25	-2.58	Price, 1976
2-Ethyltoluene [611-14-3]	25.0	-3.21	Tewari et al., 1982
Ethyl trimethylacetate [3938-95-2]	25.0	-1.38	Stephenson and Stuart, 1986*
Etofenprox [80844-07-1]	25	-5.58	Worthing and Hance, 1991
Etridiazole [2593-15-9]	25	-3.69	Worthing and Hance, 1991
Etrimfos [38260-54-7]	23	-3.86	Worthing and Hance, 1991
Fenabutatin oxide [13356-08-6]	23	-8.32	Worthing and Hance, 1991
Fenarimol [60168-88-9]	25	-4.38	Worthing and Hance, 1991
Fenchlorazole [103112-36-3]	20	-5.62	Worthing and Hance, 1991
Fenchlorim [3740-92-9]	20	-4.95	Worthing and Hance, 1991
(1R)-endo-Fenchyl alcohol [2219-02-9]	25	-2.27	Fichan et al., 1999
Fenchlorphos [299-88-4]	20	-5.11	Fühner and Geiger, 1977
	20	-5.51	Ellgehausen et al., 1981
Fenfuram [24691-80-3]	20	-3.30	Worthing and Hance, 1991
Fenitropan [65934-95-4]	20	-3.97	Worthing and Hance, 1991
Fenitrothion [122-14-5]	20	-4.12	Worthing and Hance, 1991
Fenobucarb [3766-81-2]	20	-2.50	Worthing and Hance, 1991
Fenothiocarb [62580-32-2]	20	-3.93	Worthing and Hance, 1991
Fenoxaprop-ethyl [66441-23-4]	20	-2.60	Humburg et al., 1989
Fenoxycarb [79127-80-3]	25	-4.72	Worthing and Hance, 1991
Fenpiclonil [74738-17-3]	20	-5.07	Worthing and Hance, 1991
Fenpropathrin [39515-41-8]	25	-6.02	Worthing and Hance, 1991
Fenpropidin [67306-00-7]	25	-2.89	Worthing and Hance, 1991
Fenpropimorph [67306-03-0]	25	-4.85	Worthing and Hance, 1991
Fensulfothion [115-90-2]	25	-2.30	Worthing and Hance, 1991
Fenthion [55-38-9]	20	-5.14	Worthing and Hance, 1991
Fenuron [101-42-8]	20	-1.65	Fühner and Geiger, 1977
	20	-1.65	Ellgehausen et al., 1981
Fenuron-TCA [4482-55-7]	25	-1.83	Humburg et al., 1989
Ferimzone [89269-64-7]	30	-3.20	Worthing and Hance, 1991
Fluazifop-butyl [69806-50-4]	25	-5.28	Humburg et al., 1989
Fluazifop-P-butyl [79241-46-6]	25	-5.58	Humburg et al., 1989
	25	-1.93	Worthing and Hance, 1991
Fluchloralin [33245-39-5]	20	<-5.55	Humburg et al., 1989
Flucythrinate [70124-77-5]	21	-5.96	Worthing and Hance, 1991
Fluometuron [2164-17-2]	20	-3.41	Humburg et al., 1989
Fluorobenzene [462-06-6]	0	-1.77	Stephenson, 1992*
	9.5	-1.76	Stephenson, 1992*
	19.2	-1.75	Stephenson, 1992*
	25.0	-1.87	Andrews and Keefer, 1950
	29.7	-1.79	Stephenson, 1992*
	30	-1.80	Freed et al., 1979
	39.6	-1.80	Stephenson, 1992*
	47.7	-1.74	Stephenson, 1992*
	60.1	-1.70	Stephenson, 1992*
	70.0	-1.70	Stephenson, 1992*

Compound [CAS No.]	Temp. °C	Log S mol/L	Reference
	80.0	-1.71	Stephenson, 1992*
m-Fluorobenzyl chloride [456-42-8]	25.0	-2.54	Tewari et al., 1982
	30.00	-1.79	Gross et al., 1933
o-Fluorobenzyl chloride [345-35-7]	25.0	-2.54	Tewari et al., 1982
Fluorodifen [15457-05-3]	20	-5.22	Ellgehausen et al., 1981
Fluoroglycofen [77501-60-1]	25	<-5.62	Humburg et al., 1989
Fluoromid [41205-21-4]	10	-2.49	Worthing and Hance, 1991
Fluprednisolone [53-34-9]	37	-2.56	Hamlin et al., 1965
Flupropanate [756-09-3]	20	1.31	Worthing and Hance, 1991
Flurenol-butyl [2314-09-2]	20	-3.89	Worthing and Hance, 1991
Fluridone [59756-60-4]	20	-4.44	Humburg et al., 1989
Flurochloridone [61213-25-0]	20	-4.05	Worthing and Hance, 1991
Fluoroxypyr-methyl [81406-37-3]	27.7	-5.61	Worthing and Hance, 1991
Flurtamone [96525-23-4]	20	-3.96	Worthing and Hance, 1991
Flusulfamide [106917-52-6]	25	-5.16	Worthing and Hance, 1991
Flutoanil [66332-96-5]	20	-4.53	Worthing and Hance, 1991
Flutriafol [76674-21-0]	20	-3.37	Worthing and Hance, 1991
Fluxofenim [88485-37-4]	20	-4.01	Worthing and Hance, 1991
Folpet [133-07-3]	20	-5.47	Fühner and Geiger, 1977
Fomesafen [72178-02-0]	20	-3.94	Humburg et al., 1989
Formothion [2540-82-1]	24	-2.00	Worthing and Hance, 1991
4-Formyl-2-nitrophenol [3011-34-5]	20.0	-2.95	Schwarzenbach et al., 1988
Fosamine-ammonium [25954-13-6]	25	1.02	Humburg et al., 1989
Fosmethilan [83733-82-8]	20	-5.20	Worthing and Hance, 1991
Furalaxyl [57646-30-7]	20	-3.12	Worthing and Hance, 1991
Furathiocarb [65907-30-4]	20	-4.58	Worthing and Hance, 1991
Glycine [56-40-6]	25	0.43	pH 7.0, Carta and Tola, 1996
Glyphosate [1071-83-6]	25	-1.03	Humburg et al., 1989
Guaiacol [90-05-1]	25	-0.70	Tam et al., 1994
Haloxyfop [69806-34-4]	25	-3.92	Worthing and Hance, 1991
Haloxyfop-methyl [69806-40-2]	25	-4.61	Humburg et al., 1989
Harmane [486-84-0]	15	-4.22	pH 13, Burrows et al., 1996
	16	-4.20	pH 13, Burrows et al., 1996
	17	-4.17	pH 13, Burrows et al., 1996
	20	-4.08	pH 13, Burrows et al., 1996
	37	-3.87	pH 13, Burrows et al., 1996
	38	-3.84	pH 13, Burrows et al., 1996
	45	-3.79	pH 13, Burrows et al., 1996
Harmine [442-51-3]	15	-4.73	pH 13, Burrows et al., 1996
	16	-4.71	pH 13, Burrows et al., 1996
	17	-4.70	pH 13, Burrows et al., 1996
	37	-4.55	pH 13, Burrows et al., 1996
	38	-4.56	pH 13, Burrows et al., 1996
	45	-4.53	pH 13, Burrows et al., 1996
Heptenophos [23560-59-0]	20	-2.06	Worthing and Hance, 1991
Heptopargil [73886-28-9]	20	-2.31	Worthing and Hance, 1991
Hexaflumuron [86479-06-3]	18	-7.23	Worthing and Hance, 1991
2,2′,3,3′,4,4′,6-Heptachlorobiphenyl [52663-71-5]	25	-8.26	Miller et al., 1984
2,2′,3,4,4′,5,5′-Heptachlorobiphenyl [35065-29-3]	25	-8.71	Dulfer et al., 1995
1,2,3,4,6,7,8-Heptachlorodibenzo-p-dioxin [35822-46-9]	20.0	-11.25	Friesen et al., 1985
	26.0	-11.22	Li and Andren, 1994
1,2,3,4,6,7,8-Heptachlorodibenzofuran [67462-39-4]	22.7	-11.48	Friesen et al., 1985
Heptadecanoic acid [506-12-7]	20.0	-4.81	Ralston and Hoerr, 1942
1,6-Heptadiene [3070-53-9]	25	-3.34	McAuliffe, 1966
1,6-Heptadiyne [2396-63-6]	25	-1.75	McAuliffe, 1966
Heptanal [111-71-7]	0	-1.55	Stephenson, 1993b*
	10.0	-1.80	Stephenson, 1993b*
	20.0	-1.85	Stephenson, 1993b*
	30.0	-1.91	Stephenson, 1993b*
	40.0	-1.98	Stephenson, 1993b*
	50.0	-1.94	Stephenson, 1993b*
	60.0	-1.94	Stephenson, 1993b*
	70.0	-1.94	Stephenson, 1993b*
	80.0	-1.91	Stephenson, 1993b*

Compound [CAS No.]	Temp. °C	Log S mol/L	Reference
	90.0	-1.85	Stephenson, 1993b*
Heptanoic acid [111-14-8]	20.0	-1.73	Ralston and Hoerr, 1942
	30	-1.85	Bell, 1971
1-Heptanol [111-70-6]	0	-1.69	Stephenson et al., 1984*
	10.5	-1.66	Stephenson et al., 1984b
	20.0	-1.84	Hommelen, 1979
	20.2	-1.80	Stephenson et al., 1984*
	25	-1.89	Li and Andren, 1994
	25.0	-1.95	Tewari et al., 1982
	25.0	-1.55	Booth and Everson, 1948
	25	-1.81	Butler et al., 1933
	30.6	-1.88	Stephenson et al., 1984*
	39.8	-1.85	Stephenson et al., 1984*
	50.1	-1.86	Stephenson et al., 1984*
	60.0	-1.81	Stephenson et al., 1984*
	70.1	-1.76	Stephenson et al., 1984*
	80.1	-1.73	Stephenson et al., 1984*
	90.5	-1.68	Stephenson et al., 1984*
2-Heptanol [543-49-7]	0	-1.25	Stephenson et al., 1984*
	10.2	-1.40	Stephenson et al., 1984*
	19.5	-1.44	Stephenson et al., 1984*
	30.7	-1.52	Stephenson et al., 1984*
	40.0	-1.55	Stephenson et al., 1984*
	50.0	-1.60	Stephenson et al., 1984*
	60.3	-1.61	Stephenson et al., 1984*
	70.3	-1.56	Stephenson et al., 1984*
	80.0	-1.56	Stephenson et al., 1984*
	90.2	-1.53	Stephenson et al., 1984*
3-Heptanol [589-82-2]	20.0	-1.39	Hommelen, 1959
	25	-1.46	Crittenden and Hixon, 1954
	25.0	-1.40	Stephenson and Stuart, 1986*
4-Heptanol [589-55-9]	20.0	-1.39	Hommelen, 1959
	25.0	-1.42	Stephenson and Stuart, 1986*
4-Heptanone [123-19-3]	0	-1.21	Stephenson, 1992*
	9.6	-1.33	Stephenson, 1992*
	19.5	-1.40	Stephenson, 1992*
	29.7	-1.41	Stephenson, 1992*
	39.7	-1.52	Stephenson, 1992*
	50.0	-1.48	Stephenson, 1992*
	60.6	-1.55	Stephenson, 1992*
	70.3	-1.58	Stephenson, 1992*
	80.9	-1.55	Stephenson, 1992*
	90.2	-1.56	Stephenson, 1992*
1-Heptene [592-76-7]	20	-3.50[a]	Natarajan and Venkatachalam, 1972
	25	-3.50[a]	Natarajan and Venkatachalam, 1972
	25.0	-3.73	Tewari et al., 1982
	30	-3.61[a]	Natarajan and Venkatachalam, 1972
2-Heptene [592-77-8]	15	-3.47[a]	Natarajan and Venkatachalam, 1972
	20	-3.49[a]	Natarajan and Venkatachalam, 1972
	25	-3.57[a]	Natarajan and Venkatachalam, 1972
	25	-3.69[b]	Natarajan and Venkatachalam, 1972
	30	-3.73[b]	Natarajan and Venkatachalam, 1972
	35	-3.78[b]	Natarajan and Venkatachalam, 1972
Heptyl acetate [112-06-1]	25.0	-2.88	Stephenson and Stuart, 1986*
Heptylamine [111-68-2]	0	-1.27	Stephenson, 1993a*
	3.0	-1.48	Stephenson, 1993a*
	6.0	-1.48	Stephenson, 1993a*
	10.0	-1.54	Stephenson, 1993a*
	15.0	-1.52	Stephenson, 1993a*
	16.0	-1.52	Stephenson, 1993a*
	18.0	-1.54	Stephenson, 1993a*
	20.0	-1.60	Stephenson, 1993a*
	30.0	-1.58	Stephenson, 1993a*
	40.0	-1.58	Stephenson, 1993a*

Compound [CAS No.]	Temp. °C	Log S mol/L	Reference
	50.0	-1.58	Stephenson, 1993a*
	60.0	-1.54	Stephenson, 1993a*
	70.0	-1.63	Stephenson, 1993a*
	80.0	-1.60	Stephenson, 1993a*
	90.0	-1.54	Stephenson, 1993a*
Heptyl butyrate [5870-93-9]	10.3	-2.86	Stephenson, 1992*
	19.8	-2.82	Stephenson, 1992*
	29.8	-2.71	Stephenson, 1992*
	39.7	-2.87	Stephenson, 1992*
	49.9	-2.95	Stephenson, 1992*
	60.2	-2.82	Stephenson, 1992*
	70.3	-2.93	Stephenson, 1992*
	80.1	-2.97	Stephenson, 1992*
	90.5	-3.16	Stephenson, 1992*
Heptyl formate [112-23-2]	25.0	-1.98	Stephenson and Stuart, 1986*
1-Heptyne [628-71-7]	25	-3.01	McAuliffe, 1966
2,2',3,3',4,4'-Hexachlorobiphenyl [38380-07-3]	25	-9.01	Dunnivant and Elzerman, 1988
	25	-9.11	Miller et al., 1984
	25	-8.87	Dunnivant and Elzerman, 1988
	25	-8.41	Dulfer et al., 1995
2,2',3,3',4,5-Hexachlorobiphenyl [55215-18-4]	25	-7.79	Dunnivant and Elzerman, 1988
	25	-8.59	Dunnivant and Elzerman, 1988
2,2',3,3',6,6-Hexachlorobiphenyl [38411-22-2]	25	-7.90	Dickhut et al., 1986
	25	-7.78	Miller et al., 1984
	25	-8.35	Dulfer et al., 1995
2,2',4,4',5,5'-Hexachlorobiphenyl [35065-27-1]	4.0	-7.89	Doucette and Andren, 1988a
	22.0	-8.50	Opperhuizen et al., 1988
	24	-8.58	Chiou et al., 1977
	24	-8.58	Freed et al., 1979
	25.0	-7.63	Doucette and Andren, 1988a
	25	-8.62	Dunnivant and Elzerman, 1988
	40.0	-7.45	Doucette and Andren, 1988a
2,2',4,4',6,6'-Hexachlorobiphenyl [33979-03-2]	22.0	-8.52	Opperhuizen et al., 1988
	25	-8.20	Dunnivant and Elzerman, 1988
	25	-8.03	van Haelst et al., 1996
	25.0	-8.04	Li et al., 1993
	25	-8.10	Li and Andren, 1994
	25	-8.04	Li and Doucette, 1993
	25	-8.95	Miller et al., 1984
	25	-8.56	Dunnivant and Elzerman, 1988
	25	-8.46	Dulfer et al., 1995
1,2,3,4,7,8-Hexachlorodibenzo-p-dioxin [39227-28-6]	20.0	-10.95	Friesen et al., 1985
	26.0	-10.69	Li and Andren, 1994
1,2,3,4,7,8-Hexachlorodibenzofuran [70658-26-9]	22.7	-10.66	Friesen et al., 1990
Hexacosane [630-01-3]	25.0	-8.33	Sutton and Calder, 1974
Hexadecane [544-76-3]	25.0	-8.40	Sutton and Calder, 1974
	25	-7.56	Franks, 1966
Hexadecanoic acid [57-10-3]	20.0	-4.55	Ralston and Hoerr, 1942
	25	-5.57	pH 5.7, Robb, 1966
1-Hexadecanol [36653-82-4]	25	-6.77	Robb, 1966
1,5-Hexadiene [592-42-7]	25	-2.69	McAuliffe, 1966
Hexamethylenediamine [124-09-4]	30	1.92	Du Pont, 1999c*
Hexanal [66-25-1]	0	-1.00	Stephenson, 1993b*
	10.0	-1.19	Stephenson, 1993b*
	20.0	-1.28	Stephenson, 1993b*
	30.0	-1.36	Stephenson, 1993b*
	40.0	-1.38	Stephenson, 1993b*
	50.0	-1.46	Stephenson, 1993b*
	60.0	-1.55	Stephenson, 1993b*
	70.0	-1.51	Stephenson, 1993b*
	80.0	-1.47	Stephenson, 1993b*
	90.0	-1.40	Stephenson, 1993b*
2,5-Hexane dinitrate [42730-17-6]	25	-2.68	Fischer and Ballschmiter, 1998, 1998a
Hexanoic acid [142-62-1]	20.0	-1.08	Ralston and Hoerr, 1942

Compound [CAS No.]	Temp. °C	Log S mol/L	Reference
	30	-1.19	Bell, 1971
1-Hexanol [111-27-3]	0	-1.03	Stephenson et al., 1984*
	10.2	-1.13	Stephenson et al., 1984*
	20	-1.31	Laddha and Smith, 1948
	20.0	-1.19	Stephenson et al., 1984*
	20.0	-1.22	Hommelen, 1959
	25	-1.23	Crittenden and Hixon, 1954
	25	-1.28	Li and Andren, 1994
	25.0	-1.38	Tewari et al., 1982
	25	-1.21	Booth et al., 1933
	29.7	-1.26	Stephenson et al., 1984*
	39.8	-1.30	Stephenson et al., 1984*
	50.0	-1.31	Stephenson et al., 1984*
	60.0	-1.30	Stephenson et al., 1984*
	70.3	-1.25	Stephenson et al., 1984*
	80.3	-1.21	Stephenson et al., 1984*
	90.3	-1.20	Stephenson et al., 1984*
2-Hexanol [623-93-7]	0	-0.70	Stephenson et al., 1984*
	10.1	-0.78	Stephenson et al., 1984*
	19.8	-0.90	Stephenson et al., 1984*
	20	-0.83	Ginnings and Webb, 1938
	20.0	-0.99	Hommelen, 1959
	25	-0.87	Ginnings and Webb, 1938
	29.9	-0.94	Stephenson et al., 1984*
	30	-0.90	Ginnings and Webb, 1938
	40.0	-0.97	Stephenson et al., 1984*
	50.0	-1.03	Stephenson et al., 1984*
	60.2	-1.05	Stephenson et al., 1984*
	70.0	-1.02	Stephenson et al., 1984*
	80.1	-0.96	Stephenson et al., 1984*
	90.2	-1.04	Stephenson et al., 1984*
3-Hexanol [623-37-0]	0	-0.57	Stephenson et al., 1984*
	10.1	-0.72	Stephenson et al., 1984*
	20.0	-0.79	Stephenson et al., 1984*
	25	-0.80	Ginnings and Webb, 1938
	30.0	-0.87	Stephenson et al., 1984*
	39.8	-0.92	Stephenson et al., 1984*
	50.0	-0.97	Stephenson et al., 1984*
	60.1	-1.00	Stephenson et al., 1984*
	70.2	-1.00	Stephenson et al., 1984*
	80.2	-0.98	Stephenson et al., 1984*
	90.3	-0.97	Stephenson et al., 1984*
3-Hexanone [589-38-8]	25	-0.83	Ginnings et al., 1940
Hexazinone [51235-04-2]	25	-0.88	Humburg et al., 1989
trans-2-Hexenal [6728-26-3]	0	-1.03	Stephenson, 1993*
	10.0	-1.06	Stephenson, 1993*
	20.0	-1.09	Stephenson, 1993*
	30.0	-1.13	Stephenson, 1993*
	40.0	-1.14	Stephenson, 1993*
	50.0	-1.12	Stephenson, 1993*
	60.0	-1.14	Stephenson, 1993*
	70.0	-1.03	Stephenson, 1993*
	80.0	-1.05	Stephenson, 1993*
	90.0	-1.01	Stephenson, 1993*
2-Hexene [592-43-8]	20	-3.06[a]	Natarajan and Venkatachalam, 1972
	25	-3.10[a]	Natarajan and Venkatachalam, 1972
	25	-3.15[b]	Natarajan and Venkatachalam, 1972
	30	-3.18[a]	Natarajan and Venkatachalam, 1972
	30	-3.22[b]	Natarajan and Venkatachalam, 1972
	35	-3.29[b]	Natarajan and Venkatachalam, 1972
1-Hexen-3-ol [4798-44-1]	20	-0.57	Ginnings et al., 1939
	25	-0.60	Ginnings et al., 1939
	30	-0.63	Ginnings et al., 1939
4-Hexen-3-ol [4798-58-7]	20	-0.39	Ginnings et al., 1939

Compound [CAS No.]	Temp. °C	Log S mol/L	Reference
	25	-0.42	Ginnings et al., 1939
	30	-0.45	Ginnings et al., 1939
n-Hexyl acetate [142-92-7]	25.0	-2.48	Stephenson and Stuart, 1986*
Hexylamine [111-26-2]	30.0	-1.02	Stephenson, 1993a*
	40.0	-1.11	Stephenson, 1993a*
	50.0	-1.04	Stephenson, 1993a*
	60.0	-1.03	Stephenson, 1993a*
	70.0	-1.05	Stephenson, 1993a*
	80.0	-1.03	Stephenson, 1993a*
	90.0	-1.08	Stephenson, 1993a*
Hexylbenzene [1077-16-3]	10.00	-5.25	May et al., 1983[c]
	12.00	-5.24	May et al., 1983[c]
	19.00	-5.24	May et al., 1983[c]
	22.00	-5.23	May et al., 1983[c]
	23.00	-5.22	May et al., 1983[c]
	25.00	-5.21	May et al., 1983[c]
	25.0	-5.20	Tewari et al., 1982
	25.0	-5.25	Owens et al., 1986
Hexyl butyrate [2639-63-6]	11.9	-3.18	Stephenson, 1992*
	29.0	-2.90	Stephenson, 1992*
	40.0	-2.57	Stephenson, 1992*
	50.4	-2.60	Stephenson, 1992*
	60.1	-2.72	Stephenson, 1992*
	70.3	-2.59	Stephenson, 1992*
Hexyl ether [112-58-3]	0	-2.97	Stephenson, 1992*
	20.0	-2.99	Stephenson, 1992*
	39.4	-2.99	Stephenson, 1992*
	70.3	-3.12	Stephenson, 1992*
	80.2	-3.12	Stephenson, 1992*
	90.3	-2.99	Stephenson, 1992*
Hexyl formate [629-33-4]	25.0	-1.97	Stephenson and Stuart, 1986*
Hexyl propionate [2445-76-3]	25.0	-1.94	Stephenson and Stuart, 1986*
Hexyne [693-02-7]	25.0	-2.08	Tewari et al., 1982
	25	-2.36	McAuliffe, 1966
Hexythiazox [78587-05-0]	20	-5.85	Worthing and Hance, 1991
HFC-134a [811-97-2]	25	-1.83	Du Pont, 1999d*
4-Hydroxyacetanilide [103-90-2]	30	-0.94	Granberg and Rasmuson, 1999
p-Hydroxyacetophenone [99-93-4]	25	-1.27	Jin et al., 1998
Hydroxyatrazine [2163-68-0]	26.0	-2.94	pH 3.0, Ward and Weber, 1968
	26.0	-4.52	pH 7.0, Ward and Weber, 1968
	26.0	-4.48	pH 10.0, Ward and Weber, 1968
3-Hydroxybenzaldehyde [100-83-4]	25	-1.16	Jin et al., 1998
4-Hydroxybenzoic acid [99-96-7]	20	-1.46	Corby and Elworthy, 1971
5-Hydroxyindan [1470-94-6]	25	-1.59	Southworth and Keller, 1986
4-Hydroxy-3-methoxybenzaldehyde [121-33-5]	25	-1.30	Jin et al., 1998
8-Hydroxyquinoline sulfate [134-31-6]	20	-0.11	Worthing and Hance, 1991
Hymexazol [10004-44-1]	25	-0.07	Worthing and Hance, 1991
Imazapyr-isopropylammonium [81510-83-0]	15	-1.52	Worthing and Hance, 1991
	25	-1.45	Worthing and Hance, 1991
Imazaquin [81335-37-7]	25	-3.72	Humburg et al., 1989
Imazethapyr [81335-77-5]	25	-2.32	Humburg et al., 1989
Imibenconazole [86598-92-7]	20	-5.38	Worthing and Hance, 1991
Inabenfide [82211-24-3]	30	-5.53	Worthing and Hance, 1991
Iodine [7553-56-2]	25.0	-2.89	Vesala, 1974
Iodobenzene [591-50-4]	25.0	-2.95	Vesala, 1974
	25.0	-3.11	Andrews and Keefer, 1950
	30	-2.78	Freed et al., 1979
1-Iodoheptane [4282-40-0]	25.0	-4.81	Tewari et al., 1982
α-Ionone [127-41-3]	25.0	-3.20	Etzweiler et al., 1995
β-Ionone [79-77-6]	25	-3.06	Fichan et al., 1999
Ioxynil [1689-83-4]	20	-3.87	Führer and Geiger, 1977
	25	-3.46	Humburg et al., 1989
Ipatone [3004-70-4]	26.0	-2.87	pH 3.0, Ward and Weber, 1968
	26.0	-3.44	pH 7.0, Ward and Weber, 1968

Compound [CAS No.]	Temp. °C	Log S mol/L	Reference
	26.0	-3.43	pH 10.0, Ward and Weber, 1968
Ipazine [1912-25-0]	26.0	-3.94	pH 3.0, Ward and Weber, 1968
	26.0	-3.95	pH 7.0, Ward and Weber, 1968
	26.0	-3.98	pH 10.0, Ward and Weber, 1968
Iprobenfos [26087-47-8]	20	-2.83	Worthing and Hance, 1991
Iprodione [36734-19-7]	20	-4.40	Worthing and Hance, 1991
Isazofos [42509-80-8]	20	-3.32	Ellgehausen et al., 1981
Isobutyl butyrate [539-90-2]	25.0	-2.46	Stephenson and Stuart, 1986*
	25.0	-3.01	Tewari et al., 1982
	30.00	-2.78	Gross et al., 1933
Isobutyl isobutyrate [97-85-8]	25	-2.04	Amoore and Buttery, 1978
Isobutyl propionate [540-42-1]	0	-1.62	Stephenson, 1992*
	9.3	-1.65	Stephenson, 1992*
	19.4	-1.76	Stephenson, 1992*
	31.1	-1.85	Stephenson, 1992*
	40.3	-1.87	Stephenson, 1992*
	51.1	-1.90	Stephenson, 1992*
	61.0	-1.93	Stephenson, 1992*
	71.0	-1.94	Stephenson, 1992*
	81.3	-1.94	Stephenson, 1992*
	91.3	-1.96	Stephenson, 1992*
Isobutyraldehyde [78-84-2]	0	0.03	Stephenson, 1993b*
	10.0	-0.04	Stephenson, 1993b*
	25.0	-0.11	Stephenson, 1993b*
	25	0.09	Amoore and Buttery, 1978
	30.0	-0.20	Stephenson, 1993b*
	40.0	-0.26	Stephenson, 1993b*
	50.0	-0.30	Stephenson, 1993b*
	60.0	-0.35	Stephenson, 1993b*
Isofenphos [25311-71-1]	20	-4.28	Worthing and Hance, 1991
Isononyl acetate [40379-24-6]	25.0	-2.99	Stephenson and Stuart, 1986*
Isopentyl acetate [123-92-2]	25.0	-1.79	Stephenson and Stuart, 1986*
Isopentyl butyrate [106-27-4]	25.0	-2.80	Stephenson and Stuart, 1986*
Isopentyl propionate [105-68-0]	25.0	-2.37	Stephenson and Stuart, 1986*
Isopropalin [33820-53-0]	20	-6.59	Humburg et al., 1989
	25.0	-3.76	Banerjee et al., 1980
Isopropyl butyrate [638-11-9]	25.0	-0.68	Stephenson and Stuart, 1986*
4-Isopropylpyridine [696-30-0]	0	-0.89	Stephenson, 1993*
	10.0	-0.94	Stephenson, 1993*
	20.0	-1.00	Stephenson, 1993*
	30.0	-1.05	Stephenson, 1993*
	40.0	-1.08	Stephenson, 1993*
	50.0	-1.10	Stephenson, 1993*
	60.0	-1.12	Stephenson, 1993*
	70.0	-1.10	Stephenson, 1993*
	80.0	-1.05	Stephenson, 1993*
	90.0	-1.05	Stephenson, 1993*
2-Isopropyltoluene [527-84-4]	25	-3.44	Lun et al., 1997
3-Isopropyltoluene [535-77-3]	25	-3.50	Lun et al., 1997
4-Isopropyltoluene [99-87-6]	25	-3.42	Lun et al., 1997
Isoprothiolane [50512-35-1]	20	-3.78	Worthing and Hance, 1991
Isoproturon [34123-59-6]	20	-3.57	Worthing and Hance, 1991
Isouron [55861-78-4]	25	-2.43	Worthing and Hance, 1991
Isovaleraldehyde [590-86-3]	0	-0.38	Stephenson, 1993b*
	10.0	-0.56	Stephenson, 1993b*
	20.0	-0.63	Stephenson, 1993b*
	30.0	-0.68	Stephenson, 1993b*
	40.0	-0.73	Stephenson, 1993b*
	50.0	-0.79	Stephenson, 1993b*
	60.0	-0.83	Stephenson, 1993b*
	70.0	-0.86	Stephenson, 1993b*
	80.0	-0.84	Stephenson, 1993b*
	90.0	-0.95	Stephenson, 1993b*
Isoxaben [82558-50-7]	25	-5.52	Humburg et al., 1989

Compound [CAS No.]	Temp. °C	Log S mol/L	Reference
Isoxapyrifop [87757-18-4]	25	-4.59	Worthing and Hance, 1991
Isoxathion [18854-01-8]	25	-5.22	Worthing and Hance, 1991
Karbutilate [4849-32-5]	20	-2.93	Worthing and Hance, 1991
Kasugamycin [19408-46-9]	25	-0.54	Worthing and Hance, 1991
Lactofen [77501-63-4]	22	-6.76	Humburg et al., 1989
Lauraldehyde [112-54-9]	0	-2.01	Stephenson, 1993b*
	20.0	-2.79	Stephenson, 1993b*
	30.0	-2.66	Stephenson, 1993b*
	40.0	-2.79	Stephenson, 1993b*
Lenacil [2164-08-1]	25	-4.59	Worthing and Hance, 1991
Leptophos [21609-90-5]	20	-7.94	Chiou et al., 1977
L-Leucine [61-90-5]	25	-0.77	pH 7.0, Carta and Tola, 1996
R-(+)-Limonene [5989-27-5]	25	-3.82	Fichan et al., 1999
(+)-Limonene oxide [470-82-6]	25	-2.34	Fichan et al., 1999
(±)-Linalool [78-70-6]	25	-2.00	Fichan et al., 1999
Linuron [330-55-2]	20	-3.52	Humburg et al., 1989
Maleic hydrazide [123-33-1]	25	-1.27	Humburg et al., 1989
MCPA [94-74-6]	25	-2.39	Worthing and Hance, 1991
MCPA-thioethyl [25319-90-8]	25	-5.03	Worthing and Hance, 1991
Mecoprop [93-65-2]	20	-2.54	Humburg et al., 1989
Mefenacet [73250-68-7]	20	-4.87	Worthing and Hance, 1991
Mefluidide [53780-34-0]	22	-3.24	Humburg et al., 1989
Menadione [58-27-5]	33	-3.05	Dubbs and Gupta, 1998
Mepanipyrim [110235-47-7]	20	-4.60	Worthing and Hance, 1991
Mephosfolan [950-10-7]	25	-3.67	Worthing and Hance, 1991
Mepronil [55814-41-0]	20	-4.22	Worthing and Hance, 1991
Mercuric chloride [7487-94-7]	20	-0.59	Worthing and Hance, 1991
Mercuric oxide [21908-53-2]	25	-3.64	Worthing and Hance, 1991
Mercurous chloride [7546-30-7]	18	-4.42	Worthing and Hance, 1991
Metalaxyl [57837-19-1]	20	-1.67	Ellgehausen et al., 1980
	20	-1.60	Ellgehausen et al., 1981
Metaldehyde [9002-91-9]	17	-2.94	Worthing and Hance, 1991
Methacrolein [75-85-3]	0	-0.20	Stephenson, 1993b*
	10.0	-0.21	Stephenson, 1993b*
	20.0	-0.24	Stephenson, 1993b*
	30.0	-0.21	Stephenson, 1993b*
	40.0	-0.32	Stephenson, 1993b*
	50.0	-0.31	Stephenson, 1993b*
	60.0	-0.34	Stephenson, 1993b*
Metamitron [41394-05-2]	20	-2.08	Worthing and Hance, 1991
Metazachlor [67129-08-2]	20	-4.21	Worthing and Hance, 1991
Methabenzthiazuron [18691-97-9]	20	-3.57	Worthing and Hance, 1991
Methacrifos [30864-28-9]	20	-2.78	Worthing and Hance, 1991
Methane [74-82-8]	25	-2.79	McAuliffe, 1966
	25.0	-2.79	McAuliffe, 1963
Methazole [20354-26-1]	25	-5.24	Humburg et al., 1989
Methidathion [950-37-8]	20	-3.08	Führer and Geiger, 1977
Methiocarb [2032-65-7]	20	-3.92	Worthing and Hance, 1991
Methomyl [16752-77-5]	25	-0.45	Worthing and Hance, 1991
Methoprotryne [841-06-5]	20	-2.93	Führer and Geiger, 1977
4-Methoxy-2-nitrophenol [1568-70-3]	20.0	-2.84	Schwarzenbach et al., 1988
9-Methylanthracene [779-02-2]	25.0	-5.56	Vadas et al., 1991
	25	-5.87	Mackay and Shiu, 1977
1-Methylbenz[a]anthracene [2498-77-3]	27	-6.64	Davis et al., 1942
9-Methylbenz[a]anthracene [2381-16-0]	24	-6.82	Hollifield, 1979
	27	-6.56	Davis et al., 1942
10-Methylbenz[a]anthracene [2381-15-9]	24	-7.34	Hollifield, 1979
	27	-6.64	Davis et al., 1942
Methyl benzoate [93-58-3]	25.0	-1.74	Stephenson and Stuart, 1986*
4-Methylbiphenyl [644-08-6]	4.9	-4.96	Doucette and Andren, 1988a
	25.0	-4.62	Doucette and Andren, 1988a
	40.0	-4.38	Doucette and Andren, 1988a
	25	-5.15	Li et al., 1993
3-Methylbutanoic acid [503-74-2]	25	-0.33	Amoore and Buttery, 1978

Compound [CAS No.]	Temp. °C	Log S mol/L	Reference
2-Methyl-1-butanol [137-32-6]	0.5	-0.35	Stephenson et al., 1984*
	9.7	-0.42	Stephenson et al., 1984*
	19.6	-0.49	Stephenson et al., 1984*
	20	-0.49	Ginnings and Baum, 1937
	25	-0.47	Ginnings and Baum, 1937
	25	-0.47	Crittenden and Hixon, 1954
	29.6	-0.55	Stephenson et al., 1984*
	30	-0.54	Ginnings and Baum, 1937
	39.3	-0.57	Stephenson et al., 1984*
	49.6	-0.60	Stephenson et al., 1984*
	59.3	-0.63	Stephenson et al., 1984*
	69.5	-0.60	Stephenson et al., 1984*
	79.7	-0.58	Stephenson et al., 1984*
	90.8	-0.55	Stephenson et al., 1984*
2-Methyl-2-butanol [75-85-4]	0.5	0.25	Stephenson et al., 1984*
	9.8	0.22	Stephenson et al., 1984*
	20.8	0.12	Stephenson et al., 1984*
	20	0.15	Ginnings and Baum, 1937
	25	0.10	Ginnings and Baum, 1937
	29.5	0.05	Stephenson et al., 1984*
	30	0.07	Ginnings and Baum, 1937
	39.5	-0.03	Stephenson et al., 1984*
	49.0	-0.09	Stephenson et al., 1984*
	60.0	-0.15	Stephenson et al., 1984*
	70.2	-0.17	Stephenson et al., 1984*
	80.1	-0.19	Stephenson et al., 1984*
	90.2	-0.21	Stephenson et al., 1984*
3-Methyl-2-butanol [598-75-4]	0	-0.02	Stephenson et al., 1984*
	10.1	-0.09	Stephenson et al., 1984*
	20	-0.16	Ginnings and Baum, 1937
	20.0	-0.17	Stephenson et al., 1984*
	25	-0.20	Ginnings and Baum, 1937
	30	-0.23	Ginnings and Baum, 1937
	30.0	-0.24	Stephenson et al., 1984*
	40.0	-0.30	Stephenson et al., 1984*
	50.0	-0.34	Stephenson et al., 1984*
	60.0	-0.42	Stephenson et al., 1984*
	70.0	-0.38	Stephenson et al., 1984*
	79.5	-0.39	Stephenson et al., 1984*
	90.0	-0.39	Stephenson et al., 1984*
3-Methyl-2-butanone [563-80-4]	0	0.06	Stephenson, 1992*
	9.5	-0.03	Stephenson, 1992*
	18.3	-0.11	Stephenson, 1992*
	25	-0.15	Ginnings et al., 1940
	30.8	-0.17	Stephenson, 1992*
	39.6	-0.24	Stephenson, 1992*
	50.0	-0.25	Stephenson, 1992*
	60.8	-0.29	Stephenson, 1992*
	69.8	-0.31	Stephenson, 1992*
	80.0	-0.34	Stephenson, 1992*
	89.0	-0.38	Stephenson, 1992*
2-Methyl-2-butene [513-35-9]	15	-2.29[a]	Natarajan and Venkatachalam, 1972
	15	-2.37[d]	Natarajan and Venkatachalam, 1972
	20	-2.31[a]	Natarajan and Venkatachalam, 1972
	20	-2.38[d]	Natarajan and Venkatachalam, 1972
	25	-2.34[a]	Natarajan and Venkatachalam, 1972
	25	-2.40[d]	Natarajan and Venkatachalam, 1972
N-Methylbutylamine [110-68-9]	43.0	0.13	Stephenson, 1993a*
	44.0	-0.10	Stephenson, 1993a*
	50.0	-0.06	Stephenson, 1993a*
	60.0	-0.26	Stephenson, 1993a*
	70.0	-0.27	Stephenson, 1993a*
	80.0	-0.29	Stephenson, 1993a*
	90.0	-0.37	Stephenson, 1993a*

Compound [CAS No.]	Temp. °C	Log S mol/L	Reference
(1-Methylbutyl)amine [625-30-9]	87.0	0.34	Stephenson, 1993a*
	88.0	0.13	Stephenson, 1993a*
	89.0	0.13	Stephenson, 1993a*
	90.0	0.06	Stephenson, 1993a*
Methyl tert-butyl ether [1634-04-4]	0	-0.03	Stephenson, 1992*
	9.7	-0.24	Stephenson, 1992*
	19.8	-0.32	Stephenson, 1992*
	25	-0.24	McBain and Richards, 1946
	29.6	-0.45	Stephenson, 1992*
	39.3	-0.55	Stephenson, 1992*
	48.6	-0.67	Stephenson, 1992*
2-Methylbutyraldehyde [96-17-3]	0	-0.52	Stephenson, 1993b*
	10.0	-0.67	Stephenson, 1993b*
	20.0	-0.81	Stephenson, 1993b*
	30.0	-0.79	Stephenson, 1993b*
	40.0	-0.85	Stephenson, 1993b*
	50.0	-0.90	Stephenson, 1993b*
	60.0	-0.94	Stephenson, 1993b*
	70.0	-0.97	Stephenson, 1993b*
	80.0	-0.97	Stephenson, 1993b*
	90.0	-1.10	Stephenson, 1993b*
Methyl butyrate [623-42-7]	25.0	-0.76	Stephenson and Stuart, 1986*
Methyl chloroacetate [96-34-4]	25.0	-0.33	Stephenson and Stuart, 1986*
Methyl 2-chlorobutyrate [3153-37-5]	26.0	-1.09	Stephenson and Stuart, 1986*
Methyl 2-chloropropionate [17639-93-9]	25.0	-0.79	Stephenson and Stuart, 1986*
3-Methylcholanthrene [56-49-5]	25	-7.92	Means et al., 1980
	25	-7.97	Mackay and Shiu, 1977
5-Methylchrysene [3697-24-3]	27	-5.69	Davis et al., 1942
6-Methylchrysene [1705-85-7]	27	-6.57	Davis et al., 1942
2-Methylcyclohexanol [583-59-5]	25.0	-0.83	Stephenson and Stuart, 1986*
3-Methylcyclohexanol [591-23-1]	25.0	-0.93	Stephenson and Stuart, 1986*
4-Methylcyclohexanol [589-91-3]	25.0	-0.97	Stephenson and Stuart, 1986*
4-Methylcyclohexanone [589-92-4]	0	-0.50	Stephenson, 1992*
	9.6	-0.60	Stephenson, 1992*
	20.1	-0.66	Stephenson, 1992*
	29.7	-0.72	Stephenson, 1992*
	40.3	-0.76	Stephenson, 1992*
	50.1	-0.77	Stephenson, 1992*
	61.0	-0.78	Stephenson, 1992*
	72.0	-0.77	Stephenson, 1992*
	80.5	-0.76	Stephenson, 1992*
	91.8	-0.75	Stephenson, 1992*
1-Methylcyclohexene [591-47-9]	25	-3.27	McAuliffe, 1966
Methyl decanoate [110-42-9]	25.0	-4.69	Tewari et al., 1982
Methyl dichloroacetate [116-54-1]	25.0	-1.64	Stephenson and Stuart, 1986*
6-Methyl-2,4-dinitrophenol [534-52-1]	20.0	-3.00	Schwarzenbach et al., 1988
Methyldymron [42609-73-4]	20	-3.35	Worthing and Hance, 1991
Methyl enanthate [106-73-0]	25.0	-2.20	Stephenson and Stuart, 1986*
1-Methylfluorene [1730-37-6]	25	-4.96	Billington et al., 1988
Methyl hexanoate [106-70-7]	25.0	-1.89	Stephenson and Stuart, 1986*
2-Methyl-2-hexanol [625-23-0]	20	-1.03	Ginnings and Hauser, 1938
	25	-1.08	Ginnings and Hauser, 1938
	30	-1.12	Ginnings and Hauser, 1938
3-Methyl-3-hexanol [597-96-6]	20	-0.93	Ginnings and Hauser, 1938
	25	-0.99	Ginnings and Hauser, 1938
	30	-1.03	Ginnings and Hauser, 1938
5-Methyl-2-hexanol [627-59-8]	25.0	-1.38	Stephenson and Stuart, 1986*
5-Methyl-2-hexanone [110-12-3]	0	-1.15	Stephenson, 1992*
	9.0	-1.24	Stephenson, 1992*
	19.1	-1.33	Stephenson, 1992*
	29.7	-1.40	Stephenson, 1992*
	39.5	-1.43	Stephenson, 1992*
	49.8	-1.45	Stephenson, 1992*
	59.5	-1.45	Stephenson, 1992*

Compound [CAS No.]	Temp. °C	Log S mol/L	Reference
	70.0	-1.44	Stephenson, 1992*
	79.9	-1.41	Stephenson, 1992*
	89.7	-1.44	Stephenson, 1992*
5-Methyl-3-hexanone [623-56-3]	0	-1.17	Stephenson, 1992*
	9.5	-1.28	Stephenson, 1992*
	19.6	-1.39	Stephenson, 1992*
	31.0	-1.46	Stephenson, 1992*
	39.6	-1.50	Stephenson, 1992*
	50.5	-1.53	Stephenson, 1992*
	61.3	-1.54	Stephenson, 1992*
	70.6	-1.54	Stephenson, 1992*
	81.2	-1.55	Stephenson, 1992*
	91.4	-1.55	Stephenson, 1992*
(1-Methylhexyl)amine [123-82-0]	3.0	-1.08	Stephenson, 1993a*
	4.0	-1.07	Stephenson, 1993a*
	6.0	-1.11	Stephenson, 1993a
	10.0	-1.18	Stephenson, 1993a*
	20.0	-1.36	Stephenson, 1993a*
	30.0	-1.42	Stephenson, 1993a*
	40.0	-1.46	Stephenson, 1993a*
	50.0	-1.46	Stephenson, 1993a*
	60.0	-1.45	Stephenson, 1993a*
	70.0	-1.48	Stephenson, 1993a*
	80.0	-1.49	Stephenson, 1993a*
	90.0	-1.50	Stephenson, 1993a*
Methyl isobutyrate [547-63-7]	25.0	-0.75	Stephenson and Stuart, 1986*
Methyl isothiocyanate [556-61-6]	20	-0.95	Worthing and Hance, 1991
Methyl p-methoxybenzoate [121-98-2]	20	-2.41	Corby and Elworthy, 1971
1-Methylnaphthalene [90-12-0]	8.6	-3.85	Schwarz, 1977
	10	-3.80	Schwarz and Wasik, 1977
	14.0	-3.80	Schwarz, 1977
	14	-3.70	Schwarz and Wasik, 1977
	17.1	-3.79	Schwarz, 1977
	20	-3.70	Schwarz and Wasik, 1977
	20.0	-3.75	Schwarz, 1977
	22	-3.68	Schwarz and Wasik, 1977
	23.0	-3.71	Schwarz, 1977
	25.0	-3.74	Eganhouse and Calder, 1976
	25	-3.68	Schwarz and Wasik, 1977
	25.0	-3.68	Schwarz, 1977
	25	-3.70	Mackay and Shiu, 1977
	26.1	-3.67	Schwarz, 1977
	29.2	-3.63	Schwarz, 1977
	31.7	-3.59	Schwarz, 1977
3-Methyl-2-nitrophenol [4920-77-8]	20.0	-1.64	Schwarzenbach et al., 1988
3-Methyl-4-nitrophenol [2581-34-2]	20.0	-2.11	Schwarzenbach et al., 1988
4-Methyl-2-nitrophenol [119-33-5]	20.0	-2.55	Schwarzenbach et al., 1988
5-Methyl-2-nitrophenol [700-838-9]	20.0	-2.75	Schwarzenbach et al., 1988
Methyl nonanoate [1731-84-6]	25.0	-3.88	Tewari et al., 1982
Methyl pentanoate [624-24-8]	25.0	-6.05	Price, 1976
2-Methyl-1-pentanol [105-30-6]	0	-0.90	Stephenson et al., 1984*
	10.0	-0.99	Stephenson et al., 1984*
	19.6	-1.07	Stephenson et al., 1984*
	25	-1.23	Crittenden and Hixon, 1954
	30.8	-1.11	Stephenson et al., 1984*
	40.3	-1.12	Stephenson et al., 1984*
	50.0	-1.13	Stephenson et al., 1984*
	60.3	-1.13	Stephenson et al., 1984*
	70.1	-1.12	Stephenson et al., 1984*
	80.3	-1.09	Stephenson et al., 1984*
	90.7	-1.05	Stephenson et al., 1984*
2-Methyl-2-pentanol [590-36-3]	20	-0.45	Ginnings and Webb, 1938
	25	-0.50	Ginnings and Webb, 1938
	30	-0.53	Ginnings and Webb, 1938

Compound [CAS No.]	Temp. °C	Log S mol/L	Reference
2-Methyl-3-pentanol [565-67-3]	20	-0.66	Ginnings and Webb, 1938
	25	-0.71	Ginnings and Webb, 1938
	30	-0.75	Ginnings and Webb, 1938
3-Methyl-1-pentanol [589-35-5]	25.0	-1.04	Stephenson and Stuart, 1986*
3-Methyl-2-pentanol [565-60-6]	20	-0.69	Ginnings and Webb, 1938
	25	-0.72	Ginnings and Webb, 1938
	30	-0.75	Ginnings and Webb, 1938
3-Methyl-3-pentanol [77-74-7]	9.8	-0.35	Stephenson et al., 1984*
	19.5	-0.46	Stephenson et al., 1984*
	20	-0.32	Ginnings and Webb, 1938
	25	-0.38	Ginnings and Webb, 1938
	29.8	-0.47	Stephenson et al., 1984*
	30	-0.42	Ginnings and Webb, 1938
	39.8	-0.57	Stephenson et al., 1984*
	49.7	-0.62	Stephenson et al., 1984*
	59.5	-0.70	Stephenson et al., 1984*
	70.1	-0.73	Stephenson et al., 1984*
	80.1	-0.74	Stephenson et al., 1984*
	90.4	-0.78	Stephenson et al., 1984*
4-Methyl-2-pentanol [108-11-2]	0	-0.56	Stephenson et al., 1984*
	9.7	-0.69	Stephenson et al., 1984*
	20	-0.75	Ginnings and Webb, 1938
	20.0	-0.80	Stephenson et al., 1984*
	25	-0.79	Ginnings and Webb, 1938
	25.0	-0.81	Ginnings and Hauser, 1938
	25	-0.78	Crittenden and Hixon, 1954
	30	-0.82	Ginnings and Webb, 1938
	30.0	-0.82	Stephenson et al., 1984*
	40.3	-0.90	Stephenson et al., 1984*
	50.0	-0.93	Stephenson et al., 1984*
	60.1	-0.96	Stephenson et al., 1984*
	70.2	-0.96	Stephenson et al., 1984*
	80.2	-0.92	Stephenson et al., 1984*
	90.2	-0.94	Stephenson et al., 1984*
2-Methyl-3-pentanone [565-69-5]	25	-0.82	Ginnings et al., 1940
3-Methyl-2-pentanone [565-61-7]	25	-0.68	Ginnings et al., 1940
4-Methyl-1-penten-3-ol [4798-45-2]	20	-0.48	Ginnings et al., 1939
	25	-0.51	Ginnings et al., 1939
	30	-0.54	Ginnings et al., 1939
Methyl tert--pentyl ether [994-05-8]	0	-0.62	Stephenson, 1992*
	9.7	-0.81	Stephenson, 1992*
	19.9	-0.97	Stephenson, 1992*
	31.0	-1.08	Stephenson, 1992*
	40.3	-1.20	Stephenson, 1992*
	48.8	-1.25	Stephenson, 1992*
	59.2	-1.34	Stephenson, 1992*
	69.2	-1.36	Stephenson, 1992*
	79.2	-1.45	Stephenson, 1992*
3-Methylphenol [108-39-4]	25.0	-1.59	Tewari et al., 1982
N-Methylpiperidine [626-67-5]	44.0	0.18	Stephenson, 1993*
	45.0	0.09	Stephenson, 1993*
	46.0	0.03	Stephenson, 1993*
	47.0	-0.04	Stephenson, 1993*
	48.0	-0.06	Stephenson, 1993*
	49.0	-0.12	Stephenson, 1993*
	50.0	-0.11	Stephenson, 1993*
	51.0	-0.14	Stephenson, 1993*
	52.0	-0.15	Stephenson, 1993*
	54.0	-0.21	Stephenson, 1993*
	56.0	-0.25	Stephenson, 1993*
	70.0	-0.33	Stephenson, 1993*
	80.0	-0.44	Stephenson, 1993*
	90.0	-0.50	Stephenson, 1993*
2-Methylpiperidine [109-05-7]	70.0	0.26	Stephenson, 1993*

Compound [CAS No.]	Temp. °C	Log S mol/L	Reference
	71.0	0.17	Stephenson, 1993*
	72.0	0.12	Stephenson, 1993*
	73.0	0.07	Stephenson, 1993*
	74.0	0.05	Stephenson, 1993*
	74.0	0.02	Stephenson, 1993*
	76.0	-0.01	Stephenson, 1993*
	77.0	0.00	Stephenson, 1993*
	78.0	-0.05	Stephenson, 1993*
	79.0	-0.06	Stephenson, 1993*
	80.0	-0.08	Stephenson, 1993*
	81.0	-0.09	Stephenson, 1993*
	82.0	-0.09	Stephenson, 1993*
	84.0	-0.10	Stephenson, 1993*
	86.0	-0.12	Stephenson, 1993*
	88.0	-0.14	Stephenson, 1993*
	90.0	-0.14	Stephenson, 1993*
	92.5	-0.16	Stephenson, 1993*
3-Methylpiperidine [626-56-2]	67.0	0.14	Stephenson, 1993*
	68.0	0.14	Stephenson, 1993*
	70.0	0.01	Stephenson, 1993*
	80.0	-0.16	Stephenson, 1993*
	90.0	-0.19	Stephenson, 1993*
4-Methylpiperidine [626-58-4]	86.0	0.14	Stephenson, 1993*
	87.0	0.09	Stephenson, 1993*
	88.0	0.06	Stephenson, 1993*
	90.0	0.00	Stephenson, 1993*
	95.0	-0.05	Stephenson, 1993*
Methyl propionate [554-12-1]	25.0	-0.17	Stephenson and Stuart, 1986*
Methyl salicylate [119-36-8]	25.0	-2.10	Stephenson and Stuart, 1986*
2-Methyltetrahydrofuran [96-47-9]	0	0.39	Stephenson, 1992*
	9.5	0.32	Stephenson, 1992*
	19.3	0.22	Stephenson, 1992*
	29.5	0.12	Stephenson, 1992*
	39.6	0.03	Stephenson, 1992*
	50.1	-0.04	Stephenson, 1992*
	60.7	-0.12	Stephenson, 1992*
	70.6	-0.16	Stephenson, 1992*
3-Methyltetrahydropyran [26093-63-0]	0	-0.44	Stephenson, 1992*
	9.7	-0.67	Stephenson, 1992*
	29.8	-0.74	Stephenson, 1992*
	40.8	-0.81	Stephenson, 1992*
	50.3	-0.85	Stephenson, 1992*
	60.5	-0.89	Stephenson, 1992*
	70.5	-0.91	Stephenson, 1992*
	81.6	-0.92	Stephenson, 1992*
	92.0	-0.98	Stephenson, 1992*
Methyl trichloroacetate [598-99-2]	25.0	-2.29	Stephenson and Stuart, 1986*
Methyl trimethylacetate [598-98-1]	25.0	-1.74	Stephenson and Stuart, 1986*
2-Methylvaleraldehyde [123-15-9]	0	-1.26	Stephenson, 1993b*
	10.0	-1.25	Stephenson, 1993b*
	20.0	-1.33	Stephenson, 1993b*
	30.0	-1.31	Stephenson, 1993b*
	40.0	-1.42	Stephenson, 1993b*
	50.0	-1.38	Stephenson, 1993b*
	60.0	-1.44	Stephenson, 1993b*
	70.0	-1.41	Stephenson, 1993b*
	80.0	-1.44	Stephenson, 1993b*
	90.0	-1.44	Stephenson, 1993b*
Metolachlor [51218-45-2]	20	-2.81	Ellgehausen et al., 1980
	20	-2.73	Ellgehausen et al., 1981
Metolcarb [1129-41-5]	30	-1.80	Worthing and Hance, 1991
Metoxuron [19937-59-8]	20	-2.58	Ellgehausen et al., 1981
Metribuzin [21087-64-9]	20	-2.24	Humburg et al., 1989
Metsulfuron-methyl [74223-64-6]	25	-3.15	pH 4.6, Humburg et al., 1989

Compound [CAS No.]	Temp. °C	Log S mol/L	Reference
	25	-2.34	pH 5.4, Humburg et al., 1989
	25	-1.60	pH 6.7, Humburg et al., 1989
Mibolerone [3704-09-4]	37	-3.82	Hamlin et al., 1965
Mirex [2385-85-5]	24	-6.44	Hollifield, 1979
Molinate [2212-67-1]	20	-2.37	Humburg et al., 1989
Monalide [7287-36-7]	23	-4.02	Worthing and Hance, 1991
Monocrotophos [6923-22-4]	20	0.65	Worthing and Hance, 1991
Monolinuron [1746-81-2]	25	-2.47	Worthing and Hance, 1991
Monuron [150-68-5]	20	-2.94	Ellgehausen et al., 1980
	20	-3.00	Ellgehausen et al., 1981
	20	-3.00	Fühner and Geiger, 1977
Myclobutanil [88671-89-0]	25	-3.31	Worthing and Hance, 1991
Myrcene [123-35-3]	25	-3.66	Fichan et al., 1999
1-Naphthalenemethanol [4780-79-4]	25	-2.12	Southworth and Keller, 1986
1-Naphthol [90-15-3]	23	-2.23	Fu and Luthy, 1986
	25	-2.22	Hassett et al., 1981
	25	-2.10	Southworth and Keller, 1986
Napropamide [5299-99-7]	20	-3.57	Humburg et al., 1989
	25	-3.60	Gerstl and Yaron, 1983
Naptalam [132-66-1]	20	-3.16	Humburg et al., 1989
Naptalam-sodium [132-67-2]	25	-0.10	Humburg et al., 1989
2-(1-Naphthyl)acetamide [86-86-2]	40	-3.68	Worthing and Hance, 1991
2-(1-Naphthyl)acetic acid [86-87-3]	20	-2.65	Worthing and Hance, 1991
2-(2-Naphthyloxy)propionanilide [52570-16-8]	27	-5.60	Worthing and Hance, 1991
Napropamide [15299-99-7]	20	-3.57	Worthing and Hance, 1991
Neburon [555-37-3]	20	-4.76	Fühner and Geiger, 1977
Niclosamide [50-65-7]	20	-5.31	pH 6.4, Worthing and Hance, 1991
	20	-3.47	pH 9.1, Worthing and Hance, 1991
2-Nitroanisole [91-23-6]	30.00	-1.96	Gross et al., 1933
4-Nitroanisole [100-02-7]	30.00	-2.41	Gross et al., 1933
3-Nitrophenol [554-84-7]	20	-1.08	Hashimoto et al., 1984
Nitrothal-isopropyl [10552-74-6]	20	-5.88	Worthing and Hance, 1991
2,2',3,3',4,4',5,5',6-Nonachlorobiphenyl [40186-72-9]	22.0	-9.77	Opperhuizen et al., 1988
	25	-10.26	Dickhut et al., 1986
2,2',3,3',4,5,5',6,6'-Nonachlorobiphenyl [52663-77-1]	25	-10.41	Miller et al., 1984
1,8-Nonadiyne [2396-65-8]	25	-2.98	McAuliffe, 1966
Nonanal [124-19-6]	20.0	-2.68	Stephenson, 1993b*
	30.0	-2.37	Stephenson, 1993b*
	40.0	-2.55	Stephenson, 1993b*
	50.0	-2.31	Stephenson, 1993b*
	60.0	-2.68	Stephenson, 1993b*
	70.0	-2.37	Stephenson, 1993b*
	80.0	-2.45	Stephenson, 1993b*
	90.0	-2.45	Stephenson, 1993b*
Nonanoic acid [112-05-0]	20.0	-2.78	Ralston and Hoerr, 1942
1-Nonanol [143-08-8]	20.0	-3.03	Hommelen, 1959
	25.0	-3.13	Tewari et al., 1982
	25.0	-2.68	Stephenson and Stuart, 1986*
2-Nonanol [628-99-9]	25.0	-2.59	Stephenson and Stuart, 1986*
2-Nonanone [821-55-6]	0	-2.43	Stephenson, 1992*
	9.8	-2.52	Stephenson, 1992*
	25.0	-2.58	Tewari et al., 1982
	30.3	-2.65	Stephenson, 1992*
	39.8	-2.62	Stephenson, 1992*
	50.0	-2.75	Stephenson, 1992*
	60.0	-2.66	Stephenson, 1992*
	70.4	-2.62	Stephenson, 1992*
	81.2	-2.72	Stephenson, 1992*
	91.2	-2.57	Stephenson, 1992*
3-Nonanone [925-78-0]	0	-2.46	Stephenson, 1992*
	9.5	-2.49	Stephenson, 1992*
	29.7	-2.41	Stephenson, 1992*
	39.8	-2.50	Stephenson, 1992*
	50.0	-2.42	Stephenson, 1992*

Compound [CAS No.]	Temp. °C	Log S mol/L	Reference
	60.6	-2.39	Stephenson, 1992*
	80.2	-2.49	Stephenson, 1992*
5-Nonanone [502-56-7]	0	-2.30	Stephenson, 1992*
	9.5	-2.35	Stephenson, 1992*
	19.8	-2.42	Stephenson, 1992*
	30.1	-2.47	Stephenson, 1992*
	40.8	-2.55	Stephenson, 1992*
	50.0	-2.69	Stephenson, 1992*
	60.6	-2.61	Stephenson, 1992*
	70.3	-2.66	Stephenson, 1992*
	80.0	-2.69	Stephenson, 1992*
	90.6	-2.57	Stephenson, 1992*
1-Nonene [124-11-8]	25.0	-5.05	Tewari et al., 1982
Nonyl formate [5451-92-3]	0	-2.87	Stephenson, 1992*
	9.5	-3.16	Stephenson, 1992*
	30.7	-3.46	Stephenson, 1992*
	40.1	-3.39	Stephenson, 1992*
	60.1	-2.77	Stephenson, 1992*
	70.2	-2.80	Stephenson, 1992*
	80.1	-2.74	Stephenson, 1992*
	90.3	-2.65	Stephenson, 1992*
4-Nonylphenol [104-40-5]	20.5	-4.61	Ahel and Giger, 1993
1-Nonyne [3452-09-3]	25	-4.24	McAuliffe, 1966
Norflurazon [27314-13-2]	25	-4.08	Humburg et al., 1989
Norharmane [244-63-3]	16	-3.86	pH 13, Burrows et al., 1996
	17	-3.80	pH 13, Burrows et al., 1996
	37	-3.60	pH 13, Burrows et al., 1996
	38	-3.59	pH 13, Burrows et al., 1996
	45	-3.54	pH 13, Burrows et al., 1996
Noruron [2163-79-3]	25	-3.17	Humburg et al., 1989
Nuarimol [63284-71-9]	25	-4.08	Worthing and Hance, 1991
2,2',3,3',4,4',5,5'-Octachlorobiphenyl [35694-08-7]	22.0	-9.54	Opperhuizen et al., 1988
2,2',3,3',5,5',6,6'-Octachlorobiphenyl [2136-99-4]	25	-9.47	Dickhut et al., 1986
	25	-9.04	Miller et al., 1984
Octachlorodibenzo-p-dioxin [3268-87-9]	20.0	-12.06	Friesen et al., 1985
	20	-12.06	Webster et al., 1985
	25	-12.79	Shiu et al., 1988
	40.0	-12.17	Doucette and Andren, 1988a
	60.0	-11.40	Doucette and Andren, 1988a
	80.0	-11.77	Doucette and Andren, 1988a
Octachlorodibenzofuran [39001-02-0]	39.5	-11.06	Doucette and Andren, 1988a
	58.6	-10.50	Doucette and Andren, 1988a
	80.0	-10.86	Doucette and Andren, 1988a
Octadecane [593-45-3]	25.0	-8.08	Sutton and Calder, 1974
Octadecan acid [57-11-4]	20.0	-4.99	Ralston and Hoerr, 1942
	25	-5.70	pH 5.7, Robb, 1966
Octanal [124-13-0]	0	-2.11	Stephenson, 1993b*
	10.0	-2.11	Stephenson, 1993b*
	20.0	-2.11	Stephenson, 1993b*
Octanoic acid [127-04-2]	20.0	-2.33	Ralston and Hoerr, 1942
	30	-2.46	Bell, 1971
1-Octanol [111-87-5]	20.5	-2.42	Stephenson et al., 1984*
	20.0	-2.43	Hommelen, 1959
	22	-2.46	Jepsen and Lick, 1999
	25	-2.42	Shinoda et al., 1959
	25.0	-2.39	Sanemassa et al., 1987
	25	-2.41	Crittenden and Hixon, 1954
	25	-2.49	Li and Andren, 1994
	25	-2.35	Butler et al., 1933
	30.6	-2.31	Stephenson et al., 1984*
	40.1	-2.30	Stephenson et al., 1984*
	50.0	-2.09	Stephenson et al., 1984*
	60.3	-2.17	Stephenson et al., 1984*
	70.3	-2.23	Stephenson et al., 1984*

Compound [CAS No.]	Temp. °C	Log S mol/L	Reference
	80.1	-2.18	Stephenson et al., 1984*
	90.3	-2.18	Stephenson et al., 1984*
2-Octanol [123-96-6]	20.0	-2.07	Hommelen, 1959
	25	-1.51	Crittenden and Hixon, 1954
	25.0	-2.07	Stephenson and Stuart, 1986*
	25	-2.01	Mitchell, 1926
3-Octanol [589-98-0]	25.0	-1.97	Stephenson and Stuart, 1986*
2-Octanone [111-13-7]	0	-1.75	Stephenson, 1992*
	9.7	-1.78	Stephenson, 1992*
	19.2	-1.87	Stephenson, 1992*
	25.0	-2.05	Tewari et al., 1982
	29.7	-2.10	Stephenson, 1992*
	40.4	-2.10	Stephenson, 1992*
	50.0	-2.06	Stephenson, 1992*
	60.4	-2.08	Stephenson, 1992*
	80.6	-2.10	Stephenson, 1992*
	91.0	-2.14	Stephenson, 1992*
3-Octanone [106-68-3]	0	-1.85	Stephenson, 1992*
	20.1	-1.97	Stephenson, 1992*
	29.7	-1.98	Stephenson, 1992*
	39.6	-2.03	Stephenson, 1992*
	50.0	-2.07	Stephenson, 1992*
	60.0	-2.12	Stephenson, 1992*
	70.5	-2.11	Stephenson, 1992*
	80.3	-2.04	Stephenson, 1992*
	91.0	-2.08	Stephenson, 1992*
2-Octene [111-67-1]	15	-3.53[a]	Natarajan and Venkatachalam, 1972
	20	-3.60[a]	Natarajan and Venkatachalam, 1972
	25	-3.67[a]	Natarajan and Venkatachalam, 1972
	25	-3.74[b]	Natarajan and Venkatachalam, 1972
	30	-3.80[b]	Natarajan and Venkatachalam, 1972
	35	-3.86[b]	Natarajan and Venkatachalam, 1972
Octyl acetate [112-14-1]	0	-2.91	Stephenson, 1992*
	19.2	-2.94	Stephenson, 1992*
	29.7	-2.98	Stephenson, 1992*
	40.0	-2.98	Stephenson, 1992*
	50.5	-3.03	Stephenson, 1992*
	71.2	-3.01	Stephenson, 1992*
	80.6	-3.09	Stephenson, 1992*
	92.1	-3.16	Stephenson, 1992*
Octylamine [111-86-4]	40.0	-1.94	Stephenson, 1993a*
	50.0	-1.71	Stephenson, 1993a*
	60.0	-1.68	Stephenson, 1993a*
	70.0	-1.57	Stephenson, 1993a*
	80.0	-1.70	Stephenson, 1993a*
	90.0	-1.91	Stephenson, 1993a*
2-Octylbenzene [777-22-0]	25	-5.80	Deno and Berkheimer, 1960
Octyl formate [112-32-3]	25.0	-2.48	Stephenson and Stuart, 1986*
1-Octyne [629-05-0]	25	-3.66	McAuliffe, 1966
	25.0	-9.09	Doucette and Andren, 1988a
	40.0	-8.75	Doucette and Andren, 1988a
Ofurace [58810-48-3]	21	-3.30	Worthing and Hance, 1991
Orbencarb [34622-58-7]	27	-4.03	Worthing and Hance, 1991
Oryzalin [19044-88-3]	25	-5.12	Humburg et al., 1989
Oxabetrinil [74782-23-3]	20	-4.06	Worthing and Hance, 1991
Oxadiazon [19666-30-9]	20	-5.69	Humburg et al., 1989
Oxadixyl [77732-09-3]	25	-1.91	Worthing and Hance, 1991
Oxamyl [23135-22-0]	25	0.11	Worthing and Hance, 1991
Oxolinic acid [14698-29-4]	25	-7.94	Worthing and Hance, 1991
Oxycarboxin [5259-88-1]	25	-2.43	Worthing and Hance, 1991
Oxyfluorfen [42874-03-3]	25	-6.56	Humburg et al., 1989
Parathion-methyl [298-00-0]	20	-3.68	Worthing and Hance, 1991
Pebulate [1114-71-2]	20	-3.53	Humburg et al., 1989
Penconazole [66246-88-6]	20	-3.61	Worthing and Hance, 1991

Compound [CAS No.]	Temp. °C	Log S mol/L	Reference
Pencycuron [66063-05-6]	20	-6.04	Worthing and Hance, 1991
Pendimethalin [40487-42-1]	25	-2.01	Humburg et al., 1989
2,2',4,5,5'-Pentachlorobiphenyl [37680-72-3]	20	-7.91	Swann et al., 1983
	24	-7.51	Freed et al., 1979
	25	-7.23	Miller et al., 1984
2,2',4,6,6'-Pentachlorobiphenyl [56558-16-8]	25	-7.32	Dunnivant and Elzerman, 1988
2,3,4,5,6-Pentachlorobiphenyl [18259-05-7]	22.0	-7.38	Opperhuizen et al., 1988
	25	-7.91	Dunnivant and Elzerman, 1988
	25	-7.77	Miller et al., 1984
	25	-7.68	Dunnivant and Elzerman, 1988
	25	-7.91	Dulfer et al., 1995
1,2,3,4,7-Pentachloro-p-dioxin [39227-61-7]	20.0	-9.47	Friesen et al., 1985
	26.0	-9.33	Li and Andren, 1994
2,3,4,7,8-Pentachlorodibenzofuran [57117-31-4]	22.7	-9.16	Friesen et al., 1990
Pentadecanoic acid [1002-84-2]	20.0	-4.30	Ralston and Hoerr, 1942
1-Pentadecanol [629-76-5]	25	-6.35	Robb, 1966
1,5-Pentane dinitrate [3457-92-9]	25	-2.19	Fischer and Ballschmiter, 1998, 1998a
2,4-Pentanedione [123-54-6]	0	0.15	Stephenson, 1992*
	9.6	0.17	Stephenson, 1992*
	19.8	0.21	Stephenson, 1992*
	29.8	0.25	Stephenson, 1992*
	39.7	0.28	Stephenson, 1992*
	50.1	0.32	Stephenson, 1992*
	60.6	0.39	Stephenson, 1992*
	70.5	0.43	Stephenson, 1992*
	80.3	0.51	Stephenson, 1992*
	90.5	0.60	Stephenson, 1992*
Pentanoic acid [109-52-4]	30	-0.60	Bell, 1971
1-Pentanol [71-41-0]	0	-0.43	Stephenson et al., 1984*
	10.2	-0.53	Stephenson et al., 1984*
	20	-0.57	Ginnings and Baum, 1937
	20	-0.77	Laddha and Smith, 1948
	20.2	-0.59	Stephenson et al., 1984*
	25	-0.64	Li and Andren, 1994
	25	-0.54	Hansen et al., 1949
	25	-0.60	Ginnings and Baum, 1937
	30	-0.63	Ginnings and Baum, 1937
	25.0	-0.88	Tewari et al., 1982
	25	-0.60	Butler et al., 1933
	25	-0.60	Crittenden and Hixon, 1954
	25.0	-0.50	Booth and Everson, 1948
	25	-0.54	Donahue and Bartell, 1952
	30.6	-0.64	Stephenson et al., 1984*
	37	-0.67	Evans et al., 1978
	40.2	-0.67	Stephenson et al., 1984*
	50.0	-0.68	Stephenson et al., 1984*
	60.3	-0.68	Stephenson et al., 1984*
	70.0	-0.66	Stephenson et al., 1984*
	80.0	-0.65	Stephenson et al., 1984*
	90.7	-0.60	Stephenson et al., 1984*
2-Pentanol [6032-29-7]	0	-0.08	Stephenson et al., 1984*
	10.1	-0.18	Stephenson et al., 1984*
	19.5	-0.28	Stephenson et al., 1984*
	20	-0.25	Ginnings and Baum, 1937
	25	-0.30	Ginnings and Baum, 1937
	30	-0.32	Ginnings and Baum, 1937
	30.6	-0.35	Stephenson et al., 1984*
	40.0	-0.39	Stephenson et al., 1984*
	50.0	-0.42	Stephenson et al., 1984*
	60.0	-0.45	Stephenson et al., 1984*
	70.1	-0.46	Stephenson et al., 1984*
	79.9	-0.45	Stephenson et al., 1984*
	90.3	-0.46	Stephenson et al., 1984*
3-Pentanol [584-02-1]	0	-0.03	Stephenson et al., 1984*

Compound [CAS No.]	Temp. °C	Log S mol/L	Reference
	10.2	-0.10	Stephenson et al., 1984*
	20.0	-0.18	Stephenson et al., 1984*
	20	-0.19	Ginnings and Baum, 1937
	25	-0.23	Ginnings and Baum, 1937
	25	-0.24	Crittenden and Hixon, 1954
	30.0	-0.22	Stephenson et al., 1984*
	30	-0.26	Ginnings and Baum, 1937
	40.0	-0.33	Stephenson et al., 1984*
	50.0	-0.32	Stephenson et al., 1984*
	60.0	-0.40	Stephenson et al., 1984*
	70.0	-0.41	Stephenson et al., 1984*
	80.0	-0.43	Stephenson et al., 1984*
	90.0	-0.41	Stephenson et al., 1984*
2-Pentanone [107-87-9]	0	0.00	Stephenson, 1992*
	9.7	-0.10	Stephenson, 1992*
	19.7	-0.16	Stephenson, 1992*
	31.0	-0.24	Stephenson, 1992*
	39.6	-0.27	Stephenson, 1992*
	49.8	-0.31	Stephenson, 1992*
	60.1	-0.33	Stephenson, 1992*
	70.2	-0.33	Stephenson, 1992*
	80.0	-0.36	Stephenson, 1992*
	90.5	-0.40	Stephenson, 1992*
3-Pentanone [96-22-0]	0	-0.05	Stephenson, 1992*
	9.7	-0.14	Stephenson, 1992*
	19.3	-0.21	Stephenson, 1992*
	25	-0.25	Ginnings et al., 1940
	30.6	-0.31	Stephenson, 1992*
	40.3	-0.35	Stephenson, 1992*
	50.0	-0.38	Stephenson, 1992*
	60.1	-0.40	Stephenson, 1992*
	70.1	-0.42	Stephenson, 1992*
	80.2	-0.44	Stephenson, 1992*
	25.0	-0.28	Tewari et al., 1982
2-Pentenenitrile [13284-42-9]	20	-0.91	Du Pont, 1999e*
1-Penten-3-ol [616-25-1]	25	-0.02	Ginnings et al., 1939
3-Penten-2-ol [1569-50-2]	25	0.02	Ginnings et al., 1939
4-Penten-1-ol [821-09-0]	25	-0.18	Ginnings et al., 1939
Pentyl acetate [628-63-7]	25.0	-1.84	Stephenson and Stuart, 1986*
Pentylbenzene [528-68-1]	25.0	-4.59	Tewari et al., 1982
	25.0	-4.64	Owens et al., 1986
tert-Pentylbenzene [2049-95-8]	25.0	-4.25	Andrews and Keefer, 1950a
Pentyl butyrate [540-18-1]	25.0	-2.80	Stephenson and Stuart, 1986*
1-Pentyl nitrate [1002-16-0]	25	-2.57	Hauff et al., 1998
2-Pentyl nitrate [21981-48-6]	25	-2.48	Hauff et al., 1998
1-Pentyne [627-19-0]	25.0	-1.81	Tewari et al., 1982
	25	-1.64	McAuliffe, 1966
Perfluidone [37924-13-3]	22	-3.80	Worthing and Hance, 1991
Permethrin [52645-53-1]	30	-6.29	Worthing and Hance, 1991
Perylene [198-55-0]	25	-8.80	Mackay and Shiu, 1977
sec-Phenethyl alcohol [98-85-1]	25	-1.80	Southworth and Keller, 1986
Phenetole [103-73-1]	25.0	-2.33	Vesala, 1974
	25	-2.35	McGowan et al., 1966
Phenmedipham [13684-63-4]	20	-5.48	Führer and Geiger, 1977
Phenothrin [26002-80-2]	30	-2.24	Worthing and Hance, 1991
Phenoxyacetic acid [122-59-8]	10	-1.10	Freed et al., 1979
Phenthoate [2597-03-7]	24	-4.46	Worthing and Hance, 1991
Phenyl acetate [122-79-2]	0	-1.34	Stephenson, 1992*
	9.7	-1.35	Stephenson, 1992*
	20.0	-1.36	Stephenson, 1992*
	29.7	-1.39	Stephenson, 1992*
	40.0	-1.34	Stephenson, 1992*
	51.0	-1.36	Stephenson, 1992*
	60.7	-1.28	Stephenson, 1992*

Compound [CAS No.]	Temp. °C	Log S mol/L	Reference
	70.5	-1.29	Stephenson, 1992*
	80.2	-1.23	Stephenson, 1992*
	90.8	-1.18	Stephenson, 1992*
Phenylacetic acid [103-82-2]	20	-0.93	Freed et al., 1979
Phenylbutazone [50-33-9]	20	-4.08	Hafkenscheid and Tomlinson, 1983
2-Phenyldecane [4537-13-7]	25.0	-7.59	Sherblom et al., 1992
3-Phenyldecane [4621-36-7]	25.0	-7.42	Sherblom et al., 1992
4-Phenyldecane [4537-12-6]	25.0	-7.44	Sherblom et al., 1992
5-Phenyldecane [4537-11-5]	25.0	-7.46	Sherblom et al., 1992
2-Phenyldodecane [2719-61-1]	25.0	-8.40	Sherblom et al., 1992
3-Phenyldodecane [2400-00-2]	25.0	-8.15	Sherblom et al., 1992
4-Phenyldodecane [2719-64-4]	25.0	-8.30	Sherblom et al., 1992
5-Phenyldodecane [2719-63-6]	25.0	-8.30	Sherblom et al., 1992
6-Phenyldodecane [2719-62-2]	25.0	-8.40	Sherblom et al., 1992
m-Phenylenediamine [108-45-2]	25	0.51	Du Pont, 1999f*
	40	0.92	Du Pont, 1999f*
o-Phenylenediamine [95-54-5]	35	-0.43	Du Pont, 1999g*
Phenyl ether [101-84-8]	25.0	-3.96	Vesala, 1974
2-Phenylethanol [60-12-8]	25.0	-0.67	Stephenson and Stuart, 1986*
4-Phenyl-2-nitrophenol [885-82-5]	20.0	-4.41	Schwarzenbach et al., 1988
2-Phenylphenol [90-43-7]	25	-2.39	Worthing and Hance, 1991
1-Phenyl-1-propanol [93-54-9]	25.0	-1.12	Stephenson and Stuart, 1986*
2-Phenyltetradecane [4534-59-2]	25.0	-8.40	Sherblom et al., 1992
3-Phenyltetradecane [4534-58-1]	25.0	-8.30	Sherblom et al., 1992
4-Phenyltetradecane [4534-57-0]	25.0	-8.40	Sherblom et al., 1992
5-Phenyltetradecane [4534-56-9]	25.0	-8.30	Sherblom et al., 1992
6-Phenyltetradecane [4534-55-8]	25.0	-8.40	Sherblom et al., 1992
2-Phenyltridecane [4534-53-6]	25.0	-8.40	Sherblom et al., 1992
3-Phenyltridecane [4534-52-5]	25.0	-8.40	Sherblom et al., 1992
4-Phenyltridecane [4534-51-4]	25.0	-8.40	Sherblom et al., 1992
5-Phenyltridecane [4534-50-3]	25.0	-8.40	Sherblom et al., 1992
6-Phenyltridecane [4534-49-0]	25.0	-8.40	Sherblom et al., 1992
2-Phenylundecane [4536-88-3]	25.0	-8.10	Sherblom et al., 1992
3-Phenylundecane [4536-87-2]	25.0	-7.92	Sherblom et al., 1992
4-Phenylundecane [4536-86-1]	25.0	-8.05	Sherblom et al., 1992
5-Phenylundecane [4537-57-0]	25.0	-8.00	Sherblom et al., 1992
6-Phenylundecane [4537-14-8]	25.0	-7.96	Sherblom et al., 1992
Phosalone [2310-17-0]	20	-5.23	Chiou et al., 1977
Phosdiphen [36519-00-3]	20	-5.77	Worthing and Hance, 1991
Phosmet [732-11-6]	25	-4.16	Worthing and Hance, 1991
Phoxim [14816-18-3]	20	-5.30	Worthing and Hance, 1991
o-Phthalic acid [88-99-3]	27.00	-2.97	Han et al., 1999
m-Phthalic acid [121-91-5]	27.00	-1.38	Han et al., 1999
p-Phthalic acid [88-99-3]	27.00	-3.55	Han et al., 1999
Picloram [1918-02-1]	25	-2.75	Humburg et al., 1989
Pimaricin [7681-93-8]	20-22	-2.21	Worthing and Hance, 1991
Pindone [83-26-1]	25	-4.11	Worthing and Hance, 1991
(−)-α-Pinene [7785-26-4]	25	-4.43	Fichan et al., 1999
(−)-β-Pinene [18172-67-3]	25	-4.09	Fichan et al., 1999
α-Pinene oxide [1686-14-2]	25	-2.59	Fichan et al., 1999
Piperophos [24151-93-7]	20	-4.15	Worthing and Hance, 1991
Pirimicarb [23103-98-2]	25	-1.95	Worthing and Hance, 1991
Pirimiphos-methyl [29232-93-7]	30	-4.49	Worthing and Hance, 1991
Prallethrin [23031-36-9]	25	-4.57	Worthing and Hance, 1991
Prednisolone [50-24-8]	37	-3.01	Hamlin et al., 1965
Pretilachlor [51218-49-6]	20	-3.80	Worthing and Hance, 1991
Primisulfuron [86209-51-0]	20	-3.83	pH 7, Worthing and Hance, 1991
	20	-2.02	pH 9, Worthing and Hance, 1991
Prochloraz [67747-09-5]	25	-4.04	Worthing and Hance, 1991
Procymidone [32809-16-8]	25	-4.80	Worthing and Hance, 1991
Prodiamine [29091-21-2]	25	-3.43	Humburg et al., 1989
Profenofos [41198-08-7]	20	-4.27	Ellgehausen et al., 1980, 1981
Profluralin [26399-36-0]	20	-6.54	Ellgehausen et al., 1981
Proglinazine-ethyl [68228-18-2]	25	-2.56	Worthing and Hance, 1991

Compound [CAS No.]	Temp. °C	Log S mol/L	Reference
Prohexadione-calcium [124537-28-6]	20	-3.17	Worthing and Hance, 1991
Promecarb [2631-37-0]	25	-3.36	Worthing and Hance, 1991
Prometone [1610-18-0]	20	-2.56	Fühner and Geiger, 1977
	20	-2.48	Humburg et al., 1989
	26.0	-1.73	pH 2.0, Ward and Weber, 1968
	26.0	-2.35	pH 3.0, Ward and Weber, 1968
	26.0	-2.58	pH 5.0, Ward and Weber, 1968
	26.0	-2.52	pH 7.0, Ward and Weber, 1968
	26.0	-2.53	pH 10.0, Ward and Weber, 1968
Prometryn [7287-19-6]	20	-3.78	Fühner and Geiger, 1977
	20	-3.70	Humburg et al., 1989
	25	-4.68	Somasundaram et al., 1991
	26.0	-1.87	pH 2.0, Ward and Weber, 1968
	26.0	-3.07	pH 3.0, Ward and Weber, 1968
	26.0	-3.76	pH 5.0, Ward and Weber, 1968
	26.0	-3.78	pH 7.0, Ward and Weber, 1968
	26.0	-3.76	pH 10.0, Ward and Weber, 1968
Pronamide [23950-58-5]	24	-1.82	Humburg et al., 1989
Propachlor [1918-16-7]	20	-2.56	Humburg et al., 1989
	25	-2.54	Humburg et al., 1989
Propamocarb hydrochloride [25606-41-1]	25	0.59	Worthing and Hance, 1991
1,3-Propane dinitrate [3457-90-7]	25	-1.15	Fischer and Ballschmiter, 1998, 1998a
Propanil [709-98-8]	20	-3.22	Fühner and Geiger, 1977
Propaphos [7292-16-2]	25	-3.39	Worthing and Hance, 1991
Propaquizafop [111479-05-1]	25	-5.37	Worthing and Hance, 1991
Propargite [2312-35-8]	29	-5.26	Worthing and Hance, 1991
Propazine [139-40-2]	20	-4.66	Fühner and Geiger, 1977
	26.0	-3.21	pH 1.0, Ward and Weber, 1968
	26.0	-4.08	pH 2.0, Ward and Weber, 1968
	26.0	-4.68	pH 3.0, Ward and Weber, 1968
	26.0	-4.57	pH 5.0, Ward and Weber, 1968
	26.0	-4.70	pH 7.0, Ward and Weber, 1968
	26.0	-4.66	pH 10.0, Ward and Weber, 1968
Propene [115-07-1]	25	-2.32	McAuliffe, 1966
Propetamphos [31218-83-4]	23	-3.41	Worthing and Hance, 1991
Propham [122-42-9]	24	-2.86	Humburg et al., 1989
Propioconazole [60207-90-1]	20	-3.49	Worthing and Hance, 1991
Propionaldehyde [123-38-6]	12.3	0.91	Stephenson, 1993b*
	15.6	0.80	Stephenson, 1993b*
	20.0	0.73	Stephenson, 1993b*
	30.0	0.67	Stephenson, 1993b*
	40.0	0.61	Stephenson, 1993b*
	50.0	0.53	Stephenson, 1993b*
Propiophenone [93-55-0]	19.4	-1.62	Stephenson, 1992*
	40.8	-1.65	Stephenson, 1992*
	60.3	-1.83	Stephenson, 1992*
	71.2	-1.66	Stephenson, 1992*
	80.2	-1.75	Stephenson, 1992*
	90.5	-1.62	Stephenson, 1992*
Propoxur [114-26-1]	20	-2.04	Worthing and Hance, 1991
Propyl bromide [106-94-5]	30	-1.73	Gross and Saylor, 1931
Propyl butyrate [105-66-8]	25.0	-1.88	Stephenson and Stuart, 1986*
Propyl ether [111-43-3]	25	-1.61	Bennett and Philip, 1928
Propyl propionate [106-36-5]	25.0	-1.29	Stephenson and Stuart, 1986*
Propyne [74-99-7]	25	-1.06	McAuliffe, 1966
Prosulfocarb [52888-80-9]	20	-4.28	Worthing and Hance, 1991
Prothiofos [34643-46-4]	20	-6.84	Worthing and Hance, 1991
Prothoate [2275-18-5]	20	-2.06	Worthing and Hance, 1991
Pyraclofos [77458-01-6]	20	-4.04	Worthing and Hance, 1991
Pyrazolynate [58011-68-0]	25	-6.89	Worthing and Hance, 1991
Pyrazon [58858-18-7]	20	-2.74	Humburg et al., 1989
Pyrazophos [13457-18-6]	20	-4.95	Worthing and Hance, 1991
Pyrazosulfuron-ethyl [93697-74-6]	20	-4.46	Worthing and Hance, 1991
Pyrazoxyfen [71561-11-0]	20	-5.65	Worthing and Hance, 1991

Compound [CAS No.]	Temp. °C	Log S mol/L	Reference
Pyridaben [96489-71-3]	20	-7.48	Worthing and Hance, 1991
Pyridate [55512-33-9]	20	-5.40	Humburg et al., 1989
Pyrifenox [88283-41-4]	20	-3.41	Worthing and Hance, 1991
Pyriproxyfen [95737-68-1]	20-25	0.08	Worthing and Hance, 1991
Proquilon [57369-32-1]	20	-1.64	Worthing and Hance, 1991
Quinalphos [13593-03-8]	23-24	-4.13	Worthing and Hance, 1991
Quinclorac [84087-01-4]	20	-3.52	Humburg et al., 1989
	20	-3.50	Worthing and Hance, 1991
Quinmerac [90717-03-6]	20	-4.00	Worthing and Hance, 1991
Quinoline [91-22-5]	0	-1.26	Stephenson, 1993*
	10.0	-1.22	Stephenson, 1993*
	20.0	-1.19	Stephenson, 1993*
	23	-1.28	Fu and Luthy, 1986
	30.0	-1.29	Stephenson, 1993*
	40.0	-1.21	Stephenson, 1993*
	50.0	-1.19	Stephenson, 1993*
	60.0	-1.22	Stephenson, 1993*
	70.0	-1.21	Stephenson, 1993*
	80.0	-1.14	Stephenson, 1993*
	90.0	-1.00	Stephenson, 1993*
Quizalofop-ethyl [76578-14-8]	20	-5.09	Humburg et al., 1989
Quizalofop-P-ethyl [100646-51-3]	20	-5.97	Worthing and Hance, 1991
Rotenone [83-79-4]	100	-4.42	Worthing and Hance, 1991
Salicyaldehyde [90-02-8]	0	-1.62	Stephenson, 1993b*
	10.0	-1.61	Stephenson, 1993b*
	20.0	-1.61	Stephenson, 1993b*
	30.0	-1.57	Stephenson, 1993b*
	40.0	-1.43	Stephenson, 1993b*
	50.0	-1.41	Stephenson, 1993b*
	60.0	-1.41	Stephenson, 1993b*
	70.0	-1.33	Stephenson, 1993b*
	80.0	-1.17	Stephenson, 1993b*
	90.0	-1.13	Stephenson, 1993b*
Salicylic acid [69-72-7]	20	-1.88	Freed et al., 1979
Secbumeton [26269-45-0]	1	-2.53	pH 6, Gaynor and Van Volk, 1981
	8	-2.49	pH 6, Gaynor and Van Volk, 1981
	20	-2.51	pH 6, Gaynor and Van Volk, 1981
	20	-2.57	Fühner and Geiger, 1977
	29	-2.48	pH 6, Gaynor and Van Volk, 1981
Siduron [1982-49-6]	25	-4.11	Worthing and Hance, 1991
Simatone [673-04-1]	26.0	-1.88	pH 3.0, Ward and Weber, 1968
	26.0	-1.92	pH 7.0, Ward and Weber, 1968
	26.0	-1.89	pH 10.0, Ward and Weber, 1968
Simazine [122-34-9]	0	-5.00	Humburg et al., 1989
	20	-4.76	Fühner and Geiger, 1977
	26.0	-3.24	pH 1.0, Ward and Weber, 1968
	26.0	-4.11	pH 2.0, Ward and Weber, 1968
	26.0	-4.54	pH 3.0, Ward and Weber, 1968
	26.0	-4.60	pH 5.0, Ward and Weber, 1968
	26.0	-4.60	pH 7.0, Ward and Weber, 1968
	26.0	-4.60	pH 10.0, Ward and Weber, 1968
	85	-3.38	Humburg et al., 1989
Simetryn [1014-70-6]	26.0	-2.50	pH 3.0, Ward and Weber, 1968
	26.0	-2.68	pH 7.0, Ward and Weber, 1968
	26.0	-2.67	pH 10.0, Ward and Weber, 1968
Sodium chlorate [7775-09-9]	0	0.87	Humburg et al., 1989
	100	1.33	Humburg et al., 1989
Sodium dimethyl isophthalate-5-sulfonate [3965-55-7]	19.2	-2.17	Jiang et al., 2000*
	23.7	-2.01	Jiang et al., 2000*
	29.7	-1.90	Jiang et al., 2000*
	34.7	-1.79	Jiang et al., 2000*
	39.7	-1.72	Jiang et al., 2000*
	50.8	-1.54	Jiang et al., 2000*
	60.3	-1.41	Jiang et al., 2000*

Compound [CAS No.]	Temp. °C	Log S mol/L	Reference
	70.2	-1.19	Jiang et al., 2000*
	78.7	-0.05	Jiang et al., 2000*
	85.0	0.07	Jiang et al., 2000*
Sodium fluoride [7681-49-4]	18	0.00	Worthing and Hance, 1991
Sodium hexafluorosilicate [16893-85-9]	17.5	-1.46	Worthing and Hance, 1991
Sodium sulfate [7757-82-6]	10	-0.23	Okorafor, 1999
	20	0.05	Okorafor, 1999
	30	0.31	Okorafor, 1999
	40	0.36	Okorafor, 1999
	50	0.35	Okorafor, 1999
trans-Stilbene [103-03-0]	25.0	-5.93	Andrews and Keefer, 1950a
Streptomycin [57-92-1]	20-25	-1.46	Worthing and Hance, 1991
Sulfometuron-methyl [74222-97-2]	25	-4.56	pH 5, Humburg et al., 1989
	25	-3.08	pH 7, Humburg et al., 1989
Sulfur dioxide [7446-09-5]	25.0	0.04	Vesala, 1974
Sulfur hexafluoride [2551-62-4]	74.1	0.81	Mroczek, 1997
Sulfuryl fluoride [2699-79-8]	25	-2.13	Worthing and Hance, 1991
Sulprofos [35400-43-2]	20	-6.02	Worthing and Hance, 1991
Tebuconazole [107534-96-3]	20	-3.98	Worthing and Hance, 1991
Terbufos [13071-79-9]	20	-4.72	Bowman and Sans, 1982
Tebutam [35256-85-0]	25	-5.47	Worthing and Hance, 1991
Tebuthiuron [34014-18-1]	25	-1.96	Humburg et al., 1989
Tecloftalam [76280-91-6]	26	-4.51	Worthing and Hance, 1991
Teflubenzuron [83121-18-0]	20-23	-7.28	Worthing and Hance, 1991
Tefluthrin [79538-32-2]	20	-7.32	Worthing and Hance, 1991
Temephos [3383-96-8]	25	-7.19	Worthing and Hance, 1991
Terbacil [5902-51-2]	25	-2.48	Humburg et al., 1989
Terbumeton [33693-04-8]	20	-3.24	Worthing and Hance, 1991
Terbuthylazine [5915-41-3]	20	-4.66	Führner and Geiger, 1977
Terbutryn [886-50-0]	1	-3.96	pH 6, Gaynor and Van Volk, 1981
	8	-3.96	pH 6, Gaynor and Van Volk, 1981
	20	-3.84	pH 6, Gaynor and Van Volk, 1981
	20	-3.62	Ellgehausen et al., 1980
	20	-3.98	Ellgehausen et al., 1981
	20	-3.98	Führner and Geiger, 1977
	29	-3.78	pH 6, Gaynor and Van Volk, 1981
m-Terphenyl [92-06-8]	25.0	-5.18	Akiyoshi et al., 1987
o-Terphenyl [84-15-1]	25.0	-5.27	Akiyoshi et al., 1987
p-Terphenyl [92-94-4]	25.0	-7.11	Akiyoshi et al., 1987
α-Terpineol [10482-56-1]	25	-1.91	Fichan et al., 1999
2,3,4,5-Tetrachloroanisole [938-86-3]	25	-5.26	Lun et al., 1995
2,3,5,6-Tetrachloroanisole [6936-40-9]	25	-5.13	Lun et al., 1995
2,2',3,3'-Tetrachlorobiphenyl [38444-93-8]	25	-7.27	Dunnivant and Elzerman, 1988
	25	-7.01	Dulfer et al., 1995
2,2',4,4'-Tetrachlorobiphenyl [2437-79-8]	22	-6.81	Jepsen and Lick, 1999
	22.0	-6.73	Opperhuizen et al., 1988
2,2',4,5-Tetrachlorobiphenyl [41464-40-8]	25	-7.25	Miller et al., 1984
2,2',5,5'-Tetrachlorobiphenyl [35693-99-3]	22.0	-7.28	Opperhuizen et al., 1988
	25	-6.43	Dunnivant and Elzerman, 1988
2,2',5,6'-Tetrachlorobiphenyl [41464-41-9]	25	-6.79	Dunnivant and Elzerman, 1988
2,2',6,6'-Tetrachlorobiphenyl [15968-05-5]	22.0	-8.03	Opperhuizen et al., 1988
	25	-7.39	Dunnivant and Elzerman, 1988
2,3,4,5-Tetrachlorobiphenyl [33284-53-6]	25	-7.32	Dunnivant and Elzerman, 1988
	25.0	-7.33	Li et al., 1993
	25	-7.33	Li and Doucette, 1993
	25	-7.14	Miller et al., 1984
	25	-7.18	Dunnivant and Elzerman, 1988
	25	-7.26	Dulfer et al., 1995
2,4,4',6-Tetrachlorobiphenyl [32598-12-2]	25	-6.51	Dunnivant and Elzerman, 1988
3,3',4,4'-Tetrachlorobiphenyl [32598-13-3]	22.0	-8.21	Opperhuizen et al., 1988
	25	-8.73	Dunnivant and Elzerman, 1988
	25	-8.71	Dickhut et al., 1986
	25	-8.59	Dunnivant and Elzerman, 1988
3,3',5,5'-Tetrachlorobiphenyl [33284-52-5]	25	-8.37	Dunnivant and Elzerman, 1988

Compound [CAS No.]	Temp. °C	Log S mol/L	Reference
1,2,3,4-Tetrachloro-*p*-dioxin [30746-58-8]	4.0	-9.45	Doucette and Andren, 1988a
	5	-8.97	Shiu et al., 1988
	15	-8.85	Shiu et al., 1988
	25	-8.92	Santl et al., 1994
	25	-8.77	Shiu et al., 1988
	25.0	-8.84	Doucette and Andren, 1988a
	35	-8.45	Shiu et al., 1988
	45	-8.19	Shiu et al., 1988
	40.0	-8.44	Doucette and Andren, 1988a
1,2,3,7-Tetrachloro-*p*-dioxin [67028-18-6]	20.0	-8.87	Friesen et al., 1985
	26.0	-8.65	Li and Andren, 1994
1,3,6,8-Tetrachlorodibenzo-*p*-dioxin [33423-92-6]	20.0	-9.00	Friesen et al., 1985
	20	-9.00	Webster et al., 1985
2,3,7,8-Tetrachlorodibenzofuran [51207-31-9]	22.7	-8.86	Friesen et al., 1990
2,4,5,6-Tetrachloro-3-methylphenol [10460-33-0]	25	-4.60	pH 5.1, Blackman et al., 1955
2,3,4,5-Tetrachloronitrobenzene [879-39-0]	20	-4.55	Eckert, 1962
2,3,5,6-Tetrachloronitrobenzene [117-18-0]	20	-5.10	Eckert, 1962
2,3,4,6-Tetrachlorophenol [58-90-2]	25	-3.10	pH 5.1, Blackman et al., 1955
4,5,6,7-Tetrachlorophthalide [27355-22-2]	25	-5.04	Worthing and Hance, 1991
Tetrachloroveratrole [944-61-6]	25	-5.24	Lun et al., 1995
Tetrachlorvinphos [22248-79-9]	20	-4.52	Worthing and Hance, 1991
Tetraconazole [112281-77-3]	20	-3.39	Worthing and Hance, 1991
Tetradecane [629-59-4]	23	-8.78	Coates et al., 1985
	25.0	-7.96	Sutton and Calder, 1974
	25	-7.46	Franks, 1966
Tetradecanoic acid [544-63-8]	20.0	-4.06	Ralston and Hoerr, 1942
	25	-5.43	Robb, 1966
1-Tetradecanol [112-72-1]	25	-5.84	Robb, 1966
Tetradifon [116-29-0]	10	-6.85	Worthing and Hance, 1991
1,2,3,4-Tetrahydronaphthalene [119-64-2]	20.0	-3.49	Burris and MacIntyre, 1986
Tetrahydropyran [142-68-7]	0	0.18	Stephenson, 1992*
	9.4	0.07	Stephenson, 1992*
	19.9	0.00	Stephenson, 1992*
	31.0	-0.10	Stephenson, 1992*
	39.6	-0.15	Stephenson, 1992*
	50.5	-0.22	Stephenson, 1992*
	60.7	-0.27	Stephenson, 1992*
	71.2	-0.28	Stephenson, 1992*
	81.3	-0.30	Stephenson, 1992*
Tetramethrin [7696-12-0]	30	-4.86	Worthing and Hance, 1991
1,1,3,3-Tetramethylbutylamine [107-45-9]	0	-0.84	Stephenson, 1993a*
	10.0	-0.99	Stephenson, 1993a*
	20.0	-1.11	Stephenson, 1993a*
	30.0	-1.20	Stephenson, 1993a*
	40.0	-1.29	Stephenson, 1993a*
	50.0	-1.35	Stephenson, 1993a*
	60.0	-1.39	Stephenson, 1993a*
	70.0	-1.42	Stephenson, 1993a*
	80.0	-1.43	Stephenson, 1993a*
	90.0	-1.47	Stephenson, 1993a*
Thiazafluron [25366-23-8]	20	-2.20	Ellgehausen et al., 1980
	20	-2.06	Ellgehausen et al., 1981
Thicyofen [116170-30-0]	20	-3.01	Worthing and Hance, 1991
Thidiazuron [51707-55-2]	25	-3.85	Worthing and Hance, 1991
Thifensulfuron-methyl [79277-27-3]	25	-4.21	Worthing and Hance, 1991
Thioanisole [110-68-5]	25	-2.39	Hine and Weimar, 1965
Thiobencarb [28249-77-6]	20	-3.52	Humburg et al., 1989
Thiabendazole [148-79-8]	25	-1.30	pH 2, Worthing and Hance, 1991
Thiocyclam [31895-21-3]	23	-0.51	Worthing and Hance, 1991
Thiodicarb [59669-26-0]	25	-4.01	Worthing and Hance, 1991
Thiofanox [39196-18-4]	22	-1.62	Worthing and Hance, 1991
Thiometon [640-15-3]	25	-3.09	Worthing and Hance, 1991
Thiophanate-methyl [23564-05-8]	25	-4.11	Worthing and Hance, 1991
Thiophene [110-02-1]	25	-1.45	Price, 1976

Compound [CAS No.]	Temp. °C	Log S mol/L	Reference
Thiophenol [108-98-5]	25	-2.12	Hine and Weimar, 1965
Tiocarbazil [36756-79-3]	30	-5.05	Worthing and Hance, 1991
α-Tocopherol [59-02-9]	33	-4.32	Dubbs and Gupta, 1998
o-Tolualdehyde [529-20-4]	0	-1.76	Stephenson, 1993b*
	10.0	-1.88	Stephenson, 1993b*
	20.0	-1.93	Stephenson, 1993b*
	30.0	-1.90	Stephenson, 1993b*
	40.0	-1.88	Stephenson, 1993b*
	50.0	-1.72	Stephenson, 1993b*
	60.0	-1.65	Stephenson, 1993b*
	70.0	-1.59	Stephenson, 1993b*
	80.0	-1.60	Stephenson, 1993b*
	90.0	-1.58	Stephenson, 1993b*
p-Tolualdehyde [104-87-0]	0	-1.90	Stephenson, 1993b*
	10.0	-1.85	Stephenson, 1993b*
	20.0	-1.82	Stephenson, 1993b*
	40.0	-1.80	Stephenson, 1993b*
	50.0	-1.70	Stephenson, 1993b*
	70.0	-1.72	Stephenson, 1993b*
	80.0	-1.68	Stephenson, 1993b*
	90.0	-1.60	Stephenson, 1993b*
p-Toluidine [106-49-0]	20	-1.21	Hashimoto et al., 1984
p-Tolunitrile [104-85-8]	20	-1.89	McGowan et al., 1966
Tralkoxydim [87820-88-0]	20	-4.82	Worthing and Hance, 1991
Triadimefon [43121-43-3]	20	-3.05	Worthing and Hance, 1991
Tridimenol [55219-65-3]	20	-3.68	Worthing and Hance, 1991
Triallate [2303-17-5]	25	-4.88	Humburg et al., 1989
Triallylamine [102-70-5]	0	-1.41	Stephenson, 1993a*
	10.0	-1.50	Stephenson, 1993a*
	20.0	-1.68	Stephenson, 1993a*
	30.0	-1.78	Stephenson, 1993a*
	40.0	-1.82	Stephenson, 1993a*
	50.0	-1.80	Stephenson, 1993a*
	60.0	-1.86	Stephenson, 1993a*
	70.0	-1.82	Stephenson, 1993a*
	80.0	-1.88	Stephenson, 1993a*
	90.0	-1.76	Stephenson, 1993a*
Triasulfuron [82097-50-5]	20	-2.43	Worthing and Hance, 1991
Triazoxide [72459-58-6]	20	-3.92	Worthing and Hance, 1991
1,2,4-Tribromobenzene [615-54-3]	25	-4.50	Yalkowsky et al., 1979
Tri-n-butylamine [102-82-9]	0	-2.29	Stephenson, 1993a*
	10.0	-2.31	Stephenson, 1993a*
	20.0	-2.29	Stephenson, 1993a*
	25.0	-3.12	Vesala, 1974
	30.0	-2.25	Stephenson, 1993a*
	40.0	-2.26	Stephenson, 1993a*
	50.0	-2.57	Stephenson, 1993a*
	60.0	-2.58	Stephenson, 1993a*
	70.0	-2.68	Stephenson, 1993a*
	80.0	-2.68	Stephenson, 1993a*
S,S,S-Tributyl phosphorotrithioate [78-48-8]	20	-5.14	Worthing and Hance, 1991
Trichlorfon [52-68-6]	20	-0.33	Worthing and Hance, 1991
Trichloroacetic acid [76-03-9]	25	0.87	Bailey and White, 1965
2,3,4-Trichloroanisole [54135-80-7]	25	-4.29	Lun et al., 1995
2,4,6-Trichloroanisole [87-40-1]	25	-4.20	Lun et al., 1995
2,2',5-Trichlorobiphenyl [37680-65-2]	25	-5.70	Dunnivant and Elzerman, 1988
	25	-5.60	Dunnivant and Elzerman, 1988
2,3',5-Trichlorobiphenyl [38444-81-4]	25	-6.01	Dunnivant and Elzerman, 1988
2,4,4'-Trichlorobiphenyl [7012-37-5]	22.0	-6.58	Opperhuizen et al., 1988
	25	-6.34	Dunnivant and Elzerman, 1988
	25	-6.35	Chiou et al., 1986
	25	-6.00	Dunnivant and Elzerman, 1988
2,4,5-Trichlorobiphenyl [15862-07-4]	25	-6.20	Miller et al., 1984
	25	-6.08	Dulfer et al., 1995

Compound [CAS No.]	Temp. °C	Log S mol/L	Reference
2,4,6-Trichlorobiphenyl [35693-92-6]	4.0	-6.51	Doucette and Andren, 1988a
	25.0	-6.14	Doucette and Andren, 1988a
	25	-6.01	Dunnivant and Elzerman, 1988
	25	-6.03	Li et al., 1993
	25	-6.06	Miller et al., 1984
	25	-5.83	Dulfer et al., 1995
	40.0	-5.77	Doucette and Andren, 1988a
1,2,4-Trichlorodibenzo-p-dioxin [39227-58-2]	5	-8.12	Shiu et al., 1988
	15	-7.78	Shiu et al., 1988
	25	-7.62	Santl et al., 1994
	25	-7.53	Shiu et al., 1988
	35	-7.24	Shiu et al., 1988
	45	-7.01	Shiu et al., 1988
2,4,6-Trichloro-3,5-dimethylphenol [6972-47-0]	25	-4.66	pH 5.1, Blackman et al., 1955
3,4,5-Trichloroguaiacol [57057-83-7]	5	-3.30	Tam et al., 1994
	15	-3.11	Tam et al., 1994
	25	-2.86	Tam et al., 1994
	35	-2.80	Tam et al., 1994
	45	-2.63	Tam et al., 1994
4,5,6-Trichloroguaiacol [2668-24-8]	7	-3.88	Tam et al., 1994
	15	-3.73	Tam et al., 1994
	25	-3.66	Tam et al., 1994
	35	-3.45	Tam et al., 1994
	45	-3.19	Tam et al., 1994
2,4,6-Trichloro-3-methylphenol [551-76-8]	25	-3.28	pH 5.1, Blackman et al., 1955
2,3,4-Trichloronitrobenzene [17700-09-3]	20	-3.94	Eckert, 1962
2,4,5-Trichloronitrobenzene [89-69-0]	20	-3.89	Eckert, 1962
3,4,5-Trichloroveratrole [16766-29-3]	25	-4.37	Lun et al., 1995
Triclopyr [55335-06-3]	20	-3.52	Humburg et al., 1989
Tricyclazole [41814-78-2]	25	-2.07	Worthing and Hance, 1991
Tridecanoic acid [638-53-9]	20.0	-3.81	Ralston and Hoerr, 1942
Tridecylbenzene [123-02-4]	25	-9.05	Sherblom et al., 1992
Tridiphane [51528-03-1]	21	-5.25	Humburg et al., 1989
Trietazine [1912-26-1]	26.0	-3.91	pH 3.0, Ward and Weber, 1968
	26.0	-3.90	pH 7.0, Ward and Weber, 1968
	26.0	-3.83	pH 10.0, Ward and Weber, 1968
4-Trifluoromethyl-2-nitrophenol [400-99-7]	20.0	-2.50	Schwarzenbach et al., 1988
Triflumizole [68694-11-1]	20	-1.44	Worthing and Hance, 1991
Triflumuron [64628-44-0]	20	-7.16	Worthing and Hance, 1991
3,4,5-Trimethoxybenzaldehyde [86-81-7]	25	-2.12	Jin et al., 1998
Trimethylamine [75-50-3]	25	0.84	Amoore and Buttery, 1978
2,3,3-Trimethyl-2-butanol [594-83-2]	40	-0.72	Ginnings and Hauser, 1938
1,1,3-Trimethylcyclopentane [4516-69-2]	25	-4.48	Price, 1976
3,5,5-Trimethylhexanal [5435-64-3]	0	-1.95	Stephenson, 1993b*
	10.0	-2.37	Stephenson, 1993b*
	20.0	-2.55	Stephenson, 1993b*
	30.0	-2.37	Stephenson, 1993b*
	40.0	-2.45	Stephenson, 1993b*
	50.0	-2.45	Stephenson, 1993b*
	60.0	-2.55	Stephenson, 1993b*
	70.0	-2.37	Stephenson, 1993b*
	80.0	-2.37	Stephenson, 1993b*
	90.0	-2.20	Stephenson, 1993b*
3,3,5-Trimethyl-1-hexanol [1484-87-3]	20.0	-2.51	Hommelen, 1959
	25.0	-1.99	Stephenson and Stuart, 1986*
1,4,5-Trimethylnaphthalene [213-41-1]	25	-4.91	Mackay and Shiu, 1977
2,6,8-Trimethyl-4-nonanone [123-18-2]	9.6	-3.19	Stephenson, 1992*
	39.7	-3.19	Stephenson, 1992*
	50.1	-3.12	Stephenson, 1992*
	60.8	-3.04	Stephenson, 1992*
	50.1	-3.12	Stephenson, 1992*
	60.8	-3.03	Stephenson, 1992*
	80.0	-3.12	Stephenson, 1992*
2,2,3-Trimethyl-3-pentanol [7294-83-2]	20	-1.24	Ginnings and Coltrane, 1939

Compound [CAS No.]	Temp. °C	Log S mol/L	Reference
	25	-1.28	Ginnings and Coltrane, 1939
	30	-1.31	Ginnings and Coltrane, 1939
2,3,5-Trimethylphenol [697-82-5]	25	-2.30	Southworth and Keller, 1986
2,4,6-Trinitropyridine [108-75-8]	6.5	0.03	Stephenson, 1993*
	7.5	-0.10	Stephenson, 1993*
	8.2	-0.30	Stephenson, 1993*
	20.0	-0.53	Stephenson, 1993*
	30.0	-0.67	Stephenson, 1993*
	40.0	-0.76	Stephenson, 1993*
	50.0	-0.82	Stephenson, 1993*
	60.0	-0.85	Stephenson, 1993*
	70.0	-0.86	Stephenson, 1993*
	80.0	-0.90	Stephenson, 1993*
	90.0	-0.87	Stephenson, 1993*
2,4,6-Trinitrotoluene [118-96-7]	6	-3.64	pH 3.7, Ro et al., 1996
	6	-3.65	pH 6.9, Ro et al., 1996
	12	-3.55	pH 6.9, Ro et al., 1996
	13	-3.50	pH 3.7, Ro et al., 1996
	13	-3.55	pH 6.9, Ro et al., 1996
	20	-3.42	pH 4.2, Ro et al., 1996
	20	-3.41	pH 7.3, Ro et al., 1996
	20	-3.37	pH 9.2, Ro et al., 1996
	20	-3.38	pH 9.3, Ro et al., 1996
	21	-3.48	pH 3.5, Ro et al., 1996
	21	-3.44	pH 6.8, Ro et al., 1996
	21	-3.41	pH 9.1, Ro et al., 1996
	25	-3.35	pH 3.5, Ro et al., 1996
	25	-3.35	pH 6.8, Ro et al., 1996
	25	-3.31	pH 9.1, Ro et al., 1996
	42	-3.04	pH 4.0, Ro et al., 1996
	42	-3.05	pH 6.8, Ro et al., 1996
	42	-3.13	pH 9.3, Ro et al., 1996
Tri-n-propylamine [102-69-2]	0	-2.30	Stephenson, 1993a*
	10.0	-2.31	Stephenson, 1993a*
	20.0	-2.81	Stephenson, 1993a*
	25	-3.00	Luke et al., 1989
	25.0	-2.28	Vesala, 1974
Triphenylene [217-59-4]	8.00	-7.88	May et al., 1983[c]
	12.00	-7.88	May et al., 1983[c]
	14.80	-7.83	May et al., 1983[c]
	20.50	-7.67	May et al., 1983[c]
	25.0	-6.74	Akiyoshi et al., 1987
	25	-6.73	Mackay and Shiu, 1977
	25	-6.73	Klevens, 1950
	27	-6.78	Davis et al., 1942
	27.30	-7.48	May et al., 1983[c]
	28.20	-7.45	May et al., 1983[c]
Tris(1,3-dichloroisopropyl) phosphate [13674-87-8]	24	-4.79	Hollifield, 1979
Tris(2-ethylhexyl) phosphate [126-72-7]	24	-4.94	Hollifield, 1979
L-Tyrosine [60-18-4]	25	-2.57	pH 5.39, Carta and Tola, 1996
	25	-2.55	pH 7.00, Carta and Tola, 1996
	25	-2.57	pH 9.12, Carta and Tola, 1996
1-Undecanol [112-42-5]	25.0	-2.63	Stephenson and Stuart, 1986*
Uniconazole [83657-22-1]	25	-4.54	Worthing and Hance, 1991
Vernolate [1929-77-7]	20	-3.35	Humburg et al., 1989
Vinclozolin [50471-44-8]	20	-4.93	Worthing and Hance, 1991
4-Vinyl-1-cyclohexene [100-43-3]	25	-3.33	McAuliffe, 1966
Xylylcarb [2425-10-7]	20	-2.49	Worthing and Hance, 1991
Ziram [137-30-4]	20	-4.88	Fühner and Geiger, 1977

[a] In 0.001M HNO_3; [b] In 0.05M HCl. HNO_3; [c] Converted from mole fraction solubility units; [d] In 0.1M HCl; * Solubility originally reported as wt%. In converting the units to mol/L, it was assumed the density of the solution containing the solute was equal to the density of water at 0 °C or 1.000 g/cm³.

Table 7. Henry's Law Constants of Inorganic and Organic Compounds

Compound [CAS No.]	Temp. °C	Log K_H atm·m³/mol	Reference
Acetophenone [98-86-2]	14.9	-5.24	Betterton, 1991[a]
	15	-5.58	Allen et al., 1997
	17.4	-5.45	Allen et al., 1997
	20	-5.34	Allen et al., 1997
	25	-4.98	Allen et al., 1997
	25	-4.97	Shiu and Mackay, 1997
	25.1	-5.04	Betterton, 1991[a]
	35.0	-4.69	Betterton, 1991[a]
	45.0	-4.40	Betterton, 1991[a]
Alachlor [15972-60-8]	23	-8.08	Fendinger and Glotfelty, 1988
tert-Amyl methyl ether [994-05-8]	23	-2.72	Miller and Stuart, 2000
Androst-16-en-3-ol [1153-51-1]	25	-3.53	Amoore and Buttery, 1978
Argon [7440-37-1]	25	-0.15	Morrison and Johnstone, 1954
Benzaldehyde [100-52-7]	5	-5.26	Allen et al., 1997
	10	-5.09	Allen et al., 1997
	15	-4.83	Betterton and Hoffman, 1988[b]
	15	-4.93	Allen et al., 1997
	20	-4.75	Allen et al., 1997
	25	-4.62	Zhou and Mopper, 1990
	25	-4.51	Allen et al., 1997
	25	-4.57	Betterton and Hoffman, 1988[b]
	35	-4.34	Betterton and Hoffman, 1988[b]
	45	-4.11	Betterton and Hoffman, 1988[b]
Benzyl chloride [100-44-7]	20	-3.45	Hovorka and Dohnal, 1997
Bis(hydroxymethyl)peroxide [17088-73-2]	25	-8.65	Zhou and Lee, 1992
	25	<-10.00	Staffelbach and Kok, 1993
Bromine [7726-95-6]	20	-2.99	Jenkins and King, 1965
	25	-2.85	Kelley and Tartar, 1956
	25	-2.85	Hill et al., 1968
Bromine chloride [13863-41-7]	25.0	-2.97	Bartlett and Margerum, 1999
Bromoacetic acid [79-08-3]	25	-8.18	Bowden et al., 1998
1-Bromobutane [109-65-9]	25	-1.67	Hoff et al., 1993
4-Bromophenol [106-41-2]	25	-6.85	Parsons et al., 1971
Butanal [123-72-8]	25	-3.98	Zhou and Mopper, 1990
	25	-3.94	Buttery et al., 1969
1,3-Butane dinitrate [6423-44-5]	25	-4.76	Fischer and Ballschmiter, 1998, 1998a
1,4-Propane dinitrate [3457-91-8]	25	-5.20	Fischer and Ballschmiter, 1998, 1998a
2,3-Butanedione [431-03-8]	14.9	-5.15	Betterton, 1991[a]
	25.0	-4.87	Betterton, 1991[a]
	25	-4.76	Snider and Dawson, 1985
	35.1	-4.69	Betterton, 1991[a]
	44.9	-4.30	Betterton, 1991[a]
Butanoic acid [107-92-6]	25	-6.67	Khan et al., 1995
	25	-6.28	Butler and Ramchandani, 1935
3-Buten-2-one [78-94-4]	5	-5.18	Allen et al., 1997
	10	-4.89	Allen et al., 1997
	15	-4.72	Allen et al., 1997
	20	-4.54	Allen et al., 1997
	25	-4.33	Allen et al., 1997
	25	-4.61	Iraci et al., 1998
Butyl mercaptan [109-79-5]	25	-2.34	Przyjazny et al., 1983
1-Butyl nitrate [928-45-0]	25	-3.00	Kames and Schurath, 1992
	25	-3.00	Luke et al., 1989
	25	-2.81	Hauff et al., 1998
4-tert-Butylphenol [98-54-4]	25	-5.95	Parsons et al., 1972
Butyronitrile [109-74-0]	25	-4.28	Butler and Ramchandani, 1935
	25	-4.16	Hawthorne et al., 1985
	25.0	-4.13	Ramachandran et al., 1996
Carbon dioxide [124-38-9]	25	-1.53	Morgan and Maas, 1931
	25	-1.56	Zheng et al., 1997
Carbon monoxide [630-08-0]	25	-0.87	Meadows and Spedding, 1974
Carbon tetrafluoride [75-73-0]	25	0.70	Morrison and Johnstone, 1954
Carbonyl chloride [353-50-4]	25	-4.54	Mirabel et al., 1996
Carbonyl sulfide [463-58-1]	25	-1.34	De Bruyn et al., 1995
Chlornitrofen [1836-77-7]	25	<-5.91	Kawamoto and Urano, 1989

1123

Compound [CAS No.]	Temp. °C	Log K_H atm·m³/mol	Reference
Chloroacetic acid [79-11-8]	25	-8.04	Bowden et al., 1998
Chloroacetone [78-95-5]	14.9	-5.08	Betterton, 1991[a]
	25.0	-4.77	Betterton, 1991[a]
	34.9	-4.51	Betterton, 1991[a]
	44.9	-4.30	Betterton, 1991[a]
Chloroamine [10599-90-3]	25	-4.97	Holzwarth et al., 1984
1-Chlorobutane [109-69-3]	23.0	-1.82	Leighton and Calo, 1981
2-Chlorobutane [78-86-4]	24.9	-1.65	Leighton and Calo, 1981
Chlorodifluoroacetic acid [76-04-0]	25	-7.40	Bowden et al., 1998
Chlorodifluoromethane [75-45-6]	25	-1.57	Zheng et al., 1997
1-Chlorohexane [544-10-5]	23.0	-1.64	Leighton and Calo, 1981
Chloroiodomethane[593-71-5]	0	-3.56[c]	Moore et al., 1995
	10	-3.30[c]	Moore et al., 1995
	20	-3.07[c]	Moore et al., 1995
1-Chloronaphthalene [90-13-1]	25	-3.45	Shiu and Mackay, 1997
1-Chloropentane [543-59-9]	24.3	-1.63	Leighton and Calo, 1981
1-Chloro-2-propanone[78-95-5]	25	-4.77	Betterton, 1991[a]
Chlorothalonil [1897-45-6]	25	--6.71	Kawamoto and Urano, 1989
Cyclohexylamine [108-98-8]	25	-4.98	Amoore and Buttery, 1978
Cyclopentanone	25	-4.92	Hawthorne et al., 1985
Decanal [112-31-2]	25	-2.79	Zhou and Mopper, 1990
Dibromoacetic acid [631-64-1]	25	-8.36	Bowden et al., 1998
Dibromomethane [74-95-3]	0	-3.56[c]	Moore et al., 1995
	10	-3.32[c]	Moore et al., 1995
	20	-3.09[c]	Moore et al., 1995
	20	-3.15	Tse et al., 1992
	25	-3.04	Wright et al., 1992
	30	-2.96	Tse et al., 1992
	35	-2.85	Tse et al., 1992
	40	-2.77	Tse et al., 1992
Dichloroacetic acid [79-43-6]	25	-8.08	Bowden et al., 1998
Dibutylamine [111-92-2]	25	-4.04	Christie and Crisp, 1967
Dichloroamine [3400-09-7]	25	-4.46	Holzwarth et al., 1984
2,2'-Dichlorobiphenyl [13029-08-8]	25	-3.47	Dunnivant and Elzerman, 1988
	25	-3.64	Brunner et al., 1990
2,3-Dichlorobiphenyl [16605-91-7]	25	-3.64	Brunner et al., 1990
2,3'-Dichlorobiphenyl [25569-80-6]	25	-3.60	Brunner et al., 1990
2,4-Dichlorobiphenyl [33284-50-3]	25	-3.55	Brunner et al., 1990
2,4'-Dichlorobiphenyl [34883-43-7]	23	-3.02	Atlas et al., 1982
	23	-2.72	Atlas et al., 1982[d]
2,5-Dichlorobiphenyl [34833-39-1]	25	-3.41	Dunnivant and Elzerman, 1988
3,3'-Dichlorobiphenyl [2050-60-1]	25	-3.63	Dunnivant and Elzerman, 1988
3,4-Dichlorobiphenyl [2974-92-7]	25	-3.85	Brunner et al., 1990
	25	-3.69	Dunnivant and Elzerman, 1988
4,4'-Dichlorobiphenyl [2050-68-2]	25	-3.70	Dunnivant and Elzerman, 1988
2,7-Dichlorodibenzo-p-dioxin [33857-26-0]	25	-4.23	Santl et al., 1994
cis-1,2-Dichloroethylene [156-59-2]	10	-2.57	Ashworth et al., 1988
	15	-2.49	Ashworth et al., 1988
	20	-2.44	Ashworth et al., 1988
	20	-2.50	Hovorka and Dohnal, 1997
	20	-2.49	Tse et al., 1992
	25	-2.43	Gossett, 1987
	25	-2.38	Wright et al., 1992
	25	-2.34	Ashworth et al., 1988
	30	-2.24	Ashworth et al., 1988
	30	-2.31	Tse et al., 1992
	40	-2.14	Tse et al., 1992
1,3-Dichloropropylene [542-75-6]	25	-2.81	Wright et al., 1992
Diethyl disulfide [110-81-6]	20	-2.67	Vitenberg et al., 1975
	25	-2.81	Przyjazny et al., 1983
Diethyl peroxide [628-37-5]	25	-5.53	O'Sullivan et al., 1996
Diethyl sulfide [352-93-2]	25	-2.75	Przyjazny et al., 1983
Difluoroacetic acid [381-73-7]	25	-7.48	Bowden et al., 1998
1,1-Difluoroethane [75-37-6]	25	-1.73	Zheng et al., 1997

Compound [CAS No.]	Temp. °C	Log K_H atm·m³/mol	Reference
Diiodomethane [75-11-6]	0	-4.07[c]	Moore et al., 1995
	10	-3.76[c]	Moore et al., 1995
	20	-3.50[c]	Moore et al., 1995
3,4-Dimethylaniline\ [95-64-7]	25	-5.73	Jayasinghe et al., 1992
Dimethyl disulfide [624-92-0]	20	-2.92	Vitenberg et al., 1975
	25	-2.98	Przyjazny et al., 1983
1,5-Dimethylnaphthalene [571-61-9]	25	-3.45	Shiu and Mackay, 1997
2,6-Dimethylphenol [87-65-0]	25.	-5.18	Hawthorne et al., 1985
2,2-Dimethylpropanoic acid [75-98-9]	25	-6.54	Khan et al., 1995
2,6-Dimethylpyrazine [108-50-9]	25	-5.00	Chaintreau et al., 1995
2,3-Dimethylpyridine[583-61-9]	25	-5.15	Andon et al., 1954
2,4-Dimethylpyridine[108-47-4]	25	-5.18	Andon et al., 1954
	25	-5.00	Hawthorne et al., 1985
2,5-Dimethylpyridine [589-93-5]	25	-5.08	Andon et al., 1954
2,6-Dimethylpyridine [108-48-5]	25	-4.98	Andon et al., 1954
	25	-4.82	Hawthorne et al., 1985
3,4-Dimethylpyridine [583-58-4]	25	-5.43	Andon et al., 1954
3,5-Dimethylpyridine [591-22-0]	25	-5.15	Andon et al., 1954
Dimethyl sulfide [75-18-3]	20	-2.79	Vitenberg et al., 1975
	25	-2.68	De Bruyn et al., 1995
	25	-2.74	Przyjazny et al., 1980
	25	-2.75	Dacey et al., 1984
1,3-Dioxolane [646-06-0]	25	-4.60	Cabani et al., 1971
Dipropylamine [142-84-7]	25	-4.28	Christie and Crisp, 1967
Dipropyl sulfide [111-47-7]	25	-2.52	Przyjazny et al., 1983
Ethanedial [107-22-2]	25	-8.56	Zhou and Mopper, 1990
1,2-Ethane dinitrate [628-96-6]	20	-5.81	Kames and Schurath, 1992
Ethanol [64-17-5]	25	-5.28	Snider and Dawson, 1985
	25	-5.34	Burnett, 1963
	60	-4.37	Chai and Zhu, 1998
Ethyl *n*-butyl ether [628-81-9]	23	-2.81	Miller and Stuart, 2000
Ethyl *tert*-butyl ether [637-92-3]	23	-2.62	Miller and Stuart, 2000
Ethylene glycol [107-21-1]	25	-9.60	Bone et al., 1983
Ethyl formate [109-94-4]	25	-2.30	Hartkopf and Karger, 1973
2-Ethyl-3-methoxypyrazine [25680-58-4]	25	-4.83	Buttery et al., 1971
Ethyl nitrate [625-58-1]	25	-3.20	Kames and Schurath, 1992
Ethyl peroxide [3031-74-1]	25	-5.53	O'Sullivan et al., 1996
2-Ethylpyrazine [13925-00-3]	25	-5.60	Buttery et al., 1971
2-Ethylpyridine [100-71-0]	25	-4.79	Andon et al., 1954
3-Ethylpyridine [536-78-7]	25	-4.98	Andon et al., 1954
4-Ethylpyridine [536-75-4]	25	-5.08	Andon et al., 1954
Fenitrothion [122-14-5]	20	-6.03	Metcalf et al., 1980
Fluoroacetic acid [144-49-0]	25	-7.48	Bowden et al., 1998
Fluorobenzene [462-06-6]	2.0	-2.70	Dewulf et al., 1999
	6.0	-2.59	Dewulf et al., 1999
	10.0	-2.16	Dewulf et al., 1999
	18.0	-2.30	Dewulf et al., 1999
	25.0	-2.18	Dewulf et al., 1999
2,2′,3,3′,4,4′,5-Heptachlorobiphenyl [35065-30-6]	25	-5.05	Brunner et al., 1990
2,2′,3,4,4′,5,5′-Heptachlorobiphenyl [35065-29-3]	10	-6.04	Koelmans et al., 1999
	25	-5.00	Brunner et al., 1990
2,2′,3,3′,4,5,5′-Heptachlorobiphenyl [52663-74-8]	25	-4.89	Brunner et al., 1990
2,2′,3,3′,4,5,6-Heptachlorobiphenyl [68194-16-1]	25	-4.85	Brunner et al., 1990
2,2′,3,4,5,5′,6-Heptachlorobiphenyl [51712-05-7]	25	-4.80	Brunner et al., 1990
2,2′,3,3′4,5,6′-Heptachlorobiphenyl [38411-25-5]	25	-4.85	Brunner et al., 1990
2,2′,3,3′,5,5′,6-Heptachlorobiphenyl [52663-67-9]	25	-4.64	Brunner et al., 1990
2,2′,3,3′,5,6,6′-Heptachlorobiphenyl [52663-64-6]	25	-4.62	Brunner et al., 1990
Heptanal [111-71-7]	25	-3.52	Zhou and Mopper, 1990
	25	-3.57	Buttery et al., 1969
1-Heptanol [111-70-6]	25	-4.26	Shiu and Mackay, 1997
2,2′,3,3′,4,4′-Hexachlorobiphenyl [38380-07-3]	25	-4.89	Brunner et al., 1990
	25	-4.52	Dunnivant and Elzerman, 1988
2,2′,3,3′,4,5-Hexachlorobiphenyl [55215-18-4]	25	-4.54	Brunner et al., 1990
2,2′,3,3′,4,5′-Hexachlorobiphenyl [52663-66-8]	25	-4.43	Brunner et al., 1990

Compound [CAS No.]	Temp. °C	Log K_H atm·m³/mol	Reference
2,2′,3,3′,4,6-Hexachlorobiphenyl [61798-70-7]	25	-4.41	Brunner et al., 1990
2,2′,3,3′,4,6′-Hexachlorobiphenyl [38380-05-1]	25	-4.36	Brunner et al., 1990
2,2′,3,3′,5,6-Hexachlorobiphenyl [52704-70-8]	25	-4.31	Brunner et al., 1990
2,2′,3,3′,5,6′-Hexachlorobiphenyl [52744-13-5]	25	-4.25	Brunner et al., 1990
2,2′,3,3′,6,6′-Hexachlorobiphenyl [38411-22-2]	25	-4.06	Brunner et al., 1990
2,2′,3,4,4′,5-Hexachlorobiphenyl [35065-28-2]	10	-4.80	Koelmans et al., 1999
	25	-4.68	Brunner et al., 1990
2,2′,3,4,5,5′-Hexachlorobiphenyl [52712-04-6]	25	-4.64	Brunner et al., 1990
2,2′,3,4′,5,5′-Hexachlorobiphenyl [51908-16-8]	25	-4.60	Brunner et al., 1990
2,2′,3,4′,5,6-Hexachlorobiphenyl [68194-13-8]	25	-4.29	Brunner et al., 1990
2,2′,3,5,5′,6-Hexachlorobiphenyl [52663-63-5]	25	-4.23	Brunner et al., 1990
2,2′,4,4′,5,5′-Hexachlorobiphenyl [35065-27-1]	10	-4.80	Koelmans et al., 1999
	25	-3.88	Dunnivant and Elzerman, 1988
	25	-4.64	Brunner et al., 1990
2,2′,4,4′,6,6′-Hexachlorobiphenyl [33979-02-2]	25	-3.12	Dunnivant and Elzerman, 1988
2,3,3′,4,5,5′-Hexachlorobiphenyl [39635-35-3]	25	-4.70	Brunner et al., 1990
2,2′,3,3′,4,5,6-Hexachlorobiphenyl [41411-62-5]	25	-4.70	Brunner et al., 1990
2,2′,3,3′,4′,5,6-Hexachlorobiphenyl [74472-44-9]	25	-4.82	Brunner et al., 1990
2,2′,3,3′,5,5′,6-Hexachlorobiphenyl [74472-46-1]	25	-4.54	Brunner et al., 1990
trans-2-Hexadienal [142-83-6]	25	-5.00	Buttery et al., 1971
1,1,1,3,3,3-Hexafluoro-2-propanol [920-66-1]	25	-4.38	Rochester and Symonds, 1973
Hexamethyleneimine [111-49-9]	25	-5.20	Cabani et al., 1971a
Hexanal [66-25-1]	25	-3.69	Zhou and Mopper, 1990
	25	-3.67	Buttery et al., 1969
Hexanoic acid [142-62-1]	25	-6.15	Khan et al., 1995
1-Hexanol [111-27-3]	25	-4.77	Buttery et al., 1969
3-Hexanone [589-38-8]	2.0	-3.05	Dewulf et al., 1999
	6.0	-2.90	Dewulf et al., 1999
	10.0	-2.90	Dewulf et al., 1999
	18.0	-2.92	Dewulf et al., 1999
	25.0	-2.89	Dewulf et al., 1999
trans-2-Hexenal [6728-26-3]	25	-4.30	Buttery et al., 1971
Hexylamine [111-26-2]	25	-4.57	Christie and Crisp, 1967
Hydrazoic acid [7782-79-8]	25	-4.08	Betterton and Robinson, 1997
Hydrogen peroxide [7722-84-1]	3	-8.71	Hwang and Dasgupta, 1985
	5.0	-8.70	O'Sullivan et al., 1996
	10	-8.47	Hwang and Dasgupta, 1985
	18.0	-8.17	O'Sullivan et al., 1996
	20	-8.13	Hwang and Dasgupta, 1985
	20.0	-8.15	Yoshizumi et al., 1984
	24	-7.96	O'Sullivan et al., 1996
	25	-7.92	O'Sullivan et al., 1996
	25	-7.93	Zhou and Lee, 1992
	25	-8.00	Lind and Kok, 1994
	25	-8.04	Staffelbach and Kok, 1993
	30	-7.58	Hwang and Dasgupta, 1985
Hydrogen sulfide [7783-06-4]	25	-1.94	De Bruyn et al., 1995
Hydroxyacetaldehyde [141-46-8]	25	-7.62	Betterton and Hoffman, 1988[b]
	45	-7.19	Betterton and Hoffman, 1988[b]
4-Hydroxybenzaldehyde [123-08-0]	25	-9.28	Parsons et al., 1971
Hydroxymethyl hydroperoxide [15932-89-5]	25	-9.23	O'Sullivan et al., 1996
	25	-9.20	Staffelbach and Kok, 1993
	25	-8.68	Zhou and Lee, 1992
Hypobromous acid [13517-11-8]	25	<-6.28	Blatchley et al., 1992
Hypochlorous acid [7790-92-3]	25	-5.68	Hanson and Ravishankara, 1991
	25	-5.86	Holzwarth et al., 1984
	25	-5.97	Blatchley et al., 1992
Isoamyl nitrate [543-87-3]	25	-2.65	Hauff et al., 1998
Isobutyk isobutyrate [97-85-8]	25	-3.01	Amoore and Buttery, 1978
3-Isobutyl-3-methoxypyrazine [24683-00-9]	25	-4.30	Buttery et al., 1971
Isobutyl nitrate [543-29-3]	25	-2.81	Kames and Schurath, 1992
	25	-2.81	Luke et al., 1989
	25	-2.64	Hauff et al., 1998
Isobutryaldehyde [78-84-2]	25	-3.71	Amoore and Buttery, 1978

Compound [CAS No.]	Temp. °C	Log K_H atm·m³/mol	Reference
Isopropyl nitrate [1712-64-7]	25	-2.90	Kames and Schurath, 1992
Krypton [7439-90-9]	25	-0.38	Morrison and Johnstone, 1954
Methacrolein [75-85-3]	5	-4.19	Allen et al., 1997
	10	-4.04	Allen et al., 1997
	15	-3.89	Allen et al., 1997
	20	-3.78	Allen et al., 1997
	25	-3.63	Allen et al., 1997
4-Methylaniline [106-49-0]	25	-5.65	Jayasinghe et al., 1992
3-Methylbutanoic acid [503-74-2]	25	-6.08	Khan et al., 1995
2-Methyl-1-butanol [137-32-6]	25	-4.85	Butler et al., 1935
2-Methyl-2-butanol [75-85-4]	25	-4.86	Butler et al., 1935
Methyl tert-butyl ether [1634-04-4]	23	-3.20	Miller and Stuart, 2000
	25	-3.28	Robbins et al., 1993
	30	-2.92	Robbins et al., 1993
	40	-2.65	Robbins et al., 1993
	45	-2.44	Robbins et al., 1993
	50	-2.39	Robbins et al., 1993
Methyl butyrate [623-42-7]	25	-3.69	Buttery et al., 1969
Methylglyoxal [78-98-8]	25	-7.51	Zhou and Mopper, 1990
	25	-6.57	Betterton and Hoffman, 1988[b]
Methyl hexanoate[106-70-7]	25	-3.43	Buttery et al., 1969
1-Methylnaphthalene [90-12-0]	25	-3.59	Mackay and Shiu, 1981
	25	-3.62	Shiu and Mackay, 1997
Methyl nitrate [598-58-3]	25	-3.30	Kames and Schurath, 1992
Methyl octanoate [111-11-5]	25	-3.11	Buttery et al., 1969
Methyl parathion [298-00-0]	20	-7.42	Rice et al., 1997a
Methyl pentanoate [624-24-8]	25	-3.43	Buttery et al., 1969
Methyl peroxide [3031-73-0]	25	-5.48	Lind and Kok, 1994
	25	-5.49	O'Sullivan et al., 1996
Methyl phenyl ether [100-66-3]	2.0	-4.02	Dewulf et al., 1999
	6.0	-3.88	Dewulf et al., 1999
	10.0	-3.77	Dewulf et al., 1999
	18.0	-3.58	Dewulf et al., 1999
	25.0	-3.43	Dewulf et al., 1999
2-Methylpropanoic acid [79-31-2]	25	-6.04	Khan et al., 1995
	25	-6.76	Servant et al., 1991
2-Methylpropenal [78-85-3]	25	-3.81	Iraci et al., 1998
2-Methyl-2-propenoic acid [79-41-4]	25	-6.41	Khan et al., 1992
Methyl propionate [554-12-1]	25	-3.76	Buttery et al., 1969
N-Methylpiperidine [626-67-5]	25	-4.46	Cabani et al., 1971a
2-Methylpyrazine [109-08-0]	25	-5.65	Buttery et al., 1971
2-Methylpyridine [109-06-8]	25	-5.00	Andon et al., 1954
	25	-4.92	Hawthorne et al., 1985
3-Methylpyridine [108-99-6]	25	-5.11	Andon et al., 1954
	25	-4.62	Chaintreau et al., 1995
4-Methylpyridine[108-89-4]	25	-5.23	Andon et al., 1954
N-Methylpyrrolidine [120-94-5]	25	-4.52	Cabani et al., 1971a
2-Methylthiophene [554-14-3]	25	-2.62	Przyjazny et al., 1983
Metolachlor [51218-45-2]	20	-7.11	Rice et al., 1997a
Mirex [2385-85-5]	22	-3.29	Yin and Hassett, 1986
Nitric monoxide[10102-43-9]	25	-0.15	Zafiriou and McFarland, 1980
Nitrogen dioxide [10102-44-0]	22	-0.85	Lee and Schwartz, 1981
Nitrogen trichloride [10025-85-1]	25	-2.00	Holzwarth et al., 1984
Nitrogen trioxide [12033-49-7]	25	-3.26	Thomas et al., 1998
Nitrosyl chloride [2696-92-6]	25	<-1.70	Scheer et al., 1997
Nitrous acid [7782-77-6]	25	-4.69	Park and Lee, 1988
	25	-4.70	Becker et al., 1996
Nonanal [124-19-6]	25	-3.00	Zhou and Mopper, 1990
	25	-3.11	Buttery et al., 1969
2-Nonanone [821-55-6]	25	-3.43	Buttery et al., 1969
2,2',3,3',4,4',5,5'-Octachlorobiphenyl [35694-08-7]	25	-5.00	Brunner et al., 1990
2,2',3,3',4,4',5,6-Octachlorobiphenyl [52663-78-2]	25	-4.96	Brunner et al., 1990
2,2',3,3',4,4',5,6'-Octachlorobiphenyl [42740-50-1]	25	-5.00	Brunner et al., 1990
2,2',3,3',4,5,5',6-Octachlorobiphenyl [68194-17-2]	25	-4.85	Brunner et al., 1990

Compound [CAS No.]	Temp. °C	Log K_H atm·m³/mol	Reference
2,2′,3,3′,4,5,5′,6′-Octachlorobiphenyl [52663-75-9]	25	-5.00	Brunner et al., 1990
2,2′,3,3′,4,5′,6,6′-Octachlorobiphenyl [40186-71-8]	25	-4.77	Brunner et al., 1990
2,2′,3,3′,5,5′,6,6′-Octachlorobiphenyl [2136-99-4]	25	-4.74	Brunner et al., 1990
Octamethylcyclotetrasiloxane [556-67-2]	20	-1.09	Hamelink et al., 1996
Octanal [124-13-0]	25	-3.32	Zhou and Mopper, 1990
	25	-3.28	Buttery et al., 1969
1-Octanol [111-87-5]	25	-4.60	Buttery et al., 1969
2-Octanone [111-13-7]	25	-3.73	Buttery et al., 1969
trans-2-Octenal [2548-87-0]	25	-4.11	Buttery et al., 1971
Oxygen [7782-44-7]	25	-0.08	Carpenter, 1966
Ozone [10028-15-6]	0.0	-1.30	Kosak-Channing and Helz, 1983
	10.0	-1.23	Kosak-Channing and Helz, 1983
	15.0	-1.17	Kosak-Channing and Helz, 1983
	20.0	-1.11	Kosak-Channing and Helz, 1983
	20.0	-1.03	Rischbieter et al., 2000
	25	-1.11	Briner and Perrottet, 1939
	25	-1.06	Kosak-Channing and Helz, 1983
	25.0	-1.04	Rischbieter et al., 2000
	30	-1.00	Kosak-Channing and Helz, 1983
2,2′,3,4,4′-Pentachlorobiphenyl [65510-45-4]	25	-4.18	Brunner et al., 1990
2,2′,3,4,5′-Pentachlorobiphenyl [38380-02-8]	25	-4.13	Brunner et al., 1990
2,2′,3′,4,5-Pentachlorobiphenyl [41464-51-1]	25	-4.13	Brunner et al., 1990
2,2′,3,5′,6-Pentachlorobiphenyl [38379-99-6]	25	-3.92	Brunner et al., 1990
2,2′,4,4′,5-Pentachlorobiphenyl [38380-01-7]	10	-4.80	Koelmans et al., 1999
	25	-4.11	Brunner et al., 1990
2,2′,4,5,5′-Pentachlorobiphenyl [37680-72-3]	25	-3.60	Dunnivant and Elzerman, 1988
	25	-4.05	Brunner et al., 1990
2,2′,4,5,6′-Pentachlorobiphenyl [68194-06-9]	25	-4.05	Brunner et al., 1990
2,2′,4,6,6′-Pentachlorobiphenyl [56558-16-8]	25	-3.05	Dunnivant and Elzerman, 1988
2,3,4,4′,5′-Pentachlorobiphenyl [31508-00-6]	10	-5.25	Koelmans et al., 1999
2,3′,4,5,5′-Pentachlorobiphenyl [68194-12-7]	25	-4.25	Brunner et al., 1990
2,2,3,3,3-Pentafluoro-1-propanol [422-05-9]	25	-4.65	Rochester and Symonds, 1973
Pentanal [110-62-3]	25	-3.81	Zhou and Mopper, 1990
	25	-3.83	Buttery et al., 1969
Pentanoic acid [109-52-4]	25	-6.34	Khan et al., 1995
1-Pentanol [71-41-0]	25	-4.89	Butler et al., 1935
2-Pentanol [6032-29-7]	25	-4.83	Butler et al., 1935
Pentylamine [110-58-7]	25	-4.61	Christie and Crisp, 1967
1-Pentyl nitrate [1002-16-0]	20	-3.08	Kames and Schurath, 1992
	25	-2.78	Hauff et al., 1998
2-Pentyl nitrate [21981-48-6]	25	-2.57	Kames and Schurath, 1992
	25	-2.53	Hauff et al., 1998
Pernitric acid [26404-66-0]	25	-6.60	Amels et al., 1996
Peroxyacetic acid [79-21-0]	25	-5.92	O'Sullivan et al., 1996
	25	-5.83	Lind and Kok, 1994
Peroxyacetyl nitrate [2278-22-0]	25	-3.45	Kames et al., 1991
	25	-3.61	Kames and Schurath, 1995
	25	-3.70	Holden et al., 1984
Peroxyacetyl radical [36709-10-1]	25	>-2.00	Villalta et al., 1996
Piperidine [110-89-4]	25	-5.34	Cabani et al., 1971a
	25	-5.31	Amoore and Buttery, 1978
Propanal [123-38-6]	25	-4.11	Zhou and Mopper, 1990
	25	-4.13	Buttery et al., 1969
1,2-Propane dinitrate [6423-43-4]	20	-5.26	Kames and Schurath, 1992
	25	-4.51	Fischer and Ballschmiter, 1998, 1998a
1,3-Propane dinitrate [3457-90-7]	25	-5.11	Fischer and Ballschmiter, 1998, 1998a
1,3-Propanediol [504-63-2]	20	-8.96	Bone et al., 1983
1,2,3-Propanetriol [56-81-5]	25	-7.78	Butler and Ramchandani, 1935
Propanoic acid [79-09-4]	25	-6.76	Khan et al., 1995
	25	-6.79	Servant et al., 1991
	25	-6.36	Butler and Ramchandani, 1935
2-Propanol [67-63-0]	25	-5.11	Snider and Dawson, 1985
	25	-5.08	Butler et al., 1935
Propionitrile [107-12-0]	25	-4.43	Butler and Ramchandani, 1935

Compound [CAS No.]	Temp. °C	Log K_H atm·m³/mol	Reference
	25	-4.28	Hawthorne et al., 1985
Propylamine [107-10-8]	25	-4.90	Butler and Ramchandani, 1935
	25	-4.83	Christie and Crisp, 1967
Propyl ether [111-43-3]	25	-2.36	Hartkopf and Karger, 1973
Propyl mercaptan [107-03-9]	25	-2.40	Przyjazny et al., 1983]
1-Propyl nitrate [627-13-4]	25	-3.04	Kames and Schurath, 1992
Pyrrole [109-97-7]	25	-4.74	Hawthorne et al., 1985
Pyrrolidine [123-75-1]	25	-5.62	Cabani et al., 1971a
	25	-5.63	Amoore and Buttery, 1978
3-Pyrroline [109-96-6]	25	-5.70	Amoore and Buttery, 1978
Pyruvaldehyde [78-98-8]	15	-6.93	Betterton and Hoffman, 1988[b]
	25	-6.57	Betterton and Hoffman, 1988[b]
	25	-7.51	Zhou and Mopper, 1990
	35	- 6.11	Betterton and Hoffman, 1988[b]
	45	-5.89	Betterton and Hoffman, 1988[b]
Pyruvic acid [127-17-3]	25	-8.49	Khan et al., 1992, 1995
Quintozene [82-68-8]	25	-5.43	Kawamoto and Urano, 1989
Radon [10043-92-2]	25	-0.97	Morrison and Johnstone, 1954
2,2',3,3'-Tetrachlorobiphenyl [38444-93-8]	25	-4.00	Brunner et al., 1990
2,2',3,4-Tetrachlorobiphenyl [52663-59-9]	25	-3.85	Brunner et al., 1990
2,2',3,4'-Tetrachlorobiphenyl [36559-22-5]	25	-3.85	Brunner et al., 1990
	25	-3.69	Dunnivant and Elzerman, 1988
2,2',3,5'-Tetrachlorobiphenyl [41464--39-5]	23	-3.11	Atlas et al., 1982
	23	-2.27	Atlas et al., 1982[d]
	25	-3.72	Brunner et al., 1990
2,2',4,4'-Tetrachlorobiphenyl [58194--04-7]	23	-3.12	Atlas et al., 1982
	23	-2.31	Atlas et al., 1982[d]
2,2',4,5'-Tetrachlorobiphenyl [41464-40-8]	25	-3.68	Brunner et al., 1990
2,2',5,5'-Tetrachlorobiphenyl [35693-99-3]	10	-4.33	Koelmans et al., 1999
	23	-3.03	Atlas et al., 1982
	23	-2.31	Atlas et al., 1982[d]
	25	-3.47	Dunnivant and Elzerman, 1988
	25	-3.70	Brunner et al., 1990
2,2',5,6'-Tetrachlorobiphenyl [41464-41-9]	25	-3.47	Dunnivant and Elzerman, 1988
2,2',6,6'-Tetrachlorobiphenyl [15968-05-5]	23	-3.12	Atlas et al., 1982
	23	-2.31	Atlas et al., 1982[d]
	25	-3.26	Dunnivant and Elzerman, 1988
	25	-3.70	Brunner et al., 1990
2,3,4,4'-Tetrachlorobiphenyl [33025--41-1]	23	-3.08	Atlas et al., 1982
	23	-2.34	Atlas et al., 1982[d]
2,3',4,5-Tetrachlorobiphenyl [73575-53-8]	25	-4.00	Brunner et al., 1990
2,3',4',5-Tetrachlorobiphenyl [32598-11-1]	25	-4.00	Brunner et al., 1990
2,3,4,6--Tetrachlorobiphenyl [54230-22-7]	25	-3.68	Brunner et al., 1990
2,4,4',5-Tetrachlorobiphenyl [32598-12-2]	25	-4.00	Brunner et al., 1990
3,3',4,4'-Tetrachlorobiphenyl [32598-13-3]	25	-4.03	Dunnivant and Elzerman, 1988
1,2,3,4-Tetrachlorodibenzo-p-dioxin [30746-58-8]	25	-4.70	Santl et al., 1994
1,1,1,2-Tetrachloroethane[630-20-6]	20	-2.77	Tse et al., 1992
	30	-2.55	Tse et al., 1992
	35	-2.44	Tse et al., 1992
	40	-2.34	Tse et al., 1992
	25	-2.60	Wright et al., 1992
4,5,6,7-Tetrachlorophthalide [27355-22-2]	25	-6.27	Kawamoto and Urano, 1989
1,1,1,2-Tetrafluoroethane [811-97-2]	25	-1.26	Zheng et al., 1997
2,2,3,3-Tetrafluoro-1-propanol [76-37-9]	25	-5.20	Rochester and Symonds, 1973
Tetrahydro-2,5-dimethylfuran [1003-38-9]	25	-3.76	Cabani et al., 1971
Tetrahydro-2-methylfuran [96-47-9]	25	-4.04	Cabani et al., 1971
1,2,3,4-Tetrahydronaphthalene [119-64-2]	10	-3.12	Ashworth et al., 1988
	15	-2.98	Ashworth et al., 1988
	20	-2.87	Ashworth et al., 1988
	25	-2.73	Ashworth et al., 1988
	30	-2.57	Ashworth et al., 1988
Tetrahydropyran [142-68-7]	25	-3.90	Cabani et al., 1971
Thiobencarb [28249-77-6]	25	-5.69	Kawamoto and Urano, 1989
Thiophene [109-75-5]	25	-2.63	Przyjazny et al., 1980

Compound [CAS No.]	Temp. °C	Log K_H atm·m³/mol	Reference
Tribromoacetic acid [75-96-7]	25	-8.48	Bowden et al., 1998
Trichlorfon [52-68-6]	25	-7.91	Kawamoto and Urano, 1989
Trichloroacetaldehyde [75-87-6]	15	-8.70	Betterton and Hoffman, 1988[b]
	25	-8.54	Betterton and Hoffman, 1988[b]
	35	-8.39	Betterton and Hoffman, 1988[b]
	45	-8.19	Betterton and Hoffman, 1988[b]
Trichloroacetic acid [76-03-9]	25	-7.87	Bowden et al., 1998
Trichloroacetyl chloride [76-02-8]	25	-3.30	Mirabel et al., 1996
2,2',3-Trichlorobiphenyl [38444-78-9]	23	-3.10	Atlas et al., 1982
	23	-2.50	Atlas et al., 1982[d]
	25	-3.70	Brunner et al., 1990
2,2',5-Trichlorobiphenyl [37680-65-2]	23	-3.00	Atlas et al., 1982
	23	-2.57	Atlas et al., 1982[d]
	25	-3.60	Brunner et al., 1990
2,2',6-Trichlorobiphenyl [38444-73-4]	25	-3.64	Brunner et al., 1990
2,3,3'-Trichlorobiphenyl [38444-84-7]	23	-3.10	Atlas et al., 1982
	23	-2.41	Atlas et al., 1982[d]
	25	-3.80	Brunner et al., 1990
2,3,4'-Trichlorobiphenyl [38444-85-8]	25	-3.85	Brunner et al., 1990
2,3',5-Trichlorobiphenyl [38444-81-4]	25	-3.49	Dunnivant and Elzerman, 1988
	25	-3.70	Brunner et al., 1990
2,3,6-Trichlorobiphenyl [55702-45-9]	25	-3.66	Brunner et al., 1990
	25	-3.70	Brunner et al., 1990
2,4,5-Trichlorobiphenyl [15862-07-4]	25	-3.70	Brunner et al., 1990
2,4',5-Trichlorobiphenyl [16606-02-3]	23	-3.03	Atlas et al., 1982
	23	-2.44	Atlas et al., 1982[d]
	25	-3.72	Brunner et al., 1990
2,4,6-Trichlorobiphenyl [35693-92-6]	25	-3.19	Dunnivant and Elzerman, 1988
3,3',5-Trichlorobiphenyl [38444-87-0]	25	-3.77	Brunner et al., 1990
3,4,4'-Trichlorobiphenyl [38444-90-5]	23	-3.08	Atlas et al., 1982
	23	-2.34	Atlas et al., 1982[d]
	25	-4.00	Brunner et al., 1990
1,2,4-Trichlorodibenzo-p-dioxin [39227-58-2]	25	-4.44	Santl et al., 1994
Trichloroethanal [115-20-8]	25	-8.53	Betterton and Hoffman, 1988[b]
1,2,3-Trichloropropane [96-18-4]	25	-0.68	Tancréde and Yanagisawa, 1990
1,1,2-Trichloro-1,2,2-trifluoroethane [76-13-1]	2.0	-0.99	Dewulf et al., 1999
	6.0	-0.88	Dewulf et al., 1999
	10.0	-0.82	Dewulf et al., 1999
	18.0	-0.58	Dewulf et al., 1999
	25.0	-0.48	Dewulf et al., 1999
Trifluoroacetic acid [76-05-1]	25	-6.95	Bowden et al., 1996
1,1,1-Trifluoroacetone [421-50-1]	15.1	-5.56	Betterton, 1991[a]
	25.1	-5.14	Betterton, 1991[a]
	35.0	-4.60	Betterton, 1991[a]
	45.0	-4.34	Betterton, 1991[a]
Trifluoroacetyl chloride[354-32-5]	25	-3.30	Mirabel et al., 1996
Trifluoroacetyl fluoride [354-32-5]	25	-3.48	Mirabel et al., 1996
2,2,2-Trifluoroethanol [75-89-8]	25	-4.77	Rochester and Symonds, 1973
Trifluoromethane [75-46-7]	25	-1.15	Zheng et al., 1997
1,1,1-Trifluoro-2-propanol [374-01-6]	25	-4.65	Rochester and Symonds, 1973
1,1,1-Trifluoro-2-propanone [421-50-1]	25	-5.15	Betterton, 1991[a]
Trimethylamine [75-50-3]	25	-3.98	Christie and Crisp, 1967
	25	-3.89	Amoore and Buttery, 1978
2,4,5-Trimethylaniline [137-17-7]	25	-5.59	Jayasinghe et al., 1992
2-Undecanone	25	-3.20	Buttery et al., 1969
Xenon [7440-63-3]	25	-0.63	Morrison and Johnstone, 1954

[a] Values originally reported as effective Henry's law constant, $H^* = \{[RR'CO]aq + [RR'C(OH)2]\}/[RR'CO]g$ in units of M/atm. In this equation, [RR'HO]aq is the total concentration of the aldehyde in the aqueous phase and [RR'C(OH)₂] is the concentration of ketone present in the *gem*-diol form; [b] Values originally reported as effective Henry's law constant, $H^* = \{[RCHO]aq + [RCH(OH)_2]\}/[RCHO]g$ in units of M/atm. In this equation, [RCHO]aq is the total concentration of the aldehyde in the aqueous phase and [RCH(OH)₂] is the concentration of aldehyde present in the *gem*-diol form; [c] Henry's law constant determined using seawater (salinity 30.4‰); [d] Henry's law constant determined using seawater (salinity 36‰).

Table 8. Organic Compounds Detected in Water-Soluble Fractions of Regular Gasoline, Super (Unleaded) Gasoline, Gasohol, and Four Middle Distillate Fuels

Compound	Regular Gasoline[a]	Super Gasoline[b]	Gasohol	#2 Fuel Oil	Jet Fuel A	Diesel Fuel	JP-4[c]
			Concentration (mg/L)				
Amyl alcohol	ND	ND	2.67	--[d]	--	--	--
Aniline	0.55	ND	0.20	ND	ND	ND	ND
Benzene	24.0-56.7	80.7	32.3	0.50	0.23	0.20-0.28	17.6
t-Butyl alcohol	ND	3.72	ND	--	--	--	--
C_8-alkylphenol	3.00	0.13	0.72	7.71	1.69	2.51	1.78
C_8-alkylphenol	1.25	ND	0.40	6.36	0.65	1.23	0.72
C_9-alkylphenol	--	--	--	0.83	0.26	0.19	0.20
C_9-alkylphenol	--	--	--	ND	0.08	ND	0.08
C_9-alkylphenol	--	--	--	3.42	0.85	1.24	0.50
C_9-alkylphenol	--	--	--	2.37	0.70	0.79	0.42
m+p-Cresol	6.03	0.60	1.76	1.84	0.43	1.31	0.92
o-Cresol	6.61	0.57	1.17	2.64	0.72	1.36	1.51
Cumene	ND	0.14	0.15	ND	ND	ND	ND
2,3-Dimethylphenol	2.46	0.10	0.47	0.22	0.75	1.08	0.69
2,6-Dimethylphenol	0.47	ND	ND	1.38	0.32	ND	0.41
3,4-Dimethylphenol	1.10	ND	0.22	1.24	0.14	0.51	0.20
EGMME[e]	--	--	--	ND	ND	ND	60.6
Ethanol	ND	ND	8,400	--	--	--	--
Ethylbenzene	2.38-3.8	7.42	3.54	0.21	0.41	0.10-0.17	1.57
2-Ethylphenol	0.74	ND	0.08	0.67	0.27	0.25	0.32
2-Ethyltoluene	0.32	0.62	0.44	0.05	0.15	ND-0.09	0.07
3+4-Ethyltoluene	1.23	2.23	2.17	0.17	0.27	0.11	0.21
Indan	0.40	0.23	0.50	0.05	0.15	0.06	ND
Isocumene	0.18	0.51	0.40	ND	0.05	ND	0.05
2-Isopropylphenol	--	--	--	0.30	ND	ND	0.14
3+4-Isopropylphenol	--	--	--	3.78	1.29	1.23	0.96
Methyl t-butyl ether	ND	137	ND	--	--	--	--
1-Methylnaphthalene	--	--	--	0.21	0.12	0.16	0.05
2-Methylnaphthalene	--	--	--	0.42	0.17	0.27	0.07
Naphthalene	0.24	0.21	0.29	0.60	0.34	0.25	0.18
Phenol	1.53	0.19	0.33	0.09	0.09	0.07	0.22
N-Propylbenzene	--	--	--	--	--	0.023	--
Sulfolane	ND	ND	1.41	ND	ND	ND	ND
Toluene	25.9-30.8	86.9	60.8	1.54	1.05	0.55-0.86	32.0
o+p-Toluidine	0.80	ND	0.19	ND	ND	ND	ND
m-Toluidine	0.51	ND	0.19	ND	ND	ND	ND
1,2,3-Trimethylbenzene	0.30	0.81	0.80	0.22	ND	0.09-0.12	0.19
1,2,4-Trimethylbenzene	1.11	3.11	2.90	0.51	0.44	0.13-0.39	0.39
1,3,5-Trimethylbenzene	0.34	1.29	0.48	0.08	0.09	0.01-0.03	0.09
m+p-Xylene	7.00-13.0	20.1	14.6	--	--	0.23	--
o-Xylene	3.83-6.2	11.4	8.49	1.73	0.87	0.17-1.75	1.99
p-Xylene	3.9-4.38	--	--	1.11	1.23	0.56	5.48

Data obtained from API (1985), Potter (1996), and Thomas and Delfino (1991).

[a] 87 octane; [b] 94 octane; [c] Military jet fuel; [d] Unclear from Potter (1996) whether compound was targeted for analysis or compound was ND; [e] Ethylene glycol monomethyl ether; ND – According to Potter (1996), not detected at compound-specific method detection limit which ranged from 0.005 to 0.100 mg/L.

Table 9. Concentrations of Organic Compounds Detected in Neat Diesel Fuels and Diesel-Powered Medium-Duty Truck Exhaust

Compound	Diesel Fuel Composition		Gas-Phase Emissions
	mg/L	μg/g	μg/km
Acenaphthene	--	--	19.3
Acephenanthrylene	--	--	12.0
Acenaphthylene	--	--	70.1
Acetaldehyde	--	41,800	--
Acetone	--	22,000	--
Acetophenone	--	5,100	--
Acrolein	--	3,400	--
Anthracene	0.026-40	5	12.5
Benaldehyde	--	3,800	--
Benz[a]anthracene	0.018-5.9	--	2.98
Benzene	--	--	2,740
Benzo[b&k]fluoranthene	0.0027-1.7	--	--
Benzo[ghi]fluoranthene	0.0022-3.1	--	5.82
Benzo[a]fluorene	0.0047-11	--	--
Benzofuran	--	53.2	
Benzoic acid	--	1,260	--
Benzo[ghi]perylene	0.0080-0.35	--	--
Benzo[e]pyrene	0.047-2.1	--	
Boacetyl	--	900	
Butanal	--	1,300	--
n-Butane	--	--	3,830
Butanone	--	7,500	--
cis-2-Butene	--	--	260
trans-2-Butene	--	--	520
C_2-fluorene	--	190	65.2
C_2-naphthalenes	--	2,050	542
C_3-naphthalenes	--	1,360	240
C_4-naphthalenes	--	760	97.3
Chrysene/triphenylene	0.0074-4.3	--	--
Crotonaldehyde	--	13,400	--
Cyclohexane	--	--	210
Cyclopentane	--	--	410
Cyclopentene	--	--	
Cyclopenta[cd]pyrene	0.014-3.2	--	2.06
Decanal	--	2,800	--
Decanoic acid	--	72.9	--
Decylcyclohexane	--	420	38.2
Dibenzofuran	--	29	28.7
Dibenzothiophene	--	1.98	--
Dibenzothiozole	--	251	--
Didecylcyclohexane	--	200	16.8
2,5-Dimethylbenzaldehyde	--	4,100	--
2,2-Dimethylbutane	--	--	310
2,3-Dimethylbutane	--	--	570
8β,13α-Dimethyl-14β-n-butylpodocarpane	--	2.1	44.0
2,3-Dimethylhexane	--	--	160
2,4-Dimethylhexane	--	--	50
2,5-Dimethylhexane	--	--	50
8β,13α-Dimethyl-14β-[3'-methylbutyl]-podocarpene	--	0.6	13.8
2,3-Dimethylpentane	--	--	720
2,4-Dimethylpentane	--	--	410
n-Docosane	--	4,340	--
Dodecanal	--	1,200	--
n-Dodecane	--	15,500	503
Dodecanoic acid	--	13.1	--
n-Eicosane	--	6,530	206
Ethene	--	--	8,560
Ethylbenzene	--	--	470
3-Ethylhexane	--	--	210
3-Ethyltoluene	--	--	210
4-Ethyltoluene	--	--	520
Ethyne	--	--	4,600

Compound	Diesel Fuel Composition		Gas-Phase Emissions
	mg/L	μg/g	μg/km
Farnesane	--	9,220	434
Fluoranthene	0.060-9.8	--	53.0
Fluorene	--	52	34.6
Fluorenone	--	34.6	--
Formaldehyde	--	22,300	--
Glyoxal	--	2,100	--
n-Heneicosane	--	5,220	65.8
n-Heptacosane	--	180	--
n-Heptadecane	--	5,700	614
Heptanal	--	3,200	--
n-Heptane	--	--	470
Heptylcyclohexane	--	730	20.0
n-Hexacosane	--	290	--
n-Hexadecane	--	10,100	711
Hexanal	--	2,200	--
cis-2-Hexene	--	--	100
trans-2-Hexene	--	--	160
Hexylcyclohexane	--	830	14.9
Indanone	--	69.5	--
Indeno[1,2,3-cd]pyrene	0.0056-0.85	--	--
Isobutene	--	--	1,140
Isopentane	--	--	2,740
Methacrolein	--	4,000	--
2-Methylanthracene	0.13-160	6	10.4
Methylbenzoic acids	--	772	--
2-Methyl-1-butene	--	--	260
3-Methyl-1-butene	--	--	160
Methylcyclohexane	--	--	520
Methylcyclopentane	--	--	620
Methylglyoxal	--	1,700	--
2-Methylheptane	--	--	100
2-Methylhexane	--	--	570
3-Methylhexane	--	--	310
1-Methyl-7-isopropylphenanthrene	0.013-35	--	--
1-Methylnaphthalene	--	580	378
2-Methylnaphthalene	--	980	511
2-Methylpentane	--	--	930
3-Methylpentane	--	--	670
2-Methyl-2-pentene	--	--	210
1-Methylphenanthrene	0.10-210	28	17.0
2-Methylphenanthrene	--	45	42.0
3-Methylphenanthrene	0.11-99	51	30.3
4&9-Methylphenanthrene	0.11-300	--	--
9-Methylphenanthrene	--	35	22.9
1-Methylpyrene	0.021-12	--	--
2-Methylpyrene	0.032-9.3	--	--
Naphthalene	--	600	617
n-Nonadecane	--	7,020	411
Nonanal	--	4,400	--
n-Nonane	--	--	160
Nonanoic acid	--	240	--
Nonylcyclohexane	--	490	24.7
Norfarnesane	--	16,300	360
Norpristane	--	8,670	566
n-Octacosane	--	36	--
n-Octadecane	--	9,212	601
Octanal	--	3,100	--
n-Octane	--	--	260
Octanoic acid	--	125	--
Octylcyclohexane	--	500	26.2
n-Pentacosane	--	730	--
n-Pentadecane	--	10,500	398
n-Pentane	--	--	1,860

Compound	Diesel Fuel Composition		Gas-Phase Emissions
	mg/L	μg/g	μg/km
trans-2-Pentene	--	--	50
Pentylcyclohexane	--	--	83.9
Pentadecylcyclohexane	--	150	12.8
Phenanthrene	0.17-110	57	93.1
Phytane	--	5,770	439
Picene	0.0031-0.12	--	--
Pristane	--	5,840	443
Propanal	--	14,000	--
n-Propylbenzene	--	--	100
Pyrene	0.16-24	64	71.9
n-Tetracosane	--	1,680	--
Tetradecanoic acid	--	5.3	--
n-Tetradecane	--	13,500	629
Tetradecylcyclohexane	--	160	15.9
Toluene	--	--	3,980
n-Tricosane	--	2,670	--
Tridecanal	--	2,000	--
n-Tridecane	--	15.700	477
Tridecanoic acid	--	13.1	--
Tridecylcyclohexane	--	170	16.5
1,2,4-Trimethylbenzene	--	--	880
1,3,5-Trimethylbenzene	--	--	260
2,2,4-Trimethylpentane	--	--	1,240
2,3,4-Trimethylhexane	--	--	310
2,6,10-Trimethyltridecane	--	8,830	367
Undecanal	--	2,600	--
Undecanoic acid	--	206	--
Undecylcyclohexane	--	430	23.9
Xanthone	--	12,4	--
o-Xylene	--	--	830
m&p-Xylene	--	--	2,330

References: Adapted from Schauer et al. (1999) and Westerholm and Li (1994).

Table 10. Water-Soluble Concentrations of Organic Compounds and Metals in New and Used Motor Oil and Organic Compounds in Kerosene

Compound	Concentration, μg/L		
	New Motor Oil	Used Motor Oil	Kerosene
Aromatic Compounds			
Benzene	0.37-0.40	195-198	73-349
Ethylbenzene	0.15-0.17	117-124	171-314
2-Ethyltoluene	---	---	113-179
3&4-Ethyltoluene	---	---	231
Isopropylbenzene	---	---	22-28
n-Propylbenzene	---	---	57-82
Toluene	16.3-16.9	781-814	448-1,065
1,2,3-Trimethylbenzene	---	---	291-405
1,2,4-Trimethylbenzene	---	---	401-478
1,3,5-Trimethylbenzene	---	---	86-91
m+p-Xylene	0.26-0.29	302-339	360-658
o-Xylene	16.2-17.5	294-308	232-382
Base Neutral & Acid Extractable Compounds			
Acenaphthene	D	D	2
Acenaphthylene	D	4.5-4.6	---
Anthracene	D	1.1-1.3	12
Bis(2-ethylhexyl) phthalate	17-21	D-1.2	---
Butyl benzyl phthalate	8.6-13	14-17	---
Chrysene	D	D	---
Di-n-butyl phthalate	38-43	15-23	---
2,6-Dimethylnaphthalene	---	---	27
2,4-Dimethylphenol	---	---	99
Dimethyl phthalate	D	ND	---
Di-n-octyl phthalate	1.3-1.4	73-78	---
Fluoranthene	D	1.3-1.5	---
Fluorene	---	---	3
1-Methylnaphthalene	0.68-0.82	26-34	147-285
2-Methylnaphthalene	0.42-0.66	46-54	211-354
Naphthalene	D	116-117	278-644
Phenanthrene	1.9-2.1	2.1-2.2	ND
Phenol	---	---	8
Pyrene	D	D	---
Heavy Metals			
Cadmium	ND	Not determined	Not determined
Copper	ND	Not determined	Not determined
Lead	ND	Not determined	Not determined
Nickel	80	120	Not determined
Zinc	2,190	4,550	Not determined

Adapted from Chen et al (1994) and Thomas and Delfino (1991).

Concentrations of targeted analytes were determined by gently stirring Pennzoil® SAE 10W-40 motor oil with distilled water at various time intervals and oil to water ratios. The concentrations listed above were determined after mixing oil and water (1:10 ratio) for 24 hours at 20 °C; D – Detected but not quantified; ND – Not detected.

Table 11. Toxicity of Inorganic and Organic Chemicals to Various Species

Compound [CAS No.]	Species	Results log, mg/L		Reference
Acephate [30560-19-1]	Chironomus plumosus	EC_{50} (48-h)	>1.70	Mayer and Ellersieck, 1986
	Pteronarcella badia	LC_{50} (24-h)	1.47	Mayer and Ellersieck, 1986
	Gammarus pseudolimnaeus	LC_{50} (24-h)	>1.70	Mayer and Ellersieck, 1986
	Salmo gairdneri	LC_{50} (24-h)	>3.00	Mayer and Ellersieck, 1986
		LC_{50} (96-h)	3.04	Mayer and Ellersieck, 1986
4-Acetaminophenol [103-90-2]	Daphnia pulex	EC_{50} (24-h)	2.13	Lilius et al., 1995
Acetochlor [34256-82-1]	Daphnia magna	EC_{50} (48-h)	1.20	Humburg et al., 1989
		LC_{50} (48-h)	1.89	Humburg et al., 1989
	Lepomis machrochirus	LC_{50} (96-h)	0.11	Humburg et al., 1989
	Salmo gairdneri	LC_{50} (96-h)	-0.35	Humburg et al., 1989
Acetophenone [98-86-2]	Spirostomum ambiguum	EC_{50} (24-h)	2.45	Nałecz-Jawecki and Sawicki, 1999
		EC_{50} (48-h)	2.36	Nałecz-Jawecki and Sawicki, 1999
		LC_{50} (24-h)	2.95	Nałecz-Jawecki and Sawicki, 1999
		LC_{50} (48-h)	2.87	Nałecz-Jawecki and Sawicki, 1999
	Pimephales promelas	LC_{50} (96-h)	2.21	Veith et al., 1983
Acifluorfen [50594-66-6]	Lepomis machrochirus	LC_{50} (96-h)	1.49	Humburg et al., 1989
	Salmo gairdneri	LC_{50} (96-h)	1.73	Humburg et al., 1989
Akton [1757-18-2]	redear sunfish	LC_{50} (96-h)	-0.42	Mayer and Ellersieck, 1986
	channel catfish	LC_{50} (24-h)	0.62	Mayer and Ellersieck, 1986
		LC_{50} (96-h)	-0.40	Mayer and Ellersieck, 1986
	Lepomis machrochirus	LC_{50} (24-h)	-0.23	Mayer and Ellersieck, 1986
		LC_{50} (96-h)	-0.77	Mayer and Ellersieck, 1986
Alachlor [15972-60-8]	channel catfish	EC_{50} (48-h)	0.32	Humburg et al., 1989
	Chironomus plumosus	EC_{50} (48-h)	0.51	Mayer and Ellersieck, 1986
	crayfish	EC_{50} (48-h)	2.51	Humburg et al., 1989
	Daphnia magna	EC_{50} (48-h)	1.32	Mayer and Ellersieck, 1986
	Salmo gairdneri	LC_{50} (24-h)	0.63	Mayer and Ellersieck, 1986
		LC_{50} (96-h)	0.38	Mayer and Ellersieck, 1986
	Lepomis machrochirus	LC_{50} (24-h)	1.06	Mayer and Ellersieck, 1986
		LC_{50} (96-h)	0.63	Mayer and Ellersieck, 1986
	rats	LD_{50}[b]	2.97	Humburg et al., 1989
Aldicarb [116-06-3]	Gammarus italicus	LC_{50} (96-h)	-0.38	Pantani et al., 1997
	Echinogammarus tibaldii	LC_{50} (96-h)	-0.66	Pantani et al., 1997
	Paramecium multimicro- nucleatum	LC_{50} (9-h)[a]	2.16	Edmiston et al., 1985
		LC_{50} (13-h)[a]	2.09	Edmiston et al., 1985
		LC_{50} (17-h)[a]	2.02	Edmiston et al., 1985
		LC_{50} (24-h)[a]	1.97	Edmiston et al., 1985
	Lepomis machrochirus	LC_{50} (72-h)	2.00	Day, 1991
	Salmo gairdneri	LC_{50} (24-h)	-0.05	Mayer and Ellersieck, 1986
		LC_{50} (96-h)	-0.21	Mayer and Ellersieck, 1986
Allethrin [584-79-2]	Daphnia pulex	EC_{50} (48-h)	-1.68	Mayer and Ellersieck, 1986
	Simocephalus serrulatus	EC_{50} (48-h)	-1.25	Mayer and Ellersieck, 1986
	Lepomis machrochirus	LC_{50} (24-h)	-1.19	Mayer and Ellersieck, 1986
		LC_{50} (96-h)	-1.25	Mayer and Ellersieck, 1986
	Salmo gairdneri	LC_{50} (24-h)	-1.70	Mayer and Ellersieck, 1986
		LC_{50} (96-h)	-1.72	Mayer and Ellersieck, 1986
	Gammarus fasciatus	LC_{50} (24-h)	-1.42	Mayer and Ellersieck, 1986
		LC_{50} (96-h)	-1.96	Mayer and Ellersieck, 1986
Allyl isothiocyanate [57-06-7]	Oryzias latipes	LC_{50} (96-h)	-1.11	Holcombe et al., 1995
Aluminum chloride [7446-70-0]	Neantheses arenaceodentata	LC_{50} (96-h)	>0.30	Petrich and Reish, 1979
	Capitella capitata	LC_{50} (96-h)	0.30	Petrich and Reish, 1979
	Ctenodrilus serratus	LC_{50} (96-h)	-0.32	Petrich and Reish, 1979
Aluminum nitrate [13473-90-0]	Pimephales promelas	LC_{50} (96-h)	0.63	Mayer and Ellersieck, 1986
Aluminum sulfate [10043-01-3]	Pimephales promelas	LC_{50} (96-h)	0.64	Mayer and Ellersieck, 1986
Amdro [67485-29-4]	Salmo gairdneri	LC_{50} (96-h)	-1.12	Mayer and Ellersieck, 1986
Ametryn [834-12-8]	Lepomis machrochirus	LC_{50} (96-h)	0.61	Humburg et al., 1989
	Salmo gairdneri	LC_{50} (96-h)	0.94	Humburg et al., 1989
Aminocarb [2032-59-9]	Gammarus pseudolimnaeus	EC_{50} (24-h)	0.55	Mayer and Ellersieck, 1986
		EC_{50} (48-h)	0.36	Mayer and Ellersieck, 1986
	Daphnia pulex	EC_{50} (48-h)	-0.49	Mayer and Ellersieck, 1986

Compound [CAS No.]	Species	Results log, mg/L		Reference
	Caecidolea racovitzai racovitzai	LC_{50} (96-h)	1.56	12 °C, Richardson et al., 1983
		LC_{50} (96-h)	1.08	20 °C, Richardson et al., 1983
	Pteronarcella badia	LC_{50} (24-h)	-1.46	Mayer and Ellersieck, 1986
		LC_{50} (96-h)	-1.59	Mayer and Ellersieck, 1986
	Chironomus pseudolimnaeus	LC_{50} (48-h)	1.13	Mayer and Ellersieck, 1986
	Lepomis machrochirus	LC_{50} (24-h)	0.51	Mayer and Ellersieck, 1986
		LC_{50} (96-h)	0.52	Mayer and Ellersieck, 1986
	Salmo gairdneri	LC_{50} (24-h)	-1.64	Mayer and Ellersieck, 1986
		LC_{50} (96-h)	-1.70	Mayer and Ellersieck, 1986
4-Aminopyridine [504-24-5]	*Coturnix coturnix* (male)	LD_{50}[b]	2.65	Schafer et al., 1975
	Coturnix coturnix (female)	LD_{50}[b]	2.75	Schafer et al., 1975
Amitriptyline [50-48-6]	*Daphnia magna*	EC_{50} (24-h)	0.06	Lilius et al., 1995
	Daphnia pulex	EC_{50} (24-h)	0.02	Lilius et al., 1995
Amitrole [61-82-5]	*Gammarus fasciatus*	LC_{50} (24-h)	>1.00	Mayer and Ellersieck, 1986
		LC_{50} (96-h)	>1.00	Mayer and Ellersieck, 1986
	male rats	LD_{50}[b]	3.70	Humburg et al., 1989
Ammonium chloride as NH_4-N [12125-02-9]	*Pseudacris regilla*	LC_{50} (4-d)	1.78	Schuytema and Nebeker, 1999
		LC_{50} (10-d)	1.48	Schuytema and Nebeker, 1999
	Xenopus laevis	LC_{50} (4-d)	1.76	Schuytema and Nebeker, 1999
		LC_{50} (5-d)	1.75	Schuytema and Nebeker, 1999
Ammonium fluoride [12125-01-8]	*Pimephales promelas*	LC_{50} (24-h)	2.63	Curtis et al., 1978
		LC_{50} (48-h)	2.62	Curtis et al., 1978
		LC_{50} (96-h)	2.56	Curtis et al., 1978
Ammonium nitrate as NH_4-N [6484-52-2]	*Pseudacris regilla*	LC_{50} (4-d)	1.61	Schuytema and Nebeker, 1999
		LC_{50} (10-d)	1.40	Schuytema and Nebeker, 1999
	Xenopus laevis	LC_{50} (4-d)	1.51	Schuytema and Nebeker, 1999
		LC_{50} (5-d)	1.64	Schuytema and Nebeker, 1999
Ammonium perfluorononanoate [4149-60-4]	male rats	LC_{50} (4-h)[c]	-0.09	Kinney et al., 1989
Ammonium perfluorooctanoate [3825-26-1]	male rats	LC_{50} (4-h)[c]	-0.01	Kennedy et al., 1986
Ammonium sulfate [7783-20-2]	*Helisoma trivolvis*	LC_{50} (24-h)	2.85	Tchounwou et al., 1991
		LC_{100} (48-h)	3.10	Tchounwou et al., 1991
	Biomphalaria havanensis	LC_{50} (24-h)	2.82	Tchounwou et al., 1991
		LC_{100} (48-h)	3.00	Tchounwou et al., 1991
	Heteropneustes fossilis	LC_{50} (96-h)	3.60	Banerjee and Paul, 1993
Amphetamine sulfate [60-13-9]	*Daphnia magna*	EC_{50} (24-h)	1.78	Lilius et al., 1995
	Daphnia pulex	EC_{50} (24-h)	0.09	Lilius et al., 1995
Anilazine [101-05-3]	*Salmo gairdneri*	LC_{50} (24-h)	-0.84	Mayer and Ellersieck, 1986
		LC_{50} (96-h)	-0.85	Mayer and Ellersieck, 1986
	Gammarus fasciatus	LC_{50} (24-h)	0.10	Mayer and Eliersieck, 1986
		LC_{50} (96-h)	-0.57	Mayer and Ellersieck, 1986
	Lepomis machrochirus	LC_{50} (24-h)	-0.22	Mayer and Ellersieck, 1986
		LC_{50} (96-h)	-0.49	Mayer and Ellersieck, 1986
Antimony [7440-36-0]	*Daphnia magna*	LC_{50} (24-h)	>2.72	LeBlanc, 1980
		LC_{50} (24-h)	>2.72	LeBlanc, 1980
Apholate [52-46-0]	*Salmo gairdneri*	LC_{50} (24-h)	>1.60	Mayer and Ellersieck, 1986
		LC_{50} (96-h)	>1.60	Mayer and Ellersieck, 1986
	Lepomis machrochirus	LC_{50} (24-h)	>1.48	Mayer and Ellersieck, 1986
		LC_{50} (96-h)	>1.48	Mayer and Ellersieck, 1986
Aramite [140-57-8]	*Daphnia pulex*	EC_{50} (48-h)	-0.80	Mayer and Ellersieck, 1986
	Simocephalus serrulatus	EC_{50} (48-h)	-0.64	Mayer and Ellersieck, 1986
	Salmo gairdneri	LC_{50} (24-h)	-0.14	Mayer and Ellersieck, 1986
		LC_{50} (96-h)	-0.49	Mayer and Ellersieck, 1986
	Gammarus fasciatus	LC_{50} (24-h)	-0.46	Mayer and Ellersieck, 1986
		LC_{50} (96-h)	-1.22	Mayer and Ellersieck, 1986
	Lepomis machrochirus	LC_{50} (24-h)	-0.32	Mayer and Ellersieck, 1986
		LC_{50} (96-h)	-0.46	Mayer and Ellersieck, 1986
Arsenic [7440-38-2]	*Tigriopus brevicornis*	LC_{50} (96-h)	-1.56	Forget et al., 1998
Arsenic sulfide [98-88-4]	*Pimephales promelas*	LC_{50} (48-h)	2.81	Curtis et al., 1978

Compound [CAS No.]	Species	Results log, mg/L		Reference
		LC$_{50}$ (96-h)	2.13	Curtis et al., 1978
Arsenic trioxide [1327-53-3]	*Daphnia pulex*	EC$_{50}$ (24-h)	0.14	Lilius et al., 1995
	Tanytarsus dissimilis	LC$_{50}$ (48-h)	1.99	Holcombe et al., 1983
Aspirin [50-78-2]	*Daphnia magna*	EC$_{50}$ (24-h)	3.17	Lilius et al., 1995
	Daphnia pulex	EC$_{50}$ (24-h)	2.56	Lilius et al., 1995
Atrazine [1912-24-9]	*Scenedesmus subspicatus*	EC$_{10}$ (96-h)	-1.40	Geyer et al., 1985
		EC$_{50}$ (96-h)	-0.96	Geyer et al., 1985
	Chironomus riparius	LC$_{50}$ (10-d)	1.28	Taylor et al., 1991
	Chironomus tentans 1st instar	LC$_{50}$ (48-h)	-0.14	Macek et al., 1976b
	Daphnia magna	LC$_{50}$ (48-h)	0.84	Macek et al., 1976b
	Gammarus fasciatus 1st moult	LC$_{50}$ (48-h)	0.76	Macek et al., 1976b
	Pimephales promelas	LC$_{50}$ (96-h)	1.18	Macek et al., 1976b
	Gammarus italicus	LC$_{50}$ (96-h)	1.00	Pantani et al., 1997
	Gammarus pulex	LC$_{50}$ (96-h)	1.17	Taylor et al., 1991
		LC$_{50}$ (5-d)	1.13	Taylor et al., 1991
		LC$_{50}$ (10-d)	0.64	Taylor et al., 1991
	Cyprinus carpio	LC$_{50}$ (96-h)	1.27	Neskovic et al., 1993
	Echinogammarus tibaldii	LC$_{50}$ (96-h)	0.52	Pantani et al., 1997
	Tigriopus brevicornis	LC$_{50}$ (96-h)	-0.81	Forget et al., 1998
	Salvelinus fontinalis	LC$_{50}$ (96-h)	0.80	Macek et al., 1976b
Atropine sulfate [55-48-1]	*Daphnia magna*	EC$_{50}$ (24-h)	2.40	Lilius et al., 1995
Azadirachtin [11141-17-6]	*Bufo quercicus* (stage 12)	LC$_{50}$ (96-h)	0.64	Punzo, 1997
	Bufo quercicus (stage 16)	LC$_{50}$ (96-h)	0.64	Punzo, 1997
	Bufo quercicus (stage 20)	LC$_{50}$ (96-h)	0.79	Punzo, 1997
	Bufo quercicus (stage 24)	LC$_{50}$ (96-h)	1.05	Punzo, 1997
	Bufo quercicus (stage 30)	LC$_{50}$ (96-h)	1.29	Punzo, 1997
Azamethiphos [35575-96-3]	*Homarus americanus* adults	LC$_{50}$ (48-h)	0.14	Burridge, et al., 1999
Azinphos ethyl [2642-71-9]	*Artemia* sp.	EC$_{50}$ (24-h)	0.52	Guzzella et al., 1997
	Brachionus plicatilis	EC$_{50}$ (24-h)	>0.72	Guzzella et al., 1997
	Daphnia magna	EC$_{50}$ (48-h)	-2.40	Mayer and Ellersieck, 1986
	Daphnia pulex	EC$_{50}$ (48-h)	-2.49	Mayer and Ellersieck, 1986
	Simocephalus serrulatus	EC$_{50}$ (48-h)	-2.38	Mayer and Ellersieck, 1986
	Salmo gairdneri	LC$_{50}$ (24-h)	-1.30	Mayer and Ellersieck, 1986
		LC$_{50}$ (96-h)	-1.70	Mayer and Ellersieck, 1986
	Lepomis machrochirus	LC$_{50}$ (24-h)	-2.43	Mayer and Ellersieck, 1986
		LC$_{50}$ (96-h)	-2.96	Mayer and Ellersieck, 1986
Azinphos methyl [86-50-0]	*Artemia* sp.	EC$_{50}$ (24-h)	1.36	Guzzella et al., 1997
	Brachionus plicatilis	EC$_{50}$ (24-h)	1.93	Guzzella et al., 1997
	Salmo gairdneri	LC$_{50}$ (24-h)	>3.00	Mayer and Ellersieck, 1986
		LC$_{50}$ (96-h)	3.04	Mayer and Ellersieck, 1986
	Gammarus italicus	LC$_{50}$ (96-h)	-2.15	Pantani et al., 1997
	Echinogammarus tibaldii	LC$_{50}$ (96-h)	-2.41	Pantani et al., 1997
	Salmo gairdneri	LC$_{50}$ (24-h)	-1.81	Mayer and Ellersieck, 1986
		LC$_{50}$ (96-h)	-2.28	Mayer and Ellersieck, 1986
	Gammarus fasciatus	LC$_{50}$ (24-h)	-3.30	Mayer and Ellersieck, 1986
		LC$_{50}$ (96-h)	-3.90	Mayer and Ellersieck, 1986
	Ambystoma gracile larvae	LC$_{50}$ (96-h)	0.22	Nebeker et al., 1998
	Ambystoma maculatum larvae	LC$_{50}$ (96-h)	0.28	Nebeker et al., 1998
	Cyprinodon variegatus	LC$_{50}$ (96-h)	-2.70	Morton et al., 1997
	Mysidopsis bahia	LC$_{50}$ (96-h)	-3.54	Morton et al., 1997
	Pseudacris regilla	LC$_{50}$ (96-h)	0.56	Nebeker et al., 1998
	Mus musculus	LC$_{50}$ (10-d)	2.44	Meyers and Wolff, 1994
	Microtus canicaudus	LC$_{50}$ (10-d)	2.47	Meyers and Wolff, 1994
	Peromyscus maniculatus	LC$_{50}$ (10-d)	3.07	Meyers and Wolff, 1994
	Mus musculus	LD$_{50}$[b]	1.04	Meyers and Wolff, 1994
	Microtus canicaudus	LD$_{50}$[b]	1.51	Meyers and Wolff, 1994
	Peromyscus maniculatus	LD$_{50}$[b]	1.68	Meyers and Wolff, 1994
Azulene [275-51-4]	*Daphnia pulex*	EC$_{50}$ (48-h)	0.20	Passino-Reader et al., 1997
Barban [101-27-9]	*Daphnia magna*	EC$_{50}$ (48-h)	-0.37	Mayer and Ellersieck, 1986
	Chironomus plumosus	EC$_{50}$ (48-h)	-0.49	Mayer and Ellersieck, 1986
	Gammarus fasciatus	LC$_{50}$ (24-h)	0.91	Mayer and Ellersieck, 1986
		LC$_{50}$ (48-h)	0.60	Mayer and Ellersieck, 1986
		LC$_{50}$ (96-h)	0.04	Mayer and Ellersieck, 1986

Compound [CAS No.]	Species	Results log, mg/L		Reference
	Lepomis machrochirus	LC$_{50}$ (96-h)	0.07	Humburg et al., 1989
	Salmo gairdneri	LC$_{50}$ (96-h)	-0.21	Humburg et al., 1989
Barium [7440-39-3]	*Cyprinodon variegatus*	LC$_{50}$ (24-h)	2.70	Heitmuller et al., 1981
		LC$_{50}$ (48-h)	2.70	Heitmuller et al., 1981
		LC$_{50}$ (72-h)	2.70	Heitmuller et al., 1981
		LC$_{50}$ (96-h)^d	2.70	Heitmuller et al., 1981
	Daphnia magna	LC$_{50}$ (24-h)	>2.72	LeBlanc, 1980
		LC$_{50}$ (48-h)	2.61	LeBlanc, 1980
Barium nitrate [10022-31-8]	*Daphnia magna*	EC$_{50}$ (24-h)	2.13	Lilius et al., 1995
Benazolin [3813-05-6]	dogs	LD$_{50}$^{b}	>3.00	Humburg et al., 1989
	mice	LD$_{50}$^{b}	>3.60	Humburg et al., 1989
	rats	LD$_{50}$^{b}	>3.70	Humburg et al., 1989
Bendiocarb [22781-23-3]	*Gammarus italicus*	LC$_{10}$ (24-h)	-1.37	Pantani et al., 1997
	Echinogammarus tibaldii	LC$_{50}$ (24-h)	-1.96	Pantani et al., 1997
	Lepomis machrochirus	LC$_{50}$ (24-h)	0.76	Mayer and Ellersieck, 1986
		LC$_{50}$ (96-h)	0.39	Mayer and Ellersieck, 1986
Benfluralin [1861-40-1]	*Carassius auratus*	LC$_{50}$ (24-h)	-0.06	Mayer and Ellersieck, 1986
		LC$_{50}$ (48-h)	-0.07	Mayer and Ellersieck, 1986
		LC$_{50}$ (96-h)	-0.09	Mayer and Ellersieck, 1986
	dogs	LD$_{50}$^{b}	>3.30	Humburg et al., 1989
	mice	LD$_{50}$^{b}	>4.00	Humburg et al., 1989
	rats	LD$_{50}$^{b}	>3.70	Humburg et al., 1989
Benomyl [17804-35-2]	*Daphnia magna*	EC$_{50}$ (48-h)	0.45	Mayer and Ellersieck, 1986
	Chironomus plumosus	EC$_{50}$ (48-h)	0.85	Mayer and Ellersieck, 1986
	Salmo gairdneri	LC$_{50}$ (24-h)	0.08	Mayer and Ellersieck, 1986
		LC$_{50}$ (96-h)	-0.56	Mayer and Ellersieck, 1986
	Lepomis machrochirus	LC$_{50}$ (24-h)	-0.28	Mayer and Ellersieck, 1986
		LC$_{50}$ (96-h)	-0.41	Mayer and Ellersieck, 1986
Bensulfuron-methyl [83055-99-6]	*Salmo gairdneri*	LC$_{50}$ (96-h)	>2.81	Humburg et al., 1989
Bensulide [741-58-2]	*Salmo gairdneri*	LC$_{50}$ (24-h)	-0.02	Mayer and Ellersieck, 1986
		LC$_{50}$ (96-h)	-0.14	Mayer and Ellersieck, 1986
	rats	LD$_{50}$^{b}	2.89	Humburg et al., 1989
Bentazon [25057-89-0]	*Lepomis machrochirus*	LC$_{50}$ (96-h)	2.79	Humburg et al., 1989
	Salmo gairdneri	LC$_{50}$ (96-h)	2.28	Humburg et al., 1989
	cats	LD$_{50}$^{b}	2.70	Humburg et al., 1989
	mice	LD$_{50}$^{b}	2.60	Humburg et al., 1989
	rabbits	LD$_{50}$^{b}	2.88	Humburg et al., 1989
	rats	LD$_{50}$^{b}	3.04	Humburg et al., 1989
Benthiocarb [408-27-5]	*Salmo gairdneri*	LC$_{50}$ (24-h)	0.29	Mayer and Ellersieck, 1986
		LC$_{50}$ (96-h)	0.06	Mayer and Ellersieck, 1986
	Cyprinodon variegatus	LC$_{50}$ (96-h)	0.14	Schimmel et al., 1983
	Mysidopsis bahia	LC$_{50}$ (96-h)	-0.48	Schimmel et al., 1983
Benzalazine [64896-26-0]	rats	LD$_{50}$^{b}	>4.00	Herzog and Leuschner, 1994
Benzonitrile [100-47-0]	*Spirostomum ambiguum*	EC$_{50}$ (24-h)	2.70	Nałecz-Jawecki and Sawicki, 1999
		EC$_{50}$ (48-h)	2.60	Nałecz-Jawecki and Sawicki, 1999
		LC$_{50}$ (24-h)	3.10	Nałecz-Jawecki and Sawicki, 1999
		LC$_{50}$ (48-h)	2.98	Nałecz-Jawecki and Sawicki, 1999
Benzophenone [119-61-9]	*Pimephales promelas*	LC$_{50}$ (96-h)	1.17	Veith et al., 1983
Benzoyl chloride [100-44-7]	*Pimephales promelas*	LC$_{50}$ (24-h)	1.63	Curtis et al., 1978
		LC$_{50}$ (48-h)	1.54	Curtis et al., 1978
		LC$_{50}$ (96-h)	1.54	Curtis et al., 1978
Benzyl acetate [140-11-4]	*Oryzias latipes*	LC$_{50}$ (96-h)	0.60	Holcombe et al., 1995
Beryllium [7440-41-7]	*Daphnia magna*	LC$_{50}$ (24-h)	0.28	LeBlanc, 1980
		LC$_{50}$ (48-h)	0.00	LeBlanc, 1980
Bicyclohexyl [92-51-3]	*Daphnia pulex*	EC$_{50}$ (48-h)	-1.46	Passino-Reader et al., 1997
Bifenox [42576-02-3]	rats	LD$_{50}$^{b}	>3.70	Humburg et al., 1989
Binapacryl [485-31-4]	*Lepomis machrochirus*	LC$_{50}$ (24-h)	-1.38	Mayer and Ellersieck, 1986
		LC$_{50}$ (96-h)	-1.40	Mayer and Ellersieck, 1986
Bioban P-1487 [37304-88-4]	*Platynereis dumerilii*	EC$_{50}$ (1-h)	-0.49	Hutchinson et al., 1995

Compound [CAS No.]	Species	Results log, mg/L		Reference
		EC_{50} (48-h)	-0.54	Hutchinson et al., 1995
		LC_{50} (48-h)	-0.49	Hutchinson et al., 1995
Bis(β-chloroethyl)sulfide [505-60-2]	mice	LC_{50} (14-d)[c]	-1.37	Vijayaraghavan, 1997
	female mice	LC_{50} (1-h)[c]	1.63	Kumar and Vijayaraghavan, 1998
Bis[2-(dimethylamino)ethyl] ether [3033-62-3]	rats	LD_{50}[b]	3.08	Ballantyne, 1997
		LC_{50} (6-d)[c]	2.22	Ballantyne, 1997
Bromacil [314-40-9]	carp	LC_{50} (48-h)	2.21	Humburg et al., 1989
	Lepomis machrochirus	LC_{50} (48-h)	1.85	Humburg et al., 1989
	Salmo gairdneri	LC_{50} (48-h)	1.88	Humburg et al., 1989
	rats	LD_{50}[b]	>3.70	Humburg et al., 1989
Boron trifluoride [7637-07-2]	rats	LC_{50} (4-h)	0.08	Rusch et al., 1986
Bromine [7726-95-6]	Daphnia magna	LC_{50} (24-h)	0.18	LeBlanc, 1980
		LC_{50} (48-h)	0.00	LeBlanc, 1980
Bromine chloride [13863-41-7]	Palaemonetes pugio	EC_{50} (48-h)	-0.30	Burton and Margrey, 1978
		EC_{50} (96-h)	-0.40	Burton and Margrey, 1978
		LC_{50} (24-h)	0.04	Burton and Margrey, 1978
		LC_{50} (48-h)	-0.10	Burton and Margrey, 1978
		LC_{50} (96-h)	-0.22	Burton and Margrey, 1978
	Callinectes sapidus	LC_{50} (48-h)	0.08	Burton and Margrey, 1978
		LC_{50} (96-h)	-0.10	Burton and Margrey, 1978
2-Bromopropane [75-26-3]	6 mice (3 males, 3 females)	LC_{50} (4-h)[a]	4.49	Kim et al., 1996
		LC_{100} (4-h)[a]	4.52	Kim et al., 1996
Butachlor [23184-66-9]	carp	LC_{50} (96-h)	-0.49	Humburg et al., 1989
	crayfish	LC_{50} (96-h)	1.41	Humburg et al., 1989
	Daphnia magna	LC_{50} (96-h)	0.38	Humburg et al., 1989
	Lepomis machrochirus	LC_{50} (96-h)	-0.36	Humburg et al., 1989
	Salmo gairdneri	LC_{50} (96-h)	-0.28	Humburg et al., 1989
	bobwhite quail	LD_{50}[b]	4.00	Humburg et al., 1989
	mallard duck	LD_{50}[b]	3.67	Humburg et al., 1989
	rats	LD_{50}[b]	3.30	Humburg et al., 1989
1,3-Butadiene diepoxide [1464-53-5]	Poecilia reticulata	LC_{50} (14-d)	0.43	Deneer et al., 1988
		LC_{50} (7-d)	2.99	Könemann, 1981
Butylate [2008-41-5]	Salmo gairdneri	LC_{50} (24-h)	0.60	Mayer and Ellersieck, 1986
		LC_{50} (96-h)	0.32	Mayer and Ellersieck, 1986
	Lepomis machrochirus	LC_{50} (24-h)	-0.05	Mayer and Ellersieck, 1986
		LC_{50} (96-h)	-0.68	Mayer and Ellersieck, 1986
n-Butyl ether [142-96-1]	Cyprinodon variegatus	LC_{50} (24-h)	0.38	Heitmuller et al., 1981
		LC_{50} (48-h)	0.38	Heitmuller et al., 1981
		LC_{50} (72-h)	0.38	Heitmuller et al., 1981
		LC_{50} (96-h)[d]	0.38	Heitmuller et al., 1981
	Daphnia magna	LC_{50} (24-h)	1.51	LeBlanc, 1980
		LC_{50} (48-h)	1.41	LeBlanc, 1980
	Pimephales promelas	LC_{50} (96-h)	1.51	Veith et al., 1983
n-Butyl nitrite [544-16-1]	Sprague-Dawley rats	LC_{50} (4-h)[a]	2.62	Klonne et al., 1987a
Butyric acid [107-92-6]	Hyale plumulosa	LC_{50} (48-h)[d]	2.40	Onitsuka et al., 1989
	Oryzias latipes	LC_{50} (48-h)	1.95	Onitsuka et al., 1989
		LC_{50} (48-h)[d]	2.36	Onitsuka et al., 1989
Cacodylic acid [75-60-5]	male rat	LD_{50}[b]	2.86	Stevens et al., 1979
	female rat	LD_{50}[b]	2.72	Stevens et al., 1979
	male and female albino rats	LD_{50}[b]	2.92	Humburg et al., 1989
Cadmium [7440-43-9]	Daphnia galeata mendotae	EC_{50} (96-h)	-1.52	Marshall, 1979
	Acrobeloides buetschlii	LC_{50} (24-h)	2.00	Kammenga et al., 1994
		LC_{40} (48-h)	1.97	Kammenga et al., 1994
		LC_{50} (72-h)	1.77	Kammenga et al., 1994
	Aedes aegypti	LC_{50} (24-h)	1.22	Rayms-Keller et al., 1998
	Aporcelaimellus obtus-caudatus	LC_{50} (48-h)	1.38	Kammenga et al., 1994
	Aphelenchus avenae	LC_{50} (24-h)	>1.95	Kammenga et al., 1994
		LC_{50} (72-h)	1.17	Kammenga et al., 1994
	Caenorhabditis elegans	LC_{50} (24-h)	1.47	Kammenga et al., 1994
		LC_{40} (48-h)	1.18	Kammenga et al., 1994

Compound [CAS No.]	Species	Results log, mg/L		Reference
		LC_{50} (72-h)	1.17	Kammenga et al., 1994
	Cephalobus persegnis	LC_{50} (24-h)	1.29	Kammenga et al., 1994
		LC_{40} (48-h)	1.08	Kammenga et al., 1994
		LC_{50} (72-h)	0.97	Kammenga et al., 1994
	Dorylaimus stagnalis	LC_{50} (24-h)	1.58	Kammenga et al., 1994
		LC_{50} (48-h)	1.24	Kammenga et al., 1994
		LC_{50} (96-h)	1.01	Kammenga et al., 1994
	Diplogasteritus sp.	LC_{50} (24-h)	0.62	Kammenga et al., 1994
		LC_{40} (48-h)	0.52	Kammenga et al., 1994
		LC_{50} (72-h)	0.52	Kammenga et al., 1994
	Homarus americanus	LC_{50} (96-h)	-1.11	Johnson and Gentile, 1979
	Mytilopsis sallei	LC_{50} (96-h)	-0.15	Uma Devi, 1996
	Ranatra elongata	LC_{50} (24-h)	-0.30	Shukla et al., 1983
		LC_{50} (48-h)	-0.36	Shukla et al., 1983
		LC_{50} (72-h)	-0.45	Shukla et al., 1983
		LC_{50} (96-h)	-0.54	Shukla et al., 1983
	Poecilia reticulata	LC_{50} (48-h)	1.75	Miliou et al., 1998
	Plectus acuminatus	LC_{50} (24-h)	1.65	Kammenga et al., 1994
		LC_{40} (48-h)	1.17	Kammenga et al., 1994
		LC_{50} (72-h)	1.08	Kammenga et al., 1994
	Prionchulus punctatus	LC_{50} (24-h)	1.39	Kammenga et al., 1994
	Rhabditis sp.	LC_{50} (24-h)	1.48	Kammenga et al., 1994
		LC_{40} (48-h)	1.28	Kammenga et al., 1994
		LC_{50} (72-h)	1.15	Kammenga et al., 1994
	Tigriopus brevicornis	LC_{50} (96-h)	-1.32	Forget et al., 1998
	Tobrilus gracilis	LC_{50} (24-h)	1.36	Kammenga et al., 1994
		LC_{40} (48-h)	1.30	Kammenga et al., 1994
		LC_{50} (72-h)	1.13	Kammenga et al., 1994
	Tylenchus elegans	LC_{50} (24-h)	>1.95	Kammenga et al., 1994
Cadmium chloride [10108-64-2]	*Spirostomum teres*	LC_{50} (0.5-h)	-0.92	Twagilimana et al., 1998
		LC_{50} (24-h)	-0.29	Twagilimana et al., 1998
	Brachydanio rerio	LC_{50} (48-h)	0.40	Slooff, 1979
	Scylla seratta Forskall	LC_{50} (48-h)	-1.11	Ramachandran et al., 1997
	Gammarus italicus	LC_{10} (24-h)	-0.04	Pantani et al., 1997
	Echinogammarus tibaldii	LC_{50} (24-h)	0.04	Pantani et al., 1997
	Drosophila melanogaster	LC_{50}	1.67	Atkins et al., 1991
	Xenopus laevis	LC_{50} (96-h)	0.77	Sunderman et al., 1991
Cadmium sulfate [10124-36-4]	*Porcellio laevis*	LD_{50} (96-h)[b]	4.43	Odendaal and Reinecke, 1999
Caffeine [58-08-2]	*Daphnia magna*	EC_{50} (24-h)	2.83	Lilius et al., 1995
Calciferol [50-14-6]	rats	LD_{50}[b]	1.75	Worthing and Hance, 1991
Camphene [79-92-5]	*Cyprinodon variegatus*	LC_{50} (24-h)	0.26	Heitmuller et al., 1981
		LC_{50} (48-h)	0.30	Heitmuller et al., 1981
		LC_{50} (72-h)	0.30	Heitmuller et al., 1981
		LC_{50} (96-h)[d]	0.28	Heitmuller et al., 1981
	Daphnia magna	LC_{50} (24-h)	1.66	LeBlanc, 1980
		LC_{50} (48-h)	1.34	LeBlanc, 1980
Captafol [2425-06-1]	*Salmo gairdneri*	LC_{50} (24-h)	-1.40	Mayer and Ellersieck, 1986
		LC_{50} (96-h)	-1.51	Mayer and Ellersieck, 1986
	Lepomis machrochirus	LC_{50} (24-h)	-1.23	Mayer and Ellersieck, 1986
		LC_{50} (96-h)	-1.47	Mayer and Ellersieck, 1986
Captan [133-06-2]	*Salmo gairdneri*	LC_{50} (24-h)	-1.12	Mayer and Ellersieck, 1986
		LC_{50} (96-h)	-1.14	Mayer and Ellersieck, 1986
	Lepomis machrochirus	LC_{50} (24-h)	-0.84	Mayer and Ellersieck, 1986
		LC_{50} (96-h)	-0.85	Mayer and Ellersieck, 1986
(S)-a-Cedrene [469-61-4]	*Daphnia pulex*	EC_{50} (48-h)	-1.36	Passino-Reader et al., 1997
Cetyltrimethylammonium bromide [57-09-0]	*Gammarus italicus*	LC_{10} (24-h)	-0.14	Pantani et al., 1997
	Echinogammarus tibaldii	LC_{50} (24-h)	-0.62	Pantani et al., 1997
Chloramben [133-90-4]	rats	LD_{50}[b]	~3.54	Humburg et al., 1989
Chloramphenicol [56-75-7]	*Daphnia magna*	EC_{50} (24-h)	2.74	Lilius et al., 1995
Chlorfenac [85-34-7]	*Daphnia magna*	EC_{50} (48-h)	-1.53	Mayer and Ellersieck, 1986
	Simocephalus serrulatus	EC_{50} (48-h)	0.82	Mayer and Ellersieck, 1986
	mice	LD_{50}[b]	3.02	Humburg et al., 1989
	rats	LD_{50}[b]	3.25	Humburg et al., 1989

Compound [CAS No.]	Species	Results log, mg/L		Reference
Chlorfenethol [80-06-8]	*Daphnia magna*	EC_{50} (48-h)	-0.57	Mayer and Ellersieck, 1986
Chlorfenvinphos [470-90-6]	male Fischer rats	LC_{50} (4-h)[a]	-0.89	Takahashi et al., 1994
Chlorflurenol [2464-31-4]	*Lepomis machrochirus*	LC_{50} (96-h)	0.86	Humburg et al., 1989
	Salmo gairdneri	LC_{50} (96-h)	0.51	Humburg et al., 1989
	bobwhite quail	LD_{50}[b]	4.00	Humburg et al., 1989
Chlorimuron [99283-00-8]	rats	LD_{50}[b]	>3.60	Humburg et al., 1989
Chlormephos [24934-91-6]	*Gammarus italicus*	LC_{10} (24-h)	-0.11	Pantani et al., 1997
	Echinogammarus tibaldii	LC_{50} (24-h)	-1.40	Pantani et al., 1997
Chloroamine [10599-90-3]	*Chironomidae* larvae	LC_{50} (75-min)	1.51	Brozaet al., 1998
Chlorobenzilate [510-15-6]	*Daphnia pulex*	EC_{50} (48-h)	-0.06	Mayer and Ellersieck, 1986
	Simocephalus serrulatus	EC_{50} (48-h)	-0.26	Mayer and Ellersieck, 1986
1-Chlorobutane [109-69-3]	*Poecilia reticulata*	LC_{50} (7-d)	1.99	Könemann, 1981
4-Chloro-*o*-cresol [1570-64-5]	male rat	LD_{50}[b]	3.08	Hattula et al., 1979
4-Chlorodiphenyl ether [7005-72-3]	*Salvelinus fontinalis*	LC_{50} (96-h)	-0.14	Chui et al., 1990
2-Chloroethanol [107-07-3]	*Oryzias latipes*	LC_{50} (96-h)	1.48	Holcombe et al., 1995
Chloromethyl methyl ether [107-30-2]	rats	LC_{50} (14-d)[a]	1.74	Drew et al., 1975
	hamsters	LC_{50} (14-d)[a]	1.81	Drew et al., 1975
1-Chloronaphthalene [90-13-1]	*Cyprinodon variegatus*	LC_{50} (24-h)	0.53	Heitmuller et al., 1981
		LC_{50} (48-h)	0.40	Heitmuller et al., 1981
		LC_{50} (72-h)	0.38	Heitmuller et al., 1981
		LC_{50} (96-h)[d]	0.38	Heitmuller et al., 1981
	Daphnia magna	LC_{50} (48-h)	0.86	LeBlanc, 1980
4-Chlorophenol [106-48-9]	*Cyprinodon variegatus*	LC_{50} (24-h)	0.76	Heitmuller et al., 1981
		LC_{50} (48-h)	0.73	Heitmuller et al., 1981
		LC_{50} (72-h)	0.73	Heitmuller et al., 1981
		LC_{50} (96-h)[d]	0.73	Heitmuller et al., 1981
	Daphnia magna	LC_{50} (24-h)	0.94	LeBlanc, 1980
		LC_{50} (48-h)	0.61	LeBlanc, 1980
	Hydra viridissima	LC_{50} (96-h)	1.65	Pollino and Holdway, 1999
	Hydra vulgaris	LC_{50} (96-h)	1.51	Pollino and Holdway, 1999
4-Chloro-2-methylphenoxy acetic acid [94-74-6]	*Salmo trutta*	LD_{50}[b]	2.17	Hattula et al., 1978
	mice	LD_{50}[b]	2.90	Humburg et al., 1989
Chlorothalonil [1897-45-6]	*Oncorhynchus mykiss*	LC_{50} (96-h)	-1.12	Ernst et al., 1991
	Mytilus edulis	LC_{50} (96-h)	0.77	Ernst et al., 1991
	Mya arenaria	LC_{50} (96-h)	1.54	Ernst et al., 1991
4-Chlorotoluene [106-43-4]	*Poecilia reticulata*	LC_{50} (14-d)	0.77	Könemann, 1981
Chloroxuron [1982-47-4]	*Salmo gairdneri*	LC_{50} (24-h)	0.57	Mayer and Ellersieck, 1986
		LC_{50} (96-h)	-0.37	Mayer and Ellersieck, 1986
	dogs	LD_{50}[b]	>4.00	Humburg et al., 1989
	female rats	LD_{50}[b]	3.73	Humburg et al., 1989
	male rats	LD_{50}[b]	3.57	Humburg et al., 1989
Chlorpropham [101-21-3]	albino rats	LD_{50}[b]	3.58	Humburg et al., 1989
	mallard ducks	LD_{50}[b]	>3.30	Humburg et al., 1989
	rabbits	LD_{50}[b]	3.70	Humburg et al., 1989
Chlorsulfuron [64902-72-3]	*Lepomis machrochirus*	LC_{50} (96-h)	>2.40	Humburg et al., 1989
	Salmo gairdneri	LC_{50} (96-h)	>2.40	Humburg et al., 1989
	male rats	LD_{50}[b]	3.74	Humburg et al., 1989
	quail and duck	LD_{50}[b]	>3.00	Humburg et al., 1989
Cinmethylin [87818-31-3]	rats	LD_{50}[b]	3.65	Humburg et al., 1989
Clethodim [99129-21-2]	*Lepomis machrochirus*	LC_{50} (96-h)	>2.00	Humburg et al., 1989
	Salmo gairdneri	LC_{50} (96-h)	1.83	Humburg et al., 1989
	bobwhite quail	LD_{50}[b]	3.30	Humburg et al., 1989
	female rats	LD_{50}[b]	3.13	Humburg et al., 1989
	male rats	LD_{50}[b]	3.21	Humburg et al., 1989
Clomazone [81777-89-1]	Atlantic silverside	LC_{50} (96-h)	0.80	Humburg et al., 1989
	Daphnia magna	LC_{50} (96-h)	0.72	Humburg et al., 1989
	Eastern oysters	LC_{50} (96-h)	0.72	Humburg et al., 1989
	Lepomis machrochirus	LC_{50} (96-h)	1.53	Humburg et al., 1989
	Salmo gairdneri	LC_{50} (96-h)	1.28	Humburg et al., 1989
	Sheepshead minnow	LC_{50} (96-h)	1.61	Humburg et al., 1989
	bobwhite quail	LD_{50}[b]	>3.40	Humburg et al., 1989

Compound [CAS No.]	Species	Results log, mg/L		Reference
	female rats	LD_{50}[b]	3.19	Humburg et al., 1989
	male rats	LD_{50}[b]	3.41	Humburg et al., 1989
Clonitralide [1420-04-8]	Daphnia magna	EC_{50} (48-h)	-0.72	Mayer and Ellersieck, 1986
	Chironomus plumosus	EC_{50} (48-h)	0.20	Mayer and Ellersieck, 1986
	Schistosoma mansoni			
	miracidia	LC_{50} (2-h)[a]	-1.22	Tchounwou et al., 1991a
		LC_{50} (4-h)[a]	-1.52	Tchounwou et al., 1991a
		LC_{50} (6-h)[a]	-1.70	Tchounwou et al., 1991a
Clopyralid [1702-17-6]	Lepomis machrochirus	LC_{50} (96-h)	2.10	Humburg et al., 1989
	Salmo gairdneri	LC_{50} (96-h)	2.01	Humburg et al., 1989
	rats	LD_{50}[b]	3.63	Humburg et al., 1989
Cobalt chloride [7646-79-9]	Xenopus laevis	EC_{50} (96-h)	0.51	Plowman et al., 1991
		LC_{50} (96-h)	3.13	Plowman et al., 1991
Copper [7440-50-8]	Acroneuria lycorias larva	LC_{50} (96-h)	0.92	Warnick and Bell, 1969
	Aedes aegypti	LC_{50} (24-h)	1.52	Rayms-Keller et al., 1998
	Chironomus decorus	LC_{50} (48-h)	-0.13	Kosalwat and Knight, 1987
	Chironomus riparius	LC_{50} (48-h)	0.08	Taylor et al., 1991
		LC_{50} (96-h)	-0.15	Taylor et al., 1991
		LC_{50} (10-d)	-0.70	Taylor et al., 1991
	Daphnia magna	LC_{50} (72-h)	-1.05	Winner and Farrell, 1976
	Gammarus pulex	LC_{50} (48-h)	-1.33	Taylor et al., 1991
		LC_{50} (96-h)	-1.43	Taylor et al., 1991
		LC_{50} (10-d)	-1.48	Taylor et al., 1991
	Gammarus pseudolimnaeus	LC_{50} (96-h)	-1.70	Arthur and Leonard, 1970
	Hydra viridissima	LC_{50} (96-h)	-2.07	Pollino and Holdway, 1999
	Pimephales promelas	LC_{50} (96-h)	0.34	Brungs et al., 1976
	Salmo gairdneri	LC_{50} (72-h)	-0.40	Brown, 1968
	Hydra vulgaris	LC_{50} (96-h)	-1.59	Pollino and Holdway, 1999
	Scyliorhinus canicula	LC_{50} (24-h)	1.20	Torres et al., 1987
		LC_{50} (48-h)	0.60	Torres et al., 1987
	Homarus americanus	LC_{50} (96-h)	-1.32	Johnson and Gentile, 1979
	Corophium sp.	LC_{50} (96-h)	-1.08	Hyne and Everett, 1998
Copper acetate [142-71-2]	Pimephales promelas	LC_{50} (24-h)	-0.32	Curtis et al., 1978
		LC_{50} (48-h)	-0.38	Curtis et al., 1978
		LC_{50} (96-h)	-0.41	Curtis et al., 1978
Copper sulfate [7758-98-7]	Daphnia magna	EC_{50} (24-h)	0.19	Lilius et al., 1995
	Daphnia pulex	EC_{50} (24-h)	0.09	Lilius et al., 1995
	Spirostomum teres	LC_{50} (0.5-h)	-0.37	Twagilimana et al., 1998
		LC_{50} (24-h)	-1.43	Twagilimana et al., 1998
	Daphnia magna	LC_{50} (24-h)	-0.42	Ferrando et al., 1992
	Brachionus calyciflorus	LC_{50} (24-h)	-1.12	Ferrando et al., 1992
	Salmo gairdneri	LC_{50} (24-h)	-0.82	Mayer and Ellersieck, 1986
		LC_{50} (96-h)	-0.87	Mayer and Ellersieck, 1986
	Gammarus fasciatus	LC_{50} (48-h)	-0.72	Judy, 1979
	rats	LD_{50}[b]	2.67	Humburg et al., 1989
Correx [2235-25-8]	Salmo gairdneri	LC_{50} (24-h)	0.20	Mayer and Ellersieck, 1986
Coumaphos [56-72-4]	Simocephalus serrulatus	EC_{50} (48-h)	-3.00	Mayer and Ellersieck, 1986
	Salmo gairdneri	LC_{50} (24-h)	0.48	Mayer and Ellersieck, 1986
		LC_{50} (96-h)	-0.05	Mayer and Ellersieck, 1986
Crotoxyphos [7700-17-6]	Salmo gairdneri	LC_{50} (24-h)	-1.00	Mayer and Ellersieck, 1986
		LC_{50} (96-h)	-1.14	Mayer and Ellersieck, 1986
Cryloite [15096-52-3]	Daphnia pulex	EC_{50} (48-h)	1.00	Mayer and Ellersieck, 1986
	Simocephalus serrulatus	EC_{50} (48-h)	0.70	Mayer and Ellersieck, 1986
	Salmo gairdneri	LC_{50} (24-h)	-0.80	Mayer and Ellersieck, 1986
		LC_{50} (96-h)	-1.33	Mayer and Ellersieck, 1986
Cupric chloride [7758-89-6]	Gammarus italicus	LC_{10} (24-h)	-0.77	Pantani et al., 1997
	Echinogammarus tibaldii	LC_{50} (24-h)	-0.23	Pantani et al., 1997
Cyanazine [21725-46-2]	Salmo gairdneri	LC_{50} (24-h)	1.08	Mayer and Ellersieck, 1986
		LC_{50} (96-h)	0.95	Mayer and Ellersieck, 1986
	rats	LD_{50}[b]	2.52	Humburg et al., 1989
Cycloate [1134-23-2]	Salmo gairdneri	LC_{50} (96-h)	0.65	Humburg et al., 1989
	female rats	LD_{50}[b]	3.50	Humburg et al., 1989
	male rats	LD_{50}[b]	3.30	Humburg et al., 1989
Cyclohexylbenzene [827-52-1]	Daphnia pulex	EC_{50} (48-h)	-0.26	Passino-Reader et al., 1997

Compound [CAS No.]	Species	Results log, mg/L		Reference
Cyhexatin [13171-70-5]	*Daphnia magna*	EC$_{50}$ (48-h)	-3.77	Mayer and Ellersieck, 1986
p-Cymene [99-87-6]	*Cyprinodon variegatus*	LC$_{50}$ (24-h)	0.26	Heitmuller et al., 1981
		LC$_{50}$ (48-h)	0.30	Heitmuller et al., 1981
		LC$_{50}$ (72-h)	0.30	Heitmuller et al., 1981
		LC$_{50}$ (96-h)[d]	0.28	Heitmuller et al., 1981
	Daphnia magna	LC$_{50}$ (24-h)	0.97	LeBlanc, 1980
		LC$_{50}$ (48-h)	0.81	LeBlanc, 1980
Cypermethrin [52315-07-8]	*Labeo rohita*	LC$_{50}$ (96-h)	-2.28	Philip et al., 1995
Daimuron [42609-52-9]	rats	LD$_{50}$[b]	>3.70	Worthing and Hance, 1991
Dalapon [75-99-0]	*Daphnia pulex*	EC$_{50}$ (48-h)	1.04	Mayer and Ellersieck, 1986
	Simocephalus serrulatus	EC$_{50}$ (48-h)	1.20	Mayer and Ellersieck, 1986
	chicks (mixed)	LD$_{50}$[b]	3.75	Humburg et al., 1989
	female guinea pigs	LD$_{50}$[b]	3.59	Humburg et al., 1989
	female rabbits	LD$_{50}$[b]	3.59	Humburg et al., 1989
	female rats	LD$_{50}$[b]	3.88	Humburg et al., 1989
	male rats	LD$_{50}$[b]	3.97	Humburg et al., 1989
	mallard ducks	LD$_{50}$[b]	>3.70	Worthing and Hance, 1991
Dazomet [533-74-4]	male albino mice	LD$_{50}$[b]	2.81	Humburg et al., 1989
	rats	LD$_{50}$[b]	2.72	Worthing and Hance, 1991
2,4-DB-dimethylammonium [2758-42-1]	*Salmo gairdneri*	LC$_{50}$ (96-h)	0.605	Humburg et al., 1989
2,4-D butoxylethyl ester [1929-73-3]	*Daphnia magna*	EC$_{50}$ (48-h)	0.61	Mayer and Ellersieck, 1986
	Cypridopsis vidua	EC$_{50}$ (48-h)	0.34	Mayer and Ellersieck, 1986
	Chironomus plumosus	EC$_{50}$ (48-h)	-0.10	Mayer and Ellersieck, 1986
DCPA [1861-32-1]	*Daphnia magna*	EC$_{50}$ (48-h)	>2.00	Mayer and Ellersieck, 1986
	Chironomus plumosus	EC$_{50}$ (48-h)	>2.00	Mayer and Ellersieck, 1986
	male albino mice	LD$_{50}$[b]	>4.00	Humburg et al., 1989
2,4-D dimethylamine [2008-39-1]	*Daphnia magna*	EC$_{50}$ (48-h)	>2.00	Mayer and Ellersieck, 1986
	Cypridopsis vidua	EC$_{50}$ (48-h)	0.90	Mayer and Ellersieck, 1986
	Chironomus plumosus	EC$_{50}$ (48-h)	>2.00	Mayer and Ellersieck, 1986
Decanoic acid [334-48-5]	*Hyale plumulosa*	LC$_{50}$ (48-h)[d]	1.61	Onitsuka et al., 1989
	Oryzias latipes	LC$_{50}$ (48-h)[d]	1.49	Onitsuka et al., 1989
1-Decanol [112-30-1]	*Pimephales promelas*	LC$_{50}$ (96-h)	0.38	Veith et al., 1983
2-Decanone [693-54-9]	*Pimephales promelas*	LC$_{50}$ (96-h)	0.76	Veith et al., 1983
Deltamethrin [52918-63-5]	*Poecilia reticulata*	LC$_{50}$[a]	-1.80	Mittal et al., 1994
	ducks	LD$_{50}$[b]	>3.67	Worthing and Hance, 1991
Demeton [8065-48-3]	*Daphnia pulex*	EC$_{50}$ (48-h)	-1.85	Mayer and Ellersieck, 1986
Diallate [2303-16-4]	Harlequin fish	LC$_{50}$ (48-h)	0.91	Humburg et al., 1989
	Daphnia magna	LC$_{50}$ (48-h)	0.88	Humburg et al., 1989
	Lepomis machrochirus	LC$_{50}$ (96-h)	0.40	Humburg et al., 1989
	Salmo gairdneri	LC$_{50}$ (96-h)	0.51	Humburg et al., 1989
	rats	LD$_{50}$[b]	3.02	Humburg et al., 1989
Diazepam [439-14-5]	*Daphnia magna*	EC$_{50}$ (24-h)	0.63	Lilius et al., 1995
	Daphnia pulex	EC$_{50}$ (24-h)	1.08	Lilius et al., 1995
Diazinon [333-41-5]	*Photobacterium phosphoreum*	EC$_{50}$ (5-min)	1.01	Somasundaram et al., 1990
	Artemia sp.	EC$_{50}$ (24-h)	1.28	Guzzella et al., 1997
	Brachionus plicatilis	EC$_{50}$ (24-h)	1.45	Guzzella et al., 1997
	Daphnia pulex	EC$_{50}$ (48-h)	-3.10	Mayer and Ellersieck, 1986
	Simocephalus serrulatus	EC$_{50}$ (48-h)	-2.80	Mayer and Ellersieck, 1986
	Daphnia magna	LC$_{50}$ (24-h)	-3.05	Fernandez-Casalderry et al., 1994
	Brachionus calyciflorus	LC$_{50}$ (24-h)	1.47	Fernandez-Casalderry et al., 1992
	Micropterus salmoides	LC$_{50}$ (24-h)	-0.05	Pan and Dutta, 1998
	Anguilla anguilla	LC$_{50}$ (24-h)	-0.80	Sancho et al., 1992
		LC$_{50}$ (48-h)	-0.96	Sancho et al., 1992
		LC$_{50}$ (72-h)	-1.05	Sancho et al., 1992
		LC$_{50}$ (96-h)	-1.10	Sancho et al., 1992
		LC$_{50}$ (96-h)	-1.08	Ceron et al., 1996
	Oryzias latipes	LC$_{50}$ (48-h)	0.64	Tsuda et al., 1997
Diazinon oxon [962-58-3]	*Oryzias latipes*	LC$_{50}$ (48-h)	-0.66	Tsuda et al., 1997
1,4-Dibromobenzene [106-37-6]	*Gambusia affinis*	LC$_{50}$ (96-h)	-2.52	Chaisuksant et al., 1998
Dibutylhexamethylenediamine [4835-11-4]	rats	LC$_{50}$ (48-h)	-0.66	Kennedy and Chen, 1984

Compound [CAS No.]	Species	Results log, mg/L		Reference
Dicamba [1918-00-9]	*Daphnia magna*	EC$_{50}$ (48-h)	>2.00	Mayer and Ellersieck, 1986
	mallard duck	LD$_{50}$[b]	3.30	Humburg et al., 1989
	rats	LD$_{50}$[b]	3.23	Humburg et al., 1989
Dichlobenil [1194-65-6]	*Daphnia pulex*	EC$_{50}$ (48-h)	0.57	Mayer and Ellersieck, 1986
	Simocephalus serrulatus	EC$_{50}$ (48-h)	0.76	Mayer and Ellersieck, 1986
	Scenedesmus subspicatus	EC$_{10}$ (96-h)	-0.52	Geyer et al., 1985
		EC$_{50}$ (96-h)	0.43	Geyer et al., 1985
Dichlone [117-80-6]	*Cypridopsis vidua*	EC$_{50}$ (48-h)	-1.49	Mayer and Ellersieck, 1986
Dichlormid [37764-25-3]	female rats	LD$_{50}$[b]	3.30	Humburg et al., 1989
2′,4′-Dichloroacetophenone [2234-16-4]	*Pimephales promelas*	LC$_{50}$ (96-h)	1.04	Veith et al., 1983
2,4-Dichloroaniline [554-00-7]	*Poecilia reticulata*	LC$_{50}$ (14-d)	1.07	Könemann, 1981
3,4-Dichloroaniline [95-76-1]	*Brachionus calyciflorus*	LC$_{50}$ (24-h)	1.79	Ferrando et al., 1992
	Chironomus riparius	LC$_{50}$ (48-h)	1.17	Taylor et al., 1991
		LC$_{50}$ (96-h)	0.87	Taylor et al., 1991
		LC$_{50}$ (5-d)	0.74	Taylor et al., 1991
		LC$_{50}$ (10-d)	0.62	Taylor et al., 1991
	Daphnia magna	LC$_{50}$ (24-h)	-0.70	Ferrando et al., 1992
		LC$_{50}$ (96-h)	0.00	Adema and Vink, 1981
	Dreissena polymorpha	LC$_{50}$ (48-h)	1.34	Adema and Vink, 1981
	Gammarus pulex	LC$_{50}$ (24-h)	1.55	Taylor et al., 1991
		LC$_{50}$ (48-h)	1.24	Taylor et al., 1991
		LC$_{50}$ (10-d)	0.70	Taylor et al., 1991
	Hydrozetes lacustris	LC$_{50}$ (96-h)	0.67	Schmitz and Nagel, 1995
	Poecilia reticulata	LC$_{50}$ (96-h)	0.94	Adema and Vink, 1981
	Pristina longiseta	LC$_{50}$ (96-h)	0.40	Schmitz and Nagel, 1995
2,4-Dichlorobenzyl chloride [94-99-5]	*Poecilia reticulata*	LC$_{50}$ (14-d)	-0.62	Könemann, 1981
trans-1,4-Dichloro-2-butene [110-57-6]	*Poecilia reticulata*	LC$_{50}$ (7-d)	<1.60	Könemann, 1981
1,1-Dichloro-1-fluoroethane [1717-00-6]	rats	LC$_{50}$ (4-h)	4.79	Brock et al., 1995
		LC$_{50}$*	>3.70	Brock et al., 1995
1,5-Dichloropentane [628-76-2]	*Poecilia reticulata*	LC$_{50}$ (7-d)	<1.05	Könemann, 1981
3,5-Dichlorophenol [591-35-5]	*Platynereis dumerilii*	EC$_{50}$ (1-h)	0.28	Hutchinson et al., 1995
		EC$_{50}$ (48-h)	0.33	Hutchinson et al., 1995
		LC$_{50}$ (48-h)	0.55	Hutchinson et al., 1995
1,1-Dichloropropane [78-99-9]	*Daphnia magna*	LC$_{50}$ (24-h)	1.36	LeBlanc, 1980
		LC$_{50}$ (48-h)	1.48	LeBlanc, 1980
1,3-Dichloropropane [142-28-9]	*Cyprinodon variegatus*	LC$_{50}$ (24-h)	1.94	Heitmuller et al., 1981
		LC$_{50}$ (48-h)	1.94	Heitmuller et al., 1981
		LC$_{50}$ (72-h)	1.94	Heitmuller et al., 1981
		LC$_{50}$ (96-h)[d]	1.94	Heitmuller et al., 1981
	Poecilia reticulata	LC$_{50}$ (7-d)	1.92	Könemann, 1981
1,3-Dichloropropene [542-75-6]	*Cyprinodon variegatus*	LC$_{50}$ (24-h)	0.83	Heitmuller et al., 1981
		LC$_{50}$ (48-h)	0.52	Heitmuller et al., 1981
		LC$_{50}$ (72-h)	0.34	Heitmuller et al., 1981
		LC$_{50}$ (96-h)[d]	0.26	Heitmuller et al., 1981
	Daphnia magna	LC$_{50}$ (24-h)	0.86	LeBlanc, 1980
		LC$_{50}$ (48-h)	0.79	LeBlanc, 1980
2,3-Dichloropropene [78-88-6]	*Poecilia reticulata*	LC$_{50}$ (7-d)	<1.05	Könemann, 1981
Dichlorprop [120-36-5]	mice	LD$_{50}$[b]	2.60	Humburg et al., 1989
	rats	LD$_{50}$[b]	2.90	Humburg et al., 1989
Dichlorosilane [4109-96-0]	male mice	LC$_{50}$ (4-h)[a]	2.16	Nakashima et al., 1996
2,4-Dichlorotoluene [95-73-8]	*Poecilia reticulata*	LC$_{50}$ (14-d)	0.67	Könemann, 1981
3,4-Dichlorotoluene [95-75-0]	*Pimephales promelas*	LC$_{50}$ (96-h)	0.46	Veith et al., 1983
α,α′-Dichloro-*m*-xylene [626-16-4]	*Poecilia reticulata*	LC$_{50}$ (14-d)	-0.92	Könemann, 1981
Diclofop-methyl [51338-27-3]	bobwhite quail	LD$_{50}$[b]	3.64	Humburg et al., 1989
	dogs	LD$_{50}$[b]	>3.20	Humburg et al., 1989
	rats	LD$_{50}$[b]	2.75	Humburg et al., 1989
Dicrotophos [141-66-2]	*Simocephalus serrulatus*	EC$_{50}$ (48-h)	-0.57	Mayer and Ellersieck, 1986
Dicyclopentadiene [77-73-6]	*Daphnia pulex*	EC$_{50}$ (48-h)	0.62	Passino-Reader et al., 1997
Diethanolamine [111-42-2]	*Cyprinodon variegatus*	LC$_{50}$ (24-h)	>2.73	Heitmuller et al., 1981

Compound [CAS No.]	Species	Results log, mg/L		Reference
		LC_{50} (48-h)	>2.73	Heitmuller et al., 1981
		LC_{50} (72-h)	>2.73	Heitmuller et al., 1981
		LC_{50} (96-h)[d]	>2.73	Heitmuller et al., 1981
	Daphnia magna	LC_{50} (24-h)	2.23	LeBlanc, 1980
		LC_{50} (48-h)	1.74	LeBlanc, 1980
1,2,7,8-Diepoxyoctane [2426-07-5]	*Poecilia reticulata*	LC_{50} (14-d)	0.83	Deneer et al., 1988
Diethatyl ethyl [58727-55-8]	bobwhite quail	LD_{50}[b]	>4.00	Humburg et al., 1989
	female albino mice	LD_{50}[b]	3.61	Humburg et al., 1989
	female albino rats	LD_{50}[b]	3.57	Humburg et al., 1989
	male albino mice	LD_{50}[b]	3.22	Humburg et al., 1989
	male albino rats	LD_{50}[b]	3.37	Humburg et al., 1989
	mallard duck	LD_{50}[b]	>4.00	Humburg et al., 1989
Di(ethylene glycol) [111-46-6]	*Poecilia reticulata*	LC_{50} (7-d)	4.79	Könemann, 1981
Diethylene glycol butyl ether [112-34-5]	*Poecilia reticulata*	LC_{50} (7-d)	3.14	Könemann, 1981
Difenzoquat methyl sulfate [43222-48-6]	female mice	LD_{50}[b]	1.64	Humburg et al., 1989
	female rats	LD_{50}[b]	3.03	Humburg et al., 1989
	male mice	LD_{50}[b]	1.49	Humburg et al., 1989
	male rats	LD_{50}[b]	1.73	Humburg et al., 1989
Diflubenzuron [35367-38-5]	*Daphnia magna*	EC_{50} (48-h)	1.18	Mayer and Ellersieck, 1986
	Chironomus plumosus	EC_{50} (48-h)	2.75	Mayer and Ellersieck, 1986
	Hyalella azteca	LC_{50} (96-h)	-2.74	Fischer and Hall, 1992
	Salmo gairdneri	LC_{50} (96-h)	>2.18	McKague and Pridmore, 1978
	Skwala sp.	LC_{50} (96-h)	>2.00	Fischer and Hall, 1992
	Oncorhynchus kisutch	LC_{50} (96-h)	>2.18	McKague and Pridmore, 1978
	Lepomis machrochirus	LC_{50} (96-h)	2.13	Madder and Lockhart, 1978
	Fundulus heteroclitus	LC_{50} (96-h)	2.41	Madder and Lockhart, 1978
	Clistorinia magnifica	LC_{50} (30-d)	-1.00	Fischer and Hall, 1992
1,1-Difluoroethane [75-37-6]	rats	LC_{50} (4-h)[a]	>5.69	Keller et al., 1996
Digoxin [20830-75-5]	*Daphnia magna*	EC_{50} (24-h)	1.38	Lilius et al., 1995
9,10-Dihydroanthracene [613-31-0]	*Daphnia pulex*	EC_{50} (48-h)	-0.25	Passino-Reader et al., 1997
1,2-Dihydronaphthalene [447-53-0]	*Daphnia pulex*	EC_{50} (48-h)	0.63	Passino-Reader et al., 1997
9,10-Dihydrophenanthrene [776-35-2]	*Daphnia pulex*	EC_{50} (48-h)	-0.38	Passino-Reader et al., 1997
Diisopropyl ether [108-20-3]	*Pimephales promelas*	LC_{50} (96-h)	1.96	Veith et al., 1983
Dimethoate [60-51-5]	*Artemia* sp.	EC_{50} (24-h)	2.48	Guzzella et al., 1997
	Brachionus plicatilis	EC_{50} (24-h)	2.39	Guzzella et al., 1997
	Gammarus italicus	LC_{10} (24-h)	0.58	Pantani et al., 1997
	Echinogammarus tibaldii	LC_{50} (24-h)	0.61	Pantani et al., 1997
p-Dimethoxybenzene [150-78-7]	*Pimephales promelas*	LC_{50} (96-h)	2.07	Veith et al., 1983
2,6-Dimethoxytoluene [5673-07-4]	*Pimephales promelas*	LC_{50} (96-h)	1.31	Veith et al., 1983
3,5-Dimethylaniline [108-69-0]	*Oryzias latipes*	LC_{50} (24-h)	1.54	Tonogai et al., 1982
		LC_{50} (48-h)	1.23	Tonogai et al., 1982
	T. pyriformis	LC_{100} (48-h)	2.44	Schultz et al., 1978
Dimethylarsinic acid [124-65-2]	*Lepomis machrochirus*	LC_{50} (24-h)	1.32	Mayer and Ellersieck, 1986
		LC_{50} (96-h)	1.23	Mayer and Ellersieck, 1986
3,3-Dimethyl-2-butanone [75-97-8]	*Pimephales promelas*	LC_{50} (96-h)	1.94	Veith et al., 1983
N,N-Dimethylethanolamine [108-01-0]	Wistar rats	LC_{50} (4-h)[a]	3.22	Klonne et al., 1987
	rats	LC_{50} (4-h)[a]	3.16	Ballantyne and Leung, 1996
2,6-Dimethylquinoline [877-43-0]	*Xenopus laevis*	LC_{50} (96-h)	0.81	Dumont et al., 1979
Dimethyl sulfoxide [67-68-5]	*Palaemonetes pugio* embryos	LC_{50} (4-d)	4.35	Rayburn and Fisher, 1997
		LC_{50} (12-d)	4.09	Rayburn and Fisher, 1997
2,3-Dinitrotoluene [602-01-7]	*Cyprinodon variegatus*	LC_{50} (24-h)	>0.88	Heitmuller et al., 1981
		LC_{50} (48-h)	0.70	Heitmuller et al., 1981
		LC_{50} (72-h)	0.46	Heitmuller et al., 1981
		LC_{50} (96-h)[d]	0.36	Heitmuller et al., 1981

Compound [CAS No.]	Species	Results log, mg/L		Reference
	Daphnia magna	LC$_{50}$ (24-h)	0.45	LeBlanc, 1980
		LC$_{50}$ (48-h)	-0.18	LeBlanc, 1980
Dinoseb [88-85-7]	rats	LD$_{50}$[b]	1.76	Humburg et al., 1989
Dioxathion [78-34-2]	*Daphnia magna*	EC$_{50}$ (48-h)	-3.46	Mayer and Ellersieck, 1986
Diphenamide [957-51-7]	*Daphnia magna*	EC$_{50}$ (48-h)	1.76	Mayer and Ellersieck, 1986
	Cypridopsis vidua	EC$_{50}$ (48-h)	1.71	Mayer and Ellersieck, 1986
Dipentyl ether [693-65-2]	*Pimephales promelas*	LC$_{50}$ (96-h)	0.51	Veith et al., 1983
Dipropetryn [4147-51-7]	rats	LD$_{50}$[b]	3.70	Humburg et al., 1989
Diquat dibromide [85-00-7]	*Roccus saxatilis*	LC$_{50}$ (96-h)	-0.60	Wellborn, 1969
Disulfoton [298-04-4]	male mice	LD$_{50}$ (24-h)[b]	0.76	Pawar and Fawade, 1978
	female mice	LD$_{50}$ (24-h)[b]	0.43	Pawar and Fawade, 1978
	male rat	LD$_{50}$ (24-h)[b]	0.86	Pawar and Fawade, 1978
	female rat	LD$_{50}$ (24-h)[b]	0.51	Pawar and Fawade, 1978
Dithiopyr [97886-45-8]	*Daphnia magna*	LC$_{50}$ (48-h)	1.23	Humburg et al., 1989
	common carp	LC$_{50}$ (96-h)	-0.14	Humburg et al., 1989
	Lepomis machrochirus	LC$_{50}$ (96-h)	-0.15	Humburg et al., 1989
	rainbow trout	LC$_{50}$ (96-h)	-0.32	Humburg et al., 1989
	bobwhite quail	LD$_{50}$[b]	3.35	Humburg et al., 1989
Diuron [330-54-1]	rats	LD$_{50}$[b]	3.53	Humburg et al., 1989
n-Docosane [629-97-0]	*Cyprinodon variegatus*	LC$_{50}$ (24-h)	>2.70	Heitmuller et al., 1981
		LC$_{50}$ (48-h)	>2.70	Heitmuller et al., 1981
		LC$_{50}$ (72-h)	>2.70	Heitmuller et al., 1981
		LC$_{50}$ (96-h)[d]	>2.70	Heitmuller et al., 1981
	Daphnia magna	LC$_{50}$ (24-h)	>2.72	LeBlanc, 1980
1-Dodecanol [112-53-8]	*Pimephales promelas*	LC$_{50}$ (96-h)	0.00	Veith et al., 1983
Dodecylbenzene sulfonic acid, sodium salt [25155-30-0]	*Gammarus italicus*	LC$_{10}$ (24-h)	1.31	Pantani et al., 1997
Doxefazepam [40762-15-0]	mice, rats, and dogs	LD$_{50}$[b]	>3.30	Bertoli et al., 1989
	Echinogammarus	LC$_{50}$ (24-h)	1.22	Pantani et al., 1997
Ebrotidine [100981-43-9]	mice	LD$_{50}$[b]	2.56	Grau et al., 1997
	rats	LD$_{50}$[b]	2.50	Grau et al., 1997
Endosulfan [115-29-7]	*Brachionus calyciflorus*	LC$_{50}$ (24-h)	0.71	Fernandez-Casalderry et al., 1992
	Daphnia magna	LC$_{50}$ (24-h)	-0.21	Fernandez-Casalderry et al., 1994
	Atalophlebia australis nymphs	LC$_{50}$ (48-h)	-3.22	Leonard et al., 1999
	Cheumatopsyche sp. larvae	LC$_{50}$ (48-h)	-3.40	Leonard et al., 1999
	Jappa kutera nymphs	LC$_{50}$ (48-h)	-2.89	Leonard et al., 1999
	Puntius conchonius	LC$_{50}$ (48-h)	1.33	Gill et al., 1991
	Channa punctatus	LC$_{50}$ (96-h)	0.49	Haider and Inbaraj, 1986
	Hydra viridissima	LC$_{50}$ (96-h)	-0.17	Pollino and Holdway, 1999
	Hydra vulgaris	LC$_{50}$ (96-h)	-0.09	Pollino and Holdway, 1999
Epibromohydrin [3132-64-7]	*Poecilia reticulata*	LC$_{50}$ (14-d)	-0.01	Deneer et al., 1988
1,2-Epoxybutane [106-88-7]	*Poecilia reticulata*	LC$_{50}$ (14-d)	1.52	Deneer et al., 1988
1,2-Epoxydecane [2404-44-6]	*Poecilia reticulata*	LC$_{50}$ (14-d)	0.52	Deneer et al., 1988
1,2-Epoxydodecane [2855-19-8]	*Poecilia reticulata*	LC$_{50}$ (14-d)	0.04	Deneer et al., 1988
1,2-Epoxyhexane [1436-34-6]	*Poecilia reticulata*	LC$_{50}$ (14-d)	1.27	Deneer et al., 1988
1,2-Epoxyoctane [2984-50-1]	*Poecilia reticulata*	LC$_{50}$ (14-d)	1.02	Deneer et al., 1988
EPTC [759-94-4]	blue crab	LC$_{50}$ (24-h)	>1.30	Humburg et al., 1989
	Lepomis machrochirus	LC$_{50}$ (96-h)	1.43	Humburg et al., 1989
	Salmo gairdneri	LC$_{50}$ (96-h)	1.28	Humburg et al., 1989
	male albino mice	LD$_{50}$[b]	3.50	Humburg et al., 1989
	male albino rats	LD$_{50}$[b]	3.22	Humburg et al., 1989
Ethanol [64-17-5]	*Daphnia magna*	EC$_{50}$ (24-h)	4.14	Lilius et al., 1995
	Daphnia pulex	EC$_{50}$ (24-h)	4.06	Lilius et al., 1995
	Spirostomum ambiguum	EC$_{50}$ (24-h)	4.08	Nałecz-Jawecki and Sawicki, 1999
		EC$_{50}$ (48-h)	4.08	Nałecz-Jawecki and Sawicki, 1999
		LC$_{50}$ (24-h)	4.43	Nałecz-Jawecki and Sawicki, 1999
		LC$_{50}$ (48-h)	4.31	Nałecz-Jawecki and Sawicki, 1999
	Pimephales promelas	LC$_{50}$ (96-h)	4.17	Veith et al., 1983
	Palaemonetes pugio embryos	LC$_{50}$ (4-d)	4.08	Rayburn and Fisher, 1997

Compound [CAS No.]	Species	Results log, mg/L		Reference
	Poecilia reticulata	LC_{50} (7-d)	4.04	Könemann, 1981
	Palaemonetes pugio embryos	LC_{50} (12-d)	3.56	Rayburn and Fisher, 1997
Ethepon [16672-87-0]	albino mice	LD_{50}[b]	3.63	Humburg et al., 1989
	bobwhite quail	LD_{50}[b]	3.00	Humburg et al., 1989
Ethion [563-12-2]	*Daphnia magna*	EC_{50} (48-h)	-4.25	Mayer and Ellersieck, 1986
	Daphnia pulex	EC_{50} (48-h)	-2.55	Mayer and Ellersieck, 1986
	Simocephalus serrulatus	EC_{50} (48-h)	-2.33	Mayer and Ellersieck, 1986
Ethofumesate [26225-79-6]	*Lepomis machrochirus*	LC_{50} (96-h)	<2.51	Humburg et al., 1989
	Salmo gairdneri	LC_{50} (96-h)	<2.26	Humburg et al., 1989
	bobwhite quail	LD_{50}[b]	<3.94	Humburg et al., 1989
	mallard duck	LD_{50}[b]	3.55	Humburg et al., 1989
2-(2-Ethoxyethoxy)ethanol [111-90-0]	*Pimephales promelas*	LC_{50} (96-h)	4.42	Veith et al., 1983
Ethylene glycol [107-21-1]	*Daphnia magna*	EC_{50} (24-h)	4.69	Lilius et al., 1995
	Daphnia pulex	EC_{50} (24-h)	4.78	Lilius et al., 1995
	Poecilia reticulata	LC_{50} (7-d)	4.69	Könemann, 1981
2-Ethylhexanol [104-76-7]	*Chironomus plumosus*	EC_{50} (48-h)	1.53	Mayer and Ellersieck, 1986
Ethyl nitrite [109-95-5]	Sprague-Dawley rats	LC_{50} (4-h)[a]	2.20	Klonne et al., 1987a
Fenamiphos [22224-92-6]	*Gammarus italicus*	LC_{10} (24-h)	-1.70	Pantani et al., 1997
	Echinogammarus tibaldii	LC_{50} (24-h)	-1.96	Pantani et al., 1997
Fenitrothion [122-14-5]	*Daphnia magna*	EC_{50} (48-h)	-1.96	Mayer and Ellersieck, 1986
	Chironomus plumosus	EC_{50} (48-h)	-2.15	Mayer and Ellersieck, 1986
	Daphnia magna	LC_{50} (48-h)	-4.17	Ferrando et al., 1996
	Oryzias latipes	LC_{50} (48-h)	0.54	Tsuda et al., 1997
	Callinectes sapidus	LC_{50} (96-h)	0.93	Johnston and Corbett, 1995
Fenoxaprop-ethyl [66441-23-4]	*Daphnia magna*	LC_{50}	1.06	Humburg et al., 1989
	Lepomis machrochirus	LC_{50}	0.52	Humburg et al., 1989
	Salmo gairdneri	LC_{50}	0.53	Humburg et al., 1989
	female rats	LD_{50}[b]	3.53	Humburg et al., 1989
	male rats	LD_{50}[b]	3.52	Humburg et al., 1989
Fenoxycarb [79127-80-3]	*Gambusia affinis*	LC_{50} (24-h)	0.02	Tietze et al., 1991
Fenthion [55-38-9]	*Daphnia pulex*	EC_{50} (48-h)	-3.10	Mayer and Ellersieck, 1986
	Cypridopsis vidua	EC_{50} (48-h)	-1.74	Mayer and Ellersieck, 1986
	Simocephalus serrulatus	EC_{50} (48-h)	-3.21	Mayer and Ellersieck, 1986
	Gambusia affinis	LC_{50} (24-h)	0.47	Tietze et al., 1991
	Ceriodaphnia dubia	LC_{50} (48-h)	-2.76	Roux et al., 1995
	Daphnia pulex	LC_{50} (48-h)	-2.89	Roux et al., 1995
	Poecilia reticulata	LC_{50} (96-h)	-2.67	Roux et al., 1995
	Cyprinus carpio	LC_{50} (96-h)	-2.60	Roux et al., 1995
	Tilapia rendalli	LC_{50} (96-h)	-2.53	Roux et al., 1995
	Oreochromis mossambicus	LC_{50} (96-h)	-2.77	Roux et al., 1995
Fenticonazole [72479-26-6]	mice and rats	LD_{50}[b]	3.48	Graziani et al., 1981
Fenuron-TCA [4482-52-7]	albino rats	LD_{50}[b]	3.60	Humburg et al., 1989
Fenvalerate [51630-58-1]	*Daphnia magna*	EC_{50} (48-h)	-2.68	Mayer and Ellersieck, 1986
	Chironomus plumosus	EC_{50} (48-h)	-3.37	Mayer and Ellersieck, 1986
	Atherinops affinis	LC_{50} (96-h)	-3.18	Goodman et al., 1992
	Pimephales promelas	LC_{50} (96-h)	-3.07	Jarvinen et al., 1989
	Cyprinodon variegatus	LC_{50} (96-h)	-2.30	Schimmel et al., 1983
	Menidia menidia	LC_{50} (96-h)	-3.51	Schimmel et al., 1983
	Mugil cephalus	LC_{50} (96-h)	-3.24	Schimmel et al., 1983
	Mysidopsis bahia	LC_{50} (96-h)	-5.10	Schimmel et al., 1983
	Opsanus beta	LC_{50} (96-h)	-2.27	Schimmel et al., 1983
	Penaeus duorarum	LC_{50} (96-h)	-3.08	Schimmel et al., 1983
	steelhead trout	LC_{50} (96-h)	-4.06	Curtis et al., 1985
	Colinus virginianus adults	LD_{50}[b]	>3.60	Bradbury and Coats, 1982
Ferrous sulfate [7720-78-7]	*Daphnia magna*	EC_{50} (24-h)	1.16	Lilius et al., 1995
	Daphnia pulex	EC_{50} (24-h)	2.00	Lilius et al., 1995
Fluazifop-P-butyl [79241-46-6]	*Salmo gairdneri*	LC_{50} (96-h)	0.73	Humburg et al., 1989
	mallard duck	LD_{50}[b]	3.55	Humburg et al., 1989
Fluchloralin [33245-39-5]	*Lepomis machrochirus*	LC_{50} (96-h)	-1.80	Humburg et al., 1989
	Salmo gairdneri	LC_{50} (96-h)	-1.92	Humburg et al., 1989
	bobwhite quail	LD_{50}[b]	3.85	Humburg et al., 1989
	dogs	LD_{50}[b]	>3.81	Humburg et al., 1989
	mallard duck	LD_{50}[b]	4.11	Humburg et al., 1989

Compound [CAS No.]	Species	Results log, mg/L		Reference
	mice	$LD_{50}{}^b$	2.86	Humburg et al., 1989
	rabbits	$LD_{50}{}^b$	3.90	Humburg et al., 1989
	rats	$LD_{50}{}^b$	3.19	Humburg et al., 1989
Flucythrinate [70124-77-5]	*Cyprinodon variegatus*	LC_{50} (96-h)	-2.96	Schimmel et al., 1983
	Mysidopsis bahia	LC_{50} (96-h)	-5.10	Schimmel et al., 1983
	Penaeus duorarum	LC_{50} (96-h)	-3.66	Schimmel et al., 1983
Flumethrin [62924-70-3]	*Damalinia caprae*	LC_{50}	2.08	Gard et al., 1998
Fluometuron [2164-17-2]	*Daphnia magna*	EC_{50} (24-h)	>1.00	Mayer and Ellersieck, 1986
	catfish	LC_{50} (96-h)	1.74	Humburg et al., 1989
	Lepomis machrochirus	LC_{50} (96-h)	1.98	Humburg et al., 1989
	Salmo gairdneri	LC_{50} (96-h)	1.67	Humburg et al., 1989
	rats	$LD_{50}{}^b$	3.26	Humburg et al., 1989
Fluridone [59756-60-4]	*Daphnia magna*	EC_{50} (24-h)	0.64	Mayer and Ellersieck, 1986
	dogs	$LD_{50}{}^b$	>2.70	Humburg et al., 1989
	mice	$LD_{50}{}^b$	>4.00	Humburg et al., 1989
	rats	$LD_{50}{}^b$	>4.00	Humburg et al., 1989
1-Fluoro-4-nitrobenzene [350-46-9]	*Pimephales promelas*	LC_{50} (96-h)	1.45	Veith et al., 1983
Fomesafen [72178-02-0]	female rats	$LD_{50}{}^b$	3.82	Humburg et al., 1989
	male rats	$LD_{50}{}^b$	3.91	Humburg et al., 1989
Fonofos [944-22-9]	*Photobacterium phosphoreum*	EC_{50} (5-min)	0.72	Somasundaram et al., 1990
	Artemia sp.	EC_{50} (24-h)	0.96	Guzzella et al., 1997
	Brachionus plicatilis	EC_{50} (24-h)	0.94	Guzzella et al., 1997
Fosamine ammonium [25954-13-6]	*Daphnia magna*	EC_{50} (48-h)	>2.00	Mayer and Ellersieck, 1986
	guinea pigs	$LD_{50}{}^b$	3.87	Humburg et al., 1989
	rats	$LD_{50}{}^b$	4.39	Humburg et al., 1989
Furan [110-00-9]	*Pimephales promelas*	LC_{50} (96-h)	1.79	Veith et al., 1983
3-Furanemethanol [4412-91-3]	*Pimephales promelas*	LC_{50} (96-h)	2.71	Veith et al., 1983
Germanium [7440-56-4]	Wistar rats	LC_{50} (4-h)^c	>3.73	Arts et al., 1990
Germanium dioxide [1310-53-8]	Wistar rats	LC_{50} (4-h)^c	3.15	Arts et al., 1994
Glyphosate [1071-83-6]	*Daphnia magna*	EC_{50} (24-h)	0.72	Mayer and Ellersieck, 1986
		EC_{50} (48-h)	0.47	Mayer and Ellersieck, 1986
	Daphnia magna	LC_{50} (48-h)	2.89	Humburg et al., 1989
	Atlantic oyster	LC_{50} (96-h)	>1.00	Humburg et al., 1989
	Fiddler crab	LC_{50} (96-h)	2.97	Humburg et al., 1989
	Harlequin fish	LC_{50} (96-h)	2.23	Humburg et al., 1989
	Lepomis machrochirus	LC_{50} (96-h)	2.08	Humburg et al., 1989
	Salmo gairdneri	LC_{50} (96-h)	1.93	Humburg et al., 1989
	Shrimp	LC_{50} (96-h)	2.45	Humburg et al., 1989
Haloxyfop-methyl [69806-40-2]	*Daphnia* sp.	LC_{50} (48-h)	0.79	Humburg et al., 1989
	Lepomis machrochirus	LC_{50} (96-h)	-0.70	Humburg et al., 1989
	Salmo gairdneri	LC_{50} (96-h)	-0.40	Humburg et al., 1989
	mallard duck	$LD_{50}{}^b$	3.33	Humburg et al., 1989
1,2,3,4,7,8-HCDD [70648-26-9]	*Oryzias latipes*	LC_{50} (3-d)	-2.54	Wisk and Cooper, 1990
HCFC-123 [306-83-2]	rats	LC_{50} (4-h)^a	4.51	Trochimowicz, 1989
HCFC-141b [1717-00-6]	rats	LC_{50} (4-h)^a	>4.79	Seckar, 1989
HCFC-225ca [422-56-0]	male and female rats	LC_{50} (4-h)^a	4.57	Brock et al., 1999
HCFC-225cb[507-55-1]	male and female rats	LC_{50} (4-h)^a	4.57	Brock et al., 1999
Heptachloronorbornene [2440-02-0]	*Pimephales promelas*	LC_{50} (30-d)	-1.22	Spehar et al., 1979
		LC_{50} (96-h)	-1.07	Spehar et al., 1979
1-Heptanol [111-70-6]	*Xenopus* sp.	LC_{50} (5-d)	2.24	Bernardini et al., 1994
Hexachloronorbornadiene [28680-44-6]	*Pimephales promelas*	LC_{50} (30-d)	-0.91	Spehar et al., 1979
		LC_{50} (96-h)	-0.73	Spehar et al., 1979
Hexachlorophene [70-30-4]	*Daphnia magna*	EC_{50} (24-h)	-0.70	Lilius et al., 1995
Hexafluoroisobutylene	rats	LC_{50} (4-h)^c	3.15	Gad et al., 1986
Hexanoic acid [142-62-1]	*Hyale plumulosa*	LC_{50} (48-h)^d	2.35	Onitsuka et al., 1989
	Oryzias latipes	LC_{50} (48-h)^d	2.35	Onitsuka et al., 1989
1-Hexanol [111-27-3]	*Pimephales promelas*	LC_{50} (96-h)	1.99	Veith et al., 1983
Hexavalent chromium [18520-29-9]	*Colisa fasciatus*	LC_{50} (96-h)	1.78	Nath and Kumar, 1988

Compound [CAS No.]	Species	Results log, mg/L		Reference
HFC-134a [811-97-2]	rats	LC$_{50}$ (4-h)[a]	>5.70	Millischer, 1989
Hydramethylnon [67485-29-4]	*Chironomus plumosus*	EC$_{50}$ (48-h)	-0.89	Mayer and Ellersieck, 1986
	Daphnia magna	EC$_{50}$ (48-h)	-0.85	Mayer and Ellersieck, 1986
	Lepomis machrochirus	LC$_{50}$ (96-h)	-0.92	Mayer and Ellersieck, 1986
Hydrocyanic acid [74-90-8]	*Asellus communis*	LC$_{50}$ (96-h)	0.36	Oseid and Smith, 1979
	Gammarus pseudolimnaeus	LC$_{50}$ (96-h)	-0.77	Oseid and Smith, 1979
Hydrogen sulfide [7783-06-4]	rats	LC$_{10}$ (2-h)[a]	2.74	Prior et al., 1988
		LC$_{10}$ (4-h)[a]	2.63	Prior et al., 1988
		LC$_{10}$ (6-h)[a]	2.77	Prior et al., 1988
		LC$_{50}$ (2-h)[a]	2.77	Prior et al., 1988
		LC$_{50}$ (4-h)[a]	2.70	Prior et al., 1988
		LC$_{50}$ (6-h)[a]	2.85	Prior et al., 1988
2'-Hydroxy-4'-methoxy-acetophenone [552-41-0]	*Pimephales promelas*	LC$_{50}$ (96-h)	1.74	Veith et al., 1983
Imazethapyr [81335-77-5]	bobwhite quail	LD$_{50}$[b]	>3.33	Humburg et al., 1989
	channel catfish	LC$_{50}$ (96-h)	2.38	Humburg et al., 1989
	Lepomis machrochirus	LC$_{50}$ (96-h)	2.62	Humburg et al., 1989
	Salmo gairdneri	LC$_{50}$ (96-h)	2.53	Humburg et al., 1989
	female rabbits	LD$_{50}$[b]	>3.70	Humburg et al., 1989
	male mice	LD$_{50}$[b]	>3.70	Humburg et al., 1989
	male mice	LD$_{50}$[b]	>3.70	Humburg et al., 1989
	mallard duck	LD$_{50}$[b]	>3.33	Humburg et al., 1989
Ioxynil [1689-83-4]	rats	LD$_{50}$[b]	2.04	Humburg et al., 1989
Ioxynil octanoate [3861-47-0]	mallard duck	LD$_{50}$[b]	3.08	Humburg et al., 1989
	pheasant	LD$_{50}$[b]	3.00	Humburg et al., 1989
	rats	LD$_{50}$[b]	2.11	Humburg et al., 1989
Ioxynil-sodium [2961-62-8]	harlequin fish	LC$_{50}$ (48-h)	0.52	Humburg et al., 1989
	rats	LD$_{50}$[b]	2.08	Humburg et al., 1989
Irloxacin [91524-15-1]	mice and rats	LD$_{50}$[b]	3.30	Guzman et al., 1999
Isoamyl nitrite [110-46-3]	Sprague-Dawley rats	LC$_{50}$ (4-h)[a]	2.85	Klonne et al., 1987a
Isobutyl nitrite [542-56-3]	Sprague-Dawley rats	LC$_{50}$ (4-h)[a]	2.89	Klonne et al., 1987a
Isofenphos [25311-71-1]	*Photobacterium phosphoreum*	EC$_{50}$ (5-min)	1.99	Somasundaram et al., 1990
Isopropalin [33820-53-0]	bobwhite quail	LD$_{50}$[b]	>3.30	Humburg et al., 1989
	rats	LD$_{50}$[b]	>3.70	Humburg et al., 1989
2-Isopropoxyethanol [109-59-1]	*Poecilia reticulata*	LC$_{50}$ (7-d)	3.74	Könemann, 1981
Isoxaben [82558-50-7]	bobwhite quail	LD$_{50}$[b]	>3.70	Humburg et al., 1989
Ivermectin [70288-86-7]	*Culicoides variipennis*	LC$_{50}$ (48-h)	-0.46	Holbrook and Mullens, 1994
Lactofen [77501-63-4]	rats	LD$_{50}$[b]	>3.70	Humburg et al., 1989
Lead acetate [301-04-2]	*Drosophila melanogaster*	LC$_{50}$	3.47	Atkins et al., 1991
Lead nitrate [10099-74-8]	*Spirostomum teres*	LC$_{50}$ (0.5-h)	-1.05	Twagilimana et al., 1998
		LC$_{50}$ (24-h)	1.03	Twagilimana et al., 1998
	Rana ridibunda	LC$_{50}$ (96-h)	2.14	Vogiatzis and Loumbourdis, 1999
Leptophos [21609-90-5]	*Leiostomus xanthurus*	LC$_{50}$ (96-h)	-2.39	Schimmel et al., 1979
	Mysidopsis bahia	LC$_{50}$ (96-h)	-2.50	Schimmel et al., 1979
	Penaeus duorarum	LC$_{50}$ (96-h)	-2.73	Schimmel et al., 1979
Levofloxacin [100986-85-4]	female mice	LD$_{50}$[b]	3.26	Kato et al., 1992
	male mice	LD$_{50}$[b]	3.27	Kato et al., 1992
	female rats	LD$_{50}$[b]	3.18	Kato et al., 1992
	male rats	LD$_{50}$[b]	3.17	Kato et al., 1992
	female monkeys	LD$_{50}$[b]	2.40	Kato et al., 1992
Linuron [330-55-2]	*Daphnia magna*	EC$_{50}$ (48-h)	-0.57	Mayer and Ellersieck, 1986
	Lepomis machrochirus	LC$_{50}$ (96-h)	1.20	Humburg et al., 1989
	rainbow trout	LC$_{50}$ (96-h)	1.20	Humburg et al., 1989
	female rats	LD$_{50}$[b]	3.08	Humburg et al., 1989
	male rats	LD$_{50}$[b]	3.10	Humburg et al., 1989
Lithium sulfate [10377-48-7]	*Daphnia magna*	EC$_{50}$ (24-h)	2.29	Lilius et al., 1995
	Daphnia pulex	EC$_{50}$ (24-h)	2.16	Lilius et al., 1995
Longicyclene [1137-12-8]	*Daphnia pulex*	EC$_{50}$ (48-h)	-1.19	Passino-Reader et al., 1997
Longifolene [475-20-7]	*Daphnia pulex*	EC$_{50}$ (48-h)	-1.10	Passino-Reader et al., 1997
Malaoxon [1634-78-2]	*Oryzias latipes*	LC$_{50}$ (48-h)	-0.55	Tsuda et al., 1997
Maleic hydrazide, sodium salt [28330-26-9]	rats	LD$_{50}$[b]	3.84	Humburg et al., 1989

Compound [CAS No.]	Species	Results log, mg/L		Reference
Mefluidide [53780-34-0]	*Lepomis machrochirus*	LC_{50} (4-d)	3.20	Humburg et al., 1989
	trout	LC_{50} (4-d)	>3.08	Humburg et al., 1989
	mice	LD_{50}[b]	3.28	Humburg et al., 1989
	rats	LD_{50}[b]	>3.60	Humburg et al., 1989
Mercuric acetate [1600-27-7]	*Pimephales promelas*	LC_{50} (24-h)	-0.28	Curtis et al., 1978
		LC_{50} (48-h)	-0.38	Curtis et al., 1978
		LC_{50} (96-h)	-0.72	Curtis et al., 1978
Mercuric chloride [7487-94-7]	*Gammarus italicus*	LC_{10} (24-h)	-0.17	Pantani et al., 1997
	Echinogammarus tibaldii	LC_{50} (24-h)	-0.30	Pantani et al., 1997
	Spirostomum teres	LC_{50} (0.5-h)	-0.96	Twagilimana et al., 1998
		LC_{50} (24-h)	-2.40	Twagilimana et al., 1998
	Channa punctatus	LC_{50} (96-h)	0.26	Sastry and Agrawal, 1979
Mercuric thiocyanate [592-85-8]	*Pimephales promelas*	LC_{50} (24-h)	-0.41	Curtis et al., 1978
		LC_{50} (48-h)	-0.41	Curtis et al., 1978
		LC_{50} (96-h)	-0.82	Curtis et al., 1978
Mercury [7439-97-6]	*Aedes aegypti*	LC_{50} (24-h)	0.49	Rayms-Keller et al., 1998
	Homarus americanus	LC_{50} (96-h)	-1.70	Johnson and Gentile, 1979
	Mytilopsis sallei	LC_{50} (96-h)	-0.59	Devi, 1996
Methazole [20354-26-1]	albino rats	LD_{50}[b]	3.40	Humburg et al., 1989
Methomyl [16752-77-5]	*Daphnia magna*	EC_{50} (24-h)	-1.92	Mayer and Ellersieck, 1986
	Daphnia magna	EC_{50} (48-h)	-2.09	Mayer and Ellersieck, 1986
	Gammarus italicus	LC_{10} (24-h)	-1.33	Pantani et al., 1997
	Echinogammarus tibaldii	LC_{50} (24-h)	-0.60	Pantani et al., 1997
Methoprene [40596-69-8]	*Salmo gairdneri*	LC_{50} (96-h)	2.03	McKague and Pridmore, 1978
	Oncorhynchus kisutch	LC_{50} (96-h)	1.93	McKague and Pridmore, 1978
	Ictalurus punctatus	LC_{50} (96-h)	>2.00	Madder and Lockhart, 1978
	trout	LC_{50} (96-h)	0.52	Madder and Lockhart, 1978
2-Methoxyethanol [109-86-4]	*Poecilia reticulata*	LC_{50} (7-d)	4.24	Könemann, 1981
3-Methyl-2-butanone [563-80-4]	*Pimephales promelas*	LC_{50} (96-h)	2.94	Veith et al., 1983
Methyl *t*-butyl ether [1634-04-4]	*Pimephales promelas*	LC_{50} (96-h)	2.85	Veith et al., 1983
6-Methyl-5-hepten-2-one [110-93-0]	*Pimephales promelas*	LC_{50} (96-h)	1.93	Veith et al., 1983
5-Methyl-2-hexanone [110-12-3]	*Pimephales promelas*	LC_{50} (96-h)	2.20	Veith et al., 1983
Methyl nitrite [626-91-9]	Sprague-Dawley rats	LC_{50} (4-h)[a]	2.25	Klonne et al., 1987a
Methyl parathion [298-00-0]	*Daphnia magna*	EC_{50} (48-h)	-3.82	Mayer and Ellersieck, 1986
	Simocephalus serrulatus	EC_{50} (48-h)	-3.43	Mayer and Ellersieck, 1986
	Mysidopsis bahia	LC_{50} (96-h)	-3.11	Schimmel et al., 1983
	Penaeus duorarum	LC_{50} (96-h)	-2.92	Schimmel et al., 1983
2-Methyl-2,4-pentanediol [107-41-5]	*Pimephales promelas*	LC_{50} (96-h)	4.03	Veith et al., 1983
2-Methylquinoline [91-63-4]	*Xenopus laevis*	LC_{50} (96-h)	1.42	Dumont et al., 1979
Metolachlor [51218-45-2]	*Daphnia magna*	EC_{50} (48-h)	1.37	Mayer and Ellersieck, 1986
	Chironomus plumosus	EC_{50} (48-h)	0.58	Mayer and Ellersieck, 1986
	channel catfish	LC_{50} (96-h)	0.69	Humburg et al., 1989
	Lepomis machrochirus	LC_{50} (96-h)	1.00	Humburg et al., 1989
	rainbow trout	LC_{50} (96-h)	0.30	Humburg et al., 1989
Metribuzin [21087-64-9]	*Daphnia magna*	EC_{50} (48-h)	>2.00	Mayer and Ellersieck, 1986
	Lepomis machrochirus	LC_{50} (96-h)	>2.00	Humburg et al., 1989
	rainbow trout	LC_{50} (96-h)	>2.00	Humburg et al., 1991
	female rats	LD_{50}[b]	3.04	Humburg et al., 1989
	male rats	LD_{50}[b]	3.08	Humburg et al., 1989
Metsulfuron-methyl [74223-64-6]	male rats	LD_{50}[b]	3.70	Humburg et al., 1989
Mexacarbate [315-18-4]	*Daphnia pulex*	EC_{50} (48-h)	-2.00	Mayer and Ellersieck, 1986
	Paramecium multimicro-nucleatum	LC_{50} (7-h)	1.92	Edmiston et al., 1985
		LC_{50} (9-h)	1.76	Edmiston et al., 1985
		LC_{50} (13-h)	1.54	Edmiston et al., 1985
		LC_{50} (17-h)	1.40	Edmiston et al., 1985
		LC_{50} (24-h)	1.28	Edmiston et al., 1985
	Simocephalus serrulatus	EC_{50} (48-h)	-1.89	Mayer and Ellersieck, 1986
Mirex [2385-85-5]	*Daphnia magna*	EC_{50} (48-h)	>0.00	Mayer and Ellersieck, 1986
	Daphnia pulex	EC_{50} (48-h)	>-1.00	Mayer and Ellersieck, 1986
	Chironomus plumosus	EC_{50} (48-h)	>0.00	Mayer and Ellersieck, 1986

Compound [CAS No.]	Species	Results log, mg/L		Reference
	Simocephalus serrulatus	EC_{50} (48-h)	>-1.00	Mayer and Ellersieck, 1986
	Hydra sp.	LC_{50} (96-h)	0.61	Lue and Cruz, 1978
		LC_{50} (72-h)	1.37	Lue and Cruz, 1978
		LC_{50} (48-h)	2.83	Lue and Cruz, 1978
Molinate [2212-24-9]	*Cypridopsis vidua*	EC_{50} (48-h)	-0.74	Mayer and Ellersieck, 1986
	Gammarus italicus	LC_{10} (24-h)	0.34	Pantani et al., 1997
	Echinogammarus tibaldii	LC_{50} (24-h)	0.26	Pantani et al., 1997
	Lepomachro chirus	LC_{50} (96-h)	1.46	Humburg et al., 1989
	goldfish	LC_{50} (96-h)	1.48	Humburg et al., 1989
	rainbow trout	LC_{50} (96-h)	0.11	Humburg et al., 1989
	Oryzias latipes embryo	LC_{50} (5-d)	2.09	Wolfe et al., 1993
	male albino mice	LD_{50}[b]	2.90	Humburg et al., 1989
	male albino rats	LD_{50}[b]	2.86	Humburg et al., 1989
Monocrotophos [2157-98-4]	*Oreochromis niloticus*	LC_{50} (96-h)	0.69	Thangnipon et al., 1995
Naled [300-76-5]	mice	LD_{50}[a]	2.35	Berteau and Deen, 1978
1-Naphthol [90-15-3]	*Spirostomum ambiguum*	EC_{50} (24-h)	1.12	Nałecz-Jawecki and Sawicki, 1999
		EC_{50} (48-h)	1.13	Nałecz-Jawecki and Sawicki, 1999
		LC_{50} (24-h)	1.48	Nałecz-Jawecki and Sawicki, 1999
		LC_{50} (48-h)	1.41	Nałecz-Jawecki and Sawicki, 1999
Napropamide [15299-99-7]	goldfish	LC_{50} (96-h)	>1.00	Humburg et al., 1989
	male and female rats	LD_{50}[b]	>3.70	Humburg et al., 1989
Naptalam [132-66-1]	rats	LD_{50}[b]	3.91	Humburg et al., 1989
Naptalam-sodium [132-67-12	rats	LD_{50}[b]	3.25	Humburg et al., 1989
Nefiracetam [77191-36-7]	beagle dogs	LD_{50}[b]	2.70	Sugawara et al., 1994
	female mice	LD_{50}[b]	3.29	Sugawara et al., 1994
	male mice	LD_{50}[b]	3.30	Sugawara et al., 1994
	female rats	LD_{50}[b]	3.15	Sugawara et al., 1994
	male rats	LD_{50}[b]	3.07	Sugawara et al., 1994
Nickel chloride [7718-54-9]	*Neantheses arenaceodentata*	LC_{50} (7-d)	1.23	Petrich and Reish, 1979
		LC_{50} (96-h)	1.49	Petrich and Reish, 1979
	Capitella capitata	LC_{50} (96-h)	>1.70	Petrich and Reish, 1979
	Ctenodrilus serratus	LC_{50} (96-h)	1.23	Petrich and Reish, 1979
	Xenopus laevis	LC_{50} (96-h)	1.64	Hopfer et al., 1990
Nicotine [54-11-5]	*Daphnia magna*	EC_{50} (24-h)	0.75	Lilius et al., 1995
	Daphnia pulex	EC_{50} (24-h)	0.14	Lilius et al., 1995
Nitralin [4726-14-1]	*Daphnia magna*	EC_{50} (48-h)	-2.40	Mayer and Ellersieck, 1986
Nitrapyrin [4684-94-0]	*Daphnia magna*	EC_{50} (48-h)	0.20	Mayer and Ellersieck, 1986
4-Nitrobenzyl bromide [100-11-8]	*Poecilia reticulata*	LC_{50} (14-d)	-0.97	Hermens et al., 1985
Nitrogen dioxide [10102-44-0]	mice (strain ICR)	LC_{50} (16-h)[a]	1.58	Ichinose et al., 1982
	mice (strain BALB/c)	LC_{50} (16-h)[a]	1.69	Ichinose et al., 1982
	mice (strain ddy)	LC_{50} (16-h)[a]	1.71	Ichinose et al., 1982
	mice (strain C57BL/6)	LC_{50} (16-h)[a]	1.84	Ichinose et al., 1982
Nitrogen trichloride [10025-85-1]	rats	LC_{50} (1-h)[a]	2.05	Barbee et al., 1983
p-Nitrophenyl phenyl ether [620-88-2]	*Pimephales promelas*	LC_{50} (96-h)	0.43	Veith et al., 1983
1-Nonanol [143-08-8]	*Pimephales promelas*	LC_{50} (96-h)	0.76	Veith et al., 1983
5-Nonanone [502-56-7]	*Pimephales promelas*	LC_{50} (96-h)	1.49	Veith et al., 1983
Norbornane [279-23-2]	*Daphnia pulex*	EC_{50} (48-h)	0.72	Passino-Reader et al., 1997
Norflurazon [27314-13-2]	rats	LD_{50}[b]	>3.90	Humburg et al., 1989
Noruron [18530-56-8]	dogs	LD_{50}[b]	3.57	Humburg et al., 1989
	Wistar rats	LD_{50}[b]	3.17	Humburg et al., 1989
	Sprague-Dawley rats	LD_{50}[b]	3.83	Humburg et al., 1989
Nuarimol [63284-71-9]	male rats	LD_{50}[b]	3.10	Worthing and Hance, 1991
Octanoic acid [124-07-2]	*Hyale plumulosa*	LC_{50} (48-h)[d]	2.11	Onitsuka et al., 1989
	Oryzias latipes	LC_{50} (48-h)[d]	2.18	Onitsuka et al., 1989
1-Octanol [111-87-5]	*Pimephales promelas*	LC_{50} (96-h)	1.13	Veith et al., 1983
	Spirostomum ambiguum	EC_{50} (24-h)	-1.47	Nałecz-Jawecki and Sawicki, 1999

Compound [CAS No.]	Species	Results log, mg/L		Reference
		EC_{50} (48-h)	-1.47	Nałecz-Jawecki and Sawicki, 1999
		LC_{50} (24-h)	-1.30	Nałecz-Jawecki and Sawicki, 1999
		LC_{50} (48-h)	-1.40	Nałecz-Jawecki and Sawicki, 1999
2-Octanone [111-13-7]	*Pimephales promelas*	LC_{50} (96-h)	1.56	Veith et al., 1983
Omethoate [1113-02-6]	*Artemia* sp.	EC_{50} (24-h)	2.40	Guzzella et al., 1997
	Brachionus plicatilis	EC_{50} (24-h)	2.47	Guzzella et al., 1997
Orphenadrine HCl [341-69-5]	*Daphnia magna*	EC_{50} (24-h)	1.47	Lilius et al., 1995
Oryzalin [19044-88-3]	*Lepomis macrochirus*	LC_{50} (96-h)	0.46	Humburg et al., 1989
	rainbow trout	LC_{50} (96-h)	0.51	Humburg et al., 1989
	bobwhite quail	LD_{50}[b]	2.70	Humburg et al., 1989
	mice	LD_{50}[b]	>3.70	Humburg et al., 1989
	rats	LD_{50}[b]	>3.70	Humburg et al., 1989
Oxabetrinil [74782-23-3]	*Lepomis macrochirus*	LC_{50} (96-h)	1.08	Humburg et al., 1989
	Salmo gairdneri	LC_{50} (96-h)	0.85	Humburg et al., 1989
	bobwhite quail	LD_{50}[b]	>3.30	Humburg et al., 1989
	hamsters	LD_{50}[b]	>3.48	Humburg et al., 1989
	mallard duck	LD_{50}[b]	>3.30	Humburg et al., 1989
	mice	LD_{50}[b]	>3.70	Humburg et al., 1989
	rats	LD_{50}[b]	>3.70	Humburg et al., 1989
Oxadiazon [19666-30-9]	bobwhite quail	LD_{50}[b]	3.78	Humburg et al., 1989
	mallard ducks	LD_{50}[b]	>3.00	Humburg et al., 1989
	rats	LD_{50}[b]	>3.70	Humburg et al., 1989
Oxalyl chloride [79-37-8]	rats	LC_{50} (1-h)[a]	3.26	Barbee et al., 1995
Oxamyl [23135-22-0]	*Daphnia magna*	EC_{50} (48-h)	-0.38	Mayer and Ellersieck, 1986
	Chironomus plumosus	EC_{50} (48-h)	-0.74	Mayer and Ellersieck, 1986
	male rats	LC_{50} (1-h)	-0.77	Kennedy, 1986
	female rats	LC_{50} (1-h)	-0.92	Kennedy, 1986
	rats	LC_{50} (4-h)	-1.19	Kennedy, 1986
	Gammarus italicus	LC_{10} (24-h)	-0.66	Pantani et al., 1997
	Echinogammarus tibaldii	LC_{50} (24-h)	-0.52	Pantani et al., 1997
	fasted rats	LD_{50}[b]	0.40	Kennedy, 1986
	fasted mice	LD_{50}[b]	0.36	Kennedy, 1986
	guinea pigs	LD_{50}[b]	0.85	Kennedy, 1986
Oxyfluorfen [42874-03-3]	*Oreochromis niloticus*	LC_{50} (96-h)	0.48	Hassanein et al., 1999
	dogs	LD_{50}[b]	>3.70	Humburg et al., 1989
	rats	LD_{50}[b]	>3.70	Humburg et al., 1989
Ozone [10028-15-6]	*Daphnia magna*	LC_{50} (1-h)	-0.85	Leynen et al., 1998
		LC_{50} (2-h)	-1.03	Leynen et al., 1998
		LC_{50} (4-h)	-1.24	Leynen et al., 1998
		LC_{50} (24-h)	-1.68	Leynen et al., 1998
		LC_{50} (48-h)	-1.96	Leynen et al., 1998
	Clarias gariepinus larvae	LC_{50} (48-h)	-1.46	Leynen et al., 1998
	Cyprinus carpio larvae	LC_{50} (48-h)	-1.51	Leynen et al., 1998
	Leuciscus idus larvae	LC_{50} (48-h)	-1.44	Leynen et al., 1998
Paraquat [1910-42-5]	*Daphnia magna*	EC_{50} (24-h)	1.07	Lilius et al., 1995
	Daphnia pulex	EC_{50} (24-h)	1.20	Lilius et al., 1995
		EC_{50} (48-h)	0.60	Mayer and Ellersieck, 1986
	Simocephalus serrulatus	EC_{50} (48-h)	0.57	Mayer and Ellersieck, 1986
	rats	LD_{50}[b]	2.14	Humburg et al., 1989
Parathion methyl [298-00-0]	*Artemia* sp.	EC_{50} (24-h)	1.30	Guzzella et al., 1997
	Brachionus plicatilis	EC_{50} (24-h)	1.83	Guzzella et al., 1997
	Gammarus italicus	LC_{10} (24-h)	-2.24	Pantani et al., 1997
	Echinogammarus tibaldii	LC_{50} (24-h)	-2.80	Pantani et al., 1997
	Spirostomum teres	LC_{50} (0.5-h)	-0.37	Twagilimana et al., 1998
		LC_{50} (24-h)	0.91	Twagilimana et al., 1998
Pebulate [1114-71-2]	mice	LD_{50}[b]	3.22	Humburg et al., 1989
	rats	LD_{50}[b.]	2.96	Humburg et al., 1989
1,2,3,7,8-PeCDD [57117-41-6]	*Oryzias latipes*	LC_{50} (3-d)	-4.57	Wisk and Cooper, 1990
Pendimethalin [40487-42-2]	male and female rats	LD_{50}[b]	>3.70	Humburg et al., 1989
Pentachloroethane [76-01-7]	*Cyprinodon variegatus*	LC_{50} (24-h)	2.08	Heitmuller et al., 1981
		LC_{50} (48-h)	2.08	Heitmuller et al., 1981

Compound [CAS No.]	Species	Results log, mg/L		Reference
		LC_{50} (72-h)	2.08	Heitmuller et al., 1981
		LC_{50} (96-h)[d]	2.06	Heitmuller et al., 1981
2-Pentanedione [123-54-6]	female rats	LC_{50} (4-h)[a]	3.09	Ballantyne et al., 1986
1-Pentanol [71-41-0]	*Spirostomum ambiguum*	EC_{50} (24-h)	3.09	Nałecz-Jawecki and Sawicki, 1999
		EC_{50} (48-h)	3.05	Nałecz-Jawecki and Sawicki, 1999
		LC_{50} (24-h)	3.34	Nałecz-Jawecki and Sawicki, 1999
		LC_{50} (48-h)	3.32	Nałecz-Jawecki and Sawicki, 1999
3-Pentanol [584-02-1]	*Poecilia reticulata*	LC_{50} (7-d)	3.00	Könemann, 1981
3-Pentanone [96-22-0]	*Spirostomum ambiguum*	EC_{50} (24-h)	3.27	Nałecz-Jawecki and Sawicki, 1999
		EC_{50} (48-h)	3.26	Nałecz-Jawecki and Sawicki, 1999
		LC_{50} (24-h)	3.49	Nałecz-Jawecki and Sawicki, 1999
		LC_{50} (48-h)	3.47	Nałecz-Jawecki and Sawicki, 1999
	Pimephales promelas	LC_{50} (96-h)	3.19	Veith et al., 1983
Permethrin [52645-53-1]	*Daphnia magna*	EC_{50} (48-h)	-2.90	Mayer and Ellersieck, 1986
	Chironomus plumosus	EC_{50} (48-h)	-3.25	Mayer and Ellersieck, 1986
	Hydropsyche spp.	LC_{50} (1-h)	0.66	Sibley and Kaushik, 1991
	Isonychia bicolor	LC_{50} (1-h)	1.13	Sibley and Kaushik, 1991
	Simulium vittatum	LC_{50} (1-h)	0.43	Sibley and Kaushik, 1991
	Daphnia magna	LC_{50} (72-h)	-3.22	Sibley and Kaushik, 1991
	Cyprinodon variegatus	LC_{50} (96-h)	-2.11	Schimmel et al., 1983
	Daphnia magna	LC_{50} (96-h)	-2.17	Sibley and Kaushik, 1991
	Menidia menidia	LC_{50} (96-h)	-2.66	Schimmel et al., 1983
	Mugil cephalus	LC_{50} (96-h)	-2.26	Schimmel et al., 1983
	Mysidopsis bahia	LC_{50} (96-h)	-4.70	Schimmel et al., 1983
	Penaeus duorarum	LC_{50} (96-h)	-3.08	Schimmel et al., 1983
	Procambarus clarkii	LC_{50} (96-hr)	-3.36	Jarboe and Romaire, 1991
Phenmedipham [13684-63-4]	chickens	LD_{50}[b]	>3.48	Humburg et al., 1989
	dogs	LD_{50}[b]	>3.60	Humburg et al., 1989
	guinea pigs	LD_{50}[b]	>3.60	Humburg et al., 1989
	mice	LD_{50}[b]	>3.90	Humburg et al., 1989
	rats	LD_{50}[b]	>3.90	Humburg et al., 1989
Phenthoate [2597-03-7]	*Channa punctatus*	LC_{50} (48-h)	-0.33	Rao et al., 1985
2-Phenoxyethanol [122-99-6]	*Pimephales promelas*	LC_{50} (96-h)	2.54	Veith et al., 1983
Phorate [298-02-2]	*Artemia* sp.	EC_{50} (24-h)	>1.70	Guzzella et al., 1997
	Brachionus plicatilis	EC_{50} (24-h)	>1.70	Guzzella et al., 1997
Phosmet [732-11-6]	*Daphnia magna*	EC_{50} (48-h)	-2.25	Mayer and Ellersieck, 1986
Phosphamidon [13171-21-6]	*Daphnia pulex*	EC_{50} (48-h)	-2.00	Mayer and Ellersieck, 1986
	Simocephalus serrulatus	EC_{50} (48-h)	-2.03	Mayer and Ellersieck, 1986
	Puntius conchonius	LC_{50} (48-h)	2.65	Gill et al., 1991
Phthalic acid[88-99-3]	*Chironomus plumosus*	EC_{50} (48-h)	>1.86	Mayer and Ellersieck, 1986
Picloram [19180-02-1]	*Daphnia magna*	EC_{50} (48-h)	1.88	Mayer and Ellersieck, 1986
	cattle	LD_{50}[b]	2.88	Humburg et al., 1989
	rats	LD_{50}[b]	3.91	Humburg et al., 1989
	sheep	LD_{50}[b]	>3.00	Humburg et al., 1989
α-Pinene[7785-26-4]	*Daphnia magna*	LC_{50} (24-h)	1.83	LeBlanc, 1980
		LC_{50} (48-h)	1.61	LeBlanc, 1980
Pirimicarb [23103-98-2]	*Daphnia magna*	EC_{50} (48-h)	-1.80	Kusk, 1996
Pirimiphos-methyl [29232-93-7]	*Pseudomugil signifer*	LC_{50} (96-h)	-1.04	Brown et al., 1998
Polyoxyethylene (4) lauryl ether [9002-92-0]	*Gammarus italicus*	LC_{10} (24-h)	0.88	Pantani et al., 1997
	Echinogammarus tibaldii	LC_{50} (24-h)	0.56	Pantani et al., 1997
Potassium azide [20762-60-1]	*Daphnia pulex*	EC_{50} (48-h)	0.88	Mayer and Ellersieck, 1986
	Gammarus fasciatus	LC_{50} (24-h)	1.18	Mayer and Ellersieck, 1986
		LC_{50} (96-h)	0.81	Mayer and Ellersieck, 1986
	Lepomis machrochirus	LC_{50} (24-h)	0.43	Mayer and Ellersieck, 1986
		LC_{50} (96-h)	-0.11	Mayer and Ellersieck, 1986

Compound [CAS No.]	Species	Results log, mg/L		Reference
Potassium chloride [7447-40-7]	*Daphnia magna*	EC_{50} (24-h)	3.05	Lilius et al., 1995
Potassium cyanide [151-50-8]	*Daphnia magna*	EC_{50} (24-h)	-0.22	Lilius et al., 1995
	Daphnia pulex	EC_{50} (24-h)	-0.25	Lilius et al., 1995
	Brachydanio rerio	LC_{50} (48-h)	-0.36	Slooff, 1979
Potassium dichromate [7778-50-9]				
	Gammarus italicus	LC_{10} (24-h)	0.64	Pantani et al., 1997
	Echinogammarus tibaldii	LC_{50} (24-h)	-0.34	Pantani et al., 1997
	Macrobrachium lamarrei	LC_{50} (24-h)	0.74	Murti et al., 1983
		LC_{50} (48-h)	0.57	Murti et al., 1983
		LC_{50} (72-h)	0.39	Murti et al., 1983
		LC_{50} (96-h)	0.26	Murti et al., 1983
	Spirostomum teres	LC_{50} (0.5-h)	0.96	Twagilimana et al., 1998
		LC_{50} (24-h)	0.95	Twagilimana et al., 1998
Prodiamine [29091-21-2]	rats	LD_{50}[b]	>3.70	Humburg et al., 1989
Profenofos [41198-08-7]	*Daphnia magna*	EC_{50} (48-h)	-2.85	Mayer and Ellersieck, 1986
	Chironomus plumosus	EC_{50} (48-h)	-3.00	Mayer and Ellersieck, 1986
Prometon [7287-19-6]	rainbow trout	LC_{50} (96-h)	1.30	Humburg et al., 1989
	rats	LD_{50}[b]	3.36	Humburg et al., 1989
Prometryn [7287-19-6]	*Lepomis machrochirus*	LC_{50} (96-h)	1.00	Humburg et al., 1989
	rainbow trout	LC_{50} (96-h)	0.40	Humburg et al., 1989
Pronamide [23950-58-5]	female rats	LD_{50}[b]	3.75	Humburg et al., 1989
	male rats	LD_{50}[b]	3.92	Humburg et al., 1989
	mongrel dogs	LD_{50}[b]	4.00	Humburg et al., 1989
Propachlor [1918-16-7]	*Daphnia magna*	EC_{50} (48-h)	0.84	Mayer and Ellersieck, 1986
	Chironomus plumosus	EC_{50} (48-h)	-0.10	Mayer and Ellersieck, 1986
	Lepomis machrochirus	LC_{50} (96-h)	>0.15	Humburg et al., 1989
	rainbow trout	LC_{50} (96-h)	-0.77	Humburg et al., 1989
	bobwhite quail	LD_{50}[b]	1.96	Humburg et al., 1989
	rats	LD_{50}[b]	3.26	Humburg et al., 1989
Propanil [709-98-8]	dog	LD_{50}[b]	3.09	Ambrose et al., 1972
	sheep	LD_{50}[b]	2.40	Palmer, 1964
2-Propanol [67-63-0]	*Spirostomum ambiguum*	EC_{50} (24-h)	3.84	Nałecz-Jawecki and Sawicki, 1999
		EC_{50} (48-h)	3.85	Nałecz-Jawecki and Sawicki, 1999
		LC_{50} (24-h)	4.35	Nałecz-Jawecki and Sawicki, 1999
		LC_{50} (48-h)	4.33	Nałecz-Jawecki and Sawicki, 1999
	Pimephales promelas	LC_{50} (96-h)	4.02	Veith et al., 1983
	Poecilia reticulata	LC_{50} (7-d)	3.85	Könemann, 1981
	Sprague-Dawley female rats	LC_{50} (8-h)[a]	4.28	Laham et al., 1980
Propazine [139-40-2]	*Lepomis machrochirus*	LC_{50} (96-h)	>2.00	Humburg et al., 1989
	rainbow trout	LC_{50} (96-h)	1.26	Humburg et al., 1989
	Sprague-Dawley male rats	LC_{50} (8-h)[a]	4.35	Laham et al., 1980
Propham [122-42-9]	*Daphnia pulex*	EC_{50} (48-h)	0.91	Mayer and Ellersieck, 1986
	Simocephalus serrulatus	EC_{50} (48-h)	1.00	Mayer and Ellersieck, 1986
	mice	LD_{50}[b]	3.48	Humburg et al., 1989
	rats	LD_{50}[b]	3.95	Humburg et al., 1989
Propoxur [114-26-1]	*Gammarus italicus*	LC_{10} (24-h)	-1.30	Pantani et al., 1997
	Echinogammarus tibaldii	LC_{50} (24-h)	-1.82	Pantani et al., 1997
Propoxyphene HCl [1639-60-7]	*Daphnia magna*	EC_{50} (24-h)	1.32	Lilius et al., 1995
	Daphnia pulex	EC_{50} (24-h)	1.33	Lilius et al., 1995
Propranolol HCl [318-99-9]	*Daphnia magna*	EC_{50} (24-h)	0.49	Lilius et al., 1995
	Daphnia pulex	EC_{50} (24-h)	0.64	Lilius et al., 1995
n-Propyl nitrite [543-67-9]	Sprague-Dawley rats	LC_{50} (4-h)[a]	2.48	Klonne et al., 1987a
Pyrazon [58858-18-7]	mice	LD_{50}[b]	3.48	Humburg et al., 1989
	rabbits	LD_{50}[b]	3.10	Humburg et al., 1989
	rats	LD_{50}[b]	3.56	Humburg et al., 1989
Pyridate [55512-33-9]	male and female rats	LD_{50}[b]	3.67	Humburg et al., 1989
Pyriproxyfen [95737-68-1]	*Aedes aegypti*	LC_{50} (6-h)	-4.67	Loh and Yap, 1989
	Daphnia carinata sensu lato	LC_{50} (48-h)	-1.10	Trayler and Davis, 1996
Quizalofop-ethyl [76578-14-8]	*Lepomis machrochirus*	LC_{50} (96-h)	-0.34	Humburg et al., 1989
	rainbow trout	LC_{50} (96-h)	-0.06	Humburg et al., 1989

Compound [CAS No.]	Species	Results log, mg/L		Reference
	female rats	LD_{50}[b]	3.17	Humburg et al., 1989
	male rats	LD_{50}[b]	3.22	Humburg et al., 1989
	mallard duck	LD_{50}[b]	3.30	Humburg et al., 1989
Quinoline [91-22-5]	*Xenopus laevis*	LC_{50} (96-h)	1.42	Dumont et al., 1979
Resmethrin [10453-86-8]	*Gambusia affinis*	LC_{50} (24-h)	-2.15	Tietze et al., 1991
	Toxorhynchites splendens	LC_{50} (24-h)	-2.54	Tietze et al., 1993
	mice	LD_{50}[a]	3.14	Berteau and Deen, 1978
Resorcinol [108-46-3]	*Pimephales promelas*	LC_{50} (24-h)	1.95	Curtis et al., 1978
		LC_{50} (48-h)	1.86	Curtis et al., 1978
		LC_{50} (96-h)	1.75	Curtis et al., 1978
Rotenone [83-79-4]	*Daphnia pulex*	EC_{50} (48-h)	-1.00	Mayer and Ellersieck, 1986
	Simocephalus serrulatus	EC_{50} (48-h)	-0.51	Mayer and Ellersieck, 1986
Selenium [7782-49-2]	*Cyprinodon variegatus*	LC_{50} (24-h)	1.75	Heitmuller et al., 1981
		LC_{50} (48-h)	1.41	Heitmuller et al., 1981
		LC_{50} (72-h)	1.11	Heitmuller et al., 1981
		LC_{50} (96-h)[d]	0.83	Heitmuller et al., 1981
	Daphnia magna	LC_{50} (24-h)	-0.37	LeBlanc, 1980
		LC_{50} (48-h)	-0.18	LeBlanc, 1980
Sertaconazole [99592-32-2]	mice and rats	LD_{50}[b]	>3.90	Grau et al., 1992
Sethoxydim [74051-80-2]	rats	LD_{50}[b]	3.51	Worthing and Hance, 1991
Siduron [1982-49-6]	male and female rats	LD_{50}[b]	>3.88	Humburg et al., 1989
Silver [7440-22-4]	*Cyprinodon variegatus*	LC_{50} (24-h)	1.73	Heitmuller et al., 1981
		LC_{50} (48-h)	1.81	Heitmuller et al., 1981
		LC_{50} (96-h)[d]	1.76	Heitmuller et al., 1981
	Daphnia magna	LC_{50} (24-h)	-2.82	LeBlanc, 1980
		LC_{50} (48-h)	-2.82	LeBlanc, 1980
Silver nitrate [7761-88-8]	*Pimephales promelas*	LC_{50} (96-h)	-2.17	Holcombe et al., 1983
Simazine [122-34-9]	*Lepomis machrochirus*	LC_{50} (96-h)	1.20	Humburg et al., 1989
	oyster	LC_{50} (96-h)	>0.00	Humburg et al., 1989
	rainbow trout	LC_{50} (96-h)	0.45	Humburg et al., 1989
	mallard duck	LD_{50}[b]	4.71	Humburg et al., 1989
	rats	LD_{50}[b]	>3.70	Humburg et al., 1989
Sodium arsenite [7784-46-5]	*Daphnia pulex*	EC_{50} (48-h)	0.48	Mayer and Ellersieck, 1986
	Simocephalus serrulatus	EC_{50} (48-h)	0.15	Mayer and Ellersieck, 1986
Sodium azide [26628-22-8]	*Daphnia pulex*	EC_{50} (48-h)	0.62	Mayer and Ellersieck, 1986
	Gammarus fasciatus	LC_{50} (24-h)	1.15	Mayer and Ellersieck, 1986
		LC_{50} (96-h)	0.70	Mayer and Ellersieck, 1986
	Lepomis machrochirus	LC_{50} (24-h)	0.32	Mayer and Ellersieck, 1986
		LC_{50} (96-h)	-0.17	Mayer and Ellersieck, 1986
Sodium chlorate [7775-09-9]	rats	LD_{50}[b]	3.70	Humburg et al., 1989
Sodium chloride [7647-14-5]	*Daphnia magna*	EC_{50} (24-h)	3.34	Lilius et al., 1995
	Daphnia pulex	EC_{50} (24-h)	3.52	Lilius et al., 1995
Sodium cyanide [143-33-9]	*Anas platyrhynchos*	LD_{50}[b]	0.43	Henny et al., 1994
	Canis latrans	LD_{50}[b]	0.61	Sterner, 1979
	Coragyps atratus	LD_{50}[b]	0.68	Wiemeyer et al., 1986
	Coturnix japonica	LD_{50}[b]	0.97	Wiemeyer et al., 1986
	Falco sparverius	LD_{50}[b]	0.60	Wiemeyer et al., 1986
	Gallus domesticus	LD_{50}[b]	1.32	Wiemeyer et al., 1986
	Otus asio	LD_{50}[b]	0.93	Wiemeyer et al., 1986
	Sturnus vulgaris	LD_{50}[b]	1.23	Wiemeyer et al., 1986
Sodium fluoride [7681-49-4]	*Cyprinodon variegatus*	LC_{50} (24-h)	0.96	Heitmuller et al., 1981
		LC_{50} (48-h)	0.96	Heitmuller et al., 1981
		LC_{50} (72-h)	0.96	Heitmuller et al., 1981
		LC_{50} (96-h)[d]	0.96	Heitmuller et al., 1981
	Daphnia magna	EC_{50} (24-h)	2.48	Lilius et al., 1995
		LC_{50} (24-h)	2.83	LeBlanc, 1980
		LC_{50} (48-h)	2.53	LeBlanc, 1980
	Daphnia pulex	EC_{50} (24-h)	2.69	Lilius et al., 1995
	rats	LD_{50}[b]	2.26	Worthing and Hance, 1991
Sodium methanearsonate [2163-80-6]	*Peromyscus leucopus*	LD_{50}[b]	-3.48	Judd, 1979
Sodium nitrate [7731-99-4]	*Pseudacris regilla*	LC_{50} (10-d)	2.76	Schuytema and Nebeker, 1999
Sodium oleate [143-19-1]	*Oryzias latipes*	LC_{50} (48-h)[d]	2.34	Onitsuka et al., 1989
Sodium oxalate [62-76-0]	*Daphnia magna*	EC_{50} (24-h)	2.60	Lilius et al., 1995

Compound [CAS No.]	Species	Results log, mg/L		Reference
Sodium pentachlorophenate				
[131-52-2]	*Palaemonetes pugio*	LC_{50} (96-h)	-0.36	Conklin and Rao, 1978
	Spirostomum teres	LC_{50} (0.5-h)	0.00	Twagilimana et al., 1998
		LC_{50} (24-h)	0.13	Twagilimana et al., 1998
Sodium selenite [10102-18-8]	*Daphnia magna*	EC_{50} (48-h)	0.40	Mayer and Ellersieck, 1986
	Chironomus plumosus	EC_{50} (48-h)	1.64	Mayer and Ellersieck, 1986
Sodium stearate [822-16-2]	*Oryzias latipes*	LC_{50} (48-h)[d]	2.10	Onitsuka et al., 1989
Styrene oxide [96-09-3]	*Poecilia reticulata*	LC_{50} (14-d)	0.85	Deneer et al., 1988
Sulfometuron-methyl				
[74222-97-2]	*Daphnia magna*	LC_{50} (96-h)	>1.10	Humburg et al., 1989
	Lepomis machrochirus	LC_{50} (96-h)	>1.10	Humburg et al., 1989
	oyster	LC_{50} (96-h)	>1.10	Humburg et al., 1989
	mallard duck	LD_{50}[b]	3.70	Humburg et al., 1989
Temephos [3383-96-8]	*Gambusia affinis*	LC_{50} (24-h)	0.75	Tietze et al., 1991
	Pseudomugil signifer	LC_{50} (96-h)	-0.23	Brown et al., 1998
Terbufos [13071-79-9]	*Daphnia magna*	EC_{50} (48-h)	-3.40	Mayer and Ellersieck, 1986
	Chironomus plumosus	EC_{50} (48-h)	-2.85	Mayer and Ellersieck, 1986
	Poecilia reticulata	LC_{50}	1.53	Mittal et al., 1994
Terbutryn [886-50-0]	channel catfish	LC_{50} (96-h)	0.48	Humburg et al., 1989
	Lepomis machrochirus	LC_{50} (96-h)	0.60	Humburg et al., 1989
	rainbow trout	LC_{50} (96-h)	0.48	Humburg et al., 1989
	bobwhite quail	LD_{50}[b]	>4.30	Humburg et al., 1989
	mallard duck	LD_{50}[b]	>3.67	Humburg et al., 1989
1,1,1,2-Tetrachloroethane				
[630-30-6]	*Daphnia magna*	LC_{50} (24-h)	1.43	LeBlanc, 1980
		LC_{50} (48-h)	1.38	LeBlanc, 1980
α,α-2,6-Tetrachlorotoluene				
[81-19-6]	*Pimephales promelas*	LC_{50} (96-h)	-0.01	Veith et al., 1983
Tetraethoxysilane [78-10-4]	male ICR mice	LC_{50} (4-h)[a]	>3.00	Nakashima et al., 1994
Thallium [7440-28-0]	*Cyprinodon variegatus*	LC_{50} (24-h)	1.65	Heitmuller et al., 1981
		LC_{50} (96-h)[d]	1.32	Heitmuller et al., 1981
	Daphnia magna	LC_{50} (24-h)	0.56	LeBlanc, 1980
		LC_{50} (48-h)	0.34	LeBlanc, 1980
Thallium sulfate [7446-18-6]	*Daphnia magna*	EC_{50} (24-h)	0.91	Lilius et al., 1995
	Daphnia pulex	EC_{50} (24-h)	0.67	Lilius et al., 1995
Thanite [115-31-1]	*Daphnia magna*	EC_{50} (48-h)	-0.94	Mayer and Ellersieck, 1986
Theophylline [58-55-9]	*Daphnia magna*	EC_{50} (24-h)	2.19	Lilius et al., 1995
	Daphnia pulex	EC_{50} (24-h)	2.52	Lilius et al., 1995
Thifenuron-methyl [79277-27-3]	*Daphnia magna*	LC_{50} (48-h)	>3.00	Humburg et al., 1989
	Lepomis machrochirus	LC_{50} (96-h)	>2.00	Humburg et al., 1989
	Salmo gairdneri	LC_{50} (96-h)	>2.00	Humburg et al., 1989
	mallard duck	LD_{50}[b]	>3.40	Humburg et al., 1989
	rats	LD_{50}[b]	>3.70	Humburg et al., 1989
Thiobencarb [28249-77-6]	*Anguilla anguilla*	LC_{50} (96-h)	1.12	Fernandez-Vega et al., 1999
		LC_{50} (72-h)	1.23	Fernandez-Vega et al., 1999
		LC_{50} (48-h)	1.34	Fernandez-Vega et al., 1999
		LC_{50} (24-h)	1.41	Fernandez-Vega et al., 1999
	Oryzias latipes embryo	LC_{50} (6-d)	1.28	Wolfe et al., 1993
	bobwhite quail	LD_{50}[b]	>3.89	Humburg et al., 1989
	mallard ducks	LD_{50}[b]	>4.00	Humburg et al., 1989
	mice	LD_{50}[b]	3.44	Humburg et al., 1989
	rats	LD_{50}[b]	2.96	Humburg et al., 1989
Thioridazine HCl [130-61-0]	*Daphnia magna*	EC_{50} (24-h)	-0.12	Lilius et al., 1995
	Daphnia pulex	EC_{50} (24-h)	-1.50	Lilius et al., 1995
Thiourea [62-56-6]	*Scenedesmus subspicatus*	EC_{10} (96-h)	-0.38	Geyer et al., 1985
		EC_{50} (96-h)	0.83	Geyer et al., 1985
Thiram [137-26-8]	*Spirostomum teres*	LC_{50} (0.5-h)	-0.77	Twagilimana et al., 1998
		LC_{50} (24-h)	-0.36	Twagilimana et al., 1998
m-Toluidine [95-53-4]	*Spirostomum ambiguum*	EC_{50} (24-h)	2.68	Nałecz-Jawecki and Sawicki, 1999
		EC_{50} (48-h)	2.67	Nałecz-Jawecki and Sawicki, 1999
		LC_{50} (24-h)	2.92	Nałecz-Jawecki and Sawicki, 1999

Compound [CAS No.]	Species	Results log, mg/L		Reference
		LC_{50} (48-h)	2.83	Nałecz-Jawecki and Sawicki, 1999
p-Toluidine [99-94-5]	Spirostomum ambiguum	EC_{50} (24-h)	2.29	Nałecz-Jawecki and Sawicki, 1999
		EC_{50} (48-h)	2.00	Nałecz-Jawecki and Sawicki, 1999
		LC_{50} (24-h)	2.49	Nałecz-Jawecki and Sawicki, 1999
		LC_{50} (48-h)	2.21	Nałecz-Jawecki and Sawicki, 1999
Triallate [2303-17-5]	Daphnia magna	EC_{50} (48-h)	-1.10	Mayer and Ellersieck, 1986
		LC_{50} (48-h)	-0.37	Humburg et al., 1989
	Lepomis machrochirus	LC_{50} (96-h)	0.11	Humburg et al., 1989
	rainbow trout	LC_{50} (96-h)	0.08	Humburg et al., 1989
	bobwhite quail	LD_{50}[b]	>3.35	Humburg et al., 1989
Triasulfuron [82097-50-5]	rats	LD_{50}[b]	>3.70	Humburg et al., 1989
Triazophos [24017-47-8]	dogs	LD_{50}[b]	>2.51	Worthing and Hance, 1991
Tribenuron [106040-48-6]	Daphnia magna	EC_{50} (48-h)	2.86	Humburg et al., 1989
	Lepomis machrochirus	LC_{50} (96-h)	>3.00	Humburg et al., 1989
	Salmo gairdneri	LC_{50} (96-h)	>3.00	Humburg et al., 1989
	bobwhite quail	LD_{50}[b]	>3.70	Humburg et al., 1989
1,2,4-Tribromobenzen [615-54-3]	Gambusia affinis	LC_{50} (96-h)	-3.22	Chaisuksant et al., 1998
S,S,S- Tributyl phosphoro-trithioate [78-48-8]	Daphnia magna	EC_{50} (48-h)	-2.17	Mayer and Ellersieck, 1986
	Chironomus plumosus	EC_{50} (48-h)	-1.40	Mayer and Ellersieck, 1986
Tributyltin [688-73-3]	Mytilus edulis	LC_{50} (24-h)	-3.60	Stenalt et al., 1998.
Trichlorfon [52-68-6]	Daphnia pulex	EC_{50} (48-h)	-3.74	Mayer and Ellersieck, 1986
	Simocephalus serrulatus	EC_{50} (48-h)	-3.29	Mayer and Ellersieck, 1986
2′,3′,4′-Trichloroaceto-phenone [13608-87-2]	Pimephales promelas	LC_{50} (96-h)	0.30	Veith et al., 1983
2,2,2-Trichloroethanol [115-20-8]	Pimephales promelas	LC_{50} (96-h)	2.48	Veith et al., 1983
Triclopyr [55335-06-3]	Lepomis machrochirus	LC_{50} (96-h)	2.17	Humburg et al., 1989
	Salmo gairdneri	LC_{50} (96-h)	2.07	Humburg et al., 1989
	mallard duck	LD_{50}[b]	3.23	Humburg et al., 1989
Tricyclene [508-32-7]	Daphnia pulex	EC_{50} (48-h)	0.15	Passino-Reader et al., 1997
Tridiphane [51528-03-1]	Eastern oyster	LC_{50} (96-h)	-1.44	Humburg et al., 1989
	fathead minnow	LC_{50} (96-h)	0.11	Humburg et al., 1989
	fiddler crab	LC_{50} (96-h)	>3.00	Humburg et al., 1989
	pink shrimp	LC_{50} (96-h)	-0.85	Humburg et al., 1989
	Salmo gairdneri	LC_{50} (96-h)	2.07	Humburg et al., 1989
	Salmo gairdneri	LC_{50} (96-h)	2.07	Humburg et al., 1989
	Lepomis machrochirus	LC_{50} (96-h)	-0.43	Humburg et al., 1989
	Salmo gairdneri	LC_{50} (96-h)	-0.28	Humburg et al., 1989
	mallard duck	LD_{50}[b]	>3.40	Humburg et al., 1989
Triethylene glycol [112-27-6]	Pimephales promelas	LC_{50} (96-h)	4.84	Veith et al., 1983
	Poecilia reticulata	LC_{50} (7-d)	4.80	Könemann, 1981
Triethyllead chloride [1067-14-7]	Drosophila melanogaster	LC_{50}	1.47	Atkins et al., 1991
1,1,1-Trifluoroethane [420-46-2]	rats	LC_{50} (4-h)[a]	>5.73	Brock et al., 1996
2′,3′,4′-Trimethoxyaceto-phenone [13909-27-6]	Pimephales promelas	LC_{50} (96-h)	2.24	Veith et al., 1983
Tri(dimethylamino)silane [15112-89-7]	male and female rats	LC_{50} (4-h)[a]	1.58	Ballantyne et al., 1989
Tris-BP [126-72-7]	Scenedesmus subspicatus	EC_{10} (96-h)	-0.77	Geyer et al., 1985
		EC_{50} (96-h)	0.49	Geyer et al., 1985
1-Undecanol [112-42-5]	Pimephales promelas	LC_{50} (96-h)	0.02	Veith et al., 1983
Urea [57-13-6]	Helisoma trivolvis	LC_{50} (24-h)	4.48	Tchounwou et al., 1991
	Biomphalaria havanensis	LC_{50} (24-h)	4.42	Tchounwou et al., 1991
Valproic acid [77-66-1]	Xenopus	LC_{50} (96-h)	2.97	Dawson et al., 1992
Vernolate [1929-77-7]	Cypridopsis vidua	EC_{50} (48-h)	-0.60	Mayer and Ellersieck, 1986
	fingerling trout	LC_{50} (96-h)	0.98	Humburg et al., 1989
White phosphorus [7723-14-0]	Cygnus olor	LD_{50}[b]	0.56	Sparling et al., 1999

Compound [CAS No.]	Species	Results log, mg/L		Reference
Zinc chloride [7646-85-7]	*Gammarus italicus*	LC$_{10}$ (24-h)	0.94	Pantani et al., 1997
	Echinogammarus tibaldii	LC$_{50}$ (24-h)	1.41	Pantani et al., 1997
	Heteropneustes fossilis	LC$_{50}$ (96-h)	1.88	Hemalatha and Banerjee, 1997
Zinc phosphide [1314-84-7]	*Gallus domesticus*	LD$_{50}$[a]	1.40	Shivanandappa et al., 1979
Zinc sulphate [7733-02-0]	*Pimephales promelas*	LC$_{50}$ (96-h)	0.49	Judy and Davies, 1979
	Spirostomum teres	LC$_{50}$ (0.5-h)	-0.52	Twagilimana et al., 1998
		LC$_{50}$ (24-h)	0.95	Twagilimana et al., 1998

[a] Values for non-aquatic species (e.g., mice, rats, hamsters) were determined from inhalation studies and are given in ppm; [b] Acute oral value in mg/kg; [c] Units are in mg/m^3; [d] In seawater.

Table 12. Freundlich and Soil Adsorption Coefficients of Miscellaneous Organic Chemicals and Metals

Compound [CAS No.]	Adsorbent	% sand	% silt	% clay	CEC[a]	pH	K_F	K_F units	$1/n$	C units	x/m units	f_{oc} %	log K_{oc}[b]	Reference
Acenaphthene [83-32-9]	Filtrasorb 300	--	--	--	--	5.3/7.4	190	L/g	0.36	mg/L	mg/g	--	--	Dobbs and Cohen, 1980
	Filtrasorb 300	--	--	--	--	5.0	140	L/g	0.43	mg/L	mg/g	--	--	Burks, 1981
	Filtrasorb 400	--	--	--	--		626	L/g	0.457	mg/L	mg/g	--	--	Walters and Luthy, 1982
Acenaphthylene [208-96-8]	Filtrasorb 400	--	--	--	--	5.3/7.4	115	L/g	0.37	mg/L	mg/g	--	--	Dobbs and Cohen, 1980
	Filtrasom 300	--	--	--	--		266	L/g	0.302	mg/L	mg/g	--	--	Walters and Luthy, 1982
	Filtrasorb 400	--	--	--	--		0.013	L/g	0.911	mg/L	mg/g	--	--	Chanda et al., 1985
Acetic acid [64-19-7]	Poly(4-vinyl pyridine)	--	--	--	--		0.446	L/g	0.64	mg/L	mg/g	--	--	Chanda et al., 1985
	Polybenzimidazole	--	--	--	--		74	L/g	0.44	mg/L	mg/g	--	--	Dobbs and Cohen, 1980
Acetophenone [98-86-2]	Filtrasorb 300	--	--	--	--	5.3/7.4	9.63	L/mg	0.291	μg/L	μg/g	--	--	DiGiano et al., 1980
	Hydrodarco 3000	4	10	86	76	4.5	--	unitless	--	--	--	0.11	2.27	Southworth and Keller, 1986
	Apison, TN soil	2	38	60	129	4.2	--	unitless	--	--	--	1.2	2.02	Southworth and Keller, 1986
	Dormont, WV soil	11	21	68	64	4.4	--	unitless	--	--	--	0.05	2.43	Southworth and Keller, 1986
	Fullerton, TN soil	3.0	41.8	55.2	23.72	7.79	0.89	mL/g	--	μg/mL	μg/g	2.07	1.63	Khan et al., 1979a
	River sediment	33.6	35.4	31.0	19.00	7.44	0.56	mL/g	--	μg/mL	μg/g	2.28	1.38	Khan et al., 1979a
	River sediment	0.3	31.2	68.6	33.01	7.83	0.68	mL/g	--	μg/mL	μg/g	0.72	1.98	Khan et al., 1979a
	River sediment	82.4	10.7	6.8	3.72	8.32	0.07	mL/g	--	μg/mL	μg/g	0.15	1.68	Khan et al., 1979a
	River sediment	7.1	75.6	17.4	12.40	8.34	0.09	mL/g	--	μg/mL	μg/g	0.11	1.91	Khan et al., 1979a
	Iowa sediment	2.1	34.4	63.6	18.86	8.45	0.12	mL/g	--	μg/mL	μg/g	0.48	1.40	Khan et al., 1979a
	WV sediment	15.6	48.7	35.7	15.43	7.76	0.27	mL/g	--	μg/mL	μg/g	0.95	1.45	Khan et al., 1979a
	Ohio River sediment	34.6	25.8	39.5	15.43	7.76	0.30	mL/g	--	μg/mL	μg/g	0.66	1.66	Khan et al., 1979a
	River sediment	0.0	71.4	28.6	8.50	5.50	0.29	mL/g	--	μg/mL	μg/g	1.30	1.34	Khan et al., 1979a
	Il soil	50.2	42.7	7.1	8.33	7.60	0.85	mL/g	--	μg/mL	μg/g	1.88	1.65	Khan et al., 1979a
	Il sediment	26.2	52.7	21.2	23.72	7.55	0.53	mL/g	--	μg/mL	μg/g	1.67	1.49	Khan et al., 1979a
	Illinois river sediment	17.3	13.6	69.1	31.15	6.70	0.68	mL/g	--	μg/mL	μg/g	2.38	1.46	Khan et al., 1979a
	Illinois river sediment	1.6	55.4	42.9	20.86	7.75	0.66	mL/g	--	μg/mL	μg/g	1.48	1.65	Khan et al., 1979a
	River sediment	67.6	13.9	18.6	3.72	6.35	0.44	mL/g	--	μg/mL	μg/g	1.21	1.56	Khan et al., 1979a
	Stream sediment													
2-Acetylaminofluorene [53-96-3]	Filtrasorb 300	--	--	--	--	5.3/7.4	318	L/g	0.12	mg/L	mg/g	--	--	Dobbs and Cohen, 1980
9-Acetylanthracene [784-04-3]	Apison, TN soil	4	10	86	76	4.5	--	unitless	--	--	--	0.11	3.92	Southworth and Keller, 1986
	Dormont, WV soil	2	38	60	129	4.2	--	unitless	--	--	--	1.2	3.24	Southworth and Keller, 1986
	Fullerton, TN soil	11	21	68	64	4.4	--	unitless	--	--	--	0.05	3.07	Southworth and Keller, 1986
4-Acylbiphenyl [92-91-1]	Apison, TN soil	4	10	86	76	4.5	--	unitless	--	--	--	0.11	3.38	Southworth and Keller, 1986
	Dormont, WV soil	2	38	60	129	4.2	--	unitless	--	--	--	1.2	3.28	Southworth and Keller, 1986
	Fullerton, TN soil	11	21	68	64	4.4	--	unitless	--	--	--	0.05	2.85	Southworth and Keller, 1986
2-Acetylnaphthalene [93-08-3]	Apison, TN soil	4	10	86	76	4.5	--	unitless	--	--	--	0.11	3.08	Southworth and Keller, 1986
	Dormont, WV soil	2	38	60	129	4.2	--	unitless	--	--	--	1.2	2.98	Southworth and Keller, 1986

Chemical [CAS]	Sample	11	21	68	64	4.4								Reference
Acid blue 113 [3351-05-1]	Fullerton, TN soil	--	--	--	--	--	--	--	unitless	--	--	0.05	2.61	Southworth and Keller, 1986
Acifluorfen [50594-66-6]	MLSS	--	--	--	--	--	6.7	1.2	L/g	mg/L	mg/g	--	--	Shaul et al., 1986
	Filtrasorb 400	--	--	--	--	6.9	60.200	0.198	L/kg	μg/L	μg/g	--	--	Speth and Miltner, 1998
	Dundee loam	--	--	13.1	12.1[c]	5.59	0.48	--	L/kg	μmol/L	μmol/kg	0.747	1.81	Locke et al., 1997
	Dundee sicl	--	--	35.2	19.5[c]	5.08	0.56	--	L/kg	μmol/L	μmol/kg	0.748	1.87	Locke et al., 1997
	Dundee sl (CT)	--	--	22.0	14.3[c]	5.29	1.30	--	L/kg	μmol/L	μmol/kg	1.19	2.04	Locke et al., 1997
	Dundee sl (NT)	--	--	22.0	16.7[c]	5.13	3.15	--	L/kg	μmol/L	μmol/kg	2.24	2.15	Locke et al., 1997
	Humic acid A	--	--	--	--	3.1	148.7	0.93	L/kg	mg/L	mg/kg	--	--	Celi et al., 1996
	Humic acid A	--	--	--	--	2.7	346.0	0.91	L/kg	mg/L	mg/kg	--	--	Celi et al., 1996
	Humic acid B	--	--	--	--	3.4	84.1	0.95	L/kg	mg/L	mg/kg	--	--	Celi et al., 1996
	Humic acid C	--	--	--	--	3.4	94.4	0.93	L/kg	mg/L	mg/kg	--	--	Celi et al., 1996
	Humic acid D	--	--	--	--	3.2	87.3	1.02	L/kg	mg/L	mg/kg	--	--	Celi et al., 1996
	Humic acid D	--	--	--	--	3.0	107.0	1.00	L/kg	mg/L	mg/kg	--	--	Celi et al., 1996
	Humic acid E	--	--	--	--	3.4	51.4	0.99	L/kg	mg/L	mg/kg	--	--	Celi et al., 1996
	Humic acid F	--	--	--	--	3.4	126.8	0.98	L/kg	mg/L	mg/kg	--	--	Celi et al., 1996
	Humic acid G	--	--	--	--	3.4	87.2	0.99	L/kg	mg/L	mg/kg	--	--	Celi et al., 1996
	Humic acid H	--	--	--	--	3.4	89.6	0.95	L/kg	mg/L	mg/kg	--	--	Celi et al., 1996
	Lafitte muck	--	--	20.0	78.0[c]	4.10	89.6	--	L/kg	μmol/L	μmol/kg	19.13	2.67	Locke et al., 1997
	Mahan lfs	--	--	4.7	3.8[c]	4.20	1.66	--	L/kg	μmol/L	μmol/kg	1.20	2.14	Locke et al., 1997
	Mahan fsl	--	--	14.8	4.1[c]	4.40	0.86	--	L/kg	μmol/L	μmol/kg	0.010	3.93	Locke et al., 1997
	Miami sl (CT)	--	--	40.0	14.1[c]	6.16	1.10	--	L/kg	μmol/L	μmol/kg	1.90	1.76	Locke et al., 1997
	Miami sl (NT)	--	--	40.0	15.6[c]	6.36	1.80	--	L/kg	μmol/L	μmol/kg	3.50	1.71	Locke et al., 1997
	Sharkey clay	--	--	61.0	43.7[c]	6.00	2.16	--	L/kg	μmol/L	μmol/kg	1.69	2.11	Locke et al., 1997
	Ships clay	--	--	50.0	40.1[c]	7.50	0.61	--	L/kg	μmol/L	μmol/kg	0.834	1.86	Locke et al., 1997
	Weswood sl	--	--	21.0	16.4[c]	7.70	0.30	--	L/kg	μmol/L	μmol/kg	0.311	1.98	Locke et al., 1997
Acridine [260-94-6]	River sediment	3.0	41.8	55.2	23.72	7.79	334	--	mL/g	μg/mL	μg/g	2.07	4.21	Banwart et al., 1982
	River sediment	33.6	35.4	31.0	19.00	7.44	278	--	mL/g	μg/mL	μg/g	2.28	4.09	Banwart et al., 1982
	River sediment	0.3	31.2	68.6	33.01	7.83	132	--	mL/g	μg/mL	μg/g	0.72	4.26	Banwart et al., 1982
	River sediment	82.4	10.7	6.8	3.72	8.32	33	--	mL/g	μg/mL	μg/g	0.15	4.34	Banwart et al., 1982
	Iowa sediment	7.1	75.6	17.4	12.40	8.34	34	--	mL/g	μg/mL	μg/g	0.11	4.49	Banwart et al., 1982
	WV sediment	2.1	34.4	63.6	18.86	4.45	142	--	mL/g	μg/mL	μg/g	0.48	4.47	Banwart et al., 1982
	Ohio River sediment	15.6	48.7	35.7	15.43	7.76	165	--	mL/g	μg/mL	μg/g	0.95	4.24	Banwart et al., 1982
	River sediment	34.6	25.8	39.5	15.43	7.76	101	--	mL/g	μg/mL	μg/g	0.66	4.18	Banwart et al., 1982
	Illinois soil	0.0	71.4	28.6	8.50	5.50	193	--	mL/g	μg/mL	μg/g	1.30	4.17	Banwart et al., 1982
	Illinois river sediment	50.2	42.7	7.1	8.33	7.60	207	--	mL/g	μg/mL	μg/g	1.88	4.04	Banwart et al., 1982
	Illinois river sediment	26.2	52.7	21.2	23.72	7.55	92	--	mL/g	μg/mL	μg/g	1.67	3.74	Banwart et al., 1982
	Illinois river sediment	17.3	13.6	69.1	31.15	6.70	287	--	mL/g	μg/mL	μg/g	2.38	4.08	Banwart et al., 1982
	River sediment	1.6	55.4	42.9	20.86	7.75	213	--	mL/g	μg/mL	μg/g	1.48	4.16	Banwart et al., 1982
	Stream sediment	67.6	13.9	18.6	3.72	6.35	185	--	mL/g	μg/mL	μg/g	1.21	4.18	Banwart et al., 1982
Acridine orange [494-38-2]	Filtrasorb 300	--	--	--	--	5.3/7.4	180	0.29	L/g	mg/L	mg/g	--	--	Dobbs and Cohen, 1980
Acridine yellow [92-26-2]	Filtrasorb 300	--	--	--	--	5.3/7.4	230	0.12	L/g	mg/L	mg/g	--	--	Dobbs and Cohen, 1980

Compound [CAS No.]	Adsorbent	% sand	% silt	% clay	CEC[a]	pH	K_F	K_F units	$1/n$	C units	x/m units	f_{oc} %	log K_{oc}[b]	Reference
Acrolein [107-02-8]	Filtrasorb 300	--	--	--	--	5.3/7.4	1.2	L/g	0.65	mg/L	mg/g	--	--	Dobbs and Cohen, 1980
Acrylonitrile [107-03-1]	Filtrasorb 300	--	--	--	--	5.3/7.4	1.4	L/g	0.51	mg/L	mg/g	--	--	Dobbs and Cohen, 1980
Adipic acid [124-04-9]	Filtrasorb 300	--	--	--	--	5.3/7.4	20	L/g	0.47	mg/L	mg/g	--	--	Dobbs and Cohen, 1980
Alachlor [15972-60-8]	Filtrasorb 400	--	--	--	--	--	482	L/g	0.26	mg/L	mg/g	--	--	Miltner et al., 1989
	Cellulose	--	--	--	--	--	6.45	mL/g	--	--	--	--	1.16	Torrents et al., 1997
	Chitin	--	--	--	--	--	5.40	mL/g	--	--	--	--	1.08	Torrents et al., 1997
	Collagen	--	--	--	--	--	257	mL/g	--	--	--	--	2.66	Torrents et al., 1997
	Lignin	--	--	--	--	--	402	mL/g	--	--	--	--	2.85	Torrents et al., 1997
Aldicarb [116-06-3]	Arredondo sand	93.8	3.0	3.2	--	6.8	0.20	μgmL/g	0.94[d]	--	--	0.80	1.40	Bilkert and Rao, 1985
	Batcombe sl	--	--	--	--	--	--	μgmL/g	0.95[d]	--	--	2.05	1.63	Briggs, 1981
	Cecil sal	65.8	19.5	14.7	--	5.6	0.18	L/kg	0.86	μM/L	μM/kg	0.90	1.30	Bilkert and Rao, 1985
	Clarion soil	37	42	21	21.02	5.00	0.78	L/kg	0.85	μM/L	μM/kg	2.64	1.47	Felsot and Dahm, 1979
	Harps soil	21	55	24	37.84	7.30	1.13	mL/g	--	--	--	3.80	1.47	Felsot and Dahm, 1979
	Palmyra sal	53	37	10	8.2[c]	4.9	0.07	mL/g	--	--	--	1.17	0.78	pH 4.9, Lemley et al., 1988
	Palmyra sal	53	37	10	8.2[c]	4.9	0.06	mL/g	--	--	--	1.17	0.71	pH 7.1, Lemley et al., 1988
	Peat	42	39	19	77.34	6.98	4.16	L/kg	0.89	μM/L	μM/kg	18.36	1.36	Felsot and Dahm, 1979
	Sarpy fsl	77	15	8	5.71	7.30	0.19	L/kg	0.93	μM/L	μM/kg	0.51	1.57	Felsot and Dahm, 1979
	Thurman fsl	83	9	8	6.01	6.83	0.22	L/kg	0.95	μM/L	μM/kg	1.07	1.31	Felsot and Dahm, 1979
	Webster sicl	18.4	45.3	38.3	--	7.3	0.76	μgmL/g	0.97[d]	--	--	3.97	1.30	Bilkert and Rao, 1985
Aldicarb sulfone [1646-88-4]	Cottenham sal	--	--	--	--	--	0.004	mL/g	--	--	--	1.17	0.50	Briggs, 1981
	Palmyra sal	53	37	10	8.2[c]	4.9	--	--	--	--	--	2.05	0.08	pH 4.9, Lemley et al., 1988
Aldrin [309-00-2]	Batcombe sl	--	--	--	--	--	--	--	--	--	--	--	--	Briggs, 1981
	Filtrasorb 300	--	--	--	--	5.3/7.4	651	L/g	0.92	mg/L	mg/g	--	4.69	Dobbs and Cohen, 1980
Ametryn [834-12-8]	Altura loam	--	--	--	27.6	8.0	2.55	mL/g	--	μg/mL	μg/g	2.13	2.08	Liu et al., 1970
	Coto clay	--	--	--	14.0	7.7	2.51	mL/g	--	μg/mL	μg/g	1.84	2.13	Liu et al., 1970
	Humata sandy clay	--	--	--	10.1	4.5	4.70	mL/g	--	μg/mL	μg/g	0.98	2.68	Liu et al., 1970
2-Aminobenzoic acid [118-92-3]	Activated charcoal	--	--	--	--	--	2.22	unitless	0.369	mmol/g	mol/kg	--	--	20 °C, Hartman et al., 1946
2-Aminoanthracene [613-13-8]	EPA-B2 sediment	67.5	18.6	13.9	3.72	6.35	321.6	mL/g	--	ng/mL	ng/g	1.21	4.42	Means et al., 1982
	EPA-4 sediment	3.0	55.2	41.8	23.72	7.79	329.2	mL/g	--	ng/mL	ng/g	2.07	4.20	Means et al., 1982
	EPA-5 sediment	33.6	31.0	35.4	19.00	7.44	304.1	mL/g	--	ng/mL	ng/g	2.28	4.14	Means et al., 1982
	EPA-6 sediment	0.2	68.6	31.2	33.01	7.83	259.5	mL/g	--	ng/mL	ng/g	0.72	4.56	Means et al., 1982
	EPA-8 sediment	82.4	6.8	10.7	3.72	8.32	79.0	mL/g	--	ng/mL	ng/g	0.15	4.72	Means et al., 1982
	EPA-9 soil	7.1	17.4	75.6	12.40	8.34	103.7	mL/g	--	ng/mL	ng/g	0.11	4.97	Means et al., 1982
	EPA-14 soil	2.1	63.6	34.4	18.86	4.54	145.1	mL/g	--	ng/mL	ng/g	0.48	4.48	Means et al., 1982
	EPA-15 sediment	15.6	35.7	48.7	11.30	7.79	391.9	mL/g	--	ng/mL	ng/g	0.95	4.62	Means et al., 1982
	EPA-18	34.6	39.5	25.8	15.43	7.76	283.0	mL/g	--	ng/mL	ng/g	0.66	4.63	Means et al., 1982
	EPA-20 sediment	0.0	28.6	71.4	8.50	5.50	458.7	mL/g	--	ng/mL	ng/g	1.30	4.55	Means et al., 1982

Compound / Sorbent													Reference
EPA-21	50.2	7.1	42.7	8.33	7.60	531.9	mL/g	--	ng/mL	ng/g	1.88	4.45	Means et al., 1982
EPA-22	26.1	21.2	52.7	8.53	7.55	502.1	mL/g	--	ng/mL	ng/g	1.67	4.48	Means et al., 1982
EPA-23	17.3	69.1	13.6	31.15	6.70	875.2	mL/g	--	ng/mL	ng/g	2.38	4.57	Means et al., 1982
EPA-26	1.6	42.9	55.4	20.86	7.75	688.7	mL/g	--	ng/mL	ng/g	1.48	4.67	Means et al., 1982
4-Aminobiphenyl [92-67-1]													
Activated charcoal	--	--	--	--	--	2.01	unitless	0.369	mmol/L	mol/kg	--	--	30 °C, Hartman et al., 1946
Activated charcoal	--	--	--	--	--	1.85	unitless	0.369	mmol/L	mol/kg	--	--	40 °C, Hartman et al., 1946
Activated charcoal	--	--	--	--	--	1.67	unitless	0.369	mmol/L	mol/kg	--	--	50 °C, Hartman et al., 1946
6-Aminochrysene [2642-98-0]													
Filtrasorb 300	--	--	--	--	5.3/7.4	200	L/g	0.26	mg/L	mg/g	--	--	Dobbs and Cohen, 1980
EPA-B2 sediment	67.5	18.6	13.9	3.72	6.35	1,735.5	mL/g	--	ng/mL	ng/g	1.21	5.16	Means et al., 1982
EPA-4 sediment	3.0	55.2	41.8	23.72	7.79	3,115.7	mL/g	--	ng/mL	ng/g	2.07	5.18	Means et al., 1982
EPA-5 sediment	33.6	31.0	35.4	19.00	7.44	3,972.5	mL/g	--	ng/mL	ng/g	2.28	5.24	Means et al., 1982
EPA-6 sediment	0.2	68.6	31.2	33.01	7.83	1,078.7	mL/g	--	ng/mL	ng/g	0.72	5.18	Means et al., 1982
EPA-8 sediment	82.4	6.8	10.7	3.72	8.32	573.3	mL/g	--	ng/mL	ng/g	0.15	5.58	Means et al., 1982
EPA-9 soil	7.1	17.4	75.6	12.40	8.34	686.4	mL/g	--	ng/mL	ng/g	0.11	5.80	Means et al., 1982
EPA-14 soil	2.1	63.6	34.4	18.86	4.54	924.3	mL/g	--	ng/mL	ng/g	0.48	5.28	Means et al., 1982
EPA-15 sediment	15.6	35.7	48.7	11.30	7.79	1,292.2	mL/g	--	ng/mL	ng/g	0.95	5.13	Means et al., 1982
EPA-18	34.6	39.5	25.8	15.43	7.76	1,424.5	mL/g	--	ng/mL	ng/g	0.66	5.33	Means et al., 1982
EPA-20 sediment	0.0	28.6	71.4	8.50	5.50	871.9	mL/g	--	ng/mL	ng/g	1.30	4.83	Means et al., 1982
EPA-21	50.2	7.1	42.7	8.33	7.60	2,616.0	mL/g	--	ng/mL	ng/g	1.88	5.14	Means et al., 1982
EPA-22	26.1	21.2	52.7	8.53	7.55	1,459.0	mL/g	--	ng/mL	ng/g	1.67	4.94	Means et al., 1982
EPA-23	17.3	69.1	13.6	31.15	6.70	3,923.3	mL/g	--	ng/mL	ng/g	2.38	5.22	Means et al., 1982
EPA-26	1.6	42.9	55.4	20.86	7.75	1,688.8	mL/g	--	ng/mL	ng/g	1.48	5.06	Means et al., 1982
3-Aminonitrobenzene [100-01-6] Batcombe sl	--	--	--	--	--	--	--	--	--	--	--	1.73	Briggs, 1981
3-Aminophenol [591-27-5]													
Poly(4-vinyl pyridine)	--	--	--	--	--	0.12	L/g	0.881	mg/L	mg/g	--	--	Chanda et al., 1985
Polybenzimidazole	--	--	--	--	--	0.044	L/g	0.95	mg/L	mg/g	--	--	Chanda et al., 1985
Anethole [104-46-1] Filtrasorb 300	--	--	--	--	5.3/7.4	300	L/g	0.42	mg/L	mg/g	--	--	Dobbs and Cohen, 1980
Aniline [62-53-3]													
H-montmorillonite	--	--	--	73.5	9.0	1,300	L/g	0.81	μmol/L	μmol/g	--	--	Bailey et al., 1968
Na-montmorillonite	--	--	--	87.0	9.0	130	L/g	0.90	μmol/L	μmol/g	--	--	Bailey et al., 1968
2-Anisidine [90-04-0]													
Filtrasorb 300	--	--	--	--	5.3/7.4	50	L/g	0.34	mg/L	mg/g	--	--	Dobbs and Cohen, 1980
Hydrodarco 1030	--	--	--	--	7	4.06	L/mg	0.46	μg/L	mg/g	--	--	Weber and Pirbazari, 1981
Norit carbon	--	--	--	--	7	13.72	L/mg	0.27	μg/L	mg/g	--	--	Weber and Pirbazari, 1981
Nuchar WV-G	--	--	--	--	7	15.41	L/mg	0.25	μg/L	mg/g	--	--	Weber and Pirbazari, 1981
100/200 mesh GAC	--	--	--	--	7	3.44	L/mg	0.664	μg/L	mg/g	--	--	Weber and Pirbazari, 1981
Anthracene [120-12-7]													
Filtrasorb 300	--	--	--	--	5.3/7.4	376	L/g	0.70	mg/L	mg/g	--	--	Dobbs and Cohen, 1980
Filtrasorb 400	--	--	--	--	--	330	L/g	0.62	mg/L	mg/g	--	--	Walters and Luthy, 1982
Anthracene-9-carboxylic acid [723-62-6]													
EPA-B2 sediment	67.5	18.6	13.9	3.72	6.35	5.27	mL/g	--	ng/mL	ng/g	1.21	2.64	Means et al., 1982
EPA-4 sediment	3.0	55.2	41.8	23.72	7.79	5.49	mL/g	--	ng/mL	ng/g	2.07	2.42	Means et al., 1982
EPA-5 sediment	33.6	31.0	35.4	19.00	7.44	7.96	mL/g	--	ng/mL	ng/g	2.28	2.54	Means et al., 1982
EPA-6 sediment	0.2	68.6	31.2	33.01	7.83	5.47	mL/g	--	ng/mL	ng/g	0.72	2.88	Means et al., 1982
EPA-8 sediment	82.4	6.8	10.7	3.72	8.32	1.84	mL/g	--	ng/mL	ng/g	0.15	3.09	Means et al., 1982

Compound [CAS No.]	Adsorbent	% sand	% silt	% clay	CEC[a]	pH	K_F	K_F units	1/n	C units	x/m units	f_{oc} %	log K_{oc}[b]	Reference
	EPA-9 soil	7.1	17.4	75.6	12.40	8.34	2.82	mL/g	--	ng/mL	ng/g	0.11	3.41	Means et al., 1982
	EPA-14 soil	2.1	63.6	34.4	18.86	4.54	10.03	mL/g	--	ng/mL	ng/g	0.48	3.32	Means et al., 1982
	EPA-15 sediment	15.6	35.7	48.7	11.30	7.79	2.66	mL/g	--	ng/mL	ng/g	0.95	2.45	Means et al., 1982
	EPA-18	34.6	39.5	25.8	15.43	7.76	1.78	mL/g	--	ng/mL	ng/g	0.66	2.43	Means et al., 1982
	EPA-20 sediment	0.0	28.6	71.4	8.50	5.50	13.27	mL/g	--	ng/mL	ng/g	1.30	3.01	Means et al., 1982
	EPA-21	50.2	7.1	42.7	8.33	7.60	6.45	mL/g	--	ng/mL	ng/g	1.88	2.54	Means et al., 1982
	EPA-22	26.1	21.2	52.7	8.53	7.55	5.59	mL/g	--	ng/mL	ng/g	1.67	2.53	Means et al., 1982
	EPA-23	17.3	69.1	13.6	31.15	6.70	9.88	mL/g	--	ng/mL	ng/g	2.38	2.62	Means et al., 1982
	EPA-26	1.6	42.9	55.4	20.86	7.75	7.50	mL/g	--	ng/mL	ng/g	1.48	2.71	Means et al., 1982
9-Anthracenemethanol [1468-95-7]	Apison, TN soil	4	10	86	76	4.5	--	unitless	--	--	--	0.11	3.90	Southworth and Keller, 1986
	Dormont, WV soil	2	38	60	129	4.2	--	unitless	--	--	--	1.2	3.43	Southworth and Keller, 1986
	Fullerton, TN soil	11	21	68	64	4.4	--	unitless	--	--	--	0.05	3.18	Southworth and Keller, 1986
Aroclor 1016 [12674-11-2]	Doe Run Pond	56.0	44.0	<1.0	--	6.1	1,290	L/mg	--	mg/g	mg/g	1.4	5.21	Steen et al., 1978
	Filtrasorb 400	--	--	--	--	7	6,650	L/mg	0.36	µg/L	mg/g	--	--	Weber and Pirbazari, 1981
	Hickory Hill Pond	55.0	45.0	<1.0	--	6.3	1,300	unitless	--	mg/g	mg/g	2.4	5.51	Steen et al., 1978
	Oconee River	93.0	6.0	1.0	--	6.5	620	unitless	--	mg/g	mg/g	0.4	4.41	Steen et al., 1978
	USDA Pond	--	--	--	--	6.4	1,370	unitless	--	mg/g	mg/g	0.8	4.99	Steen et al., 1978
Aroclor 1221 [11104-28-2]	Filtrasorb 300	--	--	--	--	5.3/7.4	242	L/g	0.70	mg/g	mg/g	--	--	Dobbs and Cohen, 1980
Aroclor 1232 [11141-16-5]	Filtrasorb 300	--	--	--	--	5.3/7.4	630	L/g	0.73	mg/L	mg/g	--	--	Dobbs and Cohen, 1980
Aroclor 1242 [53469-21-9]	Doe Run Pond	56.0	44.0	<1.0	--	6.1	1.09	unitless	--	mg/g	mg/g	1.4	5.13	Steen et al., 1978
	Hickory Hill Pond	55.0	45.0	<1.0	--	6.3	1.25	unitless	--	mg/g	mg/g	2.4	5.50	Steen et al., 1978
	Oconee River	93.0	6.0	1.0	--	6.5	0.54	unitless	--	mg/g	mg/g	0.4	4.35	Steen et al., 1978
	USDA Pond	--	--	--	--	6.4	1.21	unitless	--	mg/g	mg/g	0.8	4.41	Steen et al., 1978
Aroclor 1254 [11097-69-1]	Filtrasorb	--	--	--	--	7	0.73	L/mg	1.14	µg/L	µg/kg	0.9	4.88	Weber and Pirbazari, 1981
	Glendale clay	--	--	--	--	--	687	--	1.45	µg/L	µg/kg	0.7	4.47	Fairbanks and O'Connor, 1984
	Harvey fsl	--	--	--	--	--	205	--	1.53	µg/L	µg/kg	--	--	Fairbanks and O'Connor, 1984
	Illite lay	--	--	--	--	--	63.1	--	1.1[d]	µg/L	ng/g	--	--	Haque et al., 1974
	Lea sal	--	--	--	--	--	227	--	1.22	µg/L	µg/kg	0.9	4.40	Fairbanks and O'Connor, 1984
	Woodburn sl	--	--	--	--	--	26.3	L/kg	0.81[d]	µg/L	ng/g	1.80	3.16	Haque et al., 1974
Arsenic [7440-38-2], as AsO_4^{3-}	Alligator soil	5.9	39.4	54.7	30.2[c]	4.8	47.8	L/kg	0.636[d]	mg/L	mg/kg	1.54	3.49	Buchter et al., 1989
	Calciorthid sediment	70.0	19.3	10.7	14.7[c]	8.5	8.87	L/kg	0.554[d]	mg/L	mg/kg	0.44	3.30	Buchter et al., 1989
	Cecil soil	78.8	12.9	8.3	2.0[c]	5.7	19.8	L/kg	0.618[d]	mg/L	mg/kg	0.61	3.51	Buchter et al., 1989
	Lafitte soil	60.7	21.7	17.6	26.9[c]	3.9	71.0	L/kg	0.747[d]	mg/L	mg/kg	11.6	2.79	Buchter et al., 1989
	Molokai soil	25.7	46.2	28.2	11.0[c]	6.0	156	L/kg	0.561[d]	mg/L	mg/kg	1.67	3.97	Buchter et al., 1989
	Norwood soil	79.2	18.1	2.8	4.1[c]	6.9	8.53	L/kg	0.510[d]	mg/L	mg/kg	0.21	3.61	Buchter et al., 1989
	Olivier soil	4.4	89.4	6.2	8.6[c]	6.6	46.0	L/kg	0.548[d]	mg/L	mg/kg	0.83	3.74	Buchter et al., 1989
	Spodosol sediment	90.2	6.0	3.8	2.7[c]	4.3	18.8	L/kg	0.797[d]	mg/L	mg/kg	1.98	2.98	Buchter et al., 1989

Compound	Soil												Reference
Atratone [1610-17-9]	Webster soil	27.5	48.6	23.9	48.1[c]	7.6	23.6	L/kg	0.648[d]	mg/L	mg/kg	4.39	2.73	Buchter et al., 1989
	Windsor soil	76.8	20.5	2.8	2.0[c]	5.3	105	L/kg	0.601[d]	mg/L	mg/kg	2.03	3.71	Buchter et al., 1989
	H-montmorillonite	–	–	–	73.5	9.0	1,300	L/g	0.81	μmol/L	μmolg	–	–	Bailey et al., 1968
	Na-montmorillonite	–	–	–	87.0	9.0	440	L/g	0.48	μmol/L	μmolg	–	–	Bailey et al., 1968
Atrazine [15972-60-8]	Alluvium 9	93	6	6	–	7.61	0.202	L/kg	–	–	μg/kg	0.09	2.35	Roy and Krapac, 1994
	Alluvium 10	92	7	1	–	7.61	0.185	L/kg	–	–	μg/kg	0.15	2.09	Roy and Krapac, 1994
	Bates sl	1	26	73	9.3	6.5	0.8	L/kg	–	–	–	0.81	1.99	Talbert and Fletchall, 1965
	Baxter csl	9	72	19	11.2	6.0	2.3	L/kg	–	–	–	1.22	2.28	Talbert and Fletchall, 1965
	Begbroke soil	66.0	18.4	15.6	–	7.1	1.0	mL/g	–	–	μg/g	1.93	1.71	Grover and Hance, 1969
	Cecil sal	65.8	19.5	14.7	6.8	5.6	0.89	–	1.04[d]	μg/mL	–	0.90	2.00	Rao and Davidson, 1979
	Clarksville sl	20	67	13	5.7	5.7	1.7	L/kg	–	–	–	0.81	2.32	Talbert and Fletchall, 1965
	Cumberland csl	20	70	17	6.5	6.4	1.4	L/kg	–	–	–	0.70	2.30	Talbert and Fletchall, 1965
	Eldon sl	8	72	20	12.9	5.9	2.5	L/kg	–	–	–	1.74	2.16	Talbert and Fletchall, 1965
	Eustis fine sand	93.8	3.0	3.2	5.2	5.6	0.62	–	0.79[d]	–	–	0.56	2.04	Rao and Davidson, 1979
	Filtrasorb 400	–	–	–	–	–	283	L/g	0.29	mg/L	mg/g	–	–	Miltner et al., 1989
	Gerald sl	1	26	73	11.0	4.7	3.2	L/kg	–	–	–	1.57	2.31	Talbert and Fletchall, 1965
	Grundy sicl	3	67	30	13.5	5.6	4.8	L/kg	–	–	–	2.09	2.36	Talbert and Fletchall, 1965
	Hickory Hill silt	–	–	–	–	–	7.07	mL/g	–	μg/mL	μg/g	3.27	2.33	Brown and Flagg, 1981
	Knox sl	4	72	24	18.8	5.4	3.6	L/kg	–	–	–	1.68	2.33	Talbert and Fletchall, 1965
	Lebanon sl	13	70	17	7.7	4.9	2.2	L/kg	–	–	–	1.04	2.33	Talbert and Fletchall, 1965
	Lindley loam	30	44	26	6.9	4.7	2.6	L/kg	–	–	–	0.87	2.48	Talbert and Fletchall, 1965
	Lintonia loamy sand	84	11	5	3.2	5.3	0.6	L/kg	–	–	–	0.35	2.23	Talbert and Fletchall, 1965
	Marian sl	9	74	17	9.9	4.6	2.2	L/kg	–	–	–	0.81	2.43	Talbert and Fletchall, 1965
	Marshall sicl	4	66	30	21.3	5.4	4.5	L/kg	–	–	–	2.44	2.27	Talbert and Fletchall, 1965
	Menfro sl	4	85	11	9.1	5.3	1.7	L/kg	–	–	–	1.39	2.09	Talbert and Fletchall, 1965
	Na-montmorillonite	–	–	–	87.0	9.0	15	L/g	0.85	μmol/L	μmolg	–	–	Bailey et al., 1968
	Netonia sl	11	75	14	8.8	5.2	1.8	L/kg	–	–	–	0.93	1.87	Talbert and Fletchall, 1965
	Oswego sicl	5	67	28	21.0	6.4	2.7	L/kg	–	–	–	1.68	2.21	Talbert and Fletchall, 1965
	Putnam sl	6	74	20	12.3	5.3	1.9	L/kg	–	–	–	1.10	2.24	Talbert and Fletchall, 1965
	Salix loam	32	50	18	17.9	6.3	2.3	L/kg	–	–	–	1.22	2.28	Talbert and Fletchall, 1965
	Sand 1	88	4	8	–	7.53	0.201	L/kg	–	–	μg/kg	0.18	2.05	Roy and Krapac, 1994
	Sand 2	89	6	5	–	7.10	0.315	L/kg	–	–	μg/kg	0.12	2.42	Roy and Krapac, 1994
	Sand 3	87	11	2	–	7.20	0.727	L/kg	–	–	μg/kg	0.14	2.72	Roy and Krapac, 1994
	Sand 5	88	6	6	–	5.58	1.68	L/kg	–	–	μg/kg	0.01	3.23	Roy and Krapac, 1994
	Sand 7	85	8	7	–	6.05	0.900	L/kg	–	–	μg/kg	0.1	2.95	Roy and Krapac, 1994
	Sand 8	88	4	8	–	6.91	0.348	L/kg	–	–	μg/kg	0.09	2.59	Roy and Krapac, 1994
	Sandy loam	40	–	–	9.0	6.6	0.607	mL/g	–	–	–	1.24	1.69	Xu et al., 1999
	Sarpy loam	40	41	19	14.3	7.1	2.2	L/kg	–	–	–	0.75	2.47	Talbert and Fletchall, 1965
	Shelby loam	26	43	31	20.1	4.3	3.2	L/kg	–	–	–	2.09	2.19	Talbert and Fletchall, 1965
	Sharkey clay	25	30	45	28.2	5.0	3.1	L/kg	–	–	–	1.45	2.33	Talbert and Fletchall, 1965
	Summit silty clay	5	48	47	35.1	4.8	5.6	L/kg	–	–	–	0.40	3.15	Talbert and Fletchall, 1965

Compound [CAS No.]	Adsorbent	% sand	% silt	% clay	CEC[a]	pH	K_F	K_F units	1/n	C units	x/m units	f_oc %	log K_oc[b]	Reference
	Till 4	31	52	17	--	8.15	2.58	L/kg	--	μg/L	μg/kg	0.32	2.91	Roy and Krapac, 1994
	Till 6	33	56	11	--	7.65	2.83	L/kg	--	μg/L	μg/kg	0.33	2.93	Roy and Krapac, 1994
	Union sl	1	79	19	6.8	5.4	4.1	L/kg	--	--	--	1.04	2.60	Talbert and Fletchall, 1965
	Wabash clay	1	36	63	40.3	5.7	3.7	L/kg	--	--	--	1.28	2.46	Talbert and Fletchall, 1965
	Waverley sl	14	16	20	12.8	6.4	3.0	L/kg	--	--	--	1.16	2.41	Talbert and Fletchall, 1965
	Webster sicl	18.4	45.3	38.3	54.7	7.3	6.03	--	0.73[d]	mg/L	μg/g	3.87	2.19	Rao and Davidson, 1979
Azinphos-methyl [86-50-0]	Cohansey sand	90.0	8.0	2.0	--	--	96.7	L/kg	--	mg/L	μg/g	2.60	3.57	Reducker et al., 1988
	Sandy loam	--	--	--	9.0	6.6	6.07	mL/g	--	--	--	1.24	2.69	Xu et al., 1999
Azobenzene [103-33-3]	Batcombe sl	--	--	--	--	--	--	--	--	--	--	--	3.13	Briggs, 1981
Benzene [71-43-2]	Filtrasorb 300	--	--	--	--	5.3	1.0	L/g	1.6	mg/L	mg/g	--	--	Dobbs and Cohen, 1980
	Filtrasorb 400	--	--	--	--	7	1.12	L/mg	0.39	μg/L	mg/g	--	--	Weber and Pirbazari, 1981
	Filtrasorb 400	--	--	--	--		0.036	L/g	0.48	mg/L	mg/g	--	--	El-Dib and Badawy, 1979
	Hydrodarco 1030	--	--	--	--	7	1.18	L/g	0.36	mg/L	mg/g	--	--	Weber and Pirbazari, 1981
	Norit peat carbon	--	--	--	--	7	0.73	L/g	0.61	mg/L	mg/g	--	--	Weber and Pirbazari, 1981
	Nuchar WV-G	--	--	--	--	7	1.07	L/g	0.48	mg/L	mg/g	--	--	Weber and Pirbazari, 1981
	Agricultural soil	65.2	25.6	9.2	9.0	7.4	0.96	L/kg	--	mol/L	mol/kg	2.2	1.64	Seip et al., 1986
	Cohansey Sand	90.0	8.0	2.0	5.1	3.8	8.76	L/kg	0.999	mg/L	μg/g	2.55	2.53	Uchrin and Mangels, 1987
	Forest soil	69.5	20.5	10.1	2.9	4.2	1.98	L/kg	--	mol/L	mol/kg	3.7	1.73	Seip et al., 1986
	Forest soil	97.3	2.2	0.5	0.48	5.6	0.076	L/kg	--	mol/L	mol/kg	0.2	1.58	Seip et al., 1986
	Hastings sicl	1	68	31	17	5.6	2.4	mL/g	0.89	ng/mL	ng/g	2.6	1.97	Rogers et al, 1980
	Leie River clay	0	0	100	--	--	2.822	L/kg	--	--	--	4.12	1.84	2.3 °C, Dewulf et al., 1999
	Leie River clay	0	0	100	--	--	3.00	L/kg	--	--	--	4.12	1.86	3.8 °C, Dewulf et al., 1999
	Leie River clay	0	0	100	--	--	3.05	L/kg	--	--	--	4.12	1.87	6.2 °C, Dewulf et al., 1999
	Leie River clay	0	0	100	--	--	3.10	L/kg	--	--	--	4.12	1.88	8.0 °C, Dewulf et al., 1999
	Leie River clay	0	0	100	--	--	3.31	L/kg	--	--	--	4.12	1.90	13.5 °C, Dewulf et al., 1999
	Leie River clay	0	0	100	--	--	3.06	L/kg	--	--	--	4.12	1.87	18.6 °C, Dewulf et al., 1999
	Leie River clay	0	0	100	--	--	3.31	L/kg	--	--	--	4.12	1.90	25.0 °C, Dewulf et al., 1999
	PRM sal	70.4	24.0	5.6	5.6	5.5	2.67	L/kg	0.905	mg/L	μg/g	1.28	2.31	Uchrin and Mangels, 1987
	Al-montmorillonite	0	0	100	80	4.2	30.9	mL/g	1.08	ng/mL	ng/g	0	--	Rogers et al., 1980
	Ca-montmorillonite	0	0	100	80	6.6	4.4	mL/g	0.99	ng/mL	ng/g	0	--	Rogers et al., 1980
	Overton sicl	15	51	34	29	7.8	1.8	mL/g	0.94	ng/mL	ng/g	1.8	2.00	Rogers et al., 1980
1,2-Benzenediol [120-80-9]	Brookston clay loam	--	--	--	22.22	5.7	3.18	mL/g	0.36	μmol/mL	μmol/g	2.96	2.07	Boyd, 1982
1,3-Benzenediol [108-46-3]	Brookston clay loam	--	--	--	22.22	5.7	0.28	mL/g	0.40	μmol/mL	μmol/g	2.96	1.02	Boyd, 1982
Benzidine [92-87-5]	Chalmers sl	17.5	64.3	18.2	25.0[c]	5.4	9,460	L/kg	0.842	nmol/L	nmol/kg	--	5.68	Graveel et al., 1986
	Milford sicl	4.3	58.7	37.0	45.4[c]	6.4	21,000	L/kg	0.604	nmol/L	nmol/kg	--	5.91	Graveel et al., 1986
	River sediment	3.0	41.8	55.2	23.72	7.79	570.5	mL/g	0.513	nmol/mL	nmol/g	2.07	4.44	Zierath et al., 1980
	River sediment	33.6	35.4	31.0	19.00	7.44	589.4	mL/g	0.468	nmol/mL	nmol/g	2.28	4.41	Zierath et al., 1980
	River sediment	0.3	31.2	68.6	33.01	7.83	1,657.8	mL/g	0.568	nmol/mL	nmol/g	0.72	5.36	Zierath et al., 1980

Chemical	Matrix					pH		units		units	units			Reference
	River sediment	82.4	10.7	6.8	3.72	8.32	86.2	mL/g	0.496	nmol/mL	nmol/g	0.15	4.76	Zierath et al., 1980
	Iowa sediment	7.1	75.6	17.4	12.40	8.34	551.7	mL/g	0.372	nmol/mL	nmol/g	0.11	5.70	Zierath et al., 1980
	WV sediment	2.1	34.4	63.6	18.86	4.45	3,941.3	mL/g	0.664	nmol/mL	nmol/g	0.48	5.91	Zierath et al., 1980
	Ohio River sediment	15.6	48.7	35.7	15.43	7.76	1,705.4	mL/g	0.266	nmol/mL	nmol/g	0.95	5.25	Zierath et al., 1980
	River sediment	34.6	25.8	39.5	15.43	7.76	564.6	mL/g	0.413	nmol/mL	nmol/g	0.66	4.93	Zierath et al., 1980
	Illinois soil	0.0	71.4	28.6	8.50	5.50	2,332.9	mL/g	0.426	nmol/mL	nmol/g	1.30	5.25	Zierath et al., 1980
	Illinois river sediment	50.2	42.7	7.1	8.33	7.60	49.6	mL/g	0.694	nmol/mL	nmol/g	1.88	3.42	Zierath et al., 1980
	Illinois river sediment	26.2	52.7	21.2	23.72	7.55	73.9	mL/g	0.640	nmol/mL	nmol/g	1.67	3.65	Zierath et al., 1980
	Illinois river sediment	17.3	13.6	69.1	31.15	6.70	1,072.7	mL/g	0.569	nmol/mL	nmol/g	2.38	4.65	Zierath et al., 1980
	River sediment	1.6	55.4	42.9	20.86	7.75	108.2	mL/g	0.656	nmol/mL	nmol/g	1.48	3.95	Zierath et al., 1980
	Russell sl	18.4	69.2	12.4	14.7[c]	5.4	8,110	L/kg	0.834	nmol/L	nmol/kg	--	5.94	Graveel et al., 1986
	Stream sediment	67.6	13.9	18.6	3.72	6.35	500.4	mL/g	0.423	nmol/mL	nmol/g	1.21	4.62	Zierath et al., 1980
Benzidine hydrochloride [531-85-1]														
Benz[b]fluoranthene [205-99-2]	Filtrasorb 300	--	--	--	--	5.3/7.4	220	L/g	0.37	mg/L	mg/g	--	--	Dobbs and Cohen, 1980
Benz[k]fluoranthene [207-08-9]	Filtrasorb 300	--	--	--	--	5.3/7.4	57	L/g	0.37	mg/L	mg/g	--	--	Dobbs and Cohen, 1980
Benz[a]anthracene [56-55-3]	Filtrasorb 300	--	--	--	--	5.3/7.4	181	L/g	0.57	mg/L	mg/g	--	--	Dobbs and Cohen, 1980
Benzoic acid [65-85-0]	Filtrasorb 400	--	--	--	--	--	216	L/g	0.50	mg/L	mg/g	--	--	Walters and Luthy, 1982
	Carbon (200 mesh)	--	--	--	--	--	1.15	unitless	0.368	mmol/g	mol/kg	--	--	20 °C, Hartman et al., 1946
	Carbon (200 mesh)	--	--	--	--	--	1.06	unitless	0.368	mmol/g	mol/kg	--	--	30 °C, Hartman et al., 1946
	Carbon (200 mesh)	--	--	--	--	--	0.78	unitless	0.368	mmol/g	mol/kg	--	--	40 °C, Hartman et al., 1946
	Carbon (200 mesh)	--	--	--	--	--	0.56	unitless	0.368	mmol/g	mol/kg	--	--	40 °C, Hartman et al., 1946
	Filtrasorb 300	--	--	--	--	5.3/7.4	0.76	L/g	1.8	mg/L	mg/g	--	--	Dobbs and Cohen, 1980
	Esrum sandy till	78	4	18	9.1	4.71	0.23	L/kg	0.95[d]	mg/L	mg/kg	0.06	2.58	Lokke, 1984
	Gribskov, A-horizon	87	6	3	4.8	3.23	1.17	L/kg	0.93[d]	mg/L	mg/kg	1.41	1.92	Lokke, 1984
	Gribskov, B-horizon	82	4	7	9.6	3.59	1.97	L/kg	0.84[d]	mg/L	mg/kg	2.58	1.88	Lokke, 1984
	Gribskov, C-horizon	88	3	5	7.0	4.07	1.39	L/kg	0.95[d]	mg/L	mg/kg	1.82	1.88	Lokke, 1984
	Roskilde soil	66	18	12	14.0	5.40	1.46	L/kg	0.89[d]	mg/L	mg/kg	1.64	1.95	Lokke, 1984
	Strødam, AB-horizon	84	5	4	13.0	3.88	1.53	L/kg	0.91[d]	mg/L	mg/kg	5.11	1.48	Lokke, 1984
	Strødam, C-horizon	92	3	3	1.6	4.95	0.45	L/kg	0.97[d]	mg/L	mg/kg	0.09	2.70	Lokke, 1984
	Tisvilde, C-horizon	96	1	2	1.3	4.21	0.32	L/kg	0.79[d]	mg/L	mg/kg	0.15	2.32	Lokke, 1984
Benzophenone [119-61-9]	Apison, TN soil	4	10	76	--	4.5	--	unitless	--	--	--	0.11	2.76	Southworth and Keller, 1986
	Dormont, WV soil	2	38	60	129	4.2	--	unitless	--	--	--	1.2	2.64	Southworth and Keller, 1986
	Fullerton, TN soil	11	21	68	64	4.4	--	unitless	--	--	--	0.05	2.72	Southworth and Keller, 1986
Benzo[ghi]perylene [191-24-2]	Filtrasorb 300	--	--	--	--	5.3/7.4	10.7	L/g	0.37	mg/L	mg/g	--	--	Dobbs and Cohen, 1980
Benzo[a]pyrene [50-32-8]	Filtrasorb 300	--	--	--	--	--	33.6	L/g	0.44	mg/L	mg/g	--	--	Dobbs and Cohen, 1980
Benzothiazole [95-16-9]	Filtrasorb 300	--	--	--	--	5.3/7.4	120	L/g	0.27	mg/L	mg/g	--	--	Dobbs and Cohen, 1980
Benzyl alcohol [100-51-6]	Apison, TN soil	4	10	86	76	4.5	--	unitless	--	--	--	0.11	<0.70	Southworth and Keller, 1986
	Dormont, WV soil	2	38	60	129	4.2	--	unitless	--	--	--	1.2	<0.70	Southworth and Keller, 1986
	Fullerton, TN soil	11	21	68	64	4.4	--	unitless	--	--	--	0.05	<0.70	Southworth and Keller, 1986
α-BHC [319-84-9]	Filtrasorb 300	--	--	--	--	5.3/7.4	303	L/g	0.43	mg/L	mg/g	--	--	Dobbs and Cohen, 1980
	Pokkali soil	--	--	--	--	--	70.79	L/kg	0.94	mg/L	μg/g	--	--	Wahid and Sethunathan, 1979

Compound [CAS No.]	Adsorbent	% sand	% silt	% clay	CEC[a]	pH	K_F	K_F units	1/n	C units	x/m units	f_oc %	log K_oc[b]	Reference
β-BHC [319-85-7]	Kari soil	--	--	--	--	--	501.20	L/kg	1.16	mg/L	μg/g	--	--	Wahid and Sethunathan, 1979
	Filtrasorb 300	--	--	--	--	5.3/7.4	220	L/g	0.49	mg/L	mg/g	--	--	Dobbs and Cohen, 1980
	Ca-bentonite	--	--	--	--	--	2.92	mL/g	0.886	μg/mL	μg/g	--	--	20 °C, Mills and Biggar, 1969
	Ca-bentonite	--	--	--	--	--	2.71	mL/g	0.911	μg/mL	μg/g	--	--	20 °C, Mills and Biggar, 1969
	Ca-Staten peaty muck	--	--	--	--	--	456	mL/g	0.950	μg/mL	μg/g	12.76	3.55	20 °C, Mills and Biggar, 1969
	Ca-Staten peaty muck	--	--	--	--	--	437	mL/g	0.990	μg/mL	μg/g	12.76	3.54	20 °C, Mills and Biggar, 1969
	Ca-Venado clay	--	--	--	--	--	62.8	mL/g	0.861	μg/mL	μg/g	3.48	3.12	20 °C, Mills and Biggar, 1969
	Ca-Venado clay	--	--	--	--	--	60.2	mL/g	0.883	μg/mL	μg/g	3.48	3.07	30 °C, Mills and Biggar, 1969
	Pokkali soil	--	--	--	--	--	79.43	L/kg	0.80	mg/L	μg/g	--	--	Wahid and Sethunathan, 1979
	Kari soil	--	--	--	--	--	158.50	L/kg	0.80	mg/L	μg/g	--	--	Wahid and Sethunathan, 1979
	Silica gel	--	--	--	--	--	2.13	mL/g	0.971	μg/mL	μg/g	--	--	19.8 °C, Mills and Biggar, 1969
	Silica gel	--	--	--	--	--	1.64	mL/g	0.985	μg/mL	μg/g	--	--	30 °C, Mills and Biggar, 1969
Biphenyl [92-52-4]	Apison, TN soil	4	10	86	76	4.5	--	unitless	--	--	--	0.11	3.52	Southworth and Keller, 1986
	Clay loam	--	--	--	12.4	5.91	15.6	L/g	1.00	mg/L	mg/g	1.42	3.04	Kishi et al., 1990
	Clay loam	--	--	--	35.0	4.89	124	L/g	0.775	mg/L	mg/g	10.40	3.08	Kishi et al., 1990
	Dormont, WV soil	2	38	60	129	4.2	--	unitless	--	--	--	1.2	2.94	Southworth and Keller, 1986
	Fullerton, TN soil	11	21	68	64	4.4	--	unitless	--	--	--	0.05	2.95	Southworth and Keller, 1986
	Light clay	--	--	--	13.2	5.18	31.8	L/g	1.00	mg/L	mg/g	1.51	3.32	Kishi et al., 1990
	Light clay	--	--	--	28.3	5.26	56.8	L/g	0.98	mg/L	mg/g	3.23	3.26	Kishi et al., 1990
	Sandy loam	--	--	--	35.0	5.41	90.0	L/g	1.00	mg/L	mg/g	7.91	3.04	Kishi et al., 1990
4-Biphenylmethanol [3597-91-9]	Apison, TN soil	4	10	86	76	4.5	--	unitless	--	--	--	0.11	2.85	Southworth and Keller, 1986
	Dormont, WV soil	2	38	60	129	4.2	--	unitless	--	--	--	1.2	2.69	Southworth and Keller, 1986
	Fullerton, TN soil	11	21	68	64	4.4	--	unitless	--	--	--	0.05	2.02	Southworth and Keller, 1986
2,2'-Biquinoline [119-91-5]	River sediment	3.0	41.8	55.2	23.72	7.79	61	mL/g	--	μg/mL	μg/g	2.07	4.75	Banwart et al., 1982
	River sediment	0.3	31.2	68.6	33.01	7.83	172	mL/g		μg/mL	μg/g	0.72	4.38	Banwart et al., 1982
	River sediment	82.4	10.7	6.8	3.72	8.32	57	mL/g		μg/mL	μg/g	0.15	4.58	Banwart et al., 1982
	Iowa sediment	7.1	75.6	17.4	12.40	8.34	61	mL/g		μg/mL	μg/g	0.11	4.75	Banwart et al., 1982
	WV sediment	2.1	34.4	63.6	18.86	4.45	88	mL/g		μg/mL	μg/g	0.48	4.26	Banwart et al., 1982
	Ohio River sediment	15.6	48.7	35.7	15.43	7.76	135	mL/g		μg/mL	μg/g	0.95	4.15	Banwart et al., 1982
	River sediment	34.6	25.8	39.5	15.43	7.76	106	mL/g		μg/mL	μg/g	0.66	4.21	Banwart et al., 1982
	Illinois soil	0.0	71.4	28.6	8.50	5.50	93	mL/g		μg/mL	μg/g	1.30	3.85	Banwart et al., 1982
	Illinois river sediment	50.2	42.7	7.1	8.33	7.60	216	mL/g		μg/mL	μg/g	1.88	4.06	Banwart et al., 1982
	Illinois river sediment	26.2	52.7	21.2	23.72	7.55	181	mL/g		μg/mL	μg/g	1.67	4.03	Banwart et al., 1982
	Illinois river sediment	17.3	13.6	69.1	31.15	6.70	243	mL/g		μg/mL	μg/g	2.38	4.01	Banwart et al., 1982
	River sediment	1.6	55.4	42.9	20.86	7.75	180	mL/g		μg/mL	μg/g	1.48	4.09	Banwart et al., 1982
	Stream sediment	67.6	13.9	18.6	3.72	6.35	105	mL/g		μg/mL	μg/g	1.21	3.94	Banwart et al., 1982
Bis(2-chloroethoxy)methane [111-91-1]	Filtrasorb 300	--	--	--	--	5.3/7.4	11	L/g	0.65	mg/L	mg/g	--	--	Dobbs and Cohen, 1980

Chemical [CAS]	Sorbent	Sand	Silt	Clay	OC/CEC	pH	value	units	value	units	value	units	1/n	Reference
Bis(2-chloroethyl) ether [111-44-4]														
Bis(2-ethylhexyl) phthalate [117-81-7]	Filtrasorb 300					5.3	0.086	L/g	1.84	mg/L	—	mg/g	—	Dobbs and Cohen, 1980
Boron [7440-42-8]	Filtrasorb 300					5.3	1.13×10^{-1}	L/g	1.5	μM	—	μmol/g	—	Dobbs and Cohen, 1980
	Soil			3.4	1.6	7.03	0.087	L/g	0.935	μM	0.17	μmol/g	—	Elrashidi and O'Connor, 1982
	Soil			5.0	6.2	8.00	0.421	L/g	1.19	μM	0.02	μmol/g	—	Elrashidi and O'Connor, 1982
	Soil			5.6	7.8	7.82	0.125	L/g	0.947	μM	0.04	μmol/g	—	Elrashidi and O'Connor, 1982
	Soil			7.7	8.1	7.89	0.162	L/g	0.843	μM	0.04	μmol/g	—	Elrashidi and O'Connor, 1982
	Soil			10	5.5	6.02	0.409	L/g	0.666	μM	0.45	μmol/g	—	Elrashidi and O'Connor, 1982
	Soil			13.7	14.0	7.42	2.16	L/g	0.645	μM	0.43	μmol/g	—	Elrashidi and O'Connor, 1982
	Soil			27.3	18.5	7.62	2.53	L/g	0.618	μM	1.10	μmol/g	—	Elrashidi and O'Connor, 1982
	Soil			25	16.2	6.02	1.93	L/g	0.644	μM	1.00	μmol/g	—	Elrashidi and O'Connor, 1982
	Soil			57	35.2	7.57	3.99	L/g	0.572	μM	0.97	μmol/g	—	Elrashidi and O'Connor, 1982
Boron [7440-42-8], as BO₃.	Alligator soil	5.9	39.4	54.7	30.2[c]	4.8	1.49	L/kg	0.363[d]	mg/L	1.54	mg/kg	1.99	Buchter et al., 1989
	Calciorthid sediment	70.0	19.3	10.7	14.7[c]	8.5	0.85	L/kg	0.787[d]	mg/L	0.44	mg/kg	2.29	Buchter et al., 1989
	Molokai soil	25.7	46.2	28.2	11.0[c]	6.0	1.39	L/kg	0.518[d]	mg/L	1.67	mg/kg	1.92	Buchter et al., 1989
	Webster soil	27.5	48.6	23.9	48.1[c]	7.6	1.60	L/kg	0.641[d]	mg/L	4.39	mg/kg	1.56	Buchter et al., 1989
Bromacil [314-40-9]	Adkins loamy sand	84	13	3		7.3	0.51	Lkg	0.91[d]	μmol/L	0.40	μmol/kg	2.11	4 °C, Madhun et al., 1986
	Adkins loamy sand	84	13	3		7.3	0.30	L/kg	0.97[d]	μmol/L	0.40	μmol/kg	1.88	25 °C, Madhun et al., 1986
	Basinger fine sand			1.9	1.52[c]	5.75	0.57	mL/g			0.61	—	1.97	Reddy et al., 1992
	Bet Dagan I soil			13.7		7.9	0.10	mL/g			0.40	—	1.40	Gerstl and Yaron, 1983
	Bet Dagan II soil			42.5		7.8	0.32	mL/g			1.01	—	1.51	Gerstl and Yaron, 1983
	Boca fine sand			3.5	10.79[c]	7.05	0.76	mL/g			1.67	—	1.66	Reddy et al., 1992
	Chobee fsl			16.1	8.10[c]	7.18	0.76	mL/g			1.39	—	1.74	Reddy et al., 1992
	Gilat soil			23.1		7.8	0.16	mL/g			0.55	—	1.46	Gerstl and Yaron, 1983
	Holopaw fine sand			1.3	2.03[c]	6.10	0.33	mL/g			0.50	—	1.82	Reddy et al., 1992
	Hula-2 peat				74.0[c]	6.9	5.05	L/kg	1.04	mg/L	7.85	mg/kg	1.81	Angemar et al., 1984
	Hula-1 peat				95.0[c]	6.3	19.75	L/kg	0.93	mg/L	29.88	mg/kg	1.82	Angemar et al., 1984
	Mivtachim soil			7.5		8.5	0.03	mL/g			0.06	—	1.70	Gerstl and Yaron, 1983
	Oxidized Hula-2 peat				30.0[c]	6.9	1.71	L/kg	1.05	mg/L	3.08	mg/kg	1.74	Angemar et al., 1984
	Pineda fine sand			0.5	8.40[c]	7.11	0.57	mL/g			1.22	—	1.67	Reddy et al., 1992
	Neve Ya'ar soil			70.0		7.7	0.39	mL/g			1.18	—	1.52	Gerstl and Yaron, 1983
	Newe Ya'ar loess	20	25	55		7.3	1.12	L/kg	0.78	mg/L	1.22	mg/kg	1.95	Angemar et al., 1984
	Riviera fine sand			5.1	4.11[c]	6.20	0.47	mL/g			0.94	—	1.70	Reddy et al., 1992
	Sa'ad sal	36	31	33	18.0[c]	7.6	0.63	L/kg	0.79	mg/L	0.56	mg/kg	2.05	Angemar et al., 1984
	Semiahmoo mucky peat	19	65	16		5.4	26.73	L/kg	0.86[d]	μmol/L	27.8	μmol/kg	1.98	4 °C, Madhun et al., 1986
	Semiahmoo mucky peat	19	65	16		5.4	21.33	L/kg	0.89[d]	μmol/L	27.8	μmol/kg	1.89	25 °C, Madhun et al., 1986
	Shefer soil			70.0		7.2	0.24	mL/g			0.72	—	1.52	Gerstl and Yaron, 1983
	Wabasso sand			2.4	2.54[c]	6.62	0.57	mL/g			0.61	—	1.97	Reddy et al., 1992
4-Bromoaniline [106-40-1]	Batcombe sl						—						1.96	Briggs, 1981
Bromodichloromethane [75-27-4]	Filtrasorb 300					5.3	7.9	L/g	0.61	mg/L	—	mg/g	—	Dobbs and Cohen, 1980

Compound [CAS No.]	Adsorbent	% sand	% silt	% clay	CEC[a]	pH	K_F	K_F units	1/n	C units	x/m units	f_{oc} %	log K_{oc}[b]	Reference
Bromoform [75-25-2]	Filtrasorb 300	--	--	--	--	5.3/7.4	19.6	L/g	0.52	mg/L	mg/g	--	--	Dobbs and Cohen, 1980
	Filtrasorb F-400	--	--	--	--	6.0	1,802	L/g	0.56	μg/L	μg/g	--	--	10-20 °C, Crittenden et al., 1985
	Filtrasorb F-400	--	--	--	--	6.0	436.6	L/g	0.69	μg/L	μg/g	--	--	20-22 °C, Crittenden et al., 1985
	Hydrodarco 3000	--	--	--	--	6.0	632	L/g	0.56	μg/L	μg/g	--	--	20-22 °C, Crittenden et al., 1985
	Westvaco WV-W	--	--	--	--	6.0	475	L/g	0.65	μg/L	μg/g	--	--	20-22 °C, Crittenden et al., 1985
4-Bromo-3-methylaniline [6933-10-4]	Batcombe sl	--	--	--	--	5.3/7.4	--	--	--	--	--	--	1.96	Briggs, 1981
4-Bromo-3-methylphenyl urea [78508-46-0]	Batcombe sl	--	--	--	--	6.0	--	--	--	--	--	--	2.37	Briggs, 1981
4-Bromonitrobenzene [586-78-7]	Batcombe sl	--	--	--	--	6.0	--	--	--	--	--	--	2.42	Briggs, 1981
4-Bromophenol [106-41-2]	Batcombe sl	--	--	--	--	6.0	--	--	--	--	--	1.46	2.41	Briggs, 1981
	Filtrasorb 400	--	--	--	--	--	0.130	L/mg	0.225	mg/L	mg/g	--	--	Sheindorf et al., 1982
4-Bromophenyl phenyl ether [101-55-3]	Filtrasorb 300	--	--	--	--	5.3/7.4	144	L/g	0.68	mg/L	mg/g	--	--	Dobbs and Cohen, 1980
3-Bromophenyl urea [2989-98-2]	Batcombe sl	--	--	--	--	--	--	--	--	--	--	--	2.06	Briggs, 1981
4-Bromophenyl urea [1967-25-5]	Batcombe sl	--	--	--	--	5.3/7.4	--	--	--	--	--	--	2.12	Briggs, 1981
5-Bromouracil [51-20-7]	Filtrasorb 300	--	--	--	--	5.3/7.4	44	L/g	0.47	mg/L	mg/g	--	--	Dobbs and Cohen, 1980
1-Butanol [71-36-3]	Dowex-1 resin	--	--	--	--	--	0.94	unitless	--	--	--	--	--	Small and Bremer, 1964
	Dowex-1-X2 resin	--	--	--	--	--	4.43	unitless	--	--	--	--	--	Small and Bremer, 1964
	Fly ash	--	--	--	--	--	15.7	L/kg	0.71	mg/L	μg/g	--	--	Banerjee et al., 1988
	GAC from wood	--	--	--	--	--	0.547	L/g	0.638	mg/L	mg/g	--	--	Abe et al., 1983
	GAC from coal	--	--	--	--	--	3.20	L/g	0.509	mg/L	mg/g	--	--	Abe et al., 1983
	GAC from coconut shell	--	--	--	--	--	8.12	L/g	0.393	mg/L	mg/g	--	--	Abe et al., 1983
2-Butanol [78-92-2]	GAC from coal	--	--	--	--	--	2.49	L/g	0.490	mg/L	mg/g	--	--	Abe et al., 1983
2-Butanone [78-93-3]	Filtrasorb 400	--	--	--	--	8.0	2.530	L/g	0.295	μg/L	μg/g	--	--	Speth and Miltner, 1998
tert-Butyl alcohol [75-65-0]	GAC from wood	--	--	--	--	--	0.077	L/g	0.813	mg/L	mg/g	--	--	Abe et al., 1983
	GAC from coal	--	--	--	--	--	1.48	L/g	0.471	mg/L	mg/g	--	--	Abe et al., 1983
	GAC from coconut	--	--	--	--	--	0.97	L/g	0.601	mg/L	mg/g	--	--	Abe et al., 1983
Butyl benzyl phthalate [85-68-7]	Filtrasorb 300	--	--	--	--	5.3/7.4	1,520	L/g	1.26	mg/L	mg/g	--	--	Dobbs and Cohen, 1980
2-Butylphenol [3180-09-4]	Filtrasorb 400	--	--	--	--	7	192	L/g	0.149	mg/L	mg/g	--	--	Belfort et al., 1983
Butyraldehyde [123-72-8]	Fly ash	--	--	--	--	--	65.3	L/kg	0.56	mg/L	μg/g	--	--	Banerjee et al., 1988
n-Butyric acid [107-92-6]	Poly(4-vinyl pyridine)	--	--	--	--	--	0.042	L/g	0.895	mg/L	mg/g	--	--	Chanda et al., 1985
	Polybenzimidazole	--	--	--	--	--	1.14	L/g	0.606	mg/L	mg/g	--	--	Chanda et al., 1985
Cadmium [7440-43-9], as Cd^{2+}	Alligator soil	5.9	39.4	54.7	30.2[c]	4.8	52.5	L/kg	0.902[d]	mg/L	mg/kg	1.54	3.53	Buchter et al., 1989
	Calciorthid sediment	70.0	19.3	10.7	14.7[c]	8.5	288	L/kg	0.568[d]	mg/L	mg/kg	0.44	4.82	Buchter et al., 1989

Compound	Medium													Reference
	Cecil soil	78.8	12.9	8.3	2.0[c]	5.7	13.9	L/kg	0.768[d]	mg/L	mg/kg	0.61	3.36	Buchter et al., 1989
	Lafitte soil	60.7	21.7	17.6	26.9[c]	3.9	52.7	L/kg	0.850[d]	mg/L	mg/kg	11.6	2.66	Buchter et al., 1989
	Molokai soil	25.7	46.2	28.2	11.0[c]	6.0	91.2	L/kg	0.773[d]	mg/L	mg/kg	1.67	3.74	Buchter et al., 1989
	Norwood soil	79.2	18.1	2.8	4.1[c]	6.9	28.8	L/kg	0.668[d]	mg/L	mg/kg	0.21	4.14	Buchter et al., 1989
	Olivier soil	4.4	89.4	6.2	8.6[c]	6.6	98.0	L/kg	0.658[d]	mg/L	mg/kg	0.83	4.07	Buchter et al., 1989
	Soil	--	--	--	33.8	5.2	6.8	L/g	0.66	μmol/L	μmol/g	9.45	1.86	Garcia-Miragaya, 1980
	Soil	--	--	--	60.0	8.4	23.2	L/g	0.72	μmol/L	μmol/g	0.42	3.74	Garcia-Miragaya, 1980
	Soil	--	--	--	23.8	5.8	3.6	L/g	0.63	μmol/L	μmol/g	0.87	2.82	Garcia-Miragaya, 1980
	Soil	--	--	--	25.0	6.0	8.3	L/g	0.95	μmol/L	μmol/g	1.04	2.90	Garcia-Miragaya, 1980
	Spodosol sediment	90.2	6.0	3.8	2.7[c]	4.3	5.47	L/kg	0.840[d]	mg/L	mg/kg	1.98	2.44	Buchter et al., 1989
	Webster soil	27.5	48.6	23.9	48.1[c]	7.6	755	L/kg	0.569[d]	mg/L	mg/kg	4.39	4.24	Buchter et al., 1989
	Windsor soil	76.8	20.5	2.8	2.0[c]	5.3	14.4	L/kg	0.782[d]	mg/L	mg/kg	2.03	2.85	Buchter et al., 1989
Captafol [2425-06-1]	Batcombe sl	--	--	--	--	--	--	--	--	--	--	2.05	3.32	Briggs, 1981
Captan [133-06-2]	Batcombe sl	--	--	--	--	--	--	--	--	--	--	2.05	2.30	Briggs, 1981
Carbaryl [63-25-2]	Batcombe sl	--	--	--	--	--	--	--	--	--	--	2.05	2.02	Briggs, 1981
	Ca-bentonite	--	--	--	--	--	0.100	L/g	2.597	μmol/L	μmol/g	--	--	5 °C, Aly et al., 1980
	Ca-bentonite	--	--	--	--	--	0.050	L/g	2.538	μmol/L	μmol/g	--	--	18 °C, Aly et al., 1980
	Ca-bentonite	--	--	--	--	--	0.023	L/g	2.545	μmol/L	μmol/g	--	--	30 °C, Aly et al., 1980
	Beverly sal	56	30	14	--	6.8	4	mL/g	0.98	pmol/mL	pmol/g	1.45	2.44	Sharom et al., 1980
	Big Creek sediment	71	22	7	--	6.6	8	mL/g	0.96	pmol/mL	pmol/g	1.62	2.69	Sharom et al., 1980
	Borg El-Arab	51.6	21.6	27.0	11.6	7.95	1.000	L/g	1.072	μmol/L	μmol/g	0.24	2.62	18 °C, Aly et al., 1980
	Borg El-Arab	51.6	21.6	27.0	11.6	7.95	0.855	L/g	1.120	μmol/L	μmol/g	0.24	2.55	30°C, Aly et al., 1980
	Organic soil	52	34	14	--	6.1	173	mL/g	0.97	pmol/mL	pmol/g	43.67	2.60	Sharom et al., 1980
	Plainfield sand	91.5	1.5	7	--	7.0	1	mL/g	1.08	pmol/mL	pmol/g	0.41	2.39	Sharom et al., 1980
Carbofuran [1563-66-2]	Filtrasorb 400	--	--	--	--	--	275	mL/g	0.41	pmol/mL	pmol/g	--	--	Miltner et al., 1989
	Beverly sal	56	30	14	--	6.8	1.6	mL/g	1.07	pmol/mL	pmol/g	1.45	2.04	Sharom et al., 1980
	Big Creek sediment	71	22	7	--	6.6	2	mL/g	0.98	pmol/mL	pmol/g	1.62	2.09	Sharom et al., 1980
	Organic soil	52	34	14	--	6.1	27	mL/g	1.08	pmol/mL	pmol/g	43.67	1.79	Sharom et al., 1980
	Plainfield sand	91.5	1.5	7	--	7.0	0.1	mL/g	0.88	pmol/mL	pmol/g	0.41	1.39	Sharom et al., 1980
Carbon tetrachloride [56-23-5]	Filtrasorb 300	--	--	--	--	5.3	11.1	L/g	0.83	mg/L	mg/g	--	--	Dobbs and Cohen, 1980
	Filtrasorb 400	--	--	--	--	--	6.80	L/g	0.469	mg/L	mg/g	--	--	Amy et al., 1987
	Filtrasorb 400	--	--	--	--	7	0.23	L/mg	0.74	μg/L	μg/g	--	--	Weber and Pirbazari, 1981
	Hydrodarco 1030	--	--	--	--	7	0.13	L/mg	0.68	μg/L	μg/g	--	--	Weber and Pirbazari, 1981
	Leie River clay	0	0	100	--	--	6.47	L/kg	--	--	--	4.12	2.20	2.3 °C, Dewulf et al., 1999
	Leie River clay	0	0	100	--	--	7.15	L/kg	--	--	--	4.12	2.24	3.8 °C, Dewulf et al., 1999
	Leie River clay	0	0	100	--	--	7.10	L/kg	--	--	--	4.12	2.24	6.2 °C, Dewulf et al., 1999
	Leie River clay	0	0	100	--	--	7.33	L/kg	--	--	--	4.12	2.25	8.0 °C, Dewulf et al., 1999
	Leie River clay	0	0	100	--	--	7.83	L/kg	--	--	--	4.12	2.28	13.5 °C, Dewulf et al., 1999
	Leie River clay	0	0	100	--	--	7.63	L/kg	--	--	--	4.12	2.27	18.6 °C, Dewulf et al., 1999
	Leie River clay	0	0	100	--	--	8.87	L/kg	--	--	--	4.12	2.33	25.0 °C, Dewulf et al., 1999
	Michawye soil	--	--	--	--	--	0.542	--	1.04	μg/L	--	0.13	2.62	Chin et al., 1988

Compound [CAS No.]	Adsorbent	% sand	% silt	% clay	CEC[a]	pH	K_F	K_F units	1/n	C units	x/m units	f_{oc} %	log K_{oc}[b]	Reference
Chloral hydrate [302-17-0]	Norit carbon	--	--	--	--	7	0.16	L/mg	0.75	μg/L	mg/g	--	--	Weber and Pirbazari, 1981
	Nuchar WV-G	--	--	--	--	7	0.22	L/mg	0.69	μg/L	mg/g	--	--	Weber and Pirbazari, 1981
Chloramben [133-90-4]	Filtrasorb 400	--	--	--	--	5.8	18.900	L/g	0.051	μg/L	μg/g	--	--	Speth and Miltner, 1998
	Anselmo sal	--	--	--	7.0	7.0	0.10	mL/g	--	μg/mL	μg/g	0.098	1.01	Schliebe et al., 1965
	Fargo clay	--	--	--	30.1	7.9	0.30	mL/g	--	μg/mL	μg/g	4.56	0.82	Schliebe et al., 1965
	Keith sal	--	--	--	11.6	6.2	0.30	mL/g	--	μg/mL	μg/g	1.67	1.25	Schliebe et al., 1965
	Monona sal	--	--	--	17.5	5.8	0.50	mL/g	--	μg/mL	μg/g	2.42	1.32	Schliebe et al., 1965
Chloranocryl [2164-09-2]	H-montmorillonite	--	--	--	73.5	9.0	30	L/g	0.62	μmol/L	μmol g	--	--	Bailey et al., 1968
Chlorbromuron [13360-45-7]	Batcombe sl	--	--	--	--	--	--	--	--	--	--	--	2.58	Briggs, 1981
Chlordane [57-74-9]	Filtrasorb 300	--	--	--	--	5.3/7.4	245	L/g	--	--	--	--	--	Dobbs and Cohen, 1980
cis-Chlordane [5103-74-2]	Lucustrine sediment	--	--	--	--	--	5,216	--	1.00	mg/L	mg/g	1.42	5.57	Chin et al., 1988
	Michawye soil	--	--	--	--	--	182.7	--	0.93	μg/L	--	0.13	5.15	Chin et al., 1988
Chlorfenvinphos [470-90-6]	Batcombe sl	--	--	--	--	--	--	--	--	--	--	2.05	2.47	Briggs, 1981
4-Chloroaniline [106-47-8]	Alfisol	12.9	64.3	19.6	--	7.45	9.5	mL/g	1.19	μg/mL	μg/g	0.76	3.10	Rippen et al., 1982
	Alumina	--	--	--	--	--	8.8	mL/g	0.64	μg/mL	μg/g	--	--	Rippen et al., 1982
	Cellulose	--	--	--	--	--	18	mL/g	1.10	μg/mL	μg/g	--	--	Rippen et al., 1982
	Entisol	8.5	68.3	20.6	--	7.9	17	mL/g	0.92	μg/mL	μg/g	1.11	3.18	Rippen et al., 1982
	Fuller's earth	--	--	--	--	--	2.5	mL/g	1.44	μg/mL	μg/g	--	--	Rippen et al., 1982
	Silica gel	--	--	--	--	--	2.6	mL/g	1.09	μg/mL	μg/g	--	--	Rippen et al., 1982
	Speyer soil 2.1	--	--	--	--	7.0	9.3	mL/g	1.23	μg/mL	μg/g	0.69	3.13	Rippen et al., 1982
	Speyer soil 2.2	--	--	--	--	5.8	2.5	mL/g	1.10	μg/mL	μg/g	2.24	2.05	Rippen et al., 1982
	Spodosol	81.5	10.0	7.2	--	3.88	3.4	mL/g	1.15	μg/mL	μg/g	3.56	1.98	Rippen et al., 1982
3-Chloro-p-anisidine [5345-54-0]	Batcombe sl	--	--	--	--	--	--	--	--	--	--	--	1.93	Briggs, 1981
Chlorobenzene [108-90-7]	Filtrasorb 300	--	--	--	--	7.4	91	L/g	0.99	mg/L	mg/g	--	--	Dobbs and Cohen, 1980
	KS1 sediment	--	--	--	--	--	1.2	mL/g	--	--	--	0.73	2.22	Schwarzenbach and Westall, 1981
	KS1H sediment	--	--	--	--	--	0.4	mL/g	--	--	--	0.08	2.70	Schwarzenbach and Westall, 1981
	Kaolin	--	--	--	--	--	0.6	mL/g	--	--	--	0.06	3.00	Schwarzenbach and Westall, 1981
	γ-Alumina	--	--	--	--	--	0.6	mL/g	--	--	--	<0.01	--	Schwarzenbach and Westall, 1981
	Silica	--	--	--	--	--	4.2	mL/g	--	--	--	<0.01	--	Schwarzenbach and Westall, 1981
o-Chlorobenzoic acid [118-91-2]	Carbon (200 mesh)	--	--	--	--	--	1.66	unitless	0.406	mmol/g	mol/kg	--	--	20 °C, Hartman et al., 1946
	Carbon (200 mesh)	--	--	--	--	--	1.53	unitless	0.406	mmol/g	mol/kg	--	--	30 °C, Hartman et al., 1946
	Carbon (200 mesh)	--	--	--	--	--	1.35	unitless	0.406	mmol/g	mol/kg	--	--	40 °C, Hartman et al., 1946
	Carbon (200 mesh)	--	--	--	--	--	1.24	unitless	0.406	mmol/g	mol/kg	--	--	50 °C, Hartman et al., 1946
2-Chlorobiphenyl [2051-60-7]	MLSS	--	--	--	--	--	20.5	L/g	0.8	μg/L	μg/g	--	--	20 °C, Bell and Tsezos, 1987
p-Chloro-m-cresol [59-50-7]	Filtrasorb 300	--	--	--	--	5.3/7.4	124	L/g	0.16	mg/L	mg/g	--	--	Dobbs and Cohen, 1980
Chlorodibromomethane [124-48-1]	Filtrasorb 300	--	--	--	--	5.3/7.4	4.8	L/g	0.34	mg/L	mg/g	--	--	Dobbs and Cohen, 1980

Compound	Adsorbent/Soil				pH	K		C		q		1/n	Reference
Chloroethane [75-00-3]	Filtrasorb F-400	--	--	--	6.0	1,266	L/g	0.52	µg/L	--	µg/g	--	10-20 °C, Crittenden et al., 1985
	Hydrodarco 3000	--	--	--	6.0	281	L/g	0.59	µg/L	--	µg/g	--	20-22 °C, Crittenden et al., 1985
	Filtrasorb 300	--	--	--	5.3	0.59	L/g	0.95	mg/L	--	mg/g	--	Dobbs and Cohen, 1980
2-Chloroethyl vinyl ether [110-75-8]		--	--	--	--	--	--	--	--	--	--	--	--
Chloroform [67-66-3]	Filtrasorb 300	--	--	--	5.3	3.9	L/g	0.80	mg/L	--	mg/g	--	Dobbs and Cohen, 1980
	Ambersorb XE-340	--	--	--	--	18.2	L/g	0.81	mg/L	--	mg/g	--	Oulman, 1981
	Filtrasorb 300	--	--	--	5.3	2.6	L/g	0.73	mg/L	--	mg/g	--	Dobbs and Cohen, 1980
	Filtrasorb F-400	--	--	--	6.0	285	L/g	0.532	µg/L	--	µg/g	--	10-20 °C, Crittenden et al., 1985
	Filtrasorb F-400	--	--	--	6.0	39.2	L/g	0.756	µg/L	--	µg/g	--	20-22 °C, Crittenden et al., 1985
	Hydrodarco 3000	--	--	--	6.0	92.5	L/g	0.67	µg/L	--	µg/g	--	20-22 °C, Crittenden et al., 1985
	Westvaco WV-W	--	--	--	6.0	55.7	L/g	0.738	µg/L	--	µg/g	--	20-22 °C, Crittenden et al., 1985
	Leie River clay	0	100	--	--	1.74	L/kg	--	--	4.12	--	1.63	2.3 °C, Dewulf et al., 1999
	Leie River clay	0	100	--	--	1.84	L/kg	--	--	4.12	--	1.65	3.8 °C, Dewulf et al., 1999
	Leie River clay	0	100	--	--	1.74	L/kg	--	--	4.12	--	1.63	6.2 °C, Dewulf et al., 1999
	Leie River clay	0	100	--	--	1.88	L/kg	--	--	4.12	--	1.66	8.0 °C, Dewulf et al., 1999
	Leie River clay	0	100	--	--	2.00	L/kg	--	--	4.12	--	1.69	13.5 °C, Dewulf et al., 1999
	Leie River clay	0	100	--	--	1.85	L/kg	--	--	4.12	--	1.65	18.6 °C, Dewulf et al., 1999
	Leie River clay	0	100	--	--	2.08	L/kg	--	--	4.12	--	1.70	25.0 °C, Dewulf et al., 1999
3-Chloro-4-bromonitrobenzene [29682-39-1]	Batcombe sl	--	--	--	--	--	--	--	--	--	--	2.60	Briggs, 1981
3-Chloro-4-methoxyphenyl urea [25277-05-8]	Batcombe sl	--	--	--	--	--	--	--	--	--	--	2.00	Briggs, 1981
2-Chloronaphthalene [91-58-7]	Filtrasorb 300	--	--	--	5.3/7.4	280	L/g	0.46	mg/L	--	mg/g	--	Dobbs and Cohen, 1980
2-Chloronitrobenzene [88-73-3]	Filtrasorb 300	--	--	--	5.3/7.4	130	L/g	0.46	mg/L	--	mg/g	--	Dobbs and Cohen, 1980
2-Chlorophenol [95-57-8]	Brookston clay loam	--	--	22.22	5.7	1.37	mL/g	0.80	µmol/mL	2.96	µmol/g	1.71	Boyd, 1982
	Filtrasorb 300	--	--	--	4.8-5.6	8	L/g	0.30	mg/L	--	mg/g	--	Knettig et al., 1986
	Filtrasorb 300	--	--	--	5.3/7.4	51.0	L/g	0.41	mg/L	--	mg/g	--	Dobbs and Cohen, 1980
	Filtrasorb 400	--	--	--	7	240	L/g	0.098	mg/L	--	mg/g	--	Peel and Benedek, 1980
3-Chlorophenol [108-43-0]	Brookston clay loam	--	--	22.22	5.7	1.78	mL/g	0.83	µmol/mL	2.96	µmol/g	1.82	Boyd, 1982
	Kootwijk humic sand	--	--	--	3.4	7.9	L/kg	1.25	µg/L	0.017	µg/kg	--	Lagas, 1988
	Holten humic-rich sand	--	--	--	4.7	31.6	L/kg	1.25	µg/L	0.032	µg/kg	--	Lagas, 1988
	Maasdijk light loam	--	--	--	7.5	12.6	L/kg	1.67	µg/L	0.009	µg/kg	--	Lagas, 1988
	Opjinen heavy loam	--	--	--	7.1	10.0	L/kg	1.43	µg/L	0.017	µg/kg	--	Lagas, 1988
	Rolde humic sand	--	--	--	4.9	15.9	L/kg	1.25	µg/L	0.022	µg/kg	--	Lagas, 1988
	Schipluiden peat	--	--	--	4.6	316.2	L/kg	1.43	µg/L	0.298	µg/kg	--	Lagas, 1988
4-Chlorophenol [106-48-9]	Amberlite XAD-2	--	--	--	--	1.98	L/g	0.65	mg/L	--	mg/g	--	Aguwa et al., 1984
	Amberlite XAD-4	--	--	--	--	10.50	L/g	0.49	mg/L	--	mg/g	--	Aguwa et al., 1984
	Brookston clay loam	--	--	22.22	5.7	1.88	mL/g	0.70	µmol/mL	2.96	µmol/g	1.85	Boyd, 1982
	Filtrasorb 400	--	--	--	--	126	dm³/g	0.25	mg/dm³	--	mg/g	--	McKay et al., 1985
	Poly(4-vinyl pyridine)	--	--	--	--	2.05	L/g	0.850	mg/L	--	mg/g	--	Chanda et al., 1985
	Polybenzimidazole	--	--	--	--	0.20	L/g	0.927	mg/L	--	mg/g	--	Chanda et al., 1985

Compound [CAS No.]	Adsorbent	% sand	% silt	% clay	CEC[a]	pH	K_F	K_F units	$1/n$	C units	x/m units	f_{oc} %	log K_{oc}[b]	Reference
4-Chlorophenyl phenyl ether [7005-72-3]	Filtrasorb 300	--	--	--	--	5.3/7.4	111	L/g	0.26	mg/L	mg/g	--	--	Dobbs and Cohen, 1980
2-Chlorophenyl urea [114-38-5]	Batcombe sl	--	--	--	--		--		--		--	--	1.61	Briggs, 1981
3-Chlorophenyl urea [1967-27-7]	Batcombe sl	--	--	--	--		--		--		--	--	2.01	Briggs, 1981
Chloropicrin [76-06-0]	Filtrasorb 400	--	--	--	--	4.7	30,200	L/g	0.155	μg/L	μg/g	--	--	Speth and Miltner, 1998
Chlorothalonil [1897-45-6]	Cohansey sand	90.0	8.0	2.0	--	--	25	L/kg	--	mg/L	μg/g	2.60	2.98	Reducker et al., 1988
Chlorotoluron [15545-48-9]	Adkins loamy sand	84	13	3	--	7.3	1.47	L/kg	0.86[d]	μmol/L	μmol/kg	0.40	2.56	4 °C, Madhun et al., 1986
	Adkins loamy sand	84	13	3	--	7.3	1.08	L/kg	0.87[d]	μmol/L	μmol/kg	0.40	2.43	25 °C, Madhun et al., 1986
	Batcombe sl	--	--	--	--		--		--		--	--	2.02	Briggs, 1981
5-Chlorouracil [1820-81-1]	Semiahmoo mucky peat	19	65	16	--	5.4	157.4	L/kg	0.85[d]	μmol/L	μmol/kg	27.8	2.75	4 °C, Madhun et al., 1986
	Semiahmoo mucky peat	19	65	16	--	5.4	116.41	L/kg	0.87[d]	μmol/L	μmol/kg	27.8	2.63	25 °C, Madhun et al., 1986
Chlorpropham [101-21-3]	Filtrasorb 300	--	--	--	--	5.3/7.4	25	L/g	0.58	mg/L	mg/g	--	--	Dobbs and Cohen, 1980
	H-montmorillonite	--	--	--	73.5	9.0	30	L/g	0.92	μmol/L	μmolg	--	--	Bailey et al., 1968
	Na-montmorillonite	--	--	--	87.0	9.0	27	L/g	1.08	μmol/L	μmolg	--	--	Bailey et al., 1968
Chlorpyrifos [2921-88-2]	Beverly sal	56	30	14	--	6.8	118	mL/g	0.99	pmol/mL	pmol/g	1.45	3.91	Sharom et al., 1980
	Big Creek sediment	71	22	7	--	6.6	139	mL/g	0.98	pmol/mL	pmol/g	1.62	3.93	Sharom et al., 1980
	Clarion soil	37	42	21	21.02	5.00	161.81	L/kg	0.91	μM/L	μM/kg	2.64	3.79	Felsot and Dahm, 1979
	Harps soil	21	55	24	37.84	7.30	397.19	L/kg	0.98	μM/L	μM/kg	3.80	4.02	Felsot and Dahm, 1979
	Organic soil	52	34	14	--	6.1	1,862	mL/g	1.09	pmol/mL	pmol/g	43.67	3.63	Sharom et al., 1980
	Plainfield sand	91.5	1.5	7	5.71	7.0	18	mL/g	0.98	pmol/mL	pmol/g	0.41	3.65	Sharom et al., 1980
	Sarpy fsl	77	15	8	6.01	7.30	28.28	L/kg	0.86	μM/L	μM/kg	0.51	3.74	Felsot and Dahm, 1979
	Thurman fsl	83	9	8	--	6.83	46.88	L/kg	0.77	μM/L	μM/kg	1.07	3.64	Felsot and Dahm, 1979
Chromium [7440-47-3], as Cr^{3+}	Filtrasorb 400	--	--	--	--	--	2.3	dm³/g	0.44	mg/dm³	mg/g	--	--	McKay et al., 1985
Chromium [7440-47-3], as Cr^{6+}	MLSS	--	--	--	--	--	2.60	L/g	0.524	μg/L	μg/g	--	--	20 °C, Lee et al., 1989a
Chromium [7440-47-3], as CrO_4^{2-}	Alligator soil	5.9	39.4	54.7	30.2[d]	4.8	3.41	L/kg	0.504[d]	mg/L	mg/kg	1.54	2.35	Buchter et al., 1989
	Lafitte soil	60.7	21.7	17.6	26.9[c]	3.9	30.3	L/kg	0.374[d]	mg/L	mg/kg	11.6	2.42	Buchter et al., 1989
	Molokai soil	25.7	46.2	28.2	11.0[c]	6.0	6.41	L/kg	0.607[d]	mg/L	mg/kg	1.67	2.58	Buchter et al., 1989
	Spodosol sediment	90.2	6.0	3.8	2.7[c]	4.3	5.47	L/kg	0.394[d]	mg/L	mg/kg	1.98	2.44	Buchter et al., 1989
	Windsor soil	76.8	20.5	2.8	2.0[c]	5.3	8.47	L/kg	0.521[d]	mg/L	mg/kg	2.03	2.62	Buchter et al., 1989
Chrysene [218-01-9]	Filtrasorb 300	--	--	--	--	5.0	6.07	L/g	0.50	mg/L	mg/g	--	--	Burks, 1981
	Filtrasorb 400	--	--	--	--	--	716	L/g	0.458	mg/L	mg/g	--	--	Walters and Luthy, 1982
Ciprofloxacin [85721-33-1]	Germany soil	80.1	16.7	2.5	12.0	5.3	427.0	mL/g	--	mg/mL	mg/g	0.70	4.79	Nowara et al., 1997
Clopyralid [1702-17-6]	Rolla NT	--	--	--	15.2[c]	5.09	1.00	L/kg	--	--	--	2.37	1.63	Shang and Arshad, 1998
	Rolla CT	--	--	--	12.4[c]	5.82	0.32	L/kg	--	--	--	2.27	1.15	Shang and Arshad, 1998
	Rycroft NT	--	--	--	28.6[c]	4.94	1.57	L/kg	--	--	--	4.70	1.52	Shang and Arshad, 1998
	Rycroft CT	--	--	--	27.7	5.38	2.06	L/kg	--	--	--	4.45	1.67	Shang and Arshad, 1998

Compound	Material													Reference
Cobalt [7440-48-4], as Co²⁺	Alligator soil	5.9	39.4	54.7	30.2[c]	4.8	35.7	L/kg	0.953[d]	mg/L	1.54	mg/kg	3.37	Buchter et al., 1989
	Calciorthid sediment	70.0	19.3	10.7	14.7[c]	8.5	251	L/kg	0.546[d]	mg/L	0.44	mg/kg	4.76	Buchter et al., 1989
	Cecil soil	78.8	12.9	8.3	2.0[c]	5.7	6.56	L/kg	0.745[d]	mg/L	0.61	mg/kg	3.03	Buchter et al., 1989
	Lafitte soil	60.7	21.7	17.6	26.9[c]	3.9	33.9	L/kg	1.009[d]	mg/L	11.6	mg/kg	2.47	Buchter et al., 1989
	Molokai soil	25.7	46.2	28.2	11.0[c]	6.0	92.5	L/kg	0.621[d]	mg/L	1.67	mg/kg	3.74	Buchter et al., 1989
	Norwood soil	79.2	18.1	2.8	4.1[c]	6.9	27.4	L/kg	0.627[d]	mg/L	0.21	mg/kg	4.12	Buchter et al., 1989
	Olivier soil	4.4	89.4	6.2	8.6[c]	6.6	67.0	L/kg	0.584[d]	mg/L	0.83	mg/kg	3.91	Buchter et al., 1989
	Spodosol sediment	90.2	6.0	3.8	2.7[c]	4.3	2.55	L/kg	0.811[d]	mg/L	1.98	mg/kg	2.11	Buchter et al., 1989
	Webster soil	27.5	48.6	23.9	48.1[c]	7.6	363	L/kg	0.782[d]	mg/L	4.39	mg/kg	3.93	Buchter et al., 1989
	Windsor soil	76.8	20.5	2.8	2.0[c]	5.3	6.28	L/kg	0.741[d]	mg/L	2.03	mg/kg	2.49	Buchter et al., 1989
Copper [7440-50-8], as Cu²⁺	Alligator soil	5.9	39.4	54.7	30.2[c]	4.8	258	L/kg	0.544[d]	mg/L	1.54	mg/kg	4.22	Buchter et al., 1989
	Calciorthid sediment	70.0	19.3	10.7	14.7[c]	8.5	2,624	L/kg	1.140[d]	mg/L	0.44	mg/kg	5.78	Buchter et al., 1989
	Cecil soil	78.8	12.9	8.3	2.0[c]	5.7	53.7	L/kg	0.546[d]	mg/L	0.61	mg/kg	3.94	Buchter et al., 1989
	Lafitte soil	60.7	21.7	17.6	26.9[c]	3.9	221	L/kg	0.987[d]	mg/L	11.6	mg/kg	3.28	Buchter et al., 1989
	Molokai soil	25.7	46.2	28.2	11.0[c]	6.0	368	L/kg	0.516[d]	mg/L	1.67	mg/kg	4.34	Buchter et al., 1989
	Norwood soil	79.2	18.1	2.8	4.1[c]	6.9	89.1	L/kg	0.471[d]	mg/L	0.21	mg/kg	4.63	Buchter et al., 1989
	Olivier soil	4.4	89.4	6.2	8.6[c]	6.6	218	L/kg	0.495[d]	mg/L	0.83	mg/kg	4.42	Buchter et al., 1989
	Spodosol sediment	90.2	6.0	3.8	2.7[c]	4.3	56.2	L/kg	0.602[d]	mg/L	1.98	mg/kg	3.45	Buchter et al., 1989
	Webster soil	27.5	48.6	23.9	48.1[c]	7.6	6,353	L/kg	1.420[d]	mg/L	4.39	mg/kg	5.16	Buchter et al., 1989
	Windsor soil	76.8	20.5	2.8	2.0[c]	5.3	77.1	L/kg	0.567[d]	mg/L	2.03	mg/kg	4.13	Buchter et al., 1989
2-Cresol [95-48-7]	Brookston clay loam	–	–	–	22.22	5.7	0.59	µmol/g	0.66	µmol/mL	2.96	µmol/g	1.34	Boyd, 1982
	Filtrasorb 400	–	–	–	–	7	73.4	L/mg	0.188	mg/L	–	mg/g	–	Belfort et al., 1983
	Hydrodarco 3000	–	–	–	–	–	25.9	L/µg	0.199	µg/L	–	mg/g	–	DiGiano et al., 1980
3-Cresol [108-39-4]	Brookston clay loam	–	–	–	22.22	5.7	0.92	µmol/g	0.87	µmol/mL	2.96	µmol/g	1.54	Boyd, 1982
	Filtrasorb 300	–	–	–	–	4.8-5.6	56	mL/g	0.25	mg/L	–	mg/g	–	Knettig et al., 1986
4-Cresol [106-44-5]	Poly(4-vinyl pyridine)	–	–	–	–	–	0.324	L/g	0.948	mg/L	–	mg/g	–	Chanda et al., 1985
	Polybenzimidazole	–	–	–	–	–	0.11	L/g	0.935	mg/L	–	mg/g	–	Chanda et al., 1985
	Apison, TN soil	4	10	86	76	4.5	–	unitless	–	–	0.11	unitless	3.53	Southworth and Keller, 1986
	Brookston clay loam	–	–	–	22.22	5.7	1.31	mL/g	0.68	µmol/mL	2.96	µmol/g	1.69	Boyd, 1982
	Dormont, WV soil	2	38	60	129	4.2	–	unitless	–	–	1.2	unitless	2.06	Southworth and Keller, 1986
	Fullerton, TN soil	11	21	68	64	4.4	–	unitless	–	–	0.05	unitless	3.53	Southworth and Keller, 1986
Cyanazine [21725-46-2]	Hickory Hill silt	–	–	–	–	–	5.98	µg/g	–	µg/mL	3.27	µg/g	2.26	Brown and Flagg, 1981
Cyclohexanol [108-93-0]	GAC from wood	–	–	–	–	–	1.31	L/g	0.619	mg/L	–	mg/g	–	Abe et al., 1983
	GAC from coal	–	–	–	–	–	7.93	L/g	0.445	mg/L	–	mg/g	–	Abe et al., 1983
	GAC from coconut	–	–	–	–	–	15.3	L/g	0.386	mg/L	–	mg/g	–	Abe et al., 1983
Cyclohexanone [108-94-1]	Filtrasorb 300	–	–	–	–	5.3/7.4	6.2	L/g	0.75	mg/L	–	mg/g	–	Dobbs and Cohen, 1980
Cyclopentanol [96-41-3]	GAC from wood	–	–	–	–	–	0.441	L/g	0.677	mg/L	–	mg/g	–	Abe et al., 1983
	GAC from coal	–	–	–	–	–	4.69	L/g	0.406	mg/L	–	mg/g	–	Abe et al., 1983
	GAC from coconut	–	–	–	–	–	5.68	L/g	0.469	mg/L	–	mg/g	–	Abe et al., 1983
Cytosine [71-30-7]	Filtrasorb 300	–	–	–	–	5.3/7.4	1.1	L/g	1.6	mg/L	–	mg/g	–	Dobbs and Cohen, 1980
2,4-D [94-75-7]	Activated charcoal	–	–	–	–	9.4	1,850	µg/g	0.74[d]	µg/g	–	mg/g	–	Grover and Smith, 1974

Compound [CAS No.]	Adsorbent	% sand	% silt	% clay	CEC[a]	pH	K_F	K_F units	1/n	C units	x/m units	f_{oc} %	log K_{oc}[b]	Reference
	Alfisol	12.9	64.3	19.6	--	7.45	1.1	mL/g	0.69	µg/mL	µg/g	0.76	2.16	Rippen et al., 1982
	Alumina	--	--	--	--	--	2.2	mL/g	1.04	µg/mL	µg/g	--	--	Rippen et al., 1982
	Anion-exchange resin	--	--	--	--	3.7	2,450	µg/g	0.99[d]	--	--	--	--	Grover and Smith, 1974
	Bernard-Fagne sl	10.9	61.7	27.4	8.75	3.60	13.18	mL/g	0.88	µg/mL	µg/g	2.42	2.74	Moreale and Van Bladel, 1980
	Bullingen sl	23.8	59.8	6.3	8.23	3.55	1.83	mL/g	0.91	µg/mL	µg/g	2.06	1.95	Moreale and Van Bladel, 1980
	Cation-exchange resin	--	--	--	--	5.4	0.01	µg/g	2.68[d]	--	--	--	--	Grover and Smith, 1974
	Cellulose	--	--	--	--	--	0.6	mL/g	0.92	µg/mL	µg/g	--	--	Rippen et al., 1982
	Cellulose powder	--	--	--	--	5.9	0.68	µg/g	1.17[d]	--	--	--	--	Grover and Smith, 1974
	Cellulose triacetate	--	--	--	--	4.9	13.0	µg/g	1.25[d]	--	--	--	--	Grover and Smith, 1974
	FA-montmorillonite	--	--	100	--	3.5	0.815	--	0.76[d]	ppm	µmol/g	--	--	5 °C, Khan, 1974
	FA-montmorillonite	--	--	100	--	3.5	0.716	--	0.83[d]	ppm	µmol/g	--	--	25 °C, Khan, 1974
	Fleron sicl	3.0	62.9	34.1	12.29	3.75	2.04	mL/g	0.94	µg/mL	µg/g	3.24	1.80	Moreale and Van Bladel, 1980
	Gribskov, A-horizon	87	6	3	4.8	3.23	2.21	L/kg	0.92[d]	mg/L	mg/kg	1.41	2.20	Lokke, 1984
	Gribskov, B-horizon	82	4	7	9.6	3.59	5.45	L/kg	0.91[d]	mg/L	mg/kg	2.58	2.32	Lokke, 1984
	Gribskov, C-horizon	88	3	5	7.0	4.07	2.70	L/kg	0.92[d]	mg/L	mg/kg	1.82	2.18	Lokke, 1984
	Heverlee sal	76.0	17.1	6.9	10.70	5.84	0.85	mL/g	0.92	µg/mL	µg/g	1.45	1.77	Moreale and Van Bladel, 1980
	Kaolinite	--	--	--	--	5.1	0.00	µg/g	--	--	--	--	--	Grover and Smith, 1974
	Lubbeek sand	89.0	5.2	5.8	2.37	6.46	0.09	mL/g	0.88	µg/mL	µg/g	0.07	2.11	Moreale and Van Bladel, 1980
	Lubbeek sand	91.3	2.1	6.6	2.30	6.43	0.05	mL/g	0.88	µg/mL	µg/g	0.02	2.33	Moreale and Van Bladel, 1980
	Lubbeek sal	71.3	18.9	9.8	4.51	6.71	0.31	mL/g	0.91	µg/mL	µg/g	0.53	1.77	Moreale and Van Bladel, 1980
	Lubbeek sl	32.6	55.2	12.2	7.02	6.91	0.43	mL/g	0.91	µg/mL	µg/g	0.65	1.82	Moreale and Van Bladel, 1980
	Lubbeek sl	15.0	69.9	15.1	9.52	6.62	0.77	mL/g	0.92	µg/mL	µg/g	1.15	1.83	Moreale and Van Bladel, 1980
	Meerdael sl	18.5	73.0	8.5	11.74	4.00	7.05	mL/g	0.88	µg/mL	µg/g	3.59	2.29	Moreale and Van Bladel, 1980
	Montmorillonite	--	--	--	--	9.3	0.00	µg/g	--	--	--	--	--	Grover and Smith, 1974
	Nodebais sl	19.3	72.9	7.8	8.40	6.20	0.39	mL/g	0.89	µg/mL	µg/g	0.73	1.73	Moreale and Van Bladel, 1980
	Peat	--	--	--	--	3.4	110	µg/g	1.64[d]	--	--	--	--	Grover and Smith, 1974
	Silica gel	--	--	--	--	7.5	0.40	µg/g	1.86[d]	--	--	--	--	Grover and Smith, 1974
	Soignes sl	31.0	68.5	0.5	16.94	3.40	16.18	mL/g	0.76	µg/mL	µg/g	4.94	2.52	Moreale and Van Bladel, 1980
	Spa sicl	6.6	59.6	33.8	12.01	3.25	23.89	mL/g	0.86	µg/mL	µg/g	1.89	3.10	Moreale and Van Bladel, 1980
	Speyer soil 2.2	--	--	--	--	5.8	2.9	mL/g	0.91	µg/mL	µg/g	2.24	2.11	Rippen et al., 1982
	Spodosol	81.5	10.0	7.2	--	3.88	4.3	mL/g	0.70	µg/mL	µg/g	3.56	2.05	Rippen et al., 1982
	Stavelot sl	8.2	69.7	22.1	5.61	3.90	7.60	mL/g	0.92	µg/mL	µg/g	2.53	2.48	Moreale and Van Bladel, 1980
	Stookrooie loamy sand	83.9	10.3	5.8	2.90	5.64	1.81	mL/g	0.93	µg/mL	µg/g	1.07	2.23	Moreale and Van Bladel, 1980
	Strødam, AB-horizon	84	5	4	13.0	3.88	2.38	L/kg	0.97[d]	mg/L	mg/kg	5.11	1.70	Lokke, 1984
	Tisvilde, C-horizon	96	1	2	1.3	4.21	0.14	L/kg	0.65[d]	mg/L	mg/kg	0.15	1.95	Lokke, 1984
	Zolder sand	94.6	2.1	3.3	1.66	3.84	10.36	mL/g	0.86	µg/mL	µg/g	1.86	2.75	Moreale and Van Bladel, 1980
	Zolder sand	96.6	1.3	2.1	0.68	4.73	0.42	mL/g	0.93	µg/mL	µg/g	0.07	2.78	Moreale and Van Bladel, 1980
	Zolder sand	96.8	2.6	0.6	0.45	4.23	0.95	mL/g	0.88	µg/mL	µg/g	0.19	2.70	Moreale and Van Bladel, 1980

2,4-D ammonium salt [2307-55-3]													
Wheat straw	--	--	--	--	5.6	1.50	μg/g	1.27d	--	--	--	--	Grover and Smith, 1974
Cecil sal	65.8	19.5	14.7	6.8	5.6	0.65	μg/g	0.83d	--	--	0.90	1.86	Rao and Davidson, 1979
Eustis fine sand	93.8	3.0	3.2	5.2	5.6	0.76	--	0.73d	--	--	0.56	2.13	Rao and Davidson, 1979
Webster sicl	18.4	45.3	38.3	54.7	7.3	4.62	--	0.70d	--	--	3.87	2.08	Rao and Davidson, 1979
Weyburn Oxbow loam	53.3	27.5	19.2	--	6.5	0.45	μg/g	1.05d	--	--	3.75	1.08	Grover and Smith, 1974
2,4-D dimethylammonium [3599-58-4]													
Activated charcoal	--	--	--	--	9.4	32,000	μg/g	1.25d	--	--	--	--	Grover and Smith, 1974
Anion-exchange resin	--	--	--	--	3.7	1,350	μg/g	0.87d	--	--	--	--	Grover and Smith, 1974
Cation-exchange resin	--	--	--	--	5.7	0.16	μg/g	1.25d	--	--	--	--	Grover and Smith, 1974
Cellulose powder	--	--	--	--	5.9	0.01	μg/g	2.05d	--	--	--	--	Grover and Smith, 1974
Cellulose triacetate	--	--	--	--	4.4	23.8	μg/g	0.54d	--	--	--	--	Grover and Smith, 1974
Indian Head loam	69.3	12.3	18.5	--	7.8	0.53	μg/g	0.97d	--	--	2.36	1.35	Grover and Smith, 1974
Kaolinite	--	--	--	--	6.0	2.40	μg/g	0.78d	--	--	--	--	Grover and Smith, 1974
Melfort loam	47.5	33.2	20.3	--	5.9	0.99	μg/g	1.00d	--	--	6.08	1.21	Grover and Smith, 1974
Montmorillonite	--	--	--	--	9.4	0.00	μg/g	--	--	--	--	--	Grover and Smith, 1974
Peat	--	--	--	--	3.3	29.5	μg/g	0.96d	--	--	--	--	Grover and Smith, 1974
Regina heavy clay	5.3	25.3	69.5	--	6.5	0.19	μg/g	1.22d	--	--	2.41	0.90	Grover and Smith, 1974
Silica gel	--	--	--	--	6.5	0.17	μg/g	0.97d	--	--	--	--	Grover and Smith, 1974
Weyburn Oxbow loam	53.3	27.5	19.2	--	6.5	0.45	μg/g	1.05d	--	--	3.75	1.08	Grover and Smith, 1974
Wheat straw	--	--	--	--	6.7	2.75	μg/g	1.16d	--	--	--	--	Grover and Smith, 1974
p,p'-DDE [72-55-9]													
Filtrasorb 300	--	--	--	--	5.3/7.4	232	L/g	0.37	mg/L	mg/g	--	--	Dobbs and Cohen, 1980
p,p'-DDT [50-29-3]													
Filtrasorb 300	--	--	--	--	5.3/7.4	322	L/g	0.50	mg/L	mg/g	--	--	Dobbs and Cohen, 1980
Diazinon [333-41-5]													
Batcombe sl	--	--	--	--	--	--	--	--	--	--	2.05	2.36	Briggs, 1981
Beverly sal	56	30	14	--	6.8	6	mL/g	0.99	pmol/mL	pmol/g	1.45	2.62	Sharom et al., 1980
Big Creek sediment	71	22	7	--	6.6	7	mL/g	1.07	pmol/mL	pmol/g	1.62	2.63	Sharom et al., 1980
MLSS	--	--	--	--	--	0.4	L/g	1.0	μg/L		--	--	20 °C, Bell and Tsezos, 1987
Organic soil	52	34	14	--	6.1	325	mL/g	1.00	pmol/mL	pmol/g	43.67	2.87	Sharom et al., 1980
Plainfield sand	91.5	1.5	7	--	7.0	2	mL/g	1.08	pmol/mL	pmol/g	0.41	2.69	Sharom et al., 1980
Dibenz[a,h]anthracene [53-70-3]													
Filtrasorb 300	--	--	--	--	5.3/7.4	69.3	L/g	0.75	mg/L	mg/g	--	--	Dobbs and Cohen, 1980
EPA-B2 sediment	67.5	18.6	13.9	3.72	6.35	20,461	mL/g	--	ng/mL	ng/g	1.21	6.23	Means et al., 1980
EPA-4 sediment	3.0	55.2	41.8	23.72	7.79	34,929	mL/g	--	ng/mL	ng/g	2.07	6.23	Means et al., 1980
EPA-5 sediment	33.6	31.0	35.4	19.00	7.44	18,361	mL/g	--	ng/mL	ng/g	2.28	5.91	Means et al., 1980
EPA-6 sediment	0.2	68.6	31.2	33.01	7.83	18,882	mL/g	--	ng/mL	ng/g	0.72	6.42	Means et al., 1980
EPA-8 sediment	82.4	6.8	10.7	3.72	8.32	1,759	mL/g	--	ng/mL	ng/g	0.15	6.07	Means et al., 1980
EPA-9 soil	7.1	17.4	75.6	12.40	8.34	2,506	mL/g	--	ng/mL	ng/g	0.11	6.36	Means et al., 1980
EPA-14 soil	2.1	63.6	34.4	18.86	4.54	14,497	mL/g	--	ng/mL	ng/g	0.48	6.48	Means et al., 1980
EPA-15 sediment	15.6	35.7	48.7	11.30	7.79	25,302	mL/g	--	ng/mL	ng/g	0.95	6.43	Means et al., 1980
EPA-18	34.6	39.5	25.8	15.43	7.76	20,192	mL/g	--	ng/mL	ng/g	0.66	6.49	Means et al., 1980
EPA-20 sediment	0.0	28.6	71.4	8.50	5.50	7,345	mL/g	--	ng/mL	ng/g	1.30	5.75	Means et al., 1980
EPA-21	50.2	7.1	42.7	8.33	7.60	55,697	mL/g	--	ng/mL	ng/g	1.88	6.47	Means et al., 1980

Compound [CAS No.]	Adsorbent	% sand	% silt	% clay	CEC[a]	pH	K_F	K_F units	1/n	C units	x/m units	f_{oc} %	log K_{oc}[b]	Reference
	EPA-22	26.1	21.2	52.7	8.53	7.55	39,809	mL/g	--	ng/mL	ng/g	1.67	6.38	Means et al., 1980
	EPA-23	17.3	69.1	13.6	31.15	6.70	19,254	mL/g	--	ng/mL	ng/g	2.38	5.91	Means et al., 1980
	EPA-26	1.6	42.9	55.4	20.86	7.75	39,840	mL/g	--	ng/mL	ng/g	1.48	6.43	Means et al., 1980
1,2,7,8-Dibenzocarbazole [239-64-5]	River sediment	3.0	41.8	55.2	23.72	7.79	258	mL/g	--	μg/mL	μg/g	2.07	6.10	Banwart et al., 1982
	River sediment	33.6	35.4	31.0	19.00	7.44	146	mL/g	--	μg/mL	μg/g	2.28	5.81	Banwart et al., 1982
	River sediment	82.4	10.7	6.8	3.72	8.32	31	mL/g	--	μg/mL	μg/g	0.15	6.31	Banwart et al., 1982
	Iowa sediment	7.1	75.6	17.4	12.40	8.34	10	mL/g	--	μg/mL	μg/g	0.11	5.95	Banwart et al., 1982
	WV sediment	2.1	34.4	63.6	18.86	4.45	81	mL/g	--	μg/mL	μg/g	0.48	6.23	Banwart et al., 1982
	Ohio River sediment	15.6	48.7	35.7	15.43	7.76	135	mL/g	--	μg/mL	μg/g	0.95	6.15	Banwart et al., 1982
	River sediment	34.6	25.8	39.5	15.43	7.76	141	mL/g	--	μg/mL	μg/g	0.66	6.33	Banwart et al., 1982
	Illinois soil	0.0	71.4	28.6	8.50	5.50	155	mL/g	--	μg/mL	μg/g	1.30	6.08	Banwart et al., 1982
	Illinois river sediment	50.2	42.7	7.1	8.33	7.60	170	mL/g	--	μg/mL	μg/g	1.88	5.96	Banwart et al., 1982
	Illinois river sediment	26.2	52.7	21.2	23.72	7.55	180	mL/g	--	μg/mL	μg/g	1.67	6.03	Banwart et al., 1982
	Illinois river sediment	17.3	13.6	69.1	31.15	6.70	139	mL/g	--	μg/mL	μg/g	2.38	5.76	Banwart et al., 1982
	River sediment	1.6	55.4	42.9	20.86	7.75	216	mL/g	--	μg/mL	μg/g	1.48	6.16	Banwart et al., 1982
	Stream sediment	67.6	13.9	18.6	3.72	6.35	296	mL/g	--	μg/mL	μg/g	1.21	6.39	Banwart et al., 1982
Dibromochloromethane [124-48-1]	Filtrasorb 300	--	--	--	--	5.3	4.8	L/g	0.34	mg/L	mg/g	--	--	Dobbs and Cohen, 1980
1,2-Dibromo-3-chloropropane [96-12-8]	Filtrasorb 300	--	--	--	--	5.3/7.4	53	L/g	0.47	mg/L	mg/g	--	--	Dobbs and Cohen, 1980
	Filtrasorb 300	--	--	--	--	3.0	220	L/g	0.45	mg/L	mg/g	--	--	Dobbs and Cohen, 1980
Di-n-butyl phthalate [84-74-2]	Activated charcoal	--	--	--	--	8.7	110,000	μg/g	1.86[d]	--	--	--	--	Grover and Smith, 1974
Dicamba [1918-00-9]	Anion-exchange resin	--	--	--	--	3.6	480	μg/g	0.78[d]	--	--	--	--	Grover and Smith, 1974
	Cation-exchange resin	--	--	--	--	5.5	0.34	μg/g	0.97[d]	--	--	--	--	Grover and Smith, 1974
	Cellulose powder	--	--	--	--	6.4	0.00	μg/g	--	--	--	--	--	Grover and Smith, 1974
	Cellulose triacetate	--	--	--	--	4.5	4.00	μg/g	1.01[d]	--	--	--	--	Grover and Smith, 1974
	Kaolinite	--	--	--	--	6.1	0.00	μg/g	--	--	--	--	--	Grover and Smith, 1974
	Melfort loam	47.5	33.2	20.3	--	5.9	0.11	μg/g	0.72[d]	--	--	6.08	0.26	Grover and Smith, 1974
	Montmorillonite	--	--	--	--	9.3	0.00	μg/g	--	--	--	--	--	Grover and Smith, 1974
	Peat	--	--	--	--	3.5	8.5	μg/g	0.99[d]	--	--	--	--	Grover and Smith, 1974
	Rolla NT	--	--	--	15.2[c]	5.09	0.23	L/kg	--	--	--	2.37	0.99	Shang and Arshad, 1998
	Rycroft NT	--	--	--	28.6[c]	4.94	1.43	L/kg	--	--	--	4.70	1.48	Shang and Arshad, 1998
	Rycroft CT	--	--	--	27.7[c]	5.38	1.61	L/kg	--	--	--	4.45	1.56	Shang and Arshad, 1998
	Silica gel	--	--	--	--	7.0	0.00	μg/g	1.01[d]	--	--	--	--	Grover and Smith, 1974
	Wheat straw	--	--	--	--	6.3	4.10	μg/g	--	--	--	--	--	Grover and Smith, 1974
Dicamba dimethylammonium [2300-66-5]	Activated charcoal	--	--	--	--	9.1	11,200	μg/g	1.17[d]	--	--	--	--	Grover and Smith, 1974

Compound	Adsorbent					pH	Kf		C				Reference
	Anion-exchange resin	—	—	—	—	3.6	1.00[d]	μg/g	1,100	μg/L	—	—	Grover and Smith, 1974
	Cation-exchange resin	—	—	—	—	5.9	—	μg/g	0.00	μg/L	—	—	Grover and Smith, 1974
	Cellulose powder	—	—	—	—	5.2	—	μg/g	0.00	μg/L	—	—	Grover and Smith, 1974
	Cellulose triacetate	—	—	—	—	4.4	0.77[d]	μg/g	1.18	μg/L	—	—	Grover and Smith, 1974
	Kaolinite	—	—	—	—	5.8	—	μg/g	0.00	μg/L	—	—	Grover and Smith, 1974
	Melfort loam	47.5	33.2	20.3	—	5.9	1.30[d]	μg/g	0.08	μg/L	6.08	0.12	Grover and Smith, 1974
	Montmorillonite	—	—	—	—	9.4	—	μg/g	0.00	μg/L	—	—	Grover and Smith, 1974
	Peat	—	—	—	—	3.5	1.04[d]	μg/g	6.30	μg/L	—	—	Grover and Smith, 1974
	Silica gel	—	—	—	—	7.1	—	μg/g	0.00	μg/L	—	—	Grover and Smith, 1974
	Weyburn Oxbow loam	53.3	27.5	19.2	—	6.5	1.27[d]	μg/g	0.07	μg/L	3.75	0.27	Grover and Smith, 1974
	Wheat straw	—	—	—	—	6.9	0.88[d]	μg/g	1.75	μg/L	—	—	Grover and Smith, 1974
Dichloroacetic acid [79-43-6]	Filtrasorb 400	—	—	—	—	5.7	0.462	L/g	1.630	μg/L	—	—	Speth and Miltner, 1998
Dichloroacetonitrile [3018-12-0]	Filtrasorb 400	—	—	—	—	4.3	0.232	L/g	261.000	μg/L	—	—	Speth and Miltner, 1998
3,4-Dichloroaniline [95-76-1]	Agricultural soil	13	67	20	—	—	—	mg/g	0.03	mg/L	1.5	0.30	Beyerle-Pfnür and Lay, 1990
	Agricultural soil	13	67	20	—	—	—	mg/g	0.07	mg/L	1.5	0.67	Beyerle-Pfnür and Lay, 1990
	Batcombe sl	—	—	—	—	7.9	—	—	—	—	—	2.29	Briggs, 1981
	Filtrasorb 400	—	—	—	—	—	0.630	L/g	5.910	μg/L	—	—	Speth and Miltner, 1998
1,2-Dichlorobenzene [95-50-1]	Pond sediment	37	46	17	—	5.5	—	mg/g	0.07	mg/L	3.0	0.37	Beyerle-Pfnür and Lay, 1990
	Pond sediment	37	46	17	—	5.1	—	mg/g	0.15	mg/L	3.0	0.70	Beyerle-Pfnür and Lay, 1990
	Sandy soil	79	12	9	—	5.1	1.43	mg/g	0.027	mg/L	1.2	0.35	Beyerle-Pfnür and Lay, 1990
	Filtrasorb 300	—	—	—	—	—	0.43	mg/g	129	mg/L	—	—	Dobbs and Cohen, 1980
1,3-Dichlorobenzene [541-73-1]	Filtrasorb 300	—	—	—	—	—	0.45	mg/g	118	mg/L	—	—	Dobbs and Cohen, 1980
1,4-Dichlorobenzene [106-46-7]	Filtrasorb 300	—	—	—	—	—	0.47	mg/g	121	mg/L	—	—	Dobbs and Cohen, 1980
	Filtrasorb 400	—	—	—	—	7	0.37	L/mg	17.1	μg/L	—	—	Weber and Pirbazari, 1981
	KS1 sediment	4	—	—	—	—	—	mL/g	4.4	mg/L	0.73	2.78	Schwarzenbach and Westall, 1981
	Apison, TN soil	10	86	76	—	4.5	—	unitless	—	—	0.11	2.82	Southworth and Keller, 1986
	Dormont, WV soil	2	38	60	129	4.2	—	unitless	—	—	1.2	2.45	Southworth and Keller, 1986
	Fullerton, TN soil	11	21	68	64	4.4	—	unitless	—	—	0.05	2.93	Southworth and Keller, 1986
	KS1H sediment	—	—	—	—	—	—	mL/g	1.1	—	0.08	3.14	Schwarzenbach and Westall, 1981
	Kaolin	—	—	—	—	—	—	mL/g	1.1	—	0.06	3.26	Schwarzenbach and Westall, 1981
	γ-Alumina	—	—	—	—	—	—	mL/g	0.9	—	<0.01	—	Schwarzenbach and Westall, 1981
	Silica	—	—	—	—	—	—	mL/g	6.0	—	<0.01	—	Schwarzenbach and Westall, 1981
3,3'-Dichlorobenzidine [91-94-1]	Filtrasorb 300	—	—	—	—	5.3/7.4	0.20	L/g	300	mg/L	—	—	Dobbs and Cohen, 1980
2,4'-Dichlorobiphenyl [34883-43-7]	Catlin Ap	0	69	21	18.1	6.1	—	—	—	—	—	4.55	Girvin and Scott, 1997
	Cloudland E	41	40	19	3	4.7	—	—	—	—	—	4.67	Girvin and Scott, 1997
	Kenoma Ap	8	70	22	11.8	6.2	—	—	—	—	—	4.61	Girvin and Scott, 1997
	Kenoma Bt2	5	42	53	33.2	6.0	—	—	—	—	—	1.12	Girvin and Scott, 1997
	Kenoma C	8	45	37	24.6	7.2	—	—	—	—	—	4.68	Girvin and Scott, 1997
	Norborne Ap	23	59	18	15.8	7.3	—	—	—	—	—	4.53	Girvin and Scott, 1997

Compound [CAS No.]	Adsorbent	% sand	% silt	% clay	CEC[a]	pH	K_F	K_F units	1/n	C units	x/m units	f_{oc} %	$\log K_{oc}$[b]	Reference
	Norborne Bt1	44	39	17	13.8	7.2	--	--	--	--	--	--	4.56	Girvin and Scott, 1997
	Norborne Bt2	47	37	16	12.8	7.2	--	--	--	--	--	--	4.58	Girvin and Scott, 1997
4,4′-Dichlorobiphenyl [2050-68-2]														
1,1-Dichloroethane [75-34-3]	Lucustrine sediment	--	--	--	--	--	2,668	--	1.09	μg/L	--	1.42	5.27	Chin et al., 1988
	Filtrasorb 300	--	--	--	--	5.3	1.79	--	0.53	mg/L	mg/g	--	--	Dobbs and Cohen, 1980
	Leie River clay	0	0	100	--	--	1.12	L/kg	--	--	--	4.12	1.43	2.3 °C, Dewulf et al., 1999
	Leie River clay	0	0	100	--	--	1.18	L/kg	--	--	--	4.12	1.46	3.8 °C, Dewulf et al., 1999
	Leie River clay	0	0	100	--	--	1.11	L/kg	--	--	--	4.12	1.43	6.2 °C, Dewulf et al., 1999
	Leie River clay	0	0	100	--	--	1.25	L/kg	--	--	--	4.12	1.48	8.0 °C, Dewulf et al., 1999
	Leie River clay	0	0	100	--	--	1.29	L/kg	--	--	--	4.12	1.50	13.5 °C, Dewulf et al., 1999
	Leie River clay	0	0	100	--	--	1.26	L/kg	--	--	--	4.12	1.49	18.6 °C, Dewulf et al., 1999
	Leie River clay	0	0	100	--	--	1.45	L/kg	--	--	--	4.12	1.55	25.0 °C, Dewulf et al., 1999
1,2-Dichloroethane [107-06-2]	Filtrasorb 300	--	--	--	--	5.3	1.57	L/g	0.83	mg/L	mg/g	--	--	Dobbs and Cohen, 1980
	Leie River clay	0	0	100	--	--	1.80	L/kg	--	--	--	4.12	1.64	2.3 °C, Dewulf et al., 1999
	Leie River clay	0	0	100	--	--	1.82	L/kg	--	--	--	4.12	1.65	3.8 °C, Dewulf et al., 1999
	Leie River clay	0	0	100	--	--	1.78	L/kg	--	--	--	4.12	1.64	6.2 °C, Dewulf et al., 1999
	Leie River clay	0	0	100	--	--	1.95	L/kg	--	--	--	4.12	1.68	8.0 °C, Dewulf et al., 1999
	Leie River clay	0	0	100	--	--	2.07	L/kg	--	--	--	4.12	1.70	13.5 °C, Dewulf et al., 1999
	Leie River clay	0	0	100	--	--	1.84	L/kg	--	--	--	4.12	1.65	18.6 °C, Dewulf et al., 1999
	Leie River clay	0	0	100	--	--	1.97	L/kg	--	--	--	4.12	1.68	25.0 °C, Dewulf et al., 1999
1,1-Dichloroethylene [75-35-4]	Filtrasorb 300	--	--	--	--	5.3	1.91	L/g	0.54	mg/L	mg/g	--	--	Dobbs and Cohen, 1980
cis-1,2-Dichloroethylene [156-59-2]	Filtrasorb F-400	--	--	--	--	6.0	151	L/g	0.70	μg/L	μg/g	--	--	15 °C, Crittenden et al., 1985
	Westvaco WV-W	--	--	--	--	6.0	180	L/g	0.64	μg/L	μg/g	--	--	21 °C, Crittenden et al., 1985
trans-1,2-Dichloroethylene [156-60-5]														
2,4-Dichlorophenol [120-83-2]	Filtrasorb 300	--	--	--	--	5.3	3.05	L/g	0.51	mg/L	mg/g	--	--	Dobbs and Cohen, 1980
	Brookston clay loam	--	--	--	22.22	5.7	3.38	mL/g	0.67	μmol/mL	μmol/g	2.96	2.10	Boyd, 1982
	Filtrasorb 300	--	--	--	--	5.3/7.4	157	L/g	0.15	mg/L	mg/g	--	--	Dobbs and Cohen, 1980
	Filtrasorb 300	--	--	--	--	4.8-5.6	131	L/g	0.24	mg/L	mg/g	--	--	Knettig et al., 1986
	Agricultural soil	65.2	25.6	9.2	9.0	7.4	7.8	L/kg	--	mol/L	mol/kg	2.2	2.55	Seip et al., 1986
	Forest soil	69.5	20.5	10.1	2.9	4.2	>17	L/kg	--	mol/L	mol/kg	3.7	>2.68	Seip et al., 1986
	Forest soil	97.3	2.2	0.5	0.48	5.6	1.39	L/kg	--	mol/L	mol/kg	0.2	2.84	Seip et al., 1986
3,4-Dichlorophenol [95-77-2]	Kootwijk humic sand	--	--	--	--	3.4	25.1	L/kg	1.11	μg/L	μg/kg	0.017	--	Lagas, 1988
	Holten humic-rich sand	--	--	--	--	4.7	79.4	L/kg	1.25	μg/L	μg/kg	0.032	--	Lagas, 1988
	Maasdijk light loam	--	--	--	--	7.5	12.6	L/kg	1.25	μg/L	μg/kg	0.009	--	Lagas, 1988
	Opijnen heavy loam	--	--	--	--	7.1	31.6	L/kg	1.43	μg/L	μg/kg	0.017	--	Lagas, 1988
	Rolde humic sand	--	--	--	--	4.9	50.1	L/kg	1.11	μg/L	μg/kg	0.022	--	Lagas, 1988
	Schipluiden peat	--	--	--	--	4.6	316.2	L/kg	1.43	μg/L	μg/kg	0.298	--	Lagas, 1988

Chemical	Soil/sediment					pH								Reference
3,4-Dichloronitrobenzene [99-54-7]	Batcombe sl	--	--	--	--	--	--	--	--	--	--	--	2.53	Briggs, 1981
3,4-Dichlorophenyl urea [2327-02-8]	Batcombe sl	--	--	--	--	--	--	--	--	--	--	--	2.49	Briggs, 1981
1,2-Dichloropropylene [563-54-2]	--	--	--	--	--	--	--	--	--	--	--	--	--	--
Dieldrin [60-57-1]	Filtrasorb 300	--	--	--	--	5.3/7.4	8.21	L/g	0.46	mg/L	mg/g	2.05	--	Dobbs and Cohen, 1980
	Batcombe sl	--	--	--	--	--	--	--	--	--	--	--	4.11	Briggs, 1981
	Beverly sal	56	30	14	--	6.8	338	mL/g	0.89	pmol/mL	pmol/g	1.45	4.37	Sharom et al., 1980
	Big Creek sediment	71	22	7	--	6.6	445	mL/g	0.91	pmol/mL	pmol/g	1.62	4.44	Sharom et al., 1980
	Clay loam	--	--	--	12.4	5.91	195	L/g	0.954	mg/L	mg/g	1.42	4.15	Kishi et al., 1990
	Filtrasorb 300	--	--	--	--	5.3/7.4	606	L/g	0.51	mg/L	mg/g	--	--	Dobbs and Cohen, 1980
	Filtrasorb 400	--	--	--	--	7	2.74	L/g	0.56	μg/L	mg/g	--	--	Weber and Pirbazari, 1981
	Organic soil	52	34	14	--	6.1	4,246	mL/g	1.08	pmol/mL	pmol/g	43.67	3.99	Sharom et al., 1980
	Plainfield sand	91.5	1.5	7	--	7.0	106	mL/g	0.88	pmol/mL	pmol/g	0.41	4.42	Sharom et al., 1980
Diethyl phthalate [84-66-2]	Filtrasorb 300	--	--	--	--	5.3/7.4	110	L/g	0.27	mg/L	mg/g	2.05	--	Dobbs and Cohen, 1980
Dimethoate [60-51-5]	Batcombe sl	--	--	--	--	--	--	--	--	--	--	--	0.96	Briggs, 1981
	AL-01 soil	--	--	13.1	13.1c	8.5	4.21	L/kg	0.64d	mg/L	mg/kg	1.17	2.56	Valverde-Garcia et al., 1988
	AL-02 soil	--	--	14.3	11.9c	8.5	3.15	L/kg	0.65d	mg/L	mg/kg	1.48	2.33	Valverde-Garcia et al., 1988
	AL-03 soil	--	--	7.8	6.3c	8.5	2.35	L/kg	0.69d	mg/L	mg/kg	0.65	2.56	Valverde-Garcia et al., 1988
	AL-04 soil	--	--	4.1	10.0c	8.5	1.81	L/kg	0.78d	mg/L	mg/kg	1.65	2.04	Valverde-Garcia et al., 1988
	AL-05 soil	--	--	5.9	3.8c	8.5	1.06	L/kg	0.59d	mg/L	mg/kg	0.37	2.46	Valverde-Garcia et al., 1988
	AL-06 soil	--	--	12.2	4.4c	8.5	2.21	L/kg	0.70d	mg/L	mg/kg	0.33	2.83	Valverde-Garcia et al., 1988
	AL-07 soil	--	--	14.1	25.6c	8.5	8.94	L/kg	0.54d	mg/L	mg/kg	2.06	2.64	Valverde-Garcia et al., 1988
	AL-08 soil	--	--	11.7	9.4c	8.5	1.51	L/kg	0.60d	mg/L	mg/kg	0.91	2.22	Valverde-Garcia et al., 1988
Dimethylamine [120-40-3]	Asquith sal	81.6	10.4	8.0	--	7.5	4.5	μg/g	0.99d	--	--	1.03	2.64	Grover and Smith, 1974
	Indian Head loam	69.3	12.3	18.5	--	7.8	9.2	μg/g	0.99d	--	--	2.36	2.59	Grover and Smith, 1974
	Melfort loam	47.5	33.2	20.3	--	5.9	32.6	μg/g	1.00d	--	--	6.08	2.72	Grover and Smith, 1974
	Regina heavy clay	5.3	25.3	69.5	--	6.5	12.0	μg/g	1.00d	--	--	2.41	2.70	Grover and Smith, 1974
	Weyburn Oxbow loam	53.3	27.5	19.2	--	6.5	11.7	μg/g	1.00d	--	--	3.75	2.49	Grover and Smith, 1974
Dimethylaminoazobenzene [60-11-7]	Filtrasorb 300	--	--	--	--	5.3/7.4	249	L/g	0.24	mg/L	mg/g	--	--	Dobbs and Cohen, 1980
7,12-Dimethylbenz[a]anthracene [57-97-6]	EPA-B2 sediment	67.5	18.6	13.9	3.72	6.35	2,371	mg/g	--	ng/mL	ng/g	1.21	5.29	Means et al., 1980
	EPA-4 sediment	3.0	55.2	41.8	23.72	7.79	2,646	mg/g	--	ng/mL	ng/g	2.07	5.11	Means et al., 1980
	EPA-5 sediment	33.6	31.0	35.4	19.00	7.44	5,210	mg/g	--	ng/mL	ng/g	2.28	5.36	Means et al., 1980
	EPA-6 sediment	0.2	68.6	31.2	33.01	7.83	1,346	mg/g	--	ng/mL	ng/g	0.72	5.27	Means et al., 1980
	EPA-8 sediment	82.4	6.8	10.7	3.72	8.32	611	mg/g	--	ng/mL	ng/g	0.15	5.61	Means et al., 1980
	EPA-9 soil	7.1	17.4	75.6	12.40	8.34	1,028	mg/g	--	ng/mL	ng/g	0.11	5.97	Means et al., 1980
	EPA-14 soil	2.1	63.6	34.4	18.86	4.54	562	mg/g	--	ng/mL	ng/g	0.48	5.07	Means et al., 1980
	EPA-15 sediment	15.6	35.7	48.7	11.30	7.79	3,742	mg/g	--	ng/mL	ng/g	0.95	5.60	Means et al., 1980
	EPA-18	34.6	39.5	25.8	15.43	7.76	1,895	mg/g	--	ng/mL	ng/g	0.66	5.46	Means et al., 1980

Compound [CAS No.]	Adsorbent	% sand	% silt	% clay	CEC[a]	pH	K_F	K_F units	1/n	C units	x/m units	f_{oc} %	log K_{oc}[b]	Reference
	EPA-20 sediment	0.0	28.6	71.4	8.50	5.50	1,617	mL/g	--	ng/mL	ng/g	1.30	5.09	Means et al., 1980
	EPA-21	50.2	7.1	42.7	8.33	7.60	5,576	mL/g	--	ng/mL	ng/g	1.88	5.47	Means et al., 1980
	EPA-22	26.1	21.2	52.7	8.53	7.55	2,679	mL/g	--	ng/mL	ng/g	1.67	5.21	Means et al., 1980
	EPA-23	17.3	69.1	13.6	31.15	6.70	6,777	mL/g	--	ng/mL	ng/g	2.38	5.45	Means et al., 1980
	EPA-26	1.6	42.9	55.4	20.86	7.75	3,740	mL/g	--	ng/mL	ng/g	1.48	5.40	Means et al., 1980
1,1-Dimethyl-3-(3-fluorophenyl)urea [330-39-2]	Batcombe sl	--	--	--	--	--	--	--	--	--	--	--	1.73	Briggs, 1981
1,1-Dimethyl-3-(4-fluorophenyl)urea [332-33-2]	Batcombe sl	--	--	--	--	--	--	--	--	--	--	--	1.43	Briggs, 1981
1,1-Dimethyl-3-(3-methoxy-phenyl)urea [28170-54-9]	Batcombe sl	--	--	--	--	--	--	--	--	--	--	--	1.72	Briggs, 1981
2,4-Dimethylphenol [105-67-9]	Filtrasorb 300	--	--	--	--	5.3/7.4	70	L/g	0.44	mg/L	mg/g	--	--	Dobbs and Cohen, 1980
	Filtrasorb 300	--	--	--	--	5	184	L/g	0.09	mg/L	mg/g	--	--	Burks, 1981
3,5-Dimethylphenol [108-68-9]	Apison, TN soil	4	10	86	76	4.5	--	unitless	--	--	--	0.11	2.66	Southworth and Keller, 1986
	Dormont, WV soil	2	38	60	129	4.2	--	unitless	--	--	--	1.2	2.28	Southworth and Keller, 1986
	Fullerton, TN soil	11	21	68	64	4.4	--	unitless	--	--	--	0.05	3.15	Southworth and Keller, 1986
Dimethylphenylcarbinol [617-94-7]	Filtrasorb 300	--	--	--	--	5.3/7.4	210	L/g	0.74	mg/L	mg/g	--	--	Dobbs and Cohen, 1980
Dimethyl phthalate [131-11-3]	Filtrasorb 300	--	--	--	--	5.3/7.4	97	L/g	0.41	mg/L	mg/g	--	--	Dobbs and Cohen, 1980
	Agricultural soil	65.2	25.6	9.2	9.0	7.4	0.94	L/kg	--	mol/L	mol/kg	2.2	1.63	Seip et al., 1986
	Forest soil	69.5	20.5	10.1	2.9	4.2	2.57	L/kg	--	mol/L	mol/kg	3.7	1.84	Seip et al., 1986
	Forest soil	97.3	2.2	0.5	0.48	5.6	0.015	L/kg	--	mol/L	mol/kg	0.2	0.88	Seip et al., 1986
2,2-Dimethyl-1-propanol [75-84-3]	GAC from wood	--	--	--	--	--	0.500	L/g	0.690	mg/L	mg/g	--	--	Abe et al., 1983
	GAC from coal	--	--	--	--	--	3.66	L/g	0.481	mg/L	mg/g	--	--	Abe et al., 1983
	GAC from coconut	--	--	--	--	--	5.05	L/g	0.494	mg/L	mg/g	--	--	Abe et al., 1983
4,6-Dinitro-o-cresol [534-52-1]	Filtrasorb 300	--	--	--	--	5.3/7.4	169	L/g	0.35	mg/L	mg/g	--	--	Dobbs and Cohen, 1980
2,4-Dinitrophenol [51-28-5]	Filtrasorb 300	--	--	--	--	5.3/7.4	33	L/g	0.61	mg/L	mg/g	--	--	Dobbs and Cohen, 1980
2,4-Dinitrotoluene [121-14-2]	Filtrasorb 300	--	--	--	--	5.3/7.4	146	L/g	0.31	mg/L	mg/g	--	--	Dobbs and Cohen, 1980
2,6-Dinitrotoluene [606-20-2]	Filtrasorb 300	--	--	--	--	5.3/7.4	145	L/g	0.32	mg/L	mg/g	--	--	Dobbs and Cohen, 1980
Diphenylamine [122-39-4]	Filtrasorb 300	--	--	--	--	5.3/7.4	120	L/g	0.31	mg/L	mg/g	--	--	Dobbs and Cohen, 1980
	Batcombe sl	--	--	--	--	--	--	--	--	--	--	--	2.78	Briggs, 1981
1,1-Diphenylhydrazine [530-50-7]	Filtrasorb 300	--	--	--	--	5.3/7.4	135	L/g	0.16	mg/L	mg/g	--	--	Dobbs and Cohen, 1980
	Hydrodarco KB	--	--	--	--	--	92	L/g	0.257	mg/L	mg/g	--	--	Fochtman and Eisenberg, 1979
1,2-Diphenylhydrazine [122-66-7]	Filtrasorb 300	--	--	--	--	5.3/7.4	1.6×10^{-1}	L/g	0.16	mg/L	mg/g	--	--	Dobbs and Cohen, 1980
Diuron [330-54-1]	Adkins loamy sand	84	13	3	--	7.3	2.63	L/kg	0.86[d]	μmol/L	μmol/kg	0.40	2.82	4 °C, Madhun et al., 1986

Compound [CAS]	Sorbent	Sand	Silt	Clay	CEC	pH	K	units	1/n	conc. units	sorbed units	%OC	log Koc	Reference
	Adkins loamy sand	84	13	3	--	7.3	1.91	L/kg	0.86[d]	µmol/L	µmol/kg	0.40	2.68	25 °C, Madhun et al., 1986
	Alluvial soil (OS-26)	--	--	--	--	--	13.5	L/kg	0.85[d]	mg/L	µg/g	--	--	Yuen and Hilton, 1962
	Alluvial soil (K-47)	--	--	--	--	--	9.6	L/kg	0.84[d]	mg/L	µg/g	--	--	Yuen and Hilton, 1962
	Batcombe sl	--	--	--	--	--	--	--	--	--	--	--	2.21	Briggs, 1981
	Begbroke soil	66.0	18.4	15.6	--	7.1	2.7	mL/g	--	µg/mL	µg/g	1.93	2.14	Grover and Hance, 1969
	Latosol (W-A-Op 9)	--	--	--	--	--	1.65	L/kg	0.94[d]	mg/L	µg/g	--	--	Yuen and Hilton, 1962
	Latosol (OS-27A)	--	--	--	--	--	1.35	L/kg	0.94[d]	mg/L	µg/g	--	--	Yuen and Hilton, 1962
	Latosol (OS-27A)	--	--	--	--	--	0.94	L/kg	0.92[d]	mg/L	µg/g	--	--	Yuen and Hilton, 1962
	H-montmorillonite	--	--	--	73.5	9.0	70	L/g	1.05	µmol/L	µmol/g	--	--	Bailey et al., 1968
	Na-montmorillonite	--	--	--	87.0	9.0	23	L/g	0.93	µmol/L	µmol/g	--	--	Bailey et al., 1968
	Regosol soil (K-4)	--	--	--	--	--	1.7	L/kg	0.94[d]	µmol/L	µmol/kg	--	--	Yuen and Hilton, 1962
	Semiahmoo mucky peat	19	65	16	--	5.4	300.61	L/kg	0.87[d]	µmol/L	µmol/kg	27.8	3.04	4 °C, Madhun et al., 1986
	Semiahmoo mucky peat	19	65	16	--	5.4	244.34	L/kg	0.87[d]	µmol/L	µmol/kg	27.8	2.95	25 °C, Madhun et al., 1986
	Wenster sandy clay loam	--	--	--	--	5.3/7.4	24.4	mL/g	0.75[d]	µg/mL	µg/g	3.34	2.87	Nkedi-Kizza et al., 1983
Edetic acid [60-00-4]	Filtrasorb 300	--	--	--	--	5.3/7.4	0.86	L/g	1.5	mg/L	mg/g	--	--	Dobbs and Cohen, 1980
α-Endosulfan [959-98-8]	Filtrasorb 300	--	--	--	--	5.3/7.4	194	L/g	0.50	mg/L	mg/g	--	--	Dobbs and Cohen, 1980
β-Endosulfan [33213-65-9]	Filtrasorb 300	--	--	--	--	5.3/7.4	615	L/g	0.83	mg/L	mg/g	--	--	Dobbs and Cohen, 1980
Endosulfan sulfate [1031-07-8]	Filtrasorb 300	--	--	--	--	5.3/7.4	688	L/g	0.81	mg/L	mg/g	--	--	Dobbs and Cohen, 1980
Endothall [145-73-3]	Filtrasorb 400	--	--	--	--	7.1	2.280	L/g	0.329	µg/L	µg/g	--	--	Speth and Miltner, 1998
	Pat Mayse Lake	62	12	26	16	6.3	0.937	L/kg	--	mg/L	mg/kg	0.683	2.14	Reinert and Rodgers, 1984
	Roselawn Cemetery	7	32	60	34.2	8.4	1.42	L/kg	--	mg/L	mg/kg	1.29	2.04	Reinert and Rodgers, 1984
Endrin [72-20-8]	Beverly sal	56	30	14	--	6.8	112	mL/g	1.12	pmol/mL	pmol/g	1.45	3.89	Sharom et al., 1980
	Big Creek sediment	71	22	7	--	6.6	258	mL/g	0.99	pmol/mL	pmol/g	1.62	4.20	Sharom et al., 1980
	Filtrasorb 300	--	--	--	--	5.3/7.4	666	L/g	0.80	mg/L	mg/g	--	--	Dobbs and Cohen, 1980
	Organic soil	52	34	14	--	6.1	3,404	mL/g	1.08	pmol/mL	pmol/g	43.67	3.89	Sharom et al., 1980
	Plainfield sand	91.5	1.5	7	--	7.0	58	mL/g	1.03	pmol/mL	pmol/g	0.41	4.15	Sharom et al., 1980
Enrofloxacin [93106-60-6]	Brazil soil	27.2	28.8	41.7	13.6	4.9	3,037	mL/g	--	mg/mL	mg/g	1.63	5.27	Nowara et al., 1997
	France soil	39.1	36.6	23.4	19.1	7.5	260	mL/g	--	mg/mL	mg/g	1.58	4.22	Nowara et al., 1997
	Germany soil	80.1	16.7	2.5	12.0	5.3	496.0	mL/g	--	mg/mL	mg/g	0.70	4.85	Nowara et al., 1997
	Illite	--	--	--	--	6.1	4,670	mL/g	--	mg/mL	mg/g	--	--	Nowara et al., 1997
	Kaolinite	--	--	--	--	6.0	3,548	mL/g	--	mg/mL	mg/g	--	--	Nowara et al., 1997
	Montmorillonite	--	--	--	--	5.9	6,310	mL/g	--	mg/mL	mg/g	--	--	Nowara et al., 1997
	Philippines soil	39.4	42.9	17.2	16.9	5.3	5,612	mL/g	--	mg/mL	mg/g	0.73	5.89	Nowara et al., 1997
	Sweden soil	84.0	8.4	7.2	7.4	6.0	1,230	mL/g	--	mg/mL	mg/g	1.23	5.00	Nowara et al., 1997
	Vermiculite	--	--	--	--	5.8	5,986	mL/g	--	mg/mL	mg/g	--	--	Nowara et al., 1997
Ethanol [64-17-5]	Dowex-1 resin	--	--	--	--	--	0.831	unitless	--	--	--	--	--	Small and Bremer, 1964
	Dowex-1-X2 resin	--	--	--	--	--	0.753	unitless	--	--	--	--	--	Small and Bremer, 1964
Ethion [563-12-2]	Beverly sal	56	30	14	--	6.8	125	mL/g	0.99	pmol/mL	pmol/g	1.45	3.94	Sharom et al., 1980
	Big Creek sediment	71	22	7	--	6.6	158	mL/g	1.03	pmol/mL	pmol/g	1.62	3.99	Sharom et al., 1980
	Organic soil	52	34	14	--	6.1	2,818	mL/g	0.96	pmol/mL	pmol/g	43.67	3.81	Sharom et al., 1980

Compound [CAS No.]	Adsorbent	% sand	% silt	% clay	CEC[a]	pH	K_F	K_F units	1/n	C units	x/m units	f_{oc} %	log K_{oc}[b]	Reference
Ethylbenzene [100-41-4]	Plainfield sand	91.5	1.5	7	–	7.0	41	mL/g	1.01	pmol/mL	pmol/g	0.41	4.00	Sharom et al., 1980
	Filtrasorb 400	–	–	–	–	–	0.100	L/g	0.40	mg/L	mg/g	–	–	El-Dib and Badawy, 1979
	Filtrasorb 300	–	–	–	–	7.4	53	L/g	0.79	mg/L	mg/g	–	–	Dobbs and Cohen, 1980
	Leie River clay	0	0	100	–	–	12.6	L/kg	–	–	–	4.12	2.49	2.3 °C, Dewulf et al., 1999
	Leie River clay	0	0	100	–	–	21.9	L/kg	–	–	–	4.12	2.73	3.8 °C, Dewulf et al., 1999
	Leie River clay	0	0	100	–	–	18.2	L/kg	–	–	–	4.12	2.65	6.2 °C, Dewulf et al., 1999
	Leie River clay	0	0	100	–	–	22.1	L/kg	–	–	–	4.12	2.73	8.0 °C, Dewulf et al., 1999
	Leie River clay	0	0	100	–	–	24.4	L/kg	–	–	–	4.12	2.77	13.5 °C, Dewulf et al., 1999
	Leie River clay	0	0	100	–	–	21.9	L/kg	–	–	–	4.12	2.73	18.6 °C, Dewulf et al., 1999
	Leie River clay	0	0	100	–	–	22.8	L/kg	–	–	–	4.12	2.74	25.0 °C, Dewulf et al., 1999
Ethylene dibromide [106-93-4]	Filtrasorb F-400	–	–	–	–	6.0	1,795	L/g	0.48	µg/L	µg/g	–	–	15 °C, Crittenden et al., 1985
Ethylene thiourea [96-45-7]	Filtrasorb 400	–	–	–	–	5.4	0.716	L/g	0.669	µg/L	µg/g	–	–	Speth and Miltner, 1998
4-Ethylphenol [123-07-9]	Filtrasorb 400	–	–	–	–	7	123	L/g	0.166	mg/L	mg/g	–	–	Belfort et al., 1983
Fenamiphos [22224-92-6]	Arredondo sand	93.8	3.0	3.2	–	6.8	1.18	µgmL/g	0.78[d]	–	–	0.80	2.17	Bilkert and Rao, 1985
	Batcombe sl	–	–	–	–	–	–	–	–	–	–	2.05	2.52	Briggs, 1981
	Cecil sal	65.8	19.5	14.7	–	5.6	1.77	µgmL/g	0.89[d]	–	–	0.90	2.29	Bilkert and Rao, 1985
	Webster sicl	18.4	45.3	38.3	–	7.3	9.62	µgmL/g	0.83[d]	–	–	3.97	2.40	Bilkert and Rao, 1985
Fenitrothion [122-14-5]	Lake Superior sediment	–	–	–	–	7.1	15.5	mL/g	1.20[d]	µg/mL	µg/g	1.68	2.97	Baarschers et al., 1983
	Ontario Gleysol	–	–	–	–	4.9	17.0	mL/g	0.98[d]	µg/mL	µg/g	0.52	3.51	Baarschers et al., 1983
	Ontario Gleysol	–	–	–	–	4.4	57.5	mL/g	0.89[d]	µg/mL	µg/g	3.54	3.21	Baarschers et al., 1983
	Ontario Luvisol	–	–	–	–	4.5	100.0	mL/g	0.63[d]	µg/mL	µg/g	3.19	3.50	Baarschers et al., 1983
	Potting soil	–	–	–	–	5.3	354.8	mL/g	1.82[d]	µg/mL	µg/g	19.20	3.27	Baarschers et al., 1983
	Sandy loam	–	–	–	9.0	6.6	15.6	mL/g	–	–	–	1.24	3.10	Xu et al., 1999
Fenuron [101-42-8]	Batcombe sl	–	–	–	–	–	–	–	–	–	–	–	1.12	Briggs, 1981
	H-montmorillonite	–	–	–	73.5	9.0	115	L/g	0.82	µmol/L	µmolg	–	–	Bailey et al., 1968
	Na-montmorillonite	–	–	–	87.0	9.0	14	L/g	1.00	µmol/L	µmolg	–	–	Bailey et al., 1968
Fiponil [120068-37-3]	Banizoumbou soil	79.4	18.2	2.4	0.7	5.8	4.4	L/kg	1.4[d]	mg/L	mg/kg	0.17	3.41	Bobé et al., 1997
	Montpellier soil	54.4	23	22.6	23.5	8.3	45.5	L/kg	1.0[d]	mg/L	mg/kg	3.77	3.08	Bobé et al., 1997
	Saguia soil	98	1.4	0.6	0.7	5.3	4.3	L/kg	1.0[d]	mg/L	mg/kg	0.06	3.86	Bobé et al., 1997
Fluometuron [2164-17-2]	Batcombe sl	–	–	–	–	–	–	–	–	–	–	–	1.82	Briggs, 1981
	Lexington sl CT	19	71	10	–	5.2	1.31	L/kg	0.61	mg/L	mg/g	1.42	1.96	Suba and Essington, 1999
	Lexington sl NT	11	76	13	–	5.1	1.96	L/kg	0.38	mg/L	mg/g	2.45	1.90	Suba and Essington, 1999
	Lexington sl SUB	8	75	16	–	5.3	0.48	L/kg	0.314	mg/L	mg/g	0.37	2.11	Suba and Essington, 1999
Fluoranthene [206-44-0]	Filtrasorb 300	–	–	–	–	5.3/7.4	664	L/g	0.28	mg/L	mg/g	–	–	Dobbs and Cohen, 1980
	Filtrasorb 300	–	–	–	–	5	88.1	L/g	–	mg/L	mg/g	–	–	Burks, 1981
	Filtrasorb 300	–	–	–	–	–	242	L/g	–	mg/L	mg/g	–	–	Walters and Luthy, 1982
Fluorene [86-73-7]	Filtrasorb 300	–	–	–	–	5.3/7.4	330	L/g	0.57	mg/L	mg/g	–	–	Dobbs and Cohen, 1980
	Filtrasorb 300	–	–	–	–	5	196	L/g	–	mg/L	mg/g	–	–	Burks, 1981

Chemical [CAS]	Material	Sand (%)	Silt (%)	Clay (%)	OM (%)	pH	K	Cₑ	x/m	Koc	Reference
Fluoridone [59756-60-4]	Filtrasorb 300	--	--	--	--	--	674 L/g	0.604 mg/L	mg/g	--	Walters and Luthy, 1982
	Hydrosoil	12	36	52	--	7.1	--	--	9.2	3.01	Muir et al., 1980
	Hydrosoil	5	37	58	--	6.8	--	--	3.0	2.97	Muir et al., 1980
	Hydrosoil	3	56	41	--	7.7	--	--	5.9	3.36	Muir et al., 1980
	Hydrosoil	0	23	77	--	7.8	--	--	1.9	3.34	Muir et al., 1980
	Hydrosoil	1	24	75	--	7.6	--	--	3.7	3.39	Muir et al., 1980
	Hydrosoil	7	45	48	--	7.7	--	--	2.3	2.95	Muir et al., 1980
2-Fluorophenyl urea [656-31-5]	Batcombe sl	--	--	--	--	--	--	--	--	1.31	Briggs, 1981
3-Fluorophenyl urea [770-19-4]	Batcombe sl	--	--	--	--	--	--	--	--	1.77	Briggs, 1981
4-Fluorophenyl urea [659-30-3]	Batcombe sl	--	--	--	--	--	--	--	--	1.52	Briggs, 1981
5-Fluorouracil [51-21-8]	Filtrasorb 300	--	--	--	--	5.3/7.4	5.5 L/g	1.0 mg/L	mg/g	2.55	Dobbs and Cohen, 1980
Fuberidazole [3878-19-1]	Sandy loam soil	--	--	--	9.0	6.6	4.40 mL/g	--	1.24	3.27	Xu et al., 1999
Folpet [133-07-3]	Batcombe sl	--	--	--	--	6.1	--	--	--	2.05	Briggs, 1981
Hexachlorobenzene [118-74-1]	Alfisol	12.9	64.3	19.6	--	7.45	3.8 mL/g	0.99 μg/mL	0.76 μg/g	2.70	Rippen et al., 1982
	Batcombe sl	--	--	--	--	6.1	--	--	--	1.53	Briggs, 1981
	Cellulose	--	--	--	--	--	2.2 mL/g	0.079 μg/mL	μg/g	4.49	Rippen et al., 1982
	Speyer soil 2.2	--	--	--	--	5.8	8.2 mL/g	1.56 μg/mL	2.24 μg/g	2.56	Rippen et al., 1982
2,2',4,4',5,5'-Hexachloro-biphenyl [35065-27-1]	Catlin Ap	0	69	21	18.1	6.1	--	--	--	6.45	Girvin and Scott, 1997
	Cloudland E	41	40	19	3	4.7	--	--	--	6.81	Girvin and Scott, 1997
	Kenoma Ap	8	70	22	11.8	6.2	--	--	--	6.48	Girvin and Scott, 1997
	Kenoma Bt2	5	42	53	33.2	6.0	--	--	--	6.54	Girvin and Scott, 1997
	Norborne Ap	23	59	18	15.8	7.3	--	--	--	6.22	Girvin and Scott, 1997
	Norborne Bt1	44	39	17	13.8	7.2	--	--	--	6.55	Girvin and Scott, 1997
	Norborne Bt2	47	37	16	12.8	7.2	--	--	--	6.72	Girvin and Scott, 1997
Formic acid [64-18-6]	Poly(4-vinyl pyridine)	--	--	--	--	--	0.057 L/g	1.003 mg/L	mg/g	--	Chanda et al., 1985
	Polybenzimidazole	--	--	--	--	--	3.04 L/g	0.443 mg/L	mg/g	--	Chanda et al., 1985
Geosmin [19700-21-1]	Aqua Nuchar	--	--	--	--	--	4.3 L/mg	0.71 ng/L	--	--	Lalezary et al., 1986
	Calgon WPH	--	--	--	--	--	915 L/mg	0.39 ng/L	--	--	Lalezary et al., 1986
Glyphosphate [1071-83-6]	Soil	15	17	65	12.6[c]	--	30.9 L/kg	0.778[d] μg/mL	1.7	3.46	de Jonge and de Jonge, 1999
	Peat (HA-1)	--	--	--	--	--	454 mL/g	0.55 μg/mL	--	--	Piccolo et al., 1996
	Volcanic soil (HA-2)	--	--	--	--	--	179 mL/g	0.54 μg/mL	--	--	Piccolo et al., 1996
	Oxidized coal (HA-3)	--	--	--	--	--	98 mL/g	0.51 μg/mL	--	--	Piccolo et al., 1996
	Lignite (HA-4)	--	--	--	--	--	7 mL/g	0.97 μg/mL	--	--	Piccolo et al., 1996
Guanine [73-40-5]	Filtrasorb 300	--	--	--	--	5.3/7.4	120 L/g	0.40 mg/L	--	--	Dobbs and Cohen, 1980
Heptahlor [76-44-8]	Filtrasorb 300	--	--	--	--	5.3/7.4	1,220 L/g	0.95 mg/L	--	--	Dobbs and Cohen, 1980
Heptachlor epoxide [1024-57-3]	Filtrasorb 300	--	--	--	--	5.3/7.4	1,040 L/g	0.70 mg/L	--	--	Dobbs and Cohen, 1980
Hexachlorobenzene [118-74-1]	Filtrasorb 300	--	--	--	--	5.3/7.4	450 L/g	0.60 mg/L	--	--	Dobbs and Cohen, 1980
2,2',4,4',5,5'-Hexachloro-biphenyl [35065-27-1]	Saginaw Bay sediment	--	--	--	--	--	15.7 L/g	0.97 ng/L	0.23 ng/g	3.83	Horzempa and Di Toro, 1983
	Saginaw Bay sediment	--	--	--	--	--	14.1 L/g	0.93 ng/L	3.13 ng/g	2.65	Horzempa and Di Toro, 1983
	Saginaw Bay sediment	--	--	--	--	--	9.3 L/g	1.14 ng/L	2.84 ng/g	2.52	Horzempa and Di Toro, 1983

Compound [CAS No.]	Adsorbent	% sand	% silt	% clay	CEC[a]	pH	K_F	K_F units	1/n	C units	x/m units	f_{oc} %	log K_{oc}[b]	Reference
Hexachlorobutadiene [87-68-3]	Saginaw Bay sediment	15	--	--	--	--	11.3	L/g	0.91	ng/L	ng/g	2.67	2.63	Horzempa and Di Toro, 1983
	Saginaw Bay sediment	85	--	--	--	--	12.6	L/g	0.94	ng/L	ng/g	0.70	3.26	Horzempa and Di Toro, 1983
	Saginaw River soil	18	--	--	--	--	10.6	L/g	0.95	ng/L	ng/g	2.84	2.57	Horzempa and Di Toro, 1983
Hexachlorocyclopentadiene [77-47-4]	Filtrasorb 300	--	--	--	--	5.3/7.4	258	L/g	0.45	mg/L	mg/g	--	--	Dobbs and Cohen, 1980
Hexachloroethane [67-72-1]	Filtrasorb 300	--	--	--	--	5.3/7.4	370	L/g	0.17	mg/L	mg/g	--	--	Dobbs and Cohen, 1980
	Filtrasorb 300	--	--	--	--	5.3/7.4	96.5	L/g	0.38	mg/L	mg/g	--	--	Dobbs and Cohen, 1980
1-Hexanol [111-27-3]	GAC from wood	--	--	--	--	--	5.91	L/g	0.593	mg/L	mg/g	--	--	Abe et al., 1983
	GAC from coal	--	--	--	--	--	25.6	L/g	0.384	mg/L	mg/g	--	--	Abe et al., 1983
	GAC from coconut	--	--	--	--	--	58.9	L/g	0.288	mg/L	mg/g	--	--	Abe et al., 1983
Humic acids [1415-93-6]	Filtrasorb 400	--	--	--	--	--	37.5	L/g	0.56	mg/L	mg/g	--	--	Oulman, 1981
	Filtrasorb 400	--	--	--	--	7	13.4	L/g	0.42	mg/L	mg/g	--	--	Weber and Pirbazari, 1981
	Hydrodarco 1030	--	--	--	--	7	9.84	L/g	0.486	mg/L	mg/g	--	--	Weber and Pirbazari, 1981
	Norit carbon	--	--	--	--	7	12.4	L/mg	0.44	μg/L	mg/g	--	--	Weber and Pirbazari, 1981
	Nuchar WV-G	--	--	--	--	7	6.85	L/mg	0.62	μg/L	mg/g	--	--	Weber and Pirbazari, 1981
	Westvaco WV-G	--	--	--	--	--	27.6	L/g	0.67	mg/L	mg/g	--	--	Oulman, 1981
Hydroquinone [123-31-9] 2-Hydroxybenzoic acid [69-72-7]	Filtrasorb 300	--	--	--	--	5.3/7.4	970	L/g	0.25	mg/L	mg/g	--	--	Dobbs and Cohen, 1980
	Carbon (200 mesh)	--	--	--	--	--	1.86	unitless	0.384	mg/L	mg/kg	--	--	20 °C, Hartman et al., 1946
	Carbon (200 mesh)	--	--	--	--	--	1.73	unitless	0.384	mg/L	mg/kg	--	--	30 °C, Hartman et al., 1946
	Carbon (200 mesh)	--	--	--	--	--	1.62	unitless	0.384	mg/L	mg/kg	--	--	40 °C, Hartman et al., 1946
	Carbon (200 mesh)	--	--	--	76	--	1.53	unitless	0.384	mg/L	mg/kg	--	--	50 °C, Hartman et al., 1946
5-Hydroxyindan [1470-94-6]	Apison, TN soil	4	10	86	76	4.5	--	unitless	--	--	--	0.11	4.06	Southworth and Keller, 1986
	Dormont, WV soil	2	38	60	129	4.2	--	unitless	--	--	--	1.2	2.38	Southworth and Keller, 1986
	Fullerton, TN soil	11	21	68	64	4.4	--	unitless	--	--	--	0.05	4.00	Southworth and Keller, 1986
Imazamethabenz [100728-84-5]	Loamy sand	47.6	24.7	27.7	--	7.76	0.99	L/kg	0.63	mg/L	mg/kg	0.35	2.45	5 °C, Cartón et al., 1997
	Loamy sand	47.6	24.7	27.7	--	7.76	0.88	L/kg	0.64	mg/L	mg/kg	0.35	2.40	15 °C, Cartón et al., 1997
	Loamy sand	47.6	24.7	27.7	--	7.76	0.68	L/kg	0.65	mg/L	mg/kg	0.35	2.29	25 °C, Cartón et al., 1997
	Loamy sand	47.6	24.7	27.7	--	7.76	0.53	L/kg	0.80	mg/L	mg/kg	0.35	2.18	35 °C, Cartón et al., 1997
	Clay loamy sand	39.9	30.0	30.1	--	7.84	1.19	L/kg	0.72	mg/L	mg/kg	0.75	2.20	5 °C, Cartón et al., 1997
	Clay loamy sand	39.9	30.0	30.1	--	7.84	0.83	L/kg	0.85	mg/L	mg/kg	0.75	2.04	15 °C, Cartón et al., 1997
	Clay loamy sand	39.9	30.0	30.1	--	7.84	0.79	L/kg	0.82	mg/L	mg/kg	0.75	2.02	25 °C, Cartón et al., 1997
	Clay loamy sand	39.9	30.0	30.1	--	7.84	0.70	L/kg	0.76	mg/L	mg/kg	0.75	1.97	35 °C, Cartón et al., 1997
	Clay loamy sand	43.0	24.0	33.0	--	7.98	0.94	L/kg	0.67	mg/L	mg/kg	0.87	2.03	5 °C, Cartón et al., 1997
	Clay loamy sand	43.0	24.0	33.0	--	7.98	0.83	L/kg	0.69	mg/L	mg/kg	0.87	1.98	15 °C, Cartón et al., 1997
	Clay loamy sand	43.0	24.0	33.0	--	7.98	0.82	L/kg	0.67	mg/L	mg/kg	0.87	1.97	25 °C, Cartón et al., 1997
	Clay loamy sand	43.0	24.0	33.0	--	7.98	0.66	L/kg	0.72	mg/L	mg/kg	0.87	1.88	35 °C, Cartón et al., 1997
Imazapyr [81334-34-1]	Torba clay loam	5.1	24.9	30.7	53.32	4.43	33.24	L/kg	0.63	μM	μmol/kg	14.85	2.35	Pusino et al., 1997

Compound / Soil														Reference
Imidacloprid [105827-78-9]														
Coazze loamy sand	72.6	25.2	2.2	--	4.72	11.63	4.56	L/kg	0.50	μM	μmol/kg	3.20	2.15	Pusino et al., 1997
Macomer clay loam	32.9	26.2	14.6	--	6.05	17.62	1.91	L/kg	0.81	μM	μmol/kg	9.28	1.31	Pusino et al., 1997
Vercelli sal	48.6	36.8	10.8	--	6.28	7.68	1.68	L/kg	0.71	μM	μmol/kg	0.84	2.30	Pusino et al., 1997
Cadriano clay loam	32.6	38.7	23.6	--	5.24	17.92	1.47	L/kg	0.72	μM	μmol/kg	0.83	2.25	Pusino et al., 1997
Monte loamy sand	70.1	16.8	9.1	--	5.24	4.03	0.84	L/kg	0.65	μM	μmol/kg	1.39	1.78	Pusino et al., 1997
Verndale sal	--	28	--	--	6.1	--	6.1	mL/g	0.81	μg/mL	μg/g	1.4	2.64	Cox et al., 1997
Verndale sal	--	28	--	--	6.1	--	6.70	mL/g	0.66	μg/mL	μg/g	1.4	2.68	Cox et al., 1997
Verndale sal	--	28	--	--	6.1	--	5.00	mL/g	0.73	μg/mL	μg/g	1.4	2.55	Cox et al., 1997
Waukegan sl	--	22	--	--	5.5	--	6.7	mL/g	0.82	μg/mL	μg/g	1.4	2.57	Cox et al., 1997
Waukegan sl	--	22	--	--	5.5	--	7.39	mL/g	0.69	μg/mL	μg/g	1.4	2.61	Cox et al., 1997
Waukegan sl	--	22	--	--	5.5	--	5.81	mL/g	0.75	μg/mL	μg/g	1.4	2.51	Cox et al., 1997
Webster clay loam	--	35	--	--	6.7	--	18	mL/g	0.76	μg/mL	μg/g	4.1	2.64	Cox et al., 1997
Webster clay loam	--	35	--	--	6.7	--	16.7	mL/g	0.70	μg/mL	μg/g	4.1	2.61	Cox et al., 1997
Webster clay loam	--	35	--	--	6.7	--	15.5	mL/g	0.72	μg/mL	μg/g	4.1	2.58	Cox et al., 1997
Ipazine [1912-25-0]														
Begbroke soil	66.0	18.4	15.6	--	7.1	--	47.7	mL/g	--	μg/mL	μg/g	1.93	3.39	Grover and Hance, 1969
Hickory Hill silt	--	--	--	--	--	--	26.6	mL/g	--	μg/mL	μg/g	3.27	2.91	Brown and Flagg, 1981
Isobutyl alcohol [78-83-1]														
GAC from wood	--	--	--	--	--	--	0.251	L/g	0.736	mg/L	mg/g	--	--	Abe et al., 1983
GAC from coal	--	--	--	--	--	--	2.75	L/g	0.465	mg/L	mg/g	--	--	Abe et al., 1983
GAC from coconut	--	--	--	--	--	--	4.06	L/g	0.476	mg/L	mg/g	--	--	Abe et al., 1983
Isopentyl alcohol [123-51-3]														
GAC from wood	--	--	--	--	--	--	1.54	L/g	0.603	mg/L	mg/g	--	--	Abe et al., 1983
GAC from coal	--	--	--	--	--	--	9.58	L/g	0.403	mg/L	mg/g	--	--	Abe et al., 1983
GAC from coconut	--	--	--	--	--	--	17.4	L/g	0.384	mg/L	mg/g	--	--	Abe et al., 1983
2-Isobutyl-3-methoxypyrazine [24683-00-9]														
Aqua Nuchar	--	--	--	--	--	--	4.3	L/mg	0.78	ng/L	ngm/g	--	--	Lalezary et al., 1986
Isophorone [78-59-1]														
Filtrasorb 300	--	--	--	--	5.3/7.4	--	32	L/g	0.39	mg/L	mg/g	--	--	Dobbs and Cohen, 1980
2-Isopropyl-3-methoxypyrazine [25733-40-4]														
Aqua Nuchar	--	--	--	--	--	--	4.7	L/mg	0.55	ng/L	ngm/g	--	--	Lalezary et al., 1986
Isoproturon [34123-59-6]														
Cultivated plot soil	30.8	44.1	25.1	--	5.8	--	1.8	L/kg	--	mg/L	mg/kg	0.79	2.35	Benoit et al., 1999
Grass strip (0–2 cm)	--	--	--	--	5.7	--	5.0	l?kg	--	mg/L	mg/kg	2.56	2.32	Benoit et al., 1999
Grass strip (2–6 cm)	32.9	40.4	26.7	--	6.0	--	3.5	l?kg	--	mg/L	mg/kg	2.18	2.20	Benoit et al., 1999
Grass strip (6–13 cm)	--	--	--	--	6.5	--	3.1	l?kg	--	mg/L	mg/kg	1.91	2.20	Benoit et al., 1999
Lead [7439-91-1], as Pb^{2+}														
Alligator soil	5.9	39.4	54.7	30.2[c]	4.8	--	1,807	L/kg	0.853[d]	mg/L	mg/kg	1.54	5.07	Buchter et al., 1989
Cecil soil	78.8	12.9	8.3	2.0[c]	5.7	--	236	L/kg	0.662[d]	mg/L	mg/kg	0.61	4.59	Buchter et al., 1989
Lafitte soil	60.7	21.7	17.6	26.9[c]	3.9	--	918	L/kg	0.558[d]	mg/L	mg/kg	11.6	3.90	Buchter et al., 1989
Molokai soil	25.7	46.2	28.2	11.0[c]	6.0	--	8,166	L/kg	1.678[d]	mg/L	mg/kg	1.67	5.69	Buchter et al., 1989
Norwood soil	79.2	18.1	2.8	4.1[c]	6.9	--	385	L/kg	0.741[d]	mg/L	mg/kg	0.21	5.26	Buchter et al., 1989
Olivier soil	4.4	89.4	6.2	8.6[c]	6.6	--	16,406	L/kg	0.998[d]	mg/L	mg/kg	0.83	6.30	Buchter et al., 1989
Spodosol sediment	90.2	6.0	3.8	2.7[d]	4.3	--	136	L/kg	0.743[d]	mg/L	mg/kg	1.98	3.84	Buchter et al., 1989
Windsor soil	76.8	20.5	2.8	2.0[c]	5.3	--	472	L/kg	0.743[d]	mg/L	mg/kg	2.03	4.37	Buchter et al., 1989
Leptophos [21609-90-5]														
Big Creek sediment	71	22	7	--	6.6	--	1,905	mL/g	0.94	pmol/mL	pmol/g	1.62	5.07	Sharom et al., 1980
Plainfield sand	91.5	1.5	7	--	7.0	--	501	mL/g	1.08	pmol/mL	pmol/g	0.41	5.09	Sharom et al., 1980

Compound [CAS No.]	Adsorbent	% sand	% silt	% clay	CEC[a]	pH	K_F	K_F units	1/n	C units	x/m units	f_{oc} %	log K_{oc}[b]	Reference
Levofloxacin [100986-85-4]	Germany soil	80.1	16.7	2.5	12.0	5.3	309.0	mL/g	--	mg/mL	mg/g	0.70	4.65	Nowara et al., 1997
Lindane [58-89-9]	Alfisol	12.9	64.3	19.6	--	7.45	7.7	mL/g	0.8	μg/mL	μg/g	0.76	3.00	Rippen et al., 1982
	Alligator soil	5.9	39.4	54.7	30.2[c]	4.8	108	L/kg	0.741[d]	mg/L	mg/kg	1.54	3.85	Buchter et al., 1989
	Beverly sal	56	30	14	--	6.8	16	mL/g	0.97	pmol/mL	pmol/g	1.45	3.04	Sharom et al., 1980
	Big Creek sediment	71	22	7	--	6.6	24	mL/g	0.96	pmol/mL	pmol/g	1.62	3.17	Sharom et al., 1980
	Ca-Staten peaty muck	--	--	--	--	--	377	mL/g	0.956	μg/mL	μg/g	12.76	3.47	10 °C, Mills and Biggar, 1969
	Ca-Staten peaty muck	--	--	--	--	--	331	mL/g	0.969	μg/mL	μg/g	12.76	3.41	20 °C, Mills and Biggar, 1969
	Ca-Staten peaty muck	--	--	--	--	--	269	mL/g	0.981	μg/mL	μg/g	12.76	3.32	30 °C, Mills and Biggar, 1969
	Ca-Staten peaty muck	--	--	--	--	--	197	mL/g	0.983	μg/mL	μg/g	12.76	3.19	40 °C, Mills and Biggar, 1969
	Ca-Venado clay	--	--	--	--	--	53.8	mL/g	0.866	μg/mL	μg/g	3.48	3.19	10 °C, Mills and Biggar, 1969
	Ca-Venado clay	--	--	--	--	--	45.7	mL/g	0.841	μg/mL	μg/g	3.48	3.12	20 °C, Mills and Biggar, 1969
	Ca-Venado clay	--	--	--	--	--	41.3	mL/g	0.845	μg/mL	μg/g	3.48	3.07	30 °C, Mills and Biggar, 1969
	Ca-Venado clay	--	--	--	--	--	36.1	mL/g	0.846	μg/mL	μg/g	3.48	3.02	40 °C, Mills and Biggar, 1969
	Clay	18	41	41	--	--	9.247	L/kg	0.989	mg/?l	mg/kg	0.85	3.04	Albanis et al., 1989
	Clay loam	--	--	--	12.4	5.91	9.82	L/g	0.794	mg/L	mg/g	1.42	2.84	Kishi et al., 1990
	Clay loam	--	--	--	35.0	4.89	79.4	L/g	0.938	mg/L	mg/g	10.40	2.88	Kishi et al., 1990
	Light clay	--	--	--	13.2	5.18	20.0	L/g	0.905	mg/L	mg/g	1.51	3.11	Kishi et al., 1990
	Light clay	--	--	--	28.3	5.26	37.4	L/g	0.805	mg/L	mg/g	3.23	3.08	Kishi et al., 1990
	MLSS	--	--	--	--	--	1.5	L/g	1.0	μg/L	μg/g	--	--	20 °C, Bell and Tsezos, 1987
	Organic soil	52	34	14	--	6.1	899	mL/g	0.98	pmol/mL	pmol/g	43.67	3.31	Sharom et al., 1980
	Plainfield sand	91.5	1.5	7	--	7.0	8	mL/g	0.88	pmol/mL	pmol/g	0.99	3.29	Sharom et al., 1980
	Sandy clay loam	50	25	25	--	--	6.180	L/kg	0.988	mg/L	mg/kg	1.07	2.76	Albanis et al., 1989
	Sandy loam	--	--	--	35.0	5.41	75.8	L/g	1.00	mg/L	mg/g	7.91	2.98	Kishi et al., 1990
	Silica gel	--	--	--	--	--	6.88	mL/g	0.938	μg/mL	μg/g	--	--	10 °C, Mills and Biggar, 1969
	Silica gel	--	--	--	--	--	4.62	mL/g	0.957	μg/mL	μg/g	--	--	20 °C, Mills and Biggar, 1969
Linuron [330-55-2]	Batcombe sl	--	--	--	--	--	--	--	--	--	--	--	2.43	Briggs, 1981
	Begbroke soil	66.0	18.4	15.6	--	7.1	10.8	mL/g	--	μg/mL	μg/g	1.93	2.74	Grover and Hance, 1969
	Cellulose	--	--	--	--	--	17	mL/g	1.1	μg/mL	μg/g	--	--	Rippen et al., 1982
	Filtrasorb 300	--	--	--	--	5.3/7.4	256	L/g	0.49	mg/L	mg/g	--	--	Dobbs and Cohen, 1980
	Medisaprist peat	--	--	--	73.9	6.77	371.53	L/kg	0.61	--	--	43.6	2.93	Franco et al., 1997
	Sandy loam	48.4	32.9	18.7	17.6	6.1	12.02	L/kg	0.62	--	--	1.72	2.84	Franco et al., 1997
	Sphagnofibrist peat	--	--	--	176.4	3.5	269.15	L/kg	0.53	--	--	41.3	2.81	Franco et al., 1997
Malathion [121-75-5]	MLSS	--	--	--	--	--	403	L/g	0.6	μg/L	μg/g	--	--	20 °C, Bell and Tsezos, 1987
	Pokkali soil	--	--	--	--	--	31.62	L/kg	0.80	mg/L	μg/g	--	--	Wahid and Sethunathan, 1979
MCPA [94-74-6]	Calciorthid sediment	70.0	19.3	10.7	14.7[c]	8.5	19.6	L/kg	0.313[d]	μg/mL	mg/kg	0.44	3.65	Buchter et al., 1989
	Cecil soil	78.8	12.9	8.3	2.0[c]	5.7	81.3	L/kg	0.564[d]	μg/kg	mg/kg	0.61	4.12	Buchter et al., 1989
	Lafitte soil	60.7	21.7	17.6	26.9[c]	3.9	190	L/kg	0.751[d]	mg/L	mg/kg	11.6	3.21	Buchter et al., 1989
	Molokai soil	25.7	46.2	28.2	11.0[c]	6.0	120	L/kg	0.960[d]	mg/L	mg/kg	1.67	3.86	Buchter et al., 1989

Chemical [CAS]	Sorbent					pH	K_f	units	C	units	x/m	units	K_{oc}	Reference
Mercury [7439-97-6], as Hg^{2+}	Norwood soil	79.2	18.1	2.8	4.1c	6.9	113	L/kg	0.582d	mg/L	0.21	mg/kg	4.73	Buchter et al., 1989
	Olivier soil	4.4	89.4	6.2	8.6c	6.6	129	L/kg	1.122d	mg/L	0.83	mg/kg	4.19	Buchter et al., 1989
	Rolla NT	—	—	—	15.2c	5.09	2.02	L/kg	—	—	2.37	—	1.93	Shang and Arshad, 1998
	Rolla CT	—	—	—	12.4c	5.82	0.73	L/kg	—	—	2.27	—	1.51	Shang and Arshad, 1998
	Rycroft NT	—	—	—	28.6c	4.94	7.35	L/kg	—	—	4.70	—	2.19	Shang and Arshad, 1998
	Rycroft CT	—	—	—	27.7	5.38	10.45	L/kg	—	—	4.45	—	2.37	Shang and Arshad, 1998
	Spodosol sediment	90.2	6.0	3.8	2.7c	4.3	86.3	L/kg	0.513d	mg/L	1.98	mg/kg	3.64	Buchter et al., 1989
	Webster soil	27.5	48.6	23.9	48.1c	7.6	299	L/kg	2.158d	mg/L	4.39	mg/kg	3.83	Buchter et al., 1989
	Windsor soil	76.8	20.5	2.8	2.0c	5.3	130	L/kg	0.681d	mg/L	2.03	mg/kg	3.81	Buchter et al., 1989
	Filtrasorb 400	—	—	—	—	—	20	dm^3/g	0.46	mg/dm^3	—	mg/g	—	McKay et al., 1985
	Kari soil	—	—	—	—	—	446.70	L/kg	1.10	mg/L	—	μg/g	—	Wahid and Sethunathan, 1979
	Ferric hydroxide	—	—	—	—	5-8	90.8	L/g	0.76	μmol/L	—	μmol/g	—	Lockwood and Chen, 1974
Metalaxyl [57837-19-1]	Bradford muck	76.9	17.0	6.1	—	—	31.62	nmol/kg	—	—	36.42	—	1.94	Sharom and Edgington, 1982
	Fox sal	24.1	53.0	23.0	—	—	0.45	nmol/kg	—	—	0.99	—	1.66	Sharom and Edgington, 1982
	Guelph loam	32.4	46.8	20.8	—	—	1.25	nmol/kg	—	—	3.31	—	1.58	Sharom and Edgington, 1982
	Honeywood loam	—	—	—	—	—	0.89	nmol/kg	—	—	2.61	—	1.53	Sharom and Edgington, 1982
Metamitron [41394-05-2]	Medisaprist peat	48.4	32.9	18.7	73.9	6.77	54.95	L/kg	0.38	mg/L	43.6	L/kg	2.10	Franco et al., 1997
	Sandy loam	—	—	—	17.6	6.1	0.061	L/kg	2.06	mg/L	1.72	L/kg	1.55	Franco et al., 1997
	Sphagnofibrist peat	—	—	—	176.4	3.5	23.99	L/kg	0.50	mg/L	41.3	L/kg	1.76	Franco et al., 1997
Methanol [72-43-5]	Dowex-1 resin	—	—	—	—	—	0.822	unitless	—	—	—	—	—	Small and Bremer, 1964
	Dowex-1-X2 resin	—	—	—	—	—	0.535	unitless	—	—	—	—	—	Small and Bremer, 1964
Methiocarb [2032-65-7]	Sandy loam soil	—	—	—	9.0	6.6	2.20	mL/g	—	—	—	—	2.25	Xu et al., 1999
	Various soils	—	—	—	—	—	—	—	—	—	—	—	2.32	Briggs, 1981
Methomyl [16752-77-5]	Filtrasorb 400	—	—	—	2.8	—	4.780	L/g	0.290	μg/L	1.24	μg/g	—	Speth and Miltner, 1998
2-Methoxyphenol [489-86-1]	Brookston clay loam	—	—	—	22.22	5.7	1.07	mL/g	0.56	μmol/mL	2.96	μmol/g	1.60	Boyd, 1982
3-Methoxyphenol [150-19-6]	Brookston clay loam	—	—	—	22.22	5.7	0.94	mL/g	0.50	μmol/mL	2.96	μmol/g	1.55	Boyd, 1982
4-Methoxyphenol [150-76-5]	Brookston clay loam	—	—	—	22.22	5.7	1.50	mL/g	0.80	μmol/mL	2.96	μmol/g	1.75	Boyd, 1982
Methyl bromide [74-83-9]	Carsetas loam sand	11.0	—	—	—	7.3	0.09	mL/g	—	—	1.46	mg/g	0.79	Gan and Yates, 1996
	Greenfield sal	9.5	—	—	—	7.4	0.07	mL/g	—	—	0.53	mg/g	1.12	Gan and Yates, 1996
	Linne clay loam	25.1	—	—	—	7.2	0.10	mL/g	—	—	1.73	mg/g	0.76	Gan and Yates, 1996
2-Methyl-1-butanol [137-32-6]	GAC from wood	—	—	—	—	—	1.06	L/g	0.658	mg/L	—	mg/g	—	Abe et al., 1983
	GAC from coal	—	—	—	—	—	8.98	L/g	0.407	mg/L	—	mg/g	—	Abe et al., 1983
	GAC from coconut	—	—	—	—	—	16.9	L/g	0.375	mg/L	—	mg/g	—	Abe et al., 1983
2-Methyl-2-butanol [75-85-4]	GAC from wood	—	—	—	—	—	0.456	L/g	0.703	mg/L	—	mg/g	—	Abe et al., 1983
	GAC from coal	—	—	—	—	—	6.91	L/g	0.391	mg/L	—	mg/g	—	Abe et al., 1983
	GAC from coconut	—	—	—	—	—	11.1	L/g	0.374	mg/L	—	mg/g	—	Abe et al., 1983
3-Methyl-2-butanol [598-75-4]	GAC from wood	—	—	—	—	—	0.843	L/g	0.661	mg/L	—	mg/g	—	Abe et al., 1983
	GAC from coal	—	—	—	—	—	4.76	L/g	0.492	mg/L	—	mg/g	—	Abe et al., 1983
	GAC from coconut	—	—	—	—	—	9.61	L/g	0.432	mg/L	—	mg/g	—	Abe et al., 1983
3-Methylcholanthrene [56-49-5]	EPA-B2 sediment	67.5	18.6	13.9	3.72	6.35	15,140	mL/g	—	ng/mL	1.21	ng/g	6.10	Means et al., 1980
	EPA-4 sediment	3.0	55.2	41.8	23.72	7.79	30,085	mL/g	—	ng/mL	2.07	ng/g	6.16	Means et al., 1980

Compound [CAS No.]	Adsorbent	% sand	% silt	% clay	CEC[a]	pH	K_F	K_F units	1/n	C units	x/m units	f_oc %	log K_oc[b]	Reference
	EPA-5 sediment	33.6	31.0	35.4	19.00	7.44	8,273	mL/g	--	ng/mL	ng/g	2.28	5.56	Means et al., 1980
	EPA-6 sediment	0.2	68.6	31.2	33.01	7.83	15,820	mL/g	--	ng/mL	ng/g	0.72	6.34	Means et al., 1980
	EPA-8 sediment	82.4	6.8	10.7	3.72	8.32	2,257	mL/g	--	ng/mL	ng/g	0.15	6.18	Means et al., 1980
	EPA-9 soil	7.1	17.4	75.6	12.40	8.34	2,694	mL/g	--	ng/mL	ng/g	0.11	6.39	Means et al., 1980
	EPA-14 soil	2.1	63.6	34.4	18.86	4.54	30,627	mL/g	--	ng/mL	ng/g	0.48	6.80	Means et al., 1980
	EPA-15 sediment	15.6	35.7	48.7	11.30	7.79	23,080	mL/g	--	ng/mL	ng/g	0.95	6.39	Means et al., 1980
	EPA-18	34.6	39.5	25.8	15.43	7.76	20,642	mL/g	--	ng/mL	ng/g	0.66	6.50	Means et al., 1980
	EPA-20 sediment	0.0	28.6	71.4	8.50	5.50	16,231	mL/g	--	ng/mL	ng/g	1.30	6.10	Means et al., 1980
	EPA-21	50.2	7.1	42.7	8.33	7.60	24,506	mL/g	--	ng/mL	ng/g	1.88	6.12	Means et al., 1980
	EPA-22	26.1	21.2	52.7	8.53	7.55	20,972	mL/g	--	ng/mL	ng/g	1.67	6.10	Means et al., 1980
	EPA-23	17.3	69.1	13.6	31.15	6.70	17,127	mL/g	--	ng/mL	ng/g	2.38	5.86	Means et al., 1980
	EPA-26	1.6	42.9	55.4	20.86	7.75	37,364	mL/g	--	ng/mL	ng/g	1.48	6.40	Means et al., 1980
Methylene chloride [75-09-2]	Filtrasorb 300	--	--	--	--	5.8	1.30	L/g	1.16	mg/L	mg/g	--	--	Dobbs and Cohen, 1980
Methyl iodide [74-88-4]	Carsetas loam sand	--	--	11.0	--	7.3	0.16	mL/g	--	--	--	1.46	1.04	Gan and Yates, 1996
	Greenfield sal	--	--	9.5	--	7.4	0.09	mL/g	--	--	--	0.53	1.23	Gan and Yates, 1996
	Linne clay loam	--	--	25.1	--	7.2	0.15	mL/g	--	--	--	1.73	0.94	Gan and Yates, 1996
3-Methyl-4-nitrophenol [2581-34-2]	Lake Superior sediment	--	--	--	--	7.1	3.0	mL/g	1.10[d]	µg/mL	µg/g	1.68	2.25	Baarschers et al., 1983
	Ontario Gleysol	--	--	--	--	5.5	2.1	mL/g	1.07[d]	µg/mL	µg/g	0.52	2.61	Baarschers et al., 1983
	Ontario Gleysol	--	--	--	--	4.7	39.8	mL/g	0.80[d]	µg/mL	µg/g	3.54	3.05	Baarschers et al., 1983
	Ontario Luvisol	--	--	--	--	4.7	43.7	mL/g	0.51[d]	µg/mL	µg/g	3.19	3.14	Baarschers et al., 1983
	Potting soil	--	--	--	--	5.3	147.8	mL/g	0.56[d]	µg/mL	µg/g	19.20	2.89	Baarschers et al., 1983
Methyl parathion [298-00-0]	Cecil sal	65.8	19.5	14.7	6.8	5.6	3.95	--	0.85[d]	--	--	0.90	2.64	Rao and Davidson, 1979
	Eustis fine sand	93.8	3.0	3.2	5.2	5.6	2.72	--	0.86[d]	--	--	0.56	2.69	Rao and Davidson, 1979
	Webster sicl	18.4	45.3	38.3	54.7	7.3	13.39	--	0.75[d]	--	--	3.87	2.54	Rao and Davidson, 1979
4-Methyl-2-pentanone [108-10-1]	El-Nahda calcareous soil	--	--	71.9	40.0	--	0.0430	L/g	1.155	µM/L	µM/g	1.00	0.63	Kishk et al., 1979
	Filtrasorb 400	--	--	--	--	9.1	8.850	L/g	0.279	µg/L	µg/g	--	--	Speth and Miltner, 1998
	Tahreer desert sand	--	--	5.1	4.25	--	0.0120	L/g	1.150	µM/L	µM/g	0.25	0.68	Kishk et al., 1979
	Tel El-Kabeer alluvium	--	--	55.9	24.5	--	0.0056	L/g	1.225	µM/L	µM/g	0.37	0.18	Kishk et al., 1979
Metobromuron [3060-89-7]	Batcombe sl	61	28	11	10.2	--	--		--	--	--	--	2.02	Briggs, 1981
Metolachlor [51218-45-2]	Cape Fear sal	--	--	--	--	--	8.52	mL/g	--	µg/mL	µg/g	5.34	2.20	Kozak et al., 1983
	Cellulose	--	--	--	--	--	460	mL/g	--	--	--	--	2.90	Torrents et al., 1997
	Chitin	--	--	--	--	--	225	mL/g	--	--	--	--	2.60	Torrents et al., 1997
	Collagen	--	--	--	--	--	3.20	mL/g	--	--	--	--	0.86	Torrents et al., 1997
	Fulvic acid	--	--	--	--	3.46	0.071	mL/g	--	--	--	--	--	Mathew and Khan, 1996
	Fulvic acid	--	--	--	--	7.00	0.055	mL/g	--	--	--	--	--	Mathew and Khan, 1996
	Goethite	--	--	--	--	4.41	0.007	mL/g	--	--	--	--	--	Mathew and Khan, 1996

Chemical / Medium	%sand	%silt	%clay	CEC	pH	K	(units)	C	(units)	S	(units)	log Koc	Reference
Goethite	--	--	--	--	7.00	0.002	mL/g	--	--	--	--	--	Mathew and Khan, 1996
Kaolinite	--	--	--	--	5.54	0.043	mL/g	--	--	--	--	--	Mathew and Khan, 1996
Kaolinite	--	--	--	--	7.00	0.025	mL/g	--	--	--	--	--	Mathew and Khan, 1996
Lignin	--	--	--	--	--	3.50	mL/g	--	--	--	--	0.90	Torrents et al., 1997
Montmorillonite	--	--	--	--	5.47	0.047	mL/g	--	--	--	--	--	Mathew and Khan, 1996
Montmorillonite	--	--	--	--	7.00	0.030	mL/g	--	--	--	--	--	Mathew and Khan, 1996
Norfolk fsl	87	11	2	2.3	--	2.20	mL/g	1.19	µg/mL	0.99	µg/g	2.35	Kozak et al., 1983
Rains fine loamy sand	52	41	7	7.1	--	3.20	mL/g	1.03	µg/mL	1.45	µg/g	2.34	Kozak et al., 1983
Taloka sl	--	--	7.3	--	6.2	0.657	L/kg	0.95	mg/L	0.64	mg/kg	2.01	Barnes et al., 1992
Taloka sl	--	--	6.6	--	6.4	0.347	L/kg	0.97	mg/L	0.58	mg/kg	1.78	Barnes et al., 1992
Taloka sl	--	--	6.7	--	6.1	0.285	L/kg	1.00	mg/L	0.58	mg/kg	1.69	Barnes et al., 1992
Taloka sl	--	--	7.3	--	5.4	0.221	L/kg	--	mg/L	0.46	mg/kg	1.68	Barnes et al., 1992
Taloka sl	--	--	8.8	--	4.9	0.134	L/kg	--	mg/L	0.46	mg/kg	1.46	Barnes et al., 1992
Metoxuron [108-44-1] Batcombe sl	--	--	--	--	--	--	--	--	--	--	--	1.74	Briggs, 1981
Mevinphos [26718-65-0] Organic soil	52	34	14	--	6.1	19	mL/g	0.95	pmol/mL	43.67	pmol/g	1.64	Sharom et al., 1980
Molybdenum [7439-98-7], as Mo₇O₂₄⁶⁻ ($Mo_7O_{24}^{6-}$) Alligator soil	5.9	39.4	54.7	30.2[c]	4.8	57.5	L/kg	0.882[d]	mg/L	1.54	mg/kg	3.57	Buchter et al., 1989
Cecil soil	78.8	12.9	8.3	2.0[c]	5.7	18.0	L/kg	0.617[d]	mg/L	0.61	mg/kg	3.47	Buchter et al., 1989
Lafitte soil	60.7	21.7	17.6	26.9[c]	3.9	81.5	L/kg	0.607[d]	mg/L	11.6	mg/kg	2.85	Buchter et al., 1989
Molokai soil	25.7	46.2	28.2	11.0[c]	6.0	118	L/kg	0.664[d]	mg/L	1.67	mg/kg	3.85	Buchter et al., 1989
Spodosol sediment	90.2	6.0	3.8	2.7[c]	4.3	25.6	L/kg	0.451[d]	mg/L	1.98	mg/kg	3.11	Buchter et al., 1989
Monolinuron [1746-81-2] Batcombe sl	--	--	--	--	--	--	--	--	--	--	--	1.84	Briggs, 1981
Monuron [150-68-5] Windsor soil	76.8	20.5	2.8	2.0[c]	5.3	43.8	L/kg	0.544[d]	mg/L	2.03	mg/kg	3.33	Buchter et al., 1989
Alluvial soil (OS-26)	--	--	--	--	--	6.0	L/kg	0.70[d]	mg/L	--	--	--	Yuen and Hilton, 1962
Alluvial soil (K-47)	--	--	--	--	--	4.8	L/kg	0.70[d]	mg/L	--	--	--	Yuen and Hilton, 1962
Batcombe sl	--	--	--	--	--	--	--	--	--	--	--	1.70	Briggs, 1981
Regosol soil (K-4)	--	--	--	--	--	1.2	L/kg	0.86[d]	mg/L	--	--	--	Yuen and Hilton, 1962
Latosol (WA-Op 9)	--	--	--	--	--	1.0	L/kg	0.80[d]	mg/L	--	--	--	Yuen and Hilton, 1962
Latosol (OS-27A)	--	--	--	--	--	0.9	L/kg	0.88[d]	mg/L	--	--	--	Yuen and Hilton, 1962
Latosol (OS-27A)	--	--	--	--	--	0.7	L/kg	0.94[d]	mg/L	--	--	--	Yuen and Hilton, 1962
H-montmorillonite	--	--	--	73.5	9.0	100	L/g	0.98	µmol/mL	--	µmol/g	--	Bailey et al., 1968
Na-montmorillonite	--	--	--	87.0	9.0	24	L/g	0.48	µmol/mL	--	µmol/g	--	Bailey et al., 1968
Naphthalene [91-20-3] Alfisol	12.9	64.3	19.6	--	7.45	9.8	mL/g	0.88	µg/mL	0.76	µg/g	3.11	Rippen et al., 1982
Alumina	--	--	--	--	--	1	mL/g	0.86	µg/mL	--	--	--	Rippen et al., 1982
Apison, TN soil	4	10	86	76	4.5	--	unitless	--	--	0.11	mg/kg	3.00	Southworth and Keller, 1986
Bjodstrup clayey till	25	34	41	40.5	7.64	2.1	L/kg	0.72	mg/L	0.13	µg/g	3.20	Lokke, 1984
Cellulose	--	--	--	--	--	125	mL/kg	1.42	µg/mL	--	--	--	Rippen et al., 1982
Clay loam	--	--	--	12.4	5.91	6.28	L/g	0.912	mg/L	1.42	mg/g	2.64	Kishi et al., 1990
Clay loam	--	--	--	35.0	4.89	56.2	L/g	0.996	mg/L	10.40	mg/g	2.73	Kishi et al., 1990
Diatomaceous earth	--	--	--	--	--	0.8	mL/g	0.86	µg/mL	--	--	--	Rippen et al., 1982
Dormont, WV soil	2	38	60	129	4.2	--	unitless	--	--	1.2	µg/g	2.60	Southworth and Keller, 1986

Compound [CAS No.]	Adsorbent	% sand	% silt	% clay	CEC[a]	pH	K_F	K_F units	1/n	C units	x/m units	f_{oc} %	log K_{oc}[b]	Reference
	Entisol	8.5	68.3	20.6	--	7.9	16	mL/g	0.61	μg/mL	μg/g	1.11	3.16	Rippen et al., 1982
	Esrum sandy till	78	4	18	9.1	4.71	0.32	L/kg	0.77	mg/L	mg/kg	0.06	2.72	Løkke, 1984
	Fuller's earth	--	--	--	--	--	6.2	mL/g	0.85	μg/mL	μg/g	--	--	Rippen et al., 1982
	Fullerton, TN soil	11	21	68	64	4.4	--	unitless	--	--	--	0.05	2.98	Southworth and Keller, 1986
	Filtrasorb 300	--	--	--	--	5.6	132	L/g	0.42	mg/L	mg/g	--	--	Dobbs and Cohen, 1980
	Filtrasorb 300	--	--	--	--	5.0	123	L/g	0.41	mg/L	mg/g	--	--	Burks, 1981
	Filtrasorb 400	--	--	--	--	--	277	L/g	0.43	mg/L	mg/g	--	--	Walters and Luthy, 1982
	Gribskov, A-horizon	87	6	3	4.8	3.23	18	L/kg	0.82	mg/L	mg/kg	1.41	3.11	Løkke, 1984
	Gribskov, B-horizon	82	4	7	9.6	3.59	14	L/kg	0.76	mg/L	mg/kg	2.58	2.72	Løkke, 1984
	Gribskov, C-horizon	88	3	5	7.0	4.07	9	L/kg	0.84	mg/L	mg/kg	1.82	2.72	Løkke, 1984
	Hydrodarco KB	--	--	--	--	--	58	L/g	0.276	mg/L	mg/g	--	--	Fochtman and Eisenberg, 1979
	Light clay	--	--	--	13.2	5.18	12.6	L/g	0.842	mg/L	mg/g	1.51	2.92	Kishi et al., 1990
	Light clay	--	--	--	28.3	5.26	24.0	L/g	0.810	mg/L	mg/g	3.23	2.87	Kishi et al., 1990
	Liposorb LP 650	--	--	--	--	--	300	mL/g	0.80	μg/mL	μg/g	--	--	Rippen et al., 1982
	Roskilde soil	66	18	12	14.0	5.40	14	L/kg	0.84	mg/L	mg/kg	1.64	2.93	Løkke, 1984
	Sandy loam	--	--	--	35.0	5.41	33.1	L/g	0.785	mg/L	mg/g	7.91	2.62	Kishi et al., 1990
	Silica gel	--	--	--	--	--	3	mL/g	0.88	μg/mL	μg/g	--	--	Rippen et al., 1982
	Speyer soil 2.1	--	--	--	--	7.0	22	mL/g	1.04	μg/mL	μg/g	0.69	3.50	Rippen et al., 1982
	Speyer soil 2.2	--	--	--	--	5.8	610	mL/g	1.05	μg/mL	μg/g	2.24	4.43	Rippen et al., 1982
	Speyer soil 2.3	--	--	--	--	7.1	18	mL/g	0.97	μg/mL	μg/g	1.12	3.21	Rippen et al., 1982
	Strodam, AB-horizon	84	5	4	13.0	3.88	30	L/kg	0.88	mg/L	mg/kg	5.11	2.76	Løkke, 1984
	Strodam, C-horizon	92	3	3	1.6	4.95	1.9	L/kg	0.73	mg/L	mg/kg	0.09	3.32	Løkke, 1984
	Tirstrup subsoil	90	7	3	1.4	6.14	0.92	L/kg	0.87	mg/L	mg/kg	0.05	3.26	Løkke, 1984
	Tisvilde, C-horizon	96	1	2	1.3	4.21	2.4	L/kg	0.57	mg/L	mg/kg	0.15	3.20	Løkke, 1984
	Various soils	--	--	--	--	--	--	--	--	--	--	--	2.62	Briggs, 1981
1-Naphthalenemethanol [4780-79-4]	Apison, TN soil	4	10	86	76	4.5	--	unitless	--	--	--	0.11	2.22	Southworth and Keller, 1986
	Dormont, WV soil	2	38	60	129	4.2	--	unitless	--	--	--	1.2	2.43	Southworth and Keller, 1986
	Fullerton, TN soil	11	21	68	64	4.4	--	unitless	--	--	--	0.05	<0.70	Southworth and Keller, 1986
1-Naphthol [90-15-3]	Filtrasorb 300	--	--	--	--	5.3/7.4	180	L/g	0.32	mg/L	mg/g	--	--	Dobbs and Cohen, 1980
2-Naphthol [135-19-3]	Filtrasorb 300	--	--	--	--	5.3/7.4	200	L/g	0.26	mg/L	mg/g	--	--	Dobbs and Cohen, 1980
1-Naphthylamine [134-32-7]	Filtrasorb 300	--	--	--	--	5.3/7.4	160	L/g	0.34	mg/L	mg/g	--	--	Dobbs and Cohen, 1980
	Milford sicl	4.3	58.7	37.0	45.4[c]	6.4	49.7	L/kg	0.992	nmol/L	nmol/kg		3.58	Graveel et al., 1986
	Morocco silty sand	76.6	21.3	2.1	16.9[c]	4.7	34.3	L/kg	1.062	nmol/L	nmol/kg		3.43	Graveel et al., 1986
	Oakville silty sand	83.2	12.7	4.1	10.2[c]	5.2	98.2	L/kg	1.061	nmol/L	nmol/kg		3.50	Graveel et al., 1986
2-Naphthylamine [91-59-8]	Filtrasorb 300	--	--	--	--	5.3/7.4	150	L/g	0.30	mg/L	mg/g	--	--	Dobbs and Cohen, 1980
	Hydrodarco KB	--	--	--	--	--	67	L/g	0.395	mg/L	mg/g	--	--	Fochtman and Eisenberg, 1979
Napropamide [5299-99-7]	Bet Dagan I soil	--	--	13.7	--	7.9	1.40	mL/g	--	--	--	0.40	2.54	Gerstl and Yaron, 1983

Chemical	Soil	% sand	% silt	% clay	CEC	pH	K	units	1/n	units	units	% OC	log Koc	Reference
	Bet Dagan II soil	--	--	42.5	--	7.8	2.96	mL/g	--			1.01	2.47	Gerstl and Yaron, 1983
	Gilat soil	--	--	23.1	--	7.8	1.92	mL/g	--			0.55	2.54	Gerstl and Yaron, 1983
	Al-montmorillonite	--	--	--	--	6.4	1.03	L/kg	0.59			--	--	Lee et al., 1990
	Cu-montmorillonite	--	--	--	--	6.4	0.84	L/kg	0.57			--	--	Lee et al., 1990
	Na-montmorillonite	--	--	--	--	6.4	0.33	L/kg	0.48			--	--	Lee et al., 1990
	Neve Yaar soil	--	--	70.0	--	7.7	2.94	mL/g	--			1.18	2.40	Gerstl and Yaron, 1983
	Mivtachim soil	--	--	7.5	--	8.5	0.27	mL/g	--			0.06	2.65	Gerstl and Yaron, 1983
	Shefer soil	--	--	70.0	--	7.2	2.35	mL/g	--			0.72	2.51	Gerstl and Yaron, 1983
	Tujunga loamy sand	--	--	--	--	6.7	1.14	L/kg	--			--	--	Lee et al., 1990
Nickel [7440-02-0], as Ni²⁺	Alligator soil	5.9	39.4	54.7	30.2[c]	4.8	37.8	L/kg	0.939[d]	mg/kg	mg/L	1.54	3.39	Buchter et al., 1989
	Calciorthid sediment	70.0	19.3	10.7	14.7[c]	8.5	206	L/kg	0.504[d]	mg/kg	mg/L	0.44	4.67	Buchter et al., 1989
	Cecil soil	78.8	12.9	8.3	2.0[c]	5.7	6.84	L/kg	0.688[d]	mg/kg	mg/L	0.61	3.05	Buchter et al., 1989
	Lafitte soil	60.7	21.7	17.6	26.9[c]	3.9	50.1	L/kg	0.903[d]	mg/kg	mg/L	11.6	2.64	Buchter et al., 1989
	Molokai soil	25.7	46.2	28.2	11.0[c]	6.0	44.9	L/kg	0.720[d]	mg/kg	mg/L	1.67	3.43	Buchter et al., 1989
	Norwood soil	79.2	18.1	2.8	4.1[c]	6.9	20.9	L/kg	0.661[d]	mg/kg	mg/L	0.21	4.00	Buchter et al., 1989
	Olivier soil	4.4	89.4	6.2	8.6[c]	6.6	50.5	L/kg	0.646[d]	mg/kg	mg/L	0.83	3.78	Buchter et al., 1989
	Spodosol sediment	90.2	6.0	3.8	2.7[c]	4.3	3.44	L/kg	0.836[d]	mg/kg	mg/L	1.98	2.24	Buchter et al., 1989
	Webster soil	27.5	48.6	23.9	48.1[c]	7.6	337	L/kg	0.748[d]	mg/kg	mg/L	4.39	3.89	Buchter et al., 1989
	Windsor soil	76.8	20.5	2.8	2.0[c]	5.3	8.43	L/kg	0.741[d]	mg/kg	mg/L	2.03	2.62	Buchter et al., 1989
Nicosulfuron [111991-09-4]	Canisteo loam	--	--	25.3	32.1[c]	7.5	4.98	L/kg	--	mg/kg	mg/L	2.76	2.26	Gonzalez and Ukrainczyk, 1999
	Crippen sicl	--	--	29.9	35.6[c]	7.8	3.70	L/kg	--	mg/kg	mg/L	2.75	2.13	Gonzalez and Ukrainczyk, 1999
	Fruitfield sand	--	--	2.1	4.2[c]	7.8	0.21	L/kg	--	mg/kg	mg/L	0.33	1.80	Gonzalez and Ukrainczyk, 1999
	Fruitfield sand	--	--	3.7	4.4[c]	6.2	0.36	L/kg	--	mg/kg	mg/L	0.20	2.26	Gonzalez and Ukrainczyk, 1999
	Fruitfield sand	--	--	3.9	4.1[c]	6.1	0.90	L/kg	--	mg/kg	mg/L	0.18	2.70	Gonzalez and Ukrainczyk, 1999
	Galva sl	--	--	51.1	29.9[c]	5.5	8.78	L/kg	--	mg/kg	mg/L	2.29	2.58	Gonzalez and Ukrainczyk, 1999
	Keomah sl	--	--	23.1	17.2[c]	6.0	3.12	L/kg	--	mg/kg	mg/L	2.36	2.12	Gonzalez and Ukrainczyk, 1999
	Webster clay loam	--	--	37.8	27.0[c]	8.2	4.00	L/kg	--	mg/kg	mg/L	1.47	2.43	Gonzalez and Ukrainczyk, 1999
Nitrapyrin [1929-82-4]	Cottenham sal	--	--	--	--	--	--		--			--	2.24	Briggs, 1981
4-Nitroaniline [100-01-6]	Filtrasorb 300	--	--	--	--	5.3/7.4	140		0.37	mg/g	mg/L	--	--	Dobbs and Cohen, 1980
Nitrobenzene [98-95-3]	Filtrasorb 300	--	--	--	--	5.3/7.4	68		0.43	mg/g	mg/L	--	--	Dobbs and Cohen, 1980
	Agricultural soil	65.2	25.6	9.2	9.0	7.4	0.61		--	mol/kg	mol/L	2.2	1.49	Seip et al., 1986
	Forest soil	69.5	20.5	10.1	2.9	4.2	3.81		--	mol/kg	mol/L	3.7	2.01	Seip et al., 1986
	Forest soil	97.3	2.2	0.5	0.48	5.6	0.94		--	mol/kg	mol/L	0.2	1.63	Seip et al., 1986
	Gribskov, B-horizon	82	4	7	9.6	3.59	5.5	L/kg	0.90	mg/kg	mg/L	2.58	2.32	5 °C, Løkke, 1984
	Gribskov, B-horizon	82	4	7	9.6	3.59	4.4	L/kg	0.92	mg/kg	mg/L	2.58	2.23	21 °C, Løkke, 1984
	Gribskov, C-horizon	88	3	5	7.0	4.07	3.1	L/kg	0.82	mg/kg	mg/L	1.82	2.23	5 °C, Løkke, 1984
	Gribskov, C-horizon	88	3	5	7.0	4.07	6.7	L/kg	0.66	mg/kg	mg/L	1.82	2.57	21 °C, Løkke, 1984
4-Nitrobiphenyl [92-93-3]	Filtrasorb 300	--	--	--	--	5.3/7.4	370		0.27	mg/g	mg/L	--	--	Dobbs and Cohen, 1980
2-Nitrophenol [88-75-5]	Brookston clay loam	--	--	--	22.22	5.7	3.05		0.89	μmol/g	μmol/mL	2.96	2.06	Boyd, 1982
3-Nitrophenol [554-84-7]	Filtrasorb 300	--	--	--	--	5.3/7.4	99		0.34	mg/g	mg/L	--	--	Dobbs and Cohen, 1980
	Brookston clay loam	--	--	--	22.22	5.7	1.42		0.73	μmol/g	μmol/mL	2.96	1.72	Boyd, 1982

Compound [CAS No.]	Adsorbent	% sand	% silt	% clay	CEC[a]	pH	K_F	K_F units	1/n	C units	x/m units	f_{oc} %	log K_{oc}[b]	Reference
4-Nitrophenol [100-02-7]	Bjødstrup clayey till	25	34	41	40.5	7.64	0.68	L/kg	0.99	mg/L	mg/kg	0.13	2.72	Løkke, 1984
	Brookston clay loam	-	-	-	22.22	5.7	1.48	mL/g	0.72	μmol/mL	μmol/g	2.96	1.74	Boyd, 1982
	Esrum sandy till	78	4	18	9.1	4.71	0.32	L/g	0.96	mg/L	mg/kg	0.06	2.72	Løkke, 1984
	Filtrasorb 300	-	-	-	-	5.3/7.4	76.2	L/g	0.25	-	mg/g	-	-	Dobbs and Cohen, 1980
	Gribskov, A-horizon	87	6	3	4.8	3.23	1.56	L/kg	0.93	mg/L	mg/kg	1.41	2.04	Løkke, 1984
	Gribskov, B-horizon	82	4	7	9.6	3.59	2.69	L/kg	0.91	mg/L	mg/kg	2.58	2.00	Løkke, 1984
	Gribskov, C-horizon	88	3	5	7.0	4.07	1.02	L/kg	0.86	mg/L	mg/kg	1.82	1.75	Løkke, 1984
	Roskilde soil	66	18	12	14.0	5.40	2.05	L/kg	0.91	mg/L	mg/kg	1.64	2.11	Løkke, 1984
	Strødam, AB-horizon	84	5	4	13.0	3.88	3.28	L/kg	0.91	mg/L	mg/kg	5.11	1.81	Løkke, 1984
	Strødam, C-horizon	92	3	3	1.6	4.95	0.25	L/kg	0.99	mg/L	mg/kg	0.09	2.45	Løkke, 1984
	Tirstrup subsoil	90	7	3	1.4	6.14	0.12	L/kg	0.79	mg/L	mg/kg	0.05	2.38	Løkke, 1984
	Tisvilde, C-horizon	96	1	2	1.3	4.21	0.16	L/kg	0.73	mg/L	mg/kg	0.15	2.00	Løkke, 1984
N-Nitrosodimethylamine [62-75-9]	Filtrasorb 300	-	-	-	-	5.3/7/4	6.8×10^{-5}	L/g	6.6	mg?l	mg/g	-	-	Dobbs and Cohen, 1980
	Hydrodarco KB	-	-	-	-	-	1.4×10^{-6}	L/g	8.15	mg/L	mg/g	-	-	Fochtman and Eisenberg, 1979
N-Nitrosodiphenylamine [86-30-6]	Filtrasorb 300	-	-	-	-	5.3/7/4	220	L/g	0.37	mg?l	mg/g	-	-	Dobbs and Cohen, 1980
4-Nonylphenol [104-40-5]	Filtrasorb 300	-	-	-	-	5.3/7/4	250	L/g	0.37	mg/L	mg/g	-	-	Dobbs and Cohen, 1980
Norflurazon [27314-13-2]	Lexington sl CT	19	71	10	-	5.2	6.47	L/kg	-	mg/L	mg/g	1.42	2.66	Suba and Essington, 1999
	Lexington sl NT	11	76	13	-	5.1	12.15	L/kg	-	mg/L	mg/g	2.45	2.70	Suba and Essington, 1999
	Lexington sl SUB	8	75	16	-	5.3	2.04	L/kg	-	mg/L	mg/g	0.37	2.74	Suba and Essington, 1999
Oxamyl [23135-22-0]	Arredondo sand	93.8	3.0	3.2	-	6.8	0.06	μgmL/g	0.94[d]	-	-	0.80	0.90	Bilkert and Rao, 1985
	Cecil sal	65.8	19.5	14.7	-	5.6	0.05	μgmL/g	0.95[d]	-	-	0.90	0.78	Bilkert and Rao, 1985
	Cottenham sal	-	-	-	-	-	-	-	-	-	-	-	0.71	Briggs, 1981
	Webster sicl	18.4	45.3	38.3	-	7.3	0.40	μgmL/g	0.97[d]	-	-	3.97	1.00	Bilkert and Rao, 1985
2,2'-Oxybis(1-chloropropane) [108-60-1]	Filtrasorb 300	-	-	-	-	5.3/7.4	24	L/g	0.57	mg/L	mg/g	-	-	Dobbs and Cohen, 1980
Parathion [56-38-2]	Batcombe sl	-	-	-	-	-	-	-	-	-	-	2.05	3.02	Briggs, 1981
	Beverly sal	56	30	14	-	6.8	14	mL/g	1.01	pmol/mL	pmol/g	1.45	2.98	Sharom et al., 1980
	Big Creek sediment	71	22	7	-	6.6	22	mL/g	0.99	pmol/mL	pmol/g	1.62	3.13	Sharom et al., 1980
	Clarion soil	37	42	21	21.02	5.00	33.81	L/kg	0.88	μM/L	μM/kg	2.64	3.11	Felsot and Dahm, 1979
	Harps soil	21	55	24	37.84	7.30	41.12	L/kg	0.80	μM/L	μM/kg	3.80	3.03	Felsot and Dahm, 1979
	Organic soil	52	34	14	-	6.1	741	mL/g	0.95	pmol/mL	pmol/g	43.67	3.23	Sharom et al., 1980
	Plainfield sand	91.5	1.5	7	-	7.0	7	mL/g	0.98	pmol/mL	pmol/g	0.41	3.24	Sharom et al., 1980
	Peat	42	39	19	77.34	6.98	254.68	L/kg	0.81	μM/L	μM/kg	18.36	3.14	Felsot and Dahm, 1979
	Sarpy fsl	77	15	8	5.71	7.30	5.68	L/kg	0.83	μM/L	μM/kg	0.51	3.05	Felsot and Dahm, 1979
	Thurman fsl	83	9	8	6.01	6.83	11.07	L/kg	0.83	μM/L	μM/kg	1.07	3.01	Felsot and Dahm, 1979
Pencycuron [66063-05-6]	Sandy loam soil	-	-	-	9.0	6.6	26.5	mL/g	-	-	-	1.24	3.33	Xu et al., 1999

Compound	Adsorbent					pH	K_F	units	$1/n$	units	units	value	log	Reference
Pentachlorophenol [87-86-5]	Filtrasorb 300					5.3/7.4	150	L/g	0.42	mg/L	mg/g	–	–	Dobbs and Cohen, 1980
	Agricultural soil	65.2	25.6	9.2	9.0	7.4	6.4	L/kg	–	mol/L	mol/kg	2.2	2.47	Seip et al., 1986
	Forest soil	69.5	20.5	10.1	2.9	4.2	>17	L/kg	–	mol/L	mol/kg	3.7	>2.68	Seip et al., 1986
	Forest soil	97.3	2.2	0.5	0.48	5.6	1.84	L/kg	–	mol/L	mol/kg	0.2	2.96	Seip et al., 1986
	Holten humic-rich sand					4.7	398.1	L/kg	1.00	µg/L	µg/kg	0.032	–	Lagas, 1988
	Kootwijk humic sand					3.4	158.5	L/kg	1.11	µg/L	µg/kg	0.017	–	Lagas, 1988
	Maasdijk light loam					7.5	12.6	L/kg	1.11	µg/L	µg/kg	0.009	–	Lagas, 1988
	Opijnen heavy loam					7.1	31.6	L/kg	1.25	µg/L	µg/kg	0.017	–	Lagas, 1988
	Rolde humic sand					4.9	158.5	L/kg	1.11	µg/L	µg/kg	0.022	–	Lagas, 1988
	Schipluiden peat					4.6	316.2	L/kg	1.25	µg/L	µg/kg	0.298	–	Lagas, 1988
	H-montmorillonite				73.5	9.0	58	L/g	0.17	µmol/L	µmolg			Bailey et al., 1968
	Na-montmorillonite				87.0	9.0	130	L/g	0.90	µmol/L	µmolg			Bailey et al., 1968
Pentanochlor [2307-68-8]	Dowex-1 resin						1.38	unitless		–	–			Small and Bremer, 1964
	Dowex-1-X2 resin						10.8	unitless		–	–			Small and Bremer, 1964
1-Pentanol [71-41-0]	GAC from wood						2.13	L/g	0.561	mg/L	mg/g			Abe et al., 1983
	GAC from coal						10.5	L/g	0.454	mg/L	mg/g			Abe et al., 1983
2-Pentanol [6032-29-7]	GAC from coconut					4.5	25.6	L/g	0.331	mg/L	mg/g			Abe et al., 1983
3-Pentanol [584-02-1]	GAC from coal					4.2	9.89	L/g	0.400	mg/L	mg/g			Abe et al., 1983
	GAC from coal					4.4	6.67	L/g	0.469	mg/L	mg/g			Abe et al., 1983
Permethrin [52645-53-1]	Lake St. George	18	34	48				L/kg		µg/L	µg/kg	24.94	3.19	Sharom and Solomon, 1981
Phenanthrene [85-01-8]	Filtrasorb 300					5	389	L/g	0.45	mg/L	mg/g			Burks, 1981
	Filtrasorb 300					5.3/7.4	135	L/g	0.44	mg/L	mg/g			Dobbs and Cohen, 1980
	Filtrasorb 300						215	L/g	0.406	mg/L	mg/g			Walters and Luthy, 1982
	Apison, TN soil	4	10	86	76	4.5		unitless		–	–	0.11	3.77	Southworth and Keller, 1986
	Dormont, WV soil	2	38	60	129	4.2		unitless		–	–	1.2	3.76	Southworth and Keller, 1986
	Fullerton, TN soil	11	21	68	64	4.4		unitless		–	–	0.05	3.15	Southworth and Keller, 1986
sec-Phenethyl alcohol [98-85-1]	Apison, TN soil	4	10	86	76	4.5		unitless		–	–	0.11	1.57	Southworth and Keller, 1986
	Dormont, WV soil	2	38	60	129	4.2		unitless		–	–	1.2	1.72	Southworth and Keller, 1986
	Fullerton, TN soil	11	21	68	64	4.4		unitless		–	–	0.05	<0.70	Southworth and Keller, 1986
Phenol [108-95-2]	Amberlite XAD-4						273	L/g	0.76	mg/L	mg/g			Aguwa et al., 1984
	Batcombe sl					6.7	0.91	mL/g	0.79	µmol/mL	µmol/g	1.46	1.72	Briggs, 1981
	Brookston clay loam				22.22	5.7	0.48	L/g	0.54	mg/L	mg/g	2.96	1.21	Boyd, 1982
	Filtrasorb 300					5.3/7.4	21	L/g	0.53	mg/L	mg/g			Dobbs and Cohen, 1980
	Filtrasorb 300					4.8-5.6	29	dm³/g	0.156	mg/dm³	mg/g			Knettig et al., 1986
	Filtrasorb 400					7	49.8	L/mg	0.26	mg/L	mg/g			Belfort et al., 1983
	Filtrasorb 400						50	L/g	0.371	mg/L	mg/g			McKay et al., 1985
	Filtrasorb 400					7	0.037	L/g	0.212	mg/L	mg/g			Sheindorf et al., 1982
	Filtrasorb 400					7	78.1	L/g	0.894	mg/L	mg/g			Peel and Benedek, 1980
	Poly(4-vinyl pyridine)						0.223	L/g	0.917	mg/L	mg/g			Chanda et al., 1985
	Polybenzimidazole						0.079	L/g		mg/L	mg/g			Chanda et al., 1985
	Apison, TN soil	4	10	86	76	4.5		unitless		–	–	0.11	1.74	Southworth and Keller, 1986

Compound [CAS No.]	Adsorbent	% sand	% silt	% clay	CEC[a]	pH	K_F	K_F units	1/n	C units	x/m units	f_{oc} %	log K_{oc}[b]	Reference
4-Phenoxyphenyl urea [78508-44-8]	Dormont, WV soil	2	38	60	129	4.2	--	unitless	--	--	--	1.2	0.85	Southworth and Keller, 1986
	Fullerton, TN soil	11	21	68	64	4.4	--	unitless	--	--	--	0.05	2.85	Southworth and Keller, 1986
3-Phenyl-1-cyclohexyl urea [886-59-9]	Batcombe sl	--	--	--	--	--	--	--	--	--	--	--	2.56	Briggs, 1981
3-Phenyl-1-cyclopropyl urea [886-59-9]	Batcombe sl	--	--	--	--	--	--	--	--	--	--	--	2.07	Briggs, 1981
3-Phenyl-1-cyclopropyl urea [13140-86-8]	Batcombe sl	--	--	--	--	--	--	--	--	--	--	--	1.72	Briggs, 1981
Phenylmercuric acetate [62-38-4]	Filtrasorb 300	--	--	--	--	5.3/7.4	270	L/g	0.44	mg/L	mg/g	--	--	Dobbs and Cohen, 1980
Phenyl urea [64-10-8]	Batcombe sl	--	--	--	--	--	--	--	--	--	--	--	1.35	Briggs, 1981
Phorate [298-02-2]	Batcombe sl	--	--	--	--	--	--	--	--	--	--	2.05	2.82	Briggs, 1981
	Clarion soil	37	42	21	21.02	5.00	9.62	L/kg	0.92	μM/L	μM/kg	2.64	2.56	Felsot and Dahm, 1979
	Harps soil	21	55	24	37.84	7.30	16.14	L/kg	0.88	μM/L	μM/kg	3.80	2.63	Felsot and Dahm, 1979
	Peat	42	39	19	77.34	6.98	73.79	L/kg	1.01	μM/L	μM/kg	18.36	2.60	Felsot and Dahm, 1979
	Sarpy fsl	77	15	8	5.71	7.30	2.32	L/kg	0.94	μM/L	μM/kg	0.51	2.66	Felsot and Dahm, 1979
	Thurman fsl	83	9	8	6.01	6.83	5.48	L/kg	0.91	μM/L	μM/kg	1.07	2.71	Felsot and Dahm, 1979
Phosphorus [7723-14-0], as PO_4^{3+}	Alligator soil	5.9	39.4	54.7	30.2[c]	4.8	34.9	L/kg	0.508[d]	mg/L	mg/kg	1.54	3.36	Buchter et al., 1989
	Calciorthid sediment	70.0	19.3	10.7	14.7[c]	8.5	13.0	L/kg	0.829[d]	mg/L	mg/kg	0.44	3.47	Buchter et al., 1989
	Cecil soil	78.8	12.9	8.3	2.0[c]	5.7	35.2	L/kg	0.250[d]	mg/L	mg/kg	0.61	3.76	Buchter et al., 1989
	Lafitte soil	60.7	21.7	17.6	26.9[c]	3.9	149	L/kg	0.466[d]	mg/L	mg/kg	11.6	3.11	Buchter et al., 1989
	Molokai soil	25.7	46.2	28.2	11.0[c]	6.0	173	L/kg	0.303[d]	mg/L	mg/kg	1.67	4.02	Buchter et al., 1989
	Olivier soil	4.4	89.4	6.2	8.6[c]	6.6	44.6	L/kg	0.321[d]	mg/L	mg/kg	0.83	3.73	Buchter et al., 1989
	Spodosol sediment	90.2	6.0	3.8	2.7[c]	4.3	173	L/kg	0.247[d]	mg/L	mg/kg	1.98	3.94	Buchter et al., 1989
	Webster soil	27.5	48.6	23.9	48.1[c]	7.6	38.5	L/kg	0.506[d]	mg/L	mg/kg	4.39	2.94	Buchter et al., 1989
	Windsor soil	76.8	20.5	2.8	2.0[c]	5.3	146	L/kg	0.293[d]	mg/L	mg/kg	2.03	3.86	Buchter et al., 1989
Picloram [1918-02-1]	H-montmorillonite	--	--	--	73.5	9.0	37	L/g	0.78	μmol/L	μmol/g	--	--	Bailey et al., 1968
Picloram - K+ salt [11562-68-2]	Fiddletown sl	--	--	18	20	--	0.976	mL/g	0.849	μg/mL	μg/g	2.44	1.60	Farmer and Aochi, 1974
	Palouse sl	--	--	27	19	--	0.553	mL/g	0.816	μg/mL	μg/g	2.09	1.42	Farmer and Aochi, 1974
	Molokai clay	--	--	83	14	--	0.310	mL/g	0.829	μg/mL	μg/g	1.39	1.35	Farmer and Aochi, 1974
	Linne clay loam	--	--	33	41	--	0.409	mL/g	0.743	μg/mL	μg/g	1.39	1.47	Farmer and Aochi, 1974
	Kentwood sand loam	--	--	9	12	--	0.118	mL/g	0.838	μg/mL	μg/g	0.93	1.10	Farmer and Aochi, 1974
	Ephrata sand loam	--	--	8	8	--	0.070	mL/g	0.596	μg/mL	μg/g	0.55	1.11	Farmer and Aochi, 1974
Prochloraz [67747-09-5]	Soil	15	17	65	12.6[c]	--	62.7	L/kg	0.830[d]	--	--	1.7	4.13	de Jonge and de Jonge, 1999
	NK1 sl	2	80	18	--	7.1	73	L/kg	--	μg/L	μg/kg	0.97	3.88	Rütters et al., 1999
	NK2 sl	3	69	28	--	7.8	56	L/kg	--	μg/L	μg/kg	0.77	3.86	Rütters et al., 1999
	NW1 ls	76	19	5	--	5.3	247	L/kg	--	μg/L	μg/kg	1.52	4.21	Rütters et al., 1999

Compound	Soil	(a)	(b)	(c)	(d)	pH	K	Units	1/n	Units	Units	(k)	(l)	Reference
Prometon [1610-18-0]	NW2 sand	86	11	3	--	5.8	78	L/kg	--	μg/L	μg/kg	0.53	4.17	Rütters et al., 1999
	NW3 ls	77	15	8	--	6.8	322	L/kg	--	μg/L	μg/kg	2.52	4.11	Rütters et al., 1999
	NW4 ls	81	13	6	--	5.9	552	L/kg	--	μg/L	μg/kg	4.44	4.09	Rütters et al., 1999
	Bates sl	1	26	73	9.3	6.5	0.6	L/kg	--	--	--	0.81	1.87	Talbert and Fletchall, 1965
	Baxter csl	9	72	19	11.2	6.0	1.7	L/kg	--	--	--	1.22	2.14	Talbert and Fletchall, 1965
	Clarksville sl	20	67	13	5.7	5.7	2.2	L/kg	--	--	--	0.81	2.43	Talbert and Fletchall, 1965
	Cumberland csl	20	70	17	6.5	6.4	0.5	L/kg	--	--	--	0.70	1.85	Talbert and Fletchall, 1965
	Eldon sl	8	72	20	12.9	5.9	1.3	L/kg	--	--	--	1.74	1.87	Talbert and Fletchall, 1965
	Gerald sl	1	26	73	11.0	4.7	6.0	L/kg	--	--	--	1.57	2.58	Talbert and Fletchall, 1965
	Grundy sicl	3	67	30	13.5	5.6	6.3	L/kg	--	--	--	2.09	2.48	Talbert and Fletchall, 1965
	Knox sl	4	72	24	18.8	5.4	6.4	L/kg	--	--	--	1.68	2.58	Talbert and Fletchall, 1965
	Lebanon sl	13	70	17	7.7	4.9	7.8	L/kg	--	--	--	1.04	2.88	Talbert and Fletchall, 1965
	Lindley loam	30	44	26	6.9	4.7	4.7	L/kg	--	--	--	0.87	2.73	Talbert and Fletchall, 1965
	Lintonia loamy sand	84	11	5	3.2	5.3	0.7	L/kg	--	--	--	0.35	2.30	Talbert and Fletchall, 1965
	Marian sl	9	74	17	9.9	4.6	14.9	L/kg	--	--	--	0.81	3.26	Talbert and Fletchall, 1965
	Marshall sicl	4	66	30	21.3	5.4	8.8	L/kg	--	--	--	2.44	2.56	Talbert and Fletchall, 1965
	Menfro sl	4	85	11	9.1	5.3	1.2	L/kg	--	--	--	1.39	1.94	Talbert and Fletchall, 1965
	Na-montmorillonite	--	--	--	87.0	9.0	150	L/g	0.64	μmol/L	μmol/g	--	--	Bailey et al., 1968
	Netonia sl	11	75	14	8.8	5.2	2.4	L/kg	--	--	--	0.93	1.99	Talbert and Fletchall, 1965
	Oswego sicl	5	67	28	21.0	6.4	3.4	L/kg	--	--	--	1.68	2.31	Talbert and Fletchall, 1965
	Putnam sl	6	74	20	12.3	5.3	2.8	L/kg	--	--	--	1.10	2.41	Talbert and Fletchall, 1965
	Salix loam	32	50	18	17.9	6.3	4.6	L/kg	--	--	--	1.22	2.58	Talbert and Fletchall, 1965
	Sarpy loam	40	41	19	14.3	7.1	1.5	L/kg	--	--	--	0.75	2.30	Talbert and Fletchall, 1965
	Sharkey clay	25	30	45	28.2	5.0	55.2	L/kg	--	--	--	1.45	3.58	Talbert and Fletchall, 1965
	Shelby loam	26	43	31	20.1	4.3	22.3	L/kg	--	--	--	2.09	3.03	Talbert and Fletchall, 1965
	Summit silty clay	5	48	47	35.1	4.8	11.6	L/kg	--	--	--	0.40	3.46	Talbert and Fletchall, 1965
	Union sl	1	79	19	6.8	5.4	6.1	L/kg	--	--	--	1.04	2.77	Talbert and Fletchall, 1965
	Wabash clay	14	36	63	40.3	5.7	17.0	L/kg	--	--	--	1.28	3.12	Talbert and Fletchall, 1965
	Waverley sl	1	16	20	12.8	6.4	3.8	L/kg	--	--	--	1.16	2.52	Talbert and Fletchall, 1965
Prometryn [7287-19-6]	Bates sl	1	26	73	9.3	6.5	1.6	L/kg	--	--	--	0.81	2.30	Talbert and Fletchall, 1965
	Baxter csl	9	72	19	11.2	6.0	4.3	L/kg	--	--	--	1.22	2.55	Talbert and Fletchall, 1965
	Cape Fear sal	61	28	11	10.2	--	844.0	mL/g	--	μg/mL	μg/g	5.34	4.20	Kozak et al., 1983
	Clarksville sl	20	67	13	5.7	5.7	5.1	L/kg	--	--	--	0.81	2.80	Talbert and Fletchall, 1965
	Cumberland csl	20	70	17	6.5	6.4	1.4	L/kg	--	--	--	0.70	2.30	Talbert and Fletchall, 1965
	Eldon sl	8	72	20	12.9	5.9	3.6	L/kg	--	--	--	1.74	2.32	Talbert and Fletchall, 1965
	Gerald sl	1	26	73	11.0	4.7	9.4	L/kg	--	--	--	1.57	2.78	Talbert and Fletchall, 1965
	Grundy sicl	3	67	30	13.5	5.6	9.2	L/kg	--	--	--	2.09	2.64	Talbert and Fletchall, 1965
	Knox sl	4	72	24	18.8	5.4	8.4	L/kg	--	--	--	1.68	2.70	Talbert and Fletchall, 1965
	Lebanon sl	13	70	17	7.7	4.9	9.0	L/kg	--	--	--	1.04	2.35	Talbert and Fletchall, 1965
	Lindley loam	30	44	26	6.9	4.7	7.9	L/kg	--	--	--	0.87	2.96	Talbert and Fletchall, 1965
	Marian sl	9	74	17	9.9	4.6	14.2	L/kg	--	--	--	0.81	3.24	Talbert and Fletchall, 1965

Compound [CAS No.]	Adsorbent	% sand	% silt	% clay	CEC[a]	pH	K_F	K_F units	1/n	C units	x/m units	f_{oc} %	log K_{oc}[b]	Reference
	Marshall sicl	4	66	30	21.3	5.4	12.3	L/kg	--	--	--	2.44	2.70	Talbert and Fletchall, 1965
	Menfro sl	4	85	11	9.1	5.3	3.3	L/kg	--	--	--	1.39	2.38	Talbert and Fletchall, 1965
	Netonia sl	11	75	14	8.8	5.2	3.5	L/kg	--	--	--	0.93	2.16	Talbert and Fletchall, 1965
	Norfolk fsl	87	11	2	2.3	--	87	mL/g	--	µg/mL	µg/g	0.99	3.94	Kozak et al. 1983
	Oswego sicl	5	67	28	21.0	6.4	5.0	L/kg	--	--	--	1.68	2.47	Talbert and Fletchall, 1965
	Putnam sl	6	74	20	12.3	5.3	3.8	L/kg	--	--	--	1.10	2.54	Talbert and Fletchall, 1965
	Rains fine loamy sand	52	41	7	7.1	--	139.0	mL/g	--	µg/mL	µg/g	1.45	3.98	Kozak et al., 1983
	Salix loam	32	50	18	17.9	6.3	6.4	L/kg	--	--	--	1.22	2.72	Talbert and Fletchall, 1965
	Sarpy loam	40	41	19	14.3	7.1	2.9	L/kg	--	--	--	0.75	2.59	Talbert and Fletchall, 1965
	Sharkey clay	25	30	45	28.2	5.0	43.4	L/kg	--	--	--	1.45	3.48	Talbert and Fletchall, 1965
	Shelby loam	26	43	31	20.1	4.3	22.1	L/kg	--	--	--	2.09	3.02	Talbert and Fletchall, 1965
	Summit silty clay	5	48	47	35.1	4.8	17.7	L/kg	--	--	--	0.40	3.65	Talbert and Fletchall, 1965
	Union sl	1	79	19	6.8	5.4	8.6	L/kg	--	--	--	1.04	2.92	Talbert and Fletchall, 1965
	Wabash clay	1	36	63	40.3	5.7	17.3	L/kg	--	--	--	1.28	3.13	Talbert and Fletchall, 1965
	Waverley sl	14	16	20	12.8	6.4	5.8	L/kg	--	--	--	1.16	2.70	Talbert and Fletchall, 1965
Propachlor [1918-16-7]	Cellulose	--	--	--	--	--	140	mL/g	--	--	--	--	2.39	Torrents et al., 1997
	Chitin	--	--	--	--	--	22.9	mL/g	--	--	--	--	1.61	Torrents et al., 1997
	Collagen	--	--	--	--	--	1.45	mL/g	--	--	--	--	0.11	Torrents et al., 1997
	Lignin	--	--	--	--	--	0.57	mL/g	--	--	--	--	2.85	Torrents et al., 1997
Propanil [709-98-8]	H-montmorillonite	--	--	--	73.5	9.0	65	L/g	0.59	µmol/L	µmolg	--	--	Bailey et al., 1968
	Na-montmorillonite	--	--	--	87.0	9.0	16	L/g	1.11	µmol/L	µmolg	--	--	Bailey et al., 1968
	Fresno, CA aquifer soil	86.6	6.2	7.2	--	7.7	0.06	L/kg	--	mg/L	mg/kg	0.02	2.48	Deeley et al., 1991
	Fresno, CA aquifer soil	83.9	4.1	12.0	--	7.3	0.07	L/kg	--	mg/L	mg/kg	0.02	2.54	Deeley et al., 1991
1-Propanol [71-23-8]	Dowex-1 resin	--	--	--	--	--	0.945	unitless	--	--	--	--	--	Small and Bremer, 1964
	Dowex-1-X2 resin	--	--	--	--	--	1.8	unitless	--	--	--	--	--	Small and Bremer, 1964
Propazine [139-40-0]	Bates sl	1	26	73	9.3	6.5	0.7	L/kg	--	--	--	0.81	1.94	Talbert and Fletchall, 1965
	Baxter csl	9	72	19	11.2	6.0	1.9	L/kg	--	--	--	1.22	2.19	Talbert and Fletchall, 1965
	Clarksville sl	20	67	13	5.7	5.7	2.1	L/kg	--	--	--	0.81	2.41	Talbert and Fletchall, 1965
	Cumberland csl	20	70	17	6.5	6.4	0.7	L/kg	--	--	--	0.70	2.00	Talbert and Fletchall, 1965
	Eldon sl	8	72	20	12.9	5.9	1.8	L/kg	--	--	--	1.74	2.01	Talbert and Fletchall, 1965
	Gerald sl	1	26	73	11.0	4.7	1.8	L/kg	--	--	--	1.57	2.06	Talbert and Fletchall, 1965
	Grundy sicl	3	67	30	13.5	5.6	2.8	L/kg	--	--	--	2.09	2.13	Talbert and Fletchall, 1965
	Hickory Hill silt	--	--	--	--	--	11.9	mL/g	--	µg/mL	µg/g	3.27	2.56	Brown and Flagg, 1981
	Knox sl	4	72	24	18.8	5.4	2.7	L/kg	--	--	--	1.68	2.21	Talbert and Fletchall, 1965
	Lebanon sl	13	70	17	7.7	4.9	2.0	L/kg	--	--	--	1.04	2.28	Talbert and Fletchall, 1965
	Lindley loam	30	44	26	6.9	4.7	2.2	L/kg	--	--	--	0.87	2.40	Talbert and Fletchall, 1965
	Lintonia loamy sand	84	11	5	3.2	5.3	0.1	L/kg	--	--	--	0.35	1.46	Talbert and Fletchall, 1965
	Marian sl	9	74	17	9.9	4.6	2.1	L/kg	--	--	--	0.81	2.41	Talbert and Fletchall, 1965

Soil/Sediment													Reference
Marshall sicl	4	66	30	21.3	5.4	3.0	L/kg	--	--	--	2.44	2.09	Talbert and Fletchall, 1965
Menfro sl	4	85	11	9.1	5.3	1.8	L/kg	--	--	--	1.39	2.11	Talbert and Fletchall, 1965
Na-montmorillonite	11	--	--	87.0	9.0	18	L/g	1.12	µmol/L	µmol/g	--	--	Bailey et al., 1968
Netonia sl	5	75	14	8.8	5.2	1.4	L/kg	--	--	--	0.93	2.09	Talbert and Fletchall, 1965
Oswego sicl	6	67	28	21.0	6.4	1.9	L/kg	--	--	--	1.68	2.05	Talbert and Fletchall, 1965
Putnam sl	32	74	20	12.3	5.3	1.1	L/kg	--	--	--	1.10	2.00	Talbert and Fletchall, 1965
Salix loam	40	50	18	17.9	6.3	1.9	L/kg	--	--	--	1.22	2.19	Talbert and Fletchall, 1965
Sarpy loam	25	41	19	14.3	7.1	1.2	L/kg	--	--	--	0.75	2.20	Talbert and Fletchall, 1965
Sharkey clay	26	30	45	28.2	5.0	3.0	L/kg	--	--	--	1.45	2.32	Talbert and Fletchall, 1965
Shelby loam	5	43	31	20.1	4.3	2.8	L/kg	--	--	--	2.09	2.13	Talbert and Fletchall, 1965
Summit silty clay	1	48	47	35.1	4.8	3.4	L/kg	--	--	--	0.40	2.93	Talbert and Fletchall, 1965
Union sl	1	79	19	6.8	5.4	2.4	L/kg	--	--	--	1.04	2.36	Talbert and Fletchall, 1965
Wabash clay	14	36	63	40.3	5.7	3.1	L/kg	--	--	--	1.28	2.38	Talbert and Fletchall, 1965
Waverley sl	--	16	20	12.8	6.4	2.0	L/kg	--	--	--	1.16	2.24	Talbert and Fletchall, 1965
Propham [122-42-9]													
Batcombe sl	--	--	--	--	--	--	--	--	--	--	--	1.95	Briggs, 1981
Propanoic acid [79-09-4]													
H-montmorillonite	--	--	--	73.5	9.0	30	L/g	0.64	µmol/L	µmol/g	--	--	Bailey et al., 1968
Poly(4-vinyl pyridine)	--	--	--	--	--	0.020	L/g	0.905	mg/L	mg/g	--	--	Chanda et al., 1985
Polybenzimidazole	--	--	--	--	--	0.620	L/g	0.628	mg/L	mg/g	--	--	Chanda et al., 1985
Pyrene [129-00-0]													
Filtrasorb 300	--	--	--	--	5.0	65.6	L/g	0.24	mg/L	mg/g	--	--	Burks, 1981
Filtrasorb 400	--	--	--	--	--	389	L/g	0.386	mg/L	mg/g	--	--	Walters and Luthy, 1982
EPA-B2 sediment	67.5	18.6	13.9	3.72	6.35	760	mL/g	--	ng/mL	ng/g	1.21	4.80	Means et al., 1980
EPA-4 sediment	3.0	55.2	41.8	23.72	7.79	1,065	mL/g	--	ng/mL	ng/g	2.07	4.71	Means et al., 1980
EPA-5 sediment	33.6	31.0	35.4	19.00	7.44	1,155	mL/g	--	ng/mL	ng/g	2.28	4.70	Means et al., 1980
EPA-6 sediment	0.2	68.6	31.2	33.01	7.83	614	mL/g	--	ng/mL	ng/g	0.72	4.93	Means et al., 1980
EPA-8 sediment	82.4	6.8	10.7	3.72	8.32	101	mL/g	--	ng/mL	ng/g	0.15	4.83	Means et al., 1980
EPA-9 soil	7.1	17.4	75.6	12.40	8.34	71	mL/g	--	ng/mL	ng/g	0.11	4.81	Means et al., 1980
EPA-14 soil	2.1	63.6	34.4	18.86	4.54	277	mL/g	--	ng/mL	ng/g	0.48	4.76	Means et al., 1980
EPA-15 sediment	15.6	35.7	48.7	11.30	7.79	783	mL/g	--	ng/mL	ng/g	0.95	4.92	Means et al., 1980
EPA-18	34.6	39.5	25.8	15.43	7.76	504	mL/g	--	ng/mL	ng/g	0.66	4.88	Means et al., 1980
EPA-20 sediment	0.0	28.6	71.4	8.50	5.50	723	mL/g	--	ng/mL	ng/g	1.30	4.78	Means et al., 1980
EPA-21	50.2	7.1	42.7	8.33	7.60	1,119	mL/g	--	ng/mL	ng/g	1.88	4.77	Means et al., 1980
EPA-22	26.1	21.2	52.7	8.53	7.55	806	mL/g	--	ng/mL	ng/g	1.67	5.69	Means et al., 1980
EPA-23	17.3	69.1	13.6	31.15	6.70	1,043	mL/g	--	ng/mL	ng/g	2.38	4.64	Means et al., 1980
EPA-26	1.6	42.9	55.4	20.86	7.75	994	mL/g	--	ng/mL	ng/g	1.48	4.83	Miltner et al., 1989
Simazine [122-34-9]													
Filtrasorb 400	--	--	--	151	5.75	0.73	L/g	0.23	--	--	--	--	Reddy et al., 1992
Basinger fine sand	--	--	--	1.52[c]	--	--	--	--	--	--	0.61	2.08	Reddy et al., 1992
Batcombe sl	--	--	1.9	--	--	--	--	--	--	--	2.05	1.68	Briggs, 1981
Bates sl	1	26	73	9.3	6.5	1.0	L/kg	--	--	--	0.81	2.09	Talbert and Fletchall, 1965
Baxter csl	9	72	19	11.2	6.0	2.3	L/kg	--	--	--	1.22	2.28	Talbert and Fletchall, 1965
Boca fine sand	--	--	3.5	10.79[c]	7.05	0.85	mL/g	--	--	--	1.67	1.71	Reddy et al., 1992
Chobee fsl	--	--	16.1	8.10[c]	7.18	1.06	mL/g	--	--	--	1.39	1.88	Reddy et al., 1992

Compound [CAS No.]	Adsorbent	% sand	% silt	% clay	CEC[a]	pH	K_F	K_F units	1/n	C units	x/m units	f_{oc} %	log K_{oc}[b]	Reference
	Clarksville sl	20	67	13	5.7	5.7	1.4	L/kg	--	--	--	0.81	2.24	Talbert and Fletchall, 1965
	Cumberland csl	20	70	17	6.5	6.4	1.2	L/kg	--	--	--	0.70	2.23	Talbert and Fletchall, 1965
	Eldon sl	8	72	20	12.9	5.9	2.9	L/kg	--	--	--	1.74	2.22	Talbert and Fletchall, 1965
	Gerald sl	1	26	73	11.0	4.7	4.2	L/kg	--	--	--	1.57	2.43	Talbert and Fletchall, 1965
	Grundy sicl	3	67	30	13.5	5.6	6.5	L/kg	--	--	--	2.09	2.49	Talbert and Fletchall, 1965
	Holopaw fine sand	--	--	1.3	2.03c	6.10	0.29	mL/g	--	μg/mL	μg/g	0.50	1.76	Reddy et al., 1992
	Hickory Hill silt	--	--	--	--	--	7.04	mL/g	--			3.27	2.33	Brown and Flagg, 1981
	Knox sl	4	72	24	18.8	5.4	5.1	L/kg	--	--	--	1.68	2.48	Talbert and Fletchall, 1965
	Lintonia loamy sand	84	11	5	3.2	5.3	1.0	L/kg	--	--	--	0.35	2.46	Talbert and Fletchall, 1965
	Marian sl	9	74	17	9.9	4.6	3.5	L/kg	--	--	--	0.81	2.64	Talbert and Fletchall, 1965
	Marshall sicl	4	66	30	21.3	5.4	7.2	L/kg	--	--	--	2.44	2.47	Talbert and Fletchall, 1965
	Menfro sl	4	85	11	9.1	5.3	2.5	L/kg	--	--	--	1.39	2.25	Talbert and Fletchall, 1965
	Netonia sl	11	75	14	8.8	5.2	3.0	L/kg	--	--	--	0.93	2.09	Talbert and Fletchall, 1965
	Oswego sicl	5	67	28	21.0	6.4	3.9	L/kg	--	--	--	1.68	2.37	Talbert and Fletchall, 1965
	Pineda fine sand	--	--	0.5	8.40c	7.11	0.55	mL/g	--	--	--	1.22	1.65	Reddy et al., 1992
	Putnam sl	6	74	20	12.3	5.3	2.2	L/kg	--	--	--	1.10	2.30	Talbert and Fletchall, 1965
	Riviera fine sand	--	--	5.1	4.11c	6.20	0.59	mL/g	--	--	--	0.94	1.80	Reddy et al., 1992
	Sharkey clay	25	30	45	28.2	5.0	7.0	L/kg	--	--	--	1.45	2.68	Talbert and Fletchall, 1965
	Shelby loam	26	43	31	20.1	4.3	5.1	L/kg	--	--	--	2.09	2.39	Talbert and Fletchall, 1965
	Union sl	1	79	19	6.8	5.4	3.8	L/kg	--	--	--	1.04	2.56	Talbert and Fletchall, 1965
	Wabasso sand	--	--	2.4	2.54c	6.62	0.80	mL/g	--	--	--	0.61	2.12	Reddy et al., 1992
	Waverley sl	14	16	20	12.8	6.4	1.0	L/kg	--	--	--	1.16	1.94	Talbert and Fletchall, 1965
Simeton [673-04-1]	Na-montmorillonite	--	--	--	87.0	9.0	2,200	L/g	0.31	μmol/L	μmolg	--	--	Bailey et al., 1968
Strychnine [57-24-9]	Ascalon sal	76	12	12	11.5	8.3	95	mL/g	0.97	μg/mL	μg/g	0.68	4.14	Starr et al., 1996
	Kim loam	54	21	25	16.4	8.0	169	mL/g	0.93	μg/mL	μg/g	1.05	4.20	Starr et al., 1996
	Orovada fsl	--	--	--	6.8	8.0	8.0	L/mg	0.41	mg/L	μg/g	0.46	3.24	Miller et al., 1983
	Rose Creek Variant fsl	--	--	--	13.5	7.6	53.9	L/mg	0.58	mg/L	μg/g	0.31	4.24	Miller et al., 1983
	Table Mountain scl	51	29	20	15.3	7.7	119	mL/g	0.82	μg/mL	μg/g	1.26	3.97	Starr et al., 1996
	Truckee sl	--	--	--	16.2	7.7	83.0	L/mg	0.67	mg/L	μg/mg	1.14	3.86	Miller et al., 1983
	Valent loamy sand	89	2	9	2.3	6.7	40	mL/g	0.75	μg/mL	μg/g	0.25	4.20	Starr et al., 1996
Styrene [100-42-5]	Filtrasorb 300	--	--	--	--	5.3/7.4	120	L/g	0.56	mg/L	mg/g	--	--	Dobbs and Cohen, 1980
Sulfur [7704-34-9], as SO$_4^{2-}$	Molokai soil	25.7	46.2	28.2	11.0c	6.0	0.42	L/kg	1.108d	mg/L	mg/kg	1.67	1.40	Buchter et al., 1989
2,4,5-T [93-76-5]	H-montmorillonite	--	--	--	73.5	9.0	105	L/g	0.42	μmol/L	μmolg	--	--	Bailey et al., 1968
Tebuconazole [107534-96-3]	Sandy loam soil	--	--	--	9.0	6.6	5.80	mL/g	0.99d	--	--	1.24	2.67	Xu et al., 1999
Terbacil [5902-51-2]	Cecil sal	65.8	19.5	14.7	6.8	5.6	0.38	--	0.99d	--	--	0.90	1.63	Rao and Davidson, 1979
	Eustis fine sand	93.8	3.0	3.2	5.2	5.6	0.12	--	0.88d	--	--	0.56	1.33	Rao and Davidson, 1979
	Webster sicl	18.4	45.3	38.3	54.7	7.3	2.46	--	0.88d	--	--	3.87	1.80	Rao and Davidson, 1979
Terbufos [13071-79-9]	Clarion soil	37	42	21	21.02	5.00	8.30	L/kg	0.96	μM/L	μM/kg	2.64	2.50	Felsot and Dahm, 1979

Compound	Sample												Reference
1,2,3,4-Tetrachlorobenzene [634-66-2]	Harps soil	21	55	24	37.84	7.30	21.13	L/kg	0.97	μM/L	μM/kg	3.80	2.75	Felsot and Dahm, 1979
	Peat	42	39	19	77.34	6.98	52.72	L/kg	0.97	μM/L	μM/kg	18.36	2.46	Felsot and Dahm, 1979
	Sarpy fsl	77	15	8	5.71	7.30	3.34	L/kg	0.95	μM/L	μM/kg	0.51	2.82	Felsot and Dahm, 1979
	Thurman fsl	83	9	8	6.01	6.83	11.40	L/kg	0.94	μM/L	μM/kg	1.07	3.03	Felsot and Dahm, 1979
1,2,4,5-Tetrachlorobenzene [95-94-3]	Clay loam	--	--	--	12.4	5.91	47.0	L/g	0.893	mg/L	mg/g	1.42	3.52	Kishi et al., 1990
	Clay loam	--	--	--	35.0	4.89	341	L/g	1.00	mg/L	mg/g	10.40	3.52	Kishi et al., 1990
	Light clay	--	--	--	13.2	5.18	122	L/g	1.00	mg/L	mg/g	1.51	3.91	Kishi et al., 1990
	Light clay	--	--	--	28.3	5.26	180	L/g	0.866	mg/L	mg/g	3.23	3.75	Kishi et al., 1990
	Sandy loam	--	--	--	35.0	5.41	315	L/g	0.883	mg/L	mg/g	7.91	3.48	Kishi et al., 1990
2,2',4,5-Tetrachlorobiphenyl [41464-40-8]	KS1 sediment	--	--	--	--	--	37.9	mL/g	--	--	--	0.73	3.72	Schwarzenbach and Westall, 1981
	KS1H sediment	--	--	--	--	--	6.2	mL/g	--	--	--	0.08	3.89	Schwarzenbach and Westall, 1981
	Kaolin	--	--	--	--	--	4.9	mL/g	--	--	--	0.06	3.91	Schwarzenbach and Westall, 1981
	γ-Alumina	--	--	--	--	--	2.2	mL/g	--	--	--	<0.01	--	Schwarzenbach and Westall, 1981
	Silica	--	--	--	--	--	12.1	mL/g	--	--	--	<0.01	--	Schwarzenbach and Westall, 1981
2,2',5,5'-Tetrachlorobiphenyl [35693-99-3]	Lucustrine sediment	--	--	--	--	--	6,193	--	1.15	μg/L	--	1.42	5.64	Chin et al., 1988
	Catlin Ap	0	69	21	18.1	6.1	--	--	--	--	--	--	5.37	Girvin and Scott, 1997
	Cloudland E	41	40	19	3	4.7	--	--	--	--	--	--	5.64	Girvin and Scott, 1997
	Kenoma Ap	8	70	22	11.8	6.2	--	--	--	--	--	--	5.31	Girvin and Scott, 1997
	Kenoma C	8	45	37	24.6	7.2	--	--	--	--	--	--	5.49	Girvin and Scott, 1997
	Norborne Ap	23	59	18	15.8	7.3	--	--	--	--	--	--	5.32	Girvin and Scott, 1997
	Kenoma Bt2	5	42	53	33.2	6.0	--	--	--	--	--	--	5.38	Girvin and Scott, 1997
	Norborne Bt1	44	39	17	13.8	7.2	--	--	--	--	--	--	5.31	Girvin and Scott, 1997
	Norborne Bt2	47	37	16	12.8	7.2	--	--	--	--	--	--	5.38	Girvin and Scott, 1997
2,2',6,6'-Tetrachlorobiphenyl [15968-05-5]	Doe Run Pond	56.0	44.0	<1.0	--	6.1	1,080	unitless	--	--	mg/g	1.4	4.89	Steen et al., 1978
	Hickory Hill Pond	55.0	45.0	<1.0	--	6.3	1,270	unitless	--	--	mg/g	2.4	4.72	Steen et al., 1978
	Oconee River	93.0	6.0	1.0	--	6.5	510	unitless	--	--	mg/g	0.4	5.11	Steen et al., 1978
	USDA Pond	--	--	--	--	6.4	650	unitless	--	--	mg/g	0.8	4.91	Steen et al., 1978
2,3',4',5-Tetrachlorobiphenyl [32598-11-1]	Doe Run Pond	56.0	44.0	<1.0	--	6.1	990	unitless	--	--	mg/g	1.4	4.85	Steen et al., 1978
	Hickory Hill Pond	55.0	45.0	<1.0	--	6.3	1,180	unitless	--	--	mg/g	2.4	4.69	Steen et al., 1978
	Oconee River	93.0	6.0	1.0	--	6.5	420	unitless	--	--	mg/g	0.4	5.02	Steen et al., 1978
	USDA Pond	--	--	--	--	6.4	580	unitless	--	--	mg/g	0.8	4.86	Steen et al., 1978
1,1,2,2-Tetrachloroethane [79-34-5]	Filtrasorb 300	--	--	--	--	5.3	10.6	L/g	0.37	mg/L	mg/g	--	--	Dobbs and Cohen, 1980
	Filtrasorb 300	--	--	--	--	5.3	50.8	L/g	0.56	mg/L	mg/g	--	--	Dobbs and Cohen, 1980
	Filtrasorb F-400	--	--	--	--	6.0	10,389	L/g	0.458	μg/L	μg/g	--	--	11 °C, Crittenden et al., 1985
Tetrachloroethylene [127-18-4]														

Compound [CAS No.]	Adsorbent	% sand	% silt	% clay	CEC[a]	pH	K_F	K_F units	1/n	C units	x/m units	f_{oc} %	log K_{oc}[b]	Reference
	Westvaco WV-C	--	--	--	--	6.0	7,524	L/g	0.502	µg/L	µg/g	--		21 °C, Crittenden et al., 1985
	Agricultural soil	65.2	25.6	9.2	9.0	7.4	4.51	L/kg	--	mol/L	mol/kg	2.2	2.31	Seip et al., 1986
	Forest soil	97.3	2.2	0.5	0.48	5.6	0.35	L/kg	--	mol/L	mol/kg	0.2	2.25	Seip et al., 1986
	Leie River clay	0	0	100	--	--	17.3	L/kg	--	--	--	4.12	2.62	2.3 °C, Dewulf et al., 1999
	Leie River clay	0	0	100	--	--	25.6	L/kg	--	--	--	4.12	2.79	3.8 °C, Dewulf et al., 1999
	Leie River clay	0	0	100	--	--	22.7	L/kg	--	--	--	4.12	2.74	6.2 °C, Dewulf et al., 1999
	Leie River clay	0	0	100	--	--	26.1	L/kg	--	--	--	4.12	2.80	8.0 °C, Dewulf et al., 1999
	Leie River clay	0	0	100	--	--	29.1	L/kg	--	--	--	4.12	2.85	13.5 °C, Dewulf et al., 1999
	Leie River clay	0	0	100	--	--	25.1	L/kg	--	--	--	4.12	2.78	18.6 °C, Dewulf et al., 1999
	Leie River clay	0	0	100	--	--	27.9	L/kg	--	--	--	4.12	2.83	25.0 °C, Dewulf et al., 1999
1,2,3,4-Tetrahydronaphthalene [119-64-2]	Filtrasorb 300	--	--	--	--	5.3/7.4	74	L/g	0.81	mg/L	mg/g	--	--	Dobbs and Cohen, 1980
2,3,4,6-Tetrachlorophenol [58-90-3]	Agricultural soil	65.2	25.6	9.2	9.0	7.4	3.38	L/kg	--	mol/L	mol/kg	2.2	2.19	Seip et al., 1986
	Forest soil	69.5	20.5	10.1	2.9	4.2	>17	L/kg	--	mol/L	mol/kg	3.7	>2.68	Seip et al., 1986
	Forest soil	97.3	2.2	0.5	0.48	5.6	1.50	L/kg	--	mol/L	mol/kg	0.2	2.88	Seip et al., 1986
	Holten humic-rich sand	--	--	--	--	4.7	251.2	L/kg	1.00	µg/L	µg/kg	0.032	--	Lagas, 1988
	Kootwijk humic sand	--	--	--	--	3.4	79.4	L/kg	1.00	µg/L	µg/kg	0.017	--	Lagas, 1988
	Maasdijk light loam	--	--	--	--	7.5	2.5	L/kg	1.11	µg/L	µg/kg	0.009	--	Lagas, 1988
	Opijnen heavy loam	--	--	--	--	7.1	10.0	L/kg	1.25	µg/L	µg/kg	0.017	--	Lagas, 1988
	Rolde humic sand	--	--	--	--	4.9	100.0	L/kg	1.11	µg/L	µg/kg	0.022	--	Lagas, 1988
	Schipluiden peat	--	--	--	--	4.6	1,259	L/kg	1.25	µg/L	µg/kg	0.298	--	Lagas, 1988
Thiram [137-26-8]	AL-01 soil	--	--	13.1	13.1[c]	8.5	12.00	L/kg	0.42[d]	mg/L	mg/kg	1.17	3.01	Valverde-García et al., 1988
	AL-02 soil	--	--	14.3	11.9[c]	8.5	12.93	L/kg	0.47[d]	mg/L	mg/kg	1.48	2.94	Valverde-García et al., 1988
	AL-03 soil	--	--	7.8	6.3[c]	8.5	9.15	L/kg	0.56[d]	mg/L	mg/kg	0.65	3.15	Valverde-García et al., 1988
	AL-04 soil	--	--	4.1	10.0[c]	8.5	11.17	L/kg	0.32[d]	mg/L	mg/kg	1.65	2.83	Valverde-García et al., 1988
	AL-05 soil	--	--	5.9	3.8[c]	8.5	4.81	L/kg	0.58[d]	mg/L	mg/kg	0.37	3.11	Valverde-García et al., 1988
	AL-06 soil	--	--	12.2	4.4[c]	8.5	8.08	L/kg	0.60[d]	mg/L	mg/kg	0.33	3.39	Valverde-García et al., 1988
	AL-07 soil	--	--	14.1	25.6[c]	8.5	13.73	L/kg	0.42[d]	mg/L	mg/kg	2.06	2.82	Valverde-García et al., 1988
	AL-08 soil	--	--	11.7	9.4[c]	8.5	11.96	L/kg	0.46[d]	mg/L	mg/kg	0.91	3.12	Valverde-García et al., 1988
Thymine [65-71-4]	Filtrasorb 300	--	--	--	--	5.3/7.4	27	L/g	0.51	mg/L	mg/g	--	--	Dobbs and Cohen, 1980
Toluene [108-88-3]	Filtrasorb 300	--	--	--	--	5.6	26.1	L/g	0.44	mg/L	mg/g	--	--	Dobbs and Cohen, 1980
	Filtrasorb 300	--	--	--	--	5.0	40.2	L/g	0.35	mg/L	mg/g	--	--	Burks, 1981
	Filtrasorb 400	--	--	--	--	--	0.090	L/g	0.30	mg/L	mg/g	--	--	El-Dib and Badawy, 1979
	Agricultural soil	65.2	25.6	9.2	9.0	7.4	2.08	L/kg	0.982	mol/L	mol/kg	2.2	1.97	Seip et al., 1986
	Cohansey Sand	90.0	8.0	2.0	5.1	3.8	19.3	L/kg	--	mg/L	µg/g	2.55	2.87	Uchrin and Mangels, 1987
	Forest soil	69.5	20.5	10.1	2.9	4.2	4.95	L/kg	--	mol/L	mol/kg	3.7	2.13	Seip et al., 1986
	Forest soil	97.3	2.2	0.5	0.48	5.6	0.11	L/kg	--	mol/L	mol/kg	0.2	1.74	Seip et al., 1986

Chemical [CAS]	Medium					pH	K	units	1/n	units	units			Reference
	Leie River clay	0	0	100	--	--	7.04	L/kg	--	--	--	4.12	2.23	2.3 °C, Dewulf et al., 1999
	Leie River clay	0	0	100	--	--	8.51	L/kg	--	--	--	4.12	2.32	3.8 °C, Dewulf et al., 1999
	Leie River clay	0	0	100	--	--	8.88	L/kg	--	--	--	4.12	2.33	6.2 °C, Dewulf et al., 1999
	Leie River clay	0	0	100	--	--	8.92	L/kg	--	--	--	4.12	2.34	8.0 °C, Dewulf et al., 1999
	Leie River clay	0	0	100	--	--	10.3	L/kg	--	--	--	4.12	2.40	13.5 °C, Dewulf et al., 1999
	Leie River clay	0	0	100	--	--	8.47	L/kg	--	--	--	4.12	2.31	18.6 °C, Dewulf et al., 1999
	Leie River clay	0	0	100	--	--	9.08	L/kg	--	--	--	4.12	2.34	25.0 °C, Dewulf et al., 1999
o-Toluic acid [118-90-1]	PRM sal	70.4	24.0	5.6	--	5.5	143.7	L/kg	0.998	mg/L	μg/g	1.28	4.04	Uchrin and Mangels, 1987
	Carbon (200 mesh)	--	--	--	--	--	0.95	unitless	0.388	mg/L	mg/kg	--	--	20 °C, Hartman et al., 1946
	Carbon (200 mesh)	--	--	--	--	--	0.85	unitless	0.388	mg/L	mg/kg	--	--	30 °C, Hartman et al., 1946
	Carbon (200 mesh)	--	--	--	--	--	0.56	unitless	0.388	mg/L	mg/kg	--	--	40 °C, Hartman et al., 1946
	Carbon (200 mesh)	--	--	--	--	--	0.30	unitless	0.388	mg/L	mg/kg	--	--	50 °C, Hartman et al., 1946
m-Toluic acid [99-04-7]	Carbon (200 mesh)	--	--	--	--	--	0.98	unitless	0.358	mg/L	mg/kg	--	--	20 °C, Hartman et al., 1946
	Carbon (200 mesh)	--	--	--	--	--	0.87	unitless	0.358	mg/L	mg/kg	--	--	30 °C, Hartman et al., 1946
	Carbon (200 mesh)	--	--	--	--	--	0.66	unitless	0.358	mg/L	mg/kg	--	--	40 °C, Hartman et al., 1946
	Carbon (200 mesh)	--	--	--	--	--	0.42	unitless	0.358	mg/L	mg/kg	--	--	50 °C, Hartman et al., 1946
p-Toluic acid [99-94-5]	Carbon (200 mesh)	--	--	--	--	--	1.36	unitless	0.370	mg/L	mg/kg	--	--	20 °C, Hartman et al., 1946
	Carbon (200 mesh)	--	--	--	--	--	1.19	unitless	0.370	mg/L	mg/kg	--	--	30 °C, Hartman et al., 1946
	Carbon (200 mesh)	--	--	--	--	--	0.93	unitless	0.370	mg/L	mg/kg	--	--	40 °C, Hartman et al., 1946
	Carbon (200 mesh)	--	--	--	--	--	0.65	unitless	0.370	mg/L	mg/kg	--	--	50 °C, Hartman et al., 1946
m-Toluidine [108-44-1]	Batcombe sl	--	--	--	--	--	--	--	--	--	--	--	1.65	Briggs, 1981
p-Toluidine [106-49-0]	Batcombe sl	--	--	--	--	--	--	--	--	--	--	--	1.90	Briggs, 1981
Triadimenol [55219-65-3]	Milford sicl	4.3	58.7	37.0	45.4[c]	6.4	13.21	L/kg	0.992	nmol/L	nmol/kg	--	2.71	Graveel et al., 1986
	Morocco silty sand	76.6	21.3	2.1	16.9[c]	4.7	5.97	L/kg	1.062	nmol/L	nmol/kg	--	2.51	Graveel et al., 1986
	Oakville silty sand	83.2	12.7	4.1	10.2[c]	5.2	5.36	L/kg	1.061	nmol/L	nmol/kg	--	2.70	Graveel et al., 1986
Trichloroacetic acid [76-03-9]	Sandy loam soil	--	--	--	9.0	6.6	1.10	mL/g	--	--	--	1.24	1.95	Xu et al., 1999
2,3,6-Trichloroanisole [50375-10-5]	Filtrasorb 400	--	--	--	--	5.6	11.700	--	0.216	μg/L	μg/g	--	--	24 °C, Speth and Miltner, 1998
1,2,3-Trichlorobenzene [87-61-6]	Aqua Nuchar	--	--	--	--	--	4.6	L/mg	1.53	ng/L	ng/g	--	--	Lalezary et al., 1986
	Clay loam	--	--	--	12.4	5.91	10.7	L/g	0.854	mg/L	mg/g	1.42	3.18	Kishi et al., 1990
	Light clay	--	--	--	35.0	4.89	180	L/g	1.00	mg/L	mg/g	10.40	3.23	Kishi et al., 1990
	Light clay	--	--	--	13.2	5.18	36.5	L/g	1.00	mg/L	mg/g	1.51	3.38	Kishi et al., 1990
	Sandy loam	--	--	--	28.3	5.26	85.7	L/g	0.965	mg/L	mg/g	3.23	3.43	Kishi et al., 1990
	Sandy loam	--	--	--	35.0	5.41	143	L/g	0.979	mg/L	mg/g	7.91	3.26	Kishi et al., 1990
1,2,4-Trichlorobenzene [120-82-1]	Filtrasorb 300	--	--	--	--	5.3	157	L/g	0.31	mg/L	mg/g	--	--	Dobbs and Cohen, 1980
	KS1 sediment	--	--	--	--	--	14.5	mL/g	--	--	--	0.73	3.30	Schwarzenbach and Westall, 1981
	KS1H sediment	--	--	--	--	--	2.5	mL/g	--	--	--	0.08	3.49	Schwarzenbach and Westall, 1981
	Kaolin	--	--	--	--	--	2.4	mL/g	--	--	--	0.06	3.60	Schwarzenbach and Westall, 1981
	γ-Alumina	--	--	--	--	--	1.5	mL/g	--	--	--	<0.01	--	Schwarzenbach and Westall, 1981
	Apison, TN soil	4	10	86	76	4.5	--	unitless	--	--	--	0.11	3.32	Southworth and Keller, 1986

Compound [CAS No.]	Adsorbent	% sand	% silt	% clay	CEC[a]	pH	K_F	K_F units	1/n	C units	x/m units	f_{oc} %	log K_{oc}[b]	Reference
	Dormont, WV soil	2	38	60	129	4.2	--	unitless	--	--	--	1.2	2.95	Southworth and Keller, 1986
	Fullerton, TN soil	11	21	68	64	4.4	--	unitless	--	--	--	0.05	3.11	Southworth and Keller, 1986
	Michawye soil	--	--	--	--	--	12.45	mL/g	1.06	μg/L	--	0.13	3.98	Chin et al., 1988
	Silica	--	--	--	--	--	7.6	--	--	--	--	<0.01	--	Schwarzenbach and Westall, 1981
1,3,5-Trichlorobenzene [108-70-3]	Agricultural soil	65.2	25.6	9.2	9.0	7.4	>9.0	L/kg	--	mol/L	mol/kg	2.2	>2.61	Seip et al., 1986
	Forest soil	69.5	20.5	10.1	2.9	4.2	>13	L/kg	--	mol/L	mol/kg	3.7	>2.55	Seip et al., 1986
	Forest soil	97.3	2.2	0.5	0.48	5.6	1.40	L/kg	--	mol/L	mol/kg	0.2	2.85	Seip et al., 1986
2,2',5-Trichlorobiphenyl [37680-65-2]	Lucustrine sediment	--	--	--	--	--	2,295	--	1.14	μg/L	--	1.42	5.21	Chin et al., 1988
	Michawye soil	--	--	--	--	--	37.70	--	1.18	μg/L	--	0.13	4.46	Chin et al., 1988
1,1,1-Trichloroethane [71-55-6]	Filtrasorb 300	--	--	--	--	5.3	2.48	L/g	0.34	mg/L	mg/g	--	1.95	Dobbs and Cohen, 1980
	Leie River clay	0	0	100	--	--	3.63	L/kg	--	--	--	4.12	1.98	2.3 °C, Dewulf et al., 1999
	Leie River clay	0	0	100	--	--	3.92	L/kg	--	--	--	4.12	1.98	3.8 °C, Dewulf et al., 1999
	Leie River clay	0	0	100	--	--	3.92	L/kg	--	--	--	4.12	1.98	6.2 °C, Dewulf et al., 1999
	Leie River clay	0	0	100	--	--	4.01	L/kg	--	--	--	4.12	1.99	8.0 °C, Dewulf et al., 1999
	Leie River clay	0	0	100	--	--	4.22	L/kg	--	--	--	4.12	2.01	13.5 °C, Dewulf et al., 1999
	Leie River clay	0	0	100	--	--	3.98	L/kg	--	--	--	4.12	1.98	18.6 °C, Dewulf et al., 1999
	Leie River clay	0	0	100	--	--	4.41	L/kg	--	--	--	4.12	2.03	25.0 °C, Dewulf et al., 1999
1,1,2-Trichloroethane [79-00-5]	Filtrasorb 300	--	--	--	--	5.3	5.81	L/g	0.60	mg/L	mg/g	--	--	Dobbs and Cohen, 1980
	Agricultural soil	65.2	25.6	9.2	9.0	7.4	1.40	L/kg	--	mol/L	mol/kg	2.2	1.80	Seip et al., 1986
	Forest soil	69.5	20.5	10.1	2.9	4.2	4.00	L/kg	--	mol/L	mol/kg	3.7	2.03	Seip et al., 1986
	Forest soil	97.3	2.2	0.5	0.48	5.6	0.12	L/kg	--	mol/L	mol/kg	0.2	1.78	Seip et al., 1986
Trichloroethylene [79-01-6]	Filtrasorb 300	--	--	--	--	5.3	28.0	L/g	0.62	mg/L	mg/g	--	--	Dobbs and Cohen, 1980
	Filtrasorb F-400	--	--	--	--	6.0	3,390	L/g	0.42	μg/L	μg/g	--	--	11 °C, Crittenden et al., 1985
	Hydrodarco 3000	--	--	--	--	6.0	713	L/g	0.47	μg/L	μg/g	--	--	21 °C, Crittenden et al., 1985
	Westvaco WV-C	--	--	--	--	6.0	2,847	L/g	0.394	μg/L	μg/g	--	--	11 °C, Crittenden et al., 1985
	Westvaco WV-C	--	--	--	--	6.0	1,062	L/g	0.50	μg/L	μg/g	--	--	21 °C, Crittenden et al., 1985
	Agricultural soil	65.2	25.6	9.2	9.0	7.4	2.11	L/kg	--	mol/L	mol/kg	2.2	1.98	Seip et al., 1986
	Forest soil	69.5	20.5	10.1	2.9	4.2	5.24	L/kg	--	mol/L	mol/kg	3.7	2.15	Seip et al., 1986
	Forest soil	97.3	2.2	0.5	0.48	5.6	0.14	L/kg	--	mol/L	mol/kg	0.2	1.86	Seip et al., 1986
	Leie River clay	0	0	100	--	--	7.05	L/kg	--	--	--	4.12	2.23	2.3 °C, Dewulf et al., 1999
	Leie River clay	0	0	100	--	--	8.71	L/kg	--	--	--	4.12	2.32	3.8 °C, Dewulf et al., 1999
	Leie River clay	0	0	100	--	--	9.15	L/kg	--	--	--	4.12	2.35	6.2 °C, Dewulf et al., 1999
	Leie River clay	0	0	100	--	--	9.08	L/kg	--	--	--	4.12	2.34	8.0 °C, Dewulf et al., 1999
	Leie River clay	0	0	100	--	--	8.93	L/kg	--	--	--	4.12	2.34	13.5 °C, Dewulf et al., 1999
	Leie River clay	0	0	100	--	--	9.53	L/kg	--	--	--	4.12	2.36	18.6 °C, Dewulf et al., 1999
	Leie River clay	0	0	100	--	--	10.6	L/kg	--	--	--	4.12	2.41	25.0 °C, Dewulf et al., 1999

Chemical [CAS]	Medium													Reference
Trichlorofluoromethane [75-69-4]	Filtrasorb 300	--	--	--	--	5.3	5.6	L/g	0.24	mg/L	mg/g	--	--	Dobbs and Cohen, 1980
4,5,6-Trichloroguaiacol [2668-24-8]	Agricultural soil	65.2	25.6	9.2	9.0	7.4	>10.2	L/kg	--	mol/L	mol/kg	2.2	>2.67	Seip et al., 1986
	Forest soil	69.5	20.5	10.1	2.9	4.2	>17	L/kg	--	mol/L	mol/kg	3.7	>2.68	Seip et al., 1986
	Forest soil	97.3	2.2	0.5	0.48	5.6	1.94	L/kg	--	mol/L	mol/kg	0.2	2.99	Seip et al., 1986
2,4,5-Trichlorophenol [95-95-4]	Brookston clay loam	--	--	--	22.22	5.7	9.75	mL/g	0.71	µmol/mL	µmol/g	2.96	2.56	Boyd, 1982
	Holten humic-rich sand	--	--	--	--	4.7	199.5	L/kg	1.25	µg/L	µg/kg	0.032	--	Lagas, 1988
	Kootwijk humic sand	--	--	--	--	3.4	63.1	L/kg	1.11	µg/L	µg/kg	0.017	--	Lagas, 1988
	Maasdijk light loam	--	--	--	--	7.5	12.6	L/kg	1.25	µg/L	µg/kg	0.009	--	Lagas, 1988
	Opijnen heavy loam	--	--	--	--	7.1	31.6	L/kg	1.25	µg/L	µg/kg	0.017	--	Lagas, 1988
	Rolde humic sand	--	--	--	--	4.9	100.0	L/kg	1.11	µg/L	µg/kg	0.022	--	Lagas, 1988
	Schipluiden peat	--	--	--	--	4.6	1,259	L/kg	1.25	µg/L	µg/kg	0.298	--	Lagas, 1988
2,4,6-Trichlorophenol [88-06-2]	Agricultural soil	65.2	25.6	9.2	9.0	7.4	2.01	L/kg	--	mol/L	mol/kg	2.2	1.96	Seip et al., 1986
	Forest soil	69.5	20.5	10.1	2.9	4.2	>17	L/kg	--	mol/L	mol/kg	3.7	>2.68	Seip et al., 1986
	Forest soil	97.3	2.2	0.5	0.48	5.6	1.13	L/kg	--	mol/L	mol/kg	0.2	2.74	Seip et al., 1986
Trietazine [1912-26-1]	Na-montmorillonite	--	--	--	87.0	9.0	58	L/g	1.00	µmol/L	µmolg	3.27	2.74	Bailey et al., 1968
Trifluralin [1582-09-8]	Hickory Hill silt	--	--	--	--	--	17.9	mL/g	--	µg/mL	µg/g	3.27	4.49	Brown and Flagg, 1981
	Hickory Hill silt	--	--	--	--	--	999	mL/g	--	µg/mL	µg/g	--	--	Brown and Flagg, 1981
3-(Trifluoromethyl)aniline [98-16-8]	Batcombe sl	--	--	--	--	--	--	--	--	--	--	--	2.36	Briggs, 1981
3-(Trifluoromethylphenyl) urea [13114-87-9]	Batcombe sl	--	--	--	--	--	--	--	--	--	--	--	1.96	Briggs, 1981
2,3,5-Trimethylphenol [697-82-5]	Apison, TN soil	4	10	86	76	4.5	--	unitless	--	--	--	0.11	3.76	Southworth and Keller, 1986
	Dormont, WV soil	2	38	60	129	4.2	--	unitless	--	--	--	1.2	2.41	Southworth and Keller, 1986
	Fullerton, TN soil	11	21	68	64	4.4	--	unitless	--	--	--	0.05	3.81	Southworth and Keller, 1986
Vanadium [7440-62-2], as VO_3^-	Alligator soil	5.9	39.4	54.7	30.2[c]	4.8	142	L/kg	0.592[d]	mg/L	mg/kg	1.54	3.96	Buchter et al., 1989
	Calciorthid sediment	70.0	19.3	10.7	14.7[c]	8.5	10.8	L/kg	0.857[d]	mg/L	mg/kg	0.44	3.39	Buchter et al., 1989
	Cecil soil	78.8	12.9	8.3	2.0[c]	5.7	39.7	L/kg	0.629[d]	mg/L	mg/kg	0.61	3.81	Buchter et al., 1989
	Lafitte soil	60.7	21.7	17.6	26.9[c]	3.9	103	L/kg	0.679[d]	mg/L	mg/kg	11.6	2.95	Buchter et al., 1989
	Molokai soil	25.7	46.2	28.2	11.0[c]	6.0	505	L/kg	0.847[d]	mg/L	mg/kg	1.67	4.48	Buchter et al., 1989
	Norwood soil	79.2	18.1	2.8	4.1[c]	6.9	18.6	L/kg	0.877[d]	mg/L	mg/kg	0.21	3.95	Buchter et al., 1989
	Olivier soil	4.4	89.4	6.2	8.6[c]	6.6	91.2	L/kg	0.607[d]	mg/L	mg/kg	0.83	4.04	Buchter et al., 1989
	Spodosol sediment	90.2	6.0	3.8	2.7[c]	4.3	90.8	L/kg	0.483[d]	mg/L	mg/kg	1.98	3.66	Buchter et al., 1989
	Webster soil	27.5	48.6	23.9	48.1[c]	7.6	80.7	L/kg	0.762[d]	mg/L	mg/kg	4.39	3.26	Buchter et al., 1989
	Windsor soil	76.8	20.5	2.8	2.0[c]	5.3	153	L/kg	0.647[d]	mg/L	mg/kg	2.03	3.88	Buchter et al., 1989
o-Xylene [95-47-6]	Filtrasorb 400	--	--	--	--	--	0.120	L/g	0.22	mg/L	mg/g	--	--	El-Dib and Badawy, 1979
	Fly ash	--	--	--	--	--	31.0	µg/g	0.91	mg/L	µg/g	--	--	Banerjee et al., 1988
	Leie River clay	0	100	--	--	--	20.8	L/kg	--	--	--	4.12	2.70	3.8 °C, Dewulf et al., 1999
	Leie River clay	0	100	--	--	--	15.5	L/kg	--	--	--	4.12	2.58	6.2 °C, Dewulf et al., 1999

Compound [CAS No.]	Adsorbent	% sand	% silt	% clay	CEC[a]	pH	K_F	K_F units	1/n	C units	x/m units	f_{oc} %	log K_{oc}[b]	Reference
m-Xylene [108-38-3]	Leie River clay	0	0	100	--	--	19.8	L/kg	--	--	--	4.12	2.68	8.0 °C, Dewulf et al., 1999
	Leie River clay	0	0	100	--	--	21.9	L/kg	--	--	--	4.12	2.73	13.5 °C, Dewulf et al., 1999
	Leie River clay	0	0	100	--	--	20.4	L/kg	--	--	--	4.12	2.69	18.6 °C, Dewulf et al., 1999
	Leie River clay	0	0	100	--	--	19.8	L/kg	--	--	--	4.12	2.68	25.0 °C, Dewulf et al., 1999
	Filtrasorb 400	--	--	--	--	6.3	4.930	L/g	0.614	μg/L	μg/g	--	--	Speth and Miltner, 1998
	Agricultural soil	65.2	25.6	9.2	9.0	7.4	3.47	L/kg	--	mol/L	mol/kg	2.2	2.20	Seip et al., 1986
	Forest soil	69.5	20.5	10.1	2.9	4.2	10.7	L/kg	--	mol/L	mol/kg	3.7	2.46	Seip et al., 1986
p-Xylene [106-42-3]	Filtrasorb 300	--	--	--	--	7.3	85	L/g	0.19	mg/L	mg/g	--	--	Dobbs and Cohen, 1980
	Leie River clay	0	0	100	--	--	12.7	L/kg	--	--	--	4.12	2.49	2.3 °C, Dewulf et al., 1999
	Leie River clay	0	0	100	--	--	23.1	L/kg	--	--	--	4.12	2.75	3.8 °C, Dewulf et al., 1999
	Leie River clay	0	0	100	--	--	18.3	L/kg	--	--	--	4.12	2.65	6.2 °C, Dewulf et al., 1999
	Leie River clay	0	0	100	--	--	23.5	L/kg	--	--	--	4.12	2.76	8.0 °C, Dewulf et al., 1999
	Leie River clay	0	0	100	--	--	25.3	L/kg	--	--	--	4.12	2.79	13.5 °C, Dewulf et al., 1999
	Leie River clay	0	0	100	--	--	24.0	L/kg	--	--	--	4.12	2.77	18.6 °C, Dewulf et al., 1999
	Leie River clay	0	0	100	--	--	24.7	L/kg	--	--	--	4.12	2.78	25.0 °C, Dewulf et al., 1999
Zinc [7440-66-6], as Zn^{2+}	Alligator soil	5.9	39.4	54.7	30.2[c]	4.8	28.1	L/kg	1.01[d]	mg/L	mg/kg	1.54	3.26	Buchter et al., 1989
	Calciorthid sediment	70.0	19.3	10.7	14.7[c]	8.5	420	L/kg	0.510[d]	mg/L	mg/kg	0.44	4.98	Buchter et al., 1989
	Cecil soil	78.8	12.9	8.3	2.0[c]	5.7	11.2	L/kg	0.724[d]	mg/L	mg/kg	0.61	3.26	Buchter et al., 1989
	Lafitte soil	60.7	21.7	17.6	26.9[c]	3.9	20.1	L/kg	0.891[d]	mg/L	mg/kg	11.6	2.24	Buchter et al., 1989
	Molokai soil	25.7	46.2	28.2	11.0[c]	6.0	80.4	L/kg	0.675[d]	mg/L	mg/kg	1.67	3.68	Buchter et al., 1989
	Norwood soil	79.2	18.1	2.8	4.1[c]	6.9	42.1	L/kg	0.515[d]	mg/L	mg/kg	0.21	4.30	Buchter et al., 1989
	Olivier soil	4.4	89.4	6.2	8.6[c]	6.6	89.1	L/kg	0.625[d]	mg/L	mg/kg	0.83	4.03	Buchter et al., 1989
	Spodosol sediment	90.2	6.0	3.8	2.7[c]	4.3	2.12	L/kg	0.962[d]	mg/L	mg/kg	1.98	2.03	Buchter et al., 1989
	Webster soil	27.5	48.6	23.9	48.1[c]	7.6	774	L/kg	0.697[d]	mg/L	mg/kg	4.39	4.25	Buchter et al., 1989
	Windsor soil	76.8	20.5	2.8	2.0[c]	5.3	9.68	L/kg	0.792[d]	mg/L	mg/kg	2.03	2.68	Buchter et al., 1989

[a] meq/100 g; [b] Average values reported are followed by a number in parentheses indicating the number of adsorbents used in the calculation. Sharom and Solomon (1981) reported a sediment organic matter content of 43%. Log K_{oc} value was calculated assuming 0.58 represents the fraction of carbon present in the organic matter. Similarly, Briggs (1981) reported sorption coefficients as log K_{om} and Talbert and Fletchall (1965) reported carbon as organic matter. Reported values were determined using the relationship $K_{oc} = 1.72K_{om}$; [c] cmol(+)/kg which is defined as the sum of exchangeable cations that the adsorbent can adsorb at a specific pH; [d] Value of exponent = n; csl: clayey silt loam; CT: conventional tillage; fsl: fine sandy loam; GAC: granular activated carbon; lfs: loamy fine sand; ls: loamy sand; MLSS: mixed liquor suspended solids; NT: no tillage; scl: sandy clay loam; sicl: silty clay loam; sl: sandy loam; sl: silt loam; SUB: chisel plow subsoil.

123-31-9	Hydroquinone	565-75-3	2,3,4-Trimethylpentane
123-42-2	Diacetone alcohol	583-60-8	*o*-Methylcyclohexanone
123-51-3	Isoamyl alcohol	584-84-9	2,4-Toluene diisocyanate
123-86-4	*n*-Butyl acetate	589-34-4	3-Methylhexane
123-91-1	1,4-Dioxane	589-81-1	3-Methylheptane
123-92-2	Isoamyl acetate	591-49-1	1-Methylcyclohexene
124-18-5	Decane	591-76-4	2-Methylhexane
124-40-3	Dimethylamine	591-78-6	2-Hexanone
124-48-1	Dibromochloromethane	591-93-5	1,4-Pentadiene
126-73-8	Tributyl phosphate	592-41-6	1-Hexene
126-99-8	Chloroprene	600-25-9	1-Chloro-1-nitropropane
127-18-4	Tetrachloroethylene	606-20-2	2,6-Dinitrotoluene
127-19-5	*N,N*-Dimethylacetamide	608-93-5	Pentachlorobenzene
129-00-0	Pyrene	613-12-7	2-Methylanthracene
131-11-3	Dimethyl phthalate	621-64-7	*N*-Nitrosodi-*n*-propylamine
132-64-9	Dibenzofuran	624-83-9	Methyl isocyanate
134-32-7	1-Naphthylamine	626-38-0	*sec*-Amyl acetate
135-98-8	*sec*-Butylbenzene	626-39-1	1,3,5-Tribromobenzene
137-26-8	Thiram	627-13-4	*n*-Propyl nitrate
140-88-5	Ethyl acrylate	627-20-3	*cis*-2-Pentene
141-43-5	Ethanolamine	628-63-7	*n*-Amyl acetate
141-78-6	Ethyl acetate	634-66-2	1,2,3,4-Tetrachlorobenzene
141-79-7	Mesityl oxide	634-90-2	1,2,3,5-Tetrachlorobenzene
142-29-0	Cyclopentene	636-28-2	1,2,4,5-Tetrabromobenzene
142-82-5	Heptane	646-04-8	*trans*-2-Pentene
143-50-0	Kepone	691-37-2	4-Methyl-1-pentene
144-62-7	Oxalic acid	763-29-1	2-Methyl-1-pentene
151-56-4	Ethylenimine	832-69-9	1-Methylphenanthrene
156-60-5	*trans*-1,2-Dichloroethylene	872-55-9	2-Ethylthiophene
191-24-2	Benzo[*ghi*]perylene	959-98-8	α-Endosulfan
192-97-2	Benzo[*e*]pyrene	1024-57-3	Heptachlor epoxide
193-39-5	Indeno[1,2,3-*cd*]pyrene	1031-07-8	Endosulfan sulfate
205-99-2	Benzo[*b*]fluoranthene	1563-66-2	Carbofuran
206-44-0	Fluoranthene	1582-09-8	Trifluralin
207-08-9	Benzo[*k*]fluoranthene	1640-89-7	Ethylcyclopentane
208-96-8	Acenaphthylene	1746-01-6	TCDD
218-01-9	Chrysene	1929-82-4	Nitrapyrin
287-92-3	Cyclopentane	2040-96-2	Propylcyclopentane
291-64-5	Cycloheptane	2104-64-5	EPN
299-84-3	Ronnel	2207-01-4	*cis*-1,2-Dimethylcyclohexane
300-76-5	Naled	2216-34-4	4-Methyloctane
309-00-2	Aldrin	2234-13-1	Octachloronaphthalene
319-84-6	α-BHC	2698-41-1	*o*-Chlorobenzylidenemalononitrile
319-85-7	β-BHC	2921-88-2	Chlorpyrifos
319-86-8	δ-BHC	3073-66-3	1,1,3-Trimethylcyclohexane
330-54-1	Diuron	3522-94-9	2,2,5-Trimethylhexane
463-82-1	2,2-Dimethylpropane	3689-24-5	Sulfotepp
479-45-8	Tetryl	3741-00-2	Pentylcyclopentane
496-11-7	Indan	4170-30-3	Crotonaldehyde
496-15-1	Indoline	4516-69-2	1,1,3-Trimethylcyclopentane
504-29-0	2-Aminopyridine	5103-71-9	*trans*-Chlordane
509-14-8	Tetranitromethane	5103-74-2	*cis*-Chlordane
526-73-8	1,2,3-Trimethylbenzene	6443-92-1	*cis*-2-Heptene
528-29-0	1,2-Dinitrobenzene	6876-23-9	*trans*-1,2-Dimethylcyclohexane
532-27-4	α-Chloroacetophenone	7005-72-3	4-Chlorophenyl phenyl ether
534-52-1	4,6-Dinitro-*o*-cresol	7421-93-4	Endrin aldehyde
538-93-2	Isobutylbenzene	7664-41-7	Ammonia
540-84-1	2,2,4-Trimethylpentane	7786-34-7	Mevinphos
540-88-5	*tert*-Butyl acetate	8001-35-2	Toxaphene
541-73-1	1,3-Dichlorobenzene	10061-01-5	*cis*-1,3-Dichloropropylene
541-85-5	5-Methyl-3-heptanone	10061-02-6	*trans*-1,3-Dichloropropylene
542-88-1	*sym*-Dichloromethyl ether	11096-82-5	PCB-1260
542-92-7	Cyclopentadiene	11097-69-1	PCB-1254
556-52-5	Glycidol	11104-28-2	PCB-1221
562-49-2	3,3-Dimethylpentane	11141-16-5	PCB-1232
563-45-1	3-Methyl-1-butene	12672-29-6	PCB-1248
565-59-3	2,3-Dimethylpentane	12674-11-2	PCB-1016

Empirical Formula Index

Synonym Index

A 1068, see Chlordane
AA, see Allyl alcohol
AAF, see 2-Acetylaminofluorene
2-AAF, see 2-Acetylaminofluorene
Aahepta, see Heptachlor
Aalindan, see Lindane
AAT, see Parathion
Aatack, see Thiram
AATP, see Parathion
AC 3422, see Parathion
ACC 3422, see Parathion
Accelerator thiuram, see Thiram
Acenaphthalene, see Acenaphthene
2-Acetamidofluorene, see 2-Acetylaminofluorene
2-Acetaminofluorene, see 2-Acetylaminofluorene
Acetdimethylamide, see *N,N*-Dimethylacetamide
Acetic acid amyl ester, see *n*-Amyl acetate
Acetic acid anhydride, see Acetic anhydride
Acetic acid (aqueous soln), see Acetic acid
Acetic acid 2-butoxy ester, see *sec*-Butyl acetate
Acetic acid butyl ester, see *n*-Butyl acetate
Acetic acid *sec*-butyl ester, see *sec*-Butyl acetate
Acetic acid *tert*-butyl ester, see *tert*-Butyl acetate
Acetic acid dimethylamide, see *N,N*-Dimethylacetamide
Acetic acid 1,3-dimethylbutyl ester, see *sec*-Hexyl acetate
Acetic acid 1,1-dimethylethyl ester, see *tert*-Butyl acetate
Acetic acid ethenyl ester, see Vinyl acetate
Acetic acid 2-ethoxyethyl ester, see 2-Ethoxyethyl acetate
Acetic acid ethyl ester, see Ethyl acetate
Acetic acid ethylene ester, see Vinyl acetate
Acetic acid isobutyl ester, see Isobutyl acetate
Acetic acid isopentyl ester, see Isoamyl acetate
Acetic acid isopropyl ester, see Isopropyl acetate
Acetic acid 2-methoxyethyl ester, see Methyl cellosolve acetate
Acetic acid 3-methylbutyl ester, see Isoamyl acetate
Acetic acid methyl ester, see Methyl acetate
Acetic acid 1-methylethyl ester, see Isopropyl acetate
Acetic acid 1-methylpropyl ester, see *sec*-Butyl acetate
Acetic acid 2-methylpropyl ester, see Isobutyl acetate
Acetic acid pentyl ester, see *n*-Amyl acetate
Acetic acid propyl ester, see *n*-Propyl acetate
Acetic acid *n*-propyl ester, see *n*-Propyl acetate
Acetic acid vinyl ester, see Vinyl acetate
Acetic aldehyde, see Acetaldehyde
Acetic ether, see Ethyl acetate
Acetic oxide, see Acetic anhydride
Acetidin, see Ethyl acetate
Acetoaminofluorene, see 2-Acetylaminofluorene
3-(Acetonylbenzyl)-4-hydroxycoumarin, see Warfarin
3-(α-Acetonylbenzyl)-4-hydroxycoumarin, see Warfarin
Acetosol, see 1,1,2,2-Tetrachloroethane
Aceto TETD, see Thiram
Acetoxyethane, see Ethyl acetate
1-Acetoxyethylene, see Vinyl acetate
2-Acetoxypentane, see *sec*-Amyl acetate
1-Acetoxypropane, see *n*-Propyl acetate
2-Acetoxypropane, see Isopropyl acetate
2-(Acetylamino)fluorene, see 2-Acetylaminofluorene
N-Acetyl-2-aminofluorene, see 2-Acetylaminofluorene
Acetyl anhydride, see Acetic anhydride
Acetylene dibromide, see Ethylene dibromide
Acetylene dichloride, see *trans*-1,2-Dichloroethylene
trans-Acetylene dichloride, see *trans*-1,2-Dichloroethylene
Acetylene tetrabromide, see 1,1,2,2-Tetrabromoethane

Acetylene tetrachloride, see 1,1,2,2-Tetrachloroethane
Acetylene trichloride, see Trichloroethylene
Acetyl ether, see Acetic anhydride
Acetyl oxide, see Acetic anhydride
Acquinite, see Chloropicrin
Acraldehyde, see Acrolein
Acritet, see Acrylonitrile
Acrylaldehyde, see Acrolein
Acrylamide monomer, see Acrylamide
Acrylic acid ethyl ester, see Ethyl acrylate
Acrylic acid methyl ester, see Methyl acrylate
Acrylic aldehyde, see Acrolein
Acrylon, see Acrylonitrile
Acrylonitrile monomer, see Acrylonitrile
Acutox, see Pentachlorophenol
Acylic amide, see Acrylamide
Adakane 12, see Dodecane
Adronal, see Cyclohexanol
Aether, see Ethyl ether
Aerothene, see 1,1,1-Trichloroethane
Aerothene MM, see Methylene chloride
Aerothene TT, see 1,1,1-Trichloroethane
Aethylis, see Chloroethane
Aethylis chloridum, see Chloroethane
AF 101, see Diuron
Aficide, see Lindane
AGE, see Allyl glycidyl ether
Agreflan, see Trifluralin
Agricide maggot killer (F), see Toxaphene
Agriflan 24, see Trifluralin
Agrisol G-20, see Lindane
Agritan, see *p,p'*-DDT
Agroceres, see Heptachlor
Agrocide, see Lindane
Agrocide 2, see Lindane
Agrocide 6G, see Lindane
Agrocide 7, see Lindane
Agrocide III, see Lindane
Agrocide WP, see Lindane
Agronexit, see Lindane
Agrotect, see 2,4-D
Aldehyde, see Acetaldehyde
Aldifen, see 2,4-Dinitrophenol
Aldrec, see Aldrin
Aldrex, see Aldrin
Aldrex 30, see Aldrin
Aldrite, see Aldrin
Aldrosol, see Aldrin
Algofrene type 1, see Trichlorofluoromethane
Algofrene type 2, see Dichlorodifluoromethane
Algofrene type 5, see Dichlorofluoromethane
Algylen, see Trichloroethylene
Alkron, see Parathion
Alleron, see Parathion
Alltex, see Toxaphene
Alltox, see Toxaphene
Allyl al, see Allyl alcohol
Allyl aldehyde, see Acrolein
Allylene, see Propyne
Allyl-2,3-epoxypropyl ether, see Allyl glycidyl ether
Allylic alcohol, see Allyl alcohol
1-Allyloxy-2,3-epoxypropane, see Allyl glycidyl ether
Allyl trichloride, see 1,2,3-Trichloropropane
Altox, see Aldrin
Alvit, see Dieldrin
Amarthol fast orange R base, see 3-Nitroaniline
Amatin, see Hexachlorobenzene
Ameisenatod, see Lindane

Ameisenmittel merck, see Lindane
American Cyanamid 3422, see Parathion
American Cyanamid 4049, see Malathion
Amfol, see Ammonia
Amidox, see 2,4-D
Amine 2,4,5-T for rice, see 2,4,5-T
Aminic acid, see Formic acid
4-Aminoaniline, see *p*-Phenylenediamine
p-Aminoaniline, see *p*-Phenylenediamine
2-Aminoanisole, see *o*-Anisidine
4-Aminoanisole, see *p*-Anisidine
o-Aminoanisole, see *o*-Anisidine
p-Aminoanisole, see *p*-Anisidine
Aminobenzene, see Aniline
p-Aminobiphenyl, see 4-Aminobiphenyl
1-Aminobutane, see *n*-Butylamine
1-Amino-4-chloroaniline, see 4-Chloroaniline
1-Amino-4-chlorobenzene, see 4-Chloroaniline
1-Amino-*p*-chlorobenzene, see 4-Chloroaniline
4-Aminochlorobenzene, see 4-Chloroaniline
p-Aminochlorobenzene, see 4-Chloroaniline
4-Aminodiphenyl, see 4-Aminobiphenyl
p-Aminodiphenyl, see 4-Aminobiphenyl
Aminoethane, see Ethylamine
1-Aminoethane, see Ethylamine
2-Aminoethanethiol, see Ethyl mercaptan
2-Aminoethanol, see Ethanolamine
β-Aminoethyl alcohol, see Ethanolamine
Aminoethylene, see Ethylenimine
Aminomethane, see Methylamine
1-Amino-2-methoxybenzene, see *o*-Anisidine
1-Amino-4-methoxybenzene, see *p*-Anisidine
p-Aminomethoxybenzene, see *p*-Anisidine
1-Amino-2-methylbenzene, see *o*-Toluidine
2-Amino-1-methylbenzene, see *o*-Toluidine
p-Aminomethylphenyl ether, see *p*-Anisidine
1-Aminonaphthalene, see 1-Naphthylamine
2-Aminonaphthalene, see 2-Naphthylamine
3-Aminonitrobenzene, see 3-Nitroaniline
m-Aminonitrobenzene, see 3-Nitroaniline
4-Aminonitrobenzene, see 4-Nitroaniline
p-Aminonitrobenzene, see 4-Nitroaniline
1-Amino-2-nitrobenzene, see 2-Nitroaniline
1-Amino-3-nitrobenzene, see 3-Nitroaniline
1-Amino-4-nitrobenzene, see 4-Nitroaniline
Aminophen, see Aniline
2-Aminopropane, see Isopropylamine
Amino-2-pyridine, see 2-Aminopyridine
α-Aminopyridine, see 2-Aminopyridine
o-Aminopyridine, see 2-Aminopyridine
2-Aminotoluene, see *o*-Toluidine
o-Aminotoluene, see *o*-Toluidine
Ammonia anhydrous, see Ammonia
Ammonia gas, see Ammonia
Amoxone, see 2,4-D
AMS, see α-Methylstyrene
Amyl acetate, see *n*-Amyl acetate
Amyl acetic ester, see *n*-Amyl acetate
Amyl acetic ether, see *n*-Amyl acetate
α-*n*-Amylene, see 1-Pentene
Amyl ethyl ketone, see 5-Methyl-3-heptanone
Amyl hydride, see Pentane
Amyl methyl ketone, see 2-Heptanone
n-Amyl methyl ketone, see 2-Heptanone
An, see Acrylonitrile
Anaesthetic ether, see Ethyl ether
Anamenth, see Trichloroethylene

Anesthenyl, see Methylal
Anesthesia ether, see Ethyl ether
Anesthetic ether, see Ethyl ether
Anhydrous ammonia, see Ammonia
Aniline oil, see Aniline
Anilinobenzene, see 4-Aminobiphenyl
Anilinomethane, see Methylaniline
4-Anisidine, see *p*-Anisidine
2-Anisylamine, see *o*-Anisidine
p-Anisylamine, see *p*-Anisidine
Ankilostin, see Tetrachloroethylene
Annulene, see Benzene
Anodynon, see Chloroethane
Anofex, see *p,p'*-DDT
Anol, see Cyclohexanol
Anone, see Cyclohexanone
Anozol, see Diethyl phthalate
Anthracin, see Anthracene
Anticarie, see Hexachlorobenzene
Antinonin, see 4,6-Dinitro-*o*-cresol
Antinonnon, see 4,6-Dinitro-*o*-cresol
Antisal 1a, see Toluene
Antisol 1, see Tetrachloroethylene
Anturat, see ANTU
Anyvim, see Aniline
Aparasin, see Lindane
Apavap, see Dichlorvos
Apavinphos, see Mevinphos
Aphamite, see Parathion
Aphtiria, see Lindane
Aplidal, see Lindane
Aptal, see *p*-Chloro-*m*-cresol
Aqua-kleen, see 2,4-D
Aqualin, see Acrolein
Aqualine, see Acrolein
Arasan, see Thiram
Arasan 70, see Thiram
Arasan 75, see Thiram
Arasan-M, see Thiram
Arasan 42-S, see Thiram
Arasan-SF, see Thiram
Arasan-SF-X, see Thiram
Arbitex, see Lindane
Arborol, see 4,6-Dinitro-*o*-cresol
Arcton 6, see Dichlorodifluoromethane
Arcton 7, see Dichlorofluoromethane
Arcton 9, see Trichlorofluoromethane
Arctuvin, see Hydroquinone
Areginal, see Ethyl formate
Arkotine, see *p,p'*-DDT
Arochlor 1221, see PCB-1221
Arochlor 1232, see PCB-1232
Arochlor 1242, see PCB-1242
Arochlor 1248, see PCB-1248
Arochlor 1254, see PCB-1254
Arochlor 1260, see PCB-1260
Aroclor 1016, see PCB-1016
Aroclor 1221, see PCB-1221
Aroclor 1232, see PCB-1232
Aroclor 1242, see PCB-1242
Aroclor 1248, see PCB-1248
Aroclor 1254, see PCB-1254
Aroclor 1260, see PCB-1260
Arthodibrom, see Naled
Artic, see Methyl chloride
Artificial ant oil, see Furfural
Artificial oil of ants, see Furfural

Arylam, see Carbaryl
ASP 47, see Sulfotepp
Aspon-chlordane, see Chlordane
Astrobot, see Dichlorvos
Atgard, see Dichlorvos
Atgard C, see Dichlorvos
Atgard V, see Dichlorvos
Athrombine-K, see Warfarin
Athrombin-K, see Warfarin
Attac 4-2, see Toxaphene
Attac 4-4, see Toxaphene
Attac 6, see Toxaphene
Attac 6-3, see Toxaphene
Attac 8, see Toxaphene
Atul fast yellow R, see *p*-Dimethylaminoazobenzene
Aules, see Thiram
Avlothane, see Hexachloroethane
Avolin, see Dimethyl phthalate
. Azabenzene, see Pyridine
Azacyclopropane, see Ethylenimine
1-Azaindene, see Indole
Azine, see Pyridine
Azirane, see Ethylenimine
Aziridine, see Ethylenimine
Azoamine Red ZH, see 4-Nitroaniline
Azobase MNA, see 3-Nitroaniline
Azoene fast orange GR base, see 2-Nitroaniline
Azoene fast orange GR salt, see 2-Nitroaniline
Azofix orange GR salt, see 2-Nitroaniline
Azofix red GG salt, see 4-Nitroaniline
Azogene fast orange GR, see 2-Nitroaniline
Azoic diazo component 6, see 2-Nitroaniline
Azoic diazo component 37, see 4-Nitroaniline
Azoic diazo component 112, see Benzidine
Azotox, see *p,p'*-DDT
B 404, see Parathion
BA, see Benzo[*a*]anthracene
B(*a*)A, see Benzo[*a*]anthracene
Baker's P and S liquid and ointment, see Phenol
Baktol, see *p*-Chloro-*m*-cresol
Baktolan, see *p*-Chloro-*m*-cresol
Baltana, see 1,1,1-Trichloroethane
Banana oil, see *n*-Amyl acetate, Isoamyl acetate
Bantu, see ANTU
Basaklor, see Heptachlor
BASF ursol D, see *p*-Phenylenediamine
Bay 19149, see Dichlorvos
Bay 70143, see Carbofuran
Bay E-393, see Sulfotepp
Bay E-605, see Parathion
Bayer-E 393, see Sulfotepp
BBC 12, see 1,2-Dibromo-3-chloropropane
BBP, see Benzyl butyl phthalate
BBX, see Lindane
BCEXM, see Bis(2-chloroethoxy)methane
BCF-bushkiller, see 2,4,5-T
BCIE, see Bis(2-chloroisopropyl) ether
BCME, see *sym*-Dichloromethyl ether
BCMEE, see Bis(2-chloroisopropyl) ether
BDCM, see Bromodichloromethane
BEHP, see Bis(2-ethylhexyl) phthalate
Belt, see Chlordane
Benfos, see Dichlorvos
Benhex, see Lindane
Bentox 10, see Lindane
Benxole, see Benzene
1,2-Benzacenaphthene, see Fluoranthene

Benz[*e*]acephenanthrylene, see Benzo[*b*]fluoranthene
3,4-Benz[*e*]acephenanthrylene, see Benzo[*b*]fluoranthene
Benzal alcohol, see Benzyl alcohol
Benzanthracene, see Benzo[*a*]anthracene
Benz[*a*]anthracene, see Benzo[*a*]anthracene
1,2-Benzanthracene, see Benzo[*a*]anthracene
1,2-Benz[*a*]anthracene, see Benzo[*a*]anthracene
2,3-Benzanthracene, see Benzo[*a*]anthracene
1,2:5,6-Benzanthracene, see Dibenz[*a,h*]anthracene
Benzanthrene, see Benzo[*a*]anthracene
1,2-Benzanthrene, see Benzo[*a*]anthracene
1-Benzazole, see Indole
Benzenamine, see Aniline, *N*-Nitrosodiphenylamine
Benzeneazodimethylaniline, see *p*-Dimethylaminoazobenzene
Benzene carbinol, see Benzyl alcohol
Benzenecarboxylic acid, see Benzoic acid
Benzene chloride, see Chlorobenzene
1,4-Benzenediamine, see *p*-Phenylenediamine
p-Benzenediamine, see *p*-Phenylenediamine
1,2-Benzenedicarboxylic acid anhydride, see Phthalic anydride
1,2-Benzenedicarboxylic acid bis(2-ethylhexyl) ester, see Bis(2-ethylhexyl) phthalate
1,2-Benzenedicarboxylic acid butyl phenylmethyl ester, seeBenzyl butyl phthalate
1,2-Benzenedicarboxylic acid dibutyl ester, see Di-*n*-butylphthalate
o-Benzenedicarboxylic acid dibutyl ester, see Di-*n*-butyl phthalate
Benzene-*o*-dicarboxylic acid di-*n*-butyl ester, see Di-*n*-butyl phthalate
1,2-Benzenedicarboxylic acid diethyl ester, see Diethyl phthalate
1,2-Benzenedicarboxylic acid dimethyl ester, see Dimethyl phthalate
1,2-Benzenedicarboxylic acid dioctyl ester, see Di-*n*-octyl phthalate
1,2-Benzenedicarboxylic acid di-*n*-octyl ester, see Di-*n*-octyl phthalate
o-Benzenedicarboxylic acid dioctyl ester, see Di-*n*-octyl phthalate
1,4-Benzenediol, see Hydroquinone
p-Benzenediol, see Hydroquinone
Benzeneformic acid, see Benzoic acid
Benzene hexachloride, see Lindane
Benzene hexachloride-α-isomer, see α-BHC
Benzene-*cis*-hexachloride, see β-BHC
Benzene-γ-hexachloride, see Lindane
α-Benzene hexachloride, see α-BHC
β-Benzene hexachloride, see β-BHC
δ-Benzene hexachloride, see δ-BHC
γ-Benzene hexachloride, see Lindane
trans-α-Benzene hexachloride, see β-BHC
Benzene hexahydride, see Cyclohexane
Benzene methanoic acid, see Benzoic acid
Benzene methanol, see Benzyl alcohol
Benzene tetrahydride, see Cyclohexene
Benzenol, see Phenol
2,3-Benzfluoranthene, see Benzo[*b*]fluoranthene
3,4-Benzfluoranthene, see Benzo[*b*]fluoranthene
8,9-Benzfluoranthene, see Benzo[*k*]fluoranthene
Benzidam, see Aniline
Benzidene base, see Benzidine
p-Benzidine, see Benzidine
2,3-Benzindene, see Fluorene

Benzinoform, see Carbon tetrachloride

Benzinol, see Trichloroethylene

Benzoanthracene, see Benzo[*a*]anthracene

1,2-Benzoanthracene, see Benzo[*a*]anthracene

Benzoate, see Benzoic acid

Benzo[*d,e,f*]chrysene, see Benzo[*a*]pyrene

Benzoepin, see α-Endosulfan, β-Endosulfan

2,3-Benzofluoranthene, see Benzo[*b*]fluoranthene

3,4-Benzo[*b*]fluoranthene, see Benzo[*b*]fluoranthene

11,12-Benzofluoranthene, see Benzo[*k*]fluoranthene

Benzo[*e*]fluoranthene, see Benzo[*b*]fluoranthene

11,12-Benzo[*k*]fluoranthene, see Benzo[*k*]fluoranthene

3,4-Benzofluoranthene, see Benzo[*b*]fluoranthene

8,9-Benzofluoranthene, see Benzo[*k*]fluoranthene

Benzo[*jk*]fluorene, see Fluoranthene

Benzofur D, see *p*-Phenylenediamine

Benzohydroquinone, see Hydroquinone

Benzol, see Benzene

Benzole, see Benzene

Benzolene, see Benzene

1,12-Benzoperylene, see Benzo[*ghi*]perylene

Benzo[*a*]phenanthrene, see Benzo[*a*]anthracene, Chrysene

Benzo[*α*]phenanthrene, see Chrysene

Benzo[*b*]phenanthrene, see Benzo[*a*]anthracene

Benzo[*def*]phenanthrene, see Pyrene

1,2-Benzophenanthrene, see Chrysene

2,3-Benzophenanthrene, see Benzo[*a*]anthracene

1,2-Benzopyrene, see Benzo[*a*]pyrene, Benzo[*e*]pyrene

3,4-Benzopyrene, see Benzo[*a*]pyrene

4,5-Benzopyrene, see Benzo[*e*]pyrene

6,7-Benzopyrene, see Benzo[*a*]pyrene

Benzo[*α*]pyrene, see Benzo[*a*]pyrene

4,5-Benzo[*e*]pyrene, see Benzo[*e*]pyrene

Benzopyrrole, see Indole

2,3-Benzopyrrole, see Indole

Benzo[*b*]pyrrole, see Indole

1-Benzo[*b*]pyrrole, see Indole

1,4-Benzoquine, see *p*-Quinone

Benzoquinol, see Hydroquinone

Benzoquinone, see *p*-Quinone

1,4-Benzoquinone, see *p*-Quinone

p-Benzoquinone, see *p*-Quinone

Benzoyl alcohol, see Benzyl alcohol

1,12-Benzperylene, see Benzo[*ghi*]perylene

Benz[*a*]phenanthrene, see Chrysene

1,2-Benzphenanthrene, see Chrysene

2,3-Benzphenanthrene, see Benzo[*a*]anthracene

Benz[*a*]pyrene, see Benzo[*a*]pyrene

1,2-Benzpyrene, see Benzo[*a*]pyrene, Benzo[*e*]pyrene

3,4-Benzpyrene, see Benzo[*a*]pyrene

3,4-Benz[*a*]pyrene, see Benzo[*a*]pyrene

Benzyl *n*-butyl phthalate, see Benzyl butyl phthalate

Benzyl chloride anhydrous, see Benzyl chloride

3,4-Benzypyrene, see Benzo[*a*]pyrene

Beosit, see α-Endosulfan, β-Endosulfan

Betaprone, see β-Propiolactone

Bexol, see Lindane

B(*b*)F, see Benzo[*b*]fluoranthene

B(*k*)F, see Benzo[*k*]fluoranthene

BFV, see Formaldehyde

BHC, see Lindane

γ-BHC, see Lindane

BH 2,4-D, see 2,4-D

4,4′-Bianiline, see Benzidine

N,N′-Bianiline, see 1,2-Diphenylhydrazine

p,p′-Bianiline, see Benzidine

Bibenzene, see Biphenyl

Bibesol, see Dichlorvos

Bicarburet of hydrogen, see Benzene

1,2-Bichloroethane, see 1,2-Dichloroethane

Bicyclo[4.4.0]decane, see Decahydronaphthalene

Biethylene, see 1,3-Butadiene

Bihexyl, see Dodecane

Biisopropyl, see 2,3-Dimethylbutane

2,3,1′,8′-Binaphthylene, see Benzo[*k*]fluoranthene

Binitrobenzene, see 1,3-Dinitrobenzene

Bio 5462, see α-Endosulfan, β-Endosulfan

Biocide, see Acrolein

Bioflex 81, see Bis(2-ethylhexyl) phthalate

Bioflex DOP, see Bis(2-ethylhexyl) phthalate

1,1′-Biphenyl, see Biphenyl

Biphenylamine, see 4-Aminobiphenyl

4-Biphenylamine, see 4-Aminobiphenyl

p-Biphenylamine, see 4-Aminobiphenyl

[1,1′-Biphenyl]-4-amine, see 4-Aminobiphenyl

(1,1′-Biphenyl)-4,4′-diamine, see Benzidine

4,4′-Biphenyldiamine, see Benzidine

p,p′-Biphenyldiamine, see Benzidine

4,4′-Biphenylenediamine, see Benzidine

p,p′-Biphenylenediamine, see Benzidine

o-Biphenylenemethane, see Fluorene

Biphenylene oxide, see Dibenzofuran

Biphenyl ether, see Phenyl ether

o-Biphenylmethane, see Fluorene

Biphenyl oxide, see Phenyl ether

Birnenoel, see *n*-Amyl acetate

2,2-Bis(*p*-anisyl)-1,1,1-trichloroethane, see Methoxychlor

S-1,2-Bis(carbethoxy)ethyl-*O,O*-dimethyl dithiophosphate, see Malathion

Bis(β-chloroethyl)acetal ethane, see Bis(2-chloroethoxy)methane

Bis(β-chloroethyl) ether, see Bis(2-chloroethyl) ether

Bis(2-chloroethyl)formal, see Bis(2-chloroethoxy)methane

Bis(β-chloroethyl)formal, see Bis(2-chloroethoxy)methane

Bis(β-chloroisopropyl) ether, see Bis(2-chloroisopropyl) ether

Bis(chloromethyl) ether, see *sym*-Dichloromethyl ether

Bis(2-chloro-1-methylethyl) ether, see Bis(2-chloroisopropyl) ether

Bis(2-chloro-3-methylethyl) ether, see Bis(2-chloroisopropyl) ether

1,1-Bis(4-chlorophenyl)-2,2-dichloroethane, see *p,p*′-DDD

1,1-Bis(*p*-chlorophenyl)-2,2-dichloroethane, see *p,p*′-DDD

2,2-Bis(4-chlorophenyl)-1,1-dichloroethane, see *p,p*′-DDD

2,2-Bis(*p*-chlorophenyl)-1,1-dichloroethane, see *p,p*′-DDD

2,2-Bis(4-chlorophenyl)-1,1-dichloroethene, see *p,p*′-DDE

2,2-Bis(*p*-chlorophenyl)-1,1-dichloroethene, see *p,p*′-DDE

1,1-Bis(*p*-chlorophenyl)-2,2,2-trichloroethane, see *p,p*′-DDT

2,2-Bis(4-chlorophenyl)-1,1,1-trichloroethane, see *p,p*′-DDT

2,2-Bis(*p*-chlorophenyl)-1,1,1-trichloroethane, see *p,p*′-DDT

α,α-Bis(*p*-chlorophenyl)-β,β,β-trichloroethane, see *p,p*′-

DDT

Bis-cme, see *sym*-Dichloromethyl ether

Bis-*O,O*-diethylphosphoric anhydride, see Tetraethyl pyrophosphate

Bis-*O,O*-diethylphosphorothionic anhydride, see Sulfotepp

Bis(dimethylamino)carbonothioyldisulfide, see Thiram

Bis(dimethylthiocarbamoyl)disulfide, see Thiram

Bis(dimethylthiocarbamyl)disulfide, see Thiram

S-1,2-Bis(ethoxycarbonyl)ethyl-*O,O*-dimethyl phosphorodithioate, see Malathion

S-1,2-Bis(ethoxycarbonyl)ethyl-*O,O*-dimethyl thiophosphate, see Malathion

Bis(2-ethylhexyl)-1,2-benzenedicarboxylate, see Bis(2-ethylhexyl) phthalate

1,1-Bis(*p*-methoxyphenyl)-2,2,2-trichloroethane, see Methoxychlor

2,2-Bis(*p*-methoxyphenyl)-1,1,1-trichloroethane, see Methoxychlor

Bivinyl, see 1,3-Butadiene

Black and white bleaching cream, see Hydroquinone

Blacosolv, see Trichloroethylene

Bladafum, see Sulfotepp

Bladafume, see Sulfotepp

Bladafun, see Sulfotepp

Bladan, see Parathion, Tetraethyl pyrophosphate

Bladan F, see Parathion

Blancosolv, see Trichloroethylene

Blue oil, see Aniline

Bonoform, see 1,1,2,2-Tetrachloroethane

Borer sol, see 1,2-Dichloroethane

2-Bornanone, see Camphor

Bosan Supra, see *p,p*'-DDT

Bovidermol, see *p,p*'-DDT

BP, see Benzo[*a*]pyrene

3,4-BP, see Benzo[*a*]pyrene

B(*a*)P, see Benzo[*a*]pyrene

B(*e*)P, see Benzo[*e*]pyrene

B(*ghi*)P, see Benzo[*ghi*]perylene

BPL, see β-Propiolactone

Bran oil, see Furfural

Brentamine fast orange GR base, see 2-Nitroaniline

Brentamine fast orange GR salt, see 2-Nitroaniline

Brevinyl, see Dichlorvos

Brevinyl E50, see Dichlorvos

Brilliant fast oil yellow, see *p*-Dimethylaminoazobenzene

Brilliant fast spirit yellow, see *p*-Dimethylaminoazobenzene

Brilliant fast yellow, see *p*-Dimethylaminoazobenzene

Brilliant oil yellow, see *p*-Dimethylaminoazobenzene

Brocide, see 1,2-Dichloroethane

Brodan, see Chlorpyrifos

Bromchlophos, see Naled

Bromex, see Naled

Bromic ether, see Ethyl bromide

4-Bromodiphenyl ether, see 4-Bromophenyl phenyl ether

p-Bromodiphenyl ether, see 4-Bromophenyl phenyl ether

Bromoethane, see Ethyl bromide

Bromofluoroform, see Bromotrifluoromethane

Bromofume, see Ethylene dibromide

Brom-o-gaz, see Methyl bromide

Bromomethane, see Methyl bromide

1-Bromo-4-phenoxybenzene, see 4-Bromophenyl phenyl ether

1-Bromo-*p*-phenoxybenzene, see 4-Bromophenyl phenyl ether

4-Bromophenyl ether, see 4-Bromophenyl phenyl ether

p-Bromophenyl ether, see 4-Bromophenyl phenyl ether

p-Bromophenyl phenyl ether, see 4-Bromophenyl phenyl ether

Brumolin, see Warfarin

Brush-off 445 low volatile brush killer, see 2,4,5-T

Brush-rhap, see 2,4-D, 2,4,5-T

Brushtox, see 2,4,5-T

B-Selektonon, see 2,4-D

Bunt-cure, see Hexachlorobenzene

Bunt-no-more, see Hexachlorobenzene

Butadiene, see 1,3-Butadiene

α,γ-Butadiene, see 1,3-Butadiene

Buta-1,3-diene, see 1,3-Butadiene

1-Butanamine, see *n*-Butylamine

n-Butane, see Butane

Butanethiol, see *n*-Butyl mercaptan

1-Butanethiol, see *n*-Butyl mercaptan

n-Butanethiol, see *n*-Butyl mercaptan

Butanol-2, see *sec*-Butyl alcohol

2-Butanol, see *sec*-Butyl alcohol

n-Butanol, see 1-Butanol

sec-Butanol, see *sec*-Butyl alcohol

t-Butanol, see *tert*-Butyl alcohol

tert-Butanol, see *tert*-Butyl alcohol

Butan-1-ol, see 1-Butanol

Butan-2-ol, see *sec*-Butyl alcohol

1-Butanol acetate, see *n*-Butyl acetate

2-Butanol acetate, see *sec*-Butyl acetate

Butanone, see 2-Butanone

2-Butenal, see Crotonaldehyde

cis-Butenedioic anhydride, see Maleic anhydride

2-Butenoic acid 3-[(dimethoxyphosphinyl)oxy]methyl ester, see Mevinphos

Butter yellow, see *p*-Dimethylaminoazobenzene

Butyl acetate, see *n*-Butyl acetate

1-Butyl acetate, see *n*-Butyl acetate

2-Butyl acetate, see *sec*-Butyl acetate

t-Butyl acetate, see *tert*-Butyl acetate

Butyl alcohol, see 1-Butanol

n-Butyl alcohol, see 1-Butanol

t-Butyl alcohol, see *tert*-Butyl alcohol

2-Butyl alcohol, see *sec*-Butyl alcohol

sec-Butyl alcohol acetate, see *sec*-Butyl acetate

Butylamine, see *n*-Butylamine

Butylbenzene, see *n*-Butylbenzene

t-Butylbenzene, see *tert*-Butylbenzene

Butyl benzyl phthalate, see Benzyl butyl phthalate

n-Butyl benzyl phthalate, see Benzyl butyl phthalate

Butyl cellosolve, see 2-Butoxyethanol

α-Butylene, see 1-Butene

γ-Butylene, see 2-Methylpropene

Butylene hydrate, see *sec*-Butyl alcohol

Butylene oxide, see Tetrahydrofuran

Butyl ethanoate, see *n*-Butyl acetate

n-Butyl ethanoate, see *n*-Butyl acetate

Butylethylene, see 1-Hexene

Butyl ethyl ketone, see 3-Heptanone

n-Butyl ethyl ketone, see 3-Heptanone

Butyl glycol, see 2-Butoxyethanol

Butyl glycol ether, see 2-Butoxyethanol

Butyl hydroxide, see 1-Butanol

t-Butyl hydroxide, see *tert*-Butyl alcohol

Butyl ketone, see 2-Hexanone

Butyl mercaptan, see *n*-Butyl mercaptan

Butyl methyl ketone, see 2-Hexanone

n-Butyl methyl ketone, see 2-Hexanone

Butyl oxitol, see 2-Butoxyethanol

2,2'-Dichloroisopropyl ether, see Bis(2-chloroisopropyl) ether

Dichloromethane, see Methylene chloride

Dichloromethyl ether, see *sym*-Dichloromethyl ether

Dichloromonofluoromethane, see Dichlorofluoromethane

3-(3,4-Dichlorophenol)-1,1-dimethylurea, see Diuron

Dichlorophenoxyacetic acid, see 2,4-D

(2,4-Dichlorophenoxy)acetic acid, see 2,4-D

3-(3,4-Dichlorophenyl)-1,1-dimethylurea, see Diuron

N'-(3,4-Dichlorophenyl)-*N,N*-dimethylurea, see Diuron

Dichlorophos, see Dichlorvos

α,β-Dichloropropane, see 1,2-Dichloropropane

cis-1,3-Dichloropropene, see *cis*-1,3-Dichloropropylene

trans-1,3-Dichloropropene, see *trans*-1,3-Dichloropropylene

cis-1,3-Dichloro-1-propene, see *cis*-1,3-Dichloropropylene

trans-1,3-Dichloro-1-propene, see *trans*-1,3-Dichloropropylene

1,3-Dichloroprop-1-ene, see *cis*-1,3-Dichloropropylene, *trans*-1,3-Dichloropropylene

(*E*)-1,3-Dichloropropene, see *trans*-1,3-Dichloropropylene

(*E*)-1,3-Dichloro-1-propene, see *trans*-1,3-Dichloropropylene

(*Z*)-1,3-Dichloropropene, see *cis*-1,3-Dichloropropylene

(*Z*)-1,3-Dichloro-1-propene, see *cis*-1,3-Dichloropropylene

cis-1,3-Dichloro-1-propylene, see *cis*-1,3-Dichloropropylene

trans-1,3-Dichloro-1-propylene, see *trans*-1,3-Dichloropropylene

2,2-Dichlorovinyl dimethyl phosphate, see Dichlorvos

2,2-Dichlorovinyl dimethyl phosphoric acid ester, see Dichlorvos

Dichlorovos, see Dichlorvos

Dicophane, see *p,p'*-DDT

Dicopur, see 2,4-D

Dicotox, see 2,4-D

β,β-Dicyano-*o*-chlorostyrene, see *o*-Chlorobenzylidenemalononitrile

Didakene, see Tetrachloroethylene

Didigam, see *p,p'*-DDT

Didimac, see *p,p'*-DDT

Dieldrine, see Dieldrin

Dieldrite, see Dieldrin

Dieldrix, see Dieldrin

Diethamine, see Diethylamine

1,2-Di(ethoxycarbonyl)ethyl-*O,O*-dimethyl phosphorodithioate, see Malathion

S-1,2-Di(ethoxycarbonyl)ethyl dimethyl phosphorothiolothionate, see Malathion

Diethyl, see Butane

N,N-Diethylamine, see Diethylamine

Diethylaminoethane, see Triethylamine

Diethylaminoethanol, see 2-Diethylaminoethanol

β-Diethylaminoethanol, see 2-Diethylaminoethanol

N-Diethylaminoethanol, see 2-Diethylaminoethanol

2-*N*-Diethylaminoethanol, see 2-Diethylaminoethanol

2-Diethylaminoethyl alcohol, see 2-Diethylaminoethanol

β-Diethylaminoethyl alcohol, see 2-Diethylaminoethanol

Diethyl (dimethoxyphosphinothioylthio) butanedioate, see Malathion

Diethyl (dimethoxyphosphinothioylthio) succinate, see Malathion

Diethylene dioxide, see 1,4-Dioxane

1,4-Diethylene dioxide, see 1,4-Dioxane

Diethylene ether, see 1,4-Dioxane

Diethylene oxide, see 1,4-Dioxane, Tetrahydrofuran

Diethylene oximide, see Morpholine

Diethyleneimid oxide, see Morpholine

Diethyleneimide oxide, see Morpholine

Diethylenimide oxide, see Morpholine

N,N-Diethylethanamine, see Triethylamine

Diethylethanolamine, see 2-Diethylaminoethanol

N,N-Diethylethanolamine, see 2-Diethylaminoethanol

Diethyl ether, see Ethyl ether

Di(2-ethylhexyl) orthophthalate, see Bis(2-ethylhexyl) phthalate

Di(2-ethylhexyl) phthalate, see Bis(2-ethylhexyl) phthalate

Diethyl-(2-hydroxyethyl)amine, see 2-Diethylaminoethanol

N,N-Diethyl-*N*-(β-hydroxyethyl)amine, see 2-Diethylaminoethanol

Diethyl mercaptosuccinic acid *O,O*-dimethyl phosphorodithioate, see Malathion

Diethyl mercaptosuccinate, *O,O*-dimethyl phosphorodithioate, see Malathion

Diethyl mercaptosuccinate, *O,O*-dimethyl thiophosphate, see Malathion

Diethylmethane, see Pentane

Diethylmethylmethane, see 3-Methylpentane

Diethyl-4-nitrophenyl phosphorothionate, see Parathion

O,O-Diethyl-*O*-4-nitrophenyl phosphorothioate, see Parathion

O,O-Diethyl *O-p*-nitrophenyl phosphorothioate, see Parathion

Diethyl-*p*-nitrophenyl thionophosphate, see Parathion

O,O-Diethyl-*O*-4-nitrophenyl thionophosphate, see Parathion

O,O-Diethyl-*O-p*-nitrophenyl thionophosphate, see Parathion

Diethyl-*p*-nitrophenyl thiophosphate, see Parathion

O,O-Diethyl-*O-p*-nitrophenyl thiophosphate, see Parathion

Diethyl oxide, see Ethyl ether

Diethylparathion, see Parathion

Diethyl-*o*-phthalate, see Diethyl phthalate

O,O-Diethyl-*O*-3,5,6-trichloro-2-pyridyl phosphorothioate, see Chlorpyrifos

Difluorodibromomethane, see Dibromodifluoromethane

Difluorodichloromethane, see Dichlorodifluoromethane

1,1-Difluoro-1,2,2,2-tetrachloroethane, see 1,1-Difluorotetrachloroethane

1,2-Difluoro-1,1,2,2-tetrachloroethane, see 1,2-Difluorotetrachloroethane

2,2-Difluoro-1,1,1,2-tetrachloroethane, see 1,1-Difluorotetrachloroethane

Digermin, see Trifluralin

Dihexyl, see Dodecane

1,2-Dihydroacenaphthylene, see Acenaphthene

Dihydroazirine, see Ethylenimine

Dihydro-1*H*-azirine, see Ethylenimine

2,3-Dihydro-2,2-dimethyl-7-benzofuranol methyl carbamate, see Carbofuran

1,3-Dihydro-1,3-dioxoisobenzofuran, see Phthalic anhydride

2,3-Dihydroindene, see Indan

2,3-Dihydro-1*H*-indene, see Indan

2,3-Dihydroindole, see Indoline

2,3-Dihydro-1*H*-indole, see Indoline

Dihydroxybenzene, see Hydroquinone

4,4′-DMDT, see Methoxychlor
p,p′-DMDT, see Methoxychlor
DMF, see N,N-Dimethylformamide
DMFA, see N,N-Dimethylformamide
DMK, see Acetone
DMN, see N-Nitrosodimethylamine
DMNA, see N-Nitrosodimethylamine
DMP, see Dimethyl phthalate
2,4-DMP, see 2,4-Dimethylphenol
DMTD, see Methoxychlor
DMS, see Dimethyl sulfate
DMU, see Diuron
DN, see 4,6-Dinitro-o-cresol
DNC, see 4,6-Dinitro-o-cresol
DN-dry mix no. 2, see 4,6-Dinitro-o-cresol
DNOC, see 4,6-Dinitro-o-cresol
DNOP, see Di-n-octyl phthalate
DNP, see 2,4-Dinitrophenol
2,4-DNP, see 2,4-Dinitrophenol
DNT, see 2,4-Dinitrotoluene
2,4-DNT, see 2,4-Dinitrotoluene
2,6-DNT, see 2,6-Dinitrotoluene
DNTP, see Parathion
Dodat, see p,p′-DDT
n-Dodecane, see Dodecane
Dolco mouse cereal, see Strychnine
Dolen-pur, see Hexachlorobutadiene
Dol granule, see Lindane
Dolochlor, see Chloropicrin
DOP, see Bis(2-ethylhexyl) phthalate, Di-n-octyl
    phthalate
Dormone, see 2,4-D
Dowanol EB, see 2-Butoxyethanol
Dowanol EE, see 2-Ethoxyethanol
Dowanol EM, see Methyl cellosolve
Dowcide 7, see Pentachlorophenol
Dowco-163, see Nitrapyrin
Dowco-179, see Chlorpyrifos
Dow ET 14, see Ronnel
Dow ET 57, see Ronnel
Dowfume, see Methyl bromide
Dowfume 40, see Ethylene dibromide
Dowfume EDB, see Ethylene dibromide
Dowfume MC-2, see Methyl bromide
Dowfume MC-2 soil fumigant, see Methyl bromide
Dowfume MC-33, see Methyl bromide
Dowfume W-8, see Ethylene dibromide
Dowfume W-85, see Ethylene dibromide
Dowfume W-90, see Ethylene dibromide
Dowfume W-100, see Ethylene dibromide
Dowicide 2, see 2,4,5-Trichlorophenol
Dowicide 2S, see 2,4,6-Trichlorophenol
Dowicide 7, see Pentachlorophenol
Dowicide B, see 2,4,5-Trichlorophenol
Dowicide EC-7, see Pentachlorophenol
Dowicide G, see Pentachlorophenol
Dowklor, see Chlordane
Dow pentachlorophenol DP-2 antimicrobial, see Penta-
    chlorophenol
Dow-per, see Tetrachloroethylene
Dowtherm E, see 1,2-Dichlorobenzene
Dow-tri, see Trichloroethylene
DPH, see 1,2-Diphenylhydrazine
DPN, see N-Nitrosodi-n-propylamine
DPNA, see N-Nitrosodi-n-propylamine
DPP, see Parathion
Dracylic acid, see Benzoic acid

Drexel parathion 8E, see Parathion
Drinox, see Aldrin and Heptachlor
Drinox H-34, see Heptachlor
Dukeron, see Trichloroethylene
Duodecane, see Dodecane
Duo-kill, see Dichlorvos
Durafur black R, see p-Phenylenediamine
Duraphos, see Mevinphos
Duravos, see Dichlorvos
Durene, see 1,2,4,5-Tetramethylbenzene
Durol, see 1,2,4,5-Tetramethylbenzene
Durotox, see Pentachlorophenol
Dursban, see Chlorpyrifos
Dursban F, see Chlorpyrifos
Dutch liquid, see 1,2-Dichloroethane
Dutch oil, see 1,2-Dichloroethane
Dykol, see p,p′-DDT
Dynex, see Diuron
E 393, see Sulfotepp
E 605, see Parathion
E 3314, see Heptachlor
EAK, see 5-Methyl-3-heptanone
Eastern states duocide, see Warfarin
EB, see Ethylbenzene
Ecatox, see Parathion
ECH, see Epichlorohydrin
Ectoral, see Ronnel
EDB, see Ethylene dibromide
EDB-85, see Ethylene dibromide
E-D-BEE, see Ethylene dibromide
EDC, see 1,2-Dichloroethane
Edco, see Methyl bromide
Effusan, see 4,6-Dinitro-o-cresol
Effusan 3436, see 4,6-Dinitro-o-cresol
Egitol, see Hexachloroethane
EGM, see Methyl cellosolve
EGME, see Methyl cellosolve
EI, see Ethylenimine
Ekagom TB, see Thiram
Ekasolve EE acetate solvent, see 2-Ethoxyethyl acetate
Ekatin WF & WF ULV, see Parathion
Ekatox, see Parathion
Ektasolve, see 2-Butoxyethanol, Methyl cellosolve
Ektasolve EE, see 2-Ethoxyethanol
EL 4049, see Malathion
Elancolan, see Trifluralin
Elaol, see Di-n-butyl phthalate
Eldopaque, see Hydroquinone
Eldoquin, see Hydroquinone
Electro-CF 11, see Trichlorofluoromethane
Electro-CF 12, see Dichlorodifluoromethane
Elgetol, see 4,6-Dinitro-o-cresol
Elgetol 30, see 4,6-Dinitro-o-cresol
Elipol, see 4,6-Dinitro-o-cresol
Embafume, see Methyl bromide
Emmatos, see Malathion
Emmatos extra, see Malathion
Emulsamine BK, see 2,4-D
Emulsamine E-3, see 2,4-D
Endocel, see α-Endosulfan, β-Endosulfan
Endosol, see α-Endosulfan, β-Endosulfan
Endosulfan, see α-Endosulfan, β-Endosulfan
Endosulfan I, see α-Endosulfan
Endosulfan II, see β-Endosulfan
Endosulphan, see α-Endosulfan, β-Endosulfan
Endrex, see Endrin
Endrine, see Endrin

Enial yellow 2G, see *p*-Dimethylaminoazobenzene
ENT 54, see Acrylonitrile
ENT 154, see 4,6-Dinitro-*o*-cresol
ENT 262, see Dimethyl phthalate
ENT 987, see Thiram
ENT 1506, see *p,p'*-DDT
ENT 1656, see 1,2-Dichloroethane
ENT 1716, see Methoxychlor
ENT 1860, see Tetrachloroethylene
ENT 4225, see *p,p'*-DDD
ENT 4504, see Bis(2-chloroethyl) ether
ENT 4705, see Carbon tetrachloride
ENT 7796, see Lindane
ENT 8538, see 2,4-D
ENT 9232, see α-BHC
ENT 9233, see β-BHC
ENT 9234, see δ-BHC
ENT 9735, see Toxaphene
ENT 9932, see Chlordane
ENT 15108, see Parathion
ENT 15152, see Heptachlor
ENT 15349, see Ethylene dibromide
ENT 15406, see 1,2-Dichloropropane
ENT 15949, see Aldrin
ENT 16225, see Dieldrin
ENT 16273, see Sulfotepp
ENT 16391, see Kepone
ENT 17034, see Malathion
ENT 17251, see Endrin
ENT 17298, see EPN
ENT 18771, see Tetraethyl pyrophosphate
ENT 20738, see Dichlorvos
ENT 22324, see Mevinphos
ENT 23284, see Ronnel
ENT 23969, see Carbaryl
ENT 23979, see α-Endosulfan, β-Endosulfan
ENT 24988, see Naled
ENT 25552-X, see Chlordane
ENT 25584, see Heptachlor epoxide
ENT 27164, see Carbofuran
ENT 27311, see Chlorpyrifos
ENT 50324, see Ethylenimine
Entomoxan, see Lindane
Entsufon, see 2,4,6-Trinitrotoluene
Envert 171, see 2,4-D
Envert DT, see 2,4-D
Envert-T, see 2,4,5-T
EP 30, see Pentachlorophenol
α-Epichlorohydrin, see Epichlorohydrin
(*dl*)-α-Epichlorohydrin, see Epichlorohydrin
Epichlorophydrin, see Epichlorohydrin
Epihydric alcohol, see Glycidol
Epihydrin alcohol, see Glycidol
EPN 300, see EPN
1,2-Epoxy-3-allyloxypropane, see Allyl glycidyl ether
1,2-Epoxy-3-chloropropane, see Epichlorohydrin
Epoxy heptachlor, see Heptachlor epoxide
Epoxypropane, see Propylene oxide
1,2-Epoxypropane, see Propylene oxide
2,3-Epoxypropanol, see Glycidol
2,3-Epoxy-1-propanol, see Glycidol
Epoxypropyl alcohol, see Glycidol
2,3-Epoxypropyl chloride, see Epichlorohydrin
Equigard, see Dichlorvos
Equigel, see Dichlorvos
Eradex, see Chlorpyrifos
Ergoplast FDO, see Bis(2-ethylhexyl) phthalate

ESEN, see Phthalic anhydride
Eskimon 11, see Trichlorofluoromethane
Eskimon 12, see Dichlorodifluoromethane
Essence of mirbane, see Nitrobenzene
Essence of myrbane, see Nitrobenzene
Estercide T-2 and T-245, see 2,4,5-T
Esteron, see 2,4-D, 2,4,5-T
Esteron 44 weed killer, see 2,4-D
Esteron 76 BE, see 2,4-D
Esteron 99, see 2,4-D
Esteron 99 concentrate, see 2,4-D
Esteron 245 BE, see 2,4,5-T
Esteron brush killer, see 2,4-D, 2,4,5-T
Esterone 4, see 2,4-D
Esterone 245, see 2,4,5-T
Estol 1550, see Diethyl phthalate
Estonate, see *p,p'*-DDT
Estone, see 2,4-D
Estonox, see Toxaphene
Estrosel, see Dichlorvos
Estrosol, see Dichlorvos
ET 14, see Ronnel
ET 57, see Ronnel
Ethanal, see Acetaldehyde
Ethanamine, see Ethylamine
1,2-Ethanediamine, see Ethylenediamine
Ethane dichloride, see 1,2-Dichloroethane
Ethanedioic acid, see Oxalic acid
1,2-Ethanediol dipropanoate, see Crotonaldehyde
Ethanedionic acid, see Oxalic acid
Ethane hexachloride, see Hexachloroethane
Ethanenitrile, see Acetonitrile
Ethane pentachloride, see Pentachloroethane
Ethane tetrachloride, see 1,1,2,2-Tetrachloroethane
Ethanethiol, see Ethyl mercaptan
Ethane trichloride, see 1,1,2-Trichloroethane
Ethanoic acid, see Acetic acid
Ethanoic anhydrate, see Acetic anhydride
Ethanoic anhydride, see Acetic anhydride
β-Ethanolamine, see Ethanolamine
Ethene dichloride, see 1,2-Dichloroethane
Ethenyl acetate, see Vinyl acetate
Ethenyl ethanoate, see Vinyl acetate
Ether, see Ethyl ether
Ether chloratus, see Chloroethane
Ether hydrochloric, see Chloroethane
Ether muriatic, see Chloroethane
Ethinyl trichloride, see Trichloroethylene
Ethiolacar, see Malathion
Ethlon, see Parathion
Ethoxyacetate, see 2-Ethoxyethyl acetate
Ethoxycarbonylethylene, see Ethyl acrylate
Ethoxyethane, see Ethyl ether
Ethoxyethyl acetate, see 2-Ethoxyethyl acetate
2-Ethoxyethyl acetate, see 2-Ethoxyethyl acetate
β-Ethoxyethyl acetate, see 2-Ethoxyethyl acetate
Ethoxy-4-nitrophenoxy phenylphosphine sulfide, see
    EPN
Ethyenyl benzene, see Styrene
Ethyl acetic ester, see Ethyl acetate
Ethyl acetone, see 2-Pentanone
Ethyl aldehyde, see Acetaldehyde
Ethyl amyl ketone, see 5-Methyl-3-heptanone
Ethyl *sec*-amyl ketone, see 5-Methyl-3-heptanone
Ethylbenzol, see Ethylbenzene
Ethyl butyl ketone, see 3-Heptanone
Ethyl carbinol, see 1-Propanol

Ethyl cellosolve, see 2-Ethoxyethanol
Ethyl chloride, see Chloroethane
Ethyldimethylmethane, see 2-Methylbutane
Ethylene aldehyde, see Acrolein
Ethylene bromide, see Ethylene dibromide
Ethylene bromide glycol dibromide, see Ethylene dibromide
Ethylene carboxamide, see Acrylamide
Ethylene chlorhydrin, see Ethylene chlorohydrin
Ethylene chloride, see 1,2-Dichloroethane
1,2-Ethylenediamine, see Ethylenediamine
1,2-Ethylene dibromide, see Ethylene dibromide
Ethylene dichloride, see 1,2-Dichloroethane
1,2-Ethylene dichloride, see 1,2-Dichloroethane
Ethylene dipropionate, see Crotonaldehyde
Ethylene glycol dipropionate, see Crotonaldehyde
Ethylene glycol ethyl ether, see 2-Ethoxyethanol
Ethylene glycol ethyl ether acetate, see 2-Ethoxyethyl acetate
Ethylene glycol methyl ether, see Methyl cellosolve
Ethylene glycol methyl ether acetate, see Methyl cellosolve acetate
Ethylene glycol monobutyl ether, see 2-Butoxyethanol
Ethylene glycol mono-*n*-butyl ether, see 2-Butoxyethanol
Ethylene glycol monoethyl ether, see 2-Ethoxyethanol
Ethylene glycol monoethyl ether acetate, see 2-Ethoxyethyl acetate
Ethylene glycol monomethyl ether, see Methyl cellosolve
Ethylene glycol monomethyl ether acetate, see Methyl cellosolve acetate
Ethylene hexachloride, see Hexachloroethane
Ethylene monochloride, see Vinyl chloride
Ethylenenaphthalene, see Acenaphthene
1,8-Ethylenenaphthalene, see Acenaphthene
Ethylene propionate, see Crotonaldehyde
Ethylene tetrachloride, see Tetrachloroethylene
Ethylene trichloride, see Trichloroethylene
N-Ethylethanamine, see Diethylamine
Ethyl ethanoate, see Ethyl acetate
Ethylethylene, see 1-Butene
Ethyl formic ester, see Ethyl formate
Ethylhexyl phthalate, see Bis(2-ethylhexyl) phthalate
2-Ethylhexyl phthalate, see Bis(2-ethylhexyl) phthalate
Ethyl hydrosulfide, see Ethyl mercaptan
Ethylic acid, see Acetic acid
Ethylidene chloride, see 1,1-Dichloroethane
Ethylidene dichloride, see 1,1-Dichloroethane
1,1-Ethylidene dichloride, see 1,1-Dichloroethane
Ethylimine, see Ethylenimine
Ethylisobutylmethane, see 2-Methylhexane
Ethyl methanoate, see Ethyl formate
Ethylmethyl carbinol, see *sec*-Butyl alcohol
Ethyl methyl ketone, see 2-Butanone
N-Ethylmorpholine, see 4-Ethylmorpholine
Ethyl nitrile, see Acetonitrile
Ethyl *p*-nitrophenyl benzenethionophosphate, see EPN
Ethyl *p*-nitrophenyl benzenethiophosphonate, see EPN
Ethyl *p*-nitrophenyl ester, see EPN
Ethyl *p*-nitrophenyl phenylphosphonothioate, see EPN
O-Ethyl *O*-4-nitrophenyl phenylphosphonothioate, see EPN
O-Ethyl *O*-*p*-nitrophenyl phenylphosphonothioate, see EPN
Ethyl *p*-nitrophenyl thionobenzenephosphate, see EPN
Ethyl *p*-nitrophenyl thionobenzenephosphonate, see EPN
Ethylolamine, see Ethanolamine
Ethyl oxide, see Ethyl ether

Ethyl parathion, see Parathion
O-Ethyl phenyl *p*-nitrophenyl phenylphosphorothioate, see EPN
O-Ethyl phenyl *p*-nitrophenyl thiophosphonate, see EPN
Ethyl phthalate, see Diethyl phthalate
Ethyl propenoate, see Ethyl acrylate
Ethyl-2-propenoate, see Ethyl acrylate
Ethyl pyrophosphate, see Tetraethyl pyrophosphate
Ethyl sulfhydrate, see Ethyl mercaptan
Ethyl thioalcohol, see Ethyl mercaptan
Ethyl thiopyrophosphate, see Sulfotepp
Etilon, see Parathion
Etiol, see Malathion
Etrolene, see Ronnel
Eviplast 80, see Bis(2-ethylhexyl) phthalate
Eviplast 81, see Bis(2-ethylhexyl) phthalate
Evola, see 1,4-Dichlorobenzene
Exagama, see Lindane
Experimental insecticide 269, see Endrin
Experimental insecticide 4049, see Malathion
Experimental insecticide 7744, see Carbaryl
Extermathion, see Malathion
Extrar, see 4,6-Dinitro-*o*-cresol
F 11, see Trichlorofluoromethane
F 12, see Dichlorodifluoromethane
F 112, see 1,2-Difluorotetrachloroethane
F 13B1, see Bromotrifluoromethane
FA, see Formaldehyde
FAA, see 2-Acetylaminofluorene
2-FAA, see 2-Acetylaminofluorene
Falitram, see Thiram
Falkitol, see Hexachloroethane
Fannoform, see Formaldehyde
Farmco, see 2,4-D
Farmco diuron, see Diuron
Farmco fence rider, see 2,4,5-T
Fasciolin, see Carbon tetrachloride and Hexachloroethane
Fasco fascrat powder, see Warfarin
Fasco-terpene, see Toxaphene
Fast corinth base B, see Benzidine
Fast garnet base B, see 1-Naphthylamine
Fast garnet B base, see 1-Naphthylamine
Fast oil yellow B, see *p*-Dimethylaminoazobenzene
Fast orange base GR, see 2-Nitroaniline
Fast orange base GR salt, see 2-Nitroaniline
Fast orange base JR, see 2-Nitroaniline
Fast orange base R, see 3-Nitroaniline
Fast orange GR base, see 2-Nitroaniline
Fast orange M base, see 3-Nitroaniline
Fast orange MM base, see 3-Nitroaniline
Fast orange O base, see 2-Nitroaniline
Fast orange O salt, see 2-Nitroaniline
Fast orange R base, see 3-Nitroaniline
Fast orange R salt, see 3-Nitroaniline
Fast orange salt JR, see 2-Nitroaniline
Fast red 2G base, see 4-Nitroaniline
Fast red base GG, see 4-Nitroaniline
Fast red base 2J, see 4-Nitroaniline
Fast red 2G salt, see 4-Nitroaniline
Fast red GG base, see 4-Nitroaniline
Fast red GG salt, see 4-Nitroaniline
Fast red MP base, see 4-Nitroaniline
Fast red P base, see 4-Nitroaniline
Fast red P salt, see 4-Nitroaniline
Fast red salt GG, see 4-Nitroaniline
Fast red salt 2J, see 4-Nitroaniline

Fast scarlet base B, see 2-Naphthylamine
Fast yellow, see *p*-Dimethylaminoazobenzene
Fat yellow, see *p*-Dimethylaminoazobenzene
Fat yellow A, see *p*-Dimethylaminoazobenzene
Fat yellow AD OO, see *p*-Dimethylaminoazobenzene
Fat yellow ES, see *p*-Dimethylaminoazobenzene
Fat yellow ES extra, see *p*-Dimethylaminoazobenzene
Fat yellow extra conc., see *p*-Dimethylaminoazobenzene
Fat yellow R, see *p*-Dimethylaminoazobenzene
Fat yellow R (8186), see *p*-Dimethylaminoazobenzene
FC 11, see Trichlorofluoromethane
FC 12, see Dichlorodifluoromethane
Fecama, see Dichlorvos
Fedal-UN, see Tetrachloroethylene
Fence rider, see 2,4,5-T
Fenchchlorphos, see Ronnel
Fenchlorfos, see Ronnel
Fenchlorophos, see Ronnel
Fenoxyl carbon n, see 2,4-Dinitrophenol
Fermentation amyl alcohol, see Isoamyl alcohol
Fermentation butyl alcohol, see Isobutyl alcohol
Fermide, see Thiram
Fermine, see Dimethyl phthalate
Fernacol, see Thiram
Fernasan, see Thiram
Fernasan A, see Thiram
Fernesta, see 2,4-D
Fernide, see Thiram
Fernimine, see 2,4-D
Fernoxone, see 2,4-D
Ferxone, see 2,4-D
Fleck-flip, see Trichloroethylene
Fleet-X, see 1,3,5-Trimethylbenzene
Fleximel, see Bis(2-ethylhexyl) phthalate
Flexol DOP, see Bis(2-ethylhexyl) phthalate
Flexol plasticizer DOP, see Bis(2-ethylhexyl) phthalate
Flock-flip, see Trichloroethylene
Flo pro T, seed protectant, see Thiram
Fluate, see Trichloroethylene
Flukoids, see Carbon tetrachloride
N-Fluoren-2-acetylacetamide, see 2-Acetylaminofluorene
9H-Fluorene, see Fluorene
2-Fluorenylacetamide, see 2-Acetylaminofluorene
N-2-Fluorenylacetamide, see 2-Acetylaminofluorene
N-Fluorenyl-2-acetamide, see 2-Acetylaminofluorene
N-9*H*-Fluoren-2-ylacetamide, see 2-Acetylaminofluorene
Fluorocarbon 11, see Trichlorofluoromethane
Fluorocarbon 12, see Dichlorodifluoromethane
Fluorocarbon 21, see Dichlorofluoromethane
Fluorocarbon 113, see 1,1,2-Trichlorotrifluoroethane
Fluorodichloromethane, see Dichlorofluoromethane
Fluorotrichloromethane, see Trichlorofluoromethane
Fly-die, see Dichlorvos
Fly fighter, see Dichlorvos
FMC 5462, see α-Endosulfan, β-Endosulfan
Folidol, see Parathion
Folidol E605, see Parathion
Folidol E & E 605, see Parathion
Foredex 75, see 2,4-D
Forlin, see Lindane
Formal, see Malathion, Methylal
Formaldehyde bis(β-chloroethylacetal), see Bis(2-chloro-
    ethoxy)methane
Formaldehyde dimethylacetal, see Methylal
Formalin, see Formaldehyde
Formalin 40, see Formaldehyde
Formalith, see Formaldehyde

Formic acid ethyl ester, see Ethyl formate
Formic acid methyl ester, see Methyl formate
Formic aldehyde, see Formaldehyde
Formic ether, see Ethyl formate
Formol, see Formaldehyde
Formosa camphor, see Camphor
Formula 40, see 2,4-D
N-Formyldimethylamine, see *N,N*-Dimethylformamide
Formylic acid, see Formic acid
Formyl trichloride, see Chloroform
Forron, see 2,4,5-T
Forst U 46, see 2,4,5-T
Fortex, see 2,4,5-T
Forthion, see Malathion
Fosdrin, see Mevinphos
Fosfermo, see Parathion
Fosferno, see Parathion
Fosfex, see Parathion
Fosfive, see Parathion
Fosfothion, see Malathion
Fosfotion, see Malathion
Fosova, see Parathion
Fostern, see Parathion
Fostox, see Parathion
Fosvex, see Tetraethyl pyrophosphate
Fouramine D, see *p*-Phenylenediamine
Fourrine 1, see *p*-Phenylenediamine
Fourrine D, see *p*-Phenylenediamine
Four thousand forty-nine, see Malathion
Freon 10, see Carbon tetrachloride
Freon 11, see Trichlorofluoromethane
Freon 11A, see Trichlorofluoromethane
Freon 11B, see Trichlorofluoromethane
Freon 12, see Dichlorodifluoromethane
Freon 12-B2, see Dibromodifluoromethane
Freon 13-B1, see Bromotrifluoromethane
Freon 20, see Chloroform
Freon 21, see Dichlorofluoromethane
Freon 30, see Methylene chloride
Freon 112, see 1,2-Difluorotetrachloroethane
Freon 112A, see 1,1-Difluorotetrachloroethane
Freon 113, see 1,1,2-Trichlorotrifluoroethane
Freon 150, see 1,2-Dichloroethane
Freon F-12, see Dichlorodifluoromethane
Freon HE, see Trichlorofluoromethane
Freon MF, see Trichlorofluoromethane
Frigen 11, see Trichlorofluoromethane
Frigen 12, see Dichlorodifluoromethane
Frigen 113 TR-T, see 1,1,2-Trichlorotrifluoroethane
Fruitone A, see 2,4,5-T
Fumagon, see 1,2-Dibromo-3-chloropropane
Fumazone, see 1,2-Dibromo-3-chloropropane
Fumazone 86, see 1,2-Dibromo-3-chloropropane
Fumazone 86E, see 1,2-Dibromo-3-chloropropane
Fumigant-1, see Methyl bromide
Fumigrain, see Acrylonitrile
Fumo-gas, see Ethylene dibromide
Fungifen, see Pentachlorophenol
Fungol, see Pentachlorophenol
Furadan, see Carbofuran
Fural, see Furfural
2-Furaldehyde, see Furfural
Furale, see Furfural
2-Furanaldehyde, see Furfural
2-Furancarbinol, see Furfuryl alcohol
2-Furancarbonal, see Furfural
2-Furancarboxaldehyde, see Furfural

2,5-Furanedione, see Maleic anhydride
Furanidine, see Tetrahydrofuran
2-Furanmethanol, see Furfuryl alcohol
Fur black 41867, see *p*-Phenylenediamine
Fur brown 41866, see *p*-Phenylenediamine
Furfural alcohol, see Furfuryl alcohol
Furfuraldehyde, see Furfural
Furfurol, see Furfural
Furfurole, see Furfural
Furole, see Furfural
α-Furole, see Furfural
Furro D, see *p*-Phenylenediamine
Fur yellow, see *p*-Phenylenediamine
Furyl alcohol, see Furfuryl alcohol
Furylcarbinol, see Furfuryl alcohol
2-Furylcarbinol, see Furfuryl alcohol
α-Furylcarbinol, see Furfuryl alcohol
2-Furylmethanal, see Furfural
2-Furylmethanol, see Furfuryl alcohol
Fusel oil, see Isoamyl alcohol
Futramine D, see *p*-Phenylenediamine
Fyde, see Formaldehyde
Fyfanon, see Malathion
G 25, see Chloropicrin
Gallogama, see Lindane
Gamacid, see Lindane
Gamaphex, see Lindane
Gamene, see Lindane
Gamiso, see Lindane
Gammahexa, see Lindane
Gammalin, see Lindane
Gammexene, see Lindane
Gammopaz, see Lindane
Gamonil, see Carbaryl
GC-1189, see Kepone
Gearphos, see Parathion
Gemalgene, see Trichloroethylene
General chemicals 1189, see Kepone
Genetron 11, see Trichlorofluoromethane
Genetron 12, see Dichlorodifluoromethane
Genetron 21, see Dichlorofluoromethane
Genetron 112, see 1,2-Difluorotetrachloroethane
Geniphene, see Toxaphene
Genithion, see Parathion
Genitox, see *p,p'*-DDT
Genklene, see 1,1,1-Trichloroethane
Geranium crystals, see Phenyl ether
Germain's, see Carbaryl
Germalgene, see Trichloroethylene
Gesafid, see *p,p'*-DDT
Gesapon, see *p,p'*-DDT
Gesarex, see *p,p'*-DDT
Gesarol, see *p,p'*-DDT
Gesfid, see Mevinphos
Gestid, see Mevinphos
Gettysolve-B, see Hexane
Gettysolve-C, see Heptane
Gexane, see Lindane
Glacial acetic acid, see Acetic acid
Glazd penta, see Pentachlorophenol
Glycol dichloride, see 1,2-Dichloroethane
Glycerin trichlorohydrin, see 1,2,3-Trichloropropane
Glycerol epichlorohydrin, see Epichlorohydrin
Glycerol trichlorohydrin, see 1,2,3-Trichloropropane
Glyceryl trichlorohydrin, see 1,2,3-Trichloropropane
Glycide, see Glycidol
Glycidyl alcohol, see Glycidol

Glycidyl allyl ether, see Allyl glycidyl ether
Glycinol, see Ethanolamine
Glycol bromide, see Ethylene dibromide
Glycol chlorohydrin, see Ethylene chlorohydrin
Glycol dibromide, see Ethylene dibromide
Glycol ether EE, see 2-Ethoxyethanol
Glycol ether EE acetate, see 2-Ethoxyethyl acetate
Glycol ether EM, see Methyl cellosolve
Glycol ether EM acetate, see Methyl cellosolve acetate
Glycol ethylene ether, see 1,4-Dioxane
Glycol methyl ether, see Methyl cellosolve
Glycol monobutyl ether, see 2-Butoxyethanol
Glycol monochlorohydrin, see Ethylene chlorohydrin
Glycol monoethyl ether, see 2-Ethoxyethanol
Glycol monoethyl ether acetate, see 2-Ethoxyethyl
    acetate
Glycol monomethyl ether, see Methyl cellosolve
Glycol monomethyl ether acetate, see Methyl cellosolve
    acetate
Goodrite GP 264, see Bis(2-ethylhexyl) phthalate
GP-40-66:120, see Hexachlorobutadiene
GPKh, see Heptachlor
Granox NM, see Hexachlorobenzene
Graphlox, see Hexachlorocyclopentadiene
Grasal brilliant yellow, see *p*-Dimethylaminoazobenzene
Green oil, see Anthracene
Grisol, see Tetraethyl pyrophosphate
Grundier arbezol, see Pentachlorophenol
Guesapon, see *p,p'*-DDT
Gum camphor, see Camphor
Gy-phene, see Toxaphene
Gyron, see *p,p'*-DDT
H-34, see Heptachlor
Halane, see 1,3-Dichloro-5,5-dimethylhydantoin
Hallucinogen, see *N,N*-Dimethylacetamide
Halocarbon 11, see Trichlorofluoromethane
Halocarbon 112, see 1,2-Difluorotetrachloroethane
Halocarbon 112a, see 1,1-Difluorotetrachloroethane
Halocarbon 113, see 1,1,2-Trichlorotrifluoroethane
Halocarbon 13B1, see Bromotrifluoromethane
Halon, see Dichlorodifluoromethane
Halon 21, see Dichlorofluoromethane
Halon 104, see Carbon tetrachloride
Halon 122, see Dichlorodifluoromethane
Halon 1001, see Methyl bromide
Halon 1011, see Bromochloromethane
Halon 1202, see Dibromodifluoromethane
Halon 1301, see Bromotrifluoromethane
Halon 2001, see Ethyl bromide
Halon 10001, see Methyl iodide
Halowax 1051, see Octachloronaphthalene
Hatcol DOP, see Bis(2-ethylhexyl) phthalate
Havero-extra, see *p,p'*-DDT
HCB, see Hexachlorobenzene
HCBD, see Hexachlorobutadiene
HCCH, see Lindane
HCCP, see Hexachlorocyclopentadiene
HCCPD, see Hexachlorocyclopentadiene
HCE, see Heptachlor epoxide, Hexachloroethane
HCH, see Lindane
α-HCH, see α-BHC
β-HCH, see β-BHC
δ-HCH, see δ-BHC
γ-HCH, see Lindane
HCPD, see Hexachlorocyclopentadiene
HCS 3,260, see Chlordane
Heclotox, see Lindane

Hylene TM, see 2,4-Toluene diisocyanate
Hylene TM-65, see 2,4-Toluene diisocyanate
Hylene TRF, see 2,4-Toluene diisocyanate
Hytrol O, see Cyclohexanone
IBA, see Isobutyl alcohol
Idryl, see Fluoranthene
IG base, see 4-Nitroaniline
Illoxol, see Dieldrin
Indenopyren, see Indeno[1,2,3-*cd*]pyrene
1*H*-Indole, see Indole
Inexit, see Lindane
Inhibisol, see 1,1,1-Trichloroethane
Insecticide 497, see Dieldrin
Insecticide 4049, see Malathion
Insectophene, see α-Endosulfan, β-Endosulfan
Inverton 245, see 2,4,5-T
Iodomethane, see Methyl iodide
IP, see Indeno[1,2,3-*cd*]pyrene
Ipaner, see 2,4-D
IPE, see Isopropyl ether
Isceon 11, see Trichlorofluoromethane
Isceon 122, see Dichlorodifluoromethane
Iscobrome, see Methyl bromide
Iscobrome D, see Ethylene dibromide
Isoacetophorone, see Isophorone
α-Isoamylene, see 3-Methyl-1-butene
Isoamyl ethanoate, see Isoamyl acetate
Isoamylhydride, see 2-Methylbutane
Isoamylol, see Isoamyl alcohol
1,3-Isobenzofurandione, see Phthalic anhydride
Isobutane, see 2-Methylpropane
Isobutanol, see Isobutyl alcohol
Isobutene, see 2-Methylpropene
Isobutenyl methyl ketone, see Mesityl oxide
Isobutyl carbinol, see Isoamyl alcohol
Isobutylene, see 2-Methylpropene
Isobutyl ketone, see Diisobutyl ketone
Isobutyl methyl ketone, see 4-Methyl-2-pentanone
Isobutyltrimethylmethane, see 2,2,4-Trimethylpentane
Isocumene, see *n*-Propylbenzene
Isocyanatomethane, see Methyl isocyanate
Isocyanic acid methyl ester, see Methyl isocyanate
Isocyanic acid methylphenylene ester, see 2,4-Toluene-
diisocyanate
Isocyanic acid 4-methyl-*m*-phenylene ester, see 2,4-Tolu-
ene diisocyanate
Isodrin epoxide, see Endrin
Isoforon, see Isophorone
Isoforone, see Isophorone
Isoheptane, see 2-Methylhexane
Isohexane, see 2-Methylpentane
β-Isomer, see β-BHC
γ-Isomer, see Lindane
Isonitropropane, see 2-Nitropropane
Isooctane, see 2,2,4-Trimethylpentane
Isooctaphenone, see Isophorone
Isopentane, see 2-Methylbutane
Isopentanol, see Isoamyl alcohol
Isopentene, see 3-Methyl-1-butene
Isopentyl acetate, see Isoamyl acetate
Isopentyl alcohol, see Isoamyl alcohol
Isopentyl alcohol acetate, see Isoamyl acetate
Isophoron, see Isophorone
Isoprene, see 2-Methyl-1,3-butadiene
Isopropenylbenzene, see α-Methylstyrene
2-Isopropoxypropane, see Isopropyl ether
Isopropylacetone, see 4-Methyl-2-pentanone

Isopropylbenzol, see Isopropylbenzene
Isopropylcarbinol, see Isobutyl alcohol
Isopropyldimethylmethane, see 2,3-Dimethylbutane
Isopropylethylene, see 3-Methyl-1-butene
Isopropylidene acetone, see Mesityl oxide
1-Isopropyl-2-methylethene, see 4-Methyl-1-pentene
1-Isopropyl-2-methylethylene, see 4-Methyl-1-pentene
Isotox, see Lindane
Isotron 2, see Dichlorodifluoromethane
Isotron 11, see Trichlorofluoromethane
Isovalerone, see Diisobutyl ketone
Ivalon, see Formaldehyde
Ivoran, see *p,p'*-DDT
Ixodex, see *p,p'*-DDT
Jacutin, see Lindane
Japan camphor, see Camphor
Jeffersol EB, see 2-Butoxyethanol
Jeffersol EE, see 2-Ethoxyethanol
Jeffersol EM, see Methyl cellosolve
Julin's carbon chloride, see Hexachlorobenzene
K III, see 4,6-Dinitro-*o*-cresol
K IV, see 4,6-Dinitro-*o*-cresol
Kaiser chemicals 11, see 1,1,2-Trichlorotrifluoroethane
Kaiser chemicals 12, see Dichlorodifluoromethane
Kamfochlor, see Toxaphene
Kanechlor, see PCB-1260
Karbaspray, see Carbaryl
Karbatox, see Carbaryl
Karbofos, see Malathion
Karbosep, see Carbaryl
Karlan, see Ronnel
Karmex, see Diuron
Karmex diuron herbicide, see Diuron
Karmex DW, see Diuron
Karsan, see Formaldehyde
Kayafume, see Methyl bromide
Kelene, see Chloroethane
Ketohexamethylene, see Cyclohexanone
Ketole, see Indole
Ketone propane, see Acetone
β-Ketopropane, see Acetone
2-Keto-1,7,7-trimethylnorcamphane, see Camphor
Killax, see Tetraethyl pyrophosphate
Kilmite 40, see Tetraethyl pyrophosphate
Kodaflex DOP, see Bis(2-ethylhexyl) phthalate
Kokotine, see Lindane
Kolphos, see Parathion
Kopfume, see Ethylene dibromide
Kopsol, see *p,p'*-DDT
KOP-thiodan, see α-Endosulfan, β-Endosulfan
Kop-thion, see Malathion
Korax, see 1-Chloro-1-nitropropane
Korlan, see Ronnel
Korlane, see Ronnel
Krecalvin, see Dichlorvos
Kregasan, see Thiram
Kresamone, see 4,6-Dinitro-*o*-cresol
p-Kresol, see 4-Methylphenol
Krezotol 50, see 4,6-Dinitro-*o*-cresol
Krotiline, see 2,4-D
Krysid, see ANTU
Krystallin, see Aniline
Kumader, see Warfarin
Kumadu, see Warfarin
Kwell, see Lindane
Kwik-kil, see Strychnine
Kypchlor, see Chlordane

Kyanol, see Aniline
Kypfarin, see Warfarin
Kypfos, see Malathion
Kypthion, see Parathion
L-36352, see Trifluralin
β-Lactone, see β-Propiolactone
Lanadin, see Trichloroethylene
Lanstan, see 1-Chloro-1-nitropropane
Larvacide 100, see Chloropicrin
Laurel camphor, see Camphor
Lauxtol, see Pentachlorophenol
Lauxtol A, see Pentachlorophenol
Lawn-keep, see 2,4-D
Ledon 11, see Trichlorofluoromethane
Ledon 12, see Dichlorodifluoromethane
Lemonene, see Biphenyl
Lendine, see Lindane
Lentox, see Lindane
Lethalaire G-52, see Tetraethyl pyrophosphate
Lethalaire G-54, see Parathion
Lethalaire G-57, see Sulfotepp
Lethurin, see Trichloroethylene
Lidenal, see Lindane
Lilly 36352, see Trifluralin
Lindafor, see Lindane
Lindagam, see Lindane
Lindagrain, see Lindane
Lindagranox, see Lindane
Lindan, see Dichlorvos
α-Lindane, see α-BHC
β-Lindane, see β-BHC
δ-Lindane, see δ-BHC
γ-Lindane, see Lindane
Lindapoudre, see Lindane
Lindatox, see Lindane
Lindosep, see Lindane
Line rider, see 2,4,5-T
Lintox, see Lindane
Lipan, see 4,6-Dinitro-o-cresol
Liroprem, see Pentachlorophenol
Liqua-tox, see Warfarin
Liquefied petroleum gas, see 2-Methylpropane
Lirohex, see Tetraethyl pyrophosphate
Lirothion, see Parathion
Lorexane, see Lindane
Lorsban, see Chlorpyrifos
LPG ethyl mercaptan 1010, see Ethyl mercaptan
Lyddite, see Picric acid
Lysoform, see Formaldehyde
M 140, see Chlordane
M 410, see Chlordane
M 5055, see Toxaphene
MA, see Methylaniline
MAAC, see sec-Hexyl acetate
Mace, see α-Chloroacetophenone
Macrondray, see 2,4-D
Mafu, see Dichlorvos
Mafu strip, see Dichlorvos
Magnacide H, see Acrolein
Malacide, see Malathion
Malafor, see Malathion
Malagran, see Malathion
Malakill, see Malathion
Malamar, see Malathion
Malamar 50, see Malathion
Malaphele, see Malathion
Malaphos, see Malathion

Malasol, see Malathion
Malaspray, see Malathion
Malathion E50, see Malathion
Malathion LV concentrate, see Malathion
Malathion ULV concentrate, see Malathion
Malathiozoo, see Malathion
Malathon, see Malathion
Malathyl LV concentrate & ULV concentrate, see Malathion
Malatol, see Malathion
Malatox, see Malathion
Maldison, see Malathion
Maleic acid anhydride, see Maleic anhydride
Malix, see α-Endosulfan, β-Endosulfan
Malmed, see Malathion
Malphos, see Malathion
Maltox, see Malathion
Maltox MLT, see Malathion
Maralate, see Methoxychlor
Marfrin, see Warfarin
Marlate, see Methoxychlor
Marlate 50, see Methoxychlor
Marmer, see Diuron
Maroxol-50, see 2,4-Dinitrophenol
Martin's mar-frin, see Warfarin
Marvex, see Dichlorvos
Matricaria camphor, see Camphor
Maveran, see Warfarin
MB, see Methyl bromide
M-B-C fumigant, see Methyl bromide
MBK, see 2-Hexanone
MBX, see Methyl bromide
MCB, see Chlorobenzene
ME-1700, see p,p'-DDD
MEA, see Ethanolamine
MEBR, see Methyl bromide
MECS, see Methyl cellosolve
Meetco, see 2-Butanone
MEK, see 2-Butanone
Melinite, see Picric acid
Melipax, see Toxaphene
Mendrin, see Endrin
Meniphos, see Mevinphos
Menite, see Mevinphos
1-Mercaptobutane, see n-Butyl mercaptan
Mercaptomethane, see Methyl mercaptan
Mercaptosuccinic acid diethyl ester, see Malathion
Mercaptothion, see Malathion
Mercuram, see Thiram
Mercurialin, see Methylamine
Merex, see Kepone
Mesitylene, see 1,3,5-Trimethylbenzene
Metafume, see Methyl bromide
Methacide, see Toluene
Methacrylic acid methyl ester, see Methyl methacrylate
Methanal, see Formaldehyde
Methanamine, see Methylamine
Methanecarbonitrile, see Acetonitrile
Methanecarboxylic acid, see Acetic acid
Methane dichloride, see Methylene chloride
Methane tetrachloride, see Carbon tetrachloride
Methanethiol, see Methyl mercaptan
Methane trichloride, see Chloroform
6,9-Methano-2,4,3-benzodioxathiepin, see Endosulfan sulfate
Methanoic acid, see Formic acid
Methenyl chloride, see Chloroform

Methenyl tribromide, see Bromoform
Methenyl trichloride, see Chloroform
Methogas, see Methyl bromide
Methoxcide, see Methoxychlor
Methoxo, see Methoxychlor
2-Methoxy-1-aminobenzene, see *o*-Anisidine
4-Methoxy-1-aminobenzene, see *p*-Anisidine
2-Methoxyaniline, see *o*-Anisidine
4-Methoxyaniline, see *p*-Anisidine
o-Methoxyaniline, see *o*-Anisidine
p-Methoxyaniline, see *p*-Anisidine
2-Methoxybenzenamine, see *o*-Anisidine
4-Methoxybenzenamine, see *p*-Anisidine
p-Methoxybenzenamine, see *p*-Anisidine
Methoxycarbonylethylene, see Methyl acrylate
2-Methoxycarbonyl-1-methylvinyl dimethyl phosphate,
    see Mevinphos
cis-2-Methoxycarbonyl-1-methylvinyl dimethyl phos-
    phate, see Mevinphos
1-Methoxycarbonyl-1-propen-2-yl dimethyl phosphate,
    see Mevinphos
4,4′-Methoxychlor, see Methoxychlor
p,p′-Methoxychlor, see Methoxychlor
Methoxy-DDT, see Methoxychlor
2-Methoxyethanol, see Methyl cellosolve
2-Methoxyethanol acetate, see Methyl cellosolve acetate
2-Methoxyethyl acetate, see Methyl cellosolve acetate
Methoxyhydroxyethane, see Methyl cellosolve
4-Methoxyphenylamine, see *p*-Anisidine
o-Methoxyphenylamine, see *o*-Anisidine
p-Methoxyphenylamine, see *p*-Anisidine
Methyl acetone, see 2-Butanone
Methyl acetylene, see Propyne
β-Methylacrolein, see Crotonaldehyde
Methyl alcohol, see Methanol
Methyl aldehyde, see Formaldehyde
(Methylamino)benzene, see Methylaniline
1-Methyl-2-aminobenzene, see *o*-Toluidine
2-Methyl-1-aminobenzene, see *o*-Toluidine
N-Methylaminobenzene, see Methylaniline
Methylamyl acetate, see *sec*-Hexyl acetate
Methyl amyl ketone, see 2-Heptanone
Methyl *n*-amyl ketone, see 2-Heptanone
2-Methylaniline, see *o*-Toluidine
N-Methylaniline, see Methylaniline
o-Methylaniline, see *o*-Toluidine
β-Methylanthracene, see 2-Methylanthracene
2-Methylbenzenamine, see *o*-Toluidine
N-Methylbenzenamine, see Methylaniline
o-Methylbenzenamine, see *o*-Toluidine
Methylbenzene, see Toluene
Methylbenzol, see Toluene
β-Methylbivinyl, see 2-Methyl-1,3-butadiene
2-Methylbutadiene, see 2-Methyl-1,3-butadiene
2-Methyl-4-butanol, see Isoamyl alcohol
3-Methylbutanol, see Isoamyl alcohol
3-Methyl-1-butanol, see Isoamyl alcohol
3-Methylbutan-1-ol, see Isoamyl alcohol
3-Methyl-1-butanol acetate, see Isoamyl acetate
1-Methylbutyl acetate, see *sec*-Amyl acetate
3-Methylbutyl acetate, see Isoamyl acetate
3-Methyl-1-butyl acetate, see Isoamyl acetate
3-Methylbutyl ethanoate, see Isoamyl acetate
Methyl *n*-butyl ketone, see 2-Hexanone
Methyl carbamate-1-naphthalenol, see Carbaryl
Methyl carbamate-1-naphthol, see Carbaryl
Methyl carbamic acid 2,3-dihydro-2,2-dimethyl-7-benzo-

furanyl ester, see Carbofuran
Methyl carbamic acid 1-naphthyl ester, see Carbaryl
Methyl chloroform, see 1,1,1-Trichloroethane
3-Methyl-4-chlorophenol, see *p*-Chloro-*m*-cresol
Methyl cyanide, see Acetonitrile
2-Methylcyclohexanone, see *o*-Methylcyclohexanone
1-Methyl-1-cyclohexene, see 1-Methylcyclohexene
Methyl 3-(dimethoxyphosphinyloxy)crotonate, see
    Mevinphos
1-Methyl-2,4-dinitrobenzene, see 2,4-Dinitrotoluene
2-Methyl-1,3-dinitrobenzene, see 2,6-Dinitrotoluene
2-Methyl-4,6-dinitrophenol, see 4,6-Dinitro-*o*-cresol
6-Methyl-2,4-dinitrophenol, see 4,6-Dinitro-*o*-cresol
Methylene bichloride, see Methylene chloride
Methylenebiphenyl, see Fluorene
2,2′-Methylenebiphenyl, see Fluorene
1,1′-[Methylenebis(oxy)]bis(2-chloroethane), see Bis(2-
    chloroethoxy)methane
1,1′-[Methylenebis(oxy)]bis(2-chloroformaldehyde), see
    Bis(2-chloroethoxy)methane
Methylene chlorobromide, see Bromochloromethane
Methylene dichloride, see Methylene chloride
Methylene dimethyl ether, see Methylal
Methylene glycol, see Formaldehyde
Methylene oxide, see Formaldehyde
Methyl ethanoate, see Methyl acetate
(1-Methylethenyl)benzene, see α-Methylstyrene
Methyl ethoxol, see Methyl cellosolve
1-Methylethylamine, see Isopropylamine
(1-Methylethyl)benzene, see Isopropylbenzene
Methylethylcarbinol, see *sec*-Butyl alcohol
Methyl ethylene oxide, see Propylene oxide
Methyl ethyl ketone, see 2-Butanone
Methylethylmethane, see Butane
N-(1-Methylethyl)-2-propanamine, see Diisopropylamine
Methyl formal, see Methylal
Methyl glycol, see Methyl cellosolve
Methyl glycol acetate, see Methyl cellosolve acetate
Methyl glycol monoacetate, see Methyl cellosolve
    acetate
5-Methylheptane, see 3-Methylheptane
3-Methyl-5-heptanone, see 5-Methyl-3-heptanone
4-Methylhexane, see 3-Methylhexane
1-Methylhydrazine, see Methylhydrazine
Methyl hydroxide, see Methanol
1-Methyl-4-hydroxybenzene, see 4-Methylphenol
2-Methylhydroxybenzene, see 2-Methylphenol
4-Methylhydroxybenzene, see 4-Methylphenol
o-Methylhydroxybenzene, see 2-Methylphenol
p-Methylhydroxybenzene, see 4-Methylphenol
Methylisoamyl acetate, see *sec*-Hexyl acetate
Methyl isobutenyl ketone, see Mesityl oxide
Methylisobutylcarbinol acetate, see *sec*-Hexyl acetate
Methyl isobutyl ketone, see 4-Methyl-2-pentanone
Methyl ketone, see Acetone
Methyl methacrylate monomer, see Methyl methacrylate
N-Methylmethanamine, see Dimethylamine
Methyl methanoate, see Methyl formate
Methyl-α-methylacrylate, see Methyl methacrylate
Methyl-2-methyl-2-propenoate, see Methyl methacrylate
β-Methylnaphthalene, see 2-Methylnaphthalene
N-Methyl-1-naphthyl carbamate, see Carbaryl
N-Methyl-α-naphthyl carbamate, see Carbaryl
N-Methyl-α-naphthylurethan, see Carbaryl
1-Methyl-2-nitrobenzene, see 2-Nitrotoluene
1-Methyl-3-nitrobenzene, see 3-Nitrotoluene
1-Methyl-4-nitrobenzene, see 4-Nitrotoluene

2-Methylnitrobenzene, see 2-Nitrotoluene
3-Methylnitrobenzene, see 3-Nitrotoluene
4-Methylnitrobenzene, see 4-Nitrotoluene
m-Methylnitrobenzene, see 3-Nitrotoluene
o-Methylnitrobenzene, see 2-Nitrotoluene
p-Methylnitrobenzene, see 4-Nitrotoluene
n-Methyl-*N*-nitrosomethanamine, see *N*-Nitrosodimeth-
   ylamine
5-Methyloctane, see 4-Methyloctane
Methylol, see Methanol
Methylolpropane, see 1-Butanol
Methyloxirane, see Propylene oxide
Methyl oxitol, see Methyl cellosolve
4-Methyl-2-pentanol acetate, see *sec*-Hexyl acetate
2-Methyl-2-pentanol-4-one, see Diacetone alcohol
2-Methyl-4-pentanone, see 4-Methyl-2-pentanone
2-Methylpentene, see 2-Methyl-1-pentene
2-Methyl-2-penten-4-one, see Mesityl oxide
4-Methyl-2-pentene-2-one, see Mesityl oxide
4-Methyl-2-pentyl acetate, see *sec*-Hexyl acetate
α-Methylphenanthrene, see 1-Methylphenanthrene
o-Methylphenol, see 2-Methylphenol
p-Methylphenol, see 4-Methylphenol
Methylphenylamine, see Methylaniline
N-Methylphenylamine, see Methylaniline
4-Methylphenylene diisocyanate, see 2,4-Toluene diisocyanate
4-Methylphenylene isocyanate, see 2,4-Toluene diisocyanate
1-Methyl-1-phenylethylene, see α-Methylstyrene
o-Methylphenylol, see 2-Methylphenol
2-Methyl-1-phenylpropane, see Isobutylbenzene
2-Methyl-2-phenylpropane, see *tert*-Butylbenzene
Methyl phthalate, see Dimethyl phthalate
2-Methylpropanol, see Isobutyl alcohol
2-Methylpropanol-1, see Isobutyl alcohol
2-Methyl-1-propanol, see Isobutyl alcohol
2-Methyl-1-propan-1-ol, see Isobutyl alcohol
2-Methyl-2-propanol, see *tert*-Butyl alcohol
Methyl propenate, see Methyl acrylate
Methylpropene, see 2-Methylpropene
2-Methyl-1-propene, see 2-Methylpropene
Methyl propenoate, see Methyl acrylate
Methyl-2-propenoate, see Methyl acrylate
1-Methylpropyl acetate, see *sec*-Butyl acetate
2-Methylpropyl acetate, see Isobutyl acetate
2-Methyl-1-propyl acetate, see Isobutyl acetate
2-Methylpropyl alcohol, see Isobutyl alcohol
(1-Methylpropyl)benzene, see *sec*-Butylbenzene
(2-Methylpropyl)benzene, see Isobutylbenzene
2-Methylpropylene, see 2-Methylpropene
1-Methyl-1-propylethene, see 2-Methyl-1-pentene
1-Methyl-1-propylethylene, see 2-Methyl-1-pentene
β-Methylpropyl ethanoate, see Isobutyl acetate
Methyl propyl ketone, see 2-Pentanone
Methyl *n*-propyl ketone, see 2-Pentanone
Methyl sulfate, see Dimethyl sulfate
Methyl sulfhydrate, see Methyl mercaptan
N-Methyl-*N*,2,4,6-tetranitroaniline, see Tetryl
N-Methyl-*N*,2,4,6-tetranitrobenzenamine, see Tetryl
Methyl thioalcohol, see Methyl mercaptan
Methyl thiram, see Thiram
Methyl thiuramdisulfide, see Thiram
m-Methyltoluene, see *m*-Xylene
o-Methyltoluene, see *o*-Xylene
p-Methyltoluene, see *p*-Xylene
Methyl tribromide, see Bromoform

Methyl trichloride, see Chloroform
Methyltrichloromethane, see 1,1,1-Trichloroethane
1-Methyl-2,4,6-trinitrobenzene, see 2,4,6-Trinitrotoluene
2-Methyl-1,3,5-trinitrobenzene, see 2,4,6-Trinitrotoluene
Methyl tuads, see Thiram
Methyl yellow, see *p*-Dimethylaminoazobenzene
Metox, see Methoxychlor
MIBK, see 4-Methyl-2-pentanone
Microlysin, see Chloropicrin
Mighty 150, see Naphthalene
Mighty RD1, see Naphthalene
MIK, see 4-Methyl-2-pentanone
Mil-B-4394-B, see Bromochloromethane
Milbol 49, see Lindane
Miller's fumigrain, see Acrylonitrile
Mineral naphthalene, see Benzene
Mipax, see Dimethyl phthalate
Miracle, see 2,4-D
Mirbane oil, see Nitrobenzene
MLT, see Malathion
MME, see Methyl methacrylate
MMH, see Methylhydrazine
MNA, see 3-Nitroaniline
MNBK, see 2-Hexanone
MNT, see 3-Nitrotoluene
Mole death, see Strychnine
Mole-nots, see Strychnine
Mollan 0, see Bis(2-ethylhexyl) phthalate
Mondur TD, see 2,4-Toluene diisocyanate
Mondur TD-80, see 2,4-Toluene diisocyanate
Mondur TDS, see 2,4-Toluene diisocyanate
Monobromobenzene, see Bromobenzene
Monobromoethane, see Ethyl bromide
Monobromomethane, see Methyl bromide
Monobromotrifluoromethane, see Bromotrifluoromethan-
   ane
Monobutylamine, see *n*-Butylamine
Mono-*n*-butylamine, see *n*-Butylamine
Monochlorbenzene, see Chlorobenzene
Monochlorethane, see Chloroethane
Monochloroacetaldehyde, see Chloroacetaldehyde
Monochlorobenzene, see Chlorobenzene
Monochlorodiphenyl oxide, see 4-Chlorophenyl phenyl
   ether
Monochloroethane, see Chloroethane
2-Monochloroethanol, see Ethylene chlorohydrin
Monochloroethene, see Vinyl chloride
Monochloroethylene, see Vinyl chloride
Monochloromethane, see Methyl chloride
"Monocite" methacrylate monomer, see Methyl meth-
   acrylate
Monoethanolamine, see Ethanolamine
Monoethylamine, see Ethylamine
Monofluorotrichloromethane, see Trichlorofluorometh-
   ane
Monohydroxybenzene, see Phenol
Monohydroxymethane, see Methanol
Monoisopropylamine, see Isopropylamine
Monomethylamine, see Methylamine
Monomethylaniline, see Methylaniline
N-Monomethylaniline, see Methylaniline
Monomethylhydrazine, see Methylhydrazine
Monosan, see 2,4-D
Monsanto penta, see Pentachlorophenol
Moosuran, see Pentachlorophenol
Mopari, see Dichlorvos
Morbicid, see Formaldehyde

Mortopal, see Tetraethyl pyrophosphate
Moscardia, see Malathion
Moth balls, see Naphthalene
Moth flakes, see Naphthalene
Motor benzol, see Benzene
Motox, see Toxaphene
Mottenhexe, see Hexachloroethane
Mouse pak, see Warfarin
Mouse-rid, see Strychnine
Mouse-tox, see Strychnine
Moxie, see Methoxychlor
Moxone, see 2,4-D
MPK, see 2-Pentanone
Mszycol, see Lindane
Murfos, see Parathion
Muriatic ether, see Chloroethane
Muthmann's liquid, see 1,1,2,2-Tetrabromoethane
Mutoxin, see *p,p'*-DDT
MVC, see Vinyl chloride
NA 1120, see 1-Butanol
NA 1230, see Methanol
NA 1247, see Methyl methacrylate
NA 1583, see Chloropicrin
NA 1648, see Acetonitrile
NA 2757, see Carbaryl
NA 2761, see Aldrin, Dieldrin, Endrin, Heptachlor, *p,p'*-DDD, Kepone, Lindane, Toxaphene
NA 2762, see Aldrin, Chlordane
NA 2765, see 2,4-D, 2,4,5-T
NA 2767, see Diuron
NA 2771, see Thiram
NA 2783, see Chlorpyrifos, Dichlorvos, Malathion, Parathion, Mevinphos, Naled, Tetraethyl pyrophosphate
NA 2821, see Phenol
NA 9037, see Hexachloroethane
NA 9094, see Benzoic acid
NAC, see Carbaryl
Nacconate 100, see 2,4-Toluene diisocyanate
Nadone, see Cyclohexanone
Nako H, see *p*-Phenylenediamine
Nanchor, see Ronnel
Nanker, see Ronnel
Nankor, see Ronnel
2-Naphthalamine, see 2-Naphthylamine
Naphthalane, see Decahydronaphthalene
1-Naphthalenamine, see 1-Naphthylamine
2-Naphthalenamine, see 2-Naphthylamine
1,2-(1,8-Naphthalenediyl)benzene, see Fluoranthene
1-Naphthalenol methyl carbamate, see Carbaryl
1-Naphthalenylthiourea, see ANTU
Naphthalidam, see 1-Naphthylamine
Naphthalidine, see 1-Naphthylamine
Naphthalin, see Naphthalene
Naphthaline, see Naphthalene
Naphthane, see Decahydronaphthalene
Naphthanthracene, see Benzo[*a*]anthracene
Naphthene, see Naphthalene
Naphtolean orange R base, see 3-Nitroaniline
Naphtolean red GG base, see 4-Nitroaniline
1-Naphthol-*N*-methyl carbamate, see Carbaryl
6-Naphthylamine, see 2-Naphthylamine
α-Naphthylamine, see 1-Naphthylamine
β-Naphthylamine, see 2-Naphthylamine
2-Naphthylamine mustard, see 2-Naphthylamine
1,2-(1,8-Naphthylene)benzene, see Fluoranthene
1-Naphthyl methyl carbamate, see Carbaryl

1-Naphthyl-*N*-methyl carbamate, see Carbaryl
α-Naphthyl-*N*-methyl carbamate, see Carbaryl
α-Naphthylthiocarbamide, see ANTU
1-(1-Naphthyl)-2-thiourea, see ANTU
N-1-Naphthylthiourea, see ANTU
α-Naphthylthiourea, see ANTU
Narcogen, see Trichloroethylene
Narcotil, see Methylene chloride
Narcotile, see Chloroethane
Narkogen, see Trichloroethylene
Narkosoid, see Trichloroethylene
Natasol fast orange GR salt, see 2-Nitroaniline
Naugard TJB, see *N*-Nitrosodiphenylamine
Naxol, see Cyclohexanol
NBA, see 1-Butanol
NCI-C00044, see Aldrin
NCI-C00099, see Chlordane
NCI-C00113, see Dichlorvos
NCI-C00124, see Dieldrin
NCI-C00157, see Endrin
NCI-C00180, see Heptachlor
NCI-C00191, see Kepone
NCI-C00204, see Lindane
NCI-C00215, see Malathion
NCI-C00226, see Parathion
NCI-C00259, see Toxaphene
NCI-C00442, see Trifluralin
NCI-C00464, see *p,p'*-DDT
NCI-C00475, see *p,p'*-DDD
NCI-C00497, see Methoxychlor
NCI-C00500, see 1,2-Dibromo-3-chloropropane
NCI-C00511, see 1,2-Dichloroethane
NCI-C00522, see Ethylene dibromide
NCI-C00533, see Chloropicrin
NCI-C00555, see *p,p'*-DDE
NCI-C00566, see α-Endosulfan, β-Endosulfan
NCI-C01854, see 1,2-Diphenylhydrazine
NCI-C01865, see 2,4-Dinitrotoluene
NCI-C02039, see 4-Chloroaniline
NCI-C02200, see Styrene
NCI-C02664, see PCB-1254
NCI-C02686, see Chloroform
NCI-C02799, see Formaldehyde
NCI-C02880, see *N*-Nitrosodiphenylamine
NCI-C02904, see 2,4,6-Trichlorophenol
NCI-C03054, see 1,3-Dichloro-5,5-dimethylhydantoin
NCI-C03361, see Benzidine
NCI-C03554, see 1,1,2,2-Tetrachloroethane
NCI-C03601, see Phthalic anhydride
NCI-C03689, see 1,4-Dioxane
NCI-C03714, see TCDD
NCI-C03736, see Aniline
NCI-C04535, see 1,1-Dichloroethane
NCI-C04546, see Trichloroethylene
NCI-C04579, see 1,1,2-Trichloroethane
NCI-C04580, see Tetrachloroethylene
NCI-C04591, see Carbon disulfide
NCI-C04604, see Hexachloroethane
NCI-C04615, see Allyl chloride
NCI-C04626, see 1,1,1-Trichloroethane
NCI-C04637, see Trichlorofluoromethane
NCI-C06111, see Benzyl alcohol
NCI-C06224, see Chloroethane
NCI-C06360, see Benzyl chloride
NCI-C07272, see Toluene
NCI-C50044, see Bis(2-chloroisopropyl) ether
NCI-C50099, see Propylene oxide

NCI-C50102, see Methylene chloride
NCI-C50124, see Phenol
NCI-C50135, see Ethylene chlorohydrin
NCI-C50384, see Ethyl acrylate
NCI-C50533, see 2,4-Toluene diisocyanate
NCI-C50602, see 1,3-Butadiene
NCI-C50680, see Methyl methacrylate
NCI-C52733, see Bis(2-ethylhexyl) phthalate
NCI-C52904, see Naphthalene
NCI-C53894, see Pentachloroethane
NCI-C54262, see 1,1-Dichloroethylene
NCI-C54375, see Benzyl butyl phthalate
NCI-C54853, see 2-Ethoxyethanol
NCI-C54886, see Chlorobenzene
NCI-C54933, see Pentachlorophenol
NCI-C54944, see 1,2-Dichlorobenzene
NCI-C54955, see 1,4-Dichlorobenzene
NCI-C55005, see Cyclohexanone
NCI-C55107, see α-Chloroacetophenone
NCI-C55118, see o-Chlorobenzylidenemalononitrile
NCI-C55130, see Bromoform
NCI-C55141, see 1,2-Dichloropropane
NCI-C55209, see Oxalic acid
NCI-C55243, see Bromodichloromethane
NCI-C55254, see Dibromochloromethane
NCI-C55276, see Benzene
NCI-C55301, see Pyridine
NCI-C55345, see 2,4-Dichlorophenol
NCI-C55367, see tert-Butyl alcohol
NCI-C55378, see Pentachlorophenol
NCI-C55481, see Ethyl bromide
NCI-C55492, see Bromobenzene
NCI-C55549, see Glycidol
NCI-C55607, see Hexachlorocyclopentadiene
NCI-C55618, see Isophorone
NCI-C55834, see Hydroquinone
NCI-C55845, see p-Quinone
NCI-C55947, see Tetranitromethane
NCI-C55992, see 4-Nitrophenol
NCI-C56155, see 2,4,6-Trinitrotoluene
NCI-C56177, see Furfural
NCI-C56224, see Furfuryl alcohol
NCI-C56279, see Crotonaldehyde
NCI-C56326, see Acetaldehyde
NCI-C56393, see Ethylbenzene
NCI-C56428, see Dimethylaniline
NCI-C56655, see Pentachlorophenol
NCI-C56666, see Allyl glycidyl ether
NCI-C60048, see Diethyl phthalate
NCI-C60082, see Nitrobenzene
NCI-C60220, see 1,2,3-Trichloropropane
NCI-C60402, see Ethylenediamine
NCI-C60537, see 4-Nitrotoluene
NCI-C60560, see Tetrahydrofuran
NCI-C60571, see Hexane
NCI-C60786, see 4-Nitroaniline
NCI-C60822, see Acetonitrile
NCI-C60866, see n-Butyl mercaptan
NCI-C60913, see N,N-Dimethylformamide
NCI-C61187, see 2,4,5-Trichlorophenol
NDMA, see N-Nitrosodimethylamine, see N-Nitrosodiphenylamine
NDPA, see N-Nitrosodi-n-propylamine
NDPhA, see N-Nitrosodiphenylamine
Neantine, see Diethyl phthalate
Necatorina, see Carbon tetrachloride
Necatorine, see Carbon tetrachloride

Nema, see Tetrachloroethylene
Nemabrom, see 1,2-Dibromo-3-chloropropane
Nemafume, see 1,2-Dibromo-3-chloropropane
Nemagon, see 1,2-Dibromo-3-chloropropane
Nemagon 20, see 1,2-Dibromo-3-chloropropane
Nemagon 90, see 1,2-Dibromo-3-chloropropane
Nemagon 206, see 1,2-Dibromo-3-chloropropane
Nemagon 20G, see 1,2-Dibromo-3-chloropropane
Nemagon soil fumigant, see 1,2-Dibromo-3-chloropropane
Nemanax, see 1,2-Dibromo-3-chloropropane
Nemapaz, see 1,2-Dibromo-3-chloropropane
Nemaset, see 1,2-Dibromo-3-chloropropane
Nematocide, see 1,2-Dibromo-3-chloropropane
Nematox, see 1,2-Dibromo-3-chloropropane
Nemazon, see 1,2-Dibromo-3-chloropropane
Nendrin, see Endrin
Neocid, see p,p'-DDT
Neohexane, see 2,2-Dimethylbutane
Neopentaene, see 2,2-Dimethylpropane
Neopentane, see 2,2-Dimethylpropane
Neoprene, see Chloroprene
Neo-scabicidol, see Lindane
Nephis, see Ethylene dibromide
Nerkol, see Dichlorvos
Netagrone, see 2,4-D
Netagrone 600, see 2,4-D
Nexen FB, see Lindane
Nexit, see Lindane
Nexit-stark, see Lindane
Nexol-E, see Lindane
NIA 5462, see α-Endosulfan, β-Endosulfan
NIA 10242, see Carbofuran
Niagara 5462, see α-Endosulfan, β-Endosulfan
Nialk, see Trichloroethylene
Niax TDI, see 2,4-Toluene diisocyanate
Niax TDI-P, see 2,4-Toluene diisocyanate
Nicochloran, see Lindane
Nifos, see Tetraethyl pyrophosphate
Nifos T, see Tetraethyl pyrophosphate
Nifost, see Tetraethyl pyrophosphate
Nipar S-20 solvent, see 2-Nitropropane
Nipar S-30 solvent, see 2-Nitropropane
Niran, see Chlordane, Parathion
Niran E-4, see Parathion
Nitrador, see 4,6-Dinitro-o-cresol
Nitramine, see Tetryl
Nitran, see Trifluralin
Nitranilin, see 3-Nitroaniline
4-Nitraniline, see 4-Nitroaniline
m-Nitraniline, see 3-Nitroaniline
o-Nitraniline, see 2-Nitroaniline
p-Nitraniline, see 4-Nitroaniline
Nitrapyrine, see Nitrapyrin
Nitration benzene, see Benzene
Nitrazol 2F extra, see 4-Nitroaniline
Nitric acid propyl ester, see n-Propyl nitrate
Nitrile, see Acrylonitrile
3-Nitroaminobenzene, see 3-Nitroaniline
m-Nitroaminobenzene, see 3-Nitroaniline
m-Nitroaniline, see 3-Nitroaniline
o-Nitroaniline, see 2-Nitroaniline
p-Nitroaniline, see 4-Nitroaniline
2-Nitrobenzenamine, see 2-Nitroaniline
3-Nitrobenzenamine, see 3-Nitroaniline
m-Nitrobenzenamine, see 3-Nitroaniline
4-Nitrobenzenamine, see 4-Nitroaniline

Omchlor, see 1,3-Dichloro-5,5-dimethylhydantoin
OMS 1, see Malathion
OMS 14, see Dichlorvos
OMS 18, see Dieldrin
OMS 29, see Carbaryl
OMS, see Ronnel
OMS 193, see Heptachlor
OMS 194, see Aldrin
OMS 197, see Endrin
OMS 219, see EPN
OMS 466, see Methoxychlor
OMS-570, see α-Endosulfan, β-Endosulfan
OMS 629, see Carbaryl
OMS 864, see Carbofuran
OMS 971, see Chlorpyrifos
OMS 1078, see *p,p'*-DDD
OMS 1437, see Chlordane
Omnitox, see Lindane
ONA, see 2-Nitroaniline
ONP, see 2-Nitrophenol
ONT, see 2-Nitrotoluene
Optal, see 1-Propanol
Orange base CIBA II, see 2-Nitroaniline
Orange base IRGA I, see 3-Nitroaniline
Orange base IRGA II, see 2-Nitroaniline
Orange GRS salt, see 2-Nitroaniline
Orange salt CIBA II, see 2-Nitroaniline
Orange salt IRGA II, see 2-Nitroaniline
Organol yellow ADM, see *p*-Dimethylaminoazobenzene
Orient oil yellow GG, see *p*-Dimethylaminoazobenzene
Orsin, see *p*-Phenylenediamine
Ortho 4355, see Naled
Orthocresol, see 2-Methylphenol
Orthodibrom, see Naled
Orthodibromo, see Naled
Orthodichlorobenzene, see 1,2-Dichlorobenzene
Orthodichlorobenzol, see 1,2-Dichlorobenzene
Orthoklor, see Chlordane
Orthomalathion, see Malathion
Orthonitroaniline, see 2-Nitroaniline
Orthophos, see Parathion
Orvinylcarbinol, see Allyl alcohol
OS 1987, see 1,2-Dibromo-3-chloropropane
OS 2046, see Mevinphos
Osmosol extra, see 1-Propanol
Ottafact, see *p*-Chloro-*m*-cresol
Ovadziak, see Lindane
Owadziak, see Lindane
1-Oxa-4-azacyclohexane, see Morpholine
Oxacyclopentane, see Tetrahydrofuran
2-Oxetanone, see β-Propiolactone
Oxidation base 10, see *p*-Phenylenediamine
Oxiranemethanol, see Glycidol
Oxiranylmethanol, see Glycidol
Oxitol, see 2-Ethoxyethanol
2-Oxobornane, see Camphor
Oxolane, see Tetrahydrofuran
Oxomethane, see Formaldehyde
Oxybenzene, see Phenol
1,1'-Oxybisbenzene, see Phenyl ether
1,1'-Oxybis(2-chloroethane), see Bis(2-chloroethyl) ether
Oxybis(chloromethane), see *sym*-Dichloromethyl ether
2,2'-Oxybis(1-chloropropane), see Bis(2-chloroisopropyl) ether
1,1'-Oxybis(ethane), see Ethyl ether
2,2'-Oxybis(propane), see Isopropyl ether

Oxy DBCP, see 1,2-Dibromo-3-chloropropane
Oxymethylene, see Formaldehyde
Oxytol acetate, see 2-Ethoxyethyl acetate
o-Oxytoluene, see 2-Methylphenol
p-Oxytoluene, see 4-Methylphenol
Pac, see Parathion
Palatinol A, see Diethyl phthalate
Palatinol AH, see Bis(2-ethylhexyl) phthalate
Palatinol BB, see Benzyl butyl phthalate
Palatinol C, see Di-*n*-butyl phthalate
Palatinol M, see Dimethyl phthalate
Panam, see Carbaryl
Panoram 75, see Thiram
Panoram D-31, see Dieldrin
Panthion, see Parathion
Para, see *p*-Phenylenediamine
Parachlorocidum, see *p,p'*-DDT
Paracide, see 1,4-Dichlorobenzene
Para-cresol, see 4-Methylphenol
Para crystals, see 1,4-Dichlorobenzene
Paradi, see 1,4-Dichlorobenzene
Paradichlorobenzene, see 1,4-Dichlorobenzene
Paradichlorobenzol, see 1,4-Dichlorobenzene
Paradow, see 1,4-Dichlorobenzene
Paradust, see Parathion
Paraflow, see Parathion
Paraform, see Formaldehyde
Paramar, see Parathion
Paramar 50, see Parathion
Paramethylphenol, see 4-Methylphenol
Paraminodiphenyl, see 4-Aminobiphenyl
Paramoth, see 1,4-Dichlorobenzene
Paranaphthalene, see Anthracene
Paranuggetts, see 1,4-Dichlorobenzene
Paraphos, see Parathion
Paraspray, see Parathion
Parathene, see Parathion
Parathion-ethyl, see Parathion
Parawet, see Parathion
Parazene, see 1,4-Dichlorobenzene
Parmetol, see *p*-Chloro-*m*-cresol
Parodi, see 1,4-Dichlorobenzene
Parol, see *p*-Chloro-*m*-cresol
PCC, see Toxaphene
PCE, see Tetrachloroethylene
PCL, see Hexachlorocyclopentadiene
PCMC, see *p*-Chloro-*m*-cresol
PCNB, see *p*-Chloronitrobenzene
PCP, see Pentachlorophenol
PD 5, see Mevinphos
PDAB, see *p*-Dimethylaminoazobenzene
PDB, see 1,4-Dichlorobenzene
PDCB, see 1,4-Dichlorobenzene
Pear oil, see Isoamyl acetate, *n*-Amyl acetate
PEB1, see *p,p'*-DDT
Pedraczak, see Lindane
Pelagol D, see *p*-Phenylenediamine
Pelagol DR, see *p*-Phenylenediamine
Pelagol grey D, see *p*-Phenylenediamine
Peltol D, see *p*-Phenylenediamine
Penchlorol, see Pentachlorophenol
Pennamine, see 2,4-D
Pennamine D, see 2,4-D
Penphene, see Toxaphene
Penphos, see Parathion
Penta, see Pentachlorophenol
Pent-acetate, see *n*-Amyl acetate

Pent-acetate 28, see *n*-Amyl acetate
Pentachlorfenol, see Pentachlorophenol
Pentachlorin, see *p,p'*-DDT
Pentachlorofenol, see Pentachlorophenol
Pentachlorofenolo, see Pentachlorophenol
Pentachlorophenate, see Pentachlorophenol
2,3,4,5,6-Pentachlorophenol, see Pentachlorophenol
Pentachlorophenyl chloride, see Hexachlorobenzene
Pentacon, see Pentachlorophenol
Penta-kil, see Pentachlorophenol
Pentalin, see Pentachloroethane
Pentamethylene, see Cyclopentane
n-Pentane, see Pentane
tert-Pentane, see 2,2-Dimethylpropane
1-Pentanol acetate, see *n*-Amyl acetate
2-Pentanol acetate, see *sec*-Amyl acetate
Pentasol, see Pentachlorophenol
Pentech, see *p,p'*-DDT
cis-Pentene-2, see *cis*-2-Pentene
(*E*)-2-Pentene-2, see *trans*-2-Pentene
(*Z*)-2-Pentene-2, see *cis*-2-Pentene
trans-Pentene-2, see *trans*-2-Pentene
Pentole, see Cyclopentadiene
1-Pentyl acetate, see *n*-Amyl acetate
2-Pentyl acetate, see *sec*-Amyl acetate
n-Pentyl acetate, see *n*-Amyl acetate
Penwar, see Pentachlorophenol
Peratox, see Pentachlorophenol
Perawin, see Tetrachloroethylene
PERC, see Tetrachloroethylene
Perchlor, see Tetrachloroethylene
Perchlorethylene, see Tetrachloroethylene
Perchlorobenzene, see Hexachlorobenzene
Perchlorobutadiene, see Hexachlorobutadiene
Perchlorocyclopentadiene, see Hexachlorocyclopentadiene
Perchloroethane, see Hexachloroethane
Perchloroethylene, see Tetrachloroethylene
Perchloromethane, see Carbon tetrachloride
Perclene, see Tetrachloroethylene
Perclene D, see Tetrachloroethylene
Percosolv, see Tetrachloroethylene
Perhydronaphthalene, see Decahydronaphthalene
Periethylenenaphthalene, see Acenaphthene
Peritonan, see *p*-Chloro-*m*-cresol
Perk, see Tetrachloroethylene
Perklone, see Tetrachloroethylene
Perm-a-chlor, see Trichloroethylene
Permacide, see Pentachlorophenol
Perm-a-clor, see Trichloroethylene
Permaguard, see Pentachlorophenol
Permasan, see Pentachlorophenol
Permatox DP-2, see Pentachlorophenol
Permatox Penta, see Pentachlorophenol
Permite, see Pentachlorophenol
Persec, see Tetrachloroethylene
Persia-Perazol, see 1,4-Dichlorobenzene
Pertite, see Picric acid
Pestmaster, see Ethylene dibromide, Methyl bromide
Pestmaster EDB-85, see Ethylene dibromide
Pestox plus, see Parathion
Pethion, see Parathion
Petrol yellow WT, see *p*-Dimethylaminoazobenzene
Petzinol, see Trichloroethylene
Pflanzol, see Lindane
Phenachlor, see 2,4,5-Trichlorophenol, 2,4,6-Trichlorophenol

Phenacide, see Toxaphene
Phenacyl chloride, see α-Chloroacetophenone
Phenanthren, see Phenanthrene
Phenantrin, see Phenanthrene
Phenatox, see Toxaphene
Phene, see Benzene
Phenethylene, see Styrene
Phenic acid, see Phenol
Phenoclor DP6, see PCB-1260
Phenohep, see Hexachloroethane
Phenol carbinol, see Benzyl alcohol
Phenol trinitrate, see Picric acid
Phenox, see 2,4-D
Phenoxybenzene, see Phenyl ether
Phenphane, see Toxaphene
3-(1'-Phenyl-2'-acetylethyl)-4-hydroxycoumarin, see
    Warfarin
3-α-Phenyl-β-acetylethyl-4-hydroxycoumarin, see Warfarin
Phenylamine, see Aniline
4-Phenylaniline, see 4-Aminobiphenyl
p-Phenylaniline, see 4-Aminobiphenyl
Phenylbenzene, see Biphenyl
Phenyl bromide, see Bromobenzene
Phenyl-4-bromophenyl ether, see 4-Bromophenyl phenyl
Phenyl-*p*-bromophenyl ether, see 4-Bromophenyl phenyl
1-Phenylbutane, see *n*-Butylbenzene
2-Phenylbutane, see *sec*-Butylbenzene
Phenyl carbinol, see Benzyl alcohol
Phenylcarboxylic acid, see Benzoic acid
Phenyl chloride, see Chlorobenzene
Phenyl chloromethyl ketone, see α-Chloroacetophenone
1,4-Phenylenediamine, see *p*-Phenylenediamine
1,10-(1,2-Phenylene)pyrene, see Indeno[1,2,3-*cd*]pyrene
1,10-(*o*-Phenylene)pyrene, see Indeno[1,2,3-*cd*]pyrene
2,3-Phenylene-*o*-pyrene, see Indeno[1,2,3-*cd*]pyrene
2,3-Phenylenepyrene, see Indeno[1,2,3-*cd*]pyrene
3,4-(*o*-Phenylene)pyrene, see Indeno[1,2,3-*cd*]pyrene
o-Phenylenepyrene, see Indeno[1,2,3-*cd*]pyrene
Phenylethane, see Ethylbenzene
Phenylethene, see Styrene
Phenylethylene, see Styrene
Phenylformic acid, see Benzoic acid
Phenyl hydrate, see Phenol
Phenyl hydride, see Benzene
Phenyl hydroxide, see Phenol
Phenylic acid, see Phenol
Phenylic alcohol, see Phenol
Phenylmethane, see Toluene
Phenyl methanol, see Benzyl alcohol
Phenyl methyl alcohol, see Benzyl alcohol
N-Phenylmethylamine, see Methylaniline
4-Phenylnitrobenzene, see 4-Nitrobiphenyl
p-Phenylnitrobenzene, see 4-Nitrobiphenyl
Phenyl perchloryl, see Hexachlorobenzene
Phenyl phosphate, see Triphenyl phosphate
Phenylphosphonothioic acid *O*-ethyl *O*-*p*-nitrophenyl
    ester, see EPN
1-Phenylpropane, see *n*-Propylbenzene
2-Phenylpropane, see Isopropylbenzene
2-Phenylpropene, see α-Methylstyrene
β-Phenylpropene, see α-Methylstyrene
2-Phenylpropylene, see α-Methylstyrene
β-Phenylpropylene, see α-Methylstyrene
o-Phenylpyrene, see Indeno[1,2,3-*cd*]pyrene
Philex, see Trichloroethylene
Phortox, see 2,4,5-T

Phosdrin, see Mevinphos

cis-Phosdrin, see Mevinphos

Phosfene, see Mevinphos

Phosflex 179-C, see Tri-*o*-cresyl phosphate

Phoskil, see Parathion

Phosphemol, see Parathion

Phosphenol, see Parathion

Phosphonothioic acid *O,O*-diethyl *O*-(3,5,6-trichloro-2-pyridinyl) ester, see EPN

Phosphoric acid 1,2-dibromo-2,2-dichloroethyl dimethyl ester, see Naled

Phosphoric acid 2,2-dichloroethenyl dimethyl ester, see Dichlorvos

Phosphoric acid 2,2-dichlorovinyl dimethyl ester, see Dichlorvos

Phosphoric acid (1-methoxycarboxypropen-2-yl) dimethyl ester, see Mevinphos

Phosphoric acid tributyl ester, see Tributyl phosphate

Phosphoric acid tri-*o*-cresyl ester, see Tri-*o*-cresyl phosphate

Phosphoric acid triphenyl ester, see Triphenyl phosphate

Phosphoric acid tris(2-methylphenyl) ester, see Tri-*o*-cresyl phosphate

Phosphoric acid tri-2-tolyl ester, see Tri-*o*-cresyl phosphate

Phosphorothioic acid *O,O*-diethyl *O*-(4-nitrophenyl)-ester, see Parathion

Phosphorothioic acid *O,O*-dimethyl *O*-(2,4,5-trichlorophenyl) ester, see Ronnel

Phosphorothionic acid *O,O*-diethyl *O*-(3,5,6-trichloro-2-pyridyl) ester, see Chlorpyrifos

Phosphostigmine, see Parathion

Phosphothion, see Malathion

Phosvit, see Dichlorvos

Phthalandione, see Phthalic anhydride

1,3-Phthalandione, see Phthalic anhydride

Phthalic acid anhydride, see Phthalic anhydride

Phthalic acid bis(2-ethylhexyl) ester, see Bis(2-ethylhexyl) phthalate

Phthalic acid dibutyl ester, see Di-*n*-butyl phthalate

Phthalic acid dimethyl ester, see Dimethyl phthalate

Phthalic acid dioctyl ester, see Bis(2-ethylhexyl) phthalate

Phthalic acid methyl ester, see Dimethyl phthalate

Phthalol, see Diethyl phthalate

Picclor, see Chloropicrin

Picfume, see Chloropicrin

Picride, see Chloropicrin

Picronitric acid, see Picric acid

Picrylmethylnitramine, see Tetryl

Picrylnitromethylamine, see Tetryl

Pied piper mouse seed, see Strychnine

Pielik, see 2,4-D

Pimelic ketone, see Cyclohexanone

Pin, see EPN

Pirofos, see Sulfotepp

Pittsburgh PX-138, see Bis(2-ethylhexyl) phthalate

Pivacin, see Pindone

Pival, see Pindone

Pivaldione, see Pindone

2-Pivaloylindane-1,3-dione, see Pindone

2-Pivaloyl-1,3-indanedione, see Pindone

Pivalyl, see Pindone

Pivalyl indandione, see Pindone

2-Pivalyl-1,3-indandione, see Pindone

Pivalyl valone, see Pindone

Placidol E, see Diethyl phthalate

Planotox, see 2,4-D

Plant dithio aerosol, see Sulfotepp

Plantfume 103 smoke generator, see Sulfotepp

Plantgard, see 2,4-D

Platinol AH, see Bis(2-ethylhexyl) phthalate

Platinol DOP, see Bis(2-ethylhexyl) phthalate

PNA, see 4-Nitroaniline

PNB, see 4-Nitrobiphenyl

PNCB, see *p*-Chloronitrobenzene

PNP, see 4-Nitrophenol

PNT, see 4-Nitrotoluene

Polychlorcamphene, see Toxaphene

Polychlorinated camphenes, see Toxaphene

Polychlorocamphene, see Toxaphene

Polycizer 162, see Di-*n*-octyl phthalate

Polycizer DBP, see Di-*n*-butyl phthalate

Polyoxymethylene glycols, see Formaldehyde

Polyram ultra, see Thiram

Poly-solv EE, see 2-Ethoxyethanol

Poly-solv EE acetate, see 2-Ethoxyethyl acetate

Poly-solv EM, see Methyl cellosolve

Pomarsol, see Thiram

Pomasol, see Thiram

Pomersol forte, see Thiram

PPD, see *p*-Phenylenediamine

PPzeidan, see *p,p'*-DDT

Preventol CMK, see *p*-Chloro-*m*-cresol

Preventol I, see 2,4,5-Trichlorophenol

Priltox, see Pentachlorophenol

Primary amyl acetate, see *n*-Amyl acetate

Primary isoamyl alcohol, see Isoamyl alcohol

Primary isobutyl alcohol, see Isoamyl alcohol

Prioderm, see Malathion

Prist, see Methyl cellosolve

Profume A, see Chloropicrin

Profume R 40B1, see Methyl bromide

Prokarbol, see 4,6-Dinitro-*o*-cresol

2-Propanamine, see Isopropylamine

Propanol, see 1-Propanol

n-Propanol, see 1-Propanol

Propanolide, see β-Propiolactone

Propanone, see Acetone

2-Propanone, see Acetone

Propellant 12, see Dichlorodifluoromethane

Propenal, see Acrolein

Prop-2-en-1-al, see Acrolein

2-Propenal, see Acrolein

Propenamide, see Acrylamide

2-Propenamide, see Acrylamide

Propenenitrile, see Acrylonitrile

2-Propenenitrile, see Acrylonitrile

Propene oxide, see Propylene oxide

2-Propenoic acid ethyl ester, see Ethyl acrylate

Propenoic acid methyl ester, see Methyl acrylate

2-Propenoic acid methyl ester, see Methyl acrylate

2-Propenoic acid 2-methyl methyl ester, see Methyl methacrylate

Propenol, see Allyl alcohol

1-Propenol-3, see Allyl alcohol

Propen-1-ol-3, see Allyl alcohol

1-Propen-3-ol, see Allyl alcohol

2-Propenol, see Allyl alcohol

2-Propen-1-ol, see Allyl alcohol

Propenol-3, see Allyl alcohol

2-Propen-1-one, see Acrolein

Propenyl alcohol, see Allyl alcohol

2-Propenyl alcohol, see Allyl alcohol

[(2-Propenyloxy)methyl]oxirane, see Allyl glycidyl ether
Propine, see Propyne
Propiolactone, see β-Propiolactone
1,3-Propiolactone, see β-Propiolactone
3-Propiolactone, see β-Propiolactone
β-Propionolactone, see β-Propiolactone
3-Propionolactone, see β-Propiolactone
β-Proprolactone, see β-Propiolactone
Propyl acetate, see *n*-Propyl acetate
1-Propyl acetate, see *n*-Propyl acetate
2-Propyl acetate, see Isopropyl acetate
Propylacetone, see 2-Hexanone
Propyl alcohol, see 1-Propanol
1-Propyl alcohol, see 1-Propanol
n-Propyl alcohol, see 1-Propanol
sec-Propylamine, see Isopropylamine
2-Propylamine, see Isopropylamine
Propylbenzene, see *n*-Propylbenzene
Propylcarbinol, see 1-Butanol
Propylene aldehyde, see Crotonaldehyde
Propylene chloride, see 1,2-Dichloropropane
Propylene dichloride, see 1,2-Dichloropropane
α,β-Propylene dichloride, see 1,2-Dichloropropane
1,2-Propylene oxide, see Propylene oxide
Propylethylene, see 1-Pentene
Propyl hydride, see Propane
Propylic alcohol, see 1-Propanol
Propyl iodide, see 1-Iodopropane
n-Propyl iodide, see 1-Iodopropane
Propylmethanol, see 1-Butanol
Propyl methyl ketone, see 2-Pentanone
Propyl nitrate, see *n*-Propyl nitrate
1-Propyne, see Propyne
Prothromadin, see Warfarin
PS, see Chloropicrin
Pseudobutylbenzene, see *tert*-Butylbenzene
Pseudocumene, see 1,2,4-Trimethylbenzene
Pseudocumol, see 1,2,4-Trimethylbenzene
Puralin, see Thiram
PX 104, see Di-*n*-butyl phthalate
PX 138, see Di-*n*-octyl phthalate
Pyranton, see Diacetone alcohol
Pyranton A, see Diacetone alcohol
β-Pyrene, see Pyrene
2-Pyridinamine, see 2-Aminopyridine
α-Pyridinamine, see 2-Aminopyridine
2-Pyridylamine, see 2-Aminopyridine
α-Pyridylamine, see 2-Aminopyridine
β-Pyrine, see Pyrene
Pyrinex, see Chlorpyrifos
Pyroacetic acid, see Acetone
Pyroacetic ether, see Acetone
Pyrobenzol, see Benzene
Pyrobenzole, see Benzene
Pyroligneous acid, see Acetic acid
Pyromucic aldehyde, see Furfural
Pyropentylene, see Cyclopentadiene
Pyrophosphoric acid tetraethyl ester, see Tetraethyl pyro-
    phosphate
Pyrophosphorodithioic acid *O,O,O,O*-tetraethyl dithiono-
    pyrophosphate, see Sulfotepp
Pyrophosphorodithioic acid tetraethyl ester, see Sulfo-
    tepp
Pyroxylic spirit, see Methanol
Pyrrolylene, see 1,3-Butadiene
QCB, see Pentachlorobenzene
Quellada, see Lindane

Quinol, see Hydroquinone
β-Quinol, see Hydroquinone
Quinone, see Hydroquinone, *p*-Quinone
4-Quinone, see *p*-Quinone
Quintox, see Dieldrin
R 10, see Carbon tetrachloride
R 12, see Dichlorodifluoromethane
R 20, see Chloroform
R 20 (refrigerant), see Chloroform
R 113, see 1,1,2-Trichlorotrifluoroethane
R 717, see Ammonia
Rafex, see 4,6-Dinitro-*o*-cresol
Rafex 35, see 4,6-Dinitro-*o*-cresol
Raphatox, see 4,6-Dinitro-*o*-cresol
Raschit, see *p*-Chloro-*m*-cresol
Raschit K, see *p*-Chloro-*m*-cresol
Rasenanicon, see *p*-Chloro-*m*-cresol
Rat-a-way, see Warfarin
Rat-b-gon, see Warfarin
Rat-gard, see Warfarin
Rat-kill, see Warfarin
Rat & mice bait, see Warfarin
Rat-mix, see Warfarin
Rat-o-cide #2, see Warfarin
Rat-ola, see Warfarin
Ratorex, see Warfarin
Ratox, see Warfarin
Ratoxin, see Warfarin
Ratron, see Warfarin
Ratron G, see Warfarin
Rats-no-more, see Warfarin
Rattrack, see ANTU
Rat-trol, see Warfarin
Rattunal, see Warfarin
Ravyon, see Carbaryl
Rax, see Warfarin
RB, see Parathion
RC plasticizer DOP, see Bis(2-ethylhexyl) phthalate
RCRA waste number F027, see 2,4,6-Trichlorophenol
RCRA waste number P001, see Warfarin
RCRA waste number P003, see Acrolein
RCRA waste number P004, see Aldrin
RCRA waste number P005, see Allyl alcohol
RCRA waste number P016, see *sym*-Dichloromethyl
    ether
RCRA waste number P022, see Carbon disulfide
RCRA waste number P023, see Chloroacetaldehyde
RCRA waste number P024, see 4-Chloroaniline
RCRA waste number P028, see Benzyl chloride
RCRA waste number P037, see Dieldrin
RCRA waste number P047, see 4,6-Dinitro-*o*-cresol
RCRA waste number P048, see 2,4-Dinitrophenol
RCRA waste number P050, see α-Endosulfan,
    β-Endosulfan
RCRA waste number P051, see Endrin
RCRA waste number P054, see Ethylenimine
RCRA waste number P059, see Heptachlor
RCRA waste number P064, see Methyl isocyanate
RCRA waste number P077, see 4-Nitroaniline
RCRA waste number P082, see *N*-Nitrosodimethylamine
RCRA waste number P089, see Parathion
RCRA waste number P108, see Strychnine
RCRA waste number P109, see Sulfotepp
RCRA waste number P111, see Tetraethyl pyro-
    phosphate
RCRA waste number P112, see Tetranitromethane
RCRA waste number P123, see Toxaphene

Refrigerant 112, see 1,2-Difluorotetrachloroethane
Refrigerant 112a, see 1,1-Difluorotetrachloroethane
Refrigerant 113, see 1,1,2-Trichlorotrifluoroethane
Refrigerant 13B1, see Bromotrifluoromethane
Refrigerant 21, see Dichlorofluoromethane
Renal PF, see p-Phenylenediamine
Reomol DOP, see Bis(2-ethylhexyl) phthalate
Reomol D 79P, see Bis(2-ethylhexyl) phthalate
Resinol yellow GR, see p-Dimethylaminoazobenzene
Resoform yellow GGA, see p-Dimethylaminoazo-
    benzene
Retarder AK, see Phthalic anhydride
Retarder BA, see Benzoic acid
Retarder ESEN, see Phthalic anhydride
Retarder J, see N-Nitrosodiphenylamine
Retarder PD, see Phthalic anhydride
Retardex, see Benzoic acid
Rezifilm, see Thiram
Rhodia, see 2,4-D
Rhodiachlor, see Heptachlor
Rhodiasol, see Parathion
Rhodiatox, see Parathion
Rhodiatrox, see Parathion
Rhothane, see p,p'-DDD
Rhothane D-3, see p,p'-DDD
Rodafarin, see Warfarin
Ro-deth, see Warfarin
Rodex, see Strychnine, Warfarin
Rodex blox, see Warfarin
Rosex, see Warfarin
Rothane, see p,p'-DDD
Rotox, see Methyl bromide
Rough & ready mouse mix, see Warfarin
Royal TMTD, see Thiram
R-pentine, see Cyclopentadiene
Rubinate TDI 80/20, see 2,4-Toluene diisocyanate
Rukseam, see p,p'-DDT
Rylam, see Carbaryl
S 1, see Chloropicrin
Sadofos, see Malathion
Sadophos, see Malathion
Sadoplon, see Thiram
Salvo, see 2,4-D
Salvo liquid, see Benzoic acid
Salvo powder, see Benzoic acid
Sanaseed, see Strychnine
Sandolin, see 4,6-Dinitro-o-cresol
Sandolin A, see 4,6-Dinitro-o-cresol
Sanocide, see Hexachlorobenzene
Santicizer 160, see Benzyl butyl phthalate
Santobane, see p,p'-DDT
Santobrite, see Pentachlorophenol
Santochlor, see 1,4-Dichlorobenzene
Santoflex IC, see p-Phenylenediamine
Santophen, see Pentachlorophenol
Santophen 20, see Pentachlorophenol
Santox, see EPN
SBA, see sec-Butyl alcohol
Scintillar, p-Xylene
Sconatex, see 1,1-Dichloroethylene
SD 1750, see Dichlorvos
SD 1897, see 1,2-Dibromo-3-chloropropane
SD 5532, see Chlordane
Seedrin, see Aldrin
Seedrin liquid, see Aldrin
Seffein, see Carbaryl
Selephos, see Parathion

Selinon, see 4,6-Dinitro-o-cresol
Septene, see Carbaryl
Sevimol, see Carbaryl
Sevin, see Carbaryl
Sextone, see Cyclohexanone
Sextone B, see Methylcyclohexane
SF 60, see Malathion
Shell MIBK, see 4-Methyl-2-pentanone
Shell SD-5532, see Chlordane
Shellsol 140, see Nonane
Shimose, see Picric acid
Shinnippon fast red GG base, see 4-Nitroaniline
Sicol 150, see Bis(2-ethylhexyl) phthalate
Sicol 160, see Benzyl butyl phthalate
Silotras yellow T2G, see p-Dimethylaminoazobenzene
Silvanol, see Lindane
Sinituho, see Pentachlorophenol
Sinox, see 4,6-Dinitro-o-cresol
Siptox I, see Malathion
Sixty-three special E.C. insecticide, see Parathion
Slimicide, see Acrolein
Smut-go, see Hexachlorobenzene
Snieciotox, see Hexachlorobenzene
SNP, see Parathion
Soilbrom-40, see Ethylene dibromide
Soilbrom-85, see Ethylene dibromide
Soilbrom-90, see Ethylene dibromide
Soilbrom-90EC, see Ethylene dibromide
Soilbrom-100, see Ethylene dibromide
Soilfume, see Ethylene dibromide
Sok, see Carbaryl
Solaesthin, see Methylene chloride
Soleptax, see Heptachlor
Solfarin, see Warfarin
Solfo black B, see 2,4-Dinitrophenol
Solfo black 2B supra, see 2,4-Dinitrophenol
Solfo black BB, see 2,4-Dinitrophenol
Solfo black G, see 2,4-Dinitrophenol
Solfo black SB, see 2,4-Dinitrophenol
Solmethine, see Methylene chloride
Solvanol, see Diethyl phthalate
Solvanom, see Dimethyl phthalate
Solvarone, see Dimethyl phthalate
Solvent ether, see Ethyl ether
Solvent III, see 1,1,1-Trichloroethane
Somalia yellow A, see p-Dimethylaminoazobenzene
Soprathion, see Parathion
Special termite fluid, see 1,2-Dichlorobenzene
Spirit of Hartshorn, see Ammonia
Spontox, see 2,4,5-T
Spotrete, see Thiram
Spotrete-F, see Thiram
Spotrete 75WDG, see Thiram
Spray-trol brand roden-trol, see Warfarin
Spritz-hormin/2,4-D, see 2,4-D
Spritz-hormit/2,4-D, see 2,4-D
Spritz-rapidin, see Lindane
Spruehpflanzol, see Lindane
SQ 1489, see Thiram
Stabilized ethyl parathion, see Parathion
Staflex DBP, see Di-n-butyl phthalate
Staflex DOP, see Bis(2-ethylhexyl) phthalate
Stathion, see Parathion
Stear yellow JB, see p-Dimethylaminoazobenzene
Strathion, see Parathion
Strobane-T, see Toxaphene
Strobane T-90, see Toxaphene

Tetralite, see Tetryl
sym-1,2,4,5-Tetramethylbenzene, see 1,2,4,5-Tetramethylbenzene
Tetramethyldiurane sulphite, see Thiram
Tetramethylene oxide, see Tetrahydrofuran
Tetramethylenethiuram disulfide, see Thiram
Tetramethylmethane, see 2,2-Dimethylpropane
Tetramethylthiocarbamoyl disulfide, see Thiram
Tetramethylthioperoxydicarbonic diamide, see Thiram
Tetramethylthiuram bisulfide, see Thiram
Tetramethylthiuram bisulphide, see Thiram
Tetramethylthiuram disulfide, see Thiram
N,N-Tetramethylthiuram disulfide, see Thiram
N,N,N′,N′-Tetramethylthiuram disulfide, see Thiram
Tetramethylthiuran disulfide, see Thiram
Tetramethylthiurane disulphide, see Thiram
Tetramethylthiurum disulfide, see Thiram
Tetramethylthiurium disulphide, see Thiram
Tetra olive N2G, see Anthracene
Tetraphene, see Benzo[a]anthracene
Tetrapom, see Thiram
Tetrasipton, see Thiram
Tetrasol, see Carbon tetrachloride
Tetrastigmine, see Tetraethyl pyrophosphate
Tetrasulphur black PB, see 2,4-Dinitrophenol
Tetrathiuram disulfide, see Thiram
Tetrathiuram disulphide, see Thiram
Tetravec, see Tetrachloroethylene
Tetravos, see Dichlorvos
Tetril, see Tetryl
Tetroguer, see Tetrachloroethylene
Tetron, see Tetraethyl pyrophosphate
Tetron-100, see Tetraethyl pyrophosphate
Tetropil, see Tetrachloroethylene
Tetrosulphur PBR, see 2,4-Dinitrophenol
2,4,6-Tetryl, see Tetryl
Texaco lead appreciator, see tert-Butyl acetate
Texadust, see Toxaphene
THF, see Tetrahydrofuran
Thiacyclopentadiene, see Thiophene
Thiaphene, see Thiophene
Thifor, see α-Endosulfan, β-Endosulfan
Thillate, see Thiram
Thimer, see Thiram
Thimul, see α-Endosulfan, β-Endosulfan
Thiobutyl alcohol, see n-Butyl mercaptan
Thiodan, see α-Endosulfan, β-Endosulfan
Thiodiphosphoric acid tetraethyl ester, see Sulfotepp
Thioethanol, see Ethyl mercaptan
Thioethyl alcohol, see Ethyl mercaptan
Thiofalco M-50, see Ethanolamine
Thiofor, see α-Endosulfan, β-Endosulfan
Thiofos, see Parathion
Thiofuram, see Thiophene
Thiofuran, see Thiophene
Thiofurfuran, see Thiophene
Thiole, see Thiophene
Thiomethanol, see Methyl mercaptan
Thiomethyl alcohol, see Methyl mercaptan
Thiomul, see α-Endosulfan, β-Endosulfan
Thionex, see α-Endosulfan, β-Endosulfan
Thiophen, see Thiophene
Thiophos 3422, see Parathion
Thiophosphoric acid tetraethyl ester, see Sulfotepp
Thiopyrophosphoric acid tetraethyl ester, see Sulfotepp
Thiosan, see Thiram
Thiosulfan, see α-Endosulfan, β-Endosulfan

Thiotepp, see Sulfotepp
Thiotetrole, see Thiophene
Thiotex, see Thiram
Thiotox, see Thiram
Thiram 75, see Thiram
Thiramad, see Thiram
Thiram B, see Thiram
Thirame, see Thiram
Thirasan, see Thiram
Thiulix, see Thiram
Thiurad, see Thiram
Thiuram, see Thiram
Thiuram D, see Thiram
Thiuram M, see Thiram
Thiuram M rubber accelerator, see Thiram
Thiuramin, see Thiram
Thiuramyl, see Thiram
Thompson's wood fix, see Pentachlorophenol
Threthylen, see Trichloroethylene
Threthylene, see Trichloroethylene
Thylate, see Thiram
Tionel, see α-Endosulfan, β-Endosulfan
Tiophos, see Parathion
Tiovel, see α-Endosulfan, β-Endosulfan
Tippon, see 2,4,5-T
Tirampa, see Thiram
Tiuramyl, see Thiram
TJB, see N-Nitrosodiphenylamine
TL 314, see Acrylonitrile
TL 337, see Ethylenimine
TL 1450, see Methyl isocyanate
TLA, see tert-Butyl acetate
TM-4049, see Malathion
TMB, see 1,3,5-Trimethylbenzene
TMTD, see Thiram
TMTDS, see Thiram
TNM, see Tetranitromethane
TNT, see 2,4,6-Trinitrotoluene
α-TNT, see 2,4,6-Trinitrotoluene
TNT-tolite, see 2,4,6-Trinitrotoluene
TOCP, see Tri-o-cresyl phosphate
TOFK, see Tri-o-cresyl phosphate
Tolit, see 2,4,6-Trinitrotoluene
Tolite, see 2,4,6-Trinitrotoluene
Toluene 2,4-diisocyanate, see 2,4-Toluene diisocyanate
Toluene hexahydride, see Methylcyclohexane
α-Toluenol, see Benzyl alcohol
2-Toluidine, see o-Toluidine
Toluol, see Toluene
2-Toluol, see 2-Methylphenol
4-Toluol, see 4-Methylphenol
o-Toluol, see 2-Methylphenol
p-Toluol, see 4-Methylphenol
Tolu-sol, see Toluene
Toluene-2,4-diisocyanate, see 2,4-Toluene diisocyanate
p-Tolyl alcohol, see 4-Methylphenol
o-Tolylamine, see o-Toluidine
Tolyl chloride, see Benzyl chloride
Tolylene-2,4-diisocyanate, see 2,4-Toluene diisocyanate
2,4-Tolylene diisocyanate, see 2,4-Toluene diisocyanate
m-Tolylene diisocyanate, see 2,4-Toluene diisocyanate
o-Tolyl phosphate, see Tri-o-cresyl phosphate
Topanel, see Crotonaldehyde
Topichlor 20, see Chlordane
Topiclor, see Chlordane
Topiclor 20, see Chlordane
Tormona, see 2,4,5-T

TOTP, see Tri-*o*-cresyl phosphate

Tox 47, see Parathion

Toxadust, see Toxaphene

Toxadust 10, see Toxaphene

Toxakil, see Toxaphene

Toxan, see Carbaryl

Tox-hid, see Warfarin

Toxichlor, see Chlordane

Toxilic anhydride, see Maleic anhydride

Toxon 63, see Toxaphene

Toyo oil yellow G, see *p*-Dimethylaminoazobenzene

Toxyphen, see Toxaphene

TPP, see Triphenyl phosphate

Trametan, see Thiram

Transamine, see 2,4-D, see 2,4,5-T

Trefanocide, see Trifluralin

Treficon, see Trifluralin

Treflam, see Trifluralin

Treflan, see Trifluralin

Treflanocide elancolan, see Trifluralin

Trethylene, see Trichloroethylene

Tri, see Trichloroethylene

Tri-6, see Lindane

Triad, see Trichloroethylene

Trial, see Trichloroethylene

Triasol, see Trichloroethylene

Tri-Ban, see Pindone

sym-Tribromobenzene, see 1,3,5-Tribromobenzene

Tribromomethane, see Bromoform

Tributon, see 2,4-D, 2,4,5-T

Tri-*n*-butyl phosphate, see Tributyl phosphate

Tricarnam, see Carbaryl

Trichloran, see Trichloroethylene

Trichloren, see Trichloroethylene

unsym-Trichlorobenzene, see 1,2,4-Trichlorobenzene

sym-Trichlorobenzene, see 1,3,5-Trichlorobenzene

1,1,1-Trichloro-2,2-bis(*p*-anisyl)ethane, see
    Methoxychlor

1,1,1-Trichlorobis(4-chlorophenyl)ethane, see *p,p'*-DDT

1,1,1-Trichlorobis(*p*-chlorophenyl)ethane, see *p,p'*-DDT

1,1,1-Trichloro-2,2-bis(*p*-chlorophenyl)ethane, see *p,p'*-
    DDT

1,1,1-Trichloro-2,2-bis(*p*-methoxyphenol)ethanol, see
    Methoxychlor

1,1,1-Trichloro-2,2-bis(*p*-methoxyphenyl)ethane, see
    Methoxychlor

1,1,1-Trichloro-2,2-di(4-chlorophenyl)ethane, see *p,p'*-
    DDT

1,1,1-Trichloro-2,2-di(*p*-chlorophenyl)ethane, see *p,p'*-
    DDT

1,1,1-Trichloro-2,2-di(4-methoxyphenyl)ethane, see
    Methoxychlor

β-Trichloroethane, see 1,1,2-Trichloroethane

1,2,2-Trichloroethane, see 1,1,2-Trichloroethane

α-Trichloroethane, see 1,1,1-Trichloroethane

Trichloroethene, see Trichloroethylene

1,1,2-Trichloroethene, see Trichloroethylene

1,2,2-Trichloroethene, see Trichloroethylene

1,1,2-Trichloroethylene, see Trichloroethylene

1,2,2-Trichloroethylene, see Trichloroethylene

1,1'-(2,2,2-Trichloroethylidene)bis(4-chlorobenzene),
    see *p,p'*-DDT

1,1'-(2,2,2-Trichloroethylidene)bis(4-methoxybenzene),
    see Methoxychlor

Trichloroform, see Chloroform

Trichlorohydrin, see 1,2,3-Trichloropropane

Trichlorometafos, see Ronnel

Trichloromethane, see Chloroform

Trichloromonofluoromethane, see Trichlorofluorometh-
    ane

Trichloronitromethane, see Chloropicrin

(2,4,5-Trichlorophenoxy)acetic acid, see 2,4,5-T

Trichlorotrifluoroethane, see 1,1,2-Trichlorotrifluoroeth-
    ane

1,1,2-Trichloro-1,2,2-trifluoroethane, see 1,1,2-Trichlor-
    otrifluoroethane

Triclene, see Trichloroethylene

Triclor, see Chloropicrin

Tricresyl phosphate, see Tri-*o*-cresyl phosphate

Tridipam, see Thiram

Trielene, see Trichloroethylene

Trieline, see Trichloroethylene

Tri-ethane, see 1,1,1-Trichloroethane

Trifina, see 4,6-Dinitro-*o*-cresol

Trifluoralin, see Trifluralin

Trifluorobromomethane, see Bromotrifluoromethane

α,α,α-Trifluoro-2,6-dinitro-*N,N*-dipropyl-*p*-toluidine, see
    Trifluralin

Trifluoromonobromomethane, see Bromotrifluorometh-
    ane

1,1,2-Trifluorotrichloroethane, see 1,1,2-Trichlorotri-
    fluoroethane

Trifluraline, see Trifluralin

Triflurex, see Trifluralin

Trifocide, see 4,6-Dinitro-*o*-cresol

Trikepin, see Trifluralin

Triklone, see Trichloroethylene

Trilen, see Trichloroethylene

Trilene, see Trichloroethylene

Triline, see Trichloroethylene

Trilit, see 2,4,6-Trinitrotoluene

Trim, see Trifluralin

Trimar, see Trichloroethylene

asym-Trimethylbenzene, see 1,2,4-Trimethylbenzene

sym-Trimethylbenzene, see 1,3,5-Trimethylbenzene

Trimethylbenzol, see 1,3,5-Trimethylbenzene

1,7,7-Trimethylbicyclo[2.2.1]heptan-2-one, see Camphor

Trimethylcarbinol, see *tert*-Butyl alcohol

1,3,3-Trimethylcyclohexane, see 1,1,3-Trimethylcyclo-
    hexane

1,1,3-Trimethyl-3-cyclohexene-5-one, see Isophorone

Trimethylcyclohexenone, see Isophorone

3,5,5-Trimethyl-2-cyclohexen-1-one, see Isophorone

1,3,3-Trimethylcyclopentane, see 1,1,3-Trimethylcyclo-
    pentane

2,5,5-Trimethylhexane, see 2,2,5-Trimethylhexane

Trimethylmethane, see 2-Methylpropane

Trimethyl methanol, see *tert*-Butyl alcohol

Trimethylphenylmethane, see *tert*-Butylbenzene

Tri-2-methylphenyl phosphate, see Tri-*o*-cresyl phos-
    phate

1,3,5-Trinitrophenol, see Picric acid

2,4,6-Trinitrophenol, see Picric acid

Trinitrophenylmethylnitramine, see Tetryl

2,4,6-Trinitrophenylmethylnitramine, see Tetryl

2,4,6-Trinitrophenyl-*N*-methylnitramine, see Tetryl

Trinitrotoluene, see 2,4,6-Trinitrotoluene

sym-Trinitrotoluene, see 2,4,6-Trinitrotoluene

Trinitrotoluol, see 2,4,6-Trinitrotoluene

α-Trinitrotoluol, see 2,4,6-Trinitrotoluene

sym-Trinitrotoluol, see 2,4,6-Trinitrotoluene

Trinoxol, see 2,4-D, 2,4,5-T

Triol, see Trichloroethylene

Triorthocresyl phosphate, see Tri-*o*-cresyl phosphate

USAF P-220, see *p*-Quinone
USAF XR-42, see Diuron
VAC, see Vinyl acetate
Valerone, see Diisobutyl ketone
VAM, see Vinyl acetate
Vampirin III, see Warfarin
Vampirinip II, see Warfarin
Vancida TM-95, see Thiram
Vancide TM, see Thiram
Vapona, see Dichlorvos
Vaponite, see Dichlorvos
Vapophos, see Parathion
Vapora II, see Dichlorvos
Vapotone, see Tetraethyl pyrophosphate
VC, see 1,1-Dichloroethylene
VC, see Vinyl chloride
VCM, see Vinyl chloride
VCN, see Acrylonitrile
VDC, see 1,1-Dichloroethylene
Vegfru malatox, see Malathion
Velsicol 104, see Heptachlor
Velsicol 1068, see Chlordane
Velsicol 53-CS-17, see Heptachlor epoxide
Velsicol heptachlor, see Heptachlor
Ventox, see Acrylonitrile
Veon, see 2,4,5-T
Veon 245, see 2,4,5-T
Verdican, see Dichlorvos
Verdipor, see Dichlorvos
Vergemaster, see 2,4-D
Vermoestricid, see Carbon tetrachloride
Versneller NL 63/10, see Dimethylaniline
Vertac 90%, see Toxaphene
Vertac toxaphene 90, see Toxaphene
Verton, see 2,4-D
Verton D, see 2,4-D
Verton 2D, see 2,4-D
Verton 2T, see 2,4,5-T
Vertron 2D, see 2,4-D
Vestinol 80, see Bis(2-ethylhexyl) phthalate
Vestrol, see Trichloroethylene
Vetiol, see Malathion
Vidon 638, see 2,4-D
Vinegar acid, see Acetic acid
Vinegar naphtha, see Ethyl acetate
Vinicizer 85, see Di-*n*-octyl phthalate
Vinyl acetate H.Q., see Vinyl acetate
Vinyl alcohol 2,2-dichlorodimethyl phosphate, see Dichlorvos
Vinyl A Monomer, see Vinyl acetate
Vinyl benzene, see Styrene
Vinyl benzol, see Styrene
Vinyl carbinol, see Allyl alcohol
Vinyl chloride monomer, see Vinyl chloride
Vinyl 2-chloroethyl ether, see 2-Chloroethyl vinyl ether
Vinyl β-chloroethyl ether, see 2-Chloroethyl vinyl ether
Vinyl C monomer, see Vinyl chloride
Vinyl cyanide, see Acrylonitrile
Vinylethylene, see 1,3-Butadiene
Vinylofos, see Dichlorvos
Vinylophos, see Dichlorvos
Vinyl trichloride, see 1,1,2-Trichloroethane
Vinylidene chloride, see 1,1-Dichloroethylene
Vinylidene chloride (II), see 1,1-Dichloroethylene
Vinylidene dichloride, see 1,1-Dichloroethylene
Vinylidine chloride, see 1,1-Dichloroethylene
Viozene, see Ronnel

Visko-rhap, see 2,4-D
Visko-rhap drift herbicides, see 2,4-D
Visko-rhap low volatile ester, see 2,4,5-T
Visko-rhap low volatile 4L, see 2,4-D
Viton, see Lindane
Vitran, see Trichloroethylene
Vitrex, see Parathion
Vonduron, see Diuron
Vulcafor TMTD, see Thiram
Vulcalent A, see *N*-Nitrosodiphenylamine
Vulcard, see *N*-Nitrosodiphenylamine
Vulcatard A, see *N*-Nitrosodiphenylamine
Vulkacit MTIC, see Thiram
Vulkanox 4020, see *p*-Phenylenediamine
Vultrol, see *N*-Nitrosodiphenylamine
VyAc, see Vinyl acetate
WARF 42, see Warfarin
Waran, see Warfarin
Warfarat, see Warfarin
Warfarine, see Warfarin
Warfarin plus, see Warfarin
Warfarin Q, see Warfarin
Warf compound 42, see Warfarin
Warficide, see Warfarin
Waxoline yellow AD, see *p*-Dimethylaminoazobenzene
Waxoline yellow ADS, see *p*-Dimethylaminoazobenzene
Weddar, see 2,4,5-T
Weddar-64, see 2,4-D
Weddatul, see 2,4-D
Weedar, see 2,4-D
Weed-b-gon, see 2,4-D
Weedez wonder bar, see 2,4-D
Weedone, see Pentachlorophenol, 2,4-D, 2,4,5-T
Weedone LV4, see 2,4-D
Weedone 2,4,5-T, see 2,4,5-T
Weed-rhap, see 2,4-D
Weed tox, see 2,4-D
Weedtrol, see 2,4-D
Weeviltox, see Carbon disulfide
Westron, see 1,1,2,2-Tetrachloroethane
Westrosol, see Trichloroethylene
White tar, see Naphthalene
Winterwash, see 4,6-Dinitro-*o*-cresol
Witicizer 300, see Di-*n*-butyl phthalate
Witicizer 312, see Bis(2-ethylhexyl) phthalate
Witophen P, see Pentachlorophenol
Wood alcohol, see Methanol
Wood naphtha, see Methanol
Wood spirit, see Methanol
Xenylamine, see 4-Aminobiphenyl
1,2-Xylene, see *o*-Xylene
1,3-Xylene, see *m*-Xylene
1,4-Xylene, see *p*-Xylene
1,3,4-Xylenol, see 2,4-Dimethylphenol
2,4-Xylenol, see 2,4-Dimethylphenol
m-Xylenol, see 2,4-Dimethylphenol
o-Xylol, see *o*-Xylene
p-Xylol, see *o*-Xylene
Yellow G soluble in grease, see *p*-Dimethylaminoazobenzene
Zeidane, see *p,p'*-DDT
Zerdane, see *p,p'*-DDT
Zeset T, see Vinyl acetate
Zithiol, see Malathion
Zoba black D, see *p*-Phenylenediamine
Zytox, see Methyl bromide

References

Abd-El-Bary, M.F., Hamoda, M.F., Tanisho, S., and Wakao, N. Henry's constants for phenol over its diluted aqueous solution, *J. Chem. Eng. Data*, 31(2):229-230, 1986.

Abdelmagid, H.M. and Tabatabai, M.A. Decomposition of acrylamide in soils, *J. Environ. Qual.*, 11(4):701-704, 1982.

Abdul, S.A. and Gibson, T.L. Equilibrium batch experiments with six polycyclic aromatic hydrocarbons and two aquifer materials, *Haz. Waste Haz. Mater.*, 3(2):125-137, 1986.

Abdul, S.A., Gibson, T.L., and Rai, D.N. Statistical correlations for predicting the partition coefficient for nonpolar organic contaminants between aquifer organic carbon and water, *Haz. Waste Haz. Mater.*, 4(3):211-222, 1987.

Abe, I., Hayashi, K., and Hirashima, T. Prediction of adsorption isotherms of organic compounds from water on activated carbon, *J. Colloid Interface Sci.*, 94(1):201-206, 1983.

Abel, P.D. Toxicity of gamma-hexachlorocyclohexane (lindane) to *Gammarus pulex*: mortality in relation to concentration and duration of exposure, *Freshwater Biol.*, 10:251-259, 1980.

Abraham, M.H. Thermodynamics of solution of homologous series of solutes in water, *J. Chem. Soc., Faraday Trans. 1*, 80:153-181, 1984.

Abraham, T., Bery, V., and Kudchadker, A.P. Densities of some organic substances, *J. Chem. Eng. Data*, 16(3):355-356, 1971.

Abram, F.S.H. and Sims, I.R. The toxicity of aniline to rainbow trout, *Water Res.*, 16(8):1309-1312, 1982.

Abram, F.S.H. and Wilson, P. The acute toxicity of CS rainbow trout, *Water Res.*, 13(7):631-635, 1979.

ACGIH. Documentation of the Threshold Limit Values and Biological Exposure Indices (Cincinnati, OH: American Conference of Governmental Industrial Hygienists, 1986), 744 p.

Achard, C., Jaoui, M., Schwing, M., and Rogalski, M. Aqueous solubilities of phenol derivatives by conductivity measurements, *J. Chem. Eng. Data*, 41(3):504-507, 1996.

Adams, A.J. and Blaine, K.M. A water solubility determination of 2,3,7,8-TCDD, *Chemosphere*, 15(9-12):1397-1400, 1986.

Adams, A.J., DeGraeve, G.M., Sabourin, T.D., Cooney, J.D., and Mosher, G.M. Toxicity and bioconcentration of 2,3,7,8-TCDD to fathead minnows (*Pimephales promelas*), *Chemosphere*, 15(9-12):1503-1511, 1986.

Adams, P.M., Hanlon, R.T., and Forsythe, J.W. Toxic exposure to ethylene dibromide and mercuric chloride: effects on laboratory-reared octopuses, *Neurotoxicol. Teratol.*, 19(6):519-523, 1988.

Adams, R.S. and Li, P. Soil properties influencing sorption and desorption of lindane, *Soil Sci. Soc. Am. Proc.*, 35:78-81, 1971.

Adema, D.M.M. and Vink, G.L. A comparative study of the toxicity of 1,1,2-trichloroethane, dieldrin, pentachlorophenol and 3,4-dichloroaniline for marine and freshwater organisms, *Chemosphere*, 10:533-554, 1981.

Adeniji, S.A., Kerr, J.A., and Williams, M.R. Rate constants for ozone-alkene reactions under atmospheric conditions, *Int. J. Chem. Kinet.*, 13:209-217, 1981.

Adewuyi, Y.G. and Carmichael, G.R. Kinetics of hydrolysis and oxidation of carbon disulfide by hydrogen peroxide in alkaline medium and application to carbonyl sulfide, *Environ. Sci. Technol.*, 21(2):170-177, 1987.

Adhya, T.K., Sudhakar-Barik, and Sethunathan, N. Fate of fenitrothion, methyl parathion, and parathion in anoxic sulfur-containing soil systems, *Pestic. Biochem. Physiol.*, 16(1):14-20, 1981.

Adhya, T.K., Sudhakar-Barik, and Sethunathan, N. Stability of commercial formulation of fenitrothion, methyl parathion, and parathion in anaerobic soils, *J. Agric. Food Chem.*, 29(1):90-93, 1981a.

Adrian, P., Lahaniatis, E.S., Andreux, F., Mansour, M., Scheunert, I., and Korte, F. Reaction of the soil pollutant 4-chloroaniline with the humic acid monomer catechol, *Chemosphere*, 18(7/8):1599-1609, 1989.

Affens, W.A. and McLaren, G.W. Flammability properties of hydrocarbon solutions in air, *J. Chem. Eng. Data*, 17(4):482-488, 1972.

Afghan, B.K. and Mackay, D., Eds. *Hydrocarbons and Halogenated Hydrocarbons in the Environment*, Volume 16, (New York: Plenum Press, 1990).

Agrawal, A. and Tratnyek, P.G. Reduction of nitro aromatic compounds by zero-valent iron metal, *Environ. Sci. Technol.*, 30(1):153-160, 1996.

Aguwa, A.A., Patterson, J.W., Haas, C.N., and Noll, K.E. Estimation of effective intraparticle diffusion coefficients with differential reactor columns, *J. Water Pollut. Control Fed.*, 56(5):442-448, 1984.

Ahel, M. and Giger, W. Aqueous solubility of alkylphenols and alkylphenol polyethoxylates, *Environ. Sci. Technol.*, 26(8):1461-1470, 1993.

Ahling B., Bjorseth, A., and Lunde, G. Formation of chlorinated hydrocarbons during combustion of polyvinyl chloride, *Chemosphere*, 10:799-806, 1978.

Ahmad, N., Walgenbach, D.D., and Sutter, G.R. Degradation rates of technical carbofuran and a granular formation in four soils with known insecticide use history, *Bull. Environ. Contam. Toxicol.*, 23(4/5):572-574, 1979.

Akimoto, H. and Takagi, H. Formation of methyl nitrite in the surface reaction of nitrogen dioxide and methanol. 2. Photoenhancement, *Environ. Sci. Technol.*, 20(4):387-393, 1986.

Akiyoshi, M., Deguchi, T., and Sanemasa, I. The vapor saturation method for preparing aqueous solutions of solid aromatic hydrocarbons, *Bull. Chem. Soc. Jpn.*, 60(11):3935-3939, 1987.

Aksnes, G. and Sandberg, K. On the oxidation of benzidine and *o*-dianisidine with hydrogen peroxide and acetylcholine in alkaline solution, *Acta Chem. Scand.*, 11:876-880, 1957.

Alaee, M., Whittal, R.M., and Strachan, W.M.J. The effect of water temperature and composition on Henry's law constant for various PAHs, *Chemosphere*, 32(6):1153-1164, 1996.

Albanis, T.A., Pomonis, P.J., and Sdoukos, A.T. The influence of fly ash on pesticide fate in the environment. I. Hydrolysis, degradation and adsorption of lindane in aqueous mixtures of soil with fly ash, *Toxicol. Environ. Chem.*, 19(3+4):161-169, 1989.

Aldrich. Catalog Handbook of Fine Chemicals (Milwaukee, WI: Aldrich Chemical, 1990), 2105 p.

Alexander, D.M. The solubility of benzene in water, *J. Phys. Chem.*, 63(6):1021-1022, 1959.

Alexander, H.C., McCarty, W.M., and Bartlett, E.A. Toxicity of perchloroethylene, trichloroethylene, 1,1,1-trichloroethane, and methylene chloride to fathead minnows, *Bull. Environ. Contam. Toxicol.*, 20(3):344-352, 1978.

Alexander, M. Biodegradation of chemicals of environmental concern, *Science (Washington, D.C.)*, 211(4478):132-138, 1981.

Alexander, M. and Lustigman, B.K. Effects of chemical structure on microbial degradation of substituted benzenes, *J. Agric. Food Chem.*, 14:410-413, 1966.

Alexeeff, G.V., Kilgore, W.W., Munoz, P., and Watt, D. Determination of acute toxic effects in mice following exposure to methyl bromide, *J. Toxicol. Environ. Health*, 15(1):109-123, 1985.

Ali, S. Degradation and environmental fate of endosulfan isomers and endosulfan sulfate in mouse, insect, and laboratory ecosystem, Ph.D. Thesis, University of Illinois, Ann Arbor, MI, 1978.

Allen, J.M., Balcavage, W.X., and Ramachandran, B.R. Measurement of Henry's law constants using a new *in-situ* optical absorbance method, in *American Chemical Society - Division of Environmental Chemistry, Preprints of Extended Abstracts*, 37(1):56-58, 1997.

Alley, E.G., Layton, B.R., and Minyard, J.P., Jr. Identification of the photoproducts of the insecticides mirex and kepone, *J. Agric. Food Chem.*, 22(3):442-445, 1974.

Almgren, M., Grieser, F., Powell, J.R., and Thomas, J.K. A correlation between the solubility of aromatic hydrocarbons in water and micellar solutions, with their normal boiling points, *J. Chem. Eng. Data*, 24(4):285-287, 1979.

Altom, J.D. and Stritzke, J.F. Degradation of dicamba, picloram, and four phenoxy herbicides in soils, *Weed Sci.*, 21:556-560, 1973.

Altshuller, A.P. Measurements of the products of atmospheric photochemical reactions in laboratory studies and in ambient air-relationships between ozone and other products, *Atmos. Environ.*, 17(12):2383-2427, 1983.

Altshuller, A.P. Chemical reactions and transport of alkanes and their products in the troposphere, *J. Atmos. Chem.*, 12:19-61, 1991.

Altshuller, A.P. and Everson, H.E. The solubility of ethyl acetate in water, *J. Am. Chem. Soc.*, 75(7):1727, 1953.

Altshuller, A.P., Kopczynski, S.L., Lonneman, W.A., Sutterfield, F.D., and Wilson, D.L. Photochemical reactivities of aromatic hydrocarbon-nitrogen oxide and related systems, *Environ. Sci. Technol.*, 4(1):44-49, 1970.

Aly, M.I., Bakry, N., Kishk, F., and El-Sebae, A.H. Carbaryl adsorption on calcium-bentonite and soils, *Soil Sci. Soc. Am. J.*, 44(6):1213-1215, 1980.

Aly, O.M. and El-Dib, M.A. Studies on the persistence of some carbamate insecticides in the aquatic environment. I. Hydrolysis of sevin, baygon, pyrolam and dimetilan in waters, *Water Res.*, 5(12):1191-1205, 1971.

Aly, O.M. and Faust, S.D. Studies on the fate of 2,4-D and ester derivatives in natural surface waters, *J. Agric. Food Chem.*, 12(6):541-546, 1964.

Ambrose, A.M., Larson, P.S., Borzelleca, J.F., and Henniger, G.R., Jr. Toxicologic studies on 3′,4′-dichloropropion-anilide, *Toxicol. Appl. Pharmacol.*, 23(4):650-659, 1972.

Ambrose, D., Ellender, J.H., Lees, E.B., Sprake, C.H.S., and Townsend, R. Thermodynamic properties of organic compounds. XXXVIII. Vapor pressures of some aliphatic ketones, *J. Chem. Thermodyn.*, 7(5):453-472, 1975.

Ambrose, D. and Sprake, H.S. The vapor pressure of indane, *J. Chem. Thermodyn.*, 8:601-602, 1976.

Amels, P., Elias, H., Götz, U., Steingens, U., and Wannowius, K.J. Kinetic investigation of the stability of peroxonitric acid and its reaction with sulfur (IV) in aqueous solution, in *Heterogeneous and Liquid-Phase Processes*, Warneck, P., Ed. (Berlin: Springer Verlag, 1996), pp. 77-88.

Amidon, G.L., Yalkowsky, S.H., Anik, S.T., and Valvani, S.C. Solubility of non-electrolytes in polar solvents. V. Estimation of the solubility of aliphatic monofunctional compounds in water using a molecular surface area approach, *J. Phys. Chem.*, 79(21):2239-2246, 1975.

Aminabhavi, T.M. and Banerjee, K. Density, viscosity, refractive index, and speed of sound in binary mixtures of 1-chloronaphthalene with benzene, methylbenzene, 1,4-dimethylbenzene, 1,3,5-trimethylbenzene, and methoxybenzene at (298.15, 303.15, and 308.15) K, *J. Chem. Eng. Data*, 44(3):547-552, 1999.

Amoore, J.E. and Buttery, R.G. Partition coefficients and comparative olfactometry, *Chem. Sens. Flavour*, 3(1):57-71, 1978.

Amoore, J.E. and Hautala, E. Odor as an aide to chemical safety: odor thresholds compared with threshold limit values and volatilities for 214 industrial chemicals in air and water dilution, *J. Appl. Toxicol.*, 3(6):272-290, 1983.

Amy, G.L., Narbaitz, R.M., and Cooper, W.J. Removing VOCs from groundwater containing humic substances by means of coupled air stripping and adsorption, *J. Am. Water Works Assoc.*, 79(1):49-54, 1987.

Anbar, M. and Neta, P. A compilation of specific bimolecular rate and hydroxyl radical with inorganic and organic compounds in aqueous solution, *Int. J. Appl. Rad. Isotopes*, 18:493-523, 1967.

Anderson, B.D., Rytting, J.H., and Higuchi, T. Solubility of polar organic solutes in nonaqueous systems: role of specific interactions, *J. Pharm. Sci.*, 69(6):676-680, 1971.

Anderson, M.A. Influence of surfactants on vapor-liquid partitioning, *Environ. Sci. Technol.*, 26(11):2186-2191, 1992.

Anderson, R.L. and DeFoe, D.L. Toxicity and bioaccumulation of endrin and methoxychlor in aquatic invertebrates and fish, *Environ. Pollut. (Series A)*, 22(2):111-121, 1980.

Andersson, H.F., Dahlberg, J.A., and Wettstrom, R. On the formation of phosgene and trichloroacetylchloride in the non-sensitized photooxidation of perchloroethylene in air, *Acta Chem. Scand.*, A29:473-474, 1975.

Andersson, J.T., Häussler, R., and Ballschmiter, K. Chemical degradation of xenobiotics. III. Simulation of the biotic transformation of 2-nitrophenol and 3-nitrophenol, *Chemosphere*, 15(2):149-152, 1986.

Andersson, J.T. and Schräder, W. A method for measuring 1-octanol-water partition coeficients, *Anal. Chem.*, 71(16):3610-3614, 1999.

Ando, M. and Sayato, Y. Studies on vinyl chloride migrating into drinking water from polyvinyl chloride pipe and reaction between vinyl chloride and chlorine, *Water Res.*, 18(3):315-318, 1984.

Andon, R.J.L., Biddiscombe, D.P., Cox, F.D., Handley, R., Harrop, D., Herington, E.F.G., and Martin, J.F. Thermodynamic properties of organic oxygen compounds. Part 1. Preparation of physical properties of pure phenol, cresols and xylenols, *J. Chem. Soc. (London)*, pp. 5246-5254, 1960.

Andon, R.J.L., Counsell, J.F., Herington, E.F.G., and Martin, J.F. Thermodynamic properties of organic oxygen compounds, *Trans. Faraday Soc.*, 59:830-835, 1963.

Andon, R.J.L., Counsell, J.F., Lees, E.B., Martin, J.F., and Mash, M.J. Thermodynamic properties of organic oxygen compounds. Part 17. Low-temperature heat capacity and entropy of the cresols, *Trans. Faraday Soc.*, 63:1115-1121, 1967.

Andon, R.J.L., Cox, J.D., and Herington, E.F.G. Phase relationships in the pyridine series. Part V. The thermodynamic properties of dilute solutions of pyridine bases in water at 25 °C and 40 °C, *J. Chem. Soc. (London)*, pp. 3188-3196, 1954.

Andreozzi, R., Insola, A., Caprio, V., and D'Amore, M.G. Ozonation of pyridine in aqueous solution: mechanistic and kinetic aspects, *Water Res.*, 25(6):655-659, 1991.

Andrews, L.J. and Keefer, R.M. Cation complexes of compounds containing carbon-carbon double bonds. IV. The argentation of aromatic hydrocarbons, *J. Am. Chem. Soc.*, 71(11):3644-3647, 1949.

Andrews, L.J. and Keefer, R.M. Cation complexes of compounds containing carbon-carbon double bonds. VI. The argentation of substituted benzenes, *J. Am. Chem. Soc.*, 72(7):3113-3116, 1950.

Andrews, L.J. and Keefer, R.M. Cation complexes of compounds containing carbon-carbon double bonds. VII. Further studies on the argentation of substituted benzenes, *J. Am. Chem. Soc.*, 72(11):5034-5037, 1950a.

Andrews, L.J., Lynn, G., and Johnston, J. The heat capacities and heat of crystallization of some isomeric aromatic compounds, *J. Am. Chem. Soc.*, 48:1274-1287, 1926.

Angemar, Y., Rebhun, M., and Horowitz, M. Adsorption, phytotoxicity, and leaching of bromacil in some Israeli soils, *J. Environ. Qual.*, 13(2):321-326, 1984.

Anliker, R. and Moser, P. The limits of bioaccumulation of organic pigments in fish: their relation to the partition coefficient and the solubility in water and octanol, *Ecotoxicol. Environ. Saf.*, 13(1):43-52, 1987.

Antelo, J.M., Arce, F., Barbadillo, F., Casado, J., and Varela, A. Kinetics and mechanism of ethanolamine chlorination, *Environ. Sci. Technol.*, 15(8):912-917, 1981.

API. Laboratory Study on Solubilities of Petroleum Hydrocarbons in Groundwater, API Publication No. 4395 (Washington, D.C.: American Petroleum Institute, 1985).

Appelman, L.M., ten Berge, W.F., and Reuzel, P.G. Acute inhalation toxicity study of ammonia in rats with variable exposure periods, *Am. Ind. Hyg. Assoc. J.*, 43(9):662-665, 1982.

Appleton, H.T. and Sikka, H.C. Accumulation, elimination, and metabolism of dichlorobenzidine in the bluegill sunfish, *Environ. Sci. Technol.*, 14(1):50-54, 1980.

Aquan-Yuen, M., Mackay, D., and Shiu, W.-Y. Solubility of hexane, phenanthrene, chlorobenzene, and *p*-dichloro-benzene in aqueous electrolyte solutions, *J. Chem. Eng. Data*, 24(1):30-34, 1979.

Aralaguppi, M.I., Jadar, C.V., and Aminabhavi, T.M. Density, refractive index, viscosity, and speed of sound in bi-nary mixtures of cyclohexanone with hexane, heptane, octane, nonane, decane, dodecane, and 2,2,4-trimethylpent-ane, *J. Chem. Eng. Data*, 44(3):435-440, 1999.

Aralaguppi, M.I., Jadar, C.V., and Aminabhavi, T.M. Density, viscosity, refractive index, and speed of sound in binary mixtures of acrylonitrile with methanol, ethanol, propan-1-ol, butan-1-ol, pentan-1-ol, hexan-1-ol, heptan-1-ol, and butan-2-ol, *J. Chem. Eng. Data*, 44(2):216-221, 1999a.

Arcand, Y., Hawari, J., and Guiot, S.R. Solubility of pentachlorophenol in aqueous solutions: the pH effect, *Water Res.*, 29(1):131-136, 1995.

Archer, T.E. Malathion residues on ladino clover seed screens exposed to ultraviolet irradiation, *Bull. Environ. Contam. Toxicol.*, 6(2):142-143, 1971.

Archer, T.E., Nazer, I.K., and Crosby, D.G. Photodecomposition of endosulfan and related products in thin films by ultraviolet light irradiation, *J. Agric. Food Chem.*, 20(5):954-956, 1972.

Archer, T.E., Stokes, J.D., and Bringhurst, R.S. Fate of carbofuran and its metabolites on strawberries in the environment, *J. Agric. Food Chem.*, 25(3):536-541, 1977.

Arey, J., Zielinska, B., Atkinson, R., Winer, A.M., Randahl, T., and Pitts, J.N., Jr. The formation of nitro-PAH from the gas-phase reactions of fluoranthene and pyrene with the OH radical in the presence of NO$_x$, *Atmos. Environ.*, 20(12):2339-2345, 1986.

Arijs, E. and Brasseur, G. Acetonitrile in the stratosphere and implications for positive ion composition, *J. Geophys. Res.*, 91D:4003-4016, 1986.

Arisoy, M. Biodegradation of chlorinated organic compounds by white-rot fungi, *Bull. Environ. Contam. Toxicol.*, 60(6):872-876, 1998.

Arrebola, M.L., Navaro, M.C., Jimenez, J., and Ocana, F.A. Variations in yield and composition of the essential oil of *Satureja obovata*, *Phytochemistry*, 35(1):83, 1994.

Arthur, J.W. and Leonard E.N. Effects of copper on *Gammarus pseudolimnaeus*, *Physa integra* and *Campeloma decisum* in soft water, *J. Fish Res. Board Can.*, 27:1227-1283, 1970.

Arthur, M.F. and Frea, J.I. 2,3,7,8-Tetrachloro-*p*-dioxin: aspects of its important properties and its potential biodegradation in soils, *J. Environ. Qual.*, 18(1):1-11, 1989.

Arts, J.H., Reuzel, P.G., Falke, H.E., and Beems, R.B. Acute and sub-acute inhalation toxicity of germanium metal powder in rats, *Food Chem. Toxicol.*, 28(8):571-579, 1990.

Arts, J.H., Til, H.P., Kuper, C.F., de Neve, R., and Swennen, B. Acute and subacute inhalation toxicity of germanium dioxide powder in rats, *Food Chem. Toxicol.*, 32(11):1037-1046, 1994.

Arunachalam, K.D. and Lakshmanan, M. Microbial uptake and accumulation of (^{14}C-carbofuran) 1,3-dihydro-2,2-dimethyl-7 benzofuranylmethyl carbamate in twenty fungal strains isolated by miniecosystem studies, *Bull. Environ. Contam. Toxicol.*, 41(1):127-134, 1988.

Arunachalam, S., Jeyalakshmi, K., and Aboobucker, S. Toxic and sublethal effects of carbaryl on a freshwater catfish, *Mystus vittatus* (Bloch), *Arch. Environ. Contam. Toxicol.*, 9(3):307-316, 1980.

Ashton, F.M. and Monaco, T.J. *Weed Science: Principles and Practices* (New York: John Wiley & Sons, 1991), 466 p.

Ashworth, R.A., Howe, G.B., Mullins, M.E., and Rogers, T.N. Air-water partitioning coefficients of organics in dilute aqueous solutions, *J. Haz. Mater.*, 18:25-36, 1988.

Aston, J.G., Kennedy, R.M., and Schumann, S.C. The heat capacity and entropy, heats of fusion and vaporization and the pressure of isobutane, *J. Am. Chem. Soc.*, 62:2059-2063, 1940.

Atkins, J.M., Schreoder, J.A., Brower, D.L., and Aposhian, H.V. Evaluation of *Drosophila melanogaster* as an alternative animal for studying heavy metal toxicity, *Toxicologist*, 11(1):39, 1991.

Atkinson, R. Kinetics and mechanisms of the gas-phase reactions of hydroxyl radical with organic compounds under atmospheric conditions, *Chem. Rev.*, 85:69-201, 1985.

Atkinson, R. Structure-activity relationship for the estimation of rate constants for the gas phase reactions of OH radicals with organic compounds, *Int. J. Chem. Kinet.*, 19:799-828, 1987.

Atkinson, R. Estimation of OH radical reaction rate constants and atmospheric lifetimes for polychlorinated biphenyls, dibenzo-*p*-dioxins, and dibenzofurans, *Environ. Sci. Technol.*, 21(3):305-307, 1987a.

Atkinson, R. Kinetics and mechanisms of the gas-phase reactions of the hydroxyl radical with organic compounds, *J. Phys. Chem. Ref. Data*, Monograph 1, pp. 1-246, 1989.

Atkinson, R. Gas-phase tropospheric chemistry of organic compounds: a review, *Atmos. Environ.*, 24A(1):1-41, 1990.

Atkinson, R. Kinetics and mechanisms of the gas-phase reactions of the NO$_3$ radical with organic compounds, *J. Phys. Chem. Data*, 20(3):459-507, 1991.

Atkinson, R., Ashmann, S.M., and Arey, J. Reactions of OH and NO$_3$ radicals with phenol, cresols, and 2-nitrophenol at 296 ± 2 K, *Environ. Sci. Technol.*, 26(7):1397-1403, 1992.

Atkinson, R., Ashmann, S.M., Carter, P.L., and Pitts, J.N., Jr. Effects of ring strain on gas-phase rate constants. 1. Ozone reactions with cycloalkenes, *Int. J. Chem. Kinet.*, 15:721-731, 1983.

Atkinson, R., Ashmann, S.M., Fitz, D.R., Winer, A.M., and Pitts, J.N., Jr. Rate constants for the gas-phase reactions of O$_3$ with selected organics at 296 K, *Int. J. Chem. Kinet.*, 14:13-18, 1982.

Atkinson, R., Ashmann, S.M., and Pitts, J.N., Jr. Rate constants for the gas-phase reactions of the NO$_3$ radicals with a series of organic compounds at 296 ± 2 K, *J. Phys. Chem.*, 92:3454-3457, 1988.

Atkinson, R., Ashmann, S.M., Winer, A.M., and Carter, W.P.L. Rate constants for the gas-phase reactions of NO$_3$ radicals with furan, thiophene, and pyrrole at 295 ± 1 K, *Environ. Sci. Technol.*, 19(1):87-90, 1985.

Atkinson, R. and Carter, W.P.L. Kinetics and mechanisms of gas-phase ozone reaction with organic compounds under atmospheric conditions, *Chem. Rev.*, 84:437-470, 1984.

Atkinson, R. and Carter, W.P.L. Kinetics and mechanisms of gas-phase ozone with organic compounds under atmospheric conditions, *Chem. Rev.*, 85:69-201, 1985.

Atkinson, R., Darnall, K.R., Lloyd, A.C., Winer, A.M., and Pitts, J.N., Jr. Kinetics and mechanisms of reaction of hydroxyl radicals with organic compounds in the gas phase, *Adv. Photochem.*, 11:375-488, 1979.

Atkinson, R., Darnall, K.R., Lloyd, A.C., Winer, A.M., Pitts, J.N., Jr. Kinetics and mechanisms of reaction of hydroxyl radicals with a series of dialkenes, cycloalkenes, and monoterpenes at 295 ± 1 K, *Environ. Sci. Technol.*, 18(5):370-375, 1984a.

Atkinson, R. and Lloyd, A.C. Evaluation of kinetic and mechanistic data for modeling of photochemical smog, *J. Chem. Kinet.*, 13:315-444, 1984.

Atkinson, R., Perry, R.A., and Pitts, J.N., Jr. Rate constants for the reactions of the OH radical with (CH$_3$)$_2$NH, (CH$_3$)$_3$N, and C$_2$H$_5$NH$_2$ over the temperature range 298-426°K, *J. Chem. Phys.*, 68(4):1850-1853, 1978.

Atkinson, R., Plum, C.N., Carter, W.P.L., Winer, A.M., and Pitts, J.N., Jr. Rate constants for the gas-phase reactions of the nitrate radicals with a series of organics in air at 298 ± 1 K, *J. Chem. Phys.*, 88:1210-1215, 1988.

Atkinson, R., Tuazon, E.C., Wallington, T.J., Aschmann, S.M., Arey, J., Winer, A.M., and Pitts, J.N., Jr. Atmospheric chemistry of aniline, *N,N*-dimethylaniline, pyridine, 1,3,5-triazine, and nitrobenzene, *Environ. Sci. Technol.*, 21(1):64-72, 1987.

Atlas, E., Foster, R., and Giam, C.S. Air-sea exchange of high molecular weight organic pollutants: laboratory studies, *Environ. Sci. Technol.*, 16(5):283-286, 1982.

ATSDR. Agency for Toxic Substances and Disease Registry (Atlanta, GA, 1995).

Attaway, H.H., Camper, N.D., and Paynter, M.J.B. Anaerobic microbial degradation of diuron by pond sediment, *Pestic. Biochem. Physiol.*, 17:96-101, 1982.

Attaway, H.H., Paynter, M.J.B., and Camper, N.D. Degradation of selected phenylurea herbicides by anaerobic pond sediment, *J. Environ. Sci. Health*, B17:683-700, 1982a.

Autian, J. Toxicity and health threats of phthalate esters: review of the literature, *Environ. Health Perspect.*, 4:3-26, 1973.

Awasthi, N., Manickam, N., and Kumar, A. Biodegradation of endosulfan by a bacterial coculture, *Bull. Environ. Contam. Toxicol.*, 59(6):928-934, 1997.

Ayanaba, A., Verstraete, W., and Alexander, M. Formation of dimethylnitrosamine, a carcinogen and mutagen, in soils treated with nitrogen compounds, *Soil Sci. Soc. Am. Proc.*, 37:565-568, 1973.

Baarschers, W.H., Elvish, J., and Ryan, S.P. Adsorption of fenitrothion and 3-methyl-4-nitrophenol on soils and sediment, *Bull. Environ. Contam. Toxicol.*, 30(5):621-627, 1983.

Babers, F.H. The solubility of DDT in water determined radiometrically, *J. Am. Chem. Soc.*, 77(17):4666, 1955.

Bachmann, A., Wijnen, W.P., de Bruin, W., Huntjens, J.L.M., Roelofsen, W., and Zehnder, A.J.B. Biodegradation of *alpha*- and *beta*-hexachlorocyclohexane in a soil slurry under different redox conditions, *Appl. Environ. Microbiol.*, 54(1):143-149, 1988.

Baek, N.H. and Jaffé, P.R. The degradation of trichloroethylene in mixed methanogenic cultures, *J. Environ. Qual.*, 18(4):515-518, 1989.

Bahadur, N.P., Shiu, W.-Y., Boocock, D.G.B., and Mackay, D. Temperature dependence of octanol-water partition coefficient for selected chlorobenzenes, *J. Chem. Eng. Data*, 42(4):685-688, 1997.

Bailey, G.W. and White, J.L. Herbicides: a compilation of their physical, chemical, and biological properties, *Residue Rev.*, 10:97-122, 1965.

Bailey, G.W., White, J.L., and Rothberg, T. Adsorption of organic herbicides by montmorillonite: role of pH and chemical character of adsorbate, *Soil Sci. Am. Proc.*, 32:222-234, 1968.

Bailey, H.D., Liu, D.H.W., and Javitz, H.A. Time/toxicity relationships in short-term static, dynamic and plug-flow bioassays, in *Aquatic Toxicology and Hazard Assessment: Eighth Symposium*, ASTM Special Technical Publication 891, Bahner, R.C. and Hansen, D.J., Eds. (Philadelphia, PA: American Society for Testing and Materials, 1985), pp. 193-212.

Baird, R., Carmona, L., and Jenkins, R.L. Behavior of benzidine and other aromatic amines in aerobic wastewater treatment, *J. Water Pollut. Control Fed.*, 49(7):1609-1615, 1977.

Baker, E.G. Origin and migration of oil, *Science (Washington, D.C.)*, 129(3353):871-874, 1959.

Baker, E.G. A hypothesis concerning the accumulation of sediment hydrocarbons to form crude oil, *Geochim. Cosmochim. Acta*, 19:309-317, 1980.

Baker, M.D. and Mayfield, C.I. Microbial and nonbiological decomposition of chlorophenols and phenols in soil, *Water, Air, Soil Pollut.*, 13:411-424, 1980.

Baker, M.D., Mayfield, C.I., and Inniss, W.E. Degradation of chlorophenol in soil, sediment and water at low temperature, *Water Res.*, 14(7):765-771, 1980.

Ballantyne, B. The acute toxicity and irritancy of bis[2-(dimethylamino)ethyl] ether, *Vet. Hum. Toxicol.*, 39(5):290-295, 1997.

Ballantyne, B., Dodd, D.E., Myers, R.C., and Nachreiner, D.J. The acute toxicity and irritancy of 2,4-pentanedione, *Drug. Chem. Toxicol.*, 9(2):133-146, 1986.

Ballantyne, B., Dodd, D.E., Muers, R.C., Nachreiner, D.J., and Pritts, L.M. The acute toxicity of tris(dimethylamino)-silane, *Toxicol. Ind. Health*, 5(1):45-54, 1989.

Ballantyne, B., Dodd, D.E., Pritts, I.M., Nachreiner, D.J., and Fowler, E.H. Acute vapour inhalation toxicity of acrolein and its influence as a trace contaminant in 2-methoxy-3,4-dihydro-2*H*-pyran, *Hum. Toxicol.*, 8(3):229-235, 1989a.

Ballantyne, B. and Leung, H.W. Acute toxicity and primary irritancy of alkylalkanolamines, *Vet. Hum. Toxicol.*, 38(6):422-426, 1996.

Ballschiter, K. and Scholz, C. Mikrobieller Abbau von chlorier Aromaten: VI. Bildung von Dichlorphenolen in mikromolarer Lösung durch *Pseudomonas* sp., *Chemosphere*, 9(7/8):457-467, 1980.

Balson, E.W. Studies in vapour pressure measurement, Part III. - An effusion manometer sensitive to 5 x 10⁻⁶ millimetres of mercury: vapour pressure of DDT and other slightly volatile substances, *Trans. Faraday Soc.*, 43:54-60, 1947.

Bamford, H.A., Baker, J.E., and Poster, D.L. Temperature dependence of Henry's law constants of thirteen polycyclic aromatic hydrocarbons between 4 °C and 31 °C, *Environ. Toxicol. Chem.*, 18(9):1905-1912, 1998.

Banerjee, S. Solubility of organic mixtures in water, *Environ. Sci. Technol.*, 18(8):587-591, 1984.

Banerjee, K., Horng, P.Y., Cheremisinoff, P.N., Sheih, M.S., and Cheng, S.L. Sorption of selected organic pollutants on fly ash in *Proceedings of the 43ʳᵈ Purdue Industrial Waste Conference*, May 1998, pp. 397-408.

Banerjee, P., Piwoni, M.D., and Ebeid, K. Sorption of organic contaminants to a low carbon subsurface core, *Chemosphere*, 14(8):1057-1067, 1985.

Banerjee, S., Howard, P.H., and Lande, S.S. General structure-vapor pressure relationships for organics, *Chemosphere*, 21(10/11):1173-1180, 1990.

Banerjee, S., Howard, P.H., Rosenburg, A.M., Dombrowski, A.E., Sikka, H., and Tullis, D.L. Development of a general kinetic model for biodegradation and its application to chlorophenols and related compounds, *Environ. Sci. Technol.*, 18(6):416-422, 1984.

Banerjee, S., Sikka, H.C., Gray, R., and Kelly, C.M. Photodegradation of 3,3'-dichlorobenzidine, *Environ. Sci. Technol.*, 12(13):1425-1427, 1978.

Banerjee, S., Yalkowsky, S.H., and Valvani, S.C. Water solubility and octanol/water partition coefficients of organics: limitations of the solubility-partition coefficient correlation, *Environ. Sci. Technol.*, 14(10):1227-1229, 1980.

Banerjee, T.K. and Paul, V.I. Estimation of acute toxicity of ammonium sulphate to the fresh-water catfish, *Heteropneustes fossilis*. II. A histopathological analysis of the epidermis, *Biomed. Environ. Sci.*, 6(1):45-58, 1993.

Banwart, W.L., Hassett, J.J., Wood, S.G., and Means, J.C. Sorption of nitrogen-heterocyclic compounds by soils and sediments, *Soil Sci.*, 133(1):42-47, 1982.

Barbash, J.E. and Reinhard, M. Abiotic dehalogenation of 1,2-dichloroethane and 1,2-dibromoethane in aqueous solution containing hydrogen sulfide, *Environ. Sci. Technol.*, 23(11):1349-1358, 1989.

Barbee, S.J., Stone, J.J., and Hilaski, R.J. Acute inhalation toxicology of oxalyl chloride, *Am. Ind. Hyg. Assoc. J.*, 56(1):74-76, 1995.

Barbee, S.J., Thackara, J.W., and Rinehart, W.E. Acute inhalation toxicology of nitrogen trichloride, *Am. Ind. Hyg. Assoc. J.*, 44(2):145-146, 1983.

Barbeni, M., Pramauro, E., and Pelizzetti, E. Photodegradation of pentachlorophenol catalyzed by semiconductor particles, *Chemosphere*, 14(2):195-208, 1985.

1260 References

Barbeni, M., Pramauro, E., Pelizzetti, E., Vincenti, M., Borgarello, E., and Serpone, N. Sunlight photodegradation of 2,4,5-trichlorophenoxyacetic acid and 2,4,5-trichlorophenol on TiO_2. Identification of intermediates and degradation pathway, *Chemosphere*, 16(6):1165-1179, 1987.

Barnes, D.W., Sanders, V.M., White, K.L., Shopp, G.M., and Munson, A.E. Toxicology of *trans*-1,2-dichloroethylene in the mouse, *Drug Chem. Toxicol.*, 8(5):373-392, 1985.

Barber, L.B. Dichlorobenzene in ground water: evidence for long-term persistence, *Ground Water*, 26(6):696-702, 1988.

Barber, L.B., Thurman, E.M., and Runnells, D.D. Geochemical heterogeneity in a sand and gravel aquifer: effect of sediment mineralogy and particle size on the sorption of chlorobenzene, *J. Contam. Hydrol.*, 9:35-54, 1992.

Barham, H.N. and Clark, L.W. The decomposition of formic acid at low temperatures, *J. Am. Chem. Soc.*, 73(10):4638-4640, 1951.

Barlas, H. and Parlar, H. Reactions of naphthaline with nitrogen dioxide in UV-light, *Chemosphere*, 16(2/3):519-520, 1987.

Barnes, C.J., Lavy, T.L., and Talbert, R.E. Leaching, dissipation, and efficacy of metolachlor applied by chemigation or conventional methods, *J. Environ. Qual.*, 21(2):232-236, 1992.

Barone, G., Crescenzi, V., Liquori, A.M., and Quadrifoglio, F. Solubilization of polycyclic aromatic hydrocarbons in poly(methacrylic acid) aqueous solutions, *J. Phys. Chem.*, 71(7):2341-2345, 1967.

Barone, G., Crescenzi, V., Pispisa, B., and Quadrifoglio, F. Hydrophobic interactions in polyelectrolyte solutions. II. Solubility of some C_3-C_6 alkanes in poly(methacrylic acid) aqueous solutions, *J. Macromol. Chem.*, 1(4):761-771, 1966.

Barrio-Lage, G.A., Parsons, F.Z., Narbaitz, R.M., Lorenzo, P.A., and Archer, H.E. Enhanced anaerobic biodegradation of vinyl chloride in ground water, *Environ. Toxicol. Chem.*, 9(1):403-415, 1990.

Barrio-Lage, G.A., Parsons, F.Z., Nassar, R.S., and Lorenzo, P.A. Sequential dehalogenation of chlorinated ethenes, *Environ. Sci. Technol.*, 20(1):96-99, 1986.

Barrio-Lage, G.A., Parsons, F.Z., Nassar, R.S., and Lorenzo, P.A. Biotransformation of trichloroethylene in a variety of subsurface materials, *Environ. Toxicol. Chem.*, 6(8):571-578, 1987.

Barrows, M.E., Petrocelli, S.R., and Macek, K.J. Bioconcentration and elimination of selected water pollutants by bluegill sunfish (*Lepomis macrochirus*), in *Dynamic, Exposure, Hazard Assessment Toxic Chemicals*, Haque, R., Ed. (Ann Arbor, MI: Ann Arbor Science Publishers, 1980).

Bartlett, W.P. and Margerum, D.W. Temperature dependencies of the Henry's law constant and the aqueous phase dissociation constant of bromine chloride, *Environ. Sci. Technol.*, 33(19):3410-3414, 1999.

Bartley, T. and Gangstad, E.O. Environmental aspects of aquatic plant control, *J. Irrig. Drain. Div.-ASCE*, 100:231-244, 1974.

Baser, K.H.C., Ozek, T., Akgul, A., and Tumen, G. Composition of the essential oil of *Nepeta racemosa* Lam, *J. Ess. Oil Res.*, 5:215-217, 1993.

Baser, K.H.C., Ozek, T., Kurkuoglu, M., and Tumen, G. Composition of the essential oil of *Origanum sipyleum* of Turkish origin, *J. Ess. Oil Res.*, 4:139-142, 1992.

Baser, K.H.C., Tumen, G., and Sezik, E. The essential oil of *Origanum minutiflorum*, *J. Ess. Oil Res.*, 3:445-446, 1991.

Battersby, N.S. and Wilson, V. Survey of the anaerobic biodegradation potential of organic compounds in digesting sludge, *Appl. Environ. Microbiol.*, 55(2):433-439, 1989.

Baulch, D.L., Cox, R.A., Hampson, R.F., Jr., Kerr, J.A., Troe, J., and Watson, R.T. Evaluated kinetic and photochemical data for atmospheric chemistry (Supplement II), *J. Phys. Chem. Ref. Data*, 13:1255-1380, 1984.

Baur, J.R. and Bovey, R.W. Ultraviolet and volatility loss of herbicides, *Arch. Environ. Contam. Toxicol.*, 2(3):275-288, 1974.

Baxter, R.M. Reductive dechlorination of certain chlorinated organic compounds by reduced hematin compared with their behaviour in the environment, *Chemosphere*, 21(4/5):451-458, 1990.

Baxter, R.M. and Sutherland, D.A. Biochemical and photochemical processes in the degradation of chlorinated biphenyls, *Environ. Sci. Technol.*, 18(8):608-610, 1984.

Bayona, J.M., Fernandez, P., Porte, C., Tolosa, I., Valls, M., and Albaiges, J. Partitioning of urban wastewater organic microcontaminants among coastal compartments, *Chemosphere*, 23(3):313-326, 1991.

Beard, J.E. and Ware, G.W. Fate of endosulfan on plants and glass, *J. Agric. Food Chem.*, 17(2):216-220, 1969.

Beaumont, R.H., Clegg, B., Gee, G., Herbert, J.B.M., Marks, D.J., Roberts, R.C., and Sims, D. Heat capacities of propylene oxide and of some polymers of ethylene and propylene oxide, *Polymer*, 7:401-416, 1966.

Beck, J. and Hansen, K.E. The degradation of quintozene, pentachlorobenzene, hexachlorobenzene and pentachloroaniline in soil, *Pestic. Sci.*, 5(1):41-48, 1974.

Becker, K.H., Kleffmann, J., Kurtenbach, R., and Wiesen, P. Solubility of nitrous acid (HONO) in sulfuric acid solutions, *J. Phys. Chem.*, 100:14984-14990, 1996.

Bedard, D.L., Wagner, R.E., Brennan, M.J., Haberl, M.L., and Brown, J.F., Jr. Extensive degradation of Aroclors and environmentally transformed polychlorinated biphenyls by *Alcaligenes eutrophus* H850, *Appl. Environ. Microbiol.*, 53(5):1094-1102, 1987.

Beeman, R.W. and Matsumura, F. Metabolism of *cis*- and *trans*-chlordane by a soil microorganism, *J. Agric. Food Chem.*, 29(1):84-89, 1981.

Behymer, T.D. and Hites, R.A. Photolysis of polycyclic aromatic hydrocarbons adsorbed on simulated atmospheric particulates, *Environ. Sci. Technol.*, 19(10):1004-1006, 1985.

Beland, F.A., Farwell, S.O., Robocker, A.E., and Geer, R.D. Electrochemical reduction and anaerobic degradation of lindane, *J. Agric. Food Chem.*, 24(4):753-756, 1976.

Belay, N. and Daniels, L. Production of ethane, ethylene, and acetylene from halogenated hydrocarbons by methanogenic bacteria, *Appl. Environ. Microbiol.*, 53(7):1604-1610, 1987.

Belfort, G., Altshuler, G.L., Thallam, K.K., Feerick, C.P., Jr., and Woodfield, K.L. Office of Research and Development, U.S. EPA Report 600/2-83-047, 1983.

Bell, G.H. The action of monocarboxylic acids on *Candida tropicalis* growing on hydrocarbon substrates, *Antonie van Leeuwenhoek*, 37:385-400, 1971.

Bell, G.R. Photochemical degradation of 2,4-dichlorophenoxyacetic acid and structurally related compounds, *Botan. Gaz.*, 118:133-136, 1956.

Bell, J.P. and Tsezos, M. Removal of hazardous organic pollutants by biomass adsorption, *J. Water Pollut. Control Fed.*, 59(4):191-198, 1987.

Bell, R.P. and McDougall, A.O. Hydration equilibria of some aldehydes and ketones, *Trans. Faraday Soc.*, 56:1281-1285, 1960.

Bellin, C.A., O'Connor, G.A., and Jin, Y. Sorption and degradation of pentachlorophenol in sludge-amended soils, *J. Environ. Qual.*, 19(3):603-608, 1990.

Beltrame, P., Beltrame, P.L., and Cartini, P. Influence of feed concentration on the kinetics of biodegradation of phenol in a continuous stirred reactor, *Water Res.*, 18(4):403-407, 1984.

Beltrame, P., Beltrame, P.L., and Cartini, P. Lipophilic and electrophilic factors in the inhibiting action of substituted phenols on the biodegradation of phenol, *Chemosphere*, 19(10/11):1623-1627, 1989.

Beltran, F.J., Encinar, J.M., and Garcia-Araya, J.F. Ozonation of *o*-cresol in aqueous solutions, *Water Res.*, 24(11):1309-1316, 1990.

Bender, M.E. The toxicity of the hydrolysis and breakdown products of malathion to the fathead minnow (*Pimephales promelas*, Rafinesque), *Water Res.*, 3(8):571-582, 1969.

Benezet, H.I. and Matsumura, F. Isomerization of γ-BHC to a α-BHC in the environment, *Nature (London)*, 243(5408):480-481, 1973.

Benkelberg, H.J., Hamm, S., and Warneck, P. Henry's law coefficients for aqueous solutions of acetone, acetalde-hyde and acetonitrile, and equilibrium constants for the addition compounds of acetone and acetaldehyde with bi-sulfite, *J. Atmos. Chem.*, 20(1):17-34, 1995.

Ben-Naim, A. and Wilf, J. Solubilities and hydrophobic interactions in aqueous solutions of monoalkylbenzene molecules, *J. Phys. Chem.*, 84(6):583-586, 1980.

Bennett, D. and Canady, W.J. Thermodynamics of solution of naphthalene in various water-ethanol mixtures, *J. Am. Chem. Soc.*, 106(4):910-915, 1984.

Bennett, G.M. and Philip, W.G. The influence of structure on the solubilities of ethers. Part I. Aliphatic ethers, *J. Chem. Soc. (London)*, 131:1930-1937, 1928.

Benoit, P., Barriuso, E., Vidon, Ph., and Réal, B. Isoproturon sorption and degradation in a soil from grassed buffer strip, *J. Environ. Qual.*, 28(1):121-129, 1999.

Benson, J.M., Brooks, A.L., Cheng, Y.S., Henderson, T.R., and White, J.E. Environmental transformation of 1-nitropyrene on glass surfaces, *Atmos. Environ.*, 19(7):1169-1174, 1985.

Benson, W.R. Photolysis of solid and dissolved dieldrin, *J. Agric. Food Chem.*, 19(1):66-72, 1971.

Benson, W.R., Lombado, P., Egry, I.J., Ross, R.D., Jr., Barron, R.P., Mastbrook, D.W., and Hansen, E.A. Chlordane photoalteration products: their preparation and identification, *J. Agric. Food Chem.*, 19(5):857-862, 1971.

Benter, T. and Schindler, R.N. Absolute rate coefficients for the reaction of NO_3 radicals with simple dienes, *Chem. Phys. Lett.*, 145:67-70, 1988.

Berg, G.L., Ed. *The Farm Book* (Willoughby, OH: Meister Publishing, 1983), 440 p.

Bernardini, G., Vismara, C., Boracchi, P., and Camatini, M. Lethality, teratogenicity and growth inhibition of heptanol in *Xenopus* assayed by a modified frog embryo teratogenesis assay-*Xenopus* (FETAX) procedure, *Sci. Total Environ.*, 151(1):1-8, 1994.

Berry, D.F., Madsen, E.L., and Bollag, J.-M. Conversion of indole to oxindole under methanogenic conditions, *Appl. Environ. Microbiol.*, 53(1):180-182, 1987.

Berteau, P.E. and Deen, W.A. A comparison of oral and inhalation toxicities of four insecticides to mice and rats, *Bull. Environ. Contam. Toxicol.*, 19(1):113-120, 1978.

Berthod, A., Han, Y.I., and Armstrong, D.W. Centrifugal partition chromatography. V. Octanol-water partition coefficients, direct and indirect determination, *J. Liq. Chromatogr.*, 11:1441-1456, 1988.

Bertoli, D., Borelli, G., and Carazzone, M. Toxicological evaluations of the benzodiazepine doxefazepam, *Arzneim. Forsch.*, 39(4):480-484, 1989.

Betterton, E.A. The partitioning of ketones between the gas and aqueous phases, *Atmos. Environ.*, 25A:1473-1477, 1991.

Betterton, E.A. and Hoffmann, M.R. Henry's law constants of some environmentally important aldehydes, *Environ. Sci. Technol.*, 22(12):1415-1418, 1988.

Betterton, E.A. and Robinson, J.L. Henry's law coefficient of hydrazoic acid, *J. Air & Waste Manage. Assoc.*, 47:1216-1219, 1991.

Beurskens, J.E.M., Dekker, C.G.C., van der Heuvel, H., Swart, M., de Wolf, and Dolfing, J. Dehalogenation of chlorinated benzenes by an anaerobic microbial consortium that selectively mediates the thermodynamic and most favorable reactions, *Environ. Sci. Technol.*, 28(4):701-706, 1993.

Bevenue, A. and Beckman, H. Pentachlorophenol: a discussion of its properties and its occurrence as a residue in human and animal tissues, *Residue Rev.*, 19:83-134, 1967.

Beyerle-Pfnür, R. and Lay, J.-P. Adsorption and desorption of 3,4-dichloroaniline on soil, *Chemosphere*, 21(9):1087-1094, 1990.

Bhatia, S.R. and Sandler, S.I. Temperature dependence of infinite dilution activity coefficients in octanol and octanol/water partition coefficients of some volatile halogenated organic compounds, *J. Chem. Eng. Data*, 40(6):1196-1198, 1995.

Bhavnagary, H.M. and Jayaram, M. Determination of water solubilities of lindane and dieldrin at different temperatures, *Bull. Grain Technol.*, 12(2):95-99, 1974.

Bidleman, T.F. Estimation of vapor pressures for nonpolar organic compounds by capillary gas chromatography, *Anal. Chem.*, 56(13):2490-2496, 1984.

Bielaszczyk, E., Czerwińska, E., Janko, Z., Kotarski, A., Kowalik, R., Kwiatkowski, M., and Żoledziowska, J. Aerobic reduction of some nitrochloro substituted benzene compounds by microorganisms, *Acta Microbiol. Pollut.*, 16:243-248, 1967.

Biggar, J.W., Nielson, D.R., and Tillotson, W.R. Movement of DBCP in laboratory soil columns and field soils to groundwater, *Environ. Geol.*, 5(3):127-131, 1984.

Biggar, J.W. and Riggs, I.R. Apparent solubility of organochlorine insecticides in water at various temperatures, *Hilgardia*, 42(10):383-391, 1974.

Bilkert, J.N. and Rao, P.S.C. Sorption and leaching of three nonfumigant nematicides in soils, *J. Environ. Sci. Health*, B20(1):1-26, 1985.

Billington, J.W., Huang, G.-L., Szeto, F., Shiu, W.Y., and Mackay, D. Preparation of aqueous solutions of sparingly soluble organic substances: I. Single component systems, *Environ. Toxicol. Chem.*, 7:117-124, 1988.

Birge, W.J., Black, J.A., Hudson, J.E., and Bruser, D.M. Embyro-larval toxicity tests with organic compounds, in *Aquatic Toxicology, ASTM STP 667*, Marking, L.L. and Kimberle, R.A. (Philadelphia, PA: American Society for Testing and Materials, 1979), pp. 131-147.

Bjørseth, A. *Handbook of Polycyclic Aromatic Hydrocarbons* (New York: Marcel Dekker, 1983).

Black, J.A., Birge, W.J., Westerman, A.G., and Francis, P.C. Comparative aquatic toxicology of aromatic hydrocarbons, *Fund. Appl. Toxicol.*, 3:353-358, 1983.

Blackman, G.E., Parke, M.H., and Garton, G. The physiological activity of substituted phenols. I. Relationships between chemical structure and physiological activity, *Arch. Biochem. Biophys.*, 54(1):55-71, 1955.

Blackmann, R.A.A. Toxicity of oil-sinking agents, *Mar. Pollut. Bull.*, 5(8):116-118, 1974.

Blais, C. and Hayduk, W. Solubility of butane and isobutane in butanol, chlorobenzene, and carbon tetrachloride, *J. Chem. Eng. Data*, 28(2):181-183, 1983.

Blatchley, E.R., III, Johnson, R.W., Alleman, J.E., and McCoy, W.F. Effective Henry's law constants for free chlorine and free bromine, *Water Res.*, 26(1):99-106, 1992.

Blazquez, C.H., Vidyarthi, A.D., Sheehan, T.D., Bennett, M.J., and McGrew, G.T. Effect of pinolene (β-pinene polymer) on carbaryl foliar residues, *J. Agric. Food Chem.*, 18(4):681-684, 1970.

Blockwell, S.J., Maund, S.J., and Pascoe, D. The acute toxicity of lindane to *Hyalella azteca* and the development of a sublethal bioassay based on precopulatory guarding behavior, *Arch. Environ. Contam. Toxicol.*, 35(3):432-440, 1998.

Blum, D.J. W., Suffet, I.H., and Duguet, J.P. Quantitative structure-activity relationship using molecularconnectivity for the activated carbon adsorption of organic chemicals in water, *Water Res.*, 28:687-699, 1994.

Blume, C.W., Hisatsune, I.C., and Heicklen, J. Gas phase ozonolysis of *cis*- and *trans*-dichloroethylene, *Int. J. Chem. Kinet.*, 8:235-258, 1976.

Bobé, A., Coste, C.M., and Cooper, J.-F. Factors influencing the adsorption of fipronil on soils, *J. Agric. Food Chem.*, 45(12):4861-4865, 1997.

Bobra, A., Mackay, D., and Shiu, W.-Y. Distribution of hydrocarbons among oil, water, and vapor phases during oil dispersant toxicity tests, *Bull. Environ. Contam. Toxicol.*, 23(4/5):558-565, 1979.

Bobra, A., Shiu, W.-Y., and Mackay, D. A predictive correlation for the acute toxicity of hydrocarbons and chlorinated hydrocarbons to the water flea, *Chemosphere*, 12:1121-1129, 1983.

Boddington, J.M. Priority substances list assessment report No. 1: Polychlorinated dibenzodioxins and polychlorinated dibenzofurans, Environment Canada, 56 p., 1990.

Boehm, P.D. and Quinn, J.G. Solubilization of hydrocarbons by the dissolved organic matter in sea water, *Geochim. Cosmochim. Acta*, 37(11):2459-2477, 1973.

Boehncke, A., Martin, K., Müller, M.G., and Cammenga, H.K. The vapor pressure of lindane (γ-1,2,3,4,5,6-Hexachlorocyclohexane) - A comparison of Knudsen effusion measurements with data from other techniques, *J. Chem. Eng. Data*, 41(3):543-545, 1996.

Boersma, D.C., Ellenton, J.A., and Yagminas, A. Investigation of the hepatic mixed-function oxidase system in herring gull embryos in relation to environmental contaminants, *Environ. Toxicol. Chem.*, 5:309-318, 1986.

Boese, B.L., Winsor, M., Lee, H., II, Specht, D.T., and Rukavina, K.C. Depuration kinetics of hexachlorobenzene in the clam (*Macoma nasuta*), *Comp. Biochem. Physiol. C*, 96(2):327-331, 1990.

Boesten, J.J.T.I., van der Pas, L.J.T., Leistra, M., Smelt, J.H., and Houx, N.W.H. Transformation of ^{14}C-labelled 1,2-dichloropropane in water-saturated subsoil materials, *Chemosphere*, 24(8):993-1011, 1992.

Boethling, R.S. and Alexander, M. Effect of concentration of organic chemicals on their biodegradation by natural microbial communities, *Appl. Environ. Microbiol.*, 37(6):1211-1216, 1979.

Boggs, J.E. and Buck, A.E., Jr. The solubility of some chloromethanes in water, J. Phys. Chem., 62:1459-1461, 1958.

Bohon, R.L. and Claussen, W.F. The solubility of aromatic hydrocarbons in water, *J. Am. Chem. Soc.*, 73(4):1571-1578, 1951.

Bollag, J.-M. and Liu, S.-Y. Hydroxylations of carbaryl by soil fungi, *Nature (London)*, 236(5343):177-178, 1972.

Bone, R., Cullis, P., and Wolfenden, R. Solvent effects on equilibria of addition of nucleophiles to acetaldehyde and the hydrophilic character of diols, *J. Am. Chem. Soc.*, 105:1339-1343, 1983.

Bonnet, P., Morele, Y., Raoult, G., Zissu, D., and Gradiski, D. Determination of the medial lethal concentration of the main aromatic hydrocarbons in the rat, *Arch. Mal. Prof.*, 43(4):461-465, 1982.

Bonnet, P., Raoult, G., and Gradiski, D. Lethal concentration of 50 main aromatic hydrocarbons, *Arch. Mal. Prof.*, 40(8-9):805-810, 1979.

Bonoli, L. and Witherspoon, P.A. Diffusion of aromatic and cycloparaffin hydrocarbons in water from 2 to 60 °C, *J. Phys. Chem.*, 72:2532-2534, 1968.

Booth, H.S. and Everson, H.E. Hydrotrophic solubilities. I. Solubilities in forty percent xylene-sulfonate, *Ind. Eng. Chem.*, 40(8):1491-1493, 1948.

Booth, H.S. and Everson, H.E. Hydrotrophic solubilities. I. Solubilities in aqueous sodium arylsulfonate solutions, *Ind. Eng. Chem.*, 41(11):2627-2628, 1949.

Bopp, R.F., Simpson, H.J., Olsen, C.R., Trier, R.M., and Kostyk, N. Chlorinated hydrocarbons and radionuclide chronologies in sediments of the Hudson River and estuary, New York, *Environ. Sci. Technol.*, 16(10):666-676, 1982.

Borello, R., Minero, C., Pramauro, E., Pelizzetti, E., Serpone, N., and Hidaka, H. Photocatalytic degradation of DDT mediated in aqueous semiconductor slurries by simulated sunlight, *Environ. Toxicol. Chem.*, 8(11):997-1002, 1989.

Borsetti, A.P. and Roach, J.A. Identification of kepone alteration products in soil and mullet, *Bull. Environ. Contam. Toxicol.*, 20(2):241-247, 1978.

Bossert, I., Kachel, W.M., and Bartha, R. Fate of hydrocarbons during oil sludge disposal in soil, *Appl. Environ. Microbiol.*, 47(4):763-767, 1984.

Bossert, I.D. and Bartha, R. Structure-biodegradability relationships of polycyclic aromatic-hydrocarbon systems in soil, *Bull. Environ. Contam. Toxicol.*, 37(4):490-495, 1986.

Boto, P., Moreau, G., and Vandycke, C. Molecular structures: Perception, autocorrelation descriptor and SAR studies. System of atomic contributions for the calculation of *n*-octanol/water partition coefficients, *Eur. J. Med. Chem.-Chim. Ther.*, 19:71-78, 1984.

Botré, C., Memoli, A., and Alhaique, F. TCDD solubilization and photodecomposition in aqueous solutions, *Environ. Sci. Technol.*, 12(3):335-336, 1978.

Boublik, T., Fried, V., and Hala, E. *The Vapor Pressures of Pure Substances*, 2nd ed. (Amsterdam: Elsevier, 1973).

Boule, P., Guyon, C., and Lemaire, J. Photochemistry and environment. IV. Photochemical behaviour of monochlorophenols in dilute aqueous solution, *Chemosphere*, 11(12):1179-1188, 1982.

Bouwer, E.J. and McCarty, P.L. Transformations of 1- and 2-carbon halogenated aliphatic organic compounds under methanogenic conditions, *Appl. Environ. Microbiol.*, 45(4):1286-1294, 1983.

Bove, J.L. and Arrigo, J. Formation of polycyclic aromatic hydrocarbon via gas phase benzyne, *Chemosphere*, 14(1):99-101, 1985.

Bove, J.L. and Dalven, P. Pyrolysis of phthalic acid esters: their fate, *Sci. Total Environ.*, 36(1):313-318, 1984.

Bowden, D.J., Clegg, S.L., and Brimblecombe, P. The Henry's law constant of trifluoroacetic acid and its partitioning into liquid water in the atmosphere, *Chemosphere*, 32(2):405-420, 1996.

Bowden, D.J., Clegg, S.L., and Brimblecombe, P. The Henry's law constant of the haloacetic acids, *J. Atmos. Chem.*, 29:85-107, 1998.

Bowman, B.T. and Sans, W.W. Adsorption of parathion, fenitrothion, methyl parathion, aminoparathion and paraoxon by Na^+, Ca^{2+}, and Fe^{3+} montmorillonite suspensions, *Soil Sci. Soc. Am. J.*, 41:514-519, 1977.

Bowman, B.T. and Sans, W.W. The aqueous solubility of twenty-seven insecticides and related compounds, *J. Environ. Sci. Health*, B14(6):625-634, 1979.

Bowman, B.T. and Sans, W.W. Adsorption, desorption, soil mobility, aqueous persistence and octanol-water partitioning coefficients of terbufos, terbufos sulfoxide and terbufos sulfone, *J. Environ. Sci. Health*, B17(5):447-462, 1982.

Bowman, B.T. and Sans, W.W. Further water solubility determininations of insecticidal compounds, *J. Environ. Sci. Health*, B18(2):221-227, 1983.

Bowman, B.T. and Sans, W.W. Determination of octanol-water partitioning coefficients (K_{ow}) of 61 organophosphorus and carbamate insecticides and their relationship to respective water solubility (S) values, *J. Environ. Sci. Health*, B18(6):667-683, 1983a.

Bowman, B.T. and Sans, W.W. Effect of temperature on the water solubility of insecticides, *J. Environ. Sci. Health*, B20(6):625-631, 1985.

Bowman, F.J., Borcelleca, J.F., and Munson, A.E. The toxicity of some halomethanes in mice, *Toxicol. Appl. Pharmacol.*, 44:213-215, 1978.

Bowman, M.C., Acree, F., Jr., and Corbett, M.K. Solubility of carbon-14 DDT in water, *J. Agric. Food Chem.*, 8(5):406-408, 1960.

Bowman, M.C., King, J.R., and Holder, C.R. Benzidine and congeners: analytical chemical properties and trace analysis in five substrates, *Int. J. Environ. Anal. Chem.*, 4:205-223, 1976.

Bowman, M.C., Schecter, M.S., and Carter, R.L. Behavior of chlorinated insecticides in a broad spectrum of soil types, *J. Agric. Food Chem.*, 13:360-365, 1965.

Bowmer, K.H. and Higgins, M.L. Some aspects of the persistence and fate of acrolein herbicide in water, *Arch. Environ. Contam. Toxicol.*, 5(1):87-96, 1976.

Bowmer, K.H., Lang, A.R.G., Higgins, M.L., Pillay, A.R., and Tchan, Y.T. Loss of acrolein from water by volatilization and degradation, *Weed Res.*, 14:325-328, 1974.

Boyd, S.A. Adsorption of substituted phenols by soil, *Soil Sci.*, 134(5):337-343, 1982.

Boyd, S.A., Kao, C.W., and Suflita, J.M. Fate of 3,3′-dichlorobenzidine in soil: Persistence and binding, *Environ. Toxicol. Chem.*, 3:201-208, 1984.

Boyd, S.A. and King, R. Adsorption of labile organic compounds by soil, *Soil Sci.*, 137(2):115-119, 1984.

Boyle, T.P., Robinson-Wilson, E.F., Petty, J.D., and Weber, W. Degradation of pentachlorophenol in simulated lenthic environment, *Bull. Environ. Contam. Toxicol.*, 24(2):177-184, 1980.

Bradbury, S.P. and Coats, J.R. Toxicity of fenvalerate to bobwhite quail (*Colinus virginianus*) including brain and liver residues associated with mortality, *J. Toxicol. Environ. Health*, 10(2):307-319, 1982.

Bradley, P.M. and Chapelle, F.H. Role for acetotrophic methanogens in methanogenic biodegradation of vinyl chloride, *Environ. Sci. Technol.*, 3319):3473-3476, 1999.

Bradley, R.S. and Cleasby, T.G. The vapour pressure and lattice energy of some aromatic ring compounds, *J. Chem. Soc. (London)*, pp. 1690-1692, 1953.

Brady, A.P. and Huff, H. The vapor pressure of benzene over aqueous detergent solutions, *J. Phys. Chem.*, 62(6).644-649, 1958.

Braker, W. and Mossman, A.L. *Matheson Gas Data Book* (East Rutherford, NJ: Matheson Gas Products, 1971), 574 p.

Brannon, J.M., Pennington, J.C., McFarland, V.A., and Hayes, C. The effects of sediment contact time on K_{oc} of nonpolar organic contaminants, *Chemosphere*, 31(6):3465-3473, 1995.

Branson, D.R. A new capacitor fluid - A case study in product stewardship, in *Aquatic Toxicology and Hazard Evaluation*, ASTM Special Technical Publication 657, Mayer, F.L. and Hamelink, J.L., Eds. (Philadelphia, PA: American Society for Testing and Materials, 1977), pp. 44-61.

Branson, D.R. Predicting the fate of chemicals in the aquatic environment from laboratory data, in *Estimating the Hazard of Chemical Substances to Aquatic Life*, Cairns, J., Jr., Dickson, K.L., and A.W. Maki, Eds. (Philadelphia, PA: American Society for Testing and Materials, 1978), pp. 55-70.

Branson, D.R., Takahasi, I.T., Parker, W.M., and Blau, G.E. Bioconcentration kinetics of 2,3,7,8-tetrachloro-*p*-dioxin in rainbow trout, *Environ. Toxicol. Chem.*, 4:779-788, 1985.

Brett, C.H. and Bowery, T.G. Insecticide residues on vegetables, *J. Econ. Entomol.*, 51:818-821, 1958.

Bridié, A.L., Wolff, C.J.M., Winter, M. BOD and COD of some petrochemicals, *Water Res.*, 13(8):627-630, 1979.

Briggs, G.G. Theoretical and experimental relationships between soil adsorption, octanol-water partition coefficients, water solubilities, bioconcentration factors, and the parachor, *J. Agric. Food Chem.*, 29(5):1050-1059, 1981.

Bright, N.F.H. The vapor pressure of diphenyl, dibenzyl and diphenylethane, *J. Chem. Soc. (London)*, pp. 624-626, 1951.

Bright, N.F.H., Cuthill, J.C., and Woodbury, N.H. The vapour pressure of parathion and related compounds, *J. Sci. Food Agric.*, 1:344-348, 1950.

Briner, E. and Perrottet, E. Détermination des solubilités de l'ozone dans l'eau et dans une solution aqueuse de chlorure de sodium: calcul des solubilités de l'ozone atmosphérique dans les eaux, *Helv. Chim. Acta*, 22:397-303, 1939.

Bringmann, G. and Kühn, G. Comparison of the toxicity thresholds of water pollutants to bacteria, algae and protozoa in the cell multiplication inhibition test, *Water Res.*, 14(3):231-241, 1980.

Broadhurst, M.G. Use and replaceability of polychlorinated biphenyls, *Environ. Health Perspect.*, (October 1972), pp. 81-102.

Brock, W.J., Shin-Ya, S., Rusch, G.M., Hardy, C.J., and Trochimowicz, H.J. Inhalation toxicity and genotoxicity of hydrochlorofluorocarbon HCFC-225ca and HCFC-225cb, *J. Appl. Toxicol.*, 19(2):101-112, 1999.

Brock, W.J., Trochimowicz, H.J., Farr, C.H., Millischer, R.J., and Rusch, G.M. Acute, subchronic, and developmental toxicity and genotoxicity of 1,1,1-trifluoroethane (HFH-143a), *Fundam. Appl. Toxicol.*, 31(2):200-209, 1996.

Brock, W.J., Trochimowicz, H.J., Millischer, R.J., Farr, C., Kawano, T., and Rusch, G.M. Acute and subchronic toxicity of 1,1-dichloro-1-fluoroethane (HCFC-141b), *Food Chem. Toxicol.*, 33(6):483-490, 1995.

Broholm, K. and Feenstra, S. Laboratory measurements of the aqueous solubility of mixtures of chlorinated solvents, *Environ. Toxicol. Chem.*, 14(1):9-15, 1995.

Broholm, M.M., Broholm, K., and Arvin, E. Sorption of heterocyclic compounds on natural clayey till, *J. Contam. Hydrol.*, 39:183-200, 1999.

Broman, D., Naf, C., Rolff, C., and Zebuhr, Y. Occurrence and dynamics of polychlorinated dibenzo-*p*-dioxins and dibenzofurans and polycyclic aromatic hydrocarbons in the mixed surface layer of remote coastal and offshore waters of the Baltic, *Environ. Sci. Technol.*, 25(11):1850-1864, 1991.

Brooke, D.N., Dobbs, A.J., and Williams, N. Octanol:water partition coefficients (P): measurement, estimation, and interpretation, particularly for chemicals with P >10^5, *Ecotoxicol. Environ. Saf.*, 11(3):251-260, 1986.

Brooke, D., Nielsen, I., de Bruijn, J., and Hermens, J. An interlaboratory evaluation of the stir-flask method for the determination of octanol-water partition coefficients (log P_{ow}), *Chemosphere*, 21(1/2):119-133, 1990.

Brooker, P.J. and Ellison, M. The determination of the water solubility of organic compounds by a rapid turbidimetric method, *Chem. Ind.*, (October 1974), pp. 785-787.

Brookman, G.T., Flanagan, M., and Kebe, J.O. Literature survey: hydrocarbon solubilities and attenuation mechanisms, API Publication 4414, (Washington, D.C.: American Petroleum Institute, 1985), 101 p.

Brooks, G.T. *Chlorinated Insecticides, Volume I, Technology and Applications* (Cleveland, OH: CRC Press, 1974), 249 p.

Brophy, J.J., Goldsack, R.J., and Clarkson, J.R. The essential oil of *Ocimum tenuiflorum* L. (Lamiacae) growing in northern Australia, *J. Ess. Oil Res.*, 5:459-461, 1993.

Brouwer, D.H., De Haan, M., Leenheers, L.H., de Vreede, S.A.F., and van Hemmen, J.J. Half-lives of pesticides on greenhouse crops, *Bull. Environ. Contam. Toxicol.*, 58(6):976-984, 1997.

Brown, A.C., Canosa-Mas, C.E., and Wayne, R.P. A kinetic study of the reactions of OH with CH_3I and CF_3I, *Atmos. Environ.*, 24A(2):361-367, 1990.

Brown, D. and Thompson, R.S. Phthalates and the aquatic environment: Part I. The effect of di-2-ethylhexyl phthalate (DEHP) and diisodecyl phthalate (DIDP) on the reproduction of *Daphnia magna* and observations on their bioconcentration, *Chemosphere*, 11(4):417-426, 1982.

Brown, D. and Thompson, R.S. Phthalates and the aquatic environment: Part II. The bioconcentration and depuration of di-2-ethylhexyl phthalate (DEHP) and diisodecyl phthalate (DIDP) in mussels (*Mytilus edulis*), *Chemosphere*, 11(4):427-435, 1982a.

Brown, D.S. and Flagg, E.W. Empirical prediction of organic pollutant sorption in natural sediments, *J. Environ. Qual.*, 10(3):382-386, 1981.

Brown, L., Bancroft, K.C.C., and Rhead, M.M. Laboratory studies on the adsorption of acrylamide monomer by sludge, sediments, clays, peat and synthetic resins, *Water Res.*, 14(7):779-781, 1980.

Brown, L. and Rhead, M. Liquid chromatographic determination of acrylamide monomer in natural and polluted aqueous environments, *Analyst*, 104(1238):391-399, 1979.

Brown, M.A., Petreas, M.T., Okamoto, H.S., Mischke, T.M., and Stephens, R.D. Monitoring of malathion and its impurities and environmental transformation products on surfaces and in air following an aerial application, *Environ. Sci. Technol.*, 27(2):388-397, 1993.

Brown, M.D., Thomas, D., and Kay, B.H. Acute toxicity of selected pesticides to the Pacific blue-eye, *Pseudomugil signifer* (Pisces), *J. Am. Mosq. Control Assoc.*, 14(4):463-466, 1998.

Brown, R.L. and Wasik, S.P. A method of measuring the solubilities of hydrocarbons in aqueous solutions, *J. Res. Nat. Bur. Stand.*, 78A(4):453-460, 1974.

Brown, V.M. The calculation of the acute toxicity of mixtures of poisons to rainbow trout, *Water Res.*, 2(10):723-733, 1968.

Brownawell, B.J. The role of colloid organic matter in the marine geochemistry of PCBs, Ph.D. Dissertation, Woods Hole-MIT Joint Program, Cambridge, MA, 1986.

Broza, M., Halpern, M., Teltsch, B., Porat, R., and Gasith, A. Shock chloramination: Potential treatment for *Chironomidae* (Diptera) larvae nuisance abatement in water supply systems, *J. Econ. Entomol.*, 91(4):834-840, 1998.

Bruggeman, W.A., Van Der Steen, J., and Hutzinger, O. Reversed-phase thin-layer chromatography of polynuclear aromatic hydrocarbons and chlorinated biphenyls. Relationship with hydrophobicity as measured by aqueous solubility and octanol-water partition coefficient, *J. Chromatogr.*, 238(2):335-346, 1982.

Brunelle, D.J. and Singleton, D.A. Chemical reaction of polychlorinated biphenyls on soils with poly(ethylene glycol)/KOH, *Chemosphere*, 14(2):173-181, 1985.

Brungs, W.A., Geckler, J.R., and Gast, M. Acute and chronic toxicity of copper to the fathead minnow in a surface water of variable quality, *Water Res.*, 10(1):37-43, 1976.

Brunner, S., Hornung, E., Santl, H., Wolf, E., Piringer, O.G., Altschuh, J., and Brüggemann, R. Henry's law constants for polychlorinated biphenyls: experimental determination and structure-property relationships, *Environ. Sci. Technol.*, 24(11):1751-1754, 1990.

Brusseau, M.L., Jessup, R.E., and Rao, P.S.C. Sorption kinetics of organic chemicals: evaluation of gas-purge and miscible-displacement techniques, *Environ. Sci. Technol.*, 24(5):727-735, 1990.

Brusseau, M.L. and Rao, P.S.C. Influence of sorbent structure on nonequilibrium sorption of organic compounds, *Environ. Sci. Technol.*, 25(8):1501-1506, 1991.

Buccafusco, R.J., Ells, S.J., and LeBlanc, G.A. Acute toxicity of priority pollutants to bluegill (*Lepomis macrochirus*), *Bull. Environ. Contam. Toxicol.*, 26(4):446-452, 1981.

Buchter, B., Davidoff, B., Amacher, M.C., Hinz, C., Iskandar, K., and Selim, H.M. Correlation of Freundlich K_d and n retention parameters with soils and elements, *Soil Sci.*, 148(5):370-379, 1989.

Bufalini, J.J. and Altshuller, A.P. Kinetics of vapor-phase hydrocarbon-ozone reactions, *Can. J. Chem.*, 43:2243-2250, 1965.

Bufalini, J.J., Gay, B.W., and Kopczynski, S.L. Oxidation of *n*-butane by the photolysis of NO_2, *Environ. Sci. Technol.*, 5(4):333-336, 1971.

Bull, D.L. and Ridgway, R.L. Metabolism of trichlorfon in animals and plants, *J. Agric. Food Chem.*, 27(2):268-272, 1979.

Bulla, C.D., III, and Edgerley, E. Photochemical degradation of refractory organic compounds, *J. Water Pollut. Control Fed.*, 40:546-556, 1968.

Bumpus, J.A. and Aust, S.D. Biodegradation of DDT [1,1,1-trichloro-2,2-bis(4-chlorophenyl)ethane] by the white rot fungus *Phanerochaete chrysosporium*, *Appl. Environ. Microbiol.*, 53(9):2001-2008, 1987.

Bumpus, J.A., Tien, M., Wright, D., and Aust, S.D. Oxidation of persistent environmental pollutants by a white rot fungus, *Science (Washington, D.C.)*, 228(4706):1434-1436, 1985.

Bunce, N.J., Nakai, J.S., and Yawching, M. A model for estimating the rate of chemical transformation of a VOC in the troposphere by two pathways: photolysis by sunlight and hydroxyl radical attack, *Chemosphere*, 22(3/4):305-315, 1991.

Bunton, C.A., Fuller, N.A., Perry, S.G., and Shiner, V.J. The hydrolysis of carboxylic anhydrides. Part III. Reactions in initially neutral solution, *J. Chem. Soc. (London)*, pp. 2918-2926, 1963.

Burczyk, L., Walczyk, K., and Burczyk, R. Kinetics of reaction of water addition to the acrolein double-bond in dilute aqueous solution, *Przem. Chem.*, 47(10):625-627, 1968.

Burdick, G.E., Dean, H.J., and Harris, E.J. Toxicity of aqualin to fingerling brown trout and bluegills, *NY Fish Game J.*, 11(2):106-114, 1964.

Burge, W.D. Anaerobic Decomposition of DDT in soil. Acceleration by volatile components of alfalfa, *J. Agric. Food Chem.*, 19(2):375-378, 1971.

Burkhard, L.P., Armstrong, D.E., and Andren, A.W. Henry's law constants for the polychlorinated biphenyls, *Environ. Sci. Technol.*, 19(7):590-596, 1985.

Burkhard, L.P. and Kuehl, D.W. *n*-Octanol/water partition coefficients by reverse phase liquid chromatography/mass spectrometry for eight tetrachlorinated planar molecules, *Chemosphere*, 15(2):163-167, 1986.

Burkhard, L.P., Kuehl, D.W., and Veith, G.D. Evaluation of reverse phase liquid chromatography/mass spectrometry for estimation of *n*-octanol/water partition coefficients for organic chemicals, *Chemosphere*, 14(10):1551-1560, 1985a.

Burks, S.L. Evaluation of the effectiveness of granular activated carbon adsorption and aquaculture for removing toxic compounds from treated petroleum refinery effluents, Robert S. Kerr Environmental Research Laboratory, U.S. EPA Report 600/2-81-067, 1981.

Burlinson, N.E., Lee, L.A., and Rosenblatt, D.H. Kinetics and products of hydrolysis of 1,2-dibromo-3-chloropropane, *Environ. Sci. Technol.*, 16(9):627-632, 1982.

Burnett, M.G. Determination of partition coefficients at infinite dilution by the gas chromatographic analysis of the vapor above dilute solutions, *Anal. Chem.*, 35:1567-1570, 1963.

Burridge, L.E., Haya, K., Zitko, V., and Waddy, S. The lethality of salmosan (azamethiphos) to American lobster (*Homarus americanus*) larvae, postlarvae, and adults, *Ecotoxicol. Environ. Saf.*, 43(2):165-169, 1999.

Burris, D.R. and MacIntyre, W.G. A thermodynamic study of solutions of liquid hydrocarbon mixtures in water, *Geochim. Cosmochim. Acta*, 50(7):1545-1549, 1986.

Burrows, H.D., Miguel, M.M., Varela, A.P., and Becker, R.S. The aqueous solubility and thermal behaviour of some β-carbolines, *Thermochim. Acta*, 279:77-82, 1996.

Burton, D.T. and Margrey, S.L. An evaluation of acute bromine chloride toxicity to grass shrimp (*Palaemonetes* spp.) and juvenile blue crabs (*Callinectes sapidus*), *Bull. Environ. Contam. Toxicol.*, 19(2):131-138, 1978.

Burton, W.B. and Pollard, G.E. Rate of photochemical isomerization of endrin in sunlight, *Bull. Environ. Contam. Toxicol.*, 12(1):113-116, 1974.

Buser, H.R. Rapid photolytic decomposition of brominated and brominated/chlorinated dibenzodioxins and dibenzofurans, *Chemosphere*, 17(5):899-903, 1988.

Butler, J.A.V. and Ramchandani, C.N. The solubility of non-electrolytes. Part II. The influence of the polar group on the free energy of hydration of aliphatic compounds, *J. Chem. Soc. (London)*, pp. 952-955, 1935.

Butler, J.A.V., Ramchandani, C.N., and Thompson, D.W. The solubility of non-electrolytes. Part I. The free energy of hydration of some aliphatic alcohols, *J. Chem. Soc. (London)*, pp. 280-285, 1935.

Butler, J.A.V., Thomson, D.W., and Maclennan, W.H. The free energy of the normal aliphatic alcohols in aqueous solution. Part I. The partial vapour pressures of aqueous solutions of methyl, *n*-propyl, and *n*-butyl alcohols. Part II. The solubilities of some normal aliphatic alcohols in water. Part III. The theory of binary solutions, and its application to aqueous-alcoholic solutions, *J. Chem. Soc. (London)*, 136:674-686, 1933.

Butler, J.C. and Webb, W.P. Upper explosive limits of cumene, *J. Chem. Eng. Data*, 2(1):42-46, 1957.

Buttery, R.G., Bomben, J.L., Guadagni, D.G., and Ling, L.C. Some considerations of the volatilities of organic flavor compounds in foods, *J. Agric. Food Chem.*, 19:1045-1048, 1971.

Buttery, R.G., Ling, L.C., and Guadagni, D.G. Volatilities of aldehydes, ketones, and esters in dilute water solution, *J. Agric. Food Chem.*, 17(2):385-389, 1969.

Buttery, R.G., Seifert, R.M., Guadagni, D.G., and Ling, L.C. Characterization of some volatile constituents of bell peppers, *J. Agric. Food Chem.*, 17(6):1322-1327, 1969a.

Butz, R.G., Yu, C.C., and Atallah, Y.H. Photolysis of hexachlorocyclopentadiene in water, *Ecotoxicol. Environ. Saf.*, 6(4):347-357, 1982.

Byast, T.H. and Hance, R.J. Degradation of 2,4,5-T by South Vietnamese soils incubated in the laboratory, *Bull. Environ. Contam. Toxicol.*, 14(1):71-76, 1975.

Cabani, S., Conti, G., and Lepori, L. Thermodynamic study on aqueous dilute solutions of organic compounds. Part 2. - Cyclic ethers, *Trans. Faraday Soc.*, 67:1943-1950, 1971.

Cabani, S., Conti, G., and Lepori, L. Thermodynamic study on aqueous dilute solutions of organic compounds. Part 1. - Cyclic amines, *Trans. Faraday Soc.*, 67:1933-1942, 1971a.

Cadle, R.D. and Schadt, C. Kinetics of the gas-phase reaction of olefins with ozone, *J. Am. Chem. Soc.*, 74:6002-6004, 1952.

Calamari, D., Galassi, S., Setti, F., and Vighi, M. Toxicity of selected chlorobenzenes to aquatic organisms, *Chemosphere*, 12(2):253-262, 1983.

Caldwell, R.S., Calderone, E.M., and Mallon, M.H. Effects of a seawater fraction of cook inlet crude oil and its major aromatic components on larval stages of the Dungeness crab, *Cancer magister dana*, in *Fate and Effects of Petroleum Hydrocarbons in Marine Organisms and Ecosystems*, (New York: Pergoman, 1976), pp. 210-220.

Call, D.J., Brooke, L.T., Harting, S.L., Poirer, S.H., and McCauley, D.J. Toxicity of phenanthrene to several freshwater species, Center for Lake Superior Environmental Studies, University of Wisconsin, Superior, WI, 1986.

Call, D.J., Brooke, L.T., and Lu, P.Y. Uptake, elimination and metabolism of three phenols by fathead minnows, *Arch. Environ. Contam. Toxicol.*, 9:699-714, 1980.

Call, F. The mechanism of sorption of ethylene dibromide on moist soils, *J. Sci. Food Agric.*, 8:630-639, 1957.

Callahan, M.A., Slimak, M.W., Gable, N.W., May, I.P., Fowler, C.F., Freed, J.R., Jennings, P., Durfee, R.L., Whitmore, F.C., Maestri, B., Mabey, W.R., Holt, B.R., and Gould, C. Water-related environmental fate of 129 priority pollutants, Office of Research and Development, U.S. EPA Report 440/4-79-029, 1979, 1160 p.

Calvert, J.G., Kerr, J.A., Demerjian, K.L., and McQuigg, R.D. Photolysis of formaldehyde as a hydrogen atom source in the lower atmosphere, *Science (Washington, D.C.)*, 175(4023):751-752, 1972.

Calvert, J.G. and Pitts, J.N., Jr. *Photochemistry* (New York: John Wiley & Sons, 1966), 899 p.

Camilleri, P., Watts, S.A., and Boraston, J.A. A surface area approach to determine partition coefficients, *J. Chem. Soc., Perkin Trans. 2*, pp. 1699-1707, 1988.

Campbell, A.N. and Campbell, A.J.R. The system naphthalene-*p*-nitrophenol: an experimental investigation of all the variables in an equation of the freezing point curve, *Can. J. Res.*, B19:73-79, 1941.

Campbell, I.M., McLaughlin, D.F., and Handy, B.J. Rate constants for reactions of hydroxyl radicals with alcohol vapors at 292 K, *Chem. Phys. Lett.*, 38:362-364, 1976.

Campbell, J.R. and Luthy, R.G. Prediction of aromatic solute partition coefficients using the UNIFAC group contribution model, *Environ. Sci. Technol.*, 19(10):980-985, 1985.

Camper, N.D. Effects of pesticide degradation products on soil microflora, in *Pesticide Transformation Products. Fate and Significance in the Environment*, ACS Symposium Series 429, Somasundaram, L. and Coats, J.R., Eds. (New York: American Chemical Society, 1991), pp. 205-216.

Canton, J.H., Sloof, W., Kool, H.J., Struys, J., Pouw, Th. J.M., Wegman, R.C.C., and Piet, G.J. Toxicity, biodegradability, and accumulation of a number of Cl/N-containing compounds for classification and establishing water quality criteria, *Reg. Toxicol. Pharm.*, 5:123-131, 1985.

Canton, J.H., van Esch, G.J., Greve, P.A., and van Hellemond, A.B.A.M. Accumulation and elimination of hexachloro-cyclohexane (HCH) by the marine algae *Chlamydomonas* and *Dunaliella*, *Water Res.*, 11(1):111-115, 1977.

Canton, J.H., Wegman, C.C., Vulto, T.J.A., Verhoef, C.H., and van Esch, G.J. Toxicity, accumulation, and elimination studies of α-hexachlorocyclohexane (α-HCH) with saltwater organisms of different trophic levels, *Water Res.*, 12(9):687-690, 1978.

Capel, P.D., Giger, W., Reichert, P., and Wanner, O. Accidental input of pesticides into the Rhine River, *Environ. Sci. Technol.*, 22(9):992-997, 1988.

Carlberg, G.E., Martinson, K., Kringstad, A., Gjessing, E., Grande, M., Källqvist, T., and Skåre, J.U. Influence of aquatic humus on the bioavailability of chlorinated micropollutants in Atlantic salmon, *Arch. Environ. Contam. Toxicol.*, 15(5):543-548, 1986.

Carlisle, P.J. and Levine, A.A. Stability of chlorohydrocarbons. I. Methylene chloride, *Ind. Eng. Chem.*, 24(2):146-147, 1932.

Carlisle, P.J. and Levine, A.A. Stability of chlorohydrocarbons. II. Trichloroethylene, *Ind. Eng. Chem.*, 24(10):1164-1168, 1932a.

Carlson, A.R., Carlson, R., and Kopperman, H.L. Determination of partition coefficients by liquid chromatography, *J. Chromatogr.*, 107:211-223, 1975.

Carlson, A.R. and Kosian, P.A. Toxicity of chlorinated benzenes to fathead minnows (*Pimephales promelas*), *Arch. Environ. Contam. Toxicol.*, 16(2):129-135, 1987.

Caro, J.H., Freeman, H.P., Glotfelty, D.E., Turner, B.C., and Edwards, W.M. Dissipation of soil incorporated carbofuran in the field, *J. Sci. Food Agric.*, 21(5):1010-1015, 1973.

Caro, J.H., Freeman, H.P., and Turner, B.C. Persistence in soil and losses in runoff in soil-incorporated carbaryl in a small watershed, *J. Agric. Food Chem.*, 22(5):860-863, 1974.

Caron, G., Suffet, I.H., and Belton, T. Effect of dissolved organic carbon on the environmental distribution of nonpolar organic compounds, *Chemosphere*, 14(8):993-1000, 1985.

Carothers, W.H., Williams, I., Collins, A.M., and Kirby, J.E. Acetylene polymers and their derivatives II. A new synthetic rubber: chloroprene and its polymers, *J. Am. Chem. Soc.*, 53:4203-4225, 1931.

Carpenter, C.P., Smyth, H.F., and Pozzani. The assay of acute vapor toxicity, and the grading and interpretation of results on 96 chemical compounds, *J. Ind. Hyg. Toxicol.*, 31:343-346, 1949.

Carpenter, J.H. New measurements of oxygen solubility in pure and natural water, *Limnol. Oceanogr.*, 11:264-277, 1966.

Carswell, T.G. and Nason, H.K. Properties and uses of pentachlorophenol, *Ind. Eng. Chem.*, 30(6):622-626, 1938.

Carta, R. and Tola, G. Solubilities of L-cystine, L-tyrosine, L-leucine, and glycine in aqueous solutions at various pHs and NaCl concentrations, *J. Chem. Eng. Data*, 41(3):414-417, 1996.

Carter, C.W. and Suffet, I.H. Interactions between dissolved humic and fulvic acids and pollutants in aquatic environments, in *Fate of Chemicals in the Environment*, ACS Symposium Series 225, Swann, R.L. and Eschenroeder, A., Eds. (Washington, D.C.: American Chemical Society, 1983), pp. 215-229.

Carter, W.P.L., Winer, A.M., and Pitts, J.N., Jr. Major atmospheric sink for phenol and the cresols. Reaction with the nitrate radical, *Environ. Sci. Technol.*, 15(7):829-831, 1981.

Cartón, A., Isla, T., and Alvarez-Benedí, J. Sorption-desorption of imazamethabenz on three Spanish soils, *J. Agric. Food Chem.*, 45(4):1454-1458, 1997.

Casellato, F., Vecchi, C., Girelli, A., and Casu, B. Differential calorimetric study of polycyclic aromatic hydrocarbons, *Thermochim. Acta*, 6:361-368, 1973.

Casida, J.E., Gatterdam, P.E., Getzin, Jr., L.W., and Chapman, R.K. Residual properties of the systemic insecticide *o,o*-dimethyl 1-carbomethoxy-1-propen-2-yl phosphate, *J. Agric. Food Chem.*, 4(3):236-243, 1956.

Casida, J.E., Holmstead, R.L., Khalifa, S., Knox, J.R., Ohsawa, T., Palmer, K.J., and Wong, R.Y. Toxaphene insecticide: a complex biodegradable mixture, *Science (Washington, D.C.)*, 183(4124):520-521, 1974.

Casserly, D.M., Davis, E.M., Downs, T.D., and Guthrie, R.K. Sorption of organics by *Selenastrum capricornutum*, *Water Res.*, 17(11):1591-1594, 1983.

Castro, C.E. The rapid oxidation of iron(II) porphyrins by alkyl halides. A possible mode of intoxication of organisms by alkyl halides, *J. Am. Chem. Soc.*, 86(2):2310-2311, 1964.

Castro, C.E. and Belser, N.O. Hydrolysis of *cis*- and *trans*-1,3-dichloropropene in wet soil, *J. Agric. Food Chem.*, 14(1):69-70, 1966.

Castro, C.E. and Belser, N.O. Biodehalogenation. Reductive dehalogenation of the biocides ethylene dibromide, 1,2-dibromo-3-chloropropane, and 2,3-dibromobutane in soil, *Environ. Sci. Technol.*, 2(10):779-783, 1968.

Castro, C.E. and Belser, N.O. Photohydrolysis of methyl bromide and chloropicrin, *J. Agric. Food Chem.*, 29(5):1005-1009, 1981.

Castro, C.E. and Belser, N.O. Biodehalogenation: oxidative and reductive metabolism of 1,1,2-trichloroethane by

Pseudomonas putida - Biogeneration of vinyl chloride, *Environ. Toxicol. Chem.*, 9(6):707-714, 1990.

Castro, C.E., Wade, R.S., and Belser, N.O. Biodehalogenation. The metabolism of chloropicrin by *Pseudomonas* sp., *J. Agric. Food Chem.*, 31(6):1184-1187, 1983.

Castro, C.E., Wade, R.S., Riebeth, D.M., Bartnicki, E.W., and Belser, N.O. Biodehalogenation: rapid metabolism of vinyl chloride by a soil *Pseudomonas* sp. - direct hydrolysis of a vinyl C-Cl bond, *Environ. Toxicol. Chem.*, 11(6):757-764, 1992.

Castro, C.E. and Yoshida, T. Degradation of organochlorine insecticides in flooded soils in the Philippines, *J. Agric. Food Chem.*, 19:1168-1170, 1971.

Caswell, R.L., DeBold, K.J., and Gilbert, L.S., Eds. *Pesticide Handbook*, 29th ed. (College Park, MD: Entomological Society of America, 1981), 286 p.

Caturla, F., Martin-Martinez, J.M., Molina-Sabio, M., Rodriguez-Reinoso, F., and Torregrosa, R. Adsorption of substituted phenols on activated carbon, *J. Colloid Interface Sci.*, 124(2):528-534, 1988.

Cavalieri, E. and Rogan, E. Role of radical cations in aromatic hydrocarbon carcinogenesis, *Environ. Health Perspect.*, 64:69-84, 1985.

Celi, L., Nègre, M., and Gennari, M. Adsorption of the herbicide acifluorfen on soil humic acids, *J. Agric. Food Chem.*, 44(10):3388-3392, 1996.

Celorie, J.A., Woods, S.L., Vinson, T.S., and Istok, J.D. A comparison of sorption equilibrium distribution coefficients using batch and centrifugation methods, *J. Environ. Qual.*, 18(3):307-313, 1989.

Cepeda, E.A., Gomez, B., and Diaz, M. Solubility of anthracene and anthraquinone in some pure and mixed solvents, *J. Chem. Eng. Data*, 34(3):273-275, 1989.

Cerniglia, C.E. and Crow, S.A. Metabolism of aromatic hydrocarbons by yeasts, *Arch. Microbiol.*, 129:9-13, 1981.

Cerniglia, C.E. and Gibson, D.T. Fungal oxidation of benzo[a]pyrene: evidence for the formation of a BP-7,8-diol 9,10-epoxide and BP-9,10-diol 7,8-epoxides, *Abstr. Ann. Mtg. Am. Soc. Microbiol.*, 1980, p. 138.

Cerón, J.J., Ferrando, M.D., Sancho, E., Gutiérrez-Panizo, C., and Andreu-Moliner, E. Effects of diazinon exposure on cholinesterase activity in different tissues of European eel (*Anguilla anguilla*), *Ecotoxicol. Environ. Saf.*, 35(3):222-225, 1996.

Cerón, J.J., Gutiérrez-Panizo, C., Barba, A., and Cámara, M.A. Endosulfan isomers and metabolite residue degradation in carnation (*Dianthus caryophyllus*) by-product under different environmental conditions, *J. Environ. Sci. Health*, B30(2):221-232, 1995.

Cessna, A.J. and Grover, R. Spectroscopic determination of dissociation constants of selected acidic herbicides, *J. Agric. Food Chem.*, 26(1):289-292, 1978.

Chacko, C.I., Lockwood, J.L., and Zabik, M. Chlorinated hydrocarbon pesticides: degradation by microbes, *Science (Washington, D.C.)*, 154(3751):893-895, 1966.

Chai, X.S. and Zhu, J.Y. Simultaneous measurements of solute concentration and Henry's constant using multiple headspace extraction gas chromatography, *Anal. Chem.*, 70(16):3481-3487, 1998.

Chaigneau, M., LeMoan, G., and Giry, L. Décomposition pyrogénée du bromuredeméthyle en l'absence d'oxygene et à 550 °C, *Comp. Rend. Ser. C.*, 263:259-261, 1966.

Chaintreau, A., Grade, A., and Muñoz-Box, R. Determination of partition coefficients and quantification of headspace volatile compounds, *Anal. Chem.*, 67(18):3300-3304, 1995.

Chaisuksant, Y., Yu, Q., and Connell, D.W. Bioconcentration of bromo- and chlorobenzenes by fish (*Gambusia affinis*) *Water Res.*, 31(1):61-68, 1997.

Chaisuksant, Y., Yu, Q., and Connell, D.W. Effects of halobenzenes on growth rate of fish (*Gambusia affinis*) *Ecotoxicol. Environ. Saf.*, 39(2):120-130, 1998.

Chambers, C.W., Tabak, H.H., and Kabler, P.W. Degradation of aromatic compounds by phenol-adapted bacteria, *J. Water Pollut. Control. Fed.*, 35(12):1517-1528, 1963.

Chan, W.F. and Larson, R.A. Formation of mutagens from the aqueous reactions of ozone and anilines, *Water Res.*, 25(12):1529-1538, 1991.

Chan, W.F. and Larson, R.A. Mechanisms and products of ozonolysis of aniline in aqueous solution containing nitrite ion, *Water Res.*, 25(12):1549-1544, 1991a.

Chanda, M., O'Driscoll, K.F., and Rempbel, G.L. Sorption of phenolics and carboxylic acids on polybenzimidazole, *Reactive Polymers*, 4(1):39-48, 1985.

Chao, J., Gadalla, N.A.M., Gammon, B.E., Marsh, K.N., Rodgers, A.S., Somayajulu, G.R., and Wilhoit, R.C. Thermodynamic and thermophysical properties of organic nitrogen compounds. Part I. Methanamine, ethanamine, 1- and 2-propanamine, benzenamine, 2-, 3-, and 4-methylbenzenamine, *J. Phys. Chem. Ref. Data*, 19(6):1547-1615, 1990.

Chao, J., Hall, K.R., and Yao, J.M. Thermodynamic properties of simple alkenes, *Thermochim. Acta*, 64(3):285-303, 1983.

Chapman, P.J. An outline of reaction sequences used for the bacterial degradation of phenolic compounds, in *Degradation of Synthetic Organic Molecules in the Biosphere: Natural, Pesticidal, and Various Other Man-Made Compounds* (Washington, D.C.: National Academy of Sciences, 1972), pp. 17-55.

Chapman, R.A. and Chapman, P.C. Persistence of granular and ec formulations of chlorpyrifos in a mineral and organic soil incubated in open and closed containers, *J. Environ. Sci. Health*, B21(6):447-456, 1986.

Chapman, R.A. and Cole, C.M. Observations on the influence of water and soil pH on the persistence of insecticides, *J. Environ. Sci. Health*, B17(5):487-504, 1982.

Chapman, R.A. and Harris, C.R. Persistence of four pyrethroid insecticides in a mineral and an organic soil, *J. Environ. Sci. Health*, B15(1):39-46, 1980.

Chapman, R.A., Harris, C.R., Svec, H.J., and Robinson, J.R. Persistence and mobility of granular insecticides in an organic soil following furrow application for onion maggot control, *J. Environ. Sci. Health*, B19(3):259-270, 1984.

Chapman, R.A., Tu, C.M., Harris, C.R., and Cole, C. Persistence of five pyrethroid insecticides in sterile and natural, mineral and organic soil, *Bull. Environ. Contam. Toxicol.*, 26(4):513-519, 1981.

Chellman, G.J., White, R.D., Norton, R.M., and Bus, J.S. Inhibition of the acute toxicity of methyl chloride in male B6C3F1 mice by glutathione depletion, *Toxicol. Appl. Pharmacol.*, 86(1):93-104, 1986.

Chen, C.S.-H., Delfino, J.J., and Rao, P.S.C. Partitioning of organic and inorganic components from motor oil, *Chemosphere*, 28(7):1385-1400, 1994.

Chen, P.N., Junk, G.A., and Svec, H.J. Reactions of organic pollutants. I. Ozonation of acenaphthylene and acenaphthene, *Environ. Sci. Technol.*, 13(4):451-454, 1979.

Chen, P.R., Tucker, W.P., and Dauterman, W.C. Structure of biologically produced malathion monoacid, *J. Agric. Food Chem.*, 17(1):86-90, 1969.

Cheung, H.M. and Kurup, S. Sonochemical destruction of CFC 11 and CFC 13 in dilute aqueous solution, *Environ. Sci. Technol.*, 28(9):1619-1622, 1994.

Chey, W. and Calder, G.V. Method for determining solubility of slightly soluble organic compounds, *J. Chem. Eng. Data*, 17(2):199-200, 1972.

Chin, Y.-P., Peven, C.S., and Weber, W.J. Estimating soil/sediment partition coefficients for organic compounds by high performance reverse phase liquid chromatography, *Water Res.*, 22(7):873-881, 1988.

Chin, Y.-P. and Weber, W.J., Jr. Estimating the effects of dispersed organic polymers on the sorption of contaminants by natural solids. 1. A predictive thermodynamic humic substance-organic solute interaction model, *Environ. Sci. Technol.*, 23(8):978-984, 1989.

Chin, Y.-P., Weber, W.J., Jr., and Voice, T.C. Determination of partition coefficients and aqueous solubilities by reverse phase chromatography - II. Evaluation of partitioning and solubility models, *Water Res.*, 20(11):1443-1451, 1986.

Chiodini, G., Rindone, B., Cariati, F., Polesello, S., Restilli, G., and Hjorth, J. Comparison between the gas-phase and the solution reaction of the nitrate radical and methylarenes, *Environ. Sci. Technol.*, 27(8):1659-1664, 1993.

Chiou, C.T. Partition coefficients of organic compounds in lipid-water systems and correlations with fish bioconcentration factors, *Environ. Sci. Technol.*, 19(1):57-62, 1985.

Chiou, C.T., Freed, V.H., Peters, W., and Kohnert, R.L. Evaporation of solutes from water, *Environ. Int.*, 3:231-236, 1980.

Chiou, C.T., Freed, V.H., Schmedding, D.W., and Kohnert, R.L. Partition coefficients and bioaccumulation of selected organic chemicals, *Environ. Sci. Technol.*, 11(5):475-478, 1977.

Chiou, C.T., Peters, L.J., and Freed, V.H. A physical concept of soil-water equilibria for nonionic organic compounds, *Science (Washington, D.C.)*, 206(4420):831-832, 1979.

Chiou, C.T., Porter, P.E., and Schmedding, D.W. Partition equilibria of nonionic organic compounds between soil organic matter and water, *Environ. Sci. Technol.*, 17(4):227-231, 1983.

Chiou, C.T., Malcolm, R.L., Brinton, T.I., and Kile, D.E. Water solubility enhancement of some organic pollutants and pesticides by dissolved humic and fulvic acids, *Environ. Sci. Technol.*, 20(5):502-508, 1986.

Chiou, C.T., Schmedding, D.W., and Manes, M. Partitioning of organic compounds in octanol-water systems, *Environ. Sci. Technol.*, 16(1):4-10, 1982.

Chiou, C.T., Shoup, T.D., and Porter, P.E. Mechanistic roles of soil humus and minerals in the sorption of nonionic organic compounds from aqueous and organic solutions, *Org. Geochem.*, 8(1):9-14, 1985.

Chopra, N.M. and Mahfouz, A.M. Metabolism of endosulfan I, endosulfan II and endosulfan sulfate in tobacco leaf, *J. Agric. Food Chem.*, 25(1):32-36, 1977.

Chou, S.-F.J. and Griffin, R.A. Soil, clay and caustic soda effects on solubility, sorption and mobility of hexachlorocyclopentadiene, Environmental Geology Notes 104, Illinois Department of Energy and Natural Resources, 1983, 54 p.

Chou, J.T. and Jurs, P.C. Computer assisted computation of partition coefficients from molecular structures using fragment constants, *J. Chem. Info. Comp. Sci.*, 19:172-178, 1979.

Chou, S.-F.J., Griffin, R.A., Chou, M.-I.M., and Larson, R.A. Products of hexachlorocyclopentadiene (C-56) in aqueous solution, *Environ. Toxicol. Chem.*, 6(5):371-376, 1987.

Choudhry, G.G. and Hutzinger, O. Acetone-sensitized and nonsensitized photolyses of tetra-, penta-, and hexachlorobenzenes in acetonitrile-water mixtures: Photoisomerization and formation of several products including polychlorobiphenyls, *Environ. Sci. Technol.*, 18(4):235-241, 1984.

Choudhry, G.G., Sundström, G., Ruzo, L.O., and Hutzinger, O. Photochemistry of chlorinated diphenyl ethers, *J. Agric. Food Chem.*, 25(6):1371-1376, 1977.

CHRIS Hazardous Chemical Data, Vol. 2, U.S. Department of Transportation, U.S. Coast Guard, U.S. Government Printing Office (November, 1984).

Christensen, J.J., Hansen, L.D., and Izatt, R.M. *Handbook of Proton Ionization Heats* (New York: John Wiley & Sons, 1975), 269 p.

Christiansen, V.O., Dahlberg, J.A., and Andersson, H.F. On the nonsensitized photooxidation of 1,1,1-trichloroethane vapor in air, *Acta Chem. Scand.*, A26:3319-3324, 1972.

Christie, A.O. and Crisp, D.J. Activity coefficients on the *n*-primary, secondary and tertiary aliphatic amines in aqueous solution, *J. Appl. Chem.*, 17:11-14, 1967.

Chu, I., Secours, V., Marino, I., and Villeneuve, D.C. The acute toxicity of four trihalomethanes in male and female rats, *Toxicol. Appl. Pharmacol.*, 52:351-353, 1980.

Chu, W. and Jafvert, C.T. Photodechlorination of polychlorobenzene congeners in surfactant micelle solutions, *Environ. Sci. Technol.*, 28(13):2415-2422, 1994.

Chui, Y.C., Addison, R.F., and Law, P.C. Acute toxicity and toxicokinetics of chlorinated diphenyl ethers in trout, *Xenobiotica*, 20(5):489-499, 1990.

Chung, D.Y. and Boyd, S.A. Mobility of sludge-borne 3,3'-dichlorobenzidine in soil columns, *J. Environ. Qual.*, 16(2):147-151, 1987.

Clarke, F.H. and Cahoon, N.M. Ionization constants by curve fitting: determination of partition and distribution coefficients of acids and bases and their ions, *J. Pharm. Sci.*, 76:611-620, 1987.

Claus, D. and Walker, N. The decomposition of toluene by soil bacteria, *J. Gen. Microbiol.*, 36:107-122, 1964.

Claussen, W.F. and Polglase, M.F. Solubilities and structures in aqueous aliphatic hydrocarbon solutions, *J. Am. Chem. Soc.*, 74(19):4817-4819, 1952.

Clayton, G.D. and Clayton, F.E., Eds. *Patty's Industrial Hygiene and Toxicology*, 3rd ed. (New York: John Wiley & Sons, 1981), 2878 p.

Clegg, S.L. and Brimblecombe, P. Solubility of ammonia in pure aqueous and multicomponent solutions, *J. Phys. Chem.*, 93:7237-7238, 1989.

Cleland, J.G. Project summary - Environmental hazard rankings of pollutants generated in coal gasification processes, Office of Research and Development, U.S. EPA Report 600/S7-81-101, 1981, 19 p.

Cleland, J.G. and Kingsbury, G.L. Multimedia environmental goals for environmental assessment, Volume II. MEG charts and background information, Office of Research and Development, U.S. EPA Report-600/7-77-136b, 1977, 454 p.

Coates, H. The chemistry of phosphorus insecticides, *Am. Appl. Biol.*, 36:156-159, 1949.

Coates, M., Connell, D.W., and Barron, D.M. Aqueous solubility and octan-1-ol to water partition coefficients of aliphatic hydrocarbons, *Environ. Sci. Technol.*, 19(7):628-632, 1985.

Coffee, R.D., Vogel, Jr., P.C., and Wheeler, J.J. Flammability characteristics of methylene chloride (dichloromethane), *J. Chem. Eng. Data*, 17(1):89-93, 1972.

Collander, R. The partition of organic compounds between higher alcohols and water, *Acta Chem. Scand.*, 5:774-780, 1951.

Collett, A.R. and Johnston, J. Solubility relations of isomeric organic compounds. VI. Solubility of the nitroanilines in various liquids, *J. Phys. Chem.*, 30(1):70-82, 1926.

Collin, J. and Lossing, F.P. Ionization potentials of some olefins, diolefins and branched paraffins, *J. Am. Chem. Soc.*, 81(5):2064-2066, 1959.

Conklin, P.J. and Rao, K.R. Toxicity of sodium pentachlorophenate (Na-PCP) to the grass shrimp, *Palaemonetes pugio*, at different stages of the molt cycle, *Bull. Environ. Contam. Toxicol.*, 20(2):275-279, 1978.

Connors, T.F., Stuart, J.D., and Cope, J.B. Chromatographic and mutagenic analyses of 1,2-dichloropropane and 1,3-dichloropropylene and their degradation products, *Bull. Environ. Contam. Toxicol.*, 44(2):288-293, 1990.

Cooney, R.V., Ross, P.D., Bartolini, G.L., and Ramseyer, J. *N*-Nitrosamine and *N*-nitroamine formation: factors influencing the aqueous reactions of nitrogen dioxide with morpholine, *Environ. Sci. Technol.*, 21(1):77-83, 1987.

Cooper, W.J., Mehran, M., Riusech, D.J., and Joens, J.A. Abiotic transformations of halogenated organics. 1. Elimination reaction of 1,1,2,2-tetrachloroethane and formation of 1,1,2-trichloroethane, *Environ. Sci. Technol.*, 21(11):1112-1114, 1987.

Coover, M.P. and Sims, R.C.C. The effects of temperature on polycyclic aromatic hydrocarbon persistence in an unacclimated agricultural soil, *Haz. Waste Haz. Mater.*, 4:69-82, 1987.

Cope, V.W. and Kalkwarf, D.R. Photooxidation of selected polycyclic aromatic hydrocarbons and pyrenequinones coated on glass surfaces, *Environ. Sci. Technol.*, 21(7):643-648, 1987.

Corbett, M.D. and Corbett, B.R. Metabolism of 4-chloronitrobenzene by the yeast *Rhodosporidium* sp., *Appl. Environ. Microbiol.*, 41(4):942-949, 1981.

Corbin, F.T. and Upchurch, R.P. Influence of pH on detoxification of herbicides in soils, *Weeds*, 15:370-377, 1967.

Corby, T.C. and Elworthy, P.H. The solubility of some compounds in hexadecylpolyoxyethylene monoethers, polyethylene glycols, water and hexane, *J. Pharm. Pharmac.*, 23:39S-48S, 1971.

Corless, C.E., Reynolds, G.L., Graham, N.J.D., and Perry, R. Ozonation of pyrene in aqueous solution, *Water Res.*, 24(9):1119-1123, 1990.

Corruccini, R.J. and Ginnings, D.C. The enthalpy, entropy and specific heat of *p*-xylene from 0 to 300°. The heat of fusion, *J. Am. Chem. Soc.*, 69:2291-2294, 1947.

Cottalasso, D., Barisione, G., Fontana, L., Domenicotti, C., Pronzato, M.A., and Nanni, G. Impairment of lipoglycoprotein metabolism in rat liver cells induced by 1,2-dichloropropane, *Occup. Environ. Med.*, 51(4):281-285, 1994.

Coutant, R.W. and Keigley, G.W. An alternative method for gas chromatographic determination of volatile organic compounds in water, *Anal. Chem.*, 60(22):2536-2537, 1988.

Cowen, W.F. and Baynes, R.K. Estimated application of gas chromatographic headspace analysis for priority pollutants, *J. Environ. Sci. Health*, A15:413-427, 1980.

Cox, D.P. and Goldsmith, C.D. Microbial conversion of ethylbenzene to 1-phenethanol and acetophenone by *Nocardia tartaricans* ATCC 31190, *Appl. Environ. Microbiol.*, 38(3):514-520, 1979.

Cox, L., Koskinen, W.C., and Yen, P.Y. Sorption-desorption of imidacloprid and its metabolites in soils, *J. Agric. Food Chem.*, 45(4):1468-1472, 1997.

Cox, R.A., Derwent, R.G., Eggleton, A.E.J., and Lovelock, J.E. Photochemical oxidation of halocarbons in the troposphere, *Atmos. Environ.*, 10:305-308, 1976.

Cox, R.A., Derwent, R.G., and Williams, M.R. Atmospheric photooxidation reactions. Rates, reactivity, and mechanism for reaction of organic compounds with hydroxyl radicals, *Environ. Sci. Technol.*, 14(1):57-61, 1980.

Cox, R.A., Patrick, K.F., and Chant, S.A. Mechanism of atmospheric photooxidation of organic compounds. Reactions of alkoxy radicals in oxidation of *n*-butane and simple ketones, *Environ. Sci. Technol.*, 15(5):587-592, 1981.

Cox, R.A. and Penkett, S.A. Aerosol formation from sulfur dioxide in the presence of ozone and olefinic hydrocarbons,

J. Chem. Soc., Faraday Trans. 1., 68:1735-1753, 1972.

Coyle, G.T., Harmon, T.C., and Suffet, I.H. Aqueous solubility depression for hydrophobic organic chemicals in the presence of partially miscible organic solvents, *Environ. Sci. Technol.*, 31(2):384-389, 1997.

Cozzarelli, I.M., Eganhouse, R.P., and Baedecker, M.G. Transformation of monoaromatic hydrocarbons to organic acids in anoxic groundwater environment, *Environ. Geol. Water Sci.*, 16(2):135-141, 1990.

Cremlyn, R.J. *Agrochemicals - Preparation and Mode of Action* (New York: John Wiley & Sons, 1991), 396 p.

Crittenden, E.D., Jr. and Hixon, A.N. Extraction of hydrogen chloride from aqueous solutions, *Ind. Eng. Chem.*, 46:265-268, 1954.

Crittenden, J.C., Luft, P., Hand, D.W., Oravitz, J.F., Loper, S.W., and Ari, M. Prediction of multi-component adsorption equilibria using ideal adsorbed solution theory, *Environ. Sci. Technol.*, 19(11):1037-1044, 1985.

Crosby, D.G. and Hamadmad, N. The photoreduction of pentachlorobenzenes, *J. Agric. Food Chem.*, 19(6):1171-1174, 1971.

Crosby, D.G. and Leitis, E. The photodecomposition of trifluralin in water, *Bull. Environ. Contam. Toxicol.*, 10(4):237-241, 1973.

Crosby, D.G. and Moilanen, K.W. Vapor-phase photodecomposition of aldrin and dieldrin, *Arch. Environ. Contam. Toxicol.*, 2(1):62-74, 1974.

Crosby, D.G., Moilanen, K.W., and Wong, A.S. Environmental generation and degradation of dibenzodioxins and dibenzofurans, *Environ. Health Perspect.*, Experimental Issue No. 5 (September 1973), pp. 259-266.

Crosby, D.G. and Tutass, H.O. Photodecomposition of 2,4-dichlorophenoxyacetic acid, *J. Agric. Food Chem.*, 14(6):596-599, 1966.

Crosby, D.G. and Wong, A.S. Photodecomposition of 2,4,5-trichlorophenoxyacetic Acid (2,4,5-T) in water, *J. Agric. Food Chem.*, 21(6):1052-1054, 1973.

Crosby, D.G., Wong, A.S., Plimmer, J.R., and Woolson, E.A. Photodecomposition of chlorinated dibenzo-*p*-dioxins, *Science (Washington, D.C.)*, 173(3998):748-749, 1971.

Crossland, N.O. and Wolff, C.J.M. Fate and biological effects of pentachlorophenol in outdoor ponds, *Environ. Toxicol. Chem.*, 4:73-86, 1985.

Crowe, R.W. and Smyth, C.P. Heat capacities, dielectric constants and molecular rotational freedom in solid trichloroethanes and disubstituted propanes, *J. Am. Chem. Soc.*, 72:4009-4015, 1950.

Crum, J.A., Bursian, S.J., Aulerich, R.J., Polin, D., and Braselton, W.E. The reproductive effects of dietary heptachlor in mink (*Mustela vison*), *Arch. Environ. Contam. Toxicol.*, 24(2):156-164, 1993.

Crummett, W.B. and Stehl, R.H. Determination of chlorinated dibenzo-*p*-dioxins and dibenzofurans in various materials, *Environ. Health Perspect.* (September 1973), pp. 15-25.

Cserjsei, A.J. and Johnson, E.L. Methylation of pentachlorophenol by *Trichoderma virgatum*, *Can. J. Microbiol.*, 18:45-49, 1972.

Cullimore, D.R. Interaction between herbicides and soil microorganisms, *Residue Rev.*, 35:65-80, 1971.

Cupitt, L.T. Fate of toxic and hazardous materials in the air environment, Office of Research and Development, U.S. EPA Report-600/3-80-084, 1980, 28 p.

Curtice, S., Felton, E.G., and Prengle, H.W., Jr. Thermodynamics of solutions. Low-temperature densities and excess volumes of *cis*-pentene-2 and mixtures, *J. Chem. Eng. Data*, 17(2):192-194, 1972.

Curtis, L.R., Seim, W.K., and Chapman, G.A. Toxicity of fenvalerate to developing steelhead trout following continuous or intermittent exposure, *J. Toxicol. Environ. Health*, 15(3-4):445-457, 1985.

Curtis, M.W., Copeland, T.L., and Ward, C.H. Aquatic toxicity of substances proposed for spill prevention regulation, in *Proceedings of the 1978 National Conference on Control of Hazardous Material Spills* (Miami, FL: Oil Spill Control Association of America, 1978), pp. 99-103.

Cushman, J.R., Rausina, G.A., Cruzan, G., Gilbert, J., Williams, E., Harrass, M.C., Sousa, J.V., Putt, A.V., Garvey, N.A., St. Laurent, J.P., Hoberg, J.R., and Machado, M.W. Ecotoxicity hazard assessment of styrene, *Ecotoxicol. Environ. Saf.*, 37(2):173-180, 1997.

Dacey, J.W.H., Wakeham, S.G., and Howes, B.L. Henry's law constants for dimethyl sulfide in freshwater and seawater, *Geophys. Res. Letters*, 11:991-994, 1984.

Dagaut, P., Wallington, T.J., Liu, R., and Kurylo, M.J. The gas-phase reactions of hydroxyl radicals with a series of carboxylic acids over the temperature range 240-440 K, *Int. J. Chem. Kinet.*, 20:331-338, 1988.

Dagley, S. Microbial degradation of stable chemical structures: general features of metabolic pathways, in *Degradation of Synthetic Organic Molecules in the Biosphere: Natural, Pesticidal, and Various Other Man-Made Compounds* (Washington, D.C.: National Academy of Sciences, 1972), pp. 1-16.

Dalich, G.M., Larson, R.E., and Gingerich, W.H. Acute and chronic toxicity studies with monochlorobenzene in rainbow trout, *Aquat. Toxicol.*, 2:127-142, 1982.

Daniels, S.L., Hoerger, F.D., and Moolenar, R.J. Environmental exposure assessment: experience under the Toxic Substance Control Act, *Environ. Toxicol. Chem.*, 4:107-117, 1985.

Darnall, K.R., Atkinson, R., and Pitts, J.N., Jr. Rate constants for the reaction of the OH radical with selected alkanes at 300 K, *J. Phys. Chem.*, 82:1581-1584, 1978.

Darnall, K.R., Lloyd, A.C., Winer, A.M., and Pitts, J.N. Reactivity scale for atmospheric hydrocarbons based on reaction with hydroxyl radicals, *Environ. Sci. Technol.*, 10(7):692-696, 1976.

Dasgupta, P.G. and Dong, S. Solubility of ammonia in liquid water and generation of trace levels of standard gaseous ammonia, *Atmos. Environ.*, 20(3):565-570, 1986.

Dauble, D.D., Carlile, D.W., and Hanf, R.W., Jr. Bioaccumulation of fossil fuel components during single-compound and complex-mixture exposures of *Daphnia magna*, *Bull. Environ. Contam. Toxicol.*, 37(1):125-132, 1986.

David, D.J. Determination of specific heat and heat of fusion by differential thermal analysis. Study of theory and operating parameters, *Anal. Chem.*, 36:2162-2166, 1964.

Davidson, J.M., Rao, P.S.C., Ou, L.T., Wheeler, W.B., and Rothwell, D.F. Adsorption, movement, and biological degradation of large concentrations of selected pesticides in soils, U.S. EPA Report 600/2-80-124, 1980, 110 p.

Davies, R.P. and Dobbs, A.J. The prediction of bioconcentration in fish, *Water Res.*, 18(10):1253-1262, 1984.

Davis, A.C. and Kuhr, R.J. Dissipation of chlorpyrifos from muck soil and onions, *J. Econ. Entomol.*, 69:665-666, 1976.

Davis, J.B. and Raymond, R.L. Oxidation of alkyl-substituted cyclic hydrocarbons by a *Nocardia* during growth on alkanes, *Appl. Microbiol.*, 9:383-388, 1961.

Davis, J.T. and Hardcastle, W.S. Biological assay of herbicides for fish toxicity, *Weeds*, 7:397-404, 1959.

Davis, J.W. and Carpenter, C.L. Aerobic biodegradation of vinyl chloride in groundwater samples, *Appl. Environ. Microbiol.*, 56(12):3878-3880, 1990.

Davis, J.W. and Madsen, S.S. The biodegradation of methylene chloride in soils, *Environ. Toxicol. Chem.*, 10:465-474, 1991.

Davis, K.R., Schultz, T.W., and Dumont, J.N. Toxic and teratogenic effects of selected aromatic amines on embryos of the amphibian *Xenopus laevis*, *Arch. Environ. Contam. Toxicol.*, 10(3):371-391, 1981.

Davis, S.N. and DeWiest, R.J.M. *Hydrogeology* (New York: John Wiley & Sons, 1966).

Davis, W.W., Krahl, M.E., and Clowes, G.H.A. Solubility of carcinogenic and related hydrocarbons in water, *J. Am. Chem. Soc.*, 64(1):108-110, 1942.

Davis, W.W. and Parke, T.V., Jr. A nephelometric method for determination of solubilities of extremely low order, *J. Am. Chem. Soc.*, 64(1):101-107, 1942.

Dawson, D.A., Schultz, T.W., Wilke, T.S., and Bryant, S. Developmental toxicity of valproic acid: studies with *Xenopus*, *Teratology*, 45(5):493, 1992.

Day, K.E. Pesticide transformation products in surface waters, in *Pesticide Transformation Products, Fate and Significance in the Environment*, ACS Symposium Series 459, Somasundaram, L. and Coats, J.R., Eds. (Washington, D.C.: American Chemical Society, 1991), pp. 217-241.

Dean, J.A., Ed. *Lange's Handbook of Chemistry*, 11th ed. (New York: McGraw-Hill, 1973), 1570 p.

Dean, J.A. *Handbook of Organic Chemistry* (New York: McGraw-Hill, 1987).

Dearden, J.C. Partitioning and lipophilicity in quantitative structure-activity relationships, *Environ. Health Perspect.* (September 1985), pp. 203-228.

de Bruijn, J., Busser, F., Seinen, W., and Hermens, J. Determination of octanol/water partition coefficients for hydrophobic organic chemicals with the 'slow-stirring' method, *Environ. Toxicol. Chem.*, 8:499-512, 1989.

De Bruyn, W.J., Swartz, E., Hu, J.H., Shorter, J.A., Davidovits, P., Worsnop, D.R., Zahniser, M.S., and Kolb, C.E. Henry's Law solubilities and Śetchenow coefficients for biogenic reduced sulfur species obtained from gas-liquid uptake measurements, *J. Geophys. Res.*, 100D:7245-7251, 1995.

Deeley, G.M., Reinhard, M., and Stearns, S.M. Transformation and sorption of 1,2-dibromo-2-chloropropane in subsurface samples collected at Fresno, California, *J. Environ. Qual.*, 20(3):547-556, 1991.

DeFoe, D.L., Holcombe, G.W., Hammermeister, D.E., and Biesinger, K.E. Solubility and toxicity of eight phthalate esters to four aquatic organisms, *Environ. Toxicol. Chem.*, 9(5):623-636, 1990.

DeGraeve, G.M., Elder, R.G., Woods, D.C., and Bergmann, H.L. Effects of naphthalene and benzene on fathead minnows and rainbow trout, *Arch. Environ. Contam. Toxicol.*, 11(4):487-490, 1982.

Dehn, W.M. Comparative solubilities in water, in pyridine and in aqueous pyridine, *J. Am. Chem. Soc.*, 39(7):1399-1404, 1917.

de Jonge, H. and de Jonge, L.W. Influence of pH and solution composition on the sorption of glyphosate and prochloraz to a sandy loam soil, *Chemosphere*, 39(5):753-763, 1999.

DeKock, A.C. and Lord, D.A. A simple procedure for determining octanol-water partition coefficients using reverse phase high performance liquid chromatography (RPHPLC), *Chemosphere*, 16(1):133-142, 1987.

de Kruif, C.G. Enthalpies of sublimation and vapour pressures of 11 polycyclic hydrocarbons, *J. Chem. Thermodyn.*, 12:243-248, 1980.

DeLassus, P.T. and Schmidt, D.D. Solubilities of vinyl chloride and vinylidene chloride in water, *J. Chem. Eng. Data*, 26(3):274-276, 1981.

DeLaune, R.D., Gambrell, R.P., and Reddy, K.S. Fate of pentachlorophenol in estuarine sediment, *Environ. Pollut. Series B*, 6:297-308, 1983.

Dellinger, B., Taylor, P.H., and Tirey, D.A. Minimization and control of hazardous combustion by-products, Risk Reduction Laboratory, U.S. EPA Report 600/S2-90-039, 1991.

DeLoos, Th.W., Penders, W.G., and Lichtenthaler, R.N. Phase equilibria and critical phenomena in fluid (*n*-hexane + water) at high pressures and temperatures, *J. Chem. Thermodyn.*, 14:83-91, 1982.

De Lorenzi, L., Fermeglia, M., and Torriano, G. Density, refractive index, and kinematic viscosity of diesters and triesters, *J. Chem. Eng. Data*, 42(5):919-923, 1997.

De Maagd, P.G.-J., ten Hulscher, D.th.E.M., van der Heuvel, H., Opperhuizen, A., and Sijm, D.T.H.M. Physico-chemical properties of polycyclic aromatic hydrocarbons: aqueous solubilities, *n*-octanol/water partition coefficients, and Henry's law constants, *Environ. Toxicol. Chem.*, 17(2):251-257, 1998.

Demond, A.H. and Lindner, A.S. Estimation of interfacial tension between organic liquids and water, *Environ. Sci. Technol.*, 27(12):2318-2331, 1993.

DeMore, W.B. and Bayes, K.D. Rate constants for the reactions of hydroxyl radical with several alkanes, cycloalkanes, and dimethyl ether, *J. Phys. Chem. A*, 103(15):2649-2654, 1999.

Deneer, J.W., Sinnige, T.L., Seinen, W., and Hermens, J.L.M. A quantitative structure-activity relationship for the acute toxicity of some epoxy compounds to the guppy, *Aquat. Toxicol.*, 13(3):195-204, 1988.

Deng, B., Campbell, T.J., and Burris, D.R. Kinetics of vinyl chloride reduction by metallic iron in zero-headspace systems, in *American Chemical Society - Division of Environmental Chemistry, Preprints of Extended Abstracts*, 37(1):81-83, 1997.

Dennis, W.H., Jr., Chang, Y.H., and Cooper, W.J. Catalytic dechlorination of organochlorine compounds-Aroclor 1254, *Bull. Environ. Contam. Toxicol.*, 22(6):750-753, 1979.

Dennis, W.H., Jr. and Cooper, W.J. Catalytic dechlorination of organochlorine compounds. II. Heptachlor and chlordane, *Bull. Environ. Contam. Toxicol.*, 16(4):425-430, 1976.

Deno, N.C. and Berkheimer, H.E. Activity coefficients as a function of structure and media, *J. Chem. Eng. Data*, 5(1):1-5, 1960.

Derbyshire, M.K., Karns, J.S., Kearney, P.C., and Nelson, J.O. Purification and characterization of an *N*-methyl-carbamate pesticide hydrolyzing enzyme, *J. Agric. Food Chem.*, 35(6):871-877, 1987.

Derlacki, Z.J., Easteal, A.J., Edge, A.V.J., Woolf, L.A., and Roksandic, Z. Diffusion coefficients of methanol and water and the mutual diffusion coefficient in methanol-water solutions at 278 and 298 K, *J. Phys. Chem.*, 89(24):5318-5322, 1985.

De Santis, R., Marrelli, L., and Muscetta, P.N. Liquid-liquid equilibria in water-aliphatic alcohol systems in the presence of sodium chloride, *Chem. Eng. J.*, 11:207-214, 1976.

Devi, V.U. Changes in oxygen consumption and biochemical composition of the marine fouling dreissinid bivalve *Mytilopsis sallei* (Reclux) exposed to mercury, *Ecotoxicol. Environ. Saf.*, 33(2):168-174, 1996.

Devillers, J., Bintein, S., and Domine, D. Comparison of BCF models based on log P, *Chemosphere*, 33(6):1047-1065, 1996.

DeVoe, H., Miller, M.M., and Wasik, S.P. Generator columns and high pressure liquid chromatography for determining aqueous solubilities and octanol-water partition coefficients of hydrophobic substances, *J. Res. Nat. Bur. Stand.*, 86(4):361-366, 1981.

Dewulf, J., Drijvers, D., and Van Langenhove, H. Measurement of Henry's law constant as function of temperature and salinity for the low temperature range, *Atmos. Environ.*, 29(3):323-331, 1995.

Dewulf, J., Van Langenhove, H., and Everaert, P. Determination of Henry's law coefficients by combination of the equilibrium partitioning in closed systems and solid-phase microextraction techniques, *J. Chromatogr. A*, 830:353-363, 1999.

Dewulf, J., Van Langenhove, H., and Graré, S. Sediment/water and octanol/water equilibrium partitioning of volatile organic compounds: temperature dependence in the 2-25 °C range, *Water Res.*, 33(10:):2424-2436, 1999a.

Dexter, R.N. and Pavlou, S.P. Mass solubility and aqueous activity coefficients of stable organic chemicals in the marine environment: Polychlorinated biphenyls, *Mar. Sci.*, 6:41-53, 1978.

Dickenson, W. The vapour pressure of 1:1:*p:p'*-dichlorodiphenyl trichloroethane (D.D.T.), *Trans. Faraday Soc.*, pp. 31-35, 1956.

Dickhut, R.M., Andren, A.W., and Armstrong, D.E. Aqueous solubilities of six polychlorinated biphenyl congeners at four temperatures, *Environ. Sci. Technol.*, 20(8):807-810, 1986.

DiGeronimo, M.J. and Antoine, A.D. Metabolism of acetonitrile and propionitrile by *Nocardia rhodochrous* LL 100-21, *Appl. Environ. Microbiol.*, 31(6):900-906, 1976.

DiGiano, F.A., Frye, W.H., and Baxter, C.W. A rational approach to utilization of carbon beds in reducing microorganic contamination in drinking water, Water Resources Research Center, University of Massachusetts, Amherst, MA, January 1980.

Dilling, W.L. Interphase transfer processes. II. Evaporation rates of chloro methanes, ethanes, ethylenes, propanes, and propylenes from dilute aqueous solutions. Comparisons with theoretical predictions, *Environ. Sci. Technol.*, 11(4):405-409, 1977.

Dilling, W.L., Lickly, L.C., Lickly, T.D., Murphy, P.G., and McKellar, R.L. Organic photochemistry. 19. Quantum yields for *O,O*-diethyl *O*-(3,5,6-trichloro-2-pyridinyl) phosphorothioate (chlorpyrifos) and 3,5,6-trichloro-2-pyridinol in dilute aqueous solutions and their environmental phototransformation rates, *Environ. Sci. Technol.*, 18(7):540-543, 1984.

Dilling, W.L., Tefertiller, N.B., and Kallos, G.J. Evaporation rates and reactivities of methylene chloride, chloroform, 1,1,1-trichloroethane, trichloroethylene, tetrachloroethylene, and other chlorinated compounds in dilute aqueous solutions, *Environ. Sci. Technol.*, 9(9):833-837, 1975.

Ding, J.Y. and Wu, S.C. Partition coefficients of organochlorine pesticides on soil and on the dissolved organic matter in water, *Chemosphere*, 30(12):2259-2266, 1995.

Dobbs, A.J. and Cull, M.R. Volatilisation of chemicals - Relative loss rates and the estimation of vapour pressures, *Environ. Pollut. (Series B)*, 3(4):289-298, 1982.

Dobbs, R.A. and Cohen, J.M. Carbon adsorption isotherms for toxic organics, Municipal Environmental Research Laboratory, U.S. EPA Report-600/8-80-023, 1980.

Dodd, D.E., Fowler, E.H., Snellings, W.M., Pritts, I.M., and Baron, R.L. Acute inhalation studies with methyl isocyanate vapor. I. Methodology and LC_{50} determinations in guinea pigs, rats, and mice, *Fundam. Appl. Toxicol.*, 6(4):747-755, 1986.

Dodd, D.E., Frank, F.R., Fowler, E.H., Troup, C.M., and Milton, R.M. Biological effects of short-term, high-concentration exposure to methyl isocyanate. I. Study objectives and inhalation exposure design, *Environ. Health Perspect.*, 72:13-19, 1987.

Dodd, D.E., Snellings, W.M., Maronpot, R.R., and Ballantyne, B. Ethylene glycol monobutyl ether: acute 9-day and 90-day vapor inhalation studies in Fischer 344 rats, *Toxicol. Appl. Pharmacol.*, 68(3):405-414, 1983.

Dodge R.H., Cerniglia, C.E., and Gibson, D.T. Fungal metabolism of biphenyl, *Biochem. J.*, 178:223-230, 1979.

Dogger, J.R. and Bowery, T.G. A study of residues of some commonly used insecticides on alfalfa, *J. Econ. Entomol.*,

51:392-394, 1958.

Dogra, R.K.S., Chandra, K., Chandra, S., Khanna, S., Srivastava, S.N., Shukla, L., Katiyar, J.C., and Shanker, R. Di-*n*-octyl phthalate induced altered host resistance: viral and protozoal models in mice, *Ind. Health*, 27:83-87, 1989.

Dojlido, J.R. Investigations of biodegradability and toxicity of organic compounds. Final Report 1975-79, Municipal Environmental Research Laboratory, U.S. EPA Report-600/2-79-163, 1980.

D'Oliveira, J.-C., Al-Sayyed, G., and Pichat, P. Photodegradation of 2- and 3-chlorophenol in TiO_2 aqueous suspensions, *Environ. Sci. Technol.*, 24(7):990-996, 1990.

Domalski, E.S. and Hearing, E.D. Condensed phase heat capacity data, in *NIST Standard Reference Database Number 69*, Mallard, W.G. and Linstrom, P.J., Eds. (Gaithersburg, MD: National Institute of Standards and Technology, 1998) (http://webbook.nist.gov).

Donahue, D.J. and Bartell, F.E. The boundary tension at water-organic liquid interfaces, *J. Phys. Chem.*, 56(4):480-484, 1952.

Donberg, P.A., Odelson, D.A., Klečka, G.M., and Markham, D.A. Biodegradation of acrylonitrile in soil, *Environ. Toxicol. Chem.*, 11:1583-1594, 1992.

Dong, S. and Dasgupta, P.K. Solubility of gaseous formaldehyde in liquid water and generation of trace standard gaseous formaldehyde, *Environ. Sci. Technol.*, 20(6):637-640, 1986.

Dorough, H.W., Skrentny, R.F., and Pass, B.C. Residues in alfalfa and soils following treatment with technical chlordane and high purity chlordane (HCS 3260) for alfalfa weevil control, *J. Agric. Food Chem.*, 20(1):42-47, 1972.

Doucette, W.J. and Andren, A.W. Estimation of octanol/water partition coefficients: evaluation of six methods for highly hydrophobic aromatic hydrocarbons, *Chemosphere*, 17(2):345-359, 1988.

Doucette, W.J. and Andren, A.W. Aqueous solubility of biphenyl, furan, and dioxin congeners, *Chemosphere*, 17(2):243-252, 1988a.

Douslin, D.R. and Huffman, H.M. Low-temperature thermal data on the five isomeric hexanes, *J. Am. Chem. Soc.*, 68:1704-1708, 1946.

Draper, W.M. and Crosby, D.G. Solar photooxidation of pesticides in dilute hydrogen peroxide, *J. Agric. Food Chem.*, 32(2):231-237, 1984.

Dreher, R.M. and Podratzki, B. Development of an enzyme immunoassay for endosulfan and its degradation products, *J. Agric. Food Chem.*, 36(5):1072-1075, 1988.

Dreisbach, R.R. *Pressure-Volume-Temperature Relationships of Organic Compounds* (Sandusky, OH: Handbook Publishers, 1952), 349 p.

Dreisbach, R.R. *Physical Properties of Chemical Compounds. Adv. Chem. Ser.* 15 (Washington, D.C.: American Chemical Society, 1955).

Dreisbach, R.R. *Physical Properties of Chemical Compounds II. Adv. Chem. Ser.* 22 (Washington, D.C.: American Chemical Society, 1959).

Drew, R.T. and Fouts, J.R. The lack of effects of pretreatment with phenobarbital and chlorpromazine on the acute toxicity of benzene in rats, *Toxicol. Appl. Pharmacol.*, 27:183-193, 1974.

Drew, R.T., Laskin, S., Kuschner, M., and Nelson, N. Inhalation carcinogenicity of alpha halo ethers. I. The acute inhalation toxicity of chloromethyl methyl ether and bis(chloromethyl)ether, *Arch. Environ. Health*, 30(2):61-69, 1975.

Drinking Water Health Advisory. Pesticides. (Chelsea, MI: Lewis Publishers, 1989), 819 p.

Dubbs, M.D. and Gupta, R.B. Solubility of vitamin E (α-tocopherol) and vitamin K_3 (menadione) in ethanol-water mixture, *J. Chem. Eng. Data*, 43(4):590-591, 1998.

Dugan, P.R. *Biochemical Ecology of Water Pollution* (New York: Plenum Press, 1972), 159 p.

Duke, J.A. *Handbook of Phytochemical Constituents of Grass, Herbs, and Other Economic Plants* (Boca Raton, FL: CRC Press, 1992).

Duke, S.O., Moorman, T.B., and Bryson, C.T. Phytotoxicity of pesticide degradation products, in *Pesticide Transformation Products, Fate and Significance in the Environment*, ACS Symposium Series 459, Somasundaram, L. and Coats, J.R., Eds. (Washington, D.C.: American Chemical Society, 1991), pp. 188-204.

Dulfer, W.J., Bakker, M.W.C., and Govers, H.A.J. Micellar solubility and micelle/water partitioning of polychlorinated biphenyls in solutions of sodium dodecyl sulfate, *Environ. Sci. Technol.*, 29(4):985-992, 1995.

Dulin, D., Drossman, H., and Mill, T. Products and quantum yields for photolysis of chloroaromatics in water, *Environ. Sci. Technol.*, 20(1):72-77, 1986.

Dumont, J.N., Schultz, T.W., and Jones, R.D. Toxicity and teratogenicity of aromatic amines to *Xenopus laevis*, *Bull. Environ. Contam. Toxicol.*, 22(1/2):159-166, 1979.

Dunnivant, F.M. and Elzerman, A.W. Aqueous solubility and Henry's law constant for PCB congeners for evaluation of quantitative structure-property relationships (QSPRs), *Chemosphere*, 7(3):525-531, 1988.

Du Pont. *Solubility Relationships of the Freon Fluorocarbon Compounds*, Technical Bulletin B-7, (Wilmington, DE: Du Pont de Nemours, 1966).

Du Pont. Product Description: Adiponitrile (Wilmington, DE: Du Pont de Nemours, 1999).

Du Pont. Product Description: Dimethyl sulfate (Wilmington, DE: Du Pont de Nemours, 1999a).

Du Pont. Product Description: Dodecanedioic acid (Wilmington, DE: Du Pont de Nemours, 1999b).

Du Pont. Product Description: Hexamethylenediamine (Wilmington, DE: Du Pont de Nemours, 1999c).

Du Pont. Product Description: HFC-134a (Wilmington, DE: Du Pont de Nemours, 1999d).

Du Pont. Product Description: 2-Pentenenitrile (Wilmington, DE: Du Pont de Nemours, 1999e).

Du Pont. Product Description: *m*-Phenylenediamine (Wilmington, DE: Du Pont de Nemours, 1999f).

Du Pont. Product Description: *o*-Phenylenediamine (Wilmington, DE: Du Pont de Nemours, 1999g).

Du Pont. Product Description: *p*-Phenylenediamine (Wilmington, DE: Du Pont de Nemours, 1999h).

Eadie, B.J., Morehead, N.R., and Landrum, P.F. Three-phase partitioning of hydrophobic organic compounds in Great

Lakes waters, *Chemosphere*, 20(1/2):161-178, 1990.

Eadsforth, C.V. Application of reverse-phase H.P.L.C. for the determination of partition coefficients, *Pestic. Sci.*, 17(3):311-325, 1986.

Eckert, J.W. Fungistatic and phytotoxic properties of some derivatives of nitrobenzene, *Phytopathology*, 52:642-650, 1962.

Eder, G. and Weber, K. Chlorinated phenols in sediments and suspended matter of the Weser estuary, *Chemosphere*, 9(2):111-118, 1980.

Edney, E.O., Kleindienst, T.E., and Corse, E.W. Room temperature rate constants for the reaction of OH with selected chlorinated and oxygenated halocarbons, *Int. J. Chem. Kinet.*, 18:1355-1371, 1986.

Edmiston, C.E., Jr., Goheen, M., Malaney, G.W., and Mills, W.L. Evaluation of carbamate toxicity: acute toxicity in a culture of *Paramecium multimicronucleatum* upon exposure to aldicarb, carbaryl, and mexacarbate as measured by a Warburg respirometry and acute plate assay, *Environ. Res.*, 36(2):338-350, 1985.

Edwards, C.A. Nature and origins of pollution of aquatic systems by pesticides, in *Pesticides in Aquatic Environments*, Khan, M.A.Q., Ed. (New York: Plenum Press, 1977).

Edwards, D.A., Luthy, R.G., and Liu, Z. Solubilization of polycyclic aromatic hydrocarbons in micellar nonionic surfactant solutions, *Environ. Sci. Technol.*, 25(1):127-133, 1991.

Eganhouse, R.P. and Calder, J.A. The solubility of medium weight aromatic hydrocarbons and the effect of hydrocarbon co-solutes and salinity, *Geochim. Cosmochim. Acta*, 40(5):555-561, 1976.

Eiceman, G.A. and Vandiver, V.J. Adsorption of polycyclic aromatic hydrocarbons on fly ash from a municipal incinerator and a coal-fired power plant, *Atmos. Environ.*, 17(3):461-465, 1983.

Eichelberger, J.W. and Lichtenberg, J.J. Persistence of pesticides in river water, *Environ. Sci. Technol.*, 5(6):541-544, 1971.

Eisenhauer, H.R. The ozonization of phenolic wastes, *J. Water Pollut. Control Fed.*, 40(11):1887-1899, 1968.

Eisenreich, S.J., Looney, B.B., and Thornton, J.D. Airborne organic contaminants in the Great Lakes ecosystem, *Environ. Sci. Technol.*, 15(1):30-38, 1981.

Eisler, R. Acute toxicities of insecticides to marine decapod crustaceans, *Crustaceana*, 16(3):302-310, 1969.

Eklund, G., Pedersen, J.R., and Strömberg, B. Formation of chlorinated organic compounds during combustion of propane in the presence of HCl, *Chemosphere*, 16(1):161-166, 1987.

El-Dib, M.A. and Aly, O.S. Persistence of some phenylamide pesticides in the aquatic environment-I. Hydrolysis, *Water Res.*, 10(12):1047-1050, 1976.

El-Dib, M.A. and Badawy, M.I. Adsorption of soluble aromatic hydrocarbons on granular activated carbon, *Water Res.*, 13(3):255-258, 1979.

Ellgehausen, H., D'Hondt, C., and Fuerer, R. Reversed-phase chromatography as a general method for determining octan-1-ol/water partition coefficients, *Pestic. Sci.*, 12:219-227, 1981.

Ellgehausen, H., Guth, J.A., and Esser, H.O. Factors determining the bioaccumulation potential of pesticides in the individual compartments of aquatic food chains, *Ecotoxicol. Environ. Saf.*, 4(2):134-157, 1980.

Elliott, S. Effect of hydrogen peroxide on the alkaline hydrolysis of carbon disulfide, *Environ. Sci. Technol.*, 24(2):264-267, 1990.

Ellington, J.J. Octanol/water partition coefficients and water solubilities of phthalate esters, *J. Chem. Eng. Data*, 44(6):1414-1418, 1999.

Ellington, J.J. and Floyd, T.F. Octanol water partition coefficients for eight phthalate esters, Environmental Research Laboratory, Athens, GA., U.S. EPA Report-600/S-96/006, 1996.

Ellington, J.J., Karickhoff, S.W., Kitchens, B.E., Kollig, H.P., Long, J.M., Weber, E.J., and Wolfe, N.L. Environmental fate constants for organic chemicals under consideration for EPA's hazardous waste identification projects, Office of Research and Development, U.S. EPA Report 600/R-93/132, 1993.

Ellington, J.J. and Stancil, F.E., Jr. Octanol/water partition coefficients for evaluation of hazardous waste land disposal: selected chemicals, Office of Research and Development, U.S. EPA Report 600/M-88/010, 1988.

Ellington, J.J., Stancil, F.E., Jr., and Payne, W.D. Measurement of hydrolysis rate constants for evaluation of hazardous waste land disposal. Volume 1, Office of Research and Development, U.S. EPA Report 600/3-86/043, 1986, 122 p.

Ellington, J.J., Stancil, F.E., Jr., Payne, W.D., and Trusty, C.D. Measurement of hydrolysis rate constants for evaluation of hazardous waste land disposal. Volume 2. Data on 54 chemicals, Office of Research and Development, U.S. EPA Report 600/3-87/019, 1987, 152 p.

Ellington, J.J., Stancil, F.E., Payne, W.D., and Trusty, C.D. Measurement of hydrolysis rate constants for evaluation of hazardous waste land disposal. Volume 3. Data on 70 chemicals, Office of Research and Development, U.S. EPA Report 600/3-88/028, 1988, 23 p.

Elliott, S. The solubility of carbon disulfide vapor in natural aqueous systems, *Atmos. Environ.*, 23(9):1977-1980, 1989.

Ellis, P.A. and Camper, N.D. Aerobic degradation of diuron by aquatic microorganisms, *J. Environ. Sci. Health*, B17(3):277-289, 1982.

Elonen, G.F., Spehar, R.L., Holcombe, G.W., Johnson, R.D., Fernandez, J.D., Erickson, R.J., Tietge, J.E., and Cook, P.M. Comparative toxicity of 2,3,7,8-tetrachlorodibenzo-*p*-dioxin to seven freshwater fish species during early life-stage development, *Environ. Toxicol. Chem.*, 17(3):472-483, 1997.

Elrashidi, M.A. and O'Connor, G.A. Boron sorption and desorption in soils, *Soil Sci. Soc. Am. J.*, 46:87-93, 1982.

El-Refai, A. and Hopkins, T.L. Parathion absorption, translocation, and conversion to paraoxon in bean plants, *J. Agric. Food Chem.*, 14(6):588-592, 1966.

Elsner, E., Bieniek, D., Klein, W., and Korte, F. Verteilung und Umwandlumg von Aldrin, Heptachlor, und Lindan in der Grünalge *Chlorella pyrenoidosa*, *Chemosphere*, 1(6):247-250, 1972.

El Tayar, N., van de Waterbeemd, H., Grylaki, M., Testa, B., and Trager, W.F. The lipophilicity of deuterium atoms. A comparison of shake-flask and HPLC (high performance liquid chromatography) methods, *Int. J. Pharmaceut.*, 19:271-281, 1984.

El Zorghani, G.A. and Omer, M.E.H. Metabolism of endosulfan isomers by *Aspergillus niger*, *Bull. Environ. Contam. Toxicol.*, 12(2):182-185, 1974.

Emmerling, U., Figurski, G., and Rasmussen, P. Densities and kinematic viscosities for the systems benzene + methyl formate, benzene + ethyl formate, benzene + propyl formate, and benzene + butyl formate, *J. Chem. Eng. Data*, 43(3):289-292, 1998.

Enfield, C.G., Bengtsson, G., and Lindqvist, R. Influence of macromolecules on chemical transport, *Environ. Sci. Technol.*, 23(10):1278-1286, 1989.

Enfield, C.G. and Yates, S.R. Organic chemical transport in groundwater, in *Pesticides in the Soil Environment: Processes, Impacts, and Modeling*, Soil Science of Society of America, Cheng, H.H., Ed. (Madison, WI: SSSA, 1990), pp. 271-302.

Engelhardt, G., Wallnöfer, P.R., and Hutzinger, O. The microbial metabolism of di-*n*-butyl phthalate and related dialkyl phthalates, *Bull. Environ. Contam. Toxicol.*, 13(3):342-347, 1975.

Environment Canada. Priority Substances List Assessment Report No. 4. Toluene, 26 p., 1993.

Environment Canada. Priority Substances List Assessment Report: bis(2-chloroethyl) ether, 15 p., 1993.

Environment Canada. Priority Substances List Assessment Report: bis(2-chloromethyl) ether and chloromethyl methyl ether, 1993a, 17 p.

Eon, C., Pommier, C., and Guiochon, G. Vapor pressures and second virial coefficients of some five-membered heterocyclic derivatives, *J. Chem. Eng. Data*, 16(4):408-410, 1971.

Epling, G.A., McVicar, W.M., and Kumar, A. Borohydride-enhanced photodehalogenation of aroclor 1232, 1242, 1254, and 1260, *Chemosphere*, 17(7):1355-1362, 1988.

EPRI. *Land and Water Quality News*. Electric Power Research Institute, 5(3):1-10, 1990.

Ernst, W. Determination of the bioconcentration potential of marine organisms. A steady state approach, *Chemosphere*, 6(11):731-740, 1977.

Ernst, W., Doe, K., Jonah, P., Young, J., Julien, G., and Hennigar, P. The toxicity of chlorothalonil to aquatic fauna and the impact of its operational use on a pond system, *Arch. Environ. Contam. Toxicol.*, 21(1):1-9, 1991.

Erstfeld, K.M., Simmons, M.S., and Atallah, Y.H. Sorption and desorption characteristics of chlordane onto sediments, *J. Environ. Sci. Health*, B31(1):43-58, 1996.

Etnier, E. Water quality criteria for 2,4-dinitrotoluene and 2,6-dinitrotoluene. Final report, Oak Ridge National Laboratory, Oak Rudge, TN, Report ORNL-6312.

Ettre, L.S., Welter, C., and Kolb, B. Determination of gas-liquid partition coefficients by automatic equilibrium headspace-gas chromatography utilizing the phase ratio variation method, *Chromatographia*, 35:73-84, 1993.

Etzweiler, F., Senn, E., and Schmidt, H.W.H. Method for measuring aqueous solubilities of organic compounds, *Anal. Chem.*, 67(3):655-658, 1995.

Evans, B.K., James, K.C., and Luscombe, D.K. Quantitative structure-activity relationships and carminative activity, *J. Pharm. Sci.*, 67:277-278, 1978.

Evmorfopoulos, E. and Glavas, S. Formation of nitrogenous compounds in the photooxidation of *n*-butane under atmospheric conditions, *Chem. Monthly*, 129(11):1151-1159, 1998.

Exxon. Tables of Useful Information (Houston, TX: Exxon, 1974), 70 p.

Fairbanks, B.C. and O'Connor, G.A. Effect of sewage sludge on the adsorption of polychlorinated biphenyls by three New Mexico soils, *J. Environ. Qual.*, 13(2):297-300, 1984.

Fairbanks, B.C., Schmidt, N.E., and O'Connor, G.A. Butanol degradation and volatilization in soils amended with spent acid or sulfuric acid, *J. Environ. Qual.*, 14(1):83-86, 1985.

Farm Chemicals Handbook (Willoughby, OH: Meister Publishing, 1988).

Farmer, W.J. and Aochi, Y. Picloram sorption by soils, *Soil Sci. Soc. Am. Proc.*, 38:418-423, 1974.

Farmer, W.J., Yang, M.S., Letey, J., and Spencer, W.F. Hexachlorobenzene: its vapor pressure and vapor phase diffusion in soil, *Soil Sci. Soc. Am. J.*, 44:676-680, 1980.

Farrell, P.G. and Newton, J. Ionization potentials of primary aromatic amines and azo-hydrocarbons, *Tetrahedron Lett.*, p. 5517, 1966.

Farrell, P.G., Shahidi, F., Casellato, F., Vecchi, C., and Girelli, A. DSC studies of aromatic picrates, *Thermochim. Acta*, 33:275-280, 1979.

Fathepure, B.Z., Tiedje, J.M., and Boyd, S.A. Reductive dechlorination of hexachlorobenzene to tri- and dichlorobenzenes in anaerobic sewage sludge, *Appl. Environ. Microbiol.*, 54(2):327-330, 1988.

Fatiadi, A.J. Effects of temperature and ultraviolet radiation on pyrene adsorbed on garden soil, *Environ. Sci. Technol.*, 1(7):570-572, 1967.

Faulkner, J.K. and Woodstock, D. Metabolism of 2,4-dichlorophenoxyacetic acid (2,4-D) by *Aspergillus niger*, *Nature (London)*, 203:865, 1964.

Felsot, A. and Dahm, P.A. Sorption of organophosphorus and carbamate insecticides by soil, *J. Agric. Food Chem.*, 27(3):557-563, 1979.

Felsot, A.J., Maddox, J.V., and Bruce, W. Dissipation of pesticide from soils with histories of furadan use, *Bull. Environ. Contam. Toxicol.*, 26(6):781-788, 1981.

Fendinger, N.J. and Glotfelty, D.W. A laboratory method for the experimental determination of air-water Henry's law constants for several pesticides, *Environ. Sci. Technol.*, 22(11):1289-1293, 1988.

Fendinger, N.J. and Glotfelty, D.W. Henry's law constants for selected pesticides, PAHs and PCBs, *Environ. Toxicol. Chem.*, 9(6):731-735, 1990.

Fendinger, N.J., Glotfelty, D.W., and Freeman, H.P. Comparison of two experimental techniques for determining air/water Henry's law constants, *Environ. Sci. Technol.*, 23(12):1528-1531, 1989.

Fernandez-Casalderry, A., Ferrando, M.D., and Andreu-Moliner. Endosulfan and diazinon toxicity to the freshwater rotifer *Brachionus calyciflorus*, *J. Environ. Sci. Health*, B27(2):155-164, 1992.

Fernandez-Casalderry, A., Ferrando, M.D., and Andreu-Moliner. Effect of sublethal concentrations of pesticides on the feeding behavior of *Daphnia magna*, *Ecotoxicol. Environ. Saf.*, 27(1):82-89, 1994.

Fernández-Vega, C., Sancho, E., Ferrando, M.D., and Andreu-Moliner, E. Rhiobencarb toxicity and plasma AchE inhibition in the European eel, *J. Environ. Sci. Health*, B34(1):61-73, 1999.

Ferrando, M.D. and Andreu-Moliner. Acute lethal toxicity of some pesticides to *Brachionus calyciforus* and *Brachionus plicatilis*, *Bull. Environ. Contam. Toxicol.*, 47(3):479-484, 1991.

Ferrando, M.D., Andreu-Moliner, E., and Fernandez-Casalderrey, A. Relative sensitivity of *Daphnia magna* and *Brachionus calyciflorus* to five pesticides, *J. Environ. Sci. Health*, B27(5):511-522, 1992.

Ferrando, M.D., Sancho, E., and Andreu-Moliner, E. Chronic toxicity of fenitrothion to an alage (*Nannochloris oculata*), a rotifer (*Brachionus calyciflorus*), and the cladoceran (*Daphnia magna*), *Ecotoxicol. Environ. Saf.*, 35(2):112-122, 1996.

Ferreira, G.A.L. and Sieber, J.N. Volatilization and exudation losses of three *n*-methylcarbamate insecticides applied systemically to rice, *J. Agric. Food Chem.*, 29(1):93-99, 1981.

Ferris, I.G. and Lichtenstein, E.P. Interactions between agricultural chemicals and soil microflora and their effects on the degradation of [^{14}C]parathion in a cranberry soil, *J. Agric. Food Chem.*, 28(5):1011-1019, 1980.

Feung, C.-S., Hamilton, R.H., and Mumma, R.O. Metabolism of 2,4-dichlorophenoxyacetic acid. V. Identification of metabolites in soybean callus tissue cultures, *J. Agric. Food Chem.*, 21(4):637-640, 1973.

Feung, C.-S., Hamilton, R.H., and Witham, F.H. Metabolism of 2,4-dichlorophenoxyacetic acid by soybean cotyledon callus tissue cultures, *J. Agric. Food Chem.*, 19(3):475-479, 1971.

Feung, C.-S., Hamilton, R.H., Witham, F.H., and Mumma, R.O. The relative amounts and identification of some 2,4-dichlorophenoxyacetic acid metabolites isolated from soybean cotyledon callus cultures, *Plant Physiol.*, 50:80-86, 1972.

Fewson, C.A. Biodegradation of aromatics with industrial relevance, in *FEMS Symp. No. 12. Microbial Degradation of Xenobiotics and Recalcitrant Compounds*, 12:141-179, 1981.

Fichan, I., Larroche, C., and Gros, J.B. Water solubility, vapor pressure, and activity coefficients of terpenes and terpenoids, *J. Chem. Eng. Data*, 44(1):56-62, 1999.

Finger, S.E., Little, E.F., Henry, M.G., Fairchild, J.F., and Boyle, T.P. Comparison of laboratory and field assessment of fluorene. Part I: effects of fluorene on the survival, growth, reproduction and behavior of aquatic organisms in laboratory tests, in *Validation and Predictability of Laboratory Methods for Assessing the Fate and Effects of Contaminants in Aquatic Environments, ASTM STP 865*, Boyle, T.P., Ed. (Philadelphia, PA: American Society for Testing and Materials, 1985), pp. 120-133.

Finke, H.L., Gross, M.E., Waddington, G., and Huffman, H.M. Low temperature thermal data for the nine normal paraffin hydrocarbons from octane to hexadecane, *J. Am. Chem. Soc.*, 76:333-341, 1954.

Finke, H.L., Messerly, J.F., Lee, S.H., Osborn, A.G., and Douslin, D.R. Comprehensive thermodynamic studies of seven aromatic hydrocarbons, *J. Chem. Thermodyn.*, 9:937-956, 1977.

Finlayson-Pitts, B.J.F. and Pitts, J.N., Jr. *Atmospheric Chemistry: Fundamentals and Experimental Techniques* (New York: John Wiley & Sons, 1986), 1098 p.

Fischer, R.G. and Ballschmiter, K. Prediction of the environmental distribution of alkyl dinitrates - Chromatographic determination of vapor pressure p°, water solubility S_{H2O}, gas-water partition coefficient K_{GW} (Henry's law constant) and octanol-water partition coefficient K_{ow}, *Fresenius J. Anal. Chem.*, 360:769-776, 1998.

Fischer, R.G. and Ballschmiter, K. Determination of vapor pressure, water solubility, gas-water partition coefficient P_{GW}, Henry's law constant, and octanol-water partition coefficient P_{ow} of 26 alkyl dinitrates, *Chemosphere*, 36(14):2891-2901, 1998a.

Fischer, I. and Ehrenberg, L. Studies of the hydrogen bond. II. Influence of the polarizability of the heteroatom, *Acta Chem. Scand.*, 2:669-677, 1948.

Fischer, S.A. and Hall, L.W., Jr. Environmental concentrations and aquatic toxicity data on diflubenzuron (dimilin), *Crit. Rev. Toxicol.*, 22(1):45-79, 1992.

Fishbein, L. Potential halogenated industrial carcinogenic and mutagenic chemicals. III. Alkane halides, alkanols, and ethers, *Sci. Total Environ.*, 11(3):223-257, 1979.

Fishbein, L. An overview of environmental and toxicological aspects of aromatic hydrocarbons. III. Xylene, *Sci. Total Environ.*, 43(1/2):165-183, 1985.

Fishbein, L. and Albro, P.W. Chromatographic and biological aspects of the phthalate esters, *J. Chromatogr.*, 70(2):365-412, 1972.

Fleeker, J. and Steen, R. Hydroxylation of 2,4-D in several weed species, *Weed Sci.*, 19:507, 1971.

Fluka. Fluka Catalog 1988/89 - Chemika-Biochemika (Ronkonkoma, NY: Fluka Chemical, 1988), 1536 p.

Fluka. Fluka Catalog 1989/90 - Chemika-Biochemika (Ronkonkoma, NY: Fluka Chemical, 1990), 1480 p.

Fochtman, E.G. and Eisenberg, W. Treatability of carcinogenic and other hazardous organic compounds, Municipal Environmental Research Laboratory, U.S. EPA Report-600/2-79-097, 1979.

Fogel, S., Lancione, R., Sewall, A., and Boething, R.S. Enhanced biodegradation of methoxychlor in soil under enhanced environmental conditions, *Appl. Environ. Microbiol.*, 44(1):113-120, 1982.

Forbis, D.A. Acute toxicity of methyl methacrylate to *Selenastrum capricornutum printz*, Analytical Biochemistry Laboratories, Inc. Report 37329, 1990.

Fordyce, C.R. and Meyer, L.W.A. Plasticizers for cellulose acetate and cellulose acetate butyrate, *Ind. Eng. Chem.*, 32(8):1053-1060, 1940.

Fordyce, R.G. and Chapin, E.C. Copolymerization. I. The mechanism of emulsion copolymerization of styrene and acrylonitrile, *J. Am. Chem. Soc.*, 69(3):581-583, 1947.

Foreman, W.T. and Bidleman, T.F. Vapor pressure estimates of individual polychlorinated biphenyls and commercial fluids using gas chromatographic retention data, *J. Chromatogr.*, 330(2):203-216, 1985.

Forget, J., Pavillon, J.F., Menasria, M.R., and Bocquene, G. Motality and LC₅₀ values for several stages of the marine copepod *Tigriopus brevicornis* (Muller) exposed to the metals arsenic and cadmium and the pesticides atrazine, carbofuran, dichlorvos, and malathion, *Ecotoxicol. Environ. Saf.*, 40(3):239-244, 1998.

Forrest, M., Lord, K.A., and Walker, N. The influence of soil treatments on the bacterial degradation of diazinon and other organophosphorus insecticides, *Environ. Pollut. Series A*, 24:93-104, 1981.

Forziati, A.F., Norris, W.R., and Rossini, F.D. Vapor pressures and boiling points of sixty API-NBS hydrocarbons, *J. Res. Nat. Bur. Stand.*, 43:555-563, 1949.

Foster, R.K. and McKercher, R.B. Laboratory incubation studies of chlorophenoxyacetic acids in chernozemic soils, *Soil Biol. Biochem.*, 5:333-337, 1973.

Fowkes, F.M. Surface effects of anisotropic london dispersion forces in *n*-alkanes, *J. Phys. Chem.*, 84(5):510-512, 1980.

Fowler, L., Trump, W.N., and Vogler, C.E. Vapor pressure of naphthalene - New measurements between 40° and 180 °C, *J. Chem. Eng. Data*, 13(2):209-210, 1968.

Francis, A.J., Spanggord, R.J., and Ouchi, G.I. Degradation of lindane by *Escherichia coli*, *Appl. Environ. Microbiol.*, 29(4):567-568, 1975.

Francis, A.W. and Robbins, G.W. The vapor pressures of propane and propylene, *J. Am. Chem. Soc.*, 55(11):4339-4342, 1933.

Franco, I., Vischetti, C., Baca, M.T., De Nobili, M., Mondini, C., and Leita, L. Adsorption of linuron and metamitron on soil and peats at two different decomposition stages, *J. Soil Contam.*, 6(3):307-315, 1997.

Frandsen, M. Cryoscopic constant, heat of fusion, and heat capacity of camphor, *Bur. Stand. J. Res.*, 7:477-483, 1931.

Frank, R., Braun, H.E., Van Hove Holdrimet, M., Sirons, G.J., and Ripley, B.D. Agriculture and water quality in the Canadian Great Lakes basin: V. Pesticide use in 11 agricultural watersheds and presence in stream water, 1975-1977, *J. Environ. Qual.*, 11(3):497-505, 1982.

Frankel, L.S., McCallum, K.S., and Collier, L. Formation of bis(chloromethyl) ether from formaldehyde and hydrogen chloride, *Environ. Sci. Technol.*, 8(4):356-359, 1974.

Franklin, J.L., Dillard, J.G., Rosenstock, H.M., Herron, J.T., Draxl, K., and Field, F.H. Ionization potentials, appearance potentials and heats of formation of gaseous positive ions, National Bureau of Standards Report NSRDS-NBS 26, U.S. Government Printing Office, 1969, 289 p.

Franks, F. Solute-water interactions and the solubility behaviour of long-chain paraffin hydrocarbons, *Nature (London)*, 210(5031):87-88, 1966.

Franks, F., Gent, M., and Johnson, H.H. The solubility of benzene in water, *J. Chem. Soc. (London)*, pp. 2716-2723, 1963.

Freed, V.H., Chiou, C.T., and Haque, R. Chemodynamics: transport and behavior of chemicals in the environment - A problem in environmental health, *Environ. Health Perspect.*, 20:55-70, 1977.

Freed, V.H., Chiou, C.T., and Schmedding, D.W. Degradation of selected organophosphate pesticides in water and soil, *J. Agric. Food Chem.*, 27(4):706-708, 1979.

Freed, V.H., Schmedding, D., Kohnert, R., and Haque, R. Physical chemical properties of several organophosphates: some implication in environmental and biological behavior, *Pestic. Biochem. Physiol.*, 10:203-211, 1979a.

Freese, E., Levin, B.C., Pearce, R., Sreevalson, T., Kaufman, J.J., Koski, W.S., and Semo, N.M. Correlation between the growth inhibitory effects, partition coefficients and teratogenic effects of lipophilic acids, *Teratology*, 20(3):413-440, 1979.

Freitag, D., Ballhorn, L., Geyer, H., and Korte, F. Environmental hazard profile of organic chemicals, *Chemosphere*, 14(10):1589-1616, 1985.

Freidig, A.P., Verhaar, H.J.M., and Hermens, J.L.M. Quantitative structure-property relationships for the chemical reactivity of acrylates and methacrylates, *Environ. Toxicol. Chem.*, 18(6):1133-1139, 1998.

Freitag, D., Geyer, H., Kraus, A., Viswanathan, R., Kotzias, D., Attar, A., Klein, W., and Korte, F. Ecotoxicological profile analysis. VII. Screening chemicals for their environmental behavior by comparative evaluation, *Ecotoxicol. Environ. Saf.*, 6(1):60-81, 1982.

Freitag, D., Scheunert, I., Klein, W., and Korte, F. Long-term fate of 4-chloroaniline-¹⁴C in soil and plants under outdoor conditions. A contribution to terrestrial ecotoxicology of chemicals, *J. Agric. Food Chem.*, 32(2):203-207, 1984.

Freundt, K.J., Liebalt, G.P., and Liberwirth, E. Toxicity studies of *trans*-1,2-dichloroethane, *Toxicology*, 7:141-153, 1977.

Friant, S.L. and Suffet, I.H. Interactive effects of temperature, salt concentration, and pH on head space analysis for isolating volatile trace organics in aqueous environmental samples, *Anal. Chem.*, 51(13):2167-2172, 1979.

Frías-Espericueta, M.G., Harfush-Melendez, M., Osuna-López, J.I., and Páez-Osuna, F. Acute toxicity of ammonia to juvenile shrimp *Penaeus vannamei* Boone, *Bull. Environ. Contam. Toxicol.*, 62(5):646-652, 1999.

Friberg, L., Kylin, B., and Nystrom, A. Toxicities of trichloroethylene and tetrachloroethylene and Fujiwara's pyridine-alkali reaction, *Acta Pharmacol. Toxicol.*, 9:303-312, 1953.

Fries, G.F. Degradation of chlorinated hydrocarbons under anaerobic conditions, in *Fate of Organic Pesticides in the Aquatic Environment, Advances in Chemistry Series*, Gould, R.F., Ed. (Washington, D.C.: American Chemical Society, 1972), pp. 256-270.

Fries, G.R., Marrow, G.S., and Gordon, C.H. Metabolism of *o,p'*- and *p,p'*-DDT by rumen microorganisms, *J. Agric.*

Food Chem., 17(4):860-862, 1969.

Friesel, P., Milde, G., and Steiner, B. Interactions of halogenated hydrocarbons with soils, *Fresenius Z. Anal. Chem.*, 319:160-164, 1984.

Friesen, K.J., Sarna, L.P., and Webster, G.R.B. Aqueous solubility of polychlorinated dibenzo-*p*-dioxins determined by high pressure liquid chromatography, *Chemosphere*, 14(9):1267-1274, 1985.

Friesen, K.J., Vilk, J., and Muir, D.C.G. Aqueous solubilities of selected 2,3,7,8-substituted polychlorinated dibenzofurans (PCDFs), *Chemosphere*, 20(1/2):27-32, 1980.

Fu, J.-K. and Luthy, R.G. Aromatic compound solubility in solvent/water mixtures, *J. Environ. Eng.*, 112(2):328-345, 1986.

Fu, M.H. and Alexander, M. Biodegradation of styrene in samples of natural environments, *Environ. Sci. Technol.*, 26(8):1540-1544, 1992.

Fühner, H. Water-solubility in homologous series, *Ber. Dtsch. Chem. Ges.*, 57B:510-515, 1924.

Fühner, R. and Geiger, M. A simple method of determining the aqueous solubility of organic substances, *Pestic. Sci.*, 8(4):337-344, 1977.

Fuhremann, T.W. and Lichtenstein, E.P. A comparative study of the persistence, movement and metabolism of six C^{14}-insecticides in soils and plants, *J. Agric. Food Chem.*, 28:446-452, 1980.

Fujisawa, S. and Masuhara, E. Determination of partition coefficients of acrylates, methacrylates and vinyl monomers using high performance liquid chromatography (HPLC), *J. Biomed. Mater. Res.*, 15:787-793, 1981.

Fujita, T., Ishii, A., and Sakagami, Y. Photodecomposition of endrin, *J. Hyg. Chem.*, 15(1):9-12, 1969.

Fujita, T., Iwasa, J., and Hansch, C. A new substituent constant, π, derived from partition coefficients, *J. Am. Chem. Soc.*, 86(23):5175-5180, 1964.

Fukano, I. and Obata, Y. Solubility of phthalates in water [Chemical abstract 120601u, 86(17):486, 1977]: *Purasuchikkusu*, 27(7):48-49, 1976.

Fukuda, K., Yasushi, I., Maruyama, T., Kojima, H.I., and Yoshida, T. On the photolysis of alkylated naphthalenes in aquatic systems, *Chemosphere*, 17(4):651-659, 1988.

Fukui, S., Hirayama, T., Shindo, H., and Nohara, M. Photochemical reaction of biphenyl (BP) and *o*-phenylphenol (OPP) with nitrogen monoxide (1), *Chemosphere*, 9(12):771-775, 1980.

Fuller, B.B. *Air Pollution Assessment of Tetrachloroethylene* (McLean, VA: The Mitre Corp., 1976), 87 p.

Funaski, N., Hada, S., and Neya, S. Partition coefficients of aliphatic ethers-molecular surface area approach, *J. Phys. Chem.*, 89:3046-3049, 1985.

Fung, H.-L. and Higuchi, T. Molecular interactions and solubility of polar nonelectrolytes in nonpolar solvents, *J. Pharm. Sci.*, 60(12):1782-1788, 1971.

Furukawa, G.T., McCoskey, R.E., and King, G.J. Calorimetric properties of benzoic acid from 0 to 410K, *J. Res. Nat. Bur. Stand.*, 47:256-261, 1951.

Furukawa, K., Arai, T., and Hashimoto, S. Volatile components of *Telosma cordata* Merrill flowers, *Flavour Fragr. J.*, 8:221-223, 1993.

Gäb, S., Parlar, H., Nitz, S., Hustert, K., and Korte, F. Beitrage zur okologischen Chemie. LXXXI. Photochemischer Abbau von Aldrin, Dieldrin und Photodieldrin als Festkorper im Sauerstoffstrom, *Chemosphere*, 3(5):183-186, 1974.

Gäb, S., Šaravanja, V., and Korte, F. Irradiation studies of aldrin and chlordene adsorbed on a silica gel surface, *Bull. Environ. Contam. Toxicol.*, 13(3):301-306, 1975.

Gabler, T., Paschke, A., and Schüürmann, G. Diffusion coefficients of substituted benzenes at high dilution in water, *J. Chem. Eng. Data*, 41(1):33-36, 1996.

Gad, S.C., Rusch, G.M., Darr, R.W., Cramp, A.L., Hoffman, G.M., and Peckham, J.C. Inhalation toxicity of hexafluoroisobutylene, *Toxicol. Appl. Pharmacol.*, 86(3):327-340, 1986.

Gaffney, J.S., Streit, G.E., Spall, W.D., and Hall, J.H. Beyond acid rain, *Environ. Sci. Technol.*, 21(6):519-524, 1987.

Gaines, T.B. Acute toxicity of pesticides, *Toxicol. Appl. Pharmacol.*, 14:515-534, 1969.

Galassi, S., Mingazzini, M., Viganò, L., Cesareo, D., and Tosato, M.L. Approaches to modeling toxic responses of aquatic organisms to aromatic hydrocarbons, *Ecotoxicol. Environ. Saf.*, 16(2):158-169, 1988.

Gälli, R. and McCarty, P.L. Biotransformation of 1,1,1-trichloroethane, trichloromethane, and tetrachloromethane by a *Clostridium* sp., *Appl. Environ. Microbiol.*, 55(4):837-844, 1989.

Gambrell, R.P., Taylor, B.A., Reddy, K.S., and Patrick, W.H., Jr. Fate of selected toxic compounds under controlled redox potential and pH conditions in soil and sediment-water systems, U.S. EPA Report 600/3-83-018, 1984.

Gan, J. and Yates, S.R. Degradation and phase partition of methyl iodide in soil, *J. Agric. Food Chem.*, 44(12):4001-4008, 1996.

Gandhi, N.N., Kumar, M.D., and Sathyamurthy, N. Effect of hydrotropes on solubility and mass-transfer coefficient of butyl acetate, *J. Chem. Eng. Data*, 43(5):695-699, 1998.

Gannon, N. and Decker, G.C. The conversion of heptachlor to its epoxide on plants, *J. Econ. Entomol.*, 51(1):3-7, 1958.

Garbarini, D.R. and Lion, L.W. Influence of the nature of soil organics on the sorption of toluene and trichloroethylene, *Environ. Sci. Technol.*, 20(12):1263-1269, 1986.

Garcia-Miragaya, J. Specific sorption of trace amounts of cadmium by soils, *Comm. Soil Sci. Plant Anal.*, 11:1157-1166, 1980.

Gardner, D.R. and Vincent, S.R. Insecticidal and neuroexitant actions of DDT analogs on the cockroach, *Periplaneta americana*, *Bull. Environ. Contam. Toxicol.*, 20(3):294-302, 1978.

Garg, S.K., Katoch, R., and Bhushan, C. Efficacy of flumethrin pour-on against *Damalinia caprae* of goats (*Capra hircus*), *Trop. Anim. Health Prod.*, 30(5):273-278, 1998.

Garraway, J. and Donovan, J. Gas phase reaction of hydroxyl radical with alkyl iodides, *J. Chem. Soc. Chem. Comm.*, 23:1108, 1979.

Garst, J.E. Accurate, wide-range, automated, high-performance liquid chromatographic method for the estimation of octanol/water partition coefficients II: equilibrium in partition coefficient measurements, additivity of substituent constants, and correlation of biological data, *J. Pharm. Sci.*, 73(11):1623-1629, 1984.

Garst, J.E. and Wilson, W.C. Accurate, wide-range, automated, high-performance liquid chromatographic method for the estimation of octanol/water partition coefficients I: effects of chromatographic conditions and procedure variables on accuracy and reproducibility of the method, *J. Pharm. Sci.*, 73(11):1616-1622, 1984.

Garten, C.T. and Trabalka, J.R. Evaluation of models for predicting terrestrial food chain behavior of xenobiotics, *Environ. Sci. Technol.*, 17(10):590-595, 1983.

Gauthier, T.D., Selfz, W.R., and Grant, C.L. Effects of structural and compositional variations of dissolved humic materials on K_{oc} values, *Environ. Sci. Technol.*, 21(5):243-248, 1987.

Gauthier, T.D., Shane, E.C., Guerin, W.F., Seitz, W.R., and Grant, C.L. Fluorescence quenching method for determining equilibrium constants for polycyclic aromatic hydrocarbons binding to humic materials, *Environ. Sci. Technol.*, 20(11):1162-1166, 1986.

Gay, B.W., Jr., Hanst, P.L., Bufalini, J.J., and Noonan, R.C. Atmospheric oxidation of chlorinated ethylenes, *Environ. Sci. Technol.*, 10(1):58-67, 1976.

Gaynor, J.D. and Van Volk, V. *s*-Triazine solubility in chloride salt solutions, *J. Agric. Food Chem.*, 29(6):1143-1146, 1981.

Geissbühler, H., Martin, H., and Voss, G. The substituted ureas, in *Herbicides: Chemistry, Degradation, and Mode of Action*, Kearney, P.C. and Kaufman, D.D., Eds. (New York: Marcel Dekker, 1975), pp. 209-292.

Georgacakis, E. and Khan, M.A.Q. Toxicity of the photoisomers of cyclodiene insecticides to freshwater animals, *Nature (London)*, 233(5315):120-121, 1971.

Gerade, H.W. and Gerade, D.F. Industrial experience with 3,3′-dichlorobenzidine: an epidemiological study of a chemical manufacturing plant, *J. Occup. Med.*, 16:322-344, 1974.

Gersich, F.M. and Milazzo, D.P. Chronic toxicity of aniline and 2,4-dichlorophenol to *Daphnia magna* Straus, *Bull. Environ. Contam. Toxicol.*, 40(1):1-7, 1988.

Gerstl, Z. and Helling, C.S. Evaluation of molecular connectivity as a predictive method for the adsorption of pesticides in soils, *J. Environ. Sci. Health*, B22(1):55-69, 1987.

Gerstl, Z. and Mingelgrin, U. Sorption of organic substances by soils and sediments, *J. Environ. Sci. Health*, B19(3):297-312, 1984.

Gerstl, Z., Nye, P.H., and Yaron, B. Diffusion of a biological pesticide: II. As affected by microbial decomposition, *Soil Sci. Am. Proc. J.*, 43(5):843-848, 1979.

Gerstl, Z. and Yaron, B. Behavior of bromacil and napropamide in soils: I. Adsorption and degradation, *Soil Sci. Soc. Am. J.*, 47(3):474-478, 1983.

Getzin, L.W. Persistence and degradation of carbofuran in soil, *Environ. Entomol.*, 2:461-467, 1973.

Getzin, L.W. Degradation of chlorpyrifos in soil: influence of autoclaving, soil moisture, and temperature, *J. Econ. Entomol.*, 74(2):158-162, 1981.

Getzin, L.W. Dissipation of chlorpyrifos from dry soil surfaces, *J. Econ. Entomol.*, 74(6):707-713, 1981a.

Getzin, L.W. Factors influencing the persistence and effectiveness of chlorpyrifos in soil, *J. Econ. Entomol.*, 78:412-418, 1985.

Getzin, L.W. and Rosefield, I. Organophosphorus insecticide degradation by heat-labile substances in soil, *J. Agric. Food Chem.*, 16(4):598-601, 1968.

Geyer, H., Kraus, A.S., Klein, W., Richter, E., and Korte, F. Relationship between water solubility and bioaccumulation potential of organic chemicals in rats, *Chemosphere*, 9(5/6):277-294, 1980.

Geyer, H., Politzki, G., and Freitag, D. Prediction of ecotoxicological behaviour of chemicals: relationship between *n*-octanol/water partition coefficient and bioaccumulation of organic chemicals by alga *Chlorella*, *Chemosphere*, 13(2):269-284, 1984.

Geyer, H., Scheunert, I., Brüggemann, R., Matthies, M., Steinberg, C.E.W., Zitko, V., Kettrup, A., and Garrison, W. The relevance of aquatic organisms' lipid content to the toxicity of lipophilic chemicals: toxicity of lindane to different fish species, *Ecotoxicol. Environ. Saf.*, 28(1):53-70, 1994.

Geyer, H.J., Scheunert, I., and Korte, F. The effects of organic environmental chemicals on the growth of the alga *Scenedesmus subspicatus*: a contribution to environmental biology, *Chemosphere*, 14(9):1355-1369, 1985.

Geyer, H.J., Scheunert, I., and Korte, F. Correlation between the bioconcentration potential of organic environmental chemicals in humans and their *n*-octanol/water partition coefficients, *Chemosphere*, 16(1):239-252, 1987.

Geyer, H., Sheehan, P., Kotzias, D., Freitag, D., and Korte, F. Prediction of ecotoxicological behaviour of chemicals: relationship between physico-chemical properties and bioaccumulation of organic chemicals in the mussel *Mytilus edulis*, *Chemosphere*, 11(11):1121-1134, 1982.

Gherini, S.A., Summers, K.V., Munson, R.K., and Mills, W.B. *Chemical Data for Predicting the Fate of Organic Compounds in Water, Volume 2: Database* (Lafayette, CA: Tetra Tech, 1988), 433 p.

Giam, C.S., Atlas, E., Chan, H.S., and Neff, G.S. Phthalate esters, PCB and DDT residues in the Gulf of Mexico atmosphere, *Atmos. Environ.*, 14(1):65-69, 1980.

Gibbard, H.F. and Creek, J.L. Vapor pressure of methanol from 288.15 to 337.65K, *J. Chem. Eng. Data*, 19(4):308-310, 1974.

Gibson, D.T. Microbial degradation of aromatic compounds, *Science (Washington, D.C.)*, 161(3846):1093-1097, 1968.

Gibson, D.T., Mahadevan, V., Jerina, D.M., Yagi, H., and Yeh, H.J.C. Oxidation of the carcinogens benzo[*a*]pyrene and benzo[*a*]anthracene to dihydrols by a bacterium, *Science (Washington, D.C.)*, 189(4199):295-297, 1975.

Gibson, H.W. Linear free energy relationships. VIII. Ionization potentials of aliphatic compounds, *Can. J. Chem.*, 55(14):2637-2641, 1977.

Gibson, W.P. and Burns, R.G. The breakdown of malathion in soil and soil components, *Microbial Ecol.*, 3:219-230, 1977.

Gile, J.D. and Gillett, J.W. Fate of selected fungicides in a terrestrial laboratory ecosystem, *J. Agric. Food Chem.*, 27(6):1159-1164, 1979.

Gill, T.S., Pande, J., and Tewari, H. Individual and combined toxicity of common pesticides to teleost *Puntius conchonius* Hamilton, 29(2):145-148, 1991.

Gillette, L.A., Miller, D.L., and Redman, H.E. Appraisal of a chemical waste problem by fish toxicity tests, *Sewage Ind. Wastes*, 24(11):1397-1401, 1952.

Gingell, R., Boatman, R.J., and Lewis, S. Acute toxicity of ethylene glycol mono-*n*-butyl ether in the guinea pig, *Food Chem. Toxicol.*, 36(9-10):825-829, 1998.

Ginnings, P.M. and Baum, R. Aqueous solubilities of the isomeric pentanols, *J. Am. Chem. Soc.*, 59(6):1111-1113, 1937.

Ginnings, P.M. and Coltrane, D. Aqueous solubility of 2,2,3-trimethylpentanol-3, *J. Am. Chem. Soc.*, 61(2):525, 1939.

Ginnings, P.M. and Furukawa, G.T. Heat capacity standards for the range 14 to 1200°K, *J. Am. Chem. Soc.*, 75:522-527, 1953.

Ginnings, P.M. and Hauser, M. Aqueous solubilities of some isomeric heptanols, *J. Am. Chem. Soc.*, 60(11):2581-2582, 1938.

Ginnings, P.M., Herring, E., and Coltrane, D. Aqueous solubilities of some unsaturated alcohols, *J. Am. Chem. Soc.*, 61(4):807-808, 1939.

Ginnings, P.M., Plonk, D., and Carter, E. Aqueous solubilities of some aliphatic ketones, *J. Am. Chem. Soc.*, 62(8):1923-1924, 1940.

Ginnings, P.M. and Webb, R. Aqueous solubilities of some isomeric hexanols, *J. Am. Chem. Soc.*, 60(8):1388-1389, 1938.

Girifalco, L.A. and Good, R.J. A theory for the estimation of surface and interfacial energies: I. Derivation and application to interfacial tension, *J. Phys. Chem.*, 61(7):904-909, 1957.

Girvin, D.C. and Scott, A.J. Polychlorinated biphenyl sorption by soils: measurement of soil-water partition coefficients at equilibrium, *Chemosphere*, 35(9):2007-2025, 1997.

Gladis, G.P. Effects of moisture on corrosion in petrochemical environments, *Chem. Eng. Prog.*, 56(10):43-51, 1960.

Glass, B.L. Relation between the degradation of DDT and the iron redox system in soils, *J. Agric. Food Chem.*, 20:324-327, 1972.

Gledhill, W.E., Kaley, R.G., Adams, W.J., Hicks, O., Michael, P.R., and Saeger, V.W. An environmental safety assessment of butyl benzyl phthalate, *Environ. Sci. Technol.*, 14(3):301-305, 1980.

Glew, D.N. and Moelwyn-Hughes, E.A. Chemical studies of the methyl halides in water, *Disc. Faraday Soc.*, 15:150-161, 1953.

Glew, D.N. and Robertson, R.E. The spectrophotometric determination of the solubility of cumene in water by a kinetic method, *J. Phys. Chem.*, 60(3):332-337, 1956.

Glooschenko, V., Holdrinet, M., Lott, J.N.A., and Frank, R. Bioconcentration of chlordane by the green alga *Scenedesmus quadricauda*, *Bull. Environ. Contam. Toxicol.*, 21(4/5):515-520, 1979.

Gobas, F.A.P.C., Lahittete, J.M., Garofalo, G., Shiu, W.Y., and Mackay, D. A novel method for measuring membrane-water partition coefficients of hydrophobic organic compounds: comparison with 1-octanol-water partitioning, *J. Pharm. Sci.*, 77(3):265-272, 1988.

Godsy, E.M., Goerlitz, D.F., and Ehrlich, G.G. Methanogenesis of phenolic compounds by a bacterial consortium from a contaminated aquifer in St. Louis Park, Minnesota, *Bull. Environ. Contam. Toxicol.*, 30(3):261-268, 1983.

Goebel, H. Chemical and physical properties of endosulfan isomers and its degradation products, *Residue Rev.*, 83:8-28, 1982.

Golab, T., Althaus, W.A., and Wooten, H.L. Fate of [14]trifluralin in soil, *J. Agric. Food Chem.*, 27(1):163-179, 1979.

Golab, T., Herberg, R.J., Parka, S.J., and Tepe, J.B. Metabolism of carbon-14 trifluralin in carrots, *J. Agric. Food Chem.*, 15(4):638-641, 1967.

Golovanova, Yu.G. and Kolesov, V.P. Enthalpies of melting, melting temperatures, and cryoscopic constants of some haloorganic compounds, *Vestn. Mosk Univ., Ser. 2: Khim*, 25(3):244-248, 1984.

Golovleva, L.A., Polyakova, A.B., Pertsova, R.N., and Finkelstein, Z.I. The fate of methoxychlor in soils and transformation by soil microorganisms, *J. Environ. Sci. Health*, B19:523-538, 1984.

Gomez, J., Bruneau, C., Soyer, N., and Brault, A. Identification of thermal degradation products from diuron and iprodione, *J. Agric. Food Chem.*, 30(1):180-182, 1982.

Gonzalez, J. and Ukrainczyk, L. Transport of nicosulfuron in soil columns, *J. Environ. Qual.*, 28(1):101-107, 1999.

González, V., Ayala, J.H., and Afonso, A.M. Degradation of carbaryl in natural waters: enhanced hydrolysis rate in micellar solution, *Bull. Environ. Contam. Toxicol.*, 42(2):171-178, 1992.

Gooch, J.A. and Hamdy, M.K. Uptake and concentration factor of arochlor 1254 in aquatic organisms, *Bull. Environ. Contam. Toxicol.*, 31(4):445-452, 1983.

Good, D.A., Hanson, J., Francisco, J.S., Li, Z., and Jeong, G.-R. Kinetics and reaction mechanism of hydroxyl radical reaction with methyl formate, *J. Phys. Chem. A*, 103(50):10893-10898, 1999.

Goodman, L.R., Hemmer, M.J., Middaugh, D.P., and Moore, J.C. Effects of fenvalerate on the early life stages of topsmelt (*Atherinops affinis*), *Environ. Toxicol. Chem.*, 11(3):409-414, 1992.

Goodman, M.A. Vapor pressure of agrochemicals by the Knudsen effusion method using a quartz crystal microbalance, *J. Chem. Eng. Data*, 42(6):1227-1231, 1997.

Gorder, G.W., Dahm, P.A., and Tollefson, J.J. Carbofuran persistence in cornfield soils, *J. Econ. Entomol.*, 75(4):637-642, 1982.

Gordon, A.J. and Ford, R.A. *The Chemist's Companion* (New York: John Wiley & Sons, 1972), 551 p.

Gordon, J.E. and Thorne, R.L. Salt effects on the activity coefficient of naphthalene in mixed aqueous electrolyte solutions. I. Mixtures of two salts, *J. Phys. Chem.*, 71(13):4390-4399, 1967.

Gore, R.C., Hannah, R.W., Pattacini, S.C., and Porro, T.J. Infared and ultraviolet spectra of seventy-six pesticides, *J. Assoc. Off. Anal. Chem.*, 54(5):1040-1082, 1971.

Gossett, J.M. Measurement of Henry's law constants for C_1 and C_2 chlorinated hydrocarbons, *Environ. Sci. Technol.*, 21(2):202-208, 1987.

Goursot, P., Girdhar, H.L., and Westrum, E.F., Jr. Thermodynamics of polynuclear aromatic molecules. III. Heat capacities and enthalpies of fusion of anthracene, *J. Phys. Chem.*, 74:2538-2541, 1970.

Gowda, T.K.S. and Sethunathan, N. Persistence of endrin in Indian rice soils under flooded conditions, *J. Agric. Food Chem.*, 24(4):750-753, 1976.

Gowda, T.K.S. and Sethunathan, N. Endrin decomposition in soils as influenced by aerobic and anaerobic conditions, *Soil Sci.*, 125(1):5-9, 1977.

Graham, J.L., Hall, D.L., and Dellinger, B. Laboratory investigation of thermal degradation of a mixture of hazardous organic compounds, *Environ. Sci. Technol.*, 20(7):703-710, 1986.

Graham, R.E., Burson, K.R., Hammer, C.F., Hansen, L.B., and Kenner, C.T. Photochemical decomposition of heptachlor epoxide, *J. Agric. Food Chem.*, 21(5):824-834, 1973.

Granberg, R.A. and Rasmuson, A.C. Solubility of paracetamol in pure solvents, *J. Chem. Eng. Data*, 44(6):1391-1395, 1999.

Grant, B.F. Endrin toxicity and distribution in freshwater: a review, *Bull. Environ. Contam. Toxicol.*, 15(3):283-290, 1976.

Grathwohl, P. Influence of organic matter from soils and sediments from various origins on the sorption of some chlorinated aliphatic hydrocarbons: implications on K_{oc} correlations, *Environ. Sci. Technol.*, 24(11):1687-1693, 1990.

Grau, M.T., Romero, A., Sacristan, A., and Ortiz, J.A. Acute toxicity studies of sertaconazole, *Arzneim. Forsch.*, 42(5a):725-7261, 1992.

Grau, M.T., Romero, A., Villamayor, F., Sacristan, A., and Ortiz, J.A. Acute toxicity studies of ebrotidine, *Arzneim. Forsch.*, 47(4a):490-491, 1997.

Graveel, J.G., Sommers, L.E., and Nelson, D.W. Decomposition of benzidine, α-naphthylamine, and *p*-toluidine in soils, *J. Environ. Qual.*, 15(1):53-59, 1986.

Grayson, B.T. and Fosbraey, L.A. Determination of vapor pressures of pesticides, *Pestic. Sci.*, 13:269-278, 1982.

Graziani, G., Cazzulani, P., and Barbadoro, E. Toxicological and pharmacological properties of fenticonazole, a new topical antimycotic, *Arzneim. Forsch.*, 31(12):2145-2151, 1981.

Green, D.W.J., Williams, K.A., and Pascoe, D. Studies on the acute toxicity of pollutants to freshwater macroinvertebrates. 4. Lindane (gamma-hexachlorocyclohexane), *Arch. Hydrobiol.*, 106:263-273, 1986.

Green, W.J., Lee, G.F., Jones, R.A., and Palit, T. Interaction of clay soils with water and organic solvents: implications for the disposal of hazardous wastes, *Environ. Sci. Technol.*, 17(5):278-282, 1983.

Greene, S., Alexander, M., and Leggett, D. Formation of *N*-nitrosodimethylamine during treatment of municipal waste water by simulated land application, *J. Environ. Qual.*, 10(3):416-421, 1981.

Greenhalgh, R. and Belanger, A. Persistence and uptake of carbofuran in a humic mesisol and the effects of drying and storing soil samples on residue levels, *J. Agric. Food Chem.*, 29:231-235, 1981.

Greiner, N.R. Hydroxyl-radical kinetics by kinetic spectroscopy. II. Reactions with C_2H_6, C_3H_8, and *iso*-C_4H_{10} at 300 K, *J. Chem. Phys.*, 46:3389-3392, 1967.

Greiner, N.R. Hydroxyl-radical kinetics by kinetic spectroscopy. VI. Reactions with alkanes in the range 300-500 K, *J. Chem. Phys.*, 53:1070-1076, 1970.

Greve, P.A. and Wit, S.L. Endosulfan in the Rhine River, *J. Water Pollut. Control Fed.*, 43(12):2338-2348, 1971.

Grimes, D.J. and Morrison, S.M. Bacterial bioconcentration of chlorinated hydrocarbon insecticides from aquatic systems, *Microbial Ecol.*, 2:43-59, 1975.

Grisak, G.E., Pickens, J.F., and Cherry, J.A. Solute transport through fractured media. 2. Column study of fractured till, *Water Resourc. Res.*, 16(4):731-739, 1980.

Griswold, J., Chen, J.-N., and Klecka, M.E. Pure hydrocarbons from solvents, *Ind. Eng. Chem.*, 42(6):1246-1251, 1950.

Groenewegen, D. and Stolp, H. Microbial breakdown of polycyclic aromatic hydrocarbons, *Zentralbl. Bakteriol. Parasitenkd. Infekitionskr. Hyg. Abt., Abt. 1:Orig., Reihe, B.*, 162:225-232, 1976.

Grosjean, D. Atmospheric reactions of ortho cresol: gas phase and aerosol products, *Atmos. Environ.*, 19(8):1641-1652, 1984.

Grosjean, D. Photooxidation of methyl sulfide, ethyl sulfide, and methanethiol, *Environ. Sci. Technol.*, 18(6):460-468, 1984a.

Grosjean, D. Atmospheric reactions of styrenes and peroxybenzoyl nitrate, *Sci. Total Environ.*, 50:41-59, 1985.

Grosjean, D. Atmospheric chemistry of alcohols, *J. Braz. Chem. Soc.*, 8(4):433-442, 1997.

Gross, F.C. and Colony, J.A. The ubiquitous nature and objectionable characteristics of phthalate esters in aerospace industry, *Environ. Health Perspect.* (January 1973), pp. 37-48.

Gross, P. The determination of the solubility of slightly soluble liquids in water and the solubilities of the dichloro-ethanes and propanes, *J. Am. Chem. Soc.*, 51(8):2362-2366, 1929.

Gross, P.M. and Saylor, J.H. The solubilities of certain slightly soluble organic compounds in water, *J. Am. Chem. Soc.*, 53(5):1744-1751, 1931.

Gross, P.M., Saylor, J.H., and Gorman, M.A. Solubility studies. IV. The solubilities of certain slightly soluble organic compounds in water, *J. Am. Chem. Soc.*, 55(2):650-652, 1933.

Grover, R. and Hance, R.J. Adsorption of some herbicides by soil and roots, *Can. J. Plant Sci.*, 49:378-380, 1969.

Grover, R. and Smith, A.E. Adsorption studies with the acid and dimethylamine forms of 2,4-D and dicamba, *Can. J. Soil Sci.*, 54:179-186, 1974.

Groves, F., Jr. Solubility of cycloparaffins in distilled water and salt water, *J. Chem. Eng. Data*, 33(2):136-138, 1988.

Grunwell, J.R. and Erickson, R.H. Photolysis of parathion (*O,O*-diethyl-*O*-(4-nitrophenyl)thiophosphate). New products, *J. Agric. Food Chem.*, 21(5):929-931, 1973.

Guadagni, D.G., Buttery, R.G., and Kana, S. Odour threshold of some organic compounds associated with food flavours, *J. Agric. Food Chem.*, 14:761-765, 1963.

Gückel, W., Kästel, R., Lwewrenz, J., and Synnatschke, G. A method for determining the volatility of active ingredients used in plant production. Part III: The temperature relationship between vapour pressure and evaporation rate, *Pestic. Sci.*, 13:161-168, 1982.

Guenzi, W.D. and Beard, W.E. Anaerobic biodegradation of DDT to DDD in soil, *Science (Washington, D.C.)*, 156(3778):1116-1117, 1967.

Guenzi, W.D. and Beard, W.E. DDT degradation in soil as related to temperature, *J. Environ. Qual.*, 5:391-394, 1976.

Guenzi, W.D., Beard, W.E., and Viets, F.G., Jr. Influence of soil treatment on persistence of six chlorinated hydrocarbon insecticides in the field, *Soil Sci. Soc. Am. Proc.*, 35:910-913, 1971.

Guisti, D.M., Conway, R.A., and Lawson, C.T. Activated carbon adsorption of petrochemicals, *J. Water Pollut. Control Fed.*, 46(5):947-965, 1974.

Gunther, F.A., Westlake, W.E., and Jaglan, P.S. Reported solubilities of 738 pesticide chemicals in water, *Residue Rev.*, 20:1-148, 1968.

Gupta, G.D., Misra, A., and Agarwal, D.K. Inhalation toxicity of furfural vapours: an assessment of biochemical response in rats, *J. Appl. Toxicol.*, 11(5):343-347, 1991.

Guseva, A.N. and Parnov, E.I. The solubility of aromatic hydrocarbons in water, *Vestn. Mosk. Univ., Ser. II Khim.*, 18(1):76-79, 1963.

Guseva, A.N. and Parnov, E.I. Solubility of cyclohexane in water, *Zh. Fiz. Khim.*, 37(12):2763, 1963a.

Guseva, A.N. and Parnov, E.I. Isothermal cross-sections of the systems cyclanes-water, *Vestn. Mosk. Univ., Ser. II Khim.*, 19:77-78, 1964.

Güsten, H., Filby, W.G., and Schoop, S. Prediction of hydroxyl radical reaction rates with organic compounds in the gas phase, *Atmos. Environ.*, 15:1763-1765, 1981.

Guttman, L., Westrum, E.F., Jr., and Pitzer, K.S. The thermodynamics of styrene (phenylethylene), including equilibrium of formation from ethylbenzene, *J. Am. Chem. Soc.*, 65:1246-1247, 1943.

Guzman, A., Garcia, C., and Demestre, I. Acute and subchronic toxicity studies of the new quinoline antibacterial agent irloxacin in rodents, *Arzneim. Forsch.*, 49(5):448-456, 1999.

Guzzella, L., Gronda, L., and Colombo, L. Acute toxicity of organophosphorus insecticides to marine invertebrates, *Bull. Environ. Contam. Toxicol.*, 59(2):313-320, 1997.

Haag, W.R. and Yao, D.C.C. Rate constants for the reaction of hydroxyl radicals with several drinking water contaminants, *Environ. Sci. Technol.*, 26(5):1005-1013, 1992.

Haderlein, S.B. and Schwarzenbach, R.P. Adsorption of substituted nitrobenzenes and nitrophenols to mineral surfaces, *Environ. Sci. Technol.*, 27(2):316-326, 1993.

Hafkenscheid, T.L. and Tomlinson, E. Estimation of aqueous solubilities of organic non-electrolytes using liquid chromatographic retention data, *J. Chromatogr.*, 218:409-425, 1981.

Hafkenscheid, T.L. and Tomlinson, E. Isocratic chromatographic retention data for estimating aqueous solubilities of acidic, basic and neutral drugs, *Int. J. Pharm.*, 17:1-21, 1983.

Hageman, W.J.M., Van der Weijden, C.W., and Loch, J.P.G. Sorption of benzo[*a*]pyrene and phenanthrene on suspended harbor sediment as a function of suspended sediment concentration and salinity: a laboratory study using the cosolvent partition coefficient, *Environ. Sci. Technol.*, 29(2):363-371, 1995.

Häggblom, M.M. and Young, L.Y. Chlorophenol degradation coupled with sulfate reduction, *Appl. Environ. Microbiol.*, 56(11):3255-3260, 1990.

Hagin, R.D., Linscott, D.L., and Dawson, J.E. 2,4-D metabolism in resistant grasses, *J. Agric. Food Chem.*, 18:848-850, 1970.

Haider, K., Jagnow, G., Kohnen, R., and Lim, S.U. Degradation of chlorinated benzenes, phenols, and cyclohexane derivatives by benzene- and phenol-utilizing bacteria under aerobic conditions, in *Decomposition of Toxic and Nontoxic Organic Compounds in Soil*, Overcash, V.R., Ed. (Ann Arbor, MI: Ann Arbor Science Publishers, 1981), pp. 207-223.

Haider, S. and Inbaraj, R.M. Relative toxicity of technical material and commercial formulation of malathion and endosulfan to a freshwater fish, *Channa punctatus*, *Ecotoxicol. Environ. Saf.*, 11(3):347-351, 1986.

Haight, G.P., Jr. Solubility of methyl bromide in water and in some fruit juices, *Ind. Eng. Chem.*, 43(8):1827-1828, 1951.

Haigler, B.E., Nishino, S.F., and Spain, J.C. Degradation of 1,2-dichlorobenzene by a *Pseudomonas* sp., *Appl. Environ. Microbiol.*, 54(2):294-301, 1988.

Hake, C.L. and Rowe, V.K. Ethers, in *Industrial Hygiene and Toxicology*, Patty, F.A., Ed. (New York: John Wiley & Sons, 1963).

Hakuta, T., Negishi, A., Goto, T., Kato, J., and Ishizaka, S. Vapor-liquid equilibria of some pollutants in aqueous and saline solutions, *Desalination*, 21(1):11-21, 1977.

Hales, J.M. and Drewes, D.R. Solubility of ammonia in water at low concentrations, *Atmos. Environ.*, 13:1133-1147, 1979.

Halfon, E. and Reggiani, M.G. On ranking chemicals of environmental concern, *Environ. Sci. Technol.*, 20(11):1173-

1179, 1986.

Hall, P.G. Thermal decomposition and phase transitions in solid nitramines, *Trans. Faraday Soc.*, 67(3):556-562, 1971.

Hallas, L.E. and Alexander, M. Microbial transformation of nitroaromatic compounds in sewage effluent, *Appl. Environ. Microbiol.*, 45(4):1234-1241, 1983.

Hamelink, J.L., Simon, P.B., and Silberhorn, E.M. Henry's law constant, volatilization rate, and aquatic half-life of octamethylcyclotetrasiloxane, *Environ. Sci. Technol.*, 30(6):1946-1952, 1996.

Hamlin, W.E., Northam, J.I., and Wagner, J.G. Relationship between *in vitro* dissolution rates and solubilities of numerous compounds representative of various chemical species, *J. Pharm. Sci.*, 54:1651-1653, 1965.

Hamm, S., Hahn, J., Helas, G., and Warneck, P. Acetonitrile in the troposphere: residence time due to rainout and uptake by the ocean, *Geophys. Res. Lett.*, 11:1207-1210, 1984.

Hammers, W.E., Meurs, G.J., and de Ligny, C.L. Correlations between liquid chromatographic capacity ratio data on Lichrosorb RP-18 and partition coefficients in the octanol-water system, *J. Chromatogr.*, 247:1-13, 1982.

Han, N., Zhu, L., Wang, L., and Fu, R. Aqueous solubility of *m*-phthalic, *o*-phthalic acid and *p*-phthalic acid from 298 to 483 K, *Separ. Purifica. Technol.*, 16:175-180, 1999.

Hance, R.J. Decomposition of herbicides in the soil by nonbiological chemical processes, *J. Sci. Food Agric.*, 18:544-547, 1967.

Hansch, C. and Anderson, S.M. The effect of intramolecular hydrophobic bonding on partition coefficients, *J. Org. Chem.*, 32(8):2583-2586, 1967.

Hansch, C. and Fujita, T. ρ-σ-π Analysis. A method for the correlation of biological activity and chemical structure, *J. Am. Chem. Soc.*, 86(8):1616-1617, 1964.

Hansch, C. and Leo, A. *Substituent Constants for Correlation Analysis in Chemistry and Biology* (New York: John Wiley & Sons, 1979), 339 p.

Hansch, C. and Leo, A. *Medchem Project Issue No. 26* (Claremont, CA: Pomona College, 1985).

Hansch, C., Leo, A., and Nikaitani, D. On the additive-constitutive character of partition coefficients, *J. Org. Chem.*, 37(20):3090-3092, 1972.

Hansch, C., Quinlan, J.E., and Lawrence, G.L. The linear free-energy relationship between partition coefficients and aqueous solubility of organic liquids, *J. Org. Chem.*, 33(1):347-350, 1968.

Hansch, C., Vittoria, A., Silipo, C., and Jow, P.Y.C. Partition coefficients and the structure-activity relationship of the anesthetic gases, *J. Med. Chem.*, 18(6):546-548, 1975.

Hansen, D.A., Atkinson, R., and Pitts, J.N., Jr. Rates constants for the reaction of OH radicals with a series of hydrocarbons, *J. Phys. Chem.*, 79(17):1763-1766, 1975.

Hansen, D.J., Goodman, L.R., Cripe, G.M., and Macauley, S.F. Early life-stage toxicity test methods for gulf toadfish (*Opsanus beta*) and results using chlorpyrifos, *Ecotoxicol. Environ. Saf.*, 11(1):15-22, 1986.

Hansen, K.C., Zhou, Z., Yaws, C.L., and Aminabhavi, T.M. Determination of Henry's law constants of organics in dilute aqueous solutions, *J. Chem. Eng. Data*, 38:546-550, 1993.

Hansen, K.C., Zhou, Z., Yaws, C.L., and Aminabhavi, T.M. A laboratory method for the determination of Henry's law constants of volatile organic chemicals, *J. Chem. Educ.*, 72(1):93-96, 1995.

Hansen, R.S., Fu, Y., and Bartell, F.E. Multimolecular adsorption from binary liquid solutions, *J. Phys. Chem.*, 53:769-785, 1949.

Hanson, D.R. and Ravishankara, A.R. The reaction probabilities of $ClONO_2$ and N_2O_5 on 40 to 75% sulfuric acid solutions, *J. Geophys. Res.*, 96D:17307-17314, 1991.

Hanst, P.L. and Gay, B.W., Jr. Photochemical reactions among formaldehyde, chlorine, and nitrogen dioxide in air, *Environ. Sci. Technol.*, 11(12):1105-1109, 1977.

Hanst, P.L. and Gay, B.W., Jr. Atmospheric oxidation of hydrocarbons: formation of hydroperoxides and peroxyacids, *Atmos. Environ.*, 17(11):2259-2265, 1983.

Hanst, P.L., Spence, J.W., and Miller, M. Atmospheric chemistry of *N*-nitrosodimethylamine, *Environ. Sci. Technol.*, 11(4):403-405, 1977.

Hao, L. and Leaist, D.G. Binary mutual diffusion coefficients of aqueous alcohols. Methanol to 1-heptanol, *J. Chem. Eng. Data*, 41(2):210-213, 1996.

Haque, R., Schmedding, D.W., and Freed, V.H. Aqueous solubility, adsorption, and vapor behavior of polychlorinated biphenyl Aroclor 1254, *Environ. Sci. Technol.*, 8(2):139-142, 1974.

Harbison, K.G. and Belly, R.T. The biodegradation of hydroquinone, *Environ. Toxicol. Chem.*, 1(1):9-15, 1982.

Hardy, J.T., Dauble, D.D., and Felice, L.J. Aquatic fate of synfuel residuals: bioaccumulation of aniline and phenol by the freshwater phytoplankten *Scenedesmus quadricauda*, *Environ. Toxicol. Chem.*, 4:29-35, 1985.

Hargrave, B.T., Phillips, G.A., Vass, W.P., Bruecker, P., Welch, H.E., and Siferd, T.D. Seasonality in bioaccumulation of organochlorines in lower tropic level arctic marine biota, *Environ. Sci. Technol.*, 34(6):980-987, 2000.

Harkins, W.D., Clark, G.L., and Roberts, L.E. The orientation of molecules in surfaces, surface energy, adsorption, and surface catalysis. V. The adhesional work between organic liquids and water, *J. Am. Chem. Soc.*, 42(4):700-713, 1920.

Harnsich, M., Möckel, H.J., and Schulze, G. Relationship between Log P_{ow} shake-flask values and capacity values derived from reversed-phase HPLC for *n*-alkylbenzenes and some OECD reference substances, *J. Chromatogr.*, 282:315-332, 1983.

Harris, C.I. Movement of pesticides in soil, *J. Agric. Food Chem.*, 17(1):80-82, 1969.

Harris, C.I., Chapman, R.A., Harris, C., and Tu, C.M. Biodegradation of pesticides in soil: rapid induction of carbamate degrading factors after carbofuran treatment, *J. Environ. Sci. Health*, B19:1-11, 1984.

Harris, C.I. and Lichtenstein, E.P. Factors affecting the volatilization of insecticidal residues from soils, *J. Econ. Entomol.*, 54:1038-1045, 1961.

Harris, G.W., Kleindienst, T.E., and Pitts, J.N., Jr. Rate constants for the reaction of OH radicals with CH₃CN, CH₂H₅CN and CH₂=CH-CN in the temperature range 298-424 K, *Chem. Phys. Lett.*, 80:479-483, 1981.

Harrison, R.B., Holmes, D.C., Roburn, J., and Tatton, J.O'G. The fate of some organochlorine pesticides on leaves, *J. Sci. Food Agric.*, 18:10-15, 1967.

Harrison, R.M., Perry, R., and Wellings, R.A. Polynuclear aromatic hydrocarbons in raw, potable and waste waters, *Water Res.*, 9(4):331-346, 1975.

Hartkopf, A. and Karger, B.L. Study of the interfacial properties of water by gas chromatography, *Acc. Chem. Res.*, 6:209-216, 1973.

Hartley, D.M. and Johnston, J.B. Use of the freshwater clam *Corbicula manilensis* as a monitor for organochlorine pesticides, *Bull. Environ. Contam. Toxicol.*, 31(1):33-40, 1983.

Hartley, D. and Kidd, H., Eds. *The Agrochemicals Handbook*, 2nd ed. (Cambridge: Royal Society of Chemistry, 1987).

Hartman, R.J., Kern, R.A., and Bobalek, E.G. Adsorption isotherms of some substituted benzoic acids, *J. Colloid Sci.*, 1:271-276, 1946.

Hashimoto, Y., Tokura, K., Kishi, H., and Strachan, W.M.J. Prediction of seawater solubility of aromatic compounds, *Chemosphere*, 13(8):881-888, 1984.

Hashimoto, Y., Tokura, K., Ozaki, K., and Strachan, W.M.J. A comparison of water solubilities by the flask and micro-column methods, *Chemosphere*, 11(10):991-1001, 1982.

Hassanein, H.M., Banhawy, M.A., Soliman, F.M., Abdel-Rehim, S.A., Muller, W.E., and Schroder, H.C. Induction of hsp70 by the herbicide oxyfluorfen (Goal) in the Egyptian Nile fish, *Oreochromis niloticis*, *Ecotoxicol. Environ. Saf.*, 37(1):78-84, 1999.

Hassett, J.J., Means, J.C., Banwart, W.L., and Wood, S.G. Sorption properties of sediments and energy-related pollutants, Office of Research and Development, U.S. EPA Report-600/3-80-041, 1980, 150 p.

Hassett, J.J., Means, J.C., Banwart, W.L., and Wood, S.G. Sorption of α-naphthol: implications concerning the limits of hydrophobic sorption, *Soil Sci. Soc. Am. J.*, 45(1):38-42, 1981.

Hastings, S.H. and Matsen, F.A. The photodecomposition of nitrobenzene, *J. Am. Chem. Soc.*, 70(10):3514-3515, 1948.

Hatakeyama, S., Ohno, M., Weng, J., Takagi, H., and Akimoto, H. Mechanism for the formation of gaseous and particulate products from ozone-cycloalkene reactions in air, *Environ. Sci. Technol.*, 21(1):52-57, 1987.

Hatch, L.F. and Kidwell, Jr., L.E. Preparation and properties of 1-bromo-1-propyne, 1,3-dibromopropyne and 1-bromo-3-chloro-1-propyne, *J. Am. Chem. Soc.*, 76(1):289-290, 1954.

Hattula, M.L., Reunanen, H., and Arstila, A.U. The toxicity of MCPA to fish, light and electron microscopy and the chemical analysis of the tissue, *Bull. Environ. Contam. Toxicol.*, 19(4):465-470, 1978.

Hattula, M.L., Reunanen, H., and Arstila, A.U. Toxicity of 4-chloro-*o*-cresol to rat: light microscopy and chemical observations, *Bull. Environ. Contam. Toxicol.*, 21(4/5):492-497, 1979.

Hattula, M.L., Wasenius, V.-M., Reunanen, H., and Arstila, A.U. Acute toxicity of some chlorinated phenols, catechols and cresols in trout, *Bull. Environ. Contam. Toxicol.*, 26(3):295-298, 1981.

Hauff, K., Fischer, R.G., and Ballschmiter, K. Determination of C₁-C₅ alkyl nitrates in rain, snow, white frost, lake, and tap water by a combined codistillation headspace chromatography technique. Determination of Henry's law constants by head-space GC, *Chemosphere*, 37(13):2599-2615, 1998.

Hawari, J., Demeter, A., and Samson, R. Sensitized photolysis of polychlorobiphenyls in alkaline 2-propanol: dechlorination of Aroclor 1254 in soil samples by solar radiation, *Environ. Sci. Technol.*, 26(10):2022-2027, 1992.

Hawari, J., Tronczynski, J., Demeter, A., Samson, R., and Mourato, D. Photodechlorination of chlorobiphenyls by sodium methyl siliconate, *Chemosphere*, 22(1/2):189-199, 1991.

Hawker, D.W. and Connell, D.W. Influence of partition coefficient of lipophilic compounds on bioconcentration kinetics with fish, *Water Res.*, 22(6):701-707, 1988.

Hawkins, M.D. Hydrolysis of phthalic and 3,6-dimethylphthalic anhydrides, *J. Chem. Soc., Perkin Trans. 2*, pp. 282-284, 1975.

Hawley, G.G. *The Condensed Chemical Dictionary* (New York: Van Nostrand Reinhold, 1981), 1135 p.

Hawthorne, S.B., Sievers, R.E., and Barkley, R.M. Organic emissions from shale oil wastewaters and their implications for air quality, *Environ. Sci. Technol.*, 19(10):992-997, 1985.

Hayashi, M. and Sasaki, T. Measurements of solubilities of sparingly soluble liquids in water and aqueous detergent solutions using nonionic surfactant, *Bull. Chem. Soc. Jpn.*, 29:857-859, 1956.

Hayduk, W., Asatani, H., and Miyano, Y. Solubilities of propene, butane, isobutane and isobutene gases in *n*-octane, chlorobenzene and *n*-butanol solvents, *Can. J. Chem. Eng.*, 66:466-473, 1988.

Hayduk, W. and Laudie, H. Prediction of diffusion coefficients for nonelectrolytes in dilute aqueous solution, *Am. Inst. Chem. Eng.*, 20(3):611-615, 1974.

Hayduk, W. and Laudie, H. Vinyl chloride gas compressibility and solubility in water and aqueous potassium laurate solutions, *J. Chem. Eng. Data*, 19(3):253-257, 1974a.

Hayduk, W. and Minhas, B.S. Diffusivity and solubility of 1,3-butadiene in heptane and other properties of heptane-styrene and aqueous potassium laurate solutions, *J. Chem. Eng. Data*, 32(3):285-290, 1987.

Hayes, W.J. *Pesticides Studied in Man* (Baltimore, MD: Williams and Wilkens, 1982), pp. 234-247.

Hayes, J.R., Condie, L.W., Egle, J.L., and Borzelleca, J.F. The acute and subchronic toxicity in rats of *trans*-1,2-dichloroethene in drinking water, *J. Am. Coll. Toxicol.*, 6(4):471-478, 1987.

Healy, J.B. and Young, L.Y. Anaerobic biodegradation of eleven aromatic compounds to methane, *Appl. Environ. Microbiol.*, 38(1):85-89, 1979.

Heitkamp, M.A., Freeman, J.P., and Cerniglia, C.E. Naphthalene biodegradation in environmental microcosms: estimates of degradation rates and characterization of metabolites, *Appl. Environ. Microbiol.*, 53(1):129-136, 1987.

Heitkamp, M.A., Freeman, J.P., Miller, D.W., and Cerniglia, C.E. Pyrene degradation by a *Mycobacterium* sp.:

identification of ring oxidation and ring fission products, *Appl. Environ. Microbiol.*, 54(10):2556-2565, 1988.

Heitmuller, P.T., Hollister, T.A., and Parrish, P.R. Acute toxicity of 54 industrial chemicals to sheepshead minnows (*Cyprinodon variegatus*), *Bull. Environ. Contam. Toxicol.*, 27(5):596-604, 1981.

Helling, C.S., Isensee, A.R., Woolson, E.A., Ensor, P.D.J., Jones, G.E., Plimmer, J.R., and Kearney, P.C. Chlorodioxins in pesticides, soils, and plants, *J. Environ. Qual.*, 2(2):171-178, 1973.

Hemalatha, S. and Banerjee, T.K. Histopathological analysis of sublethal toxicity of zinc chloride to the respiratory organs of the air breathing catfish *Heteropneustes fossilis* (Bloch), *Biol. Res.*, 30(1):11-121, 1997.

Henderson, G.L. and Crosby, D.G. The photodecomposition of dieldrin residues in water, *Bull. Environ. Contam. Toxicol.*, 3(3):131-134, 1968.

Henderson, C., Pickering, Q.H., and Tarzwell, C.M. Relative toxicity of ten chlorinated hydrocarbon insecticides to four species of fish, *Trans. Am. Fish Soc.*, 88(1):23-32, 1959.

Hendry, D.G. and Kenley, R.A. Atmospheric reaction products of organic compounds, Office of Toxic Substances, U.S. EPA Report-560/5-79-001, 1974.

Henny, C.J., Hallock, R.J., and Hill, E.F. Cyanide and migratory birds at gold mines in Nevada, USA, *Ecotoxicology*, 3:45-58, 1994.

Henry, T.R., Spitsbergen, J.M., Hornung, M.W., Abnet, C.C., and Peterson, R.E. Early life stage toxicity of 2,3,7,8-tetrachlorodibenzo-*p*-dioxin in zebrafish (*Danio rerio*), *Toxicol. Appl. Pharmacol.*, 142(1):56-68, 1997.

Herbes, S.E. and Schwall, L.R. Microbial transformation of polycyclic aromatic hydrocarbons in pristine and petroleum contaminated sediments, *Appl. Environ. Microbial.*, 35(2):306-316, 1978.

Heritage, A.D. and MacRae, I.C. Identification of intermediates formed during the degradation of hexachlorocyclo-hexanes by *Clostridium sphenoides*, *Appl. Environ. Microbiol.*, 33(6):1295-1297, 1977.

Heritage, A.D. and MacRae, I.C. Degradation of lindane by cell-free preparations of *Clostridium sphenoides*, *Appl. Environ. Microbiol.*, 34(2):222-224, 1977a.

Herman, D.C., Inniss, W.E., and Mayfield, C.I. Impact of volatile aromatic hydrocarbons, alone and in combination, on growth of the freshwater alga *Selenastrum capricornutum*, *Aquat. Toxicol.*, 18(2):87-100, 1990.

Herman, D.C., Mayfield, C.I., and Innis, W.E. The relationship between toxicity and bioconcentration of volatile aromatic hydrocarbons by the alga *Selenastrum capricornutum*, *Chemosphere*, 22(7):665-676, 1991.

Hermann, R.B. Theory of hydrophobic bonding. II. The correlation of hydrocarbon solubility in water with solvent cavity surface area, *J. Phys. Chem.*, 76(19):2754-2759, 1972.

Hermanutz, R.O. Endrin and malathion toxicity to flagfish (*Jordanella floridae*), *Arch. Environ. Contam. Toxicol.*, 7(2):159-168, 1978.

Hermens, J., Busser, F., Leeuwangh, P., and Musch, A. Quantitative correlation studies between the acute lethal toxicity of 15 organic halides to the guppy *Poecilia reticulata* and chemical reactivity towards 4-nitrobenzylpyridine, *Toxicol. Environ. Chem.*, 9:219-236, 1985.

Hermens, J., Canton, H., Janssen, P., and de Jong, R. Quantitative structure-activity relationships and toxicity studies of mixtures of chemicals with an anaesthetic potency: acute lethal and sublethal toxicity to *Daphnia magna*, *Aquat. Toxicol.*, 5:143-154, 1984.

Heron, G., Christensen, T.H., and Enfield, C.G. Henry's law constants for trichloroethylene between 10 and 95 °C, *Environ. Sci. Technol.*, 32(10):1433-1437, 1998.

Herzel, F. and Murty, A.S. Do carrier solvents enhance the water solubility of hydrophobic compounds? *Bull. Environ. Contam. Toxicol.*, 32(1):53-58, 1984.

Herzog, R. and Leuschner, J. Single and subacute local and systemic toxicity studies of benzalazine, *Arzneim. Forsch.*, 44(12):1353-1356, 1994.

Heukelekian, H. and Rand, M.C. Biochemical oxygen demand of pure organic compounds, *Sewage Ind. Wastes*, 27:1040-1053, 1955.

Hiatt, M.H. Bioconcentration factors for volatile organic compounds in vegetation, *Anal. Chem.*, 70(5):851-856, 1998.

Hickey, C.L. and Martin, M.L. Chronic toxicity of ammonia to the freshwater bivalve *Sphaerium novaezelandiae*, *Arch. Environ. Contam. Toxicol.*, 36(1):38-46, 1999.

Hilarides, R.J., Gray, K.A., Guzzetta, J., Cortellucci, N., and Sommer, C. Radiolytic degradation of 2,3,7,8-TCDD in artificially contaminated soils, *Environ. Sci. Technol.*, 28(13):2249-2258, 1994.

Hill, A.E. The system, silver perchlorate-water-benzene, *J. Am. Chem. Soc.*, 44(6):1163-1193, 1922.

Hill, A.E. The mutual solubility of liquids. I. The mutual solubility of ethyl ether and water. II. The solubility of water in benzene, *J. Am. Chem. Soc.*, 46(5):1143-1155, 1924.

Hill, A.E. and Macy, R. Ternary systems. II. Silver perchlorate, aniline and water, *J. Am. Chem. Soc.*, 46:1132-1143, 1924.

Hill, A.E. and Malisoff. The mutual solubility of liquids. III. The mutual solubility of phenol and water. IV. The mutual solubility of normal butyl alcohol and water, *J. Am. Chem. Soc.*, 48:918-927, 1926.

Hill, D.W. and McCarty, P.L. Anaerobic degradation of selected chlorinated hydrocarbon pesticides, *J. Water Pollut. Control Fed.*, 39:1259-1277, 1967.

Hill, G.D., McGahen, J.W., Baker, H.M., Finnerty, D.W., and Bingeman, C.W. The fate of substituted urea herbicides in agricultural soils, *Agron. J.*, 47:93-104, 1955.

Hill, J., Kollig, H.P., Paris, D.F., Wolfe, N.L., and Zepp, R.G. Dynamic behavior of vinyl chloride in aquatic eco-systems, U.S. EPA Report-600/3-76-001, 1976, 59 p.

Hill, J.O., Worsley, I.G., and L.G. Hepler. Calorimetric determination of the distribution coefficient and thermodynamic properties of bromine in water and carbon tetrachloride, *J. Phys. Chem.*, 72:3695-3697, 1968.

Hinckley, D.A., Bidleman, T.F., Foreman, W.T., and Tuschall, J.R. Determination of vapor pressures for nonpolar and semipolar organic compounds from gas chromatographic retention data, *J. Chem. Eng. Data*, 35(3):232-237, 1990.

Hine, J., Haworth, H.W., and Ramsey, O.B. Polar effects on rates and equilibria. VI. The effect of solvent on the transmission of polar effects, *J. Am. Chem. Soc.*, 85(10):1473-1475, 1963.

Hine, J. and Mookerjee, P.K. The intrinsic hydrophilic character of organic compounds. Correlations in terms of structural contributions, *J. Org. Chem.*, 40(3):292-298, 1975.

Hine, J. and Weimar, Jr., R.D. Carbon basicity, *J. Am. Chem. Soc.*, 87(15):3387-3396, 1965.

Hinga, K.R. and Pilson, M.E.Q. Persistence of benz[a]anthracene degradation products in an enclosed marine ecosystem, *Environ. Sci. Technol.*, 21(7):648-653, 1987.

Hinga, K.R., Pilson, M.E.Q., Lee, R.F., Farrington, J.W., Tjessem, K., and Davis, A.C. Biogeochemistry of benzanthracene in an enclosed marine ecosystem, *Environ. Sci. Technol.*, 14(9):1136-1143, 1980.

Hirsch, M. and Hutzinger, O. Naturally occurring proteins from pond water sensitize hexachlorobenzene photolysis, *Environ. Sci. Technol.*, 23(10):1306-1307, 1989.

Hirzy, J.W., Adams, W.J., Gledhill, W.E., and Mieure, J.P. Phthalate esters: the environmental issues, unpublished seminar document, Monsanto Industrial Chemicals, 1978, 54 p.

HNU. Instruction Manual - Model ISP1 101: intrinsically Safe Portable Photoionization Analyzer (Newton, MA: HNU Systems, 1986), 86 p.

Hoben, H.J., Ching, S.A., and Casarett. A study of inhalation of pentachlorophenol by rats: Part III. Inhalation toxicity study, *Bull. Environ. Contam. Toxicol.*, 15(4):463-465, 1976.

Hochstein, J.R., Bursian, S.J., and Aulerich, R.J. Effects of dietary exposure to 2,3,7,8-tetrachlorodibenzo-*p*-dioxin in adult female mink (*Mustela vison*), *Arch. Environ. Contam.Toxicol.*, 35(2):348-353, 1998.

Hodgman, C.D., Weast, R.C., Shankland, R.S., and Selby, S.M. *Handbook of Chemistry and Physics* (Cleveland, OH: Chemical Rubber, 1961).

Hodson, J. and Williams, N.A. The estimation of the adsorption coefficient (K_{oc}) for soils by high performance liquid chromatography, *Chemosphere*, 19(1):67-77, 1988.

Hodson, P.V., Dixon, D.G., and Kaiser, K.L.E. Measurement of median lethal dose as a rapid indication of contaminant toxicity to fish, *Environ. Toxicol. Chem.*, 3(2):243-254, 1984.

Hoff, J.T., Mackay, D., Gillham, R., and Shiu, W.-Y. Partitioning of organic chemicals at the air-water interface in environmental systems, *Environ. Sci. Technol.*, 27(10):2174-2180, 1993.

Hoffman, J. and Eichelsdoerfer, D. Effect of ozone on chlorinated-hydrocarbon-group insecticides in water, *Vom. Wasser*, 38:197-206, 1971.

Hogfeldt, E. and Bolander, B. On the extraction of water and nitric acid by aromatic hydrocarbons, *Arkiv Kemi*, 21(16):161-186, 1963.

Hoigné, J. and Bader, H. Rate constants of reactions of ozone with organic and inorganic compounds in water - II. Dissociation organic compounds, *Water Res.*, 17(2):185-194, 1983.

Holbrook, F.R. and Mullens, B.A. Effects of ivermectin on survival, fecundity, and egg fertility in *Culicoides variipennis* (Diptera: Certaopogonidae), *J. Am. Mosq. Control Assoc.*, 10(1):70-73, 1994.

Holcombe, G.W., Benoit, D.A., Hammermeister, D.E., Leonard, E.N., and Johnson, R.D. Acute and long-term effects of nine chemicals on the Japanese medaka (*Oryzias latipes*), *Arch. Environ. Contam. Toxicol.*, 28(3):287-297, 1995.

Holcombe, G.W., Phipps, G.L., and Fiandt, J.T. Toxicity of selected priority pollutants to various aquatic organisms, *Ecotoxicol. Environ. Saf.*, 7(4):400-409, 1983.

Holcombe, G.W., Phipps, G.L., Sulaiman, A.H., and Hoffman, A.D. Simultaneous multiple species testing: acute toxicity of 13 chemicals to 12 diverse freshwater amphibian, fish and invertebrate families, *Arch. Environ. Contam. Toxicol.*, 16(6):697-710, 1987.

Holdren, M.W., Spicer, C.W., and Hales, J.M. Peroxyacetyl nitrate solubility and decomposition rate in acidic water, *Atmos. Environ.*, 18(6):1171-1173, 1984.

Hollifield, H.C. Rapid nephelometric estimate of water solubility of highly insoluble organic chemicals of environmental interest, *Bull. Environ. Contam. Toxicol.*, 23(4/5):579-586, 1979.

Holzwarth, G., Balmer, R.G., and Soni, L. The fate of chlorine and chloramines in cooling towers, *Water Res.*, 18(11):1421-1427, 1984.

Hommelen, J.R. The elimination of errors due to evaporation of the solute in the determination of surface tensions, *J. Colloid Sci.*, 34:385-400, 1959.

Honma, T., Miyagawa, M., Sato, M., and Hasegawa, H. Neurotoxicity and metabolism of methyl bromide in rats, *Toxicol. Appl. Pharmacol.*, 81(2):183-191, 1985.

Hopfer, S.M., Plowman, M.C., Bantle, J.A., and Sunderman, F.W., Jr. Teratogenicity of $NiCl_2$ in *Xenopus laevis*, assayed by the "FETAX" procedure, *Ann. Clin. Lab. Sci.*, 20(4):295, 1990.

Horowitz, A., Shelton, D.R., Cornell, C.P., and Tiedje, J.M. Anaerobic degradation of aromatic compounds in sediment and digested sludge, *Dev. Ind. Microbiol.*, 23:435-444, 1982.

Horvath, A.L. *Halogenated Hydrocarbons. Solubility-Miscibility with Water* (New York: Marcel Dekker, 1982), 889 p.

Horvath, R.S. Microbial cometabolism of 2,4,5-trichlorophenoxyacetic acid, *Bull. Environ. Contam. Toxicol.*, 5(6):537-541, 1970.

Horvath, R.S. Co-metabolism of methyl- and chloro-substituted catechols by an *Achromobacter* sp. possessing a new *meta*-cleavage oxygenase, *Biochem. J.*, 119:871-876, 1970a.

Horzempa, L.M. and Di Toro, D.M. The extent of reversibility of polychlorinated biphenyl adsorption, *Water Res.*, 17(8):851-859, 1983.

Hou, C.T. Microbial transformation of important industrial hydrocarbons, in *Microbial Transformations of Bioactive Compounds*, Rosazza, J.P., Ed. (Boca Raton, FL: CRC Press, 1982), pp. 81-107.

Houser, J.J. and Sibbio, B.A. Liquid-phase photolysis of dioxane, *J. Org. Chem.*, 42(12):2145-2151, 1977.

Hovorka, S. and Dohnal, V. Determination of air-water partitioning of volatile halogenated hydrocarbons by the inert gas stripping method, *J. Chem. Eng. Data*, 42(5):924-933, 1997.

Howard, C.J. Rate constants for the gas-phase reactions of OH radicals with ethylene and halogenated ethylene compounds, *J. Chem. Phys.*, 65:4771-4777, 1976.

Howard, C.J. and Evenson, K.M. Rate constants for the reaction of OH with ethane and some halogen substituted ethanes at 296 °K, *J. Chem. Phys.*, 64(11):4303-4306, 1976.

Howard, P.H. *Handbook of Environmental Fate and Exposure Data for Organic Chemicals - Volume I. Large Production and Priority Pollutants* (Chelsea, MI: Lewis Publishers, 1989), 574 p.

Howard, P.H. *Handbook of Environmental Fate and Exposure Data for Organic Chemicals - Volume II. Solvents* (Chelsea, MI: Lewis Publishers, 1990), 546 p.

Howard, P.H. *Handbook of Environmental Fate and Exposure Data for Organic Chemicals - Volume III. Pesticides* (Chelsea, MI: Lewis Publishers, 1991), 684 p.

Howard, P.H., Banerjee, S., and Robillard, K.H. Measurement of water solubilities, octanol/water partition coefficients and vapor pressures of commercial phthalate esters, *Environ. Toxicol. Chem.*, 4:653-661, 1985.

Howard, P.H. and Doe, P.G. Degradation of aryl phosphates in aquatic environments, *Bull. Environ. Contam. Toxicol.*, 22(3):337-344, 1979.

Howard, P.H. and Durkin, P.R. Sources of contamination, ambient levels, and fate of benzene in the environment, Office of Toxic Substances, U.S. EPA Report-560/5-75-005, 1974, 73 p.

Hoy, K.L. New values of the solubility parameters from vapor pressure data, *J. Paint Technol.*, 42(541):76-118, 1970.

HSDB. U.S. Department of Health and Human Services. *Hazardous Substances Data Bank*. National Library of Medicine, Toxnet File (Bethesda, MD: National Institutes of Health, 1989).

Hsu, C.C. and McKetta, J.J. Pressure-volume-temperature properties of methyl chloride, *J. Chem. Eng. Data*, 9(1):45-51, 1964.

Hu, Q., Wang, X., and Brusseau. Quantitative structure-activity relationships for evaluating the influence of sorbate structure on sorption of organic compounds by soil, *Environ. Toxicol. Chem.*, 14(7):1133-1140, 1995.

Hückelhoven, R., Schuphan, I., Thiede, B., and Schmidt, B. Biotransformation of pyrene by cell cultures of soybean (*Glycine max* L.), wheat (*Triticum aestivum* L.), jimsonweed (*Datura stramonium* L.), and purple foxglove (*Digitalis purpurea* L.), *J. Agric. Food Chem.*, 45(1):263-269, 1997.

Huckins, J.N., Fairchild, J.F., and Boyle, T.P. Role of exposure mode in the bioavailability of triphenyl phosphate to aquatic organisms, *Arch. Environ. Contam. Toxicol.*, 21(4):481-485, 1991.

Huddleston, E.W. and Gyrisco, G.G. Residues of phosdrin on alfalfa and its effectiveness on the insect complex, *J. Econ. Entomol.*, 54:209-210, 1961.

Huffman, J.M., Parks, G.S., and Barmore, M. Thermal data on organic compounds. X. Further studies on the heat capacities, entropies and free energies of hydrocarbons, *J. Am. Chem. Soc.*, 53:3876-3888, 1931.

Huffman, J.M., Parks, G.S., and Daniels, A.C. Thermal data on organic compounds. VII. The heat capacities, entropies and free energies of twelve aromatic hydrocarbons, *J. Am. Chem. Soc.*, 52:1547-1558, 1930.

Huggenberger, F., .Letey, J., and Farmer, W.J. Observed and calculated distribution of lindane in soil columns as influenced by water movement, *Soil Sci. Soc. Am. Proc.*, 36:544-548, 1972.

Hull, L.A., Hisatsune, I.C., and Heicklen, J. Reaction of ozone with 1,1-dichloroethylene, *Can. J. Chem.*, 51:1504, 1973.

Humburg, N.E., Colby, S.R., Lym, R.G., Hill, E.R., McAvoy, W.J., Kitchen, L.M., and Prasad, R., Eds. *Herbicide Handbook of the Weed Science Society of America*, 6th ed. (Champaign, IL: Weed Science Society of America, 1989), 301 p.

Hunter-Smith, R.J., Balls, P.W., and Liss, P.S. Henry's law constants and the air-sea exchange of various low molecular weight halocarbon gases, *Tellus*, 35B:170-176, 1983.

Huntress, E.H. and Mulliken, S.P. *Identification of Pure Organic Compounds - Tables of Data on Selected Compounds of Order I* (New York: John Wiley & Sons, 1941), 691 p.

Hurd, C.D. and Carnahan, F.L. The action of heat on ethylamine and benzylamine, *J. Am. Chem. Soc.*, 52(10):4151-4158, 1930.

Hussam, A. and Carr, P.W. Rapid and precise method for the measurement of vapor/liquid equilibria by headspace gas chromatography, *Anal. Chem.*, 57(4):793-801, 1985.

Hustert, K., Kotzias, D., and Korte, F. Photokatalytischer Abbau von Chlornitrobenzolen an TiO_2 in wässriger Phase, *Chemosphere*, 16(4):809-812, 1987.

Hustert, K., Mansour, M., Parlar, H., and Korte, F. The EPA test - A method to determine the photochemical degradation of organic compounds in aquatic systems, *Chemosphere*, 10:995-998, 1981.

Hustert, K. and Moza, P.N. Photokatalytischer Abbau von Phthalaten an Titandioxid in wässriger Phase, *Chemosphere*, 17(9):1751-1754, 1988.

Hutchinson, T.C., Hellebust, J.A., Tam, D., Mackay, D., Mascarenhas, R.A., and Shiu, W.Y. The correlation of the toxicity to algae of hydrocarbons and halogenated hydrocarbons with their physical-chemical properties, *Environ. Sci. Res.*, 16:577-586, 1980.

Hutchinson, T.H., Jha, A.N., and Dixon, D.R. The polychaete *Platynereis dumerilii* (Audouin and Milne Edwards): a new species for assessing the hazardous potential of chemicals in the marine environment, *Ecotoxicol. Environ. Saf.*, 31(3):271-281, 1995.

Hutzinger, O., Safe, S., and Zitko, V. *The Chemistry of PCB's* (Boca Raton, FL: CRC Press, 1974), 269 p.

Hutzler, N.J., Crittenden, J.C., and Gierke, J.S. Transport of organic compounds with saturated groundwater flow: experimental results, *Water Resourc. Res.*, 22(3):285-295, 1986.

Hwang, H. and Dasgupta, P.K. Thermodynamics of the hydrogen peroxide-water system, *Environ. Sci. Technol.*, 19(3):255-258, 1985.

Hwang, H.-M., Hodson, R.E., and Lee, R.F. Degradation of phenol and chlorophenols by sunlight and microbes in estuarine water, *Environ. Sci. Technol.*, 20(10):1002-1007, 1986.

Hyne, R.V. and Everett, D.A. Application of a benthic euryhaline amphipod, *Corophium* sp., as a sediment toxicity testing organism for both freshwater and estuarine systems, *Arch. Environ. Contam. Toxicol.*, 34(1):26-33, 1998.

IARC. 1973. *IARC Monographs on the Evaluation of Carcinogenic Risk of Chemicals to Man. Certain Polycyclic Aromatic Hydrocarbons and Heterocyclic Compounds*, Volume 3 (Lyon, France: International Agency for Research on Cancer), 271 p.

IARC. 1974. *IARC Monographs on the Evaluation of Carcinogenic Risk of Chemicals to Man. Some Organochlorine Pesticides*, Volume 5 (Lyon, France: International Agency for Research on Cancer), 241 p.

IARC. 1978. *IARC Monographs on the Evaluation of Carcinogenic Risk of Chemicals to Man. Some N-Nitroso Compounds*, Volume 17 (Lyon, France: International Agency for Research on Cancer), 365 p.

IARC. 1979. *IARC Monographs on the Evaluation of Carcinogenic Risk of Chemicals to Man. Some Halogenated Hydrocarbons*, Volume 20 (Lyon, France: International Agency for Research on Cancer), 609 p.

IARC. 1983. *IARC Monographs on the Evaluation of the Carcinogenic Risk of Chemicals to Humans. Polynuclear Aromatic Compounds, Part 1, Chemical, Environmental and Experimental Data*, Volume 32 (Lyon, France: International Agency for Research on Cancer), 477 p.

Ibusuki, T. and Takeuchi, K. Toluene oxidation on U.V.-irradiated titanium dioxide with and without O_2, NO_2 or H_2O at ambient temperature, *Atmos. Environ.*, 29(9):1711-1715, 1986.

Ichinose, T., Suzuki, A.K., Tsubone, H., and Sagai, M. Biochemical studies on strain differences of mice in the susceptibility to nitrogen dioxide, *Life Sci.*, 31(18):1963-1972, 1982.

Inman, J.C., Strachan, S.D., Sommers, L.E., and Nelson, D.W. The decomposition of phthalate esters in soil, *J. Environ. Sci. Health*, B19:245-257, 1984.

Inokuchi, H., Shiba, S., Handa, T., and Akamatsu, H. Heats of sublimation of condensed polynuclear aromatic hydrocarbons, *Bull. Chem. Soc. Jpn.*, 25:299-302, 1952.

Iraci, L.T., Baker, B.M., Tyndall, G.S., and Orlando, J.J. Measurements of the Henry's law coefficients of 2-methyl-3-butenol, methacrolein, and methylvinyl ketone, *J. Atmos. Chem.*, submitted, 1998.

Isaac, R.A. and Morris, J.C. Transfer of active chlorine from chloramine to nitrogenous organic compounds. 1. Kinetics, *Environ. Sci. Technol.*, 17(12):738-742, 1983.

Isaacson, P.J. and Frink, C.R. Nonreversible sorption of phenolic compounds by sediment fractions: the role of sediment organic matter, *Environ. Sci. Technol.*, 18(1):43-48, 1984.

Isensee, A.R., Holden, E.R., Woolson, E.A., and Jones, G.E. Soil persistence and aquatic bioaccumulation potential of hexachlorobenzene, *J. Agric. Food Chem.*, 24(6):1210-1214, 1976.

Isensee, A.R. and Jones, G.E. Distribution of 2,3,7,8-tetrachlorodibenzo-*p*-dioxin (TCDD) in aquatic model ecosystem, *Environ. Sci. Technol.*, 9(7):668-672, 1975.

Ishikawa, S. and Baba, K. Reaction of organic phosphate esters with chlorine in aqueous solution, *Bull. Environ. Contam. Toxicol.*, 41(1):143-150, 1988.

Isnard, S. and Lambert, S. Estimating bioconcentration factors from octanol-water partition coefficient and aqueous solubility, *Chemosphere*, 17(1):21-34, 1988.

ITII. *Toxic and Hazardous Industrial Chemicals Safety Manual for Handling and Disposal with Toxicity and Hazard Data* (Tokyo, Japan: International Technical Information Institute, 1986), 700 p.

Ivie, G.W. and Casida, J.E. Enhancement of photoalteration of cyclodiene insecticide chemical residues by rotenone, *Science (Washington, D.C.)*, 167(3935):1620-1622, 1970.

Ivie, G.W. and Casida, J.E. Sensitized photodecomposition and photosensitizer activity of pesticide chemicals exposed to sunlight of silica gel chromatoplates, *J. Agric. Food Chem.*, 19(3):405-409, 1971.

Ivie, G.W. and Casida, J.E. Photosensitizers for the accelerated degradation of chlorinated cyclodienes and other insecticide chemicals exposed to sunlight on bean leaves, *J. Agric. Food Chem.*, 19(3):410-416, 1971a.

Ivie, G.W., Knox, J.R., Khalifa, S., Yamamoto, I., and Casida, J.E. Novel photoproducts of heptachlor epoxide, *trans*-chlordane, and *trans*-nonachlor, *Bull. Environ. Contam. Toxicol.*, 7(6):376-383, 1972.

Iwasaki, M., Ibuki, T., and Takezaki, Y. Primary processes of the photolysis of ethylenimine at Xe and Kr resonance lines, *J. Chem. Phys.*, 59(12):6321-6327, 1973.

Iwata, Y., Westlake, W.E., and Gunther, F.A. Varying persistence of polychlorinated biphenyls in six California soils under laboratory conditions, *Bull. Environ. Contam. Toxicol.*, 9(4):204-211, 1973.

Iyengar, L. and Rao, A.V.S.P. Metabolism of chlordane and heptachlor by *Aspergillus niger*, *J. Gen. Appl. Microbiol.*, 19(4):321-324, 1973.

Jacobson, S.N. and Alexander, M. Enhancement of the microbial dehalogenation of a model chlorinated compound, *Appl. Environ. Microbiol.*, 42(6):1062-1066, 1981.

Jacoby, W.A., Nimios, M.R., Blake, D.M., Noble, R.D., and Koval, C.A. Products, intermediates, mass balances, and reaction pathways for the oxidation of trichloroethylene in air via heterogeneous photocatalysts, *Environ. Sci. Technol.*, 28(9):1661-1668, 1994.

Jafvert, C.T., Westall, J.C., Grieder, E., and Schwarzenbach, R.P. Distribution of hydrophobic ionogenic organic compounds between octanol and water: organic acids, *Environ. Sci. Technol.*, 24(12):1795-1803, 1990.

Jafvert, C.T. and Wolfe, N.L. Degradation of selected halogenated ethanes in anoxic sediment-water systems, *Environ. Toxicol. Chem.*, 6(11):827-837, 1987.

James, T.L. and Wellington, C.A. Thermal decomposition of β-propiolactone in the gas phase, *J. Am. Chem. Soc.*, 91(27):7743-7746, 1969.

Jana, T.K. and Das, B. Sorption of carbaryl (1-naphthyl *N*-methyl carbamate) by soil, *Bull. Environ. Contam. Toxicol.*, 59(1):65-71, 1997.

Jańczuk, B., Wójcik, W., and Zdziennicka, A. Determination of the components of the surface tension of some liquids from interfacial liquid-liquid tension measurements, *J. Colloid Interface Sci.*, 157(2):384-393, 1993.

Janini, G.M. and Attari, S.A. Determination of partition coefficient of polar organic solutes in octanol/micellar solutions, *Anal. Chem.*, 55(4):659-661, 1983.

Janssen, D.B., Jager, D., and Wilholt, B. Degradation of *n*-haloalkanes and α,ω-dihaloalkanes by wild type and mutants of *Acinetobacter* sp. strain GJ70, *Appl. Environ. Microbiol.*, 53(3):561-566, 1987.

Japar, S.M., Wu, C.H., and Niki, H. Rate constants for the reaction of ozone with olefins in the gas phase, *J. Phys. Chem.*, 23:2318-2320, 1974.

Jarboe, H.H. and Romaire, R.P. Acute toxicity of permethrin to four sizes of red swamp crayfish (*Procambarus clarkii*) and observations of post-exposure effects, *Arch. Environ. Contam. Toxicol.*, 20(3):337-342, 1991.

Jarvinen, A.W., Tanner, D.K., and Kline, E.R. Toxicity of chlorpyrifos, endrin, or fenvalerate to fathead minnows following episodic or continuous exposure, *Ecotoxicol. Environ. Saf.*, 15(1):78-95, 1988.

Jayasinghe, D.S., Brownawell, B.J., Chen, H., and Westall, J.C. Determination of Henry's law constants of organic compounds of low volatility, *Environ. Sci. Technol.*, 26(11):2275-2281, 1992.

Jeffers, P.M., Ward, L.M., Woytowitch, L.M., and Wolfe, N.L. Homogeneous hydrolysis rate constants for selected chlorinated methanes, ethanes, ethenes, and propanes, *Environ. Sci. Technol.*, 23(8):965-969, 1989.

Jenkins, J. and King, M.B. Vapor-liquid equilibria for the system bromine/water at low bromine concentrations, *Chem. Eng. Sci.*, 20:921-922, 1965.

Jensen, L.D. and Gaufin, A.R. Acute and long-term effects of organic insecticides on two species of stonefly naiads, *J. Water Pollut. Control Fed.*, 38(8):1273-1286, 1966.

Jensen, S., Göthe, R., and Kindstedt, M.-O. Bis(*p*-chlorophenyl)acetonitrile (DDN), a new DDT derivative formed in anaerobic digested sewage sludge and lake sediment, *Nature (London)*, 240(5381):421-422, 1972.

Jepsen, R. and Lick, W. Nonlinear and interactive effects in the sorption of hydrophobic organic chemicals by sediments, *Environ. Toxicol. Chem.*, 18(8):1627-1636, 1999.

Jerina, D.M., Daly, J.W., Jeffrey, A.M., and Gibson, D.T. *cis*-1,2-Dihydroxy-1,2-dihydronaphthalene: a bacterial metabolite from naphthalene, *Arch. Biochem. Biophys.*, 142:394-396, 1971.

Jewett, D. and Lawless, J.G. Formate esters of 1,2-ethanediol: major decomposition products of *p*-dioxane during storage, *Bull. Environ. Contam. Toxicol.*, 25(1):118-121, 1980.

Jiang, Q., Gao, G.-H., Yu, Y.-X., and Qin, Y. Solubility of sodium dimethyl isophthalate-5-sulfonate in water and in water + methanol containing dosium sulfate, *J. Chem. Eng. Data*, 45(2):292-294, 2000.

Jianlong, W., Ping, L., and Yi, Q. Microbial degradation of di-*n*-butyl phthalate, *Chemosphere*, 31(9):4051-4056, 1995.

Jigami, Y., Omori, T., and Minoda, Y. The degradation of isopropylbenzene and isobutylbenzene by *Pseudomonas* sp., *Agric. Biol. Chem.*, 39(9):1781-1788, 1975.

Jin, L.-J., wei, Z., Dai, J.-Y., Guo, P., and Wang, L.-S. Prediction of partitioning properties for benzaldehydes by various molecular descriptors, *Bull. Environ. Contam. Toxicol.*, 61(1):1-7, 1998.

Johansen, C.A., Westlake, W.E., Butler, L.I., and Bry, R.E. Residual action and toxicity of methoxychlor and parathion to the cherry fruit fly, *J. Econ. Entomol.*, 47:746-749, 1954.

Johnsen, R.E. DDT Metabolism in microbial systems, *Residue Rev.*, 61:1-28, 1976.

Johnsen, S., Gribbestad, I.S., and Johansen, S. Formation of chlorinated PAH - A possible health hazard from water chlorination, *Sci. Total Environ.*, 81/82:231-238, 1989.

Johnson, B.J., Betterton, E.A., and Craig, D. Henry's law coefficients of formic and acetic acids, *J. Atmos. Chem.*, 24:113-119, 1996.

Johnson, C.A. and Westall, J.C. Effect of pH and KCl concentration on the octanol-water distribution of methylanilines, *Environ. Sci. Technol.*, 24(12):1869-1875, 1990.

Johnson, M.W. and Gentile, J.H. Acute toxicity of cadmium, copper, and mercury to larval american lobster *Homarus americanus*, *Bull. Environ. Contam. Toxicol.*, 22(1/2):258-264, 1979.

Johnson-Logan, L.R., Broshears, R.E., and Klaine, S.J. Partitioning behavior and the mobility of chlordane in groundwater, *Environ. Sci. Technol.*, 26(11):2234-2239, 1992.

Johnston, H.S. and Heicklen, J. Photochemical oxidations. IV. Acetaldehyde, *J. Am. Chem. Soc.*, 86(20):4254-4256, 1964.

Johnston, J.J. and Corbett, M.D. The effects of temperature, salinity and a simulated tidal cycle on the toxicity of fenitrothion to *Callinectes sapidus*, *Comp. Biochem. Physiol. C*, 80(1):145-149, 1985.

Joiner, R.L. and Baetcke, K.P. Parathion: persistence on cotton and identification of its photoalteration products, *J. Agric. Food Chem.*, 21(3):391-396, 1973.

Joiner, R.L., Chambers, H.W., and Baetcke, K.P. Toxicity of parathion and several of its photoalteration products to boll weevils, *Bull. Environ. Contam. Toxicol.*, 6(3):220-224, 1971.

Jones, B.K. and Hathway, D.E. Differences in metabolism of vinylidene chloride between mice and rats, *Br. J. Cancer*, 37:411-417, 1978.

Jones, C.J., Hudson, B.C., and Smith, A.J. The leaching of some halogenated organic compounds from domestic waste, *J. Haz. Mater.*, 2:227-233, 1977.

Jones, D.C. The systems of *n*-butyl alcohol-water and *n*-butyl alcohol-water-acetone, *J. Chem. Soc. (London)*, pp. 799-816, 1929.

Jones, D.M. and Wood, N.F. The mechanism of vinyl ether hydrolysis, *J. Chem. Soc. (London)*, pp. 5400-5403, 1964.

Jones, J.H. and McCants, J.F. Ternary solubility data. 1-Butanol-methanol 1-butyl ketone-water, 1-butyraldehyde-ethyl acetate-water, 1-hexane-methyl ethyl ketone-water, *Ind. Eng. Chem.*, 46(9):1956-1958, 1954.

Jones, W.M. and Giauque, W.F. The entropy of nitromethane. Heat capacity of solid and liquid. Vapor pressure, heats of fusion and vaporization, *J. Am. Chem. Soc.*, 69:983-987, 1947.

Jönsson, J.A., Vejrosta, J., and Novak, J. Air/water partition coefficients for normal alkanes (*n*-pentane to *n*-nonane), *Fluid Phase Equil.*, 9:279-286, 1982.

Jordan, T.E. *Vapor Pressures of Organic Compounds* (New York: Interscience Publishers, 1954), 266 p.

Joschek, H.I. and Miller, S.T. Photooxidation of phenol, cresols, and dihydroxybenzenes, *J. Am. Chem. Soc.*, 88:3273-3281, 1966.

Judd, F.W. Acute toxicity and effects of sublethal dietary exposure of monosodium methanearsonate herbicide to *Peromyscus leucopus* (Rodentia: Cricetidae), *Bull. Environ. Contam. Toxicol.*, 22(1/2):143-150, 1979.

Judy, R.D., Jr. The acute toxicity of copper to *Gammarus fasciatus* say, a freshwater amphipod, *Bull. Environ. Contam. Toxicol.*, 21(1/2):219-224, 1979.

Judy, R.D., Jr. and Davies, P.H. Effects of calcium addition as $Ca(NO_3)_2$ on zinc toxicity to fathead minnows, *Pimephales promelas*, Rafinesque, *Bull. Environ. Contam. Toxicol.*, 28(1/2):88-94, 1979.

Jury, W.A., Focht, D.D., and Farmer, W.J. Evaluation of pesticide pollution potential from standard indices of soil-chemical adsorption and biodegradation, *J. Environ. Qual.*, 16(4):422-428, 1987.

Jury, W.A., Russo, D., Streile, G., and Abd, H.E. Evaluation of volatilization by organic chemicals residing below the soil surface, *Water Resourc. Res.*, 26(1):13-20, 1990.

Jury, W.A., Spencer, W.F., and Farmer, W.J. Use of models for assessing relative volatility, mobility, and persistence of pesticides and other trace organics in soil systems, in *Hazard Assessment of Chemicals, Volume 2*, Saxena, J., Ed. (New York: Academic Press, 1983), pp. 1-43.

Jury, W.A., Spencer, W.F., and Farmer, W.J. Behavior assessment model for trace organics in soil: IV. Review of experimental evidence, *J. Environ. Qual.*, 13(4):580-586, 1984.

Kaars Sijpesteijn, A., Dekhuijzen, H.M., and Vonk, J.W. Biological conversion of fungicides in plants and microorganisms, in *Antifungal Compounds* (New York: Marcel Dekker, 1977), pp. 91-147.

Kaczmar, S.W., D'Itri, F.M., Zabik, M.J. Volatilization rates of selected haloforms from aqueous environments, *Environ. Toxicol. Chem.*, 3(1):31-35, 1984.

Kaiser, K.L. and Wong, P.T.S. Bacterial degradation of polychlorinated biphenyls. I. Identification of some metabolic products from Aroclor 1242, *Bull. Environ. Contam. Toxicol.*, 11(3):291-296, 1974.

Kaiser, K.L.E. and Valdmanis, I. Apparent octanol/water partition coefficients of pentachlorophenol as a function of pH, *Can. J. Chem.*, 60(16):2104-2106, 1982.

Kale, S.P., Murthy, N.B.K., and Raghu, K. Effect of carbofuran, carbaryl, and their metabolites on the growth of *Rhizobium* sp. and *Azotobacter chroococcum*, *Bull. Environ. Contam. Toxicol.*, 42(5):769-772, 1989.

Kallos, G.J. and Tou, J.C. Study of photolytic oxidation and chlorination reactions of dimethyl ether and chlorine in ambient air, *Environ. Sci. Technol.*, 11(12):1101-1105, 1977.

Kalsch, W., Nagel, R. and Urich, K. Uptake, elimination, and bioconcentration of ten anilines in zebrafish (*Brachydanio rerio*), *Chemosphere*, 22(3/4):351-363, 1991.

Kames, J. and Schurath, U. Alkyl nitrates and bifunctional nitrates of atmospheric interest: Henry's law constants and their temperatures dependencies, *J. Atmos. Chem.*, 15:79-95, 1992.

Kames, J., Schweighoefer, S., and Schurath, U. Henry's law constant and hydrolysis of peroxyacetyl nitrate (PAN), *J. Atmos. Chem.*, 12:169-180, 1991.

Kamlet, M.J., Abraham, M.H., Doherty, R.M., and Taft, R.W. Solubility properties in polymers and biological media. 4. Correlation of octanol/water partition coefficients with solvatochromic parameters, *J. Am. Chem. Soc.*, 106(2):464-466, 1984.

Kamlet, M.J., Doherty, R.M., Abraham, M.H., Carr, P.W., Doherty, R.F., and Taft, R.W. Linear solvation energy relationships. 41. Important differences between aqueous solubility relationships for aliphatic and aromatic solutes, *J. Phys. Chem.*, 91(7):1996-2004, 1987.

Kammenga, J.E., Van Gestel, C.A.M., and Bakker, J. Patterns of sensitivity to cadmium and pentachlorophenol among nematode species from different taxonomic and ecological groups, *Arch. Environ. Contam. Toxicol.*, 27(1):88-94, 1994.

Kanazawa, J. Relationship between the soil sorption constants for pesticides and their physicochemical properties, *Environ. Toxicol. Chem.*, 8(6):477-484, 1989.

Kaneda, Y., Nakamura, K., Nakahara, H., and Iwaida, M. Degradation of organochlorine pesticides with calcium hypochlorite [Chemical Abstracts 82:94190e]: *I. Eisei Kagaku*, 20:296-299, 1974.

Kannan, S. and Kishore, K. Absolute viscosity and density of trisubstituted phosphoric esters, *J. Chem. Eng. Data*, 44(4):649-655, 1999.

Kanno, S. and Nojima, K. Studies on photochemistry of aromatic hydrocarbons. V. Photochemical reaction of chlorobenzene with nitrogen oxides in air, *Chemosphere*, 8(4):225-232, 1979.

Kanno, S., Nojima, K., and Ohya, T. Formation of cyanide ion or cyanogen chloride through the cleavage of aromatic rings by nitrous acid or chlorine. IV. On the reaction of aromatic hydrocarbons with hypochlorous acid in the presence of ammonium ion, *Chemosphere*, 11(7):663-667, 1982.

Kapoor, I.P., Metcalf, R.L., Hirwe, A.S., Coats, J.R., and Khalsa, M.S. Structure activity correlations of biodegradability of DDT analogs, *J. Agric. Food Chem.*, 21(2):310-315, 1973.

Kapoor, I.P., Metcalf, R.L., Nystrom, R.F., and Sanghua, G.K. Comparative metabolism of methoxychlor, methiochlor, and DDT in mouse, insects, and in a model ecosystem, *J. Agric. Food Chem.*, 18(6):1145-1152, 1970.

Kappeler, T. and Wuhrmann, K. Microbial degradation of the water-soluble fraction of gas oil - II. Bioassays with pure strains, *Water Res.*, 12(5):335-342, 1978.

Karabaev, M.K., Abduzhaminov, T.P., Kenisarin, M.M., and Saidov, A.A. Thermodynamics of the crystal-liquid phase transition in acrylates and methacrylates, *Ser. Fiz.-Mat. Nauk*, 5:74-77, 1985.

Karickhoff, S.W. Correspondence - On the sorption of neutral organic solutes in soils, *J. Agric. Food Chem.*,

29(2):425-426, 1981.

Karickhoff, S.W. Semi-empirical estimation of sorption of hydrophobic pollutants on natural sediments and soils, *Chemosphere*, 10(8):833-846, 1981a.

Karickhoff, S.W., Brown, D.S., and Scott, T.A. Sorption of hydrophobic pollutants on natural sediments, *Water Res.*, 13(3):241-248, 1979.

Karickhoff, S.W. and Morris, K.R. Sorption dynamics of hydrophobic pollutants in sediment suspensions, *Environ. Toxicol. Chem.*, 4:469-479, 1985.

Karickhoff, S.W. and Morris, K.R. Impact of tubificid oligochaetes on pollutant transport in bottom sediments, *Environ. Sci. Technol.*, 19(1):51-56, 1985a.

Karinen, J.F., Lamberton, J.G., Stewart, N.E., and Terriere, L.C. Persistence of carbaryl in the marine estuarine environment. Chemical and biological stability in aquarium systems, *J. Agric. Food Chem.*, 15(1):148-156, 1967.

Kato, M., Furuhama, K., Yoshida, M., Akahane, K., and Takayama, S. Acute oral toxicity of the new quinoline antibacterial agent levofloxacin in mice, rats, and monkeys, *Arzneim. Forsch.*, 42(3a):365-366, 1992.

Katz, M. Acute toxicity of some organic insecticides to three species of salmonids and to threespine stickleback, *Trans. Am. Fish Soc.*, 90(3):264-268, 1961.

Kawaguchi, H. and Furuya, M. Photodegradation of monochlorobenzene in titanium dioxide aqueous suspensions, *Chemosphere*, 21(12):1435-1440, 1990.

Kawamoto, K. and Urano, K. Parameters for predicting fate of organochlorine pesticides in the environment (I) octanol-water and air-water partition coefficients, *Chemosphere*, 18(9/10):1987-1996, 1989.

Kawasaki, M. Experiences with the test scheme under the Chemical Control Law of Japan: an approach to structure-activity correlations, *Ecotoxicol. Environ. Saf.*, 4(4):444-454, 1980.

Kay, B.D. and Elrick, D.E. Adsorption and movement of lindane in soils, *Soil Sci.*, 104:314, 1967.

Kayal, S.I. and Connell, D.W. Partitioning of unsubstituted polycyclic aromatic hydrocarbons between surface sediments and the water column in the Brisbane River estuary, *Aust. J. Mar. Freshwater Res.*, 41:443-456, 1990.

Kazano, H., Kearney, P.C., and Kaufman, D.D. Metabolism of methylcarbamate insecticides in soils, *J. Agric. Food Chem.*, 20(5):975-979, 1972.

Kearney, P.C., Isensee, A.R., Helling, C.S., Woolson, E.A., and Plimmer, J.R. Environmental significance of chlorodioxins, in *Chlorodioxins - Origin and Fate, Adv. Chem. Ser.*, 120:105-111, 1971.

Kearney, P.C. and Kaufman, D.D. *Herbicides: Chemistry, Degradation and Mode of Action* (New York: Marcel Dekker, 1976), 1036 p.

Keck, J., Sims, R.C., Coover, M., Park, K., and Symons, B. Evidence for cooxidation of polynuclear aromatic hydrocarbons in soil, *Water Res.*, 23(12):1467-1476, 1989.

Keeley, D.F., Hoffpauir, M.A., and Meriwether, J.R. Solubility of aromatic hydrocarbons in water and sodium chloride solutions of different ionic strengths: benzene and toluenes, *J. Chem. Eng. Data*, 33(2):87-89, 1988.

Keen, R. and Baillod, C.R. Toxicity to *Daphnia* of the end products of wet oxidation of phenol and substituted phenols, *Water Res.*, 19(6):767-772, 1985.

Keith, L.H. and Walters, D.B. *The National Toxicology Program's Chemical Data Compendium - Volume II. Chemical and Physical Properties* (Chelsea, MI: Lewis Publishers, 1992), 1642 p.

Keller, A.E. Acute toxicity of several pesticides, organic compounds, and a wastewater effluent to the freshwater mussel, *Anodonta imbecilis, Ceriodaphnia dubia*, and *Pimephales promelas, Bull. Environ. Contam. Toxicol.*, 51(5):696-702, 1993.

Keller, D.A., Roe, D.C., and Lieder, P.H. Fluoroacetate-mediated toxicity of fluorinated ethanes, *Fundam. Appl. Toxicol.*, 30(2):213-219, 1995.

Kelley, C.M. and Tartar, H.V. On the system: bromine-water, *J. Am. Chem. Soc.*, 78:5752-5756, 1956.

Kelley, J.D. and Rice, F.O. The vapor pressures of some polynuclear aromatic hydrocarbons, *J. Phys. Chem.*, 68(12):3794-3796, 1964.

Kelley, K.K. The heat capacity of toluene from 14K to 298K. The entropy and free energy of formation, *J. Am. Chem. Soc.*, 51:2738-2741, 1929.

Kelly, T.J., Mukund, R., Spicer, C.W., and Pollack, A.J. Concentrations and transformations of hazardous air pollutants, *Environ. Sci. Technol.*, 28(8):378A-387A, 1994.

Kenaga, E.E. *Environmental Dynamics of Pesticides* (New York: Plenum Press, 1975), 243 p.

Kenaga, E.E. Correlation of bioconcentration factors of chemicals in aquatic and terrestrial organisms with their physical and chemical properties, *Environ. Sci. Technol.*, 14(5):553-556, 1980.

Kenaga, E.E. Predicted bioconcentration factors and soil sorption coefficients of pesticides and other chemicals, *Ecotoxicol. Environ. Saf.*, 4(1):26-38, 1980a.

Kenaga, E.E. and Goring, C.A.I. Relationship between water solubility, soil sorption, octanol-water partitioning and concentration of chemicals in biota, in *Aquatic Toxicology, ASTM STP 707*, Eaton, J.G., Parrish, P.R., and Hendricks, A.C., Eds. (Philadelphia, PA: American Society for Testing and Materials, 1980), pp. 78-115.

Kenley, R.A., Davenport, J.E., and Hendry, D.G. Hydroxyl radical reactions in the gas phase. Products and pathways for the reaction of OH with toluene, *J. Phys. Chem.*, 82(9):1095-1096, 1978.

Kennedy, G.D., Jr. Acute toxicity studies with oxamyl, *Fundam. Appl. Toxicol.*, 6(3):423-429, 1986.

Kennedy, G.L., Jr. and Chen, H.C. Inhalation toxicity of dibutylhexamethylenediamine in rats, *Food Chem. Toxicol.*, 22(6):425-429, 1984.

Kennedy, G.L., Jr., Hall, G.T., Brittelli, M.R., Barnes, J.R., and Chen, H.C. Inhalation toxicity of ammonium perfluorooctanoate, *Food Chem. Toxicol.*, 24(12):1325-1329, 1986.

Kennedy, G.L., Jr. and Sherman, H. Acute subchronic toxicity of dimethylformamide and dimethylacetamide following various routes of administration, *Drug Chem. Toxicol.*, 9(2):147-170, 1986.

Kennedy, M.V., Stojanovic, B.J., and Shuman, F.L., Jr. Analysis of decomposition products of pesticides, *J. Agric. Food Chem.*, 20(2):341-343, 1972.

Kennedy, M.V., Stojanovic, B.J., and Shuman, F.L., Jr. Chemical and thermal aspects of pesticide disposal, *J. Environ. Qual.*, 1(1):63-65, 1972a.

Ketelaar, J.A.A. and Bloksma, A.H. The rate of hydrolysis and composition of tetraethyl pyrophosphate, *Recueil.*, 67:665-676, 1948.1991.

Key, P.B. and Scott, G.I. Lethal and sublethal effects of chlorine, phenol, and chlorine-phenol mixtures on the mud crab, *Panopeus herbstii, Environ. Health Perspect.*, 60:307-312, 1986.

Khalifa, S., Holmstead, R.L., and Casida, J.E. Toxaphene degradation by iron(II) protoporphyrin systems, *J. Agric. Food Chem.*, 24(2):277-282, 1976.

Khalil, M.A.K. and Rasmussen, R.A. Global sources, lifetimes and mass balances of carbonyl sulfide (OCS) and carbon disulfide (CS_2) in the earth's atmosphere, *Atmos. Environ.*, 18(9):1805-1813, 1984.

Khan, A., Hassett, J.J., and Banwart, W.L. Sorption of acetophenone by sediments and soils, *Soil Sci.*, 128(5):297-302, 1979a.

Khan, I. and Brimblecombe, P. Henry's law constants of low molecular weight (<130) organic acids, *J. Aerosol Sci.*, 23(Suppl. 1):897-900, 1992.

Khan, I., and Brimblecombe, P., and Clegg, S.L. Solubilities of pyruvic acid and the lower (C_1-C_6) carboxylic acids. Experimental determination of equilibrium vapour pressures above pure aqueous and salt solutions, *J. Atmos. Chem.*, 22:285-302, 1995.

Khan, M.A.Q., Feroz, M., and Sudershan, P. Metabolism of cyclodiene insecticides by fish, in *Pesticide and Xenobiotic Metabolism in Aquatic Organisms* ACS Symposium Series 99 (Washington, D.C.: American Chemical Society, 1979).

Khan, S.U. Adsorption of 2,4-D from aqueous solution by fulvic acid-clay complex, *Environ. Sci. Technol.*, 8(3):236-238, 1974.

Kieatiwong, S., Nguyen, L.V., Hebert, V.R., Hackett, M., Miller, G.C., Miille, M.J., and Mitzel, R. Photolysis of chlorinated dioxins in organic solvents and on soils, *Environ. Sci. Technol.*, 24(10):1575-1580, 1990.

Kieckbusch, T.G. and King, C.J. An improved method of determining vapor-liquid equilibria for dilute organics in aqueous solution, *J. Chromatogr. Sci.*, 17:273-276, 1979.

Kile, D.E., Chiou, C.T., Zhou, H., Li, H., and Zu, O. Partition of nonpolar organic pollutants from water to soil and sediment organic matters, *Environ. Sci. Technol.*, 29(5):1401-1406, 1995.

Kilzer, L., Scheunert, I., Geyer, H., Klein, W., and Korte, F. Laboratory screening of the volatilization rates of organic chemicals from water and soil, *Chemosphere*, 8(10):751-761, 1979.

Kim, H.Y., Chung, Y.H., Yi, K.H., Kim, J.G., and Yu, I.J. LC_{50} of 2-bromopropane, *Ind. Health*, 34(4):403-407, 1996.

Kim, I.-Y. and Saleh, F.W. Aqueous solubilities and transformations of tetrahalogenated benzenes and effects of aquatic fulvic acids, *Bull. Environ. Contam. Toxicol.*, 44(6):813-818, 1990.

Kim, Y.-H., Woodrow, J.E., and Seiber, J.N. Evaluation of a gas chromatographic method for calculating vapor pressures with organophosphorus pesticides, *J. Chromatogr.*, 314:37-53, 1984.

Kimura, E.T., Ebert, D.M., and Dodge, P.W. Acute toxicity and limits of solvent residue for sixteen organic solvents, *Toxicol. Appl. Pharmacol.*, 19:699-704, 1971.

Kincannon, D.F and Lin, Y.S. Microbial degradation of hazardous wastes by land treatment, *Proc. Indust. Waste Conf.*, 40:607-619, 1985.

Kinney, L.A., Chromey, N.C., and Kennedy, G.L., Jr. Acute inhalation toxicity of ammonium perfluorononanoate, *Food Chem. Toxicol.*, 27(7):465-468, 1989.

Kinoshita, M. The absorption spectra of the molecular complexes of aromatic compounds with *p*-bromanil, *Bull. Chem. Soc. Jpn.*, 35:1609, 1962.

Kirchnerová, J. and Cave, G.C.B. The solubility of water in low-dielectric solvents, *Can. J. Chem.*, 54(24):3909-3916, 1976.

Kirimer, N., Baser, K.H.C., Ozek, T., and Kurkcuoglo, M. Composition of the essential oil of *Calamintha nepeta subsp. Glandulosa, J. Ess. Oil Res.*, 4:189-190, 1992.

Kirk, P.W.W., Rogers, H.R., and Lester, J.N. The fate of chlorobenzenes and permethrins during anaerobic sewage sludge digestion, *Chemosphere*, 18:1771-1784, 1989.

Kiser, R.W., Johnson, G.D., and Shetlar, M.D. Solubilities of various hydrocarbons in methanol, *J. Chem. Eng. Data*, 6(3):338-341, 1961.

Kishi, H. and Hashimoto, Y. Evaluation of the procedures for the measurement of water solubility and *n*-octanol/water partition coefficient of chemicals. Results of a ring test in Japan, *Chemosphere*, 18(9/10):1749-1759, 1989.

Kishi, H., Kogure, N., and Hashimoto, Y. Contribution of soil constituents in adsorption coefficient of aromatic compounds, halogenated alicyclic and aromatic compounds to soil, *Chemosphere*, 21(7):867-876, 1990.

Kishino, T. and Kobayashi, K. Relation between the chemical structures of chlorophenols and their dissociation constants and partition coefficients in several solvent-water systems, *Water Res.*, 28(7):1547-1552, 1994.

Kishk, F.M., Abu-Shara, T.M., Bakry, N., and Abou-Donia, M.B. Adsorption of methyl parathion by soils, *Bull. Environ. Contam. Toxicol.*, 22(6):733-738, 1979.

Kjeldsen, P., Kjølholt, J., Schultz, B., Christensen, T.H., and Tjell, J.C. Sorption and degradation of chlorophenols, nitrophenols and organophosphorus pesticides in the subsoil under landfills, *J. Contam. Hydrol.*, 6(2):165-184, 1990.

Klaassen, C. and Plaa, G. Relative effect of various chlorinated hydrocarbons on kidney and liver function in mice, *Toxicol. Appl. Pharmacol.*, 9:131-151, 1966.

Klečka, G.M., Gonsior, S.J., and Markham, D.A. Biological transformations of 1,1,1-trichloroethane in subsurface soils and groundwater, *Environ. Toxicol. Chem.*, 9(12):1437-1451, 1990.

Klein, A.W., Harnish, M., Porenski, H.J., and Schmidt-Bleek, F. OECD chemicals testing programme physico-chemical tests, *Chemosphere*, 10:153-207, 1981.

Klein, E., Weaver, J.W., and Weber, B.G. Solubility of acrylonitrile in aqueous bases and alkali salts, *Chem. Eng. Data Ser.*, 2:72-75, 1957.

Klein, R.G. Calculations and measurements on the volatility of *n*-nitrosoamines and their aqueous solutions, *Toxicology*, 23:135-147, 1982.

Klein, W., Geyer, H., Freitag, D., Rohleder, H. Sensitivity of schemes for ecotoxicological hazard ranking of chemicals, *Chemosphere*, 13(1):203-211, 1984.

Klein, W., Kohli, J., Weisgerber, I., and Korte, F. Fate of aldrin-^{14}C in potatoes and soil under outdoor conditions, *J. Agric. Food Chem.*, 21(2):152-156, 1973.

Klein, W., Kördel, W., Weiß, M., and Poremski, H.J. Updating of the OECD test guideline 107 'partition coefficient-octanol/water': OECD laboratory intercomparison test of the HPLC method, *Chemosphere*, 17(2):361-386, 1988.

Kleopfer, R.D., Easley, D.M., Haas, Jr., B.B., Deihl, T.G., Jackson, D.E., and Wurrey, C.J. Anaerobic degradation of trichloroethylene in soil, *Environ. Sci. Technol.*, 19(3):277-280, 1985.

Klemenc, A. and Löw, M. Die Löslichkeit in Wasser und ihr Zusammenhang der drei Dichlorbenzole. Eine Methode zur Bestimmung der Löslichkeit sehr wenig löslicher und zugleich sehr flüchtiger Stoffe, *Rec. Trav. Chim. Pays-Bas*, 49(4):629-640, 1930.

Klevens, H.B. Solubilization of polycyclic hydrocarbons, *J. Phys. Colloid Chem.*, 54(2):283-298, 1950.

Klonne, D.R., Dodd, D.E., Pritts, I.M., Nachreiner, D.J., Fowler, E.H., Troup, C.M., Homan, E.R., and Ballantyne, B. Dimethylethanolamine: acute, 2-week, and 13-week inhalation toxicity studies in rats, *Fundam. Appl. Toxicol.*, 9(3):512-521, 1987.

Klonne, D.R., Ulrich, C.E., Weissmann, J., and Morgan, A.K. Acute inhalation toxicity of aliphatic (C1-C5) nitrites in rats, *Fundam. Appl. Toxicol.*, 8(1):101-106, 1987a.

Klöpffer, W., Haag, F., Kohl, E.-G., and Frank, R. Testing of the abiotic degradation of chemicals in the atmosphere: the smog chamber approach, *Ecotoxicol. Environ. Saf.*, 15(3):298-319, 1988.

Klöpffer, W., Kaufman, G., Rippen, G., and Poremski, H.-P. A laboratory method for testing the volatility from aqueous solution: first results and comparison with theory, *Ecotoxicol. Environ. Saf.*, 6(6):545-559, 1982.

Knettig, E., Thomson, B.M., and Hrudey, S.E. Competitive activated carbon adsorption of phenolic compounds, *Environ. Pollut. (Series B)*, 12(4):281-299, 1986.

Knight, E.V., Novick, N.J., Kaplan, D.L., and Meeks, J.R. Biodegradation of 2-furaldehyde under nitrate-reducing and methanogenic conditions, *Environ. Toxicol. Chem.*, 9(6):725-730, 1990.

Knobloch, K., Szedzikowski, S., and Slusarcyk-Zablobona. Acute and subacute toxicity of acenaphthene and acenaphthylene, *Med. Pracy*, 20:210-222, 1969.

Knoevenagel, K. and Himmelreich, R. Degradation of compounds containing carbon atoms by photo-oxidation in the presence of water, *Arch. Environ. Contam. Toxicol.*, 4(3):324-333, 1976.

Knoph, M.B. Acute toxicity of ammonia to Atlantic salmon (*Salmo salar*) parr, *Comp. Biochem. Physiol. C*, 101(2):275-282, 1992.

Knowlton, M.F. and Huckins, J.N. Fate of radiolabeled sodium pentachlorophenate in littoral microcosms, *Bull. Environ. Contam. Toxicol.*, 30(2):206-213, 1983.

Knuutinen, J., Palm, H., Hakala, H., Haimi, J., Huhta, V., and Salminen, J. Polychlorinated phenols and their metabolites in soil and earthworms of sawmill environment, *Chemosphere*, 20(6):609-623, 1990.

Kobayashi, H. Metabolism of pentachlorophenol in fish, in *Pesticide and Xenobiotic Metabolism in Aquatic Organisms* ACS Symposium Series 99 (Washington, D.C.: American Chemical Society, 1979).

Kobayashi, H. and Rittman, B.E. Microbial removal of hazardous organic compounds, *Environ. Sci. Technol.*, 16(3):170A-183A, 1982.

Kobrinsky, P.C. and Martin, M.R. High-energy methyl radicals; the photolysis of methyl bromide at 1850 Å, *J. Chem. Phys.*, 48(12):5728-5729, 1968.

Koelmans, A.A., van der Woude, H., Hattink, J., and Niesten, D.J.M. Long-term bioconcentration kinetics of hydrophobic chemicals in *Selenastrum capricornutum* and *Microcystis aeruginosa*, *Environ. Toxicol. Chem.*, 18(6):1164-1172, 1999.

Koester, C.J. and Hites, R.A. Calculated physical properties of polychlorinated dibenzo-*p*-dioxins and dibenzofurans, *Chemosphere*, 17(12):2355-2362, 1988.

Kohring, G.-W., Rogers, J.E., and Wiegel, J. Anaerobic biodegradation of 2,4-dichlorophenol in freshwater lake sediments at different temperatures, *Appl. Environ. Microbiol.*, 55(2):348-353, 1989.

Kolb, B., Welter, C., and Bichler, C. Determination of partition coefficients by automatic equilibrium headspace gas chromatography by vapor phase calibration, *Chromatographia*, 34:235-240, 1992.

Kollig, H.P. Environmental fate constants for organic chemicals under consideration for EPA's hazardous waste identification projects, Office of Research and Development, U.S. EPA Report 600/R-93/132, 1993, 172 p.

Kollig, H.P. Environmental fate constants for additional 27 organic chemicals under consideration for EPA's hazardous waste identification projects, Office of Research and Development, U.S. EPA Report 600/R-95/039, 1995, 18 p.

Kollig, H.P., Ellington, J.J., Weber, E.J., and Wolfe, N.L. Environmental research brief - Pathway analysis of chemical hydrolysis for 14 RCRA chemicals, Office of Research and Development, U.S. EPA Report 600/M-89/009, 1990, 6 p.

Könemann, H. Quantitative structure-activity relationships in fish toxicity studies. Part 1: Relationship for 50 industrial pollutants, *Toxicology*, 19:209-221, 1981.

Könemann, H. and van Leeuwen, K. Toxicokinetics in fish: accumulation and elimination of six chlorobenzenes by guppies, *Chemosphere*, 9(1):3-19, 1980.

Könemann, H., Zelle, R., and Busser, F. Determination of log P_{oct} values of chloro-substituted benzenes, toluenes, and anilines by high-performance liquid chromatography on ODS-silica, *J. Chromatogr.*, 178:559-565, 1979.

Konietzko, H. Chlorinated ethanes: sources, distribution, environmental impact, and health effects, in *Hazard Assessment of Chemicals, Volume 3*, Saxena, J., Ed. (New York: Academic Press, 1984), pp. 401-448.

Konrad, J.G., Chesters, G., and Armstrong, D.E. Soil degradation of malathion, a phosphorothioate insecticide, *Soil Sci. Soc. Am. Proc.*, 33:259-262, 1969.

Kopczynski, S.L., Altshuller, A.P., and Sutterfield, F.D. Photochemical reactivities of aldehyde-nitrogen oxide systems, *Environ. Sci. Technol.*, 8(10):909-918, 1974.

Kopinke, F.-D., Pörschmann, J., and Stottmeister, U. Sorption of organic pollutants on anthropogenic humic matter, *Environ. Sci. Technol.*, 29(4):941-950, 1995.

Korenman, I.M. and Aref'eva, R.P. Determination of the solubility of liquid hydrocarbons in water, *Otkrytiya Izobret. Prom. Obraztsy Tovarnye Znaki*, 54(13):101, 1977.

Korenman, Ya I., Nefedova, T.E., and Byukova, R.I. Extraction of picramic acid, *Int. J. Pharmacol. Ther. Toxicol.*, 51:734-735, 1977.

Korenman, I.M. and Nikolaev. Determination of the protonation constants of weak diacidic bases by an extraction methods, *Zh. Phys. Chem.*, 48:2545-2549, 1974.

Korfmacher, W.A., Wehry, E.L., Mamantov, G., and Natusch, D.F.S. Resistance to photochemical decomposition of polycyclic aromatic hydrocarbons vapor-adsorbed on coal fly ash, *Environ. Sci. Technol.*, 14(9):1094-1099, 1980.

Korman, S. and La Mer, V.K. Deuterium exchange equilibria in solution and the quinhydrone electrode, *J. Am. Chem. Soc.*, 58:1396-1403, 1936.

Korte, F., Freitag, D., Geyer, H., Klein, W., Kraus, A.G., and Lahaniatis, E. Ecotoxicologic profile analysis - A concept for establishing ecotoxicologic priority lists for chemicals, *Chemosphere*, 1(1):79-102, 1978.

Korte, F., Ludwig, G., and Vogel, J. Umwandlung von Aldrin-[^{14}C] und dieldrin-[^{14}C] durch Mikroorganismen, Leberhomogenate, und Moskitolarven, *Liebigs Ann. Chem.*, 656:135-140, 1962.

Kosak-Channing, L.F. and Helz, G.R. Solubility of ozone in aqueous solutions of 0-0.6 M ionic strength at 5-30 °C, *Environ. Sci. Technol.*, 17(3):145-149, 1983.

Kosalwat, P. and Knight, A.W. Acute toxicity of aqueous and bound substrate bound copper to the midge, *Chironomus decorus*, *Arch. Environ. Contam. Toxicol.*, 16(3):275-282, 1987.

Koskinen, W.C., Leffler, H.R., Oliver, J.E., Kearney, P.C., and McWhorter, C.G. Effect of trifluralin soil metabolites on cotton boll components and fiber and seed properties, *J. Agric. Food Chem.*, 33(5):958-961, 1985.

Koskinen, W.C., Oliver, J.E., Kearney, P.C., and McWhorter, C.G. Effect of trifluralin soil metabolites on cotton growth and yield, *J. Agric. Food Chem.*, 32(6):1246-1248, 1984.

Kotronarou, A., Mills, G., and Hoffmann, M.R. Decomposition of parathion in aqueous solution by ultrasonic irradiation, *Environ. Sci. Technol.*, 26(7):1460-1462, 1992.

Kozak, J., Weber, J.B., and Sheets, T.J. Adsorption of prometryn and metolachlor by selected soil organic matter fractions, *Soil Sci.*, 136(2):94-101, 1983.

Kramer, C.R. and Henze, U. Partitioning properties of benzene derivatives. I. Temperature dependence of the partitioning of monosubstituted benzenes and nitrobenzenes in the n-octanol/water system, *Z. Phys. Chem.*, 271(3):503-513, 1990.

Krasnoshchekova, R.Ya. and Gubergrits, M. Solubility of paraffin hydrocarbons in fresh and salt water, *Neftekhimiya*, 13(6):885-888, 1973.

Krasnoshchekova, R.Ya. and Gubergrits, M. Solubility of n-alkylbenzene in fresh and salt waters, *Vodn. Resur.*, 2:170-173, 1975.

Kresheck, G.C., Schneider, H., and Scheraga, H.A. The effect of D_2O on the thermal stability of proteins. Thermodynamic parameters for the transfer of model compounds from H_2O to D_2O, *J. Phys. Chem.*, 69(9):3132-3144, 1965.

Kretschmer, C.B. and Wiebe, R. Solubility of gaseous paraffins in methanol and isopropyl alcohol, *J. Am. Chem. Soc.*, 74(5):1276-1277, 1952.

Kriebel, D. The dioxins: toxic and still troublesome, *Environment*, 23(1):6-13, 1981.

Kriegman-King, M.R. and Reinhard, M. Transformation of carbon tetrachloride in the presence of sulfide, biotite, and vermiculite, *Environ. Sci. Technol.*, 26(11):2198-2206, 1992.

Krien, G., Licht, H.H., and Zierath, J. Thermochemical investigation of nitramines, *Thermochim. Acta*, 6:465-472, 1973.

Krijgsheld, K.R. and van der Gen, A. Assessment of the impact of the emission of certain organochlorine compounds on the aquatic environment, *Chemosphere*, 15(7):861-880, 1986.

Krijgsheld, K.R. and van der Gen, A. Assessment of the impact of the emission of certain organochlorine compounds on the aquatic environment - Part I: monochlorophenols and 2,4-dichlorophenol, *Chemosphere*, 15(7):825-860, 1986a.

Krishna, B. and Gupta. ῶ-Type calculations on π-electron systems with inclusion of overlap charges. I. Ionization potential of some alternant hydrocarbons, *J. Am. Chem. Soc.*, 92(25):7247-7248, 1970.

Krishnamurthy, T. and Wasik, S.P. Fluorometric determination of partition coefficients of naphthalene homologues in octanol-water mixtures, *J. Environ. Sci. Health*, A13(8):595-602, 1978.

Kronberger, H. and Weiss, J. Formation and structure of some organic molecular compounds. III. The dielectric polarization of some solid crystalline molecular compounds, *J. Chem. Soc. (London)*, pp. 464-469, 1944.

Krzyzanowska, T. and Szeliga, J. A method for determining the solubility of individual hydrocarbons, *Nafta*, 28:414-417, 1978.

Ku, H.-C. and Tu, C.-H. Density and viscosity of binary mixtures of propan-2-ol, 1-chlorobutane, and acetonitrile,

J. Chem. Eng. Data, 43(3):465-468, 1998.

Kubáň, V. Determination of octa-1-ol-water partition coefficients by flow-injection extraction without phase separation, *Anal. Chim. Acta*, 248:493-499, 1991.

Kuchta, J.M., Furno, A.L., Bartkowiak, A., and Martindill, G.H. Effect of pressure and temperature on flammability limits of chlorinated hydrocarbons in oxygen-nitrogen and nitrogen tetroxide-nitrogen atmospheres, *J. Chem. Eng. Data*, 13(3):421-428, 1968.

Kucklick, J.R., Bidleman, T.F., McConnel, L.L., Walla, M.D., and Ivanov, G.P. Organochlorines in the water and biota of Lake Baikal, Siberia, *Environ. Sci. Technol.*, 28(1):31-37, 1994.

Kudchadker, A.P., Kudchadker, S.A., Shukla, R.P., and Patnaik, P.R. Vapor pressures and boiling points of selected halomethanes, *J. Phys. Chem. Ref. Data*, 8(2):499-517, 1979.

Kuhr, R.J. Metabolism of methylcarbamate insecticide chemicals in plants, *J. Agric. Food Chem. Suppl.*, 44-49, 1968.

Kumar, O. and Vijayaraghavan, R. Effect of sulphur mustard inhalation exposure on some urinary variables in mice, *J. Ppl. Toxicol.*, 18(4):257-259, 1998.

Kuo, P.P.K., Chian, E.S.K., and Chang, B.J. Identification of end products resulting from ozonation and chlorination of organic compounds commonly found in water, *Environ. Sci. Technol.*, 11(13):1177-1181, 1977.

Kurane, R., Suzuki, T., and Takahara, Y. Isolation of microorganisms growing on phthalate esters and degradation of phthalate ester by *Pseudomonas acidovorans*, *Agric. Biol. Chem.*, 41:2119-2123, 1977.

Kurihara, N., Uchida, M., Fujita, T., and Nakajima, M. Studies on BHC isomers and related compounds. V. Some physicochemical properties of BHC isomers (1), *Pestic. Biochem. Physiol.*, 2(4):383-390, 1973.

Kurlo, M.J. and Knable, G.L. A kinetic investigation of the gas-phase reactions of atomic chlorine (^{2}P) and hydroxyl (X^2) with acetonitrile: atmospheric significance and evidence of decreased reactivity between strong electrophiles, *J. Phys. Chem.*, 88:3305-3308, 1984.

Kusk, K.O. Bioavailability and effect of pirimicarb on *Daphnia magna* in a laboratory freshwater/sediment system, *Arch. Environ. Contam. Toxicol.*, 31(2):252-255, 1996.

Kwok, E.S.C., Atkinson, R., and Arey, J. Kinetics of the gas-phase reactions of indan, indene, fluorene, and 9,10-dihydro-anthracene with OH radicals, NO_3 radicals and O_3, *Int. J. Chem. Kinet.*, 29:4299-4310, 1997.

Laanio, T.L., Kearney, P.C., and Kaufman, D.D. Microbial metabolism of dinitramine, *Pestic. Biochem. Physiol.*, 3:271-277, 1973.

Lacorte, S., Lartiges, S.B., Garrigues, P., and Barceló, D. Degradation of organophosphorus pesticides and their transformation products in estuarine waters, *Environ. Sci. Technol.*, 29(2):431-438, 1995.

Laddha, G.S. and Smith, J.M. The systems: glycol-*n*-amyl alcohol-water and glycol-*n*-hexyl alcohol-water, *Ind. Eng. Chem.*, 40:494-496, 1948.

Laffort, P. and Dravnieks, A. An approach to a physico-chemical model of olfactory stimulation in vertebrates by single compounds, *J. Theor. Biol.*, 38:335-345, 1973.

LaFleur, K.S. Sorption of pesticides by model soils and agronomic soils: rates and equilibria, *Soil Sci.*, 127(2):94-101, 1979.

Lafrance, P., Marineau, L., Perreault, L., and Villenueve, J.-P. Effect of natural dissolved organic matter found in groundwater on soil adsorption and transport of pentachlorophenol, *Environ. Sci. Technol.*, 28(13):2314-2320, 1994.

Lagas, P. Sorption of chlorophenols in the soil, *Chemosphere*, 17(2):205-216, 1988.

Laha, S. and Luthy, R.G. Oxidation of aniline and other primary aromatic amines by manganese dioxide, *Environ. Sci. Technol.*, 24(3):363-373, 1990.

Laham, S., Potvin, M., Schrader, K., and Marino, I. Studies on inhalation toxicity of 2-propanol, *Drug Chem. Toxicol.*, 3(4):343-360, 1980.

Lalezary, S., Pirbazari, M., and McGuire, M.J. Evaluating activated carbons for removing low concentrations of taste and odor producing organics, *J. Am. Water Works Assoc.*, 78(11):76-82, 1986.

Lamai, S.L., Warner, G.F., and Walker, C.H. Effects of dieldrin on life stages of the African catfish, *Clarias gariepinus* (Burchell), *Ecotoxicol. Environ. Saf.*, 42(1):22-29, 1999.

Lamparski, L.L., Stehl, R.H., and Johnson, R.L. Photolysis of pentachlorophenol-treated wood. Chlorinated dibenzo-*p*-dioxin formation, *Environ. Sci. Technol.*, 14(2):196-200, 1980.

Lande, S.S., Hagen, D.F., and Seaver, A.E. Computation of total molecular surface area from gas phase ion mobility data and its correlation with aqueous solubilities of hydrocarbons, *Environ. Toxicol. Chem.*, 4:325-334, 1985.

Landrum, P.F., Nihart, S.R., Eadie, B.J., and Gardner, W.S. Reverse-phase separation method for determining pollutant binding to aldrich humic acid and dissolved organic carbon of natural waters, *Environ. Sci. Technol.*, 18(3):187-192, 1984.

Lane, D.A., Bunce, N.J., Liu, L., and Zhu, J. Products of the reaction of naphthalene with the OH radical, in *American Chemical Society - Division of Environmental Chemistry, Preprints of Extended Abstracts*, 37(1):309-310, 1997.

Lane, W.H. Determination of the solubility of styrene in water and of water in styrene, *Ind. Eng. Chem. Anal. Ed.*, 18:295-296, 1946.

Langlois, B.E. Reductive dechlorination of DDT by *Escherichia coli*, *J. Dairy Sci.*, 50:1168-1170, 1967.

Lao, R.C., Thomas, R.S., and Monkman, J.L. Computerized gas chromatographic-mass spectrometric analysis of polycyclic aromatic hydrocarbons in environmental samples, *J. Chromatogr.*, 112:681-700, 1975.

Laplanche, A., Martin, G., and Tonnard, F. Ozonation schemes of organophosphorous pesticides. Application in drinking water treatment, *Ozone: Sci. Eng.*, 6:207-219, 1984.

Larachi, F., Leroux, M., Hamoudi, S., Bernis, A., and Sayari, A. Solubility and infinite dilution activity coefficient for 5-chlorovanillin and 4-chloroguaiacol in water over the temperature range 280 to 363 K, *J. Chem. Eng. Data*, 45(2):404-408, 2000.

Larsen, T., Kjeldsen, P., and Christensen, T.H. Correlation of benzene, 1,1,1-trichloroethane, and naphthalene distribution coefficients to the characteristics of aquifer materials with low organic carbon content, *Chemosphere*, 24(8):979-991, 1992.

Lau, Y.L., Oliver, B.G., and Krishnappan, B.G. Transport of some chlorinated contaminants by the water, suspended sediments, and bed sediments in the St. Clair and Detroit Rivers, *Environ. Toxicol. Chem.*, 8(4):293-301, 1989.

Laughlin, R.B., Neff, J.M., Hrung, Y.C., Goodwin, T.C., and Gian, C.S. The effects of three phthalates esters on the larval development of the grass shrimp *Palaemonetes pugio* (Holthius), *Water, Air, Soil Pollut.*, 9:323-336, 1978.

Leadbetter, E.R. and Foster, J.W. Oxidation products formed from gaseous alkanes by the bacterium *Pseudomonas methanica*, *Arch. Biochem. Biophys.*, 82:491-492, 1959.

Leavitt, D.D. and Abraham, M.A. Acid-catalyzed oxidation of 2,4-dichlorophenoxyacetic acid by ammonium nitrate in aqueous solution, *Environ. Sci. Technol.*, 24(4):566-571, 1990.

Lebedev, B.V., Lebedev, N.K., Smirnova, N.N., Kozyreva, N.M., Kirillin, A., and Korshak. The isotope effect in the thermodynamic parameters of polymerization of styrene, *Dokl. Akad. Nauk*, SSSR 281:379-383, 1985.

Lebedev, B.V., Lityagov, V.Ya., Krentsina, T.I., and Milov, V.I. Thermodynamic properties of tetrahydrofuran in the range 8-322 K, *Zhur. Fiz. Khim.*, 53:264-265, 1979.

Lebedev, B.V. and Rabinovich, I.B. Heat capacities and thermodynamic functions of methyl methacrylate and poly(methyl methacrylate), *Tr. Khim. Khim. Teknol.*, 1:8-11, 1971.

Lebedev, B.V. and Rabinovich, I.B. Heat capacities and thermodynamic functions of α-methylstyrene and poly(α-methylstyrene), *Tr. Khim. Khim. Teknol.*, 1:12-15, 1971a.

Lebedev, B.V., Rabinovich, I.B., and Budarina, V.A. Heat capacity of vinyl chloride, polyvinyl chloride and polyvinylidene chloride in the region of 60-300°K, *Polymer Sci.*, USSR 9A:545-552, 1967.

Lebedev, B.V., Rabinovich, I.B., Milov, V.I., and Lityagov, V.Ya. Thermodynamic properties of tetrahydrofuran from 8 to 322 K, *J. Chem. Thermodyn.*, 10:321-329, 1978.

Lebedev, B.V. and Yevstropov, A.A. Thermodynamics of β-propiolactone, *t*-butyrolactone, *d*-valerolactone, and *ε*-caprolactone from 13.8 to 340 K, *J. Chem. Thermodyn.*, 15:115-128, 1983.

LeBlanc, G.A. Acute toxicity of priority pollutants to water flea (*Daphnia magna*), *Bull. Environ. Contam. Toxicol.*, 24(5):684-691, 1980.

Lee, D.-Y., Farmer, W.J., and Aochi, Y. Sorption of napropamide on clay and soil in the presence of dissolved organic matter, *J. Environ. Qual.*, 19(3):567-573, 1990.

Lee, J.F., Crum, J.R., and Boyd, S.A. Enhanced retention of organic contaminants by soils exchanged with organic cations, *Environ. Sci. Technol.*, 23(11):1365-1372, 1989.

Lee, L.S., Hagwall, M., Delfino, J.J., and Rao, S.C. Partitioning of polycyclic aromatic hydrocarbons from diesel fuel into water, *Environ. Sci. Technol.*, 26(11):2104-2110, 1992.

Lee, M.C., Chian, E.S.K., and Griffin, R.A. Solubility of polychlorinated biphenyls and capacitor fluid in water, *Water Res.*, 13(12):1249-1258, 1979.

Lee, M.D., Wilson, J.T., and Ward, C.H. Microbial degradation of selected aromatics in a hazardous waste site, *Dev. Ind. Microbiol.*, 25:557-565, 1984.

Lee, R.F. Oil spill conference, *Am. Petrol. Instit.*, pp. 611-616, 1977.

Lee, R.F., Sauerheber, R., and Benson, A.A. Petroleum hydrocarbons: uptake and discharge by the marine mussel *Mytilus edulis*, *Science (Washington, D.C.)*, 177:344-345, 1972.

Lee, S.E., Shin, H.S., and Paik, B.C. Treatment of Cr(VI)-containing wastewater by addition of powdered activated carbon to the activated sludge process, *Environ. Sci. Technol.*, 23(11):67-72, 1989a.

Lee, S.-L. and Tu, C.-H. Densities and viscosities of four alkyl esters with nitromethane systems at (293.15, 303.15, and 313.15) K, *J. Chem. Eng. Data*, 44(1):108-111, 1999.

Lee, Y.-N. and Schwartz, S.E. Reaction kinetics of nitrogen dioxide with liquid water at low partial pressure, *J. Phys. Chem.*, 85:840-848, 1981.

Leenheer, J.A., Malcolm, R.L., and White, W.R. Investigation of the reactivity and fate of certain organic components of an industrial waste after deep-well injection, *Environ. Sci. Technol.*, 10(5):445-451, 1976.

Leffingwell, J.T. The Photolysis of DDT in Water, Ph.D. Thesis, University of California, Davis, 1975.

Legierse, K.C.H.M., Sijm, D.T.H.M., van Leeuwen, C.J., Seinen, W., and Hermens, J.L.M. Bioconcentration kinetics of chlorobenzenes and the organophosphorus pesticide chlorthion in the pond snail *Lymnaea stagnalis* - a comparison with the guppy *Poecilia reticulata*, *Aquat. Toxicol.*, 41:301-323, 1998.

Lei, Y.D., Wania, F., and Shiu, W.Y. Vapor pressure of the polychlorinated naphthalenes, *J. Chem. Eng. Data*, 44(3):577-582, 1999.

Leigh, W. Degradation of selected chlorinated hydrocarbon insecticides, *J. Water Pollut. Control Fed.*, 41(11):R450-R460, 1969.

Leighton, D.T., Jr. and Calo, J.M. Distribution coefficients of chlorinated hydrocarbons in dilute air-water systems for groundwater contamination applications, *J. Chem. Eng. Data*, 26(4):382-385, 1981.

Leinonen, P.J. and Mackay, D. The multicomponent solubility of hydrocarbons in water, *Can. J. Chem. Eng.*, 51:230-233, 1973.

Leinster, P., Perry, R., and Young, R.J. Ethylenebromide in urban air, *Atmos. Environ.*, 12:2382-2398, 1978.

Leistra, M. Distribution of 1,3-dichloropropene over the phases of soil, *J. Agric. Food Chem.*, 18(6):1124-1126, 1970.

Leitis, E. and Crosby, D.G. Photodecomposition of trifluralin, *J. Agric. Food Chem.*, 22(5):842-848, 1974.

Leland, H.V., Bruce, W.N., and Shimp, N.F. Chlorinated hydrocarbon insecticides in sediments in southern Lake Michigan, *Environ. Sci. Technol.*, 7(9):833-838, 1973.

Lemley, A.T., Wagenet, R.J., and Zhong, W.Z. Sorption and degradation of aldicarb and its oxidation products in a soil-water flow system as a function of pH and temperature, *J. Environ. Qual.*, 17(3):408-414, 1988.

Lenchitz, C. and Velicky, R.W. Vapor pressure and heat of sublimation of three nitrotoluenes, *J. Chem. Eng. Data*,

15(3):401-403, 1970.

Leo, A., Hansch, C., and Church, C. Comparison of parameters currently used for the study of structure-activity relationships, *J. Med. Chem.*, 12:766-771, 1969.

Leo, A., Hansch, C., and Elkins, D. Partition coefficients and their uses, *Chem. Rev.*, 71(6):525-616, 1971.

Leo, A., Jow, P.Y.C., Silipo, C., and Hansch, C. Calculation of hydrophobic constant from π and f constants, *J. Med. Chem.*, 18(9):865-868, 1975.

Leonard, A.W., Hyne, R.V., Lim, R.P., and Chapman, J.C. Effect of endosulfan runoff from cotton fields on macroinvertebrates in the Naomi River, *Ecotoxicol. Environ. Saf.*, 42(2):125-134, 1999.

Leoni, V., Hollick, C.B., D'Alessandro, D.E., Collison, R.J., and Merolli, S. The soil degradation of chlorpyrifos and the significance of its presence in the superficial water in Italy, *Agrochimica*, 25:414-426, 1981.

Lépine, F., Milot, S., and Vincent, N. Formation of toxic PCB congeners and PCB-solvent adducts in a sunlight irradiated cyclohexane solution of Aroclor 1254, *Bull. Environ. Contam. Toxicol.*, 48(1):152-156, 1992.

Lesage, S., Jackson, R.E., Priddle, M.W., and Riemann, P.G. Occurrence and fate of organic solvent residues in anoxic groundwater at the Gloucester landfill, Canada, *Environ. Sci. Technol.*, 24(4):559-566, 1990.

Leuck, D.B., Jones, R.L., and Bowman, M.C. Chlorpyrifos-methyl insecticide residues: their analysis and persistence in coastal bermuda grass and forage corn, *J. Econ. Entomol.*, 69:287-290, 1975.

Leuenberger, C., Ligocki, M.P., and Pankow, J.F. Trace organic compounds in rain. 4. Identities, concentrations, and scavenging mechanisms for phenols in urban air and rain, *Environ. Sci. Technol.*, 19(11):1053-1058, 1985.

Lewis Publishers. *Drinking Water Health Advisory - Pesticides* (Chelsea, MI: Lewis Publishers, 1989), 819 p.

Lewis, R.J., Sr. *Rapid Guide to Hazardous Chemicals in the Workplace* (New York: Van Nostrand Reinhold, 1990), 286 p.

Leyder, F. and Boulanger, P. Ultraviolet absorption, aqueous solubility and octanol-water partition for several phthalates, *Bull. Environ. Contam. Toxicol.*, 30(2):152-157, 1983.

Leynen, M., Duvivier, L., Girboux, P., and Ollevier, F. Toxicity of ozone to fish larvae and *Daphnia magna*, *Ecotoxicol. Environ. Saf.*, 41(2):176-179, 1998.

Li, A. and Andren, A.W. Solubility of polychlorinated biphenyls in water/alcohol mixtures. 1. Experimental data, *Environ. Sci. Technol.*, 28(1):47-52, 1994.

Li, A. and Doucette, W.J. The effect of cosolutes on the aqueous solubilities and octanol/water partition coefficients of selected polychlorinated biphenyl congeners, *Environ. Toxicol. Chem.*, 12:2031-2035, 1993.

Li, C.F. and Bradley, R.L. Degradation of chlorinated hydrocarbon pesticides in milk and butter oil by ultraviolet energy, *J. Dairy Sci.*, 52:27-30, 1969.

Li, J., Dallas, A.J., Eikens, D.I., Carr, P.W., Bergmann, D.L., Hait, M.J., and Eckert, C.A. Measurement of large infinite dilution activity coefficients of nonelectrolytes in water by inert gas stripping and gas chromatography, *Anal. Chem.*, 65(22):3212-3218, 1993.

Liao, Y.-Y., Wang, Z.-T., Chen, J.-W., Han, S.-K., Wang, L.-S., Lu, G.-Y., and Zhao, T.-N. The prediction of soil sorption coefficients of heterocyclic nitrogen compounds by octanol/water partition coefficient, water solubility, and by molecular connectivity indices, *Bull. Environ. Contam. Toxicol.*, 56(5):711-716, 1996.

Lias, S.G. Ionization Energy Evaluation, in *NIST Standard Reference Database Number 69*, Mallard, W.G. and Linstrom, P.J., Eds. (Gaithersburg, MD: National Institute of Standards and Technology, 1998) (http://webbook.nist.gov).

Lias, S.G., Levin, R.D., and Kafafi, S.A. Ion Energetics Data, in *NIST Standard Reference Database Number 69*, Mallard, W.G. and Linstrom, P.J., Eds. (Gaithersburg, MD: National Institute of Standards and Technology, 1998) (http://webbook.nist.gov).

Lias, S.G. and Liebman, J.F. Ion Energetics Data, in *NIST Standard Reference Database Number 69*, Mallard, W.G. and Linstrom, P.J., Eds. (Gaithersburg, MD: National Institute of Standards and Technology, 1998) (http://webbook.nist.gov).

Librando, V. and Lane, D.A. Degradation of chemicals with significant environmental impact, in *American Chemical Society - Division of Environmental Chemistry, Preprints of Extended Abstracts*, 37(1):283-284, 1997.

Lichtenstein, E.P. 'Bound' residues in soils and transfer of soil residues in crops, *Residue Rev.*, 76:147-153, 1980.

Lichtenstein, E.P., DePew, L.J., Eshbaugh, E.L., and Sleesman, J.P. Persistence of DDT, aldrin, and lindane in some midwestern soils, *J. Econ. Entomol.*, 53:136-142, 1960.

Lichtenstein, E.P., Fuhremann, T.W., and Schulz, K.R. Persistence and vertical distribution of DDT, lindane, and aldrin residues, 10 and 15 years after a single soil application, *J. Agric. Food Chem.*, 19(4):718-721, 1971.

Lichtenstein, E.P., Liang, T.T., and Koeppe, M.K. Effects of fertilizers, captafol, and atrazine on the fate and translocation of [^{14}C]fonofos and [^{14}C]parathion in a soil-plant microcosm, *J. Agric. Food Chem.*, 30(5):871-878, 1982.

Lichtenstein, E.P. and Schulz, K.R. Breakdown of lindane and aldrin in soils, *J. Econ. Entomol.*, 52:118-124, 1959.

Lieberman, M.T. and Alexander, M. Microbial and nonenzymatic steps in the decomposition of dichlorvos (2,2-dichlor-ovinyl *O,O*-dimethyl phosphate), *J. Agric. Food Chem.*, 31(2):265-267, 1983.

Lide, C.R. *CRC Handbook of Chemistry and Physics*, 70th ed. (Boca Raton, FL: CRC Press, 1990).

Lilius, H., Hästbacka, T., and Isomaa, B. A comparison of the toxicity of 30 reference chemicals to *Daphnia magna* and *Daphnia pulex*, *Environ. Toxicol. Chem.*, 14(12):2085-2088, 1995.

Lin, S. and Carlson, R.M. Susceptibility of environmentally important heterocycles to chemical disinfection: reactions with aqueous chlorine, chlorine dioxide, and chloramine, *Environ. Sci. Technol.*, 18(10):743-748, 1984.

Lin, S., Lukasewycz, M.T., Liukkonen, R.J., and Carlson, R.M. Facile incorporation of bromine into aromatic systems under conditions of water chlorination, *Environ. Sci. Technol.*, 18(12):985-986, 1984.

Lin, W.S. and Kapoor, M. Induction of aryl hydrocarbon hydroxylase in *Neurospora crassa* by benzo[*a*]pyrene, *Curr.*

Microbiol., 3:177-180, 1979.

Lind, J.A. and Kok, G.L. Henry's law determination for aqueous solutions of hydrogen peroxide, methyl hydroperoxide, and peroxyacetic acid, *J. Geophys. Res.*, 91D:7889-7895, 1986.

Linscott, D.L. and Hagin, R.D. Additions to the aliphatic moiety of chlorophenoxy compounds, *Weed Sci.*, 18:197-198, 1970.

Lipczynska-Kochany, E. Degradation of aqueous nitrophenols and nitrobenzene by means of the Fenton reaction, *Chemosphere*, 22(5/6):529-536, 1991.

Liss, P.S. and Slater, P.G. Flux of gases across the air-sea interface, *Nature (London)*, 247(5438):181-184, 1974.

Liu, D. Biodegradation of aroclor 1221 type PCBs in sewage wastewater, *Bull. Environ. Contam. Toxicol.*, 27(5):695-703, 1981.

Liu, D., Strachan, W.M.J., Thomson, K., and Kwasniewska, K. Determination of the biodegradability of organic compounds, *Environ. Sci. Technol.*, 15(7):788-793, 1981.

Liu, K.F. and Ziegler, W.T. Heat capacity from 80° to 300°K, melting point and heat of fusion of nitroethane, *J. Chem. Eng. Data*, 11:187-189, 1966.

Liu, L.C., Cibes-Viade, M., and Koo, F.K.S. Adsorption of ametryne and diuron by soils, *Weed Sci.*,, 18:470-474, 1970.

Liu, S.-Y. and Bollag, J.M. Carbaryl decomposition to 1-naphthyl carbamate by *Aspergillus terreus*, *Pestic. Biochem. Physiol.*, 1:366-372, 1971.

Liu, S.-Y. and Bollag, J.M. Metabolism of carbaryl by a soil fungus, *J. Agric. Food Chem.*, 19(3):487-490, 1971a.

Lloyd, A.C., Atkinson, R., Lurmann, F.W., and Nitta, B. Modeling potential ozone impacts from natural hydrocarbons - I. Development and testing of a chemical mechanism for the NO_x-air photooxidations of isoprene and α-pinene under ambient conditions, *Atmos. Environ.*, 17(10):1931-1950, 1983.

Lo, J.M., Tseng, C.L., and Yang, J.Y. Radiometric method for determining solubility of organic solvents in water, *Anal. Chem.*, 58(7):1596-1597, 1986.

Loch, J.P.G., Kool, H.J., Lagas, P., and Verheul, J.H.A.M. Removal and retention of volatile chlorinated hydrocarbons in the soils' unsaturated zone, in *Contaminated Soil*, Assink J.W. and van der Brink, W.J., Eds. (Dordrecht, the Netherlands: Matinus Nijhoff Publishers, 1986), pp. 63-77.

Locke, M.A., Gaston, L.A., and Zablotowicz, R.M. Acifluorfen sorption and sorption kinetics in soil, *J. Agric. Food Chem.*, 45(1):286-293, 1997.

Lockwood, R.A. and Chen, K.Y. Adsorption of Hg(II) by ferric hydroxide, *Environ. Letters*, 6:151-161, 1974.

Lodge, K.B. Solubility studies using a generator column for 2,3,7,8-tetrachlorodibenzo-*p*-dioxin, *Chemosphere*, 18(1-6):933-940, 1989.

Lodge, K.B. and Cook, P.M. Partition studies of dioxin between sediment and water: the measurement of K_{oc} for Lake Ontario sediment, *Chemosphere*, 18(1-6):439-444, 1989.

Loehr, R.C. and Matthews, J.E. Loss of organic chemicals in soil: pure compound treatability studies, *J. Soil Contam.*, 1(4):339-360, 1992.

Loh, P.Y. and Yap, H.H. Laboratory studies on the efficacy and sublethal effects of an insect growth regulator, pyripoxyfen (S-31183) against *Aedes aegypti* (Linnaeus), *Trop. Biomed.*, 6:7-12, 1989.

Løkke, H. Sorption of selected organic pollutants in danish soils, *Ecotoxicol. Environ. Saf.*, 8(5):395-409, 1984.

Lombardo, P., Pomerantz, I.H., and Egry, I.J. Identification of photoaldrin chlorohydrin as a photoalteration product of dieldrin, *J. Agric. Food Chem.*, 20(6):1278-1279, 1972.

Lopez, C.E. and Kirkwood, J.I. Isolation of microorganisms from a texas soil capable of degrading urea derivative herbicides, *Soil Sci. Soc. Am. Proc.*, 38:309-312, 1974.

Lopez-Gonzales, J.De D. and Valenzuela-Calahorro, C. Associated decomposition of DDT to DDE in the diffusion of DDT on homoionic clays, *J. Agric. Food Chem.*, 18(3):520-523, 1970.

Lord, K.A., Briggs, G.G., Neale, M.C., and Manlove, R. Uptake of pesticides from water and soil by earthworms, *Pestic. Sci.*, 11(4):401-408, 1980.

Low, G.K.-C., McEvoy, S.R., and Matthews, R.W. Formation of nitrate and ammonium ions in titanium dioxide mediated photocatalytic degradation of organic compounds containing nitrogen atoms, *Environ. Sci. Technol.*, 25(3):460-467, 1991.

Lu, P.-Y. and Metcalf, R.L. Environmental fate and biodegradability of benzene derivatives as studied in a model ecosystem, *Environ. Health Perspect.*, 10(1-7):269-284, 1975.

Lu, P.-Y., Metcalf, R.L., Hirwe, A.S., and Williams, J.W. Evaluation of environmental distribution and fate of hexachlorocyclopentadiene, chlordene, heptachlor, and heptachlor epoxide in a laboratory model ecosystem, *J. Agric. Food Chem.*, 23(5):967-973, 1975.

Lu, P.-Y., Metcalf, R.L., Plummer, N., and Mandel, D. The environmental fate of three carcinogens: benzo[*a*]pyrene, benzidine, and vinyl chloride evaluated in laboratory model ecosystems, *Arch. Environ. Contam. Toxicol.*, 6(2/3):129-142, 1977.

Ludzack, F.J. and Ettinger, M.B. Chemical structures resistant to aerobic biochemical stabilization, *J. Water Pollut. Control. Fed.*, 32:1173-1200, 1960.

Lue, K.Y. and de laCruz, A.A. Mirex incorporation in the environment: toxicity in *Hydra*, *Bull. Environ. Contam. Toxicol.*, 19(4):412-416, 1978.

Luke, W.T., Dickerson, R.R., and Nunnermacker, L.J. Direct measurements of the photolysis rate coefficients and Henry's law constants of several alkyl nitrates, *J. Geophys. Res.*, 94D:14905-14921, 1989.

Lun, R., Shiu, W.-Y., and Mackay, D. Aqueous solubilities and octanol-water partition coefficients of chloroveratroles and chloroanisoles, *J. Chem. Eng. Data*, 40(4):959-962, 1995.

Lun, R., Varhanickova, D., Shiu, W.-Y., and Mackay, D. Aqueous solubilities and octanol-water partition coefficients of cymenes and chlorocymenes, *J. Chem. Eng. Data*, 42(5):951-953, 1997.

Lyman, W.J., Reehl, W.F., and Rosenblatt, D.H. *Handbook of Chemical Property Estimation Methods: Environmental Behavior of Organic Compounds* (New York: McGraw-Hill, 1982).

Lyons, C.D., Katz, S., and Bartha, R. Mechanisms and pathways of aniline elimination from aquatic environments, *Appl. Environ. Microbiol.*, 48(3):491-496, 1984.

Lyons, C.D., Katz, S.E., and Bartha, R. Fate of herbicide-derived aniline residues during ensilage, *Bull. Environ. Contam. Toxicol.*, 35(5):704-710, 1985.

Ma, K.C., Shiu, W.Y., and Mackay, D. Aqueous solubility of chlorophenols at 25 °C, *J. Chem. Eng. Data*, 38:364-366, 1993.

Mabey, W. and Mill, T. Critical review of hydrolysis of organic compounds in water under environmental conditions, *J. Phys. Chem. Ref. Data*, 7(2):383-415, 1978.

Mabey, W.R., Smith, J.H., Podoll, R.T., Johnson, H.L., Mill, T., Chou, T.-W., Gates, J., Partridge, I.W., Jaber, H., and Vandenberg, D. Aquatic fate process data for organic priority pollutants - Final report, Office of Regulations and Standards, U.S. EPA Report-440/4-81-014, 1982, 407 p.

Mabury, S.A. and Crosby, D.G. Pesticide reactivity toward hydroxyl and its relationship to field persistence, *J. Agric. Food Chem.*, 44(7):1920-1924, 1996.

Macalady, D.L. and Wolfe, N.L. New perspectives on the hydrolytic degradation of the organophosphorothioate insecticide chlorpyrifos, *J. Agric. Food Chem.*, 31(6):1139-1147, 1983.

Macalady, D.L. and Wolfe, N.L. Effects of sediment sorption and abiotic hydrolyses. 1. Organophosphorothioate esters, *J. Agric. Food Chem.*, 33(2):167-173, 1985.

Macek, K.J., Buxton, K.S., Deer, S.K., Dean, J.W., and Sauter, S. Chronic toxicity of lindane to selected aquatic invertebrates and fish, U.S. EPA Report 600/3-76-046, 1976a.

Macek, K.J., Buxton, K.S., Sauter, S., Gnilka, S., and Dean, J.W. Chronic toxicity of atrazine to selected aquatic invertebrates and fish, U.S. EPA Report 600/3-76-047, 1976b.

Macek, K.J. and McAllister, W.A. Insecticide susceptibility of some common fish family representatives, *Trans. Am. Fish Soc.*, 99(1):20-27, 1970.

MacIntyre, W.G. and deFur, P.O. The effect of hydrocarbon mixtures on adsorption of substituted naphthalenes by clay and sediment from water, *Chemosphere*, 14(1):103-111, 1985.

Mackay, D. Correlation of bioconcentration factors, *Environ. Sci. Technol.*, 16(5):274-278, 1982.

Mackay, D., Bobra, A., Chan, D.W., and Shiu, W.-Y. Vapor pressure correlations for low-volatility environmental chemicals, *Environ. Sci. Technol.*, 16(10):645-649, 1982.

Mackay, D., Bobra, A., Shiu, W.-Y., and Yalkowsky, S.H. Relationships between aqueous solubility and octanol-water partition coefficients, *Chemosphere*, 9:701-711, 1980.

Mackay, D. and Leinonen, P.J. Notes - Rate of evaporation of low-solubility contaminants from water bodies to atmosphere, *Environ. Sci. Technol.*, 9(13):1178-1180, 1975.

Mackay, D. and Paterson, S. Calculating fugacity, *Environ. Sci. Technol.*, 15(9):1006-1014, 1981.

Mackay, D. and Shiu, W.-Y. The determination of the solubility of hydrocarbons in aqueous sodium chloride solutions, *Can. J. Chem. Eng.*, 53:239-242, 1975.

Mackay, D. and Shiu, W.-Y. Aqueous solubility of polynuclear aromatic hydrocarbons, *J. Chem. Eng. Data*, 22(4):399-402, 1977.

Mackay, D. and Shiu, W.-Y. A critical review of Henry's law constants for chemicals of environmental interest, *J. Phys. Chem. Ref. Data*, 10(4):1175-1199, 1981.

Mackay, D., Shiu, W.-Y., and Sutherland, R.P. Determination of air-water Henry's law constants for hydrophobic pollutants, *Environ. Sci. Technol.*, 13(3):333-337, 1979.

Mackay, D. and Wolkoff, A.W. Rate of evaporation of low-solubility contaminants from water bodies to atmosphere, *Environ. Sci. Technol.*, 7(7):611-614, 1973.

Mackay, D. and Yeun, A.T.K. Mass transfer coefficient correlations for volatilization of organic solutes from water, *Environ. Sci. Technol.*, 17(4):211-217, 1983.

Macknick, A.B. and Prausnitz, J.M. Vapor pressures of high molecular weight hydrocarbons, *J. Chem. Eng. Data*, 24(3):175-178, 1979.

MacLeod, H., Jourdain, L., Poulet, G., and LeBras, G. Kinetic study of reactions of some organic sulfur compounds with OH radicals, *Atmos. Environ.*, 18(12):2621-2626, 1984.

MacMichael, G.J. and Brown, L.R. Role of carbon dioxide in catabolism of propane by *Nocardia paraffinicum* (*Rhodococcus rhodochrous*), *Appl. Environ. Microbiol.*, 53(1):65-69, 1987.

MacRae, I.C. Microbial metabolism of pesticides and structurally related compounds, *Rev. Environ. Contam. Toxicol.*, 109:2-87, 1989.

MacRae, I.C., Raghu, K., and Bautista, E.M. Anaerobic degradation of the insecticide lindane by *Clostridium* sp., *Nature (London)*, 221(5183):859-860, 1969.

MacRae, I.C., Raghu, K., and Castro, T.F. Persistence and biodegradation of four common isomers of benzene hexachloride in submerged soils, *J. Agric. Food Chem.*, 15(5):911-914, 1967.

Madder, D.J. and Lockhart, W.L. A preliminary study of the effects of diflubenzuron and methoprene on rainbow trout (*Salmo gairdneri* Richardson), *Bull. Environ. Contam. Toxicol.*, 20(1):66-70, 1978.

Mader, B.T., Uwe-Goss, K., and Eisenreich, S.J. Sorption of nonionic, hydrophobic organic chemicals to mineral surfaces, *Environ. Sci. Technol.*, 31(4):1079-1086, 1997.

Madhun, Y.A. and Freed, V.H. Degradation of the herbicides bromacil, diuron and chlortoluron in soil, *Chemosphere*, 16(5):1003-1011, 1987.

Madhun, Y.A., Freed, V.H., Young, J.L., and Fang, S.C. Sorption of bromacil, chlortoluron, and diuron by soils, *Soil Sci. Soc. Am. J.*, 50:1467-1471, 1986.

Maeda, N., Ohya, T., Nojima, K., and Kanno, S. Formation of cyanide ion or cyanogen chloride through the cleavage of aromatic rings by nitrous acid or chlorine. IX. On the reactions of chlorinated, nitrated, carboxylated or methylated benzene derivatives with hypochlorous acid in the presence of ammonium ion, *Chemosphere*, 16(10-12):2249-2258, 1987.

Magee, B.R., Lion, L.W., and Lemley, A.T. Transport of dissolved organic macromolecules and their effect on the transport of phenanthrene in porous media, *Environ. Sci. Technol.*, 25(2):323-331, 1991.

Mahaffey, W.R., Gibson, D.T., and Cerniglia, C.E. Bacterial oxidation of chemical carcinogens: formation of polycyclic aromatic acids from benz[*a*]anthracene, *Appl. Environ. Microbiol.*, 54(10):2415-2423, 1988.

Mailhot, H. and Peters, R.H. Empirical relationships between the 1-octanol/water partition coefficient and nine physiochemical properties, *Environ. Sci. Technol.*, 22(12):1479-1488, 1988.

Majewski, H.S., Klaverkamp, J.F., and Scott, D.P. Acute lethality and sublethal effects of acetone, ethanol and propylene glycol on the cardiovascular and respiratory system of rainbow trout, *Water Res.*, 13(2):217-221, 1978.

Majka, J.T. and Lavy, T.L. Adsorption, mobility, and degradation of cyanazine and diuron in soils, *Weed Sci.*, 25(5):401-406, 1977.

Makela, T.P. and Oikari, A.O. Pentachlorophenol accumulation in the freshwater mussels *Anodonta anatina* and *Pseudanodonta complanata*, and some physiological consequences of laboratory maintenance, *Chemosphere*, 31(7):3651-3662, 1995.

Makino, M., Kamiya, M., and Matsushita, H. Photodegradation mechanism of 2,3,7,8-tetrachlorodibenzo-*p*-dioxin as studied by the PM3-MNDO method, *Chemosphere*, 24(3):291-307, 1992.

Maksimov, Y.Y. Vapor pressures of aromatic nitro compounds at various temperatures, *Zh. Fiz. Khim.*, 42(11):2921-2925, 1968.

Mallard, W.G. and Linstrom, P.J., Eds. *NIST Chemistry WebBook, NIST Standard Reference Database Number 69*, (Gaithersburg, MD: National Institute of Standards and Technology, 1998) (http://webbook.nist.gov).

Mallik, M.A.B. and Tesfai, K. Transformation of nitrosamines in soil and *in vitro* by soil microorganisms, *Bull. Environ. Contam. Toxicol.*, 27(1):115-121, 1981.

Mallon, B.J. and Harrison, F.L. Octanol-water partition coefficient of benzo[*a*]pyrene: measurement, calculation and environmental implications, *Bull. Environ. Contam. Toxicol.*, 32(3):316-323, 1984.

Manilal, V.B. and Alexander, M. Factors affecting the microbial degradation of phenanthrene in soil, *Appl. Microbiol. Biotechnol.*, 35:401-405, 1991.

Mansour, M., Feicht, E., and Méallier, P. Improvement of the photostability of selected substances in aqueous medium, *Toxicol. Environ. Chem.*, 20-21:139-147, 1989.

Mansour, M., Thaller, S., and Korte, F. Action of sunlight on parathion, *Bull. Environ. Contam. Toxicol.*, 30(3):358-364, 1983.

Maraqa, M.A., Zhao, X., Wallace, R.B., and Voice, T.C. Retardation coefficients of nonionic organic compounds determined by batch and column techniques, *Soil Sci. Soc. Am. J.*, 62:142-152, 1998.

Marchidan, D.I. and Ciopec, M. Relative enthalpies and related thermodynamic funstions of some organic compounds by drop calorimetry, *J. Therm. Anal.*, 14:131-150, 1978.

Marple, L., Berridge, B., and Throop, L. Measurement of the water-octanol partition coefficient of 2,3,7,8-tetrachlorodibenzo-*p*-dioxin, *Environ. Sci. Technol.*, 20(4):397-399, 1986.

Marple, L., Brunck, R., and Throop, L. Water solubility of 2,3,7,8-tetrachlorodibenzo-*p*-dioxin, *Environ. Sci. Technol.*, 20(2):180-182, 1986.

Marshall, J.S. Cadmium toxicity to laboratory and field populations of *Daphnia galeata mendotae*, *Bull. Environ. Contam. Toxicol.*, 21(4/5):453-457, 1979.

Martens, D.A. and Frankenberger, W.T., Jr. Enhanced degradation of polycyclic aromatic hydrocarbons in soil treated with an advanced oxidative process - Fenton's reagent, *J. Soil Contam.*, 4(2):175-190, 1995.

Martens, R. Degradation of [8,9-^{14}C] endosulfan by soil microorganisms, *Appl. Environ. Microbiol*, 31(6):853-858, 1976.

Martens, R. Degradation of endosulfan-8,9-^{14}C in soil under different conditions, *Bull. Environ. Contam. Toxicol.*, 17(4):438-446, 1977.

Martin, E., Yalkowsky, S.H., and Wells, J.E. Fusion of disubstituted benzenes, *J. Pharm. Sci.*, 68(5):565-568, 1979.

Martin, G., Laplanche, A., Morvan, J., Wei, Y., and LeCloirec, C. Action of ozone on organo-nitrogen products, in *Proceedings Symposium on Ozonization: Environmental Impact and Benefit* (Paris, France: International Ozone Association, 1983), pp. 379-393.

Martin, H., Ed. *Pesticide Manual*, 3rd ed. (Worcester, England: British Crop Protection Council, 1972).

Maruya, K.A., Risebrough, R.W., and Horne, A.J. Partitioning of polynuclear aromatic hydrocarbons between sediments from San Francisco Bay and their porewaters, *Environ. Sci. Technol.*, 39(10):2942-2948, 1996.

Masi, J.F. and Scott, R.B. Some thermodynamic properties of bromobenzene from 0 to 1500K, *J. Res. Nat. Bur. Stand.*, 79(A):619-628, 1975.

Masten, L.W., Boeri, R.L., and Walker, J.D. Strategies employed to determine the aquatic toxicity of ethylbenzene, a highly volatile, poorly water-soluble chemical, *Ecotoxicol. Environ. Saf.*, 27(3):335-348, 1994.

Masterton, W.L. and Lee, T.P. Effect of dissolved salts on water solubility of lindane, *Environ. Sci. Technol.*, 6(10):919-921, 1972.

Mastrangelo, S.V.R. Adiabatic calorimeter for determination of cryoscopic data, *Anal. Chem.*, 29(5):841-845, 1957.

Masunaga, S., Urushigawa, Y., and Yonezawa, Y. Biodegradation pathway of *o*-cresol by heterogeneous culture, *Water Res.*, 20(4):477-484, 1986.

Mateson, J.F. Olfactometry: its techniques and apparatus, *J. Air Pollut. Control Assoc.*, 5:167-170, 1955.

Matheson, L.J. and Tratnyek, P.G. Reductive dehalogenation of chlorinated methanes by iron metal, *Environ. Sci.*

Technol., 28(12):2045-2053, 1994.

Mathew, R. and Khan, S.U. Photodegradation of metolachlor in water in the presence of soil mineral and organic constituents, *J. Agric. Food Chem.*, 44(12):3996-4000, 1996.

Mathias, E., Sanhueza, E., Hisatsune, I.C., and Heicklen, J. Chlorine atom sensitized oxidation and the ozonolysis of tetrachloroethylene, *Can. J. Chem.*, 52:3852-3862, 1974.

Mathur, S.P. Mailton, H.A., Greenhalgh, R., Macmillan, K.A., and Khan, S.U. Effect of microorganisms and persistence of field-applied carbofuran and dyfonate in a humic meisol, *Can. J. Soil Sci.*, 56:89-96, 1976.

Mathur, S.P. and Rouatt, J.W. Utilization of the pollutant di-2-ethylhexyl phthalate by a bacterium, *J. Environ. Qual.*, 4(2):273-275, 1975.

Mathur, S.P. and Saha, J.G. Microbial degradation of lindane-C^{14} in a flooded sandy loam soil, *Soil Sci.*, 120(4):301-307, 1975.

Mathur, S.P. and Saha, J.G. Degradation of lindane-^{14}C in a mineral soil and in an organic soil, *Bull. Environ. Contam. Toxicol.*, 17(4):424-430, 1977.

Matsumura, F. and Benezet, H.J. Studies on the bioaccumulation and microbial degradation of 2,3,7,8-tetrachloro-*p*-dioxin, *Environ. Health Perspect.*, 5:253-258, 1973.

Matsumura, F. and Boush, G.M. Malathion degradation by *Trichoderma viride* and a *Pseudomonas* species, *Science (Washington, D.C.)*, 153(3741):1278-1280, 1966.

Matsumura, F. and Boush, G.M. Degradation of insecticides by a soil fungus, *Trichoderma viride*, *J. Econ. Entomol.*, 61:610-612, 1968.

Matsumura, F., Patil, K.C., and Boush, G.M. DDT metabolized by microorganisms from Lake Michigan, *Nature (London)*, 230(5292):324-325, 1971.

Matthews, R.W. Photo-oxidation of organic material in aqueous suspensions of titanium dioxide, *Water Res.*, 20(5):569-578, 1986.

Mattson, A.M., Spillane, J.T., and Pearce, G.W. Dimethyl 2,2-dichlorovinyl phosphate (DDVP), an organic phosphorus compound highly toxic to insects, *J. Agric. Food Chem.*, 3(4):319-321, 1955.

Mattson, V.R., Arthur, J.W., and Walbridge, C.T. Acute toxicity of selected organic compounds to fathead minnows, Office of Research and Development, U.S. EPA Report 600/3-76-097, 1976.

Maule, A., Plyte, S., and Quirk, A.V. Dehalogenation of organochlorine insecticides by mixed anaerobic microbial populations, *Pestic. Biochem. Physiol.*, 277:229-236, 1987.

Maund, S.J., Peither, A., Taylor, E.J., Juttner, I., Byerle-Pfnur, R., Lay, J.P., and Pascoe, D. Toxicity of lindane to freshwater insect larvae in compartments of an experimental pond, *Ecotoxicol. Environ. Saf.*, 23(1):76-88, 1992.

May, W.E., Wasik, S.P., and Freeman, D.H. Determination of the aqueous solubility of polynuclear aromatic hydrocarbons by a coupled column liquid chromatographic technique, *Anal. Chem.*, 50(1):175-179, 1978.

May, W.E., Wasik, S.P., and Freeman, D.H. Determination of the solubility behavior of some polycyclic aromatic hydrocarbons in water, *Anal. Chem.*, 50(7):997-1000, 1978a.

May, W.E., Wasik, S.P., Miller, M.M., Tewari, Y.B., Brown-Thomas, J.M., and Goldberg, R.N. Solution thermodynamics of some slightly soluble hydrocarbons in water, *J. Chem. Eng. Data*, 28(2):197-200, 1983.

Mayer, F.L. and Ellersieck, M.R. Manual of acute toxicity: interpretation and database for 410 chemicals and 66 species of freshwater animals, U.S. Department of the Interior, U.S. Fish and Wildlife Service, Resource Publication 160, 1986.

Mayer, F.L. and Sanders, H.O. Toxicology of phthalate acid esters in aquatic organisms, *Environ. Health Perspect.*, 3:153-157, 1973.

Mazzocchi, P.H. and Bowen, M.W. Photolysis of dioxane, *J. Org. Chem.*, 40(18):2689-2690, 1975.

McAuliffe, C. Solubility in water of C_1-C_9 hydrocarbons, *Nature (London)*, 200(4911):1092-1093, 1963.

McAuliffe, C. Solubility in water of paraffin, cycloparaffin, olefin, acetylene, cycloolefin, and aromatic compounds, *J. Phys. Chem.*, 70(4):1267-1275, 1966.

McAuliffe, C. Solubility in water of normal C_9 and C_{10} alkane hydrocarbons, *Science (Washington, D.C.)*, 163(3866):478-479, 1969.

McAuliffe, C. GC Determination of solutes by multiple phase equilibrium, *Chem. Technol.*, 1:46-51, 1971.

McBain, J.W. and Lissant, K.J. The solubilization of four typical hydrocarbons in aqueous solution by three typical detergents, *J. Am. Chem. Soc.*, 73:655-662, 1951.

McBain, J.W. and Richards, P.H. Solubilization of insoluble organic liquids by detergents, *Ind. Eng. Chem.*, 38(6):642-644, 1946.

McCall, P.J., Swann, R.L., Laskowski, D.A., Vrona, S.A., Unger, S.M., and Dishburger, H.J. Prediction of chemical mobility in soil from sorption coefficients, in *Aquatic Toxicology and Hazard Assessment, Fourth Conference, ASTM STP 737*, Branson, D.R. and K.L. Dickson, Eds. (Philadelphia, PA: American Society for Testing and Materials, 1981), pp. 49-58.

McCall, P.J., Vrona, S.A., and Kelley, S.S. Fate of uniformly carbon-14 ring labeled 2,4,5-trichlorophenoxyacetic acid and 2,4-dichlorophenoxy acid, *J. Agric. Food Chem.*, 29(1):100-107, 1981a.

McCants, J.F., Jones, J.H., and Hopson, W.H. Ternary solubility data for systems involving isopropanol and water, *Ind. Eng. Chem.*, 45(2):454-456, 1953.

McCarthy, J.F. Role of particulate organic matter in decreasing accumulation of polynuclear aromatic hydrocarbons by *Daphnia magna*, *Arch. Environ. Contam. Toxicol.*, 12(5):559-568, 1983.

McCarthy, J.F. and Jimenez, B.D. Reduction in bioavailability to bluegills of polycyclic aromatic hydrocarbons bound to dissolved humic material, *Environ. Toxicol. Chem.*, 4:511-521, 1985.

McCloskey, S.E., Gershanik, J.J., Lertora, J.L., White, L. and George, W.J. Toxicity of benzyl alcohol in adult and neonatal mice, *J. Pharm. Sci.*, 75(7):702-705, 1986.

McConnell, G., Ferguson, D.M., and Pearson, C.R. Chlorinated hydrocarbons and the environment, *Endeavour*, 34(121):13-18, 1975.

McConnell, G. and Schiff, H.I. Methyl chloroform: impact on stratospheric ozone, *Science (Washington, D.C.)*, 199:174-177, 1978.

McCrady, J.K., Johnson, D.E., and Turner, L.W. Volatility of ten priority pollutants from fortified avian toxicity test diets, *Bull. Environ. Contam. Toxicol.*, 34(5):634-644, 1985.

McCullough, J.P., Finke, H.L., Gross, M.E., Messerly, J.F., and Waddington, G. Low temperature calorimetric studies of seven 1-olefins: effect of orientational disorder in the solid state, *J. Phys. Chem.*, 61:289-301, 1957.

McCullough, J.P., Finke, H.L., Messerly, J.F., Kincheloe, T.C., and Waddington, G. The low temperature thermodynamic properties of naphthalene, 1-methylnaphthalene, 2-methylnaphthalene, 1,2,3,4-tetrahydronaphthalene, *trans*-decahydronaphthalene and *cis*-decahydronaphthalene, *J. Phys. Chem.*, 61:1105-1116, 1957a.

McDevit, W.F. and Long, F.A. The activity coefficient of benzene in aqueous salt solutions, *J. Am. Chem. Soc.*, 74(7):1773-1777, 1952.

McDuffie, B. Estimation of octanol/water partition coefficients for organic pollutants using reversed phase HPLC, *Chemosphere*, 10(1):73-78, 1981.

McGuire, R.R., Zabik, M.J., Schuetz, R.D., and Flotard, R.D. Photochemistry of bioactive compounds. Photochemical reactions of heptachlor: kinetics and mechanisms, *J. Agric. Food Chem.*, 20(4):856-861, 1972.

McGovern, E.W. Chlorohydrocarbon solvents, *Ind. Eng. Chem.*, 35(12):1230-1239, 1943.

McGowan, J.C., Atkinson, P.N., and Ruddle, L.H. The physical toxicity of chemicals. V. Interaction terms for solubilities and partition coefficients, *J. Appl. Chem.*, 16:99-102, 1966.

McKague, A.B. and Pridmore, R.B. Toxicity of altosid and dimilin to juvenile rainbow trout and coho salmon, *Bull. Environ. Contam. Toxicol.*, 20(2):167-169, 1978.

McKay, G., Bino, M.J., and Altamemi, A.R. The adsorption of various pollutants from aqueous solutions onto activated carbon, *Water Res.*, 19(4):491-495, 1985.

McMurry, P.H. and Grosjean, D. Photochemical formation of organic aerosols: growth laws and mechanisms, *Atmos. Environ.*, 19(9):1445-1451, 1985.

McNally, M.E. and Grob, R.L. Determination of the solubility limits of organic priority pollutants by gas chromatographic headspace analysis, *J. Chromatogr.*, 260:23-32, 1983.

McNally, M.E. and Grob, R.L. Headspace determination of solubility limits of the base neutral and volatile organic components from the Environmental Protection Agency's list of priority pollutants, *J. Chromatogr.*, 284:105-116, 1984.

McVeety, B.D. and Hites, R.A. Atmospheric deposition of polycyclic aromatic hydrocarbons to water surfaces: a mass balance approach, *Atmos. Environ.*, 22(3):511-536, 1988.

Meadows, R.W. and Spedding, D.J. The solubility of very low concentrations of carbon monoxide in aqueous solution, *Tellus*, 26:143-149, 1974.

Means, J.C., Hassett, J.J., Wood, S.G., and Banwart, W.L. Sorption properties of energy-related pollutants and sediments, in *Polynuclear Aromatic Hydrocarbons, Third International Symposium on Chemistry, Biology, Carcinogenesis and Mutagenesis*, Jones, P.W. and Leber, P., Eds. (Ann Arbor, MI: Ann Arbor Science Publishers, 1979), pp. 327-340.

Means, J.C., Wood, S.G., Hassett, J.J., and Banwart, W.L. Sorption of polynuclear aromatic hydrocarbons by sediments and soils, *Environ. Sci. Technol.*, 14(2):1524-1528, 1980.

Means, J.C., Wood, S.G., Hassett, J.J., and Banwart, W.L. Sorption of amino- and carboxy-substituted polynuclear aromatic hydrocarbons by sediments and soils, *Environ. Sci. Technol.*, 16(2):93-98, 1982.

Medley, D.R. and Stover, E.L. Effects of ozone on the biodegradability of biorefractory pollutants, *J. Water Pollut. Control Fed.*, 55(5):489-494, 1983.

Meier, U., Grotheer, H.H., Riekert, G., and Just, T. Temperature dependence and branching ratio of the $C_2H_5OH + OH$ reaction, *Chem. Phys. Lett.*, 115:221-225, 1985.

Meikle, R.W., Kurihara, N.H., and DeVries, D.H. The photocomposition rates in dilute aqueous solution and on a surface, and the volatilization rate from a surface, *Arch. Environ. Contam. Toxicol.*, 12(2):189-193, 1983.

Meikle, R.W. and Youngson, C.R. The hydrolysis rate of chlorpyrifos, *O,O*-diethyl *O*-(3,5,6-trichloro-2-pyridyl) phosphorothioate, and its dimethyl analog, chloropyrifos-methyl in dilute aqueous solution, *Arch. Environ. Contam. Toxicol.*, 7(1):13-22, 1978.

Meites, L., Ed. *Handbook of Analytical Chemistry*, 1st ed. (New York: McGraw-Hill, 1963), 1782 p.

Melancon, M.J. and Lech, J.J. Uptake metabolism and elimination of 14C-labeled 1,2,4-trichlorobenzene in rainbow trout and carp, *J. Toxicol. Environ. Health*, 6(3):645-658, 1980.

Melnikov, N.N. *Chemistry of Pesticides* (New York: Springer-Verlag, 1971), 480 p.

Mehrle, P.M., Buckler, D.R., Little, E.E., Smith, L.M., Petty, J.D., Peterman, P.H., Stalling, D.L., DeGraeve, G.M., Coyle, J.J., and Adams, W.J. Toxicity and bioconcentration of 2,3,7,8-tetrachlorodibenzodioxin and 2,3,7,8-tetrachlorodibenzofuran in rainbow trout, *Environ. Toxicol. Chem.*, 7:47-62, 1988.

Melcer, H. and Bedford, W.K. Removal of pentachlorophenol in municipal activated sludge systems, *J. Water Pollut. Control Fed.*, 60(5):622-626, 1988.

Menges, R.A., Bertrand, G.L., and Armstrong, D.W. Direct measurement of octanol-water partition coefficient chromatograph with a back-flushing technique, *J. Liq. Chromatogr.*, 13(15):3061-3077, 1990.

Mercer, J.W., Skipp, D.C., and Giffin, D. Basics of pump-and-treat groundwater remediation technology, U.S. EPA Report 600/8-90-003, 1990, 60 p.

Messerly, J.F., Finke, H.L., Good, W.D., and Gammon, B.E. Condensed-phase heat capacities and derived thermodynamic properties for 1,4-dimethylbenzene, 1,2-diphenylethane, and 2,3-dimethylnaphthalene, *J. Chem.*

Thermodyn., 20:485-501, 1988.

Messerly, J.F., Guthrie, G.B., Todd, S.S., and Finke, H.L. Low-temperature thermal data for *n*-pentane, *n*-heptadecane, and *n*-octadecane, *J. Chem. Eng. Data*, 12:338-346, 1967.

Messerly, J.F. and Kennedy, R.M. The heat capacity and entropy, heats of fusion, and vaporization and the vapor pressure of *n*-pentane, *J. Am. Chem. Soc.*, 62:2988-2991, 1940.

Messerly, J.F., Todd, S.S., Finke, H.L., Lee-Bechtold, S.H., Guthrie, G.B, Steele, W.V., and Chirico, R.D. Heat capacities of pent-1-ene (10K to 320K), *cis*-hex-2-ene (10K to 330K), non-1-ene (10K to 400K), and hexadec-1-ene (10K to 400K), *J. Chem. Thermodyn*, 22:1107-1128, 1990.

Messerly, J.F., Todd, S.S., and Guthrie, G.B. Chemical thermodynamic properties of the pentadienes, *J. Chem. Eng. Data*, 15:227-232, 1970.

Metcalf, C.D., McLeese, D.W., and Zitko, V. Role of volatilization of fenitrothion from fresh water, *Chemosphere*, 9(3):151-155, 1980.

Metcalf, R.L. A century of DDT, *J. Agric. Food Chem.*, 21(4):511-519, 1973.

Metcalf, R.L., Booth, G.M., Schuth, C.K., Hansen, D.J., and Lu, P.-Y. Uptake and fate of di-2-ethylhexyl phthalate in aquatic organisms and in a model ecosystem, *Environ. Health Perspect.* (June 1973), pp. 27-34.

Metcalf, R.L., Fukuto, T.R., Collins, C., Borck, K., Abd El-Aziz, S., Munoz, R., and Cassil, C.C. Metabolism of 2,2-dimethyl-2,3-dihydrobenzofuranyl-7 *n*-methylcarbamate (furadan) in plants, insects, and mammals, *J. Agric. Food Chem.*, 16(2):300-311, 1968.

Metcalf, R.L., Kapoor, I.P., Lu, P.-Y., Schuth, C.K., and Sherman, P. Model ecosystem studies of the environmental fate of six organochlorine pesticides, *Environ. Health Perspect.* (June 1973a), pp. 35-44.

Metcalf, R.L., Sanborn, J.R., Lu, P.-Y., and Nye, D. Laboratory model ecosystem studies of the degradation and fate of radiolabeled tri-, tetra-, and pentachlorobiphenyl compared with DDE, *Arch. Environ. Contam. Toxicol.*, 3(2):151-165, 1975.

Metcalf, R.L., Sangha, G.K., and Kapoor, I.P. Model ecosystem for the evaluation of pesticide biodegradability and ecological magnification, *Environ. Sci. Technol.*, 5(8):709-713, 1971.

Meva'a, L.M. and Lichanot, A. Proprietes thermodynamiques en phase condensee des ortho, meta et para fluorotoluene, cresol et toluidine, *Thermochim. Acta*, 158:335-345, 1990.

Meyers, N.L., Ahlrichs, J.L., and White, J.L. Adsorption of insecticides on pond sediments and watershed soils, *Proc. Indiana Acad. Sci.*, 79:432-437, 1970.

Meyers, S.M. and Wolff, J.O. Comparative toxicity of azinphos-methyl to house mice, laboratory mice, deer mice, and gray-tailed voles, *Arch. Environ. Contam. Toxicol.*, 26(4):478-482, 1994.

Meylan, W., Howard, P.H., and Boethling, R.S. Molecular topology/fragment contribution method for predicting soil sorption coefficients, *Environ. Sci. Technol.*, 26(8):1560-1567, 1992.

Mick, D.L. and Dahm, P.A. Metabolism of parathion by two species of *Rhizobium*, *J. Econ. Entomol.*, 63:1155-1159, 1970.

Mihelcic, J.R. and Luthy, R.G. Microbial degradation of acenaphthene and naphthalene under denitrification conditions in soil-water systems, *Appl. Environ. Microbiol.*, 54(5):1188-1198, 1988.

Mikami, Y., Fukunaga, Y., Arita, M., Obi, Y., and Kisaki, T. Preparation of aroma compounds by microbial transformation of isophorone by *Aspergillus niger*, *Agric. Biol. Chem.*, 45(3):791-793, 1981.

Mikesell, M.D. and Boyd, S.A. Reductive dechlorination of the pesticides 2,4-D, 2,4,5-T, and pentachlorophenol in anaerobic sludges, *J. Environ. Qual.*, 14(3):337-340, 1985.

Milano, J.C., Guibourg, A., and Vernet, J.L. Nonbiological evolution, in water, of some three- and four-carbon atoms organohalogenated compounds: hydrolysis and photolysis, *Water Res.*, 22(12):1553-1562, 1988.

Miles, C.J. and Takashima, S. Fate of malathion and *O,O,S*-trimethyl phosphorothioate by-product in Hawaiian soil and water, *Arch. Environ. Contam. Toxicol.*, 20(3):325-329, 1991.

Miles, C.J., Trehy, M.L., and Yost, R.A. Degradation of *N*-methylcarbamate and carbamoyl oxime pesticides in chlorinated water, *Bull. Environ. Contam. Toxicol.*, 41(6):838-843, 1988.

Miles, J.R.W., Harris, C.R., and Tu, C.M. Influence of temperature on the persistence of chlorpyrifos and chlorfenvinphos in sterile and natural mineral and organic soils, *J. Environ. Sci. Health*, B18(6):705-712, 1983.

Miles, J.R.W., Harris, C.R., and Tu, C.M. Influence of moisture on the persistence of chlorpyrifos and chlorfenvinphos in sterile and natural mineral and organic soils, *J. Environ. Sci. Health*, B19(2):237-243, 1984.

Miles, J.R.W. and Moy, P. Degradation of endosulfan and its metabolites by a mixed culture of soil microorganisms, *Bull. Environ. Contam. Toxicol.*, 23(1/2):13-19, 1979.

Miles, J.R.W., Tu, C.M., and Harris, C.R. Metabolism of heptachlor and its degradation products by soil microorganisms, *J. Econ. Entomol.*, 62:1334, 1969.

Miles, J.R.W., Tu, C.M., and Harris, C.R. Degradation of heptachlor epoxide and heptachlor by a mixed culture of soil microorganisms, *J. Econ. Entomol.*, 64:839-841, 1971.

Miles, J.R.W., Tu, C.M., and Harris, C.R. Persistence of eight organophosphorus insecticides in sterile and nonsterile mineral and organic soils, *Bull. Environ. Contam. Toxicol.*, 23(3):312-318, 1979.

Miliou, H., Zaboukas, N., and Moraitou-Apostolopoulou, M. Biochemical composition, growth, and survival of the guppy, *Poecilia reticulata*, during chronic sublethal exposure to cadmium, *Arch. Environ. Contam. Toxicol.*, 35(1):58-63, 1998.

Mill, T. Hydrolysis and oxidation processes in the environment, *Environ. Toxicol. Chem.*, 1:135-141, 1982.

Mill, T., Mabey, W.R., Lan, B.Y., and Baraze, A. Photolysis of polycyclic aromatic hydrocarbons in water, *Chemosphere*, 10(11/12):1281-1290, 1981.

Millemann, R.E., Binje, W.J., Black, J.A., Curham, R.M., Daniels, K.L., Franco, P.J., Giddings, J.M., McCarthy, J.F., and Stewart, A.J. Comparative acute toxicity to aquatic organisms of components of coal-derived synthetic fuels,

Trans. Am. Fish Soc., 113:74-85, 1984.

Miller, C.T. and Weber, W.J., Jr. Sorption of hydrophobic organic pollutants in saturated soil systems, *J. Contam. Hydrol.*, 1(1/2):243-261, 1986.

Miller, G.C. and Crosby, D.G. Photooxidation of 4-chloroaniline and *N*-(4-chlorophenyl)benzenesulfonamide to nitroso- and nitro-products, *Chemosphere*, 12(9/10):1217-1227, 1983.

Miller, G.C., Miller, W.W., Hanks, L., and Warren, J. Soil sorption and alfalfa uptake of strychnine, *J. Environ. Qual.*, 12(4):526-529, 1983.

Miller, G.C. and Zepp, R.G. Photoreactivity of aquatic pollutants sorbed on suspended sediments, *Environ. Sci. Technol.*, 13(7):860-863, 1979.

Miller, L.L. and Narang, R.S. Induced photolysis of DDT, *Science (Washington, D.C.)*, 169(3943):368-370, 1970.

Miller, M.E. and Stuart, J.D. Measurement of aqueous Henry's law constants for oxygenates and aromatics found in gasolines by the static headspace method, *Anal. Chem.*, 72(3):622-625, 2000.

Miller, M.M., Ghodbane, S., Wasik, S.P., Tewari, Y.B., and Martire, D.E. Aqueous solubilities, octanol/water partition coefficients, and entropies of melting of chlorinated benzenes and biphenyls, *J. Chem. Eng. Data*, 29(2):184-190, 1984.

Miller, M.M., Wasik, S.P., Huang, G.-L., Shiu, W.-Y., and Mackay, D. Relationships between octanol-water partition coefficient and aqueous solubility, *Environ. Sci. Technol.*, 19(6):522-529, 1985.

Milligan, L.H. The solubility of gasoline (hexane and heptane) in water at 25 °C, *J. Phys. Chem.*, 28(5):495-497, 1924.

Millischer, R.J. The toxicity of HFC 134a, *J. Am. Coll. Toxicol.*, 8(6):1220, 1989.

Mills, A.C. and Biggar, J.W. Solubility-temperature effect on the adsorption of gamma- and beta-BHC from aqueous and hexane solutions by soil materials, *Soil Sci. Soc. Am. Proc.*, 33:210-216, 1969.

Mills, A.L. and Alexander, M. Factors affecting dimethylnitrosamine formation in samples of soil and water, *J. Environ. Qual.*, 5(4):437-440, 1976.

Mills, E.J., Jr. and Stack, V.T., Jr. Biological oxidation of synthetic organic chemicals, in *Proceedings 8th Industrial Waste Conference*, Purdue University, 81:492-517, 1953.

Mills, G. and Hoffman, M.R. Photocatalytic degradation of pentachlorophenol on TiO_2 particles: identification of intermediates and mechanism of reaction, *Environ. Sci. Technol.*, 27(8):1681-1689, 1993.

Mills, W.B., Porcella, D.B., Ungs, M.J., Gherini, S.A., Summers, K.V., Mok, L., Rupp, G.L., and Bowie, G.L. Water quality assessment: a screening procedure for toxic and conventional pollutants in surface and groundwater - Part I, Office of Research and Development, U.S. EPA Report 600/6-85-002a, 1985, 638 p.

Miltner, R.J., Baker, D.B., Speth, T.F., and Fronk, C.A. Treatment of seasonal pesticides in surface waters, *J. Am. Water Works Assoc.*, 81(1):43-52, 1989.

Minard, R.D., Russel, S., and Bollag, J.-M. Chemical transformation of 4-chloroaniline to a triazene in a bacterial culture medium, *J. Agric. Food Chem.*, 25(4):841-844, 1977.

Mingelgrin, U. and Gerstl, Z. Reevaluation of partitioning as a mechanism of nonionic chemicals adsorption in soils, *J. Environ. Qual.*, 12(1):1-11, 1983.

Mingelgrin, U., Saltzman, S., and Yaron, B. A possible model for the surface-induced hydrolysis of organophosphorus pesticides on kaolinite clays, *Soil Sci. Soc. Am. J.*, 41:519-523, 1977.

Mirabel, P., George, C., Magi, L., and Ponche, J.L. Gas-liquid interactions, in *Heterogeneous and Liquid-Phase Processes*, Warneck, P., Ed. (Berlin: Springer-Verlag, 1996), pp. 175-181.

Mirrlees, M.S., Moulton, S.J., Murphy, C.T., and Taylor, P.J. Direct measurement of octanol-water partition coefficient by high-pressure liquid chromatography, *J. Med. Chem.*, 19(5):615-619, 1976.

Mirvish, S.S., Issenberg, P., and Sornson, H.C. Air-water and ether-water distribution of *N*-nitroso compounds: implications for laboratory safety, analytic methodology, and carcinogenicity for the rat esophagus, nose, and liver, *J. Nat. Cancer Instit.*, 56(6):1125-1129, 1976.

Mitchell, J.W., Vratsanos, M.S., Hanley, B.F., and Parekh, V.S. Experimental flash points of industrial amines, *J. Chem. Eng. Data*, 44(2):209-211, 1999.

Mitchell, S. A method for determining the solubility of sparingly soluble substances, *J. Chem. Soc. (London)*, 1333-1337, 1926.

Mittal, P.K., Adak, T., and Sharma, V.P. Comparative toxicity of certain mosquitocidal compounds to larvivorous fish, *Poecilia reticulata*, *Indian J. Malariol.*, 31(2):43-47, 1994.

Miyano, Y. and Hayduk, W. Solubility of butane in several polar and nonpolar solvents and in an acetone-butanol solvent solution, *J. Chem. Eng. Data*, 31(1):77-80, 1986.

Moilanen, K.W., Crosby, D.G., Humphrey, J.R., and Giles, J.W. Vapor-phase photodecomposition of chloropicrin (trichloronitromethane), *Tetrahedron*, 34(22):3345-3349, 1978.

Moles, A., Bates, S., Rice, S.D., and Korn, S. Reduced growth of coho salmon fry exposed to two petroleum components, toluene and naphthalene, in fresh water, *Trans. Am. Fish Soc.*, 110:430-436, 1981.

Moles, A., Rice, S.D., and Korn, S. Sensitivity of Alaskan freshwater and anadromous fishes to Prudhoe Bay crude oil and benzene, *Trans. Am. Fish Soc.*, 108:408-414, 1979.

Monsanto Industrial Chemicals Company. PCBs-Aroclors, Technical Bulletin O/PL 306A, 1974.

Montgomery, J.H. Unpublished results, 1989.

Moore, J.W. and Ramamoorthy, S. *Organic Chemicals in Natural Waters - Applied Monitoring and Impact Assessment* (New York: Springer-Verlag, 1984), 289 p.

Moore, R.M., Green, C.E., and Tait, V.K. Determination of Henry's law constants for a suite of naturally occurring halogenated methanes in seawater, *Chemosphere*, 30(6):1183-1191, 1995.

Moreale, A. and Van Bladel, R. Soil interactions of herbicide-derived aniline residues: a thermodynamic approach, *Soil Sci.*, 127(1):1-9, 1979.

Moreale, A. and Van Bladel, R. Behavior of 2,4-D in Belgian soils, *J. Environ. Qual.*, 9(4):627-633, 1980.

Morehead, N.R., Eadie, B.J., Lake, B., Landrum, P.F., and Berner, D. The sorption of PAH onto dissolved organic matter in Lake Michigan waters, *Chemosphere*, 15(4):403-412, 1986.

Morgan, O.M. and Maas, O. An investigation of the equilibria existing in gas-water systems forming electrolytes, *Can. J. Res.*, 5:162-169, 1931.

Morgulis, M.S., Oliveira, G.H., Dagli, M.L., and Palmero-Neto, J. Acute 2,4-dichlorophenoxyacetic acid intoxication in broiler chicks, *Poult. Sci.*, 77(4):509-515, 1998.

Mori, Y., Goto, S., Onodera, S., Naito, S., and Matsushita, H. Aqueous chlorination of tetracyclic aromatic hydrocarbons: reactivity and product distribution, *Chemosphere*, 22(5/6):495-501, 1991.

Moriguchi, I. Quantitative structure-activity studies on parameters related to hydrophobicity, *Chem. Pharm. Bull.*, 23:247-257, 1975.

Morita, M., Nakagawa, J., and Rappe, C. Polychlorinated dibenzofuran (PCDF) formation from PCB mixture by heat and oxygen, *Bull. Environ. Contam. Toxicol.*, 19(6):665-670, 1978.

Morris, D.A. and Johnson, A.I. Summary of hydrologic and physical properties of rocks and soil materials as analyzed by the hydrologic laboratory of the U.S. Geological Survey, 1948-1960, U.S. Geological Survey Water-Supply Paper 1839-D, 1967, 42 p.

Morris, E.D., Jr. and Niki, H. Mass spectrometric study of the reaction of hydroxyl radical with formaldehyde, *J. Chem. Phys.*, 55:1991-1992, 1971.

Morris, E.D., Jr., Stedman, D.H., and Niki, H. Mass spectrometric study of the reactions of the hydroxyl radical with ethylene, propylene, and acetaldehyde in a discharge-flow system, *J. Am. Chem. Soc.*, 93:3570-3572, 1971.

Morrison, R.T. and Boyd, R.N. *Organic Chemistry* (Boston, MA: Allyn & Bacon, 1971), 1258 p.

Morrison, T.J. and Billett, F. The salting-out of non-electrolytes. Part II. The effect of variation in non-electrolytes, *J. Chem. Soc. (London)*, pp. 3819-3822, 1952.

Morrison, T.J. and Johnstone, N.B. Solubilities of the inert gases in water, *J. Chem. Soc. (London)*, pp. 3441-2446, 1954.

Mortland, M.M. and Raman, K.V. Catalytic hydrolysis of some organic phosphate pesticides by copper(II), *J. Agric. Food Chem.*, 15(1):163-167, 1967.

Morton, M.G., Mayer, F.L., Jr., Dickson, K.L., Waller, W.T., and Moore, J.C. Acute and chronic toxicity of azinphos-methyl to two estuarine species, *Mysidopsis bahia* and *Cyprinodon variegatus*, *Arch. Environ. Contam. Toxicol.*, 32(4):436-441, 1997.

Mosher, D.R. and Kadoum, A.M. Effects of four lights on malathion residues on glass beads, sorghum grain, and wheat grain, *J. Econ. Entomol.*, 65:847-850, 1972.

Mosier, A.R., Guenzi, W.D., and Miller, L.L. Photochemical decomposition of DDT by a free-radical mechanism, *Science (Washington, D.C.)*, 164(3883):1083-1085, 1969.

Mostafa, I.Y., Bahig, M.R.E., Fakhr, I.M.I., and Adam, Y. Metabolism of organophosphorus pesticides. XIV. Malathion breakdown by soil fungi, *Z. Naturforsch. B.*, 27:1115-1116, 1972.

Moussavi, M. Effect of polar substituents on autoxidation of phenols, *Water Res.*, 13(12):1125-1128, 1979.

Moza, P.N. and Feicht, E. Photooxidation of aromatic hydrocarbons as liquid film on water, *Toxicol. Environ. Chem.*, 20-21:135-138, 1989.

Moza, P.N., Fytianos, K., Samanidou, V., and Korte, F. Photodecomposition of chlorophenols in aqueous medium in presence of hydrogen peroxide, *Bull. Environ. Contam. Toxicol.*, 41(5):678-682, 1988.

Mroczek, E.K. Henry's law constants and distribution coefficients of sulfur hexafluoride in water from 25 °C to 230 °C, *J. Chem. Eng. Data*, 42(1):116-119, 1997.

Muccini, M., Layton, A.C., Sayler, G.S., and Schultz, T.W. Aquatic toxicities of halogenated benzoic acids to *Tetrahymena pyriformis*, *Bull. Environ. Contam. Toxicol.*, 62(5):616-622, 1999.

Muir, D.C.G., Grift, N.P., Blouw, A.P., and Lockhart, W.L. Persistence of fluridone in small ponds, *J. Environ. Qual.*, 9(1):151-156, 1980.

Mulla, M.S., Mian, L.S., and Kawecki, J.A. Distribution, transport and fate of the insecticides malathion and parathion in the environment, *Residue Rev.*, 81:116-125, 1981.

Mulley, B.A. and Metcalf, A.D. Solubilization of phenols by non-ionic surface active reagents, *Sci. Pharm.*, 6:481-488, 1966.

Munakata, K. and Kuwahara, M. Photochemical degradation products of pentachlorophenol, *Residue Rev.*, 25:13-23, 1969.

Munnecke, D.M. and Hsieh, D.P.H. Pathways of microbial metabolism of parathion, *Appl. Environ. Microbiol.*, 31(1):63-69, 1976.

Munshi, J.D., Dutta, H.M., Singh, N.K., Roy, P.K., Adhikari, S., Dogra, J.V., and Ali, M.M. Effect of malathion, an organophosphorous pesticide, on the serum proteins of *Heteropneustes fossilis* (BLOCH), *J. Environ. Pathol. Toxicol. Oncol.*, 18(1):79-83, 1999.

Munshi, H.B., Rama Rao, K.V.S., and Iyer, R.M. Characterization of products of ozonolysis of acrylonitrile in liquid phase, *Atmos. Environ.*, 23(9):1945-1948, 1989.

Munshi, H.B., Rama Rao, K.V.S., and Iyer, R.M. Rate constants of the reactions of ozone with nitriles, acrylates and terpenes in gas phase, *Atmos. Environ.*, 23(9):1971-1976, 1989a.

Munz, C. and Roberts, P.V. Air-water phase equilibria of volatile organic solutes, *J. Am. Water Works Assoc.*, 79(5):62-69, 1987.

Murphy, N.F., Lastovica, J.E., and Fallis, J.G. Correlation of interfacial tension of two-phase three-component systems, *Ind. Eng. Chem.*, 49(6):1035-1042, 1957.

Murphy, T.J., Mullin, M.D., and Meyer, J.A. Equilibrium of polychlorinated biphenyls and toxaphene with air and water, *Environ. Sci. Technol.*, 21(2):155-162, 1987.

Murray, J.M., Pottie, R.F., and Pupp, C. The vapor pressures and enthalpies of sublimation of five polycyclic aromatic hydrocarbons, *Can. J. Chem.*, 52(4):557-563, 1974.

Murthy, N.B.K., Kaufman, D.D., and Fries, G.F. Degradation of pentachlorophenol (PCP) in aerobic and anaerobic soil, *J. Environ. Sci. Health*, B14(1):1-14, 1979.

Murti, R., Omkar, and Shukla, G.S. Chromium toxicity to a freshwater prawn *Macrobrachium lamarrei* (H.M. Edwards), *Toxicol. Lett.*, 18(3):257-261, 1983.

Musoke, G.M.S., Roberts, D.J., and Cooke, M. Heterogeneous hydrodechlorination of chlordan, *Bull. Environ. Contam. Toxicol.*, 28(4):467-472, 1982.

Nahum, A. and Horvath, C. Evaluation of octanol-water partition coefficients by using high-performance liquid chromatography, *J. Chromatogr.*, 192:315-322, 1980.

Naik, M.N., Jackson, R.B., Stokes, J., and Swabay, R.J. Microbial degradation and phytotoxicity of picloram and other substituted pyridines, *Soil Biol. Biochem.*, 4:13-23, 1972.

Nakagawa, T., Sato, Y., Watabe, A., Kawamura, T., and Morita, M. Determination of benzo(*a*)pyrene in liquid paraffin by high performance liquid chromatography (HPLC), *Bull. Environ. Contam. Toxicol.*, 19(6):703-706, 1978.

Nakajima, S., Nakato, N., and Tani, T. Microbial transformation of 2,4-D and its analog, *Chem. Pharm. Bull.*, 21:671-673, 1973.

Nakashima, H., Omae, K., Sakai, T., Yamazaki, K., and Sakurai, H. Acute and sunchronic inhalation toxicity of tetraethoxysilane (TEOS) in mice, *Arch. Toxicol.*, 68(5):277-283, 1994.

Nakashima, H., Omae, K., Takebayashi, T., Ishizuka, C., Sakurai, H., Yamazaki, K., Nakaza, M., Shibata, T., Kudo, M., and Koshi, S. Acute and subacute inhalation toxicity of dichlorosilane in male ICR mice, *Arch. Toxicol.*, 70(3-4):218-223, 1996.

Nałecz-Jawecki, G. and Sawicki, J. Spirotox - A new tool for testing the toxicity of volatile compounds, *Chemosphere*, 38(14):3211-3218, 1999.

Namiot, Yu. and Beider, S.Ya. The water solubility of *n*-pentane and *n*-hexane, *Khim. Tekhnol. Topl. Masel*, 5(7):52-55, 1960.

NAS. 1977. Drinking Water and Health (Washington, D.C.: National Academy of Sciences), 939 p.

Nash, R.G. Comparative volatilization and dissipation rates of several pesticides from soil, *J. Agric. Food Chem.*, 31(2):210-217, 1983.

Nash, R.G. and Woolson, E.A. Persistence of chlorinated insecticides in soils, *Science (Washington, D.C.)*, 157(3791):924-927, 1967.

Natarajan, G.S. and Venkatachalam, K.A. Solubilities of some olefins in aqueous solutions, *J. Chem. Eng. Data*, 17(3):328-329, 1972.

Nath, K. and Kumar, N. Hexavalent chromium: toxicity and its impact on certain aspects of carbohydrate metabolism of the freshwater teleost, *Colisa fasciatus*, *Sci. Total Environ.*, 72:175-181, 1988.

Nathan, M.F. Choosing a process for chloride removal, *Chem. Eng.*, 85:93-100, 1978.

Nathwani, J.S. and Phillips, C.R. Adsorption-desorption of selected hydrocarbons in crude oil on soils, *Chemosphere*, 6(4):157-162, 1977.

Nauman, C.H. and Schaum, J.L. Human exposure estimation for 2,3,7,8-TCDD, *Chemosphere*, 16(8/9):1851-1856, 1987.

Nebeker, A.V. and Schuytema, G.S. Chronic effects of the herbicide diuron on freshwater cladocerans, amphipods, midges, minnows, worms, and snails, *Arch. Environ. Contam. Toxicol.*, 35(3):441-446, 1998.

Nebeker, A.V., Schuytema, G.S., Griffis, W.L., and Cataldo, A. Impact of guthion on survival and growth of the frog *Pseudacris regilla* and the salamanders *Ambystoma gracile* and *Ambystoma maculatum*, *Arch. Environ. Contam. Toxicol.*, 35(1):48-51, 1998.

Neely, W.B. and Blau, G.E., Eds. *Environmental Exposure from Chemicals. Volume 1* (Boca Raton, FL: CRC Press, 1985), 245 p.

Neely, W.B., Branson, D.R., and Blau, G.E. Partition coefficient to measure bioconcentration potential of organic pesticides in fish, *Environ. Sci. Technol.*, 6(7):629-632, 1974.

Nelson, H.D. and DeLigny, C.L. The determination of the solubilities of some *n*-alkanes in water at different temperatures by means of gas chromatography, *Rec. Trav. Chim. Pays-Bays*, 87(6):528-544, 1968.

Nelson, N.H. and Faust, S.D. Acidic dissociation constants of selected aquatic herbicides, *Environ. Sci. Technol.*, 3(11):1186-1188, 1969.

Nelson, P.E. and Hoff, J.E. Food volatiles: gas chromatographic determination of partition coefficient in water-lipid systems, *J. Food Sci.*, 33:479-482, 1968.

Nelson, E.C., Ghoshal, S., Edwards, J.C., Marsh, G.X., and Luthy, R.G. Chemical characterization of coal tar-water interfacial films, *Environ. Sci. Technol.*, 30(3):1014-1022, 1996.

Nesbitt, H.J. and Watson, J.R. Degradation of the herbicide 2,4-D in river water. II. The role of suspended sediment. Nutrients and water temperature, *Water Res.*, 14:1689-1694, 1980.

Neskovic, N.K., Elezovic, I., Karan, V., Poleksic, V., and Budimir, M. Acute and subacute toxicity of atrazine to carp (*Cyprinus carpio* L.), *Ecotoxicol. Saf.*, 25(2):173-182, 1993.

Nestrick, T.J., Lamparski, L.L., and Townsend, D.I. Identification oftetrachlorodibenzo-*p*-dioxin isomers at the 1-ng level by photolytic degradation and pattern recognition techniques, *Anal. Chem.*, 52(12):1865-1874, 1980.

Neudorf, S. and Khan, M.A.Q. Pick up and metabolism of DDT, dieldrin, and photodieldrin by a freshwater alga (*Ankistrodesmus amalloides*) and a microcrustacean (*Daphnia pulex*), *Bull. Environ. Contam. Toxicol.*, 13(4):443-450, 1975.

Neuhauser, E.F., Loehr, R.C., Malecki, M.R., Milligan, D.L., and Durkin, P.R. The toxicity of selected organic chemicals to the earthworm *Eisenia fetida*, *J. Environ. Qual.*, 14(3):383-388, 1985.

Newland, L.W., Chesters, G., and Lee, G.B. Degradation of γ-BHC in simulated lake impoundments as affected by aeration, *J. Water Pollut. Control Fed.*, 41(5):R174-R188, 1969.

Newsted, J.L. and Giesy, G.P. Predictive models for photoinduced acute toxicity of polycyclic aromatic hydrocarbons to *Daphnia magna*, Strauss (Cladocera, Crustacea), *Environ. Toxicol. Chem.*, 6(6):445-461, 1987.

Newton, M. and Snyder, S.P. Exposure of forest herbivores to 2,3,7,8-tetrachlorodibenzo-p-dioxin (TCDD) in areas sprayed with 2,4,5-T, *Bull. Environ. Contam. Toxicol.*, 20(6):743-750, 1978.

NFPA. 1984. Fire Protection Guide on Hazardous Materials (Quincy, MA: National Fire Protection Association), 443 p.

Ngabe, B., Bidleman, T.F., and Falconer, R.L. Base hydrolysis of α- and β-hexachlorocyclohexanes, *Environ. Sci. Technol.*, 27(9):1930-1933, 1993.

Nicholson, B.C., Maguire, B.P., and Bursill, D.B. Henry's law constants for the trihalomethanes: effects of water composition and temperature, *Environ. Sci. Technol.*, 18(7):518-521, 1984.

Nielsen, F., Olsen, E., and Fredenslund, A. Henry's law constants and infinite dilution activity coefficients for volatile organic compounds in water by a validated batch air stripping method, *Environ. Sci. Technol.*, 28(12):2133-2138, 1994.

Nielsen, P.H., Bjerg, P.L., Nielsen, P., Smith, P., and Christensen, T.H. *In situ* and laboratory determined first-order degradation rate constants of specific organic compounds in an aerobic aquifer, *Environ. Sci. Technol.*, 30(1):31-37, 1996.

Nielsen, T., Ramdahl, T., and Bjørseth, A. The fate of airborne polycyclic organic matter, *Environ. Health Perspect.*, 47:103-114, 1983.

Niessen, R., Lenoir, D., and Boule. P. Phototransformation of phenol induced by excitation of nitrate ions, *Chemosphere*, 17(10):1977-1984, 1988.

Nigg, H.N., Reinert, J.A., and Stamper, J.H. Disappearance of acephate, methamidophos, and malathion from citrus foliage, *Bull. Environ. Contam. Toxicol.*, 26(2):267-272, 1981.

Nikam, P.S., Jagdale, B.S., Sawant, A.B., and Hasan, M. Densities and viscosities for binary mixtures of benzonitrile with methanol, ethanol, propan-1-ol, butan-1-ol, pentan-1-ol, and 2-methypropan-2-ol at (303.15, 308.15, and 313.15) K, *J. Chem. Eng. Data*, 45(2):214-218, 2000.

Nikam, P.S., Manhale, T.R., and Hasan, M. Densities and viscosities for ethyl acetate + pentan-1-ol, + hexan-1-ol, + 3,5,5-trimethylhexan-1-ol, + heptan-1-ol, + octan-1-ol, and decan-1-ol at (298.15, 303.15, and 308.15) K, *J. Chem. Eng. Data*, 43(3):436-440, 1998.

Niki, H., Maker, P.D., Savage, C.M., and Breitenbach, L.P. Relative rate constants for the reaction of hydroxyl radical with aldehydes, *J. Phys. Chem.*, 82:132-134, 1978.

Nikolaou, K., Masclet, P., and Mouvier, G. Sources and chemical reactivity of polynuclear aromatic hydrocarbons in the atmosphere - A critical review, *Sci. Total Environ.*, 32(2):103-132, 1984.

Nilsen, O.G., Haugen, O.A., Zahlsen, K., Halgunset, J., Helseth, A., Aarset, H., and Eide, I. Toxicity of n-C9 to n-C13 alkanes in the rat on short term inhalation, *Pharmacol. Toxicol.*, 62(5):259-266, 1988.

Nimitz, J.S. and Skaggs, S.R. Estimating tropospheric lifetimes and ozone depletion potentials of one- and two-carbon hydrofluorocarbons and hydrochlorofluorocarbons, *Environ. Sci. Technol.*, 26(4):739-744, 1992.

NIOSH Pocket Guide to Chemical Hazards, U.S. Department of Health and Human Services, U.S. Government Printing Office, 1997, 440 p.

Nisbet, I.C. and Sarofim, A.F. Rates and routes of transport of PCBs in the environment, *Environ. Health Perspect.*, (April 1972), pp. 21-38.

Nkedi-Kizza, P., Rao, P.S.C., and Hornsby, A.G. Influence of organic cosolvents on sorption of hydrophobic organic chemicals by soils, *Environ. Sci. Technol.*, 19(10):975-979, 1985.

Nkedi-Kizza, P., Rao, P.S.C., and Johnson, J.W. Adsorption of diuron and 2,4,5-T on soil particle-size separates, *J. Environ. Qual.*, 12(2):195-197, 1983.

Nojima, K., Fukaya, K., Fukui, S., and Kanno, S. Studies on photochemistry of aromatic hydrocarbons, *Chemosphere*, 4(2):77-82, 1975.

Nojima, K., Ikarigawa, T., and Kanno, S. Studies on photochemistry of aromatic hydrocarbons. VI. Photochemical reaction of bromobenzene with nitrogen oxides in air, *Chemosphere*, 9(7/8):421-436, 1980.

Nojima, K. and Kanno, S. Studies on photochemistry of aromatic hydrocarbons. VII. Photochemical reaction of p-dichlorobenzene with nitrogen oxides in air, *Chemosphere*, 9(7/8):437-440, 1980.

Noulty, R.A. and Leaist, D.G. Diffusion coefficient of aqueous benzoic acid at 25 °C, *J. Chem. Eng. Data*, 32(4):418-420, 1987.

Novoselova, N.V., Rabinovich, I.B., Tsvetkova, L.Ya., Moseeva, E.M., and Babinkov, A.G. Heat capacity and thermodynamic functions of tetrachloroethylene, *Zhur. Fiz. Khim.*, 1627-1630, 1986.

Nowara, A., Burhenne, J., and M. Spiteller. Binding of fluoroquinoline carboxylic acid derivatives to clay minerals, *J. Agric. Food Chem.*, 45(4):1459-1463, 1997.

Nunes, P. and Benville, Jr., P.E. Uptake and depuration of petroleum hydrocarbons in the Manila clams, *Tapes semidecussata* Reeve, *Bull. Environ. Contam. Toxicol.*, 21(6):719-726, 1979.

Nyholm, N., Lindgaard-Jørgensen, P., and Hansen, N. Biodegradation of 4-nitrophenol in standardized aquatic degradation tests, *Ecotoxicol. Environ. Saf.*, 8(6):451-470, 1984.

Nyman, M.C., Ferra, J.J., Nyman, A.K., Kenttämaa, H.I., and Blatchley, E.R., III. Photodegradation of 3,3'-dichlorobenzidine in the aqueous environment, in *American Chemical Society - Division of Environmental Chemistry, Preprints of Extended Abstracts*, 37(1):293-294, 1997.

Nyssen, G.A., Miller, E., Glass, T.F., Quinn II, C.R., Underwood, J., and Wilson, D.J. Solubilities of hydrophobic compounds in aqueous-organic solvent mixtures, *Environ. Monitor. Assess.*, 9(1):1-11, 1987.

O'Connell, W.L. Properties of heavy liquids, *Trans. Am. Inst. Mech. Eng.*, 226(2):126-132, 1963.

Odendaal, J.P. and Reinecke, A.J. Short-term toxicological effects of cadmium on the woodlouse, *Porcellio laecis (Crustacea, Isopoda)*, *Ecotoxicol. Environ. Saf.*, 43(1):30-34, 1999.

Odeyemi, O. and Alexander, M. Resistance of *Rhizobium* strains to phygon, spergon, and thiram, *Appl. Environ. Microbiol.*, 33(4):784-790, 1977.

Oetting, F.L. Low-temperature heat capacity and related thermodynamic functions of propylene oxide, *J. Chem. Phys.*, 41:149-153, 1964.

Oetting, F.L. Absolute entropies of the methyl alkyl ketones at 298.15 K, *J. Chem. Eng. Data*, 10:122-125, 1965.

Ofstad, E.B. and Sletten, T. Composition and water solubility determination of a commercial tricresylphosphate, *Sci. Total Environ.*, 43(3):233-241, 1985.

Ogata, M., Fujisawa, K., Ogino, Y., and Mano, E. Partition coefficients as a measure of bioconcentration potential of crude oil compounds in fish and shellfish, *Bull. Environ. Contam. Toxicol.*, 33(4):561-567, 1984.

Ogata, M. and Miyake, Y. Disappearance of aromatic hydrocarbons and organic sulfer compounds from fish flesh reared in crude oil suspension, *Water Res.*, 12(11):1041-1044, 1978.

Ohe, T. Mutagenicity of photochemical reaction products of polycyclic aromatic hydrocarbons with nitrite, *Sci. Total Environ.*, 39(1/2):161-175, 1984.

Ohta, T. and Ohyama, T. A set of rate constants for the reactions of hydroxy radicals with aromatic hydrocarbons, *Bull. Chem. Soc. Jpn.*, 58:3029, 1985.

Oja, V. and Suuberg, E.M. Vapor pressures and enthalpies of sublimation of polycyclic aromatic hydrocarbons and their derivatives, *J. Chem. Eng. Data*, 43(3):486-492, 1998.

Okamoto, K., Hondo, F., Itaya, A., and Kusabayashi, S. Kinetics of dye-sensitized photodegradation of aqueous phenol, *J. Chem. Eng. Jpn.*, 15:368-368, 1982.

Okorafor, O.C. Solubility and density isotherms for the sodium sulfate-water-methanol system, *J. Chem. Eng. Data*, 44(3):488-490, 1999.

Oliver, B.G. Desorption of chlorinated hydrocarbons from spiked and anthropogenically contaminated sediments, *Chemosphere*, 14(8):1087-1106, 1985.

Oliver, B.G. and Carey, J.H. Photochemical production of chlorinated organics in aqueous solutions, *Environ. Sci. Technol.*, 11(9):893-895, 1977.

Oliver, B.G. and Charlton, M.N. Chlorinated organic contaminants on settling particulates in the Niagara River vicinity of Lake Ontario, *Environ. Sci. Technol.*, 18(12):903-908, 1984.

Oliver, B.G. and Niimi, A.J. Bioconcentration of chlorobenzenes from water by rainbow trout: correlations with partition coefficients and environmental residues, *Environ. Sci. Technol.*, 17(5):287-291, 1983.

Oliver, B.G. and Niimi, A.J. Bioconcentration factors of some halogenated organics for rainbow trout: limitations in their use for prediction of environmental residues, *Environ. Sci. Technol.*, 19(9):842-849, 1985.

Oliverira-Filho, E.C. and Paumgartten, F.J.R. Comparative study on the acute toxicities of α, β, γ, and δ isomers of hexachlorocyclohexane to freshwater fishes, *Bull. Environ. Contam. Toxicol.*, 59(6):984-988, 1997.

Ollis, D.F. Contaminant degradation in water, *Environ. Sci. Technol.*, 19(6):480-484, 1985.

Oloffs, P.C. and Albright, L.J. Transport of some organophosphorus in B.C. waters, *Proc. Int'l Conf. Trans. Persistent Chem. Aquat. Ecosys.*, 1:89-92, 1974.

Oloffs, P.C., Albright, L.J., and Szeto, S.Y. Fate and behavior of five chlorinated hydrocarbons in three natural waters, *Can. J. Microbiol.*, 18(9):1393-1398, 1972.

Oloffs, P.C., Albright, L.J., Szeto, S.Y., and Lau, J. Factors affecting the behavior of five chlorinated hydrocarbons in two natural waters and their sediments, *J. Fish Res. Board Can.*, 30(11):1619-1623, 1973.

Olsen, R.L. and Davis, A.L. Predicting the fate and transport of organic compounds in groundwater, *Haz. Mat. Control*, 3:40-64, 1990.

Omkar and Shukla, G.S. Nature of dichlorvos intoxication in a freshwater prawn, *Macrobrachium lamarrei* (H. Milne Edwards), *Environ. Res.*, 37(2):349-354, 1985.

Onitsuka, S., Kasai, Y., and Yoshimura, K. Quantitative structure-toxicity activity relationship of fatty acids and the sodium salts to aquatic organisms, *Chemosphere*, 18(7/8):1621-1631, 1989.

Opperhuizen, A., Gobas, F.A.P.C., Van der Steen, J.M.D., and Hutzinger, O. Aqueous solubility of polychlorinated biphenyls related to molecular structure, *Environ. Sci. Technol.*, 22(6):638-646, 1988a.

Opperhuizen, A., Serné, P., and Van der Steen, J.M.D. Thermodynamics of fish/water and octan-1-ol/water partitioning of some chlorinated benzenes, *Environ. Sci. Technol.*, 22(3):286-298, 1988.

O'Reilly, K.T. and Crawford, R.L. Kinetics of *p*-cresol degradation by an immobilized *Pseudomonas* sp., *Appl. Environ. Microbiol.*, 55(4):866-870, 1989.

Orth, W.S. and Gillham, R.W. Dechlorination of trichloroethene in aqueous solution using Fe^0, *Environ. Sci. Technol.*, 30(1):66-71, 1996.

Osborn, A.G. and Douslin, D.R. Vapor-pressure relationships for 15 hydrocarbons, *J. Chem. Eng. Data*, 19(2):114-117, 1974.

Osborn, A.G. and Douslin, D.R. Vapor pressures and derived enthalpies of vaporization of some condensed-ring hydrocarbons, *J. Chem. Eng. Data*, 20(3):229-231, 1975.

Osborne, A.D., Pitts, J.N., Jr., and Darley, E.F. On the stability of acrolein toward photooxidation in the near ultraviolet, *Int. J. Air and Water Pollut.*, 6:1-3, 1962.

Osborne, D.W., Garner, C.S., Doescher, R.N., and Yost, D.M. The heat capacity, entropy, heats of fusion and vaporization and vapor pressure of fluorotrichloromethane, *J. Am. Chem. Soc.*, 63:3496-3499, 1941.

Oseid, D.M. and Smith, Jr., L.L. The effects of hydrogen cyanide on *Asellus communis* and *Gammarus pseudolimnaeus* and changes in their competitive response when exposed simultaneously, *Bull. Environ. Contam. Toxicol.*,

21(4/5):439-447, 1979.

OSHA. 1976. Criteria for a recommended standard. . .Occupational exposure to phenol (Cincinnati, OH: National Institute for Occupational Safety and Health), 167 p.

OSHA. 1982. General Industry Standards for Toxic and Hazardous Substances, U.S. Code of Federal Regulations 1910, Subpart Z, Section 1910.1000.

O'Sullivan, D.W., Lee, M., Noone, B.C., and Heikes, B.G. Henry's law constant determinations for hydrogen peroxide, methyl hydroperoxide, hydroxymethyl hydroperoxide, ethyl hydroperoxide, and peroxyacetic acid, *J. Phys. Chem.*, 100(8):3241-3247, 1996.

Othmer, D.F., Bergen, W.S., Shlechter, N., and Bruins, P.F. Liquid-liquid extraction data. Systems used in butadiene manufacture from butylene glycol, *Ind. Eng. Chem.*, 37(9):890-894, 1945.

Ou, L.-T. 2,4-D degradation and 2,4-D degrading microoorganisms in soils, *Soil Sci.*, 137(2):100-1107, 1984.

Ou, L.-T., Gancarz, D.H., Wheeler, W.B., Rao, P.S.C., and Davidson, J.M. Influence of soil temperature and soil moisture on degradation and metabolism of carbofuran in soils, *J. Environ. Qual.*, 11(2):293-298, 1982.

Ou, L.-T., Rothwell, D.F., Wheeler, W.B., and Davidson, J.M. The effects of high 2,4-D concentrations on degradation and carbon dioxide evolutions in soils, *J. Environ. Qual.*, 7(2):241-246, 1978.

Ou, L.-T. and Street, J.J. Monomethylhydrazine degradation and its effect on carbon dioxide evolution and microbial populations in soil, *Bull. Environ. Contam. Toxicol.*, 41(3):454-460, 1988.

Oulman, C. Trace organics removal using activated carbon and polymeric adsorbents, Municipal Environmental Research Laboratory, U.S. EPA Report 600/2-81-079, 1981.

Overend, R. and Paraskevopoulos, G. Rates of OH radical reactions. 4. Reactions with methanol, ethanol, 1-propanol, and 2-propanol at 296 K, *J. Phys. Chem.*, 51:837-857, 1978.

Owens, J.W., Wasik, S.P., and DeVoe, H. Aqueous solubilities and enthalpies of solution of *n*-alkylbenzenes, *J. Chem. Eng. Data*, 31(1):47-51, 1986.

Oyler, A.R., Llukkonen, R.J., Lukasewycz, M.T., Heikkila, K.E., Cox, D.A., and Carlson, R.M. Chlorine 'disinfection' chemistry of aromatic compounds. Polynuclear aromatic hydrocarbons: rates, products, and mechanisms, *Environ. Sci. Technol.*, 17(6):334-342, 1983.

Ozek, T., Kirimer, N., and Baser, K.H.C. Composition of the essential oil of *Micromeria myrtifolia* Boiss. Et Hohen, *J. Ess. Oil Res.*, 4:79-80, 1992.

Pacor, P. Applicability of the DuPont 900 DTA apparatus in quantitative differential thermal analysis, *Anal. Chim. Acta*, 37:200-208, 1967.

Pal, D., Weber, J.B., and Overcash, M.R. Fate of polychlorinated biphenyls (PCBs) in soil plant systems, *Residue Rev.*, 74:45-98, 1980.

Palau-Casellas, A. and Hutchinson, T.H. Acute toxicity of chlorinated organic compounds to the embryos and larvae of the marine worm *Platynersis dumerilii* (Polychaete: Nereidae), *Environ. Toxicol. Water Qual.*, 13(2):149-155, 1998.

Palit, S.R. Electronic interpretations of organic chemistry. II. Interpretation of the solubility of organic compounds, *J. Phys. Chem.*, 51(3):837-857, 1947.

Palmer, J.S. Toxicity of carbamate, triazine, dichloropropionanilide, and diallylacetamide compounds in sheep, *Am. Vet. Med. Assoc. J.*, 145:917-920, 1964.

Pan, G. and Dutta, H.M. The inhibition of brain acetylcholinesterase activity of juvenile largemouth bass *Micropterus salmoides* by sublethal concentrations of diazinon, *Environ. Res.*, 79(2):133-137, 1998.

Pancirov, R.J. and Brown, R.A. Analytical methods for polynuclear aromatic hydrocarbonsin crude oils, heating oils and marine tissues, in *Prceedings of Joint Conference on Prevention and Control of Oil Pollution*, American Petroleum Institute, pp. 103-115, 1975.

Pankow, J.F. and Rosen, M.E. Determination of volatile compounds in water by purging directly to a capillary column with whole column cryotrapping, *Environ. Sci. Technol.*, 22(4):398-405, 1988.

Pantani, C., Pannunzio, G., De Cristofaro, M., Novelli, A.A., and Salvatori, M. Comparative acute toxicity of some pesticides, metals, and surfactants to *Gammarus italicus* Goedm. and *Echinogammarus tibaldii* pink and stock (Crustacea: amphipoda), *Bull. Environ. Contam. Toxicol.*, 59(6):963-967, 1997.

Paris, D.F. and Lewis, D.L. Chemical and microbial degradation of ten selected pesticides in aquatic systems, *Residue Rev.*, 45:95-124, 1973.

Paris, D.F., Lewis, D.L., and Barnett, J.T. Bioconcentration of toxaphene by microorganisms, *Bull. Environ. Contam. Toxicol.*, 17(5):564-572, 1977.

Paris, D.F., Steen, W.C., and Baughman, G.L. Role of the physico-chemical properties of Aroclors 1016 and 1242 in determining their fate and transport in aquatic environments, *Chemosphere*, 7(4):319-325, 1978.

Paris, D.F. and Wolfe, N.L. Relationship between properties of a series of anilines and their transformation by bacteria, *Appl. Environ. Microbiol.*, 53(5):911-916, 1987.

Park, J.-Y. and Lee, Y.-N. Solubility and decomposition kinetics of nitrous acid in aqueous solution, *J. Phys. Chem.*, 92:6294-6302, 1988.

Park, K.S. and Bruce, W.N. The determination of the water solubility of aldrin, dieldrin, heptachlor, and heptachlor epoxide, *J. Econ. Entomol.*, 61(3):770-774, 1968.

Park, K.S., Sims, R.C., Dupont, R.R., Doucette, W.J., and Matthews, J.E. Fate of PAH compounds in two soil types: influence of volatilization, abiotic loss and biological activity, *Environ. Toxicol. Chem.*, 9:187-195, 1990.

Parks, G.S. and Huffman, H.M. Thermal data on organic compounds. IX. A study on the effect of unsaturation on the heat capacities, entropies and free energies of some hydrocarbons and other compounds, *J. Am. Chem. Soc.*, 52:4381-4391, 1930.

Parks, G.S. and Huffman, H.M. Some fusion and transition data for hydrocarbons, *Ind. Eng. Chem.*, 23(10):1138-1139, 1931.

Parks, G.S., Huffman, H.M., and Thomas, S.B. Thermal data on organic compounds. VI. The heat capacities, entropies and free energies of some saturated non-benzenoid hydrocarbons, *J. Am. Chem. Soc.*, 52:1032-1041, 1930.

Parks, G.S., Shomate, C.H., Kennedy, W.D., and Crawford, B.L., Jr. The entropies of *n*-butane and isobutane, with some heat capacity data for isobutane, *J. Chem. Phys.*, 5:359-363, 1937.

Parks, G.S., Todd, S.S., and Moore, W.A. Thermal data on organic compounds. XVI. Some heat capacity, entropy and free energy data for typical benzene derivatives and heterocyclic compounds, *J. Am. Chem. Soc.*, 58:398-401, 1936.

Parks, G.S., Todd, S.S., and Shomate, C.H. Thermal data on organic compounds. XVII. Some heat capacity, entropy and free energy data for five higher olefins, *J. Am. Chem. Soc.*, 58:2505-2508, 1936a.

Parlar, H. Photoinduced reactions of two toxaphene compounds in aqueous medium and adsorbed on silica gel, *Chemosphere*, 17(11):2141-2150, 1988.

Parr, J.F. and Smith, S. Degradation of trifluralin under laboratory conditions and soil anaerobiosis, *Soil Sci.*, 115(1):55-63, 1973.

Parr, J.F. and Smith, S. Degradation of DDT in an Everglades muck as affected by lime, ferrous ion, and anaerobiosis, *Soil Sci.*, 118(1):45-52, 1974.

Parris, G.E. Covalent binding of aromatic amines to humates. 1. Reactions with carbonyls and quinones, *Environ. Sci. Technol.*, 14(9):1099-1106, 1980.

Parsons, F. and Lage, G.B. Chlorinated organics in simulated groundwater environments, *J. Am. Water Works Assoc.*, 75(5):52-59, 1985.

Parsons, G.H., Rochester, C.H., Rostron, A., and Sykes, P.C. The thermodynamics of hydration of phenols, *J. Chem. Soc., Perkin Trans. 2*, pp. 136-138, 1982.

Parsons, G.H., Rochester, C.H., and Wood, C.C. Effect of 4-substitution on the thermodynamics of hydration of phenol and the phenoxide anion, *J. Chem. Soc. B (London)*, pp. 533-536, 1971.

Paschke, A., Popp, P., and Schürmann, G. Water solubility and octanol/water-partitioning of hydrophobic chlorinated organic substances determined by using SPME/GC, *Fresenius J. Anal. Chem.*, 360:52-57, 1998.

Passino, D.R.M. and Smith, S.B. Acute bioassays and hazard evaluation of representative contaminants detected in Great Lakes fish, *Environ. Toxicol. Chem.*, 6:901-907, 1987.

Passino-Reader, D.R., Hickey, J.P., and Ogilvie, L.M. Toxicity to *Daphnia pulex* and QSAR predictions for polycyclic hydrocarbons representative of Great Lakes contaminants, *Bull. Environ. Contam. Toxicol.*, 59(5):834-840, 1997.

Patil, K.C. and Matsumura, F. Degradation of endrin, aldrin, and DDT by soil microorganisms, *Appl. Microbiol.*, 19:879-881, 1970.

Patil, K.C., Matsumura, F., and Boush, G.M. Metabolic transformation of DDT, dieldrin, aldrin, and endrin by marine microorganisms, *Environ. Sci. Technol.*, 6(7):629-632, 1972.

Patnaik, P. *A Comprehensive Guide to the Hazardous Properties of Chemical Substances* (New York: Van Nostrand Reinhold, 1992), 763 p.

Paul, M.A. The solubilities of naphthalene and biphenyl in aqueous solutions of electrolytes, *J. Am. Chem. Soc.*, 74(21):5274-5277, 1952.

Paulson, S.E. and Seinfeld, J.H. Atmospheric photochemical oxidation of 1-octene: OH, O_3, and $O(^3P)$ reactions, *Environ. Sci. Technol.*, 26(6):1165-1173, 1992.

Pavlostathis, S.G. and Mathavan, G.N. Desorption kinetics of selected volatile organic compounds from field contaminated soils, *Environ. Sci. Technol.*, 26(3):532-538, 1992.

Pawar, S.S. and Fawade, M.M. Alterations in the toxicity of thiodemeton due to the pretreatment of inducers, substrates and inhibitors of mixed function oxidase system, *Bull. Environ. Contam. Toxicol.*, 20(6):805-810, 1978.

Paya-Perez, A.B., Riaz, M., and Larsen, B.R. Soil sorption of 20 PCB congeners and six chlorobenzenes, *Ecotoxicol. Environ. Saf.*, 21(1):1-17, 1991.

Pearlman, R.S., Yalkowsky, S.H., and Banerjee, S. Water solubilities of polynuclear aromatic and heteroaromatic compounds, *J. Phys. Chem. Ref. Data*, 13(2):555-562, 1984.

Pearson, C.R. and McConnell, G. Chlorinated C_1 and C_2 hydrocarbons in the marine environment, *Proc. R. Soc. London*, B189(1096):305-332, 1975.

Peel, R.G. and Benedek, A. Attainment of equilibrium in activated carbon isotherm studies, *Environ. Sci. Technol.*, 14(1):66-71, 1980.

Peng, J. and Wan, A. Effect of ionic strength on Henry's constants of volatile organic compounds, *Chemosphere*, 36(13):2731-2740, 1998.

Pereira, W.E., Rostad, C.E., Chiou, C.T., Brinton, T.I., Barber, II, L.B., Demcheck, D.K., and Demas, C.R. Contamination of estuarine water, biota and sediment by halogenated organic compounds: a field study, *Environ. Sci. Technol.*, 22(7):772-778, 1988.

Pereira, W.E., Short, D.L., Manigold, D.B., and Roscio, P.K. Isolation and characterization of TNT and its metabolites in groundwater by gas chromatograph-mass spectrometer-computer techniques, *Bull. Environ. Contam. Toxicol.*, 21(4/5):554-562, 1979.

Pérez-Tejeda, P., Yanes, C., and Maestre, A. Solubility of naphthalene in water + alcohol solutions at various temperatures, *J. Chem. Eng. Data*, 35(3):244-246, 1990.

Peris-Cardells, E., Terol, J., Mauri, A.R., de la Guardia, M., and Pramauro, E. Continuous flow photocatalytic degradation of carbaryl in aqueous media, *J. Environ. Sci. Health*, B28(4):431-445, 1993.

Perlinger, J.A., Eisenreich, S.J., and Capel, P.D. Application of headspace analysis to the study of sorption of hydrophobic organic compounds to α-Al_2O_3, *Environ. Sci. Technol.*, 27(5):928-937, 1993.

Perrin, D.D. *Dissociation Constants of Organic Bases in Aqueous Solutions*, IUPAC Chemical Data Series; Supplement (London: Butterworth, 1972).

Perry, R.A., Atkinson, R., and Pitts, J.N., Jr. Rate constants for the reaction of OH radicals with CH_2=CHF, CH_2=CHCl, and CH$_2$=CHBr over the temperature range 299-426 °K, *J. Chem. Phys.*, 67:458-462, 1977.

Perscheid, M., Schlueter, H., and Ballschmiter, K. Aerober Abbau von Endosulfan durch Bodenmikroorganismen, *Z.. Naturforsch.*, 28:761-763, 1973.

Peterson, M.S., Lion, L.W., and Shoemaker C.A. Influence of vapor-phase sorption and dilution on the fate of trichloroethylene in an unsaturated aquifer system, *Environ. Sci. Technol.*, 22(5):571-578, 1988.

Petrasek, A.C., Kugelman, I.J., Austern, B.M., Pressley, T.A., Winslow, L.A., and Wise, R.H. Fate of toxic organic compounds in wastewater treatment plants, *J. Water Pollut. Control Fed.*, 55(10):1286-1296, 1983.

Petrich, S.M. and Reish, D.J. Effects of aluminum and nickel on survival and reproduction in polychaetous annelids, *Bull. Environ. Contam. Toxicol.*, 23(4/5):698-702, 1979.

Petrier, C., Micolle, M., Merlin, G., Luche, J.-L., and Reverdy, G. Characteristics of pentachlorophenate degradation in aqueous solution by means of ultrasound, *Environ. Sci. Technol.*, 26(8):1639-1642, 1992.

Petriris, V.E. and Geankopolis, C.J. Phase equilibrium in 1-butanol-water-lactic acid system, *J. Chem. Eng. Data*, 4:197-198, 1959.

Peyton, T.O., Steel, R.V., and Mabey, W.R. Carbon disulfide, carbonyl sulfide: literature review and environmental assessment, U.S. EPA Report PB-257-947/2, 1976, 64 p.

Pfaender, F.K. and Alexander, M. Extensive microbial degradation of DDT *in vitro* and DDT metabolism by natural communities, *J. Agric. Food Chem.*, 20(4):842-846, 1972.

Pfaender, F.K. and Alexander, M. Effect of nutrient additions on the apparent cometabolism of DDT, *J. Agric. Food Chem.*, 21(3):397-399, 1973.

Phatak, P.V. and Gaikar, V.G. Solubilities of *o*- and *p*-chlorobenzoic acids and *o*- and *p*-nitroanilines in *N,N*-dimethyl-formamide + water, *J. Chem. Eng. Data*, 41(5):1052-1054, 1996.

Philip, G.H., Reddy, P.M., and Sridevi, G. Cypermethrin-induced *in vivo* alterations in the carbohydrate metabolism of freshwater fish, *Labeo rohita*, *Ecotoxicol. Environ. Saf.*, 31(2):173-178, 1995.

Phillips, D.D., Pollard, G.E., and Soloway, S.B. Thermal isomerization of endrin and its behavior in gas chromatography, *J. Agric. Food Chem.*, 10(3):217-221, 1962.

Piacente, V., Scardala, P., Ferro, D., and Gigli, R. Vaporization study of *o*-, *m*-, and *p*-chloroaniline by torsion-weighing effusion vapor pressure measurements, *J. Chem. Eng. Data*, 30(4):372-376, 1985.

Piccolo, A., Celano, G., and Conte, P. Adsorption of glyphosphate by humic substances, *J. Agric. Food Chem.*, 44(8):2442-2446, 1996.

Pierce, R.H., Olney, C.E., and Felbeck, G.T. *p,p'*-DDT adsorption to suspended particulate matter in sea water, *Geochim. Cosmochim. Acta*, 38(7):1061-1073, 1974.

Pierotti, C., Deal, C., and Derr, E. Activity coefficient and molecular structure, *Ind. Chem. Eng. Fundam.*, 51:95-101, 1959.

Pignatello, J.J. Microbial degradation of 1,2-dibromoethane in shallow aquifer materials, *J. Environ. Qual.*, 16(4):307-312, 1987.

Pignatello, J.J. Slowly reversible sorption of aliphatic halocarbons in soils. I. Formation of residual fractions, *Environ. Toxicol. Chem.*, 9:1107-1115, 1990.

Pilipiuk, Z.I., Gorban, G.M., Solomin, G.I., and Gorshunova, A.I. Toxicity of 1,4-dioxane, *Kosm. Biol. Aviakosm. Med.*, 11(6):53-57, 1977.

Pillai, P., Helling, C.S., and Dragun, J. Soil-catalyzed oxidation of aniline, *Chemosphere*, 11(3):299-317, 1982.

Pinal, R., Lee, L.S., Rao, P.S.C. Prediction of the solubility of hydrophobic compounds in nonideal solvent mixtures, *Chemosphere*, 22(9/10):939-951, 1991.

Pinal, R., Rao, P.S.C., Lee, L.S., Cline, P.V., and Yalkowsky, S.H. Cosolvency of partially miscible organic solvents on the solubility of hydrophobic organic chemicals, *Environ. Sci. Technol.*, 24(5):639-647, 1990.

Pirbazari, M. and Weber, W.J., Jr. Removal of dieldrin by activated carbon, *J. Environ. Eng. Div. Am. Soc. Civil Eng.*, 110(3):656-669, 1984.

Pitter, P. Determination of biological degradability of organic substances, *Water Res.*, 10(3):231-235, 1976.

Pitter, P. and Chudoba, J. *Biodegradability of Organic Substances in the Aquatic Environment* (Boca Raton, FL: CRC Press, 1990), 306 p.

Pitts, J.N., Jr., Atkinson, R., Sweetman, J.A., and Zielinska, B. The gas-phase reaction of naphthalene with N$_2$O$_5$ to form nitronaphthalenes, *Atmos. Environ.*, 19(5):701-705, 1985.

Pitts, J.N., Jr., Grosjean, D., Cauwenberghe, K.V., Schmid, J.P., and Fitz, D.R. Photooxidation of aliphatic amines under simulated atmospheric conditions: formation of nitrosamines, nitramines, amides, and photochemical oxidant, *Environ. Sci. Technol.*, 12(8):946-953, 1978.

Pitts, J.N., Jr., Lokensgard, D.M., Ripley, P.S., Van Cauwenberghe, K.A., Van Vaeck, L., Schaffer, S.D., Thill, A.J., and Belser, Jr., W.L. 'Atmospheric' epoxidation of benzo[a]pyrene by ozone: formation of the metabolite benzo[a]py-rene-4,5-oxide, *Science (Washington, D.C.)*, 210(4476):1347-1349, 1980.

Pitts, J.N., Jr., Zielinska, B., Sweetman, J.A., Atkinson, R., and Winer, A.M. Reactions of adsorbed pyrene and perylene with gaseous N$_2$O$_5$ under simulated atmospheric conditions, *Atmos. Environ.*, 19(6):911-915, 1985.

Pitzer, K.S., Guttman, L., and Westrum, E.F., Jr. The heat capacity, heats of fusion and vaporization, vapor pressure, entropy vibration frequencies and barrier to internal rotation of styrene, *J. Am. Chem. Soc.*, 68:2209-2212, 1946.

Pitzer, K.S. and Scott, D.W. The thermodynamics of branched-chain paraffins. The heat capacity, heat of fusion and vaporization, and entropy of 2,3,4-trimethylpentane, *J. Am. Chem. Soc.*, 63:2419-2422, 1941.

Pitzer, K.S. and Scott, D.W. The thermodynamics and molecular structure of benzene and its methyl derivatives, *J. Am. Chem. Soc.*, 65:803-829, 1943.

Piwoni, M.D. and Banerjee, P. Sorption of volatile organic solvents from aqueous solution onto subsurface solids, *J. Contam. Hydrol.*, 4(2):163-179, 1989.

Plastourgou, M. and Hoffman, M.R. Transformation and fate of organic esters in layered-flow systems: the role of trace metal catalysis, *Environ. Sci. Technol.*, 18(10):756-764, 1984.

Platford, R.F., Carey, J.H., and Hale, E.J. The environmental significance of surface films: Part 1. Octanol-water partition coefficients for DDT and hexachlorobenzene, *Environ. Pollut. (Series B)*, 3(2):125-128, 1982.

Plimmer, J.R. The photochemistry of halogenated herbicides, *Residue Rev.*, 33:47-74, 1970.

Plimmer, J.R. Photolysis of TCDD and trifluralin on silica and soil, *Bull. Environ. Contam. Toxicol.*, 20(1):87-92, 1978.

Plimmer, J.R., Kearney, P.C., and Von Endt, D.W. Mechanism of conversion of DDT to DDD by *Aerobacter aerogenes*, *J. Agric. Food Chem.*, 16(4):594-597, 1968.

Plimmer, J.R. and Klingebiel, U.J. Photolysis of hexachlorobenzene, *J. Agric. Food Chem.*, 24(4):721-723, 1976.

Plimmer, J.R., Klingbiel, U.J., Crosby, D.G., and Wong, A.S. Photochemistry of dibenzo-*p*-dioxins, *Adv. Chem. Ser.*, 120:82-92, 1973.

Plimmer, J.R., Klingbiel, U.J., and Hummer, B.E. Photooxidation of DDT and DDE, *Science (Washington, D.C.)*, 167(3914):67-69, 1970.

Plowman, M.C., Peracha, H., Hopfer, S.M., and Sunderman, F.W., Jr. Teratogenicity of cobalt chloride in *Xenopus laevis*, assayed by the FETAX procedure, *Teratog. Carcinog. Mutagen*, 11(2):83-92, 1991.

Podoll, R.T., Irwin, K.C., and Parish, H.J. Dynamic studies of naphthalene sorption on soil from aqueous solution, *Chemosphere*, 18(11/12):2399-2412, 1989.

Podoll, R.T., Jaber, H.M., and Mill, T. Tetrachlorodioxin: Rates of volatilization and photolysis in the environment, *Environ. Sci. Technol.*, 20(5):490-492, 1986.

Poeti, G., Fanelli, E., and Braghetti, M. A differential scanning calorimetric study of some phenol derivatives, *J. Therm. Anal.*, 24(2):273-279, 1982.

Pogány, E., Wallnöfer, P.R., Ziegler, W., and Mücke, W. Metabolism of *o*-nitroaniline and di-*n*-butyl phthalate in cell suspension cultures of tomatoes, *Chemosphere*, 21(4/5):557-562, 1990.

Polak, J. and Lu, B.C.-Y. Mutual solubilities of hydrocarbons and water at 0 and 25 °C, *Can. J. Chem.*, 51(24):4018-4023, 1973.

Poleksic, V. and Karan, V. Effects of trifluralin on carp: biochemical and histological evaluation, *Ecotoxicol. Environ. Saf.*, 43(2):213-221, 1999.

Politzki, G.R., Bieniek, D., Lahaniatis, E.S., Scheunert, I., Klein, I., and Korte, F. Determination of vapor pressures of nine organic chemicals on silica gel, *Chemosphere*, 11(12):1217-1229, 1982.

Pollero, R. and dePollero, S.C. Degradation of DDT by a soil amoeba, *Bull. Environ. Contam. Toxicol.*, 19(3):345-350, 1978.

Polles, S.G. and Vinson, S.B. Effect of droplet size on persistence of malathion and comparison of toxicity of ULF and EC malathion to tobacco budworm larvae, *J. Econ. Entomol.*, 62:89-94, 1969.

Pollino, C.A. and Holdway, D.A. Potential of two hydra species as standard toxicity test animals, *Ecotoxicol. Environ. Saf.*, 43(3):309-316, 1999.

Pollock, G.A. and Kilgore, W.W. Toxaphene, *Residue Rev.*, 69:87-140, 1978.

Potera, G.T. The effects of benzene, toluene, and ethylbenzene on several important members of the estuarine ecosystem, Ph.D. Thesis, Lehigh University, Bethlehem, PA, 1975.

Pothuluri, J.V., Freeman, J.P., Evans, F.E., and Cerniglia, C.E. Fungal metabolism of acenaphthene by *Cunninghamella elegans*, *Appl. Microbiol.*, 58(11):3654-3659, 1992.

Potter, T.L. Analysis of petroleum-contaminated water by GC/FID with direct aqueous injection, *Groundwater Monitor. Remed.*, 16(2):157-162, 1996.

Poulsen, M., Lemon, L., and Barker, J.F. Dissolution of monoaromatic hydrocarbons into groundwater from gasoline-oxygenate mixtures, *Environ. Sci. Technol.*, 26(12):2483-2489, 1992.

Poluson, S.R., Harrington, R.R., and Drever, J.I. The solubility of toluene in aqueous salt solutions, *Talanta*, 48:633-641, 1999.

Pozzani, U.C., Weil, C.S., and Carpenter, C.P. The toxicological basis for threshold limit values. 5. The experimental inhalation of vapor mixtures by rats, with notes upon the relationship between single dose inhalation and single dose oral data, *Am. Ind. Hyg. Assoc. J.*, 20:364-369, 1959.

Price, L.C. Aqueous solubility of petroleum as applied to its origin and primary migration, *Am. Assoc. Pet. Geol. Bull.*, 60(2):213-244, 1976.

Price, K.S., Waggy, G.T., and Conway, R.A. Brine shrimp bioassay and seawater BOD of petrochemicals, *J. Water Pollut. Control Fed.*, 46(1):63-77, 1974.

Prior, M.G., Sharma, A.K., Yong, S., and Lopez, A. Concentration-time interactions in hydrogen sulphide toxicity in rats, *Can. J. Vet. Res.*, 52(3):375-379, 1988.

Probst, G.W., Golab, T., Herberg, R.J., Holzer, F.J., Parka, S.J., van der Schans, C., and Tepe, J.B. Fate of trifluralin in soils and plants, *J. Agric. Food Chem.*, 15(4):592-599, 1967.

Probst, G.W. and Tepe, J.B. Trifluralin and related compounds, in *Degradation of Herbicides*, Kearney, P.C. and Kaufman, D.D., Eds. (New York: Marcel Dekker, 1969), pp. 255-282.

Pruden, A.L. and Ollis, D.F. Degradation of chloroform by photoassisted heterogeneous catalysis in dilute aqueous suspensions of titanium dioxide, *Environ. Sci. Technol.*, 17(10):628-631, 1983.

Przyjazny, A., Janicki, W. Chrzanowski, and Staszewski, R. Headspace gas chromatographic determination of distribution coefficients of selected organosulphur compounds and their dependence of some parameters, *J. Chromatogr.*, 280:249-260, 1983.

Punzo, F. The effects of azadirachtin on larval stages of the oak toad, *Bufo quercicus* (Holbrook), *Florida Scientist*, 60(3):158-165, 1997.

Pupp, C., Lao, R.C., Murray, J.J., and Pottie, R.F. Equilibrium vapor concentrations of some polycyclic hydrocarbons, arsenic trioxide (As$_4$O$_6$) and selenium dioxide and the collection efficiencies of these air pollutants, *Atmos. Environ.*, 8:915-925, 1974.

Pusino, A., Petretto, S., and Gessa, C. Adsorption and desorption of imazapyr by soil, *J. Agric. Food Chem.*, 45(3):1012-1016, 1997.

Pussemier, L., Szabo, G., and Bulman, R.A. Prediction of the soil adsorption coefficient K$_{oc}$ for aromatic pollutants, *Chemosphere*, 21(10-11):1199-1212, 1990.

Putnam, T.B., Bills, D.D., and Libbey, L.M. Identification of endosulfan based on the products of laboratory photolysis, *Bull. Environ. Contam. Toxicol.*, 13(6):662-665, 1975.

Que Hee, S.S. and Sutherland, R.G. *The Phenoxyalkanoic Herbicides. Vol. 1. Chemistry, Analysis, and Environmental Pollution* (Boca Raton, FL: CRC Press, Inc, 1981), 321 p.

Que Hee, S.S., Sutherland, R.G., McKinlay, K.S., and Sara, J.G. Factors affecting the volatility of DDT, dieldrin, and dimethylamine salt of (2,4-dichlorophenoxy)acetic acid (2,4-D) from leaf and glass surfaces, *Bull. Environ. Contam. Toxicol.*, 13(3):284-290, 1975.

Quirke, J.M.E., Marei, A.S.M., and Eglinton, G. The degradation of DDT and its degradative products by reduced iron (III) porphyrins and ammonia, *Chemosphere*, 8(3):151-155, 1979.

Qureshi, A.A., Flood, K.W., Thompson, S.R., Janhurst, S.M., Inniss, C.S., and Rokosh, D.A. Comparison of a luminescent bacterial test with other bioassays for determining toxicity of pure compounds and complex effluents, in *Aquatic Toxicology and Hazard Assessment*, Pearson, J.G., Foster, R.B., and Bishop, W.E., Eds. (Philadelphia, PA: American Society for Testing and Materials, 1982), pp. 179-195.

Rabinovich, L.B., Pet'kov, V.I., Zarudaeva, S.S., and Ovchinnikov, E.Yu. Specific-heat and thermodynamic functions of triphenyl phosphate at 12-340 K, *Zhur. Fiz. Khim.*, 60:767-769, 1986.

Racke, K.D., Coats, J.R., and Titus, K.R. Degradation of chlorpyrifos and its hydrolysis product, 3,5,6-trichloro-2-pyridinol, in soil, *J. Environ. Sci. Health*, B23(6):527-539, 1988.

Radchenko, L.G. and Kitiagorodskii, A.I. Vapor pressure and heat of sublimation of naphthalene, biphenyl, octafluoro-naphthalene, decafluorobiphenyl, acenaphthene and α-nitronaphthalene, *Zhur. Fiz. Khim.*, 48:2702-2704, 1974.

Radding, S.B., Mill, T., Gould, C.W., Lia, D.H., Johnson, H.L., Bomberger, D.S., and Fojo, C.V. The environmental fate of selected polynuclear aromatic hydrocarbons, Office of Toxic Substances, U.S. EPA Report-560/5-75-009, 1976, 122 p.

Radosevich, S.R. and Winterlin, W.L. Persistence of 2,4-D and 2,4,5-T in chaparral vegetation and soil, *Weed Sci.*, 25:423-425, 1977.

Raghu, K. and MacRae, I.C. Biodegradation of the gamma isomer of benzene hexachloride in submerged soils, *Science (Washington, D.C.)*, 154(3746):263-264, 1966.

Raha, P. and Das, A.K. Photodegradation of carbofuran, *Chemosphere*, 21(1/2):99-106, 1990.

Rai, U.S. and Mandal, K.D. Chemistry of organic eutectics: *p*-phenylenediamine-*m*-nitrobenzoic acid system involving the 1:2 addition compound, *Bull. Chem. Soc. Jpn.*, 63(5):1496-1502, 1990.

Rai, U.S., Singh, O.P., and Singh, N.B. Some thermodynamic aspects of organic eutectics, succinonitrile-phenanthrene system, *Indian J. Chem.*, 26A:949-949, 1987.

Rajagopal, B.S., Brahmaprakash, G.P., Reddy, B.R., Singh, U.D., and Sethunathan, N. Effect and persistence of selected carbamate pesticides in soil, *Residue Rev.*, 93:1-199, 1984.

Rajagopal, B.S., Chendrayan, K., Reddy, B.R., and Sethunathan, N. Persistence of carbaryl in flooded soils and its degradation by soil enrichment cultures, *Plant Soil*, 73(1):35-45, 1983.

Rajagopal, B.S., Panda, S., and Sethunathan, N. Accelerated degradation of carbaryl and carbofuran in a flooded soil pretreated with hydrolysis products, 1-naphthol and carbofuran phenol, *Bull. Environ. Contam. Toxicol.*, 36(6):827-832, 1986.

Rajagopal, B.S., Rao, V.R., Nagendrappa, G., and Sethunathan, N. Metabolism of carbaryl and carbofuran by soil-enrichment and bacterial cultures, *Can. J. Microbiol.*, 30(12):1458-1466, 1984a.

Rajagopal, V.K. and Burris, D.R. Reduction of 1,2-dibromoethane in the presence of zero-valent iron, *Environ. Toxicol. Chem.*, 18(8):1779-1782, 1999.

Rajagopalan, R., Vohra, K.G., and Mohan Rao, A.M. Studies on oxidation of benzo[*a*]pyrene by sunlight and ozone, *Sci. Total Environ.*, 27(1):33-42, 1983.

Rajaram, K.P. and Sethunathan, N. Effect of organic sources on the degradation of parathion in flooded alluvial soil, *Soil Sci.*, 119(4):296-300, 1975.

Rajini, P.S., Krishnakumari, M.K., and Majumder, S.K. Cytotoxicity of certain organic solvents and organophosphorus insecticides to the ciliated protozoan *Paramecium caudatum*, *Microbios*, 59(240-241):157-163, 1989.

Ralston, A.W. and Hoerr, C.W. The solubilities of the normal saturated fatty acids, *J. Org. Chem.*, 7:546-555, 1942.

Ramachandran, B.R., Allen, J.M., and Halpern, A.M. Air-water partitioning of environmentally important organic compounds, *J. Chem. Educ.*, 73(11):1058-1061, 1996.

Ramachandran, S., Patel, T.R., and Colbo, M.H. Effect of copper and cadmium on three Malaysian tropical estuarine invertebrate larvae, *Ecotoxicol. Environ. Saf.*, 36(2):183-188, 1997.

Ramamoorthy, S. Competition of fate processes in the bioconcentration of lindane, *Bull. Environ. Contam. Toxicol.*, 34(3):349-358, 1985.

Ramanand, K., Panda, S., Sharmila, M., Adhya, T.K., and Sethunathan, N. Development and acclimatization of carbofuran-degrading soil enrichment cultures at different temperatures, *J. Agric. Food Chem.*, 36(1):200-205, 1988.

Ramanand, K., Sharmila, M., and Sethunathan, N. Mineralization of carbofuran by a soil bacterium, *Appl. Environ. Microbiol.*, 54(8):2129-2133, 1988a.

Ramdahl, T., Becher, G., and Bjørseth, A. Nitrated polycyclic aromatic hydrocarbons in urban air particles, *Environ. Sci. Technol.*, 16(12):861-865, 1982.

Randall, T.L. and Knopp, P.V. Detoxification of specific organic substances by wet oxidation, *J. Water Pollut. Control Fed.*, 52(8):2117-2130, 1980.

Randall, W.F., Dennis, W.H., and Warner, M.C. Acute toxicity of dechlorinated DDT, chlordane and lindane to bluegill (*Lepomis machrochirus*) and *Daphnia magna*, *Bull. Environ. Contam. Toxicol.*, 21(6):849-854, 1979.

Rao, K.R., Rao, K.S., Sahib, I.K., and Rao, K.V. Combined action of carbaryl and phenthoate on a freshwater fish (*Channa punctatus* bloch), *Ecotoxicol. Environ. Saf.*, 10(2):209-217, 1985.

Rao, P.S.C. and Davidson, J.M. Adsorption and movement of selected pesticides at high concentrations in soils, *Water Res.*, 13(4):375-380, 1979.

Rao, P.S.C. and Davidson, J.M. Estimation of pesticide retention and transformation parameters required in nonpoint source pollution models, in *Environmental Impact of Nonpoint Source Pollution*, Overcash, M.R. and Davidson, J.M., Eds. (Ann Arbor, MI: Ann Arbor Science Publishers, 1980), pp. 23-67.

Rao, P.S.C. and Sethunathan, N. Effect of ferrous sulfate on parathion degradation in flooded soil, *J. Environ. Sci. Health*, B14(3):335-351, 1979.

Rapaport, R.A. and Eisenreich, S.J. Chromatographic determination of octanol-water partition coefficients (K_{ow}'s) for 58 polychlorinated biphenyl congeners, *Environ. Sci. Technol.*, 18(3):163-170, 1984.

Rappe, C., Choudhary, G., and Keith, L.H. *Chlorinated Dioxins and Dibenzofurans in Perspective* (Chelsea, MI: Lewis Publishers, 1987), 570 p.

Rastogi, R.P. and Bassi, P.S. Mechanism of eutectic crystallization, *J. Phys. Chem.*, 68:2398-2406, 1964.

Rav-Acha, C. and Choshen, E. Aqueous reactions of chlorine dioxide with hydrocarbons, *Environ. Sci. Technol.*, 21(11):1069-1074, 1987.

Ravishankara, A.R. and Davis, D.D. Kinetic rate constants for the reaction of OH with methanol, ethanol, and tetrahydrofuran at 298 K, *J. Phys. Chem.*, 82:2852-2853, 1978.

Rayburn, J.R. and Fisher, W.S. Development toxicity of three carrier solvents using embryos of the grass shrimp, *Palaemonetes pugio*, *Arch. Environ. Contam.Toxicol.*, 33(2):217-221, 1997.

Rayms-Keller, A., Olson, K.E., McGaw, M., Oray, C., Carlson, J.O., and Beaty, B.J. Effects of heavy metals on *Aedes aegypti* (Diptera: Culcidae) larvae, *Ecotoxicol. Environ. Saf.*, 39(1):41-47, 1998.

Razo-Flores, E., Lettinga, G., and Field, J.A. Biotransformation and biodegradation of selected nitroaromatics under anaerobic conditions, *Biotechnol. Prog.*, 15(3):358-365, 1999.

Reddy, B.R. and Sethunathan, N. Mineralization of parathion in the rice rhizophere, *Appl. Environ. Microbiol.*, 45(3):826-829, 1983.

Reddy, K.N., Singh, M., and Alva, A.K. Sorption and leaching of bromacil and simazine in Florida flatwoods soils, *Bull. Environ. Contam. Toxicol.*, 48(5):662-670, 1992.

Reducker, S., Uchrin, C.G., and Winnett, G. Characteristics of the sorption of chlorothalonil and azinphos-methyl to a soil from a commercial cranberry bog, *Bull. Environ. Contam. Toxicol.*, 41(5):633-641, 1988.

Reed, C.D. and McKetta, J.J. Solubility of 1,3-butadiene in water, *J. Chem. Eng. Data*, 4(4):294-295, 1959.

Rehberg, C.E. and Dixon, M.B. *n*-Alkyl lactates and their acetates, *J. Am. Chem. Soc.*, 72(5):1918-1922, 1950.

Reinbold, K.A., Hassett, J.J., Means, J.C., and Banwart, W.L. Adsorption of energy-related organic pollutants: a literature review, Office of Research and Development, U.S. EPA Report-600/3-79-086, 1979, 180 p.

Reinert, K.H. and Rodgers, J.H., Jr. Influence of sediment types on the sorption of endothall, *Bull. Environ. Contam. Toxicol.*, 32(5):557-564, 1984.

Reinert, K.H. and Rodgers, J.H. Fate and persistence of aquatic herbicides, *Rev. Environ. Contam. Toxicol.*, 98:61-98, 1987.

Reish, D.J. and Kauwling, T.J. Marine and estuarine pollution, *J. Water Pollut. Control Fed.*, 50:1424-1469, 1978.

Renberg, L.O., Sundström, S.G., Rosén-Olofsson, A.-C. The determination of partition coefficients of organic compounds in technical products and waste waters for the estimation of their bioaccumulation potential using reversed phase thin layer chromatography, *Toxicol. Environ. Chem.*, 10:333-349, 1985.

Renberg, L., Tarkpea, M., and Lindén, E. The use of the bivalve *Mytilus edulis* as a test organism for bioconcentration studies, *Ecotoxicol. Environ. Saf.*, 9(2):171-178, 1982.

Renner, G. Gas chromatographic studies of chlorinated phenols, chlorinated anisoles, and chlorinated phenylacetates, *Toxicol. Environ. Chem.*, 27:217-224, 1990.

Renner, G., Hopfer, C., and Gokel, J.M. Acute toxicities of pentachlorophenol, pentachloroanisole, tetrachlorohydro-quinone, tetrachlorocatechol, tetrachlororesorcinol, tetrachlorodimethoxybenzenes, and tetrachlorodibenzenediol diacetates administered to mice, *Toxicol. Environ. Chem.*, 11:37-50, 1986.

Reshetyuk, A.I., Talakina, E.I., and En'yakova, P.A. Toxicological evaluation of acenaphthene and acenaphthylene, *Gig. Tr. Prof. Zabol.*, 14:46-47, 1970.

Reuber, M.D. Carcinogenicity of lindane, *Environ. Res.*, 19:460-491, 1979.

Rex, A. Über die Löslichkeit der Halogenderivate der Kohlenwasserstoffe in Wasser, *Z. Phys. Chem.*, 55:355-370, 1906.

Rhodes, R.C. Metabolism of [2-^{14}C]terbacil in alfalfa, *J. Agric. Food Chem.*, 25(5):1066-1068, 1977.

Rice, C.P., Chernyak, S.M., Hapeman, C.J., and Bilboulian, S. Air-water distribution of the endosulfan isomers, *J. Environ. Qual.*, 26(4):1101-1106, 1997.

Rice, C.P., S.M. Chernyak, and McConnell, L.L. Henry's law constants for pesticides measured as a function of temperature and salinity, *J. Agric. Food Chem.*, 45(6):2291-2298, 1997a.

Rice, F.O. and Murphy, M.T. The thermal decomposition of five-membered rings, *J. Am. Chem. Soc.*, 64(4):896-899, 1942.

Richard, Y. and Bréner, L. Removal of pesticides from drinking water by ozone, in *Handbook of Ozone Technology and Applications, Volume II. Ozone for Drinking Water Treatment*, Rice, A.G. and Netzer, A., Eds. (Montvale, MA:

Butterworth Publishers, 1984), pp. 77-97.

Richardson, G.M., Qadri, S.U., and Jessiman, B. Acute toxicity, uptake, and clearance of aminocarb by the aquatic isopod, *Caecidolea racovitzai racovitzai, Ecotoxicol. Environ. Saf.*, 7(6):552-557, 1983.

Richter, J.E., Peterson, S.F., and Kleiner, C.F. Acute and chronic toxicity of some chlorinated benzenes, chlorinated ethanes, and tetrachloroethylene to *Daphnia magna, Arch. Environ. Contam. Toxicol.*, 12(6):679-684, 1983.

Riddick, J.A., Bunger, W.B., and Sakano, T.K. *Organic Solvents - Physical Properties and Methods of Purification. Volume II* (New York: John Wiley & Sons, 1986), 1325 p.

Riederer, M. Estimating partitioning and transport of organic chemicals in the foliage/atmosphere system: discussion of a fugacity-based model, *Environ. Sci. Technol.*, 24(6):829-837, 1990.

Ringo, J., Jona, G., Rockwell, R., Segal, D., and Cohen, E. Genetic variation for resistance to chlorpyrifos in *Drosophila melanogaster* (Diptera: Drosophilidae) infesting grapes in Israel, *J. Econ. Entomol.*, 88(5):1158-1163, 1995.

Rippen, G., Ilgenstein, M., and Klöpffer, W. Screening of the adsorption behavior of new chemicals: Natural soils and model adsorbents, *Ecotoxicol. Environ. Saf.*, 6(3):236-245, 1982.

Rischbieter, E., Stein, H., and Schumpe, A. Ozone solubilities in water and aqueous salt solutions, *J. Chem. Eng. Data*, 45(2):338-340, 2000.

Riser-Roberts, E. *Bioremediation of Petroleum Contaminated Sites* (Boca Raton, FL: CRC Press, 1992), 496 p.

Rittman, B.E. and McCarty, P.L. Utilization of dichloromethane by suspended and fixed film bacteria, *Appl. Environ. Microbiol.*, 39:1225-1226, 1980.

Ro, K.S., Venugopal, A., Adrian, D.D., Constant, D., Qaisi, K., Valsaraj, K.T., Thibodeaux, L.J., and Roy, D. Solubility of 2,4,6-trinitrotoluene (TNT) in water, *J. Chem. Eng. Data*, 41(4):758-761, 1996.

Roark, R.C. A digest of information on chlordane, Bureau of Entomology and Plant Quarantine, U.S. Dept. of Agric. Report E-817, 1951, 132 p.

Robb, I.D. Determination of the aqueous solubility of fatty acids and alcohols, *Aust. J. Chem.*, 18:2281-2285, 1966.

Robbins, G., Wang, S., and Stuart, J.D. Using the static headspace method to determine Henry's law constants, *Anal. Chem.*, 65(21):3113-3118, 1993.

Robeck, G.G., Dostal, K.A., Cohen, J.M., and Kreissl, J.F. Effectiveness of water treatment processes in pesticide removal, *J. Am. Water Works Assoc.*, 57(2):181-200, 1965.

Roberts, A.L. and Gschwend, P.M. Mechanism of pentachloroethane dehydrochlorination to tetrachloroethylene, *Environ. Sci. Technol.*, 25(1):76-86, 1991.

Roberts, J.R., DeFrietas, A.S.W., and Bidney, M.A.J. Influence of lipid pool size on bioaccumulation of the insecticide chlordane by northern redhorse suckers (*Maxostoma macrolepidotum*), *J. Fish. Res. Board Can.*, 34(1):89-97, 1977.

Roberts, M.S., Anderson, R.A., and Swarbrick, J. Permeability of human epidermis to phenolic compounds, *J. Pharm. Pharmac.*, 29:677-683, 1977.

Roberts, P.V. Nature of organic contaminants in groundwater and approaches to treatment, in *Proceedings, AWWA Seminar on Organic Chemical Contaminants in Groundwater: Transport and Removal* (St. Louis, MO: American Water Works Association, 1981), pp. 47-65.

Roberts, P.V. and Dändliker, P.G. Mass transfer of volatile organic contaminants from aqueous solution to the atmosphere during surface aeration, *Environ. Sci. Technol.*, 17(8):484-489, 1983.

Roberts, P.V., McCarty, P.L., Reinhard, M., and Schreiner, J. Organic contaminant behavior during groundwater recharge, *J. Water Pollut. Control Fed.*, 52:161-172, 1980.

Roberts, P.V., Munz, C., and Dändliker, P. Modeling volatile organic solute removal by surface and bubble aeration, *J. Water Pollut. Control Fed.*, 56(2):157-163, 1985.

Roberts, T.R. and Stoydin, G. The degradation of (Z)- and (E)-1,3-dichloropropenes and 1,2-dichloropropane in soil, *Pestic. Sci.*, 7:325-335, 1976.

Robertson, R.E. and Sugamori, S.E. The hydrolysis of dimethyl sulfate and diethyl sulfate in water, *Can. J. Chem.*, 44(14):1728-1730, 1966.

Robinson, J., Richardson, A., Bush, B., and Elgar, K.E. A photo-isomerization product of dieldrin, *Bull. Environ. Contam. Toxicol.*, 1(4):127-132, 1966.

Robinson, K. and Novak, J.T. Fate of 2,4,6-trichloro-(^{14}C)-phenol bound to dissolved humic acid, *Water Res.*, 18(4):445-452, 1994.

Roburn, J. Effect of sunlight and ultraviolet radiation on chlorinated pesticide residues, *Chem. Ind.*, pp. 1555-1556, 1963.

Rochester, H. and Symonds, J.R. Thermodynamic studies of fluoroalcohols, *J. Chem. Soc. Faraday Trans. 1*, 69:1577-1585, 1973.

Rodorf, B.F. Thermodynamic and thermal properties of polychlorinated compounds: the vapor pressures and flow tube kinetics of ten dibenzo-*para*-dioxins, *Chemosphere*, 14(7):885-892, 1985.

Roeder, K.D. and Weiant, E.A. The site of action of DDT in the cockroach, *Science (Washington, D.C.)*, 103(2671):304-305, 1946.

Roesijadi, G., Petrocelli, S.R., Anderson, J.W., Giam, C.S., and Neff, G.E. Toxicity of polychlorinated biphenyls (aroclor 1254) to adult, juvenile, and larval stages of the shrimp *Palaemonetes pugio, Bull. Environ. Contam. Toxicol.*, 15(3):297-304, 1976.

Rogers, J.D. Rate constant measurements for the reaction of of the hydroxyl radical with cyclohexene, cyclopentene, and glutaraldehyde, *Environ. Sci. Technol.*, 23(2):177-181, 1989.

Rogers, K.S. and Cammarata, A. Superdelocalizability and charge density. A correlation with partition coefficients, *J. Med. Chem.*, 12(4):692-693, 1969.

Rogers, R.D. and McFarlane, J.C. Sorption of carbon tetrachloride, ethylene dibromide, and trichloroethylene on soil and clay, *Environ. Monitor. Assess.*, 1(2):155-162, 1981.

Rogers, R.D., McFarlane, J.C., and Cross, A.J. Adsorption and desorption of benzene in two soils and montmorillonite clay, *Environ. Sci. Technol.*, 14(4):457-461, 1980.

Rohrschneider, L. Solvent characterization by gas-liquid coefficients of selected solutes, *Anal. Chem.*, 45:1245-1247, 1973.

Rondorf, B. Thermal properties of dioxins, furans and related compounds, *Chemosphere*, 15:1325-1332, 1986.

Rontani, J.F., Rambeloarisoa, E., Bertrand, J.C., and Giusti, G. Favourable interaction between photooxidation and bacterial degradation of anthracene in sea water, *Chemosphere*, 14(11/12):1909-1912, 1985.

Rordorf, B.F. Prediction of vapor pressures, boiling points and enthalpies of fusion for twenty-nine halogenated dibenzo-*p*-dioxins and fifty-five dibenzofurans by a vapor pressure correlation method, *Chemosphere*, 18(1-6):783-788, 1989.

Rosen, J.D. and Carey, W.F. Preparation of the photoisomers of aldrin and dieldrin, *J. Agric. Food Chem.*, 16(3):536-537, 1968.

Rosen, J.D. and Sutherland, D.J. The nature and toxicity of the photoconversion products of aldrin, *Bull. Environ. Contam. Toxicol.*, 2(1):1-9, 1967.

Rosen, J.D., Sutherland, J., and Lipton, G.R. The photochemical isomerization of dieldrin and endrin and effects on toxicity, *Bull. Environ. Contam. Toxicol.*, 1(4):133-140, 1966.

Rosenberg, A. and Alexander, M. Microbial cleavage of various organophosphorus insecticides, *Appl. Environ. Microbiol.*, 37(5):886-891, 1979.

Rosenberg, A. and Alexander, M. Microbial metabolism of 2,4,5-trichlorophenoxyacetic acid in soil, soil suspensions, and axenic culture, *J. Agric. Food Chem.*, 28:297-302, 1980.

Rosenfield, C. and Van Valkenburg, W. Decomposition of (*O,O*-dimethyl-*O*-2,4,5-trichlorophenyl) phosphorothioate (Ronnel) adsorbed on bentonite and other clays, *J. Agric. Food Chem.*, 13(1):68-72, 1972.

Rosenstock, H.M., Draxl, K., Steiner, B.W., and Herron, J.T. Ion energetics data, in *NIST Standard Reference Database Number 69*, Mallard, W.G. and Linstrom, P.J., Eds. (Gaithersburg, MD: National Institute of Standards and Technology, 1998) (http://webbook.nist.gov).

Ross, R.D. and Crosby, D.G. The photooxidation of aldrin in water, *Chemosphere*, 4(5):277-282, 1975.

Ross, R.D. and Crosby, D.G. Photooxidant activity in natural waters, *Environ. Toxicol. Chem.*, 4(6):773-778, 1985.

Rossi, S.S. and Thomas, W.H. Solubility behavior of three aromatic hydrocarbons in distilled water and natural seawater, *Environ. Sci. Technol.*, 15(6):715-716, 1981.

Rothman, A.M. Low vapor pressure determination by the radiotracer transpiration method, *Anal. Chem.*, 28(6):1225-1228, 1980.

Roux, D., Jooste, S., Truter, E., and Kempster, P. An aquatic toxicological evaluation of fenthion in the context of finch control in South Africa, *Ecotoxicol. Environ. Saf.*, 31(2):164-172, 1995.

Roy, R.S. Spectrophotometric estimation of ester hydrolysis, *Anal. Chem.*, 44:2096-2098, 1972.

Roy, W.R., Ainsworth, C.C., Chou, S.F.J., Griffin, R.A., and Krapac, I.G. Development of standardized batch adsorption procedures: experimental considerations, in *Proceedings of the Eleventh Annual Research Symposium of the Solid and Hazardous Waste Research Division, April 29-May 1, 1985* (Cincinnati, OH: U.S. Environmental Protection Agency, 1985), U.S. EPA Report 600/9-85-013, pp. 169-178.

Roy, W.R. and Krapac, I.G. Adsorption and desorption of atrazine and deethylatrazine by low organic carbon geologic materials, *J. Environ. Qual.*, 23(3):549-556, 1994.

Roy, W.R., Krapac, I.G., Chou, S.F.J., and Griffin, R.A. Batch-type adsorption procedures for estimating soil attenuation of chemicals, Draft Technical Resource Document (Cincinnati, OH: U.S. Environmental Protection Agency, 1987), U.S. EPA Report 530-SW-87-006.

RTECS (Registry of Toxic Effects of Chemical Substances), U.S. Department of Health and Human Services, National Institute of Occupational Safety and Health, 1985, 2050 p.

Rubin, H.E., Subba-Rao, R.V., and Alexander, M. Rates of mineralization of trace concentrations of aromatic compounds in lake water and sewage samples, *Appl. Environ. Microbiol.*, 43(5):1133-1138, 1982.

Ruepert, C., Grinwis, A., and Govers, H. Prediction of partition coefficients of unsubstituted polycyclic aromatic hydrocarbons from C_{18} chromatographic and structural properties, *Chemosphere*, 14(3/4):279-291, 1985.

Rusch, G.M., Hoffman, G.M., McConnell, R.F., and Rinehart, W.E. Inhalation toxicity studies with boron trifluoride, *Toxicol. Appl. Pharmacol.*, 83(1):69-78, 1986.

Russell, D.J. and McDuffie, B. Chemodynamic properties of phthalate esters: partitioning and soil migration, *Chemosphere*, 15(8):1003-1021, 1986.

Ruth, J.N. Odor threshold and irritation levels of several chemical substances. A review, *Am. Ind. Myg. Assoc. J.*, 47:142-151, 1986.

Rutherford, D.W. and Chiou, C.T. Effect of water saturation in soil organic matter on the partition of organic compounds, *Environ. Sci. Technol.*, 26(5):965-970, 1992.

Rutherford, D.W., Chiou, C.T., and Kile, D.E. Influence of soil organic matter composition on the partition of organic compounds, *Environ. Sci. Technol.*, 26(2):336-340, 1992.

Rütters, H., Höllrigl-Rosta, A., Kreuzig, R., and Bahadir, M. Sorption of prochloraz in different soils, *J. Agric. Food Chem.*, 47(3):1242-1246, 1999.

Saarikoski, J., Lindström, R., Tyynelä, M., and Viliksela, M. Factors affecting the absorption of phenolics and carboxylic acids in the guppy (*Poecilia reticulata*), *Ecotoxicol. Environ. Saf.*, 11(1):158-173, 1986.

Sabbah, R. and El Watik, L. New reference materials for the calibration (temperature and energy) of differential thermal analysers and scanning calorimeters, *J. Therm. Anal.*, 38(4):855-863, 1992.

Sabljić, A. Predictions of the nature and strength of soil sorption of organic pollutants by molecular topology, *J. Agric.*

Food Chem., 32(2):243-246, 1984.

Sabljić, A. and Güsten, H. Predicting the night-time NO$_3$ radical reactivity in the troposphere, *Atmos. Environ.*, 24A(1):73-78, 1990.

Sada, E., Kito, S., and Ito, Y. Solubility of toluene in aqueous salt solutions, *J. Chem. Eng. Data*, 20(4):373-375, 1975.

Saeger, V.W., Hicks, O., Kaley, R.G., Michael, P.R., Mieure, J.P., and Tucker, E.S. Environmental fate of selected phosphate esters, *Environ. Sci. Technol.*, 13(7):840-844, 1979.

Saeger, V.W. and Tucker, E.S. Biodegradation of phthalate acid esters in river water and activated sludge, *Appl. Environ. Microbiol.*, 31(1):29-34, 1976.

Safe, S., Bunce, N.J., Chittin, B., Hutzinger, O., and Ruzo, L.O. Photodecomposition of halogenated aromatic compounds, in *Identification and Analysis of Pollutants in Water*, Keith, L.H., Ed., (Ann Arbor, MI: Ann Arbor Science Publishers, 1976), pp. 35-46.

Sahu, S.K., Patnaik, K.K., and Sethunathan, N. Dehydrochlorination of δ-isomer of hexachlorocyclohexane by a soil bacterium, *Pseudomonas* sp., *Bull. Environ. Contam. Toxicol.*, 48(2):265-268, 1992.

Sahyun, M.R.V. Binding of aromatic compounds to bovine serum albumin, *Nature (London)*, 209(5023):613-614, 1966.

Saini, M.L. and Dorough, H.W. Persistence of malathion and methyl parathion when applied as ultra-low-volume and emulsifiable concentrate sprays, *J. Econ. Entomol.*, 63:405-408, 1970.

Saito, S., Tanoue, A., and Matsuo, M. Applicability of i/o-characters to a quantitative description of bioconcentration of organic chemicals in fish, *Chemosphere*, 24(1):81-87, 1992.

Salamone, M.F. Toxicity of 41 carcinogens and noncarcinogenic analogs, *Prog. Mutat. Res.* in *Evaluation of Short-Term Tests for Carcinogens. Report of the International Corroborative Program. Progress in Mutation Research*, de Serres, F.J. and Ashby, Eds. (Holland: Elsevier, 1981), pp. 682-685.

Saleh, M.A. and Casida, J.E. Reductive dechlorination of the toxaphene component 2,2,5-*endo*-6-*exo*,8,9,10-heptachlorobornane in various chemical, photochemical, and metabolic systems, *J. Agric. Food Chem.*, 26(3):583-590, 1978.

Saleh, F.Y., Dickson, K.L., and Rodgers, Jr., J.H. Fate of lindane in the aquatic environment: rate constants of physical and chemical processes, *Environ. Toxicol. Chem.*, 1(4):289-297, 1982.

Saltzman, S., Kliger, L., and Yaron, B. Adsorption-desorption of parathion as affected by soil organic matter, *J. Agric. Food Chem.*, 20(6):1224-1226, 1972.

Sanborn, J.R., Francis, B.M., and Metcalf, R.L. The degradation of selected pesticides in soil: a review of the published literature, Office of Research and Development, U.S. EPA Report-600/9-77-022, 1977, 616 p.

Sanborn, J.R., Metcalf, R.L., Bruce, W.N., and Lu, P.-Y. The fate of chlordane and toxaphene in a terrestrial-aquatic model ecosystem, *Environ. Entomol.*, 5(3):533-538, 1976.

Sanborn, J.R. and Yu, C. The fate of dieldrin in a model ecosystem, *Bull. Environ. Contam. Toxicol.*, 10(6):340-346, 1973.

Sancho, E., Ferrando, M.D., and Andreu, E. and Gamon, M. Acute toxicity, uptake and clearance of diazinon by the European eel, *Anguilla anguilla* (L.), *J. Environ. Sci. Health*, B27(2):209-221, 1992.

Sandell, K.B. Distribution of aliphatic amines between water and mixture of organic solvents, *Naturwissenschaften*, 49:12-13, 1962.

Sanderman, W., Stockmann, H., and Casten, R. Über die Pyrolyse des Pentachlorphenols, *Chem. Ber.*, 90:690-692, 1957.

Sanders, H.O. Toxicities of some herbicides to six species of freshwater crustaceans, *J. Water Pollut. Control Fed.*, 42(8):1544-1550, 1970.

Sanders, H.O. and Cope, O.B. Toxicities of several pesticides to two species of cladocerans, *Trans. Am. Fish Soc.*, 95(2):165-169, 1966.

Sanders, H.O. and Cope, O.B. The relative toxicities of several pesticides to naiads of three species of stoneflies, *Limnol. Oceanogr.*, 13(1):112-117, 1968.

Sanders, H.O., Huckins, J., Johnson, B.T., and Skaar, D. Biological effects of kepone and mirex in freshwater invertebrates, *Arch. Environ. Contam. Toxicol.*, 10(5):531-539, 1981.

Sanders, H.O., Mayer, F.L., and Walsh, D.F. Toxicity, residue dynamics, and reproduction effects of phthalate esters in aquatic invertebrates, *Environ. Res.*, 6:84-90, 1973.

Sanemasa, I., Arakawa, S., Piao, C.Y., and Deguchi, T. Equilibrium solubilities of benzene, toluene, ethylbenzene, and propylbenzene in silver nitrate solutions, *Bull. Chem. Soc. Jpn.*, 58(10):3033-3034, 1985.

Sanemasa, I., Araki, M., Deguchi, T., and Nagai, H. Solubility measurements of benzene and the alkylbenzenes in water by making use of solute vapor, *Bull. Chem. Soc. Jpn.*, 55(4):1054-1062, 1982.

Sanemasa, I., Miyazaki, Y., Arakawa, S., Kumamaru, M., and Deguchi, T. The solubility of benzene-hydrocarbon binary mixtures in water, *Bull. Chem. Soc. Jpn.*, 60(2):517-523, 1987.

Sangster, J. Octanol-water partition coefficients of simple organic compounds, *J. Phys. Chem. Ref. Data*, 18(3):1111-1229, 1989.

Sanhueza, E., Hisatsune, I.C., and Heicklen, J. Oxidation of haloethylenes, *Chem. Rev.*, 76:801-826, 1976.

Santl, H., Brandsch, R., and Gruber, L. Experimental determination of Henry's law constants (HLC) for some lower chlorinated dibenzodioxins, *Chemosphere*, 29(9-11):2209-2214, 1994.

Saravanja-Bozanic, V., Gäb, S., Hustert, K., and Korte, F. Reaktionen von Aldrin, Chlorden, und 2,2-Dichlorobiphenyl mit O(^{3}P), *Chemosphere*, 6(1):21-26, 1977.

Sastry, K.V. and Agrawal, M.K. Mercuric chloride induced enzymological changes in kidney and ovary of a teleost fish, *Channa punctatus*, *Bull. Environ. Contam. Toxicol.*, 22(1/2):38-43, 1979.

Sastry, N.V., George, A., Jain, N.J., and Bahadur, P. Densities, relative permittivities, excess volumes, and excess molar polarizations for alkyl ester (methyl propanoate, methyl butanoate, ethyl propanoate, and ethyl butanoate) + hydrocarbons (*n*-heptane, benzene, chlorobenzene, and 1,1,2,2-tetrachloroethane) at 308.15 and 318.15 K, *J. Chem. Eng. Data*, 44(3):456-464, 1999.

Sato, A. and Nakajima, T. A structure-activity relationship of some chlorinated hydrocarbons, *Arch. Environ. Health*, 34(2):69-75, 1979.

Sax, N.I. and Lewis, R.J., Sr. *Hazardous Chemicals Desk Reference* (New York: Van Nostrand Reinhold, 1987), 1084 p.

Sax, N.I. *Dangerous Properties of Industrial Materials* (New York: Van Nostrand Reinhold, 1984), 3124 p.

Sax, N.I., Ed. *Dangerous Properties of Industrial Materials Report*, Volume 4 (New York: Van Nostrand Reinhold, 1984).

Sax, N.I., Ed. *Dangerous Properties of Industrial Materials Report*, Volume 5 (New York: Van Nostrand Reinhold, 1985).

Saxena, P.K. and Garg, M. Effects of insecticidal pollution on ovarian recrudescence in fresh water teleost *Channa punctatus* (B1), *Indian J. Exp. Biol.*, 16:689, 1978.

Saylor, J.H., Stuckey, J.M., and Gross, P.M. Solubility studies. V. The validity of Henry's law for the calculation of vapor solubilities, *J. Am. Chem. Soc.*, 60(2):373-376, 1938.

Scala, A.A. and Salomon, D. The gas phase photolysis and γ radiolysis of ethylenimine, *J. Chem. Phys.*, 65(11):4455-4461, 1976.

Schafer, E.W., Brunton, R.B., and Lockyer, N.F. The effects of subacute and chronic exposure to 4-aminopyridine on reproduction in coturnix quail, *Bull. Environ. Contam. Toxicol.*, 13(6):758-764, 1975.

Schantz, M.M. and Martire, D.E. Determination of hydrocarbon-water partition coefficients from chromatographic data and based on solution thermodynamics and theory, *J. Chromatogr.*, 391(1):35-51, 1987.

Schauer, J.J., Kleeman, M.J., Cass, G.R., and Simoneit, B.R.T. Measurement of emissions from air pollution sources. 2. C_1 through C_{30} organic compounds from medium duty diesel trucks, *Environ. Sci. Technol.*, 33(10):1578-1587, 1999.

Scheer, V., Frenzel, A., Behnke, W., Zetzsch, C., Magi, L., George, C., and Mirabel, P. Uptake of nitrosyl chloride (NOCl) by aqueous solutions, *J. Phys. Chem. A*, 101:9359-9366, 1997.

Schellenberg, K., Leuenberger, C., and Schwarzenbach, R.P. Sorption of chlorinated phenols by natural sediments and aquifer materials, *Environ. Sci. Technol.*, 18(9):652-657, 1984.

Scheunert, I., Topp, E., Attar, A., and Korte, F. Uptake pathways of chlorobenzenes in plants and their correlation with octanol/water partition coefficients, *Ecotoxicol. Environ. Saf.*, 27:90-104, 1994.

Scheunert, I., Vockel, D., Schmitzer, J., and Korte, F. Biomineralization rates of ^{14}C-labelled organic chemicals in aerobic and anaerobic suspended soil, *Chemosphere*, 16(5):1031-1041, 1987.

Schimmel, S.C. and Garnas, R.L. Results: interlaboratory comparison-bioconcentration tests using the eastern oyster, Environmental Research Laboratory, U.S. EPA Report-600/4-81-011, 1981, 199 p.

Schimmel, S.C., Garnas, R.L., Patrick, J.M., Jr., and Moore, J.C. Acute toxicity, bioconcentration, and persistence of AC 222,705, benthiocarb, chlorpyrifos, fenvalerate, methyl parathion, and permethrin in the estuarine environment, *J. Agric. Food Chem.*, 31(1):104-113, 1983.

Schimmel, S.C., Hamaker, T.L., and Forester, J. Toxicity and bioconcentration of EPN and leptophos to selected estuarine animals, U.S. EPA Report-600/J-79-086, 1979.

Schimmel, S.C., Patrick, J.M., Jr., and Forester, J. Heptachlor: toxicity to and uptake by several estuarine organisms, *J. Toxicol. Environ. Health*, 1(6):955-965, 1976.

Schimmel, S.C., Patrick, J.M., Jr., and Forester, J. Heptachlor: uptake, depuration, retention, and metabolism by spot (*Leiostomus xanthurus*), *J. Toxicol. Environ. Health*, 2(1):169-178, 1976a.

Schimmel, S.C., Patrick, J.M., Jr., and Forester, J. Uptake and bioconcentration of BHC and lindane in selected estuarine animals, *Arch. Environ. Contam. Toxicol.*, 6(2/3):355-363, 1977a.

Schimmel, S.C., Patrick, J.M., Jr., and Wilson, A.J., Jr. Acute toxicity to and bioconcentration of endosulfan by estuarine animals, in *Aquatic Toxicity and Hazard Evaluation*, Mayer F.L. and Hamelink, J.L., Eds. (Philadephia, PA: American Society for Testing and Materials, 1977b), pp. 241-252.

Schimmel, S.C. and Wilson, A.J., Jr. Acute toxicity of kepone to four estuarine animals, *Chesapeake Sci.*, 18(2):224-227, 1977.

Schliebe, K.A., Burnside, O.C., and Lavy, T.L. Dissipation of amiben, *Weeds*, 13:321-325, 1965.

Schmidt-Bleek, F., Haberland, W., Klein, A.W., and Caroli, S. Steps towards environment assessment of new chemicals, *Chemosphere*, 11:383-415, 1982.

Schmitz, A. and Nagel, R. Influence of 3,4-dichloroaniline (3,4-DCA) on benthic invertebrates in indoor experimental streams, *Ecotoxicol. Environ. Saf.*, 30(1):63-71, 1995.

Schnoor, J.L., Atchison, E.W., Kelley, S.L., Alvarez, P.J.J., Wakefield, S., Burken, J.G., and Just, C.L. Phyto-remediation of 1,4-dioxane by hybrid poplars, in *American Chemical Society - Division of Environmental Chemistry, Preprints of Extended Abstracts*, 37(1):197-199, 1997.

Schocken, M.J. and Gibson, D.T. Bacterial oxidation of the polycyclic aromatic hydrocarbons acenaphthene and acenaphthylene, *Appl. Environ. Microbiol.*, 48(1):10-16, 1984.

Schoen, S.R. and Winterlin, W.L. The effects of various soil factors and amendments on the degradation of pesticide mixtures, *J. Environ. Sci. Health*, B22(3):347-377, 1987.

Schrap, S.M., de Vries, P.J., and Opperhuizen, A. Experimental problems in determining sorption coefficients of organic chemicals; an example of chlorobenzenes, *Chemosphere*, 28(5):931-945, 1994.

Schrap, S.M., Haller, M., and Opperhuizen, A. Investigating the influence of incomplete separation of sediment and water on experimental sorption coefficients of chlorinated, *Environ. Toxicol. Chem.*, 14(2):219-228, 1995.

Schroy, J.M., Hileman, F.D., and Cheng, S.C. Physical/chemical properties of 2,3,7,8-TCDD, *Chemosphere*,

14(6/7):877-880, 1985.

Schultz, T.W., Kite, L.M., and Dumont, J.N. Structure-toxicity correlations of organic contaminants in aqueous coal-conversion effluents, *Arch. Environ. Contam. Toxicol.*, 7(4):457-463, 1978.

Schultz, T.W., Wesley, S.K., and Baker, L.L. Structure-activity relationships for di and tri alkyl and/or halogen substituted phenols, *Bull. Environ. Contam. Toxicol.*, 43(2):192-198, 1989.

Schumacher, H.G., Parlar, H., Klein, W., and Korte, F. Beiträge zur ökologischen Chemie. LV. Photochemische Reaktion von Endosulfan, *Chemosphere*, 3(2):65-70, 1974.

Schumann, S.C., Aston, J.S., and Sagenkahn, M. The heat capacity and entropy, heats of fusion and vaporization and the vapor pressures of isopentane, *J. Am. Chem. Soc.*, 64:1039-1043, 1942.

Schüth, C. and Reinhard, M. Catalytic hydrodehalogenation of some aromatic compounds using palladium on different support materials, in *American Chemical Society - Division of Environmental Chemistry, Preprints of Extended Abstracts*, 37(1):173-174, 1997.

Schuytema, G.S. and Nebeker, A.V. Comparative toxicity of diuron on survival and growth of Pacific treefrog, bullfrog, red-legged frog, and African clawed frog embryos and tadpoles, *Arch. Environ. Contam. Toxicol.*, 34(4):370-376, 1998.

Schuytema, G.S. and Nebeker, A.V. Comparative toxicity of ammonium and nitrate compounds on Pacific treefrog and African clawed frog embyyos, *Arch. Environ. Contam.Toxicol.*, 36(2):200-206, 1999.

Schuytema, G.S., Nebeker, A.V., Griffis, W.L., and Wilson, K.N. Teratogenesis, toxicity, and bioconcentration in frogs exposed to dieldrin, *Arch. Environ. Contam. Toxicol.*, 21(3):332-350, 1991.

Schwarz, F.P. Determination of temperature dependence of solubilities of polycyclic aromatic hydrocarbons in aqueous solutions by a fluorescence method, *J. Chem. Eng. Data*, 22(3):273-277, 1977.

Schwarz, F.P. Measurement of the solubilities of slightly soluble organic liquids in water by elution chromatography, *Anal. Chem.*, 52(1):10-15, 1980.

Schwarz, F.P. and Miller, J. Determination of the aqueous solubilities of organic liquids at 10.0, 20.0, and 30.0 °C by elution chromatography, *Anal. Chem.*, 52(13):2162-2164, 1980.

Schwarz, F.P. and Wasik, S.P. Fluorescence measurements of benzene, naphthalene, anthracene, pyrene, fluoranthene, and benzo[e]pyrene in water, *Anal. Chem.*, 48(3):524-528, 1976.

Schwarz, F.P. and Wasik, S.P. A fluorescence method for the measurement of the partition coefficients of naphthalene, 1-methylnaphthalene, and 1-ethylnaphthalene in water, *J. Chem. Eng. Data*, 22(3):270-273, 1977.

Schwarzenbach, R.P., Gschwend, P.M., and Imboden, D.M. *Environmental Organic Chemistry* (New York: John Wiley & Sons, 1993), 681 p.

Schwarzenbach, R.P., Giger, W., Schaffner, C., and Wanner, O. Groundwater contamination by volatile halogenated alkanes: abiotic formation of volatile sulfur compounds under anaerobic conditions, *Environ. Sci. Technol.*, 19(4):322-327, 1985.

Schwarzenbach, R.P., Giger, W., Hoehn, E., and Schneider, J.K. Behavior of organic compounds during infiltration of river water to groundwater. Field studies, *Environ. Sci. Technol.*, 17(8):472-479, 1983.

Schwarzenbach, R.P., Stierli, R., Folsom, B.R., and Zeyer, J. Compound properties relevant for assessing the environmental partitioning of nitrophenols, *Environ. Sci. Technol.*, 22(1):83-92, 1988.

Schwarzenbach, R.P. and Westall, J. Transport of nonpolar organic compounds from surface water to groundwater. Laboratory sorption studies, *Environ. Sci. Technol.*, 15(11):1360-1367, 1981.

Schwille, F. *Dense Chlorinated Solvents* (Chelsea, MI: Lewis Publishers, 1988), 146 p.

Scott, D.W., Guthrie, G.B., Messerly, J.F., Todd, S.S., Berg, W.T., Hossenlopp, I.A., and McCullough, J.P. Toluene: thermodynamic properties, molecular vibrations, and internal rotation, *J. Phys. Chem.*, 66:911-914, 1962.

Scott, H.D., Wolf, D.C., and Lavy, T.L. Adsorption and degradation of phenol at low concentrations in soil, *J. Environ. Qual.*, 12(1):91-95, 1983.

Scully, F.E., Jr. and Hoigné, J. Rate constants for reactions of singlet oxygen with phenols and other compounds in water, *Chemosphere*, 16(4):681-694, 1987.

Sears, G.W. and Hopke, E.R. Vapor pressures of naphthalene, anthracene, and hexachlorobenzene in a low pressure region, *J. Am. Chem. Soc.*, 71(5):1632-1634, 1949.

Seckar, J.A. The toxicity of HCFC-141b, *J. Am. Coll. Toxicol.*, 8(6):1221, 1989.

Sedlak, D.L. and Andren, A.W. Oxidation of chlorobenzene with Fenton's reagent, *Environ. Sci. Technol.*, 25(4):777-782, 1991.

Seiber, J.N., Madden, S.C., McChesney, M.M., and Winterlin, W.L. Toxaphene dissipation from treated cotton field measurements: component residual behavior on leaves and in air, soil, and sediments determined by gas chromatography, *J. Agric. Food Chem.*, 27:284-290, 1979.

Seip, H.M., Alstad, J., Carlberg, G.E., Martinsen, K., and Skaane, R. Measurement of mobility of organic compounds in soils, *Sci. Total Environ.*, 50:87-101, 1986.

Sepic, E., Bricelj, M., and Leskovsek, H. Degradation of fluoranthene by *Pasteurella* sp. IFA and *Mycrobacterium* sp. PYR-1: isolation and identification of metabolites, *J. Appl. Microbiol.*, 85(4):746-754, 1998.

Serrano, R., Hernández, F., López, F.J., and Peña, J.B. Bioconcentration and depuration of chlorpyrifos in the marine mollusc *Mytilus edulis*, *Arch. Environ. Contam. Toxicol.*, 33(1):47-52, 1997.

Servant, J., Kouadio, G., Cros, B., and Delmas, R. Carboxylic monoacids in the air of Mayombe forest (Congo): Role of the forest as a source or sink, *J. Atmos. Chem.*, 12:367-380, 1991.

Sethunathan, N., Siddaramappa, R., Rajaram, K.P., Barik, S., and Wahid, P.A. Parathion: residues in soil and water, *Residue Rev.*, 68:91-122, 1977.

Sethunathan, N. and Yoshida, T. A *Flavobacterium* sp. that degrades diazinon and parathion, *Can. J. Microbiol.*, 19(5):873-875, 1973.

Sethunathan, N. and Yoshida, T. Degradation of chlorinated hydrocarbons by *Clostridium* sp. isolated from lindane-amended, flooded soil, *Plant Soil*, 38:663-666, 1973a.

Sforzolini, G.S., Saviano, A., and Merletti, C. The action of chlorine on some hydrocarbons, polycyclic hydrocarbons, contributing to the study of the decontamination of water from carcinogenic compounds, *Bull. Soc. Ital. Biol. Sper.*, 46:903-906, 1970.

Shan, Z. and Asfour, A.-F.A. Viscosities and densities of nine binary 1-alkanol systems at 293.15 K and 298.15 K, *J. Chem. Eng. Data*, 44(1):118-123, 1999.

Shang, C. and Arshad, M.A. Sorption of clopyralid, dicamba and MCPA by two soils with conventional and no-till management, *Can. J. Soil Sci.*, 78:181-186, 1998.

Sharma, R.K. and Palmer, H.B. Vapor pressure of biphenyl near fusion temperature, *J. Chem. Eng. Data*, 19(1):6-8, 1974.

Sharma, R.C. and Sharma, M.M. Kinetics of alkaline hydrolysis of esters. II. Unsaturated esters and oxalic esters, *Bull. Chem. Soc. Jpn.*, 43:642-645, 1970.

Sharma, S.R., Singh, R.P., and Ahmed, S.R. Effect of different leachates on the movement of some phosphorus containing pesticides in soils using thin layer chromatography, *Ecotoxicol. Environ. Saf.*, 11(2):229-240, 1986.

Sharmila, M., Ramanand, K., and Sethunathan, N. Effect of yeast extract on the degradation of organophosphorus insecticides by soil enrichment and bacterial cultures, *Can. J. Microbiol.*, 35(12):1105-1110, 1989.

Sharom, M.S. and Edgington, L.V. The adsorption, mobility, and persistence of metalaxyl in soil and aqueous systems, *Can. J. Plant Pathol.*, 4:334-340, 1982.

Sharom, M.S., Miles, J.R.W., Harris, J.W., and McEwen, F.L. Persistence of 12 insecticides in water, *Water Res.*, 14(8):1089-1093, 1980.

Sharom, M.S., Miles, J.R.W., Harris, C.R., and McEwen, F.L. Behavior of 12 pesticides in soil and aqueous suspensions of soil and sediment, *Water Res.*, 14(8):1095-1100, 1980a.

Sharom, M.S. and Solomon, K.R. Adsorption-desorption, degradation, and distribution of permethrin in aqueous systems, *J. Agric. Food Chem.*, 29(6):1122-1125, 1981.

Shaul, G.M., Lieberman, R.J., Dempsey, C.R., and Dostal, K.A. Treatability of water soluble azo dyes by the activated sludge process, in *Proceedings of the Industrial Wastes Symposium*, 59th Water Pollution Control Conference, October 1986.

Sheindorf, C., Rebhun, M., and Sheintuch, M. Organic pollutants adsorption from multicomponent systems modeled by Freundlich type isotherm, *Water Res.*, 16(3):357-362, 1982.

Shelton, D.R., Boyd, S.A., and Tiedje, J.M. Anaerobic biodegradation of phthalic acid esters in sludge, *Environ. Sci. Technol.*, 18(2):93-97, 1984.

Sherblom, P.M., Gschwend, P.M., and Eganhouse, R.P. Aqueous solubilities, vapor pressures, and 1-octanol-water partition coefficients for C_9-C_{14} linear alkylbenzenes, *J. Chem. Eng. Data*, 37(4):394-399, 1992.

Shevchenko, M.A., Taran, P.N., and Marchenko, P.V. Technology of water treatment and demineralization, *Soviet J. Water Chem. Technol.*, 4(4):53-71, 1982.

Shiaris, M.P. and Sayler, G.S. Biotransformation of PCB by natural assemblages of freshwater microorganisms, *Environ. Sci. Technol.*, 16(6):367-369, 1982.

Shinoda, K., Yamanaka, T., and Kinoshita, K. Surface chemical properties in aqueous solutions of non-ionic surfactants: octyl glycol ether, α-octyl glyceryl ether and octyl glycoside, *J. Phys. Chem.*, 63:648-650, 1959.

Shinozuka, N., Lee, C., and Hayano, S. Solubilizing action of humic acid from marine sediment, *Sci. Total Environ.*, 62:311-314, 1987.

Shiraishi, H., Pilkington, N.H., Otsuki, A., and Fuwa, K. Occurrence of chlorinated polynuclear aromatic hydrocarbons in tap water, *Environ. Sci. Technol.*, 19(7):585-590, 1985.

Shiu, W.-Y., Doucette, W., Gobas, F.A.P.C., Mackay, D., and Andren, A.W. Physical-chemical properties of chlorinated dibenzo-*p*-dioxins, *Environ. Sci. Technol.*, 22(6):651-658, 1988.

Shiu, W.-Y. and Mackay, D. Henry's law constants of selected aromatic hydrocarbons, alcohols, and ketones, *J. Chem. Eng. Data*, 42(1):27-30, 1997.

Shiu, W.-Y., Wania, F., Hung, H., and Mackay, D. Temperature dependence of aqueous solubility of selected chlorobenzenes, polychlorinated biphenyls, and dibenzofuran, *J. Chem. Eng. Data*, 42(2):293-297, 1997.

Shivanandappa, T., Ramesh, H.P., and Krishnakumari, M.K. Rodenticidal poisoning of non-target animals: acute oral toxicity of zinc phosphide to poultry, *Bull. Environ. Contam. Toxicol.*, 23(4/5):452-455, 1979.

Shriner, C.R., Drury, J.S., Hammons, A.S., Towill, L.E., Lewis, E.B., and Opresko, D.M. Reviews of the environmental effects of pollutants: II. Benzidine, Office of Research and Development, U.S. EPA Report-600/1-78-024, 1978, 157 p.

Sibley, P.K. and Kaushik, N.K. Toxicity of microencapsulated permethrin to selected nontarget aquatic invertebrates, *Arch. Environ. Contam. Toxicol.*, 29(2):168-176, 1991.

Siddaramappa, R., Rajaram, K.P., and Sethunathan, N. Degradation of parathion by bacteria isolated from flooded soil, *Appl. Microbiol.*, 26(6):846-849, 1973.

Sidgwick, N.V. and Sutton, L.E. CLXX. - The system cyclohexanol and water, *J. Chem. Soc. (London)*, pp. 1323-1326, 1930.

Signer, R., Arm, H., and Daenicker, H. Dampfdrücke, Dichten, thermodynamische Mischfunktionen und Brechungsindices der binären Systeme Wasser-Tetrahydrofuran und Wasser-Diäthyläther bei 25 °, *Helv. Chim. Acta*, 52:2347-2351, 1969.

Sigworth, E.A. Identification and removal of herbicides and pesticides, *J. Am. Water Works Assoc.*, 57:1016-1022, 1964.

Sijm, R.T.H.M., Haller, M., and S.M. Schrap. Influence of storage on sediment characteristics and of drying sediment on sorption coefficients of organic contaminants, *Bull. Environ. Contam. Toxicol.*, 58(6):961-968, 1997.

Sijm, R.T.H.M., Wever, H., and Opperhuizen, A. Influence of biotransformation on the accumulation of PCDDs from fly-ash in fish, *Chemosphere*, 19:475-480, 1989.

Sikka, H.C., Appleton, H.T., and Banerjee, S. Fate of 3,3'-dichlorobenzidine in aquatic environments, U.S. EPA Report 600/3-78-068, 1978.

Simmon, V.F., Rosenkranz, H.S., Zeiger, E., and Poirier, L.A. Mutagenic activity of chemical carcinogens and related compounds in the intraperitoneal host-mediated assay, *J. Nat. Cancer Inst.*, 62:911-918, 1979.

Simmons, M.S. and Zepp, R.G. Influence of humic substances on photolysis of nitroaromatic compounds in aquatic systems, *Water Res.*, 20(7):899-904, 1986.

Sims, R.C., Doucette, W.C., McLean, J.E., Grenney, W.J., and DuPont, R.R. Treatment potential for 56 EPA listed hazardous chemicals in soil, U.S. EPA Report-600/6-88-001, 1988, 105 p.

Sims, G.K. and Sommers, L.E. Degradation of pyridine derivatives in soil, *J. Environ. Qual.*, 14(4):580-584, 1985.

Singer, M.E. and Finnerty, W.R. Microbial metabolism of straight-chain and branched alkanes, in *Petroleum Microbiology*, Atlas, R.M., Ed. (New York: Macmillan, 1984), pp. 1-59.

Singh, J. Conversion of heptachloride to its epoxide, *Bull. Environ. Contam. Toxicol.*, 4(2):77-79, 1969.

Singh, R.P. and Chiba, M. Solubility of benomyl in water at different pHs and its conversion to methyl 2-benzimida-zolecarbamate, 3-butyl-2,4-dioxo[1,2-*a*]-*s*-triazinobenzimidazole, and 1-(2-benzimidazolyl)-3-*n*-butyl urea, *J. Agric. Food Chem.*, 33(1):63-67, 1985.

Singh, R.N., Gupta, J.P., Singh, N., Singh, N.P., Singh, O.P., Hopkins, R.H., and Mazelsky, R. Growth conditions of organic non-linear optical crystals, *Thermochim. Acta*, 165(2):297-299, 1990.

Singh, N.B. and Kumar, P. Solidification behavior of the cinnamic acid-*p*-nitrophenol eutectic system, *J. Chem. Eng. Data*, 31:406-408, 1986.

Singh, A.K. and Seth, P.K. Degradation of malathion by microorganisms isolated from industrial effluents, *Bull. Environ. Contam. Toxicol.*, 43(1):28-35, 1989.

Singmaster, J.A., III. Environmental behavior of hydrophobic pollutants in aqueous solutions, Ph.D. Thesis, University of California, Davis, 1975.

Sinke, G.C. A method for measurement of vapor pressures of organic compounds below 0.1 torr, naphthalene as a reference substance, *J. Chem. Thermodyn.*, 6:311-316, 1974.

Siragusa, G.R. and DeLaune, D.D. Mineralization and sorption of *p*-nitrophenol in estuarine sediment, *Environ. Toxicol. Chem.*, 5(1):175-178, 1986.

Sisler, H.D. and Cox, C.E. Effects of tetramethyl thiuram disulfide on metabolism of *Fusarium roseum*, *Am. J. Bot.*, 41:338, 1954.

Sittig, M. *Handbook of Toxic and Hazardous Chemicals and Carcinogens* (Park Ridge, NJ: Noyes Publications, 1985), 950 p.

Skea, J.C., Simonin, H.A., Dean, H.J., Colquhoun, J.R., Spagnoli, J.J., and Veith, G.D. Bioaccumulation of aroclor 1016 in Hudson River fish, *Bull. Environ. Contam. Toxicol.*, 22(3):332-336, 1979.

Skrzec, A.E. and Murphy, N.F. Liquid-liquid equilibria. Acetic acid in water, with 1-butanol, methyl ethyl ketone, furfural, cyclohexanol, and nitromethane, *Ind. Eng. Chem.*, 46(10):2245-2247, 1954.

Skurlatov, Y.I., Zepp, R.G., and Baughman, G.L. Photolysis rates of (2,4,5-trichlorophenoxy)acetic acid and 4-amino-3,5,6-trichloropicolinic acid in natural waters, *J. Agric. Food Chem.*, 31(5):1065-1071, 1983.

Slater, R.M. and Spedding, D.J. Transport of dieldrin between air and water, *Arch. Environ. Contam. Toxicol.*, 10(1):25-33, 1981.

Slinn, W.G.N., Hasse, L., Hicks, B.B., Hogan, A.W., Lal, D., Liss, P.S., Munnich, K.O., Sehmel, G.A., and Vittori, O. Some aspects of the transfer of atmospheric trace constituents past the air-sea interface, *Atmos. Environ.*, 12:2055-2087, 1978.

Slonium, A.R. Acute toxicity of selected hydrazines to the common guppy, *Water Res.*, 11(10):889-895, 1977.

Slooff, W. Detection limits of biological monitoring system based on fish respiration, *Bull. Environ. Contam. Toxicol.*, 23(4/5):517-523, 1979.

Slooff, W. Benthic microinvertebrates and water quality assessment: some toxicological considerations, *Aquat. Toxicol.*, 4:73-82, 1983.

Slooff, W. and Baerselman, R. Comparison of the usefulness of the Mexican axolotl (*Ambystoma mexicanum*) and the clawed toad (*Zenopus laevis*) in toxicological bioassays, *Bull. Environ. Contam. Toxicol.*, 24(3):439-443, 1980.

Slooff, W., Canton, J.H., and Hermens, J.L.M. Comparison of the susceptibility of 22 freshwater species to 15 chemical compounds, *Aquat. Toxicol.*, 4:113-128, 1983.

Small, H. and Bremer, D.N. Sorption of nonionic solutes on ion exchange resins bearing amphiphilic counter ions, *Environ. Sci. Technol.*, 3(4):361-3367, 1964.

Smith, A.E. Identification of 2,4-dichloroanisole and 2,4-dichlorophenol as soil degradation products of ring-labelled [^{14}C]2,4-D, *Bull. Environ. Contam. Toxicol.*, 34(2):150-157, 1985.

Smith, C.T., Shaw, F., Lavigne, R., Archibald, J., Fenner, H., and Stern, D. Residues of malathion on alfalfa and in milk and meat, *J. Econ. Entomol.*, 53:495-496, 1960.

Smith, G.W. Phase behavior of some linear polyphenyls, *Mol. Cryst. Liq. Cryst.*, 49:207-209, 1979.

Smith, J.A., Chiou, C.T., Kammer, J.A., and Kile, D.E. Effect of soil moisture on sorption of trichloroethylene vapor to vadose-zone soil at Picatinny Arsenal, New Jersey, *Environ. Sci. Technol.*, 24(5):676-698, 1990.

Smith, J.G., Lee, S.-F., and Netzer, A. Model studies in aqueous chlorination: the chlorination of phenols in dilute aqueous solutions, *Water Res.*, 10(11):985-990, 1976.

Smith, J.H., Bomberger, D.C., Jr., and Haynes, D.L. Prediction of the volatilization rates of high-volatility chemicals from natural water bodies, *Environ. Sci. Technol.*, 14(11):1332-1337, 1980.

Smith, J.H., Mabey, W.R., Bohonos, N., Holt, B.R., and Lee, S.S. Environmental pathways of selected chemicals in freshwater systems. Part II. Laboratory studies, Environmental Research Laboratory, Athens, GA., U.S. EPA Report-600/7-78-074, 1978, 432 p.

Smith, L.R. and Dragun, J. Degradation of volatile chlorinated aliphatic priority pollutants in groundwater, *Environ. Int.*, 19(4):291-298, 1984.

Smith, R.V. and Rosazza, J.P. Microbial models of mammalian metabolism. Aromatic hydroxylation, *Arch. Biochem. Biophys.*, 161(2):551-558, 1974.

Smith, S., Reagan, T.E., Flynn, J.L., and Willis, G.H. Azinphosmethyl and fenvalerate runoff loss from a sugarcane-insect IPM system, *J. Environ. Qual.*, 12(4):534-537, 1983.

Smyth, H.F., Jr. and Carpenter, C.P. Further experience with the range finding test in the industrial toxicology laboratory, *J. Ind. Hyg. Toxicol.*, 30:63-68, 1948.

Snell, T. and Persoone, G. Acute toxicity bioassays using rotifers. I. A test for brackish and marine environments with *Brachionus plicatilis*, *Aquat. Toxicol.*, 14:65-80, 1989.

Snider, E.H. and Alley, F.C. Kinetics of the chlorination of biphenyl under conditions of waste treatment processes, *Environ. Sci. Technol.*, 13(10):1244-1248, 1979.

Snider, J.R. and Dawson, G.A. Tropospheric light alcohols, carbonyls, and acetonitrile: concentrations in the southwestern United States and Henry's law data, *J. Geophys. Res., D: Atmos.*, 90(D2):3797-3805, 1985.

Sobotka, H. and Kahn, J. Determination of solubility of sparingly soluble liquids in water, *J. Am. Chem. Soc.*, 53(8):2935-2938, 1931.

Socha, S.B. and Carpenter, R. Factors affecting the pore water hydrocarbon concentrations in Puget Sound sediments, *Geochim. Cosmochim. Acta*, 51(5):1273-1284, 1987.

Soderquist, C.J., Crosby, D.G., Moilanen, K.W., Seiber, J.N., and Woodrow, J.E. Occurrence of trifluralin and its photoproducts in air, *J. Agric. Food Chem.*, 23(2):304-309, 1975.

Solon, J.M. and Nair, J.H. The effect of a sublethal concentration of LAS on the acute toxicity of various phosphate pesticides to the fathead minnow *Pimephales promelas* Rafinesque, *Bull. Environ. Contam. Toxicol.*, 5(5):408-413, 1970.

Somasundaram, L. and Coats, J.R. Interactions between pesticides and their major degradation products, in *Pesticide Transformation Products. Fate and Significance in the Environment*, ACS Symposium Series 459, Somasundaram, L. and Coats, J.R., Eds. (Washington, D.C.: American Chemical Society, 1991), pp. 162-171.

Somasundaram, L., Coats, J.R., and Racke, K.D. Degradation of pesticides in soil as influenced by the presence of hydrolysis metabolites, *J. Environ. Sci. Health*, B24(5):457-478, 1989.

Somasundaram, L., Coats, J.R., Racke, K.D., and Shanbhag, V.M. Mobility of pesticides and their metabolites in soil, *Environ. Toxicol. Chem.*, 10(2):185-194, 1991.

Somasundaram, L., Coats, J.R., Racke, K.D., and Stahr, H.M. Application of the Microtox system to assess the toxicity of pesticides and their hydrolysis metabolites, *Bull. Environ. Contam. Toxicol.*, 44(2):254-259, 1990.

Sonnefeld, W.J., Zoller, W.H., and May, W.E. Dynamic coupled-column liquid chromatographic determination of ambient temperature vapor pressures of polynuclear aromatic hydrocarbons, *Anal. Chem.*, 55(2):275-280, 1983.

Sonobe, H., Kamps, L.R., Mazzola, E.P., and Roach, J.A.G. Isolation and identification of a new conjugated carbofuran metabolite in carrots: angelic acid ester of 3-hydroxycarbofuran, *J. Agric. Food Chem.*, 29(6):1125-1129, 1981.

Soriano-Cano, M.C., Sotomayor-Sanchez, J.A., Sanchez-Gomez, P., and Garcia-Vallejo, M.C. Essential oils of the *Rosemarinus eriocalyx-tomentosus* complex in southeast Spain, *J. Ess. Oil Res.*, 5:243-246, 1993.

Southworth, G.R. Transports and transformation of anthracene in natural waters, in *Aquatic Toxicology, ASTM STP 667*, Marking, L.L. and Kimerle, R.A., Eds. (Philadelphia, PA: American Society for Testing and Materials, 1977), pp. 359-380.

Southworth, G.R. The role of volatilization in removing polycyclic aromatic hydrocarbons from aquatic environments, *Bull. Environ. Contam. Toxicol.*, 21(4/5):507-514, 1979.

Southworth, G.R., Beauchamp, J.J., and Schmieders, P.K. Bioaccumulation potential of polycyclic aromatic hydrocarbons in *Daphnia pulex*, *Water Res.*, 12(11):973-977, 1978.

Southworth, G.R. and Keller, J.L. Hydrophobic sorption of polar organics by low organic carbon soils, *Water, Air, Soil Pollut.*, 28(3-4):239-248, 1986.

Spacek, W., Bauer, R., and Heisler, G. Heterogeneous and homogeneous wastewater treatment - Comparison between photodegradation with TiO_2 and photo-Fenton reaction, *Chemosphere*, 30(3):477-484, 1995.

Spacie, A., Lundrum, R.F., and Leversee, G.L. Uptake, depuration and biotransformation of anthracene and benzo[*a*]pyrene in bluegill sunfish, *Ecotoxicol. Environ. Saf.*, 7(3):330-341, 1983.

Spaght, T., Thomas, S.B., and Parks, G.S. Some heat capacity data on organic compounds obtained with a radiation calorimeter, *J. Phys. Chem.*, 36:882-888, 1932.

Sparling, D.W., Day, D., and Klein, P. Acute toxicity and sublethal effects of white phosphorus in mute swans, *Cygnus olor*, *Arch. Environ. Contam. Toxicol.*, 36(3):316-322, 1999.

Special Occupational Hazard Review for DDT (Rockville, MD: National Institute for Occupational Safety and Health, 1978), 205 p.

Spehar, R.L., Christensen, G.M., Curtis, C., Lemke, A.E., Norberg, T.J., and Pickering, Q.H. Effects of pollution on freshwater fish, *J. Water Pollut. Control Fed.*, 54(6):877-922, 1982.

Spehar, R.L., Veith, G.D., DeFoe, D.L., and Bergstedt, B.V. Toxicity and bioaccumulation of hexachlorocyclopentadiene, hexachloronorbornadiene and heptachloronorbornene in larval and early juvenile fathead minnows, *Pimephales promelas*, *Bull. Environ. Contam. Toxicol.*, 21(4/5):576-583, 1979.

Spence, J.W., Hanst, P.L., and Gay, B.W., Jr. Atmospheric oxidation of methyl chloride, methylene chloride, and chloroform, *J. Air Pollut. Control Assoc.*, 26(10):994-996, 1976.

Spencer, W.F., Adams, J.D., Shoup, T.D., and Spear, R.C. Conversion of parathion to paraoxon on soil dusts and clay minerals as affected by ozone and UV light, *J. Agric. Food Chem.*, 28(2):366-371, 1980.

Spencer, W.F. and Cliath, M.M. Vapor density of dieldrin, *Environ. Sci. Technol.*, 3(7):670-674, 1969.

Spencer, W.F. and Cliath, M.M. Vapor density and apparent vapor pressure of lindane (γ-BHC), *J. Agric. Food Chem.*, 18(3):529-530, 1970.

Spencer, W.F. and Cliath, M.M. Volatility of DDT and related compounds, *J. Agric. Food Chem.*, 20(3):645-649, 1972.

Spencer, W.F., Cliath, M.M., Jury, W.A., and Zhang, L.-Z. Volatilization of organic chemicals from soil as related to their Henry's law constants, *J. Environ. Qual.*, 17(3):504-509, 1988.

Speth, T.F. and Miltner, R.J. Technical note: adsorption capacity of GAC for synthetic organics, *J. Am. Water Works Assoc.*, 90(4):171-174, 1998.

Staffelbach, T.A. and Kok, G.L. Henry's law constants for aqueous solutions of hydrogen peroxide and hydroxymethyl hydroperoxide, *J. Geophys. Res.*, 98D:12713-12717, 1993.

Stalling, D.L. and Meyer, F.L. Toxicities of PCBs to fish and environmental residues, *Environ. Health Perspect.*, 1:159-164, 1972.

Standen, A., Ed. *Kirk-Othmer Encyclopedia of Chemical Technology, Volume 1*, 2nd ed. (New York: John Wiley & Sons, 1963), 990 p.

Standen, A., Ed. *Kirk-Othmer Encyclopedia of Chemical Technology, Volume 3*, 2nd ed. (New York: John Wiley & Sons, 1964), 927 p.

Standen, A., Ed. *Kirk-Othmer Encyclopedia of Chemical Technology, Volume 5*, 2nd ed. (New York: John Wiley & Sons, 1965), 884 p.

Standen, A., Ed. *Kirk-Othmer Encyclopedia of Chemical Technology, Volume 12*, 2nd ed. (New York: John Wiley & Sons, 1967), 905 p.

Standen, A., Ed. *Kirk-Othmer Encyclopedia of Chemical Technology, Volume 15*, 2nd ed. (New York: John Wiley & Sons, 1968), 923 p.

Standen, A., Ed. *Kirk-Othmer Encyclopedia of Chemical Technology, Volume 19*, 2nd ed. (New York: John Wiley & Sons, 1969), 839 p.

Standen, A., Ed. *Kirk-Othmer Encyclopedia of Chemical Technology, Volume 21*, 2nd ed. (New York: John Wiley & Sons, 1970), 707 p.

Stanley, J.G. and Trial, J.G. Disappearance constants of carbaryl from streams contaminated by forest spraying, *Bull. Environ. Contam. Toxicol.*, 25(5):771-776, 1980.

Starr, R.I. and Cunningham, D.J. Leaching and degradation of 4-aminopyridine-^{14}C in several soil systems, *Arch. Environ. Contam. Toxicol.*, 3(1):72-83, 1975.

Starr, R.I., Timm, R.W., Doxtader, K.G., Hurlbut, D.B., Volz, S.A., and Goodall, M. Sorption and aerobic biodegradation of strychnine alkaloid in various soil systems, *J. Agric. Food Chem.*, 44(6):1603-1608, 1996.

Stauffer, T.B. and MacIntyre, W.G. Sorption of low-polarity organic compounds on oxide minerals and aquifer material, *Environ. Toxicol. Chem.*, 5(11):949-955, 1986.

Stauffer, T.B., MacIntyre, W.G., and Wickman, D.C. Sorption of nonpolar organic chemicals on low-carbon-content aquifer materials, *Environ. Toxicol. Chem.*, 8(10):845-852, 1989.

Stearns, R.S., Oppenheimer, H., Simon, E., and Harkins, W.D. Solubilization by solutions of long-chain colloidal electrolytes, *J. Chem. Phys.*, 15(7):496-507, 1947.

Stedman, D.H. and Niki, H. Ozonolysis rates of some atmospheric gases, *Environ. Lett.*, 4:303-310, 1973.

Steele, W.V., Chirico, R.D., Knipmeyer, S.E., and Nguyen, A. Vapor pressure, heat capacity, and density along the saturation line, measurements for cyclohexanol, 2-cyclohexen-1-one, 1,2-dichloropropane, 1,4-di-*tert*-butyl benzene, (±)-2-ethylhexanoic acid, 2-(methylamino)ethanol, perfluoro-*n*-heptane, and sulfolane, *J. Chem. Eng. Data*, 42(6):1021-1036, 1997.

Steen, W.C. Microbial transformation rate constants of structurally diverse man-made chemicals, Environmental Research Laboratory, Athens, GA., U.S. EPA Report-600/S93-91/016, 1991, 2 p.

Steen, W.C. and Karickhoff, S.W. Biosorption of hydrophobic organic pollutants by mixed microbial populations, *Chemosphere*, 10(1):27-32, 1981.

Steen, W.C., Paris, D.F., and Baughman, G.L. Partitioning of selected polychlorinated biphenyls to natural sediments, *Water Res.*, 12(9):655-657, 1978.

Steenson, T.I. and Walker, N. The pathway of breakdown of 2,4-dichloro- and 4-chloro-2-methylphenoxyacetic acid by bacteria, *J. Gen. Microbiol.*, 16:146-155, 1957.

Stehly, G.R. and Hayton, W.L. Effect of pH on the accumulation kinetics of pentachlorophenol in goldfish, *Arch. Environ. Contam. Toxicol.*, 19(3):464-470, 1990.

Steinberg, S.M., Pignatello, J.J., and Sawhney, B.L. Persistence of 1,2-dibromoethane in soils: entrapment in intraparticle micropores, *Environ. Sci. Technol.*, 21(12):1201-1208, 1987.

Steinhagen, W.H., Swenberg, J.A., and Barrow, C.S. Acute inhalation toxicity and sensory irritation of dimethylamine, *Am. Ind. Hyg. Assoc. J.*, 43(6):411-417, 1982.

Steinwandter, H. Experiments on lindane metabolism in plants III. Formation of β-HCH, *Bull. Environ. Contam. Toxicol.*, 20(4):535-536, 1978.

Steinwandter, H. and Schluter, H. Experiments on lindane metabolism in Plants IV. A kinetic investigation, *Bull. Environ. Contam. Toxicol.*, 20(2):174-179, 1978.

Stenalt, E., Johansen, B., Lillienskjold, S., and Hansen, B.W. Mesocosm study of *Mytilus edulis* larvae and postlarvae, including the settlement phase, exposed to a gradient of tributyltin, *Ecotoxicol. Environ. Saf.*, 40(3):212-225, 1998.

Stephen, H. and Stephen, T. *Solubilities of Inorganic and Organic Compounds - Part 1, Volume 1* (London: Pergamon Printing & Art Services, Ltd., 1963), 960 p.

Stephenson, G.L., Kaushik, N.K., and Solomon, K.R. Acute toxicity of pure pentachlorophenol and a technical formulation to three species of *Daphnia*, *Arch. Environ. Contam. Toxicol.*, 20(1):73-80, 1991.

Stephenson, R. Mutual solubilities: Water-ketones, water-ethers, and water-gasoline-alcohols, *J. Chem. Eng. Data*, 37(1):80-95, 1992.

Stephenson, R. Mutual solubility of water and pyridine derivatives, *J. Chem. Eng. Data*, 38(3):428-431, 1993.

Stephenson, R. Mutual solubility of water and aliphatic amines, *J. Chem. Eng. Data*, 38(4):625-629, 1993a.

Stephenson, R. Mutual solubility of water and aldehydes, *J. Chem. Eng. Data*, 38(4):630-633, 1993b.

Stephenson, R. and Stuart, J. Mutual binary solubilities: water-alcohols and water-esters, *J. Chem. Eng. Data*, 31(1):56-70, 1986.

Stephenson, R., Stuart, J., and Tabak, M. Mutual solubility of water and aliphatic alcohols, *J. Chem. Eng. Data*, 29(3):287-290, 1984.

Stephenson, R.M. and Malanowski, S. *Handbook of the Thermodynamic of Organic Compounds* (New York: Elsevier Science, 1987).

Stepp, T.D., Camper, N.D., and Paynter, M.J.B. Anaerobic microbial degradation of selected 2,4-dihalogenated aromatic compounds, *Pestic. Biochem. Physiol.*, 23:256-260, 1985.

Sterner, R.T. Effects of sodium cyanide and diphacinone in coyotes (*Canis latrans*): applications as predacides in livestock toxic collars, *Bull. Environ. Contam. Toxicol.*, 23(1/2):211-217, 1979.

Stevens, A.A., Slocum, C.J., Seeger, D.R., and Robeck, G.G. Chlorination of organics in drinking water, *J. Am. Water Works Assoc.*, 68(11):615-620, 1976.

Stevens, J.T., DiPasquale, L.C., and Farmer, J.D. The acute inhalation toxicology of the technical grade organoarsenical herbicides, cacodylic acid and disodium methanearsonic acid: a route comparison, *Bull. Environ. Contam. Toxicol.*, 21(3):304-311, 1979.

Stewart, D.K.R. and Cairns, K.G. Endosulfan persistence in soil and uptake by potato tubers, *J. Agric. Food Chem.*, 22(6):984-986, 1974.

Stewart, D.K.R. and Chisholm, D. Long-term persistence of BHC, DDT, and chlordane in a sandy loam soil, *Can. J. Soil Sci.*, 51:379-383, 1971.

Stewart, D.K.R. and Gaul, S.O. Persistence of 2,4-D and 2,4,5-T and dicamba in a Dykeland soil, *Bull. Environ. Contam. Toxicol.*, 18(2):210-218, 1977.

Stewart, N.E., Millemann, R.E., and Breese, W.P. Acute toxicity of the insecticide sevin and its hydrolysis product 1-naphthol to some marine organisms, *Trans. Am. Fish. Soc.*, 96(1):25-30, 1967.

Stockdale, M. and Selwyn, M.J. Effects of rings substituents on the activity of phenols as inhibitors and uncouplers of mitochondrial respiration, *Eur. J. Biochem.*, 21:565-574, 1971.

Stockhardt, J.S. and Hull, C.M. Vapor-liquid equilibria and boiling point composition relations for systems *n*-butanol-water and isobutanol-water, *Ind. Eng. Chem.*, 23:1438-1440, 1931.

Stott, D.E., Martin, J.P., Focht, D.D., and Haider, K. Biodegradation, stabilization in humus, and incorporation into soil biomass of 2,4-d and chlorocatechol carbons, *Soil Sci. Soc. Am. J.*, 47:66-70, 1983.

Streitwieser, A., Jr. and Nebenzahl, L.L. Carbon acidity. LII. Equilibrium acidity of cyclopentadiene in water and in cyclohexylamine, *J. Am. Chem. Soc.*, 98(8):2188-2190, 1976.

Struif, B., Weil, L., and Quinn, K.-E. Verhalten herbizider Phenoxyalkancarbonäuren bei der Wasseraufbereitung mit Ozon, *Zeit. für Wasser und Abwasser-Forschung*, 3/4:118-127, 1978.

Stucki, G. and Alexander, M. Role of dissolution rate and solubility in biodegradation of aromatic compounds, *Appl. Environ. Microbiol.*, 53(2):292-297, 1987.

Stull, D.R. A semi-micro calorimeter for measuring heat capacities at low temperatures, *J. Am. Chem. Soc.*, 59:2726-2733, 1937.

Stull, D.R. Vapor pressures of pure organic compounds, *Ind. Eng. Chem.*, 39(4):517-560, 1947.

Su, F., Calvert, J.G., and Shaw, J.H. Mechanism of the photooxidation of gaseous formaldehyde, *J. Phys. Chem.*, 83(25):3185-3191, 1979.

Suba, J.D. and Essington, M.E. Adsorption of fluometuron and norflurazon: effect of tillage and dissolved organic carbon, *Soil Sci.*, 164(3):145-155, 1999.

Subba-Rao, R.V. and Alexander, M. Effect of DDT metabolites on soil respiration and on an aquatic alga, *Bull. Environ. Contam. Toxicol.*, 25(2):215-220, 1980.

Subba-Rao, R.V., Rubin, H.E., and Alexander, M. Kinetics and extent of mineralization of organic chemicals at trace levels in freshwater and sewage, *Appl. Environ. Microbiol.*, 43(5):1139-1150, 1982.

Sud, R.K., Sud, A.K., and Gupta, K.G. Degradation of sevin (1-naphthyl *N*-methylcarbamate) by *Achromobacter* sp., *Arch. Microbiol.*, 87:353-358, 1972.

Sudhakar-Barik and Sethunathan, N. Biological hydrolysis of parathion in natural ecosystems, *J. Environ. Qual.*, 7(3):346-348, 1978.

Sudhakar-Barik and Sethunathan, N. Metabolism of nitrophenols in flooded soils, *J. Environ. Qual.*, 7(3):349-352, 1978a.

Sudhakar-Barik, Siddaramappa, R., and Sethunathan, N. Metabolism of nitrophenols by bacteria isolated from parathion-amended flooded soil, *Antonie van Leeuwenhoek*, 42(4):461-470, 1976.

Sudhakar-Barik, Siddaramappa, R., Wahid, P.A., and Sethunathan, N. Conversion of *p*-nitrophenol to 4-nitrocatechol by a *Pseudomonas* sp., *Antonie van Leeuwenhoek*, 44(2):171-176, 1978a.

Sudhakar-Barik, Wahid, P.A., Ramakrishna, C., and Sethunathan, N. A change in the degradation pathway of parathion after repeated applications to flooded soil, *J. Agric. Food Chem.*, 27(6):1391-1392, 1979.

Suess, M.J. The environmental load and cycle of polycyclic aromatic hydrocarbons, *Sci. Total Environ.*, 6:239-250, 1976.

Suffet, I.H., Faust, S.D., and Carey, W.F. Gas-liquid chromatographic separation of some organophosphate pesticides, their hydrolysis products, and oxons, *Environ. Sci. Technol.*, 1(8):639-643, 1967.

Suflita, J.M., Stout, J., and Tiedje, J.M. Dechlorination of (2,4,5-trichlorophenoxy)acetic acid by anaerobic micro-organisms, *J. Agric. Food Chem.*, 32(2):218-221, 1984.

Sugawara, T., Kato, M., Furuhama, K., Inage, F., Suzuki, N., and Takayama, S. Single dose toxicity study of the new cognition-enhancing agent nefiracetam in mice, rats and dogs, *Arzneim. Forsch.*, 44(2a):211-213, 1994.

Sugiura, K., Washino, T., Hattori, M., Sato, E., and Goto, M. Accumulation of organochlorine compounds in fishes. Difference of accumulation factors by fishes, *Chemosphere*, 8(6):359-364, 1979.

Sullivan, R.G., Knoche, H.W., and Markle, J.C. Photolysis of trifluralin: characterization of azobenzene and azoxybenzene photodegradation products, *J. Agric. Food Chem.*, 28(4):746-755, 1980.

Sumer, K.M. and Thompson, A.R. Refraction, dispersion, and densities of benzene, toluene, and xylene mixtures, *J. Chem. Eng. Data*, 13(1):30-34, 1968.

Summers, L. The *alpha*-haloalkyl ethers, *Chem. Rev.*, 55:301-353, 1955.

Sun, S. and Boyd, S.A. Sorption of nonionic organic compounds in soil-water systems containing petroleum sulfonate-oil surfactants, *Environ. Sci. Technol.*, 27(7):1340-1348, 1993.

Sun, Y. and Pignatello, J.J. Photochemical reactions involved in the total mineralization of 2,4-D by $Fe^{3+}/H_2O_2/UV$, *Environ. Sci. Technol.*, 27(2):304-310, 1993.

Sunderman, F.W., Jr., Plowman, M.C., and Hopfer, S.M. Embryotoxicity and teratogenicity of calcium chloride in *Xenopus laevis*, assayed by the FETAX procedure, *Ann. Clin. Lab. Sci.*, 21(6):381-391, 1991.

Sunshine, I., Ed. *Handbook of Analytical Toxicology* (Cleveland, OH: Chemical Rubber, 1969), 1081 p.

Suntio, L.R., Shiu, W.-Y., and Mackay, D. A review of the nature and properties of chemicals present in pulp mill effluents, *Chemosphere*, 17(7):1249-1290, 1988.

Sutton, C. and Calder, J.A. Solubility of higher-molecular-weight *n*-paraffins in distilled water and seawater, *Environ. Sci. Technol.*, 8(7):654-657, 1974.

Sutton, C. and Calder, J.A. Solubility of alkylbenzenes in distilled water and seawater at 25 °C, *J. Chem. Eng. Data*, 20(3):320-322, 1975.

Suzuki, J., Hagino, T., and Suzuki, S. Formation of 1-nitropyrene by photolysis of pyrene in water containing nitrite ion, *Chemosphere*, 16(4):859-867, 1987.

Suzuki, M., Yamato, Y., and Watanabe, T. Residue in soil, organochlorine insecticide residues in field soils of the Kita Kyusha District - Japan 1970-1974, *Pestic. Monitor. J.*, 11:88-93, 1977.

Svec, H.J. and Junk, G.A. Electron-impact studies of substituted alkanes, *J. Am. Chem. Soc.*, 89(4):790-796, 1967.

Swain, C.G. and Thornton, E.R. Initial-state and transition-state isotope effects of methyl halides in light and heavy water, *J. Am. Chem. Soc.*, 84:822-826, 1962.

Swann, R.L., Laskowski, D.A., McCall, P.J., Vander Kuy, K., and Dishburger, H.J. A rapid method for the estimation of the environmental parameters octanol/water partition coefficient, soil sorption constant, water to air ratio, and water solubility, *Residue Rev.*, 85:17-28, 1983.

Swartz, R.C., Ferraro, S.P., Lamberson, J.O., Cole, F.A., Ozretich, R.J., Boese, B.L., Schults, D.W., Behrenfeld, M., and Ankley, G.T. Photoactivation and toxicity of mixtures of polycyclic aromatic hydrocarbon compounds in marine sediment, *Environ. Toxicol. Chem.*, 16(10):2151-2157, 1997.

Syunyaev, Z.I., Tumanyan, B.P., Kolesnikov, S.I., and Zhokhova, N.I. Some anomalies in melting points of binary mixtures of solid hydrocarbons, *Zhur. Prikl. Khim. (Leningrad)*, 57:666-669, 1984.

Szabo, G., Prosser, S.L., and Bulman, R.A. Determination of the adsorption coefficient (K_{oc}) of some aromatics for soil by RP-HPLC on two immobilized humic acid phases, *Chemosphere*, 21(6):777-788, 1990.

Tabak, H.H., Quave, S.A., Mashni, C.I., and Barth, E.F. Biodegradability studies with organic priority pollutant compounds, *J. Water Pollut. Control Fed.*, 53(10):1503-1518, 1981.

Taha, A.A., Grigsby, R.D., Johnson, J.R., Christian, S.D., and Affsprung, H.E. Manometric apparatus for vapor and solution studies, *J. Chem. Educ.*, 43(8):432-43435, 1966.

Takagi, H., Hatakeyama, S., Akimoto, H., and Koda, S. Formation of methyl nitrite in the surface reaction of nitrogen dioxide and methanol. 1. Dark reaction, *Environ. Sci. Technol.*, 20(4):387-393, 1986.

Takahashi, H., Yoshida, M., Murao, N., and Maita, K. Different inhalation lethality between micron-sized and submicron-sized aerosols of organophosphorus insecticide, chlorfenvinphos, in rats, *Toxicol. Lett.*, 73(2):103-111, 1994.

Takahashi, N. Ozonation of several organic compounds having low molecular weight under ultraviolet irradiation, *Ozone Sci. Eng.*, 12(1):1-18, 1990.

Takeuchi, K., Yazawa, T., and Ibusuki, T. Heterogeneous photocatalytic effect of zinc oxide on photochemical smog formation reaction of $C_4H_8-NO_2$-Air, *Atmos. Environ.*, 17(11):2253-2258, 1983.

Talbert, R.E. and Fletchall, O.H. The adsorption of *s*-triazines in soils, *Weeds*, 46-52, 1965.

Talekar, N.S., Sun, L.-T., Lee, E.-M., and Chen, J.-S. Persistence of some insecticides in subtropical soil, *J. Agric. Food Chem.*, 25:348-352, 1977.

Tam, D., Varhaníčková, D., Shiu, W.-Y., and Mackay, D. Aqueous solubility of chloroguaiacols, *J. Chem. Eng. Data*, 39(1):83-86, 1994.

Tan, A.-C., Sorai, M., and Suga, H. Low-temperature heat capacity and thermodynamic functions of 2-chloro-6-(tri-chloromethyl)pyridine, *Sci. China Ser. B*, 32(10):1194-1207, 1989.

Tan, C.K. and Wang, T.C. Reduction of trihalomethanes in a water-photolysis system, *Environ. Sci. Technol.*, 21(5):508-511, 1987.

Tanaka, F.S., Hoffer, B.L., and Wien, R.G. Detection of halogenated biphenyls from sunlight photolysis of chlorinated herbicides in aqueous solution, *Pestic. Sci.*, 16:265-270, 1985.

Tancréde, M.V. and Yanagisawa, Y. An analytical method to determine Henry's law constant for selected volatile organic compounds at concentrations and temperatures corresponding to tap water use, *J. Air Waste Manage. Assoc.*, 40:1658-1663, 1990.

Tanii, H. and Hashimoto, K. Structure-toxicity relationship of acrylates and methacrylates, *Toxicol. Lett.*, 11:125-129, 1982.

Tanii, H. and Hashimoto, K. Studies on the mechanism of acute toxicity of nitriles in mice, *Arch. Toxicol.*, 55:47-54, 1984.

Tanii, H., Tsuji, H., and Hashimoto, K. Structure-activity relationship of monoketones, *Toxicol. Letters*, 30:13-17, 1984.

Tarr, B.D., Barron, M.G., and Hayton, W.L. Effect of body size on the uptake and bioconcentration of bis(2-ethylhexyl) phthalate in rainbow trout, *Environ. Toxicol. Chem.*, 9:989-995, 1990.

Tate, R.L. III, and Alexander, M. Microbial formation and degradation of dimethylamine, *Appl. Environ. Microbiol.*, 31(3):399-403, 1976.

Taylor, E.J., Maund, S.J., and Pascoe, D. Toxicity of our common pollutants to the freshwater macroinvertebrates *Chironomus riparius Meigen* (Insecta: Diptera) and *Gammarus pulex* (L.) (Crustacea: Amphipoda), *Arch. Environ. Contam. Toxicol.*, 21(3):371-376, 1991.

Taymaz, K., Williams, D.T., and Benoit, F.M. Chlorine dioxide oxidation of aromatic hydrocarbons commonly found in water, *Bull. Environ. Contam. Toxicol.*, 23(3):398-404, 1979.

Tchounwou, P.B., Englande, A.J., Jr., and Malek, E.A. Toxicity evaluation of ammonium sulphate and urea to three developmental stages of freshwater snails, *Arch. Environ. Contam. Toxicol.*, 21(3):359-364, 1991.

Tchounwou, P.B., Englande, A.J., Jr., Malek, E.A., Anderson, A.C., and Abdelghani, A.A. The effects of baylusclde and malation on the survival of *Schistosoma mansoni miracidia*, *J. Environ. Sci. Health*, B26(1):69-82, 1991a.

ten Hulscher, Th.E.M., van der Velde, L.E., and Bruggeman, W.A. Temperature dependence of Henry's law constants for selected chlorobenzenes, polychlorinated biphenyls and polycyclic aromatic hydrocarbons, *Environ. Toxicol. Chem.*, 11(11):1595-1603, 1992.

Tesconi, M. and Yalkowsky, S.H. A novel thermogravimetric method for estimating the saturated vapor pressure of low-volatility compounds, *J. Pharm. Sci.*, 87(12):1512-1520, 1998.

Tewari, Y.B., Miller, M.M., Wasik, S.P., and Martire, D.E. Aqueous solubility and octanol/water partition coefficient of organic compounds at 25.0 °C, *J. Chem. Eng. Data*, 27(4):451-454, 1982.

Thangnipon, W., Thangnipon, W., Luangpaiboon, P., and Chinabut, S. Effects of the organophosphorous insecticide, monocrotophos, on acetylcholinesterase activity in the nile tilapia fish (*Oreochromis niloticus*) brain, *Neurochem. Res.*, 20(5):587-591, 1995.

Thom, N.S. and Agg, A.R. The breakdown of synthetic organic compounds in biological processes, *Proc. R. Soc. London*, B189(1096):347-357, 1975.

Thomas, D.B. and Delfino, J.J. A gas chromatographic/chemical indicator approach to assessing ground water contamination by petroleum products, *Groundwater Monitor. Rev.*, 11(4):90-100, 1991.

Thomas, E.W., Loughman, B.C., and Powell, R.G. Metabolic fate of some chlorinated phenoxyacetic acids in the stem tissue of *Avena sativa*, *Nature*, 204:286, 1964.

Thomas, K., Volz-Thomas, A., Mihelcic, D., Smit, H.G.J., and Kley, D. On the exchange of NO_3 radicals with aqueous solutions: solubility and sticking coefficient, *J. Atmos. Chem.*, 29:17-43, 1998.

Thompson, C.R., Kats, G., and Lennox, R.W. Phytotoxicity of air pollutants by high explosive production, *Environ. Sci. Technol.*, 13(10):1263-1268, 1979.

Thybaud, E. and Caquet, T. Uptake and elimination of lindane by *Lymnaea palustris* (mollusca: Gastropoda): a pharmacokinetic approach, *Ecotoxicol. Environ. Saf.*, 21(3):365-376, 1991.

Tierney, D.R., Blackwood, T.R., and Piana, M.R. Status assessment of toxic chemicals: vinylidine chloride, Office of Research and Development, U.S. EPA Report 600/2-79-210, 1979.

Tietze, N.S., Hester, P.G., Hallmon, C.F., Olson, M.A., and Shaffer, K.R. Acute toxicity of mosquitocidal compounds to young mosquitofish, *Gambusia affinis*, *J. Am. Mosq. Control Assoc.*, 7(2):290-293, 1991.

Tietze, N.S., Schreiber, E.T., Hester, P.G., Hallmon, C.F., Olson, M.A., and Shaffer, K.R. Susceptibility of first instar *Toxorhynchites splendens* to malathion, naled and resmethrin, *J. Am. Mosq. Control Assoc.*, 9(1):97-99, 1993.

Tinsley, L.J. *Chemical Concepts in Pollutant Behavior* (New York: John Wiley & Sons, 1979), 265 p.

Tipker, J., Groen, C.P., Van Den Bergh-Swart, J.K., and Van Den Berg, J.H.M. Contribution of electronic effects to the lipophilicity determined by comparison of values of log *P* obtained by high-performance liquid chromatography and calculation, *J. Chromatogr.*, 452:227-239, 1988.

Tjeerdema, R.S., Fan, T.W., Higashi, R.M., and Crosby, D.G. Sublethal effects of pentachlorophenol in the abalone (*Haliotis rufescens*) as measured by *in vivo* 31P NMR spectroscopy, *J. Biochem. Toxicol.*, 6(1):45-56, 1991.

Todd, S.S., Oliver, G.D., and Huffman, H.M. The heat capacities, heats of fusion and entropies of the six pentenes, *J. Am. Chem. Soc.*, 69:1519-1525, 1947.

Tokoro, R., Bilcovicz, R., and Osteryoung, J. Polarographic reduction and determination of 1,2-dibromoethane in aqueous solutions, *Anal. Chem.*, 58(9):1964-1969, 1986.

Tolocka, M.P. and Miller, J.H. Production of polycyclic hydrocarbons from underventilated hydrocarbon diffusion flames, in *Proceedings: Chemical and Physical Processes in Combustion* (Worcester, MA: Combustion Institute/Eastern States section, October 16-18, 1995), pp. 253-256.

Tomkiewicz, M.A., Groen, A., and Cocivera, M. Electron paramagnetic resonance spectra of semiquinone intermediates observed during the photooxidation of phenol in water, *J. Am. Chem. Soc.*, 93(25):7102-7103, 1971.

Tomlinson, E. Chromatographic hydrophobic parameters in correlation analysis of structure-activity relationships, *J. Chromatogr.*, 113:1-45, 1975.

Tonogai, Y., Ogawa, S., Ito, Y., and Twaida, M. Actual survey on Tl_m (median tolerance limit) values of environmental pollutants, especially on amines, aromatic nitrogen compounds and artificial dyes, *J. Toxicol. Sci.*, 7:193-203, 1982.

Torrents, A., Jayasundera, S., and Schmidt, W.J. Influence of the polarity of organic matter on the sorption of acetamide pesticides, *J. Agric. Food Chem.*, 45(8):3320-3325, 1997.

Torres, P., Tort, L., and Flos, R. Acute toxicity of copper to Mediterranean dogfish, *Comp. Biochem. Physiol. C*, 86(1):169-171, 1987.

Tou, J.C. and Kallos, G.J. Kinetic study of the stabilities of chloromethyl methyl ether and bis(chloroethyl) ether in humid air, *Anal. Chem.*, 46(12):1866-1869, 1974.

Tou, J.C. and Kallos, G.J. Study of aqueous HCl and formaldehyde mixtures for formation of bis(chloromethyl) ether, *J. Am. Ind. Hyg. Assoc.*, 35(7):419-422, 1974a.

Tou, J.C., Westover, L.B., and Sonnabend. Kinetic studies of bis(chloromethyl) ether hydrolysis by mass spectrometry, *J. Phys. Chem.*, 78(11):1096-1098, 1974.

Toyota, H. and Kuwahara, M. The study of production of PCP chemical fertilizer and its effect as herbicide and fertilizer, the solubility in water of PCP in PCP chemical fertilizer, *Nippon Dojohiryogaku Zasshi.* 38(2,3):93-97, 1967.

Tratnyek, P.G. and Hoigné, J. Oxidation of substituted phenols in the environment: a QSAR analysis of rate constants for reaction with singlet oxygen, *Environ. Sci. Technol.*, 25(9):626-631, 1991.

Travenius, S.Z.M. Formation and occurrence of bis(chloromethyl) ether and its prevention in the chemical industry, *Scand. J. Work Environ. Health*, 8(Suppl. 3):1-86, 1982.

Travis, C.C. and Arms, A.D. Bioconcentration of organics in beef, milk and vegetation, *Environ. Sci. Technol.*, 22(3):271-274, 1988.

Travis, C.C. and Hattemer-Frey, H. Human exposure to 2,3,7,8-TCDD, *Chemosphere*, 16(10-12):2331-2342, 1987.

Trayler, K.M. and Davis, J.A. Sensitivity of *Daphnia carinata sensu lato* to the insect growth regulator, pyriproxyfen, *Ecotoxicol. Environ. Saf.*, 33(2):154-156, 1996.

Trochimowicz, H.J. The toxicity of HCFC-123 (2,2-dichloro-1,1,1-trifluoroethane), *J. Am. Coll. Toxicol.*, 8(6):1220, 1989.

Trstenjak, B. and Perdih, A. Sorption of polychlorinated biphenyls by the fungus *Phanerochaete chrysosporium*, *Acta Chim. Slov.*, 46(3):307-313, 1999.

Trucco, R.G., Englehardt, F.R., and Tracey, B. Toxicity, accumulation and clearance of aromatic hydrocarbons in *Daphnia pulex*, *Environ. Poll. Ser. A*, 31:191-202, 1983.

Trujillo, D.A., Ray, L.E., Murray, H.E., and Giam, C.S. Bioaccumulation of pentachlorophenol by killifish (*Fundulus similus*), *Chemosphere*, 11:25-31, 1982.

Tse, G., Orbey, H., and Sandler, S.I. Infinite dilution activity coefficients and Henry's law coefficients of some priority water pollutants determined by a relative gas chromatographic method, *Environ. Sci. Technol.*, 25(10):2017-2022, 1992.

Tsierkezos, N.G., Palaiologou, M.M., and Molinou, I.E. Densities and viscosities of 1-pentanol binary mixtures at 293.15 K, *J. Chem. Eng. Data*, 45(2):272-275, 2000.

Tsuda, T., Aoki, S., Kojima, M., and Harada, H. Bioconcentration and excretion of diazinon, IBP, malathion and fenitrothion by willow shiner, *Toxicol. Environ. Chem.*, 24(3):185-190, 1989.

Tsuda, T., Kojima, M., Harada, H., Nakajima, A., and Aoki, S. Acute toxicity, accumulation and excretion of organophosphorous insecticides and their oxidation products in killifish, *Chemosphere*, 35(5):939-949, 1997.

Tu, C.M. Utilization and degradation of lindane by soil microorganisms, *Arch. Microbiol.*, 108(3):259-263, 1976.

Tuazon, E.C., Atkinson, R., Aschmann, S.M., Arey, J., Winer, A.M., and Pitts, J.N., Jr. Atmospheric loss processes of 1,2-dibromo-3-chloropropane and trimethyl phosphate, *Environ. Sci. Technol.*, 20(10):1043-1046, 1984.

Tuazon, E.C., Atkinson, R., Winer, A.M., and Pitts, J.N., Jr. A study of the atmospheric reactions of 1,3-dichloropropene and other selected organochlorine compounds, *Arch. Environ. Contam. Toxicol.*, 13(6):691-700, 1984.

Tuazon, E.C., Carter, W.P.L., Atkinson, R., Winer, A.M., and Pitts, J.N., Jr. Atmospheric reactions of *N*-nitrosodimethylamine and dimethylnitramine, *Environ. Sci. Technol.*, 18(1):49-54, 1984a.

Tuazon, E.C., Leod, H.M., Atkinson, R., and Carter, W.P.L. α-Dicarbonyl yields from the NO_x-air photooxidations of a series of aromatic hydrocarbons in air, *Environ. Sci. Technol.*, 20(4):383-387, 1986a.

Tuazon, E.C., Winer, A.M., Graham, R.A., Schmid, J.P., and Pitts, J.N., Jr. Fourier transform infrared detection of nitramines in irradiated amine-NO_x systems, *Environ. Sci. Technol.*, 12(8):954-958, 1978.

Tuhkanen, T.A. and Beltrán, F.J. Intermediates in the oxidation of naphthalene in water with the combination of hydrogen peroxide and UV radiation, *Chemosphere*, 30(8):1463-1475, 1995.

Tumen, G. The volatile constituents of *Satureja cuneifolia*, *J. Ess. Oil Res.*, 3:365-366, 1991.

Tundo, P., Facchetti, S., Tumiatti, W., and Fortunati, U.G. Chemical degradation of 2,3,7,8-TCDD by means of polyethylene-glycols in the presence of weak bases and an oxidant, *Chemosphere*, 14(5):403-410, 1985.

Turf & Ornamental Chemicals Reference 1991/92 (New York: Chemical and Pharmaceutical Press, 1991).

Twagilimana, L., Bohatier, J., Groliere, C.-A., Bonnemoy, F., and Sargos, D. A new low-cost microbiotest with the Protozoan *Spirostomum teres*: culture conditions and assessment of sensitivity of the ciliate to 14 pure chemicals, *Ecotoxicol Environ. Saf.*, 41(3):231-244, 1998.

Uchrin, C.G. and Mangels, G. Sorption equilibria of benzene and toluene on two New Jersey Coastal Plain ground water aquifer solids, *J. Environ. Sci. Health*, A22(8):743-758, 1987.

Uchrin, C.G. and Michaels, G. Chloroform sorption to New Jersey Coastal Plain ground water aquifer solids, *Environ. Toxicol. Chem.*, 5(4):339-343, 1986.

Ueberreiter, K. and Orthmann, H.-J. Spezifische Wärme, spezifisches Volumen, Temperatur- und Wärme-Leittähigkeit einiger disunstituierter Benzole und polycyclischer Systeme, *Z. Naturforsch*, 5(a):101-108, 1950.

Uma Devi, V. Bioaccumulation and metabolic effects of cadmium on marine fouling dressinid bivalve, *Mytilopsis sallei* (Recluz), *Arch. Environ. Contam.Toxicol.*, 31(1):47-53, 1996.

Umeyama, H., Nagai, T., and Nogami, H. Mechanism of adsorption of phenols by carbon black from aqueous solution, *Chem. Pharm. Bull.*, 19(8):1714-1721, 1971.

Unger, S.H., Cook, J.R., and Hollenberg, J.S. Simple procedure for determining octanol-aqueous partition, distribution, and ionization coefficients by reversed-phase high-pressure liquid chromatography, *J. Pharm. Sci.*, 67(10):1364-1366, 1978.

Ungnade, H.E. and McBee, E.T. The chemistry of perchlorocyclopentadienes and cyclopentadienes, *Chem. Rev.*, 58:249-320, 1957.

Urano, K. and Kato, Z. Evaluation of biodegradation ranks of priority pollutant organic compounds, *J. Haz. Mater.*, 13:135-145, 1986.

Ursin, C. Degradation of organic chemicals at trace levels in seawater and marine sediment. The effect of concentration on the initial turnover rate, *Chemosphere*, 14(10):1539-1550, 1985.

U.S. EPA. Preliminary study of selected potential environmental contaminants - Optical brighteners, methyl chloroform, trichloroethylene, tetrachloroethylene, ion exchange resins, Office of Toxic Substances, Report 560/2-75-002, 1975, 286 p.

U.S. EPA. Report on the problem of halogenated air pollutants and stratigraphic ozone, Office of Research and Development, Report-600/9-75-008, 1975a, 55 p.

U.S. EPA. A literature survey oriented towards adverse environmental effects resultant from the use of azo compounds, brominated hydrocarbons, EDTA, formaldehyde resins, and *o*-nitrochlorobenzene, Office of Toxic Substances, Report 560/2-76-005, 1976, 480 p.

U.S. EPA. Materials balance for chlorobenzenes. Level 1 - Preliminary, Office of Research and Development, Report-560/13-80-001, 1980.

U.S. EPA. Treatability manual. Volume 1, Office of Research and Development, Report-600/8-80-042a, 1980a, 1035 p.

U.S. EPA. Aquatic fate process data for organic priority pollutants, Office of Water Regulations and Standards, Report 440/4-81-014, 1982, 407 p.

U.S. EPA. Chemical, physical, and biological properties of compounds present at hazardous waste sites, Report 530/SW-89-010, 1985, 619 p.

U.S. EPA. Pesticides in ground water: background document, Office of Ground-Water Protection, 1986.

U.S. EPA. Sediment quality criteria for the protection of benthic organisms: acenaphthene, Office of Science and Technology, Report 822-R-93-013, 1993, 90 p.

U.S. EPA. Drinking water regulations and health advisories, Office of Water, Report 822-B-002, 1996.

Uyeta, M., Taue, S., Chikasawa, K., and Mazaki, M. Photoformation of polychlorinated biphenyls from chlorinated benzenes, *Nature (London)*, 264(5586):583-584, 1976.

Vadas, G.G., MacIntyre, W.G., and Burris, D.R. Aqueous solubility of liquid hydrocarbon mixtures containing dissolved solid components, *Environ. Toxicol. Chem.*, 10:633-639, 1991.

Vaishnav, D.D. and Babeu, L. Comparison of occurrence and rates of chemical biodegradation in natural waters, *Bull. Environ. Contam. Toxicol.*, 39(2):237-244, 1987.

Valsaraj, K.T. On the physio-chemical aspects of partitioning of non-polar hydrophobic organics at the air-water interface, *Chemosphere*, 17(5):875-887, 1988.

Valvani, S.C., Yalkowsky, S.H., and Roseman, T.J. Solubility and partitioning IV: aqueous solubility and octanol-water partition coefficients of liquid non-electrolytes, *J. Pharm. Sci.*, 70(5):502-506, 1981.

Valverde-García, A., González-Pradas, E., Villafranca-Sánchez, M., del Rey-Bueno, F., and García-Rodriguez, A. Adsorption of thiram and dimethoate on Almeria soils, *Soil Sci. Soc. Am. J.*, 52(6):1571-1574, 1988.

Van Arkel, A.E. and Vles, S.E. Löslichkeit von organischen Verbindungen in Wasser, *Recl. Trav. Chim. Pays-Bas*, 55:407-411, 1936.

Van Bladel, M. and Moreale, A. Adsorption of herbicide-derived *p*-chloroaniline residues in soils: a predictive equation, *J. Soil Sci.*, 28:93-102, 1977.

Van der Linden, A.C. Degradation of oil in the marine environment, *Dev. Biodegradation Hydrocarbons*, 1:165-200, 1978.

Van Duuren, B.L., Katz, C., Goldschmidt, M., Frenkel, K., and Sirak, A. Carcinogenicity of haloethers. II. Structure-activity relationships of analogs of bis(chloromethyl) ether, *J. Nat. Cancer Inst.*, 48:1431-1439, 1972.

Vane, L.M. and Giroux, E.L. Henry's law constants and micellar partitioning of volatile organic compounds in surfactant solutions, *J. Chem. Eng. Data*, 45(1):38-47, 2000.

Van Gestel, C.A.M. and Ma, W.-C. Toxicity and bioaccumulation of chlorophenols in earthworms, in relation to bioavailability in soil, *Ecotoxicol. Environ. Saf.*, 15(3):289-297, 1988.

Van Gestel, C.A.M. and Ma, W.-C. Development of QSAR's in soil ecotoxicology: earthworm toxicity and soil sorption of chlorophenols, chlorobenzenes and chloroanilines, *Water, Air Soil Pollut.*, 69(3-4):265-276, 1993.

Van Ginkel, C.G., Welten, H.G.J., and de Bont, J.A.M. Oxidation of gaseous and volatile hydrocarbons by selected alkene-utilizing bacteria, *Appl. Environ. Microbiol.*, 53(12):2903-2907, 1987.

van Haelst, A.G., Zhao, Q., van der Wielen, F.W.M., and Govers, H.A.J. Determination of aqueous solubilities of tetra-chlorobenzyltoluenes individually and in a mixture by a modified generator column technique, *Chemosphere*, 33(2):257-264, 1996.

van Hattum, B., Curto Pons, M.J., and Cid Montañés, J.F. Polycyclic aromatic hydrocarbons in freshwater isopods and field-partitioning between abiotic phases, *Arch. Environ. Contam. Toxicol.*, 35(2):257-267, 1998.

van Hoogan, G. and Opperhuizen, A. Toxicokinetics of chlorobenzenes in fish, *Environ. Toxicol. Chem.*, 7:213-219, 1988.

Van Leeuwen, Adema, D.M.M., and Hermens, J. Quantitative structure-activity relationships for fish early life stage toxicity, *Aquat. Toxicol.*, 16:321-334, 1990.

Van Meter, F.M. and Neumann, H.M. Solvation of the tris(1,10-phenanthroline)iron(II) cation as measured by solubility and nuclear magnetic resonance shifts, *J. Am. Chem. Soc.*, 98(6):1382-1388, 1976.

Van Ness, G.F., Solch, J.G., Taylor, M.L., and Tiernan, T.O. Tetrachlorodibenzo-*p*-dioxins in chemical wastes, aqueous effluents and soils, *Chemosphere*, 9(9):553-563, 1980.

Van Wijgaarden, R., Leeuwangh, P., Lucassen, W.G.H., Romijn, K., Ronday, R., van der Velde, R., and Willigenburg, W. Acute toxicity of chlorpyrifos to fish, a newt, and aquatic invertebrates, *Bull. Environ. Contam. Toxicol.*, 51(5):716-723, 1993.

Veierov, D., Fenigstein, A., Melamed-Madjar, V., and Klein, M. Effects of concentration and application method on decay and residual activity of foliar chlorpyrifos, *J. Econ. Entomol.*, 81(2):621-627, 1988.

Veith, G.D., Austin, N.M., and Morris, R.T. A rapid method for estimating log P for organic chemicals, *Water Res.*, 13(1):43-47, 1979a.

Veith, G.D., Call, D.J., and Brooke, L.T. Structure-toxicity relationships for the fathead minnow *Pimephales promelas*: Narcotic industrial chemicals, *Can. J. Fish. Aquat. Sci.*, 40(6):743-748, 1983.

Veith, G.D., DeFoe, D.L., and Bergstedt, B.V. Measuring and estimating the bioconcentration factor of chemicals in fish, *J. Fish Res. Board Can.*, 26:1040-1048, 1979.

Veith, G.D., Macek, K.J., Petrocelli, S.R., and Carroll, J. An evaluation of using partition coefficients and water solubility to estimate bioconcentration factors for organic chemicals in fish, in *Aquatic Toxicology, ASTM STP 707*, Eaton, J.G., Parrish, P.R., and Hendricks, A.C., Eds. (Philadelphia, PA: American Society for Testing and Materials, 1980), pp. 116-129.

Venkatesulu, D., Venkatesu, P., and Rao, M.V.P. Viscosities and densities of trichloroethylene or tetrachloroethylene with 2-alkoxyethanols at 303.15 and 313.15 K, *J. Chem. Eng. Data*, 42(2):365-367, 1997.

Venkateswarlu, K., Chendrayan, K., and Sethunathan, N. Persistence and biodegradation of carbaryl in soils, *J. Environ. Sci. Health*, B15(4):421-429, 1980.

Venkateswarlu, K., Gowda, T.K.S., and Sethunathan, N. Persistence and biodegradation of carbofuran in flooded soils, *J. Agric. Food Chem.*, 25(3):533-536, 1977.

Venkateswarlu, K. and Sethunathan, N. Metabolism of carbofuran in rice straw-amended and unamended rice soils, *J. Environ. Qual.*, 8(3):365-368, 1979.

Venkateswarlu, K. and Sethunathan, N. Degradation of carbofuran by *Azospirillium lipoferum* and *Streptomyces* spp. isolated from flooded alluvial soil, *Bull. Environ. Contam. Toxicol.*, 33(5):556-560, 1984.

Verschueren, K. *Handbook of Environmental Data on Organic Chemicals* (New York: Van Nostrand Reinhold, 1983), 1310 p.

Vesala, A. Thermodynamics of transfer of nonelectrolytes from light to heavy water. I. Linear free energy correlations of free energy of transfer with solubility and heat of melting of a nonelectrolyte, *Acta Chem. Scand.*, A28(8):839-845, 1974.

Vesala, A. and Lönnberg, H. Salting-in effects of alkyl-substituted pyridinium chlorides on aromatic compounds, *Acta Chem. Scand.*, A34:187-192, 1980.

Vijayaraghavan, R. Modifications of breathing pattern induced by inhaled sulphur mustard in mice, *Arch. Toxicol.*, 71(3):157-164, 1997.

Villalta, P.W., Lovejoy, E.R., and Hanson, D.R. Reaction probability of peroxyacetyl radical on aqueous surfaces, *Geophys. Res. Lett.*, 23:1765-1768, 1996.

Virtanen, M.T. and Hattula, M.L. The fate of 2,4,6-trichlorophenol in aquatic continuous-flow system, *Chemosphere*, 11:641-649, 1982.

Viswanadhan, V.N., Ghose, A.K., Revankar, G.R., and Robins, R.K. Atomic physiochemical parameters for three dimensional structure directed quantitative structure-activity relationships. 4. Additional parameters for hydrophobic and dispersive interactions and their application for an automated superposition of certain naturally occurring nucleoside antibiotics, *J. Chem. Inf. Comput. Sci.*, 29:163-172, 1989.

Vitenberg, A.G., Ioffe, B.V., Dimitrova, Z.S., and Butaeva, I.L. Determination of gas-liquid partition coefficients by means of gas chromatographic analysis, *J. Chromatogr.*, 112:319-327, 1975.

Vogel, T.M., Criddle, C.S., and McCarty, P.L. Transformations of halogenated aliphatic compounds, *Environ. Sci. Technol.*, 21(8):722-736, 1987.

Vogel, T.M. and McCarty, P.L. Biotransformation of tetrachloroethylene to trichloroethylene, dichloroethylene, vinyl chloride, and carbon dioxide under methanogenic conditions, *Appl. Environ. Microbiol.*, 49(5):1080-1083, 1985.

Vogel, T.M. and Reinhard, M. Reaction products and rates of disappearance of simple bromoalkanes, 1,2-dibromopropane, and 1,2-dibromoethane in water, *Environ. Sci. Technol.*, 20(10):992-997, 1986.

Vogiatzis, A.K. and Loumbourdis, N.S. Exposure of *Rana ridibunda* to lead. I. Study of lead accumulation in various tissues and hepatic λ-aminolevulinic acid dehydratase activity, *J. Appl. Toxicol.*, 19(1):25-29, 1999.

Voice, T.C. and Weber, W.J., Jr. Sorbent concentration effects in liquid/solid partitioning, *Environ. Sci. Technol.*, 19(9):789-796, 1985.

Vontor, T., Socha, J., and Vecera, M. Kinetics and mechanism of hydrolysis of 1-naphthyl, –methylcarbamate and *N,N*-dimethyl carbamates, *Collect. Czech. Chem. Commun.*, 37:2183-2196, 1972.

Vowles, P.D. and Mantoura, R.F.C. Sediment-water partition coefficients and HPLC retention factors of aromatic hydrocarbons, *Chemosphere*, 16(1):109-116, 1987.

Vozňáková, Z., Popl, M., and Berka, M. Recovery of aromatic hydrocarbons from water, *J. Chromatogr. Sci.*, 16:123-127, 1978.

Wahid, P.A., Ramakrishna, C., and Sethunathan, N. Instantaneous degradation of parathion in anaerobic soils, *J. Environ. Qual.*, 9(1):127-130, 1980.

Wahid, P.A. and Sethunathan, N. Sorption-desorption of α, β, and γ isomers of hexachlorocyclohexane in soils, *J.*

Agric. Food Chem., 27(5):1050-1053, 1979.

Wahner, A. and Zetsch, C. Rate constants for the addition of hydroxyl radicals to aromatics (benzene, *p*-chloroaniline, and *o*-, *m*- and *p*-dichlorobenzene) and the unimolecular decay of the adduct. kinetics into a quasi-equilibrium, *J. Phys. Chem.*, 87:4945-4951, 1983.

Waites, R.E. and Van Middelem, C.H. Residue studies of DDT and malathion on turnip tops, collards, snap beans, and lettuce, *J. Econ. Entomol.*, 51:306-308, 1958.

Wajon, J.E., Rosenblatt, D.H., and Burrows, E.P. Oxidation of phenol and hydroquinone by chlorine dioxide, *Environ. Sci. Technol.*, 16(7):396-402, 1982.

Wakeham, S.G., Canuel, E.A., Doering, P.H., Hobbie, J.E., Helfrich, J.V.K., and Lough, R.G.R. The biogeochemistry of toluene in coastal seawater: radiotracer experiments in controlled ecosystems, *Biogeochemistry*, 1:307-328, 1985.

Wakeham, S.G., Davis, A.C., and Karas, J.L. Microcosm experiments to determine the fate and persistence of volatile organic compounds in coastal seawater, *Environ. Sci. Technol.*, 17(10):611-617, 1983.

Walker, W.W. Chemical and microbiological degradation of malathion and parathion in an estuarine environment, *J. Environ. Qual.*, 5(2):210-216, 1976.

Walker, W.W., Cripe, C.R., Pritchard, P.H., and Bourquin, A.W. Biological and abiotic degradation of xenobiotic compounds in *in vitro* estuarine water and sediment/water systems, *Chemosphere*, 17(12):2255-2270, 1988.

Walker, W.W. and Stojanovic, B.J. Microbial versus chemical degradation of malathion in soil, *J. Environ. Qual.*, 2:229-232, 1973.

Walker, W.W. and Stojanovic, B.J. Malathion degradation by an *Arthrobacter* sp., *J. Environ. Qual.*, 3(1):4-10, 1974.

Wallington, T.J., Dagaut, P., Liu, R., and Kurylo, M.J. Rate constants for the gas phase reactions of OH with C_5 through C_7 aliphatic alcohols and ethers: predicted and experimental values, *Int. J. Chem. Kinet.*, 20:541-547, 1988.

Wallington, T.J., Dagaut, P., Kurylo, M.J. Correlation between gas-phase and solution-phase reactivities of hydroxyl radicals toward saturated organic compounds, *J. Phys. Chem.*, 92:5024-5028, 1988a.

Wallington, T.J., Dagaut, P., Liu, R., and Kurylo, M.J. The gas phase reactions of hydroxyl radicals with a series of esters over the temperature range 240-440 K, *Int. J. Chem. Kinet.*, 20:177-186, 1988b.

Wallington, T.J., Dagaut, P., Liu, R., and Kurylo, M.J. Gas-phase reactions of hydroxy radicals with the fuel additives methyl *t*-butyl ether and *t*-butyl alcohol over the temperature range 240-440 K, *Environ. Sci. Technol.*, 22(7):842-844, 1988c.

Wallington, T.J. and Japar, S.M. Atmospheric chemistry of diethyl ether and ethyl *tert*-butyl ether, *Environ. Sci. Technol.*, 25(3):410-415, 1991.

Wallington, T.J. and Kurylo, M.J. Flash photolysis resonance fluorescence investigation of the gas-phase reactions of OH radicals with a series of aliphatic ketones over the temperature range 240-440 K, *J. Phys. Chem.*, 92:5050-5054, 1987.

Walsh, P.R. and Hites, R.A. Dicofol solubility and hydrolysis in water, *Bull. Environ. Contam. Toxicol.*, 22(3):305-311, 1979.

Walters, R.W. and Guiseppi-Elie, A. Sorption of 2,3,7,8-tetrachlorodibenzo-*p*-dioxin to soils from water/methanol mixtures, *Environ. Sci. Technol.*, 22(7):819-825, 1988.

Walters, R.W. and Luthy, R.G. Experimental investigation and predictive modeling of the adsorption of polycyclic aromatic hydrocarbons from water onto activated carbon, U.S. DOE Report-DOE/PC/30246-1237, 1982.

Walters, R.W., Ostazeski, S.A., and Guiseppi-Elie, A. Sorption of 2,3,7,8-tetrachlorodibenzo-*p*-dioxin from water by surface soils, *Environ. Sci. Technol.*, 23(4):480-484, 1989.

Walton, B.T., Anderson, T.A., Hendricks, M.S., and Talmage, S.S. Physicochemical properties as predictors of organic chemical effects on soil microbial respiration, *Environ. Toxicol. Chem.*, 8(1):53-63, 1989.

Walton, B.T., Hendricks, M.S., Anderson, T.A., Griest, W.H., Merriweather, R., Beauchamp, J.J., and Francis, C.W. Soil sorption of volatile and semivolatile organic compounds in a mixture, *J. Environ. Qual.*, 21(4):552-558, 1992.

Walton, W.C. *Practical Aspects of Ground Water Modeling* (Worthington, OH: National Water Well Association, 1985), 587 p.

Wams, T.J. Diethylhexylphthalate as an environmental contaminant - A review, *Sci. Total Environ.*, 66:1-16, 1987.

Wang, C.X., Yediler, A., Peng, A., and Kettrup, A. Photodegradation of phenanthrene in the presence of humic substances and hydrogen peroxide, *Chemosphere*, 30(3):501-510, 1995.

Wang, J., Liu, P., Shi, H., and Qian, Y. Kinetics of phthalic acid ester degradation by acclimated sludge, *Process Biochem.*, 32(7):567-571, 1997.

Wang, L., Kong, L., and Chang, C. Photodegradation of 17 PAHs in methanol (or acetonitrile)-water solution, *Environ. Chem.*, 10(2):15-20, 1991.

Wang, L., Zhao, Y., and Hong, G. Predicting aqueous solubility and octanol/water partition coefficients of organic chemicals from molecular volume, *Environ. Toxicol. Chem.*, 11:55-70, 1992.

Wang, T.C. and Bricker, J.L. 2-Butanone and tetrahydrofuran contamination in the water supply, *Bull. Environ. Contam. Toxicol.*, 23(4/5):620-623, 1979.

Wang, T.C. and Tan, C.K. Enhanced degradation of halogenated hydrocarbons in a water-photolysis system, *Bull. Environ. Contam. Toxicol.*, 40(1):60-65, 1988.

Wang, T.C., Tan, C.K., and Liou, M.C. Degradation of bromoform and chlorodibromomethane in a catalyzed H_2-water system, *Bull. Environ. Contam. Toxicol.*, 41(4):563-568, 1988.

Wang, Y.-S., Subba-Rao, R.V., and Alexander, M. Effect of substrate concentration and organic and inorganic compounds on the occurrence and rate of mineralization and cometabolism, *Appl. Environ. Microbiol.*, 47(6):1195-1200, 1984.

Wang, Y.-S., Yen, J.-H., Hsieh, Y.-N., and Chen, Y.-L. Dissipation of 2,4-D, glyphosate and paraquat in river water, *Water, Air, Soil Pollut.*, 72(1):1-7, 1994.

Wang, Y.-T. Methanogenic degradation of ozonation products of biorefractory or toxic aromatic compounds, *Water Res.*, 24(2):185-190, 1990.

Wania, F., Shiu, W.-Y., and Mackay, D. Measurement of vapor pressure of several low-volatility organophosphorus chemicals at low temperatures with gas saturation method, *J. Chem. Eng. Data*, 39:572-577, 1994.

Wannstedt, C., Rotella, D., and Siuda, J.F. Chloroperoxidase mediated halogenation of phenols, *Bull. Environ. Contam. Toxicol.*, 44(2):282-287, 1990.

Ward, C.T. and Matsumura, F. Fate of 2,3,7,8-tetrachloro-*p*-dioxin (TCDD) in a model aquatic environment, *Arch. Environ. Contam. Toxicol.*, 7(3):349-357, 1978.

Ward, M. and Weber, J.B. Aqueous Solubility of aklylamino-*s*-triazines as a function of pH and molecular structure, *J. Agric. Food Chem.*, 16(6):959-964, 1968.

Ward, T.M. and Getzen, F.W. Influence of pH on the adsorption of aromatic acids on activated carbon, *Environ. Sci. Technol.*, 4(1):64-67, 1970.

Warfield, R.W. and Petree, M.C. Thermodynamic properties of polystyrene and styrene, *J. Polymer Sci.*, 55:497-405, 1961.

Warneck, P. and Zerbach, T. Synthesis of peroxyacetyl nitrate in air by acetone photolysis, *Environ. Sci. Technol.*, 26(1):74-79, 1992.

Warner, H.P., Cohen, J.M., and Ireland, J.C. Determination of Henry's law constants of selected priority pollutants, Office of Science and Development, U.S. EPA Report-600/D-87/229, 1987, 14 p.

Warnick, S.L. and Bell, H.L. The acute toxicity of some heavy metals to different species of aquatic insects, *J. Water Pollut. Control Fed.*, 41:280-284, 1969.

Wasik, S.P., Miller, M.M., Tewari, Y.B., May, W.E., Sonnefeld, W.J., DeVoe, H., and Zoller, W.H. Determination of the vapor pressure, aqueous solubility, and octanol/water partition coefficient of hydrophobic substances by coupled generator column/liquid chromatographic methods, *Residue Rev.*, 85:29-42, 1983.

Wasik, S.P., Schwarz, F.P., Tewari, Y.B., and Miller, M.M. A head-space method for measuring activity coefficients, partition coefficients, and solubilities of hydrocarbons in saline solutions, *J. Res. Nat. Bur. Stand.*, 89(3):273-277, 1984.

Wasik, S.P., Tewari, Y.B., Miller, M.M., and Martire, D.E. Octanol/water partition coefficients and aqueous solubilities of organic compounds, U.S. EPA Report PB82-141797, 1981, 56 p.

Wasik, S.P. and Tsang, W. Gas chromatographic determination of partition coefficients of some unsaturated hydrocarbons and their deuterated isomers in aqueous silver nitrate solutions, *J. Phys. Chem.*, 74:2970-2976, 1970.

Watanabe, I., Kashimoto, T., and Tatsukawa, R. Brominated phenol production from the chlorination of wastewater containing bromide ions, *Bull. Environ. Contam. Toxicol.*, 33(4):395-399, 1984.

Watanabe, N., Sato, E., and Ose, Y. Adsorption and desorption of polydimethylsiloxane, cadmium nitrate, copper sulfate, nickel sulfate and zinc nitrate by river surface sediments, *Sci. Total Environ.*, 41(2):151-161, 1985.

Watarai, H., Tanaka, M., and Suzuki, N. Determination of partition coefficients of halobenzenes in heptane/water and 1-octanol/water systems and comparison with the scaled particle calculation, *Anal. Chem.*, 54(4):702-705, 1982.

Watson, E. and Parrish, C.F. Laser induced decomposition of 1,4-dioxane, *J. Chem. Phys.*, 54(3):1427-1428, 1971.

Wauchope, R.D. and Getzen, F.W. Temperature dependence of solubilities in water and heats of fusion of solid aromatic hydrocarbons, *J. Chem. Eng. Data*, 17(1):38-41, 1972.

Wauchope, R.D. and Haque, R. Effects of pH, light and temperature on carbaryl in aqueous media, *Bull. Environ. Contam. Toxicol.*, 9(5):257-261, 1973.

Weast, R.C. and Astle, M.J., Eds. *CRC Handbook of Data on Organic Compounds, Volumes 1 and 2* (Boca Raton, FL: CRC Press, 1986).

Weast, R.C., Ed. *CRC Handbook of Chemistry and Physics*, 67th ed. (Boca Raton, FL: CRC Press, 1986), 2406 p.

Webb, R.F., Duke, A.J., and Smith, L.S.A. Acetals and oligoacetals. Part I. Preparation and properties of reactive oligoformals, *J. Chem. Soc. (London)*, pp. 4307-4319, 1962.

Weber, J.B. Interaction of organic pesticides with particulate matter in aquatic and soil systems, in *Fate of Organic Pesticides in the Aquatic Environment, Advances in Chemistry Series*, Gould, R.F., Ed. (Washington, D.C.: American Chemical Society, 1972), pp. 55-120.

Weber, W.J., Jr. and Pirbazari, M. Effectiveness of activated carbon for removal of toxic and/or carcinogenic compounds from water supplies, Environmental Research Laboratory, Athens, GA., U.S. EPA Report-600/1-81-057, 1981, 356 p.

Weber, W.J., Jr., Voice, T.C., Pirbazari, M., Hunt, G.E., and Ulanoff, D.M. Sorption of hydrophobic compounds by sediments, soils, and suspended solids. II. Sorbent evaluation studies, *Water Res.*, 17(10):1443-1452, 1983.

Webster, G.R.B., Friesen, K.J., Sarna, L.P., and Muir, D.C.G. Environmental fate modelling of chlorodioxins: determination of physical constants, *Chemosphere*, 14(6/7):609-622, 1985.

Wedemeyer, G.A. Dechlorination of DDT by *Aerobacter aerogenes*, *Science (Washington, D.C.)*, 152(3722):647, 1966.

Weil, L., Dure, G., and Quentin, K.E. Solubility in water of insecticide chlorinated hydrocarbons and polychlorinated biphenyls in view of water pollution, *Z. Wasser Forsch.*, 7(6):169-175, 1974.

Weiss, G. *Hazardous Chemicals Data Book* (Park Ridge, NJ: Noyes Data Corp., 1986), 1069 p.

Weiss, U.M., Scheunert, I., Klein, W., and Korte, F. Fate of pentachlorophenol-^{14}C in soil under controlled conditions, *J. Agric. Food Chem.*, 30(6):1191-1194, 1982.

Wellborn, T.L., Jr. Toxicity of nine therapeutic and herbicidal compounds to striped bass, *Prog. Fish Cult.*, 31:27-32, 1969.

Welling, W. and de Vries, J.W. Bioconcentration kinetics of the organophosphorus insecticidechlorpyrifos in guppies (*Poecilia reticulata*), *Ecotoxicol. Environ. Saf.*, 23(1):64-75, 1992.

Wescott, J.W. and Bidleman, T.F. Determination of polychlorinated biphenyl vapor pressures by capillary gas chromatography, *J. Chromatogr.*, 210(2):331-336, 1981.

Westcott, J.W., Simon, C.G., and Bidleman, T.F. Determination of polychlorinated biphenyl vapor pressures by a semimicro gas saturation method, *Environ. Sci. Technol.*, 15(11):1375-1378, 1981.

Westerholm, R. and Li, H. A multivariate statistical analysis of fuel-related polycyclic aromatic hydrocarbon emissions from heavy-duty vehicles, *Environ. Sci. Technol.*, 28(5):965-972, 1994.

Westheimer, F.H. and Ingraham, L.L. The entropy of chelation, *J. Phys. Chem.*, 60:1668-1670, 1956.

Wheeler, H.G., Smith, F.F., Yeomans, A.H., and Fields, E. Persistence of low-volume and standard formulations of malathion on lima bean foliage, *J. Econ. Entomol.*, 60:400-402, 1967.

Whitbeck, M. Photo-oxidation of methanol, *Atmos. Environ.*, 17(1):121-126, 1983.

Whitehouse, B.G. The effects of temperature and salinity on the aqueous solubility of polynuclear aromatic hydrocarbons, *Mar. Chem.*, 14:319-332, 1984.

WHO. Environmental Health Criteria 10: Carbon disulfide (Geneva: World Health Organization), 1979, 100 p.

WHO. Environmental Health Criteria 28: Acrylonitrile (Geneva: World Health Organization), 1983, 125 p.

WHO. Environmental Health Criteria 26: Styrene (Geneva: World Health Organization), 1983a, 123 p.

WHO. Environmental Health Criteria 38: Heptachlor (Geneva: World Health Organization), 1984, 81 p.

WHO. Environmental Health Criteria 45: Camphechlor (Geneva: World Health Organization), 1984a, 66 p.

Wiemeyer, S.N., Hill, E.F., Carpenter, J.W., and Krynitsky, A.J. Acute oral toxicity of sodium cyanide in birds, *J. Wildl. Dis.*, 22:538-546, 1986.

Wiese, C.S. and Griffin, D.A. The solubility of Aroclor 1254 in seawater, *Bull. Environ. Contam. Toxicol.*, 19(4):403-411, 1978.

Wierzbicki, T. and Wojcik, O. Preliminary trials on the decomposition of acrolein, allyl alcohol, and glycerol by activated sludge, *Zesz. Nauk. Politech Slaska Inz. Sanit.*, 8:173-185 [Chemical Abstracts 68:1589z, 1968].

Wilhoit, R.C. and Zwolinski, B.J. *Handbook of Vapor Pressure and Heats of Vaporization of Hydrocarbons and Related Compounds*, Publication 101 (College Station, TX: Thermodynamics Research Station, 1971), 329 p.

Willhite, C.C. Inhalation toxicology of acute exposure to aliphatic nitriles, *Clin. Toxicol.*, 18:991-1003, 1981.

Williams, E.F. Properties of *O,O*-diethyl *O-p*-nitrophenyl thiophosphate and *O,O*-diethyl *O-p*-nitrophenyl phosphate, *Ind. Eng. Chem.*, 43(4):950-954, 1951.

Williams, I.H., Brown, M.J., and Whitehead, P. Persistence of carbofuran residues in some British Columbia soils, *Bull. Environ. Contam. Toxicol.*, 15(2):242-243, 1976.

Williams, M.D., Adams, W.J., Parkerton, T.F., Biddinger, G.R., and Robillard, K.A. Sediment sorption coefficient measurements for four phthalate esters: experimental results and model theory, *Environ. Toxicol. Chem.*, 14(9):1477-1486, 1995.

Willingham, C.B., Taylor, W.J., Pignocco, J.M., and Rossinni, F.D. Vapor pressure and boiling points of some paraffin, akylcyclopentane, akylcyclohexane, and akylbenzene hydrocarbons, *J. Res. Nat. Bur. Stand.*, 34:219, 1945.

Wilson, B.H., Smith, G.B., and Rees, J.F. Biotransformations of selected alkylbenzenes and halogenated aliphatic hydrocarbons in methanogenic aquifer material: a microcosm study, *Environ. Sci. Technol.*, 20(10):997-1002, 1986.

Wilson, J.T., Enfield, C.G., Dunlap, W.J., Cosby, R.L., Foster, D.A., and Baskin, L.B. Transport and fate of selected organic pollutants in a sandy soil, *J. Environ. Qual.*, 10(4):501-506, 1981.

Wilson, R.G. and Cheng, H.H. Fate of 2,4-D in a Naff silt loam soil, *J. Environ. Qual.*, 7(2):281-286, 1978.

Windholz, M., Budavari, S., Blumetti, R.F., and Otterbein, E.S., Eds. *The Merck Index*, 10th ed. (Rahway, NJ: Merck, 1983), 1463 p.

Windholz, M., Budavari, S., Stroumtsos, L.S., and Fertig, M.N., Eds. *The Merck Index*, 9th ed. (Rahway, NJ: Merck, 1976), 1313 p.

Winer, A.M., Darnall, K.R., Atkinson, R., Pitts, J.N., Jr. Smog chamber study of the correlation of hydroxyl radical rate constants with ozone formation, *Environ. Sci. Technol.*, 13:622-626, 1979.

Winner, R.W. and Farrell, M.P. Acute and chronic toxicity of copper to four species of *Daphnia*, *J. Fish. Res. Board Canada*, 33:1685-1691, 1976.

Wise, S.A., Bonnett, W.J., Guenther, F.R., and May, W.E. A relationship between reversed phase C_{18} liquid chromatographic retention and the shape of polycyclic aromatic hydrocarbons, *J. Chromatogr. Sci.*, 19:457-465, 1981.

Wiseman, A., Lim, T.-K., and Woods, L.F.J. Regulation of the biosynthesis of cytochrome P-450 in Brewer's yeast. Role of cyclic AMP, *Biochim. Biophys. Acta*, 544:615-623, 1978.

Wisk, J.D. and Cooper, K.R. Comparison of the toxicity of several polychlorinated dibenzo-*p*-dioxins and 2,3,7,8-tetrachlorodibenzofuran in embryos of the Japanese medaka (*Oryzias latipes*), *Chemosphere*, 20(3/4):361-377, 1990.

Wisk, J.D. and Cooper, K.R. The stage specific toxicity of *2,3,7,8*-tetrachlorodibenzo-p-dioxin in embryos of the Japanese medaka (*Oryzias latipes*), *Environ. Toxicol. Chem.*, 9(9):1159-1169, 1990a.

Witherspoon, P.A. and Bonoli, L. Correlation of diffusion coefficients for paraffin, aromatic, and cycloparaffin hydrocarbons in water, *Ind. Eng. Chem. Fundam.*, 8(3):589-591, 1969.

Witte, F., Urbanik, E., and Zetzsch, C. Temperature dependence of the rate constants for the addition of hydroxyl radical to benzene and to some monosubstituted aromatics (aniline, bromobenzene, and nitrobenzene) and the unimolecular decay of the adducts. Part 2. Kinetics into a quasi-equilibrium, *J. Phys. Chem.*, 90:3251-3259, 1986.

Wofford, H.W., Wilsey, C.D., Neff, G.S., Giam, C.S., and Neff, J.M. Bioaccumulation and metabolism of phthalate esters by oysters, brown shrimp and sheepshead minnows, *Ecotoxicol. Environ. Saf.*, 5(2):202-210, 1981.

Wolfe, M.F., Villalobos, S.A., Seiber, J.N., and Hinton, D.E. A comparison of carbamate toxicity to medaka (*Oryzias latipes*) embryos and larvae, *Toxicologist*, 13(1):268, 1993.

Wolfe, N.L., Staff, D.C., Armstrong, J.F., and Comer, S.W. Persistence of parathion in soil, *Bull. Environ. Contam. Toxicol.*, 10(1):1-9, 1973.

Wolfe, N.L., Steen, W.C., and Burns, L.A. Phthalate ester hydrolysis: linear free energy relationships, *Chemosphere*, 9(7/8):403-408, 1980.

Wolfe, N.L., Zepp, R.G., Baughman, G.L., Fincher, R.C., and Gordon, J.A. Chemical and photochemical transformations of selected pesticides in aquatic systems, U.S. EPA Report 600/3-76-067, 1976, 141 p.

Wolfe, N.L., Zepp, R.G., Gordon, J.A., Baughman, G.L., and Cline, D.M. Kinetics of chemical degradation of malathion in water, *Environ. Sci. Technol.*, 11(1):88-93 1977a.

Wolfe, N.L., Zepp, R.G., and Paris, D.F. Use of structure-reactivity relationships to estimate hydrolytic persistence of carbamate pesticides, *Water Res.*, 12(8):561-563, 1978.

Wolfe, N.L., Zepp, R.G., and Paris, D.F. Carbaryl, phospham and chloropropham: a comparison of the rates of hydrolysis and photolysis with the rate of biolysis, *Water Res.*, 12(8):565-571, 1978a.

Wolfe, N.L., Zepp, R.G., Paris, D.F., Baughman, G.L., and Hollis, R.C. Methoxychlor and DDT degradation in water: rates and products, *Environ. Sci. Technol.*, 11(12):1077-1081, 1977.

Wong, A.S. and Crosby, D.G. Photodecomposition of pentachlorophenol in water, *J. Agric. Food Chem.*, 29(1):125-130, 1981.

Wood, A.L., Bouchard, D.C., Brusseau, M.L., and Rao, P.S.C. Cosolvent effects on sorption and mobility of organic contaminants in soils, *Chemosphere*, 21(4/5):575-587, 1990.

Woodburn, K.B., Doucette, W.J., and Andren, A.W. Generator column determination of octanol/water partition coefficients for selected polychlorinated biphenyl congeners, *Environ. Sci. Technol.*, 18(6):457-459, 1984.

Woodburn, K.B., Lee, L.S., Suresh, P., Rao, C., and Delfino, J.J. Comparison of sorption energetics for hydrophobic organic chemicals by synthetic and natural sorbents from methanol/water solvent mixtures, *Environ. Sci. Technol.*, 23(4):407-413, 1989.

Woodiwiss, F.S. and Fretwell, G. The toxicities of sewage effluents, industrial discharges and some chemical substances to brown trout in the Trent River Authority area, *Water Pollut. Control (Great Britain)*, 73:464, 1974.

Woodrow, J.E., Crosby, D.G., Mast, T., Moilanen, K.W., and Seiber, J.N. Rates of transformation of trifluralin and parathion vapors in air, *J. Agric. Food Chem.*, 26(6):1312-1316, 1978.

Woodrow, J.E., Crosby, D.G., and Seiber, J.N. Vapor-phase photochemistry of pesticides, *Residue Rev.*, 85:111-125, 1983.

Worley, J.D. Benzene as a solute in water, *Can. J. Chem.*, 45:2465-2467, 1967.

Worthing, C.R. and Hance, R.J., Eds. *The Pesticide Manual - A World Compendium*, 9th ed. (Great Britain: British Crop Protection Council, 1991), 1141 p.

Wright, C.G., Leidy, R.B., and Dupree, H.E., Jr. Insecticides in the ambient air of rooms following their application for control of pests, *Bull. Environ. Contam. Toxicol.*, 26(4):548-553, 1981.

Wright, D.A., Sandler, S.I., and DeVoll, D. Infinite dilution activity coefficients and solubilities of halogenated hydrocarbons in water at ambient temperatures, *Environ. Sci. Technol.*, 26(9):1828-1831, 1992.

Wu, S. and Gschwend, P.M. Sorption kinetics of hydrophobic organic compounds to natural sediments and soils, *Environ. Sci. Technol.*, 20(7):717-725, 1986.

Xie, T.M. Determination of trace amounts of chlorophenols and chloroguaicols in sediment, *Chemosphere*, 12(9/10):1183-1191, 1983.

Xie, T.M., Hulthe, B., and Folestad, S. Determination of partition coefficients of chlorinated phenols, guaiacols and catechols by shake-flask GC and HPLC, *Chemosphere*, 13(3):445-459, 1984.

Xu, F., Liang, X.-M., Su, F., Zhang, Q., Lin, B.-C., Wu, W.-Z., Yediler, A., and Kettrup, A. A column method for determination of soil organic partition coefficients of eight pesticides, *Chemosphere*, 39(5):787-794, 1999.

Yagi, O. and Sudo, R. Degradation of polychlorinated biphenyls by microorganisms, *J. Water Pollut. Control Fed.*, 52(5):1035-1043, 1980.

Yalkowsky, S.H. Solubility and partitioning. V: dependence of solubility on melting point, *J. Pharm. Sci.*, 70(8):971-973, 1981.

Yalkowsky, S.H., Orr, R.J., and Valvani, S.C. Solubility and partitioning. 3. The solubility of halobenzenes in water, *Indust. Eng. Chem. Fundam.*, 18(4):351-353, 1979.

Yalkowsky, S.H. and Valvani, S.C. Solubilities and partitioning 2. Relationships between aqueous solubilities, partition coefficients, and molecular surface areas of rigid aromatic hydrocarbons, *J. Chem. Eng. Data*, 24(2):127-129, 1979.

Yalkowsky, S.H. and Valvani, S.C. Solubility and partitioning. I. Solubility of nonelectrolytes in water, *J. Pharm. Sci.*, 69:912-922, 1980.

Yalkowsky, S.H., Valvani, S.C., and Mackay, D. Estimation of the aqueous solubility of some aromatic compounds, *Residue Rev.*, 85:43-55, 1983.

Yalkowsky, S.H., Valvani, S.C., and Roseman, T.J. Solubility and partitioning VI. Octanol solubility and octanol-water partition coefficients, *J. Pharm. Sci.*, 72(8):866-870, 1983a.

Yan, H., Ye, C., and Yin, C. Kinetics of phthalate ester biodegradation by *Chlorella pyrenoidosa*, *Environ. Toxicol. Chem.*, 14(6):931-938, 1995.

Yang, Y., Miller, D.J., and Hawthorne, S.B. Toluene solubility in water and organic partitioning from gasoline and diesel fuel into water at elevated temperatures and pressures, *J. Chem. Eng. Data*, 42(5):908-913, 1997.

Yaron, B. and Saltzman, S. Influence of water and temperature on adsorption of parathion by soils, *Soil Sci. Soc. Am. Proc.*, 36:583-586, 1972.

Yasuhara, A. and Morita, M. Formation of chlorinated compounds in pyrolysis of trichloroethylene, *Chemosphere*, 21(4/5):479-486, 1990.

Ye, D., Siddiqi, A., Maccubbin, A.E., Kumar, S., and Sikka, H.C. Degradation of polynuclear aromatic hydrocarbons by *Sphingomonas paucimobilis*, *Environ. Sci. Technol.*, 30(1):136-142, 1996.

Yin, C. and Hassett, J.P. Gas partitioning approach for laboratory and field studies of mirex fugacity in water, *Environ.*

Sci. Technol., 20(12):1213-1217, 1986.

Yokley, R.A., Garrison, A.A., Mamantov, G., and Wehry, E.L. The effect of nitrogen dioxide on the photochemical and nonphotochemical degradation of pyrene and benzo[*a*]pyrene adsorbed on coal fly ash, *Chemosphere*, 14(11/12):1771-1778, 1985.

Yonezawa, Y. and Urushigawa, Y. Chemical-biological interactions in biological purification systems. V. Relation between biodegradation rate constants of aliphatic alcohols by activated sludge and their partition coefficients in a 1-octanol-water system, *Chemosphere*, 8(3):139-142, 1979.

Yoshida, T. and Castro, T.F. Degradation of 2,4-D, 2,4,5-T, and picloram in two Philippine soils, *Soil Sci. Plant Nutr.*, 21:397-404, 1975.

Yoshida, K., Tadayoshi, S., and Yamauchi, F. Relationship between molar refraction and octanol/water partition coefficient, *Ecotoxicol. Environ. Saf.*, 7(6):558-565, 1983.

Yoshida, K., Shigeoka, T., and Yamauchi, F. Non-steady-state equilibrium model for the preliminary prediction of the fate of chemicals in the environment, *Ecotoxicol. Environ. Saf.*, 7(2):179-190, 1983a.

Yoshioka, Y., Mizuno, T., Ose, Y., and Sato, T. The estimation of toxicity of chemicals on fish by physico-chemical properties, *Chemosphere*, 15(2):195-203, 1986.

Yoshizumi, K., Aoki, K., Nouchi, I., Okita, T., Kobayashi, T., Kamakura S., and Tajima, M. Measurements of the concentration in rainwater and of the Henry's law constant of hydrogen peroxide, *Atmos. Environ.*, 18(2):395-401, 1984.

Youezawa, Y. and Urushigawa, Y. Chemico-biological interactions in biological purification systems. V. Relation between biodegradation rate constants of aliphatic alcohols by activated sludge and their partition coefficients in a 1-octanol-water system, *Chemosphere*, 8(3):139-142, 1979.

Young, L.Y. and Rivera, M.D. Methanogenic degradation of four phenolic compounds, *Water Res.*, 19(10):1325-1332, 1985.

Young, W.F., Horth, H., Crane, R., Ogden, T., and Arnott, M. Taste and odour threshold concentrations of potential potable water contaminants, *Water Res.*, 30(2):331-340, 1996.

Yu, C.-C., Hansen, D.J., and Booth, G.M. Fate of dicamba in a model ecosystem, *Bull. Environ. Contam. Toxicol.*, 13(3):280-283, 1975.

Yu, Y.S. and Bailey, G.W. Reduction of nitrobenzene by four sulfide minerals: kinetics, products, and solubility, *J. Environ. Qual.*, 21(1):86-94, 1992.

Yuan, X. and Lang, P. QSAR study of the toxicity of nitrobenzenes to river bacteria and *Photobacterium phosphoreum*, *Bull. Environ. Contam. Toxicol.*, 58(1):123-127, 1997.

Yuen, Q.H. and Hilton, H.W. The adsorption of monuron and diuron by Hawaiian sugarcane soils, *J. Agric. Food Chem.*, 10(5):386-392, 1962.

Yule, W.N., Chiba, M., and Morley, H.V. Fate of insecticide residues. Decomposition of lindane in soil, *J. Agric. Food Chem.*, 15(6):1000-1004, 1967.

Zabik, M.J., Schuetz, R.D., Burton, W.L., and Pape, B.E. Photochemistry of bioreactive compounds. Studies of a major product of endrin, *J. Agric. Food Chem.*, 19(2):308-313, 1971.

Zafiriou, O.C. Reaction of methyl halides with seawater and marine aerosols, *J. Mar. Res.*, 33(1):75-81, 1975.

Zafiriou, O.C. and McFarland, M. Determination of trace levels of nitric oxide in aqueous solution, *Anal. Chem.*, 52(11):1662-1667, 1980.

Zaidi, B.R., Murakami, Y., and Alexander, M. Predation and inhibitors in lake water affect the success of inoculation to enhance biodegradation of organic chemicals, *Environ. Sci. Technol.*, 23(7):859-863, 1989.

Zayed, S.M.A.D., Mostafa, I.Y., Farghaly, M.M., Attaby, H.S.H., Adam, Y.M., and Mahdy, F.M. Microbial degradation of trifluralin by *Aspergillus carneus*, *Fusarium oxysporum* and *Trichoderma viride*, *J. Environ. Sci. Health*, B18(2):253-267, 1983.

Zepp, R.G. and Schlotzhauer, P.F. Photoreactivity of selected aromatic hydrocarbons in water, in *Polynuclear Aromatic Hydrocarbons, 3rd International Symposium on Chemistry, Biology, Carcinogenesis and Mutagenesis*, Jones, P.W. and Leber, P., Eds. (Ann Arbor, MI: Ann Arbor Science Publishers, 1979), pp. 141-158.

Zepp, R.G. and Schlotzhauer, P.F. Influence of algae on photolysis rates of chemicals in water, *Environ. Sci. Technol.*, 17(8):462-468, 1983.

Zepp, R.G., Wolfe, N.L., Azarraga, L.V., Cox, R.H., and Pape, C.W. Photochemical transformation of the DDT and methoxychlor degradation products, DDE and DMDE, by sunlight, *Arch. Environ. Contam. Toxicol.*, 6(2/3):305-314, 1977.

Zepp, R.G., Wolfe, N.L., Baughman, G.L., Schlotzhauer, P.F., and MacAllister, J.N. Dynamics of processes influencing the behavior of hexachlorocyclopentadiene in the aquatic environment, 178th Meeting of the American Chemical Society, Washington, D.C. (September 1979).

Zepp, R.G., Wolfe, N.L., Gordon, J.A., and Fincher, R.C. Light-induced transformations of methoxychlor in aquatic systems, *J. Agric. Food Chem.*, 24(4):727-733, 1976.

Zeyer, J. and Kearney, P.C. Microbial metabolism of [^{14}C]nitroanilines to [^{14}C]carbon dioxide, *J. Agric. Food Chem.*, 31(2):304-308, 1983.

Zeyer, J. and Kearney, P.C. Degradation of *o*-nitrophenol and *m*-nitrophenol by a *Pseudomonas putida*, *J. Agric. Food Chem.*, 32(2):238-242, 1984.

Zhang, S. and Rusling, J.F. Dechlorination of polychlorinated biphenyls by electrochemical catalysis in a biocontinuous microemulsion, *Environ. Sci. Technol.*, 27(7):1375-1380, 1993.

Zhang, W. and Wang, C.-B. Rapid and complete dechlorination of TCE and PCBs by nanoscale Fe and Pd/Fe particles, in *American Chemical Society - Division of Environmental Chemistry, Preprints of Extended Abstracts*, 37(1):78-79, 1997.

Zheng, D.-Q., Guo, T.-M., and Knapp, H. Experimental and modeling studies on the solubility of CO_2, $CHClF_2$, CHF_3, $C_2H_2F_4$ and $C_2H_4F_2$ in water and aqueous NaCl solutions under low pressures, *Fluid Phase Equilib.*, 197-209, 1997.

Zhou, X. and Lee, Y.-N. Aqueous solubility and reaction kinetics of hydroxymethyl hydro. eroxide, *J. Phys. Chem.*, 96:265-272, 1992.

Zhou, X. and Mopper, K. Apparent partition coefficients of 15 carbonyl compounds between air and seawater and between air and freshwater; implications for air-sea exchange, *Environ. Sci. Technol.*, 24(12):1864-1869, 1990.

Ziegler, W.W. and Andrews, D.H. The heat capacity of benzene-d_6, *J. Am. Chem. Soc.*, 64:2482-2485, 1942.

Zierath, D.L., Hassett, J.J., and Banwart, W.L. Sorption of benzidine by sediments and soils, *Soil Sci.*, 129(5):277-281, 1980.

Zitko, V. Polychlorinated biphenyls (PCBs) solubilized in water by monoionic surfactants for study of toxicity to aquatic animals, *Bull. Environ. Contam. Toxicol.*, 5(3):279-285, 1970.

Zoeteman, B.C.J., DeGreef, E., and Brinkmann, F.J.J. Persistency of organic contaminants in groundwater. Lessons from soil pollution incidents in the Netherlands, *Sci. Total Environ.*, 21:187-202, 1981.

Zok, S., Görge, G., Kalsch, W., and Nagel, R. Bioconcentration, metabolism and toxicity of substituted anilines in the zebrafish (*Brachydanio rerio*), *Sci. Total Environ.*, 109/110:411-421, 1991.

Zoro, J.A., Hunter, J.M., and Ware, G.C. Degradation of *p,p'*-DDT in reducing environments, *Nature (London)*, 247(5438):235-237, 1974.

Cumulative Index

Note: I and II refer to this work and Montgomery (1998), respectively.

1337